A Course in

# Electrical and Electronic Measurements and Instrumentation

**A.K. Sawhney**
*M.Sc. (Engg.)*
*Formerly Professor & Head,*
*Electrical and Electronics Engineering Deptt.,*
*Thapar Institute of Engineering and Technology,*
*(Deemed University) PATIALA – 147004*

**Puneet Sawhney**
*M.Sc.*
*University of Waterloo, Canada*
*Senior Engineer, Engineering Global Solutions*
*DEARBORN (U.S.A)*

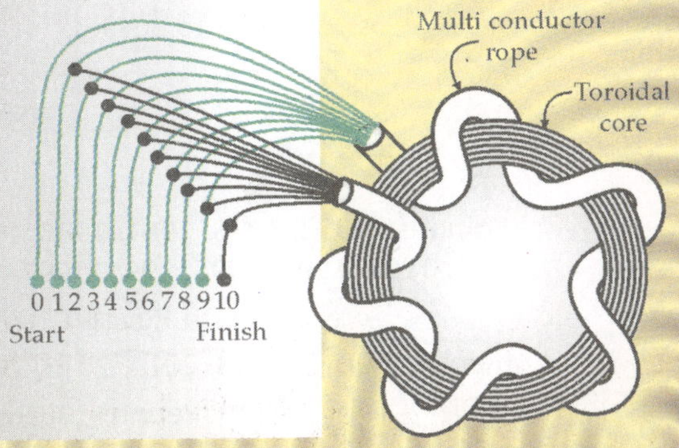


**DHANPAT RAI & Co. (Pvt.) Ltd.**
*EDUCATIONAL AND TECHNICAL PUBLISHERS*

A Course in
Electrical and Electronic Measurements and Instrumentation

Published by : **GAGAN KAPUR**
**Dhanpat Rai & Co. (P) Ltd., Delhi**

Sales Office:
1710, Nai Sarak,
Delhi-110006
Phone : 2326 5367, 2325 0251
**dhanpatrai@gmail.com**

Regd. Office:
4576/15, Agarwal Road
Darya Ganj, New Delhi-110002
Phone : 2324 7736, 37, 38
**dhanpatrai@gmail.com**


Edited by: Smt. Chandra Kanta Sawhney

## Disclaimer

Every effort has been made to avoid errors or omissions in this publication. Inspite of this, some errors might have crept in. Any mistake, error or discrepancy noted may be brought to our notice which shall be taken care ofin the next edition. It is notified that neither the publisher nor the author or seller will be responsible for any damage or loss of action to any one, of any kind, in any manner, therefrom.

Editions : 1973, 1976, 1981, 1983, 1987, 1988, 1989, 1990,
1991, 1993, 1994, 1995, 1996, 1997, 1998, 1999, 2000, 2007
Nineteenth Revised Edition: 2011
Reprint : 2012, 2013, 2014, 2015, 2016, 2017, 2018, 2019, 2020
Reprint : 2026
Price : ₹ 1250/-

Typesetting by: North Delhi Computer Convention, Delhi–110009, ndcc.in@gmail.com
Printed at : Natraj Offset, Delhi

# Preface to the Nineteenth Edition

The authors are pleased to bring out the nineteenth edition of the book 'Electrical and Electronic Measurements and Instrumentation', and are grateful to students and faculty members of different institutions in India and abroad for their warm affectionate support.

The book has been completely reproduced and care has been taken to rectify the mistakes which had cropped up in the earlier editions. The authors take this opportunity to thank all those who were kind enough to send critical suggestions for the improvement of the text.

The present edition retains all the features of the earlier editions. A new salient feature of the book is, addition of Five New Chapters (Chapters 32 to 36) at the end of the book. The authors hope that with these additions, the book covers all the topics required in the syllabus and that the students and practising engineers need to familiarise themselves with the present day technology and practice in the area of Electrical and Electronic Measurements and Instrumentation.

The authors are highly obliged to Mrs. Chandra Kanta Sawhney for editing the present edition.

Authors again request the students, teachers and practising engineers to send their suggestions for updating the text.

**A.K. Sawhney**
**Puneet Sawhney**

# Preface To The Fourth Edition

The author is pleased to bring out the Fourth Edition of the book and is thankful to both teachers and students for their affectionate and warm reception to the third edition of the book which has been sold out in a period of less than one year. The present edition retains essentially the same subject-matter as the third edition. However, the typographical and other errors, which had crept in earlier edition, have been corrected.

The book is intended as a standard text for students studying for their first degree in Electrical, Electronics and Instrumentation Engineering at Indian Universities and abroad, and also for those appearing for A.M.I.E. Section B and other professional examinations. The book is equally useful for postgraduate students as well as practising engineers involved in the field of Measurements and Instrumentation.

There have been significant changes in curriculum of almost all the universities in recent years. Electrical and Electronic Instrumentation is now offered as separate paper in many universities. This has been necessitated on account of latest technological advances which put greater emphasis and reliance on Electronic Instrumentation.

The contents of the book have been drastically modified, re-arranged and updated to acquaint the reader of modern trends in the field of Measurements and Instrumentation.

The book has been divided into *two* Parts. **Part I** deals with Measurements and Measuring Instruments and **Part II** takes care of the Instrumentation. There is an extensive coverage of Electrical and Electronic Instrumentation in this edition of the book as compared with the earlier ones. The coverage of Instrumentation portion is about 500 pages out of a total of nearly 1200 pages. The subject of Instrumentation has been developed in logical steps. Separate chapters are devoted to topics like Generalized Measurement Systems, Transducers, Signal Conditioning, Data Transmission and Telemetry, Display Devices and Recorders, Measurement of Non-electrical Quantities and Data Acquisition Systems. Also there are additional chapters on static and dynamic characteristics of Measurement Systems. There is an increased emphasis on digital instruments and instrumentation which is consonant with the present trends.

Conversions from various systems of units to S.I. units is given as Appendix at the end of the book. The significant additions to the measurements portion are Transformer ratio bridges (which are fast replacing the conventional four arm a.c. bridges), spectrum analyzer, vector impedance meter, vector voltmeter, digital maximum demand indicator and Hall effect multiplier to name a few.

The book in fact covers a very wide spectrum of the field of Electrical and Electronic Measurements and Instrumentation and is a complete reference in itself.

Another outstanding feature of the book is the inclusion of over 400 solved problems which in addition to linking the theory with actual applications gives an insight of the industrial practice. Also about 300 unsolved problems (*with answers*) have been included to give the students practice in solving problems.

The author considers the inclusion of problems on Instrumentation (both solved and unsolved) as a speciality of this book. This is a pioneering effort which is the outcome of author's experience of teaching the subject for almost two decades.

The book though voluminous, covers *two* papers, *i.e.*, first on Electrical and Electronic Measurements and Instruments and second on Electrical and Electronic Instrumentation and thus fully justifies its volume. SI units have been uniformly used in the book throughout.

The author will feel highly obliged to all readers for their constructive suggestions and healthy criticism of the book which will go a long way in the improvement of the text.

The author is thankful to his wife, Chandra, for rendering assistance in the compilation and editing of the work.

The author is grateful to his brother Ravish and Ajay for their constant help during the preparation of the text.

To author's parents who have been a source of encouragement and guidance.

**A. K. Sawhney**

# Contents

## Part I

## Electrical and Electronic Measurement and Measuring Instruments

## Chapter 3 ERRORS IN MEASUREMENTS AND THEIR STATISTICAL ANALYSIS 49



## Chapter 4 DYNAMIC CHARACTERISTICS OF INSTRUMENTS AND MEASUREMENT SYSTEMS 81

## Chapter 5 Units, Systems, Dimensions and Standards 125



## Chapter 6 Circuit Components (Resistors, Inductors and Capacitors and Their Residues) 159



## Chapter 7 Analog Instruments 177

## Chapter 8 GALVANOMETERS 197



## Chapter 9 ANALOG AMMETERS, VOLTMETERS AND OHMMETERS 237

## Chapter 10 INSTRUMENT TRANSFORMERS 313

## Chapter 11 Measurement of Power and Wattmeters 351



## Chapter 12 Measurement of Energy and Industrial Metering 379



## Chapter 13 Measurement of Phase and Frequency 405

## Chapter 14 MEASUREMENT OF RESISTANCE 421



## Chapter 15 POTENTIOMETERS 455



## Chapter 16 A.C. BRIDGES 479

## Chapter 17 HIGH VOLTAGE MEASUREMENTS AND TESTING 513



## Chapter 18 MAGNETIC MEASUREMENTS 541

## Chapter 19 OPTOELECTRONIC MEASUREMENTS 571



## Chapter 20 ELECTRONIC INSTRUMENTS 605



## Chapter 21 CATHODE RAY OSCILLOSCOPE (CRO) 641

## Chapter 22 INSTRUMENTS FOR GENERATION OF WAVEFORMS 683



## Chapter 23 SIGNAL ANALYSERS 713

## Chapter 24 HIGH FREQUENCY MEASUREMENTS 725



# Part II

# Electrical & Electronic Instrumentation

## Chapter 25 PRIMARY SENSING ELEMENTS AND TRANSDUCERS 739

## Chapter 26 SIGNAL CONDITIONING 873

## Chapter 27 DATA TRANSMISSION AND TELEMETRY 961

## Chapter 28 DISPLAY DEVICES AND RECORDERS 1005

## Chapter 29 Measurement of Non-Electric Quantities 1055

## Chapter 30 MICROPROCESSORS 1135

## Chapter 31 DATA ACQUISITION SYSTEMS 1203



## Chapter 32 MODERN SENSORS 1219



## Chapter 33 DIGITAL MEASUREMENT TECHNIQUES 1237

## Chapter 34 CHEMICAL SENSORS AND ANALYTICAL INSTRUMENTS 1249



## Chapter 35 AUTOMATIC PROCESS CONTROL 1269



## Chapter 36 BIOMEDICAL INSTRUMENTATION 1293

*Dedicated to*
Anuradha

# 1

# Measurements and Measurement Systems

## 1.1 MEASUREMENTS

The measurement of a given quantity is essentially an act or the result of comparison between the quantity (whose magnitude is unknown) and a predefined standard. Since two quantities are compared, the result is expressed in numerical values. In fact, measurement is the process by which one can convert physical parameters to meaningful numbers. The measuring process is one in which the property of an object or system under consideration is compared to an accepted standard unit, a standard defined for that particular property. The number of times the unit standard fits into the quantity being measured is the numerical measure. The numerical measure is meaningless unless followed by a unit used, since it (unit) identifies the characteristic or property measured.

In order that the results of the measurement are meaningful, there are *two* basic requirements :

(*i*) the standard used for comparison purposes must be accurately defined and should be commonly accepted, and

(*ii*) the apparatus used and the method adopted must be provable.

## 1.2 SIGNIFICANCE OF MEASUREMENTS

The importance of measurement is simply and eloquently expressed in the following statement of the famous physicist Lord Kelvin :

> *"I often say that when you can measure what you are speaking about and can express it in numbers, you know something about it; when you cannot express in it numbers your knowledge is of a meagre and unsatisfactory kind."*

The advancement of Science and Technology is dependent upon a parallel progress in measurement techniques. The reasons for this are obvious. As Science and Technology move ahead, new phenomena and relationships are discovered and these advances make new types of measurements imperative. New discoveries are not of any practical utility unless the results are backed by actual measurements. The measurements, no doubt, confirm the validity of a hypothesis but also add to its understanding. This results in an unending chain which leads to new discoveries that require more, new and sophisticated measurement techniques. Hence modern Science and Technology are associated with sophisticated methods of measurement.

There are *two* major functions of all branches of engineering :

(*i*) Design of equipment and processes, and

(*ii*) Proper operation and maintenance of equipment and processes.

Both these functions require measurements. This is because proper and economical design, operation and maintenance require a feedback of information. Measurements play a significant role in achieving goals and objectives of Engineering because of the feedback information supplied by them.

## 1.3 METHODS OF MEASUREMENTS

The methods of measurements may be broadly classified into *two* categories :

(*i*) Direct Methods and (*ii*) Indirect Methods.

### 1.3.1 Direct Methods

In these methods, the unknown quantity (also called the **measurand**) is directly compared against a standard. The result is expressed as a numerical number and a unit. The standard, in fact, is a physical embodiment of a unit.

Direct methods are quite common for the measurement of physical quantities like length, mass and time. Suppose we want to measure the length of a bar. The unit of length is metre. A bar is so many times long because that many units on our standard have the same length as the bar. A human being can make direct length comparisons with a preciseness of about 0.25 mm. Therefore, on account of human factors it is not possible to make very accurate measurements. The direct method for measurement of length can be utilized with a good degree of accuracy but when it comes to measurement of mass, the problem becomes much more intricate. It is just not possible for human beings to distinguish between wide margins of mass.

### 1.3.2 Indirect Methods

Measurement by direct methods are not always possible, feasible and practicable. These methods in most of the cases, are inaccurate because they involve human factors. They are also less sensitive. Hence direct methods are not preferred and are rarely used.

In engineering applications **Measurement Systems** are used which require need of indirect methods for measurement purposes.

## 1.4 INSTRUMENTS AND MEASUREMENT SYSTEMS

Measurements involve the use of **instruments** as a physical means of determining quantities or variables. The instrument serves as an extension of human faculties and enables the man to determine the value of unknown quantity or variable which his unaided human faculties cannot measure, a measuring instrument exists to provide information about the physical value of some variables being measured. In simple cases, an instrument consists of a single unit which gives an output reading or signal according to the unknown variable (measurand) applied to it. In more complex measurement situations, however, a measuring instrument may consist of several separate elements. These elements may consist of transducing elements which convert the measurand to an analogous form. The analogous signal is then processed by some intermediate means and then fed to the end devices to present the results of the measurement for the purposes of display and or control. The above mentioned components might be contained within one or more boxes, and the boxes holding individual measurement elements might be either close together or physically separate. Because of this modular nature of the elements within it, it is common to refer the measuring instrument as a **measurement system**.

## 1.5 MECHANICAL, ELECTRICAL AND ELECTRONIC INSTRUMENTS

The first instruments used by mankind were mechanical in nature and the principles on which these instruments worked are even in vogue today. The earliest scientific instruments used the same three essential elements as our modern instruments do. These elements are :

(*i*) a detector,

(*ii*) an intermediate transfer device, and

(*iii*) an indicator, recorder or a storage device.

The history of development of instruments encompasses *three* phases of instruments, *viz.* :

(*i*) mechanical instruments,

(*ii*) electrical instruments, and

(*iii*) electronic instruments.

### 1.5.1 Mechanical Instruments

These instruments are very reliable for static and stable conditions. But they suffer from a very major disadvantage which is because, they are unable to respond rapidly to measurements of dynamic and transient conditions. This is due to the fact that these instruments have moving parts that are rigid, heavy and bulky and consequently have a large mass. Mass presents inertia problems and hence these instruments cannot faithfully follow the rapid changes which are involved in dynamic measurements. Thus it would be virtually impossible to measure a 50 Hz voltage by using a mechanical instrument but it is relatively easy to measure a slowly varying pressure using these instruments. Another disadvantage of mechanical instruments is that most of them are a potential source of noise and cause noise pollution.

### 1.5.2 Electrical Instruments

Electrical methods of indicating the output of detectors are more rapid than mechanical methods. It is unfortunate that an electrical system normally depends upon a mechanical meter movement as indicating device. This mechanical movement has some inertia and therefore these instruments have a limited time (and hence, frequency) response. For example, some electrical recorders can give full scale response in 0.2 s, the majority of industrial recorders have responses of 0.5 to 24 s. Some galvanometers can follow 50 Hz variations, but even these are too slow for present day requirements of fast measurement.

### 1.5.3 Electronic Instruments

These days most of the scientific and industrial measurements require very fast responses. The mechanical and electrical instruments and systems cannot cope up with these requirements. The necessity to step up response time and also the detection of dynamic changes in certain parameters, which require the monitoring time of the order of ms and many a times, μs, have led to the design of today's electronic instruments and their associated circuitry. These instruments require use of semi-conductor devices. Since in electronic devices, the only movement involved is that of electrons, the response time is extremely small on account of very small inertia of electrons. For example, a Cathode ray oscilloscope (CRO) is capable of following dynamic and transient changes of the order of a few ns ($10^{-9}$ s).

Electronically controlled power supplies are used to provide stable voltages for studies in the field of chemical reactions and nuclear instrumentation. Electronic instruments are steadily becoming more reliable on account of improvements in design and manufacturing processes of semi-conductor devices. Another advantage of using electronic devices is that very weak signals can be detected by using pre-amplifiers and amplifiers. The foremost importance of the electronic instruments is the power amplification provided by the electronic amplifiers, which results in higher sensitivity. This is particularly important in the area of *Bio-instrumentation* since Bio-electric potentials are very weak *i.e.*, lower than 1 mV. Therefore, these signals are too small to operate electro-mechanical devices like recorders and they must be amplified. Additional power may be fed into the system to provide an increased power output beyond that of the input. This has been only possible through the use of electronic amplifiers, which have no important mechanical counterpart. This is particularly important where the data presentation devices use stylus type recorders, galvanometers, cathode ray oscilloscopes and magnetic tape recorders. It is a fact that hydraulic and pneumatic systems may be used for power amplification of signals. However, their use is limited to slow acting control applications like servo-systems, chemical processes and power systems. Another advantage of electronic instruments is the ability to obtain indication at a remote location which helps in monitoring inaccessible or dangerous locations. The most important use of electronic instruments is their usage in measurement of non-electrical quantities, where the non-electrical quantity is converted into electrical form through the use of transducers. Electronic instruments find extensive use in detection of electro-magnetically produced signals such as radio, video, and microwave. Electrical and electronic instruments are particularly useful in the intermediate signal modifying stage. Electronic instruments are light compact, have a high degree of reliability and their power consumption is very low.

Communications is a field which is entirely dependent upon the electronic instruments and associated apparatus. Space communications, especially, makes use of air borne transmitters and receivers and job of interpreting the signals is left entirely to the electronic instruments.

Electronic instruments make it possible to build analog and digital computers without which the modern developments in science and technology are virtually impossible. Computers require a very fast time response and it is only possible with use of electronic instruments.

Summarizing, it may be stated that in general electronic instruments have

(*i*) a higher sensitivity, (*ii*) a faster response,
(*iii*) a greater flexibility, (*iv*) lower weight,
(*v*) lower power consumption and
(*vi*) a higher degree of reliability than their mechanical or purely electrical counterparts.

## 1.6 CLASSIFICATION OF INSTRUMENTS

There are many ways in which instruments can be classified. Broadly, instruments are classified into *two* categories :

(*i*) Absolute Instruments, and
(*ii*) Secondary Instruments.

**1. *Absolute instruments.*** These instruments give the magnitude of the quantity under measurement in terms of physical constants of the instrument. The examples of this class of instruments are Tangent Galvanometer and Rayleigh's Current Balance.

**2. *Secondary instruments.*** These instruments are so constructed that the quantity being measured can only be measured by observing the output indicated by the instrument. These instruments are calibrated by comparison with an absolute instrument or another secondary instrument which has already been calibrated against an absolute instrument.

Working with absolute instruments for routine work is time consuming since every time a measurement is made, it takes a lot of time to compute the magnitude of the quantity under measurement. Therefore secondary instruments are most commonly used. Absolute instruments are seldom used except in standards institutions while secondary instruments find usage almost in every sphere of measurement. A voltmeter, a glass thermometer and a pressure gauge are typical examples of secondary instruments.

## 1.7 DEFLECTION AND NULL TYPE INSTRUMENTS

Instruments can be classified into two major categories – deflection and null types depending upon the way they present the result of measurements.

### 1.7.1 Deflection Type

The instruments of this type, the deflection of the instrument provides a basis for determining the quantity under measurement. The measured quantity produces some physical effect with deflects or produces a mechanical displacement of the moving system of the instrument. An opposing effect is built in the instrument which tries to oppose the deflection or the mechanical displacement of the moving system. The opposing effect is closely related to the deflection or mechanical displacement which can be directly observed. The opposing effect is so designed that its magnitude increases with the increase of deflection or mechanical displacement of the moving system caused by the quantity under measurement. The balance is achieved when opposing effect equals to cause producing the deflection or mechanical movement. The value of the measured quantity can then be inferred from the deflection or mechanical displacement at the point of balance.

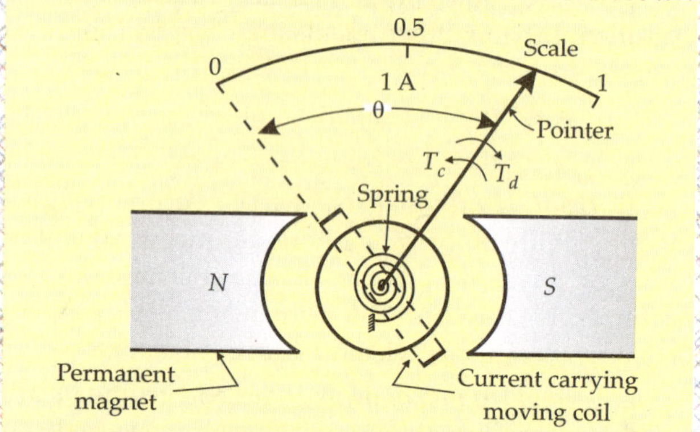


**Fig. 1.1** Deflection type instrument (PMMC type instrument).

For example, in a permanent magnet moving coil (PMMC) ammeter, the deflection of the moving coil is proportional to the current $I$, the quantity under measurement. The torque $T_d$ acting on the moving coil (*See* Fig. 1.1) is proportional to current $I$ or $T_d = GI$ where G is a constant, which is dependent upon flux density, number of turns and area of moving coil. The opposing effect is produced by a spring whose torque $T_c$ is proportional to deflection, θ or $T_c = K\theta$ where $K$ is the spring constant whose value depends upon the material and the dimensions of the spring. Under conditions of balance, $T_c = T_d$ or deflection $\theta = (G/K)I$ or current $I = (K/G)\theta$.

The value of measured quantity, current $I$, in this case depends upon the value of deflection θ, and the constants of the meter, $G$ and $K$. Thus, in order to make the instrument direct reading *i.e.*, it reads the value of current $I$ directly in terms of the deflection θ, it will have to be calibrated to take into consideration the values of $G$ and $K$.

Thus, we conclude, in a deflection type instrument, the value of measured quantity depends upon the calibration of the instrument.

### 1.7.2 Null Type

In a null type of instrument, a zero or null indication leads to determination of the magnitude of measured quantity. The null condition is dependent upon some other known conditions.

In contrast to deflection type of instruments, a null type instrument attempts to maintain the deflection at zero by suitable application of an effect opposing that generated by the measured quantity. Therefore, for the operation of a null type of instrument, the following are required :

(*i*) the effect produced by the measured quantity ;

(*ii*) the opposing effect, whose value is accurately known. This is necessary in order to determine the numerical value of the measured quantity accurately ;

(*iii*) a detector, which detects the null conditions *i.e.*, a device which indicates zero deflection (balance conditions) when the effect produced by the measured quantity is equal to the effect produced by the opposing quantity. The detector should be capable of displaying unbalance *i.e.*, a condition when the effect produced by the measured quantity is not equal to the opposing effect. Also, the detector should have means (automatic or manual) for restoring balance.

Figure 1.2 shows the elementary form of a d.c. potentiometer. It is a null type of instrument wherein an unknown emf, $E_x$, is measured. The slide wire of the potentiometer has been calibrated in terms of emf with the help of a standard emf source. The null detector is a current galvanometer whose deflection is proportional to the unbalance emf *i.e.*, the difference

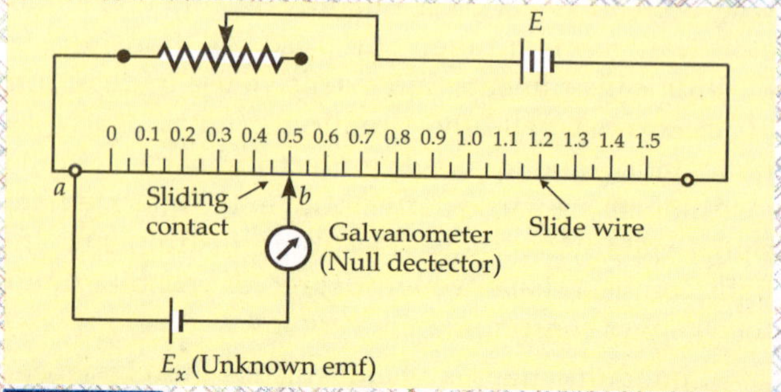


**Fig. 1.2** Null type instrument (D.C. potentiometer).

between the emf $E_{ab}$ across portion *a b* of slide wire and the unknown emf $E_x$. As soon as the two are equal, there is no current through the galvanometer and therefore it shows zero deflection thereby indicating null conditions. Therefore, the unknown emf $E_x$ is equal to $E_{ab}$, which is directly indicated by the calibrated scale placed alongside the slide wire.

### 1.7.3 Comparison of Deflection and Null Type Instruments

Comparing the advantages and disadvantages of the *two* types instruments.

(*i*) The accuracy of null type of instruments is higher than that of deflection type. This is because the opposing effect is calibrated with the help of standards which have high degree of accuracy. On the other hand, accuracy of deflection type of instruments is dependent upon their calibration which depends upon the instrument constants which are normally not known to a high degree of accuracy.

(*ii*) In the null type of instruments, the measured quantity is balanced out. This means the detector has to cover a small range around the balance (null) point and therefore can be made highly sensitive. Also the detector need not be calibrated since it has only to detect the presence and direction of unbalance and not the magnitude of unbalance. On the other hand, a deflection type of instrument must be larger in size, more rugged, and thus less sensitive if it is to measure large magnitude of unknown quantity.

(*iii*) Null type of instruments require many manipulations before null conditions are obtained and hence are apparently not suitable for dynamic measurements wherein the measured quantity changes with time. On the other hand, deflection type of instruments can follow the variations of the measured quantity more rapidly and hence are more suitable for dynamic measurements on account of their faster response. However, there are commercially automatic control instruments (such as self balancing potentiometers) that maintain a continuous null under rapidly changing conditions and thereby eliminate the need for manipulative operations.

Summarizing the above :

(*i*) Null type of instruments are more accurate than deflection type instruments.

(*ii*) Null type instruments can be highly sensitive as compared with deflection type instruments.

(*iii*) Deflection type of instruments are more suited for measurements under dynamic conditions than null type of instruments whose intrinsic response is slower.

## 1.8 ANALOG & DIGITAL MODES OF OPERATION

1

Secondary instruments work in *two* modes :

(*i*) Analog mode, and (*ii*) Digital mode.

Signals that vary in a continuous fashion and take on an infinite number of values in any given range are called **analog signals**. The devices which produce these signals are called **analog devices**.

In contrast, the signals which vary in discrete steps and thus take up only finite different values in a given range are called **digital signals**. The devices that produce such signals are called **digital devices**.

Let us elaborate further on digital and analog instruments and systems. In an analog system the function varies continuously. A typical example of variation is shown in Fig. 1.3. On the other hand, the digital values are **discrete and vary in equal steps.** Each digital number is a fixed sum of equal steps which is defined by the number.

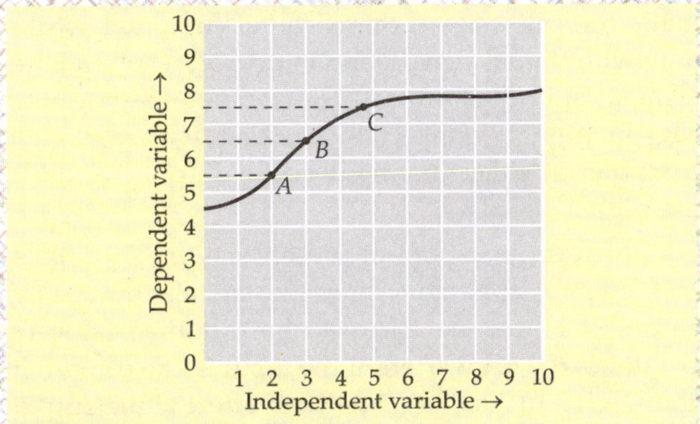


**Fig. 1.3** Representation of an analog quantity.

In order to convert an analog quantity into a digital number, the vertical displacement must be divided into equal parts. For example, in Fig. 1.3, the vertical quantities are divided into 10 equal parts and each part has a length of 1 unit. When dealing with digital numbers, a quantity between 0 to 0.5 is 0 while a quantity between 0.5 to 1.5 is 1 and a quantity between 1.5 to 2.5 is 2. For example, a point *A* on the analogue curve is 5.5 from the origin but in digital system it would be read as 5. From *A* to *B* is 6 and from *B* to *C* is 7. It apparently seems that if we adopt digital system, the errors involved will be considerable. But if we divide each of the 10 steps into 10 equal parts, we get 100 steps instead of 10. And if these 100 steps are further divided into 10 parts each, we will have 1000 steps. This gives much better accuracy in converting analog quantities into digital numbers. We can go on subdividing further and further till the desired accuracy is achieved. But it should be kept in mind that a digital number is still a sum of equal units.

In a digital system, magnitude lying within one of these steps lose their identity and are all defined by the same number. For example, if we have ten steps, numbers lying between 2.5 to 3.5 *i.e.*, 2.6, 2.7, 2.8, 2.9, 3.0, 3.1, 3.2, 3.3, 3.4, would all be read as 3.

From the above discussion we conclude that the difference between analog and digital information is that the analog output is a continuous function while the digital output is a discrete number of units. The last digit of any digital number is rounded to ± 0.5 of the last digit. It should also be marked that the magnitude of the digital quantity is measured only at the instant the reading is taken. One reading persists till another reading is taken (unlike the analog quantity which is a continuous function).

The majority of present day instruments are analog type. The importance of digital instruments is increasing, mainly because of the increasing use of digital computers in both data processing and automatic control systems. Since digital computer works only with digital signals, any information supplied to it must be in digital form. The computer's output is also in digital form. Thus working with a digital computer at either the input or the output, we must use digital signals.

However, most of our present day measurement and control apparatus produce signals of analog nature, it is thus necessary to have both **Analog to Digital (A/D) Converters** at the input to the computer and **Digital to Analog (D/A) Converters** at the output of the computer.

## 1.9 FUNCTIONS OF INSTRUMENTS AND MEASUREMENT SYSTEMS

There is another way in which instruments or measurement systems may be classified. This classification is based upon the functions they perform. The *three* main functions are explained below :

**1.** ***Indicating function.*** Instruments and systems use different kinds of methods for supplying information concerning the variable quantity under measurement. Most of the time this information is obtained as a deflection of a pointer of a measuring instrument. In this way, the instrument performs a function which is commonly known as **indicating function**. For example, the deflection of pointer of a speedometer indicates the speed of the automobile at that moment. A pressure gauge is used for indicating pressure.

**2.** ***Recording function.*** In many cases the instrument makes a written record, usually on paper, of the value of the quantity under measurement against time or against some other variable. Thus the instrument performs a **recording function**. For example, a potentiometric type of recorder used for monitoring temperature records the instantaneous values of temperatures on a strip chart recorder.

**3.** ***Controlling function.*** This is one of the most important functions especially in the field of industrial control processes. In this case, the information is used by the instrument or the system to control the original measured quantity.

Thus there are three main groups of instruments. The largest group has the indicating function. Next in line is the group of instruments which have both indicating and or recording functions. The last group falls into a special category and performs all the three functions, *i.e.*, indicating, recording and controlling.

In this text, main emphasis is laid upon instruments whose functions are mainly indicating and recording, especially those instruments which are used for engineering analysis purposes. The control function will be analyzed in those cases where controlling enters as an integral part of the indicating and recording functions of instrumentation. The examples of controlling instruments are thermostats for temperature control and floats for liquid level control.

## 1.10 APPLICATIONS OF MEASUREMENT SYSTEMS

In order to build up background for our later detailed study of measuring instruments and systems and their characteristics, it is useful to discuss, in general, the various ways these instruments are put to use. The way the instruments and measurement systems are used for different applications are as under :

- Monitoring of processes and operations,
- Control of processes and operations, and
- Experimental Engineering analysis.

**1.** ***Monitoring of processes and operations.*** There are certain applications of measuring instruments that have essentially a monitoring function. They simply indicate the value or condition of parameter under study and their readings do not serve any control functions. For example, an ammeter or a voltmeter indicates the value of current or voltage being monitored (measured) at a particular instant. Similarly, water and electric energy meters installed in homes keep track of commodity used so that later on its cost may be computed to be realized from the user.

**2.** ***Control of processes and operations.*** A very useful application of instruments is in automatic control systems. There has been a very strong association between measurement and control.

In order that process variables like temperature, pressure, humidity, etc. may be controlled, the prerequisite is that they can be measured at the desired location in the individual plants. Same is true of servo-systems, *i.e.,* systems connected with measurement of position, velocity and acceleration.

A block diagram of a simple control system is shown in Fig. 1.4. Let us assume that the output variable to be controlled is non-electrical and the control action is through electrical means. The input is reference which corresponds to the desired value of the output. The input is compared with the output with the help of a comparator. The output is a non-electrical quantity and is converted into a corresponding electrical form by a **transducer** connected in the feedback loop. In case the input and output differ, there is a resultant error signal. This error signal is amplified and then fed to an **actuator**, which produces power to drive the controlled circuitry.

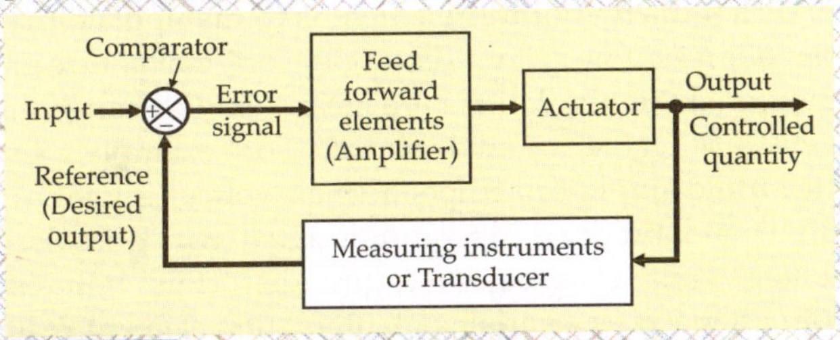


**Fig. 1.4** Block diagram of a simple control system.

The corrective action goes on till the output is at the same level as the input which corresponds to the desired output. At this stage, there is no error signal and hence there is no input to the actuator and the control action stops.

Examples of this type of application are numerous. A common one is the typical refrigeration system which employs a thermostatic control. A temperature measuring device (often a bimetallic element) senses the room temperature, thus providing the information necessary for proper functioning of the control system.

**3. *Experimental engineering analysis.*** For solution of engineering problems, theoretical and experimental methods are available. Many applications require application of both the methods. The relative affability of the method depends upon the nature of the problem. Experimental engineering analysis has many uses and some are listed below :

(*i*) Testing the validity of theoretical predictions.

(*ii*) Formulations of generalized empirical relationships in cases where no proper theoretical backing exists.

(*iii*) Determination of system parameters, variables and performance indices.

(*iv*) For development in important spheres of study where there is ample scope of study.

(*v*) Solutions of mathematical relationships with the help of analogies.

## 1.11 TYPES OF INSTRUMENTATION SYSTEMS

The advent of microprocessors has completely revolutionized the field of instrumentation and control. Microprocessor based systems are increasingly being used for dedicated applications in process instrumentation. The introduction of microprocessors results a new classification of instrumentation systems and these are :

- Intelligent Instrumentation systems, and
- Dumb Instrumentation systems.

**1. *Intelligent instrumentation.*** This term has to come mean the use of an instrumentation system to evaluate a physical variable employing a digital computer to perform all or nearly all signal and information processing. In this system after a measurement has been made of the variable, further processing whether in digital or analog form is carried out to refine the data, for the purpose of presentation to an observer or to other computers.

**2. *Dumb instrumentation.*** In this system once the measurement is made, the data must be processed by the observer.

## 1.12 INFORMATION AND SIGNAL PROCESSING

The two terms *"Information"* and *"Signals"* are usually considered as synonymous. However, they have precisely different definitions.

- **Information.** It is the data or details relating to an object or event.
- **Signals.** They carry the information about magnitude or time relating to an object or event *i.e.,* a physical quantity.

## 1.13 ELEMENTS OF A GENERALIZED MEASUREMENT SYSTEM

It is important to have a systematic organization and analysis of measurement systems. An instrument may be defined as a device or a system which is designed to maintain a functional relationship between prescribed properties of physical variables and must include ways and means of communication to a human observer. The functional relationship remains valid only as long as the static calibration of system remains constant. On the other hand, the performance of a measurement system can be described in terms of static and dynamic characteristics.

It is possible and desirable to describe the operation of a measuring instrument or a system in a generalized manner without resorting to intricate details of the physical aspects of a specific instrument or a system. The whole operation can be described in terms of functional elements.

Most of the measurement systems contain *three* main functional elements. They are :

1. Primary sensing element,
2. Variable conversion element, and
3. Data presentation element.

Each functional element is made up of a distinct component or groups of components which perform the required and definite steps in the measurement. These may be taken as basic elements, whose scope is determined by their functioning rather than their construction.

### 1.13.1 Primary Sensing Element

The quantity under measurement makes its first contact with the primary sensing element of a measurement system. In other words, the measurand is first detected by primary sensor. This act is then immediately followed by the conversion of measurand into an analogous electrical signal. This is done by a **transducer**. A transducer in general, is defined as a device which converts energy from one form to another. But in Electrical measurement systems, this definition is limited in scope. *A transducer is defined as a device which converts a physical quantity into an electrical quantity*. The physical quantity to be measured, in the first place is sensed and detected by an element which gives the output in a different analogous form. This output is then converted into an electrical signal by a transducer. This is true of most of the cases but is not true for all. In many cases the physical quantity is directly converted into an electrical quantity by a transducer. The first stage of a measurement system is known as a **detector transducer stage**.

### 1.13.2 Variable Conversion Element

The output of the primary sensing element may be electrical signal of any form. It may be a voltage, a frequency or some other electrical parameter. Sometimes this output is not suited to the system. For the instrument to perform the desired function, it may be necessary to convert this output to some other suitable form while preserving the information content of the original signal. We may cite an example. Suppose output is in analog form and the next stage of the system accepts input signals only in digital form and therefore, an A/D converter will have to be used for converting signals from analog to digital form for them to be acceptable for the next stage of the system.

Many instruments do not need any variable conversion element, while others need more than one variable conversion element.

### 1.13.3 Variable Manipulation Element

The function of this element is to manipulate the signal presented to it preserving the original nature of the signal. Manipulation here means only a change in *numerical* value of the signal. For example, an electronic amplifier accepts a small voltage signal as input and produces an output signal which is also voltage but of greater magnitude. Thus voltage amplifier acts as a variable manipulation element. It is not necessary that a variable manipulation element should follow the variable conversion element as shown in Fig. 1.5. It may precede the variable conversion element in many cases. In case, the voltage is too high, attenuators are used which lower the voltage or power for the subsequent stages of the system. Further the impedance of the attenuator to be matched to the input and output circuits and further for multi ratio attenuator these impedance ratios are constant irrespective of the ratio setting.

As discussed earlier, the output of transducer contains information needed for further processing by the system and the output signal is usually a voltage or some other form of electrical signal. The two most important properties of voltage are its magnitude and frequency though polarity may be a consideration in some cases. Many transducers develop low voltages of the order of mV and some even μV. A fundamental

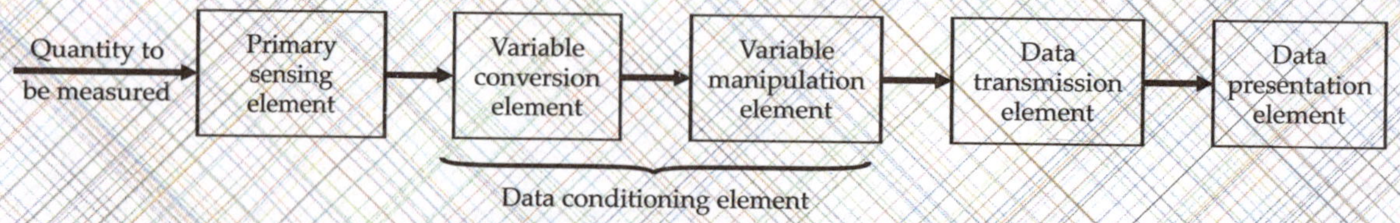


**Fig. 1.5** Functional elements of an instrumentation system.

problem is to prevent this signal being contaminated by unwanted signals like noise due to an extraneous source which may interfere with the original output signal. Another problem is that a weak signal may be distorted by processing equipment. The signal after being sensed cannot be directly transmitted to the next stage without removing the interfering sources, as otherwise highly distorted results may be obtained which are far from true. Many a times it becomes necessary to perform certain operations on the signal before it is transmitted further. These processes may be linear like amplification, attenuation, integration, differentiation, addition and subtraction. Some non-linear processes like modulation, detection, sampling, filtering, chopping and clipping etc. are also performed on the signal to bring it to the desired form to be accepted by the next stage of measurement system. This process of conversion is called **Signal Conditioning**. The term signal conditioning includes many other functions in addition to variable conversion and variable manipulation. In fact the element that follows the primary sensing element in any instrument or measurement system is called **Signal Conditioning Element.**

When the elements of an instrument are actually physically separated, it becomes necessary to transmit data from one to another. The element that performs this function is called a **Data Transmission Element**. For example, space-crafts are physically separated from the earth where the control stations guiding their movements are located. Therefore control signals are sent from these stations to space-crafts by a complicated telemetry systems using radio signals.

The signal conditioning and transmission stage is commonly known as **Intermediate Stage**.

### 1.13.4 Data Presentation Element

The information about the quantity under measurement has to be conveyed to the personnel handling the instrument or the system for monitoring, control, or analysis purposes. The information conveyed must be in a form intelligible to the personnel or to the intelligent instrumentation system. This function is done by data presentation element. In case data is to be monitored, visual display devices are needed. These devices may be analog or digital indicating instruments like ammeters, voltmeters etc. In case the data is to be recorded, recorders like magnetic tapes, high speed camera and T.V. equipment, storage type C.R.T., printers, analog and digital computers or microprocessors may be used. For control and analysis purpose microprocessors or computers may be used.

1

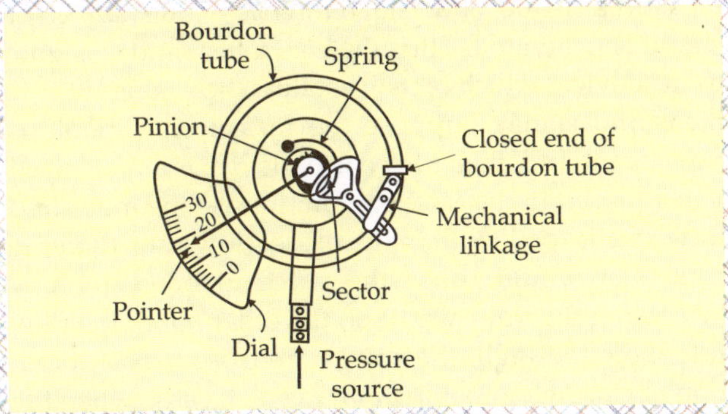


**Fig. 1.6** Bourdon tube pressure gauge.

The final stage in a measurement system is known as **terminating stage**. As an example of a measurement system, consider the simple bourdon tube pressure gauge as shown in Fig. 1.6. This gauge offers a good example of a measurement system. In this case the bourdon tube acts as the primary sensing element and a variable conversion element. It senses the input quantity (pressure in this case). On account of the pressure the closed end of the bourdon tube is displaced. Thus the pressure is converted into a small displacement. The closed end of the bourdon tube is connected through mechanical linkage to a gearing arrangement. The gearing arrangement amplifies the small displacement and makes the pointer to rotate through a large angle. The mechanical linkage thus acts as a data transmission element while the gearing arrangement acts as a data manipulation element.

The final data presentation stage consists of the pointer and dial arrangement, which when calibrated with known pressure inputs, gives an indication of the pressure signal applied to the bourdon tube. The schematic diagram of this measurement system is given in Fig. 1.7.

When a control device is used for the final measurement stage, it is necessary to apply some feedback to the input signal to accomplish the control

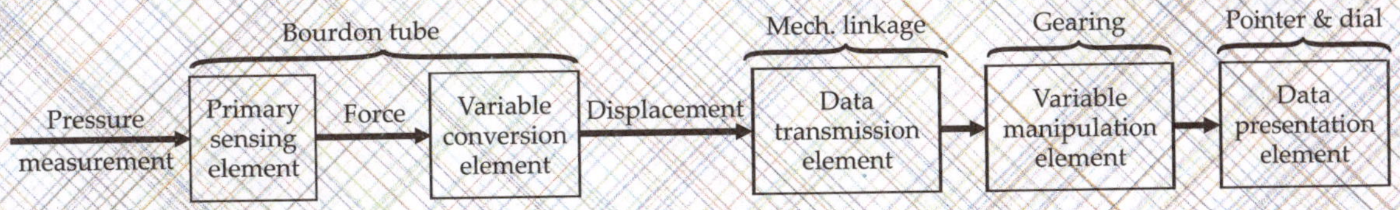


**Fig. 1.7** Schematic diagram of a Bourdon tube pressure gauge.

objectives. The control stage compares the signal representing the measured variable with a reference signal of the same form. This reference signal has a value the measured signal should have and is presented to a controller. If the measured signal agrees with the reference value, the controller does nothing. However, if there is a difference between the measured value and the reference value, an error signal is generated. Thus the controller sends a signal to a device which acts to alter the value of the measured signal. Suppose the measured variable is flow of a liquid, then the control device is a motorized valve placed in the flow system. In case the measured flow rate is too low than the preset flow rate, then the controller would cause the valve to open, thereby increasing the flow rate. If on the other hand, the flow rate were too high, the valves are closed. The operation of closing or opening of valve will cease when the output flow rate is equal to preset value of flow rate.

## 1.14 INPUT-OUTPUT CONFIGURATIONS OF MEASURING INSTRUMENTS AND MEASUREMENT SYSTEMS

A generalised configuration in instruments and measurement systems which brings out a significant input-output relationship present in them is shown in Fig. 1.8. Input quantities are classified into *three* categories :

(*i*) desired inputs, (*ii*) interfering inputs,
(*iii*) modifying inputs.

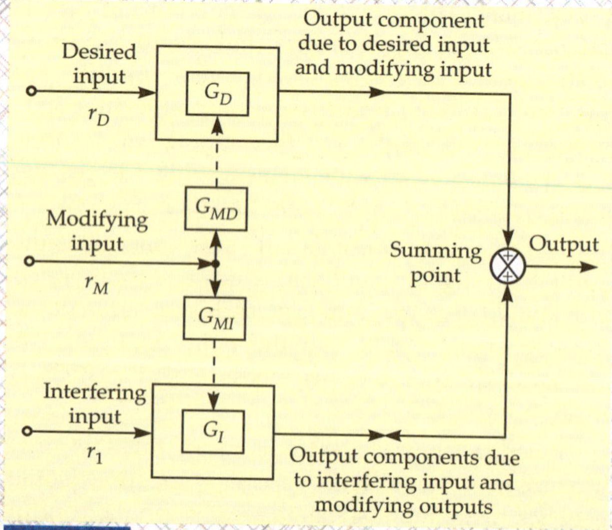


**Fig. 1.8** Generalised input-output configuration of measurement systems.

### 1.14.1 Desired Inputs

*Desired inputs* are defined as quantities for which the instrument or the measurement system is specifically designed to measure and respond. The desired input, $r_D$ produces an output, component $c_D = G_D r_D$ in accordance with an input-output relationship symbolised by a mathematical operator, $G_D$, which is defined as *Transfer Function*.

$G_D$ is necessarily a mathematical operation to get an output from a desired input. Thus if an input '$r$' is operated upon by a transfer function '$G$', the output is $c = Gr$. The transfer function may simply be a constant, $K$, which multiplies the static input, $r_D$, to get an output $c_D = K r_D$ to obtain either an amplified or an attenuated output in linear systems. It should be understood that a constant cannot be used for describing the input-output relationships for non-linear systems. For non-linear systems, the transfer function is represented by either an algebraic or a transcendental function. The input-output relationships for systems subjected to dynamic inputs are represented by differential equations.

In case, a description of the output 'scatter' or dispersion for repeated equal static inputs is desired, a statistical function is needed to represent the input-output relationship.

The transfer function, $G_D$, is therefore representative of a wide range of functions from a constant in the case of linear systems to a statistical function used for statistical measurements.

### 1.14.2 Interfering Inputs

*Interfering inputs* represent quantities to which an instrument or a measurement system becomes unintentionally sensitive. The instruments or measurement systems are not desired to respond to interfering inputs but they give an output due to interfering inputs on account of their principle of working, design and many other factors like the environments in which they are placed. The interfering input $r_I$ is operated upon by a transfer function $G_I$ to produce an output in the same manner as a desired input is operated upon by a transfer function, $G_D$, to produce an output.

### 1.14.3 Modifying Inputs

This class of inputs can be included among the interfering inputs. However, a separate classification is essential since such a classification is more significant. *Modifying inputs* are defined as inputs which cause a change in input-output relationships for either desired inputs or interfering inputs or for both. Thus, a modifying input, $r_M$, is an input that modifies $G_D$ and/or $G_I$. The symbols $G_{MD}$ and $G_{MI}$ represent the specific manner in which $r_M$ affects $G_D$ and $G_I$ respectively. These symbols, $G_{MD}$ and $G_{MI}$ are interpreted in the same general way as $G_D$ and $G_I$ are.

The block diagram shown in Fig. 1.8 depicts the concepts described above, the circle with a cross in it represents the *summing point*. The two plus signs shown indicate that the output of the summing point is the sum of the instantaneous values of its two inputs. An instrument or a measurement system may be subjected to desired, interfering and modifying inputs and therefore the output is the sum of instantaneous output components resulting from these inputs. An instrument or a measurement system may have several inputs of each of these three types and thus have several outputs. Hence, it may be necessary to draw more complex block diagrams than that shown in Fig. 1.8.

Let us consider an case of measurement of differential pressure of a fluid with the help of a *U*-tube manometer using mercury as manometric fluid as shown in Fig. 1.9. The pressures applied at the two ends of the *U*-tube are $P_1$ and $P_2$. The differential pressure is $P = P_1 - P_2$ and is given by the following expression

$$P = P_1 - P_2 = gh(\rho_m - \rho_f) \qquad ...(1.1)$$

where $\rho_m, \rho_f$ = density of mercury, and fluid respectively ; $kg/m^3$,

$g$ = gravitational constant $ms^{-2}$,

$h$ = scale regarding, m.

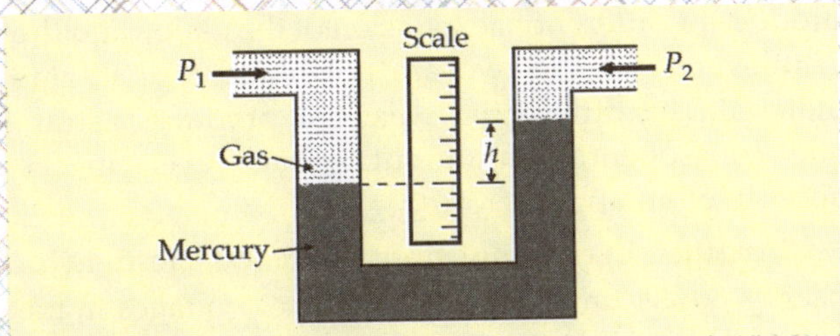


**Fig. 1.9** Measurement of differential pressure with manometer.

Let us consider that the fluid is a gas and therefore the density $\rho_f$ is negligible as compared with $\rho_m$. Therefore, Eqn. 1.1 can be written as :

$$P = P_1 - P_2 = g\rho_m h \qquad ...(1.2)$$

$P_1$ and $P_2$ are the desired inputs while scale reading, $h$, is the output. The product $g\rho_m$ relates the output with input.

In case $P_1 = P_2$, the differential pressure is zero and therefore output *i.e.*, scale reading, $h$ should be zero. However, there are examples when the scale reading, $h$ is not zero even though $P_1 = P_2$, the differential pressure, *i.e.*, desired input is zero. This is on account of presence of interfering inputs, action of two of which is shown in Figs. 1.10 and 1.11. Figure 1.10 shows a manometer which is placed on a vehicle that is accelerating. Although the pressures at the two ends of the manometer are equal, a differential pressure is created on account of acceleration and therefore the scale indicates a reading, $h$. Thus, acceleration acts as an interfering input.

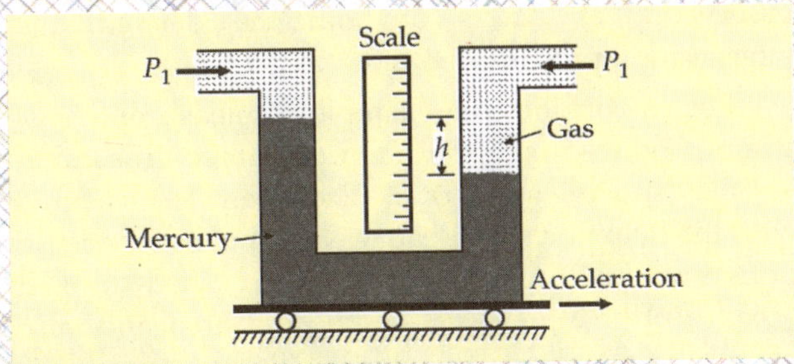


**Fig. 1.10** Acceleration as an interfering input.

Figure 1.11 shows another case of interfering input. In this case, the manometer is not properly aligned with the gravity vector because of the angle of tilt. Therefore, there is an output, $h$, even though there is no differential pressure. Hence, angle of tilt, $\theta$, acts as an interfering input.

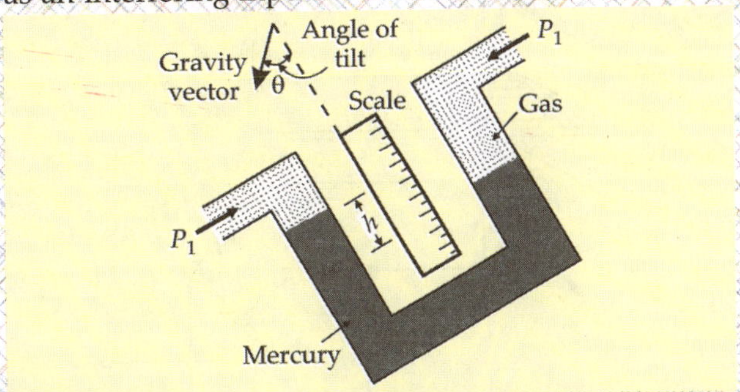


**Fig. 1.11** Angle of tilt as an interfering input.

The modifying inputs for the manometer are ambient temperature and gravitational force. Ambient temperature manifests in many ways in modifying input-output relationship as given in Eqn. 1.2. Some of these are :

(*i*) changes in ambient temperature change the length of calibrated scale thereby modifying the proportionality factor relating the input $(P_1 - P_2)$ with output $h$.

(*ii*) changes in ambient temperature change the value density of mercury, $\rho_m$ and therefore the proportionality factor relating the input with the output is changed.

It is clear from Eqn. 1.2, that the proportionality factor relating input with output depends upon the value of $g$. Now the value of $g$ is different at different locations. Therefore, a change in gravitational force due to the changes in location of manometer like using it at different heights above sea level or putting it in a space ship leads to modification of the input-output for the desired inputs. Also, since $g$ can also be interfering input due to angle of tilt as explained before, there is modification in the input-output relationship for the interfering input as well. Thus, it is observed that effects of both the desired and interfering inputs may be modified.

Let us consider the example of measurement of strain produced on application of a force on a cantilever as shown in Fig. 1.12. The strain is measured with the help of a strain gauge, which consists of grid of fine wire whose resistance changes when strained because of change in its length, and area and, also because of piezo-resistive effect. The strain gauge is firmly cemented to the cantilever at the point where the strain is to be measured. The strain gauge forms one arm of a Wheatstone bridge. Let the unstrained resistance of the strain gauge be $R_1 = R$. The other three arms of the bridge consist of resistances $R_2$, $R_3$ and $R_4$. The bridge is balanced if $R_1 / R_3 = R_2 / R_4$.

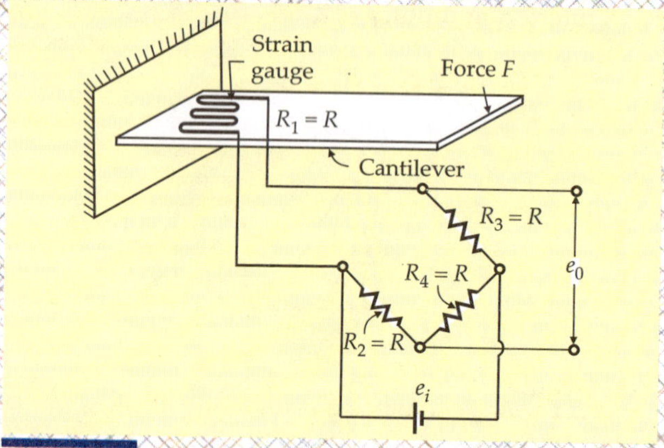


**Fig. 1.12** Measurement of strain using Wheatstone bridge.

Let $R_2 = R_3 = R_4 = R$ and therefore when no strain is applied the bridge is balanced and the output voltage, $e_0$, is zero. When strained, the resistance of the strain gauge changes according to the following relationship :

$$\Delta R = G_f \varepsilon R \quad \text{(See Eqn. 25.68 )} \qquad ...(1.3)$$

where, $\Delta R$ = change in resistance of strain gauge ; $\Omega$,

$G_f$ = gauge factor ; and $\varepsilon$ = strain.

From Eqn. 1.3, it is clear that the change in resistance of strain gauge is proportional to strain. Thus if the change in resistance of the strain gauge is measured, the value of strain can be computed. The change in resistance of strain gauge is measured with the help of a Wheatstone bridge as shown in Fig. 1.12. This bridge has four arms, the strain gauge forms one of the arms, each of rest of the three arms consists of a resistance whose value is equal to $R$, the resistance of the strain gauge when it is not strained.

When no force is applied to the cantilever, the strain is zero and the resistance of the strain gauge is $R$. Since, the resistance in each of the other three arms of the bridge is $R$, the bridge is balanced and there is no output from the bridge *i.e.*, output voltage $e_0 = 0$.

When a force, $F$, is applied to the cantilever, it causes a strain, $\varepsilon$, which in turn produces a change, $\Delta R$, in the resistance of strain gauge. This unbalances the bridge, producing an output voltage, $e_0$, which is proportional to strain, $\varepsilon$, and can be measured using a meter or an oscilloscope. The output voltage is given by :

$$e_0 = \frac{G_f}{4} \varepsilon e_i \quad \text{(See Eqn. 29.6)} \qquad ...(1.4)$$

The desired input is obviously strain, $\varepsilon$, which produces a proportional output which is voltage $e_0$.

One of the interfering inputs is temperature. Suppose there is a change in temperature at the place where the strain gauge is located. It will cause a change in resistance of strain gauge thereby producing an output voltage even if the force applied to the cantilever is zero and there is no strain.

Temperature has another interfering effect. It produces a differential expansion of the gauge and the cantilever since the two are not made of the same material. This differential expansion produces a strain and consequently an output voltage even in the absence of an applied force.

Another interfering input is the 50 Hz field of nearby power lines. This field induces voltages in the strain gauge circuit producing an output voltage even when the strain is zero.

The battery voltage is a modifying input since it changes the proportionality factor between the desired input $\varepsilon$ and the output $\varepsilon_0$ and also between the interfering input (strain gauge temperature) and output $e_0$.

## 1.15 METHODS OF CORRECTION FOR INTERFERING AND MODIFYING INPUTS

It is normal practice to try and nullify or reduce the effects of interfering and modifying inputs. There are many methods available that are adopted at the designer's level and/or user's level. Some of these methods are described below.

1. ***Method of inherent insensitivity.*** In this method the instruments or the measurement systems are so designed that they are sensitive only to desired

inputs and are insensitive to both interfering and modifying inputs. This means that it is desirable to make $G_I$ and/or $G_{MD}$ as nearly equal to zero as possible so that even though $r_I$ and/or $r_M$ may exist they cannot affect the output.

The approach adopted is not practicable in many situations, but acts as a motivating factor in many cases to use ingenuity for elimination of effects of spurious inputs. For example, temperature is an interfering input in the case of measurement of strain with the help of strain gauges. The method of inherent insensitivity motivates one for finding and using materials for the strain gauges that have a very low (almost zero) resistance temperature co-efficient. If such a material could be found, the problem of interfering temperature inputs is partially solved.

**2. *Method of high gain feedback.*** Feedback systems play an important part in modern measurement and control systems because they can easily be adopted to perform their assigned tasks automatically. The most significant advantage of using a feedback system is that the use of feedback makes the system response relatively insensitive to modifying inputs.

A non-feedback (open loop) system is shown in Fig. 1.13(*a*). The transfer function for the desired inputs is $G_D$ and the desired input is $r_D$.

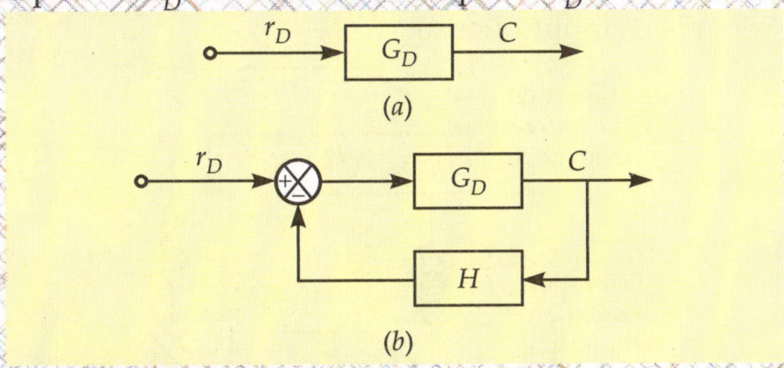


**Fig. 1.13** (*a*) Non-feedback (open loop) system. (*b*) Feedback (closed loop) system.

$\therefore$ Output due desired input $C = G_D r_D$

Suppose due to modifying input like changes in ambient temperature, the transfer function changes by $\Delta G_D$ and therefore the modified transfer function is $G_{MD} = G_D + \Delta G_D$. The output of the non-feedback systems changes to

$$C + \Delta C = (G_D + \Delta G_D) r_D = G_D r_D + \Delta G_D r_D$$

$\therefore$ Output due to change in transfer function on account of modifying input is,

$$\Delta C = \Delta G_D r_D \quad \text{...(1.5)}$$

Figure 1.13(*b*) shows a feedback (closed loop) system having $H$ as the transfer function of feedback elements.

The output corresponding to desired input is

$$C = \frac{G_D}{1 + G_D H} r_D$$

In the presence of modifying inputs, the output changes to

$$C + \Delta C = \frac{(G_D + \Delta G_D)}{1 + (G_D + \Delta G_D) H} r_D$$

$$= \left( \frac{G_D}{1 + G_D H + \Delta G_D H} + \frac{\Delta G_D}{1 + G_D H + \Delta G_D H} \right) r_D$$

$$= \frac{G_D}{1 + G_D H} r_D + \frac{\Delta G_D}{1 + G_D H} r_D \quad (\text{as } \Delta G_D << G_D)$$

$\therefore$ Output due to change in transfer function on account of modifying output is,

$$\Delta C = \frac{\Delta G_D}{1 + G_D H} r_D \quad \text{...(1.6)}$$

Hence, comparing Eqns. 1.5 and 1.6, it is clear that in comparison to the non-feedback system, the output in the feedback system, corresponding to change of transfer function because of modifying inputs is reduced by a factor $(1 + G_D H)$. Now, if the feedback systems are designed with a high gain feedback, $H$, the value of output on account of modifying inputs is considerably reduced.

**3. *Method of calculated output corrections.*** The use of this method requires the knowledge of magnitude of interfering and/or modifying inputs and also the mathematical relationships which describe the way these inputs affect the output. With this information it becomes possible to calculate the errors in output caused by the interfering and/or modifying inputs. Thus corrections can be applied to the measured output to compute the output corresponding to desired input. For example, for the case of measurement of differential pressure with the help of mercury manometer, the effects of temperature (an interfering and a modifying input) on the scale length and on the density of mercury can easily be calculated in case the temperature is known.

The gravitational force is also an interfering and modifying input for the manometers. The corrections for this spurious input can easily be computed provided the elevation and latitude of the place at which the manometer is used are known. The *method of calculated output corrections,* though applicable to any form of input, can only be used for inputs that are essentially constant.

**4. *Method of signal filtering.*** The method of signal filtering is based upon introduction of *filters* in instruments and measurement systems. The filter blocks the passage of unwanted signals in such a way that the output on account of spurious signals is either completely eliminated or is considerably reduced. The filter may be introduced into any of the three major stages of a measurement system, the choice, however, depends upon the application. Schemes involving *Input Filtering* have filters introduced into the input stage of measurement system for interfering and modifying signals. This is schematically shown in Fig. 1.14.

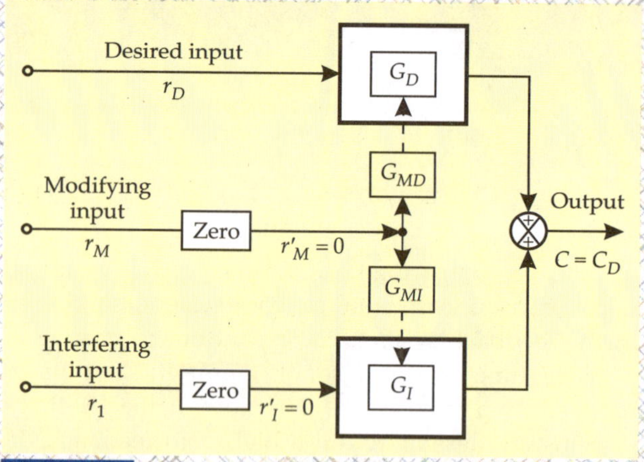


**Fig. 1.14** Schematic diagram of input filtering.

The interfering and the modifying inputs are passed through a filter whose transfer function is ideally zero and therefore the outputs $r_I'$ and $r_M'$ corresponding respectively to interfering input $r_I$ and the modifying input $r_M$ are zero. Therefore, the output that is available from the summing point is only on account of the desired input.

Figure 1.15 shows the schematic diagram for *Output Filtering*.

The output C, though a single signal, is

$$C = C_D + C_M + C_I$$

where $C_D$, $C_M$, $C_I$ are respectively the outputs corresponding to desired, modifying and interfering inputs. In many a situations, it is possible to selectively block $C_M$ and $C_I$ but allow $C_D$ to pass faithfully, the output $C_D' = C_D$, the output corresponding to the desired input. This is done by making transfer functions for $C_I$ and $C_M$ equal to zero.

Filters may take several forms. If a filter is put directly in the path of a interfering or a modifying input, it may be designed to reject completely the interfering and modifying signal. However, in most of the situations, the input to a filter contains both desired as well as the spurious inputs. In such situations the filters have to be designed so as to be selective *i.e.,* they should pass the desired signals and reject or considerably attenuate the spurious inputs.

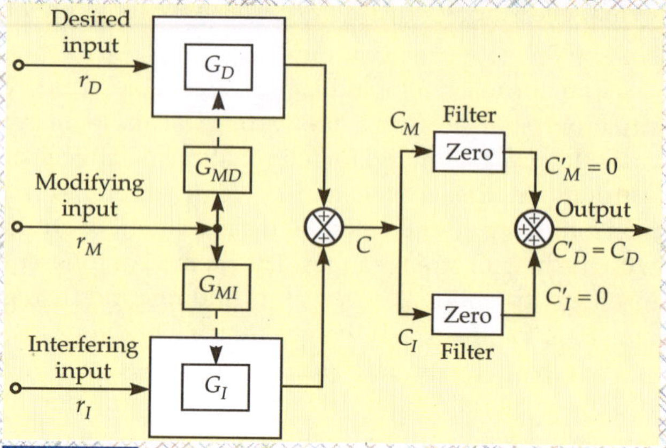


**Fig. 1.15** Schematic diagram for output filtering.

Electromechanical devices used for navigation, aircraft and missile applications should be free from any shock and vibrations since delicate instruments are attached to the structures of these devices. Figure 1.16 shows the manner in which the interfering vibration inputs may be filtered out. The mass-spring system is actually a mechanical filter which transmits only a fraction of the vibrations from the vibrating structure to the measuring instruments. This is an example of input filtering.

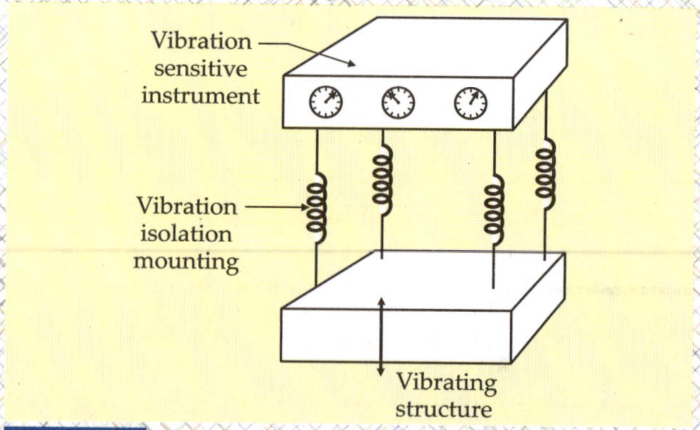


**Fig. 1.16** Isolation of vibration sensitive instruments.

A thermocouple is used for measurement of temperature as shown in Fig. 1.17. In order that the

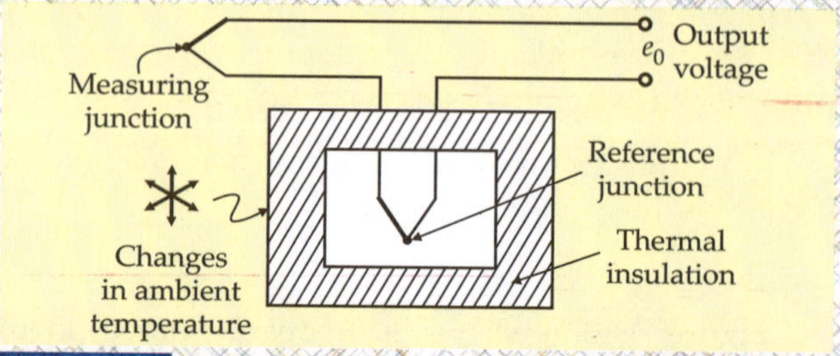


**Fig. 1.17** Isolation of thermocouple reference junction from changes in ambient temperature.

calibration of the thermocouple should hold good, any changes in the ambient temperature should not have any effect on the reference function. This needs a thermal insulation which acts as an *Input Filter*, for temperature or heat flow interfering inputs.

Figure 1.18 shows another example of input filtering. Here, the 50 Hz magnetic field tries to introduce an interfering input to a strain gauge bridge. The interfering magnetic field is prevented from the strain gauge with the help of a magnetic shield which may take the shape of a box made up of a ferromagnetic material.

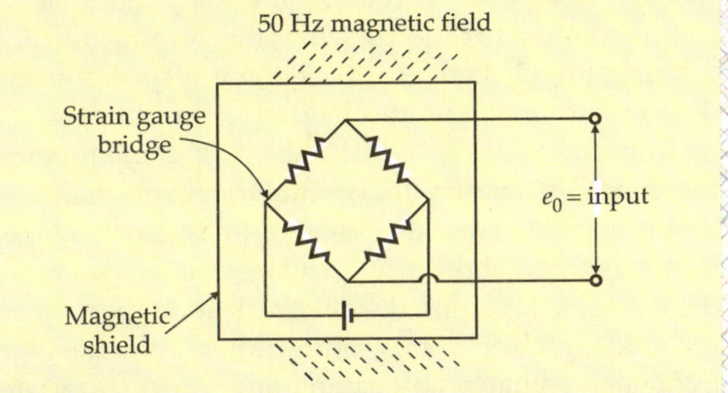


**Fig. 1.18** Magnetic shield for a strain gauge bridge for input filtering of power frequency fields.

Some examples of ***output filtering*** are given below :

Figure 1.19(*a*) shows an arrangement which is used for measurement of strains that are mainly steady and do not vary more rapidly than 1 Hz. Ideally, the output signal should vary at the same rate as the input. However, there is every likelihood that the output from the strain gauge bridge may pick up signals from 50 Hz power lines, thereby distorting the output. Therefore, it is essential that the 50 Hz signal eliminated so that there is no spurious signal applied to the measurement circuit.

Figure 1.19(*b*) shows the input signal to the measurement system which contains the desired signal arising out on account of the change in strain (at a frequency of 1 Hz) plus the 50 Hz power line interfering signal. A low-pass *RC* filter can be

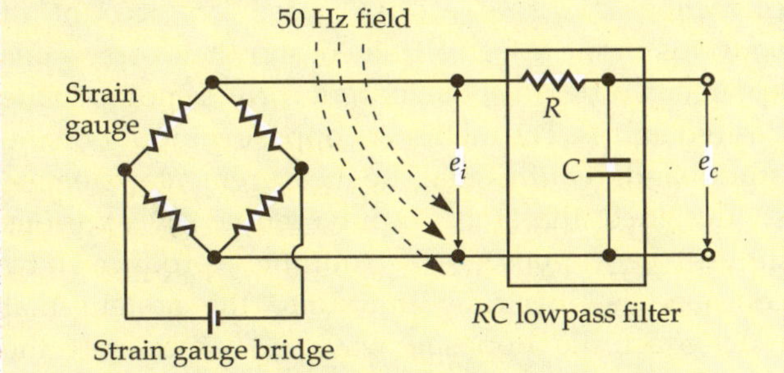


(*a*) Strain gauge circuit

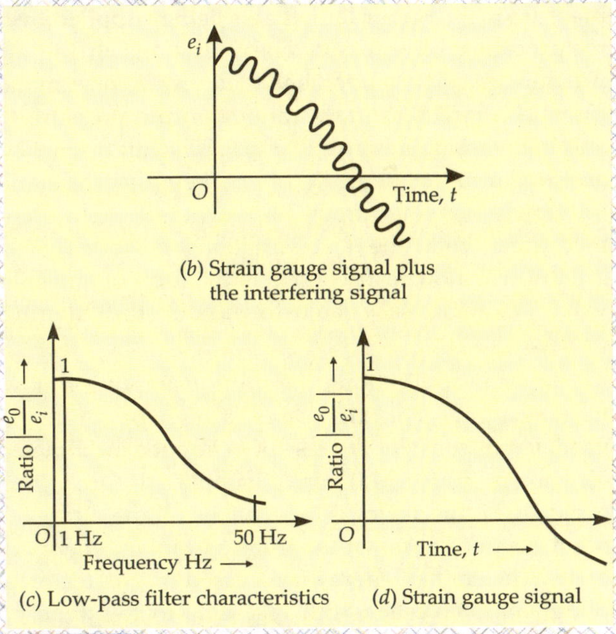


**Fig. 1.19** Output filtering of strain gauge bridge signal by a low-pass filter.

designed to have a characteristic as shown in Fig. 1.19(*c*) so that the desired signal (1 Hz in this case) is reproduced faithfully, while the high frequency signal of 50 Hz is greatly attenuated. The output, which is a faithful representation of the desired strain gauge input signal is shown in Fig. 1.19(*d*).

Figure 1.20(*a*) shows a pressure gauge whose response is modified by the insertion of a flow restriction between the source of the pressure and piston chamber as shown in Fig. 1.20(*b*). This arrangement is useful for measurement of average

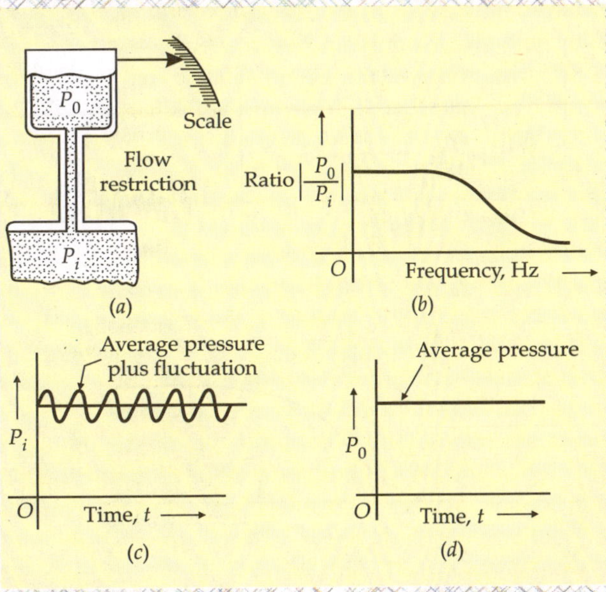


**Fig. 1.20** Measurement of average pressure.

pressure in a large air tank that is being supplied by a reciprocating compressor. The pulsations in the air pressure may be smoothened out by a pneumatic filter. The variations in the ratio of output pressure $P_0$ and in the input pressure $P_i$ *i.e.*, $|P_0 / P_i|$ with frequency are similar to that of an *RC* filter. Therefore, the static or slowly varying input pressures are accurately measured while the fast variations are attenuated greatly. The flow restriction is in the form of a needle valve allowing easy adjustments in the filtering process.

In some applications, the spurious input is a slow varying signal. For such cases, a high pass filter is used. Let us consider the case of a chopped radiometer as shown in Fig. 1.21 in a simplified form. The objective of this device is to sense the temperature, $\theta_s$, of an object in terms of the infrared radiant energy it emits. The emitted energy is focussed on a detector, a photo-emissive cell, which responds to this type of radiation. The radiant energy changes the temperature, $\theta_d$, of the photocell. Since the detector is sensitive to the changes in temperature, it will produce a voltage $e_i$ which is amplified to $e_0$ by an amplifier. Thus if $\theta_s$, changes, the temperature of the detector $\theta_d$ changes, with the result that $e_i$ changes and consequently the output voltage $e_0$ changes.

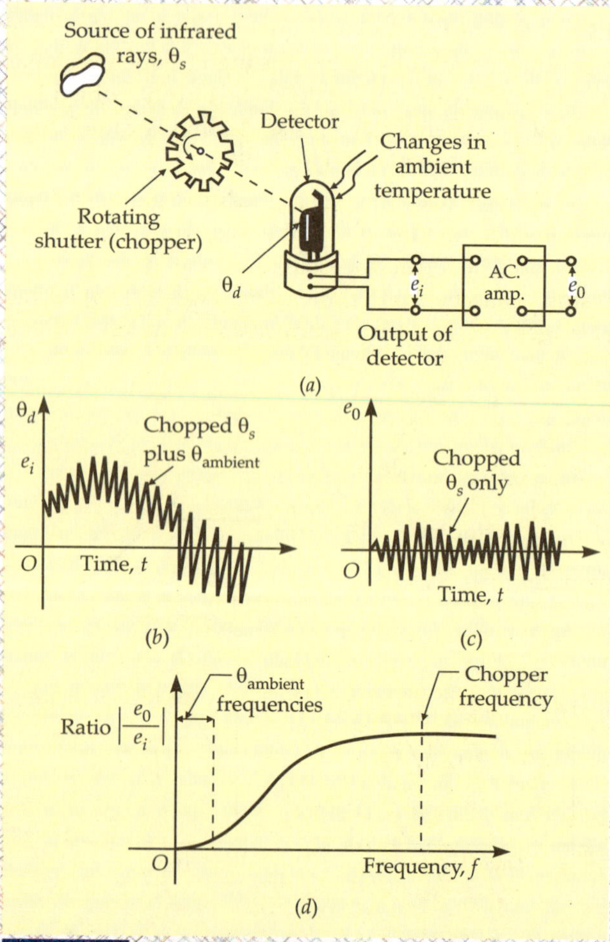


**Fig. 1.21** A highpass filter for infrared radiant energy.

The major problem with these devices is that the ambient temperature $\theta_{ambient}$ as well as the temperature of the source $\theta_s$ affect the temperature of the detector, thus affecting the output voltage $e_0$. The effect is serious because the radiant energy to be measured causes very small changes in $\theta_d$. On the other hand, even small changes in ambient temperature, $\theta_{ambient}$ may completely distort the output, thereby completely shrouding the output signal. A simple solution to this problem is the introduction of a rotating shutter between the radiant source and the detector so that the desired input is chopped or modulated, at a known frequency. The modulating frequency is chosen to be much higher than the frequencies at which the change in ambient temperature occurs. The output voltage signal $e_i$, of the detector is a superposed signal with slow changes on account of ambient temperature variations and a high frequency signal whose amplitude varies in accordance with $\theta_s$. As the desired and the interfering inputs are widely separated in frequency band, they can be conveniently and selectively filtered. For such an application, it is desired that the filter used should reject the slow varying interfering outputs on account of temperature variations, but pass fast varying signals on account of variations in $\theta_s$. This requires the use of a *high pass filter*. In fact such applications require the use of an a.c. amplifier. Since a.c. amplification, even otherwise needed, the use of an a.c. amplifier with low frequency cut off solves both the problems *i.e.*, the problem of attenuation of a low frequency signal and amplification of a high frequency a.c. signal without distortion. A wide frequency band is necessary and this is conveniently provided in the arrangement shown in Fig. 1.21.

**5. *Method of opposing inputs.*** This method is basically akin to method of calculated output corrections. The method of calculated output corrections requires errors caused on account of spurious input to be calculated and then corrections applied in order to obtain output corresponding to the desired inputs. The method of opposing inputs, on the other hand, requires that the measurement system be so designed that the outputs caused by spurious inputs be opposed by outputs produced by components physically built into the system that exactly act the opposite way and cancel out the outputs on account of spurius signals.

Let us consider the case of measurement of strain with the help of strain gauge bridge as shown in Fig. 1.22. As explained earlier temperature is an interfering input and there is an output on account of changes in ambient temperature even though there is no input strain. This effect can be eliminated by using a *dummy gauge* in the adjacent arm of the Wheatstone bridge. The arrangement is shown in Fig. 1.22.

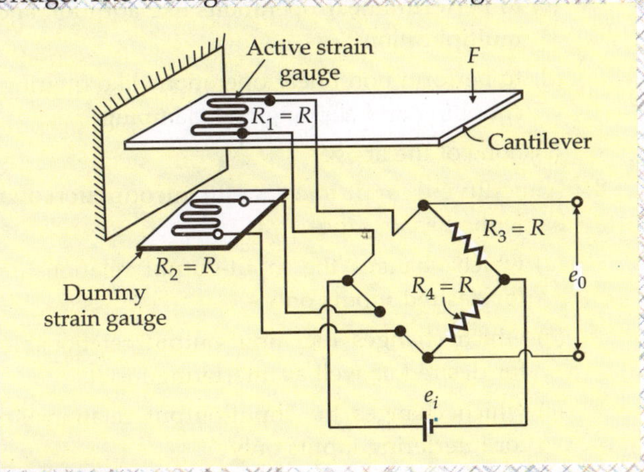


**Fig. 1.22** Use of dummy gauge for temperature compensation.

The active strain gauge is installed on the test specimen (cantilever in this case) while the dummy gauge is installed on a like piece of material and is not subjected to any strain.

The bridge is initially balanced and therefore $R_1/R_3 = R_2/R_4$. Supposing a change in temperature occurs, and the resistances of active and dummy gauges change by $\Delta R_1$ and $\Delta R_2$ respectively. In order that there should be no output the bridge should be balanced and this requires

$$\frac{R_1 + \Delta R_1}{R_3} = \frac{R_2 + \Delta R_2}{R_4}$$

or

$$\frac{\Delta R_1}{R_3} = \frac{\Delta R_2}{R_4}$$

Let $R_3 = R_4 = R$ and therefore for bridge balance $\Delta R_1$ should be equal to $\Delta R_2$. Now if the active and the dummy gauges are identical *i.e.*, $R_1 = R_2 = R$ and if the dummy gauge in placed in the same environments as the active strain gauge $\Delta R_1 = \Delta R_2$ and, therefore, there will be no output due to changes in ambient temperature.

# Review Questions

1. Distinguish between the direct and indirect methods of measurement. Cite examples to support your answer.
2. List the advantages of electronic instruments over electrical and mechanical instruments.
3. Distinguish between "Intelligent Instrumentation Systems" and "Dumb Instrumentation Systems". Cite examples to support your answer.
4. Describe the difference between deflection and null type of instruments giving suitable examples. Discuss about their accuracy, sensitivity and suitability for dynamic measurements.
5. Explain the analog and digital modes of operation of instruments. Explain how the resolution of digital instruments can be increased.
6. What are the basic blocks of a Generalized Instrumentation System ? Draw the various blocks and explain their functions.
7. What are "Desired", "Modifying" and 'Interfering" inputs for an Instrumentation System ? Give examples for each of these. Draw a block diagram for showing their influence on the output.
8. Explain how the effect of Modifying and Interfering inputs is minimized/eliminated in measurement systems. Cite suitable examples.
9. Define "Input Filtering" and "Output Filtering". Explain with suitable examples.
10. Figure 1.23 shows an electromagnetic type of balance which utilizes the modulation of light energy falling on a phototube for measurement of an unknown force. Draw a generalized block diagram of the measurement system and explain the functions of each block.

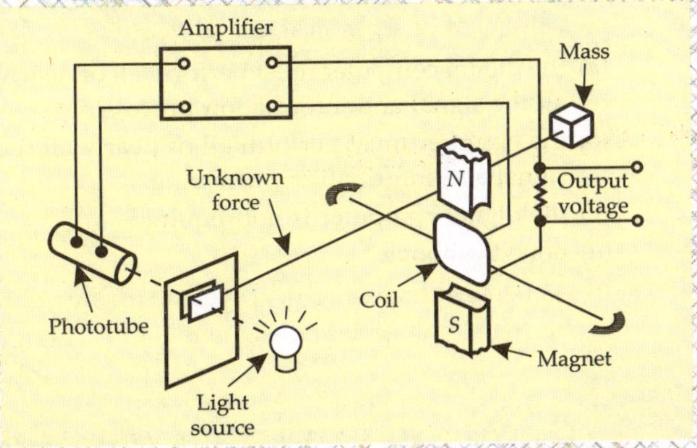


**Fig. 1.23** The electromagnetic balance.

# Objective Type Questions

**Tick (√) the most appropriate answer**

1. The measurement of a quantity
   (*a*) is an act of comparison of an unknown quantity with another quantity
   (*b*) is an act of comparison of an unknown quantity with a known quantity whose accuracy may be known or may not be known
   (*c*) is an act of comparison of an unknown quantity with a predefined acceptable standard which is accurately known
   (*d*) none of the above.
2. In present day measurement systems
   (*a*) direct methods are commonly used
   (*b*) use of direct methods is limited but indirect methods are commonly used
   (*c*) both direct and indirect methods are commonly used
   (*d*) all the above.
3. Purely mechanical instruments cannot be used for dynamic measurements because they have :
   (*a*) high inertia (*b*) large time constant
   (*c*) higher response time (*d*) all the above.
4. The usage of electronic instruments is becoming more extensive because they have
   (*a*) a high sensitivity and reliability
   (*b*) a fast response and compatibility with digital computers
   (*c*) the capability to respond to signals from remote places
   (*d*) all the above.
5. A null type of instrument as compared to a deflection type instrument has
   (*a*) a higher accuracy (*b*) a lower sensitivity
   (*c*) a faster response (*d*) all the above.
6. In an intelligent information system used for the measurement of a physical quantity
   (*a*) the digital computer must perform all or nearly all the signal and information processing
   (*b*) the observer must perform all or nearly all the signal and information processing
   (*c*) the digital computer is not required
   (*d*) none the above.
7. In a generalized measurement system, the function of the signal manipulating element is to
   (*a*) change the quantity under measurement to an analogous signal
   (*b*) change the magnitude of the input signal while retaining its identity
   (*c*) to perform linear operations like addition and multiplication
   (*d*) to perform non-linear operations like filtering, chopping and clipping and clamping
   (*e*) none of the above.
8. A modifying input to a measurement system can be defined as an input
   (*a*) which changes the input-output relationship for desired inputs only
   (*b*) which changes the input-output relationship for desired as well as interfering inputs
   (*c*) which changes the input-output relationship for interfering inputs only
   (*d*) none of the above.
9. In a measurement system, the open loop transfer function is 10 and changes by 10 percent. If a closed loop system is adopted wherein the feedback system has a transfer function of 100. What will be the change in output in the case of open loop and in closed loop systems
   (*a*) 10% , $10 \times 10^{-3}$ % (*b*) 10 %, $1 \times 10^{-3}$ %
   (*c*) 100 %, $10 \times 10^{-3}$ % (*d*) 100 %, $1 \times 10^{-3}$ %.
10. In a permanent magnet moving coil type of ammeter, the current is, $I = \dfrac{K\theta}{NBA}$
    where $\theta$ = deflection, $K$ = spring constant, $N$ = number of turns of moving coil, $A$ = area of moving coil and $B$ = flux density of field produced permanent magnet. Temperature is an interfering input to the system. It is known that strength of magnets reduces by 0.02 per cent per °C and the strength of springs reduces by 0.04 per cent per °C. Which method should be adopted to apply corrections for the interfering input ?
    (*a*) Method of inherent insensitivity
    (*b*) Method of high gain feedback
    (*c*) Method of calculated output corrections
    (*d*) Method of opposing inputs.

## *Answers*

1. (*c*) 2. (*b*) 3. (*d*) 4. (*d*) 5. (*a*) 6. (*a*) 7. (*b*) 8. (*b*) 9. (*a*) 10. (*c*)

# 2

# Characteristics of Instruments and Measurement Systems

## 2.1 MEASUREMENT SYSTEM PERFORMANCE

The treatment of instrument and measurement system characteristics can be divided into *two* distinct categories *viz.* :

(*i*) Static characteristics, and

(*ii*) Dynamic characteristics.

Some applications involve the measurement of quantities that are either constant or vary slowly with time. Under these circumstances it is possible to define a set of criteria that gives a meaningful description of quality of measurement without interfering with dynamic descriptions that involve the use of differential equations. These criteria are called **Static Characteristics.**

Normally static characteristics of a measurement system are, in general, those that must be considered when the system or instrument is used to measure a condition not varying with time.

However many measurements are concerned with rapidly varying quantities and, therefore, for such cases we must examine the dynamic relations which exist between the output and the input. This is normally done with the help of differential equations. Performance criteria based upon dynamic relations constitute the **Dynamic Characteristics.**

## 2.2 STATIC CALIBRATION

All the static performance characteristics are obtained in one form or another by a process called **static calibration.** The calibration of all instruments is important since it affords the opportunity to check the instrument against a known standard and subsequently to find errors and accuracy. Calibration procedures involve a comparison of the particular instrument with either (1) a primary standard, (2) a secondary standard with a higher accuracy than the instruments be calibrated, or (3) an instrument of known accuracy.

Actually all working instruments, *i.e.,* those instruments which are actually used for measurement work must be calibrated against some reference instruments which have a higher accuracy. Thus reference instruments in turn must be calibrated against instrument of still higher grade of accuracy, or against primary standard, or against other standards of known accuracy. It is essential that any measurement made must ultimately be traceable to the relevant primary standards.

## 2.3 STATIC CHARACTERISTICS

The main static characteristics discussed here are :

| | |
|---|---|
| (*i*) Accuracy | (*ii*) Sensitivity |
| (*iii*) Reproducibility | (*iv*) Drift |
| (*v*) Static error and | (*vi*) Dead Zone |

The qualities (*i*), (*ii*) and (*iii*) are desirable, while qualities (*iv*), (*v*) and (*vi*) are undesirable. In addition to above characteristics, definitions of many other characteristics have been given.

It must be stated, however, that there are many definitions of the above characteristics, and in some cases the definitions are unrelated. Care has been taken to select the most generally accepted definitions so as to avoid confusion.

## 2.4 ERRORS IN MEASUREMENT

Measurements done in a laboratory or at some other place always involve errors. No measurement is free from errors. If the precision of the equipment is adequate, no matter what its accuracy is, a discrepancy will always be observed between two measured results.

Since errors are a must in any measurement, it is imperative to interpret the results of a quantitative measurement in an intelligent manner. An understanding and thorough evaluation of the errors is essential.

## 2.5 TRUE VALUE

The true value of quantity to be measured may be defined as the average of an infinite number of measured values when the average deviation due to the various contributing factors tends to zero. Such an ideal situation is impossible to realise in practice and hence it is not possible to determine the **"True Value"** of a quantity by experimental means. The reason for this is that the positive deviations from the true value do not equal the negative deviations and hence do not cancel each other. Thus, normally an experimenter would never know that the value or quantity being measured by experimental means is the **"True Value"** of the quantity or not.

In fact, in practice, the term **"True Value"**', then, refers to a value that would be obtained if the quantity under consideration were measured by an **"Examplar Method"**, that is a method agreed upon by experts as being sufficiently accurate for the purposes to which the data will ultimately be put to use.

## 2.6 STATIC ERROR

The most important characteristic of an instrument or measurement system is its accuracy, which is the agreement of the instrument reading with the true value of quantity being measured. The accuracy of an instrument is measured in terms of its *error.*

We have mentioned earlier that it is impossible to measure the true value of a quantity. An approximation of the **"true value"** obtained by sufficiently extended series of measurements and also taking into account parameters and conditions to which corrections may be applied, we obtain, what is called the **best measured value** of the quantity. While it is never possible to measure the true or exact value of a quantity, it is nearly always possible to give a best measured value. **Static error** is defined as the difference between the measured value and the true value of the quantity. Then :

$$\delta A = A_m - A_t \quad ...(2.1)$$

where $\delta A$ = error,

$A_m$ = measured value of quantity,

and $A_t$ = true value of quantity

$\delta A$ is also called the absolute static error of quantity $A$.

We have $\varepsilon_0 = \delta A \quad ...(2.2)$

where $\varepsilon_0$ = absolute static error of quantity $A$ (under measurement).

The absolute value of $\delta A$ does not indicate precisely the accuracy of measurements. As an example, an error of $\pm$ 2 A is negligible when the current being measured is of the order of 1000 A while the same error of $\pm$2 A may be regarded as intolerable when the current under measurement is 10 A or so. Thus the quality of measurement is provided by the relative static error, *i.e.*, the ratio of absolute static error $\delta A$ to $A_t$, the true value of the quantity under measurement. Therefore, the relative static error $\varepsilon_r$, is given by :

$$\varepsilon_r = \frac{\text{absolute error}}{\text{true value}} = \frac{\delta A}{A_t} = \frac{\varepsilon_0}{A_t} \quad ...(2.3)$$

Percentage static error, $\%\varepsilon_r = \varepsilon_r \times 100 \quad ...(2.4)$

We have $A_t = A_m - \delta A$ (see Eqn. 2.1)

$= A_m - \varepsilon_0 = A_m - \varepsilon_r A_t$ (see Eqns. 2.2, 2.3)

$$= \frac{A_m}{1+\varepsilon_r} \quad ...(2.5)$$

However, when the absolute static error $\varepsilon_0 = \delta A$ is small, which means that the difference between measured and true values is small, $\varepsilon_r << 1$.

$\therefore$ Equation 2.5 may be written as :

$$A_t = A_m(1-\varepsilon_r) \quad ...(2.6)$$

## 2.7 STATIC CORRECTION

It is the difference between the true value and the measurement value of the quantity, or

$$\delta C = A_t - A_m \quad ...(2.7)$$

where $\delta C$ = static correction = $-\delta A$ ...(2.8)

**Example 2.1** *A meter reads 127.50 V and the true value of the voltage is 127.43 V. Determine : (a) the static error, and (b) the static correction for this instrument.*

**Solution.** From Eqn. 2.1, the error is

$$\delta A = A_m - A_t = 127.50 - 127.43 = +0.07\text{ V}$$

Static correction, $\delta C = -\delta A = -0.07$ V

**Example 2.2** *A thermometer reads 95.45°C and the static correction given in the correction curve is – 0.08°C. Determine the true value of the temperature.*

**Solution.** True value of the temperature

$$A_t = A_m + \delta C = 95.45 - 0.08 = 95.37°\text{C}$$

**Example 2.3** *A voltage has a true value of 1.50 V. An analog indicating instrument with a scale range of 0 – 2.50 V shows a voltage of 1.46 V. What are the values of absolute error and correction ? Express the error as a fraction of the true value and the full scale deflection (f.s.d.).*

**Solution.** Absolute error

$$\delta A = A_m - A_t = 1.46 - 1.50 = -0.04\text{ V}$$

Absolute correction, $\delta C = -\delta A = +0.04$ V

Relative error, $\varepsilon_r = \frac{\delta A}{A_t} = \frac{-0.04}{1.50} \times 100 = -2.67\%$

Relative error (expressed as a percentage of f.s.d.)

$$= \frac{-0.04}{2.5} \times 100 = -1.60\,\%.$$

When f.s.d. is the full scale deflection.

## 2.8 SCALE RANGE AND SCALE SPAN

In an analog indicating instrument the value of measurand is indicated on a scale by a pointer while in a recording instrument it is indicated on a chart by a pen mechanism. The choice of proper range for instruments is important in instrumentation. The **Scale Range** of an instrument is defined as the difference between the largest and the smallest reading of the instrument. Supposing the highest point of calibration is $X_{max}$ units while the lowest is $X_{min}$ units and that the calibration is continuous between the two points. Then we can say that the instrument range is between $X_{min}$ and $X_{max}$ (or many a times we say that the instrument range is $X_{max}$).

The instrument **span** is given by :

$$\text{Span} = X_{max} - X_{min} \qquad \text{...(2.9)}$$

The same definitions apply in the case of digital instruments.

For a pyrometer calibrated between 0 to 1000°C, the range is 1000°C and span is also 1000°C. For a thermometer calibrated between 200°C to 500°C, the range is 200°C to 500°C (or 500°C) but the span is $500 - 200 = 300°C$.

There is another factor that must be considered while determining the range of the instrument. This is the **Frequency Range**, which is defined as frequencies over which measurements can be performed with a specified degree of accuracy. For example, a moving iron instrument may have a 0 – 250 V range and 0 – 135 Hz frequency range with an accuracy of 0.5 or 1% of full scale reading.

**Example 2.4** *A thermometer is calibrated 150°C to 200°C. The accuracy is specified within ± 0.25 percent of instrument span. What is the maximum static error ?*

**Solution.** Span of thermometer $= 200 - 150 = 50°C$

$$\therefore \text{ Maximum static error} = \frac{\pm 0.25 \times 50}{100} = \pm 0.125°C$$

## 2.9 ERROR CALIBRATION CURVE

Error calibration means that an instrument has been calibrated against a suitable standard as discussed in Article 2.2 and its static error determined at a number of points on its scale. These data form an error curve, which can be used for correcting instrument readings. A typical error curve is shown in Fig. 2.1.

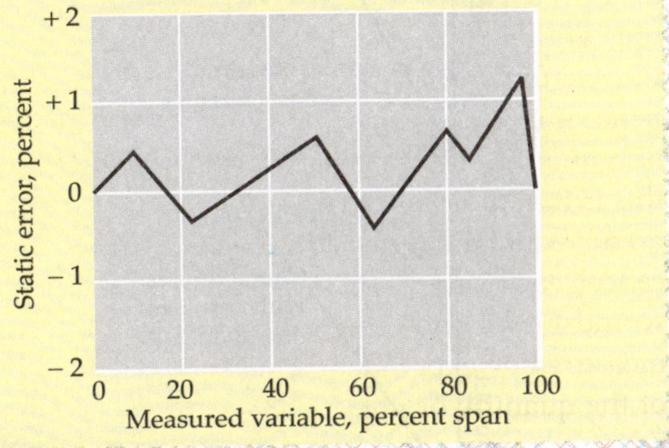


**Fig. 2.1** Error calibration curve.

## 2.10 REPRODUCIBILITY AND DRIFT

**Reproducibility.** It is the degree of closeness with which a given value may be repeatedly measured. It may be specified in terms of units for a given period of time. Perfect reproducibility means that the instrument has no **drift**.

No drift means that with a given input the measured values do not vary with time.

**Drift** may be classified into *three* categories :

(*i*) ***Zero drift.*** If the whole calibration gradually shifts due to slippage, permanent set, or due to undue warming up of electronic tube circuits, zero drift sets in. This can be prevented by zero setting. The input-output characteristics with zero drift are shown in Fig. 2.2(*a*).

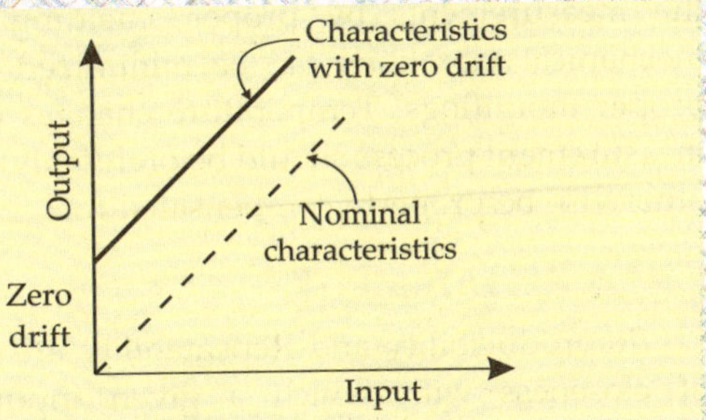


**Fig. 2.2** (*a*) Zero drift.

(*ii*) ***Span drift or Sensitivity drift.*** If there is proportional change in the indication all along the upward scale, the drift is called *span drift* or *sensitivity drift*. The characteristics with span drift are shown in Fig. 2.2(*b*) and characteristics with both zero and span drift are shown in Fig. 2.2(*c*).

(*iii*) ***Zonal drift.*** In case the drift occurs only over a portion of span of an instrument, it is called *zonal drift*.

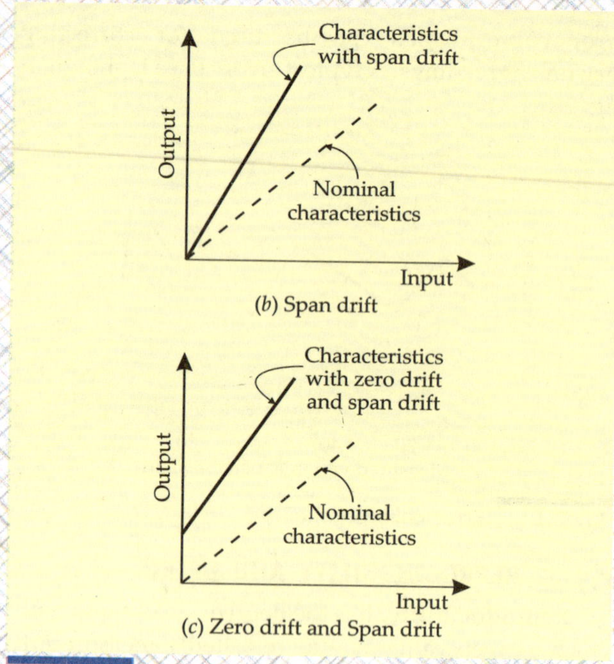


**Fig. 2.2** Input-output Characteristics with drift.

There are many environmental factors which cause drift. They may be stray electric and magnetic fields, thermal emfs, changes in temperature, mechanical vibrations, wear and tear, and high mechanical stresses developed in some parts of the instruments and systems.

**Drift** is an undesirable quality in industrial instruments because it is rarely apparent and cannot be easily compensated for. Thus it must be carefully guarded against by continuous prevention, inspection and maintenance. For example, stray electrostatic and electromagnetic fields can be prevented from affecting the measurements by proper shielding. Effect of mechanical vibrations can be minimized by having proper mountings. Temperature changes during the measurement process should be preferably avoided or otherwise be properly compensated for.

## 2.11 REPEATABILITY

Reproducibility and Repeatability are a measure of closeness with which a given input may be measured over and over again. The two terms cause confusion. Therefore, a distinction is made between the two terms. **Reproducibility** is specified in terms of scale readings over a given period of time. On the other hand, **Repeatability** is defined as the variation of scale reading and is random in nature. Figure 2.3 shows this repeatability.

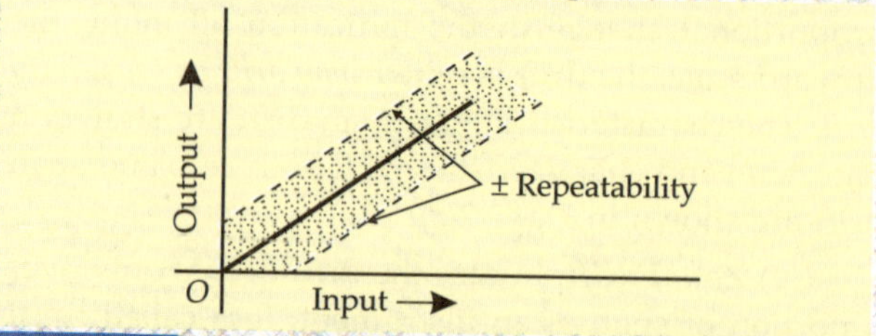


**Fig. 2.3** Input-output relationship with ± repeatability.

## 2.12 NOISE

A spurious current or voltage extraneous to the current or voltage of interest in an electrical or electronic circuit is called **Noise**. In fact, noise is a signal that does not convey any useful information. Noise may be generated external to a particular system of interest and enter the system in various ways or it may be generated inside the system of interest.

The effect of noise in a measurement system may be an annoying "hum" in the speaker of a radio receiver owing to 50 Hz power line frequency noise, to the transmission of incorrect data in telemetering systems or data communication systems, to life threatening situations caused by incorrect interpretation of waveforms related to vital human organs in biomedical instrumentation systems where the expected (desired) signals are instrinsically weak. This is because the extraneous noise signals constitute a background against which the desired signal may be read.

### 2.12.1 Signal to Noise Ratio (S/N)

Noise is an unwanted signal superimposed upon the signal of interest thereby causing a deviation of the output from its expected value. The expected value of the output is the one that is obtained only when desired signal or signal of interest is present. The extent to which noise becomes important in the measurement systems depends upon the relative value of magnitude of unwanted signal (noise) to that of signal of interest. If the magnitude of unwanted signal (noise) is small as compared with that of signal of interest, then signal to noise ratio (S/N) is large and therefore the noise becomes unimportant.

The ratio of desired signal to the unwanted noise is called **signal to noise ratio** and is expressed as

$$\frac{S}{N} = \frac{\text{signal power}}{\text{noise power}}$$

$$= \frac{(\text{signal of interest expressed in volt})^2}{(\text{unwanted noise expressed in volt})^2} \quad \ldots(2.10)$$

In any measurement system, it is desirable to have a large signal-to-noise ratio. This can be achieved by increasing the signal level without increasing the noise level or decreasing the noise level with some suitable technique.

### 2.12.2 Sources of Noise

The noise in the output is broadly of *three* types :

(*i*) Generated noise, (*ii*) Conducted noise,

(*iii*) Radiated noise.

This can be illustrated by considering the case of an amplifier which is commonly used in measurement systems to identify the different sources of noise. The block diagram of the amplifier is shown in Fig 2.4.

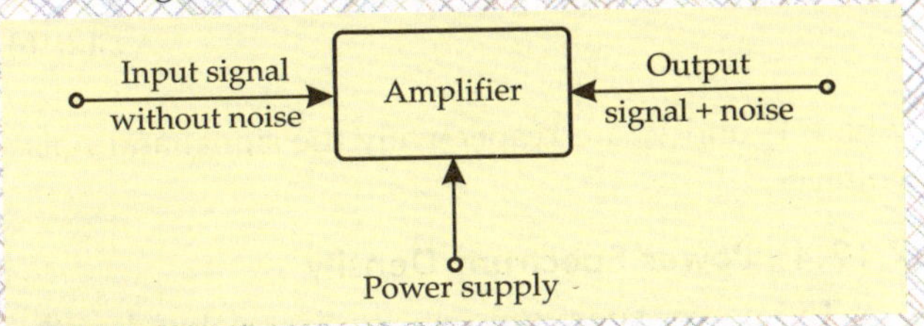


**Fig. 2.4** An amplifier for illustrating the different types of noise signals.

(*i*) ***Generated noise.*** Suppose the input signals contain no noise. The power supply serves as a source of energy for the operation of amplifier. The output signal is amplifier gain times the input signal plus a noise signal. One of the possible sources of noise is on account of internal components of the amplifier like resistors, capacitors and transistors etc. Therefore the noise in this case is generated inside the amplifier and therefore is called **Generated noise**. The internally generated noise is on account of components like resistors, capacitors, transistors etc. as stated above.

The conductive portion of a resistor consists of a regularly arranged groups of atoms that maintain the same general physical position in the conductor. These atoms contribute conduction electrons due to which a current flows. Although the atoms maintain their general physical position, they are in a state of rapid vibratory motion on account of temperature and thermal effects. This vibratory motion of atoms is transferred to the conduction electrons, thereby producing a noise component of current. Since this noise is temperature dependent, it increases with internal heating ($I^2 R$ loss) or with an increase in the ambient temperature. This noise is called **Johonson noise**.

The vibrations produced by thermal effects within a resistor cover a wide frequency range, and therefore the noise generated consists of a wide spectrum of frequencies. This wideband noise is sometimes called **White noise**.

The internally generated noise in resistors can be reduced by lowering the internal temperature. Special film and glass substrates are used for minimizing this noise.

As the noise is internally generated, external shielding cannot help in reducing this noise. Also since the noise has a wide frequency band, selective filtering is ineffective in reducing the magnitude of this noise.

A second type of noise is generated internally by short time electrical events within an active device like a transistor. In semi-conductor devices charges cross junctions within the device, and therefore move from one energy level to another. This movement results in acceleration of charges which generate electro-magnetic disturbances, which in turn produce an internally generated noise. Since, the period of transition from one energy level to another is short, the period of acceleration is also short, and hence the resulting frequency spectrum of generated noise is wideband. The same type of noise is generated as electrons pass through various electrode areas of a vacuum tube which have different electric field intensity thereby causing electrons to accelerate and cause electromagnetic disturbances.

The internally generated noise in semi-conductor and vacuum tube devices on account of random movement of charges is called **Shot Noise.** Not much can be done to reduce the internally generated noise in semi-conductor and vacuum tube devices as this noise is on account of acceleration of charges, a phenomenon so essential for operation of these devices. Selective filtering may be helpful to some extent in reduction of this noise.

Changing electric fields in the region between plates of a capacitor and changing magnetic fields around in an inductor are other sources of internally generated noise. However, in these cases, the electric and magnetic fields change in an orderly manner and therefore the noise signal has a fixed frequency and hence its magnitude can be reduced by using a filter tuned to this frequency.

(*ii*) ***Conducted noise.*** The power supply to the amplifier could be the source of noise since it may have spikes, ripples or random deviations that are conducted to the amplifier circuit through power wiring. This type of noise is called **Conducted noise**.

One of the most common sources for conducted noise is the 50 Hz power supply and the harmonics contained in it. The conducted noise can be reduced to use filters in the leads to trap out the noise.

(*iii*) ***Radiated noise.*** There may be electric or magnetic fields or disturbances in the environments around the amplifier because of which unwanted signals are radiated into the interior of the amplifier and this is called **Radiated noise**.

2

One of the common sources of radiated noise are the electromagnetic impulses radiated from the ignition wiring of spark plugs. The radiated noise can be reduced by proper shielding.

### 2.12.3 Johonson Noise

As stated earlier, Johonson noise is the thermally generated noise in a resistor.

The noise power $P_n$ generated in a conductor is

$$P_n = kT\Delta f \quad \text{...(2.11)}$$

where $k$ = Boltzmann constant

$= 1.38 \times 10^{-23}$ J/°K

$T$ = absolute temperature of resistor ; °K

$\Delta f$ = bandwidth of noise signal ; Hz

The noise generation system can be represented by a voltage source of magnitude in series with an equivalent resistance, $R_n$, as shown in Fig. 2.5.

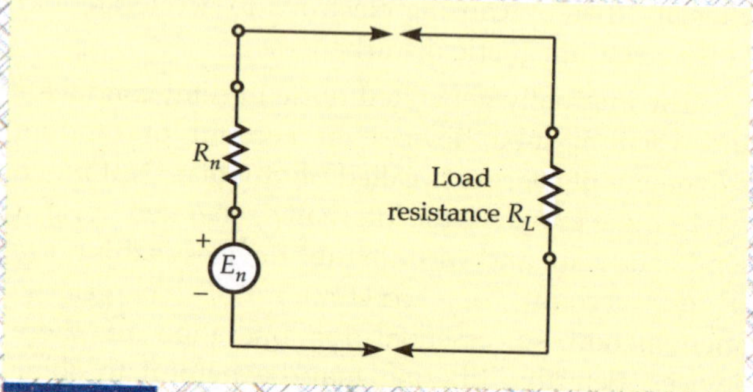


**Fig. 2.5** Circuit representation of noise signals.

If the noise generator is connected to an external load resistance, $R_L$, the noise energy will be transferred to the load. Under conditions of maximum power transfer ($R_L = R_n$), the noise power delivered to load.

$$P_{nL} = \frac{E_n^2}{4R_n} \quad \text{...(2.12)}$$

or $\quad E_n^2 = 4R_n P_{nL} = 4kT\,R_n\,\Delta f$

or noise voltage $E_n = 2\sqrt{kTR_n\Delta f}$ ...(2.13)

Suppose two resistors $R_1$ and $R_2$ are connected in series, the total noise is the sum of the noises of individual resistors.

$$E_n^2 = E_{n1}^2 + E_{n2}^2$$

$$= 4kTR_1\Delta f + 4kTR_2\Delta f = 4kTR_s\Delta f$$

or $\quad E_n = 2\sqrt{kTR_s\Delta f}$ ...(2.14)

where, $R_s = R_1 + R_2$

Similarly, the total noise voltage for parallel connected resistors is

$$E_n = 2\sqrt{kTR_p\Delta f} \quad \text{...(2.15)}$$

where $R_p$ = resultant resistance of parallel connected resistors.

$$\text{Now,} \quad \frac{S}{N} = \frac{\text{signal power}}{\text{noise power}}$$

$$= \frac{V_s^2 / R}{V_n^2 / R}$$

$$= \frac{V_s^2}{V_n^2} \quad \text{...(2.16)}$$

where $V_s$ and $V_n$ are respectively the signal and noise voltages.

### 2.12.4 Power Spectrum Density

The power spectrum density $S_n$ is defined as the noise power per unit of frequency bandwidth. Therefore from Eqn. 2.11,

$$S_n = \frac{P_n}{\Delta f} = kT \quad \text{...(2.17)}$$

Power spectrum Density represents noise energy generated per cycle of vibratory motion in the generating conductor.

### 2.12.5 Noise Factor and Noise Figure

The noise factor is defined as :

$$F = \frac{S/N \text{ at input}}{S/N \text{ at output}} \quad \text{...(2.18)}$$

Noise factor measurements are important because they are a measure of noise added to a signal by a device in the measurement system. If noise factor is expressed in decibles, it is known as **Noise figure**.

Noise figure $\quad nf = 10 \log F$ ...(2.19)

The measurement of noise figure is most meaningful measurement for amplifiers, transistors and vacuum tubes since it is a measure of the noise generated within the device. The measurement of noise figure requires the use of known source of noise.

The noise figure from Eqn. 2.19 is,

$$nf = 10\log\left[\frac{\text{noise voltage at output with noise at input}}{\text{noise voltage output with no noise at input}}\right]$$

$$= 10\log\frac{V_n}{V_0} \quad \text{...(2.20)}$$

where $V_n$ = output noise voltage with a noise source injected into the input

$V_0$ = output noise voltage without noise at input.

The difference between the two voltages represents the noise added by the device under test.

**Example 2.5** *An amplifier whose bandwidth is 100 kHz has a noise power spectrum density input of $7 \times 10^{-21}$ J. If the input resistance is 50 kΩ and the amplifier gain 100, what is the noise output voltage ?*

**Solution.** From Eqn. 2.17, power density spectrum

$$S_n = kT = 7 \times 10^{-21}$$

Input resistance $R = 50 \times 10^3 \ \Omega$,

Bandwidth $= 100 \times 10^3$ Hz

From Eqn. 2.13, noise voltage at the input is,

$$E_n = 2\sqrt{kTR\Delta f}$$

$$= 2\sqrt{7 \times 10^{-21} \times 50 \times 10^3 \times 100 \times 10^3} = 11.83 \ \mu\text{V}$$

∴ Noise voltage at the output,

$$= E_n \times \text{gain of amplifier}$$

$$= 11.83 \times 100 = 1.183 \ \text{mV}$$

**Example 2.6** *A low pass filter has an input S/N of 20. The input voltage is 3 mV. Calculate the noise voltage.*

**Solution.** From Eqn. 2.16, the signal to noise ratio is,

$$\frac{S}{N} = \frac{V_S^2}{V_n^2}$$

∴ Noise voltage

$$V_n = \frac{V_S}{\sqrt{S/N}} = \frac{3}{\sqrt{20}} = 0.67 \ \text{mV}$$

**Example 2.7** *(a) At the input an amplifier has a signal voltage level of 3 μV and a noise voltage level of 1 μV. What is the signal to noise ratio at the input ?*

*(b) If the voltage gain of the amplifier is 20, what is the S / N ratio at the output ?*

*(c) If the amplifier adds 5 μV of noise, what is S / N ratio at the output ? Calculate also the noise factor and the noise figure.*

**Solution.** (*a*) Signal to noise ratio at the input,

$$= \left(\frac{3 \times 10^{-6}}{1 \times 10^{-6}}\right)^2 = 9$$

(*b*) Voltage level of signal at the output

$$= 20 \times 3 = 60 \ \mu\text{V}$$

Voltage level of noise at the output $= 20 \times 1 = 20 \ \mu$V

∴ Signal to noise ratio at the output is,

$$= \left(\frac{60 \times 10^{-6}}{20 \times 10^{-6}}\right)^2 = 9$$

(*c*) If the amplifier adds 5 μV to the noise, then the voltage level of noise at the output

$$= 20 + 5 = 25 \ \mu\text{V}$$

Hence, signal to noise ratio at the output

$$= \left(\frac{60 \times 10^{-6}}{25 \times 10^{-6}}\right)^2 = 5.76$$

From Eqn. 2.18,

$$\text{Noise factor} \quad F = \frac{S/N \text{ at input}}{S/N \text{ at output}} = \frac{9}{5.76} = 1.56$$ 2

From Eqn. 2.19, noise figure

$$nf = 10 \log F = 10 \log 1.56 = 1.93 \ \text{db}$$

**Example 2.8** *A measurement system uses a resistance strain gauge for measurement of pressure. The strain gauge has an unstrained resistance of 120 Ω and is connected in one arm of a wheatstone bridge. The other three arms of the bridge have each a resistance of 120 Ω. The temperature is 300° K and the bandwidth is 100 kHz. The output of the bridge is a voltage signal.*

*(i) When a pressure of 7000 kN/m² is applied the output voltage is 0.12 mV. Find the ratio of the output (signal) voltage to the noise voltage generated by the resistors.*

*(ii) Find the ratio of output (signal) voltage to the noise voltage if the applied pressure is 7 kN / m².*

*The Boltzmann constant is $1.38 \times 10^{-23}$ J/° K. Comment upon the results. Consider only the noise produced in the strain gauge.*

**Solution.** (*i*) From Eqn. 2.13, the noise voltage is :

$$E_n = 2\sqrt{kTR\Delta f}$$

$$= 2\sqrt{1.38 \times 10^{-23} \times 300 \times 120 \times 100 \times 10^3} \ \text{volt}$$

$$= 0.446 \ \mu\text{V}$$

Ratio of signal voltage to noise voltage is,

$$\frac{0.12 \times 10^{-3}}{0.446 \times 10^{-6}} = 269.$$

In this case the noise voltage is negligible as compared with the signal voltage and therefore the interference due to noise is insignificant and thus does not distort the signal. Hence, the output is faithfully reproduced.

(*ii*) Assume a linear relationship between the output voltage of the bridge and the applied pressure.

∴ Output (signal) voltage, when the applied pressure is 7 kN / m²

$$= \frac{7 \times 10^3}{7000 \times 10^3} \times 0.12 \times 10^{-3} \ \text{volt}$$

$$= 0.12 \times 10^{-6} = 0.12 \ \mu\text{V}$$

∴ Ratio of signal voltage to noise voltage

$$= \frac{0.12 \times 10^{-6}}{0.446 \times 10^{-6}} = 0.27.$$

This indicates that the noise has a magnitude which is about 3.7 times that of signal, hence the signal will be completely lost in the noise. Amplification of the signal provides no solution to the problem as the signal and noise voltages will be equally amplified *i.e.*, the ratio of signal voltage to noise voltage will remain the same at the output.

## 2.13 ACCURACY AND PRECISION

In ordinary usage, the distinction between words "**Accuracy**" and "**Precision**" is usually very vague. In fact even the dictionaries invariably link the definition of one with the other. But as far as field of measurements is concerned, there is a difference between the two terms as they have sharp differences in meanings. As far as measurements are concerned, the two terms may be defined as :

### 2.13.1 Accuracy

It is the closeness with which an instrument reading approaches the true value of the quantity being measured. Thus accuracy of a measurement means conformity to truth.

The accuracy may be specified in terms of *inaccuracy or limits of errors* and can be expressed in the following ways :

**1.** ***Point accuracy.*** This is the accuracy of the instrument only at one point on its scale. The specification of this accuracy does not give any information about the accuracy at other points on the scale or in other words, does not give any information about the general accuracy of the instrument. However, the general accuracy may be given by drawing up a table that specifies the accuracy at a number of points throughout the range of instrument.

**2.** ***Accuracy as "Percentage of Scale Range".*** When an instrument has uniform scale, its accuracy may be expressed in terms of scale range. For example, the accuracy of a thermometer having a range of 500°C may be expressed as ± 0.5 percent of scale range. This means that the accuracy of the thermometer when the reading is 500°C is ± 0.5% which is negligible, but when the reading is 25°C (*i.e.*, 20 percent of scale range), the error is as high as $(500/25) \times (\pm 0.5) = 10$ percent and therefore specification of accuracy in this manner is highly misleading.

**3.** ***Accuracy as "Percentage of True Value".*** The best way to conceive the idea of accuracy is to specify it in terms of the true value of the quantity being measured *i.e.*, within ± 0.5 per cent of true value. This statement means thus as the readings get smaller so do the errors. Thus at 5 percent of full scale the accuracy of the instrument would be 20 percent better than that of an instrument which is accurate to ± 0.5 % of scale range.

### 2.13.2 Precision

It is a measure of the reproducibility of the measurements, *i.e.*, given a fixed value of a quantity, precision is a measure of the degree of agreement within a group of measurements. The term '**Precise**' means clearly or sharply defined. As an example of the difference in meaning of the two terms accuracy and precision, suppose that we have an ammeter which possesses high degree of precision by virtue of its clearly legible, finely divided, distinct scale and a knife edge pointer with mirror arrangements to remove parallax. Let us say that its readings can be taken to 1/100 of an ampere. At the same time, its zero adjustment is wrong. Now every time we take a reading, the ammeter is as **precise** as ever, we can take readings down to 1/100 of an ampere, and the readings are consistent and "*clearly defined*". However, the readings taken with this ammeter are not **accurate,** since they do not confirm to truth on account of its faulty zero adjustment.

Let us cite another example. Consider the measurement of a known voltage of 100 V with a meter. Five readings are taken, and the indicated values are 104, 103, 105, 103 and 105 V. From these values it is seen that the instrument cannot be depended on for an accuracy better than 5% (5 V in this case), while a precision of ± 1% is indicated since the maximum deviation from the mean reading of 104 V is only 1.0 V. Thus we find that the instrument can be calibrated so that it could be used to read ± 1 V dependably. This example illustrates that **accuracy** can be improved upon but not the **precision** of the instrument by calibration. Another point which is evident from above is that although the readings are close together they have a small scatter (or dispersion) and thus have a high degree of precision but the results are far from accurate. The precision of an instrument is usually dependent upon many factors and requires many sophisticated techniques of analysis.

Thus, when it is stated that a set of readings shows precision it means that the results agree among themselves. Agreement, however, is no guarantee of accuracy, as there may be some systematic disturbing effect that causes all the measured values to be in error.

## 2.14 INDICATIONS OF PRECISION

Precision is composed of *two* characteristics :

- Conformity and
- Number of significant figures.

**Precision** is used in measurements to describe the *consistency* or the *reproducibility* of results. A quantity called **precision index** defined in chapter 3 describes the spread, or dispersion of repeated result about some central value. High precision means a tight cluster of repeated results while low precision indicates a broad scattering of results. But this may not lead us to the misconception that high precision indicates high degree of accuracy since all the repetitions in result may be biased in the same way by some systematic effect that produces same deviation of results from the true value. For example, a spring scale used with a spring designed for a different spring scale may repeatedly show the same value of weight. Thus the readings may display exceptional agreement between themselves, but they all would be inaccurate values of weight since use of a wrong spring introduces a systematic shift of all readings.

We may well cite another example to illustrate another aspect of precision. Consider, for example, that a resistor whose true resistance is 1,385,692 Ω, is being measured by an ohmmeter. The ohmmeter consistently and repeatedly indicates the true value. But the observer cannot read this value from the scale. The observer's estimates from the scale reading consistently yield a value of 1.4 MΩ. This is as close to the true as he can read the scale by estimation. Although there are no deviations from the observed value, the error created by the limitation of the scale reading is a precision error.

The above example illustrates that *conformity* is necessary, but not sufficient for precision because of lack of *significant figures* obtained. Similarly, *precision* is a necessary, but not sufficient condition for *accuracy*.

In critical work, good practice requires an independent set of measurements using different instruments or different measurement techniques which are not subject to the same systematic errors. Where this is not possible, the experimenter must take steps that ensure proper functioning of instruments and to discover and eliminate any systematic disturbing factor. Calibration against a known standard may be resorted to in order to achieve the above purpose.

## 2.15 SIGNIFICANT FIGURES

An indication of the precision of the measurement is obtained from the number of significant figures in which it is expressed. Significant figures convey actual information regarding the magnitude and the measurement precision of a quantity. The more the significant figures, the greater the precision of measurement.

Let us take an example, if a voltage is specified as 256 V, its value should be taken as closer to 256 V than to either 257 V or 255 V. If the value of voltage is described as 256.0 V, it means that the voltage is closer to 256.0 V than it is to 256.1 V or 255.9 V. In 256, there are three significant figures while in 256.0 there are four. The latter, with more significant figures, expresses a measurement of greater precision than the former. 2

Frequently large numbers with zeros before a decimal point are used for approximate populations or amounts of money. For example, the population of a city is reported as 490,000. This may due to misconception, imply that the true value lies between 489,999 and 490,001, which is six significant figures. But in fact, what is meant, however, is that the population is closer to 490,000 than to 480,000 or 500,000. Since the population can be reported only to two significant figures. How else can large numbers be expressed ? A more technically correct notation uses powers of ten like $49 \times 10^4$ or $4.9 \times 10^5$. This indicates that the population figure is only accurate to two significant figures. Thus reference to population of a city as 3,000,000 would be interpreted automatically as an approximate number while reference to velocity of light as 300,000,000 metre per second creates no confusion to a person with a technical background. But expressing the velocity of light as $3 \times 10^8$ metre per second leaves no scope for confusion. Uncertainty caused by zeros to the left of decimal point is therefore usually resolved by scientific notation using powers of ten.

**Example 2.9** *State the number of significant figures in each of the following numbers :*

*(a) 302 A* *(b) 302.10 V* *(c) 0.00030 Ω*
*(d) 0.000030 MΩ* *(e) $5.01 \times 10^4$* *(f) 502000.*

**Solution.** (*a*) The number is 302. This means that it is more close to 302 than to either 301 or 303. Thus this number has 3 significant figures.

(*b*) The number involved is 302.10. Therefore it is more close to 302.10 than either to 301.09 or to 302.11. Thus it has 5 significant figures.

(*c*) The number is 0.00030. This means that it is more close to 0.00030 than either to 0.00029 or 0.00031. Thus it has 5 significant figures.

(*d*) The resistance is 0.000030 MΩ. This can be written as 30 Ω. Thus it is more close to 30 Ω than to either 29 Ω or 31 Ω. Therefore it has 2 significant figures. The zeros to the left of 3 are due to large size of unit.

(*e*) The number under consideration is $5.01 \times 10^4$. Hence it is more close to $5.01 \times 10^4$ than to either $5.00 \times 10^4$ or $5.02 \times 10^4$. Thus it has 3 significant figures.

(*f*) The number is 502000. This is a large number and a situation may arise that causes uncertainty. Strictly 502000 means that the number is more close to 502,000 than either is 501,999 or 502,001. This means that the number has 6 significant figures. However, if this number is population of a town it could be written as $501 \times 10^3$. Under this situation we can say that the population is close to $502 \times 10^3$ than to either $501 \times 10^3$ or $503 \times 10^3$. This means that number has 3 significant figures.

## 2.16 RANGE OF DOUBT OR POSSIBLE ERRORS AND DOUBTFUL FIGURES

It is customary in measurement work to record the result with all the digits of which we are sure, and final digit which is believed to be nearest to the true value. This usually implies that the reported result is good to ± 1 in the digit in the last place *i.e.*, the digit in doubt. For example, in reading wattmeter the power may be read as 22.6 W. This simply indicates that the power read by the observer to the best estimation is closer to 22.6 W, than to either 22.5 W or 22.7 W. Another way of expressing this result indicates the **'range of doubt or possible error'**. The way to express doubt regarding the last place, the digit in doubt is put in italics form. For example, 22.6 W indicates that we know the power to better than unit and that we are not sure about the tenths but all the same believe the value of digit in the tenths place is close to 6. A similar but definite method is to express the above power as 22.6 ± 0.05 W indicating that the power lies between, 22.55 W and 22.65 W giving the range of doubt or possible error as 0.1 W.

When a number of independent measurements are taken in order to obtain the **best measured value**, the result is usually expressed as **arithmetic mean** of all readings. The **range of doubt or possible error** is **largest deviation** from the mean.

**Example 2.10** *A set of independent current measurements was recorded as 10.03, 10.10, 10.11 and 10.08 A. Calculate (a) the average current, and (b) the range of error.*

**Solution.**

(*a*) Average current

$$I_{av} = \frac{I_1 + I_2 + I_3 + I_4}{4} = \frac{10.03 + 10.10 + 10.11 + 10.08}{4}$$

$$= 10.08 \text{ A}$$

(*b*) Maximum value of current $I_{max} = 10.11$ A

Range $= I_{max} - I_{av} = 10.11 - 10.08 = 0.03$ A

Minimum value of current $I_{min} = 10.03$ A

$$I_{av} - I_{min} = 10.08 - 10.03 = 0.05 \text{ A}$$

Therefore average range of error

$$= \pm \frac{0.03 + 0.05}{2} = \pm 0.04 \text{ A}$$

The number of significant figures in a quantity is one measure of precision, though not as definite as a percentage statement. Suppose the range of doubt in a 101 Ω resistor (the value has 3 significant figures) is 1 Ω. This means that the value lies between 100.5 Ω and 101.5 Ω. This range of doubt in percentage terms is 1 percent. The same range of doubt of 1 Ω in a 999 Ω resistor (again the value has 3 significant figures) creates a percentage range of doubt as only 0.1 per cent. Therefore three significant figures may cover a percentage range of doubt of 0.1 to 1 per cent and hence is indefinite as a measure of precision.

Superfluous figures are sometimes allowed to accumulate in ordinary arithmetic processes of addition, subtraction, multiplication and division. In these computational processes, the doubtful figures are written in italics, and also figures that result from their use in the processes and are thereby placed in doubt. There is another thing that is obvious, it is useless to give more than one doubtful figure. To illustrate the point we should not put the result 117.6 since the figure in the units place (*i.e.*, 7) is in doubt, and therefore, it is useless to put a figure in tenths place. The correct way to express the above result is 117.

*To illustrate further, two examples are given below :*

**Example 2.11** *Three resistors have values of 72.3, 2.73 and 0.612 Ω respectively with an uncertainty of one unit in the last figure in each case. Find the sum of three resistors connected in series.*

**Solution.** Since the three resistances are in series, their sum is : $R_S = R_1 + R_2 + R_3$

Now $R_1 = 72.3\,\Omega$

$R_2 = 2.73\,\Omega$

$R_3 = 0.612\,\Omega$

$R_S = 75.642\,\Omega$

The result cannot be expressed as 75.642 Ω, as even the figure in the tenth place *i.e.*, 6 is in doubt. Therefore the resultant resistance is 75.6 Ω with 6 as first doubtful figure.

**Example 2.12** *The voltage and current recorded in a d.c. circuit are respectively 12.16 V and 1.34 A with an uncertainty of one unit in the last figure in each case. Calculate the power.*

**Solution.** Power = 12.16 × 1.34 W.

Multiply by long hand, we have

```
      12·16
       1·34
    -------
      4864
     3648×
    1216××
    -------
    16·2944
```

The power is expressed as 16.2 W as 2 is the first doubtful figure. It would obviously be absurd to write the answer with the entire product obtained.

When two or more measurements with different degrees of accuracy are added, the result is only as accurate as the least accurate measurement. This is illustrated by the following example.

**Example 2.13** *Two resistors $R_1$ and $R_2$ are connected in series with $R_1 = 28.7\,\Omega$ and $R_2 = 3.624\,\Omega$. Calculate the total resistance to the appropriate number of significant figures.*

**Solution.**

$R_1 = 28.7\,\Omega$ (*three significant figures*)

$R_2 = 3.624\,\Omega$ (*four significant figures*)

Total resistance $R = 28.7 + 3.624$

$= 32.324\,\Omega$ (*five significant figures*)

$= 32.3\,\Omega$ (*three significant figures*)

This is done because one of the resistances is accurate only to three significant figures (or tenth of an ohm in this case) and therefore the result should also be reduced to three significant figures (or the nearest tenth of an ohm in this case) and hence the value 32.3 Ω.

The number of significant figures in multiplication may increase rapidly, but only the appropriate figures are retained in the answer, as is illustrated by the following example.

**Example 2.14** *In calculating voltage drop, a current of 4.37 A is recorded in a resistance of 31.27 Ω. Calculate the voltage drop across the resistor to the appropriate number of significant figures.*

**Solution.**

Current $I = 4.37$ A (*three significant figures*)

Resistance $R = 31.27\,\Omega$ (*four significant figures*)

Voltage drop $E = IR = 4.37 \times 31.27 = 136.6499$ V (*seven significant figures*)

Since there are three significant figures involved in the multiplication, the answer can be written only to a maximum of three significant figures *i.e.*, $E = 137$ V.

## 2.17 STATIC SENSITIVITY

The *static sensitivity* of an instrument or an instrumentation system is the ratio of the magnitude of the output signal or response to the magnitude of input signal or the quantity being measured. Its units are millimetre per micro-ampere, counts per volt etc. depending upon the type of input and output.

Sometimes the static sensitivity is expressed as the ratio of the magnitude of the measured quantity to the magnitude of the response. Thus the sensitivity expressed this way has the units of micro-ampere per millimetre or volt per count etc. as the case may be depending upon nature of input and output. Thus it is reciprocal of the sensitivity as defined above. This ratio is defined as **deflection factor** or **inverse sensitivity**. Many manufacturers define the sensitivity of their instruments in terms of inverse sensitivity and still call it sensitivity. 2

When a calibration curve is linear as in Fig. 2.6(*a*) the sensitivity of the instrument can be defined as in slope of the calibration curve. For this case the sensitivity is constant over the entire range of the instrument. However, if the curve is not nominally a straight line the sensitivity varies with the input as in Fig. 2.6(*b*).

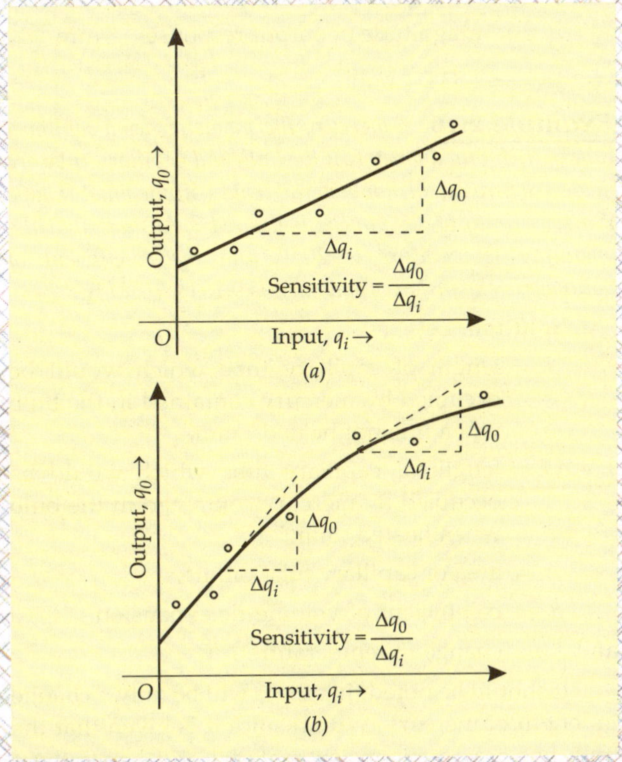


**Fig. 2.6** Definition of sensitivity.

In general, the static sensitivity at the operating point is defined as

$$\text{Static sensitivity} = \frac{\text{infinitesimal change in output}}{\text{infinitesimal change in input}} = \frac{\Delta q_0}{\Delta q_i} \quad \text{...(2.21)}$$

Similarly,

$$\text{Inverse sensitivity or deflection factor} = \frac{\Delta q_i}{\Delta q_o} \quad \text{...(2.22)}$$

The sensitivity of an instrument should be high and therefore the instrument should not have a range greatly exceeding the value to be measured. However, some margin should be kept for any accidental overloads.

**Example 2.15** *A Wheatstone bridge requires a change of 7 Ω in the unknown arm of the bridge to produce a change in deflection of 3 mm of the galvanometer. Determine the sensitivity. Also determine the deflection factor.*

**Solution.**

$$\text{Sensitivity} = \frac{\text{magnitude of output response}}{\text{magnitude of input}}$$

$$= \frac{3\,\text{mm}}{7\,\Omega} = 0.429\ \text{mm}/\Omega$$

Inverse sensitivity or scale factor

$$= \frac{\text{magnitude of input}}{\text{magnitude of output response}} = \frac{7\,\Omega}{3\,\text{mm}}$$

$$= 2.33\,\Omega/\text{mm}.$$

**Example 2.16** *A mercury thermometer has a capillary tube of 0.25 mm diameter. If the bulb and capillary tube are made of a zero expansion material what volume must it have if a sensitivity of 2.5 mm/°C is desired ? Assume that the operating temperature is 20°C and the co-efficient of volumetric expansion of mercury is 0.181× $10^{-3}$ /°C.*

**Solution.** Let :

$L_c$ = length of capillary tube which would be occupied by mercury contained in the bulb when it is not heated ; mm,

$L_c + \Delta L_c$ = length of capillary tube which would be occupied by mercury contained in the bulb when heated ; mm,

$A_c$ = area of capillary tube ; $\text{mm}^2$,

$\alpha_v$ = co-efficient of volumetric expansion ; $°\text{C}^{-1}$,

and $\Delta\theta$ = change in temperature ; °C.

It should be noted that there will be only a change in length of mercury column since it is given that the bulb has a zero expansion material and hence there will be no changes in its area and length. This is true of capillary tube as well.

$$\text{Sensitivity} \quad S = \frac{\Delta q_o}{\Delta q_i} = \frac{(L_c + \Delta L_c) - L_c}{\Delta\theta} = \frac{\Delta L_c}{\Delta\theta}$$

$$= 2.5\ \text{mm/°C}.$$

$$\text{Now} \quad A_c(L_c + \Delta L_c) = A_c(L_c + \alpha_v L_c \Delta\theta)$$

∴ Length of capillary tube

$$L_c = \frac{1}{\alpha_v} \cdot \frac{\Delta L_c}{\Delta\theta} = \frac{1}{0.181 \times 10^{-3}} \times 2.5$$

$$= 13.8 \times 10^3\ \text{mm} = 13.8\ \text{m}.$$

Hence area of bulb

$$A_b = A_c L_c = \frac{\pi}{4}(0.25)^2 \times 13.8 \times 10^3 = 677.4\ \text{mm}^3$$

## 2.18 LINEARITY

One of the best characteristics of an instrument or a measurement system is considered to be linearity, that is, the output is linearly proportional to the input. Most of the systems require a linear behaviour as it is desirable. This is because the conversion from a scale reading to the corresponding measured value of input quantity is most convenient if one merely has to multiply by a fixed constant rather than consult a non-linear calibration curve or compute from non-linear calibration equations. Also when the instrument is part of a large data or control system, linear behaviour of the part often simplifies the design and analysis of whole system. Therefore relationships to the degree of straight line relationship (*i.e.*, linearity) are common.

If for an instrument calibration curve (relating output to input) is not a straight line, it should not be concluded that the instrument is inaccurate. This is a misconception, a non-linear behaviour does not essentially lead to inaccuracy. Such an instrument may be highly accurate as ever.

However, most of the time it is necessary that measurement system components should have linear characteristics. For example, the resistance used in a potentiometer should vary linearly with displacement of the sliding contact in order that the displacement is directly proportional to the sliding contact voltage. Any departure from linearity would result in error in the read out system.

There are many definitions of linearity that exist. However, linearity defined in terms of **Independent Linearity** is the most preferable in many cases. The computation of independent linearity is done with reference to a straight line showing the relationship between output and input. This straight line is drawn by using the method of least squares from the given calibration data. This straight line is sometimes called an idealized straight line expressing the input-output relationship. The linearity is simply a measure of maximum deviation of any of the calibration points from this straight line. Figure 2.7 shows the actual calibration curve *i.e.*, a relationship between input and output and a straight line drawn from the origin using the method of least squares.

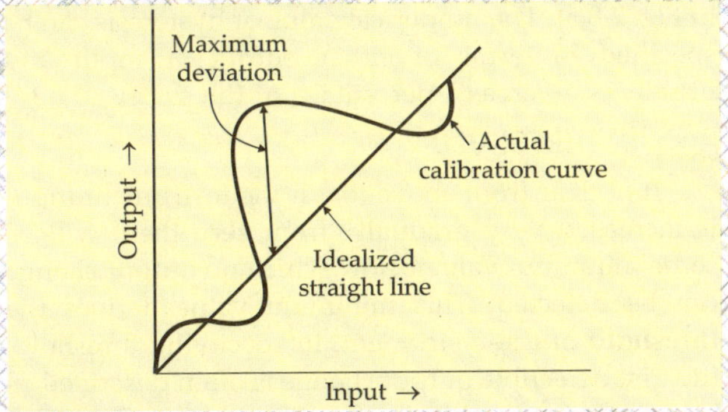


**Fig. 2.7** Linearity depicted as deviation between actual calibration curve and idealized straight line relationship.

The independent non-linearity may be defined as

$$\text{Non-linearity} = \frac{\text{(maximum deviation of output from the idealized straight line)}}{\text{actual reading}} \times 100 \quad \ldots(2.23)$$

$$\text{or Non-linearity} = \frac{\text{(maximum deviation of output from the idealized straight line)}}{\text{full scale deflection}} \times 100 \quad \ldots(2.24)$$

Equation 2.23 expresses the non-linearity in terms of the percentage of the instrument reading and therefore recognizes the desirability of a constant-percentage non-linearity. On the other hand, Eqn. 2.24 expresses, the non-linearity as a percentage of full scale reading and therefore recognizes the impossibility of testing extremely small deviations near zero since deviations approach zero as the reading approaches zero. Therefore, we will be using the definition of non-linearity as given by Eqn. 2.24. Accuracy and linearity are closely related quantities. It is desirable to keep the non-linearity as small as possible as it would in that case result in small errors. In instruments, considered essentially linear the specification of non-linearity is equivalent to a specification of overall, inaccuracy. Therefore, in many linear instruments only linearity specification is given and accuracy specification is omitted. For example, a potentiometer in a self balancing potentiometer readout system had a linearity of 0.1 percent, its accuracy would be 1 part in 1000. With a 1000 count digital encoder connected to the driving shaft, the variation in resistance of the potentiometer would produce an error equal to 1 count in 1000 counts. If the potentiometer and encoder were required to read 2000 counts, then the linearity of potentiometer would have to be less than 0.05 percent for an accuracy of 1 count in 2000 counts. Therefore, it would produce an error of 1 count in 1000 count in case a 1000 count digital encoder is connected to its shaft.

**Example 2.17** *A 10,000 Ω variable resistance has a linearity of 0.1% and the movement of contact arm is 320°. (a) Determine the maximum position deviation in degrees and the resistance deviation in ohm. (b) If this instrument is to be used as a potentiometer with a linear scale of 0 to 1.6 V determine the maximum voltage error.* 2

**Solution.** (*a*) From Eqn. 2.24,

Maximum displacement deviation

$$= \frac{\text{percent linearity} \times \text{full scale reading}}{100}$$

$$= \frac{0.1 \times 320}{100} = 0.32°$$

Similarly, maximum resistance displacement

$$= \frac{0.1 \times 10{,}000}{100} = 10\ \Omega.$$

(*b*) A displacement 320° corresponds to 1.6 V and therefore 0.32° corresponds to a voltage of

$$\left(\frac{0.32}{320}\right) \times 1.6 = 1.6 \times 10^{-3}\ \text{V}$$

$\therefore$ Maximum voltage error $= 1.6 \times 10^{-3}\ \text{V} = 1.6\ \text{mV}$.

## 2.19 HYSTERESIS

Hysteresis effects show up in any physical, chemical or electrical phenomenon. Hysteresis is a phenomenon which depicts different output effects when loading and unloading whether it is a mechanical system or an electrical system and for that matter any system. Hysteresis is non-coincidence of loading and unloading curves. Hysteresis, in a system, arises due to the fact that all the energy put into the stressed parts when loading is not recoverable upon unloading. This is because the second law of thermodynamics rules out any perfectly reversible process in the world.

Consider an instrument which has no friction due to sliding parts. When the input of this instrument is slowly varied from zero to full scale and then back to zero, its output varies as shown in Fig. 2.8(*a*). The non-coincidence of output when the input is increased and then decreased is on account of internal friction or hysteretic damping. Why we do not obtain the same output for both increasing and decreasing values of the input is explained by the fact that the energy put into the stressed parts when loading cannot be recovered back when unloading. Thus we obtain two different values of output for the same input under increasing and decreasing conditions. In case of instruments which are used on both sides of zero *i.e.*, input applied on both positive and negative sides, the variation of output is as shown in Fig. 2.8(*b*).

In case of instruments, which do not have internal friction, but have external sliding friction *i.e.*, constant coulomb friction the input-output relationships are like the ones shown in Figs. 2.8(*c*) and (*d*). Similar input-output relationships are exhibited if there is any free play or any looseness in any mechanism of the instrument.

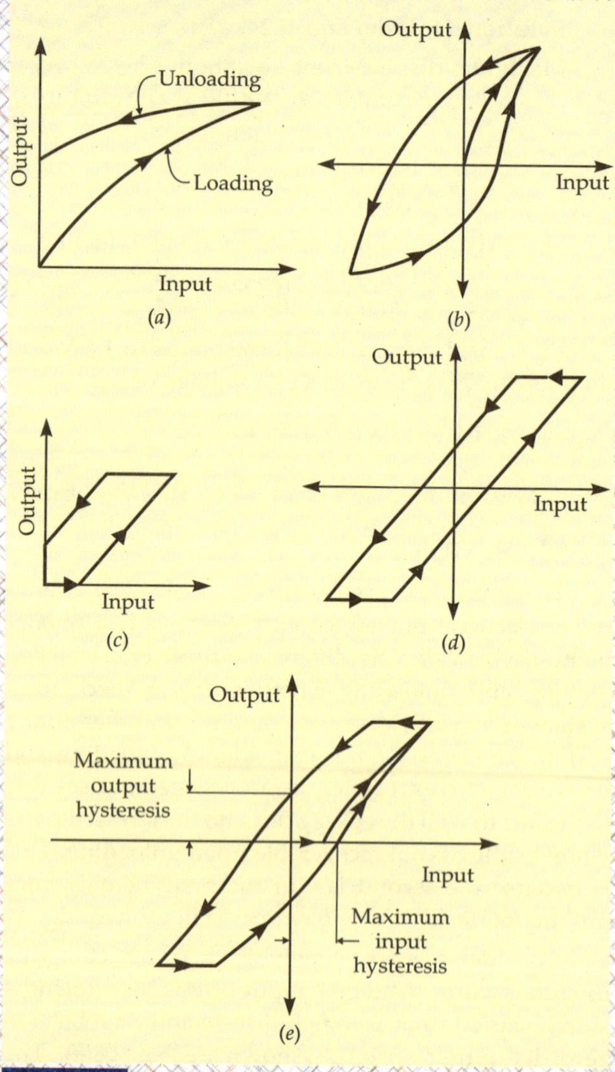


**Fig. 2.8** Hysteresis effects.

Hysteresis effects are there in electrical phenomena. One of the examples is the relationship between output voltage and field current in a d.c. generator. This is due to magnetic hysteresis. The relationship between field current and output voltage is similar to the one shown in Fig. 2.8(*b*).

In a given instrument a number of causes, such as listed above, combine to give an overall effect which may result in output-input relationship such as shown in Fig. 2.8(*e*). The numerical value of hysteresis can be specified in terms of either output or input and usually is given as a percentage of full scale.

## 2.20 THRESHOLD

It is clear from above that if the instrument input is increased very gradually from zero there will be some minimum value below which no output change can be detected. This minimum value defines the **threshold** of the instrument. In specifying threshold, the first detectible output change is often described as being any "noticeable measurable change".

This phenomenon is due to input hysteresis. In mechanical instruments, the first noticeable measurable change may not occur on account of backlash. In Fig. 2.9, which shows a gear train, the driven gear will not move *i.e.*, there will be no noticeable change in the movement of the driven gear unless the driving gear moves through a distance *x* which is the **backlash** between the gears.

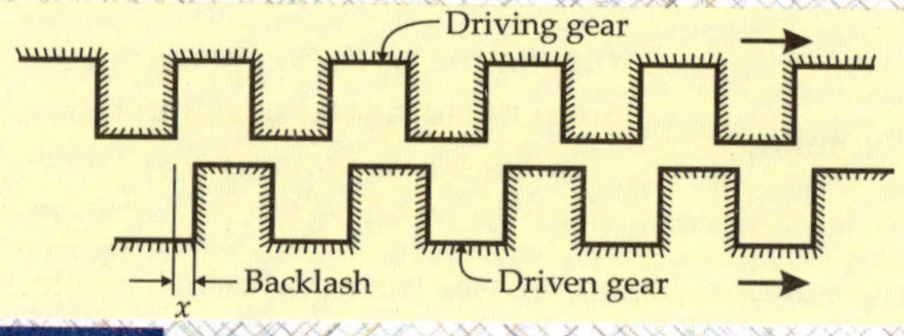


**Fig. 2.9** Threshold because of Backlash.

## 2.21 DEAD TIME

Dead time is defined as the time required by a measurement system to begin to respond to a change in the measurand. Fig. 2.10 shows the measured quantity and its value as indicated by an instrument. **Dead time**, in fact, is the time before the instrument begins to respond after the measured quantity has been changed.

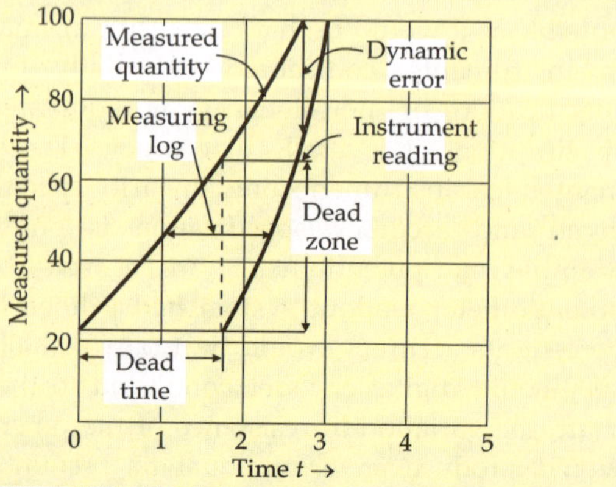


**Fig. 2.10** Dead time and dead zone.

## 2.22 DEAD ZONE

It is defined as the largest change of input quantity for which there is no output of the instrument (Fig. 2.10). For example, the input applied to the instrument may not be sufficient to overcome the friction and will, in that case, not move at all. It will only move when the input is such that it produces a driving force which can overcome friction forces. As stated in Article 2.19 and other factors which produce dead zone are backlash and hysteresis in the instrument.

The term **"dead zone"** is sometimes used interchangeably with term hysteresis. However, it may be defined as the total range of input values possible for a given output and may thus be numerically twice the hysteresis defined in Fig. 2.8 (*e*).

**Example 2.18** *The dead zone in a certain pyrometer is 0.125 percent of span. The calibration is 400°C to 1000°C. What temperature change might occur before it is detected ?*

**Solution.** Span = 1000 − 400 = 600°C.

$$\therefore \text{ Dead zone} = \frac{0.125}{100} \times 600 = 0.75°$$

A change of 0.75°C must occur before it is detected.

## 2.23 RESOLUTION OR DISCRIMINATION

If the input is slowly increased from some arbitrary (non-zero) input value, it will again be found that output does not change at all until a certain increment is exceeded. This increment is called **resolution** or **discrimination** of the instrument. Thus the smallest increment in input (the quantity being measured) which can be detected with certainty by an instrument is its **resolution** or **discrimination**. So **resolution** defines the smallest measurable input change while the threshold defines the smallest measurable input.

**Example 2.19** *A moving coil voltmeter has a uniform scale with 100 divisions, the full scale reading is 200 V and 1/10 of a scale division can be estimated with a fair degree of certainty. Determine the resolution of the instrument in volt.*

**Solution.** $1 \text{ scale division} = \frac{200}{100} = 2 \text{ V}$

$$\text{Resolution} = \frac{1}{10} \text{ scale division} = \frac{1}{10} \times 2 = 0.2 \text{ V}$$

**Example 2.20** *A digital voltmeter has a read-out range from 0 to 9,999 counts. Determine the resolution of the instrument in volt when the full scale reading is 9.999 V.*

**Solution.** The resolution of this instrument is 1 or 1 count in 9,999.

$$\therefore \text{ Resolution} = \frac{1}{9999} \text{ count} = \frac{1}{9999} \times 9.999 \text{ volt}$$
$$= 10^{-3} \text{ V} = 1 \text{ mV}$$

## 2.24 LOADING EFFECTS

The ideal situation in a measurement system is that when an element used for any purpose may be for signal sensing, conditioning, transmission or detection is introduced into the system, the original signal should remain un-distorted. This means that the original signal should not be distorted in any form by introduction of any element in the measurement system. However, under practical conditions it has been found that introduction of any element in a system results, invariably, in extraction of energy from the system there by distorting the original signal. This distortion may take the form of attenuation (reduction in magnitude), waveform distortion, phase shift and many a time all these undesirable features put together. This makes ideal measurements impossible. The incapability of the system to faithfully measure, record, or control the input signal (measurand) in undistorted form is called the **loading effect**.

2

A measurement system consists of *three* distinct stages :

(*i*) Detector-Transducer stage,

(*ii*) Signal conditioning stage (including signal transmission stage), and

(*iii*) Signal presentation stage.

The loading effects do not occur only in the first stage but may occur in any of the two subsequent stages while the first stage detector transducer loads the input signal, the second stage loads the first stage, and finally the third stage loads the second stage. In fact, the loading problem may be carried right down to the basic elements themselves.

## 2.25 INPUT AND OUTPUT IMPEDANCES

In measurement system, we deal with both electrical and mechanical quantities and elements and hence loading effects may occur on account of both electrical and mechanical elements. The loading effects are due to impedances of various elements connected in a system and hence it is desirable at this stage to analyze their effects.

In measurement systems, we talk of *generalized impedance* since impedance may take any form ; it may be electrical, mechanical, thermal or fluidic in nature. Before, we introduce the concept of generalized impedance, we will first discuss about the electrical impedance because it helps in the study of generalized impedance.

Two types of impedances *i.e.*, input impedance and output impedance are discussed as follows :

### 2.25.1 Input impedance

Figure 2.11 shows a voltage signal source and input device connected across it. The magnitude of the impedance of element connected across the signal source is called **Input Impedance.**

The magnitude of the input impedance is given by

$$Z_i = \frac{e_i}{i_i} \quad ...(2.25)$$

The instantaneous power extracted by the input device from the signal source is :

$$p = e_i i_i = \frac{e_i^2}{Z_i} \quad ...(2.26)$$

Thus it is clear from Eqns. 2.25 and 2.26 that a low input impedance device connected across the voltage signal source draws more current and drains more power from signal source than a high input impedance device. In other words, a low input impedance device connected across a voltage signal source loads the source more heavily than a high input impedance device.

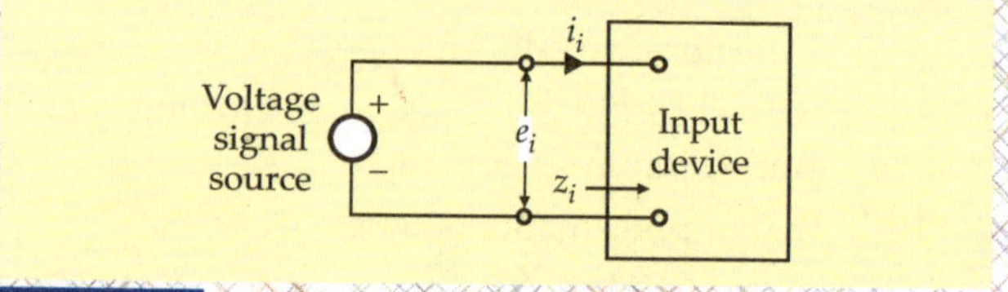


**Fig. 2.11** Voltage source and input device.

Hence, it automatically follows that when an instrument, which is to be connected across the system for measurements, is introduced into the system, the system conditions change from the original conditions on account of the power drawn from the system and the amount by which they deviate from the original conditions depend upon input impedance of the device. The lower the input impedance, higher is the power extracted and hence higher is the distortion of the original signal. Since a voltmeter is connected across the system for measurement of voltage, it should have a high input impedance. A low impedance voltmeter loads the system heavily and thus results in wrong measurements. Ideally a voltmeter when connected to the system should not draw any power from the system and this is possible only if the voltmeter has an impedance that has an infinite value.

### 2.25.2 Input Admittance

When the signal is of the form of current then series input devices are used. It is helpful to use the concept of input admittance in such cases.

Consider a constant current source and an input device connected across it as shown in Fig. 2.12.

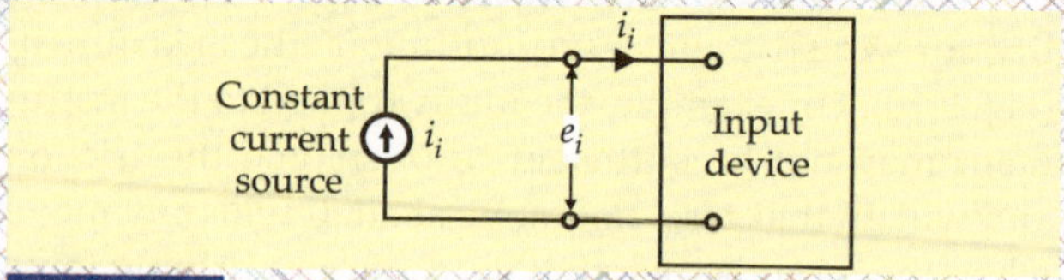


**Fig. 2.12** Current source and input device.

The magnitude of input admittance is given by :

$$Y_i = \frac{i_i}{e_i} \quad ...(2.27)$$

$$\text{Input impedance } Z_i = \frac{e_i}{i_i} = \frac{1}{Y_i} \quad ...(2.28)$$

The instantaneous power extracted from signal source is :

$$p = i_i e_i = \frac{i_i^2}{Y_i} \quad ...(2.29)$$

$$= i_i^2 Z_i \quad ...(2.30)$$

Thus it is clear from Eqns. 2.29 and 2.30 that in case of series elements the power drawn from the current signal source is small when the input admittance of the device is high *i.e.*, input impedance is low. Therefore in case of series devices, the loading effects are small when their input admittance is large (*i.e.*, when their input impedance is small). An ammeter, which is a series device, thus should be designed with a low input impedance so that the current is correctly measured. Ideally an ammeter should have an infinite input admittance or zero input impedance.

### 2.25.3 Output Impedance

The output impedance of a device is defined as its equivalent impedance as seen by the load. The definition of output device is only meaningful for an active device at a pair of terminals considered as source. Consider the circuit of 2.13(*a*). The term

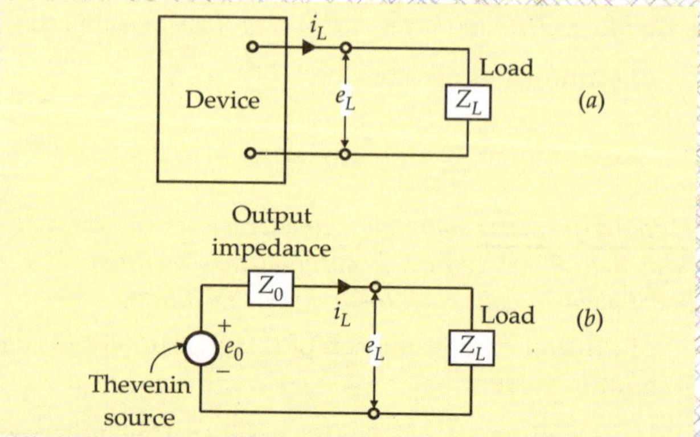


**Fig. 2.13** An active bilateral device and its Thevenin's equivalent circuit.

*equivalent impedance* implies that the device can be represented by a *Thevenin's equivalent circuit*. This circuit is shown in Fig. 2.13(*b*).

Let,

$e_0$ = voltage appearing across the output terminals of the device when the load is not connected, and

$e_L$ = voltage appearing across the output terminals when the load is connected.

The output impedance of an active device is defined as

$$Z_0 = \frac{e_0 - e_L}{i_L} \quad ...(2.31)$$

Equation 2.31 conveys the idea that the drop in the output voltage $(e_0 - e_L) = i_L Z_0$, is determined directly by the output impedance as the system is loaded. It is clear from the expression derived the lower the output impedance, the lesser is the effect of load on the output voltage.

Power loss in voltage source

$$p = (e_0 - e_L)i_L = i_L^2 Z_0 \quad ...(2.32)$$

Thus, for voltage sources, the lower the output impedance the lower is the voltage drop and also lower is the power consumption. Ideally there should not be any loading effect and this requires that output impedance $Z_0$ of the voltage source be equal to zero.

### 2.25.4 Output Admittance

When dealing with current sources it is preferable to deal with output admittance. The output admittance of current source is defined as,

$$Y_0 = \frac{i_0 - i_L}{e_L} \quad ...(2.33)$$

where $i_0$ = current delivered by the constant current source.

$i_L$ = current flowing through the load when it is connected to the source.

$$\therefore \quad Y_0 = \frac{1}{Z_0} \quad ...(2.34)$$

Power loss in the current source

$$p = (i_0 - i_L)e_L = e_L^2 Y_0 = \frac{e_L^2}{Z_0} \quad ...(2.35)$$

From Eqn. 2.35, we find that power drained from the current source is small if its output admittance is small or when its output impedance large. Ideally for no power loss in a current source, its output admittance should be zero and output impedance infinite.

**NOTE** While drawing the equivalent circuit of a voltage source, the output impedance is shown in series with the source. On the other hand, in case of current sources the output impedance is in parallel with the current source.

## 2.26 LOADING EFFECTS DUE TO SHUNT CONNECTED INSTRUMENTS



In measurement systems, voltage measuring, displaying and recording instruments like voltmeters, oscilloscopes and strip chart recorders are connected across the circuit *i.e.*, in shunt (parallel) with the circuit.

Consider a network consisting of linear bilateral impedances and generators with output terminals *A* and *B* as shown in Fig. 2.14. This is a black box with a Thevenin generator of voltage $E_0$ and an output impedance $Z_0$ in series. Suppose we are primarily concerned with the voltage of the output signal. Let $E_0$ be the open circuit voltage *i.e.*, the voltage that appears across the terminals *A* and *B* when the load (or any other measuring or recording device) which is a voltmeter in this case is not connected to the terminals.

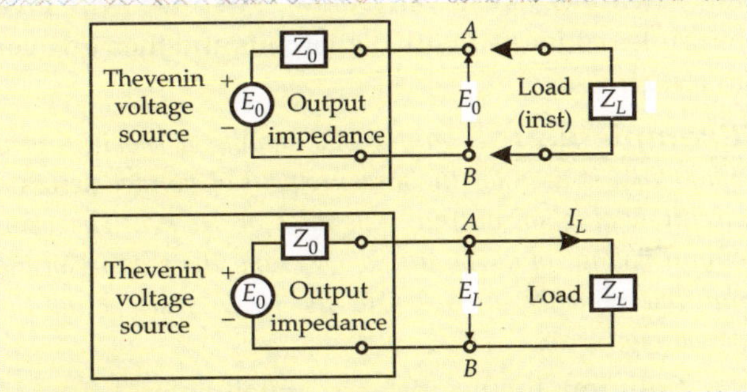


**Fig. 2.14** Voltage source & shunt connected instrument.

Ideally, when the load is connected across terminals *A* and *B* the output voltage should remain the same. However, the load impedance is not infinite and therefore when a voltmeter with an input impedance $Z_L$ is connected across *A* and *B*, a current $I_L$ flows. This causes a voltage drop $I_L Z_0$ in the output impedance of the source.

∴ Output voltage under loaded condition is

$$E_L = E_0 - I_L Z_0 = I_L Z_L \quad \text{or} \quad E_0 = I_L(Z_L + Z_0)$$

∴ Ratio of actual voltage appearing across the load (when the instrument is connected) to the voltage under no load conditions (ideal in this case) is

$$\frac{E_L}{E_0} = \frac{I_L Z_L}{I_L(Z_L + Z_0)} = \frac{1}{1 + Z_0/Z_L} \quad ...(2.36)$$

∴ Actual voltage measured,

$$E_L = \frac{E_0}{1 + Z_0/Z_L} \quad ...(2.37)$$

Thus the voltage which is measured is modified both in phase and magnitude. This means that the original voltage signal is distorted on account of connection of measuring instrument across it.

It is clear from Eqn. 2.37 that in order that the original signal $E_0$ should remain undistorted the value of input impedance of the instrument, $Z_L$, should be infinite (or the value of output impedance of the source, $Z_0$ should be equal to zero which is not attained in practice). In order to obtain as less distortion as possible the value of $Z_L$, the input impedance of instrument, should be very high as compared with $Z_0$, the output impedance of the source. In fact, it may be proved that in order to reduce error caused by the loading effects on account of shunt connected instruments to 1%, the value of input resistance of device should be atleast 100 times that of the output resistance of the source while if an error of 5% can be tolerated the input resistance of devices should be atleast 20 times the output resistance of the source.

To illustrate the loading effects of shunt devices a few examples are given below. First we will give examples connected with d.c. circuits and then go over to a.c. circuits.

**Example 2.21** *A multimeter having a sensitivity of 2,000 Ω/V is used for the measurement of voltage across a circuit having an output resistance of 10 kΩ. The open circuit voltage of the circuit is 6 V. Find the reading of the multimeter when it is set to its 10 V scale. Find the percentage error.*

**Solution.** Input resistance of voltmeter

$$Z_L = 2000 \times 10\ \Omega = 20\ \text{k}\Omega$$

Output resistance of circuit $Z_0 = 10\ \text{k}\Omega$

Open circuit voltage of circuit under measurement

$E_0 = 6\ \text{V}$.

From Eqn. 2.37, reading of voltmeter is

$$E_L = \frac{E_0}{1 + Z_0/Z_L} = \frac{6}{1 + 10/20} = 4\ \text{V}$$

∴ Percentage error in voltage reading

$$= \frac{4-6}{6} \times 100 = -33\% \text{ or } 33\% \text{ low.}$$

The loading problem given in Example 2.21 is typical of cases where a common type of voltmeter such as a multimeter whose input resistance is comparable with that of the resistance of the circuit under test is used for measuring voltage. Thus voltmeters having output resistance comparable to that of circuit under test should not be used as they seriously modify the value of test voltages. Such types of voltmeters are unsuitable for communication and electronic circuits where the impedance levels are high. However, they can be used for low impedance circuits giving a good accuracy.

If we wish to achieve 99% accuracy in voltage measurements, the input resistance of the voltmeter should be greater than 100 times the output resistance of source. For an accuracy of 95% the input resistance should be atleast 20 times the output resistance of source.

**Example 2.22** *Suppose the voltmeter of Example 2.21 is used for measurement of voltage of a circuit having an output resistance of 1000 Ω and an open circuit voltage of 6 V at its 10 V scale. Find the error in measurement.*

**Solution.** Output resistance of circuit

$$Z_0 = 1000\ \Omega = 1\ \text{k}\Omega.$$

Reading of voltmeter

$$E_L = \frac{E_0}{1 + Z_0/Z_L} = \frac{6}{1 + 1/20} = 5.71\ \text{V}$$

$$\text{Percentage Error} = \frac{5.71 - 6}{6} \times 100$$

$$= -4.8\% \text{ or } 4.8\% \text{ low.}$$

Thus when the output impedance of source is quite low as compared with input impedance of voltmeter, the error involved in measurements is quite small.

**Example 2.23** *A 50 V range voltmeter is connected across the terminals A and B of the circuit shown in Fig. 2.15. Find the reading of the voltmeter under open circuit and loaded conditions. Find the accuracy and the loading error. The voltmeter has a resistance of 1000 kΩ.*

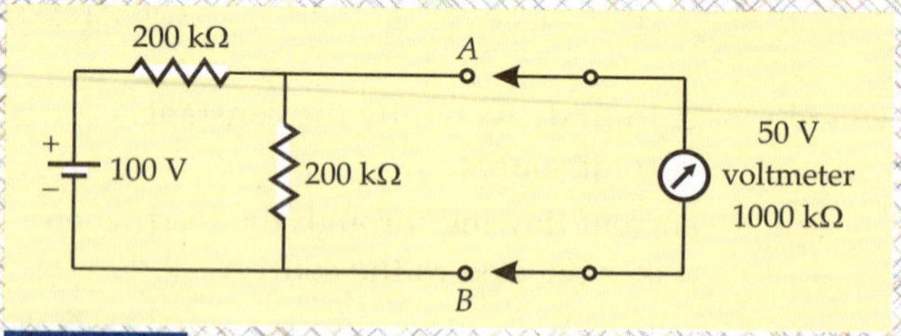


**Fig. 2.15** Diagram of Example 2.23.

**Solution.** Let us reduce the circuit to its Thevenin's equivalent circuit. The open circuit voltage, $E_0$, appearing across terminals $A$ and $B$ is

$$E_0 = \frac{100 \times 200}{400} = 50\ \text{V}.$$

The output impedance (resistance in this case) of the source as looking into terminals $A$ and $B$ is

$$Z_0 = \frac{200 \times 200}{200 + 200} = 100\ \text{k}\Omega$$

The Thevenin equivalent circuit under loaded conditions is shown in Fig. 2.16.

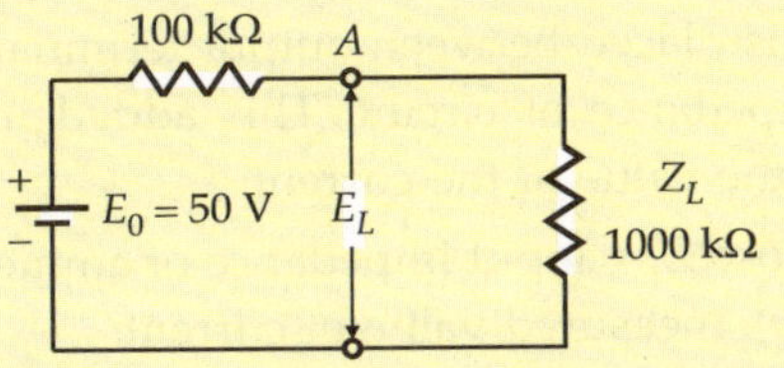


**Fig. 2.16** Thevenin Equivalent circuit of Example 2.23.

Voltage appearing across terminals *A* and *B* under loading conditions is

$$E_L = \frac{E_0}{1+Z_0/Z_L} = \frac{50}{1+100/100} = 45.45 \text{ V}.$$

$$\text{Loading error } = \frac{45.45-50}{50} \times 100$$

$$= -9.1\% \text{ or } 9.1\% \text{ low}$$

$$\text{Accuracy} = 100 - \% \text{ loading error}$$

$$= 100 - 9.1 = 90.9\%.$$

***Operation with A.C.*** The analysis of loading-effects with a.c. is not so easy. It should be borne in mind that both $Z_0$ and $Z_L$ are dependent upon frequency. Therefore the indicated voltage will depend upon the frequency of operation. On account of the input capacitance effects of the instruments, the value of input impedance of instruments, $Z_L$ becomes low at high frequencies with the result the input signal is substantially attenuated at high frequencies. It is not only the magnitude of the signal that is affected but also its phase. Worst still, as is consequence of the shunt capacitance, the non-sinusoidal signals are distorted in waveform also.

The magnitude of the measured signal becomes substantially smaller with increase in frequency as shown in Fig. 2.17.

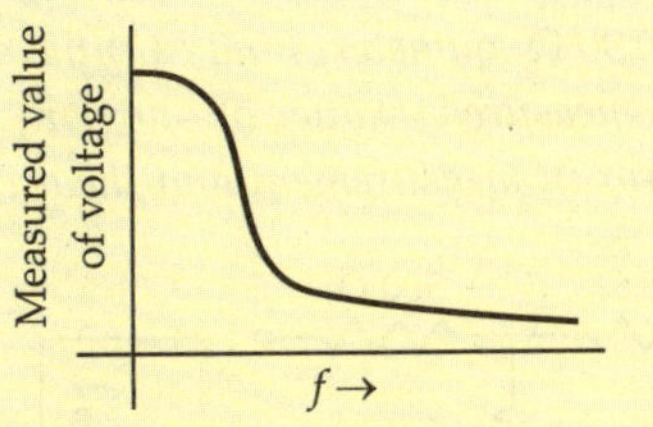


**Fig. 2.17** Effect of frequency on measured output.

The sharply changing non-sinusoidal waveforms are rounded off because of the finite time it takes to charge a capacitor. This is shown in Fig. 2.18.

The effect of frequency on magnitude and phase shift of a signal is shown in the following example.

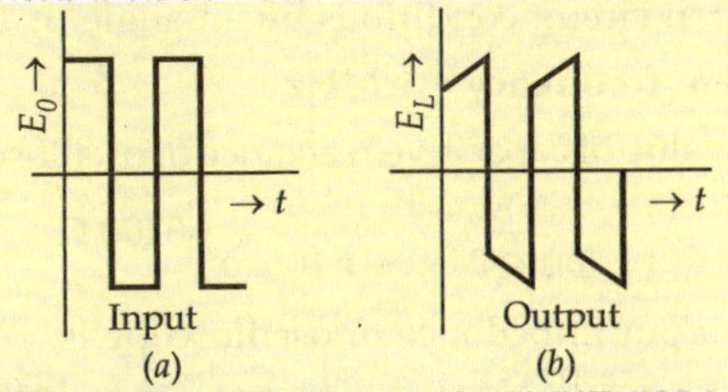


**Fig. 2.18** Waveform distortion of output of non-sinusoidal signals.

**Example 2.24** *An oscilloscope having an input resistance of 1 MΩ shunted by 50 pF capacitance is connected across a circuit having an effective output resistance of 10 kΩ. If the open circuit voltage has 1.0 V peak for a 100 kHz sine wave, what will be the voltage indicated on the oscilloscope when the frequency is (i) 100 kHz and (ii) 1 MHz ?*

**Solution.** The equivalent circuit for the measurement system is shown in Fig. 2.19.

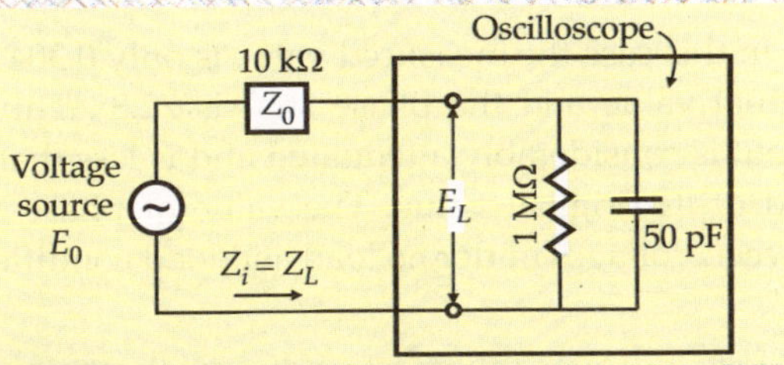


**Fig. 2.19** Diagram of Example 2.24.

**When frequency = 100 kHz ;**

The value of capacitive reactance at 100 kHz is

$$X_c = \frac{1}{2\pi f C}$$

$$= \frac{1}{2\pi \times 100 \times 1000 \times 50 \times 10^{-12}} = 32000\ \Omega$$

The input impedance of the oscilloscope is

$$Z_L = \frac{R(-jX_c)}{R-jX_c} = \frac{10^6 \times (-j32 \times 10^3)}{10^6 - j32 \times 10^3}$$

$$= -j32 \times 10^3 \Omega = 32 \times 10^3 \angle -90° \Omega$$

The voltage across the load

$$E_L = \frac{E_0}{1+Z_0/Z_L}$$

$$= 1.0\angle 0°.\frac{1}{1+\dfrac{10 \times 10^3 \angle 0°}{32 \times 10^3 \angle -90°}}$$

$$= 0.954\angle -17.4° \text{V (peak)}.$$

This means that the magnitude of voltage indicated by the oscilloscope is 0.954 of its original value.

∴ The error is (1 – 0.954) × 100 = 4.6 percent. Also the voltage under loaded conditions lags the voltage under open circuit conditions by an angle of 17.4°.

**When frequency = 1 MHz**

The value of capacitive reactance of oscilloscope is,

$$X_c = \frac{1}{2\pi\times 10^6 \times 50\times 10^{-12}} = 3200\ \Omega$$

The input impedance of oscilloscope is,

$$Z_L = \frac{R(-jX_c)}{R-jX_c}$$

$$= \frac{10^6(-j3.2\times 10^3)}{10^6 - j3.2\times 10^3} = 3.2\times 10^3\ \angle -90°\Omega$$

∴ The voltage across the load is $E_L = \dfrac{E_0}{1+Z_0/Z_L}$

$$= 10\angle 0° \times \frac{1}{1+\dfrac{10\times 10^3\angle -0°}{3.2\times 10^3\angle -90°}}$$

$$= 0.304\angle -72.3°\text{V (peak)}.$$

In this case the measured value is only 0.304 of its original value and the phase shift is 72.3°. Thus the output is considerably attenuated and is less than one third of its original value. The value of phase angle between voltages under no load and load conditions is 72.3°.

This indicates the effect of distortion of signal on account of increased shunting effect due to increase in frequency.

## 2.27 LOADING EFFECTS DUE TO SERIES CONNECTED INSTRUMENTS

Consider a network represented by a voltage source having a voltage $E_0$ and an output impedance $Z_0$. The output terminals are *A* and *B* as shown in Fig. 2.20(*a*).

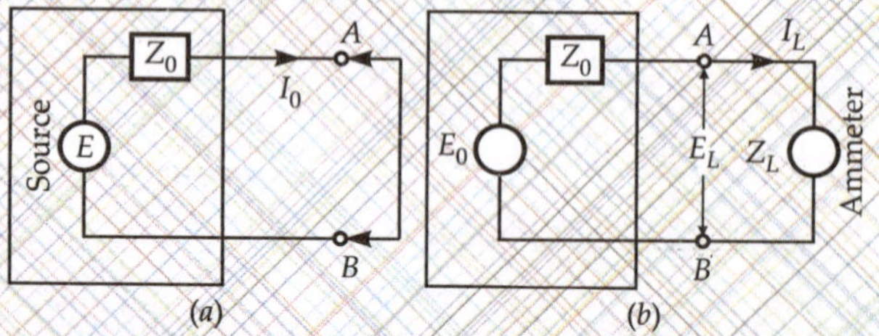


**Fig. 2.20** Loading effect of series connected instruments.

The value of current flowing between terminals *A* and *B* under ideal conditions is $I_0$. It is the current that flows when terminals *A* and *B* are shorted.

$$I_0 = \frac{E_0}{Z_0} \quad \text{or} \quad E_0 = I_0 Z_0$$

However, when we actually measure the current, a current measuring device is introduced between terminals *A* and *B*. It is usually an ammeter. When an ammeter is placed between output terminals, it adds to the impedance of circuit. This added impedance modifies the value of the current.

Suppose $Z_L$ = input impedance of ammeter. From Fig. 2.20(*b*), measured value of current

$$I_L = \frac{E_0}{Z_0+Z_L} = \frac{I_0 Z_0}{Z_0+Z_L} = \frac{I_0}{1+Z_L/Z_0} \qquad \text{...(2.38)}$$

In order that the measured value of current, be equal to the actual value of the current, $I_0$, the value of $Z_0 >> Z_L$. This means that the input impedance of the ammeter should be very small as compared with the output impedance of the source.

We can express the relationships of currents under loaded and unloaded conditions in terms of admittance. From Eqn. 2.38,

$$I_L = \frac{I_0}{1+Z_L/Z_0} = \frac{I_0 Z_0}{Z_0/Z_L}$$

$$= \frac{I_0/Y_0}{1/Y_0+1/Y_L} = \frac{I_0 Y_L}{Y_L+Y_0} \qquad \text{...(2.39)}$$

$$I_L = \frac{I_0}{1+Y_0/Y_L} \qquad \text{...(2.40)}$$

In other words, the input admittance of the series elements should be very large as compared with the output admittance of the source in order to reduce loading effect.

It may be concluded from above equations that for achieving 99% accuracy (1% error) in measurements the output resistance should be at least 100 times the resistance of the input device. In order to have 95% accuracy (5% error) the output resistance should be at least 20 times the resistance of input device.

**Example 2.25** *It is desired to measure the value of current in the 500 Ω resistor as shown in Fig. 2.21 by connecting a 100 Ω ammeter. Find : (a) the actual value of current, (b) measured value of current, and (c) the percentage error in measurement and the accuracy.*

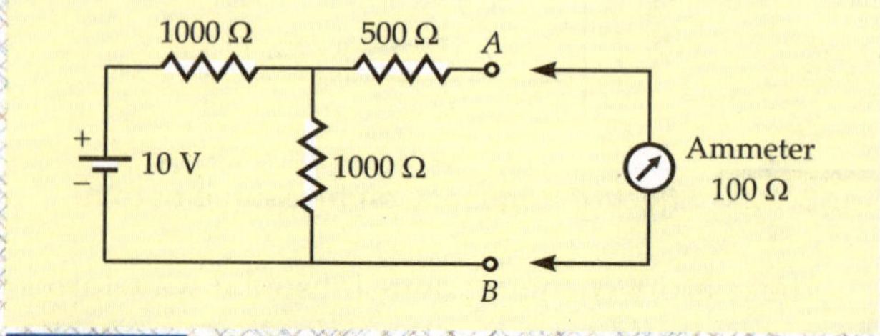


**Fig. 2.21** Diagram of Example 2.25.

**Solution.** (*a*) Let us reduce the actual circuit to an equivalent Thevenin's source.

Open circuit voltage as applying at terminals *A* and *B* is,

$$E_0 = 10 - \frac{10}{1000+1000} \times 1000 = 5 \text{ V}.$$

Output impedance of source as looking into terminals *A* and *B* is,

$$Z_0 = \frac{1000 \times 1000}{1000+1000} + 500 = 1000 \ \Omega$$

∴ The Thevenin equivalent circuit is as shown in Fig. 2.22.

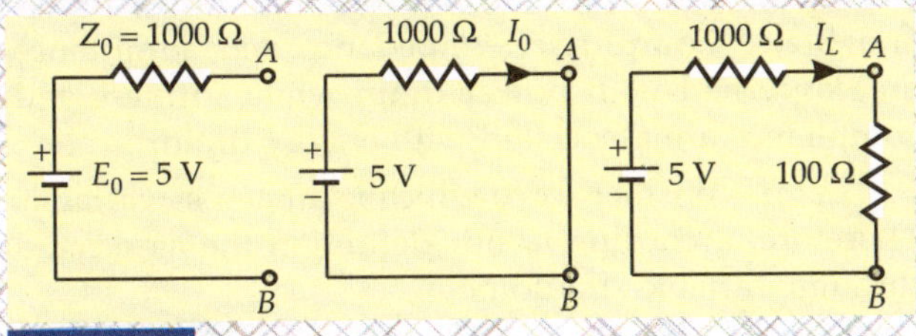


**Fig. 2.22** Thevenin Equivalent Circuit of Example 2.25.

$$\therefore \text{ Actual value of current } I_0 = \frac{E_0}{Z_0} = \frac{5}{1000} \text{ A} = 5 \text{ mA}$$

(*b*) When the ammeter is introduced into the circuit the value of current is modified.

Measured value of current

$$I_L = \frac{E_0}{Z_0 + Z_L} = \frac{5}{1000+100} \text{ A} = 4.55 \text{ mA}.$$

$$\text{(c) Error} = \frac{4.55-5}{5} \times 100 = -9\,\% = 9\,\% \text{ low}$$

Accuracy of measurement $= 100 - 9 = 91\%$.

## 2.28 GENERALIZED IMPEDANCE AND STIFFNESS CONCEPTS

The introduction of any measuring instrument in a measurement system results in extraction of energy from the system thereby resulting in "*Loading Effects*" causing a distortion of the measurand. This renders the perfect measurement of the measurand impossible. In order to assess the imperfection in measurements it becomes imperative to introduce some means of analysis whereby it would become possible to determine the extent to which the measurand has been affected so that corrective measures can be taken up to eliminate these effects if possible or to minimize these effects. In order to understand the loading effects the concepts of **Static Impedance** and **Static Stiffness** are being introduced here.

Loading effects occur due to extraction of energy from the measurand. The transfer of energy requires the specification of two variables. At the input of each element of a measurement system there exists a variable $q_{i1}$ with which we are primarily concerned. At the same point, there is an associated variable, $q_{i2}$. The dimensions of these two variables are such that their product has the dimensions of power *i.e.*,

$$p = q_{i1} \times q_{i2} \qquad ...(2.41)$$

2

The above product represents the instantaneous energy of the system.

The physical variables which determine the flow of energy in all dynamical systems can be classified as :

- Through variables and
- Across variables.

**1. *Through Variables.*** *Through variables* are those which can be specified and measured at one point in space. The through variables are extensive variables, in the sense, that their magnitude depends upon the extent of the system taking part in the energy transfer. Through variables are also called *Flow Variables* or *Pervariables*.

**2. *Across Variables.*** *Across variables* are those which can be specified and measured by two points in space. Usually one point is a reference point. The across variables are intensive variables whose magnitude is independent of the material being considered. Across variables are also known as *Effort Variables* or *Transvariables*.

Time is a variable having a spatial independence. All objects in a measurement system (for that matter in any dynamical system) involve a measured or defined relationship between an *Across Variable* (or *Transvariable*) and a *Through Variable* (or *Pervariable*). For instance, an electrical capacitor is defined in terms of the voltage (which is an across variable) and the charge (which is a through variable).

When we are primarily concerned with an input variable $q_{i1}$, and its associated variable $q_{i2}$ where $q_{i1}$ is the across variable and $q_{i2}$ is the through variable, the **generalized input impedance** is :

$$Z_{gi} = \frac{q_{i1}}{q_{i2}} \qquad ...(2.42)$$

The power drawn from the system is

$$p = \frac{q_{i1}^2}{Z_{gi}} \qquad ...(2.43)$$

Compare Eqn. 2.43 with Eqn. 2.26, where voltage is the across variable. From above we conclude that a large input impedance is necessary to keep the power drain small.

By analogy, (see Eqn. 2.37) value of measured quantity is

$$q_{i1L} = \frac{1}{1+Z_{g0}/Z_{gi}} q_{i10} \quad ...(2.44)$$

where $q_{i1L}$ = measured value of across variable,

$q_{i10}$ = undisturbed value of across variable,

$Z_{g0}$ = generalized output impedance of source,

$Z_{gi}$ = generalized input impedance of measurement device.

Of course, if we know both $Z_{g0}$ and $Z_{gi}$ we can find the value of measurand by means of Eqn. 2.44. However, this is inconvenient. Also $Z_{g0}$ is also not clearly known in non-electrical systems where definitions of both $Z_{gi}$, and $Z_{g0}$ are not always straight forward. Thus a high value of $Z_{gi}$, is desirable since then corrections are unnecessary.

However, if $q_{i1}$, is the through variable, the situation becomes quite different. In that case it is more convenient to use admittances instead of impedances.

The **generalized input admittance** is defined as

$$Y_{gi} = \frac{\text{through variable}}{\text{across variable}} = \frac{q_{i1}}{q_{i2}} \quad ...(2.45)$$

The power extracted from the measurement system is,

$$p = q_{i1} q_{i2} = \frac{q_{i1}^2}{Y_{gi}} \quad ...(2.46)$$

(Compare with Eqn. 2.29).

Thus it is clear from Eqn. 2.46 that a large value of input admittance $Y_{gi}$ is required to limit the power extracted from the measurement system in case of through variables.

The value of measurand (which is through variable in this case), under loaded conditions, can be written by analogy with Eqn. 2.40 as

$$q_{i1L} = \frac{1}{1+Y_{g0}/Y_{gi}} \times q_{i10} \quad ...(2.47)$$

where $Y_{g0}$ = generalized output admittance of preceding element,

$Y_{gi}$ = generalized input admittance of instrument,

$q_{i1L}$ = measured value of through variable,

and $q_{i10}$ = undisturbed value of through variable.

Thus it is clear from Eqn. 2.47 when a through variable is being measured, the value of the input admittance of the input element should be very large as compared with output admittance of the source.

## 2.29 STATIC STIFFNESS AND STATIC COMPLIANCE

There are some instruments which do not have a loading effect on the preceding stage under static conditions meaning thereby that when an input signal is applied to them under static conditions, the output signal is a faithful representation of the input signal. However, when the same instruments are subjected to dynamic conditions *i.e.,* they have to perform measurements where they are required to go from one steady state to another steady state, they produce loading effects because energy is required for these operations. Under these conditions, the concepts of impedance and admittance may not be useful and, therefore, it is appropriate introduce the concepts of **Static Stiffness** and **Static Compliance**. Impedance and admittance are concerned with power drain while stiffness and compliance are associated with energy drain.

Let us consider force which is an across variable. Therefore, according to our definition the associated through variable should be

$$= \frac{\text{power}}{\text{across variable}} = \frac{\text{N-m/s}}{\text{N}} = \text{m/s} = \text{velocity}.$$

Thus the through variable, associated with across variable force, is velocity.

∴ The mechanical impedance is given by

$$\frac{\text{across variable}}{\text{through variable}} = \frac{\text{force}}{\text{velocity}}$$

Suppose now we calculate the mechanical impedance of an elastic system by applying a constant force and noting the resultant velocity. The mechanical impedance will be :

$$\frac{\text{force}}{\text{velocity}} = \frac{\text{force}}{0} = \text{infinity}$$

(this is because the velocity of the system under static conditions is zero.)

The question now is, how to characterize such a system. The answer lies in using concepts of energy rather than those of power under such situations. The variables now have to be defined in terms of energy and not in terms of power associated with them. Therefore a new term **Stiffness** has to be used. This term is associated with energy and not with power. Thus, we define mechanical stiffness for this case as :

$$\frac{\text{force}}{\int(\text{velocity})dt} = \frac{\text{force}}{\text{displacement}}$$

In general the **Mechanical Stiffness** is defined as :

$$S_g = \frac{\text{across variable}}{\int(\text{through veriable})dt} \quad ...(2.48)$$

The same formulae can be used for calculating the value of the output signals as were used for impedance except that stiffness is used in place of impedance.

Thus Eqn. 2.44 becomes

$$q_{i1L} = \frac{1}{1+S_{g0}/S_{gi}} \times q_{i10} \qquad ...(2.49)$$

where $q_{i1L}$ = measured value of across variable,

$q_{i10}$ = undisturbed value of across variable,

$S_{g0}$ = generalized static output stiffness of source,

and $S_{gi}$ = generalized static stiffness of the input source.

When the measured variable is a through variable, the admittance rather than impedance is a convenient tool as seen earlier. The same analogy can be applied here. Compliance is used instead of stiffness. **Compliance** is defined as inverse of stiffness as admittance is inverse of impedance.

$$\text{Compliance } C_g = \frac{1}{S_g} = \frac{\text{through vairable}}{\int(\text{across variable})dt} \qquad ...(2.50)$$

By analogy with Eqn. 2.49,

$$q_{i1L} = \frac{1}{1+C_{g0}/C_{gi}} \times q_{i10} \qquad ...(2.51)$$

where $q_{i1L}$ = measured value of through variable,

$q_{i10}$ = undisturbed value of through variable,

$C_{gi}$ = generalized static input compliance of measuring instrument,

and $C_{g0}$ = generalized static compliance of measured system.

Table 2.1 illustrates the general applicability of the impedance, admittance, stiffness and compliance for measurement problems. The classification is divided into *two* parts :

(*i*) **Power based.** Where the output of across variable is calculated on the basis of generalized impedance and admittance concepts.

(*ii*) **Energy based.** Where the output of through variable is calculated on the basis of generalized stiffness and compliance concepts.

**Table 2.1** *Variables and their Associated Power and Energy Variables*

| S. No. | Measured Variable | Associated Variable | | Impedance Z | Admittance Y | Stiffness S | Compliance C |
|---|---|---|---|---|---|---|---|
| | | Power based | Energy based | | | | |
| 1. | Voltage (Across) | current | charge | V/A | — | V/C | — |
| 2. | Current (Through) | voltage | $\int$(voltage)$dt$ | — | A/V | — | $\frac{A}{V\text{-}s}$ |
| 3. | Force (Across) | translational velocity | translational displacement | $\frac{N}{m/s}$ | — | $\frac{N}{m}$ | — |
| 4. | Translational displacement (Through) | $d/dt$ (force) | force | — | $\frac{m}{N/s}$ | — | $\frac{m}{N}$ |
| 5 | Torque (Across) | rotational velocity | rotational displacement | $\frac{N\text{-}m}{rad/s}$ | — | $\frac{N\text{-}m}{rad/s}$ | — |
| 6 | Rotational displacement (Through) | $d/dt$ (torque) | torque | $\frac{rad}{N\text{-}m}$ | — | $\frac{rad}{N\text{-}m}$ | — |
| 7 | Translational velocity | force | $\int$(force)$dt$ | — | $\frac{m/s}{N}$ | — | $\frac{m/s}{N\text{-}s}$ |
| 8 | Rotational velocity (Through) | torque | $\int$(torque)$dt$ | — | $\frac{rad/s}{N\text{-}m}$ | — | $\frac{rad/s}{N\text{-}m}$ |
| 9 | Translation acceleration (Through) | $\int$(force)$dt$ | $\int[(\text{force})dt]dt$ | — | $\frac{m/s^2}{N\text{-}s}$ | — | $\frac{m/s^2}{N\text{-}s^2}$ |
| 10 | Rotational acceleration (Through) | $\int$(torque) | $\int[(\text{torque})dt]dt$ | — | $\frac{rad/s^2}{N\text{-}m/s}$ | — | $\frac{rad/s^2}{N\text{-}m/s^2}$ |
| 11 | Fluid pressure (Across) | volume flow rate | volume flow | $\frac{N/m^2}{m^3/s}$ | — | $\frac{N/m^2}{m^3}$ | — |
| 12 | Volume flow rate (Through) | fluid pressure | $\int$(fluid pressure)$dt$ | — | $\frac{m^3/s}{N/m^2}$ | — | $\frac{m^3/s}{(N/m^2)s}$ |
| 13 | Temperature (Across) | heat transfer rate per unit temperature difference | total heat transfer per unit temperature difference | $\frac{°C}{J/(s\text{-}°C)}$ | — | — | $\frac{°C}{J/°C}$ |

## 2.30 IMPEDANCE MATCHING AND MAXIMUM POWER TRANSFER

In many applications it is desirable to *match* the impedance of the input device to the output impedance of the signal source instead of making the impedance of the input device either too high or too low. Many low frequency cases such as audio amplifiers, feeding loudspeakers and other electromechanical transducers require impedance matching for high power transfer.

The problem of maximum power transfer is analyzed by considering the source and also the input device (load) to have only pure resistance (See Fig. 2.14).

Let $E_0$ = voltage of the source under no load conditions.

$E_L$ = voltage of the source under loaded conditions,

$R_L$ = output resistance of the source,

and $R_L$ = input resistance of the load.

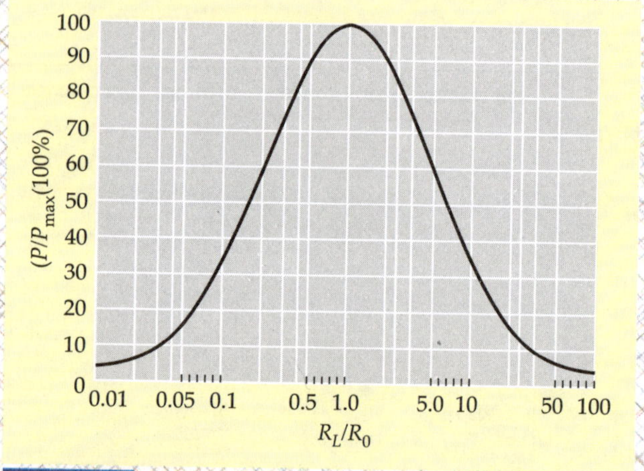


**Fig. 2.23** Maximum Power transfer.

We have, $$E_L = \frac{R_L}{R_0 + R_L} E_0 \quad \text{(See. Eqn. 2.36)}$$

The power delivered to the load is given by

$$P = \frac{E_L^2}{R_L}$$

The task now is to find out the value of load resistance so that maximum power transfer takes place between source and the load. We can write the above relationship as :

$$P = \frac{E_0^2}{R_L}\left(\frac{R_L}{R_0 + R_L}\right)^2$$

$$= \frac{E_0^2 R_L}{(R_0 + R_L)^2} \qquad \text{...(2.52)}$$

Maximum power transfer takes place when $\frac{dP}{dR_L} = 0.$

Differentiating Eqn. 2.52, we have :

$$\frac{dP}{dR_L} = \frac{(R_0 + R_L)^2 E_0^2 - E_0^2 R_L \times 2(R_0 + R_L)}{(R_0 + R_L)^4} = 0$$

which gives $R_L = R_0$ ...(2.53)

Thus in order that maximum power may be drawn from a source (when output impedance of the source is a resistance and also that of the load is a pure resistance) is when the external load resistance matches the internal resistance of the source. For maximum power transfer, $R_L = R_0$.

∴ Maximum power delivered to the load is

$$P_{max} = \frac{E_0^2}{(R_0 + R_L)^2} R_L$$

$$= \frac{E_0^2}{4R_0^2} \cdot R_0 \qquad \text{...(2.54)}$$

In case of a.c. circuits for maximum power transfer the impedance of the load should be made equal to the complex conjugate of the Thevenin equivalent impedance of the source.

Under conditions of maximum power transfer, the power supplied by source is

$$\frac{E_0^2}{R_L + R_0} = \frac{E_0^2}{2R_0}$$

while the power delivered to the load is only $E_0^2 / 4R_0$. Therefore, it is quite interesting to note that under conditions of maximum power transfer the efficiency is only 50%.

The variation power $P$ with ratio $R_L / R_0$ is shown in Fig. 2.23.

**Example 2.26.** *Measurements on a human nerve cell indicate an open circuit voltage of 80 mV, and a current of 5 nA through a 6 MΩ load. What is a maximum power available from the cell ?*

**Solution.** Let $R_0$ be the internal (output) resistance of the cell, $I_L$ the load current and $R_L$ the load resistance.

∴ Open circuit voltage of the cell

$$E_0 = I_L(R_0 + R_L)$$

or $$80 \times 10^{-3} = 5 \times 10^{-9}(R_0 + 6 \times 10^6)$$

Hence, output resistance of cell

$$R_0 = 10 \times 10^6 \Omega = 10 \text{ M}\Omega.$$

Maximum power available from the cell

$$P_{max} = \frac{E_0^2}{4R_0}$$

$$= \frac{(80 \times 10^{-3})^2}{4 \times 10 \times 10^6} = 160 \times 10^{-12}\ W = 0.16\ nW$$

**Example 2.27** *The output voltage of an audio amplifier is 10 V and 4 V when delivering a power of 25 W and 16 W respectively. Find the open circuit voltage and output resistance of the amplifier. What is the maximum power that the amplifier can give ?*

**Solution.** Let $R_0$ and $R_L$ be the output resistance of amplifier and load respectively.

Load resistance in the first case

$$R_{L1} = \frac{E_{L1}^2}{P_1} = \frac{(10)^2}{25} = 4\ \Omega$$

Load resistance in the second case

$$R_{L2} = \frac{E_{L2}^2}{P_2} = \frac{(4)^2}{4} = 1\ \Omega$$

Voltage across load $E_L = \dfrac{E_0 R_L}{R_0 + R_L}$

∴ In the first case

$$10 = \frac{E_0 \times 4}{R_0 + 4} \quad \text{or} \quad 10\,R_0 + 40 = 4\,E_0 \quad ...(i)$$

In the second case

$$4 = \frac{E_0 \times 1}{R_0 + 1} \quad \text{or} \quad 4\,R_0 + 4 = E_0 \quad ...(ii)$$

From (*i*) and (*ii*), we have,

Output resistance of amplifier $R_0 = 4\ \Omega$

and Open circuit voltage of amplifier $E_0 = 20$ V

The maximum power that a source can give is when its output resistance is equal to load resistance, *i.e.*,

$$R_0 = R_L = 4\ \Omega.$$

∴ The maximum power output of the amplifier

$$= E_0^2 / 4R_0 = (20)^2 / (4 \times 4) = 25\ W$$

2

# Review Questions

1. Define the term "True value". Explain why it is not practically possible to know the true value of a quantity. What is "Examplar Method" and why it is used in measurements ?
2. Explain the terms (*i*) Static Error (*ii*) Static Correction (*iii*) Relative Error and (*iv*) Percentage Relative Error.
3. Differentiate between the terms "Scale Range" and "Scale Span" giving suitable examples.
4. What is drift ? Explain the different types of drifts with sketches of input-output relationships in each case.
5. Define noise. Explain with examples the generated noise, conducted noise and radiated noise. Describe the techniques used for reducing the magnitude of the above mentioned categories of noise.
6. Explain the terms (*i*) Johonson noise (*ii*) Shot noise (*iii*) White noise (*iv*) Noise factor and (*v*) Noise figure.
7. Define the terms (*i*) Repeatability (*ii*) Accuracy (*iii*) Precision (*iv*) Static sensitivity (*v*) Resolution and (*vi*) Linearity.
8. Explain the phenomenon of hysteresis in measurement systems. Alongside explain the terms 'threshold', 'maximum input hysteresis', 'maximum output hysteresis', 'Dead Zone' and 'Backlash'. Draw neat diagrams to explain your answer.
9. Prove that when a shunt connected instrument is connected to a circuit, the measured voltage is given by $E_L = \dfrac{E_0}{1 + Z_0/Z_L}$

   where, $E_0$ = voltage at no load (without the instrument connected)

   $Z_0$ = output impedance of circuit

   $Z_L$ = input impedance of voltage measuring device

   Discuss the methods of reducing the loading error in the above case.
10. Prove that when a ammeter is introduced into a circuit for measurement of current, the measured value of current is given by

    $$I_L = \frac{I_0}{1 + Y_0 / Y_L}$$

    where, $I_0$ = actual value of current

    $Y_0$ = output admittance of circuit

    $Y_L$ = input admittance of ammeter.

    Prove that the loading error can be reduced to about 1 percent if the input admittance of ammeter is atleast 100 times the output admittance of the source.
11. Explain the terms "Across variable" and "Through variable". Cite examples. Define static input impedance and admittance.

12. The table given, below shows a list of power based through and across variables. Match the correct across variable with its associated through variable. Give the units of each and prove the product of pairs is power.

| Across Variable | Through Variable |
|---|---|
| Fluid pressure | Current |
| Voltage | Volume flow rate |
| Force | Rotational velocity |
| Torque | Heat transfer rate per unit temperature difference |
| Temperature | Translational velocity |

13. Explain why it is necessary to introduce terms "Static Stiffness" and " Static Compliance" in addition "Static Impedance" and "Static Admittance" for analysis of loading effects in generalized measurement systems.

14. The table given below shows a list of energy based through and across variables. Match the correct through variable with its associated across variable. Give the units of each and prove the product of pairs is energy.

| Through Variable | Across Variable |
|---|---|
| Translational displacement | $\int\int((\text{torque})dt)dt$ |
| Rotational velocity | $\int(\text{fluid pressure})dt$ |
| Translational acceleration | $\int(\text{torque } dt)dt$ |
| Volume flow rate | $\int \text{torque } dt$ |
| Rotational acceleration | force |

15. Derive the expression for maximum power transfer in a d.c. circuit. Explain why in many situations it is not advisable to work the measurement system under conditions of maximum power transfer.

## Unsolved Problems

1. An ammeter reads 6.7 A and the true value of current is 6.54 A. Determine the error and the correction for this instrument. [**Ans.** 0.16 A, – 0.16 A]
2. A voltmeter reads 109.5 V. The error taken from an error curve is – 0.37 V. Determine the true voltage. [**Ans.** 109.87 V]
3. The true value of voltage across a resistor is 50 V. The measurement finds a value of 49 V. Calculate (*a*) the absolute error (*b*) the percent error and (*c*) the percent accuracy. [**Ans.** (*a*) 1 V (*b*) 2% (*c*) 98%]
4. A 0 – 100 V voltmeter has 200 scale divisions which can be read to 1/2 division. Determine the resolution of the meter in volt. [**Ans.** 0.25 V]
5. A diaphragm type pressure measuring instrument is calibrated for absolute pressures of 6 to 760 mm of mercury. It has an accuracy of ± 1% based upon scale span. Calculate the scale range, scale span and maximum static error. [**Ans.** 760 mm, 754 mm, ± 7.54 mm]
6. State the number of significant figures in each of the following :
(*a*) 542 A, (*b*) 1.65 V, (*c*) 27.25 W, (*d*) 0.00005 Ω, (*e*) $4 \times 10^6$, (*f*) 20,002, (*g*) 0.345 kΩ
[**Ans.** (*a*) 3, (*b*) 3, (*c*) 4, (*d*) 5, (*e*) 1, (*f*) 5, (*g*) 3]
7. The following set of ten measurements was recorded during an experiment. Calculate the precision of fourth measurement. [**Ans.** 0.96]

| Measurement No. | Quantity |
|---|---|
| 1 | 98 |
| 2 | 102 |
| 3 | 101 |
| 4 | 97 |
| 5 | 100 |
| 6 | 103 |
| 7 | 98 |
| 8 | 106 |
| 9 | 107 |
| 10 | 99 |

8. A voltmeter has an input resistance of 15 MΩ. How much noise voltage will it generate for each cycle of frequency bandwidth at 17°C ? The Boltzmann constant is $1.38 \times 10^{-23}$ J/°K. [**Ans.** 0.49 μV]
9. Three 15 MΩ resistors are connected in parallel. Calculate the total noise voltage generated in the resistors at a temperature of 27°C over a 100 kHz bandwidth. The Boltzmann constant is $1.38 \times 10^{-23}$ J/°K. [**Ans.** 91 μV]
10. Calculate the noise power available in 100 kHz of bandwidth at a temperature of 300° K. The Boltzmann constant is $1.38 \times 10^{-23}$ J/°K. [**Ans.** $41.4 \times 10^{-15}$ W]

11. Calculate power spectrum density from a welding torch whose temperature is 7000°C. The Boltzmann constant is $1.38 \times 10^{-23}$ J/°K.
[**Ans.** $0.1 \times 10^{-8}$ W / Hz]

12. An amplifier has a noise figure of 30 db. If the input signal S/N is 100, what is the output S/N ? Calculate also the noise factor.
[**Ans.** 0.1, 1000]

13. A high impedance voltmeter has a 15 MΩ input impedance which is composed of three 5 MΩ resistors in series. Calculate the total noise voltage generated in the resistors at a temperature of 290°K over a 100 kHz frequency range. Boltzmann constant is $1.38 \times 10^{-23}$ J/°K. [**Ans.** 155 μV]

14. A resistor at a temperature of 290°K has a noise voltage of 2 μV for a 50 kHz bandwidth. Calculate the temperature at which the noise is 20% of its value at 290°K. [**Ans.** 11.6°K]

15. In order to measure noise figure of an amplifier, a noise voltage that is 10% of the signal voltage is applied at the input of the amplifier. The noise voltage at the output of the amplifier is 20% of the signal voltage. Calculate the noise figure of the amplifier. [**Ans.** 6.02 db]

16. A mercury thermometer has a capillary tube of 0.3 mm diameter. If the bulb is made of zero expansion material, what volume must it have if a sensitivity of 3 mm/°C is desired ? Assume operating temperature of be 20°C and co-efficient of volumetric expansion of mercury is $0.181 \times 10^{-3}$.
[**Ans.** 1171.6 mm$^3$]

17. Four resistors are connected in series. The values of resistors are 28.4 Ω, 4.25 Ω, 56.605 Ω, and 0.75 Ω, with an uncertainty of one unit, in the last digit in each. Calculate the total series resistance, giving only significant figures in the answer.
[**Ans.** 90 Ω]

18. A current of 2.56 A is flowing in a resistance of 45.73 Ω. Each quantity has an uncertainty of one unit in the last digit. Find the value of voltage drop to the appropriate number of significant figures.
[**Ans.** 117 V]

19. Determine the linearity of a potentiometer to obtain an error not to exceed 1 part in 10,000.
[**Ans.** 0.01%]

20. A recording instrument requires a current of 0.05 A to overcome initial friction and produce motion of the movement. Define this effect and list factors which produce it. [**Ans.** Dead zone = 0.05 A]

21. A self balancing potentiometer is connected to a spatial encoder required to read 2000. The linearity is 0.05%, find the accuracy of readout system.
[**Ans.** 1 part in 2000]

22. The dead zone of a certain pyrometer is 0.125 per cent of the span. The calibration is 800°C to 1800°C. What temperature change must occur before it is detected ?
[**Ans.** 1.25°C]

23. What is the true value of voltage across the 500 kΩ resistor connected between terminals *A* and *B* as shown in Fig. 2.24 ? What would a voltmeter with a sensitivity of 20 kΩ / V read on the following ranges : 50, 15, 5 V when connected across terminals *C* and *D*.
[**Ans.** 10 V, 8 V, 5.45 V, 2.86 V]

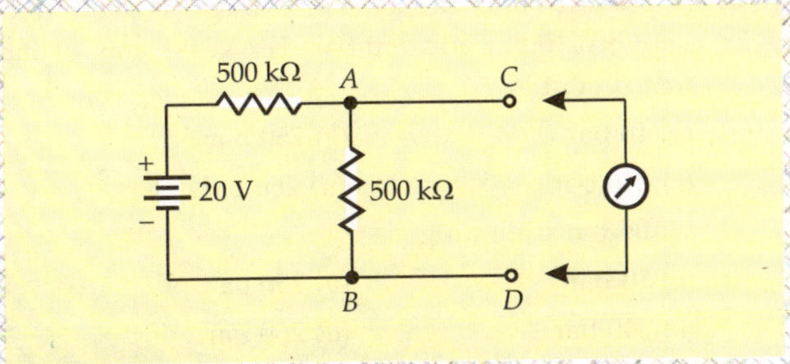


**Fig. 2.24** Diagram of Problem 23

24. What is the true value of current in the 15 kΩ resistor of Fig. 2.25 ? If an ammeter of 2 kΩ resistance is used to measure the current in the 15 kΩ resistor, what will it read ? If a loading accuracy of 99 percent is desired in measuring the current, what should the ammeter resistance be ?
[**Ans.** 100 μA, 92.6 μA, 250 Ω]

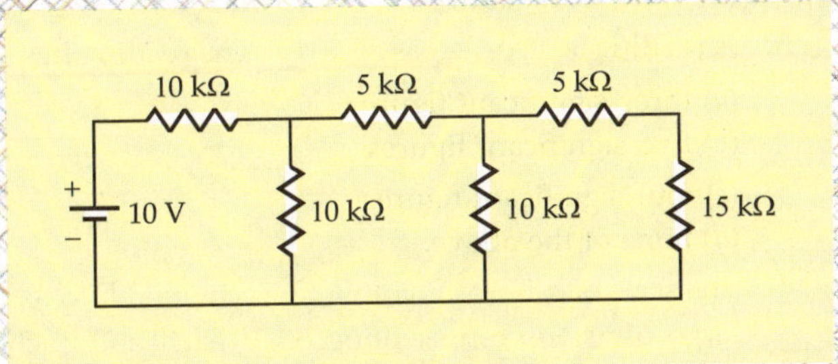


**Fig. 2.25** Diagram of Problem 24

25. A high resistance voltmeter reads 8 V across a 4 Ω resistor connected to a battery. The voltage rises to 9 V when the resistance is increased to 9 Ω. Calculate the maximum power available from the battery.
[**Ans.** 25 W]

# Objective Type Questions

**Tick ( √ ) the most appropriate answer**

1. In measurement systems, which of the following static characteristics are desirable :
   (*a*) Accuracy (*b*) Sensitivity
   (*c*) Reproducibility (*d*) All the above.
2. In measurement systems, which of the following are undesirable static characteristics :
   (*a*) Sensitivity and accuracy
   (*b*) Drift, static error and dead zone
   (*c*) Reproducibility and non-linearity
   (*d*) Drift, static error, dead zone and non-linearity.
3. A pressure measurement instrument is calibrated between 10 bar and 250 bar. The scale span of the instrument is
   (*a*) 10 bar (*b*) 250 bar
   (*c*) 240 bar (*d*) 260 bar.
   while the scale range is
   (*a*) 10 bar (*b*) 250 bar
   (*c*) 240 bar (*d*) 260 bar.
4. The output of an instrument under particular environmental conditions is given by the relationship $q_0 = a + bq_i$ where $q_i$ = input and $a$ and $b$ are constants $b \neq 1$. The theoretical input-output relationship is represented by a straight passing through zero. This is a case of
   (*a*) zero drift only
   (*b*) sensitivity drift only
   (*c*) zero and sensitivity drift together
   (*d*) none of the above.
5. A reading is recorded as 23.90 C°. The reading has
   (*a*) three significant figures
   (*b*) five significant figures
   (*c*) four significant figures
   (*d*) none of the above.
6. The scale of a 0-500 V voltmeter is divided into ten large divisions representing 50 V each and each large division is further subdivided into 10 small divisions, each representing 5 V. It is used for measurement of output voltage of a potentiometer which can be varied from 0 to 500 V. It is observed that when the sliding contact is moved from its zero position, there is no perceptible movement of pointer of the voltmeter till the sliding contact reaches a position where the output voltage should be 5 V. Therefore, it can be concluded that
   (*a*) the threshold of the voltmeter is 5 V
   (*b*) the resolution of the voltmeter is 5 V
   (*c*) the sensitivity of the voltmeter is 5 V
   (*d*) none of the above.
7. In the centre zero analog ammeter having a range of –10 A + 10 A, there is a detectable change of the pointer from its zero position on either side of the scale only if the current reaches a value of 1 A (on either side). The ammeter has a
   (*a*) resolution of 1 A (*b*) dead zone of 1 A
   (*c*) dead zone of 2 A (*d*) sensitivity of 1 A.
8. A pressure gauge is calibrated from 0 – 50 kN/m$^2$. It has a uniform scale with 100 scale divisions. One fifth of a scale division can be read with certainty. The gauge has a :
   (*a*) resolution of 0.1 kN/m$^2$
   (*b*) threshold of 0.1 kN/m$^2$
   (*c*) dead zone of 0.2 kN/m$^2$
   (*d*) a resolution of 0.5 kN/m$^2$.
9. A d.c. circuit can be represented by an internal voltage source of 50 V with an output resistance of 100 kΩ. In order to achieve 99 percent accuracy for voltage measurement across its terminals, the voltage measuring device should have
   (*a*) a resistance of atleast 10 MΩ
   (*b*) a resistance of 100 kΩ
   (*c*) a resistance of atleast 10 Ω
   (*d*) none of the above.
10. A d.c. circuit can be represented by a voltage source of 10 V in series with an output resistance of 1 kΩ. An ammeter of 50 Ω resistance is connected to the source terminals for measurement of current. The accuracy of measurement is nearly
    (*a*) – 4.8 percent (*b*) + 4.8 percent
    (*c*) 99 percent (*d*) 95.2 percent.
11. In a.c. circuits, the connection of measuring instruments cause loading errors which may affect
    (*a*) only the magnitude of quantity being measured
    (*b*) only the phase of the quantity being measured
    (*c*) both the magnitude and phase of the quantity being measured
    (*d*) magnitude, phase and also the waveform of the quantity being measured.

12. The input resistance of a cathode ray oscilloscope is of the order of

(a) tens of ohm (b) megohm
(c) kilohm (d) fraction of an ohm.

13. A resistor, at room temperature of 290 °K has a noise voltage of 2 μV for a 50 kHz bandwidth. The temperature at which the noise voltage is 20% of its value at room temperature is

(a) 58 °K (b) 11.6 °K
(c) 241.7 °K (d) none of the above.

14. An amplifier has an input signal voltage of 25 μV and a noise voltage 2.5 μV. The signal to noise ratio is

(a) 10 (b) 100
(c) 10 db (d) none of the above.

15. Johonson noise is :

(a) caused by thermal agitation of free electrons carrying current thereby modulating the current
(b) noise carried into a circuit through conductors
(c) noise of an electromagnetic origin that is radiated into a circuit
(d) shot noise which results from random emission of electrons across a p-n junction.

16. In a measurement system if a $q_1$ is an across variable and its associated through variable is $q_2$ (power based), the static impedance of the system is :

(a) $q_1 \times q_2$ (b) $q_1 / q_2$
(c) $q_2 / q_1$ (d) $q_1^2 / q_2$.

17. In a measurement system 'force' is an across variable. Its associated power based through variable is :

(a) translational displacement
(b) translational acceleration
(c) translational velocity
(d) rotational displacement.

18. The power can be controlled in a circuit by :

(a) controlling the voltage applied to load
(b) controlling the current level in the circuit
(c) using pulse modulation of power
(d) all the above.

19. The values of static stiffness and compliance in a measurement system determine the amount of

(a) power drain from a system
(b) current drain from a system
(c) energy drain from a system
(d) none the above.

20. A Wheatstone bridge is balanced with all the four resistances equal to 1 kΩ each. The bridge supply voltage is 100 V. The value of one of the resistances is changed to 1010 Ω. The output voltage is measured with a voltage measuring device of infinite resistance.

The bridge sensitivity is :

(a) 10 V/Ω (b) 25 mV/Ω
(c) 2.5 mV/Ω (d) none of the above.

## Answers

| | | | | | | | | | |
|---|---|---|---|---|---|---|---|---|---|
| 1. (d) | 2. (d) | 3. (c) | 4. (c) | 5. (c) | 6. (a) | 7. (c) | 8. (a) | 9. (a) | 10. (d) |
| 11. (d) | 12. (b) | 13. (b) | 14. (b) | 15. (a) | 16. (b) | 17. (c) | 18. (d) | 19. (c) | 20. (b) |

2

# 3

# Errors in Measurements and Their Statistical Analysis

## 3.1 LIMITING ERRORS (GUARANTEE ERRORS)

The accuracy and precision of an instrument depends upon its design, the material used and the workmanship that goes into making the instrument. The choice of an instrument for a particular application depends upon the accuracy desired. If only a fair degree of accuracy is desired, it is not economical to use expensive materials and skill for the manufacture of the instrument. But an instrument used for an application requiring a high degree of accuracy has to use expensive material and a highly skilled workmanship. The economical production of any instrument requires the proper choice of material, design and skill. In order to assure the purchaser of the quality of the instrument, the manufacturer guarantees a certain accuracy. In most instruments the accuracy is guaranteed to be within a certain percentage of full scale reading. Components are guaranteed to be within a certain percentage of the rated value. Thus the manufacturer has to specify the deviations from the **nominal value** of a particular quantity. The limits of these deviations from the specified value are defined as **Limiting Errors** or **Guarantee Errors**.

We can say that the manufacturer guarantees or promises that the error in the item he is selling is no greater than the limit set. The magnitude of a quantity having a nominal value $A_s$ and a maximum error or limiting error of $\pm\delta A$ must have a magnitude $A_a$ between the limits $A_s - \delta A$ and $A_s + \delta A$ or

$$\text{Actual value of quantity } A_a = A_s \pm \delta A \quad \text{...(3.1)}$$

For example, the nominal magnitude of a resistor is 100 Ω with a limiting error of ± 10 Ω. The magnitude of the resistance will be between the limits

$$A_a = 100 \pm 10\ \Omega \text{ or } A_a \geq 90\ \Omega \text{ and } A_a \leq 110\ \Omega.$$

In other words, the manufacturer guarantees that the value of resistance of the resistor lies between 90 Ω and 110 Ω.

## 3.2 RELATIVE (FRACTIONAL) LIMITING ERROR

The relative (fractional) error is defined as the ratio of the error to the specified (nominal) magnitude of a quantity. Therefore,

$$\text{Relative limiting error } \varepsilon_r = \frac{\delta A}{A_s} = \frac{\varepsilon_0}{A_s} \quad \text{...(3.2)}$$

or

$$\varepsilon_0 = \delta A = \varepsilon_r A_s \quad \text{...(3.3)}$$

Then from Eqns. 3.1 and 3.3 limiting values are :

$$A_a = A_s \pm \delta A = A_s \pm \varepsilon_r A_s = A_s(1 \pm \varepsilon_r) \quad \text{...(3.4)}$$

$$\text{Percentage limiting error } \%\ \varepsilon_r = \varepsilon_r \times 100 \quad \text{...(3.5)}$$

In the example considered in Eqn. 3.1, we have

$$A_s = 100\ \Omega \quad \text{and} \quad \delta A = \pm 10\ \Omega$$

∴ Relative limiting error

$$\varepsilon_r = \frac{\delta A}{A_s} = \pm\frac{10}{100} = \pm 0.1.$$

Percentage limiting error

$$\%\ \varepsilon_r = 0.1 \times 100 = \pm 10\%$$

and limiting values of resistance are :

$$A_a = A_s(1 \pm \varepsilon_r) = 100(1 \pm 0.1) = 100 \pm 10\ \Omega$$

In limiting errors the specified quantity $A_s$ is taken as the true quantity, and the quantity which has the maximum deviation from $A_a$ is taken as the erroneous quantity. Thus, we have

$$\delta A = A_a - A_s \quad \text{...(3.6)}$$

$$\therefore \text{ Relative limiting error, } \varepsilon_r = \frac{A_a - A_s}{A_s}$$

$$= \frac{\text{actual value} - \text{nominal value}}{\text{nominal value}} \quad \text{...(3.7)}$$

**Example 3.1** *The value of capacitance of a capacitor is specified as 1 μF ±5% by the manufacturer. Find the limits between which the value of the capacitance is guaranteed.*

**Solution.** The guaranteed value of the capacitance lie within the limits :

$$A_a = A_s(1 \pm \varepsilon_r) = 1 \times (1 \pm 0.05) = 0.95 \text{ to } 1.05\ \mu\text{F}.$$

**NOTE** The same idea of a guarantee limiting the worst possible case applies to electrical measurements. The measurements may involve several components, each of which may be delimited by a guarantee error. Thus the same treatment is to be followed for quantities under measurement as is followed for specified quantities.

**Example 3.2** *A 0-150 V voltmeter has a guaranteed accuracy of 1 percent of full scale reading. The voltage measured by this instrument is 75 V. Calculate the limiting error in percent. Comment upon the result.*

**Solution.** The magnitude of limiting error of instrument is

$$\delta A = \varepsilon_r A_s = 0.01 \times 150 = 1.5 \text{ V}$$

The magnitude of the voltage being measured is 75 V.

The relative error at this voltage is

$$\varepsilon_r = \frac{\delta A}{A_s} = \frac{1.5}{75} = 0.02$$

Therefore, the voltage being measured is between the limits of :

$$A_a = A_s(1 \pm \varepsilon_r)$$
$$= 75(1 \pm 0.02)\text{V} = 75 \pm 1.5 \text{ V}$$

The percentage limiting error is :

$$\%\ \varepsilon_r = \frac{(1.5)}{75} \times 100 = 2 \text{ percent.}$$

***Comments.*** It is important to note that this meter is guaranteed to have an error of less than 1 percent of full scale or the limiting error is 1 percent at the full scale deflection of 150 V. But when the meter reads 75 V *i.e.,* half the full scale value the limiting error is 2 percent (twice that at full scale deflection). The percent limiting error will be greater if a smaller voltage is measured. If the meter reads 37.5 V, the percentage limiting error is,

$$\frac{1.5}{37.5} \times 100 = 4 \text{ percent.}$$

This increase in the percentage limiting error as smaller voltages are measured occurs because the magnitude of the limiting error $\delta A$ is based upon the full scale reading of the meter and is a fixed quantity, while the actual voltage readings can be of any magnitude from 0 to 150 V. We infer from here that the percentage error increases as the voltage being measured decreases.

Thus while selecting instruments, particular care should be taken as regards the range. The values to be measured should not lie in the lower third of the range. This is particularly important if the meter accuracy is specified in terms of the full scale deflection (f.s.d.) another name for f.s.d. is fudicial value. As considerable error (as a percentage of actual value), may occur as is observed in Example 3.2, thus meters which read well up their scale should be selected as far as possible.

Another example is given below which is illustrative of the comments given above.

**Example 3.3** *A wattmeter having a range 1000 W has an error of ± 1% of full scale deflection. If the true power is 100 W, what would be the range of readings ? Suppose the error is specified as percentage of true value, what would be the range of the readings ?*

**Solution.** When the error is specified as a percentage of full scale deflection, the magnitude of limiting error at full scale $= \pm \frac{1}{100} \times 1000 = \pm 10$ W

Thus the wattmeter reading when the true reading is 100 W may be 100 ± 10 W *i.e.,* between 90 to 110 W.

$$\text{Relative error} = \frac{\pm 10}{100} \times 100 = \pm 10\%$$

Now suppose the error is specified as percentage of true value.

$$\text{The magnitude of error} = \pm \frac{1}{100} \times 100 = \pm 1 \text{ W.}$$

Therefore the meter may read $100 \pm 1$ W or between 99 to 101 W.

## 3.3 COMBINATION OF QUANTITIES WITH LIMITING ERRORS

When two or more quantities, each having a limiting error, are combined, it is advantageous to be able to compute the limiting error of the combination. The limiting error can be easily found by considering the relative increment of the function if the final result is in the form of an algebraic equation.

(*i*) ***Sum of two quantities.*** Let X be the final result which is the sum of measured quantities $x_1$ and $x_2$.

$$\therefore \quad X = x_1 + x_2$$

The relative increment of the function is given by :

$$\frac{dX}{X} = \frac{d(x_1 + x_2)}{X}$$
$$= \frac{dx_1}{X} + \frac{dx_2}{X}$$

Expressing the result in terms of relative increment of the component quantities, we have :

$$\frac{dX}{X} = \frac{x_1}{X}\frac{dx_1}{x_1} + \frac{x_2}{X}\frac{dx_2}{x_2}$$

If the limiting errors in the component quantities are represented by $\pm\delta x_1$ and $\pm\delta x_2$, then the corresponding relative limiting error in $X$ is given by :

$$\frac{\delta X}{X} = \pm\left[\frac{x_1}{X}\frac{\delta x_1}{x_1} + \frac{x_2}{X}\frac{\delta x_2}{x_2}\right] \quad ...(3.8)$$

The above equation shows that the resultant limiting relative error is equal to the sum of the products formed by multiplying the individual relative limiting errors by the ratio of each term to the function.

(ii) ***Difference of two quantities.*** Let

$$X = x_1 - x_2$$

$$\therefore \quad \frac{\delta X}{X} = \frac{dx_1}{X} - \frac{dx_2}{X}$$

Expressing the result in terms of relative increments of component quantities, we get

$$\frac{dX}{X} = \frac{x_1}{X}\frac{dx_1}{x_1} - \frac{x_2}{X}\frac{dx_2}{x_2}$$

If the errors in $x_1$ and $x_2$ are $\pm\delta x_1$ and $\pm\delta x_2$ respectively, the signs may be interpreted to give worst possible discrepancy, *i.e.*, when the error in $x_1$ is $\pm\delta x_1$, the error in $x_2$ is $-\delta x_2$ and *vice versa*, then the relative limiting error in X is given by :

$$\frac{\delta x}{X} = \pm\left(\frac{x_1}{X}\cdot\frac{\delta x_1}{x_1} + \frac{x_2}{X}\cdot\frac{\delta x_2}{x_2}\right) \quad ...(3.9)$$

Equation 3.9 is the same as Eqn. 3.8. It may, however, be mentioned that in this case when $x_1$ and $x_2$ are almost equal in magnitude, *i.e.*, when $X << x_1$ or $X << x_2$, then the relative error in X would be very large.

(iii) ***Sum of difference of more than two quantities.*** If we have

$$X = \pm x_1 \pm x_2 \pm x_3,$$

then the relative limiting error in X is given by :

$$\frac{\delta X}{X} = \pm\left(\frac{x_1}{X}\cdot\frac{\delta x_1}{x_1} + \frac{x_2}{X}\cdot\frac{\delta x_2}{x_2} + \frac{x_3}{X}\cdot\frac{\delta x_3}{x_3}\right) \quad ...(3.10)$$

(iv) ***Product of two components.*** Let

$$X = x_1 x_2.$$

$$\therefore \quad \log_e X = \log_e x_1 + \log_e x_2.$$

Differentiating the above with respect to X,

$$\frac{1}{X} = \frac{1}{x_1}\cdot\frac{dx_1}{dX} + \frac{1}{x_2}\cdot\frac{dx_2}{dX} \quad \text{or} \quad \frac{dX}{X} = \frac{dx_1}{x_1} + \frac{dx_2}{x_2}$$

Representing the errors in $x_1$ and $x_2$ as $\pm\delta x_1$ and $\pm\delta x_2$ respectively, the relative limiting error in X is given by :

$$\frac{\delta X}{X} = \pm\left(\frac{\delta x_1}{x_1} + \frac{\delta x_1}{x_2}\right) \quad ...(3.11)$$

Thus from above we conclude that the relative limiting error of product of terms is equal to the sum of relative limiting errors of terms.

(v) ***Quotient.*** Let

$$X = \frac{x_1}{x_2} \quad \text{or} \quad \log_e X = \log_e x_1 - \log_e x_2.$$

Differentiating the above expression with respect to X,

$$\frac{1}{X} = \frac{1}{x_1}\frac{dx_1}{dX} - \frac{1}{x_2}\frac{dx_2}{dX}$$

or

$$\frac{dX}{X} = \frac{dx_1}{x_1} - \frac{dx_2}{x_2}$$

Representing the limiting errors in $x_1$ and $x_2$ as $\pm\delta x_1$ and $\pm\delta x_2$ respectively and considering the worst condition when $\delta x_1/x_1$ is +ve and $\delta x_2/x_2$ is −ve or *vice versa*.

Relative limiting error in X is :

$$\frac{\delta X}{X} = \pm\left(\frac{\delta x_1}{x_1} + \frac{\delta x_2}{x_2}\right) \quad ...(3.12)$$

The above result is the same as the corresponding result for product of two quantities.

(vi) ***Product or quotient of more than two quantities.***

Let $X = x_1 x_2 x_3$ or $X = \dfrac{x_1}{x_2 x_3}$ or $X = \dfrac{1}{x_1 x_2 x_3}$

Therefore from Eqns. 3.11 and 3.12, we have relative limiting error in X as :

$$\frac{\delta X}{X} = \pm\left(\frac{\delta x_1}{x_1} + \frac{\delta x_2}{x_2} + \frac{\delta x_3}{x_3}\right) \quad ...(3.13)$$

(vii) ***Power of a factor.***

Let $\quad X = x_1^n$

$$\log_e X = n \log_e x_1$$

Differentiating above with respect to X, we have :

$$\frac{1}{X} = n\cdot\frac{1}{x_1}\frac{dx_1}{dX} \quad \text{or} \quad \frac{dX}{X} = n\cdot\frac{\delta x_1}{x_1}$$

Hence relative limiting error in X is,

$$\frac{\delta X}{X} = \pm n\frac{\delta x_1}{x_1} \quad ...(3.14)$$

It is clear from above that the relative limiting error in the result under these circumstances is magnified $n$ times.

3

*(viii)* ***Composite factors***

Let $X = x_1^n \cdot x_2^m$

$\log_e X = n\log_e x_1 + m\log_e x_2.$

Differentiating with respect to X, we have,

$$\frac{1}{X} = \frac{n}{x_1}\frac{dx_1}{dX} + \frac{m}{x_2}\frac{dx_2}{dX}$$

or
$$\frac{dX}{X} = n\frac{dx_1}{x_1} + m\frac{dx_2}{x_2}$$

Thus the limiting error in X is :

$$\frac{\delta X}{X} = \pm\left(n\frac{\delta x_1}{x_1} + m\frac{\delta x_2}{x_2}\right) \quad ...(3.15)$$

**Example 3.4** *Three resistors have the following ratings :*

$R_1 = 37\ \Omega \pm 5\%$, $R_2 = 75\ W \pm 5\%$, $R_3 = 50\ W \pm 5\%$

*Determine the magnitude and limiting error in ohm and in percent of the resistance of these resistances connected in series.*

Solution. The values of resistances are :

$$R_1 = 37 \pm \frac{5}{100} \times 37 = 37 \pm 1.85\ \Omega$$

$$R_2 = 75 \pm \frac{5}{100} \times 75 = 75 \pm 3.75\ \Omega$$

$$R_3 = 50 \pm \frac{5}{100} \times 50 = 50 \pm 2.50\ \Omega$$

The limiting value of resultant resistance,

$$R = (37 + 75 + 50) \pm (1.85 + 3.75 + 2.50)$$
$$= 162 \pm 8.10\ \Omega$$

$\therefore$ Magnitude of resistance $= 162\ \Omega$ and error in ohm $= \pm 8.1\ \Omega$.

Percent limiting error of series combination of resistances $= \pm\frac{8.1}{162} \times 100 = \pm 5\%$

**Example 3.5** *The resistance of a circuit is found by measuring current flowing and the power fed into the circuit. Find the limiting error in the measurement of resistance when the limiting errors in the measurement of power and current are respectively ± 1.5% and ± 1.0%.*

**Solution.** Resistance $R = \frac{\text{(power)}}{\text{(current)}^2} = \frac{P}{I^2} = PI^{-2}.$

From Eqn. 3.14, relative limiting error in measurement of resistance is,

$$\frac{\delta R}{R} = \pm\left(\frac{\delta P}{P} + 2\frac{\delta I}{I}\right) = \pm(1.5 + 2 \times 1.0) = \pm 3.5\%$$

**Example 3.6** *The solution for the unknown resistance for a Wheatstone bridge is :*

$$R_x = \frac{R_2 R_3}{R_1}$$

*where* $R_1 = 100 \pm 0.5\%\ \Omega$, $R_2 = 1000 \pm 0.5\%\ \Omega$, $R_3 = 842 \pm 0.5\%\ \Omega$

*Determine the magnitude of the unknown resistance and the limiting error in percent and in ohm for the unknown resistance $R_x$.*

**Solution.** Unknown resistance is,

$$R_x = \frac{R_2 R_3}{R_1} = \frac{1000 \times 842}{100} = 8420\ \Omega$$

Relative limiting error of unknown resistance is,

$$\frac{\delta R_x}{R_x} = \pm\left(\frac{\delta R_2}{R_2} + \frac{\delta R_3}{R_3} + \frac{\delta R_1}{R_1}\right)$$
$$= \pm(0.5 + 0.5 + 0.5) = \pm 1.5\%$$

Limiting error in ohm $= \pm\frac{1.5}{100} \times 8420 = \pm 126.3\ \Omega$

Guaranteed values of resistance are between :

$8420 - 126.3 = 8293.7\ \Omega$ and $8420 + 126.3 = 8546.3\ \Omega$

**Example 3.7** *A 4-dial decade box has :*

*decade a of $10 \times 1000\ W \pm 0.1\%$,*

*decade b of $10 \times 100\ W \pm 0.1\%$.*

*decade c of $10 \times 10\ \Omega \pm 0.5\%$,*

*decade d of $10 \times 1\ \Omega \pm 1.0\%$.*

*It is set at 4639 Ω. Find the percentage limiting error and the range of resistance value.*

**Solution.** Decade *a* is set at 4000 Ω and therefore,

$$\text{error} = \pm 4000 \times \frac{0.1}{100} = \pm 4\ \Omega$$

Decade *b* is set at 600 Ω and therefore,

$$\text{error} = \pm 600 \times \frac{0.1}{100} = \pm 0.6\ \Omega$$

Similarly,

$$\text{error in decade } c = \pm 30 \times \frac{0.5}{100} = \pm 0.15\ \Omega$$

and
$$\text{error in decade } d = \pm 9 \times \frac{1}{100} = \pm 0.09\ \Omega$$

Hence total error $= \pm(4 + 0.6 + 0.15 + 0.09) = 4.84\ \Omega$

Relative limiting error

$$\varepsilon_r = \pm\frac{4.84}{4639} = 1 \pm 0.00104$$

Percentage limiting error

$$\%\varepsilon_r = \pm(0.00104 \times 100) = \pm 0.104\%$$

Limiting values of resistance

$$A_a = 4639(1 \pm 0.00104) = 4639 \pm 5\ \Omega.$$

Thus we conclude from the above examples that the guarantee values are obtained by taking a direct sum of the possible errors, adopting the algebraic signs that give the worst possible case. In fact setting of guarantee limits is necessarily a pessimistic process. This is true from manufacturer's viewpoint as regards his promise to the buyer, and it is also true of the user in setting accuracy limits in results of his measurements.

## 3.4 KNOWN ERRORS

When the error of a quantity or an instrument is known the effect of this error, when combined with other errors, can be computed in a manner similar to the combinations of limiting errors. But the difference is that in case of known errors the signs of relative errors are given and must be preserved in the calculations.

**Example 3.8.** *A resistance is rated at 3200 Ω and the current flowing through this is 64 mA. (a) Compute the power loss in the resistor. (b) It was later found that the resistance of the resistor was 0.2 percent greater than the specified resistance and the ammeter read 0.75 percent more than the true current. Determine the known error in the computed power in part (a).*

**Solution.** (*a*) Power consumed

$$P = I^2R = (64\times10^{-3})^2\times3200 = 13.1\text{ W}.$$

(*b*) Relative error in power

$$\frac{\delta P}{P} = \left(\frac{2\delta I}{I} + \frac{\delta R}{R}\right) = (2\times0.75+0.2) = 1.7\%\text{ higher.}$$

**Example 3.9** *Current was measured during a test as 30.4 A, flowing in a resistor of 0.105 Ω. It was discovered later that the ammeter reading was low by 1.2 percent and the marked resistance was high by 0.3 percent. Find the true power as a percentage of the power that was originally calculated.*

**Solution.** True value of current

$$I = 30.4(1+0.012) = 30.765\text{ A}$$

True value of resistance

$$R = 0.105(1-0.003) = 0.1047\ \Omega$$

True value of power

$$= I^2R = (30.765)^2\times0.1047 = 99.097\text{ W}$$

Measured value of power

$$= (30.4)^2\times0.105 = 97.037\text{ W}.$$

$$\frac{\text{True power}}{\text{Measured power}}\times100 = \frac{99.097}{97.037}\times100 = 102.1\text{ percent.}$$

We arrive at the same results by using the following method :

Power $P = I^2R$

Total relative error

$$\frac{\delta P}{P} = 2\frac{\delta I}{I} + \frac{\delta R}{R} = 2\times(-0.12)+0.003 = -0.021$$

$$\therefore\quad \frac{\text{True power}}{\text{Measured power}} = \frac{1}{1+\text{relative error}}$$

$$= \frac{1}{1-0.021} = 1.021 = 102.1\%$$

**Example 3.10** *Three resistors of having resistances of 250 Ω, a 500 Ω and a 375 Ω are connected in parallel. The 250 Ω resistor has a +0.025 fractional error, the 500 Ω resistor has a – 0.036 fractional error, and the 375 Ω resistor has a + 0.014 fractional error. Determine (a) the total resistance neglecting errors, (b) total resistance considering the error of each resistor and (c) the fractional error of the total resistance based upon rated values.*

**Solution.** (*a*) Total resistance of resistors connected in parallel and neglecting their errors is :

$$R = \frac{1}{1/R_1+1/R_2+1/R_3}$$

$$= \frac{1}{1/250+1/500+1/375} = 115.38\ \Omega$$

(*b*) The fractional error in $R_1 = 250\ \Omega$ is +0.025

$\therefore\quad \delta R_1 = (0.025\times250) = +\,6.25\ \Omega.$

Hence $\quad R_1 = 250-6.25 = 243.75\ \Omega$

Similarly,

$\quad \delta R_1 = (-0.036\times500) = -18\ \Omega$

and $\quad R_2 = 500+18 = 518\ \Omega$

$\therefore\quad \delta R_3 = (+0.014\times375) = 5.25\ \Omega$

and $\quad R_3 = 375-5.25 = 369.75\ \Omega$

Therefore the resultant resistance of three resistances in parallel

$$R = \frac{1}{1/R_1+1/R_2+1/R_3}$$

$$= \frac{1}{1/243.75+1/518+1/369.75} = 114.45\ \Omega$$

(*c*) The fractional error of the parallel resistances based on the rated values is :

$$\frac{115.38-114.45}{115.38} = +0.008 = +0.8\%$$

**Example 3.11** *A 160± pF capacitor, an inductor of 160 ± 10% μH and a resistor of 1200 ± 10 Ω are connected in series.*

(*a*) *If all the three components are ± 0% and resonant frequency is* $f_r = \dfrac{1}{2\pi}\sqrt{\dfrac{1}{LC}}$, *compute the resonant frequency of the combination.*

(*b*) *If all the three components are +10%, compute the expected resonance frequency of the combination and the percentage error when compared to the results of part (a).*

(c) *If all the three components are – 10%, compute the expected resonant frequency and the percentage error when compared to the results of part (a).*

**Solution.** (*a*) When all the components have zero error,

$$L = 160\ \mu H = 160 \times 10^{-6}\ H$$

and $$C = 160\ pF = 160 \times 10^{-12}\ F.$$

∴ Resonant frequency

$$f_r = \frac{1}{2\pi}\sqrt{\frac{1}{LC}} = \frac{1}{2\pi}\sqrt{\frac{1}{160 \times 10^{-6} \times 160 \times 10^{-12}}}$$

$$= 1\ MHz.$$

(*b*) When the components are +10%,

$$C = 160 + 0.1 \times 160 = 176\ pF,$$

$$L = 160 + 0.1 \times 160 = 176\ \mu H$$

$$\therefore\ f_r = \frac{1}{2\pi}\sqrt{\frac{1}{176 \times 10^{-6} \times 176 \times 10^{-12}}} = 0.9\ MHz$$

Hence, $$\text{error} = \frac{0.9 - 1.0}{1.0} = -10\%$$

This is a case of known errors and can be solved by using Eqn. 3.15.

$$f_r = \frac{1}{2\pi} L^{-1/2} C^{-1/2}$$

∴ Relative error in $f_r$ is,

$$\frac{\delta f_r}{f_r} = \left(-\frac{1}{2}\frac{\delta L}{L} - \frac{1}{2}\frac{\delta C}{C}\right) = -\frac{1}{2}(0.1 + 0.1)$$

$$= -0.1 = -10\%.$$

(*c*) When the components are – 10%.

$$C = 160 - 0.1 \times 160 = 144\ pF,$$

$$L = 160 - 0.1 \times 160 = 144\ \mu H$$

$$\therefore\ f_r = \frac{1}{2\pi}\sqrt{\frac{1}{144 \times 10^{-6} \times 144 \times 10^{-12}}} = 1.1\ MHz$$

$$\text{Error} = \frac{1.1 - 1.0}{1.0} \times 100 = +10\%$$

## 3.5 TYPES OF ERRORS

No measurement can be made with perfect accuracy but it is important to find out what accuracy actually is and how different errors have entered into the measurement. A study of errors is a first step in finding ways to reduce them. Errors may arise from different sources and are usually classified as under :

- Gross Errors.
- Systematic Errors.
- Random Errors.

## 3.6 GROSS ERRORS

This class of errors mainly covers human mistakes in reading instruments and recording and calculating measurement results. The responsibility of the mistake normally lies with the experimenter. The experimenter may grossly misread the scale. For example, he may, due to an oversight, read the temperature as 31.5°C while the actual reading may be 21.5°C. He may transpose the reading while recording. For example, he may read 25.8°C and record 28.5°C instead. But as long as human beings are involved, some gross errors will definitely be committed. Although complete elimination of gross errors is probably impossible, one should try to anticipate and correct them. Some gross errors are easily detected while others may be very difficult to detect.

Gross errors may be of any amount and therefore their mathematical analysis is impossible. However, they can be avoided by adopting *two* means. They are :

1. Great care should be taken in reading and recording the data.
2. Two, three or even more readings should be taken for the quantity under measurement. These readings should be taken preferably by different experimenters and the readings should be taken at a different reading point to avoid re-reading with the same error. It should be understood that no reliance be placed on a single reading. It is always advisable to take a large number of readings as a close agreement between readings assures that no gross error has been committed.

## 3.7 SYSTEMATIC ERRORS

These types of errors are divided into *three* categories :

1. Instrumental Errors.
2. Environmental Errors.
3. Observational Errors.

### 3.7.1 Instrumental Errors

These errors arise due to *three* main reasons :

(*i*) Due to inherent shortcomings in the instrument,

(*ii*) Due to misuse of the instruments, and

(*iii*) Due to loading effects of instruments.

**1. *Inherent shortcomings of instruments.*** These errors are inherent in instruments because of their mechanical structure. They may be due to construction, calibration or operation of the instruments or measuring devices. These errors may cause the instrument to read too low or too high. For example, if the spring (used for producing controlling torque) of a permanent magnet instrument has become weak, the instrument will always read high. Errors may be caused because of friction, hysteresis or even gear backlash.

While making precision measurements, we must recognize the possibility of such errors as it is often possible to eliminate them, or at least reduce them to a great extent by using the following methods :

(*i*) The procedure of measurement must be carefully planned. Substitution methods or calibration against standards may be used for the purpose.

(*ii*) Correction factors should be applied after determining the instrumental errors.

(*iii*) The instrument may be re-calibrated carefully.

2. ***Misuse of instruments.*** There is an old saying that instruments are better than the people who use them. Too often, the errors caused in measurements are due to the fault of the operator than that of the instrument. A good instrument used in an unintelligent way may give erroneous results. Examples which may be cited for this misuse of instrument may be failure to adjust the zero of instruments, poor initial adjustments, using leads of too high a resistance etc.

No doubt the above improper practices may not cause a permanent damage to the instrument but all the same they cause errors. However, there are certain ill practices like using the instrument contrary to manufacturer's instructions and specifications which in addition to producing errors cause permanent damage to the instruments as a result of overloading and overheating that may ultimately result in failure of the instrument and sometimes the system itself.

3. ***Loading effects.*** One of the most common errors committed by beginners, is the improper use of an instrument for measurement work. For example, a well calibrated voltmeter may give a misleading voltage reading when connected across a high resistance circuit (See Example 3.12). The same voltmeter, when connected in a low resistance circuit, may give a more dependable reading (See Example 3.13). These examples illustrate that the voltmeter has a **loading effect** on the circuit, altering the actual circuit conditions by the measurement process. A detailed analysis of loading effects of instruments has been given in Chapter 2 (See Arts. 2.24, 2.25, 2.26 and 2.27 on pages nos. 33 to 39).

**Example 3.12** *A voltmeter having a sensitivity of 1000 Ω/V reads 100 V on its 150 V scale when connected across an unknown resistor in series with a multi-ammeter. When the milli-ammeter reads 5 mA, calculate :*

*(a) apparent resistance of the unknown resistor,*

*(b) actual resistance of the unknown resistor, and*

*(c) error due to the loading effect of voltmeter.*

**Solution.** Total circuit resistance

$$R_T = \frac{E_T}{I_T} = \frac{100}{5 \times 10^{-3}} = 20 \text{ k}\Omega.$$

Neglecting the resistance of milli-ammeter, the value of unknown resistor $R_x = 20$ kΩ.

(*b*) Resistance of voltmeter

$$R_V = 1000 \times 150 \ \Omega = 150 \text{ k}\Omega.$$

As the voltmeter is in parallel with the unknown resistance, we have

$$R_T = \frac{R_x R_V}{R_V + R_x}$$

or unknown resistance

$$R_x = \frac{R_T R_V}{R_V - R_T} = \frac{20 \times 150}{150 - 20} = 23.077 \text{ k}\Omega.$$

(*c*) Percentage error

$$= \frac{\text{measured value} - \text{true value}}{\text{true value}} \times 100$$

$$= \frac{20 - 23.077}{23.077} \times 100 = -13.33 \ \%.$$

**Example 3.13** *Repeat example 3.12 if the milli-ammeter reads 800 mA and the voltmeter reads 40 V on its 150 V scale.*

**Solution.** (*a*) $R_T = \dfrac{E_T}{I_T} = \dfrac{40}{800 \times 10^{-3}} = 50 \ \Omega$

(*b*) $R_V = 1000 \times 150 \ \Omega = 150 \ \Omega$

$$\therefore \quad R_x = \frac{R_T R_V}{R_V - R_T}$$

$$= \frac{50 \times 150 \times 10^3}{150 \times 10^3 - 50} = 50.017 \ \Omega.$$

(*c*) Percentage error

$$= \frac{50.0 - 50.017}{50.017} \times 100 = -0.034\%$$

Therefore errors caused by loading effects of the meters can be avoided by using them intelligently. For example, when measuring a low resistance by ammeter-voltmeter method a voltmeter having a very high value of resistance should be used.

In planning any measurement, the loading effect of instruments should be considered and corrections for these effects should be made, if needed, or more suitable instruments should be used. Preferably those methods should be used which result in negligible or no loading effects.

### 3.7.2 Environmental Errors

These errors are due to conditions external to the measuring device including conditions in the area surrounding the instrument. These may be effects of temperature pressure, humidity, dust, vibrations or of external magnetic or electrostatic fields. The corrective measures employed to eliminate or to reduce these undesirable effects are :

1. Arrangements should be made to keep the conditions as nearly as constant as possible. For example, temperature can be kept constant by keeping the equipment in a temperature controlled enclosure.
2. Using equipment which is immune to these effects. For example, variations in resistance with temperature can be minimized by using resistance materials which have a very low resistance temperature co-efficient.
3. Employing techniques which eliminate the effects of these disturbances. For example, the effect of humidity dust etc. can be entirely eliminated by hermetically sealing the equipment.
4. In case it is suspected that external magnetic or electrostatic fields can affect the readings of the instruments, magnetic or electrostatic shields may be provided.
5. *Applying computed corrections* : Efforts are normally made to avoid the use of application of computed corrections, but where these corrections are needed and are necessary, they are incorporated for the computations of the results.

### 3.7.3 Observational Errors

There are many sources of observational errors. As an example, the pointer of a voltmeter rests slightly above the surface of the scale. Thus an error on account of *PARALLAX* will be incurred unless the line of vision of the observer is exactly above the pointer. To minimize parallax errors, highly accurate meters are provided with mirrored scales, as shown in Fig. 3.1.

When the pointer's image appears hidden by the pointer, observer's eye is directly in line with the pointer. Although a mirrored scale minimizes parallax error, an error is necessarily present though it may be very small.

Since the parallax errors arise on account of pointer and the scale not being in the same plane, we can eliminate this error by having the pointer and the scale in the same plane as shown in Fig. 3.2.

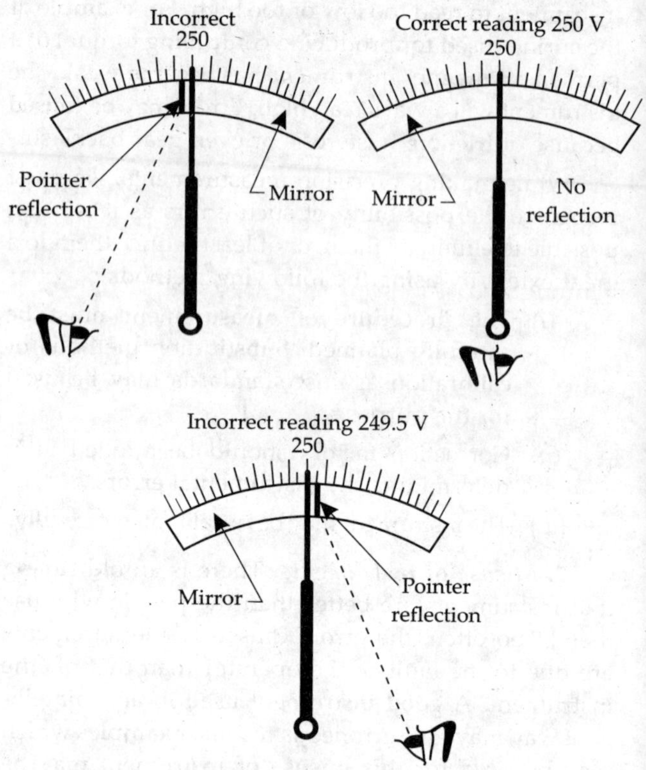


**Fig. 3.1** Errors due to Parallax.

There are human factors involved in measurement. The sensing capabilities of individual observers affect the accuracy of measurement. No two persons observe the same situation in exactly the same way where small details are concerned. For example, there are observational errors in measurements involving timing of an event. One observer may tend to anticipate the signal and read too soon. Different experimenters may produce different results, especially when sound and light measurements are involved since no two observers possess the same physical responses.

Modem electrical instruments have digital display of output which completely eliminates the errors on account of human observational or sensing powers as the output is in the form of digits.

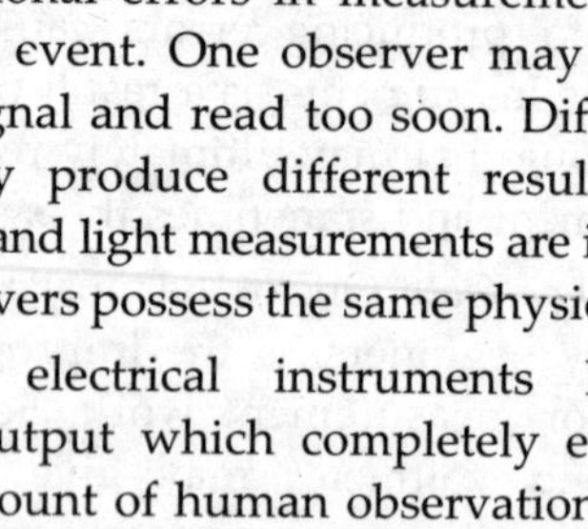


**Fig. 3.2** Arrangements showing scale and pointer in the same plane.

## 3.8 RANDOM (RESIDUAL) ERRORS

It has been consistently found that experimental results show variation from one reading to another, even after all systematic errors have been accounted for. These errors are due to a multitude of small factors which change or fluctuate from one measurement to another and are due surely to chance. The quantity being measured is affected by many happenings throughout the universe. We are aware of and account for some of the factors influencing the measurement, but about the rest we are unaware. The happenings or disturbances about which we are unaware are lumped together and called *"Random"* or *"Residual"*. Hence the errors caused by these happenings are called **Random** (or **Residual**) **Errors**. Since these errors remain even after the systematic errors have been taken care of, we call these errors as **Residual (Random) Errors.**

## 3.9 CENTRAL VALUE

As stated above, the random errors are caused by a large number of small effects, each one being a variable. These variables may be additive in some cases and subtractive in some, in their effect on the quantity being measured. In many measurements the positive and negative effects are nearly equal, so that the resultant error is small. If we make a large number of measurements and if the plus effects are equal to the minus effects, they would cancel each other and we would obtain the scatter round a **Central Value.** Since this condition is frequently met in practice, we are justified in using this concept as a basis of our study of errors which are unknown to us. Thus mathematical laws or "**Probability**" can be applied for the study of random errors. There is no other way as the random errors are unknown and only statistical study can lead us to the best approximation of the true value of the quantity under study.

## 3.10 STATISTICAL TREATMENT OF DATA

The experimental data is obtained in *two* forms of tests :

(*i*) Multisample test and (*ii*) Single-sample test.

***Multisample test.*** In this test, repeated measurement of a given quantity are done using different test conditions such as employing different instruments, different ways of measurement and by employing different observers. Simply making measurements with the same equipment, procedure, technique and same observer do not provide multisample results.

***Single-sample test.*** A single measurement (or succession of measurements) done under identical conditions excepting for time is known as single-sample test.

In order to get the exact value of the quantity under measurement, tests should be done using as many different procedures, techniques and experimenters as practicable. It should be borne in mind that the statistical means which help us to arrive at correct results are only valid for multisample tests.

### 3.10.1 Histogram

When a number of multisample observations are taken experimentally there is a scatter of the data about some central value. One method presenting test results in the form of a **Histogram.** The technique is illustrated in Fig. 3.3 representing the data given in Table 3.1. This table shows a set of fifty readings of a length measurement. The most probable or central value of length is 100 mm and the data are taken and recorded to the nearest 0.1 mm.

**Table 3.1**

| *Length (mm)* | *Number of Readings* |
|---|---|
| 99.7 | 1 |
| 99.8 | 4 |
| 99.9 | 12 |
| 100.0 | 19 |
| 100.1 | 10 |
| 100.2 | 3 |
| 100.3 | 1 |

Total number of readings = 50.

This histogram of Fig. 3.3 represents these data where the ordinate indicates the number of observed readings (frequency or occurrence) of a particular value. A histogram is also called a **frequency distribution curve.** At the central value of 100 mm is a large number of readings, 19 in this case, with other values placed almost symmetrically on either side. If smaller incremental steps, say 100 readings at 0.05 mm intervals are taken, the general form of the histogram will be almost the same but since the steps have smaller increments and we get a smoother curve.

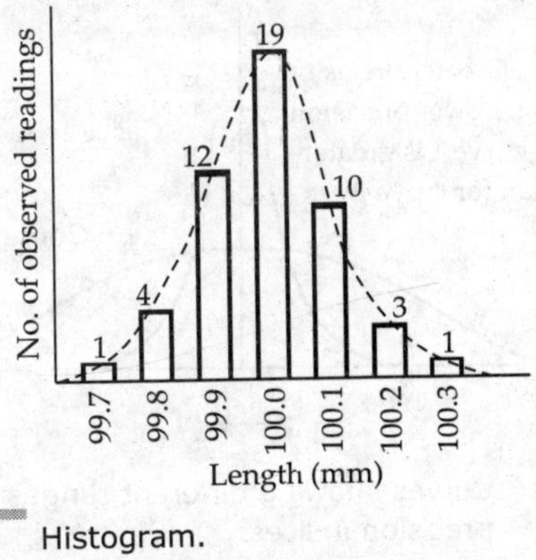


**Fig. 3.3** Histogram.

With more and more data taken at smaller and smaller increments the histogram would finally change into a smooth curve, as indicated by the dashed line in Fig. 3.3.

The smooth curve is symmetrical with respect to the central value. Many physical cases have been found which give experimental data agreeing fairly well with the smooth symmetrical curve.

### 3.10.2 Arithmetic Mean

The most probable value of measured variable (variate) is the arithmetic mean of the number of readings taken. The best approximation is made when the number of readings of the same quantity are very large. Theoretically, an infinite number of readings would give the best result, although in practice, only a finite number of measurements can be made.

The arithmetic mean is given by :

$$\overline{X} = \frac{x_1 + x_2 + x_3 + x_4 + ... + x_n}{n} = \frac{\Sigma x}{n} \quad ...(3.16)$$

where, $\overline{X}$ = arithmetic mean.

$x_1, x_2, ..., x_n$ = readings or variates or samples.

and $n$ = number of readings.

### 3.10.3 Measure of Dispersion from the Mean

*Dispersion.* The property which denotes the extent to which the values are dispersed about the central value is termed as **dispersion.** The other names used for dispersion are **spread** or **scatter**.

Figure 3.4 shows two sets of data. In one case (curve 1) the values vary from $x_1$ to $x_2$ and in other case (curve 2) the values vary from $x_3$ to $x_4$. Though their central value is the same, clearly set of data represented by curve 1 has a smaller dispersion than that of the data represented by curve 2.

It is very important to have a measure of the dispersion from the central value as it is an indication of the degree of consistency (precision) and regularity of the data.

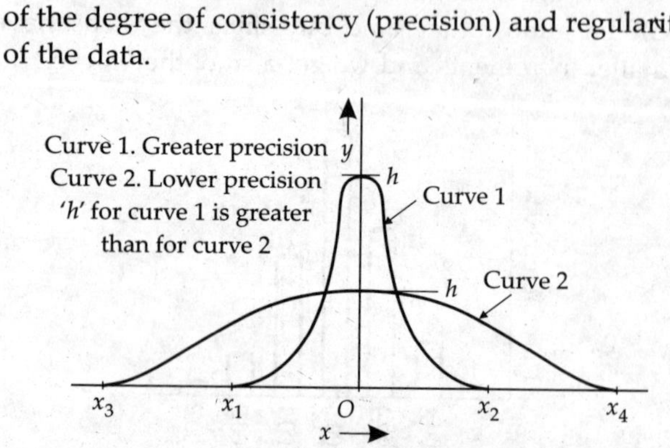


**Fig. 3.4** Curves showing different ranges and precision indices.

A large dispersion indicates that some factors involved in the measurement process are not under close control and therefore it becomes difficult to estimate the measured quantity with confidence and definiteness. For example, if we compare two sets of data and find that one set has less dispersion that the other set, we can certainly place more reliance on it and can definitely regard it as a superior set as regards random errors.

There are certain terms which must be defined as they form the basis of defining the measure of dispersion of data.

### 3.10.4 Range

The simplest possible measure of dispersion is the range which is the difference between greatest and least values of data. For example, in Fig. 3.4 the range of curve 1 is $(x_2 - x_1)$ and that of curve 2 is $(x_4 - x_3)$.

### 3.10.5 Deviation

Deviation is departure of the observed reading from the arithmetic mean of the group of readings. Let the deviation of reading $x_1$ be $d_1$ and that of reading $x_2$ be $d_2$, etc.

Then
$$d_1 = x_1 - \overline{X}$$
$$d_2 = x_2 - \overline{X}$$
..................
$$d_n = x_n - \overline{X} \quad ...(3.17)$$

and
$$\overline{X} = \frac{\Sigma(x_n - d_n)}{n} \quad ...(3.18)$$

Algebraic sum of deviations

$$= d_1 + d_2 + d_3 + ... + d_n$$
$$= (x_1 - \overline{X}) + (x_2 - \overline{X}) + (x_3 - \overline{X}) + ... + (x_n - \overline{X})$$
$$= (x_1 + x_2 + x_3 + ... + x_n) - n\overline{X} = 0$$

as $x_1 + x_2 + x_3 + ... + x_n = n\overline{X}$

Therefore the algebraic sum of deviations is zero.

### 3.10.6 Average Deviation

The average deviation is an indication of the precision of the instruments used in making the measurements. Highly precise instruments yield a low average deviation between readings.

Average deviation is defined as the sum of the absolute values of deviations divided by the number of readings. The absolute value of deviation is the value without respect to its sign. Average deviation may be expressed as :

$$\overline{D} = \frac{|-d_1| + |-d_2| + |-d_3| + ... + |-d_n|}{n} = \frac{\Sigma|d|}{n} \quad ...(3.19)$$

### 3.10.7 Standard Deviation (S.D.)

Another important term in the analysis of random errors is the **standard deviation** or the root mean square deviation. The Standard Deviation of an infinite number of data is defined as the square root of the sum of the individual deviations squared, divided by the number of readings.

Thus standard deviation is :

$$\text{S.D.} = \sigma = \sqrt{\frac{d_1^2 + d_2^2 + ... + d_n^2}{n}} = \sqrt{\frac{\Sigma d^2}{n}} \quad ...(3.20)$$

In practice, however, the number of observations is finite. When the number of observations is greater than 20, S.D. is denoted by symbol σ while if the number of observations is less than 20, the symbol used is s. The Standard Deviation of a finite number of data is given by :

$$s = \sqrt{\frac{d_1^2 + d_2^2 + d_3^2 + ... + d_n^2}{n-1}} = \sqrt{\frac{\Sigma d^2}{n-1}} \quad ...(3.21)$$

### 3.10.8 Variance

The variance is the mean square deviation, which is the same as S.D., except that square root is not extracted.

$$\text{Variance } V = (\text{Standard Deviation})^2$$

$$= (\text{S.D.})^2 = \sigma^2 = \frac{d_1^2 + d_2^2 + d_3^2 + ... + d_n^2}{n} \quad ...(3.22)$$

$$= \frac{\Sigma d^2}{n} \quad ...(3.23)$$

But when the number of observations is less than 20

$$\text{Variance } V = s^2 = \frac{\Sigma d^2}{n-1} \quad ...(3.24)$$

**Example 3.14** *A circuit was tuned for resonance by eight different students, and the values of resonant frequency in kHz were recorded as 532, 548, 543, 535, 546, 531, 543 and 536. Calculate (a) the arithmetic mean, (b) deviations from mean, (c) the average deviation, (d) the standard deviation, and (e) variance.*

**Solution.** (*a*) From Eqn. 3.16, the arithmetic mean of readings is,

$$\overline{X} = \frac{\Sigma x}{n} = \frac{532 + 548 + 543 + 535 + 546 + 531 + 543 + 536}{8}$$

$= 539.25$ kHz.

(*b*) From Eqn. 3.17, the deviations are,

$$d_1 = x_1 - \overline{X} = 532 - 539.25 = -7.25 \text{ kHz}$$
$$d_2 = x_2 - \overline{X} = 548 - 539.25 = +8.75 \text{ kHz}$$
$$d_3 = x_3 - \overline{X} = 543 - 539.25 = +3.75 \text{ kHz}$$
$$d_4 = x_4 - \overline{X} = 535 - 539.25 = -4.25 \text{ kHz}$$
$$d_5 = x_5 - \overline{X} = 546 - 539.25 = +6.75 \text{ kHz}$$
$$d_6 = x_6 - \overline{X} = 531 - 539.25 = -8.25 \text{ kHz}$$
$$d_7 = x_7 - \overline{X} = 543 - 539.25 = +3.75 \text{ kHz}$$
$$d_8 = x_8 - \overline{X} = 536 - 539.25 = -3.25 \text{ kHz}$$

(*c*) From Eqn. 3.19, average deviation is :

$$\overline{D} = \frac{\Sigma|d|}{n}$$

$$= \frac{7.25 + 8.75 + 3.75 + 4.25 + 6.75 + 8.25 + 3.75 + 3.25}{8}$$

$= 5.75$ kHz

(Note that for calculation of average deviations, the signs of deviations are disregarded).

(*d*) Since the number of readings is 8 which is less than 20, therefore Eqn. 3.24 is used for calculation of standard deviation.

$$\text{Standard deviation } s = \sqrt{\frac{\Sigma d^2}{n-1}}$$

$$= \sqrt{\frac{(-7.25)^2 + (+8.75)^2 + (+3.75) + (-4.25)^2 + (+6.75)^2 + (-8.25)^2 + (+3.75)^2 + (-3.25)^2}{(8-1)}}$$

$= 6.54$ kHz.

Variance $V = s^2 = 42.77\ (\text{kHz})^2$.

### 3.10.9 Normal or Gaussian Curve of Errors

The **Normal** or **Gaussian law** of errors is the basis for the major part of study of random effects. This type of distribution is most frequently met in practice.

The law of probability states the normal occurrence of deviations from average value of an infinite number of measurements or observations can be expressed by :

$$y = \frac{h}{\sqrt{\pi}} \exp(-h^2x^2) \quad ...(3.25)$$

**NOTE** The reader here is cautioned not to confuse $x$ with magnitude of a quantity. Here $x$ means deviation.

where $x$ = magnitude of deviation from mean,

$y$ = number of readings at any deviation $x$, (the probability of occurrence of deviation $x$).

and $h$ = a constant called precision index.

Equation 3.25 leads to curve of type shown in Fig. 3.5 and this curve showing $y$ plotted against $x$ is called "**Normal or Gaussian Probability Curve**".

3

Another, more convenient form of equation describing Gaussian curve uses standard deviation $\sigma$ and is given by :

$$y = \frac{1}{\sigma\sqrt{2\pi}} \exp(-x^2/2\sigma^2) \qquad ...(3.26)$$

The form of Eqn 3.26 is particularly useful because $\sigma$ is usually the known quantity of interest. In Fig. 3.5, the deviations from the mean value are divided in terms of $\sigma$ units, so that deviation $x = 1\sigma, 2\sigma, 3\sigma$, etc. The Gaussian curve of Fig. 3.5 may represent the error distribution of any set of data obeying the normal law and may be used as such once the value of $\sigma$ has been determined.

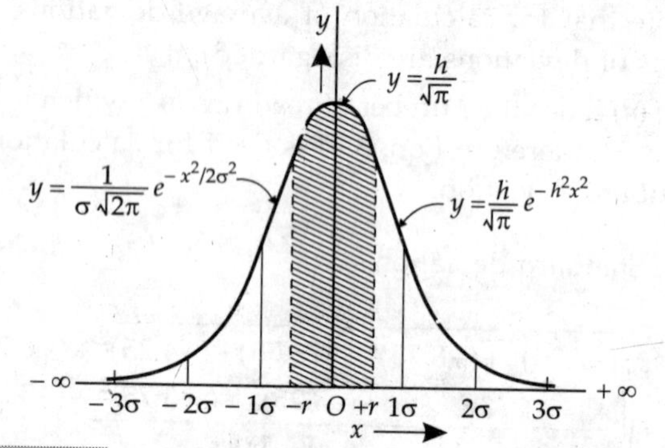


**Fig. 3.5** Normal probability curve.

This curve is symmetrical about the arithmetic mean value, and area under the curve is unity. Under the conditions specified here the total number of readings taken is represented by 1. This can be explained as follows : Suppose for the time being that we consider $h/\sqrt{\pi}$ to be unknown and replace it by the symbol $A$. If we have a large number of readings, $n$, the probable number $\Delta n$, with deviation between $x$ and $x + \Delta x$ is given by :

$$\Delta n = ny\Delta x = nA \exp(-h^2x^2)\Delta x$$

If we integrate the above expression for $-\infty$ to $+\infty$ we shall have all the cases, or

$$nA \int_{-\infty}^{\infty} \exp(-h^2x^2)dx = n$$

Thus : $$A \int_{-\infty}^{\infty} \exp(-h^2x^2)dx = 1$$

or $$\frac{h}{\sqrt{\pi}} \int_{-\infty}^{-\infty} \exp(-h^2x^2)dx = 1 \qquad ...(3.27)$$

Thus the integral of $y$ from $-\infty$ to $+\infty$ is equal to unity. The fraction of the total number of readings occurring between the values $x_1$ and $x_2$ will be equal to the area under the curve between these values of $x$.

$$n_{1-2} = \frac{h}{\sqrt{\pi}} \int_{x_1}^{x_2} \exp(-h^2x^2)dx = n \qquad ...(3.28)$$

where $n_{1-2}$ = number of readings occurring between the values $x_1$ and $x_2$.

If the area between $x_1$ and $x_2$ is 0.5, then 50 per cent of the deviations fall between $x_1$ and $x_2$.

In general the probability for finding a deviation in an interval between $x_1$ and $x_2$ becomes

$$P(x_1, x_2) = \int_{x_1}^{x_2} y\,dx = \frac{h}{\sqrt{\pi}} \int_{x_1}^{x_2} \exp(-h^2x^2)dx \qquad ...(3.29)$$

The fraction of the total number of deviations falling between zero and $x$ is :

$$n_{0-x} = \frac{h}{\sqrt{\pi}} \int_0^x \exp(-h^2x^2)dx \qquad ...(3.30)$$

This value as a function of $(x/\sigma)$ is given in table for probability integrals (See Table 3.2).

### 3.10.10 Precision Index

From Eqn. 3.25,

when $x = 0$, we have $$y = \frac{h}{\sqrt{\pi}} \qquad ...(3.31)$$

Thus it is clear from above that the maximum value of $y$ depends upon $h$. The larger the value of $h$, the sharper the curve. Thus the value of $h$ determines the sharpness of the curve since the curve drops sharply owing to the term $(-h^2)$ being in the exponent. A sharp curve evidently indicates that the deviations are more closely grouped together around deviation $x = 0$.

Figure 3.4 shows two curves having different values of $h$. Curve 1 has a large value of $h$ while curve 2 has a smaller value of $h$. Therefore curve 1, wherein the deviations are closely clustered together, is indicative of greater precision than curve 2, wherein the deviations are spread over a much wider range.

It is clear that the probability that a variate lies in a given range becomes less as the deviation of the range becomes greater. For a given deviation $x$, the probability is less greater the $h$ and vice versa. Thus the name **precision index** for $h$ is reasonable. A large value of $h$ represents high precision of the data because the probability of occurrence of variates in a given range falls off rapidly as the deviation increases because the variates tend to cluster (become closer) into a narrow range. On the other hand, a small value of $h$ represents low precision because the probability of occurrences of variates in a given range falls off gradually as the deviation increases ; this is because the variates are spread over a wide range.

It is evident therefore that curve 1 of Fig. 3.4 represents a data of greater precision than that of curve 2 since the value of $h$ for curve 1 is greater than that of curve 2.

### 3.10.11 Probable Error

We have observed above that the most probable or best value of a Gaussian distribution is obtained by taking arithmetic mean of the various values of the variate. In addition it has been indicated that the confidence in the best value (most probable value) is connected with the sharpness of the distribution curve.

Let us consider the two points $-r$ and $+r$ marked in Fig. 3.5. These points are so located that the area bounded by the curve, the $x$-axis, and the ordinates erected at $x=-r$ and $x=+r$ is equal to half of the total area under the curve. That is half the deviations lie between $x=\pm r$.

A convenient measure of precision is the quantity $r$. It is called **Probable Error** or simply **P.E.** The reason for this name is the fact mentioned above that half the observed values lie between the limits $\pm r$. If we determine $r$ as the result of $n$ measurements and then make an additional measurement, the chances are 50 – 50 percent that the new value will lie between $-r$ and $+r$ that is, the chances are even that any one reading will have an error no greater than $\pm r$.

The location of point $r$ can be found from Eqn. 3.28, by putting

$$\frac{h}{\sqrt{\pi}}\int_{-r}^{+r}\exp(-h^2x^2)dx=\frac{1}{2}$$

This gives probable error P.E.

$$r=\frac{0.4769}{h} \quad \text{...(3.32)}$$

### 3.10.12 Average Deviation for the Normal Curve

The average deviation may be computed when more than one reading is present at a given deviation by multiplying the amount of the deviation by the number of points on the deviation. Then this product is added to other similar products (without regard to sign) until all readings are taken into account ; then divide by the number of readings.

In the case of normal curve,

$$\text{Average deviation } \overline{D}=\int_{-\infty}^{+\infty}|x|\,y\,dx$$

$$=\frac{2h}{\sqrt{\pi}}\int_{-\infty}^{+\infty}\exp(-h^2x^2)|x|\,dx=\frac{1}{\sqrt{xh}} \quad \text{...(3.33)}$$

or $$h=\frac{1}{\sqrt{\pi}\,\overline{D}}$$

Substituting the value of $h$ in Eqn. 3.32, we have :

Average deviation

$$\overline{D}=\frac{r}{0.4769\sqrt{\pi}}=\frac{r}{0.8453} \quad \text{...(3.34)}$$

### 3.10.13 Standard Deviation for the Normal Curve

3

From Eqn. 3.20,

$$\text{Standard deviation } \sigma=\sqrt{\frac{\Sigma d^2}{n}}$$

Following a method similar to that followed above, we have :

$$\sigma=\frac{2h}{\sqrt{\pi}}\int_{0}^{+\infty}\exp(-h^2x^2)dx=\frac{1}{2h^2}$$

or standard deviation for normal curve

$$\sigma=\frac{1}{\sqrt{2}\,h} \quad \text{...(3.35)}$$

$$=\frac{r}{\sqrt{2}\times 0.4769}=\frac{r}{0.6745} \quad \text{...(3.36)}$$

From Eqns. 3.34 and 3.36, we have

$$\text{P.E.}=r=0.8453\,\overline{D} \quad \text{...(3.37)}$$

$$=0.6745\,\sigma. \quad \text{...(3.38)}$$

### 3.10.14 Probable Error of a Finite Number of Readings

In the analysis of the normal Gaussian error curve we have assumed that an infinite number of readings were taken. All the formulae derived above are based upon this assumption. With a finite number of readings, there is a slight difference between the computed values given above and the values obtained with a finite set of readings. For example, substitution of Eqn. 3.20 into Eqn. 3.38 gives the probable error as

$$r=0.6745\sqrt{\frac{d_1^2+d_2^2+d_3^2+...+d_n^2}{n}}$$

$$=0.6745\sqrt{\frac{\Sigma|d|^2}{n}} \quad \text{...(3.39)}$$

for an infinite number of deviations forming the normal probability curve *i.e.,* where $n$ is infinite. But for a finite number of deviations $\sigma$ is replaced by s (see Eqn. 3.21) and therefore the probable error for one reading is :

$$r=0.6745\sqrt{\frac{d_1^2+d_2^2+d_3^2+...+d_n^2}{n-1}}$$

$$=0.6745\sqrt{\frac{\Sigma|d|^2}{n-1}} \quad \text{...(3.40)}$$

This in fact means that, for a computed probable error $r$ obtained from $n$ readings, one more reading would have an even chance of being above or below $r_1$.

With a finite number of readings, the average reading has a probable error of :

$$r_m = \frac{1}{\sqrt{n}} r_1 = 0.6745\sqrt{\frac{d_1^2 + d_2^2 + d_3^2 + ... + d_n^2}{n(n-1)}} = 0.6745\sqrt{\frac{\Sigma|d|^2}{n(n-1)}} \quad ...(3.41)$$

The above equation means that for $n$ finite readings, the probable error is $r_m$.

If we have $n >> 1$ then $n - 1 \approx n$

$$r_1 \approx 0.6745\ \sigma \quad ...(3.42)$$

and

$$r_m = 0.6745\frac{\sigma}{\sqrt{n}} \quad ...(3.43)$$

### 3.10.15 Standard Deviation of Mean

When we have a multiple sample data, it is evident that the mean of various sets of data can be analyzed by statistical means. This is done by taking standard deviation of the mean.

The standard deviation of the mean is given by

$$\sigma_m = \frac{\sigma}{\sqrt{n}} \quad ...(3.44)$$

### 3.10.16 Standard Deviation of Standard Deviation

For a multiple sample data, the standard deviation of the standard deviation is :

$$\sigma_\sigma = \frac{\sigma}{\sqrt{2n}} \quad ...(3.45)$$

$$= \frac{\sigma_m}{\sqrt{2}} \quad ...(3.46)$$

## 3.11 PROBABILITY TABLES

The probability that the measurement will fall within deviations $x_1$ and $x_2$ from the mean is given by :

$$P(x_1, x_2) = \frac{h}{\sqrt{\pi}}\int_{x_1}^{x_2} \exp(-h^2x^2)dx$$

for a Normal Gaussian curve. The parameter $h = \frac{1}{\sqrt{2}\sigma}$ and the mean value $\overline{X}$ may be known from the raw data. Evaluating the integral is a formidable task hence *probability tables* have been prepared from which the value of probability can be directly read off.

Let $t = \sqrt{2}\,hx = x/\sigma$. This means deviation $x = \sigma t$.

Now $P(x_1, x_2) = P(t_1, t_2) = \frac{1}{\sqrt{2\pi}}\int_{t_1}^{t_2} \exp\left(\frac{-t^2}{2}\right)dt \quad ...(3.47)$

where $t_1 = \sqrt{2}\,hx_1$ and $t_2 = \sqrt{2}\,hx_2$.

Equation 3.47 may be written as,

$$P(t_1, t_2) = \frac{1}{\sqrt{2\pi}}\int_0^{t_2} \exp\left(\frac{-t^2}{2}\right)dt - \frac{1}{\sqrt{2\pi}}\int_0^{t_1} \exp\left(\frac{-t^2}{2}\right)dt \quad ...(3.48)$$

Each of the two integrals in Eqn. 3.48 is a function of a single parameter $t_1$ or $t_2$. Consequently, probability,

$$P(0, t) = \frac{1}{\sqrt{2\pi}}\int_0^{t} \exp\left(\frac{-t^2}{2}\right)dt \quad ...(3.49)$$

The results of this integral are given in Table 3.2 and can be used to compute the probability between limits $x_1$, and $x_2$. Tabulated values of the integral represent the area under the normalized Gaussian error curve for $t = 0$ and $t = t$ as shown in Fig. 3.6.

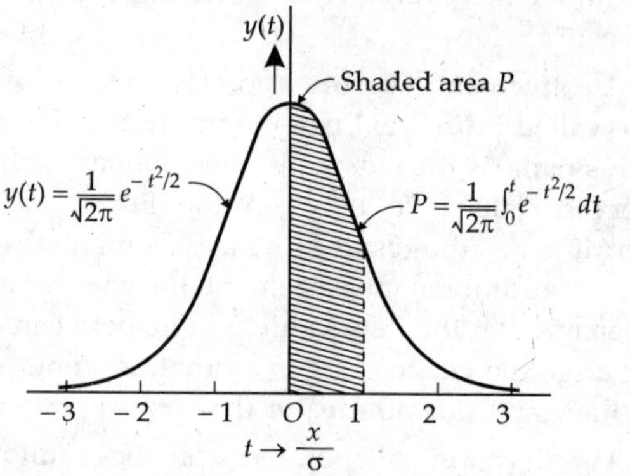


**Fig. 3.6** Graphical representation of probability.

Since the curve is symmetrical, the same table may be used for negative deviations replacing $t$ by $-t$ in Eqn. 3.49 as this substitution does not affect the value of probability.

It should be noted that there can be another table of the Gaussian error curve which tabulates twice the area as represented by Eqn. 3.49 and therefore can give the probability of occurrence between two symmetrical deviations about the mean. For such a curve, supposing it is desired to find the probability that the deviations lie between 0 and $+\sigma$. Now $x/\sigma = t = 1$. The probability is given by integral,

$$P = \frac{1}{\sqrt{2\pi}}\int_0^{1} \exp\left(\frac{-t^2}{2}\right)dt \quad ...(3.50)$$

Table 3.2 indicates a value of $P = 0.3413$ for $t = 1$. Because of the symmetry of the Gaussian curve, it follows that an equal probability exists for deviations to lie between $-\sigma$ and 0 *i.e.*, between $t = 1$ and $t = 0$.

Hence, the probability that deviations lie between $-\sigma$ to $+\sigma$ is $2 \times 0.3413 = 0.6826$.

**Table 3.2** *Probability Tables*

*Areas of the Gaussian Error Curve, Table gives values of the area under the curve between the ordinates at t = 0 and t. Example : Area = 0.1331 for t = 0.34.*

| t | 0 | .01 | .02 | .03 | .04 | .05 | .06 | .07 | .08 | .09 |
|---|---|---|---|---|---|---|---|---|---|---|
| .0 | .0000 | .0040 | .0080 | .0120 | .0160 | .0199 | .0239 | .0279 | .0319 | .0359 |
| .1 | .0398 | .0438 | .0478 | .0517 | .0557 | .0596 | .0636 | .0675 | .0714 | .0754 |
| .2 | .0793 | .0832 | .0871 | .0910 | .0948 | .0987 | .1026 | .1064 | .1103 | .1141 |
| .3 | .1179 | .1217 | .1255 | .1293 | .1331 | .1368 | .1406 | .1443 | .1480 | .1517 |
| .4 | .1554 | .1591 | .1628 | .1664 | .1700 | .1736 | .1772 | .1808 | .1844 | .1879 |
| .5 | .1915 | .1950 | .1985 | .2019 | .2054 | .2088 | .2123 | .2157 | .2190 | .2224 |
| .6 | .2258 | .2291 | .2324 | .2357 | .2389 | .2422 | .2454 | .2486 | .2518 | .2549 |
| .7 | .2580 | .2612 | .2642 | .2673 | .2704 | .2734 | .2764 | .2794 | .2823 | .2852 |
| .8 | .2881 | .2910 | .2939 | .2967 | .2996 | .3023 | .3051 | .3079 | .3106 | .3133 |
| .9 | .3159 | .3186 | .3212 | .3238 | .3264 | .3289 | .3351 | 3340 | .3365 | .3389 |
| 1.0 | .3413 | .3438 | .3461 | .3485 | .3508 | .3531 | .3554 | .3577 | .3599 | .3621 |
| 1.1 | .3643 | .3665 | .3686 | .3708 | .3729 | .3749 | .3770 | .3790 | .3810 | .3830 |
| 1.2 | .3849 | .3869 | .3888 | .3907 | .3925 | .3944 | .3962 | .3980 | .3997 | .4015 |
| 1.3 | .4032 | .4049 | .4066 | .4082 | .4099 | .4115 | .4131 | .4147 | .4162 | .4177 |
| 1.4 | .4192 | .4207 | .4222 | .4236 | .4251 | .4265 | .4279 | .4292 | .4306 | .4319 |
| 1.5 | .4332 | .4345 | .4357 | .4370 | .4382 | .4394 | .4406 | .4418 | .4430 | .4441 |
| 1.6 | .4452 | .4463 | .4474 | .4485 | .4495 | .4505 | .4515 | .4525 | .4535 | .4545 |
| 1.7 | .4554 | .4564 | .4573 | .4582 | .4591 | .4599 | .4608 | .4616 | .4625 | .4633 |
| 1.8 | .4641 | .4649 | .4656 | .4664 | .4671 | .4678 | .4686 | .4693 | .4700 | .4706 |
| 1.9 | .4713 | .4719 | .4726 | .4732 | .4738 | .4744 | .4750 | .4756 | .4762 | .4767 |
| 2.0 | .4773 | .4778 | .4783 | .4788 | .4793 | .4798 | .4803 | .4808 | .4812 | .4817 |
| 2.1 | .4821 | .4826 | .4830 | .4834 | .4838 | .4842 | .4846 | .4850 | .4854 | .4857 |
| 2.2 | .4861 | .4865 | .4868 | .4871 | .4875 | .4878 | .4881 | .4884 | .4887 | .4890 |
| 2.3 | .4893 | .4896 | .4898 | .4901 | .4904 | .4906 | .4909 | .4911 | .4913 | .4916 |
| 2.4 | .4918 | .4920 | .4922 | .4925 | .4927 | .4929 | .4931 | .4932 | .4934 | .4936 |
| 2.5 | .4938 | .4940 | .4941 | .4943 | .4945 | .4946 | .4948 | .4949 | .4951 | .4952 |
| 2.6 | .4953 | .4955 | .4956 | .4957 | .4959 | .4960 | .4961 | .4962 | .4963 | .4964 |
| 2.7 | .4965 | .4966 | .4967 | .4968 | .4969 | .4970 | .4971 | .4972 | .4973 | .4974 |
| 2.8 | .4974 | .4975 | .4976 | .4977 | .4977 | .4978 | .4979 | .4980 | .4980 | .4981 |
| 2.9 | .4981 | .4982 | .4983 | .4983 | .4984 | .4984 | .4985 | .4985 | .4986 | .4986 |
| 3.0 | .4987 | .4987 | .4987 | .4988 | .4988 | .4989 | .4989 | .4989 | .4990 | .4990 |
| 3.1 | .4990 | .4991 | .4991 | .4991 | .4992 | .4992 | .4992 | .4992 | .4993 | .4993 |
| 3.2 | .4993 | .4993 | .4994 | .4994 | .4994 | .4994 | .4994 | .4995 | .4995 | .4995 |
| 3.3 | .4995 | .4995 | .4996 | .4996 | .4996 | .4996 | .4996 | .4996 | .4996 | .4997 |
| 3.4 | .4995 | .4997 | .4997 | .4997 | .4997 | .4997 | .4997 | .4997 | .4998 | .4998 |
| 3.5 | .4998 | .4998 | .4998 | .4998 | .4998 | .4998 | .4998 | .4998 | .4998 | .4998 |
| 3.6 | .4998 | .4999 | .4999 | .4999 | .4999 | .4999 | .4999 | .4999 | .4999 | .4999 |
| 3.7 | .4999 | .4999 | .4999 | .4999 | .4999 | .4999 | .4999 | .4999 | .4999 | .4999 |
| 3.8 | .4999 | .4999 | .4999 | .4999 | .4999 | .4999 | .4999 | .5000 | .5000 | .5000 |

**Example 3.15** *The following 10 observations were recorded when measuring a voltage : 41.7, 42.0, 41.8, 42.0, 42.1, 41.9, 42.0, 41.9, 42.5 and 41.8 volt. Find (i) the mean (ii) the standard deviation (iii) the probable error of one reading (iv) the probable error of mean and (v) range.*

**Solution.** For the sake of ease in calculations, the observations are tabulated and manipulated as under :

| x | d | $d^2$ |
|---|---|---|
| 41.7 | – 0.27 | 0.0729 |
| 42.0 | + 0.03 | 0.0009 |
| 41.8 | – 0.17 | 0.0289 |
| 42.0 | + 0.03 | 0.0009 |
| 42.1 | + 0.13 | 0.0169 |
| 41.9 | – 0.07 | 0.0049 |
| 42.0 | + 0.03 | 0.0049 |
| 41.9 | – 0.07 | 0.0049 |
| 42.5 | + 0.53 | 0.2809 |
| 41.8 | – 0.17 | 0.0289 |
| $\Sigma x = 419.7$ | | $\Sigma d^2 = 0.441$ |

(*i*) Mean length $\overline{X} = \dfrac{\Sigma x}{n} = \dfrac{419.7}{10} = 41.97$ volt

(*ii*) The value of standard deviation is

$$\sigma = \sqrt{\frac{d^2}{n}} = \sqrt{\frac{0.441}{10}} = 0.21 \text{ volt}$$

(See Eqn. 3.20)

if the data is considered to be a set of infinite readings. However, the number of observations is only 10 and therefore the standard deviation is :

$$s = \sqrt{\frac{d^2}{n-1}} = \sqrt{\frac{0.441}{(10-1)}} = 0.22 \text{ volt.}$$

(See Eqn. 3.21)

(*iii*) Probable error of one reading

$$r_1 = 0.6745\ s = 0.15 \text{ volt.}$$

(See Eqn. 3.40)

(*iv*) Probable error of mean

$$r_m = \frac{r_1}{\sqrt{n-1}} = \frac{0.15}{\sqrt{9}} = 0.05 \text{ volt.}$$

(See Eqn. 3.41)

(*v*) Range $= 42.5 - 41.7 = 0.8$ volt

---

**Example 3.16** *In a test, temperature is measured 100 times with variations in apparatus and procedures. After applying the corrections, the results are :*

| ***Temperature °C*** | 397 | 398 | 399 | 400 | 401 | 402 | 403 | 404 | 405 |
|---|---|---|---|---|---|---|---|---|---|
| ***Frequency of occurrence*** | 1 | 3 | 12 | 23 | 37 | 16 | 4 | 2 | 2 |

*Calculate (a) arithmetic mean, (b) mean deviation, (c) standard deviation, (d) the probable error of one reading, (e) the standard deviation and the probable error of the mean, (f) the standard deviation of the standard deviation.*

**Solution.** The computations are done in a tabular form as under :

| ***Temperature T°C*** | ***Frequency of occurrence, f*** | ***T × f*** | ***Deviation d*** | ***f × d*** | ***d²*** | ***f × d²*** |
|---|---|---|---|---|---|---|
| 397 | 1 | 397 | – 3.78 | – 3.78 | 14.288 | 14.288 |
| 398 | 3 | 1194 | – 2.78 | – 8.34 | 7.728 | 23.185 |
| 399 | 12 | 4788 | – 1.78 | – 21.36 | 3.168 | 38.020 |
| 400 | 23 | 9200 | – 0.78 | + 17.94 | 0.608 | 13.993 |
| 401 | 37 | 14837 | + 0.22 | + 8.14 | 0.048 | 1.708 |
| 402 | 16 | 6432 | + 1.22 | + 19.52 | 1.488 | 23.814 |
| 403 | 4 | 1612 | + 2.22 | + 8.88 | 4.928 | 19.714 |
| 404 | 2 | 808 | + 3.22 | + 6.44 | 10.368 | 20.737 |
| 405 | 2 | 810 | + 4.22 | + 8.44 | 17.808 | 35.618 |
| **Total** | **100** | **40078** | | $\Sigma\lvert f \times d\rvert = 102.8$ | | $\Sigma fd^2 = 191.08$ |

(*a*) Mean temperature $= \frac{40078}{100} = 400.780°C$

(*b*) Mean deviation $\bar{D} = \frac{102.8}{100} = 1.028°C$

(*c*) Standard deviation $\sigma = \sqrt{\frac{191.08}{100}} = 1.380°C$

(*d*) Probable error of one reading

$$r_1 = 0.6745\ \sigma = 0.6745 \times 1.38 = 0.93°C$$

(*e*) Probable error of the mean

$$r_m = \frac{0.93}{\sqrt{100}} = 0.093°C$$

Standard deviation of the mean

$$\sigma_m = \frac{0.93}{\sqrt{100}} = 0.138°C$$

(*f*) Standard deviation of the standard deviation

$$\sigma_\sigma = \frac{\sigma_m}{\sqrt{2}} = \frac{0.138}{\sqrt{2}} = 0.0796°C$$

**Example 3.17** *A value $R = 92.2 \pm 0.1\,\Omega$ (where $0.1\,\Omega$ is the standard deviation) is specified for a batch of 1000 resistors. How many would you estimate have values in the range $R = 92.2 \pm 0.15\Omega$ ? Assume normal distribution. Consult probability tables.*

**Solution.** Given Deviation $x = \pm 0.15\Omega$, Standard deviation $\sigma = \pm 0.1\,\Omega$

$$\therefore \text{ Ratio } \quad t = \frac{x}{\sigma} = \frac{\pm 0.15}{\pm 0.1} = 1.5.$$

Corresponding to 1.5, the area under the Gaussian curve is 0.4332 (*See Table 3.2*).

Therefore the probable number of resistors having a value of $92.2 \pm 0.15\,\Omega = 2 \times 0.4332 \times 1000 \doteq 866.$

**Example 3.18** *An underdamped galvanometer was energized 100 times under the same carefully controlled experimental conditions and the maximum deflection was read in each case. The readings were normally distributed about a mean value of 26.3 mm and had a probable error of 2.5 mm. How many of 100 readings would you estimate exceeded 30 mm ?*

**Solution.** Deviation $x = 30 - 26.3 = 3.7$ mm.

Probable error $r = 2.5$ mm.

From Eqn. 3.36, standard deviation,

$$\sigma = \frac{r}{0.6745} = \frac{2.5}{0.6745} = 3.7\,\text{mm}$$

$$\text{Ratio} \qquad t = \frac{x}{\sigma} = \frac{3.7}{3.7} = 1$$

Corresponding to ratio $t = x/\sigma = 1$, the area under the curve is 0.3413.

Therefore readings having a deviation within ± 3.7 mm is,

$$2 \times 0.3413 \times 100 \approx 68.$$

$\therefore$ Readings exceeding a deviation of ± 3.7 mm are,

$$100 - 68 = 32.$$

Hence readings having a deviation of + 3.7 mm is $32/2 = 16$.

Thus the number of readings exceeding a maximum deflection of 30 mm, *i.e.*, (26.3 + 3.7) mm is 16.

**Example 3.19** *A machine shop manufactured 25000 steel rods of nominal length 10.0 mm which were not to exceed 10.25 mm and not be shorter than 9.50 mm. It was found that 2000 of the rods were too long to fit into a gauge set at 10.25 mm. Predict the number of the remaining 23,000 rods which will confirm to the specifications. Assume normal distribution. Consult probability tables.*

**Solution.** The total number of rods is 25000. Since the distribution is normal there is equal probability of + and – side about the mean.

$\therefore$ 12500 rods have lengths greater than 10 mm and 12500 smaller than 10 mm.

Now 2000 rods have a length greater than 10.25 mm.

Number of rods whose lengths lie between 10 mm and (10 + 0.25) mm = 12500 – 2000 = 10500.

The probability that 10500 rods have a value greater than 10 mm and less than 10 + 0.25 = (10.25 mm) is $\frac{10500}{25000} = 0.42.$

From Table 3.2, corresponding to a probability of 0.42, the value of $t = 1.41$.

As we have to find the probability of number of rods whose lengths lie between 9.5 mm and 10 mm or whose lengths lie between (10 – 0.5) mm and 10 mm.

$$\therefore \qquad t = 2 \times 1.41 = 2.82$$

(as the deviation is 0.5 mm which is twice of the earlier deviation of 0.25 mm)

Corresponding $t = 2.82$, the probability is 0.4975.

$\therefore$ Number of rods whose lengths lie between 9.5 mm and 10 mm = 0.4975 × 25000 = 12437.

Hence total number of rods whose lengths lie between specified limits *i.e.*, 9.5 mm to 10.25 mm = 10500 + 12437 = 22937.

3

## 3.12 SPECIFYING ODDS

The probability of occurrence can be stated in terms of **Odds**. Odds is the number of chances that a particular reading will occur when the error limit is specified. For example, if the error limits are specified as ± 0.6745 σ, the chances are that 50% of the observations will lie between the above limits or in other words we can say that *odds* are 1 to 1. The odds can be calculated as under

$$\text{Probability of occurrence} = \frac{\text{odds}}{\text{odds}+1} \quad ...(3.51)$$

The odds that the observation lies between $\pm\sigma$ limits are, $\frac{\text{odds}}{\text{odds}+1} = 0.6828$ or odds are 2.15 : 1 (when $\sigma = 1, t = 1$ and therefore from Table 3.2, the probability of deviations to lie between 0 and σ is 0.3414 or probability of deviations to lie between $+\sigma$ and $-\sigma$ limits is $2 \times 0.3414 = 0.6828$).

Similar calculations can be done for other deviations referring to probability table given in Table 3.2. They are summarized as given below :

**Table 3.3** *Probability and Odds*

| Deviation | Probability | Odds |
|---|---|---|
| ± 0.6745 σ | 0.5000 | 1 to 1 |
| ± σ | 0.6828 | 2.15 to 1 |
| ± 2σ | 0.9546 | 21 to 1 |
| ± 3σ | 0.9974 | 256 to 1 |

## 3.13 SPECIFYING MEASUREMENT DATA

After carrying out statistical analysis of multi-sample data, the results of measurements must be specified. The results are expressed as deviations about a mean value. The deviations are expressed as :

(*i*) ***Standard deviation.*** The result is expressed as $\overline{X} \pm \sigma$. The error limit in this case is the standard deviation. This means that 0.6828 (about 68%) of the readings are within the limits $\sigma = \pm 1$ and the odds are 2.15 to 1. Thus there is approximately a 2 to 1 possibility that a new reading will fall within this limit.

(*ii*) ***Probable error.*** The result is expressed as $\overline{X} \pm 0.6745\ \sigma$. This means that 50 percent of the readings lie within this limit and the odds are 1 to 1. This means that there is an even possibility that a new reading will lie within this limit.

(*iii*) $\pm 2\sigma$ ***limit.*** The result is expressed as $\overline{X} \pm 2\sigma$. In this case the probability range is increased. Now 0.9546 (or about 95%) of readings fall within this limit. The odds in this case are 21 to 1.

(*iv*) $+3\sigma$ ***limit.*** The result is expressed as $\overline{X} \pm 3\sigma$. The maximum or boundary error limit is $\pm 3\sigma$.

The probability in this case is 0.9974. Therefore, 99.74% of the readings will fall within this limit. In other words, it can be stated that there is a possibility of 26 readings out of 10000 will fall beyond this limit. Thus practically all the readings are included in this limit. The odds of any reading falling within this limit are 256 to 1.

## 3.14 CONFIDENCE INTERVAL AND CONFIDENCE LEVEL

It is possible to state through statistical analysis of the data that a range of deviation from the mean value within which a certain fraction of all values is expected to lie. This range is called the *Confidence Interval*.

The probability that the value of a randomly selected observation will lie in this range is called the *Confidence Level*.

If the number of observations is large and their errors are random and follow the normal Gaussian distribution, the various confidence intervals about the mean value $\overline{X}$ are given in Table 3.4.

**Table 3.4** *Confidence Intervals*

| Confidence Level | Confidence Interval | Values lying outside Confidence Interval |
|---|---|---|
| 0.500 | $\overline{X} \pm 0.674\sigma$ | 1 in 2 |
| 0.800 | $\overline{X} \pm 1.282\sigma$ | 1 in 5 |
| 0.900 | $\overline{X} \pm 1.645\sigma$ | 1 in 10 |
| 0.950 | $\overline{X} \pm 1.960\sigma$ | 1 in 20 |
| 0.990 | $\overline{X} \pm 2.576\sigma$ | 1 in 100 |
| 0.999 | $\overline{X} \pm 3.291\sigma$ | 1 in 1000 |

If the number of observations is small (say less than 20) and the standard deviation is not accurately known, the confidence interval must be broadened. Here the standard deviation is computed as :

$$s = \sqrt{\frac{\Sigma d^2}{n-1}} \quad \text{(See Eqn. 3.21)}$$

This standard is multiplied by a suitable factor to establish the confidence interval. The results are given in Table 3.5.

**Table 3.5** *Confidence Intervals (When the number of observations is small)*

| No. of Degrees of Freedom | No. of Observations | | Confidence Level 0.5 | 0.9 | 0.95 | 0.99 |
|---|---|---|---|---|---|---|
| 1 | 2 | Confidence Interval | $\overline{X} \pm 1.00s$ | $\overline{X} \pm 6.31s$ | $\overline{X} \pm 12.71s$ | $\overline{X} \pm 63.66s$ |
| 2 | 3 | | $\overline{X} \pm 0.82s$ | $\overline{X} \pm 2.92s$ | $\overline{X} \pm 4.30s$ | $\overline{X} \pm 9.92s$ |
| 3 | 4 | | $\overline{X} \pm 0.77s$ | $\overline{X} \pm 2.35s$ | $\overline{X} \pm 3.18s$ | $\overline{X} \pm 5.84s$ |
| 4 | 5 | | $\overline{X} \pm 0.74s$ | $\overline{X} \pm 2.13s$ | $\overline{X} \pm 2.78s$ | $\overline{X} \pm 4.60s$ |
| 5 | 6 | | $\overline{X} \pm 0.73s$ | $\overline{X} \pm 2.02s$ | $\overline{X} \pm 2.57s$ | $\overline{X} \pm 4.03s$ |
| 6 | 7 | | $\overline{X} \pm 0.72s$ | $\overline{X} \pm 1.94s$ | $\overline{X} \pm 2.45s$ | $\overline{X} \pm 3.71s$ |
| 7 | 8 | | $\overline{X} \pm 0.71s$ | $\overline{X} \pm 1.90s$ | $\overline{X} \pm 2.37s$ | $\overline{X} \pm 3.50s$ |
| 8 | 9 | | $\overline{X} \pm 0.71s$ | $\overline{X} \pm 1.86s$ | $\overline{X} \pm 2.31s$ | $\overline{X} \pm 3.36s$ |
| 9 | 10 | | $\overline{X} \pm 0.70s$ | $\overline{X} \pm 1.83s$ | $\overline{X} \pm 2.26s$ | $\overline{X} \pm 3.25s$ |
| 10 | 11 | | $\overline{X} \pm 0.70s$ | $\overline{X} \pm 1.81s$ | $\overline{X} \pm 2.23s$ | $\overline{X} \pm 3.17s$ |
| 15 | 16 | | $\overline{X} \pm 0.69s$ | $\overline{X} \pm 1.75s$ | $\overline{X} \pm 2.13s$ | $\overline{X} \pm 2.95s$ |

In order to obtain confidence intervals for mean of a group of observations from the corresponding intervals for an individual observation, the later is divided by $\sqrt{n}$.

Confidence interval of mean

$$= \frac{\text{confidence interval of individual observation}}{\sqrt{n}} \quad \ldots(3.52)$$

Thus the expectation that the mean of a group of observations will not differ by more than a certain amount from the theoretical mean of an infinite set of observations can also be expressed in terms of a confidence interval and a confidence level.

**Example 3.20** *The following 10 observations were recorded when measuring a voltage : 41.7, 42.0, 41.8, 42.0, 42.1, 41.9, 42.0, 41.9, 42.5 and 41.8 V. Find (a) the confidence intervals for an individual observation for confidence levels of 0.5, 0.9, 0.95 and 0.99 and (b) the confidence intervals for the group of observations for above mentioned confidence levels.*

*Consult table for confidence intervals.*

*Comment upon the results.*

**Solution.** This problem is an extension of Example 3.15.

$$\text{Standard deviation } s = \sqrt{\frac{\Sigma d^2}{n-1}} = 0.22 \text{ V}.$$

(*a*) Now $s = 0.22$ V, therefore the confidence levels and the corresponding intervals (about the mean) are for an individual observation given below. (For these calculations consult Table 3.5).

| ***Confidence Level*** | 0.5 | 0.9 | 0.95 | 0.99 |
|---|---|---|---|---|
| ***Confidence Interval*** | ± 0.7 *s*<br>± 0.15 V | ± 1.83 *s*<br>± 0.40 V | ± 2.26 *s*<br>± 0.50 V | ± 3.25 *s*<br>± 0.72 V |

(*b*) The confidence intervals for various confidence levels for the group average are calculated by dividing the confidence intervals of individual readings by $\sqrt{n}$. The confidence intervals for group average corresponding to various confidence levels are :

**Comments :**

| *Confidence level* : | 0.5 | 0.9 | 0.95 | 0.99 |
|---|---|---|---|---|
| *Confidence Interval* : | ± 0.05 | ± 0.13 | ± 0.17 | ± 0.24 |

Thus the average value found from the ten observations would be expected with a probability of 5% to be within 0.05 V of mean value of infinite number of observations.

At a confidence level of 0.99, the observed and the theoretical results differ by not more than 0.24 V.

## 3.15 REJECTION OF DATA

In most of the experiments, the experimenter finds that some of the data points are noticeably different from the majority of the data. If these data points were obtained under abnormal conditions involving gross blunders and the experimenter is sure about their dubious nature, they can be discarded straight away.

However the experimenter cannot reject a data simply because it is different from the others, he must rely on certain standard mathematical methods for rejecting any experimental data. There are many methods available for assessing whether the data be rejected or retained. The *three* commonly used methods are :

- Chauvenet's criterion
- Use of confidence intervals and
- $3\sigma$ limits.

These methods are given below.

### 3.15.1 Chauvenet's Criterion

Suppose $n$ observations are made for measurement of a quantity. We assume that $n$ is large enough that the results will follow a normal Gaussian distribution.

**Table 3.6** *Chauvenet's Criterion for rejecting Data*

| Number of readings | Ratio of maximum acceptable deviation to standard deviation, $d_{max} / \sigma$ |
|---|---|
| 2 | 1.15 |
| 3 | 1.38 |
| 4 | 1.54 |
| 5 | 1.65 |
| 6 | 1.73 |
| 7 | 1.80 |
| 10 | 1.96 |
| 15 | 2.13 |
| 25 | 2.33 |
| 50 | 2.57 |
| 100 | 2.81 |
| 300 | 3.14 |
| 500 | 3.29 |
| 1000 | 3.48 |

This distribution may be used to compute the probability that a given reading will deviate by a certain amount from the mean. Chauvenet's criterion specifies that a reading may be rejected if the probability of obtaining the particular deviation from the mean is less than $1/2\,n$. Table 3.6 gives the values of the ratio of deviation to standard deviation for various values of $n$ according to this criterion.

When applying Chauvenet's criterion, in order to eliminate any dubious data, the mean value and the standard deviation are first calculated using all data points. The deviations of individual readings are then compared with standard deviation. If the ratio of deviation of a reading to the standard deviation exceeds the limits given in Table 3.6, that reading is rejected. The mean value and the standard deviations are again calculated by excluding the rejected reading from the data.

### 3.15.2 Rejection of Data based upon Confidence Intervals

A criterion used for discarding a data point is that its deviation from the mean exceeds four times the probable error of a single reading. This results in discarding a data outside a confidence interval for a single reading at a confidence level of 0.993.

A better criterion which does not involve the evaluation of probable error when the set of data points as small and standard deviation is not accurately known, is to discard a reading that lies outside the interval corresponding to confidence level of 0.99 for a single observation. (See Tables 3.4 and 3.5). On this basis, not more than 1 reading in 100 would lie outside this range.

A still better method is to use the confidence interval corresponding to a confidence level of 0.95 in order to scrutinize the measurement procedure adopted.

### 3.15.3 Rejection of Data based upon $\pm 3\sigma$ Limits

The probability that the reading will lie within $\pm 3\sigma$ limits of central value is 0.9974 which is very high and therefore any reading not lying within this limit should be rejected.

**Example 3.21** *A laboratory experiment is conducted to measure the viscosity of a specimen of oil. A series of tests give the values as $5.30 \times 10^{-3}$, $5.73 \times 10^{-3}$, $6.77 \times 10^{-3}$, $5.26 \times 10^{-3}$, $4.33 \times 10^{-3}$, $5.45 \times 10^{-3}$, $6.09 \times 10^{-3}$, $5.64 \times 10^{-3}$, $5.81 \times 10^{-3}$ and $5.75 \times 10^{-3}$ $m^2/s$. Point out any reading that can be rejected by applying Chauvenet's criterion. The ratio of maximum deviation to standard deviation should not exceed 1.96.*

**Solution.** The results are tabulated as below :

| S.No. | $x_i$ | $d_i = x_i - \overline{X}$ | $d_i^2$ | $\lvert d_i \rvert / s$ |
|---|---|---|---|---|
| 1. | $5.30 \times 10^{-3}$ | $-0.313 \times 10^{-3}$ | $98.0 \times 10^{-9}$ | 0.50 |
| 2. | $5.73 \times 10^{-3}$ | $+0.117 \times 10^{-3}$ | $13.7 \times 10^{-9}$ | 0.19 |
| 3. | $6.77 \times 10^{-3}$ | $+1.157 \times 10^{-3}$ | $1339 \times 10^{-9}$ | 1.85 |
| 4. | $5.26 \times 10^{-3}$ | $-0.353 \times 10^{-3}$ | $125 \times 10^{-9}$ | 0.56 |
| 5. | $4.33 \times 10^{-3}$ | $-1.283 \times 10^{-3}$ | $1649 \times 10^{-9}$ | 2.05 |
| 6. | $5.45 \times 10^{-3}$ | $-0.163 \times 10^{-3}$ | $26.6 \times 10^{-9}$ | 0.26 |
| 7. | $6.09 \times 10^{-3}$ | $+0.477 \times 10^{-3}$ | $228 \times 10^{-9}$ | 0.76 |
| 8. | $5.64 \times 10^{-3}$ | $+0.027 \times 10^{-3}$ | $0.729 \times 10^{-9}$ | 0.04 |
| 9. | $5.81 \times 10^{-3}$ | $+0.197 \times 10^{-3}$ | $38.9 \times 10^{-9}$ | 0.31 |
| 10. | $5.75 \times 10^{-3}$ | $+0.137 \times 10^{-3}$ | $38.7 \times 10^{-9}$ | 0.22 |

$\Sigma x_i = 56.13 \times 10^{-3}$ ; $\Sigma d_i^2 = 3537 \times 10^{-9}$

$\therefore$ Mean value of viscosity,

$$\overline{X} = \frac{\Sigma x_i}{n} = \frac{56.13 \times 10^{-3}}{10}$$

$$= 5.613 \times 10^{-3} \text{ m}^2/\text{s}$$

Standard deviation,

$$s = \sqrt{\frac{\Sigma d_i^2}{n-1}} = \sqrt{\frac{3537 \times 10^{-9}}{(10-1)}}$$

$$= 0.627 \times 10^{-3} \text{ m}^2/\text{s}$$

It is given that for 10 readings the ratio of deviation to standard deviation is not to exceed 1.96 and therefore reading No. 5 *i.e.,* $4.33 \times 10^{-3}$ m$^2$/s should be rejected.

## 3.16 METHOD OF LEAST SQUARES

Supposing we have a set of $n$ readings $x_1, x_2, \ldots, x_n$. The sum of the squares of their deviation from some mean value $X_m$ is :

$$S = \sum_{i=1}^{n} (x_i - X_m)^2$$

Supposing we wish to minimize $S$ with respect to mean value $X_m$. We have,

$$\frac{\delta S}{\delta X_m} = 2\left(\sum_{i=1}^{n} x_i - nX_m\right) = 0$$

or $$X_m = \frac{1}{n} \sum_{i=1}^{n} x_i$$

But $$\frac{1}{n} \sum_{i=1}^{n} x_i = \overline{X} \qquad \therefore \; X_m = \overline{X}$$

Thus the mean value which minimizes the sum of the squares of the deviations is the arithmetic mean. This is a simple example of application of the method of least squares. Another application of method of least squares is given below :

Suppose that two variables $x$ and $y$ are measured over a range of values. We want to obtain a simple analytical expression for $y$ as a function of $x$. The simplest type of function is a linear one and hence we try to find the best linear function connecting $y$ and $x$.

Suppose the linear function is $y = ax + b$

Now we want to minimize the quantity,

$$S = \sum_{i=1}^{n} [y_i - (ax_i + b)]^2$$

This is done by taking derivatives of $S$ with respect to $a$ and $b$ and setting them equal to zero. This gives

$$nb + a\Sigma x_i = \Sigma y_i$$

and $$b\Sigma x_i + a\Sigma x_i^2 = \Sigma x_i y_i$$

Solving the above equations, we get,

$$a = \frac{n\Sigma x_i y_i - (\Sigma x_i)(\Sigma y_i)}{n\Sigma x_i^2 - (\Sigma x_i)^2} \quad \text{...(3.53)}$$

$$b = \frac{(\Sigma y_i)(\Sigma x_i^2) - (\Sigma x_i y_i)(\Sigma x_i)}{n\Sigma x_i^2 - (\Sigma x_i)^2} \quad \text{...(3.54)}$$

The standard deviations of $a$ and $b$ may be found as :

$$s_a = \sqrt{\frac{n}{n\Sigma x_i^2 - (\Sigma x_i)^2}}\, s_y \quad \text{...(3.55)}$$

$$s_b = \sqrt{\frac{\Sigma x_i^2}{n\Sigma x_i^2 - (\Sigma x_i)^2}}\, s_y \quad \text{...(3.56)}$$

$$s_y = \sqrt{\frac{1}{n}\Sigma(ax_i + b - y_i)^2} \quad \text{...(3.57)}$$

Standard deviation of $x$ is,

$$s_x = \sqrt{\frac{1}{n}\Sigma\left(\frac{y_i - b}{a} - x_i\right)^2} = \frac{sy}{a} \quad \text{...(3.58)}$$

***Example 3.22*** *The following data are expected to follow a linear relationship of the form $y = ax + b$. Obtain the best linear relation in accordance with a least square analysis. Calculate the standard deviation of the data from the result.*

| x | 0.9 | 2.3 | 3.3 | 4.5 | 5.7 | 6.7 |
|---|---|---|---|---|---|---|
| y | 1.1 | 1.6 | 2.6 | 3.2 | 4.0 | 5.0 |

**Solution.** The solution is of the form $y = ax + b$.

The various quantities are calculated as under :

| $x_i$ | $y_i$ | $x_iy_i$ | $x_i^2$ |
|---|---|---|---|
| 0.9 | 1.1 | 0.99 | 0.81 |
| 2.3 | 1.6 | 3.68 | 5.29 |
| 3.3 | 2.6 | 8.58 | 10.89 |
| 4.5 | 3.2 | 14.40 | 20.25 |
| 5.6 | 4.0 | 22.80 | 32.49 |
| 6.7 | 5.0 | 33.50 | 44.89 |
| $\Sigma x_i = 23.4$ | $\Sigma y_i = 17.5$ | $\Sigma x_iy_i = 83.95$ | $\Sigma x_i^2 = 114.62$ |

Number of readings, $n = 6$

From Eqns. 3.53 and 3.54,

$$a = \frac{(6)\times(83.95)-(23.4)\times(17.5)}{(6)\times(114.62)-(23.4)^2} = 0.672$$

and

$$b = \frac{(17.5)(114.62)-(83.95)(23.4)}{(6)(114.62)-(23.4)^2} = 0.296.$$

Therefore the desired relationship is

$$y = 0.672x + 0.296.$$

Figure 3.6 shows the plot of this relationship and also the actual experimental data.

Let us find out the standard deviation in $y$.

| $x_i$ | $y_i$ | $ax_i + b$ | $ax_i + b - y_i$ | $(ax_i + b - y_i)^2$ |
|---|---|---|---|---|
| 0.9 | 1.1 | 0.901 | – 0.199 | 0.0396 |
| 2.3 | 1.6 | 1.842 | + 0.242 | 0.0586 |
| 3.3 | 2.6 | 2.514 | – 0.0860 | 0.0074 |
| 4.5 | 3.2 | 3.320 | + 0.1200 | 0.0144 |
| 5.6 | 4.0 | 4.126 | + 0.1260 | 0.0159 |
| 6.7 | 5.0 | 4.798 | – 0.2020 | 0.0408 |
| | | | | $\Sigma(ax_i+b-y_i)^2 = 0.1766$ |

∴ Standard deviation of $y$ is,

$$s_y = \sqrt{\frac{1}{n}\Sigma(ax_i+b-y_i)^2} \quad \text{(See Eqn. 3.57)}$$

$$= \sqrt{\frac{0.1766}{6}} = \pm 0.1716.$$

From Eqn. 3.58, standard deviation of $x$ is,

$$s_x = \frac{s_y}{a} = \pm\frac{0.1716}{0.672} = \pm 0.255$$

From Eqns. 3.55 and 3.56 the standard deviations in $a$ and $b$ are :

$$s_a = \sqrt{\frac{n}{n\Sigma x_i^2 - (\Sigma x_i)^2}}\, s_y$$

$$= \sqrt{\frac{6}{(6\times 114.62)-(23.4)^2}} \times 0.1716 = \pm 0.0355$$

$$S_b = \sqrt{\frac{\Sigma x_i^2}{n\Sigma_i^2 - (\Sigma x_i)^2}}\, s_y$$

$$= \sqrt{\frac{(114.62)}{(6\times 114.62)-(23.4)^2}} \times 0.1716 = \pm 0.155.$$

***Example 3.23*** *The iron losses in a ferromagnetic material used in a transformer vary with frequency, f of the supply given to the transformer. For a particular transformer these iron losses were determined at various frequencies with a constant flux density in the ferromagnetic material. The results are :*

| *Frequency, f Hz* | 1100 | 1400 | 1700 | 2000 |
|---|---|---|---|---|
| *Iron losses, P, mW* | 46 | 62 | 94 | 122 |

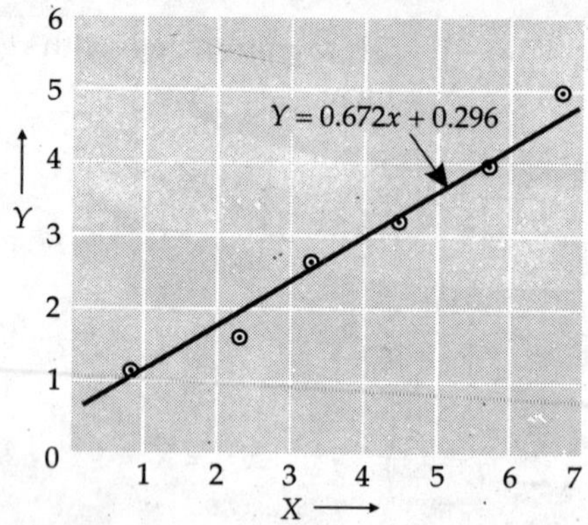


**Fig. 3.7** Method of Least Squares for linear curve fitting.

*Assuming the iron losses to have general form $P = Af^2 + Bf$, determine the constants A and B to achieve best for the four measured values, using method of least squares.*

**Solution.** Now, $P = Af^2 + Bf$

or $\quad P/f = Af + B$

∴ The above relationship can be expressed as

$$y = Ax + B$$

which is a straight line relationship between

$$y = P/f \quad \text{and} \quad x = f$$

Writing the values for finding the results for the best straight line relationship between $y = P/f$ and $x = f$.

| $x = f$ | 1100 | 1400 | 1700 | 2000 |
|---|---|---|---|---|
| $y = P_i/f$ | 0.04182 | 0.04429 | 0.05529 | 0.061 |

The various quantities are tabulated below :

| $x_i$ | $y_i$ | $x_i y_i$ | $x_i^2$ |
|---|---|---|---|
| 1100 | 0.04182 | 46 | $1.21\times10^6$ |
| 1400 | 0.04429 | 62 | $1.96\times10^6$ |
| 1700 | 0.05529 | 94 | $2.89\times10^6$ |
| 2000 | 0.061 | 122 | $4.00\times10^6$ |
| $\Sigma x_i = 6200$ | $\Sigma y_i = 0.2024$ | $\Sigma x_i y_i = 324$ | $\Sigma x_i^2 = 10.06\times10^6$ |

The values of $A$ and $B$ for the best straight line fit using the method of least squares are found by using Eqns. 3.53 and 3.54 are :

$$A = \frac{4\times324 - 6200\times0.2024}{4\times10.06\times10^6 - (6200)^2} = 22.8\times10^{-6}$$

and $$B = \frac{0.2024\times10.06\times10^6 - 324\times6200}{4\times10.06\times10^6 - (6200)^2} = 15.19\times10^{-3}$$

Hence, the best fit relationship is :

$$P = 22.8\times10^{-6} f^2 + 15.19\times10^{-3} f \text{ mW}.$$

## 3.17 VARIANCE AND STANDARD DEVIATIONS OF COMBINATION OF COMPONENTS

Suppose $X$ is a function of several component variables, each of which is subject to random effects, thus we have :

$$X = f(x_1, x_2, x_3, \ldots, x_n)$$

Now if $x_1, x_2, \ldots, x_n$, are independent variables, then for small variations in $x_1, x_2, \ldots, x_n$ from their mean value, denoted by $\delta x_1, \delta x_2, \ldots, \delta x_n$, the resulting variations of $X$ from its mean value for any one determination is given by :

$$\delta X = \frac{\partial X}{\partial x_1}\cdot \partial_{x1} + \frac{\partial X}{\partial x_2}\cdot \delta n_2 + \ldots \qquad \ldots(3.59)$$

(Eqn. 3.59, ignores the higher differentials).

Squaring Eqn. 3.59, we obtain

$$(\delta X)^2 = \left(\frac{\partial X}{\partial x_1}\right)^2 (\delta_{x1})^2 + \left(\frac{\partial X}{\partial x_2}\right)^2 (\delta_{x2})^2 + \ldots$$
$$+2\left(\frac{\partial X}{\partial x_1}\right)\left(\frac{\partial X}{\partial x_2}\right)(\delta x_1\, \delta x_2) + \ldots \qquad ..(3.60)$$

Now, if the variations of $x_1, x_2$ etc., are independent, as assumed, positive values of one increment are equally likely to be associated with positive or negative values of other increments. Thus the sum of the cross product terms tends to be zero for repeated measurements. By definition, variance $V$ is the mean square error. Thus the mean of $(\delta X)^2$ becomes the variance of X for repeated measurements. This variance of X is denoted by $V_x$ and hence we can write :

$$Vx = (\delta X)^2$$
$$= \left(\frac{\partial X}{\partial x_1}\right)^2 V_{x1} + \left(\frac{\partial X}{\partial x_1}\right)^2 V_{x2} + \ldots + \left(\frac{\partial X}{\partial x_1}\right)^2 V_{xn} \qquad \ldots(3.61)$$

as in repeated measurements $(\delta x_1)^2$ tends to become mean value of variance of $x_1$, *i.e.*, $V_{x1}$.

Eqnuation 3.61 may be written as

$$V_x = V_{x1} + V_{x2} + \ldots + V_{xn} \qquad \ldots(3.62)$$

This shows the component variances are additive with weighing factors $(\delta X/\delta x_1)^2$. The weighted variance $x_1$ can be written as

$$V_{x1} = \left(\frac{\partial X}{\partial x_1}\right)^2 V_{x1} \qquad \ldots(3.63)$$

The standard deviation of X may be found from Eqn. 3.61.

The standard deviation of X is $\sigma_x$ and is equal to

$$\sigma_x = \sqrt{V_x}$$
$$= \sqrt{\left(\frac{\partial X}{\partial x_1}\right)^2 V_{x1} + \left(\frac{\partial X}{\partial x_1}\right)^2 V_{x2} + \ldots + \left(\frac{\partial X}{\partial x_n}\right)^2 V_{xn}} \qquad \ldots(3.64)$$

but $V_{xn} = (\sigma_{xn})^2$

$$\therefore \quad \sigma_x = \sqrt{\left(\frac{\partial X}{\partial x_1}\right)^2 V_{x1} + \left(\frac{\partial X}{\partial x_2}\right)^2 V_{x2} + \ldots + \left(\frac{\partial X}{\partial x_n}\right)^2 V_{xn}}$$
$$= \sqrt{\left(\frac{\partial X}{\partial x_1}\right)^2 \sigma_{x1}^2 + \left(\frac{\partial X}{\partial x_2}\right)^2 \sigma_{x2}^2 + \ldots + \left(\frac{\partial X}{\partial x_n}\right)^2 \sigma_{xn}^2} \qquad \ldots(3.65)$$

It is clear from Eqn. 3.65 both component standard deviations are additive with weighing factors $\left(\frac{\partial X}{\partial x_1}\right)^2$, etc. which express the relative influence of the various components on the combined function.

3

Therefore we can write

$$\sigma_x = \sqrt{\sigma_{x1}^2 + \sigma_{x2}^2 + ... + \sigma_{xn}^2} \quad ...(3.66)$$

where $\sigma_{x1}$ is the weighted standard deviation of $x_1$,

$$\sigma_{x1} = \left(\frac{\partial X}{\partial x_1}\right)\sigma_{x1} \quad ...(3.67)$$

It is important to note that the above expressions are valid only if component quantities $x_1, x_2, ...$ etc. are independent of each other and also that the increments are small so that the terms of higher order than the first may be neglected. Actually in engineering applications, the increments are small as it is generally possible to keep the random effects under control.

### 3.17.1 Probable Error of Combination of Components

Suppose $X$ is a function of several component variables $x_1, x_2, ..., x_n$, each of which is an independent variate.

Then we have the standard deviation as :

$$\sigma_x = \sqrt{\left(\frac{\partial X}{\partial x_i}\right)^2 \sigma_{x1}^2 + \left(\frac{\partial X}{\partial x_2}\right)^2 \sigma_{x2}^2 + ... + \left(\frac{\partial X}{\partial x_n}\right)^2 \sigma_{xn}^2}$$

But from Eqn. 3.36 the probable error $r = 0.6745\ \sigma$, or probable error is proportional to standard deviation.

Hence we can wriie the probable error in $X$ as :

$$r_x = \sqrt{\left(\frac{\partial X}{\partial x_1}\right)^2 r_{x1}^2 + \left(\frac{\partial X}{\partial x_2}\right)^2 r_{x2}^2 + ... + \left(\frac{\partial X}{\partial x_n}\right)^2 r_{xn}^2} \quad ...(3.68)$$

where $r_{x1}, r_{x2}$ etc. are the probable errors in $x_1$, $x_2$ etc. The contribution of probable error of $x_1$ to the total error in X is $\left(\frac{\partial X}{\partial x_1}\right)^2 r_{x1}^2$ and this contribution may be written in another form as $r_{x1}^2$.

Therefore Eqn. 3.58 becomes as

$$r_x = \sqrt{r_{x1}^2 + r_{x2}^2 + ... + r_{xn}^2} \quad ...(3.69)$$

where the weighted probable error of $x_n$ becomes as

$$r_{xn} = \left(\frac{\partial X}{\partial x_n}\right) r_{xn} \quad ...(3.70)$$

**Example 3.24** *We have a parallel circuit having two branches. The current in one branch is $I_1 = 100 \pm 2$ A and in other is $I_2 = 200 \pm 5$ A. Determine the value of the total current $I = I_1 + I_2$,*

*(a) considering the errors in $I_1$ and $I_2$ as limiting errors, and*

*(b) considering the errors as standard deviations.*

*Comment upon the result.*

**Solution.** (*a*) Now $I = I_1 + I_2$

∴ Fractional error in

$$I = \frac{\delta I}{I} = \pm\left(\frac{I_1}{I}\cdot\frac{\delta I_1}{I_1} + \frac{I_2}{I}\cdot\frac{\delta I_2}{I_2}\right)$$

but $\frac{\delta I_1}{I_1} = \frac{2}{100} = 0.02$ and $\frac{\delta I_2}{I_2} = \frac{5}{200} = 0.025$

Now $I = 200 + 100 = 300$ A

Hence fractional error

$$\frac{\delta I}{I} = \pm\left(\frac{100}{300} \times 0.02 + \frac{200}{300} \times 0.025\right)$$

$$= \pm 0.0233 = 2.33 \text{ percent.}$$

Hence $I$ can be written as

$$I = 300(1 \pm 0.0233) = 300 \pm 7 \text{ A.}$$

(*b*) When the errors are standard deviation, we have

Standard deviation of $I$,

$$\sigma_1 = \sqrt{\left(\frac{\partial I}{\partial I_1}\right)^2 \sigma_{11}^2 + \left(\frac{\partial I}{\partial I_2}\right)^2 \sigma_{12}^2}$$

(See Eqns. 3.68 and 3.69)

$$= \sqrt{(2)^2 + (5)^2} = 5.38 \text{ A as } \frac{\partial I}{\partial I_1} = \frac{\partial I}{\partial I_2} = 1$$

∴ $I = 300 \pm 5.38$ A.

Standard deviation in $I$ expressed as a fraction is $5.38/300 = 0.018 = 1.8$ percent.

It is clear from above calculations that limiting errors of 2 per cent in $I_1$ and 2.5 per cent in $I_2$ combine in this case, to give a limiting error of 2.33 per cent in their sum $I$. While these very errors, when they are standard deviations, combine to give an error of only 1.8 per cent.

The use of standard deviation rather than limiting errors gives a more optimistic result. This is reasonable since the probability that both $I_1$ and $I_2$ are far from their respective means is small.

**Example 3.25** *A resistance is determined by voltmeter ammeter method. The voltmeter reads 100 V with a probable error of ± 12 V and ammeter reads 10 A with a probable error of ± 2 A. Determine the probable error in the computed value of resistance.*

**Solution.** We have resistance $R = V/I = VI^{-1}$

∴ Weighted probable error in the resistance due to voltage is,

$$r_{RV} = \frac{\partial_R}{\partial V} r_V = I^{-1} r_v = \frac{r_v}{I}$$

$$= \pm \frac{12}{10} = \pm 12\,\Omega \qquad \text{(See Eqns. 3.70)}$$

Weighted probable error in resistance due to current,

$$r_{RI} = \frac{\partial_R}{\partial I} r_1 = \frac{V}{I^2} r_1$$

$$= \frac{100}{(10)^2} \times (\pm 2) = \pm 2\,\Omega$$

From Eqn. 3.69 probable error in computed resistance is,

$$r_R = \sqrt{(r_{RV})^2 + (r_{RI})^2}$$

$$= \sqrt{(12)^2 + (2)^2} = 2.33\,\Omega.$$

**Example 3.26** *The law of deflection of a galvanometer is $I = K\theta / \cos\theta$, where I is the current ; K a constant and θ is the deflection. If the angle of deflection θ is known to be within ± 0.1° (standard deviation) of 15°, what is the percent standard deviation of current, I ?*

**Solution.** Now $I = \dfrac{K\theta}{\cos\theta} = K\theta(\cos\theta)^{-1}$

$$\therefore \quad \frac{\partial I}{\partial \theta} = K[(\cos\theta)^{-1} + \theta(\cos\theta)^{-2}\sin\theta]$$

$$= K\left[\frac{1}{0.966} + \frac{\pi}{180} \times 15 \times \frac{1}{(0.966)^2} \times 0.259\right]$$

$$= 1.108\,K$$

Standard deviation of $I$ is

$$\sigma_1 = \frac{\partial I}{\partial \theta}\sigma_\theta = \pm(1.108\,K)\left(0.1 \times \frac{\pi}{180}\right) \text{rad}$$

Percentage standard deviation of $I$

$$= \frac{\sigma I}{I} \times 100$$

$$= \pm \frac{(1.108\,K)(0.1 \times \pi/180)}{K \times (\pi/180) \times 15 \times \cos 15^\circ} \times 100$$

$$= \pm\, 0.765 \text{ percent.}$$

## 3.18 UNCERTAINTY ANALYSIS AND TREATMENT OF SINGLE SAMPLE DATA

Many a times the data available is a single sample data and therefore the statistical methods discussed earlier cannot be applied directly. On account of the single sample nature of the data, it is not possible to observe their scatter by plotting a frequency distribution curve. Hence, it becomes essential to modify the approach.

Kline and McClintock have proposed a method based upon probability and statistics which analyses the data employing *uncertainty distribution* rather than frequency distribution. They have defined the uncertainty distribution as the error distribution the experimenter believes would exist if the situation permits multisampling.

Kline and McClintock suggest that a single sample result may be expressed in terms of a *mean value* and an *uncertainty interval* based upon stated *odds.*

The result may be written as follows :

$$X = \overline{X} \pm w \ (b \text{ to } 1)$$

where $\overline{X}$ = the value if only one reading is available on the arithmetic mean of several readings.

$w$ = uncertainty interval.

$b$ = odds or the chance that the true value lies within the stated range, based upon the opinion of the experimenter.

The concept of *uncertainty* may be explained by the following example :

The results of a temperature measurement may be expressed as $\theta = 100°\text{C} \pm 1°\text{C}$.

This means that there is an uncertainty of ±1°C in the result. In other words, the experimenter is stating in precise terms the accuracy of results with which they have been made according to him. This brings about another dimension in measurements and that is, how far the experimenter is sure that his measurement falls within the specified limits. Therefore the need for a further specification arises. As mentioned earlier, Kline and McClintock proposed that the experimenters specify certain odds for the uncertainty. The aforesaid results may be given as :

$$\theta = 100°\text{C} \pm 1°\text{C} \qquad (20 \text{ to } 1)$$

Now the results expressed in the above form become more specific in nature. This is because the experimenter is willing to bet 20 to 1 odds that the temperature measurement which he has made are within ±1°C of 100°C.

This approach is of a particular value in setting up an experiment, especially when it involves expenses in terms of man-power, time and equipment. It provides essentials for establishing basis for predetemined estimates of the reliability of results through a study of propagation of uncertainties (discussed below in Art. 3.18.1). In this way evaluation of the test results can be made even before the test is physically done.

### 3.18.1 Propagation of Uncertainties

The uncertainty analysis in measurements when many variates are involved is done on the same basis as is done for error analysis when the results are expressed as standard deviations or probable errors.

Suppose X is a function of several variables,

$$X = f(x_1, x_2, x_3, \ldots, x_n).$$

where $x_1, x_2, x_3, \ldots, x_n$ are independent variables with the same degree of odds.

Let $w_x$ be the resultant uncertainty and, $w_{x1}, w_{x2}, w_{x3}, \ldots, w_{xn}$ be the uncertainties in the independent variables $x_1, x_2, x_3, \ldots, x_n$, respectively. The uncertainty in the result is given by :

$$w_x = \sqrt{\left(\frac{\partial X}{\partial x_1}\right)^2 w_{x1}^2 + \left(\frac{\partial X}{\partial x_2}\right)^2 w_{x2}^2 + \ldots + \left(\frac{\partial X}{\partial x_n}\right)^2 w_{xn}^2} \qquad \ldots(3.71)$$

**Example 3.27.** *A certain resistor has a voltage drop of 110.2 V and a current of 5.3 A. The uncertainties in the measurements are : ± 0.2 V and ± 0.06 A respectively. Calculate the power dissipated in the resistor and the uncertainty in power.*

**Solution.** Power $P = \text{voltage} \times \text{current}$

$$= VI = 110.2 \times 5.3 = 584 \text{ W}$$

Now $\quad P = VI$

$$\therefore \quad \frac{\partial P}{\partial V} = I = 5.3 \text{ and } \frac{\partial P}{\partial I} = V = 110.2$$

$$w_V = 500.2 \text{ and } w_I = \pm 0.06.$$

∴ Uncertainty in power

$$= \sqrt{\left(\frac{\partial P}{\partial V}\right)^2 w_V^2 + \left(\frac{\partial P}{\partial I}\right)^2 w_I^2}$$

$$= \sqrt{(5.3)^2 \times (0.2)^2 + (110.2)^2 + (0.06)^2}$$

$$= \pm 6.7 \text{ W} = \pm \frac{6.7}{584} \times 100 = \pm 1.15\%$$

**Example 3.28.** *Two resistors $R_1$ and $R_2$ are connected in series and then in parallel. The values of resistances are :*

$$R_1 = 100.0 \pm 0.1 \,\Omega, \quad R_2 = 50 \pm 0.03 \,\Omega$$

*Calculate the uncertainty in the combined resistance for both series and parallel arrangements.*

**Solution.** When the two resistances are connected in series the resultant resistance is :

$$R = R_1 + R_2$$

$$\therefore \quad \frac{\partial R}{\partial R_1} = 1 \text{ and } \frac{\partial R}{\partial R_2} = 1$$

Hence, uncertainty in the total resistance is

$$w_R = \pm \sqrt{\left(\frac{\partial R}{\partial R_1}\right)^2 w_{R1}^2 + \left(\frac{\partial R}{\partial R_2}\right)^2 w_{R2}^2}$$

$$= \pm \sqrt{(1)^2 \times (0.1)^2 + (1)^2 \times (0.03)^2} = \pm 0.1044 \,\Omega$$

The total resistance is $R = 100 + 50 = 150 \,\Omega$

and can be expressed as $R = 150 \pm 0.1044 \,\Omega$.

When the two resistances are connected in parallel, the resultant resistance is :

$$R = \frac{R_1 R_2}{R_1 + R_2} = \frac{100 \times 50}{100 + 50} = 33.33 \,\Omega$$

Now, $\quad R = (R_1 R_2)(R_1 + R_2)^{-1}$

$$\therefore \quad \frac{\partial R}{\partial R_1} = (R_2)(R_1 + R_2)^{-1} - R_1 R_2 (R_1 + R_2)^{-2}$$

$$= \frac{R_2}{R_1 + R_2} - \frac{R_1 R_2}{(R_1 + R_2)^2}$$

$$= \frac{50}{150} - \frac{100 \times 50}{(150)^2} = 0.111$$

$$\frac{\partial R}{\partial R_2} = \frac{R_1}{R_1 + R_2} - \frac{R_1 R_2}{(R_1 + R_2)^2}$$

$$= \frac{100}{150} - \frac{100 \times 50}{(150)^2} = 0.444.$$

Hence, uncertainty in total resistance is :

$$w_R = \pm \sqrt{\left(\frac{\partial R}{\partial R_1}\right)^2 w_{R1}^2 + \left(\frac{\partial R}{\partial R_2}\right)^2 w_{R2}^2}$$

$$= \pm \sqrt{(0.111)^2 \times (0.1)^2 + (0.444)^2 \times (0.03)^2}$$

$$= \pm 0.01734 \,\Omega.$$

The total resistance can be written as

$$R = 33.33 \pm 0.01734 \,\Omega.$$

**Example 3.29.** *A plot of land has measured dimensions of 50 by 150 m. The uncertainty in the 50 m dimension is ± 0.01 m. Calculate the uncertainty with which the 150 m dimension must be measured to ensure that the total uncertainty in the area is not greater than 150 per cent of that value it would have if 150 m dimension were exact.*

**Solution.** Let $L$ = length of the plot = 150 m,

$B$ = width of the plot = 50 m

and $\quad A$ = area of the plot

$$= L \times B = 150 \times 50 = 7500 \text{ m}^2.$$

Now $A = LB \quad \therefore \quad \frac{\partial A}{\partial L} = B$ and $\frac{\partial A}{\partial B} = L$

Uncertainty in area

$$w_A = \pm\sqrt{\left(\frac{\partial A}{\partial L}\right)^2 w_L^2 + \left(\frac{\partial A}{\partial B}\right)^2 w_R^2}$$

$$= \pm\sqrt{B^2 w_L^2 + L^2 w_B^2}$$

*Case* **I.** When there is no uncertainty in measurement of $L$, $w_L = 0$.

Uncertainty in measurement of area

$$w_A = \pm\sqrt{B^2 w_L^2 + L^2 w_B^2} = \pm\sqrt{L^2 w_B^2}$$

$$= \pm L w_B = \pm 150 \times 0.01 = \pm 1.5\,\text{m}^2$$

*Case* **II.** When there is uncertainty in measurement of $L$.

The uncertainty in area is not to exceed

$$1.5 \times 1.5 = \pm 2.25\ \text{m}^2.$$

$$w_A = \sqrt{B^2 w_L^2 + L^2 w_B^2}$$

or $$2.25 = \sqrt{(50)^2 w_L^2 + (150)^2 (0.01)^2}$$

Hence uncertainty in measurement of $L$ is $w_L = \pm 0.0335$ m

**Example 3.30** *A resistor has a nominal value of 10 Ω ± 0.1%. A voltage is applied across the resistor and the power consumed in the resistor is calculated in two ways :*

*(i) from $P = E^2 / R$, and (ii) from $P = EI$.*

*Calculate the uncertainty in the power determination in each case when the measured values of E and I are :*

$$E = 100\,\text{V} \pm 1\% \quad \text{and} \quad I = 10\,\text{A} \pm 1\%$$

*Comment upon the result.*

**Solution.** The circuit diagram is shown in Fig. 3.8.

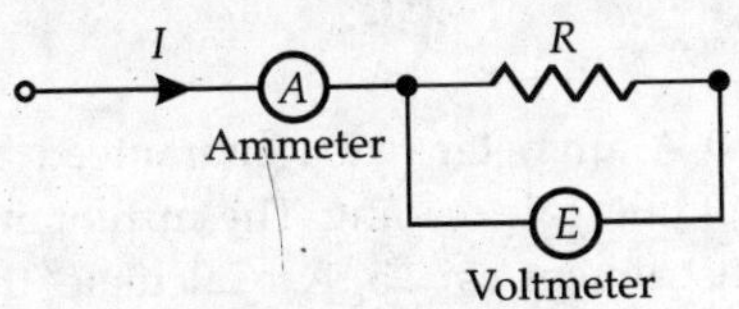


**Fig. 3.8** Figure of example 3.30.

(i) $P = \dfrac{E^2}{R}$

$$\therefore \quad \frac{\partial P}{\partial E} = \frac{2E}{R} \quad \text{and} \quad \frac{\partial P}{\partial R} = -\frac{E^2}{R^2}$$

Hence uncertainty in power measurement

$$w_P = \sqrt{\left(\frac{\partial P}{\partial E}\right)^2 w_E^2 + \left(\frac{\partial P}{\partial E}\right)^2 w_R^2}$$

$$= \sqrt{\left(\frac{2E}{R}\right)^2 w_E^2 + \left(-\frac{E^2}{R^2}\right) w_R^2}$$

∴ Percentage uncertainty in measurement of power is calculated by putting $P = E^2 / R$.

$$\frac{w_p}{P} \times 100 = \sqrt{4\left(\frac{w_E}{E}\right)^2 + \left(\frac{w_R}{R}\right)^2} \times 100$$

$$= \sqrt{4(0.01)^2 + (0.01)^2} \times 100 = \pm 2.236\,\%$$

(ii) $P = EI \quad \therefore \quad \dfrac{\partial P}{\partial E} = I$ and $\dfrac{\partial P}{\partial I} = E.$

Percentage uncertainty in power measurement

$$\frac{w_P}{P} \times 100 = \frac{I}{P}\sqrt{\left(\frac{\partial P}{\partial E}\right)^2 w_E^2 + \left(\frac{\partial P}{\partial I}\right)^2 w_I^2} \times 100$$

$$= \frac{1}{P}\sqrt{I^2 w_E^2 + E^2 w_1^2} \times 100$$

$$= \sqrt{\left(\frac{w_I}{E}\right)^2 + \left(\frac{w_I}{I}\right)^2} \times 100$$

$$= \sqrt{(0.01)^2 + (0.01)^2} \times 100$$

$$= \pm 1.414\%$$

The second method of power determination gives a much lower uncertainty than the first one even though the basic uncertainties in each quantity are the same. We conclude from here that a judicious selection of method of measurement is important in order to reduce the uncertainty in the final computed results.

# Review Questions

1. Define Limiting (Guarantee) errors. Derive the expression for relative limiting error.
2. Two quantities $x_1$ and $x_2$ have limiting errors of $\pm\,\delta x_1$ and $\pm\,\delta x_2$ respectively. Derive the expression for relative limiting error in X when

   (i) $X = x_1 + x_2$

   (ii) $X = x_1 - x_2$

   (iii) $X = x_1 x_2$

   (iv) $X = \dfrac{x_1}{x_2}$ (v) $X = x_1^n\, x_1^m$.
3. Explain the difference between limiting and known errors by citing suitable examples.

4. Errors in measurements can be classified as :
   (*i*) Gross errors (*ii*) Systematic errors
   (*iii*) Random errors.
   Explain these errors by giving suitable examples. Discuss the mean adopted to minimise these errors.
5. Systematic errors can be classified as :
   (*i*) Instrumental errors
   (*ii*) Environmental errors
   (*iii*) Observational errors.
   Discuss the above types of errors giving suitable examples. Explain the measures taken to minimize these errors.
6. Define the following terms in the context of normal frequency distribution of data :
   (*i*) mean value
   (*ii*) deviation
   (*iii*) average deviation
   (*iv*) standard deviation
   (*v*) variance.
7. Define the following for Gaussian distribution of data :
   (*i*) Precision index
   (*ii*) Probable error
   (*iii*) Standard deviation of mean
   (*iv*) Standard deviation of standard deviation
8. Explain the significance of confidence interval and confidence level in statistical analysis of data.
9. Explain the rejection of data using
   (*a*) Chauvenet's criterion
   (*b*) Confidence interval
   (*c*) $\pm 3\sigma$ limits.
10. Suppose we have a set of $n$ readings $x_1, x_2, ..., x_n$. Prove that the mean value which minimizes the sum of squares of the deviations is the arithmetic mean of the readings.
11. Suppose we have two variables $x$ and $y$. Explain how method of least squares can be used to find the best linear function connecting $y$ with $x$.
12. Suppose X is function of quantities $x_1, x_2, ..., x_n$, having standard deviations of $\sigma_{x1}, \sigma_{x2}, ..., \sigma_{xn}$. Derive the expression for standard deviation of X in terms of standard deviations of $x_1, x_2, ..., x_n$.
13. Discuss the different methods of specifying the results of measurement.
14. Define the term "ODDS". Explain odds are related to probability of occurrence of a reading within a specified limit.
15. Explain the method of treatment of single sample data with the help of uncertainty analysis by giving suitable examples. Define the terms
   (*i*) uncertainty distribution (*ii*) mean value
   (*iii*) uncertainty interval and (*iv*) odds.

# Unsolved Problems

1. The value of a resistor is specified as $500\,\Omega \pm 10\%$ by a manufacturer. Find the limits of resistance between which the value is guaranteed.
   **[Ans.** $450\,\Omega$ to $500\,\Omega$**]**
2. The hunting errors for a four dial resistance box are :
   Units : $\pm$ 0.2% Hundreds : $\pm$ 0.05 %
   Tens : $\pm$ 0.1 % Thousands : $\pm$ 0.02 %
   If the resistance value is set at $3425\,\Omega$, calculate the limiting error in the resistance value.
   **[Ans.** $\pm 0.83\,\Omega$**]**
3. A flowmeter is calibrated from 0 to 100 $m^3/s$. The accuracy is specified as within $\pm$ 0.75 per cent above 20 per cent of scale reading. What is the static error if the instrument indicates 80 $m^3/s$ ?
   **[Ans.** $\pm 0.6\ m^3/s$**]**
4. A 0-10 A ammeter has a guaranteed accuracy of 1.5% of full scale reading. The current measured by the instrument is 2.5 A. Calculate the limiting values of current and the percentage limiting error.
   **[Ans.** 2.5 $\pm$ 0.15 A, $\pm$ 6%**]**
5. A liquid flows through a pipe having a diameter of 100 mm with a velocity of 1 m/s. Calculate the flow rate. If the diameter is guaranteed within $\pm$ 1% and the velocity is known to be within $\pm$ 3% of measured value, find the limits within which rate of flow can be specified.
   **[Ans.** $7.85 \times 10^{-3}\ m^3/s, \pm 5\%$**]**
6. The resistance of an unknown resistor is determined by wheatstone bridge. The solution for the unknown resistance is stated as

$$R_4 = \frac{R_1 R_2}{R_3}$$

where limiting values of various resistances are :
$R_1 = 500\Omega \pm 1\%$, $R_2 = 615\Omega \pm 1\%$, $R_3 = 100\Omega \pm 0.5\%$.
Calculate (*a*) the nominal value of the unknown resistor, (*b*) the limiting error of the unknown resistor in ohm, and (c) the limiting error in per cent of unknown resistor.

**[Ans.** (*a*) 3075 Ω, (*b*) ± 76.88 Ω, (*c*) ± 2.5%]

**7.** The losses in a motor are calculated by measuring the input and output power of the motor and then taking their difference. In a particular motor the input is measured as 6250 W ± 2% and the output is 5000 W ± 3%. The errors are limiting errors. Calculate the losses and their percentage limiting error. Comment upon the results.

**[Ans.** 1250 W, ± 22%]

**8.** Three resistors have the following ratings :
$R_1 = 200\,\Omega \pm 5\%$, $R_2 = 100\,\Omega \pm 5\%$,
and $R_3 = 50\,\Omega \pm 5\%$,
Determine the magnitude of resultant resistance and the limiting errors in percentage and in ohm if the above resistances are connected in (*a*) series and (*b*) parallel. **[Ans.** (*a*) 350 Ω ± 5%, ± 17.5 Ω, (*b*) 28.6 Ω ± 5%, ± 1.30 Ω]

**9.** The power factor of a circuit is determined by :

$$\cos\phi = P/VI$$

where *P* is the power in watt, *V* the voltage in volt and *I* is the current in ampere. The relative errors in power, current and voltage are respectively ± 0.5%, ± 1% and ± 1%.

Calculate the relative error in power factor. Also calculate the uncertainty in the power factor if the errors were specified as uncertainties.

**[Ans.** ± 2.5%, ± 1.5%]

**10.** A 150 ± 10% Ω resistor is connected to the terminals of a power supply operating at 200 ± 0 V. What range of current would flow if the resistor varies over the range ± 10% of its expected value ? What is the range of error in the current ?

**[Ans.** 1.21 – 1.48 A, – 9% to + 11.3%]

**11.** A resistor is measured by the voltmeter-ammeter method. The voltmeter reading is 123.4 V on the 250 V scale and the ammeter reading is 283.5 mA on the 500 mA scale. Both meters are guaranteed to be accurate within ± 1 per cent of full scale reading. Calculate :

(*a*) the indicated value of resistance,

(*b*) the limits within which the result can be guaranteed.

**[Ans.** 436 Ω ± 16.5 Ω]

**12.** The four arms of a Hay's a.c. bridge are arranged as follows :

*AB* is a coil of unknown impedance.

*BC* is a non-inductance

$R_1 = 1000\,\Omega$ with an error of ± 1 part in 10,000.

*CD* is a non-reactive resistor

$R_3 = 833 \pm 0.25\,\Omega$ in series with a no loss capacitor $C = 1.43 \pm 0.001\,\mu F$.

*DA* is a non-reactive resistor

$R_2 = 16800 \pm 1$ part in 10,000.

The supply frequency is 50 ± 0.1 Hz. The bridge is balanced. Determine *L* and *R* of the coil and the limits of error. The balance conditions are :

$$L = \frac{CR_1R_3}{1+\omega^2C^2R_3^2} \quad \text{and} \quad R = \frac{R_1R_2R_3C^2\omega^2}{1+\omega^2C^2R_3^2}$$

**[Ans.** $L = 21 \pm 0.145\,H$; $R = 2480 \pm 29.5\,\Omega$]

**13.** The stress in a mild steel flat circular diaphragm is given by :

$$s = \frac{3D^2p}{16t^2}\ N/m^2$$

where *D* and *t* are respectively diameter and thickness of diaphragm : mm and *p* is the applied pressure : $N/m^2$.
A diaphragm has a diameter of 15 mm, thickness = 0.2 mm and the pressure applied is $300 \times 10^3\ N/m^2$. Calculate the stress. The known error in diameter is + 1% and in thickness is 3%. Calculate the error in the stress.

**[Ans.** $316 \times 10^6\ N/m^2$, –4%, $-12.64\ N/m^2$]

**14.** A set of independent ten measurements were made to determine the weight of a lead shot. The weights in gramms were :
1.570, 1.597, 1.591, 1.562, 1.577, 1.580, 1.564, 1.586, 1.550, 1.575.

Determine the (*a*) arithmetic mean, (*b*) average deviation, (c) standard deviation, (*d*) variance, (*e*) probable error of one reading, (*f*) probable error of the mean. **[Ans.** (*a*) 1.5752 g, (*b*) 0.01068 g, (*c*) 0.01426 g, (*d*) 2.0335 $\times 10^4\ g^2$, (*e*) 0.0024 g, (*f*) 0.00076 g]

**15.** The following values were obtained from the measurements of the value of a resistor :
147.2 Ω, 147.4 Ω, 147.9 Ω, 148.1 Ω, 147.1 Ω, 147.5 Ω, 147.6 Ω, 147.4 Ω, 147.6 Ω and 147.5 Ω. Calculate
(*a*) arithmetic mean (*b*) average deviation
(*c*) standard deviation, treating the data as finite
(*d*) standard deviation treating the data as population.
**[Ans.** (*a*) 147.53 Ω (*b*) 0.218 Ω, (*c*) 0.298 Ω (*d*) 0.283 Ω]

3

**16.** A batch of colour coded resistors of value 5.6 kΩ were measured and were found to have the following values. Determine the mean and the standard deviation ? Can any resistor be discarded on the basis of ± 3σ limits ?

5.75, 5.60, 5.65, 5.50, 5.70, 5.55, 5.80, 5.55 kΩ.

[**Ans.** 5.64 kΩ, 0.106 kΩ ; No]

**17.** The table given below lists a sample of experimental data :

| Value | Frequency of Occurrence |
|---|---|
| 3 | 1 |
| 4 | 2 |
| 5 | 3 |
| 6 | 6 |
| 7 | 7 |
| 8 | 6 |
| 9 | 4 |
| 10 | 2 |
| 11 | 1 |

Calculate (*a*) arithmetic mean (*b*) mean deviation (*c*) standard deviation (*d*) probable error of one reading (*e*) the standard deviation and the probable error of the mean (*f*) standard deviation of standard deviation. Is there any reading that can be discarded on the basis of three sigma limits ? If so, specify. [**Ans.** (*a*) 7.062 (*b*) 1.45 (*c*) 1.865 (*d*) 1.258 (*e*) 0.33, 0.226 (*f*) 0.203 ; No reading can be discarded]

**18.** A capacitor $C = 1.0 \pm 0.1 \mu F$ is charged to a voltage $20 \pm 1$ V, where errors are probable errors. Find the charge on the capacitor and its probable error.

[**Ans.** $20 \times 10^{-6}$ coulomb, $\pm 2.2 \times 10^{-6}$ coulomb]

**19.** The resistance $R$ of a copper wire is given by $R = R_0[1 + \alpha(\theta - 20)]$ where $R_0 = 4\Omega \pm 0.2\%$ is the resistance of the wire at 20°C. The value of resistance temperature co-efficient copper is $0.004\,\Omega/\Omega°C \pm 1\%$ and temperature $\theta = 25 \pm 1°C$. Find the resistance $R$ and its probable error.

[**Ans.** 4.08 Ω ± 0.3%]

**20.** The mean value of a standard resistor was determined by elaborate precision methods to be 1.000 Ω ± 0.03% (standard deviation of the mean). A year later, the resistance was measured with equal skill and care with a mean result of 0.9998 Ω. What is the probability that the standard resistance has not changed ? Consult probability tables.

[**Ans.** 0.5]

**21.** The following ten readings are taken of a certain physical length :

5.30 m, 5.73 m, 6.77 m, 5.26 m, 4.53 m,
5.45 m, 6.09 m, 5.64 m, 5.81 m, 5.75 m.

Calculate : (*a*) mean and the standard deviation, (*b*) Using the Chauvenet's criterion, test the data points for possible inconsistency (*c*) Eliminate the questional points and calculate area standard deviation for the adjusted data. In case that a data be discarded the ratio of maximum deviation to standard deviation should not exceed 1.96.

[**Ans.** (*a*) 5.613 m, 0.6266 m, (*b*) Reading No. 5 (4.33 m) should be eliminated, 0.478 m]

**22.** A certain length measurement is performed 100 times. The arithmetic mean of readings is 6.823 and the standard deviation is 0.01 m. How many readings fall within (*a*) ± 0.005 m, (*b*) ± 0.02 m, (*c*) ± 0.05 m, and (*d*) 0.001 m of the mean value ? Consult probability tables.

[**Ans.** (*a*) 38 (*b*) 95 (*c*) 100 (*d*) 8]

**23.** An oscilloscope is used for measurement of phase shift. The phase shaft is even by :

$$\phi = \sin^{-1} B/A$$

The measurements made are :

$$A = 25 \pm 0.5 \text{ mm}, \quad B = 15 \pm 0.5 \text{ mm}.$$

Calculate the nominal value of phase angle and estimate the uncertainty in measurement.

[**Ans.** 36.9°, 2.15%]

**24.** The following data were taken for the demand for electric power in a region where it is desired to build a new generating station. The data taken for 10 working days chosen at random :

| Day | Maximum Demand MW | Day | Maximum Demand MW |
|---|---|---|---|
| 1 | 2.0 | 6 | 2.9 |
| 2 | 1.2 | 7 | 1.8 |
| 3 | 2.1 | 8 | 1.6 |
| 4 | 2.3 | 9 | 2.0 |
| 5 | 3.1 | 10 | 2.6 |

(*a*) Plot a histogram of these data, (*b*) Determine the mean and standard deviation, (*c*) Find out the capacity of the plant which is able to meet the full demand on 84 per cent of the working days.

Assume that the future demand may be predicted by a normal distribution curve based on the above example.

[**Ans.** (*b*) 2.61 MW, ± 0.58 MW, (*c*) 2.8 MW]

**25.** From the following data obtain $y$ as linear function of $x$ using method of least squares.

| y | 1.2 | 2.0 | 2.4 | 3.5 | 3.5 |
|---|---|---|---|---|---|
| x | 1.0 | 1.6 | 3.4 | 4.0 | 5.2 |

[**Ans.** $y = 0.54x + 0.879$]

**26.** A load at transducer is used for measurement of weight. The output of the secondary transducer the load, cell for application of different weights for increasing and decreasing values is given below :

| Load (kg) | Output mV | |
|---|---|---|
| | Increasing | Decreasing |
| 0 | 0.08 | 0.06 |
| 5 | 0.45 | 0.88 |
| 10 | 1.02 | 2.04 |
| 15 | 1.71 | 3.10 |
| 20 | 2.55 | 4.18 |
| 25 | 3.43 | 5.13 |
| 30 | 4.48 | 6.04 |
| 35 | 5.50 | 7.02 |
| 40 | 6.53 | 8.06 |
| 45 | 7.64 | 9.35 |
| 50 | 8.70 | 10.52 |
| 55 | 9.85 | 11.80 |
| 60 | 11.01 | 12.94 |
| 65 | 12.40 | 13.86 |
| 70 | 13.32 | 14.82 |
| 75 | 14.35 | 15.71 |
| 80 | 15.40 | 16.84 |
| 85 | 16.48 | 17.92 |
| 90 | 17.66 | 18.70 |
| 95 | 18.90 | 19.51 |
| 100 | 19.93 | 20.02 |

(*a*) Determine the best fit line using method of least squares.

(*b*) Calculate the linearity.

[**Ans.** (*a*) $y = -0.6368 + 0.2079\ x$ mV,
(*b*) 5.71%, – 5.46%]

# Objective Type Questions

**Tick (√) the most appropriate answer**

**1.** An 0 - 10 A ammeter has a guaranteed accuracy of 1 percent of full scale deflection. The limiting error while reading 2.5 A is :

(*a*) 1% (*b*) 2%
(*c*) 4% (*d*) none of the above.

**2.** A 0 - 300 V voltmeter has an error of ± 2% of full scale deflection. What would be the range of readings if true voltage is 30 V ?

(*a*) 24 V – 36 V (*b*) 29.4 V – 30.6 V
(*c*) 20 V to 40 V (*d*) none of the above.

**3.** A wattmeter has a full scale range of 2500 W. It has an error ± 1% of true value. What would be range of reading if true power is 1250 W ?

(*a*) 1225 W – 1275 W (*b*) 1245 W – 1255 W
(*c*) 1200 W – 1300 W (*d*) 1237.5 W – 1262.5 W.

**4.** Power in a d.c. circuit is measured by measuring the voltage across and current through the circuit. The voltage and current measurements are made to an accuracy of ± 2 % and ± 3% respectively. The errors are limiting errors. The error in measurement of power is :

(*a*) ± 2% (*b*) ± 3%
(*c*) ± 6% (*d*) ± 5%.

**5.** The power in a circuit is measured by measuring a current through a resistor. The current is measured with an accuracy of ± 1.5% and the tolerance band of the resistor ± 0.5%. The errors are limiting or guarantee errors. The accuracy with which power is measured is :

(*a*) ± 1.125% (*b*) ± 3.5%
(*c*) ± 2% (*d*) ± 2.5%.

**6.** In a permanent magnet moving coil ammeter the deflection of the pointer is proportional to product of flux density of magnetic field produced by the permanent magnet and the current in the moving coil. If the strength of the permanent magnet becomes 95% of the original, the meter gives erraneous reading resulting into error. This error can be classified as :

(*a*) Gross error (*b*) Systematic error

(*c*) Random error (*d*) none of the above.

**7.** The voltage of a circuit is measured by a voltmeter having an input impedance comparable with the output impedance of the circuit thereby causing error in voltage measurement. This error may be called

(*a*) Gross error

(*b*) Random error

(*c*) Error caused by misuse of instrument

(*d*) Error caused by loading effect.

**8.** The mean deviation $\overline{D}$ in terms of deviations from the mean value of $n$ readings is

(*a*) $\frac{\Sigma|d|}{n}$ (*b*) $\frac{\Sigma d}{n}$

(*c*) $\frac{\sqrt{\Sigma d^2}}{n}$ (*d*) $\sqrt{\frac{\Sigma d^2}{n}}$

**9.** A set of readings has a wide range and therefore it has :

(*a*) low precision (*b*) high precision

(*c*) low accuracy (*d*) high accuracy.

**10.** The Gaussian distribution can be mathematically expressed as :

(*a*) $y = \frac{h}{\sqrt{\pi}}\exp(h^2x^2)$ (*b*) $y = -\frac{h}{\sqrt{\pi}}\exp(h^2x^2)$

(*c*) $y = \frac{h}{\pi}\exp(-h^2x^2)$ (*d*) $y = \frac{h}{\sqrt{\pi}}\exp(-h^2x^2)$

$y$ = number of readings at a deviation $x$ and $h$ is precision index.

**11.** For a Gaussian distribution, the probable error is $r$. This means that :

(*a*) area under the curve between $\pm r$ limits is 0.5

(*b*) half of the observed values lie between $\pm r$ limits

(*c*) the chances that an additional observation will lie between $\pm r$ limits are 50 percent

(*d*) all the above.

**12.** Two resistances $100\,\Omega \pm 5\,\Omega$ and $150\,\Omega \pm 15\,\Omega$ are connected in series. If the deviations are standard deviations, the resultant resistance can be expressed as :

(*a*) $250\,\Omega \pm 20\,\Omega$ (*b*) $250\Omega \pm 10\Omega$

(*c*) $250\,\Omega \pm 15.8\,\Omega$ (*d*) $250\,\Omega \pm 10.6\,\Omega$

**13.** If the result of a measurement is expressed as $\overline{X} \pm 3\sigma$ where $\overline{X}$ = mean value and $\sigma$ = standard deviation, it means that :

(*a*) approximately 99 percent of the readings lie between $\pm 3\sigma$ limit

(*b*) 26 readings out of 1000 will lie outside $\pm 3\sigma$ limit

(*c*) the odds for any readings to lie within $\pm 3\sigma$ limit

(*d*) all of the above.

**14.** A batch of resistors have a mean value of $100.00\,\Omega$ and a standard deviation $\sigma = 0.2\,\Omega$. The probability corresponding to $2\sigma$ is 0.9546. The value of odds that randomly selected resistor will lie within $100.00 \pm 0.40\,\Omega$ is :

(*a*) 1 to 1 (*b*) 2.15 to 1

(*c*) 21 to 1 (*d*) 256 to 1.

**15.** According to Chauvenet's criterion, a reading out of a set of $n$ readings should be rejected if the probability of obtaining the particular deviation from mean is :

(*a*) less than $1/2n$ (*b*) greater than $1/2n$

(*c*) less than $1/n$ (*d*) less than $1/\sqrt{2n}$

**16.** If the confidence level is 0.95, then the values lying outside the confidence interval are :

(*a*) 1 in 5 (*b*) 1 in 20

(*c*) 1 in 100 (*d*) 1 in 1000.

**17.** Uncertainty distribution is used for :

(*a*) analysis of multi-sample data

(*b*) analysis of single sample data

(*c*) analysis of both single and multi sample data

(*d*) none of the above.

## Answers

| | | | | | | | | | |
|---|---|---|---|---|---|---|---|---|---|
| **1.** (*c*) | **2.** (*a*) | **3.** (*d*) | **4.** (*d*) | **5.** (*b*) | **6.** (*b*) | **7.** (*d*) | **8.** (*a*) | **9.** (*a*) | **10.** (*d*) |
| **11.** (*d*) | **12.** (*c*) | **13.** (*d*) | **14.** (*c*) | **15.** (*a*) | **16.** (*b*) | **17.** (*b*) | | | |

# 4

# Dynamic Characteristics of Instruments and Measurement Systems

## 4.1 DYNAMIC RESPONSE

The *response* is the evaluation of the system's ability to faithfully transmit and present all the pertinent information included in the input signal and to exclude all else.

Invariably measurement systems, especially in industrial, aerospace and biological applications are subjected to inputs, which are not static but are dynamic in nature *i.e.*, the inputs that vary with time. The input varies from instant to instant, and therefore, so does the output. The behaviour of the system under such conditions is described by its **dynamic response**.

The dynamic (varying with time) inputs are of *two* types :

(*i*) transient and
(*ii*) steady state periodic.

The ***steady state periodic*** quantity is one whose magnitude has a definite repeating time cycle, whereas the time variation of a **transient** magnitude does not repeat.

All measurement systems include one or more energy storage elements. The examples of energy storage elements are electrical inductance and capacitance, mass and inertia, and thermal and fluid capacitance.

When an input is applied to a system, the energy storage elements do not allow an immediate flow of energy and therefore the measurement system does not respond to the input immediately. The measurement system goes through a *transient state* before it finally settles to its *steady state position.*

The response of a measurement system subjected to a time varying input like a sudden change in temperature can be divided into two parts, the *transient response*, and the *steady state response*. If $c(t)$ is the total time response of a system, then in general,

$$c(t) = c_t(t) + c_{ss}(t) \quad \text{...(4.1)}$$

where $c_t(t)$ and $c_{ss}(t)$ denote respectively the transient and steady state portions of the response. In measurement systems, the ***steady state response*** is simply the response when time reaches infinity.

***Transient response*** is defined as the part of response which goes to zero as time becomes large. Therefore, $c_t(t)$ has the property,

$$\lim_{t\to\infty} c_t(t) = 0 \quad \text{...(4.2)}$$

Therefore, after a time $=\infty$, the system response is,

$$c(t) = c_{ss}(t) \quad \text{...(4.3)}$$

*i.e.*, the system response is the *steady state response.*

Some measurements are made under conditions that sufficient time is available for the measurement system to settle to its final steady state conditions. Under such conditions, the study of behaviour of the system under transient state is not of much importance ; only steady state response of the system is considered. However, in many areas of measurement system applications, it becomes necessary to study the response of the system under both transient as well as steady state conditions. In many applications, the transient response of the system *i.e.*, the way system settles down to its final steady state conditions is more important than the steady state response. As an example, suppose a body is subjected to a sudden severe mechanical impact lasting for a few milliseconds. The body is accelerated and the transient response which is an acceleration-time relationship lasting for a few milliseconds during which the body is subjected to severe strains is of utmost importance.

The second type of dynamic input to a measurement system is one which has a definite periodic cycle and the input is called ***steady state periodic input.*** This input may take the form of :

(*i*) a purely sinusoidal input like the a.c. voltage with a frequency of 50 Hz or a temperature bath wherein the temperature varies cyclically with a frequency of say 0.01 Hz.

(*ii*) a complex time varying repetitive input which is a combination of the fundamental and harmonics.

Such inputs are encountered in study of vibrations.

The response of measurement systems to periodically varying inputs is not important during the first few cycles when the system goes through the transient conditions but is important when the system settles down to final steady state conditions.

The measurement systems, when subjected to periodically varying inputs exhibit in their response, a magnitude and phase relationship which is different from that of the input signal because of the energy storage elements. The output is not a faithful representation of the input in the sense that the fundamental and the harmonics in the input waveform are not reproduced in the output both in magnitude and phase relationships as in the input. The output is distorted both in terms of magnitude as well as phase relationships.

The dynamic characteristics of a measurement system are :

| | |
|---|---|
| (*i*) Speed of response | (*ii*) Measuring lag |
| (*iii*) Fidelity | (*iv*) Dynamic error. |

The characteristics listed on the left are desirable while those listed on the right are undesirable. These characteristics are defined below :

**1. *Speed of Response.*** It is defined as the rapidity with which a measurement system responds to changes in the measured quantity.

**2. *Measuring Lag.*** It is the retardation or delay in the response of a measurement system to changes in the measured quantity. The measuring lags are of two types :

(*i*) *Retardation type.* In this case the response of the measurement system begins immediately after a change in measured quantity has occurred.

(*ii*) *Time delay type.* In this case the response of the measurement system begins after a dead time after the application of the input. Dead time simply shifts the response of the system along the time scale and causes a corresponding dynamic error. The measurement lags of this type are very small and are of the order of a fraction of a second and hence can be ignored. But when the measurement system is subjected to periodically varying inputs, the performance of the systems having dead time is not satisfactory. And if the measured quantity varies at a fast rate, the dead time has a severe adverse effect on the performance of the system.

**3. *Fidelity.*** It is defined as the degree to which a measurement system indicates changes in the measured quantity without any dynamic error.

**4. *Dynamic Error.*** It is the difference between the true value of the quantity (under measurement) changing with time and the value indicated by the measurement system if no static error is assumed. It is also called *Measurement Error.*

## 4.2 DYNAMIC ANALYSIS OF MEASUREMENT SYSTEMS

The dynamic behaviour of measurement systems is studied in *two* domains *viz.* :

- Time Domain, and
- Frequency Domain.

### 1. Time Domain Analysis

In most measurement systems, time is used as an independent variable and therefore it is of interest to evaluate the time domain response of the system. The input signal is applied to the system, and the behaviour of the system is studied as a function of time. The output of a measurement system is supposed to follow the input as closely as possible and therefore it is important to compare the output and the input variables at all times.

In time domain analysis, the dynamic response of the system to different types of inputs, which are a function of time, is analysed at different intervals of time after the application of input signal. In practice, the input signals applied to a measurement systems are not known ahead of time. In most cases, the actual input signals vary in random fashion with respect to time and therefore cannot be mathematically defined. However, for the purpose of analysis and design, it is necessary to assume some basic types of input signals which can be easily defined mathematically, so that the performance of a system can be analysed with these Standard signals.

In the time domain analysis, the following standard test signals (inputs) are used :

(*i*) ***Step input.*** A step input represents an application of a sudden change. It may represent a sudden application of a force to a mechanical system like sudden rotation of a shaft or may represent the sudden closure of a switch of an electrical circuit. We will first define a unit step function. Mathematically, a unit step function $u(t)$ is defined as :

$$\left.\begin{aligned} {}^*r(t) &= 1\,;\, t \geq 0 \\ &= 0\,;\, t < 0 \end{aligned}\right\} \qquad \text{...(4.4)}$$

A unit step input is designated by $u(t)$. Therefore, a unit step input is written as,

$$r(t) = u(t) \qquad \text{...(4.5)}$$

Thus a unit step input represents a signal which changes its level from 0 to 1 in zero time and the unit magnitude is maintained thereafter. This function is plotted in Fig. 4.1($a$).

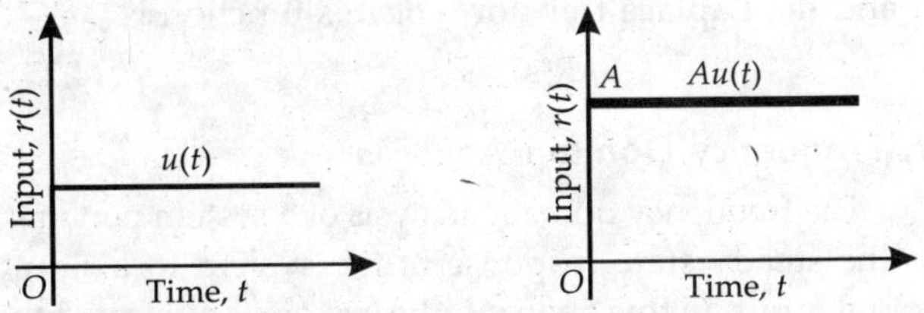


($a$) Unit step function ($b$) Step function of magnitude $A$

**Fig. 4.1** Step inputs.

The Laplace transform of a unit step input is,

$$R(s) = \int_0^\infty 1.e^{-st}\, dt = \frac{1}{s} \qquad \text{...(4.6)}$$

Figure 4.1($b$) shows a step input of magnitude $A$ wherein the magnitude changes from 0 to $A$ in zero time. This input is mathematically represented as,

$$\left.\begin{aligned} r(t) &= A\,;\, t \geq 0 \\ &= 0\,;\, t < 0 \end{aligned}\right\} \qquad \text{...(4.7)}$$

or simply $\quad r(t) = Au(t) \qquad$ ...(4.8)

The Laplace transform of a step input with magnitude $A$ is,

$$R(s) = \int_0^\infty A.e^{-st}\, dt = \frac{A}{s} \qquad \text{...(4.9)}$$

($ii$) ***Ramp input.*** It represents an input signal which changes at a constant rate with respect to time like constant velocity and therefore is called a *velocity input*.

A unit ramp signal starts at a value of zero and increases with constant slope of unit with respect to time. A unit ramp input signal is shown in Fig. 4.2($a$). It can be expressed mathematically as :

$$\left.\begin{aligned} r(t) &= t\,;\, t > 0 \\ &= 0\,;\, t < 0 \end{aligned}\right\} \qquad \text{...(4.10)}$$

or simply $\quad r(t) = t\,u(t) \qquad$ ...(4.11)

The Laplace transform of a unit ramp function is,

$$R(s) = \int_0^\infty t.e^{-st}\, dt = \frac{1}{s^2} \qquad \text{...(4.12)}$$

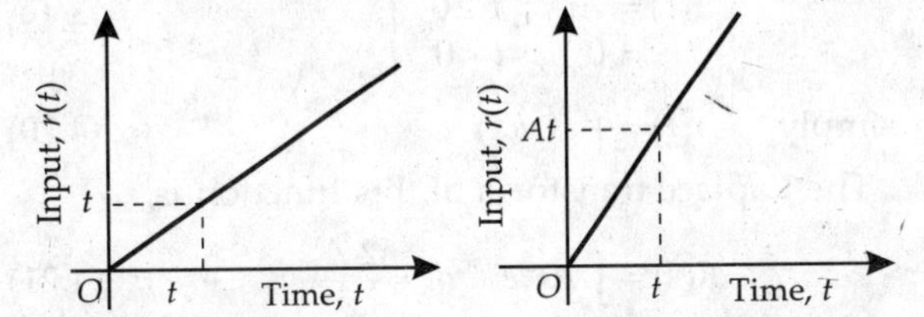


($a$) Unit-ramp function ($b$) Ramp function of magnitude $A$

**Fig. 4.2** Ramp Inputs.

A ramp input signal of magnitude $A$ is shown in Fig. 4.2($b$). Mathematically this input signal can be expressed as :

$$\left.\begin{aligned} r(t) &= At\,;\, t \geq 0 \\ &= 0\,;\ \ t < 0 \end{aligned}\right\} \qquad \text{...(4.13)}$$

or simply $\quad r(t) = At\,u(t) \qquad$ ...(4.14)

The Laplace transform of this function is :

$$R(s) = \int_0^\infty Ate^{-st}\, dt = \frac{A}{s^2} \qquad \text{...(4.15)}$$

($iii$) ***Parabolic Input.*** It represents an input signal which is proportional to the square of time and therefore represents a constant acceleration. Hence, this signal is called an *acceleration input*. The mathematical representation of a unit parabolic input is,

$$\left.\begin{aligned} r(t) &= t^2\ ;\ t \geq 0 \\ &= 0\ \ ;\ t < 0 \end{aligned}\right\} \qquad \text{...(4.16)}$$

or simply $\quad r(t) = t^2\,u(t) \qquad$ ...(4.17)

This signal is shown in Fig. 4.3($a$).

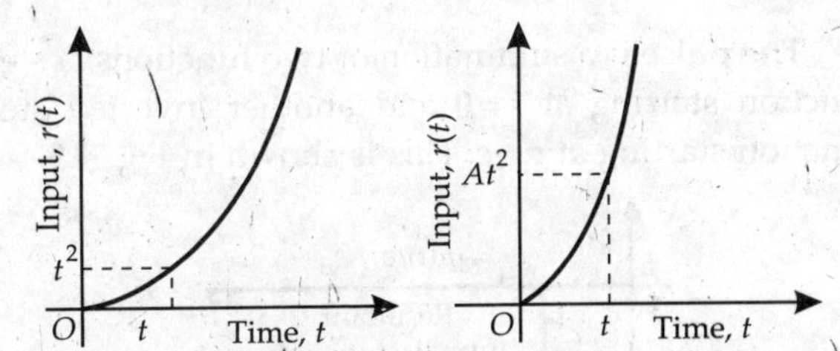


($a$) Unit parabolic function ($b$) Parabolic function of magnitude $A$

**Fig. 4.3** Parabolic inputs.

* It must be understood that a time varying input to a measurement system is designated as $r(t)$ while the output is designated as $c(t)$.

The Laplace transform of a unit parabolic function is,

$$R(s) = \int_0^\infty t^2 e^{-st}\, dt = \frac{2}{s^3} \quad \text{...(4.18)}$$

A parabolic function of magnitude $A$ is shown in Fig. 4.3($b$). Mathematically this input signal can be expressed as :

$$\left.\begin{aligned} r(t) &= At^2\ ;\ t \ge 0 \\ &= 0\ \ ;\ t < 0 \end{aligned}\right\} \quad \text{...(4.19)}$$

or simply $\quad r(t) = At^2 u(t) \quad$ ...(4.20)

The Laplace transform of this function is,

$$R(s) = \int_0^\infty At^2 e^{-st}\, dt = \frac{2A}{s^3} \quad \text{...(4.21)}$$

($iv$) ***Impulse input.*** A unit impulse is defined as a signal which has zero value everywhere except at $t = 0$ where the magnitude is finite. This function is generally called a $\delta$ (delta) function. The $\delta$ function has the following properties :

$$\left.\begin{aligned} \delta(t) &= 0\ ;\ t \ne 0 \\ \text{and} \quad \int_{-\varepsilon}^{\varepsilon} \delta(t)dt &= 1 \end{aligned}\right\} \quad \text{...(4.22)}$$

where $\varepsilon \to 0$

Since a perfect impulse function is not practically realizable, it is approximated by a pulse of small width having a unit area as shown in Fig. 4.4($a$). The diagram shows a pulse with an amplitude of $1/\varepsilon$ and duration $\varepsilon$.

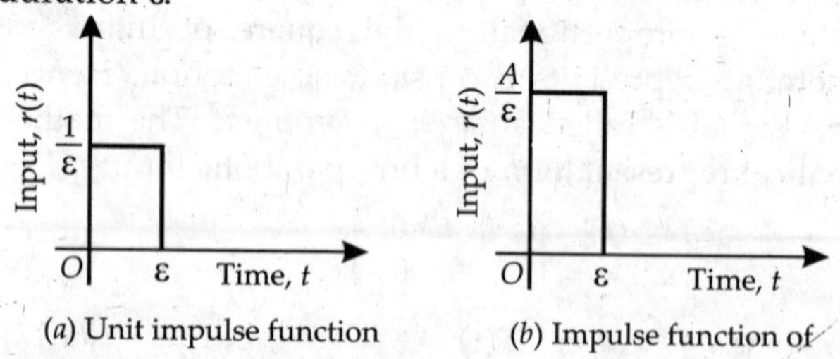


**Fig. 4.4** Impulse functions.

The pulse is a summation of two functions, a step function starting at $t = 0$ and another inverted step function starting at $t = \varepsilon$. This is shown in Fig. 4.5.

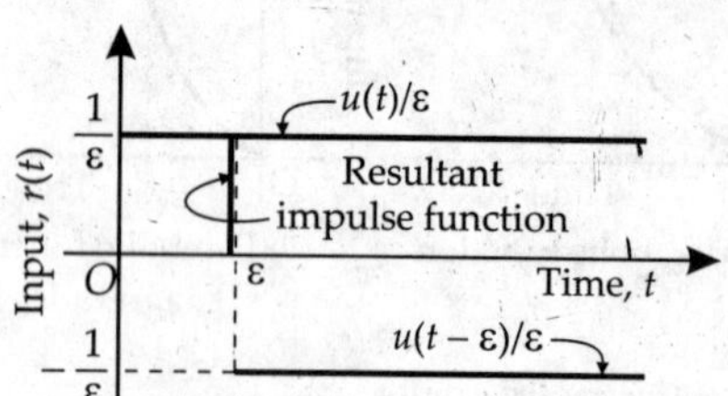


**Fig. 4.5** Impulse function as a summation of two step functions.

Therefore, a unit impulse function is expressed as :

$$\delta(t) = \lim_{\varepsilon \to 0} \frac{u(t) - u(t-\varepsilon)}{\varepsilon} \quad \text{...(4.23)}$$

The Laplace transform of a unit impulse function is,

$$R(s) = \int_0^\infty \frac{u(t) - u(t-\varepsilon)}{\varepsilon} e^{-st} dt = 1 \quad \text{...(4.24)}$$

Thus, the Laplace transform of a unit impulse function is 1.

The Laplace transform of a unit step function is 1/s and that of a unit impulse function is 1 and therefore a unit step function is a derivative of unit impulse function.

Figure 4.4($b$) shows an impulse function of strength $A$ and the Laplace transform of this function is

$$R(s) = A \quad \text{...(4.25)}$$

## 2. Frequency Domain Analysis

The frequency domain analysis of a system pertains to the steady state response of the system to a sinusoidal input. In this analysis, the system is subjected to a sinusoidal input and the system response is studied with frequency as the independent variable. The sinusoid is a unique input signal, and the resulting output signal for a linear system is sinusoidal in the steady state. However, the output signal differs from the input waveform in amplitude and phase as explained later in this chapter.

It is important to analyze the system behaviour in frequency domain through application of sinusoidal input signals because of the following reasons :

($i$) The system may be subjected to a periodically varying sinusoidal signal while in actual use.

($ii$) The system may, in practice, be subjected to a time varying signal of a complex nature. According to Fourier analysis, complex waveforms can be thought of as composed of sinusoidal signals of different frequencies. The system, therefore, can be subjected to sinusoidal signals of different amplitudes and frequencies as contained in the input waveform and the total response of the system may be obtained by superposition of the responses at different frequencies.

($iii$) A system is subjected to various types of time varying signals like step, ramp and parabolic inputs. All these time varying signals can be expressed in terms of sinusoidal signals of different amplitudes and frequencies. Therefore, the total response of the system can be obtained by superposition as explained above.

When a system is subjected to a sinusoidal input,

$$r(t) = A_i \sin \omega t \quad ...(4.26)$$

The output is,

$$c(t) = A_0 \sin(\omega t + \phi) \quad ...(4.27)$$

where $A_i$, $A_0$ are respectively the amplitudes of input and output and $\phi$ is the phase difference between input and output.

Figure 4.6 shows the variation of amplitude and phase of output with frequency for a first order system.

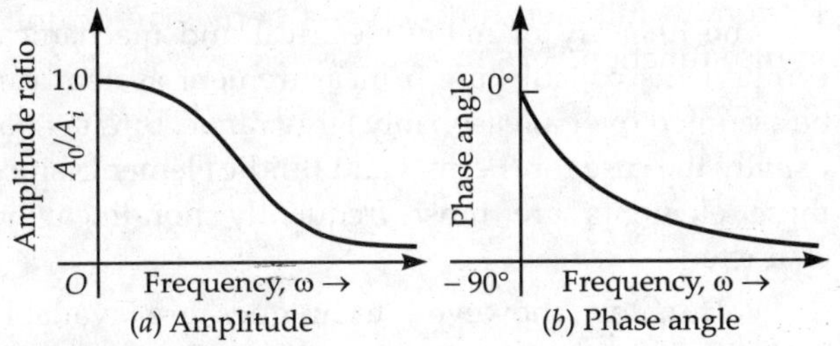


**Fig. 4.6** Frequency response of a first order system

## 4.3 MATHEMATICAL MODELS OF MEASUREMENT SYSTEMS

A measurement system is a physical system. A physical system is defined as a collection of physical objects connected together to serve a desired objective. Thus, a *measurement system* may be defined as an assemblage of diverse physical devices interconnected together to measure, analyze and control electrical, thermal, fluidic and other physical quantities.

Before one can analyze a measurement system one must be able to determine its performance. Both the ability to analyze and the precision of the results depend on how well the characteristics of each component of the system can be expressed mathematically or in other words, how well a *mathematical model* of the system can be obtained.

The first step in modelling a system is to obtain its physical model. Measurement systems are complex in nature and therefore it is impossible to obtain a physical model of a measurement system (and for that matter of any other physical system) which incorporates all the complexities of the system. However, some idealized assumptions are made in order to obtain a simplified version of the system. The model based upon idealized assumptions made keeping in view the specific problem under study and the accuracies desired is called a **physical model**.

The next step in modelling is to obtain a **mathematical model**. A **mathematical model** is a mathematical representation of a physical model and is achieved from the latter by utilizing the physical laws.

After a mathematical model of a system is obtained, the dynamic response of the system is studied for different time varying inputs.

4

## 4.4 LINEAR AND NON-LINEAR SYSTEMS

The mathematical model of a system is **linear** if it obeys the laws of *superposition* and *homogeneity*. Suppose the system is subjected to inputs,

$$x_1(t), x_2(t), ...... , x_n(t)$$

and the corresponding outputs are

$$y_1(t), y_2(t), ...... , y_n(t) \text{ respectively,}$$

then the system response to the linear combination of these inputs,

$$a_1x_1(t) + a_2x_2(t) + ...... + a_nx_n(t) \quad ...(4.28)$$

is given by the linear combination of the outputs as :

$$a_1y_1(t) + a_2y_2(t) + ...... + a_ny_n(t) \quad ...(4.29)$$

where , $a_1, a_2, ...... , a_n$ are constants.

The property of *homogeneity* means that the inputs $x_1(t), x_2(t), ...... , x_n(t)$ are of the same form and so are also the outputs $y_1(t), y_2(t), ...... , y_n(t)$.

The mathematical models describing a system are depicted by differential equations. A mathematical model is described as **linear** if the co-efficients of the differential equation describing it are either constants or are a function of the time (the independent variable).

In case the principles of superposition and homogeneity are not complied with and also in case the differential equation describing the mathematical model has co-efficients which are a function of dependent variable the system is called **non-linear**.

There are *two* further classifications of linear systems. They are :

(*i*) linear time invariant systems, and

(*ii*) linear time variant systems.

1. ***Linear time invariant system.*** In case the differential equation describing a linear system has constant co-efficients, it is called a linear time invariant system.

2. ***Linear time variant system.*** If the differential equations describing a linear system have co-efficients which are a function of time (the independent variable), the system is called a time variant system.

## 4.5 ANALYSIS OF LINEAR SYSTEMS

Powerful mathematical tools like the Laplace and Fourier transforms are available for analysis of linear systems. Unfortunately most of the measurement systems are not linear in nature and therefore assumptions have to be made to make them linear so that their analysis could be carried out. Linearizing a non-linear system, no doubt, results in simplicity but at the same time results in loss of accuracy. Hence a compromise has to be made between simplification of the model and the accuracy of results obtained from it.

The usual approach adopted is to first build a simplified model which is linear ignoring all the non-linearities and other complexities of the system which cannot be described by simple mathematics. This model is analyzed, which gives an approximate idea of the dynamic response of the system. The corrections for non-linearities and other complexities present in the system are incorporated in the solution later on to get results which are very near to the actual conditions obtaining in the system.

Summarising, the approach to dynamic system problems can be listed as follows :

1. Define the system and its components.
2. Build up the physical model and list the necessary assumptions made.
3. Formulate the mathematical model and describe the assumptions made.
4. Write down the differential equations describing the model.
5. Solve the equations for the desired output variables.
6. Examine the solutions and assumption.
7. Apply the necessary corrections to subvert the idealized corrections made in both physical and mathematical models.
8. Re-analyze and redesign till the solutions obtained are compatible with the results expected of the actual system.

## 4.6 LINEAR APPROXIMATION OF NON-LINEAR SYSTEMS

Almost all practical measurement systems are non-linear in nature. The most powerful mathematical techniques are available for analysis of linear systems and they are simple to apply. Thus it becomes partinent to make linearizing assumptions for a non-linear system for the sake of simplicity of analysis.

Suppose a system is represented by :

$$y = mx + b$$

This system is non-linear because it does not satisfy the property of homogeneity. However, this system may be considered linear for small changes $\Delta x$ and $\Delta y$ respectively, about an operating point $x_0, y_0$,

where $\quad x = x_0 + \Delta x$ and $y = y_0 + \Delta y$

$$\therefore \quad y_0 + \Delta y = m(x_0 + \Delta x) + b = mx_0 + m\Delta x + b$$

or
$$\Delta y = m\Delta x \qquad \text{...(4.30)}$$

as
$$y_0 = mx_0 + b$$

Equation 4.30 satisfies the necessary conditions of linearity for small perturbations about operating point $x_0, y_0$.

The linearity of many electrical and mechanical components constituting a measurement system can be assumed over a reasonably large range. But it is not usually the case for thermal and fluidic elements since these elements are most frequently non-linear in character.

Fortunately, however, techniques are available with the help of which it is possible to linearize non-linear elements assuming that the variations about the operating point are small. The fundamental base for this linearizing technique is that function describing the input-output by relationship for the elements can be mapped by a smooth curve and that this smooth curve deviates very little from its tangent line so long as variables describing this curve do not wander far away from the point of tangency *i.e.*, operating point. Thus if the region of operation has a restricted range, a non-linear system may be treated as linear about operating point for the purposes of analysis and hence it facilitates the solution of non-linear systems with the help of simple mathematical tools which can normally be applied for linear systems only.

Consider a general measurement system element with input $x$ and output $y$. The relationship between output and input variables is written as :

$$y = f(x).$$

The relationship may be of the form shown in Fig. 4.7. It should be noted that the relationship is no doubt non-linear but is continuous.

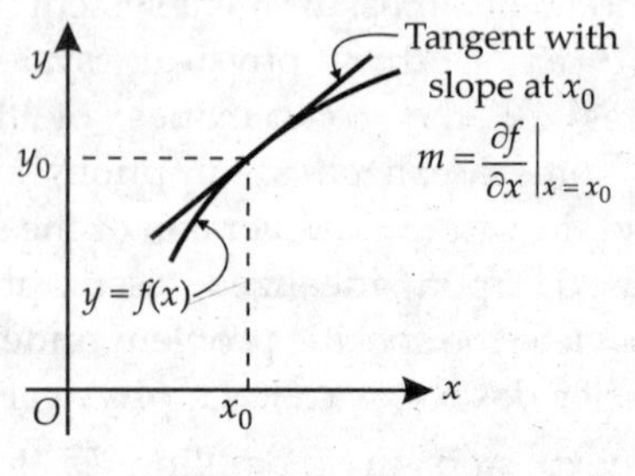


**Fig. 4.7** Linearization or a non-linear function.

The normal operating point is designated as $x_0, y_0$. Since the curve (function) is continuous over the range of interest, a Taylor Series expansion about the operating point may be used. The series gives :

$$y = f(x) = f(x_0) + \left.\frac{\partial f}{\partial x}\right|_{x=x_0} \frac{x - x_0}{1!} + \left.\frac{\partial^2 f}{\partial x^2}\right|_{x=x_0} \frac{(x-x_0)^2}{2!} + \ldots \qquad \ldots(4.31)$$

If the variation $(x - x_0)$ of the input is small about the normal operating point, the higher order terms may be neglected. Thus the slope at the operating point $m = \left.\frac{\partial f}{\partial x}\right|_{x=x_0}$ is a good approximation to the curve over the small range $(x - x_0)$. Therefore a reasonable approximation of Eqn. 4.31 becomes,

$$y = f(x_0) + \left.\frac{\partial f}{\partial x}\right|_{x=x_0} (x - x_0) \qquad \ldots(4.32)$$

$$= y_0 + m(x - x_0) \qquad \ldots(4.33)$$

where $m = \left.\frac{\partial f}{\partial x}\right|_{x=x_0}$ ...(4.34)

is the slope of the curve at $x = x_0$

Equation 4.33 may be written as :

$$y - y_0 = m(x - x_0)$$

or $$\Delta y = m \Delta x \qquad (4.35)$$

This linear approximation is accurate as long as the assumption of small signals is applicable to the specific problem.

***Example 4.1*** *Figure 4.8 shows an oscillation of a pendulum. Write down the equation of torque T acting upon it as a function of angle θ. Linearize the function for small deviations in angle θ.*

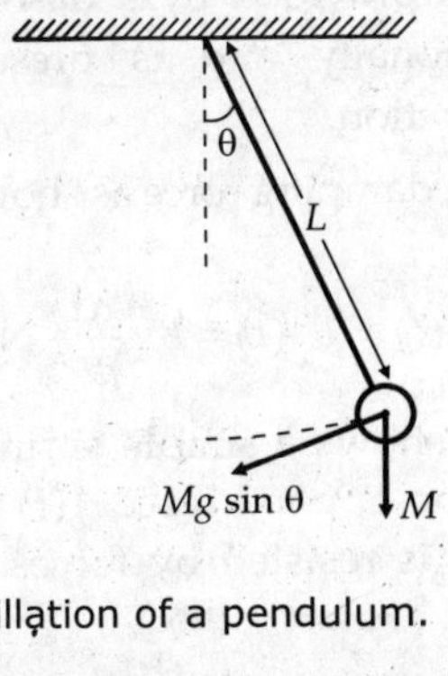


**Fig. 4.8** Oscillation of a pendulum.

**Solution.** The torque acting upon the mass when it deviates by an angle θ from its vertical axis is,

$$T = MgL \sin\theta$$

where $g$ = acceleration due to gravity.

Thus torque $T$ is proportional to sine of angle θ. Therefore a non-linear relationship exists between torque $T$ and angle θ.

The first derivative evaluated at its vertical (equilibrium) position where $\theta = \theta_0 = 0°$ provides the linear approximation. This is :

$$T = MgL \left.\frac{\partial \sin\theta}{\partial \theta}\right|_{\theta=\theta_0} (\theta - \theta_0) = MgL \cos(0°)(\theta - 0°)$$

$$= MgL\theta.$$

Thus for small deviations, there now exists a linear relationship between torque $T$ and angle θ.

## 4.7 ELECTRIC NETWORKS

An *R-L-C*-circuit is shown in Fig. 4.9. Using Kirchhoff's voltage law, the equations describing the dynamics of the circuit are written as :

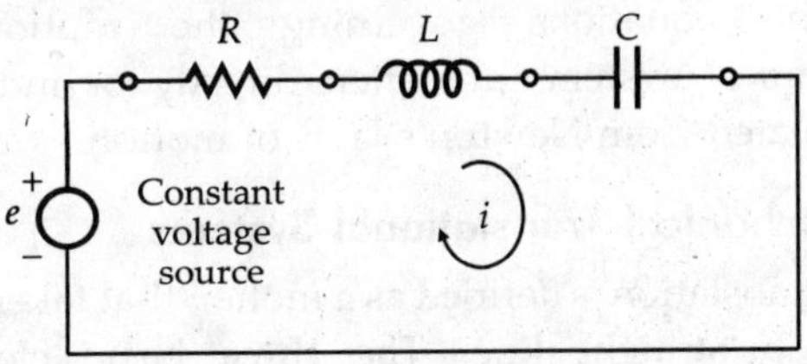


**Fig. 4.9** A series R-L-C circuit with a constant voltage source.

Input voltage

$$e(t) = L\frac{di(t)}{dt} + Ri(t) + \frac{1}{C}\int i(t)dt \qquad \ldots(4.36)$$

or $$e(t) = L\frac{d^2q(t)}{dt^2} + \frac{Rdq(t)}{dt} + \frac{q(t)}{C} \qquad \ldots(4.37)$$

where $q(t)$ = charge = $\int i(t)dt$

and $i(t)$ = current = $= \frac{dq(t)}{dt}$

Let us analyze a parallel *R-L-C* network with a constant current source as shown in Fig. 4.10.

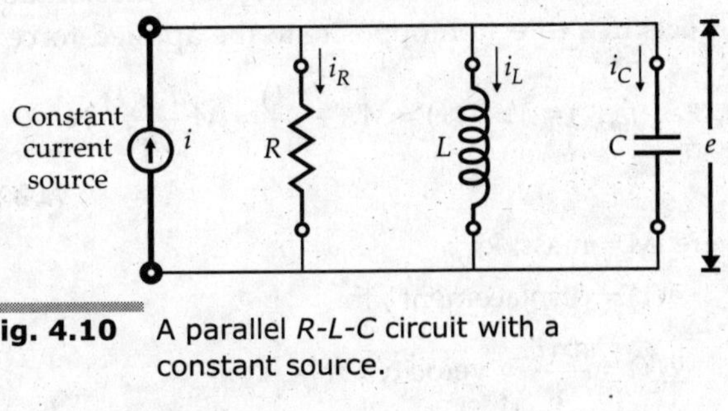


**Fig. 4.10** A parallel *R-L-C* circuit with a constant source.

Applying Kirchhoff's current law,

$$i = i_R + i_L + i_C$$

$$= \frac{e(t)}{R} + \frac{1}{L}\int e(t)dt + C\frac{de(t)}{dt} \qquad \ldots(4.38)$$

But e.m.f. $e(t) = \frac{d\psi(t)}{dt}$

where $\psi$ = flux linkages

$$\therefore \quad i = \frac{1}{R}.\frac{d\psi(t)}{dt} + \frac{\psi(t)}{L} + C\frac{d^2\psi(t)}{dt^2},$$

$$= C\frac{d^2\psi(t)}{dt^2} + \frac{1}{R}\frac{d\psi(t)}{dt} + \frac{\psi(t)}{L}. \quad ...(4.39)$$

## 4.8 MECHANICAL SYSTEMS

The motion of mechanical elements can be described in various dimensions as :

- Translational,
- Rotational and
- A combination of translational and rotational motions.

The equations governing the motions of mechanical systems are often directly or indirectly formulated from Newton's laws of motion.

### 1. Mechanical Translational Systems

Translation is defined as a motion that takes place along a straight line. The three basic elements involved in translational motion are :

(*i*) mass, (*ii*) spring and (*iii*) friction.

In this analysis, the effects of static function, coulomb friction and other friction terms which exhibit a non-linear behaviour are neglected in order to develop a linear model. Only viscous friction is considered as it exhibits a linear behaviour. Figure 4.11 shows the three elements encountered in mechanical translational systems. These elements are :

(*i*) ***Mass.*** The mass $M$ (Fig. 4.11(*a*)) is an internal element. A force applied to a mass produces an acceleration. The reaction force $f_m$ is equal and opposite to the product of the mass and acceleration and acts in a direction opposite to the applied force.

$$\therefore \quad f_m(t) = M \times a(t) = M \times \frac{dv(t)}{dt} = M\frac{d^2x(t)}{dt^2} \text{ N} \quad ...(4.40)$$

where $M$ = mass, kg ;

$x(t)$ = displacement ; m,

$v(t) = \frac{dx(t)}{dt}$ = velocity ; m/s,

$a = \frac{dv(t)}{dt} = \frac{d^2x(t)}{dt^2}$ = acceleration ; m/s$^2$.

The reaction force $f_m$ is a function of time and acts through mass $M$.

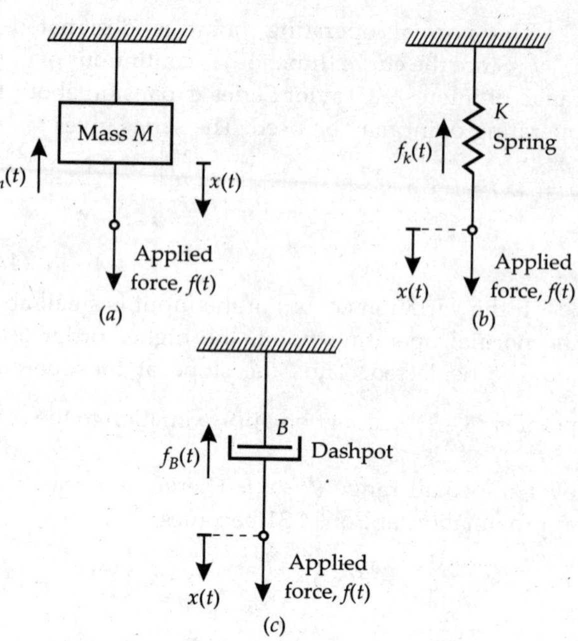


**Fig. 4.11** Elements in a mechanical translational system.

(*ii*) ***Stiffness.*** The stiffness, $K$, provides a restoring force as represented by a spring in Fig. 4.11(*b*). When a force is applied to a spring the reaction force, $f_k$ on each end of the spring is the same and is equal to the product of the stiffness $K$, and the amount of deformation, $x$, of the spring.

$$f_k(t) = Kx(t) \text{ N} \quad ...(4.41)$$

where $K$ = stiffness constant ; N/m.

(*iii*) ***Viscous Friction.*** The *co-efficient of viscous friction* or the *damping constant* $B$ characterises that the element absorbs energy or in other words, the element is an energy dissipating element. The damping may be intentional as is provided by a dashpot. But it may occur unintentionally and is present because of physical construction.

The reaction damping force as shown in Fig. 4.11(*c*) is,

$$f_B(t) = Bv(t) = B\frac{dx(t)}{dt} \text{ N} \quad ...(4.42)$$

Figure 4.12 shows a simple spring-mass-damper system where the applied force $f(t)$ is a function of time. This force is resisted by forces $f_m(t)$, $f_k(t)$ and $f_B(t)$.

We can write the equation of motion as :

$$f(t) = f_m(t) + f_B(t) + f_k(t)$$

or

$$f(t) = M\frac{d^2x(t)}{dt^2} + B\frac{dx(t)}{dt} + Kx(t) \quad ...(4.43)$$

Equation 4.43 may be written as

$$f(t) = M\ddot{x} + B\dot{x} + Kx \quad ...(4.44)$$

where $\ddot{x} = \dfrac{d^2x(t)}{dt^2}$, $\dot{x} = \dfrac{dx(t)}{dt}$ and $x = x(t)$.

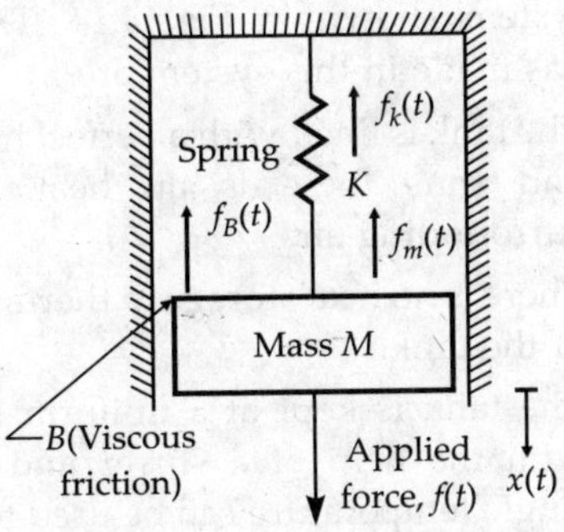


**Fig. 4.12** Spring-mass-damper system.

## 2. Mechanical Rotational Systems

The rotational motion of a body may be defined as motion about a fixed axis. Rotational systems are similar to translational systems except for the basic differences that torque equations are written in place of force equations and the displacement, velocity and acceleration terms are angular quantities. The three basic elements involved are inertia, torsional stiffness and viscous friction. An elementary mechanical rotational system is shown in Fig. 4.13. Let us analyze the system considering its basic elements.

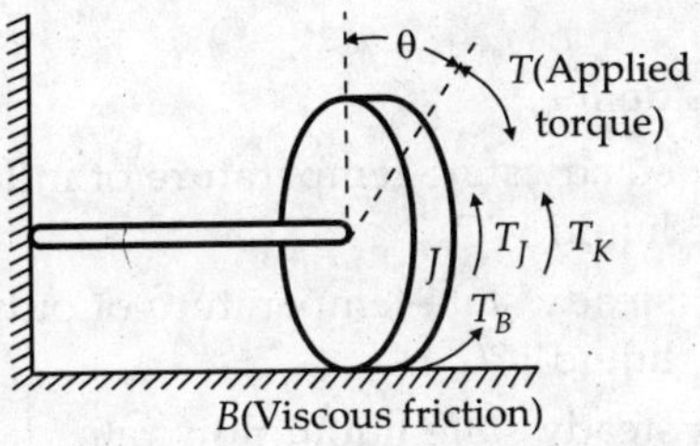


**Fig. 4.13** Mechanical rotational system.

(*i*) ***Inertia.*** The torque applied to a body having a moment of inertia, *J*, produces an angular acceleration.

The reaction torque produced by inertia is,

$$T_j(t) = J\alpha(t) = \frac{d\omega(t)}{dt} = J\frac{d^2\theta(t)}{dt^2} \quad ...(4.45)$$

where $T_j$ = intertia torque ; Nm,

$J$ = moment of inertia ; kg-m$^2$,

$\alpha$ = angular acceleration ; rad/s$^2$,

$\omega$ = angular velocity ; rad/s,

$\theta$ = angular twist ; rad.

(*ii*) ***Torsional torque.*** When a torque is applied to an elastic mass it is twisted through an angle θ. The angular torque is transmitted through the spring action and appears as a counter torque at the other end. The reaction spring or torsional torque, $T_k$, is equal to the product of the torsional constant, *K* and θ, the angular twist. Torsional torque,

$$T_k(t) = K\theta(t) \text{ Nm} \quad ...(4.46)$$

where $K$ = stiffness or torsional constant, Nm/rad.

(*iii*) ***Damping torque.*** Damping occurs whenever a body moves through a fluid. In order to produce motion of the body, a torque must be applied to overcome the reaction torque due to damping. The torque due to damping is,

$$T_B(t) = B\omega(t) = \frac{Bd\theta(t)}{dt} \text{Nm} \quad ...(4.47)$$

where $B$ = damping constant or viscous friction constant ; N-m/rad-s$^{-1}$.

The differential equation describing the dynamics of a mechanical rotational system is,

$$T(t) = J\alpha(t) + B\omega(t) + K\theta(t)$$

$$= \frac{Jd^2\theta(t)}{dt^2} + B\frac{d\theta(t)}{dt} + K\theta(t) \quad ...(4.48)$$

where $T(t)$ = applied torque ; N-m.

Eqn. 4.48 may be written in the following form :

$$T(t) = j\ddot{\theta} + B\dot{\theta} + K\theta \quad ...(4.48a)$$

where $\ddot{\theta} = \dfrac{d^2\theta(t)}{dt^2}$, $\dot{\theta} = \dfrac{d\theta(t)}{dt}$ and $\theta = \theta(t)$.

## 4.9 FORCE-VOLTAGE AND FORCE-CURRENT ANALOGIES

Comparing Eqn. 4.43 for the mechanical translational system shown in Fig. 4.12 and Eqn. 4.48 for the mechanical rotational system shown in Fig. 4.13, it is found that they have an identical form. The systems which are defined by identical differential equations are called *analogous systems.* Let us compare Eqn. 4.43 for mechanical translational system and Eqn. 4.48 for mechanical rotational system with that of a series *R-L-C* circuit (Eqn. 4.37) supplied by a constant voltage source. It is found that force, *f*, and torque *T* in the mechanical systems are analogous to voltage *e*. The analogy based upon this description is called a *force-voltage analogy.* A list of analogous variables based upon this analogy are given in Table 4.1.

**Table 4.1** *Force Voltage Analogy*

| Mechanical Translational System | Mechanical Rotational System | Electrical System |
|---|---|---|
| Force, $f$ | Torque, $T$ | Voltage, $e$ |
| Mass, $M$ | Moment of inertia, $J$ | Inductance, $L$ |
| Viscous friction constant, $B$ | Viscous friction constant, $B$ | Resistance, $R$ |
| Spring stiffness, $K$ (Elastance, $1/K$) | Torsional stiffness, $K$ (Elastance, $1/K$) | Reciprocal of capacitance, $1/C$ (Capacitance, $C$) |
| Displacement, $x$ | Angular displacement | Charge, $Q$ |
| Velocity, $\dot{x}$ | Angular velocity, $\dot{\theta}$ | Current, $i$ |

Equations 4.43 and 4.48 for mechanical translational and rotational systems respectively are analogous to Eqn. 4.39 of a parallel *R-L-C* circuit with a constant current source. In this case force, $f$ in the mechanical translational system and torque, $T$, in the mechanical rotational systems are analogous to current $i$. The analogy based upon this logic is called the *force current analogy*. A list of analogous quantities in this analogy are given in Table 4.2.

**Table 4.2** *Force-Current Analogy*

| Mechanical Translational System | Mechanical Rotational System | Electrical System |
|---|---|---|
| Force, $f$ | Torque, $T$ | Current, $i$ |
| Mass, $M$ | Moment of inertia, $J$ | Capacitance, $C$ |
| Viscous friction constant, $B$ | Viscous friction constant, $B$ | Reciprocal of resistance $1/R$ (Conductance $\sigma$) |
| Spring stiffness, $K$ (Elastance, $1/K$) | Torsional stiffness, $K$ (Elastance, $1/K$) | Reciprocal of Inductance, $1/L$ (Inductance, $L$) |
| Displacement, $x$ | Angular displacement, $\theta$ | Magnetic flux linkages, $\psi$ |
| Velocity, $\dot{x}$ | Angular velocity, $\dot{\theta}$ | Voltage, $e$ |

The concept of analogous systems is a very useful technique in the study of all types of systems which may be mechanical, thermal, fluidic or pneumatic in nature. They can be converted into analogous electrical systems and the system behaviour then can be studied and analyzed on an electrical analog computer.

## 4.10 THERMAL SYSTEMS

Thermal systems are those which involve the transfer of heat from one substance to another. Thermal systems may be analyzed in terms of resistance and capacitance. Let us consider a simple thermal system shown in Fig. 4.14. The idealized assumptions made in this system are :

(*i*) The tank is lined with a perfect heat insulator and thus there is no heat loss to the surrounding air.

(*ii*) There is no heat storage in thermal insulation of the tank.

(*iii*) The tank is kept at a uniform temperature with the help of a stirrer and therefore a single temperature can be used for the entire liquid in the tank and the outflowing liquid.

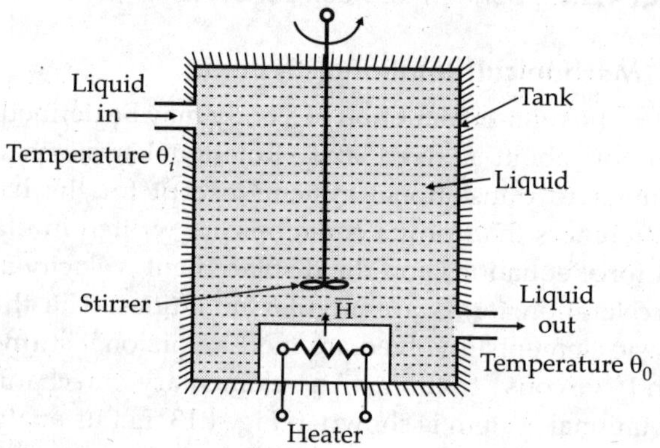


**Fig. 4.14** Thermal System.

Let us define,

$\theta_i$ = steady state temperature of inflowing liquid ; °C,

$\theta_0$ = steady state temperature of outflowing liquid ; °C,

$Q$ = steady state liquid flow rate ; kg/s,

$M$ = mass of liquid in tank ; kg,

$s$ = specific heat of liquid ; J/kg-°C,

$R$ = thermal resistance ; °C/Js$^{-1}$.

$C$ = thermal capacitance ; J/°C,

and $\overline{H}$ = steady state heat input rate ; J/s.

Assume that the temperature of the inflowing liquid is kept constant and that the heat input rate is suddenly changed from $\overline{H}$ to $\overline{H} + h_i$, where $h_i$, represents a small change in heat input rate. This is equivalent to application of a step input $h_i$. The heat outflow rate will then gradually change from $\overline{H}$ to $\overline{H} + h_0$. The temperature of the outflowing liquid will also be changed from $\theta_0$ to $\theta_0 + \theta$, where $\theta$ is the change in temperature of outflowing liquid, °C.

The steady state conditions are disturbed. A part of the increased input heat is stored in the liquid and the rest is responsible for the increase in the temperature of liquid flowing out from the tank.

Let $h_1$ = rate of heat storage in the tank ; J/s,

and $h_2$ = rate of heat outflow from the tank ; J/s.

$\therefore \quad h_i = h_1 + h_2$

Now, rate of heat storage in the tank,

$$= Ms\frac{d\theta}{dt} \quad ...(4.49)$$

$$= C\frac{d\theta}{dt} \quad ...(4.50)$$

where, $C$ = thermal capacitance = $Ms$ ; J/°C ...(4.51)

and rate of heat outflow from the tank,

$$h_2 = Qs\theta \quad ...(4.52)$$

Thus we can write :

$$h_i = C\frac{d\theta}{dt} + Qs\theta = C\frac{d\theta}{dt} + \frac{\theta}{R} \quad ...(4.53)$$

where, thermal resistance $R = 1/Qs$; °C/Js$^{-1}$ ...(4.54)

or $$h_i R = RC\frac{d\theta}{dt} + \theta$$

or $$RC\frac{d\theta}{dt} + \theta = h_i R \quad ...(4.55)$$

## 4.11 LIQUID LEVEL SYSTEMS

When analyzing liquid level systems we encounter *two* types of flows :

(*i*) laminar flow, and (*ii*) turbulent flow.

The classification of type of flow is done on the basis of Reynolds number. For the turbulent flow the Reynolds number is greater than about 3000-4000. The flow is laminar if the Reynolds number is less than about 2000. The turbulent flow is described by non-linear differential equations while laminar flow can be represented by linear differential equations. The flow is normally turbulent in industrial processes.

It is convenient to introduce the concept of resistance and capacitance to describe dynamic characteristics of liquid level systems.

Consider the flow through a short pipe connecting two tanks. The resistance for flow for such a restriction is defined as :

$$R = \frac{\text{change in level difference}}{\text{change in flow rate}} = \frac{\Delta H}{\Delta Q} \quad ...(4.56)$$

Now the relationship between the level difference and flow rate is described by different equations for laminar and turbulent flows. We shall consider both the cases.

Consider a liquid level system shown in Fig. 4.15 and assume that the flow through the load valve is laminar. For laminar flow there exists a linear relationship between liquid head and liquid flow rate and therefore,

$$Q = KH \quad ...(4.57)$$

where, $Q$ = steady state liquid flow rate ; m$^3$/s,

$H$ = steady state head ; m,

$K$ = a co-efficient ; m$^2$/s.

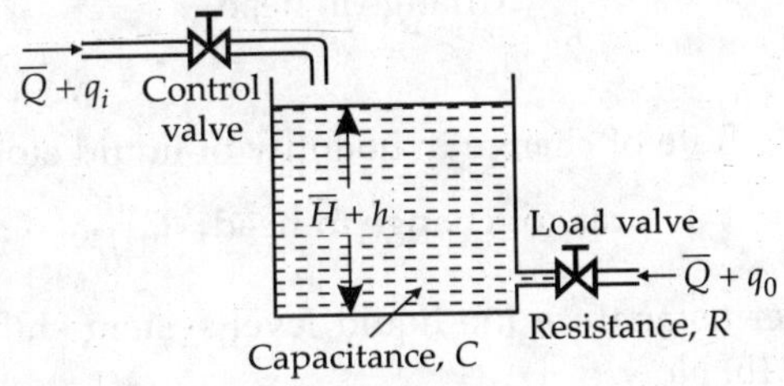


**Fig. 4.15** Liquid Level System.

It is clear from Eqn. 4.57 that the law governing laminar flow is analogous to Ohm's law which states that current is directly proportional to the potential difference, with liquid flow rate and liquid head being treated as analogous to current and voltage respectively.

For laminar flow, resistance

$$R = dH/dQ = H/Q \quad ..(4.58)$$

is constant and is analogous to electrical resistance.

Let us now consider the turbulent flow. For turbulent flow, we can write,

$$Q = K_1\sqrt{H} \quad ...(4.59)$$

$K_1$ = a coefficient : m$^{2.5}$/s

The value of $K_1 = KA\sqrt{2g}$ ...(4.60)

where $K$ = a flow coefficient = 0.6,

$A$ = area of restriction ; m$^2$,

and $g$ = acceleration due to gravity ; m/s$^2$,

Now resistance,

$$R = \frac{dH}{dQ} = \frac{2\sqrt{H}}{K_1} = \frac{2\sqrt{H}}{KA\sqrt{2g}} \quad ...(4.61)$$

There exists a non-linear relationship between head $H$, and flow rate, $Q$ and therefore the resistance $R$ is not constant but depends upon the value of head and the flow rate.

However, for turbulent flow, we can linearize the non-linear relationship between $Q$ and $H$ as given by Eqn. 4.31.

Using the technique given in Art. 4.6,

$$\text{Resistance } R = \left.\frac{dH}{dQ}\right|_{H=\bar{H}} = \left.\frac{2H}{Q}\right|_{Q=\bar{Q}} \quad ...(4.62)$$

where $\bar{H}$ and $\bar{Q}$ are respectively steady head and steady flow rate.

Such a linearization, however, is valid only for small changes in head and flow rate from their steady-state values.

The capacitance of a tank is defined as the change in quantity of stored liquid necessary to cause a unit change in the head

Capacitance C

$$= \frac{\text{Change in quantity of liquid stored}}{\text{Change in head}} ; \text{m}^2 \quad ...(4.63)$$

∴ Rate of change in quantity of liquid stored,

$$= C \times \text{rate of change in head} = C\frac{dh}{dt} \quad ...(4.64)$$

Let us analyze the liquid level system shown in Fig. 4.15. Now,

$\overline{Q}$ = steady state flow rate (before any change has occurred) ; $m^3/s$,

$q_i$ = small change in the inflow rate from its steady state value ; $m^3/s$,

$q_o$ = small change in the outflow rate from its steady state value ; $m^3/s$,

$\overline{H}$ = steady state head (before any change has occurred) ; m,

$h$ = small change in head from its steady state value ; m.

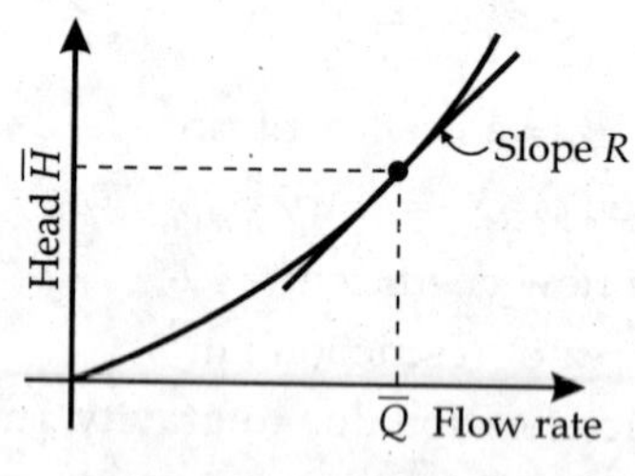


**Fig. 4.16** Head versus flow rate curve.

The inflow of liquid is equal to the sum of volume of liquid stored and the outflow of liquid.

Therefore the inflow minus the outflow during the small interval of time $dt$ is equal to the additional amount of liquid stored, we have,

$$C\,dh = (q_i - q_o)dt \quad ...(i)$$

But $\quad R = \frac{h}{q_o} \quad$ or $\quad q_o = \frac{h}{R}$

∴ Substituting the value of $R$ in (*i*), we have the differential equation for the liquid level system with a constant value of $R$ as,

$$RC\frac{dh}{dt} + h = Rq_i \quad ...(4.65)$$

## 4.12 PNEUMATIC SYSTEMS

Consider a simple pneumatic system shown in Fig. 4.17. A pneumatic source is supplying air to the pressure vessel through a pipe. If the pressure drop in the pipe line is small, the flow is laminar and hence linear differential equations may be used to describe the dynamics of the system.

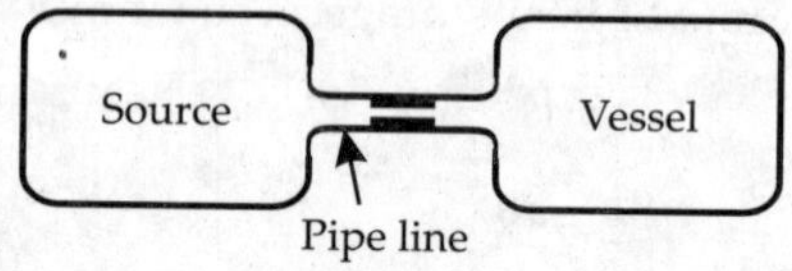


**Fig. 4.17** Pneumatic system.

Let

$P_i$ = pressure of the air of the source under steady state conditions ; $N/m^2$,

$P_0$ = pressure of air in the vessel under steady state conditions ; $N/m^2$,

$p_i$ = small change in air pressure of the source from its steady state value ; $N/m^2$,

$p_0$ = small change in air pressure of the vessel from its steady state value ; $N/m^2$,

The resistance to air flow rate into the vessel is defined as,

$$R = \frac{p_i - p_0}{q} \frac{N/m^2}{m^3/s} \quad ...(4.66)$$

where $q$ = rate of flow of air on account of differential pressure ; $m^3/s$.

The volume of air at pressure $p_i$ which gets stored in the vessel to raise the pressure inside the vessel by $p_0$ is proportional to it assuming the temperature to be constant.

∴ Volume $\quad v = Cp_0$

Constant $C$ is defined as capacitance of the vessel,

$$\therefore \text{ Capacitance } C = \frac{v}{p_0} \frac{m^3}{N/m^2} \quad ...(4.67)$$

From above, rate of storage in the vessel

$$\frac{dv}{dt} = C\frac{dp_0}{dt} \quad ...(i)$$

Thus we have, $\quad v = \int q\,dt = \int \left(\frac{p_i - p_0}{R}\right) dt$

∴ From above, rate of storage in the vessel,

$$\frac{dv}{dt} = \frac{p_i - p_0}{R} \quad ...(ii)$$

Equating (*i*) and (*ii*), we have the following differential equation describing the dynamics of a pneumatic system,

$$RC\frac{dp_0}{dt} + p_0 = p_i \quad ...(4.67a)$$

Table 4.3 gives a summary of the quantities in thermal, liquid level and pneumatic systems which are analogous to those in electrical system.

**Table 4.3** *Analogous Quantities in Thermal, Liquid Level, Pneumatic and Electrical Systems*

| S.No. | Electrical Systems | Thermal Systems | Liquid-Level Systems | Pneumatic Systems |
|---|---|---|---|---|
| 1. | Charge ; C | Heat flow ; J | Liquid flow ; $m^3$ | Air flow ; $m^3$ |
| 2. | Current ; A | Heat flow rate ; J/s | Liquid flow rate ; $m^3/s$ | Air flow rate ; $m^3/s$ |
| 3. | Voltage ; V | Temperature ; °C | Heat ; m | Pressure ; $N/m^2$ |
| 4. | Resistance ; Ω | Resistance ; $\frac{°C}{J/s}$ | Resistance ; $\frac{m}{m^3/s}$ | Resistance ; $\frac{N/m^2}{m^3/s}$ |
| 5. | Capacitance ; F | Capacitance ; J/°C | Capacitance ; $m^2$ | Capacitance ; $\frac{m^3}{N/m^2}$ |

## 4.13 TRANSFER FUNCTION

Transfer function plays an important role in the characterization of linear time invariant systems. In fact transfer function along with block diagrams and signal flow graphs forms the basis of representing the input-output relationships of linear time invariant systems. This is because in so many other areas of engineering applications like circuit theory, vibrations theory, automatic control theory, aerospace theory and control, and stability studies, the most widely used mathematical model for analysis of system dynamic response is described with the help of linear differential equations involving constant co-efficients *i.e.*, model for linear time invariant systems.

The starting point of defining the transfer function is the differential equation which defines the dynamics of a system.

Consider a linear time-invariant system which is defined by an *n*th order differential equation :

$$a_n\frac{d^n c(t)}{dt^n} + a_{n-1}\frac{d^{n-1}c(t)}{dt^{n-1}} + ... + a_1\frac{dc(t)}{dt} + a_0 c(t)$$

$$= b_m\frac{d^m r(t)}{dt^m} + b_{m-1}\frac{d^{m-1}r(t)}{dt^{m-1}} + ... + b_1\frac{dr(t)}{dt} + b_0 r(t) \quad ...(4.68)$$

where $c(t)$ = output variable, and $r(t)$ = input variable, $a_0, a_1, ....., a_{n-1}, a_n$ and $b_0, b_1, ....., b_{m-1}, b_m$ are constants which represent combiantions of system parameters with $n \geq m$.

The differential Eqn. 4.68 represents a complete description of the system between the input *r*(*t*) and the output *c*(*t*).

Any easy and convenient way of describing linear systems is the use of transfer function. The **transfer function** of a linear time invariant system is defined as the ratio of Laplace transform of the output variable to the Laplace transform of the input variable with *all initial conditions equal to zero.* Thus, in order to obtain the transfer function of a linear system represented by Eqn. 4.68 we take the Laplace transform on both sides of the equation assuming all initial conditions to be zero. Hence,

$$[a_n s^n + a_{n-1}s^{n-1} + ... + a_1 s + a_0]C(s)$$
$$= [b_m s^m + b_{m-1}s^{m-1} + ... + b_1 s + b_0]\, R(s)$$

Therefore, the transfer function of the system is,

$$G(s) = \frac{C(s)}{R(s)} = \frac{[b_m s^m + b_{m-1}s^{m-1} + ... + b_1 s + b_0]}{[a_n s^n + a_{n-1}s^{n-1} + ... + a_1 s + a_0]}\,;\, m \leq n \quad ...(4.69)$$

We can summarize the properties of a transfer function as :

1. The transfer function can be defined for a linear time invariant system.
2. The transfer function between an input variable and an output variable of system is defined as the ratio of the Laplace transform of the output to the Laplace transform of the input with all initial conditions of the system assumed to be zero.
3. All transfer functions are independent of the input excitation.

The dynamics of a mass-spring-dashpot has been described earlier by Eqn. 4.44. The differential equation describing the dynamics of the system is :

$$f(t) = M\ddot{x}(t) + B\dot{x}(t) + Kx(t)$$

where $x(t)$, the displacement is the output variable, and $f(t)$, the force is the input variable.

Taking the Laplace transform of each term of the equation and assuming that all initial conditions are zero, we have,

$$F(s) = Ms^2X(s) + BsX(s) + KX(s)$$

$$\therefore \text{ Transfer function } G(s) = \frac{X(s)}{F(s)} = \frac{1}{Ms^2 + Bs + K}$$

The highest power of the complex variable $s$ in the denominator of the transfer function determines the *order* of a system. The mass-spring-dashpot system has a transfer function whose denominator has a complex variable $s$ with highest power 2. Thus this system is a 2nd order system.

## 4.14 BLOCK DIAGRAM REPRESENTATION

Transfer function is an expression in $s$ domain which relates the output to the input for a linear time invariant system. Although it describes the input-output behaviour of the system, it does not furnish any information regarding the internal structure of the system. In order to study the dynamics of a system, the process done on the input to get the output (for the purpose of analysis), is in the form of a mathematical expression which is a representation of a physical system. Thus when transfer function of a physical system is determined, the system can be represented by a **block diagram** which is a shorthand pictorial representation of cause and effect relationship existing between input and output of the system.

The transfer function as shown by a block represents a functional operation of a sytem which can be more easily visualized through a block diagram than with the help of differential equations.

Thus a system can be represented as :

$$C(s) = G(s)\, R(s) \qquad \text{...(4.70)}$$

This representation is pictorially represented by a block diagram as shown in Fig. 4.18.

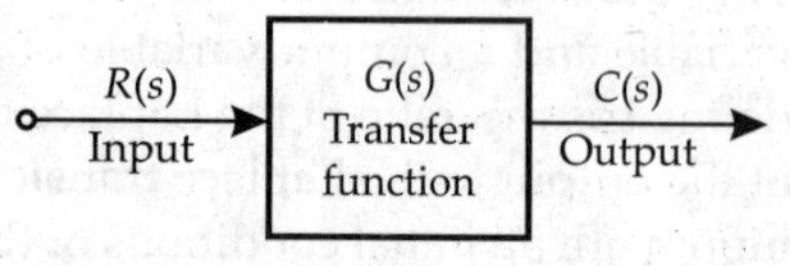


**Fig. 4.18** Block diagram representation of a system.

## 4.15 IMPULSE RESPONSE OF A LINEAR SYSTEM

The impulse response of a linear system is defined as the output response of a system when the input is a unit impulse function.

For a unit impulse, input $r(t) = \delta(t)$

Taking the Laplace transform, $R(s) = 1$.

Hence, from Eqn. 4.70, we have,

$$C(s) = G(s) \times 1 = G(s)$$

Taking the inverse Laplace transform on both the sides of the above equation, we get,

$$c(t) = g(t) \qquad \text{...(4.71)}$$

where $c(t)$ and $g(t)$ are respectively the inverse Laplace transforms of $C(s)$ and $G(s)$.

Now $g(t)$ is the impulse response of the system in time domain and thus we can state that the Laplace transform of impulse response is the transfer function $G(s)$. This function is also called the *weighing function.*

## 4.16 SINUSOIDAL TRANSFER FUNCTION

The sinusoidal transfer function is used for carrying out studies of a system in frequency domain. The study of quality of measurement systems involves the steady state response of the system when a sinusoidal input is applied to it.

The steady state response of a system to s sinusoidal input is obtained by replacing $s$ with $j\omega$ in the transfer function of the system where $\omega$ is the angular frequency in radian/s.

The general sinusoidal transfer is,

$$G(j\omega) = \frac{C(j\omega)}{R(j\omega)} = \frac{b_m(j\omega)^m + b_{m-1}(j\omega)^{m-1} + \ldots + b_1(j\omega) + b_0}{a_n(j\omega)^n + a_{n-1}(j\omega)^{n-1} + \ldots + a_1(j\omega) + a_0} \qquad \text{...(4.72)}$$

The transfer function of a mass spring-dashpot system is,

$$G(s) = \frac{X(s)}{F(s)} = \frac{1}{Ms^2 + Bs + K}$$

The sinusoidal transfer-function is written as,

$$G(j\omega) = \frac{X(j\omega)}{F(j\omega)} = \frac{1}{M(j\omega)^2 + B(j\omega) + K}$$

## 4.17 TIME DOMAIN RESPONSE

It has been stated earlier that the measurement systems may be subjected to any type of input. The type of input signals cannot be known fully ahead of time as in almost in all applications the signals are random in nature. Therefore, the time domain response of measurement systems is studied with the help of standard input signals like impulse, step, ramp and parabolic inputs.

When system studies are carried out in time domain, the dynamic behaviour of the system depends upon the system poles and not on type of input. Therefore the system behaviour to any kind of input can be predicted by studying its response to one of the standard signals. The standard input signal chosen for this purpose is a step input.

We will be studying the response of zero, first and second order systems when subjected to step inputs. The higher order systems, can, in most of the cases, be approximated as second order systems.

## 4.18 ZERO ORDER SYSTEMS

A zero order instrument is one where the highest order of the derivative describing the system behaviour is zero. Therefore from Eqn. 4.68, a zero order instrument is described by :

$$a_0 c(t) = b_0 r(t)^* \quad \text{...(4.73)}$$

Examining Eqn. 4.73, we find that it is a simple algebraic equation.

Any system that closely obeys Eqn. 4.73 over its entire operating range is defined as a **zero order system.**

We can rewrite Eqn. 4.73 as :

$$\text{Output } c(t) = (b_0 / a_0)\, r(t) \quad \text{...(4.74)}$$

$$= S r(t) \quad \text{...(4.75)}$$

where $S = b_0 / a_0$ ...(4.76)

= static sensitivity (steady state gain)

Now Eqn. 4.74 is a simple algebraic equation and therefore no matter how the input $r(t)$ varies with time, the output follows it perfectly and faithfully. This means that neither there is a distortion in magnitude nor any time lag. A linear potentiometer shown in Fig. 4.19 may be considered as a zero order instrument.

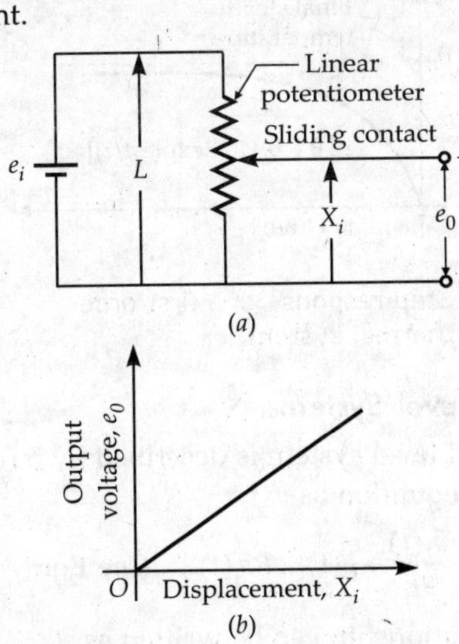


**Fig. 4.19** Linear potentiometer and its input-output characteristics.

Suppose

$e_i$ = supply voltage to potentiometer ; V,

$L$ = length of potentiometer ; m,

$x_i(t)$ = displacement of sliding contact at any time, $t$ ; m,

and $e_0(t)$ = output voltage at any time $t$ ; V.

Since the resistance of potentiometer is linear distributed along its length,

$$\text{Voltage per unit length} = \frac{e_i}{L} \text{ V/m}$$

$$\therefore \text{ Output voltage } e_0(t) = \frac{e_i}{L} x_i(t) = S x_i(t) \quad \text{...(4.77)}$$

$$\text{where static sensitivity} \quad S = \frac{e_i}{L} \quad \text{...(4.78)}$$

From Eqn. 4.78 it is clear that a linear relationship exists between the input displacement and the output voltage as shown in Fig. 4.19(*b*).

## 4.19 FIRST ORDER ELECTRICAL SYSTEMS

The highest order of derivative in first order system is one. First order systems are commonly found in measurement systems. Let us analyze a common *R-C* circuit which is a typical example of a first order system. Figure 4.20 shows a series *R-C* circuit wherein a step input voltage $E_i$ is applied to the circuit when the switch is closed.

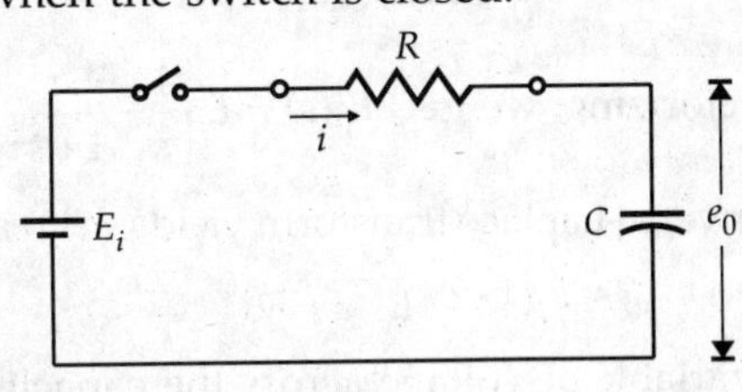


**Fig. 4.20** R-C circuit.

Let the current in the circuit by $i(t)$ at any instant $t = t$. All initial conditions are assumed to be zero.

$$E_i u(t) = R_i(t) + \frac{1}{C}\int i\,dt$$

Taking Laplace transform, we have,

$$\frac{E_i}{s} = RI(s) + \frac{1}{Cs} I(s) \quad \text{or} \quad E_i = \frac{1+RCs}{C} I(s)$$

$$\therefore \; I(s) = \frac{C E_i}{1+RCs} = \frac{E_i}{R}\left(\frac{RC}{1+RCs}\right) = \frac{E_i}{R}\left(\frac{1}{s+1/RC}\right)$$

Taking the inverse Laplace transform,

$$i(t) = \frac{E_i}{R}[\exp(-t/RC)] = \frac{E_i}{R}[\exp(-t/\tau)] \quad \text{...(4.79)}$$

where $\tau$ = time constant = $RC$ ...(4.80)

* It should be understood $c(t)$ = output, $r(t)$ = input and $a_0$ and $b_0$ are constants.

The plot of current $i$ versus time $t$ is shown in Fig. 4.21($a$). The current decays exponentially with time. The initial current is $E_i / R$ while the final current is zero which occurs only at $t = \infty$.

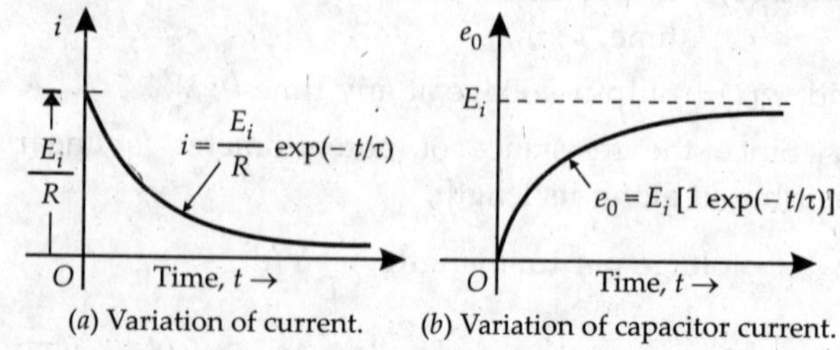


($a$) Variation of current. ($b$) Variation of capacitor current.

**Fig. 4.21** Step response of a series $R$-$C$ circuit.

Let us consider the voltage across the capacitor,

$$e_0(t) = \frac{1}{C}\int i\,dt$$

Differentiating the above equation, we have,

$$\frac{d[e_0(t)]}{dt} = \frac{1}{C}i(t) = \frac{E_i}{RC}[\exp(1-t/RC)]$$

Now take the Laplace transform of the above,

$$sE_0(s) = \frac{E_i}{(1+sRC)}$$

$$E_0(s) = \frac{E_i}{s(1+sRC)} = \frac{E_i}{s(1+s\tau)}$$

By factorizing, we get, $E_0(s) = E_i\left[\frac{1}{s} - \frac{\tau}{1+s\tau}\right]$

The Inverse Laplace transform yields the result as :

$$e_0 = E_i[1-\exp(-t/\tau)] \quad ...(4.81)$$

The variable of voltage across the capacitor with time is shown in Fig. 4.21($b$).

## 4.20 FIRST ORDER NON-ELECTRICAL SYSTEMS

### 1. Thermal System

A thermal system is described by a first order differential equation as :

$$RC\frac{d\theta(t)}{\delta\tau} + \theta(t) = h_i(t)R \quad \text{(see Eqn. 4.55)}$$

The above relationship can be written as,

$$\tau\frac{d\theta(t)}{dt} + \theta(t) = h_i(t)R \quad ...(4.82)$$

where $\tau = RC$ = thermal time constant.

Taking Laplace transform, we have,

$$\left.\begin{aligned} &\tau s\theta(s) + \theta(s) = H_i(s)R \\ \text{or} \quad &\theta(s)[(1+\tau(s)] = H_i(s)R \end{aligned}\right\} \quad ...(i)$$

Now input $r(t) = h_i(t)$ or $R(s) = H_i(s)$

and output $c(t) = \theta(t)$ or $C(s) = \theta(s)$

∴ From ($i$), transfer function,

$$G(s) = \frac{\theta(s)}{H_i(s)} = \frac{R}{1+\tau s} \quad ...(4.83)$$

For a step input,

$$r(t) = h_i u(t) \quad \text{or} \quad R(s) = \frac{1}{s}H_i(s)$$

∴ When a step input $h_i u(t)$ is applied, the Laplace transform of output is,

$$\theta(s) = G(s)H_i(s)$$

$$= \frac{sRH_i(s)}{1+\tau s} = \frac{RH_i(s)}{s}\left[\frac{1}{1+\tau s}\right]$$

$$= RH_i(s)\left[\frac{1}{s} - \frac{\tau}{1+\tau s}\right]$$

Taking inverse Laplace transform of the above relationship, we get the output in the domain,

$$\theta = Rh_i[1-\exp(-t/\tau)] \quad ...(4.84)$$

Now, when time $t$, tends to infinity, temperature tends to a value $Rh_i$ and this value of temperature is called the *final steady temperature* $\theta_0$,

where $\theta_0 = Rh_i$ ...(4.85)

∴ Equation 4.84, can be written as,

$$\theta = \theta_0[1-\exp(-t/\tau)] \quad ...(4.86)$$

Equation 4.86 indicates that temperature rises exponentially with time when heat rate is increased suddenly as shown in Fig. 4.22.

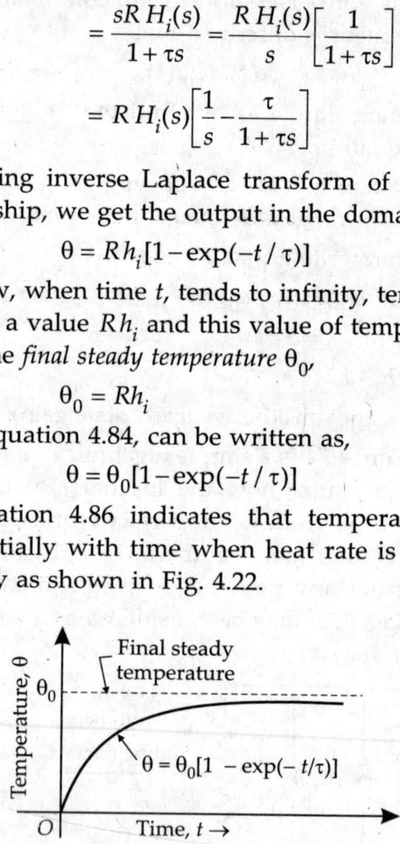


**Fig. 4.22** Step response of a first order thermal system.

### 2. Liquid Level Systems

A liquid level system is described by a first order differential equation as :

$$RC\frac{dh(t)}{dt} + h(t) = Rq_i(t) \quad \text{(see Eqn. 4.65)}$$

This relationship can be written as,

$$\tau\frac{dh(t)}{dt} + h(t) = h_i(t)R$$

where $\tau = RC$ time constant, $s$ ...(4.87)

Taking Laplace transform, we have,

$$\tau s\, H(s) + H(s) = Q_i(s)\, R$$

or $$H(s)[1+\tau s] = Q_i(s)\, R \quad ...(i)$$

Now input $r(t) = q_i(t)$ or $R(s) = Q_i(s)$

and output $c(t) = h(t)$ or $C(s) = H(s)$

∴ From (*i*), transfer function,

$$G(s) = \frac{H(s)}{Q_i(s)} = \frac{R}{1+\tau s} \quad ...(4.88)$$

For a step input,

$$r(t) = q_i\, u(t) \quad \text{or} \quad R(s) = \frac{1}{s} Q_i(s)$$

∴ When a step input $r(t) = q_i\, u(t)$ is applied, the Laplace transform of output is,

$$H(s) = G(s) Q_i(s)$$

$$= \frac{R Q_i(s)}{s(1+\tau s)} = \frac{R Q_i(s)}{s}\left[\frac{1}{1+\tau s}\right]$$

$$= R\, Q_i(s)\left[\frac{1}{s} - \frac{\tau}{1+\tau s}\right]$$

Taking inverse Laplace transform, we have the time domain response of a first order liquid level system as :

$$h = Rq_i[1 - \exp(-t/\tau)] \quad ...(4.89)$$

$$= H_0[1 - \exp(-t/\tau)] \quad ...(4.90)$$

where $H_0$ = final steady liquid head = $Rq_i$ ...(4.91)

Equation 4.90 indicates that the liquid head rises exponentially with time, when the liquid in flow rate is increased suddenly. This is shown in Fig. 4.23.

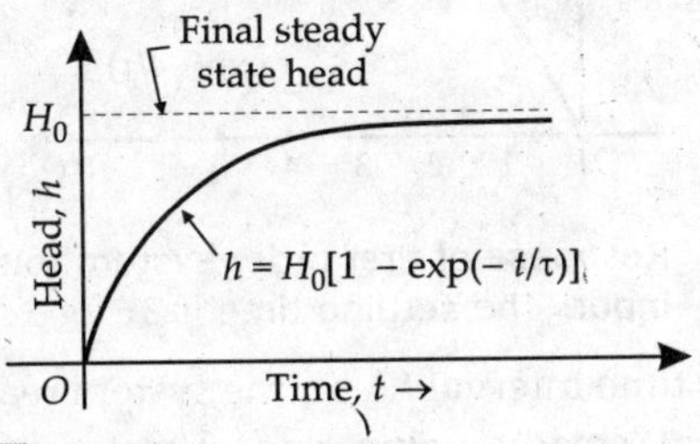


**Fig. 4.23** Step response of a first order liquid level system.

## 4.21 DIFFERENTIAL EQUATION FOR A GENERAL FIRST ORDER SYSTEM

If in Eqn. 4.68 all $a$'s and $b$'s other than $a_1, a_0$ and $b_0$ are taken as zero, we get,

$$a_1 \frac{dc(t)}{dt} + a_0\, c(t) = b_0\, r(t) \quad ...(4.92)$$

Any system that follows this equation is by definition a first order system.

Equation 4.92 has three constants *viz.* $a_0, a_1$ and $b_0$ but only two are needed. This process involves the division of the whole equation by any of these constants making one of the terms unity. Normally the equation is divided by $a_0$, giving :

$$\frac{a_1}{a_0}\frac{dc(t)}{dt} + c(t) = \frac{b_0}{a_0} r(t) \quad ...(4.93)$$

or $$\tau \frac{dc(t)}{dt} + c(t) = Sr(t) \quad ...(4.94)$$

where $S$ is the *static sentitivity* and $\tau$ is the *time constant.*

Their values are given by,

$$S = b_0 / a_0 \quad ...(4.95)$$

and $$\tau = a_1 / a_0 \quad ...(4.96)$$

The time constant $\tau$ has the dimensions of time. The static sensitivity has the dimensions of output / input. This is not only true of first order systems but of systems of any order. This is because under static conditions all derivatives are zero and the only co-efficients that remain under such conditions are $a_0$ and $b_0$. Therefore, the above definition of static sensitivity holds good for all systems.

Taking Laplace transform of Eqn. 4.94, we have,

$$\tau s\, C(s) + C(s) = SR(s)$$

$$C(s)[1+\tau s] = SR(s)$$

∴ Transfer function of a first order system is,

$$G(s) = \frac{C(s)}{R(s)} = \frac{S}{(1+\tau s)} \quad ...(4.97)$$

and the dimensionless transfer function is :

$$G(s) = \frac{1}{(1+\tau s)} \quad ...(4.98)$$

## 4.22 RESPONSE OF A FIRST ORDER SYSTEM TO A UNIT STEP INPUT

Let a unit step input $u(t)$ be applied to a first order system. Now

$$r(t) = u(t) \quad \text{or} \quad R(s) = 1/s$$

The transfer function (dimensionless) of a first order system is :

$$G(s) = \frac{1}{1+\tau s}$$

$$\therefore \text{Output } C(s) = G(s)\, R(s) = \frac{1}{s(1+\tau s)} = \frac{1}{s} - \frac{\tau}{1+\tau s}$$

Taking Inverse Laplace transform, the output is,

$$c(t) = 1 - \exp(-t/\tau) \quad ...(4.99)$$

The response is plotted in Fig. 4.24.

It is seen that the output rises exponentially from zero value to the final value of unity. The initial slope of the curve is given by :

$$\left.\frac{dc}{dt}\right|_{t=0} = \frac{1}{\tau}\exp(-t/\tau)\Big|_{t=0} = \frac{1}{\tau}$$

4

Thus, if the initial rate of change is maintained, the system will reach its final value in time $\tau$, the time constant.

The output at $t = \tau$ is,

$$c(t) = 1 - \exp(-\tau/\tau) = 0.632$$

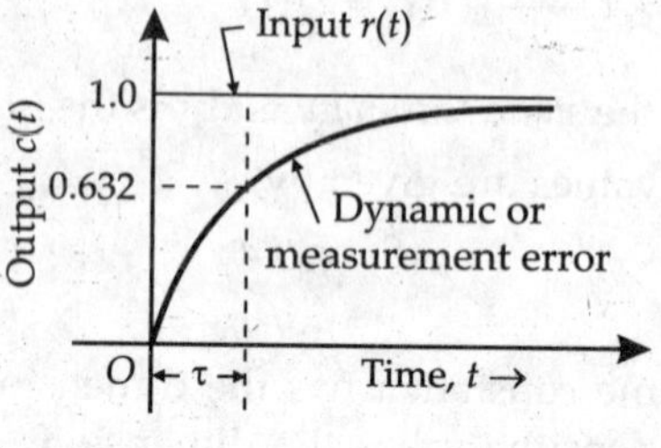


**Fig. 4.24** Response of a first order system to a step input.

Thus for a rising exponential function the *time constant $\tau$ is defined as the time to reach 63.2% of its final value.* It is obvious that all first order systems will settle to their final value at $t = \infty$.

For a decaying function the *time constant is defined as the time taken to fall to 0.368 of its initial value.*

The Dynamic or Measurement error is defined as,

$$e_m(t) = r(t) - c(t) \quad \text{...(4.100)}$$

$$= 1 - [1 - \exp(-t/\tau)]$$

$$= \exp(-t/\tau) \quad \text{...(4.101)}$$

The variation of dynamic error with time is plotted in Fig. 4.25.

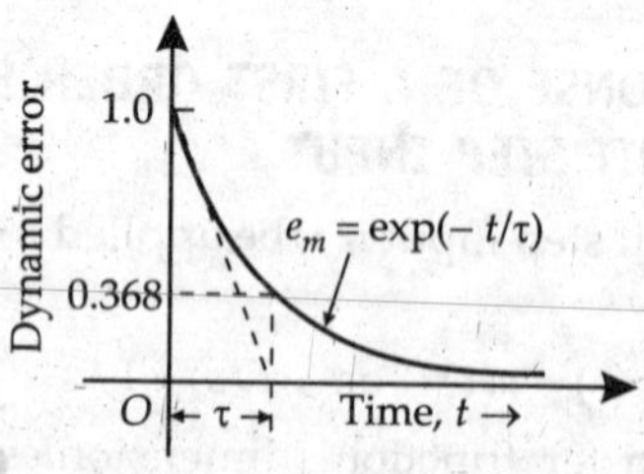


**Fig. 4.25** Dynamic (Measurement) error for a first order system when subjected to a step input.

The steady state error is,

$$e_{ss} = \lim_{t\to\infty} e_m(t) = \lim_{t\to\infty} \exp(-t/\tau) = 0 \quad \text{...(4.102)}$$

Thus the first order system tracks the unit step input with zero static error.

A dynamic characteristic useful in characterising the speed of response of any system is the *settling time* and is defined as the time (after the application of a step input) for the system to reach and stay within a plus and minus tolerance band around its steady state value. A small settling time is indicative of fast response. It is obvious that the numerical value of the settling time depends on the tolerance band used and this band must always be specified.

For a unit step input, the output reaches the values given in Table 4.4 at various intervals of time. This table also shows the per unit dynamic error at different intervals of time.

**Table 4.4** *Per Unit Output and Dynamic Error for a Step Input for a First Order System*

| $t/\tau$ | Per Unit Output | Per Unit Dynamic Error |
|---|---|---|
| 0 | 0 | 1.000 |
| 1 | 0.632 | 0.368 |
| 2 | 0.865 | 0.135 |
| 3 | 0.950 | 0.050 |
| 4 | 0.982 | 0.018 |
| 5 | 0.993 | 0.007 |
| ∞ | 1.000 | 0 |

Thus if one speaks of 5% settling time, it means that the system has reached its specified value after a time which is thrice the time constant (see Table 4.4). The system reaches 0.95 of its final steady value at $t = 3\tau$. This is shown in Fig. 4.26.

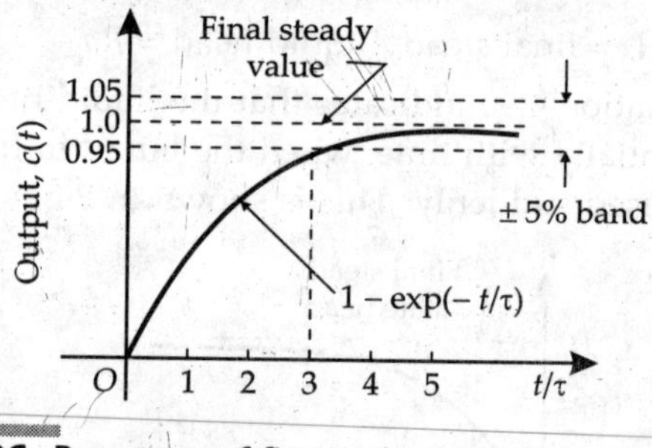


**Fig. 4.26** Response of first order system to step input. The settling time is $3\tau$ for ± 5% band.

After time interval $t = 5\tau$, the system reaches 0.993 of its final steady value and therefore it is often assumed that the instrument or system has reached its **final steady or maximum value** after an interval of $5\tau$.

The time constant is indicative of how fast the system tends to reach the final value. A large time constant corresponds to a slow system response and a small time constant represents a fast system response as shown in Fig. 4.27. It shows two curves with time constants $\tau_1$ and $\tau_2$ where $\tau_1 < \tau_2$. System with time constant $\tau_1$ is faster than the system with time constant $\tau_2$.

In some cases the initial conditions are not zero. Let us consider the case of thermometer subjected to a step input.

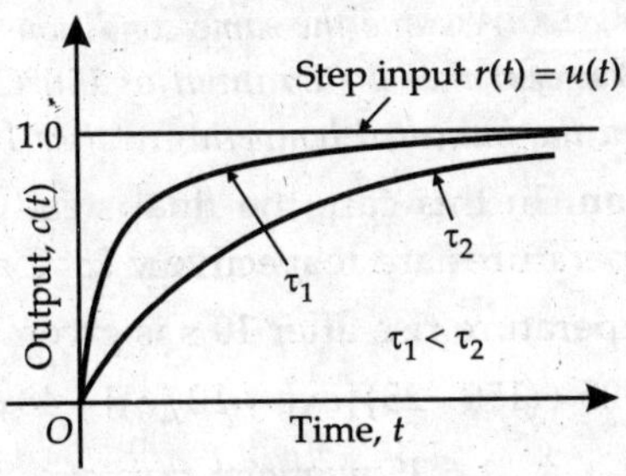


**Fig. 4.27** Response of first order systems of different time constants to a step input.

Let $\theta_i$ = initial temperature ; °C,

$\theta_0$ = final steady temperature ; °C,

and $\tau$ = time constant ; s.

The temperature at any instant is $t$ after the application of a step input is given by :

$$\theta = \theta_0[1-\exp(-t/\tau)]+\theta_i \exp(-t/\tau) \quad ...(4.103)$$

$$= \theta_0 + (\theta_i - \theta_0)[\exp(-t/\tau)] \quad ...(4.104)$$

## 4.23 MODELLING OF A THERMOMETER

A liquid in glass thermometer presents one of the finest examples of a first order system. The input quantity is the temperature $\theta_i(t)$ of the fluid surrounding the bulb of the thermometer, and the output is the displacement $x_0(t)$ of the thermometer fluid in the capillary tube (*see* Fig. 4.28). The

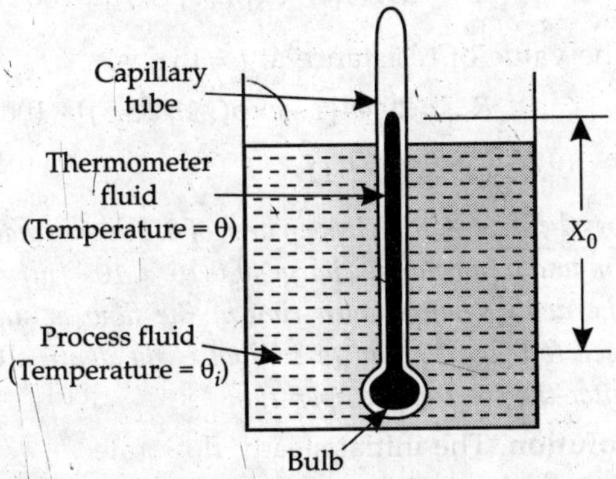


**Fig. 4.28** Liquid filled Thermometer.

assumption made is that the temperature $\theta_i(t)$ is uniform throughout the fluid at any given time but may vary in any way. The working of the meter is based upon the thermal expansion of the filling fluid which makes the liquid column rise or fall in response to the temperature variations.

Let $V_b$ = volume of bulb ; $m^3$,

$A_c$ = area of capillary tube ; $m^2$,

$\alpha$ = co-efficient of differential expansion of thermometer bulb and fluid ; $m^3/m^3$-°C

$H$ = co-efficient of heat transfer across the bulb ; $J/m^2\text{-}s^{-1}$,

$A_b$ = area of heat transfer in the bulb ; $m^2$,

$\rho$ = density of thermometer fluid ; $kg/m^3$,

$s$ = specific heat of the temperature fluid ; J/kg–°C

$\theta_i$ = process fluid temperature ; °C,

$\theta$ = temperature rise of fluid at any time $t = t$ ; °C,

$x_0$ = rise of the thermometer fluid in the capillary tube ; m.

In order to get a differential equation relating input and output in this thermometer, we consider conservation of energy over an infinitesimal interval of time $dt$ for the bulb. During this period, the temperature of thermometer fluid rises by $d\theta$.

Heat stored = heat input – heat output

Heat input = $HA_b(\theta_i - \theta)dt$

Heat output = 0 ;

assuming there is heat dissipation to outside.

Heat stored = $V_b \rho\, s\, d\theta$

$\therefore$ We have : $V_b \rho\, s\, d\theta = HA_b(\theta_i - \theta)dt$

or $$V_b \rho\, s \frac{d\theta}{dt} + HA_b\,\theta = HA_b\,\theta_i \quad ...(i)$$

But volumetric expansion

= area × linear expansion.

or $$\alpha V_b\, d\theta = A_c\, dx_0$$

or $$V_b\, d\theta = \frac{A_c}{\alpha} dx_0 \text{ and } V_b\,\theta = \frac{A_c}{\alpha} x_0$$

Substituting the values of $d\theta$ and $\theta$ in (*i*), we have,

$$\frac{\rho s A_c}{\alpha}\frac{dx_0}{dt} + \frac{HA_b A_c}{\alpha V_b} x_0 = HA_b\,\theta_i$$

Simplifying, $$\frac{\rho s V_b}{HA_b}\frac{dx_0}{dt} + x_0 = \alpha \frac{V_b}{A_c}\theta_i \quad ...(4.105)$$

Comparing Eqn. 4.105 with Eqn. 4.93, we get,

$$\frac{a_1}{a_0} = \frac{\rho s V_b}{HA_b} \text{ and } \frac{b_0}{a_0} = \alpha \frac{V_b}{A_c}$$

From Eqns. 4.95 and 4.96, the thermometer static sensitivity is,

$$S = \frac{b_0}{a_0} = \alpha \frac{V_b}{A_c} \text{ m/°C} \quad ...(4.106)$$

and time constant $$\tau = \frac{a_1}{a_0} = \frac{\rho s V_b}{HA_b} \text{ s} \quad ...(4.107)$$

Equation 4.105 can be written as,

$$\tau \frac{dx_0}{dt} + x = S\theta_i \quad ...(4.108)$$

Taking Laplace transform of Eqn. 4.108,

$$(1+s\tau)X_0(s)=S\theta_i(s)$$

∴ Transfer function $G(s)=\dfrac{X_0(s)}{\theta_i(s)}=\dfrac{S}{1+s\tau}$ ...(4.109)

The non-dimensional transfer function is,

$$G(s)=\frac{X_0(s)}{S\theta_i(s)}=\frac{1}{(1+s\tau)} \quad ...(4.110)$$

*Example 4.1* *A thermometer has a time constant of 3.5 s. It is quickly taken from a temperature 0°C to a water bath having a temperature 100°C. What temperature will be indicated after 1.5 s ?*

**Solution.** From above it is clear that the thermometer is subjected to a step input of 100°C.

From Eqn. 4.86 the output temperature after 1.5 s is given as :

$$\theta=\theta_0[1-\exp(1-t/\tau)]$$
$$=100[1-\exp(-1.5/3.5)]=34.86\,°C$$

*Example 4.2* *A temperature-sensitive transducer is subjected to a sudden temperature change. It takes 10 s for the transducer to reach equilibrium condition (five time constants). How long will it take for the transducer to read half of the temperature difference ?*

**Solution.** Time to reach equilibrium conditions

$$=5\tau=10\,\text{s}.$$

∴ Time constant $\tau=10/5=2$ s.

Now we have,

$$\theta=\theta_0[1-\exp(-t/\tau)] \text{ or } \theta/\theta_0=1-[\exp(-t/\tau)]$$

The time to read half of the temperature difference can be calculated as under :

$$0.5=1-[\exp(-t/2)] \quad \text{or} \quad t=1.39\text{ s}.$$

∴ Time to read half of the temperature difference is 1.39 s.

*Example 4.3* *A temperature sensing device can be modelled as a first order system with a time constant of 6 s. It is suddenly subjected to a step input of 25°C – 150°C. What temperature will be indicated in 10 s after the process has started ?*

**Solution.** It is given that :

Final steady state temperature $\theta_0=150°C$,

Initial temperature $\theta_i=25°C$,

and time constant $\tau=6$ s

Therefore, the temperature after 10 s can be calculated from Eqn. 4.104.

$$\theta=\theta_0+(\theta_i-\theta_0)[\exp(-t/\tau)]$$
$$=150+(25-150)[\exp(-10/6)]=126.4\,°C$$

*Example 4.4* *Assume the same conditions as given in example 4.3 except that a step input of 150°C — 25°C is applied. Find the indicated temperature after 10 s.*

**Solution.** In this case the final steady state and initial temperatures are respectively 25°C and 150°C.

∴ Temperature rise after 10 s is given by,

$$\theta=25+(150-25)[\exp(-10/6)]=48.6\,°C$$

*Example 4.5* *A 6.25 mm long resistance temperature detector (RTD) with a steady state gain of 0.3925 Ω/°C and a time constant of 5.5 s experiences a step change of 75°C in temperature. Before the temperature change, it has a stable 100 Ω resistance. Write the time-domain equation for resistance and find its value after 15 s of application of step input.*

**Solution.** The gain of RTD is 0.3925 Ω/ °C and a step input 75°C is applied to it. This is equivalent to application of $0.3925\times75=29.44\,\Omega$ step input in terms of resistance.

∴ Change in value of resistance with time

$$=29.44[1-\exp(-t/5.5)]\,\Omega$$

Hence, in order to obtain the time domain equation for resistance, the value of initial resistance must be added to it.

Therefore, the equation for resistance at any time $t$ after the application of step input is :

$$R_t=29.44[1-\exp(-t/5.5)]+100\,\Omega$$

The value of resistance at $t=15$ s is,

$$R_{15}=29.44[1-\exp(-15/5.5)]+100\,\Omega$$
$$=127.5\,\Omega.$$

*Example 4.6* *A tank has an area of 0.1 m² normal depth of 1.2 m and a normal discharge of 0.16 × 10⁻³ m³/s. How does the depth change with time if the flow is suddenly increased to 0.2 × 10⁻³ m³/s ? What is the depth after one hour after the start of the operation ?*

**Solution.** The initial steady flow rate

$$Q_{in}=0.16\times10^{-3}\,\text{m}^3/\text{s}$$

The initial head (depth)

$$H_{in}=1.2\text{ m}$$

Therefore, from Eqn. 4.59, we have,

$$K_1=\frac{Q}{\sqrt{H}}=\frac{0.16\times10^{-3}}{\sqrt{1.2}}=0.146\times10^{-3}$$

Using the same equation, the final steady head (depth)

$$H_0=\frac{Q_0}{K_1^2}=\left(\frac{0.2\times10^{-3}}{0.146\times10^{-3}}\right)^2=1.88\text{ m}$$

Linearizing the equation for flow, the hydraulic resistance to flow about the operating point is,

$$R = \frac{dH}{dQ} = \frac{1.88 - 1.2}{(0.20 - 0.16) \times 10^{-3}} = 17 \times 10^3 \text{ m}^{-2}\text{s}$$

Hydraulic capacitance $C$ = area of tank = 0.1 m$^2$.

$\therefore$ Time constant $\tau = RC = 17 \times 10^3 \times 0.1 = 1700$ s.

Writing a relationship similar to Eqn. 4.104 for thermal systems where the initial conditions are not zero, we get the head (depth) at any instant $t$ from the start of the process.

$$H = H_0 + (H_i - H_0)[\exp(-t/\tau)]$$
$$= 1.88 + (1.2 - 1.88)\ [\exp(-t/\tau)]$$
$$= 1.88 - 0.68\ [\exp(-t/1700)]$$

From the above equation we can compute the head (depth) of liquid after one hour of the start of process.

$$H = 1.88 - 0.68\ [\exp(-3600/1700)] = 1.8 \text{ m.}$$

**Example 4.7** *Figure 4.29 shows a liquid level system wherein the two tanks interact. $\overline{Q}$ is the steady state liquid in flow rate whereas $\overline{H}_1$ and $\overline{H}_2$ are the steady state heads of liquids in tank 1 and tank 2 respectively. Suppose the input rate to the tank 1 is suddenly increased by an amount $q$ which results in increased liquid flow rate $q_2$ from tank 2. Find the transfer function of the system. The capacitances of tanks 1 and 2 are $C_1$ and $C_2$ respectively. The resistances to the liquid flow are indicated in the diagram. Give an electrical analog of the system.*

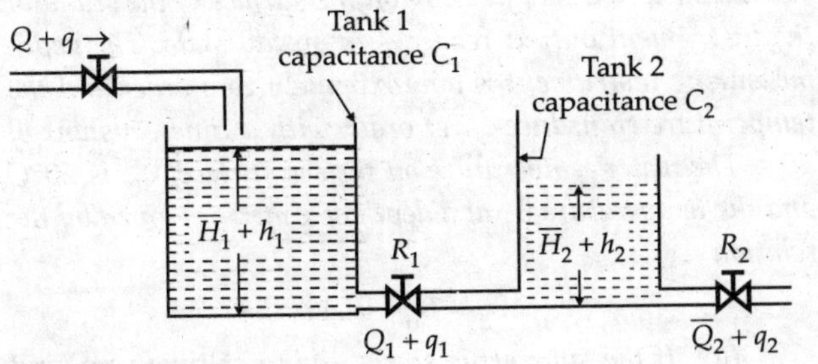


**Fig. 4.29** Two interacting tanks.

**Solution.** Let $h_1$ and $h_2$ be the increase in heads of tanks 1 and 2 respectively. Also $q_1$ and $q_2$ be the increase in liquid flow rates from tanks 1 and 2 respectively.

Therefore, we can write for the two tanks,

$$\frac{h_1 - h_2}{R_1} = q_1, \quad C_1 \frac{dh_1}{dh} = q - q_1$$

and
$$\frac{h_2}{R_2} = q_2, \quad C_2 \frac{dh_2}{dt} = q_1 - q_2$$

Solving the above differential equations, and taking Laplace transforms we get the transfer functions of the system as :

$$G(s) = \frac{Q_2(s)}{Q(s)} = \frac{1}{R_1C_1R_2C_2s^2 + (R_1C_1 + R_2C_2 + R_2C_1)s + 1}$$

The electrical analog of this system is shown in Fig. 4.30. It is cascaded $R$-$C$ network.

$e_i$ = input voltage and $e_o$ = output voltage

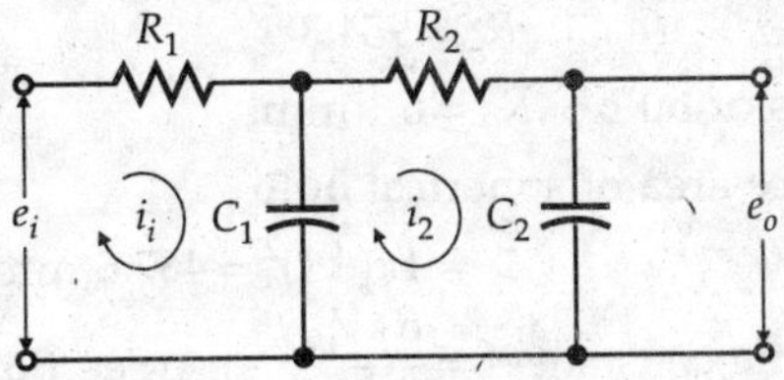


**Fig. 4.30** Electrical analog of the two interacting tanks shown in Fig. 4.29

The equations for this system are :

$$\frac{1}{C_1}\int (i_1 - i_2)dt + R_1 i_1 = e_i$$

and
$$\frac{1}{C_1}\int (i_2 - i_1)dt + R_2 i_2 = -\frac{1}{C_2}\int i_2\, dt = -e_o$$

Taking the Laplace transfoms of above equation, we obtain,

$$\frac{1}{C_1 s}[I_1(s) - I_2(s)] + R_1 I_1(s) = E_i(s)$$

$$\frac{1}{C_1 s}[I_2(s) - I_1(s)] + R_2 I_2(s) = \frac{1}{C_2 s} I_2(s) = -E_o(s)$$

Eliminating $I_1(s)$ and $I_2(s)$ from the above equations we get the transfer function between output voltage and input voltage as :

$$\frac{E_o(s)}{E_i(s)} = \frac{1}{(R_1C_1s + 1)(R_2C_2s + 1) + R_2C_1s}$$
$$= \frac{1}{R_1C_1R_2C_2s^2 + (R_1C_1 + R_2C_2 + R_2C_1)s + 1}$$

This is an equation of an electrical system which is analogous to the system shown in Fig. 4.29.

**Example 4.8** *A mercury thermometer has a capillary tube of 0.25 mm in diameter. If the bulb is made of a zero-expansion material, what volume should it have if a sensitivity 3.5 mm/°C is required ? If the bulb is spherical and is immersed in air, calculate the time constant of the thermometer. The density of mercury is $13.56 \times 10^3$ kg/m$^3$, the specific heat is 139 J/kg–°C and its volumetric thermal co-efficient of expansion is $0.18 \times 10^{-3}$ m$^3$/m$^3$-°C. The co-efficient of heat transfer is 12 J/m$^2$-°C-s.*

**Solution.** From Eqn. 4.106, sensitivity

$$S = \alpha V_b / A_c$$

$\therefore$ Volume of bulb $V_b = S\frac{A_c}{\alpha} = \frac{3.5 \times (\pi/4)(0.25)^2}{0.18 \times 10^{-3}}$

$= 954.5 \text{ mm}^3.$

The bulb is spherical and therefore

$$V_b = (4/3)\pi R_b^3$$

where $R_b$ = radius of bulb

Thus, $(4/3)\pi R_b^3 = 954.3$

or radius of bulb $R_b = 6.1$ mm.

Surface area of spherical bulb

$$A_b = 4\pi(6.1)^2 = 467.6 \text{ mm}^2.$$

$\therefore$ Time constant $\tau = \frac{\rho s V_b}{H A_b}$ (See Eqn. 4.107)

$$= \frac{13.56 \times 10^3 \times 139 \times 954.3 \times 10^{-9}}{12 \times 467.6 \times 10^{-6}} = 320.56 \text{ s}$$

## 4.24 RAMP RESPONSE OF A FIRST ORDER SYSTEM

The applied input is a unit ramp

$$r(t) = t \quad \text{or} \quad R(s) = 1/s^2$$

$\therefore$ Output $C(s) = G(s)R(s) = G(s) = \frac{1}{s^2(1+\tau s)}$

$$= \frac{1}{s^2} - \frac{\tau}{s} + \frac{\tau^2}{1+s}$$

Taking the Inverse Laplace transform, we get,

$$C(t) = t - \tau[1 - \exp(-t/\tau)] \quad ...(4.111)$$

The dynamic response of a first order system to a ramp input is shown in Fig. 4.31.

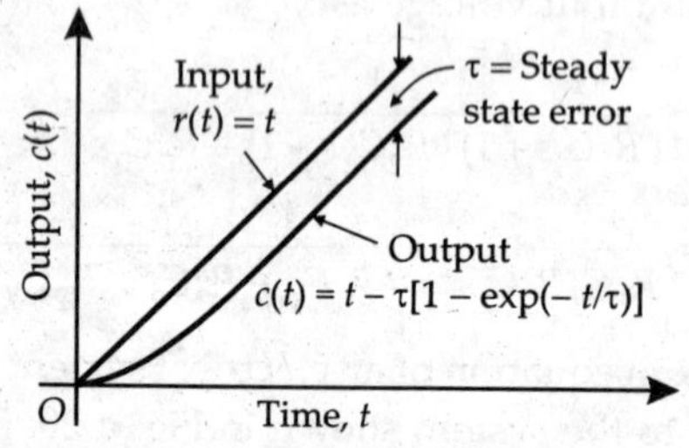


**Fig. 4.31** Ramp response of a first order system.

The dynamic error is

$$e_m(t) = r(t) - c(t)$$

$$= \tau[1 - \exp(-t/\tau)] \quad ...(4.112)$$

The final steady state error is,

$$e_{ss} = \lim_{t\to\infty} e_m(t) = \tau \quad ...(4.113)$$

Thus the first order system will track the unit ramp input with a steady state error $\tau$ which is equal to the time constant of the system.

If the system is subjected to a ramp input $r(t) = At$, then the output is,

$$c(t) = A[t - \tau(1 - \exp(-t/\tau)] \quad ...(4.114)$$

the dynamic error,

$$e_m(t) = r(t) - c(t) = A\tau[\exp(-t/\tau)] \quad ...(4.115)$$

and final steady state error $e_{ss} = A\tau$ ...(4.116)

***Example 4.9*** *The temperature of a furnace is increasing at a rate 0.1°C/s. What is the maximum permissible time constant of a first order instrument that can be used, so the temperature is read with a maximum error of 5°C ?*

**Solution.** A ramp signal of 0.1 °C/s is applied to the instrument and thus $A = 0.1$.

Maximum steady state error for a ramp signal applied to a first order instrument is given by Eqn. 4.116.

$$e_{ss} = A\tau$$

$\therefore$ Maximum allowable time constant

$$\tau = \frac{e_{ss}}{A} = \frac{5}{0.1} = 50 \text{ s}$$

***Example 4.10*** *A unmanned submarine is equipped with temperature and depth measuring instruments and has radio equipment which can transmit the output readings of these instruments back to the surface. The submarine is initially floating on the surface of the sea with the instrument output readings in steady state. The depth measuring instrument is approximately zero order and the temperature transducer first order with a time constant of 50 s. The water temperature on the sea surface, $\theta_0$, is 20°C and the temperature $\theta_x$ at a depth of x metre is given by the relation*

$$\theta_x = \theta_0 - 0.01x$$

(*a*) *If the submarine starts diving at time zero, and thereafter goes down at a velocity of 0.5 m/s, draw a table showing the temperature and depth measurements reported at intervals of 100 s over the first 500 s. Show also in the table the error in each temperature reading.*

(*b*) *What temperature does the submarine report at a depth of 1000 m ?*

**Solution.** (*a*) Let the temperature reported by the submarine at a depth of 1000 m.

Decrease in temperature

$$\Delta\theta_x = 0.01x = 0.01 \times 0.5\, t = 0.005\, t$$

This is a velocity signal.

$$\therefore \quad \Delta\theta_x(s) = \frac{0.005}{s^2}$$

Hence, $$\Delta\theta_r(s) = G(s)\cdot\Delta\theta_x(s)$$

$$= \frac{1}{1+\tau s}\times\frac{0.005}{s^2} = \frac{0.005}{s^2(1+50s)}$$

$$= 0.005\left[\frac{1}{s^2} - \frac{50}{s} + \frac{2500}{1+s}\right]$$

∴ Decrease in reported temperature,

$$\Delta\theta_r(t) = 0.005[t - 50 + 50\exp(-t/50)]$$

$$= 0.005[t - 50 + 0.25\exp(-t/50)]$$

Hence, temperature reported at time $t$ is :

$$\theta_r = \theta_0 - \Delta\theta_r$$

$$= 20 - 0.005(t-50) - 0.25\exp(-t/50)$$

Actual temperature

$$\theta_x = 20 - 0.01x = 20 - 0.005t$$

Depth $\quad x = 0.5\ t.$

The following results can be tabulated with the help of above relationships.

| *Time, t* | *Depth, x* | *Actual Temperature, $\theta_x$* | *Reported Temperature, $\theta_r$* | *Error* |
|---|---|---|---|---|
| *s* | *m* | *°C* | *°C* | *°C* |
| 0 | 0 | 20.0 | 20.0 | 0 |
| 100 | 50 | 19.5 | 19.715 | 0.216 |
| 200 | 100 | 19.0 | 19.245 | 0.245 |
| 300 | 150 | 18.5 | 18.749 | 0.249 |
| 400 | 200 | 18.0 | 18.25 | 0.25 |
| 500 | 250 | 17.5 | 17.75 | 0.25 |

(*b*) Actual temperature at 1000 m depth

$$= 20 - 0.01\times 1000 = 10°\text{C}$$

Time required to reach a depth of 1000 m is 2000 s.

Therefore, reported temperature at a depth of 1000 m

$$= 20 - 0.005(2000-50) - 0.25\exp(-2000/50)$$

$$= 10.25°\text{C}$$

## 4.25 IMPULSE RESPONSE OF A FIRST ORDER SYSTEM

Let the system be subjected to an impulse of strength $A$ as shown in Fig. 4.32.

∴ Input $r(t) = A\delta(t)$ and $R(s) = A$,

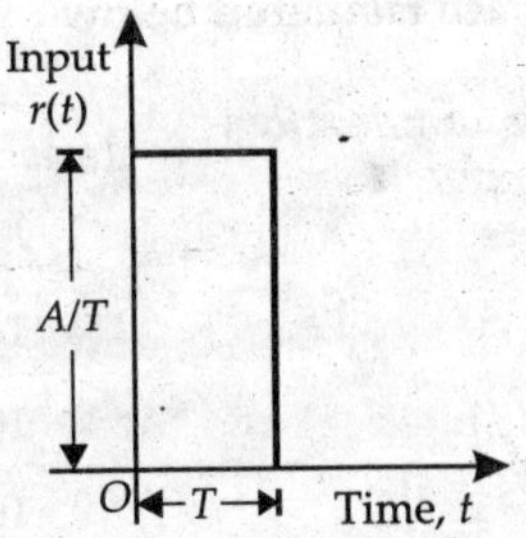


**Fig. 4.32** Impulse function of strength A.

$$\therefore \quad C(s) = G(s)\,R(s) = \frac{A}{1+\tau s}$$

Taking inverse Laplace transform, we have,

$$\text{Output} \quad c(t) = \frac{A}{\tau}[\exp(-t/\tau)] \qquad ...(4.117)$$

The response is shown in Fig. 4.33.

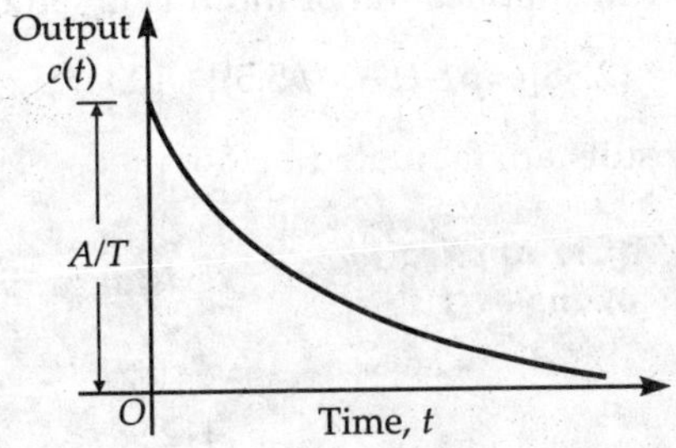


**Fig. 4.33** Impulse response a first order system.

***Example 4.11*** *A 6.25 mm long resistor temperature detector (RTD) with a steady state gain of 0.3925 Ω/°C and time constant of 5.5 s monitors the temperature of liquid flowing in pipe. A cell of liquid which is 75°C hotter than the surrounding liquid passes the RTD. It takes 3 s for the cell of superheated liquid to pass. Plot the resistance of RTD as a function of time. The initial resistance of RTD is 100 Ω.*

**Solution.** The gain of RTD is 0.3925 Ω/°C and an input of 75° C is applied to it in the form of a pulse lasting 3 s. This is equivalent to application of pulse of 0.3925 × 75 = 29.44 Ω magnitude lasting for 3 s. Therefore, it can be assumed that a step input for 3 s and is terminated thereafter. This is because the application of an impulse signal is unrealistic. (The impulse signal is defined as signal with no width and an infinite magnitude.)

Therefore, for the first 3 s of application of input the time response is given by

$$R_t = 29.44[1 - \exp(-t/5.5)] + 100 \text{ for } 0 < t \le 3\text{ s}$$

where $R_t$ is the resistance of RTD at a time after the application of input.

The results are tabulated below :

| *Time after application of input, t* s | *Resistance, $R_t$* Ω |
|---|---|
| 0 | 100 |
| 1 | 104.89 |
| 2 | 108.97 |
| 3 | 112.38 |

At $t = 3$, the pulse disappears and therefore the response decays exponentially.

The increase in the value of resistance at $t = 3$ s is $112.38 - 100 = 12.38\ \Omega$.

In order to find the response after 3 s of application of input, the origin for time axis has to be shifted by 3 s. Hence, the value of resistance at any time $t$ after the application of input is given by

$$R_t = 12.38[\exp\{-(t-3)/5.5\}] + 100 \quad t \geq 3\text{s}$$

The results are tabulated below :

| *Time after application of input, t* s | *Resistance, $R_t$* Ω |
|---|---|
| 3 | 112.38 |
| 5 | 108.60 |
| 10 | 103.47 |
| 15 | 101.40 |
| 20 | 100.56 |
| 25 | 100.23 |
| 30 | 100.09 |

The complete response is shown in Fig. 4.34.

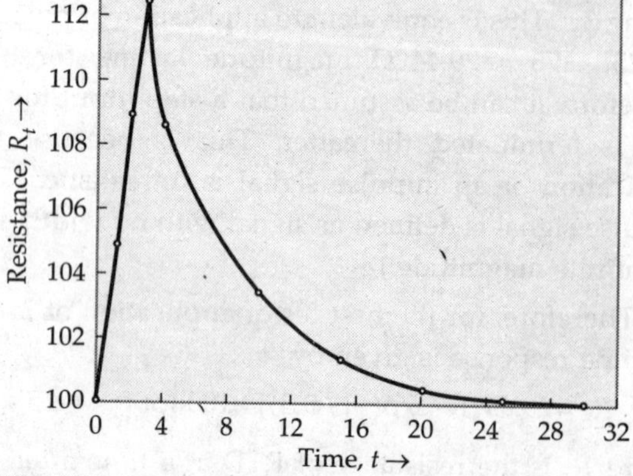


**Fig. 4.34** Response of RTD (Example 4.11).

## 4.26 RESPONSE OF SECOND ORDER SYSTEMS

A second order system may be defined by the following differential equation

$$a_2 \frac{d^2c(t)}{dt^2} + a_1 \frac{dc(t)}{dt} + a_0 c(t) = b_0 r(t) \quad ...(4.118)$$

A typical second order system is the mass damper system *i.e.*, the mechanical translational system shown in Fig. 4.12 on page 89. For this system, we have,

$$M\ddot{x}(t) + B\dot{x}(t) + Kx(t) = f(t) \quad ...(4.119)$$

where displacement $x(t)$ is the output and the force applied $f(t)$ is the input.

Taking Laplace transform, we get,

$$Ms^2X(s) + BsX(s) + KX(s) = F(s)$$

$$\therefore \text{ Transfer function } \frac{X(s)}{F(s)} = \frac{1}{Ms^2 + Bs + K} \quad ...(4.120)$$

The normalized (dimensionless) transfer function is,

$$G(s) = K\frac{X(s)}{F(s)} = \frac{K}{Ms^2 + Bs + K}$$

$$= \frac{K/M}{s^2 + (B/M)s + (K/M)}$$

$$= \frac{K/M}{s^2 + 2\left(\frac{B}{2\sqrt{KM}}\sqrt{\frac{K}{M}}\right)s + \frac{K}{M}}$$

$$= \frac{\omega_n^2}{s^2 + 2\xi\omega_n s + \omega_n^2} \quad ...(4.121)$$

where $\omega_n$ = natural frequency $= \sqrt{K/M}$ rad/s ...(4.122)

$\xi$ = damping ratio $= B/2\sqrt{KM}$ ...(4.123)

The transfer function of all second order systems is given by Eqn. 4.121.

Comparing Eqns. 4.118 and 4.119 and using Eqns. 4.122 and 4.123, we have,

Static sensitivity $S = 1/K = b_0/a_0$ ...(4.124)

Natural frequency

$$\omega_n = \sqrt{K/M} = \sqrt{a_0/a_2} \text{ rad/s} \quad ...(4.125)$$

and damping ratio

$$\xi = \frac{B}{2\sqrt{KM}} = \frac{a_1}{2\sqrt{a_0 a_2}} \quad ...(4.126)$$

The expressions for a mechanical rotational system are similar to that of a mechanical translational system with moment of inertia $J$ replacing mass $M$, the two being analogous. The expressions for a mechanical rotational system are,

Natural frequency

$$\omega_n = \sqrt{K/J} \text{ rad/s} \quad ...(4.127)$$

and damping ratio

$$\xi = B/2\sqrt{KJ} \quad ...(4.128)$$

The dimensionless transfer function of a second order system is

$$G(s) = \frac{\omega_n^2}{s^2 + 2\xi\omega_n s + \omega_n^2}$$

The *characteristic equation* of the system is,

$$s^2 + 2\xi\omega_n s + \omega_n^2 = 0 \quad ...(4.129)$$

The time domain response of a system depends upon the roots of the characteristic equation. The two roots of the characteristic equation are,

$$s_1, s_2 = -\xi\omega_n \pm \omega_n\sqrt{\xi^2 - 1} \quad ...(4.130)$$

The roots are of three different types depending upon the value of $\xi$. Hence there are three types of systems and consequently three types of response. The three types of systems are :

**1. *Overdamped systems.*** In an overdamped system $\xi > 1$ or $B^2 > 4KM$. Therefore, for this case, the roots characteristic equation are real and unequal as shown in Fig. 4.35(*a*). An overdamped system responds to any time varying input in a sluggish manner without any oscillations about the final steady state position. Figure 4.35(*b*) shows the response of an overdamped system to a step input.

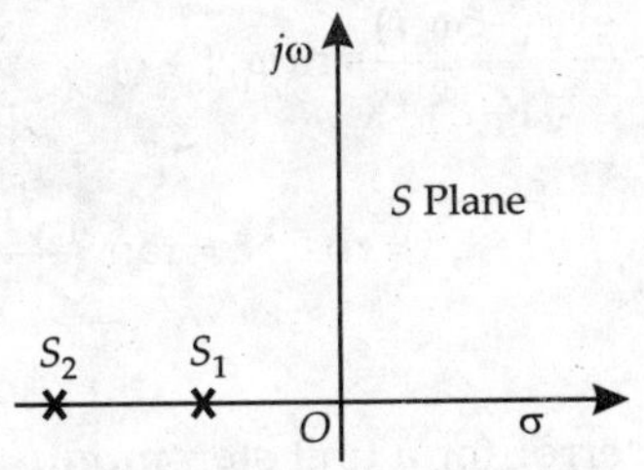


(*a*) Roots of characteristic equation.

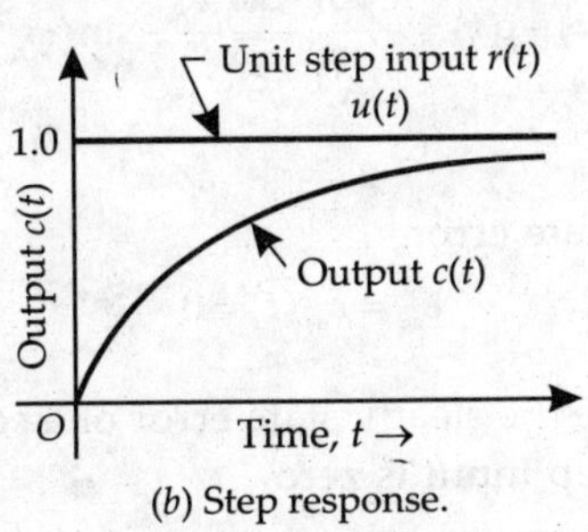


(*b*) Step response.

**Fig. 4.35** Roots and step response of an overdamped second order system.

**2. *Critically damped systems.*** For a critically damped system $\xi = 1$ or $B^2 = 4KM$. Therefore, for this case the roots are equal and real *i.e.*,

$$s_1 = s_2 = -\omega_n \quad ...(4.131)$$

The roots are shown in Fig. 4.36(*a*). The response of the system in this case is rapid and the system reaches its final steady state conditions smoothly without oscillations. The response of a critically damped system for a step input is shown in Fig. 4.36(*b*).

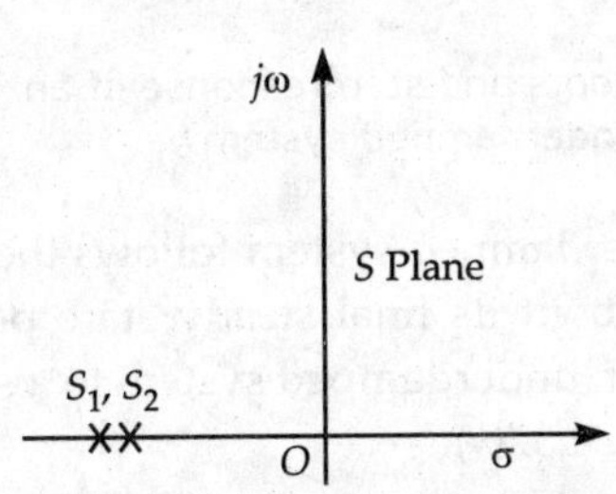


(*a*) Roots of characteristic equation

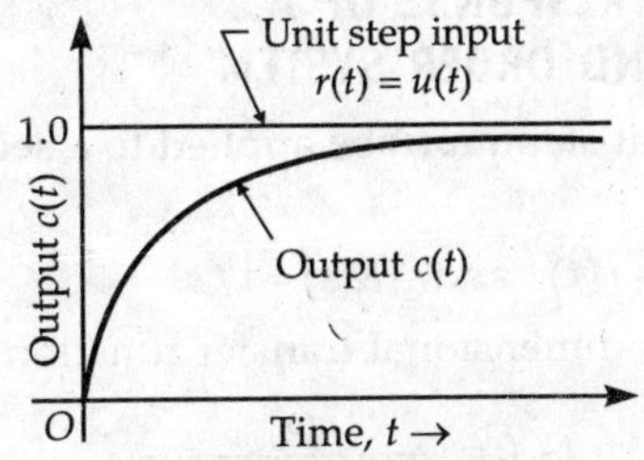


(*b*) Step response

**Fig. 4.36** Roots and step response of a critically damped system.

**3. *Underdamped systems.*** In an underdamped system $\xi < 1$ or $B^2 < 4KM$. Therefore, for this system, we get a complex conjugate pair of roots. The roots are,

$$s_1, s_2 = -\xi\omega_n \pm j\omega_n\sqrt{1-\xi^2} \quad ...(4.132)$$

$$= -\xi\omega_n \pm j\omega_d \quad ...(4.133)$$

where $\omega_d$ = frequency of damped oscillations

$$= \omega_n\sqrt{1-\xi^2} \quad ...(4.134)$$

The roots of an underdamped system are shown in Fig. 4.37(*a*).

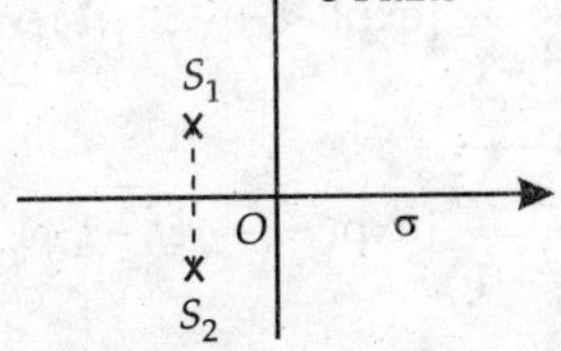


(*a*) Roots of characteristic equation.

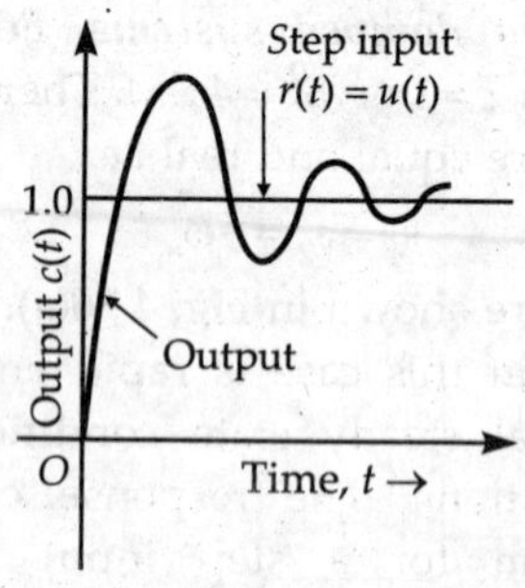


(b) Step response.

**Fig. 4.37** Roots and step response of an underdamped system.

The underdamped system follows the input with oscillations about its final steady state position. The response of an underdamped system to a step input is shown in Fig. 4.37(*b*).

## 4.27. STEP RESPONSE OF A SECOND ORDER SYSTEM

Let a unit step input be applied to a second order system.

$\therefore \quad r(t) = u(t)$ and $R(s) = 1/s$

The non-dimensional transfer function is,

$$G(s) = \frac{\omega_n^2}{s^2 + 2\xi\omega_n s + \omega_n^2}$$

$$\therefore \text{ Output } C(s) = G(s)R(s) = \frac{1}{s}\cdot\frac{\omega_n^2}{s^2 + 2\xi\omega_n s + \omega_n^2} \qquad ...(4.135)$$

As explained earlier the response of the system depends upon the poles of the system transfer function. The response is of three types depending upon the location of poles *i.e.* roots of the characteristic equation. Equation 4.135 can be solved to give the time response depending upon the location of the poles. The three different cases as outlined earlier are analyzed below.

***1. Overdamped systems.*** In this case $\xi > 1$ and the roots are real and unequal. The time response to a unit step input is,

$$c(t) = 1 - \frac{\xi + \sqrt{\xi^2 - 1}}{2\sqrt{\xi^2 - 1}}[\exp(-\xi + \sqrt{\xi^2 - 1})\omega_n t]$$

$$+ \frac{\xi - \sqrt{\xi^2 - 1}}{2\sqrt{\xi^2 - 1}}[\exp(-\xi - \sqrt{\xi^2 - 1})\omega_n t] \qquad ...(4.136)$$

The response is shown in Fig. 4.35(*b*).

***2. Critically damped systems.*** In this case $\xi = 1$ and the roots are real and equal. The equation for response to a unit step input is,

$$c(t) = 1 - (1 + \omega_n t)[\exp-(\omega_n t)] \qquad ...(4.137)$$

The response is shown in Fig. 4.36(*b*).

***3. Underdamped systems.*** For an underdamped system, $\xi < 1$ and roots of the characteristic equation are a complex conjugate pair. The response of an underdamped system to a unit step input is derived below :

$$C(s) = G(s)R(s) = \frac{1}{s}\cdot\frac{\omega_n^2}{s^2 + 2\xi\omega_n s + \omega_n^2}$$

$$= \frac{\omega_n^2}{s[s + \xi\omega_n - j\omega_n\sqrt{1-\xi^2}][s + \xi\omega_n - j\omega_n\sqrt{1-\xi^2}]}$$

In order to obtain the response in time domain, the above relationship is resolved into partial fractions and then the inverse Laplace transform is obtained. Since the roots are a complex conjugate pair the output in time domain is,

$$c(t) = \left.\frac{\omega_n^2}{s^2 + 2\xi\omega_n s + \omega_n^2}\right|_{s=0}$$

$$+ \text{Re}\left[\left.\frac{\omega_n^2}{s[s + \xi\omega_n - j\omega_n\sqrt{1-\xi^2}]}\right|_{s = -\xi\omega_n - j\omega_n\sqrt{1-\xi^2}} \exp(-\xi\omega_n t - j\omega_n\sqrt{1-\xi^2}t\right]$$

$$= 1 - \frac{\exp(-\xi\omega_n t)}{\sqrt{1-\xi^2}}\sin(\omega_n\sqrt{1-\xi^2}t + \phi) \qquad ...(4.138)$$

$$= 1 - \frac{\exp(-\xi\omega_n t)}{\sqrt{1-\xi^2}}\sin(\omega_d t + \phi) \qquad ...(4.139)$$

where $\phi = \sin^{-1}\sqrt{1-\xi^2} = \cos^{-1}\xi = \tan^{-1}\frac{\sqrt{1-\xi^2}}{\xi}$

...(4.140)

Dynamic error for a unit step input,

$$e_m(t) = r(t) - c(t) = \frac{\exp(-\xi\omega_n t)}{\sqrt{1-\xi^2}} - \sin(\omega_n\sqrt{1-\xi^2}t + \phi)$$

...(4.141)

Steady state error

$$e_{ss} = \underset{t\to\infty}{e_m}(t) = 0 \qquad \textbf{...(4.142)}$$

Therefore the steady state error of a second order system for step input is zero.

Final steady state output is,

$$c_{ss} = \lim_{t\to\infty} c(t) = 1 \qquad ...(4.143)$$

Thus a second order system tracks step input without any error.

The response of an under damped ($\xi < 1$) second order system is shown in Fig. 4.38. It is a sinusoid with a decaying amplitude.

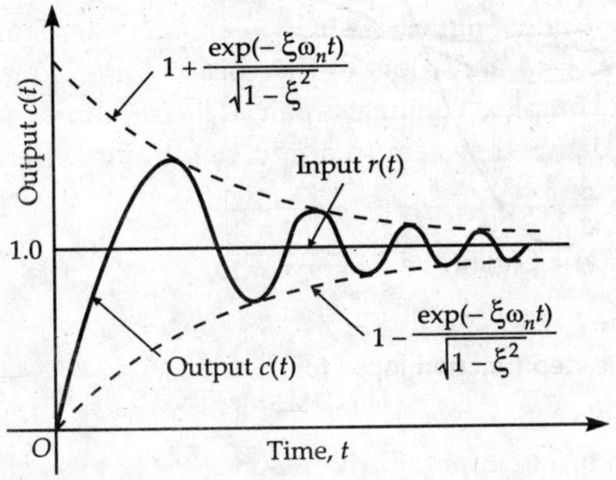


**Fig. 4.38** Unit step response of an underdamped second order system.

Let us analyse the system with various values of $\xi$. When the system is undamped *i.e.*, when $\xi = 0$, we have,

$$c(t) = 1 - \sin[\omega_n t + \pi/2]$$

$$= 1 - \cos\omega_n t \qquad \text{...(4.144)}$$

Thus the system has constant oscillations. As $\xi$ is increased, the response becomes progressively less oscillatory. When $\xi = 1$, the system is critically damped and there are no oscillations. For $\xi > 1$, the system is overdamped and hence non-oscillatory, but is highly sluggish in behaviour. The non-dimensional curves are shown in Fig. 4.40.

### 4.27.1 Time Domain Specifications

Measurement systems are designed with $\xi < 1$ with the result they exhibit an oscillatory response. Therefore the time response of second order (and for that matter of higher order systems which may be approximated as second order systems) is oscillatory when subjected to a step input. This response is underdamped in nature.

When we use an underdamped system, there are certain properties of interest. These properties put in qualitative form are,

(*i*) how fast the system moves to follow the applied input ?

(*ii*) how oscillatory is the system ?

(*iii*) how long will it take the system to practically reach its final steady state value ?

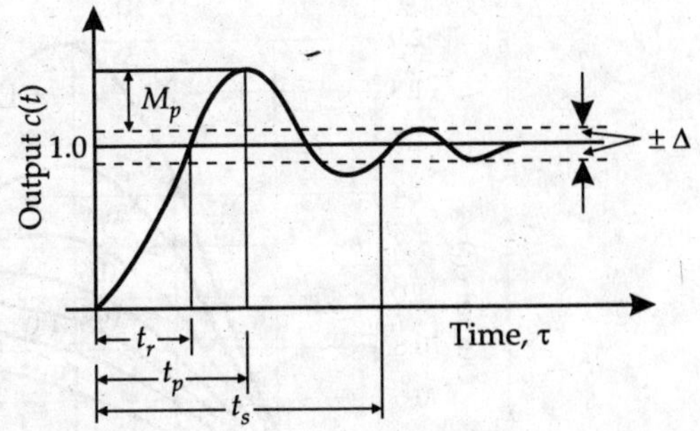


**Fig. 4.39** Time domain specifications of an under-damped second order system.

The time domain specifications cover the above properties in a quantitative form. These terms are explained with respect to a step input.

**1. *Rise time $t_r$*.** It is defined as time required for the system to rise from 0 to 100 per cent of its final value (See Fig. 4.39). Let us find the rise time for an underdamped system.

Now rise time is $t_r$ and the output at $t = t_r$ is, $c(t_r) = 1$ for a unit step input.

From Eqns. 4.138 and 4.140, we have,

$$c(t_r) = 1 - \frac{\exp(-\xi\omega_n t_r)}{\sqrt{(1-\xi^2)}} \sin[\omega_n \sqrt{(1-\xi^2)}\, t_r + \cos^{-1}\xi] = 1$$

Solving the above equation, we get,

$$\text{rise time } t_r = \frac{\pi - \cos^{-1}\xi}{\omega_n \sqrt{(1-\xi^2)}} \qquad \text{...(4.145)}$$

**2. *Peak time $t_p$*.** It is the time required for the output to reach the peak of time response of peak overshoot. In order to find the peak time, Eqn. 4.138 is differentiated and the peak time is obtained by putting the derivative equal to zero.

Following the above procedure, we have,

$$\sin[\omega_n \sqrt{(1-\xi^2)}\, t] = 0$$

Therefore the time for various peaks is given by :

$$\omega_n \sqrt{(1-\xi^2)}\, t = 0, \pi, 2\pi, 3\pi....$$

Time to reach the first peak (overshoot) is given by :

$$t_p = \frac{\pi}{\omega_n \sqrt{(1-\xi^2)}} \qquad \text{...(4.146)}$$

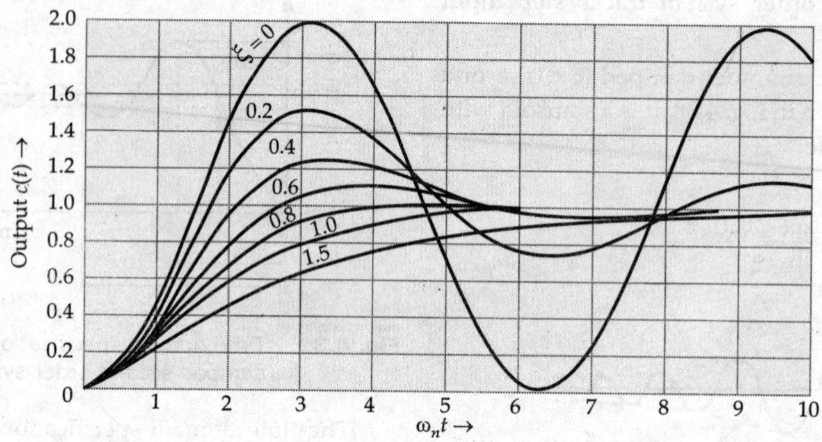


**Fig. 4.40** Transient response of a second-order system to a unit step function input for different values of damping factor ξ.

**3. *Peak overshoot* $M_p$.** The per unit peak overshoot is defined as :

$$M_p = \frac{c(t_p) - c(\infty)}{c(\infty)} \qquad \text{...(4.147)}$$

For a unit step input $c(\infty) = c_{ss} = 1$

$$\therefore \quad M_p = c(t_p) - 1$$

$$= \frac{\exp(-\xi\omega_n t_p)}{\sqrt{(1-\xi^2)}} \sin[\sqrt{(1-\xi^2)}\,\omega_n t_p + \cos^{-1}\xi]$$

$$= \exp(-\pi\xi/\sqrt{1-\xi^2}) \qquad \text{...(4.148)}$$

**4. *Settling time* $t_s$.** It is the time required for the output to reach and stay within a specified tolerance band ±Δ. This tolerance band is either ±2% or ±5% of the final steady state value. The settling time can be found by equating the exponentially varying term of Eqn. 4.138 equal to the tolerance band.

Thus for 2% tolerance band,

$$\frac{\exp(-\xi\omega_n t_s)}{\sqrt{(1-\xi^2)}} = 0.02.$$

An approximate solution of this equation is,

Settling time, $t_s = 4/(\xi\omega_n)$ ...(4.149)

For a 5% band,

Settling time, $t_s = 3/(\xi\omega_n)$ ...(4.150)

## 4.28 RAMP RESPONSE OF A SECOND ORDER SYSTEM

Let us consider a unit ramp input is applied to an underdamped second order system.

Thus, $r(t) = t$ and $R(s) = 1/s^2$

$$\therefore \text{ Output } C(s) = \frac{1}{s^2}\left[\frac{\omega_n^2}{s^2 + 2\xi\omega_n s + \omega_n^2}\right]$$

Taking partial fractions and applying inverse Laplace transform,

$$c(t) = t - \frac{2\xi}{\omega_n} + \frac{\exp(-\xi\omega_n t)}{\omega_n\sqrt{(1-\xi^2)}} \sin[\omega_n(\sqrt{1-\xi^2})t + \cos^{-1}\xi] \qquad \text{...(4.151)}$$

The response of the system is shown in Fig. 4.41 and is of the same form as is obtained with a step input. The dynamic error due to a ramp input is,

$$e_m(t) = r(t) - c(t)$$

$$= \frac{2\xi}{\omega_n} - \frac{\exp(-\xi\omega_n t)}{\omega_n\sqrt{(1-\xi^2)}} \sin[\omega_n(\sqrt{1-\xi^2})\,t + \cos^{-1}\xi]$$

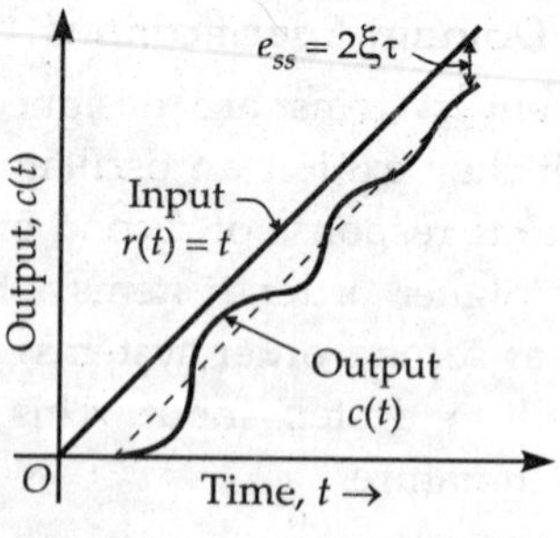


**Fig. 4.41** Response of an underdamped second order system to a unit ramp input.

∴ Steady state error

$$e_{ss} = \underset{t\to\infty}{e_m(t)} = \frac{2\xi}{\omega_n} \qquad \text{...(4.153)}$$

Time constant $\tau = 1/\omega_n$ ...(4.154)

Time lag $= 2\xi/\omega_n = 2\xi\tau.$ ...(4.155)

***Example 4.12*** *A linear second order with single degree of freedom system has a mass of $8\times10^{-3}$ kg and stiffness of 1000 N/m. Calculate the natural frequency of the system. (i) Determine the damping constant necessary to just prevent overshoot in response to a step input of force (ii) Calculate the frequency of damped oscillations if the damping ratio is reduced to 60% of its value as calculated in (i).*

**Solution.** (*i*) The given data is :

Mass $M=8\times10^{-3}$ kg and

stiffness constant $K=1000\,\text{N/m}$;

∴ From Eqn. 4.122, natural frequency

$$\omega_n=\sqrt{K/M}=\sqrt{1000/(8\times10^{-3})}=353.6 \text{ rad/s}$$

(*i*) The system should be critically damped in order that overshoots or oscillations are just prevented. For a critically damped system $\xi=1$.

$$\therefore \quad \xi=B/2\sqrt{KM}=1 \qquad \text{(see Eqn. 4.123)}$$

or damping constant

$$B=2\sqrt{KM}=2\sqrt{1000\times8\times10^{-3}}=5.66\text{ N/ms}^{-1}$$

(*ii*) New damping ratio $\xi=0.6$

∴ From Eqn. 4.134, frequency of damped oscillations

$$\omega_d=\omega_n\sqrt{1-\xi^2}=353.6\sqrt{1-(0.6)^2}=282.9 \text{ rad/s}.$$

***Example 4.13*** *A d'Arsonval galvanometer is a second order instrument. It has a spring stiffness such that a torque of 40 μ N/m gives a steady deflection of 90°. If the inertia of the rotating parts is 0.5 × $10^{-6}$ kg-m$^2$ and the damping constant is 5 μ Nm/rad s$^{-1}$, determine the damping ratio, the natural angular frequency, the frequency of damped oscillations, and the time constant of the movement. Determine of steady state dynamic error when the input voltage is increasing at a steady rate of 5 V/s.*

**Solution.** Stiffness constant

$$K=\frac{40\times10^{-6}}{\pi/2}=25.46\mu\text{N/rad}$$

From Eqn. 4.128 damping ratio

$$\xi=\frac{B}{2\sqrt{KJ}}=\frac{5\times10^{-6}}{2\sqrt{25.46\times10^{-6}\times0.5\times10^{-6}}}=0.7$$

From Eqn. 4.127 natural frequency

$$\omega_n=\sqrt{K/J}=\sqrt{\frac{25.46\times10^{-6}}{0.5\times10^{-6}}}=7.14 \text{ rad/s}$$

Frequency of damped oscillations

$$\omega_d=7.14\sqrt{1-(0.7)^2}=5.1 \text{ rad/s}$$

Time constant $\tau=1/\omega_n=1/7.14=0.14$ s

For a unit ramp input the steady state error is,

$$e_{ss}=\frac{2\xi}{\omega_n} \qquad \text{(see Eqn. 4.153)}$$

For a ramp input of 5 V/s, the steady state errors is,

$$e_{ss}=5\times2\frac{\xi}{\omega_n}=5\times2\times\frac{0.7}{7.14}=0.98\text{V}$$

From Eqn. 4.155, time lag

$$=2\xi\tau=2\times0.7\times0.14\approx0.2 \text{ s}.$$

***Example 4.14*** *A potentiometer whose moving parts weigh $5\times10^{-3}$ kg measures displacement of a spring mass system as shown in Fig. 4.42. The mass of the system is M kg. The system is subjected to a step input. The measured natural frequency is 30 Hz. If the spring constant K and mass M of the system are unknown, can the true natural frequency be found from the above data ? Suppose an additional mass of $5\times10^{-3}$ kg is attached to the potentiometer and the test is repeated. This gives a natural frequency of 25 Hz. Calculate the true natural frequency of the system.*

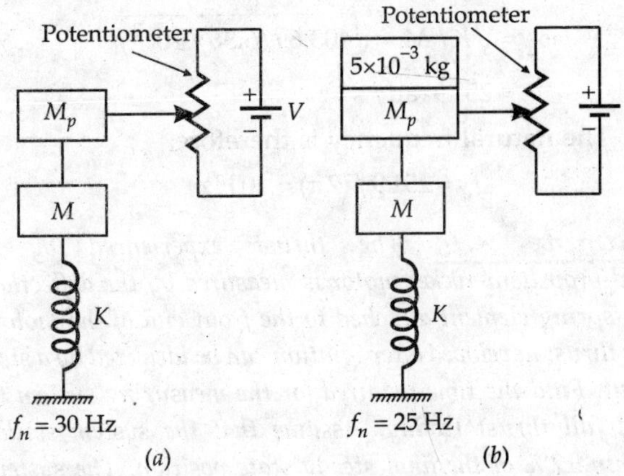


**Fig. 4.42** Determination of displacement of mass-spring system with potentiometer (Example 4.14).

**Solution.** Suppose $M_p$ is the mass of the moving parts of the potentiometer

or $$M_p=5\times10^{-3}\text{kg}.$$

Before the additional mass is attached, the total mass is

$$M+M_p \text{ or } (M+5\times10^{-3})\text{ kg}$$

This is shown in Fig. 4.42(*a*).

Now, before the additional mass is attached, the natural frequency measured is,

$$f_n=30\text{Hz} \quad \text{or} \quad \omega_n=2\pi\times30 \text{ rad/s}$$

$$\therefore \quad 2\pi\times30=\sqrt{K/(M+M_p)}=\sqrt{K/(M+5\times10^{-3})} \qquad ...(i)$$

The true natural frequency of the system

$$\omega_n = \sqrt{K/M} \quad \text{...(ii)}$$

Now we have two equations, (*i*) and (*ii*) and three unknowns *i.e.*, *K*, *M* and ω and therefore a solution cannot be obtained. In other words, with the above given data, it is not possible to find the true natural frequency of the system.

When an additional mass of $5\times10^{-3}$ kg is attached, the total mass becomes $(M+M_p+5\times10^{-3})$kg or $(M+5\times10^{-3}+5\times10^{-3})$kg as shown in Fig. 4.42(*b*). The natural frequency in this case is, $f_n = 25$ Hz or $\omega_n = 2\pi\times 25$ rad/s.

$$\therefore\ 2\pi\times25 = \sqrt{K/(M+5\times10^{-3}+5\times10^{-3})} \quad \text{...(iii)}$$

Now we have equations (*i*) and (*iii*) and therefore two unknowns *K* and *M* can be found by solving them. From (*i*) and (*iii*), we have :

$$M = 6.36\times10^{-3}\,\text{kg} \quad \text{and} \quad K = 403.6\,\text{N/m}$$

From (*ii*), natural angular frequency of the system

$$\omega_n = \sqrt{K/M} = \sqrt{403.6/(6.36\times10^{-3})}$$

$$= 251.9\,\text{rad/s}$$

The natural frequency is therefore,

$$f_n = 251.9/(2\pi) = 40\,\text{Hz}$$

***Example 4.15*** *The thrust experienced by a solid-propellent rocket motor is measured by the deflection of a spring element attached to the front end of the motor. The thrust developed after ignition can be idealized by a step input. Find the time required for the measuring system to read full thrust readily. Assume that the system settles within ±2% of the final steady state position. The system has the following parameters :*

*Mass of motor = 30 kg, spring constant = $60\times10^{-3}$ N/m and damping factor = 0.7.*

**Solution.** Natural frequency

$$\omega_n = \sqrt{K/M}$$

$$= \sqrt{60\times10^{3}/30}.$$

$$= 44.7\,\text{rad/s}$$

The system settles down within ±2% band. For a step input the settling time, for ±2% band (Eqn. 4.149) is,

$$t_s = \frac{4}{\xi\omega_n} = \frac{4}{0.7\times44.7} = 0.128\,\text{s}$$

Hence the measuring system reads the full thrust after 0.128 s.

## 4.29 IMPULSE RESPONSE OF A SECOND ORDER SYSTEM

Let a second order be subjected to impulse of strength *A*.

The non-dimensional output is,

$$c(t) = \frac{A\omega_n}{2\sqrt{(\xi^2-1)}}$$

$$[\exp(-\xi+\sqrt{\xi^2-1})\omega_n t - \exp(-\xi-\sqrt{\xi^2-1})\omega_n t] \quad \text{...(4.156)}$$

for an overdamped system.

For a critically damped system

$$c(t) = A\omega_n^2\, t[\exp(-\omega_n t)] \quad \text{...(4.157)}$$

and for an under damped system,

$$c(t) = A\omega_n\left[\frac{\exp(-\xi\omega_n t)}{\sqrt{1-\xi^2}}\sin(\sqrt{1-\xi^2}\,\omega_n t)\right] \quad \text{...(4.158)}$$

Figure 4.43 shows the response of a second order system to an impulse input in non-dimensional terms.

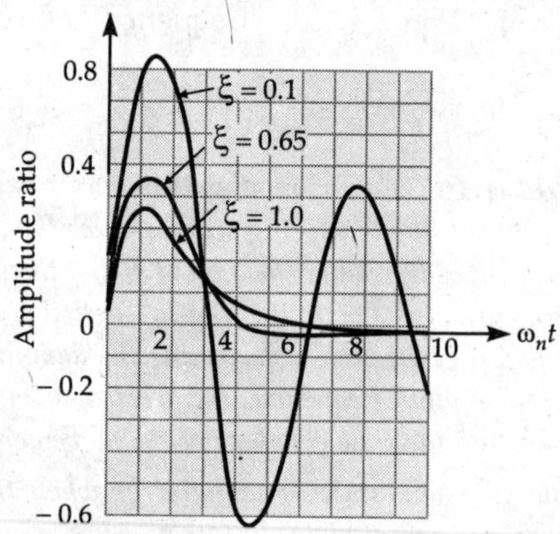


**Fig. 4.43** Non-dimensional impulse response of second order system.

## 4.30 FREQUENCY DOMAIN ANALYSIS

We have studied time response of the system with some standard inputs earlier. For studies in frequency domain, the response of the system to a sinusoidal signal is of very great importance. Steady state response to a sinusoidal signal yields a great deal of information about the system since all the standard inputs can be broken into sinusoidal signals with a wide range of frequencies starting from zero onwards.

The input standard sinusoidal signal is,

$$r(t) = A_i \sin\omega t$$

When a sinusoidal signal is applied to any system, the output is also sinusoidal with transients superimposed on it initially for a few cycles. However, in the frequency domain analysis we are not interested in the transient effects since they die out quickly and under steady state the system output is strictly sinusoidal in nature having the same frequency as the input. The amplitude of the output and its phase, however, may differ from that of the input.

The output may be represented as :

$$c(t) = A_0 \sin(\omega t + \phi) \quad ...(4.160)$$

where $A_0$ = amplitude

and $\phi$ = phase shift of the output.

The frequency of the input and the output is the same. Therefore the relation between the input and the output can be completely specified by giving their amplitude ratio and the phase shift. The amplitude ratio and the phase shift change with change in the driving frequency $\omega$.

The **frequency response** of a system consists of curves of amplitude ratio and phase shift as a function of frequency.

The sinusoidal transfer function is given by :

$$G(j\omega) = \frac{C(j\omega)}{R(j\omega)}$$

$$= \frac{b_m(j\omega)^m + b_{m-1}(j\omega)^{m-1} + ... + b_1(j\omega) + b_0}{a_n(j\omega)^n + a_{n-1}(j\omega)^{n-1} + ... + a_1(j\omega) + a_0} \quad m \le n \quad ...(4.161)$$

For any given frequency $\omega$, Eqn. 4.161 shows that $C(j\omega)/R(j\omega)$ is a complex number which can always be put in polar from $M\angle\phi$.

where
$$M = \frac{\text{amplitude of output}}{\text{amplitude of input}} = \frac{A_0}{A_i} \quad ...(4.162)$$

(This is true if $A_0$ and $A_i$ have the same dimensions).

and $\phi$= angle of phase shift between $A_0$ and $A_i$, phase shift $\phi$ is negative if the output lags behind the input.

## 4.31 FREQUENCY RESPONSE OF FIRST ORDER SYSTEMS

The dimensionless transfer function of a First order system is,

$$G(s) = 1/(1 + \tau s)$$

$\therefore$ The dimensionless sinusoidal transfer function of a First order system is :

$$G(j\omega) = \frac{C(j\omega)}{R(j\omega)} = \frac{1}{(1 + j\omega\tau)} \quad ...(4.163)$$

Suppose the system is subjected to a sinusoidal input

$$r(t) = A_i \sin \omega t \quad \text{or} \quad R(j\omega) = A_i \sin \omega t$$

$$\therefore \text{Output } C(j\omega) = G(j\omega)\, R(j\omega) = \frac{1}{(1 + j\omega\tau)} \cdot A_i \sin \omega t$$

or
$$|C(j\omega)| = A_0 = \frac{A_i}{|1 + j\omega\tau|} = \frac{A_i}{\sqrt{1 + \omega^2\tau^2}}.$$

The dimensionless ratio of output to input is,

$$\therefore \quad M = \left|\frac{A_0}{A_i}\right| = \left|\frac{1}{(1 + j\omega\tau)}\right| = \frac{1}{\sqrt{1 + \omega^2\tau^2}} \quad ...(4.164)$$

$$\therefore \text{Magnitude of output } A_0 = \frac{A_i}{\sqrt{1 + \omega^2\tau^2}} \quad ...(4.165)$$

Phase displacement of output

$$\phi = -\tan^{-1} \omega\tau \quad ...(4.166)$$

$$\text{Output} \quad C(t) = A_0 \sin(\omega t + \phi)$$

$$= \frac{A_i}{\sqrt{1 + \omega^2\tau^2}} \sin(\omega t - \tan^{-1} \omega\tau) \quad ...(4.167)$$

The phase lag is $\phi = \tan^{-1} \omega\tau$ and in terms of time this lag is called *time lag*.

$$\text{Time lag} = (1/\omega)\tan^{-1} \omega\tau \quad ...(4.168)$$

It is clear from Eqns. 4.165 and 4.166, that when,

$$\omega = 0,\ M = A_0 / A_i = 1 \text{ and } \phi = 0$$

and
$$\omega = \infty,\ M = A_0 / A_i = 0 \text{ and } \phi = -90°$$

Thus, higher the frequency, higher is the attenuation of the output and greater is the angle of lag between output and input, the highest angle of lag being – 90°. The frequency response characteristics of a *First order system* are shown in Fig. 4.44.

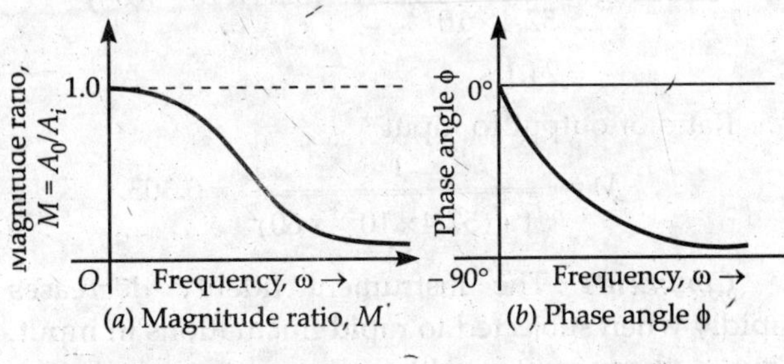


**Fig. 4.44** Frequency response characteristics of first order systems.

There is another factor which affects the frequency response of the system and that is the time constant, $\tau$, of the system. It is clear from Eqns 4.164 and 4.166 that greater the time constant, greater is the attenuation of the output and also greater is the phase shift between output and the input or in other words greater is the time lag. In order that the signal be reproduced without any attenuation or time lag, the time constant, $\tau$, should be zero. However, actual physical systems have a finite time constant and therefore there is always an attenuation of output and also a time lag between input and the output signals. Therefore, the input signal cannot be faithfully reproduced.

**Example 4.16** *(a) A therometer with time constant of 60 s is used for measuring temperature cycling with 600 s time period. Find the time lag and ratio of output to input.*

*(b) If the same instrument is used for measurement of an input with a time period of 120 s, calculate the measuring lag and the ratio of output to input.*

*Comment upon the result.*

**Solution.** (*a*) Frequency of temperature variations :

$$\omega = 2\pi/600 = 10.47 \times 10^{-3} \text{ rad/s}$$

From Eqn. 4.168,

$$\text{Time lag} = \frac{1}{\omega}\tan^{-1}\omega\tau$$

$$= \frac{1}{10.47\times 10^{-3}}\tan^{-1}(10.47\times 10^{-3}\times 60)$$

$$= 53.6 \text{ s.}$$

Ratio of output to input (See Eqn. 4.164)

$$M = \left|\frac{A_0}{A_i}\right| = \frac{1}{\sqrt{1+\omega^2\tau^2}}$$

$$= \frac{1}{\sqrt{1+(10.47\times 10^{-3}\times 60)^2}} = 0.847.$$

(*b*) Frequency of temperature variations :

$$\omega = 2\pi/120 = 52.4 \times 10^{-3} \text{ rad/s.}$$

$$\text{Time lag} = \frac{1}{52.4\times 10^{-3}}\tan^{-1}(52.4\times 10^{-3}\times 60)$$

$$= 24.1 \text{ s}$$

Ratio of output to input

$$M = \frac{1}{\sqrt{1+(52.4\times 10^{-3}\times 60)^2}} = 0.303.$$

***Comments.*** The instrument fidelity decreases rapidly when subjected to rapid fluctuations in input. Thus a thermometer with a time constant of 60 s shows 84.7% of actual input when subjected to a frequency of $10.47 \times 10^{-3}$ rad/s. But if the same instrument is subjected to the same input having a frequency which is 5 times the earlier frequency *i.e.,* 52.4 rad/s the output drops to 30.3% of the input. Thus the dynamic error is 69.7% rendering instrument readings meaningless.

**Example 4.17.** *A first order instrument must measure signals with frequency content up to 100 Hz with an amplitude inaccuracy of 5 per cent. What is the maximum allowable time constant ? What will be the shift at 50 Hz and at 100 Hz.*

**Solution.** Maximum allowable inaccuracy = 5%.

$\therefore$ Minimum allowable value of $M = 1 - 0.05 = 0.95$.

Frequency $\omega = 2\pi \times 100 = 628$ rad/s

The maximum allowable time constant can be calculated using Eqn. 4.164.

$$\therefore\ M = \frac{1}{\sqrt{1+\omega^2\tau^2}} \quad \text{or} \quad 0.95 = \frac{1}{\sqrt{1+(628\tau)^2}}$$

or time constant $\tau = 0.52 \times 10^{-3}$ s.

Phase shift at 50 Hz is,

$$\phi = -\tan^{-1}\omega\tau$$

$$= -\tan^{-1}(2\pi\times 50\times 0.52\times 10^{-3}) = -9.3°$$

Phase shift at 100 Hz is,

$$\phi = -\tan^{-1}\omega\tau$$

$$= -\tan^{-1}(2\pi\times 100\times 0.52\times 10^{-3}) = 18.1°$$

**Example 4.18.** *A liquid thermometer has its glass bulb protected by a well. The system can be represented by a double capacity system with time constants of 40 s for the well and 20 s for the bulb. The thermometer is subjected to a cyclic change of ± 10° C which occurs every 120 s. Find the maximum value of the indicated temperature and the delay time.*

**Solution.** For a double capacity system, the amplitude ratio of output to input is,

$$M = \frac{1}{\sqrt{1+(\omega\tau_1)^2}} \times \frac{1}{\sqrt{1+(\omega\tau_2)^2}}$$

and the angle of lag $= -(\tan^{-1}\omega\tau_1 + \tan^{-1}\omega\tau_2)$

$$\omega = 2\pi f = \frac{2\pi}{T} = \frac{2\pi}{120} = 52.4\times 10^{-3} \text{ rad/s.}$$

$\therefore$ Amplitude ratio,

$$M = \frac{1}{\sqrt{1+(52.4\times 10^{-3}\times 40)^2}} \times \frac{1}{\sqrt{1+(52.4\times 10^{-3}\times 20)^2}}$$

$$= \frac{1}{\sqrt{1+(2.096)^2}} \times \frac{1}{\sqrt{1+(1.048)^2}} = 0.297.$$

Maximum indicated temperature $= 0.297 \times (\pm 10°)$

$$= \pm 2.97°\text{C}$$

Angle of lag $= -(\tan^{-1}2.096 + \tan^{-1}1.048)$

$$= -110.8° = -1.93 \text{ rad}$$

$$\text{Time lag} = \frac{1.93}{52.4\times 10^{-3}} = 36.8 \text{ s.}$$

**Example 4.19.** *(i) An input of 1 sin 2t + 0.3 sin 20t is applied to a first order instrument having a time constant of 0.2 s. Find the output.*

*(ii) If the same input is applied to an instrument having a time constant of 0.002 s, what will be the output ? Comment upon the results.*

**Solution.** (*i*) The output of the instrument to an input $1\sin 2t + 0.3\sin 20t$ when the time constant is 0.2

$$= \frac{1}{\sqrt{1+(2\times0.2)^2}}\sin[(2t-\tan^{-1}(2\times0.2))]$$

$$+ \frac{0.3}{\sqrt{1+(20\times0.2)^2}}\sin[20t-\tan^{-1}(20\times0.2)]$$

$$= 0.93\sin(2t-21.8°)+0.073\sin(20t-76°).$$

(*ii*) The output of the instrument to the same input when the time constant is 0.002 s is given by :

$$= \frac{1}{\sqrt{1+(2\times0.002)^2}}\sin[(2t-\tan^{-1}(2\times0.002))]$$

$$+ \frac{0.3}{\sqrt{1+(20\times0.002)^2}}\sin[2t-\tan^{-1}(20\times0.002)]$$

$$= 1\sin(2t-0.23°)+0.3\sin(20t-2.3°)$$

***Comments.*** It is clear from the results that the output is attenuated and also the phase shift between output and input is large when the time constant of the instrument is large. However, if the time constant of the system is very small as compared with the time period of the input, the instrument reproduces the input faithfully *i.e.,* without attenuation and with an insignificant phase shift.

**Example 4.20.** *A thermocouple system has a time constant of 10 s. The system is used to measure the temperature of a furnace which fluctuates sinusoidally between 640°C and 600°C, with a periodic time of 80 s. Determine the maximum and minimum values of the temperature indicated. Calculate also the angle of phase shift and the time lag.*

**Solution.** The maximum and minimum values of temperature are respectively 640°C and 600°C.

Therefore the temperature oscillates about a mean value of $(640-600)/2 = 620°\text{C}$ with an amplitude of 20°C as shown in Fig. 4.45

Frequency of oscillation

$$\omega = 2\pi/80 = 0.0785 \text{ rad/s}$$

Amplitude of the output about the mean position

$$A_0 = \frac{A_i}{\sqrt{1+\omega^2\tau^2}} = \frac{20}{\sqrt{1+(0.0785\times10)^2}} = 15.7°\text{C},$$

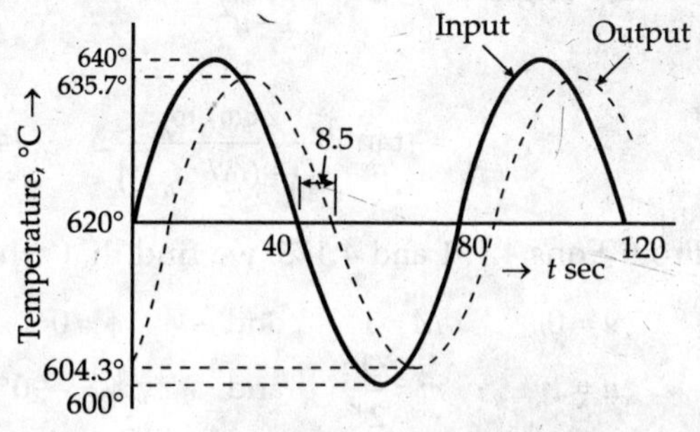


**Fig. 4.45** Variation of actual and indicated temperatures of furnace of Example 4.20.

∴ Maximum and minimum indicated temperatures are

$$(620+15.7) = 635.7°\text{C}$$

and $$(620-15.7) = 604.3°\text{C respectively.}$$

$$\text{Phase shift } \phi = -\tan^{-1}(0.0785\times10) = -0.665 \text{ rad.}$$

$$\text{Time lag} = 0.665/0.0785 = 8.5 \text{ s.}$$

## 4.32 FREQUENCY RESPONSE OF A SECOND ORDER SYSTEM

Let a second order system be subjected to a sinusoidal input,

$$r(t) = A_i \sin\omega t$$

Steady state output $c(t) = A_0\sin(\omega t + \phi)$.

The dimensionless transfer function of a second order system.

$$G(s) = \frac{\omega_n^2}{s^2 + 2\xi\omega_n s + \varepsilon_n^2}$$

The dimensionless sinusoidal transfer function is :

$$G(j\omega) = \frac{C(j\omega)}{R(j\omega)} = \frac{\omega_n^2}{(j\omega)^2 + 2\omega_n\xi(j\omega)+\omega_n^2}$$

$$= \frac{\omega_n^2}{(\omega_n^2-\omega^2)+2j\xi\omega_n\omega}$$

$$= \frac{1}{(1-u^2)+2j\xi u} \quad ...(4.169)$$

where $u$ = normalized frequency $= \omega/\omega_n$ ...(4.170)

$$\therefore \text{ Ratio } M = \left|\frac{\text{output}}{\text{input}}\right| = \left|\frac{A_0}{A_i}\right|$$

$$= \frac{1}{\sqrt{(1-u^2)+(2\xi u)^2}} \quad ...(4.171)$$

$$= \frac{1}{\sqrt{[1-(\omega/\omega_n)^2]+[2\xi\omega/\omega_n]^2}} \quad ...(4.172)$$

and phase angle $\phi = -\tan^{-1}\left(\dfrac{2\xi u}{1-u^2}\right)$ ...(4.173)

$$= -\tan^{-1}\frac{2\xi\omega/\omega_n}{[1-(\omega/\omega_n)^2]} \quad ...(4.174)$$

From Eqns 4.171 and 4.173, we find that when,

$u = 0, \quad M = 1 \quad$ and $\quad \phi = 0$

$u = 1 \quad m = \dfrac{1}{2\xi} \quad$ and $\quad \phi = -90°$

$y = \infty \quad M = 0 \quad$ and $\quad \phi = -180°.$

The magnitude ratio and the phase angle characteristics for various values of normalized frequency $u$ for a particular value of $\xi$ are shown in Figs. 4.46(*a*) and (*b*).

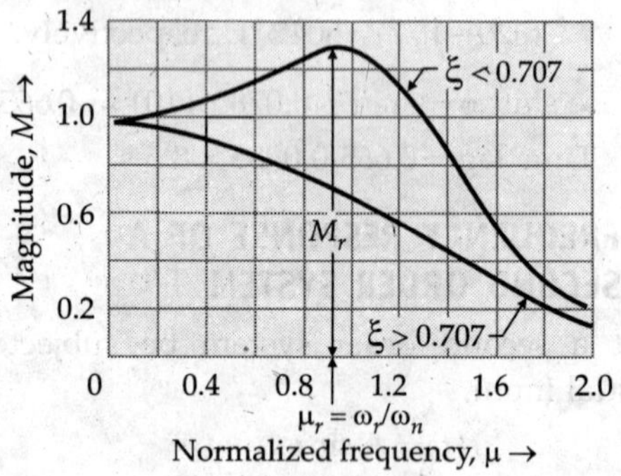


(*a*) amplitude ratio

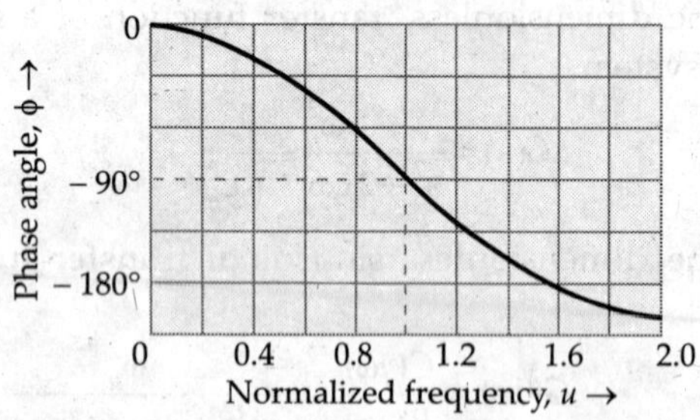


(*b*) Phase angle $\phi$

**Fig. 4.46** Frequency response characteristics of second order systems.

***1. Resonant frequency.*** The frequency at which $M$ has the maximum value is known as the resonant frequency.

Let $\omega_r$ = resonant frequency ; rad/s,

and $\quad u_r = \omega/\omega_r$ = normalized resonant frequency.

The normalized resonant frequency can be found out by differentiating the expression for $M$ with respect to $u$ and putting the differential equal to zero for $u = u_r$. Following this procedure

normalized resonant frequency,

$$u_r = \sqrt{(1-2\xi^2)} \quad ...(4.175)$$

and resonant frequency,

$$\omega_r = \omega_n\sqrt{(1-\xi^2)} \text{ rad/s} \quad ...(4.176)$$

***2. Resonant peak.*** The maximum value of $M$ occurs at $\omega = \omega_r$. This value is known as *Resonant Peak*, $M_r$. The value of $M_r$ can be found by putting $\omega = \omega_r = \omega_n\sqrt{(1-2\xi^2)}$ in Eqn. 4.172.

$$M_r = \frac{1}{2\xi\sqrt{1-\xi^2}}. \quad ...(4.177)$$

The phase shift at $\omega = \omega_r$ is,

$$\phi_r = -\tan^{-1}[(\sqrt{1-2\xi^2})/\xi] \quad ...(4.178)$$

It is clear from Eqn. 4.175 that for $\xi \geq 1/\sqrt{2}$ *i.e.*, $\xi \geq 0.707$, there is no resonant frequency and hence there is no resonant peak.

***3. Bandwidth.*** If we examine Fig. 4.46, it is found that for frequencies above $\omega > \omega_r$, the amplitude ratio $M$ decreases. The frequency at which $M$ has a value of 0.707 is called the *Cut-off frequency* $\omega_c$. Above the cut-off frequency $\omega_c$, the value of $M$ decreases below $0.707 = (1/\sqrt{2})$. Frequency $\omega_c$ represents the half power point. The band of frequencies from zero to cut-off frequency $\omega_c$ is called the *Bandwidth* of the system, $\omega_b$.

Measurement systems are low pass filters as the value of amplitude ratio is unity at $\omega = 0$. But as the frequency of the input signals increases, the output gets attenuated. Bandwidth is, therefore, indicative of the satisfactory reproduction of the input signal.

Let $u_b = \omega_b/\omega_n$ = normalized bandwidth

where $\omega_b$ = bandwidth ; rad/s.

From Eqn. 4.171, we have,

$$M = \frac{1}{\sqrt{[(1-u_b^2)^2 + (2\xi u_b)^2]}} = 0.707$$

Solving the above, we have normalized bandwidth

$$u_b = \sqrt{1-2\xi^2 + \sqrt{(2-4\xi^2+4\xi^4)}} \quad ...(4.179)$$

The de-normalised bandwidth is given by :

$$= \omega_n\sqrt{[1-2\xi^2 + \sqrt{(2-4\xi^2+2\xi^4)}]} \quad ...(4.180)$$

It is clear from Eqn. 4.179 that the normalized bandwidth is a function of damping ratio $\xi$. It is found from Eqn. 4.180 that as damping ratio $\xi$ increases, the bandwidth decreases.

***4. Frequency response curves.*** Figure 4.47 shows the variation of amplitude ratio $M$ and phase shift

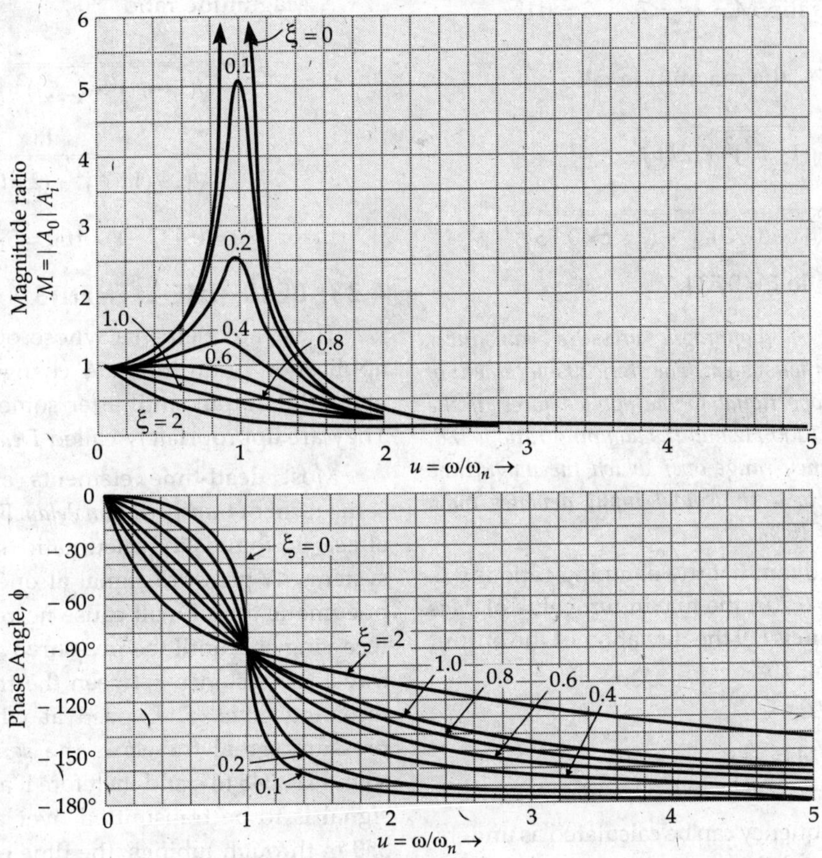


**Fig. 4.47** Frequency response curves of second order system.

with change of normalized frequency $u = \omega/\omega_n$ in a second order measurement system for various values of damping ratio $\xi$. It is clear from the diagram that if a flat frequency response is desired $u$ should be very small or the natural frequency of the system should be very high. (These are two conditions if the system has to reproduce the high frequency inputs faithfully.)

Let us examine the effect of damping ratio, $\xi$ on the frequency response curves. The damping ratio is a dominating factor in the response of a system and can be varied easily and therefore, it becomes extremely important to study its effect on the system performance. If we critically examine the magnitude ratio and phase shift curves, it is found that the optimum value of $\xi$ lies between 0.6 to 0.7, in case our criterion is the faithful response as far as magnitude is concerned, as with this value we obtain almost a flat curve for the amplitude ratio over a wide range of frequencies. However, if $\xi$ is between 0.6 and 0.7, there is a great phase shift between input and output and hence a great time lag. In situations where it is desired that the magnitude ratio should remain the same and time lag is of no consideration, a damping ratio of 0.6 to 0.7 is recommended for a second order measurement system.

***Example 4.21.*** *A torque sensing transducer is connected to the shaft of an electric motor which drives a load that has a moment of inertia of $200 \times 10^{-3}$ kg-m$^2$. What should be the damping ratio of the system when a sinusoidal torque varying at a frequency of 2 rad/s is to be reproduced with a maximum error of + 10% ? The torsional constant is 1.5 N-m/rad.*

**Solution.** Natural frequency

$$\omega_n = \sqrt{K/J}$$

$$= \sqrt{1.5/(200 \times 10^{-3})} = 2.74 \text{ rad/s.}$$

Normalized frequency

$$u = \omega/\omega_n$$

$$= 2/2.74 = 0.73.$$

The output is to be within + 10% of input.

$\therefore$ Amplitude ratio $M = \dfrac{A_0}{A_i} = \dfrac{1+0.1}{1} = 1.1.$

From Eqn. 4.171, the magnitude ratio,

$$M = \frac{1}{\sqrt{(1-u^2)^2 + (2\xi u)^2}}$$

or
$$1.1 = \frac{1}{\sqrt{[1-(0.73)^2]^2 + (2\times\xi\times 0.73)^2}}$$

$\therefore$ Damping ratio $\xi = 0.534.$

**Example 4.22.** *A diaphragm pressure transducer behaves as a second order system. The static displacement of the diaphragm is proportional to the applied force. If the natural frequency is 1000 Hz and the damping ratio is 0.6, determine the frequency range over which the magnitude ratio corresponding to a sinusoidal input deviates by a maximum amount of 10%.*

**Solution.** It is given that the damping ratio is 0.6. Since, it is less than 0.707, the maximum value of $M$ is 1.1 and the minimum is 0.9 the deviation of the output from the input being 10%.

From Eqn. 4.171,

$$M = \frac{1}{\sqrt{(1-u^2)^2 + (2\xi u)^2}}$$

The value of frequency can be calculated as under :

$$1.1 = \frac{1}{\sqrt{(1-u^2)^2 + (2\times 0.6u)^2}}$$

or $\quad u^4 - 0.5u^2 + 0.173 = 0$

The above equation gives imaginary values for frequency. Thus with $\xi = 0.6$, the output is not 1.1 of the input for any frequency.

Let us investigate, the attenuation of the input,

Now, $M = 0.9 = \dfrac{1}{\sqrt{(1-u^2)^2 + (2\times 0.6u)^2}}$

or $\quad u^4 - 0.5u^2 - 0.234 = 0$

or $\quad u = 0.916$

$\therefore \quad f = u \times f_n = 0.916 \times 1000 = 916$ Hz.

Therefore, the range over which the magnitude ratio remains within 10% of output is : 0 – 916 Hz.

**Example 4.23.** *A 2nd order instrument has a natural frequency of 4 Hz and a damping ratio of 0.66. If the excitation frequency of the system is 6 Hz, determine the error due to the proximity of excitation frequency with the natural frequency of the system.*

**Solution.** Ratio of excitation frequency to natural frequency

$$u = \omega/\omega_n = 6/4 = 1.5$$

$\therefore$ Magnitude ratio

$$M = \frac{1}{\sqrt{(1-u^2)^2 + (2\xi u)^2}}$$

$$= \frac{1}{\sqrt{[1-(1.5)^2]^2 + (2\times 0.66\times 1.5)^2}} = 0.427$$

Error $= (0.427 - 1)\times 100 = 57.3\%.$

## 4.33 DEAD-TIME ELEMENTS

These are elements whose output is delayed on application of an input. A change in input does not affect the output until after some delay or *dead time*. They are appropriately called *Dead Time Elements*.

Most dead-time elements are associated with some form of *transportation delay*. For example, such an effect is found in pneumatic signal transmission systems. A pressure signal at one end of a length of pneumatic tubing will cause no response at all at the receiving end until the pressure wave has had time to travel the distance between the transmitting and the receiving ends. The speed at which the pneumatic pressure signal travels is the same as the speed of sound *i.e.*, 345 m/s and therefore if a pneumatic pressure signal is to be transmitted over a distance of about 350 m through tubings, the time elapsed between the transmission and reception of the signal would be about 1 s and this time is treated as dead time.

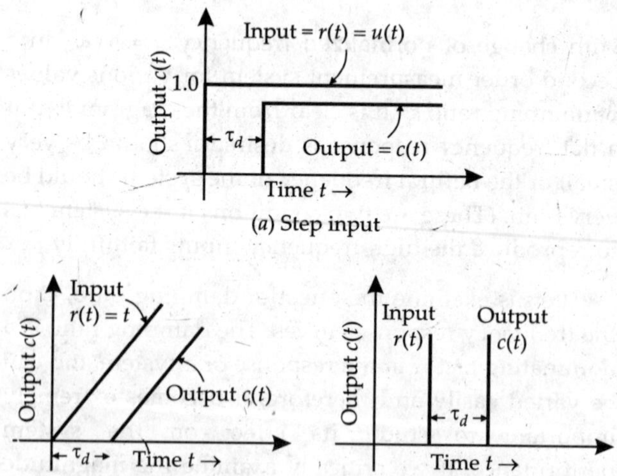


(*a*) Step input

(*b*) Ramp input

(*c*) Impulse input

**Fig. 4.48** Time response of dead-time elements to standard inputs.

A dead time element is defined as a system in which the output is of the same form as the input, but the output occurs after the delay time, $\tau_d$. The time $\tau_d$ is called the *delay time*.

Mathematically,

$$c(t) = r(t)(t - \tau_d) \quad t \geq \tau_d \qquad ...(4.181)$$

This type of element is also called a **pure delay** or **transport lag** element.

The response of dead-time elements to standard input *i.e.*, step, ramp and impulse is shown in Figs. 4.48 (*a*), 4.48 (*b*) and 4.48(*c*) respectively.

It is clear that in the dead time elements, once the time delay has expired, the output tracks the input.

However, in many practical situations, the system consisting dead time elements exhibits both dead time and first order response. Therefore, when a step input is applied to a first order plus dead time system, the output does not respond until after the transportation dealy on account of dead time is over. Then it rises exponentially, taking 5 times constants (99.3 percent of final steady state response) to complete its final response as shown in Fig. 4.49.

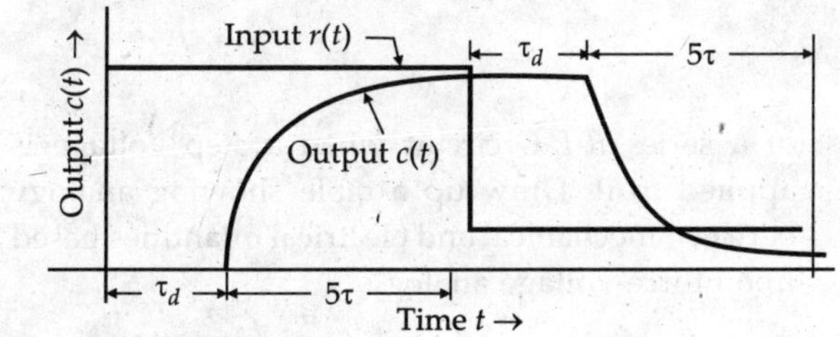


**Fig. 4.49** Step response of a first order dead time element.

For sinusoidal inputs,

$$r(t) = A_i \sin \omega t$$

The output is,

$$c(t) = A_i \sin \omega(t - \tau_d)$$
$$= A_i \sin(\omega t - \omega\tau_d) = A_i(\omega t + \phi) \qquad ...(4.182)$$

$$\text{Hence} \quad \frac{C(s)}{R(s)} = \left|\frac{A_0}{A_i}\right| \angle\phi = 1e^{-S\tau_d} \qquad ...(4.183)$$

The frequency response curves for dead time elements are shown in Fig. 4.50.

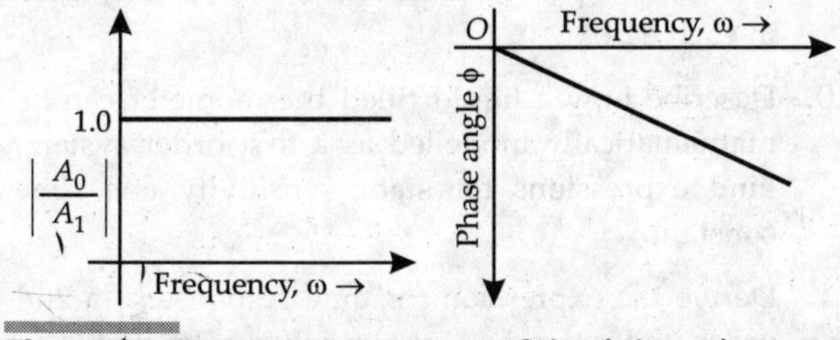


**Fig. 4.50** Frequency response of dead time elements.

## 4.34 CORRELATIONSHIP BETWEEN TIME AND FREQUENCY RESPONSE OF SECOND ORDER SYSTEMS

There exists a correlationship between time and frequency response of second order systems. Consider the step response of a second order system. The values of frequency of damped oscillations $\omega_d$ and peak overshoot $M_P$ in an underdamped second order system when subjected to a step input is given of Eqns. 4.134 and 4.148 as :

$$\omega_d = \omega_n \sqrt{1-\xi^2}$$

and $$M_P = \exp[-\pi\xi / \sqrt{(1-\xi^2)}]$$

The same system when subjected to a variable frequency sinusoidal input of unit amplitude gives the values of resonant from frequency $\omega_r$ and resonant peak (See Eqns. 4.176 and 4.177) as :

$$\omega_r = \omega_r \sqrt{1-2\xi^2}$$

$$M_r = \frac{1}{2\xi\sqrt{1-\xi^2}}$$

Comparison between $\omega_d$ and $\omega_r$, reveals that there exists a definite correlationship between them. The ratio between the two frequencies is,

$$\frac{\omega_r}{\omega_d} = \sqrt{\frac{(1-2\xi^2)}{(1-\xi^2)}} \qquad ...(4.184)$$

The ratio $\omega_r / \omega_d$ is a function of damping factor ξ plotted against it in Fig. 4.51. It should be noted that correlation ceases for values of $\xi > 0.707$ as no real value of $\omega_r$ exists for $\xi > 0.707$.

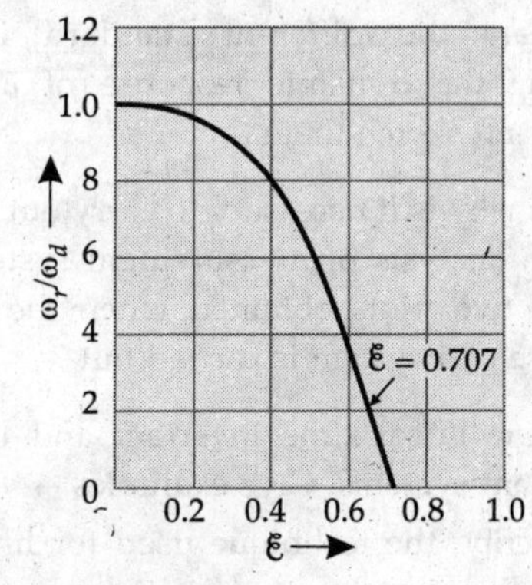


**Fig. 4.51** Correlationship between $\omega_r$ and $\omega_d$.

The values of $M_R$ and $M_r$ plotted against ξ are shown in Fig. 4.52. These plots show that the two have correlation for $0 < \xi < 0.707$ as both are functions of $\xi$. This means that a system with a particular value of $M_r$ will have some definite value of $M_P$ for

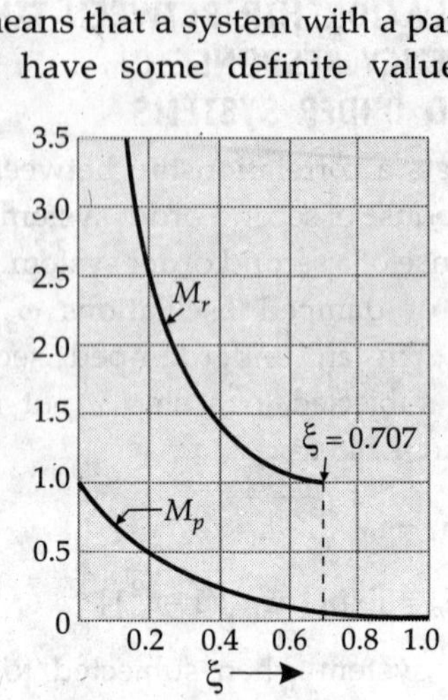


**Fig. 4.52** Correlationship between $M_r$ and $M_p$.

$0 < \xi < 0.707$ when subjected to a unit step input. The plots also show that values of both $M_P$ and $M_r$ decrease with increase in value of $\xi$ till $\xi = 0.707$ after which the correlationship ceases as there is no value of $M_r$ for $\xi > 0.707$.

The settling time in time domain is $t_s = 4/\xi\omega_n$ for $\pm$ 2% band (See Eqn. 4.149) and is a function of $\xi$ and $\omega_n$. It is indicative of speed of response in time domain. The expression for bandwidth in frequency domain is,

$$\omega_b = \omega_n [1 - 2\xi^2 + \sqrt{2 - 4\xi^2 + 4\xi^4}\,]^{1/2}$$

(see Eqn. 4.180)

Therefore bandwidth is a function of $\xi$ and $\omega_n$ and hence is indicative of speed of response in frequency domain.

# Review Questions

1. (*a*) Define the dynamic response of a system and distinguish between steady state response and transient response.

   (*b*) Define periodically varying inputs and transient inputs. Give examples.

2. Define the following terms :

   (*i*) Speed of response

   (*ii*) Measuring lag

   (*iii*) Fidelity and dynamic error.

3. What are the different standard inputs for studying the dynamic response of a system ? Define and sketch them.

4. Discuss why is it necessary to carry out Frequency Domain analysis of measurement systems. What are the two plots obtained when the frequency response of a system is carried out ?

5. (*a*) Define linear time invariant and linear time variant systems. Give examples.

   (*b*) Describe the technique used for linearizing a non-linear system for limited range about the operating point.

6. Derive the differential equation describing the dynamics of mechanical translational system subjected to a step input force. Also write the differential equation describing the time response of a series *R-L-C* circuit when a step voltage is applied to it. Draw up a table showing analogy between mechanical and electrical quantities based upon force-voltage analogy.

7. Derive the transfer function of a thermal system when the heat input rate is suddenly increased from its steady conditions. Discuss the assumptions made.

8. Derive the differential equation describing the dynamics of liquid level system when liquid inflow rate is suddenly increased from its steady state conditions. Assume turbulent flow. Is the differential equation valid for the entire range of operation or a limited range about the operating point ?

9. Derive the equations for time response of a first order system when subjected to

   (*i*) unit step input (*ii*) unit ramp input.

   Draw the response curves and find the steady error in each ease.

10. Describe how a liquid filled thermometer can be mathematically modelled as a first order system. Find expressions for static sensitivity and time constant.

11. Derive the expression for time response of a 2nd order under damped system when subjected to a

unit step input. Sketch the response. Define the following terms and find expressions for them.

(i) Rise time (ii) Peak time

(iii) Peak overshoot

(iv) Settling time.

12. Derive the expression for time response of a 2nd order under damped system when subjected to a unit ramp input. Show that the nature of the response is the same as that for a unit step input. Find expression for the steady state error.

13. Derive the expressions for magnitude and phase of a first order system when subjected to a sinusoidal input signal. Draw the magnitude and phase versus frequency plots.

Describe how the time constant of the system influences the frequency response of a first order system.

14. (a) Derive the expressions for frequency response of a 2nd order system. Draw the magnitude and phase versus frequency plots for damping ratio smaller than 0.707.

(b) Define and derive expressions for the following for a 2nd order system :

(i) Resonant frequency

(ii) Resonant peak

(iii) Bandwidth.

15. Draw the normalized frequency response curves for a 2nd order system for different values of damping factor and show that if the magnitude of the output is a faithful representation of the magnitude of input the damping factor should be between 0.6 and 0.7. Comment upon the values of phase shift and time lag under these conditions.

# Unsolved Problems

1. Figure 4.53 shows a non-linear system. Prove that the properties of superposition and homogeneity are not satisfied.

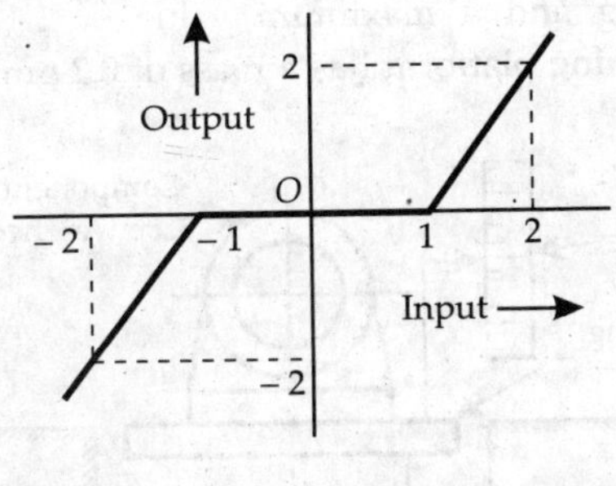


**Fig. 4.53** Response of a non-linear system.

2. A first order system when subjected to a step input has a temperature rise of 25°C after one hour and 37.5° after two hours starting from cold conditions. Calculate its final steady temperature rise and the thermal time constant. **[Ans.** 50°C ; 1.44 hour**]**

3. The flow through orifice of 30 mm diameter is turbulent. The flow co-efficient for the orifice is 0.6. The head varies from 1.5 m to 10 m. Calculate the values of discharge for the following heads.

(i) 1.5 m

(ii) 2.5 m

(iii) 5.0 m (iv) 7.5 m and

(v) 10 m.

Plot a graph between discharge and the head. Calculate the values of liquid resistance at the above heads from the graph.

**[Ans.** (i) $2.13 \times 10^{-3}$ m$^3$/s (ii) $2.75 \times 10^{-3}$ m$^3$/s (iii) $3.89 \times 10^{-3}$ m$^3$/s (iv) $4.76 \times 10^{-3}$ m$^3$/s (v) $3.5 \times 10^{-3}$ m$^3$/s
Resistance : 1410, 1820, 2572, 3150, 3638 s/m$^2$**]**

4. An *RC* circuit consists of a capacitor of 1 μF in series with a resistor of 5 kΩ. A d.c. voltage of 50 V is suddenly applied across the circuit. Calculate the value of voltage after (a) 5 ms and (b) 25 ms. **[Ans.** (a) 31.6 V ; (b) 49.6 V**]**

5. A measuring system has an exponential response to step input. The time constant of the system is 2 s. What is the time required for reaching (a) 50 per cent and (b) 80 per cent of final steady state reading ? **[Ans.** (a) 1.39 s (b) 3.2 s**]**

6. A thermometer is initially at a temperature of 70°C and is suddenly placed in liquid which is maintained at 300°C. The thermometer indicates 200°C after a time interval of 3 s. Calculate the time constant for the thermometer. Also calculate the

7. In a response test on a thermometer, it was put into a temperature-controlled bath of water maintained at 100°C, and the time was observed as the indicated temperature reached the preselected values.

The results were as under :

| *Time (s)* | *Temperature (°C)* |
|---|---|
| 0.0 | 20 |
| 1.2 | 40 |
| 3.0 | 60 |
| 5.6 | 80 |
| 8.0 | 90 |
| 11.0 | 95 |
| 15.0 | 98 |
| 18.0 | 99 |

Plot the graph of temperature versus time. Examine the graph and find out the order of system. Estimate the time constant of the thermometer.

Determine the steady state error when the same thermometer is made to measure the temperature of a liquid which is cooling at a constant rate of 0.2°C/s. **[Ans.** First order system ; 4s, + 0.8°C]

8. A balloon is equipped with temperature and altitude-measuring instruments and has radio equipment which can transmit the output readings of these instruments back to ground. The balloon is initially fixed to the ground with the instrument output readings in steady state. The altitude measuring instrument is approximately zero order and the temperature transducer first order with a time constant of 15 s. The temperature of the ground $\theta_0$ is 10°C and the temperature $\theta_x$ at an altitude of $x$ metre is given by the relation :

$$\theta_x = \theta_0 - 0.01x$$

(*a*) If the balloon is released at time zero and thereafter rises upwards at a velocity of 5 m/s, draw a table showing the temperature and altitude measurements reported at intervals of 10 s over the first 50 s of travel. Show also the error in each temperature reading.

(*b*) What does the balloon report at an altitude of 5000 m ?

| *Time, t (s)* | *Altitude, x metre* | *Actual Temperature °C* | *Reported Temperature °C* | *Error °C* |
|---|---|---|---|---|
| 0 | 0 | 10 | 10 | 0 |
| 10 | 50 | 9.5 | 9.86 | 0.36 |
| 20 | 100 | 9.0 | 9.55 | 0.55 |
| 30 | 150 | 8.5 | 9.15 | 0.65 |
| 40 | 200 | 8.0 | 8.70 | 0.70 |
| 50 | 250 | 7.5 | 8.22 | 0.72 |

**[Ans.** 8 (*a*) (*b*) –38.25° C]

9. A d'Arsonval galvanometer is a second order instrument. In the design of a d'Arsonval galvanometer's moving system, it is found that the damping ratio is 0.65 and the natural frequency of the undamped oscillations is 4 Hz.

(*a*) If the sensitivity of the movement is doubled by using springs of smaller stiffness, calculate the new damping ratio and the new natural frequency.

(*b*) If the damping ratio is now restored to its original value by altering the moment of inertia of the system, determine the natural frequency of the system.

**[Ans.** (*a*) 0.92, 2.83, Hz (*b*) 2 Hz]

10. A production line device is shown in Fig. 4.54, the masses to be weighed have a minimum value of 0.6 kg and a maximum value of 1.1. kg. The weighing platform has a mass of 0.2 gm.

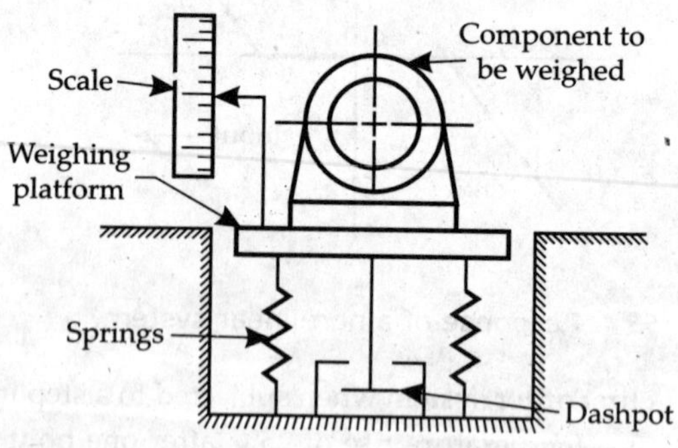


**Fig. 4.54** Weighing platform.

(*a*) Calculate the value of spring stiffness if the scale deflection is to be within ± 10 mm over the range.

(*b*) Calculate the value of damping constant provided by the dashpot when a mass of 1 kg is placed on the platform. The damping ratio is 0.7.

(c) Calculate the damping ratio with the minimum and maximum mass on the table.

(d) Determine the percentage overshoot with minimum and maximum mass on the table when they are suddenly placed on it.

[**Ans.** (a) 0.981 N/mm (b) 48 N/m-s$^{-1}$ (c) 0.73, 0.67 (d) 4%, 6%]

**11.** A first order thermometer is used to measure the temperature of air cycling at a rate of 1 cycle every five minutes. The time constant of the thermometer is 20 s. Calculate the percentage attenuation indicated by the thermometer.
If the temperature has a sinusoidal variation of ± 20°C, calculate the indicated variation in the indicated temperature. [**Ans.** 7.78%, ± 18.44°C]

**12.** Determine the ratio of output to input temperature when a first order thermometer is suddenly immersed in a tank of liquid whose temperature varies sinusoidally at a frequency of 1/300 Hz. The thermometer has a time constant of 160 s. Find also the time lag. If the temperature varies between ± 20 °C, what is the indicated value of temperature ? [**Ans.** 0.286, 61 s, ± 5.7°C]

**13.** If a cyclic change is taking place every 10 s and the time constant of the system is 10 s, what percentage of cyclic change will be registered assuming the system to be first order one ? If the time constant is doubled what would be the magnitude ratio ? [**Ans.** 15.7 percent ; 7.9 percent]

**14.** If the temperature of an oven is cycling at a true rate of ± 10°C every 40 seconds, will any thermometer show the ± 10°C variation ? If the time constant of the thermometer is 300 s, what will be the apparent cyclic variation ? [**Ans.** ± 0.212°C]

**15.** A pressure measuring gauge is first order instrument. If the pressure applied were 700± 140 kN / m$^2$ every 2 seconds (assuming a sinusoidal variation of pressure with time) find the indicated range of pressure. The time constant of the gauge is 1 s.

**16.** A first order instrument which has a time constant of 0.5 s is subjected to an input of :

$$2\sin 3t + 5\sin 30t$$

Find the response of the instrument.

[**Ans.** $1.11\sin(3t - 56°) + 0.0053\sin(30t - 86°)$]

**17.** A pressure is cycling from 50 kN/m$^2$ to 65 kN/m$^2$ with a period of 5 s. If the attenuation must be limited to 10 percent, what is the maximum time constant permissible in a pressure gauge with a first order response ? [**Ans.** 17.9 s]

**18.** In a pressure measuring transducer consisting of a number of strain gauges bonded to a diaphragm the undamped natural frequency of the system is found to be 60 kHz, and the damping ratio 0.6. Calculate the amplitude of the output signal in terms of pressure, if the input pressure to the transducer is fluctuating sinusoidally with an amplitude of 800 kN/m$^2$ at a frequency of 30 kHz. What is the percentage error in the amplitude of output ? [**Ans.** 832 kN/m$^2$ ; + 4%]

**19.** Plot the amplitude ratio for a second order system for values of the normalized frequency ratio from 0 to 1 for a damping ratio of 0.6. From the graph, determine the range of the normalized frequency for which the output amplitude is within 3% of the input amplitude. If the system has an undamped natural frequency of 3 kHz, determine the range over which this applies.

[**Ans.** up to 2.5 kHz]

**20.** A second order system has a natural frequency of 10 Hz and a damping ratio of 0.3. Calculate the value of resonant frequency and the magnitude ratio at this frequency. Also calculate the bandwidth.
What is settling time of the system for a ± 2% band when subjected to a step input ? Find the value of steady state error to unit ramp input.

[**Ans.** 9.05 Hz, 1.75, 14.5 Hz, 0.212 s, 9.5 ms.]

# Objective Type Questions

**Tick (√) the most appropriate answer**

1. A quantity whose magnitude has a definite repeating time cycle is called a :
   (a) transient
   (b) steady state periodic
   (c) steady state aperiodic
   (d) transient state periodic.

2. Dynamic response consists of :
   (a) two parts, one steady state and the other transient state response
   (b) only transient state response
   (c) only steady state response
   (d) steady state and transient frequency response.

3. The following are the desirable dynamic characteristics of a measurement system :
(a) fast response, fidelity, measuring lag and dynamic error
(b) fast response and measuring lag
(c) fidelity and measuring lag
(d) fast response and fidelity.

4. The output of a system is given by the following equation : $y = ax + b$
where $x$ is the input and $a$ and $b$ are constants.
The system exhibits :
(a) linear behaviour
(b) non-linear behaviour as it does not satisfy the property of superposition
(c) non-linear behaviour as it does not satisfy the property of homogeneity
(d) non-linear behaviour because it does not satisfy properties of both superposition and homogeneity.

5. The dynamic response of a system is given by the following differential equation :

$$a_3 \frac{d^3c(t)}{dt^3} + a_2 \frac{d^2c(t)}{dt^2} + a_1 \frac{dc(t)}{dt} + a_0 c(t) = b_2 \frac{d^2r(t)}{dt^2} + b_1 \frac{dr(t)}{dt} + b_0 r(t)$$

where $c(t)$ and $r(t)$ are respectively the output and input and $a_3, a_2, a_1, a_0, b_2, b_1$ and $b_0$ are constants. It is a :
(a) non-linear system
(b) linear system
(c) linear time variant system
(d) linear time invariant system.

6. In force-current analogy stiffness constant $K$ is analogous to :
(a) inductance
(b) reciprocal of inductance
(c) capacitance
(d) reciprocal of capacitance.

7. In force-voltage analogy viscous friction constant $B$ is analogous to :
(a) resistance
(b) reciprocal of resistance
(c) inductance
(d) capacitance.

8. In thermal systems if $M$ is mass of liquid stored, $Q$ is the liquid inflow rate and $s$ is the specific heat, the thermal resistance is given by :
(a) $Ms$ (b) $Qs$
(c) $1/Ms$ (d) $1/Qs$

9. In liquid level systems, the units of capacitance of a tank are :
(a) $m^2$ (b) $m^3$
(c) $m^2/s$ (d) $m^{-2}s$

10. The transfer function of a system is defined as :
(a) ratio of Laplace transform of input to laplace transform of output with initial conditions not equal to zero.
(b) ratio of Laplace transform of output to laplace transform of input with initial conditions not equal to zero.
(c) ratio of Laplace transform of input to laplace transform of output with all initial conditions equal to zero.
(d) ratio of Laplace transform of output to laplace transform of input with all initial conditions equal to zero.

11. A first order system is subjected to a step input. Its time constant can be defined as :
(a) time for the output to rise to 63.2% of its final steady value in the case of rising output.
(b) time for the output to drop to 36.8% of its initial value in the case of a decaying output.
(c) time for output to rise (in case of rising output) and decrease to (in case of a decaying output) reach its final steady value if the initial rate of change of output is maintained
(d) all the above.

12. A first order system has a time constant of 20 s. It is subjected to a step input. The settling time of the output is assumed to be the time it reaches 95% of the final steady state value. The settling time of the system is :
(a) 100 s (b) 60 s
(c) 20 s (d) infinity

13. An $RC$ series circuit has a resistance of 2 MΩ and a capacitance of 1 μF. It is subjected to a step input of 10 V. The dynamic error (reference to voltage across capacitor) at time $t = 4$ s of application of input is :
(a) 1.35 V (b) 0.5 V
(c) 3.68 V (d) 0 V.

14. The volume of bulb of a liquid filled thermometer is 1000 mm$^3$. Its time constant is 300 s it is desired to reduce the time constant 100 s. What is the new volume of the bulb, if all other conditions remain the same except the dimensions of the bulb :

(a) 333.3 $mm^3$ (b) 37 $mm^3$
(c) 10 $mm^3$ (d) 577 $mm^3$

15. The volume of bulb of a liquid filled thermometer is 1000 $mm^3$. Its time constant and sensitivity are respectively 300 s and 3.6 mm/°C. If the volume of the bulb is reduced to 500 $mm^3$, the sensitivity becomes :
(a) 7.2 mm/°C
(b) 10.8 mm/°C
(c) 1.8 mm/°C
(d) 1.2 mm/°C if other conditions remain the same.

16. A balloon is fitted with a first order temperature measuring device with a time constant of 20 s. It is also fitted with a zero order instrument for altitude measurements. It radios back to earth both temperature and altitude signals. The decrease in temperature is $0.02\, x\, °C$ where $x$ is the altitude in metre. The balloon travels vertically at a speed of 5 m/s. The balloon radios back altitude signal as 3000 m. The decrease in temperature radios back to earth is :
(a) 60°C (b) 58°C
(c) 62°C (d) none of the above.

17. A 2 gm mass is suspended from a simple spring. The deflection caused is 5 mm. The natural frequency of the system is :
(a) 7 Hz (b) 22 Hz
(c) 2.2 Hz (d) none of the above.

18. In a second order system, the damping factor is 0.65 and the natural frequency is 4 Hz. If the stiffness of springs is halved, the new values of damping factor and natural frequency are :
(a) 0.92, 2.83 Hz (b) 0.46, 5.66 Hz
(c) 0.92, 5.66 Hz
(d) 0.46, 2.83 Hz.

19. In a second order system, the steady state error is 0.02. If both viscous friction constant and spring constant are doubled, the steady error will be :
(a) 0.01 (b) 0.04
(c) 0.0025 (d) 0.02.

20. In second order system, the frequency of damped oscillations is 18 rad/s. The value of damping factor is 0.866. The natural frequency of oscillations is :
(a) 15.6 rad/s (b) 19.3 rad/s
(c) 36 rad/s (d) 9 rad/s.

21. In a second order system, the peak overshoot is 100%. The value of damping factor is :
(a) 1 (b) 0.707
(c) 0 (d) 0.5.

22. For a second order system, the settling time for ± 2% band is :
(a) $\frac{3}{\xi\omega_n}$ (b) $\frac{5}{\xi\omega_0}$
(c) $2\xi\omega_n$ (d) $\frac{4}{\xi\omega_n}$

where $\xi$ = damping factor and $\omega_n$ = natural frequency : rad/s.

23. In a second order under damped system, the value of peak overshoot in per unit terms is :
(a) $M_P = \exp\left(\frac{\pi\xi}{\sqrt{1-\xi^2}}\right)$
(b) $M_P = -\exp\left(\frac{\pi\xi}{\sqrt{1-\xi^2}}\right)$
(c) $M_P = \exp\left(\frac{-\pi\xi}{\sqrt{1-\xi^2}}\right)$
(d) $M_P = \left(\frac{\pi\xi}{\sqrt{1-\xi^2}}\right)$

where $\xi$ = damping factor.

24. A first order thermometer has a time constant of 50 s. It is subjected to a sinusoidal input cycling at 0.002 Hz. The time lag of the instrument is :
(a) 50 s (b) 44.6 s
(c) 500 s (d) 0.01 s.

25. In order that a first order instrument should indicate an output which is within ± 10% of the true value of the output when subjected to a sinusoidal input the product $\omega\tau$, where $\omega$ = angular frequency of input and $\tau$ time constant of instrument should be :
(a) less than 0.5
(b) greater than 0.5
(c) less than 10
(d) greater than 10.

26. A second order system when subjected to a unit step input has a peak overshoot of 10%. The same system when subjected to a sinusoidal input of unit amplitude will have a resonant peak of nearly :
(a) 103% (b) 110%
(c) 190% (d) none of the above.

27. A second order underdamped system has a damping factor of 0.8. It is subjected to a sinusoidal input of unit amplitude. It has resonant peak of :
(a) 108% (b) 92%
(c) 20% (d) it has no resonant peak.

**28.** Bandwidth, a frequency domain concept, is indicative of

(*a*) rise time in time domain

(*b*) settling time in time domain

(*c*) steady state error in the domain

(*d*) all the above.

**29.** An undamped second order instrument has a natural frequency of 500 Hz. It is subjected to a sinusoidal input of unit amplitude at 500 Hz. The amplitude of output is :

(*a*) unity (*b*) 0.5

(*c*) 2.0 (*d*) infinity.

**30.** The transfer function of a system is :

$$G(s) = \frac{100e^{-st}}{s(s+10)}$$

The system

(*a*) is a linear system

(*b*) is a non-linear system

(*c*) has a transportation lag

(*d*) none of the above.

## *Answers*

| | | | | | | | | | |
|---|---|---|---|---|---|---|---|---|---|
| **1.** (*b*) | **2.** (*a*) | **3.** (*d*) | **4.** (*c*) | **5.** (*d*) | **6.** (*b*) | **7.** (*a*) | **8.** (*d*) | **9.** (*a*) | **10.** (*d*) |
| **11.** (*d*) | **12.** (*b*) | **13.** (*a*) | **14.** (*b*) | **15.** (*c*) | **16.** (*b*) | **17.** (*a*) | **18.** (*a*) | **19.** (*d*) | **20.** (*c*) |
| **21.** (*c*) | **22.** (*d*) | **23.** (*c*) | **24.** (*b*) | **25.** (*a*) | **26.** (*a*) | **27.** (*d*) | **28.** (*b*) | **29.** (*d*) | **30.** (*c*) |

# 5

# Units, Systems, Dimensions and Standards

## 5.1 INTRODUCTION

A considerable number of systems of Units have been used at various times during human history. Some systems are of historical interest only, some are of theoretical interest while others have been used in actual experimental work and have been found acceptable in some countries. However, there are some systems of units which have been accepted throughout the world. It is thus natural for us to acquaint ourselves with important and frequently used systems of units.

## 5.2 UNIT

The result of a measurement of a physical quantity must be defined both in kind and magnitude. The standard measure of each kind of physical quantity is called a **Unit**. Measurement implies comparison with a standard value. For example, we can measure the numerical ratio ($L_1/L_0$) of a length $L_1$, to another length $L_0$, provided that $L_0$ is well defined, does not change with time and is readily reproducible to a high degree of accuracy, it can be chosen as a unit of length. In general, we can write :

Magnitude of a physical quantity

$$= \text{(Numerical ratio)} \times \text{(Unit)} \qquad ...(5.1)$$

The **Numerical Ratio** is the number of times the unit occurs in any given amount of the same quantity and, therefore, is called the **number of measures.** The numerical ratio may be called **numerical multiplier** also. From Eqn. 5.1 it is clear that the process of measurement entails estimating the numerical ratio or the number of measures. Thus when we speak of a distance of 10 metre, we know that the metre is the unit of length and the number of units in the length is ten. The physical quantity, length, in this case, is defined by the unit, **metre**. It should be understood that the numerical ratio has no physical meaning without the Unit.

## 5.3 ABSOLUTE UNITS

An absolute system of units is defined as a system in which the various units are all expressed in terms of a small number of fundamental units. Absolute measurements do not compare the measured quantity with arbitrary units of the same type but are made in terms of *Fundamental Units*.

The word **absolute** should not be confused. In its long term usage in measurements, it has been associated with relating units to the units of length, mass and time.

## 5.4 FUNDAMENTAL AND DERIVED UNITS

When we measure a physical quantity, we must express the magnitude of that quantity in terms of a unit and a numerical multiplier (as explained in Art. 5.2). If we were concerned with measurement of only one quantity we could, if we wished, select for our unit any magnitude that seemed convenient to use. However, in measurements, we are concerned with a large number of quantities which are related to each other, through established physical equations, and, therefore, we cannot choose size of units of these quantities arbitrarily and independently. This way we can avoid the use of awkward numerical constants when we express a quantity of one kind which we have derived from measurement of another quantity.

In any given piece of theoretical work the Units of the various quantities must all belong to a **consistent or coherent** system of units.

In Science and Technology, *two* kinds of units are used :

(*i*) Fundamental units and (*ii*) Derived units.

If we have $A$ kinds of quantities to evaluate and $B$ independent physical quantities expressing relationships between them, we can independently choose sizes of units of $(A - B)$ quantities. The sizes of units of remaining $B$ quantities can be derived with the help of

$B$ physical equations. A system of units so constructed is called **a consistent system**. $(A - B)$ units, which are independently chosen are called **Fundamental Units**. The remaining $B$ units are called **Derived Units**.

The **Fundamental Units** in mechanics are measures of length, mass and time. Since length, mass and time are fundamental to most other physical quantities besides those in mechanics, they are called the **Primary Fundamental Units**.

Measures of certain physical quantities in the thermal, electrical, illumination fields are also represented by fundamental units. These units are used only where these particular disciplines are involved and therefore they are called **Auxiliary Fundamental Units**.

All other units which can be expressed in terms of fundamental units with the help of physical equations are called **Derived Units**. Every derived unit originates from some physical law or equation which defines that unit. For example, the volume, $V$, of a room is equal to the product of its length $l$, width $b$ and height $h$. Therefore, $V = lbh$.

If the metre is chosen as the unit of length, then the volume of a room $6\,\text{m} \times 4\,\text{m} \times 5\,\text{m}$ is $120\ \text{m}^3$. Note that the number of measures $(6 \times 4 \times 5 = 120)$ as well the units $(\text{m} \times \text{m} \times \text{m} = \text{m}^3)$ are multiplied. The derived unit for volume is thus $\text{m}^3$. Also note that use of an awkward numerical multiplier has been avoided.

## 5.5 DIMENSIONS

Disregarding the problem of measurement and all concepts of magnitude, it is evident that every quantity has a quality which distinguishes it from all other quantities. This unique quality is called **Dimension**. The dimension is written in a characteristics notation, as for example, $[L]$ for length, $[T]$ for time etc.

A derived unit is always recognized by its **Dimensions**, which can be defined as the complete algebraic formula for the derived unit. Thus when quantity such as area $A$ of a rectangle is measured in terms of other quantities *i.e.*, length, $l$, and width, $b$, in this case, the relationship is expressed mathematically as,

$$\text{Area } A = \text{a constant} \times l \times b \qquad ...(5.2)$$

Since $l$ and $b$ each have the dimensions of a length, written $[L]$, the dimensions of area are,

$$[A] = [L][L] = [L^2] \qquad ...(5.3)$$

Since the constant is a pure numerical ratio and is, therefore, dimensionless. As discussed earlier independent units for derived quantities like volume, area etc. are unnecessary and even inconvenient, for if we take, the constant in Eqn. 5.2 as unity, and if metre (m) is unit of length, metre square $(\text{m}^2)$ can be used as unit for area.

As stated earlier, in mechanics the three fundamental units are length, mass and time. Their dimensional symbols are :

$$\text{Length} = [L]\ ;\ \text{Mass} = [M]\ ;\ \text{Time} = [T]$$

The square brackets indicate dimensional notation only. It should be understood that equality is in terms of dimensions only, and it should not be mixed up with numerical values.

## 5.6 DIMENSIONS OF MECHANICAL QUANTITIES

All mechanical quantities can be expressed in terms of the three fundamental quantities—length, mass and time. A few examples are being given below :

**1.** Velocity $= \dfrac{\text{length}}{\text{time}}$

$$\therefore \quad [v] = \frac{[L]}{[T]} = [LT^{-1}]$$

**2.** Acceleration $= \dfrac{\text{velocity}}{\text{time}}$

$$\therefore \quad [a] = \frac{[LT^{-1}]}{[T]} = [LT^{-2}]$$

**3.** Force = mass × acceleration

$$\therefore \quad [F] = [M][LT^{-2}] = [MLT^{-2}]$$

**4.** Work = force × distance

$$\therefore \quad [W] = [MLT^{-2}][L] = [ML^2T^{-2}]$$

**5.** Power $= \dfrac{\text{work}}{\text{time}}$

$$\therefore \quad [P] = \frac{[ML^2T^{-2}]}{[T]} = [ML^2T^{-3}]$$

**6.** Energy = power × time

$$[W] = [ML^2T^{-3}][T] = [ML^2T^{-2}]$$

**7.** Momentum = mass × velocity

$$= [M][LT^{-1}] = [MLT^{-1}]$$

**8.** Torque = force × distance

$$= [MLT^{-2}][L] = [ML^2T^{-2}]$$

**9.** Stiffness $= \dfrac{\text{torque}}{\text{angle}}$

$$[K] = [ML^2T^{-2}]$$

**10.** Surface Tension $= \dfrac{\text{force}}{\text{length}}$

$$[\sigma] = \frac{[MLT^{-2}]}{[L]} = [MT^{-2}]$$

**11.** Dynamic Viscosity $= \dfrac{\text{force}}{\text{area}} \div \dfrac{\text{velocity}}{\text{distance}}$

$$[\mu] = \frac{[MLT^{-2}]}{[L^2]} \div \frac{[LT^{-1}]}{[L]} = [ML^{-1}T^{-1}]$$

## 5.7 CGS SYSTEM OF UNITS

Till now the most commonly used units in electrical work were CGS units. These units involve the use of unit of a fourth quantity in addition to units of mass, length and time. Experimentally observed facts of either electrostatics or electromagnetism are used for defining CGS units for electrical work. Therefore, there are two systems of CGS units.

### 5.7.1 Electromagnetic Units (e.m. units)

Units based on electromagnetic effects are known as electromagnetic units or e.m. units and the system is known as electromagnetic system of units or e.m.u. system. This system involves the units of four quantities ; permeability ($\mu$) of the medium and the units of length, mass and time. The value of permeability of free space (vacuum) is taken as unity in this system.

### 5.7.2 Electrostatic Units (e.s. units)

Units based on electrostatic effects are known as electrostatic units or e.s. units and the system is known as electrostatic system of units or e.s.u. system. This system involves the units of four quantities ; permittivity ($\varepsilon$) of the medium and the units of length, mass and time. The value of permittivity of free space is taken as unity in this system.

The e.m.u. system is more convenient to use from the point of view of most electrical measurements, and is, much frequently used than the electrostatic system. If a quantity is expressed in C.G.S. system without any additional designation "electromagnetic" or "electrostatic" it may be taken that the electromagnetic system has been used.

## 5.8 PRACTICAL UNITS

Practical units are derived either from the absolute units or by reference to arbitrary standards. These units should be easy to handle and manage.

In working with either of the absolute system mentioned above, some most commonly used units are found to be inconveniently large or small. The practical units employed are decimal multiples, or submultiples of the absolute electromagnetic units.

The practical units, based on e.m. units, are made larger or smaller than the corresponding e.m. units by an appropriate power of 10, to bring those which are mostly used in experimental work to convenient and manageable magnitudes.

The British Association Committee on electrical units decided that the practical unit of resistance (ohm) should be $10^9$ C.G.S. electromagnetic units of resistance. It was also decided to make the practical unit of potential difference (volt) equal to $10^8$ C.G.S. electromagnetic units of potential difference. These two magnitudes being fixed, we can fix the practical units of other quantities. For example, the practical unit of current (ampere) would be

$$\text{ampere} = \frac{\text{volt}}{\text{ohm}} = \frac{10^8 \text{ e.m. units}}{10^9 \text{ e.m. units}}$$

$$= 10^{-1} \text{ e.m. unit of current.}$$

Thus the practical unit of current is 1/10 of a C.G.S. electromagnetic unit. Table 5.1 shows the symbols and magnitudes of practical units in terms of corresponding C.G.S. electromagnetic units of different quantities.

The sizes of practical units of resistance and potential difference were arbitrarily chosen to be of convenient magnitude and the other units are then derived from these units.

**Table 5.1** *Practical Units*

| No. | Quantity | Practical Unit | Symbol | Number of C.G.S. e.m. units in one practical unit | Defining Equation |
|---|---|---|---|---|---|
| 1. | Charge | coulomb | Q | $10^{-1}$ | $Q = It$ |
| 2. | Current | ampere | I | $10^{-1}$ | $E = IR$ |
| 3. | Potential Difference | volt | E | $10^8$ | |
| 4. | Resistance | ohm | R | $10^9$ | |
| 5. | Inductance | henry | L | $10^9$ | $E = L\frac{dI}{dt}$ |
| 6. | Capacitance | farad | C | $10^{-9}$ | $Q = CE$ |
| 7. | Power | watt | P | 1 watt = $10^7$ ergs per second | |
| 8. | Energy | joule | W | 1 joule = $10^7$ ergs | |

## 5.9 DIMENSIONAL EQUATIONS

All mechanical quantities can be represented in terms of the dimensions of length, mass and time, but electrical quantities need at least one other fundamental dimension.

### 5.9.1 Dimensions in Electrostatic System

In this system the dimension of permittivity $\varepsilon$ is taken as the fourth fundamental dimension. We use Coulomb's law as our starting point.

***1. Charge.*** According to Coulomb's law, the force exerted between two charges $Q_1$ and $Q_2$ is :

$$F = \frac{Q_1 Q_2}{\varepsilon d^2}$$

where $d$ is the distance between charges $Q_1$ and $Q_2$. Putting the dimensions of force and distance, we have,

$$[MLT^{-2}] = \frac{[Q^2]}{[\varepsilon][L^2]}$$

$\therefore$ Dimension of charge $[Q] = [\varepsilon^{1/2} M^{1/2} L^{3/2} T^{-1}]$

***2. Current.*** Current is charge per unit time. Let $[I]$ represent the dimensions of current.

$$[I] = \frac{[Q]}{[T]} = \frac{[\varepsilon^{1/2} M^{1/2} L^{3/2} T^{-1}]}{[T]}$$

$$= [\varepsilon^{1/2} M^{1/2} L^{3/2} T^{-2}]$$

***3. Potential difference or emf.*** Potential difference is work done per unit charge.

$$\therefore \quad [E] = \frac{[W]}{[Q]} = \frac{[ML^2T^{-2}]}{[\varepsilon^{1/2} M^{1/2} L^{3/2} T^{-1}]}$$

$$= [\varepsilon^{-1/2} M^{1/2} L^{1/2} T^{-1}]$$

where $[E]$ and $[W]$ represent the dimensions of emf and work respectively.

***4. Capacitance.*** Capacitance $C = Q/E$.

Therefore, dimensions of capacitance are :

$$[C] = \frac{[Q]}{[E]} = \frac{[\varepsilon^{1/2} M^{1/2} L^{3/2} T^{-1}]}{[\varepsilon^{-1/2} M^{1/2} L^{1/2} T^{-1}]} = [\varepsilon L]$$

***5. Resistance.*** Resistance $R = E/I$.

Therefore, dimensions of resistance are,

$$[R] = \frac{[E]}{[I]} = \frac{[\varepsilon^{-1/2} M^{1/2} L^{1/2} T^{-1}]}{[\varepsilon^{1/2} M^{1/2} L^{3/2} T^{-2}]} = [\varepsilon^{-1} L^{-1} T]$$

***6. Inductance.***

$$\text{Inductance } L = \frac{\text{emf}}{\text{rate of change of current}}$$

$$= \frac{E}{dI/dt}$$

$\therefore$ Dimensions of inductance are,

$$[L] = \frac{[E]}{[I]/[T]} = \frac{[E][T]}{[I]}$$

$$= \frac{[\varepsilon^{-1/2} M^{1/2} L^{1/2} T^{-1}][T]}{[\varepsilon^{1/2} M^{1/2} L^{3/2} T^{-2}]} = [\varepsilon^{-1} L^{-1} T^2]$$

### 5.9.2 Dimensions in Electromagnetic System

The fourth dimension in this system is the dimension of permeability $\mu$. The starting point is the law of force between poles.

***1. Pole strength.*** Force $F = \dfrac{m_1 m_2}{\mu d^2}$

where $d$ is the distance between poles of strengths $m_1$ and $m_2$. Writing the dimensions, we have,

$$\therefore \quad [MLT^{-2}] = \frac{[m^2]}{[\mu][L^2]}$$

Thus, dimensions of pole strength are,

$$[m] = [\mu^{1/2} M^{1/2} L^{3/2} T^{-1}]$$

***2. Magnetizing force.*** Magnetizing force $H$ is measured by force exerted on a unit pole. Therefore, dimensions of magnetizing force are :

$$[H] = \frac{[F]}{[m]} = \frac{[MLT^{-2}]}{[\mu^{1/2} m^{1/2} L^{3/2} T^{-1}]}$$

$$= [\mu^{-1/2} M^{1/2} L^{-1/2} T^{-1}]$$

***3. Current.*** We know that the magnetizing force at the centre of a loop of radius $r$ is,

$$H = \frac{2\pi I}{r} \quad \text{if } I \text{ is in e.m. units}$$

or

$$[H] = \frac{[I]}{[L]}$$

$\therefore$ Dimensions of current are, $[I] = [H][L]$

$$= [\mu^{-1/2} M^{1/2} L^{1/2} T^{-1}]$$

***4. Charge.*** Charge = current × time

$\therefore$ The dimensions of charge are, $[Q] = [I][T]$

$$= [\mu^{-1/2} M^{1/2} L^{1/2} T^{-1}] = [T] = [\mu^{-1/2} M^{1/2} L^{1/2}]$$

***5. Potential difference.*** Potential difference is work done per unit charge. The dimensions of potential difference are,

$$[E] = \frac{[W]}{[Q]} = \frac{[ML^2T^{-2}]}{[\mu^{-1/2} M^{1/2} L^{1/2}]}$$

$$= [\mu^{1/2} M^{1/2} L^{3/2} T^{-2}]$$

***6. Capacitance.*** The dimensions of capacitance are,

$$[C] = \frac{[Q]}{[E]} = \frac{[\mu^{-1/2} M^{1/2} L^{1/2}]}{[\mu^{1/2} M^{1/2} L^{3/2} T^{-2}]} = [\mu^{-1} L^{-1} T^2]$$

**7. Resistance.** The dimensions of resistance are,

$$[R]=\frac{[E]}{[I]}=\frac{[\mu^{1/2}M^{3/2}L^{1/2}T^{-2}]}{[\mu^{-1/2}M^{1/2}L^{1/2}T^{-1}]}=[\mu LT^{-1}]$$

**8. Inductance.** Inductance $L=\frac{E}{dI/dt}$

Putting the dimensions of $E$, $I$ and $t$, we have

$$[L]=\frac{[E]}{[I][T]}=\frac{[E][T]}{[I]}=\frac{[\mu^{1/2}M^{1/2}L^{3/2}T^{-2}][T]}{[\mu^{-1/2}M^{1/2}L^{1/2}T^{-1}]}=[\mu L]$$

Similarly, the dimensions of other quantities can be found out.

## 5.10 RELATIONSHIP BETWEEN ELECTROSTATIC AND ELECTROMAGNETIC SYSTEM OF UNITS

We observe that two different dimensional equations are obtained for the same quantity in e.m. and e.s. systems of units. Any one quantity should have the same dimensions whatever may be the system. The dimensions of charge in e.m.u. system are :

$$[Q]=[\mu^{-1/2}M^{1/2}L^{1/2}]$$

and in e.s.u. system are :

$$[Q]=[\varepsilon^{1/2}M^{1/2}L^{3/2}T^{-1}]$$

Equating the above two dimensions,

$$[\mu^{-1/2}M^{1/2}L^{1/2}]=[\varepsilon^{1/2}M^{1/2}L^{3/2}T^{-1}]$$

We have : $[\mu^{-1/2}\varepsilon^{-1/2}]=[LT^{-1}]$.

Now dimensions $[LT^{-1}]$ are those of velocity.

$$\therefore\quad \frac{1}{\sqrt{\mu\varepsilon}}=\text{a velocity.}$$

Equating the dimensional equations of other quantities, we always get the same result *i.e.*, some power of $\sqrt{\mu\varepsilon}$ is always involved. It should be understood that it is not possible to obtain separate expressions for $\mu$ and $\varepsilon$.

In any system of units the permeability of free space $\mu_0$ and permittivity of free space $\varepsilon_0$ are related by relation :

$$\mu_0\varepsilon_0=\frac{1}{v^2}$$

where $v$ is the velocity of light in the system of units considered. For example, in C.G.S. units the value of $v$ is nearly equal to $3\times10^{10}$ cm/s which is the velocity light and of electromagnetic waves in free space.

From the above relationship the dimensions of any electrical quantity can be converted from those of electrostatic system to those of the electromagnetic system, and vice versa.

For example, the dimensions of potential difference in e.s.u. system are,

$$[E]=[\varepsilon^{-1/2}M^{1/2}L^{1/2}T^{-1}]$$

and in e.m.u. system are $[E]=[\mu^{1/2}M^{1/2}L^{3/2}T^{-2}]$.

$$\therefore\quad \frac{\text{1 e.s. unit of potential difference}}{\text{1 e.m. unit of potential difference}}=\frac{[\varepsilon^{-1/2}M^{1/2}L^{1/2}T^{-1}]}{[\mu^{1/2}M^{1/2}L^{3/2}T^{-2}]}=[\varepsilon^{-1/2}\mu^{-1/2}L^{-1}T].$$

Now the units of length, mass and time are the same in both the systems. Therefore, dimensions $L$, $M$ and $T$ are neglected.

$$\therefore\quad \frac{\text{1 e.s. unit of potential difference}}{\text{1 e.m. unit of potential difference}}=[\varepsilon^{-1/2}\mu^{-1/2}]=3\times10^{10}\text{ in C.G.S. system}$$

$\therefore$ 1 e.s. unit of potential difference $=3\times10^{10}$ e.m. units of potential difference

$$=3\times10^{10}\times10^{-8}=300\text{ volt}$$

(as 1 volt $=10^8$ e.m. units of e.m.f.)

Similarly, 1 e.s. unit of capacitance

$$=\frac{1}{9\times10^{20}}\text{ e.m. units of capacitance}$$

$$=\frac{1}{9\times10^{20}}\times10^9=\frac{1}{9\times10^{11}}\text{ farad}$$

or 1 farad $=9\times10^{11}$ e.s. units of capacitance.

Since the farad is too large a unit for many practical cases, the microfarad (μF) or pico-farad (pF) are used as more convenient units.

Similarly, milli-henry or micro-henry are often used as more convenient units of inductance.

## 5.11 M.K.S. SYSTEM (GIORGI SYSTEM)

The C.G.S. system suffers from the following disadvantages :

(*i*) There are two systems of units (e.m.u. and e.s.u.) for fundamental theoretical work and a third (practical units) for practical engineering work.

(*ii*) There are two sets of dimensional equations for the same quantity.

(*iii*) If we use C.G.S. system, factor $4\pi$ occurs in formulae where its presence seems unwarranted specially in formulae which are widely used.

Owing to the above disadvantages it was thought to make modifications in the system. It was early recognized that the so called practical units themselves formed a **coherent system** and they could become an absolute system in terms of units of mass, length and time other than centimetre, gramme and second.

In 1901, professor Giorgi introduced the M.K.S. system, wherein metre, kilogramme and second were taken as the three fundamental mechanical units. In order to connect the electrical and mechanical quantities, a fourth fundamental quantity has to be used. This fourth quantity is usually permeability. The permeability of free space is taken as $\mu_0 = 10^{-7}$. The permeability $\mu$ of any other medium is given by $\mu = \mu_r \mu_0$ where $\mu_r$ is the relative permeability and has a value numerically equal to the permeability of medium when C.G.S. units are used. The permeability of free space in C.G.S. system is unity.

$\therefore$ M.K.S. unit of permeability $= 10^7 \times$ C.G.S. unit of permeability.

Now the greatest advantage of M.K.S. system of units is that if we take the three fundamental units as metre, kilogramme and second and the value of permeability of free space as $10^{-7}$, then the magnitude of the six electrical units (charge, current, potential difference, resistance, inductance and capacitance) would be identical with their values in the practical system of units. Let us examine this.

**1. *Charge.*** The dimensions of charge in e.m.u. system are $[\mu^{-1/2} M^{1/2} L^{1/2}]$.

M.K.S. unit of length,

metre = 100 centimetre = 100 × C.G.S. units of length.

M.K.S. unit of mass,

kilogramme = 1000 gramme = 1000 × C.G.S. units of mass

M.K.S. unit of time,

second = C.G.S. unit of time, second.

$$\therefore \frac{\text{M.K.S. unit of charge}}{\text{C.G.S. e.m. unit of charge}} = \left(\frac{10^7}{1}\right)^{-1/2}\left(\frac{1000}{1}\right)^{1/2}\left(\frac{100}{1}\right)^{1/2} = 10^{-1}$$

Thus M.K.S. unit of charge $= 10^{-1}$ C.G.S. e.m. unit of charge.

But referring to Table 5.1, $10^{-1}$ C.G.S. e.m. unit of charge is equal to 1 practical unit of charge *i.e.,* one coulomb.

$\therefore$ M.K.S. unit of charge = practical unit of charge = 1 coulomb.

**2. *Current.*** The dimensions of current in e.m.u. system are $[\mu^{-1/2} M^{1/2} L^{1/2} T^{-1}]$

$$\therefore \frac{\text{M.K.S. unit of current}}{\text{C.G.S. e.m. unit of current}} = \left(\frac{10^7}{1}\right)^{-1/2}\left(\frac{1000}{1}\right)^{1/2}\left(\frac{100}{1}\right)^{1/2}\left(\frac{1}{1}\right)^{-1} = 10^{-1}$$

Thus M.K.S. unit of current $= 10^{-1} \times$ C.G.S. e.m. units of current = practical unit of current = 1 ampere.

**3. *Potential difference (EMF).*** The dimensions of potential difference are $[\mu^{1/2} M^{1/2} L^{3/2} T^{-2}]$.

$$\therefore \frac{\text{M.K.S. unit of emf}}{\text{C.G.S. e.m. unit of emf}} = \left(\frac{10^7}{1}\right)^{-1/2}\left(\frac{1000}{1}\right)^{1/2}\left(\frac{100}{1}\right)^{3/2}\left(\frac{1}{1}\right)^{-2} = 10^8$$

*i.e.,* M.K.S. unit of emf $= 10^8 \times$ C.G.S. e.m. unit of emf = practical unit of emf = 1 volt.

**4. *Resistance.*** The dimensions of resistance are $[\mu L T^{-1}]$

$$\therefore \frac{\text{M.K.S. unit of resistance}}{\text{C.G.S. e.m. unit of resistance}} = \left(\frac{10^7}{1}\right)\left(\frac{100}{1}\right)\left(\frac{1}{1}\right)^{-1} = 10^9$$

M.K.S. unit of resistance $= 10^9 \times$ C.G.S. e.m. units of resistance = practical unit of resistance = 1 ohm.

**5. *Inductance.*** The dimensions of inductance are $[\mu L]$

$$\therefore \frac{\text{M.K.S. unit of inductance}}{\text{C.G.S. e.m. unit of inductance}} = \left(\frac{10^7}{1}\right)^1\left(\frac{100}{1}\right) = 10^9$$

M.K.S. system of inductance $= 10^9 \times$ C.G.S. e.m. units of inductance

= practical unit of inductance = 1 henry.

**6. *Capacitance.*** The dimensions of capacitance are $[\mu^{-1} L^{-1} T^2]$

$$\therefore \frac{\text{M.K.S. unit of capacitance}}{\text{C.G.S. e.m. unit of capacitance}} = \left(\frac{10^7}{1}\right)^{-1}\left(\frac{100}{1}\right)^{-1}\left(\frac{1}{1}\right)^2 = 10^{-9}$$

M.K.S. unit of capacitance $= 10^{-9} \times$ C.G.S. e.m. units of capacitance

= practical unit of resistance = 1 farad.

**7. *Power.*** The dimensions of power are $[ML^2T^{-3}]$

$$\therefore \frac{\text{M.K.S. unit of power}}{\text{C.G.S. e.m. unit of power}} = \left(\frac{1000}{1}\right)\left(\frac{100}{1}\right)^2\left(\frac{1}{1}\right)^{-3} = 10^7$$

M.K.S. unit of power $= 10^7 \times$ C.G.S. e.m. unit of power

$= 10^7$ ergs per second

= practical unit of power = 1 watt.

**8. *Energy*.** The dimensions of energy are $[ML^2T^{-2}]$

$$\therefore \frac{\text{M.K.S. unit of energy}}{\text{C.G.S. e.m. unit of energy}}$$

$$= \left(\frac{1000}{1}\right)\left(\frac{100}{1}\right)^2\left(\frac{1}{1}\right)^{-2} = 10^7$$

M.K.S. unit of energy $= 10^7 \times$ C.G.S. e.m. units of energy

$= 10^7$ ergs = practical unit of energy

$= 1$ joule

Thus we see that the units of all purely electrical quantities in M.K.S. system, with units of power and energy, are identical with practical system of units comprising the ampere, volt, ohm, coulomb, farad, henry, watt, joule etc. Therefore, we conclude that if M.K.S. units are used throughout, the fundamental definitions of electrical quantities give the practical units directly, without any necessity of conversion factors.

M.K.S. unit of flux is 1 weber $= 10^8$ lines or maxwells.

M.K.S. unit of flux density is 1 weber per square metre $= 10^4$ gauss.

The ***advantages*** of M.K.S. system of unit are :

(*i*) This system connects the practical units directly, with the fundamental laws of electricity and magnetism.

(*ii*) This system gives specified formulae for expressions of electromagnetism involving only practical units.

(*iii*) There are certain formulae in C.G.S. system wherein μ is frequently omitted owing to the permeability of air being unity. This absence of μ from the formulae wrongly indicates that the permeability of medium has no effect on the quantity being considered. In M.K.S. system the presence of μ in the expression emphasises the fact that the effect depends upon the medium. For example, the force per unit length between two straight conductors is $\frac{\mu I_1 I_2}{2\pi D}$ in the M.K.S. system. This means that the formula makes it clear that the force is dependent upon the permeability of medium.

## 5.12 RATIONALISED M.K.S.A. SYSTEM

The M.K.S. system in its rationalised form, utilizes four fundamental units. These are metre, kilogramme, second and ampere. The use of ampere as a fourth fundamental unit has been recommended by International Electro-technical commission in July 1950.

The dimensional equations for various quantities in terms of *L, M, T* and *I* are shown in Table 5.2.

**Table 5.2** *Rationalized M.K.S.A. System*

| No. | Quantity | Symbol | Physical Equation | Dimension |
|---|---|---|---|---|
| 1. | Current | $I$ | | $[I]$ |
| 2. | Charge | $Q$ | $Q = It$ | $[TI]$ |
| 3. | Emf | $E$ | $E = \frac{\text{work done}}{\text{charge}}$ | $[ML^2T^{-3}I^{-1}]$ |
| 4. | Resistance | $R$ | $E = IR$ | $[ML^2T^{-3}I^{-2}]$ |
| 5. | Flux (magnetic) | $\Phi$ | $E = N\frac{d\Phi}{dt}$ | $[ML^2T^{-2}I^{-1}]$ |
| 6. | Flux Density | $B$ | $B = \frac{\text{flux}}{\text{area}}$ | $[MT^{-2}I^{-1}]$ |
| 7. | Mmf | $\mathfrak{I}$ | $\mathfrak{I} = NI$ | $[I]$ |
| 8. | Magnetizing force | $H$ | $H = \frac{\text{mmf}}{\text{length}}$ | $[L^{-1}I]$ |
| 9. | Reluctance | $R$ | $R = \frac{\mathfrak{I}}{\Phi}$ | $[M^{-1}L^{-2}T^2I^{-2}]$ |
| 10. | Inductance | $L$ | $E = L\frac{dI}{dt}$ | $[ML^2T^{-2}I^{-2}]$ |
| 11. | Electric flux | $\Psi$ | $\Psi = Q$ | $[TI]$ |
| 12. | Electric flux density | $D$ | $D = \frac{\text{electric flux}}{\text{area}}$ | $[L^{-2}TI]$ |
| 13. | Electric field strength | $\varepsilon$ | $\varepsilon$ = potential gradient | $[MLT^{-3}I^{-1}]$ |
| 14. | Capacitance | $C$ | $C = \frac{Q}{E}$ | $[M^{-1}L^{-2}T^4I^2]$ |

There are many expressions in the M.K.S. system which contain the term $4\pi$. This makes the formulae a little awkward to work with. Elimination of $4\pi$ from all the formulae is impossible but the aim is to transfer the term $4\pi$ to formulae which are not very important or where it may be reasonably expected to occur as in volume of sphere. Thus rationalization means the elimination of term $4\pi$ from frequently used expressions by redefinition of certain units. This rationalization is carried out by assuming the value of permeability of free space as $4\pi \times 10^{-7}$ in place $10^{-7}$ as in the unrationalized system.

The basis of rationalization is the conception that unit flux issues from the surface of a body containing a unit charge. This flux is $4\pi$ in the case of unrationalized system. This fundamental change redefines all the relationships.

It should be emphasized again that rationalization does not mean elimination of $4\pi$ from all formulae but only from important formulae which are frequently used. The term $4\pi$ disappears in some relationships only to appear in another place in others. The advantage of rationalization is to be judged by its effects on the relationships which are considered very important.

## 5.13 SI UNITS

An international organization of which most of the advanced and developing countries, including India, are members, called the General Conference of Weights and Measures (CGPM) has been entrusted

**Table 5.3** *Base Units*

| Unit | Name | Symbol |
|---|---|---|
| Length | metre | m |
| Mass | kilogramme | kg |
| Time | second | s |
| Intensity of electric current | ampere | A |
| Thermodynamic temperature | Kelvin | K |
| Luminous intensity | candela | cd |
| *Amount of substance | mole | mol |

**Table 5.4** *Recommended Prefixes & Symbols*

| Factor by which unit is multiplied | Prefix | Symbol |
|---|---|---|
| $10^{24}$ | $Z_e$ | Z |
| $10^{12}$ | tera | T |
| $10^{9}$ | giga | G |
| $10^{6}$ | mega | M |
| $10^{4}$ | myria | ma |
| $10^{3}$ | kilo | k |
| $10^{2}$ | hecto | h |
| $10^{1}$ | deca | da |
| $10^{-1}$ | deci | d |
| $10^{-2}$ | centi | c |
| $10^{-3}$ | milli | m |
| $10^{-6}$ | micro | μ |
| $10^{-9}$ | nano | n |
| $10^{-12}$ | pico | p |
| $10^{-15}$ | femto | f |
| $10^{-18}$ | atto | a |
| $10^{-24}$ | aus | au |

with the task of prescribing definitions for fundamental units of weights and measures which are the very basis of Science and Technology today.

These definitions are adopted all over the world for purposes of science, technology, industry and commerce. The Eleventh General Conference of Weights and Measures which met in October, 1960 recommended a unified systematically constituted, coherent system of fundamental, supplementary and derived units for international use. This system, called the **International System of Units** and designated by the abbreviation, **SI**, *System internaitonale d' units* is now legally compulsory in about twenty countries. It consists of six base units, two supplementary units and 27 derived units. Principles for the use of prefixes for forming the multiples and sub-mulitples of units were also laid down. The six base units have already been adopted in the Standards of Weights and Measures Act, 1956.

The subsequent General Conferences of Weights and Measures added more units to this system of units. Tables 5.3, 5.4 and 5.5 indicate the various units, prefixes and their symbols which form the International System of Units.

* According to British Standards Publication PD 55686 : 1972, the SI system is based on seven and not six base units. The seventh base unit is mole.

**Table 5.5** *Supplementary and Derived Units*

| | | | | |
|---|---|---|---|---|
| | | ***SUPPLEMENTARY UNITS*** | | |
| 1. | Plane angle | radian | rad. | |
| 2. | Solid angle | steradian | sr. | |
| | | ***DERIVED UNITS*** | | |
| 1. | Area | square metre | $m^2$ | |
| 2. | Volume | cubic metre | $m^3$ | |
| 3. | Frequency | hertz | Hz | 1/s |
| 4. | Density (mass density) | kilogramme per cubic metre | $kg/m^3$ | |
| 5. | Velocity | metre per second | m/s | |
| 6. | Angular velocity | radian per second | rad/s | |
| 7. | Acceleration | metre per second squared | $m/s^2$ | |
| 8. | Angular acceleration | radian per second squared | $rad/s^2$ | |
| 9. | Force | newton | N | $kg\ m/s^2$ |
| 10. | Pressure (Mechanical Tension) | newton per square metre | $N/m^2$ | |
| 11. | Dynamic viscosity | newton-second per square metre | $Ns/m^2$ | |
| 12. | Kinematic viscosity | square metre per second | $m^2/s$ | |
| 13. | Work, energy, quantity of heat | joule | J | Nm |
| 14. | Power | watt | W | J/s |
| 15. | Quantity of electricity | coulomb | C | As |
| 16. | Potential, potential difference, electromotive force | volt | V | W/A |
| 17. | Electric field strength | volt per metre | V/m | |
| 18. | Resistance | ohm | Ω | V/A |
| 19. | Capacitance | farad | F | As/V |
| 20. | Magnetic flux | weber | Wb | Vs |
| 21. | Inductance | henry | H | Vs/A |
| 22. | Magnetic flux density | tesla | T | $Wb/m^2$ |
| 23. | Magnetic field strength | ampere per metre | A/m | |
| 24. | Magnetomotive force | ampere | A | |
| 25. | Luminous flux | lumen | lm | cd. sr |
| 26. | Luminance | candela per square metre | $cd/m^2$ | |
| 27. | Illuminance | lux | lx | $lm/m^2$ |

## 5.14 BASE UNITS OF SI

The seven **base units** which form the basis of the now universally accepted International Systems of Units, in their present status are defined below :

**1. Metre (m).** It is the unit of length. The metre is the length equal to 1,650,763.73 wavelengths in vacuum of radiation corresponding to the transition between the level $2_{p10}$ and $5_{d5}$ (orange red line) of the krypton 86 atom (excited at the triple point of nitrogen, 63.15° kelvin).

**2. Kilogramme (kg).** It is the unit of mass. A kilogramme is equal to the mass of the international prototype of mass. This prototype is a cylinder of platinum iridium alloy.

**3. Second (s).** It is the unit of time. A second is defined as the duration of 9,192,631,770 periods of radiation corresponding to the transition between the two hyperfine levels of the ground state of the cesium 133 atom.

**4. Ampere (A).** It is the unit of electric current. The ampere is that constant current which, if maintained in two straight parallel conductors of infinite length, of negligible cross-section, and placed one metre apart in vacuum, would produce between them a force equal to $2 \times 10^{-7}$ newton per metre length.

**5. Kelvin (K).** It is the unit of temperature. The *kelvin, unit of thermodynamic temperature, is the fraction 1/273.16 of the thermodynamic temperature of triple point of water.

**6. Candela (cd).** It is the unit of luminous intensity. It is the luminous intensity, in a perpendicular direction, of a surface of 1/600,000 square metre of a blackbody at the temperature of freezing platinum under a pressure of 101,325 newton per square metre.

**7. Mole (mol).** The mole is the amount of substance of a system which contains as many elementary entities as there are atoms in 0.012 kg of carbon 12. It should be noted that whenever mole is used, the elementary entities must be specified, which may be atoms, molecules, ions, electrons, other particles, or specified groups of such particles.

## 5.15 SUPPLEMENTARY UNITS

Two **Supplementary units** have been added to the basic six SI units. They are

- Radian for the plane angles, and
- Steradian for the solid angles.

**1. Radian (rad).** Plane angle subtended by an arc of a circle equal in length to the radius of the circle.

**2. Steradian (sr).** Solid angle subtended at the centre of a sphere by the surface whose area is equal to the square of the radius of the sphere.

## 5.16 MULTIPLYING PREFIXES OF UNITS

In the SI Units, the use of decimals, the use of powers of ten in preference to fractions simplifies any quantity. Presentation is further facilitated by the use of prefixes.

In a **coherent system** of units such as SI, the product or quotient of any two unit quantities becomes the unit of the resultant quantity without the introduction of any numerical factors, which is less prone to mistakes. In order to maintain coherent nature of the system, only SI units and not their multiples or sub-multiples be combined to form derived units. For stating value of a particular quantity, we may involve the use of factors of the form $10^n$, where $n$ is a positive or negative integer. Thus, to save writing space and to utilize time more effectively, a series of prefixes for Units has been evolved. A list of such prefixes is given in Table 5.4. Multiples or submultiples of $10^3$ are the preferred prefixes, whereas the other multipliers tend to be a nuisance in calculations.

The basic unit of mass, the gramme, will often be used as an elementary unit (to avoid absurdity of using m kg) until such time when a new name is adopted for kilogramme.

The recommended multiple and submultiple units for the basic units are given in Table 5.6.

**Table 5.6** *Recommended Multiples and Submultiples of Basic Units*

| *Basic Unit* | *Recommended Multiple and Submultiple Units* | | | | | | | *Remarks* |
|---|---|---|---|---|---|---|---|---|
| metre (m) | — | km | — | mm | μm | nm | — | μm is also called micron |
| kilogramme (kg) | Mg | — | g | mg | μg | — | — | Mg is called tonne |
| Second (s) | — | ks | — | ms | μs | ns | — | Use of min, hour and day is also allowed |
| ampere (A) | — | kA | — | mA | μA | nA | pA | |
| kelvin (K) | — | — | *k* | — | — | — | — | Multiples are not used. |
| Nominal unit multiplier | $10^6$ | $10^3$ | $10^0$ | $10^{-3}$ | $10^{-6}$ | $10^{-9}$ | $10^{-12}$ | |

* The name kelvin (symbol °K) of the thermodynamic temperature was changed from degrees kelvin (symbol K) by the 13th Conference on Weights and Measures, Paris, Oct. 1967.

A selection of derived units is given in Table 5.7. Some of the units are given special names. The recommended multiples and submultiples are also given.

**Table 5.7** *Recommended Multiples and Submultiples of Derived Units*

| Derived Unit | Recommended Multiples and Submultiples | | | | | | |
|---|---|---|---|---|---|---|---|
| Force | — | — | kN | mN | μN | — | — |
| Work (and energy) | GJ | MJ | kJ | mJ | — | — | — |
| Power | GW | MW | kW | mW | μW | nW | pW |
| Pressure | GN/m² | MN/m² | kN/m² | — | — | — | — |
| Velocity | — | — | km/s | — | — | — | — |
| Electrical potential | — | MV | kV | mV | μV | — | — |
| Electrical resistance | — | MΩ | kΩ | — | — | — | — |
| Nominal unit Multiplier | $10^9$ | $10^6$ | $10^3$ | $10^{-3}$ | $10^{-6}$ | $10^{-9}$ | $10^{-12}$ |

There are many other units which are not part of SI but are recognised by International Committee for Weights and Measures. They are listed in Table 5.8.

**Table 5.8**

| Quantity | Name of unit | Symbol | Definition |
|---|---|---|---|
| Time | minute | min | 1 min = 60 s |
| | hour | h | 1 h = 60 min = 3600 s |
| | day | d | 1 d = 24 h |
| Plane angle | degree | ° | $1° = (\pi / 180)$ rad |
| | minute | ′ | $1' = \left(\frac{1}{60}\right)^{\circ}$ |
| | second | ″ | $1'' = \left(\frac{1}{60}\right)' = \left(\frac{1}{3600}\right)^{\circ}$ |
| Volume | litre | l | $1l = 1000\ cm^3$ |
| Mass | tonne | t | $1t = 1000$ kg |
| Pressure | pascal | Pa | $1\ pa = 1\ N/m^2$ |
| | bar | bar | $1\ bar = 10^5$ Pa |

**Example 5.1** *If we adopt new units of voltage and current such that one new unit voltage = 100 volt and one new unit of current = 100 ampere, how would the units of resistance, capacitance, inductance be affected ?*

**Solution.** Let the old units of resistance, inductance, capacitance, voltage and current be $R, L, C, E$ and $I$ respectively and the new units of resistance, inductance, capacitance, voltage and current be $R', L', C', E'$ and $I'$ respectively.

(*i*) We have resistance :

$$R = \frac{E}{I} \text{ and } R' = \frac{E'}{I'} = \frac{100\,E}{100\,I} = \frac{E}{I} = R.$$

Thus the unit of resistance is equal to the old unit of resistance and, therefore, there is no change.

(*ii*) Inductance $L = \frac{E}{dI/dt}$

and $$L' = \frac{E'}{dI'/dt} = \frac{100\,E}{100\,dI/dt} = \frac{E}{dI/dt} = L$$

Thus the new unit of inductance is equal to the old unit of inductance and, therefore, there is no change.

(*iii*) Capacitance $C = \frac{\text{charge}}{\text{emf}} = \frac{It}{E}$

and $$C' = \frac{I't}{E'} = \frac{100\,It}{100\,E} = \frac{It}{E} = C$$

Thus the new unit of capacitance is equai to the old unit of capacitance and, therefore, there is no change.

**Example 5.2** *If we adopt new units of length, mass and time such that*

*one new unit of length = 10 metre,*

*one new unit of mass = 1/10 kilogram,*

*and one new unit of time = 100 second.*

*How would the units of force, power, energy and current be affected ?*

*The units of permittivity remain the same.*

**Solution.** (*i*) The dimensions of force are $[MLT^{-2}]$.

$$\therefore \text{ New unit of force} = \left(\frac{1/10}{1}\right)\left(\frac{10}{1}\right)\left(\frac{100}{1}\right)^{-2}$$

$$= 10 \text{ old units of force} = 10 \text{ N.}$$

(*ii*) The dimensions of power are $[ML^2T^{-3}]$.

$$\therefore \text{ New unit of power} = \left(\frac{1/10}{1}\right)\left(\frac{10}{1}\right)^2\left(\frac{100}{1}\right)^{-3}$$

$$= 10^{-5} \text{ old units of power} = 10^{-5} \text{ W.}$$

(*iii*) The dimensions of energy are $[ML^2T^{-2}]$.

$$\therefore \text{ New unit of energy} = \left(\frac{1/10}{1}\right)\left(\frac{10}{1}\right)^2\left(\frac{100}{1}\right)^{-2}$$

$$= 10^{-3} \text{ old units of energy} = 10^{-3} \text{ J.}$$

(*iv*) The dimensions of current are

$$[\mu^{-1/2} M^{1/2} L^{1/2} T^{-1}].$$

$\therefore$ New unit of current

$$= \left(\frac{1}{1}\right)^{-1/2}\left(\frac{1/10}{1}\right)^2\left(\frac{10}{1}\right)^{1/2}\left(\frac{100}{1}\right)^{-1}$$

$$= 10^{-2} \text{ old units of current} = 10^{-2} \text{ A.}$$

**Example 5.3** *Derive the dimensional equations for pole strength current, e.m.f., magnetic flux, magnetic flux density, mmf, magnetizing force, reluctance, permeance in the e.m. system of units.*

**Solution.** (*i*) The force exerted between two poles is $F = \dfrac{m_1 m_2}{\mu d^2}$

Dimensions of pole strength

$$= [m] = [\text{force}]^{1/2}[\text{permeability}]^{1/2}[\text{distance}]$$

$$= [MLT^{-2}]^{1/2}[\mu]^{1/2}[L] = [\mu^{1/2} M^{1/2} L^{3/2} T^{-2}]$$

(*ii*) Force between two current carrying conductors is $F = \dfrac{\mu I_1 I_2 l}{2\pi d}$.

(where $l$ = length of each conductor and $d$ = distance between two conductors)

$\therefore$ Dimensions of curent are :

$$[I] = [\text{force}]^{1/2}[\text{permeability}]^{-1/2}$$

$$= [MLT^{-2}]^{1/2}[\mu]^{-1/2} = [\mu^{-1/2} M^{1/2} L^{1/2} T^{-1}]$$

(*iii*) Dimensions of e.m.f. are :

$$[E] = \frac{[\text{power}]}{[\text{current}]} = \frac{[ML^2T^{-3}]}{[\mu^{-1/2} M^{1/2} L^{1/2} T^{-1}]}$$

$$= [\mu^{1/2} M^{1/2} L^{3/2} T^{-2}]$$

(*iv*) Emf $e$ = rate of change of flux linkages

$$= -N\frac{d\Phi}{dt} \quad \text{or} \quad d\Phi = \frac{e\,dt}{N}$$

$\therefore$ Dimensions of magnetic flux are :

$$[\Phi] = [E][T] = [\mu^{1/2} M^{1/2} L^{3/2} T^{-2}][T]$$

$$= [\mu^{1/2} M^{1/2} L^{3/2} T^{-1}]$$

as $N$ is dimensionless.

(*v*) Flux density $= \dfrac{\text{flux}}{\text{area}}$

$\therefore$ Dimensions of flux density are :

$$[B] = \frac{[\Phi]}{[L^2]} = \frac{[\mu^{1/2} M^{1/2} L^{3/2} T^{-1}]}{[L^2]}$$

$$= [\mu^{1/2} M^{1/2} L^{-1/2} T^{-1}]$$

(*vi*) Mmf = turns × ampere

$\therefore$ Dimensions of mmf $[\Im] = [I]$

$= [\mu^{-1/2} M^{1/2} L^{1/2} T^{-1}]$ as $N$ is dimensionless.

(*vii*) Magnetizing force = m.m.f. per unit length.

$$\therefore \quad [H] = \frac{[\Im]}{[L]} = [\mu^{-1/2} M^{1/2} L^{-1/2} T^{-1}]$$

(*viii*) Reluctance = mmf/flux

$$\therefore \quad [R] = \frac{[\Im]}{[\Phi]} = \frac{[\mu^{-1/2} M^{1/2} L^{1/2} T^{-1}]}{[\mu^{1/2} M^{1/2} L^{3/2} T^{-1}]} = [\mu^{-1} L^{-1}]$$

(*ix*) Permeance $\Lambda$ = 1/reluctance

$$[\Lambda] = [R]^{-1} = [\mu L]$$

**Example 5.4** *Derive the dimensions of (i) e.m.f., (ii) magnetic flux density, (iii) electric flux density, (iv) current density, (v) permeability, (vi) permittivity, (vii) resistivity and (viii) conductivity in L, M, T, I system of dimensions.*

**Solution.** (*i*) Emf $= \dfrac{\text{work done}}{\text{charge}}$

$$[E] = \frac{[ML^2T^{-2}]}{[IT]} = [I^{-1} ML^2T^{-3}]$$

(*ii*) Magnetic flux density $B$ = flux/area = $\Phi / A$

Emf $e$ = rate of change of flux linkages $= N\dfrac{d\Phi}{dt}$

Flux $[\Phi]$ = [emf] [time]

$$\therefore \quad [B]\left[\frac{\Phi}{A}\right] = \frac{[\text{emf}]\ [\text{time}]}{[\text{area}]}$$

$$= \frac{[I^{-1} ML^2 T^{-3}][T]}{[L^2]} = [I^{-1} MT^{-2}]$$

(*iii*) Electric flux density = electric flux/area

= charge/area

$$\therefore \quad [D] = \frac{[IT]}{[L^2]} = [IL^{-2}T]$$

(*iv*) Current density $\delta = \frac{\text{current}}{\text{area}}$

$$\therefore \quad [\delta] = \frac{[I]}{[L^2]} = [IL^{-2}]$$

(*v*) Flux = flux density × area

$$\therefore \quad [\Phi] = [I^{-1}MT^{-2}][L^2] = [I^{-1}ML^2T^{-2}]$$

Mmf = turns × current

$$\therefore \quad [\Im] = [I]$$

$$\text{Reluctance} = \frac{\text{mmf}}{\text{flux}}$$

$$\therefore \quad [R] = \frac{[I]}{[I^{-1}ML^2T^{-2}]} = [I^2M^{-1}L^{-2}T^2]$$

We have, reluctance $= \frac{\text{length}}{\text{permeability} \times \text{area}}$

$$\therefore \quad \text{Permeability } \mu = \frac{\text{length}}{\text{reductance} \times \text{area}}$$

$$\text{Thus, } [\mu] = \frac{[L]}{[I^2M^{-1}L^{-2}T^2][L^2]} = [I^{-2}MLT^{-2}]$$

(*vi*) Force $= \frac{Q_1Q_2}{\varepsilon d^2}$ where ε = permittivity.

$$\therefore \quad [\varepsilon] = \frac{[\text{charge}^2]}{[\text{force}][\text{distance}^2]} = \frac{[\text{current}^2][\text{time}^2]}{[\text{force}][\text{distance}^2]}$$

$$= \frac{[I^2T^2]}{[MLT^{-2}][L^2]} = [I^2M^{-1}L^{-3}T^4]$$

(*vii*) Resistance $R$ = emf/current

$$[R] = \frac{[I^{-1}ML^2T^{-3}]}{[I]} = [I^{-2}ML^2T^{-3}]$$

$$\text{Resistance} = \frac{\text{resistivity} \times \text{length}}{\text{area}}$$

$$\therefore [\text{Resistivity}] = \frac{[\text{resistance}][\text{area}]}{[\text{length}]}$$

or
$$[\rho] = \frac{[I^{-2}ML^2T^{-3}][L^2]}{[L]} = [I^{-2}ML^3T^{-3}]$$

(*viii*) Conductivity = 1/resistivity

$$\therefore \quad [\sigma] = [\rho^{-1}] = [I^2M^{-1}L^{-3}T^3]$$

**Example 5.5** *In deriving expression for resistance in Hay's bridge, the following expression is obtained :*

$$R = \frac{\omega^2R_1R_2R_3C^2}{1+\omega^2CR_2^2}$$

*where $R$, $R_1$, $R_2$, $R_3$ are resistances ; $C$, a capacitance and $\omega = 2\pi f$, where $f$ is frequency ; Hz.*

*Find out whether the expression is dimensionally correct or incorrect. In case there is an error point out the term which needs correction.*

**Solution.** The dimensions of various terms in e.m. system are :

[Resistance] $= [R] = [\mu LT^{-1}]$

[Capacitance] $= [C] = [\mu^{-1}L^{-1}T^2]$

[Angular velocity] $= [\omega]$

= [Hz or cycles per second]

$= [T^{-1}]$.

The left hand term is dimensionally expressed as :

$$[R] = [\mu LT^{-1}]$$

and the right hand term is :

$$\frac{[\omega^2R_1R_2R_3C^2]}{[1+\omega^2CR_2^2]} = \frac{[T^{-1}]^2[\mu LT^{-1}][\mu LT^{-1}][\mu LT^{-1}][\mu^{-1}L^{-1}T^2]^2}{1+[T^{-1}]^2[\mu^{-1}L^{-1}T^2][\mu LT^{-1}]^2}$$

$$= \frac{[\mu LT^{-1}]}{1+[\mu LT^{-2}]}$$

This clearly shows that the dimensions of left hand and right hand terms do not balance and, therefore, the equation is dimensionally incorrect. The dimensions of left hand term are $[\mu LT^{-1}]$. The dimensions of the numerator of right hand term are $[\mu LT^{-1}]$ and in the denominator we have two terms *i.e.*, 1 and $\omega^2CR_2^2$. Now 1 is a dimensionless quantity and, therefore, in order that the dimensions of left hand term balance those of the right hand term, the denominator of right hand term should also be dimensionaless quantity. The dimensions of second term in the denominator are $[\mu LT^{-2}]$ and in order that it may be a dimensionless quantity it should be multiplied with a term having dimensions of $[\mu^{-1}L^{-1}T^2]$ which are the dimensions of capacitance. Therefore, the second term in the denominator should contain another capacitance term or this term should be $(\omega^2CR_2^2 \times C)$ or $\omega^2C^2R_2^2$. Therefore, the correct equation is,

$$R = \frac{\omega^2R_1R_2R_3C^2}{1+\omega^2C^2R_2^2}.$$

**Example 5.6** *In the course of a derivation, an expression of the following form was arrived at :*

$$I = E\left\{\frac{1}{Z_1} + \frac{j\omega M}{Z_2}\left(\frac{1}{R} + \frac{C}{L}\right)\right\}.$$

*Show that there must be an algebraic error, and point out the term or terms which require correction.*

**Solution.** The dimensions of various quantities in e.m. system are :

[Current] $= [I] = [\mu^{-1/2} L^{1/2} M^{1/2} T^{-1}]$,
[Emf] $= [E] = [\mu^{1/2} L^{3/2} M^{1/2} T^{-2}]$,
[Impedance] $= [Z] = [\mu L T^{-1}]$,
[Mutual inductance] $= [M] = [\mu L]$,
[Capacitance] $= [C] = [\mu^{-1} L^{-1} T^2]$,
[Resistance] $= [R] = [\mu L T^{-1}]$
[Inductance] $= [L] = [\mu L]$

The left hand term is dimensionally expressed as :

$$[I] = [\mu^{-1/2} L^{1/2} M^{1/2} T^{-1}]$$

The right hand term $E$

$\left\{\frac{1}{Z_1} + \frac{j\omega M}{Z_2}\left(\frac{1}{R} + \frac{C}{L}\right)\right\}$ is dimensionally expressed as :

$$= [\mu^{1/2} L^{3/2} M^{1/2} T^{-2}] \times \left\{\frac{1}{[\mu L T^{-1}]} + \frac{[T^{-1}][\mu L]}{[\mu L T^{-1}]}\left(\frac{1}{[\mu L T^{-1}]} + \frac{[\mu^{-1} L^{-1} T^2]}{[\mu L]}\right)\right\}$$

$$= [\mu^{1/2} L^{3/2} M^{1/2} T^{-2}] \times \{[\mu^{-1} L^{-1} T] + ([\mu^{-1} L^{-1} T] + [\mu^{-2} L^{-2} T^2])\}$$

$$= \{[\mu^{-1/2} L^{1/2} M^{1/2} T^{-1}] + [\mu^{-1/2} L^{1/2} M^{1/2} T^{-1}] + [\mu^{-3/2} L^{-1/2} M^{1/2} T^0]\}$$

In order that the two sides should balance dimensionally the dimensions of right hand side should be $[\mu^{-1/2} L^{1/2} M^{1/2} T^{-1}]$. We find that the dimensions of all the terms on the right hand side are not $[\mu^{-1/2} L^{1/2} M^{1/2} T^{-1}]$ and hence there is an algebraic error. The error is due to last term having dimensions of $[\mu^{-3/2} L^{-1/2} M^{1/2} T^0]$. The error can be eliminated if it has the dimensions $[\mu^{-1/2} L^{-1/2} M^{1/2} T^{-1}]$. In order to achieve this we must multiply the last term by a quantity which has the dimensions

$$\frac{[\mu^{-1/2} L^{1/2} M^{1/2} T^{-1}]}{[\mu^{-3/2} L^{-1/2} M^{1/2} T^0]} = [\mu L T^{-1}]$$

But $[\mu L T^{-1}]$ are the dimensions of resistance or impedance. Therefore, the last term of right hand side, $C/L$ should be multiplied by a term having the dimensions of either resistance or impedance.

**Example 5.7** *The expression for eddy current produced in a metallic former moving in the field of a permanent magnet is found as :* $I_e = K\frac{BlbA}{(2b+l)\rho}$

*where* $B$ = *flux density,* $l$ = *length of former,*
$b$ = *width of former,* $A$ = *area of former,*
$\rho$ = *resistivity of conducting former,*
*and* $K$ = *a constant.*

*It is suspected that the term angular frequency* $\omega$ *is missing. Use LMTI system for dimensional analysis to detect the error.*

**Solution.** The dimensions of L.H.S. are equal to $[I]$. The dimensions of various quantities of R.H.S. are,

$$[B] = [MT^{-2} I^{-1}],\ [l] = [L].(b) = [L]$$
$$[A] = [L^2] \quad \text{and} \quad [\rho] = [ML^3 T^{-3} I^{-2}]$$

After substituting the dimensions of various quantities, we have,

$$\text{R.H.S.} = \frac{[ML^{-2} I^{-1}][L][L][L^2]}{[L][ML^3 T^{-3} I^{-2}]} = [TI]$$

In order that the two sides are dimensionally correct the R.H.S. must be multiplied by a term having a dimensions $T^{-1}$ *i.e.,* $\omega$

$\therefore$ The correct expression is $I_e = \frac{BlbA\omega}{(2b+l)\rho}$.

**Example 5.8** *The expression for mean torque of an electrodynamometer type of wattmeter may be written as :*

$$T \propto M^p e^q Z^t$$

*where* $M$ = *mutual inductance between fixed and moving coils,* $E$ = *applied voltage,* $Z$ = *impedance of load circuit. Determine the values of* $p, q, t$ *from the dimensions of the quantities.*

**Solution.** We have $T \propto M^p E^q Z^t$

Therefore, we can write $T = KM^p E^q Z^t$.

where $K$ = a constant (a dimensionless quantity).

Writing the dimensions of various quantities :

$$\text{Torque} = \frac{\text{power}}{\text{angular velocity}}$$

$$\therefore \quad [\text{Torque}] = \frac{[ML^2T^{-3}]}{[T^{-1}]} = [ML^2T^{-2}]$$

[Mutual inductance] $= [\mu L]$,
[Emf] $= [E] = [\mu^{1/2} M^{1/2} L^{3/2} T^{-2}]$,
[Impedance] $= [Z] = [\mu L T^{-1}]$

Substituting the dimensions of various quantities in the expression for torque, we have :

$$[ML^2T^{-2}] = [\mu L]^p [\mu^{1/2} M^{1/2} L^{3/2} T^{-2}]^q [\mu L T^{-1}]^t$$

$$= \left[\mu^{p+q/2+t} + M^{q/t} + L^{p+(3/2)q+t} + T^{-2q-t}\right]$$

In order that the two sides should balance dimensionally,

$$p + \frac{q}{2} + t = 0, \qquad \frac{q}{2} = 1$$

$$p + \frac{3}{2}q + t = 2 \qquad \text{and} \quad -2q - t = -2$$

From above, $p = 1$, $q = 2$ and $t = -2$.

Therefore, the equation is $T \propto ME^2Z^{-2} \propto \frac{ME^2}{Z^2}$.

5

**Example 5.9** *The energy stored in a parallel plate capacitor per unit volume (energy density) is given by :*

$$w = K\varepsilon^a V^b d^c$$

*where* $\varepsilon$ = *permittivity of medium,* $d$ = *distance between plates.* $V$ = *voltage between plates, and* $K$ = *a constant.*

*Find the values of a, b and c.*

**Solution.** Writing the dimensions of various quantities in *LMTI* system :

Energy density $w$ = energy/volume

$$[w] = \frac{[ML^2T^{-2}]}{[L^3]} = [ML^{-1}T^{-2}]$$

Now force $F = \dfrac{Q_1Q_2}{\varepsilon d^2}$ $\therefore$ permittivity $\varepsilon = \dfrac{Q_1Q_2}{Fd^2}$

or $$[\varepsilon] = \frac{[I^2T^2]}{[MLT^{-2}][L^2]} = [I^2M^{-1}L^{-3}T^4]$$

and $[\text{voltage}] = [V] = [I^{-1}ML^2T^{-3}]$, $[d] = [L]$

We get,

$$[ML^{-1}T^{-2}] = [I^2M^{-1}L^{-3}T^4]^a[I^{-1}ML^2T^{-3}]^b[L]^c$$
$$= [I]^{2a-b}[M]^{-a+b}[L]^{-3a+2b+c}[T]^{4a-3b}$$

Balancing the dimensions on both sides, we have

$0 = 2a - b$, $1 = -a + b$, $-1 = -3a + 2b + c$, $-2 = 4a - 3b$

From above, $a = 1$, $b = 2$, $c = -2$

Hence, energy stored per unit volume is,

$$w = K\varepsilon V^2 d^{-2} = \frac{K\varepsilon V^2}{d^2}.$$

**Example 5.10** *The expression for eddy current loss p per metre length of wire may be written as*

$$p \propto f^a B_m^b d^c \rho^g$$

*where* $f$ = *frequency,* $B_m$ = *maximum flux density,* $d$ = *diameter of wire,* $\rho$ = *resistivity of wire mate and a, b, c and g are constants.*

*Determine the values of a, b, c, g from the dimensions of quantities involved. Length L, mass M, time T and current I may be taken as fundamental quantities. Derive the dimensions of each quantity from fundamentals. Do not reproduce the dimensions from memory.*

**Solution.** We can write, the expression as

$$p = K f^a B_m^b d^c \rho^g.$$

where $K$ = a constant and a dimensionless quantity.

We now derive the dimensional expressions for the quantities involved.

$p$ = power loss per unit length.

$$\therefore \quad [p] = \frac{[ML^2T^{-3}]}{[L]} = [MLT^{-3}].$$

$f$ = frequency in Hz (cycles per second)

$$\therefore \quad [f] = [T^{-1}]$$

Charge = current × time

$$\therefore \quad [Q] = [I][T] = [IT]$$

$$\text{Emf} = \frac{\text{work done}}{\text{charge}}$$

$$\therefore \quad [E] = \frac{[ML^2T^{-3}]}{[Q]} = \frac{[ML^2T^{-2}]}{[IT]} = [I^{-1}ML^2T^{-3}]$$

Emf = turns × rate of change of flux

$\therefore$ [Flux] = [emf] [time]

$$[\Phi] = [I^{-1}ML^2T^{-3}][T] = [I^{-1}ML^2T^{-2}]$$

$$\text{Flux density} = \frac{\text{flux}}{\text{area}}$$

$$\therefore \quad [B_m] = \frac{[\Phi]}{[L^2]} = \frac{[I^{-1}ML^2T^{-2}]}{[L^2]} = [I^{-1}MT^{-2}]$$

We have [diameter] = [length] $\therefore [d] = [L]$

$$\text{Resistance} = \frac{\text{emf}}{\text{current}}$$

$$\therefore \quad [R] = \frac{[I^{-1}ML^2T^{-3}]}{[I]} = [I^{-2}ML^2T^{-3}]$$

$$\text{Resistance} = \frac{\text{resistivity} \times \text{length}}{\text{area}}$$

or $$\text{resistivity} = \frac{\text{resistance} \times \text{area}}{\text{length}}$$

$$\therefore [\rho] = \frac{[R][L^2]}{[L]} = \frac{[I^{-2}ML^2T^{-3}][L^2]}{[L]} = [I^{-2}ML^3T^{-3}].$$

Substituting the dimensions of quantities involved in the expression for loss per unit length,

$$[MLT^{-3}] = [I^0\,MLT^{-3}]$$
$$= [T^{-1}]^a[I^{-1}MT^{-2}]^b[L]^c[I^{-2}ML^3T^{-3}]^g$$
$$= [I^{-b-2g}M^{b+g}L^{c+3g}T^{-a-2b-3g}].$$

For dimensional balance, we must equate the corresponding indices. Therefore,

$-b - 2g = 0$, $b + g = 1$

$c + 3g = 1$ and $-a - 2b - 3g = -3$.

$\therefore$ We have, from above,

$g = -1$, $b = 2$, $c = 4$ and $a = 2$.

Hence, the expression is

$$p \propto f^2 B_m^2 d^4 \rho^{-1} \propto \frac{f^2 B_m^2 d^4}{\rho}.$$

## 5.17 DETERMINATION OF ABSOLUTE UNITS

In order to establish a satisfactory system of measurements we must determine the absolute values of electrical quantities. Determination of absolute values involves the measurement of a unit in terms of physical quantities which are measured practically. Thus if we can determine the units of, say, current and emf by practical measurements, standards for all other electrical quantities like resistance etc. can be set up in terms of current and emf.

The determination of absolute unit of emf involves the measurement of energy which is difficult to measure and, therefore, the determination of absolute unit of emf is not usually attempted. The practice is to determine the absolute units of current and resistance and to express the units of all other quantities in terms of these two quantities. Actually it may be desirable to have, in addition, primary standards of inductance and capacitance, as values of these standards involve only physical dimensions. (It is clear from the fact that the dimensions of inductance are [μL] in e.m. system and those of capacitance are [$\varepsilon L$] in e.s. system.)

## 5.18 ABSOLUTE MEASUREMENT OF CURRENT

The dimensions of current are [$\mu^{1/2} M^{1/2} L^{1/2} T^{-1}$] or [$M^{1/2} L^{1/2} T^{-1}$] if we regard permeability as a non-dimensional quantity. The dimensions of $(\text{current})^2$ are [$MLT^{-2}$] which are the dimensions of force. Therefore, the absolute measurement of current involves the measurement of a force. This force may be exerted in two ways :

(*i*) By the current in a solenoid upon a suspended magnetic needle - as in a tangent or sine galvanometer.

(*ii*) By the current in one part of a circuit upon another part of the circuit in series with it and carrying the same current as in a current balance.

The absolute measurement of current by **Rayleigh's Current Balance** is described below.

### 5.18.1 Rayleigh's Current Balance

This instrument works on the principle that if a current carrying coil is placed with its plane parallel to that of another current carrying coil with their axes coincident, there will be a force exerted between the coils. This force is proportional to the product of currents in the two coils and if the coils carry the same current, the force is proportional to the square of the current. This force can be measured if one of the coils is movable and is suspended. In this case the force can be weighed.

Figure 5.1 shows the arrangement of coils in a Rayleigh's current balance. There are two fixed coils and a small moving coil which is suspended from the arm of a sensitive balance. The three coils are connected in series. If $I$ is the value of current through them the force $F$ exerted on the suspended coil is,

$$F = I^2 \, dM/dx \text{ newton} \qquad \text{...(5.4)}$$

where $dM/dx$ is the rate of change of mutual inductance between the fixed and moving coils. This is expressed in henry/metre. The value of $dM/dx$ can be obtained from the physical dimensions of coils and $F$ is purely a mechanical quantity which can be accurately measured by the balance.

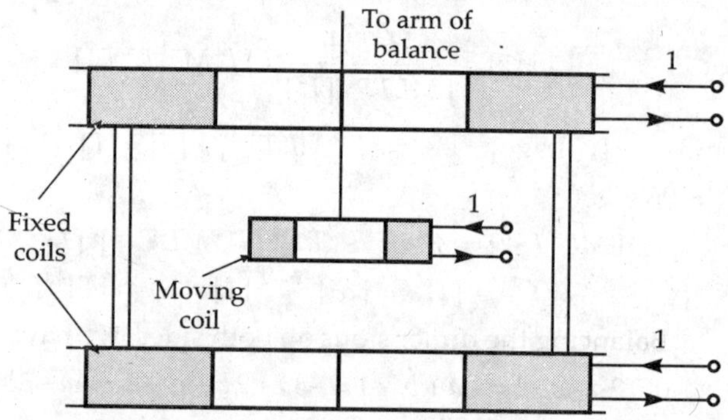


**Fig. 5.1** Rayleigh's current balance.

Thus the value of $I$ is dependent upon units of purely physical mechanical quantities and is independent of other electrical units. For measuring the value $F$, it is usual to counterbalance the suspended coil when the force on it is upwards. The current through the suspended coil is then reversed thus reversing the force. The additional counterbalance weight gives the value $2F$. It should be noted that the expression for the force is in Newton and since the weights used in counterbalancing the force are in kg, the value of '$g$' acceleration due to gravity must be known. The accuracy with which the value of '$g$' is known will determine the accuracy of current measurement. In the above measurement, we must try to minimize the errors due to:

(*i*) Inaccuracy in values of physical dimensions of the coils (on which the value of mutual inductance, $M$, depends).

(*ii*) Pull on the moving system due to leads attached to movable coil. The leads used should be flexible.

(*iii*) Heating of coils : The fixed coils should be cooled by water jackets and the moving coil by water chamber.

(*iv*) Distortion of magnetic field : For this bobbins made of non-magnetic materials should be used.

## 5.19 ABSOLUTE MEASUREMENT OF RESISTANCE

The resistance has the dimensions $[\mu LT^{-1}]$ in the e.m. system of units. Taking the permeability as dimensionless, the dimensions of resistance are $[LT^{-1}]$ *i.e.*, those of velocity. Thus the absolute measurement of resistance involves the measurement of either a velocity or the measurement of length and time. We describe here the Lorenz method of absolute measurement of resistance.

### 5.19.1 Lorenz Method

Figure 5.2 shows the experimental set up used for this method.

There is a circular metallic disc which is mounted concentrically inside a solenoid. The disc is rotated at a uniform speed. A steady current is passed through the solenoid. *R* is a low resistance whose absolute value is to be determined. The resistance *R* is connected across the disc with the help of leads and two brushes. One brush presses the edge of the disc and the other brush makes a contact with the disc near its centre. A sensitive galvanometer *G* is included in one of the leads. The magnetising current *I* for the solenoid is supplied by a battery. This current passes through resistance *R* also.

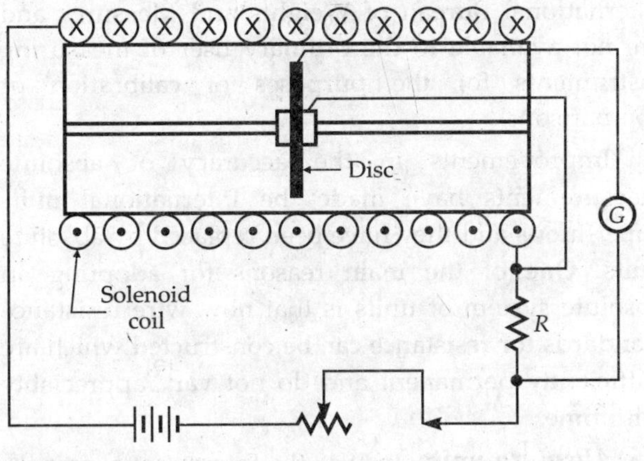


**Fig. 5.2** Lorenz method.

When we pass current through the solenoid, it produces a magnetic field. As the disc rotates an emf is induced in it. The current flowing through *R* produces a voltage drop across it. The emf generated in the disc is balanced against the voltage drop in resistance *R*, the balance being indicated by the galvanometer *G*. The speed of the disc is adjusted till a balance is obtained.

Let *M* = mutual inductance between disc and solenoid in henry. Inductance is defined as flux linkages per unit current and therefore the value of magnetic flux linkages produced by the disc when a current of *I* ampere passes through solenoid is *MI*.

But flux linkages of disc = number of turns × flux linking the disc = flux linking the disc (as the number of turns in a disc is one).

Therefore, we conclude that the value of flux passing perpendicularly through the disc surface is *MI* weber if a current *I* ampere passes through the solenoid.

Let *n* = speed of the disc in r.p.s. when the deflection of galvanometer is zero.

∴ Emf induced in the disc = rate of change of flux linkgages $= MI \times n = MIn$ volt ...(5.5)

Voltage drop in resistance = *IR* volt,

where *I* and *R* are expressed in ampere and ohm respectively.

Thus when the deflection of the galvanometer is zero,

$$MIn = IR \quad \text{or} \quad R = Mn \text{ ohm} \qquad ...(5.6)$$

The value of mutual inductance *M* is obtained from the physical dimensions of the solenoid and the disc. Hence the value of resistance *R* is determined solely in terms of linear dimensions of solenoid and disc, number of turns of solenoid and speed of rotation and is thus independent of other electrical units.

The apparatus used involves several sophisticated modifications in the construction, but the basic principle remains the same as described.

We can obtain very accurate results, if the :

(*i*) Disc and solenoid are carefully constructed and their dimensions are accurately measured.

(*ii*) Effect of earth's magnetic field upon the emf induced is made small. This is done by arranging the plane of the disc to be in the magnetic meridian and making two measurements one with current flowing in one direction and the other with the direction of current reversed.

(*iii*) Thermoelectric emfs are eliminated if two readings are taken with the current reversed in the second case.

(*iv*) Speed of rotation is accurately determined. The speed be determined by stroboscopic methods or with the help of directly driven chronograph.

**Example 5.11** *In a simple form of Lorenz apparatus for absolute measurement of resistance, the solenoid is 50 mm in diameter and 0.5 m long. It is uniformly wound with 8000 turns of wire. The rotating copper disc is 40 mm in* *diameter and the galvanometer indicates balance condition when the speed of the disc is 5952 r.p.m. Calculate the value of resistance in ohm and the mutual inductance between the disc and the coil in henry.*

**Solution.**

Number of turns $N = 8000$.

Length of solenoid $l = 0.5$ m,

Radius of solenoid $r = 25$ mm $= 0.025$ m.

Intensity of field at the centre $= H$

$$= \frac{NI}{2\sqrt{(l/2)^2 + r^2}} \text{ for a long solenoid.}$$

$$\text{Flux density at the centre } B = \mu_0 H = \frac{\mu_0 NI}{2\sqrt{(l/2)^2 + r^2}}$$

Let $r_1$ be the radius of disc.

$\therefore$ Flux linked by disc $\Phi = B \times$ area of disc

or $$\Phi = \frac{\mu_0 NI}{\sqrt{(l/2)^2 + r^2}} \times \pi r_1^2 = \frac{\pi \mu_0 r_1^2 NI}{2\sqrt{(l/2)^2 + r^2}}$$

The disc is treated as a single turn winding.

$$\therefore \text{Flux linkages of disc } \Psi = \frac{\pi \mu_0 r_1^2 NI}{2\sqrt{(l/2)^2 + r^2}}.$$

Mutual inductance between disc and solenoid is

$$M = \frac{\Psi}{I} = \frac{\pi \mu_0 r_1^2 N}{2\sqrt{(l/2)^2 + r^2}}$$

$$= \frac{\pi \times 4\pi \times 10^{-7} \times (0.02)^2 \times 8000}{2\sqrt{\left(\frac{0.5}{2}\right)^2 + (0.025)^2}} \text{ H} = 0.0251 \text{ mH.}$$

Resistance $R = Mn = 0.0251 \times 10^{-3} \times 5920/60$

$= 2.49 \text{ m}\Omega$.

## 5.20 STANDARDS AND THEIR CLASSIFICATION

A **standard** is a physical representation of a unit of measurement. The term '**standard**' is applied to a piece of equipment having a known measure of physical quantity. They are used for the purposes of obtaining the values of the physical properties of other equipment by comparison methods. In fact, a unit is realized by reference to a material standard or to a natural phenomenon including physical and atomic constants. For example, the fundamental unit of mass in the metric system (SI) is the **kilogramme**, defined as the mass of the cubic decimetre of water at its temperature of maximum density of 4°C. This unit of mass is represented by a material standard : the mass of the International Prototype kilogramme consisting of platinum iridium hollow cylinder. This unit is preserved at the International Bureau of Weights and Measures at Severes, near Paris and is the material representation of the kilogramme. Similar standards have been developed for other units of measurement, including fundamental units as well as for some of the derived mechanical and electrical units.

Just as there are fundamental and derived units of measurement we find different types of "**Standards of Measurement**" classified by their function and application in the following categories :

(*i*) International standards.

(*ii*) Primary standards.

(*iii*) Secondary standards.

(*iv*) Working standards.

### 5.20.1 International Standards

The international standards are defined on the basis of international agreement. They represent the units of measurements which are closest to the possible accuracy attainable with present day technological and scientific methods. International standards are checked and evaluated regularly against *absolute measurements* in terms of the fundamental units. The International Standards are maintained at the International Bureau of Weights and Measures and are not available to the ordinary user of measuring instruments for the purposes of calibration or comparison.

Improvements in the accuracy of absolute measurements have made the International units superfluous and they have been replaced by absolute units. One of the main reasons for adopting an absolute system of units is that now wire resistance standards for resistance can be constructed which are sufficiently permanent and do not vary appreciably with time.

***Absolute units.*** In 1948 the International ampere was superseded by the absolute ampere. The determination of absolute ampere was again made by means of a current balance, which weighs the force between two current carrying coils. Improvement in the techniques of force measurements yields a value for ampere far superior than given by earlier measurements. The relationship between the force and the current which produces the force can be calculated from fundamental electro-magnetic theory concepts. It reduces to a simple computation involving the geometric dimensions of the coils. The absolute ampere is now the fundamental unit of electric current in the SI.

### 5.20.2 Primary Standards

Primary standards are absolute standards of such high accuracy that they can be used as the ultimate reference standards. These standards are maintained by national standards laboratories in different parts of the world. The primary standards, which represent the fundamental units and some of the derived electrical and mechanical units, are independently calibrated by absolute measurements at each of the national laboratories. The results of these measurements are compared against each other, leading to a world average figure for the primary standards. Primary standards are not available for use outside the national laboratories. One of the main functions of the primary standards is the verifications and calibration of secondary standards.

The primary standards are few in number. They must have the highest possible accuracy. Also these standards must have the highest stability, *i.e.*, their values should vary as small as possible over long periods of time even if there are environmental and other changes.

In the recent past, the techniques of establishing primary standards have been drastically refined so that accuracy attainable has become of a very high level.

### 5.20.3 Secondary Standards

The secondary standards are the basic reference standards used in industrial measurement laboratories. The responsibility of maintenance and calibration of these standards lies with the particular industry involved. These standards are checked locally against reference standards available in the area. Secondary standards are normally sent periodically to the national standards laboratories for calibration and comparison against primary standards. The secondary standards are sent back to the industry by the national laboratories with a certification as regards their measured values in terms of primary standards.

### 5.20.4 Working Standards

The working standards are the major tools of a measurement laboratory. These standards are used to check and calibrate general laboratory instruments for their accuracy and performance. For example, a manufacturer of precision resistances, may use a **Standard Resistance** (which may be a working standard) in the quality control department for checking the values of resistors that are being manufactured. This way, he verifies that his measurement set up performs within the limits of accuracy that are specified.

## 5.21 STANDARDS FOR MASS AND LENGTH

***Primary standard of mass.*** We have stated before that the material representation of unit of mass is the Prototype kilogramme preserved at the International Bureau of Weights and Measures at Severes near Paris. The primary unit of mass is a Prototype kilogramme kept at National Physical Laboratories of every country. This has an accuracy of 1 part in $10^8$ and is occasionally verified against the standard kept at the International Bureau.

***Secondary standards of mass.*** The secondary standards of mass are kept by industrial laboratories. These standards have an accuracy of 1 ppm and are checked against the primary standards.

***Working standards of mass.*** The working standards of mass are available in a wide range of values so that they suit any kind of application. The accuracy of working standards is of the order of 5 ppm. The standards are verified against the secondary standards.

***International units of length.*** The international unit of length, the metre, was defined in 1960 in terms of an optical standard. This was the orange-red radiation of krypton atom. The internationally recognized krypton-86 discharge lamp when excited and observed under well defined conditions, emits orange light whose wavelength constitutes the basic unit of length. This can be measured to an accuracy of 1 ppm. The metre is equal to 1,650,76.73 wavelengths in vacuum of the orange-red light radiation of the krypton-86 atom in its trasition between levels $^2p_{10}$ and $^5d_5$. In, 1983, the metre was redefined, in terms of time which can be specified accurately using atomic standards. The metre is now the length travelled by light in vacuum in a time interval of 1/299792458 second. The speed of light therefore is now 299792458 m/s.

***Working standards of length.*** There is a wide range of working standards of length in industry. These are usually precision **Gauge Blocks** made up of steel. These blocks have two parallel surfaces and the distance between the two surfaces is specified. The accuracy tolerance of this distance is usually 0.5-0.25 micron. The most outstanding advantages gained by using of gauge blocks are their precision, low cost and accuracy which is 1 part per million (1 ppm). These features of gauge blocks have made it possible to manufacture interchangeable industrial components in a very economical manner.

## 5.22 ATOMIC FREQUENCY AND TIME STANDARDS

The recent advancements and refinements of atomic resonators has made possible control of frequency of an oscillator to a great extent. Therefore it is now very easy to construct atomic clocks of great precision and accuracy by frequency conversion (since frequency is inversely proportional to time). The transit between two energy levels $E_1$ and $E_2$ of an atom is accompanied by the emission (or absorption) by radiation given by :

$$E_2 - E_1 = h\nu$$

where $h$ = Planck's constant,

and $\nu$ = a physical constant (which is a frequency) provided that the energy states are not affected by external conditions such as magnetic fields. This physical constant depends only on the internal structure of the atom.

Since frequency is the inverse of time interval, atomic clocks can be constructed on the above mentioned principle. Investigations were done on atomic transitions of various metals. The International Committee of Weights and Measures defines the second in terms of the Caesium transitions, assigning a value of 9,192,631,770 Hz to hyperfine transition of Caesium 133 atom between two levels ($F = 4$, $m_f = 0$ and $F = 3$, $m_f = 0$) of the ground state $^2S_{1/2}$ unperturbed by external fields.

Three main atomic standards are used and they are : caesium, hydrogen, and rubidium. The caesium beam and the hydrogen maser are **primary or absolute standards** since they do not need to be referenced to any other standard for checking or calibration. The rubidium vapour standard, like a quartz oscillator, is a **secondary standard**, and it needs to be initially set up against a primary standard, and then periodically recalibrated.

Table 5.9 compares these four main types of frequency standards. Although the quartz oscillator is shown here as an independent standard, it is often used to provide feedback frequency to caesium or rubidium atomic standards, which are passive.

### 5.22.1 Caesium Beam Standard

Figure 5.3 shows the basic construction of a Caesium Beam Resonator. Caesium atoms are slowly evaporated in an oven and collimated into a beam. The state select magnet deflects each atom by an amount which depends on the energy state of that atom. Only those atoms in the $F = 4$, $m_f = 0$ state are allowed to form the caesium beam in the vacuum chamber. Here they pass through a low and uniform magnetic field (*c*-field) which is subjected to microwave excitation at a frequency of 9192.631770 MHz. This has exactly the correct energy to change the state of some of the atoms to $F = 3$, $m_f = 0$. The beam then passes a second state select magnet, identical to the first one. This deflects the atoms which have undergone the required state change onto a hot wire ionizer. The ionized caesium atoms are then filtered by a mass spectrometer, and are converted to electron current by the detector.

Figure 5.4 shows a control scheme for the caesium beam resonator, where a crystal oscillator is used to produce the microwave field frequency. The output from the detector, in the resonator, is amplified and eventually fed to a voltage controlled crystal oscillator. The driving signal from the crystal oscillator is frequency multiplied and then phase modulated by an audio oscillator. The amplitude and phase detector provides a signal magnitude and polarity, which corresponds to the difference of the oscillator frequency from the desired frequency. This signal is filtered by the integrator, to remove high-frequency components, and then used to drive the crystal oscillator, to the correct frequency required by the caesium beam resonator.

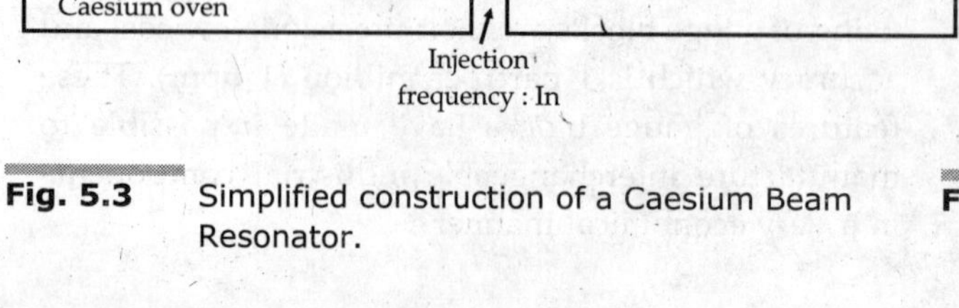


**Fig. 5.3** Simplified construction of a Caesium Beam Resonator.

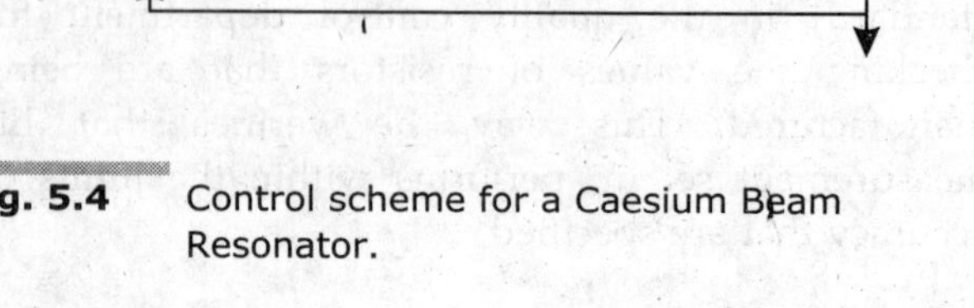


**Fig. 5.4** Control scheme for a Caesium Beam Resonator.

**Table 5.9** *Characteristics of Frequency Standards*

| Characteristics | Caesium beam | Hydrogen maser | Rubidium vapour | Quartz |
|---|---|---|---|---|
| Reproducibility | $\pm 3\times10^{-12}$ | $\pm 2\times10^{-12}$ | NA | NA |
| Stability (1 s average) | $5\times10^{-12}$ | $5\times10^{-13}$ | $5\times10^{-12}$ | $5\times10^{-12}$ |
| Drift (ageing rate) | negligible | negligible | $\pm 1\times10^{-11}$ / mon. | $\pm 5\times10^{-10}$/day |
| Resonance frequency (Hz) | 9192631770 | 1420405751 | 6834682608 | NA |
| Operating temperature | – 20 to + 60°C | 0 to 50°C | 0 to 50°C | 0 to 50°C |
| Weight (approximate) | 30 kg | 400 kg | 15 kg | 10 kg |
| Power demand (approximate) | 40 W | 200 W | 40 W | 20 W |
| Atomic interaction time | $2.5\times10^{-3}$ s | 0.5 s | $2\times10^{-3}$ s | NA |
| Atomic resonance events/s | $10^6$ | $10^{12}$ | $10^{12}$ | NA |
| Temperature of resonating atoms | 360°K | 300°K | 330°K | NA |

As indicated in Table 5.9, caesium beam standards are accurate to within 3 parts in $10^{12}$ and have good long term and short term stability. Their relatively small size and weight has also resulted in their adoption, by the majority of the world standard authorities, as their frequency reference.

### 5.22.2 Hydrogen Maser Standard

The hydrogen maser is the most stable frequency source currently known, having a frequency of 1420405751.73±0.03 Hz. However, due to its relatively large size, its use is limited to areas where stability is critical, and size is not a consideration.

Figure 5.5 shows the basic construction of a hydrogen maser standard. Unlike the caesium beam standard the hydrogen maser is an active device, and its signal can be directly amplified, or used to drive a crystal oscillator. An atomic hydrogen beam is fed past a hexapole state select magnet, which allows atoms in a higher energy state to pass into a quartz storage bulb. The bulb has teflon coated walls to minimise perturbations of the atoms, which would result in a frequency shift. The quartz bulb confines the atoms to a uniform magnetic field, in a tuned microwave cavity. This is set to a frequency equal to the transition of the hydrogen atom, between the states $F=1$, $m_f=0$ and $F=0$, $m_f=0$. The hydrogen atoms undergo many reflections off the walls of the bulb, and interact with the microwave field. This relaxes the atoms, and they give up their energy to the microwave field within the tuned cavity. This stimulates more atoms to radiate, until a maser operation is achieved.

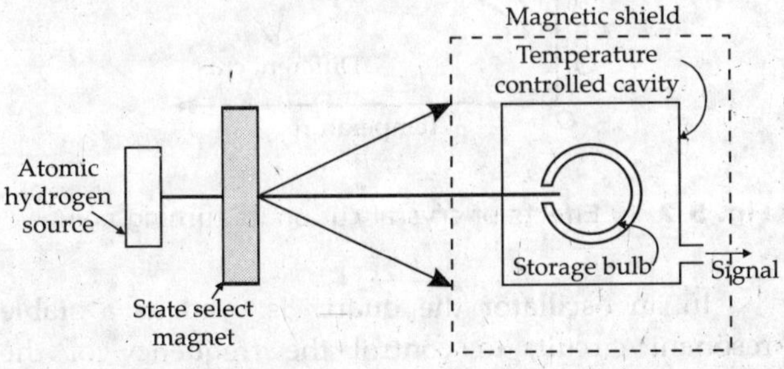


**Fig. 5.5** Simplified construction of a Hydrogen Maser Resonator.

### 5.22.3 Rubidium Vapour Standard

The rubidium vapour standard is based on the hyperfine transition in rubidium 87 gas, between the states $F=2$ and $F=1$. The transition frequency is 6834682608 Hz.

Figure 5.6 gives the basis of the Rubidium Vapour Resonator. Light from a spectral lamp is filtered, and then fed to the cell containing rubidium 87 vapour in a microwave cavity. An inert buffer gas is added to reduce Doppler broadening effects. The microwave signal is derived by multiplication of a quartz oscillator, operating in a closed loop, similar to that used for a caesium standard.

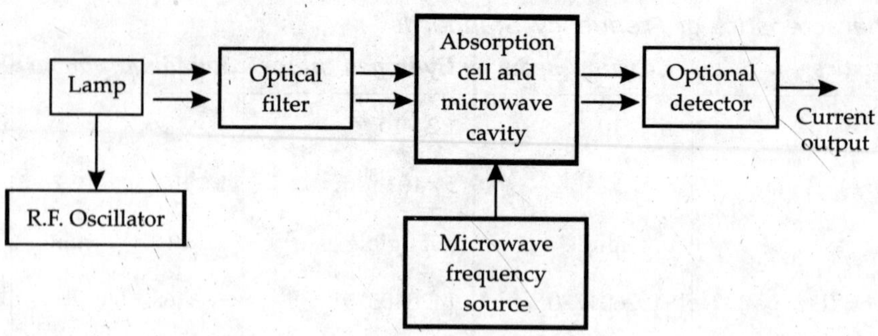


**Fig. 5.6** Simplified representation of a Rubidium Vapour Resonator.

Optical pumping of the rubidium 87 gas causes the population of the $F=2$ state to increase, at the expense of the $F=1$ state. Application of microwave energy, close to the transition frequency of the rubidium vapour states, causes transitions from the $F=2$ to the $F=1$ level, so that more light is absorbed by the cell. At a frequency corresponding to 6834682608 Hz the photodetector output reaches a minimum.

The rubidium vapour standard has good short term stability, and small size, making it a useful portable instrument. It is not self calibrating, and must be calibrated against a primary standard. The standard exhibits drift, since the resonance frequency is dependent on the gas mixture and pressure in the cell. However its drift is over 100 times less than that of the quartz standard.

### 5.22.4 Quartz Crystal Standard

The quartz oscillator relies on the piezoelectric properties of quartz, to give a frequency standard which is used in a wide range of applications. The oscillator employs the high $Q$ value of the quartz, when it is in a mechanical resonant mode. The frequency range is generally between 1 Hz-750 kHz and 1.5-200 MHz. Ranges between 750 kHz and 1.5 MHz are difficult to cover, due to strong secondary resonance. It is also preferable to use high frequency crystals, since the stability of low frequency crystals is less, and they have greater physical size.

The quartz crystal is aligned using $X$-rays, and then precision cut along one of several planes. It is then ground and lapped to exact dimensions, electrodes are vacuum deposited onto the sides and leads attached to them.

The quartz crystal has an atomic structure oriented about the $x$ (electrical), $y$ (mechanicals), and $z$ (optical) axes. Many different cuts can be made to it depending on the application. An $X$ cut gives a mechanical change in thickness, in the $x$ direction, for an electrical field applied in the $y$ direction. It has a negative temperature coefficient of frequency. A $Y$ cut gives mechanical change in length in the $y$ direction for an electrical field in the $x$ direction. It has a positive temperature coefficient of frequency. By making a cut rotated from the $x$ and $y$ axes, it is possible to combine the negative and positive temperature coefficients, to minimise the frequency shift in the crystal.

The two cuts most commonly used for quartz controlled oscillators are $AT$ and $BT$. The $AT$ cut has a lower variation of frequency with temperature, but the $BT$ cut is preferred above 10 MHz, since it is less susceptible to drive levels and load capacitances. All crystals have a frequency-temperature characteristic similar to those of Fig. 5.7. The position of the turning point is determined by the cut of the crystal, and can be chosen to occur in the range – 50°C to + 100°C. Temperature controlled ovens are used to produce a quartz crystal standard having high stability and low drift.

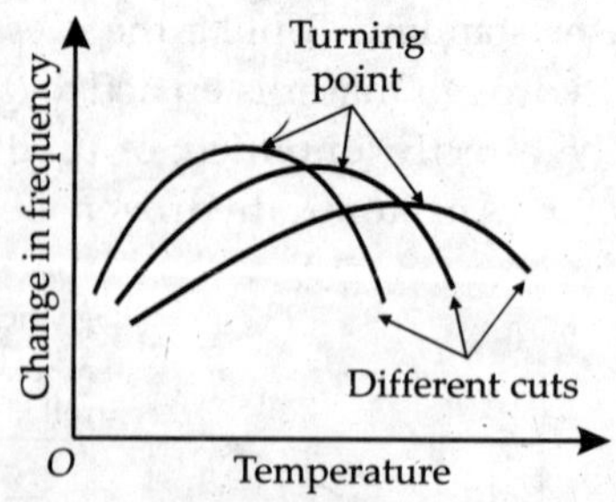


**Fig. 5.7** Effects of crystal cut on its turning point.

In an oscillator the quartz is used as a stable resonant circuit, to control the frequency of the electronic system. Figure 5.8 shows the equivalent circuit of a quartz crystal assembly. $L_1$, represents the vibrating mass of the crystal blank. $R_1$, the loading on the crystal due to ambient air and molecular friction. $C_1$, the elasticity of the quartz material, and $C_0$ the static capacitance between the electrodes plated on the blank, and the stray lead and holder capacitances.

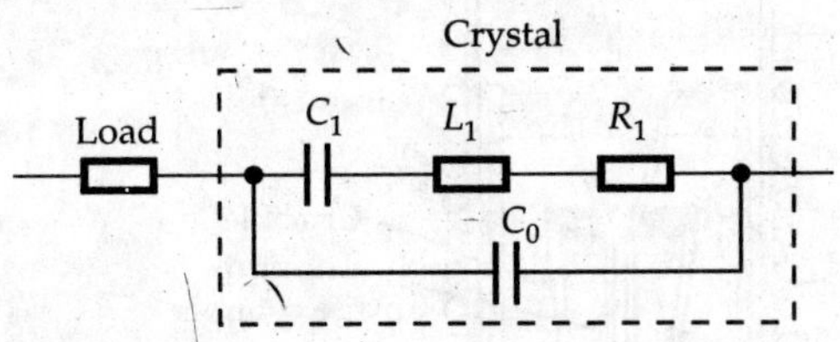


**Fig. 5.8** Equivalent circuit of a quartz crystal.

A quartz crystal can operate in a series resonant mode or in a parallel antiresonant mode. Its characteristic is similar to that shown in Fig. 5.9 where $f_s$ and $f_p$, are the series resonant and parallel antiresonant frequencies respectively. The value of the series resonance frequency and of $Q$ are given by :

$$f_s = \frac{1}{2\pi\sqrt{L_1 C_1}} \quad \text{...(5.7)}$$

$$Q = 2\pi f_s \frac{L_1}{R_1} \quad \text{...(5.8)}$$

The ability to change the operating frequency of the crystal by varying the load capacitance is called the *'pullability'* of the crystal, and is used in applications such as voltage controlled and temperature controlled oscillators.

The ageing or frequency drift of the crystal is asymptotic with time. After an initially fast rate the ageing settles down to about 5 parts in $10^{10}$ per day, in a temperature controlled oven. If an oscillarors or its oven, is switched off for a few hours, then the crystal recovers, and the ageing process starts again at its

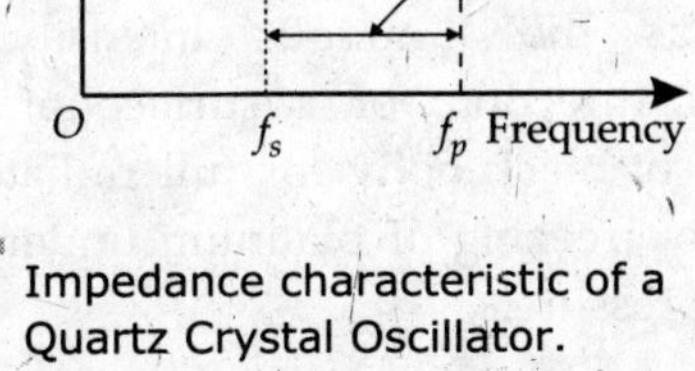


**Fig. 5.9** Impedance characteristic of a Quartz Crystal Oscillator.

initial rate; therefore a quartz standard should never be switched off. Ageing is caused by contamination in the crystal, fatigue in the mounting wires and solder, outgassing of material inside the sealed holder and leakage of the seals.

The ***advantages*** of atomic standards are :

(*i*) The accuracy of atomic standards is much higher than that of standards based upon astronomical measurements.

(*ii*) Determination of time intervals can now be made in a very short time (a few minutes at the most) and with much better accuracy than is possible with astronomical measurements that many a times took many years to complete. Some of the atomic clocks are in operation which give precision of 1 μs per day. They form the **primary frequency standard.**

(*iii*) The results with atomic standards are more uniform. Also they form a better and convenient time base.

## 5.23 TEMPERATURE STANDARDS

Thermodynarnic temperature is one of the basic **SI** quantities whose unit is kelvin. The themrodynamic kelvin scale is recognized as the basic scale to which all temperatures should be referred. The temperatures on this scale are designated as °K and denoted by the symbol *T*. The magnitude of kelvin has been fixed by defining the thermodynamic temperature of the triple point of water at exactly 273.16°K. The triple point of water is the temperature at which equilibrium between ice, liquid water, and its vapour.

Since temperature measurements on the thermodynamic scale are inherently difficult, the Seventh General Conference of Weights and Measures adopted in 1927, a practical scale which has been modified several times and is now called the **"International Practical Scale of Temperatures"**. The temperatures on this scale are designated as °C (degree Celcius) and denoted by the symbol *t*. The Celcius scale has two fundamental points the boiling point of water as 100°C and the triple point of water as 0.01°C, both points are established at atmospheric pressure. A number of primary fixed points have been established above and below the two fundamental points as given in Table 5.10.

**Table 5.10** *Fixed points on practical scale of temperature*

| *Fixed Point* | *Kelvin(°K)* | *Celcius(°C)* |
|---|---|---|
| Triple point of equilibrium of Hydrogen | 13.81 | – 259.34 |
| Boiling point of equilibrium of Hydrogen | 20.28 | – 252.87 |
| Boiling point of neon | 27.102 | – 246.048 |
| Triple point of oxygen | 54.361 | – 218.789 |
| Boiling point of oxygen | 90.188 | – 182.962 |
| Freezing point of zinc | 692.73 | 419.58 |
| Freezing point of silver | 1235.08 | 961.93 |
| Freezing point of gold | 1337.58 | 1064.43 |

The numerical values of all these points are reproducible quantities at atmospheric pressure. The conversion between kelvin scale and the Celcius scale is :

$$1(°C) = T(°K) - T_0$$

where $T_0 = 273.16$ degrees.

The primary standard for temperature is the platinum resistance thermometer. It is constructed from strain free platinum, and the values between the primary and fundamental fixed temperature points can be calculated accurately knowing the resistance properties of platinum wire.

## 5.24 LUMINOUS INTENSITY STANDARDS

The primary standard of luminous intensity in a full radiator (black body or Planckian radiator) at the temperature of solidification of platinum (2.042°K approximately) under a pressure of 101.325 kN/m$^2$. The "candela" is then defined as one sixtieth of luminous intensity per square metre of one full radiator.

Secondary standards of luminous intensity are special tungsten lamps calibrated against basic standards.

In 1948, it was decided to base the unit of luminous intensity upon the luminance (or objective brightness) of a small aperture due to light emitted from a radiator maintained at the temperature of solidification of platinum namely 2042°K. This primary standard is a full radiator at the temperature of platinum. The construction of this primary standard shown in Fig. 5.10.

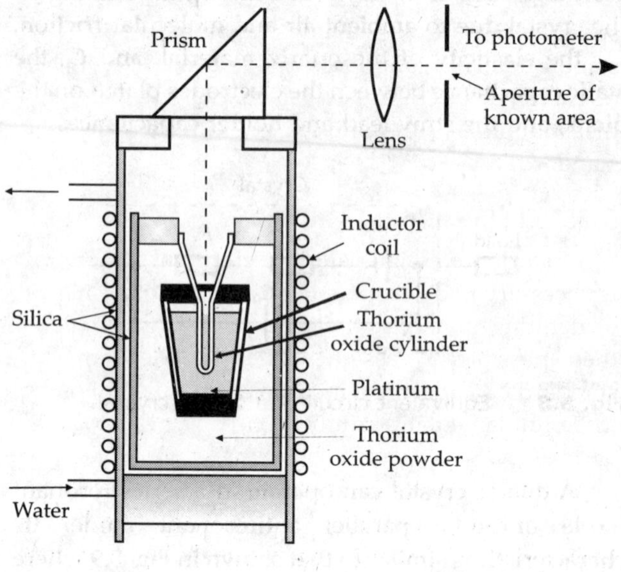


**Fig. 5.10** Primary standard of luminous intensity.

The radiator consists of a cylinder of fused thorium oxide about 45 mm long, with an internal diameter of about 2.5 mm. The bottom of the tube is packed with powdered fused thorium oxide. The cylinder is supported vertically in pure platinum contained in fused thorium oxide crucible. The latter has a lid with a small hole in the centre, about 1.5 mm in diameter, and is almost embedded in powdered fused thorium oxide in a larger refractory container having a funnel shaped opening.

The platinum is first melted by eddy currents induced in it by a high frequency coil surrounding the outer container. The coil carries alternating current at a frequency of 1.6 MHz. The platinum is allowed to cool very slowly. During the period platinum is changing from liquid to solid state, the temperature remains constant sufficiently long for measurements to be made of luminous intensity of the light passing from the aperture of known diameter. The luminance of this primary standard is found to be 598,000 international candle/m$^2$, the international candle being used prior to 1948.

In 1948 it was decided to adopt 600,000 units/m$^2$ as the luminance of primary standard. The new unit of luminous intensity was termed as "**Candela**". Candela is defined as the luminous intensity, in the perpendicular direction, of a surface of 1/600,000 square metre of black body (or full radiator) at the temperature of freezing of platinum under standard atmospheric pressure.

Secondary standards of luminous intensity are special type of tungsten lamps calibrated against basic standards. The lamp power is adjusted to give an operating temperature at which the spectral power distribution of the lamp matches that of the basic standard.

## 5.25 ELECTRICAL STANDARDS

In order to assign values to electrical standards, it is necessary to determine only two standards in terms of defining units. From electrical measurements it is then possible to assign values to the remaining standards. All electrical measurements are based upon three fundamental quantities : current, resistance and voltage. A system of measurements depends upon definitions of units of these quantities. These three electrical units are not independent but are related by Ohm's law : $E = IR$. Thus, the definition of any two units in terms of other related physical quantities is sufficient to determine the definition of the third. Accordingly, if any two units are available as samples of calibration the instruments which measure the third unit may also be calibrated.

The unit of current can be established in terms of the physical dimensions of a current balance which measures the force exerted between two current carrying coils.

The unit of resistance can also be established in an absolute manner using by Lorenz method.

Therefore national laboratories usually maintain standards of emf and resistance and base the values of other standards on these two.

In practice, an absolute volt can be obtained by passing an absolute ampere through an absolute ohm. This is possible because both absolute ampere and an absolute ohm are physically realizable.

However, there are some practical difficulties. For example : once standards of current, resistance and emf are established by definition it is necessary that samples of unit current, resistance and emf are constantly available for comparison with other samples and other associated instruments. It is possible, though not easy, to construct a standard ohm which is accurate and permanent. But it is highly impractical to maintain a standard current for the routine calibration of ammeters and voltmeters. This is because each absolute determination of current takes a long time involving precision work. Working with a current balance is a protracted and time consuming job, which would make instrument calibration very costly and also inconsistent.

Hence we conclude that from the view-point of practical considerations, the absolute determination of emf using a current balance and an absolute ohm, is not feasible. Therefore an alternative becomes imperative. Logically, the alternative should be an independent source of emf.

Fortunately for the science of measurement, it was discovered that certain types of chemical batteries developed, and maintained emf of great constancy, so that they could serve as practical reference of emfs. Therefore, the practical standard of electrical measurement is the ohm, which is a direct absolute standard, and the volt, which is originally determined from the absolute measurement of current. No working standards of current are maintained. Only working standards of resistance and emf are maintained.

## 5.26 EMF STANDARDS

The primary, secondary and working standards of emf are described below :

### 5.26.1 Primary Standard of Emf

The primary voltage standard is the saturated normal Weston cell which was developed by Edward Weston in 1892. This is a voltaic cell.

***Construction and characteristics.*** The construction of a saturated normal Weston cell is shown in Fig. 5.11. The positive electrode is mercury and the negative electrode is an amalgam of cadmium which is 1 part of cadmium and 7 parts of mercury. The electrolyte is a saturated solution of cadmium sulphate. A paste of mercurous sulphate $Hg_2SO_4$ is placed on top of mercury. This acts as the depolarizing

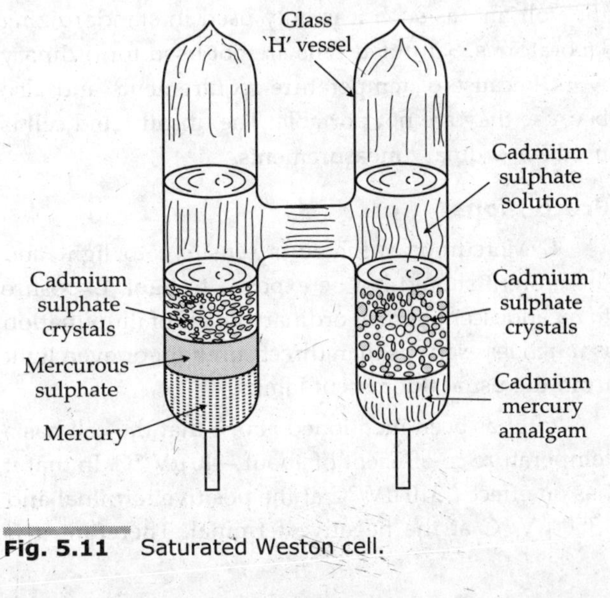


**Fig. 5.11** Saturated Weston cell.

agent. An excess of solid crystals of $CdSO_4 . 8H_2O$ (cadmium sulphate) in the electrolyte space and the depolarizer assures that the solution will be saturated at all temperatures. A simplified representation of the cell reaction is :

$$Cd + Hg_2SO_4 = CdSO_4 + 2Hg \quad ...(5.9)$$

As long as the amalgam has both a solid and liquid phase present, the emf of the cell remains absolutely constant.

The Weston standard cell is contained in an H-shaped glass vessel. The limbs of the H vessel are hermetically sealed and the connections to the external circuit are made through platinum wires. Sulphuric acid is added to act as electrolyte. Saturated cells are normally allowed to age for a year before they are considered to be of constant emf. Following this period, their life span appears indefinite excluding the possibility of accidents.

The emf of the cell is 1.01864 volt. The emf of the cell changes with temperature and the equation relating emf to temperature is :

$$\begin{aligned} E_t = {} & E_{20} - 46 \times 10^{-6}(t-20) - 0.95 \times 10^{-6}(t-20)^2 \\ & + 10 \times 10^{-9}(t-20)^3 \quad ...(5.10) \end{aligned}$$

where $E_t$ = emf of cell at any temperature $t$°C,

and $E_{20}$ = emf of cell at 20°C.

The cell voltage drops by about 40 μV/°C. The cell is kept in an oil bath to control its temperature within 0.01°C. The cell has a long life if handled carefully. It has a life span of about 10 to 20 years. The drift in value of voltage is about 1 μV per year.

The saturated cell is a standard for maintenance of the volt and as such it is only used in standardizing laboratories. Saturated cells are not used for ordinary work because of temperature requirements and also because they are not portable. The unsaturated cell is used for ordinary measurements.

## Precautions

**1.** Mercurous sulphate is sensitive to light and therefore it should not be exposed to light. Exposure to incandescent light at ordinary levels of illumination is much less serious than direct sun light or even light from the usual fluorescent lamps.

**2.** It has been mentioned above that the cell has a temperature co-efficient of about – 40 μV/°C. In that it has an effect + 310 μV/°C at the positive terminal and –350 μV/°C at the negative terminal. Therefore it is important to keep the entire cell at a uniform temperature to secure as much cancellation of positive and negative effects. Hence these cells are always used in close temperature regulation, usually an oil bath held constant to better than 0.01°C.

**3.** Saturated cells should be kept in an upright position because if they are tilted there is always the danger of mixing of electrode materials. Therefore these cells cannot be shipped and are sent through special messengers.

**4.** The principal limitation of the use of standard cells is the current drawn from the cell at any time. This should be negligibly small. Even this small current should be allowed to flow for a very short period of time. It should be understood that standard cells are potential devices and will be damaged if current is drawn from them.

It is very difficult to specify the maximum value of current that can be drawn from a standard cell since the damage to the cell is a function of both magnitude of the current and the duration of its flow. The makers specify the maximum value as 100 μA but this should be regarded as an extreme figure. This means that the current from the cell should be less than 100 μA and this current should flow momentarily.

A voltmeter should never be used for measuring the voltage of a standard cell. Firstly because of the current drain, there is a possibility of damage to the cell and secondly this measurement is meaningless since the cells have a high internal resistance (about 600 to 800 Ω). The voltage should be measured with the help of a potentiometer only because at balance the current drawn from the cell is zero.

A cell should be handled with care and never be short-circuited. The excessive current causes a change in emf which may be permanent because the cell may not be able to recover. A cell once short circuited, should be regarded with suspicion and hence is of little value as a standard.

When using the cell in potentiometric work, a resistance of 20,000 Ω should be inserted in series with the cell in order to avoid excessive drain of current when there is unbalance.

It is important that leakage resistance between the terminals of the cell be very high otherwise it may not give the standard voltage it is supposed to give. This is illustrated by the following example.

*Example 5.12 A standard cell has a voltage rating of 1.018500 V and an internal resistance of 500 Ω. The insulation resistance between its terminals is 5 MΩ. Find the current drain due to insulation resistance. Calculate also the difference between the internal voltage and the terminal voltage.*

**Solution.** Current flowing through the insulation resistance

$$= \frac{1.018500}{500 + 5 \times 10^6} = 0.204 \ \mu\text{A}.$$

Thus there is a constant drain of a current of 0.204 μA.

Voltage drop in internal resistance

$$= 0.204 \times 10^6 \times 500 \text{ V} = 102 \ \mu\text{V}.$$

∴ Difference between internal voltage and terminal voltage = 102 μV.

Terminal voltage = 1.018500 – 0.000102
= 1.018398 V.

### 5.26.2 Secondary Standard of Emf

The secondary standard of Emf is the unsaturated Weston cell.

*Unsaturated Weston Cell Construction and Characteristics.* The construction of an unsaturated cell is shown in Fig. 5.12. The unsaturated cell is similar to a saturated cell except that it does not have the cadmium sulphate crystals ($CdSO_4$). It has the same electrodes but the solution is saturated at 4°C, and, therefore, unsaturated at room temperature. In addition, a 'sputum' or retaining member (porous

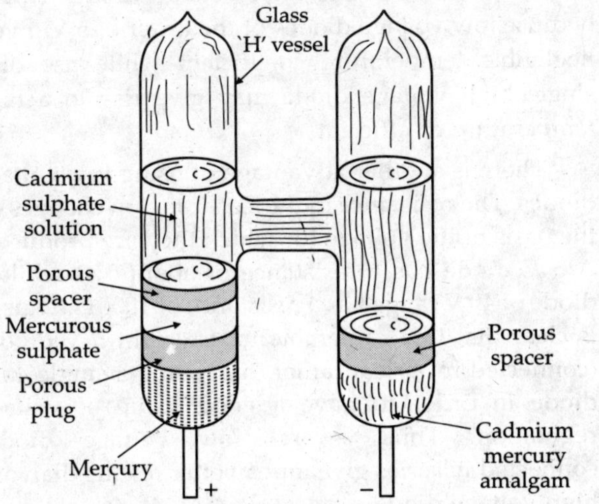


**Fig. 5.12** Unsaturated Weston cell.

plugs in this case) is used over each electrode to hold the material in place and hence these cells are considered portable. After being moved from one place to another, the cell should be allowed a few days to settle so that the results obtained from measurements involving cells are accurate.

The cells are enclosed in bakelite case which is lined with 1.5 mm thick copper to help keep all parts of the cell at a uniform temperature.

Many characteristics of unsaturated cell are different from those of a saturated cell. Not all unsaturated cells have the same emf. In fact the emf saturated cells differ from one to another. The emf of new unsaturated cells lies between 1.0180 to 1.0194 volt. Therefore each cell must be calibrated against the primary standard in order to know its exact emf. The temperature at which this emf is measured should also be specified.

The unsaturated cell becomes less stable with age ; yet it is remarkably good. The decrease in emf is about 30 to 50 μV per year. It is recommended that the cells be checked by returning them to a standardizing laboratory once a year for certification.

An unsaturated cell has a shorter life than a saturated type, yet experience has shown that some cells last even 20 years. In addition to age, the life of a cell depends on the kind of treatment the cell has received.

The temperature co-efficient of the unsaturated cell is less than of a saturated cell. It is –10 μV/°C. It has an accuracy of 0.005%. Thus it can be used for most purposes without temperature correction, if the temperature of use does not differ by a few degrees from the calibration temperature. This cell, also has a positive temperature effect at one electrode and negative at the other. Therefore it is important to keep all parts of the cell at the same temperature.

### Precautions

Same precautions as for use of saturated cells should be followed while using unsaturated cells. The only difference is that the unsaturated cells are portable and can be shipped.

Besides the precautions given for saturated cells, unsaturated cells should not be operated at temperatures below 4°C and above 40°C. Experiments made with cells beyond these limits show that the extreme upper limit of reliability occurs at 50°C. Below 4°C the emf drops rapidly and eventually the electrolyte freezes. Above 40°C the cell materials start to melt and voltage drops sharply around 55°C. It has been mentioned earlier that unsaturated cells lose emf, at an average rate of 30-50 μV per year at normal temperature of use and storage. At higher temperature the decrease is much more rapid and is about 240 μV per year at 60°C.

It is recommended that the normal operating temperature be 25°C for precision work. The cells should be mounted so that they are not exposed to extremes of temperature, to drafts, steam pipes, resistors carrying current and to light.

### 5.26.3 Laboratory Standards of Emf

In the past standard cells were used exclusively as laboratory standards but in recent years, semi-conductor devices such as Zener controlled reference sources have replaced standard cells in practically all industrial applications.

Silicon diodes have voltage current characteristics such that an extremely sharp reverse current occurs at a point on the voltage curve known as "Zener voltage". The point indicates a breakdown of the diode under reverse voltage application, but the process is reversible if safe current and heating limits are not exceeded. The Zener voltage may be controlled over a wide range by the processing techniques used during the manufacture of the Zener diodes. For voltage standard applications, Zener diodes should have a low dynamic impedance, and a low temperature co-efficient of impedance. Both alloy junction and diffused junction Zener diodes are used, the fomrer in the range 2.4 V and the latter from 6.8 - 200 V. Diffused junction diodes have a lower dynamic impedance in the region of operating point.

The circuit of Fig. 5.13 shows the Zener diode in a typical application as a laboratory standard. The supply voltage $E$ is usually much higher than the Zener voltage of diode assuring breakdown of the diode. Resistor $R$ usually has a high value and is placed in series with the diode and the battery. It serves to limit the current through the diode during breakdown to a safe value. In addition when the supply voltage varies for any reason, most of this variation is taken up in the voltage drop across $R$ with

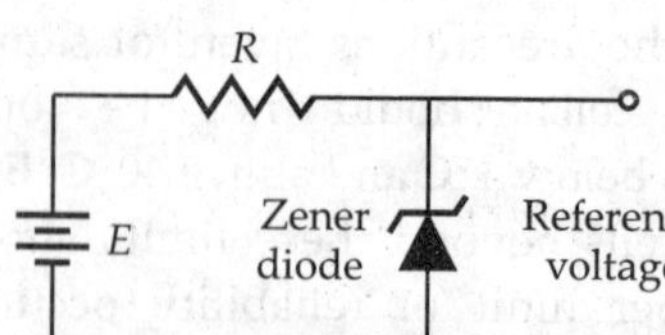


**Fig. 5.13** Zener diode laboratory standard.

a very small change in diode current and as a result the reference voltage (output voltage) remains practically the same.

Commercial voltage standards using Zener diodes essentially consist of :

(*i*) A Zener controlled voltage source placed in a temperature controlled environment to improve its long term stability.

(*ii*) A precision output voltage divider. The temperature is controlled to ± 0.03°C over an ambient temperature range of 0°C to 50°C providing an output stability of the order of 10 ppm/month. The four available outputs are :

(*a*) A 0-1000 μV source with 1 μV precision,

(*b*) A 1.000 V reference for potentiometer measurements with ratio box,

(*c*) A 1.018 +(Δ) reference for saturated Weston cell comparisons, and

(*d*) A 1.019 +(Δ) reference for unsaturated Weston cell comparisons.

*Higher voltage zener sources and temperature effects.* It is many a times convenient from the practical point of view to build higher voltage standard sources. In the following passages, problems associated with higher voltage sources and effect of temperature changes are discussed.

Zener diodes having reverse voltages in the range 4-5 V are used because they have negligible resistance temperature co-efficient. In many applications, standard sources of much higher voltage than 5 V are desired. Under these circumstances there is a distinct advantage of using several low voltage diodes in series, rather than one high voltage diode. This is because low voltage diodes of the order of 5 V give a negligible temperature co-efficient while use of a single high voltage diode may give rise to a high temperature co-efficient.

There is another advantage of using low voltage diodes. The resistance in the Zener region changes on the basis of the voltage for which they are produced. A 5 V diode has a resistance of about 5 Ω while a diode of 40 V range has a resistance of 1000 Ω. Thus, it is clear that it is preferable to use eight 5 V diodes (connected in series) rather than using a single 40 V diode in order to have a good value of voltage regulation. Thus several low voltage diodes connected in series give much better results than one high voltage diode.

It has been experienced that Zener diodes indicate permanence of value to a satisfactory level. However, it is not expected that they will replace the standard chemical cells in work involving high precision or in standardising laboratories. However, they may serve to maintain a calibrated voltage applied to potentiometer circuit for thermocouple work and recording potentiometers.

Their major ***advantages*** are :

(*i*) Commercial voltage-reference using Zener diodes plus additional circuitry to improve regulation and stability are available with voltage outputs in the vicinity of 1.0191 V ± 0.01 percent which allow them to be substituted for a standard cell.

(*ii*) The advantage of these standards over standard cell is that, unlike the standard cell. the semi-conductor device is immune to the effects of short circuit and overload conditions.

(*iii*) An attractive proposition is substitution of standard chemical cells by Zener sources in recording potentiometers which may be subjected to freezing temperatures. This is on account of the fact that the standard cells cannot work below 4°C while Zener sources give good accuracies at all temperatures if properly compensated.

(*iv*) Chemical cells can be built to voltage level of about 1 volt while Zener sources can be built to any voltage level. Thus Zener sources are preferable where higher voltage standard sources are required.

(*v*) Spring board or other applications where vibration is constantly present do not present ideal conditions for chemical cells. The Zener sources which are rugged may furnish the accuracy required under such conditions.

(*vi*) The constant voltage drop of a Zener diode may be utilized in voltage regulators, compensators, limiters, clippers, voltage reference units and voltage standards.

### 5.26.4 Josephon-Effect Standard

A current flows between two pieces of a super-conducting material which are separated by an insulator. This is known as *Josephon Effect*. Voltage standards, based upon Josephon Effect are relatively new, and have been used in only recently in some Standards Laboratories.

A voltage standard utilizing Josephon Effects, consists of a thin film of lead separated by a lead oxide which acts an insulating medium. The assembly is mounted on a glass substrate. The probe, thus formed, is then placed in a superinsulated helium Dewar system. The tunnelling junction is excited by a microwave source having a frequency of nearly 9 GHz, and is biased by a supply current source. Under these conditions the Josephon junction produces a voltage.

$$E = Nhf/2e \qquad ...(5.11)$$

where $h$ = Planck's constant,

$f$ = frequency of microwave source ; Hz.

$e$ = the electron charge and $N$ is an integer.

The value of integer $N$ can be controlled by the bias current, thereby varying the voltage produced in discrete steps.

Josephon Effect voltage standards have an accuracy of about 0.05 ppm and therefore, can be used for calibration of other emf standards using potentiometers.

## 5.27 RESISTANCE STANDARDS

The absolute value of the ohm, in SI system of units is defined in terms of fundamental units of length and mass. Absolute measurement of the ohm is done against a group of resistance standards, which are maintained at the International Bureau of Weights and Measures at Severes. This group is periodically checked against each other, and with similar groups of standards kept at other Standards Laboratories in the world.

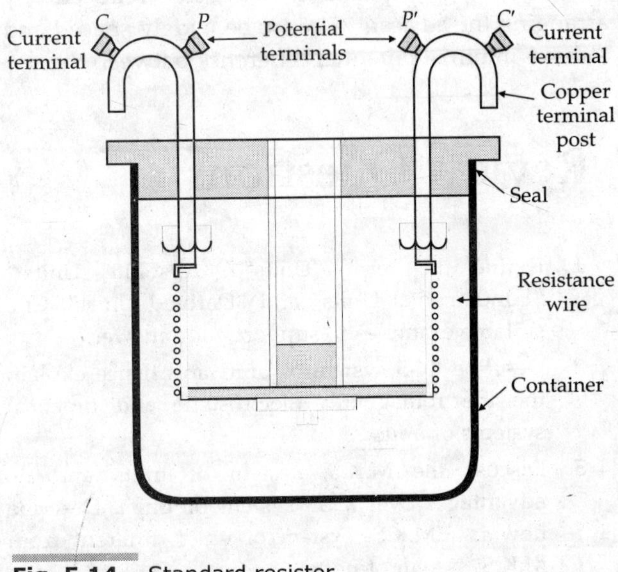


**Fig. 5.14** Standard resistor.

Figure 5.14 shows construction of a typical standard resistor. The resistance wire made from an alloy like Manganin. The constitution of Manganin is Nickel : 4%, copper : 84% and Manganese : 12%. Manganin has the following properties :

(*i*) It has a high resistivity which is $48 \times 10^{-8} \Omega$ m.

(*ii*) It has a low temperature co-efficient of resistance which is $40 \times 10^{-6}$/°C at 20°C.

(*iii*) It has a low thermal emf with copper which is below 3 μV/°C.

The resistance coil is so wound that it reduces stress. The coil is immersed in oil for temperature stabiiity. The resistor is housed in a sealed moisture free container.

The resistance of the standard at any temperature *t*°C is given by

$$R_t = R_{25} + \alpha(t-25) + \beta(t-25)^2 \quad ...(5.12)$$

where $R_{25}$ = nominal value of resistance at 25°C ; Ω, α and β are thermal co-efficients whose values are than $10 \times 10^{-6}$ and $-0.5 \times 10^{-6}$ respectively so that change in resistance for a change in temperature of 10°C is less than 50 ppm.

***Errors.*** The errors which occur in these resistance standards are :

(*i*) Due to skin effect when they are used at high frequencies. These errors are minimized by using wires of small diameter.

(*ii*) Due to stray inductance and capacitance. The stray inductance can be reduced by arranging the adjacent wires to carry current in the opposite direction so as to reduce the resultant magnetic field.

The stray capacitance can be reduced by arranging the adjacent wires to be widely spaced and to have minimal potential difference between them.

Aryton-Perry method of winding resistance wire (explained in chapter 6, page 164) achieves these objectives.

(*iii*) Due to contact resistance of the terminals in case of low value standards. The contact resistance is reduced by using a four terminal construction see page 161, in which current is passed through two outer terminals (called current terminals) and the voltage reading, for resistance value, is obtained from inner two terminals. The current drawn from the inner terminals (called potential terminals) must be kept to minimum, for accurate resistance measurement. In fact, when the value of resistance standards is determined by a potentiometer, the current drawn from the potential terminals is zero under conditions of balance.

## 5.28 CURRENT STANDARDS

It is obvious that it is impossible to build a current standard in the same sense that a resistance standard is built up. Current can be defined either in terms of an electrochemical equivalent of silver or in terms of force exterted two current carrying conductors as in Rayleigh's current balance. Such standards are rarely used in modern practice. It is much easier to measure the value of current by making it flow in a resistance standard and then measure the voltage drop across the resistance with the help of a potentiometer which has been calibrated against a voltage standard.

## 5.29 INDUCTANCE STANDARDS

These are explained in Chapter 6. (see page 166)

## 5.30 CAPACITANCE STANDARDS

These are explained in Chapter 6.(see page 171)

# Review Questions

1. Define the term "Units", "Absolute Units", "Fundamental Units" and "Derived Units". Cite suitable examples to support your answer.
2. Describe CGS system of units and hence explain the Electromagnetic, Electrostatic and Practical systems of units.
3. Discuss the M.K.S. system of units and its advantages over CGS system of units. Describe how is R.M.K.S. system of units different from M.K.S. system of units.
4. Define the seven base units of SI system.
5. (*a*) State Coulomb's inverse square law for force exerted between

   (*i*) two charges and

   (*ii*) magnetic poles.

   (*b*) Derive dimensional equation for current in both e.s. and e.m. systems.

   (*c*) Derive dimensional equations for emf in both e.m. and e.s. systems.

6. Derive the dimensions of the following quantities in *L*, *M*, *T*, *I* system

(*i*) magnetic flux (*ii*) inductance
(*iii*) capacitance (*iv*) magnetic
(*v*) magnetic flux density (*vi*) mmf
(*vii*) reluctance (*viii*) magnetising force
(*ix*) permeability (*x*) electric flux
(*xi*) permittivity (*xii*) resistance.

7. Show that product $\frac{1}{\sqrt{\mu\varepsilon}}$ has the dimensions of velocity where $u\mu$ = permeability and $\varepsilon$ = permittivity.

8. Explain the Lorentz method for determination of absolute ohm. Describe the precautions taken to minimize the error.

9. Explain how absolute value of current is determined by using Rayleigh's current balance. Describe the precautions taken to minimize the errors.

10. Distinguish between :

(*i*) International standards
(*ii*) Primary standards
(*iii*) Secondary standards and
(*iv*) Working standards.

11. Describe the International standards of Mass and Length.

12. What are' atomic standards for frequency and time ? Explain their advantages.

13. Describe the following atomic standards for frequency and time :

(*i*) Caesium beam standard
(*ii*) Hydrogen maser standard
(*iii*) Quartz crystal standard

Compare them with regards to reproducibility, stability and drift.

14. Describe the primary standard for Luminous intensity.

15. Describe why only Emf and Resistance standards are maintained in order to assign values to other electrical standards. .

16. Describe the Weston saturated and unsaturated cells used as emf standards. What are the precautions taken while using them ?

17. Describe the Zener sources used as laboratory standards for emf. Discuss their advantages.

18. What is Josephon Effect ? Describe how a standard emf source can be built using this effect.

19. Describe the construction of a Resistance standard. Describe the techniques used to minimize errors in them.

20. Explain why is it practically impossible to build a current standard on the same pattern as an emf or a resistance standard is built.

21. Describe how the value of °K has been fixed for thermodynamic scale. List some fixed points on practical scale of temperature.

5

## Unsolved Problems

1. (*a*) Sate Coulomb's inverse square law for electric charges and magnetic poles.

(*b*) Derive the dimensional equation for current in the e.m. system.

(*c*) Derive dimensional equatibn for electric potential in both e.s. and e.m. systems.

(*d*) Deduce the value of permeability of free space in R.M.K.S. system compared to its value in C.G.S. system.

2. Derive the dimensions of the following quantities in *L*, *M*, *T*, *I* system :

(*i*) potential difference
(*ii*) inductance
(*iii*) capacitance
(*iv*) magnetic flux
(*v*) flux density
(*vi*) m.m.f.
(*vii*) reluctance
(*viii*) magnetizing force
(*ix*) resistance
(*x*) current.

3. Use dimensional equations to obtain dimensions for (*i*) $\mu_0$ (*ii*) $\varepsilon_0$ in C.G.S. and e.s. systems.

4. Use dimensional equations to find out whether the following expression is dimensionally correct or not.

$$R_4 = \frac{R_2(1+\omega^2 R_3^2 C_3^2)}{\omega^2 R_1 R_3 C_3}$$

where $R_1, R_2, R_3, R_4$ are resistance $C_3$, is a capacitance, and $\omega = 2\pi f$, ($f$ being a frequency in Hz)

[**Ans.** Incorrect ; correct equation $R_4 = \dfrac{R_2(1+\omega^2 R_3^2 C_3^2)}{\omega^2 R_1 R_3 C_3^2}$ ]

**5.** In the course of a calculation, the following expression was obtained :

$$I = \frac{V\omega M}{[(\omega M + R_1 R_2)^2 \omega^2 L_1 L_2 R_1^2]^{1/2}}$$

Check whether the equation is dimensionally correct. If not, suggest any corrections.

[**Ans.** Incorrect ; $I = \dfrac{V\omega M}{[(\omega^2 M^2 + R_1 R_2)^2 + \omega^2 L_1 L_2 R_1^2]^{1/2}}$ ]

**6.** The voltage $V$ across the coil of a telephone receiver when a current $I$ is flowing through it may be written as :

$$V = I\left(Z + \frac{A^2}{z}\right)$$

where

$$Z = R + j\omega L,\ A = 2B_0\frac{N}{R},\ z = r + j\left(\omega m + \frac{s}{\omega}\right),$$

$R$ = resistance,

$L$ = inductance, and

$N$ = number of turns of the receiver coil,

$B_0$ = the flux density in air gap,

$R$ = the reluctance of the magnetic circuit,

and $r$, $m$ and $s$ are respectively the equivalent mechanical resistance, mass and stiffness of the receiver diaphragm. Suggest a suitable unit of each of the quantities involved. Find the dimensions of each quantity, and make a check of the equation for dimential equality.

**7.** Prove that the following equation is dimensionally correct :

$$e = Blv$$

where

$e$ = e.m.f. generated in a conductor,

$B$ = flux density,

$l$ = length of conductor,

$v$ = peripheral velocity of conductor.

**8.** Prove that the following equations are dimensionally correct :

(*i*) $r = \dfrac{R\omega^2 M_1 M_2}{R^2 + \omega^2 L^2}$ (*ii*) $C = \dfrac{1}{\omega^2 M}$

where $r$, $R$ = resistances,

$M_1$, $M_2$, $M$ = mutual inductances,

$L$ = self inductance,

$C$ = capacitance and $\omega$ = angular frequency.

**9.** The pull of an electromagnet is given by the expression :

$$P \propto B^a A^b \mu_0^c$$

where $B$ = flux density in the air gap of the magnet

$A$ = area of magnet, $\mu_0$ = permeability of air gap.

Determine the values of $a$, $b$ and $c$ from the dimensions of the quantities involved.

[**Ans.** $a = 2$, $b = 1$, $c = -1$]

# Objective Type Questions

***Tick (√) the most appropriate answer :***

**1.** A coherent system of units is to be evolved which has $N$ kinds of quantities to evaluate and $M$ independent physical quantities expressing relationships between them, we can independently choose the sizes of units of :

(*a*) $N + M$ quantities

(*b*) $N - M$ quantities

(*c*) $N$ quantities

(*d*) $M$ quantities.

**2.** The units whose sizes cannot be choosen independently are called :

(*a*) Derived units

(*b*) Fundamental units

(*c*) Absolute units

(*d*) Auxiliary fundamental units.

**3.** The dimensions of inductance in CGS e.s. system are :

(*a*) $[\varepsilon L^{-1} T^2]$ (*b*) $[\varepsilon^{-1} ML^{-1} T^2]$

(*c*) $[\varepsilon^{-1} LT^{-2}]$ (*d*) $[\varepsilon^{-1} L^{-1} T^2]$

**4.** The dimensions of resistance CGS e.m. system are :

(*a*) $[\mu^{-1} LT^{-1}]$ (*b*) $[\mu^{-1} MLT^{-1}]$

(*c*) $[\mu LT^{-1}]$ (*d*) $[\mu LT^{-2}]$

5. The base units in SI system are
   (a) metre, kilogramme, second
   (b) metre, kilogramme, second, ampere
   (c) metre, kilogramme, second, ampere, kelvin, candela, mole
   (d) metre, kilogramme, second, ampere, kelvin, candela.

6. The symbol nm stands for :
   (a) newton metre
   (b) nano metre
   (c) nano milli
   (d) number of moles.

7. In measuring the absolute value of ampere using Rayleigh's balance, the value of gravitational constant $g$ must be accurately known because :
   (a) the expression for the force is in newton and weights used in counterbalancing the force are in kg
   (b) value of $g$ is different at different places
   (c) corrections have to be made for the weight of moving coil
   (d) none of the above.

8. In Lorenz method for absolute measurement of ohm, two measurements are made one with current in one direction and the other with current in the opposite direction. This is done to :
   (a) eliminate the effect of thermal emfs
   (b) eliminate the effect of earth's magnetic field
   (c) eliminate the effects of both thermal emfs and earth's magnetic field
   (d) eliminate the effect of temperature on coil and disc.

9. The international standard of length is defined in terms of :
   (a) length of earth's meridian passing through Paris
   (b) distance between two lines engraved on a platinum-iridium bar
   (c) wavelength in vacuum, of radiation of krypton - 86 atom in its two specified transitions
   (d) none of the above.

10. The most stable primary atomic standard for frequency is :
   (a) Caesium beam standard
   (b) Hydrogen maser standard
   (c) Quartz standard
   (d) Rubidium vapour standard.

11. Caesium beam standard is normally used primary atomic standard for frequency because :
   (a) Caesium beam standard has an accuracy 3 parts in $10^{12}$, has short term stability, low drift and small size.
   (b) Hydrogen maser standard, though it the most stable frequency source but has a large size.
   (c) Quartz crystal's operating frequency changes with change in load capacitance and also the drift in frequency is asymptotic with time
   (d) all the above.

12. The disadvantages of atomic standards are :
   (a) their accuracy is lower than that of standards based upon astronomical measurements
   (b) the time taken for making measurements is long
   (c) the results obtained with atomic standards are not uniform
   (d) none of the above.

13. The maximum current that can be drawn from a saturated Weston cell without damaging it, is
   (a) 100 mA momentarily
   (b) 100 $\mu$A continuously
   (c) 100 $\mu$A momentarily
   (d) 1 mA momentarily.

14. The internal resistance of a saturated Weston cell is of the order of :
   (a) a few ohm
   (b) a few kilo ohm
   (c) a few hundreds of ohm
   (d) a fraction of ohm.

15. The decrease in the value of emf of an unsaturated Weston cell is :
   (a) about 1 $\mu$V per year
   (b) about 1 mV per year
   (c) about 5 – 10 $\mu$V per year
   (d) about 30 – 50 $\mu$V per year.

16. Zener sources when used as voltage standards :
   (a) are reverse biased
   (b) can be designed to cover a wide range of voltages
   (c) are immune to the effects of short circuits
   (d) all the above.

17. The output voltage Josephon effect standard :
   (a) can be changed in a continuous manner

5

(*b*) can be changed only in discrete steps

(*c*) can be changed continuously as well as in discrete steps

(*d*) cannot be changed.

**18.** Candela can be defined as luminous intensity in the perpendicular, of a surface of :

(*a*) 1/2 59000 square metre of a full radiator

(*b*) 1/600 square centimetre of a full radiator

(*c*) 1/5,89,000 square metre of a full radiator

(*d*) 1/600,000 square metre of a full radiator.

**19.** The material of wires used for making resistance standards is usually :

(*a*) Manganin

(*b*) Nichrome

(*c*) Copper

(*d*) Phosphor Bronze.

**20.** In order to reduce skin effect in resistance standards when they are used on high frequency a.c. :

(*a*) the diameter of the resistance wire should be small

(*b*) adjacent turns of the resistance coil should carry currents in the opposite direction

(*c*) they should be provided with four terminals

(*d*) all the above.

## *Answers*

| | | | | | | | | | |
|---|---|---|---|---|---|---|---|---|---|
| **1.** (*b*) | **2.** (*a*) | **3.** (*d*) | **4.** (*c*) | **5.** (*c*) | **6.** (*b*) | **7.** (*a*) | **8.** (*c*) | **9.** (*c*) | **10.** (*b*) |
| **11.** (*d*) | **12.** (*d*) | **13.** (*c*) | **14.** (*c*) | **15.** (*d*) | **16.** (*d*) | **17.** (*b*) | **18.** (*d*) | **19.** (*a*) | **20.** (*a*) |

# 6

# Circuit Components (Resistors, Inductors and Capacitors and Their Residues)

## 6.1 RESIDUES

It is almost impossible to manufacture a circuit component which has only one property *i.e.*, solely a pure resistance, an inductance or a capacitance. For example, a resistor does not have resistance only but has an associated inductance and capacitance, an inductor in addition to its inductance has resistance and interturn capacitance, and a capacitor has resistance in addition to capacitance. The impurities present in a circuit component produce unwanted quantities called *Residues*. However, a particular quantity such as resistance in a resistor, inductance in an inductor, or capacitance in a capacitor are made to dominate in order that the effect of residues is very small and it is possible to design *pure* circuit components for specified limits of accuracy.

## 6.2 RESISTORS

Resistors and resistive networks are extensively used in instrumentation and measurement work. The foremost properties of materials, used in the construction of resistors meant for precision work are : stability or permanence with time, low resistance temperature co-efficient, low thermo-electric emf with copper, high resistivity, resistance to oxidation, corrosion and moisture, ease of manufacture and low cost.

No single material possesses all the above mentioned properties and, therefore only that material is selected which is best suited for the particular application involved.

### 6.2.1 Materials Used for Resistors

The most widely used materials for precision resistors are :

***Manganin.*** It is an alloy of copper, manganese and nickel and is universally used as resistance material for precision resistors and for resistance measuring apparatus. It has a nominal composition of 84 percent copper,12 percent manganese and 4 percent nickel. It has a resistivity of 0.45 to $0.50 \times 10^{-6}\,\Omega\text{m}$ (nearly 25 times that of copper) and develops a thermal emf of 2 to 3 μV/°C against copper. When properly heat treated, it gives a stable resistance value with time *i.e.*, it shows no ageing effects.

The foremost property of manganin is that it has almost a zero temperature coefficient of resistance near about room temperatures. A representative value may be taken as 0.004 percent per °C at 20°C. Manganin is used for resistances of very high accuracy when the temperature rise is not expected to rise above 15 to 20°C.

***Constantan.*** These are a series of alloys of nickel and copper containing 40 to 60 percent nickel, with a small amount of manganese to improve their mechanical properties. All these alloys exhibit similar electrical properties. They are sold as constantan, or under various trade names for use as thermocouple materials. They have thermoelectric emfs against copper of about 40 μV per °C. However, except for their large thermoelectric emf, the electrical properties of these alloys are remarkably similar to those of manganin.

Constantan has a resistivity at ordinary temperatures of about 25 times that of copper (about the same as manganin), is corrosion resistant, inexpensive, and easy to work. This can be easily soft soldered to copper.

It finds extensive use in cases where its high thermoelectric emf against copper is not a disadvantage. For example, it is frequently used in resistors designed for a.c. operation. It also finds application in resistors of 1000 Ω and above as in voltmeter multipliers where the thermal emf generated at copper constantan junction is negligibly small as compared to the emf being measured.

***Nickel chromium alloys.*** These alloys have a somewhat higher temperature co-efficient of resistance than that of manganin and constantan. Nichrome is an example of this class of alloys. These alloys cannot be used in precision resistances. Nichrome has a very high resistivity (about 50 times that of copper) and resists corrosion even at very high temperatures. It is often used in rougher class of resistors, where small size is all important or where the operating temperatures are high. However, these alloys are difficult to solder.

***Gold chromium.*** It is an alloy of recent introduction, which appears to be very promising for some applications. It is made with slightly over 2 percent of chromium. This alloy has a resistivity at room temperature of about 20 times that of copper. Its temperature co-efficient can be made extremely small by baking it at fairly low temperatures. It has a thermo-electric emf of 7 or 8 μV per °C with copper. For many applications, the extremely small temperature co-efficient of gold chromium alloys make their use desirable especially for heat resistant standards.

### 6.2.2 Spools (Formers) for Resistance Coils

The spools (formers) made up of metals are commonly used for high quality d.c. resistors as compared with wooden spools used earlier. The advantages of metal spools are :

(*i*) They do not absorb moisture from the atmosphere unlike wooden formers. Therefore, they do not either expand or contract because of changes of humidity. Therefore, no varying stresses are produced in the resistance coil which otherwise cause change in their resistance value.

(*ii*) The metal spools readily dissipate heat produced on account of flow of current in the resistance coil. This is because the resistance wire is in intimate thermal contact with the spool and heat is readily transferred to the spools from which it is dissipated to surroundings due to convection and radiation. The outside surface of the spool is already exposed to the surroundings and hence the entire surface of the spool *i.e.*, both inside and outside is effective in dissipation of heat thereby preventing any hot spot temperatures to build up in the resistance coil.

Metal spools are ordinarily made of brass. This is because brass has a co-efficient of thermal expansion which is nearly equal to that of materials used for wires of resistance coils, and hence there is no significant differential expansion between the coil and the spool. Therefore there are no significant changes in stresses are produced in resistance wires on account of changes in temperature.

For a.c. applications, the use of metallic spools is undesirable, even out of question because of the eddy current losses produced in them. Ceramic is the universally used material for spools of a.c. resistors. The disadvantages of ceramic spools for use in resistors of high precision, are their poor thermal conductivity and their temperature co-efficient of linear expansion which is much smaller than that of the resistance wire. This leads to differential expansion with changes in temperature leading to stresses produced in the windings.

### 6.2.3 Resistance Wires

The resistance wire is generally double silk or silk and cotton covered. The wire is enamelled before these coatings are applied.

High quality resistors are wound with only one layer of wire although this requires use of smaller wire than for multilayer coils. The advantages of single layer coils are :

(*i*) heat dissipation is more satisfactory with single layer coils. On the other hand, in a multi-layer coils, the central layers have a higher temperature.

(*ii*) Single layer coils are more stable. Multi-layer coils are more sceptical to change in resistance on account change in atmospheric humidity.

(*iii*) Multi-layer coils are usually found to be less stable in resistance with time.

## 6.3 RESISTANCE STANDARDS

Figure 6.1 shows a primary resistor of 1 Ω value. The standard resistor is coil of wire made of manganin mounted on a silk insulated brass spool. The coil uses a bifilar type of arrangement (explained in Art. 6.7 on page 163) so as to reduce the residual inductance. After wound, resistance coils are impregnated with a shellac varnish and given an accelerated ageing by baking at a temperature of 150°C for about 48 hours. Ageing makes the coil resistance stable with time. The coil is supported from the hard rubber top by means of a thermometer tube in which a thermometer can be inserted from outside. The copper lead wires, which are silver soldered to the ends of resistance coil, are soft soldered to copper binding posts. The hard rubber top is screwed into the metal container which is filled with a good quality light mineral oil. The binding posts, thermometer well, and the threads by which the hard rubber top is connected to the container are all sealed with shellac which is not soluble in oil.

The purpose of the oil is to give good thermal contact between the resistor and the case in order to facilitate the dissipation heat produced in the coil. Also, oil increases the thermal capacity of the resistor so that the coil can temporarily carry large currents without getting heated.

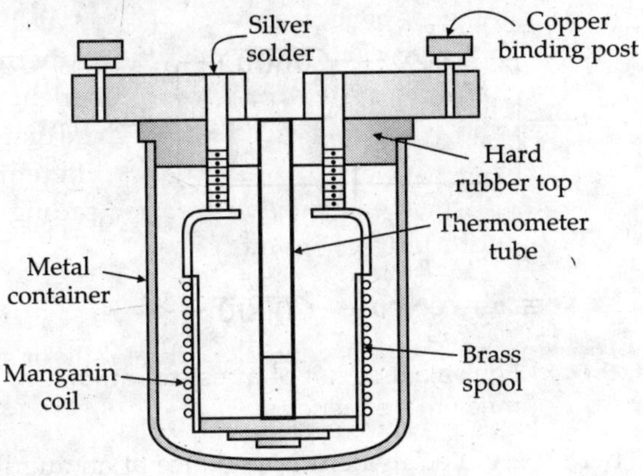


**Fig. 6.1** Primary Resistance Standard.

In some resistors, the advantage of high thermal conductivity is combined with the advantage of hermetic sealing. In this case a double walled container is used. The double walled container consists of two concentric metal co-axial cylinders with the space between cylinders being sealed. The coil is mounted in the sealed space is in good thermal contact with the inner cylinder. These double walled resistors readily give up heat to an oil bath and are not affected by changes in humidity which may otherwise cause a change in the value of resistance. The sealed space in which the coil is mounted is filled with dry air, no oil comes in physical contact with resistance coil.

In order to improve cooling, a series of holes are provided near the top just above the double walled portion.

## 6.4 LOW RESISTANCE STANDARDS

Low resistance standards have nominal value of resistance less than 1 Ω. Low resistance standards are *four terminal type*. They are provided with two current terminals $C$ and $C'$ and two potential terminals $P$ and $P'$. The advantages of four terminal resistors can be explained as under :

(*i*) Suppose a low resistance standard is provided with two terminals only as shown in Fig. 6.2(*a*).

The resistance measured ($V/I$) between the two terminals is the resistance of the coil plus the contact resistance at the current terminals. This contact resistance may be of the order of 0.001 Ω and therefore when we measure the resistance of 10 Ω resistor, the error involved is only 0.01%. However, when the resistor involved is of nominal value 0.01 Ω, the error may be as much as 10%.

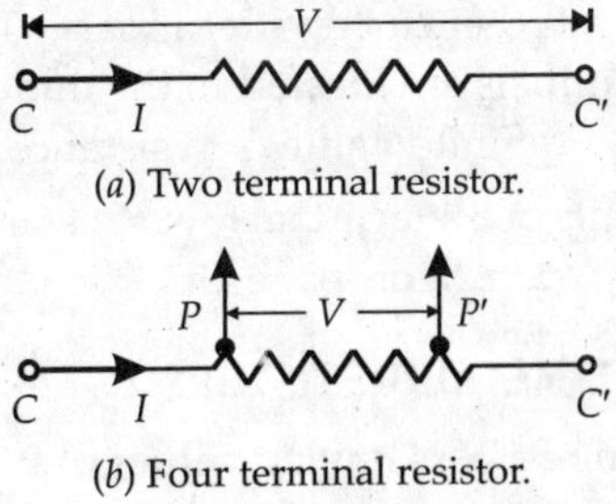


**Fig. 6.2**

The errors on account of contact resistances can be eliminated if we use four terminal resistors as shown in Fig. 6.2(*b*). The nominal value of the resistor is the resistance between potential terminals $P$ and $P'$ which are located near the middle portion of the resistor away from current terminal $C$ and $C'$. The resistance measured is the nominal value of the resistor since there is very small contact resistance drop at terminals $P$ and $P'$ owing to low value of current in the potential terminals. In fact, the value of current in potential circuit is zero under null conditions when the resistor is used for potentiometer work.

Therefore, with four terminal resistors the errors on account of contact resistance are avoided.

(*ii*) A two terminal resistor is shown in Fig. 6.3. This figure shows the current distribution along the length of the conductor. It is seen that the current distribution near the current terminals is non-uniform giving rise to non-uniform current densities while the current distribution is uniform in the middle of the resistor giving rise to uniform current density. Thus if the potential terminals are located at the middle of the resistor, they will be placed at cross-sections where the current density is uniform giving rise to a definite value of the resistance.

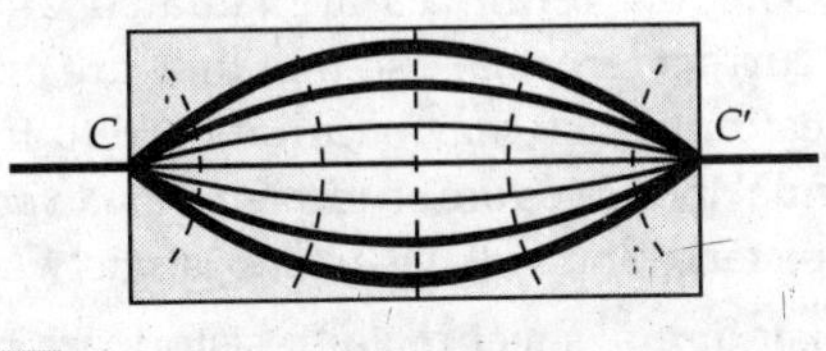


**Fig. 6.3** Current distribution in a resistor.

Also, the current distribution depends upon the type and nature terminals and connecting leads at the current terminals. Therefore, the current distribution does not remain the same and hence the value of resistance when measured between current terminals is not definite. On the other hand, the current distribution in the middle of the resistors remains the same irrespective of the type and nature of terminals and connections at the current terminals. Therefore, if potential terminals are located in the middle they will always give a definite value of resistance.

The details of a 4 terminal low resistance standard are given in Art. 5.27 on page 153.

## 6.5 SHEET METAL RESISTORS

Precision resistors having values of 0.1 Ω or lower are made of sheet manganin brazed or silver soldered to heavy copper terminal posts. A sheet metal resistor has four terminals as shown in Fig. 6.4.

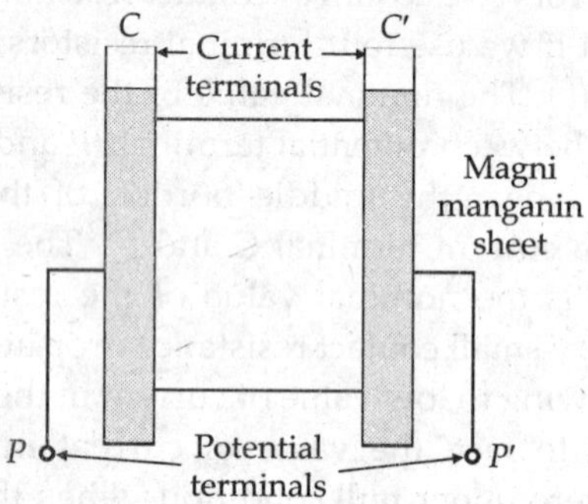


**Fig. 6.4** Sheet metal resistor.

The resistor may consist of a single sheet or several sheets in parallel. The sheets are often not straight, but are bent in *S* shape in order that greater lengths may be used as shown in Fig. 6.5.

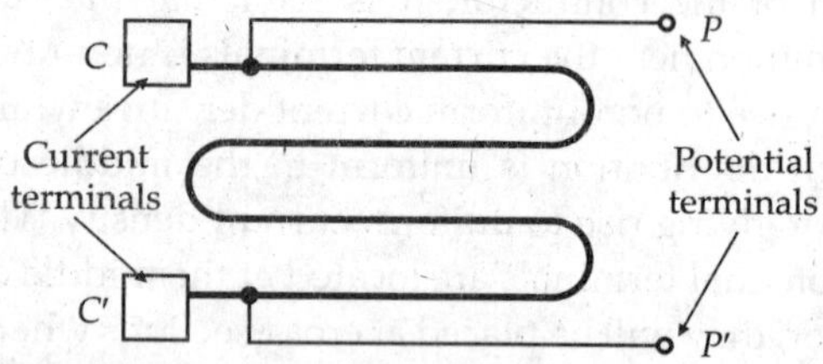


**Fig. 6.5** *S* shaped sheet metal resistors.

The potential terminals are situated at distance from the current terminals, so that they are unaffected by any heating which may occur at current terminals. The potential terminals are so located that an accurately known resistance and voltage drop exists between them.

For standards, sheet metal resistors are mounted in perforated containers with hard rubber or bakelite tops. These are often used in oil baths, in order to facilitate the dissipation of heat produced in the resistance element. An accelerated ageing of these resistors is obtained by heating them in air at 150°C for 48 hours as is done in the case of wire wound resistors.

## 6.6 FREQUENCY ERRORS IN RESISTORS

A resistor may be represented, to a first approximation, by an equivalent circuit shown in Fig. 6.6(*a*). This circuit shows an inductance *L*, in series with the d.c. resistance *R* and the combination of *R* and *L* in series is shunted by a capacitance *C*. This equivalent circuit is valid for low and medium frequencies.

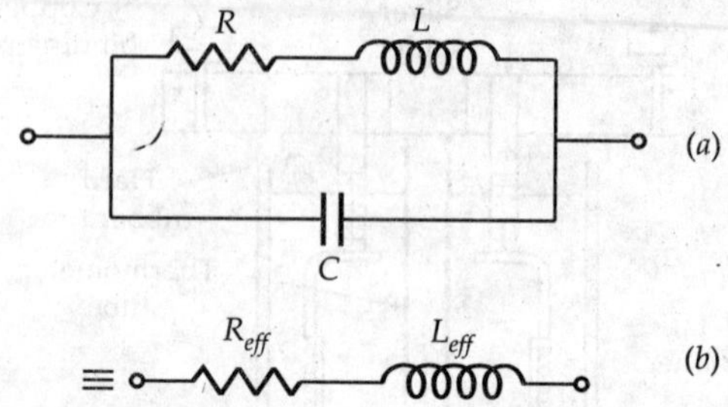


**Fig. 6.6** Equivalent circuit of a resistor at low and medium frequencies.

Let *Z* be the equivalent impedance at an angular frequency ω rad/s.

$$\therefore \quad Z = \frac{(1/j\omega C)(R + j\omega L)}{R + j\omega L + (1/j\omega C)} \qquad \ldots(6.1)$$

$$= \frac{R + j\omega(L - \omega^2 L^2 C - CR^2)}{1 + \omega^2 C^2 R^2 - 2\omega^2 LC + \omega^4 L^2 C^2} \qquad \ldots(6.2)$$

Now for a resistor that has been constructed so that its inductance and capacitance are small, $\omega^2 LC << 1$ and the term $\omega^4 C^2 L^2$ may be neglected as a square of small number. Hence Eqn. 6.2 can be simplified to :

$$Z = \frac{R + j\omega[L(1 - \omega^2 LC) - CR^2]}{1 + \omega^2 C(CR^2 - 2L)} \qquad \ldots(6.3)$$

Writing the real and imaginary terms separately, effective resistance,

$$R_{eff} = \frac{R}{1 + \omega^2 C(CR^2 - 2L)} \qquad \ldots(6.4)$$

and effective reactance,

$$X_{eff} = \frac{\omega[L(1 - \omega^2 LC) - CR^2]}{1 + \omega^2 C(CR^2 - 2L)} \qquad \ldots(6.5)$$

Now $X_{eff}$ is small and therefore the term $\omega^2 LC$ may be dropped from the numerator of Eqn. 6.5, then

$$X_{eff} = \frac{\omega[L - CR^2]}{1 + \omega^2 C(CR^2 - 2L)} \qquad \ldots(6.6)$$

Therefore, the effective inductance is,

$$L_{eff} = \frac{L - CR^2}{1 + \omega^2 C(CR^2 - 2L)} \qquad \ldots(6.7)$$

The effective inductance is also called *residual inductance*. The effect of residual inductance in resistors may be expressed by their *phase defect angle*, ϕ by which the current lags the voltage.

$$\tan\phi = \frac{X_{eff}}{R_{eff}} = \frac{\omega(L - CR^2)}{R} = \omega\left(\frac{L}{R} - CR\right) \quad \text{...(6.8)}$$

The characteristics of a resistor are sometimes expressed in terms of its *time constant* $\tau$.

$$\text{Time constant } \tau = \frac{L_{eff}}{R_{eff}} = \frac{L - CR^2}{R} = \frac{L}{R} - CR \quad \text{...(6.9)}$$

It is clear from Eqn. 6.9, that the time constant of a resistor may be either positive or negative depending upon which of the two terms $L/R$ or $CR$ predominates. The low valued resistors (1 Ω to 10 Ω) generally have a positive time constant, which means that they display a net inductive effect. The capacitance generally predominates for higher values of resistance (1000 Ω upwards) giving a negative time constant. A 100 Ω resistor is likely to have a zero time constant. The time constant of a resistor, is a useful indicator of its high frequency performance and its values range from 0.5 μs to 1 ns.

It is apparent from Eqn. 6.4, that if $CR^2 = 2L$, the d.c. resistance $R = R_{eff}$. Thus if the resistance is to remain independent of frequency, the condition is,

$$CR^2 = 2L \quad \text{...(6.10)}$$

Also, it is clear from Eqn. 6.6, if the resistance has to show no inductive effects,

$$CR^2 = L \quad \text{...(6.11)}$$

The two conditions *i.e.*, $CR^2 = 2L$ and $CR^2 = L$ cannot be met simultaneously. It is however, advisable to have zero effective inductance for which the condition is $CR^2 = 2L$. For such a resistor there will be small difference in d.c. and a.c. resistance. The effective resistance, for zero inductance is,

$$R_{eff} = \frac{R}{1 - \omega^2 LC} \quad \text{...(6.12)}$$

(This value is obtained by substituting $CR^2 = L$ in Eqn. 6.4). Now for zero effective inductance and zero phase angle

$$CR^2 = L \text{ or } R^2 = L/C \quad \text{...(6.13)}$$

This means that for small resistances, they should be so designed that inductance, $L$, is small and capacitance $C$ should be large. On the other hand, large valued resistors should be designed so that capacitance $C$ is small.

At high frequencies, the resistor is represented by an equivalent circuit having distributed parameters rather than lumped parameters as is the case with low and medium frequencies. The equivalent circuit of a resistor valid for high frequencies is shown in Fig. 6.7. This circuit is a simplification. For a complete analysis, it should include the effects of interturn capacitance, and mutual coupling between the turns. This network can be analyzed by using the transmission line theory.

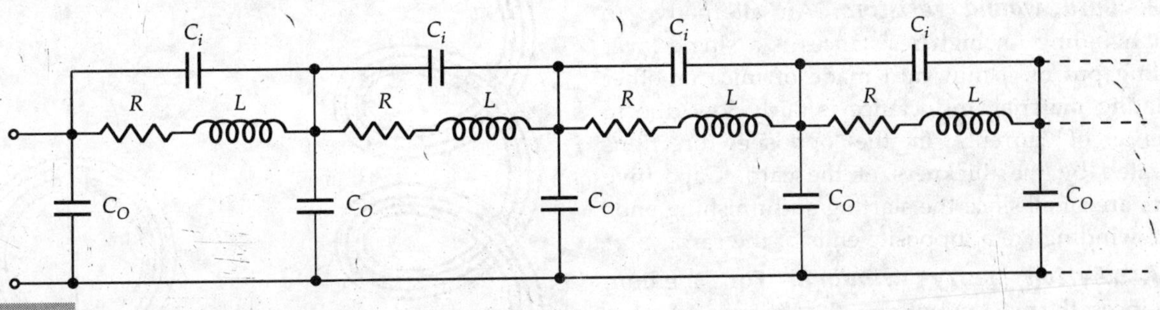


**Fig. 6.7** Equivalent circuit of resistor at high frequencies.

## 6.7 METHODS OF REDUCING RESIDUAL INDUCTANCE AND CAPACITANCE IN RESISTORS

The most commonly used methods for winding the resistance coils to reduce the residual inductance are given below. Methods of reducing residual capacitance are also explained alongside.

***1. Bifilar windings.*** This is the most commonly used method of winding non-inductive resistance coils. In this method, the wire is doubled on itself as shown in Fig. 6.8. This gives the effect of two wires, side by side, carrying currents in the opposite directions. The two wires carrying currents in the opposite direction, produce two equal and opposite magnetic fields and since the two wires are very close

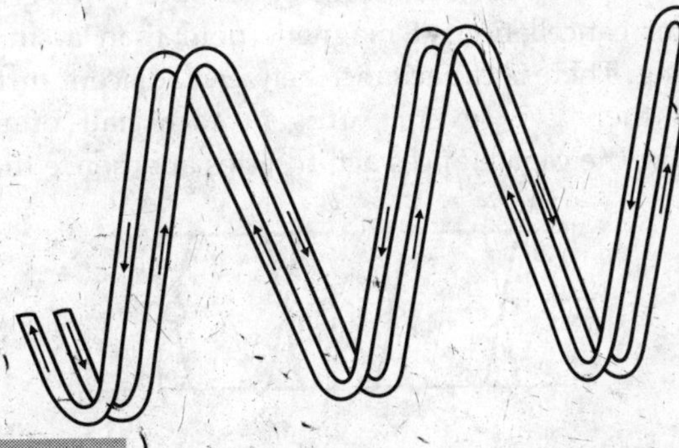

**Fig. 6.8** Bifilar Winding.

to each other, the net magnetic field is almost equal to zero. Therefore, the net inductance of the coil is almost negligible.

The bifilar type of construction is successful in reducing the inductance to a low value. However, it has the undesirable feature of a large capacitance effect if the loop is long. Thus this construction cannot be used for high resistances and is normally used for 0.1 Ω and 1 Ω resistances.

If a simple bifilar winding is used the coil capacitance may become prohibitively large in high resistances. Bifilar windings may be divided into a number of loops with all the loops connected in series (Fig. 6.9) to reduce the capacitance. If there are $n$ loops, the capacitance of each loop is $1/n$ of the capacitance of value of the whole length of wire if wound in a single loop. Now, the capacitances of these loops are in series and, therefore, the value of total capacitance is reduced to $1/n^2$ of the value if only one loop were used.

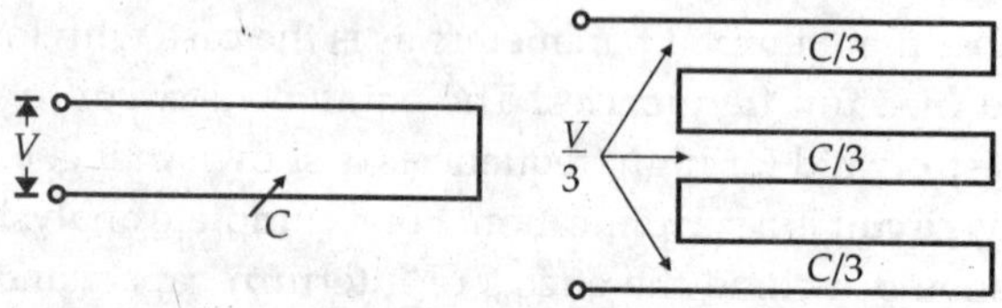


**Fig. 6.9** Sectionalizing of Bifilar Windings.

***2. Card wound resistors.*** An alternative to bifilar winding for high resistances is a single layer winding put on a thin card made of mica or other insulating material, Inductance is fairly low due to presence of currents in the opposite directions separated by the thickness of the card. Capacitive effects are small since the starting and finishing ends of the winding are at opposite ends of the card.

***3. Ayrton perry winding.*** For medium resistances, the most popular method is to use Ayrton Perry winding as shown in Fig. 6.10. This arrangement employs two windings connected in parallel and wound in opposite direction around a card with one winding spaced between turns of the other. This gives a better cancellation of magnetic field than a single winding. There is capacitance between adjacent turns, but if there are several turns in each half of the winding the capacitances act in series and hence their

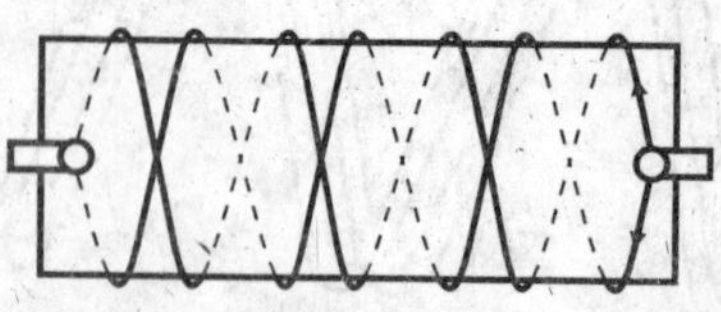

**Fig. 6.10** Ayrton Perry Winding.

resultant is small. This winding is more expensive to construct, and requires either more wire, or else finer wire, because of the two paths in parallel.

This form of winding gives extremely low time constant and is very useful for high frequency work. Ayrton Perry windings are most popular for medium resistances. Such windings are used for ohms, tens of ohm and hundreds of ohm. They are not adaptable to high resistance units.

***4. Chaperon winding.*** This is a modification of bifilar winding. In this arrangement a coil is wound in an even number of layers with the winding direction reversed at the end of each layer. Figure 6.11 shows the Chaperon method of winding. This winding is really an extension of the Bifilar principle, the current in adjacent wires neutralizing one another as regards resultant magnetic field. Both inductance and capacitance are small.

***5. Curtis and grover winding.*** It is wound on a slotted ceramic cylinder (or a slotted flat card) as shown in Fig. 6.12. Adjacent wires carry currents in the opposite direction hence magnetic effects are small giving a small inductance. Capacitive effects are also small as the voltage between adjacent wires is only the $IR$ drop which is small.

**Fig. 6.11** Chaperon Winding.

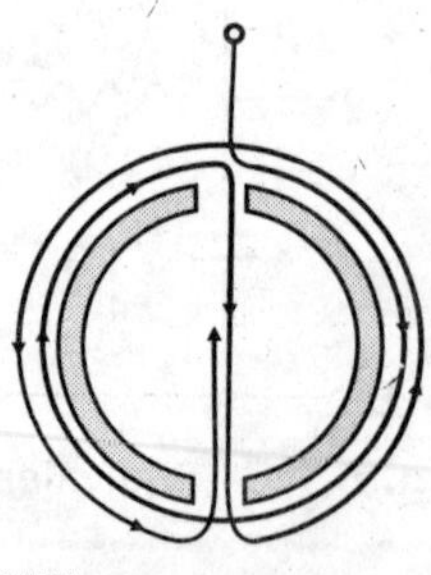

**Fig. 6.12** Curtis and Grover Winding.

***6. Woven wire resistance.*** Woven wire resistors have been constructed in which resistance wire is woven together with silk or cotton to produce a cloth in which the wire zig-zags as shown in Fig. 6.13(*i*). Part (*i*) of this diagram shows ordinary weaving. Part (*ii*) shows improved form of weaving where the adjacent pairs of wires lie close together. Part (*iii*) shows two wires in parallel similar to Ayrton Perry winding. The inductance is very small with this arrangement.

Time constants of resistors using this construction are as low as 0.4 μs for a 100 Ω resistor.

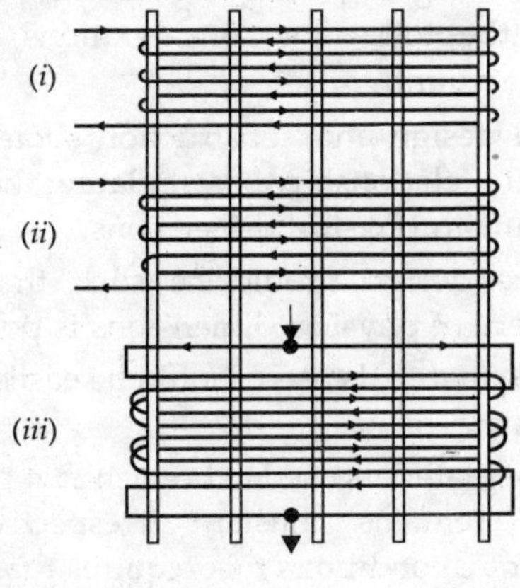


**Fig. 6.13** Woven wire resistors.

## 6.8 RESISTANCE BOXES

An assembly of resistors which can cover a wide range of values is known as a resistance box. Each resistor is constructed according to the principles described above. These resistors are mounted in the box and are so arranged that the values of resistance can be varied in steps (usually in steps of 1 Ω). The resistance is varied with the help of rotary selector switches or with the help of key plugs. In general a rotary switch is much more convenient to use. There may be more than one selector switch in a box. It is normal to arrange for one switch to select 1 to 10 Ω in steps of 1 Ω, a second switch to select 10 to 100 Ω in steps of 10 Ω, a third switch to select 100 to 1000 Ω in steps of 100 Ω, and so on. Therefore, such resistance boxes are commonly known as *decade resistance boxes.*

A decade resistance box having two dials is shown in Fig. 6.14 and is set at 22 Ω.

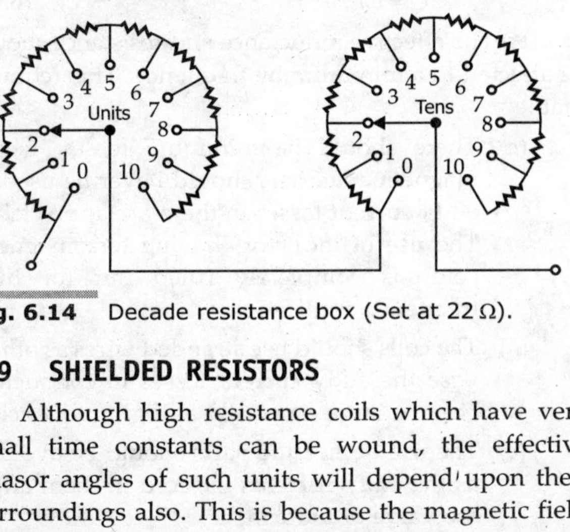


**Fig. 6.14** Decade resistance box (Set at 22 Ω).

## 6.9 SHIELDED RESISTORS

Although high resistance coils which have very small time constants can be wound, the effective phasor angles of such units will depend upon their surroundings also. This is because the magnetic field is neutralized by adopting the arrangements described earlier but an electric field exists between different parts of the resistor when it is carrying current. Also there is an electric field between the resistor and the earth. This results in distributed capacitance between one resistor and another in a decade box. Figure 6.15 shows how resistor is represented with various associated capacitances.

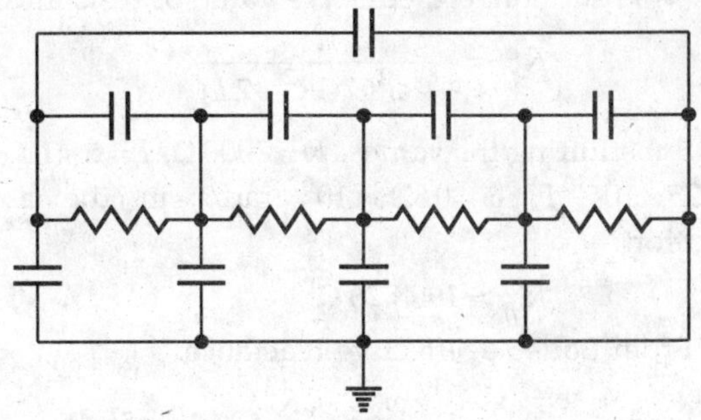

**Fig. 6.15** Distributed capacitances.

These distributed capacitances affect the operation of the resistor. They cause serious errors especially at high frequencies. This is because these distributed capacitance will vary the current from point to point along the resistor, and the potential drop along a particular section will not necessarily be exactly proportional to the resistance of the section. In order to make the effect of these distributed capacitances definite the resistor (or resistor sections) may be enclosed within metal shields which are maintained at fixed potentials with respect to resistors which they enclose.

In audio frequency working except for very high resistances, the primary need for the shield is to intercept external fields and it is not necessary to screen individual resistors of a decade.

**Example 6.1** *How much series inductance is required to render a 1000 Ω resistance coil with a shunt capacitance of 10 pF practically non-inductive over a reasonable range of frequency ? What is the effective value of resistance under the above conditions ? The frequency is 200 kHz.*

**Solution.** The condition for a resistance to remain non-inductive is given by Eqn. 6.11.

Inductance $L = CR^2 = 10 \times 10^{-12} \times (1000)^2 = 10\,\mu\text{H}$.

The effective value of resistance, when the coil is made non-inductive is given by Eqn. 6.12.

$$R_{eff} = \frac{R}{1-\omega^2 LC}$$

$$= \frac{1000}{1-(2\pi \times 200 \times 10^3)^2 \times 10 \times 10^{-6} \times 10 \times 10^{-12}}$$

$$= 1000.16\,\Omega.$$

***Example 6.2*** *A 2000 Ω resistor has a series inductance of 6 μH and a shunt capacitance of 25 pF. Find the :*

(*a*) *effective resistance and reactance*

(*b*) *phase angle and time constant of resistor if the frequency is 100 kHz.*

**Solution.** Angular frequency $\omega = 2\pi \times 100 \times 1000$
$= 0.628 \times 10^6$ rad/s.

(*a*) From Eqn. 6.4, effective value of resistance

$$R_{eff} = \frac{R}{1+\omega^2 C(CR^2 - 2L)}$$

Substituting the values, $R = 2000\,\Omega$, $L = 6\times 10^{-6}$ H, $C = 25\times 10^{-12}$ F, $\omega = 0.628\times 10^6$ rad/s in the above equation

$$R_{eff} = 1998.26\,\Omega$$

From Eqn. 6.6, effective reactance,

$$X_{eff} = \frac{\omega(L - CR^2)}{1+\omega^2 C(CR^2 - 2L)} = -59\,\Omega.$$

(*b*) Phase angle $\phi = \tan^{-1}(X_{eff}/R_{eff})$

$= \tan^{-1}(-59/1998.26) = -1.7°$

Time constant $\tau = L/R - CR$ (See Eqn. 6.9)

$= 47$ ns (negative)

***Example 6.3*** *A 100 Ω resistor has a time constant of ±30 ns. (a) Find the phase angle of the resistor (b) Calculate the value of capacitance of a capacitor to be connected in parallel with the resistor to make its phase angle equal to zero. The frequency is 1 kHz.*

**Solution.** (*a*) From Eqns. 6.8 and 6.9 phase angle and time constant are :

$$\phi = \tan^{-1}\omega(L/R - CR)$$

$$\tau = L/R - CR$$

$$\therefore\ \phi = \tan^{-1}\omega\tau = \tan^{-1}(2\pi\times 1\times 1000\times 30\times 10^{-9})$$

$$= 0.0108° = 38.9''.$$

(*b*) Suppose $C_1$ is the external resistance to be connected in parallel to the resistor to make its time constant equal to zero.

$$\therefore\quad L/R - (C + C_1)R = 0$$

or $$(L/R - CR) - (C_1 R) = 0$$

or $$\tau - C_1 R = 0$$

Hence, $C_1 = \tau/R = 30\times 10^{-9}/100 = 0.3$ nF $= 300$ pF

## 6.10 STANDARDS OF INDUCTANCE

The dimensions of inductance are [μL] in the e.m. system. Thus the dimensions of inductance are that of length only if permeability μ is considered as dimensionless. Hence, the values of standards of both self and mutual inductance depend upon their physical dimensions together with the number of turns of the inductance coil.

While designing and constructing *Inductance Standards*, the following considerations should be taken into account.

(*i*) The design and construction should be such that accurate relationship exists relating the values of inductance with physical dimensions.

(*ii*) The design should be such that accurate measurement of physical dimensions is possible.

(*iii*) The materials used should be easily workable mechanically.

(*iv*) The materials used to be such that the value of inductance remains constant irrespective of the environmental conditions this requires that :

(*a*) materials whose dimensions do not change with time, temperature and humidity,

(*b*) the inductance should be independent of the value of the current passing. This means that no ferromagnetic materials be used. When we use ferromagnetic materials, the value of permeability and hence inductance is dependent upon flux density which in turn depends upon the current flowing in the coil. Hence the inductance of an inductor constructed with ferromagnetic cores depends upon the value of the current passing. In order that inductance be independent of the current passing, air cored coils should be used for standards of inductance.

(*v*) The resistance of the windings should be very low as compared with their inductance or the inductor should have a high *L*/*R* ratio. The ratio of inductive reactance to resistance of a coil is known as its *Q* **Factor or Storage Factor**. The *Q* factor is defined as :

$$Q = \omega L/R \qquad ....(6.14)$$

(*vi*) The effective inductance and resistance should be affected to a minimum by frequency. This requires that :

(*a*) There should be minimum losses. Ferromagnetic materials should never be used as eddy current losses in them are appreciable. The use of inductors having ferromagnetic cores is completely ruled out for high frequency applications.

(*b*) The coils should use stranded wires as otherwise the eddy current losses in conductors become large especially at high frequencies.

(*c*) The use of metallic parts should be avoided as the eddy currents induced in them affect the value of inductance.

(*vii*) The inductor should be so designed that capacitive effects are negligible. This is particularly important for inductors working at high frequencies where the interturn capacitance may drastically change the effective value of inductance.

(*viii*) The inductor should be desirably unaffected by external magnetic fields and should produce a minimum interfering field of its own.

### 6.10.1 Formers for Inductance Coils

Marble is used for making formers or bobbins for the coils. The reasons for selecting marble are that it does not warp and is unaffected by atmospheric conditions. It acts as a perfect insulator and a non-magnetic material. It has a low co-efficient of thermal expansion, can be easily machined and is comparatively cheap and easy to work.

### 6.10.2 Coils of Inductance Coils

The shape of the coils should be such that a high value of inductance to resistance ratio is obtained. The $L/R$ ratio is maximum for a coil of simple circular shape with a square cross-section, with the mean diameter of coil equal to about 3 times its depth. High values of $L/R$ ratio can be obtained by using greater weight of copper or by using iron core.

## 6.11 STANDARDS OF MUTUAL INDUCTANCE

***1. Primary standards.*** These standards are always fixed standards *i.e.,* single valued. The campbell type of primary standard is shown in Fig. 6.16. It consists of a single layer primary coil of bare copper wire wound under tension in grooves machined in a cylinder of marble. The primary coil is divided into two parts. The two parts are connected in series and are separated from each other by a distance equal to three times the axial length of each. The secondary coil has many layers and is placed in a

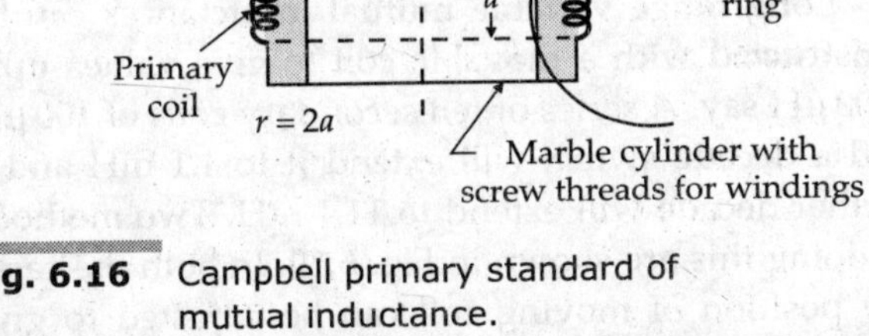


**Fig. 6.16** Campbell primary standard of mutual inductance.

channel cut in the marble cylinder. The secondary coil has a diameter of about one and half times the diameter of the primary coil and is situated midway between the two halves of primary coil. This arrangement is used in order that the secondary winding lies in a position of almost zero field. Thus the value of flux linked by secondary winding does not depend upon determination of its diameter to a great accuracy. This is desirable as it is difficult to assess the effective diameter of secondary winding because it has many layers. It is essential to use a multilayer secondary winding in place of a single layer winding in order to have a reasonably high value of inductance.

***2. Secondary standards and working-standards.*** These standards may be fixed or variable. The requirements of secondary standards are that their inductance should not vary with time and temperature and should be independent of supply frequency. Their dimensions need not be very accurately determined.

Fixed secondary standards usually consist of two coils wound on a marble or hard paraffined wood former. The coils are separated by a flange. The wire is of stranded copper in order to reduce the effect of eddy currents in conductors. After winding, the coils and former are immersed in hot paraffin wax.

Variable standards are frequently used in laboratories and are known as 'Inductometers'. They consist of two sets of coils, one fixed and other movable so that the mutual inductance could be varied.

## 6.12 STANDARDS OF SELF-INDUCTANCE

***1 Primary standards.*** The mutual inductance standards are generally regarded as primary standards of inductance as their value can be known with greater accuracy. Also the formula necessary for the calculation of the exact value of inductance for primary standards of self-inductance is very much complicated as compared with corresponding formula for mutual inductance. Hence normally primary standards of self inductance are rarely prepared and the primary standard for mutual inductance is taken as the primary standard of self-inductance.

***2. Secondary standards and working standards.*** Secondary standards of self-inductance are constructed to have a nominal value which is usually a simple fraction of 1 H. In designing and constructing these standards all the precautions taken for constructing corresponding standards of mutual inductance are followed.

Coils for use as secondary standards are wound with silk covered stranded copper wire. The formers are made of marble. After winding the coils are immersed in melted paraffin wax for some time.

## 6.13 VARIABLE INDUCTANCES

Variable inductors are very useful in many circuits. The principle adopted for most variable self-inductors is the use of two coils, one fixed and the other movable in series, mounted so that the mutual inductance between them may be varied. If $L_1$ and $L_2$ are the self-inductances of two coils and $M$ is the mutual inductance between them, the resulting self-inductance $L$ is :

$$L = L_1 + L_2 \pm 2M$$

Therefore, the inductance can be varied from

$$L = L_1 + L_2 - 2M \text{ to } L = L_1 + L_2 + 2M$$

If the self-inductances of the two coils are equal and they are closely coupled.

*i.e.,* $L_1 = L_2 = M$

In such a case the inductance can be varied from almost a small value (nearly zero) to a maximum of nearly 4 times the inductance of one coil. It is clear that all the windings are always in circuit and, therefore, the value of resistance remains the same over the complete range of variation of $L$.

***1. Ayrton Perry inductometer.*** Figure 6.17 shows the construction of a Ayrton Perry inductometer. The moving coil is mounted inside the fixed coil. The moving coil is on a spindle which also carries a pointer and a handle at the top as shown. The scale can be graduated to read values of inductance directly. This form of inductometer can be cheaply and easily constructed for use as a variable inductance in cases where the inductance must be variable but not necessarily known.

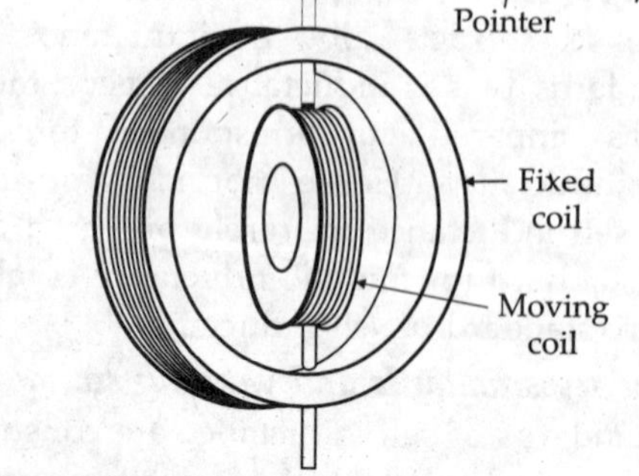


**Fig. 6.17** Ayrton Perry inductometer.

Its disadvantages are that it produces an external magnetic field which may affect the values of other units (instruments, inductors, resistors etc.) placed near it. Also its scale does not follow a straight line law. However, the shape of the scale can be improved by properly shaping the coils.

***2. Brooks and Weaver inductometer.*** A diagram of construction of this inductometer is shown in Fig. 6.18. The middle disc has two link shaped coils set in it, and the top and bottom plates each have a coil with half the turns of the moving coil. The two groups of the coils are connected with opposite polarities, that is when the flux direction is upward in the right hand coils, it is downward on the left. One coil serves as the return flux path for the other and, therefore, the magnetic field does not spread out.

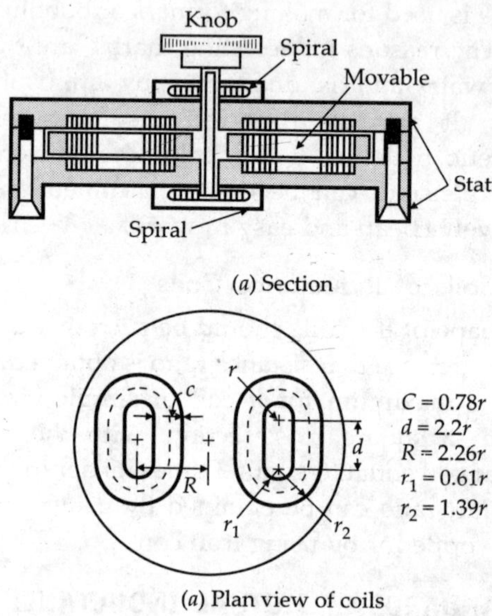


**Fig. 6.18** Brooks and Weaver inductometer.

The dimensions of the coils are specially chosen to give uniform scale and also to obtain as great inductance as possible. The Brooks inductometer is a great improvement over the Ayrton Perry type because it is compact, has high $L/R$ ratio and does not produce any external magnetic field. Also the inductance value remains constant even if the plates warp a little.

This inductometer may be used as a variable self-inductor if the fixed and moving coils are connected in series and a mutual inductor if the coils are in different circuits.

***3. Variable mutual inductometers.*** A short range variable mutual inductance can be obtained by separating the connections of fixed and moving coiis of the variable self-inductances described (*i.e.,* the Ayrton Perry and the Brooks inductometers).

Long range variable mutual inductances can be constructed with a movable coil to give values upto 1000 μH say. A series of ten secondary coils of 100 μH and a decade switch will extend it to 1.1 mH and a further decade will extend to 11.1 mH. Two methods of doing this are shown in Fig. 6.19. In both the cases the position of moving coil can be adjusted to give zero inductance or given negative values. This later feature being very useful in some of the applications.

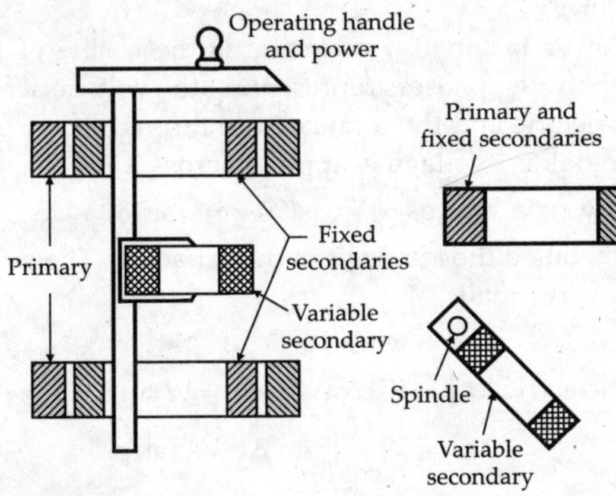


**Fig. 6.19** Variable mutual inductometers.

## 6.14 INDUCTORS FOR HIGH FREQUENCY WORK

High frequency electronic circuits frequently employ inductances of the order of μH. These inductors are usually air cored coils consisting of a large number of turns. But in case small turns are to be used to get large inductance values, high permeability cores are used. The ferromagnetic cores for these inductors should have a low eddy current loss. *Ferrite* cores, made of high resistivity ferromagnetic materials are used at high frequencies as their high resistivity makes the eddy current loss smaller.

## 6.15 INDUCTORS FOR LOW FREQUENCY WORK

Large valued inductors are made by having a large number of turns wound round a core made up of a ferromagnetic material like iron. The magnetic properties of these materials are such that the magnetic flux density is increased appreciably. In this way inductances of several hundred henry are produced. The corer of iron-core inductors are laminated in order to reduce eddy current losses.

## 6.16 FREQUENCY ERRORS IN INDUCTORS

The inductors used for measurement purposes should possess a constant value. Very small losses and small leakage. The equivalent circuit of an inductor is the same as that of a resistor as shown in Fig. 6.6. The pure inductance $L$ is in series with resistance $R$ (an impurity), with shunt capacitance $C$ (also an impurity). The increase of frequency causes a more proportionate increase in impedance due to the inevitable interturn capacitance. The arrangement becomes a resonating circuit with resonating frequency $f_0 = 1/(2\pi\sqrt{LC})$ Hz. At this frequency the inductor behaves as a pure resistor. At still higher frequencies the impedance becomes capacitive. Inductances for measurement purposes can therefore be used upto a frequency of 0.1 $f_0$. Therefore, the equivalent circuit given in Fig. 6.6 is valid for low and medium frequencies (which are far away from the resonant frequency). At these frequencies, the expression for impedance is given by :

$$Z = \frac{R + j\omega[L(1-\omega^2 LC) - CR^2]}{1+\omega^2 C^2 R^2 - 2\omega^2 LC + \omega^4 L^2 C^2} \quad \text{(See Eqn. 6.2)}$$

In the construction of an inductor, the resistance and capacitance components are kept to a minimum. Therefore, the terms $\omega^2 C^2 R^2$ and $CR^2$ which involve $R$ and $C$ can be neglected.

$$\therefore \quad Z = \frac{R + j\omega L(1-\omega^2 LC)}{1-2\omega^2 LC + \omega^4 L^2 C^2}$$

$$= \frac{R}{(1-\omega^2 LC)^2} \frac{j\omega L(1-\omega^2 LC)}{(1-\omega^2 LC)^2} \quad \text{...(6.15)}$$

$$\therefore \text{ Effective resistance } R_{eff} = \frac{R}{(1-\omega^2 LC)^2} \quad \text{...(6.16)}$$

∴ Using the binomial expansion,

$$R_{eff} = R(1+2\omega^2 LC) \quad \text{...(6.17)}$$

∴ Effective reactance is,

$$X_{eff} = \frac{\omega L(1-\omega^2 LC)}{(1-\omega^2 LC)^2} = \frac{\omega L}{1-\omega^2 LC} \quad \text{...(6.18)}$$

∴ Using the binomial expansion

$$X_{eff} = \omega L(1+\omega^2 LC) \quad \text{...(6.19)}$$

∴ The effective inductance is,

$$L_{eff} = L(1+\omega^2 LC) \quad \text{...(6.20)}$$

Eqns. 6.19 and 6.20 show that both the effective resistance and inductance of an inductor increase with frequency. The analysis is a simplification and can only be applied to low and medium frequencies

**Example 6.4** *A coil has a resistance of 200 Ω and an inductance of 1 H when measured at very low frequency. The distributed capacitance is 200 pF. Find the percentage change in effective inductance when this coil is used at a frequency of 1000 Hz. Also calculate the effective value of inductance.*

**Solution.** We have $L = 1$H

and $\omega = 2\pi \times 1000 = 2000\pi$ rad/s.

From Eqn. 6.20, effective (equivalent) inductance,

$$L_{eff} = L(1+\omega^2 LC)$$

$$\text{Percentage change} = \frac{(L_{eff} - L)}{L} \times 100 = \omega^2 LC \times 100$$

$$= (2000\pi)^2 \times 1 \times 200 \times 10^{-12} \times 100 = 0.79\%$$

Effective value of inductance $L_{eff} = L(1+\omega^2 LC)$

$$= 1 \times 1.0079 = 1.0079 \text{ H.}$$

**Example 6.5** *A coil has an effective inductance 0.102 H at 1000 Hz and 0.110 H at 3000 Hz. Find the value or inductance at low frequencies and also the distributed capacitance. Find also the effective inductance at 2000 Hz.*

**Solution.** From Eqn. 6.20,

effective inductance $L_{eff} = L(1+\omega^2 LC)$

Thus we have, $\quad 0.102 = L[1+(2\pi\times 1000)^2 LC]$

and $\quad 0.110 = L=[1+(2\pi\times 3000)^2 LC]$

From above we get, $LC = 0.25\times 10^{-9}$

Thus we have,

$$0.102 = L[1+(2\pi\times 1000)^2\times 0.25\times 10^{-9}]$$

Therefore, value of inductance at low frequencies is $L = 0.101$ H.

The distributed capacitance is

$$C = \frac{0.25\times 10^{-9}}{(0.101)} = 2.475\times 10^{-9}\text{ F} = 2475\text{ pF}.$$

Effective inductance at 2000 Hz is

$$L_{eff} = 0.101[1+(2\pi\times 2000)^2\times 0.25\times 10^{-9}] = 0.105\text{ H}.$$

## 6.17 LOSS ANGLE AND POWER FACTOR OF CAPACITORS

A capacitor fails to present a perfect and constant capacitance at all frequencies due to several effects. There are ohmic losses in the plates and connecting wires, interfacial polarization and power loss in dielectric etc. Therefore, a capacitor can be represented by an equivalent circuit as shown in Fig. 6.20(*a*). Resistance represents the losses.

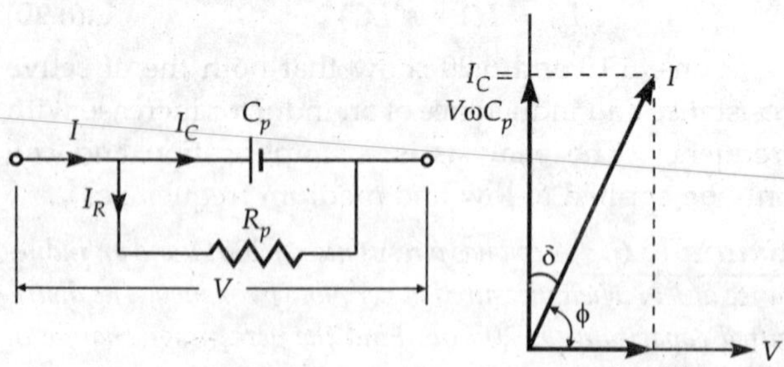

**Fig. 6.20** Capacitor equivalent circuit and phasor diagram (Parallel representation).

In the case of air capacitors and most other cases, the losses in the dielectric are very small and hence Capacitors using these dielectrics may be regarded as perfect.

Figure 6.20(*b*) shows the phasor diagram of dielectric when a voltage is impressed across it where,

$V$ = applied voltage ; $V, I_c$ = capacitive current ;

$A, I_R$ = active (loss) component of current ;

$A, C$ = capacitance for parallel representation ;

$F, R$ = resistance for parallel representation ; Ω and $\omega = 2\pi f$.

Power factor of a dielectric is the cosine of the angle between phasors representing the voltage across the dielectric and the total current through it when a sinusoidal a.c. voltage is applied across it.

Power factor $\cos\phi = \cos(90-\delta) = \sin\delta$

$\delta$ is called the *loss angle* of the capacitor. The value of $\delta$ is very small.

$$\therefore \quad \sin\sigma = \tan\delta = \delta$$

$$\text{Dielectric loss} = VI\cos(90°-\delta) = VI\sin\delta$$

$$= V\frac{I_c}{\cos\delta}\sin\delta = VI_c\tan\sigma$$

But $\quad I_c = \omega C_P V$

$$\text{Dielectric loss} = \omega CV^2\tan\delta \qquad \text{...(6.21)}$$

Also $\quad \tan\delta = \frac{I_R}{I_c} = \frac{V/R_P}{\omega C_P V} = \frac{1}{\omega C_P R_P} \qquad \text{...(6.22)}$

Euation 6.22 is very useful for calculating dielectric loss. It is clear that the loss is determined by $\tan\delta$ and therefore this item has great importance.

The capacitor may be represented as shown in Fig. 6.21(*a*). The phasor diagram for this arrangement is shown in Fig. 6.21(*b*).

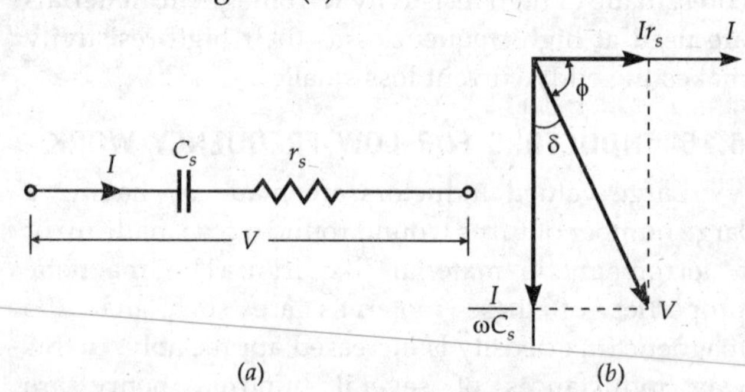

**Fig. 6.21** Capacitor equivalent circuit and phasor diagram (Series representation).

$C_s, r_s$ are respectively the equivalent series capacitance and resistance.

$$\text{Power loss} = VI\cos\phi = VI\sin\delta$$

$$= \frac{I}{\omega C_s\cos\delta} I\sin\delta = \frac{I^2}{\omega C_s}\tan\delta \qquad \text{...(6.23)}$$

and $\quad \tan\delta = \frac{Ir_s}{I/\omega C_s} = \omega C_s r_s \qquad \text{...(6.24)}$

The power factor of a capacitor is very good test of its quality. A capacitor with a small dielectric loss and small leakage will have a very low power factor, *i.e.*, a small loss angle. It is found that power factor (or tan δ) is practically constant over a wide range of frequency. tan δ is also known as *dissipation factor* of capacitor.

$$\text{Dissipation factor } D = \frac{1}{\omega C_p R_p} = \omega C_s r_s \quad \ldots(6.25)$$

## 6.18 CAPACITANCE STANDARDS

***1. Primary standards.*** The dimensions of capacitance in e.s system are [*L*]. Thus the dimensions of capacitance are those of length if the permittivity ε is taken as dimensionless. Therefore, values of primary standards of capacitance can be known by measuring lengths. The dimensions of such capacitors should be very accurately known in order that a primary standard is constructed. Air is usually used as the dielectric as it is the only dielectric whose permittivity is definitely known and which is free from absorption and dielectric loss. Three types of constructions have been used for primary standards :

(*i*) two concentric spheres
(*ii*) two concentric cylinders
(*iii*) two parallel plates with guard rings.

Parallel plate construction is normally not used. The concentric cylinder construction is shown in Fig. 6.22. The capacitance is calculated by assuming no fringing and hence a guard ring construction is necessary. Guard rings are maintained at the same potential as the outer cylinder but are not directly connected to it. Any fringing takes place between outer edges of guard rings and there is no fringing of electrostatic field between the outer and the inner cylinders. The supports are made of insulating materials like quartz.

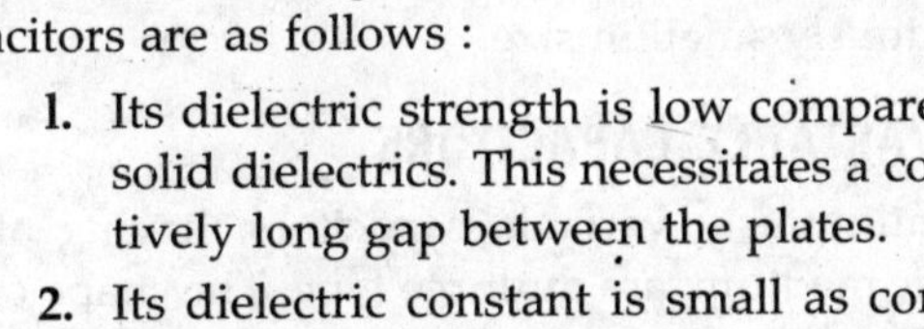


**Fig. 6.22** Primary capacitance standard.

The disadvantages of air as a dielectric in such capacitors are as follows :

1. Its dielectric strength is low compared with solid dielectrics. This necessitates a comparatively long gap between the plates.
2. Its dielectric constant is small as compared with solid dielectrics. This means that a capacitor using air as dielectric medium will be larger in size.
3. Dust particles settle in the gap between plates and cause leakage troubles and, therefore, precautions such as drying of air etc. have to be observed.
4. There is no solid dielectric between the plates to act as a spacer, but the plates must be rigidly fixed in position by supports of solid dielectric. Few solid dielectrics have been found to be suitable for the purpose owing to their tendency to warp and shift in position.

Air capacitors have extremely low losses, usually assumed zero, if the dielectric which supports the insulated electrodes is a good insulating material and is placed in a region of low field strength. Therefore, air capacitors seem at first consideration to be well suited as primary electrical standards. It has been found difficult, however, to secure the extremely high stability needed for this purpose even carefully treated metal parts warp somewhat with time and so permanency is not that much as is desired in primary standards.

***Vacuum and gas-filled capacitors.*** Vacuum and gas-filled capacitors are designed for high voltage and high frequency applications. They are not affected by variations in temperature, moisture content and humidity. They are rugged and their calibration holds good over a long range of time. They have low power dissipation and are not permanently damaged by over-voltages *i.e.,* they are self healing type. Vacuum capacitors are manufactured with capacitance values ranging from 50 pF to 500 pF. The maximum voltage level is 10 kV.

The breakdown strength of gases increases in proportion to the pressure while their dielectric consant remains unaffected. The high voltage capacitors are therefore kept in pressure vessels which are filled with nitrogen compressed to a pressure of about 16 atmospheres. These capacitors are used for voltages upto 1500 kV and have a value ranging from 20 to 200 pF. A diagrammatic section of a high voltage capacitor, fitted with guard rings, and filled with compressed gas is shown in Fig. 6.23.

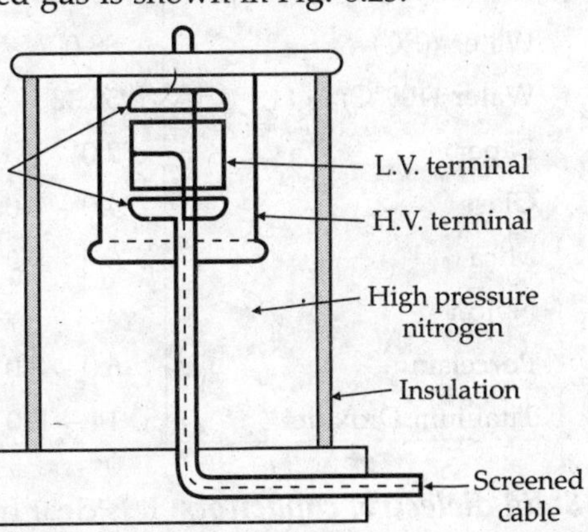


**Fig. 6.23** High voltage capacitor.

***2. Secondary standards and working standards for capacitance.*** Secondary standards and working standards of capacitance need note be provided with guard rings and their dimensions may not be accurately known as their value is fixed after comparison with primary standards. However, their capacitance should not change with time and in order to achieve this we should use materials which do not warp and so alter in dimensions. The plates should be rigidly fixed in position and should not have any appreciable change in length with change in temperature. In order to give rigid support to the capacitor plates, some solid insulating materials may have to be used in place of air. Such capacitors will not be free of dielectric loss. But by using materials such as fused silica, the loss can be decreased.

A number of plates may be used in place of two in order to get large capacitance without increasing the volume of capacitor. Capacitances upto 0.02 μF can be obtained.

These capacitors must be adequately screened to prevent the effect of stray electrostatic fields. The screen is normally connected to one of the main terminals but occasionally it is brought out to a separate terminal and the capacitor is then known as a "*three terminal*" capacitor.

Mica capacitors are used where large values of capacitance are required and where dielectric loss is of no importance.

**Table 6.1** *Dielectric Constant of Various Materials*

| *Material* | *Dielectric Constant* |
|---|---|
| Argon (0°C, 1 atm.) | 1.000545 |
| Hydrogen (0°C, 1 atm.) | 1.000264 |
| Air (19°C, 1 atm.) | 1.0006 |
| Air (19°C, 100 atm.) | 1.0548 |
| Benzene (20°C) | 2.284 |
| Methanol (20°C) | 33.62 |
| Water (0°C) | 88.00 |
| Water (100°C) | 55.33 |
| Paper | 2.0 |
| Glass | 4.0 – 6.0 |
| Mica | 6.9 – 9.2 |
| Nylon | 3.5 |
| Porcelain | 6.0 – 8.0 |
| Titanium Dioxide | 14 – 110 |

***Solid dielectric capacitors.*** It is clear from Table 6.1 that by using solid dielectric materials which have a large dielectric constant than vacuum, air or compressed gases the capacitance value can be increased. In other words, for the same value of capacitance, the size of the capacitors can be reduced. Several dielectric mediums are used in making capacitors, the most common of which are impregnated paper, mica and polystyrene. Paper is not used for capacitors meant for measurement work. Mica is used as a dielectric in the best working standards of capacitance. It has a dielectric constant of about 8, its specific volume resistivity $10^{14}$ Ω, its breakdown voltage 100 kV/mm and its co-efficient of linear expansion $5 \times 10^{-6}$/°C. However the performance of mica capacitors is effected by change in frequency.

Plastic film capacitors have plastic, (*e.g.*, polystyrene) film as the dielectric medium and electrodes are metal film plates deposited directly on the dielectric. Polystyrene has a dielectric constant of 2.4 and is not affected by moisture.

Plastic materials have very high resistivity and thus have high d.c. resistance. The dielectric losses are low even at frequencies as high as 1 MHz. The temperature co-efficient is between – 100 to $-200 \times 10^{-6}$/°C and their maximum operating temperature is 85°C.

Polystyrene capacitors are particularly suited for integrating and differentiating networks because of absence of dielectric after effect.

***Ceramic capacitors.*** These capacitors are widely used for instrumentation work as transducers. They are manufactured the same way as the polystyrene capacitors. Capacitors with different ceramic dielectrics which is usually, titanium dioxide are used for a varied type of instrumentation applications. The type of capacitor to be used for a particular application depends upon the dielectric constant of the dielectric material used and the response at various frequencies.

One type, with a dielectric constant between 6 to 16, has low dielectric loss even at frequencies as high as 100 GHz. Another type, with dielectric constant of nearly 90, can be made with a wide range of temperature co-efficients which vary from +200 to $-800 \times 10^{-6}$/°C.

There is another type which has a dielectric constant between 1000 to 5000. Thus this capacitor is tremendously small in size.

## 6.19 VARIABLE CAPACITORS

Continuously variable capacitors, having air as dielectric medium, are multiple type. The shape of the plates is usually semi-circular (Fig. 6.24).

There are two sets of plates of fixed and the other moving. The plates are arranged so that the moving plates can be rotated in the air gap between the fixed plates. The capacitance is varied owing to variation of area of moving plate interleaving with the fixed plates.

The frame, stator rods, and rotor shafts are made of alloys of aluminium which combine the mechanical strength of brass with light weight of aluminium. The plates are also of aluminium, so that all the parts have the same temperature co-efficient of linear expansion. The capacitor is provided with a slow motion worm gear device for high accuracy in setting and reading.

The capacitor may be provided with a vernier degree scale and calibration chart or may be calibrated to read directly in pico farad.

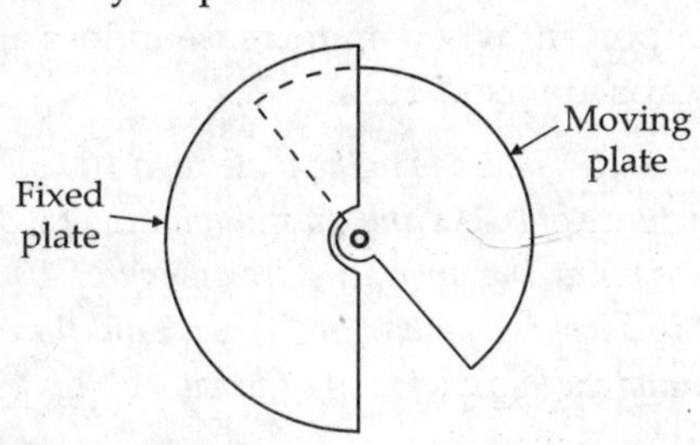


**Fig. 6.24** Variable capacitor.

## 6.20 DECADE CAPACITANCE BOXES

In practice, decade capacitance-boxes are very popular. Figure 6.25(*a*) shows a decade capacitor. There are ten fingers in the moving plate. When each group of fingers is embraced within the fixed plates there is no change in the value of capacitance for small angular movements. The capacitance angular displacement curve is shown in Fig. 6.25(*b*). This diagram clearly shows how well the condition of non-variation of capacitance for small changes in displacement is met in practice. Special compensating techniques are adopted to overcome the fringing effects. The dotted curve in Fig. 6.25(*b*) is for an ordinary variable capacitor having the same capacitance range and linear capacitance law. It is clear from the curve (with firm lines) that precise angular position of the plates is not necessary as the ten positions, each located by a click mechanism, provide a decade of capacitance.

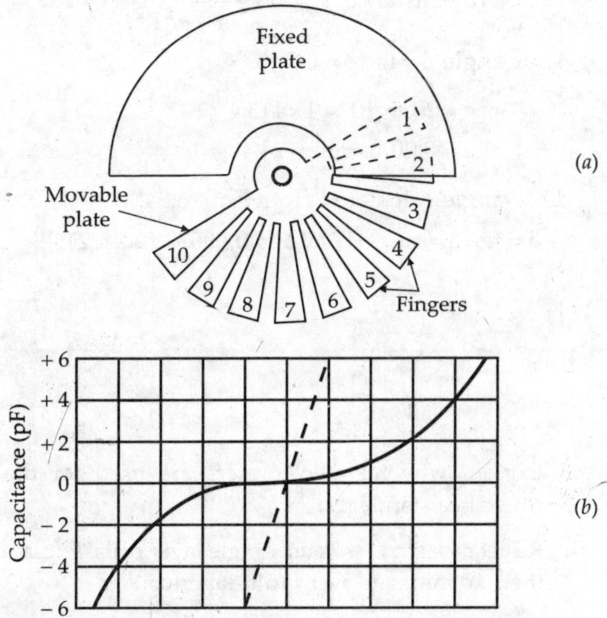


**Fig. 6.25** Decade capacitance box.

Air is used as a dielectric medium for the capacitor and hence this is a low loss capacitor. The capacitor has the advantage of being highly stable and the reading errors due to finely graduated scales (as in the case of continuously variable capacitors are eliminated). Thus a very high degree of precision is achieved in this capacitance box.

In modern practice decade boxes having three or four decades giving capacitance values upto 1.11 μF and 11.1 μF respectively in steps of 0.001 μF each are used.

## 6.21 FREQUENCY ERRORS IN CAPACITORS

A capacitor may be represented as shown in Fig. 6.26. $r$ and $L$ represent the resistance and inductance of leads, $C$ is a pure capacitance shunted by a resistance $R$ representing the dielectric loss. Thus the resistance, $R$, in the equivalent circuit will have a value which is a function of frequency. However, in the following analysis it is assumed that the capacitor is to be operated at low or medium frequencies such that $R$ may be considered as having constant magnitude.

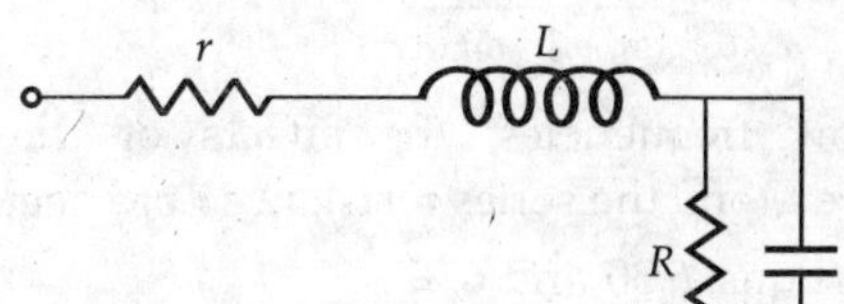


**Fig. 6.26** Equivalent Circuit of capacitor.

The impedance is :

$$Z = r + j\omega L + \frac{R/j\omega C}{R + 1/j\omega C}$$

$$= r + \frac{R}{1+\omega^2 R^2 C^2} + j\omega\left[L - \frac{CR^2}{1+\omega^2 R^2 C^2}\right] \quad ...(6.26)$$

the effective reactance is $X_{eff} = \dfrac{1}{j\omega C_{eff}}$

where $C_{eff}$ = effective capacitance

$$\therefore \quad \frac{1}{j\omega C_{eff}} = j\omega\left[L - \frac{CR^2}{1+\omega^2 R^2 C^2}\right]$$

or $$\frac{1}{\omega C_{eff}} = -\omega L + \frac{\omega CR^2}{1+\omega^2 R^2 C^2}$$

or $$\omega C_{eff} = \frac{1+\omega^2 R^2 C^2}{\omega CR^2 - \omega L(1+\omega^2 R^2 C^2)} \quad ...(6.27)$$

In a capacitor $\omega^2 C^2 R^2$ is very large as compared to 1.

$$\therefore \quad \omega C_{eff} = \frac{\omega^2 R^2 C^2}{\omega CR^2 - \omega^3 R^2 C^2 L} = \frac{\omega C}{1-\omega^2 LC}$$

or $$C_{eff} = \frac{C}{1-\omega^2 LC} \quad ...(6.28)$$

For medium frequencies, the value of effective capacitance may be written as :

$$C_{eff} = C(1+\omega^2 LC) \quad ...(6.29)$$

At a frequency $f_0 = 1/(2\pi\sqrt{LC})$, a resonant condition is produced. Under this condition the reactance vanishes and the capacitor behaves like a pure resistor. It is therefore important that the inductance of leads is kept to a minimum so that resonant frequency is high. The capacitor for measuring purposes can be used only for frequencies $f < 0.1 f_0$ if an error of less than 1% is desired.

The effective series resistance (medium frequency) is :

$$R_{eff} = r + \frac{R}{1+\omega^2 R^2 C^2} \quad ...(6.30)$$

The loss angle is $$\tan\delta = \frac{(\omega CR^2 - \omega L - \omega^3 R^2 C^2 L)}{r+\omega^2 R^2 rC + R}$$

$$= \frac{1-\omega^2 LC}{\omega r + 1/\omega CR} \quad ...(6.31)$$

At low frequencies, the effects of the series inductance $L$, and the series resistance $r$ are negligible.

From Eqns. 6.30 and 6.27

∴ The effective series resistance (low frequency) is

$$R_{eff} = \frac{R}{1+\omega^2 C^2 R^2} \quad ...(6.32)$$

and the effective capacitance is

$$C_{eff} = C + \frac{1}{\omega^2 CR^2} \quad ...(6.33)$$

The loss angle is also given by :

$$\tan\delta = \omega Cr + \frac{1}{\omega CR} \quad ...(6.34)$$

As the frequency applied to a capacitor is increased the values of capacitance and tan δ show a fall in magnitude until the effects of inductance become important when an increase in loss angle and effective capacitance occurs.

**Example 6.6** *The leads to the plates of a 100 pF capacitor have a resistance of 0.5 Ω and an inductance of 0.3 μH. The dielectric losses at the operating frequency of 10 MHz are equivalent to a leakage path of 50,000 Ω. Find the equivalent capacitance and the loss angle of the capacitor at this frequency.*

**Solution.**

Given capacitance $C = 100 \times 10^{-12}$ F

Shunt resistance $R = 50{,}000\ \Omega$,

Resistance of leads $r = 0.5\ \Omega$

Inductance of leads $L = 0.3 \times 10^{-6}$ H

$$\omega = 2\pi \times 10 \times 10^6 = 20 \times \pi \times 10^6 \text{ rad/s}$$

From Eqn. 6.28 effective capacitance,

$$C_{eff} = \frac{C}{1-\omega^2 LC}$$

Substituting the given values, $C_{eff} = 113.4$ pF.

Effective resistance, $R_{eff} = r + \dfrac{R}{1+\omega^2 R^2 C^2} = 1\ \Omega$

Loss angle $\delta = \tan\delta = \omega R_{eff} C_{eff}$

$$= 20\pi \times 10^6 \times 1 \times 113.4 \times 10^{-12}$$

$$= 0.0071 \text{ rad.} = 24.4'.$$

Loss angle calculated from Eqn. 6.34.

$$\delta = \tan\delta = \omega Cr + 1/\omega CR = 0.00632 \text{ rad} = 21.7'.$$

# Review Questions

1. Define the term 'residues' as used for circuit components.
2. Describe the desirable properties which a material should possess to be used in resistance standards.
3. List the properties of the following materials which are used for standard resistors :

   (*i*) Manganin (*ii*) Constantan (*iii*) Gold chromium.
4. Explain why metallic formers are used for d.c. resistance standards.
5. Resistance standards use single layer coils. Explain their advantages over multi-layer coils.
6. Describe the construction of a 1 Ω resistance standard for d.c.

7. Explain why is it necessary to use four terminals resistance coils for low resistance standards. Explain how the effect of contact resistance is eliminated in them and how the resistance is made to have a fixed definite value between potential terminals irrespective of the type of contacts at the current terminals.
8. Describe the construction of sheet metal resistors.
9. Derive the expressions for effective resistance, effective inductance, phase angle and time constant of a resistor used at low and medium frequencies.
10. Explain why is it not possible to design a resistor to have to resistance equal to d.c. resistance and simultaneously not to have a residual inductance ?
11. Explain the different methods used for winding of wires of resistance coils so as to render them non-inductive. How are the capacitive effects reduced in resistance coils of standard resistors ?
12. Explain the factors taken into consideration while designing and constructing standard inductors.
13. Describe the construction of a primary standard of mutual inductance.
14. Describe the construction and working of the following types of variable inductometers :

    (*i*) Aryton Perry Inductometer

    (*ii*) Brooks and Weaver Inductometer.
15. Derive the expressions for effective resistance and effective inductance of a inductor used at medium and low frequencies.
16. Derive expressions for dissipation factor for capacitors for both series and parallel equivalent circuit representations.
17. Describe the construction of high voltage capacitance standard.

    What are the advantages and disadvantages of using solid dielectric materials for standards of capacitance ?
18. Describe the construction of a decade capacitance box.
19. Derive expressions for effective resistance and effective capacitance for a capacitor used at low and medium frequencies.

6

# Unsolved Problems

1. A coil has a resistance of 2000 Ω and a shunt capacitance of 30 pF. Calculate the value of series inductance to render the coil non-inductive over a reasonable range of frequency. **[Ans.** 120 μH]
2. A 1000 Ω resistor has a series inductance of 5 μH and a shunt capacitance of 20 pF. Find the effective resistance and reactance at 20 kHz. Calculate the phase angle and time constant of resistor at this frequency. **[Ans.** 999.68 Ω, – 18.8 Ω, – 1°5′, $1.8 \times 10^{-6}$ s]
3. An inductive coil has a resistance of 400 Ω and an inductance of 3 H when measured at a very low frequency. The distributed capacitance is 100 pF. Find the percentage change in effective inductance when this coil is used at 1 kHz. **[Ans.** 1.2%]
4. A coil has effective inductance of 2.1 H and 2.15 H at frequencies of 2 kHz and 4 kHz respectively. Determine the true value of inductance. Also find the value of distributed capacitance. **[Ans.** 2.084, 25.45 pF]
5. The resistance and inductance of leads of a 100 pF capacitor are 10 Ω and 0.25 μH respectively. The leakage resistance is 100 kΩ. Find the value of effective capacitance, effective resistance and loss angle for the capacitor. The frequency is $100 \times 10^6$ rad/s. **[Ans.** 133.25 pF, 1.1 Ω, 0.8°]

# Objective Type Questions

***Tick (√) the most appropriate answer :***

1. The residues of a resistor at high frequencies are :

   (*a*) interturn capacitance (*b*) inductance

   (*c*) ground capacitance

   (*d*) all the above.
2. Materials used for precision resistors should have :

   (*a*) low resistivity

   (*b*) high resistance temperature co-efficient

   (*c*) high thermoelectric emf against copper

   (*d*) none of the above.

3. Metal spools cannot be used for resistors meant for a.c. applications because :
(a) hysteresis losses are produced in them
(b) hysteresis and eddy current losses are produced in them
(c) they produce differential expansion leading to stresses in resistance wire
(d) their use results in poor heat dissipation.

4. Accelerated ageing is done on coils for standard resistors in order to :
(a) make the resistance stable with time
(b) reduce inductive and capacitive effects
(c) reduce hot spot temperatures
(d) none of the above.

5. Four terminal resistors are used for resistance values :
(a) greater than 10 Ω (b) greater than 1 Ω
(c) less than 1 Ω (d) of the order of MΩ.

6. The condition for a resistor to have the same value of resistance at medium frequencies as with d.c. is :
(a) $CR^2 = L$ (b) $CR^2 = 2L$
(c) $CR^2 = \omega L$ (d) $CR^2 = 2\omega L$

7. A 100 Ω resistor has an effective inductance of 0.1 μH and a distributed capacitance of 20 pF. Its time constant at medium frequencies is :
(a) 1 ns (b) 2 ns
(c) – 1 ns (d) 0

8. In resistance standards, bifilar winding arrangement is used to reduce :
(a) capacitance of winding
(b) inductance of winding
(c) both capacitance and inductance of winding
(d) none of the above.

9. The resistance coil of a winding has a distributed capacitance of 9 pF when wound in a single section. The same winding is sectionalised and is wound in three sections connected in series.
The distributed capacitance of the winding with new arrangement is :
(a) 9 pF (b) 3 pF
(c) 27 pF (d) 1 pF

10. In order that value of an inductance standard should remain constant irrespective of the value of current carried by its coil :
(a) cores of ferromagnetic materials should be used
(b) air cored coils should be used
(c) solid conductors should be used
(d) all the above.

11. In order that the values of effective resistance and inductance of a standard remain unaffected due to changes in frequency :
(a) ferromagnetic materials should not be used for core.
(b) coils should use stranded conductors.
(c) use of metallic parts should be avoided.
(d) all the above.

12. The value of effective inductance standard inductors at low and medium frequencies is approximately :
(a) $L(1-\omega^2 LC)$ (b) $\omega L(1+\omega^2 LC)$
(c) $L(1+\omega^2 LC)$ (d) $L(1+2\omega^2 LC)$

13. An imperfect capacitor is represented by a capacitance $C_P$, in parallel with a resistance $R_P$. The value of its dissipation factor (tan δ) is :
(a) $\omega C_P R_P$ (b) $\omega^2 C_P R_P$
(c) $\dfrac{1}{\omega^2 C_P R_P}$ (d) $\dfrac{1}{\omega C_P R_P}$

14. Capacitors with solid dielectric materials are used because :
(a) they give a reduced dielectric loss
(b) they are smaller in size as compared with air capacitors.
(c) they have a lower capacitance value as compared with air capacitors.
(d) none of the above.

15. The effective value of capacitance of standard capacitor at low and medium frequencies is
(a) $\dfrac{C}{1-\omega^2 LC}$ (b) $\dfrac{C}{1+\omega^2 LC}$
(c) $\dfrac{C}{1-2\omega^2 LC}$ (d) $\dfrac{C}{1+2\omega^2 LC}$

## Answers

| | | | | | | | | | |
|---|---|---|---|---|---|---|---|---|---|
| 1. (d) | 2. (d) | 3. (b) | 4. (a) | 5. (c) | 6. (b) | 7. (c) | 8. (b) | 9. (d) | 10. (b) |
| 11. (d) | 12. (c) | 13. (d) | 14. (b) | 15. (a) | | | | | |

# 7

# Analog Instruments

## 7.1 ANALOG INSTRUMENTS

An analog device is one in which the output or display is a continuous function of time and bears a constant relation to its input. Analog instruments find extensive use in present day applications although digital instruments are increasing in number and applications. The areas of application which are common to both analog and digital instruments are fairly limited at present. Hence, it can safely be predicted that the analog instruments will remain in extensive use for a number of years and are not likely to be completely replaced by digital instruments for certain applications.

## 7.2 CLASSIFICATION OF ANALOG INSTRUMENTS

Broadly, the analog instruments (and for that matter digital instruments) may be classified according to the quantity they measure. *For example,* an instrument meant for measurement of current is classified as an *Ammeter* while an instrument that measures voltage is classified as a *Voltmeter*. In addition to above instruments, we have wattmeters, power factor meters, frequency meters etc.

Electrical instruments may also be classified according to the kind of current that can be measured by them. Electrical instruments may be classified as instruments for :

(*i*) direct current (d.c.)

(*ii*) alternating current (a.c.) and

(*iii*) both direct and alternating current instruments (d.c./a.c.), these instruments are appropriately called universal instruments.

Analog instruments depend for their operation on one of the many effects produced by current and voltage and thus can be classified according to which of the effects is used for their working. The various effects used are listed in Table 7.1 on page 179.

Analog instruments are also classified as :

- Indicating,
- Recording and
- Integrating.

***Indicating instruments*** are those instruments which indicate the magnitude of a quantity being measured. They generally make use of a dial and a pointer for this purpose. Ordinary voltmeters, ammeters and wattmeters belong to this category.

The analog indicating instruments may be divided into *two* groups :

(*i*) electromechanical instruments,

(*ii*) electronic instruments.

Electronic instruments are constructed by addition of electronic circuits to electromagnetic indicators in order to increase the sensitivity and input impedance.

***Recording instruments*** give a continuous record of the quantity being measured over a specified period. The variations of the quantity being measured are recorded by a pen (attached to the moving system of the instrument ; the moving system is operated by the quantity being measured) on a sheet of paper fixed or moving. For example, we may have a recording voltmeter in a sub-station which keeps record of the variations of supply voltage during the day.

***Integrating instruments*** totalize events over a specified period of time. The summation, which they give is the product of time and an electrical quantity. Ampere hour and watt hour (energy) meters are examples of this category. The integration (summation value) is generally given by a register consisting of a set of a pointers and dials.

The analog instruments may also be classified on the basis of method used for comparing the unknown quantity (measured) with the unit of measurement. The *two* categories of instruments based upon this classification are :

1. ***Direct measuring instruments.*** These instruments convert the energy of the measurand directly into energy that actuates the instrument and the value of the unknown quantity is measured or displayed or recorded directly. The examples of this class of instruments are ammeters, voltmeters, wattmeters and energy meters.

2. ***Comparison instruments.*** These instruments measure the unknown quantity by comparison with a standard. (Direct measuring instruments are the most commonly used in engineering practice because they are the most simple and inexpensive. Also their use makes the measurement possible in the shortest time.) The examples of comparison type instruments are d.c. and a.c. bridges.

Comparison type instruments are used in cases where a higher accuracy of measurement is required.

Electrical instruments may also be classified according to their *accuracy class*. The limits of intrinsic error in the measured quantity for instruments for various classes of accuracy* are :

| ***Accuracy class*** | 0.2 | 0.5 | 1.0 | 1.5 | 2.5 | 5 |
|---|---|---|---|---|---|---|
| ***Limit of error (percent)*** | ±0.2 | ±0.5 | ±1.0 | ±1.5 | ±2.5 | ±5 |

## 7.3 PRINCIPLES OF OPERATION

As mentioned earlier secondary analog instruments may be classified according to the principle of operation they utilize. The effects they utilize are :

(*i*) magnetic effect,
(*ii*) heating effect,
(*iii*) electrostatic effect,
(*iv*) electromagnetic effect and
(*v*) hall effect.

***1. Magnetic effect.*** Consider a current carrying conductor of Fig. 7.1(*a*), it produces a magnetic field in the anticlockwise direction. We now have a uniform magnetic field as shown in Fig. 7.1(*b*). Let the current carrying conductor be placed in this magnetic field. The resultant field is as shown in Fig. 7.1(*c*). This results in distortion of magnetic field causing a force *F* to act from left to right. The reversal of direction of the current will cause a force *F* in the opposite direction, *i.e.*, from right to left subject to the condition that the direction of the existing field remains the same.

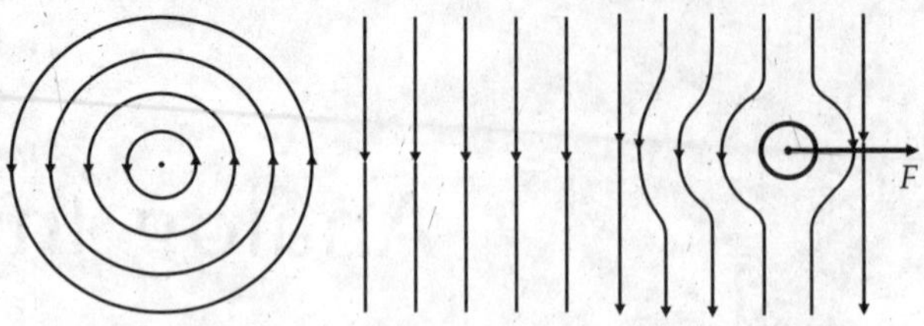


(*a*) Current carrying conductor (*b*) Magnetic field (*c*) Resultant force

**Fig. 7.1** Force produced due to a current carrying a magnetic field.

If we form the conductor into a coil, the magnetic field produced by each turn of the coil will add up and the coil will behave as an imaginary magnet as shown in Fig. 7.2.

**Fig. 7.2** Magnetic field produced by current carrying coils.

**Force of attraction or repulsion**. Consider a current carrying coil as shown in Fig. 7.2. It produces an imaginary bar magnet. When a piece of soft iron which has not been previously magnetized is brought near the end of the coil, it will be attracted by the coil. Therefore, if we pivot the soft iron on a spindle between two bearings and a coil is mounted near it, the iron piece will swing into the coil when the latter is carrying current. The effect is utilized in the *"attraction type of moving iron instruments."*

If we have two pieces of soft iron placed near the coil the two will be similarly magnetized and there will be force of repulsion between them. This effect is utilized in *"repulsion type moving iron instruments."*

**Force between a current carrying coil and a permanent magnet**. Consider the coil of Fig. 7.3. It produces an imaginary bar magnet when carrying current. When a permanent magnet is brought near it,

* IS 6236–1971

there will be either a force of attraction or repulsion. If the coil is mounted on a spindle between bearings, there will be a movement of the coil. This effect is utilized in permanent magnet moving coil instruments.

**Force between two current carrying coils.** Consider two current carrying coils shown in Fig. 7.3. For the directions of the currents shown, the two produce unlike poles near each other and thus there is a force of attraction and if one of the coils is movable and the other is fixed, there will be a motion of the movable coil.

This effect is utilised in the dynamometer type of instruments.

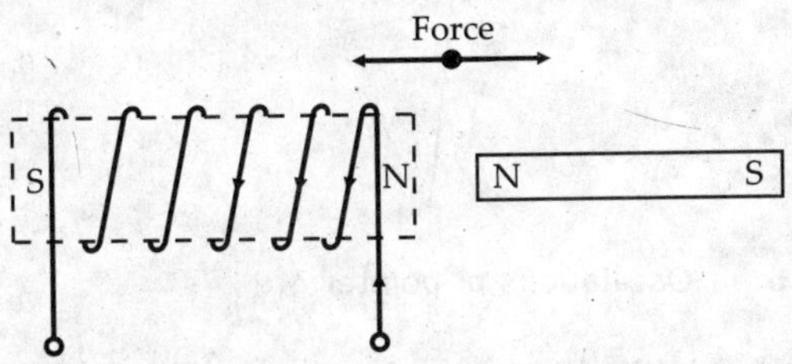


**Fig. 7.3** Force between current carrying coil and a permanent magnet.

***2. Thermal effect.*** The current to be measured is passed through a small element which heats it. The temperature rise is converted to an emf by a thermocouple attached to the element.

A thermo-couple consists of lengths of two dissimilar electric conductors joined at ends to form a closed loop. If the junctions of the two dissimilar metals are maintained at different temperatures, a current flows through the closed loop. This current can be measured and is indicative of the r.m.s. value of the current flowing through the heater elements.

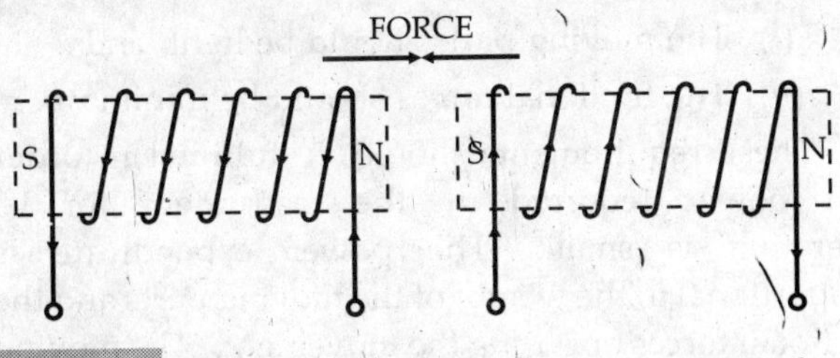


**Fig. 7.4** Force between two current carrying coils.

***3. Electrostatic effect.*** When two plates are charged, there is a force exerted between them. This force is used to move one of the plates. The instruments working on this principle are called electrostatic instruments and they are usually voltmeters.

***4. Induction effect.*** When a non-magnetic conducting pivoted disc or a drum is placed in a magnetic field produced by a system of electromagnets excited by alternating currents, an emf is induced in the disc or drum. If a closed path is provided, the emf forces a current to flow in the disc or drum. The force produced by the interaction of induced currents and the alternating magnetic fields makes the disc move. The induction effect is mainly utilized in a.c. energy meters.



***5. Hall effect.*** If a strip of conducting material carries current in the presence of a transverse magnetic field as shown in Fig. 7.5, an emf is produced between two edges of conductor. The magnitude of the voltage depends upon the current, flux density and a property of conductor called *"Hall Effect Co-efficient"*.

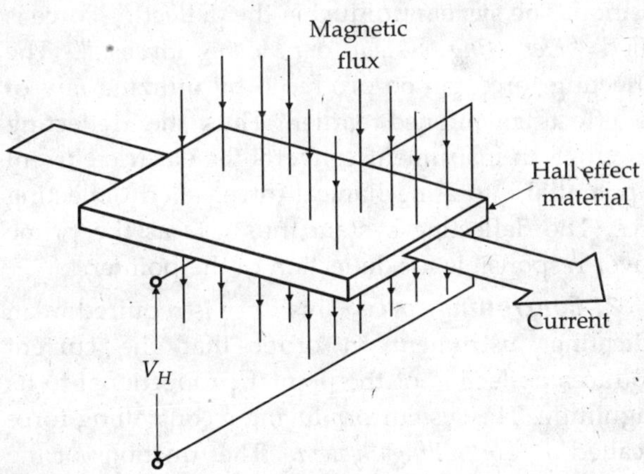


**Fig. 7.5** Hall effect element.

The emf may be measured after amplification. Hall effect elements are extensively used in magnetic measurements. They can also be used for sensing of current.

Another instrument that uses Hall effect is **Poynting Vector Wattmeter**. This wattmeter used for measuring the power loss density at the surface of a magnetic material.

Table 7.1 gives effects utilized by various types of instruments.

**Table 7.1** *Effects used in Instruments for their operations*

| *Effect* | *Instruments* |
|---|---|
| Magnetic effect | Ammeters, voltmeters, wattmeters, integrating meters. |
| Heating effect | Ammeters and voltmeters, wattmeters. |
| Electrostatic effect | Voltmeters. |
| Induction effect | A.C. ammeters, voltmeters, wattmeters, energy meters. |
| Hall effect | Flux meters, ammeters and Poynting vector wattmeter. |

## ELECTROMECHANICAL INDICATING INSTRUMENTS

### 7.4 OPERATING FORCES

Three types of forces are needed for the satisfactory operation of any indicating instrument. These are :

(*i*) Deflecting force,

(*ii*) controlling force and

(*iii*) damping force

***1. Deflecting force.*** The deflecting or operating force is required for moving the pointer from its zero position. The system producing the deflecting force is called *"Deflecting system* or *Moving System."* The deflecting force can be produced by utilizing any of the effects mentioned earlier. Thus the deflecting system of an instrument converts the electric current or potential into a mechanical force called deflecting force. The deflecting system thus acts as the prime mover responsible for deflection of the pointer.

***2. Controlling force.*** This force is required in an indicating instrument in order that the current produces deflection of the pointer proportional to its magnitude. The system producing a controlling force is called a *"Controlling System."* The functions of the controlling system are :

(*i*) To produce a force equal and opposite to the deflecting force at the final steady position of pointer in order to make the deflection of the pointer definite for a particular magnitude of current. In the absence of a controlling system, the pointer will shoot (swing) beyond the final steady position for any magnitude of current and thus the deflection will be indefinite.

(*ii*) To bring the moving system back to zero when the force causing the instrument moving system to deflect is removed. In the absence of a controlling system the pointer will not come back to zero when current is removed. Controlling force is usually provided by springs.

***3. Damping force.*** When a deflecting force is applied to the moving system, it deflects and it should come to rest at a position where the deflecting force is balanced by the controlling force. The deflecting and controlling forces are produced by systems which have inertia and, therefore, the moving system cannot immediately settle at its final position but overshoots or swings ahead of it. Consider Fig. 7.6. Suppose *O* is the equilibrium or final steady position. Because of inertia the moving system moves to position *'a'*. Now for any position *'a'* beyond the equilibrium position the controlling force is more than the deflecting force and hence the moving system swings back. Due to inertia it cannot settle at *'O'* but swings to a position say *'b'* behind the equilibrium position. At *'b'*, the deflecting force is more than the controlling force and hence the moving system again swings ahead.

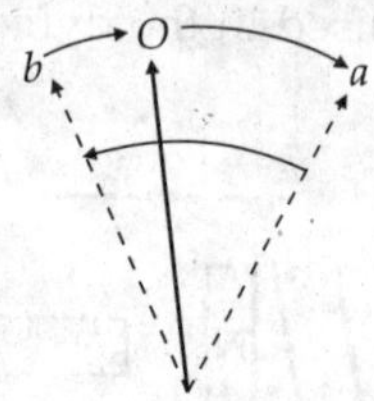


**Fig. 7.6** Oscillations of pointer.

The pointer thus oscillates about its final steady (equilibrium) position with decreasing amplitude till its kinetic energy (on account of inertia) is dissipated in friction and therefore, it will settle down at its final steady position. If extra forces are not provided to "damp" these oscillations, the moving system will take a considerable time to settle to the final position and hence time consumed in taking readings will be very large. Therefore, damping forces are necessary so that the moving system comes to its equilibrium position rapidly and smoothly without any oscillations.

### 7.5 CONSTRUCTIONAL DETAILS

***Moving system.*** The requirements of a moving system are :

(*i*) The moving parts should be light, and

(*ii*) The frictional forces should be minimum.

These requirements should be fulfilled in order that power required by the instrument for its operation is small. (The power expenditure is proportional to the weight of the moving parts and the frictional forces opposing the movement.) The moving system can be made light by using aluminium as far as possible. The frictional forces are reduced by using a spindle mounted between jewel bearings and by carefully balancing the system.

***Supporting the moving element.*** The force or torque developed by the moving element of an electrical instrument is necessarily small in order that the power consumption of the instrument be kept low so that the introduction of the instrument into a circuit may cause minimum change in the existing circuit conditions. Because of low power levels, the consideration of various methods of supporting the moving element becomes of vital importance. With the operating forces being small, the frictional forces must be kept to a minimum in order that the instrument reads correctly and is not erratic in action and is reliable.

## 7.6 TYPES OF SUPPORTS

Several types of supports are used, depending upon the sensitivity required and the operating conditions to be met. Supports may be of the following types :

- Suspension
- Taut Suspension
- Pivot and jewel bearings (double).

***1. Suspension.*** It consists of a fine, ribbon shaped metal filament for the upper suspension and coil of fine wire for the lower part (See. Fig. 8.1 page 198). The ribbon is made of a spring material like berrylium copper or phosphor bronze. This coiling of lower part of suspension is done in order to give negligible restraint on the moving system. The type of suspension requires careful levelling of the instrument, so that the moving system hangs in correct vertical position. This construction is, therefore, not suited to field use and is employed only in those laboratory applications in which very great sensitivity is required. In order to prevent shocks to the suspension during transit etc. a clamping arrangement is employed for supporting the moving system.

***2. Taut suspension.*** Suspension type of instruments can only be used in vertical position. The taut suspension has a flat ribbon suspension both above and below the moving element, with suspension kept under tension by a spring arrangement (Fig. 7.7). The advantage of this suspension is that exact levelling is not required if the moving element is properly balanced.

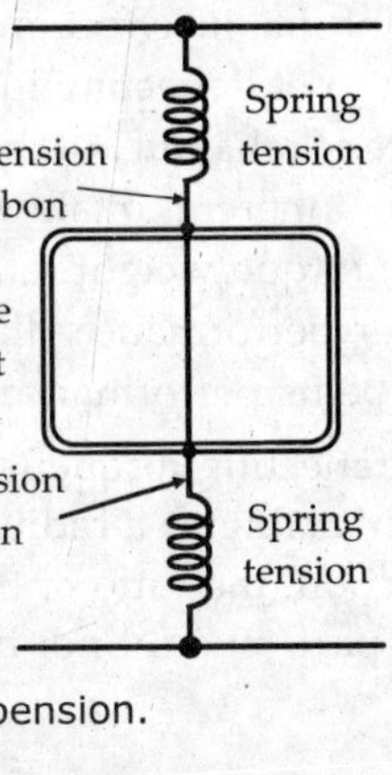


**Fig. 7.7** Taut suspension.

Suspensions and taut suspensions are customarily used in instruments of galvanometer class which require a low friction and high sensitivity mechanism. But actually there is no strict line of demarcation between galvanometers and other indicating instruments. Some sensitive wattmeters, and electrostatic voltmeters also use flexible suspension.

Ribbon suspensions, in addition to supporting the moving element, exert a controlling torque when twisted. Thus the use of suspension results in elimination of pivots, jewels and control springs and therefore, pivotless instruments are free from many defects.

***3. Pivot and jewel bearings.*** The moving system is mounted on a spindle made of hardened steel. The two ends of the spindle are made conical and then polished to form pivots. These ends fit conical holes in jewels located in the fixed parts of instruments (Fig. 7.8). These jewels, which are preferably made of sapphire form the bearings. Originally natural sapphire was used but now synthetic sapphire is being used. The combination of steel and sapphire gives lowest friction.

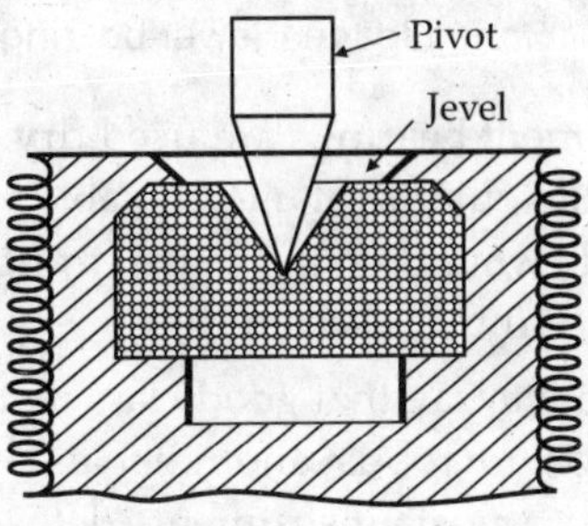


**Fig. 7.8** Jewel bearing.

It has been found that the frictional torque, for jewel bearings, is proportional to area of contact between the pivot and jewel. Thus the contact area between pivot and jewel should be small. The pivot is ground to a cone and its tip is rounded to a hemispherical surface of small area. The jewel is ground to a cone of somewhat larger angle. The pivot may have a radius at tips from 0.0125 mm to as high as 0.075 mm depending upon the weight of the mechanism and the force it will encounter. The radius of the pit of jewel is somewhat longer so that the contact is in the form of a circle. The contact area should not be too small otherwise the stress may exceed the crushing strength of the material of pivot. The pivots of many modern instruments are loaded nearly to yield point of steel. We can have an idea of the stress produced if we consider a moving element weighing 300 mg resting on the area of circle 0.005 mm diameter. The load produces a stress of about

$150 \times 10^6 \, N/m^2$ or 1/6 of the crushing limit of the steel. Moreover, if the instruments are subjected to shocks, the dynamic effect may cause the stress to be two or three times or more the static value. It is easy to visualize that rough handling of the instrument may deform the pivot and crack the jewel. Jewels are sometimes spring mounted (Fig. 7.9) to lessen the likelihood of damage because of shocks, and its bottom is also rounded.

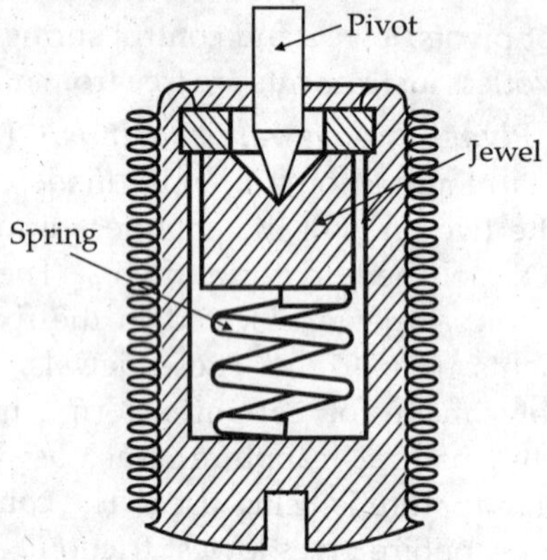


**Fig. 7.9** Spring loaded jewel bearing.

Instrument bearings are used dry. Any lubricant, such as oil, would be of questionable value in the first place and would soon cause trouble because of gumming and by collected dust.

It is true to say that good class instruments of this construction are robust and reliable in service, but it is a fact that most instrument defects are due to excessive friction caused by damaged pivots and/or dirty jewel bearings, and due to damaged or distorted control springs.

**4. *Balancing.*** In order that the wear on the bearings becomes uniform and symmetrical the centre of gravity of weight of moving parts should coincide with axis of the spindle for all positions of pointer and there are no out of balance forces. This also ensures that the deflection is unaffected by the position of the pointer and there are no reading errors.

The balance can be achieved by using balance weights carried on arms attached to the moving system. The use of balance weights is illustrated in Fig. 7.10. The balance weight whose position is adjustable compensates for the weight of pointer and any other part attached to the moving system.

A cheaper method is to replace balance weights by a wire wound round and soldered to the balance arm. Once fixed, the degree of balance is not adjustable in this case.

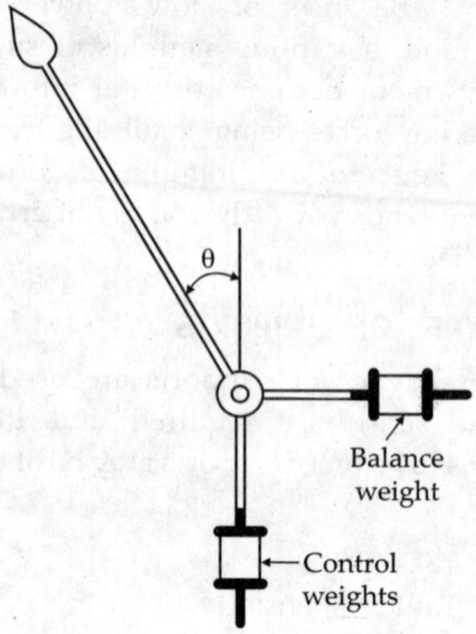


**Fig. 7.10** Control weights.

When determining the magnitude of weight and the distances at which they are placed, we must consider their effect on the moving system. A small weight at a large distance increases the moment of inertia and necessitates a large damping torque. A large weight at short distance, on the other hand, increases the weight of moving system. A compromise is usually made.

## 7.7 TORQUE/WEIGHT RATIO

The frictional torque in an instrument depends upon the weight of moving parts. If the weight of moving parts is large, the frictional torque will be large. The frictional torque exerts a considerable influence on the performance of an indicating instrument. If the frictional torque is large and is comparable to a considerable fraction of the deflecting torque the deflection of the moving system will depend upon the frictional torque to an appreciable extent. Also the deflection will depend on the direction from which the equilibrium position is approached and will be uncertain. On the other hand, if the frictional torque is very small compared with the deflecting torque, its effect on deflection is negligible. Thus the ratio of deflecting torque to frictional torque is a measure of reliability of the instrument indications and of the inherent quality of the design. Hence (deflecting) torque/weight ratio of an instrument is an index of its performance, the higher the ratio, the better will be its performance.

If the deflecting torque is expressed in terms of a force which, acting at a radius of 1 cm, produces full scale deflection, the ratio of this torque to the weight of moving parts should not be less than 0.1 as far as possible.

## 7.8 CONTROL SYSTEMS

The deflecting system of most of the commercial instruments is mounted on a pivoted spindle, the quantity being measured producing a deflecting torque proportional to its magnitude. There are two types of control systems which are used for such a mounted system :

(*i*) Gravity control, and

(*ii*) Spring control.

***1. Gravity control.*** In this type of control, a small weight is placed on an arm attached to the moving system. The position of this weight is adjustable. This weight produces a controlling torque due to gravity.

Figure 7.10 shows the pointer at zero position. In this case the control torque is zero. Suppose the system deflects through an angle θ as shown in Fig. 7.11. The weight acts at a distance *l* from the centre, the component of weight trying to restore the pointer back to zero position is *W* sin θ. Therefore, controlling torque is :

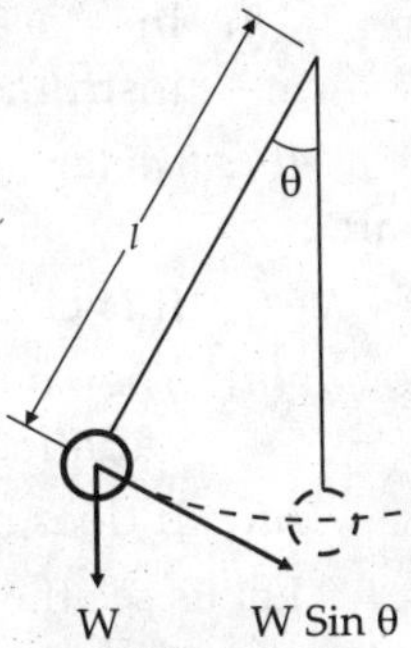


**Fig. 7.11** Gravity Control

$$T_c = W \sin\theta \times l = Wl \sin\theta \quad \text{...(7.1)}$$

$$= K_g . \sin\theta \quad \text{...(7.2)}$$

where $K_g = Wl$ = a constant.

Thus the controlling torque is proportional to sine of angle of deflection of moving system. The controlling torque can be varied by simply adjusting the position of control weight upon the arm which carries it.

It is obvious that the instruments employing gravity control must be used in vertical position in order that the control may operate. The instruments must be mounted in level position otherwise there will be a very serious zero error. For these reasons, gravity control is not suited for indicating instruments in general and portable instruments in particular. The system is normally not used now except in cases where cost and simplicity is the primary consideration.

***2. Spring control.*** A hair spring attached to the moving system (Fig. 7.12) exerts a controlling torque. The essential requirements for instrument springs are :

(*i*) They should be non-magnetic.

(*ii*) They should be proof from mechanical fatigue.

(*iii*) Where springs are used to lead current into moving system they should have a small resistance, their cross-sectional area must be sufficient to carry the current without temperature rise affecting their constant. They should also have a low resistance temperature co-efficient.

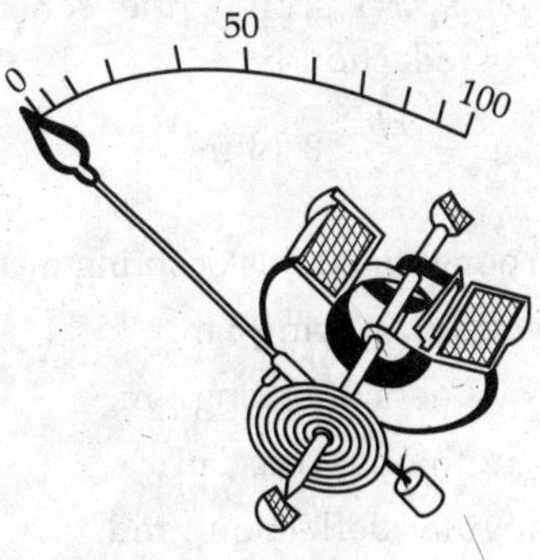


**Fig. 7.12** Spring Control.

A number of non-magnetic materials like silicon bronze, hard rolled silver or copper, platinum silver, platinum-iridium and German silver have been used but have not been found satisfactory owing to some reason or the other. For most applications phosphor bronze has been found to be the most suitable material except in instruments of low resistance (like millivoltmeters). In this case some special bronze alloys having low resistance may be used with some sacrifice in mechanical quality.

Flat spiral springs are used in almost all indicating instruments as the space required by these springs is less than for other types.

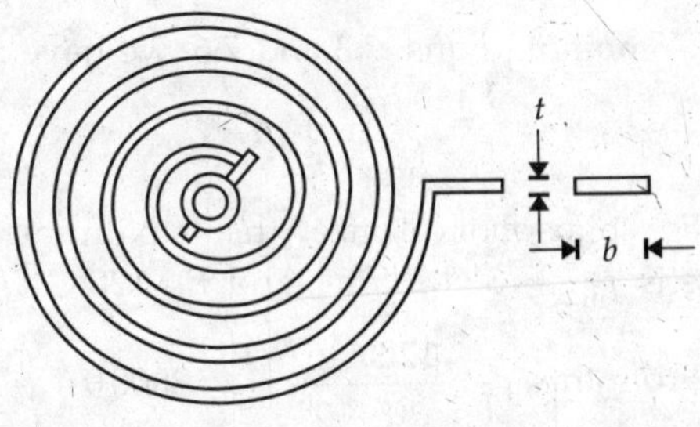


**Fig. 7.13** Flat Spiral spring.

One form of control spring mounting is shown in Fig. 7.12. The inner end of the spring is attached to the spindle and the outer end carries a spigot which engages in a circular disc surrounding the jewel screw. This disc carries an arm which is slotted and splayed out at the end. The purpose of slotted extension arm is to allow the spring to be coiled or uncoiled slightly, so that the pointer may be set at zero. The slotted arm is actuated by a set screw mounted at the front of instrument and, therefore, *zero setting* of the instrument can be done without removing the cover.

By making the number of turns large, the deformation per unit length is kept small on full scale deflection. The controlling torque is thus made proportional to the angle of deflection of the moving system.

For a flat spiral spring, the controlling torque developed by deflection is :

$$T_c = \frac{Ebt^3}{12l}\theta \text{ N-m} \quad ...(7.3)$$

where $E$ = Young's modulus of spring material ; N / m$^2$,

$b$ = width of spring ; m,

$t$ = thickness of spring ; m,

$l$ = length of spring ; m

$\theta$ = angular deflection ; rad.

$E, b, t, l$ are constant for a particular spring.

$$\therefore \quad T_c = K\theta \quad ...(7.4)$$

$$\text{where } K = \frac{Ebt^3}{12l} \text{ Nm/rad} \quad ...(7.5)$$

= a constant called "spring constant" or "control constant" or "torsion constant" or "restoring constant".

The springs should be stressed well below their elastic limit at maximum deflection of the instrument in order that there is no permanent set or that no change in deflection (or zero shift) will occur from inelastic yield.

$$\text{Maximum fibre stress } f_{max} = \frac{6T_c}{bt^2} \text{ N/m}^2 \quad ...(7.6)$$

Combining Eqns. 7.3 and 7.6, we have :

$$\frac{l}{t} = \frac{E}{f_{max}} \cdot \frac{\theta}{2} \quad ...(7.7)$$

For phosphor bronze, the maximum allowable stress is $f_{max} = 30$ MN/m$^2$, and $E = 120$ GN/m$^2$.

$$\text{From this, } \frac{l}{t} = \frac{120 \times 10^9}{30 \times 10^6} \cdot \frac{\theta}{2} = 2000\,\theta$$

Therefore, for a full scale deflection of $\theta = \frac{\pi}{2}$

and hence $\frac{l}{t} = 3000$.

Hence in order that the material is not overstressed the ratio of length to thickness of the spring strip should not be much less than 3000. Some designers use a ratio as low as 2100 (for a maximum deflection of 90°) for portable instruments. For switchboard instruments, whose indications are of lower precision, a ratio as low as 1000 is sometimes used. The ratio $b/t$ may be as high as 30/1 but is generally greater than 10/1.

Fatigue in springs may be avoided to a great extent by proper annealing and ageing during manufacture. In order to eliminate the effect of temperature variations upon the length of the spring, two springs coiled in opposite directions are used. When the moving system deflects, one spring is extended while the other is compressed.

### 7.8.1 Comparison between Spring and Gravity Control

The *disadvantages* of gravity control have been described earlier. They are

(*i*) This type of control can be used in only vertically mounted instruments and

(*ii*) In case gravity control is used, the instruments must be perfectly levelled.

Its *advantages*, when compared with spring control, are :

(*i*) It is cheap.

(*ii*) The control is independent of temperature variations.

(*iii*) It does not deteriorate with time.

Let us see the effect of the two types of controls on shape of the scale. Consider an instrument in which the deflecting torque is directly proportional to current being measured.

Therefore, deflecting torque $T_d = K_d I$

where $I$ = current being measured

and $K_d$ = a constant

If the instrument is spring controlled, the controlling torque $T_c = K\theta$

where $\theta$ = angle of deflection

and $K$ = spring constant.

At the equilibrium position, the deflecting torque is equal to the controlling torque or

$$T_c = T_d \quad \text{or} \quad K\theta = K_d I \quad \therefore \ \theta = (K_d / K)I = K_2 I$$

where $K_2$ = a constant = $K_d / K$

Therefore, the deflection is directly proportional to current (the quantity being measured) throughout the whole range of scale and therefore the scale is uniform.

If the instrument is gravity controlled $T_c = K_g \sin\theta$.

For equilibrium position $T_c = T_d$ or $K_g \sin\theta = K_d I$.

$$\text{or} \quad \sin\theta = (K_d / K_g)I = K_2' I$$

$$\therefore \quad \theta = \sin^{-1}(K_2' I)$$

Suppose a current $I_{max}$ produces a deflection of 90° in both the cases, then a current of $I_{max}/2$ would produce a deflection of 45° in the spring controlled instrument. If gravity control is employed, we have

$$\sin 90° = K_2' I_{max} \quad \text{or} \quad K_2' = 1 / I_{max}$$

$\therefore$ When $I = I_{max}/2, \qquad \theta = \sin^{-1} K_2' I$

$$= \sin^{-1}(1/I_{max})(I_{max}/2)$$

$$= \sin^{-1}(0.50) = 30°.$$

Thus, while a current $I_{max}$ gives a deflection of 90°, a current of $I_{max}/2$ gives a deflection of 30°. This means that a gravity controlled instrument does not have a uniform scale.

Hence a gravity control, taken on its merit, gives a cramped (compressed or crowded) scale at the lower end. This is a disadvantage when the pointer lies at the lower scale values, when it will not possible to read the instrument scale accurately on account of it being cramped.

**Example 7.1** *The control spring of an instrument has the following dimensions :*

*Length of strip = 370 mm ; thickness of strip = 0.073 mm, width of strip = 0.51 mm.*

*The Young's modulus is 112.8 GN/m². Estimate the torque exerted by spring when it is turned through 90°.*

**Solution.**

From Eqn. 7.3, the controlling torque

$$T_c = \frac{Ebt^3}{12t}\theta$$

$$= \frac{112.8\times10^9\times(0.51\times10^{-3})\times(0.073\times10^{-3})^3\times\pi/2}{12\times370\times10^{-3}}$$

$$= 7.9\times10^{-6}\text{ Nm}.$$

**Example 7.2** *The deflecting torque corresponding to a deflection of 60° is $1000\times10^{-6}$ Nm in an instrument. The control is exerted through two phosphorus bronze springs. Allowing a maximum stress of 65 MN/m² and taking the value of modulus of elasticity as 112.8 GN/m², calculate suitable dimensions for control springs. The width of spring strip is 1 mm.*

**Solution.** There are two springs and, therefore, each spring exerts a controlling torque which is half of the deflecting torque.

$\therefore$ Controlling torque of each spring,

$$T_c = (1000\times10^{-6})/2 = 500\times10^{-6}\text{ Nm}.$$

From Eqn. 7.6, maximum stress $f_{max} = 6\dfrac{T_c}{bt^2}$

$$bt^2 = \frac{T_c}{f_{max}} = \frac{6\times500\times10^{-6}}{65\times10^6}$$

$$= 46.15\times10^{-12}\text{ m}^3 = 46.15\times10^{-3}\text{ mm}^3$$

Width of spring $b = 1$ mm,

so $\qquad t^2 = (46.15\times10^{-3})/1 = 46.15\times10^{-3}\text{ mm}^2.$

$\therefore$ Thickness of spring $t = 0.215$ mm.

From Eqn. 7.3, controlling torque $T_c = \dfrac{Ebt^3}{12\,l}\theta$

or length of spring $l = \dfrac{Ebt^3}{12\,T_c}\theta.$

$$= \frac{112.8\times10^9\times(1\times10^{-3})\times(0.125\times10^{-3})^3}{12\times500\times10^{-6}}\times\frac{\pi}{3}$$

$$= 0.193\text{ m} = 193\text{ mm}.$$

**Example 7.3** *In a permanent magnet coil type voltmeter the deflecting torque is proportional to the applied voltage. If the deflection is 90° at a full scale value of 100 V, plot the shape of the scale if the instrument is :*

*(a) Spring controlled,*

*(b) Gravity controlled.*

*Draw the shape of scale for both cases.*

**Solution.** At equilibrium position $T_c = T_d$

For a spring controlled instrument $T_c = K\theta$

$\therefore \quad K\theta = K_d V$ or $\theta = (K/K_d)V = K_2 V.$

The deflection is directly proportion the voltage and a uniform scale is obtained as shown in Fig. 7.14(*a*).

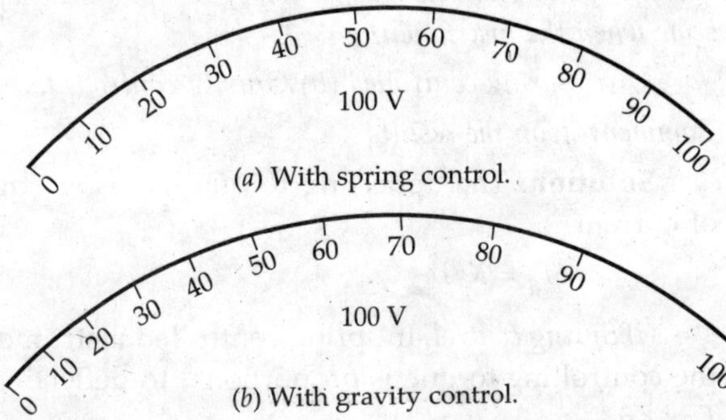


(*a*) With spring control.

(*b*) With gravity control.

**Fig. 7.14** Shape of scale of instrument when deflecting torque is directly proportional to the quantity being measured.

For a gravity controlled instrument.

$$T_c = K_g \sin\theta$$

$$\therefore \quad K_g \sin\theta = K_d V$$

or $$\theta = \sin^{-1}\left(\frac{K_d}{K_g}\right)V = \sin^{-1} K_2' V$$

At 100 V, $\theta = 90°$

$$\therefore \quad K_2' = \frac{1}{100}$$

The deflection at any voltage $V$ is $\theta = \sin^{-1}\dfrac{V}{100}.$

The deflections for various values of voltage are tabulated on next page.

| Voltage V | Deflection θ degrees |
|---|---|
| 0 | 0 |
| 10 | 5.7 |
| 20 | 11.5 |
| 30 | 17.5 |
| 40 | 23.6 |
| 50 | 30 |
| 60 | 36.9 |
| 70 | 44.4 |
| 80 | 53.1 |
| 90 | 64.2 |
| 100 | 90 |

The scale is shown in Fig. 7.14(*b*). It is clear, with gravity control, the scale is cramped at the lower end and expanded at the upper end.

**Example 7.4** *The deflecting torque of an ammeter varies as the square of the current passing through it. If a current of 5 A produces a deflection of 90°, plot the shape of scale when the instrument is :*

*(i) Spring controlled (ii) Gravity controlled.*

*Comment upon the results.*

**Solution.** The deflecting torque varies as square of current.

$\therefore \quad T_d = K_d I^2$

*(i) Spring control.* In spring controlled instruments, the controlling torque is proportional to deflection or $T_c = K\theta$.

For equilibrium position, $T_c = T_d$

or $\quad K\theta = K_d I^2 \quad$ or $\quad \theta = (K_d / K) I^2 = K_2 I^2$

The deflection is 90° when the current is 5 A :

$\therefore \quad K_2 = 90/(5)^2 = 3.6$

Hence deflection at any current $I$ is $\theta = 3.6 I^2$

*(ii) Gravity control.* For gravity controlled instruments the controlling torque is proportional to sine of angle of deflection.

$\therefore \quad T_c = K_g \sin\theta$

At equilibrium $K_g \sin\theta = K_d I^2$

or $\quad \theta = \sin^{-1}(K_g / K_d) I^2 = \sin^{-1} K_2' I^2$

$\therefore \quad K_2' = 1/25$

Hence deflection any current $I$ is

$$\theta = \sin^{-1}(I^2/25).$$

The results are tabulated below :

| Current, I (A) | Deflection θ degrees (for spring control) | Deflection θ degrees (for gravity control) |
|---|---|---|
| 0 | 0 | 0 |
| 1.0 | 3.6 | 2.3 |
| 1.5 | 8.1 | 5.2 |
| 2.0 | 14.4 | 9.2 |
| 2.5 | 22.5 | 14.5 |
| 3.0 | 32.4 | 21.1 |
| 3.5 | 44.1 | 29.3 |
| 4.0 | 57.6 | 39.8 |
| 4.5 | 72.9 | 54.4 |
| 5.0 | 9.0 | 9.0 |

The shape of scales is shown in Fig. 7.15.

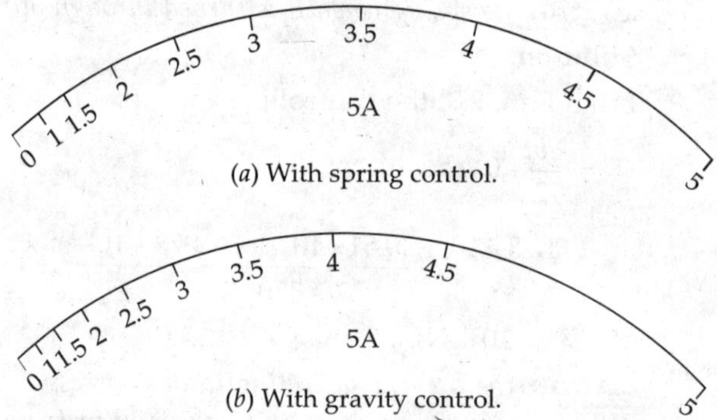


**Fig. 7.15** Shape of scale of instruments where deflecting torque is proportional to square of the quantity being measured.

The scale is non-uniform for both the cases. This is due to deflecting torque being proportional to square of quantity being measured. However, the scale is more cramped at the lower end in the case of gravity control.

**Example 7.5** *A weight of 5 g is used as the controlling weight in a gravity controlled instrument. Find its distance from spindle if the deflecting torque corresponding to a deflection of 60° is* $1.13 \times 10^{-3}$ *Nm.*

**Solution.** Deflecting torque due to a weight is

$$T_d = mg\sin\theta \times d$$

where $d$ = distance of the weight from spindle.

$$d = \frac{T_d}{mg\sin\theta} = \frac{1.13\times10^{-3}}{5\times10^{-3}\times9.81\times\sin 60°}$$

$$= 0.0266 \text{ m} = 26.6 \text{ mm}.$$

## 7.9 DAMPING SYSTEMS

The damping torque should be of such a magnitude that the pointer quickly comes to its final steady position, without overshooting. If the instrument is *underdamped*, the moving system will oscillate about the final steady position with a decreasing amplitude and will take some time before it comes to rest. When the moving system moves rapidly but smoothly to its final steady position, the instrument is said to be critically *damped or dead beat*. If the damping torque is more than what is required for critical damping, the instrument is said to be *overdamped*. In an overdamped instrument, the moving system moves slowly to its final steady position in a lethargic fashion. The readings are very tedious to take in this case.

Figure 7.16 illustrates the way an underdamped, an overdamped and a critically damped system moves to its final steady position, $\theta_F$.

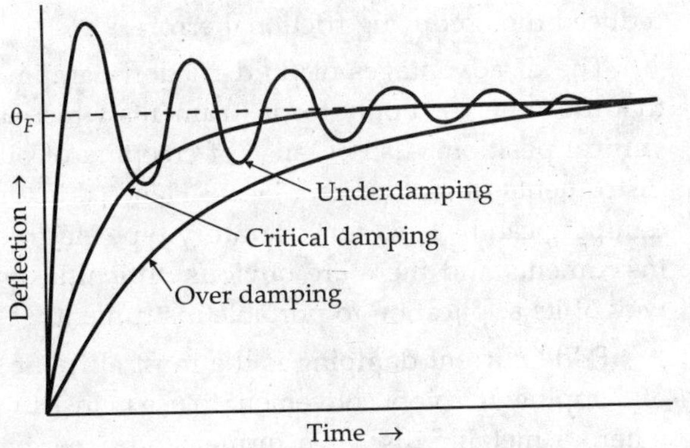


**Fig. 7.16** Setting curves for different types of damping.

The damping device should be such that it produces a damping torque only while the moving system is in motion. To be effective the damping torque should be proportional to the velocity of the moving system but independent of the operating current. It must not effect the controlling torque or increase the static friction.

The methods for producing damping torque are :

(*i*) Air friction damping,

(*ii*) Fluid friction damping,

(*iii*) Eddy current damping and

(*iv*) Electromagnetic damping.

### 7.9.1 Air Friction Damping

Two types of air friction damping devices are shown in Fig. 7.17.

The arrangement of Fig. 7.17(*a*) consists of a light aluminium piston which is attached to the moving system. This piston moves in a fixed air chamber which is closed at one end. The clearance between piston and chamber walls is uniform throughout and is very small. When there are oscillations the piston moves into and out of an air chamber. When the piston moves into the chamber, the air inside is compressed and the pressure of air, thus built up, opposes the motion of piston and hence whole of the moving system. When the piston moves out of air chamber, pressure in the closed space falls, and the pressure on the open side of piston is greater than on the other side. Thus there is again an opposition to motion.

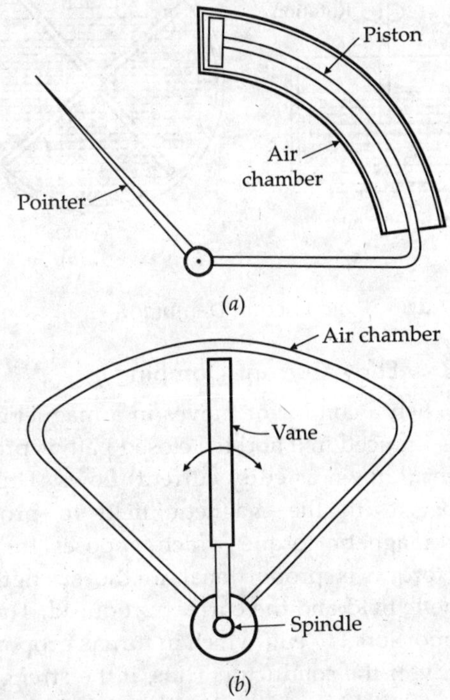


**Fig. 7.17** Air friction damping.

The arrangement of Fig. 7.17(*b*), consists of an aluminium vane which moves in a quadrant (sector) shaped air chamber. This air chamber is a recess cast in a bakelite moulding or diecasting. The chamber is completed by providing a cover plate at the top. The aluminium piston should be carefully fitted so that it does not touch the wall otherwise a serious error will be caused in readings.

### 7.9.2 Fluid Friction Damping

This form of damping is similar to air friction damping. Oil is used in place of air and as the viscosity of oil is greater, the damping force is also correspondingly greater.

A disc is attached to the moving system [Fig. 7.18(*a*)], this disc dips into an oil pot and is completely submerged in oil. When the moving system moves,

7

the disc moves in oil and a frictional drag is produced. This frictional drag always opposes the motion.

In the arrangement shown in Fig. 7.18(*b*), a number of vanes are attached to the spindle. These vanes are submerged in oil and move in a vertical plane. This arrangement gives a greater damping torque.

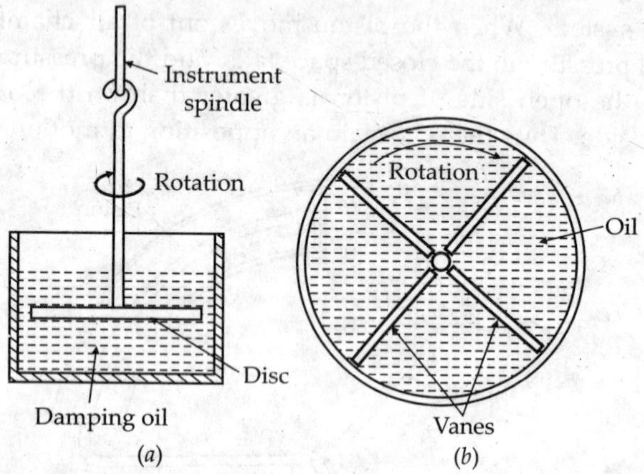


**Fig. 7.18** Fluid Friction Damping.

### 7.9.3 Eddy Current Damping

When a conductor moves in a magnetic field an emf is induced in it and if a closed path is provided, a current (known as eddy current) flows. This current interacts with the magnetic field to produce an electromagnetic torque which opposes the motion. This torque is proportional to the strength of the magnetic field and the current produced. The current is proportional to emf which in turn is proportional to velocity of the conductor. Thus, if the strength of the magnetic field is constant (if it is produced by a permanent magnet), the torque is proportional to velocity of the conductor.

### 7.9.4 Electromagnetic Damping

The movement of a coil in a magnetic field produces a current in the coil which interacts with the magnetic field to produce a torque. This torque opposes the movement of the coil and slows the response. The magnitude of the current and hence the damping torque is dependent upon the resistance of the circuit to which the instrument is connected. The electromagnetic damping is used in galvanometers.

## 7.10 COMPARISON OF METHODS OF DAMPING

Air friction damping provides a very simple and cheap method of damping. But care must be taken to see that the piston is not bent or twisted otherwise it will touch the walls of the air chamber thereby causing serious errors due to solid friction which is thus introduced. This method is used in hot wire and moving iron instruments. Air friction has the advantage, that it does not require the use of a permanent magnet whose introduction may lead to distortion of operating field. Therefore, this type of damping is used in moving iron and dynamometer type of instruments where the operating magnetic field is weak and is likely to be get distorted with the introduction of a permanent magnet.

Fluid friction damping has the advantage that the oil which is required for damping, can be used for insulation purposes in some forms of instruments which are submerged in oil. A vane moving in oil instead of air does not require the same small clearances to give effective damping and therefore this method is suitable for instruments such as electrostatic type where the movement is suspended rather than pivoted. Another advantage of fluid friction damping is that due to the upthrust of oil, the load on bearings or suspension is reduced thus reducing frictional errors.

The disadvantages of fluid friction damping are that it can be used only for instruments which are in vertical position. Also because of creeping of oil, the instruments cannot be kept clean. Hence this type of damping can be used for laboratory type electrostatic instruments and there are obvious difficulties in the way of its application to portable instruments.

Eddy current damping is the most efficient form of damping. It is very convenient to use in instruments where a metallic disc or a former and a permanent magnet already form part of the operating system. For these reasons this method is used in hot wire, moving coil and induction type instruments. This method cannot be used in instruments where introduction of a permanent magnet required for producing eddy currents will distort the existing magnetic field as in moving iron or dynamometer type of instruments.

## 7.11 METHODS OF EDDY CURRENT DAMPING

There are two common forms of damping devices :

(*i*) A metal former which carries the working coil of the instrument.

(*ii*) A thin aluminium disc attached to the moving system of the instrument. This disc moves in the field of a permanent magnet.

### 7.11.1 Eddy Current Damping Torque of Metal Former

Figure 7.19 shows a metallic former moving in the field of a permanent magnet.

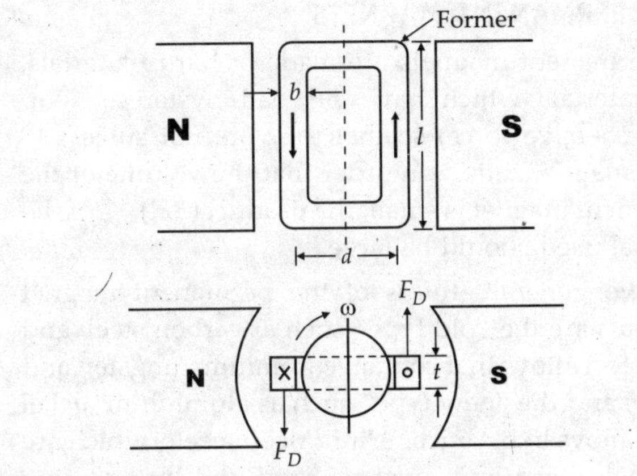


**Fig. 7.19** Eddy current damping by metal former.

Let $B$ = strength of magnetic field ; Wb/m$^2$,

$\omega$ = angular speed of former ; rad/s,

$l$ = length of former ; m,

$t$ = thickness of former ; m,

$b$ = width of former ; m,

$d$ = breadth of former ; m,

$\rho$ = resistivity of material of former, $\Omega$ m.

Linear velocity of former, $v = (d/2)\omega$

Emf generated in the former

$$E_e = 2 \times Bl(d/2)\omega = Bld\omega$$

Resistance of path of eddy currents

$$R_e = \frac{\text{resistivity} \times \text{length}}{\text{area}} = \frac{\rho \times 2(l+d)}{b \times t}$$

$$= \frac{2\rho(l+d)}{bt}$$

Eddy current

$$I_e = \frac{E_e}{R_e} = \frac{Bld\omega}{2\rho(l+d)/bt} = \frac{Blbtd\omega}{2\rho(l+d)}$$

$\therefore$ Damping force

$$F_D = BI_e l = B \times \frac{Blbtd\omega}{2\rho(l+d)} \times l = \frac{B^2l^2btd\omega}{2\rho(l+d)} \quad ...(7.8)$$

Damping torque

$$T_D = F_D \times d = \frac{B^2l^2btd\omega}{2\rho(l+d)} \times d = \frac{B^2l^2d^2bt\omega}{2\rho(l+d)} \quad ...(7.9)$$

Damping constant

$$K_D = \frac{T_D}{\omega} = \frac{B^2l^2d^2bt\omega}{2\rho(l+d)\omega} = \frac{B^2l^2d^2bt}{2\rho(l+d)} \text{ Nm/rad s}^{-1} \quad ...(7.10)$$

It is clear from above that the damping torque can be varied by changing $t$, the thickness of metal former.

### 7.11.2 Damping Torque of a Metal Disc

Let $B$ = flux density of magnetic field ; Wb/m$^2$,

$t$ = thickness of metallic disc ; m,

$b$ = width of permanent magnet ; m,

$d$ = depth of permanent magnet ; m,

$\rho$ = resistivity of material of disc ; $\Omega$m,

$\omega$ = angular speed of disc ; rad/s

It is not possible to attempt an exact solution of the problem and hence only an approximate solution is given.

Considering the emf induced in the portion under the pole face.

Let $R$ = radius measured from centre of pole to centre of disc ; m.

$\therefore$ Emf induced in the portion below the magnet

$$E_e = B/v = BdR\omega$$

Resistance of eddy current path under the pole

$$= \frac{\rho d}{bt}$$

Actual path of eddy current is not limited to the portion of the disc under the magnet but is greater than this.

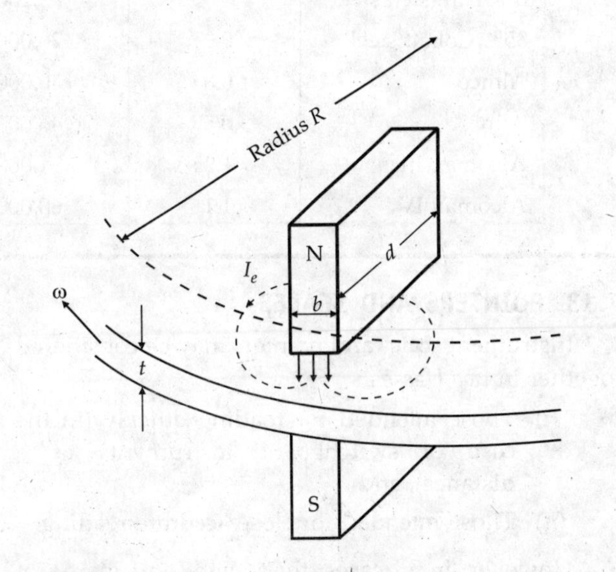


**Fig. 7.20** Eddy current damping by metallic disc.

Therefore to take this factor into account the actual resistance is taken as $K$ times $\rho d / bt$.

$\therefore$ Resistance of eddy current path $R_e = \dfrac{K\rho d}{bt}$

where $K$ = a constant which depends upon radial position of poles.

Eddy current $I_e = \frac{E_e}{R_e} = \frac{BdR\omega}{K\rho d / bt} = \frac{BRbt\omega}{K\rho}$

Damping force $F_D = B \times \frac{BRbt\omega}{K\rho} \times d$

$$= \frac{B^2 Rdbt\omega}{K\rho} \quad \text{...(7.11)}$$

Damping torque $T_D = \frac{B^2 R^2 dbt\omega}{K\rho}$ ...(7.12)

Damping constant

$$K_D = \frac{T_D}{\omega}$$

$$= \frac{B^2 R^2 dbt}{K\rho} \text{ Nm/rad s}^{-1} \quad \text{...(7.13)}$$

From Eqn. 7.12, it is clear that the damping torque can be changed by changing *R*, *i.e.*, the radial position of poles with respect to the centre of the disc.

## 7.12 PERMANENT MAGNETS

Permanent magnets are made of hard materials. *i.e.*, materials which have a broad hysteresis loop (large coercive force) so that they are not subject to self-demagnetization. In order that the volume of the permanent magnet is small, the product $(BH)_{max}$ of the material used should be large.

Two general groups of the permanent magnet materials are the "old type" such as carbon steel, and other steel alloys that contain chromium, tungsten and cobalt ; and the "new type" such as aluminium nickel cobalt alloys like Alnico, Alini, Alcomax etc. which are extremely hard and must be produced by sintering and ceramic magnets of which barium ferrite is the chief constituent. Table 7.2 gives the data for magnetic materials used for permanent magnets.

One of the major requirements of a permanent magnet is that its strength should not change with time, *i.e.*, there should be no ageing effects.

**Table 7.2** *Permanent Magnet Materials*

| *Material* | *Remenance Wb/m²* | *Coercive force A/m* | *Value of B for $(BM)_{max}$* | *Value of H for $(BM)_{max}$* | *(BM)max* |
|---|---|---|---|---|---|
| 10% carbon steel | 0.9 | 4,000 | 0.6 | 2,600 | 1,560 |
| 6% Tungsten steel | 1.05 | 5,200 | 0.7 | 3,750 | 2,625 |
| 35% Cobalt steel | 0.9 | 20,000 | 0.6 | 13,000 | 7,800 |
| Alnico | 0.8 | 40,000 | 0.56 | 24,000 | 13,500 |
| Alini | 0.55 | 46,000 | 0.36 | 28,000 | 10,000 |
| Alcomax III | 1.25 | 54,000 | 0.95 | 40,000 | 38,000 |
| Alcomax IV | 1.1 | 60,000 | 0.8 | 43,000 | 34,000 |

## 7.13 POINTERS AND SCALES

Instrument scales and pointers may be considered together in *two* classes :

(*i*) Those intended for reading quickly (in the case of switchboard instruments at a distance), and

(*ii*) Those intended for close accurate reading.

However, in all cases, the weight and inertia of pointer must be reduced so as to reduce the load on the bearing and to avoid necessity of excessive damping torque.

The pointer motion is limited by buffers or stops to a little more than the scale. These stops are constructed as very light springs so that the pointer is not bent when it strikes them sharply on a sudden overload or reversal of the operating current.

***Switchboard and panel instruments.*** Measuring instruments for switchboards have a square or a rectangular front. The pointers and scales for these instruments should be so that the position of the pointer is conspicuous, the scale markings are legible and there is contrast with the background. Pointers are usually broad or carry broad targets as their ends. Aluminium strip or tube is used for making pointers and '*A*' type of construction is used to give rigidity.

The scale is generally printed on the enamelled surface of a metal plate or on paper or card board pasted to a metal plate. Scale markings are usually broad.

***Portable instruments.*** Portable instruments usually have knife edge pointer moving over scale marked with fine lines, usually not wider than knife edge may be formed by flattening the end of a light aluminium tube or it may be a thin flat strip carried at the end of the pointer. If the pointer and scale are in different planes, observational errors may result on account of parallax. The errors due to parallax arise due to the line of sight being in an oblique plane direction. The parallax errors can be avoided by using knife edge pointers and scales mounted in the same plane as the tip of pointer.

## 7.14 SYMBOLS USED FOR ANALOG INSTRUMENTS

There are a number of symbols used on the dial of indicating instruments. These relate to the nature of measured quantity and the number of measuring elements, safety of the instruments, position of use of instruments, accuracy class of instruments, principle of working of instruments. Some of the extracts of IS 1248 (Part I)-1983 relating to indicating instruments are reproduced below.

### 7.14.1 Nature of Measured Quantity and Number of Measuring Elements

| | |
|---|---|
| Direct current circuit and/or dc responding measuring element | 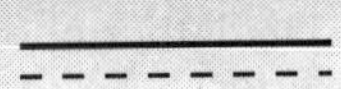 |
| Alternating current circuit and/or ac responding measuring element |  |
| Direct and/or alternating current circuit and/or dc and ac responding measuring element |  |
| Three-phase alternating current circuit (general symbol) |   |
| One measuring element for 3-wire network |   |
| One measuring element for 4-wire network |   |
| Two measuring element for 3-wire network with unbalanced loads |   |
| Two measuring elements (*E*) for 4-wire network with unbalanced loads |   |
| Three measuring elements (*E*) for 4-wire network with unbalanced loads |   |

### 7.14.2 Safety

| | |
|---|---|
| Test voltage 500 V |  |
| Test voltage above 500 V (for example, 2 kV) |   |
| Apparatus not subjected to a voltage test |   |
| High voltage flash |  |
| High voltage on accessory and/or on instrument (for example, voltmeter) | 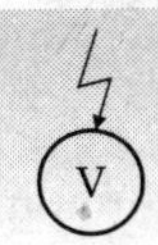  |

### 7.14.3 Position of Use

| | |
|---|---|
| Instrument to be used with the dial vertical |  |
| Instrument to be used with the dial horizontal |  |
| Instrument to be used with the dial inclined (for example 60°) from the horizontal plane | 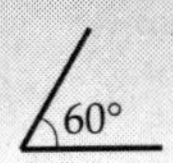  |

### 7.14.4 Accuracy Class

| | |
|---|---|
| Class index (for example, 1.5) with errors expressed as a percentage of the fiducial value except when the latter corresponds to the scale length or the span or the indicated value | 1.5 |
| Class index (for example, 1.5) when the fiducial value corresponds to the scale length |   |
| Class index (for example, 1.5) when the fiducial value corresponds to the indicated value | 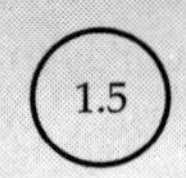  |
| Class index (for example, 1) when the fiducial value corresponds to the span |   |

### 7.14.5 Principle of Operation

| Instrument | Symbol | Instrument | Symbol |
|---|---|---|---|
| Permanent-magnet moving-coil instrument |  | Induction instrument |  |
| Permanent-magnet ratio-meter (quotientmeter) |  | Induction ratiometer (quotientmeter) | 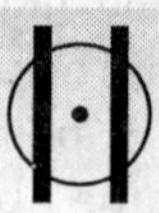 |
| Moving permanent-magnet instrument | 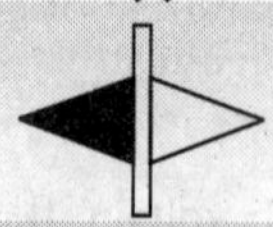 | Bimetallic instrument |  |
| Moving permanent-magnet ratiometer (quotient meter) |  | Electrostatic instrument | 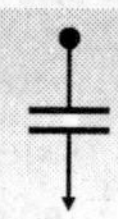 |
| Moving-iron instrument | 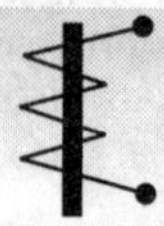 | Vibrating-reed instrument |  |
| Polarized moving-iron instrument | 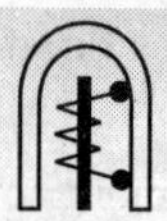 | Non-insulated thermocouple (thermal converter) |  |
| Moving-iron ratiometer | 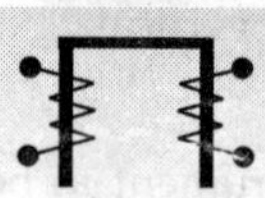 | Insulated thermocouple (thermal converter) | 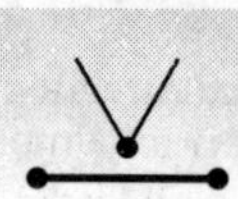 |
| Ironless electro-dynamic instrument |  | Electronic device in a measuring circuit | 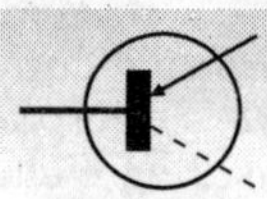 |
| Iron-cored electro-dynamic (ferrodynamic) instrument |  | Electronic device in an auxiliary circuit | 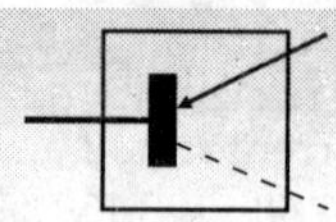 |
| Ironless electro-dynamic ratiometer (quotientmeter) | 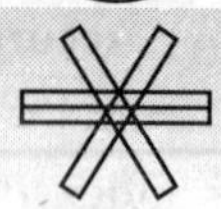 | Rectifier type instrument | 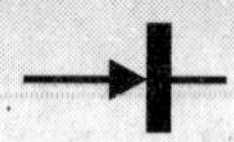 |
| Iron-cored electro-dynamic (ferrodynamic) ratiometer |  | | |

## 7.15 RECORDING INSTRUMENTS

Recording instruments are used at places where a continuous record of circuit conditions is required. This record may be used for future reference or computation work.

The indicating instruments described earlier give a continuous indication of the measurand, but they require the presence of an operator to observe variations in magnitude of readings. With the ever-increasing emphasis on automation, continuously recording instruments, in particular those designed to record permanently the variations in the quantity, are finding many applications.

To cite an example a particular type of recording instrument is obtained by modifying the construction of an indicating instruments. This is done by replacing the pointer and scale of the normal indicating instrument by a light arm carrying an ink pen. The ink pen deflects to give an ink record on a paper chart which is wound at a constant speed by a clockwork mechanism from one roller or drum to another, or by a small synchronous motor run from the mains. The path traced out by the pen gives a continuous record of the variations of deflection of the instrument.

The different types of recording instruments are described in subsequent chapter.

## 7.16 INTEGRATING INSTRUMENT

These instruments are used to measure and register total energy being supplied to a circuit in a given interval of time. This is done irrespective of the rate at which energy is consumed.

A watt-hour meter is an example of Integrating type instruments.

***Watt hour meters.*** Watt hour meters measure the energy supplied to a circuit directly in watt hour. These meters are provided with a circuit which is excited by the voltage and, therefore, the operating torque is in part, due to a current proportional to the supply voltage.

Integrating meters can be classified as :

(*i*) Clock type (*ii*) Motor type.

(*i*) ***Clock meters.*** They consist of a clock mechanism having two pendulums. There are two sets of coils, one energised by the current carried by the load and the other by the voltage across the load. The current coils are fixed but the voltage coils are carried by the pendulums. The pendulums are subjected to magnetic pull when swinging past the fixed coils. This interaction causes one pendulum to accelerate and the other to retard. The resulting difference in time period of the two pendulums is made to give an indication proportional to the energy.

Clock meters are very costly and are seldom used.

(*ii*) ***Motor meters.*** These meters are universally used for measurement of energy. The motor meters consist of three main parts which differ in detail according to the underlying principle of the motoring action utilized, which may be similar to that of a simple a.c. or d.c. motor.

These parts are :

(*i*) *Operating system.* This produces a torque proportional to the current or power and makes the moving system rotate.

(*ii*) *Braking system.* This produces a braking torque proportional to the speed of the rotation and thus causes the meter to run at a steady speed. The speed is such that the braking torque is equal to the operating torque. The braking torque is generally provided by eddy currents induced in a metallic disc moving in the field of a permanent magnet.

(*iii*) *Registering system.* The moving system is carried on a spindle. The spindle has a worm cut out on it and this worm engages a train of wheels with the help of a pinion. These wheels are caused to rotate when the spindle rotates. The spindles of the wheels in the train carry hands which move over the dials which register units, tens, hundreds and so on.

# Review Questions

1. What are the different effects used in producing deflecting torque in an analog instrument ? Cite examples, in which these effects are used.
2. Define the terms "Indicating instruments", "Recording Instruments" and "Integrating Instruments". Give examples of each case.
3. Define the classification between "Direct Measuring Instruments" and "Comparison type Instruments". Give suitable examples for each case.
4. Describe the various operating forces needed for proper operation of an analog indicating instrument.
5. Why is a controlling torque necessary in an analog indicating instrument? What would happen in the absence of a controlling torque ?
6. Describe the different methods of producing controlling torque in an analog indicating instrument. List their advantages and disadvantages.
7. Describe the various forms of supporting the moving system of an analog instrument. Draw neat sketches to support your answer.
8. Explain the significance of ratio of operating torque to weight of moving parts in an analog indicating instrument.

9. What is the use of balancing weights attached to the moving system of an analog indicating instrument ?
10. What are the different methods of damping used in analog indicating instruments ? List their advantages and disadvantages.
11. Sketch the curves showing deflection versus time for analog indicating instruments for under damping, critical damping and overdamping.

    Explain why overdamping is not used in indicating instruments. Why is it preferable to have a slightly underdamped instruments ?
12. Why is it preferable to use materials with a broad hysteresis loop for permanent magnets in instruments ?

    Prove that the energy density in permanent magnets is $\frac{1}{2}BH^2$ where $B$ = flux density in Wb/m$^2$ and $H$ = magnetizing force, A/m.
13. Describe the pointers and scales in indicating analog instruments.
14. Differentiate between Recording and Integrating instruments. Give suitable examples in each case.
15. Describe the major parts of a motor meter. Why is it necessary to have a braking system in a motor meter ?

# Unsolved Problems

1. An instrument spring made of phosphor bronze strip has a length of 400 mm, a width 0.5 mm, and a thickness of 0.08 mm. If the modulus of elasticity of phosphor bronze is taken as 120 GN/m$^2$ ; estimate the torque exerted by the spring for a deflection of 60° and 90°.

   **[Ans** (*a*) $6.7 \times 10^{-6}$ Nm (*b*) $10 \times 10^{-6}$ Nm]
2. The control torque of an indicating instrument is exerted by two flat springs. The full sale torque is $40 \times 10^{-6}$ Nm for a deflection of 90°. The modulus of elasticity and the maximum stress for the spring material are 120 GN/m$^2$ and 20 MN/m$^2$ respectively. Calculate the number of turns required and the thickness of one spring if its width is 0.75 mm and its mean diameter is 11 mm.

   **[Ans** 122, 0.884 mm]
3. A permanent magnet moving coil (PMMC) instrument has a full scale deflection of 90° for a current of 2 A. The deflecting torque in a PMMC ammeter is directly proportional to current in the moving coil. Find the value of current required for a deflection of 30° if the instrument is (*i*) spring controlled (*ii*) gravity controlled.

   **[Ans** (*i*) 0.67 A (*ii*) 1.0 A]

# Objective Type Questions

***Tick (√) the most appropriate answer***

1. A current carrying conductor is shown in Fig. 7.21(*a*). It is brought in a magnetic field shown in Fig. 7.21(*b*).

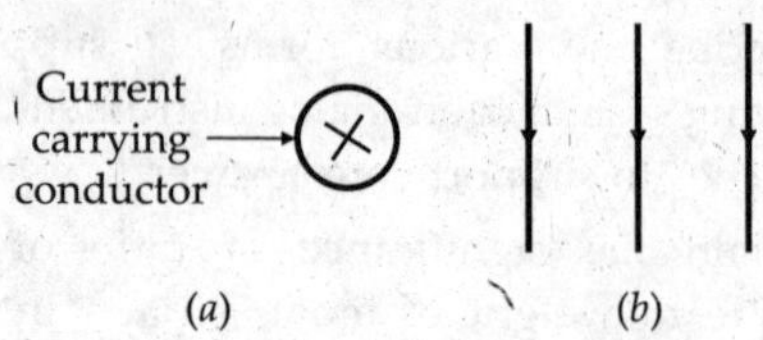


**Fig. 7.21**

   (*a*) It will experience no force
   (*b*) It will experience a force acting from left to right
   (*c*) It will experience a force acting from right to left
   (*d*) It will experience a force acting from top to bottom.
2. A current carrying conductor produces a magnetic field which in turn produces the magnetic polarities in two soft iron pieces as shown in

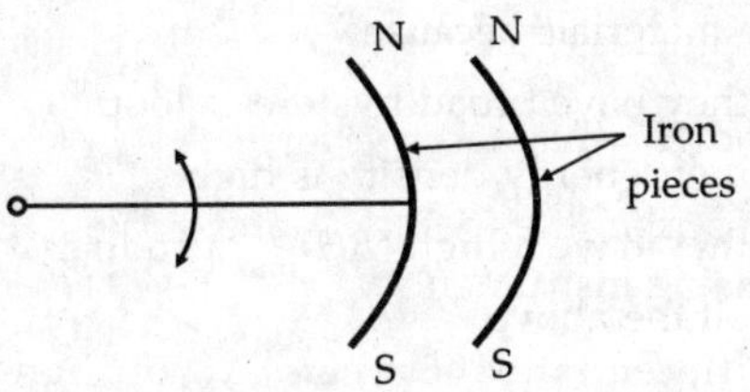


Fig. 7.22

Fig. 7.22. If the inner piece is movable and moves in the clockwise direction in what direction will it move if the direction of current in the current carrying conductor is reversed

(a) it will not move

(b) it will move in the clockwise direction

(c) it will move in the anti-clockwise direction

(d) none of the above.

3. Two coils are connected in series as shown in Fig. 7.23. One of the coils is movable and the other is stationary. With direction of current shown the movable coil moves in clockwise direction. In what direction will the movable coil move if the coils are re-connected as shown in Fig. 7.23. The current is a.c.

(a) clockwise

(b) anticlockwise

(c) does not move

(d) direction of movement is dependent upon polarity.

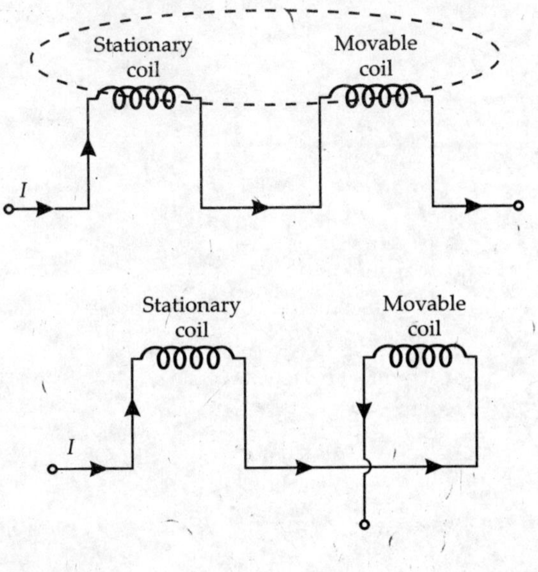


Fig. 7.23

4. If an analog PMMC 0 - 10 A ammeter is provided with no controlling mechanism and the moving parts are free to rotate, what will be the reading of the instrument if a current of 1 A (d.c.) passed through the moving coil (The torque produced is sufficient to overcome the frictional losses)

(a) 1 A

(b) 10 A

(c) the pointer will continuously rotate

(d) the pointer will remain stationary.

5. A spring controlled instrument uses phosphor bronze springs to produce controlling torque. If the ratio of length of spring to thickness of spring is 3000 for a deflection 90°, what should be this ratio if the scale is to extended to 120°

(a) 4000

(b) 2250

(c) 2000

(d) none of the above.

6. A voltage of 200 V produces a deflection of 90° in a PMMC spring controlled instrument. If the same instrument is provided with gravity control, what would be the deflection ?

(a) 90°

(b) 45°

(c) 64.2°

(d) cannot be known from the data given.

7. A spring produces a controlling torque of $16 \times 10^{-6}$ Nm for a deflection of 120°. Suppose its width and length are made two times their original value and the thickness is halved, the value of controlling torque for the same deflection will be

(a) $16 \times 10^{-6}$ Nm

(b) $8 \times 10^{-6}$ Nm

(c) $32 \times 10^{-6}$ Nm

(d) $2 \times 10^{-6}$ Nm

8. Air friction damping should not be used where the deflecting torque in the instrument is produced due to

(a) magnetic field

(b) electrostatic field

(c) thermo-electric emf

(d) none of the above.

9. Fluid friction damping can be used in
   (*a*) horizontally mounted instruments
   (*b*) vertically mounted instruments
   (*c*) both in horizontally and vertically mounted instruments
   (*d*) none of the above.

10. Permanent magnets used in instruments are hard core materials because
   (*a*) they have broad hysteresis loop
   (*b*) their energy density is high
   (*c*) they have a high $(BH)_{max}$ product
   (*d*) all the above.

## Answers

| 1. (*c*) | 2. (*b*) | 3. (*b*) | 4. (*c*) | 5. (*a*) | 6. (*d*) | 7. (*d*) | 8. (*d*) | 9. (*b*) | 10. (*d*) |
|---|---|---|---|---|---|---|---|---|---|

# 8

# Galvanometers

## 8.1 INTRODUCTION

A galvanometer is an instrument used for detecting presence of small currents or voltages in a circuit or for measuring their magnitudes. Galvanometers find their principal application in bridge and potentiometer measurements where their function is to indicate zero current. Therefore, a galvanometer in addition to being sensitive, should have a stable zero, a short periodic time and nearly critical damping.

## 8.2 D'ARSONVAL GALVANOMETER

This instrument is very commonly used in various methods of resistance measurement and also in d.c. potentiometer work.

### 8.2.1 Construction of d'Arsonval Galvanometer

The construction of a d'Arsonval Galvanometer is shown in Fig. 8.1. The description of different parts is given below.

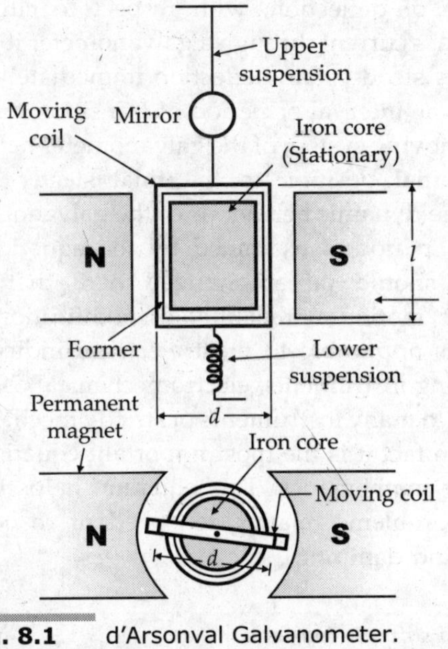


**Fig. 8.1** d'Arsonval Galvanometer.

1. ***Moving coil.*** It is the current carrying element. It is either rectangular or circular in shape and consists of a number of turns of fine wire. This coil is suspended so that it is free to turn about its vertical axis of symmetry. It is arranged in a uniform, radial, horizontal magnetic permanent magnet and iron core. The iron core is spherical in shape if the coil is circular but is cylindrical if the coil is rectangular. The iron core is used to provide a flux path of low reluctance and therefore to produce strong magnetic field for the coil to move in. This increases the deflecting torque and hence the sensitivity of the galvanometer. The length of air gap is about 1.5 mm. In some galvanometers the iron core is omitted resulting in of decreased value of flux density and the coil is made narrower to decrease the air gap. Such a galvanometer is less sensitive, but its moment of inertia is smaller on account of its reduced radius and consequently a short periodic time.

2. ***Damping.*** There is a damping torque present owing to production of eddy currents in the metal former on which the coil is mounted. Damping is also obtained by connecting a low resistance across the galvanometer terminals. Damping torque depends upon the resistance and we can obtain critical damping by adjusting the value of resistance as explained later in this chapter.

3. ***Suspension.*** The coil is supported by a flat ribbon suspension which also carries current to the coil. The other current connection in a sensitive galvanometer is a coiled wire. This is called the lower suspension and has a negligible torque effect. This type of galvanometer must be levelled carefully so that the coil hangs straight and centrally without rubbing the poles or the soft iron cylinder. Some portable galvanometers which do not require exact levelling have "*taut suspensions*" consisting of straight flat strips kept under tension for at the both top and at the bottom.

The upper suspension consists of gold or copper wire of nearly 0.0125 or 0.025 mm diameter rolled into the form of a ribbon. This is not very strong mechanically, so that the galvanometers must be handled carefully without jerks. Sensitive galvanometers are provided with coil clamps to take the strain from suspension while the galvanometer is being moved.

4. ***Indication.*** The suspension carries a small mirror upon which a beam of light is cast. The beam of light is reflected on to a scale upon which the deflection is measured. This scale is usually about 1 metre away from the instrument, although 1/2 metre may be used for greater compactness.

5. ***Zero setting.*** A torsion head is provided for adjusting the position of the coil and also for zero setting.

## 8.4 TORQUE EQUATION

Figure 8.1 shows the quantities that enter the torque equation of a galvanometer.

Let,

$l, d$ = length of respectively vertical and horizontal side (width) of coil, m,

$N$ = number of turns in the coil,

$B$ = flux density in the air gap, Wb/m$^2$,

$i$ = current through moving coil, A,

$K$ = spring* constant of suspension, Nm/rad,

$\theta_F$ = final steady state deflection of moving coil, rad.

Force on each side of coil = $N. Bil \sin \alpha$

where $\alpha$ = angle between direction of magnetic field and the conductor.

The field is radial and, therefore, $\alpha = 90°$

Hence, force on each side of coil $= NBil$ ...(8.1)

Deflecting torque $T_d$ = force × distance

$= NBild$ ...(8.2)

$= NBAi$ ...(8.3)

where $A = ld$ = area of coil, m$^2$

$N, B, A$ are constants for a galvanometer.

$\therefore$ Deflecting torque, $T_d = Gi$ ...(8.4)

where $G = NBA = NBld$ ...(8.5)

$G$ is called the *displacement constant* of the galvanometer.

Controlling torque exerted by the suspension at deflection $\theta_F$, is $T_c = K\theta_F$.

For final steady deflection,

$$T_c = T_d$$

or $$K\theta_F = Gi$$

Hence, final steady deflection

$$\theta_F = Gi/K \quad ...(8.6)$$

If the deflection is measured on a mm scale kept 1 metre away and if $\theta_F$ is small, the deflection $d$ (Fig. 8.2) is given by :

$d = 0d$ = radius × angle turned by the reflected beam

$$= 1000 \times 2\,\theta_F = 2000\; Gi/K \text{ mm} \quad ...(8.7)$$

This is because if the deflection is $\theta_F$, the mirror turns through an angle $\theta_F$ while the reflected beam turns through an angle $2\theta_F$.

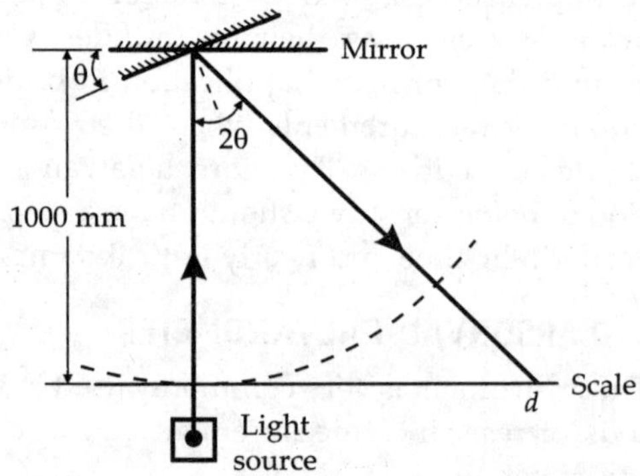


**Fig. 8.2** Measurement of deflection with lamp and scale arrangement.

## 8.5 DYNAMIC BEHAVIOUR OF GALVANOMETERS

We have considered so far the steady state relationship of deflection, with respect to current. When we pass current through a galvanometer it does not reach its steady state deflection immediately but there is a time interval or period of transition during which the moving system of the galvanometer deflects from its initial position to the final steady state position. The dynamic behaviour of the galvanometer during this period is examined by the equation of motion. It should be emphasized here that this equation of motion with suitable modifications of constants, is applicable to most electrical indicating and recording instruments, electromechanical oscillographs and to many instruments of strictly mechanical actuation. In fact it is the most important equation in engineering mathematics. This equation helps us to study the problems relating to speed of response, overshoot and damping.

* K is known by various names like control constant, restoring constant and stiffness constant.

Let us first consider the constants of a galvanometer. These constants are known as **"Intrinsic Constants"**.

(*i*) ***Displacement constant.*** The deflecting torque is given by $T_d = Gi$ (See Eqn. 8.4) where $G$ is the *displacement constant* of the galvanometer and is equal to $NBld$. The units of $G$ are Nm/A.

(*ii*) ***Inertia constant.*** A retarding torque is produced owing to inertia of moving system. This torque is dependent upon the moment of inertia of moving system and the angular acceleration.

Inertia torque $T_j = J(d^2\theta/dt^2)$ ...(8.8)

where $J$ = moment of inertia of moving system about the axis of rotation; kg-m$^2$,

$\theta$ = deflection at any time $t$ ; rad.

$J$ is also called the *"inertia constant"*.

(*iii*) ***Damping constant.*** Damping is provided by the friction due to motion of the coil in air and also by induced electrical effects if a closed circuit is provided. Damping torque is assumed to be proportional to velocity of the moving system. This is true for electromagnetic damping and is at least a fair approximation for air damping (which is usually a small part of the total). We shall study damping action in greater details in Art. 8.12, but for the present we will lump the two effects together.

Damping torque, $T_D = D(d\theta/dt)$ ...(8.9)

where $D^*$ is damping constant ; Nm/rad s$^{-1}$.

(*iv*) ***Control constant.*** A controlling torque is produced due to elasticity of the system which tries to restore the moving system back to its original position.

Controlling torque, $T_C = K\theta$,

where $K$ is *control constant*, N-m/rad.

### 8.5.1 Equation of Motion

There are four torques acting on the moving system. Deflecting torque, $T_d$, tries to accelerate the system while inertia torque, $T$ damping torque, $T_D$ and the control torque, $T_C$, try to retard the system (See Fig. 8.3).

Therefore, for any deflection $\theta$ at any instant $t$

$$T_j + T_D + T_C = T_d$$

or $$J\frac{d^2\theta}{dt^2} + D\frac{d\theta}{dt} + K\theta = Gi \qquad ...(8.10)$$

We have now a linear, second order differential equation whose solution is the sum of a *"complementary function"* representing a transient condition and a *"particular integral"* representing the steady state condition.

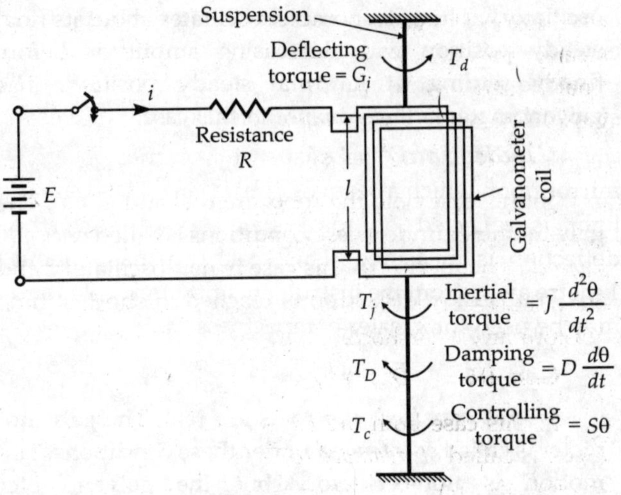


**Fig. 8.3** Torques acting on the moving system of a galvanometer.

***Complementary function.*** The auxiliary equation is

$$Jm^2 + Dm + K = 0.$$

The roots of this equation are :

$$m_1 = \frac{-D + \sqrt{D^2 - 4KJ}}{2J} \quad \text{and} \quad m_2 = \frac{-D - \sqrt{D^2 - 4KJ}}{2J}$$

Thus $\theta = A\exp^{(m_1 t)} + B\exp^{(m_2 t)}$ ...(8.11)

where $A$ and B are constants.

***Particular integral.*** We pass a steady current $i$ through the galvanometer.

Under steady state conditions

$$d^2\theta/dt^2 = 0,\ d\theta/dt = 0,\ \text{and}\ \theta = \theta_F.$$

Putting the above conditions in Eqn. 8.10, the final steady state deflection is

$$\theta_F = Gi/K \qquad ...(8.12)$$

Thus, the complete solution of differential equation is

$$\theta = A\exp(m_1 t) + B\exp(m_2 t) + \theta_F \qquad ...(8.13)$$

Now $\theta_F$ is the final steady deflection and the term $(Ae^{m_1 t} + Be^{m_2 t})$ represents a motion which may or may not be oscillatory. We can tell the type of behaviour from the form of the roots $m_1$ and $m_2$. There are three possible cases.

* The symbol for damping constant used in this chapter is $D$ whereas in chapter 4, the symbol used was $B$. This has been necessitated because in this chapter symbol $B$ is used for flux density.

**Case I** $D^2 < 4KJ$

For this case, the two roots $m_1$ and $m_2$ are imaginary. Thus under these conditions the motion is oscillatory. The galvanometer oscillates about its final steady position with decreasing amplitude before finally settling at its final steady position. The galvanometer is *underdamped* in this case.

**Case II** $D^2 = 4KJ$

In this case both the roots are real and equal. The galvanometer under these conditions is called *critically damped*. The motion in this case is non-oscillatory and the final steady deflection is reached in shortest time without any overshoot.

**Case III** $D^2 > 4KJ$

In this case both the roots are real. The galvanometer is called *overdamped* under these conditions. The motion is non-oscillatory but the galvanometer reaches its final steady position in a sluggish manner.

### 8.5.2 Underdamped Motion of a Galvanometer

The motion of a galvanometer is underdamped when $D^2 < 4KJ$.

Under these conditions both the roots $m_1$ and $m_2$ are imaginary and may be written as :

$$m_1, m_2 = \frac{-D \pm \sqrt{D^2 - 4KJ}}{2J} = \frac{-D \pm \sqrt{(-1)^2 \times 4KJ - D^2}}{2J}$$

$$= \frac{-D}{2J} + j\frac{\sqrt{4KJ - D^2}}{2J} = -\alpha + j\omega$$

and $m_2 = -\frac{-D}{2J} - j\frac{\sqrt{4KJ - D^2}}{2J} = -\alpha - j\omega.$

Putting the above values in Eqn. 8.13, we get :

$$\theta = A[\exp(-\alpha + j\omega)t] + B[\exp(-\alpha - j\omega)t] + \theta_F$$

$$= \exp(-\alpha t)[A\exp(j\omega t) + B\exp(-j\omega t)] + \theta_F.$$

Since $e^{\pm j\omega t}$ is complex and $\theta$ is real as it represents a physical quantity, it automatically follows that $A$ and $B$ must be complex.

Let $A = a + jb$ and $B = c + jd$

$$\therefore\ \theta = e^{-\alpha t}[(a + jb)\exp(j\omega t) + (c + jd)\exp(-j\omega t)] + \theta_F$$

$$= e^{-\alpha t}[(a + jb)(\cos\omega t + j\sin\omega t) + (c + jd)(\cos\omega t - j\sin\omega t)] + \theta_F$$

$$= e^{-\alpha t}[(a+c)\cos\omega t - (b+d)\sin\omega t + j(b+d)\cos\omega t + j(a-c)\sin\omega t] + \theta_F \quad ...(8.14)$$

Since $\theta$ is real and therefore the imaginary term of Eqn. 8.14 must be equal to zero for all values of $t$.

Hence, for all values of $t$ :

$$(b + d)\cos\omega t + (a - c)\sin\omega t = 0.$$

At $\omega t = 0$, $b + d = 0$ or $b = -d$

At $\omega t = \pi/2$ $a - c = 0$ or $a = c$

$\therefore$ $A = a + jb$ and $B = a - jb.$

Thus we conclude that $A$ and $B$ are complex conjugate pair.

We can therefore, write Eqn. 8.14 as

$$\theta = 2\exp(-\alpha t)[a\cos\omega t + d\sin\omega t] + \theta_F \quad ...(8.15)$$

Let $a = (F/2)\sin\phi$ and $d = (F/2)\cos\phi$

$\therefore\ F = 2\sqrt{(a^2 + d^2)}$ and $\phi = \tan^{-1}\phi$

Hence we can write

$$\theta = F\exp(-\alpha t)\sin\phi\cos\omega t + \cos\phi\sin\omega t] + \theta_F$$

$$= F\exp(-\alpha t)\sin(\omega t + \phi) + \theta_F \quad ...(8.16)$$

$$= F\exp(-Dt/2J)[\sin(\omega_d t + \phi)] + \theta_F \quad ...(8.17)$$

where $\omega_d$ = angular frequency of damped oscillation

$$= \frac{\sqrt{4KJ - D^2}}{2J}\ \text{rad/s} \quad ...(8.18)$$

Let us evaluate $\theta_F$ and $\phi$. At $t = 0$, $\theta = 0$.

Putting this condition in Eqn. 8.17, we have

$$0 = F\sin\phi + \theta_F$$

or $\sin\phi = -(\theta_F / F)$ ...(*i*)

Since $\theta_F$ and $\phi$ are positive quantities, $F$ is a negative quantity. Differentiating Eqn. 8.17, we have :

$$\frac{d\theta}{dt} = \exp(-D/2J)F\exp(-Dt/2J)[\sin(\omega_d t + \phi)] + F\exp(-Dt/2J)[\cos(\omega_d t + \phi)]\omega_d$$

At $t = 0$, $d\theta/dt = 0$

Substituting this in the above relationship

$$0 = \frac{-D}{2J}F\sin\phi + F\omega_d\cos\phi \quad ...(ii)$$

$$\tan\phi = \omega_d \cdot \frac{2J}{D} \quad ...(8.19)$$

$$= \frac{\sqrt{4KJ - D^2}}{2J} \cdot \frac{2J}{D} = \frac{\sqrt{4KJ - D^2}}{D} \quad ...(8.20)$$

Substituting the value of $\sin\phi$ in (*ii*), we have,

$$0 = \left(\frac{-D}{2J}F\right) \times \left(-\frac{\theta_F}{F}\right) + F\omega_d\cos\phi$$

or $\cos\phi = -\frac{D}{2J\omega_d} \cdot \frac{\theta_F}{F}$ ...(*iii*)

From (*i*) and (*iii*), we get

$$\left(-\frac{\theta_F}{F}\right)^2+\left(-\frac{D}{2J\omega_d}\cdot\frac{\theta_F}{F}\right)^2=1$$

or $$F=-\theta_F\left[\frac{4J^2\omega_d^2+D^2}{4J^2\omega_d^2}\right]^{1/2} \quad \text{...(iv)}$$

Now $$\omega_d=\frac{\sqrt{4KJ-D^2}}{2J}$$

$$\therefore \quad F=-\theta_F\frac{2\sqrt{KJ}}{\sqrt{4KJ-D^2}} \quad \text{...(8.21)}$$

Substituting the above values of $F$ and $\phi$ in Eqn. 8.17,

$$\theta=\theta_F\left[1-\frac{2\sqrt{KJ}}{\sqrt{2KJ-D^2}}\exp(-Dt/2J)\sin\left(\omega_d t+\tan^{-1}\frac{2J\omega_d}{D}\right)\right] \quad \text{...(8.22)}$$

$$=\theta_F\left[1-\frac{2\sqrt{KJ}}{\sqrt{2KJ-D^2}}\exp(-Dt/2J)\sin\left(\frac{\sqrt{4KJ-D^2}}{2J}t+\tan^{-1}\frac{\sqrt{4KJ-D^2}}{D}\right)\right] \quad \text{...(8.23)}$$

When a current is suddenly passed through the coil of an underdamped galvanometer, the moving system will start from its zero current position and then oscillate about its final steady state position $\theta_F$. This oscillation would be an attenuated sinusoidal motion. The angular frequency of the sinusoidal component of this motion is $\omega_d$. The frequency of this sinusoidal component (called frequency of damped oscillations) is

$$f_d=\frac{\omega_d}{2\pi}=\frac{1}{2\pi}\cdot\frac{\sqrt{2KJ-D^2}}{2J} \quad \text{...(8.24)}$$

and the time period is :

$$T_d=\frac{1}{f_d}=\frac{2\pi}{\omega_d}=2\pi.\frac{2J}{\sqrt{4KJ-D^2}} \quad \text{...(8.25)}$$

### 8.5.2 Undamped Motion of a Galvanometer

The undamped motion of a galvanometer is obtained when there are no damping forces *i.e.*, when $D=0$. Such a case is not possible under practical working conditions, but the properties of the undamped galvanometer are used in expressing its motion under actual operating conditions.

From Eqns. 8.18 and 8.21 for $D=0$, we get :

Angular frequency, $\omega_n=\sqrt{K/J}$ ...(8.26)

and $F=-\theta_F$ ...(8.27)

The frequency of undamped (called natural or free oscillation) is :

$$f_n=\frac{\omega_n}{2\pi}=\frac{1}{2\pi}\sqrt{\frac{K}{J}} \quad \text{...(8.28)}$$

and the free period of oscillations is

$$T_0=1/f_n=2\pi\sqrt{J/K} \quad \text{...(8.29)}$$

and $\phi_0=\tan^{-1}\infty=90°$ (See Eqn. 8.20)

Substituting these values in Eqn. 8.17, we have :

$$\theta=-\theta_F[\sin(\omega_n t+90°)]+\theta_F=\theta_F(1-\cos\omega_n t) \quad \text{...(8.30)}$$

Thus the motion of the undamped galvanometer is an oscillation around $\theta_F$ with a constant amplitude $\theta_F$ and a frequency $f_n$.

### 8.5.3 Critically Damped Motion of a Galvanometer

The critical damped motion of a galvanometer is obtained when $D^2=4KJ$.

Under this condition the roots $m_1$, and $m_2$ of the auxiliary equation are equal $m_1=m_2=-D/2J$.

The solution in this case is

$$\theta=\theta_F+\exp(-Dt/2J)[A+Bt] \quad \text{...(8.31)}$$

where $A$ and $B$ are constants. The values of $A$ and $B$ can be found by differentiating Eqn. 8.31.

$$\frac{d\theta}{dt}=\frac{-D}{2J}\exp(-Dt/2J)(A+Bt)+B\exp(-Dt/2j) \quad \text{...(8.32)}$$

At $t=0$, $\theta=0$ and $d\theta/dt=0$.

Putting these conditions in Eqns. 8.31 and 8.32,

$$0=\theta_F+A \text{ and } 0=-(D/2J)A+B$$

$$A=-\theta_F \quad B=-(D/2J)\theta_F$$

The solution is

$$\theta=\theta_F[1-\exp(-Dt/2J)(1+Dt/2J)] \quad \text{...(8.33)}$$

Now for critical damping

$$D=D_c=2\sqrt{KJ} \quad \text{...(8.34)}$$

where $D_c$ = damping constant for critical damping.

Under critical damping conditions

$$D/2J=2\sqrt{KJ/2J}=\sqrt{K/J}=\omega_n \quad \text{...(8.35)}$$

∴ For a critically damped galvanometer :

$$\theta=\theta_F[1-\exp(-\omega_n t)(1+\omega_n t)] \quad \text{...(8.36)}$$

For critical damping the pointer quickly moves to its final steady position without any oscillations. The curve for critical damping is shown in Fig. 7.16. It is clear from Fig 7.16 that the time taken by the coil to reach its final steady position is smaller in the case of critical damping than that for over-damping.

8

### 8.5.4 Overdamped Motion of Galvanometer

The overdamped motion of a galvanometer is obtained when $D^2 > 4KJ$. In this case the roots $m_1$ and $m_2$ are real and unequal. The complete solution of the equation of motion is :

$$\theta = \theta_F \left[ 1 + \frac{\xi + \sqrt{\zeta^2 - 1}}{2\sqrt{\zeta^2 - 1}} \exp\left\{ -\omega_n t(\zeta - \sqrt{\zeta^2 - 1}\,) \right\} - \frac{\zeta - \sqrt{\zeta^2 - 1}}{2\sqrt{\zeta^2 - 1}} \exp\left\{ -\omega_n (\zeta + \sqrt{\zeta^2 - 1}\,) \right\} \right] \quad ...(8.37)$$

The above expression represents a decaying motion without oscillations or overshoot. However, this motion is usually slow and is not desirable in indicating instruments. The value of $\xi = D/D_c$.

## 8.6 RESPONSE OF GALVANOMETERS

If we examine Fig. 7.16, we find that an underdamped galvanometer causes a great waste of time as the coil settles to its final steady position after oscillating about it. An overdamped galvanometer represents a very sluggish behaviour and loss of time. A critically damped galvanometer represents an arrangement that takes the coil to its final steady position in the minimum time that is possible without overshooting. The best operating conditions are attained with a galvanometer that is either critically damped, or slightly-underdamped which may even give a faster response.

If the damping is slightly less than the critical value the galvanometer swings a little beyond final position on the first swing, but comes to rest soon.

## 8.7 OPERATIONAL CONSTANTS

We have derived the galvanometer equations in terms of the intrinsic constants, $J$, $D$ and $K$. These are basic physical constants and are important for a designer. But unfortunately, to a user, these are not known and also their evaluation is difficult. The user is interested in the so called *"operational constants"*. These are the sensitivity, the critical damping resistance and the time period.

## 8.8 RELATIVE DAMPING

The damping of a galvanometer is expressed most conveniently with critical damping case. The damping ratio is defined as the ratio of the actual damping constant to the damping constant required for critical damping.

$\therefore$ Damping ratio $\zeta = D/D_c$ ...(8.38)

But $D_c = 2\sqrt{KJ}$

$\therefore$ $\zeta = D/2\sqrt{KJ}$ ...(8.39)

Now
$$\frac{D}{2J} = \frac{D}{2\sqrt{KJ}} \cdot \sqrt{\frac{K}{J}} = \zeta\omega_n \quad ...(8.40)$$

$$\tan\phi = \frac{\sqrt{4KJ - D^2}}{D} = \sqrt{\left(\frac{2\sqrt{KJ}}{D}\right)^2 - 1} = \frac{\sqrt{1-\zeta^2}}{\zeta} \quad ...(8.41)$$

Similarly, $\sin\theta = \sqrt{1-\zeta^2}$ and $\cos\phi = \zeta$

Now
$$\omega_d = \frac{\sqrt{4KJ - D^2}}{2J} = \sqrt{\frac{K}{J} - \frac{D^2}{4J^2}}$$
$$= \sqrt{\omega_n^2 - \zeta^2\omega_n^2}$$
$$= \omega_n\sqrt{1-\zeta^2} \quad ...(8.42)$$

Substituting the values of different quantities listed above, we can write Eqn. 8.23 as :

$$\theta = \theta_F \left[ 1 - \frac{\omega_n}{\omega_d} \exp(-\xi\omega_n t)\sin(\omega_d t + \phi) \right] \quad ...(8.43)$$

$$= \theta_F \left[ 1 - \frac{1}{\sqrt{1-\zeta^2}} \exp(-\xi\omega_n t) \sin\left( \omega_d t + \tan^{-1}\frac{\sqrt{1-\zeta^2}}{\zeta} \right) \right] \quad ...(8.44)$$

$$= \theta_F \left[ 1 - \frac{1}{\sqrt{1-\zeta^2}} \exp(-\xi\omega_n t) \sin\left( \omega_d t + \sin^{-1}\sqrt{1-\zeta^2} \right) \right] \quad ...(8.45)$$

Now $T_0/T_d = \omega_d/\omega_n$

But $\omega_d/\omega_n = \sqrt{1-\zeta^2}$

$\therefore$ $T_0/T_d = \sqrt{1-\zeta^2}$ ...(8.46)

Also, $\zeta\omega_n = \dfrac{2\pi\zeta}{T_0}$

Substituting the above values in Eqn. 8.44, we get

$$= \theta_F \left[ 1 - \frac{1}{\sqrt{1-\zeta^2}} \exp(-2\pi\xi t/T_0) \sin\left( \frac{2\pi}{T_0}\sqrt{1-\zeta^2}\,t + \sin^{-1}\sqrt{1-\zeta^2} \right) \right] \quad ...(8.47)$$

$$= \theta_F \left[ 1 - \frac{T_d}{T_0} \exp(-2\pi\xi t/T_0) \sin(2\pi t/T_d + \sin^{-1} T_0/T_1) \right] \quad ...(8.48)$$

It may be noted here that this equation describes the galvanometer motion in terms of operational constants, *i.e.*, relative damping, free period, and sensitivity (the last being implied in $\theta_F$).

## 8.9 LOGARITHMIC DECREMENT

The time required for the deflection to reach a maximum value may be obtained by putting $d\theta / dt = 0$. Differentiating Eqn. 8.43 and putting it equal to zero, we get :

$$\tan\left(\omega_n t + \tan^{-1}\frac{\sqrt{1-\zeta^2}}{\zeta}\right) = \frac{\sqrt{1-\zeta^2}}{\zeta}$$

The above can be true only if $\omega_d t = N\pi$ where $N$ is an integer.

If $N$ is an even integer, we get a minimum value of deflection since the initial value $t = 0$ represents a minimum. When $N$ is an odd integer $t$ corresponds to a maximum value.

The first maximum value of deflection occurs at $t_1$, for $N = 1$.

$\therefore$ We have $\omega_d t_1 = \pi$ or $t_1 = \dfrac{\pi}{\omega_d} = \dfrac{\pi}{\omega_n\sqrt{1-\zeta^2}}$

Substituting this value in Eqn. 8.44, the first maximum deflection is :

$$\theta_1 = \theta_F\left[1 - \frac{1}{\sqrt{1-\zeta^2}}\exp(-\pi\xi\sqrt{1-\xi^2})\right.$$

$$\left.\sin(\pi + \sin^{-1}\sqrt{1-\zeta^2})\right]$$

$$= \theta_F[1 + \exp(-\pi\xi/\sqrt{1-\xi^2})] \qquad ...(8.49)$$

or $\theta_1 - \theta_F = \theta_F \exp(-\pi\xi/\sqrt{1-\xi^2})$ ...(8.50)

Equation 8.50 gives the first over shoot.

Now, the first minimum deflection occurs at $t - t_2$ where $t_2 = 2\pi / \omega_d$.

Substituting this value of $t$ in Eqn. 8.45, we get,

$$\theta_2 = \theta_F[1 - \exp(-2\pi\xi/\sqrt{1-\xi^2})] \qquad ...(8.51)$$

The first undershoot is given by :

$$\theta_F - \theta_2 = \exp(-2\pi\xi/\sqrt{1-\xi^2}) \qquad ...(8.52)$$

Taking the ratio of successive extreme deviations from the final steady deflections we have :

$$\frac{\theta_1 - \theta_F}{\theta_F - \theta_2} = \frac{\exp(-\pi\xi/\sqrt{1-\xi^2})}{\exp(-2\pi\xi/\sqrt{1-\xi^2})}$$

$$= \exp(\pi\xi\sqrt{1-\xi^2})$$

From which we get $\log_e\left(\dfrac{\theta_1 - \theta_F}{\theta_F - \theta_2}\right) = \dfrac{\pi\zeta}{\sqrt{1-\zeta^2}}$

*Logarithmic Decrement* is defined as the Naperian logarithm of the ratio of successive swings.

$$\text{Logarithmic decrement } \lambda = \log_e\left(\frac{\theta_1 - \theta_F}{\theta_F - \theta_2}\right)$$

$$= -\log_e\left(\frac{\theta_F - \theta_2}{\theta_1 - \theta_F}\right) = \frac{\pi\zeta}{\sqrt{1-\zeta^2}} \qquad ...(8.53)$$

But $T_0 / T_d = \sqrt{1-\zeta^2}$ $\therefore$ $\lambda = \pi\zeta(T_d / T_0)$ ...(8.54)

Equation 8.48 may be expressed as

$$\theta_0 = \theta_F\left[1 - \frac{T_d}{T_0}\exp(-2\lambda t / T_d)\left\{\sin\left(\frac{2\pi t}{T_d} + \sin^{-1}\frac{\pi\zeta}{\lambda}\right)\right\}\right] \qquad ...(8.55)$$

$$= \theta_F\left[1 - \frac{\omega_n}{\omega_d}\exp(-\omega_d\lambda/\pi)\sin\left(\omega_d t + \sin^{-1}\frac{\pi\zeta}{\lambda}\right)\right] \qquad ...(8.56)$$

## 8.10 OVERSHOOT

Figure 8.4 shows an expanded scale the upper portions of curves computed from Eqn. 8.48 for $\zeta = 0.6$ and 0.8. In order to have comparison the critical curve $\zeta = 1$ is also drawn. The curve with $\zeta = 0.8$ has a very small overshoot but curve with $\zeta =$ 0.6 has an

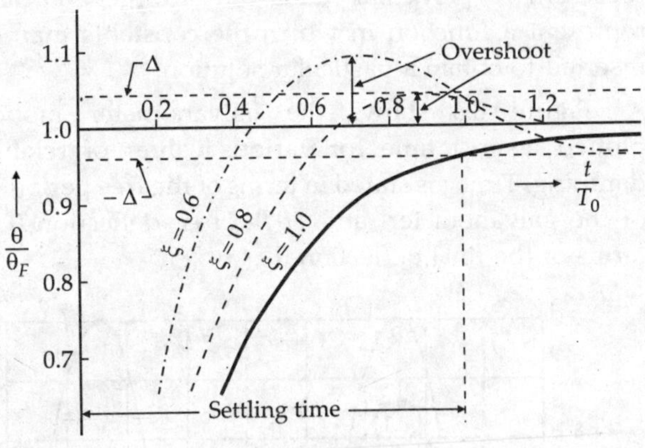


**Fig. 8.4** Overshoot and settling time.

overshoot of about 10 per cent. Actually we are interested in the time taken by the galvanometer to settle to its final position but this "time of deflection" is used in a loose way as in fact the deflection only approaches the final value asymptotically and never reaches it. If we assume that we are interested in reading the galvanometer to a precision Δ, we require only that the deflection arrives at and remains within a band $\theta_F \pm \Delta$. A band of $\pm 1$ percent is sometimes indicated. However, $\pm 2$ per cent is used in Fig. 8.4 in order to get a clear idea. The 0.8 curve enters the band at $t = 0.94\ T_0$ and stays within it while the critical curve enters the band at $t = 0.94\ T_0$. The $\zeta = 0.6$ curve cuts across both limits quite early but finally enters the band at $t = 0.94\ T_0$. The time it takes the galvanometer to reach $\theta_F \pm \Delta$ and to remain thereafter in the $\pm\Delta$ band is known as the *settling time*. If $\pm$ 1 per cent band had been used, the three curves would have nearly the same entry points. Thus it is clear that the time for deflection is small for any value of damping with $\zeta = 0.6$ to 1.0, and therefore, any damping arrangement (between $\zeta = 0.6$ to 1.0) could be selected by the operator.

Some underdamping, with $\zeta = 0.6$ to 1.0 from 0.6 to 0.8, actually has advantage over the critical condition as a small overshoot indicates to the operator that the coil is moving freely without touching the poles. Underdamping is done for all indicating instruments and the small overshoot is considered to be an advantage as it is an assurance of free movement of the metering element and good conditions of the pivots and jewels.

## 8.11 NON-DIMENSIONAL CURVES OF A GALVANOMETER MOTION

It is often advantageous to plot curves in terms of non-dimensional functions, as these curves do not represent a particular data but are general and represent a function in which the constants may be inserted to obtain a particular solution.

Figure 8.5 shows the galvanometer motion plotted against time for various values of relative damping. Time $t$ is stated in terms of the free period $T_0$ of the galvanometer, and of the free deflection $\theta$ in terms of the final deflection $\theta_F$.

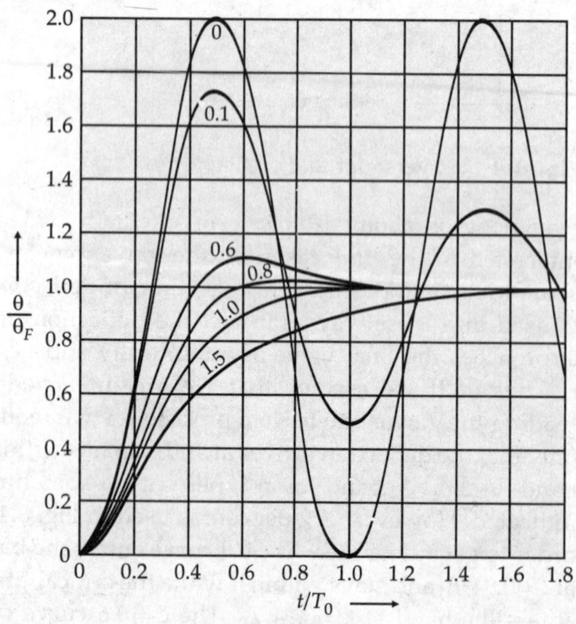


**Fig. 8.5** Non-dimensional curves of a galvanometer.

An inspection of these curves shows that :

(*i*) The actual time period, (*i.e.*, period of damped oscillations) increases slowly for damped vibrations and then more rapidly as critical damping is approached.

(*ii*) The oscillations persist for a considerable time for small values of damping but this time quickly becomes negligible as damping approaches critical value ; and

(*iii*) For damping greater than critical, the motion becomes sluggish. Thus it is clear that for overdamping ($\zeta > 1$) the time taken to reach the final steady deflection is unnecessarily long.

## 8.12 DAMPING

Damping means dissipation of energy of rotation. This dissipation of energy is due partly to mechanical effects, and usually to a greater extent, due to electromagnetic effects from the coil circuit.

(*a*) ***Mechanical damping.*** Mechanical damping is mostly due to friction produced owing to motion of coil in the air surrounding it. A small dissipation of energy is associated with flexing of suspension (elastic hysteresis).

Damping torque due to mechanical effects is

$$T_m = D_m\,(d\theta / dt) \qquad \text{...(8.57)}$$

where $D_m$ = damping constant due to mechanical damping.

(*b*) ***Electromagnetic damping.*** Electromagnetic damping is produced by the induced effects when the coil moves in the magnetic field and a closed path is provided for the currents to flow. The electromagnetic damping is because of :

(*i*) eddy currents produced in the metal former ;

(*ii*) current circulated in the coil circuit by emf generated in the coil when it rotates.

The damping may be classified as :

*Open circuit damping.* The damping effects produced by friction, elastic hysteresis, and eddy currents in metal former are always present whether the galvanometer circuit is closed or not and, therefore, the damping produced by these effects is called open circuit damping.

*Damping due to coil circuit resistance.* The damping effect produced by current flowing in the coil is only present when circuit of the coil is closed.

Let $R$ = resistance of galvanometer circuit when closed

$= R_g + R_e.$

$R_g$ = resistance of galvanometer coil,

$R_e$ = external resistance required for damping.

The voltage induced in the coil by its motion

$$e = 2N \times Blv = 2N \times Bl.(d/2)\omega$$

$$= NBld\,.(d\theta/dt)$$

The current flowing in the circuit due to emf $e$ is :

$$i = \frac{e}{R} = \frac{NBld}{R} \cdot \frac{d\theta}{dt}$$

∴ Torque produced owing to current flowing in the coil.

$T_{coil}$ = conductors × force on each conductor × distance

$$= N \times Bil \times d = NBld \times i = NBld \times \frac{NBld}{R} \cdot \frac{d\theta}{dt}$$

$$= \frac{(NBld)^2}{R}\left(\frac{d\theta}{dt}\right)$$

$$= (G^2/R)(d\theta/dt) \quad \text{...(8.58)}$$

as $G = NBld$

∴ Damping constant due to resistance of circuit

$$D_{circuit} = G^2/R \quad \text{...(8.59)}$$

Hence $T_{coil} = D_{circuit}(d\theta/dt)$ ...(8.60)

Let us consider the damping effect of former. Let $R_f$ be the resistance of the former.

∴ Proceeding as earlier, the damping torque produced by the former is :

$$T_f = \frac{(Bld)^2}{R_f} \cdot \left(\frac{d\theta}{dt}\right) \quad \text{(As } N = 1 \text{ for the former)}$$

or $$T_f = \frac{G^2}{N^2R_f}\left(\frac{d\theta}{dt}\right) = D_{former}\frac{d\theta}{dt} \quad \text{...(8.61)}$$

Damping constant due to resistance of former

$$= D_{former} = G^2/(N^2R_f) \quad \text{...(8.62)}$$

∴ Total electromagnetic damping torque

$$T_e = \left(\frac{G^2}{R} + \frac{G^2}{N^2R_f}\right)\frac{d\theta}{dt} = D_e\frac{d\theta}{dt} \quad \text{...(8.63)}$$

where $D_e$ = damping constant due to electromagnetic effects

$$= \frac{G^2}{R} + \frac{G^2}{N^2R_f} \quad \text{...(8.64)}$$

***Total damping.*** Total damping effect is made up of two components *i.e.*, damping due to mechanical effects and damping due to electromagnetic effects.

Total damping torque

$$T_D = T_m + T_e$$

$$= D_m(d\theta/dt) + D_e(d\theta/dt)$$

$$= (D_m + D_e)(d\theta/dt) \quad \text{...(8.65)}$$

$$= D.(d\theta/dt) \quad \text{...(8.66)}$$

where $D$ = constant due to combined effects of mechanical and electromagnetic damping

$$= D_m + D_e$$

***Critical resistance.*** The mechanical damping effects are very small as compared with electromagnetic effects, and therefore, $D_m = 0$.

So that we have, $D = \frac{G^2}{R} + \frac{G^2}{N^2R_f}$

The damping due to metal former is usually small and hence we can put $G^2/(N^2R_f) = 0$.

∴ $$D = G^2/R \quad \text{...(8.67)}$$

For critical damping

$$D = D_c = 2\sqrt{KJ} \quad \text{(See Eqn. 8.34)}$$

∴ For critical damping $G^2/R = 2\sqrt{KJ}$ ...(8.68)

Hence for critical damping the total resistance of the galvanometer circuit should be

$$R = G^2/2\sqrt{KJ} \quad \text{...(8.69)}$$

Thus for a particular galvanometer for which $G$, $J$ and $K$ are constants, critical damping conditions can be obtained by properly adjusting the value of the external resistance connected in the galvanometer circuit.

External series resistance required for critical damping

$$R_e = R - R_g = G^2/2\sqrt{KJ} - R_g \quad \text{...(8.70)}$$

This external resistance required for critical damping is called *CDRX* or *ECDR, critical damping resistance external.*

## 8.13 SENSITIVITY

The sensitivity of a galvanometer may be defined in several ways *i.e.*, current, voltage and Meghom sensitivities.

1. ***Current sensitivity.*** The current sensitivity of a galvanometer is defined as the deflection produced by unit current.

∴ Current sensitivity

$$S_i = \theta_F/i \quad \text{but} \quad \theta_F = G_i/K$$

∴ $$S_i = G_i/K \text{ rad/A} \quad \text{...(8.71)}$$

8

In practice the currents involved are small and the deflection is measured in terms of divisions or distance moved by spot on a scale. Thus the definition of sensitivity as given by Eqn. 8.71 is not relevant to practical measurement work. So while doing practical work, the sensitivity is expressed in terms of scale divisions per microampere or mm/μA. If the scale is placed 1 mm away and has divisions of mm, the sensitivity of current galvanometer is :

Current sensitivity,

$$S_i = \frac{d}{i \times 10^6} \text{ scale divisions/}\mu\text{A} \quad \ldots(8.72)$$

$$\therefore \quad S_i = \frac{2000\,G}{K \times 10^6} = \frac{G}{500\,K} \text{ mm/}\mu\text{A} \quad \ldots(8.73)$$

as $d = 2000\,Gi / K$ (See Eqn. 8.71)

2. ***Voltage sensitivity.*** The voltage sensitivity is the deflection in scale divisions per unit voltage impressed on the galvanometer.

Voltage sensitivity,

$S_v$ = scale division per unit voltage

$$= \frac{d}{i \times R_g \times 10^6} \text{ scale division/}\mu\text{V} \quad \ldots(8.74)$$

For a galvanometer with scale 1 m away scale division of 1 mm,

$$S_v = \frac{2000\,G}{KR_g \times 10^6} = \frac{G}{500\,K\,R_g} \text{ mm/}\mu\text{V} \quad \ldots(8.75)$$

In America it is a practice to include *CDRX* in the galvanometer circuit for the purpose of defining voltage sensitivity. Accordingly the definition of voltage sensitivity is :

$S_v$ = scale divisions per unit voltage with galvanometer and *CDRX* in series.

For a galvanometer with scale 1 m away and 1 scale division = 1 mm,

$$S_v = \frac{G}{500\,K(R_g + CDRX)} \text{ mm/}\mu\text{V} \quad \ldots(8.76)$$

3. ***Megohm sensitivity.*** The megohm sensitivity is the resistance of the circuit (in MΩ) so that the deflection will be 1 scale division with 1 volt impressed to the circuit. As the coil resistance is negligible in comparison, the megohm sensitivity represents the resistance that must be put in series with the galvanometer so that 1 volt gives a deflection of 1 scale division. In fact megohm sensitivity is the current sensitivity (expressed in microampere per scale division).

Megohm sensitivity,

$S_0$ = resistance in megohm to give a deflection of one scale division with 1 volt

$$= \frac{d}{i \times 10^{-6}} \text{ M}\Omega\text{/scale division} \quad \ldots(8.77)$$

For a galvanometer with scale 1 m away and 1 scale division = 1 mm,

$$S_0 = \frac{G}{500\,K} \text{ M}\Omega\text{/mm} \quad \ldots(8.78)$$

***Comments.*** A sensitive galvanometer is one which produces a large deflection for a small current. We have current sensitivity, $S_i = G/500\,K$.

Thus the value of $S_i$ should be large in order that a galvanometer be sensitive. Hence for a sensitive galvanometer the displacement constant *G* should be large and control constant *K* should be small.

Now, displacement constant $G = NBA$. Therefore, the sensitivity can be increased by increasing the number of turns, using high flux density and by increasing the dimensions of the coil. For a given coil area $A(= ld)$ and a constant value of flux density *B*, displacement constant *G* can be increased by increasing the number of turns. If the winding area is fixed, a large number of turns can only be wound if we use thin wire. Thus we can increase the sensitivity of a galvanometer by winding its coil with large number of turns of thin fine wire. Hence for a sensitive galvanometer the value of resistance of coil $R_g$ is large. We can also increase the sensitivity by making the control constant *K* small. Thus a suspension of small stiffness leads to increased sensitivity.

## 8.14 GALVANOMETER SHUNTS

In the case of preliminary balancing work in bridge and potentiometer measurements, the potential across the galvanometer circuit may be high owing to unbalance. If we are using galvanometer which has high sensitivity for final stages of balancing the large unbalance that probably exists at the start is likely to damage the coil or the suspension. This because the potential across the galvanometer circuit (due to unbalance) circulates a large current through the coil which results in large deflection with consequent large twisting of suspension. Large current also produces undesirable heating effects. Sometimes even if the above effects are not there, excessive currents send the light spot so far off the scale that the deflection is not readable and the operator does not know what to do to improve the conditions. Thus it is very necessary to use protective device of some sort to protect the sensitive galvanometers from the above damaging effects.

The simplest procedure is to use :

(*i*) A high resistance in series with the galvanometer to limit the current and therefore the deflection produced by the available emf under conditions of unbalance.

(*ii*) A low resistance is connected across the terminals of the galvanometer thus by-passing most of the current. This reduces the sensitivity of the galvanometer circuit so that the response of the instrument to a given current is reduced. A low resistance shunt will heavily overdamp the instrument so that its motion will be sluggish.

For both series and shunt resistances the time required to attain the final deflection is excessive. As a general rule a combination of a series and a shunt resistance is preferable which will give the desired reduction in sensitivity and at the same time the proper damping characteristics.

### 8.14.1 Ayrton Universal Shunt

Figure 8.6 shows an arrangement which incorporates combination of series and parallel resistances. It is clear from the diagram that the lower part of the resistance (*i.e.*, $bc \rightarrow c$ below the sliding contact) is in parallel while the upper part of the resistance (*i.e.*, *ab* above the sliding contact) is in series with the galvanometer. When the sliding contact is near the lower end, a small part of resistance is in parallel with the galvanometer and the major part is in series with the galvanometer. The current divides itself in two parts, the major going through the shunt and only a small part through the galvanometer, thus protecting it and giving a small sensitivity. As the sliding contact is moved up, the shunt resistance increases and the series resistance decreases. This increases the current through the galvanometer and thus the sensitivity is increased. However, there is a loss of sensitivity (as compared with galvanometer alone) with the sliding contact even at the top point. However, this is not a serious limitation if the resistance $R_{sh}$ is very large as compared with $R_g$ of the galvanometer.

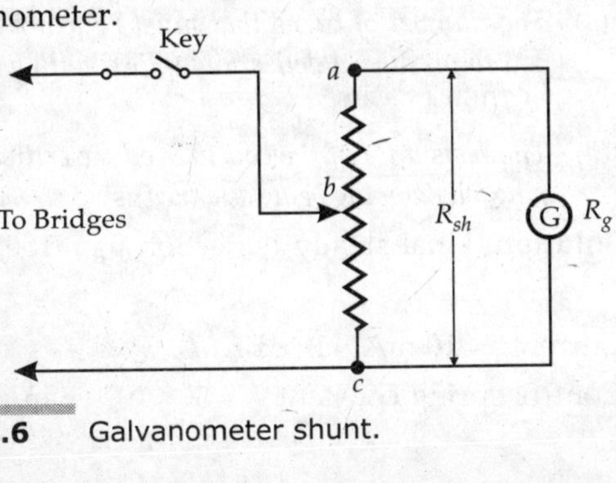


**Fig. 8.6** Galvanometer shunt.

With key *K* is open as in Fig. 8.6, there is closed circuit through the galvanometer coil and $R_{sh}$ and this provides a damping circuit. Therefore, the value of $R_{sh}$ should be equal to *CDRX* (a little higher value is satisfactory). As *CDRX* is usually several times $R_g$ the loss of sensitivity due to using a shunt of this value is not serious.

Ayrton universal shunt construction is shown in Fig. 8.7. This arrangement uses a dial type switch. The numbers on the contacts represent the relative values of galvanometer current. The resistance of each section to give the desired current can be calculated as under :

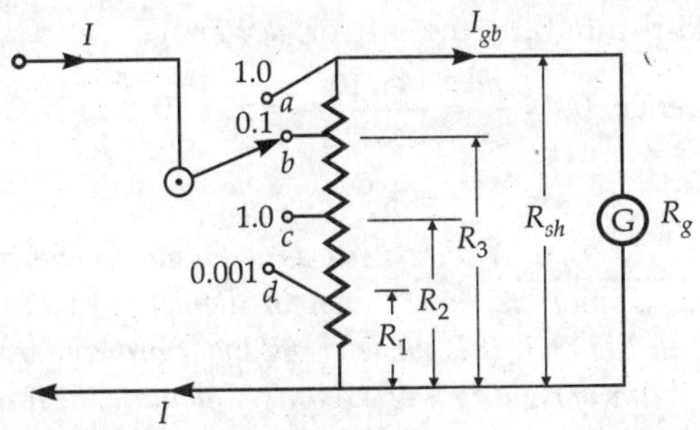


**Fig. 8.7** Ayrton universal shunt.

Current through the galvanometer at contact '*a*' or '*I*'

$$I_{ga} = I \frac{R_{sh}}{R_{sh} + R_g}$$

When the contact is at *b*, the resistance $R_3$ is in parallel with resistances $(R_{sh} - R_3)$ and $R_g$ in series.

$\therefore$ Current through the galvanometer at contact '*b*' or '0.1'

$$I_{gb} = I \frac{R_3}{R_3 + (R_{sh} - R_3 + R_g)} = I \frac{R_3}{R_{sh} + R_g}$$

The current through the galvanometer at contact '*b*' should be 0.1 of current at contact '*a*'.

We have $$\frac{I_{gb}}{I_{ga}} = \frac{IR_3/(R_{sh} + R_g)}{I R_{sh}/(R_{sh} + R_g)} = \frac{R_3}{R_{sh}}$$

$$\therefore \quad \frac{R_3}{R_{sh}} = 0.1 \text{ in order that } \frac{I_{gb}}{I_{ga}} = 0.1$$

Similarly, $R_2 / R_{sh} = 0.01$ and $R_1 / R_{sh} = 0.001$.

It should be noted that the relative values of current (through the galvanometer and through the shunt) do not depend upon the value of galvanometer resistance. The shunt thus gives the same relative current values for the various steps for all galvanometers although not the same fraction of the total current. Therefore it is called a *Universal Shunt*.

Initially we should keep the sliding contacts at the lower end. This diverts most of the current through the shunt and hence decreases the current through the galvanometer reducing its sensitivity. As balance conditions are approached the sliding contacts are then moved up increasing the current through the galvanometer thereby increasing its sensitivity.

**Example 8.1** *The scale of a galvanometer is placed at a distance of 0.4 m from the mirror. A deflection of 44 mm is observed. What is the angle through which the coil has turned ?*

**Solution.** Suppose $d$ is the deflection, $r$ is the distance between scale and mirror, and $\theta_F$ is the angle turned through by the mirror, $d = 2r\theta_F$.

Hence, $\theta_F = \dfrac{d}{2r} = \dfrac{44 \times 10^{-3}}{2 \times 0.4} = 55 \times 10^{-3}$ rad

$= 3.15° = 3°9'$.

**Example 8.2** *A wire used as a torsion member in a (small galvanometer is 0.1 mm in diameter, arranged as shown in Fig. 8.8. It is made of phosphor bronze for which the modulus of rigidity is 40 GN/m². Calculate the torsional constant and maximum allowable torque if the stress is limited to 100 MN/m².*

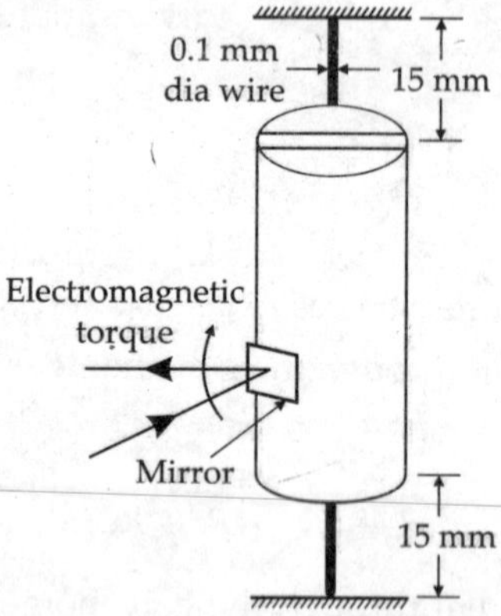


**Fig. 8.8** A galvanometer moving system.

*A collimated ray of light is reflected by the galvanometer mirror shown. The radius from the mirror to the circular scale is 200 mm. Calculate the sensitivity of the instrument in m/Nm and the maximum possible scale deflection. The polar moment of inertia for a round wire is $\pi/32\ d^4$. Where d is the diameter of wire.*

**Solution.** For a round wire $\dfrac{T}{I_p} = \dfrac{f_s}{d/2} = \dfrac{G\theta}{l}$

Hence, the deflection per unit torque is :

$$\frac{\theta}{T} = \frac{l}{GI_p}$$

where G is the modulus of rigidity.

Since there are two wires, one at the top and the other at the bottom, the deflection is half of that for single wire.

$$\therefore \frac{\theta}{T} = \frac{l}{2GI_p} = \frac{l}{2\,G \times (\pi/32)d^4} = \frac{16l}{\pi G d^4}$$

$$= \frac{16 \times (15 \times 10^{-3})}{\pi \times (40 \times 10^9) \times (0.1 \times 10^{-3})^4} = 19 \times 10^3 \text{ rad / Nm}$$

$$\text{Torsional constant } K = \frac{T}{\theta} = \frac{\pi G d^4}{16l}$$

$$= \frac{1}{19 \times 10^3} = 52.6 \times 10^{-6} \text{ Nm/rad.}$$

The maximum stress occurs at the outermost surface :

$$\therefore \text{ Maximum torque } T = \frac{f_s I_p}{d/2}$$

$$= \frac{f_s \times (\pi/32)d^4}{d/2} = \frac{\pi f_s d^3}{16}$$

Since there are two wires, the maximum allowable torque is :

$$T = \frac{2 \times \pi f_s d^3}{16} = \frac{\pi}{8} f_s d^3 = \frac{\pi}{8} \times (100 \times 10^6) \times (0.1 \times 10^{-3})^3$$

$$= 39.3 \times 10^{-6} \text{ Nm.}$$

The scale is kept at a radius of 200 mm and therefore the sensitivity of the scale is :

$$S = 2 \times (19 \times 10^3) \times 200$$

$$= 7600 \times 10^3 \text{ mm/Nm} = 7600 \text{ m/Nm.}$$

A factor of 2 has been introduced because if mirror turns through an angle $\theta$, the collimated light turns through $2\theta$.

The deflection for maximum torque

$$= 7600 \times 10^3 \times 39.3 \times 10^{-6} = 298.7 \text{ mm.}$$

**Example 8.3** *The coil of a moving coil galvanometer is wound on a non-magnetic former whose height and width are both 20 mm. It moves in a constant field of 0.12 Wb/m². The moment of Inertia of its moving parts $0.25 \times 10^{-6}$ kg-m², and the control spring constant is $30 \times 10^{-6}$ Nm/rad. Calculate :*

(*a*) *The number of turns that must be wound on the coil to produce a deflection of 150° with a current of 10 mA.*

(*b*) *The resistance of the coil to produce critical damping, all damping being assumed as electromagnetic.*

**Solution.** Final steady deflection $\theta_F = 150° = 2.62$ rad.

Current $i = 10$ mA $= 10 \times 10^{-3}$ A.

Control spring constant $K = 30 \times 10^{-6}$ Nm/rad

(*a*) Final steady deflection $\theta_F = Gi/K$.

$\therefore$ Displacement constant, $G = \frac{K\theta_F}{i}$

$$= \frac{30 \times 10^{-6} \times 2.62}{10 \times 10^{-3}} = 7.86 \times 10^{-3} \text{ Nm/A}$$

But displacement constant $G = NBld$

$\therefore$ Number of turns, $N = \frac{G}{BlD}$

$$= \frac{7.86 \times 10^{-3}}{0.12 \times 0.02 \times 0.02} = 164.$$

From Eqn. 8.68, resistance of galvanometer circuit for critical damping

$$R = \frac{G^2}{2\sqrt{KJ}} = \frac{(7.86 \times 10^{-3})^2}{2\sqrt{30 \times 10^{-6} \times 0.25 \times 10^{-6}}}$$

$$= 11.28\ \Omega.$$

**Example 8.4** *The coil of a moving coil galvanometer has 300 turns and is suspended in a uniform magnetic field of 0.1 Wb/m$^2$. The control constant is $0.2 \times 10^{-6}$ N-m/rad. The coil is 20 mm wide and 25 mm high, with a moment of inertia of $0.15 \times 10^{-6}$ kg-m$^2$.*

*If the galvanometer resistance is 200 Ω, calculate the value of the resistance which, when connected in the galvanometer terminals, will give critical damping. Assume the damping to be entirely electromagnetic.*

**Solution.** Displacement constant,

$$G = NBld = 300 \times 0.1 \times 0.025 \times 0.02$$

$$= 15 \times 10^{-3} \text{ N-m/A.}$$

For critical damping, $D = 2\sqrt{KJ}$

$$= 2\sqrt{0.2 \times 10^{-6} \times 0.15 \times 10^{-6}}$$

$$= 0.346 \times 10^{-6} \text{ Nm/rad s}^{-1}.$$

If the damping is wholly electromagnetic

$D = G^2/R$ (See Eqn. 8.67)

$\therefore$ Resistance of galvanometer circuit, $R = \frac{G^2}{D}$

$$= \frac{(15 \times 10^{-3})^2}{0.346 \times 10^{-6}} = 650\ \Omega.$$

$\therefore$ External resistance for critical damping CDRX

$$= R - R_g = 650 - 200 = 450\ \Omega.$$

**Example 8.5** *A galvanometer gives a deflection of 150 mm on a linear scale distant 2.5 m for a steady current of 1 μA. The period of oscillations is 4 s and the moment of inertia of moving system is $1 \times 10^{-6}$ kg m$^2$. Calculate the coil circuit resistance necessary to obtain critical damping, assuming damping torques produced due to other effects to be negligible. The galvanometer uses a mirror and scale arrangement with a collimated light falling on the mirror.*

**Solution.** Suppose the final steady deflection is $\theta_F$ radian. Let $r$ be the distance of scale from the galvanometer and $d$ be the deflection on the linear scale.

$\because$ $d = 2r\theta_F$

or steady deflection, $\theta_F = d/2r$

$$= 150/(2 \times 2500) = 0.03 \text{ rad.}$$

But steady deflection $\theta_F = Gi/K$.

$\therefore$ Displacement constant,

$$G = \frac{\theta_F K}{i} = \frac{0.03 \times K}{1 \times 10^{-6}} = 30\,K \times 10^3$$

From Eqn. 8.29,

Period of free oscillations, $T_0 = 2\pi\sqrt{J/K}$

or $4 = 2\pi\sqrt{1 \times 10^{-6}/K}$

$\therefore$ Control constant $K = 2.47 \times 10^{-6}$ Nm/rad.

Hence, $G = 30 \times 2.47 \times 10^{-6} \times 10^3 = 74.1 \times 10^{-3}$ Nm/A

For critical damping resistance of galvanometer circuit.

$$R = \frac{G^2}{2\sqrt{KJ}} = \frac{(74.1 \times 10^{-3})^2}{2\sqrt{2.47 \times 10^{-6} \times 1 \times 10^{-6}}} = 1747\ \Omega.$$

**Example 8.6** *A moving coil galvanometer has former of a non-conducting material. The current sensitivity of the instrument is 0.001 μA/mm at one metre. The period of undamped oscillations is 6 seconds. If the displacement constant of the instrument is $5 \times 10^{-3}$ N-m/A, calculate :*

*(i) control constant, (ii) inertia constant,*

*(iii) total circuit resistance for instrument to be dead beat.*

**Solution.** Suppose that the instrument is reflecting type. Therefore, if the final steady deflection is $\theta_F$, the spot moves through an angle $2\theta_F$.

The spot moves a distance of 1 mm at 1 m with a current of 0.001 μA or in other words, a current of 0.001 μA deflects the instrument through

$$\theta = \frac{1}{2}\frac{l}{r} = \frac{1}{2} \times \frac{1}{1000} = 0.5 \times 10^{-3} \text{ rad.}$$

$\therefore$ Control constant, $K = \frac{Gi}{\theta_F} = \frac{5 \times 10^{-3} \times 0.001 \times 10^{-6}}{0.5 \times 10^{-3}}$

$$= 10 \times 10^{-9} \text{ Nm/rad.}$$

Undamped period, $T_0 = 2\pi\sqrt{J/K}$

or inertia constant $J = \frac{KT_0^2}{4\pi^2} = \frac{10 \times 10^{-9} \times (6)^2}{4\pi^2}$

$$= 9.12 \times 10^{-9} \text{ kg m}^2.$$

Total circuit resistance for critical damping or dead beat

$$R = \frac{G^2}{2\sqrt{KJ}} = \frac{(5 \times 10^{-3})^2}{2\sqrt{10 \times 10^{-9} \times 9.12 \times 10^{-9}}} = 1309\ \Omega.$$

***Example 8.7*** *A d'Arsonval galvanometer has a rectangular coil wound on an aluminium former of resistivity $27 \times 10^{-9}$ Wm. The sides of the former, each of effective length 30 mm, move in a radial field of uniform flux density 0.15 Wb/m$^2$ against a control torque of $18 \times 10^{-6}$ Nm/rad. The width of the former is 20 mm, the ends being effectively outside the magnet field, and the moment of inertia of the moving system is $80 \times 10^{-9}$ kg-m$^2$. Determine the cross-sectional area of the aluminium former if it is to provide critical damping, other sources of damping being neglected.*

**Solution.** Total damping factor in the absence of all other damping except the damping produced by former is :

$$D = \frac{G^2}{N^2 R_f} = \frac{N^2 B^2 l^2 d^2}{N^2 R_f} = \frac{B^2 l^2 d^2}{R_f}$$

For critical damping, $D = D_c = 2\sqrt{KJ}$

$$\therefore \quad \frac{B^2 l^2 d^2}{R_f} = 2\sqrt{KJ}$$

Hence resistance of former, $R_f = \dfrac{B^2 l^2 d^2}{2\sqrt{KJ}}$

$$= \frac{(0.15)^2 \times (30 \times 10^{-3})^2 \times (20 \times 10^{-3})^2}{2\sqrt{18 \times 10^{-6} \times 80 \times 10^{-9}}}$$

$$= 3375 \times 10^{-6}\,\Omega.$$

Length of eddy current paths, $L = 2(l + d)$

$$= 2 \times (30 + 20) = 100 \text{ mm} = 0.1 \text{ m}.$$

Let $a$ be the area of cross-section of the aluminium former.

$$\therefore \quad R_f = \frac{\rho L}{a} = \frac{27 \times 10^{-9} \times 0.1}{a} = 3.375 \times 10^{-6}.$$

or area of cross-section of aluminium former

$$a = \frac{27 \times 10^{-9} \times 0.1}{3.375 \times 10^{-6}}$$

$$= 0.8 \times 10^{-6}\text{ m}^2 = 0.8 \text{ mm}^2.$$

***Example 8.8*** *The card attached to a galvanometer gives the following data :*

*Sensitivity = 0.004 μA/mm, Coil resistance = 50 Ω,*

*CDRX = 300 Ω*

*It is used in bridge measurements with an Ayrton shunt of 500 Ω resistance.*

(a) *Is the galvanometer overdamped or underdamped under this condition of use ? Explain briefly.*

(b) *What is the sensitivity of galvanometer with shunt at maximum setting ?*

(c) *Where should the shunt be tapped to give a sensitivity 0.01 as great as in (b) ?*

**Solution.** (*a*) With key $K$ open (Fig. 8.9), the galvanometer has a closed circuit for damping through its own resistance and total resistance of shunt in series.

∴ Total resistance in circuit

$$= R_{sh} + R_g = 500 + 50 = 550\,\Omega.$$

External resistance for critical damping

$$CDRX = 300\,\Omega$$

∴ Resistance of circuit for critical damping

$$= CDRX + R_g = 300 + 50 = 350\,\Omega$$

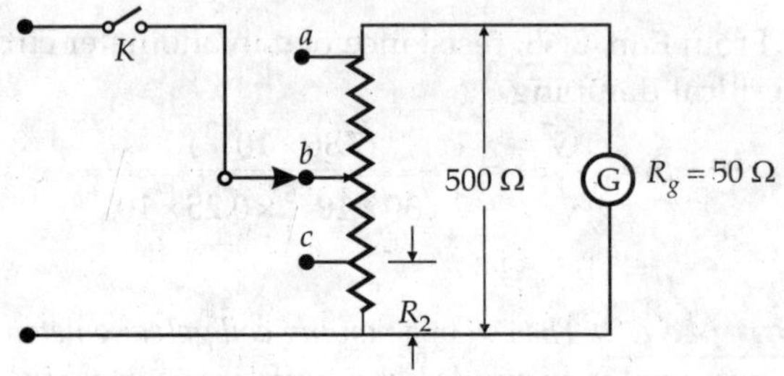


**Fig. 8.9** Ayrton shunt of Example 8.8.

The damping effect is large if the resistance of the galvanometer circuit is small. As the resistance of the galvanometer circuit (= 550 Ω) is larger than the resistance (= 350 Ω) required for critical damping, the galvanometer damping is smaller than critical. Hence the galvanometer is underdamped.

(*b*) Shunt at maximum setting means that whole of the resistance of the shunt is in parallel with the galvanometer. Therefore, the sliding contact is at '*a*' for maximum setting.

Suppose $I$ is the total circuit current in μA and $I_{ga}$ is the current through the galvanometer with sliding contact at '*a*' in μA.

$$I_{ga} = I \cdot \frac{R_{sh}}{R_{sh} + R_g} = I \cdot \frac{500}{500 + 50} = 0.909\, I$$

Now a current of 0.004 μA deflects the galvanometer through 1 mm and, therefore, deflection of galvanometer with a current 0.909 $I$ μA is :

$$d = \frac{0.909\, I}{0.004} = 227.25\, I \text{ mm}.$$

$$\therefore \text{ Sensitivity} = \frac{I}{227.25\, I} = 0.0044\ \mu\text{A/mm}.$$

(In the problem the sensitivity is expressed in μA/mm unlike our practice where sensitivity expressed in mm/μA.)

(*c*) The new value of sensitivity is 0.01 times as great as in (*b*).

$$\therefore \text{ New value of sensitivity} = 0.0044 \times \frac{1}{0.01}$$

$$= 0.44\ \mu\text{A/mm}.$$

Suppose the slider contact is at '*c*' when this sensitivity is obtained. The resistance being $R_2$ as indicated.

Current through the galvanometer,

$$I_{gc} = I.\frac{R_2}{R_2+(R_{sh}-R_2+R_g)} = I\frac{R_2}{R_{sh}+R_g}$$

$$= I\frac{R_2}{550} = 1.82\times10^{-3}\ IR_2\,\mu\text{A}.$$

$$\text{Deflection } d = 1.82\times10^{-3}\ IR_2\times\frac{1}{0.004}$$

$$= 0.455\ IR_2 \text{ mm}.$$

$$\therefore \text{ Sensitivity at '}c\text{'} = \frac{I}{d} = \frac{I}{0.455\ IR_2} = \frac{2.2}{R_2}\mu\text{A/mm}.$$

This should be equal to 0.44 μA/mm

or $\dfrac{2.2}{R_2} = 0.44 \quad \therefore \quad R_2 = \dfrac{2.2}{0.44} = 5\ \Omega.$

**Example 8.9** *The deflection of a galvanometer is zero with no current and 75 mm with a steady state current of 5.6 μA. Its first maximum deflection, after a step function of voltage which produces a steady state current is applied is 112 mm. The maximum deflection in the next cycle is 84 mm. Determine : (a) Current sensitivity, (b) The logarithmic decrement, and (c) Relative damping.*

**Solution.**

$$\text{Current sensitivity, } S_i = \frac{\theta_F}{i} = \frac{75}{5.6} = 13.4 \text{ mm/}\mu\text{A}.$$

First maximum deflection

$$\theta_1 = \theta_F\,[1+\exp(-\pi\xi/\sqrt{1-\xi^2})] \qquad ...(i)$$

(See Eqn. 8.48)

Similarly, second maximum deflection

$$\theta_3 = \theta_F\,[1+\exp(-3\pi\xi/\sqrt{1-\xi^2})] \qquad ...(ii)$$

Hence from (*i*) and (*ii*),

$$\frac{\theta_1-\theta_F}{\theta_3-\theta_F} = \exp(2\pi\xi/\sqrt{1-\xi^2})$$

But logarithmic decrement

$$\lambda = \frac{\pi\xi}{\sqrt{1-\xi^2}} \qquad \text{(See Eqn. 8.52)}$$

$$\therefore \quad \frac{\theta_1-\theta_F}{\theta_3-\theta_F} = \exp(2\lambda) \quad \text{or} \quad \frac{112-75}{84-75} = \exp\,(2\lambda)$$

or Logarithmic decrement, $\lambda = 0.707$.

$$\text{Logarithmic decrement, } \lambda = \frac{\pi\xi}{\sqrt{1-\xi^2}}$$

or $$0.707 = \frac{\pi\xi}{\sqrt{1-\xi^2}}$$

or relative damping, $\xi = 0.22$.

**Example 8.10** *The current sensitivity of a galvanometer is 100 mm/μA with the scale 1 m away. The resistance of coil is 200 Ω.*

*(a) Calculate the voltage sensitivity and the megohm sensitivity.*

*(b) What is the deflection produced by a current of 0.5 μA with the scale 0.75 m away from galvanometer ?*

*(c) A resistance of 50 Ω is placed in parallel with the coil. The scale is 1 m away from the galvanometer. Calculate the current sensitivity, voltage sensitivity, and the megohm sensitivity for the shunted galvanometer.*

*(d) The coil of the galvanometer is rewound. The area of the coil is not changed. The wire on the new coil has a diameter of one-half that of the original coil, but the total weight of the coil is the same. Calculate the current sensitivity, voltage sensitivity and megohm sensitivity.*

8

**Solution.** (*a*) The current sensitivity with scale 1 m away is 100 mm/μA.

Voltage sensitivity

$$S_v = \frac{d}{i\times R_g\times10^6} \text{ mm/}\mu\text{A.} \quad \text{(See Eqn. 8.73)}$$

$$= \frac{100}{1\times10^{-6}\times200\times10^6} = 0.5 \text{ mm/}\mu\text{V}.$$

Current required for a deflection of 1 mm is 0.01 μA.

∴ Megohm sensitivity

$$S_0 = \frac{1}{0.01\times10^{-6}} = 100 \text{ M}\Omega/\text{mm}.$$

(*b*) The deflection produced at a distance of 0.75 m is 3/4 of the deflection produced at a distance 1 m.

∴ Deflection due to a current of 0.5 μA on a scale 0.75 m away $= 0.75\times0.5\times100 = 37.5$ mm.

(*c*) The galvanometer resistance is 200 Ω and that of shunt resistance is 50 μA. When a current of 1 μA is measured, 0.2 μA passes through the galvanometer and the rest through the shunt.

∴ Current sensitivity, $S_i = 0.2\times100 = 20$ mm/μA.

The voltage sensitivity remains the same.

Now a current of 1/20 μA = 0.05 μA is required for a deflection of 1 mm.

$$\therefore \text{ Megohm sensitivity} = \frac{1}{0.05\times10^{-6}} = 20 \text{ M}\Omega\text{/mm}.$$

(*d*) Since the diameter of wire of the new coil is one half of the original, the volume of each turn of the new coil one fourth of that of original turn. Since the weight of the new coil is the same as that of original coil, the number of turns is four times that in the original coil.

Now the new coil has number of turns four times that of the original coil and area of each turn of new coil is one fourth of the original coil, the resistance of the new coil is 16 times that of the original coil.

Resistance of new coil $= 16 \times 200 = 3200\ \Omega$.

Displacement constant of new coil is four times that of the original coil.

Current sensitivity is directly proportional to the displacement constant.

$\therefore$ Current sensitivity of new coil

$$= 4 \times 100 = 400 \text{ mm}/\mu\text{A}.$$

Voltage sensitivity is inversely proportional to resistance.

$\therefore$ Voltage sensitivity of new coil

$$= 0.5/16 = 0.03125 \text{ mm}/\mu\text{V}.$$

Megohm sensitivity of the new coil $= 400\ \text{M}\Omega/\text{mm}$.

***Example 8.11*** *The following data refers to a moving coil galvanometer whose resistance is 460 Ω and a resistance of 1970 Ω is connected in series with it.*

*Number of turns = 250, flux density = 0.1 Wb/m², control constant = $0.15 \times 10^{-6}$ Nm/rad s$^{-1}$, dimensions of coil = 30 × 30 mm, moment of inertia of coil = $0.2 \times 10^{-6}$ kg m². Calculate :*

(a) *The resistance to be connected to galvanometer for critical damping,*
(b) *Relative damping,*
(c) *Logarithmic decrement,*
(d) *Frequency of free oscillations,*
(e) *Period of undamped and damped oscillations,*
(f) *First maximum deflection.*

**Solution.**

(a) Displacement constant,

$$G = NBld = 250 \times 0.1 \times 0.03 \times 0.03$$
$$= 22.5 \times 10^{-3} \text{ Nm/A}.$$

Let $R$ be the total resistance of the circuit for critical damping.

$$\therefore \quad \frac{G^2}{R} = 2\sqrt{KJ} \quad \text{or} \quad R = \frac{G^2}{2\sqrt{KJ}}$$

$$= \frac{(22.5 \times 10^{-3})}{2\sqrt{0.15 \times 10^{-6} \times 0.2 \times 10^{-6}}} = 1460\ \Omega$$

$\therefore$ External resistance required for critical damping

$$CDRX = 1460 - 460 = 1000\ \Omega.$$

(b) Relative damping, $\xi = D/D_e$

$$= \frac{G^2/\text{resistance of circuit}}{G^2/\text{resistance of circuit for critical damping}}$$

$$= \frac{G^2/(1970+460)}{G^2/(1000+460)} = 0.6$$

(c) Logarithmic decrement, $\lambda = \dfrac{\pi\xi}{\sqrt{1-\xi^2}}$

$$= \frac{\pi \times 0.6}{\sqrt{1-(0.6)^2}} = 2.36.$$

Frequency of free oscillations, $f_n = \dfrac{1}{2\pi}\sqrt{\dfrac{K}{J}}$

$$= \frac{1}{2\pi}\sqrt{\frac{0.15 \times 10^{-6}}{0.2 \times 10^{-6}}} = 0.138 \text{ Hz}.$$

Frequency of damped oscillations, $f_d = f_n\sqrt{1-\zeta^2}$

$$= 0.138 \times \sqrt{1-(0.6)^2} = 0.11 \text{ Hz}.$$

(d) Period of undamped oscillations

$$T_0 = 1/f_n = 1/0.138 = 7.24 \text{ s}.$$

Period of damped oscillations

$$T_d = 1/f_d = 1/0.11 = 9.1 \text{s}.$$

(e) First maximum deflection

$$\theta_1 = \theta_F[1 + \exp(-\pi\xi/\sqrt{1-\xi^2})]$$
$$= \theta_F[1 + \exp(-\lambda)] \qquad \text{(See Eqns. 8.48 and 8.52)}$$
$$= \theta_F[1 + \exp(-2.36)] = 1.0944\ \theta_F.$$

Thus even with relative damping $\zeta = 0.6$ the first maximum deflection is 9.44% (about 10%) more than the final steady deflection.

***Example 8.12*** *A current galvanometer has the following parameters :*

*$B = 10 \times 10^{-3}$ Wb/m², $N = 200$ turns, $l = 16$ mm, $d = 16$ mm, $K = 12 \times 10^{-9}$ Nm/rad, $J = 50 \times 10^{-9}$ kg-m², $D = 5 \times 10^{-9}$ Nm/rad s$^{-1}$.*

*The resistance of the coil is 120 Ω. Calculate :*

(a) *The deflection of the galvanometer in radian and in mm when a current of 1 μA flows through it, the scale being 1 m away,*
(b) *The current sensitivity,*
(c) *The voltage sensitivity,*
(d) *The megohm sensitivity,*
(e) *The frequency of damped oscillations,*
(f) *The period of free oscillations,*
(g) *Relative damping,*
(h) *The first maximum deflection,*
(i) *The logarithmic decrement.*

**Solution.** (*a*) Displacement constant, $G = NBld$

$$= 200 \times 10 \times 10^{-3} \times 10 \times 10^{-3} \times 16 \times 10^{-3}$$
$$= 0.512 \times 10^{-3} \text{ Nm/A}.$$

Deflection of galvanometer

$$\theta_F = \frac{Gi}{K} = \frac{0.512 \times 10^{-3} \times 1 \times 10^{-6}}{12 \times 10^{-9}} = 0.0427 \text{ rad}.$$

Deflection (in mm) $= 2000\,\theta_F = 2000 \times 0.0427 = 85.4$ mm.

(*b*) A current of 1 μA causes a deflection of 85.4 mm.

∴ Current sensitivity, $S_i = 85.4$ mm/μA.

(*c*) Voltage across the galvanometer for current of

$$1\,\mu\text{A} = 1 \times 10^{-6} \times 120 = 120 \times 10^{-6}\text{ V}.$$

The deflection corresponding to this voltage is 85.4 mm.

∴ Voltage sensitivity

$$S_v = \frac{85.4}{120 \times 10^{-6}} = 0.712 \times 10^{-6} \text{ mm/V}$$
$$= 0.712 \text{ mm/}\mu\text{V}.$$

(*d*) Megohm sensitivity

$$S_0 = \frac{d}{i \times 10^6} = \frac{85.4}{1 \times 10^{-6} \times 10^6} = 85.4 \text{ M}\Omega\text{/mm}.$$

(*e*) Frequency of damped oscillations

$$\omega_d = \frac{\sqrt{4KJ - D^2}}{2J}$$
$$= \frac{\sqrt{4 \times 12 \times 10^{-9} \times 50 \times 10^{-9} - (5 \times 10^{-9})^2}}{2 \times 50 \times 10^{-9}}$$
$$= 0.487 \text{ rad/s}$$

or $\quad f_d = \omega_d / 2\pi = 0.487/2\pi = 0.0775$ Hz.

(*f*) Frequency of natural oscillations

$$f_n = \frac{1}{2\pi}\sqrt{\frac{K}{J}} = \frac{1}{2\pi}\sqrt{\frac{12 \times 10^{-9}}{50 \times 10^{-9}}} = 0.078 \text{ Hz}.$$

Period of natural oscillations

$$T_0 = 1/f_n = 1/0.078 = 12.82 \text{ s}.$$

(*g*) Damping constant required for critical damping

$$D_0 = 2\sqrt{KJ} = 2\sqrt{12 \times 10^{-9} \times 50 \times 10^{-9}}$$
$$= 49 \times 10^{-9} \text{ Nm/rad s}^{-1}.$$

Relative damping, $\zeta = \dfrac{D}{D_c} = \dfrac{5 \times 10^{-9}}{49 \times 10^{-9}} = 0.102.$

(*h*) First maximum deflection

$$\theta_1 = \theta_F[1 + \exp(-\pi\xi/\sqrt{1-\xi^2})]$$
$$= \theta_F[1 + \exp(-\pi \times 0.102/\sqrt{1 - 0.102^2})]$$
$$= 1.72\,\theta_F = 1.72 \times 85.4 = 147.2 \text{ mm}$$

(*i*) Logarithmic decrement

$$\lambda = \frac{\pi\zeta}{\sqrt{1-\zeta^2}} = \frac{\pi \times 0.102}{\sqrt{1-(0.102)^2}} = 0.322.$$

***Example 8.13*** *Figure 8.10 shows the circuit for testing of a galvanometer. The various values are : $E = 1.5$ V, $R_1 = 1\,\Omega$ and $R_2 = 7500\,\Omega$ when $R_3 = 350\,\Omega$, the galvanometer deflection is 140 mm and when $R_3 = 1080\,\Omega$, the deflection is 50 mm.* 8

*Calculate,*

(*a*) *The resistance of galvanometer,*

(*b*) *The sensitivity of galvanometer, and*

(*c*) *The value of resistance $R_3$ when the deflection is 70 mm.*

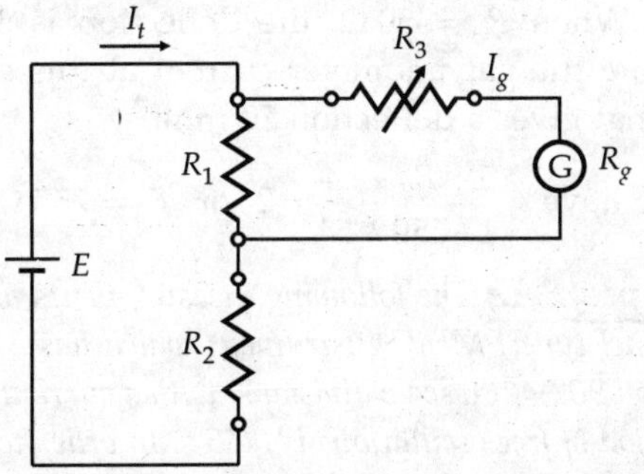


**Fig. 8.10** Figure of Example 8.13.

**Solution.** Since the variable resistor $R_3$ is in a circuit which is in parallel with $R_1$ and also $R_1$ is very small as compared with $R_2$, the variations of $R_3$ have no effect on the total current $I_t$. However, variation of R3 changes the galvanometer current.

(*a*) Galvanometer current, $I_g = I_t \dfrac{R_1}{R_1 + R_3 + R_g}$

When $R_3 = 350\,\Omega$ the galvanometer current is :

$$I_{g1} = I_t \frac{1}{1 + 350 + R_g} \qquad \text{...}(i)$$

When $R_3 = 1080\,\Omega$, the galvanometer current is

$$I_{g2} = I_t \frac{1}{1 + 1080 + R_g} \qquad \text{...}(ii)$$

The deflection of galvanometer with $I_{g1}$ is 140 mm and that $I_{g2}$ is 50 mm.

$$\therefore \quad I_{g1}/I_{g2} = 140/50 = 14/5 \qquad \text{...}(iii)$$

From (*i*), (*ii*) and (*iii*), we have :

$$\frac{I_{g1}}{I_{g2}} = \frac{1 + 1080 + R_g}{1 + 350 + R_g} = \frac{14}{5}$$

or $\quad$ galvanometer resistance, $R_g = 54\,\Omega$.

(b) The total resistance of the circuit

$$R_t = R_2 + \frac{R_1(R_3 + R_g)}{R_1 + R_2 + R_g} \approx 7500\ \Omega$$

$$\therefore \quad I_t = \frac{1.5}{7500} = 0.2 \times 10^{-3}\ \text{A}$$

For $R_3 = 350\ \Omega$,

Galvanometer current,

$$I_{g1} = 0.2 \times 10^{-3} \times \frac{1}{1 + 350 + 54}$$

$$= 0.494 \times 10^{-6}\ \text{A} = 0.494\ \mu\text{A}$$

Deflection = 140 mm.

$$\therefore \text{ Current sensitivity, } S_i = \frac{140}{0.494} = 283.4\ \text{mm}/\mu\text{A}.$$

(c) When $R_3 = 350\ \Omega$, the deflection is 140 mm. Therefore the galvanometer current at this setting is twice that gives a deflection 70 mm.

We have, $\dfrac{1 + R_3 + 54}{1 + 350 + 54} = 2$ or $R_3 = 755\ \Omega$.

**Example 8.14** *The following measurements were made on a panel type PMMC instrument which has a full scale deflection 90° = 100 scale divisions, with a current of 1 mA. The period of free oscillations is 0.55 s. In order to measure the spring constant, a small weight having a gravitational force of 98.1× $10^{-6}$ N is placed at a distance of 100 mm from the horizontal axis of rotation with the horizontal pointer acting as lever arm; the resulting deflection being 35 divisions. The first maximum deflection is 106 divisions when a current of 1 mA is suddenly passed through the meter. The value of flux density in the air gap measured with a Hall probe is 0.24 Wb/m$^2$. The length of coil is 15 mm and the average diameter of coil is 14 mm. From the above data, calculate :*

*(a) Spring constant,* *(b) Moment of inertia,*
*(c) Number of turns,* *(d) Damping ratio,*
*(e) Damping constant,*
*(f) Equivalent resistance of the short circuited winding represented by the damping action of the former of the coil.*

**Solution.** Full scale deflection = 90° = 100 scale divisions

$\therefore$ 1 scale division = 0.9° = 0.0157 rad.

(a) Restoring torque, $T_c = 98.1 \times 10^{-6} \times 100 \times 10^{-3}$

$$= 9.81 \times 10^{-6}\ \text{Nm}.$$

Deflection $\theta$ = 35 div.

$$= 35 \times 0.0157\ \text{rad} = 0.5495\ \text{rad}.$$

$$\therefore \text{ Spring constant, } K = \frac{T_c}{\theta} = \frac{9.81 \times 10^{-6}}{0.5495}$$

$$= 17.85 \times 10^{-6}\ \text{Nm/rad}.$$

(b) Period of free oscillations, $T_0 = 2\pi\sqrt{J/K}$.

$$\therefore \text{ Moment of inertia, } J = \frac{KT_0^2}{4\pi^2}$$

$$= \frac{17.85 \times 10^{-6} \times (0.55)^2}{4\pi^2} = 0.137 \times 10^{-6}\ \text{kg-m}^2.$$

(c) Control torque for a deflection of 35 scale divisions is $9.81 \times 10^{-6}$ Nm.

$\therefore$ Control torque for full scale deflection (100 div.)

$$T_c = 9.81 \times 10^{-6} \times \frac{100}{35} = 28 \times 10^{-6}\ \text{Nm}.$$

$\therefore$ Deflecting torque at full scale

$$T_d = NBldI$$

$$= N \times 0.24 \times 15 \times 10^{-3} \times 14 \times 10^{-3} \times 1 \times 10^{-3}$$

$$= 50.4\,N \times 10^{-9}\ \text{Nm}.$$

At equilibrium

$$T_d = T_c \quad \text{or} \quad 50.4\,N \times 10^{-9} = 28 \times 10^{-6}.$$

$$\therefore \text{ Number of turns, } N = \frac{28 \times 10^{-6}}{50.4 \times 10^{-9}} = 556.$$

(d) First maximum deflection

$$\theta_1 = 106 \text{ divisions} = 1.06\ \theta_F.$$

But $\theta_1 = \theta_F\,[1 + \exp(-\pi\zeta/\sqrt{1-\zeta^2})]$

or $1.06 = 1 + \exp(-\pi\xi/\sqrt{1-\xi^2})$

$\pi\xi/\sqrt{1-\xi^2} = 2.81$ or damping ratio, $\zeta = 0.667$.

(e) Damping constant, $D = \zeta \times 2\sqrt{KJ}$

$$= 0.667 \times 2 \times \sqrt{17.85 \times 10^{-6} \times 0.137 \times 10^{-6}}$$

$$= 2.09 \times 10^{-6}\ \text{Nm/rad s}^{-1}$$

(f) Considering the former to be a single turn coil, the damping constant is, $D_{former} = (Bld)^2 / R_f$.

$\therefore$ Resistance of former

$$R_f = \frac{(Bld)^2}{D_{former}} = \frac{(0.24) \times 15 \times 10^{-3} \times 14 \times (10^{-3})^2}{2.09 \times 10^{-6}}$$

$$= 1.22 \times 10^{-3}\ \Omega$$

## 8.15 BALLISTIC GALVANOMETER

A ballistic galvanometer is used for measurement of quantity of electricity (charge) passed through it. In magnetic measurements, this quantity of electricity is due to an instantaneous emf induced in a search coil connected across the ballistic galvanometer. The instantaneous emf is induced in the search coil when the flux linking with the search coil is changed. The quantity of electricity passing through the galvanometer is proportional to the emf induced and hence to the change in flux linking with the search coil. The galvanometer can therefore be calibrated to read the charge directly.

### 8.15.1 Construction of Ballistic Galvanometers

The ballistic galvanometer is of d'Arsonval type. However, it does not show a steady deflection (as is the case in a current galvanometer), owing to the transitory nature of the current passing through it, but it oscillates with decreasing amplitude, the amplitude of the first deflection or swing or throw being proportional to the charge passing.

The proportionality between the throw of the galvanometer and the charge passing holds good only if the whole charge passes through the galvanometer coil before any appreciable deflection of the coil takes place. This condition can be satisfied if the time taken by the charge to pass is small the period of the undamped oscillations of the galvanometer is large.

The time period for undamped oscillations of a galvanometer is $T_0 = 2\pi\sqrt{J/K}$.

Thus to have a long time period, (which is usually 10 to 15 seconds) the moment of inertia 'J' of the moving system should be large and the control constant of the suspension 'K' should be small.

The above conditions can be satisfied by attaching small weights to the moving system in order to increase its moment of inertia and by using suspensions of smaller stiffness so as to decrease the control constant $K$.

The damping of the galvanometer should be small in order that the amplitude of the first swing is large. After the first swing has been observed, electromagnetic damping may be used to bring the coil rapidly to rest. This can be done by having a key connected across the galvanometer terminals : this key short circuits the galvanometer coil when closed. This is because the electromagnetic torque produced on account of movement of coil is inversely proportional to resistance. When the coil is short circuited, the resistance in the circuit is only that of coil, which is small and so a large damping torque is produced which brings to the coil to rest in a short time.

Except for the above special features, the construction of a ballistic galvanometer is similar to a d'Arsonval type galvanometer.

### 8.15.2 Theory of Ballistic Galvanometer

It has been mentioned above that as long as charge passes through the meter there is no significant deflection of the moving system. The moving system deflects only after the charge has completely passed through the meter. The moving system has to deflect in order that the energy imparted to it by the charge is dissipated gradually in friction and electromagnetic damping. But during the actual motion, there is no deflecting torque as there is no current through the coil as the charge would have already passed.

Supposing $Q$ is the charge to be measured.

$$\therefore \quad Q = \int i\, dt \quad \text{or} \quad i = \frac{dQ}{dt}.$$

8

Let the instant of time during which the charge $Q$ passes be defined as the time between $t = t_0 = 0$ and $t = t_1$. Between time interval $t_0$ to $t_1$, there is no motion of the coil of the galvanometer and the equation of motion is,

$$J\frac{d^2\theta}{dt^2} + D\frac{d\theta}{dt} + K\theta = Gi$$

or
$$J\frac{d^2\theta}{dt^2} + D\frac{d\theta}{dt} + K\theta = G\frac{dQ}{dt}$$

or
$$\frac{d^2\theta}{dt^2} + \frac{D}{J}\frac{d\theta}{dt} + \frac{K}{J}\theta = \frac{G}{J}\frac{dQ}{dt}$$

Integrating above gives,

$$\left.\frac{d\theta}{dt}\right|_{t_0=0}^{t_1} + \left.\frac{D}{J}\theta\right|_{t_0=0}^{t_1} + \frac{K}{J}\int_0^{t_1}\theta\, dt = \frac{G}{J}Q \qquad \text{...(8.77)}$$

But the deflection, $\theta$, remains zero during the time interval $t_0$ to $t_1$. Then Eqn. 8.77 becomes,

$$d\theta/dt = (G/J)\theta \qquad \text{...(8.78)}$$

After passage of charge, *i.e.*, after the expiry of interval $t_1$, no current passes through the coil of the galvanometer and hence the deflecting torque is zero. Thus the equation of motion after time $t_1$ is,

$$J\frac{d^2\theta}{dt^2} + D\frac{d\theta}{dt} + K\theta = 0$$

The solution of this equation is taken from Eqn. 8.17.

$$\theta = F\exp(-Dt/2J)[\sin(\omega_d t + \phi)] + \theta_F$$

But in this case, the galvanometer does not show any steady state deflection and therefore $\theta_F = 0$.

Thus we have,

$$\theta = F\exp(-Dt/2J)[\sin(\omega_d t + \phi)] \qquad \text{...(8.79)}$$

Also since damping is small $\omega_d \approx \omega_n$.

Hence $\theta = F\exp(-D/2J)[\sin(\omega_n t + \phi)]$ ...(8.80)

Differentiating Eqn. 8.80, we have,

$$\frac{d\theta}{dt} = F\omega_n\exp(-D/2J)[\cos(\omega_n t + \phi)] - F(D/2J)\exp(-D/2J)[\sin(\omega_n t + \phi)] \qquad \text{...(8.81)}$$

The initial conditions at $t = 0$ are $\theta = 0$ and $\frac{d\theta}{dt} = (G/J)Q$.

From Eqn. 8.80 we get,

$$F \sin\phi = 0 \quad \text{or} \quad \phi = 0.$$

and from Eqn. 8.81, $F = GQ/J\omega_n$ ...(8.82)

Hence Eqn. 8.80 can be written as,

$$\theta = (G/J\omega_n)Q\exp(-Dt/J)\sin\omega_n t \quad \text{...(8.83)}$$

$$= (G/J).\sqrt{J/K}\,Q\exp(-Dt/2J)\sin\sqrt{K/J}\,t \quad \text{...(8.84)}$$

$$= AQ\exp(-Dt/2J)\sin(2\pi/T_0)t \quad \text{...(8.85)}$$

where $A = (G/J)(\sqrt{J/K})$ ...(8.86)

It is evident from Eqn. 8.83, that the charge $Q$, is proportional to the deflection, $\theta$, at any instant. Also the motion of the galvanometer is oscillatory with a decreasing amplitude. The deflection time curve for the ballistic galvanometer is shown in Fig. 8.11.

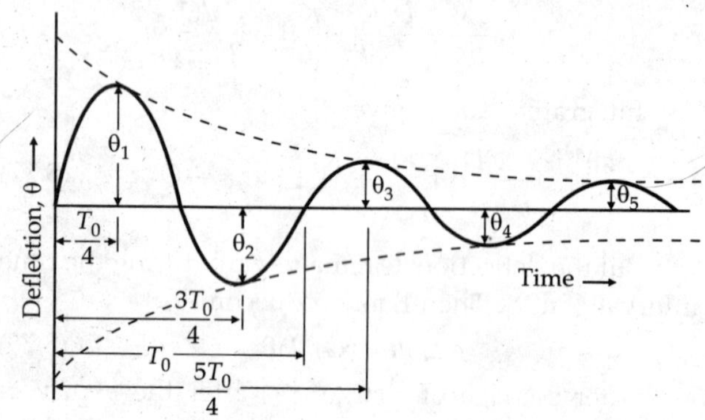


**Fig. 8.11** Variation of deflection with time for a ballistic galvanometer.

The amplitude of the oscillations decreases with time. The maxima of successive oscillations occurs at $t = T_0/4, 3T_0/4, 5T_0/4$, etc.

$$\theta_1 = KQ\exp(-D/2J.T_0/4)\sin(2\pi/T_0.T_0/4)$$

$$= KQ\exp(-D/2J.T_0/4) \quad \text{...(8.87)}$$

Putting the value of $T_0 = 2\pi\sqrt{J/K}$ in the above equation,

$$\theta_1 = AQ\exp(-D/2J.2\pi/4.\sqrt{J/K})$$

$$= AQ\exp(-\pi D/4\sqrt{JK}) \quad \text{...(8.88)}$$

Let us examine Eqn. 8.54 for logarithmic decrement.

Logarithmic decrement, $\lambda = \pi\zeta(T_d/T_0)$

For cases where damping is small, $T_d = T_0$.

$$\therefore \quad \lambda = \pi\xi = \pi D/D_c = \pi D/2\sqrt{KJ} \quad \text{...(8.89)}$$

Hence Eqn. 8.88 can be written as

$$\theta_1 = AQ\exp(-\lambda/2) \quad \text{...(8.90)}$$

Similarly, we can write

$$\theta_2 = AQ\exp(-3\lambda/2)$$

and $\theta_n = AQ\exp[-(2n-1)\lambda/2]$ ...(8.91)

Let us examine the case when the galvanometer is free from damping *i.e.*, $D = 0$. The equation of motion of an undamped galvanometer as derived from Eqn. 8.85 is :

$$\theta = AQ\sin(2\pi/T_0)t \quad \text{...(8.92)}$$

Amplitude of swings of undamped oscillations is, therefore,

$$\theta_0 = AQ\sin(2\pi/T_0 . T_0/4) = AQ \quad \text{...(8.93)}$$

The deflection-time curve for an undamped galvanometer is shown in Fig. 8.12.

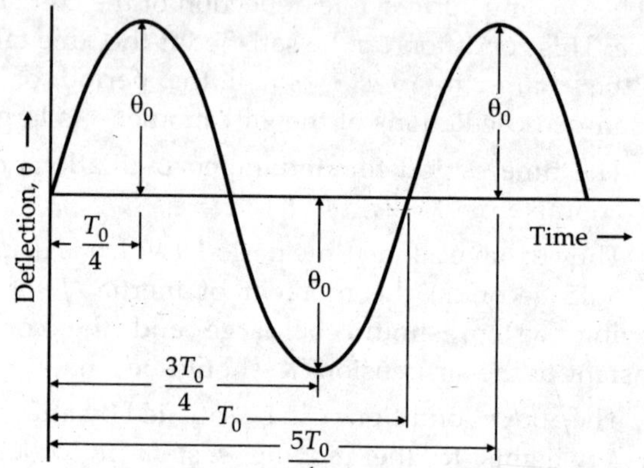


**Fig. 8.12** Undamped oscillations.

Hence, we can write

$$\theta_1 = \theta_0\exp(-\lambda/2) \quad \text{...(8.94)}$$

$$\theta_2 = \theta_0\exp(-3\lambda/2)$$

..................

$$\theta_n = \theta_0\exp[-(n-1)l/2] \quad \text{...(8.95)}$$

Logarithmic decrement is logarithm of ratio of successive swings.

$\therefore$ Logarithmic decrement

$$\lambda = \log_e(\theta_1/\theta_2) \quad \text{...(8.96)}$$

Now ratio of successive swings is :

$$\frac{\theta_1}{\theta_2} = \frac{\exp(-\lambda/2)}{\exp(-3\lambda/2)} = \exp(\lambda) \quad \text{...(8.97)}$$

Similarly, $\dfrac{\theta_2}{\theta_3} = \exp(\lambda)$

..................

$$\frac{\theta_{n-1}}{\theta_n} = \exp(\lambda) \quad \text{...(8.98)}$$

From above, we get

$$\frac{\theta_1}{\theta_2} \times \frac{\theta_2}{\theta_3} \times \frac{\theta_3}{\theta_4} \times \ldots \times \frac{\theta_{n-1}}{\theta_n} = \exp[(n-1)\lambda]$$

or

$$\frac{\theta_1}{\theta_n} = \exp[(n-1)\lambda]$$

Hence logarithmic decrement is,

$$\lambda = \frac{1}{n-1}\log_e\left(\frac{\theta_1}{\theta_n}\right) \qquad ...(8.99)$$

From Eqn. 8.96, we have

$$\theta_1 = \theta_0 \exp(-\lambda/2)$$

$$\therefore \quad \theta_0 = \theta_1 \exp(\lambda/2) \qquad ...(8.100)$$

$$\therefore \quad \theta_0 = \theta_1(1+\lambda/2) \text{ approximately} \qquad ...(8.101)$$

But $\theta_0 = AQ$

$$\text{Hence } AQ = \theta_1(1+\lambda/2) \qquad ...(8.102)$$

$$\therefore \text{Charge } Q = (\theta_1/A)(1+\lambda/2) \qquad ...(8.103)$$

Substituting the value $A = (G/J)\sqrt{J/K}$ in the above expression, we get,

$$\text{Charge, } Q = (J/G)\sqrt{K/J}\,.\,\theta_1(1+\lambda/2) \qquad ...(8.104)$$

Now $T_0 = 2\pi\sqrt{J/K}$

$$\therefore \text{Charge, } Q = \frac{J}{G}\sqrt{\frac{K}{J}}\cdot\frac{T_0}{2\pi\sqrt{J/K}}\theta_1(1+\lambda/2)$$

$$= \frac{K}{G}\frac{T_0}{2\pi}\theta_1(1+\lambda/2) \qquad ...(8.105)$$

Let us eliminate quantities $K$ and $G$ from above expression.

Suppose a steady current $I_g$ passing through the galvanometer produces a steady deflection $\theta$.

$$\therefore \quad GI_g = K\theta \quad \text{or} \quad K/G = I_g/\theta$$

$$\text{Hence} \quad Q = \frac{I_g}{\theta}\cdot\frac{T_0}{2\pi}(1+\lambda/2)\theta_1 \qquad ...(8.106)$$

$$= K_q\theta_1 \qquad ...(8.107)$$

$$\text{where} \quad K_q = \frac{I_g}{\theta}\frac{T_0}{2\pi}(1+\lambda/2) \qquad ...(8.108)$$

= constant of galvanometer.

Units of $K_q$ are coulomb per radian.

### 8.15.3 Calibration of a Ballistic Galvanometer

Calibration of a galvanometer means the experimental determination of its constants. The various methods for determining the constant are :

1. ***Using a mutual inductance.*** This method is most commonly used for the calibration purposes. The mutual inductance is usually in the form of a standard solenoid with a secondary coil. A standard solenoid consists of a long coil wound on a cylinder (Fig. 8.13). The length of the cylinder is very large as compared with its diameter. At the centre of the solenoid a secondary coil is placed. The axial length of the secondary coil is small.

The secondary coil is connected to the ballistic galvanometer. When a current is passed through the solenoid, the field strength at the centre is,

$$H = \frac{N_1 I}{l}\cos\theta = \frac{N_1 I}{l}\cdot\frac{l/2}{\sqrt{(l/2)^2 + r_1^2}}$$

$$= \frac{N_1 I}{2\sqrt{(l/2)^2 + (r_1)^2}}$$

where $N_1$ = number of turns of the solenoid,
$I$ = current through the solenoid,
$l$ = length of solenoid and
$r_1$ = mean radius of solenoid

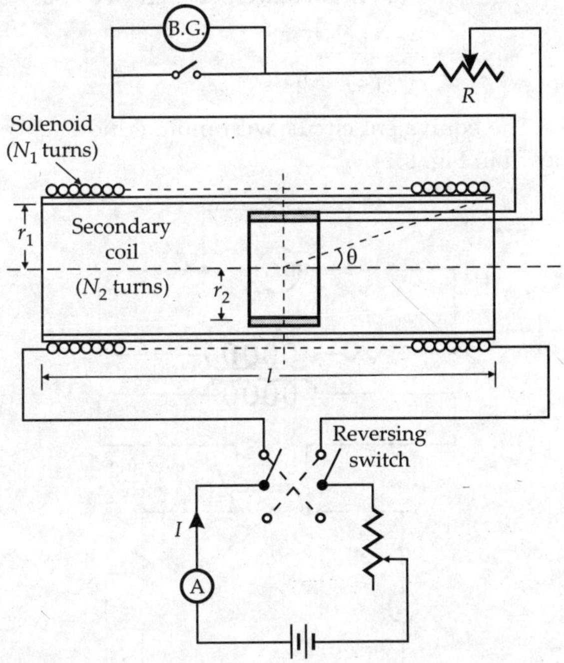


**Fig. 8.13** Calibration of ballistic galvanometer with mutual inductance.

Flux density at the centre of solenoid, $B = \mu_0 H$.

Flux linking with the secondary coil, $\Phi = B\times$ area of secondary coil $= B\times \pi r_2^2 = \pi B r_2^2 = \pi\mu_0 H r_2^2$.

where $r_2$ = mean radius of secondary coil.

Let $N_2$ be the number of turns in the secondary coil.

$\therefore$ Flux linkages of secondary coil $\Psi_2$

$$= N_2\Phi = \pi\mu_0 N_2 H r_2^2.$$

Let the current through the solenoid be reversed. This will change the value of flux linking with secondary coil from $+\Phi$ to $-\Phi$.

Change in flux $= \Phi - (-\Phi) = 2\,\Phi$.

Change in flux linkages of secondary coil $= 2\,N_2\Phi$

$$= 2\,\pi\mu_0 N_2 H r_2^2.$$

Suppose $\theta_1$ is the throw of the galvanometer

∴ Constant of galvanometer in terms of flux linkages

$$K_\phi = \frac{2\pi\mu_0 N_2 H r_2^2}{\theta_1} \text{ weber turn per radian} \quad ...(8.109)$$

$$= 2\pi\mu_0 N_2 \cdot \frac{N_1 I r_2^2}{2 \times \sqrt{(l/2)^2 + (r_1^2)}} \cdot \frac{1}{\theta_1}$$

$$= \frac{\pi\mu_0 N_1 N_2 r_2^2}{\sqrt{(l/2)^2 + (r_1^2)}} \cdot \frac{I}{\theta_1} \text{ weber turn per radian}$$

We have

$M$ = mutual inductance between two coils

$$= \frac{\pi\mu_0 N_1 N_2 r_2^2}{2 \times \sqrt{(l/2)^2 + r_1^2}}$$

The equivalent circuit with mutual inductance is shown in Fig. 8.14.

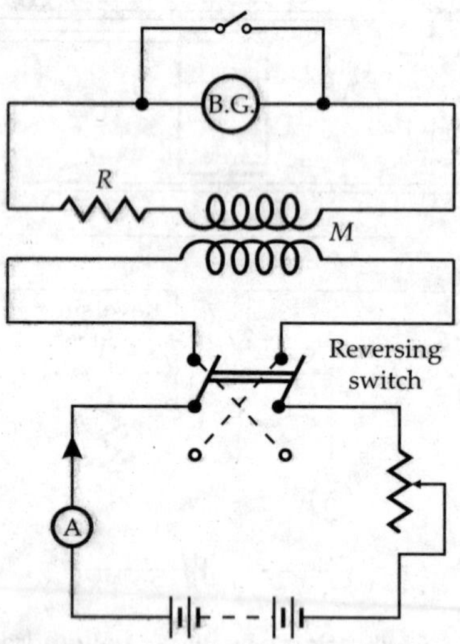


**Fig. 8.14** Calibration of B.G. with mutual inductance.

$$\therefore \quad K_\phi = \frac{2MI}{\theta_1} \text{ weber turn per radian} \quad ...(8.110)$$

Suppose the change in flux linkages takes place in time $t$.

∵ Emf induced in the secondary coil

$$= \text{rate of change in flux linkage} = -\frac{d\Psi}{dt}$$

$$= \frac{2\pi\mu_0 N_2 H r_2^2}{t}$$

This emf sends a current $i$ through the galvanometer

$$i = \frac{2\pi\mu_0 N_2 H r_2^2}{Rt}$$

where $R$ = galvanometer circuit resistance.

Charge passing in time $t$, $Q = it = \dfrac{2\pi\mu_0 N_2 H r_2^2}{R}$

But $\Theta = K_q \theta_1$

where $K_q$ = constant of galvanometer in coulomb per radian.

∴ Constant of galvanometer

$$K_q = \frac{Q}{\theta_1} = \frac{2\pi\mu_0 N_2 H r_2^2}{R} \frac{1}{\theta_1}$$

$$= \frac{\pi\mu_0 N_1 N_2 r_2^2 I}{R\sqrt{(l/2)^2 + (r_1)^2}} = \frac{2MI}{R\theta_1} \quad ...(8.111)$$

Thus the galvanometer constant can be obtained in terms of coulomb per radian.

From the above analysis, $K_q = K_\phi / R$ ...(8.112)

Sometimes terms *"Ballistic Sensitivity"* and *"Flux linkage sensitivity"* are used in place of constants of galvanometers. These terms are defined as :

Ballistic sensitivity $S_q = \theta_1 / Q$ rad/C ....(8.113)

and flux linkage sensitivity

$S_\phi = \theta_1 / \Delta\Psi$ rad/weber turn ...(8.114)

Naturally $S_q = 1/K_q$ ...(8.115)

and $S_\phi = 1/K_\phi$ ...(8.116)

Also $S_q = RS_\phi$ ...(8.117)

**2. *Using a capacitor.*** A capacitor is charged to a known emf $E$ and is then discharged through the ballistic galvanometer. The throw of the galvanometer is observed. Let this throw be $\theta_1$.

∴ Constant of galvanometer

$$K_q = Q/\theta_1 = CE/\theta_1 \text{ coulomb/radian}$$

The circuit is shown in Fig. 8.15.

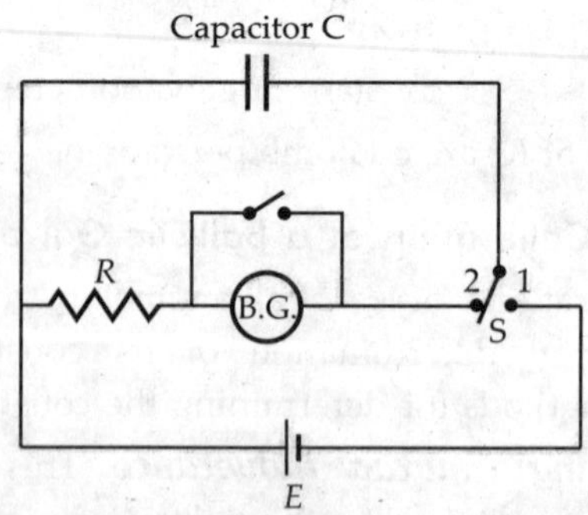


**Fig. 8.15** Calibration of Ballistic Galvanometer with capacitor.

When the switch $S$ is put at position 1, the capacitor $C$ is charged to potential $E$. The switch is then put at position 2. At this position the capacitor is discharged through a resistance.

3. ***Using Hibbert magnetic standard.*** A Hibbert magnetic standard (Fig. 8.16) consists of a circular bar magnet and an iron yoke which have a narrow annular gap between them. A brass tube carrying a coil can slide through the gap. The brass tube is taken to its top position and is then released with the help of a trigger. It falls under gravity and as it slides through the gap, its coil cuts through the magnetic field of the permanent magnet and therefore an emf is produced in the coil. The coil is connected across the terminals of the ballistic galvanometer. The throw of the B.G. (Ballistic Galvanometer) can be observed. The rate at which coil cuts through the field is constant for a particular apparatus. Thus emf per turn is constant. By the use of the standard the flux linkages which produce an observed throw of the galvanometer can be known by knowing the air gap flux and the number of turns in the coil. Hence the galvanometer constant can be evaluated.

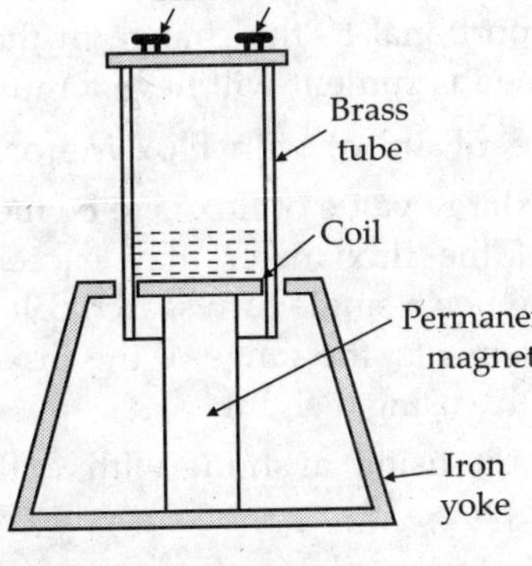


**Fig. 8.16** Hibbert magnetic standard.

## 8.16 FLUX METER

The flux meter is a special type of ballistic galvanometer in which the controlling torque is very small and the electromagnetic damping is heavy.

### 8.16.1 Construction of Flux Meter

The construction of a flux meter is shown in Fig. 8.17. In general the construction is similar to that of a moving coil milli-ammeter. A coil of small cross-section is suspended from a spring support by means of a single silk thread. The coil moves in the narrow gap of a permanent magnet. There are no control springs.

The current is led into the coil with the help of a very loose helices of very thin, annealed silver strips. The controlling torque is thus reduced to minimum.

The coil is formerless and the air friction damping is negligible.

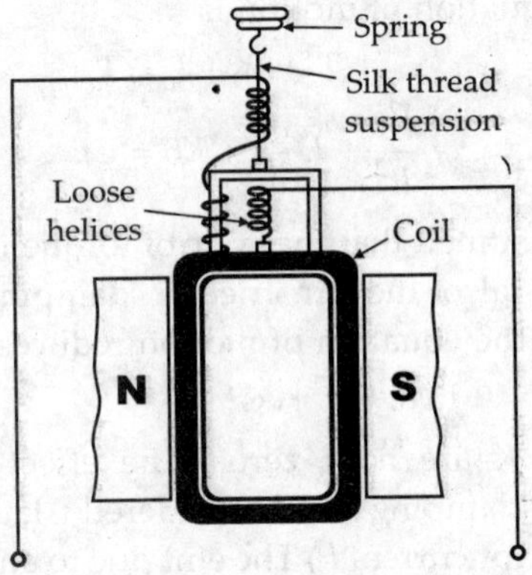


**Fig. 8.17** Flux meter.

### 8.16.2 Operation of Flux Meter

The terminals of the flux meter are connected to a search coil as shown in Fig. 8.18. The flux linking with the search coil is changed either by removing the coil from the magnetic field or by reversing the field. Due to the change in the value of flux linking with the search coil an emf is induced in it. This emf sends a current through the flux meter which deflects through an angle depending upon the change in the value of flux linkages.

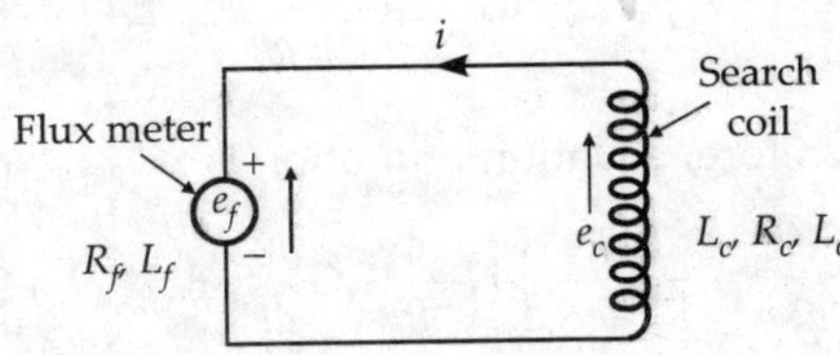


**Fig. 8.18** Flux meter with search coil.

The instrument coil deflects during the period the flux linkages change but as soon as the change ceases the coil stops, owing to high electromagnetic damping in the coil circuit. This high electromagnetic damping is obtained by having a low resistance of the circuit comprising the flux meter mid the search coil.

### 8.16.3 Theory of Flux Meter

Let

$N$ = number of turns in the search coil,

$R_c, L_c$ = resistance and inductance of the search coil respectively,

$R_f, L_f$ = resistance and inductance of flux meter respectively,

$\Phi$ = flux linking with the search coil,

$i$ = current in the circuit at any instant,

and $\theta$ = deflection of the instrument at any instant.

8

The equation of motion is

$$T_j + T_D + T_c = T_d$$

or $$J\frac{d^2\theta}{dt^2} + D\frac{d\theta}{dt} + K\theta = Gi.$$

It is assumed that the control torque is negligibly small and also the air friction damping is small. Therefore, the equation of motion reduces to :

$$J d^2\theta / dt^2 = Gi \qquad \ldots(8.118)$$

as $D$ and $K$ are both zero. (The effect of electromagnetic damping is considered later in the expression for current $i$.) The emf due to any change of flux linked with search coil : $e_c = N\Phi / dt$.

At the same time, owing to the movement of the flux meter coil in the field of the magnet, a rotational emf is induced in the coil. This emf $e_f = Gd\theta / dt$.

Also there are voltage drops in the resistance and inductance of the circuit.

Therefore, we can write

$$e_c = e_f + (L_f + L_c)\frac{di}{dt} + (R_f + R_c)i$$

or $$N\frac{d\Phi}{dt} = G\frac{d\theta}{dt} + (L_f + L_c)\frac{di}{dt} + (R_f + R_c)i$$

or $$i = \frac{N\frac{d\Phi}{dt} - G\frac{d\theta}{dt} - (L_f + L_c)\frac{di}{dt}}{R_f + R_c}$$

Therefore, equation of motion is :

$$J\frac{d^2\theta}{dt^2} = \frac{G}{R_f + R_c}\left[N\frac{d\Phi}{dt} - G\frac{d\theta}{dt} - (L_f + L_c)\frac{di}{dt}\right]$$

or $$N\frac{d\Phi}{dt} = \frac{J(R_f + R_c)}{G}\frac{d^2\theta}{dt^2} + G\frac{d\theta}{dt} + (L_f + L_c)\frac{di}{dt}$$

$$= \frac{J(R_f + R_c)}{G}\frac{d\omega}{dt} + G\frac{d\theta}{dt} + (L_f + L_c)\frac{di}{dt}$$

where $\omega$ = angular velocity of moving coil at any instant $t$.

Now if the time taken by change in flux is $T$, we have,

$$\int_0^T N\frac{d\Phi}{dt}dt = \int_0^T \frac{J(R_f + R_c)}{G}\frac{d\omega}{dt}dt + \int_0^T G\frac{d\theta}{dt}dt + \int_0^T (L_f + L_c)\frac{di}{dt}dt$$

or $$\int_{\Phi_1}^{\Phi_2} N d\Phi = +\int_{\omega_1}^{\omega_2} \frac{J(R_f + R_c)}{G}d\omega + \int_{\theta_1}^{\theta_2} G d\theta + \int_{i_1}^{i_2} (L_f + L_c)\, di$$

$$\therefore\ N(\Phi_2 - \Phi_1) = \frac{J(R_f + R_c)}{G}(\omega_2 - \omega_1) + G(\theta_2 - \theta_1) + (L_f + L_c)(i_2 - i_1) \qquad \ldots(i)$$

where $\Phi_1$ and $\Phi_2$ are the interlinking fluxes, $\omega_1$ and $\omega_2$; are the angular velocities, $\theta_1$ and $\theta_2$ are the deflections and $i_1$ and $i_2$ are the currents ; suffixes 1 and 2 indicate respectively values at the beginning and at the end of the change.

But angular velocities and currents are zero at both the beginning and the end of the change

*i.e.,* $\omega_1 = \omega_2 = 0$ and $i_1 = i_2 = 0$

Putting these values in (*i*), we have

$$N(\Phi_2 - \Phi_1) = G(\theta_2 - \theta_1) \qquad \ldots(8.118)$$

or if $\Phi$ is the change in the flux and $\theta$ the change in flux meter deflection

$$\Phi = (G/N)\theta \qquad \ldots(8.119)$$

If the flux meter permanent magnet field is uniform for all positions of the moving coil, $G$ is a constant. Thus the change in the value of flux is directly proportional to the change in the deflection and hence the instrument will have a uniform scale.

### 8.16.4 Use of Shunt with Flux Meter

If a very large value of flux is to be measured, the deflection of the flux meter may exceed the scale length even when a single turn search coil is used. For such measurements, the range of the flux meter may be extended by using a shunt.

The circuit using a shunt with a fluxmeter is shown in Fig. 8.19.

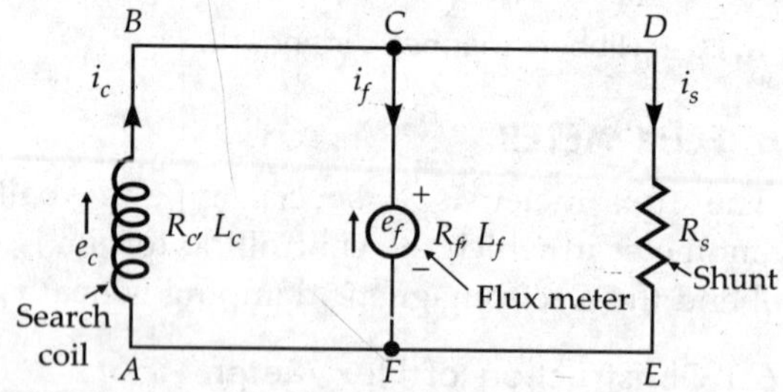


**Fig. 8.19** Flux meter with shunt.

Let,

$R_c, R_f, F_s$ = resistance of search coil, flux meter and shunt respectively,

$L_c, L_f$ = inductance of search coil and flux meter respectively,

$i_c, i_f, i_s$ = current through search coil, flux meter and shunt respectively,

$e_c$ = emf induced in the search coil because of change in flux linkages = $Nd\Phi / dt$.

$e_f$ = emf generated in the flux meter coil because of rotation = $Gd\Phi / dt$

Applying Kirchhoff's law for the mesh *ABCF*,

$$e_c = i_c R_c + L_c \frac{di_c}{dt} + i_f R_f + L_f \frac{di_f}{dt} + e_f$$

or $$N\frac{d\Phi}{dt} - L_c\frac{di_c}{dt} - L_f\frac{di_f}{dt} - G\frac{d\theta}{dt} = i_c R_c + i_f R_f.$$

Integrating as before, we have,

$$N(\Phi_2-\Phi_1) - L_c(i_{c2}-i_{c1}) - L_f(i_{f2}-i_{f2}) - G(\theta_2-\theta_1)$$
$$= R_c\int_0^T i_c\,dt + R_f\int_0^T i_f\,dt.$$

Now currents are zero at both the beginning and the end of the flux change

or $$i_{c2} = i_{c1} = 0,\ i_{f2} = i_{f1} = 0.$$

Therefore, we can write

$$N(\Phi_2-\Phi_1) - G(\theta_2-\theta_1) = R_c\int_0^T i_c\,dt + R_f\int_0^T i_f\,dt. \quad ...(ii)$$

The equation of motion is $J\frac{d^2\theta}{dt^2} = Gi_f$

or $$i_f = \frac{J}{G}\frac{d^2\theta}{dt^2} = \frac{J}{G}\frac{d\omega}{dt}$$

where $\omega$ is the angular velocity at any instant.

Hence $$\int_0^T i_f\,dt = \frac{J}{G}\int_0^T \frac{d\omega}{dt}.dt = \frac{J}{G}\int_{\omega_1}^{\omega_2} d\omega$$

or $$\int_0^T i_f\,dt = \frac{J}{G}(\omega_2-\omega_1)$$

But $\omega_2 = \omega_1 = 0$

Therefore, $\int_0^T i_f\,dt = 0$

Hence (*ii*) reduces to

$$N(\Phi_2-\Phi_1) - G(\theta_2-\theta_1) = R_c\int_0^T i_c\,dt \quad ...(iii)$$

Let us apply Kirchhoff's law to mesh *CDEF*, we have,

$$e_f + L_f\frac{di_f}{dt} + i_f R_f = i_s R_s$$

or $$G\frac{d\theta}{dt} + L_f\frac{di_f}{dt} = -i_f R_f + i_s R_s$$
$$= -i_f R_f + (i_c - i_f)R_s$$

or $$G\frac{d\theta}{dt} + L_f\frac{di_f}{dt} = -(R_f+R_s)i_f + i_c R$$

Integrating as before

$$G(\theta_2-\theta_1) + L_f(i_{f2}-i_{f1}) = -(R_f+R_s)\int_0^T i_f\,dt + R_s\int_0^T i_c\,dt$$

But $i_{f2} = i_{f1} = 0$ and $\int_0^T i_f\,dt = 0.$

Hence $$G(\theta_2-\theta_1) = R_s\int_0^T i_c\,dt$$

or $$\int_0^T i_c\,dt = \frac{G(\theta_2-\theta_1)}{R_s}$$

Substituting this value of $\int_0^T i_c\,dt$ in (*iii*), we get,

$$N(\Phi_2-\Phi_1) - G(\theta_2-\theta_1) = \frac{R_c}{R_s}G(\theta_2-\theta_1)$$

or $$N(\Phi_2-\Phi_1) = G\frac{R_c+R_s}{R_s}(\theta_2-\theta_1) \quad ...(8.121)$$

If $(\Phi_2-\Phi_1) = \Phi$ and $(\theta_2-\theta_1) = \theta$

We have $$\Phi = \frac{G}{N}\cdot\frac{R_c+R_s}{R_s}\theta \quad ...(8.122)$$

Comparison of Eqn. 8.122 with corresponding Eqn. 8.120 of unshunted flux meter shows that the multiplying factor for the shunted flux meter is :

$$m = \frac{R_c+R_s}{R_s} \quad ...(8.123)$$

**NOTE** The theory of the unshunted flux meter shows (see Eqn. 8.120) that the value of resistance of either the flux meter or the search coil does not affect the meter sensitivity. However, the value of total circuit resistance should not exceed about 20 ohm otherwise the damping will not be adequate. The theory of a shunted flux meter shows that the resistance $R_c$ of the search coil is important when the meter is shunted. The resistance $R_f$ of the meter is unimportant.

The theory also shows the inductance of search coil and the flux meter are immaterial and that the time for flux change has no effect on the indication of the flux meter. This time may be as much as 2 minutes in some eases, however, in practice, if the time exceeds about 30 seconds, serious errors are introduced because of friction, a small control torque exerted by the current leads and the effect of improper levelling etc.

### 8.16.5 Advantages and Disadvantages of Flux Meters

***Advantages :***

(*i*) The industrial form of flux meter is portable.

(*ii*) Its scale is directly calibrated in weber turns.

(*iii*) The indication of the meter is independent of the time taken by the flux change. This is of great importance in measurement of flux linking with highly inductive circuits where the flux changes may be relatively slow rendering the use of ballistic galvanometers impractical.

***Disadvantages :***

The flux meter is less sensitive and accurate than the ballistic galvanometer.

**Example 8.15** *The coil of a ballistic galvanometer has 115 turns of mean area $25 \times 40$ mm$^2$. The flux density in the air gap is 0.12 Wh/m$^2$ and the moment of inertia is $0.5 \times 10^{-6}$ kg-m$^2$. The stiffness constant of springs is $45 \times 10^{-6}$ Nm/rad. What current must be passed to give a deflection of 100° and what resistance must be added in series with the movement to give critical damping ?*

**Solution**. Displacement constant, $G = NBA$

$$= 115 \times 0.12 \times 25 \times 40 \times 10^{-6}$$

$$= 138 \times 10^{-3} \text{ N-m/A.}$$

Deflection, $\theta = 100° = 1.475$ rad.

But $$i = \frac{K\theta}{G} = \frac{4.5 \times 10^{-6}}{138 \times 10^{-3}} \times 1.475$$

$$= 4.8 \times 10^{-3} \text{ A} = 4.8 \text{ mA}$$

For critical damping, damping constant

$$D = 2\sqrt{KJ} = 2\sqrt{45 \times 10^{-6} \times 0.5 \times 10^{-6}}$$

$$= 9.48 \times 10^{-6} \text{ Nm/rad-s}^{-1}.$$

Neglecting resistance of the galvanometer coil, external resistance required for critical damping is :

$$CDRX = \frac{G^2}{D} = \frac{(138 \times 10^{-3})^2}{9.48 \times 10^{-6}}$$

$$= 2000\ \Omega.$$

**Example 8.16** *A solenoid 1 m long and wound with 800 turns has a search coil of 50 turns and cross-sectional area 300 mm$^2$ at its centre. Reversing a current of 4 A in the solenoid causes a deflection of 5 divisions in a ballistic galvanometer connected to the search coil. Calculate the galvanometer constant in flux linkages per division. Calculate also the flux linkage sensitivity.*

**Solution.** Flux density at the centre of solenoid

$$B = \mu_0 N_1 I / l$$

Flux linking with search coil, $\Phi = BA = \mu_0 N_1 IA / l$

Flux linkages of search coil, $\Psi = N_2 BA$

$$= \mu_0 N_1 N_2 IA / l$$

Change of flux linkages, $\Delta\Psi = 2\Psi = 2\mu_0 N_1 N_2 IA / l$ (since the current is reversed)

Galvanometer constant

$$K_\phi = \frac{2\Psi}{\theta_1} = \frac{2\mu_0 N_1 N_2 IA}{l\theta_1}$$

$$= \frac{2 \times 4\pi \times 10^{-7} \times 800 \times 50 \times 4 \times 300 \times 10^{-6}}{1 \times 5}$$

$$= 24.1 \times 10^{-6} \text{ weber turn per division.}$$

Flux linkage sensitivity,

$$S_\phi = \frac{1}{K_\phi} = \frac{1}{24.1 \times 10^{-6}}$$

$$= 41.5 \times 10^3 \text{ divisions per weber turn.}$$

**Example 8.17** *The mutual inductance between magnetizing winding and a search coil wound on a specimen is 9 mH. The search coil has 20 turns and the specimen has a cross-sectional area of 500 mm$^2$. Reversal of a current of 3 A in magnetising winding produces a galvanometer throw of 60 divisions. Calculate the value of the flux density in a specimen if the reversal of current in the magnetizing winding produces a galvanometer deflection of 36 divisions.*

**Solution.** Flux linkages of search coil, $\Psi = MI$.

Change in flux linkages, $\Delta\Psi = 2\Psi = 2MI$

(since the current is reversed)

Galvanometer constant

$$K_\phi = \frac{2\Psi}{\theta_1} = \frac{2MI}{\theta_1} = \frac{2 \times 9 \times 10^{-3} \times 3}{60}$$

$$= 0.9 \times 10^{-3} \text{ weber turn per division.}$$

With a galvanometer throw of 36 divisions,

Change in flux linkages

$$\Delta\Psi = 2\Psi$$

$$= 36 \times 0.9 \times 10^{-3} = 324 \times 10^{-3} \text{ weber turn.}$$

Original flux linkages, $\Psi = 162 \times 10^{-3}$ weber turn.

Flux in specimen

$$\Phi = \frac{\Psi}{N_2} = \frac{16.2 \times 10^{-3}}{20} = 0.81 \times 10^{-3} \text{ Wb.}$$

Flux density

$$B = \frac{\Phi}{A} = \frac{0.81 \times 10^{-3}}{500 \times 10^{-6}} = 1.62 \text{ Wb/m}^2.$$

**Example 8.18** *A ballistic galvanometer has a resistance of 150 Ω and an undamped period of 7.5 s. A steady emf of 3.5 mV produces a deflection of 210 mm. Determine the quantity of electricity discharged from a capacitor if the deflection produced is 750 mm. The relative damping is 0.8.*

**Solution.** Steady state current

$$I_g = 3.5 \times 10^{-3} / 150 = 23.3 \times 10^{-6} \text{ A}$$

Under steady state conditions : $K\theta = GI_g$

$$\therefore \quad \frac{K}{G}=\frac{I_g}{\theta}=\frac{23.3\times10^{-6}}{210}=0.11\times10^{-6}\text{ A/mm.}$$

Logarithmic decrement

$$\lambda=\frac{\pi\xi}{\sqrt{1-\xi^2}}=\frac{\pi\times0.8}{\sqrt{1-(0.8)^2}}=4.19.$$

From Eqn. 8.104, quantity of electricity

$$Q=\frac{K}{G}\cdot\frac{T_0}{2\pi}\theta_1(1+\lambda/2)$$

$$=0.111\times10^{-6}\times\frac{7.5}{2\pi}\times\frac{1+4.19}{2}\times750$$

$$=307.5\times10^{-6}\text{ C}=307.5\ \mu\text{C}$$

**Example 8.19** *The discharge of a capacitor through a ballistic galvanometer produces a damped frequency of 0.125 Hz and successive swings of 120, 96 and 76.8 mm, calculate the damping ratio. Also determine the logarithmic decrement and period of undamed and damped oscillations.*

**Solution.** From Eqn. 8.99, logarithmic decrement

$$\lambda=\frac{1}{n-1}\log_e\frac{\theta_1}{\theta_n}=\frac{1}{(3-1)}\log_e\frac{120}{76.8}=0.223.$$

$$\text{Period of damped oscillations, } T_d=\frac{1}{f_d}=\frac{1}{0.125}=8\text{s}$$

$$\text{From Eqn. 8.52, logarithmic decrement, } \lambda=\frac{\pi\xi}{\sqrt{1-\xi^2}}.$$

$$\text{Damping ratio, } \xi=\sqrt{\frac{\lambda^2}{\pi^2+\lambda^2}}=\sqrt{\frac{(0.223)^2}{\pi^2+(0.223)^2}}$$

$$=0.0708.$$

Frequency of undamped oscillations,

$$f_n=\frac{f_d}{\sqrt{1-\xi^2}}=\frac{0.125}{\sqrt{1-(0.0708)^2}}=0.1253\text{ Hz}$$

$$\text{Period of undamped oscillations, } T_0=\frac{1}{0.1253}=7.98\text{ s.}$$

**Example 8.20** *A moving coil ballistic galvanometer of 150 Ω resistance gives a throw of 75 divisions when the flux through a search coil to which it is connected is reversed. Find the flux density in which the reversal of the coil takes place, given that the galvanometer constant = 110 μC /div and the search coil has 1400 turns, a mean area of 5000 mm² and a resistance of 20 Ω.*

*Calculate also the ballistic and flux linkage sensitivities of the galvanometer.*

**Solution.** Resistance of galvanometer circuit

$$=20+150=170\ \Omega.$$

Galvanometer constant

$$K_q=110\ \mu\text{C/div.}$$

$$\text{Now} \quad K_q=\frac{Q}{\theta_1}=\frac{it}{\theta_1}=\frac{et}{R\theta_1}.$$

But change in flux linkages, $\Delta\Psi=2\Psi$ since the current is reversed.

$$\therefore \quad e=\frac{2\psi}{t}=\frac{2\Phi N}{t}=\frac{2BAN}{t}$$

$$\text{Hence } K_q=\frac{2BAN}{t}\cdot\frac{t}{R\theta_1}=\frac{2BAN}{RQ_1}.$$

Flux density

$$B=\frac{K_q R\theta_1}{2AN}=\frac{110\times10^{-6}\times170\times75}{2\times5000\times10^{-6}\times1400}=0.1\text{ Wb/m}^2.$$

Ballistic sensitivity

$$S_q=1/K_q=1/110=9.1\times10^{-3}\text{ div/}\mu\text{C}=9100\text{ div/C.}$$

Flux linkage sensitivity

$$S_\phi=S_q/R=9100/150=60.67\text{ weber turns/div.}$$

**Example 8.21** *A ballistic galvanometer gives a first maximum deflection of 60° for a discharge of 1000 μC. Find the quantity of electricity which when discharged through this galvanometer gives rise to a spot deflection of 10 divisions on a millimeter circular scale 1 metre away.*

**Solution.** Let us assume that angular deflection of 60° is the deflection of the reflected ray from the mirror attached to the moving coil.

$$\text{Charge} \quad Q=K_q\,\theta_1$$

$$\text{Now} \quad \theta_1=60°=\pi/3\text{ rad.}$$

$$\therefore \quad K_q=\frac{1000}{\pi/3}=\frac{3000}{\pi}\ \mu\text{C/rad.}$$

Angle turned through by reflected ray for a swing of 10 mm is : $\theta=\frac{1}{2r}=\frac{10}{2000}=0.005$ rad

∴ Charge for a swing of 0.005 radian is :

$$Q=(3000/\pi)\times0.005=4.77\ \mu\text{C.}$$

**Example 8.22** *A galvanometer with an undamped period of 9 seconds and a current sensitivity of 1 microampere per scale division is connected to a search coil in a circuit of 4000 Ω resistance. Determine the flux linkage change in the search coil to produce a first swing of 100 divisions if the logarithmic decrement is 0.2.*

**Solution.**

$$\text{Charge, } Q=\frac{T_0}{2\pi}\frac{I_g}{\theta}(1+\lambda/2)\theta_1$$

$$=(9/2\pi)\times1\times10^{-6}(1+0.2/2)\times100$$

$$=158\times10^{-6}\text{ C}$$

Let $\Delta\Psi$ be the change in flux linkages in time $t$

$$\therefore \quad \text{emf, } e=\frac{\Delta\Psi}{t},\ \text{Charge, } Q=it=\frac{e}{R}.t=\frac{\Delta\Psi}{R}.$$

$$\therefore \text{ Change in flux linkage, } \Delta\Psi=RQ$$

$$=4000\times158\times10^{-6}$$

$$=632\times10^{-3}\text{ weber turn.}$$

***Example 8.23*** *A certain flux meter has the following constants :*

*Air gap flux density* $= 0.05\ Wb/m^2$,

*turns on moving coil* $= 40$,

*area of moving coil* $= 750\ mm^2$.

*If the flux linking a 10 turn search coil of 200 mm² area connected to the flux meter is reversed in a uniform field of 0.5 Wb/m², calculate the deflection of the flux meter.*

**Solution.**

Constant of flux meter

$$G = 40 \times 0.05 \times 750 \times 10^{-6} = 1500 \times 10^{-6}$$

Flux linking with search coil

$$= 0.5 \times 200 \times 10^{-6} = 100 \times 10^{-6} \text{ Wb.}$$

Change in flux linking the coil

$$= 2 \times 100 \times 10^{-6} = 200 \times 10^{-6} \text{ Wb}$$

From Eqn. 8.120, change in flux $\Phi = (G/N)\theta$

$$\therefore \text{ Deflection, } \theta = \frac{N\Phi}{G} = \frac{10 \times 200 \times 10^{-6}}{1500 \times 10^{-6}} \text{ rad.} = 76.5°$$

***Example 8.24*** *A flux meter is connected to a search coil having 500 turns and a mean area of 500 mm². The search coil is placed at the centre of a solenoid 1 metre long, wound with 800 turns. When a current of 5 A is reversed, there is a deflection of 25 scale divisions. Calculate the calibration in flux linkages per scale division.*

**Solution.** Field strength of solenoid

$$H = \frac{N_1 I}{l} = \frac{800 \times 5}{1} = 4000 \text{ A/m}$$

(assuming it to be very long).

Flux density

$$B = \mu_0 H = 4\pi \times 10^{-7} \times 4000$$
$$= 16\pi \times 10^{-3} \text{ Wb/m}^2.$$

Flux linkages of flux meter search coil

$$= N_2 BA = 500 \times 16\pi \times 10^{-3} \times 500 \times 10^{-6}$$
$$= 0.4\pi \times 10^{-3} \text{ weber turn.}$$

Change in flux linkages

$$= 2 \times (0.4\ \pi \times 10^{-3}) = 0.8\pi \times 10^{-3} \text{ weber turn.}$$

Flux linkages per scale division

$$= 0.8\pi \times 10^{-3} / 25 = 0.1005 \times 10^{-3} \text{ weber turn.}$$

***Example 8.25*** *A flux meter when used with a single turn search coil gives full scale deflection if a given uniform field linking with the search coil is reversed. It is proposed to extend the range of flux meter to measure flux densities five times the above value using the same search coil and method of reversals. If the resistance of search coil is 1 Ω, find the resistance of the shunt to be connected in parallel with the flux meter.*

**Solution.** From Eqn. 8.123, the range of instrument is extended $(R_c + R_s)/R_s$ times by use of a shunt. Where $R_c$ = resistance of search coil and $R_s$ = resistance of shunt.

Now, we have, $\dfrac{R_c + R_s}{R_s} = 5$ or $\dfrac{1 + R_s}{R_s} = 5$

Resistance of shunt, $R_s = 0.25\ \Omega$.

***Example 8.26*** *The reluctance of a magnetic circuit excited by an mmf of 8000 A is 120,000 A/Wb. A flux meter is used to measure the flux. If the flux meter scale has 120 divisions and the flux linkages required for a deflection of 1 division are* $0.15 \times 10^{-3}$ *weber turn, calculate the resistance of the shunt required for use with the search coil. Number of turns of search coil is 1 and its resistance is 0.025 Ω. The measurement is made by switching off the excitation.*

**Solution.** Flux to be measured $= \dfrac{\text{mmf}}{\text{reluctance}}$

$$= \frac{8000}{1{,}20{,}000} = 66.7 \times 10^{-3} \text{ Wb.}$$

Value of flux when the flux meter shows full scale deflection

$$= \frac{0.15 \times 10^{-3} \times 120}{1} = 18 \times 10^{-3} \text{ Wb.}$$

$$\text{Ratio} = \frac{\text{flux to be measured}}{\text{flux for full scale deflection}}$$

$$= \frac{66.7 \times 10^{-3}}{18 \times 10^{-3}} = 3.7$$

Thus we have, $\dfrac{R_c + R_s}{R_s} = \dfrac{0.025 \times R_s}{R_s} = 3.7$

or resistance of shunt, $R_s = 0.00926\ \Omega$.

## 8.17 VIBRATION GALVANOMETERS

These galvanometers are of d'Arsonval type having a moving coil suspended between the pieces of a permanent magnet. When an alternating current is passed through the moving coil, an alternating deflecting torque is produced which makes the coil vibrate with a frequency equal to the frequency of the current passing. On account of inertia of the moving parts, the amplitude of vibrations is small. However, if the natural frequency of the moving system is made equal to the frequency of the current, mechanical resonance is obtained and the moving system vibrates with a large amplitude.

The damping is very small in these galvanometers. This is required in order to obtain a sharp resonance curve so that the galvanometer is selective *i.e.,* it gives a very small deflection for alternating currents of frequencies other than the resonant frequency.

Vibration galvanometers are suitable for use at power and low audio frequencies, but they are mainly used at power frequencies. They can be manufactured for fixed frequencies, or they can be provided with tuning devices to cover a range of frequencies.

### 8.17.1 Construction of Vibration Galvanometers

Figure 8.20 shows the construction of a Duddell's moving coil vibration galvanometer. The moving coil consists of a fine bronze or platinum silver wire. This wire passes over a small pulley at the top and is pulled tight by a spring attached to the pulley, the tension of the spring can be adjusted by turning a milled head attached to the spring. The loop of wire is stretched over two ivory bridge pieces, the distance between these pieces is adjustable. When the moving coil vibrates due to passage of a.c., the reflected beam from the mirror throws a band of light upon a scale provided for the purpose.

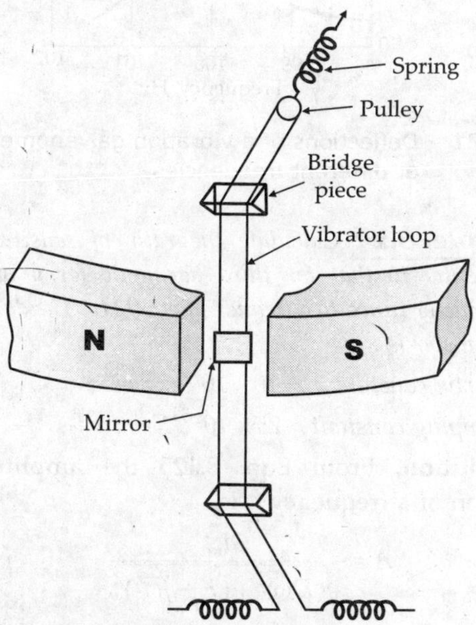


**Fig. 8.20** Vibration galvanometer.

*Tuning*. Tuning means adjustment of the natural frequency of the moving system so that it is equal to the frequency of the current passing through the coil.

In the case of Duddell's galvanometer coarse tuning is done by varying the distance between the bridge pieces. This varies the length of the loop which is free to vibrate and thus varies the natural frequency of the moving system. Fine adjustment of tuning is done by varying the tension of the spring.

When the galvanometer is tuned (*i.e.*, when it is under mechanical resonance conditions), the amplitude of vibrations is very large and consequently a very wide band of light is observed on the scale.

***Theory.*** Let the value of current passing through the moving coil at an instant $t$ be :

$$i = I_m \sin\omega t$$

$\therefore$ Deflecting torque, $T_d = Gi = GI_m \sin\omega t.$

Equation of motion is $T_j + T_D + T_c = T_d$

or
$$J\frac{d^2\theta}{dt^2} + D\frac{d\theta}{dt} + K\theta = GI_m \sin\omega t.$$

***Complementary function***. The galvanometer is underdamped and therefore the complementary function is :

$$\theta = F\exp(-Dt/2J)[\sin(\omega_d t + \phi)] \quad \text{(See Eqn. 8.17)}$$

***Particular integral.*** Writing the equation for motion we have,

$$J\frac{d^2\theta}{dt^2} + D\frac{d\theta}{dt} K\theta = GI_m \sin\omega t$$

The particular integral is of the form

$$\theta = A\sin(\omega t - \alpha)$$

where $A$ and $\alpha$ are constants.

We have, $\dfrac{d\theta}{dt} = A\omega(\cos\omega t - \alpha)$

and $\dfrac{d^2\theta}{dt^2} = -A\omega^2\sin(\omega t - \alpha)$

Substituting the above values in equation for motion, we have

$$-AJ\omega^2\sin(\omega t - \alpha) + AD\omega\cos(\omega t - \alpha)$$
$$+ AK\sin(\omega t - \alpha) = GI_m \sin\omega t$$

The above relation must hold good for all values of $t$.

When $\omega t = \alpha$, $\quad AD\omega = GI_m \sin\alpha \quad ...(i)$

When $(\omega t - \alpha) = \pi/2 \quad -AJ\omega^2 + AK = GI_m \cos\alpha \quad ...(ii)$

Squaring (*i*) and (*ii*) and adding we get :

$$A^2D^2\omega^2 + A^2(K - J\omega^2)^2 = G^2I_m^2$$

or
$$A = \frac{GI_m}{\sqrt{(D\omega)^2 + (K - J\omega^2)^2}}$$

Also from (*i*) and (*ii*),

$$\tan\alpha = \frac{D\omega}{K - J\omega^2} \quad \text{or} \quad \alpha = \tan^{-1}\frac{D\omega}{K - J\omega^2}$$

$\therefore$ Particular integral is,

$$\theta = \frac{GI_m}{\sqrt{(D\omega)^2 + (K - J\omega^2)^2}}\sin(\omega t - \alpha).$$

Hence the complete solution is

$$\theta = F\exp(-Dt/2J)[\sin(\Phi_d t + \phi)]$$
$$+ \frac{GI_m}{\sqrt{(D\omega)^2 + (K - J\omega^2)^2}}\sin(\omega t - \alpha)$$
$$...(8.124)$$

8

The first term on the right hand side of Eqn. 8.124 is a transient term and affects only the deflection for a first few vibrations after switching on and therefore it is of little importance.

Hence deflection

$$\theta = \frac{GI_m}{\sqrt{(D\omega)^2 + (K - J\omega^2)^2}} \sin(\omega t - \alpha) \quad \ldots(8.125)$$

The amplitude of vibration is

$$A = \frac{GI_m}{\sqrt{(D\omega)^2 + (K - J\omega^2)^2}} \quad \ldots(8.126)$$

Now we should make the amplitude as large as possible. For a particular value of current $I_m$ the amplitude can be made large by increasing $G$ or by decreasing $(D\omega)^2 + (K - J\omega^2)^2$.

**1. *Increasing G.*** Since $G = NBA$, it may be made large by using a coil of large area $A$ and having large number of turns $N$. But increasing the area and the number of turns, the inertia constant $J$ is increased and therefore the term $(D\omega)^2 + (K - \omega^2 J)^2$ also increases. Hence the purpose is not served. $G$ can be made large by using a high value of flux density $B$. Therefore, powerful magnets are used in vibration galvanometers to give large amplitude of vibrations, *i.e.*, to increase their sensitivity.

**2. *Decreasing* $(D\omega)^2 + (K - J\omega^2)^2$.** There are three constants involved $J$, $D$ and $K$. If we consider an existing galvanometer, $J$ and $D$ are fixed and only control constant $K$ can be changed. The control constant $K$ can be varied by adjusting the length and tension of the suspension of the moving system.

In order that the amplitude be maximum, the term $(D\omega)^2 + (K - J\omega^2)^2$ should be minimum. This requires that

$$K - J\omega^2 = 0$$

or $\omega = \sqrt{K/J}$

or supply frequency, $f = \omega/2\pi$

$$= (1/2\pi)\sqrt{K/J}$$

But $(1+2\pi)\sqrt{K/J}$ is the natural undamped frequency, $f_n$ of the moving system.

$\therefore$ For maximum amplitude $f = f_n$.

Thus the vibration galvanometer is tuned by changing the length and tension of the moving system in order that the natural frequency of moving system is equal to the supply frequency.

The deflections at various frequencies can be calculated from Eqn. 8.124, and the results plotted as shown in Fig. 8.21. It is found that for a galvanometer having a sharply peaked resonance curve a slight deviation from the resonant frequency results in a large reduction in deflection. Therefore, the galvanometer responds to only fundamental frequency (frequency for which it is turned) and the harmonics have no appreciable effect upon the deflection.

Vibration galvanometers can be used for frequencies between 100 to 1800 Hz. The current sensitivity is about 50 mm/μA with a scale 1 metre away. The sensitivity is almost inversely proportional to the frequency.

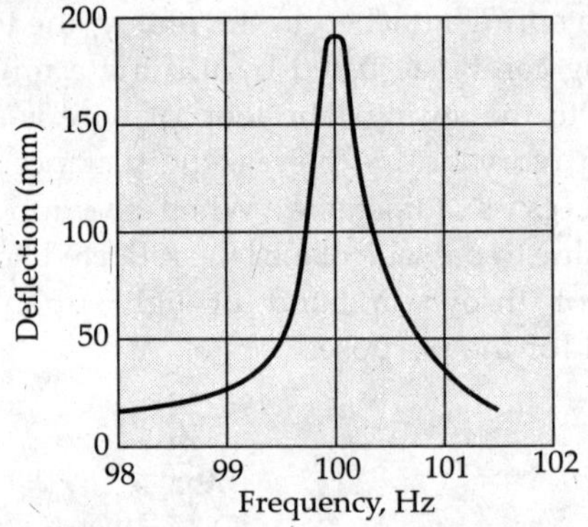


**Fig. 8.21** Deflections of a vibration galvanometer at different frequencies.

***Example 8.27*** *Calculate the ratio of sensitivity for fundamental to that for third harmonic for a vibration galvanometer tuned to a frequency of 50 Hz. The data of the galvanometer is :*

*Inertia constant* $= 10 \times 10^{-3}$ *kg-m*$^2$ ;

*damping constant* $= 25 \times 10^{-3}$ *N-m/rad-s*$^{-1}$.

**Solution.** From Eqn. 8.125 the amplitude of vibration of a frequency $\omega$ is :

$$A = \frac{GI_m}{\sqrt{(D\omega)^2 + (K - J\omega^2)^2}}$$

Therefore the ratio of amplitudes of fundamental and third harmonic is :

$$\frac{A_1}{A_3} = \frac{\sqrt{(3\,D\omega_1)^2 + (K - 9\,J\omega_1^2)^2}}{\sqrt{(D\omega_1)^2 + (K - J\omega_1^2)^2}}$$

For a tuned galvanometer

$$\omega_1 = \sqrt{K/J} \quad \text{or} \quad K = J\omega_1^2$$

Thus we have

$$\frac{A_1}{A_3} = \frac{\sqrt{(3\,D\omega_1)^2 + (J\omega_1^2 - 9\,J\omega_1^2)^2}}{D\omega_1} = \frac{\sqrt{(3\,D\omega_1)^2 + (8\,J\omega_1^2)^2}}{D\omega_1}$$

$$= \frac{\sqrt{(3 \times 25 \times 10^{-3} \times 2\pi \times 50)^2 + (8 \times 10 \times 10^{-3} \times 4\pi^2 \times 50^2)^2}}{25 \times 10^{-3} \times 2\pi \times 50} = 1000.$$

From above it is clear that for a tuned galvanometer the deflection caused by the harmonics will be insignificant as compared to that for the fundamental.

## 8.18 DUDDELL'S OSCILLOGRAPH

It consists of a single loop of thin phosphor-bronze strip which forms the moving system. It is called a vibrator loop. The vibrator loop is situated in the field of a permanent magnet but when high sensitivities are required, highly saturated electromagnets are used.

The vibrator loop (Fig. 8.22) is formed by passing the strip over an ivory pulley. A spring tension adjusting device is attached to the ivory pulley. A small mirror for reflecting light is cemented to the loop midway between the bridge pieces. A Duddell element is essentially a vibration galvanometer having a low period of vibrations. The natural frequency is about 10 kHz. The loop is immersed in oil to get a high degree of damping.

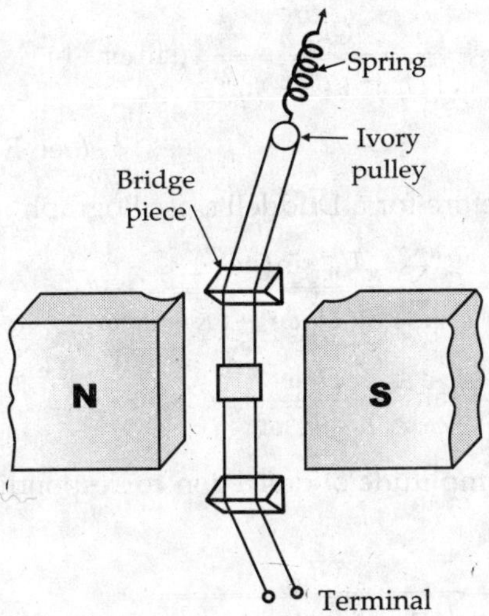


**Fig. 8.22.** Duddell's Oscillograph element.

The galvanometer units can be constructed with diameters 3 mm to 6 mm and about 75 mm length. A typical unit is shown in Fig. 8.23. These galvanometer units are often called pencil galvanometers. They comprise the moving element of the galvanometer and are set in a fixed frame of the permanent magnets. Two or more units per cm can be constructed and 50 or more galvanometers are included in a single oscillograph in order to record many waveforms simultaneously.

When an alternating current is passed through the loop, an alternating torque acts upon the loop and it vibrates. When a beam of light is cast upon the mirror, it is reflected. As the loop is vibrating, the reflected beam will move backwards and forwards. If a wave shape is to be recorded, a photographic film is passed at a constant speed in a direction perpendicular to that of the movement of the reflected beam of light. The actual speed of film depends upon the frequency of the wave being recorded.

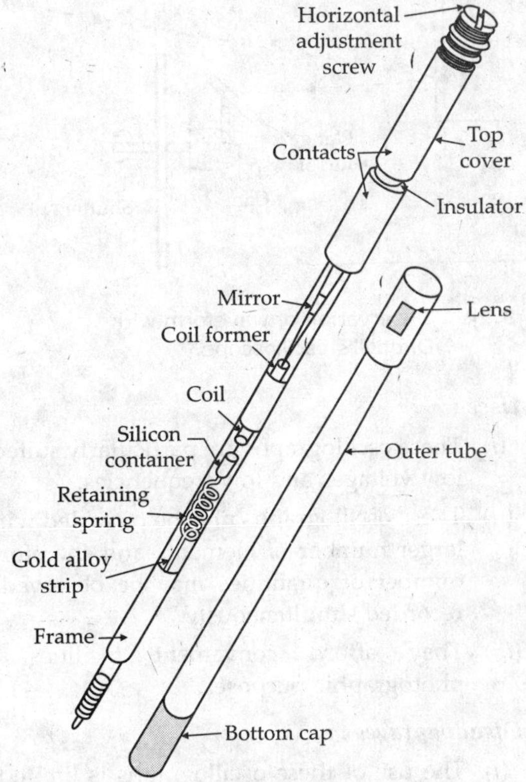


**Fig. 8.23.** Recorder Pencil Galvanometer.

The wave shape is recorded on the film as a result of the two perpendicular motions, *i.e.*, of the film and the reflected light. The oscillograph is calibrated by passing a known value of direct current in the loop and recording the deflection produced by the current.

If the waveform is not to be recorded but is only to be observed, the arrangement as shown in Fig. 8.24 is used. The reflected beam of light from the vibrator mirror falls on a plane mirror and is further reflected on to a screen. The plane mirror is rocked by a cam driven by a synchronous motor. The synchronous motor is supplied from the source whose waveform is to be observed. The plane mirror driven by the cam, gives forward motion of the reflected beam on the screen. The forward motion is carried out for 1½ cycles of the wave, and the plane mirror is returned to its original position during the next half cycle. During this cycle the light is cut off by a rotating shutter. The process is repeated continuously and the wave shape appears stationary on the screen on account of persistence of vision.

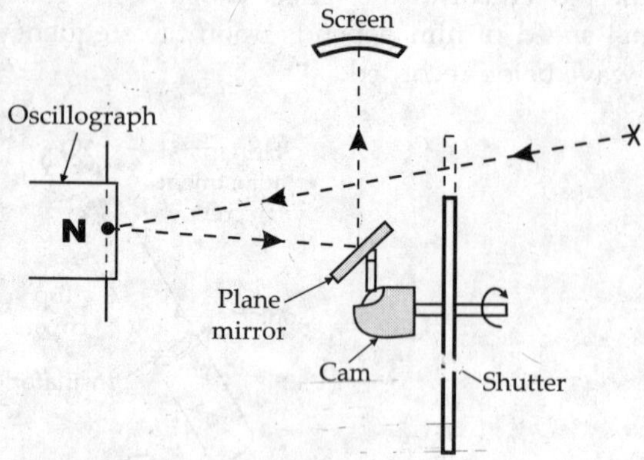


**Fig. 8.24** Observation of waveform with Duddell's oscilloscope.

*Uses :*

(*i*) These oscillographs are particularly suited at low voltages and low frequencies.

(*ii*) These oscillographs may be provided with a larger number of elements and therefore a number of quantities may be observed or recorded simultaneously.

(*iii*) They afford convenient facilities for photographic purposes.

*Disadvantages :*

(*i*) The use of these oscillographs is limited to frequencies upto 500 Hz if an accurate record of the waveform is required. Above this frequency the response to higher order harmonics is very small and therefore the harmonics are not recorded.

(*ii*) The use of these instruments is more complicated.

These oscillographs are not commonly used these days and for many purposes they have been replaced by the cathode ray oscillographs.

## 8.18.1 Theory Duddell's Oscillograph

Duddell's oscillograph is essentially a vibration galvanometer, and therefore its equation of motion is :

$$J\frac{d^2\theta}{dt^2} + D\frac{d\theta}{dt} + K\theta = Gi$$

When considering a vibration galvanometer the current $i$ was assumed to be purely sinusoidal, its equation being $i = I_m \sin \omega t$.

However, in the case of an oscillograph, this assumption cannot be made since oscillograph is used for determination of wave forms of non-sinusoidal quantities. Therefore, current $i$ in the case of an oscillograph contains harmonics and it may be written as :

$$I = \sum_{n=1}^{n=\infty} I_{mn} \sin(n\omega t + \phi_n) \quad \text{...(8.127)}$$

according to Fourier's theorem where $n$ is the order of the harmonic.

When dealing with the vibration galvanometer, the equation for deflection $\theta$ was found to contain a transient term and a term corresponding to steady vibration motion (See Eqn. 8.124). In a Duddell's oscillograph critical or nearly critical damping is provided and therefore the transient term decays after a very small interval of time and can thus be neglected.

The expression for deflection in a vibration galvanometer is

$$\theta = \frac{GI_m}{\sqrt{(D\omega)^2 + (K - J\omega^2)^2}}(\sin \omega t - \phi)$$

(See Eqn. 8.125)

Therefore for a Duddell's oscillograph :

$$\theta = G\sum_{n=1}^{n=\infty} \frac{I_{mn} \sin(n\omega t \pm \phi_n - \alpha_n)}{\sqrt{(Dn\omega)^2 + (K - Jn^2\omega^2)^2}} \quad \text{...(8.128)}$$

where $a_n = \tan^{-1}\dfrac{Dn\omega}{K - Jn^2\omega^2}$ ...(8.129)

The amplitude of deflection corresponding to $n$th harmonic

$$\theta_n = \frac{GI_{mn}}{\sqrt{(Dn\omega)^2 + (K - Jn^2\omega^2)^2}} \quad \text{...(8.130)}$$

$$= \frac{GI_{mn}/K}{\sqrt{(Dn\omega/K)^2 + (1 - Jn^2\omega^2/K)^2}} \quad \text{...(8.131)}$$

But $\dfrac{D}{K} = \dfrac{2D}{2\sqrt{KJ}} \cdot \sqrt{\dfrac{J}{K}} = \dfrac{2\xi}{\omega_n}$

as $\dfrac{D}{2\sqrt{KJ}} = \xi$ and $\sqrt{\dfrac{J}{K}} = \dfrac{1}{\omega_n}$ and $\dfrac{J}{K} = \dfrac{1}{\omega_n^2}$

Substituting the above values in Eqn. 8.131, we have :

$$\theta_n = \frac{GI_{mn}/K}{\sqrt{(1 - n^2\omega^2/\omega_n^2)^2 + (2\xi n\omega/\omega_n)^2}} \quad \text{...(8.132)}$$

But the deflection under ideal (d.c.) conditions is

$$\theta'_n = \frac{GI_{mn}}{K} \quad ...(8.133)$$

$$\theta_n = \frac{\theta'_n}{\sqrt{(1-n^2\omega^2/\omega_n^2)+(2\xi n\omega/\omega_n)^2}} \quad ...(8\ 134)$$

Hence, ratio of actual amplitude to ideal amplitude

$$\frac{\theta_n}{\theta'_n} = \frac{1}{\sqrt{(1-n^2\omega^2/\omega_n^2)^2+(2\xi n\omega/\omega_n)^2}} \quad ...(8.135)$$

$$= \frac{1}{\sqrt{(1-u^2)^2+(2\xi u)^2}} \quad ...(8.136)$$

where $u$ = frequency ratio $= n\omega/\omega_n$ ...(8.137)

In is clear from Eqn. 8.136 that the amount of damping used is of vital importance. If the galvanometer is overdamped, the amplitude will decrease as the frequency is increased while an underdamped galvanometer will produce excessive deflections for frequencies near the natural or resonant frequency. Hence a compromise must be made. The curves for deflection amplitude ratio $(\theta_n/\theta'_n)$ against frequency ratio $(n\omega/\omega_n)$ are shown in Fig. 8.25 for various values of damping ratio, $\xi$. It shows that if damping ratio $\xi = 0.64$ is used, the deflection amplitude is within ± 5 per cent of ideal amplitude for frequencies upto 60 per cent of galvanometer resonant frequency.

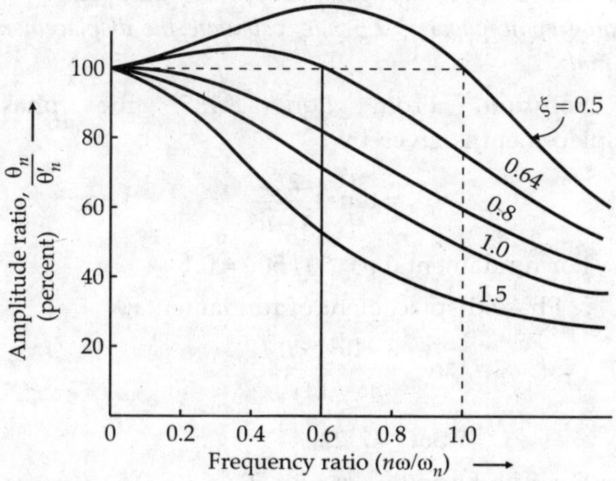


**Fig. 8.25** Amplitude *v/s* frequency ratio characteristics of Duddell's Oscillograph.

In addition to amplitude characteristics described above some phase shift, equivalent to a time delay, must inevitably occur between the input and the output on account of inertia and damping.

Let us examine the expression for phase shift

$$\alpha_n = \tan^{-1}\frac{Dn\omega}{K-Jn^2\omega^2}$$

$$= \tan^{-1}\frac{n\omega D/K}{1-n^2\omega^2 J/K}$$

$$= \tan^{-1}\frac{2\xi n\omega/\omega_n}{1-n^2\omega^2/\omega_n^2}$$

$$= \tan^{-1}\frac{2\xi u}{1-u^2} \quad ...(8.138)$$

It is clear from Eqn. 8.138 that the harmonics in the output waveshape do not have the same relative phases as those in the input current waveshape as the value of $\alpha_n$ will be different for different harmonics. Hence the oscillograph does not give a true reproduction of current wave. However, it can be made to give a reproduction which is sufficiently accurate for practical purposes.

Equation 8.138 gives the phase displacement of the output frequency with respect to its true position, measured on its own time scale of angular frequency. In order so estimate the distortion produced in the output wave, it is necessary to calculate the phase displacement of fundamental and each harmonic and to allow for the different angular frequency scales of the various harmonics. For example, a 35° phase shift for 7th harmonic on its own time scale represents $35/7 = 5°$ displacement on the angular frequency scale of the fundamental.

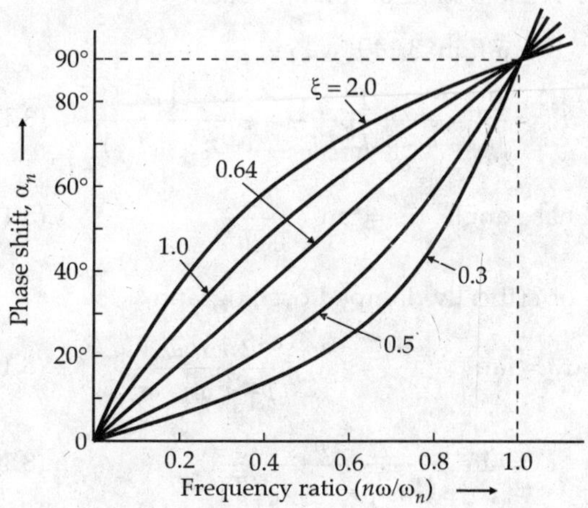


**Fig. 8.26** Phase shift *v/s* frequency ratio characteristic of a Duddell's Oscillograph.

Thus it is clear from above that the phase shift can be made to have negligible importance provided that it can he made directly proportional to frequency that all frequencies of input waveform are delayed by the same time interval.

Figure 8.26 shows the curves of phase shift against frequency ratio for different values of damping constant $\zeta$. It is clear that the relationship is linear for $\zeta = 0.64$.

Fortunately the optimum amplitude performance is also obtained with $\zeta = 0.64$.

In an ideal Duddell's oscilloscope $D = 0$ and $J = 0$

$\therefore$ From Eqn. 8.131, amplitude of deflection for $n$th harmonic

$$\theta_n = GI_{mn} / K = \theta'_n$$

From Eqn. 8.138, phase shift for $n$th harmonic

$$\alpha_n = 0$$

Hence when $D = 0$ and $J = 0$, there is no amplitude distortion and there is no phase shift of the fundamental and any of the harmonics.

### Critically Dumped Oscillograph

When the galvanometer is critically damped $\zeta = 1$. The ratio of actual amplitude to ideal amplitude is given by substituting $\zeta = 1$ in Eqn. 8.136

$$\frac{\theta_n}{\theta'_n} = \frac{1}{\sqrt{(1-u^2)^2 + (2 \times 1 \times u)^2}} = \frac{1}{1+u^2} \quad ...(8.139)$$

$$= \frac{1}{1 + n^2\omega^2 / \omega_n^2} \quad ...(8.140)$$

Suppose $T_0$ is the period of natural oscillations and $T$ is the period of fundamental.

$\therefore$ From Eqn. 8.140, we get

$$R_n = \frac{\theta_n}{\theta'_n} = \frac{1}{1 + n^2 T_0^2 / T^2} = \frac{1}{1 + n^2 (T_0 / T)^2} \quad ...(8.141)$$

$$\text{Phase angle, } \alpha_n = \tan^{-1} \frac{2\zeta u}{1-u^2} \quad ...(8.142)$$

For critically damped oscillograph,

$$\alpha_n = \tan^{-1} \frac{2u}{1-u^2} = \tan^{-1} \frac{2n\omega/\omega_n}{1 - n^2\omega^2/\omega_n^2} \quad ...(8.143)$$

$$= \tan^{-1} \frac{2n}{(T/T_0) - (n^2 T_0 / T)}. \quad ...(8.144)$$

**Example 8.28** *The relative damping in a Duddell's oscillograph is 0.6. The natural frequency is 2400 Hz, calculate the range of frequencies over which the oscillograph will have an amplitude distortion of less than ± 2%.*

**Solution.** The ratio of actual amplitude to ideal amplitude is :

$$\frac{\theta_n}{\theta'_n} = \frac{1}{\sqrt{(1-u^2)^2 + (2\zeta u)^2}} \quad \text{(See Eqn. 8.136)}$$

In order that amplitude distortion be less than ± 2 per cent, the value of $\theta_n / \theta'_n$ should be between 0.98 and 1.02.

$\therefore$ The amplitude to be –2%

$$0.98 = \frac{1}{\sqrt{(1-u^2)^2 + (2 \times 0.6u)^2}}$$

or $u = 0.79$

Hence upper frequency $= 0.79 \times 2400 = 1826$ Hz.

For amplitude to be + 2%.

$$1.02 = \frac{1}{\sqrt{(1-u^2)^2 + (2 \times 0.6u)^2}}$$

or $u = 0.29$.

Lower frequency $= 0.29 \times 2400 = 696$ Hz.

Hence the range of frequencies between which the amplitude distortion is within ± 2% is 696 Hz to 1896 Hz.

**Example 8.29** *An oscillograph has a natural frequency of 500 Hz and a damping ratio of 0.64. The fundamental frequency applied to it is 50 Hz. Calculate the phase displacement for the fundamental and 5th harmonic. Calculate the error in recording time of the 5th harmonic with regards to fundamental. Supposing the speed of the photographic paper is 2.5 m/s, calculate the displacement in mm.*

**Solution.** From Eqn. 8.137, the phase displacement is given by

$$\alpha_n = \tan^{-1} \frac{2\zeta u}{1-u^2}$$

For fundamental $u = 50/500 = 0.1$

$\therefore$ Phase displacement of fundamental,

$$\alpha_1 = \tan^{-1} \frac{2 \times 0.64 \times 0.1}{1-(0.1)^2}$$

$$= 7.366° = 7°22'$$

For fifth harmonic, $u = 5 \times 50 / 500 = 0.5$.

Phase displacement for fifth harmonic

$$\alpha_5 = \tan^{-1} \frac{2 \times 0.64 \times 0.1}{1-(0.5)^2} = 40.47° = 40°28'$$

Phase displacement of 5th harmonic on fundamental scale

$$= 40°28' / 5 = 8°5.6'.$$

Time lag for fundamental

$=7.366/360\times 50 = 0.409\times 10^{-3}$ s.

The time for a complete cycle is $1/50$ s $= 409$ μs.

Time lag for 5th harmonic $= 40.47/(360\times 5\times 50)$

$= 0.45\times 10^{-3}$ s $= 450$ μs.

Error in recording time $= 450 - 409 = 41$ μs.

Error in mm $=$ speed $\times$ time $= 2.5\times 41\times 10^{-6}$

$\approx 0.1\times 10^{-3}$ m $= 0.1$ mm.

**Example 8.30** *A Duddell's oscillograph has a moving system whose natural period of oscillation is 1/2000 second. The oscillograph is critically damped. The wave-form of a complex wave, the frequency of whose fundamental is 50 Hz is to be determined. What are the percentage errors in the production of 3rd, 5th, 7th, 11th and 13th harmonics ? Also determine the phase displacement of 13th harmonic.*

**Solution.** We have, $T_0 = 1/2000$ s and $T = 1/50$ s

We have

$$R_n = \frac{1}{1+n^2(T_0/T)^2}$$

$$\therefore \quad R_1 = \frac{1}{1+(1)^2(50/2000)^2} = 0.9994$$

$$R_3 = \frac{1}{1+(3)^2(50/2000)^2} = 0.9994$$

$$R_5 = \frac{1}{1+(5)^2(50/2000)^2} = 0.9844$$

$$R_7 = \frac{1}{1+(7)^2(50/2000)^2} = 0.9694$$

$$R_{11} = \frac{1}{1+(11)^2(50/2000)^2} = 0.9244$$

$$R_{13} = \frac{1}{1+(13)^2(50/2000)^2} = 0.9044$$

Thus the percentage errors in reproduction are :

- In 3rd harmonic – 0.56 percent
- In 5th harmonic – 1.56 percent
- In 7th harmonic – 3.06 percent
- In 11th harmonic – 7.56 percent
- In 13th harmonic – 9.56 percent.

Displacement of $n$th harmonic

$$\alpha = \tan^{-1}\frac{2n}{(T/T_0)-n^2(T_0/T)} \qquad \text{(See Eqn. 8.144)}$$

Phase displacement of 13th harmonic

$$\alpha_{13} = \tan^{-1}\frac{2\times 13}{(2000/50)-(13)^2(50/2000)}$$

$$= 36°0'$$

Phase displacement of fundamental

$$\alpha_1 = \tan^{-1}\frac{2\times 1}{(2000/50)-(1)^2(50/2000)} = 2°52'$$

On 13th harmonic scale $2°52'$ is equivalent to $13\times 2°52' = 37°16'$.

$\therefore$ 13th harmonic is shown $37°16' - 36°0' = 1°16'$ ahead of its true position.

# Review Questions

1. Describe the constructional details and principle of operation of a d'Arsonval galvanometer. Derive the expression for steady state deflection. If the instrument is provided with a mirror and a light source is placed 1 m away from the galvanometer, find the expression for the deflection of light spot.
2. Explain the intrinsic constants of a galvanometer and write down the equation for motion of a galvanometer.

   Explain the response of the galvanometer under

   (*i*) undamped conditions

   (*ii*) underdamped conditions

   (*iii*) critically damped conditions and

   (*iv*) overdamped conditions. Draw the response curves.
3. Prove for an underdamped galvanometer, the deflection θ at any time $t$ after the application of current is

$$\theta = \theta_F\left[1 - \frac{1}{\sqrt{1-\xi^2}}\exp(-\xi\omega_n t)\,\sin\left(\omega_n\sqrt{1-\xi^2}\,t + \cos^{-1}\xi\right)\right]$$

where $\theta_F$ = steady state deflection ; rad,

$\omega_n$ = natural angular frequency ; rad/s,

$\xi$ = damping factor.

4. For an undamped instrument, derive the expressions for
   (a) first overshoot
   (b) logarithmic decrement.

   Prove that logarithmic decrement
   $$\lambda = \pi\xi / \sqrt{1-\xi^2}$$
   where $\xi$ = damping factor.
5. Define the operational constants of a d'Arsonval galvanometer. Express the equation of motion of a d'Arsonval galvanometer in terms of the operational constants.
6. Discuss the effect of damping factor $\xi$ on the peak overshoot and the settling time of a galvanometers. Why are galvanometers intentionally under-damped ?
7. Derive the expression for resistance needed in a galvanometer to make it dead beat in case damping on account of all factors except electromagnetic damping are negligible.
8. Define the terms "current sensitivity", "voltage sensitivity" and "Megohm sensitivity" as applied to d'Arsonval galvanometers. Explain how current sensitivity of a galvanometer can be increased.
9. Explain the functioning of a Ayrton Universal shunt used for d'Arsonval galvanometers. Prove that the current through the galvanometer does not depend upon the resistance of the galvanometer in case an Ayrton shunt is used.
10. Describe construction and working of a ballistic galvanometer. Explain the difference in constructional details of a ballistic galvanometer and a d'Arsonval galvanometer.
11. Prove that in a ballistic galvanometer, the charge is proportional to first swing of the moving coil.
12. Describe the different methods used for calibration of a ballistic galvanometer.
13. Describe construction and working of an unshunted flux meter. Prove that the flux is proportional to the deflection of the flux meter.
14. Describe the function of a shunted flux meter. Prove that the deflection of a shunted flux meter is independent of the resistance of its coil.
15. Explain the construction and working of a vibration galvanometer. Explain the term 'tuning' and describe how a vibration galvanometer is tuned.
16. Derive the expression for steady state deflection of a vibration galvanometer when a sinusoidal voltage is applied to its coil. Explain the means employed to increase the amplitude of vibrations.
17. Explain the construction and working of Duddell's oscillograph. List its applications.
18. Derive the expressions for deflection and phase shift for $n$th harmonic in the case of Duddell's oscillograph. Sketch the curves for amplitude ratio and phase shift versus frequency ratio for different values of damping factor.

## Unsolved Problems

1. Find the steady state deflection of a moving coil galvanometer in degrees and in mm on a scale placed 1 m away when a current of 1 mA is passed through its coil. Also calculate the resistance required for critical damping if the galvanometer has a relative damping of 0.1 on open circuit. The data of the galvanometer is :

   dimensions of coil = $25 \times 20$ mm,

   number of turns = 100,

   flux density = 0.1 $\text{Wb/m}^2$.

   inertia constant = $0.1 \times 10^{-6}$ $\text{kg-m}^2$ and

   control constant = $25 \times 10^{-6}$ Nm/rad.

   **Hint** : Resistance required for critical damping
   $$= \frac{G^2}{(1-0.1) \times 2\sqrt{KJ}}$$
   **[Ans.** 11.45°, 400 mm, 6.2 Ω**]**
2. A moving coil galvanometer gives a deflection of 100 mm on a scale 2 m away from its mirror with a current of 1 μA. When swinging freely on open circuit, the time of one complete oscillation is 3 seconds. The galvanometer is dead beat when the total resistance of the circuit is 2800 Ω. Find the moment of inertia of the moving system.

   **[Ans.** $0.972 \times 10^{-9}$ $\text{kg-m}^2$**]**

3. The constants of a galvanometer are :

Control constant $= 0.23 \times 10^{-6}$ N-m/rad,

Damping constant $= 5 \times 10^{-9}$ N-m/rad-s$^{-1}$,

Moment of inertia $= 0.18 \times 10^{-6}$ kg-m$^2$.

It is proposed to increase the periodic time to 15 seconds by attaching small weights on light arms fixed to the coil spindle. Determine by how much these weights must increase the moment of inertia of the coil. [**Ans.** $1.13 \times 10^{-6}$ kg-m$^2$]

4. In a moving coil instrument, critical damping is to be provided only by eddy currents in the aluminium former carrying the moving coil. The moment of inertia of moving system is $3 \times 10^{-6}$ kg-m$^2$. The torsional constant of the control spring is $2 \times 10^{-6}$Nm/rad and the loop resistance of aluminium former is 0.006 ohm. Calculate the required coil flux linkages per radian deflection.

[**Ans.** $30.5 \times 10^{-6}$ Wb/rad]

5. A moving coil galvanometer has a coil wound upon a former of non-conducting material. The current sensitivity of the instrument is 0.001 μA/mm at 1 m and the undamped periodic time is 5 s. If the torque is 0.0632 Nm/A, calculate the control torque of the suspension, the moment of inertia of moving system and the total circuit resistance which critically damps the instrument. The effect of air damping may be neglected.

[**Ans.** $0.1264 \times 10^{-6}$ N-m/rad, $0.08 \times 10^{-6}$ kg-m$^2$, 19860 Ω]

6. A d'Arsonval galvanometer has the following data :

flux density $= 8 \times 10^{-3}$ Wb/m$^2$,

number of turns = 300,

length of coil = 15 mm,

width of coil = 30 mm,

spring constant $= 2.5 \times 10^{-9}$ Nm/rad,

moment of inertia $= 10 \times 10^{-9}$ kg-m$^2$,

damping constant $= 2 \times 10^{-9}$,

resistance of coil = 80 Ω.

Calculate :

(*a*) the deflection of galvanometer for a current of 1 μA,

(*b*) current sensitivity in mm/μA if the scale is kept 1 m away from the mirror.

Find :

(*c*) the natural frequency of oscillations

(*d*) relative damping

(*e*) frequency of damped oscillations

(*f*) first maximum deflection

(*g*) external resistance required for critical damping.

[**Ans.** (*a*) 0.432. rad (*b*) 864 mm/μA (*c*) 0.5 rad/s (*d*) 0.2 (*e*) 0.49 rad/s (*f*) 0.66 rad (*g*) 36.7 Ω]

7. The torsional constant of a galvanometer is $5 \times 10^{-6}$ Nm/rad. Calculate the energy stored in the suspension when it has been twisted through 0.4 rad.

The moment of inertia of the coil is $0.2 \times 10^{-6}$ kg-m$^2$. Calculate the angular velocity which the coil will have when it passes through its equilibrium position if damping is neglected.

**Hint :** Kinetic energy $= \frac{1}{2} J\omega_{max}^2 = \frac{1}{2} K\theta_{max}^2$

[**Ans.** 2 rad/s]

8. Figure 8.10 shows a circuit for testing of a galvanometer where $E = 1.5$ V, $R_1 = 1.0\,\Omega$, $R_2 = 2500\,\Omega$, $R_3$ is variable. When $R_3 = 450\,\Omega$, the galvanometer deflection is 1500 mm and with $R_3 = 950\Omega$, the deflection is reduced to 75 mm. Calculate

(*a*) the resistance of galvanometer,

(*b*) current sensitivity of galvanometer.

[**Ans.** (*a*) 49 Ω. (*b*) 125 mm/μA]

9. A ballistic galvanometer gives a first swing of 60° for a discharge of 1000 μC. Find the quantity of electricity to produce

(*a*) a swing of 90° in the instrument

(*b*) a spot deflection of 20 mm on a scale 1 m away.

[**Ans.** 1500 μC ; 9.55 μC]

10. When a current of 0.001 A is passed through a ballistic galvanometer, it gives a steady deflection of 50 scale divisions. When a capacitor charged to a potential of 100 V d.c. is discharged through the galvanometer, it gives a maximum throw of 220 scale divisions. The periodic time of swing is 4 seconds. Find the value of capacitance. The value of logarithmic decrement is 0.2. [**Ans.** 30.8 μF]

11. The periodic time of an undamped ballistic galvanometer is 10 s and a current of 0.15 mA gives a steady deflection of 200 scale divisions. Find the quantity of electricity which can produce or swing of 120 scale divisions. What is the quantity of electricity corresponding to this swing if the instrument has a decrement of 1.03 ?

[**Ans.** 143 μC (undamped), 216.6 μC (damped)]

12. A ballistic galvanometer of resistance 1500 Ω gives a throw of 75 divisions when the flux through the

search coil to which it is connected, is reversed. If the flux density is 0.1 $Wb/m^2$, the search coil has 1400 turns, a mean area of 5500 $mm^2$ and a resistance of 200 Ω, calculate the galvanometer constant in terms of coulomb per scale division.

[**Ans.** 12 μC/div.]

**13.** A ballistic galvanometer has an undamped free period of 8.7 s and a relative damping of 0.8. Determine the time required for the galvanometer to pass through zero after the application of charge. [**Ans.** 7.3 s]

**14.** A Duddell's oscillograph has a moving system whose natural period of oscillation is 0.1 ms. The waveform of a complex wave, the frequency of whose fundamental is 50 Hz, is to be determined. What are the percentage errors in reproduction of 3rd, 5th, 7th and 11th harmonics ?

[**Ans.** 0.0225, 0.6024, 0.1224, 0.315]

**15.** A critically damped Duddell's oscillograph is required to reproduce the 13th harmonic; of a 50 Hz wave with a relative amplitude correct to within 2 percent.

What should be the natural frequency of the movement ? What will be the relative phase angle departure of the 13th harmonic ?

[**Ans.** 4590 Hz, 16.7′]

# Objective Type Questions

### *Tick (✓) The Most Appropriate Answer*

**1.** In a d'Arsonval galvanometer, an iron core is usually used between the permanent magnet pole faces. This is used so that

(*a*) flux density in the air gap becomes high thereby a large deflecting torque is produced

(*b*) the effect of stray magnetic fields is reduced

(*c*) moment of inertia of moving parts becomes smaller.

(*d*) none of the above.

**2.** Sometimes, the d'Arsonval galvanometers, do not use ferromagnetic cores between poles of the permanent magnet. In this case

(*a*) the flux density becomes smaller resulting in low deflecting torque

(*b*) the dimensions of the moving coil can be made smaller thereby reducing the moment of inertia

(*c*) the magnetic field may not be radial resulting in a non-uniform scale even if spring control is used

(*d*) all the above.

**3.** A d'Arsonval galvanometer uses a light and scale arrangement. The light source is placed 1 m away from the moving system of galvanometer. The arrangement uses a circular scale calibrated in mm. The deflection indicated by the scale is

(*a*) 1000 mm

(*b*) 2000 mm

(*c*) 500 mm

(*d*) none of the above.

If a current of 1 μA is passed through the coil. The spring stiffness is $2 \times 10^{-6}$ Nm/rad and the displacement constant is 2 Nm/A.

**4.** The time period of free oscillations in a galvanometer having a relative damping of 0.6 is 2 s. The frequency of damped oscillations is :

(*a*) 0.5 rad/s

(*b*) 0.3 rad/s

(*c*) 0.4 rad/s

(*d*) none of the above.

**5.** The relative damping in a galvanometer is 0.8. Its logarithmic decrement is approximately

(*a*) 0.48

(*b*) 1.25

(*c*) 4.19

(*d*) – 4.19.

**6.** In a critically damped galvanometer the deflection at a time 0.9 times the time of free oscillations after a current is passed through the moving coil is approximately

(*a*) 0.986 times the final deflection

(*b*) 0.901 times the final deflection

(*c*) 0.866 times the final deflection

(*d*) none of the above.

7. The resistance required for critical damping in a circuit is 1000 Ω. The galvanometer circuit has a resistance of 800 Ω. Is the galvanometer circuit
(*a*) underdamped
(*b*) undamped
(*c*) overdamped
(*d*) none of the above ?

8. A galvanometer has a ratio of 0.9 for damped frequency oscillations to undamped frequency of oscillations. Suppose moment of inertia, stiffness constant and damping constant are made twice their original value, what would be the new ratio of damped frequency oscillations to undamped frequency oscillations,
(*a*) 0.9
(*b*) 1.11
(*c*) 4
(*d*) 2 ?

9. A current of 2 μA is passed through the moving coil of an undamped d'Arsonval galvanometer which has a displacement constant of 2 Nm/A and a control constant of $10 \times 10^{-6}$ Nm/rad. The moving oscillates with an amplitude of :
(*a*) 0.2 rad
(*b*) 0.4 rad
(*c*) 0.8 rad
(*d*) none of the above.

10. If the damping in a d'Arsonval galvanometer is only due to electromagnetic effects, the resistance required for critical damping is
(*a*) $G^2 / \sqrt{KJ}$
(*b*) $G / \sqrt{KJ}$
(*c*) $G / 2\sqrt{KJ}$
(*d*) $G^2 / 2\sqrt{KJ}$

where $G$ = displacement constant ; Nm/A, $K$ = control constant ; Nm/rad and $J$ = inertia constant ; kg-m$^2$.

11. A d'Arsonval galvanometer uses a lamp and scale arrangement. Its current sensitivity is 250 mm/μA. If the resistance of the coil is 100 Ω and the external resistance required for critical damping is 900 Ω (which is connected in the circuit), the voltage sensitivity is
(*a*) 0.25 mm/μV
(*b*) 0.25 mm/V
(*c*) 2.5 mm/μV
(*d*) 250 mm/μV.

12. Ayrton shunt is used in d'Arsonval galvanometers so as to limit the current in the galvanometer coil to its maximum permissible value. The relative value of current through the galvanometer coil and the shunt
(*a*) depends upon the value of resistance of galvanometer coil only
(*b*) depends upon the values of resistance of galvanometer coil and the shunt
(*c*) does not depend upon the value of resistance of galvanometer coil
(*d*) none of the above.

13. A Ballistic galvanometer should be designed with
(*a*) a large period of natural oscillations and a negligible damping constant
(*b*) a small period of natural oscillations and a high damping constant
(*c*) a large period of natural oscillations and a high damping factor
(*d*) a small period of natural oscillations and a low damping factor.

14. In a flux meter
(*a*) the controlling torque is produced by weights attached to moving coil
(*b*) the controlling torque is produced by springs
(*c*) there is no controlling torque
(*d*) none of the above.

15. In an unshunted flux meter, the sensitivity is dependent upon
(*a*) resistance of moving coil
(*b*) resistance of search coil
(*c*) resistance of both moving coil and search coil
(*d*) none of the above.

16. A vibration galvanometer is to be tuned to a frequency of 50 Hz. The ratio of its control constant to inertia constant should be
(*a*) 98696
(*b*) 2500
(*c*) $10.132 \times 10^{-6}$
(*d*) none of the above.

17. A vibration galvanometer is tuned
(*a*) by changing the length and tension of vibrating coil
(*b*) by attaching weight to the vibrating coil
(*c*) by changing its damping constant
(*d*) all the above.

8

**18.** A Duddell's oscillograph can be used for frequencies

(*a*) upto 50 Hz

(*b*) upto 500 Hz

(*c*) above 500 Hz

(*d*) upto 10 kHz.

**19.** A Duddell's oscillograph will give no amplitude distortion and phase displacement if its

(*a*) moment of inertia and stiffness constant are zero

(*b*) stiffness constant and damping factor are zero

(*c*) moment of inertia and damping factor are zero

(*d*) all the above.

**20.** In a Duddell's oscilloscope the phase displacements of fundamental and 13th harmonic are calculated to be 3° and 36° respectively. The oscillograph will show the 13th harmonic to be

(*a*) 39° ahead of its true position

(*b*) 38° ahead of its true position

(*c*) 3° ahead of its true position

(*d*) 3° behind its true position.

## Answers

| | | | | | | | | | |
|---|---|---|---|---|---|---|---|---|---|
| **1.** (*a*) | **2.** (*d*) | **3.** (*b*) | **4.** (*c*) | **5.** (*c*) | **6.** (*a*) | **7.** (*c*) | **8.** (*a*) | **9.** (*c*) | **10.** (*d*) |
| **11.** (*a*) | **12.** (*c*) | **13.** (*a*) | **14.** (*c*) | **15.** (*d*) | **16.** (*a*) | **17.** (*a*) | **18.** (*b*) | **19.** (*c*) | **20.** (*c*) |

# 9

# Analog Ammeters, Voltmeters and Ohmmeters

## 9.1 INTRODUCTION

Analog ammeters and voltmeters are classed together as there are no fundamental differences in their operating principles. The action of all ammeters and voltmeters, with the exception of electrostatic type of instruments, depends upon a deflecting torque produced by an electric current. In an ammeter this torque is produced by a current to be measured or by a definite fraction of it. In a voltmeter this torque is produced by a current which is proportional to the voltage to be measured. Thus all analog voltmeters and ammeters are essentially current measuring devices.

The essential requirements of a measuring instrument are :

(*i*) That its introduction into the circuit, where measurements are to be made, does not alter the circuit conditions ;

(*ii*) The power consumed by them for their operation is small.

*Ammeters* are connected in series in the circuit whose current is to be measured. The power loss in an ammeter is $I^2 R_a$, where $I$ is the current to be measured and $R_a$, is the resistance of ammeter. Therefore, ammeters should have a low electrical resistance so that they cause a small voltage drop and consequently absorb small power.

*Voltmeters* are connected in parallel with the circuit whose voltage is to be measured. The power loss in voltmeters is $V^2/R_v$, where $V$ is the voltage to be measured and $R_v$, is the resistance of voltmeter. Therefore, voltmeters should have a high electrical resistance, in order that the current drawn by them is small and consequently the power consumed is small.

*Ohmmeters* are used for measurement of resistance. They incorporate a source of emf and a current measuring device.

## 9.2 TYPES OF INSTRUMENTS

The main types of instruments used as ammeters and voltmeters are :

(*i*) Permanent magnet moving coil (PMMC)
(*ii*) Moving iron
(*iii*) Electro-dynamometer
(*iv*) Hot wire
(*v*) Thermocouple
(*vi*) Induction
(*vii*) Electrostatic
(*viii*) Rectifier.

Of these the permanent-magnet moving-coil type can be used for direct-current measurements only, and the induction type for alternating-current measurements only. The other types of meters can be used with either direct or alternating-currents.

The moving-iron and moving-coil types both depend for their action upon the magnetic effect of current. The former is the most generally used form of indicating instrument, as it is the cheapest. It can be used for either direct or alternating-current measurements and, if properly designed is very accurate. The permanent-magnet moving-coil instrument is the most accurate type for direct-current measurements, and instruments of this type are frequently constructed to have substandard* accuracy.

Electrodynamometer type of instruments are used both on a.c. as well as on d.c. Their calibration for both d.c. and a.c. is the same and hence they are very useful as "*transfer instruments*".

* The connotation, "substandard", is not used in the negative sense. It only implies a standard which is almost a secondary standard.

Thermal instruments have the advantage that their calibration is the same for both d.c. and a.c. They are particularly suited for alternating-current measurements since their deflection depends directly upon the heating effect of the alternating current, *i.e.*, upon the rms value of the current. Their readings are thus, independent of the frequency or wave-form of the current, and of any stray magnetic fields which may exist in their vicinity.

As voltmeters, electrostatic instruments have the advantage that their power consumption is exceedingly small. They can be made to cover a large range of voltage, and can be constructed to have substandard accuracy. Their main disadvantage is that the electrostatic principle is only directly applicable to voltage measurements.

The induction principle is more generally used for watt-hour meters than for ammeters and voltmeters owing to their comparatively high cost, and inaccuracy, of induction instruments of the latter types.

## 9.3 ERRORS IN AMMETERS AND VOLTMETERS

There are certain errors which occur in most types of instruments, while other errors occur only in those of particular type. The latter types of errors will be dealt with later, together with the instruments in which they occur.

Of the errors common to most types of instruments, friction and temperature errors are perhaps the most important. To reduce the effect of friction torque, and consequently the error produced by it the weight of the moving system must be made as small as possible compared with the operating force, *i.e.*, the ratio of torque to weight must be large. In high quality instruments the ratio of deflecting torque at full scale to weight of moving system is rarely much less than 0.01 Nm/kg. A vertical spindle is generally preferred to a horizontal one from the point of view of a small friction torque.

The most serious error is produced by the heat generated in the instrument, or by changes in ambient (room) temperature, is that due to a change in the resistance of the working coil. Such a change of resistance is of little importance in ammeters, but in voltmeters, in which the working current should be directly proportional to the applied voltage, it is essential that the resistance of the instrument remains as nearly constant as possible.

Thus, the power loss in the instrument should be small, and resistance coils which are likely to produce appreciable heating should be mounted, if possible, in such a position that they are well ventilated. In order to eliminate temperature errors, the working coil is wound with copper wire and is of comparatively low resistance. A high *"swamping resistance"* of material whose resistance temperature coefficient is small, is connected in series with the coil, so that, although the resistance of the coil may change considerably, the change in total resistance of circuit is small.

Other errors resulting from heating may be caused by expansion of the control spring, or of other parts of the instrument, although such errors are usually small. Lack of balance in the moving system and changes in the strength of permanent magnets (if used) are other possible sources of error which are common to several types of instruments.

## 9.4 PERMANENT MAGNET MOVING COIL INSTRUMENT (PMMC)

The permanent magnet moving coil instrument is the most accurate type for d.c. measurements. The working principle of these instruments is the same as that of the d'Arsonval type of galvanometers, the difference being that a direct reading instrument is provided with a pointer and a scale.

### 9.4.1 Construction of PMMC Instruments

The general constructional features of this instrument are shown in Fig. 9.1.

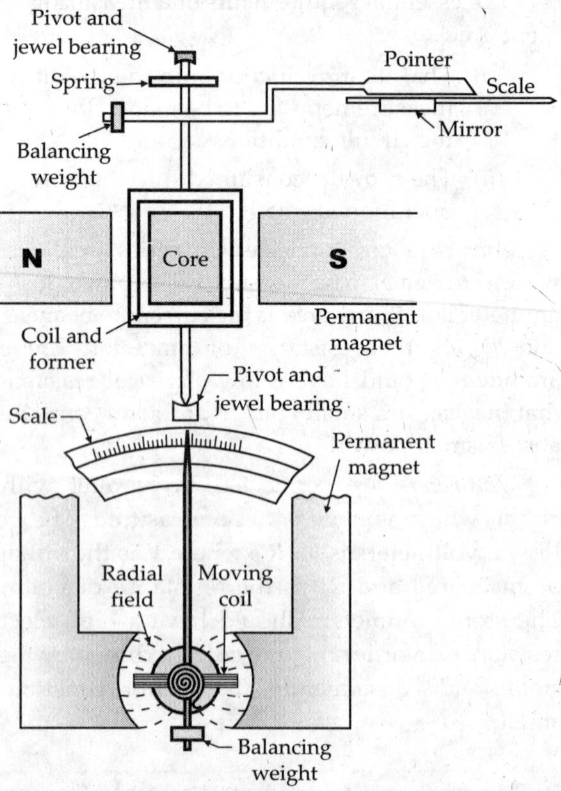


**Fig. 9.1** Permanent magnet moving coil instrument.

***Moving coil.*** The moving coil is wound with many turns of enamelled or silk covered copper wire. The coil is mounted on a rectangular aluminium former which is pivoted on jewelled bearings. The coils move freely in the field of a permanent magnet. Most voltmeter coils are wound on metal frames to provide the required electro-magnetic damping. Most ammeter coils, however, are wound on non-magnetic formers, because coil turns are effectively shorted by the ammeter shunt. The coil itself, therefore, provides electro-magnetic damping.

***Magnet systems.*** There has been considerable development in materials for permanent magnets and, therefore, magnet assemblies have undergone a lot of change in the recent past. Old style magnet system consisted of a relatively long *U* shaped permanent magnets having soft iron pole pieces. Owing to development of materials like Alcomax and Alnico, which have a high coercive force, it is possible to use smaller magnet lengths and high field intensities. The flux densities used in PMMC instruments vary from 0.1 $Wb/m^2$ to 1 $Wb/m^2$. Thus in small instruments it is possible to use a small coil having small number of turns and hence a reduction in volume is achieved. Alternatively in instruments having a large scale length it is possible to increase the air gap length to accommodate large number of turns.

The movement of the coil is restricted in the above design. This is because no actual part of the coil is allowed to reach the extreme positions near the pole tips where, there is fringing field (owing to fringing the flux density near the pole tips is smaller than that at the centre and also the field is not radial). Thus the angular span of scale is restricted to 90°. In order to obtain longer movement of the pointer and a longer angular swing of the coil a concentric magnet construction as shown in Fig. 9.2 is used. Since the magnet is concentric type it produces a radial flux pattern which extends over 250° or more. This type of construction is used for many panel type instruments and some portable instruments.

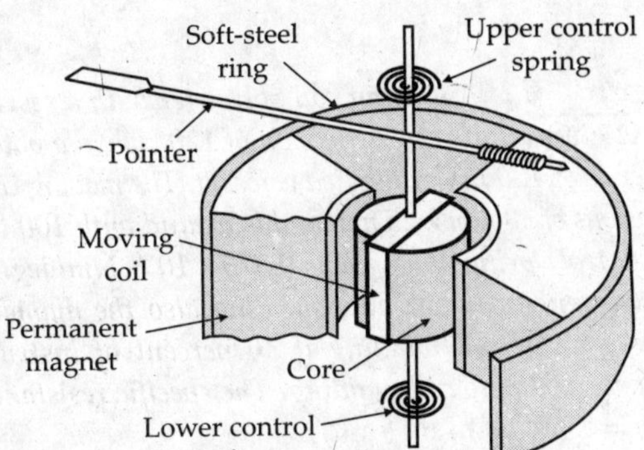


**Fig. 9.2** Concentric magnet assembly.

An air cored coil offset from the axis of rotation is used as shown in Fig. 9.3. The scale length of the instrument can be increased from 120° to 240° or even 300°, thereby giving better resolution of reading for the same scale range.

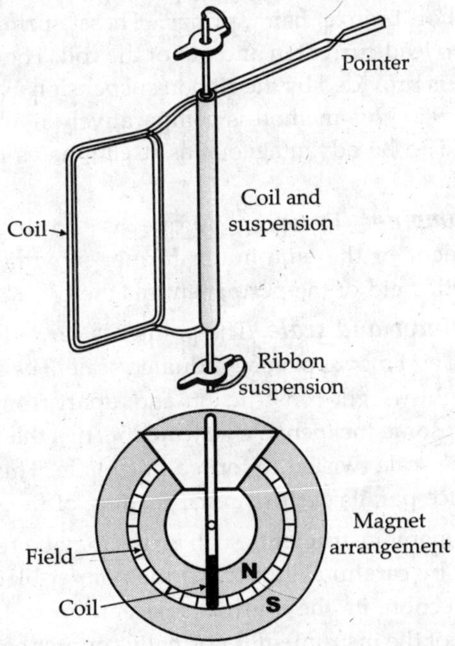


**Fig. 9.3** Long scale moving coil instrument.

In recent years, with the development of improved magnetic materials like Alnico, it has become feasible to design a magnetic system in which the magnet itself serves as the core as shown in Fig. 9.4. The moving coil moves over the magnet. The active sides of the moving coil are located in the uniform radial field between pole pieces and the steel yoke.

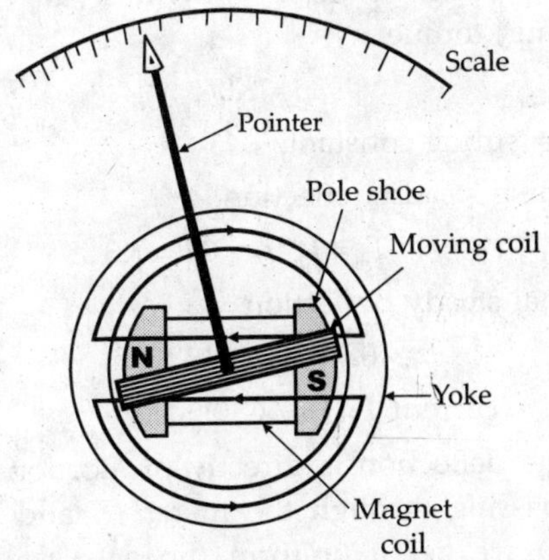


**Fig. 9.4** Core magnet construction.

9

This arrangement has the obvious advantage of being relatively unaffected by the external magnetic fields. It also eliminates the magnetic shunting effects in steel panel construction, where several meters operating side by side may affect each other's readings. The need for magnetic shielding in the form of iron cases, is also eliminated by core magnet construction.

**Control.** When the coil is supported between two jewel bearings the control torque is provided by two phosphor bronze hair springs. These springs also serve to lead current in and out of the coil. The control torque is provided by the ribbon suspension as shown in Fig. 9.2. This method is comparatively new and is claimed to be advantageous as it eliminates bearing friction.

**Damping.** Damping torque is produced by movement of the aluminium former moving in the magnetic field of the permanent magnet.

**Pointer and scale.** The pointer is carried by the spindle and moves over a graduated scale. The pointer is of light-weight construction and, apart from those used in some inexpensive instruments has the section over the scale twisted to form a fine blade. This helps to reduce parallax errors in the reading of the scale.

In many instruments such errors may be reduced further by careful alignment of the pointer blade and its reflection in the mirror adjacent to scale. The weight of the instrument is normally counter balanced by weights situated diametrically opposite and rigidly connected to it.

### 9.4.2 Torque Equation

The torque for a moving coil instrument is derived in Art. 8.4 on page 198.

Deflecting torque $T_d = NBl\,dI = GI$ ...(9.1)

where $G$ = a constant = $NBld$ ...(9.2)

The spring control provides a restoring (controlling) torque

$$T_c = K\theta \quad ...(9.3)$$

where $K$ = spring constant.

For final steady deflection

$$T_c = T_d \text{ or } GI = K\theta$$

∴ Final steady deflection

$$\theta = (G/K)\,I \quad ...(9.4)$$

or current $I = (K/G)\theta$ ...(9.5)

As the deflection is directly proportional to the current passing through the meter ($K$ and $G$ being constants) we get a uniform (linear) scale for the instrument.

In micro-ammeters and low range milli-ammeters upto about 20 mA, the entire current to be measured is sent through the moving coil. This is because instrument springs serve as current leads to the moving coil. Their current carrying capacity limits the current which can be safely carried to about 20 mA. For higher currents (usually above 20 mA) the moving coil is shunted to bypass current around the coil and the spring. D.C. ammeters are normally designed to have a voltage drop of nearly 50 mV to 100 mV for full scale deflection. Voltmeters on their own have a range of 0-50 mV or 0-100 mV. However, voltmeters having higher ranges use a moving coil together with sufficient series resistance (known as multiplier) to limit the instrument current to the desired value. Most d.c. voltmeters are designed to produce full scale deflection with a current of 20, 10, 5 or 1 mA. Normally a value of 1 mA is used.

Thus, excluding low range current measuring instruments, most d.c. ammeters are actually 50 mV (or 100 mV) millivoltmeters operated with a suitable shunt, while voltmeters are low range milli-ammeters operated with a suitable series resistance.

**Example 9.1** *A permanent magnet moving coil instrument has a coil of dimensions 15 mm × 12 mm. The flux density in the air gap is $1.8 \times 10^{-3}$ Wb/m$^2$ and the spring constant is $0.14 \times 10^{-6}$ Nm/rad. Determine the number of turns required to produce an angular deflection of 90 degrees when a current of 5 mA is flowing through the coil.*

**Solution.** Deflection, $\theta = 90° = \pi/2$ rad.

At equilibrium, $T_d = T_c$ or $NBld\,I = K\theta$.

∴ Number of turns, $N = \dfrac{K\theta}{Bld\,I}$

$$= \frac{0.14 \times 10^{-6} \times \pi/2}{1.8 \times 10^{-3} \times 15 \times 10^{-3} \times 12 \times 10^{-3} \times 12 \times 10^{-3} \times 5 \times 10^{-3}}$$

$= 136.$

**Example 9.2** *A moving coil voltmeter with a resistance of 20 Ω gives a full scale deflection of 120° when a potential difference of 100 mV is applied across it. The moving coil has dimensions of 30 mm × 25 mm and is wound with 100 turns. The control spring constant is $0.375 \times 10^{-6}$ Nm/deg. Find the flux density in the air gap. Find also the diameter of copper wire of coil winding if 30 percent of instrument resistance is due to coil winding. The specific resistance for copper = $1.7 \times 10^{-8}$ Ωm.*

**Solution.** Voltage across instrument for full scale deflection = 100 mV.

Current in instrument for full scale deflection

$$I=\frac{V}{R}=\frac{100\times10^{-3}}{20}=5\times10^{-3}\text{ A.}$$

Deflecting torque

$$T_d = NBldI$$
$$= 100\times B\times30\times10^{-3}\times25\times10^{-3}\times5\times10^{-3}$$
$$= 375\times10^{-8}\Omega\text{m.}$$

$\therefore$ Controlling torque for a deflection $\theta = 120°$,

$$T_c = K\theta = 0.375\times10^{-6}\times120$$
$$= 45\times10^{-6}\text{ N-m.}$$

At final steady position,

$$T_d = T_c \quad\text{or}\quad 375\times10^{-6}B = 45\times10^{-6}.$$

$\therefore$ Flux density in the air gap

$$B=\frac{45\times10^{-6}}{375\times10^{-6}}=0.12\text{ Wb/m}^2.$$

Resistance of coil winding

$$R_c = 0.3\times20 = 6\,\Omega$$

Length of mean turn

$$L_{mt} = 2(l+d) = 2(30+25) = 110\text{ mm.}$$

Let $a$ be the area of cross-section of wire and $\rho$ be the resistivity.

Resistance of coil, $R_c = N\rho L_{mt}/a$.

$\therefore$ Area of cross-section of wire,

$$a=\frac{100\times1.7\times10^{-8}\times110\times10^{-3}}{6}\times10^6$$

$$=31.37\times10^{-3}\text{ mm}^2$$

Diameter of wire, $d = [(4/\pi)(31.37\times10^{-3})]^{1/2} = 0.2\text{ mm}$

***Example 9.3*** *The coil of a moving coil voltmeter is 40 mm long and 30 mm wide and has 100 turns on it. The control spring exerts a torque of $240\times10^{-6}$ N-m when the deflection is 100 divisions on full scale. If the flux density of the magnetic field in the air gap is 1.0 Wb/m$^2$, estimate the resistance that must be put in series with the coil to give one volt per division. The resistance of the voltmeter coil may be neglected.*

**Solution:** Controlling torque at full scale deflection

$$T_c = 240\times10^{-6}\text{ N-m.}$$

Deflecting torque at full scale deflection

$$T_d = NBldI$$
$$= 100\times1\times40\times10^{-3}\times30\times10^{-3}I$$
$$= 120\times10^{-3}I\text{ N-m.}$$

At final steady position,

$$T_d = T_c \quad\text{or}\quad 120\times10^{-3}I = 240\times10^{-6}.$$

$\therefore$ Current at full scale deflection, $I$

$$= 2\times10^{-3}\text{ A} = 2\text{ mA.}$$

Let the resistance of the voltmeter circuit be $R$

$\therefore$ Voltage across the instrument $= 2\times10^{-3}R$.

This produces a deflection of 100 divisions

$\therefore$ Volts per division $= 2\times10^{-3}R/100$

This value should be equal to 1 in order to get 1 volt per division.

$$\therefore\quad 2\times10^{-3}R/100 = 1$$

or $\quad R = 50000\,\Omega = 50\,\text{k}\Omega$

### 9.4.3 Ammeter Shunts

The basic movement of a d.c. ammeter is a PMMC d'Arsonval galvanometer. The coil winding of a basic movement is small and light and can carry very small currents since the construction of an accurate instrument with a moving coil to carry currents greater than 100 mA is impracticable owing to the bulk and weight of the coil that would be required.

When heavy currents are to be measured, the major part of the current is bypassed through a low resistance called a "*shunt*". Figure 9.5 shows the basic movement (meter) and its shunt to produce an *ammeter*.

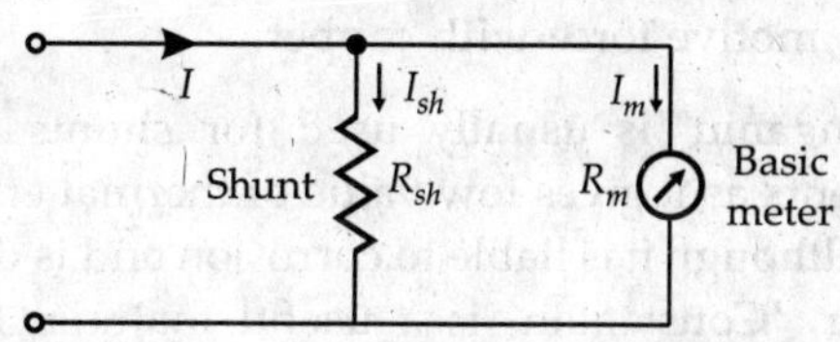


**Fig. 9.5** Basic ammeter circuit.

The resistance of the shunt can be calculated using conventional circuit analysis. See Fig. 9.5,

where $R_m$ = internal resistance of movement (*i.e.*, the coil) $\Omega$ ;

$R_{sh}$ = resistance of the shunt ; $\Omega$,

$I_m = I_{fs}$ = full scale deflection current of movement, A ;

$I_{sh}$ = shunt current, A

$I$ = current to be measured ; A.

Since the shunt resistance is in parallel with the meter movement, the voltage drops across shunt and movement must be the same.

or $\quad I_{sh}R_{sh} = I_mR_m$

$$\therefore\quad R_{sh} = I_mR_m/I_{sh} \qquad ...(9.6)$$

But $\quad I_{sh} = I - I_m$

Therefore, we can write,

$$R_{sh} = I_mR_m/(I-I_m) \qquad ...(9.7)$$

$$\therefore\quad I/I_m - 1 = R_m/R_{sh}$$

or

$$I/I_m = 1 + R_m/R_{sh} \qquad ...(9.8)$$

This ratio of total current to the current in the movement is called *multiplying power of shunt*.

∴ Multiplying power

$$m = I / I_m \quad ...(9.9)$$

$$= 1 + R_m / R_{sh} \quad ...(9.10)$$

∴ Resistance of shunt

$$R_{sh} = R_m / (m-1) \quad ...(9.11)$$

The shunt resistance used with a d'Arsonval movement may consist of a coil of resistance wire within the case of the Instrument, or it may be external shunt having a very low resistance.

***Construction of Shunts :***

The general requirements for shunts are :

(*i*) The temperature co-efficient of shunt and instrument should be low and should be as nearly as possibly the same.

(*ii*) The resistance of shunts should not vary with time,

(*iii*) They should carry the current without excessive temperature rise,

(*iv*) They should have a low thermal electromotive force with copper.

'Manganin' is usually used for shunts of d.c. instruments as it gives low value of thermal emf with copper although it is liable to corrosion and is difficult to solder. 'Coustantan' is a useful material for a.c. circuits since its comparatively high thermal emf, being unidirectional, is ineffective on these circuits.

The construction of shunts is the same as that of low resistance standards explained in Art. 6.4. Shunts for low currents are enclosed in the meter casing but for currents above 200 A, they are mounted separately (so that heat produced can be effectively dissipated).

**Example 9.4** *A 1 mA meter d'Arsonval movement with an internal resistance of 100 Ω is to be converted into a 0-100 mA ammeter. Calculate the shunt resistance required. What particulars should he specified on the shunt ?*

**Solution.** Shunt resistance

$$R_{sh} = I_m R_m / (I - I_m) = 1 \times 100 / (100-1) = 1.01\,\Omega.$$

Voltage drop across the shunt

$$= (1.0)\,(100) = 100 \text{ mV}.$$

Equivalent resistance of shunt and meter in parallel

$$= 100 \times 10^{-3} / (100 \times 10^{-3}) = 1.0\,\Omega.$$

Therefore the indications on the shunt should be :

(*i*) 100 mA, 100 mV (*ii*) 1.0 Ω, 100 mA or

(*iii*) 1.0 Ω, 100 mV.

**Example 9.5** *Find the multiplying power of a shunt of 200 Ω resistance used with a galvanometer of 1000 Ω resistance. Determine the value of shunt resistance to give a multiplying power of 50.*

**Solution.** Multiplying power

$$m = 1 + R_m / R_{sh} = 1 \times 1000 / 200 = 6.$$

We have, $m = 1 + R_m / R_{sh}$

$$\therefore \text{ Shunt resistance, } R_{sh} = \frac{R_m}{m-1} = \frac{1000}{50-1} = 20.4\,\Omega.$$

**Example 9.6** *A moving coil ammeter has a fixed shunt of 0.02 Ω. With a coil resistance of R = 1000 Ω and a potential difference of 500 mV across it, full scale deflection is obtained :*

(*a*) *To what shunted current does this correspond ?*

(*b*) *Calculate the value of R to give full scale deflection when shunted current I is*

*(i) 10 A, (ii) 75 A and*

(*c*) *With what value of R is 40% deflection obtained with I = 100 A ?*

**Solution.** (*a*) Current through shunt,

$$I_{sh} = 500 \times 10^{-3} / 0.02 = 25 \text{ A}.$$

Current through the meter to give full scale deflection

$$= 500 \times 10^{-3} / 1000 = 0.5 \times 10^{-3} \text{ A}.$$

(*b*) (*i*) Voltage across shunt for a current of 10 A

$$= 0.02 \times 10 = 0.2 \text{ V}.$$

∴ Resistance of meter for a current of 10 A to give full scale deflection

$$= 0.2 / (0.5 \times 10^{-3}) = 400\,\Omega.$$

(*ii*) Voltage across shunt for a current of 75 A

$$= 0.02 \times 75 = 1.5 \text{ V}.$$

Resistance of meter for a current of 75 A to give full scale deflection

$$= 1.5 / (0.5 \times 10^{-3}) = 3000\,\Omega.$$

(*c*) Now 40 percent deflection is obtained with 100 A.

∴ Current to give full scale deflection

$$= 100 / 0.4 = 250 \text{ A}.$$

Voltage across shunt for a current of 250 A

$$= 0.02 \times 250 = 5.0 \text{ V}$$

Resistance of meter for a current of 100 A to give 40 percent of full scale deflection

$$= 5.0 / (0.5 \times 10^{-3}) = 10{,}000\,\Omega$$

**Example 9.7** *A simple shunted ammeter using a basic meter movement with an internal resistance of 1800 Ω and a full scale deflection current of 100 μA is connected in a* *circuit and gives reading of 3.5 mA on its 5 mA scale. This reading is checked with a recently calibrated d.c. ammeter which gives a reading of 4.1 mA. The implication is that the first ammeter has a faulty shunt on its 5 mA range. Calculate :*

(*a*) *the actual value of faulty shunt ;*

(*b*) *the correct shunt for the 5 mA range.*

**Solution.** Correct multiplying power of shunt

$$m = I / I_m = 5 \times 10^{-3} / (100 \times 10^{-6}) = 50.$$

Multiplying power of faulty shunt

$$= 50 \times (4.1 / 3.5) = 58.6.$$

Resistance value of faulty shunt

$$= 1800 / (58.6 - 1) = 31.25\ \Omega.$$

Correct value of shunt resistance

$$= 1800 / (50 - 1) = 36.73\ \Omega.$$

### 9.4.4 Effect of Temperature Changes in Ammeters

The temperature error can be eliminated when the shunt and the moving-coil are made of the same material and kept at the same temperature. This method, however, is not satisfactory in practice as the temperatures of the two parts are not likely to change at the same rate. An additional disadvantage of using copper shunts is that they are likely to be bulky as the resistivity of copper is small. Copper shunts are only occasionally used in instruments with built-in shunts.

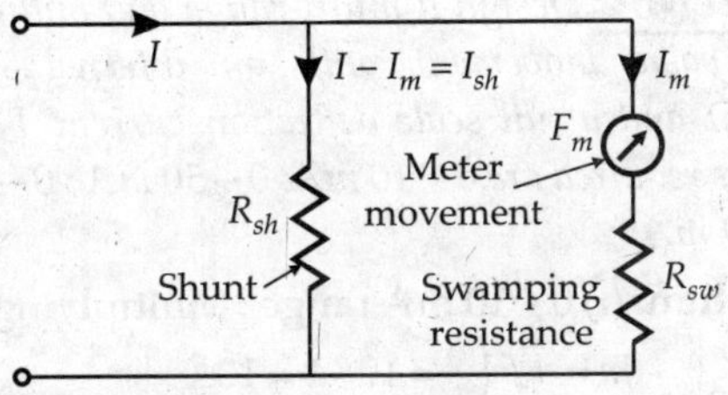


**Fig. 9.6** Meter shunt and swamping resistance.

The arrangement normally used is shown in Fig. 9.6. In this case, a *'swamping resistance'* of manganin (which has a negligible temperature co-efficient) having a resistance 20 to 30 times the coil resistance is connected in series with the coil and a shunt of manganin is connected across this combination. Since copper forms a small fraction of the series combination, the proportion in which the currents would divide between the meter and the shunt would not change appreciably with the change in temperature. This fact is illustrated with the following example.

**Example 9.8** *A moving coil instrument whose resistance is 25 Ω gives a full-scale deflection with a current of 1 mA. This instrument is to be used with a manganin shunt to extend its range to 100 mA. Calculate the error caused by a 10°C rise in temperature when :*

(*a*) *Copper moving coil is connected directly across the manganin shunt.*

(*b*) *A 75-ohm manganin resistance is used in series with the instrument moving-coil.*

*The temperature co-efficient of copper is 0.004/°C and that of manganin 0.00015/°C.*

**Solution. Case (*a*).** Refer to Fig. 9.6.

Multiplying-factor of shunt

$$m = I / I_m = 100 / 1 = 100.$$

∴ Resistance of shunt

$$R_{sh} = R_m / (m-1) = 25 / (100 - 1) = 0.2525\ \Omega.$$

Instrument resistance for 10°C rise in temperature,

$$R_{mt} = 25(1 + 10 \times 0.004) = 26\ \Omega.$$

Shunt resistance for 10°C rise in temperature,

$$R_{sht} = 0.2525(1 + 10 \times 0.00015) = 0.2529\ \Omega.$$

Current $I_{mt}$ through the meter for 100 mA in the main circuit for 10°C rise in temperature,

$$I_{mt} = 100 \times \frac{0.2529}{26 + 0.2529} = 0.963\ \text{mA}.$$

Normal meter current = 1 mA.

∴ Error due to rise in temperature

$$= (0.963 - 1) \times 100 = -3.7\%.$$

**Case (*b*).** Total resistance in the meter circuit

$$R_m + R_{sw} = 25 + 75 = 100\ \Omega.$$

Shunt resistance

$$R_{sh} = R_m / (m-1) = 100 / (100 - 1) = 1.01\ \Omega.$$

Resistance of the instrument circuit for 10°C rise in temperature,

$$R_{mt} = 25(1 + 10 \times 0.004) + 75(1 + 10 \times 0.00015) = 101.11\ \Omega.$$

Shunt resistance for 10°C rise in temperature,

$$R_{sht} = 1.01(1 + 10 \times 0.00015) = 1.0115\ \Omega$$

Instrument current for 100 mA in the main circuit for 10°C rise in temperature

$$I_{mt} = 100 \times \frac{1.0115}{101.11 + 1.0115} = 0.9905\ \text{mA}.$$

$$\therefore \quad \text{Error} = (0.9905 - 1) \times 100 = -0.95\%.$$

The improvement in the error from 3.7% to about 1% has been obtained by the use of the additional series swamping resistance of 3 times as compared to the meter resistance. We could obtain better correction by increasing the ratio of the swamping resistance as compared to the meter resistance. But by increasing this ratio, the p.d. across the meter circuit would also be increased. Thus the disadvantage of using swamping resistors is a reduction in the full scale sensitivity as a higher voltage across the instrument is necessary to sustain the full scale current.

### 9.4.5 Multi-range Ammeters

The current range of a d.c. ammeter may be further extended by a number of shunts, selected by a range switch. Such meter is called a *multirange ammeter*. Figure 9.7 shows a schematic diagram of multirange ammeter. The circuit has four shunts $R_{sh1}$, $R_{sh2}$, $R_{sh3}$ and $R_{sh4}$, which can be put in parallel with the meter movement to give four different current ranges $I_1$, $I_2$, $I_3$ and $I_4$. Let $m_1$, $m_2$, $m_3$ and $m_4$ be the shunt multiplying powers for currents $I_1$, $I_2$, $I_3$ and $I_4$.

$$\therefore \quad R_{sh1} = R_m/(m_1-1),$$

$$R_{sh2} = R_m/(m_2-1),$$

$$R_{sh3} = R_m/(m_3-1)$$

and

$$R_{sh4} = R_m/(m_4-1)$$

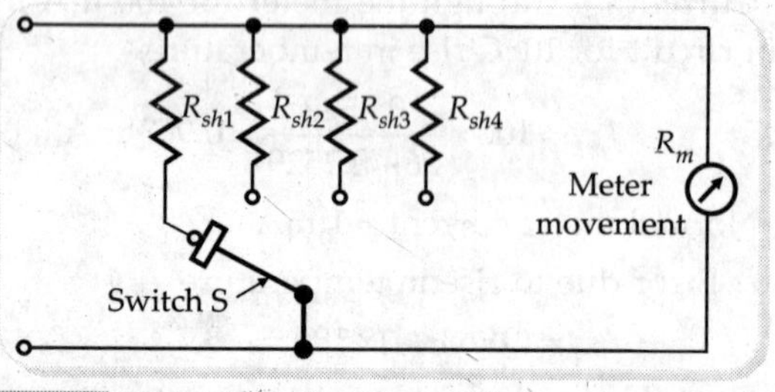


**Fig. 9.7** Multi-range ammeter.

Low range ammeters use a multiposition make-before-break switch (See in Fig. 9.7) provided on the case of the instrument. This type of switch is essential in order that the meter movement is not damaged when changing from the current range one to another. If we provide an ordinary switch, the meter remains without a shunt and as such is unprotected and therefore can be damaged when the range is changed.

Multi-range ammeters are used for ranges from 1 to 50 A. When using a multi-range ammeter, first use the highest current range, then decrease the current range until good upscale reading is obtained.

The universal shunt or Ayrton shunt shown in Fig. 9.8 is also used for multi-range ammeters. The advantage of an Ayrton shunt is that it eliminates the possibility of the meter being in the circuit without a shunt. But this advantage is gained at the cost of a higher meter resistance.

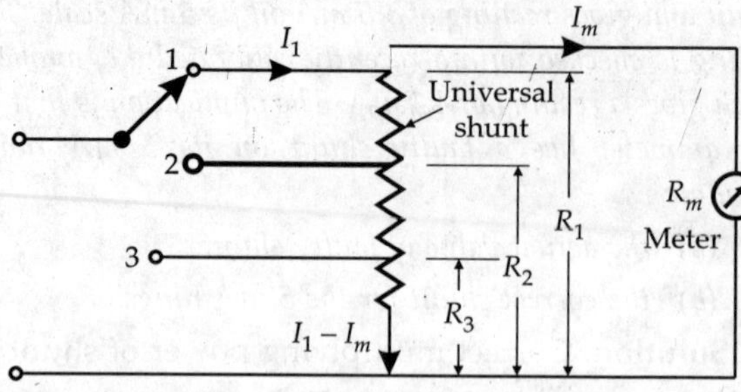


**Fig. 9.8** Multi-range ammeter using universal shunt.

***Universal or Ayrton shunt.*** Consider that the meter ranges have to be extended to $I_1$, $I_2$ and $I_3$. For the arrangement shown in Fig. 9.8 we have, for switch at position 1,

$$I_m R_m = (I_1 - I_m) R_1$$

$$\therefore \quad m_1 = I_1 / I_m = 1 + R_m / R_1$$

or

$$R_1 = R_m/(m-1)$$

For switch at position 2,

$$I_m (R_1 - R_2 + R_m) = (I_2 - I_m) R_2$$

or

$$R_2 = (R_1 + R_m)/m_2$$

For switch at position 3,

$$I_m (R_1 - R_3 + R_m) = (I_3 - I_m) R_3$$

or

$$R_3 = (R_1 + R_m)/m_3.$$

Thus the values of different sections of resistances *i.e.*, $(R_1 - R_2)$, $(R_1 - R_3)$, and $R_3$ may be found.

**Example 9.9** *Design a multi-range d.c. milli-ammeter using a basic movement with an internal resistance $R_m = 50\ \Omega$ and a full scale deflection current $I_m = 1 mA$. The ranges required are 0 – 10 mA: 0 – 50 mA ; 0 – 100 mA, and 0-500 mA.*

**Solution.** (*i*) 0 – 10 mA range : Multiplying power

$$m = I / I_m = 10/1 = 10.$$

From Eqn. 9.11, resistance of shunt

$$R_{sh1} = R_m/(m-1) = 50/9 = 5.55\ \Omega.$$

(*ii*) 0 – 50 mA range

$$m = 50/1 = 50$$

$$\therefore \quad R_{sh2} = 50/(50-1) = 1.03\ \Omega.$$

(*iii*) 0 – 100 mA range

$$m = 100/1 = 100$$

$$\therefore \quad R_{sh3} = 50/(100-1) = 0.506\ \Omega.$$

(*iv*) 0 – 500 mA range

$$m = 500/1 = 500$$

$$\therefore \quad R_{sh4} = 5/(500-1) = 0.1\ \Omega.$$

***Example 9.10*** *Design an Ayrton shunt to provide an ammeter with current ranges of 1 A, 5 A and 10 A. A basic meter with an internal resistance of 50 Ω and a full scale deflection current of 1 mA is to be used.*

**Solution.** Referring to Fig. 9.8, we have,

$$I_m = 1 \text{ mA}, \; R_m = 50\,\Omega$$

and $$I_1 = 1 \text{ A}, \; I_2 = 5 \text{ A}, \; I_3 = 10 \text{ A}.$$

Thus $$m_1 = 1/(1\times10^{-3}) = 1000$$

$$m_2 = 5/(1\times10^{-3}) = 5000.$$

$$m_3 = 10/(1\times10^{-3}) = 10{,}000.$$

Thus, $$R_1 = \frac{R_m}{m_1 - 1} = \frac{50}{100-1} \approx 0.05\,\Omega,$$

$$R_2 = \frac{R_1 + R_m}{m_2} = \frac{0.05 + 50}{5000} \approx 0.01\,\Omega$$

$$R_3 = \frac{R_1 + R_m}{m_3} = \frac{0.05 + 50}{10{,}000} \approx 0.005\,\Omega$$

Therefore the resistances of the various sections of the universal shunt are :

$$R_1 - R_2 = 0.05 - 0.01 = 0.040\,\Omega$$

$$R_2 - R_3 = 0.01 - 0.005 = 0.005\,\Omega$$

and $$R_3 = 0.005\,\Omega.$$

### 9.4.6 Voltmeter Multipliers

A d'Arsonval basic meter movement is converted into a voltmeter by connecting a series resistance with it. This series resistance is known as a *multiplier*. The combination of the meter movement and the multiplier is put across the circuit whose voltage is to be measured (See Fig. 9.9).

The multiplier limits the current through the meter so that it does not exceed the value for full scale deflection and thus prevents the movement from being damaged.

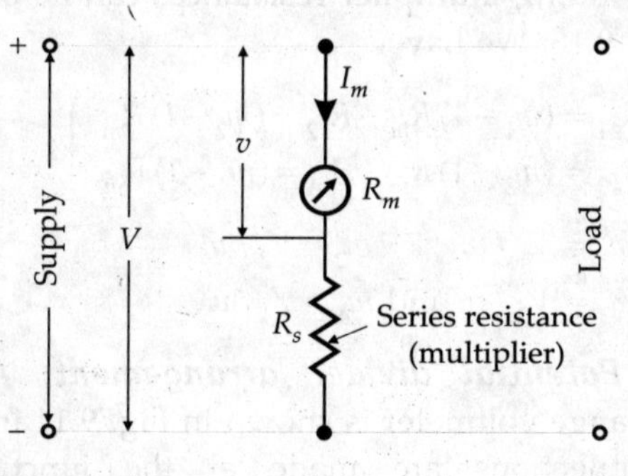


**Fig. 9.9** Meter with a multiplier.

The value of a multiplier, required to extend the voltage range, is calculated as under :

Let :

$I_m = I_{fs}$ = full scale deflection current of meter,

$R_m$ = internal resistance of meter movement,

$R_s$ = multiplier resistance,

$v$ = voltage across the meter movement for current $I_m$,

$V$ = full range voltage of instrument.

For the circuit of Fig. 9.9,

$$v = I_m R_m \qquad \text{...(9.12)}$$

$$V = I_m (R_m + R_s) \qquad \text{...(9.13)}$$

$$\therefore \quad R_s = \frac{V - I_m R_m}{I_m} = \frac{V}{I_m} - R_m \qquad \text{...(9.14)}$$

We can also express the result in terms of multiplying factor of multiplier.

Multiplying factor for multiplier,

$$m = \frac{V}{v} = \frac{I_m (R_m + R_s)}{I_m R_m} = 1 + \frac{R_s}{R_m} \qquad \text{...(9.15)}$$

$\therefore$ Resistance of multiplier, $R_s = (m-1) R_m$ ...(9.16)

Hence for the measurement of voltage $m$ times the voltage range of the instrument, the series multiplying resistance should be $(m-1)$ times the meter resistance. Thus to extend the voltage range to 10 times the instrument range, $R_s = 9\,R_m$.

### Construction of Multipliers

The essential requirements of multipliers are :

(*i*) Their resistance should not change with time,

(*ii*) The change in their resistance with temperature should be small,

(*iii*) They should be non-inductively wound for a.c. meters.

The resistance materials used for multipliers are manganin and constantan.

The general constructional details of multipliers are the same as that for medium and high resistances as explained in Chapter 5.

***Example 9.11*** *A moving coil instrument gives a full scale deflection of 10 mA when the potential difference across its terminals is 100 mV. Calculate*

(*a*) *the shunt resistance for a full scale deflection corresponding to 100 A,*

(*b*) *the series resistance for full scale reading with 1000 V.*

*Calculate the power dissipation in each case.*

**Solution.** Meter current, $I_m = 10$ mA,

∴ Meter resistance

$$R_m = 100 \times 10^{-3} / (10 \times 10^{-3}) = 10\ \Omega$$

(*a*) Shunt multiplying factor,

$$I / I_m = 100 / (10 \times 10^{-3}) = 10{,}000.$$

∴ Shunt resistance,

$$R_{sh} = 10 / (10{,}000 - 1) \approx 0.001\ \Omega.$$

Power dissipation,

$$v\,I = (100 \times 10^{-3}) \times 100 = 10 \text{ W}.$$

(*b*) Voltage multiplying factor

$$m = V / v = 1000 / (100 \times 10^{-3}) = 10{,}000.$$

∴ Multiplier resistance

$$R_s = (m-1) R_m = (10{,}000 - 1) \times 10 \approx 100{,}000\ \Omega.$$

Power dissipation

$$V I_m = (1000)(10 \times 10^{-3}) = 10 \text{ W}$$

### 9.4.7 Effect of Temperature Changes in Voltmeters

The multiplier resistance used in series with the moving-coil is usually made of manganin having a negligible resistance temperature coefficient. Since the series resistance of the multiplier is very much greater than the coil resistance (which is of copper), the variations of resistance of $R_m$ due to temperature changes are *'swamped'* by the resistance $R_s$ of multiplier. This is illustrated by the following example.

**Example 9.12** *A moving-coil instrument whose resistance is 25 Ω gives a full-scale deflection with a voltage of 25 mV. This instrument is to be used with a series multiplier to extend its range to 10 V. Calculate the error caused by 10°C rise in temperature.*

*The temperature coefficient of copper is 0.004/°C and that of manganin 0.00015/°C.*

**Solution.** Meter current for full scale deflection,

$$I_m = V / R_m = 25 \times 10^{-3} / 25 = 1 \text{ mA}.$$

Voltage multiplying factor

$$m = V / v = 10 / (25 \times 10^{-3}) = 400.$$

Multiplier resistance

$$R_s = (m-1) R_m = (400 - 1) 25 = 9975\ \Omega$$

Total resistance of voltmeter circuit,

$$R_v = 25 + 9975 = 10{,}000\ \Omega.$$

Resistance of meter with 10°C rise in temperature,

$$R_{mt} = 25(1 + 0.004 \times 10) = 26\ \Omega.$$

Resistance of multiplier with 10°C rise in temperature

$$R_{st} = 9975(1 + 0.00015 \times 10) = 9990\ \Omega.$$

Total resistance of voltmeter circuit with 10°C rise in temperature

$$R_{vt} = 26 + 9990 = 10016\ \Omega.$$

Reading of voltmeter at 10°C rise in temperature

$$= (10{,}000 / 10016) \times 10 = 9.984 \text{ V}.$$

$$\text{Percentage error} = \frac{9.984 - 10}{10} \times 100 = 0.16\% \text{ low.}$$

Thus it is clear from above that the temperature changes have no appreciable effect on the readings of voltmeters if the multipliers are made of low temperature coefficient materials, *i.e.*, the temperature changes are swamped out by the multipliers.

### 9.4.8 Multirange d.c. Voltmeters

In a multirange voltmeter, different full scale voltage ranges may be obtained by the use of individual multiplier resistors or by a potential divider arrangement.

**1. *Individual multipliers.*** We can obtain different voltage ranges by connecting different values of multiplier resistors in series with the meter. The number of these resistors is equal to the number of ranges required. Figure 9.10 shows multiplier resistors $R_{s1}$, $R_{s2}$, $R_{s3}$ and $R_{s4}$ which can be connected in series

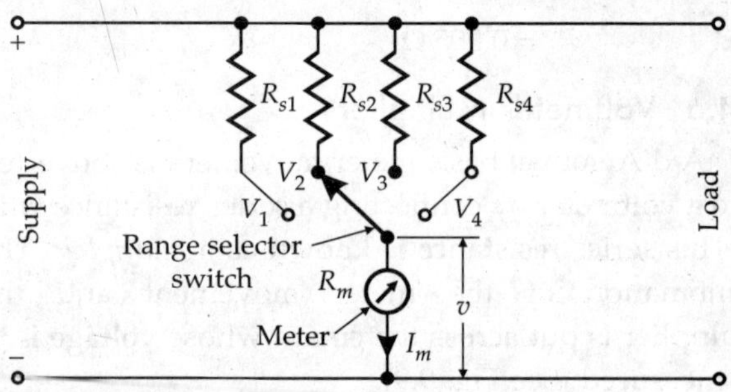


**Fig. 9.10** Multi-range voltmeter.

with the meter by a range selector switch. Consider that the ranges desired are $V_1$, $V_2$, $V_3$ and $V_4$, then the corresponding multiplier resistances can be obtained by Eqn. 9.16. We have,

$$\left.\begin{aligned} R_{s1} &= (m_1 - 1) R_m, \; R_{s2} = (m_2 - 1) R_m \\ R_{s3} &= (m_3 - 1) R_m, \; R_{s4} = (m_4 - 1) R_m \end{aligned}\right\} \quad ...(9.17)$$

where $m_1 = V_1 / v$, $m_2 = V_2 / v$, $m_3 = V_3 / v$ and $m_4 = V_4 / v$.

**2. *Potential divider arrangement.*** Another multi-range voltmeter is shown in Fig. 9.11 in which the connections are made at the junctions of resistances $R_1$, $R_2$, $R_3$ and $R_4$ in series to obtain the

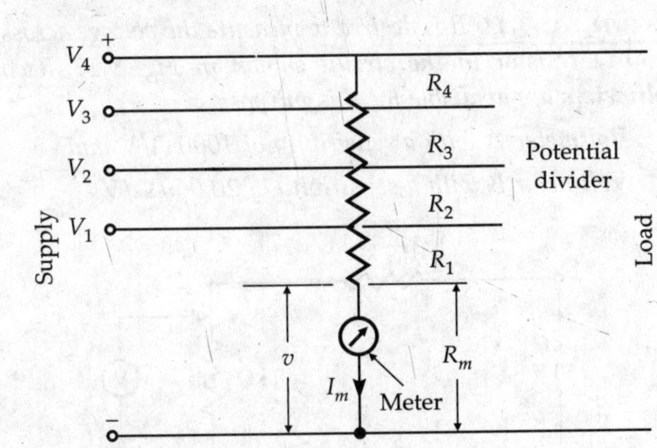


**Fig. 9.11** Multi-range voltmeter using potential divider.

voltage ranges $V_1$, $V_2$, $V_3$, and $V_4$. These connections are brought out to binding posts on the instrument, and the instrument is connected to the proper binding post for the desired voltage range. The series resistances for the voltage ranges $V_1$, $V_2$, $V_3$ and $V_4$ can be computed as follows :

$$R_1 = \frac{V_1}{I_m} - R_m = \frac{V_1}{v/R_m} - R_m$$

$$= m_1 R_m - R_m = (m_1 - 1) R_m \qquad \text{...(9.18)}$$

$$R_2 = \frac{V_2}{I_m} - R_m - R_1 = \frac{V_2}{v/R_m} - R_m - (m_1 - 1) R_m$$

$$= m_2 R_m - R_m - (m_1 - 1) R_m = (m_2 - m_1) R_m \qquad \text{...(9.19)}$$

Similarly,

$$R_3 = (V_3 / I_m) - R_m - R_1 - R_2 = (m_3 - m_2) R_m \qquad \text{...(9.20)}$$

and $$R_s = (V_4 / I_m) - R_m - R_1 - R_2 - R_3 = (m_4 - m_3) R_m \qquad \text{...(9.21)}$$

This system has the advantage that all multipliers except the first have standard resistance values and can be obtained commercially in precision tolerances. The range multiplier, $R_1$, is the only special resistor which must be manufactured to meet specific requirements.

Multi-range voltmeters are very effective for moderate range voltages. For higher-range voltages it is often desirable to use external resistors in connection with a given voltmeter.

**Example 9.13** *A basic d'Arsonval meter movement with an internal resistance $R_m$ = 100 Ω. and a full scale current of $I_m$ = 1 mA, is to be converted into a multi-range d.c. voltmeter with ranges of 0 – 10 V, 0 – 50 V, 0 – 250 V and 0 – 500 V. Find the values of various resistances using the potential divider arrangement.*

**Solution.** Voltage across the meter movement,

$$v = I_m R_m = 1 \times 100 = 100 \text{ mV}.$$

The voltage multiplying factors are :

$$m_1 = 10 / (100 \times 10^{-3}) = 100,$$

$$m_2 = 50 / (100 \times 10^{-3})$$

$$m_3 = 2500 \text{ and } m_4 = 5000$$

Referring to Fig. 9.11 and using Eqns. 9.18 to 9.21 the values of various resistances are :

$$R_1 = (m_1 - 1) R_m = (100 - 1) \times 100 = 9900 \, \Omega$$

$$R_2 = (m_2 - m_1) R_m = (500 - 100) \times 100 = 40 \text{ k}\Omega$$

$$R_3 = (m_3 - m_2) R_m = (2500 - 500) \times 100 \, \Omega = 200 \text{ k}\Omega$$

and $$R_4 = (m_4 - m_3) R_m = (5000 - 2500) \times 100 \, \Omega = 250 \text{ k}\Omega$$

### 9.4.9 Sensitivity of PMMC Voltmeters

The current sensitivity is defined as the deflection per unit current and this is equal to the ratio of constant $G$ of the electrical system divided by the control spring constant $K$. Hence for sensitivity to be large, $G$ should be large and $K$ should be small. For a given coil area $A(= ld)$ and a constant flux density, $G$ can be increased when the coil is wound with many turns of thin wire while $K$ can be made small by using a light flat spring and the coil assembly lightly pivoted. On the contrary a less sensitive instrument is wound with few turns of thick wire and has a stiffer spring. Thus as a general rule it may be stated that a sensitive instrument will have a large resistance because it is wound with many turns of fine wire.

Different d'Arsonval movements having different values of resistance are used in voltmeters of various ranges – requiring different multipliers for different ranges. Therefore, the total resistance of the voltmeter circuit is a difficult instrument rating to express. More meaningful information can be conveyed to the user via the *sensitivity rating* of the instrument. This rating, generally printed on the meter dial, specifies the resistance of the meter for a one volt range.

The sensitivity of a voltmeter is defined as

$$S_v = 1 / I_{fs} = 1 / I_m \ \Omega/\text{V} \qquad \text{...(9.22)}$$

where $I_{fs}$ = current required for full scale deflection.

The sensitivity $S_v$ of the voltmeter may be used to advantage in the sensitivity method of calculating the resistance of multiplier in a d.c. voltmeter. Consider the circuit of Fig. 9.10.

$$R_1 = S_v V_1 - R_m \qquad \text{...(9.23)}$$

$$R_2 = S_v V_2 - (R_m + R_1) \qquad \text{...(9.24)}$$

$$R_3 = S_v V_3 - (R_m + R_1 + R_2) \qquad \text{...(9.25)}$$

$$R_4 = S_v V_4 - (R_m + R_1 + R_2 + R_3) \qquad \text{...(9.26)}$$

**Example 9.14** *Solve Example 9.13 by voltage sensitivity method.*

**Solution.** Sensitivity

$S_v = 1/I_m = 1/(1 \times 10^{-3}) = 1000\ \Omega/\text{V}.$

$R_1 = S_v V_1 - R_m = 1000 \times 10 - 100 = 9900\ \Omega$

$R_2 = S_v V_2 - (R_m + R_1)$
$= 1000 \times 50 - (100 + 9900)\ \Omega = 40\ \text{k}\Omega$

$R_3 = S_v V_3 - (R_m + R_1 + R_2)$
$= 1000 \times 250 - (100 + 9900 + 40000)\ \Omega = 200\ \text{k}\Omega$

and $R_4 = S_v V_3 - (R_m + R_1 + R_2 + R_3)$
$= 1000 \times 500 - (100 + 9900 + 40{,}000 + 200{,}000)\ \Omega$
$= 250\ \text{k}\Omega$

**Example 9.15** *Which meter has a greater sensitivity : Meter A having a range of 0 – 10 V and a multiplier resistance of 18 kΩ, or meter B with a range of 0 – 300 V and a multiplier resistance of 298 kΩ ? Both meter movements have a resistance of 2 kΩ.*

**Solution. *Meter A* :**

Resistance of meter movement + resistance of multiplier
$= 2 + 18 = 20\ \text{k}\Omega.$

∴ Sensitivity $= 20/10 = 2\ \text{k}\Omega/\text{V}$

***Meter B* :**

Resistance of meter movement + resistance of multiplier
$= 2 + 298 = 300\ \text{k}\Omega$

∴ Sensitivity $= 300/300 = 1\ \text{k}\Omega/\text{V}$

Meter *A* is more sensitive than meter *B*, as it (meter *A*) needs half the current that meter *B* needs for full scale deflection.

### 9.4.10 Sensitivity of PMMC Voltmeters and their Loading Effects

The sensitivity of a d.c. voltmeter is an important factor when selecting a meter for a certain voltage measurement. A low resistance voltmeter may give correct readings when measuring voltages in low resistance circuits but is certain to produce very unreliable readings in high resistance circuits. A voltmeter, when connected across two points in a highly resistive circuit, acts as a shunt for that portion of the circuit. The meter will then give a lower voltage drop than actually existed before the meter was connected. This effect is called loading effect of an instrument and is caused principally by low sensitivity instruments.

This is illustrated by the following example.

**Example 9.16** *It is desired to measure the voltage across a 50 Ω resistor in the circuit shown in Fig. 9.12. Two voltmeters are available for this purpose :*

*Voltmeter A with a sensitivity of 1000 Ω/V and*
*Voltmeter B with a sensitivity of 20,000 Ω/V.*

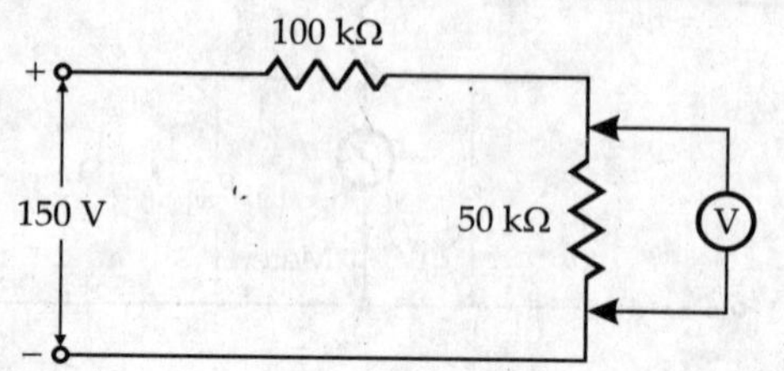


**Fig. 9.12** Circuit of Example 9.16.

*Both meters have 0 – 50 V range. Calculate :*
*(a) The reading of each voltmeter ;*
*(b) The error in each reading expressed as a percentage of true value.*

**Solution.** True value of voltage across the 50 kΩ resistor $= 50/(100 + 50) \times 150 = 50\ \text{V}.$

(a) ***Voltmeter A.*** Resistance of voltmeter

$$R_v = S_v V = 1000 \times 50\ \Omega = 50\ \text{k}\Omega.$$

Now this voltmeter is connected across the 50 kΩ resistor and therefore the resistance of parallel combination of voltmeter and resistor is

$$50 \times 50/(50 + 50) = 25\ \text{k}\Omega.$$

Voltage across the combination of voltmeter and resistor $= 25/(100 + 25) \times 150 = 30\ \text{V}.$

Hence voltmeter *A* indicates a voltage of 30 V.

***Voltmeter B.*** Resistance of voltmeter

$$R_v = S_v V = 20{,}000 \times 50\ \Omega = 1000\ \text{k}\Omega$$

Resistance of combination of voltmeter in parallel with 50 kΩ resistor

$$= 1000 \times 50/(1000 + 50) = 47.6\ \text{k}\Omega.$$

∴ Voltage across the combination of voltmeter and resistor

$$= 47.6/(100 + 47.6) \times 150 = 48.37\ \text{V}.$$

Hence voltmeter *B* indicates a voltage of 48.37 V.

(b) ***Voltmeter A.*** Percentage error

$$= \frac{\text{indicated voltage} - \text{true voltage}}{\text{true voltage}} \times 100$$

$$= \frac{30 - 50}{50} \times 100 = -40\%$$

***Voltmeter B.*** Percentage error

$$= \frac{48.36 - 50}{50} \times 100 = -3.26\%.$$

The above results indicate that the meter with higher sensitivity or ohm per volt rating gives more reliable results. This is particularly true in the cases where voltage measurements are made in high resistance circuits.

Voltage measurements in electronic circuits (where high resistances are invariably involved) are generally made with multirange voltmeters or multimeters, with sensitivities in range of 20 kΩ/V to 50 kΩ/V. In measurements of power circuits where the current is usually large voltmeters may have sensitivities as low as 100 Ω/V.

### 9.4.11 Errors in PMMC Instruments

The main sources of errors in moving coil instruments are due to :

(*i*) Weakening of permanent magnets due to ageing at temperature effects.

(*ii*) Weakening of springs due to ageing and temperature effects.

(*iii*) Change of resistance of the moving coil with temperature.

***Magnets.*** In order to have permanence in magnetism, magnets are aged by heat and vibration treatment. This process results in the loss of initial magnetism but that remains is strongly held.

***Springs.*** The weakening of springs with time can be reduced by careful use of material and pre-ageing during manufacture. However, the effect of weakening of springs on the performance of the instrument is opposite to that of ageing of magnets. The weakening of magnets tends to decrease the deflection for a particular value of current, while the weakening of springs tends to increase the deflection.

In permanent magnet moving coil instruments, a 1°C increase of temperature reduces the strength of springs by about 0.04 percent and reduces flux density in the air gap of the magnet by about 0.02 percent. Thus the net effect, on the average, is to increase the deflection by about 0.02 percent per °C.

***Moving coil.*** The moving coil of a measuring instrument is usually wound with copper wire having a temperature co-efficient of 0.004/°C. When the instrument is used as a micro-ammeter or a milli-ammeter and the moving coil is directly connected to the output terminals of the instrument, the indication of the instrument for a constant current would decrease by 0.04 percent per °C rise in temperature.

In case the moving coil instrument is used as a voltmeter a large series resistance of negligible temperature co-efficient (made of material like manganin) is used. This eliminates the error due to temperature. This is because the copper coil forms a very small fraction of the total resistance of the instrument circuit and thus any change in its resistance has a negligible effect on the total resistance.

The position when the instrument current range is extended by using a shunt is, however, different. The main source of error in this case is due to relatively larger change in the resistance of the copper moving coil as compared to that of the manganin shunt. This happens because copper has a much higher resistance temperature co-efficient as compared to manganin. To reduce the error in this account, it is usual to include in series with the moving-coil a *'swamping resistance'* of manganin so that the copper coil forms only a small fraction of the total resistance comprising the coil and the additional swamping resistance. This swamping resistance is also used as an adjustment for the final calibration of the ammeter. (See Art. 9.4.4 on page 243)

### 9.4.12 Advantages and Disadvantages of PMMC Instruments

The main *advantages* of PMMC instruments are :

1. The scale is uniformly divided.
2. The power consumption is very low as 25 μW to 200 μW.
3. The torque-weight ratio is high which gives a high accuracy. The accuracy is of the order of generally 2 percent of full scale deflection.
4. A single instrument may be used for many different current and voltage ranges by using different values for shunts and multipliers.
5. Since the operating forces are large on account of large flux densities which may be as high as 0.5 $Wb/m^2$ the errors due to stray magnetic fields are small.
6. Self-shielding magnets make the core magnet mechanism particularly useful in aircraft and aerospace applications, where a multiplicity of instruments must be mounted in close proximity to each other. An example of this type of mounting may be in the cross pointer indicator, where as many as five mechanisms are mounted in one case to form a unified display. This, obviously, results in elimination of iron cases and the corresponding weight reduction are of great advantage in aircraft and aerospace instruments.

9

The chief *disadvantages* are :

1. These instruments are useful only for d.c. The torque reverses if the current reverses. If the instrument is connected to a.c., the pointer cannot follow the rapid reversals and the deflection corresponds to mean torque, which is zero. Hence these instruments cannot be used for a.c.
2. The cost of these instruments is higher than that of moving iron instruments.

## 9.5 OHMMETERS

The ohmmeter is a convenient direct reading device for measurement of resistance. These instruments have a low degree of accuracy. The statement regarding accuracy is not intended in an unfavourable sense. There is a wide field of application for this instrument in determining the approximate value of resistance. An ohmmeter is useful for determining the approximate resistance of circuit components such as heater elements or machine field coils, measuring and sorting of resistors used in electronic circuits, checking of semi-conductor diodes and for checking continuity of circuits. It is also useful in laboratories as an aid to a precision bridge, for it can help to know the approximate value of resistance which can save time in balancing the bridge.

### 9.5.1 Series-type Ohmmeter

A circuit of a series-type ohmmeter is shown in Fig. 9.13. It consists of basic d'Arsonval movement connected in parallel with a shunting resistor $R_2$. This parallel circuit is in series with resistance $R_1$ and a battery of emf $E$. The series circuit is connected to the terminals $A$ and $B$ of the unknown resistance $R_x$.

In Fig. 9.13,

$R_1$ = current limiting resistor,

$R_2$ = zero adjusting resistor,

$E$ = emf of internal battery,

and $R_m$ = internal resistance of d'Arsonval movement.

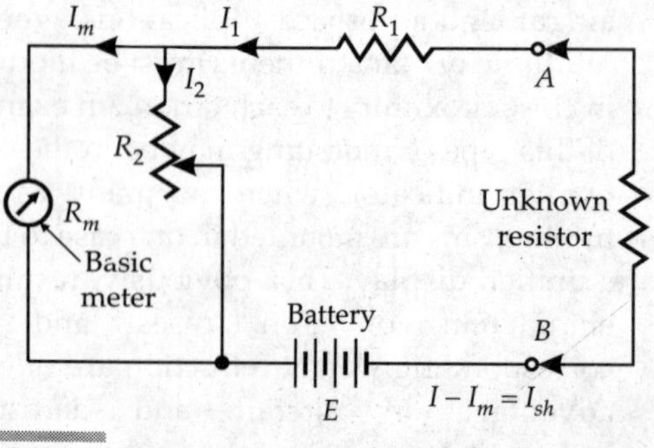


**Fig. 9.13** Series-type ohmmeter.

When the unknown resistance $R_x = 0$ (terminals $A$ and $B$ shorted) maximum current flows through the meter. Under this condition, resistor $R_2$ is adjusted until the basic movement (meter) indicates full scale current $I_{fs}$. The full-scale current position of the pointer is marked "0 Ω" on the scale. Similarly, when $R_x$ is removed from circuit, $R_x = \infty$ (that is when terminals $A$ and $B$ are open), the current in the meter drops to zero and the movement indicates zero current which is marked "∞". Thus the meter will read infinite resistance at the zero current position and zero resistance at full scale current position. Since zero resistance is indicated when the current in the meter is maximum and hence the pointer goes to the top mark. When the unknown resistance is inserted at terminals $A$, $B$ the current through the meter is reduced and hence pointer drops lower on the scale. Therefore the meter has "0" at extreme right and "∞" at the extreme left. Intermediate scale markings may be placed on the scale by different known values of resistance $R_x$ to the instrument. Figure 9.15(*a*) shows the shape of scale of series type ohmmeter.

A convenient quantity to use in the design of a series ohmmeter is the value of $R_x$ which causes the half scale deflection of the meter. At this position, the resistance across terminals $A$ and $B$ is defined as the half scale position resistance $R_h$. The design can be approached by recognizing the fact that when $R_h$ is connected across terminals $A$ and $B$ the meter current reduces to one half of its full scale value or with $R_x = R_h$, $I_m = 0.5\, I_{fs}$ where $I_m$ = current through the meter, $I_{fs}$ = current through the meter for full scale deflection. This clearly means that $R_h$ is equal to the internal resistance of the ohmmeter looking into terminals $A$ and $B$.

$$\therefore \quad R_h = R_1 + \frac{R_2 R_m}{R_2 + R_m} \qquad \text{...(9.27)}$$

The battery current at half scale deflection

$$I_h = E / 2 R_h \qquad \text{...(9.28)}$$

In order to produce full scale deflection the battery current must be doubled.

$$\therefore \quad I_1 = 2 I_h = E / R_h \qquad \text{...(9.29)}$$

Current through the shunt

$$I_2 = I_1 - I_{fs} \qquad \text{...(9.30)}$$

The voltage drop across $R_2$ is equal to voltage drop across meter

$$I_2 R_2 = I_{fs} R_m$$

or

$$R_2 = (I_{fs} / I_2) R_m \qquad \text{...(9.31)}$$

Substituting Eqn. 9.30 in Eqn. 9.31, we get

$$R_2 = \frac{I_{fs} R_m}{I_1 - I_{fs}} = \frac{I_{fs} R_m R_h}{E - I_{fs} R_h} \quad \text{...(9.32)}$$

Solving Eqn. 9.27,

$$R_1 = R_h - \frac{R_2 R_m}{R_2 + R_m} \quad \text{...(9.33)}$$

Substituting Eqn. 9.32 into Eqn. 9.33 and solving for $R_1$, we get,

$$R_1 = R_h - \frac{I_{fs} R_m R_h}{E} \quad \text{...(9.34)}$$

From Eqn. 9.32, we have,

$$R_h = \frac{ER_2}{I_{fs}(R_2 + R_m)} \quad \text{...(9.35)}$$

We observe from Eqn. 9.32, as $I_{fs} R_h$ approaches $E$, the value of $R_2$ approaches infinity. If $R_2$ is to have some finite limit say 0.5 MΩ, then from Eqn. 9.35,

$$R_h = \frac{500000\,E}{I_{fs}(500000 + R_m)}$$

If $R_2$ is to compensate for a 10% drop in battery voltage, then the maximum value of $R_h$ is

$$R_h = \frac{(500000)(0.9\,E)}{I_{fs}(500000 + R_m)} \approx \frac{0.9\,E}{I_{fs}} \quad \text{...(9.36)}$$

If the resistance $R_h$ is to be increased beyond this value, a battery with a higher emf must be used.

Hence, in terms of full scale meter current, the resistance of meter, the battery voltage, and the measured resistance at half scale, the resistances $R_1$ and $R_2$ can be determined and the maximum value of $R_h$ for a given battery e.m.f. can be computed.

The meter current for any resistance $R_x$ is

$$I_m = I_1 \frac{R_2}{R_2 + R_m}$$

But $$I_1 = \frac{E}{\text{internal resistance} + R_x} = \frac{E}{R_h + R_x}$$

or $$I_m = \frac{ER_2}{(R_h + R_x)(R_2 + R_m)} \quad \text{...(9.37)}$$

When $R_x = 0$, $I_m = I_{fs}$,

$$\therefore \quad I_{fs} = \frac{ER_2}{R_h(R_2 + R_m)} \quad \text{...(9.38)}$$

Dividing Eqn. 9.37 by Eqn. 9.38 gives

$$s = \frac{I_m}{I_{fs}} = \frac{R_h}{R_x + R_h} \quad \text{...(9.39)}$$

where $s$ is the fraction of full scale reading when measuring an unknown resistance $R_x$. Thus the scale of a given ohmmeter for a given $R_h$ can be calibrated either by direct calibration, using known values of $R_x$ or it can be calculated in terms of the fraction of the full scale readings, which can be computed from the resistance to be measured $R_x$ and the half scale resistance $R_h$.

Although the series type ohmmeter is a popular design and is used extensively in portable instruments for general service work, its use presents certain difficulties. Important among these is the internal battery whose voltage decreases gradually with time and age, so that the full scale current drops down and the meter does not read '0' when terminals $A$ and $B$ are shorted. The variable resistor $R_2$ provides an adjustment to counteract the effect of battery voltage change. Without $R_2$, it would be possible to bring the pointer back to full scale by adjusting resistance $R_1$, but this would change the calibration all along the scale, and therefore change in resistance $R_2$ is a superior solution, since the parallel resistance $R_2$ and coil resistance $R_m$ are always low as compared to $R_1$ and thus the change in $R_2$ needed for adjustment does not change calibration very much. The circuit of Fig. 9.13 does not completely compensate for ageing of the battery, but it does a reasonably good job within the expected limits of accuracy of the instrument. Some ohmmeters are constructed with a soft iron shunt across the pole pieces of the meter (permanent magnet moving coil instrument). The magnetic shunt serves to modify the air gap flux density of the magnet, and hence controls the sensitivity of the meter movement. The pointer can thus be set on the full scale mark for compensation of changes in battery voltage without changing the electrical circuit. This is a great advantage as it does not alter the electrical circuit at all.

Some ohmmeters use rectified alternating current as the power source. In this case the voltage may be regulated to give a constant voltage supply.

The range of the instrument can be extended by providing proper shunts and thus the ohmmeter may be designed to read resistance values over a wide range.

**Example 9.17** *Design a series type ohmmeter. The movement to be used requires 0.5 mA for full scale deflection and has an internal resistance of 50 Ω. The internal battery has a voltage of 3 V. The desired value of half scale resistance is 3000 Ω. Calculate*

(a) *the values of series and parallel resistances $R_1$ and $R_2$;*

(b) *the range of values of $R_2$, if the battery voltage may vary from 2.7 V to 3.1 V. Use the value of $R_1$, calculated in (a).*

9

**Solution.** The value of parallel resistance

$$R_2 = \frac{I_{fs} R_m R_h}{E - I_{fs} R_h} \quad \text{(See Eqn. 9.32)}$$

$$= \frac{0.5 \times 10^{-3} \times 50 \times 3000}{3 - 0.5 \times 10^{-3} \times 3000} = 50\ \Omega.$$

The internal resistance of ohmmeter is equal to the half scale resistance

or $$R_1 + \frac{R_2 R_m}{R_2 + R_m} = R_h$$

$$\therefore \quad R_1 + \frac{50 \times 50}{50 + 50} = 3000$$

$\therefore$ Series resistance, $R_1 = 2975\ \Omega$.

The value of resistance $R_2$ when the battery voltage is 2.7 V

$$R_2 = \frac{0.5 \times 10^{-3} \times 50 \times 3000}{2.7 - 0.5 \times 10^{-3} \times 3000} = 62.5\ \Omega.$$

When the battery voltage is 3.1 V,

$$R_2 = \frac{0.5 \times 10^{-3} \times 50 \times 3000}{3.1 - 0.5 \times 10^{-3} \times 3000} = 46.88\ \Omega.$$

### 9.5.2 Shunt Type Ohmmeter

The circuit diagram of a shunt type ohmmeter is shown in Fig. 9.14. It consists of a battery in series with an adjustable resistor $R_1$ and a basic d'Arsonval movement (meter). The unknown resistance is connected across terminals $A$ and $B$, parallel with the meter. In this circuit it is necessary to have an "off-on" switch to disconnect the battery from the circuit when the instrument is not in use. When the unknown resistor $R_x = 0\ \Omega$ ($A$ and $B$ are shorted), the meter current is zero. If the unknown resistance $R_x = \infty$ ($A$ and $B$ are open), the current finds path only through the meter and selecting a proper value for resistance $R_1$, the pointer may be made to read full scale. This ohmmeter therefore has "zero" mark on the left hand side of the scale (no current) and infinite mark on the right hand side of the scale (full scale deflection current).

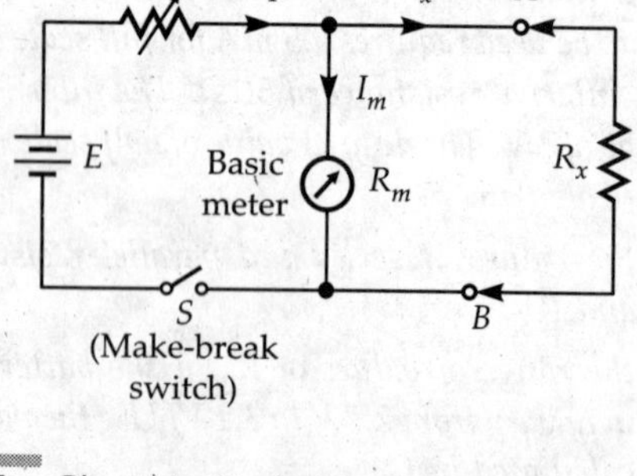


**Fig. 9.14** Shunt type ohmmeter.

The analysis of the shunt type ohmmeter is similar to that of a series type ohmmeter. In Fig. 9.14, when $R_x = \infty$, the full scale meter current is

$$I_{fs} = \frac{E}{R_1 + R_m} \quad \text{...(9.40)}$$

or $$R_1 = \frac{E}{I_{fs}} - R_m \quad \text{...(9.41)}$$

With any resistance $R_x$ to be measured, the meter current $I_m$ is :

$$I_m = \left( \frac{E}{R_1 + \frac{R_m R_x}{R_m + R_x}} \right) \left( \frac{R_x}{R_m + R_x} \right)$$

$$= \frac{E R_x}{R_1 R_m + R_x (R_1 + R_m)} \quad \text{...(9.42)}$$

At half scale reading of the meter,

$$I_m = 0.5\ I_{fs} \text{ and } R_x = R_h.$$

Then, $$0.5\, I_{fs} = \frac{E R_h}{R_1 R_m + R_h (R_1 + R_m)} \quad \text{...(9.43)}$$

Dividing Eqn. 9.42 by Eqn. 9.40 gives

$$s = \frac{I_m}{I_{fs}} = \frac{R_x (R_1 + R_m)}{R_1 R_m + R_x (R_1 + R_m)} \quad \text{...(9.44)}$$

Usually $R_1$ is very large as compared with $R_m$ and therefore,

$$s = \frac{I_m}{I_{fs}} = \frac{R_x}{R_m + R_x} \quad \text{...(9.45)}$$

Thus the meter can be calibrated by computing the fraction of full scale '$s$' for a given resistance $R_x$ in terms of $R_x$ and the meter resistance $R_m$.

In order to determine the relative scale values for a given value of $R_1$, the half scale reading may by found by dividing Eqn. 9.40 by Eqn 9.45, and solving for $R_h$ :

$$R_h = \frac{R_1 R_m}{R_1 + R_m} \quad \text{...(9.46)}$$

It is clear from the above analysis that,

(*i*) The battery emf $E$, the meter full scale current $I_{fs}$ and meter resistance $R_m$ determine the, resistance $R_1$.

(*ii*) The resistance $R_1$ and the meter resistance $R_m$ determine the half scale resistance $R_h$.

Thus the meter-resistance scale depends only on the meter and the voltage of the battery. The battery emf is 1.5-15 V, while the full scale current of the meter may vary from 0.25 mA to 1.0 A or even more. Then $R_1$ will vary from 5 to 12000 Ω. Usually $R_1$ is considerably larger than $R_m$.

Then,

$$R_1 + R_m \approx R_1 \qquad ...(9.47)$$

Substituting this in Eqn. 9.46 gives

$$R_h = R_m \qquad ...(9.48)$$

The shunt-type of ohmmeter is particularly suited for the measurement of low value resistors. It is not a commonly used instrument, but it is used in laboratories of for low resistance applications.

The shape of scales of series and shunt type of ohmmeters are given in Fig. 9.15.

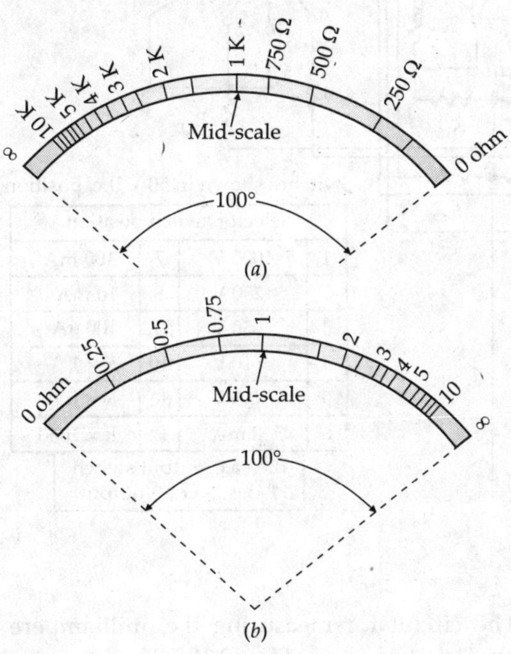


**Fig. 9.15** Shape of scale of ohmmeters.
(*a*) Series type ohmmeter with a half scale resistance value of 1 kΩ.
(*b*) Shunt type ohmmeter with a half scale resistance value of 1 Ω.

To illustrate that the shunt-type ohmmeter is particularly suited to the measurement of low resistances, an example is being given below.

***Example 9.18*** *A shunt type ohmmeter uses a 10 mA basic d'Arsonval movement with an internal resistance of 5 Ω. The battery emf is 3 V. It is desired to modify the circuit by adding appropriate shunt resistances across the movement so that its instrument indicates 0.5 Ω at the mid point on its scale.*

*Calculate :*

(*a*) *The value of shunt resistance*

(*b*) *Value of current limiting resistor $R_1$.*

**Solution.** For half scale deflection

$$I_m = 0.5\ I_{fs} = 0.5 \times 10 = 5 \text{ mA}.$$

The voltage across the movement

$$= (5 \times 10^{-3}) \times 5 = 25 \times 10^{-3} = 25 \text{ mA}.$$

(*a*) Referring to Fig. 9.14,

Voltage across unknown resistor = voltage across movement.

∴ Current through unknown resistance,

$$I_x = 25 \times 10^{-3} / 0.5 = 50 \text{ mA}.$$

Current through shunt,

$$I_{sh} = I_x - I_m = 50 - 5 = 45 \text{ mA}.$$

∴ The value of shunt resistance,

$$R_{sh} = 25 / 45 = 0.555\ \Omega.$$

(*b*) The total battery current, $I_1 = I_x + I_m + I_{sh}$

$$= 50 + 5 + 45 = 100 \text{ mA}.$$

Voltage drop across limiting resistor

$$= 3 - 25 \times 10^{-3} = 2.975 \text{ V}.$$

$$\therefore \quad R_1 = 2.975 / (100 \times 10^{-3}) = 29.75\ \Omega.$$

Thus by using proper meter and shunts, low values of resistances can be measured.

## 9.6 MULTIMETER OR VOLT-OHM-MILLI-AMMETER (V.O.M.)

The ammeter, the voltmeter and the ohmmeter all use a basic d'Arsonval movement. The difference between these instruments is the circuit in which the basic movement is used. It is therefore obvious that an instrument can be designed to perform these three measurement functions. This instrument which contains a function switch to connect the appropriate circuits to the d'Arsonval movement, is called a *"Multimeter"* or *"Volt-ohm-milli-ammeter"* (*V.O.M.*).

A representative example of a commercial multimeter is the Simpson model 260 whose complete circuit diagram is shown in Fig. 9.16. The meter is a combination of a d.c. milli-ammeter, a d.c. voltmeter, an a.c. voltmeter, a multirange ohmmeter and an output meter.

This multimeter uses a d'Arsonval movement that has a resistance of 2000 Ω and a full scale current of 50 Ω.

The instrument is provided with a selector switch which can be set for different modes of operation like measurement of voltage, current, resistance etc., and also for various ranges of these quantities.

The circuit for *d.c. voltmeter* section is shown in Fig. 9.17 when common input terminals are used for voltage ranges of 0-1.5 V to 0-1000 V. The range can be set with the help of a selector switch. An external voltage jack marked "DC 5000 V" is used for d.c.

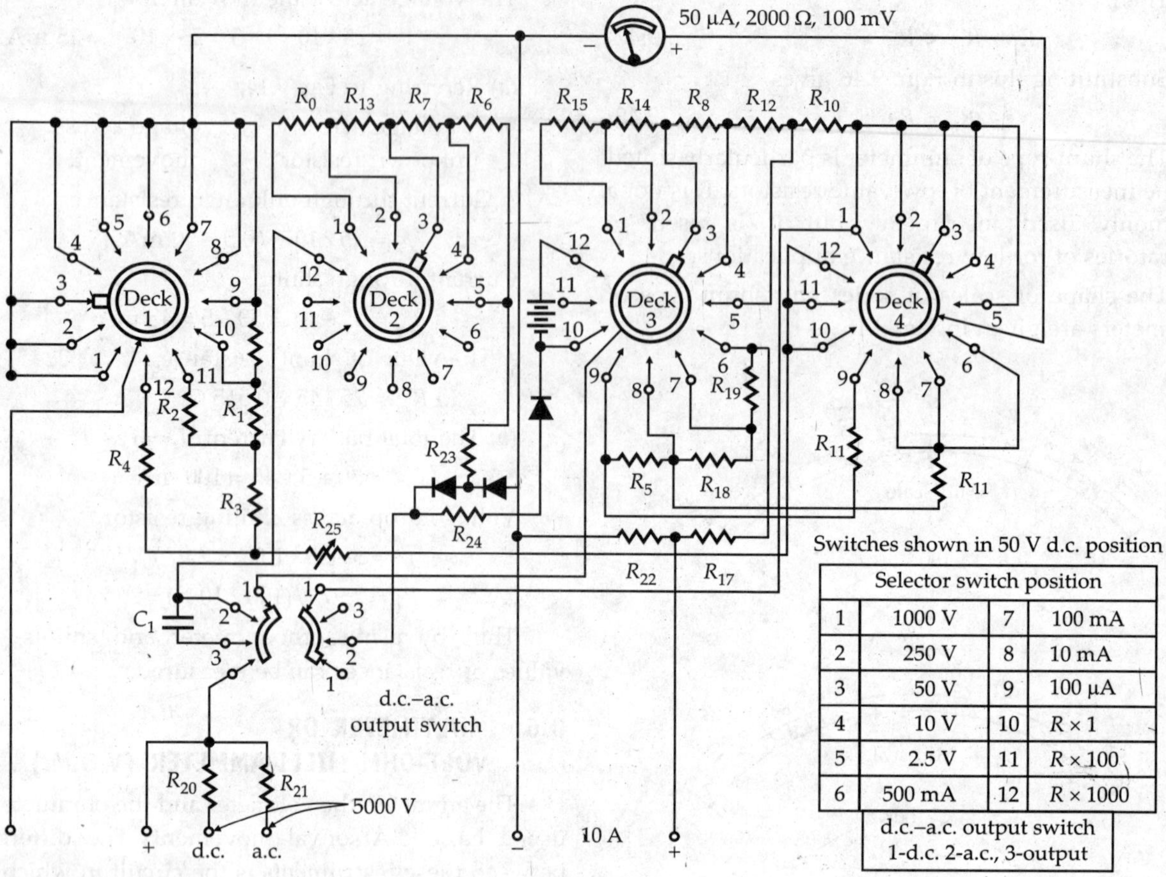


| | Selector switch position | | |
|---|---|---|---|
| 1 | 1000 V | 7 | 100 mA |
| 2 | 250 V | 8 | 10 mA |
| 3 | 50 V | 9 | 100 μA |
| 4 | 10 V | 10 | R × 1 |
| 5 | 2.5 V | 11 | R × 100 |
| 6 | 500 mA | 12 | R × 1000 |

d.c.–a.c. output switch
1-d.c. 2-a.c., 3-output

**Fig. 9.16** Circuit diagram of Simpson model 260 Multimeter.

voltage measurements upto 5000 V. The values of multiplier resistances are given in the diagram. For use on 5000 V range the selector switch should be set to 1000 V position but the test lead should be connected to the external jack marked "5000 V ". The instrument has a sensitivity of 20 kΩ/V which is fairly high and therefore the instrument is well suited to general service work in electronic measurements.

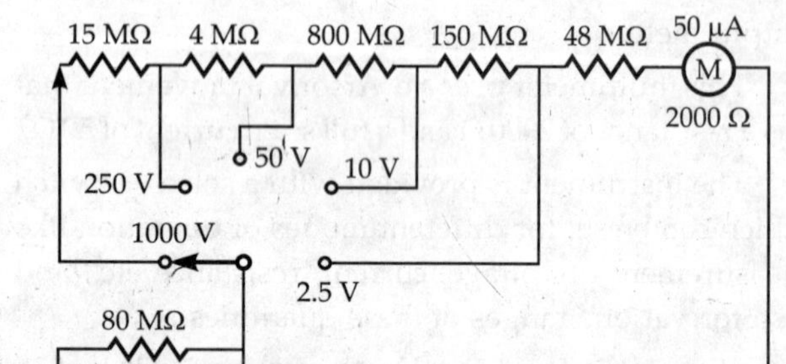


**Fig. 9.17** D.C. voltmeter section of Simpson Model 260 Multimeter.

The circuit for measuring d.c. milliampere and ampere is shown in Fig. 9.18. The common (+) terminal and negative (–) terminals are used for current measurements upto 500 mA. The jacks marked + 10 A and – 10 A are used for 0 – 10 A range.

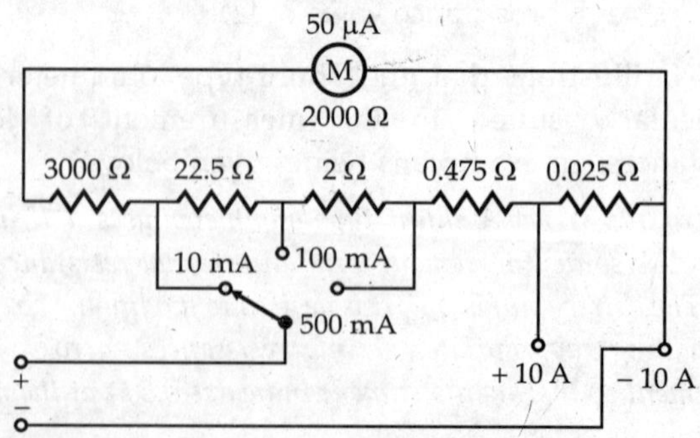


**Fig. 9.18** Ammeter section of Simpson Model 260 Multimeter.

The details of the *ohmmeter section* of the VOM meter are shown in Fig. 9.19. The circuit of Fig. 9.19(*a*) shows the ohmmeter circuit for a scale multiplication of 1 *i.e.*, $R \times 1$ scale. The instrument is a variation of shunt type ohmmeter. The instrument is short circuited and the *zero adjustment control* is varied until the meter reads zero resistance. The circuits for a scale multiplications of 100 and 10,000 are shown in Fig. 9.19(*b*) and 9.19(*c*). A.C. voltmeter readings are obtained by setting the "a.c.-d.c" switch to the a.c. position. When put on a.c. position, the input is rectified with the help of diodes and then fed to the meter.

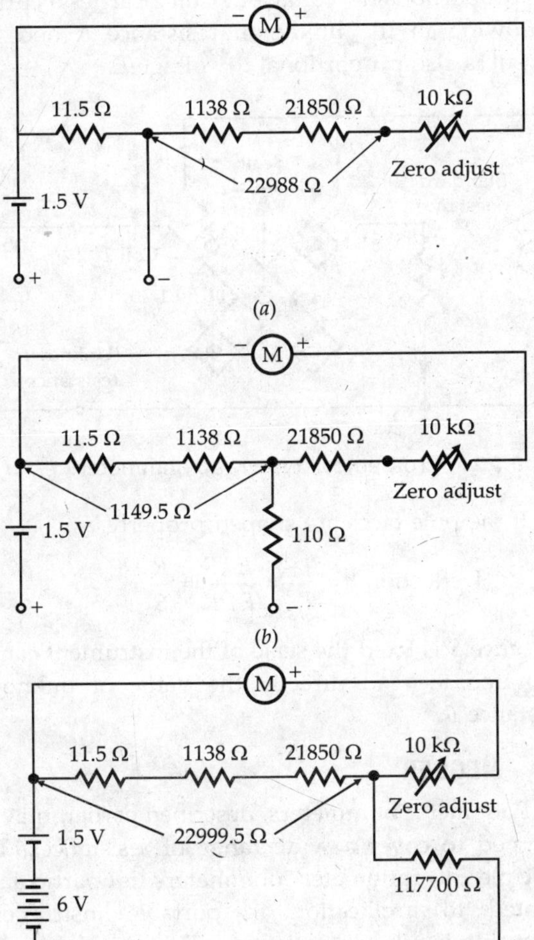


**Fig. 9.19** Ohmmeter section of Simpson Model 260 Multimeter (*a*) $R \times 1$ range (*b*) $R \times 100$ range (*c*) $R \times 10000$ range.

## 9.7 RATIOMETER

A permanent magnet ratiometer consists of two moving coils rigidly fixed together and mounted on the same spindle. They move in the magnetic field of a permanent magnet (Fig. 9.20). Current is led to the moving coils with the help of fine ligaments. There are no controlling forces. Let us consider a simple case of two coils mounted 90° apart moving in a parallel field.

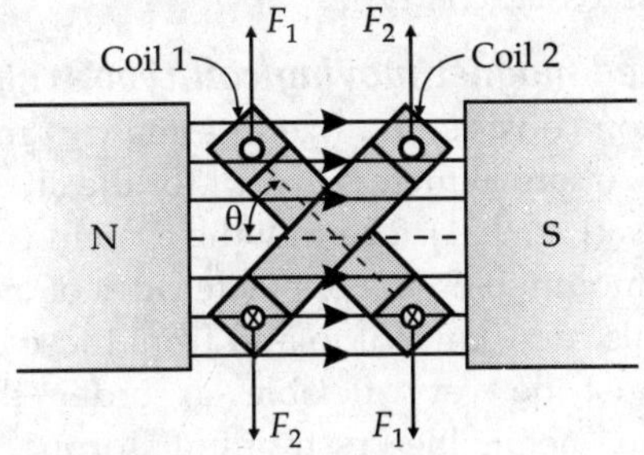


**Fig. 9.20** Ratiometer

Let

$I_1, I_2$ = current in coils 1 and 2 respectively ; A,

$N_1, N_2$ = number of turns in coil 1 and 2 respectively,

$l_1, l_2$ = length of coils 1 and 2 respectively ; m,

$d_1, d_2$ = width of coils 1 and 2 respectively ; m,

$B$ = flux density of magnetic field ; Wb/m$^2$.

Deflecting torque acting on coil 1 :

$$T_{d1} = F_1 d_1 = N_1 B l_1 d_1 I_1 \cos\theta^*$$

This torque acts in the clockwise direction

Deflecting torque acting on coil 2 :

$$T_{d2} = F_2 d_2 = N_2 B l_2 d_2 \sin\theta^*$$

This torque acts in the anti-clockwise direction. At the equilibrium position,

$$T_{d1} = T_{d2}.$$

or
$$N_1 B l_1 d_1 I_1 \cos\theta = N_2 B l_2 d_2 I_2 \sin\theta$$

or
$$\tan\theta = \frac{N_1 B l_1 d_1}{N_2 B l_2 d_2} \cdot \frac{I_1}{I_2} = k\frac{I_1}{I_2} \quad ...(9.49)$$

Thus the deflection depends upon the ratio of two currents $I_1$ and $I_2$ and hence the name *ratiometer*.

The shape of the pole faces can be modified and the angle between the two coils can be adjusted which enables the ratio $I_1 / I_2$ to be proportional to the deflection instead of tan θ. For such a meter,

$$\theta = k I_1 / I_2 \quad ...(9.50)$$

and hence the meter has a linear scale for the ratio $I_1 / I_2$.

* This is because the field is not radial but is parallel and therefore we introduce cos θ and sin θ in the expression.

### 9.7.1 Ratiometer Ohmmeters

Ratiometer ohmmeters have two forms of construction. These are :

(*i*) Fixed magnet moving coil type, and

(*ii*) Crossed coil type.

1. ***Fixed magnet moving coil type.*** This form of construction consists of a fixed permanent magnet as in most d'Arsonval meters except for the shape of the pole pieces (Fig. 9.21). The moving system consists of two coils mounted on the opposite sides of a common axis. Spirals carry current to and from these coils and spirals must be very flexible in order that they produce a negligible restraining torque on the movement and the deflection of the system is determined by the magnetic forces acting on the two coils. The central iron core is shaped as shown in Fig. 9.21 and mounted off centre so that a tapering non-uniform air gap is produced.

Currents are passed through the coils to produce torques in the opposite directions. If from initial equilibrium, the current in one coil increases relatively to the other, that coil moves to a position having a longer air gap and hence comes under the influence of a lower flux density while the other coil moves to a position of higher flux density. Equilibrium is reached when the two torques are equal. In Fig. 9.21 two resistances $R$ and $S$ connected in the circuit of the two coils are supplied by the same battery. $S$ is a standard and $R$ is the unknown resistance.

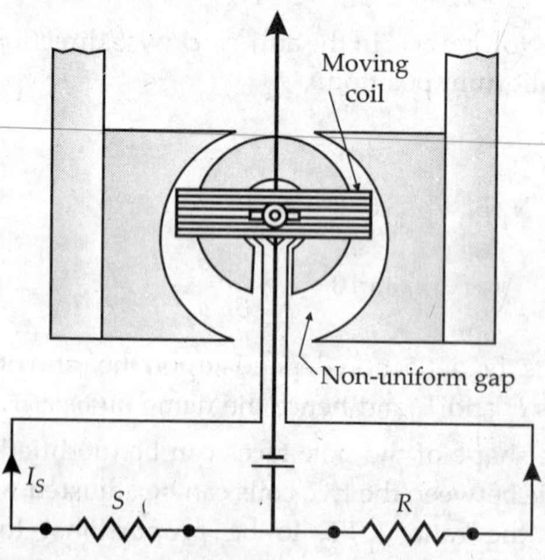


**Fig. 9.21** Fixed magnet moving coil type ratio ohmmeter.

The deflection of the moving system is a function of the relative magnitude of currents in the two coils.

$$\text{Deflection } \theta = f\left(\frac{I_S}{I_R}\right) = f\left(\frac{E/S}{E/R}\right) = f\left(\frac{R}{S}\right)$$

Thus the scale can be calibrated to read directly the value of unknown resistance $R$.

An advantage of this circuit is that a change of battery voltage affects both coils in the same proportion and hence does not affect the calibration. It is thus unnecessary to use a regulated power supply or to provide compensation for voltage changes.

2. ***Crossed coil type.*** Figure 9.22 shows a crossed coil ohmmeter. Coil 1 is connected in series with the fixed resistor to the power supply, and hence current $I_S$ is proportional to voltage $E$. Coil 2 carries a current $I_R$ flowing in the unknown resistance $R$ and this current is also proportional to voltage $E$.

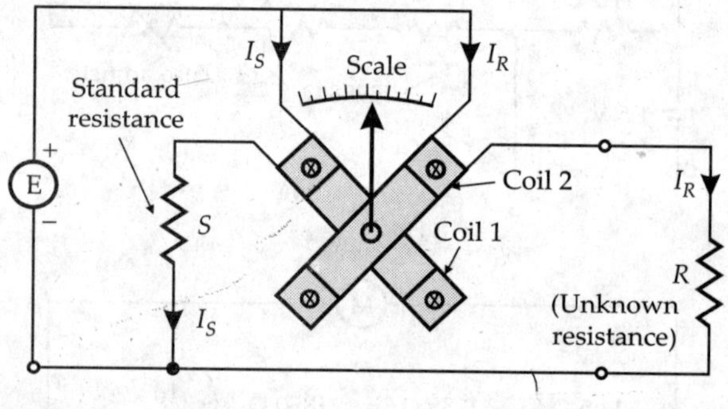


**Fig. 9.22** Crossed coil type ratio ohmmeter.

If the pole faces are shaped properly

$$\text{Deflection } \theta = \frac{I_S}{I_R} = \frac{E/S}{E/R} = \frac{R}{S}$$

Since $S$ is fixed the scale of the instrument can be calibrated to read directly the value of unknown resistance $R$.

## 9.8 MEGGER

Ratiometer ohmmeters, described earlier, may be designed to cover a wide range of resistances. The principle of ratiometer ohmmeters is particularly adapted to application in portable instruments measuring insulation resistance. This principle forms the basis of insulation testing instrument known as *Megger*. The essential parts of a Megger are shown in Fig. 9.23. The current coil is similar to that of the permanent magnet moving coil instrument. There are two voltage (potential) coils $V_1$ and $V_2$. The voltage coil $V_1$ embraces (threads over) the annular magnetic core. It is clear from Fig. 9.23 that voltage coil $V_1$ is in a weak magnetic field when the pointer is at '∞' position and hence this coil can exert very little torque.

The torque exerted by the voltage coil increases as it moves into a stronger field and this torque is maximum when it is under the pole face and under this condition the pointer is at its zero end of the resistance scale. In order to modify further the torque in the voltage circuit, another voltage coil $V_2$ is used. This coil is also so located that it moves into stronger field as the pointer moves from the '∞' position towards the zero position of the resistance scale. The coil finally embraces (threads around) the extension $H$ of the pole piece.

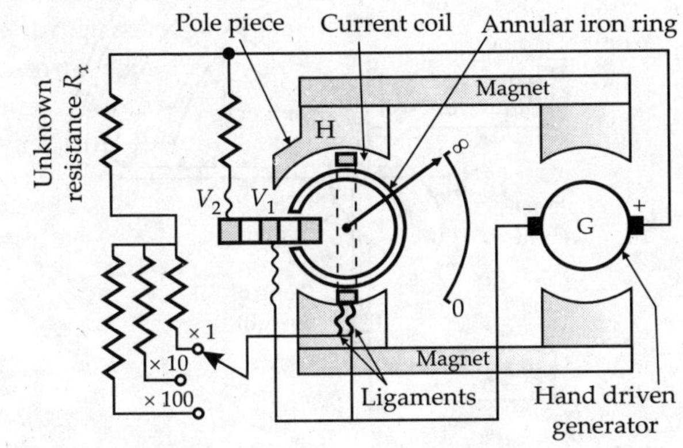


**Fig. 9.23** Megger.

The combined action of the two voltage coils $V_1$ and $V_2$ may be considered as though the coils constituted a spring of variable stiffness, being very stiff near the zero end of the scale where the current in the current coil is very large (on account of unknown resistance $R_x$ being small), and very weak near the ∞ end of the scale where the current in the current coil is very small (on account of unknown resistance $R_x$ being very large).

Thus this effect compresses the low resistance portion of the scale and opens up the high resistance portion of the scale. This is a great advantage since this instrument is meant to be used as "*insulation tester*" as the insulation resistances are quite high.

The voltage range of the instrument can be controlled by a voltage selector switch. This can be done by varying resistance '$R$' connected in series with the current coil. The test voltages, usually 500, 1000 or 2500 V are generated by a hand cranked generator $G$. A centrifugal clutch is incorporated in the generator drive mechanism which slips at a predetermined speed so that a constant voltage is applied to the insulation under test. This voltage provides a test of strength of low voltage insulation as well as a measure of its insulation resistance, since it is sufficient to cause breakdown at faults. Such breakdowns are indicated by sudden motion of the pointer off scale at zero end. As the same magnet system supplies magnetic fields for both instrument and generator, and as current and voltage coils move in a common magnetic field, the instrument indications are independent of the strength of the magnet.

## 9.9 MOVING IRON (M.I.) INSTRUMENTS

The most common ammeters and voltmeters for laboratory or switch-board use at power frequencies are the moving iron instruments. These instruments can be constructed to measure current and voltage to an accuracy needed in most engineering works and still be cheap as compared with any other type of a.c. instrument of same accuracy and ruggedness.

The general principle of working of a moving iron instrument (which is strictly relevant to attraction type of moving iron instruments) can be explained as under :

A plate or vane of soft iron or of high permeability steel forms the moving element of the system. This iron vane is so situated that it can move in a magnetic field produced by a stationary coil. The coil is excited by the current or voltage under measurement. When the coil is excited, it becomes an electromagnet and the iron vane moves in such a way so as to increase the flux of the electromagnet. This is because the vane tries to occupy a position of minimum reluctance. Thus the force produced is always in such direction so as to increase the inductance of coil (this is because inductance is inversely proportional to reluctance of magnetic circuit of the coil).

### 9.9.1 General Torque Equation of Moving Iron Instruments

An expression for the torque of a moving iron instrument may be derived by considering the energy relations when there is a small increment in current supplied to the instrument. When this happens there will be a small deflection $d\theta$ and some mechanical work will be done. Let $T_d$ be the deflecting torque.

$\therefore$ Mechanical work done $= T_d \cdot d\theta$

Alongside there will be a change in the energy stored in the magnetic field owing to change in inductance.

Suppose the initial current is $I$, the instrument inductance $L$ and the deflection $\theta$. If the current increases by $dI$ then the deflection changes by $d\theta$ and the inductance by $dL$. In order to affect an increment $dI$ in the current there must be an increase in the applied voltage given by

$$e = \frac{d}{dt}(LI) = I\frac{dL}{dt} + L\frac{dI}{dt}.$$

The electrical energy supplied

$$eIdt = I^2dL + ILdI$$

The stored energy changes from

$$= \frac{1}{2}I^2L \text{ to } \frac{1}{2}(I+dI)^2(L+dL).$$

Hence the change in stored energy,

$$= \frac{1}{2}(I^2 + 2\,IdI + dI^2)(L+dL) - \frac{1}{2}I^2L$$

Neglecting second and higher order terms in small quantities this becomes $ILdI + \frac{1}{2}I^2dL$.

From the principle of the conservation of energy,

$$\text{Electrical energy supplied} = \text{increase in stored energy} + \text{mechanical done work}$$

$$I^2dL + ILdI = ILdI = \frac{1}{2}I^2dL + T_d\,d\theta$$

Thus $\quad T_d\,d\theta = \frac{1}{2}I^2dL$

or Deflecting torque $T_d = \frac{1}{2}I^2\frac{dL}{d\theta}$ ...(9.51)

$T$ is in newton-metre, $I$ in ampere, $L$ in henry, and $\theta$ in radian.

The moving system is provided with control springs and it turns the deflecting torque $T_d$ is balanced by the controlling torque

$$T_c = K\theta.$$

where $K$ = control spring constant ; Nm/rad,

$\theta$ = deflection ; rad.

At equilibrium (or final steady) position,

$$T_c = T_d \quad \text{or} \quad K\theta = \frac{1}{2}I^2\frac{dL}{d\theta}$$

$\therefore$ Deflection $\theta = \frac{1}{2}\frac{I^2}{K}\frac{dL}{d\theta}$ ...(9.52)

Hence the deflection is proportional to square of the rms value of the operating current. The deflecting torque is, therefore, unidirectional (acts in the same direction) whatever may be the polarity of the current.

### 9.9.2 Classification of Moving Iron Instruments

Moving iron instruments are of *two* types :

(*i*) Attraction type.

(*ii*) Repulsion type.

1. ***Attraction type.*** Figure 9.24 shows the constructional details of an attraction type moving iron instrument. The coil is flat and has a narrow slot like opening. The moving iron is a flat disc or a sector eccentrically mounted. When the current flows through the coil, a magnetic field is produced and the moving iron moves from the weaker field outside the coil to the stronger field inside it or in other words, the moving iron is attracted in. The controlling torque is provided by springs but gravity control can be used for panel type of instruments which are vertically mounted.

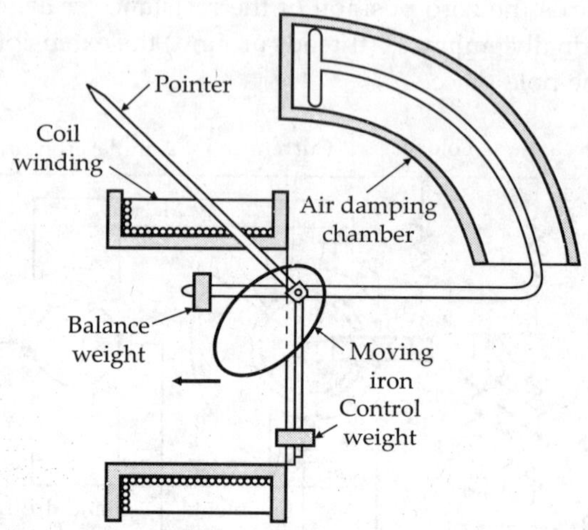


**Fig. 9.24** Attraction type moving iron instrument.

Damping is provided by air friction with the help of a light aluminium piston (attached to the moving system) which moves in a fixed chamber closed at one end as shown in Fig. 9.24 or with the help of a vane (attached to the moving system) which moves in a fixed sector shaped chamber as shown in Fig. 7.17(*b*) on page 187.

2. ***Repulsion type.*** In the repulsion type, there are two vanes inside the coil one fixed and other movable. These are similarly magnetised when the current flows through the coil and there is a force of repulsion between the two vanes resulting in the movement of the moving vane.

*Two* different designs are in common use :

(*i*) *Radial Vane Type*. In this type, the vanes are radial strips of iron. The strips are placed within the coil as shown in Fig. 9.25(*a*). The fixed vane is attached to the coil and the movable one to the spindle of the instrument.

(*ii*) *Co-axial Vane Type*. In this type of instrument, the fixed and moving vanes are sections of co-axial cylinders as shown in Fig. 9.25(*b*).

The controlling torque is provided by springs. Gravity control can also be used in vertically mounted instruments. The damping torque is produced by air friction as in attraction type instruments.

The operating magnetic field in moving iron instruments is very weak and therefore eddy current damping is not used in them as introduction of a permanent magnet required for eddy current damping would distort the operating magnetic field.

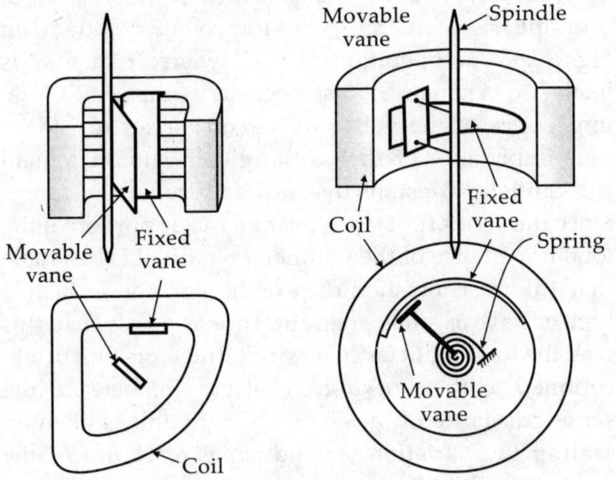


(a) Radial vane type. (b) Co-axial vane type.

**Fig. 9.25** Replusion type moving iron instruments.

It is clear that whatever may be the direction of the current in the coil of the instrument, the iron vanes are so magnetised that there is always a force of attraction in the attraction type and repulsion in the repulsion type of instruments. Thus moving iron instruments are unpolarised instruments *i.e.*, they are independent of the direction in which the current passes. Therefore, these instruments can be used on both a.c. and d.c.

### 9.9.3 Shape of Scale of Moving Iron Instruments

The deflection in a moving iron instrument is given by

$$\theta = \frac{1}{2}\frac{I^2}{K}\frac{dL}{d\theta} \qquad \text{(See Eqn. 9.52)}$$

Thus the angular deflection is proportional to the square of the operating current, and the instrument has a square law response. The deflection is in terms of rms value of current or voltage. As the deflection is proportional to square of current, it is evident that the scale of such an instrument is non-uniform.

If there is no saturation, the change of inductance with angle of deflection is uniform *i.e.*, $dL/d\theta$ = a constant. Therefore the instrument exhibits a pure square law response. For such an instrument the scale can be easily laid as the measured quantity is proportional to the square root of deflection.

In actual instruments $dL/d\theta$ is not constant and is usually a function of angular position of the moving iron and thus the scale is distorted from the square law in a manner dependent upon the way in which inductance varies with angle of deflection. This variation can be controlled by suitable design *i.e.*, by choosing proper dimensions, shape and position of iron vanes. Thus it is possible to design and construct an instrument with a scale which is very nearly uniform over a considerable part of its length. The necessary condition relating to $dL/d\theta$ against $\theta$ for linearization may be obtained from Eqn. 9.52 which gives :

$$\frac{dL}{d\theta} = \frac{2K\theta}{I^2}$$

For a linear scale $I = C\theta$ where C is a constant.

$$\therefore \quad \frac{dL}{d\theta} = \frac{2K}{C^2\theta} \quad \text{or} \quad \theta\cdot\frac{dL}{d\theta} = \text{constant.}$$

Thus for a linear scale product $\theta\frac{dL}{d\theta}$ should be a constant.

This is not possible as it requires $dL/d\theta$ to be infinite at $\theta = 0$. In practice the scale is made linear from the maximum deflection down to about 1/10th of the maximum deflection. The plot of $dL/d\theta$ against $\theta$ over the range is a rectangular hyperbola as shown in Fig. 9.26.

It is also possible to design and construct an instrument in which a small portion of the range, which is of particular interest or importance, is expanded over a large part while the remainder of the scale is compressed into a relatively small space. It is clear from the torque equation (Eqn. 9.52) that the portion of scale near zero can never be expanded or made uniform since this would require that the initial value of $dL/d\theta$ infinite.

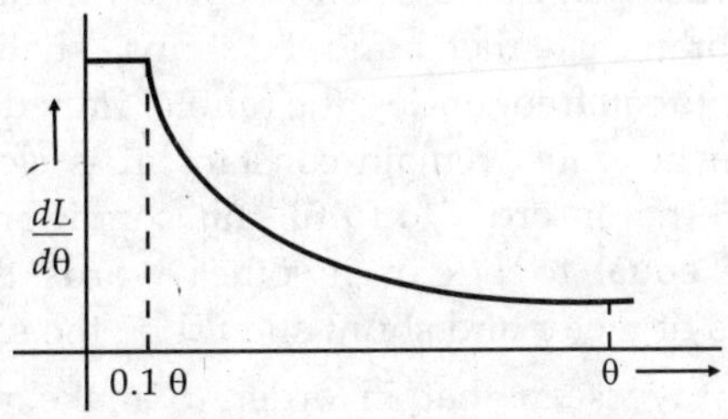


**Fig 9.26** Variation of $dL/d\theta$ v/s $\theta$.

The deflecting torque is proportional to the square of current multiplied by the rate of change of inductance with angle. The scale is compressed at its lower end because $I^2$ is small relative to full scale value and somewhat compressed at its higher end because in this range the rate of change of inductance with angle (*i.e.*, $dL/d\theta$) decreases rapidly (See Fig. 9.26). The scale is usable over about 80° of its length.

### 9.9.4 Shunts for Moving Iron Instruments

Moving iron instruments can be built for a range upto 50 A since in these instruments moving parts do not carry any current. Hence shunts are not necessary, except for very large currents. However, if a shunt is required to be used with an ammeter to extend its range, the inductances of both the meter coil and the shunt must be considered along with their resistances.

Figure 9.27 shows a shunt placed in parallel with a moving iron instrument.

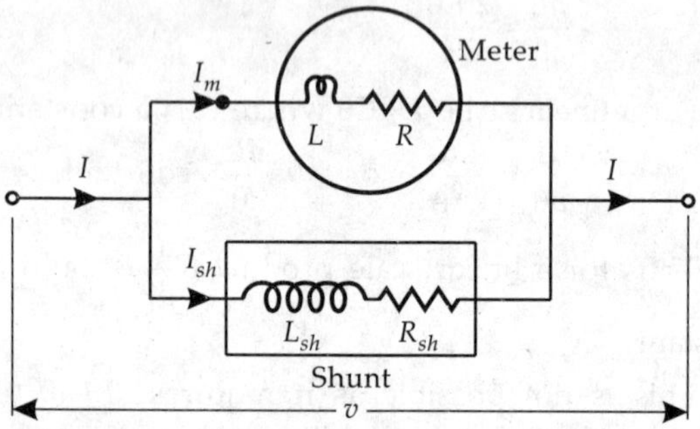


**Fig. 9.27** Shunt for moving iron instrument.

The currents in meter and shunt are in inverse ratio of their impedances.

or
$$\frac{I_{sh}}{I_m} = \sqrt{\frac{(R)^2 + (\omega L)^2}{(R_{sh})^2 + (\omega L_{sh})^2}}$$

$$= \frac{R_{sh}}{R} \cdot \frac{\sqrt{1 + (\omega L/R)^2}}{\sqrt{1 + (\omega L_{sh}/R_{sh})^2}} \quad \ldots(9.53)$$

It is clear from above that the ratio of two currents depends upon frequency.

In order that the division of current through two parallel branches - meter and the shunt - shall remain the same for all frequencies, the ratio of impedances of two branches must remain constant. It is clear from Eqn. 9.53 that in order to fulfill this condition $L_{sh}/R_{sh}$ must be equal to $L/R$ or in other words the time constants of meter and shunt should be the same.

We have seen that in order to make readings independent of the frequency the time constants of meter and shunt should be equal. But it is difficult to design a shunt with appropriate inductance, and shunts are rarely incorporated in moving iron ammeters. It has already been mentioned that coils of moving iron instruments can be designed to carry as high a current as 50 A since they are stationary and therefore shunts are normally not required. The range of moving iron ammeters for a.c. applications can be extended through use of current transformers.

### 9.9.5 Multipliers for Moving Iron Instruments

As in permanent magnet moving-coil instruments, the voltage range of moving-iron instruments may be extended by the use of a series resistance (multiplier) with the working-coil as shown in Fig. 9.28. For operation on a.c. where frequency is likely to change, it is necessary that the total impedance of the voltmeter circuit (series resistance plus impedance of the working coil) should remain substantially constant over a large frequency range. Since the series resistance forms a major portion of the total impedance of the voltmeter circuit, it is desirable that this resistor should be either of non-inductive type or have as small an inductance as possible. In this way the total inductance in the voltmeter is practically confined to the working coil itself and because the series resistance of the circuit is large, this is likely to swamp the variations of inductive reactance of the working coil with the change in frequency.

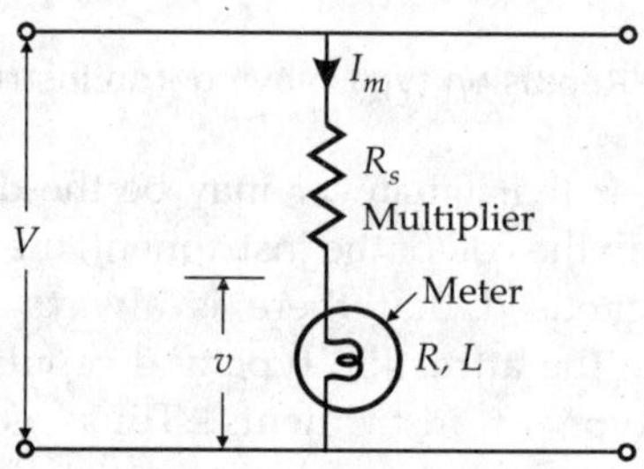


**Fig. 9.28** Voltage multipliers for moving iron instruments.

Let

$R$ = resistance of meter,

$L$ = inductance of meter,

$I_m$ = meter current for full scale deflection

$\omega$ = angular frequency.

Voltage drop across the meter for full scale deflection

$$v = I_m \sqrt{R^2 + \omega^2 L^2}$$

Let $V$ be the voltage to be measured and $R_s$ be the resistance of the non-inductive multiplier.

Total resistance of circuit $= R + R_s$

and total inductance of circuit $= L$

$\therefore$ Total impedance of circuit $= \sqrt{(R + R_s)^2 + \omega^2 L^2}$

and the current in the meter, $I_m = \dfrac{V}{\sqrt{(R + R_s)^2 + \omega^2 L^2}}$

or $\quad V = I_m \sqrt{(R + R_s)^2 + \omega^2 L^2}$

Hence, voltage multiplying power

$$m = \frac{V}{v} = \frac{\sqrt{(R+R_s)^2 + \omega^2 L^2}}{\sqrt{R^2 + \omega^2 L^2}} \quad ...(9.54)$$

It is evident from Eqn. 9.54 that the multiplying power will change with change in frequency. The multiplier may be shunted by a capacitor, in order to compensate for errors caused by change of multiplying power with change of frequency. This has been explained in Art. 9.9.7.

### 9.9.6 Comparison between Attraction and Repulsion Types of Instruments

In general it may be said that attraction-type instruments possess the same advantages, and are subject to the limitations, described for the repulsion type. An attraction type instrument will usually have a lower inductance than the corresponding repulsion type instrument, and voltmeters will therefore be accurate over a wider range of frequency and there is a greater possibility of using shunts with ammeters. On the other hand, repulsion instruments are more suitable for economical production in manufacture, and a nearly uniform scale is more easily obtained ; they are, therefore, much more common than the attraction type.

### 9.9.7 Errors in Moving Iron Instruments

There are *two* types of errors which occur in moving iron instruments — errors which occur with both a.c. and d.c. and the other which occur only with a.c. only.

#### Errors with both D.C. and A.C.

(*i*) **Hysteresis error.** This error occurs as the value of flux density is different for the same current for ascending and descending values. The value of flux density is higher for descending values of current and therefore, the instrument tends to read higher for descending values of current (and voltage) than for ascending values. This error can be minimized by making the iron parts small so that they demagnetize themselves quickly. Another method is to work the iron parts at low values of flux density so that the hysteresis effects are small.

Hysteresis may produce a 2 to 3 percent error. With the use of nickel iron alloys with narrow hysteresis loops, the error may be brought down to less than 0.05 per cent.

(*ii*) **Temperature error.** The effect of temperature changes on moving iron instruments arises chiefly from the temperature coefficient of spring. The error may be 0.02 percent per °C change in temperature. In voltmeters, errors are caused due to self-heating of coil and series resistance. The temperature of the coil may increase by 10 to 20°C for a power consumption of 1 W. Therefore, the resistance increases (by about 4 to 8%), causing a decrease in current for a given voltage. This produces a decreased deflection. Therefore, the series resistance should be made of a material like Manganin which has a small temperature co-efficient. The value of series resistance should be very large as compared with the coil resistance in order to minimize errors due to self-heating. In the case of switch board instruments, the series resistance is about 10 times the coil resistance. 9

(*iii*) **Stray magnetic fields.** The errors due to stray magnetic fields (fields other than the operating magnetic field) may be appreciable as the operating magnetic field is weak (about 0.006 to 0.0075 Wb/m$^2$ at full scale deflection) and hence can be easily distorted. Such errors depend upon the direction of the stray magnetic field relative to the held of the instrument. These errors can be minimized by using an iron case or a thin iron shield over the working parts.

#### Errors with A.C. only

(*i*) **Frequency errors.** Changes in frequency may cause errors due to changes of reactance of the working coil and also due to changes of magnitude of eddy currents set up in the metal parts of instrument.

**Reactance of instrument coil.** The change of reactance of the instrument coil is important in case of voltmeters where an additional resistance $R_s$ is used in series with the instrument coil. Let the resistance and inductance of the instrument coil be $R$ and $L$. Then the current $I$ in the instrument coil for a given applied voltage $V$ is given by :

$$I = \frac{V}{\sqrt{(R+R_s)^2 + \omega^2 L^2}}$$

The deflection of the moving-iron voltmeter depends upon the current through the coil. Therefore, the deflection for a given voltage will be less at high frequencies than at low frequencies. To some extent, compensation to this type of error is possible by connecting a capacitor $C$ across the series resistance $R_s$ as shown in Fig. 9.29.

The idea of shunting the series resistor is to make the circuit behave like a pure resistance so that the frequency changes have no effect on the readings of the instrument. Thus the compensated instrument will have a power factor nearly equal to unity. As the resistance of the meter, $R$, is considerably smaller than the series multiplier resistance $R_s$, it can be assumed that it is sufficient to ensure that the magnitude $Z$ of the impedance of the circuit formed by $L$, $R_s$ and $C$ has a value $R_s$ when used on a.c.

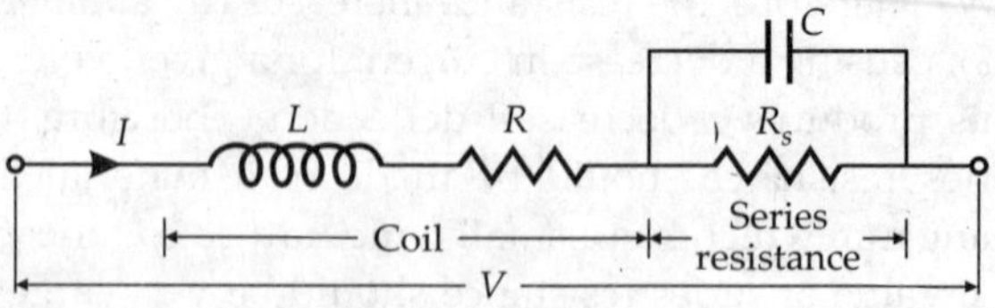


**Fig. 9.29** Frequency compensation for M.I. voltmeters.

Now $$Z = j\omega L + \frac{R_s}{1 + j\omega CR_s}$$

$$= j\omega L + \frac{R_s - j\omega CR_s^2}{1+\omega^2 C^2 R_s^2}$$

Since $\omega CR_s << 1$, we can write

$$Z = j\omega L + (R_s - j\omega CR_s^2)(1-\omega^2C^2R_s^2)$$
$$= j\omega L + R_s - j\omega CR_s^2 - \omega^2C^2R_s^3 + j\omega^3C^3R_s^4$$
$$= R_s - \omega^2C^2R_s^3 + j[\omega L - \omega CR_s^2(1-\omega^2C^2R_s^2)]$$
$$= R_s - \omega^2C^2R_s^2 + j[\omega L + \omega CR_s^2]$$
$$= R_s(1-\omega^2C^2R_s^2) + j\omega(L - CR_s^2)$$

or $$Z^2 = R_s^2(1-\omega^2C^2R_s^2)^2 + \omega^2(L-CR_s^2)^2.$$

This must equal $R_s^2$ in order that the a.c. calibration at all frequencies and d.c. calibration is the same.

$\therefore$ $$R_s^2 = R_s^2(1-\omega^2C^2R_s^2)^2 + \omega^2(L-CR_s^2)^2$$
$$= R_s^2(1-2\omega^2C^2R_s^2+\omega^4C^4R_s^4) + \omega^2(L-CR_s^2)^2$$
$$= R_s^2(1-2\omega^2C^2R_s^2) + \omega^2(L-CR_s^2) \quad \text{as } \omega^4C^4R_s^2 << 1$$
$$= R_s^2 - 2\omega^2C^2R_s^4 + \omega^2L^2 + \omega^2C^2R_s^4 - 2\omega^2LCR_s^2$$

or $$L^2 - 2LCR_s^2 - C^2R_s^4 = 0 \quad \text{or} \quad L = 2.41\ CR_s^2$$

$\therefore$ $$C = \frac{1}{2.41}\frac{L}{R_s^2} = 0.41\frac{L}{R_s^2} \qquad \text{...(9.55)}$$

It should be understood that the above analysis is valid for a limited range of frequency which in practical cases is upto 125 Hz.

***Eddy currents.*** These errors are caused by eddy currents induced in the iron parts of the instruments. Let the mutual inductance between the instrument coil and the iron parts be $M$. The induced voltage $E_e (= \omega MI)$ due to current $I$ in the instrument coil lags the current by 90° (see Fig. 9.30). As a result of this induced voltage an eddy current $I_e$ flows and its magnitude is

$$I_e = \frac{\omega MI}{\sqrt{R_e^2 + \omega^2 L_e^2}}$$

where $R_e$ and $L_e$ are respectively the resistance and inductance of this eddy current path. $I_e$ lags $E_e$ be an angle $\theta_e = \tan^{-1} \omega L_e / R_e$.

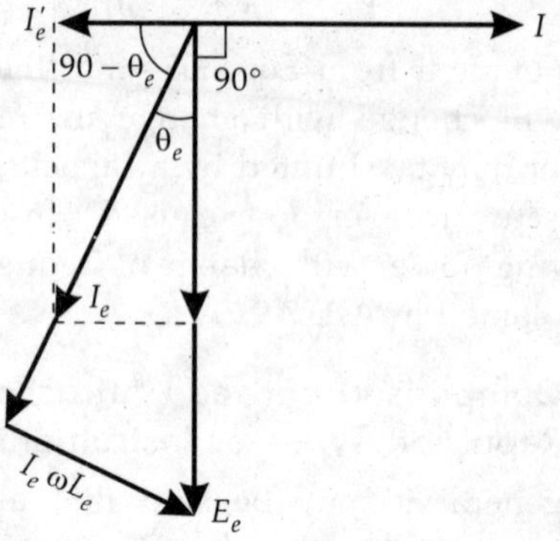


**Fig. 9.30** Effect of eddy currents in M.I. instruments.

A component of this current $I'_e = I_e \cos(90 - \theta_e) = I_e \sin\theta_e$ opposes the instrument current $I$ and sets up opposing field thus reducing the torque on the moving system. From the phasor diagram shown in Fig. 9.30, it is seen

$$I'_e = I_e \sin\theta_e = \frac{\omega MI}{\sqrt{R_e^2+\omega^2L_e^2}} \cdot \frac{\omega L_e}{\sqrt{R_e^2+\omega^2L_e^2}}$$

$$= \frac{\omega^2 ML_eI}{R_e^2+\omega^2L_e^2}$$

$\therefore$ $$I'_e \approx \frac{\omega^2 ML_eI}{R_e^2} \quad \text{when } R_e >> \omega L_e,$$

*i.e.*, when $\omega$ is small.

$$\approx \frac{MI}{L_e} \quad \text{when } \omega L_e >> R_e,$$

*i.e.*, when $\omega$ is large.

Thus at low frequencies the eddy current error increases with square of the frequency while at high frequencies the error is practically constant. For these reasons moving iron instruments are unsuitable for frequencies above 125 Hz.

### 9.9.8 Advantages and Disadvantages of Moving Iron Instruments

1. ***Universal use.*** These instruments can be used for both a.c. and d.c.

Although, moving iron instruments are responsive to d.c., the hysteresis effect causes an appreciable error in measurement. Therefore, these instruments are rarely used for d.c. applications except for very inexpensive indicators such as charging and discharging current indicators in automobiles. However, hysteresis presents no problems for use an a.c. Therefore, moving iron instruments are used extensively in industry for measurement of a.c. voltage and current where errors of the order of 5% to 10% are acceptable.

2. ***Less friction errors.*** Errors due to friction are quite small as torque-weight ratio is quite high in these instruments. This is an account of the fact that the current carrying part (heavy part of the instrument) is stationary and the moving parts are light in weight.

3. **Cheapness.** Depending upon the magnitude of the current to be measured, the coil may have a few turns of very heavy section conductor or many turns of fine wire so that the total mmf required is the same for a given displacement of the moving system irrespective of the instrument range. Hence identical moving systems may be used for an entire series of instruments from voltmeters which require a small value of current, to the highest range ammeters. Thus in a series of instruments which use an mmf of 200 A at full scale, the highest range ammeter could be for 200 A using one turn while voltmeters of the series could require 50 mA with 4000 turns for full scale deflection. The fact that a single type of moving element could cover the entire range is one reason that moving iron instruments can be built at less cost than some other types.

4. **Robustness.** The instruments are robust owing to simple construction and also that there are no current carrying moving parts.

5. ***Accuracy.*** These instruments are capable of giving an accuracy within the limits of both precision and industrial grades. Modern well designed portable moving iron instruments are expected to have a d.c. error of 2 percent or less. The initial accuracy of high grade instruments is stated to be 0.75 percent for frequencies between 25 to 125 Hz and they may be expected to be accurate within 0.2% to 0.3% at 50 Hz if carefully designed.

6. **Scale**. Moving iron instruments are now available with 240° circular scales. The greatly increased scale length being a certain advantage. The scale of moving iron instruments is not uniform and is cramped at the lower end and therefore accurate readings are not possible at this end.

7. ***Errors.*** These instruments are subjected to serious errors due to hysteresis, frequency changes and stray magnetic fields.

8. ***Waveform errors.*** The non-linearity of *B-H* curve of iron means that the deflecting torque is not exactly proportional to the square of the current. Thus, while the theory already given shows that the instrument, once calibrated in rms values, should indicate rms values whatever the waveform will not, in fact, be realized in practice.

However the precision grade instruments are so designed that the maximum flux density reached in the vanes is about 30 to 40% of the saturation value. Hence even alternating currents having highly distorted waveforms can be measured satisfactorily. This is not true of switch-board instruments meant for industrial applications where maximum flux density reaches the saturation value on account of the fact that they are designed with a high value of flux density to obtain a large deflecting torque in order to reduce the size of the instrument.

9. ***Difference between d.c. and a.c. calibrations.*** There is a difference between d.c. and a.c. calibrations on account of effect of inductance of the meter and the eddy currents when the meter is used on a.c. They must be calibrated for frequencies at which they are used.

**Example 9.19** *The inductance of a moving iron instrument is given by :*

$$L=(10+5\theta-\theta^2)\mu H$$

*where θ is the deflection in radian from zero position. The spring constant is $12\times10^{-6}$ N, m/rad. Estimate the deflection for a current of 5 A.*

**Solution.** Rate of change of inductance with deflection

$$\frac{dL}{d\theta}=\frac{d}{d\theta}(10+5\theta-\theta^2)=5-2\theta\ \mu\text{H/rad.}$$

From. Eqn. 9.52, the deflection is $\theta=\frac{1}{2}\frac{I^2}{K}\frac{dL}{d\theta}$

or $$\theta=\frac{1}{2}\times\frac{(5)^2}{12\times10^{-6}}\times(5-2\theta)\times10^{-6}$$

From above, we get $\theta=1.69$ rad $=96.8°$.

**Example 9.20** *The inductance of a moving iron ammeter with a full scale deflection of 90° at 1.5 A, is given by the expression $L=(200+40\theta-4\theta^2-\theta^3)\mu H$, where θ is the deflection in radian from the zero position. Estimate the angular deflection of the pointer for a current of 1.0 A.*

**Solution.** Inductance $L=200+40\theta-4\theta^2-\theta^3\ \mu$H.

∴ Rate of change of inductance, with deflection

$$dL/d\theta=40-8\theta-3\theta^2\ \mu\text{H/rad.}$$

$dL/d\theta$ for $\theta=\pi/2$ is :

$$40-8\times\pi/2-3(\pi/2)^2=20\ \mu\text{H/rad.}$$

Deflection, $\theta=\frac{1}{2}\frac{I^2}{K}\frac{dL}{d\theta}$

With $I = 1.5$ A, we have

$$\theta = \pi/2 \text{ and } dL/d\theta = 20 \times 10^{-6} \text{ H/rad.}$$

$$\therefore \quad \frac{\pi}{2} = \frac{1}{2}\frac{(1.5)^2}{K} \times 20 \times 10^{-6}$$

$\therefore$ Spring constant, $K = 14.32 \times 10^{-6}$ Nm/rad.

For $I = 1$ A,

$$\text{Deflection, } \theta = \frac{1}{2} \times \frac{(1)^2}{14.32 \times 10^{-6}}(40 - 8\theta - 3\theta^2) \times 10^{-6}$$

From above, we get,

$$3\theta + 36.64\theta^2 - 40 = 0 \quad \text{or} \quad \theta = 1.008 \text{ rad} = 57.8°.$$

**Example 9.21** *The law of deflection of a moving iron ammeter is given by $I = 4\theta^n$ ampere where θ is the deflection in radian and n is a constant. The self-inductance when the meter current is zero is 10 mH. The spring constant is 0.16 N-m/rad.*

(i) *Determine an expression for self-inductance of the meter as a function of θ and n.*

(ii) *With $n = 0.75$ calculate the meter current and the deflection that corresponds to a self-inductance of 60 mH.*

**Solution.**

From Eqn. 9.52, deflection in radian $\theta = \frac{1}{2}\frac{I^2}{K}\frac{dL}{d\theta}$

Rate of change of self inductance

$$\frac{dL}{d\theta} = \frac{2K\theta}{I^2} \quad \text{or} \quad dL = \frac{2K}{I^2}\theta\, d\theta.$$

Substituting the value $I = 4\theta^n$ in the above expression, we have

$$\frac{dL}{d\theta} = \frac{1}{8}K\theta^{1-2n}.$$

Integrating the above expression, we have,

$$L = \frac{K\theta^{2-2n}}{8(2-2n)} + A = \frac{K}{16(1-n)}\theta^{2-2n} + A$$

where $A$ = constant of integration.

We have $I = 4\theta^n$. Thus $\theta = 0$ when $I = 0$. It is given that when $I = 0$ the value of self-inductance $L = 10 \times 10^{-3}$ H.

Putting this in expression for $L$,

$$10 \times 10^{-3} = 0 + A \quad \text{or} \quad A = 10 \times 10^{-3}.$$

Hence the expression for self-inductance as a function of θ and $n$ is :

$$L = \frac{K}{16(1-n)}\theta^{2-2n} + 10 \times 10^{-3}$$

$$= \frac{1}{100(1-n)}\theta^{2-2n} + 10 \times 10^{-3} \text{ H}$$

as $\quad K = 0.16$ Nm/rad.

(ii) With $n = 0.75$, the expression for self-inductance is,

$$L = \frac{1}{100(1-0.75)}\theta^{2-2\times0.75} + 10 \times 10^{-3}$$

$$= 0.04\theta^{0.5} + 10 \times 10^{-3} \text{ H}$$

Putting $L = 60 \times 10^{-3}$, we have $60 \times 10^{-3}$

$$= 0.04\theta^{0.5} + 10 \times 10^{-3}.$$

$\therefore$ Deflection, $\theta = 1.56$ rad $= 89.5°$.

Current, $\quad I = 4\theta^n = 4 \times 1.56^{0.75} = 5.58$ A.

**Example 9.22** *The following figures give the relation between deflection and inductance of a moving iron instrument.*

| Deflection, Degree | Inductance, μH |
|---|---|
| 20 | 335 |
| 30 | 345 |
| 40 | 355.5 |
| 50 | 366.5 |
| 60 | 376.5 |
| 70 | 385 |
| 80 | 391.2 |
| 90 | 396 |

*Find the current and the torque to give a deflection of (a) 30°, (b) 80°. Given that control spring constant is $0.4 \times 10^{-6}$ Nm/deg.*

**Solution.** The graph between deflection and inductance is plotted as shown in Fig. 9.31.

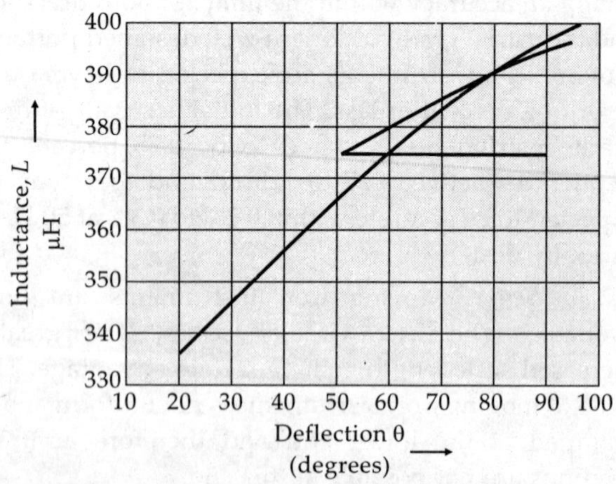


**Fig. 9.31** Plot of inductance v/s deflection for example 9.22.

The expression for deflection is

$$\theta = (1/2\,K)\,I^2(dL/d\theta)$$

Thus in order to find out current $I$ for a particular deflection, we must evaluate $dL/d\theta$ at the deflection concerned.

(*a*) **For $\theta = 30°$.**

The curve is linear in this region

$$\therefore \quad \left(\frac{dL}{d\theta}\right)_{30} = \frac{355.5-335}{40-20}$$

$$= 1.025\ \mu H/degree$$

$$= 58.7\ \mu H/rad.$$

Putting this value in the expression for θ,

$$30 = \frac{1}{2} \times \frac{1}{0.4\times10^{-6}} \times I^2 \times 58.6\times10^{-6}$$

or $\quad I = 0.65$ A.

(θ should be expressed in degrees as the value of $K$ is expressed N-m/degree).

Deflecting torque

$$T_d = T_c = K\theta = 0.4\times10^{-6}\times 30$$

$$= 12\times10^{-6}\ \text{N-m}.$$

(*b*) **For $\theta = 80°$**

A tangent is drawn to the curve at 80° and slope is found

$$\left(\frac{dL}{d\theta}\right)_{80} = \frac{396.5-375}{90-51} = 0.55/\mu H/degree$$

$$= 31.5\ \mu H/rad.$$

$$\therefore \quad 80 = \frac{1}{2} \times \frac{1}{0.4\times10^{-6}} \times I^2 \times 31.5\times10^{-6}$$

or $\quad I = 1.425$ A.

Deflecting torque

$$T_d = 0.4\times10^{-6}\times 80 = 32\times10^{-6}\ \text{N-m}.$$

***Example 9.23*** *The relationship between inductance of a moving iron ammeter, the current and the position of the pointer is as follows :*

| **Reading (A)** | 1.2 | 1.4 | 1.6 | 1.8 |
|---|---|---|---|---|
| **Deflection (degree)** | 36.5 | 49.5 | 61.5 | 74.5 |
| **Inductance (μH)** | 575.2 | 576.5 | 577.8 | 578.8 |

*Calculate the deflecting torque and the spring constant when the current is 1.5 A.*

**Solution.** The relationship between deflection and current and between deflection and inductance are plotted as shown in Fig. 9.32.

Corresponding to a current of 1.5 A, the deflection is 55.5 degrees. At this deflection, the rate of change of inductance with deflection is almost linear.

From the graph of inductance *v/s* deflection

$$\frac{dL}{d\theta} = \frac{577.65-576.5}{60-49.5} = 0.11\ \mu H/deg. = 6.3\ \mu H/rad.$$

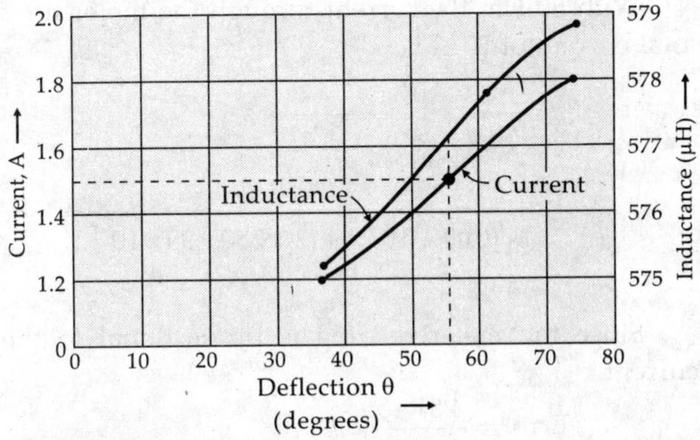


**Fig. 9.32** Plots of current and inductance *v/s* deflection for example 9.23.

Deflecting torque

$$T_d = \frac{1}{2} I^2 \frac{dL}{d\theta} = \frac{1}{2}(1.5)^2 \times 6.3\times10^{-6}$$

$$= 7.09\times10^{-6}\ \text{Nm}.$$

Deflection, $\theta = 55.5° = 0.968$ rad.

∴ Spring constant

$$K = \frac{T_d}{\theta} = \frac{7.09\times10^{6}}{0.968}$$

$$= 7.32\times10^{-6}\ \text{Nm/rad}.$$

***Example 9.24*** *Calculate the constants of a shunt to extend the range of 0-5 A moving iron ammeter to 0-50 A. The instrument constants are $R = 0.09\ \Omega$ and $L = 90\ \mu H$. If the shunt is made non-inductive and the combination is correct on d.c. find the full scale error at 50 Hz.*

**Solution.** Multiplying power of shunt

$$m = I/I_m = 50/5 = 10.$$

In order that the meter may read correctly at all frequencies the time constants of meter and shunt circuits should be equal. Under this condition multiplying power, $m = 1 + R/R_{sh}$.

∴ Resistance of shunt

$$R_{sh} = \frac{R}{m-1} = \frac{0.09}{10-1} = 0.01\ \Omega.$$

Also, $\quad \dfrac{L}{R} = \dfrac{L_{sh}}{R_{sh}}$

∴ Inductance of shunt

$$L_{sh} = \frac{L}{R} R_{sh} = \frac{90}{0.09} \times 0.01 = 10\ \mu H.$$

With d.c., the current through the meter for a total current of 50 A is,

$$I_m = \frac{R_{sh}}{R+R_{sh}} \times I = \frac{0.01}{0.09+0.01} \times 50 = 5.0\ \text{A}.$$

With 50 Hz, the current through the meter for a total current of 50 A is,

$$I_m = \frac{R_{sh}}{\sqrt{(R+R_{sh})^2+\omega^2 L^2}} \times I$$

$$= \frac{0.01}{\sqrt{(0.09+0.01)^2+(2\pi\times 50\times 90\times 10^{-6})^2}} \times 50$$

$$= 4.81 \text{ A.}$$

Since the meter reading is proportional to the current,

$$\text{Error} = \frac{4.81-5}{5} \times 100$$

$$= -3.8\% \text{ or the meter reads 3.8\% low.}$$

**Example 9.25** *The coil of a 300 V moving iron voltmeter has a resistance of 500 Ω and an inductance of 0.8 H. The instrument reads correctly at 50 Hz a.c. supply and takes 100 mA at full scale deflection.*

*What is the percentage error in the instrument reading when it is connected to 200 V d.c. supply.*

**Solution.** Reactance of coil at 50 Hz,

$$X = 2\pi\times 50\times 0.8 = 251.3\ \Omega$$

Current taken by meter at 50 Hz and 300 V,

$$I_m = 100 \text{ mA} = 0.1 \text{ A.}$$

Impedance of meter at 50 Hz,

$$Z = 300/0.1 = 3000\ \Omega.$$

∴ Resistance of meter circuit

$$= \sqrt{Z^2 - X^2} = \sqrt{(3000)^2-(251.3)^2}$$

$$= 2989\ \Omega.$$

Current taken by meter at 200 V, 50 Hz supply

$$= 200/3000 = 0.06671 \text{ A.}$$

Current taken by instrument with 200 volt d.c.

$$= 200/2989 = 0.0669 \text{ A}$$

The deflection is proportional to the operating current. Also the instrument reads correctly with 50 Hz a.c.

Hence reading of instrument with 200 V d.c.

$$= \frac{\text{current with 200 volt d.c} \times \text{reading with 200 volt 50 Hz a.c.}}{\text{current with 200 volt 50 Hz a.c.}}$$

$$= \frac{0.0669}{0.0667} \times 200 = 200.6 \text{ V.}$$

**Example 9.26** *A 0 - 100 mA moving iron ammeter is converted to a 0 - 500 V, 50 Hz voltmeter by adding a series resistance with the coil. The coil has negligible resistance and an inductance $L=(0.01+0.2\,\theta)/4\pi$ henry, where θ is the deflection in radian. The total angular span of the meter is 100°.*

*Compute*

*(a) The spring constant of the meter and*

*(b) The series resistance required.*

**Solution.** Now, $L = \dfrac{0.01+0.2\,\theta}{4\pi}$ H

$$\therefore \quad \frac{dL}{d\theta} = \frac{0.1}{2\pi} \text{ H/rad}$$

(*a*) Spring constant, $K = \dfrac{1}{2\theta} I^2 \dfrac{dL}{d\theta}$

$$= \frac{1\times(100\times 10^{-3})^2}{2\times 100\times(\pi/180)} \times \frac{0.1}{2\pi}$$

$$= 45.6\times 10^{-6} \text{ Nm/rad.}$$

(*b*) Impedance required to convert the meter to 500 V voltmeter, $Z = 500/(100\times 10^{-3}) = 5000\ \Omega$.

Inductance of meter coil full scale deflection

$$L = \frac{0.01+0.2\,\theta}{4\pi} = \frac{0.01+0.2\times 100\times \pi/180}{4\pi}$$

$$= 0.0286.$$

Reactance of meter coil at 50 Hz

$$= 2\pi\times 50\times 0.0286 = 8.98\ \Omega.$$

∴ Series resistance required

$$R_s = \sqrt{(5000)^2-(8.98)^2} = 5000\Omega.$$

**Example 9.27** *The operating coil of a 250 V moving-iron voltmeter has a resistance of 500 Ω and an inductance of 1.0 H. The series resistance is 2000 Ω. The instrument reads correctly when a direct voltage of 250 V is applied. What will it read when 250 V at 50 Hz is applied ? With what value of capacitance must the series resistance be shunted to make the meter read correctly at 50 Hz ?*

**Solution.** The impedance of the voltmeter circuit at 50 Hz,

$$Z = \sqrt{(R+R_s)^2+\omega^2 L^2}$$

$$= \sqrt{(2000+500)^2+(2\pi\times 50\times 1)^2} = 2520\ \Omega.$$

With d.c. $Z = R + R_s = 2500\ \Omega$.

The error is due to change of current on account of increase in impedance on a.c.

∴ Reading at 250 V and 50 Hz

$$= \frac{2500}{2520} \times 250 = 248 \text{ V.}$$

From Eqn. 9.55 value of capacitance required to make the circuit non-inductive

$$C = 0.41\frac{L}{R_s^2} = 0.41 \times \frac{1}{(2000)^2} \approx 0.1\ \mu\text{F.}$$

**Example 9.28** *The coil of a 150 V moving iron voltmeter has a resistance of 400 Ω and an inductance of 0.75 H. The coil is made of copper which has a resistance temperature co-efficient of 0.004/°C. The current consumed by the instrument when placed on a 150 V d.c. supply is 0.05 A. The series resistance of the voltmeter is of Manganin with a resistance temperature co-efficient 0.00015/°C. Estimate,*

(*a*) *The temperature co-efficient of the instrument,*

(*b*) *The alteration of the reading between direct current and alternating current at 100 Hz,*

(*c*) *The capacitance of the capacitor necessary to eliminate the frequency error.*

**Solution.** Total resistance of instrument circuit

$$= 150 / 0.05 = 3000\ \Omega.$$

Resistance of coil $R = 400\ \Omega$

$\therefore$ Series resistance, $R_s = 3000 - 400 = 2600\ \Omega$.

(*a*) Change in resistance of coil/°C

$$= 0.004 \times 400 \times 1 = 1.60\ \Omega.$$

Change in swamping resistance/°C

$$= 0.00015 \times 2600 \times 1 = 0.39\ \Omega$$

Total change in resistance of instrument circuit/°C

$$= 1.60 + 0.39 = 1.99\ \Omega$$

$\therefore$ Resistance temperature co-efficient of instrument

$$= \frac{\text{change in resistance/°C}}{\text{total resistance}}$$

$$= 1.99 / 3000 = 0.00066/°\text{C}.$$

(*b*) Reactance of coil at 100 Hz,

$$= 2\pi \times 100 \times 0.75 = 471.2\ \Omega.$$

Impedance of instrument at 100 Hz,

$$= \sqrt{(3000)^2 + (471.2)^2} = 3037\ \Omega.$$

Current drawn by instrument at 100 Hz,

$$= 150 / 3037 = 0.0494\ \text{A}$$

$\therefore$ Reading of instrument at 100 Hz,

$$= (0.0494 / 0.05) \times 150 = 148.2\ \text{V}.$$

$$\text{Error} = \frac{148.2 - 150}{150} \times 100 = 1.2\%\ \text{low}.$$

(*c*) In order that there is no frequency error, value of capacitance to be connected across $R_s$,

$$C = 0.41 \frac{L}{R_s^2} = 0.41 \times \frac{0.75}{(2600)} \text{F}$$

$$= 0.0455\ \mu\text{F}$$

**Example 9.29** *A moving iron instrument has full scale current of 100 mA. It is converted into a 250 V voltmeter by using a series resistance made of a material having negligible resistance temperature co-efficient. The meter has a resistance of 320 Ω at 20°C. After carrying a steady current of 100 mA for a long time, the resistance of the coil increases to 369 Ω due to self-heating. Calculate the error due to self-heating when a voltage of*

(*a*) *250 V,*

(*b*) *125 V is applied continuously.*

**Solution.** Total resistance of the circuit when the meter is converted to a 250 V voltmeter

$$R = 250 / (100 \times 10^{-3}) = 2500\ \Omega$$

Resistance of the series resistor

$$R_s = 2500 - 320 = 2180\ \Omega$$

(*a*) When a voltage of 250 V is applied continuously the total circuit resistance becomes

$$2180 + 369 = 2549\ \Omega$$

$\therefore$ Voltmeter reading

$$= (2500 / 2549) \times 250 = 245.2\ \text{V}.$$

Hence error due to self-heating

$$= \frac{245.2 - 250}{250} \times 100 = -1.92\ \%.$$

(*b*) When a voltage of 125 V is applied continuously the increase in temperature and consequently the change in resistance is 1/4 of the value with 250 V applied continuously,

$\therefore$ Change in resistance of coil

$$= (369 - 320)/4 = 12.25\ \Omega.$$

Hence the resistance of voltmeter circuit,

$$= 2500 + 12.25 = 2512.25\ \Omega$$

$\therefore$ Voltmeter reading

$$= (2500 / 2512.25) \times 125 = 124.4\ \text{V}.$$

$$\text{Hence, error} = \frac{124.4 - 125}{125} \times 100 = -0.48\%$$

## 9.10 ELECTRODYNAMOMETER (EELECTRODYNAMIC) TYPE INSTRUMENTS

The necessity for the a.c. calibration of moving iron instruments as well as other types of instruments which cannot be correctly calibrated requires the use of a transfer type of instrument. A *transfer instrument* is one that may be calibrated with a d.c. source and then used without modification to measure a.c. This requires the transfer type instrument to have same accuracy for both d.c. and a.c., which the electrodynamometer instruments have. This is necessary since all measurements, and hence the calibration of all indicating instruments, must eventually be referred to standards of voltage and resistance. These standards are precision resistors and the Weston

9

standard cell (which is a d.c. cell). It is obvious, therefore, that it would be impossible to calibrate an a.c. instrument directly against the fundamental standards.

The calibration of an a.c. instrument may be performed as follows : The transfer instrument is first calibrated on d.c. This calibration is then transferred to the a.c. instrument on alternating current, using operating conditions under which the latter operates properly.

Electrodynamic instruments are capable of service as transfer instruments. Indeed, their principal use as ammeters and voltmeters in laboratory and measurement work is for the transfer calibration of working instruments and as standards for calibration of other instruments as their accuracy is very high.

Electrodynamometer type of instruments are used as a.c. voltmeters and ammeters both in the range of power frequencies and lower part of the audio frequency range. They are used as watt-meters, varmeters and with some modification as power factor meters and frequency meters.

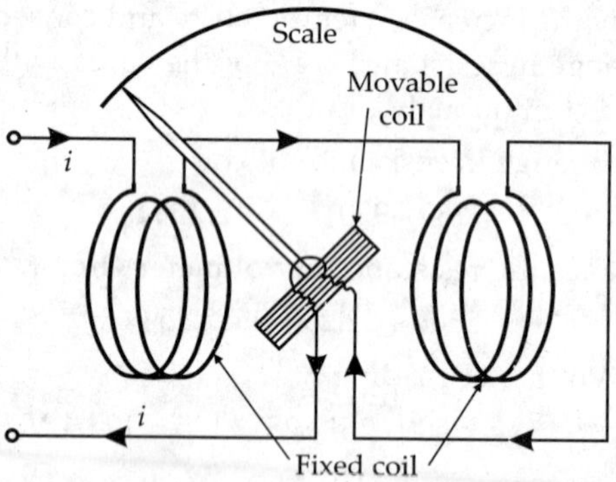


**Fig. 9.33** Electrodynamometer type instrument.

### 9.10.1 Operating Principle of Electrodynamometer Type Instrument

We can have an idea of the working principle of this instrument by taking up a permanent magnet moving coil instrument and considering how it would behave on a.c. It would have a torque in one direction during one half of the cycle and an equal effect in the opposite direction during the other half of the cycle. If the frequency were very low, the pointer would swing back and forth around the zero point. However, for an ordinary meter, the inertia is so great that on power frequencies the pointer does not go very far in either direction but merely stays (vibrates slightly) around zero. If, however, we were to reverse the direction of the flux each time the current through the movable coil reverses, a unidirectional torque would be produced for both positive and negative halves of the cycle. In electrodynamometer instruments the field can be made to reverse simultaneously with the current in the movable coil if the field (fixed) coil is connected in series with the movable coil.

### 9.10.2 Construction of Electrodynamometer Type Instrument

***Fixed coils.*** The field is produced by a fixed coil. This coil is divided into two sections to give a more uniform field near the centre and to allow passage of the instrument shaft. The instrument as shown in Fig. 9.33 may be a milliammeter, or may become a voltmeter by the addition of a series resistance. The fixed coils are wound with fine wire for such applications.

Fixed coils are usually wound with heavy wire carrying the main current in ammeters and wattmeters. The wire is stranded where necessary to reduce eddy current losses in conductors. The coils are usually varnished and baked to form a solid assembly. These are then clamped in place against the coil supports. This makes the construction rigid so that there is no shifting or change in dimensions which might effect the calibration.

The mounting supports are preferably made out of ceramic, as metal parts would weaken the field of the fixed coil on account of eddy currents.

***Moving coil.*** A single element instrument has one moving coil. The moving coil is wound either as a self-sustaining coil or else on a non-metallic former. A metallic former cannot be used as eddy currents would be induced in it by the alternating field. Light but rigid construction is used for the moving coil. It should be noted that both fixed and moving coils are air cored.

***Control.*** The controlling torque is provided by two control springs. These springs act as leads to then moving coil.

***Moving system.*** The moving coil is mounted on an aluminium spindle. The moving system also carries the counter weights and truss type pointer. Sometimes a suspension may be used in case a high sensitivity is desired.

***Damping.*** Air friction damping is employed for these instruments and is provided by a pair of aluminium vanes, attached to the spindle at the bottom. These vanes move in sector shaped chambers. Eddy current damping cannot be used in these instruments as the operating field is very weak (on account of the fact that the coils are air cored) and any introduction of a permanent magnet required for eddy current damping would distort the operating magnetic field of the instrument.

***Shielding.*** The field produced by the fixed coils is somewhat weaker than in other types of instruments. It is nearly 0.005 to 0.006 Wb/m$^2$. In d.c. measurements even the earth's magnetic field may affect the readings. Thus it is necessary to shield an electrodynamometer type instrument from the effect of stray magnetic fields. Air cored electrodynamometer type instruments are protected against external magnetic fields by enclosing them in a casing of high permeability alloy. This shunts external magnetic fields around the instrument mechanism and minimizes their effects on the indication.

***Cases and scales.*** Laboratory standard instruments are usually contained in highly polished wooden cases. These cases are so constructed as to remain dimensionally stable over long periods of time. The glass is coated with some conducting material to completely remove the electrostatic effects. The case is supported by adjustable levelling screws. A spirit level is also provided to ensure proper levelling.

The scales are hand drawn, using machine sub-dividing equipment. Diagonal lines for fine sub-division are usually drawn for main markings on the scale. Most of the high-precision instruments have a 300 mm scale with 100, 120 or 150 divisions.

### 9.10.3 Torque Equation of Electrodynamometer Instruments

Let

$i_1$ = instantaneous value of current in the fixed coils ; A,

$i_2$ = instantaneous value of current in the moving coil ; A,

$L_1$ = self-inductance of fixed coils ; H,

$L_2$ = self-inductance of moving coils ; H,

$M$ = mutual inductance between fixed and moving coils ; H.

Referring to Fig. 9.34,

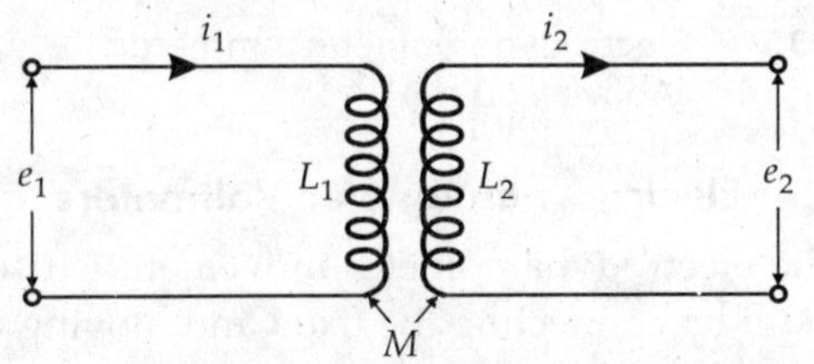


**Fig.9.34** Circuit representation of Electrodynamometer Instrument.

Flux linkages of coil 1, $\Psi_1 = L_1 i_1 + M i_2$

Flux linkages of coil 2, $\Psi_2 = L_2 i_2 + M i_1$

Electrical input energy,

$$= e_1 i_1\, dt + e_2 i_2\, dt = i_1\, d\Psi_1 + i_2 d\psi_2$$

as $e_1 = \dfrac{d\Psi_1}{dt}$ and $e_2 = \dfrac{d\Psi_2}{dt}$

$$= i_1 d(L_1 i_1 + M i_2) + i_2 d(L_2 i_2 + M i_1)$$

$$= i_1 L_1 di_1 + i_1^2 dL_1 + i_1 i_2 dM + i_1 M di_2 + i_2 L_2 di_2 + i_2^2 dL_2 + i_1 i_2 dM + i_2 M di_1 \quad ...(i)$$

Energy stored in the magnetic field

$$= \frac{1}{2} i_1^2 L_1 + \frac{1}{2} i_2^2 L_2 + i_1 i_2 M$$

Change in energy stored,

$$= d\left(\frac{1}{2} i_1^2 L_1 + \frac{1}{2} i_2^2 L_2 + i_1 i_2 M\right)$$

$$= i_1 L_1 di_1 + (i_1^2/2) dL_1 + i_2 L_2 di_2 + (i_2^2/2) dL_2 + i_1 M di_2 + i_2 M di_1 + i_1 i_2 dM \quad ...(ii)$$

From principle of conservation of energy,

Total electrical input energy = change in energy stored + mechanical energy

The mechanical energy can be obtained by subtracting (*ii*) from (*i*)

$$\therefore \text{ Mechanical energy} = \frac{1}{2} i_1^2\, dL_1 + \frac{1}{2} i_2^2\, dL_2 + i_1 i_2\, dM.$$

Now the self-inductances $L_1$ and $L_2$ are constant and therefore $dL_1$ and $dL_2$ are both equal to zero.

Hence, mechanical energy $= i_1 i_2\, dM$

Suppose $T_i$ is the instantaneous deflecting torque and $d\theta$ is the change in deflection, then,

$$\text{mechanical energy} = \text{work done} = T_i\, d\theta.$$

Thus we have,

$$T_i\, d\theta = i_1 i_2\, dM \text{ or } T_i = i_1 i_2\, dM/d\theta \quad ...(9.56)$$

***Operation with D.C.*** Let

$I_1$ = current in fixed coils ; A,

$I_2$ = current in moving coil ; A.

$\therefore$ From Eqn. 9.56 deflecting torque

$$T_d = I_1 I_2\, dM/d\theta \quad ...(9.57)$$

Equation 9.57 shows that the deflecting torque depends in general on the product of currents $I_1$ and $I_2$ and the rate of change of mutual inductance.

This deflecting torque deflects the moving coil to such a position where the controlling torque of the spring is equal to the deflecting torque.

Suppose $\theta$ = final steady deflection

∴ Controlling torque

$$T_c = K\theta$$

where $K$ = spring constant ; Nm/rad.

At final steady position, $T_d = T_c$

$$I_1 I_2 \frac{dM}{d\theta} = K\theta$$

or deflection, $\theta = \frac{I_1 I_2}{K} \cdot \frac{dM}{d\theta}$ ...(9.58)

***Operation with A.C.*** Let $i_1$ and $i_2$ be the instantaneous values of currents carried by the coils. Therefore, the instantaneous deflecting torque is :

$$T_i = i_1 i_2 \, dM/d\theta \qquad \text{(see Eqn. 9.56)}$$

The average deflection torque over a complete cycle

$$T_d = \frac{1}{T}\int_0^T T_i \, dt = \frac{dM}{d\theta}\frac{1}{T}\int_0^T i_1 i_2 \, dt \qquad ...(9.59)$$

where $T$ = time period for one complete cycle.

***Sinusoidal currents.*** If currents $i_1$ and $i_2$ are sinusoidal and are displaced by a phase angle $\phi$ *i.e.*,

$$i_1 = I_{m1} \sin \omega t \text{ and } i_2 = I_{m2} \sin(\omega t - \phi)$$

∴ The average deflecting torque

$$T_d = \frac{dM}{d\theta}\frac{1}{T}\int_0^T i_1 i_2 \, dt$$

$$= \frac{dM}{d\theta}\frac{1}{2\pi}\int_0^{2\pi} i_{m1} \sin \omega t . I_{m2} \sin(\omega t - \phi) d(\omega t)$$

$$= (I_{m1} I_m / 2) \cos \phi dM/d\theta$$

$$= I_1 I_2 \cos \phi dM/d\theta \qquad ...(9.60)$$

where $I_1$ and $I_2$ are rms values of current flowing in the coils.

At equilibrium, $T_d = T_c$

or $I_1 I_2 \cos \phi \cdot dM/d\theta = K\theta$

∴ Deflection $\theta = \frac{I_1 I_2}{K} \cos \phi \cdot \frac{dM}{d\theta}$ ...(9 61)

It follows from the above that for sinusoidal alternating currents the deflecting torque and the deflection are determined by the product of the rms values of the coil currents and cosine of the phase angle between them. It is clear that the electrodynamometer type of instruments can be used on both d.c. and a.c. When being used on a.c. the instantaneous torque is proportional to $i^2$, as the current varies so does the deflecting torque but the average torque acts in the same direction. The pointer is not able to follow the torque pulsations produced by alternating currents but takes up a position where the average deflecting torque is equal to the controlling torque. Thus, the electrodynamometer types of instruments can be used for both d.c. and a.c. applications the instrument is initially calibrated on d.c., the calibration, however, holds good for a.c. as well (due to instrument accuracy being same for both d.c. and a.c. operations). In the case of a.c. the deflecting torque is proportional to the product of r.m.s. values of currents.

### 9.10.4 Electrodynamometer Ammeters

Figure 9.33 shows that the arrangement of coils of an electrodynamometer ammeter. In this case the fixed and moving coils are connected in series and, therefore, carry the same current *i.e.*,

$$I_1 = I_2 = I \text{ and } \phi = 0.$$

∴ Deflecting torque, $T_d = I^2 \, dM/d\theta$(see Eqn. 9.60) ...(9.62)

and deflection, $\theta = \frac{I^2}{K}\frac{dM}{d\theta}$ ...(9.63)

The current through the moving coil should not exceed 100 mA and, therefore, the arrangement of Fig. 9.33 is suitable for ammeters having a range upto 100 mA.

For ammeters of higher capacity, the arrangement shown in Fig. 9.35 is used. In this case the moving coil is shunted across by a low resistance to limit the current through moving coil to 100 mA. Thus there are two separate parallel branches for fixed and moving coils. In order that the ammeter may indicate correctly at all frequencies, the time constant $L/R$ of two branches should be equal as otherwise the currents will not be independent of frequency.

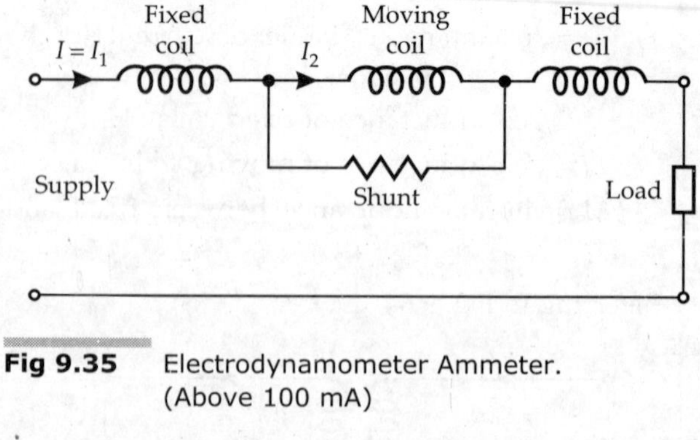


**Fig 9.35** Electrodynamometer Ammeter. (Above 100 mA)

### 9.10.5 Electrodynamometer Voltmeters

The electrodynamometer movement is used as a voltmeter by connecting the fixed and moving coils in series with a high non-inductive resistance. The arrangement is shown in Fig. 9.36.

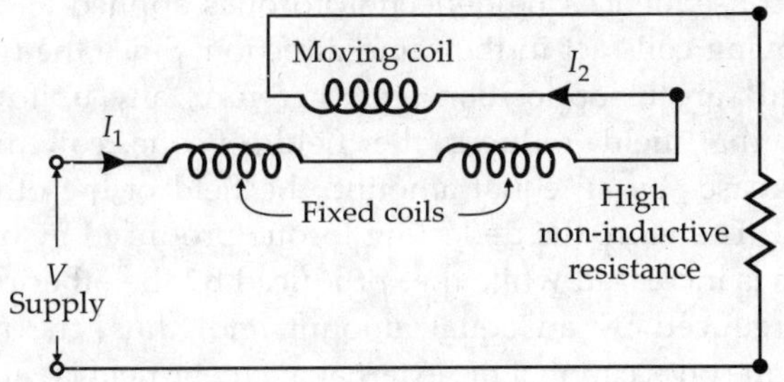


**Fig 9.36** Electrodynamometer Voltmeter.

Deflecting torque,

$$T_d = I_1 I_2 \cos\phi \cdot dM/d\theta$$

In this case,

$$I_1 = I_2 = V/Z \text{ and } \phi = 0.$$

$$\therefore \quad T_d = (V/Z).(V/Z).dM/d\theta$$

$$= (V^2/Z_1^2) \cdot dM/d\theta \quad \ldots(9.64)$$

where $V$ = voltage across the instrument.

$Z$ = impedance of the instrument circuit.

$$\text{Deflection, } \theta = \frac{V^2}{KZ^2} \cdot \frac{dM}{d\theta} \quad \ldots(9.65)$$

Electrodynamometer type voltmeter is the most accurate type of a.c. voltmeter. But the sensitivities are low in comparison with the d.c. instruments, and range from 10 to 30 Ω/V.

### 9.10.6 Errors in Electrodynamometer Instruments

The main sources of error in a dynamometer type of instruments are due to

(*i*) Low torque/weight ratio,
(*ii*) Frequency,
(*iii*) Eddy currents,
(*iv*) External magnetic fields,
(*v*) Temperature change.

(*i*) ***Torque/weight ratio.*** The magnetic field produced by air cored coils is essentially small and therefore, flux linkages per ampere in the moving coil are few (perhaps 3 to 4 percent of flux linkages per ampere in the permanent magnet moving coil instruments). Therefore, the deflecting torque is low. Thus to produce a reasonable deflecting torque, the mmf of the moving coil must be increased. The mmf of the moving coil can be increased by increasing the current through the coil or by increasing the number of turns. The current through the moving coil cannot be increased beyond 200 mA as otherwise the springs would be heated too much. On the other hand an increase the number of turns of the moving coil, will give rise to increased weight of moving parts resulting in a low torque/weight ratio. A low torque/weight ratio indicates a heavy moving system and, therefore, the friction losses in electrodynamometer type of instrument are larger than in other types. Frictional errors tend to be high.

(*ii*) ***Frequency.*** The frequency error of dynamometer type instrument is due largely as a result of variation to self-reactance of coils, with frequency. In case of voltmeters,

$$\text{Deflection, } \theta = (V^2/KZ^2) \cdot dM/d\theta$$

Now, impedance $Z = \sqrt{R^2 + (2\pi fL)^2}$ and, therefore, the value of impedance $Z$ increases with frequency and hence the voltmeters tend to read low as the frequency is increased. Therefore, in order to reduce frequency errors in voltmeters the coil winding is made a very small part of the circuit. This makes the inductive reactance, a small fraction of total impedance and, therefore, $Z = \sqrt{R^2 + (2\pi fL)^2} \approx R$.

This means that the frequency does not affect the calibration of the instrument. In fact, electrodynamometer voltmeters may be used. In general, within their guaranteed accuracy from dc to about 125 Hz.

In order to reduce frequency error in dynamometer ammeters (Fig. 9.35) the ratio of currents in fixed and moving coils *i.e.*, $I_1/I_2$ should be independent of frequency. This requires the time constants of two circuits should be the same.

Let $R_1, L_1, Z_1$ = resistance, inductance and impedance of fixed coil circuit.

$R_2, L_2, Z_2$ = resistance, inductance and impedance of moving coil circuit.

We have,

$$\frac{I_2}{I_1} = \frac{Z_1}{Z_2} = \frac{\sqrt{R_1^2 + (2\pi fL_1)^2}}{\sqrt{R_2^2 + (2\pi fL_2)^2}} \cdot \frac{R_1}{R_2}$$

$$= \frac{\sqrt{1 + (2\pi f)^2 (L_1/R_1)^2}}{\sqrt{1 + (2\pi f)^2 (L_2/R_2)^2}}$$

Now if $L_1/R_1 = L_2/R_2$, ratio of currents with a.c. becomes $I_2/I_1 = R_1/R_2$

Ratio of currents with d.c. is $I_2/I_1 = R_1/R_2$.

Thus the ratio of currents with a.c. is the same as with d.c. if the time constants (ratio $L/R$) of fixed and moving coil circuits are the same. Therefore, the frequency error in shunted ammeters can be eliminated by having equal time constants for both fixed and moving coil circuits. Also the phase difference between the two currents is zero.

(*iii*) ***Eddy currents.*** The effect of eddy currents is to produce a torque as a result of coupling between moving coil and the adjacent metal parts. Hence, metal in the coil supports and structural part is kept to the minimum possible. Any metal that is used is kept away and is selected to have high resistivity so as to reduce eddy currents induced in it. The metal present in coil supports, in the shield etc., produces a frequency error in the instrument as already seen in the case of moving iron instruments.

(*iv*) ***External magnetic fields.*** Since the operating field in electrodynamometer instruments is weak, therefore, these instruments must be protected from the external magnetic fields. Portable instruments use metal shields which serve to isolate the coils from external influences.

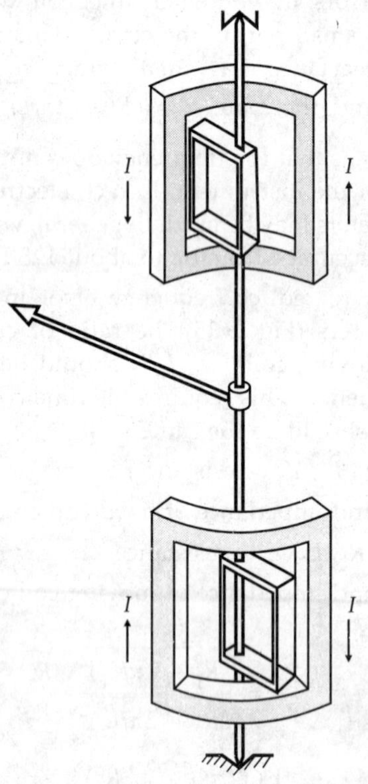


**Fig.9.37** Astatic System.

Precision type of instruments are unshielded as eddy currents are induced in the metal shields which introduce errors. In percision instruments the influence of external magnetic fields can be minimized by using *"Astatic System"*. Astatic electrodynamometer instruments are constructed with two similar sets of fixed and moving coils mounted on the same shaft (Fig. 9.37). The two pairs of fixed coils are so connected that their magnetic fields are in opposition. The currents in the moving coils are also opposed and, in consequence, the deflection torques applied to the moving coils act in the same direction. Since the two fields are in opposition and therefore, any uniform external field reduces the field of one coil and increases by an equal amount, the field of the other coil. Therefore the deflecting torque produced by one coil is increased while that produced by the other coil is reduced by an equal amount and, thus, the net torque on account of the external magnetic field is zero.

(*v*) ***Temperature changes.*** High currents are carried by coils which produce heat. Self-heating of coils produces errors. High accuracy instruments contain temperature compensating resistors which tend to neutralize the effects of temperature changes.

### 9.10.7 Shape of Scale Electrodynamometer Type Instruments

We have deflection $\theta \alpha I^2(dM/d\theta)$ for an ammeter and $\theta \alpha V^2(dM/d\theta)$ for a voltmeter. Thus it is clear from above that the scale of a dynamometer type of ammeter or a voltmeter is not uniform. The instruments have a pure *"square law"* response if $dM/d\theta =$ constant.

We have assumed the value of $(dM/d\theta)$ to be constant but it is not true. The value of $(dM/d\theta)$ is constant for a radical field but for a parallel field, which is obtained in a dynamometer type instrument, it is not constant.

Figure 9.38 shows a dynamometer type of instrument in which a parallel field is produced by the fixed coils. The mutual inductance, *M*, between fixed coils and moving coil depends upon the position of the moving coil. The maximum value of the mutual inductance $M_{max}$ occurs when $\theta = 180°$, as this position gives the maximum flux linkages. When $\theta = 0$, $M = -M_{max}$. If the flux density of the magnetic field, in which the moving coil moves, is uniform the flux linkages and hence the mutual inductance is proportional to the projected area of the coil on a plane perpendicular to the axis of the fixed coils.

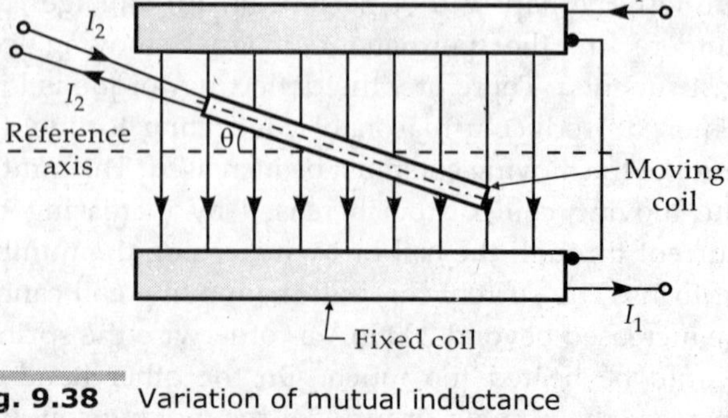


**Fig. 9.38** Variation of mutual inductance with coil position

Hence the mutual inductance $M$ at any angle $\theta$ is,

$$M = -M_{max} \times \cos\theta \quad ...(9.66)$$

or $$dM/d\theta = M_{max} \times \sin\theta \quad ...(9.67)$$

Thus the deflecting torque of a dynamometer type of instrument is given by,

$$T_d = I_1 I_2 M_{max} \cos\phi \sin\theta \quad ..(9.68)$$

and $$\theta = \frac{I_1 I_2}{K} M_{max} \cos\phi \sin\theta \quad ...(9.69)$$

It is clear from Eqn. 9.69 the instrument does not have a pure *"square law"* response.

In electrodynamometer instruments, as usually constructed, $\theta$ varies from about 45° to 135°, *i.e.*, from −45° to +45° from the position of zero mutual inductance and as changes in value of $M$ over this range are not large *i.e.*, $(dM/d\theta)$ is practically constant over this range, the instrument shows almost a square law response.

Also in practice, the field due to fixed coils is not uniform, and by careful design of the coil, $(dM/d\theta)$ may be made to constant over the range 45° to 135° with respect to reference axis shown in Fig. 9.38 and hence the instrument may show a true square law response.

### 9.10.8 Advantages and Disadvantages of Electrodynamometer Instruments

***Advantages :***

*(i)* As the coils are air cored, these instruments are free from hysteresis and eddy current errors.

*(ii)* They have a precision grade accuracy for frequencies upto 40 to 500 Hz, the lower range being decided by the oscillations of the pointer. Lower grade instruments can be used for a frequency range of 15 to 1000 Hz. Instruments using astatic movement can be used for frequencies upto 10 kHz.

*(iii)* These instruments can be used on both a.c. and d.c. and since they have a precision grade accuracy of about 0.5% of full scale deflection which is same for d.c. and a.c. operations and hence they are very useful as transfer instruments.

*(iv)* Electrodynamometer voltmeters are very useful where accurate rms values of voltage, irrespective of waveforms, are required. This is shown below.

Let the voltage applied to the voltmeter be represented as :

$$e = E_{m1}\sin(\omega t + \alpha_1) + E_{m2}\sin(2\omega t + \alpha_2) + ... + E_{mn}\sin(n\omega t + \alpha_n) + ...$$

Thus instantaneous torque in an electrodynamometer instrument is proportional to square of voltage. Therefore, we can write :

$$T_1 \,\alpha\, [E_{m1}\sin(\omega t + \alpha_1) + E_{m2}\sin(2\omega t + \alpha_2) + ... + E_{mn}\sin(n\omega t + \alpha_n) + ...]^2$$

The average deflecting torque, $T_d = \frac{1}{T}\int_0^T T_i\, dt$

$$\alpha\, \frac{1}{T}\int_0^T [E_{m1}\sin(\omega t + \alpha_1) + E_{m2}\sin(2\omega t + \alpha_2) + ... + E_{mn}\sin(n\omega t + \alpha_n) + ...]^2\, d(\omega\tau)$$

$$\alpha\, \frac{1}{2\pi}\int_0^{2\pi} [E_{m1}\sin(\alpha + \alpha_1) + E_{m2}\sin(2\alpha + \alpha_2) + ... + E_{mn}\,(n\alpha + \alpha_n) + ...]^2\,(d\alpha)$$

Now, we have

$$\frac{1}{2\pi}\int_0^{2\pi} E_{mp}^2 \sin^2(p\alpha + \alpha_p)\, d\alpha = \frac{E_{mp}^2}{2}$$

and $$\frac{1}{2\pi}\int_0^{2\pi} [E_{mp}\sin(p\alpha + \alpha_p) \times E_{mq}\sin(q\alpha + \alpha_q)]\, d\alpha = 0$$

The average deflecting torque

$$T_d \,\alpha\, \frac{1}{2}[E_{m1}^2 + E_{m2}^2 + E_{m3}^2 + ... + E_{mn}^2 + ...]$$

$$\alpha\, [E_1^2 + E_2^2 + ... + E_n^2 + ...]$$

where $E_1 = \frac{E_{m1}}{\sqrt{2}}$, $E_2 = \frac{E_{m2}}{\sqrt{2}}$ ..., $E_n = \frac{E_{mn}}{\sqrt{2}}$

But $[E_1^2 + E_2^2 + ... + E_n^2 + ...] = E^2$

where $E$ is the r.m.s. value of voltage.

$$\therefore \quad T_d \,\alpha\, E^2$$

Hence the deflecting torque in an electrodynamometer type of voltmeter is proportional to square of the rms value of applied voltage irrespective of its waveform.

***Disadvantages :***

*(i)* They have a low torque/weight ratio and hence have a low sensitivity.

*(ii)* Low torque/weight ratio gives increased frictional losses. These errors must be minimized if high accuracy is desired. This is attained by a reasonable reduction in weight of moving element, proper selection of materials for spindle and bearings and adequately polishing these parts. These measures make the instrument more costly and therefore these instruments are not used for general industrial applications where cost is a major consideration.

(*iii*) They are more expensive than either the PMMC or the moving iron type instruments.

(*iv*) These instruments are sensitive to overloads and mechanical impacts. Therefore, they must be handled with great care.

(*v*) The operating current of these instruments is large owing to the fact that they have weak magnetic field. The flux density is about 0.006 Wb/m$^2$ as against 0.1 to 0.5 Wb/m$^2$ in PMCC instruments. The low value of flux density affects the operating torque and so the meter sensitivity is low. For a voltmeter the sensitivity is about 10 Ω/V to 30 Ω/V for electrodynamometer type of instruments as compared with that of PMMC instruments which is about 20 kΩ/V. The operating current in these instruments is large and consequently they have a higher power consumption than PMMC instruments.

(*vi*) They have a non-uniform scale.

(*vii*) Their general frequency range is from 0 – 125 Hz. However, moving system tends to follow the pulsations in torque at low frequencies. Thus measurements at frequencies as low as 5 Hz to 15 Hz become impractical. Measurements at these low frequencies require special structures with abnormal damping.

The frequency range of these instruments can be increased to a few kHz by using stranded conductors and laminated supports to reduce eddy currents and through use of a compensating capacitor to for making the moving coil circuit non-inductive.

**Example 9.30** *The inductance of a 25 A electrodynamic ammeter changes uniformly at the rate of 0.0035 μH/degree. The spring constant is $10^{-6}$ N-m/degree. Determine the angular deflection at full scale.*

**Solution.** Final steady deflection, $\theta = \frac{I^2}{K} \cdot \frac{dM}{d\theta}$

(See Eqn. 9.63)

Spring constant $K$ is given in N-m/degree and therefore, we must express $(dM/d\theta)$ in H/rad if deflection is to be found in degrees.

$$\frac{dM}{d\theta} = \frac{0.0035}{\pi/180} \times 10^{-6} = 0.2 \times 10^{-6} \text{ Nm/rad}$$

$$\therefore \text{ Deflection, } \theta = \frac{(25)^2}{10^{-6}} \times 0.2 \times 10^{-6} = 125°$$

**Example 9.31** *A 10 A electrodynamic ammeter is controlled by a spring having a constant of 0.1× $10^{-6}$ N-m/degree. The full scale deflection is 110°. Determine the inductance of the instrument when measuring a current of 10 A. The mutual inductance at 0° deflection is 2 μH and the change is mutual inductance is linear with deflection.*

**Solution.** The rate of change of mutual inductance is

$$\frac{dM}{d\theta} = \frac{K\theta}{I^2} = \frac{0.1 \times 10^{-6} \times 110}{(10)^2}$$

$$= 0.11 \times 10^{-6} \text{ H/rad} = 0.00192 \text{ μH/degree.}$$

Final inductance = initial inductance

+ change in inductance

= 2.0 + 0.00192 × 110 = 2.21 μH

**Example 9.32** *In an electrodynamometer instrument the total resistance of the voltage coil circuit is 8.00 Ω and the mutual inductance changes uniformly from – 173 μH at zero deflection to + 175 μ H at full scale, the angle of full scale being 95°. If a potential difference of 100 V is applied across the voltage circuit, and a current of 3 A at a power factor of 0.75 is passed through the current coil, what will be the deflection, if the spring control constant is 4.63 × $10^{-6}$ N-m/rad ?*

**Solution.** Change in mutual inductance

$$= +175 - (-173) = 348 \text{ μH.}$$

Deflection, $\theta = 95° = 1.66$ rad.

Rate of change of mutual inductance

$$dM/d\theta = 348/1.66 = 209.6 \text{ μH/rad.}$$

This problem relates to a dynamometer type instrument where the two coils carry different values of current and also there is a phase difference between them.

Current through fixed coil

$$I_1 = 3 \text{ A.}$$

Current through moving coil

$$I_2 = 100/8200 = 12.2 \times 10^{-3} \text{ A}$$

and $\cos\phi = 0.75.$

Spring constant

$$K = 4.63 \times 10^{-6} \text{ N-m/rad}$$

$$= 0.0808 \times 10^{-6} \text{ N-m/degree.}$$

From Eqn. 9.61 deflection

$$\theta = \frac{I_1 I_2}{K} \cos\phi \frac{dM}{d\theta}$$

$$= \frac{3 \times 12.2 \times 10^{-3}}{0.0808 \times 10^{-6}} \times 0.75 \times 209.6 \times 10^{-6} = 71.2°$$

**Example 9.33** *For a certain dynamometer ammeter the mutual inductance M varies with deflection θ (expressed in degrees) as*

$$M = -6\cos(\theta + 30°) \text{ } mH.$$

*Find the deflecting torque produced by a direct current of 50 mA corresponding to a deflection of 60°.*

**Solution.** Rate of change of mutual inductance with deflection

$$\frac{dM}{d\theta} = \frac{d}{d\theta}[-6\cos(\theta+30°)] = 6\sin(\theta+30°)\text{ mH}$$

$\frac{dM}{d\theta}$ at a deflection of 60° is,

$$\left(\frac{dM}{d\theta}\right)_{\theta=60°} = 6\sin(60°+30°)\text{ mH}$$

$$= 6\times10^{-3}\text{ H/degree.}$$

Deflecting torque, $T_d$

$$= I^2\frac{dM}{d\theta} = (50\times10^{-3})^2\times6\times10^{-3}$$

$$= 15\times10^{-6}\text{ Nm} = 15\ \mu\text{Nm.}$$

**Example 9.34** *A dynamometer ammeter is fitted with two fixed coils having a total resistance of 3.0 Ω and a total inductance of 0.12 H, and a moving coil of resistance 30 Ω and an inductance of 0.003 H. Calculate the error in reading when the instrument is calibrated with d.c. and used on a.c. 50 Hz with moving coil shunted directly across the field coils. Comment upon the results.*

**Solution.** Resistance of fixed (field) coils,

$$R_1 = 3.0\ \Omega$$

Reactance of fixed coils at 50 Hz,

$$X_1 = 2\pi\times50\times0.12 = 37.7\Omega$$

Resistance of moving coil,

$$R_2 = 30\ \Omega.$$

Reactance of moving coil at 50 Hz,

$$X_2 = 2\pi\times50\times0.003 = 0.9425\ \Omega.$$

Let the current being measured be $I$.

**With D.C.** Current through fixed coils,

$$I_1 = I R_2/(R_1+R_2)$$

$$= I\times30/(3+30) = 0.909\ I.$$

Current through moving coils,

$$I_2 = I R_1/(R_1+R_2)$$

$$= I\times3/(3+30) = 0.0909\ I.$$

Deflection, $\theta \propto I_1 I_2 = K_1 I_1 I_2 = 0.0826\ K_1 I^2$.

**With A.C.** Impedance of fixed coils

$$Z_1 = \sqrt{(3)^2+(37.3)^2} = 37.8\ \Omega.$$

Phase angle, $\alpha_1 = \tan^{-1} 37.7/3 = 85.4°$.

Impedance of moving coil

$$Z_2 = \sqrt{(30)^2+(0.9425)^2} = 30\ \Omega.$$

Phase angle, $\alpha_2 = \tan^{-1} 0.09425/30 = 1.8°$.

Current through fixed coil

$$I = \frac{Z_2\angle\alpha_1}{Z_1\angle\alpha_1 + Z_2\angle\alpha_2}.\ I = 0.588\ I\angle-47.7°$$

Current through moving coil

$$I = \frac{Z_2\angle\alpha_1}{Z_1\angle\alpha_1 + Z_1\angle\alpha_2} I = 0.743\ I\angle36°.$$

Phase difference between $I_1$ and $I_2$ is

$$\phi = -47.7° - 36° = -83.7\ \therefore\ \cos\phi = 0.1097$$

Deflection with a.c.

$$\theta = K_1 I_1 I_2 \cos\phi$$

$$= K_1\times0.588\times I\times0.743\times I\times0.1097 = 0.048\ K_1 I^2$$

Percentage error,

$$= \frac{\text{reading on a.c} - \text{reading on d.c.}}{\text{reading on d.c}}\times100$$

$$= \frac{0.048\ K_1 I^2 - 0.0826\ K_1 I^2}{0.0826\ K_1 I^2}\times100 = -41.96\ \%$$

***Comments.*** The ratio of resistance to reactance in the case of fixed coils is 3/37.7 = 0.08 while in the case of moving coil, this ratio is 30/0.9425 = 31.8. The large difference in $R/X$ ratios causes a large phase shift (83.7° in this case) in the two coils thereby giving a greatly reduced deflecting torque and hence deflection under a.c. operation. If the $R/X$ (or $L/R$) ratio were same for the two coils, the difference in deflection under a.c. operating conditions and d.c. operating conditions would not have been so large (almost 42% in the present case). It is better to use an arrangement shown in Fig. 9.35 wherein the fixed and the moving coils are connected in series. The moving coil is connected across the shunt in order to limit the current through it to 100 mA.

In order to prevent errors on account of change in frequency the value of $X/R$ (or $L/R$) ratio should be the same both for fixed and the moving coil.

**Example 9.35** *A 50 volt range spring controlled, electro-dynamic voltmeter having a square law scale response takes 0.05 A on d.c. for full scale deflection of 90°. The control constant is $0.5\times10^{-6}$ N-m/degree and the initial mutual inductance of the instrument is 0.25 H. Find the true potential difference across the instrument when it reads 50 V at 50 Hz.*

**Solution.** The expression for deflection for a square law response voltmeter is :

$$\theta = \frac{V^2}{KZ^2}\frac{dM}{d\theta} \qquad \text{(See Eqn. 9.65)}$$

For d.c., impedance,

$$Z = \text{resistance } R \text{ and,}$$

therefore, $\theta = \frac{V^2}{KR^2}\frac{dM}{d\theta}$.

But $V/R = 0.05$ A, $K = 0.5 \times 10^{-6}$ and $\theta = 90°$.

Thus we have, $90 = \frac{(0.05)^2}{0.5 \times 10^{-6}} \frac{dM}{d\theta}$

or $\frac{dM}{d\theta} = 18 \times 10^{-3}$ H/rad.

Angle of deflection $= 90° = \pi/2$ rad.

Total change in mutual inductance,

$= 18 \times 10^{-3} \times \pi/2 = 28.3 \times 10^{-3}$ H.

Mutual inductance at $\theta = 90°$ is :

= initial mutual inductance

+ change in mutual inductance

$= 0.25 + 28.3 \times 10^{-3} = 0.2783$ H.

Reactance of instrument at 50 Hz,

$= 2\pi \times 50 \times 0.2783 = 87.43\ \Omega$

Resistance of instrument = 50/0.05 = 1000 Ω.

Impedance of instrument at 50 Hz,

$= \sqrt{(1000)^2 + (87.43)^2} = 1003.8\ \Omega.$

Current through the instrument when reading 50 V is 0.05 A.

∴ Actual voltage across the instrument

$= 0.05 \times 1003.8 = 50.19$ V

**Example 9.36** *A dynamometer ammeter is arranged so that 1 percent of total current passes through the moving coil and rest through the fixed coil. The mutual inductance between two coils varies with the angle of displacement of the moving coil from its zero position as follows.*

| *Angle (degree)* | *Mutual Inductance* μ*H* |
|---|---|
| 0 | – 336 |
| 15 | – 275 |
| 30 | – 192 |
| 60 | 0 |
| 90 | 192 |
| 105 | 275 |
| 120 | 336 |

*If a torque of $10.5 \times 10^{-6}$ Nm is required to give to full scale deflection of 120°, calculate the current at half and full scale deflection.*

**Solution.** The mutual inductance $M$ is plotted against deflection $\theta$ as shown in Fig. 9.39.

Half full scale deflection = 120/2 = 60°

It is clear from the data given that the $M$ varies linearly about $\theta = 60°$.

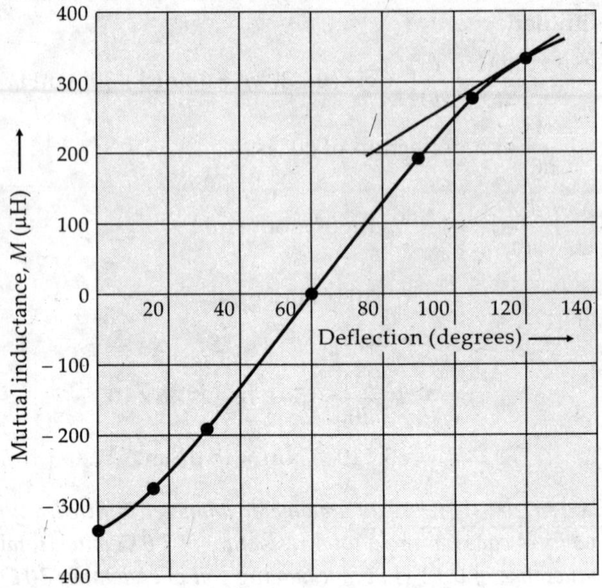


**Fig. 9.39** Variation of mutual inductance with deflection for Example 9.36.

$$\therefore \quad \left(\frac{dM}{d\theta}\right)_{60} = \frac{192 - (-192)}{90 - 30}$$

$= 6.4$ μH/degree $= 366.7$ μH/rad.

Torque at half full scale

$$T_d = \frac{1}{2} \times 10.5 \times 10^{-6} = 5.25 \times 10^{-6} \text{ Nm}$$

Suppose, $I$ = Total current

∴ Current through fixed coil, $I_1 = 0.99\ I$.

Current through moving coil, $I_2 = 0.01\ I$.

Now $T_d = I_1 I_2 \frac{dM}{d\theta}$.

$$\therefore \quad 5.25 \times 10^{-6} = 0.99\,I \times 0.01\,I \times 366.7 \times 10^{-6}$$

or $I = 1.2$ A

At full scale deflection of 120°, the rate of change of mutual inductance is determined by finding the slope of the curve

$$\left(\frac{dM}{d\theta}\right)_{120} = \frac{336 - 200}{120 - 77} = \frac{136}{43} = 3.16 \text{ μH/degree}$$

$= 181.2$ μH/rad.

Torque at full scale, $T_d = 10.5 \times 10^{-6}$ Nm.

$$\therefore \quad 10.5 \times 10^{-6} = 0.99\,I \times 0.01\,I \times 181.2 \times 10^{-6}$$

or current $I = 2.42$ A.

Full scale deflection $= 120° = 2.09$ rad.

∴ Spring constant, $K = 10.5 \times 10^{-6}/2.09$

$= 5.02 \times 10^{-6}$ Nm/rad.

## 9.11 ELECTROTHERMIC INSTRUMENTS

Electrothermic instruments depend for their indication on some property of a circuit element which is heated by flow of current. These instruments are used for measurement of current at high frequencies without any significant errors. In fact their chief field of application is in the measurement of current of frequencies above the range of moving iron and electrodynamometer type instruments. At moderate frequencies they can be used for precision measurement of voltage.

### 9.11.1 Classification of Electrothermic Instruments

There are *three* electro-thermic effects which can be utilised for measurement of current and accordingly there are three types of Electrothermic instruments.

(*i*) *Hot wire instruments*. These instruments utilize the expansion of wire heated by current being measured.

(*ii*) *Thermocouple instruments*. These instruments depend for their action on the emf developed at the junction heated by an auxiliary circuit carrying current being measured.

(*iii*) *Bolometers*. The change in resistance of a circuit element heated by current being measured is used for operation of these instruments.

## 9.12 HOT WIRE INSTRUMENTS

The constructional features of a hot wire type instrument are shown in Fig. 9.40. The current to be measured is passed through a fine platinum irridium wire. The wire is stretched between two terminals. A second wire is attached to the fine wire at one end and to a terminal at the other end. A thread is attached to the second wire. This thread passes over a pulley and is fixed to a spring.

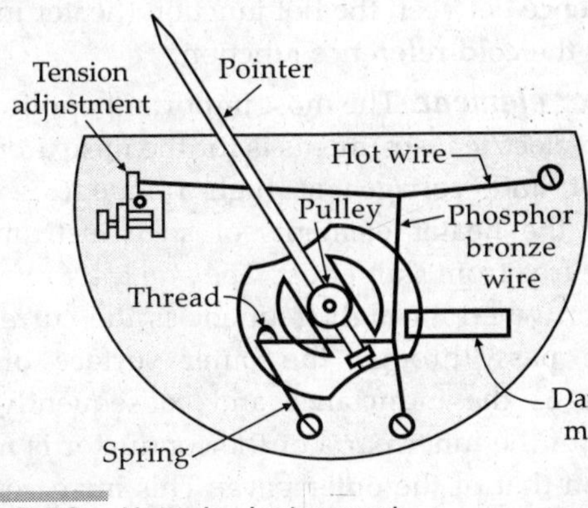


**Fig. 9.40** Hot wire instrument.

When the current is passed through the fine wire it gets heated up and expands. The sag of the wire is magnified and the expansion is taken up by the spring. This causes the pulley to rotate and the pointer to deflect, indicating the value of the current. The expansion is proportional to the heating effect of the current and hence to the square of the rms value of the current. Therefore, the meter may be calibrated to read the rms value of the current.

The instrument has many disadvantages like instability due to stretching of wire, lack of ambient temperature compensation, sluggish response, high power consumption and inability to withstand overloads and mechanical shocks. These disadvantages have made this instrument commercially unsatisfactory. Hot wire instruments are now obsolete and have been replaced by thermo-electric instruments.

## 9.13 THERMOCOUPLE INSTRUMENTS

When two metals having different work functions are placed together, a voltage is generated at the junction which is nearly proportional to the temperature. This junction is called a *Thermocouple*. This principle is used to convert heat energy to electrical energy at the junction of two conductors as shown in Fig. 9.41.

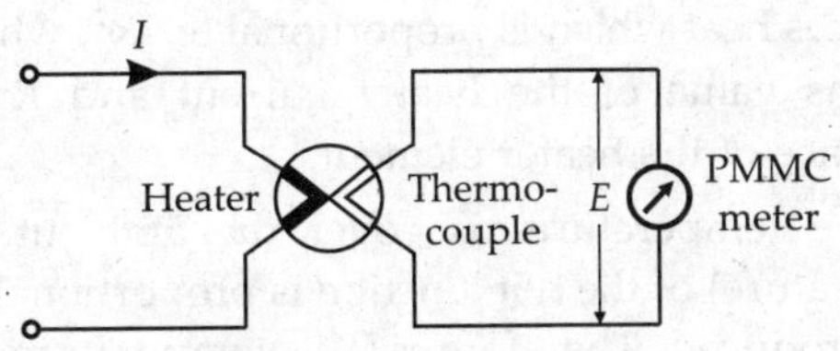


**Fig. 9.41** Circuit of a thermo-couple

The heat at the junction is produced by the electrical current flowing in the heater element while the thermocouple produces an emf at its output terminals which can be measured with the help of a PMMC instrument. The emf produced is proportional to the temperature and hence to the rms value of the current. Therefore the scale of PMMC instrument can be calibrated to read the current passing through the heater. The thermocouple type of instruments can be used for both d.c. and a.c. applications. The most attractive feature of thermocouple instruments is that they can be used for measurements of current and voltages at very high frequencies. In fact, these instruments are very accurate well above a frequency of 50 MHz.

### 9.13.1 Principle of Operation of Thermoelectric Instruments

The thermal emf developed in a circuit composed of two dissimilar metals with junctions kept at absolute temperatures $T_1$ and $T_2$ (with $T_1 > T_2$) may be approximately written as :

$$E = a(T_1 - T_2) + b(T_1 - T_2)^2 \quad ...(9.70)$$

where $a$ and $b$ are constants whose values depend upon the metals used.

Let $\Delta\theta$ = difference of temperatures of hot and cold junctions in °C $= T_1 - T_2$.

$$\therefore \quad E = a(\Delta\theta) + b(\Delta\theta)^2 \quad ...(9.71)$$

Thus emf of a thermocouple is approximately a parabolic function of the temperature difference between the junctions. The approximate values of constants $a$ and $b$ are :

$a$ = 40 to 50 μV or more per °C difference of temperature,

$b$ = few tenths or hundredths of a microvolt/(°C)$^2$

If a temperature difference of 100°C exists between the junctions, an emf of approximately 5 mV will be available. In some instruments, the hot junction has a very high temperature and an emf as high as 15 mV is produced.

The current flowing through the heater element produces heat which is proportional to $IR^2$, where $I$ is the rms value of the heater current and $R$ is the resistance of the heater element.

The temperature rise over the ambient (room temperature) of the hot junction is proportional to the heat produced. Therefore rise in temperature of hot junction $\propto I^2R$.

If the cold junction is maintained at ambient (room temperature) the rise in temperature of hot junction is equal to temperature rise of hot junction above ambient.

Hence difference in temperature $\Delta\theta$ = rise in temperature

or $\quad \Delta\theta \propto I^2R$

$\therefore \quad \Delta\theta = K_1 I^2 R$

The difference in temperature between hot and cold junctions produces an emf (Eqn. 9.71).

$$E = a(\Delta\theta) + b(\Delta\theta)^2$$

The value of $b$ is very small as compared with that of $a$ and therefore $E \approx a(\Delta\theta)$

This emf is fed to a permanent magnet moving coil millivoltmeter which gives a deflection proportional to emf $E$.

∴ Deflection of moving coil instrument

$$\theta = K_2 E = K_2(a\Delta\theta)$$

$$K_1 K_2 a I^2 R = K_3 I^2 \quad ...(9.72)$$

where $K_3 = aK_1K_2R$ = a constant.

Hence the instrument shows a square law response.

Actually we do not get an exact square law response as :

(*i*) the resistance $R$ of the heater element does not remain constant but varies with temperature rise as the element may have a positive or a negative temperature coefficient of resistance.

(*ii*) the emf is not proportional to first power of temperature difference but is a parabolic function of temperature rise as is clear from Eqn. 9.71.

These effects are usually quite small so that, if the air gap field of permanent magnet moving coil instrument is uniform, the scale of the instrument may be laid off approximately with the square of the heater current.

### 9.13.2 Construction of Thermoelectric Instruments

The thermo-electric instruments consist of the following major parts :

(*i*) Thermo-electric element and

(*ii*) Indicating instruments.

The emf produced in themro-electric instruments is d.c. in nature and is of the order of a few mV. Therefore, for the purposes of indication permanent magnet moving coil (PMMC) type of instruments are used since they have high degree of accuracy and a high sensitivity. The *thermo-elements* consist of a heater wire through which current to be measured or a current proportional to the voltage being measured is passed. The other pan of the themro-element is the thermo-couple which develops an emf proportional to the difference between the hot junction (heater in this case) and the cold reference junction.

***Heater element.*** The most important application of thermo-electric instruments is for the measurement of current and voltages at high frequencies, and therefore, the heater elements of such instruments should be free from skin effect. The *skin effect* refers to a condition where at high frequencies the current is forced to pass through the outer surface of the conductor as the inductance and consequently the reactance of the inner parts of the conductor is much larger than that of the outer parts. This increases the resistance of the conductor (since the current passes only through the outer part *i.e.* skin of the conductor and not through the inner part of the conductor whose reactance is higher than the outer part of the conductor). In order to reduce the skin effect, the heater wires are designed with a small area of cross-section. In fact, in order to minimize the skin effect, a very fine wire of non-magnetic material having a very high resistivity is used for heater elements. For example, for a 0.025 mm diameter copper wire the increase in resistance is 1% at 20 MHz, whereas the constantan wire of the same size (constantan has a much higher resistivity than that of copper), the increase in resistance is only 0.0015 % at 20 MHz and at 500 MHz the increase in resistance is 1 %.

***Thermo-elements.*** The thermo-elements used in thermocouple instruments are described below :

**1. Contact type.** The contact type thermo-element has a separate heater wire and the thermocouple has one junction (*hot junction*) in contact with the heater wire as shown in Fig. 9.42.

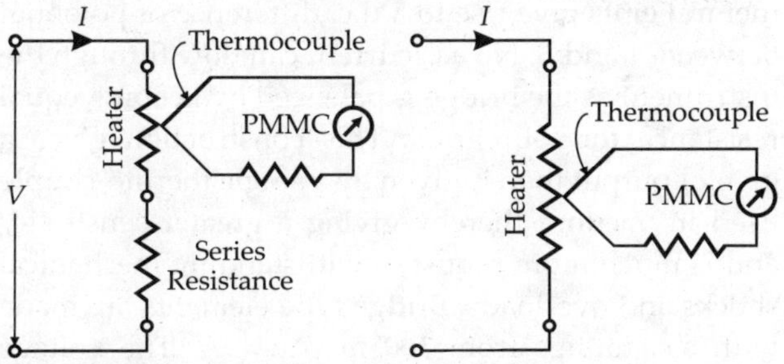


**Fig. 9.42** Contact type thermo-element

In order to ensure that the thermo-couple is responding to the temperature produced by the heater on account of current or voltage being measured, and not account of changes in the ambient temperature, a construction shown in Fig. 9.43 may be used.

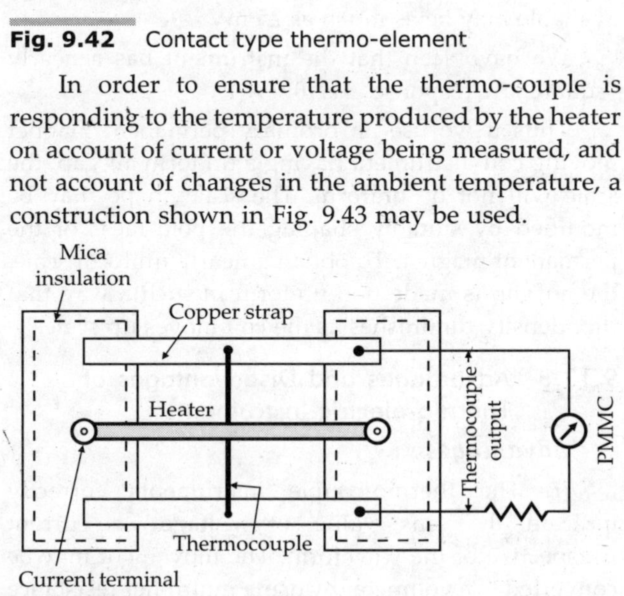


**Fig. 9.43** Arrangement of themro-element adopted for avoiding effects due to changes in ambient temperature.

The free ends of the thermo-couple are attached to the centre of copper straps, ensuring that they are at a temperature which is an average of the temperature at the ends of the heater element. The ends of the copper straps are close to, but insulated from, the ends of the heater. The temperature at the centre of the heater is greater than that at its end, and for a given current the temperature difference between the thermo-couple junctions is constant, and is not affected by the ambient temperature.

The action of the thermo-electric instruments is as follows :

(*i*) Electrical energy is converted to the thermal energy in the heater element. Most of the heat energy is dissipated away and a portion of it is transferred to the hot junction.

(*ii*) The heat energy transferred to the hot junction is converted back to electrical energy by Seebeck effect.

(*iii*) The output of the thermocouple is connected to a permanent magnet moving coil instrument. A portion of the electrical energy in the thermocouple circuit is converted to mechanical energy in deflecting the pointer and is stored as potential energy in the springs of the instrument.

The overall efficiency of the system is very low, of the order of a few hundredths of a percent under most favourable circumstances. This means that :

(*a*) the power consumption of the instrument is relatively high

(*b*) a sensitive d.c. instrument has to be used as an indicating element.

**2. Non-contact type.** In the separate heater type there is no electric contact between the heating element and the thermo-couple, the two being separated by electrical insulation. The heat is conducted from heater to junction across the insulation. This makes the instrument sluggish and less sensitive than the contact type. However, the separation between the heater and the thermo-couple becomes essential for measurement of currents in circuits operated at potentials considerably above the ground potential.

**3. Vacuum thermo-elements.** For very low-range thermo-couple instruments, the power absorption is very low. In some cases a power as little as 2 mW is used in the heater element. In order to attain a temperature rise (80°C to 100°C) needed to develop the required output in the thermo-couple instrument (about 5 mV on open circuit), we must increase the efficiency and sensitivity of the thermo-element. This is done by enclosing both the heater and thermo-couple assembly in a glass tube which is then evacuated (*see* Fig. 9.44). The efficiency is increased because there is no cooling of heater due of convection air currents.

Figure 9.44 shows an arrangement of separate type where the heater element and the thermo-couple junction are electrically isolated from each other by a bead of insulating material. This is done to minimize the effects of capacitance between the thermo-couple and the rest of the circuit at high frequencies. The insulation creates a very small capacitance between the isolated circuits and is in series with the capacitance from the instrument to ground or to other parts of the circuit. This reduces the capacitive currents which might otherwise be diverted through part of the heater element and produce errors in indication.

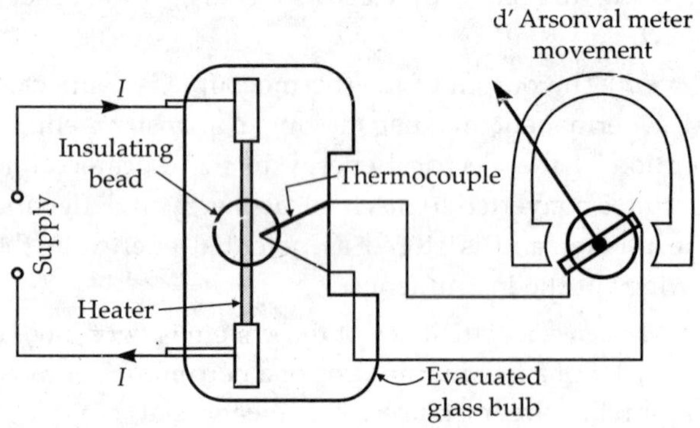


**Fig. 9.44** Vacuum type thermo-element.

The capacitive currents may be considerable part of the total current being measured if the instrument is used for measurement of very small currents. With the advent of new materials, it is now possible to use materials for bead which have very high thermal conductivities. These instruments are very sensitive and have the same response as the contact types for currents upto 100 mA. Another advantage of vacuum type is that the heater element can be operated at much higher temperatures. Vacuum type thermo-couples may the used in the range 2 mA to 500 mA.

Vacuum thermo-couples are suitable for current measurements at frequencies upto 100 MHz or more depending on size and resistivity of wire.

For voltage measurements, very great difficulties are encountered in constructing suitable series resistors for high frequency range. Voltmeters with resistors of usual type are used upto 10 kHz. Special multipliers for low voltages permit measurements at frequencies upto 100 MHz.

**4. Bridge type.** The thermo-couples connected to form a bridge circuit as shown in Fig. 9.45. No heater element is used ; instead the current to be measured passes directly through the thermocouples and raises their temperature proportional to $I^2R$.

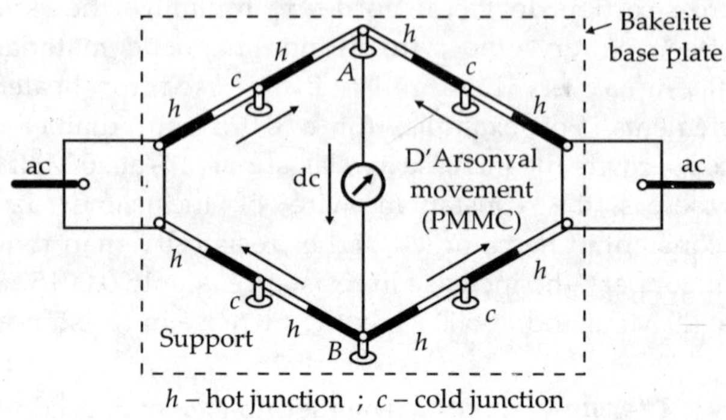


**Fig. 9.45** Bridge type thermo-couple instrument.

The cold junctions are at the pins that are embedded in the bakelite base plate and hot junctions are midway between the pins. The thermo-couples are oriented as shown in Fig. 9.45, so that the resultant themral emfs give rise to a d.c. difference of potential between $A$ and $B$. No a.c. current can flow through the instrument as the bridge is balanced by keeping equal resistance for four arms. This construction gives a greater output than is given by a single thermo-couple used in vacuum thereby giving a greater sensitivity and is much more robust in withstanding mechanical shocks and overloads. Bridge type elements are made with a.c. ratings from 100 mA to 1 A. The voltage available may be as much as 25 mV.

We have seen that the instrument has a nearly "square law response" *i.e.*, $\theta = K_3 I^2$.

Thus if we use an ordinary permanent magnet moving coil instrument having a uniform air gap, the scale will not be uniform. The scale shape may be modified by suitably shaping the pole faces of the permanent magnet. To obtain a nearly uniform scale, the air gap is made non-uniform in such a way that flux density diminishes as the coil moves up scale.

### 9.13.3 Advantages and Disadvantages of Thermo-electric Instruments

***Advantages :***

(*i*) The thermo-couple instruments correctly indicate the rms value of voltage or current irrespective of the waveform. The movement may be converted to a voltmeter by using multiplier resistance and a low current thermo-couple. Thermo-couple voltmeters are available in ranges upto 500 V with sensititives of 100-500 Ω / V.

(*ii*) These instruments are not effected by stray magnetic fields.

(*iii*) These instruments are free from frequency errors and hence they can be used over a very wide frequency range. In fact they are primarily designed for current measurements at high frequencies. In this application they are superior to any other type of instrument both in accuracy and frequency range. Their accuracy can be as high as 1% for frequencies upto 50 MHz and can be classified as *RF* instruments. Above this frequency the effective resistance of heater wire is increased on account of skin effect, thereby reducing the accuracy. Tubalar designs are used for heater wire above 3 A to reduce errors on account of skin effect produced by high frequencies. For small currents upto 3 A, the heater wire is solid and very thin.

(*iv*) These instruments can measure currents in the range of 0.5 to 20 A. For higher currents the heater element is usually situated outside the instrument case. For lower currents, a bridge type arrangement is used.

(*v*) These instruments have a high sensitivity.

(*vi*) They are very useful when used as transfer instruments to calibrate d.c. instruments by potentiometer and standard cell.

***Disadvantages*** :

At normal rated current, the heater attains a temperature of 300°C. If we pass twice the rated current the heater would give a temperature of nearly 4 times the normal temperature, *i.e.*, 1200°C. It is obvious that square law rate will bring the heater to nearly its burn out temperature. Thus it is imperative that the heater be protected against damaging overloads. The fuses do not provide any protection as due to overload the heater wire may burn out before fuse blows out.

Thus the overload capacity of thermo-couple instruments is small as compared with other instruments and is about 150% of the full scale current.

**Example 9.37** *A thermo-electric ammeter is assumed to have a perfect square law response. It gives a full scale deflection for a current of 10 A. Calculate the current which causes half scale deflection.*

**Solution.** The deflection of a thermo-electric ammeter is

$$\theta = KI^2$$

Suppose $\theta_F$ is the full scale deflection

$$\therefore\ \theta_F = K \times (10)^2 \quad \text{or} \quad \theta_F / K = (10)^2 = 100$$

The deflection at half scale is $\theta_F / 2$

$$\therefore\ \theta_F / 2 = KI^2$$

or current required to give half scale deflection

$$I = \left(\frac{\theta_F}{2K}\right)^{1/2} = \left(\frac{100}{2}\right)^{1/2} = 7.07 \text{ A}$$

**Example 9.38** *Design a basic three range 5 V, 10 V and 25 V thermo-couple voltmeter around the following specifications. The meter uses a d'Arsonval movement of 50 μA for full scale deflection and an internal resistance of 200 Ω. The heater element is designed for maximum current of 5 mA and has a resistance of 200 Ω. The characteristics of the thermo-couple are shown in Fig. 9.46.*

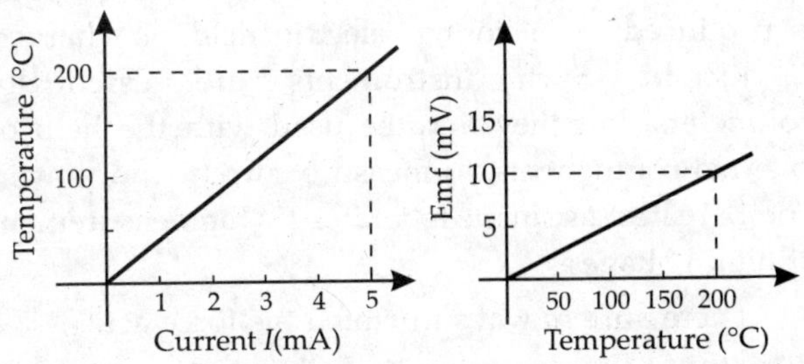


(*a*) Heater current v/s Temperature graph

(*b*) Graph of thermo-couple junction temperature v/s potential difference

**Fig. 9.46** Graphs of Example 9.38.

**Solution.** The value of multiplier resistance on each of the three ranges is :

(*a*) 5 V range $R_s = 5/(5 \times 10^{-3}) - 200 = 800\ \Omega$

(*b*) 10 V range $R_s = 10/(5 \times 10^{-3}) - 200 = 1800\ \Omega$

(*c*) 25 V range $R_s = 25/(5 \times 10^{-3}) - 200 = 4800\ \Omega$

The graph for heater current v/s temperature shown in Fig. 9.46(*a*), gives the temperature of 200°C for a current of 5 mA. The graph of thermo-couple function temperature v/s potential difference shown in Fig. 9.46(*b*) gives a potential difference of 10 mV for a temperature difference of 200°C.

When a voltage of 10 mV is applied to the d'Arsonval movement its current for full scale deflection is :

$$I = 10 \times 10^{-3} / 200 = 50 \mu\text{A}.$$

This is the rated current of the d'Arsonval movement. The schematic diagram for the circuit is shown in Fig. 9.47.

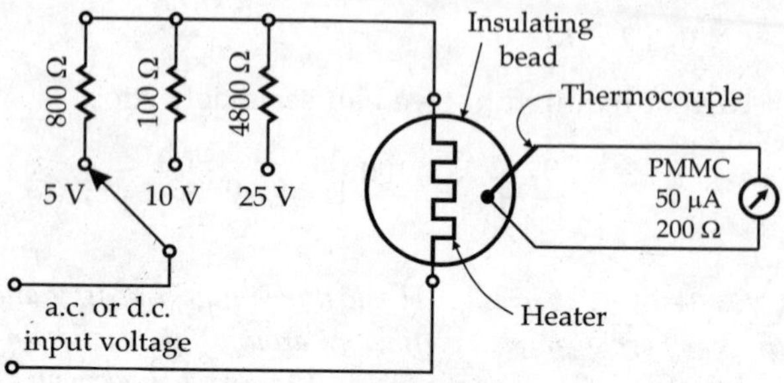


**Fig. 9.47** Multiple range voltmeter using a thermo-couple instrument of Example 9.38.

## 9.14 ELECTROSTATIC INSTRUMENTS

In electrostatic instruments, the deflecting torque is produced by action of electric field on charged conductors. Such instruments are essentially voltmeters, but they may be used with the help of external components, to measure current and power. Their greatest use in the laboratory is for measurement of high voltages.

There are *two* ways in which the force act :

(*i*) One type involves *two* oppositely charged electrodes. One of them fixed and the other is movable. Due to force of attraction, the movable electrode is drawn towards the fixed one.

(*ii*) In the other type, there are forces of attraction or repulsion or both between the electrodes which cause rotary motion of the moving electrode.

In both the cases the mechanism resembles a variable capacitor and the force or torque is due to the fact that mechanism tends to move the moving electrode to such a position where the energy stored is maximum.

### 9.14.1 Force and Torque Equations of Electrostatic Instruments

The stored energy is used as a basis for derivation of force and torque equations.

***Linear motion.*** Referring to Fig. 9.48, plate $A$ is fixed and plate $B$ is movable. The plates are oppositely charged and are restrained by a spring connected to the fixed point. Let a potential difference of $V$ volt be applied to the plates ; then a force of attraction $F$ newton exists between them. Plate $B$ moves towards $A$ until this force is balanced by that of the spring. The capacitance between the plates is then $C$ farad and the stored energy is $\frac{1}{2}CV^2$ joule.

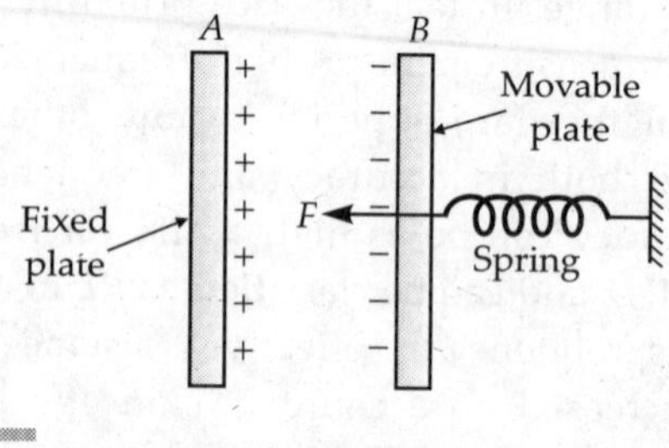


**Fig. 9.48** Illustrating theory of linear motion of electrostatic instruments.

Now let there be a small increment $dV$ in the applied voltage ; then the plate $B$ will move a small distance $dx$ towards $A$. When the voltage is being increased a capacitive current flows. This current is given by :

$$i = \frac{dq}{dt} = \frac{d}{dt}(CV) = C\frac{dV}{dt} + V\frac{dC}{dt}$$

The input energy is

$$Vi\,dt = V^2\,dC + CV\,dV$$

Change in stored energy

$$= \frac{1}{2}(C + dC)(V + dV)^2 - \frac{1}{2}CV^2$$

$$= \frac{1}{2}(C + dC)(V^2 + \frac{1}{2}V\,dV + dV^2) - \frac{1}{2}CV^2$$

$$= \frac{1}{2}V^2 dC + CV\,dV$$

neglecting the higher order terms as they are small quantities.

From the principle of the conservation of energy,

Input electrical energy = increase in stored energy + mechanical work done

or $$V^2 dC + CV\,dV = \frac{1}{2}V^2 dC + CV\,dV + F\,dx$$

$$\therefore \quad F = \frac{1}{2}V^2\frac{dC}{dx} \quad \text{...(9.73)}$$

***Rotational motion.*** The foregoing treatment can be applied to the rotational motion (Fig. 9.49) by writing an angular displacement $\theta$ in place of linear displacement $x$ and deflecting torque $T_d$ in place of force $F$ (see Eqn. 9.73).

Deflecting torque

$$T_d = \frac{1}{2}V^2\frac{dC}{d\theta} \quad \text{...(9.74)}$$

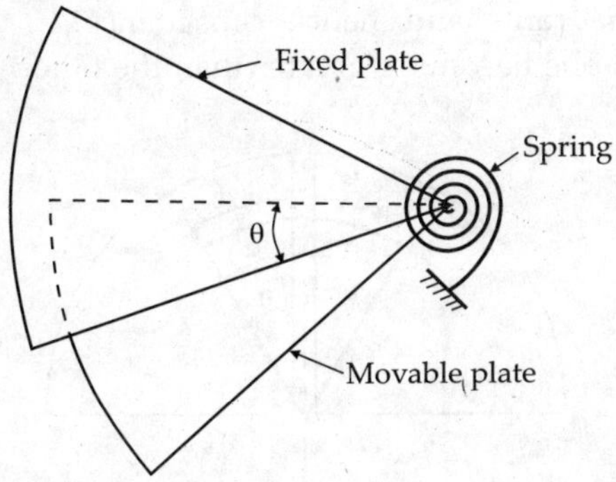


**Fig. 9.49** Illustrating theory of rotary motion of electrostatic instruments.

If the instrument is spring controlled or has a suspension,

Controlling torque, $T_c = K\theta$

where $K$ = spring constant and $\theta$ = deflection

Hence deflection, $$\theta = \frac{1}{2}\frac{V^2}{K}\frac{dC}{d\theta} \quad ...(9.75)$$

Since, the deflection is proportional to the square of voltage to be measured, the instrument can be used on both a.c. and d.c. The instrument exhibits a *square law response* and hence the scale is non-uniform is compressed at the lower end.

## 9.14.2. Quadrant Electrometer

This instrument is used for measurement of voltages ranging from about 100 V to 20 kV.

***Construction.*** There are four fixed metal quadrants arranged so as to form a shallow circular box with short air gaps between the quadrants (Fig. 9.50). A thin metal vane or needle is suspended in this partially closed box. The needle is of a double sector shape and may be suspended by means of a thread of phosphor bronze or silver quartz. The needle is equidistant from top and bottom quadrants. The needle carries a mirror and the deflection is read off by means of a lamp and scale arrangement.

An alternate arrangement may be mounting the needle on a spindle and the control torque is provided by a spring attached to the spindle. The deflecting torque moves a pointer and the deflection is read on a calibrated scale. Eddy current damping is used for these instruments.

The principle of vacuum enclosure has also been applied to deflecting electrostatic instruments. This makes it possible to manufacture extremely sensitive and accurate reflecting-type suspended-vane instruments.

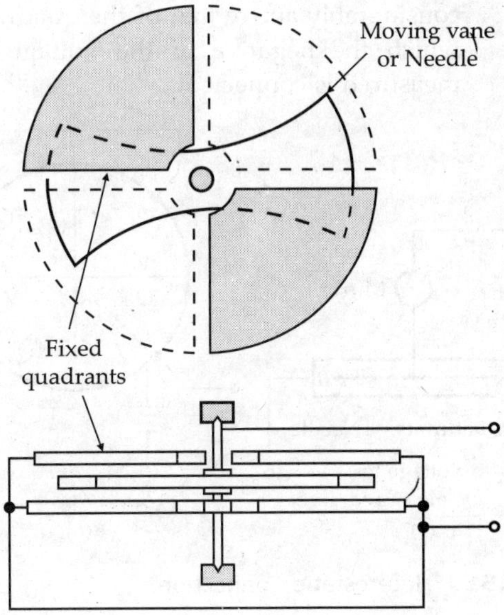


**Fig. 9.50** Quadrant Electrometer.

It is usual to connect a high resistance in series with an electrostatic voltmeter, particularly if the instrument is of the suspended-vane type. This resistance is used for limiting the current which can flow if there is accidental connection between a vane and a quadrant (*e.g.*, due to a bent vane). Without this precaution the faulty instrument would be a dead short on the supply. The voltage drop across such a resistance is normally negligible, owing to the very small current taken at the instrument. However, it may introduce errors at high frequencies. In order to avoid error, these resistances are short circuited when making high frequency measurements.

***Principle of operation.*** The fixed quadrants are connected together and the voltage to be measured is applied between the fixed quadrants and the moving needle (Fig. 9.50). Due to charge accumulated an electrostatic force are set up because of which the needle rotates. The suspension exerts a controlling torque and the needle takes up a position where the deflecting torque is equal to the controlling torque.

The deflecting torque is proportional to the square of the applied voltage and therefore, the instrument can be used for both a.c. and d.c.

### Types of Connections

The quadrant electrometer may be used in *two* ways :

1. *Heterostatic connection*. This connection is shown in Fig. 9.51. A high tension battery is

9

used for charging the needle to a potential considerably above that of the quadrants to which the negative of the voltage to be measured is connected.

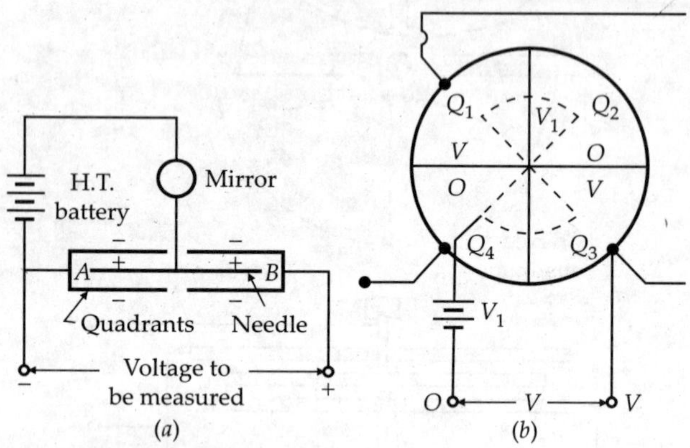


Fig. 9.51 Heterostatic Connection.

2. *Idiostatic connection.* This connection is shown in Fig. 9.52. The needle is directly connected to one pair of quadrants.

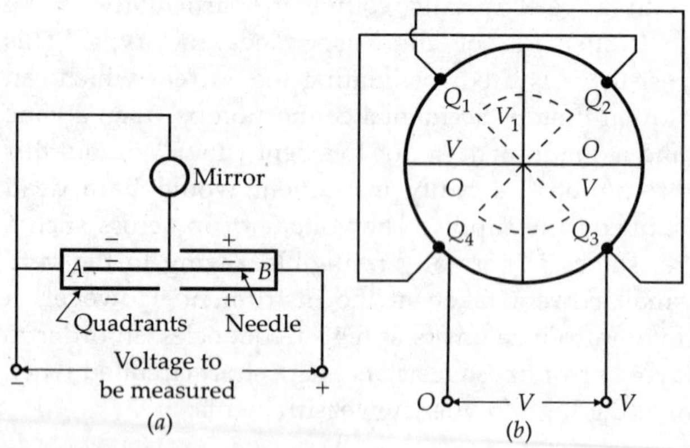


Fig. 9.52 Idiostatic Connection.

***Theory.*** In order to develop torque equation for an electrometer it is advisable to consider only one half of the needle moving in two quadrants, as shown in Fig. 9.53. The firm lines show the needle at zero position while the dotted lines show it deflected through an angle $\theta$.

The needle is considered to be a sector of a circle having a radius $r$ and spanning an angle $2\alpha$. Now we have an arrangement wherein the needle and quadrants form *two* capacitors $C_1$ and $C_2$ lying side by side.

At zero position these capacitors $C_1$ and $C_2$ have equal capacitances as the needle is symmetrically placed with respect to the two quadrants. But when the needle detlects it is no longer symmetrical with the two quadrants and hence capacitance $C_1$ and $C_2$ change, one becoming greater than the other.

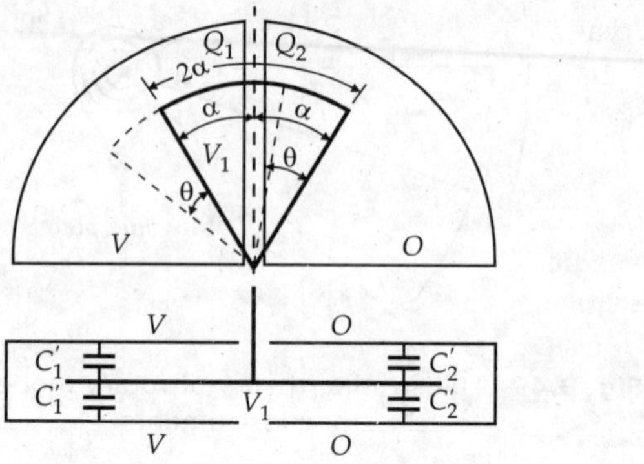

Fig. 9.53 Illustrating the theory of operation of Quadrant type Electrometers.

Let us consider that the instrument is heterostatically connected.

We have

$V_1$ = potential of needle,

$V$ = voltage being measured,

$V$ = potential of quadrant $Q_1$,

$O$ = potential of quadrant $Q_2$,

Let the needle deflect through an angle $\theta$ in the anti-clockwise direction. In this position,

$C_1$ = capacitance of left hand capacitor,

$C_2$ = capacitance of right hand capacitor.

Now $d$ is the distance of needle from either the top or the bottom plates of the quadrants.

For a parallel plate capacitor

$$\text{Capacitance, } C_1' = \frac{\varepsilon A}{d}$$

where $\varepsilon$ is permittivity of medium.

$$\therefore \; C_1' = \frac{\varepsilon\left[\frac{1}{2}(\alpha+\theta)r^2\right]}{d} = \frac{\varepsilon[(\alpha+\theta)r^2]}{2d}$$

This is because the needle spans an angle $(\alpha + \theta)$ under quadrant $Q_1$.

$$\text{Capacitance, } C_1 = 2C_1' = \frac{\varepsilon}{d}(\alpha+\theta)r^2$$

Similarly,

$$\text{capacitance } C_2 = 2C_2' = \frac{\varepsilon}{d}(\alpha-\theta)r^2$$

$$\text{Now } \frac{dC_1}{d\theta} = \frac{\varepsilon}{d}r^2 \text{ and } \frac{dC_2}{d\theta} = -\frac{\varepsilon r^2}{d}$$

From Eqn. 9.38, deflecting torque is,

$$= \frac{1}{2} V^2 \frac{dC}{d\theta}$$

$$= \frac{1}{2}(V_1 - V)^2 \frac{dC_1}{d\theta} + \frac{1}{2} V_1^2 \frac{dC_2}{dt}$$

$$= \frac{1}{2}\frac{\varepsilon r^2}{d}[(V_1 - V)^2 - V_1^2]$$

We have considered only two quadrants and half needle. Considering all the four quadrants and the whole of the needle, the deflecting torque will be doubled. Hence total deflecting torque :

$$T_d = 2 \times \frac{1}{2}\frac{\varepsilon r^2}{d}[(V_1 - V)^2 - V_1^2]$$

$$= \frac{\varepsilon r^2}{d}[V(2V_1 - V)] \quad ...(9.76)$$

The negative sign has been dropped since it only indicates that the direction of rotation is opposite to that which has been assumed.

When potential of needle $V_1$ is very large as compared to voltage being measured, the above equation may be written as :

$$T_d = \frac{2\varepsilon r^2}{d} V_1 V \quad ...(9.77)$$

$$\text{Deflection, } \theta = \frac{\varepsilon r^2}{Kd}[V(2V_1 - V^2] \approx \frac{2\varepsilon}{Kd} V_1 V \quad ...(9.78)$$

Thus heterostatic connection gives a linear scale response when the potential of needle is very high as compared to that of voltage being measured.

***Idiostatic connection.*** Figure 9.52 shows an idiostatic connection. For idiostatic connection $V_1 = 0$.

Hence from Eqn. 9.76 deflecting torque for this instrument is

$$T_d = \frac{\varepsilon r^2}{d} V^2 \quad ...(9.79)$$

$$\text{and deflection, } \theta \cong \frac{\varepsilon r^2}{Kd} V^2 \quad ...(9.80)$$

$$= K_1 V^2 \quad ...(9.81)$$

Hence with idiostatic connection, the instrument gives a square law response.

For air as dielectric

$$\varepsilon = \varepsilon_0 = 8.85 \times 10^{-12} \text{F/m}$$

$$\text{and therefore } \theta = 8.85 \times 10^{-12} \frac{r^2}{Kd} V^2$$

### 9.14.3 Kelvin Multicellular Voltmeter

The driving torque in a quadrant electrometer is small and therefore, this instrument is not suitable as such for measurement of low voltages (the force being proportional to square of voltage). In order to make the operating forces sufficiently large, we increase the number of cells and connect the fixed quadrants in parallel. A cell is quadrant electrometer consisting of four quadrants and one moving needle. As there are a number of cells in the instrument and therefore, it is known as a multicellular voltmeter. We have to increase the number of cells, as the voltage to be measured decreases ; for example, for a 150 V maximum reading voltmeter 13 cells are used, the number decreasing to one for a 3000 V instrument.

Deflecting torque for a $n$ cell instrument

$$T_d = n \times \text{torque of one cell}$$

$$= n \times \frac{\varepsilon}{d} r^2 V^2 \quad ...(9.82)$$

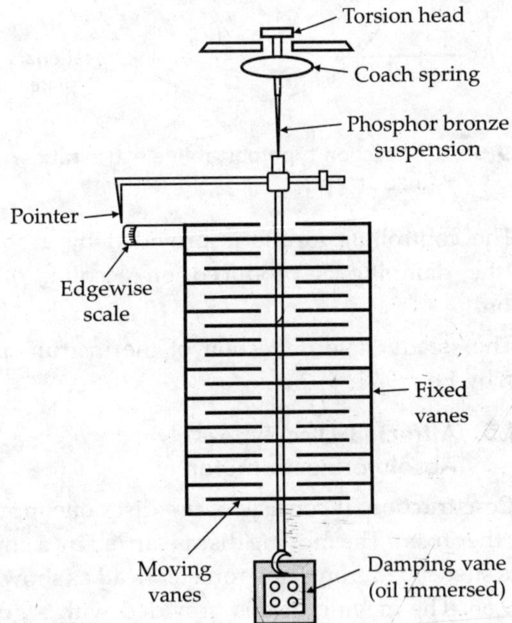


**Fig. 9.54** Kelvin multicellular voltmeter.

Figure 9.54 shows a 10 cell Kelvin multicellular instrument. There are many constructional features which distinguish it from ordinary quadrant electrometer. A coach spring is used to protect the suspension against vibrations. A torsion head actuated by worm-wheel arrangement, is provided for zero adjustment. The pointer and scale are of edgewise type.

Damping is done by using an oil dashpot.

### 9.14.4 Attraction Type Portable Electrostatic Instrument

The portable electrostatic instruments are generally of attraction type. The system consists of semi-circular plates one plate is free to move and the other is fixed as shown in Fig. 9.55. The fixed plates and the moving plates are insulated from each other and the voltage to be measured is applied across them. This results into production of an electrostatic field which produces a force of attraction between the fixed and the moving plates thereby deflecting the movable plate.

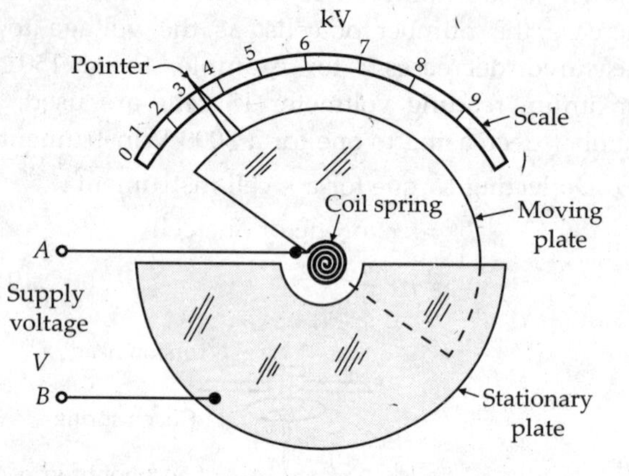


**Fig. 9.55** Attraction type portable electrostatic instrument.

The controlling torque is provided by a spring and the damping is produced on account of air friction.

The steady state deflection of the instrument is given by Eqn. 9.75.

### 9.14.5. Attracted Disc Type-Kelvin Absolute Electrometer

**Construction.** It consists of *two* discs one moving and other fixed. The moving disc is carried by a spring and is suspended from a micrometer head as shown in Fig. 9.56. The moving disc is provided with a guard ring to reduce the fringing effects.

The zero setting of the instrument is done with the help of an optical shifting system.

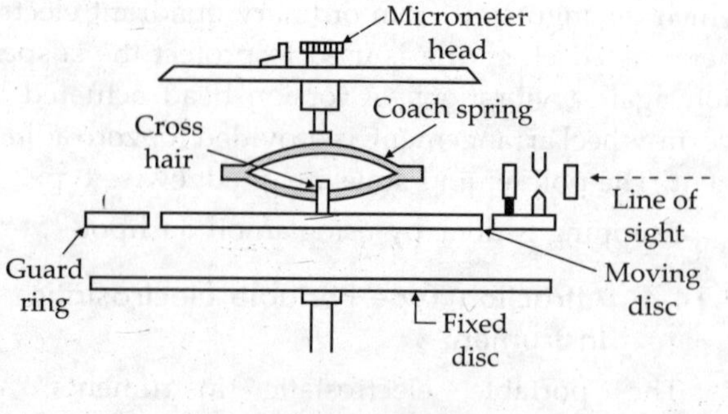


**Fig. 9.56** Kelvin absolute electrometer.

***Operation.*** The voltage to be measured is applied between the two discs. The moving disc is attracted downwards and is brought back to zero position by turning the micrometer head. The displacement is measured by the micrometer which is calibrated in terms of force. The voltage is then determined in terms of the force and the dimensions of the instrument as is clear from Eqn. 9.84.

***Theory.*** Let

$d$ = distance between plates ; m,

$A$ = area of plates ; m$^2$

= area of moving plate + $\frac{1}{2}$ area of air gap between moving plate and guard ring.

$\varepsilon$ = permittivity of dielectric ; F/m,

$V$ = voltage being measured ; V,

$F$ = force between discs ; N.

force $F = \dfrac{1}{2}V^2\dfrac{dC}{dx}$ (See Eqn. 9.73)

But for a capacitor with closely spaced plates

$$\frac{dC}{dx} = \frac{C}{d} \quad \text{and} \quad C = \frac{\varepsilon A}{d} \quad \therefore \quad F = \frac{1}{2}\varepsilon A\frac{V^2}{d^2} \quad ...(9.83)$$

Hence voltage, $V = \sqrt{2Fd^2/\varepsilon A}$ ...(9.84)

This theory shows that such an instrument will give an absolute determination of potential difference, as the p.d. is given in terms of force and linear dimensions.

The deflecting force is only adequate if the voltage is high. Special construction may be necessary to ensure good insulation, screening, and freedom from corona effects. Instruments of lower accuracy, but working on the same principle, may be arranged to read the voltage, directly on a scale being calibrated by comparison with a standard instrument.

The disadvantage of this instrument is that when the voltage being measured is small (a few hundred volt), the *two* discs should be very near together in order to get an appreciable force. In such cases the measurement of distance between the plates is difficult to carry out. The solution lies in increasing the voltage between fixed and moving plates by using heterostatic connections.

### 9.14.6. Extension of Range of Electrostatic Voltmeters

The voltage range of Electrostatic Voltmeters may be extended by using :

(*i*) resistance potential dividers,

(*ii*) capacitance potential dividers.

**1. Resistance Potential Divider**

The arrangement of a resistance potential divider for use with an electrostatic voltmeter is shown in Fig. 9.57. The voltage $V$ to be measured is connected across the total resistance $R$ while the electrostatic voltmeter of capacitance $C$ is connected across a portion of the potential divider having a resistance $r$. For operation on d.c. assuming that the electrostatic voltmeter capacitance $C$ has an infinite leakage resistance, the voltage multiplying factor $m$ of the potential divider is $R / r$.

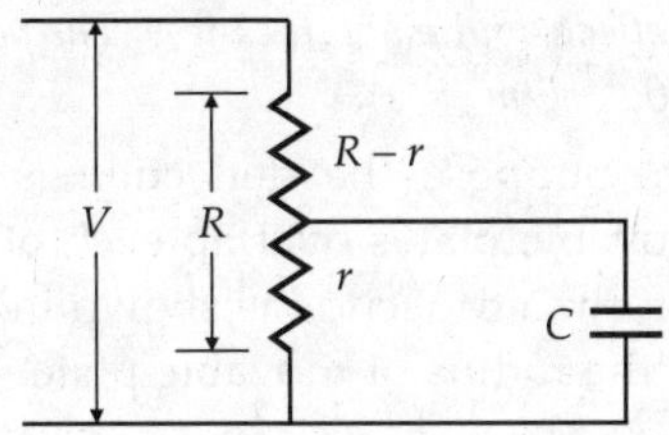


**Fig. 9.57** Resistance potential divider.

For operation on a.c. the multiplying power of the multiplier is

$$m = R / r \qquad \text{...(9.85)}$$

in case $\omega^2 C^2 r^2 \leq 1$

This can be easily analyzed.

**2. Capacitor Multipliers**

The extension of voltage range of an electrostatic voltmeter by the use of a suitable series capacitor is shown in Fig. 9.58($a$).

Let $C_v$ be the capacitance of the voltmeter and $C_s$ that of the capacitor in series with it. The impedance $Z$ of the voltmeter is $Z = 1/j\omega C_v$

The total capacitance $C_t$ due to series combination of $C_s$ and $C_v$ is :

$$C_t = \frac{C_s C_v}{C_s + C_v}$$

Total impedance of the combination is

$$Z_t = \frac{C_s + C_v}{j\omega C_s C_v} \quad \therefore \frac{Z_t}{Z} = \frac{C_s + C_v}{j\omega C_s C_v} \bigg/ \frac{1}{j\omega C_v} = \frac{C_s + C_v}{C_s}$$

Hence the multiplying factor $m$ is given by :

$$m = \frac{V}{v} = \frac{Z_t}{Z} = \frac{C_s + C_v}{C_s} = 1 + \frac{C_s}{C_v} \qquad \text{...(9.85)}$$

The voltmeter capacitance $C$, varies with deflection ($i.e.$, varies with position of the moving vane) of the voltmeter. Hence an electrostatic voltmeter should be calibrated together with its series multiplier capacitor.

In practice, the capacitor multiplier consists of a set of capacitors connected in series across the voltage to be measured and the electrostatic voltmeter is connected across one of them as shown in Fig. 9.58($b$).

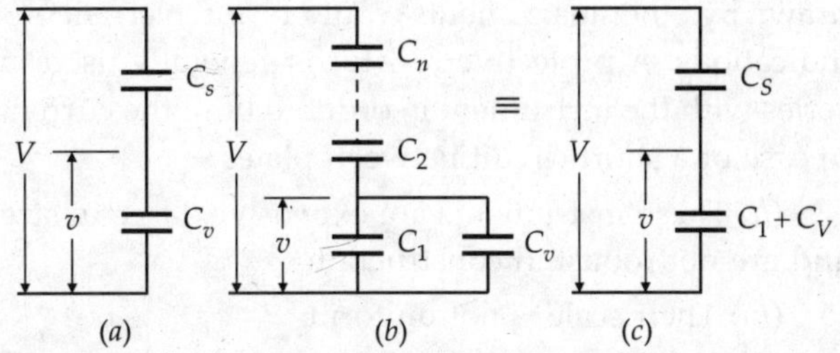


**Fig. 9.58** Extension of range by capacitor multipliers.

The resultant parallel capacitance due to $C_1$ and $C_v$ is $C_1 + C_v$. And the equivalent capacitance $C$, in series with this parallel combination is

$$\frac{1}{C_s} = \frac{1}{C_2} + \frac{1}{C_3} + ... + \frac{1}{C_v}$$

Using Eqn. 9.85, the multiplying factor $m$ is obtained by replacing $C_v$ by $C_1 + C_v$ by as in Fig. 9.58($c$).

$$\text{Hence } m = 1 + \frac{C_1 + C_v}{C_s} \qquad \text{...(9.86)}$$

Again, as $C_v$ varies with deflection, the voltmeter should be calibrated with its multiplier. However, if the capacitance $C_1$ across which $C_v$ is connected is large compared to the maximum capacitance of the voltmeter, the multiplying ratio may be taken as approximately constant.

### 9.14.7 Advantages and Disadvantages of Electrostatic Instruments

***Advantages :***

($i$) These instruments draw negligible power from the main. The initial current drawn by them when connected across d.c. mains is the charging current which dies down gradually depending upon the insulation resistance of meter which is of the order of $10^{12}\Omega$. The final current with d.c. operation to zero. On a.c. operation, current drawn depends upon the impedance of the meter circuit. The currents drawn on a.c. are fairly small and therefore the power requirements of electrostatic instruments are quite low.

($ii$) They may be used on both a.c. and d.c.

($iii$) They have no frequency and wave-form errors as the deflection is proportional to square of voltage and there is no hysteresis.

($iv$) There are no errors caused by stray magnetic fields as the instrument works on electrostatic principle.

($v$) They are particularly suited for high voltages.

***Disadvantages*** :

(*i*) The use of electrostatic instruments is limited to certain special applications, particularly in a.c. circuits of relatively high voltage, where the current drawn by other instruments would result in erraneous indications. A protective resistor is generally used in series with the instrument in order to limit the current in case of a short circuit between plates.

(*ii*) These instruments are expensive, large in size and are not robust in construction.

(*iii*) Their scale is not uniform.

(*iv*) The operating forces are small. This can be illustrated by comparing their operating forces in electromagnetic type instruments.

In electromagnetic instruments, the energy density = $1/2\ B^2/\mu$ J/m$^3$. Considering air as the medium and $B = 1$ Wb/m$^2$, the energy density is

$$\frac{1}{2}(1)^2/(4\pi\times 10^{-7}) = 40\times 10^3 \text{ J/m}^3$$

Considering an electrostatic field in air with $\varepsilon = 1$ MV/m (one third of the breakdown of air).

Energy density in electrostatic field

$$=\frac{1}{2}\times 8.85\times 10^{-12}\times(1\times 10^6) = 4.4 \text{ J/m}^3.$$

Therefore it is evident that the energy density in a magnetic field is almost $10^4$ times that in an electrostatic field. Hence, the operating torque in electrostatic instruments are much smaller than that obtained in electro-magnetic instruments.

The operating torques in electrostatic instruments can be increased by using (*i*) medium of high di-electric strength (*ii*) using plates of large area and (*iii*) smaller distance between plates, which in fact is determined by the di-electric media used.

**Example 9.39** *An electrostatic voltmeter reading upto 2000 V is controlled by a spring with a torsion constant of $5\times 10^{-6}$ Nm/rad has a full scale deflection of 90°. The capacitance at zero voltage is 15 pF. What is the capacitance when the pointer indicates 2000 V ?*

**Solution.** Deflection, $\theta = 90° = \pi/2$ rad.

From Eqn. 9.74 final steady deflection

$$\theta = \frac{1}{2}\frac{V^2}{K}\frac{dC}{d\theta}$$

or rate of change of capacitance

$$\frac{dC}{d\theta} = \frac{2\theta K}{V^2} = \frac{2\times(\pi/2)\times 5\times 10^{-6}}{(2000)^2}$$

$$= 3.93\times 10^{12} \text{ F/rad} = 3.93 \text{ pF/rad.}$$

Change in capacitance when reading from 0 to 2000 V is $3.93\times(\pi/2) = 6.17$ pF.

∴ Capacitance when reading 2000 V,

$$C = 15 + 6.17 = 21.17 \text{ pF.}$$

**Example 9.40** *An electrostatic voltmeter is constructed with six parallel, semi-circular fixed plates equi-spaced at 4 mm intervals and five interleaved semi-circular movable plates that move in planes midway between the fixed plates in air. The instrument is spring controlled. If the radius of movable plates is 40 mm, calculate the spring constant if 10 kV corresponds to full scale deflection of 100°. Neglect fringing edge effects and plate thickness. The permittivity of air is $8.85\times 10^{-12}$ F/m.*

**Solution.** Suppose the deflection is θ rad. The fixed and movable plates overlap each other over an angle θ and form a capacitor as shown in Fig. 9.59(*a*). Suppose *r* is the radius of movable plate.

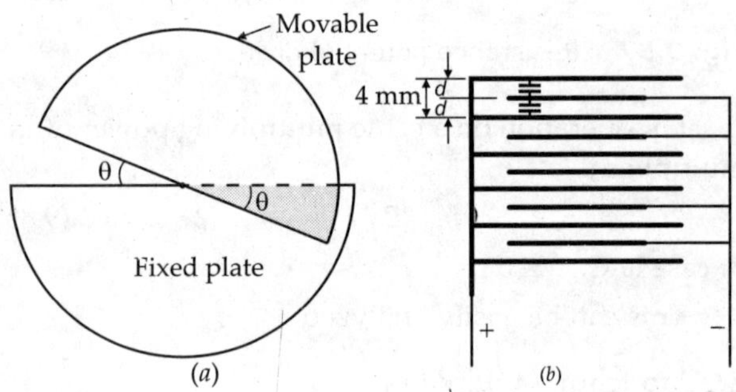


**Fig. 9.59** Electrostatic voltmeter of Example 9.40.

∴ Area of plates forming the capacitor $A = \theta r^2/2$.

Capacitance between fixed and moving plates.

$$C = 2\times(\varepsilon A/d) = 2(\varepsilon\theta r^2/2d) = \varepsilon\theta r^2/d.$$

This is because there are two capacitors, each having a capacitance of $\varepsilon A/d$ between fixed and movable plates. The capacitance are in parallel as shown in Fig. 9.59(*b*).

$$\therefore\ dC/d\theta = \varepsilon r^2/d$$

There are 6 fixed plates and 5 movable plates and hence there are 5 units exerting torque.

Torque exerted by each unit

$$=\frac{1}{2}V^2\frac{dC}{d\theta} \qquad \text{(see Eqn. 9.74)}$$

Torque exerted by *n* units

$$T_d = n\times\frac{1}{2}V^2\frac{dC}{d\theta} = \frac{n}{2}\frac{\varepsilon r^2}{d}V^2$$

$$=\frac{n}{2}\frac{\varepsilon r^2}{d}V^2.$$

Let the final steady deflection be θ.

$\therefore$ Controlling torque, $T_c = K\theta$ or $K\theta = \frac{n}{2}\frac{\varepsilon r^2}{d}V^2$

$\therefore$ Spring constant $K = \frac{n}{2\theta}\frac{\varepsilon r^2}{d}V^2$

It is given that :

number of units, $n = 5$, deflection $\theta = 100°$, permittivity, $\varepsilon = 8.85 \times 10^{-12}$ (for air)

Separation between fixed plates $2d = 4$ mm

$\therefore \quad d = 2 \text{ mm} = 2 \times 10^{-3}$ m.

Radius of movable plates = 40 mm = $40 \times 10^{-3}$ m,

Voltage, $V = 10$ kV $= 10 \times 10^3$.

Substituting these values in (*i*), we have

Spring constant,

$$K = \frac{5}{2 \times 100} \times \frac{(8.85 \times 10^{-12}) \times (40 \times 10^{-3})^2 \times (10 \times 10^3)^2}{2 \times 10^{-3}}$$

$= 17.7 \times 10^{-6}$ N-m/degree.

**Example 9.41** *In an electrostatic voltmeter the full scale deflection is obtained when the moving plate turns through 90°. Torsional constant is $10 \times 10^{-6}$ N-m/rad. The relation between the angle of deflection and capacitance between the fixed and moving plates is given by*

| *Deflection degree* | *Capacitance pF* |
|---|---|
| 0 | 81.4 |
| 10 | 121 |
| 20 | 156 |
| 30 | 189.2 |
| 40 | 220 |
| 50 | 246 |
| 60 | 272 |
| 70 | 294 |
| 80 | 316 |
| 90 | 334 |

*Find the voltage applied to the instrument when the deflection is 90°.*

**Solution.** The graph between capacitance and deflection is drawn as shown in Fig. 9.60.

The rate of change of capacitance at $\theta = 90°$ is found by drawing a tangent to the curve at that point

$$\frac{dC}{d\theta} = \frac{bc}{ab} = \frac{370 - 250}{110 - 44} = 1.82 \text{ pF/deg}$$

$= 104.2$ pF/rad.

Spring constant, $K = 10 \times 10^{-6}$ N-m/rad

$= 01.745 \times 10^{-6}$ N-m/deg.

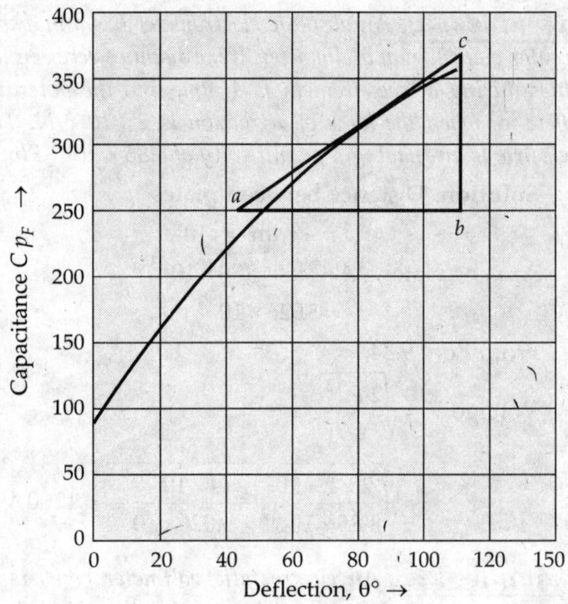


**Fig. 9.60** Plot of capacitance v/s deflection for Example 9.41.

Now $\theta = \frac{1}{2}\frac{V^2}{K}\frac{dC}{d\theta}$

or applied voltage, $V = \sqrt{\frac{2K\theta}{dC/d\theta}}$

$$= \sqrt{\frac{2 \times 0.1745 \times 10^{-6} \times 90}{104.2 \times 10^{-12}}} = 549 \text{ V.}$$

**Example 9.42** *The movable vane of a quadrant electrometer turned through 25 scale divisions when idiostatically connected to a potential difference of 60 V. When used heterostatically with the quadrants connected to a small voltage V and the needle connected to 900 V supply, the deflection was 10 scale divisions. Determine V.*

**Solution.** From Eqn. 9.79, deflection for idiostatic connection

$$\theta = \frac{\varepsilon r^2}{d}V^2 \quad \text{...(}i\text{)}$$

and from Eqn. 9.78 deflection for heterostatic connection

$$\theta = 2\frac{\varepsilon r^2}{d}V_I V \quad \text{...(}ii\text{)}$$

where $V$ = voltage being measured,

and $V_I$ = Voltage of needle.

Substituting the data given, we have

$$\frac{25}{10} = \frac{(60)^2}{2 \times 900 \times V}$$

$\therefore$ Voltage, $V = 0.8$ volt.

***Example 9.43*** *An absolute electrometer has a movable circular plate 80 mm in diameter. If the distance between the plates during a measurement is 4 mm, find the potential difference when the force of attraction is $2 \times 10^{-3}$ N. The dielectric is air, having a permittivity of $8.85 \times 10^{-3}$ F/m.*

**Solution.** Distance between plates

$$d = 4 \text{ mm} = \times 10^{-3}$$

Area of plates, $A = (\pi/4)(80 \times 10^{-3})^2$

$$= 5.026 \times 10^{-3} \text{ m}^2.$$

From Eqn. 9.84,

$$\text{Voltage } V = \sqrt{\frac{2Fd^2}{\varepsilon A}}$$

$$= \sqrt{\frac{2 \times 2 \times 10^{-3} \times (4 \times 10^{-3})^2}{8.85 \times 10^{-12} \times 5.026 \times 10^{-3}}} = 1200 \text{ V}.$$

***Example 9.44*** *An electrostatic voltmeter consists of two attracted plates, one movable, and the other fixed. Both the plates are provided with guard rings so as to make edge effects negligible. It is observed that the application of 10 kV between the plates results in a pull of $5 \times 10^{-3}$ N on the movable plate. Find the change in capacitance resulting from change in position of the movable plate by 1 mm. Diameter of movable plate is 100 mm.*

**Solution.** Area of movable plate

$$A = (\pi/4) \times (100 \times 10^{-3})^2 = 7.85 \times 10^{-3} \text{ m}^2.$$

Voltage, $V = 10 \text{ kV} = 10 \times 10^3$ V.

Force, $F = 5 \times 10^{-3}$ N.

Distance between plates

$$d = \sqrt{\frac{\varepsilon A}{2F}}.V = \frac{\sqrt{8.85 \times 10^{-12} \times 7.85 \times 10^{-3}}}{2 \times 5 \times 10^{-3}} \times 10 \times 10^3 \text{m}$$

$= 26.36$ mm.

Capacitance of a parallel plate capacitor, $C = \varepsilon A/d$.

Now $d$ changes from

$$d_1 = 26.36 \text{ mm to } d_2 = (26.36 - 1) = 25.36 \text{ mm}.$$

$\therefore$ Change in capacitance

$$= \varepsilon A\left(\frac{1}{d_2} - \frac{1}{d_1}\right)$$

$$= 8.85 \times 10^{-12} \times 7.85 \times 10^{-3}\left(\frac{1}{25.35 \times 10^{-3}} - \frac{1}{26.35 \times 10^{-3}}\right) \text{F}$$

$= 0.104$ pF.

The change in capacitance may be found as under :

Rate of change of capacitance with distance (See Eqn. 9.73)

$$\frac{dC}{dx} = \frac{2F}{V^2} = \frac{2 \times 5 \times 10^{-3}}{(10 \times 10^3)^2} \text{ F/m} = 100 \text{ pF/m}.$$

Change of capacitance

$$= (dC/dx) \times \text{distance moved}$$

$$= 100 \times 1 \times 10^{-3} = 0.1 \text{ pF}.$$

This result does not tally with the earlier one. The reason is that the rate of change of capacitance is not uniform over the small distance (1 mm) moved by the movable plate.

***Example 9.45*** *An electrostatic 250 voltmeter is composed of three square plates of 250 mm side. Two are fixed 100 mm apart in air and are electrically connected. Between this pair and the third plate which can move into the space between them, a potential difference is applied. With 12 kV, the movable plate is one quarter way in and with 32 kV, it is half way in. What potential difference is required to pull the plate three quarter way in ? Neglect fringing and assume that the displacement of the plate produces a proportional control force. The permittivity of air is $8.85 \times 10^{-12}$ F/m.*

**Solution.** The arrangement of plates is shown in Fig. 9.61 when the movable plate has moved in a distance $x$ mm.

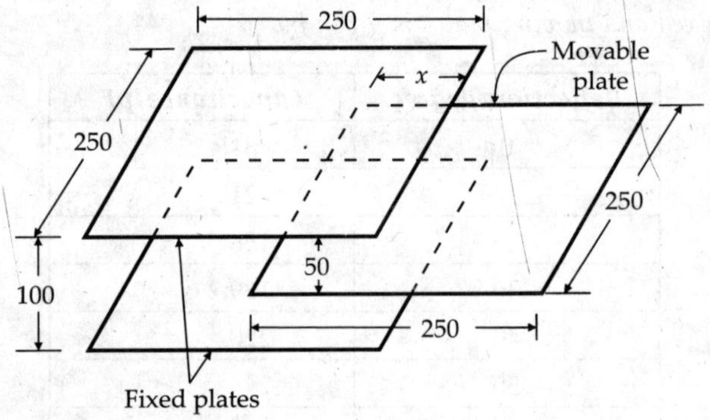


**Fig. 9.61** Diagram of Example 9.45.

Area of plates : $A = 250 \times x \text{ mm}^2$

$$= 250x \times 10^{-6} \text{ m}^2.$$

Distance between plates

$$d = 50 \text{ mm} = 50 \times 10^{-3} \text{ m}.$$

The capacitance of the arrangement is

$$C = \frac{2\varepsilon A}{d}.$$

(Since there are two capacitors in parallel)

$$= \frac{2 \times 8.85 \times 10^{-12} \times 250x \times 10^{-6}}{50 \times 10^{-3}}$$

$$= 0.0885 \times 10^{-12} x \text{ F}.$$

Rate of change of capacitance $dC/dx$

$$= 0.0885 \times 10^{-12} x \text{ F/mm}$$

$$= 88.5 \times 10^{-12} \text{ F/m}.$$

Let $x_0$ be the initial displacement in mm.

$\therefore$ Net displacement $= x - x_0$.

Hence $\quad x - x_0 = \frac{1}{2}\frac{V^2}{K}\frac{dC}{dx}$

For $\quad V = 12$ kV $= 12000$ V,

$x = 250/4 = 62.5$ mm.

$$\therefore\ 62.5 - x_0 = \frac{1}{2}\times\frac{(12000)^2}{K}\times 0.0885\times 10^{-12}$$

$$= \frac{6.37}{K}\times 10^{-6} \quad ...(i)$$

For $V = 32$ kV $= 32000$ V, $x = 250/2 = 125$ mm.

$$\therefore\ 125 - x_0 = \frac{1}{2}\times\frac{(32000)^2}{K}\times 0.0885\times 10^{-12}$$

$$= \frac{45.3}{K}\times 10^{-6} \quad ...(ii)$$

From (*i*) and (*ii*),

$$x_0 = 52.3 \text{ mm and } K = 0.623\times 10^{-6} \text{ N/m.}$$

Suppose the voltage applied for a displacement

$$x = 0.75\times 250 = 187.5 \text{ mm is } V.$$

$$\therefore\ 187.5 - 52.3 = \frac{1}{2}\times\frac{V^2}{0.623\times 10^{-6}}\times 0.0885\times 10^{-12}$$

or $\quad V = 43.63\times 10^3$ V $= 43.63$ kV

**Example 9.46** *The capacitance of a 0-2,000 V electrostatic voltmeter increases uniformly from 42 to 54 pF from zero to full scale deflection. It is required to increase the range of the instrument to 20,000 V by means of an external capacitor. Calculate the value of capacitance required. Also calculate the area of a pair of plates of an air capacitor suitable for the purpose, assuming distance between the plates = 25 mm. If the capacitor is adjusted to make the full-scale reading correct, what will be per cent error at half-scale reading ?*

**Solution.** Multiplying factor $m = 20000/2000 = 10$.

Capacitance of voltmeter at full scale $C_s = 54$ pF.

The multiplying factor with an external capacitance $C_s$, at the full-scale value (See Eqn. 9.85)

$$m = 1 + C_v / C_s \text{ or } 10 = 1 + 54/C_s$$

$\therefore$ Capacitance of series capacitor $C_s = 6$ pF.

Capacitance $C = \frac{\varepsilon A}{d}$

$$\therefore\quad 6\times 10^{-12} = \frac{8.85\times 10^{-12} A}{25\times 10^{-3}}$$

$\therefore$ Area of plates $A = 16.95\times 10^{-3}$ m$^2$ $= 16950$ mm$^2$.

The value of the capacitance of voltmeter at half-scale reading $= \frac{1}{2}(54 + 42) = 48$ pF

$\therefore$ Multiplying factor at half scale $m = 1 + 48/6 = 9$.

Reading indicated by voltmeter at half-scale

$= 20000/2 = 10000$ V.

Actual reading of voltmeter

$= 9\times 1000 = 9000$ V.

Percentage error

$$= \frac{10000 - 9000}{9000}\times 100 = 11.1\ \% \text{ high.}$$

## 9.15 RECTIFIER TYPE INSTRUMENTS

Rectifier type instruments are used for measurement of a.c. voltages and currents by employing a rectifier element which converts a.c. to a unidirectional d.c. and then using a meter responsive to d.c. to indicate the value of rectified a.c. The indicating instrument is PMMC instrument which uses a d'Arsonval movement. This method is very attractive since PMMC instruments have a higher sensitivity than the electrodynamometer or the moving iron instruments. The arrangement which employs a full wave rectifier circuit using four diodes as rectifying elements is shown in Fig. 9.62. It should be understood that rectifier type instruments are primarily used as voltmeters.

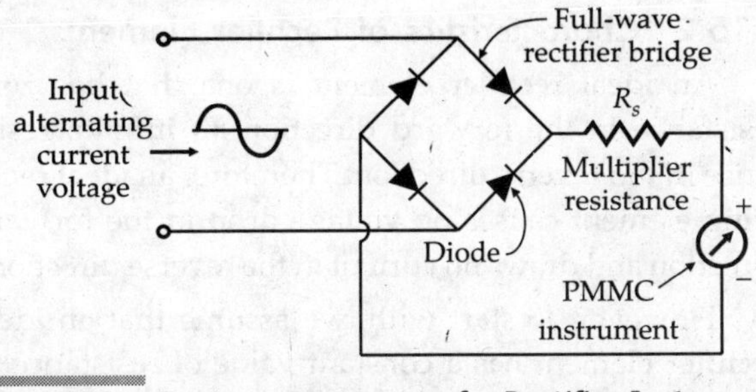


**Fig. 9.62** Basic arrangement of a Rectifier Instrument using a full wave rectifier circuit.

Rectifier instruments are particularly suited to measurements on communication circuits and for all other light current work where the voltages are low and resistances high. It is essential that for these applications that the current drawn by the voltmeters does not exceed say 1 mA in order that there are no loading effects and rectifier type instruments satisfy these requirements as their sensitivity is between 1000 Ω/V to 2000 Ω/V.

The multiplier resistance, $R_s$, is used to limit the value of the current in order that it does not exceed the current rating of PMMC instrument.

In comparison with other types of a.c. meters, the rectifier type of instrument is more widely used than either the more accurate but costly electrodynamometer type (which is used primarily at power frequencies) or the more delicate thermocouple type instruments which find extensive use at high frequencies.

### 9.15.1 Rectifier Elements

A rectifier element is used to convert a.c. to d.c. so that a unidirectional current flows through the PMMC movement. Rectifier type of instruments use a copper oxide or a Selenium cell, a Germanium or Silicon diode. Copper oxide elements have a peak inverse voltage (PIV) of about 2 V while Selenium elements have a PIV of about 10 V and also they can handle only limited amount of current and hence are becoming obsolete.

Germanium diodes have a peak inverse voltage (PIV) of the order of 300 V and a current rating of approximately 100 mA. Low current Silicon diode rectifiers have a PIV of upto 1000 V, and a current rating of the order to 500 mA. For these reasons Germanium and Silicon diodes find extensive applications in present day rectifier instruments. It may be mentioned here though Silicon diodes can carry currents as high as 85 A per rectifying element, but these high current Silicon diodes are seldom used in indicating instruments and find applications in power supply circuits.

### 9.15.2 Characteristics of Rectifier Elements

An ideal rectifier element is one that has zero resistance in the forward direction an infinite resistance in the reverse direction. Therefore, an ideal rectifying element causes no voltage drop in the forward direction and draws no current in the reverse direction.

However, to start with, we assume that an ideal rectifier element has a constant value of resistance in the forward direction and an infinite resistance in the reverse direction.

This rectifier element has a voltage - current relationship such as shown in Fig. 9.63(*a*). The relationship is linear in the forward direction while no current flows when the voltage is applied in the reverse direction.

This means that the rectifier resistance is constant in the forward direction but is infinite in the reverse direction as shown in Fig. 9.63(*b*). These ideal conditions are not realized under actual conditions. In practice, however, at low values of forward voltage the rectifier operates under highly non-linear part of its voltage-current ($v$–$i$) characteristics and its resistance is very large is compared with the resistance values with high currents. This is shown in Fig. 9.64(*b*). It should be noted that for low values of current, the resistance decreases rapidly as the current is increased. At high values of current, the forward resistance of the diode, $R_d$, is almost constant.

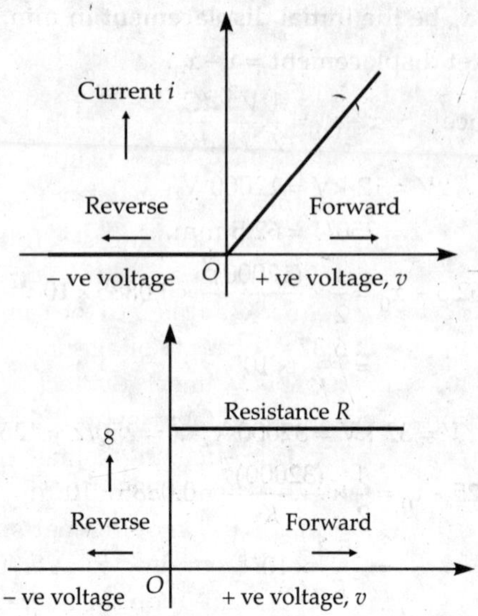


**Fig. 9.63** Characteristics of a near Ideal Rectifier Element.

It may also be noted that when a reverse voltage is applied the current is not zero (as in the case of ideal diode) but has a small finite value. For a given voltage, the forward current can be as much as 25 times the backward current or even more for the same voltage. Although, for many practical applications the reverse current may be assumed as negligible.

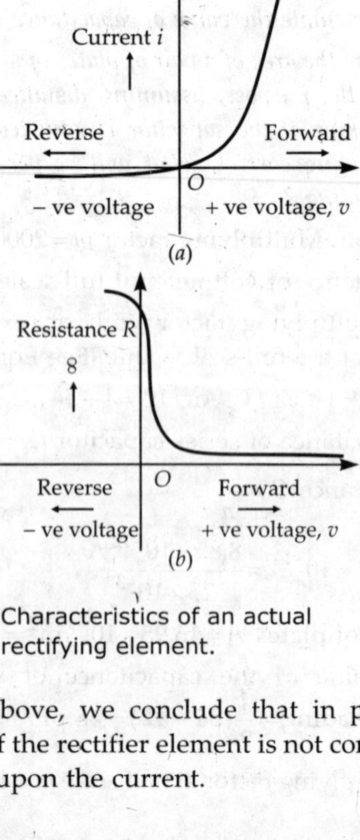


**Fig. 9.64** Characteristics of an actual rectifying element.

From above, we conclude that in practice, the resistance of the rectifier element is not constant but is dependent upon the current.

### 9.15.3 Half Wave Rectifier Circuit

A circuit of a voltmeter using a half wave rectifier is shown in Fig. 9.65. Suppose the meter resistance is $R_m$ and that of multiplier is $R_s$. Neglect the forward resistance of the diode. When a d.c. voltage $V_{dc} = V$ is applied to the circuit, the current through the meter is $I_m = V/(R_m + R_s)$ which produces full scale deflection. Now supposing an a.c. sinusoidal voltage $v = V_m \sin\omega t = \sqrt{2}\, V \sin\omega t$ is applied to the same circuit where $V_m$ = peak value of a.c. voltage and $V$ = rms value of a.c. voltage. This voltage gets rectified and a unidirectional pulsating voltage is produced at the output of the rectifier. This pulsating voltage produces a pulsating current and hence a pulsating torque. Because of the inertia of the moving parts, PMMC indicates a deflection corresponding to average value of current which is dependant upon the average value of the applied voltage.

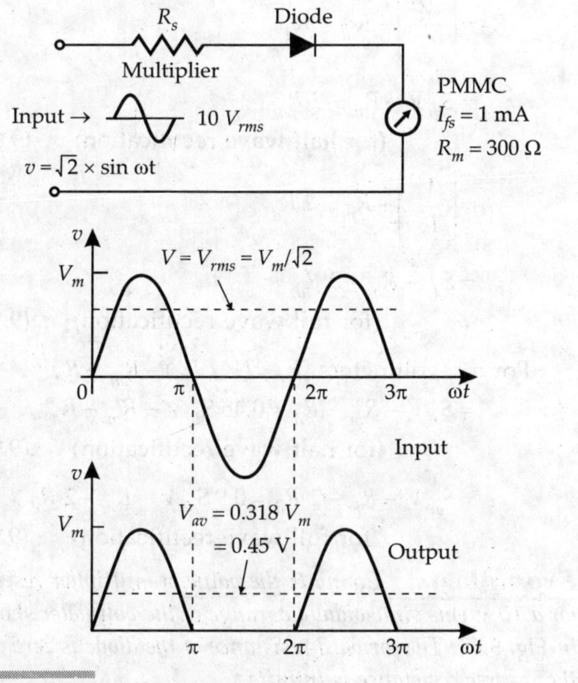


**Fig. 9.65** Half wave rectifier circuit.

Average value of voltage

$$V_{av} = \frac{1}{2\pi}\int_0^{\pi} V_m \sin\omega t\, d(\omega t) = \frac{1}{\pi} V_m$$

$$= 0.318\, V_m \quad \text{...(9.87)}$$

$$= 0.318 \times \sqrt{2}\, V = 0.45\, V \quad \text{...(9.88)}$$

Therefore, current through the meter is $0.45\, V/(R_m + R_s)$ which produces a deflection that is 0.45 times that produced with d.c. of equal magnitudes, $V$.

Hence the sensitivity of a half wave rectifier instrument with a.c. 0.45 times its sensitivity with d.c.

### 9.15.4 Full Wave Rectifier Circuit

A circuit of a voltmeter using full wave rectifier is shown in Fig. 9.66. When a d.c. voltage $V_{dc} = V$ of correct polarity is applied to the rectifier the current through the meter is $V/(R_m + R_s)$ which causes the full scale deflection of the meter. However, when a sinusoidal voltage $v = V_m \sin\omega t = \sqrt{2}\, V \sin\omega t$ is applied, the average voltage appearing across the meter is :

$$V_{av} = \frac{1}{\pi}\int_0^{\pi} V_m \sin\omega t\, d(\omega t) = \frac{2}{\pi} V_m$$

$$= 0.636\, V_m \quad \text{...(9.89)}$$

$$= 0.9\, V \quad \text{...(9.90)}$$

Therefore, the average current in meter is $0.9\, V/(R_m + R_s)$ and hence the deflection with a.c. is 0.9 times that with d.c. for the same value of voltage, $V$.

Therefore, the sensitivity of a full wave rectifier type of instruments with sinusoidal a.c. as an input is 90% of that with d.c. voltage of the same magnitude. The sensitivity of a full wave rectifier type of instrument is twice that of a half wave rectifier type.

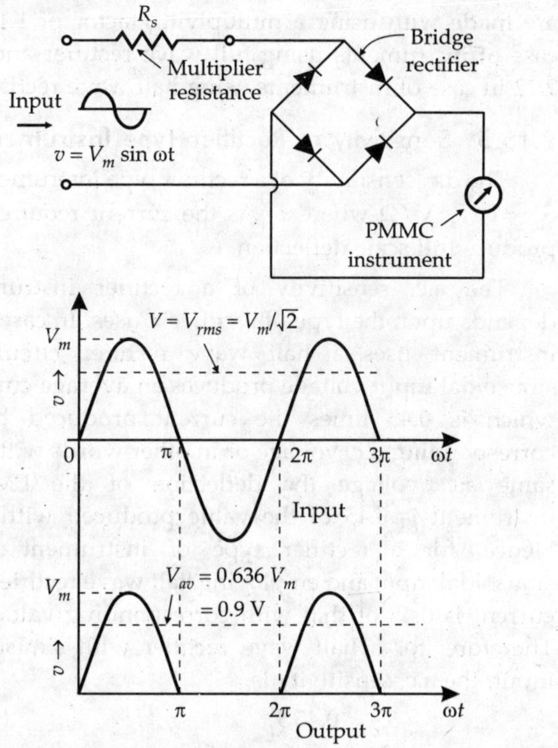


**Fig. 9.66** Voltmeter using full wave rectifier.

For practical purposes, since the currents and voltages are expressed in r.m.s. values, the meter scales are calibrated in terms of r.m.s. values of a sinusoidal function although the meter responds to average value of current. A non-sinusoidal waveform has an average value that may differ considerably from the average value of a sinusoidal wave (for which the meter is calibrated) and the indicated reading would certainly be erroneous.

The form factors relating the average value to the r.m.s. value of sinusoidal quantity are :

$$\text{Form factor, } K_f = \frac{\text{r.m.s. value of a.c. wave}}{\text{average value of a.c. wave}}$$

For a sinusoidal waveform for full wave rectification,

$$K_f = \frac{V}{V_{av}} = \frac{V_m/\sqrt{2}}{2V_m/\pi} = 1.11 \quad ...(9.91)$$

For a sinusoidal waveform for a half wave rectification,

$$K_f = \frac{V}{V_{av}} = \frac{V_m/\sqrt{2}}{V_m/\pi} = 2.22 \quad ...(9.92)$$

The rectifier type of instruments are calibrated for r.m.s. values of sinusoidal voltages or currents. In order that a rectifier type instrument should read true r.m.s. value of the sinusoidal input, the scale markings are made with using a multiplying factor of 1.11 in case of instruments using full wave rectifier and by 2.22 in case of instruments using half wave rectifiers.

### 9.15.5 Sensitivity of Rectifier Type Instruments

The d.c. sensitivity of a rectifier type instrument is $S_{dc} = 1/I_{fs}$ V/Ω where $I_{fs}$, is the current required to produce full scale deflection.

The a.c. sensitivity of a rectifier instrument depends upon the type of rectifier it uses. In case, the instrument uses a half wave rectifier circuit, a sinusoidal input voltage produces an average current which is 0.45 times the current produced by a corresponding d.c. voltage or in other words with the same a.c. voltage the deflection of the PMMC instrument is 0.45 of the value produced with d.c. Hence, for a rectifier type of instrument with sinusoidal input and employing half wave rectifier the current is 0.45 of that with corresponding value dc. Therefore, for a half wave rectifier with sinusoidal input, the a.c. sensitivity is,

$$S_{ac} = 0.45\, S_{dc} \quad ...(9.93)$$

Similarly, for a rectifier type instrument, with sinusoidal input and with full wave rectifier, the a.c. sensitivity is,

$$S_{ac} = 0.9\, S_{dc} \quad ...(9.94)$$

The above analysis is based upon that the forward resistances of diodes being zero and the reverse resistance being infinite. It should be understood that the above relationships have been derived for sinusoidal inputs. The expressions for sensitivity derived above are not valid for non-sinusoidal input waveforms such as square, triangular and sawtooth waveforms.

### 9.15.6 Extension of Range of Rectifier Instruments as Voltmeters

Suppose it is intended to extend the range of a rectifier instrument which uses a PMMC instrument having a d.c. sensitivity of $S_{dc}$.

Let

$v$ = voltage drop across the PMMC instrument,

$V$ = applied voltage.

Therefore, for d.c. operation the values of series resistance (multiplier) needed is

$$R_s = \left(\frac{V}{I_{fs}}\right) - R_m - R_d$$

$$= S_{dc}V - R_m - 2R_d$$

(for half wave rectification) ...(9.95)

$$= \left(\frac{V}{I_{fs}}\right) - R_m - 2R_d$$

$$= S_{dc} - R_m - 2R_d$$

(for full wave rectification) ...(9.96)

For a.c. voltmeters, $R_s = (V/I_{fs}) - R_m - R_d$

$$= S_{ac}V - R_m - R_d = 0.45\, S_{dc}\, V - R_m - R_d$$

(for half wave rectification) ...(9.97)

$$= S_{ac}\, V - R_m - 2R_d = 0.9\, S_{dc}\, V - R_m - 2R_d$$

(for full wave rectification) ...(9.98)

**Example 9.47** *Compute the value of multiplier resistor for a 10 V rms sinusoidal a.c. range of the voltmeter shown in Fig. 9.67. The forward resistance of the diode is zero and the reverse resistance is infinite.*

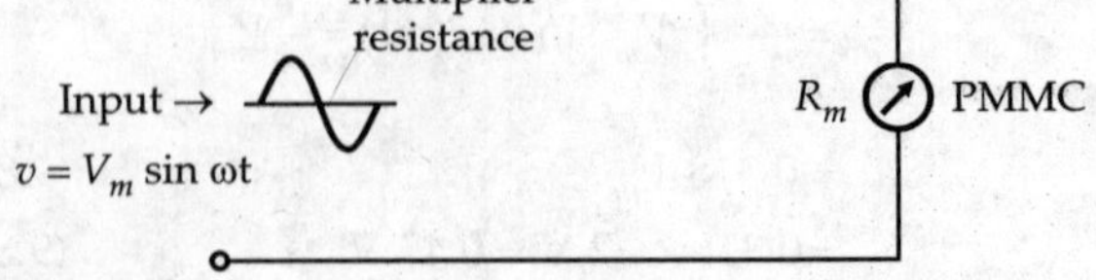


**Fig. 9.67** Rectifier type voltmeter of Example 9.47.

**Solution.** The sensitivity of meter movement for d.c.,

$$S_{dc} = 1/I_{fs} = 1/(1 \times 10^{-3}) = 1000\ \Omega/\text{V}.$$

For half wave rectifier, the a.c. sensitivity,

$$S_{ac} = 0.45 \times 1000 = 450\ \Omega / V$$

Total resistance of circuit for a.c. operation

$$= S_{ac} V = 450 \times 10 = 4500\ \Omega$$

Resistance of multiplier, $R_s = S_{ac} V - R_m - R_d$

$$= 450 \times 10 - 300 \times 10 - 300 - 0$$
$$= 4200\ \Omega.$$

**Example 9.48** *Compute the value of the multiplier resistor for a 10 V r.m.s. sinusoidal a.c. range of the voltmeter shown in Fig. 9.68. The forward resistance of the diode is zero and the reverse resistance is infinite.*

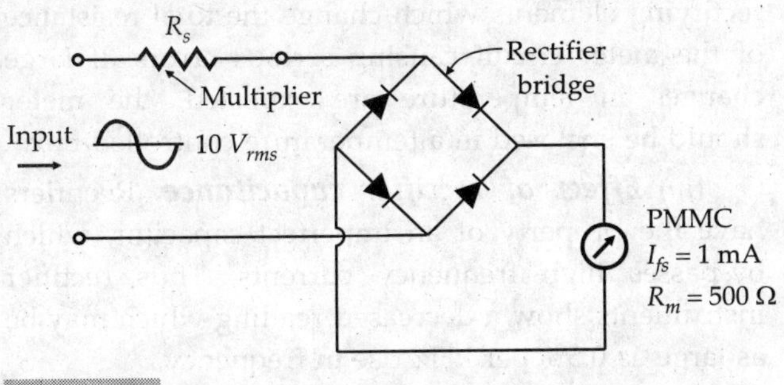


**Fig. 9.68** Rectifier type voltmeter of Example 9.48.

**Solution.** The d.c. sensitivity is :

$$S_{dc} = 1/I_{fs} = 1/(1 \times 10^{-3}) = 1000\ \Omega / V$$

For full wave rectifier circuit, a.c. sensitivity

$$S_{ac} = 0.9 \times 1000 = 900\ \Omega / V.$$

Resistance of multiplier

$$R_s = S_{ac} V - R_m - 2 R_d$$
$$= 900 \times 10 - 500 - 0 = 8500\ \Omega$$

**Example 9.49** *A rectifier type of instrument uses a basic PMMC movement of 50 μA and a resistance of 1000 Ω. It employs a full wave rectifier circuit with forward resistance of each diode being 1000 Ω. The reverse resistance of the diodes is infinite. The range of the instrument is 0 – 10 V a.c. sinusoidal. Calculate the value of*

*(a) Series multiplier* *(b) Meter sensitivity.*

**Solution.** (*a*) The d.c. sensitivity is

$$S_{dc} = 1/(5 \times 10^{-6}) = 20{,}000\ \Omega / V.$$
$$= 20\ k\Omega / V.$$

For a full wave rectifier circuit, a.c. sensitivity,

$$S_{ac} = 20{,}000 \times 0.9 = 18{,}000\ \Omega / V$$

Total resistance of the circuit,

$$= 18000 \times 10 = 180{,}000\ \Omega$$

∴ Resistance of multiplier

$$R_s = S_{ac} \times 10 - R_m - 2 R_d$$
$$= 180{,}000 - 1000 - 2000 = 177000\ \Omega$$

(*b*) Meter sensitivity on a.c.

$$S_{ac} = 180{,}000 / 10$$
$$= 18000 = 18\ k\Omega / V.$$

**Example 9.50** *A rectifier type instrument uses a bridge rectifier and has its scale calibrated in terms of rms values of a sine wave. It indicates a voltage of 2.22 V when measuring a voltage of triangular waveshape. Estimate the peak and r.m.s. values of applied voltage. Also calculate the error.*

**Solution.** The meter uses a full wave rectifier circuit and it indicates a value of 2.22 V. The form factor for full wave rectified sinusoidal waveform is 1.11.

∴ Average value of voltage

$$V_{av} = 2.22 / 1.11 = 2\ V.$$

For a triangular waveshape, peak value of voltage

$$V_m = 2\ V_{av} = 4\ V.$$

r.m.s. value of voltage

$$V = V_m / \sqrt{3} = 4\sqrt{3} = 2.31\ V$$

$$\therefore \quad \text{Error} = \frac{2.22 - 2.31}{2.31} \times 100 = -3.9\%$$

### 9.15.7 Factors Affecting the Performance of Rectifier Type Instruments

The factors which influence the performance of rectifier type instruments when used on a.c. are described below.

(*i*) ***Effect of waveform.*** The rectifier type of instruments are calibrated in terms of r.m.s. values of sinusoidal currents and voltages. The markings on the instrument are 1.11 and 2.22 times of the average current for full wave and half wave rectification respectively. It is clear at the outset that these instruments are subject to waveform errors as other waveforms may not have a form factor of 1.11 (in case of full wave rectifier) or 2.22 (in the case of half wave rectifier). This has already been explained with the help of Example 9.50.

(*ii*) ***Effect of rectifier resistance.*** The voltage-current ($v - i$) characteristics of a practical diode are non-linear and therefore its forward resistance is not constant but depends upon the operating point. The resistance is large at low values of current but is small for large values of current. The effect of the non-linear characteristics of rectifying elements is to distort the waveform of the current flowing through them. Therefore, when a sinusoidal voltage is applied, the current is not sinusoidal but has a peaked waveform (on account of low resistance at high values of current) and therefore its form factor is greater than 1.11 for full wave rectifiers (or greater than 2.22 in case of half wave rectifiers). This means that the actual r.m.s. value of current flowing is greater than the value indicated by the instrument *i.e.*, the instrument reads low. The effect of series multiplier resistance is to *swamp* the variable resistance of the rectifying elements. The greater the multiplier resistance, the greater is the swamping effect and consequently lesser the error on account of variation of diode resistance. In practice, the error is quite small for voltmeters having a range of 50 V and above on account of the fact that they have a large value of multiplier resistance and the variations in the diode resistance are able to produce negligible effect on the total value of the resistance of circuit which is made up of multiplier resistance, resistance of diodes and the resistance of the meter.

However, at low voltages, the value of forward current is small, and the rectifier operates in an extremely non-linear part of its $v-i$ characteristics where the forward is very large as compared to the resistance at higher current values (See Fig. 9.64*b*). Also, the resistance is not only large but there is a large change in its value at small currents. The problem is accentuated because for low voltage ranges, the value of multiplier resistance is small and hence it is unable to *swamp* the changes in the diode resistance. Therefore, the lower part of the a.c. scale in low range voltmeters becomes crowded. Hence, most manufacturers provide a separate scale for low voltages which is calibrated specially for this purpose. Figure 9.69 shows a separate scale for the voltage range of 0 – 25 V.

(*iii*) ***Effect of temperature changes.*** One of the major drawbacks of rectifier instruments is the errors caused by variation of the resistance of rectifying element with temperature. The compensation for this is carried out by making a part of series resistor (multiplier) of copper, which has a high positive resistance temperature co-efficient and since the rectifier element has a negative resistance temperature co-efficient, the two effects can be made to neutralize each other.

In case, there are large changes in temperature, there are wide variations in the resistance of the rectifying elements which change the total resistance of the meter circuit causing serious errors. If large changes in temperature are expected, the meter should be enclosed in a temperature controlled box.

(*iv*) ***Effect of rectifier capacitance.*** Rectifiers have the property of an imperfect capacitor, which by-passes high frequency currents. Thus rectifier instruments show a decreased reading which may be as large as 0.5% per kHz rise in frequency.

(*v*) ***Decrease in sensitivity.*** The sensitivity of rectifier type of instruments is lower on a.c. operation than on d.c. operation.

The sensitivity on a.c. when using a full wave rectifier is 90% of that with d.c. and only 45% of that with d.c. when using half wave rectifier.

In case, a meter which produces a full scale deflection for a current of 50 μA, the d.c. sensitivity

$$= 1/(50 \times 10^{-6}) = 20\,\text{k}\Omega\,/\,\text{V}$$

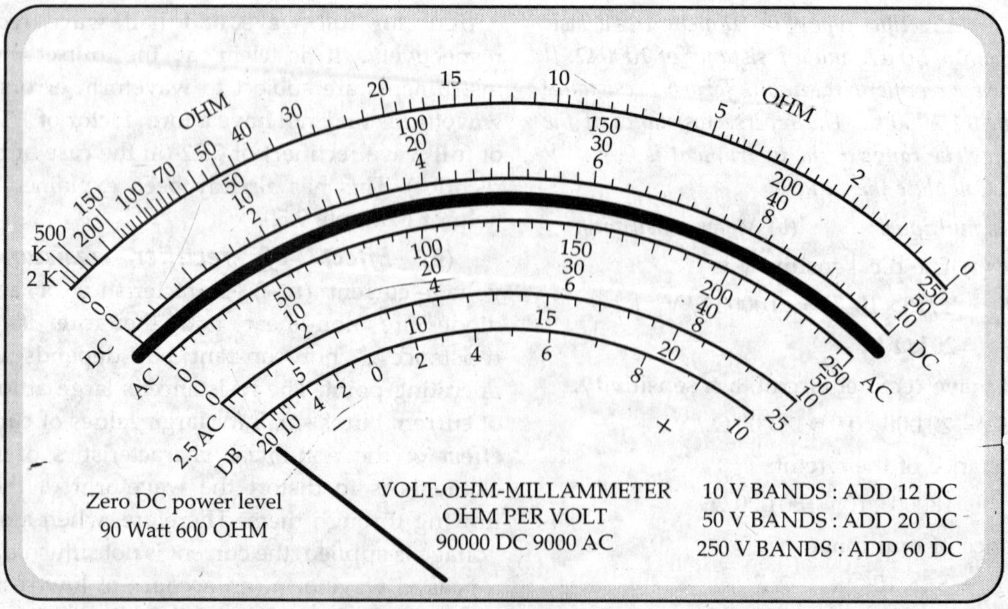


**Fig. 9.69** A multimeter using separate scale for low voltages.

In case of half wave rectifier circuits, the a.c. sensitivity

$$S_{ac} = 0.45,\ S_{dc} = 0.45 \times 20 = 9\,\text{k}\Omega\,/\,\text{V}.$$

The instrument is connected across a shunt resistance, which has almost the same value as that of meter. This lowers the value of sensitivity to half its value *i.e.*, $9/2 = 4.5\ \text{k}\Omega\,/\,\text{V}$

The actual a.c. sensitivity is normally about $5\ \text{k}\mu\,/\,\text{V}$ as against a sensitivity $20\ \text{k}\Omega\,/\,\text{V}$.

**Example 9.51** *The v–i characteristics of a diode are shown Fig. 9.70(a). This diode is used in a circuit for measurement of a.c. voltages as shown in Fig. 9.70(b).*

(*a*) *Compute the value of multiplier resistance.*

(*b*) *Suppose the scale is calibrated with the value of multiplier calculated in (a), what is the value of voltage indicated for a current 20 μA*

(*c*) *Calculate the actual value of diode resistance for current of 20 μA. Find the value of the a.c. voltage which drives this current and calculate the error. Comment upon the results.*

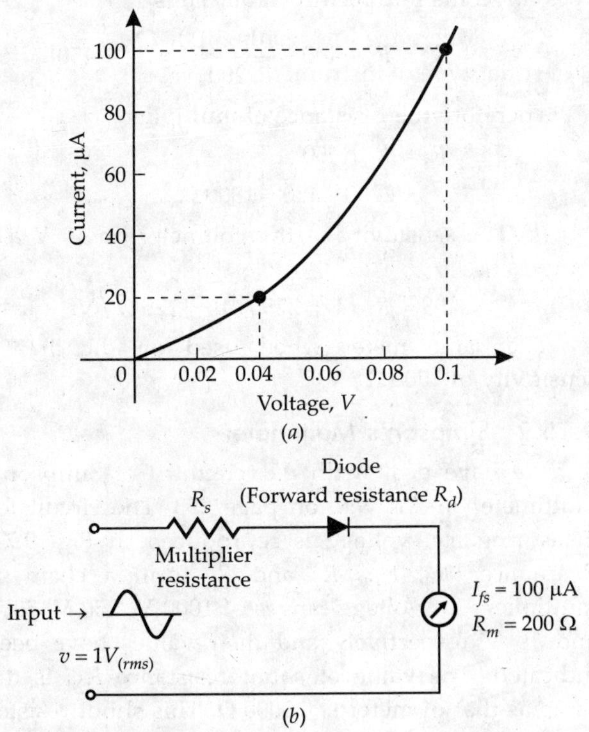


**Fig. 9.70** Characteristics of a diode and the circuit of voltmeter of Example 9.51.

**Solution.** (*a*) Resistance of the diode at 100 μA (full scale deflection)

$$R_d = 0.1/(100 \times 10^{-6}) = 1000\ \Omega$$

D.C. sensitivity of meter

$$S_{dc} = 1/(100 \times 10^{-6}) = 10{,}000\ \Omega/\text{V}.$$

∴ Resistance of multiplier for half wave rectifier circuit

$$R_s = 0.45,\ S_{dc}V - R_m - R_d$$
$$= 0.45 \times 10{,}000 \times 1 - 200 - 1000$$
$$= 3300\ \Omega$$

(*b*) If the scale is calibrated with diode having a resistance 200 μA, the indication with a current of 20 Ω will be 1/5 = 0.2 of full scale deflection *i.e.*, 0.2 V.

(*c*) The actual value of diode resistance at 20 μA is,

$$R_d = 0.04/(20 \times 10^{-6}) = 2000\ \Omega$$

The d.c. voltage required to drive a current of 20 μA is,

$$= 20 \times 10^{-6}[3300 + 200 + 2000]$$
$$= 0.11\ \text{V}$$

A.C. voltage to drive a current of 20 μA is,

$$= 0.11/0.45 = 02.44\ \text{V}$$

$$\therefore \quad \text{Error} = \frac{0.244 - 0.2}{0.2} \times 100 = 22\%$$

***Comments.*** The $v-i$ characteristics of a diode are non-linear. The forward resistance of diode is low at high values current and high at low values of current. If the scale is calibrated, considering the low value of resistance at high values of current, there are bound to be errors when low voltages (with consequent low currents) are measured because at low currents the resistance of the diode is high. Therefore, a greater voltage is needed to drive the current through the meter at low voltages *i.e.*, a value greater than what is indicated if the diode forward resistance is assumed to be what it is at high values of current (or voltage). Therefore, it becomes necessary to use a separate scale for low voltages where the multiplier resistance is not high and a change in the resistance of the diode changes the total resistance of the circuit significantly. This is not the case at higher voltages, where the multiplier resistance is high and any change in the diode resistance produces negligible effect on the total resistance of the circuit thereby undisturbing the calibration.

### 9.15.8 Multimeters

This arrangement is shown in Fig. 9.71. Two diodes (Rectifier elements) are used in this circuit, forming a full wave rectifier with the meter so. connected that it receives only half of the rectified current. Rectifier $D_1$ conducts during the positive half cycle of the input waveform and causes the meter to deflect according to the average value of this half cycle. The meter movement is shunted by a resistance

$R_{sh}$ in order to draw more current through rectifier $D_1$ and move its operating point into the linear portion of current-voltage characteristics. In the absence of rectifier $D_2$, the negative half cycle of the input voltage would apply a reverse voltage across rectifier $D_1$ causing a small leakage current in the reverse direction. The average value of the complete cycle would therefore be lower than it should be for half wave rectification. Rectifier $D_2$ deals with this problem. On the negative half cycle, $D_2$, conducts heavily and current through the measuring circuit which now is in the opposite direction, bypasses the meter.

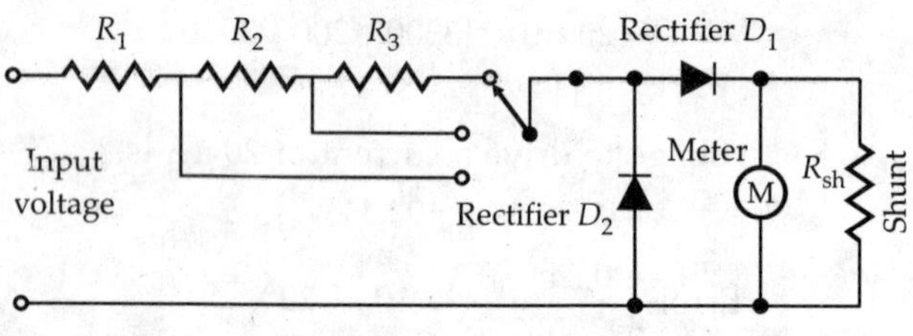


**Fig. 9.71** Multivoltmeter.

The commercial multimeter often uses the same scale markings for both its d.c. and a.c. voltage ranges. Since the d.c. component of a sine wave for half wave rectification equals 0.45 times the rms value. a problem arises immediately. In order to obtain the same deflection on corresponding d.c. and a.c. voltage ranges, the multiplier for the a.c. range must be lowered proportionately. The circuit shown in Fig. 9.72 solves this problem and is illustrated by the following example.

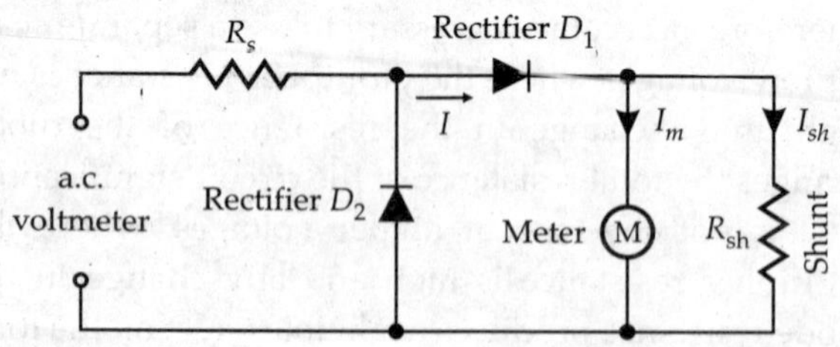


**Fig. 9.72** Single range rectifier voltmeter.

**Example 9.52** *A meter has an internal resistance of 100 Ω and requires 1 mA for full scale deflection. It is to be converted to a 10 V ac. voltmeter using circuit of Fig. 9.72. The shunt resistor, placed across the meter has a value of 100 Ω. Rectifiers $D_1$ and $D_2$ each have a forward resistance of 400 Ω and are assumed to have infinite resistance in the reverse direction. Calculate*

*(a) Value of the multiplier $R_s$ and*

*(b) Voltmeter sensitivity on the a.c. range.*

**Solution.** In order that the meter draws its full scale current of 1 mA, the source must also deliver current through the shunt resistor. Since the meter resistance $R_m$ and the shunt resistance $R_{sh}$ are equal, the total current $I$ supplied by the source must be $I = 2$ mA.

Since it is a half wave rectifier, average value of d.c. emf is :

$$V_{av} = 0.45\ V = 0.45 \times 10 = 4.5\,V.$$

Total resistance of the circuit

$$R_v = V_{av} / I$$
$$= 4.5 / (2 \times 10^{-3}) = 2250\ \Omega.$$

But $$R_v = R_s + R_{D1} + \frac{R_m R_{sh}}{R_m + R_{sh}}$$

where $R_s$ = resistance of multiplier

and $R_{D1}$ = forward resistance of rectifier $D_1$

Since we are interested only in the resistance of the circuit during the half cycle that the meter receives current, and during this period rectifier $D_2$ presents infinite resistance and hence resistance of rectifier $D_2$ does not come into picture at all. Thus :

$$R_v = R_s + 400 + \frac{100 \times 100}{200} = R_s + 450$$

Therefore the resistance of multiplier

$$R_s = R_v - 450$$
$$= 2250 - 450 = 1800\ \Omega.$$

(*b*) The sensitivity of the voltmeter on 10 V a.c. range is :

$$S = R_v / V = 2250 / 10 = 225\ \Omega / V.$$

The same meter when used on d.c. has a sensitivity of 1000 Ω / V.

### 9.15.9 Simpson's Multimeter

We have dealt with d.c. circuitry of Simpson's multimeter in Art. 9.6. on page 253. The circuit for measuring a.c. voltage is reproduced in Fig. 9.73. Resistances $R_9$, $R_{13}$, $R_7$ and $R_6$ form a chain of multipliers for voltage ranges of 1000 V, 250 V, 50 V and 10 V respectively and their values have been indicated. The value of shunt resistance $R_{24}$ is the same as that of meter *i.e.*, 2000 Ω. This shunt resistor serves to improve the performance of rectifier circuit. The forward resistance can be assumed as 500 Ω. On the 2.5 V range resistor $R_{23}$ acts as the multiplier resistance and its value is 1000 Ω. The values of multiplier resistances for other ranges are indicated in Fig. 9.73. These values give a sensitivity of 1000 Ω / V on a.c. ranges. On 2.5 V range resistor $R_{23}$ has a resistance of $R_s$.

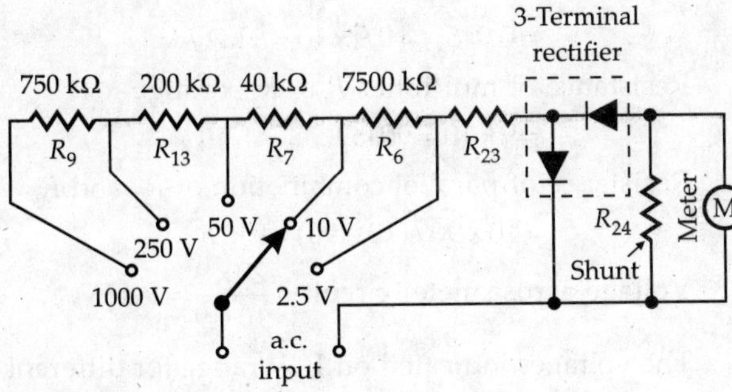


**Fig. 9.73** Simpson's multimeter circuit for measuring a.c. voltages.

### 9.15.10 Rectifier Ammeters

A rectifier ammeter consists of a bridge-connected rectifier and a moving-coil instrument connected across the secondary of a current transformer (See Fig. 9.74). It is not possible to use a shunt (as in d.c. moving-coil instruments), as the proportion of the total current passing through the movement would vary widely with the total current, owing to the change of rectifier resistance with current. The resulting scale shape would depart so far from uniformity as to be practically useless. When a current transformer is used the secondary (instrument) current is proportional to the primary current (the current to be measured) whatever the value of instrument resistance, provided this resistance is not too high. The current transformer will differ in some particulars from those described in Chapter 10, *e.g.*, its secondary will have a very large number of turns suitable for currents of the order of 1 mA. As the burden is so small, the transformer can be made very small in size and will be quite cheap; its frequency characteristics may also be made very good. The current range is 0 - 15 mA.

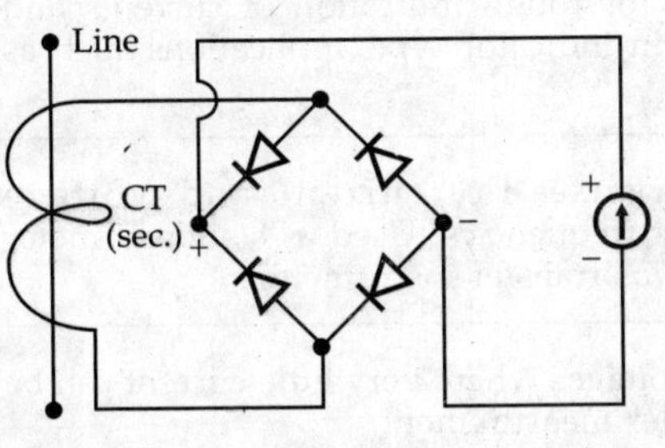


**Fig. 9.74** Rectifier ammeter.

The effects of temperature coefficient, capacitance, and form factor may be considered along the lines indicated for the case of voltmeters.

### 9.15.11 Advantages of Rectifier Instruments

The chief advantages of rectifier instruments may be summarised as :

(*i*) The frequency range-extends from about 20 Hz to high audio frequencies :

With the introduction of germanium and silicon rectifiers, rectifier instruments may be built in half wave configuration with a sensitive PMMC instrument at frequencies in the region of 3000 MHz.

(*ii*) These instruments have a much lower operating current for voltmeters (or higher value of ohm per volt) than other a.c. instruments, with the exception of electrostatic and valve voltmeters ;

(*iii*) They have a practically uniform scale for most ranges.

(*iv*) Their accuracy is about ± 5% under normal operating conditions.

The chief errors and limitations have been dealt with in the general description of rectifier instruments given earlier in Art. 9.15.7 on page 295.

### 9.15.12 Loading Effects of Rectifier Instruments

The sensitivity of rectifier type of voltmeters, using either half or full wave rectification, is less than the sensitivity of d.c. voltmeters. Therefore, the loading effects of an a.c. rectifier voltmeter is greater than that of a corresponding d.c. voltmeter. This is illustrated by the following example.

**Example 9.53** *Determine the reading obtained with a d.c. voltmeter in the circuit in Fig. 9.75 when switch S is set to position 1 and then to position 2 and determine the reading obtained with a half wave and a full wave rectifier voltmeter. All the meters use a 100 μA full scale deflection meter movement and arc set on their 10 V d.c. or a.c. rms range.*

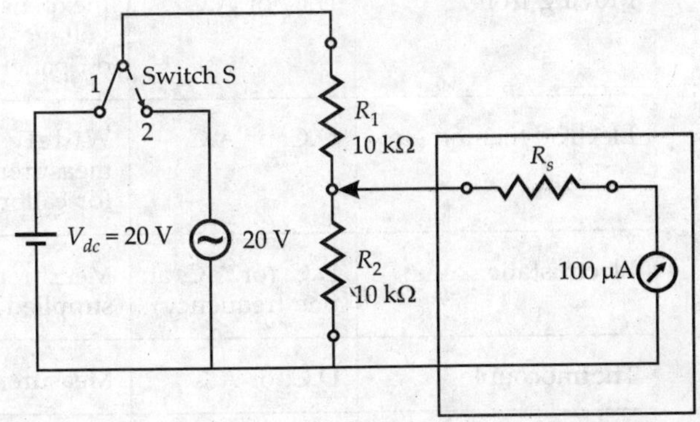


**Fig 9.75** Diagram of Example 9.53.

9

**Solution.** *When the switch S is at 1,*

D.C. sensitivity

$$S_{dc} = 1/(100 \times 10^{-6}) = 10\,\text{k}\Omega/\text{V}.$$

Resistance of multiplier considering resistance of meter to be negligible

$$R_s = S_{dc} \times \text{range} = 10 \times 10 = 100\,\text{k}\Omega$$

Resistance of parallel combination of $R_2$ and $R_s$

$$= 10 \times 100/(10+100) = 9.09\,\text{k}\Omega$$

Voltage across the meter circuit

$$= \frac{20 \times 9.09}{10 + 9.09} = 9.52\text{ V}.$$

*When the switch is at 2,*

For half wave rectification

$$S_{ac} = 0.45\,S_{dc} = 0.45 \times 10 = 4.5\,\text{k}\Omega/\text{V}$$

Resistance of multiplier

$$R_s = R_{ac} \times \text{range}$$

$$= 4.5 \times 10 = 45\,\text{k}\Omega$$

Resistance of parallel combination of $R_2$ and $R_s$

$$= 10 \times 45/(10+45) = 8.18\,\text{k}\Omega$$

Voltage across the meter circuit

$$= \frac{20 \times 8.18}{10 + 8.18} = 9\text{ V}.$$

For full wave rectification

$$S_{ac} = 0.9\,S_{dc} = 0.9 \times 10 = 9\,\text{k}\Omega/\text{V}$$

Resistance of multiplier $R_s = R_{ac} \times$ range

$$= 9 \times 10 = 90\,\text{k}\Omega.$$

Resistance of parallel combination of $R_2$ and $R_s$

$$= 10 \times 90/(10+90) = 9\,\text{k}\Omega$$

$$\text{Voltage across meter circuit} = \frac{20 \times 9}{10 + 9} = 9.47\text{ V}$$

The voltages indicated on 10 V range for different instruments are

D.C. : 9.52 V

A.C. half wave : 9.00 V

A.C. full wave : 9.47 V

Therefore, the loading effects of a.c rectifier type of voltmeters are greater than those of d.c. voltmeters. The loading effect of a.c. voltmeters using half wave rectification are greater than of those using full wave rectification.

## 9.16 SUMMARY CLASSIFICATION OF ALTERNATING AND DIRECT CURRENT METERS

A summary of a.c. and d.c. meters is given in Table 9.1, in which meter types are classified according to their suitability on d.c., a.c. or both, and also according to the areas in which they find their major use.

**Table 9.1** *Classification of Alternating Direct Current Meters*

| ***Meter type*** | ***Suitability*** | ***Major uses*** |
|---|---|---|
| PMMC (d'Arsonval) | D.C. | Most widely used meter for d.c. current and voltage and resistance measurements in low and medium impedance circuits. |
| Moving Iron | D.C. or A.C. | Inexpensive type used for rough indication of currents and voltages. Widely used in indicator type applications such as on panels. |
| Electrodynamometer | D.C. or A.C. | Widely used for precise a.c. current and voltage measurements at power frequencies. Used as standard meter for calibration and also as transfer instrument. |
| Electrostatic | D.C. (or A.C. at one frequency) | Measurement of high voltages where very little current can be supplied by circuit under measurement. |
| Thermocouple | D.C. or A.C. | Measurement of Radio frequency a.c. signals. |
| Rectifier | D.C. or A.C. | Widely used for medium sensitivity service type voltage measurements in medium impedance circuits. |

**Example 9.54** *A sinusoidal voltage of amplitude 100 V is applied across a circuit containing a rectifying device which entirely prevents current from flowing in one direction and offers a resistance of 10 Ω to the flow of current in the other direction. A hot wire type and PMMC instrument are connected in the same circuit for measurement of current. Calculate the readings of the instruments.*

**Solution.** Maximum value of voltage

$$V_m = 100\text{ V}$$

For half wave rectifier, average value of current

$$I_{av} = 0.318 \times 100/10 = 3.18\text{ A.}$$

r.m.s. value of current, $I = 100/(2 \times 10) = 5\text{ A.}$

A hot wire instrument reads rms value of current and therefore reading of this instrument

$$= 5\text{ A.}$$

A PMMC instrument reads the value of average current and therefore the reading of this instrument

$$= 3.18\text{ A.}$$

**Example 9.55** *A resistance of 50 Ω is connected in series with a rectifying device, a moving coil ammeter and a dynamometer type ammeter. The circuit is connected across a sinusoidal a.c. supply of 100 V. The resistance of the rectifying device is 50 Ω in one direction and 250 Ω in the opposite direction. Calculate the*

*(a) Readings on two ammeters*

*(b) Power supplied from mains and*

*(c) Power dissipated in rectifying element.*

**Solution.** R.M.S. value of voltage supplied

$$V = 100\text{ V}$$

Maximum value of voltage $V_m = \sqrt{2} \times 100\text{ V}$.

Instantaneous value of voltage $v = \sqrt{2} \times 100 \sin \theta$.

Resistance in the forward direction

$$R_f = 50 + 50 = 100\ \Omega$$

Resistance in the reverse direction

$$R_r = 250 + 50 = 300\ \Omega$$

Instantaneous value of current in the forward direction

$$i_f = \frac{V_m}{R_f} = \frac{\sqrt{2} \times 100}{100} \sin\theta = \sqrt{2}\sin\theta$$

Instantaneous value of current in the reverse direction

$$i_r = \frac{V_m}{R_f} = \frac{\sqrt{2} \times 100}{300} \sin\theta = \frac{\sqrt{2}}{3}\sin\theta$$

R.M.S. value of current

$$I_{rms} = \sqrt{\frac{1}{2\pi}\left[\int_0^{\pi} (\sqrt{2}\sin\theta)^2 d\theta + \int_{\pi}^{2\pi} \left(\frac{\sqrt{2}}{3}\sin\theta\right)^2 d\theta\right]}$$

$$= 0.745\text{ A.}$$

Average value of current

$$I_{av} = \frac{1}{2\pi}\left[\int_0^{\pi} \sqrt{2}\sin\theta\, d\theta + \int_{\pi}^{2\pi} \frac{\sqrt{2}}{3}\sin\theta\, d\theta\right]$$

$$= 0.3\text{ A.}$$

∴ Reading of dynamometer ammeter = 0.745 A.

Reading of moving coil ammeter = 0.3 A.

Power taken from the mains

= power supplied in the forward half cycle

+ power supplied in the backward half cycle

$$= \frac{1}{2}\left(\frac{V^2}{R_f} + \frac{V^2}{R_r}\right) = \frac{1}{2}\left[\frac{(100)^2}{100} + \frac{(100)^2}{300}\right]$$

$$= 66.7\text{ W}$$

Power consumed in 50 ohm resistor

$$= I^2R = (0.745)^2 \times 50 = 27.7\text{ W}$$

Power dissipated in rectifier

= total power supplied

– power consumed in resistor

$$= 66.7 - 27.7 = 39\text{ W.}$$

**Example 9.56** *A current of 0.5 + 0.3 sin ωt – 0.2 sin 2ωt is passed through the circuit shown in Fig. 9.76. Determine the reading of each instrument if ω = $10^6$ rad/s.*

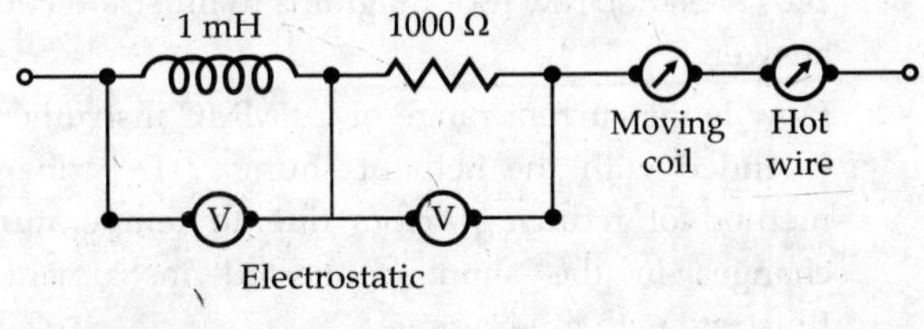


**Fig. 9.76** Circuit of Example 9.56.

**Solution.** The moving-coil reads average value of current while hot wire reads rms value of current. The electrostatic voltmeters do not take any current for their operation and they read the rms value of voltage.

Let $i$ be the instantaneous value of current

$$i = I_0 + I_{1m}\sin\omega t + I_{2m}\sin 2\omega t$$
$$= 0.5 + 0.3\sin\omega t - 0.2\sin 2\omega t$$

Average value of a sinusoidal quantity over a full number of cycles or over a time very much greater than the time period of a cycle is equal to zero.

$$\therefore \quad I_{av}' = i = 0.5 \text{ A.}$$

Hence reading of moving-coil instrument $= 0.5$ A.

Rms value of current

$$I = \sqrt{I_0^2 + \left(\frac{I_{1m}}{\sqrt{2}}\right)^2 + \left(\frac{I_{2m}}{\sqrt{2}}\right)^2}$$

$$= \sqrt{(0.5)^2 + \frac{0.09}{2} + \frac{0.04}{2}}$$

$$= 0.561 \text{ A.}$$

Hence reading of Hot-wire instrument $= 0.561$ A.

Reading of Electrostatic voltmeter across 1000 Ω resistance

$$V_R = 0.561 \times 1000 = 561 \text{ V.}$$

Instantaneous value of voltage across 1 mH inductor

$$v_L = L\frac{di}{dt} = 1 \times 10^{-3} \frac{d}{dt}(0.5 + 0.3 \sin \omega t - 0.2 \sin 2\omega t)$$

$$= 1 \times 10^{-3} \omega (0.3 \cos \omega t - 0.4 \cos 2\omega t)$$

$$= 300 \cos \omega t - 400 \cos 2\omega t \qquad (\because \omega = 10^6)$$

Hence reading of electrostatic voltmeter across 1 mH inductor

$$V_L = \sqrt{\left(\frac{300}{\sqrt{2}}\right)^2 + \left(\frac{400}{\sqrt{2}}\right)^2} = 354 \text{ V.}$$

## Review Questions

1. Describe the construction and working of PMMC instrument. Derive the equation for deflection if the instrument is spring controlled. Describe the method of damping used in these instruments.
2. Explain why PMMC instruments are the most widely used instruments. Discuss their advantages and disadvantages.
3. Explain the types of constructions used to extend the scale span of PMMC instruments to say about 240° to 300°. Draw neat diagrams to illustrate your answer.
4. How is the current range of a PMMC instrument extended with the help of shunts ? Describe a method of reducing errors due to temperature changes in the shunt connected instruments. Illustrate with an example.
5. Describe the working a universal shunt uses for multi-range ammeters. Derive expressions for resistances of different sections of a universal shunt used for a 3 range ammeter.
6. Describe bow a potential divider arrangement is used for multipliers used for multi-range voltmeters. Derive expressions for resistance of different sections for a 4 range voltmeter.
7. Describe the general requirements for a material to be used for shunts for ammeters and multipliers for voltmeters for PMMC instruments. What are the commonly used materials ? Describe their properties as regards their suitability to be used for shunts or multipliers.
8. Describe the circuit diagram of a series type ohmmeter. Explain how is it designed ? What is the significance of half scale value ? How are the adjustments done in case the battery runs down ? How are the zero adjustments made ?
9. Describe the construction and working of a shunt type ohmmeter. Write down its design equations. Why are series type ohmmeters preferred over shunt type ohmmeters ?
10. Explain the principle of working of a Ratiometer. Describe, how this principle is used in a Meggar for determination of insulation resistance.
11. Explain the working of a meggar with the help of a neat diagram.
12. Explain the working of (*i*) attraction type and (*ii*) repulsion type of moving iron instruments with the help of neat diagrams. Describe the methods of producing controlling and damping torques in them. Explain why these meters can be used on both a.c. and d.c.
13. Describe the constructional details of an attraction type moving iron instrument with the help of a neat diagram. Derive the equation for deflection if spring control is used and comment upon the shape of scale.

14. Derive a general equation for deflection for a spring controlled repulsion type moving iron instrument. Comment upon the shape of the scale. Explain the methods adopted to linearize the scale.

15. Discuss the following types of errors in moving iron instruments :
    (i) Hysteresis Error
    (ii) Temperature Error
    (iii) Error on account of stray magnetic fields
    (iv) Errors on account of change of frequency.

16. Derive the expression for capacitance to be connected across the multiplier of a moving iron voltmeter so as to make its circuit non-inductive for frequencies upto 125 Hz.

17. Describe the working and constructional details of an attraction type moving iron instrument. Discuss its advantages and disadvantages.

18. Describe the constructional details and working of an electrodynamometer type instrument.
    Derive the equation for deflection under a.c. operation if the meter is spring controlled. Discuss the shape of the scale.

19. Explain why electrodynamometer type of instruments can be used both on a.c. and d.c. Why are these instruments used as transfer instruments ?

20. What is an Astatic movement ? Explain why and how is it used in Electrodynamometer type of instruments.

21. Explain how an electrodynamometer type instrument is able to measure the true r.m.s. value of a voltage or current irrespective of its waveform.

22. Discuss the main sources of errors in electrodynamometer type instruments. Explain their advantages and disadvantages.

23. Explain the principle of working of thermocouple instruments. Draw neat diagrams to illustrate the working of
    (i) Contact type
    (ii) Non contact type and
    (iii) Bridge type thermo-elements.

24. Discuss the constructional details of a thermocouple type instrument used at very high frequencies. Describe their advantages and disadvantages.

25. Derive the expression for deflection for a rotary type electrostatic instrument using spring control. Comment upon the scale of the instrument. Since an electrostatic voltmeter does not depend upon the flow of current for its operation (it draws almost negligible current under steady state), why is it not widely used for sensitive low voltage measurements ?

26. Describe the working of an Quadrant Electrometer. Derive the expressions for deflection in the case of
    (i) Heterostatic connection and
    (ii) Idiostatic connection. If the instrument is spring controlled. Which of these instruments can be used for measurement of low voltages ?

27. Explain the construction and working of an attracted disc type Kelvin absolute electrometer. What are its advantages and disadvantages ? Can it be used for measurement of low voltages such as 100 V ? If not, why not ?

28. Why electrostatic instruments cannot be used for measurement of low voltages while electromagnetic instruments can be ? Illustrate your answer with some specific example comparing the energy densities produced in electrostatic instruments and electromagnetic instruments.

29. Comment upon the following for Electrostatic Instruments :
    (i) Current drawn
    (ii) Operating forces produced
    (iii) Frequency range.

30. Describe the working of a rectifier type of voltmeter. Discuss the sensitivity of rectifier type instruments. How does the a.c. sensitivity of rectifier type of instruments compare with their d.c. sensitivity ? Explain the reason for the fact that a voltmeter employing a 50 $\mu$A basic meter is rated at 20 k$\Omega$/V on d.c. measurement and only 5 k$\Omega$/V on a.c. measurements.

31. Discuss the effect of the following on the performance of rectifier type instruments :
    (i) Waveform of input voltage
    (ii) Resistance of diodes
    (iii) Change in ambient temperature
    (iv) Capacitance of diodes.

32. Describe the working of a rectifier type voltmeter which uses a half wave rectifier circuit. Explain the effect of using a reverse biased diode on the working of the instrument.

9

**33.** Explain the circuit of a multimeter for measurement of a.c. voltages. Why are multimeters provided with a separate scale for low voltage a.c. range ?

Explain why a rectifier type of voltmeter may have a uniform (or linear) a scale on the high voltage ranges, but generally has a non-linear scale for low voltage ranges.

**34.** Compare the effects of circuit loading when using an a.c. voltmeter with half wave rectification against those when using an a.c. voltmeter with full wave rectification.

**35.** Draw up a table to list the suitability for use on d.c. and a.c. and major uses of the following types of instruments :

(*i*) PMMC type

(*ii*) Moving iron type

(*iii*) Electrodynamometer type

(*iv*) Thermocouple type

(*iv*) Electrostatic type

(*v*) Rectifier type.

# Unsolved Problems

**1.** A moving coil instrument has the following data : number of turns = 100, width of coil = 20 mm, depth of coil = 30 mm, flux density in the gap = 0.1 Wb/m$^2$. Calculate the deflecting torque when carrying a current of 10 mA. Also calculate the deflection if the control spring constant is $2 \times 10^{-6}$ Nm/degree. **[Ans.** $60 \times 10^{-6}$ Nm, 30°]

**2.** The following data refers to a moving coil voltmeter :

Resistance, 10,000 ohm ; dimensions of coil 30 mm × 30 mm ; number of turns on coil, 100 ; flux density in gap, 0.08 Wb/m$^2$ ; spring constant, $3 \times 10^{-6}$ Nm per degree. Find the deflection produced by a voltage of 200 V. **[Ans.** 48°]

**3.** A moving coil milli-voltmeter has a resistance of 200 Ω and the full scale deflection is reached when a potential difference of 100 mV is applied across the terminals. The moving coil has effective dimensions of 30 mm × 25 mm and is wound with 100 turns. The flux density in the gap is 0.2 Wb/m$^2$. Determine the control constant of the spring if the final deflection is 100° and a suitable diameter of copper wire for the coil winding if 20 percent of the total instrument resistance is due to the coil winding. Resistivity of copper is $1.7 \times 10^{-8}$ Ωm.

**[Ans.** $0.075 \times 10^{-6}$ Nm/degree ; 0.077 mm]

**4.** The coil of a measuring instrument has a resistance of 1 Ω, and the instrument has a full scale deflection of 250 V when a resistance of 4999 Ω is connected in series with it. Find

(*a*) The current range of the instrument when used as an ammeter with the coil connected across a shunt of 1/499 Ω, and

(*b*) The value of the shunt resistance for the instrument to give a full-scale deflection of 50 A.

**[Ans.** 25 A ; 0.001001 Ω]

**5.** A moving-coil instrument has at normal temperature a resistance of 10 Ω and a current of 45 milliampere gives full scale deflection. If its resistance rises to 10.2 Ω due to temperature change, calculate the reading when a current of 2000 A is measured by means of a $0.225 \times 10^{-3}$ A shunt of constant resistance. What is the percentage error ? **[Ans.** 44.1 mA, – 1.96%]

**6.** The coil of a moving-coil voltmeter is 40 mm × 30 mm wide and has 100 turns wound on it. The control spring exerts a torque of $0.25 \times 10^{-3}$ Nm when the deflection is 50 divisions on the scale. If the flux density of the magnetic field in the air-gap is 1 Wb/m$^2$ estimate the resistance that must be put in series with the coil to give 1 volt per division. Resistance of voltmeter is 10,000 Ω.

**[Ans.** 14,000 Ω]

**7.** A direct voltage of 240 V is applied to a uniform 200 Ω resistor *abc*. A 40 Ω resistor is tapped across *ab*. Find the resistance of *ab* in order that 1 A flows in the 40 Ω resistor. **[Ans.** 71.65 Ω]

**8.** A moving-coil instrument has a resistance of 5 Ω between terminals and full scale deflection is obtained with a current of 15 mA. This instrument is to be used with a manganin shunt to measure 100 A at full scale. Calculate the error caused by a 10°C rise in temperature :

(*a*) When the internal resistance of 5 Ω is due to copper only.

(*b*) When a 4 Ω manganin swamping resistance is used in series with a copper coil of 1 Ω resistance.

The resistance temperature co-efficients of copper and manganin are 0.4%/°C and 0.015%/°C rise respectively. **[Ans.** 3.7% low : 0.8% low]

**9.** Design a multi-range ammeter with ranges of 1 A, 5 A, 25 A and 125 A employing individual shunts in each case. A d'Arsonval movement with an internal resistance of 730 Ω and a full scale current of 5 mA is available.

**[Ans.** 3.67 Ω, 0.73 Ω, 0.146 Ω, 0.0292 Ω]

**10.** A basic d'Arsonval movement with a full scale reading of 50 μA and an internal resistance of 1800 Ω is available. It is to be converted into a 0 – 1 V, 0 – 5 V, 0 – 25 V and 0 – 125 V multi-range voltmeter using individual multipliers for each range. Calculate the value of the individual multipliers.

**[Ans.** 18.2 kΩ, 98.2 kΩ, 498.2 kΩ, 2498.2 kΩ]

**11.** The Simpson multimeter model 260 uses a basic d'Arsonval movement of 50 μA with an internal resistance of 2000 Ω. The multi-voltmeter ranges of this instrument are 0 – 2.5 V, 0 – 10 V, 0 – 50 V, 0 – 250 V, 0 – 1000 V and 0 – 5000 V. The instrument utilizes the potential divider arrangement shown in Fig. 9.11. Calculate the resistances values of $R_1$, $R_2$, $R_3$, $R_4$, $R_5$ and $R_6$.

**[Ans.** 48 kΩ, 150 kΩ, 800 kΩ, 4 kΩ, 15 kΩ, 80 kΩ]

**12.** Design a series type ohmmeter (as shown in Fig. 9.13). The basic movement requires a current of 1.0 mA for full scale deflection and has an internal resistance of 50 Ω. The internal battery has a voltage of 3.0 V. The desired value of half scale resistance is 2000 Ω. Calculate

(*a*) The values resistances $R_1$ and $R_2$.

(*b*) The range of values of $R_2$, if the battery voltage may vary from 2.7 V to 3.1 V. Use value of $R_1$ as calculated in (*a*).

**[Ans.** (*a*) $R_1$ = 1966.7 Ω, 100 Ω (*b*) 142.8 – 90.9 Ω]

**13.** A shunt type ohmmeter, with circuit as shown in Fig. 9.14 uses a basic meter with a full scale current of 2 mA and an internal resistance of 25 Ω. The battery has an emf of 1.5 V. Find

(*a*) The value of resistance $R_1$,

(*b*) At what point (in percentage of full scale) will 100 Ω be marked on the scale ?

**[Ans.** (*a*) 725 Ω. (*b*) 80.5% A]

**14.** The inductance of a certain moving iron ammeter is $\left(8+4\theta-\frac{1}{2}\theta^2\right)$ μH, where θ is the deflection in radian from the zero position. The control spring torque is $12\times10^{-6}$ N-m/rad. Calculate the scale positions in radian for currents 1, 2, 3, 4 and 5 A and discuss the scale shape obtained.

**[Ans.** 0.16, 0.571, 1.09, 1.6, 2.04 rad]

**15.** The inductance of a moving iron ammeter is given by : $L=(0.01+C\theta)^2$ mH, where θ is the deflection from zero position in degree. The angular deflections of the instrument corresponding to 1.5 and 2 A are respectively 90° and 120°. Find the value of C. **[Ans.** $-0.0474\times10^{-3}$]

**16.** The relationship between inductance of a moving iron ammeter, the current and the position of the pointer is as follows :

| *Reading (A)* | *Deflection (degree)* | *Inductance (μH)* |
|---|---|---|
| 0.8 | 16.5 | 527.8 |
| 1.0 | 26 | 573.9 |
| 1.2 | 36 | 575 |
| 1.4 | 46.5 | 576.2 |
| 1.6 | 57 | 577.3 |
| 1.8 | 70 | 578.35 |
| 2.0 | 2.0 | 579.45 |

Calculate the deflecting torque when the current is 1.5 A and 2.1 A.

**[Ans.** $6.45\times10^{-6}$ Nm, $5.73\times10^{6}$ Nm]

**17.** The control spring of moving iron ammeter exerts a torque of $0.5\times10^{-6}$ N-m/degree when the deflection is 52°. The inductance of the coil varies with pointer deflection according to :

| *Deflection (degree)* | 20 | 40 | 60 | 80 |
|---|---|---|---|---|
| *Inductance (μH)* | 659 | 702 | 752 | 792 |

Determine the current passing through the meter.

**[Ans.** 0.63 A]

**18.** The copper coil of a 150 V moving iron voltmeter has a resistance of 400 ohm at 15°C and an inductance of 0.75 H. The current for full scale deflection is 0.05 A.

The temperature coefficients of resistance for copper and eureka at 15°C are 0.004/°C and 0.00001/C° respectively.

Calculate :

(*a*) The percentage increase of resistance of this instrument per degree rise in temperature

(*b*) The indication when 150 V at 100 Hz is applied, the instrument having been previously calibrated on direct current.

**[Ans.** 0.0542%, 148.17 V]

**19.** A soft-iron voltmeter for a maximum reading of 120 V has an inductance of 0.6 H and a total resistance of 2,400 Ω. It is calibrated to read correctly on a 60 Hz circuit. What series resistance would be necessary to increase its range to 600 V ? Draw up suitable workshop instructions for making up the resistance. **[Ans.** 9660 Ω]

**20.** An alternating current voltmeter, with a maximum scale reading of 50 V, has an inductance of 0.09 H and a total resistance of 500 Ω. The coil is wound with copper wire having a resistance of 50 Ω and the remainder of the voltmeter circuit consists of a non-inductive resistance in series with the coil. Find the capacitance that should be placed across the non-inductive resistor to make the instrument read correctly both on d.c. as well as a.c. **[Ans.** 0.182 μF]

**21.** The following measurements were made on a voltmeter of the dynamometer type having a range of 60 V and a resistance of 780 Ω.

| *Applied voltage(V)* | 30 | 40 | 50 | 60 |
|---|---|---|---|---|
| *Deflection (degree)* | 14 | 24 | 37 | 54 |
| *Inductance (mH)* | 74.8 | 78.5 | 82.8 | 88.6 |

Calculate the deflecting torque when the applied voltage is 45 V. **[Ans.** $64.6 \times 10^{-6}$ N-m]

**22.** The following measurements were made on a 50 V dynamometer type voltmeter having a resistance of 750 Ω.

| *Deflection (degree)* | *Mutual inductance (μH)* |
|---|---|
| 10 | 67.2 |
| 20 | 72 |
| 30 | 77 |
| 50 | 86 |
| 70 | 94.4 |
| 80 | 98 |
| 90 | 100 |

A torque of $0.5 \times 10^{-6}$ is required to give full scale deflection of 90°. Calculate the voltage at half full scale reading. **[Ans.** 23.3 V]

**23.** A thermocouple instrument reads 10 A at full scale. Determine the current when the meter reads one third of full scale. **[Ans.** 5.78 A]

**24.** The spring constant of a 3000 V electrostatic voltmeter is $7.06 \times 10^{-6}$ Nm/rad. The full scale deflection of the instrument is 80°. Assuming the rate of change of capacitance with angular deflection to be constant over the operating range, calculate the total change of capacitance from zero to full scale. **[Ans.** 3.06 pF]

**25.** A deflecting electrostatic voltmeter gives full scale deflection of 40° with 9 V applied to its terminals. Its capacitance is 10 pF at zero deflection and 30 pF at 30°, and varies uniformly with deflection. The moment of inertia of moving vanes is $0.02 \times 10^{-6}$ kg – m$^2$. Calculate the undamped period of instrument. **[Ans.** 18.9 second]

**26.** There are 7 cells in a kelvin multicellular voltmeter. The needle is in the form of two sectors of a circle of radius 50 mm. The distance between the fixed quadrants is 4 mm and the needle is hung midway between them. Application of a voltage of 1000 V gives a deflection of 90°. Calculate the spring constant. **[Ans.** $49.3 \times 10^{-6}$ Nm/rad]

**27.** The plates of an absolute electrometer, which are 250 mm in diameter and 20 mm apart are charged to a potential difference of 10 kV, calculate the force of attraction between the plates and the charge on each. **[Ans.** 0.064 N, 0.22 μC]

**28.** The movable vane of a quadrant electrometer turns through 40 scale divisions when it is idiostatically connected to a potential of 100 V. When it is used heterostatically with the quadrants connected to a small voltage *e* and the needle to a 100 V supply, the deflection is 15 scale divisions. Determine the voltage *e*. **[Ans.** 3.7 V]

**29.** The capacity of an electrostatic voltmeter reading from 0 to 2000 V increases from 80 to 90 pF as the pointer moves from zero to full scale deflection. Calculate the value of external capacitor used to increase its range to 20 kV. If the capacitor is adjusted to make the full scale reading correct, what will be the error at half-scale reading ? **[Ans.** 10 pF, 5.27% high]

**30.** A single element rectifying device is connected in series with a moving coil ammeter and a thermal ammeter and it is found that the circuit has a resistance of 100 Ω in one direction and infinity in the other direction. If a sinusoidal emf of 10 V rms is applied to this circuit what will be the reading of each instrument and what will be the power taken from the supply ?

**[Ans.** PMMC 0.045 A, Thermal 0.0707 A, 0.5 W]

**31.** In a bridge connected rectifier voltmeter each rectifier has a resistance of 20 Ω in the forward direction and infinity in the reverse direction. The moving coil indicator has resistance of 10 Ω and gives full scale deflection for 10 mA. Determine the series resistance necessary if the instrument is to read the correct rms value at 1/2 full scale when connected to a source of voltage : $v_1 = 100\sqrt{2}$ volt. If this instrument, and an accurate electrodynamic voltmeter are connected across a voltage

$$v_2 = 100\sqrt{2}\sin\omega t + 10\sqrt{2}\sin 3\omega t$$

volt determine reading of each instrument.

[**Ans.** 17970 Ω, Rectifier 103.2 V, Electrodynamic 100.5 V]

**32.** A rectifier voltmeter uses a full wave rectifier. The meter has an internal resistance of 50 Ω and requires a current of 1 mA for full scale deflection. The rectifiers each have zero forward resistance and infinite reverse resistance.

Calculate,

(*a*) the multiplier resistance required for full scale meter deflection with 10 V a.c.

(*b*) sensitivity of d.c. voltmeter,

(*c*) the sensitivity in Ω/V with a.c.

[**Ans.** (*a*) 8950 Ω, (*b*) 1000 Ω/V, (*c*) 900 Ω/V]

**33.** A commercial meter of Fig. 9.72 uses a 1 mA basic meter with an internal resistance of 100 Ω. The shunting resistance across the meter is 200 Ω. Diodes $D_1$ and $D_2$ each have a forward resistance of 200 Ω and infinite reverse resistance. Calculate the value of series multiplier so that the meter is converted to a 10 V a.c. voltmeter.

Determine the sensitivity as an a.c. voltmeter.

[**Ans.** 2733.3 Ω ; 300 Ω]

**34.** Figure 9.77 represents a meter dial for an a.c. voltmeter with full wave rectification. Calculate the values of peak to-peak voltage and the d.c. voltage and the current for the rms voltages shown if the d.c. sensitivity of the meter movement is 10 kΩ/V. Sketch the meter dial and fill in the blanks.

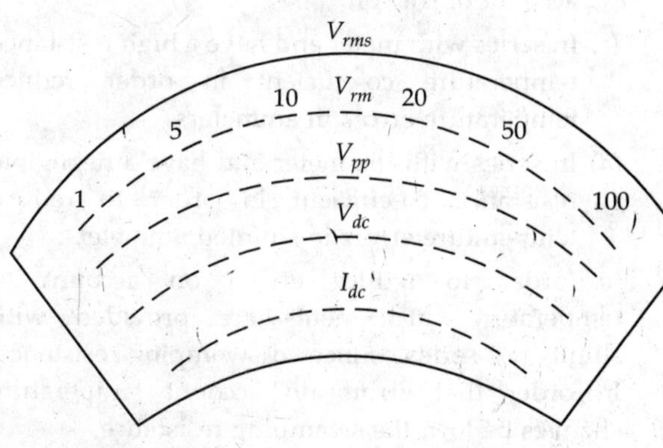


**Fig. 9.77** Diagram of problem 9.34.

**35.** A direct current of about 2 A is to be measured accurately. If the following 3 ammeters are available, which would you prefer to use for this purpose ?

Give clear reasons for your choice.

(*i*) Permanent magnet moving coil type with 0 – 5 A range and 0.5 per cent accuracy at full scale.

(*ii*) Permanent magnet moving coil with 0 – 3 A range and 1 per cent accuracy at full scale.

(*iii*) Moving iron type with 0 – 2.5 A range per cent at full scale.

**36.** What will be the reading when a voltage

$$e = 200\sin\omega t + 40\sin(3\omega t + 30°) + 30\cos(5\omega t + 21.5°) \text{ volt}$$

is applied to an electrodynamometer voltmeter.

[**Ans.** 145.8 V]

**37.** An electrodynamometer ammeter and a permanent magnet moving coil ammeter are connected in series with a 100 Ω resistor and a rectifier having a forward resistance 50 Ω and backward resistance of 500 Ω. A voltage having triangular waveform and a frequency of 50 Hz and rms value of 100 V is applied to the circuit.

Calculate

(*a*) the reading of each ammeter, and

(*b*) the effective power factor of the circuit.

[**Ans.** (*a*) dynamometer, 0.486 A ; PMCC, 0.361 A, (*b*) 0.856]

**38.** A millivoltmeter has two scales reading 0 – 20 mV and 0 – 100 mV, the resistance of two ranges being 400 Ω and 2000 Ω respectively. When used in conjunction with a thermopile (a number of thermocouple connected in series) a certain reading changed from 19.5 mV to 23.4 mV when the range was altered from the lower to the higher scale.

Calculate

(*a*) the resistance of the thermopile,

(*b*) its emf and

(*c*) the correction multiplier.

[**Ans.** (*a*) 105.4 Ω ; (*b*) 24.6 mV ; (*c*) 1.26, 1.05]

**39.** An rms reading a.c. voltmeter and a peak-to-peak reading a.c. voltmeter are used to determine the following three signals are sinusoidal or not. The readings obtained are

First Signal : peak to peak reading = 35.26 V, rms reading = 12.00 V

Second Signal : peak to peak reading
= 11.31 V, rms reading = 4.00 V

Third Signal : peak to peak reading
= 25.00 V, rms reading = 8.33 V

Determine the nature of signals.

[**Ans.** Not sinusoidal, sinusoidal, sinusoidal]

**40.** Two different a.c. voltmeters are used to measure the voltage across the 22 kΩ/V resistor shown in Fig. 9.78. Meter A has a.c. sensitivity of 10 kΩ/V, a guaranteed accuracy of 98% at full scale, and is set on its 200 V scale. Meter B has an a.c. sensitivity of 4 kΩ/V, a guaranteed accuracy of 98.5% at full scale, and is set on its 0 – 100 V range. Which meter provides more accurate results ?

[**Ans.** Meter B]

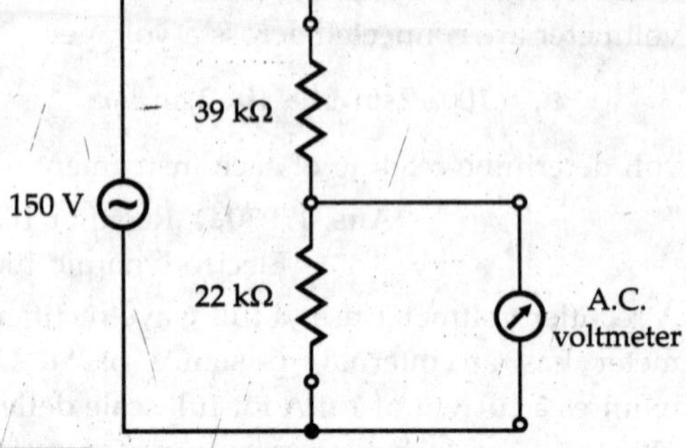


**Fig. 9.78** Diagram of problem 9.40.

# Objective Type Questions

***Tick (√) the most appropriate answer :***

**1.** A voltmeter has a resistance of 2000 Ω. When it is connected across a d.c. circuit its power consumption is 2 mW. Suppose this voltmeter is replaced by a voltmeter of 4000 Ω resistance, the power consumption will be :

(*a*) 4 mW (*b*) 1 mW
(*c*) 2 mW (*d*) none of the above.

The circuit conditions remain undisturbed.

**2.** A milllammeter of resistance 100 Ω is connected in series with a circuit. Its power consumption is 0.1 mW. Supposing it is replaced with a milliammeter of 200 Ω resistance the power consumption will be

(*a*) 0.2 mW (*b*) 0.05 mW
(*c*) 0.1 mW (*d*) none of the above.

The circuit conditions remain undisturbed.

**3.** The high torque to weight ratio in an analog indicating instrument indicates :

(*a*) High friction loss
(*b*) Low friction less
(*c*) Nothing as regards friction loss
(*d*) None of the above.

**4.** A PMMC instrument produces a deflection of 90° when a current 2 mA passes through its moving coil. The instrument is spring controlled. Supposing the instrument is redesigned with its spring constant as before, and its permanent magnet is replaced with a new permanent magnet which produces twice the flux density as before. What would be the deflection if a current of 1 mA is passed through the coil? The other features of the PMMC instrument remain the same.

(*a*) 180° (*b*) 90°
(*c*) 45° (*d*) 135°

**5.** A 1 mA ammeter has a resistance of 100 Ω. It is to be converted to a 1 A ammeter. The value of shunt resistance is :

(*a*) 0.001 Ω (*b*) 0.1001 Ω
(*c*) 100000 Ω (*d*) 100 Ω

**6.** Swamping resistance is connected

(*a*) In series with the shunt to reduce temperature error in shunted ammeters
(*b*) In series with the ammeters to reduce errors on account of friction
(*c*) In series with meter and have a high resistance temperature co-efficient in order reduce temperature errors in ammeters
(*d*) In series with the meter and have a negligible resistance co-efficient in order to reduce temperature errors in shunted ammeters.

**7.** In order to reduce errors on account of temperature, instruments are provided with shunts use series connected swamping resistance. In order that errors on account temperature changes be low, the swamping resistance

(*a*) Should be made of a material having a high resistance temperature co-efficient (*b*) Should be made of a material having a low resistance temperature co-efficient with the value of swamping resistance equal to meter resistance

(*c*) Should be made of a material having low resistance temperature co-efficient and should have a value of about 20 to 30 times that of meter resistance.

(*d*) Should have an infinite value.

**8.** A 4-range milliammeter having ranges of 0-10 mA, 0-50 mA, 0-100 mA and 0-500 mA. It is used for measurement of current whose magnitude is not known. In order to prevent damage to the instrument, the selector switch of the ammeter should be first placed at

(*a*) 0 – 50 mA range (*b*) 0 – 100 mA range

(*c*) 0 – 10 mA range (*d*) 0 – 500 mA range.

**9.** A 1 mA d'Arsonval movement has a resistance of 100 Ω. It is to be converted to a 10 V voltmeter. The value of multiplier resistance is

(*a*) 999 Ω (*b*) 9999 Ω

(*c*) 9900 Ω (*d*) 990 Ω.

**10.** A d'Arsonval movement is rated at 50 μA. Its sensitivity is

(*a*) 20000 Ω/V (*b*) 200000 Ω/V

(*c*) 200 Ω/V (*d*) Cannot be determined.

**11.** A d'Arsonal movement has a sensitivity of 40,000 Ω and its internal resistance is 4000 Ω. The resistance of multiplier to convert it to 1 V voltmeter is :

(*a*) 44000 Ω (*b*) 36000 Ω

(*c*) 3600 Ω (*d*) None of the above.

**12.** A PMMC instrument is spring controlled. The control spring stiffness decreases by about 0.04 percent per °C rise in temperature and the strength of magnets goes down by 0.02 percent per °C rise in temperature. The rise in temperature is 10°C. The deflection

(*a*) Decreases by about 0.2 percent

(*b*) Decreases by about 0.4 percent

(*c*) Increases by about 0.4 percent

(*d*) Increases by about 0.2 percent.

**13.** The power consumption PMMC instruments is typically about :

(*a*) 0.25 W to 2 W (*b*) 0.25 mW to 2 mW

(*c*) 25 μW to 200 μW (*d*) None of the above.

**14.** In a series type ohmmeter :

(*a*) 0 marking is on the left hand side of side while ∞ marking is on the right hand side

(*b*) 0 marking is on the right hand side of scale and ∞ infinity marking on the left hand side

(*c*) Any of the above two markings can be on left or right side of the scale

(*d*) 0 marking is in the middle of scale.

**15.** In a series type of ohmmeter the zero adjustment should be done

(*a*) By changing the value of series resistance

(*b*) By changing the value of shunt resistance connected across the meter movement

(*c*) By changing both series as well as the shunt resistance

(*d*) By changing the battery voltage.

**16.** A series type ohmmeter is designed with a d'Arsonval movement having full scale current of 0.5 mA and an internal resistance. It uses a battery of 3 V and has a half scale resistance of 3000 Ω. The battery voltage goes down to 2.7 V. What is the value of shunt resistance required so that full scale current is obtained if the meter terminals are short circuited ?

(*a*) 62.5 Ω (*b*) 2700 Ω

(*c*) 55 Ω (*d*) 130 Ω

**17.** A make break switch is provided to disconnect the battery when the meter is not in use in

(*a*) Both series and shunt type ohmmeters

(*b*) Only in series type ohmmeters

(*c*) Only in shunt type ohmmeters

(*d*) None of the above.

**18.** A meggar is used for measurement of

(*a*) Low valued resistances

(*b*) Medium valued resistances

(*c*) High valued resistances, particularly insulation resistance

(*d*) All the above.

**19.** The controlling torque in a meggar is provided by :

(*a*) Springs

(*b*) Weights attached to the moving system

(*c*) It does not need any controlling torque

(*d*) None of the above.

**20.** A moving iron instrument can be used for current and voltage measurements

(*a*) In a.c. circuits only

(*b*) In d.c. circuits only

(*c*) In both a.c. and d.c. circuits for any value of frequency (in case of a.c. circuits)

(*d*) In both a.c. and d.c. circuits for frequencies upto about 125 Hz (in case of a.c. circuits).

**21.** In spring controlled moving iron instruments, the scale is :

(a) Uniform

(b) Cramped at the lower end and expanded at the upper end

(c) Expanded at the lower end and cramped at the upper end

(d) Cramped both at the lower and the upper ends.

**22.** Moving iron type of instrument can be used as :

(a) Standard instruments for calibration of other instruments

(b) Transfer type instruments

(c) Indicator type instruments as on panels

(d) All the above.

**23.** Moving iron instruments when measuring voltages or currents

(a) Indicate the same values of the measurement for both ascending and descending values

(b) Indicate higher value of measurand for ascending values

(c) Indicate higher value of measurand for descending values

(d) None of the above.

**24.** The moving iron voltmeters indicate :

(a) The same value for d.c. and a.c. voltages

(b) Lower values for a.c. voltages than for corresponding d.c. voltages

(c) Higher values for a.c. voltages than for corresponding d.c. voltages

(d) None of the above.

**25.** A moving iron voltmeter reads low for a.c. voltages than for corresponding values of d.c voltages. The meter can be made to read equally for both a.c. and d.c. voltages

(a) If the resistance of the multiplier is made very high

(b) If the inductance of the coil is made small

(c) If the resistance of the coil is made very large

(d) If the multiplier resistance is shunted by a capacitor of appropriate value to make the circuit non-inductive.

**26.** Horizontally mounted moving iron instruments use :

(a) Eddy current damping

(b) Electromagnetic damping

(c) Fluid friction damping

(d) Air friction damping.

**27.** The reason why eddy current damping cannot be used in moving iron instruments is :

(a) They have a strong operating magnetic field

(b) They are not normally used in vertical position

(c) They need a large damping force which can only be provided by air friction

(d) They have a very weak operating magnetic field and introduction of a permanent magnet required for eddy current damping would distort the operating magnetic field.

**28.** The frequency range of moving iron instruments is

(a) Audio frequency band 20 Hz to 20 kHz

(b) Very low frequency band 10 Hz to 30 kHz

(c) Low frequency band 30 Hz to 300 kHz

(d) Power frequencies 0 to 125 Hz.

**29.** Spring controlled moving iron instruments exhibit a square law response resulting in a non-linear scale. The shape of the scale can be made almost linear by :

(a) Keeping rate of change of inductance, $L$, with deflection, $\frac{1}{\theta}\cdot\frac{dL}{d\theta}$, as constant

(b) Keeping $\frac{1}{K\theta}$ as constant

(c) Keeping $\frac{1}{K\theta}$ as constant where $K$ is the spring constant

(d) Keeping $\theta\cdot\frac{dL}{d\theta}$ as constant.

**30.** An electrodynamometer type of instruments finds its major use as :

(a) Standard instrument only

(b) Transfer instrument only

(c) Both as standard and transfer instrument

(d) A indicator type of instrument.

**31.** The operating magnetic held in an electrodynamometer type of instrument has a flux density typically about

(a) 1 $Wb/m^2$ (b) 0.5 $Wb/m^2$

(c) 0.05 $Wb/m^2$ (d) 0.005 $Wb/m^2$.

**32.** An Astatic movement in an electrodynamometer type of instrument is used for :

(a) Eliminating errors on account of external magnetic fields

(b) Increasing the operating torque of the instruments

(c) Providing eddy current damping

(d) None of the above.

**33.** In order that an electrodynamometer type of instrument exhibits a pure square law response, the meter range should be limited to :

(*a*) $-45°$ to $+45°$ about the position for zero mutual inductance between fixed and moving coils

(*b*) $-45°$ to $+45°$ about the position for maximum mutual inductance between fixed and moving coils

(*c*) $-22.5°$ to $22.5°$ about the position for zero inductance between fixed and moving coils

(*d*) $-90°$ to $90°$ about the position for maximum mutual inductance between fixed and moving coils.

**34.** A voltage of 200 V at 5 Hz is applied to an electrodynamometer type of instrument which is spring controlled. The indication on the instruments is

(*a*) 200 V

(*b*) 0 V

(*c*) The instrument follows the variations in voltage and does not give a steady response

(*d*) None of the above.

**35.** Thermocouple instruments can be used for a frequency range

(*a*) Upto 100 Hz (*b*) Upto 5000 Hz

(*c*) Upto 1 MHz (*d*) 50 MHz and above.

**36.** The heater wire of thermocouple instruments is made very thin in order to have

(*a*) A high value of resistance

(*b*) To reduce skin effects at high frequencies

(*c*) To reduce the weight of the instrument

(*d*) To decrease the over-ranging capacity of the instrument.

**37.** Electrostatic type instruments are primarily used as

(*a*) Ammeters (*b*) Wattmeters

(*c*) Voltmeters (*d*) Ohmmeters.

**38.** An hypothetical electrostatic voltmeter with spring control is designed with one fixed and another movable plate. Both plates are in the form of a circle with one plate over the other so as to produce a rotatry motion. If a voltage is applied between them, no rotatry motion is produced because :

(*a*) There is no change in the voltage

(*b*) There is no change in the capacitance irrespective of the relative position of the two plates

(*c*) The spring force is very large

(*d*) The applied voltage is small to produce any noticeable deflection.

**39.** An electrostatic voltmeter draws a small value of current on d.c.

(*a*) Under steady state condition respective of the applied voltage

(*b*) When switched on irrespective of the applied voltage

(*c*) When measuring low voltages

(*d*) When measuring high voltages.

**40.** An quadrant type electrostatic instrument uses two types of connections (*i*) Heterostatic and (*ii*) Idiostatic.

An external battery is used

(*a*) For Idiostatic connection

(*b*) For Heterostatic connection

(*c*) For both Idiostatic and Heterostatic connections

(*d*) None of the above.

**41.** The size of electrostatic voltmeters is large as compared with that of corresponding electromagnetic voltmeters because :

(*a*) They are used for measurement of high voltages

(*b*) Energy density in an electrostatic field is considerably smaller than that in an electromagnetic field

(*c*) Their frequency range is smaller

(*d*) They have a high dielectric loss.

**42.** The range of an electrostatic voltmeter can be extended by using :

(*a*) A capacitor in series with the voltmeter whose capacitance is greater than the capacitance of the voltmeter

(*b*) A capacitor in series with the voltmeter whose capacitance is smaller than the capacitance of the voltmeter

(*c*) A resistor in series with the voltmeter

(*d*) An inductor in series with the voltmeter.

**43.** The sensitivity of a PMMC instrument is 10 kΩ/V. If this instrument is used in a rectifier type voltmeter with half waves rectification, what would be the sensitivity ?

(*a*) 10 kΩ/V (*b*) 9 kΩ/V

(*c*) 4.5 kΩ/V (*d*) 22.2 kΩ/V

**44.** A PMMC meter rated at 50 μA is used in a rectifier type of instrument which uses full wave rectification. What is the sensitivity on sinusoidal a.c. ?

(*a*) 20 kΩ/V (*b*) 9 kΩ/V

(*c*) 22.2 kΩ/V (*d*) 18 kΩ/V

9

**45.** A reverse biased diode in addition to a forward biased diode is used in rectifier type voltmeters using half wave rectification. The use of the reverse biased diode is

(*a*) It does not allow any current to flow through the meter during negative half cycle

(*b*) It short-circuits the meter during the negative half cycle

(*c*) It does not allow the reverse leakage current to flow through the meter during the negative half cycle

(*d*) All the above.

**46.** The PMMC meters used in rectifier type voltmeters are provided with shunts even at the expense of lowering sensitivity. Why ?

(*a*) The shunting of the meter causes the circuit current to be high thereby making the rectifier work in the linear portion of its $v$ - $i$ characteristics.

(*b*) The shunt prevents the meter against sudden overloads

(*c*) The provision of shunt increases the sensitivity of the circuit

(*d*) All the above.

**47.** Why are multimeters provided with separate scale for low a.c. voltages ?

(*a*) To improve the readability of the scale.

(*b*) To have high accuracy

(*c*) To take into account the high value of resistance of rectifier at low voltages (and currents) and also the fact that at low voltages (mid currents) the value of rectifier resistance is not constant but varies considerably even for small changes in voltages (or current)

(*d*) None of the above.

**48.** Which meter has the highest accuracy in the prescribed limit of frequency range :

(*a*) PMMC (*b*) Moving Iron

(*c*) Electrodynamometer (*d*) Rectifier.

**49.** Which instrument is the cheapest disregarding the accuracy :

(*a*) PMMC (*b*) Moving Iron

(*c*) Electrodynamometer (*d*) Rectifier.

**50.** Which instrument has the highest frequency range with accuracy within reasonable limits

(*a*) Moving Iron (*b*) Electrodynamometer

(*c*) Thermocouple (*d*) Rectifier.

## *Answers*

| | | | | | | | | | |
|---|---|---|---|---|---|---|---|---|---|
| **1.** (*b*) | **2.** (*a*) | **3.** (*b*) | **4.** (*b*) | **5.** (*b*) | **6.** (*d*) | **7.** (*c*) | **8.** (*d*) | **9.** (*c*) | **10.** (*a*) |
| **11.** (*b*) | **12.** (*d*) | **13.** (*c*) | **14.** (*b*) | **15.** (*b*) | **16.** (*a*) | **17.** (*c*) | **18.** (*c*) | **19.** (*c*) | **20.** (*d*) |
| **21.** (*b*) | **22.** (*c*) | **23.** (*c*) | **24.** (*b*) | **25.** (*d*) | **26.** (*d*) | **27.** (*d*) | **28.** (*d*) | **29.** (*d*) | **30.** (*c*) |
| **31.** (*d*) | **32.** (*a*) | **33.** (*a*) | **34.** (*c*) | **35.** (*d*) | **36.** (*b*) | **37.** (*c*) | **38.** (*b*) | **39.** (*a*) | **40.** (*b*) |
| **41.** (*b*) | **42.** (*b*) | **43.** (*c*) | **44.** (*d*) | **45.** (*d*) | **46.** (*a*) | **47.** (*c*) | **48.** (*c*) | **49.** (*b*) | **50.** (*c*) |

# 10

# Instrument Transformers

## 10.1 INTRODUCTION

Transformers are used in a.c. systems for the measurement of current, voltage, power and energy. They are also used in connection with measurement of power factor, frequency and for indication of synchronism. Instrument transformers find a wide application in protection circuits of power systems for the operation of over current, under voltage, earth fault and various other types of relays.

In all the above applications, the transformer is put for measurement purposes, the actual measurements being done by measuring instruments. Transformers used in conjunction with measuring instruments for measurement purposes are called *"Instrument Transformers."* The transformer used for measurement of current is called a *"Current Transformer"* or simply *"C.T."* Transformers used for voltage measurements are called *"Voltage Transformers"* or *"Potential Transformers"* or simply *"P.T."* in short.

## 10.2 USE OF INSTRUMENT TRANSFORMERS

The extension of instrument range, so that current, voltage, power and energy can be measured with instruments or meters of moderate size is of very great importance in commercial metering. In power systems, currents and voltages handled are very large and, therefore, direct measurements are not possible as these currents and voltages are far too large for any meter of reasonable size and cost. The solution lies in stepping down these currents and voltages with the help of instrument transformers so that they could be metered with instruments of moderate sizes.

Figure 10.1 shows current being measured by a C.T. The primary winding is so connected that the current being measured passes through it and the secondary winding is connected to an ammeter. The "C.T." steps down the current to the level of ammeter.

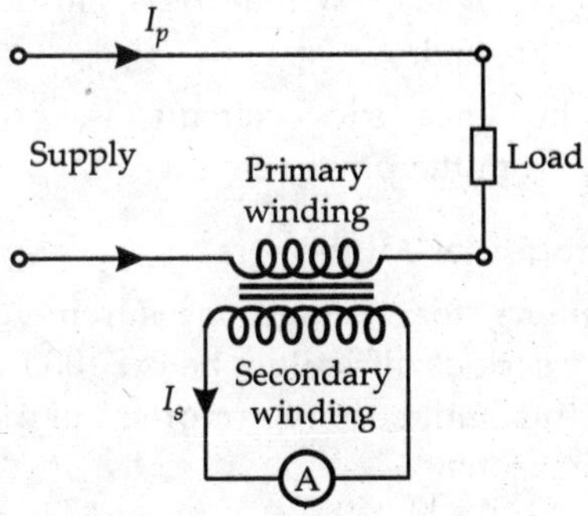


**Fig. 10.1** Current transformer.

Figure 10.2 shows voltage measurement with a P.T. The primary winding is connected to the voltage being measured and the secondary winding, to a voltmeter. The 'P.T.' *steps down* the voltage to the level of voltmeter.

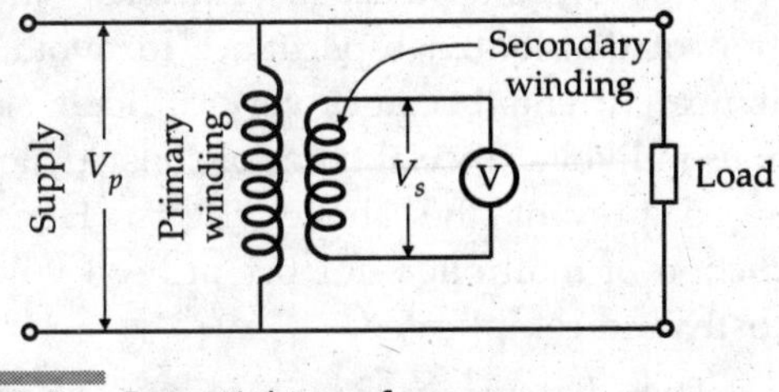


**Fig. 10.2** Potential transformer.

It might appear that extension of range could be conveniently done by the use of shunts for currents and multipliers for voltage measurements, as is done in d.c. measurements. But this method is suitable only for small values of current and voltage. There are many disadvantages with the use of shunts and multipliers.

### Disadvantages of Shunts

(*i*) It is difficult to achieve accuracy with a shunt on a.c., since division of current between a meter and shunt depends upon the ratio of reactance to resistance of two paths. For proper measurements, time constants of meter and shunt should be the same. Therefore, a separate shunt would be needed for each instrument.

Also since measurements are made over a wide frequency range it becomes difficult to obtain good accuracy with shunts.

(*ii*) The method of using shunts is limited to capacities of a few hundred ampere at most, since the power consumed by shunts at large currents would be considerably large.

(*iii*) The problem of insulation of instrument and shunt is quite difficult if measurements are done at high voltages of several hundreds or thousands of volts above ground.

(*iv*) The measuring circuit is not isolated electrically from the power circuit.

## Disadvantages of Multipliers

Multipliers for voltage measurements do not present any serious difficulties below 1000 V. But their use above this limit becomes impractical owing to the following reasons :

(*i*) The power consumed by multipliers becomes large as the voltage increases. A power dissipation of about 7.5 W is the upper limit for a self-contained instrument.

(*ii*) Care has to be exercised to keep leakage currents in high voltage multipliers down to negligible values. However, insulation of multipliers required to prevent leakage currents and the reduction of distributed capacitance to avoid shunt capacitance currents becomes very difficult above a few thousand volts. Special types of constructions are needed to prevent the above effects. Hence the construction of multipliers for use at high voltage is very costly and complicated.

(*iii*) The measuring circuit is not electrically isolated from the power circuit.

## Advantages of Instrument Transformers

The current and voltage transformers are extensively used for very precise measurements as well as for routine measurements as they have many advantages and these are :

(*i*) When instruments are used in conjunction with instrument transformers, their readings do not depend upon their constants (*R, L, C*) as is the case with shunts and multipliers. The instrument transformers produce practically the same instrument reading regardless of the constants of the instrument or, in fact, the number of instruments connected in the circuit.

(*ii*) Current transformers have been standardized at 5 A secondary winding current and the voltage transformers at from 100 to 120 V secondary winding voltage. These are very moderate ratings and the instruments for measurements are rated near these. Thus a 5 A ammeter may be used to measure 1000 A with the help of a 1000/5 A ratio current transformer or a 110 V voltmeter may be used to measure a voltage of 66 kV with the help of a 66,000/110 V potential transformer. Therefore, very cheap moderate rating instruments may be used to measure large currents and high voltages.

(*iii*) With the standardization of C.T. and P.T. secondary winding ratings, it is possible to standardize instruments around these ratings and, therefore, there is great reduction in the costs of instrument transformers and instruments. Also, replacement of instrument transformers is easy on account of the standardization of the ratings.

(*iv*) The measuring circuit is isolated from the power circuit. The fact that the meters in the secondary circuit of an instrument transformer are isolated electrically from the primary side is of very great importance in high voltage systems. Large alternators usually operate at voltages between 11 kV to 20 kV while transmission voltages as high as 750 kV are being used. Therefore, in all electrical systems it is necessary to have a measure of currents and voltages, both for metering purposes and for the operation of protective relays and other equipment. It is impossible to bring the high voltage lines directly to the switchboard to be connected to instruments as even for voltages upto a few thousand volts it would be difficult to insulate the equipment to provide safety for the operating personnel.

The situation is completely changed with the use of instrument transformers, as only the leads from the secondary windings of the instrument transformers are brought to the switchboard for connection to the metering equipment. The voltages between these leads and between leads and ground are very low and hence the operators can work safely.

In fact instrument transformers are so important for insulating and range extension purposes that it is difficult to imagine the operation of a high voltage a.c. system without them.

Summarizing the above, we can say that the use of instrument transformers for range extension has the following advantages :

1. Instruments of moderate size are used for metering *i.e.*, 5 A for current and 100 to 120 V for voltage measurements.
2. Instruments and meters can be standardized so that there is a saving in overall costs. Replacement of damaged instruments is easy.
3. Single range instruments can be used to cover large current or voltage ranges, when used with suitable multi-range instrument transformers or with several single range instrument transformers.
4. The metering circuit is isolated from the high voltage power circuits. Hence insulation is no problem and the safety is assured for the operators.
5. There is low power consumption in metering circuit.
6. Several instruments can be operated from a single instrument transformer.

## 10.3 RATIOS OF INSTRUMENT TRANSFORMERS

There are some definitions which are given below :

***Transformation ratio.*** It is the ratio of the magnitude of the primary phasor to the secondary phasor.

Transformation ratio,

$$R = \frac{|\text{primary phasor}|}{|\text{secondary phasor}|} \quad ...(10.1)$$

$$= \frac{\text{primary winding current}}{\text{secondary winding current}} \text{ for a C.T.} \quad ...(10.2)$$

$$= \frac{\text{primary winding voltage}}{\text{secondary winding voltage}} \text{ for a P.T.} \quad ...(10.3)$$

***Nominal ratio.*** It is the ratio of rated primary winding current (or voltage) to the rated secondary winding current (or voltage).

Nominal ratio,

$$K_n = \frac{\text{rated primary winding current}}{\text{rated secondary winding current}} \text{ for a C.T.} \quad ...(10.4)$$

$$= \frac{\text{rated primary winding voltage}}{\text{rated secondary winding voltage}} \text{ for a P.T.} \quad ...(10.5)$$

***Turns ratio.*** This is defined as under :

Turns Ratio,

$$n = \frac{\text{number of turns of secondary winding}}{\text{number of turns of primary winding}} \text{ for a C.T.} \quad ...(10.6)$$

$$= \frac{\text{number of turns of primary winding}}{\text{number of turns of secondary winding}} \text{ for a P.T.} \quad ...(10.7)$$

***Ratio correction factor (RCF).*** The ratio correction factor of a transformer is the transformation ratio divided by nominal ratio.

Transformation ratio = ratio correction factor × nominal ratio

or

$$R = \text{RCF} \times K_n \quad ...(10.8)$$

The ratio marked on the transformers is their nominal ratio.

## 10.4 BURDEN OF AN INSTRUMENT TRANSFORMER

It is convenient to express load across the secondary winding terminals as the output in volt-ampere at the rated secondary winding voltage. The rated burden is the volt-ampere loading which is permissible without errors exceeding the limits for the particular class of accuracy.

Total secondary winding burden

$$= \frac{(\text{secondary winding induced voltage})^2}{(\text{impedance of secondary winding circuit including impedance of secondary winding})}$$

$$= \left(\begin{array}{c}\text{secondary winding}\\ \text{current}\end{array}\right)^2 \times \left(\begin{array}{c}\text{impedance of secondary}\\ \text{winding circuit including}\\ \text{secondary winding}\end{array}\right) \quad ...(10.9)$$

Secondary Winding burden due to load

$$= \frac{(\text{secondary winding induced voltage})^2}{(\text{impedance of load on secondary winding})}$$

$$= \left(\begin{array}{c}\text{secondary winding}\\ \text{current}\end{array}\right)^2 \times \left(\begin{array}{c}\text{impedance of load on}\\ \text{secondary winding}\end{array}\right) \quad ...(10.10)$$

## 10.5 CURRENT TRANSFORMERS

The current transformer is used with its primary winding connected in series with line carrying the current to be measured and, therefore, the primary current is dependant upon the load connected to the system and is not determined by the load (burden)

10

connected on the secondary winding of the current transformer. The primary winding consists of very few turns and, therefore, there is no appreciable voltage drop across it. The secondary winding of the current transformer has larger number of turns, the exact number being determined by the turns ratio. The ammeter, or wattmeter current coil, are connected directly across the secondary winding terminals. Thus a current transformer operates its secondary winding nearly under short circuit conditions. One of the terminals of the secondary winding is earthed so as to protect equipment and personnel in the vicinity in the event of an insulation breakdown in the current transformer. Figure 10.3 shows a circuit for measurement of current and power with a current transformer.

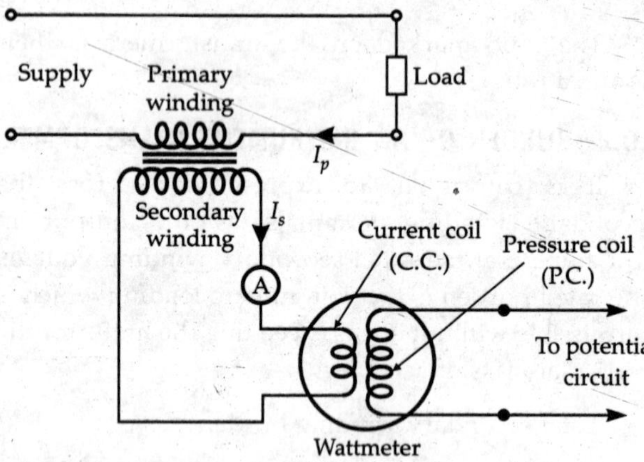


**Fig. 10.3** Use of C.T. for current and power measurement.

### 10.5.1 Relationships in a Current Transformer

Figure 10.4 represents the equivalent circuit and Fig. 10.5 the phasor diagram of a current transformer. The diagrams are same as for any other transformer.

$n$ = turns ratio

$$= \frac{\text{number of secondary winding turns}}{\text{number of primary winding turns}}$$

$r_s$ = resistance of the secondary winding,

$x_s$ = reactance of the secondary winding,

$r_e$ = resistance of external burden *i.e.*, resistance of meters, current coils etc. including leads,

$x_e$ = reactance of external burden *i.e.*, reactance of meters, current coils etc. including leads,

$E_p$ = primary winding induced voltage,

$E_s$ = secondary winding induced voltage,

$N_p$ = number of primary winding turns,

$N_s$ = number of secondary winding turns,

$V_s$ = voltage at the secondary winding terminals,

$I_s$ = secondary winding current,

$I_p$ = primary winding current,

$\theta$ = phase angle of transformer,

$\Phi$ = working flux of the transformer,

$\delta$ = angle between secondary winding induced voltage and secondary winding current,

= phase angle of total burden including impedance of secondary winding

$$= \tan^{-1}\left(\frac{x_s + x_e}{r_s + r_e}\right)$$

$\Delta$ = phase angle of secondary winding load circuit *i.e.*, of external burden $= \tan^{-1}\dfrac{x_e}{r_e}$

$I_0$ = exciting current,

$I_m$ = magnetising component of exciting current,

$I_e$ = loss component of exciting current,

$\alpha$ = angle between exciting current $I_0$ and working flux $\Phi$.

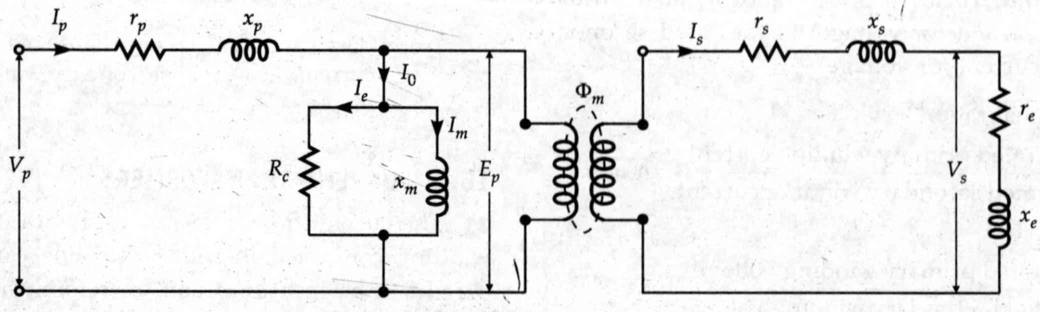


**Fig. 10.4** Equivalent circuit of a C.T.

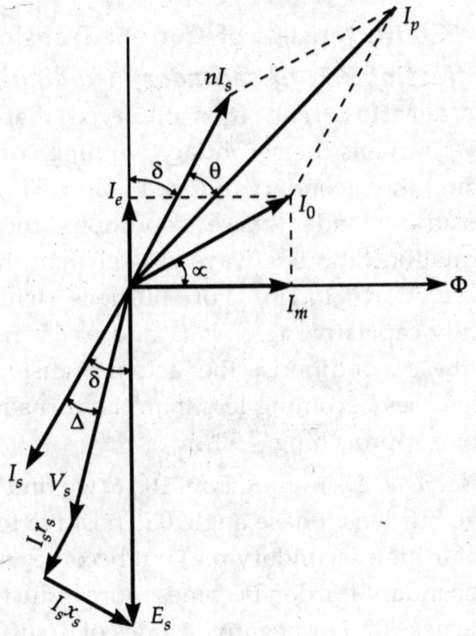


**Fig. 10.5** Phasor diagram of a C.T.

***Transformation ratio.*** Consider a small section of the phasor diagram as shown in Fig. 10.6.

We have $\angle bac = 90° - \delta - \alpha$, $ac = I_0$, $oa = nI_s$

and $oc = I_p$.

$$\therefore \quad bc = I_0 \sin(90° - \delta - \alpha) = I_0 \cos(\delta + \alpha)$$

$$ab = I_0 \cos(90° - \delta - \alpha) = I_0 \sin(\delta + \alpha)$$

Now $(oc)^2 = (oa + ab)^2 + (bc)^2$

or
$$I_p^2 = [nI_s + I_0 \sin(\delta + \alpha)]^2 + [I_0 \cos(\delta + \alpha)]^2$$
$$= n^2 I_s^2 + I_0^2 \sin^2(\delta + \alpha) + 2nI_s I_0 \sin(\delta + \alpha) + I_0^2 \cos^2(\delta + \alpha)$$
$$= n^2 I_s^2 + 2nI_s I_0 \sin(\delta + \alpha) + I_0^2$$

$$\therefore \quad I_p = [n^2 I_s^2 + 2nI_s I_0 \sin(\delta + \alpha) + I_0^2]^{1/2} \qquad \text{...(10.11)}$$

Transformation ratio,

$$R = \frac{I_p}{I_s} = \frac{\{n^2 I_s^2 + 2nI_s I_0 \sin(\delta + \alpha) + I_0^2\}^{1/2}}{I_s} \qquad \text{...(10.12)}$$

Now in a well designed current transformer $I_0 << nI_s$. Usually $I_0$ is less than 1 percent of $I_p$, and $I_p$ is, therefore, very nearly equal to $n\, I_s$,

∴ Equation 10.12 can be written as

$$R \approx \frac{[n^2 I_s^2 + 2nI_s I_0 \sin(\delta + \alpha) + I_0^2 \sin^2(\delta + \alpha)]^{1/2}}{I_s}$$

$$\approx \frac{nI_s + I_0 \sin(\delta + \alpha)}{I_s} \approx n + \frac{I_0}{I_s} \sin(\delta + \alpha) \qquad \text{...(10.13)}$$

Although only approximate, Eqn. 10.13 is sufficiently accurate for practically all purposes. The above theory is applicable to case when the secondary burden has a lagging power factor *i.e.*, when the burden is inductive which is normally the case.

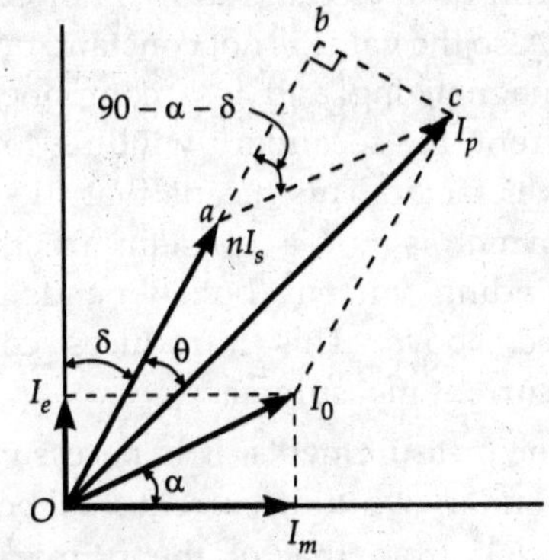


**Fig. 10.6** A section of phasor diagram of a C.T.

Equation 10.13 can be further expanded as :

$$R \approx n + \frac{I_0}{I_s}(\sin\delta\cos\alpha + \cos\delta\sin\alpha)$$

$$\approx n + \frac{I_m \sin\delta + I_e \cos\delta}{I_s} \qquad \text{...(10.14)}$$

as $I_m = I_0 \cos\alpha$ and $I_e = I_0 \sin\alpha$.

***Phase angle.*** The angle by which the secondary current phasor, when reversed, differs in phase from the primary current, is known as the *phase angle* of the transformer.

This angle is taken to be +ve if the secondary current reversed leads the primary current. The angle is taken as –ve if secondary current reversed lags behind the primary current.

The angle between $I_s$, reversed and $I_p$ is θ. Therefore, the phase angle is θ.

From the phasor diagram (Fig. 10.6)

$$\tan\theta = \frac{bc}{ob} = \frac{bc}{oa + ab} = \frac{I_0 \cos(\delta + \alpha)}{nI_s + I_0 \sin(\delta + \alpha)}$$

As θ is very small, we can write

$$\theta = \frac{I_0 \cos(\delta + \alpha)}{nI_s + I_0 \sin(\delta + \alpha)} \text{ rad.} \qquad \text{...(10.15)}$$

Now $I_0$ is very small as compared to $nI_s$, and, therefore, we can neglect the term $I_0 \sin(\delta + \alpha)$

$$\theta \approx \frac{I_0 \cos(\delta + \alpha)}{nI_s} \text{ rad} \qquad \text{...(10.16)}$$

$$\approx \frac{I_0 \cos\delta\cos\alpha - I_0 \sin\delta\sin\alpha}{nI_s}$$

$$\approx \frac{I_m \cos\delta - I_e \sin\delta}{nI_s} \text{ rad} \qquad \text{...(10.17)}$$

$$\approx \frac{180}{\pi}\left(\frac{I_m \cos\delta - I_e \sin\delta}{nI_s}\right) \text{ degree} \qquad \text{...(10.18)}$$

### 10.5.2 Errors in Current Transformers

It is clear from Eqn. 10.14 that the value of transformation ratio (actual ratio) is not equal to the turns ratio. Also the value is not constant, but depends upon the magnetizing and loss components of the exciting current, the secondary winding load current and its power factor. This means that the secondary winding current is not a constant fraction of the primary winding current but depends upon the factors listed above. This introduces considerable errors into current measurements.

In power measurements, it is necessary that the phase of secondary winding current shall be displaced by exactly 180° from that of the primary winding current. It is seen that the phase difference is different from 180° by an angle θ. Thus in power measurements, owing to use of C.T. two types of errors are introduced ; one due to actual transformation ratio being different from the turns ratio and the other due to secondary winding current not being 180° out of phase with the primary winding current.

***Ratio error.*** Ratio Error is defined as :

Percentage ratio error

$$= \frac{\text{nominal ratio} - \text{actual ratio}}{\text{actual ratio}} \times 100$$

$$= \frac{K_n - R}{R} \times 100 \qquad \text{...(10.19)}$$

***Phase angle error.*** Phase angle

$$\theta = \frac{180}{\pi}\left[\frac{I_m \cos\delta - I_e \sin\delta}{nI_s}\right] \text{degree.}$$

***Approximate formulae for errors.*** The usual instrument burden is largely resistive with some inductance and, therefore, δ is positive and is generally small.

Hence $\sin\delta \approx 0$ and $\cos\delta \approx 1$. Therefore, we can write Eqns. 10.14 and 10.18 as :

$$R \approx n + \frac{I_e}{I_s} \qquad \text{...(10.20)}$$

and

$$\theta \approx \left(\frac{180}{\pi}\right)\left(\frac{I_m}{nI_s}\right) \text{degree} \qquad \text{...(10.21)}$$

But $I_p \approx nI_s$ and, therefore, Eqns. 10.20 and 10.21 can be rewritten as

$$R \approx n + nI_e / I_p = n(1 + I_e / I_p) \qquad \text{...(10.22)}$$

and

$$\theta = \left(\frac{180}{\pi}\right)\left(\frac{I_m}{I_p}\right) \text{degree} \qquad \text{...(10.23)}$$

### 10.5.3 Characteristics of Current Transformers

1. ***Effect of P.F. of secondary winding burden on errors–ratio error.*** It is observed that for all inductive burdens the secondary winding current, $I_s$, lags behind the secondary induced voltage, $E_s$, so that δ is positive. Under these conditions the actual transformation ratio is always greater than the turns ratio (see Eqn. 10.13). For burdens which are sufficiently capacitive $I_s$ leads $E_s$ and so δ is negative. Under these conditions, the actual transformation ratio decreases becoming less than the turns ratio for values of δ approaching – 90°.

*Phase angle.* Examining Eqn. 10.16, we find that for inductive burdens, phase angle θ is positive for small values of δ (high secondary p.f.) but becomes negative as the secondary burden becomes more inductive and δ approaches 90°. For negative values of δ (sufficiently capacitive burdens) δ is always positive.

The variation of transformation ratio R and phase angle θ with δ is shown in Fig. 10.7. These conclusions are based upon the assumption that the magnitude of secondary impedance remains constant.

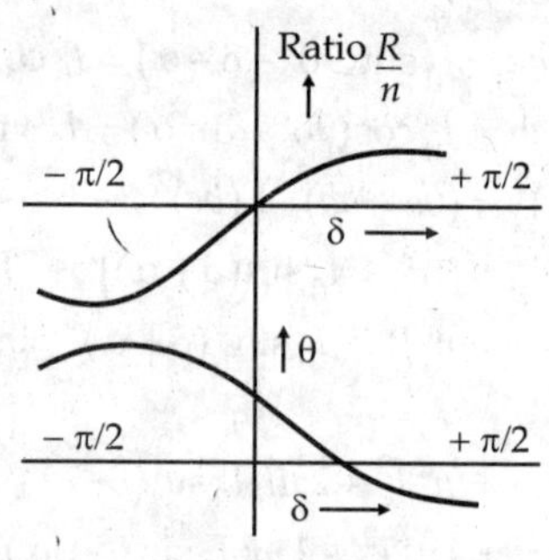


**Fig. 10.7** Variation of transformation ratio and nhase exciting angle with p.f. of secondary winding circuit.

2. ***Effect of change of primary winding current.*** If the primary winding current changes the secondary winding current changes proportionately. At low values of current $I_p$ (or $I_s$) current $I_m$ and loss component $I_e$ are a greater proportion of $I_p$ and, therefore, the errors are greater. As the current $I_p$ increases, there is an increase in $I_s$ and there is a decrease in ratio error and phase angle. This is clear from Eqns. 10.22 and 10.23. The variation of ratio error and phase angle with secondary winding current is shown in Figs. 10.8 and 10.9.

3. ***Effect of change in secondary winding circuit burden.*** An increase in secondary winding circuit burden impedance means an increase in volt ampere rating. This necessitates an increase in the secondary

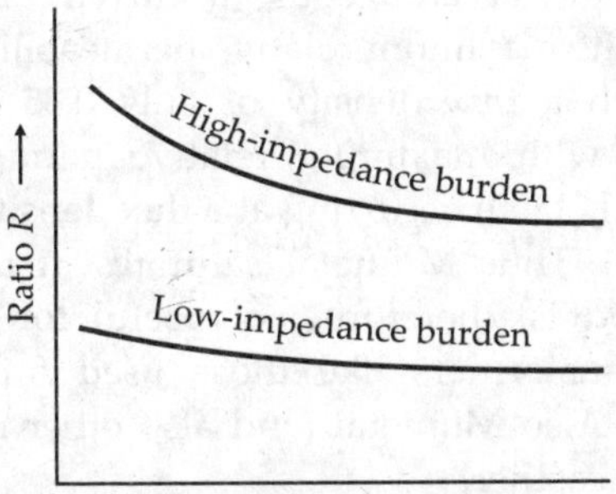


**Fig. 10.8** Variation of ratio $R$ with secondary winding current.

winding induced voltage which can be generated by an increased flux and flux density. Therefore both magnetizing component $I_m$ and loss component $I_e$ are increased. Thus it is expected that errors will increase with increase in secondary winding burden. In general, a greater burden impedance not only increases the transformation ratio, but also shifts the phase angle between primary winding current and secondary winding current reversed to more positive values as shown in Figs. 10.8 and 10.9.

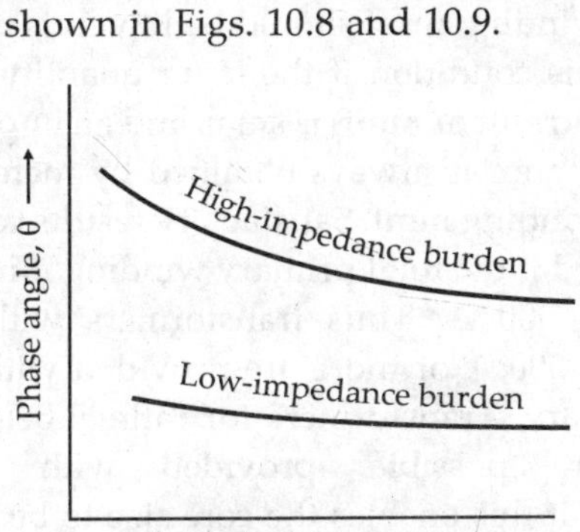


**Fig. 10.9** Variation of phase angle with secondary winding current.

4. ***Effect of change of frequency.*** The effect of increase in frequency will result in proportionate decrease in flux density. Thus, in general, the effect of increase in frequency is similar to that produced by decrease in impedance of secondary winding burden.

A current transformer is seldom used at a frequency which is very different from the one or which it is designed and, therefore, consideration of this effect is not very important.

### 10.5.4 Causes of Errors in Current Transformers

In an ideal transformer, the actual transformation ratio would be equal to the turns ratio and the phase angle would be zero. However, as a result of physical limitations inherent in electric and magnetic circuits of the transformer, there are departures from this ideal and consequently there are errors caused. The reasons are :

(*i*) There is some exciting mmf required by the primary winding to produce flux and, therefore, the transformer draws a magnetizing current $I_m$.

(*ii*) The transformer input must have a component which supplies the core losses (eddy current and hysteresis losses) and $I^2R$ losses of transformer windings due to flow of current $I_0$.

Therefore, loss component $I_e$ is required to feed the losses associated with the flux and also the associated copper loss in winding due to flow of $I_0$.

(*iii*) The flux density in the core is not a linear function of the magnetizing force, *i.e.*, the transformer core becomes saturated.

(*iv*) There is always a magnetic leakage and consequently the primary flux linkages are not equal to the secondary flux linkages.

### 10.5.5 Means to Reduce Errors in Current Transformers

It is clear from Eqns. 10.22 and 10.23 that for usual types of burdens, the difference between actual transformation ratio and the turns ratio depends largely on the loss component $I_e$ and the transformer phase angle depends largely on magnetizing current $I_m$. It is obvious that if the ratio has to be close to the turns ratio and the phase angle is to be small, $I_e$ and $I_m$ must be small as compared to $I_p$.

There are some design features which help us to minimize the errors and they are discussed below :

#### Design Features of Current Transformers

1. ***Core.*** In order to minimize the errors the magnetizing current $I_m$ and loss component $I_e$ must be kept to a low value. This means that the core must have a low reluctance and a low core loss. The reduction of reluctance of flux path can be brought about by using materials of high permeability, short magnetic paths, large cross-section of core and a low value of flux density. The current transformers are, in fact, designed for much lower flux densities than that are used for power transformers. This is especially important for current transformers used for protective relays which are frequently required to have a fair accuracy at currents many times the rated current (20 to 30 times the rated value), in order that the relay

10

operation may be correct in the event of a short circuit or a fault on the system, particularly when differential relaying schemes are used.

The number of joints in building up cores should be minimum a far as possible because joints produce air gaps which offer paths of high reluctance for the flux. The mmf consumed by joints can be reduced by properly lapping the joints and tightly binding the core. The core loss is reduced by choosing materials having low hysteresis and low eddy current losses, and by working the core at low flux densities.

Present-day magnetic materials used in current transformers are divided into three categories :

(*i*) Hot rolled silicon steel ;

(*ii*) Cold rolled grain oriented silicon steel ; and

(*iii*) Nickel iron alloys.

In current transformer practice hot rolled silicon steels (4% silicon) are used in a variety of forms. For ring type current transformers "ring" stampings are commonly used. For wound type, *T-U, L* or *E* and *I* stampings are used. In the highest grades of transformers, the core is built of ring shaped stampings stacked in cylindrical form as shown in Fig. 10.10. An alternative method employs cores that are made of strip wound in spiral form like a clock spring (Fig. 10.11). These are called *toroidal cores*. The latter method is much to be preferred when grain oriented magnetic materials are being used as it ensures that the flux path is always along the grains and hence there is minimum reluctance. Another advantage of spiral type of cores is that the joints are entirely eliminated.

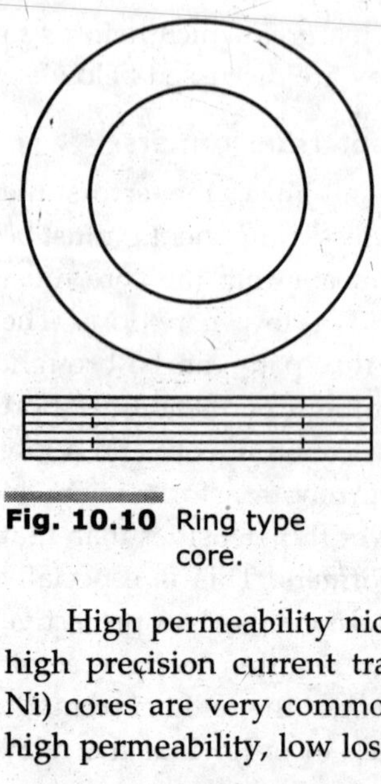

**Fig. 10.10** Ring type core.

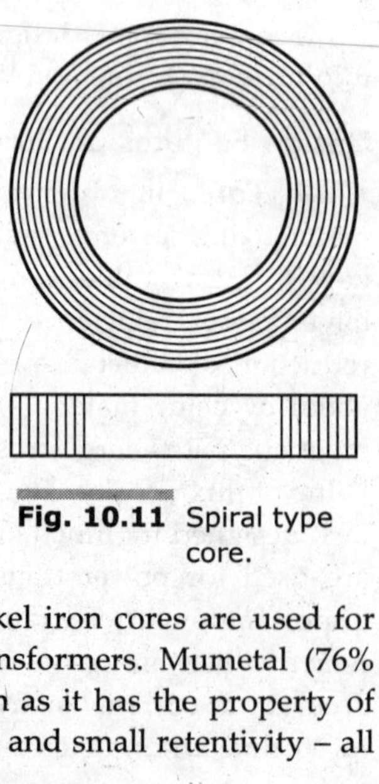

**Fig. 10.11** Spiral type core.

High permeability nickel iron cores are used for high precision current transformers. Mumetal (76% Ni) cores are very common as it has the property of high permeability, low loss and small retentivity – all of which are advantageous in current transformer work. But its maximum relative permeability (90,000) occurs with a flux density of only 0.35 Wb/m$^2$ as compared with maximum relative permeability of silicon steel (4500) occurring at a flux density of about 0.5 Wb/m$^2$. Thus Mumetal saturates at a low flux density and is, therefore, not useful for protective current transformers like those used for overload relays etc. Also Mumetal (and also other nickel iron alloys) are costlier.

Permendur (49% Co and 49% Fe) has the advantage of a very high saturation density of 2 to 4 Wb/m$^2$ as compared with 0.7 to 0.8 Wb/m$^2$ of other high permeability alloys.

Hipernik (50% Fe and 50% Ni) has high permeability at low flux densities and reasonable high saturation density and therefore it is frequently used for current transformers.

2. ***Primary winding current ratings.*** What ever equipment a C.T. is feeding, it is desirable that the ratio of exciting current to primary current should be small. This means that the ratio of excitation mmf to primary winding mmf should be low. It is difficult to achieve this condition if the latter quantity (primary winding current or mmf) is small and an improvement in performance is always obtained by increasing the primary winding mmf. Satisfactory results can usually be achieved if the total primary winding mmf at rated current is 500 A. Thus transformers with a rated current of 500 A or more are provided with a single turn primary. Transformers for ratings below 500 A are where possible, provided with multiturn windings, if this enables the core size to be reduced.

With the advent of improved magnetic materials and the development of methods for biasing the core to improve permeability, single turn primary winding can be used for even 100 A primary winding current.

3. ***Leakage reactance.*** Leakage reactance tends to increase ratio error. Therefore, the two windings, primary and secondary should be close together to reduce the secondary winding leakage reactance. Use of ring shaped cores around which toroidal windings are uniformly distributed also leads to low values of leakage reactance.

4. ***Turns compensation.*** We have, actual transformation ratio :

$$R = n + I_e / I_s \qquad \text{(Eqn. 10.20)}$$

Thus if we make the "nominal ratio" equal to the turns ratio the actual transformation ratio becomes more than the nominal ratio. Now if we reduce the turns ratio and keep the nominal ratio equal to the earlier value, the actual transformation ratio will be reduced. This would make actual transformation ratio nearly equal to the nominal ratio. Let us make it clear with the help of an example.

Let us consider a 1000/5 A current transformer with loss component equal to 0.6 percent of primary winding current.

Its nominal ratio, $K_n = 1000/5 = 200$.

Loss component, $I_e = (0.6/100) \times 1000 = 6$ A.

Let the number of primary turns, $N_p = 1$.

If the turns ratio is equal to the nominal ratio, we have $n = 200$.

$\therefore$ Secondary winding turns

$$N_s = nN_p = 200 \times 1 = 200$$

Actual ratio, $R = n + I_e / I_s = 200 + 6/5 = 201.2$.

Now suppose we do not use 200 turns for the secondary winding and instead use 199 turns.

Actual transformation ratio with turns compensation

$$R = n + I_e / I_s = 199 + 6/5 = 200/2.$$

Thus we find that by reducing the secondary winding turns slightly, the actual transformation ratio is made nearly equal to the nominal ratio.

Usually the best number of secondary winding turns is one or two less than the number which would make actual transformation ratio equal to nominal ratio of the transformer. The phase angle error is effected very little by a change of one or two turns in the secondary winding.

The correction by reduction in secondary winding turns, is exact only for a particular value of current and burden impedance. The C.T. in this case may be called *"compensated"*.

The errors can also be reduced by :

5. ***Use of shunts.*** If the secondary winding current is too large, it may be reduced by a shunt placed across the primary or the secondary winding. This method makes an exact correction only for a particular value and type of burden. It also reduces phase angle error.

6. ***Wilson compensation method.*** Reduction of one or two turns of the secondary winding, no doubt, reduces the ratio error, but it has no effect on the phase angle error. Also this method is too coarse a method for ratio adjustment and therefore we must use a method which exercises a finer control, say which is equivalent to reduction to a fraction of a turn. A compensated type of design was given by S. Wilson of the General Electric Company (Fig. 10.12). This method gives finer adjustments.

It employs a few turns of wire called auxiliary secondary turns passed through a hole in the core and connected in series with the secondary winding. A short-circuited turn is placed around one position of core to improve the phase relationships.

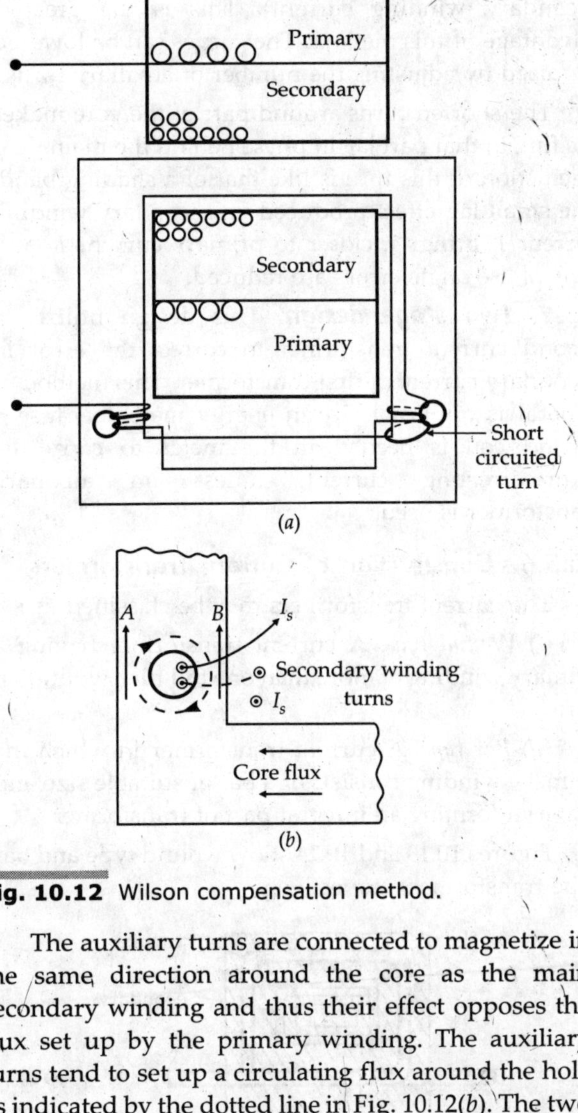


**Fig. 10.12** Wilson compensation method.

The auxiliary turns are connected to magnetize in the same direction around the core as the main secondary winding and thus their effect opposes the flux set up by the primary winding. The auxiliary turns tend to set up a circulating flux around the hole as indicated by the dotted line in Fig. 10.12(*b*). The two fluxws are addative in the section *A* of the core and subtractive in section *B*. At low flux densities the addition in flux is equal to the subtraction in flux. However, as the flux densities increase (with larger currents $I_p$ and $I_s$) section *A* tends to saturate, so that the flux in section *A* is no longer linearly proportional to the current and the increase in flux in this portion is less than linear proportional to the current. The action is equivalent to transforming some of the core flux to section *B* of the core, where it links with the auxiliary turns and gives the effect of increased secondary winding turns. An increase of secondary turns means a reduction of $I_s$ as compared with an uncompensated transformer causing an increase in ratio $R$. This action is needed to flatten the curves relating the ratio and phase angle with secondary current (Figs. 10.8 and 10.9) are flattened out so that the errors are practically constant (and of course known over a wide range of secondary winding current). This is the greatest advantage of this method. The curves can be lowered or raised by adjusting the number of auxiliary turns.

The shorted turns around part of the core makes the flux in that part lag in phase behind the main flux. The action of this turn is like that of a shading band. The small lag effect produced on secondary winding current $I_s$ brings it closer to primary current $I_p$ and thus phase angle errors are reduced.

7. ***Two stage design.*** This design utilizes a second current transformer to correct the error in secondary current of first transformer. This method in general is applicable to an energy meter because a second coil is needed in the meter to carry the error-correcting current, unless an auxiliary transformer is used.

### 10.5.6 Construction of Current Transformers

The current transformers may be classified as :

(*i*) *Wound type.* A current transformer having a primary winding of more than one full turn wound on core.

(*ii*) *Bar type.* A current transformer in which the primary winding consists of a bar of suitable size and material forming an integral part of transformer.

Figures 10.13 and 10.14 show wound type and bar type transformers respectively.

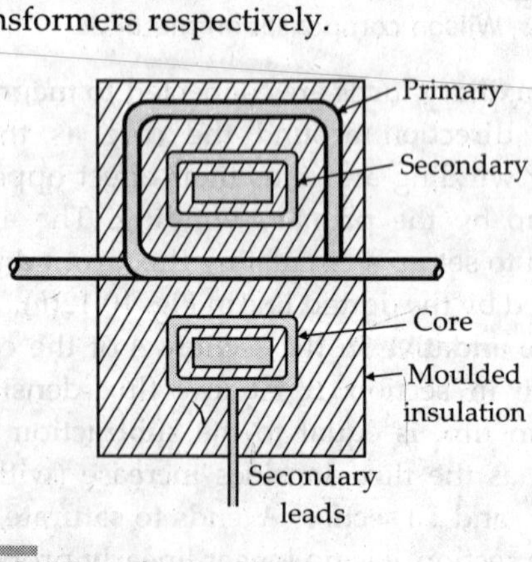


**Fig. 10.13** Wound type C.T.

The simplest form any current transformer can take, is the ring type or window type, examples of which are given in Fig. 10.15 which shows three commonly used shapes *i.e.,* stadium, circular and rectangular orifices. The core, if of a nickel-iron alloy

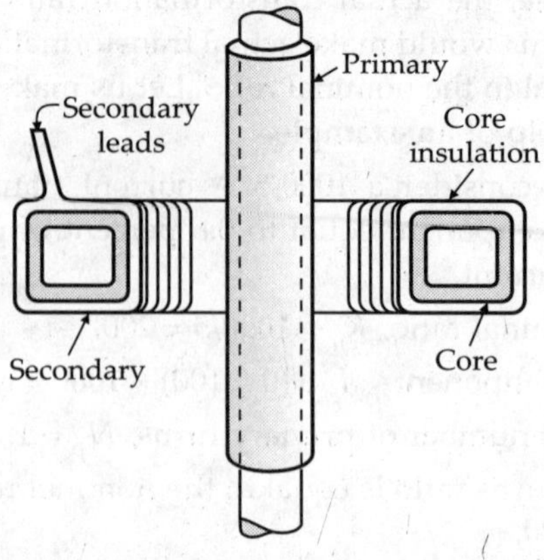


**Fig. 10.14** Bar type C.T.

or an oriented electrical steel is almost certainly of the continuously wound type. But current transformers using hot rolled steel will consist of stack of ring stampings. Before putting secondary winding on the core, the latter is insulated by means of end collars and circumferential wraps of elephantide or presspahn. These pressboards, in addition to acting as insulating medium, must also protect the secondary winding conductor from mechanical damage due to sharp comers. The secondary winding conductor is put on the core by a toroidal winding machine although hand winding is still frequently adopted if the number of secondary winding turns is small.

After the secondary winding has been placed on the core, the ring type transformer is completed by exterior taping with or without first applying exterior end collars and circumferential insulating wraps.

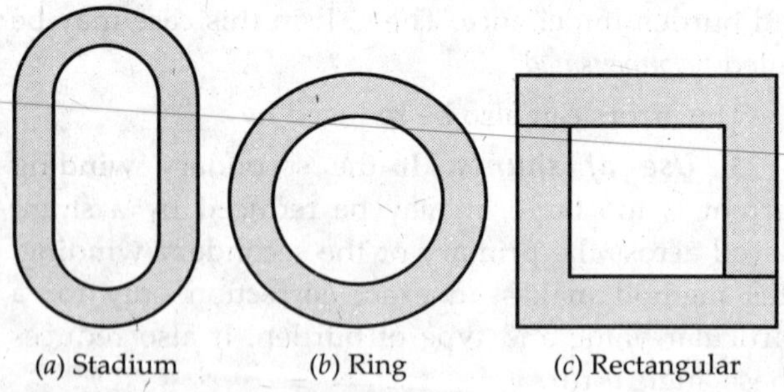


**Fig. 10.15** Stamping for window type C.Ts.

A near relative of the ring type current transformer is the so called "bushing type" transformer. This is, in fact, indistinguishable from of the ordinary ring type but the term is used when the current transformer fits over a fully insulated primary winding conductor such as over the oil end of a terminal bushing of a power transformer or an oil circuit beaker.

At very high voltages, the insulation of the current carrying conductor from the measuring circuit becomes an expensive problem. At 750 kV, cascaded current transformers are used or alternatively a coaxial shunt is used to modulate a radio frequency signal that is transmitted from the shunt placed in the high voltage line to receiving equipment on the ground, thereby overcoming the insulation problem. However. this type of system has severe limitation in its power output which has to be amplified in order to operate relays etc.

In a split core current transformer (Fig. 10.16), the core is split, each half having two finely ground or lapped gap faces. These current transformers are assembled on to the primary conductor "on site" for either permanent or temporary duty.

In a bar type current transformer, the core and secondary windings are the same as in a ring type transformer but the fully insulated bar conductor constituting the single turn primary winding is now an integral part of the current transformer. The insulation on the primary winding conductor may be bakelized paper tube or a resin directly moulded on the bar.

In a low voltage wound type current transformer the secondary winding is wound on a bakelite former or bobbin and the heavy primary conductor is either wound directly on top of secondary winding, suitable insulation being first applied over the secondary winding or the primary is wound entirely separately, taped with suitable insulating material and then assembled with the secondary winding on the core.

In the manufacture of current transformers assembly of lamination stacks dem ands somewhat greater core cross-section than ordinary transformers in order to keep down the reluctance of the inter-leaved corners as low as possible so as to minimize the magnetizing current. Sometimes cut cores are used.

Whenever possible secondary windings should utilize the whole available winding length on the core, the secondary winding turns being suitably spaced to accomplish this and the insulation between secondary winding and core and earth must be capable of withstanding the high peak voltages caused if the secondary winding is open-circuited when primary winding current is flowing. In the case of a large number of secondary winding turns, requiring more than one winding layer, the frequently adopted technique is to sectionalize the secondary winding so as to considerably reduce the peak voltage between layers.

With wound primary current transformers this particular problem is rarely met but it is of importance to try to obtain good relative positioning of primary and secondary winding coils, thus minimizing the axial forces on both coils caused by primary winding short circuit currents.

The windings should be close together to reduce the secondary winding leakage reactance as the leakage reactance increases the ratio error. Round copper wire of about 3 mm$^2$ area is frequently used for secondary windings rated at 5 A. Copper strip is used for primary winding, the dimensions of which depend upon the primary winding current.

When using bar type primary, the external diameter of the tube must be large enough to keep the voltage gradient, in the dielectric at its surface, to an acceptable value in order to avoid corona effect.

The windings must be designed to withstand, without damage, the large short circuit forces that are caused when a short circuit takes place on the system in which the current transformer is connected.

The windings are separately wound, and are insulated by tape and varnish for small line voltages. For voltages above 7 kV the transformers are oil immersed or compound filled.

### 10.5.7 Clamp on Ammeters

A current transformer with a single conductor is used in combination with a bridge rectifier and a d.c. milli-ammeter to produce a very useful service meter. The core of the transformer can be split with the help of a trigger switch (Fig. 10.16) and therefore the core can be clamped around a live conductor to measure the current. Thus this arrangement avoids the necessity of breaking the circuit in order that a current measuring device be inserted in series with it to

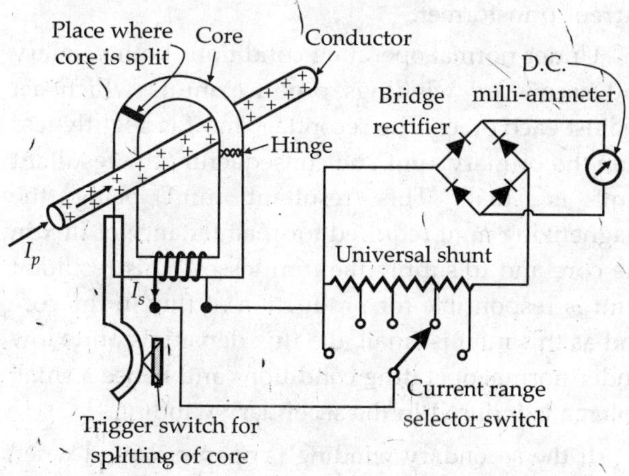


**Fig. 10.16** Clamp-on Ammeter.

measure the value of current flowing. By changing the shunt resistance of the milli-ammeter circuit ranges from 0 – 5 A to 0 – 600 A can be obtained.

The same milli-ammeter and rectifier are used with two external binding posts and a range selecting switch for a multi-range a.c. voltmeter.

This type of clamp-on transformer (or split core transformer) is used with recording ammeters. These transformers are designed for a voltage of 5 kV between the primary and the secondary windings. There are several variations of this clamp on transformer which are used for measuring maximum current in a line, real and reactive powers, and the power factor.

### 10.5.8 Effect of Secondary Open Circuit

Current transformers are always used with the secondary winding circuit closed through ammeters, wattmeter current coils, or relay coils. A precaution which should always be observed in using current transformers is the following :

*Never open the secondary winding circuit of a current transformer while its primary winding is energised.*

Failure to observe this precaution may lead to serious consequences both to the operating personnel and to the transformer. This is clear from the following :

The difference between a power transformer and a current transformer is that in a power transformer the current flowing in the primary winding is largely the reflection of that flowing in the secondary circuit, whereas in the current transformer, the primary winding is connected in series with the line whose current is being measured or indicated and this current is no way controlled or determined by the conditions of the secondary winding circuit of the current transformer.

Under normal operating conditions both primary and secondary windings produce mmfs which act against each other. The secondary mmf is slightly less than the primary mmf and consequently the resultant mmf is small. This resultant mmf being the magnetizing mmf required for maintenance of flux in the core and to supply the iron losses. This resultant mmf is responsible for production of flux in the core and as this mmf is small, the flux density is quite low under normal operating conditions and hence a small voltage is induced in the secondary winding.

If the secondary winding is open-circuited when the primary winding is carrying current, the primary winding mmf remains the same while the opposing secondary winding mmf reduces to zero. Therefore the resultant mmf is equal to the primary winding mmf $I_p N_p$ which is very large. This large mmf produces a large flux in the core till it saturates. This large flux linking the turns of the secondary winding, would induce a high voltage in the secondary winding which could be dangerous to the transformer insulation (although modem CTs are designed to withstand this voltage) and to the person who has opened the circuit. Also the eddy current and hysteresis losses would be very high under these conditions and due to this the transformer may be overheated and completely damaged. Even if it does not happen, the core may become permanently magnetized and this gives appreciable ratio and phase angle errors.

Most current transformers are provided with a short-circuiting link or a switch at the secondary winding terminals. If such a link is provided, it should always be closed before any change is made in the secondary winding circuit with primary winding excited. The secondary winding can safely be short-circuited since when used for measurement it is practically short-circuited as the impedance of the burden *i.e.*, an ammeter or a current coil of a wattmeter is very small.

### 10.5.9 Permanent Magnetization and Demagnetization of Cores of Current Transformers

The permanent magnetization of cores of current transformers may result from any of the following effects :

(*i*) When the secondary winding is open-circuited with primary winding energized. This causes a large magnetization force which produces a high value of flux density in the core. When this force is taken off, it leaves a large residual magnetism.

(*ii*) A switching transient may leave behind a residual magnetism.

(*iii*) Permanent magnetization may also result from d.c. current in either winding. This may be due to passage of d.c. current through the windings for finding resistance or for checking the polarity. Permanent magnetization may also be due to transient short circuit currents contained in the line in which it is connected. These currents contain d.c. current components in addition to a.c. currents.

The presence of permanent magnetization in the core of a C.T. may reduce its permeability at the flux densities at which it is normally operated and, therefore, results in increase of both its ratio and phase angle errors. Thus, for proper current transformer operation, this permanent magnetization should be removed and the transformer restored to its normal conditions.

***Demagnetization of core.*** There are several methods of demagnetization of core. The following *two* methods are usually adopted :

**1.** The first method is to pass a current through the primary winding equal to that which was passing when C.T. secondary winding open circuit took place. The secondary winding circuit is left open. The primary current is drawn from a motor alternator set. The motor is shut down with alternator field still excited. As the speed of the set falls the alternator voltage gradually falls to zero, and the core of the C.T. is passed through a large number of cycles of magnetization of gradually decreasing amplitude, finishing at zero magnetization.

**2.** In the second method, the primary winding is excited with full current and a very high variable resistance of several hundred ohms is connected across the secondary winding. This amounts to practically open circuiting the transformer. This resistance is then gradually reduced to zero as uniformly as possible. By this means the magnetization of the transformer core is reduced from a very high value to its normal value gradually.

***Example 10.1*** *A current transformer has a single turn primary and a 200 turns secondary winding. The secondary winding supplies a current of 5 A to a non-inductive burden of 1 Ω resistance. The requisite flux is set up in the core by an mmf of 80 A. The frequency is 50 Hz and the net cross-section of the core is 1000 mm².*

*Calculate the ratio and phase angle of the transformer. Also find the flux density in the core. Neglect the effects of magnetic leakage, iron losses and $I^2R$ losses.*

**Solution.** If we neglect the magnetic leakage, the secondary leakage reactance becomes zero. Therefore, the secondary burden is purely resistive and the impedance of burden is equal to the resistance of the secondary winding.

∴ Impedance of secondary circuit = 1 Ω.

Voltage induced in the secondary winding

$$E_s = \text{current in the secondary} \times \text{impedance of secondary winding}$$

$$= 5 \times 1 = 5 \text{ V.}$$

As the secondary burden is purely resistive, the secondary current is in phase with secondary induced voltage and the secondary power factor is unity or $\delta = 0$.

The loss component of no load current is to be neglected and, therefore,

$$I_e = 0.$$

∴ Exciting current

$$I_0 = \sqrt{I_m^2 + I_e^2} = I_m$$

$$= \text{magnetizing component and } \alpha = 0.$$

Magnetizing component of no load current

$$I_m = \frac{\text{magnetizing mmf}}{\text{primary winding turns}} = \frac{80}{1} = 80 \text{ A.}$$

Secondary winding current, $I_s = 5$ A.

Reflected secondary winding current

$$= n I_s = 200 \times 5 = 1000 \text{ A.}$$

The phasor diagram is shown in Fig. 10.17.

∴ Primary current

$$I_p = \sqrt{(nI_s)^2 + (I_m)^2}$$

$$= \sqrt{(1000)^2 + (80)^2}$$

$$= 1003.2 \text{ A.}$$

Actual transformation ratio

$$R = I_p / I_s = 1003.2/5$$

$$= 2000.64.$$

Phase angle

$$\theta = \tan^{-1} I_m / I_n I_s$$

$$= \tan^{-1} 80/1000°$$

$$= 4°34'$$

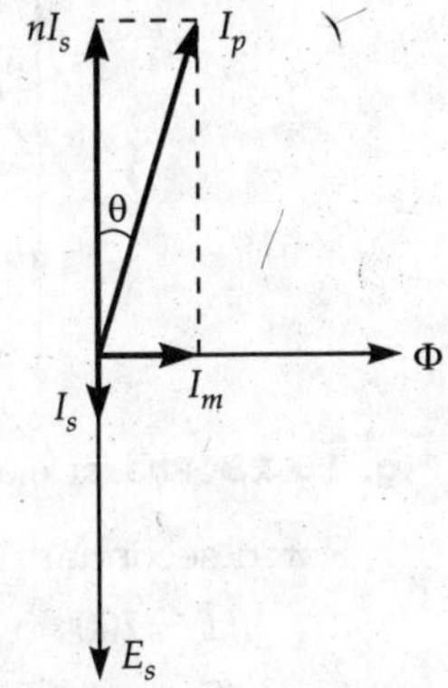


**Fig. 10.17** Phasor diagram of Example 10.1.

We know

$$E_s = 4.44 \, f \, \Phi_m \, N_s$$

∴ Maximum flux

$$\Phi_m = \frac{E_s}{4.44 \, f \, N_s}$$

$$= \frac{5}{4.44 \times 50 \times 200} = 0.1126 \times 10^{-3} \text{ Wb.}$$

Area of core $= 1000 \text{ mm}^2 = 1000 \times 10^{-6} \text{ m}^2$.

∴ Maximum flux density

$$B_m = \frac{0.1126 \times 10^{-3}}{1000 \times 10^{-6}} = 0.1126 \text{ Wb/m}^2.$$

***Example 10.2*** *The exciting current of a ring core current transformer, of ratio 1000/5 A when operating at full primary current and with a secondary burden of non-inductive resistance of 1 Ω is 1 A at a power factor of 0.4. Calculate :*

*(i) the phase displacement between primary and secondary winding currents ;*

*(ii) the ratio error at full load, assuming that there has been no compensation.*

**Solution.** The burden of secondary winding is purely resistive and, therefore, secondary winding p.f. is unity or $\delta = 0$.

The p.f. of exciting current is 0.4

$\therefore \quad \cos(90-\alpha) = 0.4 \quad \text{or} \quad \alpha = 23°35'$

Exciting current $I_0 = 1$ A.

$\therefore$ Nominal ratio $K_n = 1000/5 = 200$.

Since there is no turn compensation, the turns ratio is equal to nominal ratio or $n = K_n = 200$.

When the primary winding carries rated current of 1000 A, the secondary winding carries a current of 5 A.

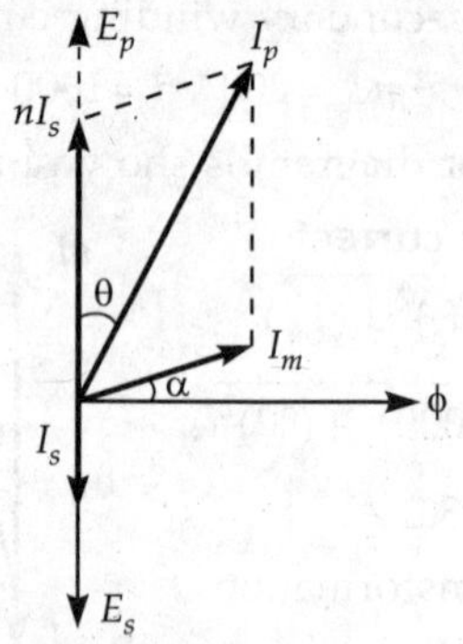


**Fig. 10.18** Phasor diagram of Example 10.2.

Rated secondary winding of current, $I_s = 5$ A

$\therefore \quad nI_s = 200 \times 5 = 1000$ A.

From Eqn. 10.13,

actual transformation ratio

$$R = n + \frac{I_0}{I_s}\sin(\delta+\alpha) = 200 + \frac{1}{5}\sin(0 + 23°35')$$

$$= 200.08.$$

$$\text{Ratio error} = \frac{\text{nominal ratio} - \text{actual ratio}}{\text{actual ratio}} \times 1000$$

$$= \frac{K_n - R}{R} \times 100 = \frac{200 - 200.08}{200.08} \times 100 = -0.04\%.$$

From Eqn. 10.16, phase angle,

$$\theta = \frac{180}{\pi}\left[\frac{I_0\cos(\delta+\alpha)}{nI_s}\right] = \frac{180}{\pi}\left[\frac{\cos 23°35'}{1000}\right]$$

$$= 0.0525° = 3'9''$$

***Example 10.3*** *A 1000/5 A, 50 Hz current transformer has a secondary burden comprising a non-inductive impedance of 1.6 Ω. The primary winding has one turn. Calculate the flux in the core and ratio error at full load. Neglect leakage reactance and assume the iron loss in the core to be 1.5 W at full load. The magnetizing mmF is 100 A.*

**Solution.** Nominal ratio, $K_n = 1000/5 = 200$.

The turns ratio is assumed equal to the nominal ratio in the absence of any other data.

$\therefore$ Turns ratio $n = K_n = 200$

Primary winding turns, $N_p = 1$

$\therefore$ Secondary winding turns, $N_s = nN_p = 200$.

Secondary impedance = 1.6 Ω.

$\therefore$ Secondary induced voltage, $E_s = 5 \times 1.6 = 8$ V

We have, $E_s = 4.44\, f\,\Phi_m\, N_s$.

or maximum flux

$$\Phi_m = \frac{E_s}{4.44\, f\, N_s} = \frac{8}{4.44 \times 50 \times 200}$$

$$= 180 \times 10^{-3} \text{ Wb.}$$

Primary induced voltage

$$E_p = \frac{E_s}{n} = \frac{8}{200} \text{ V.}$$

Loss component of exciting current

$$I_e = \frac{\text{iron loss}}{E_p} = \frac{1.5}{8/200} = 37.5 \text{ A.}$$

The magnetizing component

$$I_m = \frac{\text{magnetizing mmf}}{N_p} = \frac{100}{1} = 100 \text{ A.}$$

As the burden is purely resistive and, therefore, the secondary load angle is zero or $\delta = 0$.

From Eqn. 10.14,

$$\text{Actual ratio, } R = n + \frac{I_m\sin\delta + I_e\cos\delta}{I_s}$$

$$= 200 + \frac{37.5}{5} = 207.5$$

$$\text{Ratio error} = \frac{K_n - R}{R} \times 100 = \frac{200 - 207.5}{207.5} \times 100$$

$$= -3.61\ \%.$$

***Example 10.4*** *A current transformer with a bar primary has 300 turns in its secondary winding. The resistance and reactance of the secondary circuit are 1.5 Ω and 1.0 Ω respectively including the transformer winding. With 5 A flowing in the secondary winding, the magnetizing mmf is 100 A and the iron loss is 1.2 W. Determine the ratio and phase angle errors.*

**Solution.** Primary winding turns, $N_p = 1$,

Secondary winding turns, $N_s = 300$.

$\therefore$ Turns ratio, $n = N_s / N_p = 300$.

Secondary circuit burden impedance

$$= \sqrt{(1.5)^2 + (1.0)^2} = 1.8\ \Omega.$$

For secondary winding circuit :

$$\cos\delta = 1.5/1.8 = 0.833 \text{ and } \sin\delta = 1.0/1.8 = 0.555.$$

Secondary induced voltage

$$E_s = 5\times1.8 = 9.0\,\text{V}.$$

Primary induced voltage

$$E_p = E_s/9 = 9/300 = 0.03\ \text{V}.$$

Loss component of current referred to primary winding

$$I_e = \frac{\text{iron loss}}{E} = \frac{1.2}{0.03} = 40\ \text{A}.$$

Magnetizing current,

$$I_m = \frac{\text{magnetizing mmf}}{\text{primary winding turns}} = \frac{100}{1} = 100\ \text{A}.$$

Actual ratio,

$$R = n + \frac{I_e\cos\delta + I_m\sin\delta}{I_s}$$

$$= 300 + \frac{40\times0.833 + 100\times0.555}{5} = 317.6.$$

In the absence of any information to the contrary we can take nominal ratio to be equal to the turns ratio,

or $\quad K_n = n = 300.$

Percentage ratio error

$$= \frac{K_n - R}{R}\times100 = \frac{300-317.6}{317.6} = -5.54\%.$$

From Eqn. 10.18, we have

$$\text{Phase angle, } \theta = \frac{180}{\pi}\left(\frac{I_m\cos\delta - I_e\sin\delta}{nI_s}\right)$$

$$= \frac{180}{\pi}\left(\frac{100\times0.833 - 40\times0.555}{300\times5}\right) = 2.34°.$$

**Example 10.5** *A current transformer has a bar primary and 200 secondary winding turns. The secondary winding burden is an ammeter of resistance 1.2 Ω and reactance 0.5 Ω, the secondary winding has a resistance of 0.2 Ω and reactance 0.3 Ω. The core requires the equivalent of an mmf of 100 A for magnetization and 50 A for core losses.*

(i) *Find the primary winding current and ratio error when the ammeter in the secondary winding circuit indicates 5 A.*

(ii) *How many turns could be reduced in the secondary winding in order that the ratio error be zero for this condition ?*

**Solution.** Total secondary circuit resistance

$$= 1.2 + 0.2 = 1.4\ \Omega$$

Total secondary circuit reactance

$$= 0.5 + 0.3 = 0.8\ \Omega$$

Secondary circuit phase angle

$$\delta = \tan^{-1}0.8/1.4 = 29°42'$$

or $\quad \cos\delta = 0.8686$ and $\sin\delta = 0.4955.$

Primary winding turns, $N_p = 1,$

Secondary winding turns, $N_s = 200.$

$\therefore$ Turns ratio, $\quad n = 200.$

Magnetizing current

$$I_m = \frac{\text{magnetizing mmf}}{\text{primary turns}} = \frac{100}{1} = 100\ \text{A}.$$

Loss component

$$I_e = \frac{\text{mmf equivalent to iron loss}}{\text{primary winding turns}} = \frac{50}{1} = 50\ \text{A}.$$

$\therefore$ Actual ratio

$$R = n + \frac{I_e\cos\delta + I_m\sin\delta}{I_s}$$

$$= 200 + \frac{50\times0.8686 + 100\times0.4955}{5} = 218.6$$

Primary current

$$I_p = \text{actual transformation ratio} \times \text{secondary current}$$

$$= 218.6\times5 = 1093\ \text{A}.$$

(*ii*) In order to eliminate the ratio error, we must reduce the secondary winding turns or in other words we must reduce the turns ratio.

The nominal ratio is 200 and therefore for zero ratio error the actual transformation ratio should be equal to the nominal ratio,

Nominal ratio, $K_n = 200.$

Actual ratio, $\quad R = n + \dfrac{I_e\cos\delta + I_m\sin\delta}{I_s}$

$\therefore$ For zero ratio error

$$K_n = R.$$

or $\quad 200 = n + \dfrac{50\times0.8686 + 100\times0.4955}{5} = n + 18.6.$

Turns ratio, $n = 181.4$

Hence secondary winding turns

$$N_s = nN_p = 181.4\times1 = 181.4.$$

Reduction in secondary winding turns

$$= 200 - 181.4 \approx 19$$

**Example 10.6** *The primary winding exciting current of a current transformer with a bar primary, nominal ratio 100/1, operating on an external burden of 1.6 Ω non-inductive, the secondary winding resistance being 0.2 Ω, is 1.9 A, lagging 40.6° to the secondary voltage reversed there being 100 secondary turns. With 1 A flowing in the secondary winding, calculate (i) the actual ratio of primary winding current to secondary winding current, (ii) the phase angle between them in minutes.*

**Solution.** Primary winding turns $N_p = 1.$

Secondary winding turns $N_s = 100.$

$\therefore$ Turn ratio, $\quad n = N_s/N_p = 100.$

10

Total secondary circuit resistance

$= 1.6 + 0.2 = 1.8\,\Omega.$

Secondary circuit reactance = 0.

∴ The secondary p.f. is unity and its phase angle

$\delta = 0.$

$\alpha$ = angle between $I_0$ and $\Phi = 90° -$ (angle between $I_0$ and $E_s$ reversed) $= 90° - 40.6° = 49.4°$

∴ $\cos\alpha = 0.65$ and $\sin\alpha = 0.76.$

Actual ratio,

$$R = n + \frac{I_0}{I_s}\sin(\delta+\alpha) \quad \text{(see Eqn. 10.13)}$$

$$= 100 + \frac{1.9}{1.0}\sin(0° + 49.4°) = 101.44.$$

Phase angle,

$$\theta = \frac{180}{\pi}\cdot\frac{I_0\cos(\delta+\alpha)}{nI_s}\ \text{degree}$$

$$= \frac{180}{\pi}\times 1.9 \times \frac{\cos 49.4°}{100\times 1} = 0.708° = 42.5'.$$

***Example 10.7*** *A 100/5 A, 50 Hz current transformer has a bar primary and a rated secondary burden of 12.5 VA. The secondary winding has 196 turns and a leakage inductance of 0.96 mH. With a purely resistive burden at rated full load, the magnetization mmf is 16 A and the loss excitation requires 12 A. Find the ratio and phase angle errors.*

**Solution.** Secondary burden = 12.5 VA

Secondary winding current, $I_s = 5$ A

∴ Secondary circuit impedance $= 12.5/(5)^2 = 0.5\,\Omega$

Secondary circuit reactance

$= 2\pi \times 50 \times 96 \times 10^{-3} = 0.3\,\Omega.$

Phase angle of secondary circuit

$\delta = \sin^{-1} 0.3/0.5 = \sin^{-1} 0.6$

∴ $\sin\delta = 0.6$ and $\cos\delta = \sqrt{(1)^2 - (0.6)^2} = 0.8.$

Primary winding turns, $N_p = 1$

Secondary winding turns, $N_s = 196$

Turn ratio, $n = N_s / N_p = 196$

Nominal ratio, $K_n = 1000/5 = 200$

Magnetizing current

$$I_m = \frac{\text{magnetizing mmf}}{\text{primary winding turns}} = \frac{16}{1} = 16\text{ A}$$

Loss component

$$I_e = \frac{\text{excitation for loss}}{\text{primary winding turns}} = \frac{12}{1} = 12\text{ A}$$

Actual ratio

$$R = n + \frac{I_e\cos\delta + I_m\sin\delta}{I_s}$$

$$= 196 + \frac{12\times 0.8 + 16 + 0.6}{5} = 199.84$$

$$\text{Ratio error} = \frac{\text{nominal ratio} - \text{actual ratio}}{\text{actual ratio}} \times 100$$

$$= \frac{200 - 199.84}{199.84}\times 100 = +0.08\%.$$

Phase angle

$$\theta = \frac{180}{\pi}\left[\frac{I_m\cos\delta - I_e\sin\delta}{nI_s}\right] = 180\left[\frac{16\times 0.8 - 12\times 0.6}{196\times 5}\right]$$

$= 0.327° = 19.6'.$

***Example 10.8*** *A current transformer of turns ratio 1 : 199 is rated as 1000/5 A, 25 VA. The core loss and magnetizing component of the primary current are 4 and 7 A under rated conditions. Determine the phase angle and ratio errors for the rated burden and rated secondary current at 0.8 p.f. lagging and 0.8 p.f. leading. Neglect the resistance and leakage resistance of secondary winding.*

**Solution.** Primary winding turns, $N_p = 1$

Secondary winding turns, $N_s = 199$

∴ Turns ratio, $n = 199.$

Nominal ratio, $K_n = 1000/5 = 200$

Magnetizing current, $I_m = 7$ A

Loss component, $I_e = 4$ A.

*0.8. P.F. Lagging :*

For lagging p.f. the secondary phase angle $\delta$ is positive.

$\cos\delta$ = secondary p.f. = 0.8

$\sin\delta = \sqrt{(1)^2 - (0.8)^2} = 0.6$

From Eqn. 10.14, actual ratio,

$$R = n + \frac{I_e\cos\delta + I_m\sin\delta}{I_s}$$

$$= 199 + \frac{4\times 0.8 + 7\times 0.6}{5} = 200.48$$

∴ Ratio error

$$= \frac{K_n - R}{R}\times 100 = \frac{200 - 200.48}{200.48}\times 100 = -0.24\%$$

and Phase angle

$$\theta = \frac{180}{\pi}\left[\frac{I_m\cos\delta - I_e\sin\delta}{nI_s}\right]$$

$$= \frac{180}{\pi}\left[\frac{7\times 0.8 - 4\times 0.6}{199\times 5}\right] = +0.185° = +11'$$

*0.8. P.F. Leading :*

For leading p.f. $\delta$ is negative.

$\cos\delta = 0.8$ and $\sin\delta = -0.6$

$$\text{Actual ratio}\quad R = n\,\frac{I_e\cos\delta + I_m\sin\delta}{I_s}$$

$$= 199 + \frac{4\times 0.8 - 7\times 0.6}{5} = 198.8$$

∴ Ratio error

$$= \frac{K_n - R}{R} \times 100 = \frac{200 - 198.8}{198.8} \times 100 = +0.603\%$$

and Phase angle

$$\theta = \frac{180}{\pi} \frac{I_m \cos\delta - I_e \sin\delta}{nI_s}$$

$$= \frac{180}{\pi}\left[\frac{7\times0.8 - 4\times0.6}{199\times5}\right] = 0.46° = 27°6'.$$

***Example 10.9*** *A current transformer having a 1 turn primary is rated at 500/5 A, 50 Hz with an output of 15 VA. At rated load with non-inductive burden, the inphase and quadrature components (referred to the flux) of the exciting mmf are 8 and 10 A respectively. 'The number of turns in the secondary is 98, and the resistance and leakage reactance of the secondary winding are 0.35 Ω and 0.3 Ω respectively. Calculate the current and phase angle error.*

**Solution.** Primary winding turns, $N_p = 1$

Secondary winding turns, $N_s = 98$

∴ Turns ratio, $n = 98$

Nominal ratio, $K_n = 500/5 = 100$

Magnetizing current component is in phase while the loss component is in quadrature with the flux.

∴ Magnetizing mmf = 8 A

Mmf equivalent to loss = 10 A

Magnetizing current

$$I_m = \frac{\text{magnetizing mmf}}{\text{primary winding turns}} = \frac{8}{1} = 8 \text{ A}$$

Loss component

$$I_e = \frac{\text{loss mmf}}{\text{primary winding turns}} = \frac{10}{1} = 10 \text{ A}$$

Output volt-ampere, VA = 15.

Impedance of secondary load burden

$= \text{VA}/I_s^2 = 15/(5)^2 = 0.6\ \Omega$

It is given that the external burden is purely resistive

∴ Resistance of external burden = 0.6 Ω

Reactance of external burden = 0 Ω

Resistance of total secondary circuit burden

$= 0.6 + 0.35 = 0.95\ \Omega$

Reactance of total secondary circuit burden

$= 0 + 0.3 = 0.3\Omega$

Secondary phase angle $\delta = \tan^{-1} 0.3/0.95 = 17°30'$

So, $\cos\delta = 0.95$ and $\sin\delta = 0.3$

$$\text{Actual ratio, } R = \frac{I_e \cos\delta + I_m \sin\delta}{I_s}$$

$$= 98 + \frac{10\times0.95 + 8\times0.3}{5} = 100.38$$

Ratio (current) error

$$= \frac{100 - 100.38}{100.38} \times 100 = -0.38\%$$

Phase angle

$$\theta = \frac{180}{\pi}\left[\frac{I_m \cos\delta - I_e \sin\delta}{nI_s}\right]$$

$$= \frac{180}{\pi}\left[\frac{8\times0.95 - 10\times0.3}{98\times5}\right] = 0.537°.$$

***Example 10.10*** *At its rated load of 25 VA, a 100/5 A current transformer has an iron loss of 0.2 W and a magnetizing current of 1.5 A. Calculate its ratio error and phase angle when supplying rated output to a meter having a ratio of resistance to reactance of 5.*

**Solution.** Nominal ratio, $K_n = 100/5 = 20$

In the absence of any other data, the turns ratio is taken equal to the nominal ratio, $n = 20$.

Neglecting the burden of the secondary winding, the total burden of the secondary circuit is equal to the burden of the meter.

∴ Burden of secondary circuit = 25 VA

Voltage across primary winding

$$E_p = \text{VA}/I_p = 25/100 = 0.25 \text{ V}$$

Loss component of current

$$I_e = \text{iron loss}/E_p = 0.2/0.25 = 0.8 \text{ A}$$

Magnetising current, $I_m = 1.5$ A

Secondary circuit phase angle,

$$\delta = \tan^{-1}\frac{X}{R} \tan^{-1}\frac{1}{5} = 11°18'$$

∴ $\cos\delta = 0.98$ and $\sin\delta = 0.196$

Actual ratio

$$R = n + \frac{I_e \cos\delta + I_m \sin\delta}{I_s}$$

$$= 20 + \frac{0.8\times0.98 + 1.5\times0.196}{5} = 20.215$$

$$\text{Ratio error} = \frac{K_n - R}{R} \times 100$$

$$= \frac{20 - 20.215}{20.215} \times 100 = -1.075\%$$

Phase angle

$$\theta = \frac{180}{\pi}\left[\frac{I_m \cos\delta - I_e \sin\delta}{nI_s}\right]$$

$$= \frac{180}{\pi}\left[\frac{15\times0.98 - 0.8\times0.196}{20\times5}\right] = 0.75°$$

**Example 10.11** *The resistance and leakage reactance of the secondary winding of a 500/5 A current transformer are 0.02 and 0.03 Ω respectively. The, magnetization characteristics of the transformer are as follows :*

| Secondary emf, V | 0.25 | 0.5 | 1.0 | 1.5 | 2.5 |
|---|---|---|---|---|---|
| Primary magnetizing component, A | 0.9 | 1.5 | 2.5 | 3.2 | 4.0 |
| Primary core loss component, A | 0.5 | 1.3 | 2.7 | 3.9 | 4.6 |

*An ammeter and a wattmeter current coil are connected in series with the secondary winding. Their respective resistances are 0.04 and 0.06 Ω and reactances 0.06 and 0.03 Ω. Calculate the ratio and phase angle errors of the C.T. when a current of 5 A passes through the instruments, if there is no turns compensation. What should be the turns ratio to minimize the ratio error for this buraen ?*

**Solution.** Nominal ratio, $K_n = 500/5 = 100$

With no turn compensation, the turns ratio is equal to the nominal ratio

or $\qquad n = 100$

Total resistance of secondary circuit

$$= 0.02 + 0.04 + 0.06 = 0.12\ \Omega$$

Total reactance of secondary circuit

$$= 0.03 + 0.06 + 0.03 = 0.12\ \Omega$$

Secondary phase angle

$$\delta = \tan^{-1}\frac{X}{R} = \tan^{-1}\frac{0.12}{0.12} = 45°.$$

$\therefore \qquad \cos\delta = 0.707$ and $\sin\delta = 0.707$.

Total secondary circuit impedance

$$= \sqrt{(0.12)^2 + (0.12)^2} = 0.17\ \Omega.$$

Secondary winding induced voltage

$$E_s = 5 \times 0.17 = 0.85\ \text{V}.$$

The characteristics of the current transformer are plotted in Fig. 10.19.

Corresponding to $E_s = 0.85$, we have from Fig. 10.19,

$$I_e = 24\ \text{A} \ \text{and} \ I_m = 2.26\ \text{A}.$$

Actual ratio

$$R = n + \frac{I_e \cos\delta + I_m \sin\delta}{I_s}$$

$$= 100 + \frac{2.4 \times 0.707 + 2.26 \times 0.707}{5}$$

$$= 100 + 0.66 = 100.66$$

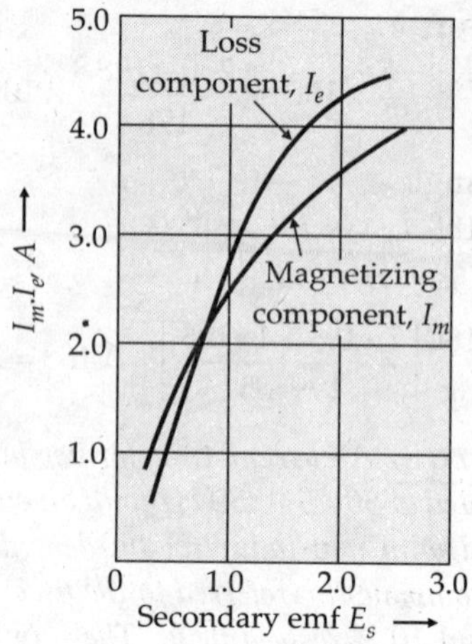


**Fig. 10.19** Diagram of Example 10.11.

Ratio error

$$= \frac{K_n - R}{R} \times 100 = \frac{100 - 100.66}{100.66} \times 100 = -0.656\%.$$

Phase angle

$$= \frac{180}{\pi}\left[\frac{I_m \cos\delta - I_e \sin\delta}{nI_s}\right]$$

$$= \frac{180}{\pi}\left[\frac{2.26 \times 0.707 - 2.4 \times 0.707}{5 \times 100}\right] \text{degree} = 0.7'$$

The ratio error can be minimized by making turns ratio, $n = 99$.

Actual ratio with, $n = 99$ is, $R = 99 + 0.66 = 99.66$

$$\therefore \ \text{Error} = \frac{100 - 99.66}{99.66} \times 100 = 0.36\%.$$

## 10.6 POTENTIAL TRANSFORMERS

Potential transformers are used to operate voltmeters, the potential coils of wattmeters and relays from high voltage lines. The primary winding of the transformer is connected across the line carrying the voltage to be measured and the voltage circuit is connected across the secondary winding.

The design of a potential transformer is quite similar to that of a power transformer but the loading of a potential transformer is always small, sometimes only a few volt-ampere. The secondary winding is designed so that a voltage of 100 to 120 V is delivered to the instrument load. The normal secondary voltage rating is 110 V.

### 10.6.1 Difference between C.T. and P.T.

There are a few differences in the operation of a current transformer and a potential transformer.

(*i*) The potential transformer may be considered as 'parallel' transformer with its secondary winding operating nearly under open circuit conditions whereas the current transformer may be thought as a 'series' transformer under virtual short circuit conditions. Thus the secondary winding of a P.T. can be open-circuited without any damage being caused either to the operator or to the transformer.

(*ii*) The primary winding current in a C.T. is independent of the secondary winding circuit conditions while the primary winding current in a P.T. certainly depends upon the secondary circuit burden.

(*iii*) In a potential transformer, full line voltage is impressed upon its terminals whereas a C.T. is connected in series with one line and a small voltage exists across its terminals. However, the C.T. carries the full line current.

(*iv*) Under normal operation the line voltage is nearly constant and, therefore, the flux density and hence the exciting current of a potential transformer varies only over a restricted range whereas the primary winding current and excitation of a C.T. vary over wide limits in normal operation.

### 10.6.2 Relationships in a Potential Transformer

The theory of a potential transformer is essentially the same as that of a power transformer. The main point of difference is that the power loading of a P.T. is very small and consequently the exciting current is of the same order as the secondary winding current while in a power transformer the exciting current is a very small fraction of secondary winding load current.

Figures 10.20 and 10.21 show the equivalent circuit and phasor diagram respectively of a potential transformer.

$\Phi$ = working flux,

$I_m$ = magnetizing component of no load (exciting) current,

$I_e$ = iron loss component of no load (exciting) current,

$I_0$ = no load (exciting) current,

$E_s$ = secondary winding induced voltage,

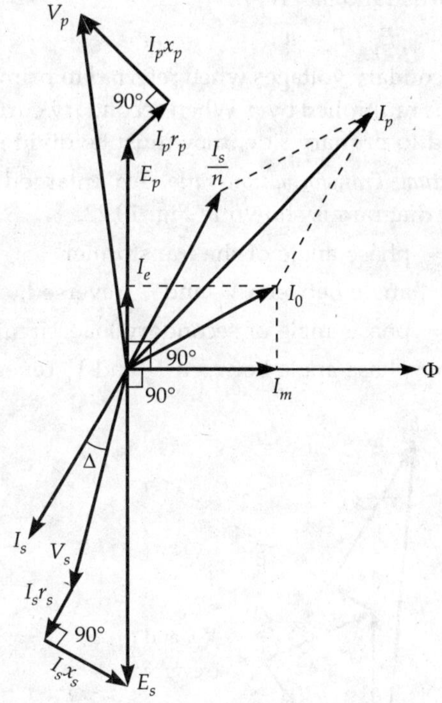


**Fig. 10.21** Phasor diagram of a potential transformer.

$V_s$ = secondary winding terminal voltage,

$N_p$ = primary winding turns,

$N_s$ = secondary winding turns,

$I_s$ = secondary winding current,

$r_s$ = resistance of secondary winding,

$x_s$ = reactance of secondary winding,

$r_e$ = resistance of secondary load circuit,

$x_e$ = reactance of secondary load circuit.

$\Delta$ = phase angle of secondary load circuit $= \tan^{-1} x_e / r_e$

$E_p$ = primary winding induced voltage,

$I_p$ = primary winding current,

$r_p$ = resistance of primary winding,

$x_p$ = reactance of primary winding.

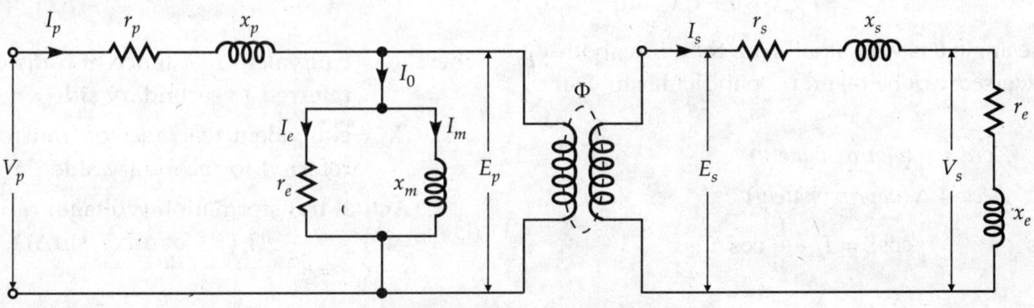


**Fig. 10.20** Equivalent circuit of a potential transformer.

Turns ratio, $n = N_p / N_s$

$\therefore \quad E_p / E_s = n$

Secondary voltages when referred to primary side are to be multiplied by *n*. When secondary currents are referred to primary side, they must be divided by *n*.

*Actual Transformation ratio.* An enlarged concise phasor diagram is shown in Fig. 10.22.

$\theta$ = phase angle of the transformer
= angle between $V_p$ and $V_s$ reversed,

$\Delta$ = phase angle of secondary load circuit,

$\beta$ = phase angle between $I_p$ and $V_s$ reversed,

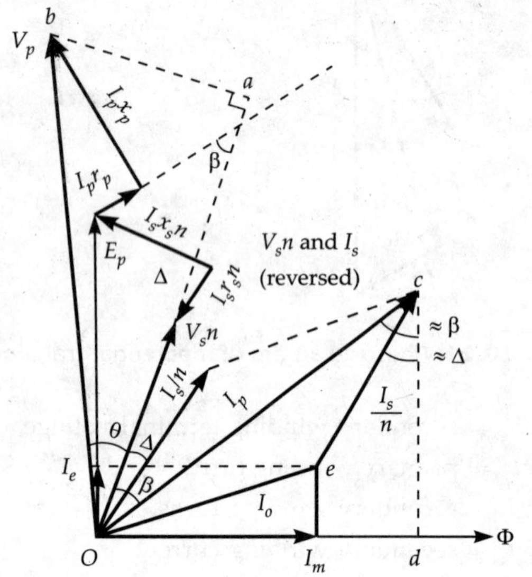


**Fig. 10.22** Enlarged and concise phasor diagram of a potential transformer.

Now, $oa = V_p \cos\theta$

From the phasor diagram

$$oa = nV_s + nI_s r_s \cos\Delta + nI_s x_s \sin\Delta + I_p r_p \cos\beta + I_p x_p \sin\beta$$

or
$$V_p \cos\theta = nV_s + nI_s r_s \cos\Delta + nI_s x_s \sin\Delta + I_p r_p \cos\beta + I_p x_p \sin\beta$$
$$= nV_s + nI_s(r_s \cos\Delta + x_s \sin\Delta) + I_p r_p \cos\beta + I_p x_p \sin\beta \quad \ldots(i)$$

Phase angle $\theta$ is very small and, therefore, both $V_p$ and $V_s$ reversed can be taken perpendicular to $\Phi$ and, hence,

$$\angle ocd = \beta \ (approximately)$$

and
$$\angle ecd = \Delta \ (approximately)$$

Thus
$$I_p \cos\beta = I_e + \frac{I_s}{n}\cos\Delta$$

and
$$I_p \sin\beta = I_m + \frac{I_s}{n}\sin\Delta$$

Now $\theta$ is very small usually less than 1° and therefore, $\cos\theta = 1$ and hence we can write :

$$V_p \cos\theta = V_p$$

Substituting the above values in (*i*), we have :

$$V_p = nV_s + nI_s(r_s \cos\Delta + x_s \sin\Delta) + \left(I_e + \frac{I_s}{n}\cos\Delta\right) r_p + \left(I_m + \frac{I_s}{n}\cos\Delta\right) x_p$$

$$= nV_s + I_s \cos\Delta\left(nr_s + \frac{r_p}{n}\right) + I_s \sin\Delta\left(nx_s + \frac{x_p}{n}\right) + I_e r_p + I_m x_p \quad \ldots(10.24)$$

$$= nV_s + \frac{I_s}{n}\cos\Delta(n^2 r_s + r_p) + \frac{I_s}{n}\sin\Delta(n^2 x_s + x_p) + I_e r_p + I_m x_p$$

$$= nV_s + \frac{I_s}{n}\cos\Delta R_p + \frac{I_s}{n}\sin\Delta X_p + I_e r_p + I_m x_p$$

$$= nV_s + \frac{I_s}{n}(R_p \cos\Delta + X_p \sin\Delta) + I_e r_p + I_m x_p. \quad \ldots(10.25)$$

Here

$R_p$ = equivalent resistance of the transformer referred to the primary side $= n^2 r_s + r_p$

and $X_p$ = equivalent reactance of the transformer referred to the primary side $= n^2 x_s + x_p$

$\therefore$ Actual transformation (voltage) ratio

$$R = V_p / V_s$$
$$= n + \frac{\frac{I_s}{n}[R_p \cos\Delta + X_p \sin\Delta] + I_e r_p + I_m x_p}{V_s} \quad \ldots(10.26)$$

Equation 10.24 may be written as :

$$V_p = nV_s + nI_s \cos\Delta\left(r_s + \frac{r_p}{n^2}\right) + nI_s \sin\Delta\left(x_s + \frac{x_p}{n^2}\right) + I_e r_p + I_m x_p$$

$$= nV_s + nI_s \cos\Delta R_s + nI_s \sin\Delta X_s + I_e r_p + I_m x_p$$

$$= nV_s + nI_s(R_s \cos\Delta + X_s \sin\Delta) + I_e r_p + I_m x_p \quad \ldots(10.27)$$

where $R_s$ = equivalent resistance of transformer referred to secondary side $= r_s + r_p / n^2$

$X_s$ = equivalent reactance of transformer referred to secondary side $= x_s + x_p / n^2$.

$\therefore$ Actual transformation (voltage) ratio

$$R = \frac{V_p}{V_s} = n + \frac{nI_s(R_s \cos\Delta + X_s \sin\Delta) + I_s r_p + I_m x_p}{V} \quad \ldots(10.28)$$

Using Eqns. 10.25 and 10.27, the difference between actual transformation ratio and turns ratio is :

$$R - n = \frac{\frac{I_s}{n}[R_p \cos\Delta + X_p \sin\Delta] + I_e r_p + I_m x_p}{V_s} \quad ...(10.29)$$

$$= \frac{n I_s [R_s \cos\Delta + X_s \sin\Delta] + I_e r_p + I_m x_p}{V_s} \quad ...(10.30)$$

*Phase Angle*

$$\tan\theta = \frac{ab}{oa}$$

$$= \frac{I_p x_p \cos\beta - I_p r_p \sin\beta + nI_s x_s \cos\Delta - nI_s r_s \sin\Delta}{nV_s + nI_s r_s \cos\Delta + nI_s x_p \sin\Delta + I_p r_p \cos\beta + I_p x_p \sin\beta}$$

The terms in the denominator involving $I_p$ and $I_s$ are small and, therefore, they can be neglected as compared with $nV_s$.

$$\therefore \quad \tan\theta = \frac{I_p x_p \cos\beta - I_p r_p \sin\beta + nI_s x_s \cos\Delta - nI_s r_s \sin\Delta}{nV_s}$$

$$= \frac{x_p\left(I_e + \frac{I_s}{m}\cos\Delta\right) - r_p\left(I_m + \frac{I_s}{m}\sin\Delta\right) + nIx_s \cos\Delta - nI_s r_s \sin\Delta}{nV_s}$$

$$= \frac{I_s \cos\Delta\left(\frac{x_p}{n} + nx_s\right) - I_s \sin\Delta\left(\frac{r_p}{n} + nr_s\right) + I_e x_p - I_m r_s}{nV_s}$$

$$= \frac{\frac{I_s \cos\Delta}{n}(x_p + n^2 x_s) - \frac{I_s \sin\Delta}{n}(r_p + n^2 r_s) + I_e x_p - I_m r_p}{nV_s}$$

$$= \frac{\frac{I_s \cos\Delta}{n} X_p - \frac{I_s \sin\Delta}{n} R_p + I_e x_p - I_m r_p}{nV_s}$$

$$= \frac{(x_p \cos\Delta - R_p \sin\Delta) + I_e x_p - I_m r_p}{nV_s}$$

Since θ is small, $\theta = \tan\theta$.

$$\therefore \quad \theta = \frac{\frac{I_s}{n}(X_p \cos\Delta - R_p \sin\Delta) + I_e x_p - I_m r_p}{nV_s} \text{ rad.} \quad ...(10.31)$$

$$= \frac{I_s}{V_s}(X_s \cos\Delta - R_s \sin\Delta) + \frac{I_e x_p - I_m r_p}{nV_s} \text{ rad.} \quad ...(10.32)$$

### 10.6.3 Errors in Potential Transformers

It is clear from above that like current transformers, the potential transformers also introduce errors into measurements.

***Ratio (voltage) error.*** The actual ratio of transformation varies with operating conditions and the error in secondary voltage may be defined as :

$$\text{Percentage ratio error} = \frac{K_n - R}{R} \times 100 \quad ...(10.33)$$

***Phase angle error.*** In an ideal voltage transformer there should not be any phase difference between primary winding voltage and the secondary winding voltage reversed. However, in an actual transformer there exists a phase difference between $V_p$ and $V_s$ reversed.

Phase angle θ

$$= \frac{\frac{I_s}{n}(X_p \cos\Delta - R_p \sin\Delta) + I_e x_p - I_m r_p}{nV_s} \text{ rad} \quad ...(10.34)$$

$$= \frac{I_s}{V_s}(X_s \cos\Delta - R_s \sin\Delta) + \frac{I_e x_p - I_m r_p}{nV_s} \quad ...(10.35)$$

The phase angle is taken as + ve when secondary winding voltage reversed leads the primary winding voltage. The angle is – ve when secondary winding voltage reversed lags the primary winding voltage.

It should be clear that while measuring voltage, ratio error is only important while for power measurements both ratio and phase angle errors are involved.

### 10.6.4 Reduction of Errors in Potential Transformers

There has been improved accuracy of potential transformers with the advent of new materials. There are some modifications in design which lead to smaller total errors. These are :

(*i*) *Reduction of magnetising and loss components.* From Eqn. 10.30 it is clear that the difference between actual ratio and the turns ratio is made up of two parts. One is dependent upon the secondary winding current and the other upon the two components of no load current. It should be understood that unlike in the case of current transformers and power transformers, the values of the two components of no load current are quite comparable with that of the load current. Hence a considerable improvement in the performance can be made by reducing the values of $I_m$ and $I_e$. Such a reduction requires short magnetic paths. good quality core material, low flux density in core and suitable precautions in the assembly and interleaving of core.

(*ii*) *Reduction of resistance and leakage reactance.* Winding resistance can be minimized by using thick conductors and by adopting the smallest length of mean turn.

The leakage reactance of windings depends upon the magnitudes of primary and secondary winding leakage fluxes and therefore, we should keep the two windings as close as possible. The spacing should be compatible with insulation requirements.

Actually we should keep the flux density in the core as high as feasible without approaching the saturation too closely. A high value of flux density results in lower cross-section of the core which means that the length of mean turn would be small and thus the resistance of the winding becomes smaller.

A high flux density means a high flux in the core and so the windings have lesser number of turns. A small number of turns naturally results in smaller leakage reactance of the windings.

It is generally agreed that the new grain oriented steels allow a flux density of 1.5 Wb/m$^2$ or even higher as compared with the older materials which allow about 1.2 Wb/m$^2$.

(*iii*) *Turns compensation.* At no load the actual ratio exceeds the trims ratio by an amount $(I_e r_p + I_m x_p)/V_s$. (See Eqn. 10.26). With an inductive or resistive load there is a further increase of ratio because of voltage drops in resistance and leakage reactance of the windings. If the trims ratio is equal to the nominal ratio, the actual ratio differs from the nominal ratio and thus there are errors.

The solution lies in making the turns ratio less than the nominal ratio This can be done by either reducing the number of primary winding turns, of increasing the number of secondary winding turns This makes it possible to make the actual ratio of transformation equal to the nominal ratio for one particular value and type of burden. At the same time there is a general reduction in the ratio error over the entire working range of burden. The phase angle θ is practically unaffected by a small change in the turns ratio.

### 10.6.5 Construction of Potential Transformers

The design and construction of potential transformers are basically the same as those of power transformers but there are few major points of difference :

(*i*) Power transformers are designed keeping in view the efficiency, regulation and cost. The cost being reduced by using small core and conductor sizes. In designing a potential transformer, economy in material is not a big consideration and the transformers are designed to give desired performance *i.e.,* constancy of ratio and smallness of phase angle. Compared to a power transformer, a potential transformer has larger core and conductor sizes. Economic designs may lead to large ratio and phase angle errors which are undesirable features.

(*ii*) The output of a potential transformer is always small and the size is quite large. Therefore, the temperature rise is small and hence there are no thermal problems caused by overloads as in power transformers. In fact, the loading of a potential transformer is limited by accuracy considerations while in a power transformer the load limitation is on heating basis. Actually, the potential transformers are able to carry loads on a thermal basis many times their rated loads. These loads range from 2 to 3 times for low voltage potential transformers and upto 30 or more times for some high voltage transformers.

***Core.*** The core may be of shell or core type of construction. Shell type of construction is normally only used for low voltage transformers. Special precautions should be taken to assemble and interleave the core laminations so that the effect of air gaps at the joints may be minimized.

***Windings.*** The primary and secondary windings are coaxial to reduce the leakage reactance to minimum. In order to simplify the insulation problems, the low voltage winding (secondary) is put next to the core. The primary winding may be a single coil in low voltage transformers but must be subdivided into a number of short coils in high voltage transformers in order to reduce the insulation needed between coil layers.

***Insulation.*** Cotton tape and varnished cambric are used as insulation for coil construction. Hard libre separators are used between coils. At low voltages, the transformers are usually filled without compound but potential transformers for use at voltages above 7 kV are oil immersed. Dry type, porcelain insulated transformers have been developed in the continent for use upto 45 kV.

***Bushings.*** Oil filled bushings are usually used for oil filled potential transformers as this minimizes the overall size of the transformer. Two bushings are used when neither side of the line is at ground potential. Some potential transformers, connected from line to neutral of grounded neutral systems, have only one high voltage bushing.

It is pertinent to point out here that a current transformer needs only one bushing as leads from the two ends of the primary winding are brought through the same insulator since there is only a small voltage between them, thus saving the expense of another high voltage insulator.

A 2-winding single phase potential transformer is shown in Fig. 10.23.

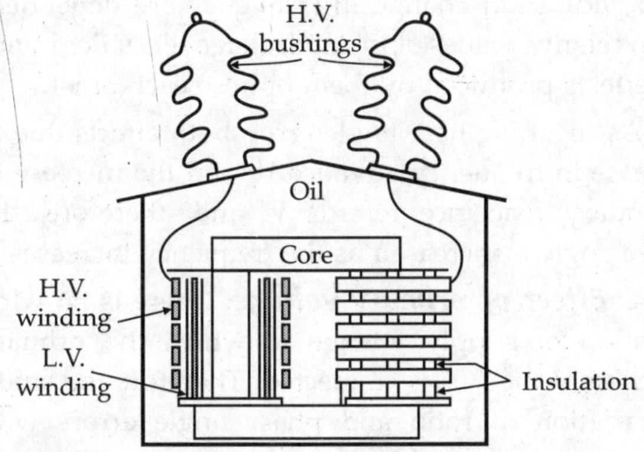


**Fig. 10.23** Single phase potential transformer.

### 10.6.6 High Voltage Potential Transformers

Conventional type potential transformers used for high voltages of 110 kV and above, are very large in size and costly to build because of insulation requirements. For example a 110 kV potential transformer has an overall height of about 7.5 metre and weighs nearly 5 tonnes. This is a very unwieldy size and also the materials utilized in the transformer construction are very uneconomically utilized.

Recently there has been a development in the design and construction of P.T.s which has resulted in considerable reduction in size and cost of transformers. Two designs have been developed which eliminate the high voltage lead in bushings. The elimination of bushings reduces the size and cost of transfor- mers. These designs are intended to measure line to ground voltages in a three phase system. The designs employ :

(*i*) ***Insulated casing.*** The transformer is built entirely in an oil filled high voltage insulator. This results in economy in space and material.

(*ii*) ***Moulded rubber potential transformer.*** Recent developments in the synthetic rubber industry have introduced the moulded rubber potential transformer, replacing the insulating oil and porcelain bushings for some applications. This unit is less expensive than the conventional oil filled P.T., and since the bushings are made of moulded rubber, the difficulties caused due to porcelain breakage are eliminated.

(*iii*) ***Cascaded transformers.*** In a cascaded arrangement, the voltage is divided among a number of transformers.

Suppose the voltage is divided among $N$ transformers and, therefore, each transformer will take $1/N$ of total voltage. In this way each requires insulation corresponding to the lower voltage, with a consequent saving in space and material. The chief application of cascade arrangement is to increase voltage rating of dry type units. This arrangement is explained in details in chapter 17 on High Voltage Measurements.

### 10.6.7 Protection of Potential Transformers

Potential transformers can be continuously operated at 1.2 times the rated voltage.

A short circuit on the secondary side of a potential transformer can lead to complete damage of the transformer. In order to protect the power system against short circuits in the potential transformers, fuses are used on the primary (high voltage) side. Fuses are used in the secondary side to protect the P.T. against faulty switching and defective earthing.

### 10.6.8 Capacitive Potential Transformers

At voltages above about 100 kV (phase) the conventional electro-magnetic type of potential transformer becomes expensive owing to insulation requirements. A less expensive alternative is capacitive voltage transformer.

This consists of a capacitance potential divider used in conjunction with a conventional auxiliary transformer (Fig. 10.24). The capacitance potential divider steps down the voltage to be measured (say to about 10 kV). The capacitance divider output voltage is further stepped down by the auxiliary transformer to the desired secondary voltage (say 110 V). The auxiliary transformer consists of an inductance $L$, which may consist wholly or partly of leakage inductance of the windings of auxiliary transformer.

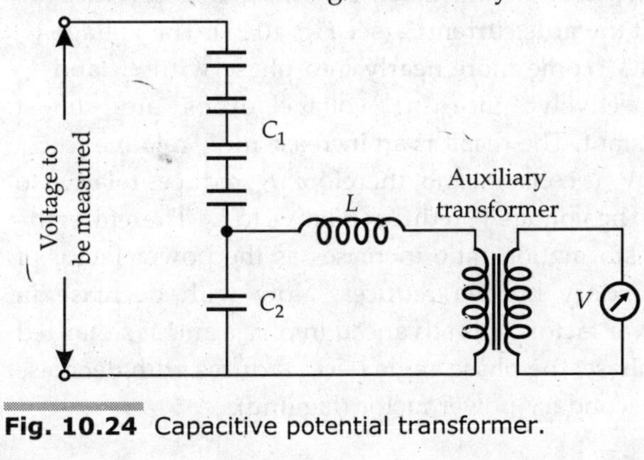


**Fig. 10.24** Capacitive potential transformer.

The value of this inductance $L$ may be adjusted to equal $\frac{1}{\omega^2(C_1+C_2)}$ so that the voltage drop due to the current drained from the divider is largely compensated. Thus the overall voltage transformation ratio is independent of burden. The overall ratio is the product of the divider and transformer ratios. This has been explained in the chapter 17 on High Voltage Measurements.

In practice, however, the compensation is not complete because of the losses in the inductance and also due to small changes in frequency which might occur. Nevertheless, the performance of capacitor voltage transformers is not much inferior to that of the electro-magnetic type of potential transformers and provided the current in the primary of the auxiliary transformer does not exceed about 10% of that in the divider.

### 10.6.9 Characteristics of Potential Transformers

1. ***Effect of secondary current or VA.*** If we increase the secondary burden. the secondary current is increased, and therefore, the primary current increases. Both primary and secondary voltage drops increase and thus for a given value of $V_p$, the value of $V_s$ decreases and hence the actual ratio increases as the burden increases. The ratio error increases becoming more negative with increase in burden. This variation of ratio error is almost linear with change in burden.

With regard to phase angle, the voltage $V_p$ is more advanced in phase because of increased voltage drops with increase in secondary burden. The phasor $V_s$ reversed is retarded in phase owing to increase in secondary winding voltage drops. Thus with increase in burden. the phase angle between $V_p$ and $V_s$ reversed increases becoming more negative.

2. ***Effect of power factor of secondary burden.*** If the power factor of secondary circuit burden is reduced. angle $\Delta$ is increased. This makes current $I_p$ to shift towards current $I_0$ (see Fig. 10.21). The voltage $V_p$ and $V_s$ come more nearly into phase with $E_p$ and $E_s$ respectively since the voltage drops are almost constant. The result is an increase in $V_p$ relative to $E_p$. But $V_p$ is constant and, therefore, $E_p$ reduces relative to $V_p$. The voltage $V_s$ reduces relative to $E_s$. Therefore, the transformation ratio increases as the power factor of secondary burden reduces. Now with decrease in power factor, $V_s$ is advanced in phase and $V_p$ retarded in phase, the phase angle (– ve) reduces with decrease in secondary power factor (lagging).

3. ***Effect of frequency.*** For a constant voltage, the flux is inversely proportional to frequency. Increase in frequency reduces the flux and, therefore. $I_m$ and $I_e$ are decreased and, therefore, the voltage ratio decreases. The decrease is not so much, as with increase in frequency the leakage reactance increases and, therefore, leakage reactance drops are increased giving an increase in ratio. Thus changes in voltage ratio because of change in frequency are dependent upon relative values of In and leakage reactances since the effects produced by them oppose each other.

As regards phase angle error, both effects due to increase in frequency advance $V_p$ and the increase in secondary reactance retards $V_s$ and, therefore, the phase angle is increased as the frequency increases.

4. ***Effect of primary voltage.*** There is no wide variation of supply voltage to which the primary winding of the P.T. is connected. Therefore the study of variation of ratio and phase angle errors with supply voltage are of no importance.

***Example 10.12*** *A potential transformer, ratio 1000 / 100 volt, has the following constants :*

*Primary resistance = 94.5 Ω*

*Secondary resistance = 0.86 Ω*

*Primary reactance = 66.2 Ω*

*Total equivalent reactance = 110 Ω*

*No load current = 0.02 A at 0.4 power factor.*

*Calculate : (i) phase angle error at no load (ii) burden in VA at unity power factor at which the phase angle will be zero.*

**Solution.** No load power factor

$$\cos\alpha = 0.4$$

$$\sin\alpha = \sqrt{(1)^2-(0.4)^2} = 0.917$$

$$\therefore \quad I_e = I_0\cos\alpha = 0.02\times 0.4 = 0.008 \text{ A}$$

$$I_m = I_0\sin\alpha = 0.02\times 0.917 = 0.01834 \text{ A}$$

Turns ratio, $n = 1000/100 = 10$

(*i*) From Eqn. 10.31, phase angle

$$\theta = \frac{\frac{I_s}{n}(X_p\cos\Delta - R_p\sin\Delta) + I_e x_p - I_m r_p}{nV_s} \text{ rad.}$$

At no load $I_s = 0$

$$\therefore \quad \theta = \frac{I_e x_p - I_m x_p}{nV_s}$$

$$= \frac{0.008\times 66.2 - 0.01834\times 94.5}{10\times 1000} \text{ rad} = -4.1'$$

(*ii*) At unity power factor $\cos\Delta = 1$ and $\sin\Delta = 0$.

$$\theta = \frac{\frac{I_s}{n}X_p - I_e x_p - I_m r_p}{nV_s}$$

For $\theta = 0$, $\frac{I_s}{n}X_p + I_e x_p - I_m r_p = 0$.

or $I_s = \frac{n}{X_p}(I_m r_p - I_e x_p)$

$$= \frac{10}{100}(0.01834 \times 94.5 - 0.008 \times 66.2) = 0.109 \text{ A}.$$

Burden $= V_s I_s = 100 \times 0.109 = 10.9$ VA.

**Example. 10.13** *A potential transformer rated 6900/115 V has 22500 turns in the primary winding and 375 turns in the secondary winding. With 6900 V applied to the primary and the secondary circuit open circuited, the primary winding current is 0.005 A lagging the voltage by 73.7°. With a particular burden connected to the secondary, the primary winding current is 0.0125 A, lagging the voltage by 53.1°.*

*Primary winding resistance = 1200 Ω*

*Primary winding reactance = 2000 Ω*

*Secondary winding resistance = 0.4 Ω*

*Secondary winding reactance = 0.7 Ω*

(*i*) *Find the secondary current and terminal voltage, using the applied primary voltage $V_p = 6900 + j0$ as reference. Find the load burden also.*

(*ii*) *Find the actual transformation ratio and also the phase angle.*

(*iii*) *If the actual ratio is equal to the nominal ratio under above conditions, what change should be made in the primary turns ?*

**Solution.** Nominal ratio, $K_n = 6900/115 = 60$

Primary winding turns, $N_p = 22500$

Secondary winding turns, $N_s = 375$

Turns ratio $n = 22500/375 = 60$

No load current $I_0 = 0.005$ A

No load p.f.

$$= \cos 73.7° = 0.28 \quad \text{and} \quad \sin 73.7° = 0.96$$

Primary current, $I_p = 0.0125$ A

Primary power factor

$$= \cos 53.1° = 0.6 \quad \text{and} \quad \sin 53.1° = 0.8$$

Now primary voltage $V_p$ is taken as reference and, therefore, we can write :

$$V_p = 6900 + j0,$$

$$I_p = 0.0125(0.6 - j0.8) = 0.0075 - j0.01$$

$$I_0 = 0.005(0.28 - j0.96) = 0.0014 - j0.0048.$$

[Phasor $\frac{I_s}{n}$ is the phasor difference of $I_p$ and $I_0$]

$$\therefore \frac{I_s}{n} = (0.0075 - j0.01) - (0.0014 - j0.0048)$$

$$= 0.0061 - j0.0052$$

$$I_s \text{ (reversed)} = n \times \frac{I_s}{n}$$

$$= 60(0.0061 - j0.0025)$$

$$= 0.366 - j0.312.$$

or $I_s = -(0.366 - j0.312) = -0.366 + j0.312$

$\therefore$ Secondary current

$$I_s = 0.48 \text{ A}.$$

Primary induced voltage

$$E_p = V_p - I_p Z_p$$

$$= 6900 - j0 - (0.0075 - j0.01)(1200 + j2000)$$

$$= 6900 - j0 - (9 + j3) = 6891 - j3 \text{ V}.$$

Secondary induced voltage $E_s$ (reversed)

$$= \frac{E_s}{n} = \frac{6891 - j3}{60} = 114.85 - j0.05 \text{ V}.$$

Secondary terminal voltage

$$V_s = E_s - I_s Z_s$$

$$= -114.85 + j0.05 - (-0.366 + j0.312)(0.4 + j0.7)$$

$$= -114.49 + j0.18 \text{ or } V_s = 114.49 \text{ V}.$$

Secondary burden $= V_s I_s = 114.49 \times 0.48 = 55$ V.

(*ii*) Actual ratio $= \frac{V_p}{V_s} = \frac{6900}{114.48} = 60.27.$

$$V_s = \text{(reversed)} = -(-114.49 + j0.18)$$

$$= 114.49 - j0.18 \text{ V}$$

Angle by which $V_s$ (reversed) lags $V_p$

$$= \tan^{-1}\frac{0.18}{114.49} \approx \frac{0.18}{114.49} \text{ rad}$$

$$= \frac{0.18}{114.49} \times \frac{180}{\pi} = 0.90° = 0.54'.$$

(*iii*) The solution to the problem lies in reducing the turns ratio (decreasing number of primary turns) so that the actual ratio equals the nominal ratio.

New turns ratio $= (60/60.27) \times 60$

$\therefore$ New value of primary turns

$$= (60/60.27) \times 60 \times N_s$$

$$= (60/60.27) \times 60 \times 375 = 22399.$$

$\therefore$ Reduction in primary turns $= 22500 - 22399 = 101$

Here we are neglecting the change in voltage drops caused by change in turns ratio and hence the solution is only approximate.

## 10.7 TESTING OF INSTRUMENT TRANSFORMERS

Methods for experimentally testing instrument transformers *i.e.,* finding their ratio and phase angle errors may be broadly classified into *two* groups.

(*i*) ***Absolute methods.*** In these methods the transformer errors are determined in terms of constants-resistance, inductance and capacitance of the testing circuit.

(*ii*) ***Comparison method***. In these methods, the errors of the transformer under test are compared with those of a standard current transformer whose errors are known.

Each of these two test methods can be classified, according to measurement technique employed as :

(*i*) *Deflection methods.* These methods use the deflections of suitable instruments like electro-dynamometer wattmeters for measuring quantities related to the phasors under consideration or to their deflection. The required ratio and phase angles are then found out from the magnitudes of deflection. These methods may be made direct reading in some cases.

(*ii*) *Null methods.* These methods make use of a network in which the appropriate phasor quantities are balanced against one another. The ratio and phase angle errors are then found out from t.he impedance elements of the network. The method may be made direct reacting in terms of calibrated scales on the adjustable elements in the network.

## 10.8 TESTING OF CURRENT TRANSFORMERS

1. ***Mutual inductance method.*** This is an absolute method using null technique. The connections are shown in Fig. 10.25.

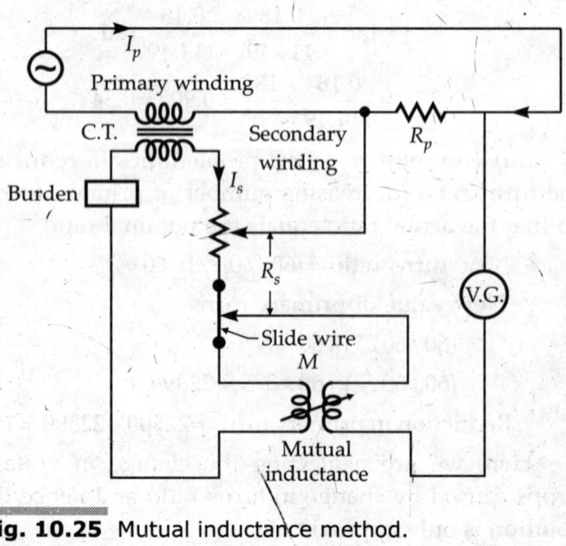


**Fig. 10.25** Mutual inductance method.

$R_p$ and $R_s$ are low resistance, non-inductive shunts, $R_s$ is variable while $R_p$ is fixed. $R_s$ has a slide wire for fine adjustment of resistance. The voltage drop across resistance $R_p$ is matched against voltage drop across $R_s$. A vibration galvanometer is put in the circuit to indicate the balance conditions. Assuming, for the moment that there is no phase difference between $I_p$ and $I_s$ the vibration galvanometer will indicate zero deflection if

$$I_pR_p = I_sR_s \quad \text{or} \quad I_p / I_s = R_s / R_p.$$

Therefore, $R_s$ and $R_p$ should be so chosen that the ratio $R_s / R_p$ is nearly equal to the nominal ratio of the current transformer.

Resistance $R_s$ is adjusted to render the two voltage drops equal. In order to obtain zero deflection the magnitude and also the phase of the voltage drops should be same. Thus a mutual inductance $M$ is put to compensate for phase difference between $I_p$ and $I_s$ as without any phase compensating device it will be impossible to obtain balance with resistance alone.

Figure 10.26 represents the phasor diagram under balance conditions.

$$\tan\theta = \frac{I_s\omega M}{I_sR_s} = \frac{\omega M}{R_s} \quad \text{and} \quad \cos\theta = \frac{I_sR_s}{I_pR_p}.$$

From these relations we have,

$$\text{Phase angle,} \quad \theta \approx \frac{\omega M}{R_s} \text{ rad.}$$

$$\text{Actual ratio,} \quad R = \frac{I_p}{I_s} = \frac{R_s}{R_p \cos\theta} \approx \frac{R_s}{R_p}$$

as $\theta$ is very small.

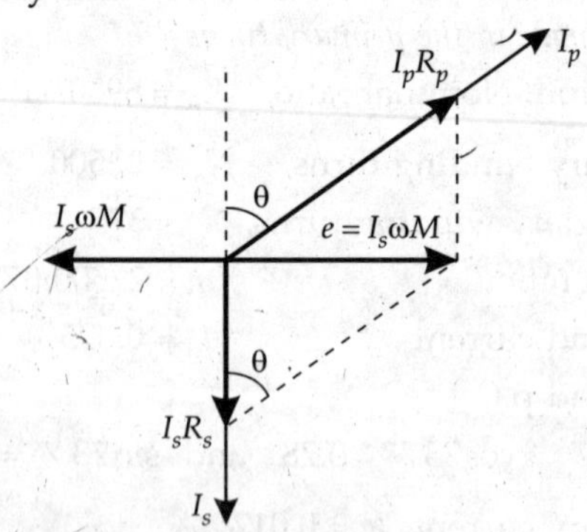


**Fig. 10.26** Phasor diagram for mutual inductance method.

It should be noted that the secondary load circuit includes the resistance $R_s$ impedance of primary winding of mutual inductance and also the impedance marked burden. This must be taken into account while stating the burden at which the errors have been measured.

2. ***Silsbee's method.*** Silsbee's method is a comparison method. There are two types of Silsbee's methods ; deflectional and null. Only the deflectional method is described here.

The arrangement for this method is shown schematically in Fig. 10.27. Here the ratio and phase angle of the test transformer $X$ are determined, in terms of that of a standard transformer $S$ having the same nominal ratio.

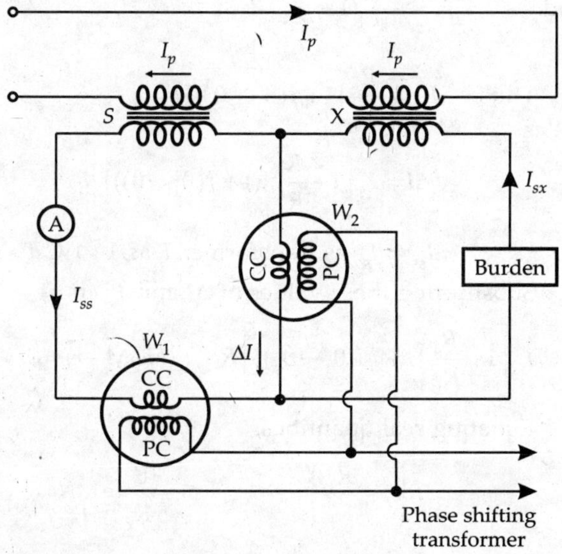


**Fig. 10.27** Silsbee's deflectional method.

The two transformers are connected with their primaries in series. An adjustable burden is put in the secondary circuit of the transformer under test.

An ammeter is included in the secondary circuit of the standard transformer so that the current may be set to the desired value. $W_1$ is a wattmeter whose current coil is connected to carry the secondary current of the standard transformer. The current coil of wattmeter $W_2$ carries a current $\Delta I$ which is the difference between the secondary currents of the standard and test transformers. The voltage circuits of the wattmeters (*i.e.*, their pressure coils) are supplied in parallel from a phase shifting transformer at a constant voltage $V$.

The phasor diagram is shown in Fig. 10.28.

**1.** The phase of the voltage is so adjusted that wattmeter $W_1$ reads zero. Under these conditions, voltage $V$ is in quadrature with current $I_{ss}$. The position of voltage phasor for this case is shown as $V_q$.

Reading of wattmeter, $W_1$,

$$W_{1q} = V_q I_{ss} \cos 90° = 0.$$

Reading of wattmeter $W_2$,

$$W_{2q} = V_q \times \text{component of current } \Delta I \text{ in phase with}$$

$$V_q = V_q I_q = V_q I_{sx} \sin(\theta_x - \theta_s).$$

where $\theta_x$ = phase angle of C.T. under test,

$\theta_s$ = phase angle of standard C.T.

**2.** The phase of voltage $V$ is shifted through 90° so that it occupies a position $V_p$ and is in phase with $I_{ss}$

Reading of wattmeter $W_1$,

$$W_{1p} = V_p I_{ss} \cos\theta = V_p I_{ss}.$$

Reading of wattmeter $W_2$,

$$W_{2p} = V_p \times \text{component of current } \Delta I \text{ in phase with } V_p.$$

$$= V_p \times \Delta I_p = V_p [I_{ss} - I_{sx} \cos(\theta_x - \theta_s)]$$

If the voltage is kept same for both sets of readings, then

$$V = V_p = V_q.$$

We have,

$$W_{2q} = VI_{sx} \sin(\theta_x - \theta_s), W_{1p} = VI_{ss}$$

$$= VI_{ss} - I_{sx} \cos(\theta_x - \theta_s)$$

$$= VI_{ss} - VI_{sx} \cos(\theta_x - \theta_s)$$

$$\approx W_{1p} - VI_{sx} \cos(\theta_x - \theta_s) = W_{1p} - VI_{sx}$$

as $(\theta_x - \theta_s)$ is very small and, therefore, $\cos(\theta_x - \theta_s) = 1$.

For above, $VI_{sx} = W_{1p} - W_{2p}$.

Actual ratio of transformer under test, $R_x = I_p / I_{sx}$.

Actual ratio of standard transformer, $R_s = I_p / I_{ss}$.

$$\therefore \quad \frac{R_x}{R_s} = \frac{I_{ss}}{I_{sx}} = \frac{VI_{ss}}{VI_{sx}} = \frac{W_{1p}}{W_{1p} - W_{2p}}$$

$$= \frac{1}{1 - (W_{2p}/W_{1p})} \approx 1 + \frac{W_{2p}}{W_{2p}}$$

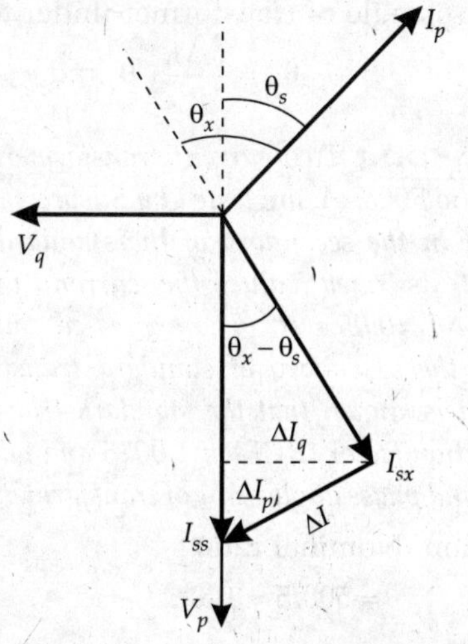


**Fig. 10.28** Phasor diagram for Silsbee's method.

$$R_x = R_s\left(1+\frac{W_{2p}}{W_{1p}}\right),$$

$$\sin(\theta_x - \theta_s) = \frac{W_{2q}}{VI_{sx}},$$

$$(\cos\theta_x - \theta_s) = \frac{VI_{ss} - W_{2p}}{VI_{sx}} = \frac{W_{1p} - W_{2p}}{VI_{sx}}$$

$$\therefore \quad \tan(\theta_x - \theta_s) = \frac{W_{2q}}{W_{1p} - W_{2p}}$$

or $$(\theta_x - \theta_s) = \frac{W_{2q}}{W_{1q} - W_{2q}} \text{ rad.}$$

or phase angle of test transformer

$$\theta = \frac{W_{2q}}{W_{1p} - W_{2p}} + \theta_s \text{ rad.}$$

$$\approx \frac{W_{2q}}{W_{1p}} + \theta_s \text{ rad as } W_{2p} \text{ is small.}$$

Hence if the ratio and phase angle errors of standard transformer are known, we can compute the errors of the test transformer.

$W_2$ must be a sensitive instrument. Its current coil may be designed for small values. It is normally designed to carry about 0.25 A for testing CTs having a secondary current of 5A.

**3. *Arnold's method.*** This is a comparison method involving null techniques.

This method is used for getting very accurate results. The errors of transformer under test $X$ (Fig. 10.29) are compared with those of standard transformer $S$. In order to isolate the measuring circuit from the secondaries of CTs, a 5/5 current transformer $T$ is used. This transformer has negligible errors.

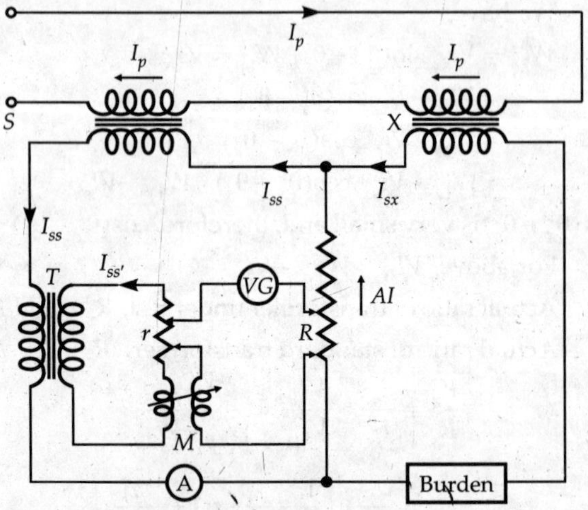


**Fig. 10.29** Arnold's method.

We use the same notations as we did in case of Silsbee's method. Balance is obtained when

$$\Delta I \times R + I'_{ss} r - I'_{ss}\, j\,\omega M = 0 \qquad \text{...}(i)$$

Now $\Delta I = I_{ss} - I_{sx}$ But $I_{ss} = -\frac{I_p}{R_x}(\cos\theta_x + j\sin\theta)$

$$= -\frac{I_p}{R_s}(1 + j\,\theta_s) \text{ as } \theta_s \text{ is very small}$$

and $$\mathbf{I}_{sx} = -\frac{I_p}{R_x}(1 + j\theta_x)$$

Thus, $$I_{sx} = \frac{R_s}{R_x}[1 + j(\theta_x - \theta_s)]\,I_{ss}$$

$$\therefore \quad \Delta I = I_{ss}\left[1 - \frac{R_s}{R_x}\{(1 + j(\theta_x - \theta))\}\right]$$

$I'_{ss} = I_{ss}$ as transformer $T$ as 1 : 1 C.T.

Substituting these values of $\Delta I$ and $I_{ss}$ in (*i*)

$$I_{ss}\left[1 - \frac{R_s}{R_x}\{1 + j\,I(\theta_x - \theta_s)\}\right]R + I_{ss}(j\omega M - r) = 0.$$

Equating real quantities,

$$R\left(1 - \frac{R_s}{R_x}\right) = r \quad \therefore \quad 1 - \frac{R_s}{R_x} = \frac{r}{R}.$$

Equating imaginary quantities,

$$\frac{R_s}{R_x} R(\theta_x - \theta_x) = \omega M$$

$$\therefore \quad \theta_x - \theta_s = \frac{\omega M}{R}\cdot\frac{R_x}{R_s} \text{ rad.}$$

But $R_x \approx R_s$ and, therefore,

$$\theta_x - \theta_s = \frac{\omega M}{R} \text{ rad.}$$

$\therefore$ Phase angle of transformer under test

$$\theta_x = \frac{\omega M}{R} + \theta_s \text{ rad.}$$

**Example 10.14** *Two current transformers of the same nominal ratio 500/5 A, are tested by Silsbee's method. With the current in the secondary of the standard transformer adjusted at its rated value, the current in the middle conductor $\Delta I = 0.05\, e^{-j126.9°}$ A expressed with respect to current in the secondary of standard transformer as the reference. It is known that the standard transformer has a ratio correction factor (RCF) of 1.0015 and phase error +8′. Find RCF and phase angle error of transformer under test.*

**Solution.** Nominal ratio

$= 500/5 = 100, \quad I_{ss} = 5$ A.

since $I_{ss}$ is reference, $I_{ss} = 5 + j0$

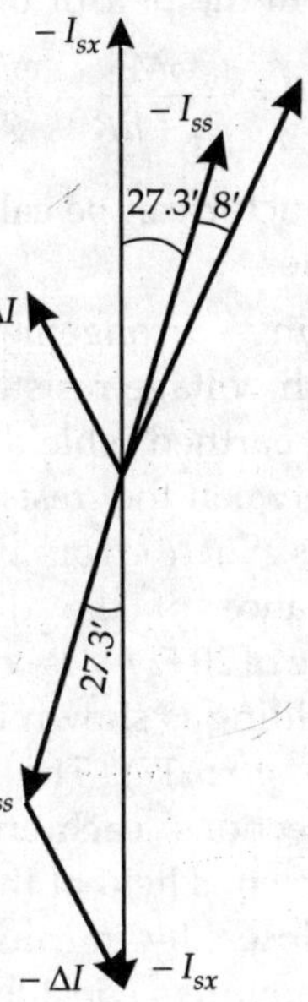


Fig. 10.30 Phasor diagram of Example 10.14.

$$\Delta \mathbf{I} = 0.05\,e^{-j126.9°}.$$
$$= 0.05[\cos 126.9° - j \sin 126.9°]$$
$$= -0.03 - j0.04.$$

Now, current in the secondary of test transformer

$$I_{sx} = I_{ss} - \Delta I = 5 + j0 - (0.03 - j0.04)$$
$$= +5.03 + j0.04$$

or $I_{sx} \approx 5.03$ A.

Angle between $I_{sx}$ and $I_{ss}$

$$= 0.04 / 5.03 \text{ rad.} = 27.3'$$

Phase angle between $I_{sx}$ reversed

and $I_p = +27.3 + 8 = +35.3'$

Ratio correction factor RCF of standard transformer = 1.0015.

∴ Actual ratio of standard transformer

$$R_s = \text{RCF} \times \text{nominal ratio}$$
$$= 1.0015 \times 100 = 100.15$$

Primary current

$$I_p = R_s \times I_{ss}$$
$$= 100.15 \times 5 = 500.75 \text{ A}$$

Actual ratio of transformer under test

$$R_x = \frac{I_p}{I_s} = \frac{500.75}{5.03} = 99.55$$

Ratio correction factor of test transformer

$$\text{RCF} = \frac{\text{actual ratio}}{\text{nominal ratio}} = \frac{99.55}{100} = 0.9955.$$

## 10.9 TESTING OF POTENTIAL TRANSFORMERS

**1. *Absolute null method.*** Figure 10.31 shows the connections for testing a potential transformer using a resistance potential divider. The burden with which the transformer is to be tested is connected across the secondary winding, and the normal primary voltage, at normal frequency is applied to the primary winding. One end of the secondary winding is connected to one end of the primary winding. A non-inductive potential divider is connected across the primary winding. The lower part of the divider consists of a Fixed resistor in series with a side-wire *S*,

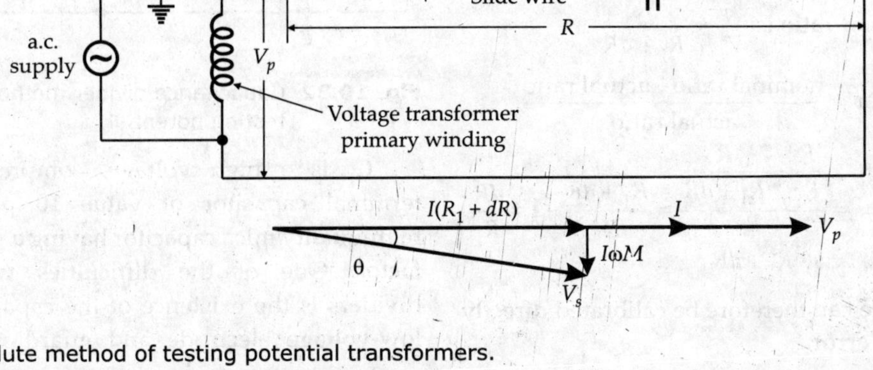


Fig. 10.31 Absolute method of testing potential transformers.

and the upper part consists of a shielded, high voltage resistor in series with the primary of a variable mutual inductor $M$. A capacitor $C$ shunts a small part $r$ of the upper part of the divider. VG is a vibration galvanometer. The resistances are chosen so that the nominal ratio of the test transformer is equal to $R/R_1$.

$L$ is the primary winding inductance of the mutual inductor and will cause a phase error in the divider. The capacitor $C$ compensates for this error in the following manner. The impedance of the whole divider is :

$$\mathbf{Z} = R - r + j\omega L + \frac{r}{1 + j\omega Cr}$$

$$= R - r + j\omega L + \frac{r}{1+\omega^2C^2r^2}(1 - j\omega Cr)$$

$$= R - r + \frac{r}{1+\omega^2C^2r^2} + j\omega\left(L - \frac{Cr^2}{1+\omega^2C^2r^2}\right)$$

By making $\frac{Cr^2}{1+\omega^2C^2r^2} = L$, the divider is non-reactive at the operating frequency, which is usually 50 Hz, and the total impedance is then :

$$Z = R - r + \frac{r}{1+\omega^2C^2r^2}$$

The term $\omega^2C^2r^2$ is normally small compared with unity and may be neglected, so that the impedance of the divider is equal to $R$. If this term is not negligible, the resistance of the upper part of the divider can be increased by a small amount to correct for it at the operating frequency.

In carrying out the test, the contacts on the slide-wire and the mutual inductor are adjusted until the vibration galvanometer gives no deflection.

The phasor diagram corresponding to the balanced condition is shown in Fig. 10.31, in which $I = \frac{V_p}{R}$ is the current in the divider. The phase angle $\theta$ of the test transformer is small so that

$$V = I(R_1 + dR) = \frac{V_p}{R}(R_1 + dR)$$

and the transformer ratio is $\frac{V_p}{V_s} = \frac{R}{R_1 + dR}$

$$\text{The ratio error} = \frac{\text{nominal ratio} - \text{actual ratio}}{\text{actual ratio}}$$

$$= \frac{\frac{R}{R_1} - \frac{R}{R_1 + dR}}{\frac{R}{R_1 + dR}} = \frac{R_1 + dR}{R_1} - 1 = \frac{dR}{R_1}$$

The slide-wire can therefore be calibrated directly in values of ratio error.

*Phase angle*. From the phasor diagram,

$$\tan\theta \approx \theta = \frac{\omega M}{R_1 + dR} \approx \frac{\omega M}{R_1}$$

and the mutual inductor can be calibrated directly in terms of phase angle.

Special shielding arrangements have to be adopted for the high voltage resistor. If it is surrounded by a simple earthed shield, the capacitance currents flowing between the resistor and shield can cause serious errors. Such errors can be reduced by keeping the resistance of the divider as low as possible, and a value of 20 Ω / V is commonly adopted. The method of shielding is shown in chapter on high voltage testing (Chapter 17). The divider is broken into a number of sections, each enclosed in its own screen, and each screen is held at the mid-potential of the section it encloses by means of an auxiliary divider. There is therefore a considerable reduction in the capacitance current from the divider to the shield.

It is not practical to construct resistance dividers suitable for voltage transformer testing at voltages much above 40 kV ; the use of a capacitance divider is preferable at higher voltages.

**2. *Methods employing capacitance dividers.*** A wide variety of methods of testing voltage transformers with the aid of capacitance dividers have been used. Figure 10.32 gives the connections of a method of testing $PT_s$ using capacitance voltage dividers.

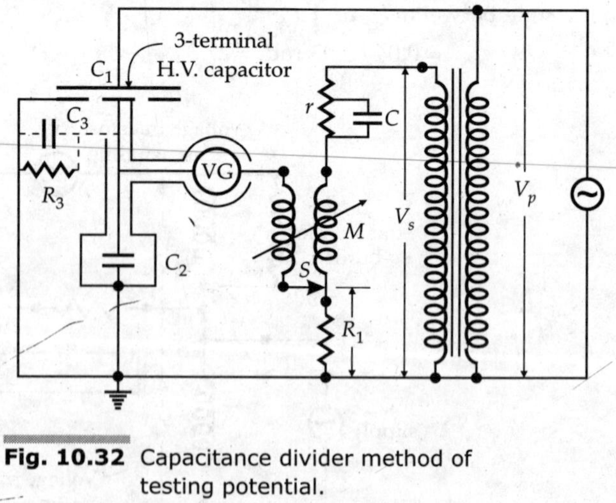


**Fig. 10.32** Capacitance divider method of testing potential.

$C_1$ is a high voltage, compressed gas, three terminal capacitor of value 10 pF and $C_2$ is a high-quality mica capacitor having a very small power factor. One of the difficulties with capacitance dividers is the existence of the capacitance $C_3$ of the low-voltage electrode and guard ring of $C_1$. This capacitance is often associated with a loss resistance $R_3$, and it is therefore important that the value of the mica capacitor $C_3$ should be large so that $C_2 + C_3$ and $R_3$ have a negligible power factor. The capacitance between the h.v. and l.v. electrodes of $C_1$ is loss-free ; hence the phase error of the divider, which should be negligible, is determined by $C_2$ shunted by $C_3$ and $R_3$. There is therefore a restriction on the minimum value of $C_2$ and on the minimum ratio of the divider. This difficulty can be overcome by the modification shown in Fig. 10.33, in which the guard ring is connected to earth through a capacitor $C_4$ which is adjusted so that $C_1 / C_2 = C_{1E} / C_4$. The guard ring is then at the same potential as the low-voltage electrode.

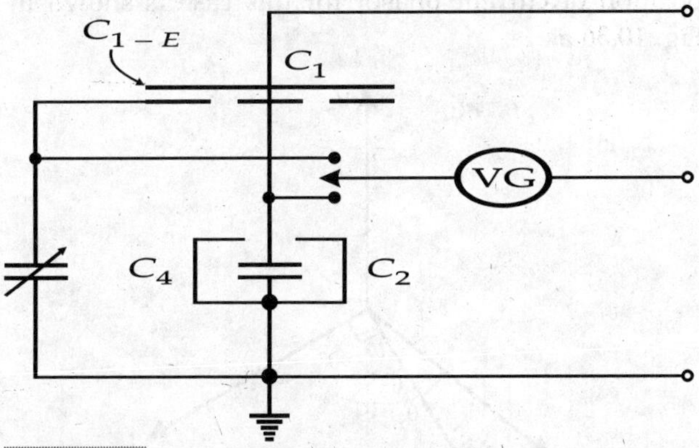


**Fig. 10.33** Guard ring connected to earth via potential capacitor $C_4$.

A resistance divider having a ratio of 10/1 is connected across the secondary winding of the potential transformer under test. This divider incorporates a variable potential slide-wire $S$ and mutual inductor $M$ for ratio and phase-angle adjustment. It is similar to the arrangement used in the resistance divider equipment previously discussed. $C$ and $r$ are used to compensate for the primary inductance of the mutual inductor $M$, so that the divider is non-reactive at the test frequency. The slide-wire $S$, and the mutual inductor can be calibrated to read ratio error and phase angle directly.

The adoption of a 10/1 ratio in the measuring resistance divider permits ratio errors of either sign to be measured. It has the added advantage that the ratio of the capacitance divider is 10 times the nominal ratio of the test transformer permitting the use of a large value of $C_2$ and thus minimizing the effect of $C_3$ and $R_3$ in the simple arrangement.

The compensation of the measuring resistance divider, and its zero setting can be very easily checked by comparing it with a simple 10/1 ratio resistance divider.

*Clothier and medina method.* This method has been developed quite recently and uses a variable- tapping transformer for the measuring device. Figure 10.34 is a simplified circuit diagram. $C_1$ and $C_2$ are three-terminal capacitors whose ratio is equal to the nominal ratio of the lest transformer 1. The windings of transformer 1 are in series aiding. The errors in transformer 1 are balanced by means of transformer 2 whose ratio can be adjusted a little above or below unity by using decade switches to select the primary turns.

Phase-angle errors are balanced by supplying a current $I_R$ through a high resistance $R$ to the junction of $C_1$ and $C_R$. $I_R$ is in phase with the secondary voltage and therefore in quadratic with the current in $C_2$. The value of $I_R$ is adjusted by a switch on transformer 2 using at tuned electronic detector.

Since the detector voltage is zero at balance, the current in $R$ is $aV / R$, where $aV$ is the voltage across $R$ with respect to earth and $V$ is the voltage across $C_2$. $aV$ can have either sign.

At balance $I_1 = I_2 + I_R$.

$$I_1 = j\omega C_1 V_p,\ I_2 = j\omega C_2 V, \text{ and } I_R = aV / R$$

$$\therefore\ j\omega C_1 V_P = j\omega C_2 V[1 + a / j\omega C_2 R]$$

or

$$\frac{V_p}{V_s} = \frac{C_2}{C_1}\frac{V}{V_s}\left[1 - \frac{ja}{\omega C_2 R}\right]$$

If the phase angle of the auxiliary transformer is negligible, and its voltage is very nearly equal to its turns ratio, then, if $N_1$ and $N_2$ are primary and secondary turns on Transformer 2, and if $N = N_2 - N_1$.

$$\frac{V}{V_s} = \frac{N_2}{N_1} = \frac{N_2}{N_2 - N} = \frac{1}{1 - N / N_2}$$

$$= \frac{1}{1 - D} \text{ where } D = N / N_2$$

The phase angle $\theta \approx \tan\theta = \dfrac{a}{\omega C_2 R}$ rad

The ratio $R_x$ of the test transformer is

$$R_x = \left|\frac{V_p}{V_s}\right| = \frac{C_2}{C_1}\frac{1}{1 - D}\sqrt{1 + \frac{a^2}{\omega^2 C_2^2 R^2}}$$

$$= \frac{C_2}{C_1}\frac{1}{1 - D}\sqrt{1 + \theta^2}.$$

If the ratio of the capacitance divider is adjusted so that $C_2 / C_1$ is equal to the nominal ratio $K_n$ of the transformer,

$$\frac{R_x}{K_n} = \frac{1}{1 - D}\sqrt{1 + \theta^2} \approx 1 + D + D^2 + \frac{1}{2}\theta^2$$

In general, $\frac{R_x}{K_n} = 1 + D$ gives sufficient accuracy.

The accuracy of the method depends upon the auxiliary transformer 2. Its ratio is nearly unity, however, and in these conditions it is not difficult to construct a transformer having very small ratio error and phase angle.

The guard rings which constitute the third terminals of capacitors $C_1$ and $C_2$ are not shown in Fig. 10.34 ; both are connected to earth.

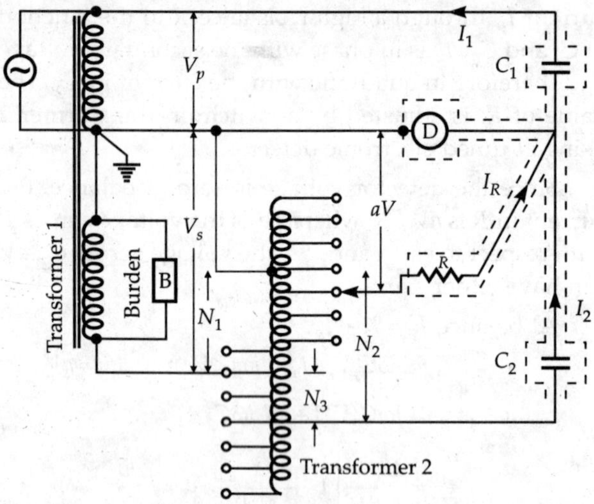


**Fig. 10.34** Clothier and Medina absolute method of testing potential transformer.

## Comparison Methods

**1. *Method using wattmeters.*** This method is analogous to Silsbee's deflectional comparison method for current transformers. The arrangement is shown in Fig. 10.35. The ratio and phase angle errors of a test transformer are determined in terms of those of a standard transformer *S* having the same nominal ratio.

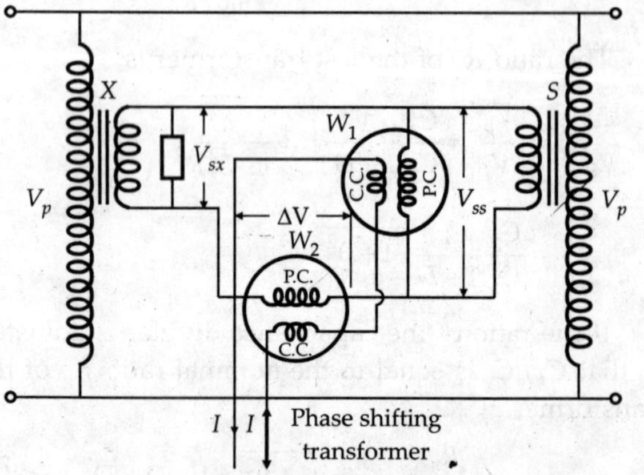


**Fig. 10.35** Comparison method using wattmeters.

The two transformers are connected with their primaries in parallel. A burden is put in the secondary circuit of test transformer. $W_1$ is a wattmeter whose potential coil is connected across the secondary of standard transformer. The pressure coil of wattmeter $W_2$ is so connected that a voltage $\Delta V$ which is the difference between secondary voltages of standard and test transformers, is impressed across it. The current coils of the two wattmeters are connected in series and are supplied from a phase shifting transformer. They carry a constant current *I*.

**1.** The phase of the current *I* is so adjusted that the wattmeter $W_1$ reads zero. Under these conditions current *I* is in $V_{ss}$ quadrature with voltage $V_{ss}$. The position of current phasor for this case is shown in Fig. 10.36 as $I_q$.

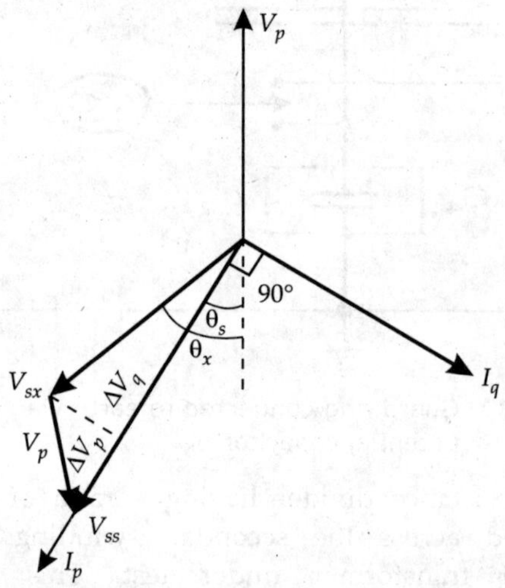


**Fig. 10.36** Phasor diagram for circuit of Fig. 10.35.

Reading of wattmeter

$$W_1, W_{1q} = V_{ss} I_q \cos 90° = 0.$$

Reading of wattmeter $W_2$ is

$W_{2q}$ = component of voltage $\Delta V$ in phase with $I_q \times I_q$.

$$= \Delta V_q I_q = V_{sx} \sin(\theta_x - \theta_s) \times I_q$$

$$= V_{sx} I_q \sin(\theta_x - \theta_s).$$

**2.** The phase of current *I* is shifted through 90° so that it occupies a position $I_p$ in phase with $V_{ss}$.

Reading of wattmeter $W_1$ is $W_{1p}$

$$= V_{ss} I_p \cos 0 = V_{ss} I_p$$

Reading of wattmeter $W_2$

$W_{2p}$ = component of voltage $\Delta V$ in phase with $I_p \times I_p$

$$= \Delta V_p \times I_p = [V_{ss} - V_{sx} \cos(\theta_x - \theta_s)] I_p$$

If the current is kept same for both the readings

$$I = I_p = I_q.$$

$$\therefore \quad W_{2q} = V_{sx} I \sin(\theta_x - \theta_s).\ W_{1p} = V_{ss} I$$

$$W_{2p} = [V_{ss} - V_{sx} \cos(\theta_x - \theta_s)] I$$

$$= V_{ss} I - V_{sx} I \cos(\theta_x - \theta_s) = W_{1p} - V_{sx} I$$

as $(\theta_x - \theta_s)$ is very small. $\therefore \quad V_{sx} I = W_{1p} - W_{2p}$

Actual ratio of test transformer

$$R_x = \frac{V_p}{V_{sx}}.$$

Actual ratio of standard transformer

$$R_s = \frac{V_p}{V_{sx}}$$

$$\therefore \quad \frac{R_x}{R_s} = \frac{V_{ss}}{V_{sx}} = \frac{V_{ss} I}{V_{sx} I} = \frac{W_{1p}}{W_{1p} - W_{2p}} \approx 1 + \frac{W_{2p}}{W_{1p}}$$

or ratio of transformer under test

$$R_x = R_s\left(1 + \frac{W_{2p}}{W_{1p}}\right)$$

We have,

$$\sin(\theta_x - \theta_s) = \frac{W_{2q}}{V_{sx} I}$$

and $$\cos(\theta_x - \theta_s) = \frac{V_{ss} I - W_{2p}}{V_{sx} I} = \frac{W_{1p} - W_{2p}}{V_{sx} I}$$

From above,

$$\tan(\theta_x - \theta_s) = \frac{W_{2q}}{W_{1p} - W_{2p}}$$

or $$\theta_x - \theta_s \approx \frac{W_{2q}}{W_{1p} - W_{2p}} \text{ rad.}$$

Phase angle of test transformer

$$\theta_x \approx \frac{W_{2q}}{W_{1p} - W_{2p}} + \theta_s \approx \frac{W_{2q}}{W_{1p}} + \theta_s \text{ rad.}$$

Wattmeter $W_2$ should have a very low voltage range since only a small voltage $\Delta V$ is impressed upon its pressure coil.

**2. *Clothier and Medina comparison method.*** A simplified diagram of the test circuit devised by Clothier and Medina is shown in Fig. 10.37. The primaries of the two transformers are connected in parallel and facilities are provided for fine adjustments of magnitude and phase of the secondary voltage of standard transformer until it equals that of the test transformer.

Ratio balance is affected by adjustment of primary turns $N$, of an autotransformer connected to the secondary of the current transformer. The phase balance is done by adjustment of a resistance $r$ through which a current flows, this current being in quadrature with the secondary voltage. The sign of the secondary voltage is selected by a reversing switch. Adjustments of resistance $r$ can be made in steps of 1 in 10.000 in ratio and 0.1′ in phase.

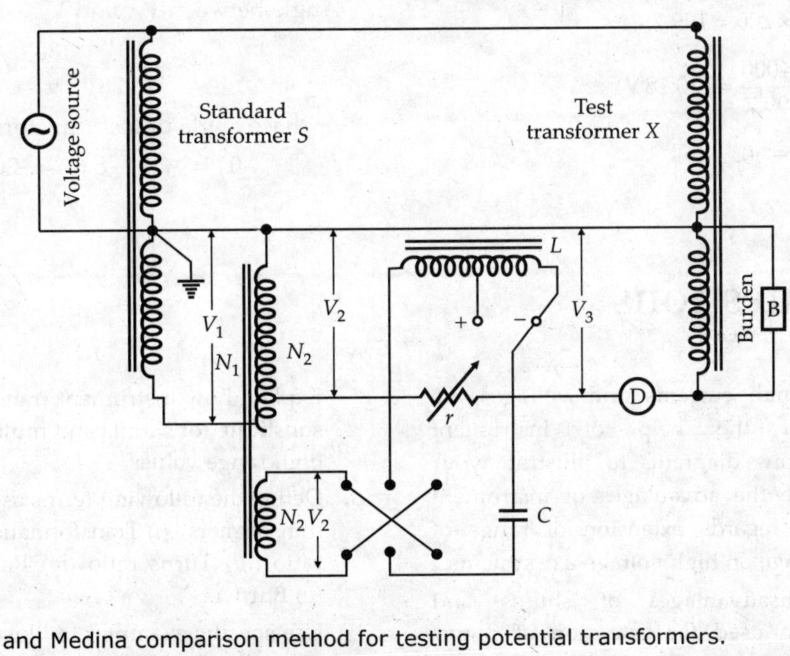


**Fig. 10.37** Clothier and Medina comparison method for testing potential transformers.

At balance :

$$\frac{R_x}{R_s} = \frac{N_1}{N_2} \text{ and } \theta_x = \theta_s \pm rC\omega$$

The circuit uses a tapped inductance $L$ in parallel with resistance $r$, which ensures that the ratio and phase relations are completely independent. This can be achieved by making $L$ equal to $2/(\omega^2 C)$ for negative and to $2/(3\omega^2 C)$ for positive phase angles.

This equipment is portable and uses an iron cored dynamometer instrument as a detector.

**Example 10.15** *An instrument potential transformer of nominal ratio 24,000/120 V is tested by comparison with a calibrated transformer of the same nominal ratio. The standard transformer is known to have an RCF of 0.9985 and a phase angle error of – 12′. The secondaries are connected in opposition and the resultant voltage ΔV (terminal voltage of test transformer – terminal voltage of standard transformer) is 0.5 ∠ 216.9° volt with respect to secondary voltage of standard transformer when the transformers are fed from rated voltage supply. Compute the RCF and phase angle of test transformer.*

**Solution.**

$V_{sx}$ = secondary voltage of test transformer,

$V_{ss}$ = secondary voltage of standard transformer,

$\theta_x$ = phase angle of test transformer,

and $\theta_s$ = phase angle of standard transformer.

Nominal ratio $= 24000/120 = 200$

Actual ratio of standard transformer

= RCF × nominal ratio

$$R_s = 0.9985 \times 200 = 199.7.$$

$$V_{ss} = \frac{V_p}{R_s} = \frac{24000}{199.7} = 120.18 \text{ V}$$

$$V_{ss} = 120.18 + j0.$$

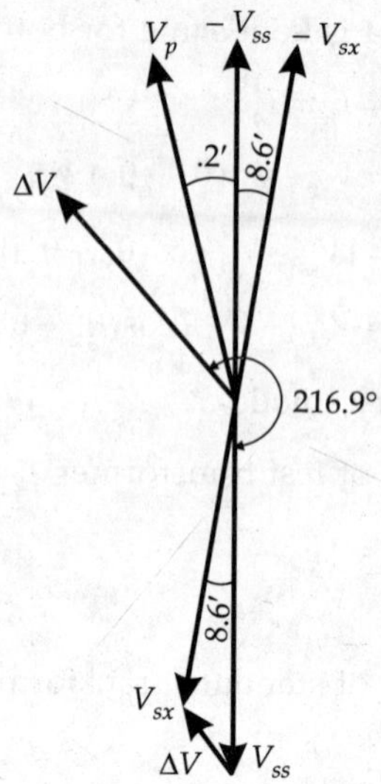


**Fig. 10.38** Phasor diagram of Example 10.15

$$\Delta V = 0.5\angle 216.9° = 0.5(\cos 216.9 + j\sin 216.9) = -0.4 - j0.3$$

It is given $V_{sx} = V_{ss} = \Delta V$

or $$V_{sx} = \Delta V + V_{ss} = -0.4 - j0.3 + 120.18 + j0 = 119.78 - j0.3.$$

∴ $$V_{sx} \approx 119.78 \text{ V}.$$

Actual ratio test transformer

$$R_x = \frac{V_p}{V_{sx}} = \frac{24000}{119.78} = 200.36$$

RCF of test transformer

$$= \frac{\text{actual ratio}}{\text{nominal ratio}} = \frac{200.36}{200} = 1.0018.$$

Angle between $V_{sx}$ and $V_{ss}$

$$= -\frac{0.3}{119.78} \times \frac{180}{\pi} \times 60 = -8.6'$$

∴ Phase angle of test transformer

$$\theta_x = -(12' + 8.6') = -20.6' \quad \text{(See Fig. 10.38).}$$

# Review Questions

1. Describe how high currents and voltages are measured with the help of instrument transformers. Draw diagrams to illustrate your answer. Describe the advantages of instrument transformers as regards extension of range of current and voltage on high voltage a.c. systems.
2. Explain the disadvantages of shunts and multipliers when used for extension of range. Explain how instrument transformers are a better substitute for shunts and multipliers especially for high range values.
3. Define the following terms as used for instrument transformers : (*i*) Transformation ratio. (*ii*) Nominal ratio (*iii*) Turns ratio (*iv*) Ratio correction factor (*v*) Burden.

   Use specific examples to illustrate your answer.

4. Draw the equivalent circuit and phasor diagram of a current transformer. Derive the expressions for ratio and phase angle errors.
5. Explain the effect of secondary burden on the ratio and phase errors of a current transformer.
6. Explain the effect of the following on the performance of current transformers :
   (i) Change of primary winding current
   (ii) Change of secondary circuit burden and
   (iii) Change of frequency.
7. Discuss the major sources of errors in current transformers. Discuss them. Describe the design and constructional features used in current transformers to reduce the errors.
8. Describe the constructional details of wound and bar types current transformers for reduction of the ratio and phase angle errors. Describe the features incorporated in the windings to reduce the leakage reactance.
9. Explain the method of turns compensation used in current transformers to reduce ratio error. The explanation should be illustrated with the help of a suitable example.
10. Explain the Wilson Compensation Method for reduction of errors in current transformers.
11. Explain the construction and working of clip-on-ammeters. Explain their utility. Are these instruments used as purely indicator type instruments or can they be used for precision measurements ? Explain.
12. Explain in details the effect of opening the secondary circuit of a current transformer when the primary winding is energized.
13. Explain the methods for demagnetization of the core of a current transformer whose secondary circuit has been accidentally opened when the primary winding was energised.
14. Draw the equivalent circuit and phasor diagram of a potential transformer. Derive the expressions for its ratio and phase angle errors. Describe the assumptions made for derivation of errors.
15. Describe the design and constructional features used in potential transformers 'for reduction of ratio and phase angle errors.
16. Describe the working of a capacitive type potential transformer. Describe how the transformation ratio can be made independent of the burden of the transformer.
17. Describe the effect of the following on the characteristics of a potential transformer :
   (i) burden (VA) of secondary winding circuit
   (ii) power factor of secondary winding circuit
   (iii) frequency
   (iv) supply voltage.
18. Describe the following methods for testing of a current transformer :
   (i) Mutual Inductance Method
   (ii) Silsbee's Method and
   (iii) Arnold's Method.
19. Explain the Absolute Null Method for testing of a potential transformer.
20. Describe the comparison method using wattmeters for testing of potential transformers.

10

## Unsolved Problems

1. A ring-core current transformer with a nominal ratio of 500/5 and a bar primary has a secondary resistance of 0.5 Ω and negligible secondary reactance. The resultant of magnetizing and iron loss components of the primary current associated with a full load secondary current of 5 A in a burden of 1.0 Ω (non-inductive) is 3 A at a power factor of 0.4. Calculate the true ratio and the phase angle error of transformer on full load. Calculate also the total flux in the core assuming a frequency of 50 Hz. **[Ans.** 100, 24 ; 0.314° ; $337 \times 10^{-6}$ Wb]
2. An 8/1 current transformer has an accurate current ratio when the secondary is short-circuited. The inductance of secondary is 60 mH and its resistance, is 0.5 Ω and the frequency is 50 Hz. Estimate the current ratio and phase angle error when the instrument load resistance is 0.4 Ω and inductance is 0.7 mH. Assume no iron loss and magnetizing current equal to 1 percent of primary current. The permeability remains constant.

   **[Ans.** 8.001 ; 0°2']
3. A current transformer with 5 primary turns has a secondary burden consisting of a resistance of 0.16Ω and an inductive resistance of 0.12 Ω. When the primary current is 200 A, the magnetizing current is 1.5 A and the iron loss current is 0.4 A.

Determine any expressions used, the number of secondary turns needed to make the current ratio 100.1 and also the phase angle under these conditions. **[Ans.** 497 ; 0.275°]

4. A current transformer of nominal ratio 1000/5 A, is operating with total secondary impedance $0.4 + j0.3\Omega$. At rated current the components of the primary current associated with the core- magnetizing and core loss effects are respectively 6 A and 1.5 A. The primary winding has 4 turns. Calculate the ratio error and phase angle at rated primary current if the secondary winding has (*a*) 800 turns (*b*) 795 turns. **[Ans.** (*a*) – 0.48% ; 13′, (*b*) + 0.14% ; 13′]

5. A bar type current transformer of toroidal construction requires 400 A to magnetize it, and 300 A to supply the iron loss for each volt per turn induced in the secondary winding and rated frequency. Across the secondary terminal is connected an impedance of 2 Ω with a phase angle φ, and the resistance of secondary winding is 0.5 Ω. The nominal ratio is 1000/5 A, and it is required to minimize both ratio and phase errors. Determine the necessary values of secondary winding turns and phase angle φ. **[Ans.** 194 ; 64°40′]

6. The resistance and reactance of the secondary of a 500/5 A current transformer are 0.02 Ω and 0.03 Ω respectively and the transformers characteristics are given by :

| *Emf., V* | 0.5 | 0.5 | 1.0 | 1.5 | 2.0 |
|---|---|---|---|---|---|
| *Magnetizing current, A* | 0.9 | 1.5 | 2.5 | 3.2 | 4.0 |
| *Core less component, A* | 0.5 | 1.3 | 2.7 | 3.9 | 4.6 |

An ammeter, a wattmeter current coil and an induction relay are connected in series with the secondary winding. Their resistances are 0.8 Ω, 0.10 Ω, 0.14 Ω and their reactances 0.09 Ω, 0.07 Ω and 0.08 Ω respectively. If the current in the instruments is 4 A, calculate the ratio and phase angle error (*a*) when all the instruments are in circuit, (*b*) when only the wattmeter is in circuit. Calculate the load VA in each case.

**[Ans.** (*a*) 1.4% 1.25′, 6.4 VA ; (*b*) – 0.601%, 2.5′, 1.96 VA]

7. A single phase potential transformer has a turns ratio of 3810/63. The nominal secondary voltage is 63 V and the total equivalent resistance and leakage reactance referred to the secondary side are 2 Ω and 1 Ω respectively. Calculate the ratio and phase angle errors when the transformer is supplying a burden of $100 + j200\,\Omega$. State the assumptions made. **[Ans.** – 0.80% ; – 0.338°]

8. A 500/100 V Potential transformer has the following constants :

Primary winding resistance = 47.25 Ω, Secondary winding resistance = 0.43 Ω, Primary winding reactance = 33.1 Ω. Secondary winding reactance is negligible ; no load primary current = 0.1 A at 0.6 p.f. Calculate

(*a*) the no load angle between primary winding and reversed secondary winding voltages,

(*b*) the value of secondary winding current at unity p.f. when the phase angle is zero.

**[Ans.** (*a*) 12°20′, (*b*) 0.0271 A]

# Objective Type Questions

## Tick (√) the most appropriate answer

1. The advantages of instrument transformers are

(*a*) The readings of instruments used in conjunction with them do not depend upon their resistance, inductance etc.

(*b*) The ratings of instrument transformers have been standardized and the rating of instruments used in conjunction used with them also get standardized. Therefore, there is reduction of cost and ease in replacements

(*c*) The metering circuit is electrically isolated from the power circuit thereby providing safety to operating personnel

(*d*) all the above.

2. The disadvantages of shunts for use at high currents are :

(*a*) it is difficult to achieve good accuracy with shunts

(*b*) power consumption of the shunts is large

(c) the metering circuit is not electrically isolated from the power circuit

(d) all the above.

**3.** The disadvantages of using multipliers with voltmeters at high voltages are :

(a) The power consumption of multipliers becomes large at large voltages

(b) The multipliers at high voltages have to be shielded in order to prevent capacitive currents

(c) The metering circuit is not electrically from the power circuit.

(d) all the above.

**4.** The nominal ratio of a current transformer is :

(a) primary winding current secondary winding current

(b) rated primary winding current/rated secondary winding current

(c) number of secondary winding turns/number of primary winding turns

(d) all the above.

**5.** The transformation ratio in the case of a potential transformer is defined as ratio of :

(a) primary winding voltage/secondary winding voltage

(b) rated primary winding voltage/rated secondary winding voltage

(c) number of turns of primary winding/number of turns of secondary winding

(d) all the above.

**6.** The burden of current transformers is expressed in terms of :

(a) secondary winding current

(b) VA rating of transformer

(c) voltage, current and power factor of secondary winding circuit

(d) none of the above.

**7.** A current transformer has a rating of 100/5 A. Its magnetizing and loss components of the exciting current are 1 A and 0.6 A respectively and secondary winding burden is purely resistive, its transformation ratio at rated current is :

(a) 20.12 (b) 20.2

(c) 200.2 (d) none of the above.

**8.** A current transformer has a rating of 1000/5 A. Its magnetizing current and loss component of exciting current are 10 A and 6 A respectively. The phase angle between secondary winding induced voltage and current is 30°. The phase angle error of the transformer at rated current is :

(a) 0.65°

(b) 0.305°

(c) 0.496°

(d) none of the these.

10

**9.** The errors in current transformers can be reduced by designing them with :

(a) high permeability and low loss core materials, avoiding any joints in the core and also keeping the flux density to a low value

(b) using primary and secondary windings as close to each other as possible

(c) using large cross-sections for both primary and secondary winding conductors

(d) all the above.

**10.** The current in the primary winding of a current transformer depends upon :

(a) burden of the secondary winding of transformer

(b) load connected to the system in which the C.T. is installed

(c) both burden of the transformer secondary winding and load connected to the system

(d) none of the above.

**11.** Turns compensation is used in current transformers primarily for reduction of :

(a) Phase angle error

(b) both ratio and phase angle errors

(c) ratio error, reduction in phase angle error is incidental

(d) none of the above.

**12.** A short-circuiting link is provided on the secondary side of a current transformer because :

(a) when the secondary winding of the C.T. is short-circuited by the link with the primary winding energized, a very high current flows on the primary side

(b) when the secondary winding of the C.T. is short-circuited by the link with the primary winding energized, it is possible to make any adjustments in the secondary winding circuit like replacing a faulty ammeter

(c) when the short-circuiting link is opened with the secondary winding open-circuited, the current on the primary side falls to almost zero

(d) all the above.

**13.** When the secondary winding of a current transformer is open-circuited with the primary winding energized

(*a*) the whole of the primary current produces large value of flux in the core (Limited only saturation) thereby producing a large voltage in the secondary winding

(*b*) the large voltage may act as safety hazard for the operators and may even rupture the insulation

(*c*) when the large magnetizing force is taken off it leaves a large value of residual magnetism

(*d*) all the above.

**14.** Clamp on meters are used because

(*a*) with their use it is possible to measure to current flowing a line without breaking the circuit

(*b*) their accuracy is high

(*c*) it is not possible to measure voltage without them

(*d*) all the above.

**15.** When the secondary winding of a potential transformer is suddenly open circuited with primary winding excited :

(*a*) large voltages are produced in the secondary winding may be a safety hazard to operating personnel

(*b*) the large voltages so produced may rupture the insulation

(*c*) the primary winding draws only the no load current

(*d*) none of the above.

**16.** The ratio and phase angle errors in potential transformers may be reduced by

(*a*) increasing the exciting current

(*b*) increasing the resistance and leakage reactance in the transformer

(*c*) by not employing turns compensation

(*d*) none of the above.

**17.** Capacitive potential transformers are used

(*a*) for primary winding phase voltages above 100 kV

(*b*) for keeping the value of transformation ratio constant irrespective of the burden by making certain adjustment

(*c*) because they are cheaper than the electromagnetic transformers above a certain voltage range

(*d*) all the above.

**18.** The size of potential transformers :

(*a*) is the same as that of power transformers of the same VA rating

(*b*) is much greater than that of power transformers of the same VA rating because they are designed for low ratio and phase angle errors which require large sized cores and winding conductors

(*c*) is smaller that that of power transformers of the same VA rating

(*d*) none of the above.

**19.** The ratio of transformation in the case of potential transformers :

(*a*) increases with increase in power factor of secondary burden

(*b*) remains constant irrespective of the power factor of secondary burden

(*c*) decreases with increase in power factor of secondary burden

(*d*) none of the above.

**20.** In case of potential transformers :

(*a*) the phase angle error is always positive

(*b*) the phase angle error is always negative

(*c*) the phase angle error is usually zero

(*d*) the phase angle error is positive when the secondary winding voltage reversed leads the primary winding voltage and is negative when the secondary winding voltage reversed lags behind the primary winding voltage.

## *Answers*

| | | | | | | | | | |
|---|---|---|---|---|---|---|---|---|---|
| **1.** (*d*) | **2.** (*d*) | **3.** (*d*) | **4.** (*b*) | **5.** (*a*) | **6.** (*b*) | **7.** (*a*) | **8.** (*b*) | **9.** (*a*) | **10.** (*b*) |
| **11.** (*c*) | **12.** (*b*) | **13.** (*d*) | **14.** (*a*) | **15.** (*c*) | **16.** (*d*) | **17.** (*d*) | **18.** (*b*) | **19.** (*c*) | **20.** (*d*) |

# 11

# Measurement of Power and Wattmeters

## 11.1 POWER IN D.C. CIRCUITS

The power taken by a load from a d.c. supply is given by the product of readings of an ammeter and a voltmeter when connected in the circuit as shown in Fig. 11.1. *i.e.*, power $P = VI$ watt.

It should be remembered that the ammeter and voltmeter require power for their operation and allowance must be made for this. Either of the two connections shown in Figs. 11.1(*a*) and (*b*) may be used. In Fig. 11.1(*a*), the ammeter is connected between load and voltmeter. Therefore, the voltmeter not only indicates the voltage $V_L$ across the load, but in addition voltage drop $V_a$ across the ammeter. If $R_a$ is the resistance of the ammeter, voltage drop across ammeter $V_a = IR_a$.

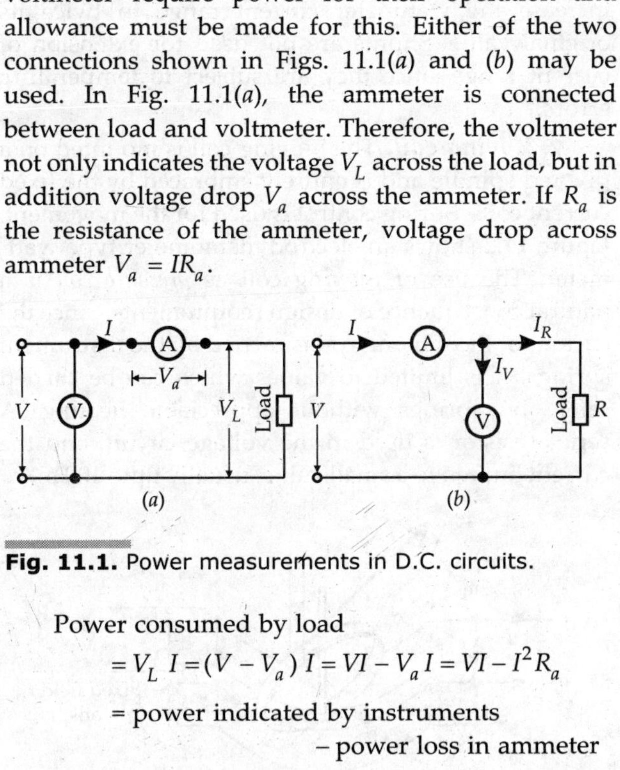

**Fig. 11.1.** Power measurements in D.C. circuits.

Power consumed by load

$$= V_L\, I = (V - V_a)\, I = VI - V_a I = VI - I^2 R_a$$

= power indicated by instruments − power loss in ammeter

∴ Power indicated by instruments

= power consumed in load + power loss in ammeter

In Fig. 11.1(*b*), voltmeter is connected between load and ammeter. Therefore, ammeter not only indicates the current through the load but in addition current in the voltmeter also.

Current through the voltmeter

$$I_V = V / R_V$$

where $R_V$ = resistance of voltmeter.

Power consumed by load

$$= VI_L = V(I - I_V) = V(I - V / R_V)$$

$$= VI - V^2 / R_V$$

= power indicated by instruments − power loss in voltmeter

∴ Power indicated by instruments

= power consumed in load + power loss in voltmeter

Thus in both the cases, the power indicated by the instruments is equal to the power consumed by the load plus the power consumed by the instrument nearer to the load terminals.

In order to obtain the true power, corrections must be applied for power loss in instruments. Under normal conditions the value of power loss in instruments is quite small as compared with the load power, and therefore, the error introduced on this account is small. However, when the output of the source to be measured is limited, the ammeter and voltmeter may load the circuit too heavily, causing serious errors.

For a permanently wired installation, when power measurements are required, it is a distinct advantage to instal a wattmeter in place of voltmeter and ammeter. Wattmeter gives direct indication of power and there is no need of multiplying two readings as is the case when voltmeter and ammeter are used. The labour involved is reduced and the accuracy is also increased.

## 11.2 POWER IN A.C. CIRCUITS

In the case of alternating currents, the instantaneous power varies continuously as the current and voltage go through a cycle. However, we are not interested in the instantaneous value of power (except where transient conditions are being studied) but in its average value over a cycle. This is because the average power multiplied by time measures the transfer of energy over a time interval, if steady state conditions exist.

The instantaneous power is given by

$$p = vi \quad \text{...(11.1)}$$

where $p$, $v$, $i$ are the instantaneous values of power, voltage and current respectively.

Thus, if both current and voltage waves are sinusoidal, the current lagging in phase by an angle $\phi$, then :

$$v = V_m \sin \omega t \quad \text{and} \quad i = i_m \sin(\omega t - \phi)$$

Therefore, the instantaneous power is given by

$$p = vi = V_m I_m \sin \omega t \sin(\omega t - \phi)$$

or writing $\theta = \omega t$, we have,

$$p = V_m I_m \sin\theta \sin(\theta - \phi)$$

$$= V_m \frac{I_m}{2}[\cos\phi - \cos(2\theta - \phi)] \quad \text{...(11.2)}$$

Average power over a cycle is

$$P = \frac{V_m I_m}{2\pi}\int_0^{2\pi}\left[\frac{\cos\phi - \cos(2\theta - \phi)}{2}\right]d\theta = \frac{V_m I_m}{2}\cos\phi$$

$$\therefore \; P = VI \cos\phi \quad \text{...(11.3)}$$

where $V$ and $I$ are rms values of voltage and current and $\cos\phi$ is the power factor of the load.

The fact the power factor ($\cos\phi$) is involved in the expression for power means that a wattmeter must be used for measurement of power in a.c. circuits instead of merely an ammeter and a voltmeter, since the later method takes no account of the power factor.

## 11.3 ELECTRODYNAMOMETER WATTMETERS

These instruments are similar in design and construction to electrodynamometer type ammeters and voltmeters described in Chapter 9 (page 267). The two coils are connected in different circuits for measurement of power. The fixed coils or *"field coils"* are connected in series with the load and so carry the current in the circuit. The fixed coils, therefore, form the *"current coil"* or simply C.C. of the wattmeter. The moving coil is connected across the voltage and, therefore, carries a current proportional to the voltage. A high non-inductive resistance is connected in series with the moving coil to limit the current to a small value. Since the moving coil carries a current proportional to the voltage, it is called the *"pressure coil"* or *"voltage coil"* or simply called P.C. of the wattmeter.

### 11.3.1. Construction of Electrodynamometer Wattmeter

1. ***Fixed coils.*** The fixed coils carry the current of the circuit. They are divided into two halves. The reason for using fixed coils as current coils is that they can be made more massive and can be easily constructed to carry considerable current since they present no problem of leading the current in or out. The fixed coils are wound with heavy wire. This wire is stranded or laminated especially when carrying heavy currents in order to avoid eddy current losses in conductors. The fixed coils of earlier wattmeters were designed to carry a current of 100 A but modern designs usually limit the maximum current ranges of wattmeters to about 20 A. For power measurements involving large load currents, it is usually better to use a 5 A wattmeter in conjunction with a current transformer of suitable range.

In the case of precision wattmeters, the two halves of the fixed coil, which are connected in series for a basic measuring range, can be connected in parallel to increase the wattmeter current range to twice its original value. Shunts are not used for extension of current range since they are subject to temperature errors.

2. ***Moving coil.*** The moving coil is mounted on a pivoted spindle and is entirely embraced by the fixed current coils. Spring control is used for the movement. Figure 11.2 shows an electrodynamometer type wattmeter. The use of moving coil as *pressure coil* is a natural consequence of design requirements. Since the current of the moving coil is carried by the instrument springs, it is limited to values which can be carried safely by springs without appreciable heating. A series resistor is used in the voltage circuit, and the current limited to a small value, usually upto 100 mA.

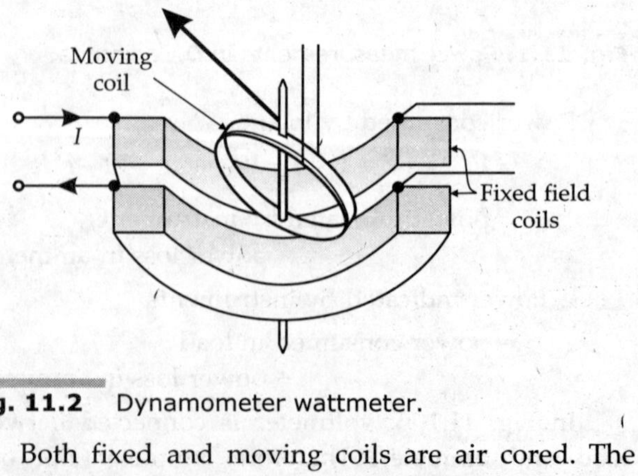


**Fig. 11.2** Dynamometer wattmeter.

Both fixed and moving coils are air cored. The voltage rating of the wattmeter is limited to about 600 V by the power requirements of the voltage circuit since most of the power is absorbed by the resistance in series with the moving coil and considerable heat is generated. For higher voltages, the pressure coil circuit is designed for 110 V, and a potential transformer is used to step down the voltage.

3. ***Control***. Spring control is used for the instrument.

4. ***Damping.*** Air friction damping is used. The moving system carries a light aluminium vane which moves in a sector shaped box. Electromagnetic or eddy current damping is not used as introduction of a permanent magnet (for damping purposes) will greatly distort the weak operating magnetic field.

5. ***Scales and pointers.*** They are equipped with mirror type scales and knife edge pointers to remove reading errors due to parallax.

### 11.3.2 Theory of Electrodynamometer Wattmeters

The instantaneous torque of an electrodynamometer instruments is given by :

$$T_i = i_1 i_2 \, dM/d\theta$$

where $i_1$ and $i_2$ are instantaneous values of currents in two coils.

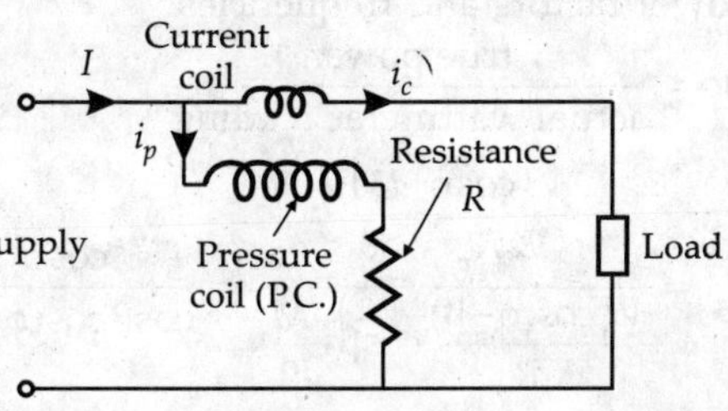


**Fig. 11.3** Circuit of electrodynamometer wattmeter.

Let $V$ and $I$ be the r.m.s. values of voltage and current being measured.

$\therefore$ Instantaneous value of voltage across the pressure coil circuit

$$v = \sqrt{2}\, V \sin \omega t.$$

If the pressure coil circuit has a very high resistance, it can be treated as purely resistive. Therefore, current $i_p$ in the pressure coil is in phase with the voltage and its instantaneous value is :

$$i_p = v / R_p = \sqrt{2}\, V/(R_p) \sin \omega t = \sqrt{2}\, I_p \sin \omega t$$

here $I_p = V_p / R_p$

= rms value of current in pressure coil circuit.

and $R_p$ = resistance of pressure coil circuit.

If the current in the current coil lags the voltage in phase by an angle $\phi$ instantaneous value of current through current coil is :

$$i_p = \sqrt{2}\, I \sin(\omega t - \phi)$$

$\therefore$ Instantaneous torque

$$T_i = \sqrt{2}\, I_p \sin \omega t \times \sqrt{2}\, I \sin(\omega t - \phi) dM/d\theta$$

$$= 2 I_p \, I \sin \omega t \sin(\omega t - \phi) \, dM/d\theta$$

$$= I_p I (\cos \phi - \cos(2\omega t - \phi)) dM/d\theta \quad ...(11.4)$$

It is clear from above that there is a component of power which varies as twice the frequency of current and voltage (mark the term containing $2\omega r$).

Average deflecting torque

$$T_d = \frac{1}{T} \int_0^T T_i \, d(\omega t)$$

$$= \frac{1}{T} \int_0^T I_p I [\cos \phi - \cos(2\omega t - \phi)] \frac{dM}{d\theta} . d(\omega t)$$

$$= I_p I \cos \phi . dM/d\theta \quad ...(11.5)$$

$$= (VI/R_p) \cos \phi . dM/d\theta \quad ...(11.6)$$

Controlling torque exerted by springs

$$T_c = K\theta$$

where $K$ = spring constant and

$\theta$ = final steady deflection.

Since the moving system of the instrument cannot follow the rapid variations in torque (the torque has a double frequency, component), it will take up a position at which the average deflection torque is equal to the restoring torque of the springs.

$\therefore$ At balance position

$$K\theta = I_p I \cos \phi . dM/d\theta$$

or deflection, $\theta = I_p I \cos \phi . (dM/d\theta)/K \quad ...(11.7)$

$$= (VI \cos \phi / R_p K) dM/d\theta \quad ...(11.8)$$

$$= K_1 VI \cos \phi . dM/d\theta \quad ...(11.9)$$

$$= (K_1 \, dM/d\theta) P \quad ...(11.10)$$

where $P$ = power being measured = $VI \cos \phi$

and $K_1 = 1/R_p K \quad ...(11.11)$

### 11.3.2. Shape of Scale of Electrodynamometer Wattmeters

It is clear from Eqn. 11.10, the deflection is directly proportional to the power being measured and the scale is essentially uniform over the range in which $(dM/d\theta)$ is constant. By suitable design, the mutual inductance between fixed and moving coil can be made to vary linearly with angle over a range of 40° to 50° on either side of zero mutual inductance position as shown in Fig. 11.4. If the position of zero mutual inductance is at the midscale, the scale will be uniform over 80° to 100° which covers almost the entire scale range.

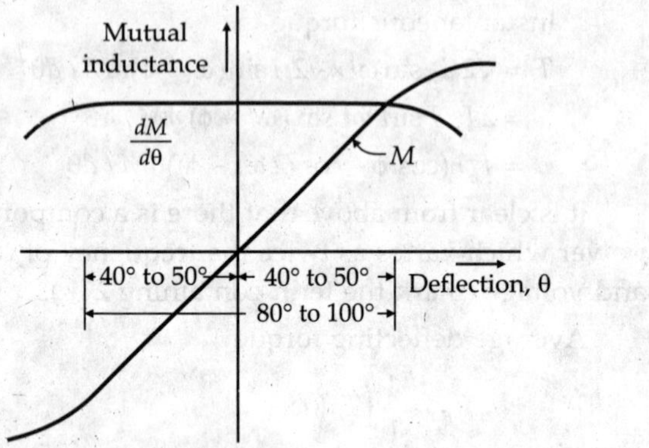


**Fig. 11.4** Variation of mutual inductance with deflection.

### 11.3.3 Errors in Electrodynamometer Wattmeters

1. ***Pressure coil inductance.*** We have,

$r_p$ = resistance of pressure coil,

$L$ = inductance of pressure coil,

$R$ = resistance in series with pressure coil,

$R_p$ = total resistance of pressure coil circuit $= r_p + R$,

$V$ = voltage applied to pressure coil circuit,

$I$ = current in the current coil circuit,

$I_p$ = current in the pressure coil circuit,

$Z_p$ = impedance of pressure coil circuit

$$= \sqrt{(R+r_p)^2+(\omega L)^2}$$

It was implied in the discussions of the idealized wattmeter that the current in the pressure coil is in phase with the applied voltage. If the pressure coils of the wattmeter has an inductance the current in it will lag the voltage by an angle β where :

$$\beta = \tan^{-1} \omega L/ R_p = \tan^{-1} \omega L/(r_p + R)$$

It will be seen from Fig. 11.5, that for lagging power factor, the angle between current in the current coil circuit and the current in the pressure coil circuit is less than ϕ, by which the load current lags the applied Voltage. The angle between the pressure coil current and the current in the current coil is :

$$\phi' = \phi - \beta$$

∴ The actual wattmeter reading is :

$$= (I_p I/K)\cos\phi' . dM/d\theta \qquad \text{(See Eqn. 11.7)}$$

$$= V/(Z_p K) . I \cos(\phi - \beta) . dM/d\theta$$

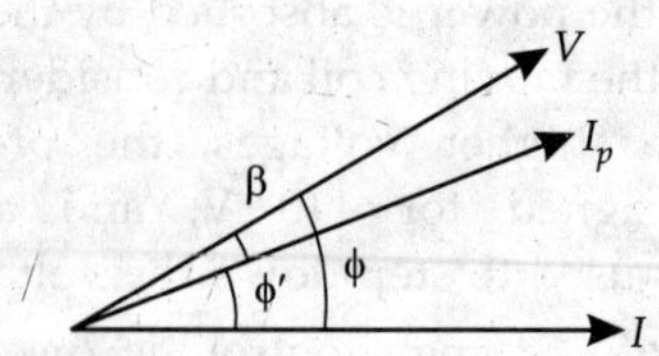


**Fig. 11.5** Wattmeter phasor diagram for lagging p.f.

Now $Z_p = R_p/\cos\beta$

∴ Actual wattmeter reading

$$= \frac{VI\cos(\phi-\beta)}{K(R_p/\cos\beta)}\frac{dM}{\theta}$$

$$= \frac{VI}{KR_p}\cos\beta\cos(\phi-\beta)\frac{dM}{d\theta} \qquad ...(11.12)$$

In the absence of inductance $Z_p = R_p$ and $\beta = 0$ and therefore the wattmeter will read true power under these conditions.

∴ True power

$$= \frac{I_p I}{K}\cos\phi\frac{dM}{d\theta} = \frac{VI\cos\phi}{KR_p}\frac{dM}{d\theta} \qquad ...(11.13)$$

It is clear from the above expression that in the absence of inductance, the wattmeter reads correctly at all power factors and frequencies.

$$\text{Ratio} = \frac{\text{true power}}{\text{actual wattmeter reading}}$$

$$= \frac{\dfrac{VI\cos\phi}{KR_p}.\dfrac{dM}{d\theta}}{\dfrac{VI\cos(\phi-\beta)}{KR_p}\cos\beta\dfrac{dM}{d\theta}} = \frac{\cos\phi}{\cos\beta\cos(\phi-\beta)}$$

∴ True power

$$= \frac{\cos\phi}{\cos\beta\cos(\phi-\beta)} \times \text{actual wattmeter reading} \qquad ...(11.14)$$

The *correction factor* is a factor by which the actual wattmeter reading is multiplied to get the true power.

∴ Correction factor

$$= \frac{\cos\phi}{\cos\beta\cos(\phi-\beta)} \text{ for lagging loads} \qquad ...(11.15)$$

It is clear from the phasor diagram (Fig. 11.5) that on lagging loads the wattmeter will read high, as the effect of the inductance of the pressure coil circuit is to bring the pressure coil current more nearly into phase with the load current than would be the case if this inductance were zero. Very serious errors may be introduced by pressure coil inductance at low power factors unless special precautions are taken.

The wattmeter will read low when the load power factor is leading as in that case the effect of pressure coil inductance is to increase the phase angle between load current and pressure coil current (See Fig. 11.6).

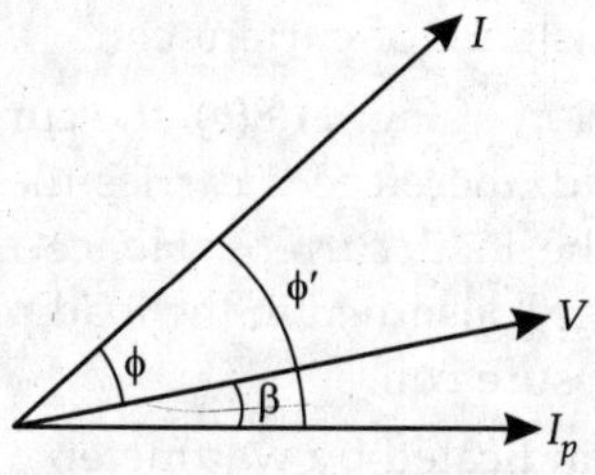


**Fig. 11.6** Wattmeter phasor diagram for leading p.f.

Correction factor for leading p.f.

$$= \frac{\cos\phi}{\cos\beta\cos(\phi+\beta)} \quad ...(11.16)$$

The errors produced on account of inductance of pressure coil in case of lagging power factor loads are discussed below :

The error in terms of instrument deflection is :

Actual wattmeter reading – true power

$$= \left[1 - \frac{\cos\phi}{\cos\beta\cos(\phi-\beta)}\right] \times \text{actual wattmeter reading} \quad ...(11.17)$$

$$\approx \left[1 - \frac{\cos\phi}{\cos(\phi-\beta)}\right] \times \text{actual wattmeter reading}$$

as $\beta$ is very small, at most a few minutes and, therefore, $\cos\beta$ is nearly equal to unity.

$$\text{Error} = \left[1 - \frac{\cos\phi}{\cos\phi\cos\beta+\sin\phi\sin\beta}\right] \times \text{actual wattmeter reading}$$

$$= \left[\frac{\sin\phi\sin\beta}{\cos\phi+\sin\phi\sin\beta}\right] \times \text{actual wattmeter reading}$$

$$= \left[\frac{\sin\beta}{\cot\phi+\sin\beta}\right] \times \text{actual wattmeter reading} \quad ...(11.18)$$

We can write Eqn. 11.14 as :

$$\frac{\text{true power}}{\text{actual wattmeter reading}} = \frac{\cos\phi}{\cos\beta\cos(\phi-\beta)}$$

$$= \frac{1+\tan^2\beta}{1+\tan\phi\tan\beta}$$

Now $\beta$ is very small and, therefore, $\tan^2\beta << 1$.

$$\therefore \quad \frac{\text{true power}}{\text{actual wattmeter reading}} \approx \frac{1}{1+\tan\phi\tan\beta}$$

or actual wattmeter reading

$$= \text{true power } (1 + \tan\phi\tan\beta) \quad ...(11.19)$$

$$\therefore \text{Error} = \text{actual wattmeter reading} - \text{true power}$$

$$= \tan\phi\tan\beta \times \text{true power} \quad ...(11.20)$$

Percentage error

$$= \frac{\text{wattmeter reading} - \text{true power}}{\text{true power}} \times 100$$

$$= \tan\phi\tan\beta \times 100 \quad ...(11.21)$$

We have, true power $= VI\cos\phi$

$$\therefore \text{Error} = \tan\phi\tan\beta \times VI\cos\phi = VI\sin\phi\tan\beta \quad ...(11.22)$$

It is clear from Eqn. 11.22 that the error is serious at low power factors.

*Compensation for inductance of pressure coil.* Many wattmeters are compensated for errors caused by inductance of pressure coil by means of a capacitor connected in parallel with a portion of multiplier (series resistance) as shown in Fig. 11.7.

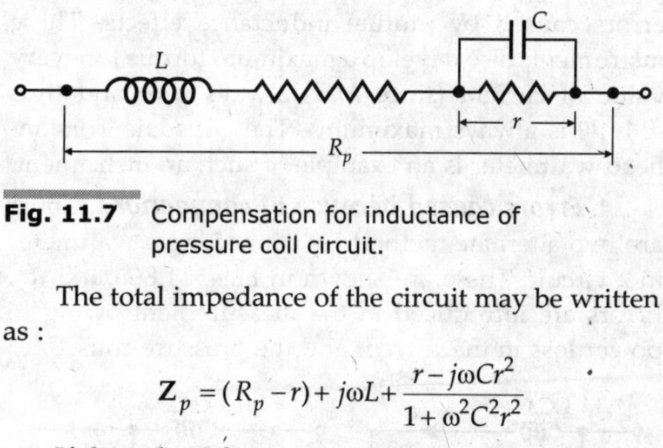


**Fig. 11.7** Compensation for inductance of pressure coil circuit.

The total impedance of the circuit may be written as :

$$\mathbf{Z}_p = (R_p - r) + j\omega L + \frac{r - j\omega Cr^2}{1+\omega^2C^2r^2}$$

If the value of circuit constants are so chosen that for power frequencies $\omega^2C^2r^2 << 1$.

We have approximately,

$$\mathbf{Z}_p \approx R_p - r + j\omega L + r - j\omega Cr^2 \approx R_p + j\omega(L - Cr^2)$$

If we make, $L = Cr^2$, then $Z_p \approx R_p$ and $\beta = 0$.

Thus the error caused by pressure coil inductance is almost completely eliminated. This type of compensation is very slightly effected by change in frequency and can be used for frequencies at which $\omega^2C^2r^2 << 1$.

The frequency range over which the above compensation holds good is 10 kHz.

2. ***Pressure coil capacitance.*** The pressure coil circuit may possess capacitance in addition to inductance. This capacitance is mainly due to

interturn capacitance of the series resistance. The effect capacitance is opposite to that produced by inductance, therefore, the wattmeter reads high on lagging power factors of the load.

The phase angle between pressure coil current and the applied voltage depends upon the reactance of the pressure coil circuit. The inductive reactance is normally greater than the capacitive reactance and, therefore, the phase angle varies with frequency, increasing with increase in frequency.

3. ***Error due to mutual inductance effects.*** Errors are caused due to mutual inductance between current and pressure coils of the wattmeter. These errors are quite low at power frequencies but they increase and become more important as the frequency is increased.

The effect of mutual inductance is to increase the phase angle for the connection in which the pressure coil is connected on the load side and to decrease the phase angle when the current coil is connected on the load side. This change in phase angle is $\tan^{-1} \omega M / R_p$.

Instruments have been developed whose coil systems are so arranged that they are always in a zero position of mutual inductance and thus are free from errors caused by mutual inductance effects. These instruments also develop a maximum torque for every value of current (since $M$ is always zero and thus $dM/d\theta$ is always maximum). The Drysdale Torsion-head wattmeter is an example of such an instrument.

4. ***Errors caused because of connections.*** There are two alternate methods of connecting a wattmeter in a circuit. These are shown in Figs. 11.8(*a*) and (*b*). Errors are introduced in the measurement owing to power loss in the current and the pressure coils.

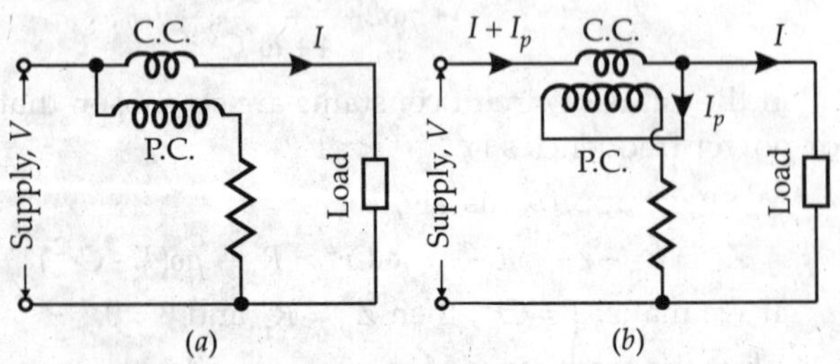


**Fig. 11.8** Wattmeter connections.

In connection of Fig. 11.8(*a*), the pressure coil is connected on the supply side and, therefore, the voltage applied to the pressure coil is the voltage across the load plus the voltage drop across the current coil. Thus the wattmeter measures the power loss in its current coil in addition to the power consumed by load.

Power indicated by wattmeter

$$= \text{power consumed by load} + \text{power loss in current coil}$$

$$= \text{power consumed by load} + I^2 / R_c$$

where $R_c$ = resistance of current coil.

In connection of Fig. 11.8(*b*), the current coil is on supply side and, therefore it carries the pressure coil current plus the load current. Hence the wattmeter reads the power consumed in the load plus the power loss in the pressure coil.

∴ Power indicated by wattmeter

$$= \text{power consumed by load} + \text{power loss in pressure coil}$$

$$= \text{power consumed by load} + V^2 / R_p,$$

where $R_p$ = resistance of potential coil.

If the load current is small, the voltage drop in the current coil is small, so that connection of Fig. 11.8(*a*) introduces a very small error. On the other hand, if the load current is large, the value of pressure coil current is very small as compared with the load current and hence power loss in pressure coil will be very small as compared with the load power and. therefore, connection of Fig. 11.8(*b*) is preferable.

It should be noted that in an uncompensated wattmeter, the reading of the wattmeter includes the power loss in the coil connected on the load side. The use of above connections for different applications [connection (*a*) for small currents and (*b*) for large currents] holds good only if the power loss in the instruments is to be neglected. However, if accuracy requires that the power loss in the instruments be taken into account, connection (*b*) is preferable. This is because for a constant voltage, the power loss $V^2 / R_p$, in the pressure coil circuit is constant and, therefore, the true power consumed by the load can be obtained by subtracting $V^2 / R_p$ from the wattmeter reading. The value of $V^2 / R_p$ can be calculated or observed directly by simply reading the wattmeter with load disconnected.

The power taken by the pressure coil circuit is constant if the voltage is constant and becomes a smaller percentage of the total as a larger amount of power is measured. However, in cases, where the load current is large and the power factor is small, connection (*b*) may result in large error since the total power measured is small. Hence in wattmeters which are designed for low power factor measurements, a compensating coil may used in the instrument to compensate for the error caused by power loss in the pressure coil circuit. The arrangement using a compensating coil is shown in Fig. 11.9.

In connection 11.8(*b*) the current coil carries a current of $(I + I_p)$, and produces a field corresponding to this current. The compensating coil is connected in series with the pressure coil circuit (Fig. 11.9) and is made as nearly as possible identical and coincident with the current coil. It is so connected that it opposes the field of the current coil. The compensating coil carries a current $I_p$ and produces a field corresponding to this current. This field acts in opposition to the current coil field. Thus the resultant field is due to current $I$ only. Hence the error caused by the pressure coil current flowing in the current coil is neutralized.

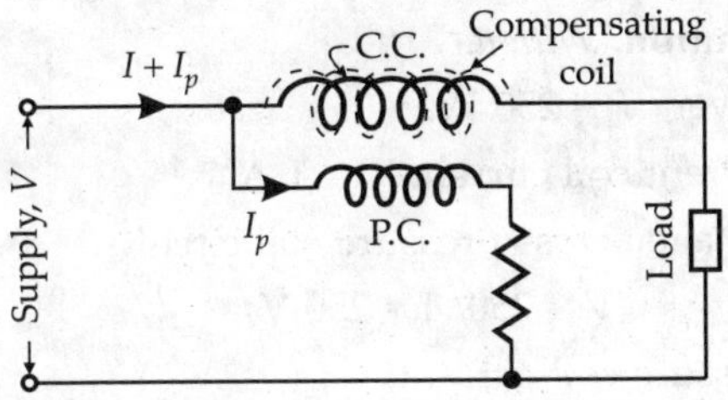


**Fig. 11.9** Compensation for power loss in pressure coil circuit.

5. ***Eddy current errors.*** Eddy currents are induced in the solid metal parts and within the thickness of the conductors by alternating magnetic field of the current coil. These currents produce a field of their own and alter the magnitude and phase of the current coil field and thus cause errors.

Refer to the phasor diagram shown in Fig. 11.10.

The phase of eddy emf $E_e$, is 90° behind the current coil flux $\Phi_c$. The eddy current $I_e$, lags eddy emf $E_e$ by a small angle. $I_e$ produces a flux $\Phi_e$ and therefore the resultant flux is $\Phi_r$. It is clear from Fig. 11.10, that the resultant flux is less than the current coil flux $\Phi_c$ and lags it by a small angle.

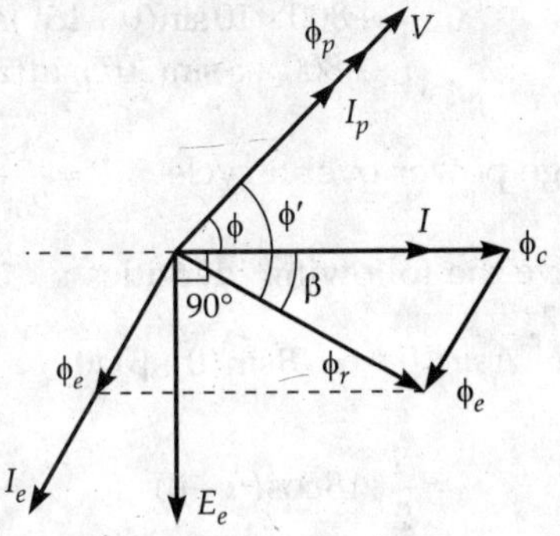


**Fig. 11.10** Phasor diagram for illustrating the effect of eddy current.

Since the resultant flux lags the current coil flux, the phase angle between the fluxes $\Phi_p$ and $\Phi_r$, which produce the torque, is increased. As the magnitude of the field is decreased and the angle between fluxes is increased, the torque produced is decreased for lagging power factors.

This error is similar in principle to the error caused by the inductance of pressure coil except that the sign of error is opposite. From Fig. 11.10, it is clear that the wattmeter will read low for lagging power factors and high for leading power factors.

The error caused by eddy currents is not easily calculable and may be serious if the metal parts are not well removed from the current coil. In fact solid metal parts should be avoided as far as possible. Stranded conductors should be used for the current coil if the current carried is large. This reduces the eddy currents induced within the thickness of the coil.

6. ***Stray magnetic field errors.*** The electrodynamometer wattmeter has a relatively weak operating field and therefore, it is particularly effected by stray magnetic fields resulting in the serious errors. Hence these instruments should be shielded against effects of stray magnetic fields.

11

Laminated iron shields are used in portable laboratory instruments while steel cases are sometimes provided to shield the switch-board instruments. Precision type of wattmeters are not provided with shields in order to keep down errors caused by eddy currents and to avoid the small d.c. errors which may be caused by permanent magnetism of the shield. Such wattmeters use an astatic system as shown in Fig. 9.37.

7. ***Errors caused by vibration of moving system.*** This error is found occasionally on alternating currents. The torque on the moving system varies cyclically with a frequency which is twice that of the voltage (see Eqn. 11.4). If the pointer, spring or some other part of the moving system has a natural frequency which is in approximate resonance with the frequency of torque pulsation, the moving system would vibrate with a considerable amplitude. These vibrations not only make the position of the pointer difficult to read but also the mean position of the pointer may be displaced from that corresponding to the average value of power. These vibrations are avoided by designing the instruments in such a way that the natural frequency of the moving system is very much away from twice the frequency of the system on which the wattmeter is intended to be used.

8. ***Temperature errors.*** The indication of a wattmeter is effected by changes in room temperature. This is because any change in room temperature changes the resistance of the pressure coil and the

stiffness of the springs. These effects are opposite in nature and would nearly neutralize each other, if the pressure coil circuit were composed of copper and of a resistance alloy having a negligible resistance temperature co-efficient in the ratio of 1 : 10.

**Example 11.1.** *In a dynamometer wattmeter the moving coil has 500 turns of mean diameter 30 mm. Estimate the torque if the axes of the field and the moving coils are at (a) 60° (b) 90° when the flux density produced by field coils is $15 \times 10^{-3}$ Wb /m², the current in moving coil is 0.05 A and the power factor is 0.866.*

**Solution.** Let us take the position of moving coil when its axis is coincident with that of fixed coil as reference. Let θ be the deflection from the reference position, *i.e.*, θ be the angle between axes of two coils.

Flux linking with moving coil

$$= \text{area} \times \text{component of flux density perpendicular to area}$$
$$= (\pi/4) D^2 \times B \times \cos\theta$$

where $D$ = mean diameter of pressure coil.

Flux linkages of moving coil

$$= N \times (\pi/4) D^2 \times B\cos\theta = (\pi/4) N D^2 B\cos\theta$$

Mutual inductance $M$

$$= \frac{\text{flux linkages of moving coil}}{\text{(current of fixed coil)}}$$

$$= \frac{\pi}{4I} N D^2 B\cos\theta = M_{max} \cos\theta$$

where $M_{max}$ = maximum mutual inductance

$$= (\pi/4I) N D^2 B.$$

Rate of change of flux linkages

$$dM/d\theta = (\pi/4I) N D^2 B\sin\theta = M_{max} \sin\theta$$

(sign disregarded)

From Eqn. 11.6, torque exerted

$$T_d = \frac{V}{R_p} I\cos\phi \frac{dM}{d\theta} = \left(\frac{VI\cos\phi}{R_p}\right) . M_{max} \sin\theta$$

$$= (I_p I\cos\phi) M_{max} \sin\theta$$

$$= I_p I \times (\pi/4I) N D^2 B\sin\theta\cos\phi$$

$$= (\pi/4) N D^2 B I_p \cos\phi \sin\theta$$

We have, $B = 15 \times 10^{-3}$ Wb / m² ;

$I_p = 0.05$ A ; $N = 500$; $D = 30 \times 10^{-3}$ m ; $\cos\phi = 0.866$.

(*a*) For $\theta = 60°$, $\sin\theta = 0.866$

∴ Deflecting torque

$$T_d = (\pi/4) \times 500 \times (30 \times 10^{-3})^2 \times 15 \times 10^{-3} \times 0.05 \times 0.866 \times 0.866$$

$$= 198.8 \times 10^{-6} \text{Nm}.$$

(*b*) When $\theta = 90°$, $\sin\theta = 1$

$$\therefore \quad T_d = (\pi/4) \times 500 \times (3 \times 10^{-2})^2 \times 15 \times 10^{-3} \times 0.05 \times 0.866 \times 1$$

$$= 229.5 \times 10^{-6} \text{Nm}.$$

**Example 11.2** *An uncompensated spring controlled dynamometer wattmeter reads 250 W with d.c. currents of 1 A and 0.05 A in its current and potential coils respectively. Calculate what this wattmeter will read when the current coil current is 10 sin (377 t + 15°) + 5 sin (1131 t) ampere and the potential coil voltage is 500 cos (377 t – 30°) + 800 sin (754 t + 45°) volt. Calculate also the resistance of the potential coil circuit assuming it to be purely resistive.*

**Solution.** *With d.c.*

Power, $P = 250$ W.

Current coil current, $I = 1$ A

Voltage across pressure coil circuit

$$V = 250/1 = 250 \text{ V}$$

Pressure coil current

$$I_p = 0.05 \text{ A}$$

∴ Resistance of potential coil circuit

$$R_p = 250/0.05 = 5000\ \Omega.$$

*With a.c.*

Instantaneous current

$$i = 10\sin(377t + 15°) + 5\sin(1131t)$$

Putting 317 $t$ = θ, we have

$$i = 10\sin(\theta + 15°) + 5\sin(3\theta)$$

Instantaneous voltage

$$v = 500\cos(377t - 30°) + 800\sin(754t + 45°)$$
$$= 500\cos(\theta - 30°) + 800\sin(2\theta + 45°)$$

Instantaneous power $p = vi$

$$= 500 \times 10\sin(\theta + 15°)\cos(\theta - 30°) + 500 \times 5\sin(3\theta)\cos(\theta - 30°) + 800 \times 10\sin(\theta + 15°)\sin(2\theta + 45°) + 800 \times 5\sin(30°)\sin(2\theta + 45°).$$

Average power over a cycle $\quad P = \frac{1}{2\pi}\int_0^{2\pi} vi\, d\theta$

We have the following identities :

(*i*) $\frac{1}{2\pi}\int_0^{2\pi} A\sin(\theta + \alpha) . B\sin(\theta + \beta) d\theta$

$$= \frac{1}{2} AB\cos(\alpha - \beta).$$

(*ii*) $\frac{1}{2\pi}\int_0^{2\pi} A\sin(\theta + \alpha) . B\sin(\theta + \beta) d\theta$

$$= \frac{1}{2} AB\sin(\alpha - \beta).$$

(*iii*) $\frac{1}{2\pi}\int_0^{2\pi} A\sin(m\theta+\alpha).B\sin(n\theta+\beta)d\theta = 0$

(*iv*) $\frac{1}{2\pi}\int_0^{2\pi} A\sin(m\theta+\alpha).B\cos(n\theta+\beta)d\theta = 0$

Therefore,

$$\frac{1}{2\pi}\int_0^{2\pi} 500\times 10\sin(\theta+15°)\cos(\theta-30°)d\theta$$

$$= \frac{1}{2}\times 500\times 10\sin 45° = 1768$$

Now $\frac{1}{2\pi}\int_0^{2\pi} 500\times 5\sin(3\theta)\cos(\theta-30°)d\theta = 0$;

$$\frac{1}{2\pi}\int_0^{2\pi} 800\times 10\sin(\theta+15°)\sin(2\theta+45°) = 0;$$

and $\frac{1}{2\pi}\int_0^{2\pi} 800\times 5\sin(3\theta)\sin(2\theta+45°) = 0$

Hence average power $P$ = 1768 W.

**Example 11.3** *A wattmeter has a current coil of 0.1 Ω resistance and a pressure coil of 6500 Ω resistance. Calculate the percentage errors, due to resistance only with each of the two methods of connection, shown in Fig. 11.8 when reading the input to an apparatus which takes (a) 12 A at 250 V with unity power factor, and (b) 12 A at 250 V and 0.4 power factor.*

**Solution.** (*a*) Power = $VI\cos\phi$

$$= 250 \times 12 \times 1 = 3000 \text{ W}.$$

The two possible ways in which the pressure coil can be connected are shown in Figs. 11.8(*a*) and (*b*).

For connection of Fig. 11.8(*a*), the error is caused by power lost in the current coil.

Power loss in current coil

$$= I^2R_c = (12)^2 \times 0.1 = 14.4 \text{ W}$$

$\therefore$ Error = (14.4 / 3000) × 100 = 0.48%

For connection of Fig. 11.8(*b*), error is caused by power lost in pressure coil.

Power loss in pressure coil

$$= V^2/R_p = (250)^2/6500 = 9.6 \text{ W}$$

$\therefore$ Error = (9.6 / 3000) × 100 = 0.32%

(*b*) Power = 250 × 12 × 0.4 = 1200 W

For connection of Fig. 11.8(*a*)

Error = (14.4 / 1200) × 100 = 1.2%

For connection of Fig. 11.8 (*b*) error

= (9.6 / 1200) × 100 = 0.8%.

**Example 11.4.** *A wattmeter has a current coil of 0.03 Ω resistance and a pressure coil of 6000 Ω resistance. Calculate the percentage error if the wattmeter is so connected that :*

*(i) the current coil is on the load side,*

*(ii) the pressure coil is on the load side*

*(a) if the load takes 20 A at a voltage of 220 V and 0.6 power factor in each case. (b) What load current would give equal errors with the two connections ?*

**Solution.** (*a*) Power consumed by load

$$= 220 \times 20 \times 0.6 = 2640 \text{ W}$$

(*i*) The wattmeter measures the loss in the current coil for this connection.

Loss in current coil

$$= I^2R_c = (20)^2 \times 0.03 = 12 \text{ W}.$$

$\therefore$ Error = (12 / 2640) × 100 = 0.45%.

(*ii*) The wattmeter measures the loss in the pressure coil circuit for this connection.

Loss in pressure coil circuit

$$= V^2/R_p = (220)^2/6000 = 8.06 \text{ W}.$$

Error = (8.06 / 2640) × 100 = 0.31%.

(*b*) For equal error for the two connections

$$I^2R_c = \frac{V^2}{R_p} \quad \text{or} \quad I^2 = \frac{V^2}{R_cR_p} = \frac{(220)^2}{0.03\times 6000}$$

$\therefore$ $I = 16.4$ A.

**Example 11.5** *A wattmeter is rated at 10 A and 25 V. The current coil has a resistance of 0.06 Ω and a reactance of 0.02 Ω. The potential coil circuit may be assumed to be purely resistive having a resistance of 6250 Ω. Find the error due to the two different connections shown in Fig. 11.8. The load is 10 A at a power factor of 0.174 lagging. The voltage across the load is 25 V.*

**Solution.** Power consumed by load

$$= 25 \times 10 \times 0.174 = 43.5 \text{ W}$$

In Fig. 11.8 (*a*), the current coil is on the load side.

Power factor

$$\cos\phi = 0.174,$$

$\therefore$ $\phi = 80°$ lagging

Current $\mathbf{I} = 10(\cos 80° - j\sin 80°)$

$$= 10(0.1736 - j0.985) = 1.736 - j9.85 \text{ A}$$

Voltage drop across current coil

$$= (1.736 - j9.86)(0.06 + j0.02)$$

$$= 0.301 - j0.5569$$

Voltage across pressure coil $V_p$

$$= 25 + 0.301 - j0.5569$$

$$= 25.316\angle -1.26° \text{ V}$$

11

Power indicated by wattmeter

$= 25.316 \times 10 \times \cos(80° - 1.26°)$

$= 49.5$ W.

(This could be calculated otherwise)

Reading of wattmeter

= power consumed in load
+ power consumed in C.C.

$= 43.5 + (10)^2 \times 0.06 = 49.6$ W.

$$\text{Error} = \frac{49.5 - 43.5}{43.5} \times 100 = 13.8\%$$

When the pressure coil is on the load side as in Fig. 11.8(*b*).

Power lost in pressure coil

$= (25)^2 / 6250 = 0.1$ W

Power indicated by wattmeter

$= 43.5 + 0.1 = 43.6$ W

$$\text{Error} = \frac{43.6 - 43.5}{43.5} \times 100 = 0.23\%.$$

**Example 11.6** *A dynamometer wattmeter is used to measure the power factor of a 20 μF capacitor. The pressure coil of the wattmeter having a resistance 1000 Ω and an inductive reactance of 15 Ω is connected across a 50 Hz supply. The current coil of the wattmeter, a variable resistor R and the capacitor are connected in series across the same supply. The wattmeter deflection is made zero by adjusting the value of R to 1.65 Ω. If the current coil resistance is 0.1 Ω and its inductance negligible : determine the power factor of the capacitor.*

**Solution.** Phase angle of pressure coil circuit

$$\beta = \tan^{-1} 15/1000 = 0.015 \text{ rad} \approx 51.6'$$

Capacitive reactance

$$X_c = 10^6 / (2\pi \times 50 \times 20) = 159.2\ \Omega.$$

Let $r$ be the series resistance of the capacitor.

Total resistance of the load circuit including resistance of current coil.

$= 0.1 + 1.65 + r = 1.75 + r$ ohm.

$\therefore \quad \phi$ = phase angle of load circuit

$= \tan^{-1} 159.2 / (1.75 + r)$

The load circuit has a leading p.f. as the reactance is capacitive.

*For leading p.f.*

Reading of wattmeter $\alpha \cos\beta \cos(\phi + \beta)$.

For a wattmeter reading to be zero,

$\cos\beta \cos(\phi + \beta) = 0$

or $\quad \cos(\phi + \beta) = 0 \quad$ or $\quad \phi + \beta = 90°$

or $\quad \phi = 90° - 51.6' = 80°8.4'$ and $\tan\phi = 66.6$

Hence, $\quad 159.2 / (1.75 + r) = 66.6$

or $\quad$ resistance of capacitor $r = 0.64\ \Omega$.

Impedance of capacitor

$$= \sqrt{(159.2)^2 + (0.64)^2} = 159.2\ \Omega$$

Power factor of capacitor $= 0.64 / 159.2 = 0.004$.

**Example 11.7** *The current coil of a wattmeter is connected in series with an ammeter and an inductive load. A voltmeter and the voltage coil are connected across a 100 Hz supply. The ammeter reading is 4.5 A and the voltmeter and wattmeter readings. are respectively, 240 V and 23 W. The inductance of voltage circuit is 10 mH and its resistance 2000 Ω. If the voltage drops across the ammeter and the current coil are negligible, what is the percentage error in the wattmeter reading ?*

**Solution.** Inductive reactance of pressure coil

$$= 2\pi \times 100 \times 10 \times 10^{-3} = 6.28\ \Omega.$$

Resistance of pressure coil circuit = 2000 Ω.

Phase angle of pressure coil circuit

$$B = \tan\beta = 6.28 / 2000 = 0.00314 \text{ rad.} = 0.18° = 10.8'.$$

*For an inductive load* : reading of wattmeter $\alpha \cos\beta \cos(\phi - \beta)$

True power $\alpha \cos\phi$

where $\cos\phi$ = load power factor

$\therefore$ True power

$$= \frac{\cos\phi}{\cos\beta\cos(\phi - \beta)} \times \text{reading of wattmeter} \quad ...(i)$$

But true power

$= I^2 R = (4.5)^2 R = (4.5)^2 Z \cos\phi,$

where $R$ = resistance of load.

Impedance of load $Z = 240 / 4.5 = 53.3\ \Omega$.

$\therefore$ True power $= (4.5)^2 \times 53.3 \cos\phi$

Reading of wattmeter = 23 W.

Substituting these values in (*i*), we have,

$$(4.5)^2 \times 53.3 \cos\phi = \frac{\cos\phi}{\cos\beta\cos(\phi - \beta)} \times 23$$

or $$\cos\beta\cos(\phi - \beta) = \frac{23}{(4.5)^2 \times 53.3} = 0.0213$$

or $$\cos 0.18° \cos(\phi - 0.18°) = 0.0213$$

or $$\phi = 88.96°$$

Percentage error

$= \tan\phi \tan\beta \times 100$

$= 55.08 \times 0.00314 \times 100 = 17.3\%$.

**Example 11.8** *A dynamometer type of wattmeter is rated at 10 A and 100 V with a full scale reading of 1000 W. The inductance of the voltage circuit is 5 mH and its* *resistance is 3000 Ω. If the voltage drop across the current coil of the wattmeter is negligible, what is the error in the wattmeter at the rated VA rating with zero power factor? Frequency is 50 Hz.*

**Solution.** Reactance of pressure coil

$$\omega L = 2\pi \times 50 \times 5 \times 10^{-3} = 1.57\Omega$$

Pressure coil phase angle

$$\beta = \tan\beta = 1.57/3000 = 0.000523 \text{ rad.}$$

Power factor of load, $\cos\phi = 0$ or $\sin\phi = 1$.

From Eqn. 11.22, error with zero p.f.

$$= VI \sin\phi \tan\beta$$

$$= 100 \times 10 \times 1 \times 0.000523 = 0.523 \text{ W.}$$

**Example 11.9** *An electrodynamometer wattmeter is used for measurement of power in a single phase circuit. The load voltage is 100 V and the load current is 9 A at a lagging power factor of 0.1. The wattmeter voltage circuit has a resistance of 3000 Ω and an inductance of 30 mH. Estimate the percentage error in the wattmeter reading when the pressure coil is connected : (i) On the supply side, and (ii) on the load side. The current coil has a resistance of 0.1 Q and negligible inductance. The frequency is 50 Hz. Comment upon the result.*

**Solution.** Load power $= 100 \times 9 \times 0.1 = 90$ W

Load power factor $\cos\phi = 0.1$

$\therefore \quad \phi = 84.26°$ and $\sin\phi = 0.995$ and $\tan\phi = 9.95$.

Resistance of pressure coil circuit

$$R_p = 3000\ \Omega$$

Reactance of pressure coil circuit

$$= 2\pi \times 50 \times 30 \times 10^{-3} = 9.42\ \Omega.$$

As the phase angle of pressure coil circuit is small,

$$\beta = \tan\beta \approx \frac{9.42}{3000} = 0.00314 \text{ rad.}$$

(*i*) When the pressure coil is connected on the load side, the wattmeter measures power loss in pressure coil circuit in addition to load power.

True power $= VI\cos\phi = 100 \times 9 \times 0.1 = 90$ W.

Let us consider only the effect of inductance.

From Eqn. 11.19 reading of wattmeter

$$= \text{true power } (1 + \tan\phi \tan\beta)$$

$$= 90(1 + 9.95 \times 0.00314) = 92.81 \text{ W.}$$

Power loss in pressure coil circuit

$$= \frac{V^2}{R_p} = \frac{(100)^2}{3000} = 3.33 \text{ W}$$

$\therefore$ Reading of wattmeter considering the power loss in pressure coil circuit

$$= 92.81 + 3.33 = 96.14 \text{ W.}$$

$$\text{Percentage error} = \frac{96.14 - 90}{90} \times 100 = 6.82\%.$$

(*ii*) When the current coil is on the load side, the wattmeter measures the power in the load plus the power loss in the current coil. In fact, the current coil acts as a load.

$\therefore$ Total power = power consumed in load $+ I^2 R_c$

$$= 90 + (9)^2 \times 0.1 = 98.1 \text{ W.}$$

Impedance of load = 100/9 = 11.1 Ω

Resistance of load $= 11.1 \times 0.1 = 1.11\ \Omega$

Reactance of load $= 11.1 \times 0.995 = 11.05\ \Omega$

Resistance of load plus resistance of current coil

$$= 1.11 + 0.1 = 1.21\ \Omega$$

Reactance of load plus reactance of current coil

$$= 11.05\ \Omega$$

Impedance of load including current coil

$$= \sqrt{(1.21)^2 + (11.05)^2} = 11.1\ \Omega$$

Power factor of load including current coil

$$= 1.21/11.1 = 0.109$$

$\therefore \quad \phi = 83.74°$ and $\tan\phi = 9.12$

Reading of wattmeter

$$= 98.1\ (1 + 9.2 \times 0.00314) = 100.9 \text{ W.}$$

$$\text{Percentage error} = \frac{100.9 - 90}{90} \times 100 = 12.1\%.$$

It is clear from above that in low power factor circuits we should not use a connection wherein the current coil is on the load side as it results in greater errors.

**Example 11.10** *The current coil of a dynamometer wattmeter is connected to a 24 V d.c. source in series with a 6 Ω resistor. The potential circuit is connected through an ideal rectifier in series with a 50 Hz source of 100 V. The inductance of pressure circuit and current coil resistance are negligible. Compute the reading of the wattmeter.*

**Solution.** Current through the current coil

$$= 24/6 = 4 \text{ A.}$$

The pressure coil is energised by an ideal rectifier. Therefore, the pressure coil carries current during one half cycle and in the other half cycle there is no current in it. This means that there is a deflecting torque on the meter during one half cycle.

$\therefore$ Reading of wattmeter

$$= \text{average power over a cycle}$$

$$= \frac{1}{2\pi}\left[\int_0^{2\pi} vi\, d\theta\right]$$

$$= \frac{1}{2\pi}\left[\int_0^{\pi} \sqrt{2} \times 100 \sin\theta \times 4\, d\theta\right] = 181 \text{ W.}$$

11

## 11.4. FERRODYNAMIC WATTMETERS

The operating torque can be considerably increased by using iron cores for the coils. Ferrodynamic wattmeters employ cores of low loss iron so that there is a large increase in the flux density and consequently an increase in operating torque with little loss in accuracy.

Figure 11.11 shows a ferrodynamic wattmeter. The fixed coil is wound on a laminated core having pole pieces designed to give a uniform radial field throughout the air gap. The moving coil is asymmetrically pivoted and is placed over a hook shaped pole piece.

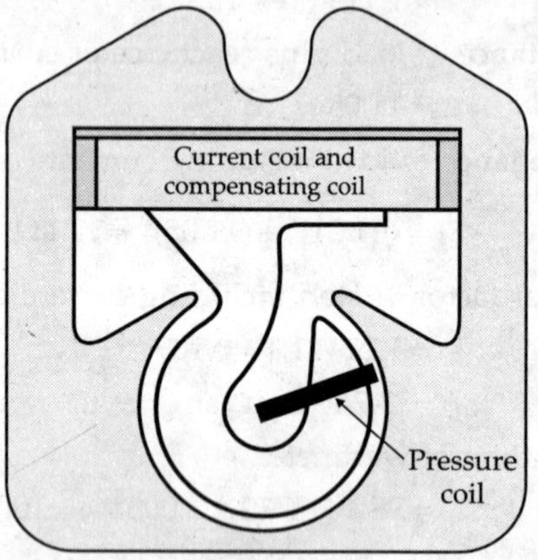


**Fig. 11.11** Ferrodynamic wattmeter.

This type of construction permits the use of a long scale upto about 270° and gives a deflecting torque which is almost proportional to the average power.

With this construction there is a tendency on the part of the pressure coil to creep (move further on the hook) when only the pressure coil is energised. This is due to the fact that coil tries to take up a position where it links with maximum flux. The creep causes errors and a compensating coil is put to compensate for this voltage creep.

The use of ferromagnetic core makes it possible to employ a robust construction for the moving element. Also the instrument is less sensitive to external magnetic fields. On the other hand, this construction introduces complementary errors due to non-linearity of magnetization curve and introduction of large eddy current and hysteresis losses in the core.

The iron losses cause the phase angle of flux $\Phi_c$ (produced by the current coil) to lag behind current $I$ in this coil by an angle $\alpha$ (see Fig. 11.12). Therefore the deflection is :

$$\theta = I_p I \frac{dM}{d\theta} \cos(\phi + \alpha - \beta)$$

$$= K_1 \frac{dM}{d\theta} . VI \cos(\phi + \alpha - \beta) \quad ...(11.23)$$

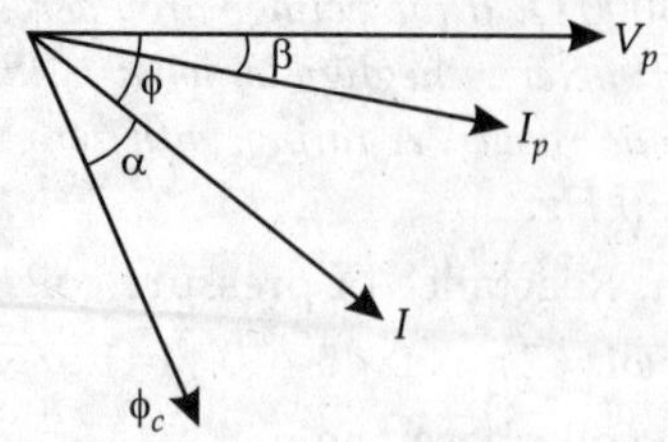


**Fig. 11.12** Phasor diagram of Ferrodynamic wattmeter.

Fulfilling the condition $\alpha = \beta$, we have

$$\theta = K_1 \frac{dM}{d\theta} VI \cos\phi = K_2 P \quad ...(11.24)$$

Thus the phase angle error of the wattmeter will be zero.

In order to fulfil the condition or $\alpha = \beta$ at comparatively high values of $\alpha$, angle $\beta$ is increased. This requires the inductance of pressure coil circuit to be increased which is done by using unifilar winding for the series resistor instead of a bifilar winding. Owing to high iron losses and a high inductance the calibration is useful over a very narrow range of frequency.

A capacitance may be used to shunt the moving coil and a part of its series resistance so as to compensate for the effects of inductance of coil and the iron losses.

## 11.5 LOW POWER FACTOR WATTMETERS (ELECTRODYNAMOMETER TYPE)

Measurement of power in circuits having low power factor by ordinary electrodynamometer wattmeters is difficult and inaccurate because :

(*i*) the deflecting torque on the moving system is small (owing to low power factor) even when the current and pressure coils are fully excited ;

(*ii*) errors introduced because of inductance of pressure coil tend to be large at low power factors.

Special features are incorporated in an electrodynamometer wattmeter to make it a low power factor type of wattmeter. These features are discussed in details below :

1. ***Pressure coil current.*** The pressure coil circuit is designed to have a low value of resistance, so that the current, flowing through it, is increased to give an increased operating torque. The pressure coil current in a low power factor wattmeter may be as much as 10 times the value employed for high power factor wattmeters.

2. ***Compensation for pressure coil current.*** The power being measured in a low power factor circuit is small and current is high on account of low power factor, connection of Fig. 11.8(*a*) cannot be used because owing to large load current there would be a large power loss in the current coil and, therefore, the wattmeter will give a large error.

If we use connection of Fig. 11.8(*b*), the power loss in the pressure coil circuit is included in the reading given by the wattmeter. Thus with this connection also the wattmeter will give a serious error as the power loss in the pressure coil may be a large percentage of the power being measured.

Therefore, it is absolutely necessary to compensate for the pressure coil current in a low power factor wattmeter. This compensation is the same as is shown in Fig. 11.9.

3. ***Compensation for inductance of pressure coil.*** The error caused by pressure coil inductance is :

$$VI \sin \phi \tan \beta \qquad \text{(Eqn. 11.22)}$$

Now, with low power factor, the value of $\phi$ is large and, therefore, the error is correspondingly large. Hence in a low power factor wattmeter we must compensate for the error caused by inductance of the pressure coil. This is done by connecting a capacitor across a part of series resistance in the pressure coil circuit as shown in Fig. 11.7.

If we incorporate the features discussed above, the low factor wattmeter can be represented as shown in Fig. 11.13.

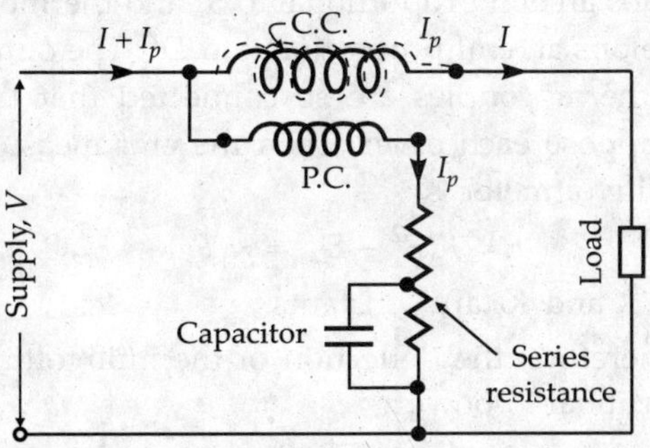


**Fig. 11.13** Low power factor wattmeter.

4. ***Small control torque.*** Low power factor wattmeters are designed with to have a small control torque so that they give full scale deflection for power factors as low as 0.1.

## 11.6 THERMOCOUPLE WATTMETER (THERMAL WATT CONVERTER)

The arrangement for measuring power with thermocouples is shown in Fig. 11.14. This method uses two similar thermocouples whose outputs are connected in opposition with a galvanometer in between.

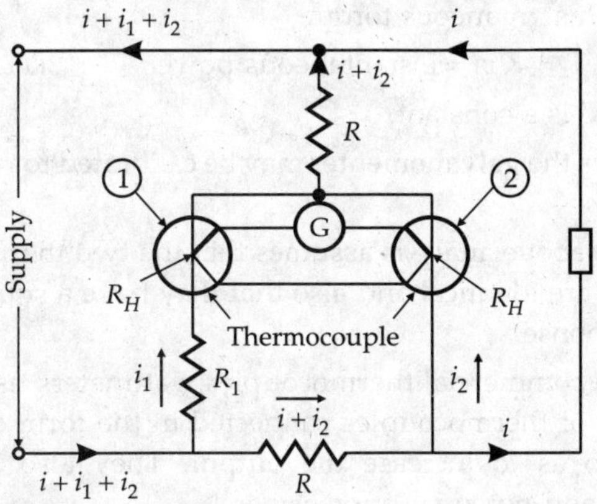


**Fig. 11.14** Thermocouple wattmeter (thermal watt-converter).

$R_H$ is the resistance of each thermocouple heating element.

Emf of thermocouple 1,

$$e_1 \ \alpha \ \text{heat produced by heating element} \ \alpha \ i_1^2 R_H$$

Emf of thermocouple 2, $e_2 \ \alpha \ i_2^2 R_H$

Emf across the galvanometer $= e_1 - e_2$

$\therefore$ Instantaneous deflecting torque of galvanometer

$$T_i = K_1(e_1 - e_2) = K_2(i_1^2 - i_2^2)$$

where $K_1$ and $K_2$ are constants.

Let $v$ be the instantaneous voltage at the load, and $i$ the instantaneous load current.

Therefore, we can write :

$$i_1(R_H + R_1) = (i_1 + i_2)R_2 + i_2 R_H$$

and $$i_2 R_H + (i_1 + i_2)R = v$$

From these two two equations we get,

$$i_1 = \frac{v(R_H + R_2) + iR_2(R_H + R)}{(R_H + R)(R_H + R_1) + R(R_H + R_2)};$$

$$i_2 = \frac{v(R_H + R_1) - iR_2 R}{(R_H + R)(R_H + R_1) + R(R_H + R_2)}$$

Substituting the values of $i_1$ and $i_2$ in the expression for instantaneous torque and assuming $R_1 = R_2$ for simplicity, we have :

$$T_1 \ \alpha \left\{ \frac{2viR_1(R_H + 2R)}{(R_H + R_1)[(R_H + R_1)^2 + R^2]} + \frac{i^2 R_1^2 [(R_H + R)^2 - R^2]}{(R_H + R_1)^2 [(R_H + R_1)^2 + R^2]} \right\}$$

$$= K(Avi + Bi^2) \qquad \text{...(11.25)}$$

where $A$ and $B$ are constants involving the resistances.

If $R_H + R$ is not very different from $R$, then $B$ is almost equal to zero and $Bi^2$ may be neglected as compared to $Avi$.

∴ Instantaneous torque

$$T_i = K'vi = \text{instantaneous power} \quad ...(11.26)$$

where $K'$ is a constant.

Thus the galvanometer may be calibrated to read the power.

The above analysis assumes that the two thermocouples are identical and also that they have a square law response.

The commercial thermocouples wattmeters use a number of thermocouples connected in the form of a chain so as to increase the output. They also use current and potential transformers.

A thermal watt converter used for a.c circuits is described below.

The power in an a.c. circuit is given by $P = VI\cos\phi$. Figure 11.15 shows a phasor diagram where current $I$ lags behind voltage $V$ by an angle $\phi$.

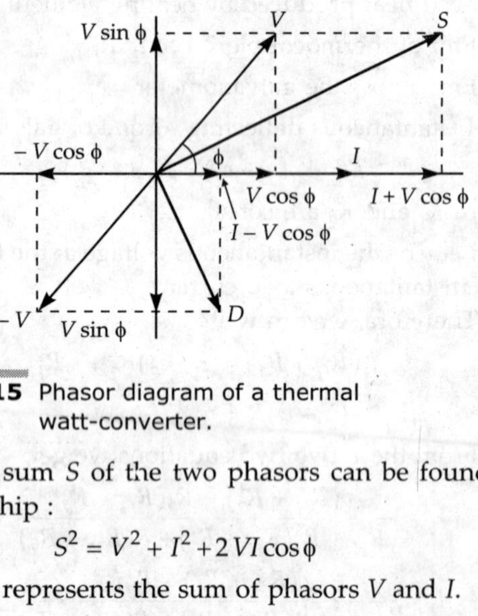


**Fig. 11.15** Phasor diagram of a thermal watt-converter.

The sum $S$ of the two phasors can be found by relationship :

$$S^2 = V^2 + I^2 + 2VI\cos\phi$$

where $S$ represents the sum of phasors $V$ and $I$. ...(*i*)

Similarly, the difference $D$, between the two phasors is found from

$$D^2 = V^2 + I^2 - 2VI\cos\phi \quad ...(ii)$$

$$S^2 - D^2 = 4VI\cos\phi$$

or Power $P = VI\cos\phi = 1/4(S^2 - D^2)$ ...(11.27)

Thus a circuit arrangement capable of measuring quantity $S^2 - D^2$ can measure a quantity proportional to $VI\cos\phi$ representing power. A thermal watt converter using instrument transformers can be used for the purpose.

Figure 11.16 shows the schematic diagram of the elementary circuit of the thermal watt-converter. The plain arrows show the direction of the current from the current transformer, and flagged arrows show the direction of the current in potential circuit at a given instant of time.

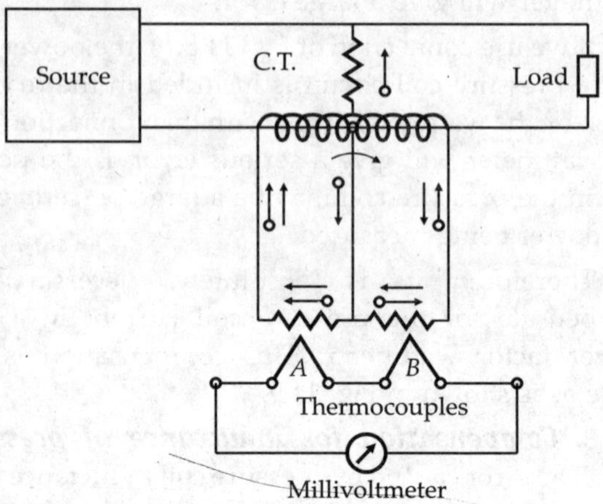


**Fig. 11.16** Elementary circuit of a thermal watt converter.

The heating element of thermocouple $A$ receives the sum of the currents produced by the current transformer and the potential circuit. The heating element of thermocouple $B$, however, receives the difference of these currents. Through proper design, the heat generated in the thermocouples and hence the developed emf, is proportional to the square of the current in the heater. Therefore, thermocouple $A$ develops an emf proportional to $S^2$ and thermocouple $B$ develops an emf proportional to $D^2$. The outputs of these thermocouples are so connected that the two emfs oppose each other. Then the emf measured by the millivoltmeter is :

$$E = K(S^2 - D^2) = K_1 P \quad ...(11.28)$$

where $K$ and $K_1$ are constants.

Therefore, the deflection of the millivoltmeter is proportional to power.

In practice, chains of thermocouples are used instead of single thermocouples in order to obtain greater emf and hence greater sensitivity. The thermocouples are self-heating and are similar to those of the bridge type element discussed on page 280. The practical circuit is shown in Fig. 11.17.

Thermal watt converters are widely used for measurement of power in different circuits, and the sum of their outputs can be applied to a recording potentiometer for graphic record of total power consumed.

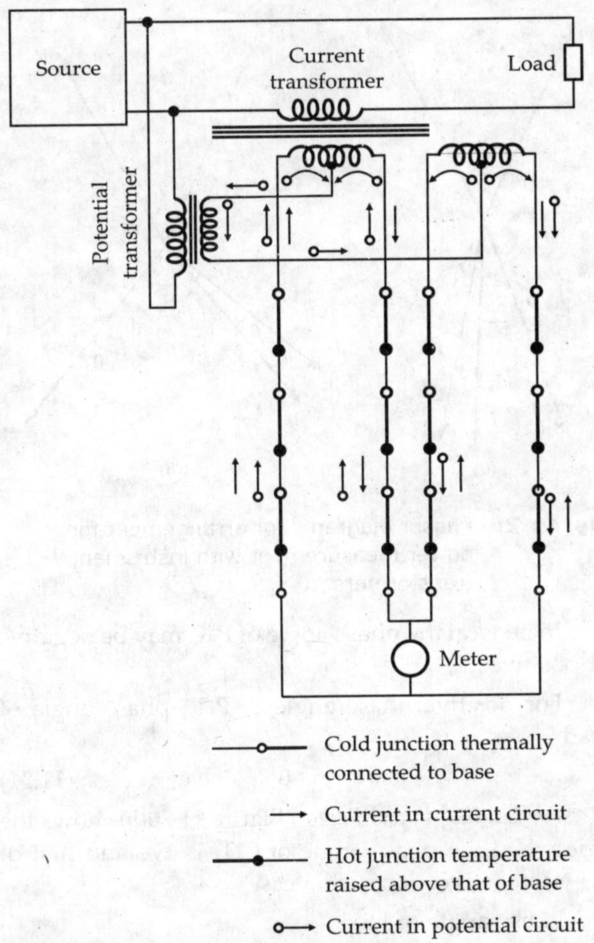


**Fig. 11.17** Circuit diagram of thermal watt converter.

They are also used for calibration of both d.c. as well as a.c. instruments and for instrumentation process monitoring.

## 11.7. HALL EFFECT MULTIPLIER

In applications where *M* the power is to be controlled or processed further, Hall Effect Multipliers are used. A Hall effect multiplier is shown in Fig. 11.18. The Hall effect multiplier uses a Hall effect element as explained on page 179.

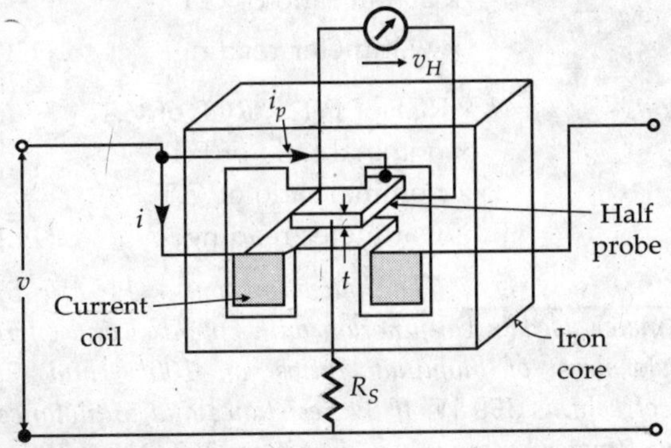


**Fig. 11.18** Hall effect multiplier.

The current is passed through the current coil which produces a magnetic field proportional to the current *i*. This field is perpendicular to the Hall effect element. A current $i_p$, proportional to the voltage, is passed through the Hall effect element in a direction perpendicular to field as shown. The current is limited by the multiplier resistance $R_s$. The output voltage of the Hall effect multiplier is :

$$v_H = K_H \, i_p . B / t \qquad \text{...(11.29)}$$

where $K_H$ = Hall co-efficient ; V - mA - Wbm$^{-2}$,

$B$ = flux density ; Wb/m$^2$,

and $t$ = thickness of Hall element ; m.

Now $B \alpha i$ and $i_p = (v / R_s) \alpha v$.

Hence $v_H \; \alpha \; v_i$,

∴ The output voltage of the Hall effect multiplier is proportional to instantaneous power.

Hence the voltmeter connected at the output terminals can be calibrated in terms of power. The Hall effect voltage which is representative of the power, can be processed further for control and other purposes. This is the major advantage of Hall effect multiplier over electrodynamometer wattmeters the output of the later being the deflection of a pointer which cannot be processed further.

## 11.8 MEASUREMENT OF POWER USING INSTRUMENT TRANSFORMERS

Power measurements are made in high voltage circuits connecting the wattmeter to the circuit through current and potential transformers as shown in Fig. 11.19.

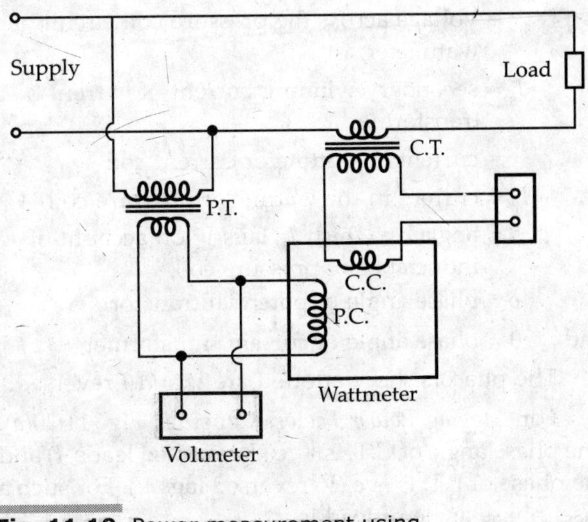


**Fig. 11.19** Power measurement using instrument transformers.

The primary winding of the C.T. is connected in series with the load and the secondary winding is connected in series with an ammeter and the current coil of a wattmeter.

The primary winding of the potential transformer is connected across the supply lines and a voltmeter and the potential coil circuit of the wattmeter are connected in parallel with the secondary winding of the transformer. One secondary terminal of voltmeter each transformer and the casings are earthed.

We have seen that both are subject to ratio and phase angle errors. The errors in good modern instrument transformers are small and may be ignored for many purposes. However, they must be considered in precision work. Also in some power measurements these errors, if not taken into account, may lead to very inaccurate results.

Voltmeters and ammeters are effected by only ratio errors while wattmeters are influenced in addition by phase angle errors. Corrections can be made for these errors if test information is available about the instrument transformers and their burdens.

Figure 11.20 gives the phasor diagrams for the current and voltages of load, and in the wattmeter coils. Fig. 11.20(*a*) refers to a load having lagging power factor while Fig. 11.20(*b*) to a load with leading power factor.

Let

$V$ = voltage across the load,

$I$ = load current,

$\phi$ = phase angle between current and voltage,

$\alpha$ = phase angle between currents in the current and pressure coils of wattmeter,

$V_s$ = voltage across secondary winding of the potential transformer,

= voltage across the pressure coil circuit of wattmeter,

$I_s$ = secondary winding current of current transformer

= current in wattmeter current coil,

$I_p$ = current in the wattmeter pressure coil,

$\beta$ = angle by which $I_p$ lags $V_s$ on account of inductance of pressure coil,

$\delta$ = phase angle of potential transformer.

and $\theta$ = phase angle of current transformer,

The phasors shown dotted are $V$ and $I$ reversed.

For *Lagging Power Factor*. Consider Fig. 11.20(*a*). The phase angle of C.T. is + ve ($I_s$ reversed leads $I$) and the phase of P.T. is –ve ($V_s$ reversed lags $V$). For such a case, phase angle of load is :

$$\phi = \alpha + \theta + \delta + \beta \quad ...(11.30)$$

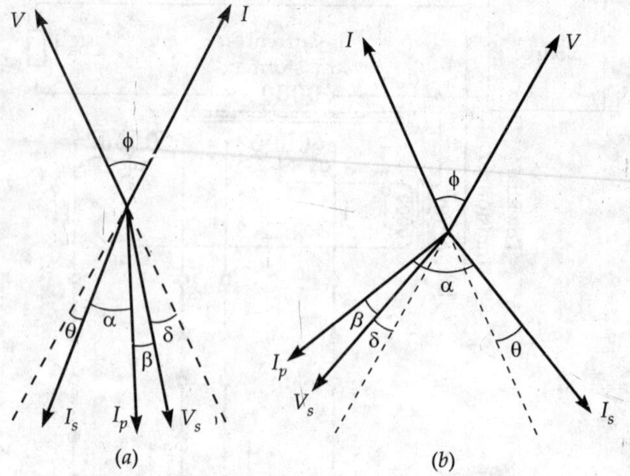


**Fig. 11.20** Phasor diagrams for arrangement for power measurement with instrument transformers.

In general the phase angle of P.T. may be negative or positive.

For positive phase angle of P.T., phase angle of load

$$\phi = \alpha + \theta + \delta + \beta \quad ...(11.31)$$

*For Leading Power Factor*. Figure 11.20(*b*) shows the case when the phase angle of C.T. is +ve and that of P.T. is negative. For such a case,

Phase angle of load

$$\phi = \alpha - \theta - \delta - \beta \quad ...(11.32)$$

If the phase angle of P.T. is +ve,

$$\phi = \alpha - \theta + \delta - \beta \quad ...(11.33)$$

***Correction factors.*** The correction factors-neglecting for the present, the ratio of transformer is :

$$K = \frac{\cos\phi}{\cos\beta\cos\alpha} \quad ....(11.34)$$

where $\alpha$ can be calculated from Eqns. 11.30 to 11.34 depending upon the conditions of the circuit.

$\therefore$ Power = $K \times$ actual ratio of P.T.
$\times$ actual ratio of C.T.
$\times$ wattmeter reading ...(11.35)

= $K \times$ RCF of P.T. $\times$ RCF of C.T.
$\times$ nominal ratio of P.T.
$\times$ nominal ratio of C.T.
$\times$ wattmeter reading. ...(11.36)

**Example 11.11.** *The indication on a 110 V, 5 A wattmeter used in conjunction with potential and current transformers of nominal ratios of 100/1 and 20/1 respectively, is 350 W. If the resistance and inductance of the wattmeter pressure coil circuit are 362 Ω and 100 mH* *respectively and the ratio errors and phase angles of potential and current transformers at particular working conditions are + 0.8%, −45′ and −0.2% +90′, what is the true value of the power being measured ? The load phase angle is 50° lagging. Frequency is 50 Hz.*

**Solution.** Reactance of pressure coil circuit

$$= 2\pi \times 50 \times 10 \times 10^{-3} = 3.14\,\Omega$$

Resistance of pressure coil circuit = 362 Ω

Phase angle of pressure coil circuit

$$\beta = \tan^{-1} 3.14/362 \approx 30'$$

Phase angle of current transformer $\theta = +90'$

Phase angle of potential transformer $\delta = -45'$

Phase angle of load = 50°

Phase angle between pressure coil current $I_p$ and a current $I_s$ of wattmeter current coil (See Fig. 11.21)

$$\alpha\, \phi - \theta - \beta - \delta = 50° - 90' - 30' - 45' = 47°15'$$

Correction factor

$$K = \frac{\cos\phi}{\cos\beta\cos\alpha}$$

$$= \frac{\cos 50°}{\cos 30' \times \cos 47°15'} = 0.947$$

∴ Percentage ratio error

$$= \frac{K_n - R}{R} \times 100$$

∴ Actual ratio

$$R = \frac{K_n \times 100}{100 + \text{percentage error ratio}}$$

Actual ratio of C.T. $= \dfrac{20 \times 100}{100 - 0.2} = 20.04$

Actual ratio of P.T. $= \dfrac{100 \times 100}{100 + 0.8} = 99.2$

From Eqn. 11.35,

Power of load $= 0.947 \times 20.04 \times 99.2 \times 350 \times 10^{-3}$

$$= 658.9\,\text{kW}.$$

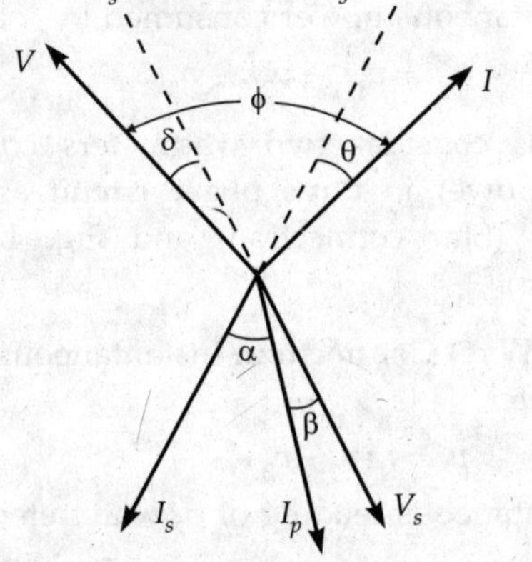


**Fig. 11.21** Phasor diagram of Example 11.11.

***Example 11.12*** *The power in a single phase high voltage circuit is measured by using instrument transformers with voltmeter, ammeter and wattmeter. Observed readings of the instruments (assuming no errors) are 115 V, 4.5 A and 200 watt. Characteristics of the transformers are*

*P.T. : nominal ratio : 11500/115 V ; ratio correction factor 0.995 ; phase angle, −25′.*

*C.T. : nominal ratio; 25/5 A ; ratio correction factor 0.997, phase angle ; +15′.*

*Neglecting the voltage phase angle in the voltmeter calculate the true power.*

**Solution.** The phasor diagram is drawn in Fig. 11.22. Phase angle between secondary current and votlage = α

as, $\beta = 0$

Since inductance of pressure coil circuit is negligible.

or $\cos\alpha = \dfrac{\text{wattmeter reading}}{V_s I_s}$

$$= \frac{200}{115 \times 4.5} = 0.3865$$

∴ $\alpha = 67°16'$

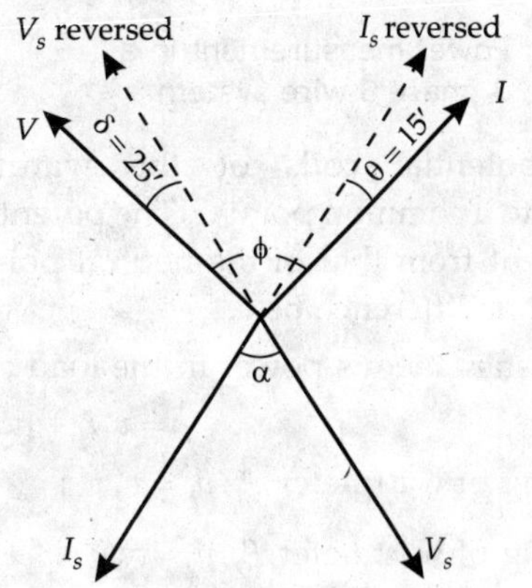


**Fig. 11.22** Phasor diagram of example 11.12.

From the phasor diagram 11.22, phase angle of load

$$\phi = \alpha + \theta + \delta = 67°16' + 15' + 25' = 67°56'$$

∴ $\cos\phi = 0.3757$

Correction factor, $K = \dfrac{\cos\phi}{\cos\beta\cos\alpha}$

∴ $= \dfrac{0.3757}{1 \times 0.3865} = 0.972$

From Eqn. 11.36, actual power

$$= 0.972 \times 0.995 \times 0.997 \times \frac{11500}{115} \times \frac{25}{5} \times 200$$

$$= 96430\text{ W}.$$

11

## 11.9 POWER IN POLY-PHASE SYSTEMS

**Blondel's Theorem.** If a network is supplied through $n$ conductors, the total power is measured by summing the readings of $n$ wattmeters so arranged that a current element of a wattmeter is in each line and the corresponding voltage element is connected between that line and a common point. If the common point is located on one of the lines, then the power may be measured by $n-1$ wattmeters.

Consider the simple case of a 3 phase, wattmeter wire system and load as shown in Fig. 11.23.

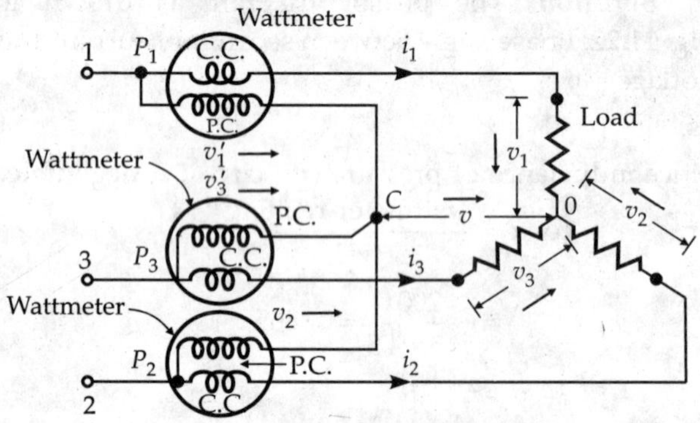


**Fig. 11.23** Power measurement in a 3 phase 3 wire system.

The potential coils of the wattmeters are connected to a common point $C$. The potential of point $C$ is different from that of the neutral point $O$ of the load. Let this difference be $v$.

The instantaneous power in the load :

$$p = v_1 i_1 + v_2 i_2 + v_3 i_3$$

Reading of wattmeter $P_1, p_1 = v_1' i_1$ ;

Reading of wattmeter $P_2, p_2 = v_2' i_3$

Reading of wattmeter $P_3, p_3 = v_3' i_3$

Now $v_1 = v + v_1', v = v + v_2',$ and $v_3 = v + v_3'$

$\therefore\ p_1 = (v_1 - v)i_1,\ p_2 = (v_2 - v)i_2$ and $p_3 = (v_3 - v)i_3$.

Sum of the wattmeter readings $= p_1 + p_2 + p_3$

$$= (v_1 - v)i_1 + (v_2 - v)i_2 + (v_3 - v)i_3$$

$$= v_1 i_1 + v_2 i_2 + v_3 i_3 - v(i_1 + i_2 + i_3)$$

Applying Kirchhoff's current law, $i_1 + i_2 + i_3 = 0$

$\therefore$ Sum of the wattmeter readings

$$= v_1 i_1 + v_2 i_2 + v_3 i_3 = p \qquad ...(11.37)$$

Equation 11.37 represents the total instantaneous power measured by the three wattmeters. Hence the sum of the readings is equal to the total power in the load.

## 11.10 MEASUREMENT OF POWER IN THREE PHASE CIRCUITS

**1. *Three wattmeter method.*** The connections as employed for a 3 phase 4 wire system are shown in Fig. 11.24.

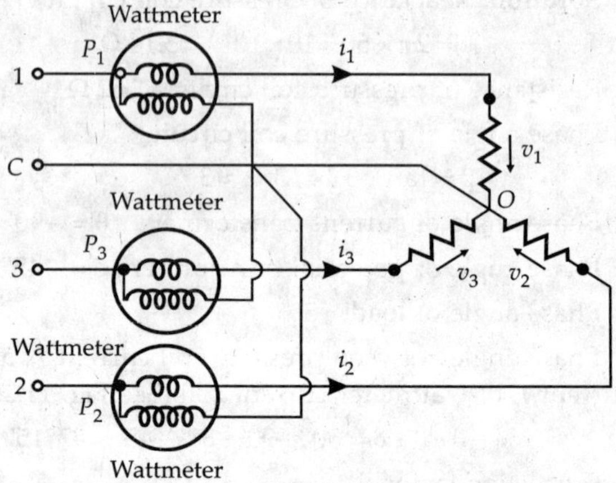


**Fig. 11.24** Three wattmeter method for measurement of power in 3 phase 4 wire circuit.

In this case, the common point $C$ of pressure coils and the neutral $O$ of the load coincide and therefore,

$$v = 0$$

and $\qquad v_1 = v_1',\ v_2 = v_2',\ v_3 = v_3'$.

Sum of the instantaneous readings of the wattmeters

$$p = p_1 + p_2 + p_3 = v_1 i_1 + v_2 i_2 + v_3 i_3$$

Instantaneous power of load

$$= v_1 i_1 + v_2 i_2 + v_3 i_3$$

Hence these three wattmeters measure the power of the load.

**2. *Two wattmeter method*.** In a three phase three wire system we require 3 elements, but if we make the common points of the pressure coils coincide with one of the lines, then we will require only $n-1=2$ elements.

Instantaneous power consumed by load

$$= v_1 i_1 + v_2 i_2 + v_3 i_3$$

Let us consider two wattmeters connected to measure power in three phase circuit as shown in Fig. 11.25 (Star connection) and Fig. 11.26 (Delta connection).

**Star (Wye) Connection.** Instantaneous reading of $P_1$ wattmeter,

$$p_1 = i_1(v_1 - v_3)$$

Instantaneous reading of $p_2$ wattmeter,

$$p_2 = i_2(v_2 - v_3)$$

Sum of instantaneous readings of two wattmeters

$$= p_1 + p_2$$
$$= i_1(v_1 - v_3) + i_2(v_2 - v_3)$$
$$= v_1 i_1 + v_2 i_2 - v_3(i_1 + i_2)$$

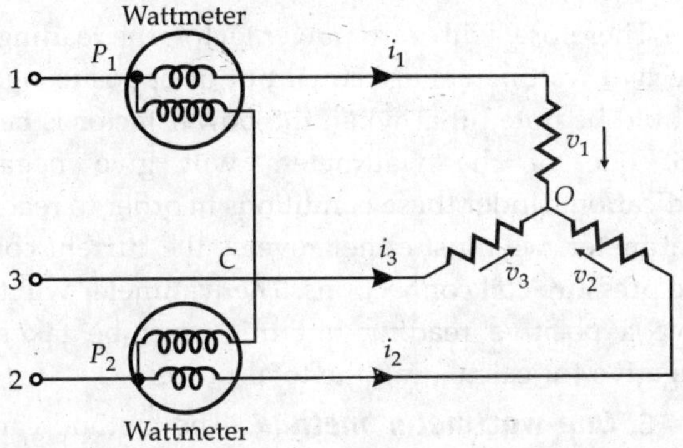


**Fig. 11.25** Two wattmeter method (star connection).

From Kirchhoff's law (Fig. 11.25)

$$i_1 + i_2 + i_3 = 0 \quad \text{or} \quad i_3 = -(i_1 + i_2)$$

$\therefore$ Sum of instantaneous readings of two wattmeters $= v_1 i_1 + v_2 i_2 + v_3 i_3$

Therefore, the sum of the two wattmeter reading is equal to the power consumed by the load. This is irrespective of whether the load is balanced or unbalanced.

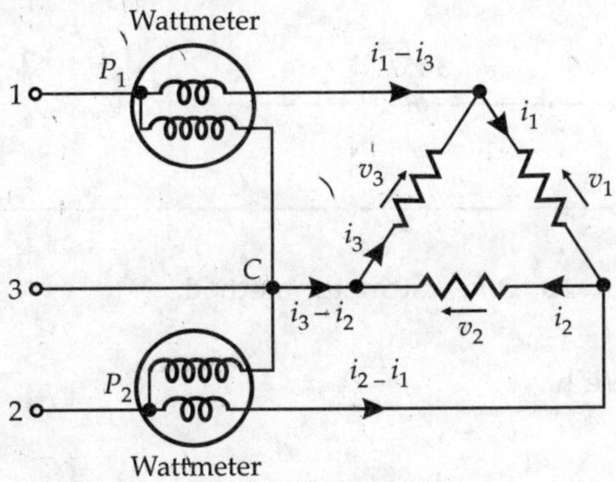


**Fig. 11.26** Two wattmeter method (Delta connection).

**Delta Connection.** Instantaneous reading of $P_1$ wattmeter

$$p_1 = -v_3(i_3 - i_3)$$

Instantaneous reading of $P_2$ wattmeter

$$p_2 = v_2(i_2 - i_1)$$

$\therefore$ Sum of instantaneous readings of wattmeters $P_1$ and $P_2$ ;

$$p_1 + p_2 = -v_3(i_1 - i_3) + v_2(i_2 - i_1)$$
$$= v_2 i_2 + v_3 i_3 - i_1(v_2 + v_3)$$

From Kirchhoff's voltage law,

$$v_1 + v_2 + v_3 = 0$$
$$\therefore \quad v_1 = -(v_2 + v_3)$$

Hence sum of instantaneous readings of two wattmeters

$$= v_2 i_2 + v_3 i_3 - i_1(-v_1) = v_1 i_1 + v_2 i_2 + v_3 i_3$$

Therefore, the sum of the two wattmeter readings is equal to the power consumed by the load. This is irrespective of whether the load is balanced or unbalanced.

Figure 11.27 shows the phasor diagram for a balanced star connected load of Fig. 11.25.

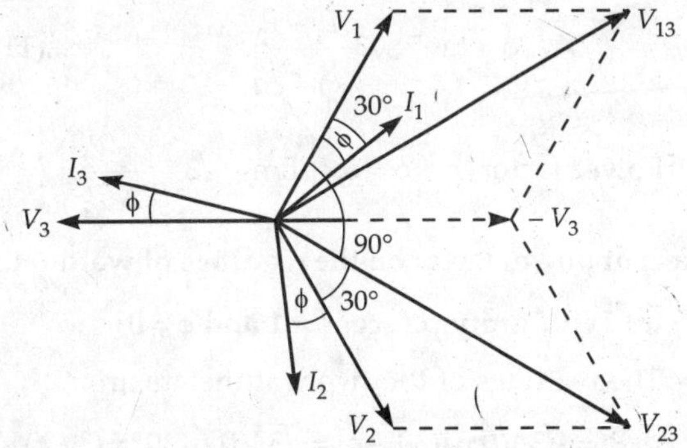


**Fig. 11.27** Phasor diagram for balanced star connected load.

11

Let $V_1, V_2, V_3$ be the rms values of phase voltage and $I_1, I_2, I_3$ be the rms values of phase currents.

The load is balanced, therefore,

Phase voltages, $V_1 = V_2 = V_3 = V$ (say)

Line voltages, $V_{13} = V_{23} = V_{12} = \sqrt{3}\,V$

Phase currents, $I_1 = I_2 = I_3 = I$ (say)

Line currents, $I_1 = I_2 = I_3 = I$

Power factor $= \cos\phi$

The phase currents lag the corresponding phasor voltages by an angle $\phi$.

The current through wattmeter $P_1$ is $I_1$ and voltage across its pressure coil is $V_{13}$. $I_1$ leads $V_{13}$ by an angle $(30° - \phi)$.

$\therefore$ Reading of $P_1$ wattmeter,

$$P_1 = V_{13} I_1 \cos(30° - \phi) = \sqrt{3}\,VI\cos(30° - \phi) \quad ...(11.38)$$

The current through wattmeter $P_2$ is $I_2$ and voltage across its pressure coil is $V_{23}$. $I_2$ lags $V_{23}$ by an angle $(30° + \phi)$.

$\therefore$ Reading of $P_2$ wattmeter,

$$P_2 = V_{23} I_2 \cos(30° + \phi)$$
$$= \sqrt{3}\,VI\cos(30° + \phi) \quad ...(11.39)$$

Sum of reading of two wattmeters :

$$P_1 + P_2 = \sqrt{3}\, VI[\cos(30° - \phi) - \cos(30° - \phi)]$$

$$= 3\, VI \cos\phi \qquad ...(11.40)$$

This is the total power consumed by load.

∴ Total power consumed by load

$$P = P_1 + P_2 \qquad ...(11.41)$$

Difference of readings of two wattmeters

$$P_1 - P_2 = \sqrt{3}\, VI[\cos(30° - \phi) - \cos(30° + \phi)] = \sqrt{3}\, VI \sin\phi \qquad ...(11.42)$$

$$\therefore \quad \frac{P_1 - P_2}{P_1 + P_2} = \frac{\sqrt{3}\, VI \sin\phi}{3\, VI \cos\phi} = \frac{\tan\phi}{\sqrt{3}}$$

or

$$\phi = \tan^{-1}\sqrt{3}\,\frac{P_1 - P_2}{P_1 + P_2} \qquad ...(11.43)$$

Power factor $\cos\phi = \cos\left(\tan^{-1}\sqrt{3}\,\frac{P_1 - P_2}{P_1 + P_2}\right)$

**Effect of power factor on the readings of wattmeters :**

**(*a*) With unity p.f. $\cos\phi = 1$ and $\phi = 0$**

The readings of the two wattmeters are :

$$P_1 = \sqrt{3}\, VI \cos(30° - \phi) = \sqrt{3}\, VI \cos 30° = (3/2)VI$$

and $P_2 = \sqrt{3}\, VI \cos(38° - \phi) = \sqrt{3}\, VI \cos 30° = (3/2)VI$

$$P_1 + P_2 = 3VI$$

At unity power factor, total power

$$= P = 3VI\cos\phi = 3VI$$

Thus at unity power factor, the readings of the two wattmeters are equal, each wattmeter reads half of the total power.

**(*b*) When p.f. = 0.5, we have $\phi = 60°$.**

Therefore,

$$P_1 = \sqrt{3}\, VI \cos(30° - \phi)$$

$$= \sqrt{3}\, VI \cos(30° - 60°) = (3/2)V$$

and $P_2 = \sqrt{3}\, VI \cos(30° + \phi)$

$$= \sqrt{3}\, VI \cos(30° + 60°) = 0$$

$$P_1 + P_2 = (3/2)VI + 0 = (3/2)VI$$

Total power $P = 3VI\cos\phi = (3/2)VI$

Therefore, when the power factor is 0.5, one of the wattmeters reads zero and the other reads total power.

**(*c*) When p.f. = 0, we have $\phi = 90°$**

Therefore,

$$P_1 = \sqrt{3}\, VI \cos(30° - \phi)$$

$$= \sqrt{3}\, VI \cos(30° - 90°) = (\sqrt{3}/2)VI$$

and $P_2 = \sqrt{3}\, VI \cos(30° + \phi)$

$$= \sqrt{3}\, VI \cos(30° + 90°) = -(\sqrt{3}/2)VI$$

or $P_1 + P_2 = 0$

Total power $P = 3VI\cos\phi = 0$

Therefore, with zero power factor, the readings of the two wattmeters are equal but of opposite sign. It should be noted that when the power factor is below 0.5, one of the wattmeters will give negative indication. Under these conditions in order to read the wattmeter, we must either reverse the current coil or the pressure coil connections. The wattmeter will then give a positive reading but this must be taken as negative for calculating the total power.

**3. *One wattmeter method.*** The method can be used only when the load is balanced. The connections are shown in Fig. 11.28. The current coil is connected in one of the lines and one end of the pressure coil to the same line, other end being connected alternately to the other two lines. 'I'he phasor diagram is shown in Fig. 11.29.

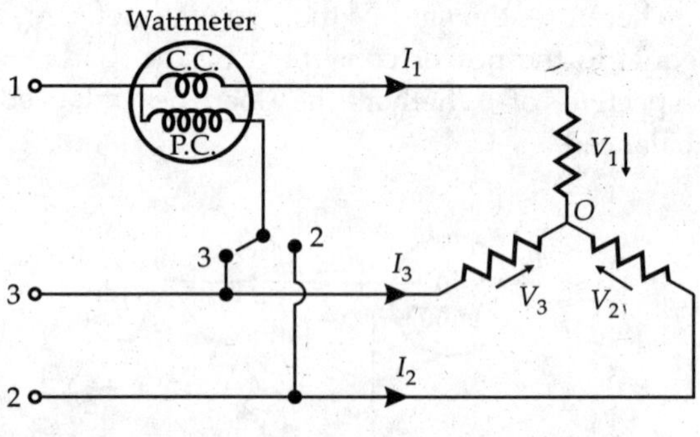


**Fig. 11.28** One wattmeter method.

We have,

$$V_1 = V_2 = V_3 = V,$$

$$I_1 = I_2 = I_3 = I$$

and $V_{13} = V_{12} = \sqrt{3}\, V.$

Reading of wattmeter when the switch is at 3 :

$$P_1 = V_{13} I_1 \cos(30° - \phi) = \sqrt{3}\, VI \cos(30° - \phi)$$

Reading of wattmeter when the switch is at 2 :

$$P_2 = V_{12} I_1 \cos(30° + \phi) = \sqrt{3}\, VI \cos(30° + \phi)$$

Sum of the two wattmeter readings,

$$P_1 + P_2 = \sqrt{3}\, VI[\cos(30° - \phi) + \cos(30° + \phi)] = 3VI\cos\phi$$

Thus the sum of the two readings is the power consumed by the load, or $P = P_1 + P_2$.

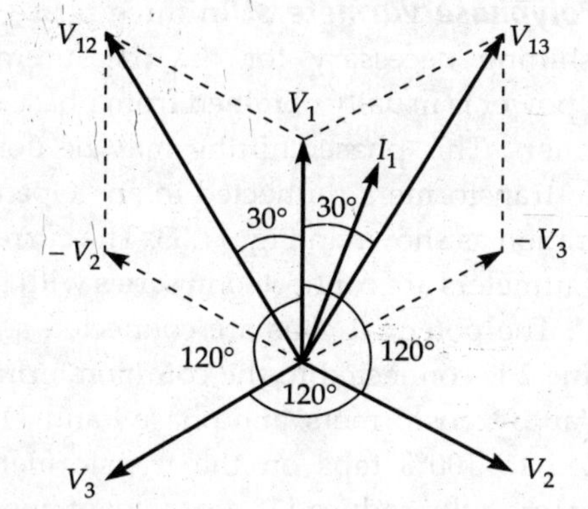


**Fig. 11.29** Phasor diagram of one wattmeter method.

Proceeding as in the case of two wattmeter method,

$$\tan\phi = \sqrt{3}\,\frac{P_1 - P_2}{P_1 + P_2} \qquad \text{...(11.44)}$$

and power factor

$$\cos\phi = \cos\left[\tan^{-1}\sqrt{3}\,\frac{P_1 - P_2}{P_1 + P_2}\right].$$

## 11.11 THREE PHASE WATTMETERS

A dynamometer type three-phase wattmeter consists of two separate wattmeter movements mounted together in one case with the two moving coils mounted on the same spindle. The arrangement is shown in Fig. 11.30.

There are two current coils and two pressure coils.

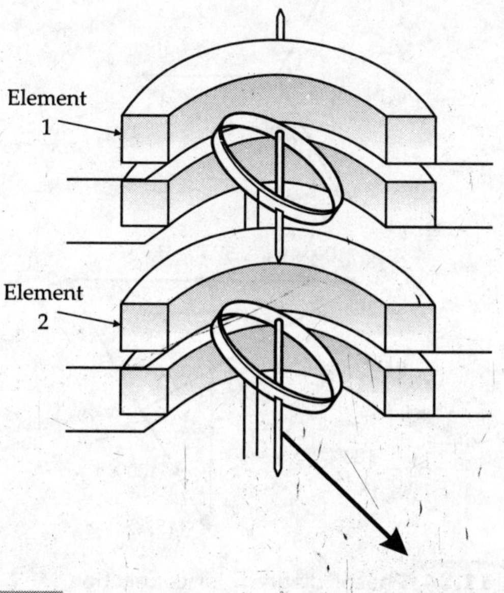


**Fig. 11.30** Three phase two element wattmeter.

A current coil together with its pressure coil is known as an element. Therefore, a three phase wattmeter has two elements.

The connections of two elements of a 3 phase wattmeter are the same as that for two wattmeter method using two single phase wattmeters.

The torque on each element is proportional to the power being measured by it. The total torque deflecting the moving system is the sum of the deflecting torque of the two elements.

Deflecting torque of element 1 $\alpha P_1$ : Deflecting torque of element 2 $\alpha P_2$.

$\therefore$ Total deflecting torque $\alpha (P_1 + P_2) \alpha P$.

Hence the total deflecting torque on the moving system is proportional to the total power.

In order that a 3 phase wattmeter read correctly, there should not be any mutual interference between the two elements. A laminated iron shield may be placed between the two elements to eliminate the mutual effects.

Compensation for mutual effects can be done by using Weston's method. The arrangement is shown in Fig. 11.31. Resistance $R$ may be adjusted to compensate for errors caused by mutual interference.

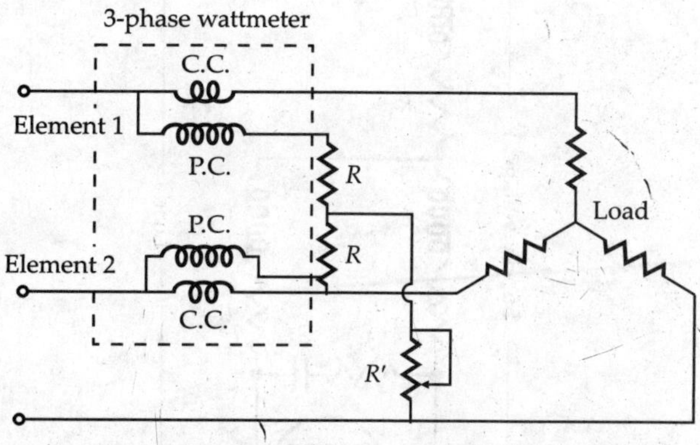


**Fig. 11.31** Compensation for mutual effects between two elements of a 3 phase wattmeter.

## 11.12 MEASUREMENT OF REACTIVE POWER

The reactive power in a circuit is

$$Q = VI \sin\phi$$

It is often convenient and even essential that the reactive power be measured. For example, in load monitoring, such a measurement gives the operator or load despatcher information concerning the nature of the load. Also the reactive power serves as a check on power factor measurements, since ratio of reactive and active power is

$$\tan\phi = Q/P$$

where $Q$ and $P$ are respectively the reactive and active powers.

Also the apparent power $S = VI$, which determines the line and generator capacity may be determined from measurements of active and reactive power.

$$\text{Apparent power } S = \sqrt{P^2 - Q^2}$$

1. ***Single phase varmeters.*** In a single phase circuit reactive power can be measured by a varmeter (volt ampere reactive meter). This is an electrodynamic wattmeter in whose pressure coil circuit a large inductive reactance is substituted for the series resistance so that the pressure coil current $I_p$ is in quadrature with the voltage. Under these conditions the wattmeter reads :

$$VI\cos(90° - \phi) = VI\sin\phi = \text{reactive power.}$$

The circuit of a varmeter is shown in Fig. 11.32(*a*) with its phasor diagram in Fig. 11.32 (*b*).

It should be noted that varmeters do not read correctly it harmonics are present or if the frequency is different from that used when calibrating the instrument.

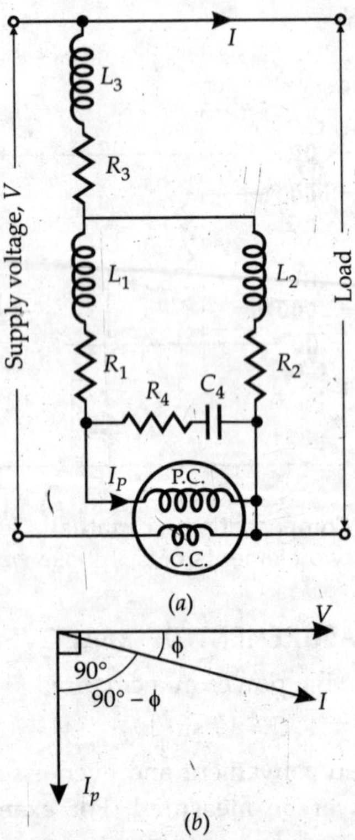


**Fig. 11.32** Single phase voltmeter and its phasor diagram.

2. ***Polyphase varmeters.*** In three phase circuits phase shifting necessary for the measurement. of reactive power is usually obtained from phase shifting transformers. This phase shifting may be done with two auto-transformers connected in an "Open Delta" configuration as shown in Fig. 11.33. The current coils of the wattmeters are connected in series with the lines as usual. The potential coils are connected as shown Phase-line 2 is connected to the common terminals of the two auto-transformers, and phase 1 and 3 lines are connected to 100% taps on the transformers. Both transformers will produce 115.4 percent of line voltage

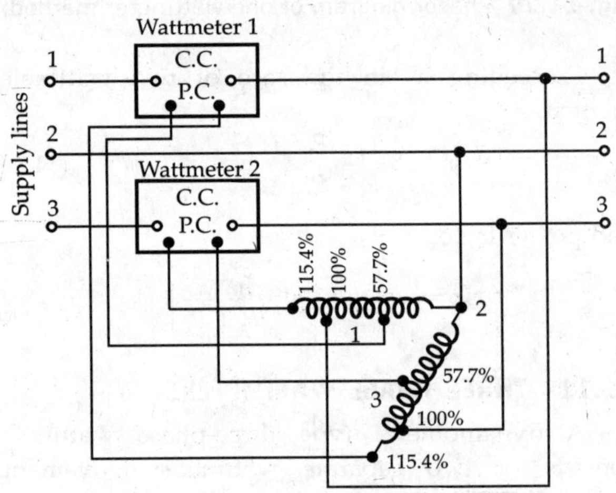


**Fig. 11.33** Reactive power measurement with two auto-transformers.

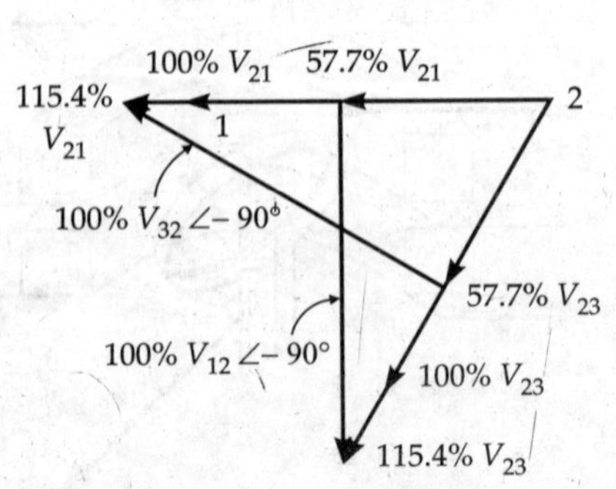


**Fig. 11.34** Phasor diagram for connection of Fig. 11.33.

across the total winding. The pressure coil of wattmeter 1 is connected from 57.7 percent tap on transformer 1 to the 115.4 percent tap on transformer 2, which produces a voltage equal to line voltage but shifted by 90°. This is shown in phasor diagram of Fig. 11.34. The pressure coil of wattmeter 2 is connected in a similar way. Since both the coils receive a voltage equal to the line voltage but displaced by 90°, the wattmeters read the reactive power consumed by the load. The arithmetic sum of the two wattmeters gives the total reactive power supplied to the load.

3. ***Reactive power measurement in three phase circuits.*** In the case of balanced three phase circuits, it is simple to use a single wattmeter to read the reactive power. The current coil of the wattmeter is connected in one line and the pressure coil is connected across the other two lines as shown in Fig. 11.35(*a*). Referring to phasor diagram of this circuit shown in Fig. 11.35(*b*).

Current through the current coil $= I_2$

Voltage across the pressure coil $= V_{13}$

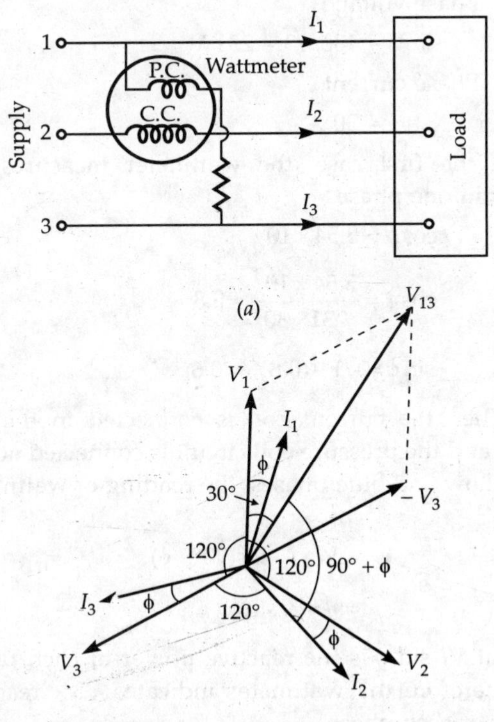


**Fig. 11.35** (*a*) Reactive power measurement with one wattmeter
(*b*) Phasor diagram for connections of Fig. 11.35(*a*)

∴ Reading of wattmeter

$$= V_{13} I_2 \cos(90°+\phi)$$

$$= \sqrt{3}\, VI \cos(90°+\phi)$$

$$= \sqrt{3}\, VI \sin\phi$$

Total reactive volt amperes of the circuit

$$Q = 3VI\sin\phi$$

$$= (-\sqrt{3}) \times \text{reading of wattmeter}$$

Phase angle

$$\phi = \tan^{-1}\frac{Q}{P}.$$

**Example 11.13** *Two wattmeters connected to measure the input to a balanced 3 phase circuit indicate 2000 W and 500 W respectively.*

*Find the power factor of circuit :*

(*a*) *when both the readings are positive and*

(*b*) *when the latter reading is obtained after reversing the connections to the current coil of first instrument.*

**Solution.** (*a*) $P_1 = 2000\,\text{W}$ and $P_2 = 500\,\text{W}$.

From Eqn. 11.44, the phase angle between current and voltage is :

$$\phi = \tan^{-1}\left(\sqrt{3}\,\frac{P_1 - P_2}{P_1 + P_2}\right)$$

$$= \tan^{-1}\left(3\sqrt{\frac{2000-500}{2000+500}}\right) = 45°54'.$$

Power factor

$$\cos\phi = \cos 45°54' = 0.696$$

(*b*) $P_1 = 2500\,\text{W}$ and $P_2 = -500\,\text{W}$ ;

$$\phi = \tan^{-1}\sqrt{3}\,\frac{2000-(-500)}{2000+(-500)} = 70°54'$$

Power factor $= \cos\phi = \cos 70°54' = 0.327.$

**Example 11.14** *A 3 phase 500 V motor load has a power factor of 0.4. Two wattmeters connected to measure the input. They show the input to be 30 kW. Find the reading of each instrument.*

**Solution.** We have

$$P_1 + P_2 = 30\,\text{kW} \qquad ...(i)$$

Now $\cos\phi = 0.4$

∴ $\phi = 66°\,24'$ and $\tan\phi = 2.289$

From Eqn. 11.44

$$\tan\phi = \sqrt{3}\frac{P_1 - P_2}{P_1 + P_2}$$

$$\therefore \quad 2.289 = \sqrt{3}\frac{P_1 - P_2}{30}$$

or $$P_1 - P_2 = 39.7 \text{ kW} \quad ...(ii)$$

From (*i*) and (*ii*), we get

$$P_1 = 34.85 \text{ kW} \quad \text{and} \quad P_2 = -4.85 \text{ kW}.$$

**Example 11.15** *The power flowing in a 3 phase, 3 wire balanced load system is measured by two wattmeter method. The reading of wattmeter A is 7500 W and of wattmeter B is – 1500 W.*

(*a*) *What is the power factor of the system ?*

(*b*) *If the voltage of the circuit is 400 V, what is the value of capacitance which must be introduced in each phase to cause the whole of the power measured to appear on wattmeter A ? The frequency is 50 Hz.*

**Solution.** (*a*) $P_1 = 7500$ W and $P_2 = -1500$ W.

Total power

$$P = P_1 + P_2$$

$$= 7500 - 1500 = 6000 \text{ W}.$$

$$\phi = \tan^{-1}\sqrt{3}\frac{P_1 - P_2}{P_1 + P_2}$$

$$= \tan^{-1}\sqrt{3}\frac{7500 - (-1500)}{7500 + (-1500)} = 69°$$

Power factor $\cos\phi = \cos 69° = 0.358$.

(*b*) Power consumed by each phase (6000/3)

$= 2000$ W.

Voltage of each phase

$$= \left(\frac{400}{\sqrt{3}}\right) = 231 \text{ V}.$$

Current in each phase

$$= \frac{2000}{231 \times 0.358} = 24.1 \text{ A}.$$

Impedance of each phase

$= 231/24.1 = 9.55\ \Omega$.

Resistance of each phase

$= 2000/(24.1)^2 = 3.44\ \Omega$.

Reactance of each phase

$$= \sqrt{(9.55)^2 - (3.44)^2} = 8.9\ \Omega$$

In order that one of the wattmeters should read zero, the power factor should be 0.5.

$$\therefore \quad \cos\phi = 0.5 \quad \text{and} \quad \tan\phi = 1.73.$$

Now $\tan\phi = X/R$

$\therefore$ Reactance of circuit

$$X = R\tan\phi = 3.44 \times 1.73 = 5.9\ \Omega.$$

$\therefore$ Capacitive reactance required

$$= 8.9 - 5.9 = 3.0\ \Omega.$$

and capacitance, $$C = \frac{1}{2\pi \times 50 \times 3} \text{ F}$$

$$= 1060\ \mu\text{F}.$$

**Example 11.16** *A wattmeter reads 5.54 kW when its current coil is connected in the red phase and its voltage coil is connected between the neutral and red phase of a symmetrical 3 phase system supplying a balanced load of 30 A at 400 V. What will be the reading of the instrument if the current coil remains unchanged and voltage coil be connected between blue and yellow phases ? The phase sequence is RYB. What does this figure represent ?*

**Solution.** Let $V$ and $I$ be the phase voltage and phase current respectively. $\cos\phi$ is the power factor of the load.

Phase voltage

$$V = 400/\sqrt{3} = 231 \text{ V}$$

and Phase current

$$I = 30 \text{ A}.$$

In the first case, the wattmeter measures the power in one phase.

$$\therefore \quad VI\cos\phi = 5.54 \times 10^3$$

or $$\cos\phi = \frac{5.54 \times 10^3}{231 \times 30} = 0.8$$

and $$\sin\phi = \sqrt{1 - (0.8)^2} = 0.6$$

When the current coil is connected in the red phase and the pressure coil circuit is connected across the yellow and blue phases, the reading of wattmeter is :

$$= V_{YB} I_R \cos(90° - \phi)$$

$$= \sqrt{3}\, VI \sin\phi$$

But $VI \sin\phi$ is the reactive power of each phase and therefore the wattmeter indicates $\sqrt{3}\times$ reactive power of each phase.

Magnitude of reading

$$= \sqrt{3} \times 131 \times 30 \times 0.6 \times 10^{-3}$$

$$= 7.2 \text{ kVAr}.$$

# Review Questions

1. Describe the constructional details of an Electrodynamometer type wattmeter. Derive the expression for torque when the instrument is used on a.c. Explain why it is necessary to make the potential coil circuit purely resistive ?
2. Derive the torque equation for an Electrodynamometer type of wattmeter. Comment upon the shape of scale if spring control is used. How is it that a uniform scale is obtained when the scale span is about $-45°$ to $45°$ of the position where there is zero mutual inductance between fixed and moving coils ?
3. Prove that the true power

$$= \frac{\cos\phi}{\cos\phi\cos(\phi-\beta)} \times \text{actual wattmeter reading}$$

for electrodynamometer type of wattmeters where $\cos\phi$ = power factor of the circuit $\beta = \tan^{-1}\omega L$ where $L$ and $R$ are the inductance and resistance of the pressure coil of the circuit. Explain why errors are large-when the power factor is low.
4. Derive the expression for the capacitance to be connected across the resistor in the pressure coil circuit so as to neutralize the effect of inductance of pressure coil circuit.
5. There are two types of connections which can be used for an electrodynamometer type wattmeter, one where the current coil is on the load side and second where the pressure coil is on the load side. Discuss the errors caused on account of the two connections. Also, explain under what conditions the each of the two types of connections should be used.
6. Explain the special features incorporated in an electrodynamometer type of wattmeter so that it can be used for low power factor applications.
7. Describe the errors in Electrodynamometer type wattmeters.
8. Explain the constructional details and working of Ferrodynamic wattmeters. What are their advantages and disadvantages ? Explain.
9. Explain the principle of working of a Thermal Watt Converter. Explain how power is measured in a thermal watt converter with the help of its circuit diagram and phasor diagram.
10. Explain the construction and working of a Hall Effect Multiplier for measurement of power.
11. Explain how power can be measured in a 3 phase circuit with the help of two wattmeters. Illustrate your answer with the help of a phasor diagram for a balanced star (Wye) connected load.
12. Explain the working of a 3 phase wattmeter. Draw a neat sketch of the wattmeter and also its connections. Describe how the mutual effects between the two elements of the wattmeter are eliminated.
13. When two wattmeter method is used for measurement of power in a three phase balanced circuit, comment upon the readings of the two wattmeters under following conditions. Support your answer by drawing phasor diagrams.

    Assume that the system is star (Wye) connected :

    (*i*) When the power factor is unity

    (*ii*) When the power factor is zero lagging

    (*iii*) When the power factor is 0.5 lagging

    (*iv*) When the power factor is 0.3 lagging.
14. Describe the method for measurement of reactive power in single phase circuits.
15. Describe the following methods of measurement of reactive power in three phase circuits :

    (*i*) Using two auto-transformers

    (*ii*) Using a single electrodynamometer type wattmeter.

11

# Unsolved Problems

1. An electrodynamometer type wattmeter has a field system which may be considered as long compared with the diameter of moving coil. the mean diameter of moving coil is 30 mm, and is wound with 500 turns. If the current through the moving coil is 50 mA and the wattmeter is measuring power flowing in a circuit having a power factor of 0.7, estimate the torque, if the axes of the field and the moving coils are at (*a*) 45° and (*b*) 90°. **[Ans.** $87.4 \times 10^{-6}$ Nm ; $120 \times 10^{-6}$ **Nm]**

2. The pressure coil of an electrodynamometer wattmeter has a resistance of 6600 Ω. When the voltage applied to the pressure coil is 120 V and a current of 20 A flows in the series coil, the deflection is 160°. What additional resistance must be connected in the pressure coil circuit to make the constant of meter equal to 20 W per degree ? **[Ans.** 2200 Ω]

3. A 250 V, 10 A electrodynamometer wattmeter has a resistance of current and potential coils of 0.5 Ω and 12,500 Ω respectively. Find the percentage error due to each of the two methods of connection shown in Figs. 11.8(*a*) and 11.8 (*b*) with unity power factor loads at 250 V with currents

(*a*) 4 A and

(*b*) 12 A.

Neglect error due to pressure coil inductance.

**[Ans.** (*a*) 0.8%, 0.5% (*b*) 2.4%, 0.166%]

4. If the reactance of the pressure coil circuit of a wattmeter is 1 per cent of its resistance, calculate the percentage error due to this cause at power-factors of 0.8, 0.5 and 0.1 respectively. **[Ans.** 0.75%. 1.7%, 9.95%]

5. If the current in the pressure coil of a wattmeter lags 2° behind the voltage, and instrument is accurate when $\cos\phi = 1$, find the percentage error when $\cos\phi = 0.8$, 0.6 and 0.4 respectively. **[Ans.** 2.56%, 4.59%, 7.94%]

6. The voltage circuit of an electrodynamic wattmeter has an inductance of 8 mH and the resistance of 2000 Ω. What is the percentage error of the instrument when measuring an inductive load having a phase angle of 89° at 50 Hz. Neglect the impedance of the current coil and assume the potential circuit current is negligible as compared with the load current. **[Ans.** 7.2%]

7. A certain circuit takes 10 A at 200 V and the power absorbed is 1000 W. If the wattmeter's current coil has a resistance of 0.15 Ω and its pressure coil a resistance of 5000 Ω and an inductance of 0.3 H, find

(*a*) the error due to resistance for each of the two possible methods of connection ;

(*b*) the error due to inductance if the frequency is 50 Hz ;

(*c*) the total error in each case.

**[Ans.** (*a*) 15 W, 8 W, (*b*) 3.3 W (*c*) 48 W, 41 W]

8. A dynamometer wattmeter measures power in a 50 Hz, single phase circuit without error, at all power factors. The resistance of the voltage coil and its series resistance are 400 Ω and 10,000 Ω respectively. The series resistance has a distributed self-capacitance equivalent to a shunt capacity of 20 pF. What is the self-inductance of the pressure coil ? **[Ans.** 2 mH]

9. A voltage :

$$100\sin\omega t + 40\cos(3\omega t - 30°) + 50\sin(5\omega t + 45°)$$

$V$ is applied to the pressure circuit of a wattmeter and through the current coil is passed a current of $8\sin\omega t + 6\cos(5\omega t - 120°)A$. What will be the reading on the wattmeter ? What percentage of this power is due to fundamental ?

**[Ans.** 439 W ; 91.2%]

10. The inductive reactance of the pressure coil circuit of dynamometer wattmeter is 0.4 percent of its resistance at normal frequency and the capacitance is negligible. Calculate the percentage error and correction factor due to reactance for loads at (*i*) 0.707 p.f. lagging (*ii*) 0.5 p.f. lagging : **[Ans.** (*a*) 0.996, 0.4% ; (*b*) 0.993, 0.7%]

11. The total resistance of the pressure circuit of a wattmeter is 4000 Ω and the inductance of the pressure coil circuit is 6.5 mH. Give the shunted capacitor method of compensating the inductance errror and determine across what portion of the series resistance a 0.1 pF capacitor should be shunted to effect compensation. **[Ans.** 225 Ω]

12. The resistance of the pressure circuit of a wattmeter is 200 times its reactance, the phase error of the potential transformer is 0.25° and the phase error of the current transformer 0.3° at full load, double this at 1/5th load non-inductive, and 50 per cent more on loads of power-factor 0.5. Find the percentage errors of the wattmeter at full and at 1/5th load of power-factor 0.5 inductive.

**[Ans.** 1.69%, 2.44%]

13. A 100 V, 5 A range wattmeter is used in conjunction with instrument transformers to measure power consumed by a load at 6 kV taking 100 A at 0.5 lagging power factor. The instrument transformers have the following data :

P.T. nominal ratio = 60 : 1 ;

ratio error = + 1% ;

phase angle = – 1°.

C.T. nominal ratio = 20 : 1 ;

ratio error = – 0.5% ;

phase angle = + 2°.

Assuming the wattmeter reads correctly, find the error in the indicated power due to transformer errors. **[Ans.** 9.5% high]

14. In a particular measurement, the wattmeter readings were 5000 and 1000 W.
Calculate the power and power factor if
(a) both meter reads direct and
(b) one of the meters has to be reversed.
[**Ans.** (a) 6000 W, 0.656 ; (b) 4000 W : 0.359]

15. Two wattmeters connected to measure the power in a 440 V, 3 phase balanced system gave readings of 5000 W and 1000 W, the latter reading being obtained after reversing the current coil connection of wattmeter. Determine what value of capacitance which will cause the whole power to be read by the first wattmeter. Assume delta connection. The frequency is 50 Hz. [**Ans.** 497 pF]

# Objective Type Questions

***Tick (√) the most appropriate answer***

1. The power in a d.c. circuit is measured with the help of ammeter and a voltmeter. The voltmeter is connected on the load side. The power indicated by the product of readings of two instruments (*VI*) is :
(a) the power consumed in the load
(b) the sum of power consumed by load and the voltrneter
(c) the sum of power consumed by load and the ammeter
(d) none of the above.

2. In an electrodynamometer type of wattmeter :
(a) the current coil is made fixed
(b) the pressure coil is fixed
(c) any of the two coils i.e. current coil or pressure coil can be made fixed
(d) both the coils should be movable.

3. In electrodynamometer type wattmeters, current coils designed for carrying heavy currents use stranded wire or laminated conductors :
(a) to reduce iron losses
(b) to reduce hysteresis losses
(c) to reduce eddy current losses in conductors
(d) all the above.

4. In electrodynamometer type of wattmeters, the inductance of pressure coll circuit produces error :
(a) which is constant irrespective of the power factor of the load
(b) which is higher at low power factors
(c) which is lower at low power factors
(d) none of the above.

5. A capacitor is connected across a portion of resistance of the multiplier in order to make the pressure coil circuit non-inductive. The value of this resistance is r while the total resistance and inductance of pressure circuit are respectively $R_p$ and $L$.
The value of capacitance C is :
(a) $0.41 L/r^2$
(b) $0.41 L/R_p^2$
(c) $L/R_p^2$
(b) $L/r^2$

6. When measuring power with an electrodynamometer wattmeter in a circuit where the load current is small
(a) the current coil should be connected on the load side
(b) the pressure coil should be connected on the load side
(c) it is immaterial whether current coil or pressure coil is connected on the load side.

7. When measuring power with an electrodynamometer wattmeter in a circuit where the load current is large :
(a) the current coil should be connected on the load side
(b) the pressure coil should be connected on the load side
(c) the pressure coil should be connected on the supply side
(d) it is immaterial whether the pressure coil or the current coil is on the load side.

8. When measuring power with an electrodynamometer wattmeter in a circuit having a low power factor :
(a) the current coil should be connected on the load side
(b) the current coil should be connected on the supply side

11

(c) the pressure coil should be connected on the load side

(d) a compensated wattmeter with pressure coil connected on the load side should be used.

9. The power in a 3 phase four wire clrcult can be measured by using

(a) 2 wattmeters

(b) 4 wattmeters

(c) 3 wattmeters

(d) 1 wattmeter.

10. The power in a 3 phase circuit is measured with the help of 2 wattmeters. The readings of one of wattmeters is positive and that of the other is negative. The magnitude of readings is different. It can be concluded that the power factor of the circuit is :

(a) unity

(b) zero (lagging)

(c) 0.5 (lagging)

(d) less than 0.5 (lagging)

## Answers

1. (b) 2. (a) 3. (c) 4. (b) 5. (d) 6. (a) 7. (b) 8. (d) 9. (c) 10. (d)

# 12

# Measurement of Energy and Industrial Metering

## 12.1 GENERAL

Energy is the total power delivered or consumed over a time interval, that is,

energy = power × time

Electrical energy developed as work or dissipated as heat over an interval of time $t$ may be expressed as :

$$W = \int_0^t vi\,dt \qquad ...(12.1)$$

If $v$ is expressed in V, $i$ in A and $t$ in s, the unit of energy is joule or watt second which is 1 watt over an interval of one second. If the unit of time is taken as hour, energy is then expressed in watt hours. In many instances, this unit is of inconvenient size and a larger unit is used–the kilowatt hour, *i.e.*, energy consumed when power is delivered at an average rate of 1000 watt for one hour.

## 12.2 MOTOR METERS

Motor meters are used for measurement of energy in both d.c. and a.c. circuits. For d.c. circuits, the meter may be an ampere hour meter or a watt hour meter.

In motor meters the moving system revolves continuously unlike in the indicating instruments where it deflects through a fraction of a revolution. The speed of rotation is proportional to power in the case of watt hour meters and ampere hours in the case of ampere hour meters. Thus the total number of revolutions made by a watt hour meter in a given interval of time is proportional to the energy supplied (and in the case of ampere hour meters, to the total quantity of electricity supplied). In this connection a term called meter constant is used. *Meter Constant* is defined as the number of revolutions made per kilowatt-hour (kWh). The value of meter constant is usually marked on the meter.

## 12.3 BRAKING

In a motor meter the speed of the moving system is controlled by a braking system. The braking system consists of a permanent magnet (braking magnet) so placed that it induces eddy currents in some part of the moving system. These eddy currents produce a braking (retarding) torque which is proportional to the speed of moving system. The part in which eddy currents are produced is usually an aluminium disc. The disc is mounted on the moving system and, therefore, when the moving system revolves this disc cuts through the field of the permanent magnet.

Emf generated in the disc

$$e = K_1 \Phi n$$

where $\Phi$ = flux of the permanent magnet,

$n$ = speed of rotation,

and $K_1$ = a constant.

Let $r$ be the resistance of the eddy current paths. Therefore, eddy current produced is :

$$i = e/r = K_1 \Phi n/r$$

The braking torque is produced by the interaction of the eddy current and the field of the permanent magnet. This torque is directly proportional to the product of flux of the magnet, magnitude of eddy current and the effective radius $R$ from the axis of the disc.

∴ Braking torque

$$T_B = K_2 \Phi i R = K_1 K_2 \Phi^2 n R/r$$
$$= K_3 \Phi^2 n R/r \qquad ...(12.2)$$

where $K_2$ and $K_3$ are constants.

If the radius $R$ of the disc is constant

$$T_B = K_4 \Phi^2 n/r \qquad ...(12.3)$$

where $K_4$ = a constant = $K_3 R$

The moving system attains a steady speed when the driving torque is equal to the braking torque.

Braking torque at steady speed $N$ is :

$$T_B = K_3\Phi^2 NR / r = K_5 N \quad ...(12.4)$$

where $K_5$ = a constant = $K_3\Phi^2 R / r$

Let $T_d$ be the driving torque at the steady speed $N$.

At steady speed,

$$T_B = T_d$$

$$\therefore \quad K_3\Phi^2 NR / r = T_d$$

or steady speed $$N = (r / K_3\Phi^2 R)T_d \quad ...(12.5)$$

$$= K_6 T_d$$

where $K_6$ = a constant = $r / K_3\Phi^2 R$ ...(12.6)

Hence the steady speed attained by a meter for a constant value of driving torque $t_d$ is directly proportional to the resistance, $r$, of the eddy current paths, inversely proportional to square of flux $\Phi$ of permanent magnet and inversely proportional, to radius $R$. In order that the calibration of the meter remains the same, it is essential that the value of $\Phi$ and $r$ should remain constant. Thus it is very important that the strength of braking magnet shall remain constant throughout the life of the meter. This constancy is assured by the careful design and proper treatment of the magnet during manufacture. Temperature changes affect the value of resistance $r$ and thus cause errors. An increase in temperature increases the resistance $r$ and thus the braking torque is reduced.

It is necessary that the steady speed of the meter should be low and in order to achieve this the disc resistance be low and that flux of the magnet and the effective radius of the disc be large. An aluminium disc is generally preferred to a copper disc since the resistance per unit weight of aluminium is smaller. Frictional forces are limiting factor on the weight and the size of the disc and thus the radius of the disc cannot be increased unduly. Therefore in order to obtain low steady speeds, strong magnets, having large pole areas are used. The air gap is kept as small as mechanical considerations permit.

The braking torque can be adjusted by using a magnetic shunt whose distance from the magnet can be changed. If the magnetic shunt is brought near the poles, it will bypass a larger amount of flux and so the disc will cut through a smaller flux resulting in reduced braking torque. On the other hand, if it is moved away from the magnet, it will bypass lesser flux giving a larger braking torque. The braking torque can also be varied by varying the radius $R$ (See Eqn. 12.2). If the magnet is placed at a larger radius, the braking torque would also be larger.

## 12.4 FRICTION

Errors caused by friction are considerably more important in motor meters than the corresponding errors in the indicating instruments. This is because the friction forces are always present in a motor meter and therefore affect the speed of the moving system for any value of current or power.

***Static friction.*** Friction forces which exist when the moving system is just about to start revolving may prevent it from starting at all if the load is small. This friction force will also cause serious errors at low loads as the friction force is comparable to load at low loads. This will make the meter registration lower than the actual consumption. Static friction may be taken as constant when the meter is running, and may be compensated for by providing a small constant driving torque. This constant torque is applied to the moving system independent of the load.

***Running friction.*** When the meter is running normally, a friction torque is applied to the moving system and the magnitude of this torque is proportional to the speed of the moving system. This friction torque adds to the braking action and may cause serious errors and therefore has to be compensated for.

The friction forces in a meter, are unavoidable. However, they must be reduced to a minimum and must remain practically constant over long periods of time. Therefore the moving parts should be made light. The jewels and pivots should be kept in good order. Spring loaded bearings should be used as their use results in lower static friction and higher life for the jewel.

The counting mechanism should have the minimum possible friction, and the worm and worm wheel connection to the spindle should be properly adjusted. The driving torque of the meter should be high so that the irregularities in friction may not cause errors.

## 12.5 ENERGY METERS FOR A.C. CIRCUITS

Induction type of energy meters are universally used for measurement of energy in domestic and industrial a.c. circuits. Induction type of meters possess lower friction and higher torque/weight ratio. Also induction type meters are inexpensive and accurate and retain their accuracy over a wide range of loads and temperature conditions.

## 12.6 THEORY OF INDUCTION TYPE METERS

In all induction instruments we have two fluxes produced by currents flowing in the windings of the instrument. These fluxes are alternating in nature and so they produce emfs in a metallic disc or a drum provided for the purpose. These emfs in turn circulate eddy currents in the metallic disc or the drum.

Figure 12.1 shows a thin circular aluminium disc free to rotate about an axis through its centre. There are two poles $P_1$ and $P_2$ producing alternating fluxes $\phi_1$ and $\phi_2$ which cut the disc. Any annular portion of the disc with centre on the axis $P_1$ will be linked by alternating flux $\phi_1$, hence it will have an induced emf and this induced emf will produce an eddy current $i_1$. Similarly, flux $\phi_2$ will produce an eddy current $i_2$.

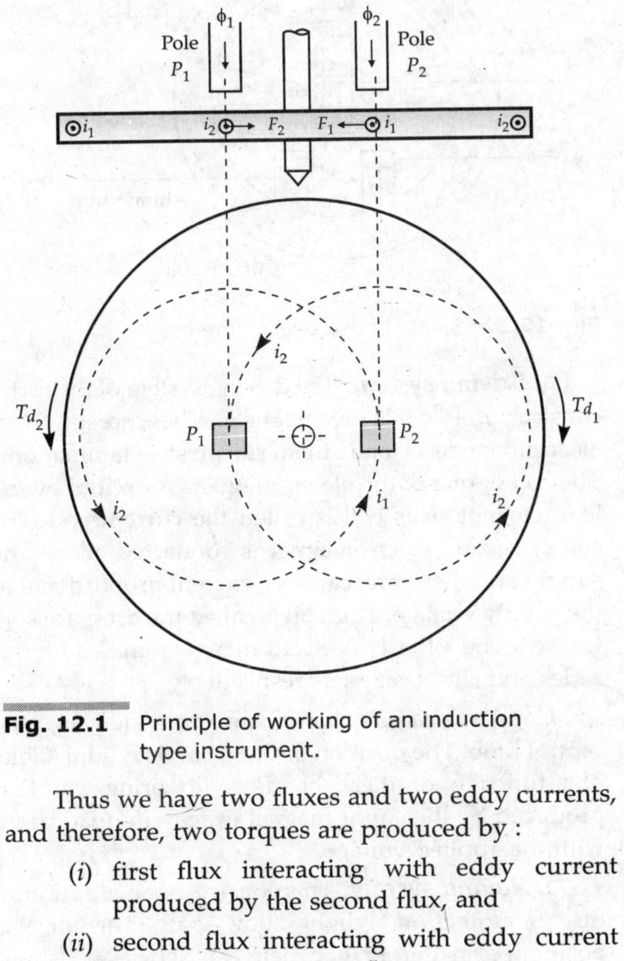


**Fig. 12.1** Principle of working of an induction type instrument.

Thus we have two fluxes and two eddy currents, and therefore, two torques are produced by :

(i) first flux interacting with eddy current produced by the second flux, and

(ii) second flux interacting with eddy current produced by the first flux.

Total torque is the sum of these two torques. We now develop an expression for torque :

Let $\phi_1$ and $\phi_2$ be the instantaneous values of two fluxes having a phase difference of β and, therefore, we can write :

$$\phi_1 = \Phi_{m1} \sin \omega t,$$

$$\phi_2 = \Phi_{m2} \sin(\omega t - \beta)$$

Their rms values are

$$\Phi_1 = \Phi_{m1}\sqrt{2} \quad \text{and} \quad \Phi_2 = \Phi_{m2}/\sqrt{2}$$

Flux $\Phi_1$ produces an emf in the disc by transformer action.

The instantaneous value of this emf is,

$$e_1 \propto \left[-\frac{d(\Phi_1)}{dt}\right], \propto \left[-\frac{d}{dt}(\Phi_{m1} \sin \omega t)\right],$$

$$\propto [-\omega \Phi_{m1} \cos \omega t], \quad \propto [-f\Phi_{m1} \cos \omega t].$$

It is clear from above that $e_1$ lags the flux by 90°

Let $E_1$ = rms value of emf $e_1$,

$$\therefore \quad E_1 \propto f \frac{\Phi_{m1}}{\sqrt{2}} \propto f\Phi_1 \qquad \text{...(12.7)}$$

This emf circulates an eddy current having an rms value $I_1$. If the impedance of eddy current path is $Z$

$$I_1 = \frac{E_1}{Z} \propto \frac{f}{Z}\Phi_1$$

$I_1$ lags the voltage $E_1$ by an angle α where tan α is the ratio of reactance to resistance of path of eddy currents in the disc.

Referring to phasor diagram shown in Fig. 12.2 the phase angle between $\Phi_2$ and $I_1$ is (90°−β+α).

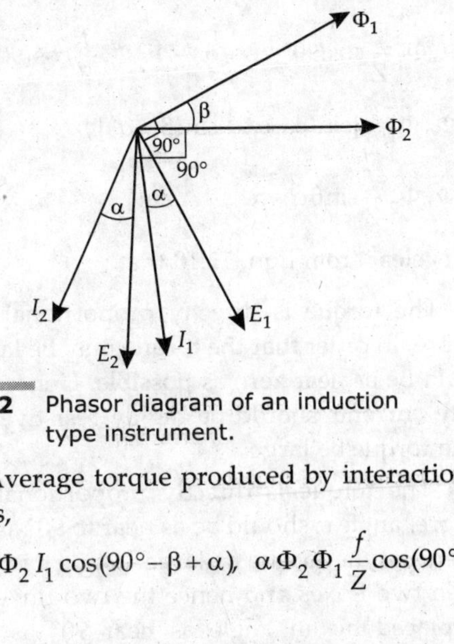


**Fig. 12.2** Phasor diagram of an induction type instrument.

∴ Average torque produced by interaction of $\Phi_2$ and $I_1$ is,

$$T_{d1} \propto \Phi_2 I_1 \cos(90° - \beta + \alpha), \propto \Phi_2 \Phi_1 \frac{f}{Z} \cos(90° - \beta + \alpha) \qquad \text{...(12.8)}$$

Now take flux $\Phi_2$ and proceed in a similar manner. It produces an eddy emf of rms value is,

$$E_2 \propto f\Phi_2.$$

$E_2$ lags the flux $\Phi_2$ by 90°, $E_2$ circulates eddy currents in the same disc. The rms value of eddy current produced by emf $E_2$ is :

$$I_2 \propto \frac{f}{Z}\Phi_2$$

This current lags the emf $E_2$ by an angle α and, therefore, the angle between flux $\Phi_1$ and current $I_2$ is (90°+β+α). (Refer to Fig. 12.2).

12

Average torque produced by interaction of flux, $\Phi_2$ and current $I_2$ is :

$$T_{d2} \propto \Phi_1 I_2 \cos(90°+\alpha+\beta), \propto \Phi_1 \Phi_2 \frac{f}{Z}\cos(90°+\alpha+\beta)$$
...(12.9)

Let the positive directions of flux be as shown and assume that at the instant under consideration, both fluxes are increasing. Application of Lenz's law shows that the directions of currents $i_1$ and $i_2$ will be those as indicated in Fig. 12.1.

Applying the left hand rule, we find that the flux $\Phi_2$ interacting with current $i_1$ produces a force $F_1$ acting from right to left.

Similarly, flux $\Phi_1$ interacting with current $i_2$ produces a force $F_2$ acting from left to right and since the two forces are in opposite direction, they produce two torques $T_{d1}$ and $T_{d2}$ acting in different directions.

Thus the total torque is the difference of the two torques or

Total deflecting torque $T_d = T_{d1} - T_{d2}$

$$\propto \Phi_2 \Phi_1 \frac{f}{Z}\cos(90°-\beta+\alpha) - \Phi_2 \Phi_1 \frac{f}{Z}\cos(90°+\beta+\alpha),$$

$$\propto \Phi_2 \Phi_1 \frac{f}{Z}[\sin(\beta-\alpha)+\sin(\beta+\alpha)],$$

$$\propto \Phi_1 \Phi_2 \frac{f}{Z}\sin\beta\cos\alpha. \quad \text{...(12.10)}$$

It is clear from Eqn. 12.10 that :

(*i*) The torque is directly proportional to $\cos\alpha$. Therefore, in order that the torque may be large angle $\alpha$ should be as near zero as possible. Hence, the path of eddy currents should be highly resistive in order that the torque be large.

(*ii*) The torque is directly proportional to $\sin\beta$. Therefore, angle $\alpha$ should be as near to 90° as possible in order that the torque be large. Angle $\beta$ is the angle between two fluxes and hence the two fluxes should be displaced by an angle as near 90° as possible. Maximum torque is achieved when $\beta = 90°$ *i.e.*, when the two fluxes are displaced by 90°.

## 12.7 SINGLE PHASE INDUCTION TYPE METERS

The construction and principle of operation of Single Phase Energy Meters is explained below :

### 12.7.1 Construction of Induction Type Energy Meters

The construction varies in details from one manufacturer's product to the next. However, the differences are very minor in nature.

There are *four* main parts of the operating mechanism :

(*i*) Driving system,
(*ii*) Moving system,
(*iii*) Braking system, and
(*iv*) Registering system.

Figure 12.3 shows the construction of a single phase induction type energy meter.

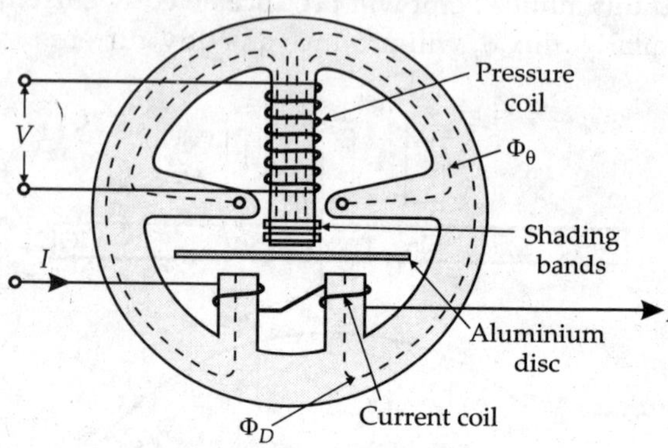


**Fig. 12.3** Single phase energy meter.

1. ***Driving system.*** The driving system of the meter consists of two electro-magnets. The core of these electromagnets is made up of silicon steel laminations. The coil of one of the electromagnets is excited by the load current. This coil is called the current coil. The coil of second electromagnet is connected across the supply and therefore, carries a current proportional to the supply voltage. This coil is called the pressure coil. Consequently the two electromagnets are known as series and shunt magnets respectively.

Copper shading bands are provided on the central limb. The position of these bands is adjustable. The function of these bands is to bring the flux produced by the shunt magnet exactly in quadrature with the applied voltage.

2. ***Moving system.*** This consists of an aluminium disc mounted on a light alloy shaft. This disc is positioned in the air gap between series and shunt magnets. The upper bearing of the rotor (moving system) is a steel pin located in a hole in the bearing cap fixed to the top of the shaft. The rotor runs on a hardened steel pivot, screwed to the foot of the shaft. The pivot is supported by a jewel bearing. A pinion engages the shaft with the counting or registering mechanism.

A unique design for the suspension of the disc is used in the floating-shaft energy meter. Here the rotating shaft has a small magnet at each end, where the upper magnet of the shaft is attracted to a magnet in the upper bearing and the lower magnet of the shaft is attracted to a magnet in the lower bearing. The moving system thus floats without touching either bearing surface, and the only contact with the movement is that of the gear connecting the shaft with the gear of the train, thus the friction is drastically reduced.

3. ***Braking system.*** A permanent magnet positioned near the edge of the aluminium disc forms the braking system. The aluminium disc moves in the field of this magnet and thus provides a braking torque. The position of the permanent magnet is adjustable, and therefore, braking torque can be adjusted by shifting the permanent magnet to different radial positions as explained earlier.

4. ***Registering (counting) mechanism.*** The function of a registering or counting mechanism is to record continuously a number which is proportional to the revolutions made by the moving system. By a suitable system, a train of reduction gears the pinion on the rotor shaft drives a series of five or six pointers.

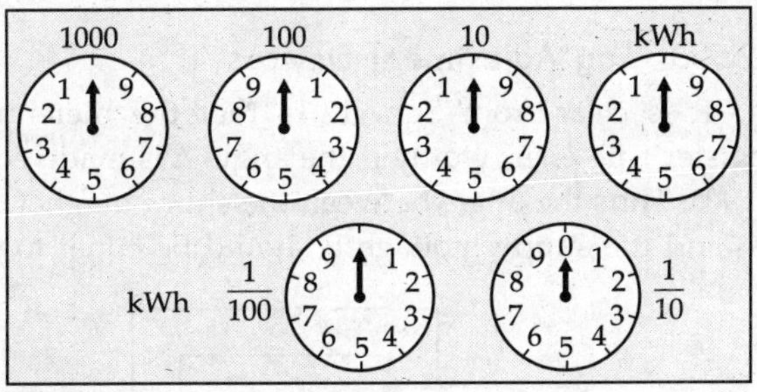


**Fig. 12.4** Pointer type of register.

These rotate on round dials which are marked with ten equal divisions. The pointer type of register is shown in Fig. 12.4. Cyclo-meter register as shown in Fig. 12.5 can also be used.

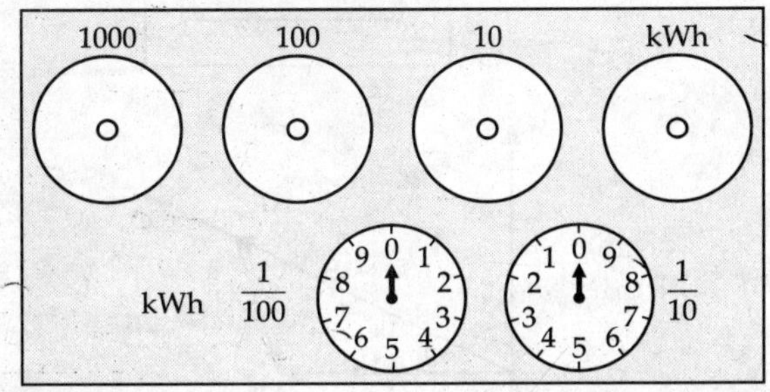


**Fig. 12.5** Cyclometer register.

## 12.7.2 Theory and Operation of Single Phase Energy Meters

A simple functional diagram of the driving system of the meter is shown in Fig. 12.6. The corresponding phasor diagram is shown in Fig. 12.7.

The supply voltage is applied across the pressure coil. The pressure coil winding is highly inductive as it has very large number of turns and the reluctance of its magnetic circuit is very small owing to presence of air gaps of very small length. Thus the current $I_p$ through the pressure coil is proportional to the supply voltage and lags it by a few degrees less than 90°. This is because the winding has a small resistance and there are iron losses in the magnetic circuit.

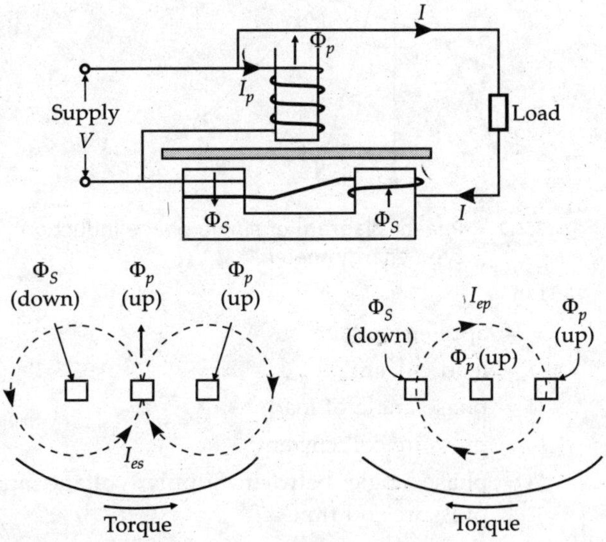


**Fig. 12.6** Working of a single phase induction type energy meter.

Current $I_p$ produces a flux $\Phi_{pt}$. This flux divides itself into two parts $\Phi_g$ and $\Phi_p$. The major portion $\Phi_g$ flows across the side gaps reluctance of this path is small. The reluctance to the path of flux $\Phi_p$ is large and hence its magnitude is small. This flux $\Phi_p$ goes across aluminium disc and hence is responsible for production of driving torque. Flux $\Phi_p$ is in phase with current $I_p$ and is proportional to it. Therefore flux $\Phi_p$ is proportional to voltage $V$ and lags it by an angle a few degrees less than 90°. Since flux $\Phi_p$ is alternating in nature, it induces an eddy emf $E_{ep}$ in the disc which in turn produces eddy current, $I_{ep}$.

The load current $I$ flows through the current coil and produces a flux $\Phi_s$. This flux is proportional to the load current and is in phase with it. This flux produces eddy current $I_{es}$ in the disc. Now the eddy current $I_{es}$ interacts with flux $\Phi_p$ to produce a torque and eddy current $I_{ep}$ interacts with $\Phi_s$ to produce another torque. These two torques are in the opposite direction (as shown in Fig. 12.6) and the net torque is the difference of these. The theory of induction type instruments has been dealt within Art. 12.6.

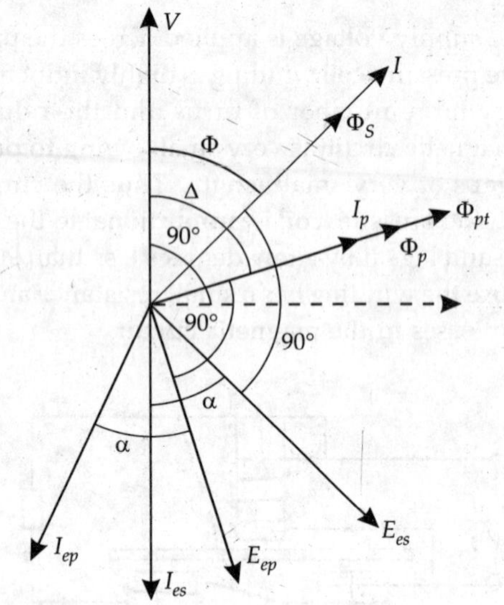


**Fig. 12.7** Phasor diagram of single phase induction type energy meter.

Let

$V$ = applied voltage

$I$ = load current

$\phi$ = phase angle of load

$I_p$ = pressure coil current

$\Delta$ = phase angle between supply voltage and pressure coil flux

$f$ = frequency.

$Z$ = impedance of eddy current paths

$\alpha$ = phase angle of eddy current paths

$E_{ep}$ = eddy emf induced by flux $\Phi_p$

$I_{ep}$ = eddy current due to flux $\Phi_p$

$E_{ev}$ = eddy emf induced by flux $\Phi_s$

$I_{es}$ = eddy current due to flux $\Phi_s$

From Eqn. 12.10, net driving torque

$$T_d \propto \Phi_1 \Phi_2 \frac{f}{Z} \sin\beta \cos\alpha = K_1 \Phi_1 \Phi_2 \frac{f}{Z} \sin\beta \cos\alpha$$

where $K_1$ = a constant

Now $\beta$ = phase angle between fluxes $\Phi_1$ and $\Phi_2$

In our case, the two fluxes are $\Phi_p$ and $\Phi_s$.

$\therefore$ $\beta$ = Phase angle between fluxes $\Phi_p$ and $\Phi_s = (\Delta - \phi)$ (See Fig. 12.7)

$\therefore$ Driving torque, $T_d = K_1 \Phi_p \Phi_s \frac{f}{Z} \sin(\Delta - \phi) \cos\alpha$

But $\Phi_p \propto V$ and $\Phi_s \propto I$.

$$\therefore \quad T_d = K_2 VI \frac{f}{Z} \sin(\Delta - \phi) \cos\alpha$$

If $f$, $Z$ and $\alpha$ are constants,

$$T_d = K_3 V I \sin(\Delta - \phi) \quad \text{...(12.11)}$$

If $N$ is the steady speed, braking torque

$$T_B = K_4 N$$

At steady speed the driving torque must equal the braking torque

$$\therefore \quad K_4 N = K_3 VI \sin(\Delta - \phi)$$

$$\text{or} \quad N = KVI \sin(\Delta - \phi) \quad \text{...(12.12)}$$

If $\Delta = 90°$,

$$\text{Speed,} \quad N = KVI \sin(90° - \phi) = KVI \cos\phi \quad \text{...(12.13)}$$

$$= K \times (\text{power})$$

Thus in order that the speed of rotation is proportional to power, angle $\Delta$ should be equal to 90°. Hence the flux $\Phi_p$, must be made to lag the supply voltage by exactly 90°.

Total number of revolutions

$$= \int N\,dt = K \int VI \sin(\Delta - \phi)\,dt \quad \text{...(12.14)}$$

If $\Delta = 90°$, total number of revolutions

$$= K \int (VI \cos\phi)\,dt \quad \text{...(12.15)}$$

$$= K \int (\text{power})\,dt = K \times (\text{energy}) \quad \text{...(12.16)}$$

### 12.7.3 Lag Adjustment Devices

It is clear from Eqn. 12.14 that the meter will register true energy only if the angle $\Delta$ is made equal to 90°. Thus the angle between the shunt magnet flux $\Phi_p$ and the supply voltage $V$ should be equal to 90°.

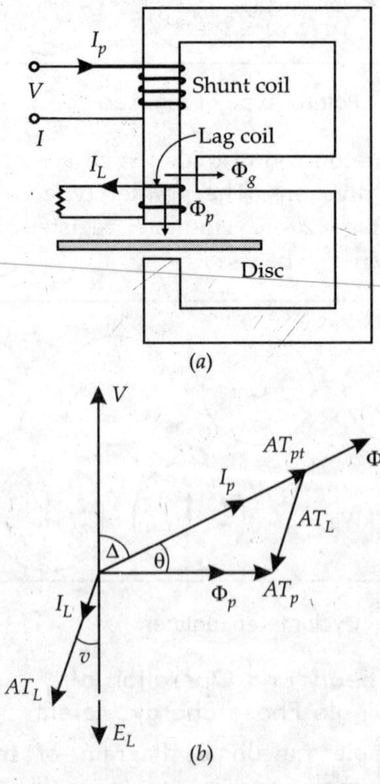


**Fig. 12.8** Adjustment of lag angle with lag coils.

This requires that the pressure coil winding should be so designed that it is highly inductive and has a low resistance and the iron losses in the core are small but even with this the phase of flux $\Phi_p$ is not 90° with respect to voltage $V$ but a few degrees less than 90°.

However, by introducing a magnetic shunt circuit which allows the main portion of the shunt magnet flux to bypass the gap in which the disc is located, it is possible to introduce an mmf in the proper phase relation to bring the shunt magnet flux in the disc air gap in exact quadrature with the voltage. This is illustrated in Fig. 12.8.

The required mmf is obtained from a 'lag coil' which is located on the central limb of the shunt magnet close to the disc gap and links with the flux that cuts the disc.

The pressure coil is excited by voltage $V$ and carries a current $I_p$ which produces an mmf $AT_{pt}$, which in turn produces a flux $\Phi_{pt}$ lagging the voltage by an angle $90° - \Delta$. The flux $\Phi_{pt}$ divides itself into two parts $\Phi_g$ and $\Phi_p$. Flux $\Phi_p$ cuts the disc and also links with the lag coil. A voltage $E_L$ is induced in this coil lagging $\Phi_p$ by 90°. This voltage circulates a current $I_L$ through the lag coil. $I_L$ lags behind $E_L$ by an angle $\gamma$ which depends upon the reactance and resistance of the lag coil. The lag coil current produces an mmf $AT_L$. The flux $\Phi_p$ in the disc air gap will be created by the combined action of the main mmf $AT_{pt}$ in phase with $I_p$ and the lag coil mmf $AT_L$ in phase with $I_L$. Thus the flux $\Phi_p$ will be in phase with the resultant mmf $AT_p$. Hence it is clear that the phase of flux $\Phi_p$ can be adjusted by varying the mmf of the shading coil either in magnitude or in phase or both.

The arrangements for adjusting the mmf of the lag coil are :

1. ***Adjustable resistance.*** A few turns of fairly thick wire are placed around the central limb of the shunt magnet and the circuit is closed through a low adjustable resistance (Fig. 12.9). The resistance of this circuit is altered to adjust the lag angles of flux $\Phi_p$. An

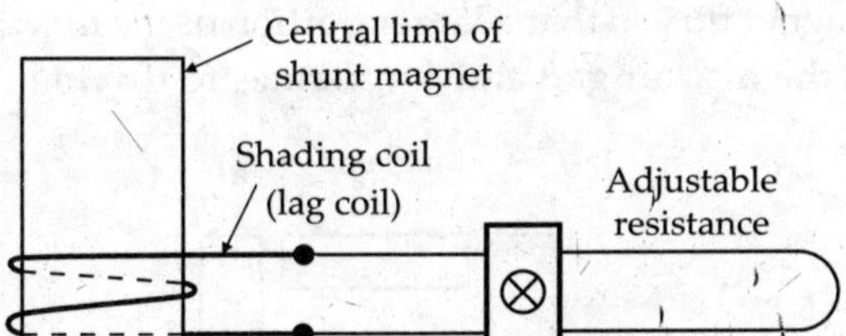


**Fig. 12.9** Lag adjustment with adjustable resistance.

increase in resistance decreases the current and mmf in the lag coil and therefore the value of lag angle $\theta$ is decreased. The value of lag angle $\theta$ is increased by decreasing the resistance of the lag coil circuit. The resistance of the lag coil is so adjusted that $\Delta$ becomes equal to 90°.

2. ***Shading bands.*** In this arrangement copper shading bands $L_1$ are placed around the central limb of shunt magnet instead of a lag coil with adjustable resistance (Fig. 12.10). The adjustment can be done by moving the shading bands along the axis of the limb. As the shading bands are moved up the limb, they embrace more flux. This results in greater values for induced emf, current and the mmf $AT_L$ produced by the shading bands and therefore the value of lag angle $\theta$ increases. When the shading bands are moved down limb, mmf $AT_L$ decreases and the lag angle is reduced. The adjustment is so done that '$\theta$' is such that it makes $\Delta = 90°$.

12

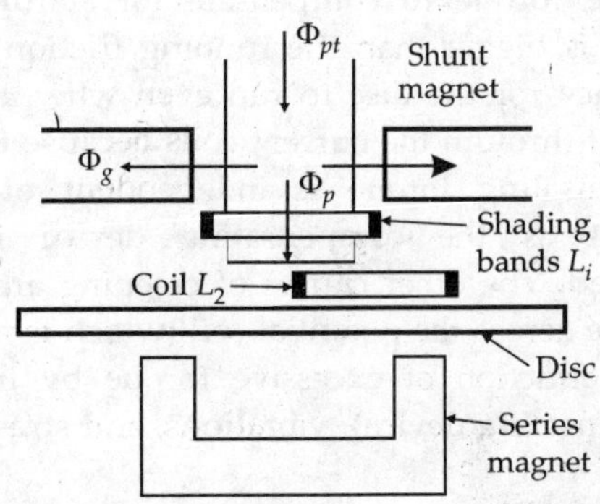


**Fig. 12.10** Lag adjustment with change of position of shading bands.

Hence by providing any of the above arrangements, the shunt magnet flux $\Phi_p$ can be brought exact quadrature with applied voltage $V$. This adjustment is known as "**lag adjustment**". Sometimes it is referred to as *"power factor"*, *"quadrature"* or *"inductive load adjustment"*.

### 12.7.4 Light Load or Friction Compensation

Despite every care taken in the design of both the jewelled-pivot bearing, which forms the lower bearing for the spindle, and of the simple sleeve pin-type bearing at the top of the spindle, friction errors are liable to be serious, particularly at light loads. In order to ensure accurate registration at low loads, it is therefore necessary to arrange for small torque, practically independent of the load on the meter which acts in the direction of rotation and which is nearly as possible equal in magnitude to the friction torque. This is usually obtained by means of a small shading loop (or vane) situated between the centre pole of the shunt magnet and the disc and slightly the one side of the center-line of the pole. This is shown as $L_2$ in Fig. 12.10. The interaction between the portions of the flux which are shaded and unshaded by this loop and the currents they induce in the disc results in a small driving torque whose value can be adjusted by lateral movement of the loop. The tests for correct adjustment are freedom from serious errors at light loads and the value of the load necessary to start the meter. Other arrangements of non-symmetrical shading loops are also used.

### 12.7.5 Creep

In some meters a slow but continuous rotation is obtained even when there is no current flowing through the current coil and only pressure coil is energized. This is called *creeping*. The major cause for creeping is over-compensation for friction. If the friction compensating device is adjusted to give a driving torque to compensate for starting friction which is bigger than the running friction there is a tendency for the disc to run even when there is no current through the current coils because the friction compensating torque is independent of the load current as the compensating device is voltage actuated. The other causes of creeping are excessive voltage across the potential coil (which is responsible for production of excessive torque by the friction compensating device), vibrations, and stray magnetic fields.

In order to prevent this creeping two diametrically opposite holes are drilled in the disc ; the disc will come to rest with one of the holes under the edge of a pole of the shunt magnet the rotation being thus limited to a maximum of half a revolution. The action may be understood by reference to Fig. 12.11. If a hole is under the edge of a pole, the circular eddy-current paths in the disc will be distorted as shown. The effective centre of the eddy-current paths is then at $A'$ and this point is the central point of the equivalent magnetic pole produced by the currents. Examination of polarities shows that there is a resulting force on the disc, tending to move $A'$ away from the pole axis $A$. Thus the disc may creep until the hole reaches a position near the edge of a pole, but further movement is opposed by the torque produced in the manner just described. The magnitude of this torque is not sufficient to affect the action of the meter on load.

In some cases a small piece of iron is attached to the edge of the disc. The force of attraction exerted by the brake magnet on the iron piece is sufficient to prevent creeping of disc.

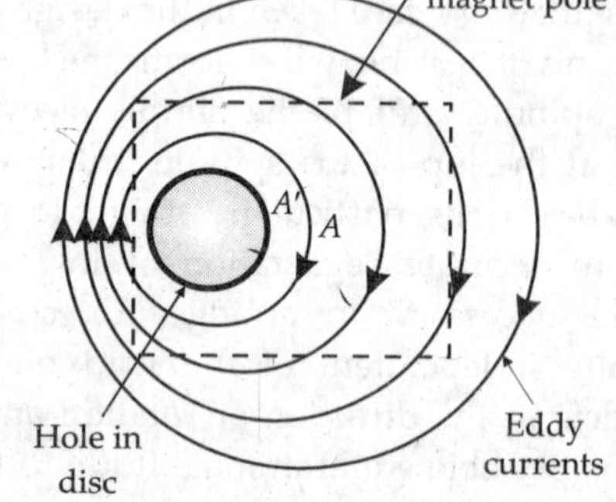


**Fig. 12.11** Prevention of creep by drilling holes in disc.

### 12.7.6 Over-Load Compensation

The disc revolves continuously in the field of the series magnet under load conditions and, therefore, there is a dynamically induced emf in the disc because of this rotation. This emf causes eddy currents which interact with the field of the series magnet to produce a braking torque. This self-braking torque is obviously proportional to the square of the load current and it assumes increasing importance as the load rises. Thus at high values of load current the registration tends to be lower than the actual.

In order to minimize the self-braking action, the full load speed of the disc is kept as low as possible. This speed is nearly 40 rpm. The current coil flux $\Phi_s$ is made small as compared pressure with coil flux $\Phi_p$. Thus the dynamically induced emf (which causes the braking torque) is very small as compared with statically induced emf which is produced by flux alterations at the rate of 50 Hz and which is responsible for production of driving torque.

It is customary to add an overload compensating device. This usually takes the form of a magnetic shunt for the series magnet core as shown in Fig. 12.12. The magnetic shunt approaches saturation and so its permeability decreases at overloads. Thus at large currents the magnetic shunt diverts less of series magnet flux so that a larger portion of the flux appears in the disc air gap and contributes to driving torque.

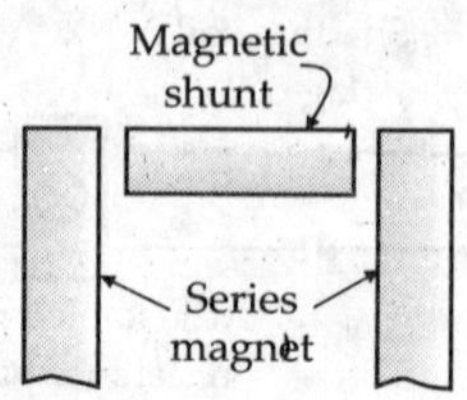


**Fig. 12.12** Overload compensation device.

### 12.7.7 Voltage Compensation

A certain amount of variation is permitted in the declared voltage of supply. Therefore, energy meters must be compensated for this variation. Voltage variations cause errors owing to two reasons :

(*i*) The relationship between shunt magnet flux and the supply voltage is not linear owing to saturation in iron parts :

(*ii*) The shunt magnet flux $\Phi_p$ produces a dynamically induced emf in the disc which in turn results in a self-braking torque which is proportional to square of the supply voltage.

Compensation for voltage variations is provided by using a saturable magnetic shunt which diverts a greater proportion of the flux into the active path when the voltage rises. The compensation can be conveniently provided by increasing the reluctance of the side limbs of the shunt magnet. This is done by providing holes in the side limbs shown in Fig. 12.2.

### 12.7.8 Temperature Compensation

An increase in temperature is accompanied by a rise in resistance of all copper and aluminium parts and results in :

(*i*) A small decrease in the potential coil flux and a reduction in angle of lag between $V$ and $\Phi_p$,

(*ii*) A decrease in torques produced by all shading bands,

(*iii*) An increase in the resistance of the eddy current paths, and

(*iv*) A decrease in the angle of lag '$\alpha$' of the eddy currents.

If we examine the above effects we find that the errors introduced by variations of temperature of an induction meter are usually small since the various effects produced tend to neutralize each other. An increase in temperature increases the resistance of the pressure coil so that it takes less current resulting in reduced flux of the shunt magnet. This reduces the driving torque. But the flux of the braking magnet is also decreased owing to increase in temperature thereby reducing the braking torque. Again an increase in temperature increases the resistance of the eddy current paths in the disc and this reduces both the driving and the braking torques. Hence these effects tend to neutralize each other. The temperature errors are certainly negligible with non-inductive loads but these errors are quite serious at low power factor loads.

The effect of increase of temperature, in general, is to cause the meter to run fast and to register high. Therefore, the temperature effects can be compensated for by simply using a temperature shunt on the brake magnet. Special magnetic materials, such as 'Mutemp' are available which show a considerable decrease in permeability with increase in temperature.

### 12.7.9 Errors in Single Phase Energy Meters

The errors caused by the driving system are :

(*i*) *Incorrect magnitude of fluxes.* This may be due to abnormal values of current or voltage. The shunt magnet flux may be in error due to changes in resistance of coil or due to abnormal frequencies.

(*ii*) *Incorrect phase angles.* There may not be proper relationship between the various phasors. This may be due to improper lag adjustments, abnormal frequencies, change in resistance with temperature, etc.

(*iii*) *Lack of symmetry in magnetic circuit.* In case the magnetic circuit is not symmetrical, a driving torque is produced which makes the meter creep.

The errors caused by the braking system are :

(*i*) Changes in strength of brake magnet,

(*ii*) Changes in disc resistance,

(*iii*) Self-braking effect of series magnet flux, and

(*iv*) Abnormal friction of moving parts.

### 12.7.10 Adjustments in Single Phase Energy Meters

Some adjustments are carried out in energy meters so that they read correctly and their errors are within allowable limits. The sequence of these adjustments is :

**1.** ***Preliminary light load adjustment.*** The disc is so positioned that the holes are not underneath the electromagnets. Rated voltage is applied to the potential coil with no current through the current coil. The light load device is adjusted until the disc just fails to start.

**2.** ***Full load unity factor adjustment.*** The pressure coil is connected across the rated supply voltage and rated full load current at unity power factor is passed through the current coils. The position of the brake magnet is adjusted to vary the braking torque so that the meter revolves at the correct speed within required limits or errors.

**3.** ***Lag adjustment (low power factor adjustment).*** The pressure coil is connected across rated supply voltage and rated full load current is passed through the current coil at 0.5 p.f. lagging. The lag device is adjusted till the meter runs at correct speed.

**4.** With rated supply voltage, rated full load current and unity power factor, the speed of the meter is checked and full load unity p.f. and low p.f. adjustments are repeated until the desired accuracy limits are reached for both the conditions.

**5.** ***Light load adjustment.*** Rated supply voltage is applied across the pressure coil and a very low current (about 5 per cent of full load) is passed through the meter at unity power factor. The light load adjustment is done so that the meter runs at correct speed.

**6.** Full load-unity power factor and light load adjustments are again done until speed is correct for both loads i.e., full load as well as light load.

**7.** The performance is rechecked at 0.5 p.f. lagging.

**8.** ***Creep adjustment.*** As a final check on light load adjustment the pressure coil is excited by 110 per cent of rated voltage with zero load current. If the light load adjustment is correct the meter should not creep under these conditions.

## 12.8 POLYPHASE ENERGY METERS

Energy in polyphase circuits can be measured by a group of single phase energy meters connected as required by Blondel's Theorem. The total energy is the sum of the readings of all energy meters (a negative sign is used for the reading of any meter which runs backwards). However, in commercial measurements the above arrangement is not used but instead polyphase energy meters are used.

Similar to the case of wattmeters for measurement of power in polyphase circuits the electrical energy in a '$n$' conductor system requires $(n-1)$ measuring elements for measurement of total energy. Thus a 3 phase, 3 wire system requires a 2 element energy meter and a 3 phase 4 wire system requires a 3 element energy meter.

In polyphase wattmeters the elements are mounted on the same spindle which drives the registering mechanism. Thus the registering mechanism registers the net effect of all the elements.

Polyphase wattmeters may be 'multi-disc type' or single disc type. In the multi-disc type each element drives a separate disc. In the single disc type meter all the elements drive the same disc. This disc is slotted or laminated in sectors to prevent interaction between eddy currents produced by one element with the flux produced by another element.

### 12.8.1 Two Element Energy Meter

Figure 12.13 shows a 2 element energy meter which is used for 3 phase 3 wire systems. The meter shown in the diagram is provided with 2 discs one for each element. It is essential that the driving torque of the two elements be exactly equal for equal amounts of power passing through each. Thus in addition to normal compensating devices attached to each element an adjustable magnetic shunt is provided on one or both the elements to balance the torques of the two. The necessary adjustment is made with the coils energized from a single phase supply. The pressure coils are connected in parallel and the current coils in series in such a manner that the torques produced by the two elements oppose each other. The magnetic shunt is adjusted to a position where the two torques are exactly equal and opposite and therefore there is no rotation of disc. This way the two elements are rendered exactly similar.

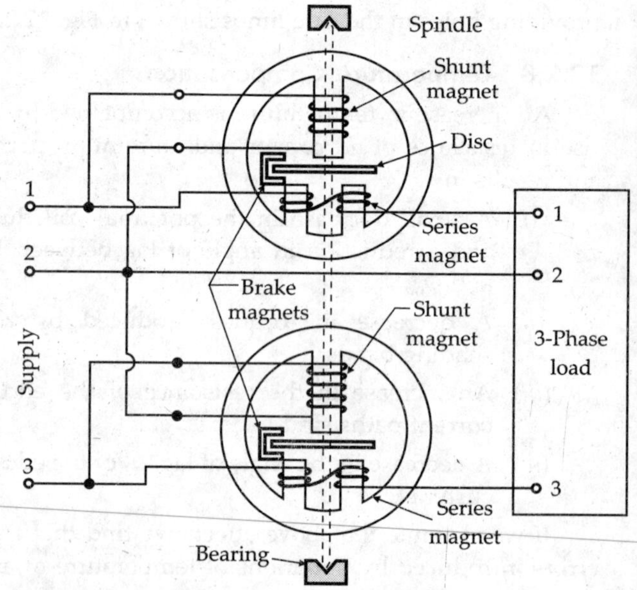


**Fig. 12.13** Two element energy meter.

The usual lag, light load and unity power factor adjustments are made independently for each element as explained earlier in Art. 12.7.9.

## 12.9 INDUSTRIAL METERING AND TARIFFS

The rating of a power generating plant must be equal to the simultaneous maximum demand it has to meet. This means that a fixed sum proportional to the maximum demand to be met has to be spent for the plant required. This type of expenses which include the cost of plant, buildings, transmission and distribution equipment, part of staff salaries are called

*"Fixed Charges"*. In addition to above, expenses have to be incurred for fuel and also for maintaining the plant. These expenses are proportional to the number of units generated and are called *"Running Charges"*.

Thus if the total cost of electric supply is to be shared equitably by various consumers, every consumer should pay a fixed sum proportional to his maximum demand in addition to a sum proportional to the energy consumed by him.

In a.c. systems the size of the plant is determined not only by the kW capacity required but also by the power factor at which the plant is required to work. Consumers should be encouraged to draw power at high power factors because a system working at low power factor has the following disadvantages :

(*i*) Cost of power station and distribution equipment is more for low power factor loads.

(*ii*) Low power factor makes the regulation poor so that it becomes increasingly difficult to maintain the supply voltage within specified limits.

(*iii*) The current drawn from the supply by low power factor loads is high. Therefore the loss in the cables, lines and other equipment will be higher at low power factors. Or if the losses are to be kept the same, the sizes of conductors and other equipment will have to be increased.

Tariffs for industrial users should take the above factors into account. There are various types of tariffs which try to balance the costs incurred with the nature of the consumer's load. Some of the industrial tariffs are :

1. ***Kilo watt-hour tariff.*** The most straight-forward but not the most usual form, is a charge per kilowatt-hour consumed. Thus the consumer's bill is directly proportional to the energy used. It is also known as *"Flat Rate"* charge.

2. ***Maximum demand tariff.*** It is a two part tariff. The total charges are divided into two parts one being simple charge per kilowatt-hour of energy consumed with another fixed charge which is based upon the maximum demand. This maximum demand is actually the average over a predetermined period of time. This period is usually 15 to 30 minutes.

3. ***Power factor tariffs.*** Power factor tariffs are devised to make a distinction between overall charge per kilowatt-hour to be recovered from *two* types of consumers ; (*i*) Those having good power factors (who should be charged less) and (*ii*) Those having poor power factors (who should be charged more). Power factor Tariffs are two part type. There are *three* different types of power factor tariffs :

(*a*) **kVA maximum demand tariffs.** The maximum demand is measured in kVA and a demand charge per kVA may be recovered in addition to the charges levelled for the energy consumption.

(*b*) **Average power factor tariffs.** In this case, an average power factor say 0.8 lagging may be assumed as reference and a surcharge for each 0.01 by which the average p.f. falls below this figure (0.8 in our case) may be made. Also, if the average p.f. is above this reference figure, a discount may be allowed in a similar manner.

(*c*) **kWh and kV Arh tariffs.** In this type of tariff both kilowatt-hour (kWh) and reactive kilovolt-ampere-hour (kV Arh) are separately charged.

The above brief consideration of tariffs is sufficient to show that special meters like maximum demand indicators, reactive kilovolt-ampere hour meters and kilovolt-ampere hour meters have to be installed at industrial establishments in order to have balance between costs of power production with the money recovered from different types of consumers.

## 12.10 MAXIMUM DEMAND INDICATORS

The chief requirement of these indicators is that they shall record the maximum power taken by a consumer during a particular period. The maximum demand indicators shall be so designed that any sudden momentary increase in loads such as due to short circuits on the system or due to high starting currents of motors are not taken into account. Therefore, a maximum demand indicator must be designed to record the average power over successive predetermined periods. These periods may be of 15 to 30 minutes duration.

Maximum demand indicators are of *four* types :

(*i*) Recording demand indicators.

(*ii*) Average demand indicators (Merz Price maximum demand indicator).

(*iii*) Thermal type maximum demand indicators.

(*iv*) Digital maximum demand indicators. Only Merz Price Maximum Demand Indicator is described here.

### 12.10.1 Average Demand Indicator (Merz Price Maximum Demand Indicator)

This type of demand indicator is not a separate instrument but is a fitting which can be attached as a unit with any type of energy meter. Thus the maximum demand indicator together with the energy meter registers the total energy consumption and the maximum value of the average power over equal intervals of time which may be of 15 or 30 minutes duration.

***Construction and operation.*** It consists of a special dial mechanism with a pointer driven forward by a gear train which is normally coupled to the energy meter spindle. This dial system is engaged to the meter spindle for a predetermined time interval say half an hour. After the expiry of this time interval, a reset device comes into operation which brings the mechanism back to zero position. The pointer, however, does not return to zero but is lightly held by a special friction device and continues indicating the energy consumed during the previous half hour. The pointer will not move forward unless the energy consumed in some subsequent time interval exceeds the one recorded by the pointer. In this way the maximum demand expressed in energy consumed per half hour for any given period of the time, is obtained.

The reset device for returning the driving mechanism of the demand indicator to zero at the end of every half hour, is operated by a switch actuated by either a small synchronous motor of the electric clock-type or by an ordinary spring driven clock-mechanism. In some cases the spring of the clock is wound electrically. Duration of the time intervals depends upon the setting of the switch which may be altered as desired.

Figure 12.14 shows a Merz-Price maximum demand indicator. A pin drives the pointer forward for a period say half hour. The energy consumed during this period is indicated on the dial. At the end of this period a cam controlled by a timing gear momentarily disengages the pinion (as indicated) by means of a bell crank. This allows the driving mechanism and the pin to return to zero position under the action of a spring. The pointer, however, remains at its position and continues recording the energy consumed during the previous half hour period. During the next half hour period the pin is again driven forward, but the pointer is only moved forward if the energy consumed during a subsequent period exceeds that consumed during all previous periods.

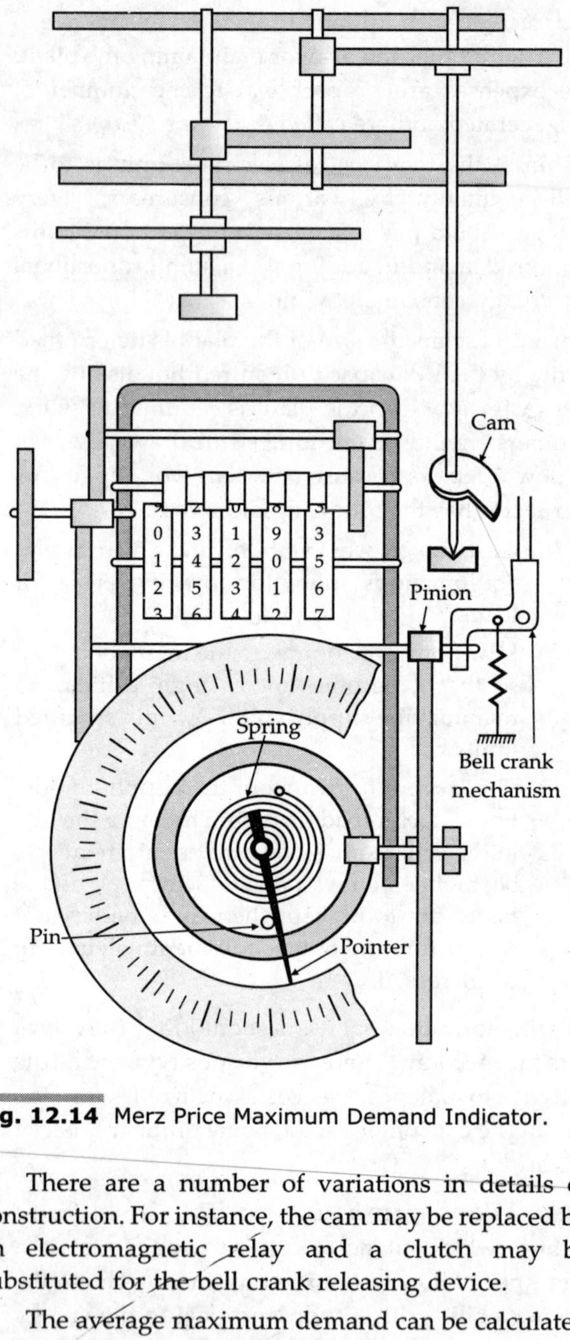


**Fig. 12.14** Merz Price Maximum Demand Indicator.

There are a number of variations in details of construction. For instance, the cam may be replaced by an electromagnetic relay and a clutch may be substituted for the bell crank releasing device.

The average maximum demand can be calculated by the following simple formula :

$$\text{Average maximum demand in kW} = \frac{\text{maximum energy recorded over a time in interval kWh}}{\text{time interval in hours}} \quad \ldots(12.17)$$

Maximum demand meters of this type can also be used to measure the maximum demand in terms of kVAh or kVArh by attaching to a suitable meter element which will measure such quantities.

***Advantages*** :

The instrument is more accurate than the thermal type. Its scale is also uniform.

***Disadvantages*** :

(*i*) This meter is costly since it involves mechanisms of great complexity.

(*ii*) The meter suffers from a serious disadvantage. If the maximum demand occurs some time after the start of one time interval and continues over only a part of the next interval the real maximum is not indicated since it is split up into two different time intervals. The thermal lagged meters whose indication does not depend upon preselected times but whose integration is continuous do not suffer from this disadvantage as they have a very long response time.

## 12.11 MEASUREMENT OF Vah AND VArh

For a circuit with a supply voltage *V*, load current *I* and power factor cos $\phi$, we can write :

$$(VI)^2 = (VI\cos\phi)^2 + (VI\sin\phi)^2$$

Now if both sides of the above equation are multiplied by the quantity $h^2$, where *h* is the time in hours, we have :

$$(VIh)^2 = (VIh\cos\phi)^2 + (VI\sin\phi)^2 \qquad ...(12.18)$$

or

$$(\text{VAh})^2 = (\text{Wh})^2 + (\text{VArh})^2 \qquad ...(12.19)$$

where VAh = volt-ampere hours,

Wh = watt-hours,

and VArh = volt-ampere hours reactive.

Expressing the above equation in the units encountered in big systems

$$(\text{kVAh})^2 = (\text{kWh})^2 + (\text{kVArh})^2 \qquad ...(12.20)$$

Equations 12.18, 12.19 and 12.20 hold good only for the condition that cos$\phi$ is maintained constant throughout the time period of *h*-hours. In practice this condition is usually not met. Let us consider a circuit where the power factor cos$\phi$ varies irregularly from time to time. If a graph is plotted with watt-hour readings horizontally and reactive volt ampere vertically, the readings being taken at hourly intervals then at the end of several hours the graph obtained would be somewhat as shown in Fig. 12.15(*b*).

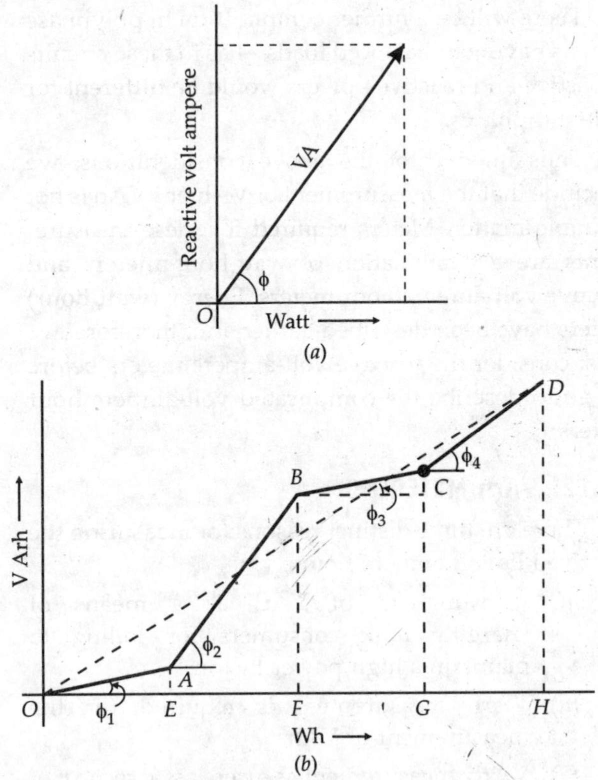


**Fig. 12.15** Relationships between Apparent, active reactive components of Power and Energy.

At the end of first hour the energy in Wh is *OE* and VArh is equal to *AE* and therefore the VAh, provided that p.f. cos $\phi_2$ has remained constant during first hour, is given by *OA* as :

$$OA = \sqrt{(OE^2) + (AE^2)}$$

At the end of 2nd hour the total Wh recorded is *OF* and the total VArh is *BF*. But the total Vah is not represented by line *OB*, since the p.f. during second hour is cos $\phi_1$, which differs from cos $\phi_2$. The true VAh is given by (*OA* + *AB*) which is, of course, greater than *OB*. Thus over a period of four hours [as represented in Fig. 12.15(*b*)], the total Wh recorded is *OH* and the total VArh is *DH*. If we apply Eqn. 12.19, the total VAh is *OD*. But the true VAh is not *OD* but is (*OA* + *AB* + *BC* + *CD*) which is larger than *OD*.

It is clear that whatever be the shape of lines, between O and D its length can never be less than the straight line *OD*. Thus if the power factor varies from one instant to the next, even by a small amount the true total VAh is always greater than the value given by expression

$$\text{VAh} = \sqrt{(\text{Wh})^2 + (\text{VArh})^2}$$

There will be a further complication in polyphase circuits having unbalanced loads — as in these circuits the active and reactive powers would be different for different phases.

Thus in view of the above considerations, we conclude that the measurement of VAh or kVAh is not a simple matter. Meters required for these measurements are a combination of watt-hour meters and reactive volt-ampere hour meters. Energy (watt-hour) meters have been described earlier and, therefore, we must consider the reactive volt-ampere meters before we go to describe the complicated volt-ampere hour meters.

## 12.12 VArh METERING

There are three distinct reasons for measuring the reactive kilovolt ampere hours, viz.

(*i*) Measurement of VArh is a means of charging bulk consumers for failing to maintain a high power factor.

(*ii*) VArh measurement is required for the measurement of VAh.

(*iii*) VArh measurement is used as a record of operating conditions when there is an exchange of energy between two or more interconnected power stations.

Integrating meters for registering VArh are usually based upon induction type energy meter element.

Consider Eqn. 12.14 of a single phase energy meter.

The total number of revolutions made in a specified time

$$= K\int VI\sin(\Delta-\phi)dt$$

In case of energy meters angle $\Delta$ is made nearly 90° the meter registers the energy consumed. However if, we make 0 or 180° (by giving an additional quadrature shift) the total number of revolutions in a specified time

$$= K\int VI\sin(0-\phi)dt = K\int VI\sin\phi dt$$

$$= K\int(\text{VAr})\,dt$$

Thus an induction type energy meter can be converted into a reactive volt-ampere hour meter if the pressure coil flux is brought into phase with the voltage. Hence we conclude that a VArh meter is an ordinary induction type energy meter with an additional quadrature shift. Several methods are available to achieve the required additional 90° phase shift.

The methods employed are :

(*i*) Special devices are used so as to achieve the requisite phase shifts in the current and voltage circuits.

(*ii*) Normal energy meters are used but voltages applied have an effective displacement of 90° from those normally applied for energy measurements. This arrangement is particularly suited to polyphase systems.

Normal induction type of energy meters can be used to register VArh with the help of following schemes. These schemes apply to 3-phase 3 wire systems.

(*i*) *Single element system.* A normal single phase induction type energy meter is used. The current coil is connected in series with one of the phases while the pressure coil is connected across the other two phases as shown in Fig. 12.16. This connection is the same as used for measurement of reactive power.

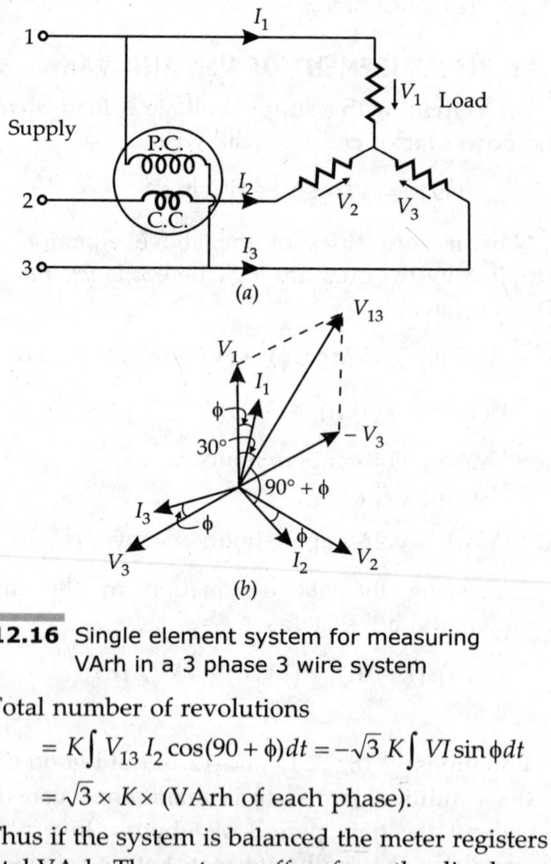


**Fig. 12.16** Single element system for measuring VArh in a 3 phase 3 wire system

Total number of revolutions

$$= K\int V_{13} I_2 \cos(90+\phi)dt = -\sqrt{3}\,K\int VI\sin\phi dt$$

$$= \sqrt{3}\times K\times (\text{VArh of each phase}).$$

Thus if the system is balanced the meter registers the total VArh. The system suffers from the disadvantage that it can only be used for balanced loads.

(*ii*) *Crossed phase method.* This method can be used for unbalanced currents in a 3 phase 3 wire system. In this case a 2 element polyphase energy meter is used. A neutral point is obtained by means of a choke as shown in Fig. 12.17(*a*). Each current is associated with a voltage differing in phase from the voltage normally used for power or energy measurements as shown in Fig. 12.17(*b*). In this method we must assure that torques acting on the two elements are opposite in direction. Evidently the registration requires a multiplier of value $\sqrt{3}$, since the voltage across the pressure coil is $1/\sqrt{3}$ of that used for power or energy measurements.

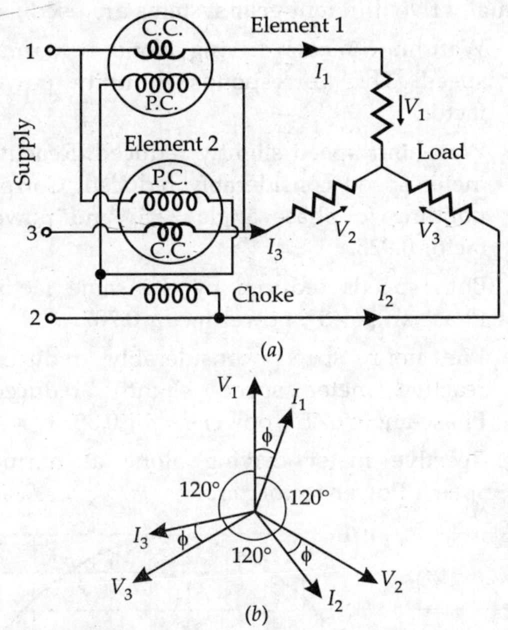


**Fig. 12.17** Crossed phase method of metering VArh in a 3 phase 3 wire system.

This is clear from the following :

Total number of revolutions

$$= K\int V_3 I_1 \cos(120°+\phi) - V_1 I_3 \cos(120°-\phi)\, dt$$

Taking a case of balanced currents

$$= K\int VI[\cos(120°+\phi) - \cos(120°-\phi)]dt$$

Total number of revolutions

$$= K\int (\sqrt{3}\, VI \sin\phi)dt$$

(*iii*) *Autotransformer method.* This method uses a two element watt hour meter with a phase shifting arrangement employing two open delta connected auto-transformers as shown in Fig. 11.33 on page 372. The meter will thus register VArh.

## 12.13 MEASUREMENT OF VAh

Volt-ampere hours are sometimes measured where the added expense for such measurement is justified. Figure 12.18 illustrates the two alternative methods of metering volt-ampere hours. *OI* represents the active power W and *OV* represents the apparent power VA. In Fig 12.18(*a*) meter *A* is the watt hour meter while meter *B* registers the volt-ampere hours. This meter *B* may be maximum demand indicator which gives VA demand.

$$\text{VAh} = \text{reading of meter } B$$

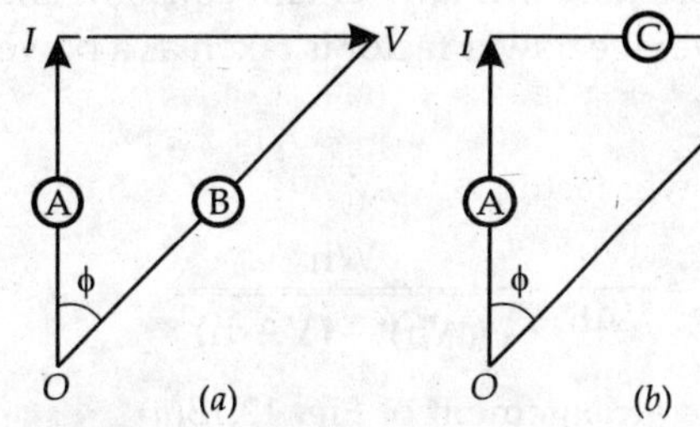


**Fig. 12.18** Methods of VAh metering.

In Fig. 12.18(*b*), *A* is a meter which records watt hours and *C* is a reactive volt-ampere hour meter. The volt-ampere hours may be calculated from these two readings :

$$\text{VAh} = \sqrt{(\text{Wh})^2 + (\text{VArh})^2}$$

$$= \sqrt{(\text{reading of meter } A)^2 + (\text{reading of meter } C)^2}$$

This principle of vector addition is utilised for the construction of a meter which measures VAh directly. The action of this meter is based upon an ingenious form of ball and disc friction gearing used to couple three shafts together as shown in Fig. 12.19. Two of these shafts are driven one by a watt hour meter and the other by a reactive volt-ampere hour meter. Due to the peculiar action of the friction gearing unit, the third shaft takes up a speed proportional to the square root of the sum of the squares of the speeds of the two shafts. Hence the friction gearing sums up the two speeds vectorially in accordance with the following expression :

$$\text{VA} = \sqrt{(\text{W})^2 + (\text{VAr})^2}$$

Therefore, the speed of this third shaft is proportional to the VA in the circuit. This shaft is connected to VAh registering mechanism which records the VAh.

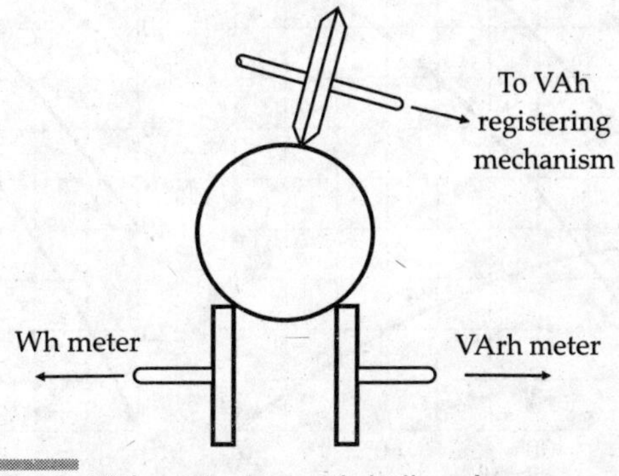


**Fig. 12.19** VAh metering with ball and disc friction gearing.

12

Actually it is not very necessary to meter volt ampere hours since the information required is merely for the establishment of an acceptable power factor index for estimating the cost of energy supplied. The power factor does not normally remain constant and, therefore, the average power factor is taken as a power factor index.

Average power factor

$$\cos\phi_a = \frac{\text{Wh}}{\text{VAh}} = \frac{\text{Wh}}{\sqrt{(\text{Wh})^2 + (\text{VArh})^2}}$$

Thus for the arrangement of Fig. 12.18(*a*)

$$\cos\phi_a = \frac{\text{reading of meter } A}{\text{reading of meter } B}$$

For the arrangement of Fig 12.18(*b*)

$$\cos\phi_a = \frac{\text{reading of meter } A}{\sqrt{(\text{reading of meter } A)^2 + (\text{reading of meter } C)^2}}$$

Average power factor is also known as "average energy factor".

**1. *Trivector meter*.** This meter measures kVAh and also kVA of maximum demand. The meter consists of a kWh meter and a reactive kVAh meter in the case with a special summator mounted between them. Both the meters drive the summator through a complicated gearing which arranges for the summator to register kVAh correctly at all power factors.

Actually five different gear systems are used :

1. Watt-hour meter driving alone at normal speed. This corresponds to unity power factor.
2. Watt-hour speed slightly reduced. Reactive meter speed considerably reduced. Corresponding to phase angle 22.5° and power factor 0.925.
3. Both speeds reduced by the same factor. Phase angle 45°, power factor 0.707.
4. Watt-hour speed considerably reduced, reactive meter speed slightly reduced. Phase-angle 67.5°, power-factor 0.38.
5. Reactive meter driving alone at normal speed. Power-factor zero.

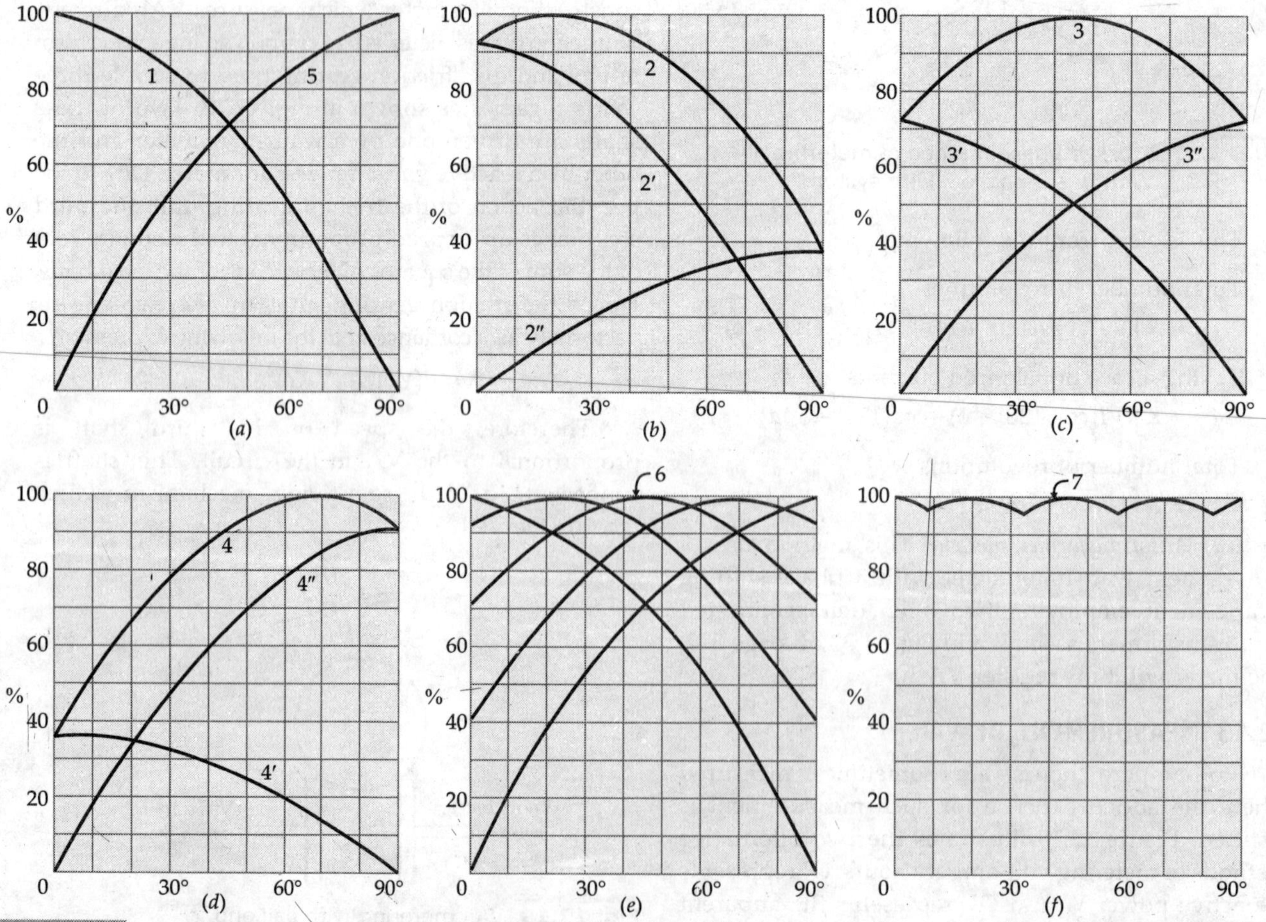


**Fig. 12.20** Diagrams related to Trivector Meter.

The final drive from each gear system is taken to a ratchet-coupling linked to a common register shaft. By this means the shaft is always driven by the planet wheel (or direct drive) which has the maximum speed, while all four of the slower shafts merely idle on the ratchets. If the power-factor changes by an appreciable amount then the drive shifts to a different ratchet as a new shaft takes up the fastest running speed. This is shown by the speed/power-factor curves for the two component meters. For a given *VI* product the speed of the watt-hour meter varies as cos $\phi$, consequently the curve of 'speed/phase-angle' over a range of 0 – 90° lagging takes the form (1) of Fig. 12.20 while the speed variations of the reactive meter follow curve (5). Thus, if only the two ungeared drives were available, then the speed of the final spindle would follow curve (1) for a range of phase angle variation 0 – 45°, but would move to curve (5) for angles 45 – 90°. Now taking conditions at the first differential gear the fundamental speeds are reduced by means of suitable gears to give the curves (2′) and (2″) for the drive speeds from the two component meters. The plant wheel speed is the sum of these two speed-curves (2). Similarly curve (3) applies to the second differential gear where the speed reduction ratios are equal, and curve (4) to the third differential where the speed ratios are the same as for curve (2) but in the reverse order. Continuing the resultant speed curves, and bearing in mind the fact that the register drive takes up the fastest ratchet speed, the overall curve (6) is obtained. If the gearing is modified a little to give slightly fast running at the peaks of the curve then the maximum effective error is not more than one or two per cent [curve 7].

**2. *VAh meter using bridge connected rectifiers.*** This meter utilizes a 3 phase bridge connected rectifier for summing up the 3 line currents. This meter, therefore, operates satisfactorily on unbalanced loads. The circuit is shown in Fig. 12.21. The three current transformers supply the bridge connected rectifier with currents proportional to the line currents The components of the direct current $I_a$ due to each of the currents $I_1'$, $I_2'$ and $I_3'$ are each independent of the phase of the corresponding current so that the total current $I_b$ is a measure of the arithmetic sum of the fine currents of the system. This current is passed through the control winding of a transductor, and the alternating current $I_c$ in the output circuit of the transductor (fed from the auxiliary transformer) is proportional to $I_b$. $I_c$ is passed through the current coil of a conventional induction meter ; the voltage circuit is fed from the lines, as shown. The phase angle between pressure coil current $I_p$ and current coil $I_c$ will be constant, so that the meter speed will be proportional to the product of line voltage and $I_c$, and hence to the product of line voltage and line current also summed by the rectifier.

12

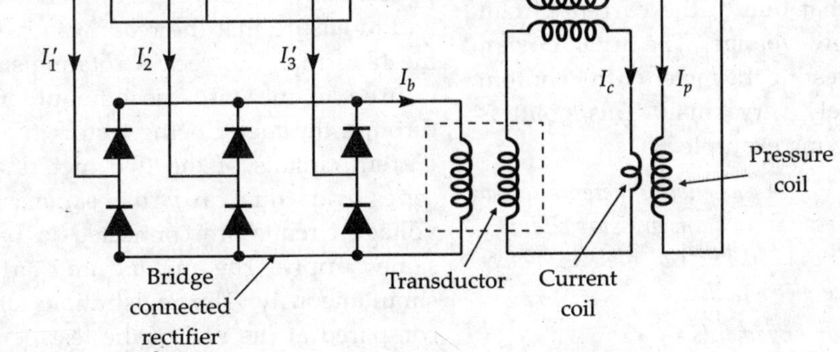


**Fig. 12.21** Bridge connected rectifier VAh meter.

## 12.14 TESTING OF ENERGY METERS

The term 'testing' includes the checking of the actual registration of the meter as well as the adjustments done to bring the errors of the meter within the prescribed limits.

**A.C. Energy meters.** Meters should be tested for the following conditions :

1. At 5 per cent of marked current with unity power factor.
2. At 100 or 125 per cent of marked current with unity power factor.
3. At one intermediate load with unity power factor.
4. At marked current and 0.5 lagging power factor.

In addition meters should be subjected to the following tests :

**1.** *Creep test.* With an applied voltage of 110 percent of its marked value and with the current circuit open, the meter should not revolve through more than one revolution. *i.e.*, it should not creep.

**2.** *Starting test.* At 0.5 per cent of marked value of current and with normal voltage the meter should start and run. Accuracy is not checked at this current.

### 12.14.1 Phantom Loading

When the current rating of a meter under test is high a test with actual loading arrangements would involve a considerable waste of power. In order to avoid this "Phantom" or "Fictitious" loading is done.

Phantom loading consists of supplying the pressure circuit from a circuit of required normal voltage, and the current circuit from a separate low voltage supply. It is possible to circulate the rated current through the current circuit with a low voltage supply as the impedance of this circuit is very low. With this arrangement the total power supplied for the test is that due to the small pressure coil current at normal voltage, plus that due to the current circuit current supplied at low voltage. The total power, therefore, required for testing the meter with Phantom loading is comparatively very small. This point is illustrated by the following example.

**Example 12.1** *A 220 V, 5 A d.c. energy meter is tested at its marked ratings. The resistance of the pressure circuit is 8800 Ω and that of current coil is 0.1 Ω. Calculate the power consumed when testing the meter with*

(i) *Direct loading arrangements.*

(ii) *Phantom loading with current circuit excited by a 6 V battery.*

**Solution.** The two testing arrangements are shown in Fig. 12.22.

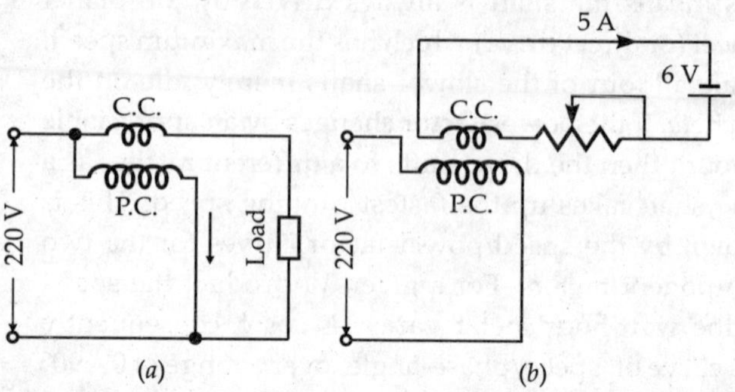


**Fig. 12.22** Diagram of Example 12.1.

(i) *Direct loading* :

Power consumed in pressure circuit

$$= (220)^2 / 8800 = 5.5 \text{ W}$$

Power consumed in current circuit

$$= 220 \times 5 = 1100 \text{ W.}$$

Total power consumed

$$= 5.5 + 1100 = 1105.5 \text{ W.}$$

(ii) *Phantom loading* :

Power consumed in pressure circuit

$$= (220)^2 / 8800 = 5.5 \text{ W.}$$

Power consumed in current circuit

$$= 6 \times 5 = 30 \text{ W.}$$

Total power consumed

$$= 5.5 + 30 = 35.5 \text{ W}$$

This power consumption is considerably smaller than the one with direct loading.

***Testing methods.*** Three main methods of testing motor meters are given below.

**Method A.** This method involves a long period Dial test.

*Long Period Dial Test.* This test is called the "long period dial test" and is performed to check the gearing and to ensure that the readings given by registration mechanism are correct. A rotating substandard meter is used to measure the amount of energy passing through the meter being tested in a given time. The current circuits of the two meters are connected in series and thus carry the same current while the voltage circuits are connected in parallel across the same supply. The meters are started and stopped simultaneously. The registrations of two meters are compared at the end of the test period (which is an hour or even more) and the error is calculated as below.

Let

$D_x$ = registration of meter under test in kWh.

$D_s$ = registration of sub-standard meter in kWh.

$\therefore$ Percentage error $= \dfrac{D_x - D_s}{D_s} \times 100.$

These tests (as they extend over long periods) give an opportunity of discovering errors due to self-heating. After the end of a long period dial test, a short time test at full load is done to detect the errors caused by heating.

**Method B.** A rotating substandard meter is used. The current coils of the substandard meter and the meter under test are connected in series, and their pressure coils are connected in parallel. Arrangements are provided to start and stop the two meters simultaneously. This test extends over a short period and therefore it is known as a short period test.

For short periods test

(*a*) either the revolutions of the test meter disc are compared with those made by the disc of substandard meter during the same time.

(*b*) or the speeds of the discs of test and substandard meters are directly compared

Let

$K_x$ = number of revolutions per kWh for the meter under test,

$K_s$ = number of revolutions per kWh for the substandard meter,

$R_x$ = number of revolutions made by the meter under test,

$R_s$ = number of revolutions made by the substandard meter during the test interval,

$N_x$ = speed of disc of meter under test,

$N_s$ = speed of disc of substandard meter,

$t$ = time interval in hours.

(*a*) *When revolutions are compared* :

Energy recorded by meter under test during test interval

$$= R_x / K_x$$

Energy recorded by substandard meter during test interval

$$= R_s / K_s$$

$$\therefore \text{ Percentage error} = \frac{(R_x / K_x - R_s / K_s)}{R_s / K_s} \times 100$$

(*b*) *When speeds are compared* :

Energy recorded by meter under test $= N_x t / K_x$

Energy recorded by substandard meter under test

$$= N_s t / K_s$$

$$\therefore \text{ Percentage error} = \left(\frac{N_x K_s}{K_x N_s} - 1\right) \times 100.$$

An additional requirement of method *B* is one *"long period dial test"* extending over an hour at full load. This is done in order to check the registering mechanism and also the effect of time. The meter is then again subjected to a short time test at full load to check for any effect due to heating.

**Method C.** In this method precision grade indicating instruments are used as reference standard. These indicating instruments are connected in the circuit of meters under test. The current and voltage are held constant during the test. The number of revolutions made by the meter disc during the test are recorded. The time taken is also measured.

Energy recorded by meter under test

$$= R_s / K_x \text{ kWh.}$$

Energy computed from the readings of the indicating instrument

$$= \text{kW} \times t$$

where

$R_x$ = number of revolutions made by disc of meter under test.

$K_x$ = number of revolutions per kWh for meter under test,

kW = power in kilowatt as computed from readings of indicating instruments, and

$t$ = time in hours.

$$\therefore \text{ Percentage error} = \frac{(R_x / K_x - \text{kW} \times t)}{\text{kW} \times t} \times 100$$

In addition to this test, a 'long period dial test' is required to check the registration mechanism.

Before conducting any of these tests on a watt hour meter its potential circuit must be connected to the supply for one hour in order to enable the self-heating of the potential coil to stabilise.

### 12.14.2 Meter Testing Circuits

These circuits incorporate Phantom loading arrangements. Therefore separate power supplies are used for potential and current circuits. In addition provision must be made for varying the phase of either current or voltage so that the meter can be tested at the desired power factor. A simple circuit for testing single phase induction type meters is shown in Fig. 12.23.

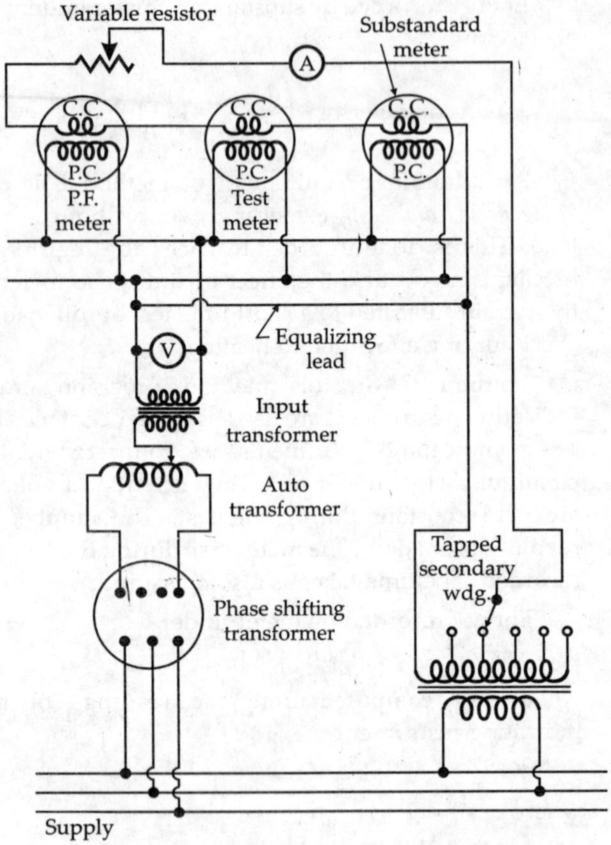


**Fig. 12.23** Testing of a.c. meters by Phantom loading (Phase shifting device in the pressure circuit).

***Pressure circuit supply.*** The potential circuit is supplied with the help of an input transformer having an output voltage of 250 V. The primary side of this transformer is connected to an autotransformer which provides fine control of output voltage from 0 to 250 V.

***Current circuit supply.*** The current circuit is supplied separately through a transformer. The primary of this transformer is connected to the same supply to which the pressure circuit is connected. The transformer has a tapped secondary and a switch is used to select the tapping which gives the required current. Fine adjustments of current can be done by a variable resistor included in the current circuit.

***Phase shifting arrangement.*** Two types of phase shifting arrangements are employed.

(*i*) The potential circuit is supplied from the secondary of a phase shifting transformer. The primary of this phase shifting transformer is connected to the same supply which feeds the current circuit. The phase of the potential circuit can be varied by rotating the rotor of the phase shifting transformer through an appropriate angle. Arrangement of Fig. 12.23 uses a phase shifting transformer.

(*ii*) Two similar alternators are coupled together and driven by a motor. One of the alternators supplies the potential circuit and the other supplies the current circuit. One of the alternators is mounted on a cradle so that its stator can be rocked around. Therefore, the position of the stator of this alternator can be changed with reference to the stator of the other alternator. Figure 12.24 shows an arrangement for phase shifting of alternator supplying the current, the phase of current in the current circuit can be altered with respect to phase of voltage across the pressure coil circuit. The angle through which the alternator stator is moved gives a measure of the phase angle between current and voltage. The frequency of the two currents is the same as the two alternators are identical and are mounted on the same shaft.

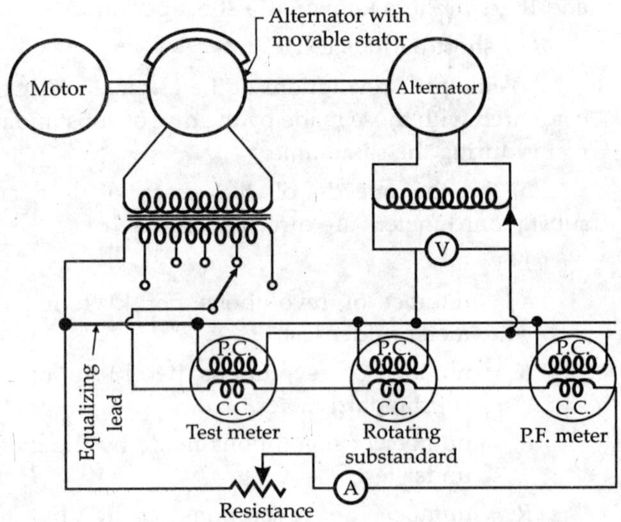


**Fig. 12.24** Testing of a.c. meters by Phantom loading (Phase shifting device in the current circuit).

When Phantom loading is used an equalizing lead is necessary (Fig. 12.24) in order to ensure that the potential difference between the current and the pressure coils of the meters is zero as it is when they are in service.

**Example 12.2** *The meter constant of a 230 V, 10 A watthour meter is 1800 revolutions per kWh. The meter is tested at half load and rated voltage and unity power factor. The meter is found to make 80 revolutions in 138 s. Determine the meter error at half load.*

**Solution.** Actual energy consumed at half load during 69 s.

$$= VI\cos\phi \times t \times 10^{-3} = 230 \times 5 \times 1 \times (138/3600) \times 10^{-3}$$

$$= 44.08 \times 10^{-3} \text{ kWh.}$$

$$\text{Energy recorded} = \frac{\text{number of revolutions made}}{\text{revolutions / kWh}}$$

$$= \frac{80}{1800} = 44.4 \times 10^{-3} \text{ kWh.}$$

$$\therefore \quad \text{Error} = \frac{44.44 - 44.08}{44.08} \times 100 = 0.817\% \text{ fast.}$$

**Example 12.3** *An energy meter is designed to make 100 revolutions of disc for one unit of energy. Calculate the number of revolutions made by it when connected to load carrying 40 A at 230 V and 0.4 power factor for an hour. If it actually makes 360 revolutions, find the percentage error.*

**Solution.** Energy actually consumed

$$= VI \cos\phi \times t \times 10^{-3} = 239 \times 40 \times 0.4 \times 1 \times 10^{-3}$$

$$= 3.68 \text{ kWh.}$$

The meter makes 100 revolutions for one unit *i.e.*, one kWh of energy consumed.

$\therefore$ Number of revolutions made $= 3.68 \times 100 = 368$.

In case the meter makes 360 revolutions,

$$\text{error} = \frac{360 - 368}{368} \times 100 = 2.17\% \text{ low.}$$

**Example 12.4** *The constant for a three phase, 3 element integrating wattmeter is 0.12 revolution of disc per kWh. If the meter is normally used with a potential transformer of ratio 22,000/110 V and a current transformer of ratio 500/5 A; find the error expressed as a percentage of the correct reading from the following test figures for the instrument only :*

*Line Voltage = 100 V; current = 5.25 A; power factor = 1*

*Time to complete 40 revolutions = 61 s.*

**Solution.** Actual energy consumed during the test period

$$= \sqrt{3} \times \text{ ratio of P.T.} \times \text{ratio of C.T} \times V_s I_s \cos\phi \times t$$

$$= \sqrt{3} \times \frac{22000}{110} \times \frac{500}{5} \times 110 \times 5.25 \times 1 \times \frac{61}{3600} \times 10^{-3}$$

$$= 339 \text{ kWh.}$$

Energy recorded by meter during the test period

$$= \frac{\text{number of revolutions made}}{\text{number of revolutions per kWh}}$$

$$= \frac{40}{0.12} = 333.3 \text{ kWh.}$$

$$\text{Percentage error} = \frac{333.3 - 339}{339} \times 100 = 1.68\% \text{ low.}$$

**Example 12.5** *A 230 V, single phase, watt hour meter has a constant load of 4 A passing through it for 6 hours at unity power factor. If the meter disc makes 2208 revolutions during this period what is the meter constant in revolutions per kWh ? Calculate the power factor of the load if the number of revolutions made by the meter are 1472 when operating at 230 V and 5 A for 4 hours.*

**Solution.** Energy supplied

$$= VI \cos\phi \times t \times 10^{-3}$$

$$= 230 \times 4 \times 1 \times 6 \times 10^{-3} = 5.52 \text{ kWh.}$$

Meter constant

$$= \frac{\text{revolutions}}{\text{kWh}} = \frac{2208}{5.52} = 400 \text{ rev/kWh}$$

Energy consumed when the meter makes 1472 revolutions

$$= 1472 / 400 = 3.68 \text{ kWh.}$$

Now energy consumed $= VI \cos\phi \times t \times 10^{-3}$

or $\quad 230 \times 5 \times \cos\phi \times 4 \times 10^{-3} = 3.68.$

$\therefore$ Power factor, $\cos\phi = 0.8$.

**Example 12.6** *A correctly adjusied, single phase, 240 V induction watthour meter has a meter constant of 600 revolutions per kWh. Determine the speed of the disc, for a current of 10 A at a power factor of 0.8 lagging.*

*If the lag adjustment is altered so that the phase angle between voltage coil flux and applied voltage is 86°. Calculate the error introduced at (a) unity p.f. (b) 0.5 p.f. lagging. Comment upon the results.*

**Solution.** Energy consumed in one minute with rated current and 0.8 p.f. lagging.

$$= 240 \times 10 \times 0.8 \times (1/60) \times 10^{-3} = 0.032 \text{ kWh.}$$

$\therefore$ Revolutions made in one minute

$$= 0.032 \times 600 = 19.2.$$

$\therefore$ Speed of disc = 19.2 rpm.

When lag adjustment is altered :

Steady speed, $N = KVI \sin(\Delta - \phi)$.

If the lag adjustment is correctly done $\Delta = 90°$.

Under this condition, steady speed

$$N = KVI \sin(90° - \phi) = KVI \cos\phi.$$

$\therefore$ Error introduced because of incorrect lag adjustment

$$= \frac{KVI[\sin(\Delta - \phi) - \cos\phi]}{KVI \cos\phi} \times 100$$

$$= \frac{\sin(\Delta - \phi) - \cos\phi}{\cos\phi} \times 100 \text{ percent.}$$

We have, $\Delta = 86°$.

(*a*) at unity p.f. $\phi = 0$.

$$\therefore \text{ Error} = \frac{\sin(86° - 0°) - 1}{1} \times 100 = -0.24\%.$$

(*b*) At 0.5 p.f. lagging, $\phi = 60°$.

$$\therefore \text{ Error} = \frac{\sin(86° - 60°) - \cos 60°}{\cos 60°} \times 100$$

$$= -12.3\%.$$

12

**Comments.** Thus it is evident that at low power factors the errors due to voltage flux not being in quadrature with applied voltage is very serious even though the phase displacement differs from 90° by a small value of 4°.

**Example 12.7** *A single phase induction type energy meter is adjusted to read correctly at unity power factor it is observed that at 1/4 full load current at 0.5 lagging power factor the effective voltage magnet flux lags behind the current magnet flux by 27°. Will it introduce any error in the measurement ? If so, calculate the percentage error introduced.*

**Solution.** Power factor $= \cos\phi = 0.5$ lagging.

Phase angle, $\phi = 60°$.

Thus the phase angle between applied voltage and load current is 60°. Under ideal conditions, the voltage magnet flux $\Phi_p$ should lag behind the applied voltage $V$ by 90° and the current magnet flux $\Phi_s$ should be in phase with load current $I$. This is shown in Fig) 12.25(*a*).

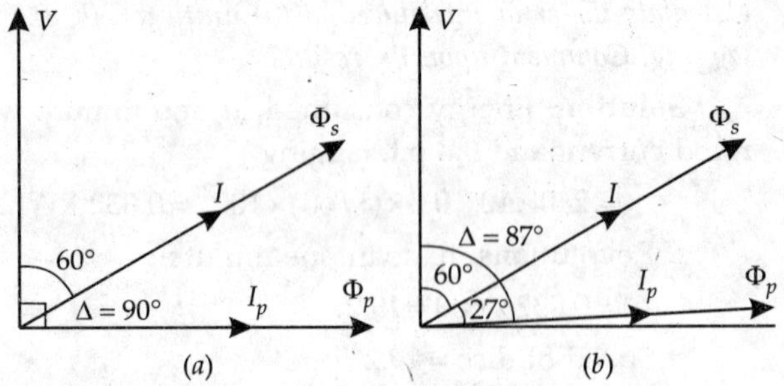


**Fig. 12.25** Phasor diagrams of Example 12.7.

Energy registered $= K\int VI\sin(\Delta-\phi)dt$

$\alpha \sin(\Delta-\phi)$

Now under ideal conditions $\Delta = 90°$.

$\therefore$ Energy registered under ideal conditions

$\alpha \sin(90°-\phi)\,\alpha \cos\phi\,\alpha \cos 60°.$

The phasor diagram for actual working conditions is shown in Fig. 12.25 (*b*).

Now $\Delta = 87°$.

$\therefore$ Energy registered under actual working conditions,

$\alpha \sin(\Delta-\phi)\,\alpha \sin(87°-60°)\,\alpha \sin 27°.$

It is clear from the above that the energy registered under actual working conditions (existing in the meter) is not the same as under ideal conditions. Therefore, the meter does not read correctly.

$$\text{Error} = \frac{[\sin 27° - \cos 60°]}{\cos 60°} \times 100$$

$$= \frac{0.454 - 0.5}{0.5} \times 100 = -9.2\%.$$

**Example 12.8** *On investigation a power company found that the polyphase meter through which an industrial plant was being served was incorrectly connected. The error resulted from incorrect polarity markings of one of the potential transformers. During two months the indicated demand had been 400 and 600 kW and the watt hour meter had registered 17000 and 25000 kWh. A careful test showed that power factor was 72 percent most of the time.*

(*a*) *By what factor should the demand meter and watthour meter readings be multiplied to obtain the actual readings ?*

(*b*) *The demand charge was Rs. 40 per kW per month and the energy charged 50 paise per kWh. Calculate the loss suffered by the company because of the error.*

**Solution.** If the meters are correctly connected :

Torque on element 1

$= KVI\cos(30°+\phi)$

and torque on element 2

$= \sqrt{3}\,K\cos(30°-\phi).$

The resultant torque is thus

$$T_d = KVI[\cos(30°+\phi)+\cos(30°-\phi)]$$
$$= \sqrt{3}\,KVI\cos\phi$$

where $K$ = a constant.

In case one of the P.Ts. is wrongly connected (*i.e.*, with its polarity reserved) the torque on that element is reserved. Let us consider that the connections on P.T. feeding element 1 are reversed. Therefore, deflecting torques are :

on element 1 $= K[-VI\cos(30°+\phi)]$

on element 2 $= KVI\cos(30°-\phi)$

Therefore, the resultant torque on the meter,

$$T_d = KVI[-\cos(30°+\phi)+\cos(30°-\phi)]$$
$$= KVI\sin\phi.$$

From above it is clear that if one of the P.Ts. is connected with reserved polarity, the meter reads corresponding to a power $VI\sin\phi$ instead of $\sqrt{3}\,VI\cos\phi$.

$$\therefore \text{True reading} = \sqrt{3}\,\frac{\cos\phi}{\sin\phi} \times \text{indicated reading}$$

$$= \sqrt{3}\cot\phi \times \text{indicated reading.}$$

Hence in order to get true readings the readings of the meter must be multiplied by a factor $\sqrt{3}\cot\phi$.

Now

$\cos\phi = 0.72$ or $\phi = 43.950$ and $\cot\phi = 1.0375.$

Thus the meter readings must be multiplied by a factor $\sqrt{3} \times 1.0375 = 1.8$ in order to get true readings.

Bill of the consumer for two months

$$= 40[400+600]+\frac{50}{100}[17000+26000]$$

$$= \text{Rs. } 56500.$$

True bill $= 1.8 \times 56500 = $ Rs. 101700

Loss suffered by company

$$= 101700 - 56500 = \text{Rs } 45200.$$

**Example 12.9** *A large consumer has a kVA demand and a kVAh tariff measured by 'sine' and 'cosine' watthour type meters each equipped with a Merz Price demand indicator. The tariff is Rs. 40 per month per kVA of demand plus 30 paise per kVAh. Determine the monthly bill for 30 days based upon the following readings :*

*'sine' meter advances by 90,000, reactive kVAr demand indicator 150 kVAr, 'cosine' meter advances by 120,000 kWh, and demand indicator by 200 kW. What is the average monthly power factor and the total cost per unit ?*

**Solution.** In a month, kVArh = 90,000

and kWh = 120,000.

∴ Total kVAh in a month

$$= \sqrt{(\text{kWh})^2 + (\text{kVArh})^2} = \sqrt{(20{,}000)^2 + (90{,}000)^2}$$

$$= 150{,}000.$$

$$\text{Average p.f., } \cos\phi_a = \frac{\text{kWh}}{\text{kVAh}} = \frac{120{,}000}{150{,}000} = 0.8$$

Now, maximum kVAr demand = 150

and maximum kW demand = 200

Maximum demand in kVA

$$= \sqrt{(\text{kW})^2 + (\text{kVAr})^2} = \sqrt{(200)^2 + (150)^2}$$

$$= 250$$

$$\text{Monthly bill} = 250 \times 40 + \frac{30}{100} \times 150{,}000 = \text{Rs. } 55000$$

$$\therefore \text{ Cost per unit} = \frac{55000}{120{,}000} \times 100 \approx 46 \text{ paise.}$$

**Example 12.10** *The following readings are obtained for one moth of 30 days : kVAhr meter = 38,830, kWh meter = 291,940, Demand indicator = 1400 kW.*

*Find out the average monthly load factor and power factor.*

**Solution.** Average load over a month

$$= \frac{\text{units consumed in a month}}{\text{number of hours in a month}}$$

$$= \frac{291940}{30 \times 24} = 405.5 \text{ kW}.$$

Average monthly load factor

$$= \frac{\text{average load}}{\text{maximum demand}} = \frac{405.5}{14000} = 0.289.$$

Total kVAh in a month

$$= \sqrt{(\text{kWh})^2 + (\text{kVArh})^2} = \sqrt{(29140)^2 + (83830)^2}$$

$$= 303{,}737.$$

$$\text{Average p.f., } \cos\phi_a = \frac{\text{kWh}}{\text{kVAh}} = \frac{291{,}940}{303{,}737} = 0.961.$$

**Example 12.11** *A 240 V, 5 A, single phase energy meter has a registration constant of 1200 revolutions per kWh. It is tested by means of a 240 V, 5 A wattmeter having 500 scale divisions which can be read to 0.1 division and a stop watch which can be read to 0.01 second and which has negligible error. When tested at full load, the meter makes 40 revolutions in 99.8 s. If the human error in timing be taken as ± 0.05 s, estimate the limits within which the error of the meter may lie. The wattmeter is accurate to within 0.05 percent of its full scale reading.*

**Solution.** Energy registered by the energy meter in 40 revolutions

$$= 40/1200 \text{ kWh} = (40/1200) \times 1000 \times 3600$$

$$= 120 \times 10^3 \text{ Ws}.$$

Power registered by wattmeter at 240 V and 5 A

$$= 240 \times 5 = 1200 \text{ W}.$$

Range of wattmeter = 1200 W

Scale divisions = 500

∴ 1 scale division = 1200/50 = 2.4 W

Observational error of wattmeter

$$= \pm 0.1 \times 2.4 = \pm 0.24 \text{ W}$$

Constructional error of wattmeter

$$= \left(\frac{0.05}{100} \times 1200\right) = \pm 0.6 \text{ W}.$$

Total error of wattmeter $= \pm 0.24 \pm 0.6 = \pm 0.84$ W.

∴ Wattmeter reading $= 1200 \pm 0.84$ W.

Observational error of stop watch $= \pm 0.05$ s.

Constructional error of stop watch $= \pm 0.01$ s.

Total error of stop watch $= \pm 0.05 \pm 0.01 = \pm 0.06$ s.

∴ Reading of stop watch $= 99.8 \pm \times 0.06 s$.

Energy obtained from readings of stop watch and wattmeter $= (1200 \pm 0.84)(99.8 \pm 0.06)$ Ws.

∴ Limits of error of energy meter

$$= \frac{120 \times 10^3 - \{(1200 \pm 0.84)(99.8 \pm 0.06)\}}{120 \times 10^3} \times 100$$

Using the positive sign — Percentage error

$$= \frac{120 \times 10^3 - \{(1200 - 0.84)(99.8 - 0.06)\}}{120 \times 10^3} \times 100 = +0.33\%.$$

Using the negative sign — Percentage error

$$= \frac{120 \times 10^3 - \{(1200 + 0.84)(99.8 + 0.06)\}}{120 \times 10^3} \times 100 = +0.07\%.$$

12

# Review Questions

1. Derive the expression for deflecting torque in single phase induction type meters. Show that the deflection is maximum when the phase angle between two fluxes is 90° and when the disc is purely non-inductive.
2. Describe the functions of the following in a single phase induction type energy meter :
   (*i*) Shunt and series magnets (*ii*) Moving disc
   (*iii*) Permanent magnet
   (*iv*) Registering mechanism (*v*) Shading bands
   (*vi*) Holes in the disc
3. Describe the constructional details of a single phase induction type energy meter. Explain, why the phase of shunt flux is made exactly in quadrature with that of applied voltage so as to produce a deflecting torque exactly proportional to power.
4. Explain how the following adjustments are made in a single phase induction type energy meter :
   (*i*) Lag adjustment
   (*ii*) Adjustment for friction compensation
   (*iii*) Creep
   (*iv*) Overload compensation
   (*v*) Temperature compensation
5. Explain the sources of errors in single phase induction type Energy Meters.
6. Describe the construction and working of a two element Induction type energy meter.
7. Explain the different types of tariffs used. Explain the underlying principle for using different types of tariffs for different types of consumers.
8. Describe the construction and working of a Merz Price Maximum Demand Indicator.
9. Explain the different types of methods used for VAh metering.
10. Describe in details the working of a Trivector meter.
11. What is Phantom loading ? Explain with an example how is it more advantageous than testing with direct loading.
12. Describe a circuit for testing of a single phase induction type energy meter at different loads and power factors.
13. Describe a VAR meter using bridge type rectifiers.
14. Enumerate the different types of tests that are carried out on single phase Induction type watt-hour meters. Describe the Long Period Dial Test.
15. Describe the crossed Phase Method for measurement of VArh in 3 phase 3 wire circuits.

# Unsolved Problems

1. The declared constant of a 5 A, 220 V d.c. watthour meter is 3275 revolutions per kWh. Calculate the speed of the disc at full load.

   In a test run at half load, the meter takes 59.5 s to complete 30 revolutions. Calculate the error of the meter. **[Ans.** 1 r.p.s., 0.84% fast]
2. A 50 A, 230 V meter on full load test makes 61 revolutions in 37 s. If the normal disc speed is 520 revolutions per kWh, find the percentage error. **[Ans.** 0.75% low]
3. A watthour meter is calibrated to measure energy on a 250 V supply. On test a steady current of 15 A is passed through it for 5 hours at unity power factor. If the meter readings before and after the test are 8234.21 kWh and 8253.13 kWh respectively, calculate the percentage error.

   If the spindle turns through 290 revolutions during 5 minutes when a current of 20 A is passing through the meter at 250 V and 0.87 power factor, calculate the meter constant.

   **[Ans.** 0.9% high ; 800 revolutions/kWh]
4. The current and flux produced by series magnet of an induction watthour meter are in phase but there is an angular departure of 3° from quadrature between voltage and the shunt magnet flux. The speed of disc at full load and unity power factor is 40 rpm. Assuming the meter to register correctly under this condition, calculate its speed at 1/4 full load and 0.5 power factor lagging. Also find the percentage error. **[Ans.** 4.51 rpm, 9.2%]
5. Two single phase watthour meters are connected through current and potential transformers, to a three phase system. The ratios of the transformers

may be considered to be correct, but the phase angle of potential transformer is 15′ (leading) and that of current transformer 3′ (leading). The power factor of the system is 0.75 lagging. Determine the relative speeds of the disc. **[Ans.** 3.2 : 1]

6. A single phase induction watthour meter, tested at its full load rating of 240 V, 10 A, is 1% slow at unity power factor and correct at a power factor 0.5 lagging. Assuming that the friction error is compensated at all power factors, estimate the error at rated VA when the power factor of the system is (*i*) 0.8 lagging and (*ii*) 0.8 leading. **[Ans.** 0.57% slow ; (*ii*) 1.43% slow]

7. Calculate the monthly bill in the case of a consumer whose maximum demand is 120 kW, average monthly load factor is 30 percent, and where tariff in use is Rs. 40 per kW of maximum demand and 30 paise per kWh consumed. **[Ans.** Rs. 12576]

8. A balanced, three-phase delta-connected load on a 400-V, 50-Hz supply has an impedance of $5 + j7\Omega$/plane. The current coil of watthour meter $W_1$ is connected in the line *R* and that of a watthour meter $W_2$ in line *Y*. The volt coils of the two instruments have a common connection on line *B*, but have their other terminals cross-connected, *i.e.*, the volt coil of $W_1$ is connected to line *Y* and that of $W_2$ to line *R*. Show that the reactive kVAlr of the load may be determined from the reading of either watthour meter, and calculate the watthour meter reading and the total reactive kV·Ah if the time interval is 1 hour. Would the method described apply to an unbalanced load ? **[Ans.** 26.22 kWh ; 45.4 kVAh]

9. A rotating substandard energy meter is tested by a wattmeter of the same VA rating having a 500 division scale which can be read to 0.1 division, and a stop watch having a negligible error with a scale which can be read to 0.01 s. The wattmeter is accurate to within 0.05 percent of its full scale reading of all parts of the scale, and the total error in starting and stopping the stop watch does not exceed 0.1 s. Determine the duration necessary for a test of the energy meter at 100 percent load to enable the measured accuracy of the meter to be guaranteed to within 0.1 percent. For test of the same duration at 10 percent load, calculate limits of accuracy of the determination. **[Ans.** 370 s ; 0.73%]

10. A large consumer has a kVA demand and a kVAh tariff measured by 'sine' and 'cosine' watthour type meters, each equipped with a Merz Price Demand Indicator. The tariff is Rs. 15 per kVA + Re 0.1 per kVAh. Determine the consumer's monthly bill for 30 days based on the following readings : sine meter registers 100 MVArh, demand indicator 125 kVAr and cosine meter registers 125 MWh and demand indicator 180 kW. What are the monthly power factor, load factor and average cost per unit ? **[Ans.** Rs. 19280, 0.781, 0.964, Re. 0.154]



# Objective Type Questions

***Tick (√) the most appropriate answer***

1. The braking torque provided by a permanent magnet in a single phase energy meter is proportional to the
   (*a*) Square of the flux of the permanent magnet
   (*b*) Speed of the meter
   (*c*) Distance of the permanent magnet from the centre of the revolving disc
   (*d*) All the above.

2. The braking torque provided by a permanent magnet in a single phase energy meter can be changed by :
   (*a*) Providing a magnetic shunt and changing its position
   (*b*) Changing the distance of the permanent magnet from the centre of the revolving disc
   (*c*) Both (*a*) and (*b*)
   (*d*) None of the above.

3. In the single phase induction meter, in order to obtain true value of energy, the shunt magnet flux should lag behind the applied voltage by :
   (*a*) 90° (*b*) 0°
   (*c*) 45° (*d*) None of the above.

4. In an induction type of meter, maximum torque is produced when the phase angle, between the two fluxes is :
   (*a*) 0° (*b*) 45°
   (*c*) 60° (*d*) 90°.

**5.** In an induction type meter, maximum torque is obtained when the parameters of rotating disc are :

(*a*) $R = 0$ (*b*) $X = 0$

(*c*) $R = X$ (*d*) None of the above.

where $R$ and $X$ are respectively the resistance and reactance of eddy current paths in the disc.

**6.** In a single phase induction type energy meter, the lag adjustment is done

(*a*) To make the current coil flux to lag 90° behind the applied voltage

(*b*) To make the pressure coil flux to lag 90° behind the applied voltage

(*c*) To bring the pressure coil flux in phase with the applied voltage

(*d*) None of the above.

**7.** In a circuit of a single phase induction energy meter, the pressure coil lags the voltage by 88°, the errors while measuring power in two circuits having power factors of unity and 0.5 lagging are respectively are :

(*a*) – 0.061%, + 6.1% (*b*) + 0.061%, – 6.1%

(*c*) – 0.061%, – 6.1% (*d*) – 6.1%, – 6.1%.

**8.** Phantom loading for testing of energy meters is used :

(*a*) To isolate the current and potential circuits

(*b*) To reduce power loss during testing

(*c*) For meters having low current ratings

(*d*) To test meters having a large current rating for which loads may not be available in the laboratory. This also reduces power losses during testing.

**9.** In an induction type energy meter, compensation for static friction is provided by

(*a*) Shading bands which are actuated by to provide a constant torque irrespective of load

(*b*) Lag circuits

(*c*) Drilling holes in the disc

(*d*) None of the above.

**10.** In a household single phase single phase induction type wattmeter, the meter can be reversed by

(*a*) Reversing the supply terminals

(*b*) Reversing the load terminals

(*c*) Opening the meter connections and reversing either the potential coil terminals or current coil terminals

(*d*) Opening the meter and reversing connections of both current and potential coil circuits.

**11.** If an induction type Energy meter runs fast, it can be slowed down by :

(*a*) Lag adjustment

(*b*) Light load adjustment

(*c*) By adjusting the position of braking magnet and making it come closer to the centre of the disc

(*d*) By adjusting the position of braking magnet and making it move away from the centre of the disc.

**12.** Creeping in a single phase induction type energy meter may be due to :

(*a*) Overcompensation for friction

(*b*) Overvoltage

(*c*) Vibrations

(*d*) All the above.

**13.** A Merz Price Maximum Demand Indicator indicates :

(*a*) Maximum demand

(*b*) Average maximum demand over a specified period of time

(*c*) Maximum energy consumption

(*d*) All the above.

**14.** VAh metering can be done by using :

(*a*) A ball and disc friction gearing

(*b*) Trivector meter

(*c*) Bridge connected Rectifiers

(*d*) All the above.

**15.** Light load adjustments for induction type energy meters are usually done at :

(*a*) 10% of full load current

(*b*) 5% of full load current

(*c*) 50% of full load current

(*d*) 1% of full load current.

## Answers

**1.** (*d*) **2.** (*c*) **3.** (*a*) **4.** (*d*) **5.** (*b*) **6.** (*b*) **7.** (*c*) **8.** (*d*) **9.** (*a*) **10.** (*c*)
**11.** (*d*) **12.** (*d*) **13.** (*b*) **14.** (*d*) **15.** (*b*)

# 13

# Measurement of Phase and Frequency

*(Including Synchroscopes and Phase Sequence Indicators)*

## 13.1 POWER FACTOR METERS

On measuring the current, voltage and power in an a.c. circuit, its power factor can be calculated from the relationship $\cos\phi = P/VI$. This method of determining the power factor of an electric circuit, is however, of low accuracy, has a number of disadvantages and is rarely used in practice. It is obviously desirable to have an instantaneous indication of the power factor of an a.c. circuit, especially where this is varying continuously, without having resource to mathematical calculations of readings of several instruments. *Power factor meters* indicate directly, by a single reading, the power factor of the circuit to which they are connected. The accuracy obtained with the use of power factor meters is sufficient for most purposes other than high precision testing.

Power factor meters – like wattmeters – have a current circuit and a pressure circuit. The current circuit carries the current (or definite fraction of this current) in the circuit whose power factor is to be measured. The pressure circuit is connected across the circuit whose power factor is to be measured and is usually split up into two parallel paths – one inductive and the other non-inductive. The deflection of the instrument depends upon the phase difference between the main current and currents in the two paths of the pressure circuit, *i.e.*, upon the phase angle or power factor of the circuit. The deflection is indicated by a pointer.

The moving system of power factor meters is perfectly balanced at equilibrium by two opposing forces and therefore there is no need for a controlling force. Hence when a power factor meter is disconnected from a circuit the pointer remains at the position which it occupied at the instant of disconnection.

There are *two* types of power factor meters :

(*i*) Electrodynamometer type, and

(*ii*) Moving Iron type.

These instruments are described in details below :

### 13.1.1 Single Phase Electrodynamometer Power Factor Meter

The construction of a single phase electrodynamometer type power factor meter is shown in Fig. 13.1. It consists of a fixed coil which acts as the current coil. This coil is split up into two parts and carries the current of the circuit under test. Therefore, the magnetic field produced by this coil is proportional to the main current. Two identical pressure coils *A* and *B* pivoted on a spindle constitute the moving system. Pressure coil *A* has a non-inductive resistance *R* connected in series with it, and coil *B* has a highly inductive choke coil *L* connected in series with it. The two coils are connected across the voltage of the

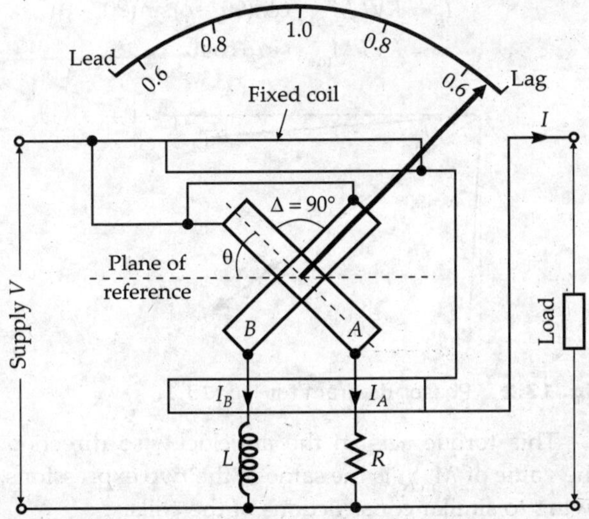


**Fig. 13.1** Single phase electrodynamometer type power factor meter.

circuit. The values of $R$ and $L$ are so adjusted that the two coils carry the same value of current abnormal frequency, *i.e.*, $R = \omega L$. The current through coil $A$ is in phase with the circuit voltage while that through coil $B$ lags the voltage by an angle $A$ which is nearly equal to 90°. The angle between the planes of coils is also made equal to $\Delta$. There is no controlling device. Connections to moving coils are made through thin silver or gold ligaments which are extremely flexible and thus give a minimum control effect on the moving system.

In order to simplify the problem, we assume that the current through coil $B$ lags the voltage by exactly 90°. Also that the angle between planes of coils is exactly 90°, (*i.e.*, $\Delta = 90°$).

Now, there will be two deflecting torques, one acting on coil $A$ and the other on coil $B$. The coil windings are so arranged that the torques due to the two coils are opposite in direction. Therefore the pointer will take up a position where these two torques are equal.

Let us consider the case of a lagging power factor of $\cos\phi$.

Deflecting torque acting on coil $A$ is :

$$T_A = KVI\,M_{max} \cos\phi \sin\theta$$

(see chapter 9, page 273)

$\theta$ = angular deflection from the plane of reference,

and $M_{max}$ = maximum value of mutual inductance between the two coils.

This torque say acts in the clockwise direction.

Deflection torque acting on coil $B$ is :

$$I_B = KVI\,M_{max} \cos(90° - \phi)\sin(90° + \theta)$$
$$= KVI\,M_{max} \sin\phi \cos\theta.$$

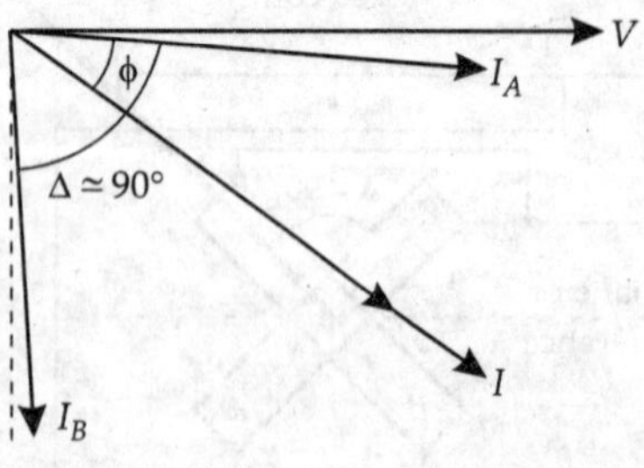


**Fig. 13.2** Phasor diagram for Fig. 13.1.

This torque acts in the anticlockwise direction. The value of $M_{max}$ is the same in the two expressions, owing to similar constructions of the coils.

The coils will take up such a position that the two torques are equal.

Hence at equilibrium

$$T_A = T_B.$$

or $$KVI\,M_{max} \cos\phi \sin\theta = KVI\,M_{max} \sin\phi \cos\theta$$

or $$\theta = \phi$$

Therefore the deflection of the instrument is a measure of phase angle of the circuit. The scale of the instrument can be calibrated in directly in terms of power factor. The instrument must be designed for, and calibrated at, the frequency of the supply on which it is to be used. In case the meter is used for any other frequency or if the supply contains harmonics it will give rise to serious errors in the indication on account change in the value of reactance of choke coil.

### 13.1.2 Three Phase Electrodynamometer Power Factor Meter

Figure 13.3 shows the construction and connections of a 3 phase electrodynamometer type power factor meter. This meter is only useful for balanced loads.

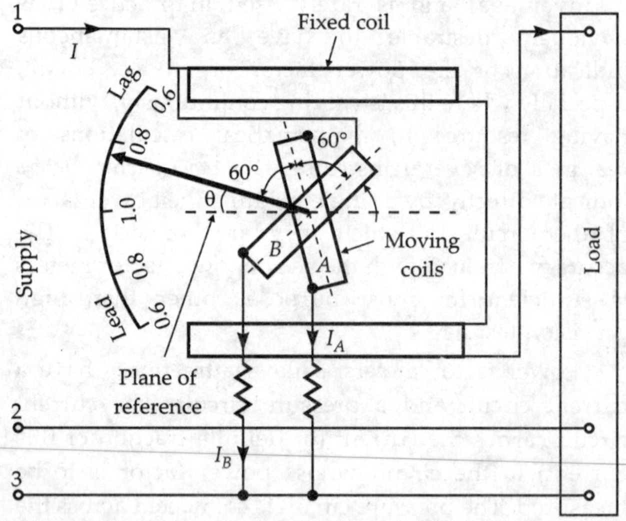


**Fig. 13.3** Three phase dynamometer type factor meter.

The two moving coils are so placed that the angle between their planes is 120°. They are connected across two different phases of the supply circuit. Each coil has a series resistance. There is no necessity for phase splitting by artificial means, since the required phase displacement between currents $I_A$ and $I_B$ in the two moving coils can be obtained from the supply itself as shown.

Voltage applied across coil $A$ is $V_{12}$ and as its circuit is resistive, current $I_A$ is in phase with $V_{12}$. Voltage applied across coil $B$ is $V_{13}$ and current $I_B$ is in phase with $V_{13}$ as the circuit of coil is resistive.

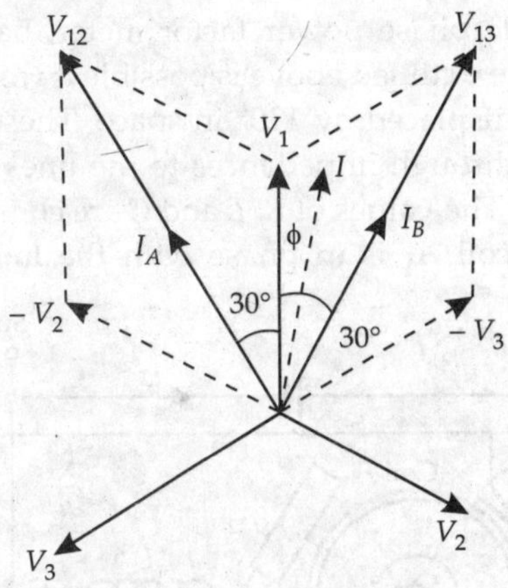


**Fig. 13.4** Phasor diagram for Fig. 13.3.

Let $\phi$ = phase angle of the circuit,

and $\theta$ = angular deflection from the plane of reference.

Now, $V_1 = V_2 = V_3 = V$.

Torque acting on coil $A$ is :

$$T_A = KV_{12}\, I\, M_{max} \cos(30°+\phi)\sin(60°+\theta)$$
$$= \sqrt{3}\, KVI\, M_{max} \cos(30°+\phi)\sin(60°+\theta)$$

Torque acting on coil $B$ is :

$$T_B = KV_{13}\, I\, M_{max} \cos(30°-\phi)\sin(120°+\theta)$$
$$= \sqrt{3}\, KVI\, M_{max} \cos(30°-\phi)\sin(120°+\theta)$$

Torques $T_A$ and $T_B$ act in the opposite directions and the moving system takes up a position where $T_A = T_B$.

$\therefore \cos(30°+\phi)\sin(60°+\theta) = \cos(30°-\phi)\sin(120°+\theta)$

Solving the above expression, we have, $\theta = \phi$

Thus the angular deflection of the pointer from the plane of reference is equal to the phase angle of the circuit to which the meter is connected.

The three phase power factor meter gives indications which are independent of waveform and frequency of supply, since the currents in the two moving coils are equally affected by any change of frequency.

For measurement of power factor in 3 phase unbalanced systems a two element power factor meter (where two sets of fixed coils and two sets of moving coils mounted on the spindle) has to be used.

### 13.1.3 Moving Iron Power Factor Meters

These instruments may be divided into two categories, according to whether the operation of the instrument depends upon a rotating magnetic field or a number of alternating fields.

***Rotating field power factor meter.*** The essential features of a rotating field type of instrument are shown in Fig. 13.5.

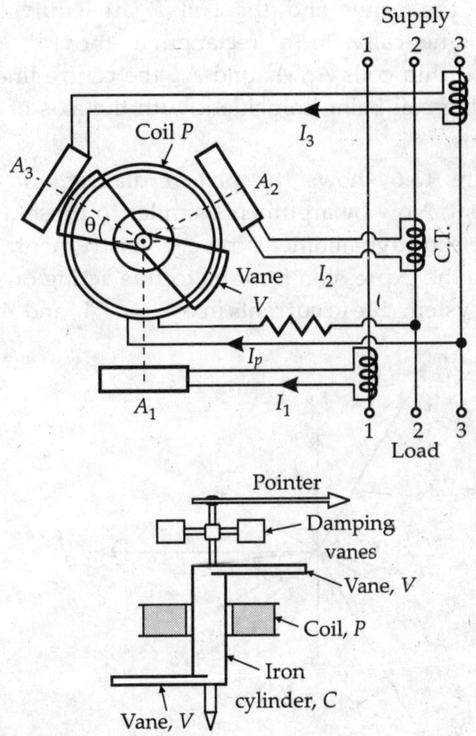


**Fig. 13.5** Rotating field moving iron power factor meter.

$A_1$, $A_2$, $A_3$ are three fixed coils, with their axes displaced 120° from each other and intersecting on the centre line of the instrument. These three coils are connected respectively in lines 1, 2 and 3 of a three phase supply. Usually current transformers are used for the purpose. $P$ is a fixed coil connected in series with a high resistance across one pair of lines say 2 and 3. There is an iron cylinder $C$ inside coil $P$. Two sector shaped iron vanes $V$ are fixed to this cylinder. The two vanes are 180° apart in space. The spindle also carries damping vanes and a pointer. There are no control springs.

Coil $P$ and the iron system produce an alternating flux, which interacts with the fluxes produced by coils $A_1$, $A_2$ and $A_3$. This causes the moving system to take up an angular position determined by the phase angle of the current.

The theory of the moving iron instrument may be developed in a similar manner to that of the electrodynamometer type instruments if we consider the cylinder $C$ and the vanes $V$, $V$ to be magnetized by current $I_p$ in coil $P$ which is in phase with and proportional to the line voltage of the system. (This is very nearly true as coil $P$ has a large resistance connected in series with it). Then if the effects of hysteresis and eddy currents are ignored, the iron cylinder, the vanes and the coil $P$ are equivalent electromagnetically to a rectangular moving coil pivoted within coils $A_1$, $A_2$ and $A_3$, the centre line of the moving coil being coincident with the axis of the iron vanes.

Figure 13.6 shows the phasor diagram of the instrument. Now by arguments similar to those used for the electrodynamometer type of instrument, we can write the expression for total torque acting on the moving system due to currents in coils $A_1$, $A_2$ and $A_3$.

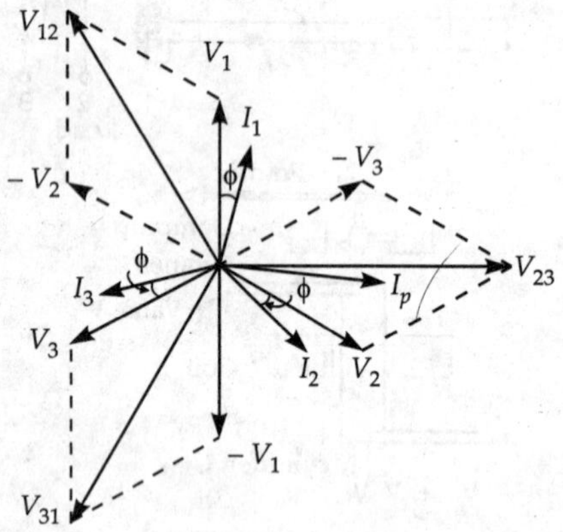


**Fig. 13.6** Phasor diagram for Fig. 13.5.

The total deflecting torque

$$T_d \propto [I_1 I_p \cos(90°-\phi)\sin(90°+\theta) + I_2 I_p \cos(330°-\phi) \sin(210°+\theta) + I_3 I_p \cos(210°-\phi)\sin(330°+\theta)]$$

For a steady deflection, the total torque must be zero. Also considering the system to be balanced *i.e.*, $I_1 = I_2 = I_3$, we have,

$$\cos(90°-\phi)\sin(90°+\theta) + \cos(330°-\phi)\sin(210°+\phi) + \cos(210°-\phi)\sin(330°+\theta) = 0.$$

Solving the above expression, we have, $\theta = \phi$.

Therefore the deflection of iron vane from the reference axis is a direct measure of the phase angle between each line current and the corresponding phase voltage.

It may be noted that the three fixed coils $A_1$, $A_2$, $A_3$ produce a rotating magnetic field and therefore owing to this there will be an induction motor action tending to drag the moving system continuously in the direction of the rotating magnetic field. This effect can be made negligibly small by using high resistivity metal for the moving irons so as to reduce the values of induced currents.

A single phase power factor meter, based upon the principle outlined above is possible if we provide 3 fixed coils displaced by 120° in space. These coils are connected through impedances to the lines as shown in Fig. 13.7. The values of $R$, $L$ and $C$ are chosen so that current in coil $A_1$ is in phase with the line voltage,

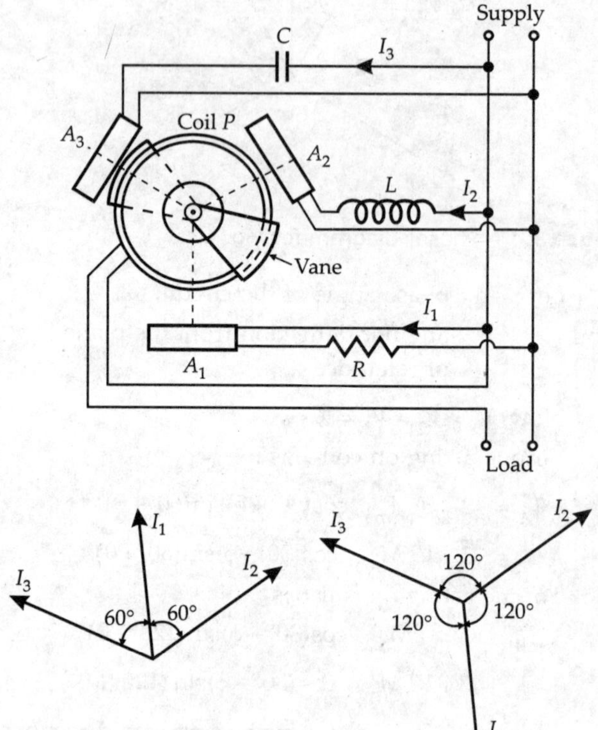


**Fig. 13.7** Single phase moving iron power factor meter.

current in coil $A_2$ lags by 60°, and current in coil $A_3$ leads by 60°. The connections of coil $A_1$ are reversed with respect to the connections of other coils so that the currents in the three coils are 120° out of phase with each other. Coil $P$ is excited by the line, current. The rest of the arrangement is similar to that described for the three phase power factor meter. The fixed coil assembly consisting of coils $A_1$, $A_2$ and $A_3$ has equal currents in each phase displaced 120° in time and since the coil axes are displaced 120° in space a revolving field is produced which makes the system behave similarly to that for a balanced 3 phase supply. *Alternating Field Power Factor Meter (Nalder Lipman Type.)* Figure 13.8 shows the construction of a Nalder Lipman type instrument used for balanced currents.

The moving system comprises of three pairs of iron vanes and cylinders which are fixed to a common spindle. This spindle is pivoted in jewel bearing and carries damping vanes and a pointer. The iron vanes are sector shaped, the arc subtending an angle of 120°. The vanes forming each pair (which are magnetically connected to the same iron cylinder) are fixed 180° apart as in the rotating field instrument. The cylinders are separated on the spindle by distance pieces *S*. These distance pieces are made of a non-magnetic material. The axes of symmetry of the three pairs of vanes are displaced by 120° from each other.

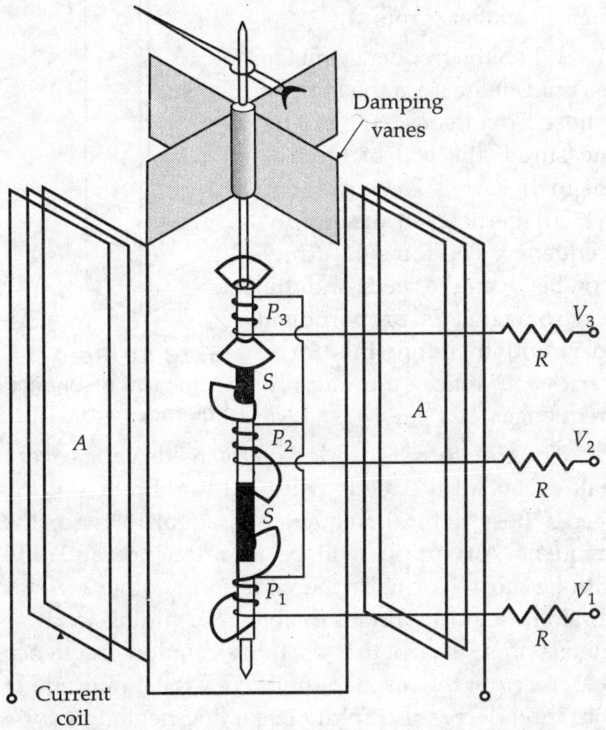


**Fig. 13.8** Nalder Lipman power factor meter.

The iron cylinders and the vanes are magnetized by three fixed co-axial pressure coils $P_1$, $P_2$, $P_3$. These pressure coils are mounted co-axially with the spindle and are excited by currents proportional to the phase voltages of the three phase system.

The current coil *A*, is wound in two equal parts which are mounted parallel to each other on opposite sides of the spindle. Coil *A* is supplied with current proportional to the current in one of the lines of the three phase system. There are no control forces.

The angular position of the moving system is determined by the phase angle of the line current with respect to the phase voltage. The operation of this instrument can be analyzed in a similar manner as used for rotating field type of instrument. The rotating field type instrument has three current coils and one pressure coil while the alternating type has three pressure coils and one current coil. In one case there are three fluxes due to three line currents, displaced from each other by 120° in space as well as in time ; together with a flux due to line voltage. In the second case there are three fluxes due to three phase voltages, displaced by 120° in space as well as in time together with a flux due to a line current. Thus the relationship between angular deflection of the moving element and phase angle of the system is the same in both the cases.

The moving system deflects into such a position that the mean torque on one pair of vanes is neutralized by the other two torques, so that the resultant torque is zero. In this steady portion, the deflection of the iron vane which is magnetized by the same phase as the current coil, is equal to the phase angle of the circuit. (provided effect of iron losses and the pressure coil inductance is neglected). The instrument as shown is used for balanced currents, but it can be modified for use on an unbalanced three phase circuit and for two phase and single phase circuits.

It may be noted here that the three coils together with their vanes and cylinders lie in three different planes and therefore no rotating magnetic field is produced. Hence in this instrument there is no tendency for the moving system to rotate continuously as is the case in rotating field type instrument.



### 13.1.4 Advantages and Disadvantages of Moving Iron P.F. Meters

***Advantages :***

1. The working forces are very large as compared with those in electrodynamometer type.
2. All the coils in a moving iron instrument are fixed and before the use of ligaments is eliminated.
3. The scale extends over 360°.
4. These instruments are simple and robust in construction. They are comparatively cheap also.

***Disadvantages :***

1. Errors are introduced in these meters owing to losses in iron parts. These losses are dependent upon the load and the frequency. They are less accurate than the electrodynamometer type.
2. The calibration of these instruments is appreciably affected by variations in supply frequency, voltage and waveform.

## 13.2 FREQUENCY METERS

The different types of frequency meters are :

1. Mechanical resonance type ;
2. Electrical resonance type ;
3. Electrodynamometer type ;
4. Weston type ;
5. Ratiometer type ;
6. Saturable core type.

The frequency can also be measured and compared by other arrangements like electronic counters, frequency bridges, stroboscopic methods and cathode ray oscilloscope.

### 13.2.1 Mechanical Resonance Type Frequency Meter (Vibrating Reed Type)

***Construction.*** This meter consists of a number of thin steel strips called reeds. These reeds are placed in a row alongside and close to an electromagnet as shown in Fig. 13.9. The electromagnet has a laminated iron core and its coil is connected in series with a resistance, across the supply whose frequency is to be measured. The reeds are approximately about 4 mm wide and 0.5 mm thick. All the reeds are not exactly similar to each other. They have either slightly different dimensions or carry different weights or flags at their tops.

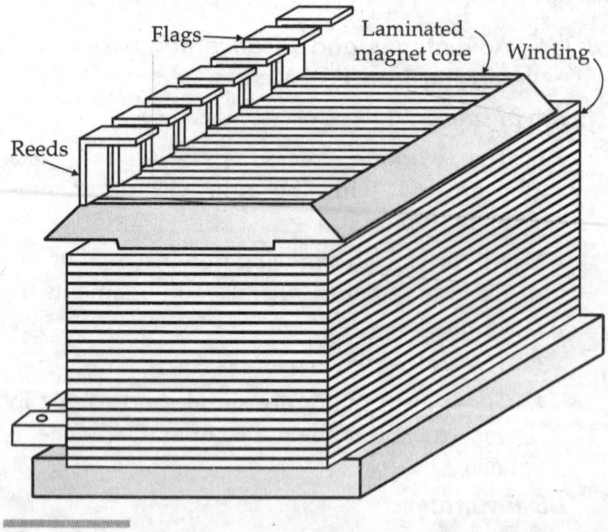


**Fig. 13.9** Vibrating reed type frequency meter.

The natural frequency of vibration of the reeds depends upon their weights and dimensions. Since the reeds have different weights and sizes, their natural frequencies of vibration are different. The reeds are arranged in ascending order of natural frequency, the difference in frequency is usually 0.5 Hz. Thus the natural frequency of first reed may be 47 Hz, of the second 47.5 Hz, of the next 48 Hz as and so on.

The reeds are fixed at the bottom end and are free at the top end. Since the reeds on a frequency meter are arranged to be viewed end on, they have a portion bent over at the free end to serve as a flag as shown in Fig. 13.10. The flags are painted white to afford maximum contrast against their black background.

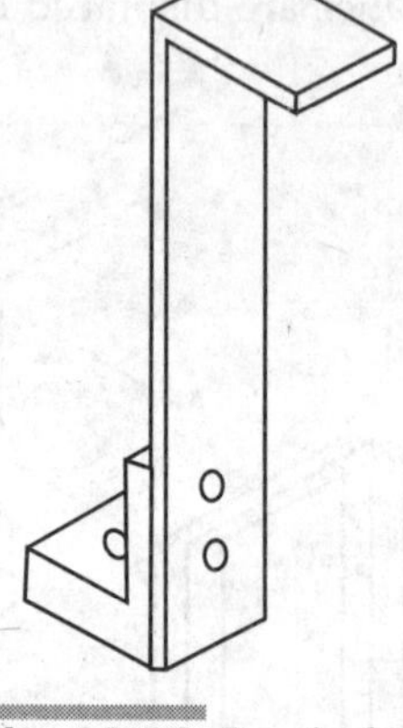

**Fig. 13.10** Reed of mechanical resonance frequency meter.

When the frequency meter is connected across the supply whose frequency is to be measured, the coil of electromagnet carries a current $i$ which alternates at the supply frequency. The force of attraction between the reeds and the electromagnet is proportional to $i^2$ and therefore this force varies at twice the supply frequency.

Thus the force is exerted on the reeds varies every half cycle. All the reeds will tend to vibrate, but the reed whose natural frequency is equal to twice the frequency of supply will be in resonance and will vibrate most. Normally the vibration of other reeds is so slight as to be unobservable. The tuning in these meters is so sharp that as the excitation frequency departs from the resonant frequency the amplitude of vibration decreases rapidly becoming negligible for a frequency perhaps 1 to 2 percent away from resonance. This is clear from Fig. 13.11.

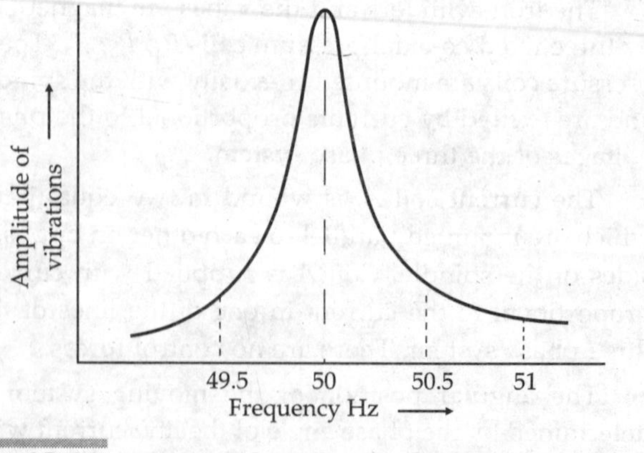


**Fig. 13.11** Variation of amplitude of vibrations with frequency (Natural frequency = 50 Hz).

When the 50 Hz reed is vibrating with its maximum amplitude (when it is in resonance) some vibrations of 49.5 Hz 50.5 Hz reeds may be observed as shown in Fig. 13.12(*b*) but very little vibrations will be observed on 49 Hz and 51 Hz reeds. For a frequency exactly midway between that of the reeds, both will vibrate with amplitudes which are equal in magnitude, but considerably less than the amplitude which is at resonance. Figure 13.12(*c*) shows the condition of vibrating reeds when the frequency is exactly midway between 49.5 Hz and 50 Hz. Figure 13.12(*a*) shows the condition of the reeds when the frequency meter is unexcited *i.e.*, is not connected to the supply.

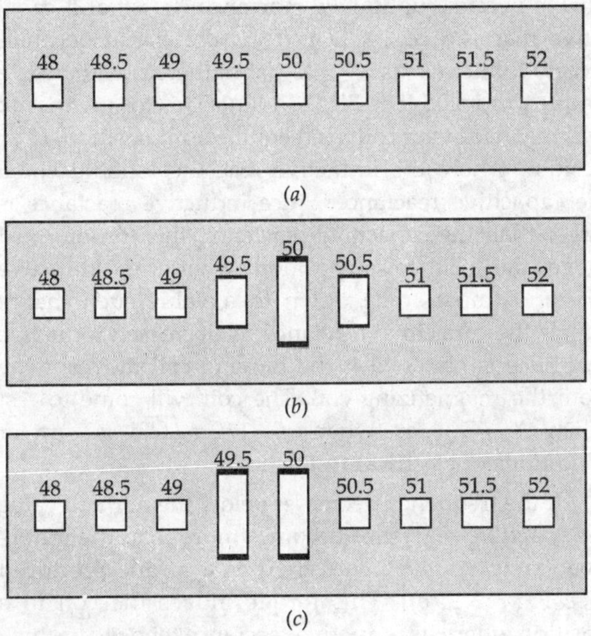


**Fig. 13.12** Indications from vibrating reeds.

The usual frequency span of these meters is six Hz say 47 Hz to 53 Hz. The frequency range of a set of reeds may be doubled in a simple manner. In the presence of an alternating flux alone, the reeds are attracted two times in a cycle and the reed whose frequency is twice the frequency of supply responds. If however, the electromagnet is polarized by a direct flux superposed on the alternating flux and is equal in magnitude, the fields (d.c. and a.c.) will cancel each other in one half cycle and reinforce during the other half cycle and so the reeds will be attracted only once in a cycle. Thus a reed whose natural frequency is 100 Hz will respond to 50 Hz when the electromagnet is unpolarized and to 100 Hz when the electromagnet is polarized. It is clear from above that the range of the frequency meter will be doubled with polarization.

The polarization may be accomplished by using a d.c. winding in addition to the a.c. winding or by using a permanent magnet.

An advantage of the reed type of frequency meter is that the indication is virtually independent of the waveform of the supply voltage. The indication is independent of magnitude of applied voltage also provided the voltage is not too low, as at low voltages the amplitude of vibrations will not be sufficient and thus the readings will not be reliable.

The disadvantage is that such instruments cannot be read much closer than half the frequency difference between adjacent reeds. Thus they cannot be used for precision measurements. The reliability of readings also depends upon the accuracy with which the meter reeds have been tuned.

### 13.2.2 Electrical Resonance Type Frequency Meters

Two types of electrical resonance frequency meters being described here.

1. ***Ferrodynamic type frequency meter.*** Figure 13.13 shows the construction of an electrical resonance frequency meter. It consists of a fixed coil which is connected across the supply whose frequency is to be measured. This coil is called magnetizing coil. The magnetizing coil is mounted on a laminated iron core. The iron core has a cross-section which varies gradually over the length, being maximum near the end where the magnetizing coil is mounted and minimum at the other end. A moving coil is pivoted over this iron core. A pointer is attached to the moving coil. The terminals of the moving coil are connected to a suitable capacitor C. There is no provision for a controlling force.



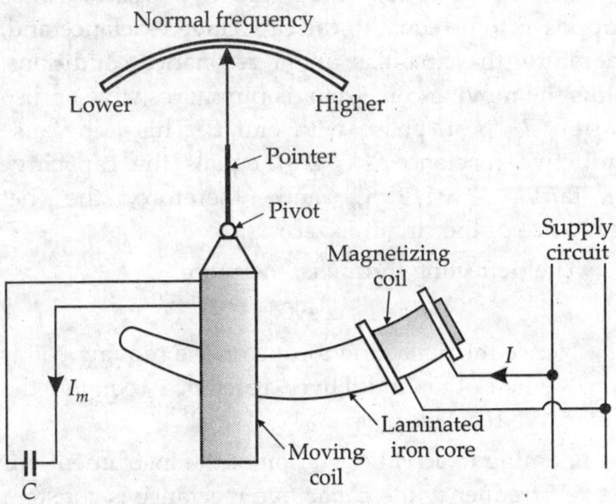


**Fig. 13.13** Electrical resonance frequency meter (Ferro-dynamic).

The operation of the instrument can be understood from the three phasor diagrams shown in Figs. 13.14(*a*), (*b*) and (*c*). The magnetizing coil carries a current $I$ and this current produces a flux $\Phi$. If we neglect the resistance of the coil and the iron losses in the core, flux Φ is in phase with current $I$. Flux Φ, being alternating in nature, induces and emf $E$ in the moving coil. This emf lags behind the flux by 90°. The emf induced circulates a current $I_m$ in the moving coil. The phase of this current $I_m$ depends upon the inductance $L$ of the moving coil and the capacitance $C$.

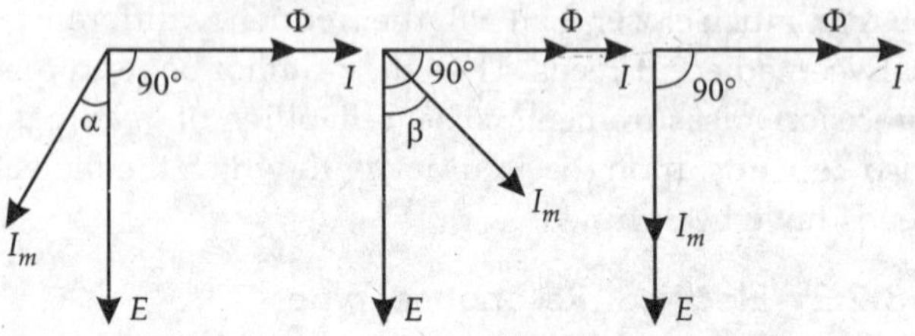


**Fig. 13.14** Phasor diagrams for electrical resonance (ferrodynamic type) frequency meter.

In Fig. 13.14(*a*), the circuit of the moving coil is assumed to be inductive and therefore current $I_m$ lags behind the emf $E$ by an angle α. The torque acting on the moving coil is thus :

$$T_d \propto I_m \cos(90°+\alpha)$$

In Fig. 13.14 (*b*), the moving coil circuit is assumed to be largely capacitive and therefore current $I_m$ leads the emf $E$ by an angle β and therefore the deflecting torque is :

$$T_d \propto I_m \cos(90°-\beta)$$

This torque is obviously opposite to what was in the case represented in Fig. 13.14(*a*).

In Fig. 13.14(*c*), the inductive reactance is supposed to be equal to the capacitive reactance and, therefore, the circuit is under resonance conditions. Thus the moving coil circuit is purely resistive and so current $I_m$ is in phase with emf $E$. This is because inductive reactance $X_L = 2\pi fL$ equals the capacitive reactance $X_c = 1/2\pi fC$, and, therefore, the net reactance of the circuit is zero.

The deflecting torque is, therefore,

$$T_d \propto I_m \cos 90° = 0.$$

Hence the deflecting torque on the moving coil is zero when the inductive reactance equals the capacitive reactance.

Coming to actual operation of the instrument, for a fixed frequency the capacitive reactance is constant but the inductive reactance of moving coil is not constant. This is because the inductance of moving coil is dependent upon the position which the moving coil occupies on the iron core. This inductance and hence inductive reactance is maximum when the moving coil occupies a position close to the magnetizing coil and minimum when it is at the other end.

The value of capacitance $C$ is so chosen that the moving coil occupies a convenient mean position on the iron core when the frequency is at its normal value. Figure 13.13 shows the position of the moving coil for normal frequency. This means that at this position, the value of inductance is such that for normal frequency the inductive reactance is equal to capacitive reactance. Now suppose that the frequency increases above the normal value. Under these conditions the inductive reactance becomes larger than the capacitive reactance. This is because unductive reactance $X_L \propto f$ and capacitive reactance $X_C \propto 1/f$. Thus the circuit becomes largely inductive (as $X_L > X_C$) and, therefore, there is a torque produced [See Fig. 13.14(*a*)]. This torque tries to pull the in moving coil to an equilibrium position, *i.e.*, a position where the inductive reactance should equal the capacitive reactance. Since inductive reactance is larger than the capacitive reactance, this torque must move the coil to a position where its inductive reactance must come down to a value such that it equals the capacitive reactance. A decreased inductive reactance is obtained if the moving coil moves away from the magnetizing coil. The coil will come to rest when $X_L = X_C$ or when $f = 1/2\pi\sqrt{LC}$ *i.e.*, under conditions of electrical resonance.

If the frequency decreases below the normal, value of capacitive reactance becomes more than the inductive reactance and hence there is a torque produced [See Fig. 13.14(*b*)]. This torque moves the coil to a position where its inductive reactance tends to become larger and thus equals capacitive reactance. Therefore, the moving coil deflects towards the magnetizing coil.

An advantage of this instrument is that great sensitivity is achieved with its use as the inductance of moving coil changes gradually with variation of its position on the iron core.

**2. *Electrodynamometer type frequency meter.*** The schematic diagram of this meter is shown in Fig. 13.15. The fixed coil is divided into two parts 1 and 2. The two parts of the fixed coil form two separate resonant circuits. Fixed coil 1 is in series with an inductance, $L_1$ and a capacitance, $C_1$ forming a resonant circuit of frequency, $f_1$ slightly below the lower end of the instrument scale. Fixed coil 2 is in series with inductance $L_2$ and capacitance $C_2$ forming a resonant circuit of frequency $f_2$ slightly higher than the upper end of the instrument scale. In the case of instruments for power frequency measurements, the circuits may be timed to frequencies of 40 Hz and 60 Hz respectively with 50 Hz in the middle of the scale.

The two parts of fixed coil are arranged as shown in the diagram, their return circuits being through the movable coil. The torque on the movable element is proportional to the current through the moving coil. This current is the sum of the currents in the two parts of the fixed coil. For applied frequency, within the limits of the frequency range of the instrument, the circuit of fixed coil 1 operates above resonant frequency (as $X_{L1} > X_{C1}$) with current $i_1$ through it, lagging the applied voltage. The circuit of fixed coil 2 operates below the resonant frequency (as $X_{C2} > X_{L2}$) with current $i_2$ leading the applied voltage. One fixed coil circuit is inductive and the other is capacitive and therefore the torques produced by the two currents $i_1$ and $i_2$ act in opposition on the moving coil. The resultant torque is a function of frequency of the applied voltage and therefore the meter scale can be calibrated in terms of frequency. The instrument scale spreads over an angle of about 90°.

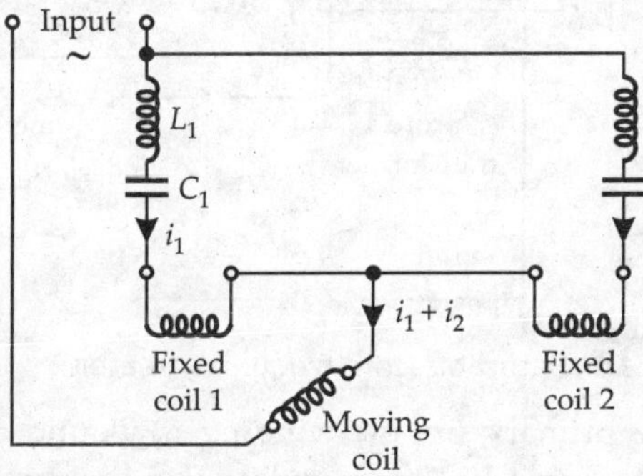


**Fig. 13.15** Electrodynamometer type frequency meter.

The controlling torque is provided by a small iron vane mounted on the moving system.

This meter is used only for power frequency measurements and is used in power systems for monitoring the frequency.

### 13.2.3 Weston Frequency Meter

This frequency meter consists of two coils mounted perpendicular to each other. Each coil is divided into two sections. The connections are as shown in Fig. 13.16. The branch circuit of coil $A$ has a resistor $R_A$ connected in series with it while coil $B$ has a reactance coil $L_B$ in series. Circuit of coil $A$ is in parallel with a reactance coil $L_A$ while that of coil $B$ is in parallel with resistance $R_B$. The moving element is a soft iron needle. This needle is pivoted on a spindle which also carries a pointer and damping varies. There is no controlling force. The series reactance coil $L$ acts to suppress higher harmonics in the current of instrument and therefore tends to minimize the waveform errors in its indication.

The meter is connected across the supply and the two coils carry currents. These currents set up two magnetic fields which are at right angles to each other. The magnitude of the field depends upon the value of current flowing in the coil. Both these fields act upon the soft iron needle and the needle takes up a position which depends upon the relative magnitudes of the two fields and hence of the currents.

The meter is so designed that the values of various resistances and inductances are such that for normal frequency of supply the value of voltage drops across reactance $L_A$ and resistance $R_B$ send equal currents through coils $A$ and $B$. Therefore, the needle takes up a position which is at 45° to both the coils and the pointer is at the centre of the scale as shown in Fig. 13.16.

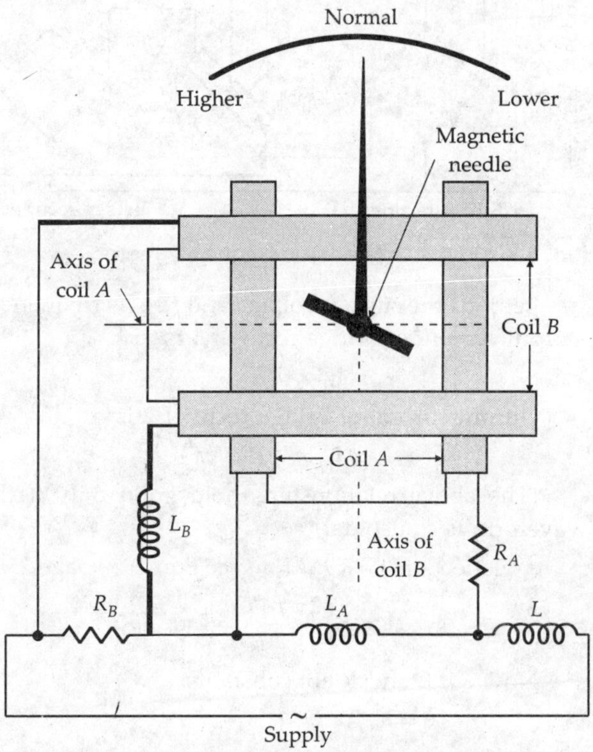


**Fig. 13.16** Weston frequency meter.

Now if the frequency increases above its normal value, reactances of $L_A$ and $L_B$ increase while resistances $R_A$ and $R_B$ remain the same. This means that with an increase in frequency, the voltage impressed upon coil $A$ increases as compared with that across coil $B$. Hence the current in coil $A$ increases while it decreases in coil $B$ (owing to increase in reactance of $L_B$). Thus the field of coil $A$ becomes stronger than that of coil $B$. The tendency of the needle is to deflect towards the stronger field and therefore, it tends to set itself in line with axis of coil $A$. Thus the pointer deflects to the left.

When the frequency decreases an opposite action takes place and the pointer deflects to the right.

### 13.2.4 Ratiometer Type Frequency Meter

A ratiometer type frequency meter consists of a ratiometer which gives a linear relationship between the current ratio and the deflection. The two coils of this ratiometer are fed with rectified output currents of two separate bridge rectifiers as shown in Fig. 13.17. The input sides of the two bridge rectifiers are connected to alternating current supply whose frequency is to be measured. Input side of one of the bridge rectifiers has a series capacitance C and the other has a series resistance R.

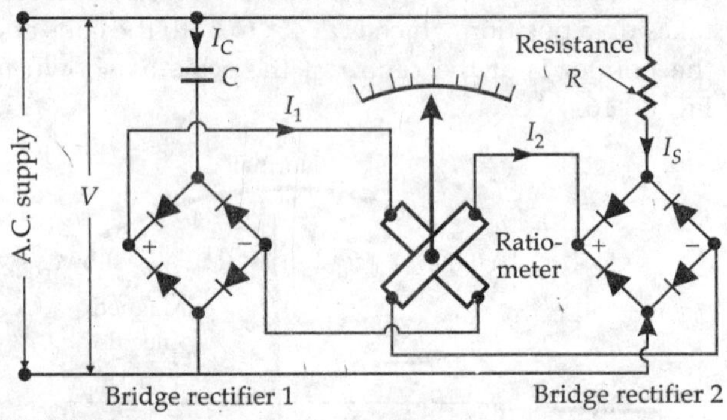


**Fig. 13.17** Ratiometer type frequency meter.

Let $V$ be the supply voltage and $f$ be its frequency. Output current of bridge rectifier 1 is :

$$I_1 \propto I_C \propto fVC$$

Output current of bridge rectifier 2 is :

$$I_2 \propto I_R \propto V/R$$

(The above relationships hold good only if the waveform is sinusoidal)

$\therefore$ Deflection $\theta = K I_1 / I_2$ (see Eqn. 9.50 page 255)

$$= K_1 \frac{2\pi f VC}{V/R} = 2\pi K_1 CRf$$

Now $K_1$, $C$ and $R$ are constants.

$$\therefore \quad \theta = K_2 f,$$

where $K_2$ = a constant $= 2\pi K_1 CR$.

Thus the instrument has a linear scale of frequency in case the ratiometer is so designed that the deflection is directly proportional to the ratio of two currents.

It is clear from above that the supply voltage $V$ does not appear in the expression for deflection. Thus the instrument may be used for a fairly wide range of voltage below the maximum specified. However, the voltage should not be too low otherwise distortions are introduced which make the meter read wrongly.

This meter may be used upto a frequency of 5000 Hz.

### 13.2.5 Saturable Core Frequency Meter

A saturable core frequency meter is shown in Fig. 13.18. This meter has a saturable core transformer as its primary detector. The core assembly consists of three parts *viz* :

(*i*) An outer core which is made of magnetic material and has a sufficiently large cross-section so that it does not saturate ;

(*ii*) A leg $A$ made of non-magnetic material ;

(*iii*) A leg $B$ which is made of magnetic material, the cross-section of leg $A$ being the same as that of $B$. The leg $B$ saturates at low values of mmfs.

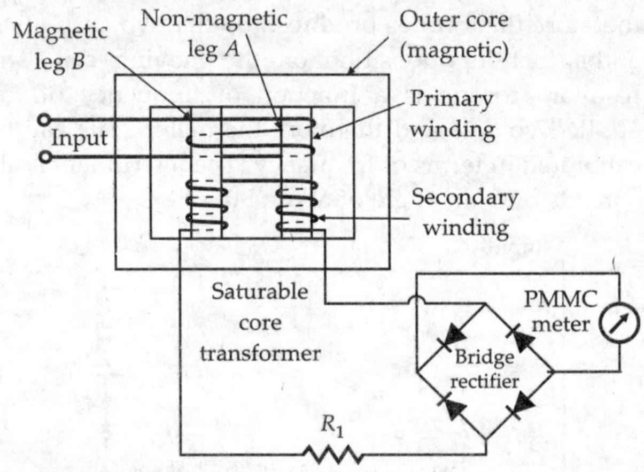


**Fig. 13.18** Saturable core frequency meter.

The primary (input) winding is wound around both legs $A$ and $B$. The secondary winding consists of two coils, one around leg $A$ and the other around $B$. The two coils are connected in series. The emfs induced in these two coils oppose each other. When there is saturation, the rate of increase of induced voltage in the secondary coil over leg $A$ will equal the rate of increase of induced voltage in secondary coil wound over magnetic leg $B$. Thus the rate of increase in these two coils will cancel, and the secondary output voltage will not be a function of the primary voltage but will be a function of frequency only.

Let $\Phi$ be the difference between the flux of leg $A$ and leg $B$. Then the induced emf in the secondary winding is ;

$$e = N\frac{d\Phi}{dt}$$

and the average secondary voltage per half cycle is :

$$E_{av} = \frac{\omega}{\pi}\int_0^{\pi/\omega} e\,dt = \frac{\omega}{\pi} N \int_{-\Phi_m}^{+\Phi_m} d\Phi = 4N f\Phi_m$$

or frequency $f = \dfrac{E_{av}}{4N\Phi_m}$

where $K$ is a constant after the magnetic leg $B$ becomes saturated.

The output of this circit is rectified and the current in the d.c. meter is :

$$I_{av} = E_{av} / R$$

where $R$ includes the resistance of the bridge rectifier, the meter and the resistant $R_1$. Therefore

$$f = KR\,I_{av}$$

The deflection θ is directly proportional to current $I_{av}$ as current is detected by a PMMC meter or deflection $\theta \propto I_{av} \propto f$.

Thus the deflection of meter is directly proportional to the frequency.

This type of frequency meter has the advantage that it can measure frequencies over a wide range and is especially well suited for use in tachnometer systems.

## 13.3 SYNCHROSCOPES

A synchroscope is used to determine the correct instant for closing the switch which connects an alternator to the power station busbars. This process of connecting at the correct instant or *synchronizing* is necessary when an unloaded "incoming" machine is to be connected to the busbars in order to share the load.

The correct instant of synchronizing is when the busbar and the incoming machine voltages

(*i*) are equal in magnitude, (*ii*) are in phase, and

(*iii*) have the same frequency.

For a 3 phase machine the phase sequence of the two should be the same. This condition is verified by a phase sequence indicator.

The voltages can be checked with the help of a voltmeter. The function of the synchroscope is to indicate the difference in phase and frequency of voltages of the busbar and the incoming machine.

Synchroscopes may either be of the electro-dynamometer type or the moving iron type. Both types are special forms of respective power factor meters.

### 13.3.1 Electro-dynamometer (Weston) Type Synchroscope

Figure 13.19 shows a simple circuit of a Weston type synchroscope. It consists of a three-limbed transformer. The winding on one of the outer limbs is excited from busbars and that on the other outer limb by the incoming machine. The winding on the central limb is connected to a lamp.

The windings on the outer limbs produce two fluxes which are forced through the central limb. The resultant flux through the central limb is equal to the phasor sum of these fluxes. This resultant flux induces an emf in the winding of the central limb. The two outer limb windings are so arranged that when the busbar and the incoming machine voltages are in phase, the two fluxes through the central limb are addative and thus emf induced in the central limb winding is maximum.

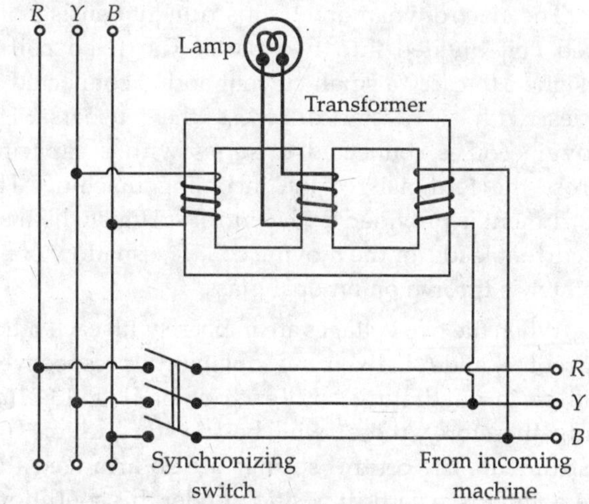


**Fig. 13.19** Weston type synchroscope.

Hence under these conditions the lamp glows with maximum brightness. When the two voltages are 180° out of phase with each other the resultant flux is zero and hence no emf is induced in the central limb winding, with the result the lamp does not glow at all and is dark. If the frequency of the incoming machine is different from that of the busbars, the lamp will be alternately bright and dark or in other words the lamp flickers. The frequency of flickering is equal to the difference in frequencies of the busbar and the incoming machine.

The correct instant of synchronizing is when the lamp is flickering at a very slow rate and is at its maximum brightness.

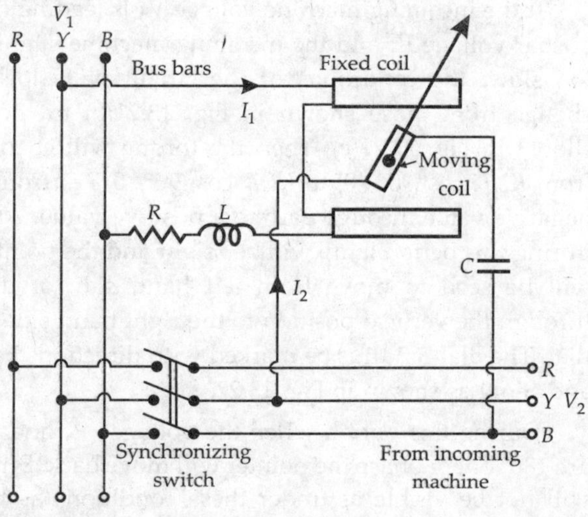


**Fig. 13.20** Weston type synchroscope.

13

One of the defects of this simple circuit is that it does not indicate whether the incoming machine is too fast or too slow. This defect can be corrected by introducing an electrodynamometer type instrument into the circuit shown in Fig. 13.20.

The electrodynamometer instrument consists of a fixed coil divided into two parts. The fixed coil is designed to carry a small current and is connected in series with a resistance across the busbars. The moving coil is connected in series with a capacitor across the terminals of the incoming machine. The instrument is provided with control springs which act as current leads for the moving coil. The shadow of the pointer is thrown on an opal glass.

When the two voltages are in phase with each other, currents $I_1$ and $I_2$ in fixed and moving coils respectively will be in quadrature with each other [Fig. 13.21(*a*)] and, therefore, there will be no torque on the instrument. The control springs are so arranged that the pointer is in vertical position under this condition.

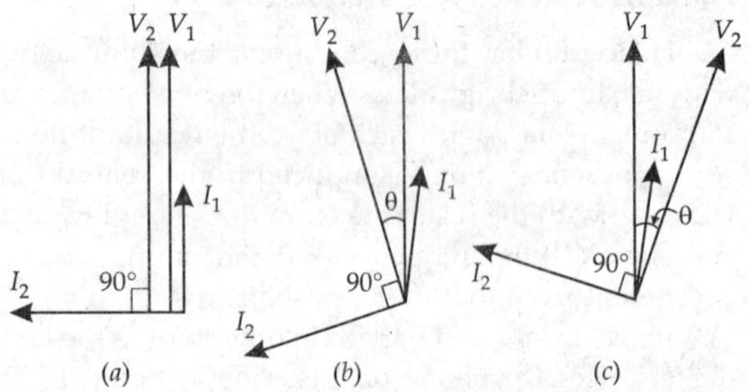


**Fig. 13.21** Phasor diagrams for different conditions of busbar and incoming machine voltages.

Also the lamp is at its maximum brightness and the pointer is silhoueted against the opal glass.

If the incoming machine voltage $V_2$ is leading the busbar voltage $V_1$ and the incoming machine slightly too slow, the conditions of the circuit will slowly change from those shown in Fig. 13.21(*b*) to those shown in Fig. 13.21(*c*). Then the torque will change from $KI_1I_2 \cos(90°+\theta)$ to $KI_1I_2 \cos(90°-\theta)$ *i.e.*, from a negative value through zero to a positive value. And during this period lamp will be bright and the pointer will be seen to move from left hand side of dial through the vertical position to the right hand side of dial. The dial can thus be marked with directions **Fast** and **Slow** as shown in Fig. 13.22.

During this period when the voltages $V_1$ and $V_2$ are 180° out of phase the pointer will move back. But it will not be visible as under these conditions as the lamp is dark.

The visible movement of the pointer is therefore a series of traverses on the dial in one direction. If the incoming machine is too fast the visible traverses will be in the other direction. The correct instant of synchronizing is when the pointer is visible at its central position and is moving very slowly.

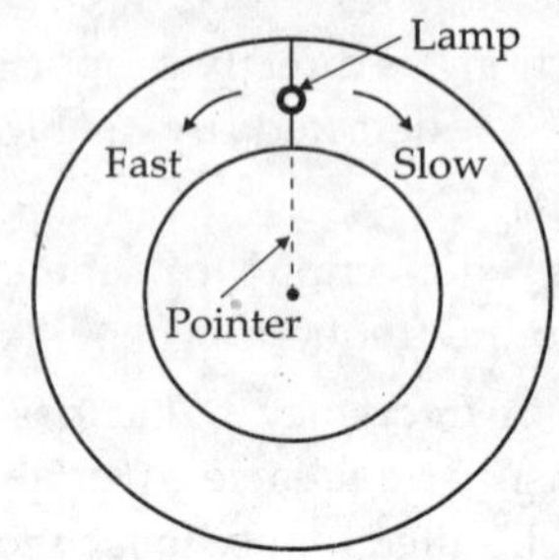


**Fig. 13.22** Dial of Weston synchroscope.

It may be observed that in order to have an exact quadrature relationship between currents $I_1$ and $I_2$ wher. voltages $V_1$ and $V_2$ are in phase, is obtained only if small inductance $L$ is introduced in the fixed coil circuit.

### 13.3.2 Moving Iron Synchroscope

Figure 13.23 shows the construction of a moving iron synchroscope which is due to Lipman. It has a fixed coil divided into two parts. This fixed coil $A$ is designed for a small value of current and is connected in series with a resistance across two phases of the busbar. There are two iron cylinders $C_1$ and $C_2$ mounted on the spindle. Each iron cylinder is provided with two iron vanes whose axes are 180° out with each other. The iron cylinders are excited by two pressure coils $P_1$ and $P_2$ which are connected to two phases of the incoming machine. One of the coils has a series resistance and the other has a series inductance. This is done in order to create an artificial phase difference of 90° between the currents of two pressure coils. There are no control springs. The instrument is provided with a pointer which moves over a dial marked *Fast* and *Slow*.

When the frequency of incoming machine is the same as that of the busbars, the instrument behaves exactly like the corresponding form of the power factor meter. The deflection of the pointer from the plane of reference is equal to phase difference between the two voltages.

However if the frequencies of the two voltages are different, the pointer rotates continuously at a speed corresponding to difference in frequency of the two voltages. The direction of rotation depends whether the incoming machine is too fast or too slow.

Let $V_1$ and $V_2$ be the voltages of busbar and incoming machine respectively. Let the frequencies of the two voltages be equal.

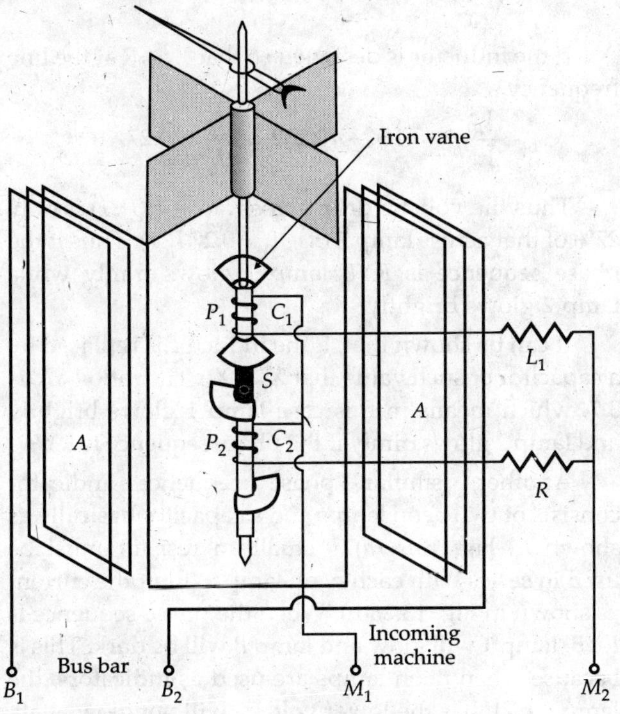


**Fig. 13.23** Nalder Lipman type synchroscope.

Torques produced by coils $P_1$ and $P_2$ are :

$$T_1 = KV_1V_2 \sin\theta \cos(\pm\alpha)$$

$$T_2 = KV_1V_2 \sin(90°-\theta)\cos(90°\pm\alpha)$$

$$= KV_1V_2 \cos\theta \sin(\pm\alpha)$$

where $\theta$ = deflection of the pointer from plane of reference.

This plane of reference is the vertical position of the pointer.

$\alpha$ = phase angle between the two voltages.

The two torques are in opposite direction, therefore, under equilibrium

$$T_1 = T_2.$$

or $KV_1V_2 \sin\theta\cos(\pm\alpha) = KV_1V_2\cos\theta\sin(\pm\alpha)$ or $\theta = \pm\alpha$.

Thus the pointer is stationary and its deflection from plane of reference equals the phase difference between the two voltages.

If the frequency of the incoming machine differs from that of the busbars, the torques are :

$$T_1 = KV_1V_2 \sin\theta\cos(\pm\pi f't \pm\alpha)$$

and $$T_2 = KV_1V_2 \sin(90°-\cos[90°-(-2\pi f't \pm\alpha)])$$

$$= KV_1V_2 \cos\theta \sin(\pm 2\pi f't \pm\alpha)$$

where $f' = f_2 - f_1$ = difference in two frequencies.

Hence at equilibrium :

$$\sin\theta\cos(\pm 2\pi f' \pm\alpha) = \cos\theta\sin(\pm 2\pi f't \pm\alpha)$$

or $$\theta = \pm 2\pi f't \pm\alpha.$$

Thus the moving system rotates with a frequency corresponding to difference in the two frequencies. The direction of rotation depends upon whether the frequency of incoming machine is higher ($f'$ positive) or lower ($f'$ negative) than the frequency at the busbars.

The instant of synchronizing is when the pointer is stationary ($f' = 0$) and when it is at its vertical position.

Moving iron synchroscopes are more common in use. They are cheap and their operation is simple and also they have a 360° scale.

## 13.4 PHASE SEQUENCE INDICATORS

These instruments are used to determine the phase sequence of three phase supplies. There are *two* types of phase sequence indicators :

(*i*) Rotating type. (*ii*) Static type.

(*i*) ***Rotating type***. The principle of working of these meters is similar to that of 3 phase induction motors.

They consist of three coils mounted 120° apart in space. The three ends of the coils are brought out and connected to three terminals marked *RYB* as shown in Fig. 13.24. The coils are star connected and are excited by the supply whose phase sequence is to be determined. An aluminium disc is mounted on the top of the coils. The coils produce a rotating magnetic field and eddy emfs are induced in the disc. These emfs cause eddy currents to flow in the aluminium disc. A torque is produced with the interaction of the eddy currents with the field. The disc revolves because of the torque and the direction of rotation depends upon the phase sequence of the supply.

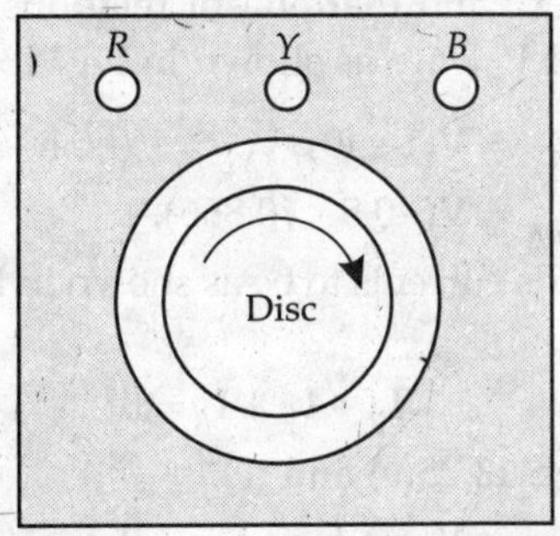


**Fig. 13.24** Phase sequence indicator.

An arrow indicates the direction of the rotation of the disc. If the direction of the rotation is the same as that indicated by the arrow head, the phase sequence of the supply is the same as marked on the terminals of the instrument. However, if the disc revolves opposite to the direction indicated by the arrowhead, the sequence of the supply is opposite to that marked on the terminals.

(*ii*) ***Static type.*** One arrangement consists of two lamps and an inductor as shown in Fig. 13.25(*a*). When the phase sequence is *RYB*, lamp 2 will be dim and lamp 2 will glow brightly. If the phase sequence is *RBY*, lamp 1 will glow brightly and lamp 2 will be dim.

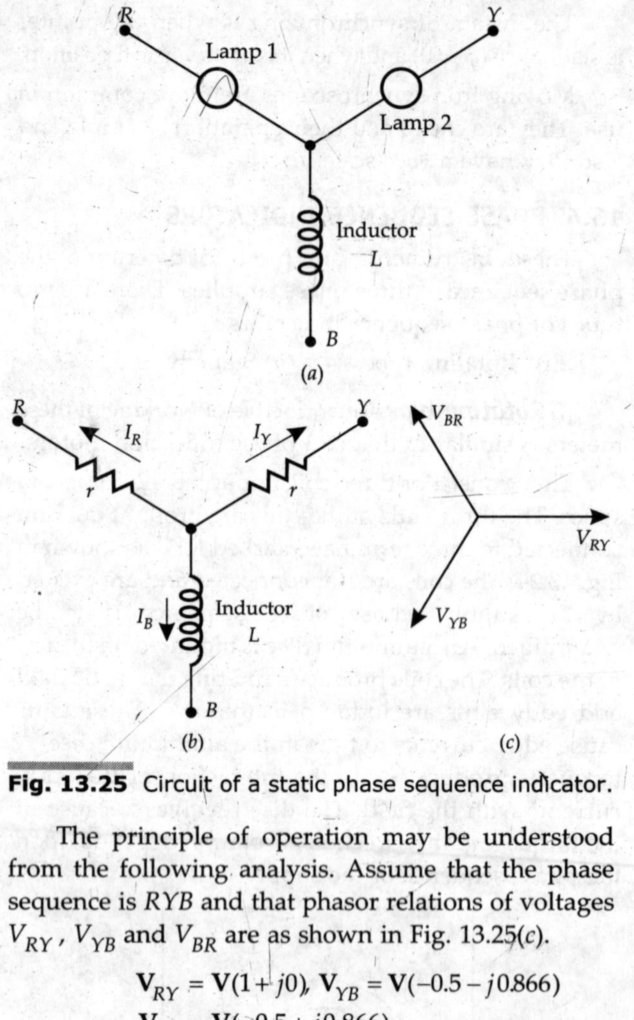


**Fig. 13.25** Circuit of a static phase sequence indicator.

The principle of operation may be understood from the following analysis. Assume that the phase sequence is *RYB* and that phasor relations of voltages $V_{RY}$, $V_{YB}$ and $V_{BR}$ are as shown in Fig. 13.25(*c*).

$$\mathbf{V}_{RY} = \mathbf{V}(1+j0), \mathbf{V}_{YB} = \mathbf{V}(-0.5-j0.866)$$
$$\mathbf{V}_{BR} = \mathbf{V}(-0.5+j0.866).$$

Assume the currents to be as shown in Fig. 13.25(*b*), that is :

$$\mathbf{I}_R + \mathbf{I}_Y + \mathbf{I}_B = 0$$

From Figs. 13.25(*b*) and (*c*),

$$\mathbf{V}_{RY} + \mathbf{I}_Y r - \mathbf{I}_R r = 0$$
$$\mathbf{V}_{YB} + \mathbf{I}_B jX_L - \mathbf{I}_Y r = 0$$

Solving for $\mathbf{I}_R$ and $\mathbf{I}_Y$, we have,

$$\mathbf{I}_R = \frac{\mathbf{V}_{RY}}{r} + \mathbf{I}_Y \text{ and } \mathbf{I}_Y = \frac{\left(\mathbf{V}_{YB} - \frac{\mathbf{V}_{RY} j\mathbf{X}_L}{r}\right)}{(r+2jX_L)}$$

then
$$\frac{\mathbf{I}_R}{\mathbf{I}_Y} = 1 + \frac{(1+j2X_L/r)}{(\mathbf{V}_{YB}/\mathbf{V}_{RY}) - jX_L/r}$$

But
$$\frac{\mathbf{V}_{YB}}{\mathbf{V}_{RY}} = -0.5 - j0.866.$$

If the inductor is designed so that $X_L = R$ at the line frequency,

$$\frac{\mathbf{I}_R}{\mathbf{I}_Y} = -0.134 + j0.232 \text{ or } \frac{I_R}{I_Y} = 0.27.$$

Thus the voltage drop across lamp 1 ($I_R\, r$) is only 27% of that across lamp 2 (*i.e.*, $I_Y = 0.27\, I_R\, r$). Thus if the phase sequence is *RYB* lamp 1 glows dimly while lamp 2 glows brightly.

It can be shown that, if the inductor is replaced by a capacitor of such value that $X_C = X_L$. The ratio $I_R/I_Y$ is 3.7, which means in this case lamp 1 glows brightly and lamp 2 glows dimly if the phase sequence is *RYB*.

Another, similar phase sequence indicator consists of two neon lamps and a capacitor basically as shown in Fig. 13.26(*a*). Actually a resistor must be used in series with each neon lamp to limit the current as shown in Fig. 13.26(*b*). When the phase sequence is *RYB*, lamp 1 will glow and lamp *B* will be dark. This is because when neon lamps are used as indicators, the lamp which has the lower voltage will not glow at all, as the voltage across it is lower than the breakdown voltage of the lamp.

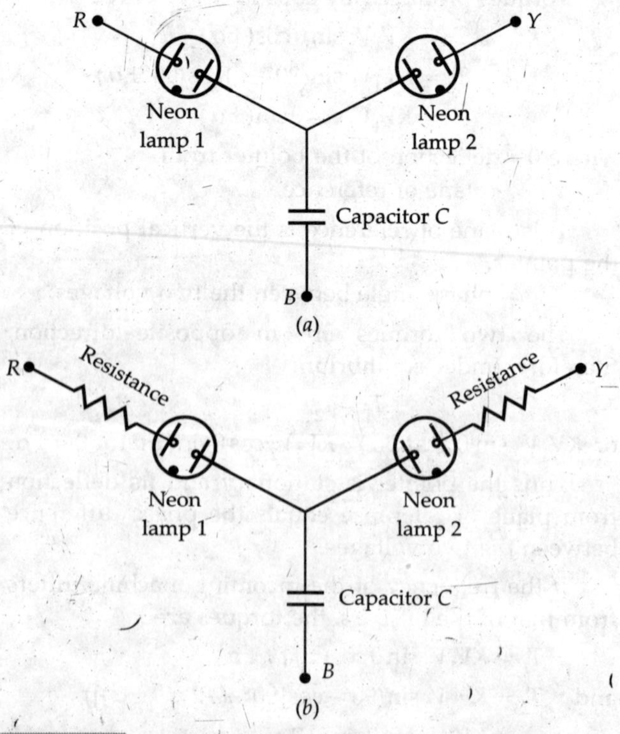


**Fig. 13.26** Phase sequence indicators using neon lamp and capacitor.

***Example 13.1*** *In a deflectional frequency meter working on the principle of electrical resonance, there are two parallel circuits each consisting of an inductance and a capacitance in series. One circuit has $C_1 = 1\,\mu F$ and is tuned to a frequency, $f_1 = 60$ Hz. The other has $C_2 = 1.5\,\mu F$ and is tuned to a frequency $f_2$, below 50 Hz. The resistance of each circuit is $R_1 = R_2 = 100\,\Omega$. What must be the inductance of the second circuit, and to what frequency must it be tuned, in order that the current in both the circuits shall be same at a frequency of 50 Hz ?*

**Solution.** For resonant circuit frequency,

$$f_1 = \frac{1}{2\pi\sqrt{L_1 C_1}}$$

$\therefore$ Inductance of circuit 1

$$L_1 = \frac{1}{4\pi^2 f_1^2 C_1}$$

$$= \frac{1}{4\pi^2 \times (60)^2 \times 1 \times 10^{-6}}$$

$$= 7.05 \text{ H.}$$

Impedance of circuit 1 at 50 Hz,

$$\mathbf{Z}_1 = R_1 + j\left(\omega L_1 - \frac{1}{\omega C_1}\right)$$

$$= 100 + j\left(2\pi \times 50 \times 7.05 - \frac{1}{2\pi \times 50 \times 1 \times 10^{-6}}\right)$$

$$= 100 + j971\,\Omega$$

Impedance of circuit 2 at 50 Hz

$$\mathbf{Z}_2 = R_2 + j\left(\omega L_2 - \frac{1}{\omega C_2}\right)$$

$$= 100 + j\left(2\pi \times 50 \times L_2 - \frac{1}{2\pi \times 50 \times 1.5 \times 10^{-6}}\right)$$

$$= 100 + j(314\,L_2 - 2120).$$

For equal currents in two circuits,

$\mathbf{Z}_1 = \mathbf{Z}_2$, or $314\,L_2 - 2120 = 971$.

inductance of circuit 2, $L_2 = 9.82$ H.

Resonant frequency of circuit 2,

$$f = \frac{1}{2\pi}\sqrt{\frac{1}{9.82 \times 1.5 \times 10^{-6}}} = 41.15 \text{ Hz.}$$

13

# Review Questions

1. Describe the constructional details and working of a single phase electrodynamometer type of power factor meter. Prove that the special displacement of moving system is equal to the phase angle of the system.
2. Describe the constructional details and working a 3 phase electrodynamometer type of power factor meter. Describe why phase splitting is not necessary in this case while in a single phase power factor phase splitting has to be done by using $R$ in one circuit and $L$ in another circuit of the moving coils.
3. Explain the construction and working of a 3 phase rotating field power factor meter. Prove that the deflection of moving system is equal to the phase angle of the system.

   What modifications are done to use this meter in single phase circuits ?
4. Describe the construction and working 3 phase alternating field power factor meter.

   Explain the advantages and disadvantages of moving iron type power factor meters.
5. What are the different methods of measurement of frequency in the power frequency range ? Explain the working of a mechanical resonance type frequency meter.
6. Explain the working of electro-resonance type power factor meters. Draw the phasor diagrams under different power factor conditions.
7. Explain the construction and working of the following :

   (*i*) Weston type frequency meter

   (*ii*) Ratiometer type frequency meter

   (*iii*) Saturable core frequency meter.
8. What is synchronizing ? Describe under what condition a 3 phase alternator can be synchronized to 3 busbars ?
9. Describe the construction and working of a Weston type synchroscope. How is it assured that the

   (*i*) incoming machine has the same voltage as that of the busbars and also whether they are in phase with each other or not

   (*ii*) incoming machine has the same phase sequence as the busbars to which it has been connected

   (*iii*) frequency of the incoming machine is the same as that of the busbars

   (*iv*) incoming machine is faster or slower than the busbars ?
10. Describe the construction and working of rotating and static type phase sequence indicators.

# Objective Type Questions

***Tick (√) the most appropriate answer :***

**1.** The power factor in a single phase circuit is $\cos \phi = P/VI$, where $P$ is the power read by a wattmeter, $V$ is the voltage read by a voltmeter and $I$ is the current read by an ammeter. All the three instruments have a limiting error of ±1%. The power factor of the circuits can be known if only one meter *i.e.*, an electrodynamometer type power factor having an limiting error ±1% used. The preferred method is to use a single instrument like a power factor rather than using 3 instruments because the error with using wattmeter, voltmeter and ammeter is

(*a*) ± 2% (*b*) ± 3%
(*c*) ± 1% (*d*) −1%

**2.** The controlling torque in single phase power factor meter is provided by :

(*a*) spring control (*b*) gravity control
(*c*) stiffness of suspension (*d*) none of the above.

**3.** There will be serious errors, if power factor of non-sinusoidal waveform is measured by electrodynamometer power factor meters. This is true of :

(*a*) only single phase meters
(*b*) only 3 phase meters
(*c*) both single and three phase meters
(*d*) none of the above.

**4.** In a reed type frequency meters all the reeds :

(*a*) have the same natural frequency
(*b*) have different natural frequency
(*c*) have different natural frequencies but the difference in natural frequencies of adjacent reeds is ± 0.5 Hz
(*d*) none of the above.

**5.** A 53 Hz reed type frequency meter is polarized with d.c. The new range of frequency meter is :

(*a*) 106 Hz (*b*) 26.5 Hz
(*c*) 53 Hz (*d*) none of the above.

**6.** For a 3 phase, moving iron synchroscope

$f_2$ = frequency of incoming machine
$f_1$ = frequency of busbars

At equilibrium

$$\theta = \pm 2\pi f't + \alpha \;;\; \text{where } f' = f_2 - f_1.$$

If $f'$ is positive

(*a*) the incoming machine is faster
(*b*) the incoming machine is slower
(*c*) the incoming machine can be synchronized to the busbars
(*d*) none of the above.

**7.** An electrodynamometer type of instrument is used in Weston type synchroscope while synchronizing an incoming machine to busbars

(*a*) for checking the power factor of the two circuits
(*b*) for checking the voltages of the two circuits
(*c*) for checking whether the incoming machine is fast or slow
(*d*) for checking the phase sequence of the two circuits.

**8.** An alternator is being synchronized to the busbars whose frequency is 50 Hz. The bulb of the synchronoscope flickers at a frequency of 10 Hz. The frequency of the alternator is :

(*a*) 60 Hz (*b*) 40 Hz
(*c*) 60 Hz or 40 Hz (*d*) none of the above.

**9.** A phase sequence indicator rotates clockwise for phase sequence of *RYB*. If the phase sequence is changed to *BRY* it will rotate

(*a*) anticlockwise
(*b*) clockwise
(*c*) clockwise or anticlockwise
(*d*) none of the above.

**10.** In a static type phase sequence indicator which uses two identical lamps of pure resistances in the circuit of phases *R* and *Y* and inductor in phase *B* having inductive reactance = resistance of bulb with circuit being connected in star. The reactance is equal to the bulb resistance. If bulb connected to phase *R* glows brightly and lamp connected to phase *Y* glows dimly, the phase sequence is :

(*a*) RBY (*b*) RYB
(*c*) BRY (*d*) none of the above.

## Answers

**1.** (*b*) **2.** (*d*) **3.** (*a*) **4.** (*c*) **5.** (*a*) **6.** (*a*) **7.** (*c*) **8.** (*c*) **9.** (*b*) **10.** (*a*)

# 14

# Measurement of Resistance

## 14.1 CLASSIFICATION OF RESISTANCES

The classification of resistances, from the point of view of measurement, is as follows :

(i) *Low resistances.* All resistances of the order of 1 Ω and under may be classified as low resistances.

(ii) *Medium resistances.* This class includes resistances from 1 Ω upwards to about 0.1 MΩ.

(iii) *High resistances.* Resistances of the order of 0.1 MΩ and upwards are classified as high resistances.

The classification outlined above is not rigid, but forms a basis for techniques, followed or measurement, which may be different for different classes.

## 14.2 MEASUREMENT OF MEDIUM RESISTANCES

The different methods used for measurement of medium resistances are :

(i) Ammeter-Voltmeter method.

(ii) Substitution method.

(iii) Wheatstone bridge method.

(iv) Ohmmeter method.

### 14.2.1 Ammeter-Voltmeter Method

This method is very popular since the instruments required for this test are usually available in the laboratory. The *two* types of connections employed for ammeter-voltmeter method are shown in Figs. 14.1(*a*) and (*b*). In both the cases, if readings of ammeter and voltmeter are taken, then the measured value of resistance is given by :

$$R_m = \frac{\text{voltmeter reading}}{\text{ammeter reading}} = \frac{V}{I} \quad ...(14.1)$$

The measured value of resistance $R_m$, would be equal to the true value, $R$, if the ammeter resistance is zero and the voltmeter resistance is infinite, so that the conditions in the circuit are not disturbed. However, in practice this is not possible and hence both the methods give inaccurate results.

Consider circuit of Fig. 14.1(*a*). In this circuit the ammeter measures the true value of the current through the resistance but the voltmeter does not measure the true voltage across the resistance. The voltmeter indicates the sum of the voltages across the ammeter and the measured resistance.

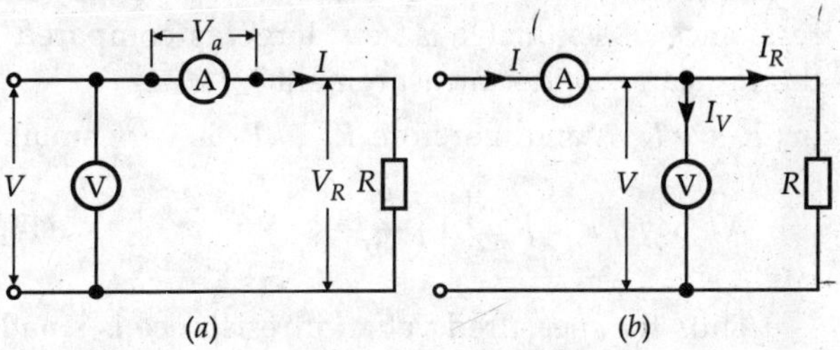


**Fig. 14.1** Measurement of resistance by ammeter-voltmeter method.

Let $R_a$ be the resistance of the ammeter.

∴ Voltage across the ammeter, $V_a = IR_a$

Now, measured value of resistance,

$$R_{m1} = \frac{V}{I} = \frac{V_R + V_a}{I} = \frac{IR + IR_a}{I} = R + R_a \quad ...(14.2)$$

∴ True value of resistance,

$$R = R_{m1} - R_a \quad ...(14.3)$$

$$= R_{m1}\left(1 - \frac{R_a}{R_{m1}}\right) \quad ...(14.4)$$

Thus the measured value of resistance is higher than the true value. It is also clear from above that the true value is equal to the measured only if the ammeter resistance, $R_a$, is zero.

$$\text{Relative error, } \varepsilon_r = \frac{R_{m1} - R}{R} = \frac{R_a}{R} \quad ...(14.5)$$

It is clear from Eqn. 14.5 that the error in measurements would be small if the value of resistance under measurement is large as compared to the internal resistance of the ammeter. Therefore the circuit of Fig. 14.1(*a*) should be used when measuring high resistance values.

**Consider the circuit of Fig. 14.1(*b*).** In this circuit the voltmeter measures the true value of voltage but the ammeter measures the sum of currents through the resistance and the voltmeter.

Let $R_v$ be the resistance of the voltmeter.

Current through the voltmeter, $I_v = V / I_v$

Measured value of resistance,

$$R_{m2} = \frac{V}{I} = \frac{V}{I_R + I_v} = \frac{V}{V/R + V/R_v} = \frac{R}{1 + R/R_v}$$

True value of resistance,

$$R = \frac{R_{m2} R_v}{R_v - R_{m2}} = R_{m2}\left(\frac{1}{1 - R_{m2}/R_v}\right) \quad ...(14.6)$$

From Eqn. 14.6 it is clear that the true value of resistance is equal to the measured value only if the resistance of voltmeter, $R_v$, is infinite. However, if the resistance of voltmeter is very large as compared to the resistance under measurement :

or $R_v >> R_{m2}$, and therefore $R_{m2} / R_v$ is very small.

We have, $R = R_{m2}\left(1 + \frac{R_{m2}}{R_v}\right)$ ...(14.7)

Thus the measured value of resistance is smaller than the true value.

Relative error, $\varepsilon_r = \frac{R_{m2} - R}{R} = \frac{R_{m2}^2}{R_v R}$ ...(14.8)

The value of $R_{m2}$ is approximately equal to R.

$\therefore \quad \varepsilon_r = -\frac{R}{R_v}$ ...(14.9)

It is clear from Eqn. 14.9 that the error in measurement would be small if the value of resistance under measurement is very small as compared to the resistance of the voltmeter. Hence the circuit of Fig. 14.1(*b*) should be used when measuring low resistance values.

The Ammeter-Voltmeter method, in the two forms explained above, is a simple method but is, essentially a comparatively rough method, the accuracy being limited by accuracy of ammeter and voltmeter used, even if corrections are made for the voltage drop across the ammeter for connections of Fig. 14.1(*a*) and for the shunting effect of voltmeter for connections of Fig. 14.1(*b*). If the two instruments are of 0.5% accuracy and are read near full scale, the instrumental error in the result may be any thing from 0 to 1%. If read near half scale, the percentage error may be twice as great and for lower readings may mount considerably higher. With less accurate instruments the possible error, of course is increased. It is difficult to obtain accuracy much better than 1% in a resistance value under usual conditions and the error sometimes may be considerably higher. However, the method is useful in some laboratory work in which high accuracy is not required.

The suitability of a particular method for resistance measurement depends upon the relative values of resistance under measurement and the resistance of the meters.

The division point between the two methods is at the resistance for which the relative errors given by the two methods are equal.

The relative errors for the two cases are equal when :

$$\frac{R_a}{R} = \frac{R}{R_v} \quad \text{(See Eqns. 14.5 and 14.9)}$$

or when true value of resistance

$$R = \sqrt{R_a R_v} \quad ...(14.10)$$

For resistances greater than the value given by Eqn. 14.10 the method of Fig. 14.1(*a*) is used while for lower resistances method of 14.1(*b*) is used.

***Example 14.1*** *In the measurement of a resistance, R, by the Ammeter-voltmeter method, connections as in Figs. 14.1(a) and 14.1(b) are used. The resistance of ammeter is 0.01 Ω and that of voltmeter, 2000 Ω. In case of (b) the current measured is 2 A and the voltage 180 V. Find the percentage error in calculating resistance R as a quotient of the readings and the true value of R. Also find the reading of the voltmeter in case of (a) if the current indicated by the ammeter is 2 A.*

**Solution.** Case (*b*) See Fig. 14.1(*b*)

Measured value of resistance

$$R_{m2} = V / I = 180/2 = 90\ \Omega.$$

Current through the voltmeter

$$I_v = V / R_v = 180/2000 = 0.09\ \text{A}.$$

Current through the resistance

$$I_R = I - I_v = (2.0 - 0.09)\ \text{A}.$$

True value of resistance

$$R = \frac{V}{I_R} = \frac{180}{(2.0 - 0.09)} = 94.24\ \Omega.$$

Percentage error

$$= \frac{R_{m2} - R}{R} \times 100 = \frac{90 - 94.24}{94.24} = -4.5\%.$$

Case (*a*) See Fig. 14.1(*a*)

Reading of voltmeter

$$V = V_a + V_R = I(R_a + R) = 2(0.01 + 94.24)$$
$$= 188.50 \text{ V}.$$

***Example 14.2*** *A resistance of approximate value of 80 Ω is to be measured by voltmeter-ammeter method using a 1 A ammeter having a resistance of 2 Ω and a 50 V voltmeter having a resistance of 5000 Ω.*

(*a*) *Suggest which one of the two methods should be used ?*

(*b*) *Supposing in the suggested method the following measurements are made :*

$$I = 0.42 \text{ A and } V = 35.5 \text{ V}.$$

*What is the resulting error if the accuracy of the instruments is ± 0.5% at full scale and the errors are standard deviations.*

**Solution.** (*a*) Value of resistance for which the errors are equal for the two types of connections

$$R = \sqrt{R_A R_v} = \sqrt{2 \times 5000} = 100\ \Omega$$

Since the resistance to be measured has a value less than 100 Ω, the method of Fig. 14.1 (*b*) should be used as it results in smaller error.

Measured value of resistance

$$R_{m2} = V/I = 35.5/0.42 = 84.52\ \Omega$$

True value of resistance

$$R = R_{m2}\left(\frac{1}{1 - R_{m2}/R_v}\right)$$

$$= 84.52\left(\frac{1}{1 - 84.52/5000}\right) = 86\ \Omega.$$

(*b*) Error in ammeter reading

$$= (0.5/100) \times 1 = 0.005 \text{ A}$$

∴ Percentage error at 0.42 A reading

$$= (0.005/0.42) \times 100 = 1.19\%$$

Error in voltmeter reading

$$= 0.5 \times (50/100) = 0.25 \text{ V}$$

∴ Percentage error at 35.5 V reading

$$= (0.25/35.5) \times 100 = 0.704\%$$

Since the errors correspond to standard deviations, error due to ammeter and voltmeter

$$= \sqrt{(1.19)^2 + (0.704)^2} = \pm 1.38\%$$

Absolute error due to ammeter and voltmeter

$$= (1.38/100) \times 86 \approx \pm 1.2\ \Omega.$$

∴ The resistance is specified as 86 ± 1.2 Ω.

***Example 14.3*** *A resistance R is measured using the connections of Fig. 14.1(a). The current measured is 10 A on range 10 A and the voltage measured is 125 V on 150 V range. The scales of the ammeter and voltmeter are uniform, the total scale divisions of ammeter are 100 and that of voltmeter are 150. The scales of these instruments are such that 1/10 of a scale division can be distinguished. The constructional error of the ammeter is ± 0.3% and that of voltmeter ± 0.4%. The resistance of the ammeter is 0.25 Ω. Calculate the value of R and the limits of possible error in the results.*

**Solution.** Reading error of ammeter

$$= \pm \frac{1}{10 \times 100} \times 100 = \pm 0.1\%$$

Reading error of voltmeter

$$= \pm \frac{1}{10 \times 150} \times 100 = \pm 0.087\%$$

Total error of ammeter $\delta I = \pm 0.3 \pm 0.1 = \pm 0.4\%$

Total error of voltmeter

$$\delta V = \pm 0.4 \pm 0.067 = \pm 0.467\%$$

Now, resistance $R = V/I$ and therefore total systematic error in measurement

$$= \pm \delta V \pm \delta A = \pm 0.467 \pm 0.4 = \pm 0.867\%$$

Measured value of resistance

$$R_{m2} = 125/10 = 12.5\ \Omega.$$

True value of resistance

$$R = R_{m1}\left(1 - \frac{R_a}{R_{m1}}\right) \quad \text{(See Eqn. 14.4)}$$

$$= 12.5\left(1 - \frac{0.25}{12.5}\right) = 12.25\ \Omega.$$

Therefore the value of $R$ is specified as

$$12.25\ \Omega \pm 0.867\% = 12.25 \pm 0.11\ \Omega.$$

### 14.2.2 Substitution Method

The connection diagram for this method is shown in Fig. 14.2. *R* is the unknown resistance while *S* is a standard variable resistance. '*A*' is an ammeter and '*r*' is a regulating resistance. There is a switch for putting *R* and *S* into the circuit alternately.

The switch is put at position '*1*' and resistance *R* is connected in the circuit. The regulating resistance *r* is adjusted till the ammeter pointer is at a chosen scale mark. Now, the switch is thrown to position '2' putting the standard variable resistance *S* in the circuit. The value of *S* is varied till the same deflection as was obtained with *R* in the circuit is obtained. The settings of the dials of *S* are read. Since the substitution of one resistance for another has left the current unaltered, and provided that the emf of battery and

14

the position of $r$ are unaltered, the two resistances must be equal. Thus the value of unknown resistance $R$ is equal to the dial settings of resistance $S$.

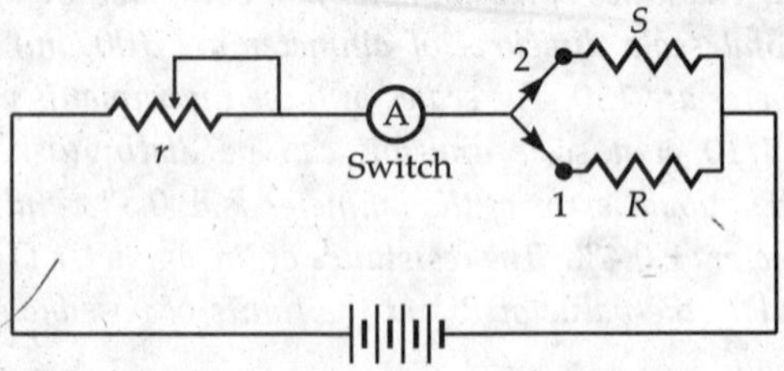


**Fig. 14.2** Substitution method.

This is a more accurate method than the Ammeter-Voltmeter method, as it is not subject to the errors encountered in the latter method. However, the accuracy of this method is greatly affected if there is any change in the battery emf during the time the readings on the two settings are taken. Thus in order to avoid errors on this account, a battery of ample capacity should be used so that its emf remains constant.

The accuracy of the measurement naturally depends upon the constancy of the battery emf and of the resistance of the circuit excluding $R$ and $S$, upon the sensitivity of the instrument, and upon the accuracy with which standard resistance $S$ is known.

This method is not widely used for simple resistance measurements and is used in a modified form for the measurement of high resistances. The substitution principle, however, is very important and finds many applications in bridge methods and in high frequency a.c. measurements.

**Example 14.4** *In a measurement of resistance by substitution method a standard 0.5 MΩ resistor is used. The galvanometer has a resistance of 10 kΩ and gives deflections as follows :*

*(i) With standard resistor, 41 divisions,*

*(ii) With unknown resistance, 51 divisions.*

*Find the unknown resistance.*

**Solution.** The deflection of the galvanometer is directly proportional to the current passing through the circuit and hence is inversely proportional to the total resistance of the circuit. Let $S$, $R$ and $G$ be respectively the resistances of standard resistor, unknown resistor and the galvanometer. Also let $\theta_1$ be the deflection with standard resistor in circuit and $\theta_2$ with unknown resistor in circuit.

$$\therefore \quad \frac{\theta_1}{\theta_2} = \frac{R+G}{S+G}$$

Hence, unknown resistance

$$R = (S+G)\frac{\theta_1}{\theta_2} - G$$

$$= (0.5 \times 10^6 + 10 \times 10^3) \times (41/51) - 10 \times 10^3$$

$$= 0.4 \times 10^6 \Omega = 0.4 \text{ M}\Omega.$$

### 14.2.3 Wheatstone Bridge

A very important device used in the measurement of medium resistances is the Wheatstone bridge. A Wheatstone bridge has been in use longer than almost any electrical measuring instrument. It is still an accurate and reliable instrument and is extensively used in industry. The Wheatstone bridge is an instrument for making *comparison measurements* and operates upon a *null indication* principle. This means the indication is independent of the calibration of the null indicating instrument or any of its characteristics. For this reason, very high degrees of accuracy can be achieved using Wheatstone bridge. Accuracy of 0.1% is quite common with a Wheatstone bridge as opposed to accuracies of 3% to 5% with ordinary ohmmeter for measurement of medium resistances. Figure 14.3 shows the basic circuit of a Wheatstone bridge. It has four resistive arms, consisting of resistances $P$, $Q$, $R$ and $S$ together with a source of emf (a battery) and a null detector, usually a galvanometer $G$ or other sensitive current meter. The current through the galvanometer depends on the potential difference between points $c$ and $d$. The bridge is said to be balanced when there is no current through the galvanometer or when the potential difference across the galvanometer is zero. This occurs when the voltage from point '$b$' to point '$a$' equals the voltage from point '$d$' to point '$b$'; or, by referring to the other battery terminal, when the voltage from point '$d$' to point '$c$' equals the voltage from point '$b$' to point '$c$'.

For bridge balance, we can write,

$$I_1 P = I_2 R \qquad \text{...(14.11)}$$

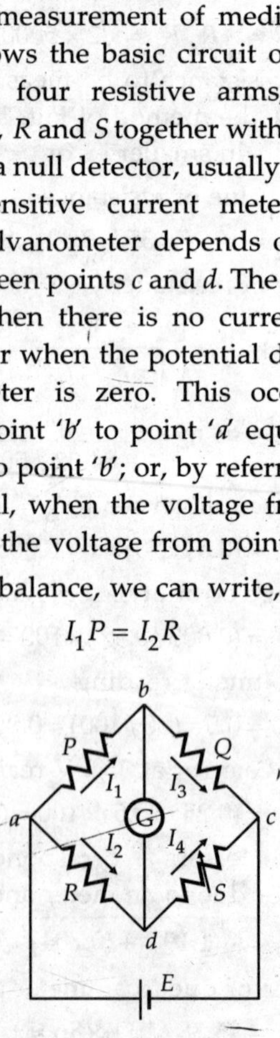


**Fig. 14.3** Wheatstone bridge.

For the galvanometer current to be zero, the following conditions also exist :

$$I_1 = I_3 = \frac{E}{P+Q} \quad ...(14.12)$$

and

$$I_2 = I_4 = \frac{E}{R+S} \quad ...(14.13)$$

where $E$ = emf of the battery.

Combining Eqns. 14.11, 14.12 and 14.13 and simplifying, we obtain :

$$\frac{P}{P+Q} = \frac{R}{R+S} \quad ...(14.14)$$

from which $QR = PS$ ...(14.15)

Equation 14.15 is the well known expression for the balance of Wheatstone bridge. If three of the resistances are known, the fourth may be determined from Eqn. 14.15 and we obtain :

$$R = S\frac{P}{Q} \quad ...(14.16)$$

where $R$ is the unknown resistance, $S$ is called the *'standard arm'* of the bridge and $P$ and $Q$ are called the *'ratio arms'*.

In the industrial and laboratory form of the bridge, the resistors which make up $P$, $Q$ and $S$ are mounted together in a box, the appropriate values being selected by dial switches. Battery and galvanometer switches are also included together with a galvanometer and a dry battery in the portable sets. $P$ and $Q$ normally consist of four resistors each, the values being 10, 100, 1000 and 10,000 Ω respectively $S$ consists of a 4 dial or 5 dial decade arrangement of resistors. Figure 14.4 shows the commercial form of Wheatstone bridge.

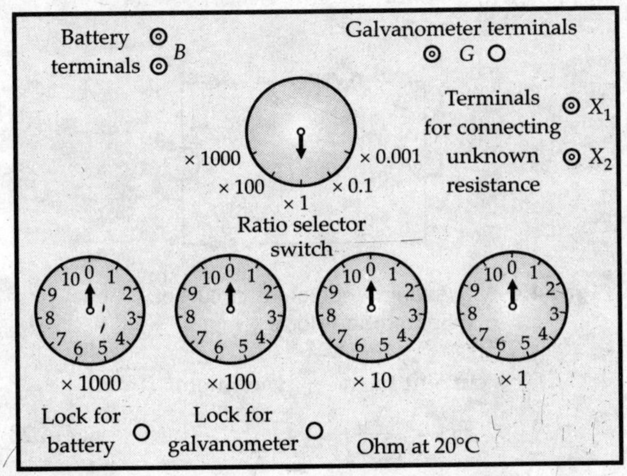


**Fig. 14.4** Commercial form of Wheatstone bridge.

### 14.2.4 Sensitivity of Wheatstone Bridge

It is frequently desirable to know the galvanometer response to be expected in a bridge which is slightly unbalanced so that a current flows in the galvanometer branch of the bridge network. This may be used for :

(*i*) selecting a galvanometer with which a given unbalance may be observed in a specified bridge arrangement,

(*ii*) determining the minimum unbalance which can be observed with a given galvanometer in the specified bridge arrangement, and

(*iii*) determining the deflection to be expected for a given unbalance.

The sensitivity to unbalance can be computed by solving the bridge circuit for a small unbalance. The solution is approached by converting the Wheatstone bridge of Fig. 14.3 to its *"Thevenin Equivalent"* circuit. Assume that the bridge is balanced when the branch resistances are $P, Q, R, S$ so that $P/Q = R/S$. Suppose the resistance $R$ is changed to $R + \Delta R$ creating an unbalance. This will to cause an emf $e$ to appear across the galvanometer branch. With galvanometer branch open, the voltage drop between points $a$ and $b$ is :

$$E_{ab} = I_1 P = \frac{EP}{P+Q}$$

Similarly, $E_{ad} = I_2(R+\Delta R) = \dfrac{E(R+\Delta R)}{R+\Delta R+S}$

Therefore voltage difference between points $d$ and $b$ is :

$$e = E_{ad} - E_{ab} = E\left[\frac{R+\Delta R}{R+\Delta R+S} - \frac{P}{P+Q}\right] \quad ...(14.17)$$

and since $\dfrac{P}{P+Q} = \dfrac{R}{R+S}$

$$\therefore \quad e = E\left[\frac{R+\Delta R}{R+\Delta R+S} - \frac{P}{P+S}\right]$$

$$= \frac{ES\Delta R}{(R+S)^2 + \Delta R(R+S)}$$

$$\approx \frac{ES\Delta R}{(R+S)^2} \quad ...(14.18)$$

as $\Delta R(R+S) << (R+S)^2$

Let $S_v$ be the voltage sensitivity of galvanometer. Therefore, deflection of galvanometer is

$$\theta = S_v e = S_v \frac{ES\Delta R}{(R+S)^2} \quad ...(14.19)$$

The bridge sensitivity $S_B$ is defined as the deflection of the galvanometer per unit fractional change in unknown resistance.

Bridge sensitivity

$$S_B = \frac{\theta}{\Delta R / R} \quad ...(14.20)$$

$$= \frac{S_v ESR}{(R+S)^2} \quad ...(14.21)$$

From Eqn. 14.21, it is clear that the sensitivity of the bridge is dependent upon bridge voltage, bridge parameters and the voltage sensitivity of the galvanometer. Rearranging the terms in the expression for sensitivity,

$$S_B = \frac{S_v E}{(R+S)^2 / SR} = \frac{S_v E}{\frac{R}{S} + 2 + \frac{S}{R}}$$

$$= \frac{S_v E}{\frac{P}{Q} + 2 + \frac{Q}{P}} \quad ...(14.22)$$

From Eqn. 14.22, it is apparent that maximum sensitivity occurs where $R/S = 1$. As the ratio becomes either larger or smaller, the sensitivity decreases. Since the accuracy of measurement is dependent upon sensitivity a limit can be seen to the usefulness for a given bridge, battery and galvanometer combination.

For a bridge with equal arms, $R = S = P = Q$,

Bridge sensitivity $\quad S_B = \frac{S_v E}{4} \quad ...(14.23)$

As explained above the sensitivity is maximum when the ratio is unity. The sensitivity with ratio $P/Q = R/S = 1000$ would be about 1/250 of that for unity ratio. The sensitivity with $P/Q = R/S = 1000$ would similarly be about 1/250 of that for unity ratio.

Thus the sensitivity decreases considerably if the ratio $P/Q = R/S$ is greater or smaller than unity. This reduction in sensitivity is accompanied by a reduction in accuracy with which a bridge can be balanced.

***Galvanometer current.*** The current through the galvanometer can be found out by finding the Thevenin equivalent circuit. The Thevenin or open circuit voltage appearing between terminals $b$ and $d$ with galvanometer circuit open circuited is,

$$E_0 = E_{ad} - E_{ac} = I_2(R + \Delta R) - I_1 P$$

$$= \frac{E(R+\Delta R)}{R+\Delta R+S} - \frac{EP}{P+Q}$$

$$= E\left[\frac{R+\Delta R}{R+\Delta R+S} - \frac{P}{P+Q}\right] \quad ...(14.24)$$

For a bridge with equal arms, $R = S = P = Q$.

$$E_0 = E\left[\frac{R+\Delta R}{R+\Delta R+R} - \frac{R}{R+R}\right]$$

$$= E\left[\frac{R+\Delta R}{2R+\Delta R} - \frac{1}{2}\right]$$

$$\approx E\left(\frac{\Delta R}{4R}\right) \text{ as } \Delta R << R. \quad ...(14.25)$$

The resistance of the Thevenin equivalent circuit is found by looking back into terminals $c$ and $d$ (Fig. 14.3) and replacing the battery by its internal resistance. In most cases, however, the extremely low resistance of the battery can be neglected and this simplifies the solution as we can assume that terminals $a$ and $b$ are shorted. The Thevenin equivalent resistance can be calculated by referring to Fig. 14.5.

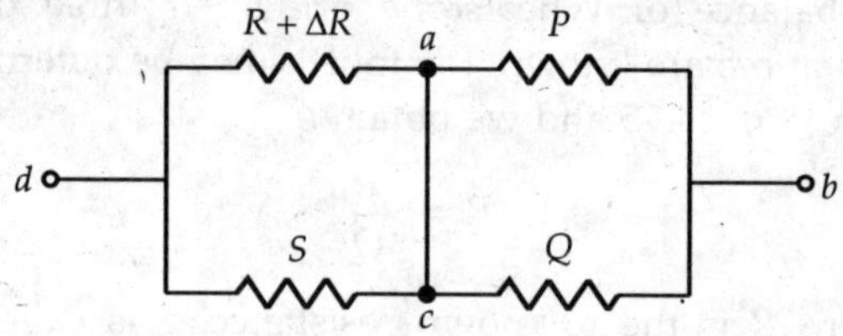


**Fig. 14.5** Finding resistance of bridge looking into terminals $d$ and $b$.

Thevenin equivalent resistance of bridge

$$R_0 = \frac{RS}{R+S} + \frac{PQ}{P+Q} \quad ...(14.26)$$

Considering $\Delta R << R$

For a bridge with equal arms,

$$P = Q = S = R,$$

$$R_0 = R \quad ...(14.27)$$

The Thevenin equivalent of the bridge circuit therefore reduces to a Thevenin generator with an emf $E_0$ and an internal resistance $R_0$. This circuit is shown in Fig. 14.6.

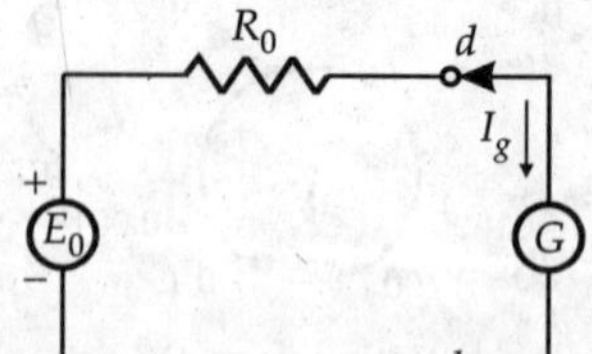


**Fig. 14.6** Thevenin equivalent circuit of Wheatstone bridge.

∴ The current in the galvanometer circuit

$$I_g = \frac{E_0}{R_0 + G} \quad ...(14.28)$$

where $G$ = resistance of the galvanometer circuit.

For a bridge with equal arms,

$$I_g = \frac{E(\Delta R/4R)}{(R+G)} \quad \text{...(14.29)}$$

The deflection of the galvanometer for a small change in resistance in the unknown arm is,

$$\theta = \frac{S_v E S \Delta R}{(R+S)^2}$$

But $$S_v = \frac{S_i}{R_0 + G}$$

where $S_i$ = current sensitivity of the galvanometer.

$$\therefore \quad \theta = \frac{S_i E S \Delta R}{(R_0+G)(R+S)^2} \quad \text{...(14.30)}$$

For a bridge with equal arms,

$$\theta = \frac{S_i E \Delta R}{4R(R+G)} \quad \text{...(14.31)}$$

Also bridge sensitivity,

$$S_B = \frac{\theta}{\Delta R/R} = \frac{S_i E S R}{(R_0+G)(R+S)^2} \quad \text{...(14.32)}$$

For a bridge with equal arms,

Bridge sensitivity,

$$S_B = \frac{S_i E}{4(R+G)} \quad \text{...(14.33)}$$

### 14.2.5 Precision Measurement of Medium Resistances with Wheatstone Bridge

It is sometimes necessary to measure resistances to a precision of 1 part in 10,000 or even greater by comparing them with standard resistances. In such cases more than ordinary precautions are necessary in order to secure the required accuracy. The following factors should be taken into consideration.

1. ***Resistance of connecting leads.*** A lead of 22 SWG wire having a length of 25 cm has a resistance of about 0.012 Ω and this represents more than 1 part in 1000 for a 10 Ω resistance or more than one part in 10,000 for a 100 Ω resistance.

2. ***Thermo-electric effects.*** Thermoelectric emfs are often present in the measuring circuit and they must be taken into account since they affect the galvanometer deflection in the same way as an emf occurring because of unbalance. The effect of thermoelectric emfs and other parasite emfs on the measurement may be eliminated by reversing the battery connections to the bridge through a quick acting switch and adjusting the balance until no change in galvanometer deflection can be observed on reversal. The results are obtained by averaging the two readings. This way the effect of thermo-electric emfs is eliminated with the added advantage of doubling the sensitivity of the bridge.

3. ***Temperature effects.*** The errors caused by change of resistance due to change of temperature produces serious errors in measurements especially in the case of resistances made up of materials having a large value of resistance temperature coefficient. For example in the case of copper, which has a resistance temperature coefficient of 0.004/°C, a change in temperature of 1°C will cause an error of 0.4% or 1 part in 250.

4. ***Contact resistance.*** Serious errors may be caused by contact resistances of switches and binding posts. A dial may have a contact resistance of about 0.003 Ω and thus a 4 dial resistance box has a contact resistance of about 0.01 Ω. This value is quite high especially when low resistance measurements are being done. Another aspect of the contact resistance is that error caused by it is difficult to account for since its magnitude *i.e.*, magnitude of contact resistance is uncertain.

14

In precision resistance measurements, the most accurate comparisons are made on an equal ratio bridge with a fixed standard resistance nominally equal to resistance under test. Then with equal leads, equal currents and equal heating of the ratio arms, the possible errors are minimized. The problem is then reduced to that of determining the exact ratio of the unknown resistance, $R_1$ to the standard resistance $S$ or the difference between them. The different methods used for this purpose are :

(*i*) *Change in ratio arms.* Small known changes are made in the ratio arms and the exact balance is obtained.

(*ii*) *Using a high resistance shunt.* In this case the bridge is balanced by means of an adjustable high resistance put in parallel with one of the bridge arms. A decade resistance box is used for this purpose. Suppose the test resistance, $R$, is slightly lower in value than the standard resistance, $S$ and let the balance be obtained with a resistance $xS$, put in parallel with $S$.

The resistance of this arm is then

$$xS^2/(S+xS), \ i.e., \ S/(1+1/x).$$

∴ Value of unknown resistance

$$R = \frac{P}{Q} \cdot \frac{S}{1+1/x}.$$

Since $x$ is very large and therefore $1/x$ is very small and hence we can write,

$$R = \frac{P}{Q} \cdot S(1 - 1/x) \quad ...(14.34)$$

For an equal arm bridge, $P = Q$ and therefore,

$$R = S(1 - 1/x) \quad ...(14.35)$$

The fractional difference, $1/x$ between $S$ and $R$ can be known. Since this method determines the fractional difference and $x$ has a large value, the high resistance put in parallel with $S$ need not be particularly accurate.

*(iii) Carey foster slide wire bridge.* A slide wire bridge is used for the purpose of determining the difference between the standard and the unknown resistances. The details are explained below.

### 14.2.6 Carey-Foster Slide-wire Bridge

The connections of this bridge are shown in Fig. 14.7, a slide-wire of length $L$ being included between $R$ and $S$ as shown. This bridge is specially suited for the comparison of two nearly equal resistances.

Resistances $P$ and $Q$ are first adjusted so that the ratio $P/Q$ is approximately equal to the ratio $R/S$. Exact balance is obtained by adjustment of the sliding contact on the slide-wire. Let $l_1$ be the distance of the sliding contact form the left-hand end of the slide wire. The resistances $R$ and $S$ are then interchanged and balance again obtained. Let the distance now be $l_2$.

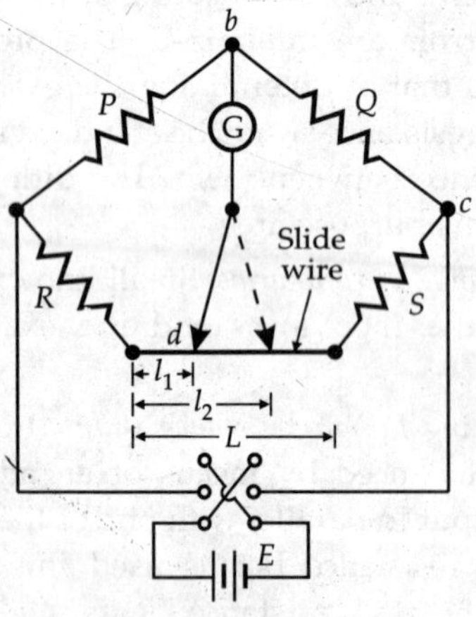


**Fig. 14.7** Carey-Foster Slide wire bridge.

Then for the first balance

$$\frac{P}{Q} = \frac{R + l_1 r}{S + (L - l_1)r}$$

where $r$ is the resistance per unit length of the slide-wire.

For the second balance, $\frac{P}{Q} = \frac{S + l_2 r}{R + (L - l_2)}$

Now, for first balance,

$$\frac{P}{Q} + 1 = \frac{R + l_1 r + S + (L - l_1)r}{S + (L - l_1)r} = \frac{R + S + Lr}{S + (L - l_1)r} \quad ...(i)$$

also, for second balance,

$$\frac{P}{Q} + 1 = \frac{S + l_2 r + R + (L - l_2)r}{R + (L - l_2)r} = \frac{S + R + Lr}{R + (L - l_2)r} \quad ...(ii)$$

From (*i*) and (*ii*), we have,

$$S + (L - l_1)r = R + (L - l_2)r$$

Hence $\quad S - R = (l_1 - l_2)r \quad ...(14.36)$

Thus the difference between $S$ and $R$ is obtained from the resistance per unit length of the slide-wire together with the difference $(l_1 - l_2)$ between the two slide-wire lengths at balance.

The slide-wire is calibrated *i.e.*, $r$ is obtained by shunting either $S$ or $R$ by a known resistance and again determining the difference in length $(l_1' - l_2')$.

Suppose that $S$ is known and that $S'$ is its value when shunted by a known resistance ; then

$$S - R = (l_1 - l_2)r \quad \text{and} \quad S' - R = (l_1' - l_2')r$$

$$\therefore \quad \frac{S - R}{l_1 - l_2} = \frac{S' - R}{l_1' - l_2'}$$

from which $\quad R = \frac{S(l_1' - l_2') - S(l_1 - l_2)}{(l_1' - l_2' - l_1 + l_2)} \quad ...(14.37)$

Equation 14.37 shows that this method gives a direct comparison between $S$ and $R$ in terms of lengths only, the resistances of $P$ and $Q$ contact resistances, and the resistances of connecting leads being eliminated.

As it is important that the two resistors $R$ and $S$ shall not be handled or disturbed during the measurement, a special switch is used to affect the interchanging of these two resistors during the test.

### 14.2.7 Kelvin-Varley Slide

A Kelvin-Varley slide is used for voltage division. This method is very precise and finds extensive applications. A Kelvin-Varley slide is shown in Fig. 14.8. It consists of several decades of resistors which are interconnected. The voltage division is carried out successively. Each voltage division decade is made up of eleven equal resistors with successive division decades having a total resistance equal to twice the value of a unit resistor in the previous decade. For example, in the Kelvin Varley Slide shown in Fig. 14.8, there are four decade dividers. This decade is constructed using 11 resistance coils having a resistance of 10 kΩ each. The second decade divider has 11 resistors of 2 kΩ) each*. Similarly, the 3rd decade has 11 resistors of 400 Ω each and the fourth and final decade has 10 resistors of 80 Ω each.

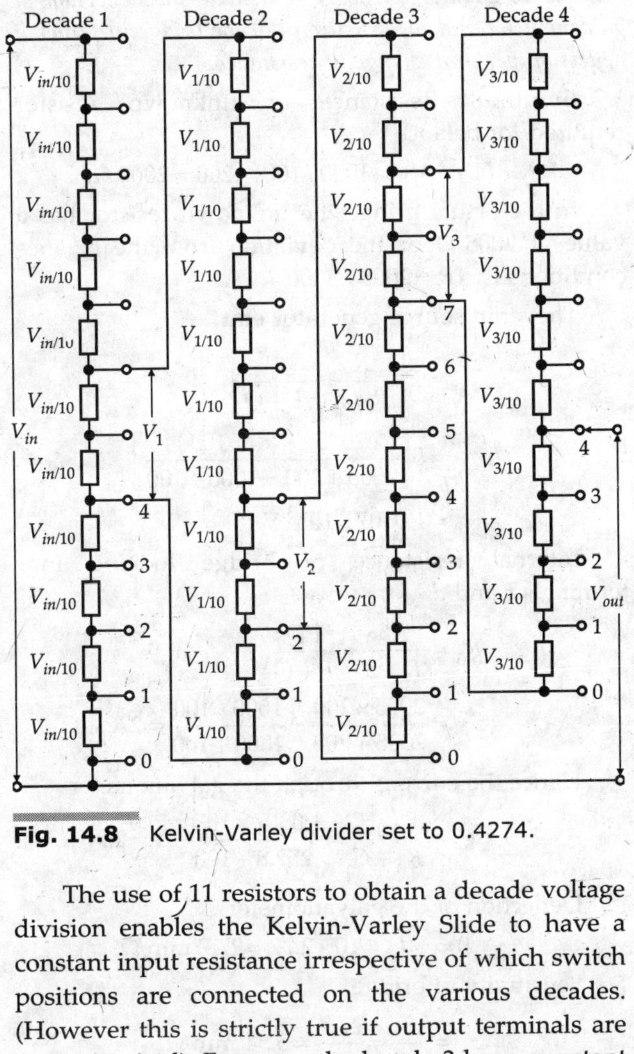

**Fig. 14.8** Kelvin-Varley divider set to 0.4274.

The use of 11 resistors to obtain a decade voltage division enables the Kelvin-Varley Slide to have a constant input resistance irrespective of which switch positions are connected on the various decades. (However this is strictly true if output terminals are open circuited). For example decade 3 has a constant input resistance of :

$$9 \times 400 + \frac{1 \times 400 + 10 \times 80}{800 + 800} = 4000\ \Omega = 4\ \text{k}\Omega.$$

Decade 2 has an input resistance of

$$9 \times 2000 + \frac{1 \times 2000 \times 4000}{4000 + 4000} = 20{,}000\ \Omega = 20\ \text{k}\Omega.$$

The input terminal impedance is 100 kΩ irrespective of the switch position if it is assumed that the output current is negligible.

The advantage of using the Kelvin-Varley Slide is the reduction of errors which arise on account of switch contact resistance due to current sharing within the device. The disadvantages of the Kelvin-Varley Slide are its calibration and the errors on account of temperature. The errors on account of temperature are dut to the fact that the resistors carry different accounts and the changes in the value of $r$ ohm resistance is different on account of different self-heating conditions. The temperature effects can be reduced to negligible proportions by using resistors made of materials having a low resistance temperature co-efficient. The error may be reduced to as low a value as ± 0.1 ppm. The principle of Kelvin-Varley Slides is used with advantage in potentio- meters and universal shunts.

Figure 14.9 shows the use of Kelvin-Varley slide in a Wheatstone bridge. The device is used to replace the simple slide wire with the advantage that it gives greater accuracy.

For the case shown in Fig. 14.9

$$\frac{R}{S} = \frac{7554}{5346}$$

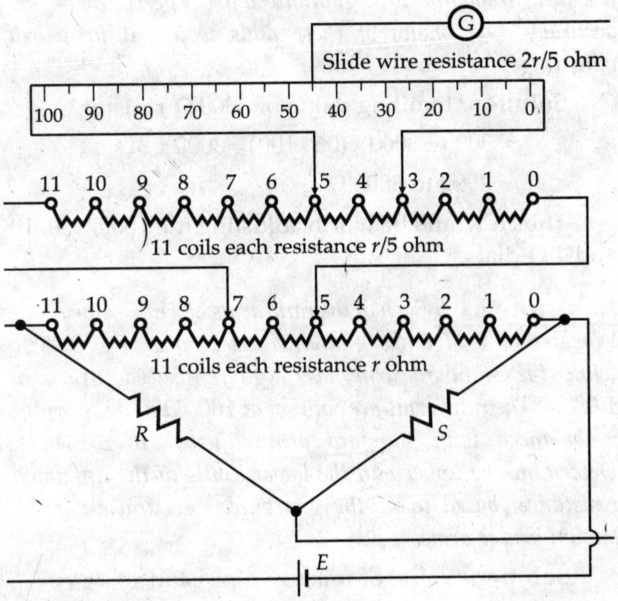


**Fig. 14.9** Kelvin-Varley slide.

### 14.2.8 Limitations of Wheatstone Bridge

The use of Wheatstone bridge is limited to the measurement of resistances ranging from a few ohm to several megohms. The upper limit is set by the reduction in sensitivity to unbalance caused by high resistance values. The upper limit can be extended to a

* Resistancc of each unit in first decade is 10 kΩ. The second decade has a total resistance of 2 × 10 = 20 kΩ. Therefore, each unit in the second decade has a resistance of 20/10 = 2 kΩ. In other words, we can say that if the first decade consists of 11 coils of resistance $r$ ohm each, the next decade has 11 coils of $r/5$ ohm each and the third decade has 11 coils of $2r/5$ ohm each.

14

certain extent by increasing the emf applied to the bridge but in this case care has to be taken to avoid overheating of any arm of the bridge. Inaccuracy may also be introduced on account of leakage over insulation of the bridge arms when measuring very high resistances. For measurement of very high resistances, a megohm bridge (described on page 441) is preferable.

The lower limit for measurement is set by the resistance of the connecting leads and by contact resistance at the binding posts. The error caused by leads may be corrected fairly well, but contact resistance presents a source of uncertainty that is difficult to overcome. The lower limit for accurate measurement is in the neighbourhood of 1 to 5 ohm. For low resistance measurements, therefore, a Kelvin bridge (described in Art. 14.3.2 on page 434) is generally preferred.

**Example 14.5** *A resistance of approximately 3000 Ω is needed to balance a bridge. It is obtained on a 5 dial resistance box having steps of 1000, 100, 10, 1 and 0.1 Ω. The measurement is to be guaranteed to 0.1 percent. For this accuracy, how many of these dials would it be worth adjusting ?*

**Solution.** Limiting value of 3000 Ω resistor

$$= 3000 \pm 3000 \times (0.1/100) = 3000 \pm 3\Omega$$

$$= 2997 \text{ to } 3003\ \Omega$$

Thus it would be worth adjusting the 1000, 100, 10 and 1 Ω dials.

**Example 14.6** *Each of the ratio arms of a laboratory type Wheatstone bridge has guaranteed accuracy of ± 0.05%, while the standard arm has a guaranteed accuracy of ± 0.1%. The ratio arms are both set at 1000 Ω and the bridge is balanced with standard arm adjusted to 3,154 Ω. Determine the upper and the lower limits of the unknown resistance, based upon the guaranteed accuracies of the known bridge arms.*

**Solution.** Value of unknown resistance

$$R = (P/Q) \times S = (1000/1000) \times 3154$$

$$= 3154\ \Omega.$$

∴ Percentage error in determination of $R$.

$$\frac{\delta R}{R} = \pm\frac{\delta P}{P} \pm \frac{\delta Q}{Q} \pm \frac{\delta S}{S}$$

$$= \pm 0.05 \pm 0.05 \pm 0.01 = \pm 0.2\ \%.$$

∴ Limiting values of

$$R = 3154 \pm 0.2\% = 3091 \text{ to } 3217\ \Omega.$$

**Example 14.7** *In the Wheatstone bridge of Fig. 14.3, the values of resistances of various arms are $P = 1000\ \Omega$, $Q = 100\ \Omega$, $R = 2{,}005\ \Omega$ and $S = 200\ \Omega$. The battery has an emf of 5 V and negligible internal resistance. The galvanometer has a current sensitivity of 10 mm/μA and an internal resistance of 100 Ω. Calculate the deflection of galvanometer and the sensitivity of the bridge in terms of deflection per unit change in resistance.*

**Solution.** Resistance of unknown resistor required for balance

$$R = (P/Q)S = (1000/100) \times 200 = 2000\ \Omega.$$

In the actual bridge the unknown resistor has a value of 2005 Ω or the deviation from the balance conditions is $\Delta R = 2005 - 2000 = 5\ \Omega$.

Thevenin source generator emf

$$E_0 = E\left[\frac{R}{R+S} - \frac{P}{P+Q}\right]$$

$$= 5\left[\frac{2005}{2005+200} - \frac{1000}{1000+100}\right]$$

$$= 1.0307 \times 10^{-3}\ \text{V}.$$

Internal resistance of bridge looking into terminals $b$ and $d$.

$$R_0 = \frac{RS}{R+S} + \frac{PQ}{P+Q}$$

$$= \frac{2005 \times 200}{2005+200} + \frac{1000 \times 100}{1000+100} = 272.8\ \Omega$$

Hence the current through the galvanometer

$$I_g = \frac{E_0}{R_0 + G} = \frac{1.0307 \times 10^{-3}}{272.8 + 100}\ \text{A} = 2.77\ \mu\text{A}.$$

Deflection of the galvanometer

$$\theta = S_i I_g = 10 \times 2.77 = 27.7\ \text{mm}/\Omega.$$

Sensitivity of bridge

$$S_B = \frac{\theta}{\Delta R} = \frac{27.7}{5} = 5.54\ \text{mm}/\Omega.$$

**Example 14.8** *A Wheatstone bridge has ratio arms of 1000 Ω and 100 Ω and is being used to measure an unknown resistance of 25 Ω. Two galvanometers are available. Galvanometer 'A' has a resistance of 50 Ω and a sensitivity of 200 mm/μA and galvanometer 'B' has values of 600 Ω and 500 mm/μA. Which of the two galvanometers more is sensitive to a small unbalance on the above bridge, and what is the ratio of sensitivities ? The galvanometer is connected from the junction of the ratio arms to the opposite comers. Comment upon the results.*

**Solution.** The arrangement of this bridge is shown in Fig. 14.3. Value of standard resistance under balance conditions

$$S = R \cdot \frac{Q}{P} = 25 \times \frac{1000}{100} = 250\ \Omega.$$

Internal resistance of bridge looking into terminals $b$ and $d$.

$$R_0 = \frac{RS}{R+S} + \frac{PQ}{P+Q}$$

$$= \frac{25 \times 250}{25+250} + \frac{100 \times 1000}{100+1000} = 113.6\ \Omega.$$

From Eqn. 14.30, the deflection for a small change in resistance

$$\theta = \frac{S_i ES\Delta R}{(R_0+G)(R+S)^2}$$

∴ Ratio of deflections for two galvanometers

$$\frac{\theta_A}{\theta_B} = \frac{(S_i)_A ES\Delta R}{(R_0+G_A)(R+S)^2} \Big/ \frac{(S_i)_B ES\Delta R}{(R_0+G_B)(R+S)^2}$$

$$= \frac{(S_i)_A}{(S_i)_B} \cdot \frac{R_0+G_B}{R_0+G_A}$$

$$= \frac{200}{500} \times \frac{(113.6+600)}{(113.6+50)} = 1.75$$

Thus galvanometer $A$ has a sensitivity of 1.75 times that of galvanometer $B$ as far as this bridge is concerned, even though on its own galvanometer $A$ is less sensitive than galvanometer $B$.

**Example 14.9** *A highly sensitive galvanometer can detect a current as low as 0.1 nA. This galvanometer is used in a Wheatstone bridge as a detector. The resistance of galvanometer is negligible. Each arm of the bridge has a resistance of 1 kΩ. The input voltage applied to the bridge is 20 V. Calculate the smallest change in resistance which can be detected. The resistance of the galvanometer can be neglected as compared with the internal resistance of bridge.*

**Solution.** Now $P = Q = R = S = 1\ k\Omega = 1000\ \Omega$.

It is a bridge with equal arms.

∴ The internal resistance of bridge

$$R_0 = R = 1000\Omega \qquad \text{(See Eqn. 14.27)}$$

Let the change in resistance be $\Delta R$.

∴ Output voltage of bridge due to unbalance

$$E_0 = E\Delta R / 4R$$

$$= 20 \times \Delta R / (4 \times 1000) = 5 \times 10^{-3}\ \Delta R$$

Current through the galvanometer

$$= \frac{E_0}{R_0} = \frac{5 \times 10^{-3}\ \Delta R}{1000} = 0.1 \times 10^{-9}.$$

∴ The smallest change in resistance which can be detected

$$\Delta R = 20 \times 10^{-6}\Omega = 20\ \mu\Omega.$$

**Example 14.10** *A regular Wheatstone bridge is used to measure high resistances (in the megohm range). The bridge has ratio arms of 10,000 Ω and 10 Ω.*

*The adjustable arm has a maximum value of 10,000 Ω. A battery of 10 V emf and negligible resistance is connected from the junction of ratio aims to the opposite corner ?*

*(a) What is the maximum resistance that can be measured by this arrangement ?*

*(b) If the galvanometer has a sensitivity of 200 mm/μA and a resistance of 50 Ω, how much unbalance is needed to give a galvanometer deflection of 1 mm for the maximum resistance of part (a) ?*

*(c) If the galvanometer of part (b) is replaced by a galvanometer of sensitivity of 1000 mm/μA and a resistance of 1000 Ω, calculate the change in resistance to cause a deflection of 1 mm.*

**Solution.** The arrangement of the bridge is shown in Fig. 14.3.

Now we have,

unknown resistance, $R = (P/Q) \times S$.

The maximum value of

$$R = (\text{maximum value of } P/Q) \times (\text{maximum value of } S)$$

$$= \frac{10000}{10} \times 10000 = 10\ M\Omega.$$

(*b*) Internal resistance of bridge

$$R_0 = \frac{RS}{R+S} + \frac{PQ}{P+Q}$$

$$= \frac{10 \times 10^6 \times 10 \times 10^3}{10 \times 10^6 + 10 \times 10^3} + \frac{10 \times 10^3 \times 10}{10 \times 10^3 + 10}$$

$$\approx 10{,}000\ \Omega$$

From Eqn 14.30, change in resistance needed for 1 mm deflection

$$\Delta R = \frac{\theta(R_0+G)(R+S)^2}{S_i ES},$$

$$= \frac{1 \times (10000+50)(10 \times 10^6 + 10 \times 10^3)^2}{200 \times 10^6 \times 10 \times 10^3}$$

$$\approx 0.5\ M\Omega.$$

(*c*) The change in resistance to cause 1 mm deflection with galvanometer having a resistance,

$$G = 1000\ \Omega$$

and current sensitivity of $S_i = 1000$ mm/μA is

$$\Delta R = \frac{1 \times (10{,}000+1000)(10 \times 10^6 + 10 \times 10^3)^2}{1000 \times 10^6 \times 10 \times 10^3}$$

$$\approx 0.1\ M\Omega$$

14

***Example 14.11*** *A Wheatstone bridge is shown in Fig. 14.10*

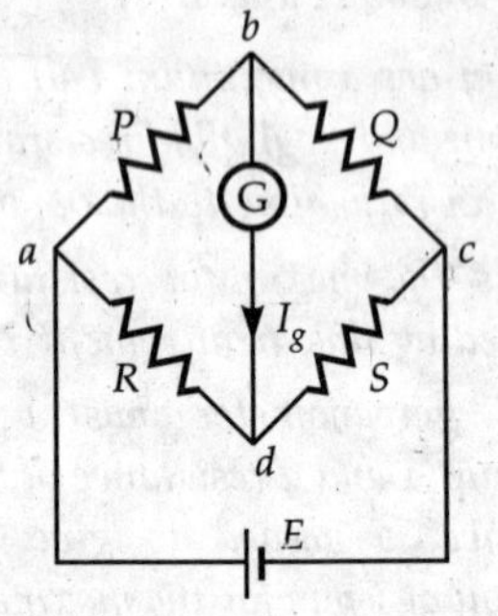


**Fig. 14.10** Diagram of Example 14.11.

*The values of resistances are :*

$$P = 1\ k\Omega,\ R = 1\ k\Omega,\ S = 5\ k\Omega,\ G = 100\ \Omega.$$

*The Thevenin source generator voltage $E_0 = 24$ mV and the galvanometer current is 13.6 µA. Calculate the value of Q.*

**Solution.** The Thevenin equivalent circuit of the bridge is shown in Fig 14.11.

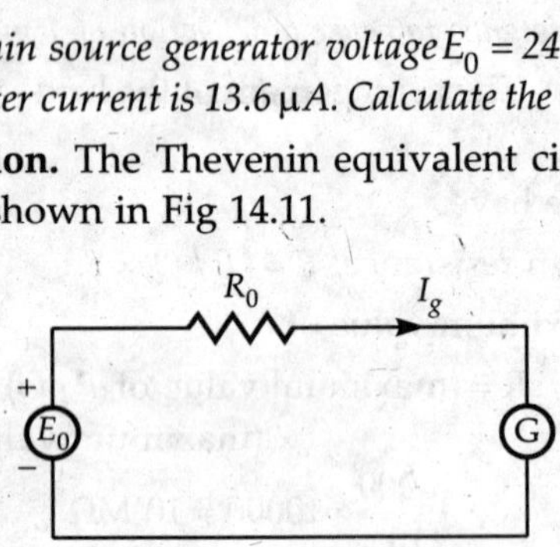


**Fig. 14.11** Thevenin source generator of Example 14.11.

$R_0$ = resistance of circuit looking into terminals *d* & *c* with terminals *a* & *b* short-circuited.

$$= \frac{RS}{R+S} + \frac{PQ}{P+Q} = \frac{1\times 5}{1+5} + \frac{1\times Q}{1+Q}$$

$$= 0.833 + \frac{Q}{1+Q}\ k\Omega.$$

Now, $R_0 + G = 24\times 10^{-3}/13.6\times 10^{6} = 1.765\ k\Omega$

or $\qquad R_0 = 1765 - 100 = 1665\ \Omega = 1.665\ k\Omega.$

$\therefore \qquad 0.833 + \dfrac{Q}{1+Q} = 1.665$

or $\qquad Q = 4.95\ k\Omega.$

***Example 14.12*** *A modified form of Wheatstone bridge is shown in Fig 14.12. Calculate the value of unknown resistance, $R_x$, if*

$$R_a = 1200\ \Omega, \qquad R_a = 1600\ R_b,$$

$$R_1 = 800\ R_b,\ R_1 = 1.25\ R_2 \text{ and } R_3 = 0.5\ R_b$$

*are the resistance values under balanced conditions.*

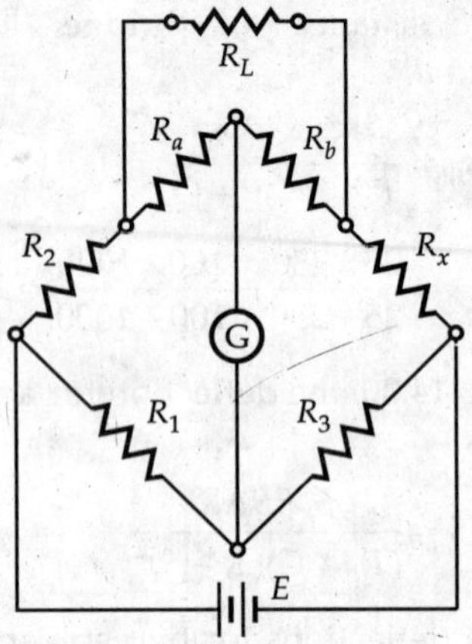


**Fig. 14.12** Wheatstone bridge of Example 14.12.

**Solution.** The calculated values are as under :

$$R_a = 1200\ \Omega,$$

$$R_b = (1/1600)\times 1200 = 0.75\ \Omega$$

$$R_1 = 800\ R_b = 800\times 0.75 = 600\ \Omega$$

$$R_2 = R_1/1.25 = 600/1.25 = 480\ \Omega$$

$$R_3 = 0.5\ R_b = 0.5\times 0.75 = 0.375\Omega.$$

Delta-star transformation has to be used for the analysis.

This is shown in Fig. 14.13.

$$R_{10} = \frac{R_a R_b}{R_a + R_b + R_c} = \frac{1200\times 0.75}{1200 + 0.75 + R_L} = \frac{900}{1200.75 + R_L}$$

$$R_{20} = \frac{R_b R_L}{R_a + R_b + R_L} = \frac{0.75\ R_L}{1200 + 0.75 + R_L} = \frac{0.75\ R_L}{1200.75 + R_L}$$

$$R_{30} = \frac{R_a R_L}{R_a + R_b + R_L} = \frac{1200\ R_L}{1200 + 0.75 + R_L} = \frac{1200\ R_L}{1200.75 + R_L}$$

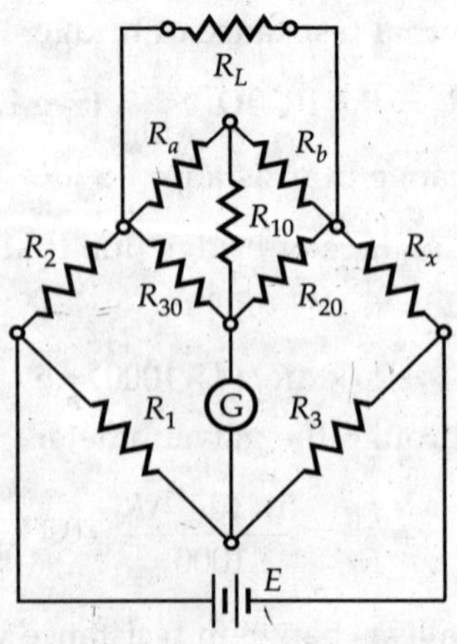


**Fig. 14.13** Delta-star transformation for analysis of Fig. 14.12.

The modified circuit is shown in Fig. 14.14.

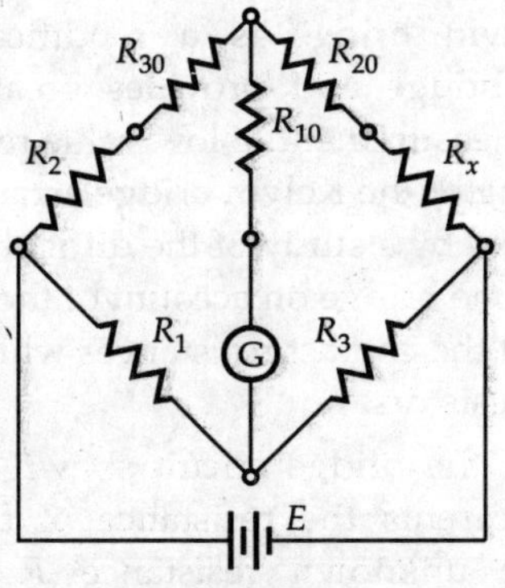


**Fig. 14.14** Configuration of bridge of Fig. 14.12 after Delta-star transformation.

For balance,

$$\frac{R_2+R_{30}}{R_1}=\frac{R_x+R_{20}}{R_3}$$

or
$$\frac{480+\dfrac{1200\,R_L}{1200.75+R_L}}{600}=\frac{R_x+\dfrac{0.75\,R_L}{1200.75+R_L}}{0.375}$$

or $R_x = 0.3\ \Omega.$

***Example 14.13*** *A Wheatstone bridge is used for measuring the value of change of resistance of a strain gauge which forms one of the arms of the bridge. All the arms of the bridge including the strain gauge have a resistance of 100 Ω each. The maximum allowable power dissipation from the strain gauge is 250 mW. Determine the value of maximum permissible current through the strain gauge and maximum allowable value of bridge supply voltage. Suppose a source of 20 V is available, find the value of series resistance to be connected between the source and the bridge to limit the input voltage of the bridge to permissible level.*

**Solution.** The resistance of strain gauge $R = 100\ \Omega$

Suppose $I$ is the current through each arm under balanced conditions.

$\therefore\ I^2R = P,$

where $P$ = power dissipation.

Hence, maximum permisible current

$$I=\sqrt{P/R}=\sqrt{\frac{250\times10^{-3}}{100}}=0.05\text{ A}=50\text{ mA}.$$

The maximum allowable voltage which can be applied to the bridge

$$=2\times50\times10^{-3}\times100=10\text{ V}.$$

$\therefore$ Voltage across the series resistor

$$=20-10=10\text{ V}.$$

Current through the series resistor

$$=2\times50\times10^{-3}=100\times10^{-3}\text{ A}.$$

$\therefore$ Resistance of series resistor

$$R_s=\frac{10}{100\times10^{-3}}=100\ \Omega.$$

***Example 14.14*** *In a Carey-Foster's bridge a resistance of 1.0125 Ω is compared with a standard resistance of 1.0000 Ω, the slide wire has a resistance of 0.250 Ω in 100 divisions. The ratio arms nominally each 10 Ω, are actually 10.05 and 9.95 Ω respectively.*

*How far (in scale divisions) are the balance positions from those which would obtain of ratio arms were true to their nominal value ? The slide wire is 100 cm long.*

**Solution.** *Balance with ratio arms equal to nominal value.* Let $l_1$ be the distance of balance point on slide wire from the unknown resistance end in cm of slide wire. Let $r$ be the resistance per cm length of slide wire.

$$\therefore\ r=\frac{0.0250}{100}$$

$$=0.00025\ \Omega/\text{cm as length of wire is 100 cm.}$$

In this case

$$P=Q=10\Omega,\ S=1.0000\ \Omega\ \text{and}\ R=1.0125\ \Omega.$$

Under balance conditions :

$$\frac{P}{Q}=\frac{R+l_1r}{S+(100-l_1)r}$$

or
$$\frac{10}{10}=\frac{1.0125+0.00025\,l_1}{1.0000+0.025-0.00025\,l_1}\ \text{or}\ l_1=25\text{ cm}.$$

Thus the balance is obtained at 25 and 75 scale divisions.

*Balance arms equal to true values :*

Now in this case, $P = 9.05\ \Omega$ and $Q = 10.05\ \Omega.$

Under balance conditions :

$$\frac{9.95}{10.05}=\frac{1.0125+0.00025\,l_1}{1.0000+0.025-0.000025\,l_1}$$

or
$$\frac{1-0.005}{1+0.005}=\frac{1.0125+0.00025\,l_1}{1.0250-0.00025\,l_1}$$

or $l_1 = 25$ cm.

Thus the balance is obtained at 5 and 95 cm.

## 14.3 MEASUREMENT OF LOW RESISTANCE

The methods used for measurement of medium resistances are unsuitable for measurement of low resistances *i.e.*, resistances having a value under 1Ω. The reason is that the resistance of leads and contacts, though small, are appreciable in comparison in the case of low resistances. For example, a contact resistance of 0.002 Ω causes a negligible error when a resistance of 100 Ω is being measured but the same contact resistance would cause an error of 10% if a low resistance of the value of 0.02 Ω is measured. Hence special type of construction and techniques have to be used for the measurement of low resistances in order to avoid serious errors occurring on account of the factors mentioned above.

Low resistances are constructed with four terminals as shown in Fig. 14.15. One pair of terminals CC′ (called the current terminals) is used to lead current to and from the resistor. The voltage drop is measured between the other two terminals PP′, called the potential terminals.

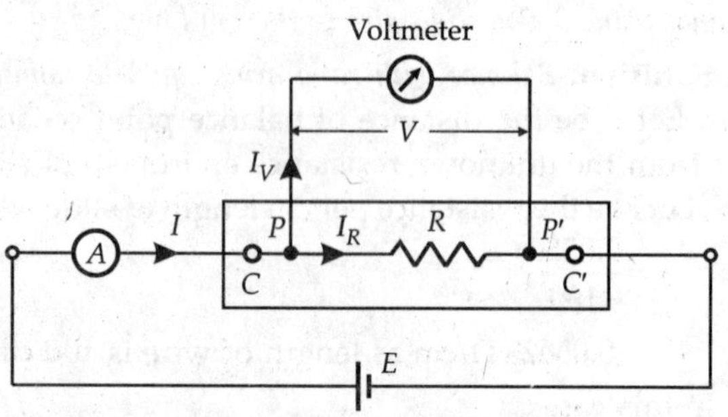


**Fig. 14.15** Ammeter voltmeter method for measuring 4 terminal resistances.

The voltage $V$, indicated in Fig. 14.15, is thus $I_R$ times the resistance $R$ between terminals $PP'$ and does not include any contact resistance drop that may be present at the current terminals $CC'$.

Resistors of low values are thus measured in terms of resistance, between potential terminals, which becomes perfectly and precisely definite in value and is independent of the contact resistance drop at the current terminals. Contact resistance drop at the potential terminals need not be a source of error, as current crossing at these terminals is usually extremely small or even zero for null methods. Also this contact resistance now becomes a part of the potential circuit and is, therefore, a negligible part of the total resistance of the potential circuit since potential circuits have a high value of resistance.

### 14.3.1 Methods for Measurement of Low Resistance

The methods for measurement of low resistance are :

1. Ammeter voltmeter method. (This method has been explained in Art. 14.2.1 on page 421)
2. Kelvin's double bridge method.
3. Potentiometer method (This is explained in Chapter 15).

### 14.3.2 Kelvin Double Bridge Method of Measurement of Low Resistances

The Kelvin bridge is a modification of the Wheatstone bridge and provides greatly increased accuracy in measurement of low value resistances. An understanding of the Kelvin bridge arrangement may PV be obtained by a study of the difficulties that arise in a Wheatstone bridge on account of the resistance of the leads and the contact resistances while measuring low valued resistors.

Consider the bridge circuit shown in Fig. 14.15 where $r$ represents the resistance of the lead that connects the unknown resistance $R$ to standard resistance $S$. Two galvanometer connections indicated by dotted lines, are possible. The connection may he either to point '$m$' or to point '$n$'. When the galvanometer is connected to point $m$, the resistance, $r$, of the connecting leads is added to the standard resistance, $S$, resulting in indication of too low an indication for unknown resistance $R$. When the connection is made to point $n$, the resistance, $r$, is added to the unknown resistance resulting in indication of too high a value for $R$.

Suppose that instead of using point $m$, which gives a low result, or $n$, which makes the result high, we, make the galvanometer connection to any intermediate point '$d$' as shown by full line in Fig. 14.16. If at point '$d$' the resistance $r$ is divided into two parts, $r_1$ and $r_2$, such that

$$\frac{r_1}{r_2} = \frac{P}{Q} \qquad \text{...(14.36)}$$

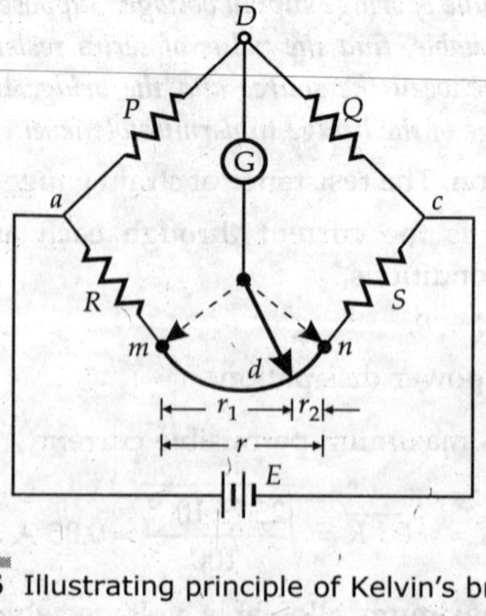


**Fig. 14.16** Illustrating principle of Kelvin's bridge.

Then the presence of $r_1$ the resistance of connecting leads, causes no error in the result. We have,

$$R + r_1 = \frac{P}{Q}\cdot(S + r_2) \quad \text{but} \quad \frac{r_1}{r_2} = \frac{P}{Q} \qquad \text{...(14.37)}$$

or $$\frac{r_1}{r_1+r_2}=\frac{P}{P+Q}$$

or $$r_1=\frac{P}{P+Q}\cdot r$$

as $r_1+r_2=r$ and $r_2=\dfrac{Q}{P+Q}\cdot r$

∴ We can write Eqn. 14.37 as

$$\left(R+\frac{P}{P+Q}r\right)=\frac{P}{Q}\left(S+\frac{Q}{P+Q}r\right)$$

or $$R=\frac{P}{Q}\cdot S \qquad \text{...(14.38)}$$

Therefore we conclude that making the galvanometer connection as at $c$, the resistance of leads does not affect the result.

The process described above is obviously not a practical way of achieving the desired result, as there would certainly be a trouble in determining the correct point for galvanometer connections. It does, however, suggest the simple modification, that two actual resistance units of correct ratio be connected between points $m$ and $n$, the galvanometer be connected to the junction of the resistors. This is the actual Kelvin bridge arrangement, which is shown in Fig. 14.17.

The Kelvin double bridge incorporates the idea of a second set of ratio arms – hence the name double bridge – and the use of four terminal resistors for low resistance arms. Figure 14.17 shows the schematic diagram of the Kelvin bridge. The first of ratio arms is $P$ and $Q$. The second set of ratio arms, $p$ and $q$ is used to connect the galvanometer to a point $d$ at the appropriate potential between points $m$ and $n$ to eliminate the effect of connecting lead of resistance $r$ between the known resistance, $R$, and the standard resistance, $S$.

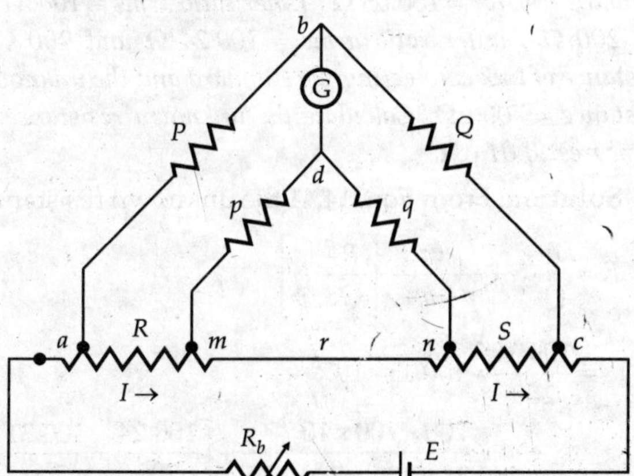


**Fig. 14.17** Kelvin double bridge

The ratio $p/q$ is made equal to $P/Q$. Under balance conditions there is no current through the galvanometer, which means that the voltage drop between $a$ and $b$, $E_{ab}$ is equal to the voltage drop $E_{amd}$ between $a$ and $c$.

Now $$E_{ab}=\frac{P}{P+Q}E_{ac}$$

and $$E_{ac}=I\left[R+S+\frac{(p+q)r}{p+q+r}\right] \qquad \text{...(14.39)}$$

and $$E_{amd}=I\left[R+\frac{p}{p+q}\left\{\frac{(p+q)r}{p+q+r}\right\}\right]$$

$$=I\left[R+\frac{pr}{p+q+r}\right] \qquad \text{...(14.40)}$$

For zero galvanometer deflection,

$$E_{ab}=E_{amd}$$

or $$\frac{P}{P+Q}I\left[R+S+\frac{(p+q)r}{p+q+r}\right]=I\left[R+\frac{pr}{p+q+r}\right]$$

or $$R=\frac{P}{Q}\cdot S+\frac{qr}{p+q+r}\left[\frac{P}{Q}-\frac{p}{q}\right] \qquad \text{...(14.41)}$$

Now, if $P/Q=p/q$, Eqn. 14.41 becomes,

$$R=\frac{P}{Q}\cdot S. \qquad \text{...(14.42)}$$

Equation 14.42 is the usual working equation for the Kelvin bridge. It indicates that the resistance of connecting lead, $r$, has no effect on the measurement, provided that the two sets of ratio arms have equal ratios. Equation 14.41 is useful, however, as it shows the error that is introduced in case the ratios are not exactly Ω equal. It indicates that it is desirable to keep $r$ as small as possible in order to minimize the errors in case there is a difference between ratios $P/Q$ and $p/q$.

The effect of thermo-electric emfs can be eliminated by making another measurement with the battery connections reversed. The true value of $R$ being the mean of the two readings.

In a typical Kelvin bridge, the range of resistance covered is 0.1 μΩ to 1.0 Ω.

The accuracies are as under :

From 1000 μΩ to 1.0 Ω : 0.05%.

From 100 μΩ to 1000 μΩ : 0.2% to 0.05%.

From 10 μΩ to 100 μΩ : 0.5%. to 0.2,

limited by thermoelectric emfs.

In this bridge there are four internal resistance standards of 1 Ω, 0.1 Ω, 0.01 Ω and 0.001 Ω respectively.

### 14.3.3 Kelvin Bridge Ohmmeter

This is a modified form of the Kelvin bridge and is intended for the rapid measurement of the winding resistances of machines and transformers, and for the measurement of contact and earth conductor resistances. The accuracy is of the order of ± 0.2%. This instrument is direct reading and the balance is obtained by rotating a single dial.

Figure 14.18 gives the circuit diagram of a typical Kelvin bridge ohmmeter. The ratio arms $P/Q$ and $p/q$ of Fig. 14.17 are replaced by a combination of fixed resistors $P$ and $p$ and a double slide wire enabling the bridge ratio to be varied continuously between values 10/1 to 200/1.

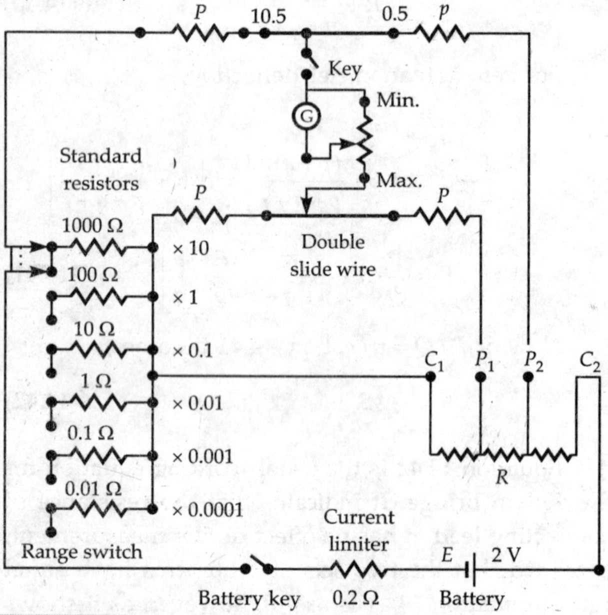


**Fig. 14.18** Kelvin bridge ohmmeter.

Six standard resistors are used and these range from 0.01 Ω to 1000 Ω. The ratio dial is calibrated from 0.5 Ω to 10.5 Ω on the × 1 range using 100 Ω standard resistor. The over all range of the instrument is from 0.00005 Ω to 105 Ω.

### 14.3.3 Unbalanced Kelvin Bridge

The galvanometer current of unbalanced Kelvin bridge can be found by the same Thevenin method as was used for Wheatstone bridge. The Thevenin source generator voltage, referring to Fig. 14.17 is :

$$E_0 = E_{ab} - E_{amd}$$

where $$E_{ab} = \frac{P}{P+Q} \cdot E_{ac} \qquad \text{...(14.43)}$$

$$E_{amd} = \frac{R + \dfrac{pr}{p+q+r}}{R + S + \dfrac{(p+q)r}{p+q+r}} E_{ab} \qquad \text{...(14.44)}$$

The circuit for the Thevenin equivalent resistance as seen from galvanometer terminals $c$ and $d$ is shown in Fig. 14.19($a$) where $R_b$ is the resistance in the battery circuit.

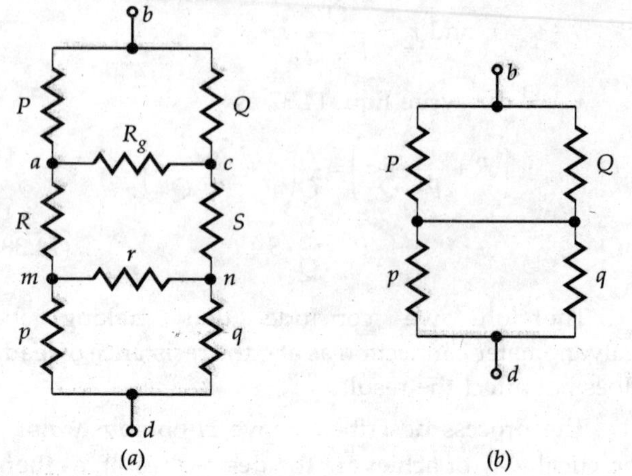


**Fig. 14.19** Equivalent circuit for calculation of resistance of Kelvin bridge.

However, $R, S, r$ and $R_b$, are usually very small and may be neglected with no appreciable-loss, in accuracy. The equivalent circuit then reduces to the one as shown in Fig. 14.19($b$). Therefore, Thevenin equivalent resistance

$$R_0 = \frac{PQ}{P+Q} + \frac{pq}{p+q} \qquad \text{...(14.45)}$$

The galvanometer current

$$I_g = \frac{E_0}{R_0 + G} \qquad \text{...(14.46)}$$

***Example 14.15*** *A 4 terminal resistor of approximately 50 μΩ resistance was measured by means of a Kelvin bridge having the following component resistances :*

*Standard resistor = 100.03 Ω ; Inner ratio arms = 100.31 Ω and 200 Ω ; outer ratio arms = 100.24 Ω and 200 Ω ; resistance of link connecting the standard and the unknown resistance = 700 μΩ. Calculate the unknown resistance to the nearest 0.01 μΩ.*

**Solution.** From Eqn. 14.41 the unknown resistance

$$R = \frac{P}{Q} \cdot S + \frac{qr}{p+q+r}\left[\frac{P}{Q} - \frac{p}{p}\right]$$

$$= \frac{100.24}{200} \times 100.03 \times 10^{-6}$$

$$+ \frac{200 \times 700 \times 10^{-6}}{100.31 + 200 + 700 \times 10^{-6}}\left[\frac{100.24}{200} - \frac{100.31}{200}\right]$$

$$= 49.97 \times 10^{-6}\,\Omega = 49.97\ \mu\Omega.$$

***Example 14.16*** *A Kelvin double bridge (shown in Fig. 14.17) each of the ratio arms $P = Q = p = q = 1000\ \Omega$. The emf of the battery is 100 V and a resistance of $5\ \Omega$ is included in the battery circuit. The galvanometer has a resistance of 500 Ω and the resistance of the link connecting the unknown resistance to the standard resistance may be neglected. The bridge is balanced when the standard resistance $S = 0.001\ \Omega$.*

*(a) Determine the value of unknown resistance.*

*(b) Determine the current (approximate value) through the unknown resistance R at balance.*

*(c) Determine the deflection of the galvanometer when the unknown resistance, R, is changed by 0.1 percent from its value at balance. The galvanometer has a sensitivity of 200 mm/μA.*

**Solution.** (a) At balance, the value of unknown resistance

$$R = \frac{P}{Q} \cdot S = \frac{1000}{1000} \times 0.001 = 0.001\ \Omega.$$

(b) If we examine the Kelvin bridge circuit, we find the resistors P, Q and p, q are in parallel with the resistance of link, r. Since r is negligible and P, Q, p and q have large values, the effect of ratios arms can be neglected for the purpose of calculation of current.

∴ Current under balance conditions

$$I = \frac{E}{R_b + R + S} = \frac{100}{5 + 0.01 + 0.0001} \approx 20\text{ A}.$$

where $R_b$ = resistance in the battery circuit.

(c) The value R is changed by 0.1 per cent.

∴ New value of $R = 1.001 \times 0.001 = 0.001001\ \Omega$.

Voltage between points a and c (Fig. 14.17)

$$E_{ad} = \frac{R+S+r}{R_b + R + S + r} \cdot E = \frac{R+S}{R_b+R+S} E \text{ as } r = 0$$

$$= \frac{R+S}{R_b} = \frac{0.001 + 0.001001}{5} \times 100 = 40 \times 10^{-3}$$

From Eqns. 14.43 and 14.44,

$$E_0 = E_{ab} - E_{amd}$$

$$= \left(\frac{P}{P+Q} - \frac{R}{R+S}\right) E_{ac}$$

$$= \left(\frac{1000}{1000+1000} - \frac{0.001001}{0.001 + 0.001001}\right) \times 40 \times 10^{-3}$$

$$= 0.01 \times 10^{-3}\text{ V} = 0.01\text{ mV. (considering } r = 0).$$

Since $R, S, r$ and $R_b$ are quite small as compared to $P, Q, p$ and $q$, we can use circuit of Fig. 14.19(b). For the calculation of internal resistance, $R_0$ as viewed from terminals d and b.

$$R_0 = \frac{PQ}{P+Q} + \frac{pq}{p+q}$$

$$= \frac{1000 \times 1000}{1000 + 1000} + \frac{1000 \times 1000}{1000 + 1000} = 1000\ \Omega$$

Galvanometer current

$$I_g = \frac{E_0}{R_0 + G} = \frac{0.01 \times 10^{-3}}{1000 + 500}$$

$$= 0.0067 \times 10^{-6}\text{ A} = 0.0067\ \mu\text{A}.$$

∴ Deflection of galvanometer

$$\theta = S_i I_g = 200 \times 0.0067 = 1.34\text{ mm}.$$

## 14.4 MEASUREMENT OF HIGH RESISTANCE

High resistance of the order of hundreds or thousands of megohm are often encountered in electrical equipment, and frequently must be measured.

*Common examples are :*

(i) Insulation resistance of components and built up electrical equipment like machines and cables ;

(ii) Resistance of high resistance circuit elements like in vacuum tube circuits ;

(iii) Leakage resistance of capacitors ;

(iv) Volume resistivity of a material, *i.e.*, the resistance between two faces of unit area separated by unit distance with all conduction from face to face being through the body of the material ;

(v) Surface resistivity, *i.e.*, the resistance between two lines of unit length and unit distance apart, the lines being on the surface of the material and all conduction being on the surface.

### 14.4.1 Difficulties in Measurement of High Resistances

High accuracy is rarely required in such measurements, hence simple circuits are used. Since the resistances under measurement have high values, very small currents are encountered in the measurement circuits. This aspect leads to several difficulties :

(i) The insulation resistance of the resistor may be comparable with the actual value of the resistor. Thus leakage currents are produced. These leakage currents are of comparable magnitude to the current being measured and must be eliminated from the measurement. Leakage currents no doubt introduce errors, but they generally vary from day to day, depending upon the humidity conditions and therefore cause additional unpredictable complications.

14

(ii) Due to electrostatic effect, stray changes can appear in the measuring circuit causing errors. Alternating fields can also effect the measurements considerably. Therefore. critical points of the measuring circuit must be carefully screened.

(iii) In order to obtain definite ratios in the potential distribution with respect to surroundings, one point of the circuit may be connected to earth for accuracy in measurements.

(iv) In measurement of insulation resistance the specimen often has considerable capacitance. On application of a direct voltage a large charging current flows initially which gradually decays down after a short interval. Further, insulating materials possess the property of dielectric absorption. *i.e.*, after the main charging current has decayed down, further charge is slowly absorbed over a considerable period of time, perhaps for minutes or even hours. Thus measurement of true conduction current should be delayed until after the cessation of the charging and absorbing currents. But since the absorbing currents take a considerably long time to decay, it is usually inevitable that the conduction current measured includes some absorption in current. The testing conditions, including the time between the application of voltage and observation of the current, must be specified.

(v) When measuring the resistance of low conductivity conductors, insulating materials and products, the effect of various factors upon their resistance should be taken into account. Thus, a change in the temperature of cardboard from 20° to 40°C is accompanied by a 13 fold change in its resistance, changes in humidity form 10 to 60 percent cause a 30 fold change in resistance of porcelain. Besides temperature and humidity, the kind of current employed for measurement, the magnitude and duration of the applied voltage, and other factors also effect the resistance being measured.

(vi) Fairly high voltages are used in tests in order to raise the currents to reasonable values in order to be measured. So normally a sensitive galvanometer or micro-ammeter is required and adequate steps have to be taken to prevent damage to these delicate instruments.

A voltage supply of 100 V upto a few kV is often used depending upon the nature and breakdown voltage of the test object. The powe supply unit is d.c. transistorized source. Proper smoothing and stabilization circuits are used to ensure constancy of voltage with time.

### 14.4.2 Use of Guard Circuit

Some form of guard circuits are generally used to eliminate the errors caused by leakage currents over insulation. Figure 14.20 illustrates the operation of a guard circuit. In Fig. 14.20(*a*), a high resistance mounted on a piece of insulating material is measured by the ammeter-voltmeter method. The micro-ammeter measures the sum of the current through the resistor ($I_R$) and the current through the leakage path around the resistor ($I_L$). The measured value of resistance computed from the readings indicated on the voltmeter and the micro-ammeter, will not be true value but will be in error. In Fig. 14.20(*b*) guard terminal has been added to resistance terminal block. The guard terminal surrounds the resistance terminal entirely and is connected to the battery side of the micro-ammeter. The leakage current $I_L$, now bypasses the micro-ammeter which then indicates the current $I_R$ through the resistor and thus allows the correct determination of the resistance value from the readings of voltmeter and micro-ammeter. The guard terminal and resistance terminal are almost at the same potential and thus there will be no flow of current between them.

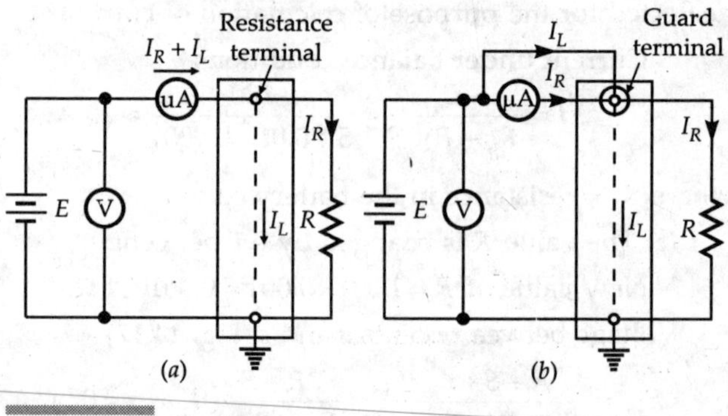


**Fig. 14.20** Application of guard circuit for measurement of high resistance.

### 14.4.3 Methods for Measurement of High Resistance

The different methods employed are :

1. Direct deflection method.
2. Loss of charge method:
3. Megohm bridge.
4. Meggar (This is explained in Art. 9.8, page 256).

***Direct deflection method.*** The direct deflection method is basically that of Fig. 14.20(*b*). For high resistances, such as insulation resistance of cables, a sensitive galvanometer of '*d*' Arsonval type (usually having a current sensitivity of at 1000 mm/μA at a scale distance of 1 metre) is used in place of the microammeter. In fact many sensitive type of galvanometers can detect currents from 0.1 — 1 nA. Therefore, with an applied voltage of 1 kV, resistances as high as $10^{12}$ to $10 \times 10^{12}$ Ω can be measured.

An illustration of the direct deflection method used for measuring insulation resistance of a cable is shown in Fig. 14.21. The galvanometer G, measures the current $I_R$ between the conductor and the metal Sheath. The leakage current $I_L$, over the insulating material is carried by the guard wire wound on the insulation and therefore does not flow through the galvanometer.

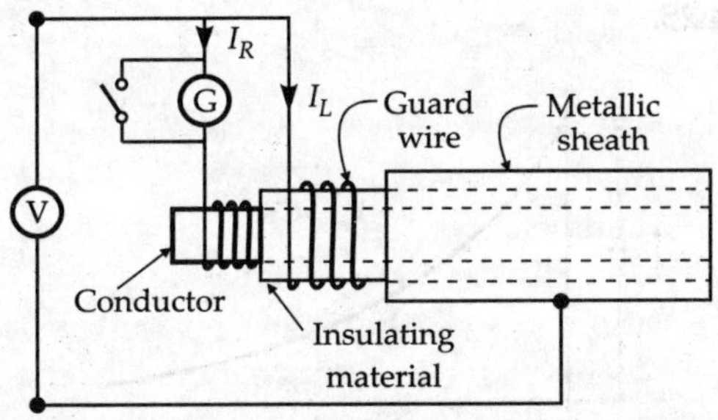


**Fig. 14.21** Measurement of insulation resistance of cable having sheath.

Cables without metal sheaths can be tested in a similar way if cable, except the end or ends on which corrections are made, is immersed in water in a tank. The water and the tank then form the return path for the current. The cable is immersed in slightly saline water for about 24 hours and the temperature is kept constant (at about 20°C) and then the measurement is taken as in Fig. 14.22.

The insulation resistance of the cable $R = V / I_R$.

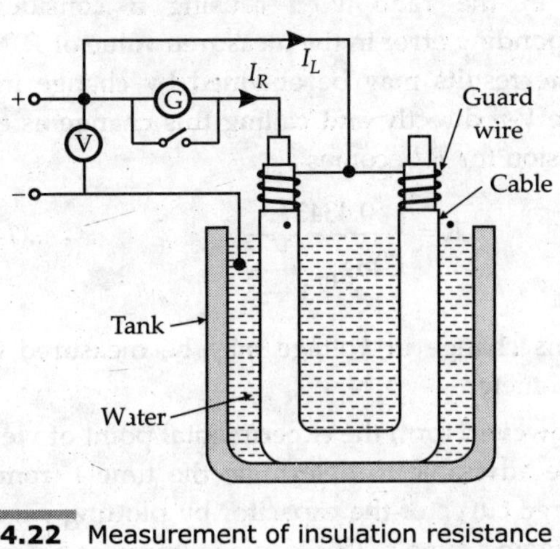


**Fig.14.22** Measurement of insulation resistance of cable having no conducting sheath.

In some cases, the deflection of the galvanometer is observed and its scale is afterwards calibrated by replacing the insulation by a standard high resistance (usually 1 MΩ), the galvanometer shunt being varied, as required to give a deflection of the same order as before.

In tests on cables the galvanometer should be short-circuited before applying the voltage. The short circuiting connection is removed only after sufficient time has elapsed so that charging and absorption currents cease to flow. The galvanometer should be well shunted during the early stages of measurement, and it is normally desirable to include a protective series resistance (of several megohm) in the galvanometer circuit. The value of this resistance should be subtracted from the observed resistance value in order to determine the true resistance. A high voltage battery of 500 V emf is required and its emf should remain constant throughout the test.

***Measurement of volume and surface resistivity.*** The direct deflection method is often used for measurement of insulation resistance of insulating material samples available in sheet form. In such cases we are interested in the measurement of volume resistivity and the surface resistivity of the material.

Figure 14.23 shows the schematic diagram for measurement of volume and surface resistivities of a specimen of insulating material. The specimen is provided with tin foil or colloidal graphite electrodes ; the upper electrode having a guard ring. For measurement of volume resistivity (which in fact is the specific resistance) readings of voltage applied and the current through the galvanometer are taken. Leakage currents over the edge of the specimen will flow between the guard ring and the lower electrode and hence will not introduce error into the measurement. The volume resistivity, ρ, can be calculated as follows :

Let $d_1$ = diameter of the upper electrode $d_1$,

$r$ = thickness of the specimen sheet,

$V_1$ = reading of voltmeter,

and $I_1$ = current through galvanometer $G_1$.

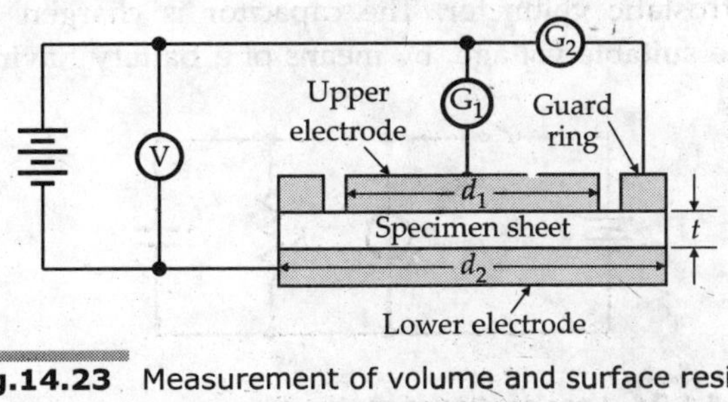


**Fig.14.23** Measurement of volume and surface resistivities of insulating material specimen.

14

∴ Resistance of specimen

$$R = V_1 / I_1. \quad \text{But } R = \rho t / \pi d_1^2$$

∴ Volume resistivity of specimen,

$$\rho = \frac{\pi d_1^2}{t} = \frac{\pi d_1^2 V_1}{t I_1} \qquad ...(14.47)$$

The resistivity of a thin layer of dielectric materials is different from volume resistivity, not only because of an adherent humidity layer but also because of contamination, chemical alterations, absorption of gases, or structural modification. The resistance $R_t$ between two electrodes embedded in or attached to a dielectric medium is composed of volume resistance $R_v$ and surface resistance $R_s$ with $1/R_t = 1/R_v + 1/R_s$.

The volume resistance, $R_v$ can be measured separately from surface resistance $R_s$ by means of guard rings as shown in Fig 14.20.

If we want to measure surface resistivity, the galvanometer is placed in position $G_2$. In this position the galvanometer measures the leakage current and current flowing between upper and lower electrodes will be eliminated from measurement. Let

$d_2$ = diameter of lower electrode disc,

$V_2$ = reading of voltmeter

and $I_2$ = current through galvanometer $G_2$.

∴ Surface resistance

$$R_s = V_2 / I_2.$$

The leakage current flows along a path of length $t$ and width $\pi d_2$ and therefore, surface resistivity,

$$\rho_s = \frac{R_s \times \pi d_2}{t} = \frac{\pi d_2}{t} \cdot \frac{V_2}{I_2} \qquad ...(14.48)$$

Other forms of specimen and electrodes are also used. For example, the electrodes and guard ring may be mercury, either placed in specially machined recesses, in moulded insulating materials, or retained by metal rings on the surface of sheet materials.

***Loss of charge method.*** In this method, (Fig. 14.24) the insulation resistance $R$ to be measured is connected in parallel with a capacitor $C$ and a electrostatic voltmeter. The capacitor is charged to some suitable voltage, by means of a battery having

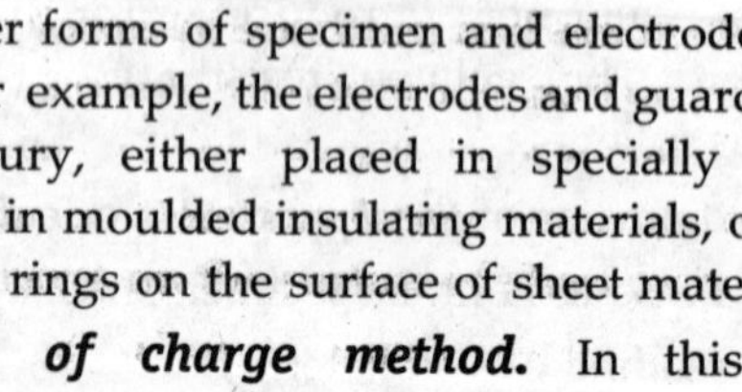


**Fig. 14.24** Loss of charge method.

voltage $V$ and is then allowed to discharge through the resistance. The terminal voltage is observed over a considerable period of time during discharge.

The voltage across the capacitor at any instant $t$ after the application of voltage is

$$V = V \exp(-t/CR)$$

or $\quad V/v = \exp(-t/CR)$

or Insulation resistance

$$R = \frac{t}{C \log_e V/v} = \frac{0.4343\, t}{C \log_{10} V/v} \qquad ...(14.49)$$

The variation of voltage $v$ with time shown in Fig. 14.25.

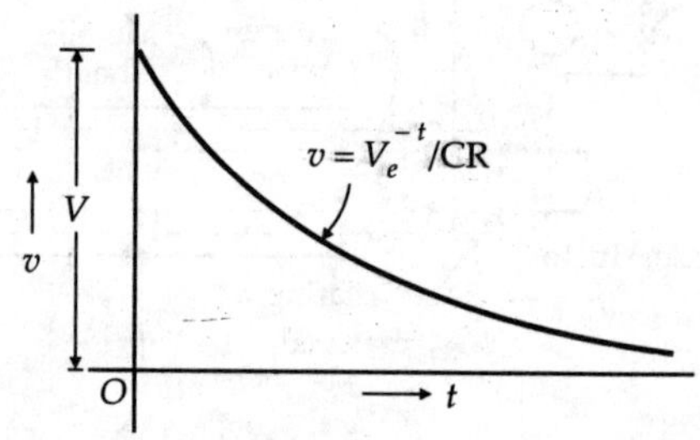


**Fig. 14.25** Variation of voltage with time.

From Eqn. 14.49 it follows that if $V, v, C$ and $t$ are known the value of $R$ can be computed.

If the resistance $R$ is very large the time for an appreciable fall in voltage is very large and thus this process may become time-consuming. Also the voltage-time curve will thus be very flat and unless great care is taken in measuring voltages at the beginning and end of the time $r$, a serious error may be made in the ratio $V/v$ causing a considerable corresponding error in the measured value of $R$. More accurate results may be obtained by change in the voltage $V$-$v$ directly and calling this change as $e$, the expression for $R$ becomes :

$$R = \frac{0.4343\, t}{C \log_{10} \dfrac{V}{V-e}} \qquad ...(14.50)$$

This change in voltage may be measured by a galvanometer.

However, form the experimental point of view, it may be advisable to determine the time $t$ from the discharge curve of the capacitor by plotting curve of $\log_e v$ against time $t$. This curve is linear as shown in Fig. 14.26 and thus determination of time $t$ from this curve for the voltage to fall from $V$ to $v$ yields more accurate results.

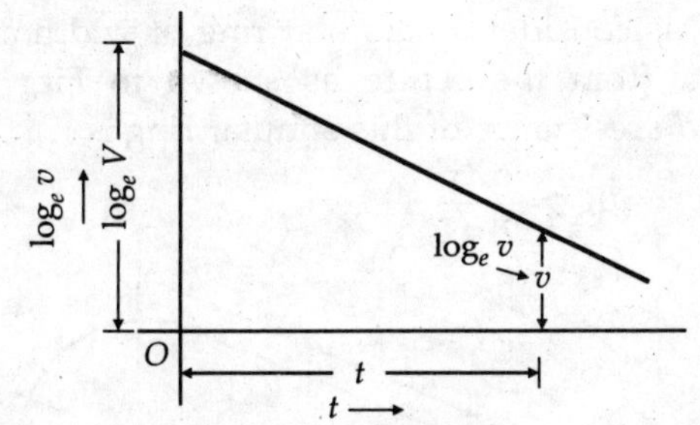


**Fig. 14.26** Plot of $\log_e v$ versus time.

This method is applicable to some high resistances, but it requires a capacitor of a very high leakage resistance as high as the resistance being measured. The method is very attractive if the resistance being measured is the leakage resistance of a capacitor as in this case auxiliary R and C units are not required.

Actually in this method we do not measure the true value of resistance since we assume here that the value of resistance of electrostatic voltmeter and the leakage resistance of the capacitor have infinite value. But in practice corrections must be applied to take into consideration the above two resistances. Figure 14.27 shows the actual circuit of the test where $R_1$ represents the leakage resistance of capacitor. Then if $R'$ is the resistance of $R_1$ and $R$ in parallel the discharge equation for capacitance gives

$$R' = \frac{0.4343\, t}{C \log_{10} V/v} \qquad \text{...(14.51)}$$

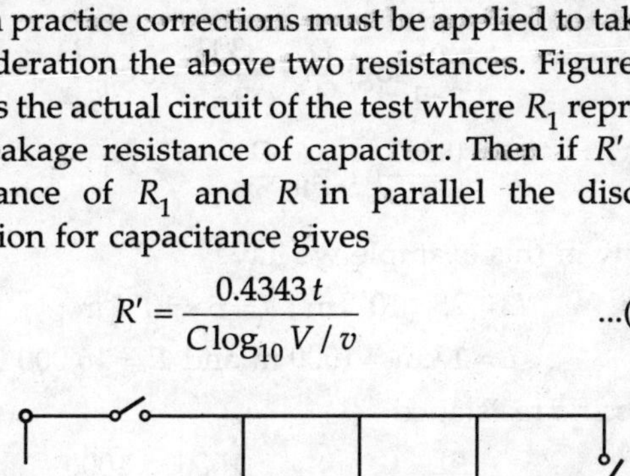


**Fig. 14.27** Loss of charge method considering effects of leakage resistance of capacitor.

The test is then repeated with the unknown resistance R, disconnected and the capacitor discharging through $R_1$. The value of $R_1$ obtained from this second test and substituted into the expression

$$R' = \frac{R R_1}{R + R_1} \qquad \text{...(14.52)}$$

in order to get value of R.

The leakage resistance of the voltmeter, unless very high should also be taken into consideration.

***Megohm bridge method.*** Figure 14.28(*a*) shows a very high resistance *R* with its two main terminals *A* and *B*, and a guard terminal, which is put on the insulation. This high resistance may be diagramatically represented as in Fig. 14.28(*b*). The resistance *R* is between main terminals *A* and *B* and the leakage resistances $R_{AG}$ and $R_{BG}$ between the main terminals *A* and *B* of from a "Three terminal resistance".

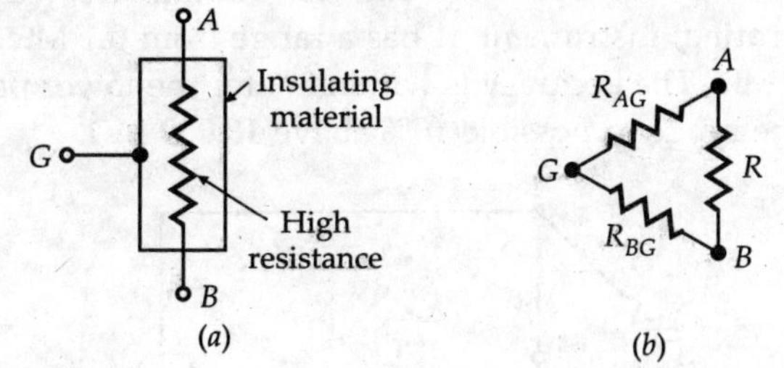


**Fig. 14.28** Three Terminal Resistances.

Let us consider the hypothetical case of a 100 MΩ resistance. We assume that each of the leakage resistances is 100 MΩ *i.e.*, $R_{AG} = R_{BG} = 100$ MΩ. Let this resistance be measured by an ordinary Wheatstone bridge as shown in Fig. 14.29(*a*). It is clear that the Wheatstone bridge will measure a resistance of $\frac{100 \times 200}{100 + 200} = 67$ MΩ instead of 100 MΩ thus giving an error of 33 percent.

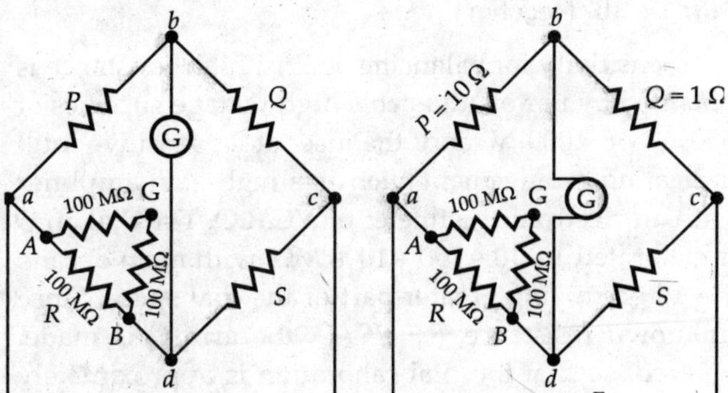


**Fig. 14.29** Use of guard circuit for measurement of high resistances.

However if the same resistance is measured by a modified Wheatstone bridge as shown in Fig. 14.29(*b*) with the guard connection G connected as indicated, the error in measurement is considerably reduced. For the arrangement shown in Fig. 14.29(*b*) resistance $R_{BG}$ is put in parallel with the galvanometer and thus it has no effect on the balance and only effects the sensitivity of the galvanometer slightly. The resistance $R_{AG} = 100$ MΩ is put in parallel with a resistance $P = 100$ kΩ and therefore for the arrangement shown the measured value has an error of only 0.01 percent and this error is entirely negligible for measurements of this type.

The arrangement of Fig. 14.30 illustrates the operation of a Megohm bridge.

Figure 14.30 shows the circuit of a completely self-contained Megohm bridge which includes power supplies, bridge a members, amplifiers, and indicating, instrument. It has a range from 0.1 MΩ to $10^6$ MΩ. The accuracy is within 3% for the lower part of the range to possible 10% above 10,000 MΩ.

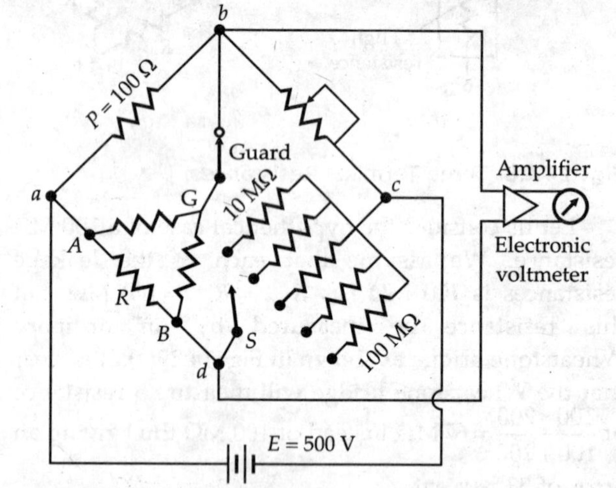


**Fig. 14.30** Megohm bridge.

Sensitivity for balancing against high resistance is obtained by use of adjustable high voltage supplies of 500 V or 1000 V and the use of a sensitive null indicating arrangement such as a high gain amplifier with an electronic voltmeter or a C.R.O. The dial on $Q$ is calibrated 1 – 10 – 100 – 1000 MΩ, with main decade 1 – 10 occupying greater part of the dial space. Since unknown resistance $R = PS/Q$, the arm $Q$ is made, tapered, so that the dial calibration is approximately logarithmic in the main decade, 1 – 10. Arm $S$ gives five multipliers, 0.1, 1, 10, 100 and 1000.

The junction of ratio arms $P$ and $Q$ is brought on the main panel and is designated as 'Guard' terminal.

***Example 14.17*** *Derive an expression for insulation resistance of single core cable. The conductor of a cable has a diameter of 5 mm and the overall diameter of the cable is 25 mm. If the insulation resistance of the cable is 16,000 Ω/km, calculate the specific resistance of insulating material.*

**Solution.** Let

$d =$ diameter of conductor,

$D =$ diameter of insulated cable,

$L =$ length of cable, and

$\rho =$ resistivity of insulating material.

Let us consider an annular ring of width $dx$ at a radius $x$ from the centre as shown in Fig. 14.31. Insulation resistance of this annular ring is :

$$dR_x = \frac{\rho\, dx}{2\pi x L}$$

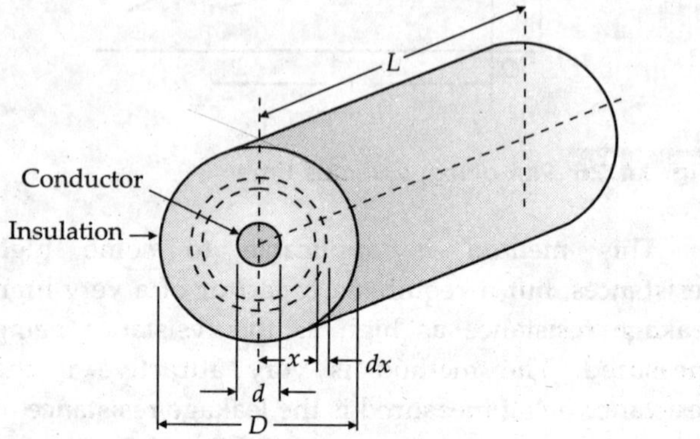


**Fig. 14.31** Insulation resistance of cable.

Total insulation resistance of cable

$$R = \int_{d/2}^{D/2} \frac{\rho\, dx}{2\pi x L}$$

$$= \frac{\rho}{2\pi L}\log_e\frac{D}{d} = \frac{2.303\rho}{2\pi L}\log_{10}\frac{D}{d}$$

$$= \frac{0.367}{L}\rho\log_{10}\frac{D}{d}$$

Now in this example we have,

$$D = 25\times10^{-3}\text{ m} ;\ d = 5\times10^{-3}\text{ m} ;$$

$$L = 1\text{ km} = 1000\text{ m and } R = 16{,}000\ \Omega.$$

Specific resistance

$$\rho = \frac{RL}{0.367\log_{10} D/d} = \frac{16000\times1000}{0.367(\log_{10}2.5/0.5)}$$

$$= 62.5\times10^{-6}\ \Omega\text{m}.$$

***Example 14.18*** *A test voltage is applied for several minutes between the conductor of a 400 metre length of cable and earth. The galvanometer connected in series reads 250 divisions, the value of universal shunt being 2.5 with a standard resistance of 1 MΩ in circuit, the scale reading is 350, the value of shunt being 1000. Calculate the insulation resistance of the cable. What would be the insulation resistance of the same cable of length 100 metre ?*

**Solution.** Since we have a universal shunt with the galvanometer and therefore the current through the circuit is proportional to the deflection of the galvanometer times universal shunt multiplier, Deflection of galvanometer with unknown resistance '$R$' is circuit 350 divisions and the shunt multiplier is 2.5.

∴ Current through the circuit with unknown resistance connected is :

$$I_R K \times 250 \times 2.5 = 625\ \text{K}$$

where $K$ is a constant.

Deflection of galvanometer with standard resistance '$S$' is circuit is 250 divisions and the shunt multiplier is 1000.

∴ Current through the circuit with standard resistance connected is :

$$I_s = K \times 350 \times 1000 = 350000\ \text{K}.$$

Now $I_R R = I_s S = E,$

where $E$ = voltage of the source.

∴ Insulation resistance of 400 metre long cable.

$$R = \frac{I_S}{I_R} \cdot S = \frac{3,50,000\ K}{625\ K} \times 1 = 560\ \text{M}\Omega.$$

The insulation resistance is inversely proportional to length of cable and therefore insulation resistance of 1000 metre long cable

$$= 560 \times (400/1000) = 224\ \text{M}\Omega.$$

**Example 14.19** *A length of cable is tested for insulation resistance by the loss of charge method. An electrostatic voltmeter of infinite resistance is connected between the cable conductor and earth, forming therewith a joint capacitance of 600 pF. It is observed that after charging the voltage falls from 250 V to 92 V in 1 minute. Calculate the insulation resistance of the cable.*

**Solution.** From Eqn. 14.49 insulation resistance of cable,

$$R = \frac{0.4343\, t}{C(\log_{10} V/v)} = \frac{0.4343 \times 60}{600 \times 10^{-12} (\log_{10} 250/92)}$$

$$= 100 \times 10^9\ \Omega = 100,000\ \text{M}\Omega.$$

**Example 14.20** *A cable is tested by loss of charge method using a ballistic galvanometer, with following results : Discharged immediately after electrification. deflection 200 divisions. Discharged after 30 s and after electrification : (i) deflection 126 divisions, (ii) when in parallel with a resistance of 10 MΩ, deflection 100 divisions. Calculate the insulation resistance of the cable.*

**Solution.** Suppose $R$ is the insulation resistance of the cable and let $R'$ be the resultant resistance of parallel combination of insulation resistance, $R$ and the 10 MΩ resistance.

$$\therefore\ R = \frac{0.4343 \times 30}{C \log_{10} 200/126} = \frac{0.4343 \times 30}{0.201 C}$$

and $$R' = \frac{0.4343 \times 30}{C \log_{10} 200/100} = \frac{0.4343 \times 30}{0.301 C}$$

Thus we have,

$$\frac{R'}{R} = \frac{0.201}{0.301} = 0.667 \quad \text{or} \quad R' = 0.667\ R.$$

Now $$R' = \frac{R \times 10}{R + 10} \quad \text{or} \quad 0.667\ R = \frac{R \times 10}{R + 10}$$

$$\therefore\ R = 5\ \text{M}\Omega$$

## 14.5 MEASUREMENT OF EARTH RESISTANCE

The provision of an earth electrode for an electrical system is necessitated by the following reasons :

**1.** All the parts of electrical equipment, like casings of machines. switches and circuit breakers, lead sheathing and armouring of cables, tanks of transformers, etc. which have to be at earth potential, must be connected to an earth electrode. The purpose of this is to protect the various parts of the installation, as well as the persons working against damage in case the insulation of a system fails at any point. By connecting V these parts to an earthed electrode a continuous low resistance path is available for leakage currents to flow to earth. This current operates the protective devices and thus the faulty circuit is isolated in case a fault occurs. 14

**2.** The earth electrode ensures that in the event of over voltage on the system due to lightning discharges or other system faults, those parts of equipment which are normally dead as far as voltages are concerned, do not attain dangerously high potentials.

**3.** In a three phase circuit the neutral of the system is earned in order to stabilize the potential of the circuit with respect to earth.

An earth electrode will only be effective so long it has a low resistance to the earth and can carry large currents without deteriorating. Since the amount of current which an earth electrode will carry is difficult to measure, the resistance value of the earth electrode is taken as sufficiently reliable indication of its effectiveness. The resistance of earth electrode should be low to give good protection and it must be measured.

The main factors on which the resistance of any earthing system depends are :

**1.** Shape and material of earth electrode or electrodes used.

**2.** Depth in the soil at which the electrodes are burried.

**3.** Specific resistance of soil surrounding and in the neighbourhood of electrodes. The specific resistance of the soil is not constant but varies from one type of soil to another. The amount of moisture present in the soil affects its specific resistance and hence the resistance of earth electrode is not a constant factor but suffers seasonal variations. This calls for periodic testing to ensure that the earthing system remains reasonably effective.

The specific resistance of soils varies between wide limits and is very much dependent upon its moisture content. Approximate figures for specific resistance of soil are $80 \times 10^3$ Ωm in for moist clay to $80 \times 10^6$ Ωm for sand of normal moisture content. A decrease of moisture content of 30% is capable of producing an increase of 300% to 400% in specific resistance. Thus it is necessary to make regular checks for earth resistance during the year round.

### 14.5.1 Methods of Measuring Earth Resistance

1. ***Fall of potential method.*** Figure 14.32 shows the circuit for measurement of earth resistance with fall of potential method. A current is passed through earth electrode an auxiliary electrode *B* (which is usually an iron spike) inserted in earth at a distance away from the earth electrode. A second auxiliary electrode *A* is inserted in earth between *E* and *B*. The potential difference *V* between *E* and *A* is measured for a given current *I*. The flow of ground currents is shown in Fig. 14.33(*c*).

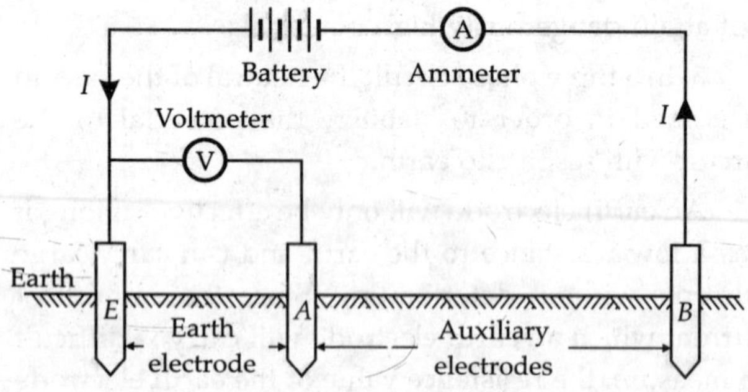


Fig. 14.32 Measurement of earth resistance by fall of potential method.

The lines of the first electrode current diverge and those of the second electrode current converge. As a result the current density is much greater in the vicinity of the electrodes than at a distance from them. The potential distribution between the electrodes is shown in Fig. 14.33(*b*). It is apparent from this curve that the potential rises in the proximity of electrodes *E* and *B* and is constant along the middle section. The resistance of earth, therefore, is

$$R_E = V / I \quad \text{or} \quad V_{EA} / I.$$

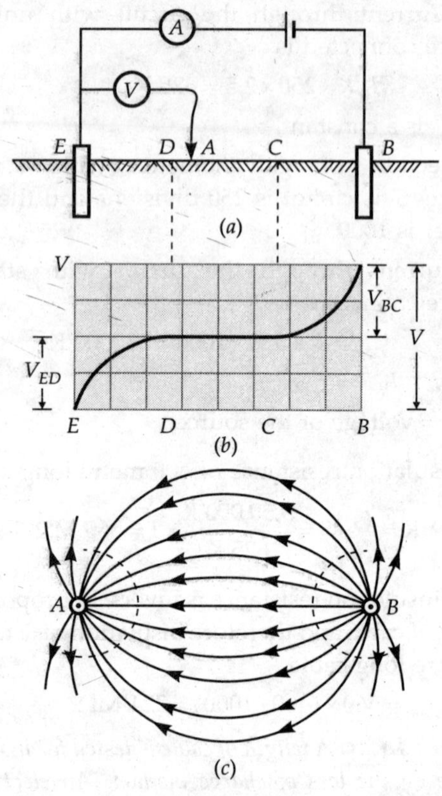


Fig. 14.33 Distribution of potentials between two earthing electrodes.

The position of electrodes *E* and *B* is fixed and the position of electrode *A* is changed and resistance measurements are done for various positions of electrode *A*.

A graph is plotted between earth resistance against the distance between electrode *E* and *A*. This graph is shown in Fig. 14.34.

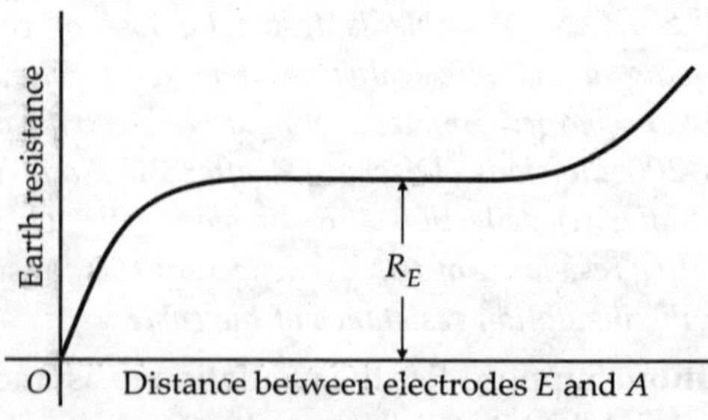


Fig. 14.34 Variation of earth resistance with distance between electrode *E* and *A*.

From Fig. 14.34, it is clear that the measured value of earth resistance depends upon the position of the auxiliary electrode *A*. The earth resistance rises rapidly initially. When the distance between earth electrode *E* and auxiliary electrode *A* is increased, it then becomes constant, and when the auxiliary electrode $A$ approaches the auxiliary electrode $B$, the resistance rises again. The placing of electrodes is thus very important and serious error may be caused by incorrect placing of the electrodes. The correct value of resistance of earth, $R_E$, is when the auxiliary electrode $A$ is at such a distance that the resistance lies on the flat part of curve of Fig. 14.30.

The spacing between the earth electrode $E$ and the auxiliary electrodes $A$, $B$ should be large so as to get proper results. The distance may be a few hundred metres in case the earth resistance is low.

**2. Earth tester.** The resistance of earth can be measured by an earth tester shown in Fig. 14.35. The "Earth Tester" is a special type of Meggar (See Art. 9.8 page 256) and it has some additional constructional features additional constructional features and they are :

(i) a rotating current reverser, and

(ii) a rectifier

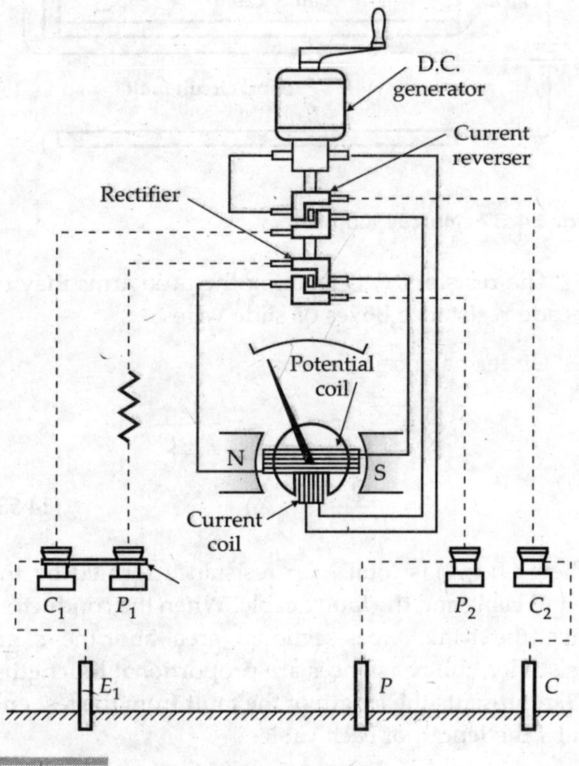


**Fig. 14.35** Earth tester.

Both these additional features consist of simple commutators made up of 'L' shaped segments. They are mounted on the shaft of the hand driven generator. Each commutator has four fixed brushes. One pair of each set of brushes is so positioned that they make contact alternately with one segment and then with the other as the commutator rotates. The second pair of each of set of brushes is positioned on the commutator so that continuous contact is made with one segment whatever the position of the commutator.

The earth tester has four terminals $P_1$, $P_2$. Two terminals $P_1$ and $C_1$ are shorted to form a common point to be connected to the earth electrode. The other two terminals $P_2$ and $C_2$ are connected to auxiliary electrodes $P$ and $C$ respectively.

The indication of the earth tester depends upon the ratio of the voltage across the pressure coil and the current through the coil. The deflection of its pointer indicates the resistance of earth directly. Although the "Earth Tester", which is a permanent magnet moving coil instrument and can operate on d.c. only, yet by including the reverser and the rectifying device it is possible to make measurements with a.c. flowing in the soil.

The sending of a.c. current through the soil has many advantages and therefore this system is used. The use of a.c. passing through the soil eliminates unwanted effects due to production of a back emf in the soil on account of electrolytic action. Also the instrument is free from effects of alternating or direct currents presents in the soil.

**Example 14.21** *On a 250 V supply a fault having a resistance of 20 Ω develops between the unearthed end of the winding of an electric heater and the frame. If the resistance of the substation earth electrode is 4 Ω that of human body 2000 Ω, and the safe maximum current through the body is 25 mA, what is the safe maximum resistance of consumer's earth electrode ?*

**Solution.** Figure 14.36 shows the diagrammatic representation of the problem.

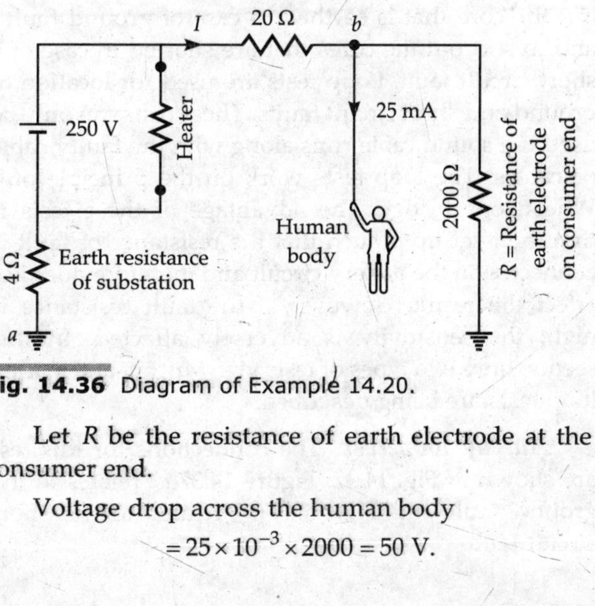


**Fig. 14.36** Diagram of Example 14.20.

Let $R$ be the resistance of earth electrode at the consumer end.

Voltage drop across the human body

$$= 25 \times 10^{-3} \times 2000 = 50 \text{ V}.$$

14

$\therefore$ Voltage between points $a$ and $b$

$$= 250 - 50 = 200 \text{ V}.$$

Current $\quad I = \dfrac{200}{20+4} = \dfrac{200}{24}$ A.

Current through human body

$$= I \frac{R}{R+2000} = 25 \times 10^{-3} \text{ A}.$$

or $$\frac{200}{24} \cdot \frac{R}{R+2000} = 25 \times 10^{-3}$$

or $$200\, R = 0.6 + 1200$$

or resistance of earth electrode on consumer end $R \approx 6\,\Omega$.

## 14.6 LOCALIZATION OF CABLE FAULTS

In this section, faults occurring in cables which are in use on lower distribution voltages are considered. The common faults which are likely to occur in such cables are :

1. *Ground fault.* The insulation of the cable may breakdown causing a flow of current from the core of the cable to the lead sheath or to the earth. This is called "Ground Fault" .
2. *Short circuit fault.* If the insulation between two conductors is faulty, a current flows between them. This is called a "Short Circuit Fault".

### 14.6.1 Methods Used for Localizing Ground and Short Circuit Faults

The methods used localizing ground and short circuit faults differ from those used for localizing open circuit faults.

In the case of multicore cables it is advisable, first of all, to measure insulation resistance of each core to earth and also between cores. This enables us to sort out the core that is earthed in case of ground fault ; and to sort out the cores that are shorted in case of a short circuit fault. Loop tests are used for location of ground and short circuit faults. These tests can only be used if a sound cable runs along with the faulty cable or cables. The loop tests work on the principle of a Wheatstone bridge. The advantage of these tests is that their set up is such that the resistance of fault is connected in the battery circuit and therefore does not effect the result. However, if the fault resistance is high, the sensitivity is adversely affected. In this section only two types of tests viz., Murray and Varley loop tests are being described.

***Murray loop test.*** The connections for this test are shown in Fig. 14.37. Figure 14.37(*a*) relates to the ground fault and Fig. 14.37(*b*) relates to the short circuit fault.

In both cases, the loop circuit formed by the cable conductors is essentially a Wheatstone bridge consisting of resistances $P$, $Q$, $R$ and $X$. $G$ is a galvanometer for indication of balance.

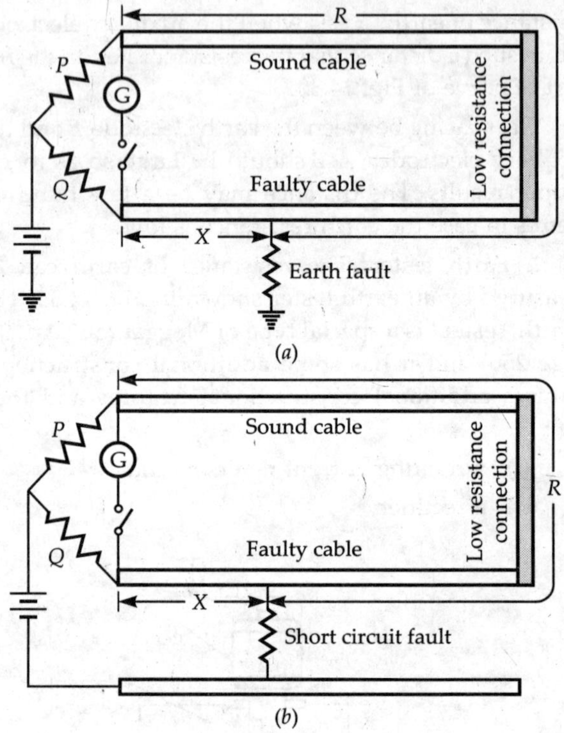


**Fig. 14.37** Murray loop test.

The resistors $P$, $Q$ forming the ratio arms may be decade resistance boxes or slide wires.

Under balance conditions :

$$\frac{X}{R} = \frac{Q}{P} \quad \text{or} \quad \frac{X}{R+X} = \frac{Q}{P+Q}$$

$$\therefore \qquad X = \frac{Q}{P+Q}(R+X) \qquad \text{...(14.53)}$$

where $(R+X)$ is total loop resistance formed by the sound cable and the faulty cable. When the conductors have the same cross-sectional area and the same resistivity, the resistances are proportional to lengths. If $l_1$ represents the length of the fault from the test end and $l$ the length of each cable.

Then

$$l_1 = \frac{Q}{P+Q} \cdot 2l \qquad \text{...(14.54)}$$

The above relation shows that the position of the fault may be located when the length of the cable is known. Also, the fault resistance does not alter the balance condition because its resistance enters the battery circuit hence affects only the sensitivity of the bridge circuit. However, if the magnitude of the fault resistance is high, difficulty may be experienced in obtaining the balance condition on account of decrease in sensitivity and hence accurate determination of the position of the fault may not be possible. In such a case, the resistance of the fault may be reduced by applying a high direct or alternating voltage – in consistance with the insulation rating of the cable – on the line so as to carbonize the insulation at the point of the fault.

***Varley loop test.*** In this test we can determine experimentally the total loop resistance instead of calculating it from the known lengths of the cable and its resistance per unit length. The necessary connections for the ground fault are shown in Fig. 14.38(*a*) and for the short circuit fault in Fig. 14.38(*b*). The treatment of the problem, in both cases, is identical.

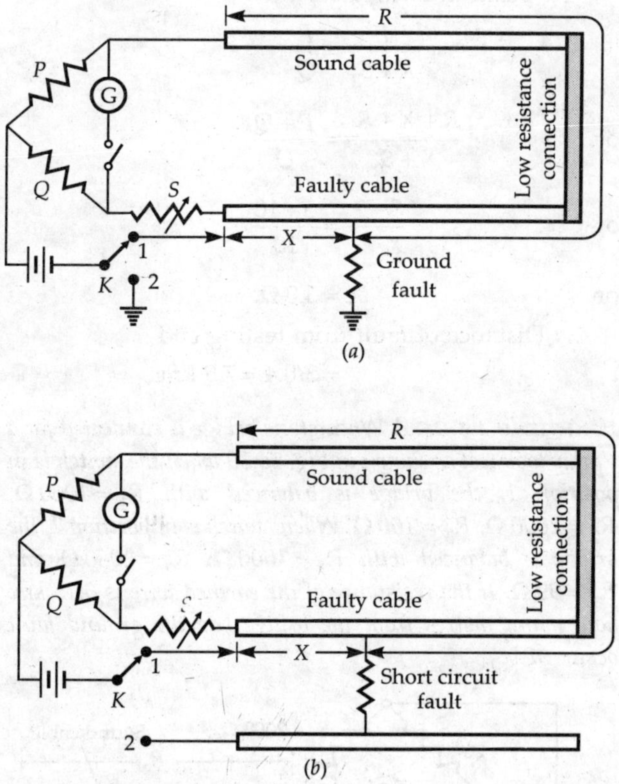


**Fig. 14.38** Varley loop test.

A single pole double throw switch $K$ is used in this circuit. Switch $K$ is first thrown to position '*1*' and the resistance $S$ is varied and balance obtained.

Let the value of $S$ for balance be $S_1$. The four arms of the Wheatstone bridge are $P, Q, R+X, S_1$.

At balance :

$$\frac{R+X}{S_1} = \frac{P}{Q} \quad \text{...(14.55)}$$

This determines $R+X$ *i.e.*, the total loop resistance as $P$, $Q$ and $S_1$ are known.

The switch $K$ is then thrown to position '2' and the bridge is rebalanced. Let the new value of $S$ for balance be $S_2$. The four arms of the bridge now are $P, Q, R, X+S_2$.

At balance,

$$\frac{R}{X+S_2} = \frac{P}{Q}.$$

or

$$\frac{R+X+S_2}{X+S_2} = \frac{P+Q}{Q}$$

or

$$X = \frac{(R+X)Q - S_2 P}{P+Q} \quad \text{...(14.56)}$$

Hence $X$ is known from the known values of $P, Q, S_2$ from this equation and $R+X$ (the total resistance of 2 cables) as determined from Eqn. 14.55. Knowing the value of $X$, the position of the fault is determined.

Now

$$\frac{X}{R+X} = \frac{l_1}{2l} \text{ or } l_1 = \frac{X}{R+X} 2l \quad \text{...(14.57)}$$

where $l_1'$ = length of fault from the test end

and $l$ = total length of conductor.

Equations 14.54 (for Murray loop test) and 14.56 (for Varley loop test) are valid only when the cable sections are uniform throughout the loop. Corrections must be applied in case the cross-sections of faulty and sound cables are different or when the cross-section of the faulty cable is not uniform over its entire length. Since temperature affects the value of resistance, corrections must be applied on this account if the temperatures of the two cables are different. Corrections may also have to be applied in case the cables have a large number of joints.

**Example 14.22** *In a test for a fault to earth by Murray loop test, the faulty cable has a length of 5.2 km. The faulty cable is looped with a sound cable of the same length and cross-section. The resistances of ratio arms are 100 Ω and 41.2 Ω at balance. Calculate the distance of the fault from the test end.*

*If the decade resistance boxes forming the ratio arms have limits of error of ± 0.5% (standard deviation) of the dial reading, what is the limit of error in the above calculated result ?*

**Solution.** Let X be the resistance of cable from the test end to place of fault and $R+X$ be the total resistance of loop. P and Q are the ratio arms.

$\therefore \quad X/R = Q/P$

or $\quad X = \dfrac{Q}{P+Q}(X+R)$

$= \dfrac{41.2}{141.2}(X+R) = 0.292\,(X+R)$

Let $l_1$ be the distance of fault from test end $l$ be the length of each cable and $r$ resistance per unit length.

$\therefore \quad X = rl_1$ and $X + R = 2rl$

$\therefore$ Distance of fault from test end :

$l_1 = 0.292 \times 2 \times 5.2 = 3.03$ km.

Limiting fractional error of unknown resistance

$$\frac{\delta X}{X} = \sqrt{\left(\frac{\delta P}{P}\right)^2 + \left(\frac{\delta P + \delta Q}{P+Q}\right)^2} = \sqrt{\left(\frac{\pm 0.5}{100}\right)^2 + \left(\frac{\pm 0.5}{100}\right)^2}$$

$= 0.707\%$.

Therefore limiting error of distance

$= 3.03 \times 0.707/100 \times 1000 = 21.4$ m.

**Example 14.23** *In a test by Murray loop method for a fault to earth on a 520 metre length of cable having a resistance of 1.1 Ω per 1000 metre, the faulty cable is looped with a sound cable of the same length but having a resistance 2.29 Ω per 1000 metre. The resistances of the other two arms of the testing network, at balance, are in the ratio of 2.7 : 1. Calculate the distance of fault from the testing end of test cable.*

**Solution.** Suppose $r_1$ and $r_2$ are the resistances per unit length of faulty and sound cable respectively.

Let $l$ be the length of each cable and $l_1$ be the distance of fault from the test end.

Referring to Fig. 14.37(*a*),

$\dfrac{X}{R} = \dfrac{Q}{P}$

or $\quad X = \dfrac{Q}{P+Q}(X+R) \quad ...(i)$

Now $\quad X = r_1 l_1 = \dfrac{1.1}{1000} l$

and $\quad X + R = r_1 l + r_2 l$

$= \left(\dfrac{1.1}{1000} + \dfrac{2.92}{1000}\right) \times 520 = 1.76\ \Omega.$

Substituting these values in (*i*), we have

$\dfrac{1.1}{1000} l_1 = \dfrac{l}{2.7+1} \times 1.76.$

$\therefore \quad l_1 = 432$ m.

**Example 14.24** *A short circuit fault is located by Varley loop test. The circuit of Fig. 14.38 (b) is used for the purpose. The ratio arms are set at P = 5 Ω and Q = 10 Ω and the values of variable resistance S are 16 Ω for position 1 of switch K and 7 Ω for position 2. The sound and faulty cables are identical and have a resistance of 0.4 Ω/km. Determine the length of each cable and the distance of fault from the test end.*

**Solution.** Let $S_1$ be the value of resistance $S_2$ for position 1 and $S_2$ for position 2.

$\therefore \quad \dfrac{R+X}{S_1} = \dfrac{P}{Q}$

Hence resistance of loop

$R + X = \dfrac{P}{Q} S_1 = \dfrac{5}{10} \times 16 = 8\ \Omega.$

Resistance of each cable = 8/2 = 4 Ω

$\therefore$ Length of each cable = 4/0.4 = 10 km.

At positron 2 we have,

$\dfrac{P}{X+S_2} = \dfrac{P}{Q}$

or $\quad \dfrac{R+X+S_2}{X+S_2} = \dfrac{P+Q}{Q}$

or $\quad \dfrac{8+7}{X+7} = \dfrac{5+10}{10}$

or $\quad X = 3.0\ \Omega.$

$\therefore$ Distance of fault from testing end

= 3/0.4 = 7.5 km.

**Example 14.25** *A Wheatstone bridge is connected for a Varley loov test as shown in Fig. 14.39 when the switch is in position 1, the bridge is balanced with $R_1 = 1000\ \Omega$, $R_2 = 2000\ \Omega$, $R_3 = 100\ \Omega$. When switch is in position 2, the bridge is balanced with $R_1 = 1000\ \Omega$, $R_2 = 2000\ \Omega$ and $R_3 = 99\ \Omega$. If the resistance of the earthed wire is 0.15 km, how many metres from the bridge has the ground fault occurred ?*

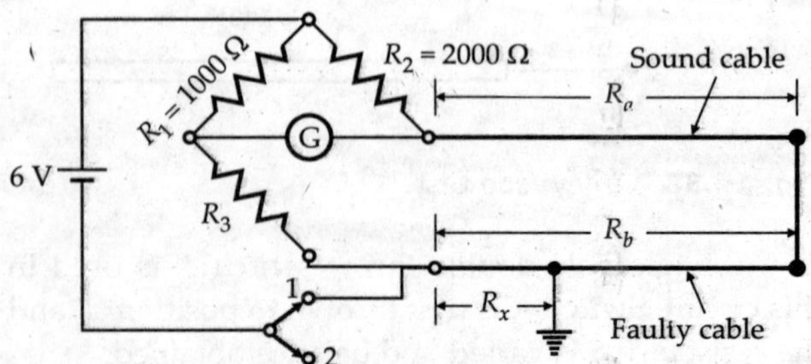


**Fig. 14.39** Diagram of Example 14.23.

**Solution.** When the switch is at position 1

$$2000/1000 = (R_a + R_b)/100$$

or $$R_a + R_b = 200\ \Omega$$

When the switch is at position 2

$$\frac{2000}{1000} = \frac{R_a + R_b - R_x}{99 + R_x}$$

$$198 + 2R_x = 200 - R_x$$

or $$R_x = 0.67\ \Omega.$$

∴ Resistance of fault from bridge

$$= \frac{0.67}{0.15} \times 1000$$

$$= 4466.7 \text{ m}.$$

# Review Questions

1. (*a*) Classify the resistances from the point of view of measurements.

   (*b*) Describe in brief the different methods used for measurement of medium resistances.

2. Describe the ammeter-voltmeter method of measurement of resistances. There are two ways in which the circuit of ammeter voltmeter method can be used

   (*i*) ammeter connected to the side of unknown resistance and

   (*ii*) voltmeter connected to the side of unknown resistance.

   Derive the condition which decides which circuit is to be used for a particular set of ammeter, voltmeter and unknown resistance. Assume equal relative error in both the cases.

3. Describe the substitution method of measurement of medium resistances. List the factors on which the accuracy of the method depends.

4. Draw the circuit of a Wheatstone bridge and derive the conditions of balance.

5. Derive the expression for bridge sensitivity for a Wheatstone bridge with equal arms. Find also the expression for current through the galvanometer for a small unbalance.

6. What are the different factors which affect the precision measurement of medium resistances with Wheatstone bridge ? Explain how their effects are minimized/eliminated.

7. Describe the working of a Carey-Foster Slide-wire bridge.

8. Draw the circuit of a Kelvin-Varley slide and explain its working and advantages.

9. What are the different problems associated with measurement of low resistances ? Explain the principle of working a Kelvin's Double Bridge and explain how the effect of contact resistance and resistance of leads is eliminated.

10. Draw the circuit of a Kelvin's Double Bridge used for measurement of low resistances. Derive the condition for balance.

11. What are the different difficulties encountered in the measurement of high resistances ? Explain how these difficulties are overcome.

12. Explain the loss of charge method for measurements of insulation resistance of cables.

13. Differentiate between volume and surface resistivities. Explain the method of measuring them.

14. Describe the working of a Megohm bridge.

15. What is the importance of the value of Earth's resistance. What are the factors which influence its value ? Describe the fall of potential method for measurement of earth resistance.

16. Describe the construction and working of an Earth Tester. Explain how it can be used for measurement of resistance of an earthing electrode.

17. Describe the Murray loop test for localization of ground and short circuit faults in cables.

18. Describe the Varley loop test for localization of ground and short circuit faults in cables.

# Unsolved Problems

1. A voltmeter of resistance 500 Ω and a milliammeter of 1.0 Ω resistance are used to measure a resistance by ammeter voltmeter method. If the voltmeter reads 20 V and milliammeter 100 mA.

   Calculate the value of measured resistance

   (*i*) if the voltmeter is put across the resistance and the milliammeter connected in series with the unknown resistance

   (*ii*) if the voltmeter is put across the unknown resistance with ammeter connected on the supply side.

   **[Ans.** 199 Ω ; 333 Ω]

2. In a laboratory a voltmeter of 200 Ω resistance and an ammeter of 0.02 Ω resistance are available. Calculate the value of resistance that can be measured by the ammeter voltmeter method for which the two different circuit measurements give equal errors. **[Ans.** 2 Ω]

3. In a measurement of resistance by the substitution method a standard resistance of 100 kΩ is used. The galvanometer has a resistance of 2000 Ω, and gives the following deflections :

   (*i*) With unknown resistance : 46 divisions,

   (*ii*) With standard resistance : 40 divisions.

   Find the value of unknown resistance

   **[Ans.** 86.7 kΩ]

4. The four arms of a Wheatstone bridge are as follows :

   $AB = 100\,\Omega$ ; $BC = 10\,\Omega$ ;

   $CD = 4\,\Omega$ and $DA = 50\,\Omega$.

   The galvanometer has a resistance of 20 Ω and is connected across *BD*. A source of 10 V d.c. is connected across *AC*. Find the current through the galvanometer. What should be the resistance in the arm *DA* for no current through the galvanometer ?

   **[Ans.** 5.15 mA ; 40 Ω]

5. The four arms of a Wheatstone bridge are as follows :

   $AB = 100\,\Omega$ ;

   $BC = 1000\,\Omega$ ;

   $CD = 4000\,\Omega$ and

   $DA = 400\,\Omega$.

   The galvanometer has a resistance of 100 Ω, a sensitivity of 100 mm/μA and is connected across *AC*. A source of 4 V d.c. is connected across *BD*. Calculate the current through the galvanometer and its deflection if the resistance of ami *DA* is changed from 400 Ω to 401 Ω.

   **[Ans.** 1.63 μA ; 16.3 mm]

6. In a Wheatstone bridge, the ratio arms $AB = 10\,\Omega$ and $BC = 100\Omega$ ; standard-resistance across $CD = 10\,\Omega$. The shunt across 10 Ω ratio arm has to be changed from 22310 to 27670 Ω, when the resistor $R_2$ was changed for $R_1$ in the arm *DA*. Calculate in magnitude the difference between the resistances of $R_1$ and $R_2$. The bridge is balanced in both the cases. **[Ans.** 86.8 μΩ]

7. A modified Wheatstone bridge network is constituted as follows :

   *AB* is a resistance *P* in parallel with resistance *p* ;

   *BC* is a resistance *Q* in parallel with a resistance *q* ;

   *CD* and *DA* are resistances *R* and *S* respectively.

   The nominal values of *P, Q* and *S* are each 10 Ω. With resistance *R* in circuit, balance is obtained with $p = 30{,}000\,\Omega$ and $q = 25{,}000$. With *R* replaced by a standard resistance of 10 Ω, balance is obtained when p = 15,000 Ω and q = 40,000 Ω. Calculate the value of *R*. **[Ans.** 9.99952 Ω]

8. A Wheatstone bridge is used to measure the resistance of a resistor. The bridge was balanced with the values shown in Fig. 14.40. It is found that, due to presence of chemical impurities emfs are set

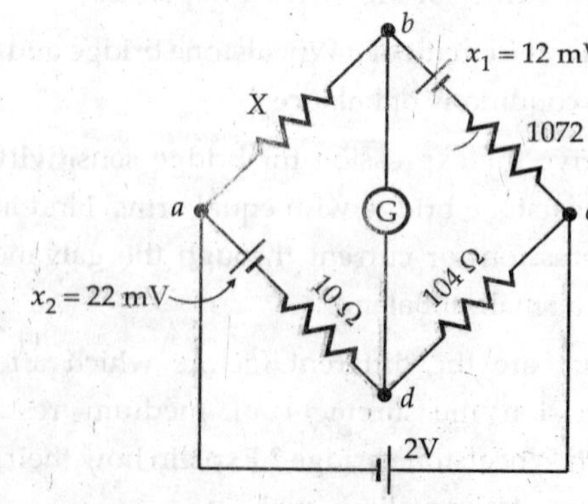


**Fig. 14.40** Diagram of Problem 14.8.

up at points $x_1$ and $x_2$ of respective values 12 mV and 22 mV with polarities being shown in the diagram. Calculate ;

(*i*) the apparent value of X,

(*ii*) the true of X,

(*iii*) the % error in the apparent value of X.

[**Ans.** 103 Ω ; 115.7 Ω ; – 12.3%]

**9.** A Kelvin bridge is balanced with the following constants :

Outer ratio arm 100 Ω and 1000 Ω ;

Inner arms, ratio 99.92 Ω and 1000.6 Ω ;

Resistance of link, 0.1 Ω ;

Standard resistance, 0.00377 Ω.

Calculate the value of unknown resistance.

[**Ans.** 0.000467 Ω]

**10.** The ratio arms of a kelvin bridge are 100 Ω each. The galvanometer has an internal resistance of 500 Ω and a current sensitivity of 200 mm/μA. The unknown resistance is 0.1002 and the standard resistance is set at 0.1000 Ω. A d.c. current of 10 A is passed through the standard and the unknown resistance from a 2.2 V battery in series with a rheostat.

Calculate the deflection of the galvanometer. Neglect the resistance of the link. Find also the resistance unbalance to produce a deflection of 1 mm.

[**Ans.** 3320 mm ; 0.6 μΩ]

**11.** Calculate insulation resistance of a cable in which the voltage falls from 100 to 80 V in 20 s. The capacitance is 300 pF. [**Ans.** 29800 MΩ]

**12.** A cable immersed in a testing tank is charged to a voltage of 200 V with an electrostatic voltmeter connected between core and tank. After one minute's electrification, the cable and voltmeter are isolated and is found that in 20 s the voltage falls to 150 V. The test is repeated with a resistance of 20 MΩ between core and tank and the voltage is found to fall to 100 V in the same time. Calculate the insulation resistance of the cable, that of the voltmeter is given as 50 MΩ.

**13.** The following observations were made for a loss of charge method for the determination of a high resistance *R*. The charged capacitor of capacitance 12.5 μF was connected across an electrostatic voltmeter and *R* in parallel and the voltage measured after intervals of time.

| *time, s* | *voltage, V* |
|---|---|
| 0 | 150 |
| 100 | 121 |
| 200 | 97 |
| 300 | 83 |
| 400 | 65 |
| 500 | 57 |
| 600 | 340 |

The further set of readings was taken with resistor *R* removed from circuit, as follows :

| *time, s* | *voltage, V* |
|---|---|
| 0 | 150 |
| 200 | 143 |
| 400 | 133 |
| 600 | 121 |

The readings of voltmeter were subject to large random errors. Calculate the value of resistance *R*.

[**Ans.** 52.5 MΩ]

**14.** A high resistance of 200 MΩ has a leakage resistance of 400 MΩ between each of its main terminals and the guard terminal. Find the percentage error in measurement if the above resistance is measured by an ordinary Wheatstone bridge without providing guard circuit.

[**Ans.** 20% low]

**15.** Two mains are working at potential difference of 220 V. A 250 V voltmeter having 10,000 Ω/V when connected between positive main and earth reads 149 V and the reading is 42 V when connected between negative main the earth. Calculate the insulation resistance of each main with respect to earth. [**Ans.** 1.73 and 0.486 MΩ]

**16.** A feeder cable 250 metre long has a fault to earth. The fault is localized by the following resistance measurements between earth and one end of cable (*a*) distant end insulated, 6.95 Ω (*b*) distant end earthed, 1.71 Ω. The cable has a total resistance of 1.80 Ω. Find the resistance of fault and its distance from test end.

**17.** In a Murray loop test for ground fault on a 500 metre long cable having a resistance of 1.6 Ω/km, the faulty cable is looped with a sound cable of same length and cross-section. If resistances of ratio arms are 3 : 1, calculate the distance of the fault from the test end. [**Ans.** 250 m]

18. A telephone line, 5 km long, has an earth fault 2.3 km from test end. If the resistance of the lines per km is 4.0 Ω, what value of variable resistance will give balance in a Varley loop test ? The ratio arms are equal. **[Ans. 21.6 Ω]**

19. A telephone wire having a resistance of 14.6 Ω km, develops a fault to earth. When looped with a sound wire of the same length, the total resistance is found to be 56 Ω. If the value of variable resistance is 16.3 at balance in a Varley loop test with ratio arms equal, calculate the distance of fault. **[Ans. 1.36 km]**

20. A Wheatstone bridge is connected for a Varley loop test as shown in Fig. 14.41. When the switch is in position *a*, the bridge is balanced with $R_1 = 1000\,\Omega$, $R_2 = 100\,\Omega$, $R_3 = 53\,\Omega$. When switch *S* is in position *b*, the bridge is balanced with

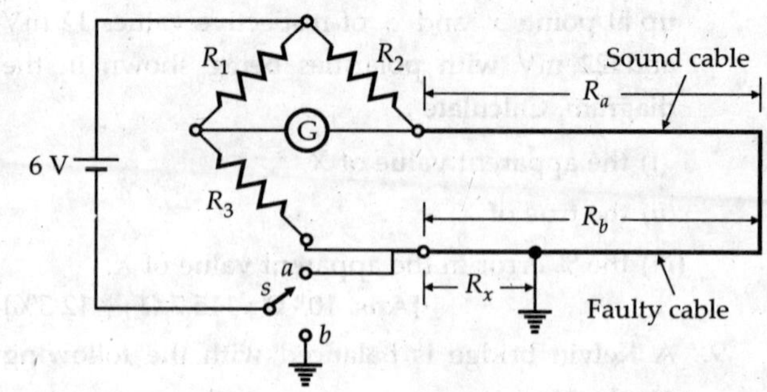


**Fig. 14.41** Diagram of Problem 14.20

$R_1 = 1000\,\Omega$, $R_2 = 100\,\Omega$ and $R_3 = 52.9\,\Omega$. If the resistance of the shorted wire is 0.015 Ω/m. What is the distance between the place of fault and the bridge ? **[Ans. 2353 m]**

# Objective Type Questions

***Tick (√) the most appropriate answer*** :

1. A resistance of value 10 Ω approximately is to be measured by ammeter-voltmeter method with resistance of ammeter is 0.02 Ω and that of voltmeter is 5000 Ω. The resistance should be measured
   (*a*) by connecting the ammeter on the side of unknown resistance as this connection gives better accuracy
   (*b*) by connecting the voltmeter on the side of unknown resistance as this connection gives better accuracy
   (*c*) by any of the two connections, as both of them give equal accuracy
   (*d*) none of the above.

2. A unknown resistance is measured by substitution method. First a standard known resistance of 100 Ω is connected in series with a circuit having a rheostat and a galvanometer. The battery voltage is 10 V and the setting of the rheostat is 500 Ω and the galvanometer shows a deflection of 60°. After this test, the battery voltage goes down to 9 V and when the unknown resistance is substituted for the known resistance, the galvanometer again shows a deflection of 60° with the same setting of the rheostat. The value of unknown resistance
   (*a*) 100 Ω
   (*b*) 54 Ω
   (*c*) 90 Ω
   (*d*) none of the above.

3. Equal resistances of 100 Ω each are connected in each arm of a Wheatstone bridge which is supplied by a 2 V battery source. The galvanometer of negligible resistance connected to the bridge can sense as low current as 1 μA. The smallest value of resistance that can be measured is :
   (*a*) 20 μΩ
   (*b*) 2 μΩ
   (*c*) 20 μΩ
   (*d*) none of the above.

4. A Wheatstone bridge has ratio arms of 1000 Ω and 100 Ω resistance, the standard resistance arms consists 4 decade resistance boxes of 1000, 100, 10, 1 Ω steps. The maximum and minimum values of unknown resistance which can be determined with this set up is
   (*a*) 111100 Ω, 1 Ω

(b) 11110 Ω, 10 Ω

(c) 111100 Ω, 10 Ω

(d) none of the above.

**5.** A Wheatstone bridge cannot be used for precision measurements because errors are introduced into on account of

(a) resistance of connecting leads

(b) thermo-electric emfs

(c) contact resistances

(d) all the above.

**6.** A Kelvin-Varley Slide consists of 4 decade dividers. The first decade is constructed by having 11 coils of 10 kΩ resistance each. The subsequent decades will have coils of :

(a) 11 coils of 20 kΩ each, 11 coils of 40 kΩ each, 11 coils of 80 kΩ each

(b) 11 coils of 10 kΩ each, 11 coils of 5 kΩ each, 10 coils of 1 kΩ each

(c) 11 coils of 2 kΩ each, 11 coils of 400 Ω each, 11 coils of 80 Ω each

(d) 11 coils of 2 kΩ each, 11 coils of 400 Ω each, 10 coils of 80 Ω each.

**7.** Low resistances are provided with four terminals

(a) to facilitate the connection of current and potential circuits

(b) in order that the resistance value becomes definite irrespective of the nature of contacts at the current terminals

(c) to eliminate the effect of thermo-electric emfs

(d) to eliminate the effect of leads.

**8.** In a Kelvin's Double Bridge two sets of readings are taken when measuring a low resistance, one with the current in one direction and the other with direction of current reversed.

This is done to

(a) eliminate the effect of contact resistance

(b) eliminate the effect of resistance of leads

(c) correct for changes in battery voltage

(d) eliminate the effect of thermo-electric emfs.

**9.** High resistances are provided with a guard terminal.

This guard terminal is used to :

(a) bypass the leakage current

(b) guard the resistance against stray electrostatic fields

(c) guard the resistance against overloads

(d) none of the above.

**10.** When measuring insulation resistance of cables using d.c. source, the galvanometer used should be initially short circuited because,

(a) cables have a low value of initial resistance

(b) cables have a high value of capacitance which draws a high value of charging current

(c) cables have a low value of capacitance which draws a high value of charging current

(d) none of the above.

**11.** A circular piece of specimen has a surface resistance $R_s$. Its diameter is $d$ and the thickness is $t$. The surface resistivity ps of the specimen is given by :

(a) $\frac{\pi d^2 R_s}{t}$

(b) $\frac{\pi d R_s}{t}$

(c) $\frac{R_s t}{\pi d^2}$

(d) $\frac{R_s t}{\pi d}$

**12.** The value of resistance of an earthing electrode depends upon :

(a) shape and material of electrode

(b) depth to which electrode is driven into earth.

(c) specific resistance of soil

(d) all the above.

**13.** From the point of view of safety, the resistance of earthing electrode should be :

(a) low

(b) high

(c) medium

(d) the value of resistance of earth electrodes does not affect the safety.

**14.** The advantage of Varley loop tests over Murray loop tests is

(a) they can be used for localizing of short circuit faults

(b) they can be used for localizing of earth fault

14

(c) the loop resistance can be experimentally determined

(d) their accuracy is higher.

15. When localizing ground fault with the help of loop tests, the resistance of the fault :

(a) affects the balance conditions

(b) affects the value of cable resistance

(c) affects the sensitivity of the bridge

(d) all the above.

## Answers

| | | | | | | | | | |
|---|---|---|---|---|---|---|---|---|---|
| **1.** (c) | **2.** (b) | **3.** (a) | **4.** (c) | **5.** (d) | **6.** (d) | **7.** (b) | **8.** (d) | **9.** (a) | **10.** (b) |
| **11.** (b) | **12.** (d) | **13.** (a) | **14.** (c) | **15.** (c) | | | | | |

# 15

# Potentiometers

## D.C. POTENTIOMETERS

### 15.1 INTRODUCTION

A potentiometer is an instrument designed to measure an unknown voltage by comparing it with a known voltage. The known voltage may be supplied by a standard cell or any other known voltage-reference source (See pages 149-153). Measurements using comparison methods are capable of a high degree of accuracy because the result obtained does not depend upon the actual deflection of a pointer, as is the case in deflectional methods, but only upon the accuracy with which the voltage of the reference source is known.

Another advantage of the potentiometers is that since a potentiometer makes use of a balance or null condition, no current flows and hence no power is consumed in the circuit containing the unknown emf when the instrument is balanced. Thus the determination of voltage by a potentiometer is quite independent of the source resistance.

Since a potentiometer measures voltage, it can also be used to determine current simply by measuring the voltage drop produced by the unknown current passing through a known standard resistance.

The potentiometer is extensively used for a calibration of voltmeters and ammeters and has in fact become the standard for the calibration of these instruments. For the above mentioned advantages the potentiometer has become very important in the field of electrical measurements and calibration.

#### 15.1.1 Basic Potentiometer Circuit

The principle of operation of all potentiometers is based on the circuit of Fig. 15.1, which shows the schematic diagram of the basic slide wire potentiometer.

With switch '*S*' in the "*operate*" position and the galvanometer key *K* open, the battery supplies the "*working current*" through the rheostat *R* and the slide wire. The working current through the slide wire may be varied by changing the rheostat setting. The method of measuring the unknown voltage, *E*, depends upon finding a position for the sliding contact such the galvanometer shows zero deflection, *i.e.*, indicates null condition, when the galvanometer key, *K*, is closed. Zero galvanometer deflection or a null means that the unknown voltage, *E*, is equal to the voltage drop $E_1$, across portion *ac* of the slide wire. Thus determination of the value of unknown voltage now becomes a matter of evaluating the voltage drop $E_1$ along the portion *ac* of the slide wire.

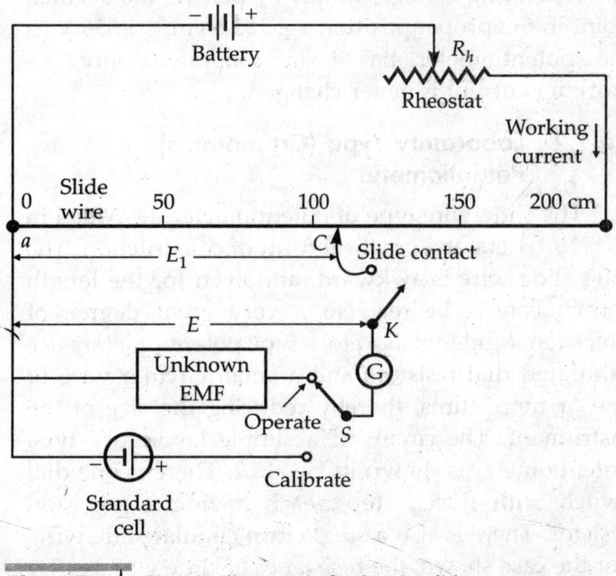


**Fig. 15.1** Circuit diagram of a basic slide wire potentiometer.

The slide wire has a uniform cross-section and hence uniform resistance along its entire length. A calibrated scale in cm and fractions of cm, is placed along the slide wire so that the sliding contact can be placed accurately at any desired position along the slide wire. Since the resistance of slide wire is known accurately, the voltage drop along the slide wire can be controlled by adjusting the value of working current. The process of adjusting the working current so as to match the voltage drop across a portion of sliding wire against a standard reference source is known as *"Standardisation"*.

***Standardisation.*** The procedure for standardisation of the potentiometer is illustrated by the following example :

The slide wire of Fig. 15.1 has a total length of 200 cm and a resistance of 200 Ω. The emf of the standard cell is 1.0186 V. Switch *'S'* is thrown to *"calibrate"* position and the sliding contact is placed at 101.86 cm mark on the slide wire scale. The rheostat $R_h$ is now adjusted so as to vary the working current. This adjustment is carried on till the galvanometer shows no deflection when key *'K'* is pressed. Under these conditions, the voltage drop along the 101.86 cm portion of the slide wire is equal to standard cell voltage of 101.86 V. Since the 101.86 cm portion of the slide wire has a resistance of 101.86 Ω, the working current in fact has been adjusted to a value ;

$$\frac{1.0186}{101.86} \times 1000 = 10 \text{ mA}.$$

The voltage at any point along the slide wire is proportional to the length of slide wire. This voltage is obtained by converting the calibrated length into the corresponding voltage, simply by placing the decimal point in the proper position *e.g.* 153.6 cm. = 1.536 V. If the potentiometer has been calibrated once, its working current is never changed.

### 15.1.2 Laboratory Type (Crompton's) Potentiometer

The slide-wire type of potentiometer described in Art. 15.1.1 is not a practical form of construction. The long slide wire is awkward, and even for the length shown cannot be read to a very great degree of precision. Modern laboratory type potentiometers use calibrated dial resistors and a small circular wire of one or more turns, thereby reducing the size of the instrument. The circuit of a simple laboratory type potentiometer is shown in Fig. 15.2. There is one dial switch with fifteen steps, each having a precision resistor. There is also a single turn circular slide wire. For the case shown, the resistance of slide wire is 10 Ω and the dial resistors have a value of 10 Ω each. Thus

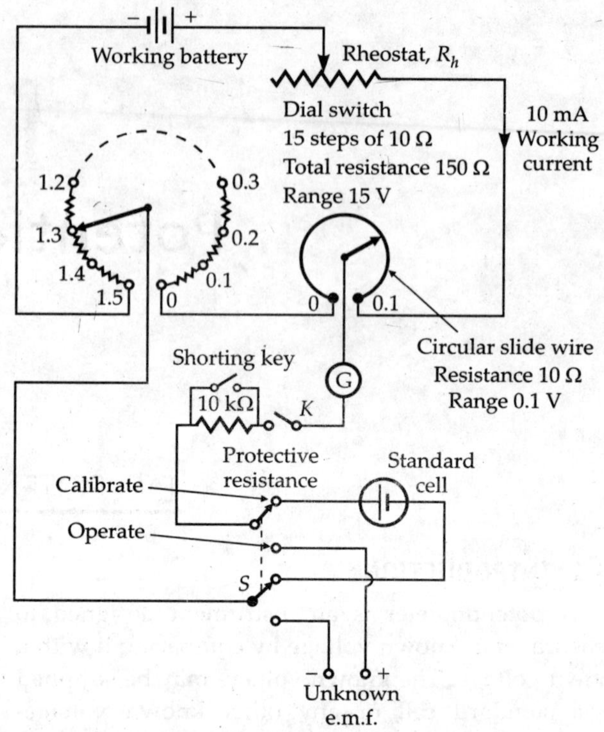


**Fig. 15.2** Laboratory type potentiometer.

the dial has a total resistance of 150 Ω and in addition the slide wire has a resistance of 10 Ω. The working current of the potentiometer is 10 mA and therefore each step of dial switch corresponds to 0.1 V. The slide wire is provided with 200 scale divisions and since the total resistance of slide wire corresponds to a voltage drop of 0.1 V, each division of slide wire corresponds to 0.1/200 = 0.0005 V. It is quite comfortable to interpolate readings upto 1/5 of a scale division and therefore with this potentiometer it is possible to estimate the readings upto 0.0001 V.

This potentiometer is provided with a double throw switch which allows the connection of either the standard cell or the unknown emf to be applied to the working circuit. A key and a protective resistance (usually about 10 kΩ) is used in the galvanometer circuit. In order to operate the galvanometer at its maximum sensitivity provision is made to short the protective resistance when near the balance conditions.

The following steps are used when making measurements with the above potentiometer :

1. The combination of dial resistors and the slide wire is set to the standard cell voltage. Supposing the value of emf of standard cell is 1.0186 V, the dial resistor is put at 1.0 V and the slide wire is put at 0.0186 setting.

2. The switch $S$ is thrown to the calibrate position and the galvanometer key is tapped while the rheostat is adjusted for zero deflection on the galvanometer. The protective resistance is kept in the circuit in the initial stages so as to protect the galvanometer from being getting damaged.
3. As the balance or null point is approached, the protective resistance is shorted so as to increase the sensitivity of the galvanometer. Final adjustments are made for zero deflection with the help of the rheostat.

   This completes the standardisation process for the potentiometer.
4. After completion of standardisation, the switch '$S$' is thrown to operate position thereby connecting the unknown emf into the potentiometer circuit. With the protective resistance in the circuit, the potentiometer is balanced by means of the main dial and the slide wire.
5. As the balance is approached, the protective resistance is shorted, and final adjustments are made to obtain true balance.
6. The value of unknown emf is read off directly from the settings of the dial adjust slide wire.
7. The standardisation of the potentiometer is checked again by returning the switch $S$ to the calibrate position. The dial settings are kept exactly the same as in the original standardisation process. If the new reading does not agree with the old one, a second measurement of unknown emf must be made. The standardisation should be again checked after the completion of measurement. This potentiometer is a form of **Crompton's Potentiometer.**

### 15.1.3 Multiple-Range Potentiometer

The single-range potentiometer shown in Fig. 15.2 is frequently constructed to cover a range of 1.6 V—though, of course, it can be designed for any desired voltage within reasonable and practical limits. For example, a Crompton's potentiometer is designed to measure voltages upto 1.9 V by simply adding three more resistance steps to the main dial.

The circuit of a single potentiometer may be modified in a simple way to add a second range, which is usually by a second factor, such as 0.1 or 0.01, in order that the direct reading features of the original circuit may still be utilized.

Figure 15.3 shows the schematic diagram of a duo-range (two range) potentiometer. The two ranges are obtained by using two resistors $R_1$ and $R_2$ and a range selecting switch $S$. The operation of duo-range potentiometer of Fig. 15.3 may be more easily understood and analyzed by redrawing in its simplified form by omitting the galvanometer and calibration (standardizing) circuit. The simplified diagram is shown in Fig. 15.4.

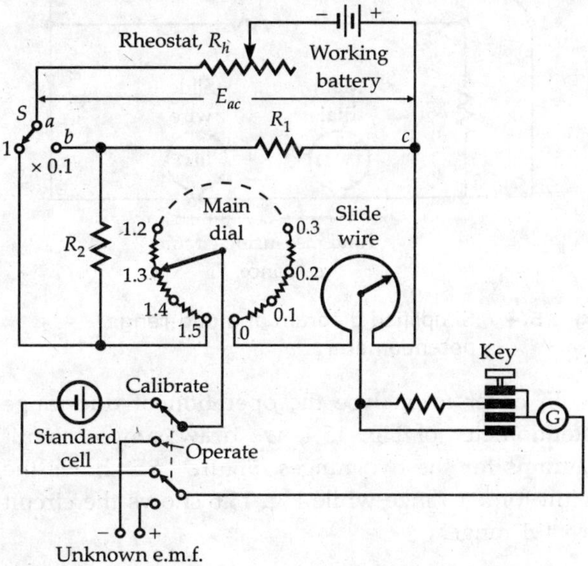


**Fig. 15.3** Duo-range type potentiometer.

In Fig. 15.4 the total resistance of measuring of circuit, $R_m$, consists of resistance of main dial in series with that of slide wire. The main dial has fifteen steps of 10 Ω each and therefore it has a total resistance of 150 Ω. The resistance of slide wire is 10 Ω. The measuring circuit current $I_m$ must be equal to 10 mA in order to produce a voltage drop of 1.6 V across the measuring circuit resistance $R_m$ *i.e.*, resistance of the main dial and the slide wire in series. The diagrams 15.3 and 15.4, with the range switch $S$ in × 1 position, represent an identical arrangement as shown in Fig. 15.2. When the range switch $S$ is thrown to position × 0.1, the measuring current $I_m$ must be reduced to 1/10 of its original value *i.e.*, 1 mA in order to produce a voltage drop of 0.16 V across the measuring circuit resistance $R_m$.

The design of circuit of a duo-range potentiometer should be such that it is possible to change the measuring ranges without re-adjusting the rheostat or changing the value of working voltage of the battery. This is essential so that once the instrument has been calibrated on × 1 range, calibration of the × 0.1 range is not necessary. The above requirement means that the

15

voltage $E_{ac}$ in Fig. 15.4 remains the same for both positions of range switch $S$. This condition is satisfied only when the total battery current has the same value for each measuring range.

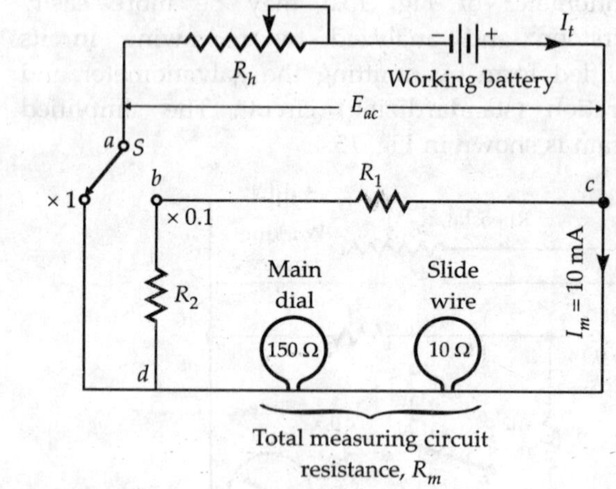


**Fig. 15.4** Simplified diagram of a duo-range potentiometer.

In order to analyse the operation of duo-range potentiometer of Fig. 15.4, we draw simple circuit diagrams for the two ranges. Figure 15.5 shows the circuit for × 1 range while Fig. 15.6 shows the circuit for × 0.1 range.

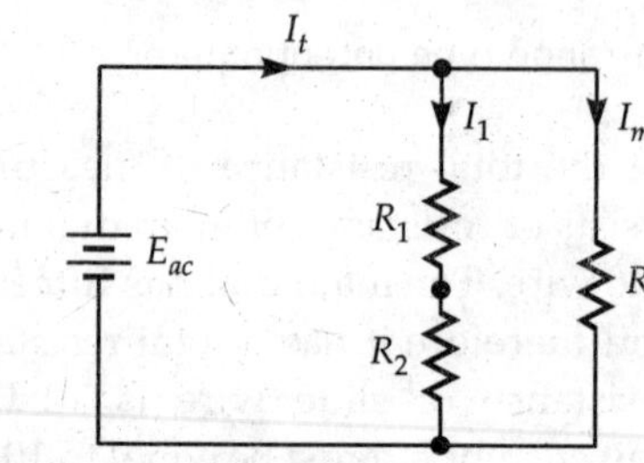


**Fig. 15.5** Circuit for × 1 range.

On range × 1 (Fig. 15.5), the range resistors $R_1$ and $R_2$ in series, are in parallel with total measuring resistance $R_m$. On the × 0.1 range (Fig. 15.6), the range resistor $R_1$, is in parallel with the series of combination $R_2$ and $R_m$.

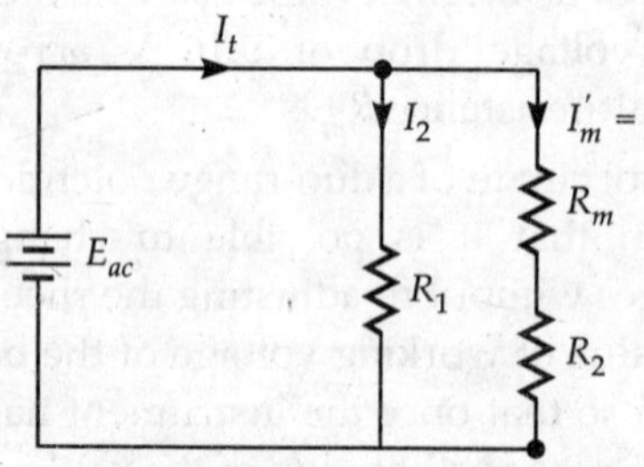


**Fig. 15.6** Circuit for × 0.1 range.

In order that the total current $I_t$ drawn from the battery is the same, we must have

$$\frac{R_m(R_1+R_2)}{R_1+R_2+R_m} = \frac{R_1(R_m+R_2)}{R_1+R_2+R}$$

or $$R_2 R_m = R_1 R_2 \quad \text{or} \quad R_1 = R_m$$

This means that range resistance $R_1$ must equal the resistance of the measuring circuit in order that the total current drawn from the battery is the same for both the ranges.

Now current through the measuring circuit on × 0.1 range should be 1/10 of the current in the measuring circuit on × 1 range.

This means that

$$I'_m = 0.1\, I_m$$

Now $$I_m = \frac{E_{ac}}{R_m} \qquad ...(i)$$

and $$I'_m = \frac{E_{ac}}{R_m + R_2} \quad \text{or} \quad 0.1\, I_m = \frac{E_{ac}}{R_m + R_2} \qquad ...(ii)$$

From (*i*) and (*ii*), we have,

$$0.1(R_m + R_2) = R_m$$

$$R_2 = 9\, R_m = 9 R_1.$$

For the circuit of Fig. 15.4, where measuring circuit resistance $R_m = 160\ \Omega$, we have,

$$R_1 = R_m = 160\ \Omega \quad \text{and} \quad R_2 = 9\, R_m = 1440\ \Omega$$

With this arrangement the total current drawn from the battery is 11 mA for each of the two ranges.

The above mentioned potentiometer measures voltages upto 1.6 V on × 1 range and upto 0.16 V on the × 0.1 range. On the lower range the dial readings are simply multiplied by the range factor of 0.1 to get the voltage being measured. If the slide wire has 100 scale divisions and each division can be interpolated to one-fifth of a division, the resolution of the potentiometer is :

$$\frac{1}{5} \times \frac{1}{100} \times 0.01\text{ V} = 20\ \mu\text{ V, on the} \times 0.1 \text{ range}$$

The duo-range potentiometer has two distinct advantages in having the low range for small voltages :

(*i*) precision of reading is increased by one decimal place, and

(*ii*) a greater part of reading is made on the dial resistors which have inherently a greater accuracy than the slide wire.

The duo-range potentiometer given in Fig. 15.3 is constructed for a voltage ratio of 10 : 1. A similar arrangement may be used for any other ratio by proper selection of range resistors $R_1$ and $R_2$. Precision laboratory type of potentiometers often have three voltage ranges usually of 1.0, 0.1 and 0.01.

### 15.1.4 Constructional Details of Potentiometers

All the resistors in a potentiometer, with the exception of slide wires are made of manganin. This is because manganin has a high stability, a low temperature co-efficient and has freedom from thermo-electric effects against copper.

The slide wire is usually made of platinum-silver alloy and the sliding contact are of a copper-gold-silver alloy. This combination of materials for slide wire and sliding contacts results in a good contact, freedom from thermo-electric emfs and minimum wear of slide wire.

The current controlling rheostat is usually a combination of a stud dial and a multiturn slide wire.

It is very important that internal thermo-electric emfs in a potentiometer are minimum. The use of manganin resistors helps in this direction. It is desirable that all the parts work at the same temperature. Therefore, all the parts are covered in a single case. This has the added advantage of protecting the contacts from fumes and dust which may cause corrosion and appearance of volatic emfs at the joints. Potentiometers designed specially for thermocouple measurements have copper terminals.

In order to prevent leakage, all the parts must be enclosed, so as to protect them from moisture. The working parts are normally mounted on ebonite or keramot panels.

### 15.1.5 Precision Type Potentiometers

The potentiometers described so far are simple instruments capable of a precision of 100 μV for readings upto 1.6 V. These difficulties to get accurate results mainly arise because of lack of uniformity of slide wire and the difficulties in maintaining the contacts in good shape. However, for some applications higher precision and accuracy are needed, requiring changes in design features. Some of the features are explained below :

### 15.1.6 Vernier Potentiometer

The limitations imposed on performance of ordinary potentiometers by slide wire are eliminated in a vernier potentiometer. This instrument has two ranges : the normal range of 1.6 V down to 10 μV and a lower range of 0.16 V down to 1 μV.

This potentiometer uses the Kelvin Varley arrangement shown in Fig. 15.7. However, in this potentiometer there are three measuring dials. The first dial measures upto 1.5 V (on the XI range) in steps 0.1 V : the middle dial has 102 studs and reads upto 0.1 V in steps of 0.001 V ; the third dial also has 102 studs and reads from – 0.0001 V to 0.001 V in steps of 0.00001 V (*i.e.*, 10 μV). There is no slide wire. The resistances of the middle dial shunt two of the coils of first dial. The moving arm of middle dial carries two arms spaced two studs apart. In actual practice the resistance of second dial is greater than that between two studs in the main (first) dial, so that the voltage drop across the second dial is greater than 0.1 V. If this is not done the voltage drop in switch contact resistances and leads would render the converge of middle dial to less than 0.1 V. The third dial is obtained from a shunt circuit which permits a true zero and a small negative setting to be obtained.

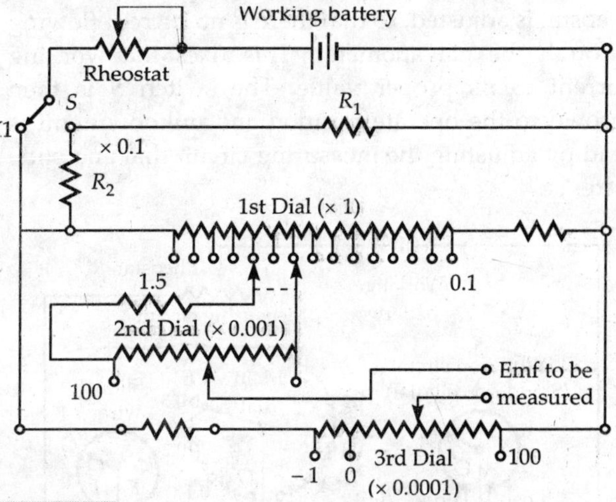


**Fig. 15.7** A duo-range vernier potentiometer.

The vernier potentiometer reads to increment of 0.00001 V (10 μV) on range XI and has readability of 1 μ V on × 0.1 range. If a third range of × 0.01 is provided, the readability becomes 0.1 μ V. This does not mean, unfortunately, that small voltages can be read with assurance to 0.1 μV. Measurements are subject to stray thermal and contact emfs in the potentiometer, galvanometer and the measuring circuits. These emfs may be of the order of one to several microvolts and are difficult to locate and control. These emfs can be minimized only by special construction *i.e.*, proper selection of metals for resistors, terminals and connecting leads and also by the use of thermal shields.

### 15.1.7 Standard Cell Dial

All modern potentiometers incorporate a separate standard cell dial circuit (calibrating circuit).

15

This provides a means of standard cell balance resistance to suit the emf value of the particular standard cell used. Also a separate standard cell dial permits the operator to check the standard cell balance at any time during the course of measurement without disturbing the potentiometer setting. Figure 15.8 shows a separate standard cell dial incorporated in a single range potentiometer.

This potentiometer is provided with an independent standardizing circuit *AD* which can be set on any range of standard cell emf from 1.016 V to 1.020 V. A drop of 1.016 V is provided by resistance $R_s$ and the remaining drop of 0.004 V is provided by the slide wire *AC*. Thus a considerable change in standard cell emf owing to temperature changes can be allowed for the operation of circuit, Fig. 15.8. The slide wire of calibrating circuit is set to read the emf of the standard cell which is connected to the potentiometer. The selector switch *S* is put at the calibrate position and the rheostat is adjusted, so that there is no current flowing through the galvanometer. This fixes the working current to its proper value. The switch *S* is then thrown to the operate position and unknown emf is read by adjusting the measuring circuit dial and slide wire.

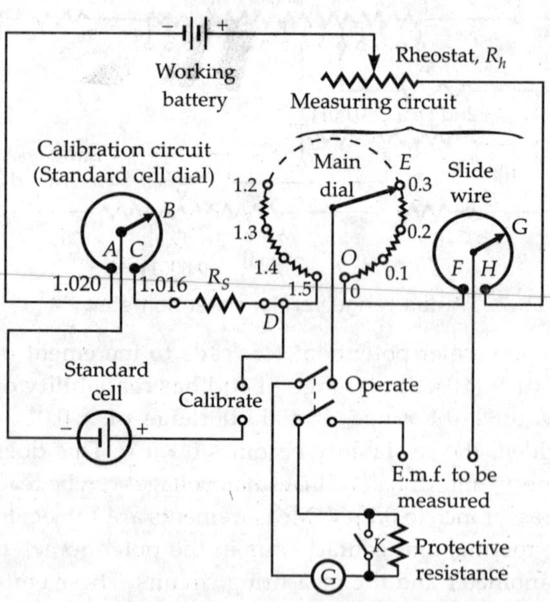


**Fig. 15.8** Single range potentiometer using independent calibrating circuit.

The standardizing circuit may be checked for constancy of working current, any time during the measurements by simply throwing the switch *S* back to calibrate position. This process does not disturb the measuring circuit while convenience and speed of measurements is increased.

### 15.1.8 True Zero

In the student type potentiometers described in Art. 15.2 it is impossible to obtain true zero because the two contacts cannot coincide absolutely. This drawback is eliminated in the simplified circuit of Fig. 15.9. The slide wire *BC* is provided with a shunt resistor which is tapped at *D*. This tapping is made 0 on the main dial. When the contact is in a position such that $\frac{r_1}{r_3} = \frac{r_2}{r_4}$ = zero, the slider can travel a little lower than zero position giving a small negative reading. The movement of slider above zero gives positive reading. The range of slide wire is usually from – 0.005 V to + 0.105 V.

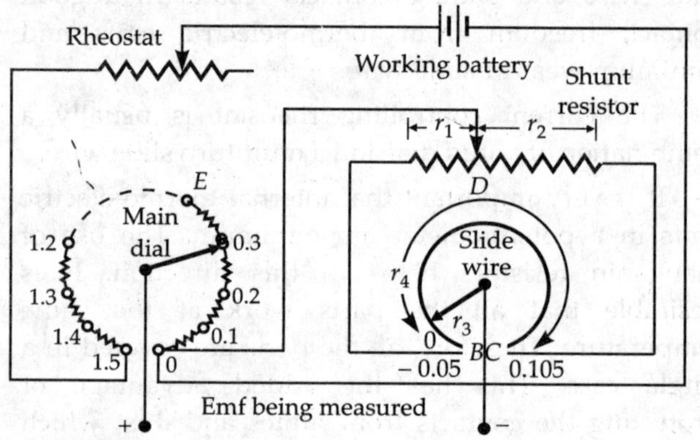


**Fig. 15.9** Potentiometer with true zero.

### 15.1.9 Brooks Deflectional Potentiometer

This potentiometer is used for applications where the voltage to be measured is continuously changing. With conventional type of potentiometers it becomes impossible to obtain exact balance even when the changes are slow since it is difficult to follow the changes by manipulation of the dials.

Figure 15.10 shows a deflectional potentiometer. In this potentiometer only one or two main dials,

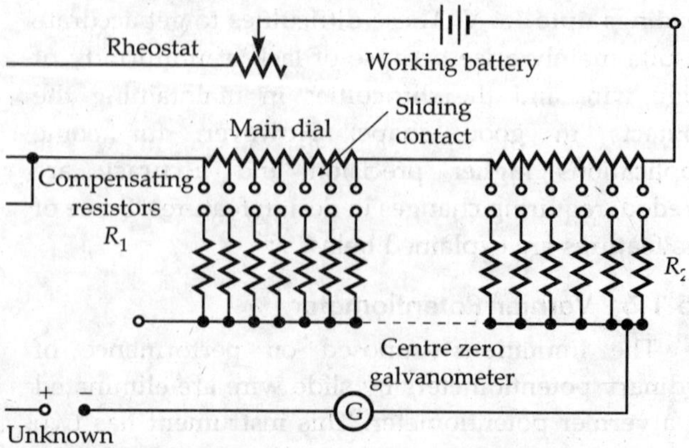


**Fig. 15.10** Deflectional Potentiometer.

consisting of decade resistance boxes, are used. A centre zero type galvanometer is included in the circuit to indicate deflection. The galvanometer circuit includes a set of compensating resistors $R_1$, $R_2$. The value of the compensating resistors is such that the resistance of the potentiometer circuit, as viewed from terminals where the unknown emf is applied, remains constant irrespective of the position of the sliding contacts. This means that current through the galvanometer will always be proportional to the out of balance current whatever may be the setting of the main dials. Thus the galvanometer scale can be calibrated to read the out of balance emf directly. The value of the unknown emf is obtained by adding the galvanometer reading to the main dial setting. The main dial setting is kept nearly equal to the emf being measured.

The potentiometer is used to monitor the value of temperature where a thermocouple is employed. The output of the thermocouple is fed to the unknown emf terminals.

Reasonably rapid changes in galvanometer indication can be followed with ease which are impossible to take with conventional potentiometers.

**Example 15.1** *A simple slide wire is used for measurement of current in a circuit. The voltage drop across a standard resistor of 0.1 Ω is balanced at 75 cm. Find the magnitude of the current if the standard cell emf of 1.45 V is balanced at 50 cm.*

**Solution.** Voltage drop per unit length

$$= 1.45/50 = 0.029 \text{ V/cm}.$$

Voltage drop across 75 cm length

$$= 0.029 \times 75 = 2.175 \text{ V}$$

∴ Current through resistor

$$I = 2.175/0.1 = 21.75 \text{ A}.$$

**Example 15.2** *A basic slide wire potentiometer has a working battery voltage of 3.0 V with negligible internal resistance. The resistance of slide wire is 400 Ω and its length is 200 cm. A 200 cm scale is placed along the slide wire. The slide wire has 1 mm scale divisions and it is possible to read upto 1/5 of a division. The instrument is standardized with 1.018 V standard cell with sliding contact at the 101.8 cm mark on scale. Calculate*

(a) *working current,*

(b) *the resistance of series rheostat,*

(c) *the measurement range and*

(d) *the resolution of instrument.*

**Solution.** (a) Since the instrument is standardized with an emf of 1.018 V with sliding contact at 101.8 cm, it is obvious that a length 101.8 cm represents a voltage of 1.018.

Resistance of 101.8 cm length of wire

$$= (101.8/200) \times 400 = 203.6 \ \Omega.$$

∴ Working current

$$I_m = 1.018/203.6 = 0.005 \text{ A} = 5 \text{ mA}.$$

(b) Total resistance of the battery circuit

= resistance of rheostat + resistance of slide wire

∴ Resistance of rheostat

$R_h$ = total resistance − resistance of slide wire

$$= \frac{3}{5 \times 10^{-3}} - 400 = 600 - 400 = 200 \ \Omega.$$

(c) The measurement range is the total voltage across the slide wire.

$$\therefore \quad \text{Range of voltage} = (5 \times 10^{-3}) \times 400 = 2.0 \text{ V}$$

(d) A length of 200 cm represents 2.0 volt and therefore 1 mm represents a voltage of

$$(2/200) \times (1/10) = 1 \text{ mV}.$$

Since it is possible to read 1/5 of 1 mm. Resolution of instrument is 1/5 × 1 = 0.2 mV.

**Example 15.3** *A single range student type potentiometer has a 18 step dial switch where each step represents 0.1 V. The dial resistors are 10 Ω. The slide wire of the potentiometer is circular and has 11 turns and a resistance of 11 Ω each. The slide wire has 100 divisions and interpolation can be done to one-fourth of a division. The working battery has a voltage of 6.0 V and negligible internal resistance. Calculate*

(a) *the measuring range of potentiometer,*

(b) *the resolution,*

(c) *working current and*

(d) *setting of rheostat.*

**Solution.** (a) Total resistance of measuring circuit

$R_m$ = resistance of dial + resistance of slide wire

$$= 18 \times 10 + 11 = 191 \ \Omega$$

Since each step of 10 Ω represents a voltage drop of 0.1 V, range of instrument

$$= 191 \times (0.1/10) = 1.91 \text{ V}$$

(b) The slide wire has a resistance of 11 Ω and therefore voltage drop across slide wire

$$= 11 \times (0.1/10) = 0.11 \text{ V}$$

The slide wire has 11 turns and therefore voltage drop across each turn

$$= (0.11/11) = 0.01 \text{ V}$$

Each turn is divided into 100 divisions and therefore each division represents a voltage drop of

$$0.01/100 = 0.0001 \text{ V}$$

15

Since each turn can be interpolated to 1/4 of a division,

∴ Resolution of instrument

$$= \frac{1}{4} \times 0.0001\text{V} = 0.000025\text{ V} = 25\ \mu\text{V}.$$

(*c*) The voltage drop across working circuit of potentiometer is 1.91 V and the resistance is 191 Ω.

∴ Working current

$$I_m = 1.91/191 = 0.001\text{ A} = 10\text{ mA}.$$

(*d*) Total resistance across battery circuit

$$= 6/(10 \times 10^{-3}) = 600\ \Omega$$

The resistance of potentiometer working circuit is 191 Ω.

∴ Resistance of series rheostat

$$R_h = 600 - 191 = 409\ \Omega$$

**Example 15.4** *A duo-range potentiometer is designed for two ranges : X 1 and X 0.1. The circuit of Fig. 15.3 is used. The main dial consists of 15 steps of 20 Ω each, and the slide wire has a resistance of 30 Ω. The standard cell voltage is 1.019 V. The potentiometer is designed to have a measuring range of 1.65 V d.c. on the X 1 range. Calculate*

(*a*) *the value of measuring current on each range,*

(*b*) *the resistance of range resistors $R_1$ and $R_2$,*

(*c*) *the current supplied by battery for each range and*

(*d*) *the resistance of rheostat if the battery has a voltage of 6.0 V.*

**Solution.** (*a*) An examination of the data shows that if we assume each step of dial resistor to represent 0.1 V, the voltage drop across the main dial is $0.1 \times 15 = 1.5\text{V}$. Now each step has a resistance of 20 Ω and therefore voltage drop across the 30 Ω slide wire

$$= (30/20) \times 0.1 = 0.15\text{V}.$$

∴ Total voltage drop across the measuring circuit

$$= 1.5 + 0.15 = 1.65\text{V}.$$

Thus to design a 1.65 V potentiometer, each step of main dial should represent 0.1 V.

Total resistance of measuring circuit

$$R_m = 15 \times 20 + 30 = 330\ \Omega$$

Measuring circuit current on X 1 range,

$$I_m = 1.65/330 = 0.005\text{ A} = 5\text{ mA}.$$

(*b*) The simplified circuits of X 1 and X 0.1 range potentiometers are given in Figs. 15.5 and 15.6. The analysis is also given along side which can be referred to

∴ $R_1 = R_m = 330\ \Omega$

and $R_2 = 9R_1 = 9 \times 330 = 2970\ \Omega$

(*c*) **Range : X 1**

Referring to Fig. 15.5,

$$\text{Current } I_1 = \frac{I_m R_m}{R_1 + R_2} = \frac{(5 \times 10^{-3}) \times 330}{330 + 2970}$$

$$= 0.5 \times 10^{-3}\text{ A} = 0.5\text{ mA}.$$

Total battery current

$$I_t = I_m + I_1 = 5 + 0.5 = 5.5\text{ mA}.$$

**Range : X 0.1**

Referring to Fig. 15.6, working current

$$I'_m = 0.1\ I_m = 0.1 \times 5 = 0.5\text{mA}$$

$$\text{Current } I_2 = \frac{I'_m (R_m + R_2)}{R_1}$$

$$= \frac{0.5 \times 10^{-3}(330 + 2970)}{330}$$

$$= 5 \times 10^{-3}\text{ A} = 5\text{mA}$$

Total battery current

$$I_t = I'_m + I_2 = 0.5 + 5 = 5.5\text{mA}.$$

(*d*) Voltage drop across working circuit = 1.65 V

Voltage of battery = 6.00 V

Voltage drop across rheostat = 6.00 – 1.65 = 4.35 V

Battery current = $5.5 \times 10^{-3}$ A.

∴ Resistance of rheostat

$$R_h = 4.35/(5.5 \times 10^{-3}) = 791\ \Omega$$

### 15.1.10 Volt-Ratio Box

General purpose potentiometers cover a maximum voltage range of approximately 0 – 1.6 V in case of American Instruments and 0 – 1.8 V in case of British Instruments. If higher voltage ranges have to be measured a precision potential divider, called a *volt-ratio box*, is used.

The volt-ratio box provides multiple ranges as shown in Fig. 15.11.

The voltage to be measured is connected to the appropriate binding post. Sometimes the volt-ratio box is provided with selector switch for voltage ranges. Supposing the voltage to be measured is connected between common terminal and 300 V switch. If the reading of the potentiometer is 0.826, then the value of unknown voltage is

$$0.826 \times (300/1.5) = 165.2.$$

It was stated earlier that the potentiometer consumes no power from the source under balance conditions. However, when a volt-ratio box is used, this statement is no longer true. An examination of Fig. 15.11 indicates that a current must flow through resistance $R$, in order to produce a voltage drop so that a voltage measurement be made by the potentiometer in order that the unknown voltage be determined. Therefore, the resistance of the volt-ratio box represents a load on the unknown voltage source and thus consumes power. The value of volt-ratio box resistance should be as high as possible in order that the power consumption is low. However, low values of resistance are preferred since they have a greater degree of stability. Also low values of resistance give greater galvanometer sensitivity and also minimize the effects of high resistance leakage paths around the terminals.

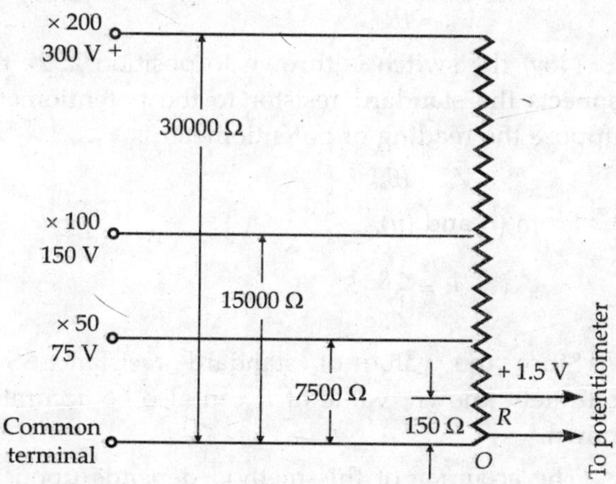


**Fig. 15.11** Volt-Ratio box.

Thus the choice of volt-ratio box resistance involves a compromise between low and high values. Volt-ratio boxes are usually constructed with resistances of 100 Ω/V or 200 Ω/V of the nominal voltage range, allowing a maximum current of 10 mA or 5 mA respectively from the measured source. Figure 15.11 shows a volt-ratio box with a resistance of 100 Ω /V and a maximum current of 10 mA.

**Example 15.5** *A volt ratio box is shown in Fig. 15.12. The volt ratio box should be so designed that when an unknown voltage of 100 V is applied to the input terminals, an output voltage of 2 V is available at the box output terminals. An additional requirement is that the total input resistance ($R_1 + R_2$) of the box is 10 MΩ. Determine the volt-box ratio and the relationship between $R_1$ and $R_2$.*

**Solution.** $$\frac{\text{Input (unknown) voltage}}{\text{Output voltage}} = \frac{V_x}{V_2} = \frac{100}{2} = 50.$$

Also $$\frac{R_1 + R_2}{R_2} = \frac{V_x}{V_2} = 50$$

or $$R_1 = 49\,R_2$$

But $$R_1 + R_2 = 100\ \text{M}\Omega$$

or $$49\,R_2 + R_2 = 100\ \text{M}\Omega$$

$$\therefore \quad R_2 = 2\ \text{M}\Omega \quad \text{and} \quad R_1 = 98\ \text{M}\Omega.$$

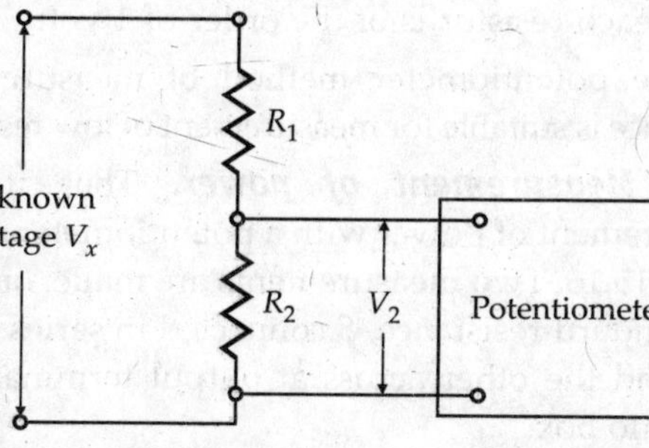


**Fig. 15.12** Diagram of Example 15.5.

### 15.1.11 Applications of D.C. Potentiometers

In addition to measurement of voltage, the potentiometer is the usual basis for calibration of all voltmeters, ammeters and wattmeters. The potentiometer may also be used for measurement of current, power and resistance. Since the potentiometer is a d.c. device, the instruments to be calibrated must be d.c. moving iron or electrodynamometer types.

**1. *Calibration of voltmeter.*** Figure 15.13 shows the circuit for calibration of a voltmeter. The foremost requirement in this calibration process is that a suitable stable d.c. voltage supply is available since any changes in the supply voltage will cause a corresponding change in the voltmeter calibration.

Figure 15.13 shows a potential divider network, consisting of two rheostats, one for coarse and the other for fine control of calibrating voltage. These controls are connected to the supply source and with the help of these controls it is possible to adjust the voltage so that the pointer coincides exactly with a major division of the voltmeter. The voltage across the voltmeter is stepped down to a value suitable for application to a potentiometer with the help of a volt-ratio box. For accuracy of measurements, it is necessary to measure voltages near the maximum range of the potentiometer, as far as possible.

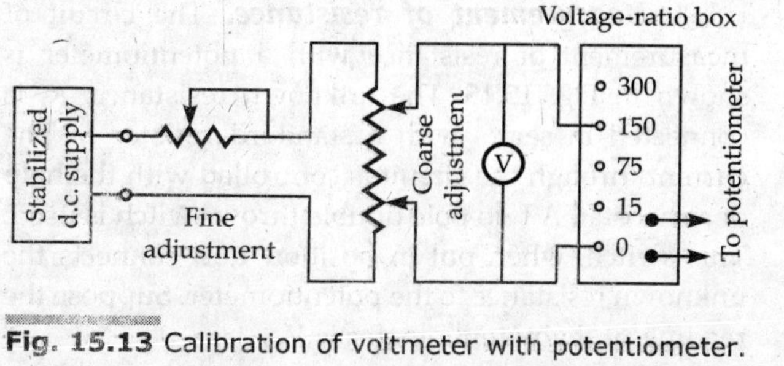


**Fig. 15.13** Calibration of voltmeter with potentiometer.

Thus if a potentiometer has a maximum range of 1.6 V, to achieve high accuracy we will have to use low, voltage ranges for voltages less than 1.6 V and use appropriate tappings on volt-ratio box for voltages higher than 1.6 V.

The potentiometer measures the true value of voltage. If the potentiometer reading does not agree with the voltmeter reading, a negative or positive error is indicated. A calibration curve may be drawn with the help of the readings of voltmeter and potentiometer.

2. ***Calibration of ammeter.*** Figure 15.14 shows the circuit for calibrating an ammeter. A standard resistance of suitable value and sufficient current carrying capacity is placed in series with the ammeter under calibration.

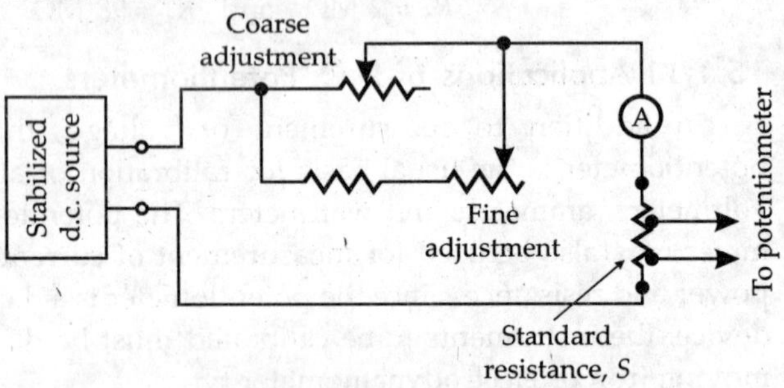


**Fig. 15.14** Calibration of an ammeter with potentiometer.

The voltage across the standard resistor is measured with the help of potentiometer and the current through the standard resistance (and hence the ammeter) can be computed.

Current $I = V_S/S$

where

$V_S$ = voltage across the standard resistor as indicated by the potentiometer,

and $S$ = resistance of standard resistor.

Since the resistance of the standard resistor is accurately known and the voltage across the standard resistor is measured by a potentiometer, this method of calibrating an ammeter is very accurate. A calibration curve indicating the errors at various scale readings of the ammeter may be plotted.

3. ***Measurement of resistance.*** The circuit of measurement of resistance with a potentiometer is shown in Fig. 15.15. The unknown resistance, $R$, is connected in series with a standard resistor $S$. The current through the circuit is controlled with the help of a rheostat. A two pole double throw switch is used. This switch, when put in position 1, 1′ connects the unknown resistance to the potentiometer. Suppose the reading of the potentiometer is $V_R$.

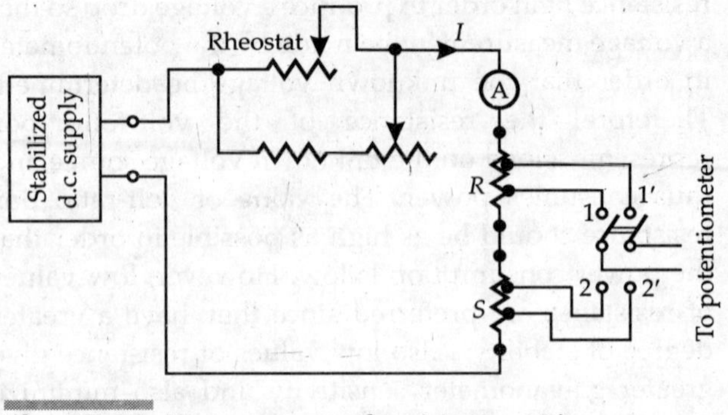


**Fig. 15.15** Measurement of resistance with potentiometer.

$$\therefore \quad V_R = IR \qquad \ldots(i)$$

Now the switch is thrown to position 2, 2′, this connects the standard resistor to the potentiometer. Suppose the reading of potentiometer is $V_S$.

$$V_S = IS \qquad \ldots(ii)$$

From (*i*) and (*ii*),

$$R = \frac{V_R}{V_S} \cdot S$$

Since the value of standard resistance $S$ is accurately known, value of $R$ can also be accurately known.

The accuracy of this method depends upon the assumption that there is no change in the value of current when the two different measurements are taken. Therefore a stable d.c. supply is absolutely necessary. The difficulty of ensuring this condition is the chief disadvantage of this method.

The resistance of the standard resistor, $S$, which must be accurately known, should be of the same order as the resistance, $R$, under measurement. The ammeter inserted in the circuit is merely for indicating whether the current flowing through the circuit is within the capacity of the resistors or not otherwise the exact value of current flowing need not be known. It is desirable that the current flowing through the circuit be so adjusted that the value of voltage drop across each resistor is of the order of 1 volt.

The potentiometer method of measurement of resistance is suitable for measurement of low resistances.

4. ***Measurement of power.*** The circuit for measurement of power with a potentiometer is shown in Fig. 15.16. Two measurements are made, one across the standard resistance, $S$ connected in series with the load and the other across at output terminals of the volt-ratio box.

The current in the circuit can be computed from the voltage drop across the standard resistance and the voltage across the load can be computed from the potentiometer reading across the output terminals of volt-ratio box.

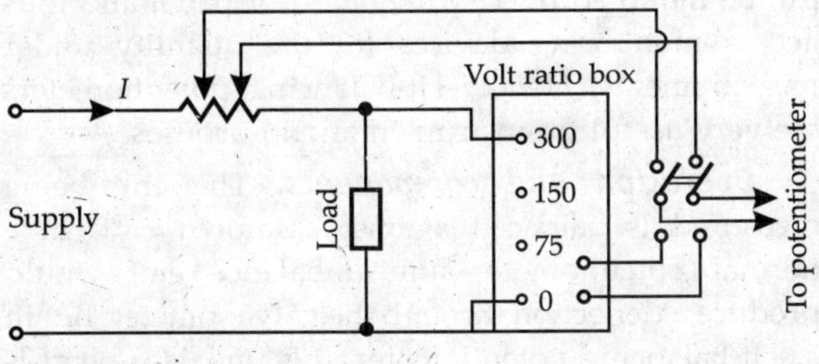


**Fig. 15.16** Measurement of power with potentiometer.

Let

$V_S$ = reading of potentiometer when connected across $S$.

$V_R$ = reading of potentiometer when connected across volt-ratio box.

∴ Current through the load, $I = V_S/S$.

Voltage across the load $V = KV_R$,

where $K$ = multiplying factor of volt-ratio box.

∴ Power consumed $= VI = KV_R = V_S / S$

**5. *Calibration of wattmeter.*** For the calibration of a wattmeter, a circuit similar to Fig. 15.16 may be used. Such an arrangement, however, results in a considerable consumption of power especially when the wattmeter has a large rating. In order to save expenditure of power, the arrangement of Fig. 15.17 is used. The current coil of wattmeter is supplied from a low voltage supply and a series rheostat is inserted to adjust the value of current.

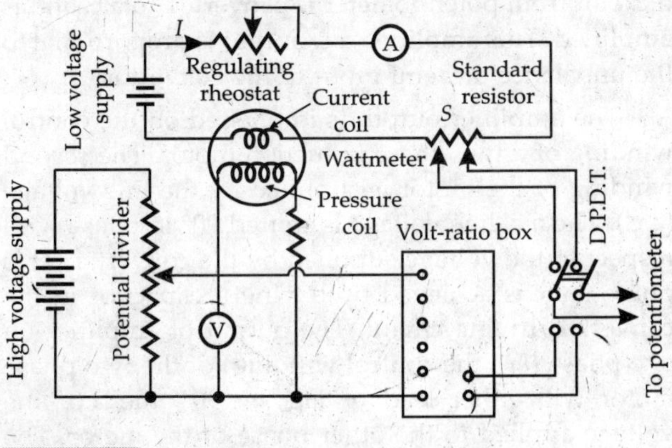


**Fig. 15.17** Calibration of wattmeter with a d.c. potentiometer.

The potential circuit is supplied from the supply. A volt-ratio box is used to step down the voltage for the potentiometer to read. As explained earlier on page 396, this type of arrangement is known as *"Phantom Loading"*. The voltage, $V$, and the current, $I$, are measured in turn with the potentiometer employing a double pole double throw (D.P.D.T.) switch. The true power is then $VI$ and the wattmeter reading may be compared with this value.

**Example 15.6** *During the measurement of a low resistance using a potentiometer the following readings were obtained :*

*Voltage drop across the low resistance under test = 0.4221 V,*

*Voltage drop across a 0.1 Ω standard resistance = 1.0235 V.*

*Calculate the value of unknown resistance, current and the power lost in it.*

**Solution.** Resistance of unknown resistor

$$R = \frac{V_R}{V_S} \cdot S = \frac{0.4221}{1.0235} \times 0.1 = 0.041208\ \Omega$$

Current through the resistor

$$= \frac{V_S}{S} = \frac{1.0235}{0.1} = 10.235\ \text{A}.$$

Power loss in unknown resistance

$$= I^2R = (10.235)^2 \times 0.041208 = 4.316\ \text{W}.$$

**Example 15.7** *In the measurement of low resistance by means of a potentiometer, the following readings were obtained :*

*Voltage drop across low resistance under test = 0.83942 V,*

*Voltage drop across a 0.1 Ω standard resistance connected in series with the unknown resistance = 1.01575 V.*

*The resistance of standard resistor at the temperature of test is 1.00014 Ω. Upon setting the potentiometer dials to zero and breaking the current passing through the unknown resistance, the thermal emf of latter produced a galvanometer deflection equivalent to 23 μV, the direction of the deflection being the same as that produced by an increase of the potentiometer reading during the voltage measurement. Calculate the resistance of the unknown.*

**Solution.** Indicated voltage drop across the unknown resistance

$$= 0.83942\ \text{V}.$$

Thermal emf with unknown resistance

$$= 23\mu\ \text{V} = 0.000023\ \text{V}.$$

∴ Actual voltage drop across unknown resistance

$$V_R = 0.839420 - 0.000023 = 0.839397\ \text{V}.$$

∴ Unknown resistance

$$R = \frac{V_R}{S} \times S = \frac{0.839397}{1.01575} \times 0.10014$$

$$= 0.08275\ \Omega$$

**Example 15.8** *A potentiometer that is accurate to ± 0.0001 V (Standard deviation) is used to measure current through a standard resistance of 0.1±0.1Ω % (Standard Deviation). The voltage across the resistance is measured to be 0.2514 V. What is the current and to what accuracy it has been determined ?*

**Solution.** Current, $I = V/S = 0.2514/0.1 = 2.514$ A.

The fractional standard deviation of current is

$$\frac{\delta I}{I} = \sqrt{\left(\frac{\delta V}{V}\right)^2 + \left(\frac{\delta S}{S}\right)^2}$$

$$= \sqrt{\left(\frac{0.0001}{0.2514}\right)^2 + (0.001)^2} = 0.00108.$$

Percentage accuracy of measurement of current

$= \pm\, 0.00108 \times 100 = \pm\, 0.108\%.$

### 15.1.12 Protection of Standard Cell and Galvanometer

When the potentiometer is not balanced, the galvanometer circuit carries current. This current may be excessively high if the working point is far away from balance conditions. Another delicate component in the potentiometer set up is the standard cell, in which the current must be limited to 0.1 mA otherwise it will be damaged. The usual practice to protect a galvanometer against over currents is to shunt it with a low resistance. But if this practice is followed in a potentiometer, it will not only fail to protect the standard cell but on the other hand increase the current in the standard cell during the standardizing process. This is because the resistance of the circuit decreases considerably on account of a low resistance put in parallel with the galvanometer.

Therefore a series resistance is used to limit the current in both the standard cell and the galvanometer. This resistance should be sufficiently large to protect the circuit if the polarity of standard cell should accidentally be reversed. The current in the galvanometer circuit is to be limited to 0.1 mA and assuming a voltage of 1.02 V for the standard cell, the total series resistance of the circuit should be

$$R = \frac{1.02 + 1.02}{0.1 \times 10^{-3}} = 20{,}400\ \Omega$$

Thus a series resistance of at least 20,000 Ω be inserted in the standard cell circuit in order to protect it against overcurrents.

## 15.2 SELF-BALANCING POTENTIOMETERS

Self-balancing potentiometers are widely used in industry because their action is automatic and thus eliminates the constant attention of an operator. In addition to their automatic self-balancing features, they also draw a curve of the quantity being measured with the help of recording mechanism. Besides they can be mounted on a switchboard or panel and thus act as monitoring devices for the quantity under measurements. Hence self-balancing potentiometers are very useful for instrumentation purposes.

***Operation and components.*** The emf being measured is balanced against a known emf. In a normal potentiometer any unbalance emf would produce a deflection through the galvanometer. But in a self-balancing potentiometer, this unbalance emf is applied to the input of an amplifier which amplifies the signal. This amplified output emf is made to drive a motor that moves a sliding contact which balances the potentiometer.

However, there are some inherent disadvantages in the above set-up. The unbalance emf is d.c. and this is fed to a d.c. amplifier. Now d.c. amplifiers are not stable and therefore their output is subject to a drift, and hence working of the system with a d.c. amplifier is not satisfactory.

To overcome the above difficulty, a convertor is placed between the potentiometer and the amplifier. The convertor consists of a vibrating reed which is excited with alternating current. The reed operates a double throw switch which reverses the flow of current through the primary winding of a transformer for each vibration cycle of the reed. Since the current through the primary winding reverses an alternating voltage is induced in the secondary winding of the transformer. The secondary winding of the transformer is connected to an a.c. amplifier. Thus the unbalance d.c. emf from potentiometer is converted to a.c. and is amplified. This amplified a.c. output is proportional to the unbalance d.c. emf input to the convertor.

The amplifier output is impressed on the control winding of a two phase induction motor. The second winding of the motor is supplied by the line voltage (a.c.). The a.c. line voltage is shifted 90° in phase with respect to the voltage supplied by the converter. This phase shift is achieved by having a capacitor in the converter driving circuit. The output of amplifier on one phase (*i.e.*, the control winding) of the two phase motor will either lead or lag, by 90°, the a.c. line voltage applied to the other phase of the motor. The direction of rotation of the motor depends upon the phase of the amplifier output voltage which, in turn, depends upon the polarity of the unbalance d.c. emf fed to the converter. Thus if the emf being measured is greater than the balancing voltage of the potentio-meter, the motor will rotate in one direction. If the emf

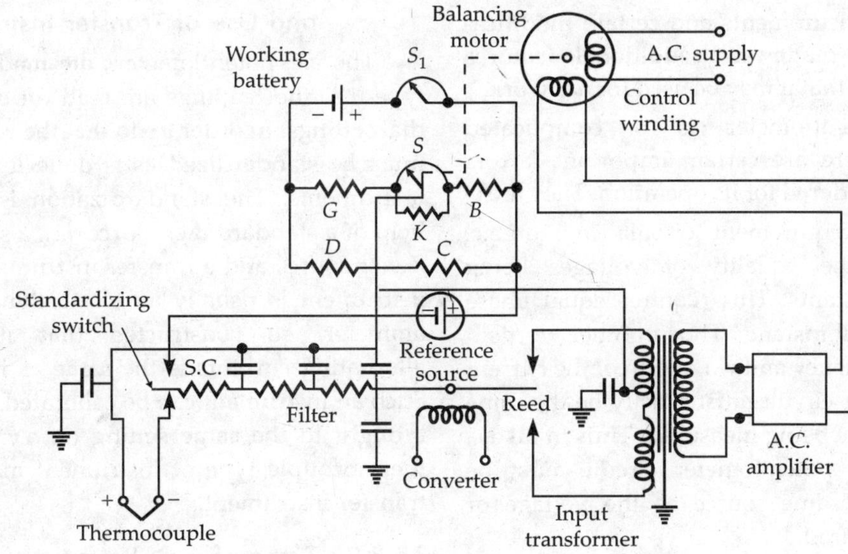


**Fig. 15.18** Self-balancing potentiometer.

being measured is smaller than the balancing voltage, the polarity of unbalance voltage is reversed and thus the output voltage of the amplifier is shifted by 180° and the motor revolves in the opposite direction.

The shaft of the motor is connected mechanically to the slide wire in such a way, that the rotation of the motor decreases the unbalance in the potentiometer circuit. When the emf being measured is equal to the potentiometer voltage, the amplifier output voltage is zero and the motor thus stops rotating. Therefore under any condition of unbalance, the amplifier output voltage will cause the motor to move and cause the sliding contact to come to balance conditions.

The circuit diagram of a Leeds Northrup spead O-Max self-balancing potentiometer is shown in Fig. 15.18. This circuit is used to record the temperature of thermocouples. The emf produced by a thermocouple is a function of the temperature difference between its hot and cold junctions. The variation in temperature of the reference junction is compensated by an electrical compensating circuit. The drop across resistor *D*, which is made of nickel copper alloy, compensates for the change in temperature of reference junction. Resistor *G* balances the voltage drop across *D* at the desired base temperature. Zero suppression may also be included in resistor *G*. Resistance *K* and slide wire *S* form the actual measuring circuit and resistor *B* produces correct voltage drop for the calibration of the circuit with the reference voltage. The reference voltage is provided by a Zener source. The working current can be adjusted by a rheostat $S_1$.

The signal supplied to the potentiometer is filtered with a low pass filter. The filter capacitors across the input circuit have no effect upon the d.c. emf supplied to the potentiometer but variations in the input signal and any stray a.c. signals that may be impressed upon the input signal are smoothened out by the filter.

The motor, which moves the slide wire contact to maintain potentiometer balance, is mechanically coupled to a pen mechanism. The pen, actuated by the mechanism, moves over a chart which is driven by a separate clock motor. Thus the temperature is recorded on the strip chart.

We have already mentioned some of the applications of self-balancing potentiometers. There are many more applications. They are even used in multiplying and dividing networks in analogue computers and in special function generators.

## 15.3 A.C. POTENTIOMETERS

The d.c. potentiometer is an accurate and versatile instrument and thus it is obvious that the potentiometer principle be applied to measurement of alternating currents and voltages. The principle of alternating current potentiometer is the same as that of the direct current potentiometer. The most important difference between a d.c. and an a.c. potentiometer is that, whereas in a d.c. potentiometer only the magnitudes of the unknown emf and potentiometer voltage drop have to be made equal to obtain balance, in the a.c. instrument both magnitudes and phases of the two have to be same to obtain

15

balance. Thus an ordinary d.c. potentiometer cannot be used for a.c. measurements and certain modifications have to be made and additional features incorporated in it so that it may be used for a.c. work.

The a.c. potentiometer is a complicated instrument and there are certain important factors which must be considered for its operation. They are :

**1.** A necessary requirement for balance in an a.c. potentiometer is the equality of voltages being compared at all instants. This requires equal phase and magnitude at all instants. Thus in other words it means that the frequency and waveform of the current in the potentiometer circuit must exactly be the same as that of the voltage being measured. Thus in all a.c. potentiometers the potentiometer circuit must be supplied from the same source as the voltage or current being measured.

**2.** A vibration galvanometer, which is a tuned device, is usually used as a detector in a.c. potentiometers. The presence of harmonics in one or both of the voltages being compared, the balance point may not be the same if a tuned detector (responding to only one frequency) is used. As an alternative arrangement, if an average indicating detector is used, it may not show the same balance point as an rms indicating device. In the presence of harmonics, it may be possible that a balance can never be achieved, the detector showing only a minimum balance. In such a situation, the accuracy of measurements is seriously affected. Therefore, the source of a.c. supply is made as sinusoidal as possible.

**3.** The ratio of two voltages (*i.e.*, the unknown voltage and the voltage across the potentiometer) may be determined with a high degree of precision, the accuracy with which the value in volt can be stated is determined by the accuracy with which the reference voltage (*i.e.*, voltage across the potentiometer) is known or the accuracy with which a reference (working) current can be known. There being no a.c. reference source (the reference source in d.c. being a standard cell or a Zener source), the absolute accuracy with which an a.c. voltage can be measured in an a.c. potentiometer cannot be comparable with corresponding type of d.c. measurement.

**4.** Extraneous or stray emfs picked up from stray fields or couplings between portions of the potentiometer circuit seriously effect the result. These emfs must be eliminated, compensated for or measured since they may add vectorially to the emf being measured.

### 15.3.1 Standardizing of A.C. Potentiometers and Use of Transfer Instruments

The a.c. potentiometers are made direct reading type, *i.e.*, the readings are read off directly from the dial settings. In order to do that the a.c. potentiometer must be standardized as is done in the case of d.c. instruments. The standardization is done with the help of a standard d.c. source, *i.e.*, a standard cell or a Zener source and a transfer instrument. This transfer instrument is usually an electrodynamometer milli-ammeter, so constructed that its response to alternating currents is the same as its d.c. response. Such an instrument can be calibrated on d.c. and then brought to the same setting on a.c. Alternatively a thermocouple type of instrument may be used as a transfer instrument.

### 15.3.2 Types of A.C. Potentiometers

A.C. potentiometers may be classified according to the manner in which the value of unknown voltage may be measured by the instrument dials and scales. The a.c. potentiometers may be broadly classified as :

**1. *Polar type.*** In these instruments, the magnitude of the unknown voltage is read from one scale and its phase angle, with respect to some reference phasor, is read directly from a second scale. Provision is made to read phase angles up to 360°. The voltage is read in the form $V\angle\theta$ (See Fig. 15.19).

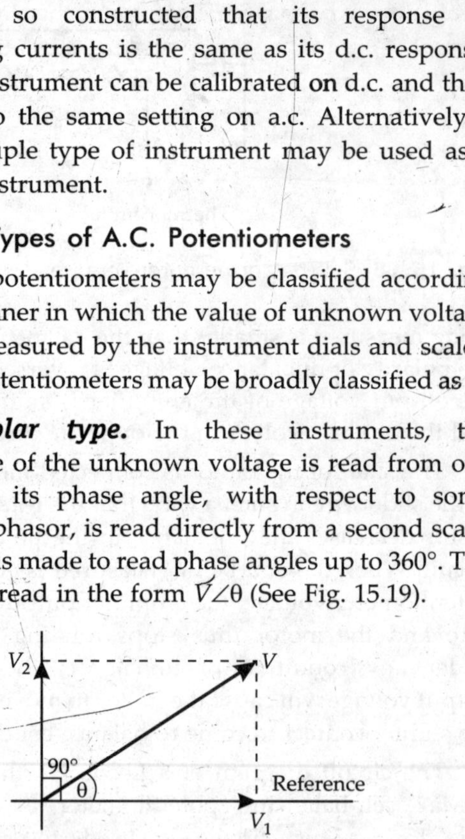


**Fig. 15.19** Phasor diagram for polar type potentiometer.

**2. *Co-ordinate type.*** These instruments are provided with two scales to read respectively the **inphase** component '$V_1$' and the **quadrature** component '$V_2$' of the unknown voltage $V$. These components are 90° out of phase with each other.

Then voltage $V = \sqrt{V_1^2 + V_2^2}$ and its phase angle with respect to current in the 'inphase' portion of the potentiometer is $\theta = \tan^{-1}(V_2/V_1)$. Provision is made in these potentiometers to read both positive and negative values of $V_1$ and $V_2$ so that all angles up to 360° are covered.

### 15.3.3 Drysdale Polar Potentiometer

This instrument consists of a potentiometer of ordinary d.c. type, except that the resistance coils are non-inductively wound and the following auxiliary apparatus is used :

1. ***Drysdale phase shifting transformer.*** The phase shifting transformer consists of a laminated silicon steel stator which houses a two or a three phase winding. The rotor is also a laminated structure having slots in which a winding is provided. There is a small air gap between the rotor and the stator. When current flows in the stator winding, a rotating field is produced, thereby inducing an emf in the rotor winding. The rotor can be adjusted at will through any required angle, the phase displacement of rotor emf being equal to the angle through which the rotor has been moved from its zero position. A scale and pointer are provided on top of the instrument to indicate the angle through which rotor has been moved from its zero position. The scale is graduated both in degrees and cosines of the angles.

A phase shifter operated from a single phase supply may also be used by means of a phase splitting device. This device is shown in Fig. 15.20. In this phase shifter it is only necessary to have two separate windings displaced by 90° in space on the stator. One winding is fed directly from the supply while the other winding is connected in series with a variable resistance $R$ and a variable capacitor $C$. The values resistance $R$ and capacitance $C$ are adjusted until the currents through the two windings are equal and are 90° displaced from each creating a uniform revolving field. The quadrature adjustment is explained in Art. 15.3.4.

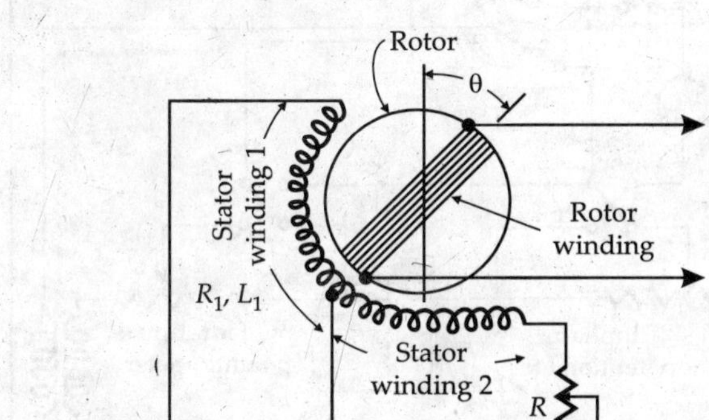


**Fig. 15.20** Drysdale phase shifting transformer.

For a particular position of the rotor the induced emf due to each of the stator winding is proportional to the cosine of the angular displacement from position of maximum emf. The emf induced in rotor winding due to stator winding 1,

$$e_1 = KI \sin \omega t \cos \theta$$

and that due to stator winding 2 is

$$e_2 = KI \sin(\omega t + 90°) \cos(\theta + 90°)$$
$$= -KI \cos \omega t \sin \theta$$

∴ The resultant emf in the rotor

$$e = KI[\sin \omega t \cos \theta - \cos \omega t \sin \theta]$$
$$= KI \sin(\omega t - \theta)$$

The above relationship shows that the rotor emf has a constant amplitude but the phase angle is given by rotor position θ.

When a three phase supply is available, it is usual to employ a three phase shifting transformer as it is less sensitive to small changes in frequency than the phase splitting circuit using single phase supply. 15

2. ***Transfer instrument.*** A precision type electro-dynamometer ammeter is required for standardization purposes. In order to standardize the a.c. potentiometer the slide wire circuit is switched on to a direct-current supply, and the standard current is obtained in the ordinary way, using a standard cell. This standard current, required to make the potentiometer direct reading, is measured by a precision ammeter which is included in the battery supply circuit of the potentiometer. During operation on alternating current, the ammeter is still included in the supply circuit, and the rms value of the slide-wire current is maintained at the same value as was required on direct current. Since the coils of the potentiometer are non-inductively wound, the potentiometer remains direct reading when used with an alternating current supply. A change over switch, to enable the potentiometer to be used on either direct or alternating current, is also included in the auxiliary apparatus.

***Operation with alternating current.*** A simplified diagram of connections of the potentiometer for use with alternating currents is given in Fig. 15.21. The Kelvin-Varley slide principle is employed in the slide-wire circuit as shown. $V.G$ is a vibration galvanometer — used as a detector for measurements at commercial frequencies. This must be carefully tuned to give resonance at the frequency of the circuit under test, $r$ is a shunting resistor for the reduction of the range of the potentiometer. When this shunt is put in circuit by the switch $S_1$, the resistor $R$ is simultaneously connected in series with the slide wire circuit in order that the resistance of the working portion of the potentiometer may be maintained constant. $R$ is a rheostat for adjustment of the slide-wire current. $A$ is the precision ammeter as mentioned above. The phase-shifting transformer whose connections are given in Fig. 15.20 is omitted for the sake of clearness.

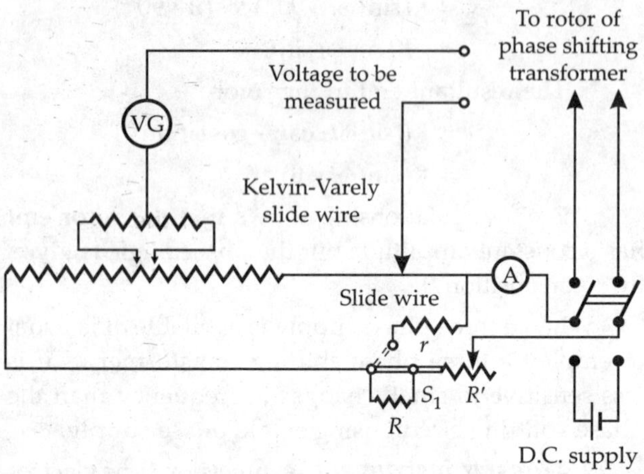


**Fig. 15.21** Drysdale (Polar type) A.C. Potentiometer.

The potentiometer is first standardized by adjusting rheostat and the standard current is noted, the switch $S$ being thrown over to the battery side for this standardization, the vibration galvanometer being replaced by a d'Arsonval galvanometer.

The switch $S$ is then thrown over to the alternating supply side, the standard cell and d'Arsonval galvanometer being previously replaced by the alternating voltage to be measured and the vibration galvanometer, respectively. The stator windings of the phase-shifter are then adjusted to exact quadrature by means of the variable resistor and capacitor, these being adjusted until the alternating current in the slide-wire is constant for all positions of the rotor.

Balance of the potentiometer is obtained by successive adjustment of the sliding contacts of the slide-wire circuit and of the rotor of the phase-shifter. The readings of potentiometer dials and the slide-wire, at balance, give the magnitude of the voltage being measured, as in the case of direct current measurements, while the reading on the scale of the phase-shifter gives the phase of the voltage being measured relative to the supply voltage. If the voltage measured is that across a standard resistance through which the current in the circuit under test is flowing, the magnitude of this current is obtained by dividing the measured value of the voltage by the value of the standard resistance, while its phase relative to the voltage of the circuit is read off from the scale of the phase shifting transformer. For accurate results it is necessary that the voltage and frequency of the supply are steady and that the waveform of the voltage is reasonably sinusoidal.

### 15.3.4 Gall-Tinsley (Co-ordinate type) A.C. Potentiometer

This potentiometer consists of two separate potentiometer circuits enclosed in a common case. One is called the *"in-phase"* potentiometer and the other the *"quadrature"* potentiometer. The slide-wire circuits are supplied with currents which have a phase difference of 90°. On the first of these potentiometers, that component of the "unknown" voltage which is in phase with the current in the slide-wire circuit of the potentiometer is measured. On the other potentiometer the component of the "unknown" voltage in phase quadrature with the current in its slide-wire circuit is measured. Since the two slide-wire currents are in quadrature, the two measured values are the quadrature components of the unknown voltage. If these measured values are $V_1$ and $V_2$ respectively, then the unknown voltage is given by $V = \sqrt{V_1^2 + V_2^2}$, and its phase difference from the current in the "in-phase" potentiometer slide-wire circuit is given by the angle $\theta$ where $\tan \theta = V_2 / V_1$.

Figure 15.22 shows the connections of the potentiometer, simplified somewhat for the purpose of clearness. The in-phase and quadrature potentiometer circuit are shown, with their sliding

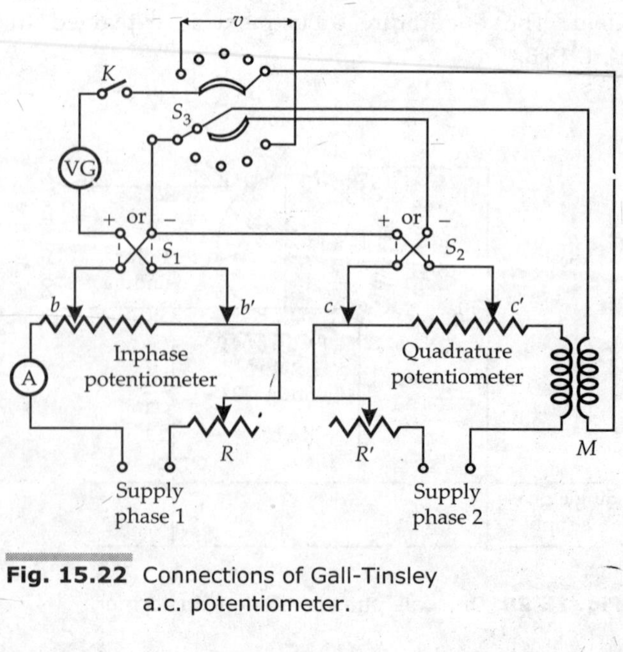


**Fig. 15.22** Connections of Gall-Tinsley a.c. potentiometer.

contacts $bb'$ and $cc'$ and rheostats $R$ and $R'$ for current adjustment. The supplies to the potentiometer are obtained from a single-phase supply by means of the arrangement shown in Fig. 15.23.

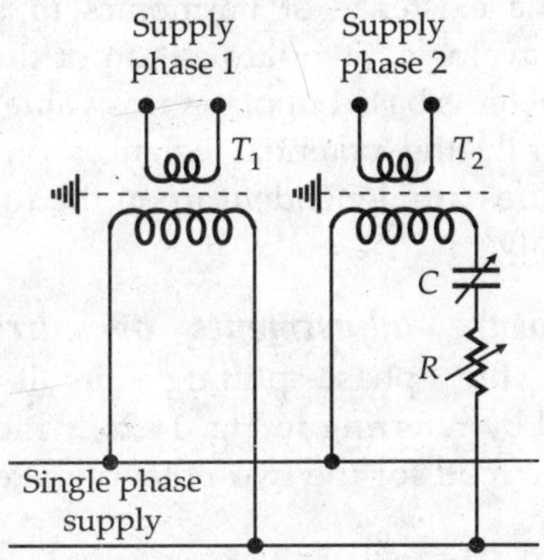


**Fig. 15.23** Phase-splitting circuit.

$T_1$ and $T_2$ are two step-down transformers which supply about 6 V to the potentiometer circuits. They also serve to isolate the potentiometer from the line and are usually provided with earthed screens between the windings. The supply to $T_2$ is obtained through a variable resistor $R$ and variable capacitor $C$ for the purpose of phase splitting. Quadrature phase displacement is obtained by adjusting $C$ and $R$.

Referring again to Fig. 15.22, V.G. is a vibration galvanometer (tuned to the supply frequency) and $K$ is its key. $A$ is a reflecting dynamometer instrument for maintaining the current in the two slide-wires at the standard value (50 mA). $S_1$ and $S_2$ are two "*sign-changing*" switches which may be necessary to reverse the direction of the unknown emf applied to the slide-wires. The necessity of these switches depends on the relative phases of the unknown and slide wire-voltages. $S_3$ is a selector switch by which the unknown voltages to be measured are placed in the circuit. There are four pairs of terminals for the application of such voltages, the connections to only one pair—to which an unknown voltage $V$ is applied—being shown in the figure. The selector switch, when in the position shown in the figure—called the "test position"—allows the current in the quadrature potentiometer slide-wire to be compared with that in the in-phase potentiometer wire, utilizing the mutual inductance $M$ for the purpose.

The current in the in-phase potentiometer wire is first adjusted to its standard value by means of a direct current supply and a standard cell, the vibration galvanometer being replaced by a galvanometer of the d'Arsonval type for this purpose. The dynamometer ammeter is of the torsion head type, and the torsion head is turned to give zero deflection on direct current. This setting is left untouched during the calibration with alternating current, the slide-wire current being adjusted to give zero deflection again. The vibration galvanometer is then placed in circuit and the direct current supply is replaced by the alternating supplies.

Now, the magnitude of the current in the quadrature potentiometer wire must be the same as that in the in-phase potentiometer—namely, the standard value of 50 mA. These two currents must also be exactly in quadrature. Rheostat $R$ is adjusted until the current in the in-phase potentiometer wire is the standard value (as indicated on $A$). The selector switch $S_3$ is then switched on to the test position (shown in Fig. 15.22). The emf induced in the secondary winding of the mutual inductor $M$—assuming $M$ to be free from eddy current effects—will lag 90° in phase behind the current in the primary winding, *i.e.*, in the quadrature potentiometer slide-wire. Also if $i$ is the primary current, then the emf induced in the secondary is $2\pi \times \text{frequency} \times M \times i$ where $M$ is the value of the mutual inductance. Therefore, for given values of frequency and mutual inductance, the induced emf when $i$ has the standard value 50 mA, can easily be calculated. Now, if $f = 50\text{Hz}$ and $M = 0.0636$ H, the secondary induced emf = $2\pi \times 50 \times 0.0636 \times 0.05 = 1.0\,\text{V}$ when $i$ has the standard value.

15

The slide-wire of the in-phase potentiometer is thus set to this calculated value of induced emf in the secondary of $M$ (the slide-wire current being maintained at its standard value), and rheostat $R$ and capacitor $C$ (See Fig 15.23) are adjusted until exact balance is obtained. For balance the current in the quadrature potentiometer slide-wire must be both equal to the standard value and also must be exactly 90° out of phase with the current in the in-phase slide-wire. This latter condition follows from the fact that the emf in the secondary of $M$ lags 90° in-phase behind the primary current, and, therefore, for this emf, to be *in-phase* with the voltage drop across a portion of the in-phase slide-wire, the current in the primary of $M$ must be in *exact quadrature* with the current in this in-phase slide-wire. Any difference in polarity between the two circuits is corrected for by the sign-changing switches $S_1$ and $S_2$.

These adjustments having been made, the unknown voltage is switched in circuit by means of the selector switch $S_3$. In this position of $S_3$, the two slide-wire circuits are in series with one another and with the vibration galvanometer. Balance is obtained by adjusting both pairs of contacts (*bb*′ and *cc*′) together with the reversal of switches $S_1$ and $S_2$, if necessary. At balance, the reading of the slide-wire of the in-phase potentiometer, together with the position of $S_1$, gives the magnitude and sign of the in-phase component of the unknown voltage, while the reading of the quadrature potentiometer, with the position of $S_2$, gives the magnitude and sign of the quadrature component.

For example, if both $S_1$ and $S_2$ are in the positive position and $V_1$ and $V_2$ are the in-phase and quadrature components respectively of the unknown voltage $V$, then the phase of $V$ is as shown in Fig. 15.24, while its magnitude is $\sqrt{V_1^2 + V_2^2}$.

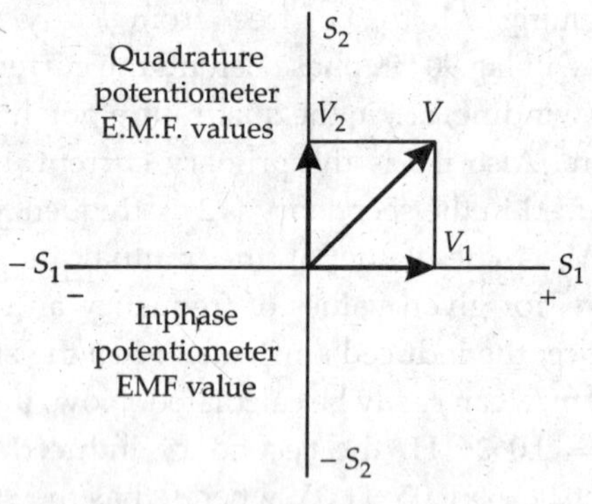


**Fig. 15.24** Phasor diagram for co-ordinate type potentiometer.

***Errors.*** The errors which may occur in using this potentiometer may be due to :

(*a*) Slight differences in the reading of the reflecting dynamometer instrument on a.c. as compared with the reading on d.c. Such errors may cause the standard current value on a.c. to be slightly incorrect.

(*b*) Mutual inductance between the various parts of the circuit. An error in the nominal value of the mutual inductance *M* would cause the current in the quadrature slide-wire circuit to be somewhat different from the standard value.

(*c*) Inaccuracy of the method of measuring the frequency, which again would cause an error in the quadrature slide-wire standard current value.

(*d*) The fact that inter-capacitance, earth capacitance and mutual inductance effects are present in the slide-wire coils and affect the potential gradient.

(*e*) The existence of harmonics in the supply waveform. Standardization of the potentiometer is based upon an rms value of current, while the potential balances on the slide-wires are dependent upon the fundamental only.

***Quadrature adjustments of currents.*** The action of the phase-splitting circuit may be understood by referring to Fig. 15.25, which gives an equivalent circuit for the two potentiometers. $R_1$ and

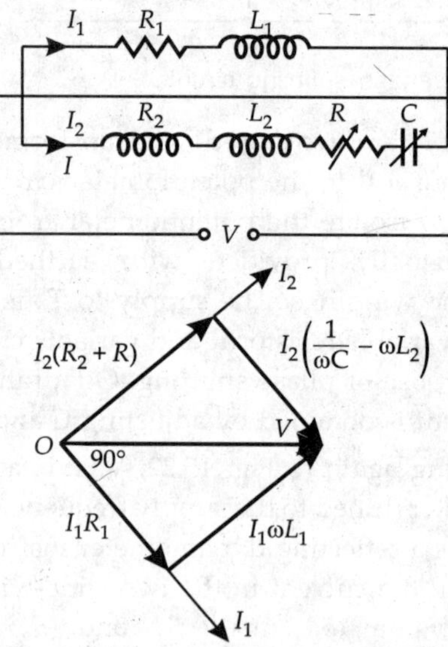


**Fig. 15.25** Phasor diagram for phase-splitting circuit.

$L_1$ are the equivalent resistance and inductance of the in-phase potentiometer circuit, and $R_2$ and $L_2$ those of the quadrature potentiometer. When the potentiometer currents are equal and in quadrature.

$$I_2 = j\,I_1$$

or
$$\frac{V}{R + R_2 + j(\omega L_2 - 1/\omega C)} = j\frac{V}{R_1 + j\omega L_1} \quad \text{...(15.1)}$$

$$\therefore \quad R_1 + j\omega L_1 = j(R + R_2) - (\omega L_2 - 1/\omega C)$$

By separating real and imaginary parts, we obtain the conditions for phase splitting as :

$$R_1 + \omega L_2 = 1/\omega C \quad \text{...(15.2)}$$

and
$$\omega L_1 - R_2 = R \quad \text{...(15.3)}$$

The phase splitting is adjusted by means of *R* and *C*.

This phase splitting has been used in later forms of the a.c. potentiometer in place of the original quadrature device described by D.C. Gall, which used a transformer and variable resistor.

### 15.3.6 Applications of A.C. Potentiometers

The applications of a.c. potentiometers are numerous and only a limited number of applications can be given in the space available here.

1. ***Voltmeter calibration.*** Low voltages up to 1.5 V or thereabouts can be measured directly. Higher voltages can be measured by using a volt-box (for medium voltages) or two capacitors in series (for high voltages) in conjunction with the potentiometers.

2. ***Ammeter calibration.*** The measurement of various alternating currents required for such calibration may be made by the use of non-inductive standard resistors with the potentiometer the method being similar to that adopted when the calibration is to be carried out with direct current.

3. ***Wattmeter and energy-meter testing.*** The testing circuit for wattmeters and energy meters is the same as that used in the case of d.c. measurements (See Fig. 15.17). A phase shifting transformer is included in the potential circuit to vary the phase of voltage with respect to current so that the wattmeters and energy meter are tested at various power factors.

4. ***Measurement of self reactance of a coil.*** A standard resistance $S$ is connected in series with the coil whose reactance is to be measured (Fig. 15.26).

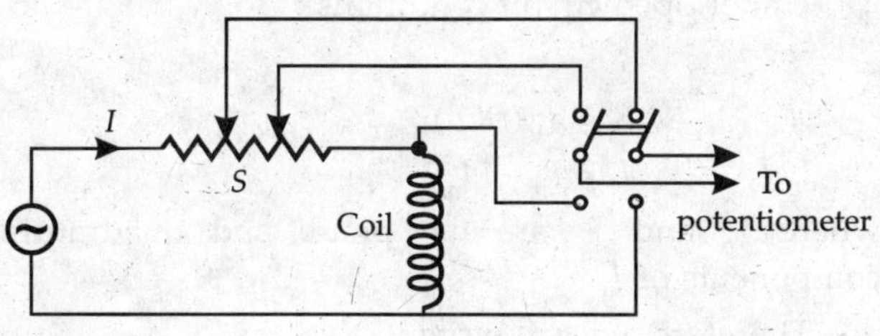


**Fig. 15.26** Measurement of self reactance of a coil.

Two voltage measurements are done, one across the standard resistance and the other across the coil. Supposing we are using a polar type of potentiometer and the readings are :

Voltage across standard resistor

$$'S' = V_s \angle \theta_s$$

Voltage across the coil

$$V_c = V_c \angle \theta_c$$

∴ Current through coil

$$I = V_s / S \angle \theta_s$$

Impedance of coil

$$Z = \frac{V_c}{I} = \frac{SV_c \angle \theta_c}{V_s \angle \theta_s} = \frac{SV_c}{V_s} \angle \theta_c - \theta_s$$

∴ Resistance of coil

$$R = Z\cos(\theta_c - \theta_s) = \frac{SV_c}{V_s}\cos(\theta_c - \theta_s)$$

Resistance of coil

$$X = Z\sin(\theta_c - \theta_s) = \frac{SV_c}{V_s}\sin(\theta_c - \theta_s)$$

5. ***Other applications.*** The practical field of usefulness of a.c. potentiometers is in engineering measurements in which an accuracy of 0.5 to 1% is adequate, and in cases where the potentiometer method may be more convenient and simpler than other types of voltage determination. The potentiometer method is indispensible when one is concerned with accurate measurements of ratio of two voltages but when one does not need to know accurately the precise magnitude of either of them. Another type of measurement in which a.c. potentiometers are used is that in which a voltage must be resolved into its two components. The a.c. potentiometer gives excellent results in magnetic testing (See chapter 18) and precise testing of instrument transformers.

15

**Example 15.9** *Calculate the inductance of a coil from the following measurement on an a.c. potentiometer :*

*Voltage drop across a 0.1 Ω standard resistor connected in series with the coil = 0.613 ∠12°6′.*

*Voltage across the test coil through a 100/1 volt-ratio box = 0.781 ∠50°48′ V.*

*Frequency is 50 Hz.*

**Solution.** Current through the coil $I$

$$= \frac{0.613 \angle 12°6'}{0.1} = 6.13 \angle 12°6' \text{ A.}$$

Voltage drop across the coil $V$

$$= 100 \times 0.781 \angle 50°48'$$

$$= 7.81 \angle 50°48' \text{ V.}$$

∴ Impedance of coil Z

$$= \frac{V}{I} = \frac{78.1 \angle 50°48'}{6.13 \angle 12°6'}$$

$$= 12.74 \angle 38°42' \Omega.$$

Reactance of coil $R$

$$= 12.74 \cos 38° 42' = 9.94 \Omega.$$

Resistance of coil X

$$= 12.74 \sin 38°42' = 7.95 \Omega$$

Inductance of coil $L$

$$= \frac{X}{2\pi f} = \frac{7.95}{2\pi \times 50} \text{ H} = 25.3 \text{ mH}.$$

**Example 15.10** *A co-ordinate type potentiometer is used for determination of impedance of a coil and the results obtained are :*

*Voltage across a 1.0 Ω resistor in series with the coil : + 0.238 V on in-phase dial and – 0.085 V on quadrature dial.*

*Voltage across a 10 : 1 potential divider used with the coil : + 0.3375 V on in-phase dial and + 0.232 V on quadrature dial.*

*Calculate the resistance and reactance of the coil.*

**Solution.** Current through the coil $I$

$$= \frac{(+0.238 - j0.085)}{1.0}$$

$$= 0.238 - j0.085 \text{ A}.$$

Voltage across the coil $V$

$$= 10(0.3375 + j0.232)$$

$$= 3.375 + j2.32 \text{ V}.$$

∴ Impedance of coil $Z$

$$= \frac{V}{I} = \frac{3.375 + j2.32}{0.238 - j0.085}$$

$$= 9.49 + j13.13\Omega.$$

∴ Resistance of coil, $R = 9.49\ \Omega$.

Reactance of coil, $X = 13.13\ \Omega$.

**Example 15.11** *In the measurement of power by a polar potentiometer, the following readings were obtained :*

*Voltage across a 0.2 Ω standard resistance in series with the load = 1.46 ∠32°.*

*Voltage across a 200 : 1 potential divider across the line = 1.37 ∠56° V.*

*Estimate the current, voltage, power and power factor of the load.*

**Solution.** Current through the load

$$I = \frac{1.46\angle 32°}{0.2} = 7.3\angle 32° \text{ A}.$$

Magnitude of current $I = 7.3$ A.

Voltage across the load $V$

$$= 200(1.37\angle 56°) = 274\angle 56° V$$

Magnitude of voltage $V$

$$= 274 \text{ V}.$$

Phase angle of the load

$$= 56° - 32° = 24°.$$

Power factor of load = cos 24° = 0.9135 (lagging).

Power consumed by load

$$= VI\cos\phi = 274 \times 7.3 \times 0.9135 = 1820 \text{ W}$$

**Example 15.12** *A current of 10 A at frequency of 50 Hz was passed through the primary of a mutual inductor having a negligible phase defect. The voltages of primary and secondary terminals were measured on a co-ordinate potentiometer and are given below :*

*With secondary open circuited :*

*Secondary voltage = – 2.72 + j1.57 V.*

*Primary voltage = – 0.211 + j0.352 V.*

*With secondary short circuited :*

*Primary voltage = – 0.051 + j0.329 V.*

*The phase of primary current relative to the potentiometer current was same in both the tests.*

*Determine the resistances and self-inductances of the two windings. Find also the mutual inductance.*

**Solution.** The voltage equations in the phasor form can be written as :

$$E_1 = I_1 (R_1 + jX_1) + I_2 (jX_m),$$
$$E_2 = I_2 (R_2 + jX_2) + I_1 (jX_m)$$

where $E_1$ = voltage of primary winding,

$E_2$ = voltage of secondary winding,

$I_1$ = current of primary winding,

$I_2$ = current of secondary winding,

$R_1$ = resistance of primary winding,

$R_2$ = resistance of secondary winding,

$X_1$ = self-reactance of primary winding,

$X_2$ = self-reactance of secondary winding,

and $X_m$ = mutual reactance.

Under open circuit conditions :

$$I_2 = 0$$

∴ $$E_2 = I_1 (jX_m)$$

Let $$I_1 = I_p + jI_q$$

where $I_p$ and $I_q$ are the phase and quadrature components of $I_1$.

Therefore, we can write :

$$E_2 = (I_p + jI_q)(jX_m)$$
$$= (-I_q + jI_p)X_m$$

From the data given, we have,

$$-2.72 + j1.57 = -I_q X_m + jI_p X_m$$

or $$I_p X_m = 1.57 \text{ V} \text{ and } I_q X_m = 272 \text{ V}.$$

or $$X_m\sqrt{I_p^2 + I_q^2} = 3.14 \text{ V}.$$

But $$\sqrt{I_p^2 + I_q^2} = I = 10 \text{ A}.$$

∴ $$X_m = \frac{3.14}{10} = 0.314\ \Omega.$$

Hence mutual inductance

$$M = \frac{X_m}{2\pi f} = \frac{0.314}{2\pi \times 50} \text{H} = 1.0 \text{ mH}.$$

$$I_p = \frac{1.57}{0.314} = 5.0 \text{ A}$$

and $$I_q = \frac{2.72}{0.314} = 8.66 \text{ A}$$

At open circuit,

$$E_1 = I_1 (R_1 + jX_1) = (I_p + jI_q)(R_1 + jX_1)$$

$$= (I_p R_1 - I_q X_1) + j(I_p X_1 + I_q R_1).$$

Putting the numerical values, we get,

$$-0.211 + j0.352 = (5R_1 - 8.66X_1) + j(5X_1 + 8.66R_1)$$

Equating real and the imaginary terms, we have

$$5R_1 - 8.66X_1 = -0.211 \quad \text{...(i)}$$

and $$5X_1 + 8.66R_1 = 0.352 \quad \text{...(ii)}$$

Solving (*i*) and (*ii*), we get

$$R_1 = 0.02\ \Omega \text{ and } X_1 = 0.0359\ \Omega.$$

∴ Self-inductance of primary winding

$$L_1 = \frac{0.0359}{2\pi \times 50} \text{H} = 0.114 \text{H}$$

At short circuit of secondary winding

$$E_2 = 0,$$

$$\therefore \quad E_2 - I_2 (R_2 + jX_2) + I_1 (jX_m) = 0.$$

or $$I_2 = -\frac{jX_m}{R_2 + jX_2} I_1$$

Thus $$E_1 = I_1 (R_1 + jX_1) + I_2 (jX_m)$$

$$= I_1 (R_1 + jX_1) + I_1 \frac{X_m^2}{R_2 + jX_2}$$

Now at short circuit :

$$E_1 = 0.051 + j0.329$$

Also $$I_1 (R_1 + jX_1) = -0.211 + j0.352$$

$$\therefore \quad -0.051 + j0.329 = -0.211 + j0.352 + \frac{(5 + j8.66) \times (0.314)^2}{R_2 + jX_2}$$

or $$R_2 + jX_2 = \frac{0.494 + j\,0.856}{0.16 - j0.023}$$

$$= 2.27 + j5.66\Omega$$

$$\therefore X_2 = 5.66\ \Omega$$

Self-reactance of secondary winding

$$L_2 = \frac{5.66}{2\pi \times 50} \text{H} = 18 \text{ mH}.$$

Resistance of secondary winding

$$R_2 = 2.27\ \Omega.$$

15

# Review Questions

1. Describe the basic principle of operation of a d.c. potentiometer. Explain why a potentiometer does not load the voltage source whose voltage is being determined.
2. Explain the term "standardization", of a potentiometer. Describe the procedure of standardization of a d.c. potentiometer.
3. Draw the circuit diagram of a Crompton's potentiometer and explain its working. Describe the steps used when measuring an unknown resistance.
4. Describe the working of a duo-range potentiometer by drawing its circuit diagram. Explain its advantages.
5. Explain how *"true zero"* is obtained in a Crompton's Potentiometer.
6. Explain the reasons why a separate *"Standard Cell Dial Circuit"* is provided in modern d.c. potentiometers.
7. How are the limitations imposed on the performance of ordinary potentiometers on account of use of slide wire eliminated through use of a vernier potentiometer ? Describe the circuit of a vernier potentiometer and explain how is it possible to measure as low voltages as 0.1 μV. Explain even though it is theoretically possible to measure voltages of the order of fraction of a μV, practically it is not possible because of certain reasons. Explain these reasons.
8. Describe the working of a Brooks' Deflectional potentiometer. Explain its circuit and its advantages.
9. What is a volt-ratio box ? Explain its construction, working and applications.
10. Describe with the help of suitable diagrams, how a d.c. potentiometer can be used for :

    (*i*) calibration of a voltmeter

(ii) calibration of an ammeter

(iii) calibration of a wattmeter.

(iv) determination of an unknown resistance.

11. Describe the working of a self-balancing potentiometer with the help of a diagram for measurement of temperature using a thermocouple.

12. Explain the reasons why d.c. potentiometers cannot be used for a.c. measurements straightaway. Explain the modifications that are needed in a d.c. potentiometer to be used for a.c. applications.

13. Describe the construction and working of a polar type potentiometer. How is it standardized ? What are the functions of the transfer instrument and the phase shifting transformer ?

Explain how an unknown voltage can be measured by using this potentiometer.

14. Describe the construction and working of a co-ordinate type a.c. potentiometer. How is it standardized ? Explain how an unknown voltage can be measured with it.

Discuss the sources of error in these instruments.

15. Explain with the help of suitable diagrams, how a.c. potentiometers can be used for :

(i) calibration of voltmeters

(ii) calibration of ammeters

(iii) calibration of wattmeters and energy meters

(iv) measurement of reactance of a coil.

# Unsolved Problems

1. The emf of a standard cell is measured with a potentiometer which gives a reading of 1.01892 V. When in 1 MΩ resistor is connected across the standard cell terminals, the potentiometer reading drops to 1.01874 V. Calculate the internal resistance of the cell. **[Ans.** 176.6 Ω]

2. A standard cell of 1.0185 V used with a simple potentiometer balances at 50 cm. Calculate

(a) the emf of the cell that balances at 72 cm,

(b) the percentage error in voltmeter which balances at 64.5 cm when reading 1.33 V,

(c) the percentage error in an ammeter that reads 0.43 A when balanced at 43.2 cm with voltage drop across a 2 Ω resistor in the ammeter circuit. **[Ans.** (a) 1.467 V ; (b) 1.23% high ; (c) 2.27% low]

3. A slide wire potentiometer is used to measure the voltage between two points of a certain d.c. circuit. The potentiometer reading is 1.0 volt. Across the same two points when a 10,000 Ω/V voltmeter is connected, the indicated reading on the voltmeter is 0.5 V on its 5 V range.

Calculate the input resistance between two points. **[Ans.** 50,000 Ω]

4. A slide wire potentiometer has a battery of 4 V and negligible internal resistance. The resistance of slide wire is 100 Ω and its length 200 cm. A standard cell of 1.018 V is used for standardizing the potentiometer and the rheostat is adjusted so that balance is obtained when the sliding contact is at 101.8 cm.

(a) Find the working current of the slide wire and the rheostat setting.

(b) If the slide wire has divisions marked in mm and each division can be interpolated to one fifth, calculate the resolution of the instrument. **[Ans.** 20 mA ; 100 Ω ; 0.2 mV]

5. In the circuit of Fig. 15.8, the working voltage of battery is 3.03 V. The current in the working circuit is 2 mA. Determine

(a) the resistance of main dial in the measuring circuit,

(b) the resistance of the slide wire in the measuring circuit,

(c) the resistance $R_s$, and

(d) the resistance of slide wire in the calibrating circuit. **[Ans.** 750 Ω ; 50 Ω ; 508 Ω ; 2 Ω]

6. Design a volt-ratio box with a resistance of 20 Ω/V and ranges 3 V, 10 V, 30 V, 100 V. The volt ratio box is to be used with a potentiometer having a measuring range of 1.6 V.

**Hint.** The output of the volt ratio box should be designed for a voltage of 1.5 V.

**[Ans.** Resistance at output terminals = 30 Ω ; The total resistances at input terminals are : 60 Ω, 200 Ω, 600 Ω and 2000 Ω.]

7. Measurements for the determination of the impedance of a coil are made on a coordinate type of potentiometer. The results are : Voltage across 1 Ω standard resistance in series with the coil + 0.952 V on inphase dial and – 0.340 V on quadrature dial ;

Voltage across 10 : 1 potential divider connected to the terminals of the coil : + 1.35 V on inphase dial and + 1.28 V on quadrature dial. Calculate the resistance and reactance of the coil.

[**Ans.** $R = 8.32\ \Omega$ and $X = 16.41\ \Omega$]

**8.** Power is measured with an a.c. potentiometer. The voltage across a 0.1 Ω standard resistance connected in series with the load is 0.35 – *j*0.10 V. The voltage across 300 : 1 potential divider connected to the supply is $0.8 + j0.15$ V. Determine the power consumed by the load and the power factor. [**Ans.** 801 W ; 0.8945]

**9.** The current taken by a small iron core choke coil is measured by a co-ordinate type a.c. potentiometer. A 1.0 Ω resistance is connected in series with the choke coil. The voltage measurements across the resistance and the choke coil are made. The data obtained :

Voltage across 1.0 Ω resistance $= +0.8 - j0.75$ V

Voltage across coil $= +1.2 + j0.3$ V.

Assuming sinusoidal voltage and current, determine the core loss and the magnetizing current of the coil. [**Ans.** 0.730 W ; 0.923 A]!

**10.** A resistor of 100 Ω is connected in series with a 1000 turn iron cored coil across a 50 Hz supply. The voltages across the resistor and across the coil measured by an a.c. potentiometer are 0.0587 – *j*0.0650 V and 0.8593 + *j*0.5010 V respectively. The voltage measured across a 500 turn search coil wound over the core is – 0.4175 – *ij*0.2565 V. Find the iron loss in the core, the $I^2R$ loss in the coil and the leakage flux considering the circuit as a transformer.

[**Ans.** 0.155 mW ; 0.024 mW ; 0.07 μWb]

**11.** A non-reactive resistor of 1000 Ω is corrected in series with a coil and a capacitor to a 50 Hz supply. If the voltages across each, measured by an a.c. potentiometer are + 0.6 – j0.24 V, + 0.6 + j0.4 V and – 0.1 V – *j*0.4 V respectively, find the power dissipated and the mean stored energy in each component.

[**Ans.** 0.418, 0.264, 0.036 mW and 0.061, 0.421 μJ]

15

# Objective Type Questions

***Tick (√) the most appropriate answer*** :

**1.** A potentiometer is basically a :

(*a*) deflectional type instrument

(*b*) null type instrument

(*c*) deflectional as well as null type instrument

(*d*) a digital instrument.

**2.** When a potentiometer is used for measurement of voltage of an unknown source, the power consumed in the circuit of the unknown source under null conditions

(*a*) is very high (*b*) is high

(*c*) is small (*d*) is ideally zero.

**3.** Standardization of potentiometers is done in order that, they become

(*a*) accurate

(*b*) precise

(*c*) accurate and direct reading

(*d*) accurate and precise.

**4.** A Crompton's Potentiometer is provided with a dial resistor having 15 steps of 10 Ω each and a slide wire of 10 Ω resistance The slide wire is divided into 100 divisions and one fifth of a division can be read with certainty. The working current of potentiometer is 10 mA. The range and resolution of the potentiometer are respectively :

(*a*) 1.6 V, 0.2 mV (*b*) 1.6 V, 0.5 mV

(*c*) 1.5 V, 0.1 mV

(*d*) 1.6 V, 0.1 mV.

**5.** A duo-range Crompton's potentiometer is provided with a dial resistor having 15 steps of 10 Ω each and a slide wire of 10 Ω resistance. The ranges are X 1 and X 0.1. The current in the potentiometer measuring circuit 10 mA on the X 1 scale. The slide wire has 100 divisions and one fifth of a scale division can be read with certainty. What is the current drawn from the battery and the resolution of the instrument on X 0.1 scale range ?

(*a*) 10 mA, 20 μV

(*b*) 11 mA, 10 μV

(*c*) 11 mA, 2 μV

(*d*) 11 mA, 20 μV.

**6.** Brooks' deflectional potentiometer is used when the unknown voltage :

(*a*) is constant

(*b*) is varying at a slow rate

(*c*) is varying very rapidly

(*d*) all the above.

7. A volt-ratio box is designed with a resistance 100 Ω/V. It has input terminals of 300 V, 150 V and 75 V, and an output terminal of 1.5 V which is connected to the potentiometer. Supposing a voltage of 270 V is connected to the 300 V terminal, what is voltage indicated by the potentiometer and what is the power consumption under null conditions ?

(*a*) 1.5 V, 3 W (*b*) 1.35 V, 3 W

(*c*) 1.5 V, 2.43 W (*d*) 1.35 V, 2.43 W.

8. The standardization of a.c. potentiometers is done by

(*a*) directly using a.c. standard voltage sources

(*b*) using d.c. standard sources and transfer instruments

(*c*) using d.c. standard sources and d'Arsonval galvanometer

(*d*) using a.c. standard sources and transfer instruments.

9. The readings of a polar type a.c. potentiometer when measuring the reactance of a coil are

$$I = 12\angle 13.8°, \quad V = 27.8\angle 29.7°.$$

The reactance of the coil is :

(*a*) 2.317 Ω (*b*) 0.634 Ω

(*c*) 2.22 Ω (*d*) – 2.22 Ω.

10. A coordinate type potentiometer is used for determination of impedance of a coil and the results obtained are : voltage across 1 Ω resistor in series with coil, 0.24 V on in phase dial and – 0.09 V on quadrature dial,

Voltage across 10 : 1 potential divider used with coil + 0.37 V on in-phase dial and +0.32 on quadrature dial.

The resistance of coil is :

(*a*) 9.14 Ω (*b*) 2.34 Ω

(*c*) 16.76 Ω (*d*) none of the above.

## *Answers*

| 1. (*b*) | 2. (*d*) | 3. (*c*) | 4. (*a*) | 5. (*d*) | 6. (*b*) | 7. (*d*) | 8. (*b*) | 9. (*b*) | 10. (*a*) |
|---|---|---|---|---|---|---|---|---|---|

# 16

# A.C. Bridges

## 16.1 INTRODUCTION

Alternating current bridge methods are of outstanding importance for measurement of electrical quantities. Measurement of inductance, capacitance, storage factor, loss factor may be made conveniently and accurately by employing a.c. bridge networks.

The a.c. bridge is a natural outgrowth of the Wheatstone bridge. An a.c. bridge, in its basic form, consists of four arms, a source of excitation, and a balance detector. In an a.c. bridge each of the four arms is an impedance, and the battery and the galvanometer of the Wheatstone bridge are replaced respectively by an a.c. source and a detector sensitive to small alternating potential differences.

The usefulness of a.c. bridge circuits is not restricted to the measurement of unknown impedances and associated parameters like inductance, capacitance, storage factor, dissipation factor etc. These circuits find other applications in communication systems and complex electronic circuits. Alternating current bridge circuits are commonly used for phase shifting, providing feedback paths for oscillators and amplifiers, filtering out undesirable signals and measuring the frequency of audio signals.

## 16.2 SOURCES AND DETECTORS

For measurements at low frequencies, the power line may act as the source of supply to the bridge circuits. For higher frequencies electronic oscillators are universally used as bridge source supplies. These oscillators have the advantage that the frequency is constant, easily adjustable, and determinable with accuracy. The waveform is very close to a sine wave, and their power output is sufficient for most bridge measurements. A typical oscillator has a frequency range of 40 Hz to 125 kHz with a power output of 7 W.

The detectors commonly used for a.c. bridges are :

(*i*) Head phones,

(*ii*) Vibration galvanometers, and

(*iii*) Tuneable amplifier detectors.

***Head phones*** are widely used as detectors at frequencies of 250 Hz and over upto 3 or 4 kHz. They are most sensitive detectors for this frequency range.

When working at a single frequency a tuned detector normally gives the greatest sensitivity and discrimination against harmonics in the supply.

***Vibration galvanometers*** are extremely useful for power and low audio frequency ranges. Vibration galvanometers are manufactured to work at various frequencies ranging from 5 Hz to 1000 Hz but are most commonly used below 200 Hz as below this frequency they are more sensitive than the head phones.

***Tuneable amplifier detectors*** are the most versatile of the detectors. The transistor amplifier can be tuned electrically and thus can be made to respond to a narrow bandwidth at the bridge frequency. The output of the amplifier is fed to a pointer type of instrument. This detector can be used, over a frequency range of 10 Hz to 100 kHz.

For ordinary a.c. bridge measurements of inductance and capacitance, a fixed frequency oscillator of 1000 Hz and output of about 1 W is adequate. For more specialised work continuously variable oscillators are preferable with outputs upto 5 W. The high power may be necessary on some occasions, but in practice it is better to limit the power supplied to the bridge. Another practice which is usually followed is to use an untuned amplifier detector. The balance detection is sensed both orally by head phones, and visually by a pointer galvanometer having a logarithmic deflection (to avoid damage to the galvanometer which may be caused by unbalance).

## 16.3 GENERAL EQUATION FOR BRIDGE BALANCE

Figure 16.1 shows a basic a.c. bridge. The four arms of the bridge are impedances $Z_1$, $Z_2$, $Z_3$ and $Z_4$.

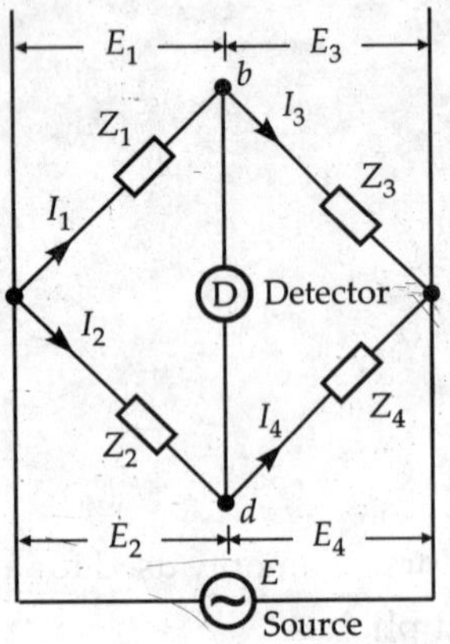


**Fig. 16.1** Basic a.c. bridge network.

The conditions for balance of bridge require that there should be no current through the detector. This requires that the potential difference between points *b* and *d* should be zero. This will be the case when the voltage drop from *a* to *b* equals to voltage drop from *a* to *d*, both in magnitude and phase. In complex notation we can, thus, write :

$$E_1 = E_2 \quad ...(16.1)$$

or

$$I_1 Z_1 = I_2 Z_2 \quad ...(16.2)$$

Also at balance,

$$I_1 = I_3 = \frac{E}{Z_1 + Z_3} \quad ...(16.3)$$

and

$$I_2 = I_4 = \frac{E}{Z_2 + Z_4} \quad ...(16.4)$$

Substitution of Eqns. 16.3 and 16.4 into Eqn. 16.2 gives,

$$Z_1 Z_4 = Z_2 Z_3 \quad ...(16.5)$$

or when using admittances instead of impedances

$$Y_1 Y_4 = Y_2 Y_3 \quad ...(16.6)$$

Equations 16.5 and 16.6 represent the basic equations for balance of an a.c. bridge. Equation 16.5 is convenient to use when dealing with series elements of a bridge while Eqn. 16.6 is useful when dealing with parallel elements.

Equation 16.5 states that the product of impedances of one pair opposite arms must equal the product of impedances of the other pair of opposite arms expressed in complex notation. This means that both magnitudes and the phase angles of the impedances must be taken into account.

Considering the polar form, the impedance can be written as $\mathbf{Z} = Z\angle\theta$, where $\mathbf{Z}$ represents the magnitude and $\theta$ represents the phase angle of the complex impedance. Now Eqn. 16.5 can be re-written in the form

$$(Z_1 \angle\theta_1)(Z_4 \angle\theta_4) = (Z_2 \angle\theta_2)(Z_3 \angle\theta_3) \quad ...(16.7)$$

Thus for balance, we must have :

$$Z_1 Z_4 \underline{/\theta_1 + \theta_4} = Z_2 Z_3 \underline{/\theta_2 + \theta_3} \quad ...(16.8)$$

Equation 16.8 shows that two conditions must be satisfied simultaneously when balancing an a.c. bridge. The first condition is that the magnitude of impedances satisfy the relationship :

$$Z_1 Z_4 = Z_2 Z_3 \quad ...(16.9)$$

The second condition is that the phase angles of impedances satisfy the relationship :

$$\angle\theta_1 + \angle\theta_4 = \angle\theta_2 + \angle\theta_3 \quad ...(16.10)$$

The phase angles are positive for an inductive impedance and negative for capacitive impedance.

If we work in terms of rectangular co-ordinates, we have

$$\mathbf{Z}_1 = R_1 + jX_1 \,; \qquad \mathbf{Z}_2 = R_2 + jX_2 \,;$$
$$\mathbf{Z}_3 = R_3 + jX_3 \text{ and } \mathbf{Z}_4 = R_4 + jX_4.$$

Thus from Eqn. 16.5, for balance,

$$\mathbf{Z}_1 \mathbf{Z}_4 = \mathbf{Z}_2 \mathbf{Z}_3$$

or $(R_1 + jX_1)(R_4 + jX_4) = (R_2 + jX_2)(R_3 + jX_3)$

or $R_1 R_4 - X_1 X_4 + j(X_1 R_4 + X_4 R_1)$

$$= R_2 R_3 - X_2 X_3 + j(X_2 R_3 + X_3 R_2) \quad ...(16.11)$$

Equation 16.11 is a complex equation and a complex equation is satisfied only if real and imaginary parts of each side of the equation are separately equal. Thus, for balance,

$$R_1 R_4 - X_1 X_4 = R_2 R_3 - X_2 X_3 \quad ...(16.12)$$

and

$$X_1 R_4 + X_4 R_1 = X_2 R_3 + X_3 R_2 \quad ...(16.13)$$

Thus there are two independent conditions for balance and both of them must be satisfied for the bridge to be balanced.

**Example 16.1** *The four impedances of an a.c. bridge shown in Fig. 16.1 are :*

$$\mathbf{Z}_1 = 400\,\Omega \angle 50° \,; \quad \mathbf{Z}_2 = 200\,\Omega \angle 40° \,;$$
$$\mathbf{Z}_3 = 800 \angle -50° \,; \quad \mathbf{Z}_4 = 400\,\Omega \angle 20°$$

*Find out whether the bridge is balanced under these conditions or not.*

**Solution.** Applying the first condition of balance for magnitudes,

$$Z_1 Z_4 = Z_2 Z_3$$

Now, $Z_1Z_4 = 400 \times 400 = 1{,}60{,}000$

and $Z_2Z_3 = 200 \times 800 = 1{,}60{,}000.$

$\therefore \quad Z_1Z_4 = Z_2Z_3.$

Thus the first condition is satisfied.

Applying the second condition for balance required for phase,

$$\angle\theta_1 + \angle\theta_4 = \angle\theta_2 + \angle\theta_3$$

Now $\angle\theta_1 + \angle\theta_4 = 50° + 20° = 70°$

and $\angle\theta_2 + \angle\theta_3 = 40° - 50° = -10°.$

This indicates that the condition for phase relationship is not satisfied and therefore the bridge is unbalanced even though the condition for equality of magnitudes is satisfied.

**Example 16.2** *An a.c. bridge circuit working at 1000 Hz is shown in Fig. 16.1, Arm ab is a 0.2 μF pure capacitance ; arm bc is a 500 Ω pure resistance ; arm cd contains an unknown impedance and arm da has a 300 Ω resistance in parallel with a 0.1 μF capacitor. Find the R and C or L constants of arm cd considering it as a series circuit.*

**Solution.** Impedance of arm *ab* is :

$$Z_1 = \frac{1}{2\pi f C_1} = \frac{1}{2\pi \times 1000 \times 0.2 \times 10.6} = 800\ \Omega.$$

and $Z_1 = 800\angle{-90°}\ \Omega$ since it is a pure capacitance.

Impedance of arm *bc* is

$$Z_3 = 500\ \Omega$$

or $Z_3 = 500\angle 0°$ since it is a pure resistance.

Arm *da* contains a 300 Ω resistance in parallel with a 0.1 μF capacitance.

$$Z_2 = \frac{R_2}{1 + j\omega C_2 R_2} = \frac{300}{1 + j(2\pi \times 1000 \times 0.1 \times 10^{-6} \times 300)}$$

$$= 294.8 - j55.4 = 300\angle{-10.6°}$$

For balance, $Z_1Z_4 = Z_2Z_3$

$\therefore$ Impedance of *cd* required for balance is,

$$Z_4 = \frac{Z_2Z_3}{Z_1} = \frac{360 \times 500}{800}\angle{-10.6° + 0° + 90°}$$

$$= 187.5\angle 78.4°.$$

The positive angle for impedance indicates that the branch consists of a series *R-L* circuit.

Resistance, $R_4 = 187.5 \cos 78.4° = 37.7\ \Omega.$

Inductive reactance

$$X_4 = 187.5 \sin 78.4° = 83.6\ \Omega.$$

Inductance, $L_4 = 183.6/2\pi(1000)\text{H} = 29.2\ \text{mH}.$

## 16.4 GENERAL FORM OF AN A.C. BRIDGE

As an example let us consider the bridge circuit of Fig. 16.2. $R_3$ and $R_4$ are non-inductive resistances, $L_1$ and $L_2$ are inductances of the negligible resistance and $R_1$ and $R_2$ are non-inductive resistors.

Therefore at balance,

$$Z_1Z_4 = Z_2Z_3$$

or $$(R_1 + j\omega L_1)R_4 = (R_2 + j\omega L_2)R_3$$

Equating the real and imaginary parts separately, we have,

$$R_1R_4 = R_2R_3 \quad \text{or} \quad R_1 = \frac{R_3}{R_4}\cdot R_2 \qquad \text{...(16.14)}$$

and $$j\omega L_1R_4 = j\omega L_2R_3 \quad \text{or} \quad L_1 = \frac{R_3}{R_4}L_2 \qquad \text{...(16.15)}$$

Thus if $L_1$ and $R_1$ are unknown, the above bridge may be used to measure these quantities in terms of $R_2, R_3, R_4$ and $L_2$. We may deduce several important conclusions from the above simple example. They are :

**1.** Two balance equations are always obtained for an a.c. bridge circuit. This follows from the fact that for balance in an a.c. bridge, both magnitude and phase relationships must be satisfied. This requires that real and imaginary terms must be separated, which give two equations to be satisfied for balance.

**2.** The two balance equations enable us to know two unknown quantities. The two quantities are usually a resistance and an inductance or a capacitance.

**3.** In order to satisfy both conditions for balance and for convenience of manipulation, the bridge must contain two variable elements in its configuration. For greatest convenience, each of the balance equations must contain one variable element, and one only. The equations are then said to be *independent*. In the bridge of Fig. 16.2, $R_2$ and $L_2$ are obvious choice as variable elements since $L_2$ does not appear in the expression for

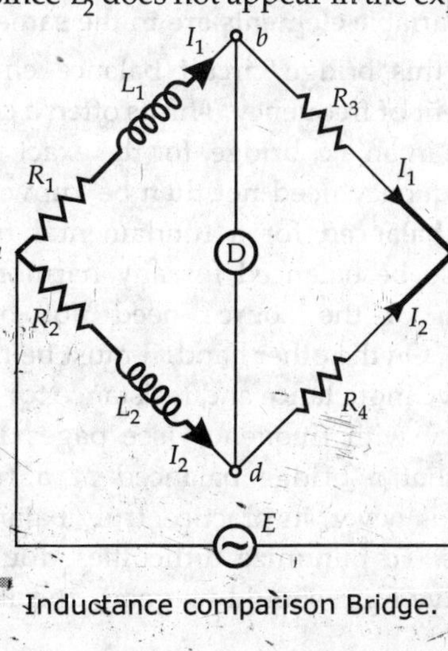


**Fig. 16.2** Inductance comparison Bridge.

16

$R_1$ and $R_2$ does not figure in the expression for $L_1$ and hence the two balance equations are independent. The technique of balancing is to adjust $L_2$ till a minimum indication is obtained on the detector, then to adjust $R_2$ until a new smaller minimum indication is obtained. Then $L_2$ and $R_2$ are alternately adjusted until the detector shows no indication.

The process of alternate manipulation of two variable elements is rather typical of the general balancing procedures adopted in most a.c. bridges. When two variables are chosen such that the two balance equations are no longer independent, the bridge has a very poor convergence of balance and gives the effect of sliding balance. The term *Sliding Balance* describes a condition of interaction between the two controls. Thus when we balance with $R_2$, then go to $R_3$ and back to $R_2$ for adjustment, we find a new apparent balance. Thus the balance point appears to move, or slide and settles only gradually to its final point after many adjustments. It may be emphasised here that in case the two balance conditions are independent, not more than two or three adjustments of the variable elements would be necessary to obtain balance. In case we choose the two variable components such that the two equations are not independent the balance procedure becomes laborious and time-consuming. For example, if we choose $R_2$ and $R_3$ as variable elements, the two equations are no longer independent since $R_3$ appears in both the equations. Thus $R_3$ is adjusted to satisfy Eqn. 16.15, its value may be becoming farther removed from that which satisfies Eqn. 16.14. There are two adjustments, one resistive and the other reactive that must be made to secure balance. For the usual magnitude responsive detector, these adjustments must be made alternately until they converge on the balance point.

The convergence to balance point is best when both the variable elements are in the same arms.

**4.** In this bridge circuit balance equations are independent of frequency. This is often a considerable advantage in an a.c. bridge, for the exact value of the source frequency need not then be known. Also, if a bridge is balanced for a fundamental frequency it should also be balanced for any harmonic and the wave-form of the source need not be perfectly sinusoidal. On the other hand, it must be realized that the effective inductance and resistance for example, of a coil, vary with frequency (see pages 162-163 and 169), so that a bridge balanced at a fundamental frequency is never, in practice, truly balanced for the harmonics. To minimize difficulties due to this the source wave form should be good, and it is often an advantage to use a detector tuned to the fundamental frequency. Further while the disappearance of the frequency factor is of advantage in many bridges, some bridges derive their usefulness from the presence of a frequency factor ; such bridges must then be supplied from a source with very good wave-form and high frequency stability.

Alternatively, they may be used to determine frequency.

## 16.5 MEASUREMENT OF SELF-INDUCTANCE

### 16.5.1 Maxwell's Inductance Bridge

This bridge circuit measures an inductance by comparison with a variable standard self-inductance. The connections and the phasor diagrams for balance conditions are shown in Fig. 16.3.

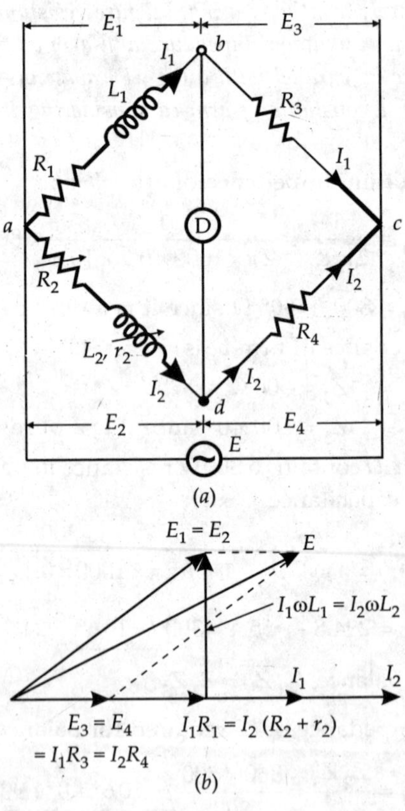


**Fig. 16.3** Maxwell's inductance Bridge.

Let

$L_1$ = unknown inductance of resistance $R_1$,

$L_2$ = variable inductance of fixed resistance $r_2$,

$R_2$ = variable resistance connected in series with inductor $L_2$,

and $R_3$, $R_4$ = known non-inductive resistances.

The theory of this bridge has been dealt with in Art. 16.4. At balance,

$$L_1 = \frac{R_3}{R_4} L_2 \qquad \text{...(16.16)}$$

and

$$R_1 = \frac{R_3}{R_4}(R_2 + r_2) \qquad \text{...(16.17)}$$

Resistors $R_3$ and $R_4$ are normally a selection of values from 10, 100, 1000 and 10,000 Ω. $r_2$ is a decade resistance box. In some cases, an additional known resistance may have to be inserted in series with unknown coil in order to obtain balance.

### 16.5.2. Maxwell's Inductance

**Capacitance bridge.** In this bridge, an inductance is measured by comparison with a standard variable capacitance. The connections and the phasor diagram at the balance conditions are given in Fig. 16.4.

Let $L_1$ = unknown inductance,

$R_1$ = effective resistance of inductor $L_1$,

$R_2, R_3, R_4$ = known non-inductive resistances,

and $C_4$ = variable standard capacitor.

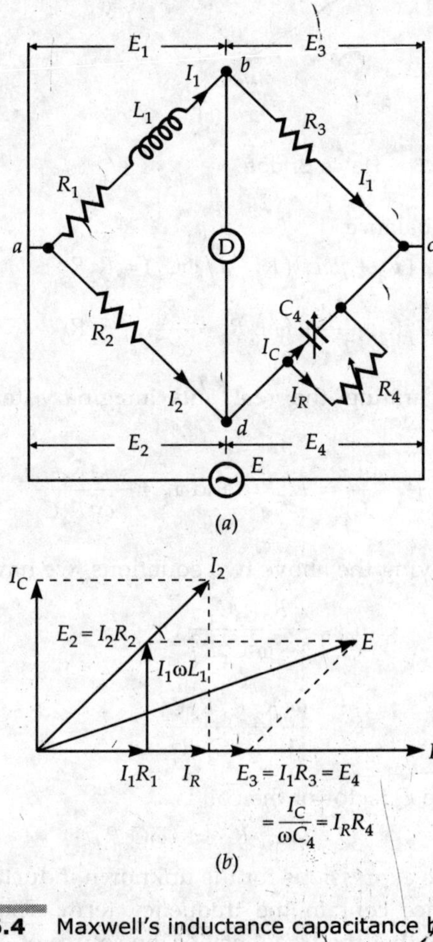


**Fig. 16.4** Maxwell's inductance capacitance bridge.

Writing the equation for balance

$$(R_1 + j\omega L_1)\left(\frac{R_4}{1 + j\omega C_4 R_4}\right) = R_2 R_3$$

or

$$R_1 R_4 + j\omega L_1 R_4 = R_2 R_3 + j\omega R_2 R_3 C_4 R_4$$

Separating the real and imaginary terms, we have

$$R_1 = \frac{R_2 R_3}{R_4} \qquad \text{...(16.18)}$$

and

$$L_1 = R_2 R_3 C_4 \qquad \text{...(16.19)}$$

Thus we have two variables $R_4$ and $C_4$ which appear in one of the two balance equations and hence the two equations are independent.

The expression for $Q$ factor

$$Q = \omega L_1 / R_1 = \omega C_4 R_4 \qquad \text{...(16.20)}$$

***Advantages :***

The advantages of this bridge are :

1. The two balance equations are independent if we choose $R_4$ and $C_4$ as variable elements.
2. The frequency does not appear in any of the two equations.
3. This bridge yields simple expression for unknowns $L_1$ and $R_1$ in terms of known bridge elements.

   Physically $R_2$ and $R_3$ are each, say, 10, 100, 1000 or 10,000 Ω and their value is selected to give suitable value of product $R_2 R_3$ which appears in both the balance equations ; $C_4$ is decade capacitor and $R_4$ a decade resistor.

   The simplicity of the bridge can be appreciated by the following example. Suppose the product $R_2 R_3$ is $10^6$. Therefore, inductance is $L_1 = C_4 \times 10^6$ (see Eqn. 16.19). Thus when balance is achieved the value of $C_4$ in μF directly gives the value of inductance in $H$.
4. The Maxwell's inductance – capacitance bridge is very useful for measurement of a wide range of inductance at power and audio frequencies.

***Disadvantages :***

The main disadvantages of this bridge are :

1. This bridge requires a variable standard capacitor which may be very expensive if calibrated to a high degree of accuracy. Therefore sometimes a fixed standard capacitor is used, either because a variable capacitor is not available or because fixed capacitors have a higher degree of accuracy and are

less expensive than the variable ones. The balance adjustments are then done by :

(a) either varying $R_2$ and $R_4$ and since $R_2$ appears in both the balance equations, the balance adjustments become difficult ; or

(b) putting an additional resistance in series with the inductance under measurement and then varying this resistance and $R_4$.

**2.** The bridge is limited to measurement of low $Q$ coils, $(1 < Q < 10)$. It is clear from Eqn. 16.20 that the measurement of high $Q$ coils demands a large value for resistance $R_4$, perhaps $10^5$ or $10^6\,\Omega$. The resistance boxes of such high values are very expensive. Thus for values of $Q > 10$, the Maxwell's bridge is unsuitable.

The Maxwell's bridge is also unsuited for coils with a very low value of $Q$ (*i.e.*, $Q < 1$). $Q$ values of this magnitude occur in inductive resistors, or in an R.F. coil if measured at low frequencies. The difficulty in measurement occurs on account of labour involved in obtaining balance since nominally a fixed capacitor is used and balance is obtained by manipulating resistances $R_2$ and $R_4$ alternately. This difficulty is explained as below :

From Eqns. 16.18 and 16.19, it is clear that $R_2$ enters into both the expressions. A preliminary inductive balance is made with $R_2$ and then $R_4$ is varied to give a resistive balance which is dependent on the $R_2$ setting. Accordingly when $R_2$ is changed for a second inductive balance, the resistive balance is disturbed and moves to a new value giving slow *"convergence"* to balance. This is particularly true of a low $Q$ coil, for which resistance is prominent (as $Q = \omega L/R$). Thus a sliding balance condition prevails and it takes many manipulations to achieve balance for low $Q$ coils with a Maxwell's bridge.

From the above discussions we conclude that a Maxwell's bridge is suited for measurements of only medium $Q$ coils.

### 16.5.3 Hay's Bridge

The Hay's bridge is a modification of Maxwell's bridge. The connection diagram and the phasor diagram for this bridge are shown in Fig. 16.5. This bridge uses a resistance in series with the standard capacitor (unlike the Maxwell's bridge which uses a resistance in parallel with the capacitor).

Let

$L_1$ = unknown inductance having a resistance $R_1$.

$R_2, R_3, R_4$ = known non-inductive resistance,

and $C_4$ = standard capacitor.

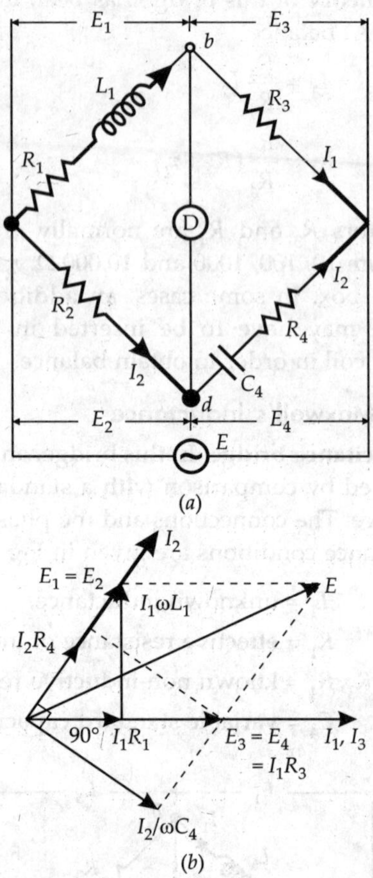


**Fig. 16.5** Hay's Bridge.

At balance,

$$(R_1 + j\omega L_1)(R_4 - j/\omega C_4) = R_2 R_3$$

or
$$R_1 R_4 + \frac{L_1}{C_4} + j\omega L_1 R_4 - \frac{jR_1}{\omega C_4} = R_2 R_3$$

Separating the real and imaginary terms, we obtain,

$$R_1 R_4 + \frac{L_1}{C_4} = R_2 R_3 \quad \text{and} \quad L_1 = -\frac{R_1}{\omega^2 R_4 C_4}$$

Solving the above two equations, we have

$$L_1 = \frac{R_2 R_3 C_4}{1 + \omega^2 C_4^2 R_4^2} \qquad ...(16.21)$$

and
$$R_1 = \frac{\omega^2 R_2 R_3 R_4 C_4^2}{1 + \omega^2 C_4^2 R_4^2} \qquad ...(16.22)$$

The $Q$ factor of the coil is,

$$Q = \omega L_1 / R_1 = 1/\omega C_4 R_4 \qquad ...(16.23)$$

The expressions for the unknown inductance and resistance contain the frequency term. Therefore it appears that the frequency of the source of supply to the bridge must be accurately known. This is not true for the inductance when a high $Q$ coil is being measured, as is explained below :

Now $L_1 = \dfrac{R_2 R_3 C_4}{1+\omega^2 C_4^2 R_4^2}$ but $Q = \dfrac{1}{\omega C_4 R_4}$

and therefore $L_1 = \dfrac{R_2 R_3 C_4}{1+(1/Q)^2}$ ...(16.24)

For a value of $Q$ greater than 10, the term $(1/Q)^2$ will be smaller than 1/100 and can be neglected.

Therefore Eqn. 16.24 reduces to

$$L_1 = R_2 R_3 C_4 \quad ...(16.25)$$

which is the same as for a Maxwell's bridge.

***Advantages :***

1. This bridge gives very simple expression for unknown inductance for high $Q$ coils, and is suitable for coils having $Q > 10$.
2. This bridge also gives a simple expression for $Q$ factor.
3. If we examine the expression for $Q$ factor :

   $$Q = 1/\omega C_4 R_4$$

   we find that the resistance $R_4$ appears in the denominator and hence for high $Q$ coils its value should be small. Thus this bridge requires only a low value resistor for $R_4$, whereas the Maxwell's bridge requires a parallel resistor, $R_4$, of a very high value.

***Disadvantages :***

The Hay's bridge is suited for the measurement of high $Q$ inductors, especially those inductors having a $Q$ greater than 10. For inductors having $Q$ values smaller than 10, the term $(1/Q)^2$ in the expression for inductance $L_1$ (Eqn. 16.24) becomes rather important and thus cannot be neglected. Hence this bridge is not suited for measurement of coils having $Q$ less than 10 and for these applications a Maxwell's bridge is more suited.

### 16.5.4 Anderson's Bridge

This bridge, in fact, is a modification of the Maxwell's inductance capacitance bridge. In this method, the self-inductance is measured in terms of a standard capacitor. This method is applicable for precise measurement of self-inductance over a very wide range of values.

Figure 16.6 shows the connections and the phasor diagram of the bridge for balanced conditions.

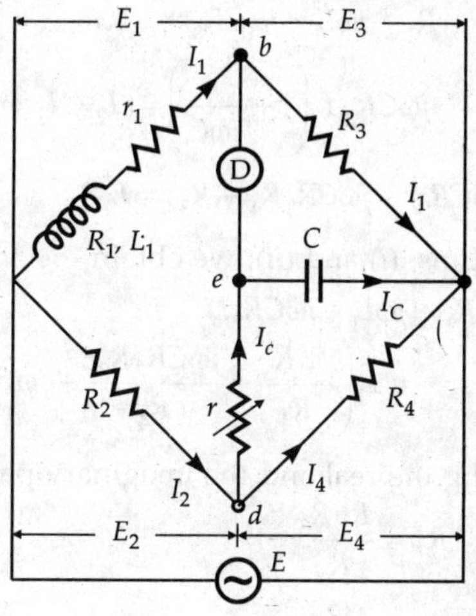

(a)

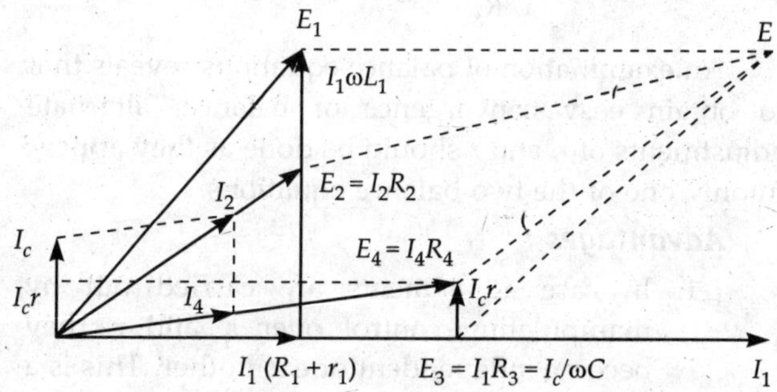

(b)

**Fig. 16.6** Anderson's Bridge.

Let

$L_1$ = self-inductance to be measured,

$R_1$ = resistance of self-inductor,

$r_1$ = resistance connected in series with self-inductor,

$r, R_2, R_3, R_4$ = known non-inductive resistances,

and $C$ = fixed standard capacitor.

At balance,

$$I_1 = I_3 \text{ and } I_2 = I_c + I_4$$

Now, $I_1 R_3 = I_c \times \dfrac{1}{j\omega C}$

$\therefore \quad I_c = I_1 j\omega C R_3.$

Writing the other balance equations

$$I_1(r_1 + R_1 + j\omega L_1) = I_2 R_2 + I_c r$$

and $$I_c\left(r + \frac{1}{j\omega C}\right) = (I_2 - I_c) R_4.$$

Substituting the value of $I_c$ in the above equations, we have

$$I_1(r_1 + R_1 + j\omega L_1) = I_2 R_2 + I_1 j\omega C R_3 r$$

or $$I_1(r+R_1+j\omega L_1-j\omega CR_3r)=I_2R_2 \quad ...(i)$$

and $$j\omega CR_3\,I_1\left(r+\frac{1}{j\omega C}\right)=(I_2-I_1 j\omega CR_3)\,R_4$$

or $$I_1(j\omega CR_3r+j\omega CR_3R_4+R_3)=I_2R_4 \quad ...(ii)$$

From Eqns. (*i*) and (*ii*), we obtain

$$I_1(r_1+R_1+j\omega L_1-j\omega CR_3r)=I_1\left(\frac{R_2R_3}{R_4}+\frac{j\omega CR_2R_3r}{R_4}+j\omega CR_3R_2\right)$$

Equating the real and the imaginary parts,

$$R_1=\frac{R_2R_3}{R_4}-r_1 \quad ...(16.26)$$

and $$L_1=C\frac{R_3}{R_4}[r(R_4+R_2)+R_2R_4] \quad ...(16.27)$$

An examination of balance equations reveals that to obtain easy convergence of balance, alternate adjustments of $r_1$ and $r$ should be done as they appear in only one of the two balance equations.

***Advantages :***

1. In case adjustments are carried out by manipulating control over $r_1$ and $r$, they become independent of each other. This is a marked superiority over sliding balance conditions met with low $Q$ coils when measuring with Maxwell's bridge. A study of convergence conditions would reveal that it is much easier to obtain balance in the case of Anderson's bridge than in Maxwell's bridge for low $Q$-coils.
2. A fixed capacitor can be used instead of a variable capacitor as in the case of Maxwell's bridge.
3. This bridge may be used for accurate determination of capacitance in terms of inductance.

***Disadvantages :***

1. The Anderson's bridge is more complex than its prototype Maxwell's bridge. The Anderson's bridge has more parts and is more complicated to set up and manipulate. The balance equations are not simple and in fact are much more tedious.
2. An additional junction point increases the difficulty of shielding the bridge.

Considering the above complication of the Anderson's bridge, in all the cases where a variable capacitor is permissible the more simple Maxwell's bridge is used instead of Anderson's bridge.

### 16.5.5 Owen's Bridge

This bridge may be used for measurement of an inductance in terms of capacitance. Figure 16.7 shows the connections and phasor diagrams, for this bridge, under balance conditions.

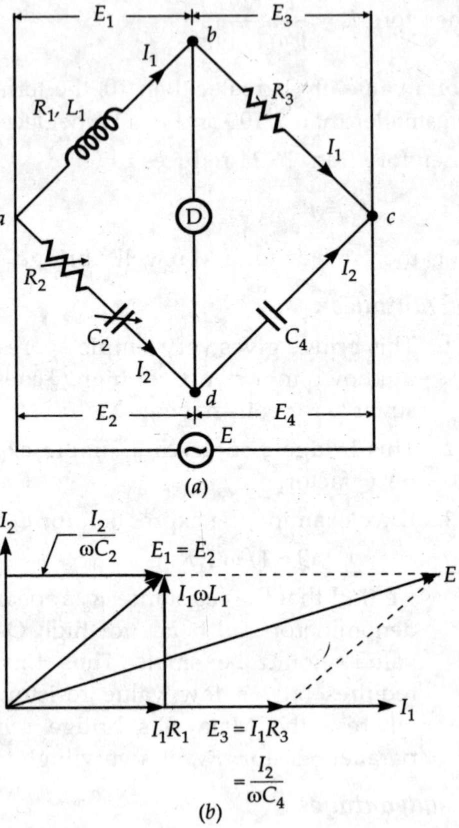


**Fig. 16.7** Owen's Bridge.

Let

$L_1$ = unknown self-inductance of resistance $R_1$,
$R_2$ = variable non-inductive resistance,
$R_3$ = fixed non-inductive resistance,
$C_2$ = variable standard capacitor,
and $C_4$ = fixed standard capacitor.

At balance,

$$(R_1+j\omega L_1)\left(\frac{1}{j\omega C_4}\right)=\left(R_2+\frac{1}{j\omega C_2}\right)R_3$$

Separating the real and imaginary terms, we obtain,

$$L_1=R_2R_3C_4 \quad ...(16.28)$$

and $$R_1=R_3\frac{C_4}{C_2} \quad ...(16.29)$$

***Advantages :***

1. Examining the equations for balance, we find that we obtain two independent equations in case $C_2$ and $R_2$ are made variable. Since $R_2$ and $C_2$, the variable elements, are in the same arm, convergence to balance conditions is much easier.
2. The balance equations are quite simple and do not contain any frequency component.
3. The bridge can be used over a wide range of measurement of inductances.

***Disadvantages :***

1. This bridge requires a variable capacitor which is an expensive item and also its accuracy is about 1 percent.
2. The value of capacitance $C_2$ tends to become rather large when measuring high $Q$ coils.

### 1.6.5.6 Measurements of Incremental Inductance

The magnetic flux linking an ion-cored coil is not in direct proportion to the current flowing in the coil, but varies in a manner usually indicated by a magnetization curve. The flux usually increases fairly rapidly when the current build-up process first begins. However, the flux increment corresponding to a particular size of current increment becomes smaller with continual increase of current as the core approaches the condition of magnetic "*saturation*". An induced voltage in the coil depends on a change of flux and hence becomes smaller for a given current change at higher values of current.

The basic expression for induced voltage in terms of a changing flux is usually replaced, for purposes of circuit study, by the "*coefficient of self-inductance*" multiplied by the rate of change of current. The new concept must be examined with care when applied to the case of an iron-cored coil. In terms of a flux change, the induced voltage is $e = -Nd\phi/dt$.

In terms of inductance, $e = -Ldi/dt$,

where $L$ = coefficient of self-inductance.

Comparison of the two expressions gives

$$L = Nd\phi/di$$

which helps to explain the nature of this quantity. Inductance may be represented by the slope of the $\phi$ - $i$ curve, or magnetization curve, of a reactor. For an air-cored coil there is a linear relationship between flux and current and, accordingly, $L$ is a constant. An iron-cored coil does not have a linear flux-current relationship, and $L$ has different values, depending on the portion of the magnetization curve being used and on the manner of defining $L$. That is, for some purpose we may be interested in the average inductance, represented by the slope of a line from the origin to a particular point on the $\phi$ - $i$ curve. For other purposes we are connected with small variations in the neighbourhood of a point on the curve (for very small variations, $L$ equals $N$ times the slope of the curve at the point in question). It is thus evident that the coefficient of self-inductance is ambiguous for iron-cored coils unless the conditions are carefully defined.

Many iron-cored coils are used as filter reactors in rectifier circuits and in other applications in which a combination of a d.c. component and a superposed a.c. are encountered. The inductance of the reactor to the a.c. in such a case may vary over a wide range, depending on the magnitude of the d.c. term. An air gap in the magnetic core tends to straighten the $\phi$ - $i$ curve, and make $L$ more nearly constant. An iron-cored coil without an air gap gives an extreme variation of $L$ if the d.c. component becomes high enough to produce a considerable degree of saturation. 16

The d.c. component of the current fixes the operating point around which changes occur on account of the alternating current. The incremental inductance at this point may be defined as turns time the slope of $\phi$ - $i$ curve.

$\therefore$ Increment inductance, $L = Nd\phi/di = N\Delta\phi/\Delta i$

Similarly, incremental permeability may be defined as the slope of the $B$ - $H$ curve at the operating point.

Incremental permeability, $\mu = dB/dH = \Delta B/\Delta H$.

The magnitude of the a.c. component also has an effect on the apparent inductance of the reactor, though usually not to so great an extent as the d.c. component in the common type of reactor applications. To make inductance measurements on an iron-cored coil definite, the data should include the amount of d.c. present, and also the frequency and magnitude of the a.c.

The non-linear character of the flux-current curve enters in the measurement problem in another way, by producing distortion of the current and voltage waveforms. That is, a sinusoidal voltage does not produce a sinusoidal current, or vice-versa, in an iron-cored coil. Even though the applied bridge voltage is sinusoidal, the detector voltage is distorted and hence may be analyzed by Fourier method into the fundamental frequency-component plus harmonics. The harmonics cause difficulties in the balancing process if the detector consists of an amplifier, a tuned circuit can be incorporated in it to pass the fundamental frequency and suppress the harmonics. The situation is a little more difficult when a telephone head set is used but a person can train himself to listen for the null point of the fundamental frequency and disregard the harmonics. Due to the difficulties of defining inductance in the first place, and of determining balance in the second place, we cannot expect to measure parameters of an iron-cored coil with the same accuracy as those of an air-cored coil. Fortunately, in most uses of the iron-cored coils, as in filter chokes, a low degree of accuracy can be tolerated.

Many bridge circuits can be modified to permit the simultaneous application of d.c. and a.c. to the reactor. One thing that must be kept in mind in arranging the circuit is the fact that the amount of d.c. specified for the test reactor must also pass through one of the bridge arms, and care must be taken not to overheat the precision resistor.

The incremental inductance can be measured with an Owen's bridge. The original circuit, however, has to be modified in order that the coil under measurement is fed from both d.c. and a.c. This circuit is shown in Fig. 16.8.

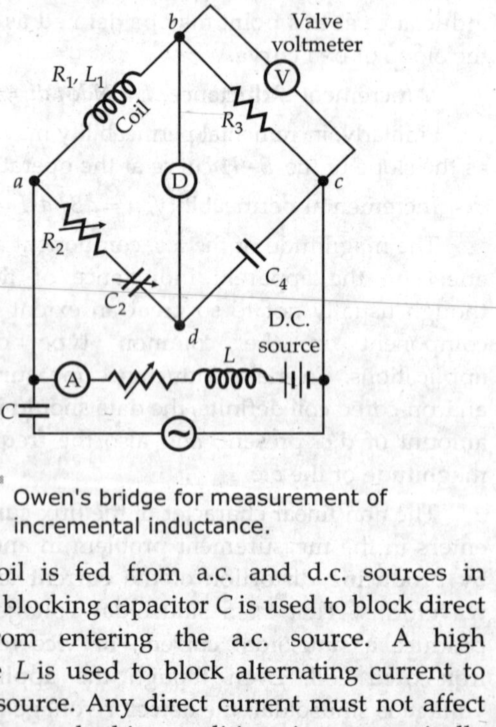


**Fig. 16.8** Owen's bridge for measurement of incremental inductance.

The coil is fed from a.c. and d.c. sources in parallel. A blocking capacitor C is used to block direct current from entering the a.c. source. A high inductance L is used to block alternating current to enter d.c. source. Any direct current must not affect the balance and this condition is automatically satisfied in Owen's bridge because capacitors $C_2$ and $C_4$ block any d.c. current flowing through the detector.

As has been mentioned earlier, it is necessary to know the magnetization conditions under which the coil is being worked. The d.c. component of current is measured by a moving coil ammeter A connected in the d.c. supply circuit. The a.c. component of current may be easily obtained from the reading of a valve voltmeter (not sensitive to d.c.) connected across the known resistance $R_3$. The value of current calculated from this reading is a.c. current through $R_3$, but this is also, at balance, the a.c. current through the coil.

At balance, incremental inductance

$$L_1 = R_2 R_3 C_4 \quad \text{...(16.30)}$$

Now inductance, $L_1 = N^2 / (l / \mu A)$

∴ Incremental permeability

$$\mu = L_1 l / N^2 A \quad \text{...(16.31)}$$

where $N$ = number of turns,

$A$ = area of flux path,

$l$ = length of flux path,

and $L_1$ = incremental inductance.

## 16.6 MEASUREMENT OF CAPACITANCE

### 16.6.1 De Sauty's Bridge

The bridge is the simplest method of comparing two capacitances. The connections and the phasor diagram of this bridge are shown in Fig. 16.9.

Let $C_1$ = capacitor whose capacitance is to be measured,

$C_2$ = a standard capacitor,

and $R_3, R_4$ = non-inductive resistors.

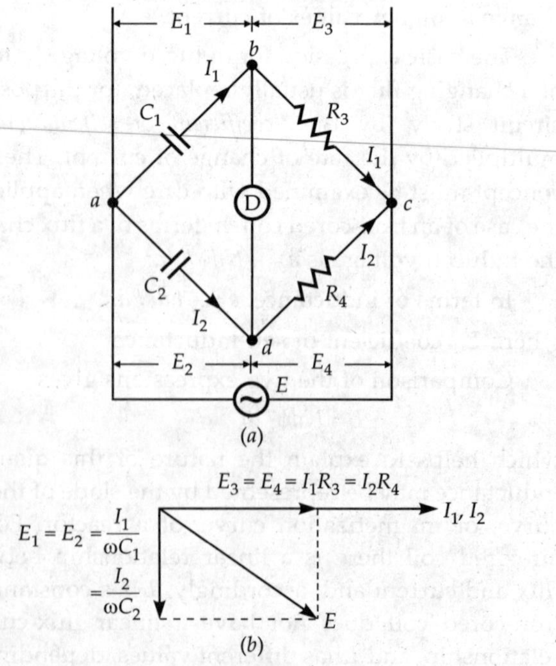


**Fig. 16.9** De Sauty's bridge.

At balance,

$$\left(\frac{1}{j\omega C_1}\right)R_4 = \left(\frac{1}{j\omega C_2}\right)R_3$$

or $$C_1 = C_2 R_4 / R_3 \quad ...(16.23)$$

The balance can be obtained by varying either $R_3$ or $R_4$. The advantage of this bridge is its simplicity. But this advantage is nullified by the fact that it is impossible to obtain balance if both the capacitors are not free from dielectric loss. Thus with this method only loss-less capacitors like air capacitors can be compared.

In order to make measurements on imperfect capacitors (*i.e.*, capacitors having dielectric loss), the bridge is modified as shown in Fig. 16.10. This modification is due to Grover.

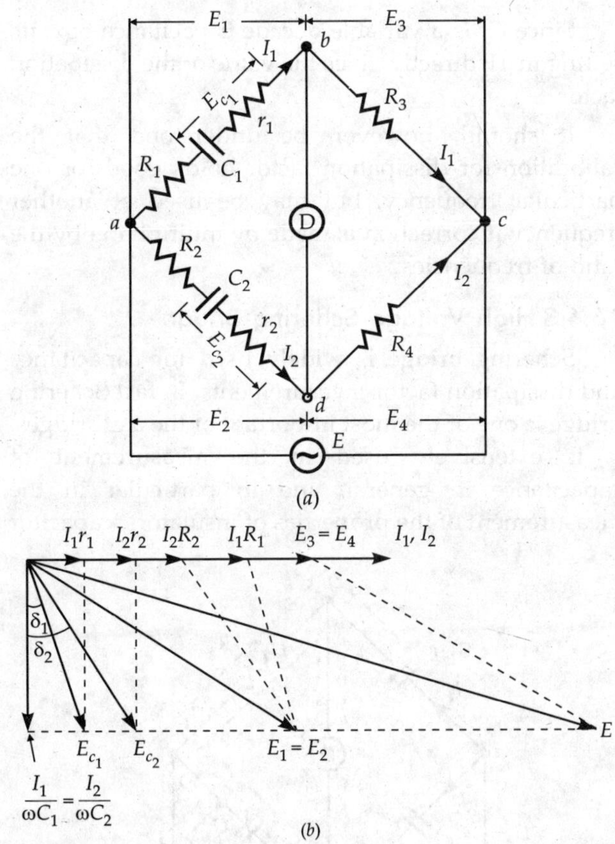


**Fig. 16.10** Modified De Sauty's bridge.

Resistors $R_1$ and $R_2$ are connected in series with $C_1$ and $C_2$ respectively. $r_1$ and $r_2$ are resistances representing the loss component of the two capacitors.

At balance,

$$\left(R_1 + r_1 + \frac{1}{j\omega C_1}\right)R_4 = \left(R_2 + r_2 + \frac{1}{j\omega C_2}\right)R_3$$

From which we have

$$\frac{C_1}{C_2} = \frac{R_2 + r_2}{R_1 + r_1} = \frac{R_4}{R_3} \quad ...(16.33)$$

The balance may be obtained by variation of resistances $R_1, R_2, R_3, R_4$.

Figure 16.10(*b*) shows the phasor diagram of the bridge under balance conditions. The angles $\delta_1$ and $\delta_2$ are the phase angles of capacitors $C_1$ and $C_2$ respectively.

Dissipation factors for the capacitors are :

$$D_1 = \tan\delta_1 = \omega C_1 r_1 \text{ and } D_2 = \tan\delta_2 = \omega C_2 r_2.$$

From Eqn. 16.33, we have

$$\frac{C_1}{C_2} = \frac{R_2 + r_2}{R_1 + r_1}$$

or $$C_2 r_2 - C_1 r_1 = C_1 R_1 - C_2 R_2$$

or $$\omega C_2 r_2 - \omega C_1 r_1 = \omega(C_1 R_1 - C_2 R_2)$$

$\therefore$ $$D_2 - D_1 = \omega(C_1 R_1 - C_2 R_2)$$

But $$C_1 / C_2 = R_4 / R_3$$

$\therefore$ $$C_1 = C_2 R_4 / R_3$$

Hence $$D_2 - D_1 = \omega C_2 \left(\frac{R_1 R_4}{R_3} - R_2\right) \quad ...(16.34)$$

Therefore, if the dissipation factor of one of the capacitors is known, the dissipation factor for the other can be determined.

This method does not give accurate results for dissipation factor since its value depends on difference of quantities $R_1 R_4 / R_3$ and $R_2$. These quantities are moderately large and their difference is very small and since this difference cannot be known with a high degree accuracy, the dissipation factor cannot be determined accurately.

### 16.6.2 Schering Bridge

The connection and phasor diagram of the bridge under balance conditions are shown in Fig. 16.11.

Let

$C_1$ = capacitor whose capacitance is to be determined,

$r_1$ = a series resistance representing the loss in the capacitor $C_1$,

$C_2$ = a standard capacitor. This capacitor is either an air or a gas capacitor and hence is loss-free. However, if necessary, a correction may be made for the loss angle of this capacitor,

$R_3$ = a non-inductive resistance,

$C_4$ = a variable capacitor,

and $R_4$ = a variable non-inductive resistance in parallel with variable capacitor $C_4$.

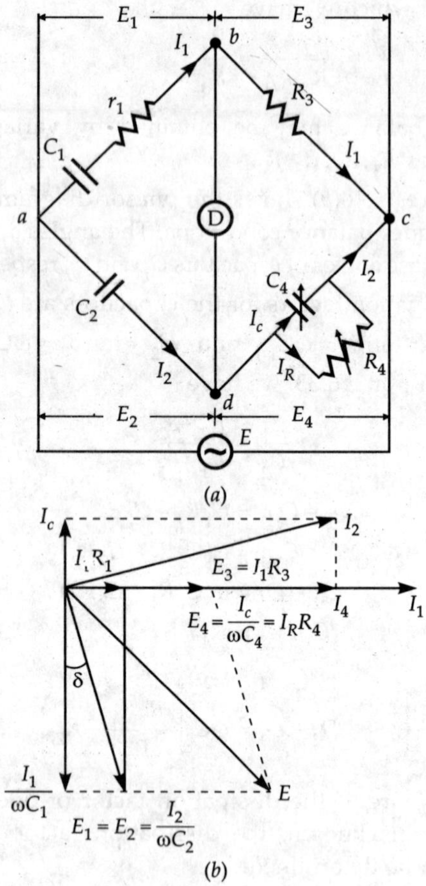


**Fig. 16.11** Low voltage Schering bridge.

At balance,

$$\left(r_1 + \frac{1}{j\omega C_1}\right)\left(\frac{R_4}{1 + j\omega C_4 R_4}\right) = \frac{1}{j\omega C_2} \cdot R_3$$

or
$$\left(r_1 + \frac{1}{j\omega C_1}\right) R_4 = \frac{R_3}{j\omega C_2}(1 + j\omega C_4 R_4)$$

or
$$r_1 R_4 - \frac{jR_4}{\omega C_1} = -j\frac{R_3}{\omega C_2} + \frac{R_3 R_4 C_4}{C_2}$$

Equating the real and imaginary terms, we obtain

$$r_1 = R_3 C_4 / C_2 \quad ...(16.35)$$

and
$$C_1 = C_2(R_4 / R_3) \quad ...(16.36)$$

Two independent balance equations are obtained if $C_4$ and $R_4$ are chosen as the variable elements.

Dissipation factor,

$$D_1 = \tan\delta = \omega C_1 r_1 = \omega(C_2 R_4 / R_3) \times (R_3 C_4 / C_2)$$
$$= \omega C_4 R_4 \quad ...(16.37)$$

Therefore values of capacitance $C_1$, and its dissipation factor are obtained from the values of bridge elements at balance.

Permanently set up Schering bridges are sometimes arranged so that balancing is done by adjustment of $R_2$ and $C_4$ with $C_2$ and $R_4$ remaining fixed. Since $R_3$ appears in both the balance equations and therefore there is some difficulty in obtaining balance but it has certain advantages as explained below :

The equation for capacitance is $C_1 = (R_4 / R_3)C_2$ and since $R_4$ and $C_2$ are fixed, the dial of resistor $R_3$ may be calibrated to read the capacitance directly.

Dissipation factor $D_1 = \omega C_4 R_4$ and in case the frequency is fixed the dial of capacitor $C_4$ can be calibrated to read the dissipation factor directly.

Let us say that the working frequency is 50 Hz and the value of $R_4$ is kept fixed at 3,180 Ω.

$\therefore$ Dissipation factor $D_1$

$$= 2\pi \times 50 \times 3180 \times C_4 = C_4 \times 10^6.$$

Since $C_4$ is a variable decade capacitance box, its setting in μF directly gives the value of the dissipation factor.

It should, however, be understood that the calibration for dissipation factor holds good for one particular frequency, but may be used at another frequency if correction is made by multiplying by the ratio of frequencies.

### 16.6.3 High Voltage Schering Bridge

Schering bridge is widely used for capacitance and dissipation factor measurements. In fact Schering bridge is one of the most important of the a.c. bridges. It is extensively used in the measurement of capacitance in general, and in particular in the measurement of the properties of insulators, capacitor

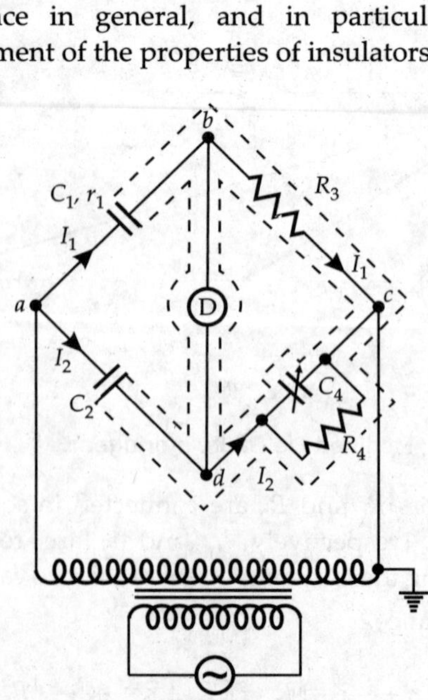

**Fig. 16.12** High Voltage Schering bridge.

bushings, insulating oil and other insulating materials. This bridge is particularly suitable for small capacitances, and is then usually supplied from a high frequency or a high voltage source. The measurements done on small capacitances suffer from many disadvantages if carried out at low voltages. High voltage Schering bridge is certainly preferable for such measurements.

The special features of a high voltage Schering bridge shown in Fig. 16.12, are explained below :

**1.** The high voltage supply is obtained from a transformer usually at 50 Hz. The detector, in this case, is a vibration galvanometer.

**2.** Arms *ab* and *ad* each contain only a capacitor and these capacitors are designed for high voltage work. The impedances of these two arms are very high in comparison with the other arms, *bc* and *dc*. Thus the major portion of potential drop will be in the arms *ab* and *ad* and very little voltage drop is there across the arms *bc* and *dc*. The point *c* is earthed. Such is the large magnitude of impedances in arms *ab* and *ad*, that even if a voltage as high as 100 kV is applied to the bridge, the voltage across arms *bc* and *dc* is a few volt above earth. This is certainly a great advantage as the controls are located in arms *bc* and *dc* and for the safety of the operator, these controls should be and are at low potential with respect to earth. For the same reason the detector is also at a low potential.

**3.** It is necessary to prevent dangerous high voltages appearing across arms *bc* and *ac* in the case of breakdown of either of the high voltage capacitors. This is done by connecting a spark gap, (set to breakdown at about 100 V) across each of the arms *bc* and *dc*.

**4.** The impedances of arms *ab* and *ad* are large and therefore the current drawn from the source is small and hence the power loss is quite small. But this small value of current also necessitates the use of a sensitive detector.

**5.** The fixed standard capacitor, $C_2$, has either air or compressed gas as dielectric. The dissipation factor of a dry and clean gas is sensibly zero, loss in the insulating supports cannot be avoided. This loss, however, can be prevented from influencing the measurements by the use of a guard ring from which both electrodes of the capacitor are supported. With this arrangement the current through the high voltage supports passes direct to earth, and as the potential difference between low voltage electrode and the guard ring is very small, the insulation of the low voltage electrode has a negligible effect. For some general applications, mica capacitors are used and in such cases, the dissipation factor of the capacitor must be accurately known.

**6.** Earthed screens are provided in order to avoid errors caused due to inter-capacitance between high and low arms of the bridge. Instead of earthing one point on the circuit as shown in Fig. 16.12, the earth capacitance effect on the galvanometer and leads is eliminated by means of a "Wagner earth device" (described later in this chapter on page 498).

### 16.6.4 Measurement of Relative Permittivity with Schering Bridge

Schering bridge is very useful for the measurement of relative permittivity of dielectric materials. The determination of relative permittivity involves the measurement of capacitance of a small capacitor with the specimen as dielectric. The capacitor with specimen as dielectric is formed by using either a parallel plate or a concentric cylinder configuration for the electrodes. Guard circuits are used in order to make the plate area definite. 16

Many different techniques are used, both in the type of specimen capacitor used and in the measuring circuit. The normal arrangement for solid materials is to use a disc specimen with metal electrodes. The electrodes may consist of thin metal foil attached to the specimen by petroleum jelly, or thin films of silver or aluminium applied by evaporation (both these arrangements normally have solid metal backing electrodes), or mercury. Mercury electrodes are obtained by floating (or supporting) the specimen on the surface of a mercury pool, the upper electrodes consisting of a smaller mercury pool held in place by a metal containing ring. Liquid specimens fill the space between the concentric cylindrical electrodes of a test cell.

The relative permittivity is calculated from the measured value of capacitance and the dimensions of the electrodes. For a parallel plate arrangement, relative permittivity

$$\varepsilon_r = \frac{C_s d}{\varepsilon_0 A} \qquad \text{...(16.38)}$$

where $C_s$ = measured value of capacitance with specimen as dielectric,

$d$ = spacing between electrodes,

$A$ = effective area of electrodes,

and $\varepsilon_0$ = permittivity of free space.

A method which avoids the necessity for close contact between electrodes and specimen uses a pair of solid electrodes (one with a guard ring) between which the specimen (thinner than the space between electrodes) is slipped. The capacitance of the arrangement is measured, the specimen is then removed, and the spacing between the electrodes is adjusted, by means of micrometer adjustments, until the capacitance is the same as before. The relative permittivity of the specimen can be calculated from the thickness of the specimen and the alteration in electrode spacing.

Let

$C$ = capacitance with specimen between electrodes,
$A$ = area of electrodes,
$d$ = thickness of specimen,
$t$ = gap between specimen and electrode,
and $x$ = reduction in separation between the two measurements.

These dimensions are shown in Fig. 16.13.

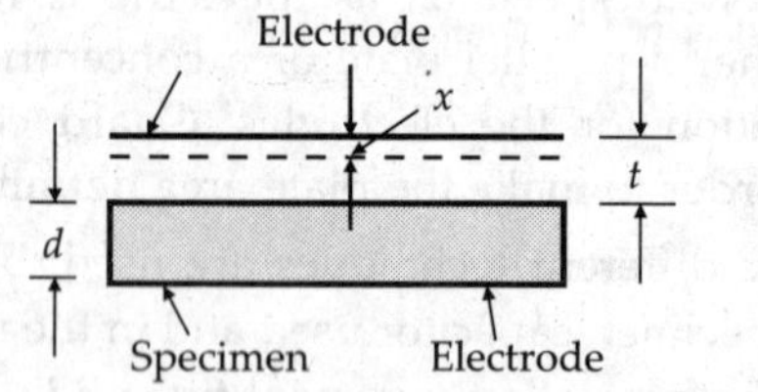


**Fig. 16.13** Measurement of relative permittivity.

Let

$C_s$ = capacitance of specimen,
$C_0$ = capacitance due to space between specimen and electrode,
and $C$ = effective capacitance of $C_s$ and $C_0$ in series.

$$\therefore\ C = \frac{C_s C_0}{C_s + C_0} = \frac{(\varepsilon_r \varepsilon_0 A/d)\cdot(\varepsilon_0 A/t)}{(\varepsilon_r \varepsilon_0 A/d)+(\varepsilon_0 A/t)} = \frac{\varepsilon_r \varepsilon_0 A}{\varepsilon_r t + d}$$

When the specimen is removed and the spacing readjusted to give the same capacitance, the expression for capacitance is :

$$C = \frac{\varepsilon_0 A}{t+d-x}.$$

Hence
$$\frac{\varepsilon_0 A}{t+d-x} = \frac{\varepsilon_r \varepsilon_0 A}{\varepsilon_r t + d}$$

$$\therefore \qquad \varepsilon_r = \frac{d}{d-x} \qquad ...(16.39)$$

## 16.7 MEASUREMENT OF MUTUAL INDUCTANCE

### 16.7.1 Uses of Mutual Inductance in Bridge Circuits

Mutual inductance has been used in a great number of bridges for a variety of purposes. Known mutual inductances are used in some circuits for the measurement of unknown mutual inductance. Variable standard mutual inductances have been used as components in bridges for the measurement of self-inductances, capacitance, and frequency. Many such bridges are found in literature, but some of the circuits represent minor modifications of other bridges ; that is, changes to achieve greater accuracy or convenience for a particular type of measurement, or a special range of unknown quantities.

We shall study only a small number of these circuits. Applications to capacitance and self-inductance determinations are of secondary interest because methods using a capacitance standard are generally more convenient and accurate for these measurements than a mutual inductor and they (capacitors) are also cheaper, compact, and more generally available. The mutual inductance circuits may have particular merit in some cases, but this enters the field of special research rather than of general measurements, so we shall not attempt to go into much detail. Much reference material is available on the subject.

The derivation of the balance equations requires a different method when we have mutual coupling between the arms. It is necessary to write equations for the voltage around the loops and then to make solution for the unknown quantities.

### 16.7.2 Mutual Inductance Measured as Self-Inductance

If the terminals of the two coils whose mutual inductance is to be measured are available for series connection, measurement may be made by one of the self-inductance bridges considered earlier. If the connections are made so that the magnetic fields of the two coils are addative as shown in Fig. 16.14(*a*), the effective inductance of the two coils in series is,

$$L_{e1} = L_1 + L_2 + 2M \qquad ...(16.40)$$

If the connections of one coil are reversed as shown in Fig. 16.14(*b*),

$$L_{e2} = L_1 + L_2 - 2M \qquad ...(16.41)$$

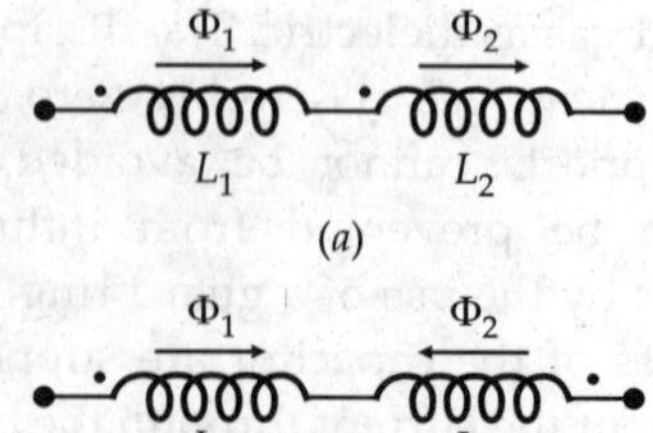


**Fig. 16.14** Positive and negative couplings for mutual inductors.

Thus from Eqns. 16.40 and 16.41, we get

Mutual inductance

$$M = \frac{1}{4}(L_{e1} - L_{e2}) \quad ...(16.42)$$

Hence mutual inductance is obtained as one-fourth of difference of self-inductance measured with series addative and series subtractive connections. This method is of advantage only when fairly high coupling between the two coils is obtained otherwise it results in poor accuracy due to nearly equal terms of Eqn. 16.42.

### 16.7.3 Heaviside Mutual Inductance Bridge

This Bridge (shown in Fig. 16.15) measures mutual inductance in terms of a known self- inductance. The same bridge, slightly modified, was used by Campbell to measure a self-inductance in terms of a known mutual inductance.

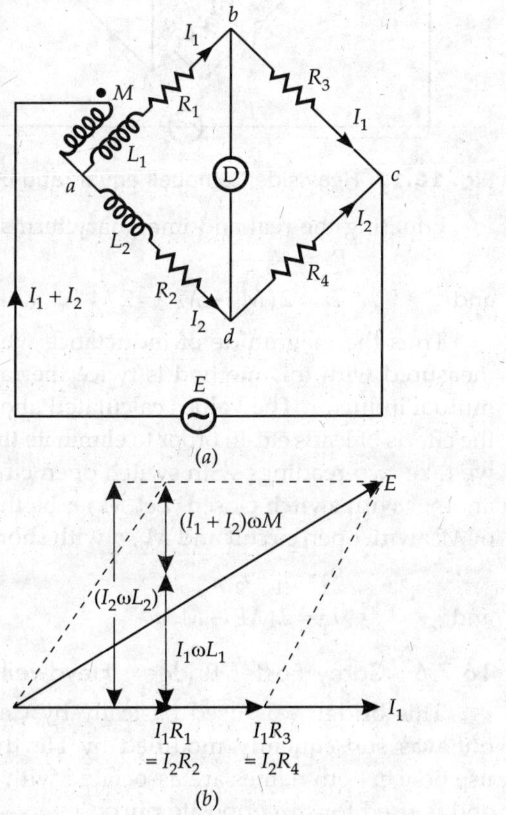


**Fig. 16.15** Heaviside mutual inductance bridge.

Let

$M$ = unknown mutual inductance,

$L_1$ = self-inductance of secondary of mutual inductance,

$L_2$ = known self-inductance, and

$R_1, R_2, R_3, R_4$ = non-inductive resistors.

At balance voltage drop between $b$ and $c$ must equal the voltage drop between $d$ and $c$. Also the voltage drop across $a$-$b$-$c$ must equal the voltage drop across $a$-$d$-$c$. Thus we have the following equations at balance :

$$I_1R_3 = I_2R_4$$

and

$$(I_1 + I_2)(j\omega M) + I_1(R_1 + R_3 + j\omega L_1) = I_2(R_2 + R_4 + j\omega L_2)$$

$$\therefore \quad I_2\left(\frac{R_4}{R_3} + 1\right)j\omega M + I_2\frac{R_4}{R_3}(R_1 + R_3 + j\omega L_1) = I_2(R_2 + R_4 + j\omega L_2)$$

or

$$j\omega M\left(\frac{R_4}{R_3} + 1\right) + \frac{R_4}{R_3}R_1 + R_4 + j\omega L_1\frac{R_4}{R_3} = R_2 + R_4 + j\omega L_2$$

Thus

$$R_1 = R_2R_3 / R_4 \quad ...(16.43)$$

and

$$M = \frac{L_2 - L_1R_4/R_3}{R_4/R_3 + 1} = \frac{R_3L_2 - R_4L_1}{R_3 + R_4} \quad ...(16.44)$$

It is clear from Eqn. 16.44, that $L_1$, the self- inductance of the secondary of the mutual inductor must be known in order that $M$ be measured by this method.

In case $R_3 = R_4$, we get

$$M = \frac{L_2 - L_1}{2} \quad ...(16.45)$$

and

$$R_1 = R_2 \quad ...(16.46)$$

This method can be used for measurement of self-inductance. Supposing $L_2$ is the self-inductance to be determined. From Eqns. 16.43 and 16.44, we get

$$L_2 = \frac{M(R_3 + R_4) + R_4L_1}{R_3}$$

$$= M\left(1 + \frac{R_4}{R_3}\right) + \frac{R_4}{R_3}L_1 \quad ...(16.47)$$

and

$$R_2 = R_1R_4 / R_3 \quad ...(16.48)$$

In case $R_4 = R_3$, we have

$$L_2 = 2M + L_1 \quad \text{and} \quad R_2 = R_1.$$

### 16.7.4 Campbell's Modification of Heaviside Bridge

Figure 16.16 shows a modified Heaviside bridge. This modification is due to Campbell. This is used to measure a self-inductance in terms of a mutual inductance. In this case an additional balancing coil $L$, $R$ is included in arm $ad$ in series with inductor under test. An additional resistance $r$ is put in arm $ab$. Balance is obtained by varying $M$ and $r$. A short circuiting switch is placed across the coil $R_2$, $L_2$ under measurement. Two sets of readings are taken one with switch being open and the other with switch being closed. Let values of $M$ and $r$, be $M_1$ and $r_1$ with switch open, and $M_2$ and $r_2$ with switch closed.

16

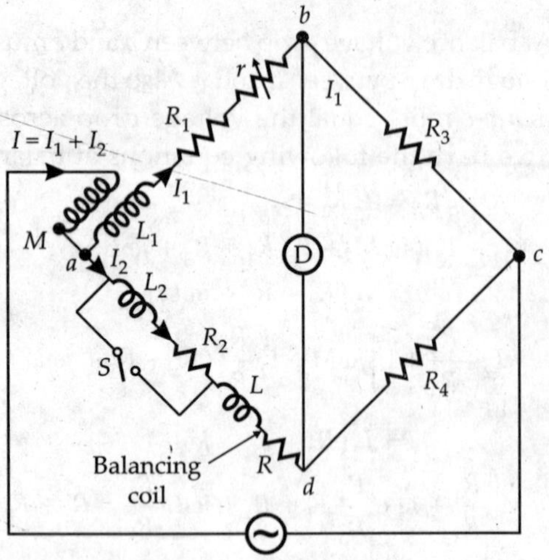


**Fig. 16.16** Compbell's modification of Heaviside bridge.

∴ With switch open, from Eqn. 16.47, we have,

$$L_2 + L = \frac{M_1(R_3 + R_4) + R_4 L_1}{R_3}$$

and with switch closed, $L = \dfrac{M_2(R_3 + R_4) + R_4 L_1}{R_3}$

Hence $L_2 = (M_1 - M_2)(1 + R_4/R_3)$ ...(16.49)

With switch open, from Eqn. 16.48, we have,

$$R_2 + R = (R_1 + r_1)R_4/R_3$$

and with switch closed

$$R = (R_1 + r_2)R_4/R_3$$

∴ $R_2 = (r_1 - r_2)R_4/R_3$ ...(16.50)

This method is a good example of the methods adopted to eliminate the effects of leads etc.

When we have equal ratio arms $R_3 = R_4$.

and therefore from Eqns. 16.49 and 16.50, we get,

$$L_2 = 2(M_1 - M_2) \quad ...(16.51)$$

and $R_2 = r_1 - r_2$ ...(16.52)

### 16.7.5 Heaviside Campbell Equal Ratio Bridge

The use of balancing coil in the above method reduces the sensitivity of the bridge. Figure 16.17 shows Heaviside Campbell equal ratio bridge. This is a better arrangement which improves sensitivity and also dispenses with the use of a balancing coil.

In this method the secondary of the mutual inductor is made up of two equal coils each having a self-inductance. One of the coils is connected in arm *ab* and the other in arm *ad*. The primary of mutual inductance reacts with both of them. $L_2$, $R_2$ is the coil whose self-inductance and resistance is to be determined. The resistances $R_3$ and $R_4$ are made equal. Balance is obtained by varying the mutual inductance and resistance *r*.

At balance, $I_1 R_3 = I_2 R_4$ but $R_3 = R_4$

and therefore $I_1 = I_2 = I/2$ as $I = I_1 + I_2$.

Writing the other equation for balance :

$$I_1(R_1 + r) + I_1 j\omega L + Ij\omega M_x = I_2 R_2 + I_2 j\omega(L_2 + L) - Ij\omega M_y$$

or $$\frac{R_1 + r}{2} + j\omega\left(\frac{L}{2} + M_x\right) = \frac{R_2}{2} + j\omega\left(\frac{L_2 + L}{2}\right) - j\omega M_y$$

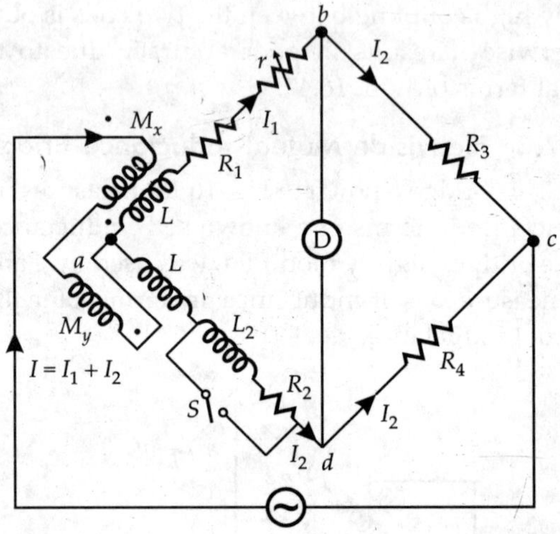


**Fig. 16.17** Heaviside Campbell equal ratio bridge.

Equating the real and imaginary terms

$$R_2 = R_1 + r \quad ...(16.53)$$

and $L_2 = 2(M_x + M_y) = 2M$ ...(16.54)

Thus the magnitude of inductance which can be measured with this method is twice the range of the mutual inductor. The values calculated above include the effects of leads etc. In order to eliminate these effects, we take two readings with switch open circuited and another with switch closed. Let $M_1$, $r_1$ be the readings of *M*, *r* with open circuit and $M_2$, $r_2$ with short circuit.

$$R_1 = r_1 - r_2 \quad ...(16.55)$$

and $L_1 = 2(M_1 - M_2)$ ...(16.56)

### 16.7.6 Carey Foster Bridge ; Heydweiller Bridge

This bridge was used basically by Carey Foster but was subsequently modified by Heydweiller for use on a.c. Both names are associated with the bridge and is used for *two* opposite purposes :

(*i*) It is used for measurement of capacitance in terms of a standard mutual inductance. The bridge in this case is known as Carey Foster's bridge.

(*ii*) It can also be used for measurement of mutual inductance in terms of a standard capacitance and is then known as Heydweiller bridge.

Figure 16.18 shows the connection diagram for the bridge under balance conditions.

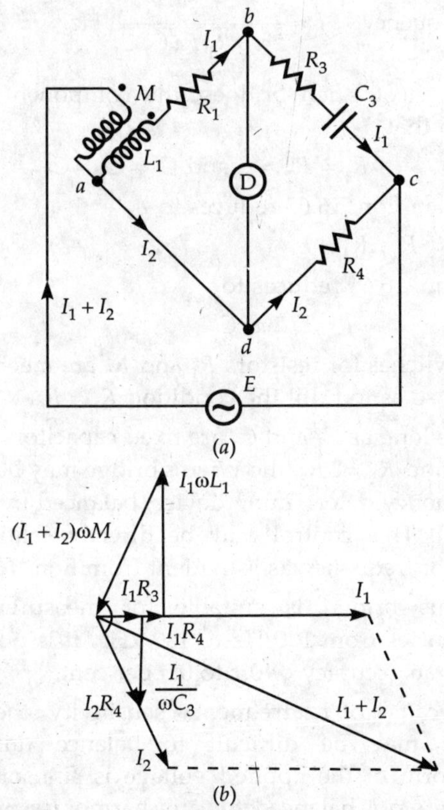


**Fig. 16.18** Carey Foster (Heydweiller) Bridge.

This bridge has an unusual feature, one of its arms, *ad*, is short circuited and therefore the potential drop across this arm is zero. In order to achieve balance, the potential drop across arm *ab* should also be equal to zero and for this reason negative coupling is needed for the mutual inductance.

At balance,

$$I_1(R_1 + j\omega L_1) - (I_1 + I_2) j\omega M = 0$$

and
$$I_1\left(R_3 + \frac{1}{j\omega C_3}\right) = I_1 R_4.$$

The solution of the above equation gives :

$$M = R_1 R_4 C_3 \qquad \text{...(16.57)}$$

and
$$L_1 = \frac{M(R_3 + R_4)}{R_4} = R_1 C_3 (R_3 + R_4) \qquad \text{...(16.58)}$$

If the bridge is used for measurement of capacitance, Eqns. 16.57 and 16.58 may be written as :

$$C_3 = M / R_1 R_4 \qquad \text{...(16.59)}$$

$$R_3 = R_4 (L_1 - M) / M \qquad \text{...(16.60)}$$

In the measurement of mutual inductance with this bridge, $R_3$ is a separate resistance while in the measurement of capacitance $R_3$ is not a separate unit but represents the equivalent series resistance of the capacitor and thus can be determined in terms of the elements of the bridge.

### 16.7.7 Campbell's Bridge

This bridge measures an unknown mutual inductance in terms of a standard mutual inductance. Figure 16.19 shows the circuit diagram for the bridge.

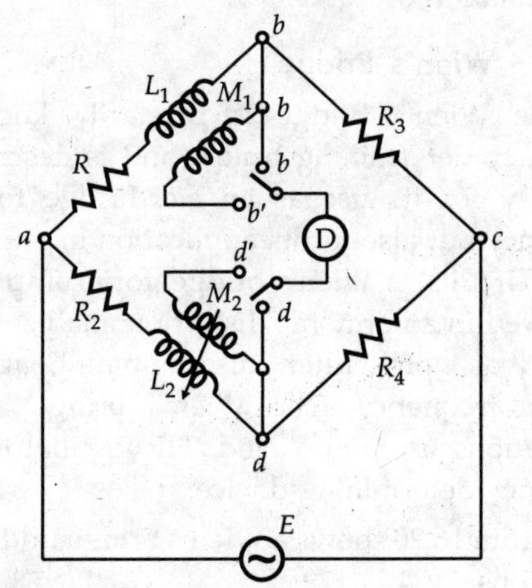


**Fig. 16.19** Campbell's Bridge.

Let

$M_1$ = unknown mutual inductance,

$L_1$ = self-inductance of secondary of mutual inductance $M_1$,

$M_2$ = variable standard mutual inductance,

$L_2$ = self-inductance of secondary of mutual inductance $M_2$, and

$R_1, R_2, R_3, R_4$ = non-inductive resistances.

There are *two* steps required in the balancing process.

1. Detector is connected between *b* and *d*. The circuit now becomes a simple self-inductance comparison bridge. The requirement for balance is :
$$\frac{L_1}{L_2} = \frac{R_1}{R_2} = \frac{R_3}{R_4} \qquad \text{...(16.61)}$$
The bridge may be balanced by adjustment of $R_3$ (or $R_4$) and $R_1$ (or $R_2$).
2. Detector is connected between *b′* and *d′*. Keeping adjustments as in step 1 above, the variable mutual inductance $M_2$ is varied to get balance. Then,
$$\frac{M_1}{M_2} = \frac{R_3}{R_4}$$
or
$$M_1 = M_2 R_3 / R_4 \qquad \text{...(16.62)}$$

## 16.8 MEASUREMENT OF FREQUENCY

Some bridges have balance equations which involve frequency directly even if the performance of individual bridge elements is independent of frequency. These bridges may be used for determination of frequency in terms of values of various bridge elements. We shall describe here the Wien's bridge, which is the most important one for determination of frequency.

### 16.8.1 Wien's Bridge

The Wien's bridge is primarily known as a frequency determining bridge and is described here not only for its use as an a.c. bridge to measure frequency but also for its application in various other useful circuits. A Wien's bridge, for example, may be employed in a harmonic distortion analyzer, where it is used as notch filter, discriminating against one specific frequency. The Wien's bridge also finds applications in audio and HF oscillators as the frequency determining device.

Figure 16.20 shows a Wien's bridge under balance conditions.

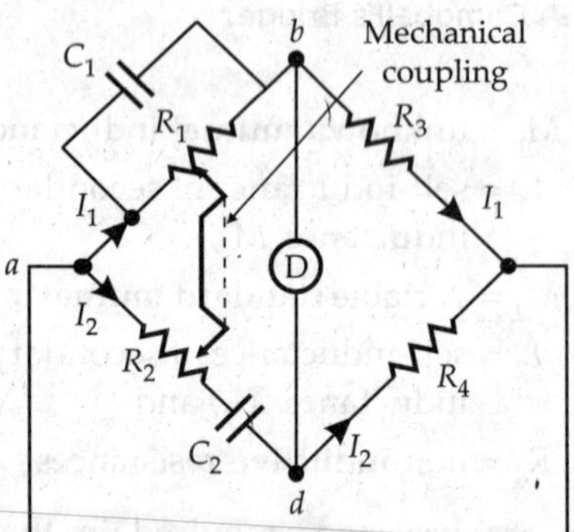


**Fig. 16.20** Wien's Bridge.

At balance

$$\left(\frac{R_1}{1+j\omega C_1 R_1}\right) R_4 = \left(R_2 - \frac{j}{\omega C_2}\right) R_3$$

or

$$\frac{R_4}{R_3} = \frac{R_2}{R_1} + \frac{C_1}{C_2} + j\left(\omega C_1 R_2 - \frac{1}{\omega C_2 R_1}\right)$$

Equating the real and imaginary parts,

or

$$\frac{R_4}{R_3} = \frac{R_2}{R_1} + \frac{C_1}{C_2} \qquad ...(16.63)$$

and

$$\omega C_1 R_2 - \frac{1}{\omega C_2 R_1} = 0$$

from which

$$\omega = \frac{1}{\sqrt{R_1 R_2 C_1 C_2}}$$

and frequency,

$$f = \frac{1}{2\pi\sqrt{R_1 R_2 C_1 C_2}} \text{ Hz.}$$

In most Wien bridges, the components are so chosen that

$$R_1 = R_2 = R \text{ and } C_1 = C_2 = C.$$

Then Eqn. 16.63 reduces to,

$$R_4 / R_3 = 2$$

and Eqn. 16.64 reduces to

$$f = 1/2\pi RC \qquad ...(16.65)$$

Switches for resistors $R_1$ and $R_2$ are mechanically linked so as to fulfil the condition $R_1 = R_2$.

As long as $C_1$ and $C_2$ are fixed capacitors equal in value and $R_4 = 2R_3$, the Wien's bridge may be used as a frequency determining device, balanced by a single control. This control may be directly calibrated in terms of frequency as is evident from Eqn. 16.65.

This bridge is suitable for measurement of frequencies from 100 Hz to 100 kHz. It is possible to obtain an accuracy of 0.1 to 0.5 per cent.

Because of its frequency sensitivity, the Wien's bridge may be difficult to balance unless the waveform of the applied voltage is sinusoidal. The bridge is not balanced for any harmonics present in the applied voltage, so that these harmonics will sometimes produce an output voltage marking the true balance point. This difficulty can be overcome by connecting a filter in series with the null detector.

A Wien's bridge may be used for measurement of capacitance also.

## 16.9 UNIVERSAL IMPEDANCE BRIDGE

One of the most useful and versatile laboratory bridges is the *Universal Impedance Bridge*. The set up of this bridge combines several of the bridge configurations described so far into a single portable instrument. This instrument is capable of measuring both d.c. and a.c. resistance, inductance and storage factor $Q$ factor of an inductor, capacitance and dissipation factor $D$ of a capacitor.

The universal bridge consists of four basic bridge circuits. It has suitable a.c. and d.c. sources, a.c. and d.c. null detectors, and impedance standards. The Wheatstone bridge is used for both d.c. and a.c. resistance measurements. Capacitance and dissipation factors are determined by modified De Sauty's bridge. The Maxwell's bridge is used for medium $Q$ inductors while Hay's bridge configuration is used for inductors having $Q$ more than 10.

For d.c. resistance measurements a suspension type galvanometer having a sensitivity 0.5 μA per scale division is used. The null indicator for a.c. measurements is usually an electron ray tube. Terminals are provided for connection of any external null detectors. High impedance head phones may also be used as a.c. detectors.

The a.c. source of the bridge consists of an oscillator having 10 kHz as the standard frequency.

## 16.10 SOURCES OF ERRORS IN BRIDGE CIRCUITS

We have assumed in our derivation of basic circuit relationships that a bridge consists of lumped impedance units connected only by the wires that are placed in the circuit for making connections. This idealized condition exists in bridge circuits to a lesser or greater extent. The idealized arrangement works fairly well if the frequency is low, if component impedances are not high and if accuracy desired is not high. But in practice there are certain factors which we have not considered yet and these complicate the behaviour or the bridge circuits considered so far.

Some of these factors are stray couplings between one bridge arm and another and from elements to ground. These stray couplings modify the balance conditions making a definite balance impossible or may lead to false balance. Thus these effects will cause incorrect values of unknown components to be attained.

***Factors causing errors.*** The various factors causing errors in a.c. bridge circuit are listed below :

(*i*) stray-conductance effects, due to imperfect insulation ;

(*ii*) mutual-inductance effects, due to magnetic coupling between various components of the bridge ;

(*iii*) stray-capacitance effects, due to electrostatic fields between conductors at different potentials ;

(*iv*) 'residues' in components – *e.g.*, the existence of small amount of series, inductance or shunt capacitance in nominally non-reactive resistors.

### 16.10.1 Precautions and Techniques used for Reducing Errors

Errors introduced by the above effects become greater in high frequency and high-voltage bridges, but precautions should always be taken to reduce them to a minimum. Even so, in high-accuracy work it is often necessary to determine the magnitude of some of the effects and allow for them in the bridge calculations. The more important methods of reducing bridge errors are now described.

**1. *Use of high-quality components.*** Good-quality bridge components will normally have the advantages of high-accuracy calibration, freedom from stray conductance effects, and a minimum of residues. Where residues exist, manufacturer supplies information giving the values of the residues.

**2. *Bridge lay-out.*** The conventional bridge circuit diagrams in the present chapter are drawn to represent the potential distribution in the circuit ; they do not indicate the best physical lay-out. In general, it is desirable that the four 'corners' of the bridge *a*, *b*, *c* and *d* (Fig. 16.21) should be closed together. Each component is then connected to the appropriate corners by its own leads. The physical lay-out of the bridge will then be as shown in Fig. 16.21. All the leads to a single corner should be brought as nearly as possible to the same point. A pair of leads to a component should not form a large loop ; if possible the leads should be co-axial. If there is more than one inductor in the bridge, leads may have to be quite long. In this manner stray mutual inductance coupling and lead self-inductances are reduced to minimum and the stray capacitances are definitely located across individual arms of bridge ; they can then be measured and compensated for relatively easily.

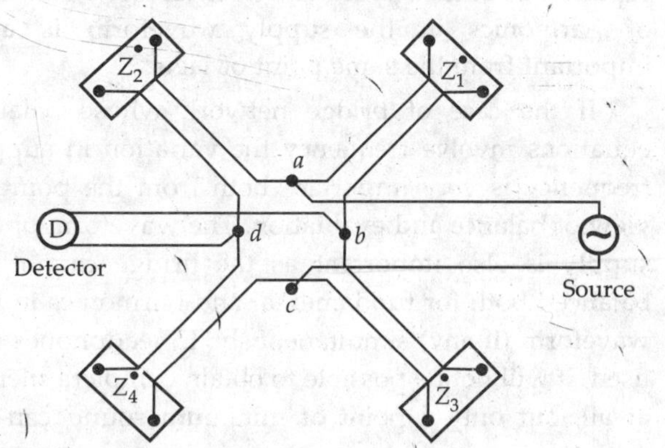


**Fig. 16.21** Bridge lay-out.

**3. *Sensitivity.*** To obtain a high degree of accuracy *e.g.*, 1 part in $10^4$ or better, it is necessary that the bridge must be operated at sufficient sensitivity which implies that the change in voltage at the detector terminals from the balance condition should be large when the detector is considered open. The treatment is similar as with the Wheatstone bridge except that with the a.c. bridge circuit we have to consider both the magnitude and phase angle of the impedance of the bridge arms.

16

**4. *Stray conductance effects.*** If the insulation between various components (elements) of a bridge circuit is not good, trouble may arise because of leakage currents from one arm to another. This is specially true in the case of high impedance bridges. To avoid this, various bridge components and other pieces of apparatus may be mounted on insulating stands.

**5. *Eddy current errors.*** Errors may result due to variation in the value of the standards, which may occur because of induced eddy currents in the standard resistors and inductors. In order to avoid such errors the presence of large conducting masses near the bridge network is avoided.

**6. *Residual errors.*** Though the resistors used are taken as non-inductive and non-capacitive but their inductance and capacitance are never zero. Residues mean small inherent inductance or capacitance of the resistors. In precise work it becomes necessary to evaluate them in order to eliminate them or compensate for errors due to these. The self-inductance is important only when the coils used are multi-turn coils and supply used is of high frequency.

**7. *Frequency and wave form errors.*** In case of bridges whose results are independent of frequency, the supply frequency is important only from the point of view of its effects on resistance, inductance and capacitance of the apparatus under test. The presence of harmonics in the supply waveform is also important from the same point of view.

If the case of bridge network whose balance equations involve frequency the variation in supply frequency is very important both from the point of view of balance and evaluation. The waveform of the supply is also important as the bridge cannot be balanced both for fundamental and harmonics in the waveform (if any) simultaneously. If headphones are used, it will not be possible to obtain complete silence at all, but only a point of minimum sound can be achieved.

This difficulty is overcome either by employing wave filters which eliminate the unwanted harmonics from the source or by employing tuned detectors in place of headphones such as vibration galvanometers which do not respond to harmonics and respond readily only to the fundamental for which they are tuned.

## 16.11 WAGNER EARTHING DEVICE

If each component in a bridge has a defining screen connected to one end, a very high accuracy in measurement is made possible by the addition of a Wagner earthing device. This device removes all the earth capacitances from the bridge network.

Figure 16.22 shows the connections of the device for use in conjunction with the general form of bridge network, in which $Z_1, Z_2, Z_3$ and $Z_4$ are the impedances of the bridge arms. $Z_5$ and $Z_6$ are the two variable impedances of the Wagner earth branch, the centre point of which is earthed as shown. These impedances may consist of variable resistances and capacitances similar to those used in the arms of the bridge proper, but not necessarily of known value. The two impedances $Z_5$ and $Z_4$, must be capable of forming a balanced bridge with $Z_1$ and $Z_3$ or $Z_2$ and $Z_4$ and can be a duplicate of either of these pairs of arms. $C_1$, $C_2$, $C_3$ and $C_4$ are the stray earth capacitances appearing at the apexes of the bridge. $D$ is the detector, which in this case is headphones.

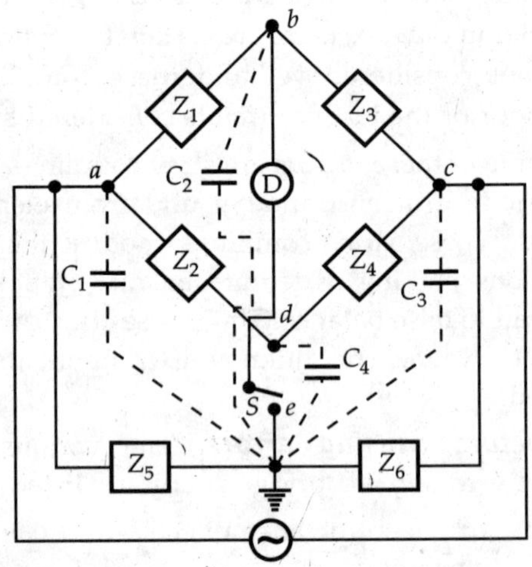


**Fig. 16.22** Wagner Earthing Device.

If the switch $S$ is on contact $d$, balance of the bridge may be obtained by adjustment of the impedances $Z_2$ and $Z_4$. The presence of the earth capacitances will prevent a true balance being obtained, but a point of minimum sound can be obtained.

After adjusting the bridge to give minimum sound, the switch $S$ is thrown to contact $e$ so that the headphones are then connected between $b$ and earth ; $Z_5$ and $Z_4$ are next adjusted until minimum sound is obtained. The headphones are next reconnected to $b$, $d$ and $Z_2$ and $Z_4$ adjusted to give minimum sound again. The process is repeated until silence is obtained with the switch on $d$, and silence or the minimum attainable sound, with the switch on $e$. Then all three points $b$, $d$ and $e$ are at earth potential. Under these conditions no current flows in the earth capacitances $C_2$ and $C_4$ and since $C_1$ and $C_3$ shunt the Wagner arms $Z_5$ and $Z_6$, these capacitances are eliminated from the bridge network $Z_1$, $Z_2$, $Z_3$ and $Z_4$.

The capacitances $C_1$ and $C_3$, shunting $Z_5$ and $Z_6$, complicate these arms, and because of this the combination may not give a true balance against $Z_2$ and $Z_4$, only a minimum being attainable during the Wagner balance. The existence of these capacitances should be borne in mind when deciding the form of the Wagner arm, which should preferably be composed of resistances and capacitances. Much of $C_1$ and $C_3$ will be due to the earth capacitance of the source transformer and a common cause of difficulty in achieving a good Wagner balance is due to the use of transformers with large and lossy earth capacitance. Poor insulation in the source transformer will cause resistance to appear shunting $C_1$ and $C_3$. Some of the disadvantages of the use of Wagner earth devices can be overcome by using double ratio a.c. bridges ; which have two sets of inductively coupled ratio arms.

**NOTE** The students must attempt these problems from fundamentals although here direct formulae have been used since analysis has been done earlier.

**Example 16.3** *A Maxwell's inductance comparison bridge is shown in Fig. 16.23. Arm ab consists of a coil with inductance $L_1$ and resistance $r_1$ in series with a non-inductive resistance R. Arm bc and coil ad are each a non-inductive resistance of 100 Ω. Arm ad consists of standard variable inductor L of resistance 32.7 Ω. Balance is obtained when $L_2$ = 47.8 mH and R = 1.36 Ω. Find the resistance and inductance of the coil in arm ab.*

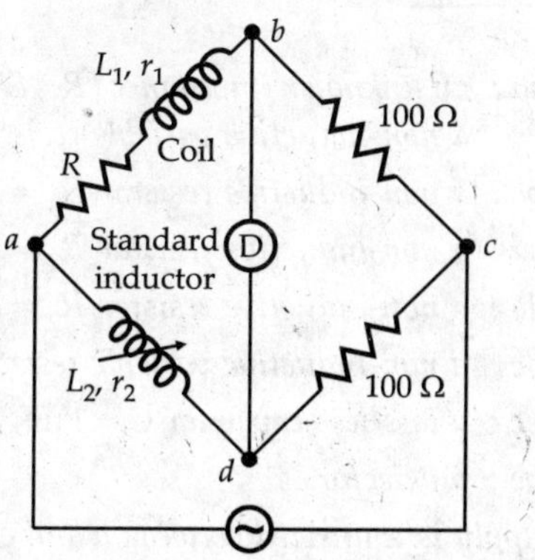


**Fig. 16.23** Diagram of Example 16.3.

**Solution.** At balance

$$[(R_1 + r_1) + j\omega L_1] \times 100 = (r_2 + j\omega L_2) \times 100$$

Equating the real and imaginary terms

$$R_1 + r_1 = r_2 \text{ and } L_2 = L_1'$$

∴ Resistance of coil :

$$r_1 = r_2 - R_1 = 32.7 - 1.36 = 31.34\ \Omega.$$

Inductance of coil,

$$L_1 = L_2 = 47.8 \text{ mH}.$$

**Example 16.4** *A Maxwell's capacitance bridge shown in Fig. 16.4 is used to measure an unknown inductance in comparison with capacitance. The various values at balance,*

$$R_2 = 400\Omega\ ;\ R_3 = 600\ \Omega\ ;\ R_4 = 1000\ \Omega\ ;\ C_4 = 0.5\ \mu F.$$

*Calculate the values of $R_1$ and $L_1$. Calculate also the value of storage (Q) factor of coil if frequency is 1000 Hz.*

**Solution.** At balance,

$$(R_1 + j\omega L_1)\left(\frac{R_4}{1 + j\omega C_4 r_4}\right) = R_2 R_3.$$

Separating the real and imaginary terms, we have

$$R_1 = \frac{R_2 R_3}{R_4} = \frac{400 \times 600}{1000} = 240\ \Omega$$

and

$$L_1 = R_2 R_3 C_4$$

$$= 400 \times 600 \times 0.5 \times 10^{-6} = 0.12 \text{ H}.$$

Storage factor,

$$Q = \frac{\omega L_1}{R_1} = \frac{2\pi \times 100 \times 0.12}{240} = 3.14.$$

**Example 16.5** *An inductance of 0.22 H and 20 Ω resistance is measured by comparison with a fixed standard inductance of 0.1 H and 40 Ω resistance. They are connected as shown in Fig. 16.24(a). The unknown inductance is in arm ab and the standard inductance is arm bc, a resistance of 750 Ω is connected in arm cd and a resistance whose amount is not known is in arm da.*

*Find the resistance of arm da and show any necessary and practical additions required to achieve both resistive and inductive balance.*

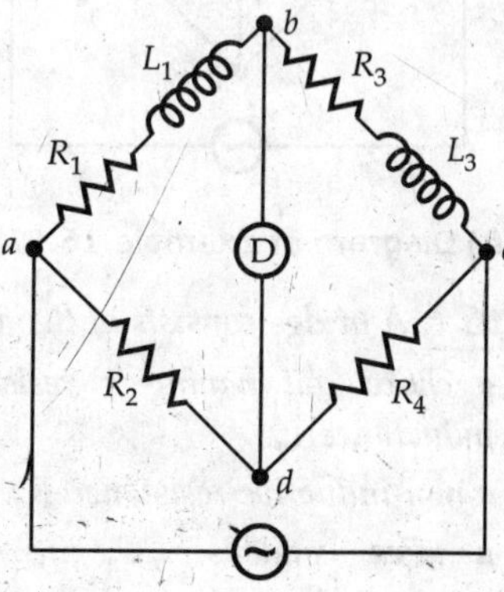


**Fig. 16.24(a)** Diagram of Example 16.5.

16

**Solution.** Now $R_1 = 20\ \Omega$ ; $L_1 = 0.22$ H ;

$R_4 = 750\ \Omega$ ; $R_3 = 40\ \Omega$ ; $L_3 = 0.1$ H.

At balance, $(R_1 + j\omega L_1)R_4 = R_2(R_3 + j\omega L_3)$

Thus the two balance equations are :

$$R_1 = \frac{R_2 R_3}{R_4} \text{ and } L_1 = \frac{R_2 L_3}{R_4}$$

From above, we have

$$\frac{L_1}{L_3} = \frac{R_2}{R_4} = \frac{R_1}{R_3}$$

$\therefore$ Value of $R_2$ required for balance

$$R_2 = R_4 \frac{L_1}{L_3} = 750 \times \frac{0.22}{0.1} = 1650\ \Omega$$

and $$\frac{L_1}{L_3} = \frac{R_2}{R_4} = 2.2.$$

Now examine the value of ratio $\frac{R_1}{R_3}$ for the existing circuit, we have

$$\frac{R_1}{R_3} = \frac{20}{40} = 0.5$$

The value of this ratio should be 2.2 for both resistive and inductive balance and therefore we must add a series resistance to arm *ab*. Let this series resistance be $r_1$. Therefore

$$\frac{R_1 + r_1}{R_3} = 2.2 \quad \text{or} \quad r_1 = 2.2 \times 40 - 20 = 68\ \Omega.$$

The modified circuit is shown in Fig. 16.24(*b*).

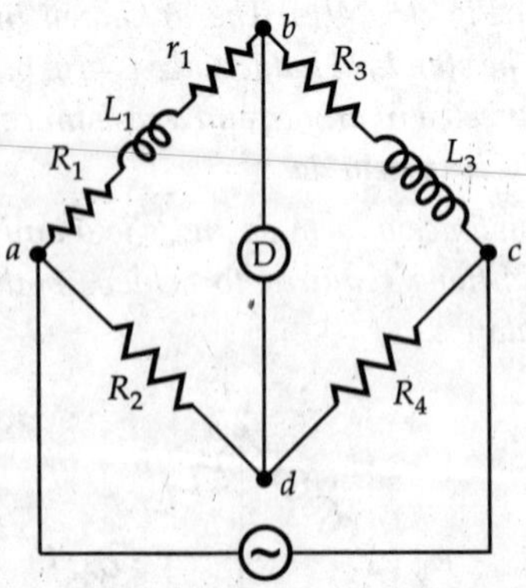


**Fig. 16.24(*b*)** Diagram of Example 16.5.

**Example 16.6** *A bridge consists of the following :*

*Arm ab : a choke coil having a resistance $R_1$ and inductance $L_1$.*

*Arm bc : a non-inductive resistance $R_3$.*

*Arm cd : a mica condenser $C_4$ in series with a non-inductive resistance $R_4$.*

*Arm da : a non-inductive resistance $R_2$*

*When this bridge is fed from a source of 500 Hz, balance is obtained under following conditions :*

*$R_2 = 2410\ \Omega$ ; $R_3 = 750\ \Omega$ ; $C_4 = 0.35\ \mu F$ ; $R_4 = 64.5\ \Omega$.*

*The series resistance of capacitor $= 0.4\Omega$. Calculate the resistance and inductance of the choke coil.*

*The supply is connected between a and c and the detector is between b and d.*

**Solution.** When we draw the sketch of this bridge, we find that it is Hay's bridge shown in Fig. 16.5. The analysis of Hay's bridge is done in Art. 16.5.3. on page 484.

Now from Eqns. 16.21 and 16.22, inductance and resistance of the coil are :

$$L_1 = \frac{R_2 R_3 C_4}{1 + \omega^2 C_4^2 R_4^2}$$

$$= \frac{2410 \times 750 \times 0.35 \times 10^{-6}}{1 + (2\pi \times 500 \times 0.35 \times 10^{-6} \times 64.9)}$$

$$= 0.63 \text{ H.}$$

and $$R_1 = \frac{\omega^2 R_2 R_3 R_4 R_4^2}{1 + \omega^2 C_4^2 R_4^2}$$

$$= \frac{(2\pi \times 500)^2 \times 2410 \times 750 \times 64.9 \times (0.35 \times 10^{-6})^2}{1 + (2\pi \times 500 \times 0.35 \times 10^{-6} \times 64.9)^2}$$

$$= 141.1\ \Omega$$

(Note that the resistance of arm *cd* is the sum of resistance of the series resistor and the series resistance of capacitor *i.e.*, $R_4 = 64.5 + 0.4 = 64.9\Omega$).

**Example 16.7** *The arms of a five node bridge are as follows :*

*arm ab : an unknown impedance $(R_1, L_1)$ in series with a non-inductive variable resistor $r_1$,*

*arm bc : a non-inductive resistor $R_3 = 100\ \Omega$,*

*arm cd : a non-inductive resistor $R_4 = 200\ \Omega$*

*arm da : a non-inductive resistor $R_2 = 250\ \Omega$,*

*arm de : a non-inductive variable resistor r*

*arm ec : a loss-less capacitor $C = 1\ \mu F$, and*

*arm be : a detector.*

*An a.c. supply is connected between a and c.*

*Calculate the resistance and inductance $R_1$, $L_1$ when under balance conditions $r_1 = 43.1\ \Omega$ and, $r = 229.7\ \Omega$.*

**Solution.** After drawing the network we find that it is the Anderson's bridge shown in Fig. 16.6 and dealt with in Art. 16.5.4 on page 485.

From Eqns. 16.26 and 16.27,

Resistance,

$$R_1 = \frac{R_2 R_3}{R_4} - r_1 = \frac{250 \times 100}{200} - 43.1 = 81.9\ \Omega.$$

Inductance,

$$L_1 = \frac{CR_3}{R_4}[r(R_4 + R_2) + R_2 R_4]$$

$$= 1 \times 10^{-6} \times \frac{100}{200}[229.7(200 + 250) + 250 \times 200]$$

$$= 0.0766 \text{ H}.$$

***Example 16.8*** *The four arms of a bridge are :*

*arm ab : an imperfect capacitor $C_1$ with an equivalent series resistance of $r_1$*

*arm bc : a non-inductive resistance $R_3$,*

*arm cd : a non-inductive resistance $R_4$,*

*arm da : an imperfect capacitor $C_2$ with an equivalent series resistance of $r_2$ series with a resistance $R_2$.*

*A supply of 450 Hz is given between terminals a and c and the detector is connected between b and d. At balance :*

$$R_2 = 4.8\ \Omega,\ R_3 = 2000\ \Omega,\ R_4 = 2850\ \Omega \text{ and}$$
$$C_2 = 0.5\ \mu\text{F and } r_2 = 0.4\ \Omega.$$

*Calculate the value of $C_1$ and $r_1$ and also of the dissipating factor for this capacitor.*

**Solution.** The bridge is shown in Fig. 16.25.

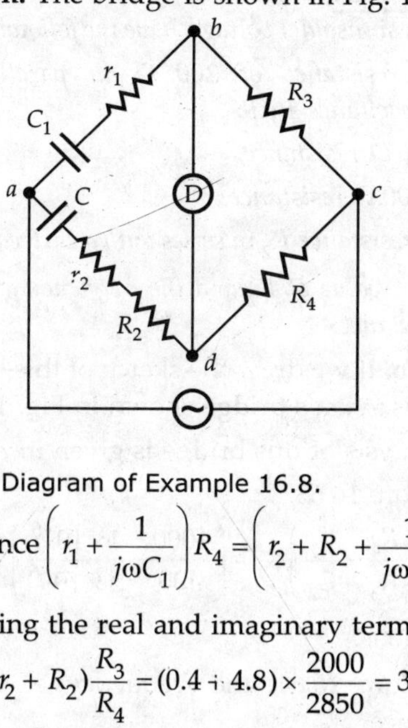

**Fig. 16.25** Diagram of Example 16.8.

$$\text{At balance}\left(r_1 + \frac{1}{j\omega C_1}\right)R_4 = \left(r_2 + R_2 + \frac{1}{j\omega C_2}\right)R_3$$

Separating the real and imaginary terms

$$r_1 = (r_2 + R_2)\frac{R_3}{R_4} = (0.4 + 4.8) \times \frac{2000}{2850} = 3.65\ \Omega.$$

and $C_1 = C_2 \dfrac{R_4}{R_3} = (0.5 \times 10^{-6}) \times \dfrac{2850}{2000} \text{F} = 0.712\ \mu\text{F}.$

Dissipating factor,

$$D_1 = \tan\delta_1 = \omega C_1 r_1 = 2\pi \times 450 \times 0.712 \times 10^{-6} \times 3.65$$
$$= 0.00734.$$

***Example 16.9*** *An Owen's bridge is used to measure the properties of a sample of sheet steel at 2 kHz. At balance, arm ab is test specimen ; arm bc is $R_3 = 100\ \Omega$ ; arm cd is $C_4 = 0.1\ \mu F$ and arm da is $R_2 = 834\ \Omega$ in series with $C = 0.124\ \mu F$. Derive balance conditions and calculate the effective impedance of the specimen under test conditions.*

**Solution.** Let $R_1$ and $L_1$ be the effective resistance and inductance of the specimen respectively.

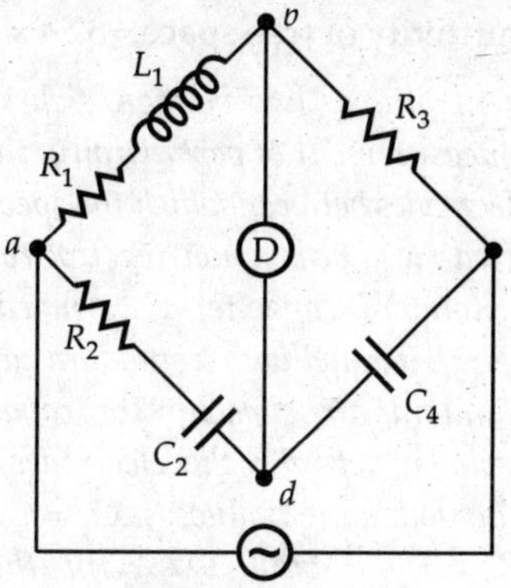

**Fig. 16.26** Diagram of Example 16.9.

$$\text{At balance } (R_1 + j\omega L_1)\frac{1}{j\omega C_4} = R_3\left(R_2 + \frac{1}{j\omega C_2}\right)$$

$$\therefore \quad L_1 = R_2 R_3 C_4 = 834 \times 100 \times 0.1 \times 10^{-6} \text{ H}$$
$$= 8.34 \text{ mH}$$

and $R_1 = R_3 \dfrac{C_4}{C_2} = 100 \times \dfrac{0.1}{0.124} = 80.7\ \Omega.$

Reactance of specimen at 2 kHz,

$$X_1 = 2\pi \times 2 \times 1000 \times 8.34 \times 10^{-3} = 104.5\ \Omega.$$

Impedance of specimen

$$Z_1 = \sqrt{(80.7)^2 + (104.5)^2} = 132\ \Omega.$$

***Example 16.10*** *A sheet of bakelite 4.5 mm thick is tested at 50 Hz between electrodes 0.12 m in diameter. The Schering bridge employs a standard air capacitor $C_2$ of 106 pF capacitance, a non-reactive resistance $R_4$ of $1000/\pi\Omega$ in parallel with a variable capacitor $C_4 = 0.5\ \mu F$, and a non-reactive variable resistance $R_3$.*

*Balance is obtained with $C_4 = 0.5\ \mu F$ and $R_2 = 260\ \Omega$. Calculate the capacitance, power factor and relative permittivity of sheet.*

**Solution.** The Schering bridge is shown in Fig. 16.11 where $C_1$ and $r_1$ represent the capacitance and series resistance of the capacitor. From Eqns. 16.35 and 16.36

$$r_1 = \frac{C_4}{C_2}R_3 = \frac{0.5 \times 10^{-6}}{106 \times 10^{-12}}260 = 1.23 \times 10^6\ \Omega$$

and $C_1 = \dfrac{R_4}{R_3}C_2 = \dfrac{1000}{\pi \times 260} \times 106 \times 10^{-12} \text{ F} = 130 \text{ pF}.$

Power factor of sheet

$= \omega C_1 r_1 = 2\pi \times 50 \times 130 \times 10^{-12} \times 1.23 \times 10^6 = 0.05.$

Now capacitance, $C_1 = \varepsilon_r \varepsilon_0 \frac{A}{d}$

$\therefore$ Relative permittivity

$$\varepsilon_r = \frac{C_1 d}{\varepsilon_0 A} = \frac{130 \times 10^{-12} \times 4.5 \times 10^{-3}}{8.854 \times 10^{-12} \times \frac{\pi}{4}(0.12)^2} = 5.9$$

where $\varepsilon_0$ = permittivity of free space $= 8.84 \times 10^{-12}$ F/m.

**Example 16.11** *In a low-voltage Schering bridge designed for the measurement of permittivity, the branch ab consists of two electrodes between which the specimen under test may be inserted; arm bc is a non-reactive resistor $R_3$ in parallel with a standard capacitor C, arm cd is a non-reactive resistor $R_4$ in parallel with a standard capacitor $C_4$; arm da is a standard air capacitor of capacitance $C_2$. Without the specimen between the electrodes, balance is obtained with the following values; $C_3 = C_4 = 120$ pF, $C_2 = 150$ pF, $R_3 = R_4 = 5000\ \Omega$. With the specimen inserted these values become $C_3 = 200$ pF, $C_4 = 1000$ pF, and $R_3 = R_4 = 5000\ \Omega$. In each test $\omega = 5000$ rad/s. Find the relative permittivity of the specimen.*

**Solution.** The voltage circuit is shown in Fig. 16.27.

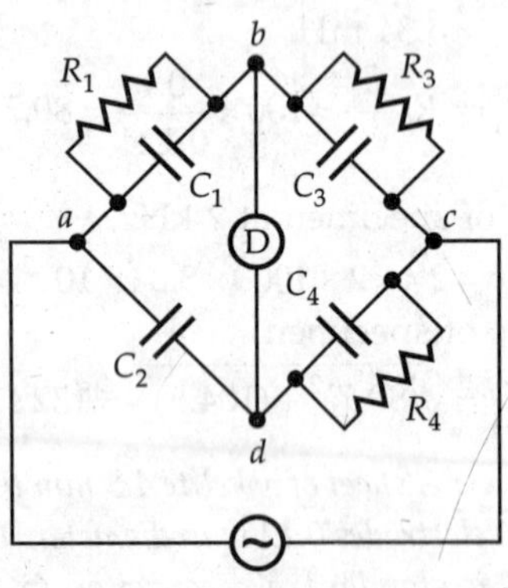


**Fig. 16.27** Diagram of Example 16.11.

For balance $Y_1 Y_4 = Y_2 Y_3$

or $$\left(\frac{1}{R_1} + j\omega C_1\right)\left(\frac{1}{R_4} + j\omega C_4\right) = (j\omega C_2)\left(\frac{1}{R_3} + j\omega C_3\right)$$

or $$\left(\frac{1}{R_1 R_4} - \omega^2 C_1 C_4\right) + j\omega\left(\frac{C_4}{R_1} + \frac{C_1}{R_4}\right) = j\omega\frac{C_2}{R_3} - \omega^2 C_2 C_3$$

Equating the real and imaginary parts, we have

$$\frac{1}{R_1 R_4} - \omega^2 C_1 C_4 = -\omega^2 C_2 C_3 \quad ...(i)$$

and $$\frac{C_4}{R_1} + \frac{C_1}{R_4} = \frac{C_2}{R_3} \quad ...(ii)$$

From (*i*) and (*ii*), we have

$$C_1 = \frac{\frac{C_2 R_4}{R_3} + \omega^2 C_2 C_3 C_4 R_4^2}{1 + \omega^2 C_4^2 R_4^2}$$

Now $$\omega^2 C_2 C_3 C_4 R_4^2 << \frac{C_2 R_4}{R_3}$$

and $$\omega^2 C_4^2 R_4^2 << 1.$$

Hence we can write

$$C_1 = C_2 \frac{R_4}{R_3}$$

When the capacitor $C_1$ is without specimen dielectric, let its capacitance be $C_0$.

$$\therefore \quad C_0 = C_2 \frac{R_4}{R_3} = 150 \times \frac{5000}{5000} = 150 \text{ pF}.$$

When the specimen is inserted as dielectric, let the capacitance be $C_s$.

$$\therefore \quad C_s = C_2 \frac{R_4}{R_3} = 900 \times \frac{5000}{5000} = 900 \text{ pF}.$$

Now $C_0 = \varepsilon_0 A/d$ and $C_s = \varepsilon_r \varepsilon_0 A/d$.

Hence relative permittivity of specimen

$$\varepsilon_r = \frac{C_s}{C_0} = \frac{900}{150} = 6.$$

**Example 16.12** *The arms of a four arm bridge abcd, supplied with sinusoidal voltage, have the following values:*

*arm ab: A resistance of 200 Ω in parallel with a capacitance 1 μF.*

*arm bc: 400 Ω resistance,*

*arm cd: 1000 Ω resistance.*

*arm da: A resistance $R_2$ in series with a 2 μF capacitance.*

*Determine the value of $R_2$ and the frequency at which the bridge will balance.*

**Solution.** If we draw the sketch of this bridge, we find that it is Wien's bridge shown in Fig. 16.20.

The analysis of this bridge is given in Art. 16.8.1.

From Eqn. 16.63,

$$R_2 = \left(\frac{R_4}{R_3} - \frac{C_1}{C_2}\right) R_1 = \left(\frac{1000}{400} - \frac{1 \times 10^{-6}}{2 \times 10^{-6}}\right) \times 200$$

$= 400\ \Omega.$

From Eqn. 16.64, the frequency at which the bridge balance,

$$f = \frac{1}{2\pi\sqrt{R_1 R_2 C_1 C_2}}$$

$$= \frac{1}{2\pi\sqrt{200 \times 400 \times 1 \times 10^{-6} \times 2 \times 10^{-6}}} = 398 \text{ Hz}.$$

***Example 16.13*** *Fig. 16.17 shows a Heaviside Campbell bridge for measurement of a self-inductance $L_2$ with the equal ratios i.e., $R_3 = R_4$, the following results were obtained :*

*With switch open $M = 15.8$ mH and $r = 25.7\ \Omega$.*

*With switch close $M = 0.2$ mH and $r = 1.2\ \Omega$.*

*Find the resistance and self-inductance of coil.*

**Solution.** The theory of this bridge is dealt in Art. 16.7.4.

From Eqns. 16.49 and 16.50, for equal ratios

Self-inductance

$$L_2 = 2(M_1 - M_2) = 2(15.8 - 0.2) = 31.2 \text{ mH}$$

Resistance

$$R_2 = r_1 - r_2 = 25.7 - 11.2 = 14.5\ \Omega.$$

***Example 16.14*** *A modified Carey Foster's bridge shown in Fig 16.18 is used for measurement of capacitance. Arm ab contains the secondary winding a mutual inductance of 18.35 mH and a total non-reactive resistance of 200 Ω. The inductance of secondary of mutual inductor in arm ab is 40.6 mH. Arm ad is short circuited. Arm bc contains the unknown capacitor in series with a resistance of 119.5 Ω. Arm cd comprises of a resistor of 100 Ω resistance. The detector is across bd. Determine the capacitance and its equivalent series resistance.*

**Solution.** From the analysis carried in Art. 16.7.6, we have

$$\text{Capacitance } C_3 = \frac{M}{R_1 R_4} = \frac{18.36 \times 10^{-3}}{200 \times 100} \text{F} = 0.918\ \mu\text{F}.$$

$$R_3 = \frac{R_4(L_1 - M)}{M} = \frac{100(40.6 - 18.36)}{18.36} = 121.2\ \Omega.$$

∴ Series resistance of capacitor

$$= 121.2 - 119.5 = 1.7\ \Omega.$$

***Example 16.15*** *In a modified Carey-Foster bridge shown in Fig. 16.18. At balance $R_4 = 10\ \Omega$, $R_3 = 25.1 \pm 0.1\ \Omega$, $R_1 = 100\ \Omega$, $L_1 = 22.82$ mH and $M = 4920 \pm 10\ \mu$H. Find the unknown capacitance $C_3$ and the limits of accuracy determined by the limits of balance measurements indicated.*

**Solution.** From Eqn. 16.59, we get

$$C_3 = \frac{M}{R_1 R_4} = \frac{4920 \pm 10}{100 \times 10} = 4.92 \pm 0.01\ \mu\text{F}.$$

$$\text{Accuracy limit} = \frac{0.01}{4.92} \times 100 = 0.2\%.$$

***Example 16.16*** *Suppose in the above example the value of $R_3$ does not include the series loss resistance of capacitor, find the loss resistance and limits of its accuracy.*

**Solution.** From Eqn. 16.60,

$$R_3 = \frac{R_4(L_1 - M)}{M} = \frac{228200}{4920(1 \pm 10/4920)}$$

$$= \frac{46.5}{1 \pm 10/4920} = 46.5\left(1 \pm \frac{10}{4920}\right)$$

$$= 46.5(1 \pm 0.00204) = 46.5 \pm 0.095\ \Omega.$$

This resistance includes the series resistance of capacitor.

Series resistance of capacitor

$$= (46.5 \pm 0.095) - (25.1 \pm 0.1)$$

$$= 21.4 \pm 0.195\ \Omega$$

$$\text{Limits of accuracy} = \frac{0.195}{21.4} \times 100 = 0.915\%.$$

## 16.12 TRANSFORMER RATIO BRIDGE

The *Transformer Ratio Bridges* are becoming increasingly popular and are being used for a wide range of applications. This is on account of versatility and accuracy of *Ratio Transformers*, which are used in the transformer ratio bridges. In fact, transformer ratio bridges are replacing the conventional a.c. bridges at a rapid rate. 16

A transformer ratio bridge consists of voltage transformer whose performance approaches that of an ideal transformer. An ideal transformer is one that has no resistance, no core loss and no leakage flux (*i.e.*, there is perfect coupling between the windings).

The ratio transformer is provided with a number of tappings in order to obtain voltage division.

Voltage appearing across the windings of a transformer is :

$$E = 4 K_f N \Phi_m f \text{ volt} \qquad ...(16.66)$$

where $N$ = number of turns,

$\Phi_m$ = maximum value of flux ; Wb,

$f$ = frequency ; Hz,

$K_f$ = form sector

(Its value is 1.11 for sinusoidal flux).

For a given value of $K_f$, flux $\Phi_m$ and frequency $f$,

$$E = K_1 N \qquad ...(16.67)$$

Figure 16.28 shows an autotransformer provided with tappings. Suppose an alternating voltage $E$ is applied across the winding. Assuming that the

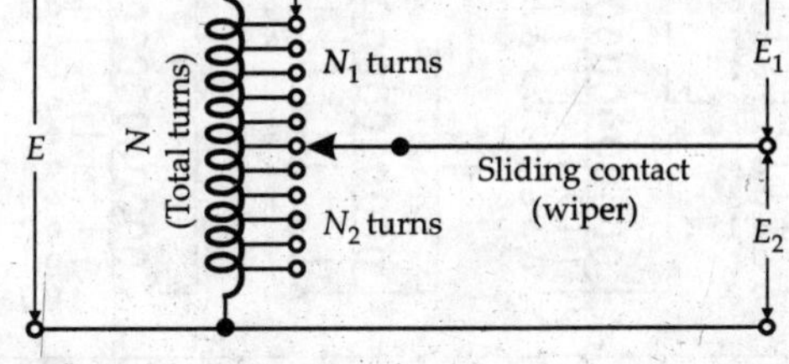


**Fig. 16.28** Tapped autotransformer.

autotransformer is ideal type, the division of applied voltage $E$ into output voltages $E_1$ and $E_2$ is :

$$E_1 = E \cdot \frac{N_1}{N} \quad \text{and} \quad E_2 = E \cdot \frac{N_2}{N}$$

Different values of $E_1$ and $E_2$ may be had by changing the position of wiper on the tappings.

However, in practice it is impossible to construct an ideal transformer. But the ideals of zero winding resistance, zero core loss and perfect coupling can be closely achieved if the design features similar to those for instrument transformers are used. The material used for construction of core should be such that it gives the smallest core losses at the desired operating frequency. The magnetizing current is reduced by using a *Toroidal Core*. The added advantage of toroidal core is that winding put on it has minimum leakage reactance giving an almost perfect coupling. The leakage reactance can be reduced further by using a special type of construction for the windings as shown in Fig. 16.29. This winding takes the form of a *Multiconductor Rope*. In order to obtain a decade of voltage division,the multicon- ductor rope has ten wires with successive sets of turns connected in series and a tapping is taken from each joint.

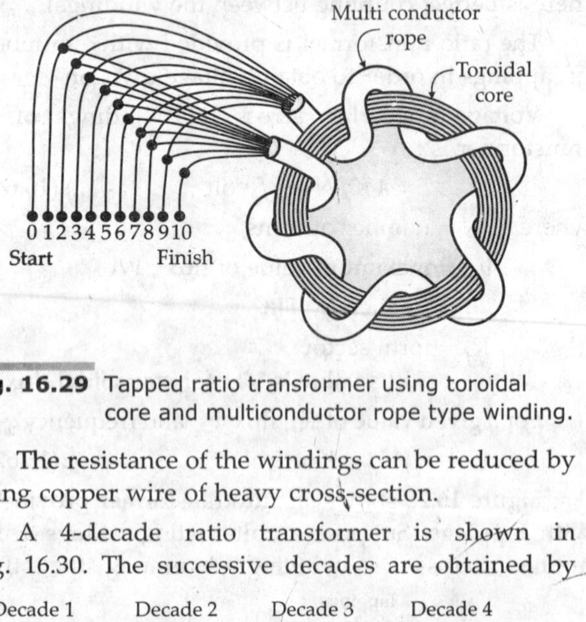


**Fig. 16.29** Tapped ratio transformer using toroidal core and multiconductor rope type winding.

The resistance of the windings can be reduced by using copper wire of heavy cross-section.

A 4-decade ratio transformer is shown in Fig. 16.30. The successive decades are obtained by using an arrangement similar to that in a Kelvin Varley slide. This transformer arrangement gives a ratio error of less than 1 part in $10^4$.

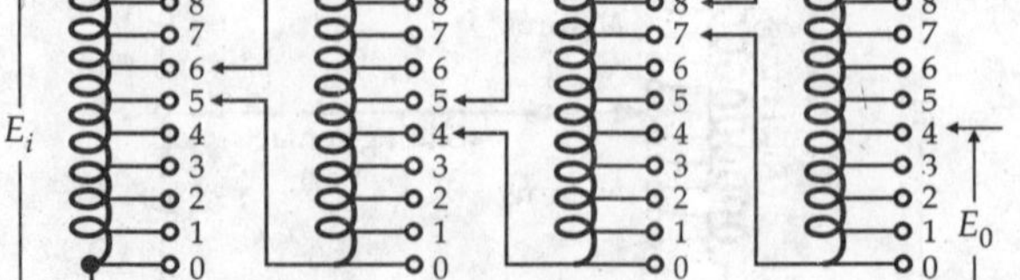


**Fig. 16.30** Four decade ratio transformer.

### 16.12.1 Applications and Features of Ratio Transformers

***Applications.*** The ratio transformers can be used for :

(*i*) Measurement of resistance capacitance and inductance in comparison with standard resistance, standard capacitance and standard inductance respectively,

(*ii*) Measurement of amplifier gain and phase shift, and

(*iii*) Measurement of transformer ratios.

***Features.*** The ratio transformers have the following features :

(*i*) They can be used on a.c. only.

(*ii*) They have very small ratio errors.

(*iii*) They have a wide frequency range extending from 50 Hz to 50 kHz.

(*iv*) They have high input impedance and low input impedance. Thus the loading effects in them are small.

### 16.12.2 Measurement of Resistance

The circuit used for measurement of an unknown resistance, $R$, in comparison with a standard resistance, $R_s$ is shown in Fig. 16.31. The position of the wiper is adjusted till the detector $D$ shows null. (The detectors used are the same as for conventional a.c. bridges.)

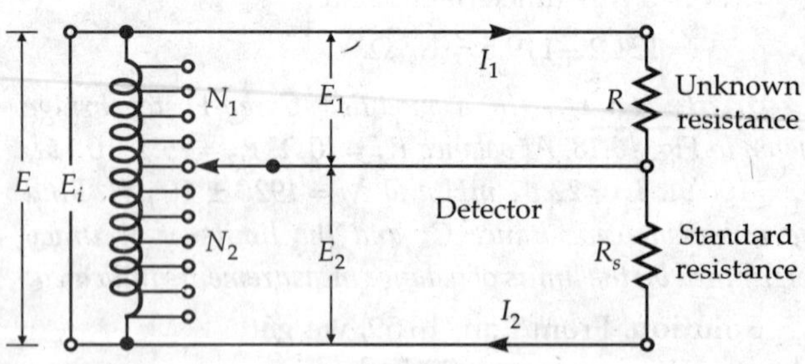


**Fig. 16.31** Measurement of resistance using ratio transformer.

Current through the unknown resistance

$$I_1 = \frac{E_1}{R} = \frac{K_1 N_1}{R}$$

Current through the standard resistance

$$I_2 = \frac{E_2}{R_s} = \frac{K_1 N_2}{R_s}$$

Under balance conditions, the current through the detector is zero *i.e.*, $I_1 = I_2$

$$\therefore \quad \frac{K_1 N_1}{R} = \frac{K_1 N_2}{R_s}$$

or
$$R = \frac{N_1}{N_2} R_s \qquad \text{...(16.68)}$$

A circuit used for measurement of low resistance is shown in Fig. 16.32. This circuit is similar to that of Kelvin's double bridge.

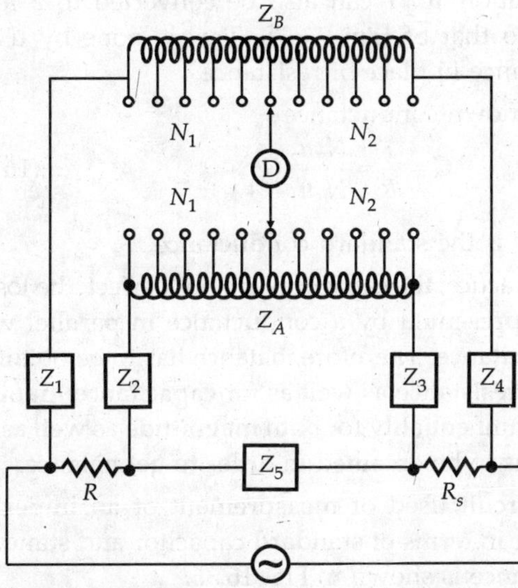


**Fig. 16.32** Measurement of low resistance.

By using two transformers a form of Kelvin's double bridge for measurement of low resistances may be devised as shown in Fig. 16.32, where

$$R = R_s\left[\frac{N_1 Z_B + Z_1}{(1-N_1)Z_B + Z_4}\right] + Z_5\left[\frac{(1-Z_2)Z_A + Z_3}{Z_1+Z_2+Z_3+Z_5}\right]$$

$$\times\left[\frac{N_1 Z_B + Z_1}{(1-N_1)Z_B + Z_4} - \frac{N_1 Z_A + Z_3}{(1-N_2)Z_A + Z_3}\right] \qquad \text{...(16.69)}$$

If the impedances of leads *i.e.*, $Z_1, Z_2, Z_3, Z_4$, and $Z_5$ are small and the resistances $R$ and $R_s$ are of the same order, the unknown resistance is given by :

$$R = \frac{N_1}{N_2} R_s \qquad \text{...(16.70)}$$

This form of Kelvin's bridge can, of course, only be used on *ac* but by plotting the values $R$ against frequency, an extrapolation may be used to find the d.c. value of $R$ with an accuracy of a few parts in a million.

### 16.12.2 Measurement of Capacitance

A circuit for measurement of capacitance is given in Fig. 16.33. An unknown capacitance $C$ is measured in comparison with a standard capacitance $C$, which is assumed to be perfect. Resistance $R$ represents the loss of the capacitor. Since for balance, the magnitude and phase of the currents passing through the detector should be the same, a variable standard resistance is connected in parallel with the standard capacitor.

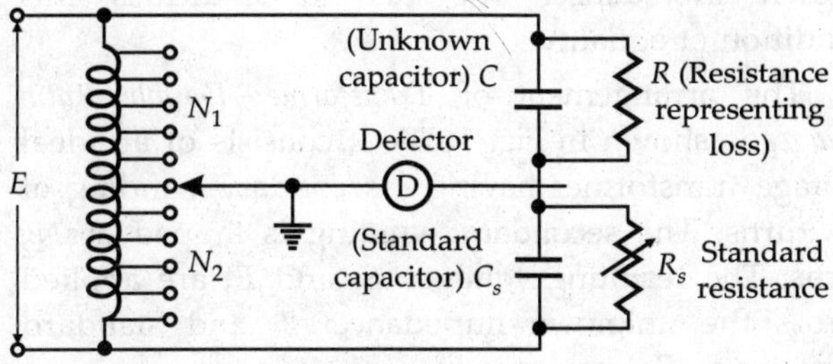


**Fig. 16.33** Measurement of capacitance using ratio transformer.

$$\text{At balance,} \quad C = \frac{N_2}{N_1} C_s \text{ and } R = \frac{N_1}{N_2} R_s \qquad \text{...(16.71)}$$

16

Dissipation factor

$$D = \tan\delta = \frac{1}{\omega CR} = \frac{1}{\omega C_s R_s} \qquad \text{...(16.72)}$$

### 16.12.3 Measurement of Phase Angle

A circuit for measurement of small phase angles with the help of ratio transformers is shown in Fig. 16.34. An *RC* circuit is used where the capacitance is variable in order to get phase shift. The value of resistance should be large in order that there are no loading effects on the ratio transformer. The capacitance is changed till the detector indicates null.

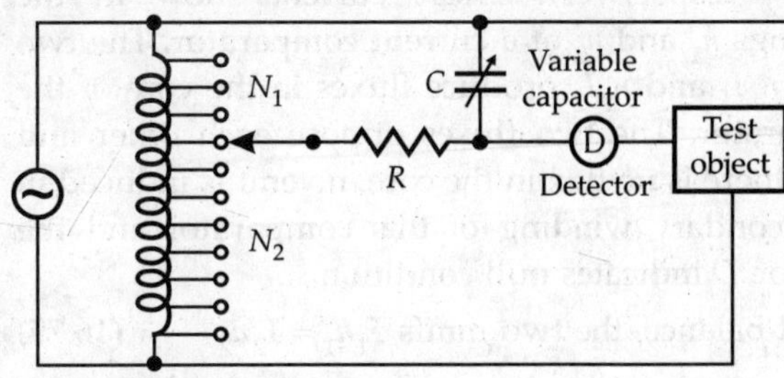


**Fig. 16.34** Measurement of phase angle using ratio transformer.

$$\text{Phase angle } \phi = \tan^{-1}(-\omega RC) \qquad \text{...(16.73)}$$

The magnitude of in-phase component is :

$$\phi_1 = \frac{N_2}{N_1 + N_2}\cos^2\phi \qquad \text{...(16.74)}$$

### 16.12.4 Transformer Double Ratio Bridges

The transformer bridges described till now in this chapter are similar to the conventional a.c. bridges in the sense that the balance conditions are indicated when the voltage across the detector is zero and therefore, the current through the detector is zero.

An alternate way of indication of balance is that the current flowing through the unknown impedance is equal and opposite to that through the standard or known impedance. The detector indicates this condition of equality.

The arrangement of *Transformer Double Ratio Bridges* is shown in Fig. 16.35. It consists of an ideal voltage transformer having a secondary winding of $N_1$ turns. The secondary winding is tapped at $N_2$ turns. The resulting voltages $E_1$ and $E_2$ are applied across the unknown impedance, $Z$, and standard impedance, $Z_s$, respectively.

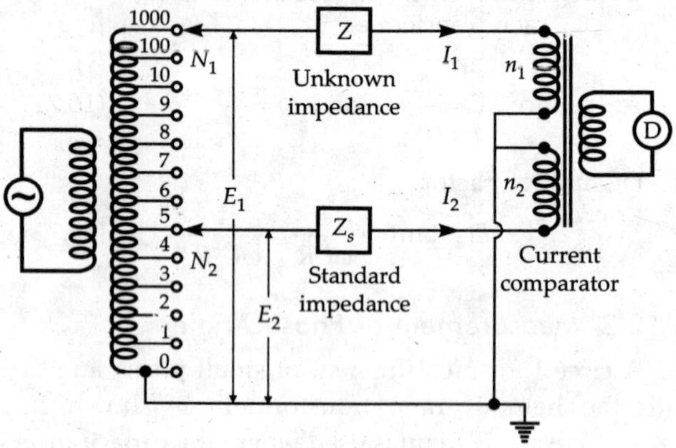


**Fig. 16.35** Basic circuit for a double ratio transformer bridge.

The two voltages $E_1$ and $E_2$ produce currents $I_1$ and $I_2$ respectively. These currents flow in the windings $n_1$ and $n_2$ of a current comparator. The two mmfs $n_1I_1$ and $n_2I_2$ produce fluxes in the core of the comparator. The two fluxes oppose each other and when there is no flux in the core, no emf is induced in the secondary winding of this comparator and the detector $D$ indicates null condition.

At balance, the two mmfs $I_1n_1 = I_2n_2$ ...(16.75)

Now $I_1 = \dfrac{E_1}{Z} = \dfrac{K_1N_1}{Z}$ and $I_2 \dfrac{E_2}{Z_s} = \dfrac{K_1N_2}{Z_s}$

$$\therefore \quad \frac{N_1n_1}{Z} = \frac{N_2n_2}{Z_s}$$

or unknown impedance

$$Z = Z_s\frac{N_1n_1}{N_2n_2} \quad ...(16.76)$$

The balance is obtained by changing the setting of $N_2$.

If the unknown impedance is a pure resistance $R$, a standard resistance $R_s$ is required for obtaining balance

$$R = \frac{N_1n_1}{N_2n_2}R_s \quad ...(16.77)$$

If the unknown impedance is a pure capacitance $C$, a standard capacitor $C_s$ is required for balance,

$$C = \frac{N_2n_2}{N_1n_1}C_s \quad ...(16.78)$$

Equation 16.77 can also be converted to a form similar to that of Eqn. 16.78. This is done by using conductance in place of resistance.

Unknown conductance

$$G = \frac{1}{R} = \frac{N_2n_2}{N_1n_1}G_s \quad ...(16.79)$$

where $G_s$ is the standard conductance.

In practice, the capacitors are not perfect, the losses being represented by a conductance in parallel with the capacitance. Therefore, balance has to be obtained both for resistance as well as for capacitance. In other words, mmf equality for both magnitude as well as for phase has to be obtained in order to get balance.

A circuit used of measurement of an imperfect capacitor in terms of standard capacitor and standard conductance is shown in Fig. 16.36.

Under balance,

$$C = \frac{n_2}{N_1n_1}(C_{s1}N_{21} + C_sN_{22}) \quad ...(16.80)$$

and $$G = \frac{n_2}{N_1n_1}(G_{s1}N'_{21} + G_{s2}N'_{22}) \quad ...(16.81)$$

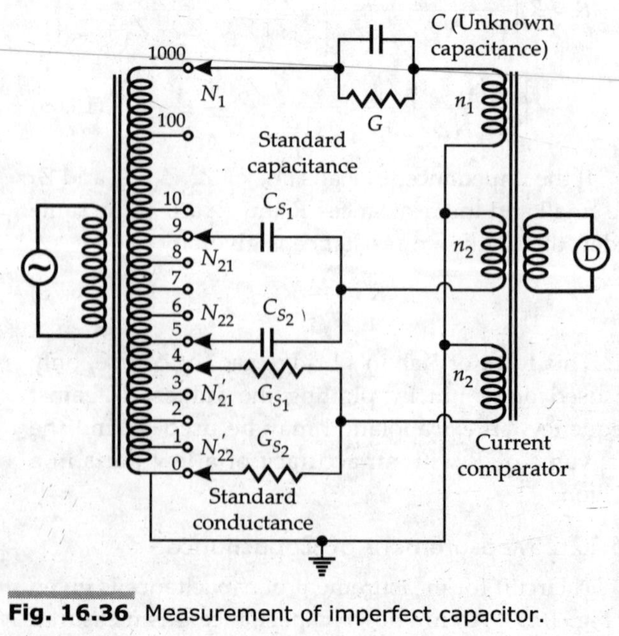


**Fig. 16.36** Measurement of imperfect capacitor.

It is clear from Eqns. 16.80 and 16.81 that the double ratio bridge allows the reactive and resistive components of an unknown to be adjusted independently of each other (capacitive balance can be had by changing $N_{21}$ and $N_{22}$ while resistive balance can be obtained by changing $N'_{21}$ and $N'_{22}$). This property of independent balance facilitates the use of this bridge in a very wide range of applications.

### 16.12.5 Measurement of Inductance

The circuit for measurement of inductance is the same as that for measurement of capacitance (Fig. 16.36) except that the unknown capacitance is replaced by the unknown inductance.

In order that independent balance of the resistive and reactive mmfs can be obtained any inductive circuit must be considered as a two terminal network, the components of which are in parallel. The mmf due to the resistive component of the unknown may then be opposed by that through the conductance standard. On the other hand, the mmf due to inductive component of the unknown must be balanced due to capacitance standard. In order to obtain mmf balance due to unknown inductance and the standard capacitance, it is necessary to reverse the comparator winding connected to the standard capacitor. This condition of bridge operation is normally indicated by addition of a negative sign to the display of capacitance $C_m$. Suppose an inductance $L_p$ in parallel with a resistance $R_p$ represents the actual inductor having an inductance $L$ in series with a resistance $R$.

At balance, $\dfrac{1}{R_p} + \dfrac{1}{j\omega L_p} = G_m - j\omega C_m$

where $G_m$ = indicated conductance, and

$C_m$ = indicated capacitance.

$\therefore$ Series resistance, $$R = \frac{G_m}{G_m^2 + \omega^2 C_m^2} \quad ...(16.82)$$

and Series inductance, $$L = \frac{C_m}{G_m^2 + \omega^2 C_m^2} \quad ....(16.83)$$

### 16.12.6 Measurement of Components in 'SITU'

One of the greatest advantages of double ratio transformer bridges is their capability to measure the values of the components while they remain connected in the circuit. Figure 16.37 shows a circuit for measurement of an unknown impedance **Z**. If the

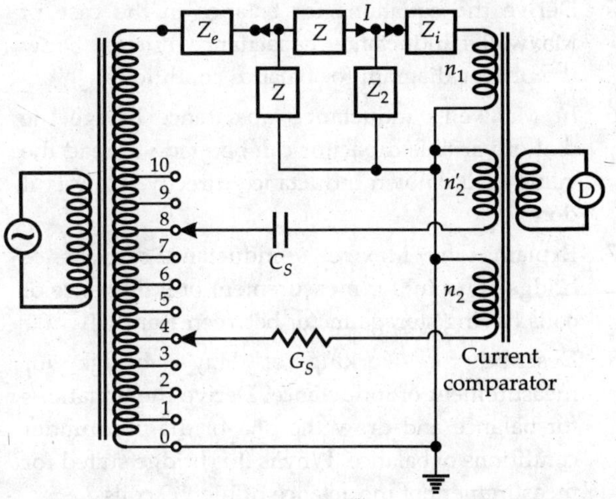


**Fig. 16.37** Measurement of components 'IN SITU' using a double ratio transformer bridge.

ratio transformers are considered ideal the values of impedances $\mathbf{Z}_e$ and $\mathbf{Z}_i$ can be assumed to be zero and the voltage applied to the unknown impedance **Z** is not affected by current flowing in impedance $\mathbf{Z}_1$. Also all the current flowing through Z would flow into $n_1$, winding as :

$$I_2 = \frac{\mathbf{Z}_i \mathbf{I}}{\mathbf{Z}_i + \mathbf{Z}_2}$$

The value of unknown impedance is approximately given by :

$$\mathbf{Z} = \left[\frac{1}{G_m + j\omega C_m}\right]\left[1 - \left(\frac{\mathbf{Z}_e}{\mathbf{Z}_1} + \frac{\mathbf{Z}_i}{\mathbf{Z}_2}\right)\right] \quad ...(16.84)$$

In practice impedances $\mathbf{Z}_e$ and $\mathbf{Z}_i$ have small finite values and corrections must be applied for this. In order that corrections be applied $\mathbf{Z}_e$ and $\mathbf{Z}_i$ must be determined.

# Review Questions

1. Derive the general equations for balance for an a.c. bridge. Prove that two conditions *i.e.,* for magnitude and phase have to be satisfied if an a.c. bridge is to be balanced unlike a d.c. bridge wherein only the magnitude condition is to be satisfied
2. Describe the sources and the null detectors that are used for a.c. bridges.
3. Explain what is meant by sliding balance. How is this condition avoided by choosing variables for

16

manipulation of balance *i.e.*, why variables are so chosen that the two equations for balance are independent of each other.

4. Why is it preferable in bridge circuits that the equations of balance are independent of frequency ? Explain.
5. Derive the equations for balance in the case of Maxwell's inductance capacitance bridge. Draw the phasor diagram for balance conditions.
6. In Maxwell's inductance-capacitance bridge the dial of variable capacitor can be made to read the value of unknown inductance directly. How is it done ?
7. Explain why Maxwell's inductance-capacitance bridge is useful for measurement of inductance of coils having storage factor between 1 and 10.
8. Describe the working of Hay's bridge for measurement of inductance. Derive the equations for balance and draw the phasor diagram under conditions of balance. Why is this bridge suited for measurement of inductance of high $Q$ coils ?
9. Derive the equations of balance for an Anderson's bridge. Draw the phasor diagram for conditions under balance.
   Discuss the advantages and disadvantages of the bridge.
10. What are incremental inductance and permeability ? Describe how are they measured with Owen's bridge.
11. Describe how an unknown capacitance can be measured with the help of D'Sauty's bridge. What are the limitations of this bridge and how are they overcome by using a modified form of D'Sauty's bridge ? Draw phasor diagrams to illustrate your answer.
12. Describe the working of a low voltage Schering bridge. Derive the equations for capacitance and dissipation factor. Draw the phasor diagram of the bridge under conditions of balance.
13. What are the modifications and additional features incorporated in a low voltage Schering bridge for it to be used on high voltages ? Explain. Describe how relative pemittivity of a specimen of insulating material can be determined using a Schering bridge.
14. Describe the working of Heaviside Mutual Inductance Bridge for determination of mutual inductance of a coil. Explain how the self-inductance of a coil can be determined with the help of a mutual inductance. Also, explain how the use of a balancing coil helps in elimination of errors on account of resistance of leads etc.
15. Explain how Wien's bridge can be used for experimental determination of frequency. Derive the expression for frequency in terms of bridge parameters.
16. What is a Universal Impedance Bridge ? What are the features incorporated in it ? Draw the circuit diagrams of the various bridges used in a Universal Impedance Bridge.
17. What are the different sources of errors in a.c. bridges ? Explain the precautions taken and the techniques used for elimination/minimization of these errors.
18. Explain the function and working of Wagner Earth Devices.
19. Describe what do you understand by the term "Ratio Transformer". Explain the construction of a Ratio transformer and describe its uses.
20. Explain how a Ratio Transformer bridge can be used for measurement of
    (*i*) resistance (*ii*) capacitance, (*iii*) phase angle.
21. What is a "Transformer Double Bridge" ? Explain how can it be used for measurement of
    (*i*) resistance, (*ii*) inductance,
    (*iii*) capacitance and loss angle of a imperfect capacitor.

## Unsolved Problems

**Note.** Students must attempt these problems from fundamentals.

1. A four arm a.c. bridge $a,b,c,d$ has the following impedances :

   Arm $ab$ : $\mathbf{Z}_1 = 200\angle 60°\Omega$ (inductive impedance).
   Arm $ad$ : $\mathbf{Z}_2 = 400\angle -60°\Omega$ (purely capacitive impedance)
   Arm $bc$ : $\mathbf{Z}_3 = 300\angle 0°\Omega$ (purely resistive),
   Arm $cd$ : $\mathbf{Z}_4 = 600\angle 30°\Omega$ (inductive impedance).

   Determine whether it is possible to balance the bridge under above conditions.

   [**Ans.** No as $\angle\theta_1 + \theta_4 \neq \angle\theta_2 + \angle\theta_3$]

2. A 1000 Hz bridge has the following constants : arm $ab$, $R_1 = 1000\ \Omega$ in parallel with $C_1 = 0.5\ \mu F$ ; arm $bc$, $R_3 = 1000\ \Omega$ in series with $C_3 = 0.5\ \mu F$ ; arm $cd$ $L_4 = 30$ mH in series with $R_4 = 200\ \Omega$. Find the constants of arm $da$ to balance the bridge. Express the result as a pure resistance $R$ in series with a pure inductance $L$ or capacitance $C$.

[**Ans.** $79.4\angle{-11.1}\,\Omega$, $R = 77.9\ \Omega$ in series with $C = 10.4\ \mu F$]

3. An a.c. bridge has in arm $ab$, a pure capacitance of 0.2 pF ; in arm $bc$, a pure resistance of $500\ \Omega$ ; in arm $cd$, a series combination of a $50\ \Omega$ resistance and of 0.1 H inductance. Arm $da$ consists of a capacitor of $0.4\ \mu F$ in series with a resistance $R_2$, $\omega = 5000$ rad/s. (*a*) Find the value of $R_2$ to give bridge balance. (*b*) Can complete balance be obtained by adjustment of $R_2$ ? If not, specify the position and the value of an adjustable resistance to complete the balance.

[**Ans.** $1000\ \Omega$, complete balance cannot be obtained by adjustment of $R_2$ only as reactive conditions are not satisfied. If an adjustable resistance of $200\ \Omega$ is connected in series in branch $c\,d$ capacitor reactive balance can be obtained]

4. The four arms of a Maxwell's capacitance bridge at balance are : arm $ab$, an unknown inductance $L_1$, having an inherent resistance $R_1$ ; arm $bc$, a non-inductive resistance of $1000\ \Omega$ ; arm $cd$, a capacitor of $0.5\ \mu F$ in parallel with a resistance of $1000\ \Omega$ ; arm $da$, a resistance of $1000\ \Omega$.

Derive the equations of balance for the bridge and determine the value of $R_1$ and $L_1$. Draw the phasor diagram of the bridge under balance conditions.

[**Ans.** $R_1 = 1000\ \Omega$ ; $L_1 = 0.5$ H]

5. In an Anderson Bridge for the measurement of inductance the arm $AB$ consists of an unknown impedance with inductance $L$ and $R$, a known variable resistance in arm $BC$, fixed resistance of $600\ \Omega$ each in arms $CD$ and $DA$, a known variable resistance in arm $DE$, and a capacitor with fixed capacitance of 1 microfarad in the arm $CE$.

The a.c. supply of 100 Hz is connected across $A$ and $C$, and the detector is connected between $B$ and $E$. If the balance is obtained with a resistance of $400\ \Omega$ in the ami $DE$ and a resistance of $800\ \Omega$ in the arm $BC$, calculate the value of unknown $R$ and $L$. Derive the conditions for balance and draw the phasor diagram under balanced conditions.

[**Ans.** $R = 800\ \Omega$, $L = 1.12$ A]

6. The four arms of a Hay's alternating current bridge are arranged as follows : $AB$ is a coil of unknown impedance ; $BC$ is a non-reactive resistor of $1000\ \Omega$ ; $CD$ is a non-reactive resistor of $833\ \Omega$ in series with a standard capacitor of $0.38\ \mu F$ : $DA$ is non-reactive resistor of $16800\ \Omega$. If the supply frequency is 50 Hz determine the inductance and the resistance at the balanced conditions. Derive the conditions for balance and draw the phasor diagram under balanced conditions.

[**Ans.** $R_1 = 210\ \Omega$, $L_1 = 6.38$ H]

7. The a.c. bridge shown in Fig. 16.38 used to measure an unknown inductance $L_x$, that has inherent resistance $R_x$. The bridge parameters are $R_1 =$

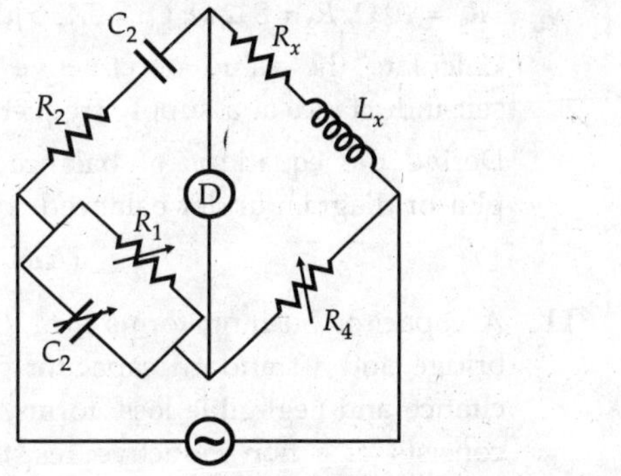


**Fig. 16.38** Diagram of Problem 7.

$20{,}000\ \Omega$, $R_2 = 50{,}000\ \Omega$, $C_2 = 0.003\ \mu F$, $\omega = 10^6$ rad/s. $C_1$ is adjustable from 10 pF to 150 pF and $R_4$ is adjustable from 0 to $10{,}000\ \Omega$.

(*a*) Show that the equations for resistive and reactive balance are independent of each other. Derive expressions for $R_x$ and $L_x$ in terms of $\omega$, $R_1$, $R_2$, $R_4$, $C_1$ and $C_2$.

(*b*) Determine the largest values of $R_x$ and $L_x$ that can be measured with the given parameters.

[**Ans.** (*a*) $R_x = R_2 R_4 / R_1 + C_1 R_4 / C_2$ ; $L_x = R_2 R_4 C_1 - R_4 / \omega^2 R_1 C_2$ ; (*b*) $2.55 \times 10^4\ \Omega$ ; 58.3 mH]

8. The four arms of a bridge network are made up as follows : $ab$, a resistor of $50\ \Omega$ in parallel with an inductor of 0.1 H ; $bc$, a resistor of $100\ \Omega$ ; $cd$, and unknown resistor $R$ in parallel with an unknown capacitor $C$ ; $da$, a resistor of $1000\ \Omega$. A 50 Hz voltage supply is applied across $ac$. Find $R$ and $C$ when a vibration galvanometer connected across $bd$ is undeflected. [**Ans.** $R = 7060\ \Omega$ ; $C = 0.72\ \mu F$]

9. A four branch bridge network $ABCD$ balanced at 1000 Hz has branches $AB$ and $BC$ of pure resistance of $1000\ \Omega$ and $1250\ \Omega$ respectively. An unknown impedance forms the arm $CD$ and the branch $DA$ consists of a standard capacitor of $0.1\ \mu F$ capacity

and negligible resistance, connected in series with a non-inductive resistance of 10 Ω to give balance. The supply voltage is 15 V and the supply is given at the points *B* and *D*. Find the components of unknown impedance and draw the necessary phasor diagram. **[Ans.** 12.5 Ω ; 0.08 μF]

**10.** The circuit for measurement of effective resistance and self-inductance of an iron cored coil is as follows : arm *ab*, the unknown impedance ; arm *bc*, a pure resistance $R_3$ ; arm *cd*, a lossless capacitor $C_2$ ; arm *da*, a capacitor $C_2$ in series with a resistance. Under balance conditions.

$R_3 = 10\,\Omega$, $R_2 = 842\,\Omega$, $C_2 = 0.135\ \mu F$ and $C_4 = 1\mu F$.

Calculate the value of effective resistance and self-inductance at a supply frequency of 100 Hz.

Derive the equations of balance and draw the phasor diagram under balanced condition.

**[Ans.** 74 Ω, 8.42 mH]

**11.** A capacitor bushing forms arm *ab* of a Schering bridge and a standard capacitor of 500 pF capacitance and negligible loss, forms arm *ad*. Arm *bc* consists of a non-inductive resistance of 300 Ω. When the bridge is balanced arm *cd* has a resis tance of 72.6 Ω in parallel with a capacitance of 0.148 μF. The supply frequency is 50 Hz. Calculate the capacitance and dielectric loss angle of capacitor.

Derive the equations for balance and draw the phasor diagram under conditions of balance.

**[Ans.** 121 pF ; 0.00338 rad]

**12.** An a.c. bridge was made up as follows : arms *AB* and *BC* equal ratio arms ; *CD* a variable capacitor *C* in series with a variable resistor *R*: *DA* a standard air capacitor of 0.001 μF in series with a fixed standard resistance of 500 Ω. The supply at 796 Hz was connected across *AC* and the detector across *BD*. Balance of the above bridge was obtained with $C = 0.001\,\mu F \pm 10$ pF and $R = 500 \pm 5\,\Omega$. When an unknown capacitor was connected across *CD*, the balance conditions changed to $C = 360 \pm 10$ pF and $R = 0$. Calculate the capacitance and power factor of the unknown capacitor and the limits of accuracy.

**[Ans.** 650 pF ± 3.2% ; 0.0039 ± 0.002%]

**13.** A Schering bridge is used for measuring the power loss in dielectrics. The specimens are in the form of discs 0.3 cm thick and have a dielectric constant of 2.3. The area of each electrode is 314 $cm^2$ and the loss angle is known to be 9 for a frequency of 50 Hz. The fixed resistor of the network has a value of 1,000 Ω and the fixed capacitance is 50 pF. Determine the values of the variable resistor and capacitor required.

Derive the equations for balance and draw the phasor diagram under balanced conditions.

**[Ans.** 4200 Ω, 0.00196 μF]

**14.** In a balanced bridge network, *AB* is a resistance of 500 Ω in series with an inductance of 0.18 H ; *BC* and *DA* are non-inductive resistances of 1000 Ω ; and *CD* consists of a resistance *R* in series with a capacitance *C*. A potential difference of 5 volt at a frequency of 5,000/2π Hz is established between the points *A* and *C*. Draw to scale a phasor diagram showing the currents and potential difference in the bridge and from it determine the values of *R* and *C*. Check the result algebraically.

**[Ans.** 472 Ω ; 0.235 μF]

**15.** A 4-arm unbalanced a.c. bridge is supplied from a source having negligible impedance. The bridge has non-reactive resistors of equal resistance in adjacent arms. The third arm has an inductor of resistance *R* and reactance *X*, where *X* is numerically equal to *R*. The fourth arm has a variable non-inductive resistor. The detector is connected between the junction of first and second arms and the junction of third and fourth arms and has a resistance *R* and negligible reactance. Determine the magnitude of the variable resistor where the detector current is in quadrature with the supply current. **[Ans.** 1.37 *R*]

**16.** An a.c. bridge consists of the following constants : arm *ab*, a resistance of 800 Ω in parallel with a capacitance 0.4 μF ; *bc*, an unknown resistance ; *cd*, a known resistance of 1200 Ω and *da*, a resistance of 500 Ω in series with a capacitance of 1 pF. Find the resistance required in arm *bc* to give balance and also the frequency for which the bridge is balanced. **[Ans.** 384 Ω ; 890 Hz]

**17.** A 4-arm bridge network, adjusted to balance conditions consists of :

Arm *AB* a standard resistor known to be within 0.1% of 100 Ω.

Arm *BC* a variable capacitor adjusted to 0.362 μF in parallel with a variable resistor adjusted to 2380 Ω.

Arm *CD* a standard resistor known to be within 0.1% of 1000 Ω.

Arm *DA* a coil of inductance *L* and series loss resistance *R*.

Evaluate *L* and *R* deriving the equations used.

Determine the maximum percentage error that can be tolerated in each of the variable components if $L$ is to be determined to within 0.5% and $R$ to within 1% of the correct value. If the variable components are known to be accurate within the limits so determined how accurately is the ratio $L/R$ known?

[**Ans.** 00.362 H ; 42 Ω ; 0.3% for capacitor ; 0.8% for resistor ; 1.1% ]

**18.** A low-resistance coil $AB$ of 500 turns is wound on a ring of high permeability magnetic alloy, and a tapping is made at a point $C$ such $AC$ = 100 turns. A secondary winding on the same is connected to an amplifier and headphones, and a supply at 796 Hz is connected between $C$ and a point $D$, so that $ACBD$ forms a bridge circuit in which the two parts $AC$ and $BC$ of the winding are the ratio arms. Balance is achieved when arm $AD$ comprises a capacitance of 0.16 pF in series with a resistance of 0.4 Ω, and arm $BD$ is the capacitor under test.

Evaluate the capacitance and the equivalent shunt loss resistance of the capacitor at supply frequency.

Mention the special advantages of this form of bridge circuit over the more conventional forms.

[**Ans.** 0.04 μF ; 976 kΩ]

**19.** A balanced Hay's bridge shown in Fig. 16.5 has
$$R_2 = R_3 = 1000\,\Omega,\quad R_4 = 8120\,\Omega,\quad C_4 = 980\text{ pF},$$
and the frequency = 4000 Hz.

Calculate $L_1$ and $R_1$.

If each of the resistors is accurate within ± 0.5% the capacitor within ± 1 pF, and the frequency within 5 Hz, determine the limits of accuracy of determination of $L$.

[**Ans.** $L = 942.3 \times 10^{-6}(1 \pm 10.00181)$ H ; accuracy limit ± 0.18%]

**20.** A resonance bridge is shown in Fig. 16.39.

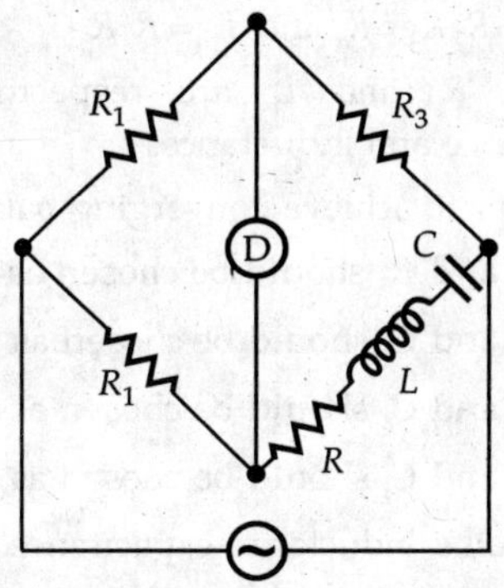


**Fig. 16.39** Resonance bridge.
(Diagram of Problem 16.20)

Prove that
$$C = \frac{1}{\omega^2 L}, \qquad L = \frac{1}{\omega^2 C}$$
and $$f = \frac{1}{2\pi\sqrt{LC}}$$

# Objective Type Questions

***Tick (√) the most appropriate answer :***

**Answer questions 1–3, referring to Fig. 16.40.**

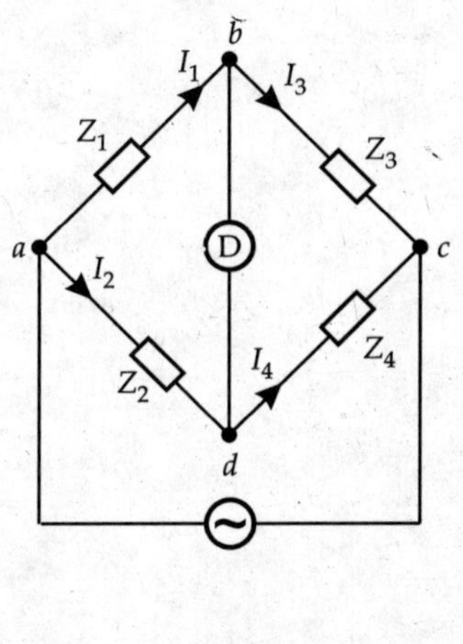


**Fig. 16.40**

**1.** In order that the bridge shown in Fig. 16.40, be balanced :

(*a*) $I_1 = I_3$ and $I_2 = I_4$

(*b*) $Z_1Z_4 = Z_2Z_3$

(*c*) $\angle\theta_1 + \angle\theta_4 = \angle\theta_2 + \angle\theta_3$

(*d*) all the above.

**2.** For the bridge shown in Fig 16.40, $Z_1 = 200\,\Omega\angle 30°$, $Z_2 = 150\,\Omega\angle 0°$, $Z_3 = 250\,\Omega\angle -40°$. In order that the bridge be balanced $Z_4$ should be :

(*a*) 187.5 Ω ∠–70° (*b*) 187.5 Ω ∠–10°

(*c*) 333.3 Ω ∠10° (*d*) 120 Ω ∠70°.

**3.** For the bridge shown in Fig. 16.40, arm *ab* consists of resistance in series with an inductance, arms *bc* and *ad* consist of pure resistance. In order to achieve balance arm *cd* should consist of

(a) a variable resistance in series with an variable inductance

(b) a variable resistance

(c) a variable resistance in series or parallel with a variable capacitance

(d) a variable capacitance.

4. The equations under balance conditions for a bridge are :

$R_1 = R_2R_3 / R_4$ and $L_1 = R_2R_3C_4$

where $R_1$ and $L_1$ are respectively unknown resistance and inductance.

In order to achieve converging balance

(a) $R_2$ and $R_3$ should be chosen as variables

(b) $R_2$ and $C_4$ should be chosen as variables

(c) $R_4$ and $C_4$ should be chosen as variables

(d) $R_3$ and $C_4$ should be chosen as variables.

5. Maxwell's inductance-capacitance bridge is used for measurement of inductance of :

(a) low $Q$ coils

(b) medium $Q$ coils

(c) high $Q$ coils

(d) low and medium $Q$ coils.

6. The advantage of Hay's bridge over Maxwell's inductance-capacitance bridge is because

(a) its equations for balance do not contain any frequency term

(b) it can be used for measurement of inductance of high $Q$ coils

(c) it can be used for measurement of inductance of low $Q$ coils

(d) none of the above.

7. In D'Sauty's bridge (unmodified form) it is :

(a) possible to obtain balance even if both the capacitors are imperfect

(b) possible to obtain balance if one of the capacitors is perfect

(c) possible to obtain balance only if both the capacitors are perfect

(d) all the above.

8. Frequency can be measured by using

(a) Maxwell's bridge

(b) Schering bridge

(c) Heaviside Campbell bridge

(d) Wien's bridge.

9. A bridge circuit works at a frequency of 2 kHz. The following can be used as detectors for detection of null conditions in the bridge

(a) vibration galvanometers and headphones

(b) headphones and tunable amplifiers

(c) vibration galvanometers and tunable amplifiers

(d) vibration galvanometers, headphones and tunable amplifiers.

10. Wagner's Earth Devices are used in a.c. bridge circuits for :

(a) eliminating the effect of earth capacitances

(b) eliminating the effect of inter-component capacitances

(c) eliminating the effect of stray electrostatic fields

(d) shielding the bridge elements.

## *Answers*

1. (d) 2. (a) 3. (c) 4. (c) 5. (b) 6. (b) 7. (c) 8. (d) 9. (b) 10. (a)

# 17

# High Voltage Measurements and Testing

## 17.1 TYPES OF TESTS

The growing extension, interconnections and use of higher voltages in electric power systems has been only possible due to extensive research work carried in the sphere of high voltage measurements and testing. Due to this research work in the high voltage field, it has been possible to reduce the size of the equipment to economic and manageable proportions. Thus it is logical that the high voltage testing of electric equipment has come into prominence and sometimes it is the sole design criterion for the determination sizes of high voltage electrical apparatus.

High voltage testing includes a large number of methods. Due to limitations of space, we will be considering a few important methods only. The high voltage tests are classified as :

- Sustained low frequency tests.
- Constant direct current tests.
- High frequency tests.
- Surge or Impulse tests.

### 1. Sustained Low Frequency Tests

These tests are most commonly used. The frequency employed is 50 Hz in India. These tests are carried out on :

(*i*) Motors, switchgear and other electrical apparatus for routine voltage testing after manufacture or in some cases after installation. The voltage normally used is 2 to 3 kV.

(*ii*) Specimens of insulation for the determination of dielectric constant and dielectric loss.

(*iii*) Supply mains for routine testing.

(*iv*) High voltage transformers, porcelain insulators and high voltage cables, etc. for works testing. The test voltage in such cases may be as high as 2000 kV.

### 2. Constant Direct Current Tests

Modern trend in electric power transmission is to use as high voltages as possible. This is because increase in trans- mission voltage results in an increase in efficiency of transmission. Overhead lines are in actual operation which employ voltages as high as 750 kV while cables for 275 kV have been made and tested.

Before high voltage lines are energised, the insulation of every part connected with the lines must withstand continuously for half an hour, a voltage as specified below :

**Table 17.1** *Test Voltages*

| ***Normal system voltage*** | ***Test voltage*** |
|---|---|
| Below 10 kV | Twice the normal voltage |
| Above 10 kV | Normal voltage + 20 kV |

From above it is clear that the transmission lines and high voltage cables would be subjected to very high voltages for a considerably long interval of time, (*i.e.*, 30 minutes). If such voltage tests are carried out with an a.c. voltage supply, the high voltage transformer required for the purpose would have to be of a very large capacity owing to the heavy charging (capacitive) currents drawn by the lines or the cables. This means that a large expenditure has to be incurred for procuring the high voltage testing transformer. Also there would be great difficulties involved in transporting the testing transformer to the site of testing owing to its large size.

Thus in order to overcome the difficulties encountered above, high voltage d.c. testing is done in place of a.c. testing.

### 3. High Voltage High Frequency Tests

The break-down and flash over of porcelain insulators used on power transmission lines is often due to high frequency disturbances in the transmission lines. These high frequency disturbances are either due to switching operations or due to some external causes and are usually damped high frequency oscillations.

It has been found that even though the porcelain insulator has satisfactorily withstood the sustained low frequency high voltage tests, it breaks down at a much lower voltage at high frequencies owing to increased dielectric loss and heating. Thus to insure reliable operation of insulators under disturbances caused by switching and arcing earths, they must be subjected to high frequency tests.

High frequency tests are also carried out for insulators used for communication purposes. Frequencies ranging from several kHz to a MHz or more are used for such purposes.

### 4. Surge or Impulse Tests

These tests are recommended with the object of determining the effect of voltage surges due to atmospheric disturbances of very short duration on electrical installations and their individual parts. The dangerous over-voltages to which power systems are exposed to are caused by lightning. Lightning may produce an over-voltage on an overhead line either by a direct stroke to the line or by an indirect stroke. An indirect stroke occurs when a charged cloud above and near to the line, electrostatically induces charges of opposite sign to that of its own charge, in the line. These charges are bound as long as the cloud remains

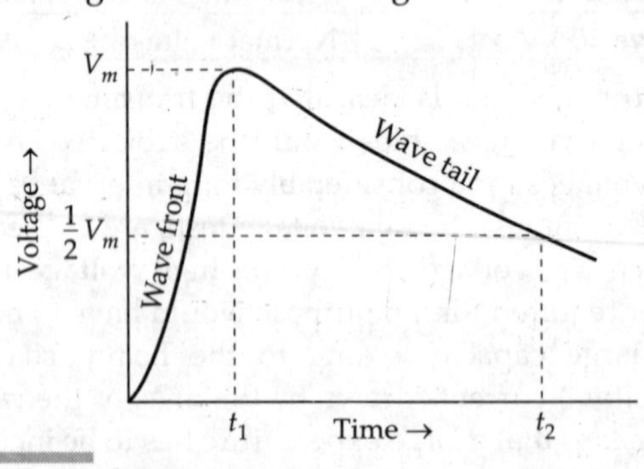


**Fig. 17.1** Surge wave.

near without discharging its electricity by lightning stroke. But if the cloud is suddenly discharged, the induced charges in the line are no longer bound, but travel with the velocity of light along the line to equalize the potential everywhere. The result is that a voltage wave travels along the line. The nature of the wave is shown in Fig. 17.1. The wave is characterised by a steep rise of voltage up to its peak value followed by a slower drop to zero voltage. The probable maximum voltage appearing on a line due to direct stroke is 10-15 MV and that by an indirect stroke is 500 kV. Fortunately direct strokes are rare.

The danger to the electric installations comes from the fact that this over-voltage acts across the insulation and that the time taken by the voltage to rise to its peak value is very small, *i.e.*, of the order of one-millionth of a second. This imposes very severe voltage stresses on the insulators, transformer windings, circuit breakers, bushings and other equipment. If the voltage stress exceeds the strength of the insulation, a flash over or a puncture results causing a short circuit and power outage. If a power arc follows, there may be disastrous results in damage to the equipment whose repair may cost considerable amount of money and time.

Since power systems are frequently exposed to lightning strokes and consequent voltage surge, impulse testing is necessary for coordination of insulation of the system in order that these surges are not allowed to damage the insulation of the important associated equipment.

## 17.2 TESTING APPARATUS

The following testing apparatus is used for sustained low frequency tests. This apparatus is usually common for all types of high voltage tests.

(*i*) high voltage testing transformer,

(*ii*) equipment for voltage regulation,

(*iii*) control gear and protective devices and

(*iv*) equipment for voltage measurement.

In addition to above, high voltage d.c. tests, high frequency tests and impulse (surge) tests require some other apparatus also which will be described along with these tests later into this chapter.

### 17.2.1 High Voltage Transformers

High voltage testing transformers are single phase core type, oil immersed and self-cooled type. They are designed for operation at the same frequency as the normal working frequency at which the test specimens operate *i.e.*, power frequency. They may also be used for higher frequencies with rated voltage, or for lower frequencies, if the voltages are reduced depending upon the frequency in order to avoid saturation of the iron core.

The kVA output and the fundamental design of the iron core and the windings, there is not much difference between a H.V. testing and a single phase power transformer as regards the thermal rating. However, H.V. testing transformers are designed with a smaller value of flux density in the core to avoid excessive magnetizing currents which contain higher harmonics that are introduced in the voltage regulator circuits supplying the test transformer. The testing transformers are also designed with compact and well insulated high voltage winding. Therefore, a single phase testing transformer may be compared with a potential transformer as regards its construction.

A single phase core type H.V. testing transformer is shown in Fig. 17.2. The primary winding is usually rated for rated voltages below 1 kV, but might often be split up in two or more winding sections which can be connected in series or parallel to increase voltage regulation capabilities. The iron core is earthed and so are the one terminal of each of the two primary and secondary windings. Two possible types of construction are shown in Figs. 17.2(*a*) and 17.2(*b*). In both cases the layout arrangement of core and windings is the same. Figure 17.2(*a*) shows a grounded metal tank, for which a h.v. bushing is necessary to bring the high voltage out of the tank. Figure 17.2(*b*) shows another arrangement, wherein the bushing is dispensed with, and a co-axial cable is used for high voltage connection which results in improved connection

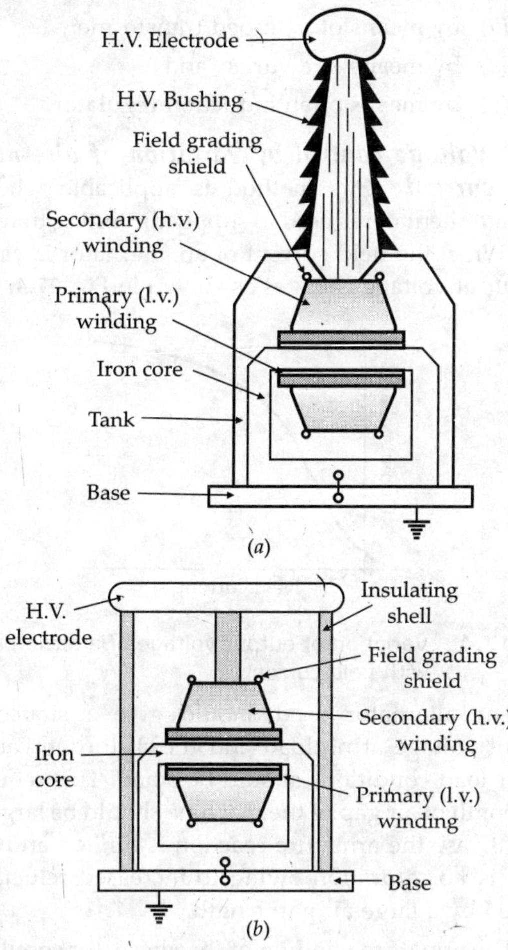


**Fig.17.2** High voltage testing transformers.

between testing transformer and the test object. In Fig. 17.2(*b*) the active part of the transformer is placed in an isolating shell, thereby avoiding the use of a bushing. This construction of Fig. 17.2(*b*) reduces the height of the transformer, however, the heat dissipation is impaired with this arrangement. In both the arrangements, tanks are filled with high quality transformer oil as most of the windings are oil-paper insulated.

The primary winding (l.v. winding) is close to the iron core and is surrounded by the h.v. winding. This arrangement reduces the leakage flux and increases the coupling between the two windings. The shape and cross-section of h.v. winding is a reference to layout of this coil : the beginning (grounded end) of the h.v. winding is located at a side close to the core, and the end close to a sliced metal shield, which prevents too high field intensities at h.v. potential. Between both ends, the single turns are arranged in layers, which are carefully insulated from each other by solid materials like kraft paper sheets. Adjacent layers, therefore, form co-axial capacitors of high values. In order to keep the potential distribution uniform under transient conditions, the capacitances have to be made equal. This is done by using reduced width of insulating layers with increasing diameters. Thus, we have trapezoidal shape for the h.v. winding.

The transformers used for insulation testing need not be of a high kVA rating because the current taken by the transformer is limited by inserting external resistances when the specimen under test breaks down. However, transformers used for cable testing are designed to deliver large currents owing to capacitive (charging) currents drawn by the cables. Attention must be paid to regulation and cooling methods employed for such transformers.

Special constructional features are incorporated in high voltage transformers to eliminate corona effects as far as possible. The high voltage terminal of the secondary winding and all other metal parts which have a high potential with respect to earth, must be constructed in such a manner that sharp corners are avoided. This is done to avoid excessive voltage gradients which cause corona. If this arrangement is not possible, corona shields (which are usually aluminium spinnings) should be provided.

For voltages upto about 500 kV, a single transformer is used. For voltages above 500 kV ; the connection of two or more transformers in cascade is more economical than the use of a single transformer because at very high voltages a single-transformer tends to be of very large size and is thus costly.

A general arrangement of two transformers connected in cascade is shown in Fig. 17.3.

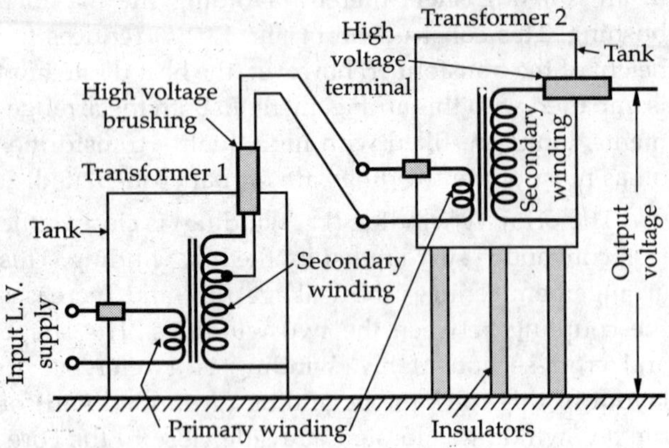


**Fig. 17.3** Cascade connection of transformers.

A low voltage supply is given to the primary winding of transformer 1. The tank of this transformer is earthed. One end of the secondary winding of this transformer is also earthed. A lead is taken from the other end of the secondary winding. This lead is taken out of the transformer through a high voltage bushing which provides insulation for full secondary voltage between this lead and the tank (earth). The secondary winding is tapped at a point and another lead is taken out of the tank through the h.v. bushing. These two leads are connected to the primary winding of transformer 2. One end of the secondary winding of transformer 2 is connected to the tank which is insulated from earth for a voltage equal to the secondary voltage of transformer 1. The other end of the secondary winding of transformer 2 is taken out with the help of a lead passing through a high voltage bushing. This lead forms the high voltage terminal of the transformer. The output voltage for test purpose, is taken between the h.v. terminal and earth. This is equal to the sum of the secondary voltages of the two transformers. The output voltage can be increased further by addition of more transformers connected in cascade.

### 17.2.2 Voltage Control

It is very important in high voltage measurements that the output voltage which is fed to the test specimen is varied smoothly. Any abrupt or crude changes in voltage will impress voltage surges on the specimen and will also affect the accuracy of measurements. It is equally important that the voltage regulating device should not distort the waveform of output voltage.

The method of voltage regulation depends upon whether a separate alternator is being used to supply the testing transformer or an available supply is being fed to the testing transformer. A separate alternator is normally used except in the case of small testing establishments which carry out routine testing work.

When a separate alternator is used, the voltage regulation is carried out by variation of the alternator field current.

In case a separate alternator is not used, the voltage regulation is done :

(*i*) by insertion of resistance in the supply circuit,

(*ii*) by insertion of inductance in the supply circuit,

(*iii*) by means of a tapped transformer,

(*iv*) by means of a variac, and

(*v*) by means of an induction regulator.

1. ***Voltage control by variation of alternator field current.*** This method is applicable when a separate alternator is used to supply the testing transformer. When the field current of an alternator is varied its output voltage changes as shown in Fig. 17.4.

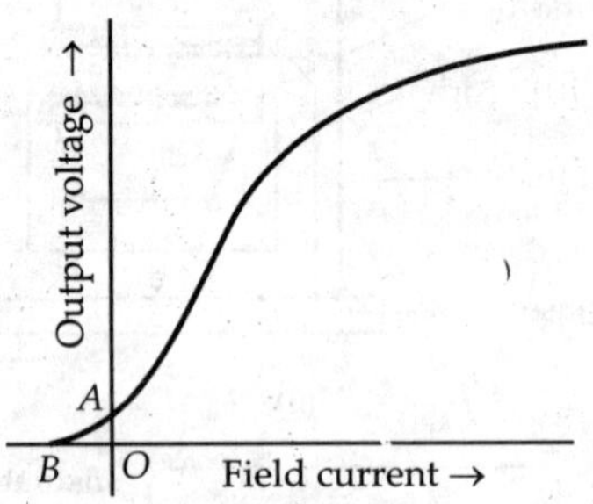


**Fig. 17.4** Variation of output voltage of alternator with field current.

The alternator used should give a sinusoidal output voltage at no load and the distortion caused under load conditions should be small. This requires the length of air gap of the machine should be large, as in that case the armature reaction is limited and also there is no saturation owing to increased reluctance offered by a large air gap length.

The armature winding of the alternator should be specially designed and also the number and shape of slots should be properly regulated so that the output voltage wave shape does not contain lower order or higher order (slot) harmonics.

The field current of the alternator is directly varied in the case of small alternators but in the case of large alternators, the field current of exciter is varied, which varies the output voltage of the exciter thus varying the alternator field current.

Figures 17.5 and 17.6 show the two connections employed. A potential divider arrangement is used in both the cases and the connections are so arranged that a small current in the reverse direction to the normal one may be obtained. Thus for this arrangement zero output voltage is obtained. Referring to Fig. 17.4, the output voltage due to residual magnetism is *OA* and a field current *OB* (in the reverse direction) is required to bring the output voltage down to zero.

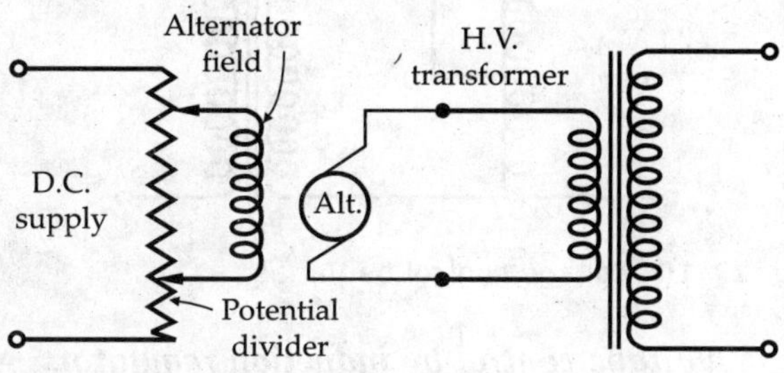


**Fig. 17.5** Voltage control by use of a potential divider in the field circuit of alternator.

The potential divider is designed to have a large number of turns so that a gradual smooth variation of output voltage is obtained.

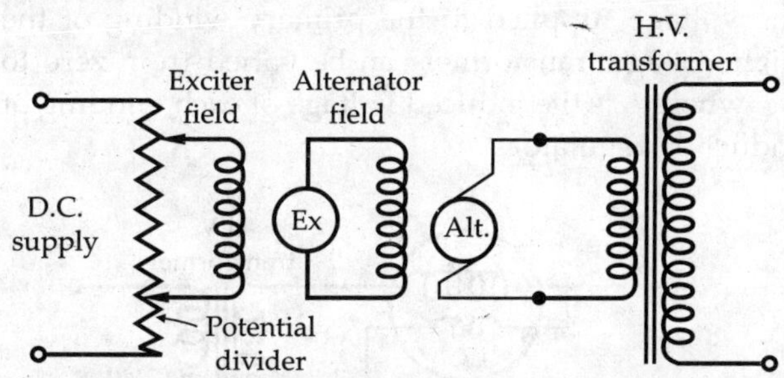


**Fig. 17.6** Voltage control by use of a potential divider in the field circuit of the exciter of alternator.

The *advantages* of this method are :

(*i*) Smooth voltage regulation from zero to full voltage is obtained.

(*ii*) There are no impedances in the circuit of supply to the h.v. transformer and therefore the input voltage waveform is not distorted.

(*iii*) The method is convenient and simple.

(*iv*) There are no disturbances caused due to frequent short circuiting of secondary winding of h.v. transformer owing to breakdown of test specimens.

2. ***Voltage control by resistance potential divider.*** This method is used when a separate alternator is not used. Potential divider arrangement as shown in Fig. 17.7 is preferred to a series resistance. This is because a smooth voltage regulation from zero to full voltage is obtained. This method is usually used for small capacity sets of about 2-3 kVA rating.

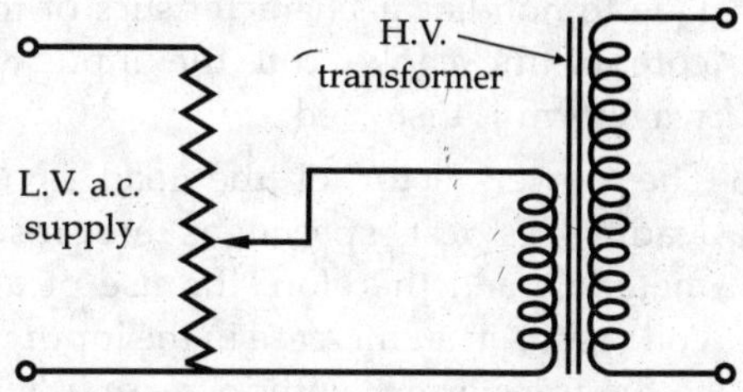


**Fig. 17.7** Voltage control by resistance potential divider.

The method has the following ***advantages*** :

(*i*) It is cheap and convenient.

(*ii*) It gives smooth voltage variation.

(*iii*) The input voltage is free from distortions. 17

However the method suffers from the following ***disadvantages*** :

(*i*) There is a large power loss in the potential divider.

(*ii*) The arrangement cannot be used for large capacity plants owing to very large size and heavy costs of the resistance potential divider.

3. ***Reactance voltage control.*** A choke coil connected in series with the primary winding of the h.v. testing transformer may be used for voltage regulation. The arrangement is shown in Fig. 17.8 wherein a choke of variable inductance is used. The reactance of the choke can be varied by withdrawal or insertion of iron core in the coil and thus the input voltage to h.v. testing transformer can be varied.

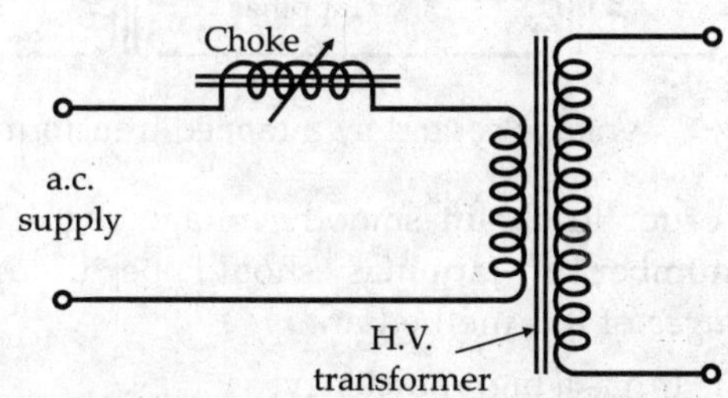


**Fig. 17.8** Variation of voltage by using series reactance.

The advantage of this method is that there is considerably smaller power loss owing to use of a choke coil in place of a resistance and consequently the efficiency is high.

The method suffers from many *disadvantages* :

(*i*) The size of the coil is large when the power to be handled is large.

(*ii*) Due to non-linear characteristics of the iron core of the choke coil the input voltage waveform is distorted.

(*iii*) The power factor of the load is usually leading as test specimens are insulating materials and therefore the use of a choke coil results in an increase in the input voltage to the transformer with increase in load.

**4. *Voltage control by tapped transformer.*** Figure 17.9 shows the arrangement for voltage regulation with the help of a tapped transformer. An inter- mediate regulating transformer is used for the purpose of tappings. The primary winding of the tapped transformer is supplied from a l.v. supply. The secondary winding of this transformer is provided with a number of tappings with the help of which the voltage fed to primary of the h.v. testing transformer can be varied. In order to avoid surges due to sudden opening of secondary circuit of regulating transformer when the tapping switch is moved, two contact brushes are used which make contacts with adjacent tapping points. A buffer resistor or a reactance coil is used between the two contact brushes to prevent the short-circuiting of a portion of the secondary winding by the brushes.

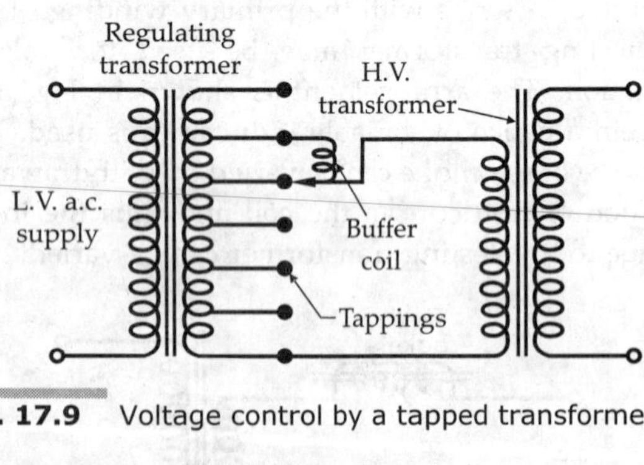


**Fig. 17.9** Voltage control by a tapped transformer.

In order to obtain smooth voltage regulation a large number of tappings should be used. The *advantages* of this method are :

(*i*) It has a high efficiency.

(*ii*) It gives a very small distortion of voltage waveform.

The *disadvantages* are :

(*i*) The voltage regulation is not smooth unless a large number of tappings is used.

(*ii*) The equipment becomes expensive if the power to be handled is large.

**5. *Voltage control by variac.*** A variac may be used to supply the h.v. transformer. The arrangement is simple and is shown in Fig. 17.10. The method has the advantage that it gives smooth voltage regulation from zero to full value of voltage.

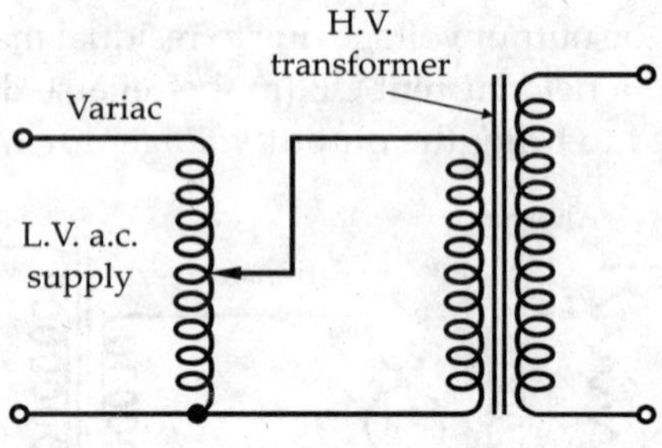


**Fig. 17.10** Voltage control by Variac.

**6. *Voltage control by induction regulators.*** An induction regulator may be used to give a smooth voltage regulation. The voltage is varied by changing the position of secondary winding with respect to primary winding. The connections are shown in Fig. 17.11. If the primary and secondary windings of an induction regulator have equal number of turns, the voltage supplied to the primary winding of the high voltage transformer can be varied from zero to 2 $E$ where $E$ is the induced voltage of each winding of induction regulator.

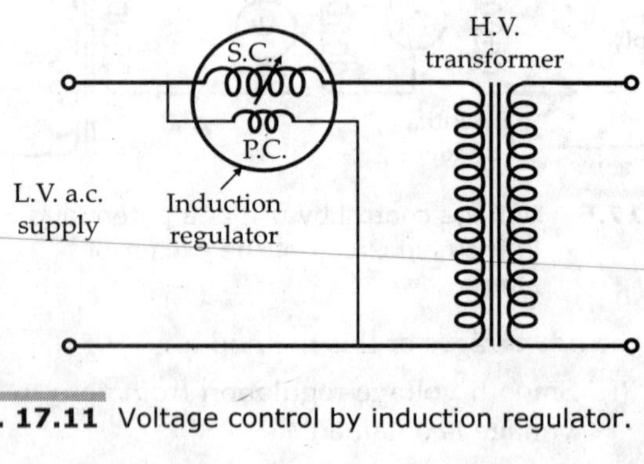


**Fig. 17.11** Voltage control by induction regulator.

The *advantages* of this method are :

(*i*) It gives a gradual and a smooth voltage regulation.

(*ii*) It can be used for all loads and power factors.

(*iii*) Voltage waveform distortion can be avoided by using a carefully designed induction regulator.

(*iv*) This method of voltage regulation is particularly suited for cable testing equipment owing to its adaptability for any load.

## 17.3 CONTROL GEAR AND PROTECTIVE DEVICES

The primary side of the high voltage transformer is provided with :

(*i*) ***Main switch.*** It is required to isolate the testing apparatus from the supply.

(*ii*) ***Fuses.***

(*iii*) ***Automatic circuit breaker.*** It is used to protect the testing apparatus from damage in case of failure of supply. It is provided with a no volt coil so that it automatically trips if the supply goes off.

(*iv*) ***Over-voltage relay.*** It is used to protect the apparatus from the effects of over voltages. This relay short circuits the no volt coil of the circuit breaker in case of over voltage and therefore the latter trips automatically isolating the apparatus from the supply.

(*v*) ***Interlocks.*** The interlocking arrangements are used :

(*a*) to ensure that the primary winding of h.v. testing transformer is switched on to a low voltage,

(*b*) to make it impossible for the operator to enter the h.v. testing enclosure while the supply is on. This is in the form of a gate switch. The supply to the primary winding of h.v. testing transformer is through a gate operated switch. When the gate is open, the switch is off and therefore the operator can safely go inside the enclosure.

(*vi*) ***Earthing.*** All the metal parts which should be at earth potential must be earthed. This includes transformer tanks, metal framework of switchgear, enclosure and one end of each transformer winding.

The protective devices on the high voltage side of the test transformer are :

(*i*) ***Protective resistances.*** When the specimen under test or sphere gap breaks down, voltage surges are produced. In order to protect the high voltage transformer from the effects of these surges a high resistance or a choke is connected in series with the secondary winding of h.v. transformer. The resistance is about 0.5 Ω/V.

(*ii*) ***Sphere gaps.*** A sphere gap is connected across the high voltage transformer. In case of over voltage, the sphere gap breaks down and thus bypasses the surge. In order to prevent pitting of sphere gaps, a resistance is used to limit the current at breakdown. This resistance is about 1 ohm per volt so that the current is limited to 1 ampere at breakdown.

Some difficulties arise in H.V. testing at power frequencies and these are primarily due to corona effect.

**Corona**. This is a phenomenon which occurs when the voltage gradient at the surface of a high voltage conductor exceeds the breakdown strength of air. This results in ionization of surrounding air accompanied by a faint glow and hissing sound. Corona results in power loss and distortion of waveform. Thus it influences the breakdown and flash-over voltages on account of surges in the circuit and because of presence of ionized air in the test room.

Steps should be taken to eliminate corona as far as possible. Since the voltage gradient at the surface decreases as the diameter increases, the conductors used in high voltage circuits should be made much more larger in size than is necessitated by current carrying considerations. Also care should be taken to avoid sharp bends and small radii of curvature.

Figure 17.12 shows a simple diagram of connections of a high voltage testing set up using a separate alternator.

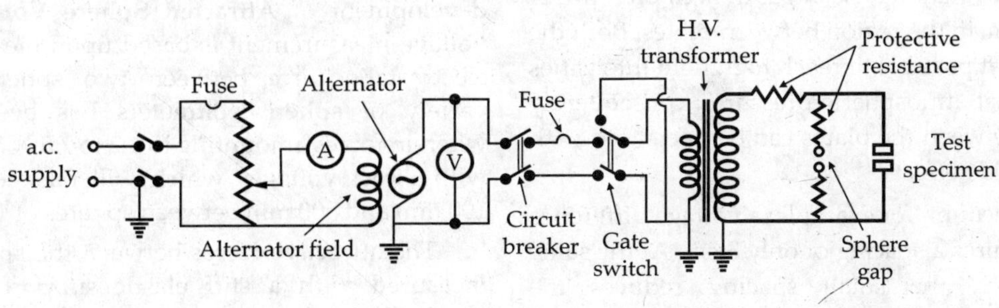


**Fig. 17.12** Circuit for h.v. testing at power frequencies.

## 17.4 EQUIPMENT FOR VOLTAGE MEASUREMENT

The high voltages encountered in practice involve a large range of frequencies extending from zero in the case of d.c. to very high frequencies of surges caused by lightning. Most methods of measurement of high voltages described in the following pages are suitable for measurement at comparatively low frequencies. However, some of them give satisfactory results even for surge frequencies.

The methods of measurement of voltage are divided into *two* categories :

(*i*) *Methods used for measurement of rms values of voltage.* We are interested in measurement of rms values of voltage because not only they have a direct bearing on the dielectric losses, but through internal heating they have a close connection with dielectric constant also.

(*ii*) *Methods of measurement of peak values of voltage.* The measurement of peak values of voltage is also necessary because of the fact that the actual breakdown or flash-over of the specimen under test is dependent upon the peak value of the applied voltage.

### 17.4.1 Measurement of R.M.S. Values of Voltage

#### 1. Electrostatic Voltmeters

These instruments have been described in details in chapter 9, page 282. The following instruments may be used for measuring rms values of high voltages.

(*i*) ***Attracted disc voltmeter.*** At high voltages, the design of the attracted disc voltmeter becomes difficult. A large spacing between plates is necessary in order to avoid flash-over and corona.

The simple expression for voltage given by Eqn. 9.84 (page 286) does not hold good and corrections have to be applied.

In order to overcome the above difficulty, an absolute high voltage attracted disc voltmeter has been evolved by making use of gas at high pressure or a high vacuum, in the region between plates. Both the designs make it possible to reach high field intensities than with air at atmospheric pressure ; consequently the spacing between the plates can be decreased with no resultant side effects.

In high vacuum, for example, a voltage difference of 300 kV requires a spacing of only 3 cm. At the same time, however, the small spacing reduces the tolerances in flatness and parallelism of the plates. Moreover the attracted electrode needs close adjustment to the plane of guard ring.

Another method of improving the uniformity of the field is to insert, between the main plates, a series of intermediate electrodes. A capacitance divider, thus formed, preserves a uniform distribution of voltage between plates.

An inherent feature of these instruments is the tendency for the electrostatic force to make the disc unstable so that an adequate elastic or electromagnetic force trying to keep the disc in coplanar position is necessary. Also when alternating voltages are being measured, care must be exercised to ensure that the forced vibrations due to pulsating component of force, are minimized.

The electrostatic force is usually balanced by an opposing gravitational or electro-magnetic force. The lack of balance can be detected by the displacement of the disc from its coplanar position and therefore very sensitive methods of measuring displacement are required in order to achieve a high degree of accuracy. Automatic adjustment of balance can be done through a servo-loop, the actuating signal is obtained from the output of the displacement detector. A disc may be loaded with a weight of approximately 90 g, or alternatively 360 g, corresponding to the attractive forces on the disc at electrostatic stresses of 5 kV/mm and 10 kV/mm and a servo-system adjusts the electrode spacing till the appropriate stress is obtained. The voltage is then read directly from a micrometer scale which indicates the spacing between the electrodes.

Alternatively, with a given electrode spacing the servo-system can be arranged to adjust the current in a coil coupled to the disc. This current produces an electromagnetic force which balances the electrostatic force and in addition stabilizes the disc powerfully in the coplanar position. The electromagnetic system is gravitationally calibrated.

(*ii*) ***Attracted sphere voltmeter.*** The tremendous classical theoretical analysis of the electrostatic field between two spherical electrodes has led to the development of **Attracted Sphere Voltmeters.** The voltage measurement is based upon measurement of electrostatic force between two spheres. A wide variety of sphere voltmeters has been produced wherein forces amounting to 4.2 to 7.5 N are produced with direct voltages which will flash over gaps of 100 mm and 500 mm between spheres of 1 m diameter.

The attractive forces between the spheres can be measured with a stiff elastic support. The errors involved are less than 1% for voltages as high as 1 MV.

Another method of measuring the electrostatic force is to observe the change produced in the period of oscillation of one of the spheres, swinging over a small arc. The voltage is proportional to

$$\sqrt{f_v^2 - f_o^2}$$

where $f_v$ = oscillating frequency with applied voltage,

and $f_o$ = oscillating frequency without applied voltage.

The values of $f_v$ and $f_o$ do not differ greatly as the restoring force is gravitational in nature. However, by using a rigid pendulum, the mass of the sphere can be counter-balanced. The value of $f_o$ is, therefore, reduced nearly to zero so that frequency of oscillations, $f_v$, is approximately proportional to the applied voltage.

(*iii*) ***Oscillating ellipsoid voltmeter.*** This is another system for which the attractive force between two electrodes can be obtained with a high degree of accuracy from the period of oscillation of an electrode.

The voltage to be measured is applied to a pair of discharge free parallel plate pair of electrodes lying in vertical planes and a small conducting electrode of accurate prolate spheroidal shape is suspended at its mid-point by a long unspun silk thread so that it hangs in the uniform field between the central parts of the plates with its long axis horizontal. When a voltage is applied an electrostatic field is produced due to which charges of opposite polarity and of calculable magnitude accumulate on the two ends of the ellipsoid. The ellipsoid, therefore, tends to align its long axis along the direction of the field. The ellipsoid oscillates about its final position on account of the moment of inertia. The controlling torque produced by the silk thread is very small. Therefore frequency of oscillations is approximately proportional to the applied voltage. The constant of proportionality can be calculated from the mass and dimensions of the ellipsoid and the geometry of the plate electrodes.

The disadvantage of the ellipsoid voltmeter is that the average electrostatic stress is much smaller than that which can be obtained when either attracted disc or attracted sphere voltmeters are used. This is on account of the concentration of the stress at the tips of the ellipsoid.

## 2. Transformer Ratio Method

In this method the primary voltage of the high voltage testing transformer is measured with a low voltage voltmeter (Fig. 17.13) and the secondary voltage is obtained as :

$$V_s = nV_p$$

where $V_s, V_p$ = terminal voltages of secondary and primary windings respectively.

$n$ = turns ratio.

This method assumes that the ratio of transformation remains the same (= $n$) irrespective of the load and power factor on the secondary side. Owing to this reason, the method though simple, gives wrong results.

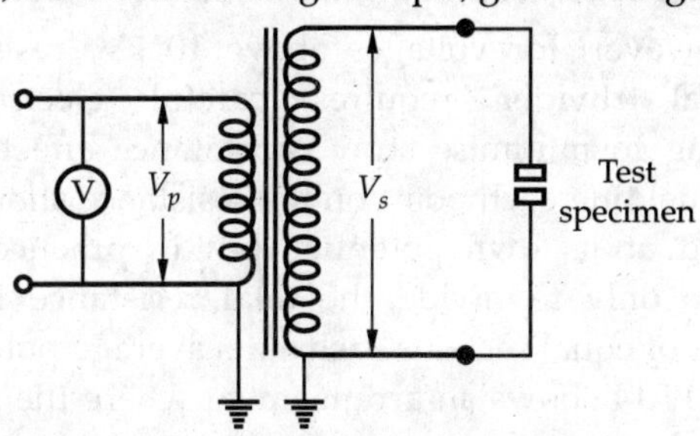


**Fig. 17.13** Measurement of rms voltage by transformer ratio method.

In some high voltage transformers, a separate voltage coil is provided whose voltage is measured with the help of a low voltage voltmeter. The secondary voltage in this case is given by :

$$V_s = (N_s / N_c) \cdot V_s$$

where $N_s$ = number of turns of the secondary winding,

$N_c$ = number of turns of the coil,

and $V_c$ = Voltage across the coil.

This method cannot be used for cascade arrangement of transformers. It gives an accuracy of 1 to 2 percent.

The advantage of this system lies in its convenience as a means of taking continuous reading after being calibrated by comparison with one of the direct systems of measurements.

## 3. Potential Divider Methods

In this method a potential divider is connected across the secondary winding (high voltage winding) of high voltage testing transformer and a definite fraction of secondary voltage is measured by a low voltage electrostatic voltmeter. A C.R.O is used to obtain records of peak values of surge voltages.

*Two* types of potential divider arrangements are used :

(*i*) Resistance potential divider, and

(*ii*) Capacitance potential divider.

Both these potential divider methods have been described in Chapter 9 (see pages 286).

We should use these methods only when either :

(*a*) the connection of the low voltage voltmeter has negligible effect on the voltage distribution, or.

(*b*) appropriate allowance for low voltage voltmeter impedance is made in computing the potential divider ratio.

17

(*i*) ***Resistance potential divider.*** Upto about 10 kV an electrostatic voltmeter can be used successfully with an unshielded resistance potential divider with resistance units wound on flat insulating sheets.

However, for voltages above 10 kV, resistance potential divider requires careful electrostatic shielding to minimise stray capacitance effects. For ideal shielding each point on the resistance should be shielded at its own potential, but in practice it is possible only to divide the total resistance into a number of equal sections each at its average potential. Figure 17.14 shows an arrangement where the shield potentials are correctly maintained by an auxiliary potential divider.

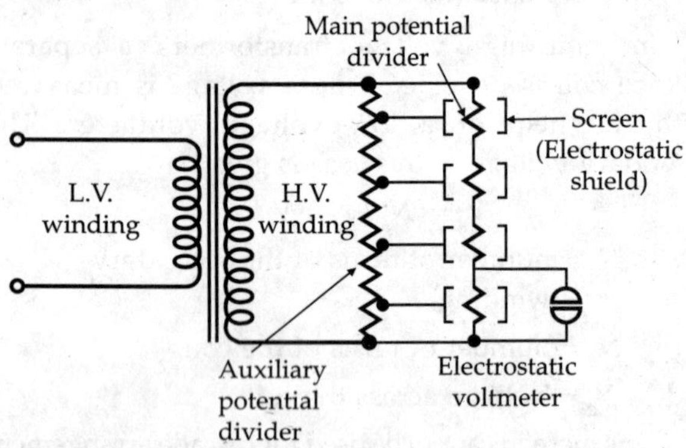


**Fig. 17.14** Resistance potential divider.

(*ii*) ***Capacitance potential divider.*** This arrangement is shown in Fig. 17.15. In this method an electrostatic voltmeter having a capacitance $C_v$ is connected in parallel with a capacitor $C_2$ and the combination is connected in series with a standard capacitor $C_1$ across the high voltage winding of testing transformer :

Let $V_s$ = voltage of the h.v. winding,

and $v$ = voltage across the voltmeter

$$V_s = \frac{C_1 + C_2 + C_v}{C_1} v$$

Now if $C_2$ is very large as compared to $C_v$,

$$V_s = \frac{C_1 + C_2}{C_1} v$$

Since the capacitance $C_v$ of voltmeter varies with deflection therefore the capacitance $C_2$ of capacitor which shunts the voltmeter is made large in order that the effects of variation of capacitance of voltmeter are swamped as is clear from the above expression.

The use of capacitance potential divider as compared to resistance potential for use at high voltage has the following ***advantages*** :

(*i*) They are simple in construction.

(*ii*) They are free from heating effects.

(*iii*) They are comparatively free from effects of stray magnetic fields and can be easily shunted.

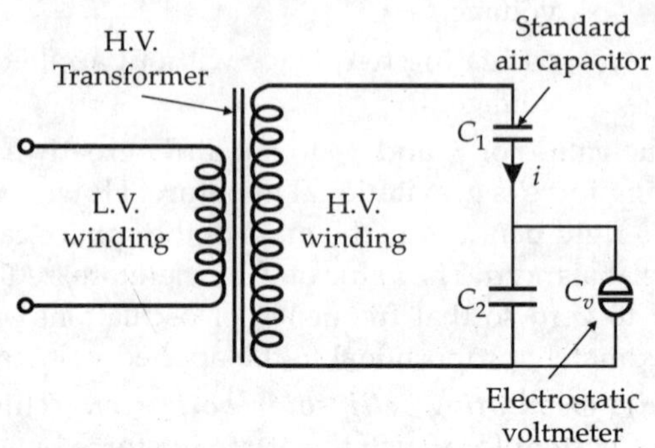


**Fig. 17.15** Capacitance potential divider.

Capacitance of standard capacitor should be accurately known and the capacitor should be free from dielectric loss. Air capacitors are usually used with capacitances ranging from 50 pF to 100 pF.

(*iii*) **Compensated capacitive potential dividers.** For any particular frequency a capacitance potential divider should have a ratio and phase angle independent of the load connected on the secondary side. This is achieved by connecting an inductance $L$ in series with the load as shown in Fig. 17.16.

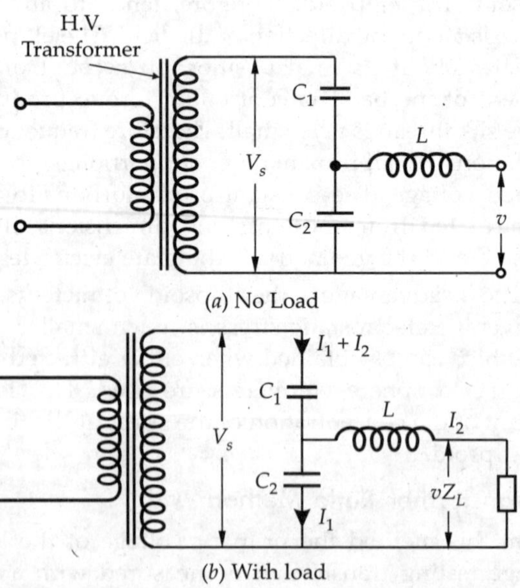


**Fig. 17.16** Compensation for constant voltage ratio irrespective of load.

(*a*) *No load conditions.* Figure 17.16(*a*) represents the no load conditions and the voltage ratio is,

$$V_s / v = (C_1 + C_2)/ C_1$$

Suppose $Z_L$ is the impedance of the load connected on the secondary side as shown in Fig. 17.16(*b*). The requirement that transformation ratio $V_s / v$ remains the same irrespective of the load on the secondary winding, the value of inductance $L$ should be such that it gives resonance with $C_1$ and $C_2$ in parallel.

or $\omega_2 L = 1/(C_1 + C_2)$

This is explained as under :

At load, $V_s = \dfrac{j(I_1 + I_2)}{\omega C_1} - \dfrac{jI_1}{\omega C_2}$

Voltage across load

$$v = V_1 - \left[\frac{-j(I_1 + I_2)}{\omega C_1} + j\omega I_2 L\right]$$

$$= -j\left[\frac{I_1}{\omega C_2} + \omega I_1 L\right]$$

$\therefore$ Voltage ratio

$$\frac{V_s}{v} = \frac{-j\left[\left(\dfrac{I_1 + I_2}{\omega C_1}\right) + \dfrac{I_1}{\omega C_2}\right]}{-j\left[\dfrac{I_1}{\omega C_2} + \omega I_2 L\right]}$$

$$= \frac{I_1(C_1 + C_2) + I_2 C_2}{C_1\,[I_1 + I_2 \omega^2 LC_2]}$$

Substituting the value of

$\omega^2 L = 1/(C_1 + C_2)$ in the above expression

we have $\dfrac{V_s}{v} = \dfrac{[I_1(C_1 + C_2) + I_2 C_2](C_1 + C_2)}{C_1[I_1(C_1 + C_2) + I_2 C_2]}$

$$= \frac{C_1 + C_2}{C_1}$$

or voltage ratio $\dfrac{V_s}{v} = \dfrac{C_1 + C_2}{C_1}$

This is the same value of ratio as we obtain on no load.

### 4. Standard Impedance Methods

A standard high impedance is connected across the terminals of the high voltage winding. The impedance may be a standard air capacitor or a high resistance. The current flowing through the impedance may be measured by a thermocouple ammeter (Fig. 17.17) by measuring voltage across a known resistance ; using an electro- static voltmeter or by using a valve voltmeter.

Secondary voltage $V_s = i \times Z$

where Z = impedance connected across the h.v. winding terminals

The impedance may be a series resistor. Series resistors are used in conjunction with moving coil, moving iron or electrodynamometer milliammeters for measurements of voltages upto 100 kV. Moving coil instrument is useful for d.c. only while the others are used for measurement of rms values of a.c. voltages.

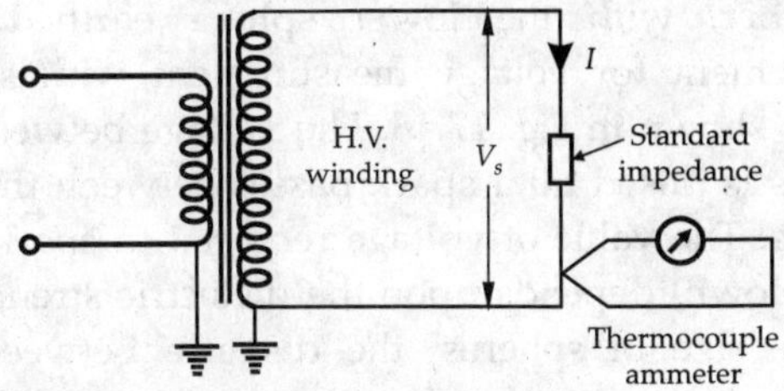


**Fig. 17.17** Measurement of voltage by standard impedance method.

### 5. Potential Transformer Method

Modern potential transformers have accuracies comparable with those of good potential dividers. In comparison with potential dividers they have the advantages that :

(*i*) the indicating instrument is isolated from the high voltage circuit,

(*ii*) within limits, the instrument impedance has negligible effect on the ratio of transformation.

For measurement of very high voltages two stage reduction by means of a combination of a potential divider and potential transformer proves very useful.

Potential transformers may be used with electrodynamometer, moving iron, or electrostatic instruments for measurement of rms values of a.c. voltages upto 200 kV.

The range may be extended by capacitor potential transformers (see Art. 10.6.7 page 335)

## 17.4.2 Measurement of Peak Values of Voltages

1. ***Transformer ratio method.*** The measurement of rms value of voltage with this method has already been explained. In order that the peak value of secondary voltage be measured, we must determine the secondary voltage waveform. From the voltage waveform the crest factor is evaluated.

$$\text{Crest factor} = \frac{\text{peak voltage}}{\text{rms voltage}}$$

or peak voltage of secondary winding

$$= \text{crest factor} \times nV_p$$

where $V_p$ = rms voltage of primary winding.

and $n$ = ratio of secondary to primary winding turns.

2. ***Sphere gaps.*** The breakdown voltage of a spark gap between two metal spheres may be used as a measure of voltages upto the highest encountered in

high voltage testing. The normal arrangement consists of two equal sized, polished aluminium or (aluminium alloy) spheres separated by an air gap. The spheres are usually mounted vertically one above the other, with the lower sphere earthed. The arrangement for voltage measurement with sphere gaps is shown in Fig. 17.18. The voltage between the spheres is raised till a spark passes between the two spheres. The value of voltage required to spark over (breakdown) depends upon the dielectric strength of air, the size of spheres, the distance between the spheres and many other factors.

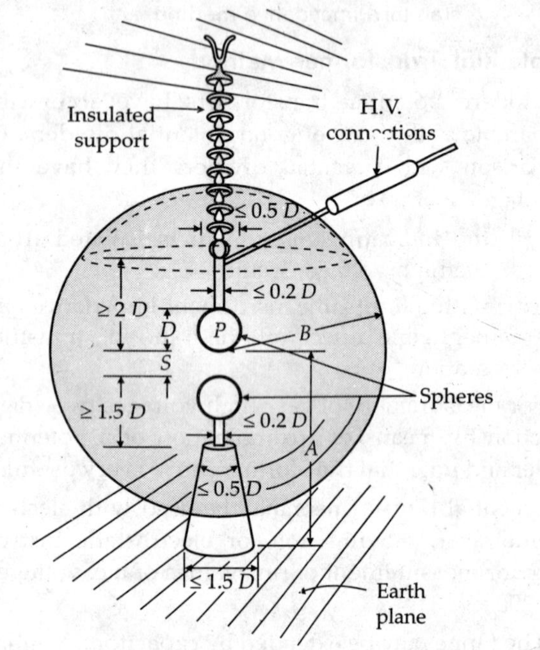


Typical vertical sphere-gap, S, showing limiting dimensions of components and clearances,

D = sphere gap diameter,

P = sparking point of H.V. sphere

A = Distance of P to earth plane,

B = radius of space around P free from external structures.

**Fig. 17.18** Measurement of voltage with sphere gaps.

Calibration tables have been compiled on the basis of experimental work with different sizes of spheres, different gap lengths at specified conditions of pressure and temperature. The breakdown voltages corresponding to actual working conditions may be directly read off from the calibration tables. However, corrections must be applied for temperature and pressure conditions.

These calibration tables give the rms values of spark-over voltages. To obtain peak voltages from results given in the tables the rms values must be multiplied by $\sqrt{2}$.

**Advantages :**

(*i*) The sphere gaps may be conveniently used for determination of peak value of a voltage wave. Their main use is for calibration of voltmeters and voltage measuring devices for high voltage tests.

(*ii*) The method is simple and peak voltages may be measured from about 2 to 2500 kV by getting a spark over and then referring to the calibration tables.

(*iii*) Sphere gaps may also be used for measurement of voltage in surge (impulse) tests.

(*iv*) Method of voltage measurement with sphere gaps is cheap, simple and reliable.

**Disadvantages :**

(*i*) The sphere gaps do not give a continuous record of voltage.

(*ii*) Voltage measurements carried out by sphere gaps can be relied upon to be accurate within 2-3% provided suitable precautions are taken. This is not a high accuracy in comparison with other electrical measurements. However even this accuracy is acceptable owing to difficult conditions of working in high voltage testing.

**Precautions :**

(*i*) Ionization of air gap depresses the spark over voltage and therefore ultra-violet or other ionization radiation should be avoided. However, in high frequency work this small ionization has an advantage and this may be provided by mild corona on high voltage cables or from a remote source of ultra-violet radiation.

(*ii*) In order to avoid pitting of spheres, a current limiting resistance of 1 ohm per volt may be inserted in the sphere gap circuit.

(*iii*) After testing, the spheres should be properly cleaned to remove any extraneous matter.

(*iv*) The spheres should not be highly polished.

(*v*) No conductor or body having a conducting surface (except the supporting shanks) shall be near the sparking-point of the h.v. sphere than the distance given by the expression 0.25 + $V$/300 m where $V$ is the voltage in kV (peak) which is to be measured.

(*vi*) The time interval between consecutive flash overs should be large enough to avoid appreciable heating of spheres.

**3. *Rectified capacitor charging current method.*** This method depends upon the fact that the average charging current of a standard air capacitor is proportional to the peak voltage. This charging current is rectified and its average value is then measured by a permanent magnet moving coil milliammeter or a d'Arsonval galvanometer.

Figure 17.19 shows the circuit for measurement of peak voltage. $C$ is a high voltage compressed gas capacitor across which practically whole of the high voltage is dropped. Specially designed rectifiers may be used for the rectification of alternating currents. $A$ is the milliammeter for measurement of average charging current.

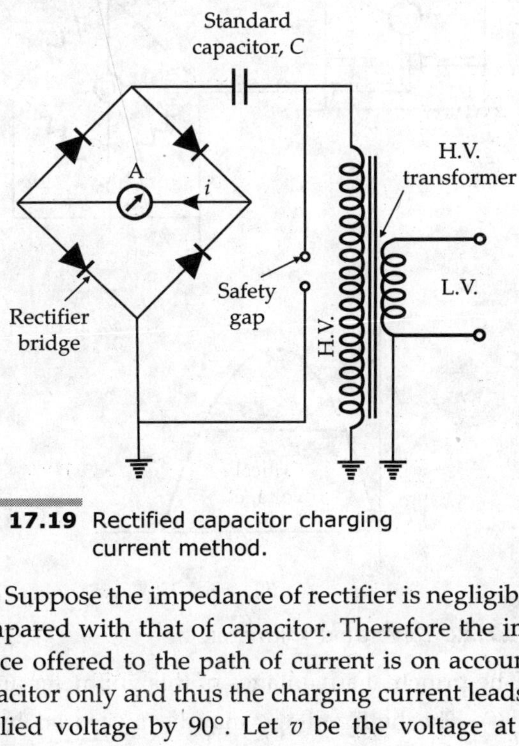


**Fig. 17.19** Rectified capacitor charging current method.

Suppose the impedance of rectifier is negligible as compared with that of capacitor. Therefore the impedance offered to the path of current is on account of capacitor only and thus the charging current leads the applied voltage by 90°. Let $v$ be the voltage at any instant.

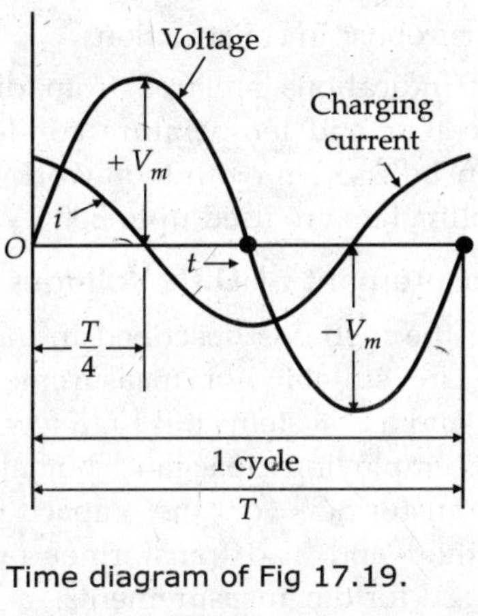


**Fig. 17.20** Time diagram of Fig 17.19.

$\therefore \quad v = V_m \sin \omega t$ where $V_m$ is the peak voltage.

Charging current at any instant

$$i = \frac{V_m}{1/\omega C} \sin(\omega t + \pi/2) = V_m \omega C \cos \omega t$$

Referring to Fig. 17.20, average charging current

$$I_{av} = \frac{1}{T/4} \int_0^{T/4} V_m \,\omega C \cos \omega t \, dt = 4 V_m fC$$

$$\left(\because T \frac{1}{f}\right)$$

Thus $\quad V_m = I_{av} / 4Cf \quad$ ...(17.1)

Equation 17.1 gives the value of peak voltage if the value of capacitance and frequency and average current are known.

The method is accurate for sine waves provided that the impedance of the rectifier is negligibly small as compared with that of the capacitor.

The method is satisfactory for all wave-shapes with the exception of :

(*i*) wave-shapes with different positive and negative maxima, and

(*ii*) wave-shapes with more than one-peak in each half cycle.

***Example 17.1*** *A permanent magnet moving coil ammeter indicates 20 mA when connected to opposite corners of a bridge rectifier. The other two corners of the bridge rectifier are connected to a secondary of a 50 Hz high voltage transformer through a 707 pF capacitor. Calculate the peak value of the voltage assuming it to be sinusoidal.*

**Solution.** From Eqn. 17.1, peak voltage of secondary of h.v. transformer,

$$V_m = \frac{i_{av}}{4Cf}$$

$$= \frac{20 \times 10^{-3}}{4 \times 707 \times 10^{-12} \times 50} \text{V} = \sqrt{2} \times 100 \,\text{kV}$$

**4. *Ryall crest voltmeter.*** Figure 17.21 shows a simplified arrangement of measuring Crest (Peak) voltage with Ryall Crest Transformer Voltmeter. It consists of two capacitors $C_1$ and $C_2$ connected in series across the high voltage winding of the test transformer. Capacitor $C_1$ is variable while $C_2$ is fixed. The capacitance of variable capacitor $C_1$ is very large as compared to that of $C_2$ with the result that almost the entire voltage is across capacitor $C_2$.

A neon lamp of special type is connected across the variable capacitor $C_1$. Neon lamps have the property that they commence to glow when the voltage applied to them reaches a definite value called a.c. striking voltage.

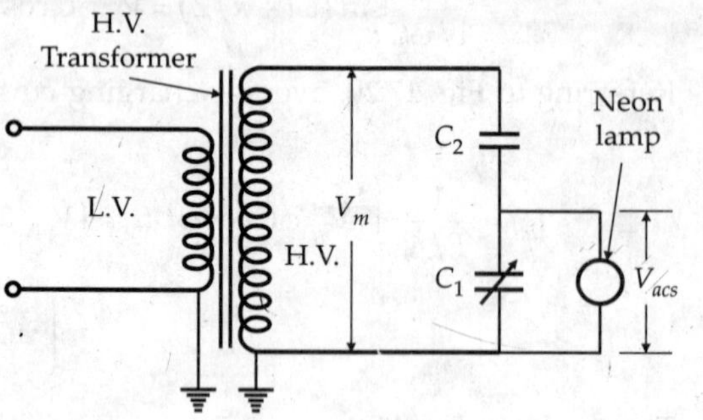


**Fig. 17.21** Ryall crest voltmeter.

In this method the capacitance of capacitor $C_1$ is gradually reduced so that there is an increased voltage across it. This voltage is increased until the lamp strikes. Then the peak voltage is given by :

$$V_m = V_{acs}\left(1+\frac{C_1}{C_2}\right),$$

where $V_{acs}$ = a.c. striking voltage of the neon lamp.

This voltmeter has the advantage that its indication is independent of the frequency. It can be designed to read upto 150 kV.

5. ***Ionic wind voltmeter.*** The principle on which this voltmeter works is given below : When the stress at a highly charged electrode in a given gas exceeds the ionization value, the ions so produced are repelled by high intensity field. During their movement the ions collide with the uncharged molecules and set them in motion away from the point, giving-rise to an ionic wind.

Figure 17.22(*a*) shows the set up of the voltmeter. A hot wire of platinum gold alloy is included in one arm of the Wheatstone bridge. This wire also forms the earthed electrode of the voltmeter. The bridge network is balanced initially. The high voltage to be measured is applied to the voltmeter and this starts an ionic wind which is directed on to the hot wire. The hot wire cools and therefore its resistance changes. This causes an appreciable out of balance voltage to appear in the bridge network and this is indicated by the galvanometer. The cooling and hence the galvanometer deflection depends upon the potential gradient (applied voltage), the temperature of the electrode, the nature and pressure of the gas and upon the frequency.

Apart from a small initial cooling effect due to convection, the main cooling action starts when the potential gradient rises above a critical value. Thus peak voltage of the measured supply must exceed the critical voltage for part of a cycle before an indication is obtained. The part of the cycle where voltage is more than the critical is only effective in the cooling [see Fig.17.22(*b*)]. It follows that the instrument is not a true peak voltmeter and a knowledge of waveform is essential in order to determine the value of peak voltage.

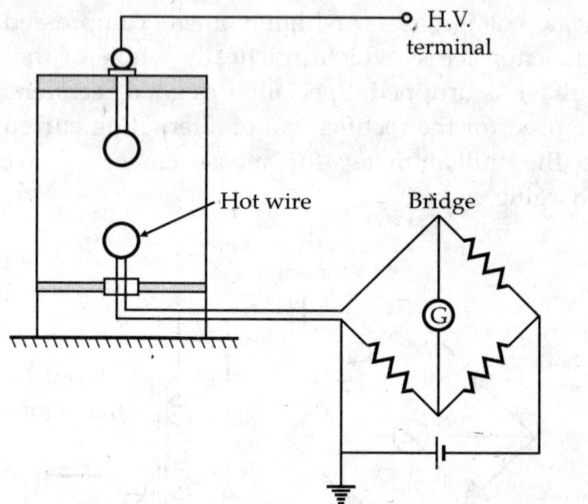


(*a*)

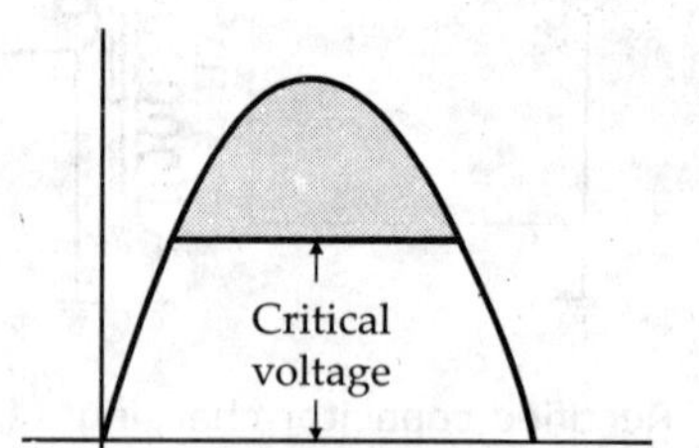


(*b*)

**Fig. 17.22** Ionic wind voltmeter.

The principal advantages of this voltmeter are :

(*i*) The high voltage may be measured by an observer at some distance from the charged conductor.

(*ii*) It is robust in construction.

(*iii*) Its indications are free from disturbances, weather and temperature conditions. So it can be used for outdoor work also. These voltmeters are used upto 300 kV.

### 17.4.3 Measurement of D.C. Voltages

Most of the methods described in the preceding paragraphs are suitable for measurement on high voltage d.c. and a.c. systems but for a few exceptions. The methods employing capacitance potential dividers, potential transformers or the capacitor charging current method and also transformer ratio method cannot be used for d.c. measurements.

Resistance potential dividers can be used. The series resistance method is also satisfactory when a high resistance is used in series with a current operated d.c. instrument.

### 17.4.4 Measurement of Instantaneous Voltage

High voltage cathode ray oscilloscope with suitable potential divider where necessary, is used for measurement of instantaneous values. A treatment of cathode ray oscilloscope is given in Chapter 21.

## 17.5 LOW FREQUENCY H.V. TESTS

Figure 17.23 shows the arrangement for sustained low frequency H.V. testing.

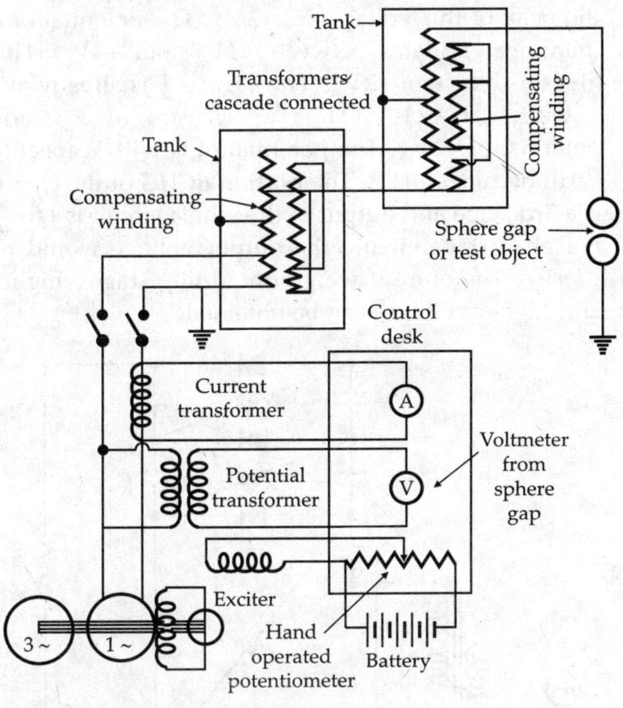


**Fig. 17.23** Sustained low frequency high voltage testing.

## 17.6 HIGH VOLTAGE D.C. TESTING

In addition to the equipment required for low frequency tests, a source for generation of d.c. is required for high voltage d.c. testing.

Generation of d.c. at high voltage by rotating electrical machines is not possible owing to limitations imposed by commutator and commutation conditions. Therefore high voltage d.c. is always obtained by rectification of alternating currents. Electronic oscillators working at 30-100 kHz may be used to feed the rectifiers upto a voltage of 50 kV as use of high frequency a.c. reduces the size of high voltage transformer. But there are some extra complications with use of high frequencies and therefore normally alternators are used to supply power at 50 Hz which is converted into d.c. by rectifiers.

The rectifying circuits commonly used are shown in Figs. 17.24 to 17.27.

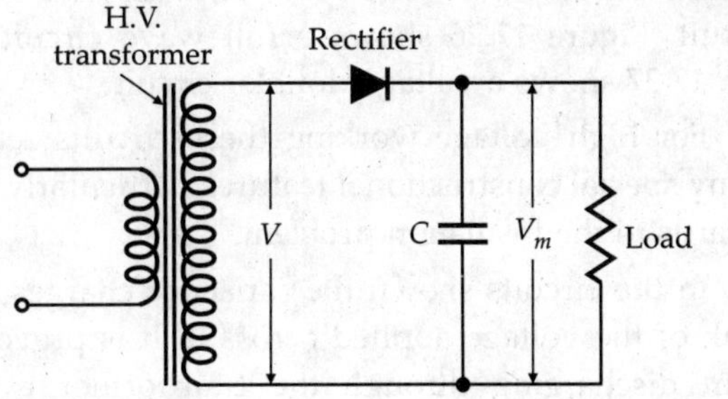


**Fig. 17.24** Rectifying circuit.

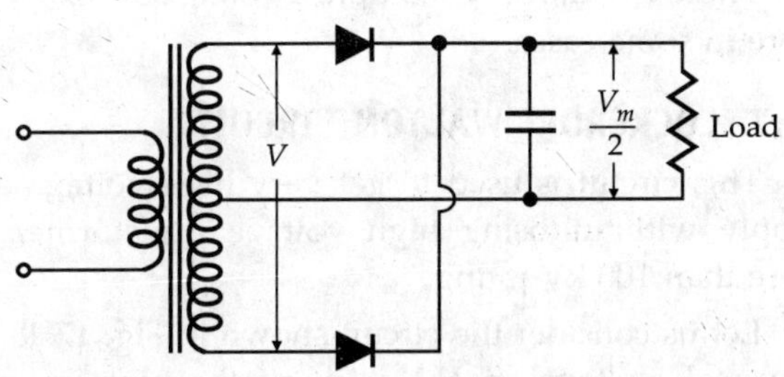


**Fig. 17.25** Rectifying circuit.

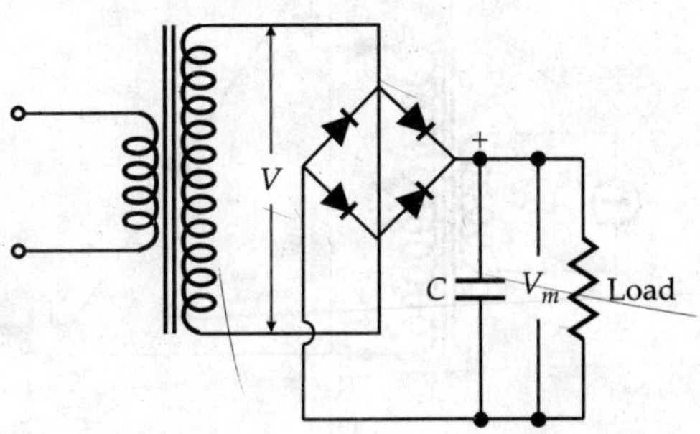


**Fig. 17.26** Rectifying circuit.

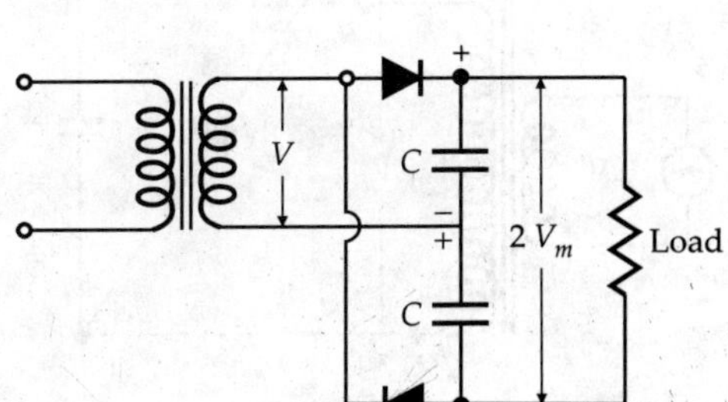


**Fig. 17.27** Rectifying circuit.

For the sake of clarity many essential parts of the rectifying equipment like smoothing circuits, protective resistors to limit the current during breakdown of test specimen have not been shown.

Figure 17.24 shows a single phase half wave circuit. Figure 17.25 shows a bi-phase half wave circuit. Figure 17.26 shows a full wave circuit and Fig. 17.27 shows a voltage doubler circuit.

For high voltage working these circuits require many special constructional features particularly with regards to the insulation problem.

In the circuits shown the capacitor charges upto peak of the voltage applied across it. It is prevented from discharging through the transformer by the rectifier but can discharge through the load (which is represented by a resistance).

These circuits are used upto a voltage of 100 kV or more in some cases.

## 17.7 COCKCROFT WALTON CIRCUIT

This circuit is used to get very high voltage a.c. supply without using high voltage transformer of more than 100 kV rating.

Let us consider the circuit shown in Fig. 17.28(*a*). The peak voltage on H.V. side of the high voltage transformer is $V_m$.

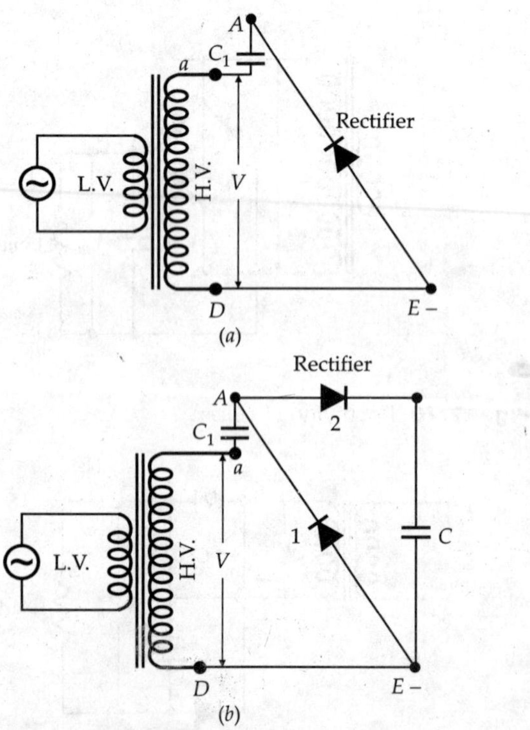


**Fig. 17.28** Basics of Cockcroft Walton circuit.

Therefore the rectifier charges the capacitor $C_1$ to a voltage + $V_m$. The voltage of the end '*a*' of transformer winding varies between $+V_m$ and $-V_m$ with respect to earth and therefore the voltage of point *A* oscillates between 0 and $+2\,V_m$ with respect to earth *E*.

Now if we connect a capacitor $C_2$ as shown in Fig. 17.28(*b*), this capacitor will be charged to a voltage $+2\,V_m$ through rectifier 2.

The complete Cockcroft-Walton circuit is shown in Fig. 17.29. It is in fact a cascade connection. As the potential of *A* varies between zero and $+2\,V_m$, *F* being at a steady potential of $+2\,V_m$ the voltage applied to capacitor $C_3$ via rectifier if $C_3$ were uncharged would vary between zero and $+2\,V_m$. Thus $C_3$ charges upto the peak of this voltage *i.e.*, $+2\,V_m$. The potential of *B* therefore oscillates between $+2\,V_m$ and $+4\,V_m$. This gives a voltage of $+2\,V_m$ across $C_4$, as *F* is already at a potential of $+2\,V_m$. Thus $C_4$ charges to a steady potential of $+2\,V_m$. The potential of *G* with respect to earth becomes $+4\,V_m$. The portion *BCHG* of the circuit is a 3rd stage and output of the whole circuit is $+6\,V_m$. For an *n* stage circuit, the output voltage would be $+2n\,V_m$. Therefore if we go on adding stages, higher and higher voltages can be obtained.

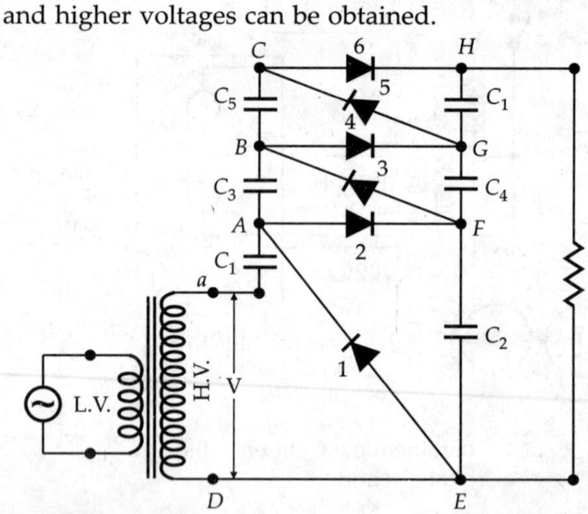


**Fig. 17.29** Cockcroft Walton circuit.

## 17.8 HIGH VOLTAGE D.C. TESTING OF CABLES

Figure 17.30 shows a circuit for cable testing. Current passes for only one half wave of the cycle. During this time the capacitor formed by cable cores and sheath is charged to a potential $V_m$ *i.e.*, the peak voltage of secondary winding of h.v. transformer. During the next half cycle, the potential of this cable capacitor remains the same and the potential of cathode rises to $V_m$ in the other direction and therefore the rectifier must be designed to withstand a voltage of $2\,V_m$.

The charging and leakage currents are indicated by a milliammeter as shown. A resistance of about 0.25 MΩ is connected in the anode circuit for protection against surge effects.

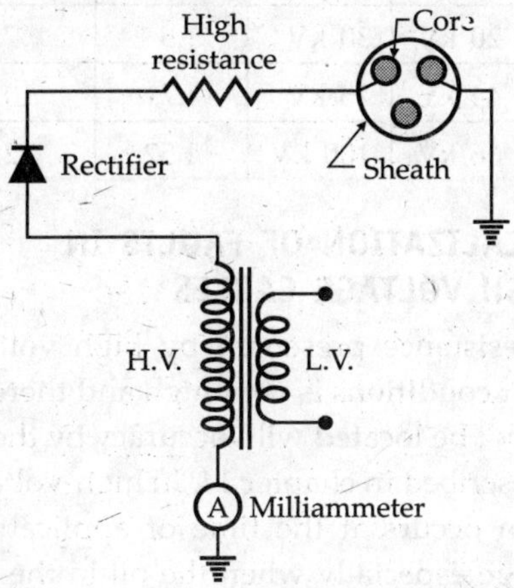


**Fig. 17.30** High voltage d.c. test on 3 core cable.

Three core cables are tested by two other connections as shown in Figs. 17.31(*a*) and 17.31(*b*).

For tests upto 200 kV, two rectifiers are used.

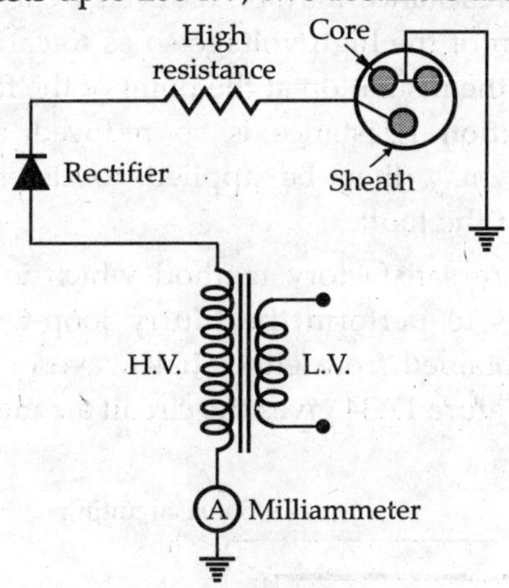


(*a*)

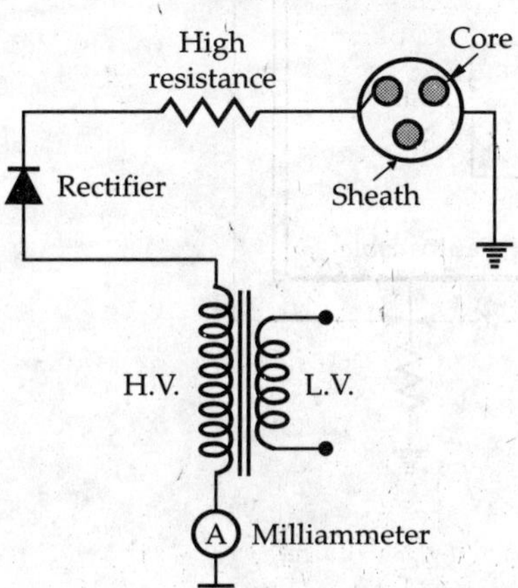


(*b*)

**Fig. 17.31** High voltage d.c. test on 3 core cable.

Figure 17.32 shows the connections for testing a 3 core cable with two rectifiers.

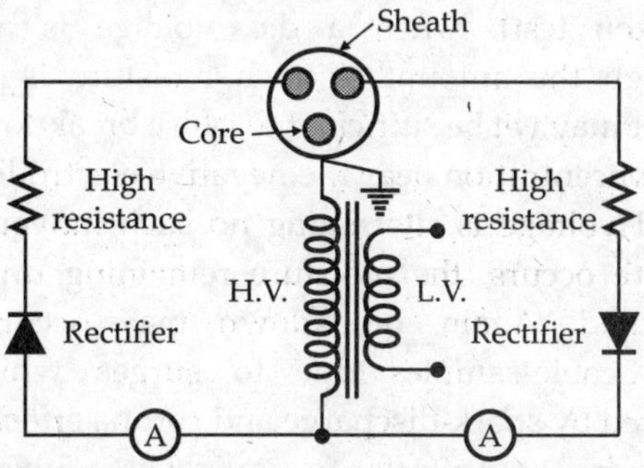


**Fig. 17.32** High voltage d.c. test on cable upto 200 kV using 2 rectifiers.

A voltage of $2V_m$ is obtained between the cores and $V_m$ between cores and earth.

For testing a single core cable, a voltage doubler circuit is used as shown in Fig. 17.33. The cables are usually tested in the field and therefore these testing sets for field duty are made portable. A small engine driven alternator is used for supplying the l.v. winding of the high voltage testing transformer. The voltage regulation is done by an auto tran[illegible]mer and fine adjustments are done by a choke coil connected in series with the primary winding (l.v.) circuit. The voltage is measured with an electrostatic voltmeter. A sphere gap is also used alongside.

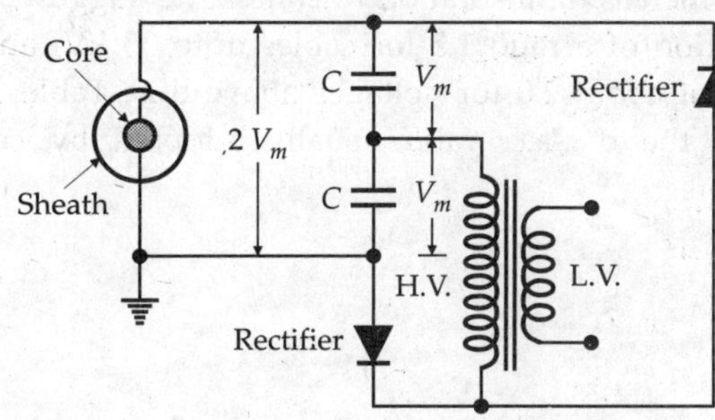


**Fig. 17.33** High voltage d.c. test on cables using voltage doubler circuit.

The testing circuit should be protected from voltage surges which cause breakdown of apparatus. Therefore care should be taken to increase the voltage gradually and to discharge the cable through a high resistance after the test. The discharging connection should be maintained for a considerably long period in order to avoid subsequent rise of voltage of cable which may result owing to dielectric absorption.

### 17.8.1 Equivalence of D.C. & A.C. Test Voltage

Owing to the "electric osmosis" effect, any moisture, which may exist within the cable dielectric, tends to move towards the negatively charged electrode (the sheath or one of the cores, depending upon the test), when a d.c. voltage is applied. Although the amount of such moisture is usually small, it may yet be sufficient to cause breakdown due to its concentration near the negative electrode. If the applied voltage is alternating no such movement of moisture occurs, the moisture remaining uniformly distributed. Again, breakdown may occur when testing cable-samples due to surges which are produced by spark-discharge and corona effects at the cable ends. These effects are more severe with a.c. than with d.c.

For these reasons it is obvious that there are other considerations beyond mere equivalence of potential gradient which must determine to what alternating voltage a given direct voltage is equivalent from the point of view of insulation breakdown.

N.A. Allen gives a table showing the ratio of d.c. to a.c. test voltages, quoted by various authorities as giving an equivalent breakdown test upon cables and dielectrics. For paper-insulated cables, although a d.c./a.c. ratio of 2.5 has been used, Allen suggests that a ratio of 1.5 to 2.0 would be more satisfactory.

Owing to the fact, also, that this ratio increases with increasing insulation thickness, he suggests the adoption of a ratio 1.5 for cables upto 33 kV, and a ratio of about 2.0 for voltages above this. Table 17.2 gives the d.c./a.c, ratio usually adopted by cable makers.

**Table 17.2** *D.C. and A.C. voltages for cables*

| *Working Voltage* | *Standard test Voltage* | | *Ratio D.C./A.C* | *Average thickness of dielectric* |
|---|---|---|---|---|
| | *A.C.* | *D.C.* | | |
| 11 kV | 20 kV | 30 kV | 1.5 | 7.62 mm |
| 22 kV | 44 kV | 75 kV | 1.71 | 10.2 mm |
| 33 kV | 66 kV | 100 kV | 1.52 | 12.7 mm |

## 17.9 LOCALIZATION OF FAULTS IN HIGH VOLTAGE CABLES

The resistance presented by high voltage cable under fault conditions is very high and therefore such faults cannot be located with accuracy by the methods already described in chapter 14. In high-voltage cables breakdown occurs at the time of application of the high voltage especially when the oil in the cable and cable joints breaks down and therefore presents a low resistance. However, when the applied voltage is withdrawn, the oil recovers its insulating property and the cable again presents high insulation resistance. The fault resistance may be reduced by a continued application of the high voltage so as to carbonize (or burn out) the insulation at the point of the fault. After the insulation resistance is so reduced ; ordinary loop-tests may then be applied to determine the position of the fault.

A more satisfactory method which is generally adopted is to perform the Murry loop-test at high voltage obtained from either half-wave or full-wave rectifier. Figure 17.34 gives the circuit arrangement for

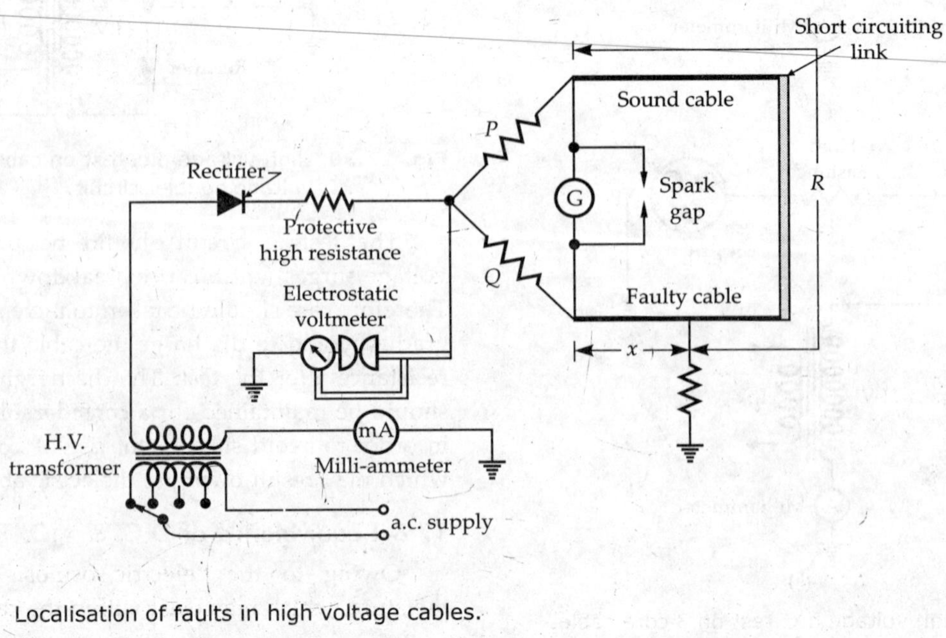


**Fig. 17.34** Localisation of faults in high voltage cables.

localization of faults in high-voltage cables. The rectified supply at about 60 kV is obtained from a high-vacuum diode. A higher voltage, if desired, may be obtained with two valves. A step up high voltage transformer is used to feed the rectifier. A current limiting high resistance and a milliammeter are included in the supply circuit for the limitation and measurement of the fault current respectively. An electrostatic voltmeter is used to measure the voltage at the junction point of the ratio arms. A loop is made connecting together the faulty and sound cables with the help of a short circuiting link. For varying the ratio arms ($P$, $Q$) a highly insulated slide-wire, having a sliding contact which can be operated by a long insulating handle is used. 'The spark gap is used for the protection of the galvanometer. The procedure adopted for the test is detailed below : The circuit is energised and the tap at the primary of the H.V. transformer is adjusted to give a low fault current of the order of 5 mA at the time of break down of the cable. The bridge is balanced by varying $P/Q$ ratio. The tap of the primary of H.V. transformer is then changed to increase the fault current to about 50 mA. The bridge is rebalanced. It may be mentioned at the time of break-down, a very large drop of voltage occurs across the protective high resistance and therefore the voltage across the cable is low. However, if the fault clears, the voltage across the cable would rise suddenly and it is on this account highly insulated slide-wire is used for the safety of the operator.

The position of the fault can be obtained using equations given for Murray loop test in Chapter 14. (see Eqns. 14.53 and 14.54, on page 446)

## 17.10 HIGH FREQUENCY HIGH VOLTAGE TESTS

It has been explained earlier that the behaviour of insulating materials working at high frequencies is different from that at ordinary power frequencies. The difference is mainly due to increased dielectric loss at high frequencies. This dielectric loss produces internal heating with the result that the breakdown strength of insulating materials is lower at high frequencies. High frequency tests are also useful for detection of in homogeneity in compound filled porcelain insulators. Two types of high frequency tests are done :

1. ***Tests with undamped high frequency oscillations.*** These tests are done on porcelain insulators used for communication purposes. Undamped high frequencies are not encountered in power systems and the insulators used therein are not subjected to these tests.

Undamped high frequency voltages upto a frequency of 30 kHz may be produced by alternators. Arc generators can also be used for the purpose but they have the disadvantage that smooth voltage regulation is not possible.

Usually electronic circuits are used for production of high frequency high voltages. Equipments have been developed which give an output of 150 kV and a frequency of 100 kHz.

2. ***Tests with dumped high frequency oscillations.*** Figure 17.35 shows a circuit for production of high frequency damped oscillations. This circuit uses a Tesla coil which is in the form of a high voltage transformer. The capacitor $C_p$ on the primary side is

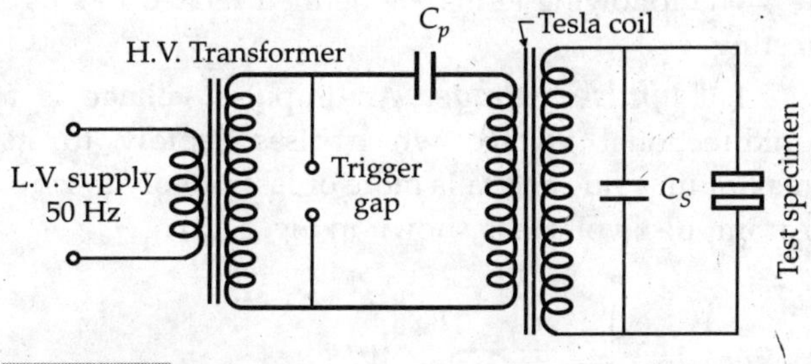


**Fig. 17.35** High frequency high voltage testing.

charged to a potential which causes a trigger gap to breakdown with the result that the capacitor discharges. Due to this discharge, a train of damped high frequency oscillations is produced in the circuit containing $C_p$, trigger gap and the primary of the Tesla coil. The charge and discharge of capacitor $C_p$ takes place two times in one voltage cycle. Therefore, there will be 100 trains of frequency oscillations (Fig. 17.36) in a second for a 50 Hz supply.

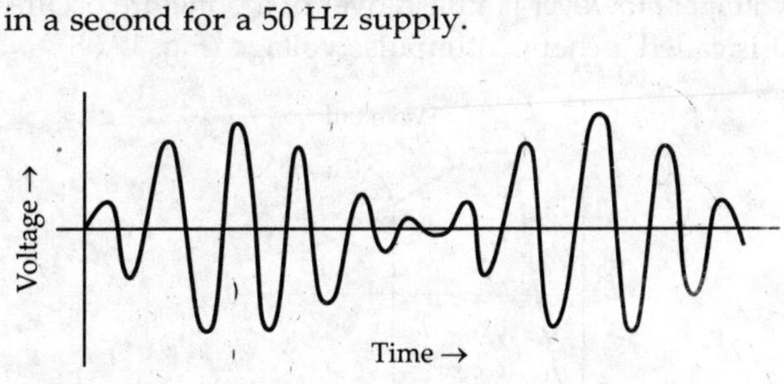


**Fig. 17.36** Damped high frequency oscillations.

The frequency of oscillations

$$f = 1/(2\pi L_p C_p)\text{Hz}$$

This frequency is of the order of 100 kHz. The highest frequency from which any trouble can be expected in a transmission system is about 80 kHz and hence 100 kHz is the maximum frequency required for testing.

Since the primary winding of the Tesla coil carries an oscillatory current, oscillations are induced in the secondary winding. If the primary and secondary windings are tuned *i.e.,* $L_pC_p = L_sC_s$ the frequency of oscillations in the secondary winding will be the same as in the primary.

These damped high frequency oscillations are impressed upon the test specimen as shown in Fig. 17.35.

A sphere gap is used for voltage measurement and a cathode ray oscilloscope may be used to record the waveform of secondary voltages of the Tesla coil.

## 17.11 SURGE (IMPULSE) TESTING

The following terms are defined related to surge testing.

1. ***Impulse voltage.*** An impulse voltage is a unidirectional voltage which rises rapidly to its maximum value and falls more or less rapidly to zero. An impulse voltage is shown in Fig. 17.37.

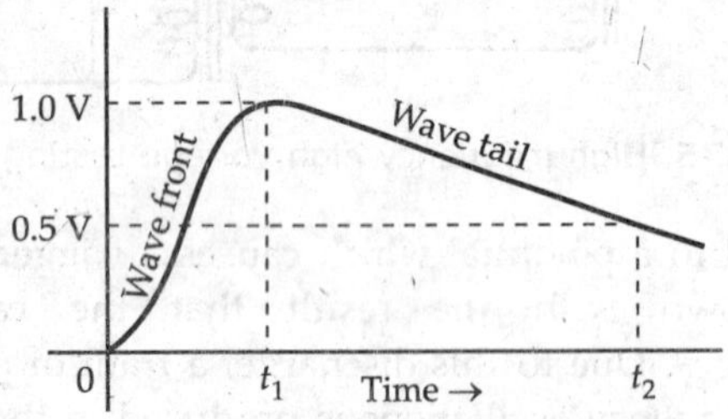


**Fig. 17.37** Full surge wave.

If an impulse voltage develops without causing a flash-over or puncture, it is called a full impulse voltage. However if a flash-over or a puncture occurs, it is called a chopped impulse voltage (Fig. 17.38).

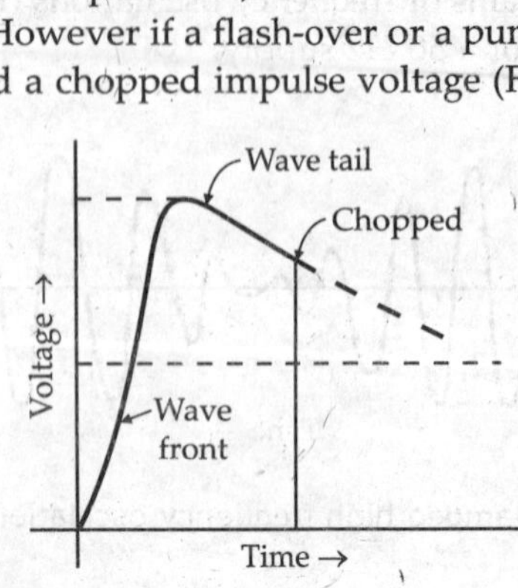


**Fig. 17.38** Chopped surge wave.

A full impulse voltage is characterized by two time intervals $t_1$ and $t_2$ defined below. The impulse voltage is conveniently expressed as a $t_1/t_2$ wave. Time interval $t_1$ is the time taken by the impulse voltage to rise to its peak value and $t_2$ is the time taken to fall to 50 percent of its peak value (see Fig. 17.37).

Normally 1/50 μs impulse voltages are used for testing. This voltage indicates that it rises to its peak value in 1 μs and falls to 50 percent of its peak value in 50 μs.

2. ***Impulse flash-over voltage.*** *50 percent impulse flash-over voltage* : It is the peak value of that impulse voltage which causes flashover of the object under test for about half the number of applied impulses.

3. ***Impulse puncture voltage.*** The impulse puncture voltage is the peak value of that impulse voltage which causes the puncture of the specimen under test when puncture occurs on wave tail and is the value of the voltage at the instant of puncture when puncture occurs on the wavefront.

4. ***Impulse ratio for flashover.*** The impulse ratio for flashover is the ratio of impulse flashover voltage to the peak value of power frequency flashover voltage.

The impulse ratio for puncture can also be defined in a similar manner.

### 17.11.1 Basic Impulse Generator Circuit

Consider the case of a fundamental impulse generator as shown in Fig. 17.39 from which all impulse generators are derived. $C$ is a capacitor charged from an independent d.c. source $E$, and discharged through inductance $L$ and resistance $R$. The impulse voltage is developed across resistance $R$. The capacitor is charged to a voltage $E$ and the switch $S$ is closed and the capacitor discharges through $R.L.$

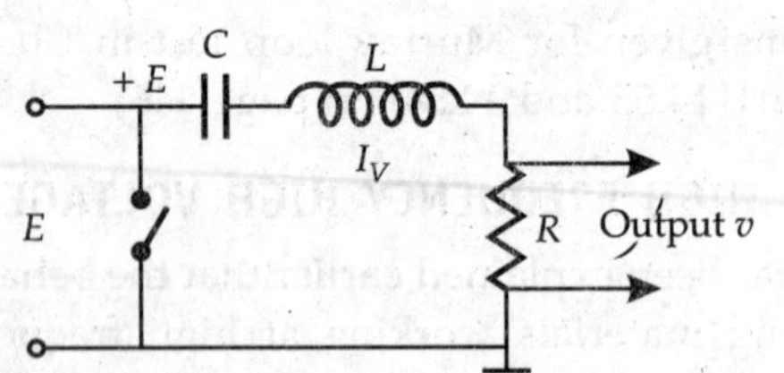


**Fig. 17.39** Basic impulse generator circuit.

We will be using Heaviside's operational calculus for the analysis.

Voltage across resistance $R$ is :

$$v = \frac{R}{R + pL + 1/pC} EI = \frac{pCR}{p^2LC + pRC + 1} EI$$

The solution of the above equation is

$$v = \frac{M(0)}{N(0)} + \sum_{k=1,2....} \frac{M(p_k)e^{pt}}{p_k N^t(p_k)},$$

Now $M(p) = EpCR$,

$\therefore \quad M(0) = 0, \quad$ and $\quad N(p) = p^2LC + pRC + 1$

Differentiating $N(p)$ we have :

$$N'(p) = 2pLC + RC$$

The roots of equation $p^2LC + pRC + 1 = 0$

are
$$p_1 p_2 = \frac{-RC \pm \sqrt{R^2C^2 - 4LC}}{2LC}$$

$$= -\frac{R}{2L} \pm \sqrt{\frac{R^2}{4L^2} - \frac{1}{LC}} = -a \pm b$$

$\therefore$ Voltage across resistance $R$.

$$v = \frac{M(0)}{N(0)} + \sum_{k=1,2....} \frac{M(p_k)e^{pt}}{p_k N'(p_k)}$$

$$= 0 + \frac{Ep_1CRe^{p_1t}}{p_1(2p_1LC + RC)} - \frac{Ep_2CRe^{p_2t}}{p_2(2p_2LC + RC)}$$

$$= E\frac{R}{L}\left(\frac{e^{p_1t}}{2p_1 + R/L} + \frac{e^{p_2t}}{2p_2 + R/L}\right)$$

$$= E\frac{R}{L}\left(\frac{e^{-(a-b)t}}{-2(a-b) + R/L} + \frac{e^{-(a+b)t}}{-2(a+b) + R/L}\right)$$

$$= E\frac{R}{L}\left[\frac{1}{b}(e^{-(a-b)t} - e^{-(a+b)t})\right]$$

$$= E\frac{a}{b}[e^{-\alpha t} - e^{-\beta t}]$$

where $\alpha = a - b$, and $\beta = a + b$.

This is a typical exponential wave of $(e^{-\alpha t} - e^{-\beta t})$ form. If values of $R$, $L$ and $C$ are inserted, the shape of wave can be found by plotting.

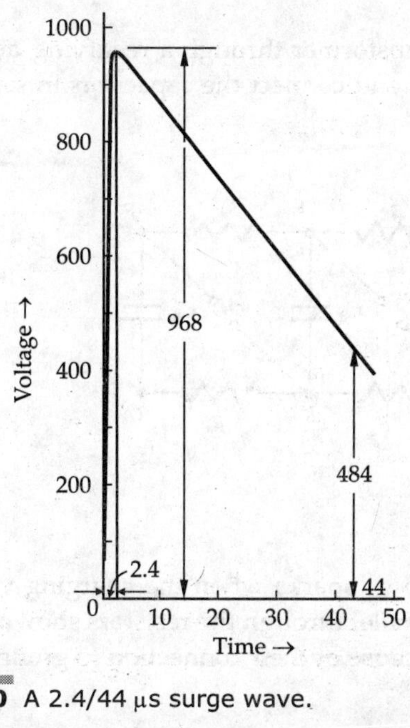


**Fig. 17.40** A 2.4/44 μs surge wave.

Let $R = 30\,\Omega$, $L = 15\,\mu$H and $C = 2\,\mu$H

With these values $a$, $b$, $\alpha$ and $\beta$ can be calculated.

$a = 10^6$, $b = 0.983 \times 10^6$, $\alpha = 0.017 \times 10^6$ and $\beta = 1.983$

$$\therefore \quad v = 1.017\,E(e^{-0.017\times10^6\,t} - e^{-1.983\times10^6\,t})$$

and if $\quad E = 1000$ V and $t$ is in μs.

$$v = 1017(e^{-0.017t} - e^{-1.983t})$$

The wave is plotted in Fig. 17.40. With these values $R$, $L$ and $C$ we obtain an approximately 2.4/44 μs wave. By adjustment of parameters $R$, $L$ and $C$ any desired wave shape may be obtained.

### 17.11.2 Single Stage Impulse Generators

Impulse generators can be single stage or of multi-stage construction. The single stage generator has a simple circuit of the type shown in Fig. 17.41. This circuit is mainly employed in low voltage applications.

The choice between a single or a multi-stage generator depends upon the output voltage desired and the type of equipment providing d.c. input to the generator.

Where expense is not the main consideration, single stage generators may be used upto a voltage output of 200 kV and even higher. However, small d.c. equipments with ratings of 4 kV and 10 kV are relatively inexpensive so that single stage generators are readily used at these voltages. For that matter multistage impulse generators with outputs of 100 kV are readily made using (say) 10 stages and a charging voltage of 10 kV.

Figure 17.41 shows a 200 V single stage impulse generator. The design is the same whether the output voltage is .10 kV or merely 200 V.

The proper single generator circuit is shown by thick lines.

A transformer supplies a rectifier which may be a value, a metal or a solid state rectifier. For rectification any of the circuits shown in Figs. 17.24 to 17.27 may be used. Resistance $R_1$ between charging equipment and generator circuit is a limiting resistance which is used to protect the rectifiers against damage when switch is closed.

The voltage across the capacitor is measured by an electrostatic voltmeter.

The recording equipment (C.R.O. etc.) may be directly connected if the generator voltage is low. In case the generator voltage is high, connections for recording purposes must be taken through a potential divider arrangement.

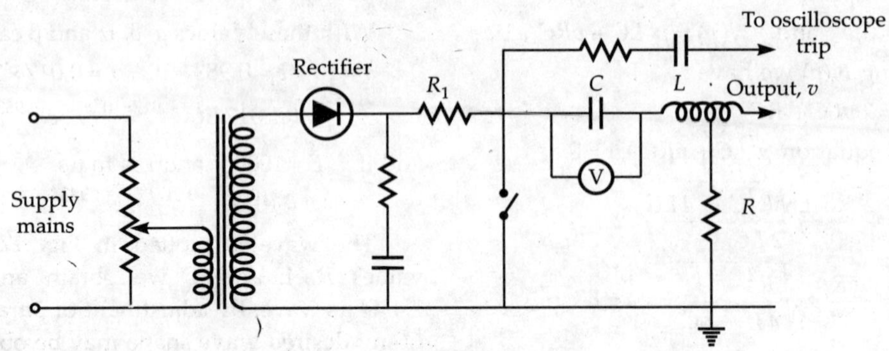


**Fig. 17.41** Single stage impulse generator.

### 17.11.3 Multi-stage Impulse Generator

The origin of the multi-stage impulse generator circuit in which capacitors are charged in parallel and discharged in series is due to Prof. Marx. A Marx circuit (with modifications by Goodlet) is shown in Fig. 17.42.

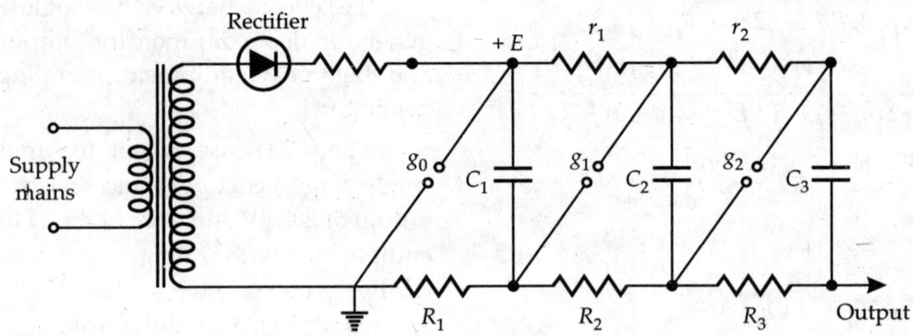


**Fig. 17.42** Marx circuit.

$C_1, C_2, C_3$... are capacitors which are connected in parallel with highest resistances $R_1, R_2$ and $r_1, r_2$ ... etc. between them $g_0, g_1, g_2$ are the trigger spark gaps.

The capacitors are charged, in parallel, from a high voltage transformer through a rectifying device. At a certain voltage depending upon their setting, the gaps break down and connect the capacitors in series.

The operation of the circuit is explained below :

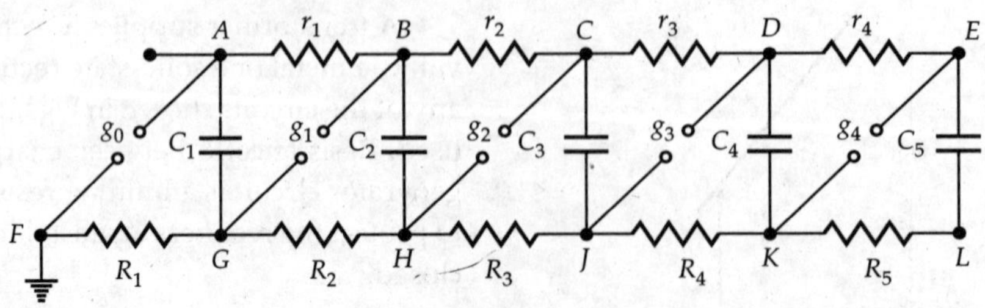


**Fig. 17.43** Charging circuit.

The gaps $g_0, g_1, g_2$ etc. are set to spark one after another. Gap $g_0$ sparks when the charging voltage has attained a voltage + $E$. All the capacitors will therefore charge in parallel through the resistors shown until they reach a potential of + $E$. Their lower plates are at earth potential because of their connection to ground through resistances.

As soon as a voltage + $E$ is attained, gap $g_0$ breaks down and there is momentary redistribution of charge potential on capacitor $C_1$. Its top plate falls to zero potential and its lower plate potential becomes $-E$ suddenly. Therefore a potential of 2 $E$ (*i.e.*, $-E$ to $+E$) suddenly appears across gap $g_1$ and this gap then instantly breaks down. This process is repeated, the potential of lower plates of capacitors of successive stages becoming $-2E, -3E, ..., -nE$, if there are $n$ stages.

Thus if there are $n$ stages, output voltage is $-nE$ if the charging (input) voltage is $+E$. Figures 17.43 and 17.44 show the charging and discharging circuits respectively.

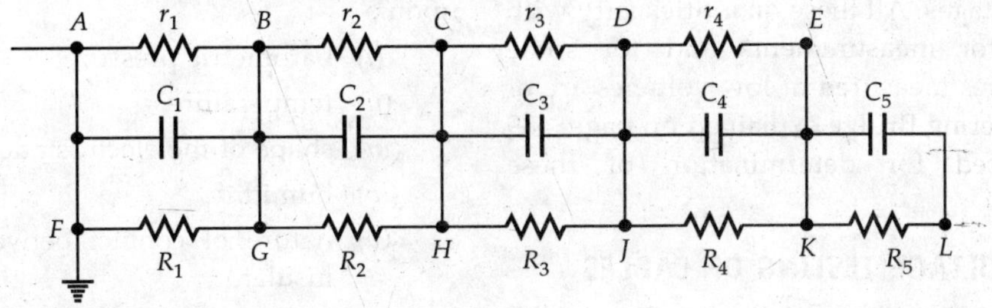


**Fig. 17.44** Discharging circuit.

### 17.11.4 Impulse Testing of Transformers

The impulse voltages at which transformers are tested are specified in the IS specifications. The impulse test voltage depends upon the rated voltage of the transformers.

The procedure for testing is as follows :

1. Prior to application of surge, the impulse generator is adjusted to give required "1/50" μs wave at 75% of required test voltage. The lightning arrester, the voltage measuring equipment and recording devices are connected in circuit. From these calibrations, the generator settings for actual tests are estimated and during tests, no change is made in any part of the circuit.
2. For full wave tests, the generator is set to give full test voltage (specified) and lightning arrester is disconnected to prevent flashover.
3. For chopped wave tests the lightning arrester is connected in circuit, and to insure that it chops on the tail of the wave, the impulse generator is set to give 15% in excess of specified test voltage. Two shots are taken.
4. Another full wave shot is taken.

During impulse test, observations are made to detect the failure of insulation (it is done by taking oscillograms of applied voltage and impulse current).

In Europe chopped wave tests are not done. Impulse tests are only type tests and are done by specific agreement between manufacturer and purchaser. Although when a transformer is made according to new specifications it has to stand an impulse test.

## 17.12 TESTING OF INSULATING MATERIALS

The dielectric strength of an insulating material may be defined as the minimum voltage gradient at which electrical failure or breakdown occurs under prescribed conditions. Though breakdown of insulation may be caused by leakage current, it is generally caused by voltage stress exceeding the dielectric strength of the material. The dielectric strength is expressed in terms of voltage per unit thickness.

There are wide variations of dielectric strength as it varies with the following factors :

(*i*) structure homogeneity and quality of sample,

(*ii*) shape of electrodes,

(*iii*) thickness of sample,

(*iv*) waveform of test voltage and duration and rate of application of voltage,

(*v*) moisture content of specimen, and

(*vi*) heat capacity of electrodes.

17

Dielectric strength is not proportional to thickness, but is relatively greater for thin sheets. It is also a function of time of application, being greater lesser is the time.

Measurements of capacitance, permittivity, dielectric loss and loss angle of insulating materials is very important. These measurements are usually made at high voltages. All these quantities vary with voltages used for measurements and for some applications values measured at low voltages are of no use at all. Schering Bridge explained on page 489 is normally used for determination of these quantities.

## 17.13 HIGH VOLTAGE TESTING OF CABLES

The tests carried out in the field after the installation of cables have been explained earlier in this chapter. Here we will describe some acceptance tests which convince the customer about the reliability of the cable when put in service. The dielectric loss and loss angle of a cable rather than its breakdown strength are an index of its reliability in service. Therefore measurements of these quantities is essential.

It has been found that the dielectric loss in a cable is not proportional to square of the voltage as is the case in most dielectrics. Also the loss angle of a cable does not remain constant. The dielectric loss increases in proportion to some power of voltage greater than second and the power factor rises with voltage.

Therefore measurements of loss angles with different voltages is made and a curve plotted between loss angle and voltage. The curve should be flat and if there are variations in the loss angle, the cable will not work satisfactorily. Also the loss angle should not change very much with temperature.

Breakdown tests of the cables with voltages applied for different intervals of time are also done and time versus breakdown voltage curves are plotted.

## 17.14 HIGH VOLTAGE TESTS ON PORCELAIN INSULATORS

The following tests are applied to porcelain insulators :

1. ***Power frequency tests.*** The frequency used is between 25 to 100 Hz of approximately sinusoidal waveform with peak value of voltage not to exceed 1.45 times the rms value.

2. ***Flashover tests.*** Porcelain insulators are designed in such a way that their puncture voltage is higher than the flash-over voltage. This means that in case of a sudden over-voltage, the air at insulator surface breaks down earlier and thus the insulator is not punctured and is saved from total destruction. Therefore flash-over tests are very important in the case of porcelain insulators.

The flash-over voltage for an insulator depends upon :

(*i*) barometric pressure,

(*ii*) temperature,

(*iii*) shape of the electrostatic field,

(*iv*) humidity,

(*v*) nature of contact between electrodes and insulator.

The various flash-over tests applied are :

(*i*) *50 percent dry impulse flash-over tests with 1/50 μs impulse voltage.*

(*ii*) *Dry flash-over and dry one minute test.* In this test the voltage is gradually raised to a specified value in approximately 10 seconds and is maintained for one minute. The voltage is then gradually raised until a flash-over occurs.

(*iii*) *Wet flash-over and one minute rain test.* During this test the insulator is sprayed constantly with water and the same procedure as for dry flash-over and dry one minute test is followed.

3. ***Sample tests.*** These include mechanical tests, temperature cycle tests, porosity tests, puncture tests etc.

4. ***Routine tests.*** These tests are applied to all the insulators. The insulators are initially subjected to a low voltage which is rapidly increased till a flash-over occurs every few seconds. The voltage is maintained at this value for a minimum of 5 minutes or if failure occurs, for five minutes after the last punctured piece has been removed. The voltage is gradually reduced to $\frac{1}{3}$rd of its value before supply is switched off.

The porcelain insulators used for communication purposes are tested with high frequency undamped oscillations.

## 17.15 TESTING OF ELECTRIC STRENGTH OF INSULATING OILS

An extract of IS specifications 6792-1972 "Method for determination of Electric strength of insulating oils" is given on next page.

***General.*** This method is primarily intended for the new insulating oils at the time of their delivery. However, in principle, it is applicable to all classes of new and used oils for transformers, circuit breakers, oil filled cables and capacitors.

The test method consists of subjecting the oil contained in a specified apparatus, to an a.c. electric field with a continuously increasing voltage till the oil breaks down.

***Test cell.*** The test cell is transparent and non-absorbent. It is made of glass or plastic and has an effective volume of 200-500 ml. It is preferably closed. Two types of cells are used and they are shown in Figs. 17.45 and 17.46.

***Electrodes.*** The electrodes may be made of copper, brass, bronze or stainless steel. The polished electrodes are either spherical (12.5 to 13.0 mm diameter) as shown in Fig. 17.45 or spherical surfaced of the shape and dimensions as shown in Fig. 17.46. The electrodes are mounted 2.5 mm apart with an accuracy of ± 0.1 mm. The axis of the electrodes is immersed to a depth of approximately 40 mm. The electrodes are replaced as soon as pitting caused by discharges is observed.

***Test method.*** The oil is poured in the dry and clean cell and is stirred so that the impurities are evenly distributed and there are no visible air bubbles. The temperature of oil is the same as that of ambient air preferably 27°C, but a temperature range of 15°C to 35°C is acceptable.

An increasing a.c. voltage of frequency 40-60 Hz is applied to the electrodes. The rate of increase of voltage is uniform and is approximately equal to 2 kV/s starting from zero to the value producing breakdown. The breakdown voltage is the voltage reached during the test at the time first spark occurs between electrodes. The test is carried out six times on the same cell filling. The electric strength is the arithmetic mean of the six results obtained.

***Electrical apparatus.*** The characteristics of the electrical apparatus comply with the following requirements.

***Transformer.*** The test voltage is obtained by using a step up transformer supplied form a.c. The low voltage side has a frequency of 40 to 62 Hz. The primary voltage is gradually increased, either manually or by an automatic control device. The voltage applied to the electrodes has approximately a



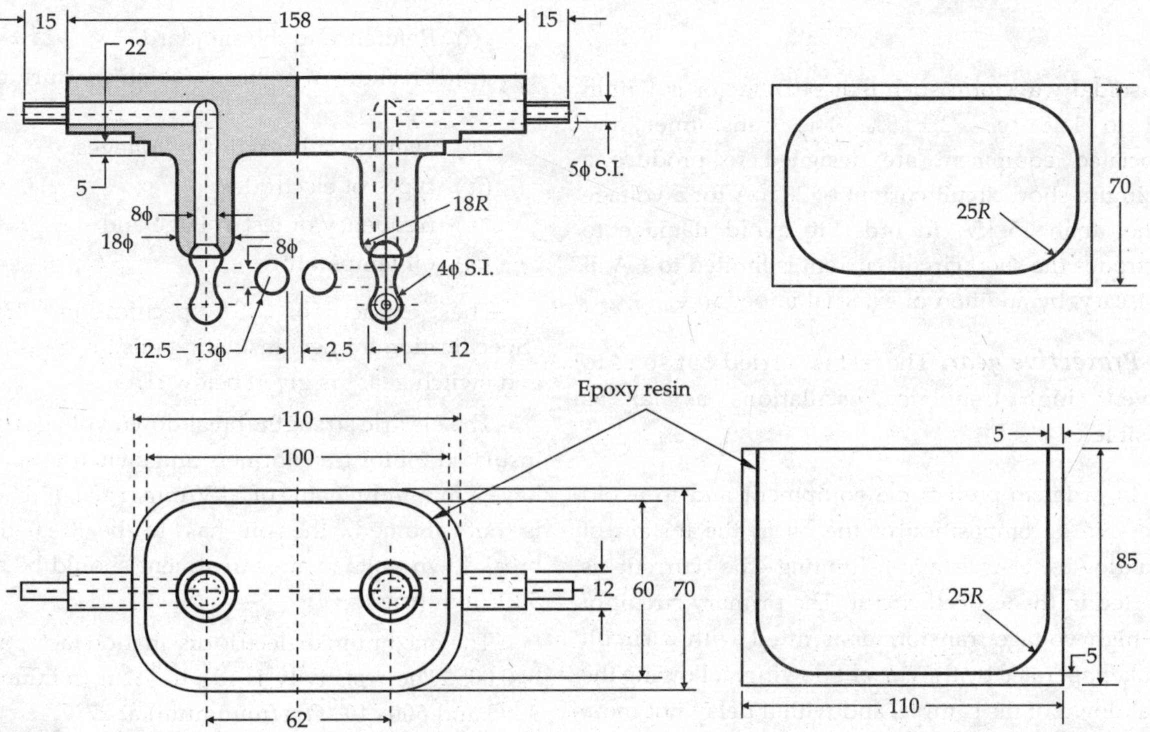


**Fig. 17.45** Test cell (All dimensions in mm)

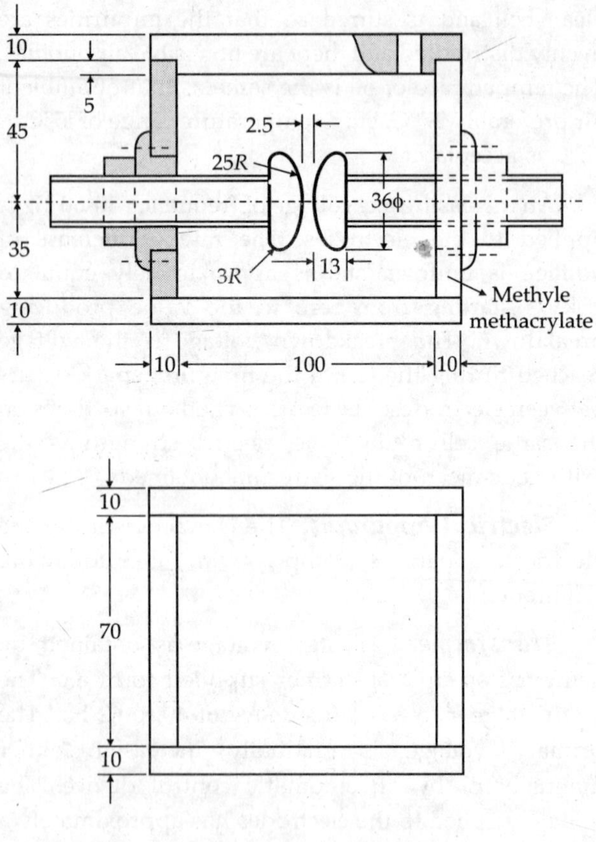


**Fig. 17.46** Test cell (All dimensions in mm).

sinusoidal waveform such that peak factor is within 1.34 to 1.48 *i.e.*, $\sqrt{2} \pm$ 5%. The transformer and associated equipment are designed to produce a minimum short circuit current of 20 mA for a voltage higher than 15 kV. In order to avoid damage to electrodes the short circuit current is limited to 1 A, if necessary, by addition of external impedance.

***Protective gear.*** The test is carried out so as to prevent high frequency oscillations as far as possible.

In order to protect the equipment and to avoid excessive decomposition of the oil at the instant of breakdown, a resistance, limiting the current is inserted in the test cell circuit. The primary circuit of the high voltage transformer is fitted with a circuit breaker operated by the current flowing, following the breakdown of the sample, and with a delay not more than 0.02 s. The circuit breaker is fitted with a no volt coil to protect equipment.

***Voltage regulation.*** Voltage regulation may be insured by one of the following methods :

(*i*) Variable ratio auto-transformer,
(*ii*) resistance type voltage regulator,
(*iii*) generator field excitation and
(*iv*) induction regulator.

Preference is given to an automatic system for increasing voltage, since it is difficult to obtain manually a reasonably uniform rate of voltage rise as a function of time.

***Measurement of test voltage.*** For the purpose of this standard, the magnitude of test voltage is defined as its peak value divided by $\sqrt{2}$. The peak value of the test voltage is measured by a peak voltmeter. The test voltage is determined by a voltmeter suitably connected to the input or output side of the testing transformer or to a voltmeter suitably connected to input or output side of the testing transformer or to a voltmeter winding provided thereon and calibrated against a sphere gap in place of the test cell.

The voltage should be measured to an accuracy better than ±4%.

**Results.** The test report should include

(*i*) Reference to the standard,
(*ii*) breakdown voltages obtained during each test,
(*iii*) average of breakdown voltages,
(*iv*) types of electrodes,
(*v*) frequency of test voltage and
(*vi*) oil temperature.

The extract of IS specification 335-1972 "Specification for new insulating oils for transformers and switch gear" is given below :

The electric strength (breakdown voltage) of new insulated oil for transformers and switch gear should have a minimum value of 3 kV/mm (rms). If this value is not attained, the oil has to be treated. The breakdown voltage after treatment should be atleast 5 kV/mm (rms).

The maximum dielectric dissipation factor at 90°C is 0.005. The resistivity is $13 \times 10^{10}\,\Omega$m (minimum) at 90°C and $500 \times 10^{10}\,\Omega$m (minimum) at 27°C.

Permittivity should be approximately 2.2 (– 0.001) at 60°C.

# Review Questions

1. Why is high voltage testing important for electrical power apparatus and insulating materials ? Describe briefly the different high voltage tests done on electrical apparatus.
2. Describe the constructional details of a H.V. transformer and explain the reasons for difference in construction of H.V. testing transformers and power transformers.
3. Explain the reasons for connecting H.V. transformers in cascade. Explain, with the help of a diagram, the cascade connections of two transformers.
4. Describe the different methods of voltage control used for high voltage test circuits. Explain their applications, advantages and disadvantages.
5. Draw the circuit diagram for high voltage testing at power frequencies. Explain the functions of the control and protective devices.
6. Describe the different electrostatic type of voltmeters used for measurement of r.m.s. value of high voltages. Explain their advantages and disadvantages.
7. Explain the following potential divider methods for measurement of r.m.s. value of high voltages,

   (*i*) resistance potential divider,

   (*ii*) capacitive potential divider and

   (*iii*) compensated capacitive potential divider.
8. Describe the use of sphere gaps for measurement of peak voltages. Discuss the advantages and disadvantages of using sphere gaps. Also describe the precautions taken when using sphere gaps for measurement of high voltages.
9. Describe the rectified capacitor charging current method for measurement of peak values of high voltages. Describe its advantages and limitations.
10. Describe the working, advantages and disadvantages of the following types of voltmeters used for measurement of peak values of high voltages.

    (*i*) Ryall crest voltmeter,

    (*ii*) Ionic wind voltmeter.
11. Describe the Cockcroft Walton circuit for production of high voltage d.c.
12. Explain why transmission lines and cables, which are used for power transmission and distribution of a.c. voltages, are tested with high voltage d.c. Describe the circuits used for high voltage d.c. te-ting of cables. Describe why is it necessary to establish an equivalence between d.c. and a.c. test voltages.
13. Describe the circuit of a single stage impulse generator. Describe how surge voltages are generated by applying a step voltage to an RLC series circuit.
14. Describe the working of Marx circuit used for impulse testing. Explain the impulse testing of transformers.
15. Describe the various high voltage tests that are carried out on porcelain insulators.
16. Explain the method for testing the dielectric strength of insulating oil according to IS specifications.

17

# Objective Type Questions

***Tick (√) the most appropriate answer***

1. A.C. transmission lines and cables are tested by using d.c.

   (*a*) they may be used for d.c. if required

   (*b*) they draw high values of charging current and hence would require a large sized h.v. transformer if tested with a.c.

   (*c*) h.v. testing transformers cannot be used for long times that are required for testing

   (*d*) all the above.
2. High frequency high voltage tests are carried out on porcelain insulators because

   (*a*) they can withstand an a.c. voltage at low frequency but fail to withstand that voltage when applied at high frequency

   (*b*) the insulators, though used on power frequency circuits, may be subjected to high frequency voltages on account of disturbances caused by switching which produce high dielectric loss and heating that may cause failure of insulator at a lower voltage (*c*) the insulators may be used for communication purposes wherein the frequencies may be of the order of MHz

(*d*) all the above.

3. If the sizes of power transformers and high voltage testing transformers of equal kVA rating are compared :
   (*a*) the size of a h.v. testing transformer is greater than that of a power transformer
   (*b*) the size of a h.v. testing transformer is smaller than that of a power transformer
   (*c*) the size of a h.v. testing transformer is the same as that of a power transformer.
4. Cascade connection of transformers is normally used when a single transformer cannot be used because the minimum test voltage :
   (*a*) is about 100 kV (*b*) is about 500 kV
   (*c*) is about 1000 kV
   (*d*) all the above.
5. The voltage control circuits do not use resistance potential dividers because :
   (*a*) they involve a large power loss
   (*b*) they cause distortion of waveform
   (*c*) they do not give a smooth variation of voltage
   (*d*) they have non-linear characteristics.
6. Sphere gaps are used for measurement of :
   (*a*) instantaneous values of voltage
   (*b*) r.m.s. values of voltage
   (*c*) peak values of voltage
   (*d*) average values of voltage.
7. A Cockcroft Walton circuit for generation of high voltage d.c. has $n$ stages. If the voltage of the secondary winding of the high voltage transformer is $V_m$, the output voltage of the circuit is :
   (*a*) $nV_m$ (*b*) $2(n-1)V_m$
   (*c*) $2(n+1)V_m$ (*d*) $2nV_m$
8. The impulse voltage wave shape is defined as $t_1/t_2$ where :
   (*a*) $t_1$ = time taken by the impulse voltage to rise to its peak and
   $t_2$ = time taken to fall to zero
   (*b*) $t_1$ = time taken by the impulse voltage to rise to its peak and
   $t_2$ = time taken to fall to 50% of its peak
   (*c*) $t_1$ = time taken by the impulse voltage to rise to 50% of its peak and
   $t_2$ = time taken to fall to zero
   (*d*) none of the above.
9. In a Marx's circuit,
   (*a*) the capacitors are charged in series and discharged in series
   (*b*) the capacitors are charged in parallel and discharged in parallel
   (*c*) are charged in series and discharged in parallel
   (*d*) charged in parallel and discharged in series.
10. The electric strength of new insulating oil should have a minimum strength (r.m.s.) of
    (*a*) 30 kV/mm (*b*) 3 kV/mm
    (*c*) 50 kV/mm (*d*) 5 kV/mm

## *Answers*

1. (*b*) 2. (*d*) 3. (*a*) 4. (*b*) 5. (*a*) 6. (*c*) 7. (*d*) 8. (*b*) 9. (*d*) 10. (*b*)

# 18

# Magnetic Measurements

## 18.1 INTRODUCTION

Electrical and magnetic phenomena are closely associated with each other. The operating characteristics of electrical machines, apparatus and instruments are greatly influenced by the properties of ferromagnetic materials used for their construction. Therefore, magnetic measurements and a thorough knowledge of characteristics of magnetic materials is of utmost importance in designing and manufacturing electrical equipment.

The principal requirements in magnetic measurements are :

(*i*) The measurement of magnetic field strength in air.

(*ii*) The determination of *B-H* curve and hysteresis loop for soft ferromagnetic materials.

(*iii*) The determination of eddy current and hysteresis losses of soft ferromagnetic materials subjected to alternating magnetic fields.

(*iv*) The testing of permanent magnets.

Magnetic measurements have some inherent inaccuracies due to which the measured values depart considerably from the true values. The inaccuracies are due to the following reasons :

(*i*) the conditions in the magnetic specimen under test are different from those assumed in calculations ;

(*ii*) the magnetic materials are not homogeneous ;

(*iii*) there is no uniformity between different batches of test specimens even if such batches are of the same composition.

## 18.2 TYPES OF TESTS

Many methods of testing magnetic materials have been devised wherein attempts have been made to eliminate the inaccuracies. However, attention will be confined to a few basic methods of testing ferromagnetic materials. They are :

(*i*) *Ballistic tests.* These tests are generally employed for the determination of *B-H* curves and hysteresis loops of ferromagnetic materials. Direct current is used to provide an adjustable mmf to the magnetic circuit, and a ballistic galvanometer or flux meter is employed for the measurement of flux density.

The ballistic methods do not directly measure the value of flux density in a magnetic material corresponding to a particular value of magnetizing force but measure instead the changes in flux brought about by changes in magnetising force. These changes in flux density are either measured by a flux meter or a ballistic galvanometer.

(*ii*) *A.C. testing.* These tests may be carried at power, audio or radio frequencies. They give information about eddy current and hysteresis losses in magnetic materials.

(*iii*) *Steady state tests.* These are performed to obtain the steady value of flux density existing in the air gap of a magnetic circuit.

## 18.3 BALLISTIC TESTS

These tests are used for determination of flux density in a specimen, determination of *B-H* curves and plotting of hysteresis loop.

### 18.3.1 Measurement of Flux Density

The measurement of flux density inside a specimen can be done by winding a search coil over the specimen. This search coil is known as a "***B* coil**". This search coil is then connected to a ballistic galvanometer or to a flux meter. Let us consider that we have to measure the flux density in a ring specimen shown in Fig. 18.1. The ring specimen is wound with a magnetizing winding which carries a current $I$. A search coil of convenient number of turns is wound on the specimen and connected through a resistance and calibrating coil, to a ballistic galvanometer as shown in Fig. 18.1.

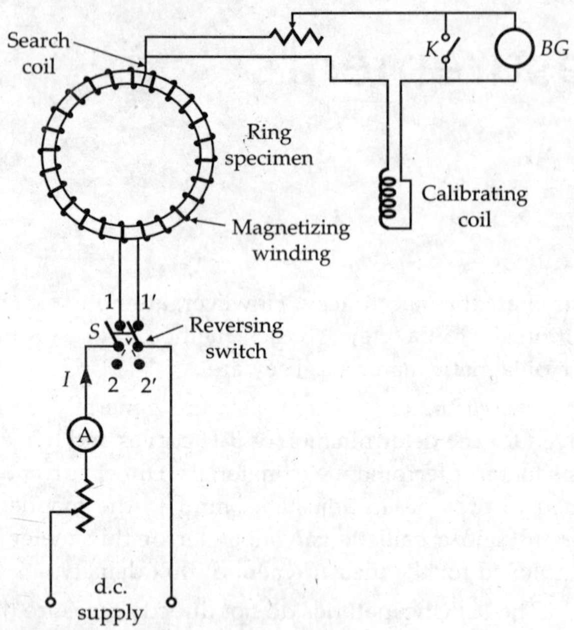


**Fig. 18.1** Measurement of flux density in ring specimens.

The current through the magnetizing coil is reversed and therefore the flux linkages of the search coil change inducing an emf in it. This emf sends a current through the ballistic galvanometer causing it to deflect.

Let

$\Phi$ = flux linking the search coil,

$R$ = resistance of the ballistic galvanometer circuit,

$N$ = number of turns in the search coil,

and $t$ = time taken to reverse the flux.

$\therefore$ Average emf induced in the search coil

$$e = Nd\Phi / dt = 2\,N\Phi / t.$$

Average current through the ballistic galvanometer is,

$$i = 2\,N\Phi / Rt$$

Charge passing is,

$$Q = it = 2\,N\Phi / R$$

Let $\theta_1$ be the throw of the galvanometer and $K_q$ be the constant of the galvanometer expressed in coulomb per unit deflection.

Charge indicated by ballistic galvanometer

$$= K_q \theta_1.$$

$$\therefore\ 2\,N\Phi / R = K_q \theta_1$$

or $\quad$ flux $\Phi = R\,K_q\,\theta_1 / 2\,N$

In a uniform field and with search coil turns at right angles to the flux density vector, we have, flux density

$$B = \frac{\text{flux}}{\text{area}} = \frac{\Phi}{A_s} = \frac{RK_q\theta_1}{2\,NA_s} \qquad \text{...(18.1)}$$

where $A_s$ = cross-sectional area of specimen.

Thus the flux density may be measured by a ballistic galvanometer by observing its throw.

***Correction for air flux.*** For the above calculations, we have assumed that the flux is uniform throughout the specimen, and that the effective area of cross-section of the search coil is equal to the cross-sectional area of the specimen. However, the search coil is usually of larger area than the specimen and thus the flux linking with the search coil is the sum of the flux existing in the specimen and the flux which is present in the air space between the specimen and the search coil.

$\therefore$ Observed value of flux,

= true value of flux in the specimen + flux in the air space between specimen and search coil.

or $$B'A_s = BA_s + \mu_0 H(A_c - A_s).$$

Hence, true value of flux density

$$B = B' - \mu_0 H\left(\frac{A_c}{A_s} - 1\right) \qquad \text{...(18.2)}$$

where $B'$ = observed (or apparent) value of flux density ; Wb/m$^2$.

$B$ = true value of flux density in specimen ; Wb/m$^2$,

$A_s$ = area of cross-section of specimen ; m$^2$,

$A_c$ = area of cross-section of coil ; m$^2$

### 18.3.2 Measurement of Value of Magnetising Force (*H*)

The magnetising force of a constant magnetic field may be measured by a ballistic galvanometer and a search coil. The value of $H$ inside a specimen can either be inferred from calculations involving data of magnetising coil and the specimen or from measurements made outside the specimen. It cannot be measured directly. If the magnetising force is to be determined in the air gap, the search coil is placed in the air gap itself. While testing ferromagnetic materials, the magnetising force, within the specimen may be determined by measuring the magnetising force on its surface, since the tangential components of the field are of equal in magnitude for both sides of the interface. The search coil, as positioned in Fig. 18.2, measures the value of flux density, $B_0$, in air. This search coil is called an **"H coil"**. While for the flux densities encountered in iron testing, there is usually no trouble in getting a good sensitivity by using $B$ coil of sufficient turns but there is some difficulty in securing adequate sensitivity in the '$H$ coil' placed at the surface. In the first instance, its cross-sectional area is much smaller than the coil surrounding the specimen (*i.e.*, $B$ coil) and then $H$ is not constant across the section. Secondly, the permeability of iron is very

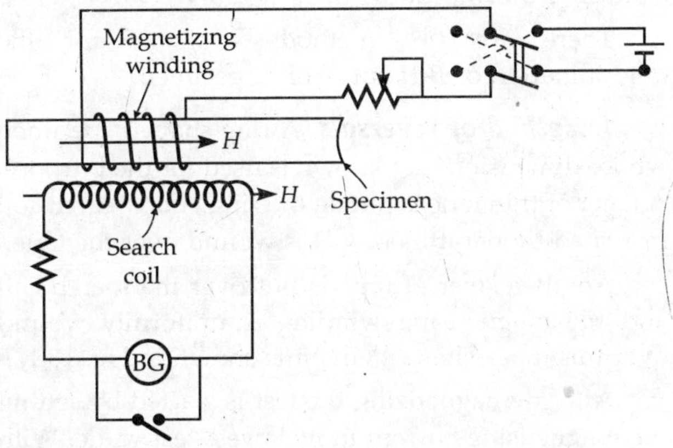


**Fig. 18.2** Determination of magnetising force $H$.

large as compared to that of air — may be some thousand times and therefore flux density $B_0$, in search coil, is very small compared to that in specimen. The value of flux density $B_0$ in $H$ coil is measured in a similar manner as described above for determination of $B$ in the specimen.

Magnetising force, $H = B_0 / \mu_0$ A/m ...(18.3)

### 18.3.3 Magnetic Potentiometer

This is a device for measurement of magnetic potential difference between two points. It can be shown that the line integral of magnetizing force $H$ produced by a coil of $N$ concentrated turns carrying a current $I$ is :

$$\int H\,dl = NI \quad ...(18.4)$$

around any closed path linking the coil. This is the circuital law of the magnetic field and forms the basis of magnetic potentiometer. A magnetic potentiometer may be used to determine the mmf around a closed path, or the magnetic potential difference between two points in a magnetic circuit.

A magnetic potentiometer consists of a one metre long flat and uniform coil made of two or four layers of thin wire wound unidirectionally on a strip of flexible non-magnetic material. The coil ends are brought out at the middle of the strip as shown in Fig. 18.3 and connected to a ballistic galvanometer. The magnetic potential difference between points $A$ and $B$ of the field is measured by placing the ends of the strip at these points and observing the throw of the ballistic galvanometer when the flux through the specimen is changed. Supposing the flux is changed by reversing the field.

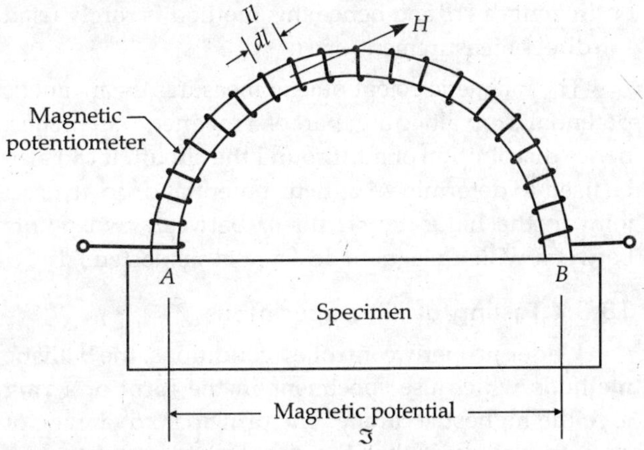


**Fig. 18.3** Magnetic potentiometer.

18

Let

$A$ = area of the strip ; $m^2$,

$n$ = number of turns per unit length of the strip,

$H_1$ = tangential component of the magnetising force ; A/m, and

$R$ = resistance of the ballistic galvanometer circuit.

∴ Flux linkages of a small infinitesimal portion of strip of length $dl$ are :

$$= \text{flux} \times \text{turns} = (\mu_0 H_1 A) n\,dl = \mu_0 H_1 A n dl.$$

Total flux linkages of the strip

$$= \int \mu_0 H_1 An\,dl = \mu_0 An \int H_1\,dl.$$

When the current in the magnetising winding is reversed, change in flux linkages

$$= 2\mu_0 An \int H_1 dl.$$

But $\int H_1\,dl = \mathfrak{I}$ = magnetic potential difference between $A$ and $B$

∴ Change in flux linkages $= 2\mu_0 An\mathfrak{I}$

$$\text{Charge} \quad Q = it = \frac{e}{R}t = \frac{2\mu_0 An\mathfrak{I}}{Rt}t = \frac{2\mu_0 An}{R}\mathfrak{I}\,\text{C} \quad ...(i)$$

Charge indicated by galvanometer

$$Q = K_q\,\theta_1 \text{ ; C} \quad ...(ii)$$

From (*i*) and (*ii*), we get,

Magnetic potential difference

$$\mathfrak{I} = \frac{R K_q \theta_1}{2 \mu_0 A n} \quad \text{...(18.5)}$$

The value of constant of galvanometer, $K_q$, can be found with the help of a calibrating circuit.

The above analysis is only valid if the portion *AB* of the specimen does not carry any part of magnetising winding. As under such condition (*i.e.*, no magnetising winding on portion *AB*), it is impossible to obtain uniform values of *H* over the entire length *AB* and hence this method is rarely used for direct measurement of *H*.

The magnetic potentiometer measures the magnetic potential drop along any part of a magnetic circuit and hence distribution of mmf round the circuit. It can also be used to determine magnetic potential drop across a joint in the magnetic circuit or between two points between which magnetic leakage is suspected.

### 18.3.4 Testing of Ring Specimens

Under properly controlled conditions, the ballistic methods which use specimens in the form of a ring give the highest accuracy. The principal advantage of ring specimens is that the end effects are absent in them and they are free from errors resulting from magnetic leakage.

The magnetizing winding should be wound uniformly on the ring otherwise the mmf distribution will be non-uniform which will result in magnetic leakage with consequent errors in the measurements. The magnetising force is usually calculated in terms of mean diameter of the ring and the mmf in the magnetising winding. However, since the inner circumference of the ring is smaller than the outer, the magnetising force and the flux density are not uniform over the cross-section of the ring. The error involved depends upon the ratio of mean diameter and the radial width of the ring. If the ratio of mean diameter to the radial width is more than 8, both magnetising force and the flux density are sufficiently uniform over the cross-section of the sepcimen and consequently the error involved is not significant.

The ring specimens may take one of the following forms :

(*i*) *Laminated.* This is the usual form for isotropic sheet materials. The ring is assembled with the directions of rolling of individual laminations distributed radially, in order that the test should yield results corresponding to mean properties of the material.

(*ii*) *Clock-spring.* A long strip of material may be wound in the form of spiral to form a ring. This form is suitable for testing anistropic materials *i.e.*, grain oriented cold rolled steel.

(*iii*) *Solids.* Solid rings prepared by either casting or forging may be used. Ring specimens are also used for testing powder or dust core materials.

The use of ring specimens in magnetic testing is limited owing to the labour involved in preparing them, since the magnetising winding has to be put with hand for each individual specimen. Also the value of magnetising force is limited by heating. Ring specimens are mainly used in the testing of small quantities of materials of very high permeability.

### 18.3.5 Determination of *B.H.* Curve

There are *two* methods available for the determination of *B-H* curve of a specimen.

**1. *Method of reversals.*** A ring shaped specimen whose dimensions are known is used for the purpose. A layer of thin tape is put on the ring and a search coil insulated by paraffined wax is wound over the tape.

Another layer of tape is put over the search coil and the magnetising winding is uniformly wound over this tape. The test circuit is shown in Fig. 18.1.

After demagnetizing the test is started by setting the magnetising current to its lowest test value. With galvanometer key *K* closed, the iron specimen is brought into a 'reproducible cyclic magnetic state' by throwing the reversing switch *S* backward and forward about twenty times. Key *K* is now opened and the value of flux corresponding to this value of *H* is measured by reversing the switch *S* and noting the throw of galvanometer. The value of flux density corresponding to this *H* can be calculated by dividing the flux by the area of the specimen.

The above procedure is repeated for various values of *H* upto the maximum testing point. The *B-H* curve may be plotted from the measured values of *B* corresponding to the various values of *H*.

**2. *Step by step method.*** The circuit for this test is shown in Fig. 18.4. The magnetising winding is supplied through a potential divider having a large number of tappings. The tappings are arranged so that the magnetising force *H* may be increased, in a number of suitable steps, upto the desired maximum value. The specimen before being tested is demagnetised.

The tapping switch $S_2$ is set on tapping 1 and the switch $S_1$ is closed. The throw of the galvanometer corresponding to this increase in flux density in the specimen, from zero to some value $B_1$ is observed.

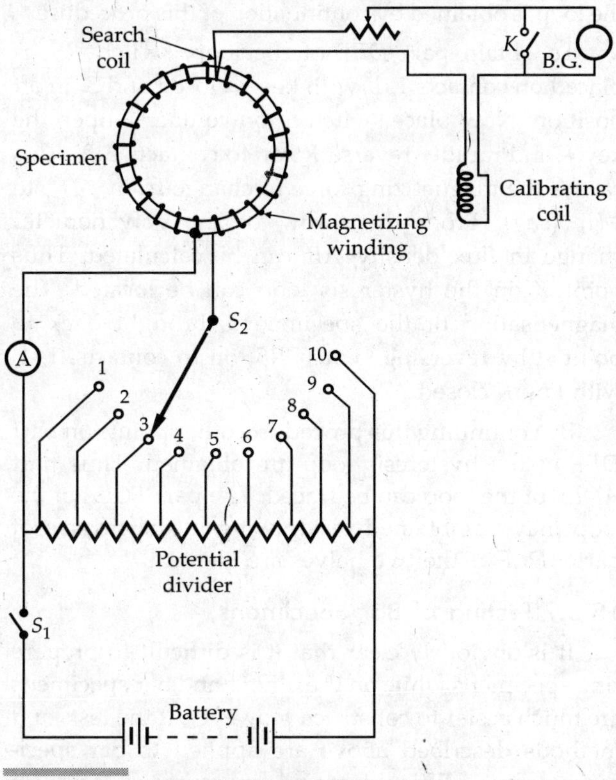


**Fig. 18.4** Determination of *B-H* curve step-by-step method.

The value of $B_1$ can be calculated from the throw of the galvanometer. The value of corresponding magnetising force $H_1$ may be calculated from the value of current flowing in the magnetising winding at tapping 1. The magnetising force is then increased to $H_2$ by switching $S_2$ suddenly to tapping 2, and the corresponding increase in flux density $\Delta B$ is determined from the throw of the galvanometer. Then flux density $B_2$ corresponding to magnetising force $H_2$ is given by $B_1 + \Delta B$. This process is repeated for other values of $H$ upto the maximum point and the complete *B-H* curve is thus obtained as shown in Fig. 18.5.

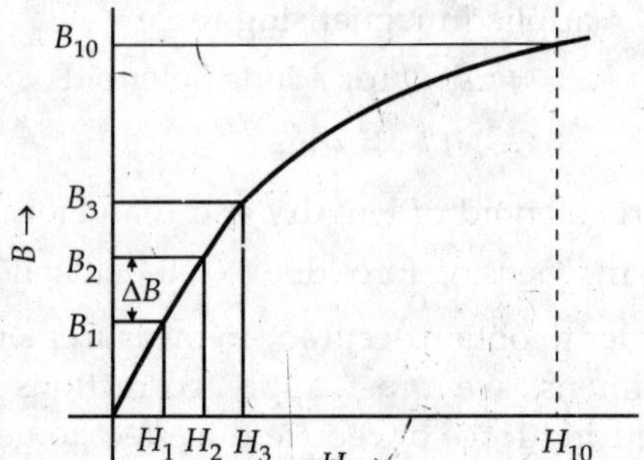


**Fig. 18.5** *B-H* curve.

### 18.3.6 Determination of Hysteresis Loop

**1. *Step by step method.*** The determination of hysteresis loop by this method is done by simply continuing the procedure described in Art. 18.3.5. for the determination of *B-H* curve. After reaching the point of maximum *H i.e.*, when switch $S_2$ is at tapping 10 (see Fig. 18.4), the magnetising current is next reduced, in steps to zero by moving switch $S_2$ down through the tapping points 9, 8, 7 .... 3, 2, 1. After reduction of magnetising force to zero, negative values of $H$ are obtained by reversing the supply to potential divider and then moving the switch $S_2$ up again in order 1, 2, 3 ,..... 7, 8, 9, 10.

**2. *Method of reversals.*** This test is done by means of a number of steps, but the change in Flux density measured at each step is the change from the maximum value + $B_m$ down to some lower value. But before the next step is commenced the iron specimen is passed through the remainder of the cycle of magnetisation back to the flux density + $B_m$. Thus the cyclic state of magnetisation is preserved.



The connections for the method of reversals are shown in Fig. 18.6. $R_1$, $R_2$ and $R_4$ are resistances in the magnetising winding and galvanometer circuits. $R_3$ is

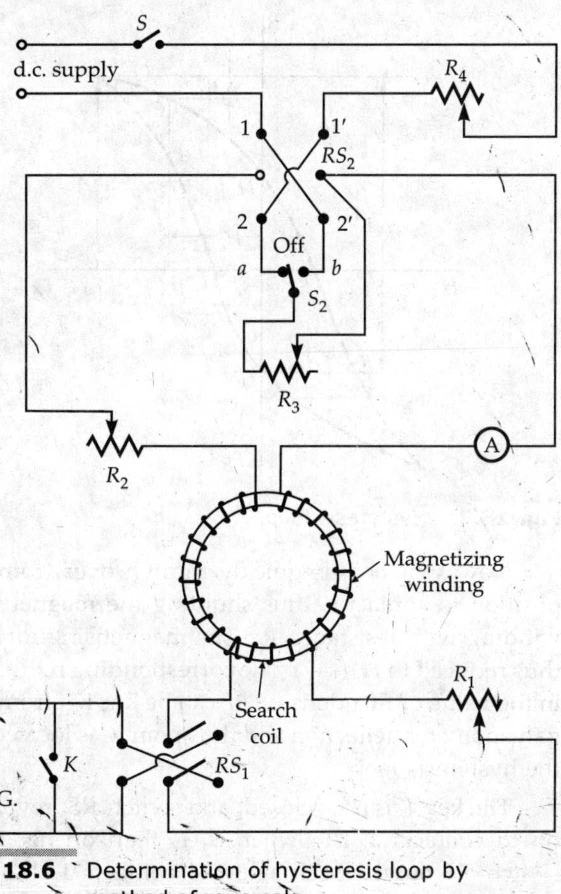


**Fig. 18.6** Determination of hysteresis loop by method of reversals.

a variable shunting resistance which is connected across the magnetising winding by moving over the switch $S_2$. Thus the current in this winding can be reduced from its maximum value down to any desired value by adjusting the value of $R_3$. The procedure for the test is :

The value of magnetising force $H_m$ required to produce flux density $B_m$ to be used during the test is obtained from the previously determined *B-H* curve of the specimen. The resistances $R_2$ and $R_1$ are then adjusted so that the magnetising current is such that this value of *H* (*i.e.*, $H_m$) is obtained when switch $S_2$ is in off position. The resistance $R_1$ is adjusted so that a convenient deflection of galvanometer is obtained when the maximum value of magnetising force is reversed. Resistance $R_3$ is adjusted to such a value that a suitable reduction of the current in the magnetising winding is obtained when the resistance is brought into the circuit.

Switch $RS_2$ is placed on contacts 1, 1′ and key *K* is opened. Since the maximum value of current is flowing in the magnetising winding, the magnetisation of the specimen corresponds to point *A* on the hysteresis loop shown in Fig. 18.7.

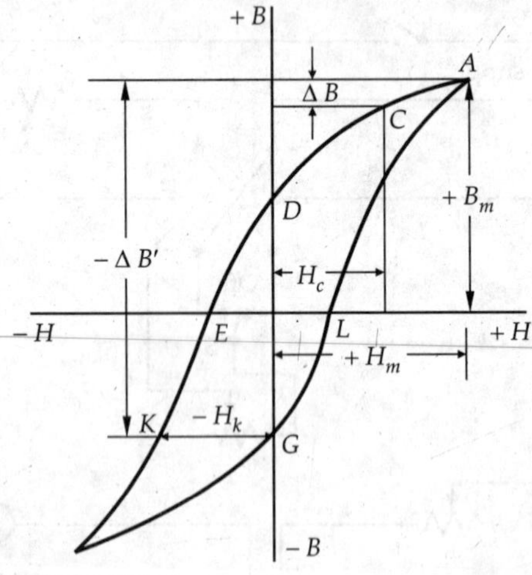


**Fig. 18.7** Hysteresis loop.

Now switch $S_2$ is quickly thrown over from off position to contact *b*, thus shunting the magnetising winding with resistance $R_3$. The magnetising force is thus reduced to $H_c$ (say). The corresponding reduction in the value of flux density *AB* can be known from the galvanometer deflection and thus point *C* is located on the hysteresis loop.

The key *K* is now closed, and switch $RS_2$ reversed on to contacts 2, 2′. Switch $S_2$ is then opened and switch *RS* moved back again to contacts 1, 1′. This procedure passes the specimen through the cycle of magnetisation and back to the point *A*. The specimen is now ready for the next step in the test. The part *AD* of the loop is obtained by continuation of this procedure.

To obtain part *DEF* of the loop, switch $RS_2$ is placed on contacts 1, 1′ with key *K* closed and $S_2$ in off position. Now place switch $S_2$ on contact *a*, open the key *K* and rapidly reverse *RS* on to contacts 2, 2′. This causes the magnetizing force to change from $+H_m$ to $-H_k$ (say). From the throw of the galvanometer change in flux density *AB*′ can be calculated. Thus point *K* on the hysteresis loop can be located. The magnetisation of the specimen is brought back to point *A* by reversing switch $RS_2$ on to contacts, 1, 1′ with key *K* closed.

By continuing this procedure, other points on part *DEF* of the hysteresis loop are obtained. Thus part *ADEF* of the loop can be traced. The part *FGLA* of the loop may be obtained by drawing in the reverse of part *ADEF* as the two halves are identical.

### 18.3.7 Testing of Bar Specimens

It is obviously clear that it is difficult to prepare ring specimens while on the other hand bar specimens are much easier to construct. However, if the tests and methods described above are applied to bar specimens, some difficulties and inaccuracies arise in testing. This is because when we use bar specimens, the return circuit for the flux is through air whose reluctance is very high. Also, when the bar is magnetised, poles are produced at its ends, and these poles produce, inside the rod a magnetising force which acts from the north pole to the south which is in opposition to the applied magnetising force. This is called self-demagnetisation (or end effects). Thus the value *H* of the magnetising force which is effective in producing the flux in the specimen is less than the applied magnetising force *H*′ as calculated from the mmf of the magnetising winding.

The true magnetising force,

$$H = H' - H_d$$

where $H'$ = applied magnetising force

$= NI/l$ for a long solenoid

$= NI/\sqrt{l^2 + d^2}$

for a short solenoid of length *l* and diameter *d*

and $H_d$ = magnetising force due to self-demagnetisation.

In order to obtain accuracy in measurements with bar specimens, we must apply corrections for the damagnetising force or reduce the effect to negligible proportions. However, the self-demagnetising force $H_d$ is somewhat an uncertain quantity.

Since the effect of demagnetising force is small when the ratio of length to diameter of bar is large, the dimensions of the specimen should be so chosen that the effect of demagnetisation is negligible. It has been found that if the ratio of length to diameter of bar specimens is 25 or more, the demagnetising has negligible effect on the value of magnetising force.

## 18.4 PERMEAMETERS

Most methods employed in magnetic testing are designed to avoid the errors and difficulties of the simpler ring or bar tests. These methods tend to combine the good qualities of both bar and ring specimens.

In general, testing is done with the help of permeameters, which make use of straight bar specimens and also provide a return path of low reluctance and thus reduce or in some cases entirely remove the effects of self demagnetisation (end effects). All permeameters use return paths of large cross-sectional area in order to make the reluctance of paths negligible as compared with that of the bar specimens. However, they differ in such details as the degree of compensation of leakage flux or the method by which the magnetisation force is determined or in the range of magnetising force, for which they can be used.

Most of the permeameters incorporate modifications of the bar and yoke arrangements first described by Hopkinson.

### 18.4.1 Hopkinson Permeameter (Bar and Yoke Method)

A test coil is wound upon the central part of a bar specimen. The bar is clamped between two halves of a massive iron yoke (*i.e.*, a yoke having a large cross-sectional area) whose reluctance is low compared to that of the bar specimen. This yoke provides a return path for the flux. The bar specimen is wound with a magnetising winding as shown in Fig. 18.8.

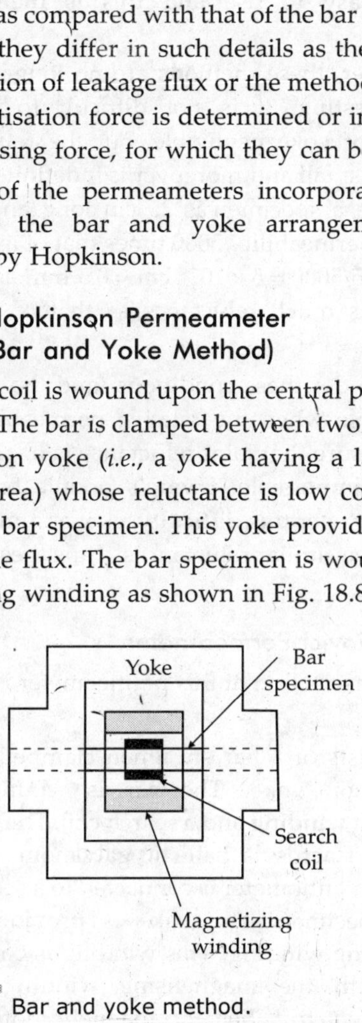


**Fig. 18.8** Bar and yoke method.

Let

$N$ = number of turns on the magnetising winding,

$I$ = current in the magnetising winding,

$l$ = length of the bar specimen between two halves of the yoke,

$A_s$ = area of cross-section of the specimen,

$\mu_s$ = permeability of the specimen when the magnetising current is $I$,

$R_y$ = reluctance of the yoke,

$R_j$ = reluctance of the joints between the bar specimen and the yoke,

and $\Phi$ = flux in the magnetic circuit.

Then,

Reluctance of specimen, $R_s = l/\mu_s A_s$.

$$\therefore \quad \text{Flux } \Phi = \frac{\text{mmf}}{\text{reluctance of magnetic circuit}} = \frac{NI}{R_y + R_j + (l/\mu_s A_s)}$$

Flux density in the specimen

$$B = \frac{\Phi}{A_s} = \frac{NI}{A_s(R_y + R_j + l/\mu_s A_s)}$$

Magnetising force

$$H = \frac{B}{\mu_s} = \frac{NI}{\mu_s A_s(R_y + R_j + l/\mu_s A_s)}$$

Let

$$m = \frac{\text{reluctance of (yoke + joints)}}{\text{reluctance of specimen}} = \frac{R_y + R_j}{l/\mu_s A_s} = \frac{\mu_s A_s}{l}(R_y + R_j) \quad \text{...(18.6)}$$

$$\therefore \quad H = \frac{NI}{l(1+m)} \quad \text{...(18.7)}$$

The value of $m$ is made small by keeping the reluctance of yoke and that joins to a small value. This can be done by carefully fitting the specimen into the yoke (so that air space between bar and yoke is negligible) and making the yoke of a large cross-section.

If $m$ is made small as described above, we have,

$$H = \frac{NI}{l}(1-m) \quad \text{...(18.8)}$$

Thus the actual value of magnetising force differs from the calculated value of magnetising force $(NI/l)$ by the amount $= mNI/l$.

The value of flux density is measured by a ballistic galvanometer in the usual way.

### 18.4.2 Ewing Double Bar Permeameter

This permeameter consists of two exactly similar bars made up of magnetic material under test. There are two pairs of magnetising coils. One pair of coils is exactly half the length of the other pair. Thus there are in all four magnetizing coils, two having full length and the two having half length. The number of turns per unit axial length is same for all the coils. Each bar is wound with two coils, one of full and the other of half length.

The bars fit tightly into holes of two yokes. The yokes are made of annealed soft iron. The position of these yokes on the bars can be adjusted. The arrangement of bars and yokes is shown in Fig. 18.9.

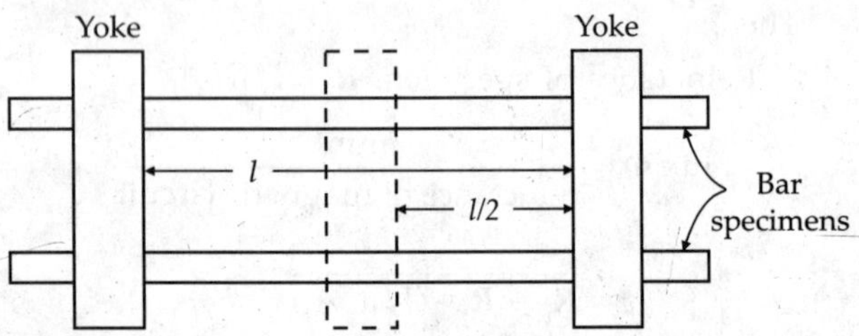


**Fig. 18.9** Ewing permeameter.

The purpose of this arrangement is the elimination of the reluctance of yoke and joints. The tests are made, one with length of specimen = $l$ and the other with length = $l/2$. It is assumed that for both positions, the reluctance of yokes and the joints is the same for a given flux density.

Let

$n$ = number of turns per unit length of magnetising coils,

$I_1$ = current in the coils when the specimen length is $l$,

$I_2$ = current in the coils when the specimen length is $l/2$,

$H_1$ = apparent magnetising force when length is $l$,

$H_2$ = apparent magnetising force when length is $l/2$,

$\Im$ = mmf required for the yokes and the joints in each case,

and $B$ = flux density in the specimen (it is the same in both the cases)

Then, $H_1 = \dfrac{nlI_1}{l} = nI_1$ and $H_2 = \dfrac{nl/2 \cdot I_2}{l/2} = nI_2$.

If $H$ is the true magnetising force in the iron for a flux density $B$, then we have,

$$H_1 l = Hl + \Im \quad \text{...(i)}$$

and $$H_2 l/2 = Hl/2 + \Im \quad \text{...(ii)}$$

This is because,

| total applied mmf | = | mmf required for iron + | mmf required for yoke and joints. |
|---|---|---|---|

From (*i*) and (*ii*), true magnetizing force

$$H = 2H_1 - H_2 \quad \text{...(18.9)}$$

The flux density $B$ corresponding to this value of $H$ is measured with the help of a ballistic galvanometer.

The disadvantages of this method are :

(*i*) the reluctance of yokes and joints is not exactly same for the two positions of the yoke,

(*ii*) the test requires two exactly similar bars, and

(*iii*) the test is quite time-consuming.

The methods which have been discussed above are very easy to use and this is their greatest advantage.

However, these methods are no longer used for magnetic testing. It is not difficult to make the reluctance of yoke negligible, but the reluctance of joints is not small and moreover is indefinite. Suppose for example a specimen is 25 cm long and is being tested at a permeability 5000 times that of air. Then an air gap of $25/5000 = 5 \times 10^{-3}$ cm = 0.5 mm at the joints will have as much reluctance as that of whole the specimen.

Thus under these conditions (and the conditions can be worse when the relative permeability is too high) to be sure that joint reluctance is negligible and definite is impossible. Ewing's double bar method does not get over this difficulty, as it assumes the two joints to have similar reluctances. Thus these methods are obsolete.

### 18.4.3 Illiovici Permeameter

The construction of this permeameter is shown in Fig. 18.10.

It consists of a bar specimen clamped against a heavy section yoke. The bar is wound with a magnetising winding and a search coil. The search coil is connected across a ballistic galvanometer $BG_1$. A magnetic potentiometer is connected to a section $AB$ of the bar specimen. The yoke is provided with a compensating winding. This winding is connected in parallel with the magnetising winding across a reversing switch. When the magnetic potentiometer indicates no magnetic potential drop, the mmf for part $AB$ is provided by the magnetizing winding while for the remaining part of the specimen, the yoke and the joints, is provided by the compensating winding.

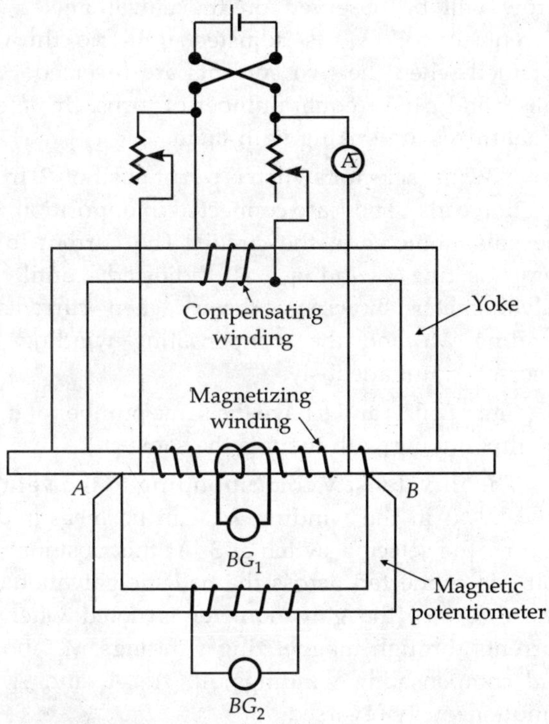


**Fig. 18.10** Illiovici permeameter

This condition is attained by adjusting the current in the main winding to the required test value and then adjusting the current in the compensating winding so that the ballistic galvanometer $BG_2$ shows no throw when the currents are reversed. Under this condition there is a no-flux through the magnetic potentiometer* and therefore no difference of magnetic potential exists between points $A$ and $B$. Thus the magnetising force for length $AB$ is given by

$$H = NI/l$$

where

$N$ = number of turns in the magnetising winding,

$I$ = current in the magnetising winding ;

and $l$ = length $AB$; m.

The flux density in the specimen is found by noting the throw of ballistic galvanometer $BG_1$ when the currents in the magnetizing and the compensating windings are reversed simultaneously.

This permeameter has the following disadvantages :

(*i*) Results obtained for this permeameter are not satisfactory because there is lack of symmetry of arrangement.

(*ii*) This permeameter does not incorporate a direct test of uniformity of magnetisation along the specimen.

(*iii*) Operation is complicated on account of readjustment of compensating winding for each reading.

(*iv*) Errors may also be caused on account of some leakage flux between yoke and specimen and the magnetic potentiometer may be affected by leakage flux.

### 18.4.4 Burrow's Permeameter

This permeameter uses a number of compensating coils which apply compensating mmf at different parts of the magnetic circuit. The mmfs of these coils are so arranged that the entire magnetic circuit is brought to a uniform distribution of magnetic potential. Under this condition there is no magnetic leakage between one part of circuit to another, and magnetising force can be calculated at any point in terms of the magnetising mmf per unit length at that point.

Figure 18.11 shows the arrangement of the magnetic circuit and the coils. $S_2$ is the bar specimen to be tested. $S_2$ is an auxiliary bar. These bars are uniform and similar in dimensions and have similar magnetic properties. The bars are surrounded by magnetising windings $M_1$ and $M_2$ which are uniformly wound over the bars. $A_1$, $A_2$, $A_3$ and $A_4$ are compensating

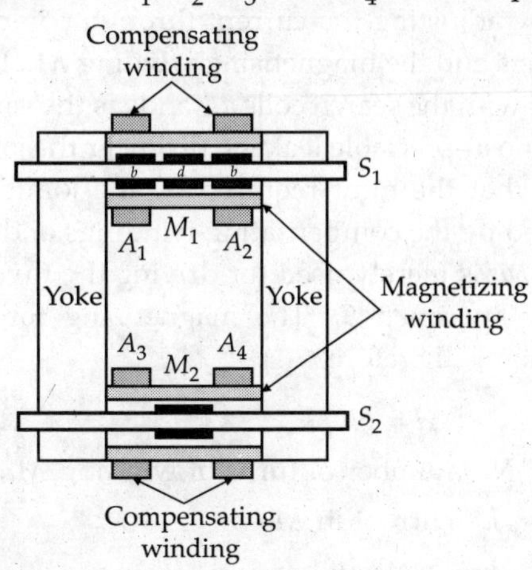


**Fig. 18.11** Burrow's permeameter.

18

* It should be noted that in general magnetic potentiometer cores should be of non-magnetic materials but in the Illiovici permeameters ferromagnetic cores are allowable since it is used as a null detector and not for actual measurement.

windings for the elimination of leakage effects at the joints between the two bars and the yoke. $a$, $b$ and $c$ are search coils and have equal number of turns. Coil $a$ surrounds the middle of the test specimen $S_1$ ; coil $b$ is divided into two equal parts which are located near the ends of bar $S_1$ ; coil $c$ is located at the middle of the auxiliary bar $S_2$.

The electrical connections of the coils are shown in Fig. 18.12. The compensating windings $A_1, A_2, A_3$ and $A_4$ are connected in series and are supplied from a separate battery. Coil $M_1$ is supplied from a separate battery and coil $M_2$ from another.

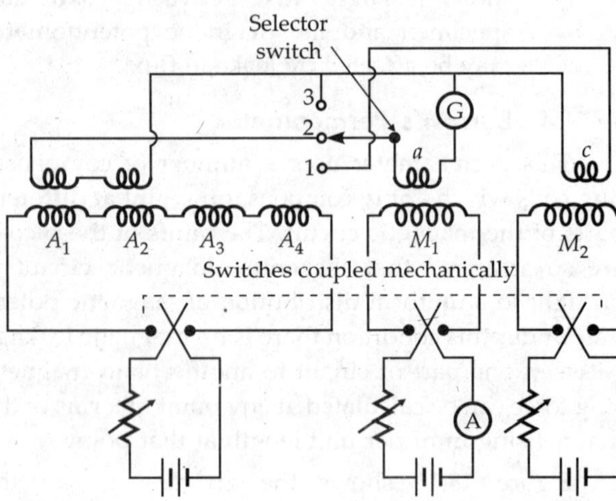


**Fig. 18.12** Electrical circuit of Burrow's permeameter.

In carrying out this test it is necessary to ensure that, for a given value of current $I$ in $M_1$, the flux linking with all the search coil $a, b, c$ is the same. This is done by adjusting the current through compensating windings and the magnetising winding $M_2$. If the flux linking with the search coils $a, b$ and $c$ is the same, there can be no appreciable leakage flux near the joints. This means that the mmf required by the joints is being supplied by the compensating windings and the mmf of coil $M_1$ is merely used for driving the flux through the bar specimen $S_1$. The magnetizing force in the specimen is therefore,

$$H = NI/l.$$

where $N$ = number of turns of winding $M_1$,

$I$ = current in $M_1$

and $l$ = length of bar $S_1$.

The following procedure is adopted to obtain equal flux linkage with all the search coils.

The test specimen is demagnetized and the value of current in $M_1$, is adjusted to the required test value.

The selector switch is put at position 1. In this position search coils $a$ and $c$ are connected in series, but in opposition to a ballistic galvanometer. The currents in coils $M_1$ and $M_2$ are simultaneously reversed. A throw will be observed on the galvanometer. The current in coil $M_2$ is adjusted until no throw is obtained when the two currents are reversed. Since coils $a$ and $c$ have equal number of turns, this means equal fluxes are linking with them.

Now the selector switch is put at position 2. In this position coils $a$ and $b$ are connected in opposition with the galvanometer in the circuit. The current in the compensating winding is adjusted until the galvanometer shows no throw when currents in winding $M_1$ and the compensating windings are reversed simultaneously.

Since coils $a$ and $b$ have the same number of turns, the flux linking with them is the same.

The flux density corresponding to the value of $(H = NI/l)$ in the winding $M_1$ can be measured by putting the selector switch at 3. At this position only coil $a$ is connected across the ballistic galvanometer. The throw of the galvanometer is noted when the currents through magnetizing windings $M_1$ and $M_2$ and compensating windings $A_1$, $A_2$, $A_3$ and $A_4$ are simultaneously reversed.

The advantages of this permeameter are :

(*i*) This method results in uniform conditions along the specimen provided that the material itself is uniform.

(*ii*) This method can be used for magnetic testing of many kinds and also for calibration of specimens that serve as reference standards in some of the simpler methods.

In fact this permeameter has been used for magnetic measurements at the Bureau of Standards (U.S.A.) and in other laboratories and has been accepted as a standard by A.S.T.M.

The main disadvantages are :

**1.** The method described above does not take into account the effects of any magnetic inhomogeneities in the specimen. Variations in permeability along the specimen result in magnetic leakage which cannot be compensated for and thus results in errors in measurements. Therefore, the method will give dependable results only when the test specimens are magnetically uniform.

**2.** In a Burrow's permeameter the necessary adjustments for carrying out compensation are very tedious and slow, therefore, it is mainly used to determine the properties of uniform bars.

### 18.4.5 Fahy's Simplex Permeameter

This permeameter is commonly used for routine testing of magnetic materials. It consists of a single specimen in the form of a bar (Fig. 18.13). The specimen is clamped against a laminated steel yoke with the help of two iron posts. The yoke carries a magnetising winding.

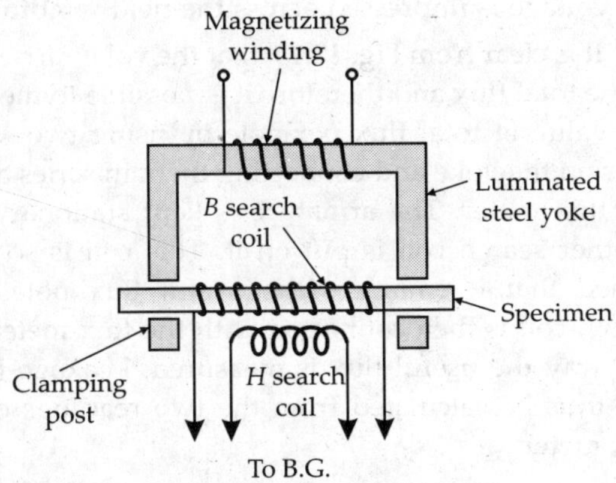


**Fig. 18.13** Fahy's permeameter.

The specimen is wound uniformly with a search coil which extends over its entire active length. This search coil is connected to a ballistic galvanometer for the measurement of flux density and therefore this coil is known as '**B coil**'. The magnetising force acting on the specimen is measured, like the flux density, by a ballistic galvanometer by connecting it across an air cored coil placed between the two clamping posts as shown. This search coil is known as '***H* coil**'. The values of magnetising force so measured are corrected by calibrating the '*H* coil' using a specimen of known magnetic characteristics in place of the test specimen.

The ***advantages*** of a Fahy's permeameter are :

(*i*) It is simpler in both construction and operation as compared with Burrow's permeameter. The data may be acquired very rapidly.

(*ii*) It requires only test specimen.

(*iii*) It is less sensitive to the effects of magnetic inhomogeneities in the specimen. This is because the *B* coil is wound uniformly over the entire length of the specimen and therefore it tends to measure the average value of flux density in the specimen. Thus the effects of local magnetic irregularities are diluted.

### 18.4.6 The National Physical Laboratory (England) form of Permeameter

This is one of the most accurate forms of permeameter and is comparatively easy to use in practice. The arrangement of the permeameter is shown schematically in Fig. 18.14. The specimen *S* may consist of a bundle of strips of 0.35 mm thick sheet iron, each strip being 250 mm long by 70 mm wide. The specimen is clamped between two similar rectangular iron yokes $Y_1$ and $Y_2$ which are in parallel in the magnetic circuit. The cross-sectional area of these yokes is large so that they present a low-reluctance path to the flux in the specimen. The yokes are often built with sheet laminations. The main magnetising winding *M* is wound on a bobbin which surrounds the specimen, and it is provided at each end with compensating windings $A_1$ and $A_2$ whose purpose is to supply the mmfs necessary to overcome the reluctance of the joints. The flux density in the specimen is measured ballistically in the usual way by means of the small search coil *a* surrounding the mid-section of the specimen. Two other coils $b_1$ and $b_2$, each have half the number of turns in *a* and are used to check the uniformity in flux in the specimen in a manner similar to that described for the Burrow's permeameter.

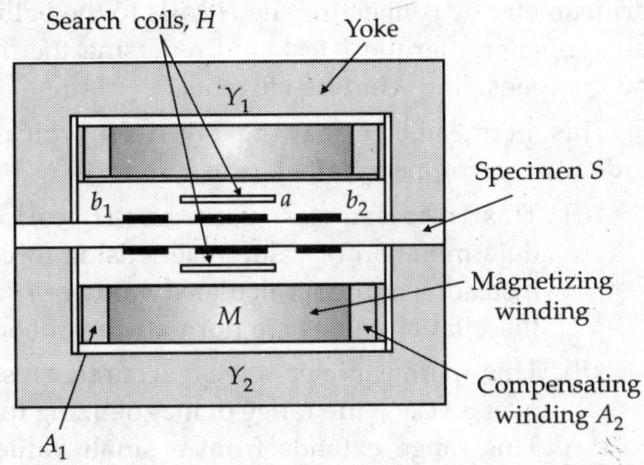


**Fig. 18.14** National Physical Laboratory permeameter.

The compensation is carried out by connecting coils $b_1$ and $b_2$ in series, and in opposition to coil *a* to the ballistic galvanometer. The main magnetising current is adjusted to give the required value of *H*, and the currents in the compensating windings are then adjusted until there is no indication on the ballistic galvanometer when the main and compensating winding currents are simultaneously reversed. When the compensation is correct, the flux density over the mid-section of the specimen is uniform. This flux density can be measured by connecting coil *a* by itself, to the ballistic galvanometer and noting the throw when the main and compensating currents are reversed simultaneously. Compensation is not necessary with specimens of low relative permeability, *i.e.*, below, 1,000.

The value of $H$ in the specimen is measured in this permeameter by means of **$H$ coils**. These coils are wound on thin rigid flat cards of insulating material, usually between 1 and 2 mm thick, of the same width as the specimen and between 50 mm and 80 mm long. The coils have a large number of turns of very fine copper wire, the number of turns being chosen to give adequate sensitivity with the ballistic galvanometer. When tests at low $H$ values are being carried out, thousands of turns, wound in several layers, are necessary. Then windings are usually impregnated with wax or varnish, to give them rigidity. Two of these coils are normally used, one placed on each side of specimen, and if the flux density is uniform in the specimen, the flux through these coils will be proportional to the $H$ in the specimen. The total flux linkages in the **$H$ coil** in a field of strength $H$ is $\mu_0 NAH$, where $N$ is the total number of turns and $A$ the mean area of the turns. It is impossible to measure the product $NA$ from the dimensions of the coil, and the $H$ coils are usually calibrated with a long solenoid.

The actual measurement of $H$ is carried out in the permeameter by connecting the **$H$ coils** to the ballistic galvanometer after the $B$ test, and reversing the main and compensating winding currents.

This permeameter may be taken as typical of modem precision permeameters as :

(*i*) This permeameter uses an **$H$ coil** for determination of value magnetising force H instead of using a calculated value of $H$ and these latter values are normally erroneous.

(*ii*) This permeameter gives accurate results over a very wide range of magnetizing force. This range extends from a small value of 10 A/m to a very high value of 300,000 A/m.

(*iii*) The field strength can be doubled by mounting a short specimen between massive soft iron pole pieces.

(*iv*) This permeameter is capable of even testing circular cross-section specimens by addition of shaped pole pieces.

## 18.5 Measurement of Leakage Factor with Flux Meter

In an electrical machine the leakage factor is defined as :

Leakage factor = total flux/useful flux ; Wb,

Useful flux = flux in the armature (flux crossing the air gap) ; Wb,

Total flux = flux in the pole bodies

= useful flux + leakage flux existing in the pole body at its root ; Wb.

Thus in order to measure the leakage factor, we must measure the flux in the pole bodies and the flux in the armature.

The flux measurements are done by a flux meter. A ballistic galvanometer cannot be used here as the field windings of electric machines are highly inductive and therefore the flux changes very slowly when the voltage is impressed across the field winding.

It is clear from Fig. 18.15, that the yoke carries half of the total flux and therefore it is possible to measure the value of total flux per pole by using two search coils on the yoke and connecting them in series across the flux meter. The armature is kept stationary and another search coil is put on it. This coil is so positioned that it embraces the useful flux pole. This search coil is then connected with the flux meter and this way the useful flux is measured. Leakage factor can thus be calculated from the two readings of the flux meter.

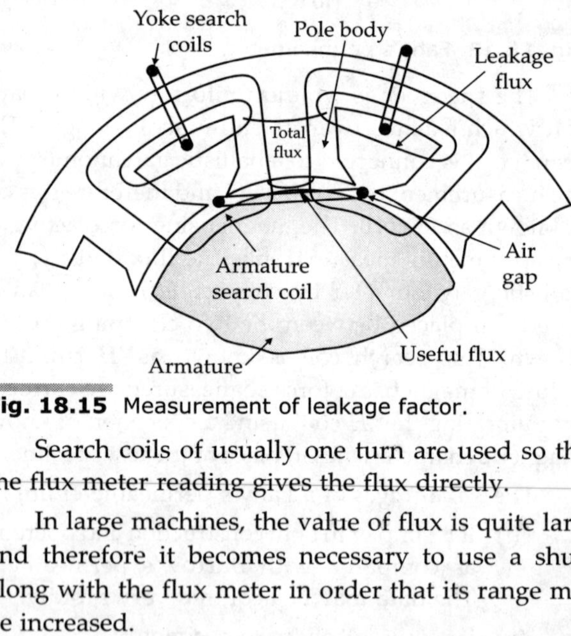


**Fig. 18.15** Measurement of leakage factor.

Search coils of usually one turn are used so that the flux meter reading gives the flux directly.

In large machines, the value of flux is quite large and therefore it becomes necessary to use a shunt along with the flux meter in order that its range may be increased.

**Example 18.1** *An iron ring of 350 mm² cross-sectional area with a mean length of 1 m is wound with a magnetising winding of 100 turns. A secondary coil with 200 turns of wire is connected to a ballistic galvanometer having a constant of 1 μC per scale division, the total resistance of the secondary circuit being 2000 Ω. On reversing a current of 10 A in the magnetising winding, the galvanometer shows a deflection of 100 scale divisions. Calculate the flux density in the specimen, and the value of permeability at this flux density.*

**Solution.** Total mmf of coil $= N_1 I_1 = 100 \times 10 = 1000$ A.

Magnetising force

$$H = N_1 I_1 / l = 1000/1 = 1000 \text{ A/m}.$$

Charge through the galvanometer

$$Q = K_q \theta_1 = 1 \times 100 = 100\ \mu C.$$

Supposing $\Phi$ is the flux through the ring.

Flux linkages of search coil

$$\Psi = N_2 \Phi = 200\,\Phi.$$

Change in flux linkages of search coil due to reversal of current $\Delta\Psi = 2\Psi = 400\,\Phi$.

Average emf induced in the search coil

$$e = \Delta\Psi / \Delta t = 400\,\Phi / \Delta t$$

where $\Delta t$ = time of reversal.

$\therefore$ Current through search coil

$$i = e/R = 400\,\Phi(\Delta t \times R)$$

Charge through the search coil circuit

$$Q = i\Delta t = \frac{400\,\Phi}{\Delta t \times R}\Delta t = \frac{400\,\Phi}{R} = \frac{400\,\Phi}{2000}$$

Thus, $400\,\Phi/2000 = 100 \times 10^{-6}$

or $\Phi = 500 \times 10^{-6}$ Wb.

Flux density in the specimen

$$B = \frac{\text{flux}}{\text{area}} = \frac{500 \times 10^{-6}}{350 \times 10^{-6}} = 1.429\ \text{Wb/m}^2$$

Permeability of specimen

$$\mu = \frac{B}{H} = \frac{1.429}{1000} = 1.429 \times 10^{-3}.$$

$\therefore$ Relative permeability of specimen

$$\mu_r = \frac{\mu}{\mu_0} = \frac{1.429 \times 10^{-3}}{4\pi \times 10^{-7}} = 1137.$$

**Example 18.2** *An iron ring has a mean diameter of 0.1 m and a cross-section of 33.5 mm². It is wound with a magnetising winding of 320 turns and a secondary winding of 220 turns. On reversing a current of 10 A in the magnetising winding, a ballistic galvanometer gives a throw of 272 scale divisions, while a Hibbert magnetic standard with 10 turns and a flux of $0.25 \times 10^{-3}$ Wb gives a reading of 102 scale divisions, other conditions remaining the same. Find the relative permeability of the specimen.*

**Solution.** In a Hibbert's magnetic standard, the coil slides through the gap of a permanent magnet.

$\therefore$ Change in flux linkages of the coil when it goes down the Hibbert's standard

= flux × turns of the Hibbert's standard's coil

$= (0.25 \times 10^{-3}) \times 10 = 2.5 \times 10^{-3}$ weber turns

Throw of galvanometer with Hibbert's standard

= 102

$\therefore$ Constant of galvanometer

$$K_\phi = 2.5 \times 10^{-3}/102$$

$= 24.5 \times 10^{-6}$ weber turns/division.

The galvanometer shows a deflection of 272 divisions when connected to search coil.

$\therefore$ Change in the flux linkages of search coil

$$\Delta\Psi = 272 \times 24.5 \times 10^{-6} = 6664 \times 10^{-6}\ \text{weber turn.}$$

This change in flux linkages is caused by reversal of flux.

$\therefore$ Flux linkages of coil

$$\Psi = \Delta\Psi/2 = 3332 \times 10^{-6}\ \text{weber turn.}$$

Flux linking with search coil

$$\Phi = \frac{\text{flux linkages of coil}}{\text{turns of coil}} = \frac{33.32 \times 10^{-6}}{220}$$

$= 15.1 \times 10^{-6}$ Wb

Flux density in specimen (ring)

$$B = \frac{15.1 \times 10^{-6}}{33.5 \times 10^{-6}} = 0.452\ \text{Wb/m}^2$$

Length of flux path in ring

$$l = \pi D = \pi \times 0.1 = 0.314\ \text{m.}$$

Total mmf of magnetising winding

$= 320 \times 10 = 3200$ A.

Magnetising force

$$H = 3200/0.314 = 10200\ \text{A/m.}$$

Relative permeability of specimen

$$\mu_r = \frac{B}{\mu_0 H} = \frac{0.452}{4\pi \times 10^{-7} \times 10{,}200} = 35.3.$$

**Example 18.3** *The constant of a given magnetic potentiometer is obtained by aid of a coil of 300 turns in which a current of 0.6 A is reversed. The resulting throw of the galvanometer is 157 scale divisions. It is then used to measure the magnetic potential difference between two points and the throw is 304 divisions. Find the magnetic potential difference.*

**Solution.** Mmf of coil $= 3000 \times 0.6 = 180$ A

Change in mmf due to reversal of current in the coil $= 2 \times 180 = 360$ A.

A deflection of 157 scale divisions is caused by an mmf of 360 A.

$\therefore$ Mmf causing a deflection of 304 divisions

$= (304/157) \times 360 = 695$ A.

Hence, magnetic potential difference = 695 A.

**Example 18.4** *A ballistic galvanometer having a circuit resistance of 5000 Ω and a constant of 0.1 μC per scale division is connected in turn with a coil of 2 turns wound round the field coil of a d.c. machine and one of 3 turns placed on the armature surface embracing total flux per pole*

18

*entering the armature. When normal field current is broken the galvanometer readings are 113 and 136 divisions respectively. Calculate the flux per pole and leakage co-efficient.*

**Solution.** Let $\Phi$ be a flux linking with a search coil of $N$ turns.

$\therefore$ Thus linkages of search coil $\Psi = N\Phi$.

If the circuit producing the flux $\Phi$ is suddenly broken, the value of flux becomes zero. Therefore, the change in the value of flux $\Phi$ is suddenly broken, the value of flux becomes zero. Therefore the change in the value of flux linkages under these conditions is $\Delta\Psi = N\Phi$.

Emf induced $e = \Delta\Psi / \Delta t = N\Phi / \Delta t$.

where $\Delta t$ = time taken by the flux to come down to zero.

Current $i = e / R = N\Phi / R\Delta t$,

where $R$ = resistance of search coil circuit.

Charge $Q = i\Delta t = N\Phi / R$.

But charge $Q = K_q\theta_1$,

where $K_q$ = constant of galvanometer in coulomb / scale division.

$\theta_1$ = deflection of galvanometer in scale division.

$\therefore K_q\theta_1 = N\Phi / R$ or Flux $\Phi = RK_q\theta_1 / N$.

**Armature Flux :**

Let $\Phi$ = armature flux per pole.

$$\therefore \quad \Phi = \frac{RK_q\theta_1}{N} = \frac{5000 \times 0.1 \times 10^{-6} \times 136}{3} = 0.0226 \text{Wb}.$$

**Pole Flux :**

Let $\Phi_i$ be the total flux in each pole.

$$\therefore \quad \Phi_t = \frac{5000 \times 0.1 \times 10^{-6} \times 113}{2} = 0.0282 \text{ Wb}.$$

$$\therefore \text{Leakage factor} = \frac{\text{total flux per pole}}{\text{armature flux per pole}}$$

$$= \frac{0.0282}{0.0226} = 1.25.$$

**Example 18.5** *A ferro-magnetic material is tested with the help of a permeameter provided with a coil wound over the specimen, this coil being known as 'B coil' and another coil wound near the surface of the coil called 'H coil'. These two coils are connected to two separate ballistic galvanometers having a constant of $25 \times 10^{-6}$ weber turn per scale division. The testing is done by reversing the current of the magnetizing winding wound over the specimen. The reading of the ballistic galvanometer connected to 'B' coil is 100 divisions and that connected to 'H' coil is 10 divisions. The data of the two coils is :*

| Coils | Turns | Area |
|---|---|---|
| B coil | 100 | $50 \times 10^{-6}$ m$^2$ |
| H coil | 10000 | $5 \times 10^{-6}$ m$^2$ |

*Find the flux density and magnetizing force in the specimen. Also find the relative permeability of the specimen.*

**Solution.** Suppose $\Phi$ and $N$ are the flux and the number of turns.

$\therefore$ Flux linkages, $\Psi = N\Phi$.

Change in flux linkages due to reversal of current

$$\Delta\Psi = 2\Psi = 2N\Phi.$$

Flux linkages indicated by galvanometer

$$= K_\phi\theta_1 \quad \therefore \ 2N\Phi = K_\phi\theta_1$$

or Flux $\Phi = K_\phi\theta_1 / 2N$

and flux density $B = K_\phi\theta_1 / 2NA$.

**'B' coil.** Since 'B' coil is wound over the specimen. it gives the flux density in the specimen.

$\therefore$ Flux density in specimen

$$B = \frac{K_\phi\theta_1}{2NA} = \frac{(25 \times 10^{-6}) \times 100}{2 \times 100 \times 50 \times 10^{-6}} = 0.25 \text{ Wb/m}^2.$$

**'H' coil.** This coil is put at the surface of the coil and hence indicates the flux density in the air.

Flux density in air

$$B_0 = \frac{25 \times 10^{-6} \times 10}{2 \times 10000 \times 5 \times 10^{-6}} = 0.0025 \text{ Wb/m}^2$$

Magnetizing force,

$$H = \frac{B_0}{\mu_0} = \frac{0.0025}{4\pi \times 10^{-7}} \approx 2000 \text{ A/m}.$$

Relative permeability,

$$\mu_r = \frac{B}{\mu_0 H} = \frac{0.25}{4\pi \times 10^{-7} \times 2000} = 100.$$

## 18.5 ALTERNATING CURRENT MAGNETIC TESTING

This testing is done in order to determine the iron losses in magnetic materials at different values of flux density and frequency, and to separate the two components of iron losses *i.e.*, eddy current losses and hysteresis losses.

### 18.5.1 Iron Loss Curves

When a magnetic material is subjected to an alternating field, loss in power occurs owing to hysteresis and eddy currents. This loss is called iron or core loss.

The hysteresis loss may be determined from the hysteresis loop obtained from d.c. tests. But this hysteresis loss may be somewhat different under actual working alternating magnetization conditions. The eddy current loss can be measured only under a.c. conditions ; and it is usually convenient to measure the combined hysteresis and eddy current loss *i.e.*, total iron loss.

Knowledge of total iron loss in ferro-magnetic materials is of considerable importance to designers. For comparison a single value of loss at a stated frequency (*e.g.*, $f = 50$ Hz) and a stated maximum flux density (*e.g.*, $B = 1.3$ Wb/m$^2$) is sometimes sufficient. But in order to give complete information, a family of curves called loss curves giving values of total iron loss per kg over a range of maximum flux density from zero to the higher value are usually given. Figure 18.16 shows loss curves for thickness $t = 0.5$ mm and $t = 0.3$ mm for a particular frequency say 50 Hz.

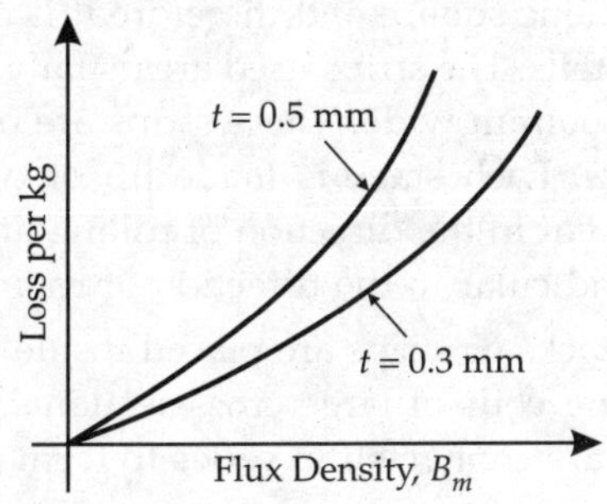


**Fig. 18.16** Loss curves.

### 18.5.2 Separation of Iron Losses

In order to analyse the effects of differences of apparatus design or of metallurgical composition of different ferro-magnetic materials, it is sometimes necessary to separate the total iron loss into hysteresis and eddy current components. Energy loss due to hysteresis in per unit volume is given by :

$$\int H\,dB = \text{area of hysteresis loop.}$$

Hysteresis loss per unit volume

= frequency × area of hysteresis loop.

Steinmetz gives the following formula for hysteresis loss :

Hysteresis loss per unit volume

$$= p_h = \eta f B_m^k \text{ watt} \quad ...(18.10)$$

where $\eta$ = hysteresis co-efficient ;

$f$ = frequency ; Hz,

$B_m$ = maximum flux density ; Wb/m$^2$,

and $k$ = steinmetz co-efficient.

The value of $k$ varies from 1.6 to 2.

Eddy current loss per unit volume for laminations,

$$p_e = \frac{4 K_f^2 f^2 B_m^2 t^2}{3\rho} \text{ watt} \quad ...(18.11)$$

where $K_f$ = form factor,

$t$ = thickness of laminations ; $m$,

and $\rho$ = resistivity of material ; $\Omega$m.

∴ Total iron loss per unit volume

$$p_i = p_h + p_e = \eta f B_m^k + \frac{4 K_f^2 f^2 B_{mi}^2 t^2}{3\rho} \quad ...(18.12)$$

Hence total iron loss in a given specimen of laminations is :

$$P_i = \text{volume} \times \left( \eta f B_m^k + \frac{4 K_f^2 f^2 B_m^2 t^2}{3\rho} \right) \quad ...(18.13)$$

For a particular specimen, volume, thickness $t$ and resistivity $p$ are constants.

$$\therefore \quad P_i = K_h f B_m^k + K_e K_f^2 f^2 B_m^2 \quad ...(18.14)$$

where $K_h$ and $K_e$ are constants.

As the hysteresis and eddy current losses have different laws of variation with both frequency and form factor, $K_f$, it is possible to separate the losses by variation of frequency or of form factor if $B_m$ can be maintained constant.

1. ***Variation of frequency.*** The form factor $K_f$, and maximum flux density $B_m$, remain constant in this test. If the frequency is varied the total iron loss may be written as :

$$P_i = K_1 f + K_2 f^2 \quad ...(18.15)$$

where $K_1 = K_h B_m^k$ ...(18.16)

and $K_2 = K_e K_f^2 B_m^2$ ...(18.17)

Both $K_1$ and $K_2$ being constants for this test.

Now we have,

$$P_i / f = K_1 + K_2 f \quad ...(18.18)$$

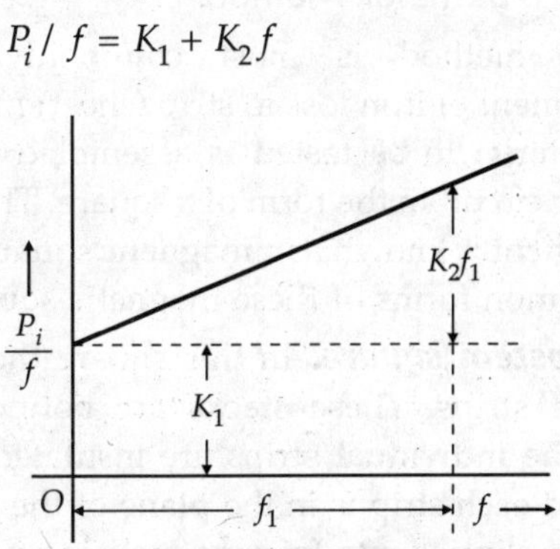


**Fig. 18.17** Variation of iron loss with frequency.

18

Thus on plotting $P_i / f$ against $f$, we obtain a straight line as shown in Fig. 18.17, the values of $K_1$ and $K_2$ can be obtained from intercept and slope of the line respectively.

2. ***Variation of form factor***. The value of maximum flux density $B_m$ and frequency $f$ are kept constant in this test. If the form factor $K_f$ is varied, the total iron losses may be written as :

$$P_i = K_3 + K_4 K_f^2 \quad ...(18.19)$$

where $K_3 = K_h f B_m^k$ ...(18.20)

and $K_4 = K_e f^2 B_m^2$ ...(18.21)

Both $K_3$ and $K_4$ being constant for this test. Thus on plotting $P_i$ against $K_f^2$, we obtain a straight line as shown in Fig. 18.18. The values of $K_3$ and $K_4$ can be obtained from its intercept and slope of the straight line respectively.

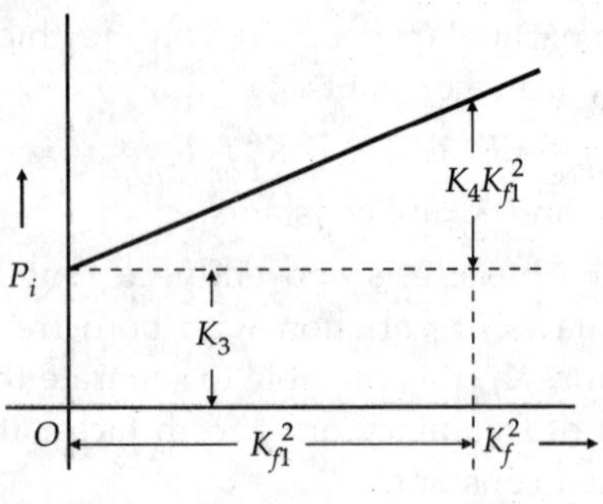


**Fig. 18.18** Variation of iron loss with $K_f^2$.

## 18.6 METHODS OF MEASUREMENT OF IRON LOSSES

There are *three* types of methods used for measurement of iron loss in ferro-magnetic materials :

(*i*) Wattmeter method

(*ii*) Bridge method and

(*iii*) Potentiometer method.

### 18.6.1 Wattmeter Method

This method is most commonly used for measurement of iron loss in strip (sheet) material. The strip material to be tested is assembled as a closed magnetic circuit in the form of a square. Therefore this arrangement is known as a magnetic square. There are two common forms of these magnetic squares.

1. ***Epstein square.*** In this square there are four stacks of strips. These stacks are bound and then taped. The individual strips are insulated from each other and each strip is in the plane of the square. The stacks are slipped into four magnetising coils with the strips projecting beyond the coils.

The ends of the four strips are interleaved (as in transfom1er core construction) and clamped at corners. The arrangement of an Epstein square is shown in Fig. 18.19.

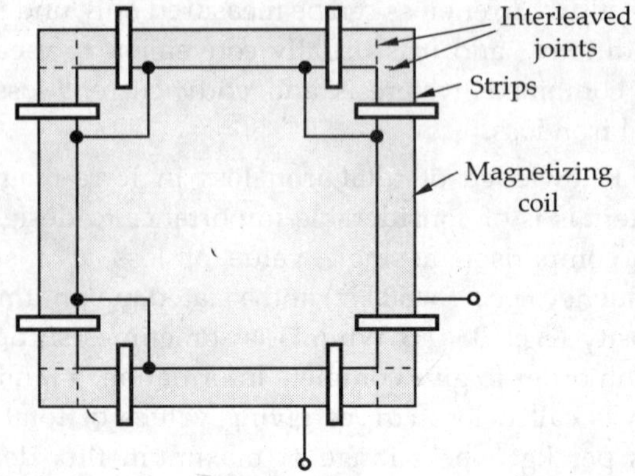


**Fig. 18.19** Epştein square.

2. ***Lloyd-Fisher square.*** This is the most commonly used magnetic square and therefore it is described in greater details. The strips used are usually 0.25 m long and 50 to 60 mm wide. These strips are built up into four stacks. Each stack is made up of two types of strips one cut in the direction of rolling and the other cut perpendicular to the direction of rolling.

The stacks or strips are placed inside four similar magnetising coils of large cross-sectional area. These four coils are connected in series to form the primary winding.

Each magnetising coil has two similar single layer coils underneath it. They are called secondary coils.

Thus in a magnetic square there are eight secondary coils. These secondary coils are connected in series in groups of four, one from each core, to form two separate secondary windings.

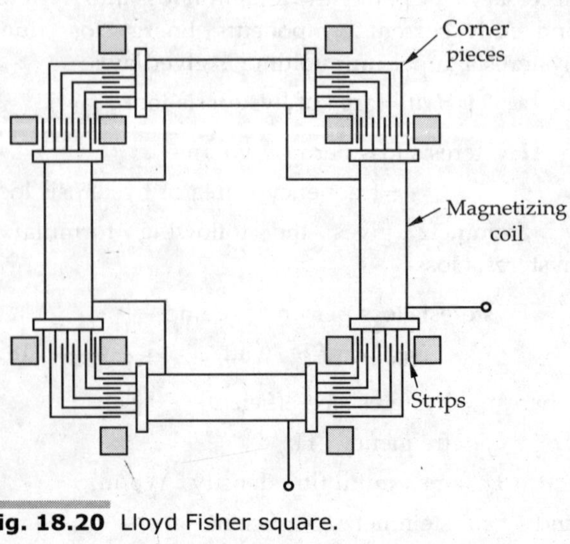


**Fig. 18.20** Lloyd Fisher square.

The ends of the strips project beyond the magnetising coils. The strips are so arranged that the plane of each strip is perpendicular to the plane of the square. The magnetic circuit is completed by bringing the four stacks together in the form of a square and joining them at the corners. The corner joints are made by a set of standard right angled corner pieces as shown in Fig. 18.20. The corner pieces are of the same material as strips or at least a material having the same magnetic properties. There is an overlapping of corner pieces and strips at the corners due to which cross-section of iron is doubled at the corners. Therefore, a correction must be applied for this. Also the measured loss has to be corrected for the loss in the corner pieces.

***Advantages :***

1. In case allowance for corner pieces is known with sufficient accuracy, the Lloyd-Fisher square gives rather more reliable results than the Epstein square.

   In the case of an Epstein square, the value of flux density in the corners is quite different from that of the square, and an allowance of this is difficult to make.

2. The Epstein square is also inferior for tests on anistropic materials as the direction of flux at the corners is partially perpendicular to the path of flux in other portions of the strip. The use of corner pieces in a Lloyd-Fisher square makes it superior for testing anistropic materials.

***Test set up.*** The test specimen is weighed before assembly and its effective cross-section is determined. The primary winding is connected to a sinusoidal voltage supply. Figure 18.21 shows the connection diagram for finding the total iron loss by wattmeter method.

The primary winding contains the current coil of the wattmeter. The pressure coil of the wattmeter is supplied from one of the secondary windings. The wattmeter is designed for low power factor operation as the power factor is usually about 0.2. The second secondary winding supplies an electrostatic voltmeter (or an electrodynamic voltmeter of very high impedance). The frequency of supply is adjusted to the correct value.

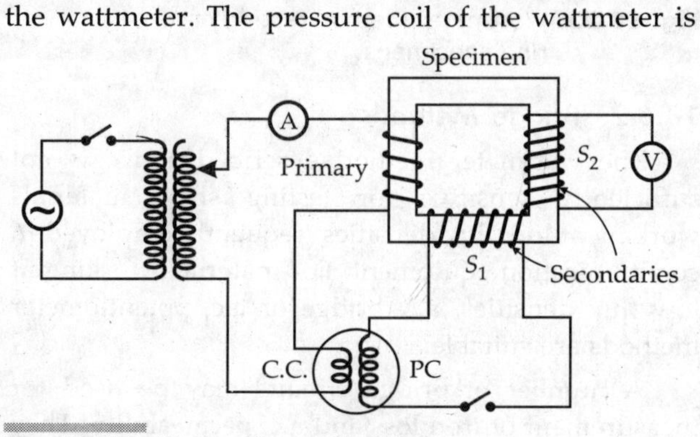


**Fig. 18.21** Wattmeter method for determining iron loss.

The voltage applied to the primary winding is adjusted, preferably with the help of a variable ratio transformer, till the magnetising current is adjusted to give the required value of $B_m$. The wattmeter and voltmeter readings are observed.

***Theory.*** The electrostatic voltmeter connected across secondary winding $S_2$ measures the rms value of induced voltage. The value of induced voltage is,

$$E = 4 K_f \Phi_m f N_2 = 4 K_f B'_m A_s f N_2$$

where

$K_f$ = form factor (1.11 for sinusoidal waves),

$\Phi_m$ = maximum value of flux ; Wb,

$B'_m$ = maximum value of flux density (apparent) ; Wb/m$^2$

$A_s$ = cross-sectional area of specimen ; m$^2$,

$f$ = frequency ; Hz,

and $N_2$ = number of turns in the secondary winding $S_2$.

∴ Maximum value of flux density

$$B'_m = \frac{E}{4 K_f A_s f N_2}$$

This value is the apparent value of the flux density.

Correction must be applied to this value of flux density as the secondary winding $S_2$ encloses the flux in the air space between specimen and the coil in addition to the flux in the specimen, as discussed earlier on page 542.

The actual value of flux density in the specimen (Eqn. 18.2) is,

$$B_m = B'_m - \mu_0 H_m \left(\frac{A_c}{A_s} - 1\right) \quad ...(18.22)$$

where

$A_c$ = cross-sectional area of coil ; m$^2$,

and $H_m$ = magnetising force corresponding to maximum flux density ; A/m. This may be taken from the *B-H* curve of the specimen

The wattmeter reading includes both the iron loss in the specimen and the copper loss in the secondary winding circuit. The copper loss in the secondary winding can be calculated and subtracted from the wattmeter readings.

Let $P_i$ = total iron loss ; W,

$P$ = wattmeter reading ; W,

$V$ = voltage applied to wattmeter pressure coil ; V,

$E$ = voltage reading = voltage induced in $S_2$ ; V,

$r_p$ = resistance of wattmeter pressure coil ; Ω,

$r_c$ = resistance of coil $S_1$ ; Ω,

and $I_p$ = current in the pressure coil circuit ; A.

Now voltage induced in $S_1$ is equal to the voltage induced in $S_2$ since both of them have equal number of turns and they link with the same flux.

∴ Voltage induced in winding $S_1 = E$

The leakage reactances of winding $S_1$ and that of pressure coil are very small as compared with resistance of the pressure coil circuit. If we neglect the leakage reactances of winding $S_1$ and pressure coil, we have,

$$E = I_p(r_p + r_c)$$

∴ Total iron loss in specimen + total copper loss in the secondary winding circuit = $PE/V$

Total copper loss in the secondary winding circuit

$$= E^2 / r_p + r_c$$

∴ Total iron loss in the specimen

$$P_i = P\frac{E}{V} - \frac{E^2}{r_p + r_c} = P\left(1 + \frac{r_e}{r_p}\right) - \frac{E^2}{r_p + r_c} W \quad \text{...(18.23)}$$

Specific iron loss *i.e.*, iron loss per kg can be calculated by dividing the total iron loss by the weight of the specimen.

The eddy current and hysteresis components of the iron loss can be determined from results of the loss measurements at different frequencies.

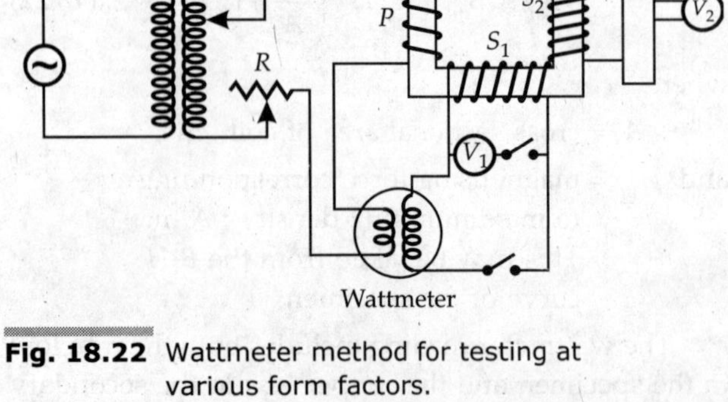


**Fig. 18.22** Wattmeter method for testing at various form factors.

Figure 18.22 shows the connections for the measurement of iron loss at different form factors. The form factor is obtained from the readings of two voltmeters $V_1$ and $V_2$. Voltmeter $V_1$ is connected across $S_1$, may be an electrostatic or an electrodynamometer type and therefore it reads the rms value of the voltage. Voltmeter $V_2$ is connected across $S_2$ a moving coil instrument supplied through a bridge rectifier and therefore it reads the average value of voltage.

$$\text{Form factor, } K_f = \frac{\text{rms value}}{\text{average value}} = \frac{\text{reading of } V_1}{\text{reading of } V_2}.$$

(as $S_1$ and $S_2$ have equal number of turns).

Variation of form factor is obtained by adjusting the values of series resistance and inductance.

3. ***Wattmeter method for testing ring specimens.*** Sheet steel in the form of rings is used for making current transformers. Also spiral cores used are for the construction of current transformers and variacs. Therefore it becomes necessary to measure iron loss in ring-specimens. The methods employed and the experimental set up is the same as that for strips. Figure 18.23 shows the connection diagram for measurement. $S_1$ and $S_2$ are two secondary windings, $V_1$ is a voltmeter which reads the rms value while voltmeter $V_2$ reads the average value of voltage.

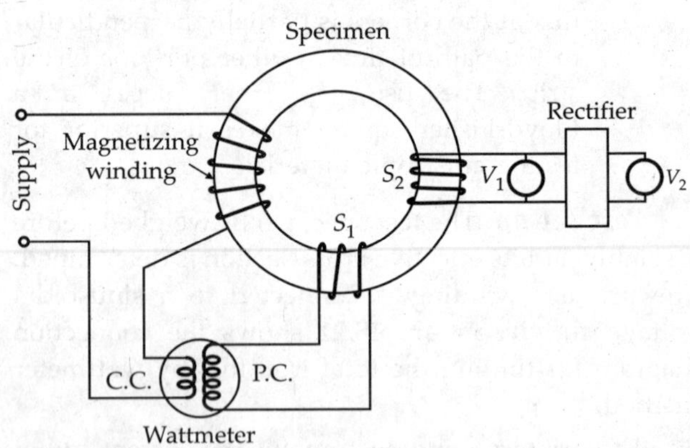


**Fig. 18.23** Wattmeter method testing of ring specimens.

### 18.6.2 Bridge Methods

The wattmeter method described above is not sufficiently sensitive for testing sheet materials working at low flux densities frequently employed in communication equipment. For materials working at low flux densities, a.c. bridge or a.c. potentiometer methods are suitable.

A number of bridge circuits may be used for measurement of iron loss and a.c. permeability. They are not only useful where materials work at low flux densities, but also where only a small quantity of material is available for testing and the test is to be carried out at commercial or audio frequencies.

**1. *Maxwell's bridge.*** The connection diagram of the Maxwell's bridge for the measurement of iron loss and permeability of a ring type specimen is given in Fig. 18.24. The theory of this bridge is given in Art. 16.5.1 on page 482. The specimen is connected in arm *ab*.

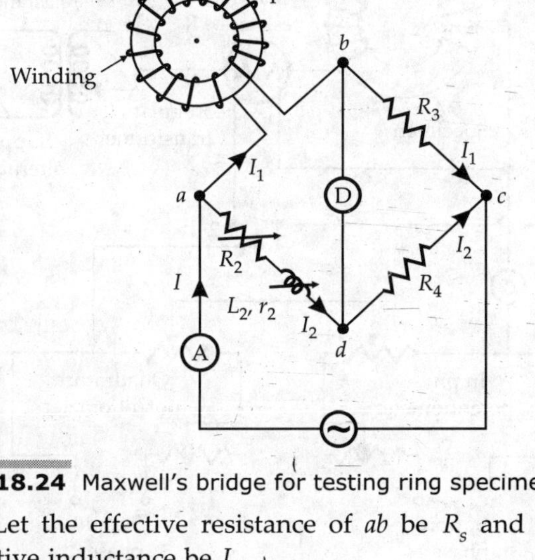


**Fig. 18.24** Maxwell's bridge for testing ring specimen.

Let the effective resistance of *ab* be $R_s$ and the effective inductance be $L_s$.

$\therefore$ From Eqn. 16.17,

$$R_s = \frac{R_3}{R_4}(R_2 + r_2).$$

Now effective resistance of arm *ab* is

$$R_s = \frac{\text{iron loss} + \text{copper loss in winding}}{(\text{current})^2}$$

$$= \frac{P_i + I_1^2 R}{I_1^2} \,;\, \Omega \qquad \text{...(18.24)}$$

where $R_w$ = resistance of winding of coil ; Ω,

$\therefore$ Iron loss $P_i = I_1^2(R_s - R_w)$W. ...(18.25)

At balance voltage drops between *bc* and *dc* are equal.

$$\therefore \qquad I_1 R_3 = I_2 R_4 - (I - I_1) R_4$$

as $\qquad I = I_1 + I_2 \quad \text{or} \quad I_1 = \frac{R_4}{R_3 + R_4} I.$

From Eqn. 18.25, we get,

$$\text{Iron loss, } P_i = I^2\left(\frac{R_4}{R_3 + R_4}\right)^2 (R_s - R_w)$$

The value of iron loss may be calculated as the value of $R_s$ is known from Eqn. 18.16.

The value of *I* can be measured and resistance, $R_w$, of the winding may be measured with d.c.

From Eqn. 18.16, inductance of specimen,

$$L_s = \frac{R_3}{R_4} L_2$$

But inductance $\quad L_s = \frac{N^2}{l_s / \mu_s A_s}$

where

$N$ = number of turns of magnetising winding,

$l_s$ = length of mean flux path in specimen,

$A_s$ = area of specimen,

and $\mu_s$ = a.c. permeability of specimen,

$$\therefore \text{ A.C. permeability, } \mu_s = \frac{l_s L_s}{N^2 A_s} = \frac{l_s R_3 L_2}{N^2 A_s R_4} \qquad \text{...(18.26)}$$

**2. *Campbell's bridge method.*** This is one of the best known methods for iron loss measurement by bridge networks. 18

The ring specimen carries two windings, one primary and one secondary, having the same cross-sectional area. *M* is a variable mutual inductance connected as shown in Fig. 18.25. $R_1$ and $R_2$ are variable resistances. The balance is obtained by varying *M* and $R_2$.

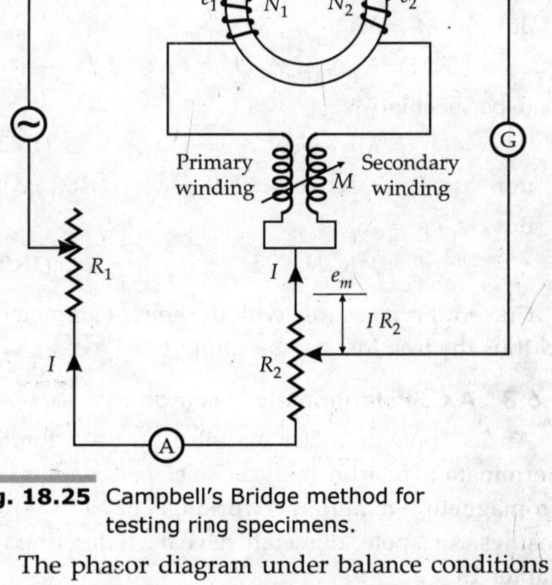


**Fig. 18.25** Campbell's Bridge method for testing ring specimens.

The phasor diagram under balance conditions is shown in Fig. 18.25. Φ is the flux in the specimen, *I* is the current in the primary winding of the specimen. On account of iron loss, *I* leads the flux by a small angle. $e_1$ and $e_2$ are respectively the induced voltages in primary and secondary windings of the specimen, $e_m$ is the induced voltage in the secondary winding of the mutual inductance *M*.

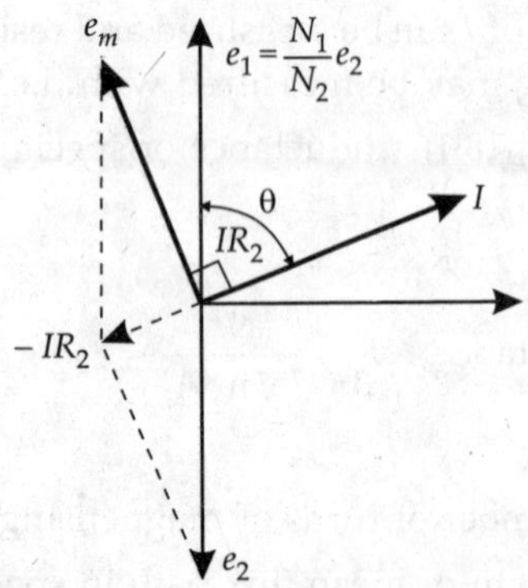


**Fig. 18.26** Phasor diagram for Campbell's Bridge.

Consider the circuit on secondary winding side. Balance is obtained when phasor sum of $e_m$ and $e_2$ is equal and opposite to the phasor $IR_2$.

Let $M'$ = mutual inductance between the primary and secondary windings of the specimen

$\therefore \quad e_2 = \omega M'I$ and $e_m = \omega MI.$

Since $IR_2$ is very small, we can write

$$e_2 \approx e_m \quad \text{or} \quad \omega M'I \approx \omega MI \quad \text{or} \quad M' \approx M.$$

Flux in the specimen

$$\Phi = N_1 I/(l_s/\mu_s A_s) = N_1 I \mu_s A_s / l_s.$$

Mutual inductance between primary and secondary windings of specimen is :

$$M = N_2 \Phi / I = N_1 N_2 \mu_s A_s / l_s$$

But $\quad M' = M$

$$\therefore \quad M = N_1 N_2 \mu_s A_s / l_s$$

or a.c. permeability

$$\mu_s = M l_s / N_1 N_2 A_s \qquad \text{...(18.27)}$$

Iron loss, $P_i = e_1 I \cos 0 - e_1 I . IR_2 / e_2 = I^2 R_2 e_1 / e_2$

But $\quad e_1 / e_2 = N_1 / N_2$

$$\therefore \quad P_i = I^2 R_2 N_1 / N_2 \qquad \text{...(18.28)}$$

Current $I$ is measured with the help of ammeter $A$ and thus the iron loss can be computed.

### 18.6.3 A.C. Potentiometer Method

An a.c. potentiometer can also be used for the determination of iron loss and a.c. permeability of ferromagnetic materials working at low flux densities. A.C. potentiometers have been described in Chapter 15.

The connections for iron loss measurement with a potentiometer are given in Fig. 18.27. The potentiometer is of Co-ordinate type.

The ring specimen carries two windings. The primary winding has $N_1$ turns and the secondary $N_2$ turns. The primary winding is supplied from an alternator through a regulator transformer. This alternator supplies the two potentiometer slide wire circuits also. The primary winding circuit has a regulating resistance. A standard resistor $R$ is connected in series with the primary winding. This resistor is introduced in the circuit for the purpose of measuring the current in the primary winding by measuring the potential drop across it. $G$ is a vibration galvanometer.

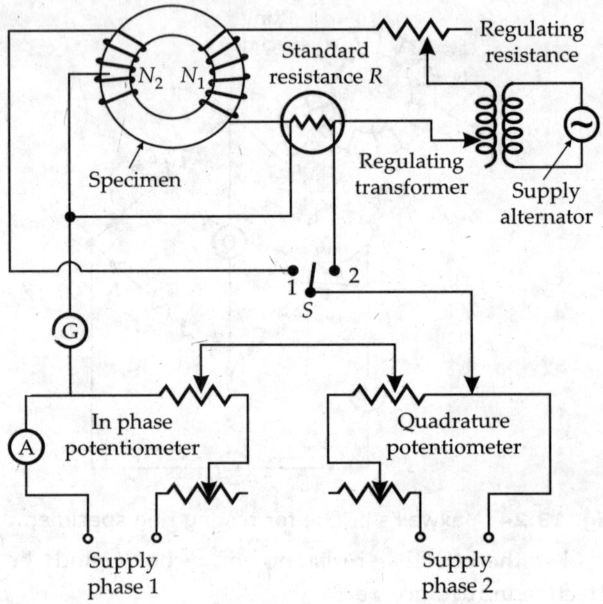


**Fig. 18.27** A.C. potentiometer method of testing ring specimens.

The potentiometer is standardized first. The supply is given to the primary winding of the specimen. A voltage $E_2$ is induced in the secondary winding.

$$E_2 = 4 K_f f \Phi_m N_2 = 4 K_f f B_m A_s N_2$$

where $K_f$ = form factor = 1.11 for sinusoidal supply,

$f$ = frequency,

$B_m$ = maximum flux density,

$A_s$ = cross-sectional area of specimen,

and $N_2$ = number of turns in the secondary winding

$\therefore$ Maximum flux density,

$$B_m = \frac{E_2}{4 K_f f A_s N_2} = \frac{E_2}{4.44 f A_s N_2}$$

for sinusoidal supply.

The voltage $E_2$ is measured by putting the switch $S$ on contact 1, setting the quadrature potentiometer at zero and adjusting the inphase potentiometer till balance is obtained. The setting of inphase potentiometer gives the value of $E_2$ directly.

Switch $S$ is then put on contact 2. At this position the potential drop in the standard resistance $R$ is matched against the inphase and quadrature potentiometers. These potentiometers are adjusted to give balance condition.

The reading of inphase potentiometer gives the value of $I_e R$ where $I_e$ is the loss component of the current in the primary winding :

$\therefore\ I_e$ = reading of inphase potentiometer/$R$

The reading of quadrature potentiometer gives the value of $I_m R$ where $I_m$ is the magnetising component of current in the primary winding.

$\therefore\ I_m$ = reading of quadrature potentiometer/$R$,

Iron loss $= I_e E_2 N_1 / N_2$

Potentiometer method is very suitable for a.c. magnetic testing at low flux densities since magnetising and iron loss components of exciting current are measured separately.

### 18.6.4 Oscillographic Method

Figure 18.28 shows arrangement used for making magnetic measurements by a cathode-ray oscillograph. Two coils are wound on the ring test specimen. Magnetising winding $N_1$ and a small non-inductive resistor $r_1$ are connected to supply through an auto-transformer.

Voltage $v_1 (= i_1 r_1)$ across resistance $r_1$ is directly proportional to magnetising current $i_1$ and, consequently, to the strength of the magnetic field in the test specimen. This voltage $v_1$ is applied to the $X$ plates through an amplifier.

Search coil $N_2$ is shunted by a capacitor $C$ and a high resistance $r_2$.

The emf induced in the search coil is given by :

$$e_2 = N_2\, d\Phi / dt = N_2 A_s\, dB / dt$$

When resistance $r_2$ is higher than that of the remaining section of the circuit, the current in the measuring circuit is,

$$i_2 = e_2 / r_2 = N_2 (A_s / r_2) \cdot (dB / dt)$$

The voltage across capacitor $C$ is :

$$v_c = \frac{1}{C}\int i_2\, dt = \frac{1}{C}\frac{N_2 A_s}{r_2}\int dB = \frac{1}{C}\frac{N_2 A_s}{r_2} B.$$

Therefore voltage $v_c$ across the capacitor $C$ is proportional to the flux density in the test specimen. This voltage $v_c$ is applied to $Y$ plates of the oscilloscope through an amplifier. The $X$-deflection is proportional to the voltage drop across resistance $r_1$ carrying the magnetising current, the deflection is therefore proportional to the magnetising force $H$.

The $Y$ deflection is proportional to voltage $v_c$ and therefore to the flux density $B$ in the specimen.

1. *Plotting of hysteresis loop.* With the arrangement shown, the $X$-deflection is proportional to $H$ and the $Y$-deflection is proportional to $B$. Thus the figure traced on the screen will be the hysteresis loop of the material.
2. *Plotting of B-H curve.* If the peak flux density in the specimen is slowly and continuously varied (*e.g.*, by varying the excitation of the alternator supplying the magnetising current), the locus of the tips of the hysteresis loops gives the *B-H* curve for the material.
3. *Determination of hysteresis loss.* The hysteresis loss for the material may be estimated from the area of the loop either by comparison with a loop for a standard specimen or by calculation from the $X$ and $Y$ sensitivities of the cathode-ray oscilloscope and the circuit constants.

Further, it must be realized that the current flowing through $r$ contains a component associated with the eddy-current loss in the specimen. Thus the loop obtained is really a 'total-loss' loop whose area represents the total iron loss. The method is normally used for such purposes as acceptance tests, where the loop obtained may be compared with a standard loop.

18

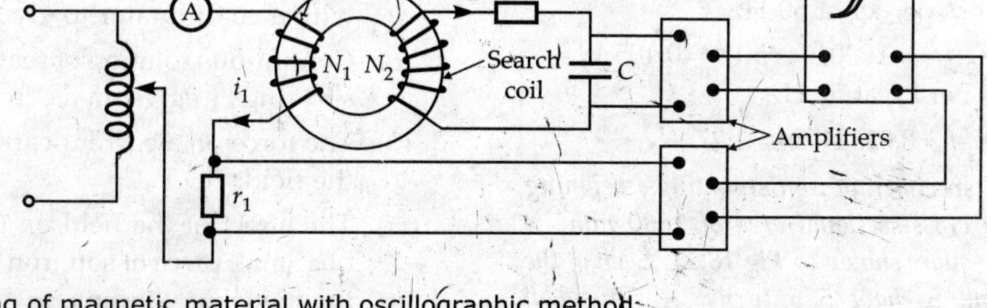


**Fig. 18.28** Testing of magnetic material with oscillographic method.

***Example 18.6*** *A manufacturer lists grain oriented steel sheet 0.3 mm thick. The resistivity of material is $50 \times 10^{-8}\,\Omega$ m. The hysteresis loop is essentially rectangular in form, with a coercive force of 12 A/m for all peak values of flux densities between 0.8 to 1.6 Wb/m$^2$. A sinusoidal flux density of 1 Wb/m$^2$ (peak) at 100 Hz is used. The density of material is 7650 kg/m$^3$. Find the total loss/kg.*

**Solution.** Base of hysteresis loop

$$= 2\,H_c = 2 \times 12 = 24\text{ A}.$$

Ordinate of hysteresis loop $= 2\,B_m = 2$ Wb/m$^2$.

Area of hysteresis loop represents the loss per cycle.

$\therefore$ Hysteresis loss per cycle $= 24 \times 2 = 48$ J.

Hysteresis loss = area of loop × frequency

$$= 48 \times 100 = 4800\text{ W}.$$

This loss is in 1 m$^3$ of volume.

$$= \frac{4\,K_f^2\,f^2\,B_m^2\,t^2}{3\rho}$$

$$= \frac{4 \times (1.11)^2 \times (100)^2 \times (1.0) \times (0.3 \times 10^{-3})^2}{3 \times 50 \times 10^{-8}} = 2970\text{ W}.$$

$\therefore$ Total loss/m$^3$ $= 4800 + 2970 = 7770$

$\therefore$ Total loss/kg $= 7770/7650 = 1.03$ W.

***Example 18.7*** *In a test on a specimen of total weight 13 kg the measured values of iron loss at a given value of peak flux density were 17.2 W at 40 Hz and 23.9 W at 60 Hz. Estimate the values of hysteresis and eddy current losses in W/kg at 50 Hz for the same value of peak flux density.*

**Solution.** Eddy current loss, $p_e = K_e\,f^2\,B_m^2$ W/kg.

Hysteresis loss, $p_h = K_h\,f\,B_m^k$ W/kg.

Since $B_m$ is constant, we can write,

$$p_h = K_1 f \text{ and } p_e = K_2 f^2.$$

$\therefore$ Total iron loss 1 kg

$$p_i = p_h + p_e = K_1 f + K_2 f^2$$

Substituting the values of iron loss per kg at 40 and 60 Hz, we have

$$7.2/13 = K_1(40) + K_2(40)^2 \qquad ...(i)$$

and $$28.9/13 = K_1(60) + K_2(60)^2 \qquad ...(ii)$$

From (*i*) and (*ii*) we get,

$$K_1 = 0.0254 \text{ and } K_2 = 0.000194.$$

Eddy current loss per kg at 50 Hz

$$= K_2 f^2 = 0.000194\,(50)^2 = 0.485\text{ W}.$$

Hysteresis loss per kg at 50 Hz

$$= K_1 f = 0.0254 \times 50 = 1.27\text{ W}.$$

***Example 18.8*** *A specimen of iron stampings weighing 10 kg and having a cross-sectional area of 1680 mm$^2$ is tested by an Epstein square shown in Fig 18.21. Each of the two windings $S_1$ and $S_2$ have 515 turns. A sinusoidal voltage supply at 50 Hz is given to the primary winding of the Epstein square and an ammeter connected in the primary circuit indicates a current of 0.35 A and the electrostatic voltmeter connected across a secondary winding indicates a voltage of 250 V. The resistance of each of the windings $S_1$ and $S_2$ is 40 Ω. The resistance of the pressure coil of the wattmeter is 80,000 Ω. Calculate the value of flux density in the specimen and the iron loss/kg at this flux density if the wattmeter connected on the primary side indicates 80 W.*

**Solution.** Maximum flux density in the specimen

$$B_m = \frac{E}{4.44\,fA_s\,N_2}$$

$$= \frac{250}{4.44 \times 50 \times 1680 \times 10^{-6} \times 5.15} = 1.3\text{ Wb/m}^2$$

Total loss $P = 80$ W.

$\therefore$ Iron loss in stampings

$$P_i = P\left(1 + \frac{r_c}{r_p}\right) - \frac{E^2}{r_p + r_c} \qquad \text{(see Eqn. 18.23)}$$

$$= 80\left(1 + \frac{40}{80{,}000}\right) - \frac{(250)^2}{40 + 80{,}000} = 79.26\text{ W}.$$

Iron loss/kg. $= \dfrac{79.26}{10} = 7.926$ W/kg.

## 18.7 METHODS OF MEASUREMENTS OF AIR GAP FLUX

Measurements described in preceding sections involve changes of magnetization. Such changes may not be always possible and measurements of an unvarying field must then be made. These measurements sometimes involve measurement of very weak fields *e.g.*, of the earth's field in geomagnetic surveys and sometimes fields of very large strength *e.g.*, fields of powerful electromagnets. Such measurements are always of the field in air gap. The range of these measurements is very large and therefore we have to select suitable method for the measurement. Some of the methods are :

(*i*) Deflection of a pivoted magnetic needle – the classical magnetometer method.

(*ii*) The rapid rotation of search coil through 180° – *e.g.*, the classical earth inductor.

(*iii*) Continuous rotation of a search coil fitted with a commutator to give a d.c. output.

(*iv*) Continuous rotation of search coil fitted with slip-rings to give an a.c. output.

(*v*) The force on a current-carrying conductor in the field.

(*vi*) The effect of the field on the magnetization of a small piece of soft iron excited by a.c. in a magnetising winding.

There are a few more methods which are described in details below :

**1.** *Measurement of field strength by withdrawal of Search Coil.* In this method a search coil is placed in a magnetic field. The terminals of the search coil are connected to a ballistic galvanometer. This search coil is suddenly withdrawn from the magnetic field, with the result the flux linking with the search coil changes. This induces an emf in the search coil which causes the galvanometer to deflect. From the throw of the galvanometer the flux density in the field may be known.

**2.** *Measurement of constant field strength by nuclear magnetic resonance.* When an alternating magnetic field of high frequency is superimposed on atomic nuclei located in a constant magnetic field, resonance absorption of energy from the high-frequency field is liable to occur at a specific ratio of the constant field strength $H$ to frequency $f$ of the alternating field.

The laws of quantum mechanics postulate that the magnetic moment of, say, the hydrogen nucleus (proton) may be oriented either in the same direction as that of the external magnetic field or in opposition to it.

The orientation of the proton magnetic moment can be changed by application of an energy quantum $hf$, where $h$ is Planck's constant and $f$ is frequency.

The difference between these two energy states of the proton amounts to $2\,m_p\,H$ where $m_p$ is the proton magnetic moment.

Evidently, $hf = 2\,m_p H$ from which it follows that :

$$\omega = 2\pi f = \frac{4\pi}{h} m_p H = v_p H$$

or

$$H = \frac{2\pi f}{v_p} \qquad \text{...(18.29)}$$

where $v_p$ is the gyromagnetic ratio of nucleus, *i.e.*, the ratio of the magnetic moment to the mechanical moment.

The gyromagnetic ratio of the nucleus of a number of substances is known to the nearest of a percent. According to this method the magnetic strength of a constant field is determined by measuring the frequency at which nuclear magnetic resonance occurs. The block diagram of the test installation used for this purpose is shown in Fig. 18.29.

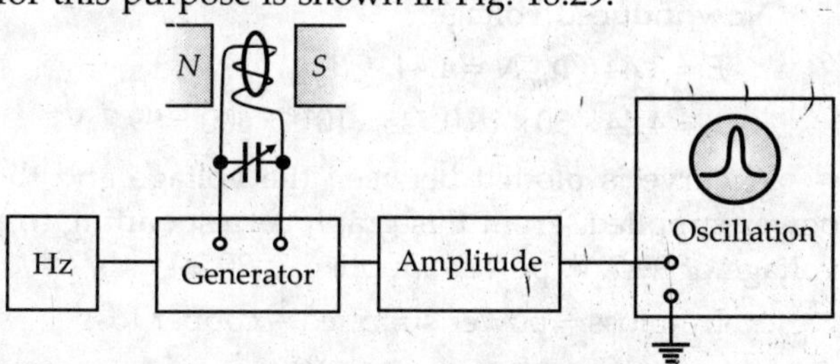


**Fig. 18.29** Measurement field strength by nuclear magnetic resonance.

A proton sample, for example, an aqueous solution of Fe $(NO_3)_3$ contained in a glass ampoule is surrounded by a small coil and inserted in the magnetic field to be measured, the axis of the coil being at right angles to the field. The coil and capacitor constitute an oscillatory circuit, which is connected to a high frequency oscillator.

Resonance is detected by a sharp change in the amplitude of the resonance curve traced on the screen of the cathode-ray oscillograph at a change in the oscillator frequency. The resonance frequency is measured by means of a frequency meter or wavemeter. This method is most suitable for measurement of fields having a magnetic strength of 800 to $1.6 \times 10^6$ A/m. The accuracy of measurement lies within $\pm$ 0.01 per cent.

## 18.8 TESTING OF PERMANENT MAGNETS

The details of the method of testing permanent magnets, when manufactured, depend upon the shape of the magnet. This varies greatly according to the purpose for which the magnet is to be used. Owing to the self-demagnetizing force in permanent magnets, it is necessary, in testing such magnets. to measure $H$ by means of search coils, of small cross-section, laid flat against the surface of the magnet. These search coils give the value of the air flux density at the surface of the magnet at different points, and search coils *i.e.*, $B$ coils in the ordinary way.

*Betteridge apparatus for magnet testing.* The principle of this apparatus for the commercial testing of permanent magnets may be followed from Fig. 18.30. The magnet to be tested is placed with its straight ends inside two magnetising coils and so that its ends press against two pole pieces, in the air gap between which a thin iron disc, plated with copper, is mounted. This disc is mounted in ball bearings and the clearances between it and the pole pieces are small. It is driven round at constant speed by a small motor, and has two small brushes making contact with its spindle and its rim. When a current flows in the magnetising windings, flux crosses the gap between the pole

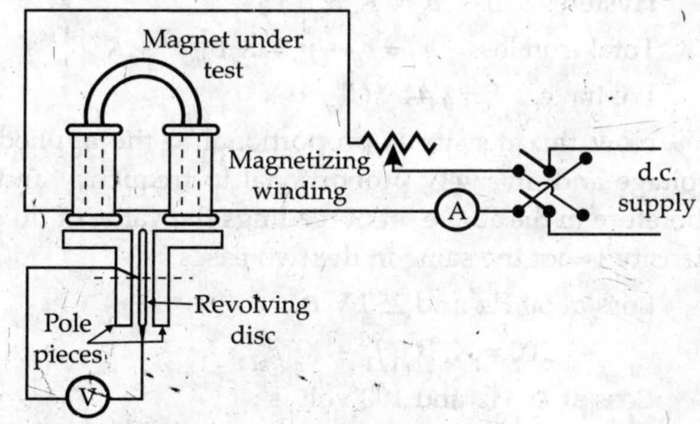


**Fig. 18.30** Betteridge apparatus.

18

pieces, and the flux density in the gap is proportional to that in the magnet. An emf is induced in the revolving disc, this emf being proportional to the gap flux density. The emf is measured by a voltmeter $V$.

Thus, the voltmeter reading gives a measure of the flux density in the specimen. The value of $H$ is obtained by the usual formula from the constants of the magnetising coils and from the current in these coils. Some leakage exists between the two arms of the magnet, but with proper design of the apparatus its effect upon the results is small.

**Example 18.9** *The iron loss in a sample is 300 W at 50 Hz with eddy current loss component 5 times as big as the hysteresis loss component. At what frequency will the iron loss be double if the flux density is kept the same ?*

**Solution.** If the value of flux density is kept constant we can write

$$p_h = K_1 f \text{ and } p_e = K_2 f^2$$

Total iron loss/kg :

$$p_i = p_h + p_e = K_1 f + K_2 f^2$$

Total iron loss at 50 Hz is

$$300 = K_1(50) + K_2(50)^2$$

but $K_2(50)^2 = 5 \times K_1(50)$

$\therefore$ $K_1 = 0.1\ K_1$.

Hence $300 = 0.1\ K_1(50)^2 + K_1(50)$ or $K_1 = 1$

Suppose iron loss is doubled at frequency $f_1$,

$\therefore$ $2 \times 300 = K_1(f_1) + K_2(f_1)^2$

or $600 = K_1 f_1 + 0.1 K_1 f_1^2$ or $f_1 = 72.6$ Hz.

**Example 18.10** *In loss tests on a sample of iron laminations the following results were recorded :*

*(i) At 60 Hz, 250 V, total iron loss = 200 W.*

*(ii) At 40 Hz, 100 V, total iron loss = 40 W.*

*Calculate the eddy current and hysteresis loss for each test. The Steinmetz index is 1.6.*

**Solution.** Eddy current loss, $p_e = K_e B_m^2 f^2$

Hysteresis loss, $p_h = K_h B_m^{1.6} f$

Total iron loss, $p_i = p_h + p_e = K_e B_m^2 f^2 + K_h B_m^{1.6} f$

We have, $E = 4.44\ f(B_m A) N$

Now flux density is proportional to the applied voltage and inversely proportional to frequency and therefore in the above set of readings the value of flux density is not the same in the two cases.

Loss at 50 Hz and 250 V is :

$$200 = K_e B_{m1}^2 f_1^2 + K_h B_{m1}^{1.6} f_1$$

Loss at 40 Hz and 100 volt is :

$$40 = K_e B_{m2}^2 f_2^2 + K_h B_{m2}^{1.6} f_2$$

But $B_{m1} f_1 \propto 250$ and $B_{m2} f_2 \propto 100$.

or $$B_{m1} = 2.5 \times \frac{40}{50} = 2 B_{m2}.$$

Hence $200 = K_e B_{m1}^2 (50)^2 + K_h (B_{m1})^{1.6} (50)$

$$40 = K_e \left(\frac{B_{m1}}{2}\right)^2 (40)^2 + K_h \left(\frac{B_{m1}}{2}\right)^{1.6} (50)$$

From above, we have,

$$K_e (B_{m1})^2 = 0.049 \text{ and } K_h (B_{m1})^{1.6} = 1.55$$

Hence, at 250 V and 50 Hz,

Eddy current loss

$$= K_e (B_{m1})^2 (50)^2 = 0.049\ (50)^2 = 122.5 \text{ W.}$$

Hysteresis loss

$$= K_h (B_{m1})^{1.6} (50) = 1.55 \times 50 = 77.5 \text{ W.}$$

At 100 volt and 40 Hz, Eddy current loss

$$= K_e \left(\frac{B_{m1}}{2}\right)^2 (40)^2 = \frac{0.049}{4} (40)^2 = 19.6 \text{ W.}$$

and Hysteresis loss

$$= K_h \left(\frac{B_{m1}}{2}\right)^{1.6} (40) = \frac{1.55}{2^{1.6}} (40) = 20.4 \text{ W.}$$

**Example 18.11** *The following test results were obtained on a sample of steel stampings at 50 Hz.*

| *Voltage V* | 4.9 | 69.3 | 91.8 | 100.5 | 110.5 | 118.0 |
|---|---|---|---|---|---|---|
| *Current A* | 0.2 | 0.3 | 0.46 | 0.52 | 0.64 | 0.77 |
| *Power W* | 9.5 | 16.8 | 27.5 | 32.5 | 39.0 | 44.8 |

*Mean width of plates = 30 mm ;*

*Mean thickness = 0.489 mm ;*

*Number of plates = 51 ;*

*Total weight = 11 kg ;*

*Number of magnetising turns in coil = 600.*

*Allowing a copper loss of 2 W in the magnetising winding, calculate iron loss in W/kg at a flux density of 1 Wb/m$_2$ and a frequency 50 Hz.*

**Solution.** Area of core $A$ = width × thickness

$$= 30 \times 10^{-3} \times 51 \times 0.489 \times 10^{-3}$$

$$= 0.748 \times 10^{-3} \text{ m}^2.$$

Now induced voltage,

$$E = 4.44\ f\Phi_m N = 4.44\ f B_m AN$$

$$= 4.44 \times 50 \times 1 \times 0.748 \times 10^{-3} \times 600 = 99.7 \text{ V.}$$

A curve is plotted between the voltage and the power supplied. From this graph corresponding to a voltage of 99.7 V, power supplied is 30.3 W.

$\therefore$ Iron loss = power supplied − copper loss

$$= 30.3 - 2 = 28.3 \text{ W.}$$

Iron loss/kg. = 28.3/11 = 2.57 W.

**Example 18.12** *In an iron loss test on a 10 kg specimen of steel laminations, the peak flux density and the form factor were maintained constant and the following results were obtained :*

| *Frequency, Hz* | 25 | 40 | 50 | 60 | 80 |
|---|---|---|---|---|---|
| *Total iron loss, W* | 18.5 | 36 | 50 | 66 | 104 |

*Evaluate the eddy current and hysteresis loss per kg at 50 Hz.*

**Solution.** If the flux density and form factor remain constant, we can write :

Total iron loss $P_i = K_1 f + K_2 f^2$

or $\frac{P_i}{f} = K_1 + K_2 f$

The values of $P_i / f$ for various values of $f$ are :

| $f$ | 25 | 40 | 50 | 60 | 80 |
|---|---|---|---|---|---|
| $P_i / f$ | 0.74 | 0.9 | 1.0 | 1.1 | 1.3 |

A graph is plotted between $f$ and $P_i / f$ as shown in Fig. 18.31.

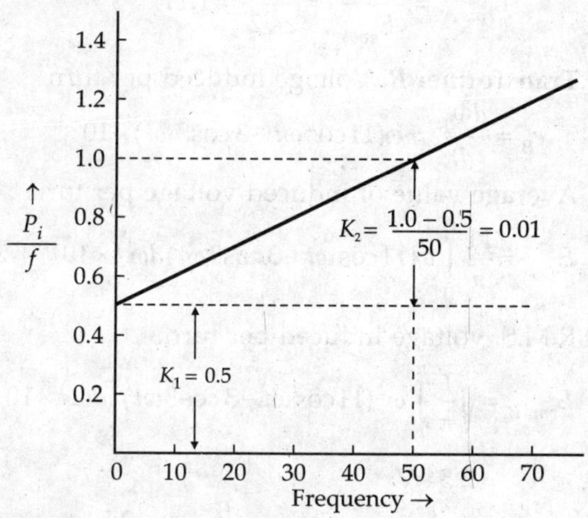


**Fig. 18.31** Diagram of Example 18.12.

From the graph

$K_1 = 0.5$ and $K_2 = 0.01$

Eddy current loss at 50 Hz

$= K_2 f^2 = 0.01(50)^2 = 25$ W.

Eddy current loss per kg $= 25/10 = 2.5$ W.

Hysteresis loss at 50 Hz $= K_1 f = 0.5 \times 50 = 25$ W.

Hysteresis loss per kg $= 25/10 = 2.5$ W.

**Example 18.13** *A test conducted on a sample of sheet laminations gave the following results at a maximum flux density of 1.0 Wb/m² with flux waveform purely sinusoidal.*

| *Frequency, Hz* | 25 | 40 | 50 | 75 |
|---|---|---|---|---|
| *Iron loss $p_i$, W/kg* | 2.32 | 4.35 | 6.0 | 10.9 |

*Determine*

(*i*) *Hysteresis and eddy current loss in W/kg at 50 Hz and at 1.0 Wb/m² and also at*

(*ii*) *a maximum flux density of 1.2 Wb/m² when the frequency is 60 Hz and the form factor of flux wave is 1.2. Assume Steinmetz index as 1.6.*

**Solution.** For a constant value of flux density and form factor :

Total iron loss $p_i = K_1 f + K_2 f^2$

or $\frac{p_i}{f} = K_1 + K_2 f$

The values of $p_i / f$ for various values of $f$ are :

| $f$ | 35 | 40 | 50 | 75 |
|---|---|---|---|---|
| $p_i / f$ | 0.093 | 0.109 | 0.12 | 0.145 |

A graph is drawn in Fig. 18.32 between $p_i / f$ and $f$. From this graph, we get : $K_1 = 0.068$ and $K_2 = 0.00108$

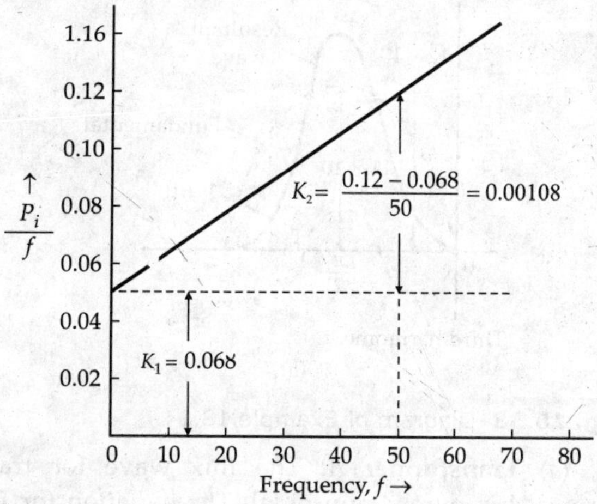


**Fig. 18.32** Diagram of Example 18.13.

∴ Hysteresis loss at 50 Hz and 1.0 W/bm²

$= K_1 f = 0.068 \times 50 = 3.4$ W/kg.

Eddy current loss at 50 Hz and 1.0 Wb/m²

$= 0.00108 \times 50^2 = 2.7$ W/kg

Now hysteresis loss $\alpha\ B_m^{1.6} f$

∴ Hysteresis loss at 60 Hz and 1.2 Wb/m²

$= 3.4 \times \left(\frac{1.2}{1.0}\right)^{1.6} \times \left(\frac{60}{50}\right) = 5.47$ Wb/m².

Eddy current loss $\alpha\ K_f^2 B_m^2 f^2$

Eddy current loss at

60 Hz, 1.2 Wb/m², and $K_f = 1.2$.

$= 2.7 \times \left(\frac{1.2}{1.11}\right)^2 \times \left(\frac{1.2}{1.0}\right)^2 \times \left(\frac{60}{50}\right)^2 = 6.5$ W/kg.

***Example 18.14*** *The core fluxes of two identical transformers A and B are $10 \times 10^{-3}$ sin ωt weber and $(11 \sin \omega t - 1 \sin 3\omega t) \times 10^{-3}$ weber respectively. Find for each transformer (a) maximum value and form factor of flux wave, (b) the mean value and form of factor induced emf per turn. (c) Also find the hysteresis and eddy current losses of transformer B if they are 40 W and 15 W respectively for transformer A. Angular frequency $\omega = 314$ rad / s.*

**Solution.** The flux waves of two transformers are shown in Fig. 18.33.

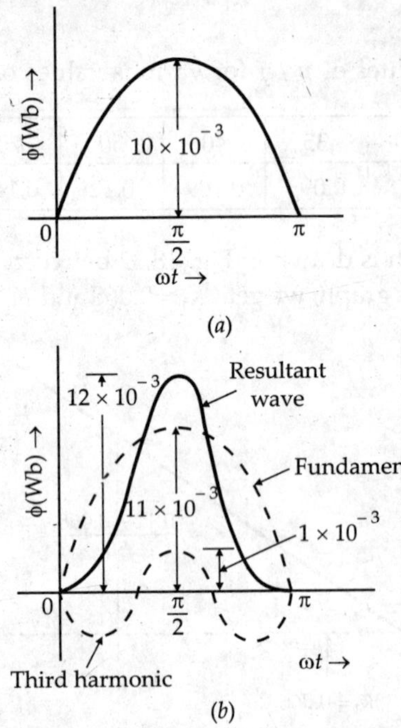


**Fig. 18.33** Diagram of Example 18.14.

(*a*) **Transformer *A*.** The flux wave for transformer *A* is purely sinusoidal. The equation for flux wave is :

$$\Phi_A = 10 \times 10^{-3} \sin \omega t$$

∴ Maximum value of flux of *A* is :

$$\Phi_{\max A} = 10 \times 10^{-3} \text{ Wb.}$$

Form factor for flux wave for transformer *A* is 1.11 as it is sinusoidal = 1.11.

**Transformer *B*.** The flux wave for transformer [Fig. 18.33(*b*)] is a combination of a fundamental and a third harmonic. The maximum value occurs at $\pi/2$ as is clear from the diagram. Maximum value of flux

$$= \Phi_{\max B} = (11+1) \times 10^{-3} = 12 \times 10^{-3} \text{ Wb.}$$

Average value of flux is

$$\Phi_{av B} = \frac{1}{\pi}\int_0^{\pi} (11 \sin \omega t - 1 \sin 3\omega t)\, d(\omega t) \times 10^{-3}$$

$$= 6.8 \times 10^{-3} \text{ Wb}$$

R.M.S. value of flux :

$$\Phi_{r.m.s B} = \sqrt{\frac{1}{\pi}\int_0^{\pi} (11 \sin \omega t - 1 \sin 3\omega t)^2 d(\omega t)} \times 10^{-3}$$

$$= 7.8 \times 10^{-3} \text{ Wb}$$

∴ Form factor of flux wave

$$= \frac{\Phi_{r.m.s A}}{\Phi_{av A}} = \frac{7.8 \times 10^{-3}}{6.8 \times 10^{-3}} = 1.15.$$

(*b*) **Transformer *A*.** Voltage induced per turn

$$e_A = \frac{d\Phi_A}{dt} = 10 \times 10^{-3} \cos \omega t \times \omega = 3140 \times 10^{-3} \cos \omega t$$

Average value of induced voltage per turn

$$E_{av A} = \frac{2}{\pi} \times 3140 \times 10^{-3} \text{ V.}$$

R.M.S. value of induced voltage per turn

$$E_{rms A} = \frac{3140 \times 10^{-3}}{\sqrt{2}}$$

∴ From factor for voltage

$$E_{rms A} = \frac{E_{rms A}}{E_{av A}} = \frac{\pi}{2\sqrt{2}} = 1.11.$$

**Transformer *B*.** Voltage induced per turn

$$e_B = \frac{d\Phi_B}{dt} = \omega(11 \cos \omega t - 3 \cos 3\omega t) \times 10^{-3}$$

Average value of induced voltage per turn

$$E_{av B} = \frac{11}{\pi}\int_0^{\pi} \omega(11 \cos \omega t - 3 \cos 3\omega t)\, d\omega t \times 10^{-3} = 2 \text{ V.}$$

R.M.S. voltage induced per turn

$$E_{rms B} = \sqrt{\frac{1}{\pi}\int_0^{\pi} \omega^2 (11 \cos \omega t - 3 \cos 3\omega t)^2\, d\omega t} \times 10^{-3}$$

$$= 2.54 \text{ V.}$$

Form factor for voltage $= \dfrac{E_{rms B}}{E_{av B}} = \dfrac{2.54}{2} = 1.27$

(*c*) Eddy current loss $P_e \propto K_f^2 B_m^2 f^2$.

Also $B_m \propto \Phi_m$

$$\therefore \quad \frac{P_{eB}}{P_{eA}} = \frac{(1.15)^2 \times (12 \times 10^{-3})^2 \times (50)^2}{(1.11)^2 \times (10 \times 10^{-3})^2 \times (50)^2} = 1.54.$$

∴ Eddy current loss in transformer *B*,

$$P_{eB} = 1.54 \times 15 = 23.1 \text{W}$$

Hysteresis loss $P_h \propto f B_m^{1.6}$

$$\therefore \quad \frac{P_{hB}}{P_{hA}} = \frac{50}{50} \times \left(\frac{12 \times 10^{-3}}{10 \times 10^{-3}}\right)^{1.6} = 1.34$$

∴ Hysteresis loss in transformer *B*,

$$P_{hB} = 1.34 \times 40 = 53.6 \text{ W.}$$

**Example 18.15** *The resistance and inductance of a test coil $L_1$, $R_1$ were determined at a frequency of $4000/2\pi$ Hz with a Maxwell's bridge shown in Fig. 16.4. The results were as follows :*

(*a*) *With air core the balance was obtained with :*

$R_2 = 550\ \Omega$, $R_3 = 18\ \Omega$,

$R_4 = 1250\ \Omega$, $C_4 = 0.5\ \mu F$.

(*b*) *With iron dust core, balance was obtained with :*

$R_1 = 550\ \Omega$, $R_3 = 18\ \Omega$,

$R_4 = 1125\ \Omega$, $C_4 = 3.85\ \mu F$.

*Determine the iron loss in the core at the test frequency. The voltage applied to the bridge is 50 V.*

**Solution.** From Eqs. 16.18 and 16.19, we have

$$R_1 = \frac{R_2 R_3}{R_4} \text{ and } L_1 = R_2 R_3 C_4.$$

(*a*) **With air core.** When there is no iron core, resistance $R_1$ represents the resistance of the winding of the coil.

∴ Resistance of winding of coil

$$R_w = \frac{550 \times 18}{1250} = 7.94\ \Omega$$

There is no use of calculating inductance with air core.

(*b*) **With iron core.** When the core is made iron. the resistance $R_1$ represents the resistance of winding plus resistance equivalent to iron loss.

∴ Effective resistance of coil

$$R/1 = \frac{550 \times 18}{1125} = 8.81\ \Omega$$

Hence resistance equivalent to iron loss in core

$$= 8.81 - 7.94 = 0.87\Omega$$

Inductance of iron cored coil

$$L_1 = 550 \times 18 \times 3.85 \times 10^{-6} = 38.2 \times 10^{-3}\ \text{H}$$

Reactance of coil

$$X_1 = 2\pi \times \frac{4000}{2\pi} \times 38.2 \times 10^{-3} = 152.8\ \Omega$$

At balance current through the coil

$$I_1 = \frac{V}{\sqrt{(R_1 + R_3)^2 + X_1^2}}$$

$$= \frac{50}{\sqrt{(8.81 + 18)^2 + (152.8)^2}} = 0.322\ \text{A}.$$

Iron loss

$$= (0.322)^2 \times 0.87 = 0.09\ \text{W}.$$

**Example 18.16** *An iron cored coil and a standard 2 Ω resistor are connected in series and the potential difference is measured across each on an a.c. potentiometer of Co-ordinate type. The readings are 3.6 + j 1.2 V for the coil and 4.8 – j 2 V for the resistor. Calculate the magnetising component and loss component of current through the coil.*

18

**Solution.** Current through the coil

$$I = \frac{4.8 - j2.0}{2} = 2.4 - j1.0$$

$$= 2.6 \angle -22°36'\ \text{A}.$$

Voltage across the coil

$$V = 3.6 + j12 = 4.07 \angle 18°18'\ \text{V}.$$

Angle between $V$ and $I$ is

$$\phi = 18°18' + 22°36' = 40°54'$$

Loss component of current

$$I_e = I\cos\phi = 2.6\cos 40°54' = 1.96\ \text{A}.$$

Magnetising component of current

$$I_m = I\sin\phi = 2.6\ \text{in}\ 40°54' = 1.72\ \text{A}.$$

# Review Questions

1. Describe briefly the different types of tests that are used for testing of magnetic materials.
2. Describe a method of experimental determination of flux density in a specimen of magnetic material using a ballistic galvanometer. Explain how the correction for flux in the air space between the specimen and the coil is applied.
3. Explain the method for experimental measurement of magnetising force acting on a specimen of magnetic material.
4. Explain the construction and working of a Magnetic Potentiometer.
5. Describe the method for determination of *B-H* curve of a magnetic material using : (*i*) Method of Reversals and (*ii*) Step by Step method.
6. Explain the Method of Reversals for experimental determination of hysteresis loop of a magnetic specimen.
7. What are Permeameters ? Explain the working of a Hopkinson Permeameter.
8. Describe the construction and working of following types of permeameters :

   (*i*) Illiovici permeameter

   (*ii*) Burrow's permeameter.

9. Explain the methods of separation of iron losses into their two components : eddy current and hysteresis losses, if the maximum value of flux density is maintained constant and
   (*i*) frequency is varied keeping the form factor constant
   (*ii*) form factor is varied keeping the frequency constant.
10. Describe the Lloyd Fisher square for measurement of iron losses in a specimen of laminations. Describe how corrections for resistance of wattmeter pressure coil resistance and resistance of secondary winding are applied.
    How is the true value of flux density obtaining in the laminations determined ?
11. Explain how the value of a.c. permeability of magnetic materials is determined through the use of :
    (*i*) Maxwell's Bridge and
    (*ii*) Campbell's Bridge.
12. Describe how magnetizing and loss components of no load current of a transformer be determined by using an a.c. potentiometer.
13. What is Nuclear Magnetic Resonance ? Explain how the strength of a magnetic field in air can be determined using this principle.
14. Explain the oscillographic method of determination of hysteresis loop of a magnetic material.
15. Explain a method for testing of permanent magnets.

## Unsolved Problems

1. A coil of 1000 turns is wound uniformly over an annular ring former of mean diameter 1 metre. A second shorter coil is mounted inside the first, co-axially with it. The coil has 50 turns and a mean diameter of 25 mm. A ballistic galvanometer is connected to this second coil. The total resistance of the second coil circuit is 2500 Ω. When a current of 5 A through the first coil is reversed, the galvanometer is deflected through 100 divisions. Neglecting damping, calculate the constant of the galvanometer. **[Ans.** $1.24 \times 10^{-9}$ coulomb/division]
2. A ring having a mean diameter of 0.3 m and a cross-sectional area 400 $\text{mm}^2$ has a primary winding of 80 turns wound uniformly. The secondary winding of 30 turns is connected to a flux meter having a constant of $0.12 \times 10^{-3}$ weber turn per division. A deflection of 46 divisions is observed when a current of 2 A is reversed in the primary winding. Calculate the relative permeability of iron. **[Ans.** 1080]
3. A ballistic galvanometer is used for determining the leakage co-efficient of a d.c. machine. The galvanometer has a constant of 0.2 μC per scale division and a circuit resistance of 8000 Ω. The galvanometer is first connected to a search coil of 3 turns wound on field coil of a machine and then to a search coil of 6 turns placed on the armature. With normal field current broken, the galvanometer readings in the two cases are 150 and 240 divisions respectively. Calculate the total flux, useful flux and leakage co-efficient of the machine. **[Ans.** 80 mWb, 64 mWb, 1.25]
4. A ring specimen of mild steel has a cross-sectional area of 600 $\text{mm}^2$ and a mean periphery of 0.25 m. It is uniformly wound with two coils, *A* and *B* having 80 and 300 turns respectively. Coil *B* is connected to a ballistic galvanometer having a constant of $12 \times 10^{-9}$ coulomb per division ; the total resistance of this circuit is 0.2 MΩ. When a current of 2.2 A through coil *A* is reversed the galvanometer gives a maximum deflection of 180 divisions. Neglecting the damping of the galvanometer calculate (*a*) flux the density in the iron (*b*) If radial saw cut of 0.5 mm wide is made in the ring, find the magnetising current in coil *A* to give the same flux density. **[Ans.** (*a*) 1.2 $\text{Wb/m}^2$. (*b*) 8.17 A]
5. The area of hysteresis loop for a certain magnetic material is 5000 $\text{mm}^2$ and the scales of the graph on which it is drawn are :
   *H* axis, 1 mm = 25 A/m ; *B* axis, 1 mm = 15 m $\text{Wb/m}^2$. Calculate the hysteresis loss per kg. The frequency is 50 Hz and the density of the material is 8500 $\text{kg/m}^3$. **[Ans.** 11 W]
6. A transformer is operated on 1000 V, 50 Hz and gives a total loss 1000 W, of which 700 W is due to hysteresis. If the transformer were to operate at 2000 V and 100 Hz, what would be losses due to hysteresis and eddy currents ? Steinmetz index is 1.6. **[Ans.** Hysteresis 1400 W ; Eddy currents = 1200 W]
7. Calculate the percentage change in the hysteresis and eddy current losses of a choke coil if other factors remain constant (*i*) the form factor of

applied voltage decreases by 20% (*ii*) the frequency of the applied voltage increases by 20%. Neglect resistance of choke coil.

[**Ans.** (*i*) hysteresis loss : no change : eddy current loss 36% decrease. (*ii*) hysteresis loss : 20% increase ; eddy current loss 44% increase]

**8.** A laminated iron ring specimen is magnetically tested and has a coil of 250 turns wound round the ring. The relationship between current and magnetic flux is :

| *Current A* | *Flux mWb* | *Current A* | *Flux mWb* |
|---|---|---|---|
| 0 | – 0.44 | 1.75 | 0.545 |
| 0.25 | – 0.37 | 2.0 | 0.56 |
| 0.5 | – 0.2 | 1.5 | 0.55 |
| 0.75 | 0.25 | 1.0 | 0.535 |
| 1.0 | 0.37 | 0.5 | 0.51 |
| 1.25 | 0.47 | 0 | 0.44 |
| 1.5 | 0.52 | | |

Calculate the hysteresis loss per cycle.

[**Ans.** 0.318 J/cycle]

**9.** The following test results refer to a sample of steel :

**1.** At a constant flux density.

| *Frequency, Hz* | 20 | 32 | 40 | 48 | 64 |
|---|---|---|---|---|---|
| *Loss, W* | 14.8 | 28.5 | 40 | 53 | 84 |

**2.** At a constant frequency of 40 Hz.

| *Peak flux density, Wb/m²* | 0.25 | 0.5 | 0.75 | 1.0 | 1.25 | 1.5 |
|---|---|---|---|---|---|---|
| *Loss, W* | 3 | 10.8 | 22.8 | 40 | 61 | 85 |

The tests were carried out on 10 kg of steel. Determine (*a*) eddy current loss/kg at $B_{max} = 1 \text{Wb}/\text{m}^2$ and frequency 40 Hz. Also find (*b*) the index $k$ in the expression $\eta B_m^k$ for the hysteresis loss per cycle.

[**Ans.** (*a*) 2.08 W/kg ; $k = 1.74$]

**10.** A ring is made of thin annular laminations of stalloy weighs 1.5 kg and carries a uniform toroidal winding. The winding is connected to a 50 Hz supply and the total power loss (due to hystersis and eddy currents) is 0.6 W at a maximum flux density of 0.5 Wb/m². Estimate power loss to be expected when the maximum flux density is increased to 1.0 Wb/m² and frequency is decreased to 20 Hz. Assume the hysteresis loss is given by the formula :

$$\eta f B_m^k \text{ W/kg.}$$

where $f$ = frequency ; $B_m$ = maximum flux density Wb/m² ; $\eta = 23 \times 10^3$ ; and $k = 1.7$ [**Ans.** 0.734 W]

**11.** The primary and secondary windings of a toroidial transformer have 330 and 25 turns respectively. A 1Ω resistor is connected in series with the primary across a 50 Hz supply. The following readings were obtained using a.c. potentiometer.

Voltage across 1 Ω resistor :

$$(0.0734 + j0.0094) \text{ V.}$$

Voltage across primary winding terminals :

$$(0.1151 + j0.1124) \text{ V}$$

Voltage across secondary winding terminals :

$$(-0.0017 - j0.0074) \text{ V.}$$

If the mean diameter of coil 10 cm and the cross-section is 2.5 × 0.3 cm, calculate :

(*i*) iron loss, (*ii*) $I^2R$ loss in the primary winding, and (*iii*) relative permeability of core.
State the assumptions made.

[**Ans.** (*i*) 2.57 mW ; (*ii*) 6.63 mW ; (*iii*) 188.3]

**12.** The cores of two identical transformers carry alternating fluxes whose instantaneous values are $10 \times 10^{-3} (\sin 314t)$
and $12 \times 10^{-3} (11 \sin 628t + 0.1 \sin 1884t)$ weber respectively.
Find the ratio of eddy current loss of transformers. Find also the ratio of hysteresis loss.

[**Ans.** Eddy current loss ratio = 5.47, hysteresis loss ratio = 2.6]

# Objective Type Questions

***Tick (√) the most appropriate answer :***

**1.** Ballistic tests are used in magnetic measurements for :

(*a*) determination of *B-H* curve of the specimen only

(*b*) determination of hysteresis loop of the specimen only

(*c*) determination of flux density, magnetizing force and *B-H* curve and hysteresis loop of the specimen

(*d*) finding out iron losses in the specimen.

2. The true value of flux density in a ring specimen is :

(a) $B = B' + \mu_0 H\left(\frac{A_c}{A_s} - 1\right)$

(b) $B = B' - \mu_0 H\left(\frac{A_c}{A_s} - 1\right)$

(c) $B = B' + \mu_0 H\left(\frac{A_s}{A_c} - 1\right)$

(d) $B = B' - \mu_0 H\left(\frac{A_s}{A_c} - 1\right)$

where $B'$ = measured value of flux density,
$H$ = magnetizing force.
$A_s$ = area of cross-section of specimen,
$A_c$ = area of cross-section of coil producing magnetizing force.

3. The reason for using ring type specimens for ballistic tests is :

(a) they are free from end effects

(b) they require less labour to prepare as compared with bar type specimens

(c) they can be used for large quantities of materials of very small permeability

(d) all the above.

4. Burrow's permeameter is preferred to other permeameters as it :

(a) incorporates arrangement for uniform distribution of mmf over the entire length of the specimen

(b) incorporates arrangements for elimination of leakage flux at the joints

(c) is very accurate and can serve as reference standard

(d) all the above.

5. In an Epstein square, used for finding the iron loss in a stack of laminations, the laminations are so arranged that

(a) plane of each lamination is perpendicular to the plane of the square

(b) plane of each lamination is parallel to the plane of the square

(c) the plane of laminations may be either parallel or perpendicular to the plane of the square

(d) the plane of laminations may be at any angle to the plane of square.

6. In a Lloyd-Fisher square used for determination of iron loss in a specimen of iron laminations, the current coil of the wattmeter is connected in the primary winding circuit, while the pressure coil is connected in the secondary winding circuit. This is done so that :

(a) the $I^2R$ losses in the current coil are not included in wattmeter reading

(b) the $I^2R$ losses in the pressure coil circuit are not included in the wattmeter reading

(c) the $I^2R$ losses both in current coil and pressure coil circuit are not included in the wattmeter reading

(d) the wattmeter reads only the iron losses in the specimen of laminations.

7. Iron losses in a specimen of iron are determined by using a co-ordinate type of potentiometer. The loss component of no load current is read by :

(a) Inphase potentiometer

(b) Quadrature potentiometer

(c) either by Inphase or Quadrature potentiometer.

(d) none of the above.

8. The iron losses when measured in a specimen of laminations at a constant maximum flux density of 1.3 Wb/m$^2$ are : 20 W at 40 Hz and 40 W at 60 Hz. What are the hysteresis and eddy current losses at 50 Hz and a flux density of 1.3 Wb/m$^2$ ?

(a) 10 W, 20 W (b) 7.38 W, 212.8 W

(c) 8.38 W, 20.8 W (d) 5 W, 25 W.

9. Magnetizing force is measured by Nuclear Magnetic Resonance method. The expression for magnetizing force is

(a) $H = f / v_p$ (b) $H = 2\pi f / h$

(c) $H = fh / 4\pi$ (d) $H = 2\pi f / v_p$

$f$ = frequency in Hz,
$h$ = Planck's constant and
$v_p$ = gyrometric ratio.

10. Permanent magnets are tested by :

(a) ballistic methods

(b) using an electric circuit having a mutual inductance

(c) potentiometric methods

(d) Betteridge apparatus.

## Answers

1. (c) 2. (b) 3. (d) 4. (d) 5. (b) 6. (a) 7. (a) 8. (c) 9. (d) 10. (d)

# 19

# Optoelectronic Measurements

## 19.1 INTRODUCTION

Optoelectronics covers that area of science which combines optical and electronic technologies. This chapter first introduces basic concepts used in optoelectronics and the associated terminology. The units of measurement are then described followed by description of measurement techniques. The chapter also introduces the concepts of **Fiber optics** and the associated parameters and the methods used for measurements.

### 19.1.1 Optical Spectrum

The electromagnetic spectrum is shown in Fig. 19.1(*a*). From the diagram it is clear that the *optical spectrum* occupies only a small part of the electro-magnetic spectrum, and the visible part of the optical spectrum the optical spectrum occupying a much smaller portion of the total electromagnetic spectrum.

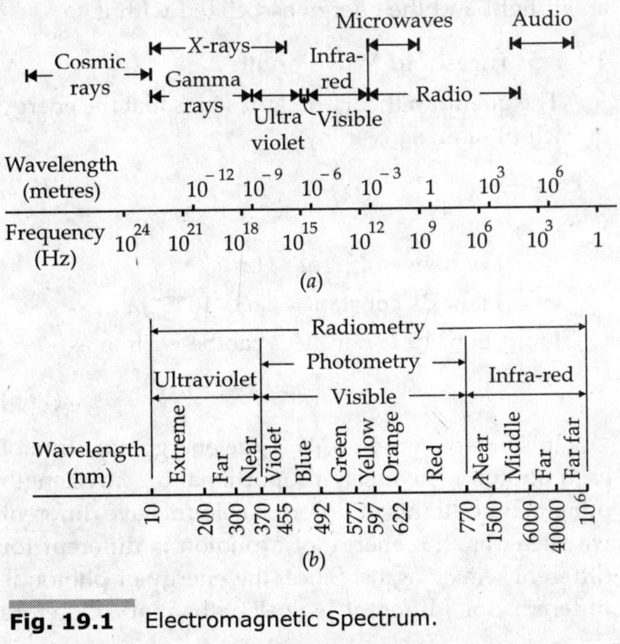


**Fig. 19.1** Electromagnetic Spectrum.

It is normal to specify the electro-magnetic spectrum in terms of wavelength rather than in terms of frequency. However, in Fig. 19.1(*a*), the range of spectrums is given both in terms of wavelength and frequency. The relationship between wavelength and frequency is :

$$f = c/\lambda \qquad ...(19.1)$$

where

$f$ = frequency ; Hz,
$\lambda$ = wavelength ; m, and
$c$ = velocity of light in m/s, which is $0.299 \times 10^9$ m/s.

The *optical spectrum* extends from 10 nm to $10^6$ nm and is divided into *three* major regions :

(*i*) Ultraviolet (UV) (10 nm – 370 nm)
(*ii*) Visible (370 nm – 770 nm)
(*iii*) Infra-red (IR) ( 770 nm – $10^6$ nm).

The ultraviolet region is further subdivided into *three* regions :

(*i*) extreme ultraviolet : 10 nm – 200 nm
(*ii*) far ultraviolet : 200 nm – 300 nm
(*iii*) near ultraviolet : 300 nm – 370 nm.

The visible region from about 370 nm – 770 nm can be seen with an average eye.

The infra-red band is subdivided into *four* regions.

(*i*) near infra-red : (770 nm – 1500 nm)
(*ii*) middle infra-red : (1500 nm – 6000 nm)
(*iii*) far infra-red : (6000 nm – 40000 nm) and
(*iv*) far far infra-red : (40000 nm – $10^6$ nm).

### 19.1.2 International Luminosity Curve

In the electromagnetic spectrum, the length of visible band is very small. Light radiation may be considered as vibrations propagated at velocity of light having wavelengths ranging from approximately 400 nm – 750 nm. The human eye responds to wavelengths between 370 nm to 770 nm.

If white light, such as given by the sun, is passed through a glass prism, it is split up into colours in the well-known spectrum-violet, indigo, blue, green, yellow, orange and red. The wavelengths of different colours are :

*Violet* : 370 nm – 455 nm, *blue* : 455 nm – 492 nm, *green* : 492 nm – 577 nm, *yellow* : 577 nm – 597 nm, *orange* : 597 nm – 662 nm and *red* : 662 nm – 770 nm.

Radiation of any wevelength outside the limits of 370 nm – 770 nm produces no sensation of light the visual apparatus of human eye.

A light source having the same rate of energy emission at each wavelength is called an equal energy source and is represented by a vertical line in the spectrum of Fig. 19.2. This diagram shows how the sensitivity or the brightness of sensation of an average human eye varies at different wavelengths for ordinary levels of illumination, the power radiated being constant for different wavelengths. This curve represents the relative luminous efficacy of different wavelengths as compared to an equal energy source. This curve is known as **"International luminosity curve"**. As far relative luminous efficacy of radiation is concerned, we define it as the ratio of the visual effect of light of a given wavelength to that of an equal energy source. It is clear from the curve that a wavelength of 555 nm in the green portion of spectrum has a relative luminous efficacy of nearly unity and thus the human eye is most sensitive to light having a wavelength of 555 nm.

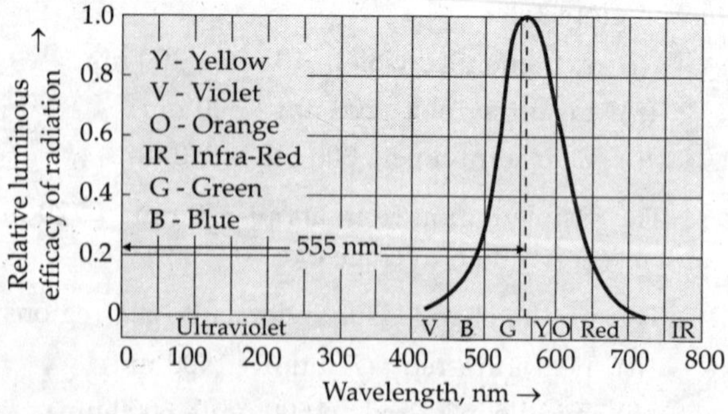


**Fig. 19.2** International luminosity curve.

The photometric evaluation of light was originally done by trained observers using visual means for matching the unknown with a calibrated and controlled source. The human eye is very proficient in detecting differences in colour and brightness but cannot make absolute measurements. This is unfortunate and is due to the fact that the colour sensitivity of the light varies from one individual to another and hence when matching lights of different colours two observers are likely to obtain different readings. To eliminate the visual observational differences in photometric measurements, the characteristics of average human eye were determined statistically and agreed by the International Commission of Illumination (CIE) and thus an International Luminosity Curve as shown in Fig. 19.2 was introduced.

### 19.1.3 Spectral Response

Spectral response is the variation of output of a transducer when light of different colours at equal illumination level is applied to it. Human eye, is in fact an optical transducer, and Fig. 19.2 depicts the spectral response of human eye.

### 19.1.4 Optical Transducers

An *optical transducer* is a light sensor *i.e.* a transducer that produces an electrical output response to light input. In instrumentation systems, measurements are not confined to visible radiation of light but are concerned with a wider spectrum of radiation including infra-red and ultraviolet. The term *light*, in fact, denotes the total range of radiation involved.The techniques used in instrumentation and process control are called **optical** because the spectrum involved is either the visible one, or infra-red or ultraviolet, the latter two are beyond the two extreme ends of the visible spectrum. Before we go in for study of optical transducers, we must know some facts about light and the energy associated with it.

### 19.1.5 Threshold Wavelength

The quantum theory of light states that the energy of each photon is given by :

$$W = h f \ J \qquad ...(19.2)$$

where

$f =$ frequency of light ; Hz,

$h =$ Planck's constant $= 6.63 \times 10^{-34}$ J/s.

From Eqn. 19.1, Eqn. 19.2 can be written as :

$$W = hc/\lambda \qquad ...(19.3)$$

It is noted from Eqn. 19.3, the energy quantum of each photon is inversely proportional to wavelength of light. Since lights of different colours have different wavelengths, the energy of a photon is different for different wavelengths. Since, the energy of photon is different for different wavelengths and therefore output of optical transducers will be different for lights of different colours although their intensities may be the same.

Let us consider the case of highly evacuated photo-emissive cell. When the light strikes the cathode, the entire energy of light quantum is converted into kinetic energy of electrons. Electrons which are imparted with greater energy that bind them together are emitted. The kinetic energy of emitted electrons is

$$W' = \frac{hc}{\lambda} e E_w \qquad ...(19.4)$$

where $E_w$ = work function ; eV,

$e$ = charge of each electron, $1.6 \times 10^{-19}$ C.

There is particular wavelength at which the energy imparted is equal to the energy required for the electrons to be liberated,

or
$$\lambda_0 = \frac{hc}{e E_w} = \frac{6.63 \times 10^{-34} \times 0.299 \times 10^9}{1.6 \times 10^{-19} E_w}$$

$$= \frac{1.24 \times 10^{-6}}{E_w} \qquad ...(19.5)$$

where $\lambda_0$ = threshold wavelength ; m.

If the frequency of any spectrum of light beyond the threshold length, there will be no response from an optical transducer.

**Example 19.1** *Show that for photoelectric emission to be possible over whole of the visible region of 370 nm — 770 nm, the work function of the photoemissive surface should be less than 1.65 eV.*

**Solution.** The threshold length is 770 nm. From Eqn. 19.5, the maximum work function is

$$E_w = \frac{1.24 \times 10^{-6}}{\lambda_0} = \frac{1.24 \times 10^{-6}}{770 \times 10^{-9}} = 1.61 \text{ eV}.$$

∴ Maximum work function is 1.61 eV which is less than 1.65 eV and hence photoelectric emission is possible over whole of the visible region.

**Example 19.2** *Determine the threshold wavelength in the case of caesium surface for which the work function is 1.8 eV.*

**Solution.** From Eqn. 19.5, the threshold wavelength is,

$$\lambda_0 = \frac{1.24 \times 10^{-6}}{1.8} = 689 \text{ nm}.$$

**Example 19.3** *Determine the maximum velocity of the emitted photoelectrons when a molybdenum surface having a work function of 4.3 eV is irridiated by mercury light having a wavelength of 0.2537 um. The e/m ratio is $0.176 \times 10^{12}$ C/kg.*

**Solution.** Energy imparted

$$= \frac{1.24 \times 10^{-6}}{0.2537 \times 10^{-6}} = 4.89 \text{ eV}.$$

The balance energy = 4.89 – 4.30 = 0.59 eV is available for acceleration of electrons.

Kinetic energy of electrons

$$= \frac{1}{2} mv^2 = e \text{ (balance energy)}$$

∴ Velocity of electrons

$$v = \sqrt{2 \times 0.59 \times 0.176 \times 10^{12}}$$
$$= 0.456 \times 10^6 \text{ m/s}.$$

## 19.2 MONOCHROMATIC LIGHT

The velocity of light through free space is constant for all types of electromagnetic radiations. However, its spectrum consists of radiations of different frequencies and wavelengths. Usually a beam of light is made up of a number of frequencies, but if the beam has a single frequency, or in practice a very narrow band of frequencies, then it is to be said to be *Monochromatic.*

## 19.3 POLARIZED WAVE SHAPES

Optical waves may be considered to move back and forth, at right angles to the direction travel of the beam of light. These beams usually are pointed in any one direction in this plane. The light in this case is called *unpolarised.* In case the beam is changed in such a way that the wave motion is only in one direction, in a plane perpendicular to the direction of motion, then it is known as *linearly polarised light.*

## 19.4 REFRACTION AND REFRACTIVE INDEX

When the light passes across the boundary of two different media it is bent provided the angle of incidence is not equal to 90°. This is called *refraction.* The *refractive index* of the medium is defined as the ratio of sine of angle of incidence to sine of angle of refraction. The refractive index varies from 1 to 3 depending on the material composition of medium and the wavelength of light.

## 19.5 REFLECTION, ABSORPTION AND TRANSMITTANCE

When a light strikes a medium, some of the light is reflected, some is absorbed and the rest is transmitted. The amount of *reflection* varies, depending upon the properties of the surface and the wavelength of incident light. The amount of reflection is about 98% for visible light on a manganese oxide surface while it is less than 1% for visible light on a lamp black surface.

The reflection is of *two* types : *viz.* specular and diffused. *Specular reflection* occurs when the surface is smooth as compared with the wavelength of light. In this case the beam is reflected with no change in divergence, and the reflected beam obeys the well known laws of reflection. In case the surface, on which the light is incident, is rough compared with the wavelength of light, *Diffused reflection* occurs. The light is scattered, and the reflection is proportional to cosine of the viewing angle. In case of a perfect diffuser, the radiance is constant and is independent of the viewing angle. The radiance is given by Lambert's law as

$$L = M / \pi \qquad ...(19.6)$$

where $L$ = radiance ; W/sr – m$^2$

and $M$ = radiant excitatance ; W/m$^2$

A part of the light falling on a surface is *absorbed*, the amount of absorption depends upon the medium. For example, pure silica glass absorbs almost zero power while the absorption is as high as 99% for lamp black.

The speed of light is highest in free space, but it decreases in other media and so does its intensity. The speed of light through a given media depends upon the wavelength of light. This is called *Dispersion*. Some of light is absorbed and some scattered. The absorbed light is converted into heat.

*Transmittance* refers to the effect when some of the light striking a surface is transmitted into the media. This transmitted light penetrates the media from zero to about 75%. The sum of the proportion of reflected, absorbed and transmitted light for any substance is equal to unity.

## 19.6 RADIOMETRY AND PHOTOMETRY

*Radiometry* refers to measurement of optical radiation covering wavelengths from 10 to $10^6$ nm. It provides an indication of optical energy measured by causing the energy to be absorbed within the instrument and then measuring the change in temperature. The instruments used should have the same response to all the frequencies within the optical band.

*Photometry* is the science concerned with measurement of light within the visible band, with respect to the response of human eye. In photometric measurements, human eye is the main the optical sensor.

The response of human eye under normal light conditions is *photopic vision* and it peaks at 555 nm as shown in the International Luminosity Curve (Fig. 19.2). The effect of different colours on the eye is clearly shown in this diagram. For example, the eye has almost ten times as much response to green as it has to blue.

As mentioned above, the average eye sees best (is at its maximum efficiency) at a wavelength of 555 nm. The conversion factor for maximum efficiency for average eye is 680 lm/W. Wavelengths, other than 555 nm within the visible range must, be converted to luminosity function in relation to CIE.

The photopic eye response curve has been developed from a sample of population under day light conditions. The response at low light level conditions is known as *scotopic vision* and has peak 507 nm as shown in Fig. 19.3.

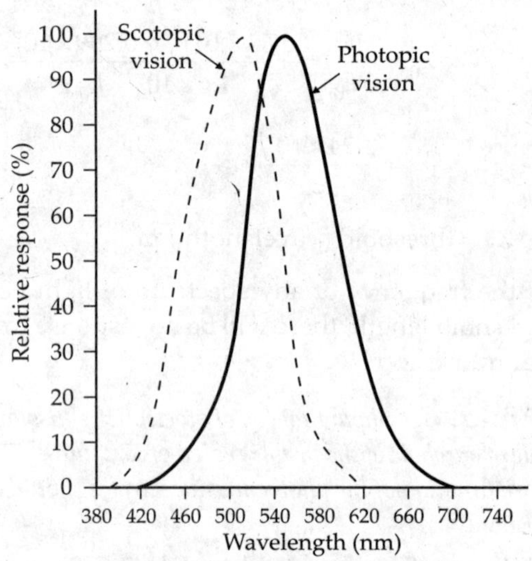


**Fig.19.3** Curves showing the response of an average eye at different wavelenghts.

## 19.7 TERMS RELATING TO PHOTOMETRY

The following terms are peculiar to measurement in the visible part of optical spectrum *i.e.*, in photometry.

1. ***Point sources and extended sources.*** Light energy is emitted in straight lines from one the following *two* types of sources :

(*i*) point source, and

(*ii*) extended source.

A *point source* of light is a source which, for photometric purposes, can be considered with sufficient accuracy to be concentrated at a point. A *uniform point source* emits light uniformally in all directions. It is also known as an *isotropic source* [Fig. 19.4(*a*)].

A point source of light is a source, which, for photometric purposes can be considered with sufficient accuracy as concentrated at one point. In practice, no source of light is a point source. However, if the source has a diameter less than 10% of the distance between it and the detector, then it can be considered to be as a point source.

The point source is idealistatically a single point ; however, even an extremely single point of origin would be finite. The extended source of light energy illuminates a point in space from several directions this is shown in Fig. 19.4 (*b*).

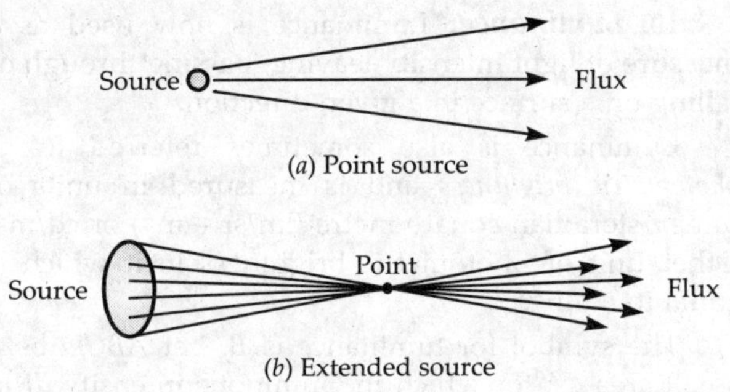


**Fig. 19.4**

2. ***Solid angle (Ω).*** The *steradian* is the unit measure of solid angle just as the radian is the lunit measure of plane angle. Figure 19.5(*a*) shows a sphere of radius *r*. A surface of area *A* on the surface of the sphere subtends a solid angle,

$$\Omega = A / r^2 \quad ...(19.7)$$

at the centre of the sphere.

In case, the area of the surface is $r^2$, it will subtend an angle of :

$\Omega = A/r^2 = r^2/r^2 = 1$ steradian at the centre of the sphere as shown in Fig. 19.5(*b*).

Therefore, a *steradian* may be defined as the solid angle subtended at the centre of a sphere of 1 m radius by a surface of area 1 $m^2$ located on the surface of the sphere.

The total surface area of a sphere of radius *r* meter is $4\pi r^2$ and therefore, a sphere contains $4\pi r^2 / r^2 = 4\pi$ steradian.

3. ***Lumen.*** The basic unit of power in photometric systems is *Lumen*. It is also called *Luminous flux* or *Luminous power*. It is defined as the rate of luminous energy emitted from a source. Like in all photometric measurements, its measurement is based on photopic response of eye, which peaks at a wavelength of 555 nm.

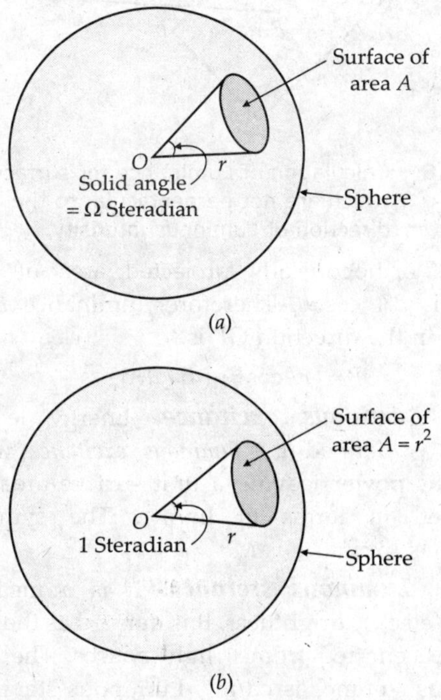


**Fig. 19.5** Illustration of solid angle.

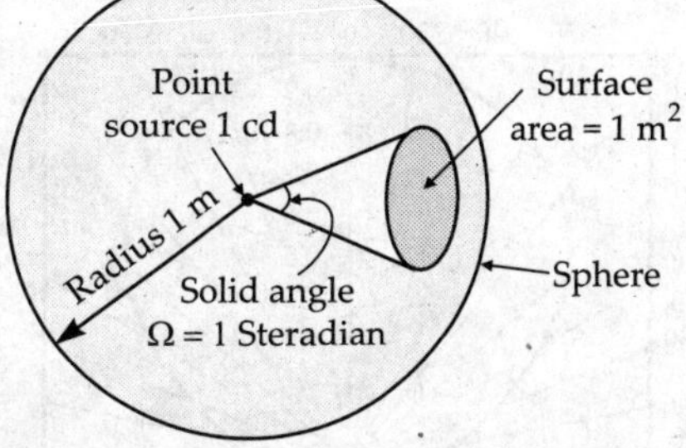


**Fig. 19.6** Relationship between candela and lumen.

The lumen can be defined as under :

If a point source of one candela is placed at the centre of a perfectly transparent sphere of radius 1 m, then the light passing through one steradian or an area of 1 $m^2$ located on the surface of the sphere is called a *"lumen"*. The symbol used for lumen is lm and the notation for luminous flux is $\Im$ or Φ.

4. ***Luminous intensity.*** This is a measure of luminous flux per unit solid angle from a point source. The unit of luminous intensity is lm/sr. The SI unit for luminous intensity is candela (cd). The symbol used for luminous intensity is *I*.

Candela (cd) is the basic unit of luminous intensity or flux per unit solid angle. The luminous intensity, in the direction of normal, of a perfect black body surface of 1/60000 $m^2$ area, at the temperature of solidification of platinum under a pressure of 101.325 $kN/m^2$.

The luminous intensity of a point source can be defined as flux per unit solid angle. Therefore,

$$\text{luminous intensity } I = d\Im / d\Omega = d\tilde{\Phi} / d\Omega \text{ cd} \quad ...(19.8)$$

A uniform point source or an isotropic point source radiates uniform energy in all directions. As there are $4\pi$ steradians in space around a point, the total flux emanating from a point source having a luminous intensity of one cd is $4\pi$ lumens.

5. ***Polar curve.*** If the source of light is not a uniform point source *i.e.* is not an isotropic source, it does not radiate light energy in all directions. In practical applications lamps and other sources of light radiate different light energies in different directions and therefore, they cannot be treated as isotropic sources. Since, these sources radiate unequal light energies in different directions, the luminous intensities produced at equidistant points in different directions are not equal. The plot of luminous intensity versus angle from a reference can be used to indicate the directional properties of the light source. This is called a polar diagram as shown in Fig. 19.7.

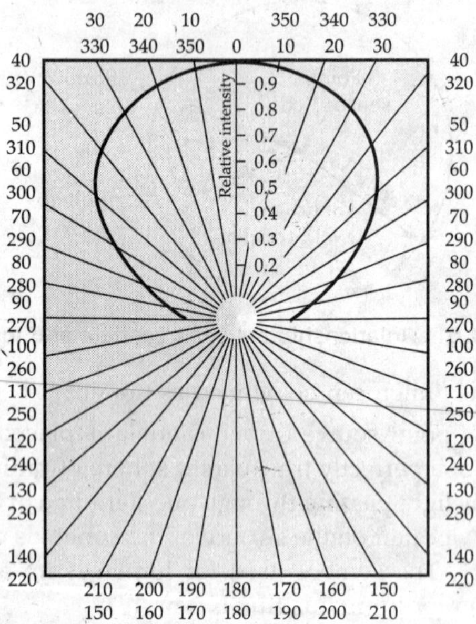


**Fig. 19.7** A polar diagram of light output from a non isotropic point source.

6. ***Mean spherical luminous intensity (M.L.S.I.).*** This is defined, for a source of light, as the mean of luminous intensities in all directions from the source of light which radiates the same total flux uniformly in all directions.

7. ***Mean horizontal luminous intensity (M.H.L.I.).*** This is defined, for a source of light, as the average value of luminous intensities in all directions in a horizontal plane passing through the light.

8. ***Mean hemispherical luminous intensity (M.H.L.I.).*** This is defined as the average value of luminous intensity in all directions within a hemisphere, either above or below the horizontal plane through the centre of source.

9. ***Illumination or Illuminance.*** The density of light energy incident on a surface is known as illumination or illuminance. This is measured in terms of lm/m$^2$ or lux. The symbol for illuminance is *E*.

Suppose a flux $d\Im$ falls on an area $dA$.

$$\therefore \text{ Illuminance } E = d\Im / dA \text{ lm/m}^2 \text{ or lux} \quad ...(19.9)$$

Equation 19.9 holds good if the area is perpendicular to the direction of flux. The symbol for illuminance is *E*.

10. ***Luminance.*** Luminance is now used as a measure of light intensity leaving, passing through or falling on a surface in a given direction.

Luminance is also sometimes referred to as *photometric brightness* and is measured in units of lumen/steradian square metre (lm/sr – m$^2$) or cd/m$^2$. Other unit of photometric brightness in it which is equal to cd/m$^2$.

The symbol for luminance is *B*. Let *ABCD* be a small area *dA* for which the luminous intensity *dI* in the direction *PQ* is inclined to the normal to the surface. The angle between normal to the surface and direction of luminous intensity being θ.

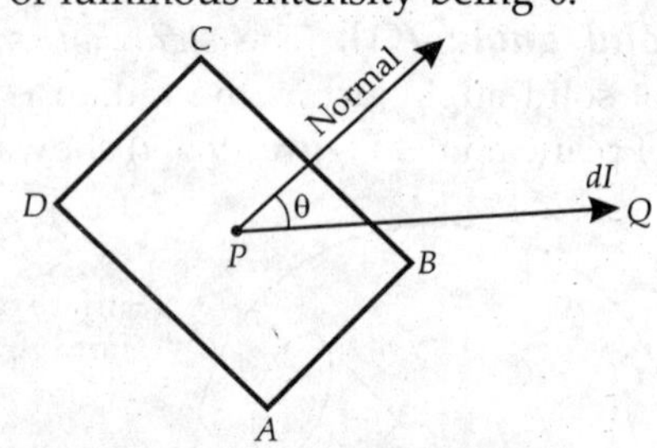


**Fig. 19.8** Calculation of Luminance for surfaces which are not perpendicular to the direction of Luminous intensity.

The orthogonically projected area of element *ABCD* is $dA \cos\theta$. Therefore, luminance *B*, when viewed in the direction *QP* is,

$$B = (1/\cos\theta)\cdot(dI/dA) \quad ...(19.10)$$

11. ***Luminous excitance.*** Energy leaving a surface is defined by *luminous excitance* which is luminous power leaving a unit surface area, and is measured in terms of lm/m$^2$. The symbol for Luminous Excitance is *M*.

12. ***Luminous sterance.*** It is a measure of photo—electric brightness. It is defined as the flux per unit area emitted from a light source. The unit for luminous sterance is cd/m$^2$. Luminous sterance was previously called as Luminance.

## 19.11 LAWS OF ILLUMINATION

1. ***Inverse square law.*** The illuminance of a surface is inversely proportional to the square of the distance of the surface from the source of light.

Consider a surface of area '$A$' situated at a distance $r$ from a point source $P$ as shown in Fig. 19.9. Let the area project a solid angle $\Omega$ at point $P$ and the flux in this solid angle be $F$.

$$\therefore \text{ Intensity } \quad I = \frac{\Im}{\Omega} = \frac{\Im}{A/r^2} \text{ and Flux } \Im = \frac{A}{r^2} I \qquad ...(19.11)$$

$$\text{Illuminance } E = \frac{\text{flux}}{\text{area}} = \frac{\Im}{A} = \frac{I}{A}\cdot\frac{A}{r^2} = \frac{I}{r^2} \qquad ...(19.12)$$

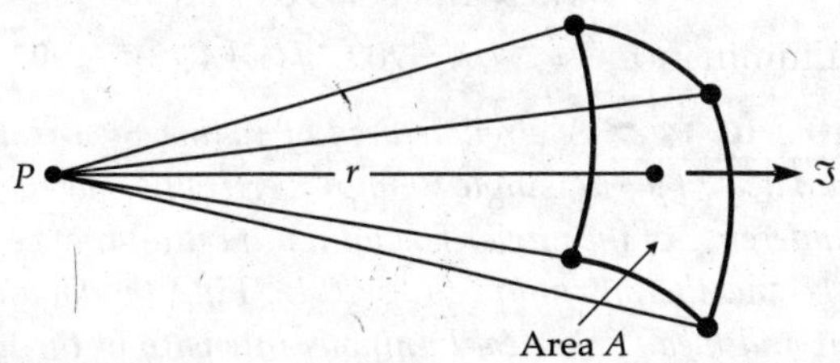


**Fig. 19.9** Depicting inverse square law.

Therefore, illuminance, $E$, of a surface from a light source of constant intensity is inversely proportional to square distance of surface from the source.

2. ***Lambert's cosine law.*** Figure 19.10 shows a surface $ABCD$ illuminated by a luminous flux in a direction $\theta$ to the normal to the direction of flux.

Let the flux in the beam be $F$ and area $ABC\,D$ be $A$.

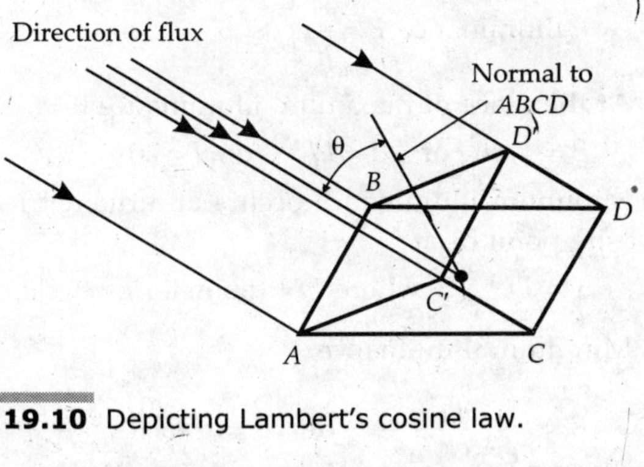


**Fig. 19.10** Depicting Lambert's cosine law.

$\therefore$ Area of $ABCD = A/\cos\theta$. The same flux falls on the two surfaces.

$$\therefore \text{ Illuminance of } ABCD = \frac{\Im}{A/\cos\theta} = \frac{\Im\cos\theta}{A} \qquad ...(19.13)$$

Thus, the illuminance of a surface is proportional to the cosine of the angle between its normal and the direction of the luminous flux. For a point source of constant $I$ in all directions, the two laws lead to the following expression for the illuminance at any point $P$ of Fig. 19.11.

$$E = \frac{I\cos\theta}{r^2} = \frac{I}{(x^2+h^2)}\cdot\frac{h}{\sqrt{x^2+h^2}}$$

$$= \frac{I\cdot h}{(x^2+h^2)^{3/2}} \qquad ...(19.14)$$

$$= \frac{Ih}{r^2} = \frac{I}{h^2}\cos^3\theta \qquad ...(19.15)$$

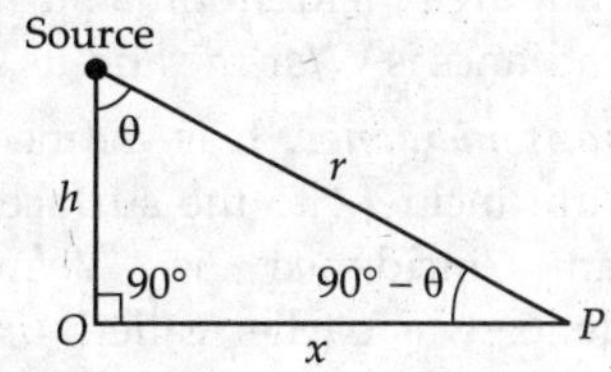


**Fig. 19.11** Depicting Lambert's cosine law.

## 19.12 TERMS RELATING TO RADIOMETRY



The following terms are peculiar to measurement in all parts of the optical spectrum and relate to radiometry. The descriptive words such as prefixes use the word radiant.

1. ***Watt (W).*** It is the basic measure of power in the radiometric system or in other words is the basic unit of radiant flux. It is an indication of energy of the wave per unit time.
2. ***Radiant energy.*** The wave energy is defined as *radiant energy* and is measured in joule. Radiant energy is the generic term used for radiant flux.
3. ***Radiant flux.*** Radiant flux is the radiometric measure for radiant energy. The unit radiant flux is W. It is also called radiant power. Its symbol is $\Phi_e$ or $\Im_e$.
4. ***Radiant intensity.*** It is a measure of radiant intensity per unit solid angle from a point source. It is defined as

$$\text{radiant intensity, } I_e = \frac{d\Im_e}{d\Omega} \text{ W/sr} \qquad ...(19.16)$$

5. ***Radiance.*** It is defined as the flux per unit solid angle per unit area of emitting surface. Its symbol is $N$.

$$\text{Radiance, } N = \frac{d\Im_e/d\Omega}{dA} \text{ W/sr-m}^2 \qquad ...(19.17)$$

6. ***Radiant sterance.*** It is defined as flux per unit solid angle per unit area of emitted surface. The unit of radiant sterance is W/sr-m$^2$ and its symbol is $B_e$. Radiant sterance was previously called as *radiance*.

7. ***Emittance.*** It is defined as radiant flux per unit area from a finite source. The unit for Emittance is W/m$^2$ and its symbol is *W*.
8. ***Radiant excitance.*** It is the measurement of radiant flux per unit area exciting from a finite source. The unit for radiant excitance is W/m$^2$. The symbol used for radiant excitance is $M_e$. It was previously known as Emittance.
9. ***Irradiance.*** It is defined as the radiant flux per unit area incident on a surface. The unit of irradiance is W/sr-m$^2$ and its symbol is *H*.
10. ***Radiant incidence.*** It is defined as the flux per unit incident to the surface. The unit of radiant incidence is W/m$^2$. Radiant incidence was earlier called *"irrudiance"*. Its symbol is $E_e$.

Figure 19.12 illustrates some of the terms relating to photometric and radiometric measurements.

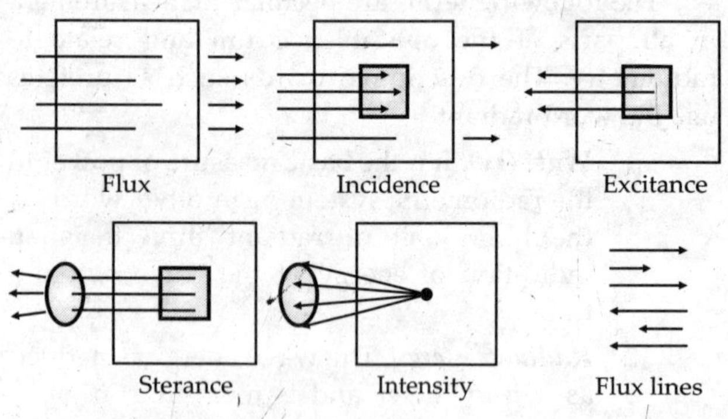


**Fig. 19.12** Illustration of terms connected with photometric and radiometric measurements.

**Example 19.4** *Six lamps are used to illuminate a certain room. If the luminous efficacy of each lamp is 11 lm/W and the lamps have to emit a total flux of 10,000 lm. Calculate, (a) M.S.L.I. per lamp, (b) case of energy consumed in 10 hours if the charges for electricity are Re 1 per unit.*

**Solution.** Luminous flux of each lamp,

$$\mathfrak{I} = 10000/6 = 1667 \text{ lm.}$$

M.H.L.I. of each lamp

$$I = \mathfrak{I}/\Omega = 1667/4\pi = 132.65 \text{ cd.}$$

Power of six lamps

$$= 10000/11 = 909.1 \text{ W}$$

Energy consumed

$$= 909.1 \times 10 \times (1/1000) = 9.1 \text{ kWh.}$$

Cost of Energy = 9.1 × 1 = ₹9.1.

**Example 19.5** *The effective area of the filament of a certain lamp when viewed from below is 2 × 10$^{-3}$ m$^2$, and the luminous intensity in the downward direction is 153 cd. Calculate the luminance of the lamp when viewed from that direction.*

**Solution.** Luminance $B = (1/\cos\theta)(dI/d\theta)$

(see Eqn. 19.10)

$$= (1/\cos 0^0)[153/2 \times 10^{-3}] = 76500 \text{ cd/m}^2.$$

**Example 19.6** *A lamp having a mean spherical luminous intensity of 80 cd has 10% of light reflected on to circular screen 3 m in diameter. Find the value of illuminance on the screen.*

**Solution.** Effective value of M.S.L.I. of lamp

$$I = 0.7 \times 80 = 56 \text{ cd.}$$

Luminous flux emitted by source

$$\mathfrak{I} = 4\pi I = 4\pi \times 56 = 703.7 \text{ lm}$$

Illuminance $E = \mathfrak{I}/A = 703.7/(\pi/4)(3)^2 = 99.5$ lx.

**Example 19.7** *A circular area of radius 6 m is to be illuminated by a single lamp vertically above the circumference of the circle. The minimum illuminance is to be 6 lx and the maximum to be 20 lx. Find the mounting height and mean spherical luminous intensity of the lamp. Assume M.S.L.I. of the lampl to be uniform in all directions.*

**Solution.** Let

$I$ = M.S.L.I. of lamp ; cd,

$h$ = height at which lamp is hung ;

$D$ = diameter of circle ; m.

The maximum illumination occurs just below the place where the lamp is hung.

From Eqn. 19.15,

$$\text{Illuminance } E = \frac{I}{h^2}\cos^3\theta$$

At the place of maximum illuminance $\theta = 0$.

$$E = I/h^2 \text{ or } 20 = I/h^2 \text{ or } I = 20h^2 \quad ...(i)$$

Minimum illuminance occurs at a diametrically-opposite point or at

$\therefore\ x = D,$ where $D$ = diameter of circle.

Minimum illuminance,

$$E = \frac{Ih}{(D^2+h^2)^{3/2}} \text{ or } 6 = \frac{20h^3}{(12^2+h^2)^{3/2}} \quad ...(ii)$$

Solving (*i*) and (*ii*), we get,

$$h = 10.82 \text{ m and } I = 2340 \text{ cd.}$$

**Example 19.8** *The illumination level of a field on account of sunlight when sun is at 45° above the horizon is 60,000 lux. How much illumination would be produced by sunlight under identical conditions, when sun reaches an altitude of 60°.*

**Solution**. From Eqn. 19.15,

$$\text{Illuminance, } E = \frac{I \cos\theta}{r^2}$$

when the sun is 45° to horizon, the angle which the direction makes with the normal to the surface is $\theta = 90° - 45° = 45°$.

$$\therefore \quad 60000 = \frac{I \cos 45°}{r^2} \text{ or } \frac{I}{r^2} = \frac{60000}{\cos 45°}$$

When the sun is 60° above horizon,

$$\theta = 90° - 60° = 30°.$$

The distance $r$, to the sun and also the luminous intensity $I$ are the same in both the cases.

Therefore, illuminance of the field when the sun is 60° above horizon is :

$$E = \frac{I}{r^2}\cos 30° = 60000 \frac{\cos 30°}{\cos 45°} = 73{,}400 \text{ lx}.$$

## 19.13 PHOTOMETRIC/RADIOMETRIC MEASUREMENT SYSTEMS

A measurement system for photometric/radiometric measurements consists of the following :

(*i*) optical sources

(*ii*) optical modulating systems and

(*iii*) optical detectors.

The system is shown in a block diagram form in Fig. 19.13.

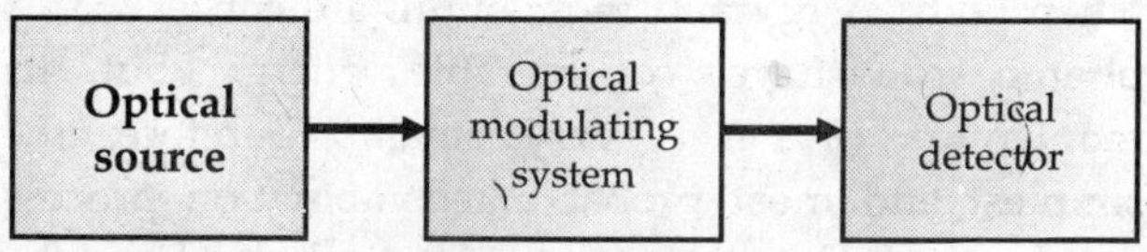


**Fig. 19.13** Generalized block diagram for Photometric/Radiometric Systems.

A photometric/radiometric measurement system is a device that measures the intensity and sterance of a light beam. Any basic photometric/radiometric measurement consists of a light source such as sunlight, LED, or a laser beam. The light is modulated with the help of light supressors, light attenuators and filters before it is made incident upon the detectors. The detectors used are photo electric transducers.

Electrical Engineers have become accustomed to measuring electrical parameters, such as voltages and currents, with an accuracy of a few parts in a million. However, optical measurements, performed even by skilled personnel and sophisticated equipment, rarely give errors below 10-20%.

Several factors contribute towards these measurement errors. Some of these are :

(*i*) The prime components for measurements are optical sources and detectors. The spectral emission of the source and the spectral response of the detector are important factors. Using a detector with the wrong spectral response for the source being using results in large errors.

For spectrometric measurements, the detector performance is specified by its closeness to the CIE photopic response curve shown in Fig. 19.14. However, these specifications are on the basis of area under the curve and can lead to large errors unless matching is done at specific frequencies. Thus as shown in Fig. 19.14, the photometric detector will give an error of about 25 percent when making measurements of a red LED source, even though it is relatively matched with CIE response curve over almost its entire range. The radiometric detector, on the other hand gives an error 5% only.

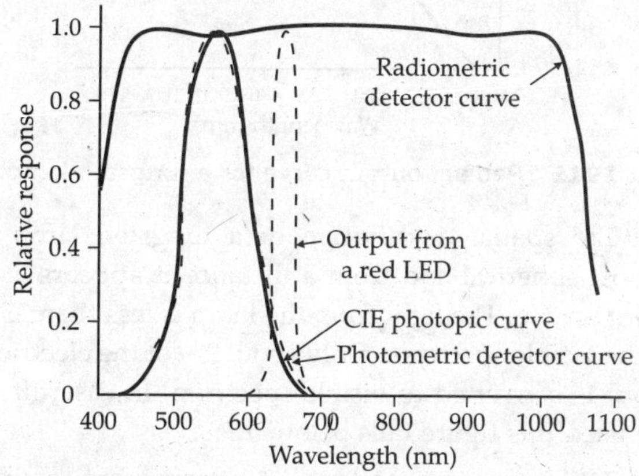


**Fig. 19.14** Comparison of detector response curves.

(*ii*) The other sources of error, apart from those arising from spectral mismatch, are ;

(*a*) ageing of detector and filters,

(*b*) stray light,

(*c*) non-linear characteristics of detectors.

## 19.14 OPTICAL SOURCES

The different optical sources used for photometric/radiometric measurements are described below.

**1. *Sunlight.*** The sunlight produces a continuous spectrum of every colour starting from ultraviolet to infra-red peaking at a wavelength of 460 nm which corresponds to blue colour.

**2. *Incandescent sources.*** Tungsten lamp is one of the most commonly used incandescent source. The tungsten lamp is used as a *secondary standard* for

19

luminous intensity. It consists of a tungsten filament enclosed in a glass envelope, and gives off light when a current is passed through it to raise it to incandescence. The filament can be designed to work on a wide range of a.c. and d.c. supplies.

The tungsten lamp has a well defined continuous spectrum as shown in Fig. 19.15. The light can be filtered to produce any colour and is of predetermined intensity. The wavelength at which peak emission takes place falls as the temperature is increased, and the intensity of output increases. It permits calibration at any given colour temperature through control of voltage applied to the lamp and the current passing through its filament.

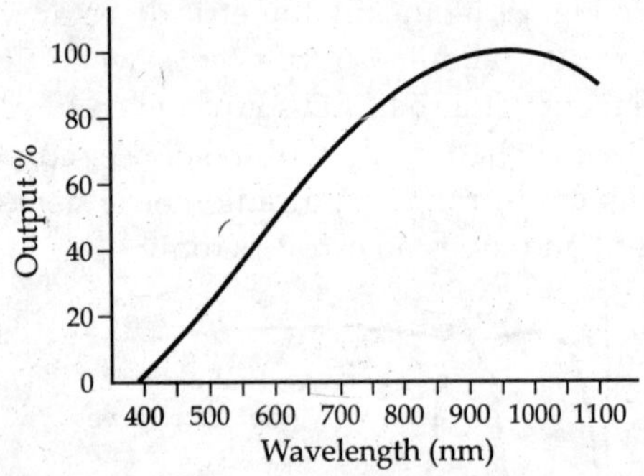


**Fig. 1915** Radiant output curve for a tungsten lamp.

The spatial distribution of a tungsten lamp is almost spherical and from a distance it appears as a point source. The light from the lamp is less than 15% in the visible spectrum. Only about 2% of the electrical input is converted to visible spectrum. In case filters are used this figure falls below 0.5%.

**3. *Flourescent sources.*** A floursecent source produces a continuous spectrum of light output, which has superimposed on it peaks at certain wavelengths. These peaks are mainly on account of discharge through mercury vapour in the lamp, and occur mainly in the ultraviolet region. The infra-red component is very low in a flourescent source. The majority part of the light spectrum produced by it is sensitive to eye. It has a higher overall efficiency as compared with incandescent lamps.

**4. *Light emitting·diodes (LEDs).*** The LED is primarily a PN junction semiconductor device. The basic mechanism of electro-magnetic radiation from the device is *injection luminescence (IL)*. This occurs in *two* steps *i.e.,*

(*i*) injection of minority carriers across the junction and

(*ii*) the radiative recombination of minority carriers.

When the diode has zero bias under thermal equilibrium conditions, the space charge (or depletion) layer potential prevents the cross-over of large concentrations of mobile conduction band electrons and the valence band holes across the junction. When a forward bias is applied to the device the magnitude of this potential barrier is reduced making it possible for diffusion of electrons and holes across the junction. The minority carrier concentrations on both sides of the junction are considerably increased which increases their rate of recombination. This is responsible for electromagnetic radiation.

Two types of radiative recombination mechanism are commonly encountered in LEDs depending upon the band gap characteristics of the semi-conductor material used. These are :

(*i*) Direct recombination and

(*ii*) Indirect recombination.

When an excess minority carrier electron concentration $\Delta N$ is injected into the $P$ side of a $PN$ junction diode, the probability of recombination with the available holes ($P$) is proportional to the concentrations of $\Delta N$ and $P$. For such a process, the radiative recombination rate is,

$$R = R_c(\Delta N)P \qquad ...(19.18)$$

where $R_c$ = recombination co-efficient.

The emission of photons as a result of recombination of electrons and holes is possible only when both energy and momentum are conserved. A photon can have considerable energy but its momentum ($h_f / c$) is very small. Therefore, the simplest and most probable recombination process will be that where the electrons and holes have the same value of momentum. This situation exists in many Group III-V compound semiconductors which are of the direct band-gap type with the conduction band minimum and valence band maximum being at zero momentum position. The recombination process in this is '*Direct type*' as shown in Fig. 19.16(*a*).

In indirect band gap semiconductors, the conduction band minimum and the valence band maximum occur at different values of momentum and therefore, in this case it is necessary to involve a third particle to conserve the momentum. Therefore, this process is known as *"indirect type"* recombination. Phonons and chemical impurities are used for this operation as shown in Fig. 19.16(*b*). The probability of recombination in this is obviously much lower than that in the direct band-gap semiconductors. As an example, the value of $R_c$ for indirect band semi-conductors, such as GaP, is about four orders of magnitude lower than that of direct band semi-conductors like GaAs, GaSb.

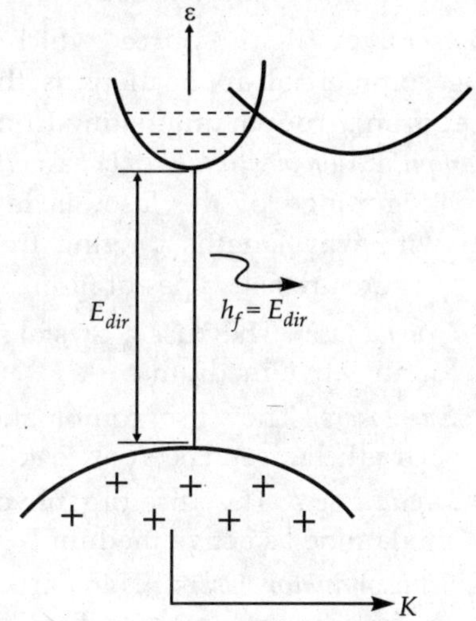


(*a*) Direct band gap type.

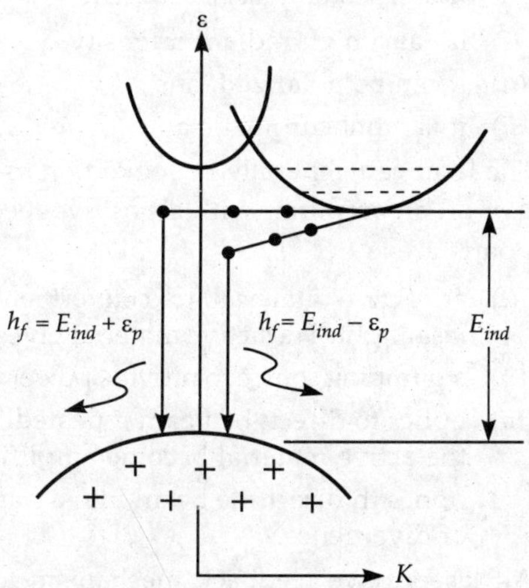


(*b*) Indirect band gap type.

**Fig. 19.16** Energy versus momentum diagrams for semi-conductors.

Referring to Fig. 19.16, for direct recombination,

$$h_f = E_{dir} \quad \text{...(19.19)}$$

and for indirect recombination,

$$h_f = E_{ind} \pm \varepsilon_p \quad \text{...(19.20)}$$

where $\varepsilon_p$ = phonon energy

and $E_{dir}, E_{ind}$ = band gap values in eV.

Usually $\varepsilon_p << E_{ind}$. Therefore, a more general expression is,

$$h_f \approx E \quad \text{...(19.21)}$$

where $E$ is the band gap semi-conductor, either direct or indirect as the case may be.

Now, wavelength of $EM$ radiation is,

$$\lambda = c/f$$
$$= hc/E \quad \text{...(19.22)}$$

The visible spectrum extends over 370 nm — 770 nm, $c = 0.299 \times 10^9$ m/s and $h = 6.63 \times 10^{-34}$ J/s.

Therefore semiconductors suitable for visible spectrum should have $E$ in the range of

$$E = \frac{hc}{\lambda}$$

or $$E = \frac{6.63 \times 10^{-34} \times 0.299 \times 10^9}{(370 \text{ to } 770) \times 10^{-9}} \text{ J}$$

or $$= \frac{6.63 \times 10^{-34} \times 0.299 \times 10^9}{(370 \text{ to } 770) \times 10^{-9} \times 1.602 \times 10^{-19}}$$

$$= 1.6 \text{ to } 3.34 \text{ eV}.$$

Several different materials have been used for LEDs, the properties of some of the commonly used materials are given in Table 19.1. Some of the considerations which govern the choice of material are :

(*i*) The gap length must be suitable for the wavelength required and

(*ii*) The efficiency must be high.

**Table 19.1** *Properties of some Led Materials*

| *Material* | *Colour* | *Peak wavelength (nm)* | *External quantum efficiency %* | *Band Structure* | *lm/W* |
|---|---|---|---|---|---|
| GaP : Zn, O | red | 695 | 15 | indirect | 20 |
| GaP : N | green | 570 | 0.5 | indirect | 600 |
| GaP : NN | yellow | 590 | 0.1 | indirect | 450 |
| Ga $As_{0.6}P_{0.4}$ | red | 649 | 0.5 | direct | 75 |
| Ga $As_{0.35}P_{0.65}$ : N | orange | 632 | 0.5 | indirect | 190 |
| Ga $As_{0.15}P_{0.85}$ : N | yellow | 589 | 0.2 | indirect | 450 |

Gallium Phosphide is mainly used for emissions in the visible spectrum. It has a gap energy of 2.26 eV and can be doped with zinc and oxygen to give red light or with nitrogen to give green light.

For infra-red emitters gallium arsenide is the most commonly used material, having a direct energy gap of 1.4 eV and a wavelength of emission about 900 nm.

Gallium arsenide and Gallium Phosphide combine to form a solid solution of Gallium arsenide phosphide. The light emission from this material peaks at about 600 nm (red) doping with nitrogen increases the conversion efficiency of phosphorus and also the wavelength of emission.

The external quantum efficiency is a measure of power output. Luminous efficiency in lm/W takes into account the sensitivity of eye. The eye is more sensitive to green than to red and hence need to the increase in luminous efficiency in this region.

Electro-luminscent diodes emit radiation in the range of green to infra-red (500 nm – 900 nm). Thus, they can be designed to give a range of outputs without the use of filters. The effect of temperature on an electro-luminescent device relates directly to its band gap. As the temperature increases the wavelength at which the light peaks also increases. This phenomenon causes shift in colours. Since this effects the response of eye, the luminous efficiency also changes. For red and green diodes the efficiency decreases with temperature, but for yellow diodes the efficiency increases because the emitted wavelength moves more into the line of response of eye.

As the spectral output of most commonly used LEDs peaks in red, they naturally present problems in measurement. In this region, the photoptically corrected detector deviates from the CIE photopic curve shown in Fig. 19.14. Therefore, photoelectric sensors have to be used which are matched in the red region of radiations.

Figure 19.17 shows the constructional details of a typical $GaAs_{0.6}P_{0.4}$ LED. It is red emitting LED and has a simple mesh construction. It uses an $N$ type alloy of $GaAs_{0.6}P_{0.4}$ which is grown epitaxially on the GaAs substrate. The $P$ region is diffused into this and capped with a comb type metal electrode to complete the $PN$ junction. The electrode is such that it distributes the diode current uniformly without masking much of the light emitted.

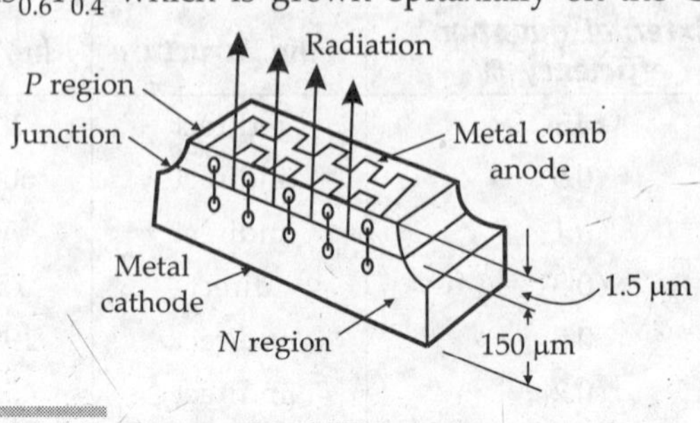


**Fig. 19.17** Ga $As_{0.6}P_{0.4}$ LED.

**5. *Lasers.*** A light source which is finding increasing commercial applications is the *laser.* The word laser is an acronym originating from the initials of *"light amplification by stimulated emission of radiation".* The operating range of the laser is in the optical spectrum with wavelengths ranging from 5 nm — $4 \times 10^6$ nm. There are *four* types of lasers.

(*i*) *Crystal laser.* They use a crystal such as ruby for an active medium.

(*ii*) *Gas lasers.* They use combination of gases such as helium and neon, as an active medium.

(*iii*) *Liquid lasers.* They use organic dyes such as rhodamine as active medium.

(*iv*) *Semiconductor lasers.* They are made from semiconductor materials like GaAs.

The properties of a laser beam are that :

(*i*) it has a high radiant intensity,

(*ii*) it can be polarized, and

(*iii*) it is monochromatic and coherent.

The laser can generally be looked on as an *optical oscillator i.e.* an amplifier with a positive feedback. Its main parts are :

(*i*) an active material to convert energy into laser light. The active material gives the gain,

(*ii*) a pumping source to provide power or energy,

(*iii*) optics to direct the beam repeatedly through the active material become amplified, and

(*iv*) optics to direct the beam into a narrow cone of divergence.

All lasers have feedback mechanism to provide continuous operation and an output coupler to transmit power out of the laser.

The most commonly used source of light is a semiconductor laser. A semiconductor laser consists of a forward biased $PN$ junction which is excited by a current. The recombination of holes and electrons takes place near the junction releasing photons. Many holes and electrons are concentrated in a small area of the junction if the bias current is large, and therefore, the photons which are released will stimulate more emission. Relatively high current densities are required to produce lasing, and many laser structures have been developed to reduce this current requirement, and to increase life.

Lasers may have d.c. or pulsed output.

## 19.15 OPTICAL DETECTORS

The optical detectors used for Photometric / Radiometric measurements are :

(i) Photo-emissive cells

(ii) Semi-Conductor Photo-electric transducers.

(iii) Thermal sensors.

### 19.15.1 Optical Transducers

A large number of different optical transducers have been developed using a variety of techniques. These transducers include :

(i) Photo-emissive cells

(ii) Semi-conductor Photoelectric transducers.

They are described below :

### Photo-emissive Cells

There are basically *two* types of photo-emissive cells :

(i) Vacuum type and

(ii) Gas filled type.

**1. Vacuum Type Photocell**

The vacuum photoelectric cell consists of a curved sheet of thin metal with its concave surface coated with a photo emissive material, which forms the cathode, and a rod anode mounted at the centre of curvature of the cathode. The whole assembly is mounted in an evacuated glass envelope as shown in Fig. 19.18.

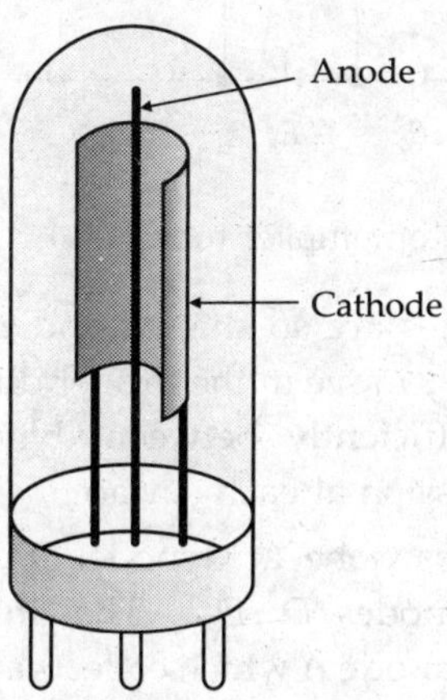


**Fig. 19.18** Photo-emissive Cell.

As explained above, the cathode is coated with emissive materials. These materials emit electrons when light radiation strikes them. The emitted electrons from the cathode are collected by a positive electrode (anode) forming an electric current.

A simple circuit for a photoemissive (or photo electric) cell is shown in Fig. 19.19. This circuit may be used for measurement of luminous flux or luminous intensity.

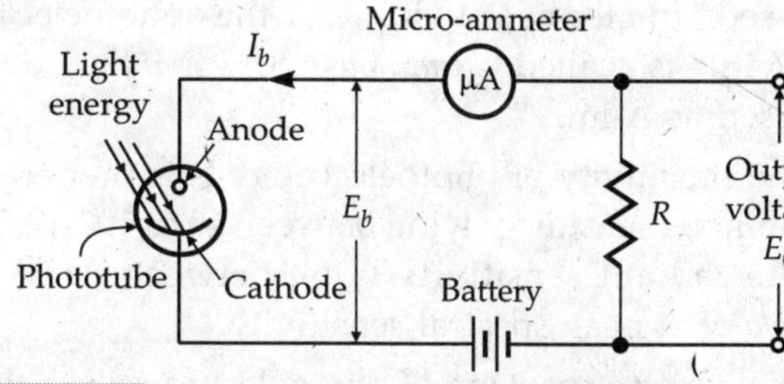


**Fig. 19.19** Circuit for photoemissive cell.

The current through the tube depends upon the following :

(i) intensity of light

(ii) colour of light or its wevelength and

(iii) voltage applied between cathode and anode.

Figure 19.20 shows the current *v/s* voltage charactensucs of a typical highly evacuated tube. A study of these curves shows that for a constant impressed voltage, the current is proportional to the luminous flux or intensity provided light of same wavelength is used.

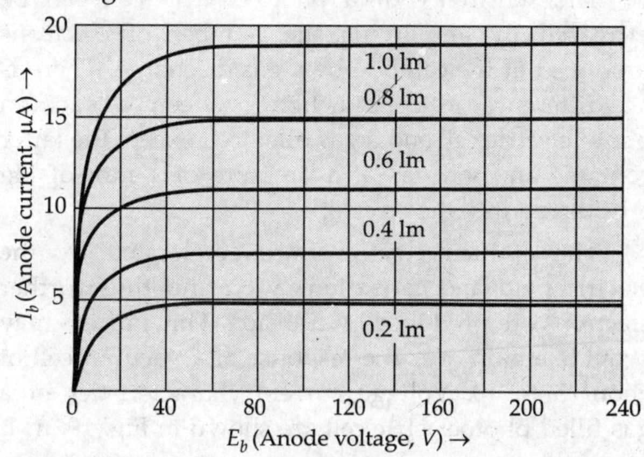


**Fig. 19.20** Current *vs.* voltage characteristics of a photo-emissive cell.

The micro-ammeter shown in Fig. 19.19 may be directly callibrated in terms of luminous flux or intensity. Else a resistor $R$ is connected in the circuit and the voltage, $E_0$, across the resistor is a measure of luminous flux or intensity. The output voltage, $E_0$, may be amplified to drive the subsequent stages of the measurement system.

The *photo-electric sensitivity* is defined as :

$$S = I_b / \Phi \quad ...(19.23)$$

where $I_b$ = photoelectric current, and

$\Phi$ = luminous flux

In case luminous flux, Φ is expressed in terms of watt, the photo electric sensitivity is called *Radiant Sensitivity* and is expressed as A/W.

On the other hand if the luminous flux is expressed in terms of lumen, the photoelectric sensitivity is called *Luminous Sensitivity* and is expressed as A/lm.

The sensitivity of photoelectric cells is moderate, the luminous sensitivity lying between 10 to 100 μA/lm and the radiant sensitivity lying between 0.002 to 0.1 μA/μW at peak spectral sensitivity.

The major advantage of photoelectric cells is that they are stable and do not change their characteristics over long periods of time provided they are operated at low voltage and are protected against excessive light.

The disadvantage of photoemissve cells is that current through the highly evacuated tubes is very small (or the order of μA) which leads low sensitivity.

The sensitivity of photoelectric transducers can be increased by using :

(*i*) gas filled tubes and
(*ii*) photomultipliers.

**2. Gas Filled Photo-emissive Tubes**

The sensitivity of a photo-electric cell can be increased by amplifying the number of electrons produced at the cathode by a gas discharge. There is no difference in construction between a vacuum photo-electric cell and a gas filled cell except the latter contains an inert gas at a pressure of 1 mm of Hg introduced into the envelop.

The emitted electrons are accelerated by the electric field and cause ionization and thus further electrons are produced by collision. This process may provide a gain over the response of a vacuum cell of about 5 to 10. Voltage current characteristics of a gas-filled photoelectric cell are shown in Fig. 19.21. If

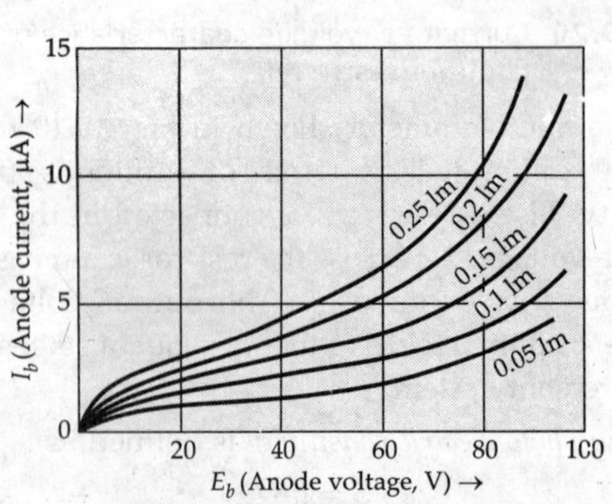


**Fig. 19.21** Characteristics of a gas filled photo-emissive cell.

the applied voltage $E_b$ exceeds a critical value (depending on tube geometry, gas filling and illumination), a glow discharge may set in which can damage the cathode. A resistor should always be connected in series with the gas-filled cell to limit the current in case of an accidental over voltage.

The luminous sensitivity lies between 40 to 150 μ A/lm and radiant sensitivity lies between 0.01 to 0.15 μA/μW.

Gas-filled photo-electric cells are not as stable as vacuum cells and their characteristics are non-linear. Also, they exhibit a time lag in response to modulated or chopped light at frequencies above about 10 kHz.

**3. Photomultipliers**

Extremely low levels of luminous intensity can be measured or detected by means of photomultiplier tubes which utilize many successive stages of secondary emission to boost up the output current from its initial very low value.

A photomultiplier tube is shown in Fig. 19.22. When light strikes a photosensitive cathode, it emits electrons. The liberated electrons are accelerated by a voltage $E_1$ and are focussed on to the next electrode, $D_1$ called *Dyanode*. These electrons striking upon the dynode cause the emission of secondary electrons.

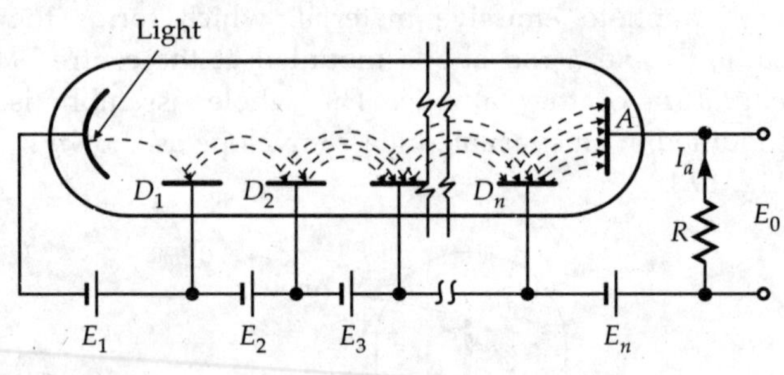


**Fig. 19.22** Photomultiplier tube.

The dynodes are so shaped and placed that the electrons always move in the proper direction and are accelerated sufficiently between dynodes to cause secondary emission at each surface.

The process done at dynode $D_1$ is repeated at subsequent dynodes $D_2, D_3$,... The final electrode in the series is an anode $A$ which collects all the electrons. The electrode $A$ is at a positive potential and the anode current $I_a$ is measured.

Let the number of electrons formed by secondary emission for each primary electron be $g$. Therefore, the gain of each stage is $g$.

If there are '$n$' dynodes, the total amplification is $A = g^n$.

The value of $g$ varies with the voltage between successive dynodes and with the surface composition and the geometry of dynodes from a value of 0.5 to 10. The value of stages of commercial multipliers is between 9 to 14. The gain $A_0$ is between $10^5$ to $10^7$. The sensitivity of a photomultiplier may be as high as 20 A/m as compared with 100 μA/1m for photoelectric cells.

The general comments about the photoemissive transducers are :

(*i*) Their disadvantage is that they are large sized and expensive and need voltage between 300 V to 2500 V for their operation..

(*ii*) Their advantages are that they have a high frequency response and a high sensitivity. Their spectral response can be varied from 100 nm to 1000 nm by changing the cathode material.

Generally these transducers are not used in general purpose electronic systems.

### 19.5.2 Semiconductor Photoelectric Transducers

Figure 19.23 illustrates the general scheme of semiconductor photoelectric transducers. Optical radiation, X-rays and certain corpuscular radiations fall on a semiconductor photoelectric transducer and

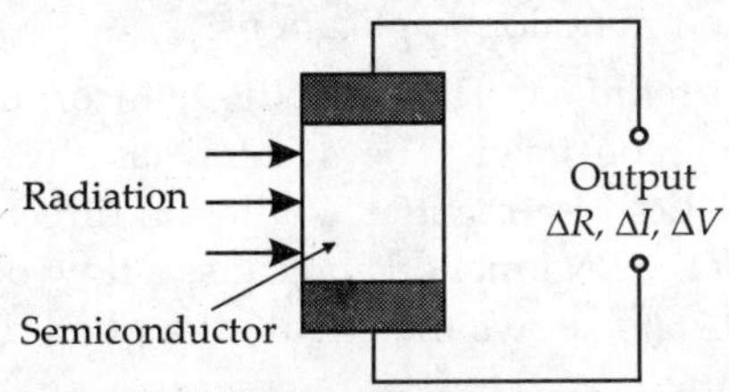


**Fig. 19.23** Schematic diagram of a semiconductor photoelectric transducer.

are absorbed. The process of energy absorption produces movable charges or a change of mobility of the charge carriers in the semiconductors there by producing one of the following effects depending upon the type of transducer used :

(*i*) change in resistance, $\Delta R$,

(*ii*) change in current output, $\Delta I$, and

(*iii*) change in voltage output, AV.

**1. *Photoconductive cells.*** Electrical conduction in semiconductor materials occurs when free charge carriers, *e.g.*, electrons, are available in the material when an electric field is applied. In certain semiconductors, light energy falling on them is of the correct order of magnitude to release charge carriers which increase flow of current produced by an applied voltage. The increase of current with increase in light intensity with the applied voltage remaining constant means that the resistance of semiconductors decreases with increase in light intensity. Therefore, these semiconductors are called *photoconductive cells* or *Photo Resistors* or sometime *Light Dependent Resistors (LDR)*, since incident light effectively varies their resistance.

The two most commonly used photoconductive semiconductor materials are cadmium sulphide (CdS) with a band gap of 2.42 eV and cadmium selenide (CdSe) with a band gap of 1.74 eV. On account of the large energy bands, both the materials have a very high resistivity at ambient temperature which gives a very high value of resistance for practical purposes. The photoconductive cells use a special type of construction which minimises resistance while providing maximum surface. Photoconductive cells are made by chemically sintering the required powder (CdS) or (CdSe) into tablets of the required shape, and enclosing them in a protective envelope of glass or plastic. Electrodes are deposited on the tablet surface and are made of materials which give an ohmic contact, but with low resistance compared with that of the photo-conductor. Gold is typically used. The electrodes are usually inter-digital *i.e.* in the form of interlocked lingers or combs as shown in Fig. 19.24. 19

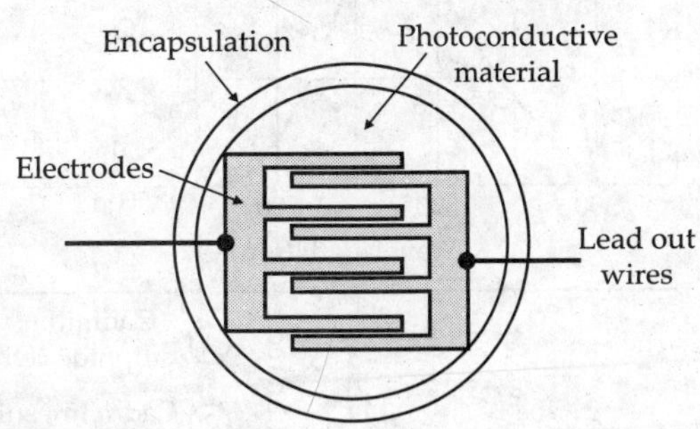


**Fig.19.24** Photoconductive cell.

### Characteristics of Photoconductive Cells

The characteristics of photoconductive cells vary considerably depending upon the type of materials used. These characteristics are given in Table 19.2.

**Table 19.2** *Characteristics of Photoconductive Cells*

| *Photoconductor* | *Time Constant* | *Spectral Band* |
|---|---|---|
| CdS | 100 μs | 0.47 – 0.71 μm |
| CdSe | 10 μs | 0.6 – 0.77 μm |
| PbS | 400 μs | 1 – 3 μm |
| PbSe | 10 μs | 1.5 – 4 μm |

When the cell is kept in darkness, its resistance is called **dark resistance.** The dark resistance may be as high as $10 \times 10^{12} \Omega$. If the cell is illuminated its resistance decreases. The resistance depends on the physical character of the photoconductive layer as well as on the dimensions of the cell and its geometric configuration. The current depends upon the d.c. voltage applied. The current is of the order of mA.

The spectral response characteristics of two commercial cells are shown in Fig. 19.25. There is almost no response to radiation of a wavelength shorter than 300 nm. Cadmium sulphide cells have a peak response in the green part of the spectrum at 510 nm and can be used in the near infra-red region up to about 750 nm. The maximum response of cadmium sulphoselenide is in the yellow orange at 615 nm. This can be used in the infra-red region up to about 1000 nm.

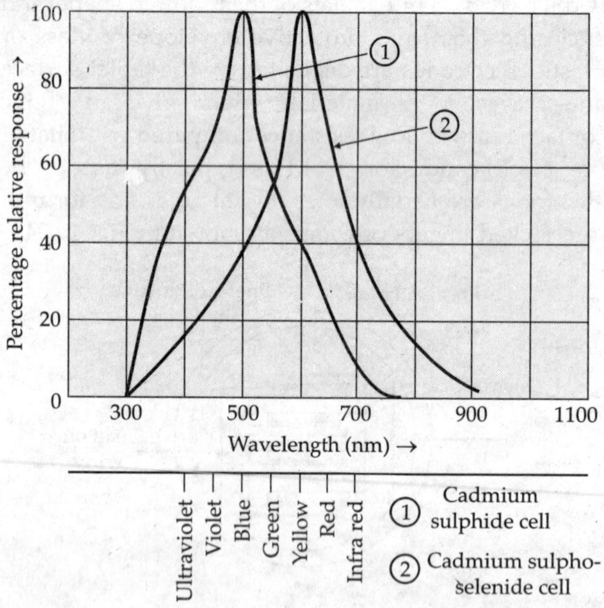


**Fig.19.25** Spectral response characteristics of photoconductive cells.

The photoconductor device described above is also called a **bulk photoconductor.** As described earlier, the photoconductor has a very high resistance at very low illumination levels, which is of the order of MΩ. The higher the intensity of light, the lower is the resistance. The resistance drops to a few kΩ when exposed to light. When using a photoresistor for a particular application it is important to select the proper dark resistance as well as the suitable sensitivity. The sensitivity of photoresistive transducer is defined as :

$$S = \frac{\Delta R}{\Delta H} \Omega/\text{W} - \text{m}^{-2} \quad ...(19.24)$$

where $\Delta R$ = change in resistance ; Ω

and $\Delta H$ = change in irradiation ; W / m$^2$.

The spectral response of the sensor must match that of the light source.

A photoconductor has a relatively large sensitive area. A small change in light intensity causes a large change in resistance. It is common for a photoconductive element to exhibit a resistance change of 1000 : 1 for a dark to light irradiance change of $5 \times 10^{-3}$ W/m$^2$ to 50 W/m$^2$. The relationship between irradiance and resistance is, however, not linear. It is closely an exponential relationship. The result may be written as :

$$R_t = R_i + (R_f - R_i)[1 - \exp(-t/\tau)] \quad ...(19.25)$$

where

$R_f$ = dark resistance ; Ω,

$R_i$ = final resistance after application of beam ; Ω,

$R_t$ = resistance at any time ; $t$

The photoconductive cell suffers from a major disadvantage that temperature changes cause substantial resistance changes for a particular light intensity. Therefore, this type of photoconductor is unsuitable for analog applications.

**2. *Photodiodes.*** The photodiode incorporates a *P* and an *N* type layer. The system has the electrical characteristics of a rectifier. Radiation directed in the vicinity of the *PN* junction and cause a flow of current. Figure 19.26(*a*) shows the circuit of a photodiode.

The photodiode is reverse biased. The reverse biased saturation current is dependent upon the intensity of the incident light. Figure 19.26(*b*) shows the typical characteristics of a photodiode. The photodiode is very useful for applications where the space is restricted. The effective area of a photodiode is about 0.2 mm$^2$ and it has a pinhead (serving as electrode) of a diameter of 0.5 mm.

The photocurrent versus light relationship is linear over a wide range. In order to maintain the linearity the bias voltage should be kept constant.

From Fig. 19.26(*b*) it is clear that the output resistance $R = \Delta V / \Delta I$, is very high and is of the order of tens of MΩ. The d.c. resistance, *V*/*I*, is the diode leakage resistance and that too is very high. This d.c. resistance depends upon the light intensity.

The frequency response of a photodiode largely dependent upon the intrinsic capacity which is typically 2 pF for a reverse bias of –10 V.

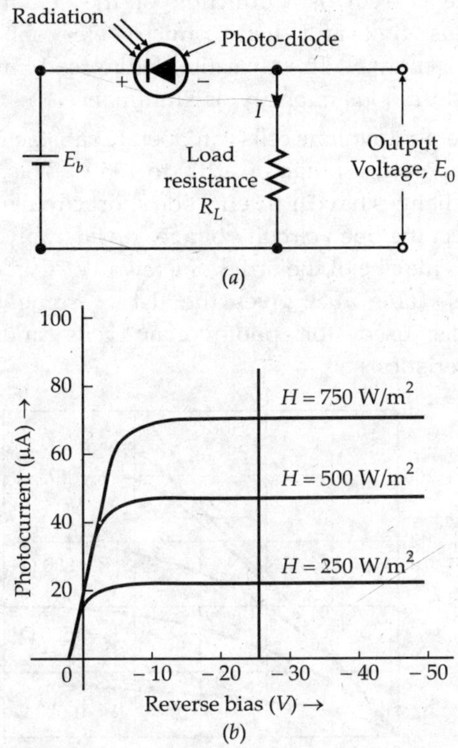


**Fig. 19.26** Circuit and characteristics of a photodiode.

The cut off frequency is given by

$$f_c = 1/2\pi R_L C$$

where $C$ is capacitance of photodiode, and

$R_L$ is the load resistance.

The cut off frequency is of the order of MHz. Even in dark there is always leakage current of the photodiode and this current is known as *dark current.* The dark current doubles about every 10°C increase in temperature for silicon and this decreases the signal to noise ratio.

Comparing photodiodes with photoconductors, the photodiode possesses considerably better frequency response, linearity, spectral response and lower noise. The disadvantages of photodiodes include small active area, rapid increase in dark current with temperature, bias voltage requirement, and the necessity of amplification at low illumination levels.

Because of their fast response time, photodiodes are used as cine film sound track readers. Similarly they can be used as detectors of modulated light in optical communication systems and also in switching circuits.

In recent years there has been considerable advance in the development of integrated circuits which combine photodiodes with circuits. The photodiodes are very small and they give rise to very small signal currents, typically as low a few picoampere. Various photodiode arrays and analogue photodetector packages are now available. Devices which contain a linear array of photodiodes on a single integrated circuit chip can be used for edge and width sensing. The photodiode centres are spaced about 25 μm apart, and 1024 diodes in a line give a chip about 27 mm in length. The diodes in the array can be sequentially addressed at a rate up to 10 MHz, so that a voltage level dependent on the light intensity falling on each diode is available in serial form. A 64 × 64 diode array containing 4096 photodiodes arranged in a square matrix is also available ; this allows area measurements. Other developments include a single photodiode coupled to an amplifier in the same encapsulation, and such an arrangement can provide high sensitivity at high frequencies. Special controlled geometry photodiodes allow measurement of position or null of position.

19

**3. *Phototransistors.*** A phototransistor is a normal transistor in which the envelope enclosing the junction is transparent to allow light to fall on the base emitter junction. At any *PN* junction hole-electron pairs are generated when light falls on the junction, so that any light falling on the base-emitter junction, produces a current which is amplified by transistor action, making the device very sensitive. The characteristics of phototransistors are shown in Fig. 19.27.

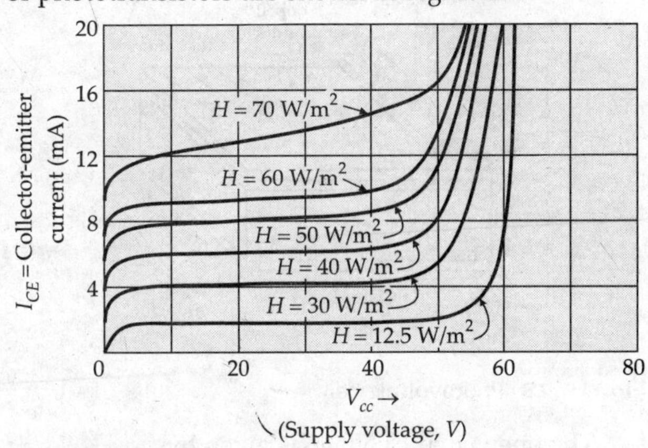


**Fig. 19.27** Characteristics of a phototransistor.

The dark current increases in a phototransistor exponentially with rise in temperature and if the temperature is sufficiently high the dark and light currents may be indistinguishable. At any given temperature the dark current can be reduced to a small value by biasing the base of the transistor which is normally left open circuited. The sensitivity is decreased but the ratio of light to dark current is increased.

The advantages of the phototransistor are : low power consumption, small size, immediate operation on switching on, low voltage operation and long life.

A phototransistor gives a high gain. This transistor is very good for digital applications because of the small rise and fall times. The rise time, which represents the response to dark-to-light irradiance is about 1 μs and the fall time which represents light-to-dark light irradiance is about 10 μs.

**4. *Photovoltaic cell.*** This is an important class of photodetectors. They generate a voltage which is proportional to EM radiation intensity. They are called photovoltaic cells because of their voltage generating characteristics. They, in fact, convert the EM energy into electrical energy. They are passive transducers *i.e.*, they do not need an external source to power them.

A photovoltaic cell is shown in Fig. 19.28. The cell is a giant diode, constructing a *PN* junction between appropriately doped semiconductors. Photons striking the cell pass through the thin *P*-doped upper layer and are absorbed by electrons in the lower *N* layer, causing formation of conduction electrons and holes. The depletion zone potential of the *PN* junction then separates these conduction holes and electrons causing a difference of potential to develop across the junction.

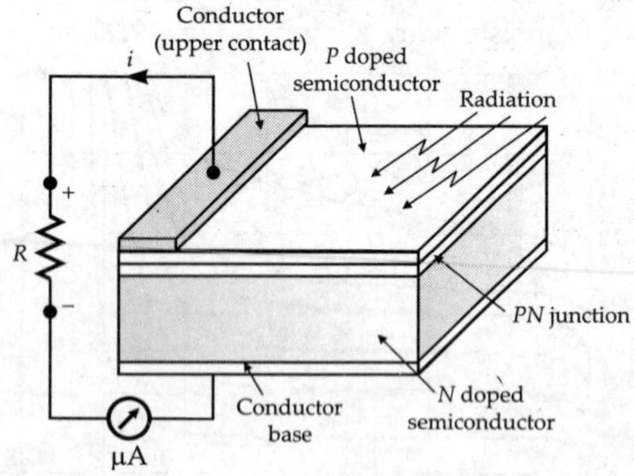


**Fig. 19.28** Photovoltaic cell.

The open circuit voltage is given by :

$$E_0 = E_c \log_e E_e : \text{V}, \qquad ...(19.26)$$

where $E_c$ = calibration voltage ; V,

and $E_e$ = radiant incidence ; W/m$^2$

All photovoltaic cells have low but finite internal resistance. When connected in a circuit having some load resistance, the cell voltage is reduced somewhat from the value given by Eqn. 19.26. Figure 19.29 shows the current *v/s* luminous flux characteristics of a typical photovoltaic cell. These characteristics show that the current as a function of the cell incident luminous flux and hence micrometer shown in Fig. 19.28 may be directly calibrated to read luminous flux or luminous intensity or illuminance.

The photovoltaic cells can operate satisfactorily in the temperature range of – 100 to 125°C. The temperature changes have little effect on short circuit current but affect the open circuit voltage considerably. These changes may be of the order of a few mV/°C in output voltage. Table 19.3 gives the list of some typical materials used for photovoltaic cells and their characteristics.

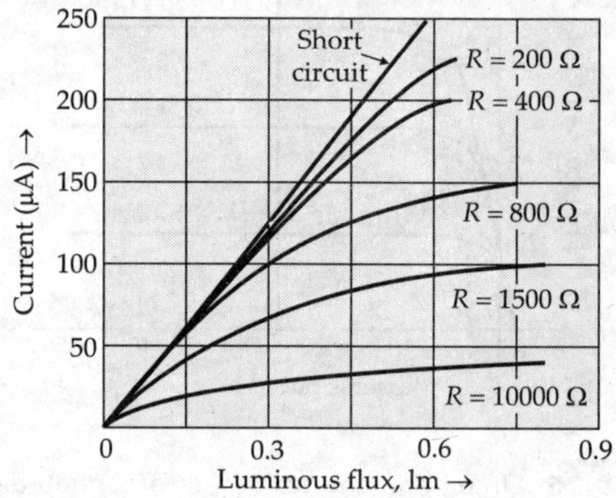


**Fig.19.29** Current *vs.* luminous flux characteristics for a Photovoltaic cell.

**Table 19.3** Typical Photovoltaic Cell Characteristics

| *Cell Material* | *Time Constant* | *Spectral Band* |
|---|---|---|
| Silicon (Si) | 20 μs | 0.44 μm – 1 μm |
| Selenium (Se) | 2 μs | 0.3 μm – 0.62 μm |
| Germanium (Ge) | 50 μs | 0.79 μm – 1.8 μm |
| Indium Arsenide (InAs) | 1 μs | 1.5 μm – 3.6 μm (cooled) |
| Indium Antimonide (InSb) | 10 μs | 2.3 μm – 7 μm (cooled) |

The tremendous advantages of the photovoltaic cell, as its name implies, are its ability to generate a voltage without any form of bias and its extremely fast response. This means that it can be used as an energy converter directly. One use of this is in photographic exposure meters which require no battery. The cell voltage operates the meter directly, or may be directly linked to the iris to control the aperture.

The advent of the space age brought about a tremendous effort in the area of direct conversion of solar energy to electrical energy. The solar cell has become the workhorse for providing power for spacecrafts and the advances in solar cell technology have found their way into many earth-based applications. In order to effectively utilise the sun as a source of power any transducer should possess a high power conversion efficiency and its spectral response should correspond to the sun's spectral output.

5. ***Photothyristors.*** A photothyristor is a *PNPN* device as shown in Fig. 19.30. It is also called a "Light activated silicon controlled rectifier" or simply an *LASCR.* When light radiation falls upon it hole-electron pairs are generated at its blocking junction. These can trigger the device truning it on. Usually the gate lead is brought out, and a bias may be applied to it to vary the threshold light level to trun it on.

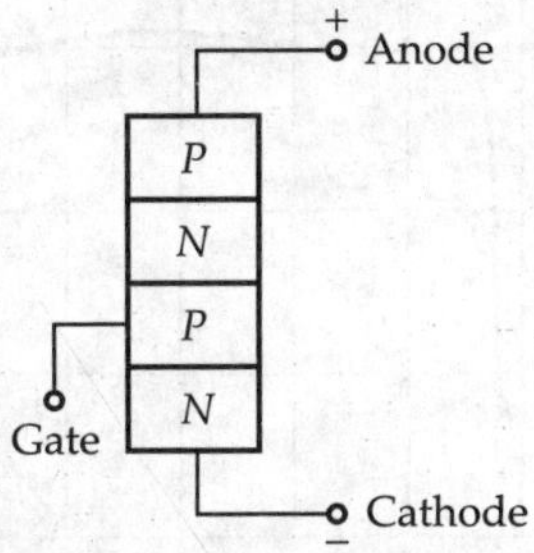


**Fig. 19.30** Light activated SCR.

### 19.15.3 Thermal Sensors

They measure the change in temperature, resulting from the absorption of radiant energy, by giving an electrical signal. These detectors have a flat response to a band of frequencies, but are suited to measuring sources which have a relatively high intensity output, and usually for wavelengths above 1000 nm. Some of the commonly used thermal sensors are described below.

1. ***Thermopile.*** A thermopile consists of a number of thermocouples connected in series as shown in Fig. 19.31. The hot junction is usually blackened to give better energy absorbtion. The spectral sensitivity

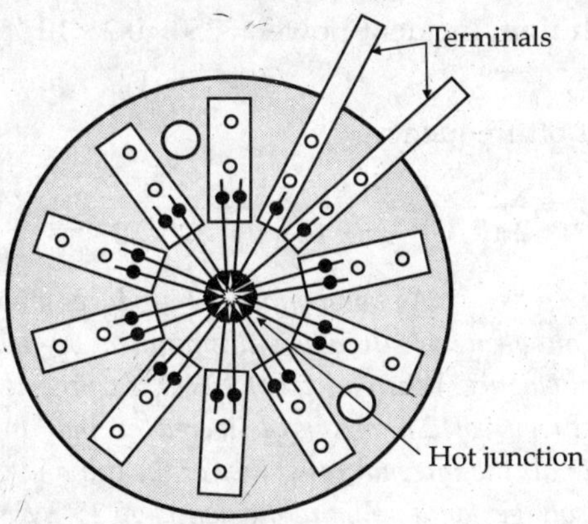


**Fig.19.31** Thermopile.

of the sensor depends on the characteristics of this black absorber. The thermopiles have a slow response and are sensitive to changes in ambient temperature. Their characteristics can be made flat over the desired portion of spectrum. They are used in radiation pyrometers and as infra-red detectors.

2. ***Bolometer.*** It consists of a strip of blackened metel sealed in galss film. When the radiant energy falls upon it, it raises its temperature and consequently its resistance. This change in resistance is measured with a wheatstone bridge. The change bridge output voltage on account of change in resistance is an indication of the radiant energy.

3. ***Pyroelectric detector.*** A pyroelectric detector consists of a crystal whose electric polarization is dependent upon temperature. It is sensitive to changes in temperature only and it can therefore, be used to measure pulsed or chopped light, and to ignore any state background light. The change in polarization results in a current variation in an external circuit. Pyroelectric detectors have a fast response and are mainly used to measure the pulsed power of lasers.

19

## 19.16 LIGHT MODULATING TECHNIQUES

Some of the light modulating techniques while making photometric/radiometric measurements are described below :

### 19.16.1 Light Suppression

When monitoring a light source, it is often necessary to utilize an external light attenuator to suppress the background light. Background light called *ambient light* may cause errors in power measurements especilly at low power levels. There are several ways for making corrections for the effect of ambient light. The following methods may be used :

(*i*) Before measuring a desired light source, a measurement of ambient light is made. Then the total source and ambient light is measured. A substraction of ambient light measurement from the total light measurement gives the true value of desired source light measurement.

(*ii*) A built in method may be used in the instruments used for power measurements. This method envisages a calibration adjustment of the power meter to account for the ambient light.

(*iii*) A light shade is placed over the detector to shut out the ambient light and only the light from the desired source is beamed upon the detector.

### 19.16.2 Light Attenuation

The optical power range of a power meter may be exceeded when measuring high powers without the operator knowing about it. This may cause extreme damage to the photoelectric detector. In order to alleviate this problem, an attenuator is placed in front of the detector, in the path of beam of light under measurement. This limits the beam power reaching the detector. The attenuator is made of diffusing material and scatters the beam within its walls. A major part of the light is absorbed by the walls of the attenuator and a small of the beam falls upon the detector. The amount of light energy falling on the detector can be changed by adjusting the distance between light detector and the attenuator diffusing material. The greater the distance between the two, the greater is the attenuation. When external attenuators are used, corrections must be made in the meter reading to account for the attenuation.

### 19.16.3 Photometric Filtering

Optical power meters are commonly used to measure light energy that is similar to that seen by human eye. In order to measure this light, the meter must respond to it the same way as a human eye would. This is accomplished by placing a photometric filter in front of the detector. Figure 19.32 shows the CIE photopic curve in relation to photodiode filters. These curves represent the wavelength response of human eye and filters. It should be noted that the response of detectors using filters and that of human eye is negligible.

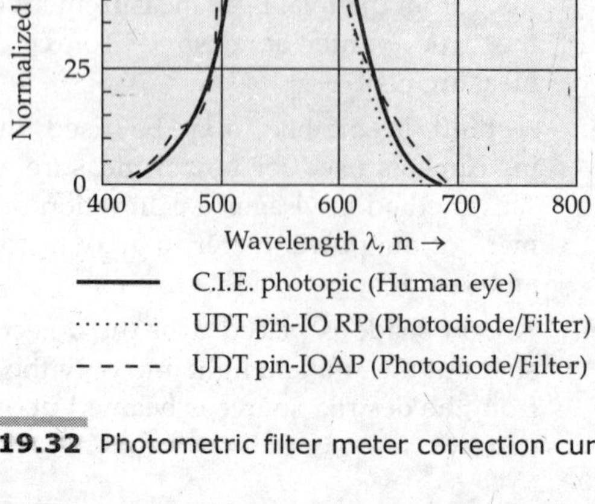


**Fig. 19.32** Photometric filter meter correction curve.

### 19.16.4 Radiometric Filters

Optical power meters are often used for measurement of light energy which does not fall within the visible spectrum. In order to measure light energy which does not fall in the visible spectrum, radiometric filters are used. The filter is placed over the detector and absorbs selected wavelengths of light while allowing other wavelengths to be incident on the detector with negligible **absorbtion**. When the filter is in place the meter responds to all wavelengths from approximately 400 to 1000 nm in the same level as shown in Fig. 19.33.

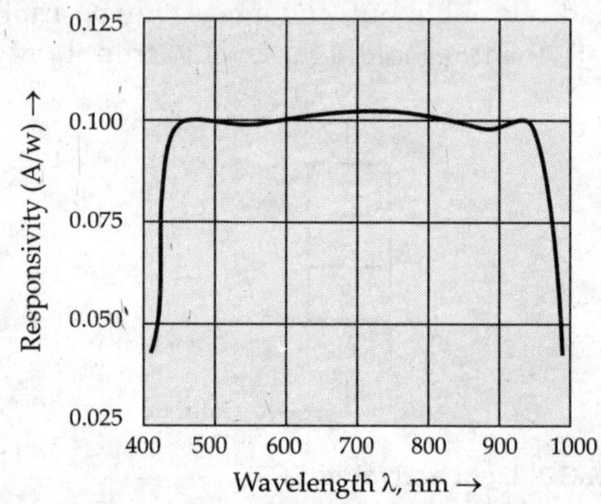


**Fig. 19.33** Radiometric filter meter correction curve.

***Example 19.9*** *A CDS cell has a dark resistance of 100 kΩ and a resistance of 30 kΩ in a day light beam. Find the resistance of the cell after 10 ms of application of a beam. The time constant is 72 ms.*

**Solution.** From Eqn. 19.25, the resistance of cell

$$R_t = R_i + (R_f - R_i)[1 - \exp(-t/\tau)]$$

$$= 30 + (100 - 30)[1 - \exp(-10/72)] = 39.1 \text{ k}\Omega.$$

***Example 19.10*** *The effective area of a photodiode is $0.2 \times 10^{-6}$ $m^2$ and the irradiance is $250 W/m^2$. Calculate the incident power. If the load resistance is 10 kΩ and the capacitance of the diode is 2 pF, find the cut off frequency.*

**Solution.** Incident power $= 250 \times 0.2 \times 10^{-6}$

$$= 50\ \mu\text{W}.$$

Cut off frequency,

$$f_c = \frac{1}{2\pi R_L C} = \frac{1}{2\pi \times 10 \times 10^3 \times 2 \times 10^{-12}} \approx 8 \text{ MHz}.$$

***Example 19.11*** *A photovoltaic cell produces a voltage of 0.33 V on open circuit when illuminated by 10 W/m² radiant incidence. When a load of 100 Ω is connected to the cell, a current of 2.2 mA is delivered at that intensity. Calculate (a) the internal resistance of the cell and (b) open circuit voltage for a radiant incidence of 25 W/m². The calibration voltage is the same for both the cases.*

**Solution.** (*a*) Let $R_i$ and $R_L$ be the internal resistance of cell and resistance of load respectively.

$$\therefore\ I = \frac{E_0}{R_i + R_L} \quad \text{or} \quad 2.2 \times 10^{-3} = \frac{0.33}{R_i + 100}$$

or internal resistance of cell $R_i = 50\ \Omega$.

(*b*) From Eqn. 19.27, open circuit voltage is,

$$E_0 = E_c \log_e E_e$$

$\therefore$ Open circuit voltage for a radiant incidence of 25 W/m$^2$ is

$$= 0.33(\log_e 25 / \log_e 10) = 0.462\ \text{V}.$$

**Example 19.12** *The characteristics of a photovoltaic cell are given in Fig. 19.29. The cell is placed 0.914 m away from a lamp having an intensity of 180 cd. The window area through which the light from the lamp is made incident upon the cell is 1935 mm$^2$. Find the value of current when a meter of a resistance 800 Ω is connected across the cell.*

**Solution.** Solid angle,

$$\Omega = \frac{A}{r^2} = \frac{1935 \times 10^{-6}}{(0.914)^2} = 2.316 \times 10^{-3}\ \text{sr.}$$

Luminous flux incident upon the cell

$$\Im = I\Omega = 180 \times 2.316 \times 10^{-3} = 0.417\ \text{lm.}$$

From Fig. 19.29, corresponding to a luminous flux of 0.417 lm and a load resistance of 800 Ω the current is 120 μA.

## 19.17 VISUAL PHOTOMETRY

Photometric measurements which depend upon human eye for judgement of equality of illuminance or luminance are termed as visual photometric measurements.

### 19.17.1 Measurements of Luminous Intensity

The principle of working of methods Standard for measurement of luminous intensity is based upon the inverse square law. A simplified set up of method for measuring luminous intensity is shown in Fig. 19.34.

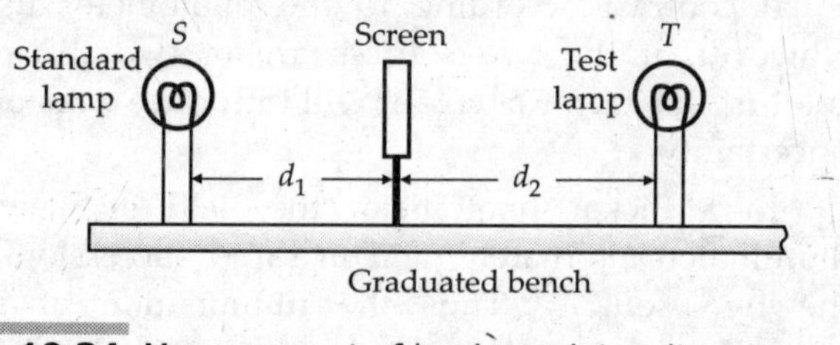


**Fig. 19.34** Measurement of luminous intensity.

Two lamps *S* and *T* are placed at a distance apart with some type of screen in between them. *S* is a standard lamp whose intensity is known and *T* is a lamp under test whose intensity is to be determined. The screen is moved in between the lamps till the illuminance (illumination) on both sides of the screen is equal.

Under this condition :

$$I_s / d_1^2 = I_T / d_2^2$$

$\therefore$ Intensity test lamp

$$I_T = I_s d_2^2 / d_1^2 \qquad \text{...(19.27)}$$

where $I_s$ = Intensity of standard lamp.

The screen used for determining the point of equal illuminance is called *"photometer Head"*. The graduated bench on which the photometer head slides is called *"Photometer Bench"*.

### 19.17.2 Photometer Heads

A photometer head consists of a device by means of which the illuminance of two surfaces, side by side may be compared under similar conditions and without moving the eye. One of the surfaces is illuminated by a standard lamp and the other by the lamp under test. 19

The various photometer heads are explained below :

1. ***Bunsen photometer head.*** It consists of a sheet of opaque white paper in the centre of which a very small spot of grease is placed. The grease spot is translucent. If the paper is illuminated from one side only then the appearance of that side will be a dark patch at the centre and a bright surrounding, while the appearance on the other side will be a bright central patch with a dark surrounding. If illuminated from both the sides (one side by standard lamp and other by test lamp) there will be a position of the photometer with respect to the sources for which the grease spot disappears when viewed simultaneously from two sides of the paper. Under this condition the illuminance of two sides of paper are equal. Therefore, intensity of test lamp under test is $I_T = I_s d_2^2 / d_1^2$.

The two sides of paper are viewed simultaneously with the help of an arrangement using two mirrors $M_1$ and $M_2$ as shown in Fig. 19.35.

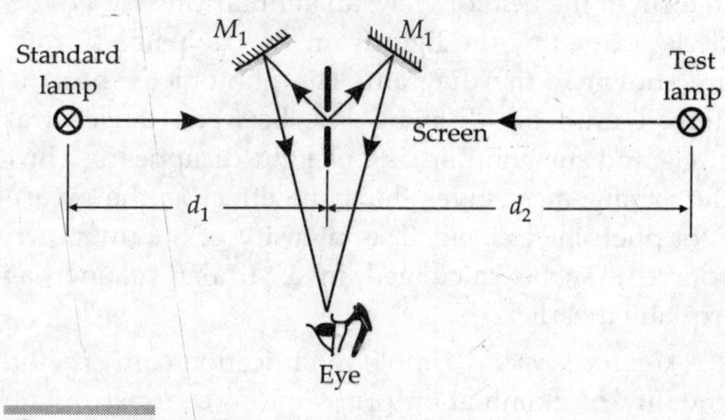


**Fig. 19.35** Bunsen photometer head.

2. ***Lummer-Brodhun photometer head.*** *Comparison type.* It is a precision head in which the precise demarcation of two halves of the visual field is achieved by optical means. The arrangement is shown in Fig. 19.36. The screen $S$ is matt white screen of plaster of paris. The two sides of this screen are illuminated by the two sources. The two sides are viewed simultaneously by means of the totally reflecting prisms $P_1$ and $P_2$ the Lummer Brodhun prism $P_3$. This is a compound prism ($P_3$) in which the hypotenuse face of one component is flat and the other

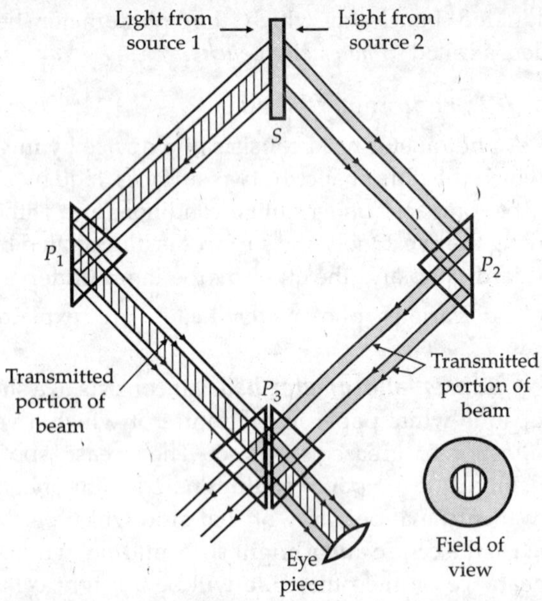


**Fig. 19.36** Lummer Brodhun photometer head (Comparison type).

spherical with a circular flat ground at the middle. Light travelling from $P_1$ to $P_3$ passes straight through the contact surface and reaches the eye piece, the resulting field of view being a circle. Light travelling from $P_2$ to $P_3$ also passes through the contact surface and therefore does not reach eye. Light which is outside this transmitted central beam is totally reflected and therefore passes through the eye piece, to form in the field of view, an annular ring round the circle formed by the light from $P_1$. The paths of rays are shown in the diagram. The photometer head is moved until the dividing line between the central circle and the annular ring of light disappears. Thus the arrangement gives the same effect as the grease spot photometer head. The intensity of the unknown source may be calculated in a similar manner as explained earlier.

*Contrast type.* A simple modification converts the equality of illumination prism into a contrast prism. The hypotenuse faces of the two component prisms are now flat and one of them is etched with trapezoidal pattern as shown in Fig. 19.37. Where the two faces are in contact, light from either side is transmitted but where the etching causes a physical separation there is total reflection. The resulting field of view is shown in Fig. 19.37. There is an additional modification in the form of two glass plates $G_1$ and $G_2$ arranged so that, because of the additional reflection set up, the illumination of the trapezoidal patch on the right hand side of the field will be somewhat less than that of surrounding field on the left hand side and *vice versa.* Denoting the difference by $x\%$ it follows that at balance, the contrast between the trapezoidal patch and its surrounding will be $x\%$ on each side.

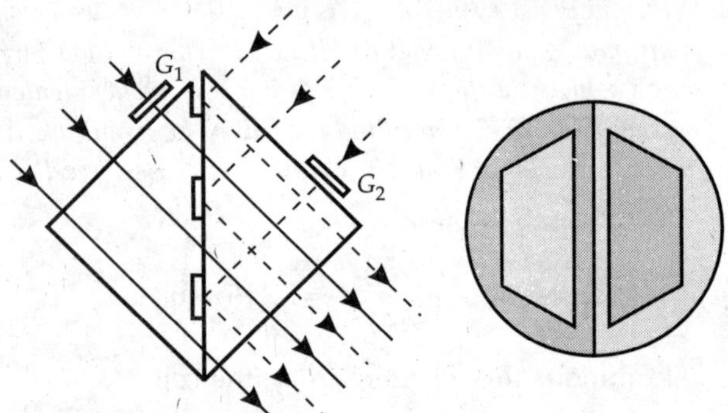


**Fig. 19.37** Lummer Brodhun Photometer Head (Contrast Type).

The contrast instrument is preferable to other because the fusion of two halves of an equality of illumination field does not lead to a high degree of accuracy of judgement as does equality of contrast.

The photometers described above are used for sources of the same colour.

3. ***Flicker photometer head.*** It will be appreciated that, when the field of view is illuminated by both sources at the same time, and the colour appearances of the two halves of field are exactly the same, the only factor calling for judgement of the part of the observer is the criterion of equal luminosity or equal contrast according to the photometer used. However, if the two sources are of two different colours, this factor of colour will introduce a factor of uncertainty.

In a flicker photometer the field of view is illuminated alternately, and in rapid succession, by the two sources. Thus the illumination at any particular instant is due to one source only. If the screen luminosities due to two sources are different, then there will be flicker due to both the causes, *i.e.,* due to difference in luminosity and colour. If the speed of the flicker is gradually raised from a low value, it is found that the flicker due to colour difference disappears before that due to luminosity difference.

The photometer can thus be used for :

(*i*) the measurement of intensity of unknown source with a standard source of same colour,

(*ii*) for the measurement of intensity of unknown source with a standard source of different colour. In this case there must first of all be adjustment of the speed till the colour difference disappears and then the normal photometric adjustments are made for getting minimum flicker.

Figure 19.38 shows a flicker photometer. One light source illuminates a white diffusing screen *S* through a prism *P*. The other source illuminates a rotating sector *F*. This sector also has a white diffusing surface.

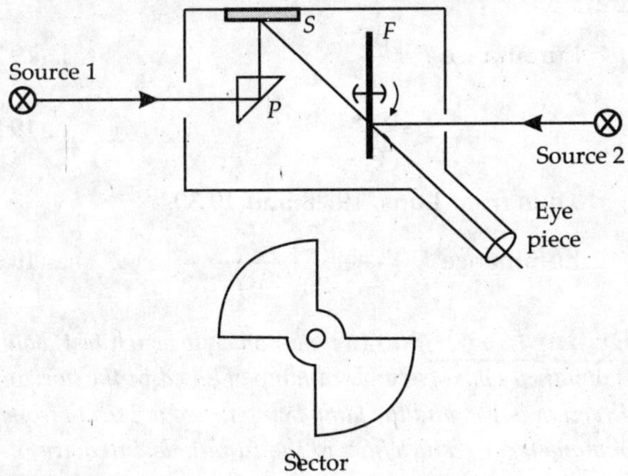


**Fig. 19.38** Flicker photometer.

When the sector revolves the eye sees the screen *S* and sector *F* alternately. The distances of sources from the photometer head are varied till minimum flicker is observed. The intensity of the test source is determined by the inverse square law.

### 19.17.3 Distribution of Luminous Intensity

In the discussions above, we have considered the intensity of the source in only one direction. However, in most lamps and sources of light, the luminous intensity is not the same in all the directions. Therefore, it is important that the intensity distribution in a number of directions round the source is known. The intensity distribution measurements are made to determine the means horizontal intensity and the mean spherical intensity.

### 19.17.4 Determination of Mean Horizontal Luminous Intensity and Polar Curve

This may be determined for any source by turning it about a vertical axis and measuring intensity in the direction of the line joining the source under test and the standard source. The readings are taken after 10° to 15° of rotation. A polar curve is plotted as shown in Fig. 19.39 from the results obtained from measurements. This curve is horizontal distribution curve of intensity. The mean horizontal luminous intensity may be obtained by plotting the intensity against the angle of rotation on rectangular axis as shown in Fig. 19.40. The mean ordinate of this curve gives the value of mean horizontal intensity.

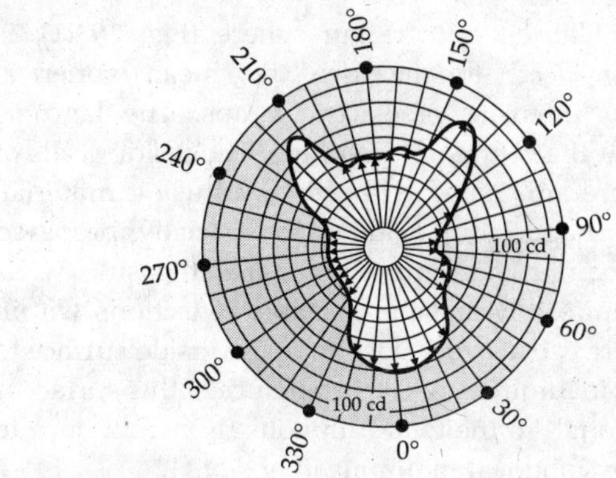


**Fig. 19.39** Polar curve.

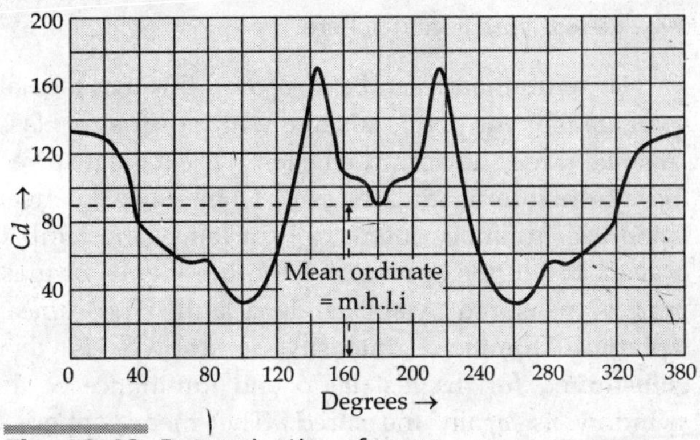


**Fig. 19.40** Determination of mean horizontal luminous intensity.

The mean horizontal intensity may also be obtained with a single measurement if the source under test is continuously revolved about its vertical axis at a speed of about 200 rpm.

### 19.17.5 Determination of Mean Spherical Luminous Intensity

The mean spherical luminous intensity is obtained from the intensity distribution curve in a vertical plane. This distribution curve is obtained from intensity measurements done by tilting the lamp in a vertical plane after intervals of 10° to 15°.

The mean spherical luminous intensity may be obtained graphically from the distribution curve with the help of Ulbricht integrating sphere.

The Ulbricht integrating sphere (Fig. 19.41), is generally used to measure the mean spherical luminous intensity of electric lamps. The lamp is located at the centre of the sphere. The inside walls of the sphere are coated with white diffusing material, such as magnesium oxide or a specially prepared sphere paint. The lamp illuminates the side walls nearly uniformly and the multiple reflections within the sphere cause every portion of the inside surface to be equal in luminance. The luminance of this surface is proportional to the total luminous flux and hence to the mean spherical luminous intensity.

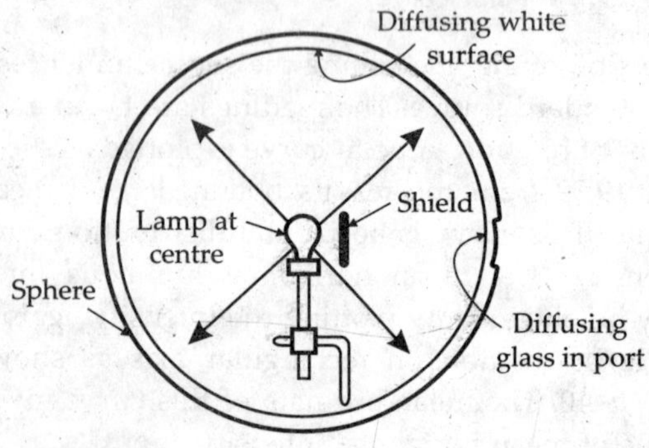


**Fig. 19.41** Integrating sphere.

This luminance can be measured through a small port, usually the port is covered with a diffusing glass. A small screen adjacent to the lamp prevents its direct light from illuminating the port. To measure the mean spherical luminous intensity of a lamp, the lamp is placed inside the sphere and the luminance of glass port is measured. A substandard lamp whose mean spherical luminous intensity is known, is then substituted for the test lamp and luminance of the window is again measured. The mean spherical luminous intensities of the two lamps are proportional to the corresponding luminances of the glass port.

This sphere can be used for measurement of total luminous flux also.

Total luminous flux

$\Im = 4\pi \times$ mean spherical luminous intensity.

*Theory of Integrating Sphere*

Let

$E$ = indirect illuminance on the surface of sphere ; lx,

$R$ = radius of sphere ; m.

$\therefore$ Total reflected flux on the inside of sphere

$= 4\pi R^2 E.$

Suppose the direct flux is $\Im$ lumen, and reflection factor is $\rho$.

Reflection factor is defined as :

Reflection factor,

$$\rho = \frac{\text{reflected luminous flux}}{\text{incident luminous flux}}$$

$\therefore$ Luminous flux absorbed $= (\Im + 4\pi R^2 E)(1-\rho)$ ...(19.28)

But luminous flux emitted = luminous flux absorbed

$$\therefore \quad \Im = (\Im + 4\pi R^2 E)(1-\rho)$$

or $$E = \frac{\rho \Im}{4\pi R^2 (1-\rho)} \quad ...(19.29)$$

Hence indirect illuminance is directly proportional to the luminous flux of the source.

We can express the above result in terms of luminance.

$$\text{Luminance } L = \frac{\rho E}{\pi} \quad ...(19.30)$$

or $$E = \frac{\pi L}{\rho} \quad ...(19.31)$$

Thus from Eqns. 19.28 and 19.30,

$$\text{Luminance } L = \frac{\rho^2 \Im}{4\pi^2 R^2 (1-\rho)} \quad ...(19.32)$$

***Example 19.13*** *(a) In a photometric bench test, balance is obtained when a standard lamp of 25 cd in the horizontal direction is 1 m and the lamp being tested is 1.25 m from the photometer screen. What is the luminous intensity of the lamp?*

*(b) If the light from the test lamp is reduced by 15 percent what will be the respective distances of the lamps from the photometric screen for balance. In this case the lamps are fixed 2.5 m apart, and the photometer screen is moved between them.*

**Solution.** (*a*) From Eqn. 19.27, luminous intensity of test lamp

$$I_T = I_s d_2^2 / d_1^2 = 25 \times (1.25)^2 / (1.0)^2 = 39.1 \text{ cd.}$$

(*b*) Intensity of test lamp

$$I_T = (1 - 0.15) \times 39.1 = 33.2 \text{ cd.}$$

Now $d_2 / d_1 = \sqrt{33.2/25} = 1.152$

or $d_2 = 1.152\, d_1$

Now $d_1 + d_2 = 25$

or $d_1 + 1.152\, d_1 = 2.5$

$\therefore$ $d_1 = 1.16$ m

and $d_2 = 1.34$ m.

Hence, the distance of standard lamp from the screen is 1.16 m while that of test lamp is 1.34 m.

***Example 19.14*** *If an Ulbricht sphere 300 mm in radius, whose inner surface has a reflection factor of 0.84, contains a lamp which produces a luminance of 1300 cd/m*$^2$ *on a portion of the sphere screened from direct light, calculate the luminous flux of the source.*

**Solution.** From Eqn. 19.32, luminous flux :

$$\Im = \frac{4\pi^2 R^2 (1-\rho) L}{\rho^2}$$

$$= \frac{4\pi^2 (300\times 10^{-3})^2 (1-0.84)\times 1300}{(0.84)^2} = 1047.4 \text{ lm.}$$

## 19.18 MODERN MEASUREMENT TECHNIQUES

The visual photometry involves the use of human eye as a detector. Modern measurement methods use special lamps, and detectors with filters to simulate the standard eye response. Errors in modern methods of measurement occur because of :

(*i*) stray light

(*ii*) non-uniform response of detectors,

(*iii*) drifts in the detectors and

(*iv*) changes in colour (wavelength) on account of changes in temperature.

A photometer normally uses a silicon photodiode, photomultiplier tube or a photoresistive device as a detector. Filters are used which correct the response to that of a standard photopic eye. The light source is coupled to the optical detector, usually with a microscope having a fixed aperture. Light is focussed onto the aperture and measured by the detector.

Radiometric instruments use photodiodes, photomultiplier tubes, photoconductive cells and thermal detectors. The sensitivity of the detector is not constant but varies with the wavelength of the incident light and therefore, the instrument should be calibrated for the wavelength of the light being measured. In case of broadband sources, filters are used if measurements are to be made in a narrow frequency band. A spectroradiometer is designed to cover a wide band of frequencies. It scans continuously over the detector, isolating all the wavelengths required and its operation is very similar to that of a filter. Spectroradiometers use many accessories like cosine receptors, LED receptors, integrating spheres, imaging optics, microscopes, telescopes and fibre optic probes. Some of the accessories used are described in Art. 19.19. Spectroradiometers are used for measurements for both sources and detectors. Standard sources are used for calibration of sources and detectors. The calibration sources should have similar intensity and optical characteristics to those of the test source.

Optical power meters give an indication of power level using specially designed chopper sensors, which convert light into an electrical signal. The sensors used with power meters are classified on the basis of power level and the wavelength of light.

Lasers, when used as sources, present many problems. The high power density of lasers causes a non-linear response in the photo detectors. The problem may be partially solved by using diffusers in front of the detectors. This, however, results in modification of the spectral response on account of reflection and absorption.

Errors are caused on account of non-uniform illumination of the detectors such as thermopiles as it result into non-uniform temperature distribution. Mirrors are commonly used in laser power measurements to reflects some of the power. Lasers have narrow beams, the reflected beam may regarded as an incident beam, and a mirror with about 90% reflection would absorb measurable amounts of heat.

Measurements of the spectrum of light sources is usually done using an optical spectrum analyzer. The light is resolved by either interference or diffraction methods. The interference method uses a Fabry-Perot or Michelson interferometer. The diffraction method resolves the spectrum using a monochromator with a diffraction grating or a prism.

## 19.19 RADIOMETER/PHOTOMETER ACCESSORY EQUIPMENT

*Microscope accessory.* The microscope accessory can measure brightness or radiance of small objects ranging from say 0.15 to 0.76 millimeters with a circular field of view.

*Spot brightness accessory.* The spot brightness accessory is a reflex viewing accessory used with an objective lens to provide a variety of working distances and resulting image sizes.

*Detector holder and aperture set.* The detector holder and aperture set provides a convenient means of varying detector area without changing detector response.

*Steradian shade.* To fix the field of view to a specific geometry, the steradian shade should be used. The geometry is fixed at say 0.155 steradians. Thus permits the measurement of surface brightness (luminance or radiance) in terms of candela per square meter, or microwatt per square metre steradian.

*Light shade.* The light shade is used in conjunction with detector/filter assemblies to limit the detector field of view in degrees thereby minimizing off-axis ambient light effects.

*LED shade.* The LED shade is a baffled tube that accurately defines the field of view, necessary when measuring LEDs.

*Neutral density filters.* The purpose of the neutral density filters is to extend the maximum power measurement capability of radiometers and photometers.

*Laser beam spreader.* The beam spreader extends laser power measurement capability.

*Laser integrating sphere.* The laser integrating sphere provides constant attenuation for high-power laser measurements.

*LED integrating sphere.* The LED integrating sphere facilitates laser power measurements. The sphere extends power measurements capability.

*Laser attenuator.* The laser attenuator provides constant attenuation for higher power laser measurements.

*Photometric filter.* When it is desired to measure light as the human eye would perceive it, the photometric filter should be used.

*Cosine diffuser.* The cosine diffuser is employed when making measurements of radiation at a surface.

*Radiometric filter.* The radiometric filter is used to provide a flat response when measuring radiation other than photometric.

*Cosine receptor.* The cosine receptor is employed to measure incident light over a wide angle.

*Narrow bandpass filters.* Optical glass interference filters with half-power bandpass at popular laser wavelengths.

*Fiber optic probes.* Probes used to make measurements in inaccessible areas. The probe is a flexible light guide constructed to glass-clad optic fibers sheathed in PCV with stainless steel end fittings.

*Objective lens.* Lens used to provide specific spot sizes.

*Close-Up objective.* Provide a variable room at a specific degree viewing angle.

*Lumilen.* Lens used with a detector head for measurement of surface brightness.

*Temperature filters.* Filters used with a radiometer / photometer to determine colour temperature in kelvin of incandescent sources.

## 19.20 FIBER OPTICS

There has been a great advancement in the technology of fiber optics sensors during the past few years. Instruments using optical fibers for measurement of various physical parameters are already available in the market and considerable research is being done related to applications of fiber optic sensors.

The fiber optic sensors provide high accuracy and sensitivity, and a broad dynamic range by simply altering the characteristics of light in response to various stimuli. In addition to accuracy, precision, sensitivity and a broad dynamic range, optical fiber sensors are being used in hostile environments, to reduce the hazards of shock in sensing applications, and to reduce the costs of remote sensing.

### 19.20.1 Optical Fibers

Optical fibers are thin, flexible threads of transparent glass or plastic that can carry visible light. Optical fibers consists of two concentric layers called the *core* and the *cladding* as shown in Fig. 19.42. The core, cladding along with the surrounding protective jacket constitute the fiber cable.

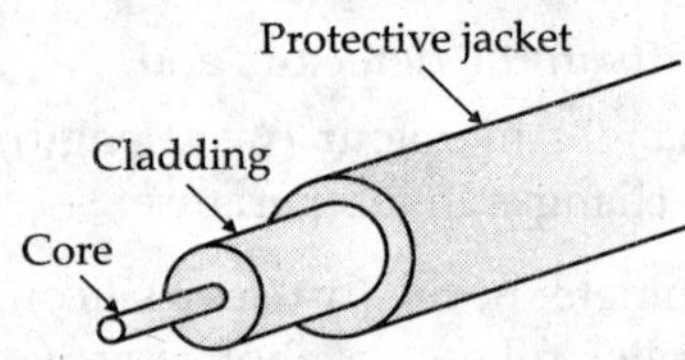


**Fig. 19.42** Fiber optic cable.

Optical fibers are classified in *two* ways :

(*i*) according to the material they are made up of,

(*ii*) according to the refractive index of the core and the number of modes by which the fiber propagates light.

*Three* different combinations of materials are used to construct optical fibers.

These are :

(*i*) Glass core and glass cladding

(*ii*) Glass core and plastic cladding

(*iii*) Plastic core and plastic cladding.

Optical fibers are also classified according to their refractive indexes into the following *three* categories :

(*i*) Step-index fiber

(*ii*) Graded-index fiber

(*iii*) Single mode fiber.

### 19.20.2 Principle of Optical Fibers

If a beam of light crosses the boundary between two materials, of refractive indices $n_1$ and $n_2$ then it will be retracted as shown in Fig. 19.43.

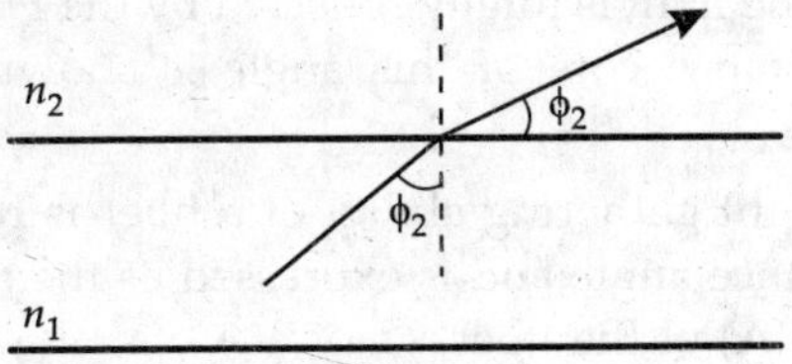


**Fig. 19.43** Illustration of refraction.

As the angle of incidence $\phi_1$ is increased, a point is reached when the beam is totally internally reflected. For this to occur inequalities 19.33 and 19.34 must both be true.

$$n_2 < n_1 \quad ...(19.33)$$

$$\phi_1 > \cos^{-1}(n_2 / n_1) \quad ...(19.34)$$

The angle at which total internal reflection first starts is called the *critical angle* and is given by :

$$\phi_c = \cos^{-1}(n_2 / n_1) \quad ...(19.35)$$

The principle of total internal reflection is used to transmit light along optical fibers as in Figs. 19.44 and 19.45(*a*). This is known as a *step index fiber* since there is an abrupt change in refractive index between the core fiber and its cladding.

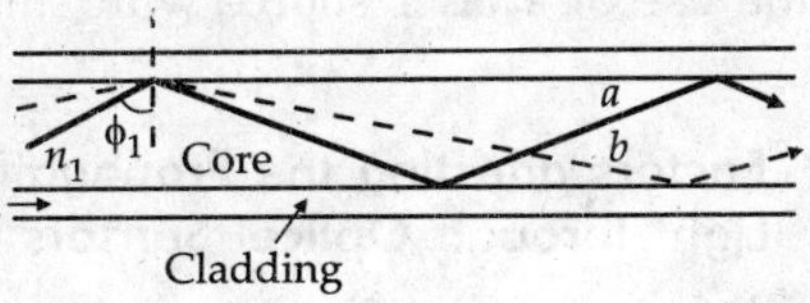


**Fig. 19.44** Light transmission in a step index fiber.

Since the path that a light wave follows as it propagates the core of an optical fiber is a function of its angle of incidence, the angle at which it strikes the core-cladding boundary, it follows naturally that there are many paths through the core. These paths are called *modes*. A *mode* is a path that a light wave can follow as it travels down the core of an optical fiber.

The number of modes for a fiber ranges from 1 to more than 100000. As explained above, light waves can travel through the core of fiber in several modes, the number of modes, *N*, is given by

$$N = \frac{2\pi^2 r^2}{\lambda^2}(n_1^2 - n_2^2) \quad ...(19.36)$$

where

$\lambda$ = wavelength of light ; m

$r$ = radius fiber core ; m, and

$n_1, n_2$ = refractive indices of core and cladding respectively.

Optical fibers used in sensors are always single mode fibers. This means that in optical sensors used in instrumentation, the light wave follows a single path as it passes through the fiber - a straight line along the longitudinal axis of fiber. The optical paths for the different classifications of opticals fibers are shown in Fig. 19.45. Single mode fibers are designed so that they have only one path for a light wave to travel. This is accomplished by making the fiber core diameter exceedingly small, typically from 5 to 10 μm. This is very close to the wavelength of propagated light wave.

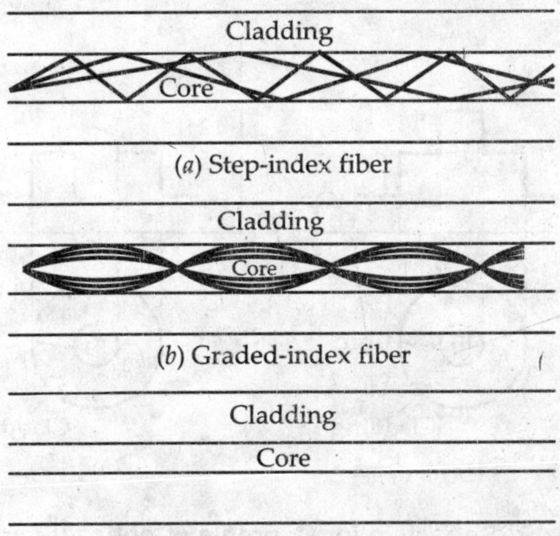


**Fig.19.45** Modes of propagation in optical fibers.

The alternate to single mode transmission is the *multi-mode transmission*. The effect of multimode transmission paths is to give signal distortion. If '*a*' and '*b*' are two modes of travel for light waves in Fig. 19.44, then it is clear that beam '*a*' takes a longer path than beam '*b*' resulting in pulse stretching as shown in Fig. 19.46.

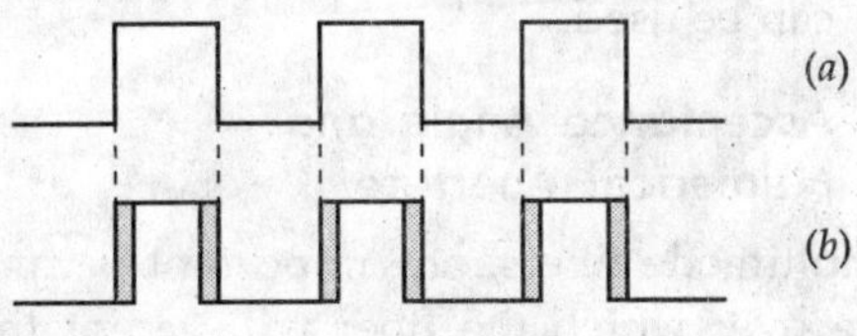


**Fig.19.46** Distortion of a light pulse passing through a fiber : (*a*) input light from source.
(*b*) output light from fiber.

This distortion can be minimized by using *Graded Index Fibers*. In this, cable, the refractive index of the core is gradually decreased from the central axis to the edge. The difference in refractive index over the core is about 1%. Figure 19.47 shows the refractive index profile for step index and graded index fibers.

The graded index fiber may be considered as a series of finite layers with refractive index changing at each boundary. The light is refracted as it crosses each boundary, but the grazing angle is successively reduced until eventually the light beam is internally reflected. This results in a parabolic path for the light as shown in Fig. 19.45(*b*).

Since the light wave travelling in the outer regions of the fiber sees lower refractive index material, it also travels faster than the light wave at the centre of the core. This partly compensates for the longer path which needs to be traversed by the outer waves, and reduces the pulse distortion as depicted in Fig.19.46

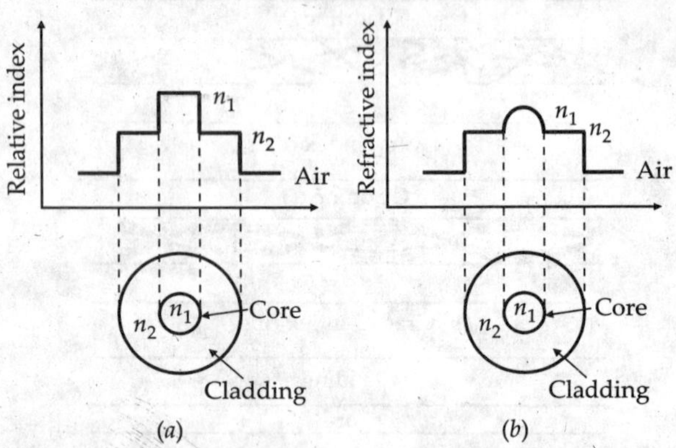


**Fig. 19.47** Refractive index profile of optical fibers : (*a*) step index, (*b*) graded index.

The disadvantages of step index fibres are :

(*i*) They use a large area so that greater light may be coupled to them.

(*ii*) their performance at low temperatures is poor

(*iii*) it is difficult to fit connectors to them

(*iv*) they suffer from pulse distortion, which reduces the bandwidth over which the fiber can be used.

### 19.20.3 Acceptance Angle and Numerical Aperture

In multimode fibers, one important parameter is the angle over which the fiber will accept the light directed towards it. This angle is called the *acceptance angle* and is defined as the half-angle of the core with

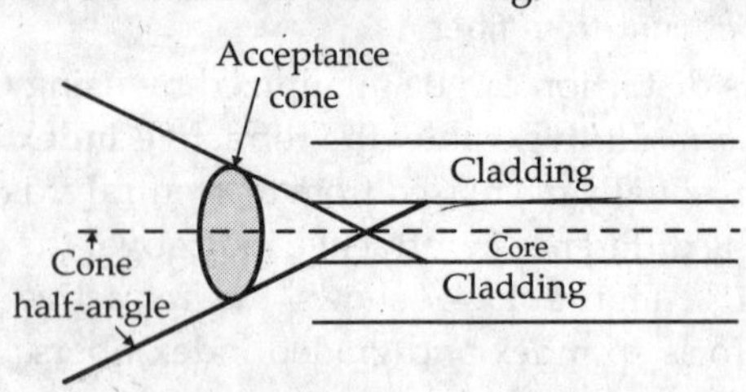


**Fig.19.48** Acceptance cone and half-angle of an optical fiber.

in which the light is totally reflected by the Fiber core. The acceptance core of this angle is shown in Fig. 19.48.

The light gathering ability of a fiber is related to its acceptance angle and is expressed as the *numerical aperture* (NA) of the fiber.

The numerical aperture of a fiber can be expressed quantitatively as :

$$\text{NA} = \sin\theta$$

or

$$\theta = \sin^{-1}\text{NA} \qquad \text{...(19.37)}$$

where $NA$ = numerical aperture, and

$\theta$ = angle of acceptance.

Manufacturers do not usually specify the value for numerical aperture for the single mode fibers marketed by them since light in single mode fibers is not reflected or refracted and therefore, does not exit the fiber at an angle through cladding. Typical values for the numerical aperture for single mode fibers is of the order of 0.2 to 0.6. The value of numerical aperture is usually not specified. The value of numerical ratio normally lies between 0.03 to 0.1 for single mode fibers. The low value for the numerical aperture means a small acceptance angle, which in turn, dictates the use of a laser source with single mode fibers.

### 19.20.4 Factors affecting the Propagation of Light through Optical Sensors

Several factors affect the propagation of light through an optical sensor. These are :

(*i*) the coherence of light source

(*ii*) the size of fiber

(*iii*) the composition of fiber

(*iv*) the numerical aperture of the source and fiber

(*v*) the amount of light injected into fiber.

The coherent light is provided by *Injection Laser Diode.* When this diode is used with single mode fibers, the coherence of light is maintained along the fiber. An injection laser diode is a semiconductor diode that spontaneously emits light from *PN* junction when a forward bias is applied. It acts as a light emitting diode below threshold current. Lasing effect takes place when the current exceeds the threshold value *i.e.,* the current at which the light becomes coherent. The coherence is very important because it permits interferometric comparison of the paths taken by two beams of light, which is a technique used to make very sensitive measurements with many optical fibers.

### 19.20.5 Fiber Optic Sensors

There are *two* basic types of fiber optic sensors which are :

(*i*) *Pure fiber sensors.* These sensors for their response depend upon on environmentally induced changes in light as it travels through the fiber.

(*ii*) *Remote optic sensors.* The use optical fibers to carry light to a separate device that responds to light stimuli.

In most pure optic sensors, one of the two effects happens to the light in response to some external effect. These are :

(*a*) The light can leak from the core into the cladding where it is eventually absorbed.

(*b*) The light can be forced to travel a different distance than light travelling an alternate optical path. This causes a phase difference to exist between two light waves. This difference can be detected by combining the two light waves by superposition and observing the resulting light and dark pattern.

Optical sensors use a single mode fiber which is a low loss fiber with a very small core and consequently a very small angle of acceptance of light. A basic scheme used for measurements involving optical fiber sensors is shown Fig. 19.49. The sensing mechanism illustrated in the diagram causes a change is some characteristic of the optical signal received at the detector. The sensing mechanism may, in fact, be a characteristics of the fiber itself, or it may be a device external to the fiber. In case tine light signal received by the detector changes as a result of some change in the fiber, then the sensor is called an *Intensity* or *Intrinsic sensor.* On the other hand, if the light signal at the detector changes as a result of some transducer type of device, then the sensor is called an *Interferometric* or *Extrinsic* sensor.

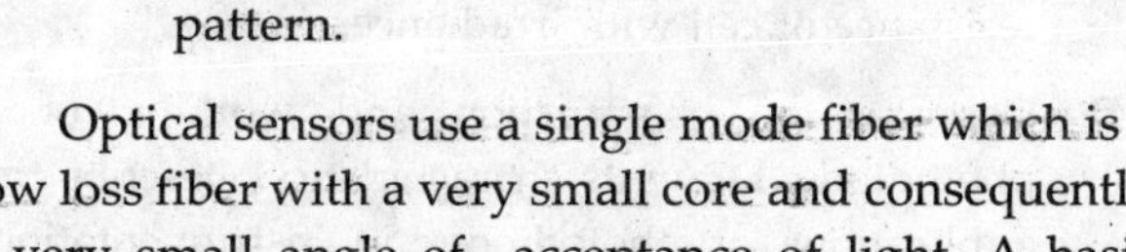


**Fig.19.49** Basic optical fiber sensor schematic.

Table 19.4 shows the list of physical parjameters that can be measured with help of Intensity/ Interferometric types of fiber optic sensors that are presently available in the market.

**Table 19.4** *Physical Parameters and the types of Optical Sensors currently available for their Measurement*

| *Physical Parameter* | *Sensor Type* | |
|---|---|---|
| | *Intensity* | *Interferometric* |
| Pressure | X | |
| Sound | X | X |
| Temperature | X | X |
| Level | X | X |
| Rotation | | X |
| Acceleration | X | |
| Displacement | | X |
| Strain | X | X |
| Electric current | X | |
| Voltage | | X |
| Electric fields | X | |
| Magnetic fields | X | |
| Acoustics | | X |
| Colour | | X |
| Pulse rate | | X |
| pH value | | X |

### 19.20.6 Advantages of Fiber Optic Sensor

The advantages of fiber optic sensors are :

(*i*) They are compatible with fiber optic communication systems in case data is being transmitted through the aforesaid medium.

(*ii*) They do not conduct any electric current and therefore optical fiber sensors are very useful in explosive environments and in high voltage equipment.

(*iii*) They are immune to electromagnetic interference and radiated signals. Optical fibers do not pick up radio frequency interference, electromagnetic interference, or interference caused by lightning, switching surges, nearby electric machinery and relays, and other sources of potential electrical noise. Therefore, the output of fiber optic sensors is free from noise of any type.

(*iv*) They are increasingly being used in measurement of physical quantities like pressure, temperature, sound, displacement and level. They can also be used for measurement of electric current, voltage, electric and magnetic fields in addition to measurements of different quantities related to the field of optics.

# Review Questions

1. Describe the electromagnetic spectrum and its major regions.
2. Explain the international luminosity curve and describe how is it related to spectral response of human eye.
3. Define threshold wavelength and derive its expression.
4. Define the following terms :
   (*i*) Monochromatic light
   (*ii*) Linearly polarized light
   (*iii*) Refraction and Refractive Index
   (*iv*) Reflection, Absorption and Transmission of Light.
5. Differentiate between the terms "Photometry" and "Radiometry". Explain the terms :
   photopic vision and scotopic vision.
6. Define the following :
   (*i*) Point source (*ii*) Extended source
   (*iii*) Solid angle.
7. Define the following Photometric terms :
   (*i*) Lumen
   (*ii*) Luminous Intensity
   (*iii*) Illuminance
   (*iv*) Luminance and
   (*v*) Luminous Excitance.
8. Describe the "Inverse Square" and "Lamberts Cosine" laws of illumination.
9. Define the following radiometric terms :
   (*i*) Radiant energy
   (*ii*) Radiant intensity
   (*iii*) Radiance
   (*iv*) Radiant excitance
   (*v*) Irradiance
   (*vi*) Radiant incidence.
10. Draw a generalized block diagram for a photometric / radiometric system. Explain the functions of source, modulating devices and detectors.
11. Explain the different types of optical sources. Describe their salient features.
12. Describe the construction and working of light emitting diodes. Explain the direct and indirect recombination modes.
13. What are lasers ? Describe different types of lasers used for instrumentation purposes along with their properties.
14. Give an overview of optical transducers.
15. Describe the construction and working of a vacuum type of photocell. Draw its characteristics and describe its advantages and disadvantages.
16. Describe the construction and working of gas filled photo-emissive tubes. Draw their characteristics and explain their advantages and disadvantages.
17. What is a photo multiplier tube ? Describe its construction and derive the expression for its gain.
18. Describe the construction and working of a photo resistive cell. Draw typical spectral response characteristics of the cells. Define "dark resistance". Write and explain the variation of resistance of cell with irradiance.
19. Describe the construction and working of a photodiode. Draw its characteristics. Describe the applications of photodiodes in instrumentation. Comment upon their frequency response.
20. Describe the construction and working of photovoltaic cells. Draw their characteristics. Describe the different materials used for these cells and draw up a table listing their time constants and the spectral band in which they can be used.
21. Describe the thermal sensors that are used in optical measurements. List their applications, advantages and disadvantages.
22. Explain the following :
    (*i*) Light Suppression
    (*ii*) Light Attenuation
    (*iii*) Photometric filters
    (*iv*) Radiometric filters.
23. Expiain the working of contrast type and flicker type of Lummer Brodhun photometric heads.
24. Explain the working and construction of an Ulbricht type of integrating sphere.
25. What are the three types of optical fibers are in use at present and which of the three is most suitable for instrumentation and why ?

26. What are the factors which affect the light propagation through an optical fiber ?

27. Describe the following :

(i) Step index fiber and

(ii) Graded index fiber.

28. Describe the conditions for total internal reflection within a fiber.

29. Define acceptance angle and numerical aperture in fiber optics. Write their expressions and explain their importance in fiber optics. Explain the difference between single and multi-mode-transmission ?

30. Describe the two basic types of optic fiber sensors. What is the difference between inrinsic and extrinsic types of sensors ?

# Unsolved Problems

1. The frequency of a given electromagnetic radiation is $6 \times 10^{14}$ Hz. Find the wavelength. Does this radiation falls within visible spectrum ?
[**Ans.** 0.5 μm ; Yes]

2. The relative efficacy of radiation or a wavelength of 0.49 μm is approximately 0.2. The international luminosity curve indicates relative efficacy of a wavelength of 0.555 μm as unity. Radiation equivalent to 20 W is produced by a wavelength of 0.49 μm. Find out the radiation (in W) at a wavelength 0.555 μm which will produce the same sensation in the human eye. [**Ans.** 4 W]

3. A metal filament gas lamp takes 0.42 A from a 230 V supply and emits 1120 lm.

Calculate

(a) the lumen per watt,

(b) the mean spherical luminous intensity (m.s.l.i.) of the lamp, and

(c) the m.s.l.i. per watt.

[**Ans.** (a) 11.6 lm/W, (b) 89.2 cd, (c) 0.923 cd/W]

4. The length of a fluorescent coating of a certain fluorescent lamp is 1460 mm and the diameter is 38 mm. If the luminous intensity in a radial direction is 340 cd, what is the luminance of the lamp ?
[**Ans.** 6130 cd/m$^2$]

5. A lamp giving 200 cd in all direction below the horizontal plane is suspended 2 m above the centre of a square table 1 m side. Calculate the maximum and minimum illuminances on the surface of the table. [**Ans.** 50 lx, 41.9 lx]

6. A 200 cd lamp emits light uniformly in all directions and is suspended 5 m above the centre of a working plane which is 7 m square. Calculate the illuminance immediately below the lamp and also at each comer of the square. If the lamp is fitted with a reflector which distributes 60% of the light emitted uniformly over a circular area 5 m in diameter, calculate the illuminance over this area.
[**Ans.** 8 lx, 2.87 lx ; 76.8 lx]

7. Four lamps are suspended 8 m above the ground at the comers of a square 4 m side. Each lamp gives 250 cd uniformly below the horizontal plane. Calculate the illuminance (a) on the ground directly under each lamp, (b) at the centre of the square. [**Ans.** 11.6 lx]

8. Two lamps of, 10 cd and 24 cd are respectively 2 m apart. A screen is placed between them 0.8 m from 16 cd lamp. Calculate the illuminance on the each side of the screen. Where must the screen be placed in order to be equally illuminated from both the sides ?
[**Ans.** 25 lx, 16.67 lx, 0.9 m from 16 cd source]

9. A certain lamp gives the following distribution of luminous intensity, in a vertical plane, the angle being measured from the axis through the centre of lamp cap :

| *Angle degree* | *Intensity, cd* |
|---|---|
| 0 | 0 |
| 15 | 84 |
| 30 | 154 |
| 45 | 200 |
| 60 | 224 |
| 90 | 254 |
| 120 | 266 |
| 150 | 276 |
| 165 | 283 |
| 180 | 288 |

Find :

(a) the mean spherical luminous intensity, and

(b) the luminous flux of the lamp.

[**Ans.** 236 cd, 2966 lm]

19

10. A 200 W tungsten lamp has a luminous efficiency of 18.3 lm/W. Calculate the luminous intensity. **[Ans.** $46 \times 10^3$ cd ]

11. A detector will just respond to an illumination of 0.14 lux. Find the maximum distance from 10 W tungsten bulb for detectibility. The luminous efficacy of a 10 W tungsten lamp is 7.8 lm/W.

12. A photomultiplier has a current gain of $3 \times 10^6$. A weak light beam produces 50 electrons per second at the photocathode. What anode to ground resistance must be used to get a voltage of 3 μV ? $1\,e = 1.6 \times 10^{-19}$ C. **[Ans.** 125 kΩ]

13. Germanium has band gap of 0.67eV. Find the maximum wavelength for resistance change by photon absorption. Planck's constant $= 6.63 \times 10^{-34}$ Js and 1 eV $= 1.6 \times 10^{-19}$ J. **[Ans.** 1.86 μm]

14. Determine the acceptance angle for a single fiber with a numerical aperture of 0.096. **[Ans.** 5.5°]

# Objective Type Questions

***Tick (√) the most appropriate answer*** :

1. The optical spectrum has wavelengths ranging from :
   (*a*) 10 nm to $10^6$ nm (*b*) 10 nm to 770 nm
   (*c*) 370 nm to 770 nm (*d*) 370 nm to $10^6$ nm.

2. The visible spectrum has wavelengths ranging from :
   (*a*) 10 nm to $10^6$ nm
   (*b*) 10 nm to 770 nm
   (*c*) 370 nm to 770 nm
   (*d*) 370 nm to $10^6$ nm.

3. A certain emissive material has a work function of 2.24 eV, the radiations emitted from it will have threshold length of approximately :
   (*a*) 900 nm (*b*) 600 nm
   (*c*) 533.6 nm (*d*) 527.6 nm.

4. In order that photoelectric emission to be possible over whole of the visible spectrum of 370 nm – 770 nm, the work function of the photoemission surface should be approximately :
   (*a*) 1.61 to 3.35 eV
   (*b*) 1.3 to 2.25 eV
   (*c*) 1.5 to 4.26 eV
   (*d*) 0.231 to 5.62 eV.

5. When alight beam is made incident on a surface which is rough as compared with the wavelength of light
   (*a*) Specular reflection takes place
   (*b*) Diffused reflection takes place
   (*c*) Absorbtion takes place
   (*d*) Transmittance takes place.

6. Radiometry refers to measurement of optical radiations
   (*a*) From a wavelength of 370 nm – 770 nm
   (*b*) From a wavelength of $10^{-12}$ nm to $10^6$ nm
   (*c*) From a wavelength of $10^2$ nm to $10^6$ nm
   (*d*) From a wavelength of 10 nm to $10^6$ nm.

7. The response of human eye under normal light conditions is called
   (*a*) Stopopic vision and it peaks at a wavelength of 555 nm
   (*b*) Stopopic vision and it peaks at a wavelength of 507 nm
   (*c*) Photopic vision and it peaks at a wavelength of 507 nm
   (*d*) Photopic vision and it peaks at a wavelength of 555 nm.

8. A light source may be considered as a point source if its diameter is
   (*a*) less than 10% of the distance between it and the detector
   (*b*) less than 1% of the distance between it and the detector
   (*c*) more than 10% of the distance between it and the detector
   (*d*) more than 1% of the distance between it and the detectors.

9. The units of luminous intensity are
   (*a*) lux (*b*) cd
   (*c*) W/sr (*d*) cd/sr.

10. The unit luminance are
   (*a*) nit (*b*) lm/sr – $m^2$
   (*c*) cd/$m^2$ (*d*) all the above.

**11.** The errors involved in photometric measurements are of the order of

(*a*) 1% (*b*) 1 – 5%

(*c*) 10 – 20% (*d*) 40 – 50%

**12.** The desirable properties of a laser beam are :

(*a*) It has a high radiant energy

(*b*) It can be polarized

(*c*) It is monochromatic and coherent

(*d*) All the above.

**13.** The current through a photoemissive tube depends upon

(*a*) Intensity of light incident upon it

(*b*) Colour and wavelength of incident light

(*c*) Voltage applied between cathode and anode of tube

(*d*) All the above.

**14.** The luminous sensitivity of gas filled photoelectric cells varies between :

(*a*) 10 to 100 μA/lm

(*b*) 40 to 150 μA/lm

(*c*) 0.002 to 0.1 μA/μW

(*d*) 0.01 to 0.15 μA/μW

**15.** A photomultiplier tube uses 10 stages. The sensitivity of each dynnode of the tube is 100 μA/lm, the overall sensitivity of the photo-multiplier tube, if the gain of each stage is 2.7, is approximately :

(*a*) 2.06 A/lm

(*b*) 270 μA/lm

(*c*) 2700 μA/lm

(*d*) None of the above.

**16.** A photoconductive cell may be called as :

(*a*) Photoresistive cell

(*b*) Photo resistor

(*c*) Light dependent resistor

(*d*) All the above.

**17.** The resistance of a photo resistive cell

(*a*) increases with increase in light intensity

(*b*) remains constant irrespective of the light intensity incident upon it

(*c*) decreases with increase in light intensity in a linear manner

(*d*) decreases with increase in light intensity in an exponential manner.

**18.** Photodiodes

(*a*) do not have any response in darkness

(*b*) have a response even in darkness

(*c*) have a leakage current even in darkness which is called dark current that doubles for every 10°C increase in the ambient temperature that which decreases the signal to noise ratio

(*d*) have a leakage current even in darkness which is called dark current that doubles for every increase of 5°C in temperature and which increases the signal to noise ratio.

**19** The sensitivity of a photovaltaic cell

(*a*) is almost constant when it is short circuited

(*b*) decreases with increase of load resistance

(*c*) is almost negligible when the load resistance is about 10000 Ω

(*d*) all the above

(*e*) none of the above.

**20.** In case of measurement of intensity of a test source, a standard source is used. The photometric head used when comparing two lights of different colours, is :

(*a*) Bunsen photometer head

(*b*) Constast type Lummer Brodhun photometer head

(*c*) Comparion type Lummer Brodhun photometer head

(*d*) Flicker photometer head.

**21.** One of the combinations of materials used for optical fibers is :

(*a*) Copper core and glass cladding

(*b*) Glass core and aluminium cladding

(*c*) Glass core and plastic cladding

(*d*) Plastic core and glass cladding.

**22.** For fiber optical sensors,

(*a*) single mode transmission is used

(*b*) multimode transmission is used

(*c*) both single and multimode transmissions are commonly used

(*d*) none of the above.

**23.** In optical fibres, if the total reflection is to take place, the following conditions must be satisfied

(*a*) $n_2 < n_1$ and $\phi_1 > \cos^{-1}(n_2/n_1)$

(*b*) $n_1 < n_2$ and $\phi_1 < \cos^{-1}(n_2/n_1)$

(*c*) $n_2 < n_1$ and $\phi < \cos^{-1}(n_2/n_1)$

(*d*) $n_2 > n_1$ and $\phi_1 > \cos^{-1}(n_2/n_1)$

where $n_1$ and $n_2$ are respectively refractive indices of core and cladding and $\phi_1$ is angle of incidence.

19

**24.** In a graded index fiber, the total reflected light takes a

(*a*) straight line path

(*b*) parabolic path

(*c*) elliptical path

(*d*) circular path.

**25.** The optical sensors may be classified as :

(*a*) Intensity and intrinsic types

(*b*) Interferometric and extrinsic types

(*c*) Intensity or Instrinsic type and interferometric or extrinsic types

(*d*) All the above.

## Answers

| | | | | | | | | | |
|---|---|---|---|---|---|---|---|---|---|
| **1.** (*a*) | **2.** (*c*) | **3.** (*c*) | **4.** (*a*) | **5.** (*b*) | **6.** (*d*) | **7.** (*d*) | **8.** (*a*) | **9.** (*b*) | **10.** (*d*) |
| **11.** (*c*) | **12.** (*d*) | **13.** (*d*) | **14.** (*b*) | **15.** (*a*) | **16.** (*d*) | **17.** (*d*) | **18.** (*c*) | **19.** (*d*) | **20.** (*d*) |
| **21.** (*c*) | **22.** (*a*) | **23.** (*a*) | **24.** (*b*) | **25.** (*c*) | | | | | |

# 20

# Electronic Instruments

## 20.1 INTRODUCTION

Electronic voltmeters, ammeters and ohmmeters use amplifiers, rectifiers and other auxiliary circuits to produce a current proportional to the quantity being measured. This current is measured by a PMMC instrument. The PMMC instruments use a taut suspension as shown in Fig. 7.7 on page 181 instead of the more conventionally used pivot and jewel arrangement shown in Fig. 7.8 on page 181. This is the analog mode of operation of electronic instruments and therefore, they are, appropriately called analog instruments. However, electronic instruments use digital read-out systems and in that case they are called digital electronic instruments. Till recently, the analog type of instruments have been predominantly in use because of the economic considerations. In addition, analog instruments are more suited for special nonlinear scales such as logarithmic scales *i.e.,* decible scales.

Although analog and digital instruments must perform the same function *i.e.,* measurement of voltage, current, power and circuit parameters, the principle of operation is different in each case. In this chapter, only analog type of instruments are being discussed.

## 20.2 ELECTRONIC VOLTMETERS AND THEIR ADVANTAGES

Almost all electronic voltmeters make use of the rectifying properties of diodes whether vacuum tubes or metal rectifiers or semi-conductor diodes. Vacuum tube diode was first used in electronic voltmeters way back in 1895 and is still quite popular as a sensing element of vacuum tube voltmeters (VTVM). With the introduction of the transistor and other semi-conductor devices, vacuum tubes are on their way out. Solid state models with junction field effect transistor (JFET) input stages are known as *Transistor Voltmeters* (TVM) and *Field Effect Voltmeters* (FETVM) are taking their place.

### Advantages of Electronic Voltmeters

**1.** ***Detection of low level signals.*** As indicated above analog instruments use PMMC movement for indication. This movement cannot be constructed with a full scale sensitivity of less than 50 μA. Any measurement using a PMMC movement must draw a current of 50 μA from the measured quantity for its operation for full scale deflection if conventional voltmeters are used. This would produce great loading effects especially in electronic and communication circuits. Electronic voltmeters avoid the loading errors by supplying power required for measurement by using external circuits like amplifiers. The amplifiers not only supply power for the operation but make it possible for low level signals, which produce a current less than 50 μA for full scale deflection, to be detected which otherwise cannot be detected in the absence of amplifiers. Let us examine the loading effect. A typical meter has a resistance of 200 Ω and its operating current at full scale is 50 μA. This means that the power consumed is only $(50 \times 10^{-6})^2 \times 200 = 0.5\,\mu\text{W}$. This is extremely a low power for power circuits but not for many electronic and communication circuits. If this power is taken from the measurand, the signal will get greatly distorted in case the power level of the circuit is very small and to offset this the power is supplied from outside through use of amplifiers.

Let us have a look at the voltage range. A meter with 200 Ω internal resistance and full scale current of 50 μA will have full scale voltage range of $50 \times 10^{-6} \times 200 = 10$ mV. Now in case lower ranges of voltages have to be measured, the use of a voltage amplifier becomes absolutely necessary. This is only possible through use of electronic voltmeters which allow the use of an amplifier. Therefore, it is possible to measure currents below 50 μA, voltages below 10 mV and keep down the power drainage below 0.5 μW by using electronic voltmeters through use of amplifiers which is otherwise not possible with conventional types of meters using PMMC movement. For the case of a.c. measurements, the use of an amplifier for detection of low level signals is even more necessary for sensitive measurements.

2. ***Low power consumption.*** A conventional PMMC voltmeter is a rugged and an accurate instrument, but it suffers from certain disadvantages. The principal problem is that it looks both high sensitivity and high input resistance. It has a sensitivity of 20 kΩ/V with a 0 – 0.5 V range and has an input resistance of only 10 kΩ at its 0.5 V range with the result it has a full scale current of 50 μA which loads the measurand considerably. In electronic and communication circuits as even this low value of current may not be tolerable on account of the fact that these circuits have very low operating currents. The *Electronic Voltmeter* (EVM), on the other hand, can have input resistances ranging from 10 MΩ to 100 MΩ with the input resistance remaining constant over all ranges instead of being different at different ranges, the EVM gives far less loading effects.

Electronic voltmeters utilize the amplifying properties of vacuum tubes and transistors and therefore the power equired for operating the deflecting element of PMMC instrument can be supplied from an auxiliary source. Thus, while the circuit whose voltage is being measured controls the sensing element of the voltmeter, the power drawn from the circuit under measurement is very small or even negligible. This can be interpreted as that the voltmeter circuit has a very high input impedance. This feature of electronic voltmeters is indispensible for voltage measurements in many high impedance circuits such as encountered in communication equipment.

3. ***High frequency range.*** The most important feature of electronic voltmeters is that their response can be made practically independent of frequency within extremely wide limits. Some electronic voltmeters permit the measurement of voltage from direct current to frequencies of the order of hundreds of MHz. The high frequency range may also be attributed to low input capacitance of most electronic devices. The capacitance may be of the order of a few pF.

## 20.3 VACUUM TUBE VOLTMETERS (VTVMs)

Vacuum tube voltmeters were considered as one of the most useful instruments for the measurement of a.c. and d.c. voltages in the past.

The description of VTVMs is being given here purely on account of their historical background and also because of the fact VTVMs are still being used in the laboratories.

The practical vacuum tube voltmeter (VTVM) circuits fall into the following categories :

1. Diode type,
2. Single triode type,
3. Balanced triode type,
4. Rectifiers amplifier type and
5. Amplifier-rectifier type.

In addition to above, many other forms of vacuum tube voltmeters have been devised. These include arrangements employing slide back vacuum tube voltmeters and vacuum tube voltmeters which use logarithmic characteristics.

### 20.3.1 Average Reading Diode Vacuum Tube Voltmeters

The simplest type of diode voltmeter is similar to a rectifier instrument. Figure 20.1(*a*) shows a schematic diagram for an average reading voltmeter using a vacuum tube diode. The arrangement includes a vacuum tube diode, a resistance *R*, and a PMMC instrument, all connected in series. The applied voltage is

$$e = E_m \sin \omega t$$

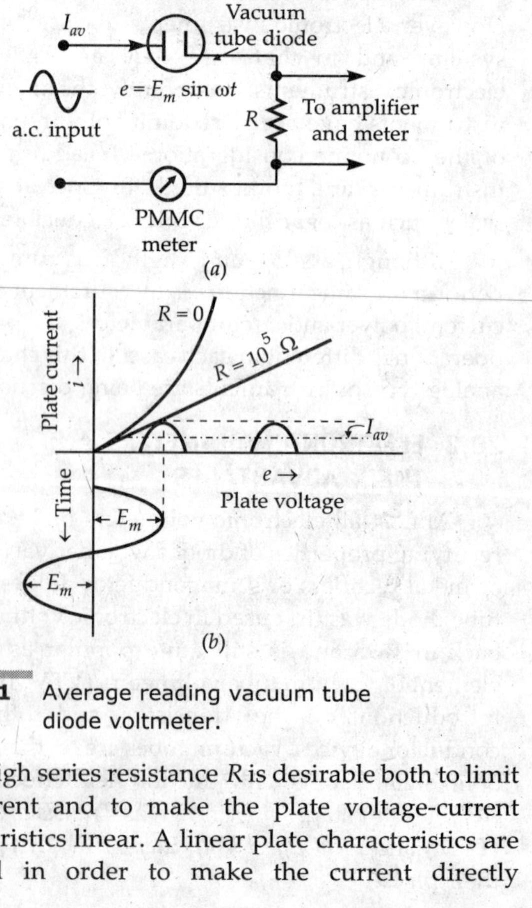


**Fig. 20.1** Average reading vacuum tube diode voltmeter.

A high series resistance *R* is desirable both to limit the current and to make the plate voltage-current characteristics linear. A linear plate characteristics are essential in order to make the current directly proportional to voltage. The plate resistance of a diode is not constant and hence in order to swamp out the variations in plate resistance, a high value of series resistance, $R$ is used. If the series resistance is sufficiently high, the resistance of the tube is negligible in comparison, and therefore we get a linear plate voltage-current characteristics thereby the scale of PMMC instrument is uniform and is independent of the variations of tube resistance. A series resistance of $10^5\ \Omega$ or more is used for the purpose of measurement of a sinusoidal input voltage as shown in Fig. 20.1(*a*).

The diode conducts during the positive half cycle and does not conduct during the negative half cycle, the average current through the meter is thus :

$$I_{av} = \frac{E_{av}}{2R} = \frac{E_{rms}}{2 \times 1.11 \times R} = 0.45\frac{E_{rms}}{R} \quad \text{...(20.1)}$$

where $E_{rms}$ and $E_{av}$ are respectively the rms and average values of applied voltage.

The factor 2 is introduced because the voltmeter works on half wave rectification.

The ***advantages*** of average reading diode voltmeters are :

**1.** *Simplicity.* The average reading diode type meter is simplest in construction.

**2.** *High input resistance and low power consumption.* The input resistance of the circuit depends upon the value of resistance, $R$ which ranges from $10^5\Omega$ to 1 M$\Omega$, and thus resistance is higher than that of multimeter circuits for a.c. voltage measurements. In most diode voltmeters, indication is obtained from the output of d.c. amplifier, the input of which is connected across the resistance $R$. This makes it possible to replace the relatively sensitive meter by a less sensitive and rugged meter. Also the input resistance, $R$ can be made greatly large say of the order of tens to hundreds of M$\Omega$. This way the power consumed by the voltmeter can be made negligibly small. (This is because the power consumed by the diode circuit is $E_{rms}^2 / 2R$ watt).

**3.** *Conversion to rms reading voltmeter.* The rms value of voltage is usually of principal interest. The average reading voltmeter can be converted into an rms reading voltmeter by calibrating the meter scale in terms of the rms value of a sine wave. Since so many waveforms in electronics are sinusoidal, the average reading voltmeter with an rms calibrated scale is an entirely satisfactory instrument. In fact an average reading voltmeter with a calibrated rms scale is much less expensive than a true rms reading voltmeter.

The rms value of a voltage wave which has similar positive and negative half cycles, is related to the average value by the form factor. For a sinusoidal voltage the form factor is 1.11. Therefore, when an average reading, voltmeter has scale markings corresponding to the rms value of the applied sinusoidal input waveform, those markings are actually corrected by a factor 1.11 from the average value of applied voltage. Non-sinusoidal waveforms, when applied to this voltmeter, will cause it to read either high or low, depending upon the form factor of the waveform. If the form factor of the applied voltage is less than that of a sinusoidal wave (*i.e.*, less than 1.11), the meter reading is high, while on the other hand if the form factor of the applied voltage is higher than 1.11, the meter readings are lower.

It should be understood that this instrument can be used to indicate d.c. voltages. Since for d.c. the diode conducts all the time, the scale markings of the instrument will have to be multiplied by $2 \times 1.11 = 2.22$ for d.c.



The ***disadvantages*** of average reading VTVM are :

**1.** *Frequency range.* The frequency range of this instrument extends well over the audio-frequency range, but at radio frequencies the valve capacitance and the distributed capacitance of the series resistance, $R$ upset the voltage division and make the instrument useless.

**2.** *Response to low voltages.* The instrument gives a deflection proportional to the mean value during the conducting half cycle provided it is not used for voltages comparable with the voltage drop across the valve. The instrument has an evenly divided scale because of linear characteristics for higher voltages but exhibits non-linearity for lower voltages and therefore for lower voltages, a graphical treatment may have to be used to determine the voltage.

### 20.3.2 Peak Reading Diode Vacuum Tube Voltmeters

The most common of the many vacuum tube arrangements that are used as vacuum tube voltmeters is the diode peak-reading voltmeter. For peak reading vacuum tube voltmeters, four types of common arrangements are used.

(*i*) Series type,
(*ii*) Shunt type,
(*iii*) Compensated shunt type and
(*iv*) Slide back type.

1. ***Series type.*** The series type of arrangement is shown in Fig. 20.2(*a*).

The series type peak reading voltmeter is an ordinary detector employing high load resistance *R* and a PMMC instrument (say a micro-ammeter) connected in series. The principle of operation is shown in Fig. 20.2(*b*). At each positive peak of the applied input voltage, the capacitor *C* is charged to a potential that is equal to the peak voltage. Between peaks a small portion of the charge on the capacitor leaks off through the resistance *R*, causing a small fall in the capacitor voltage. At the positive peak of the next cycle the charge that has been lost in this way is replenished (given back) by a pulse or current through the diode. The voltage across capacitor *C*, and hence across the resistance *R*, is thereby again built up almost to the peak of the applied voltage. Either the average voltage across *R* or the average current through *R*, can be used to indicate the peak of the applied voltage. When the applied input voltage is not too small, the relationship between peak voltage and the rectified output is linear.

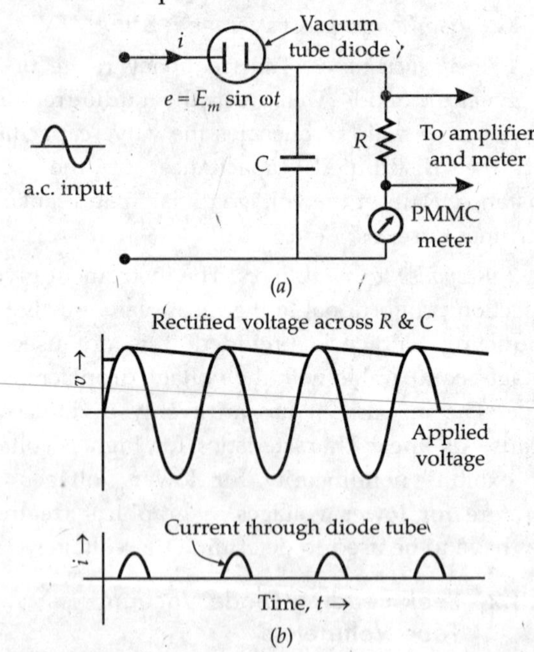


**Fig. 20.2** Series type peak reading diode vacuum tube voltmeter.

When the meter is connected as shown in Fig. 20.2(*a*), the current through the meter can be used to indicate peak applied voltage. However, the voltage across resistance *R* can be amplified by a d.c. amplifier and by using a PMMC instrument across the amplifier output the peak voltage can be indicated. The latter arrangement is preferable as explained earlier on account of the fact that the resistance *R* can be made very high, thereby reducing the power consumption and also that a less sensitive type of PMMC type of instrument could be used. The high value input resistance also results in more linear relationship between peak applied voltage and the meter indication. Also the performance of the diode with inputs consisting of pulses and modulated waves is improved.

The d.c. amplifier associated with the diode rectifier should be provided with stabilizing means to prevent drift in the indication of the output meter. In general a voltage regulated power supply combined with a compensating circuit is used.

The use of high resistance *R*, associated with a d.c. amplifier, no doubt results in a high input resistance but all the same it implies that an applied voltage of sufficient amplitude is required so that the system acts as a peak voltage device. The greatest disadvantage of this system is with regards to measurement of low voltages. If the applied voltage is too small, then some current flows throughout the cycle of the voltage as a result of the high velocity of emission of electrons, and the input resistance may be a few hundred ohm and this defeats the very purpose of using electronic instruments.

2. ***Compensated shunt type.*** The operation of shunt type and compensated shunt type VTVMs are similar to that of series type. These meters are used when it is neither possible nor desirable to have circuit under measurement as the return path for the voltmeter. Diode tubes have a characteristic which is referred to as a contact potential effect. This effect is caused by electrons striking the plate with sufficient velocity to give the plate a negative potential whenever the tube filament or heater circuit is in operation. The negative potential results in a steady deflection of the PMMC indicating meter on account of the electron emission current even in the absence of any applied voltage.

Thus the meter shows some deflection even if there is no voltage applied to it. This has to be compensated for so that the instrument can be set to

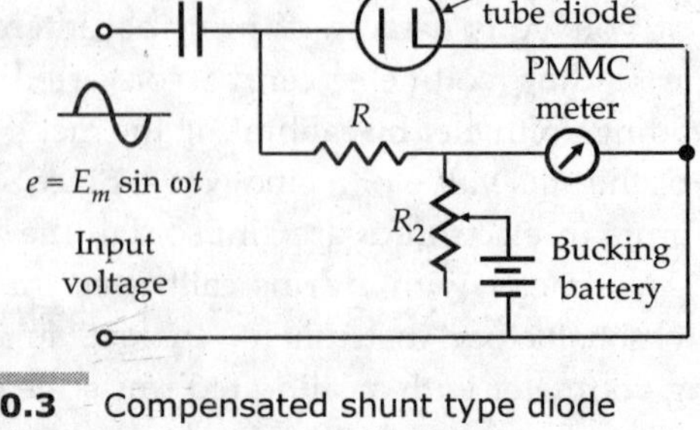


**Fig. 20.3** Compensated shunt type diode peak reading voltmeter.

zero when there is no voltage applied to the VTVM. Figure 20.3 shows the circuit for a compensated shunt type of peak reading diode VTVM.

The compensating circuit consists of a variable resistance $R_2$ and a bucking battery. The battery provides a *"bucking voltage"*. In the absence of any applied voltage, the resistance $R_2$ is adjusted so that the meter reads zero. Thus this compensating circuit is in fact a "**zero setting**" arrangement. The resistance $R_2$ must have a value greater than that of the meter in order to prevent the path formed by resistance $R_2$ and the battery forming a low impedance path for strong signals.

## 20.4 DIFFERENTIAL AMPLIFIER

A differential amplifier is the basic component of all electronic voltmeters (EVM). A block diagram of the differential amplifier is shown in Fig. 20.4(*a*). It has two input signals $v_1$ and $v_2$ as shown in Fig. 20.4(*b*) and one output signal as shown in Fig. 20.4(*d*). The output signal is an amplified version of the algebraic difference of the two input signals $v_1$ and $v_2$. If $v_1$ and $v_2$ are sinusoidal and are in phase with each other, then $(v_1 - v_2)$ is simply a new sine wave with a peak value equal to the difference in peak values of $v_1$ and $v_2$. The

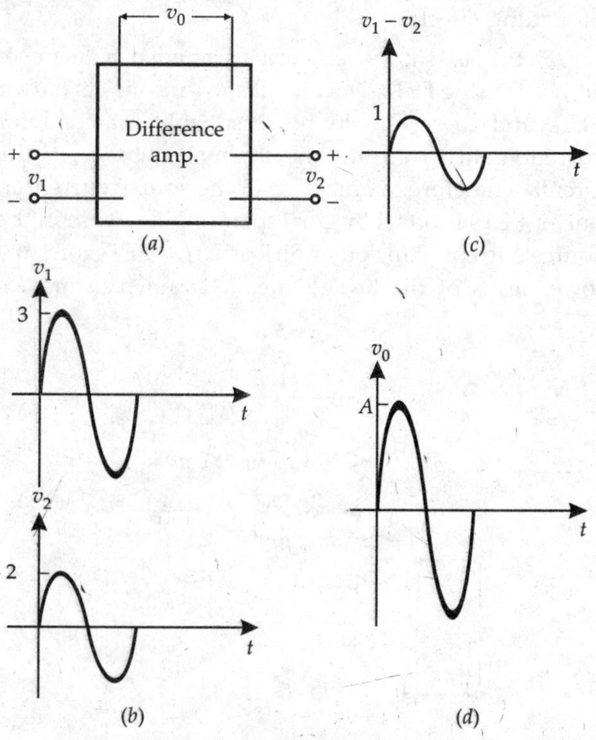


**Fig. 20.4** Difference amplifier. (*a*) Block diagram, (*b*) Input signals, (*c*) Difference of input signals, (*d*) Output of difference amplifier.

output signal is the difference of the two voltages multiplied by the gain of the amplifier. For example, $v_1 = 3\sin\omega t$ and $v_2 = 2\sin\omega t$. The difference voltage $v_1 - v_2 = 1\sin\omega t$ is shown in Fig. 20.4(*c*). The gain of the amplifier is $A$ and therefore the output is : $v_0 = A\times(v_1 - v_2) = A\sin\omega t$ which is shown in Fig. 20.4(*d*). A FET version of a difference amplifier is shown in Fig. 20.5. Let us analyze the circuit quali- tatively. Suppose $v_1$ is a 1 V-peak sine wave and $v_2$ is zero as shown in Fig. 20.6. When $v_i$ is alternating in a positive direction, the gate of left FET $Q_1$ goes positive and so the drain current increases. The voltage drop across the drain resistor increases, producing a negative going drain voltage from drain to ground as shown. The same action takes place in a single stage amplifier wherein a 180° phase shift occurs.

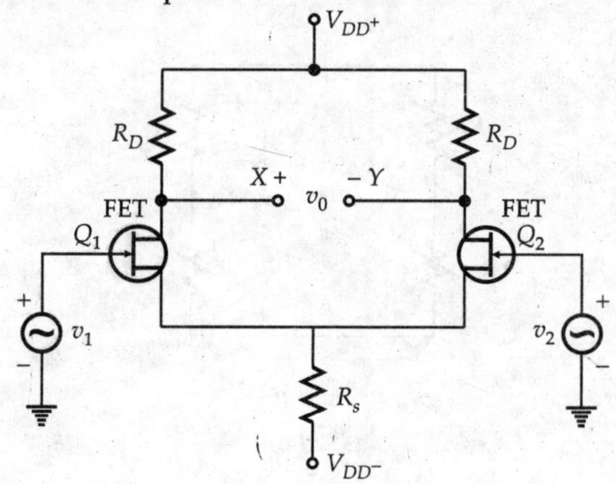


**Fig. 20.5** Difference amplifier of a field-effect transistor.

During the positive half cycle of $v_1$, the current in the source resistor increases, causing a positive going signal as shown. The source signal is such that it *reverse biases* the right FET $Q_2$. This reduces the current in the right FET, which provides a lower voltage drop across the drain resistor and therefore a positive going signal is obtained from the drain of right FET $Q_2$ to ground.

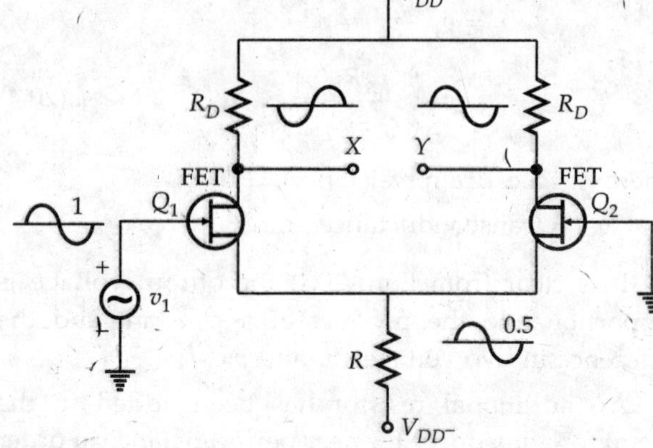


**Fig. 20.6** Difference amplifier with one input.

The signals at the drain of each FET are equal in magnitude but are 180° out of phase with each other. Therefore, output voltage, $v_0$, from terminals X and Y will be a sine wave with a peak value of twice the magnitude of each drain signal. If voltages $v_1$ and $v_2$ are sinusoidal and in phase with each other, then the output signal $v_0 = v_1 - v_2$, will be equal to zero. The difference amplifier can be represented as a bridge circuit as shown in Fig. 20.7. Each FET is represented by a variable d.c. resistance that is controlled by the input signals. When both the signals are equal, resistance of each FET is equal, and therefore, the bridge is balanced. In case, the input signals are unequal, the bridge becomes unbalanced, and a resultant output is produced (Fig. 20.8).

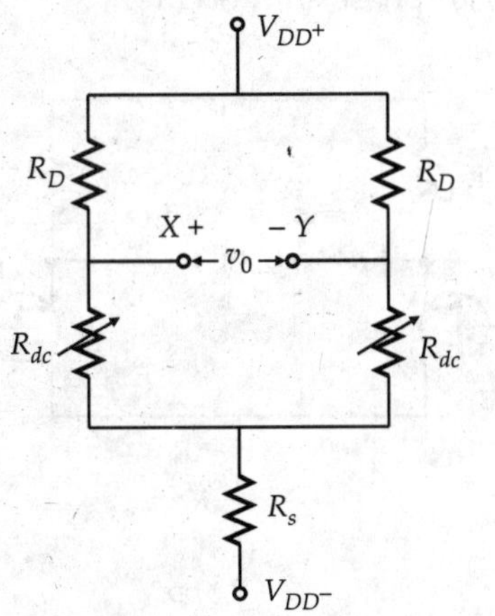


**Fig. 20.7** Interpretation of difference amplifier as a bridge.

Using equivalent circuits, it can be shown that,

$$v_x = g_m \left( \frac{r_d R_D}{r_d + R_D} \right) v_1 \quad \text{...(20.2)}$$

$$v_y = g_m \left( \frac{r_d R_D}{r_d + R_D} \right) v_2 \quad \text{...(20.3)}$$

$$v_0 = v_x - v_y$$

$$= g_m \left( \frac{r_d R_D}{r_d + R_D} \right) (v_1 - v_2) \quad \text{...(20.4)}$$

where $r_d$ = a.c. drain resistance ; Ω

$g_m$ = transconductance ; mho.

It is clear from Eqn. 20.4, the output voltage is proportional to the product of a constant and the difference in two voltages $v_1$ and $v_2$.

An additional resistor has been added in the circuit for adjustment for near perfect balance in order to compensate for the mismatch between the components. A variable resistor is used as shown in Fig. 20.8, which is adjusted till null conditions are obtained with no input signals applied to the circuit.

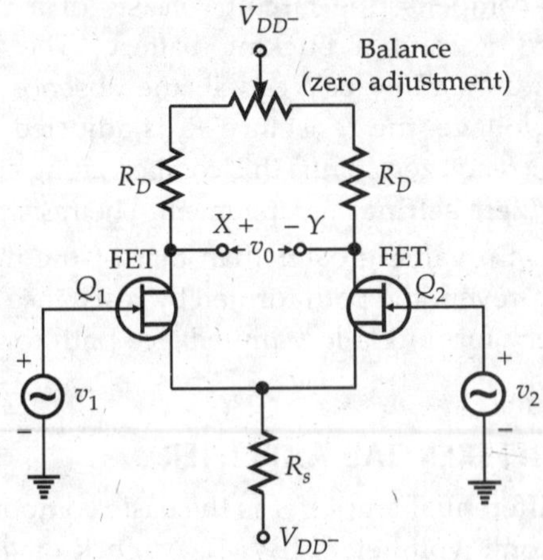


**Fig. 20.8** Difference amplifier with a balance adjustment.

## 20.5 DIFFERENCE AMPLIFIER TYPE OF ELECTRONIC VOLTMETER

Field Effect Transistors (FETs) are used for increasing the input resistance of voltmeters. The use of FETs isolates the low resistance voltmeters from the circuit under test, thereby avoiding errors on account of loading effects.

Figure 20.9 shows a schematic diagram of a difference amplifier using FETs. The circuit shown consists of two FETs and these should be reasonably matched for current gain to ensure the thermal stability of the circuit. Therefore, an increase in the source current in one of FETs is offset by corresponding decrease in the source current of the other FET. The two FETs form the lower arms of the bridge circuit. Drain resistors $R_D$

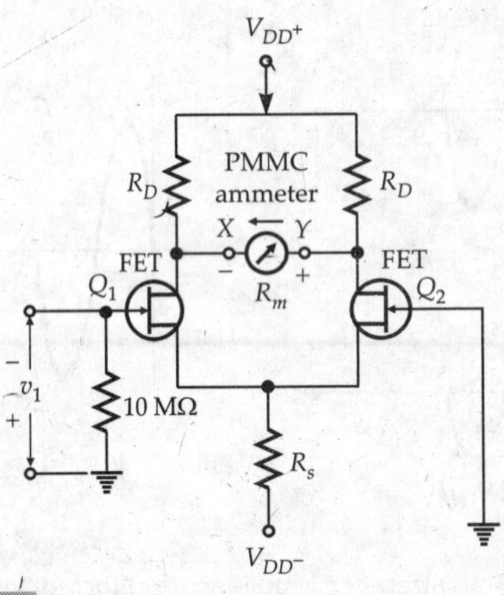


**Fig. 20.9** The difference-amplifier type of Electronic Voltmeter.

together form the upper arms. The meter is connected across the drain terminals of FETs, representing the two corners of the bridge.

The circuit is balanced when identical FETs are used so that for zero input no current flows through the ammeter. If a negative d.c. voltage is applied to the gate of left FET, a current would flow through the PMMC ammeter in the direction shown in Fig. 20.9. The magnitude of the current depends upon the magnitude of input voltage. Through proper design of the circuit, the magnitude of current can be made directly proportional to the d.c. voltage across the input. Therefore, the meter can be made to read the input voltage directly in terms of volt.

Thevenin's theorem can be used to find the relationship between the ammeter current and the input d.c. voltage with ammeter considered to be a load. With the ammeter (load in this case) removed, the circuit of Fig. 20.10($a$) is seen. The output voltage is equal to the product of gain of FET and the difference of $v_1$ and $v_2$. Since $v_2$ is zero, the voltage under open circuit conditions is,

$$v_0 = g_m \left( \frac{r_d R_D}{r_d + R_D} \right) v_1 \quad \text{...(20.5)}$$

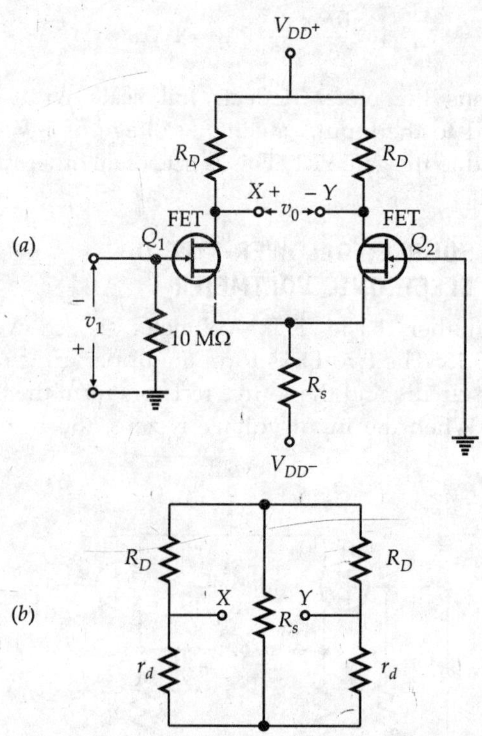


**Fig. 20.10** Applying Thevenin's theorem to the Electronic Voltmeter. ($a$) Removing the load to find open circuit voltage $V_0$ ($b$) Setting all voltages equal to zero to find Thevenin source generator resistance $R_0$.

In order to find the resistance of Thevenin source generator looking into terminals X and Y, $v_1$ and $V_{DD}$ are made zero. Under this condition both FETs have a resistance of $r_d$ as shown in Fig. 20.10($b$). From this balance circuit and assuming $R_s$ to be large, the resistance looking into terminals X and Y is,

$$R_0 = 2\frac{r_d R_D}{r_d + R_D} \quad \text{...(20.6)}$$

The Thevenin equivalent circuit with the ammeter connected as load is shown in Fig. 20.11. The resistance of ammeter is $R_m$.

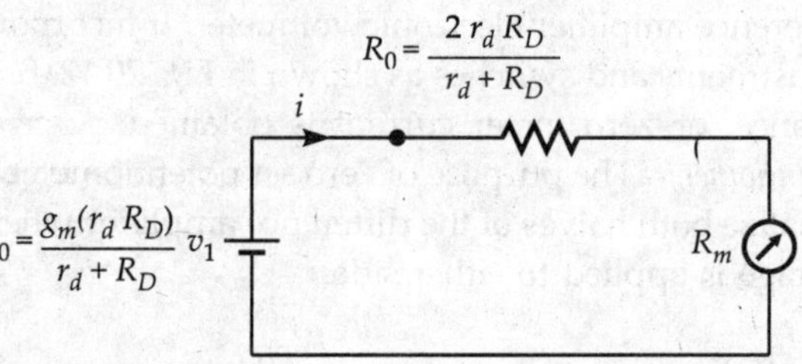


**Fig. 20.11** Thevenin Equivalent circuit for the difference-amplifier type Electronic Voltmeter.

From Fig. 20.11, the current through ammeter is,

$$i = \frac{V_0}{R_0 + R_m} = \frac{g_m r_d R_D / (r_d + R_D)}{2 r_d R_D / (r_d + R_D) + R_m} \cdot v_1 \quad \text{...(20.7)}$$

When $R_D << r_d$, Eqn. 20.7 simplifies to :

$$i = \frac{g_m R_D}{2 R_D + R_m} v_1 \quad \text{...(20.8)}$$

***Example 20.1*** *A difference amplifier type electronic voltmeter shown in Fig. 20.10 has the following data :*

*voltage applied to left FET,* $v_1 = 1$ V.

*a.c. drain resistance,* $r_d = 100$ kΩ

*transconductance,* $g_m = 0.005$ mho

$R_D = 10$ kΩ.

*Resistance of PMMC meter,* $R_m = 50\ \Omega$

*Calculate the value of current through ammeter.*

**Solution.** From Eqn. 20.7, current through the ammeter

$$= g_m \frac{r_d R_D / (r_d + R_D)}{2 r_d R_D / (r_d + R_D) + R_m} v_1$$

$$= 0.005 \times \frac{\dfrac{100 \times 10^3 \times 10 \times 10^3}{(100 \times 10^3 + 10 \times 10^3)}}{\dfrac{2 \times 100 \times 10^3 \times 10 \times 10^3}{(100 \times 10^3 + 10 \times 10^3) + 50}} \times 1$$

$= 2.5$ mA.

This result indicates that the current through the ammeter is directly proportional to the input voltage. Hence, the ammeter can be calibrated in terms of the voltage with the scale of the meter being uniform. Thus the place where the ammeter indicates 2.5 mA can be marked as 1 V for calibration as a voltmeter. The remainder of the scale is marked off in a linear manner. For example, if the full scale reading is 1 V corresponding to a current of 2.5 mA, the scale is marked as 0.5 V where the meter indicates a current of 1.25 mA.

Figure 20.12 shows a practical version of a difference amplifier electronic voltmeter. It incorporates adjustments and switches as shown in Fig. 20.12. Bridge balance, or zero meter current is obtained by *zero set potentiometer*. The purpose of zero set potentiometer is to equalize both halves of the difference amplifier when no voltage is applied to either side.

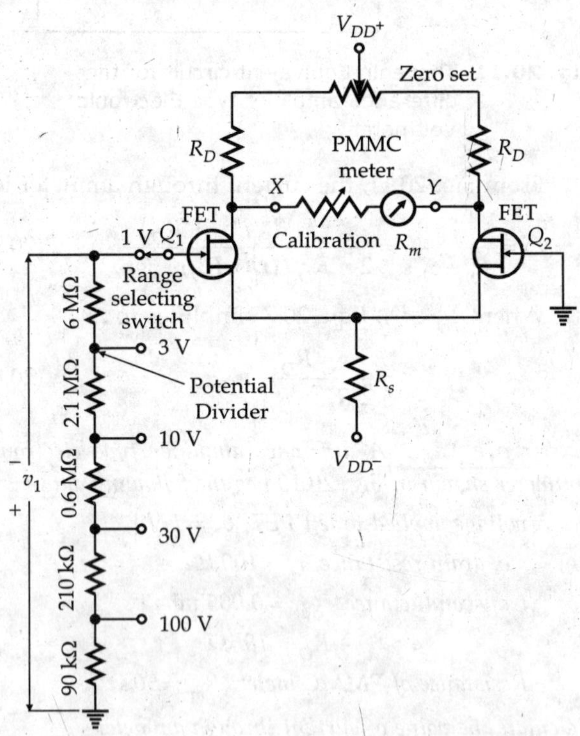


**Fig. 20.12** A practical version of the difference amplifier type of Electronic Voltmeter.

The full scale adjustment or *calibration* is affected by the potentiometer marked calibration, in series with meter movement internal resistance. This adjustment is necessary because no two FETs are identical as their transconductance $g_m$ and drain resistance $r_d$ may be different. It is clear that from Eqn. 20.7, that for a fixed value of $v_1$, different values of current will flow through the meter if $g_m$ and $r_d$ change from one FET to another. Thus if the calibration of the meter is to hold good which is marked at 1 V at full scale, the calibration potentiometer is adjusted to give a reading of 1 V exactly when the applied voltage is 1 V.

The range of a difference amplifier electronic voltmeter can be extended by using a potential divider arrangement and a range selecting switch as shown in Fig. 20.12. The potential divider along with the range selecting switch allows many full scale ranges to used. The input resistance is 9 MΩ on every switch position which allows only the permissible current to flow through the meter at full scale deflection on any meter range. The meter is calibrated to read 1 V with a current of 2.5 mA passing through the meter if the voltage is applied directly to the FET. Let us consider the case when a voltage of 10 V is to be measured and is to produce full scale deflection of the meter. The current through the meter is to be limited to 2.5 mA. For this, the range selector switch is moved to the 10 V position. The voltage applied to the gate of the FET in this case is,

$$v = \frac{(0.6+0.21+0.09)\times 10^6}{(5+2.1+0.6+0.21+0.09)\times 10^6}\times 10$$

$$= \frac{0.9\times 10^6}{9\times 10^6}\times 10 = 1 \text{ V}.$$

Thus the meter deflects full scale with 10 V applied to the input, causing a voltage of 1 V to be applied to the gate FET. This is true of all other ranges as well.

## 20.6 SOURCE FOLLOWER TYPE OF ELECTRONIC VOLTMETER

Another basic FET voltmeter is shown in Fig. 20.13. The two FETs form the upper arms of the bridge circuit, and the source resistors form the lower arms. When the input voltage is zero, the bridge is

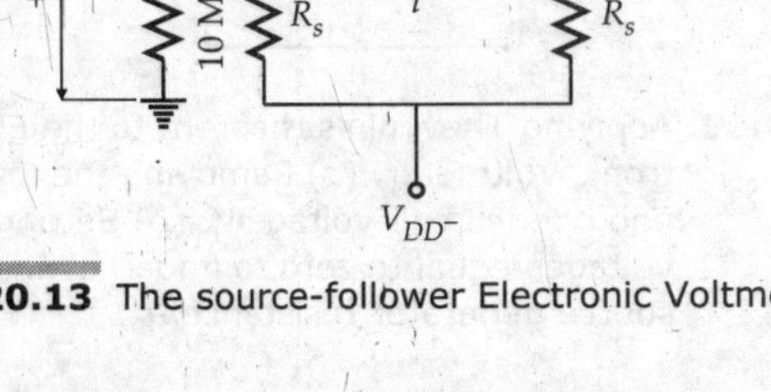


**Fig. 20.13** The source-follower Electronic Voltmeter.

balanced and no current flows through the ammeter. If $v_1$ is positive input voltage, current flows through the ammeter in the direction indicated.

The Thevenin equivalent of the circuit is first developed. The open circuit voltage is found by removing the load *i.e.,* the ammeter connected between terminals X and Y as shown in Fig. 20.14(*a*).

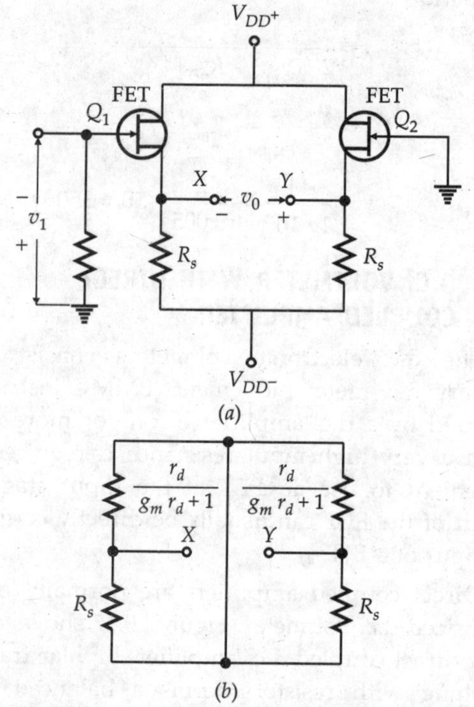


**Fig. 20.14** Applying Thevenin's theorem to the source-follower Electronic Voltmeter. (*a*) Removing the ammeter to find $v_0$. (*b*) Setting all voltages equal to zero to find $R_0$.

Since the gate of right FET is grounded, no signal is applied to its gate and therefore, no signal appears at terminal Y. The left FET, however, is a source follower. This is a *negative feedback* with a feedback factor $\beta = -1$. Therefore, the closed loop gain can be expressed as :

$$A' = \frac{A}{1 - A\beta} = \frac{A}{1 + A} = \frac{g_m (r_d R_s)/(r_d + R_s)}{1 + g_m (r_d R_s)/(r_d + R_s)}$$

$$= \frac{g_m r_d R_s}{r_d + R_s + g_m r_d R_s} \qquad \text{...(20.9)}$$

The output voltage from a source follower is,

$$v_0 = \frac{g_m r_d R_s}{r_d + R_s + g_m r_d R_s} v_1$$

$$= \frac{g_m \left(\frac{r_d}{g_m r_d + 1}\right) R_s}{\left(\frac{r_d}{g_m r_d + 1}\right) + R_s} v_1 \qquad \text{...(20.10)}$$

Comparing Eqn. 20.10 with 20.5, for the gain of common-source circuit shows that they are of the same form if $r_d / (g_m r_d + 1)$ replaces $r_d$ in common source equation.

Thus $r_d / (g_m r_d + 1)$ can be considered to be the output resistance of the FET when used in source follower circuit.

The output equation is in the simple form as that of the common source amplifier, except that $r_d$ as specified in the data sheet is replaced by

$$r_d' = r_d / (g_m r_d + 1)$$

or

$$v_0 = g_m \left(\frac{r_d' R_s}{r_d' + R_s}\right) v_1 \qquad \text{...(20.11)}$$

The Thevenin source generator voltage looking into terminals X and Y (see Fig. 20.14) is :

$$R_0 = \frac{2 R_s \cdot \frac{r_d}{g_m r_d + 1}}{R_s + r_d / (g_m r_d + 1)} = \frac{2 R_s r_d'}{R_s + r_d'} \qquad \text{...(20.12)}$$

The circuit of Fig. 20.15 shows the Thevenin Equivalent circuit with the ammeter connected between terminals X and Y.

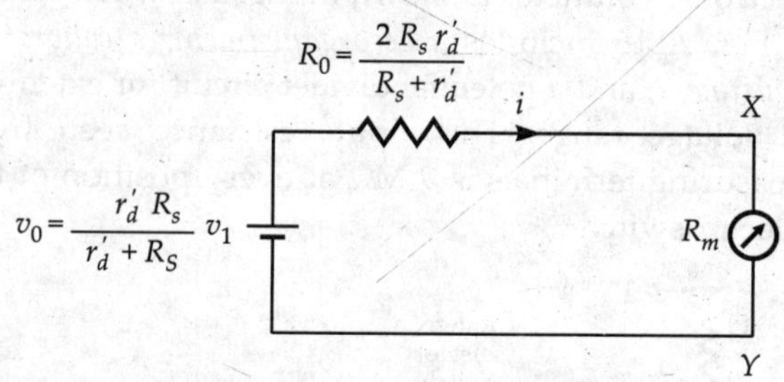


**Fig. 20.15** Equivalent circuit of a source-follower electronic voltmeter.

In order to obtain an approximate expression, we have $g_m r_d R_s >> r_d$ or $R_s$ for most FETs. Further, $R_s >> R_d / (g_m r_d + 1)$. Under these conditions, Eqn. 20.12 can be simplified as :

$$i \approx \frac{v_1}{2/(g_m) + R_m} \qquad \text{...(20.13)}$$

***Example 20.2*** *The electronic voltmeter shown in Fig. 20.14 has the following data :*

$$R_s = 1050 \text{ k}\Omega,\ r_d = 100 \text{ k}\Omega \text{ and } g_m = 0.0057 \text{ mho.}$$

*A PMMC ammeter with an internal resistance of 50 Ω is used as the detector. If the input voltage is 1 V, find the approximate value of current through the meter.*

**Solution.** The value of

$$g_m r_d R_s = 0.0057 \times 100 \times 10^3 \times 1050 \times 10^3 = 598 \text{ M}\Omega$$

which is much greater than either $r_d$ or $R_s$.

Also value of

$$\frac{r_d}{g_m r_d + 1} = \frac{100 \times 10^3}{0.0057 \times 100 \times 10^3 + 1} = 175\ \Omega$$

and the value of $R_s = 1050$ kΩ. Therefore, the value of $R_s >> r_d / (g_m r_d + 1)$ and hence Eqn. 20.13 can be used for calculating the approximate value of meter current.

From Eqn. 20.13, the meter current is

$$i = \frac{v_1}{2/g_m + R_m} = \frac{1}{2/(0.0057) + 50} = 2.5 \text{ mA}.$$

Thus, we see that the current is directly proportional to the input voltage. If an input voltage of 1 V produces a current of 2.5 mA and produces a full scale deflection, then a voltage of 0.5 V would produce a current of 1.25 mA and therefore the marking of 0.5 V would produce a current of 1.25 mA and therefore the marking of 0.5 V will be placed at midscale. Similarly, a voltage of 0.25 V would be placed at 1/2 length of scale, thereby leading to a uniform (linear) scale.

The practical circuit of a source-follower type of electronic voltmeter is shown in Fig. 20.16. The circuit of Fig. 20.16 includes *a zero adjustment, a calibration adjustment*, and a potential divider circuit for extention of voltage range. The input resistance seen from measuring terminals is 9 MΩ at every position of the selector switch.

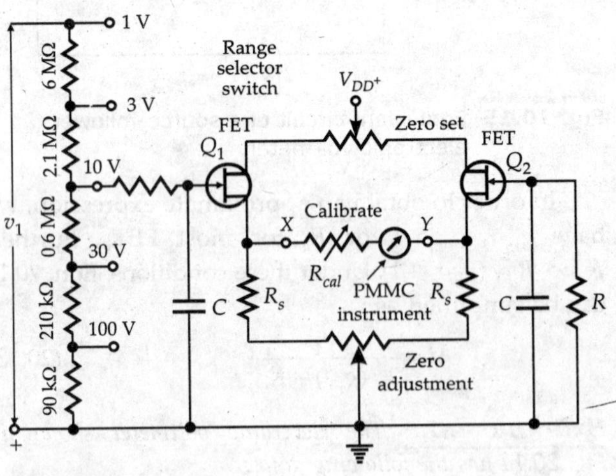


**Fig. 20.16** A practical version of the source-follower Electronic voltmeter.

***Example 20.3*** *The electronic voltmeter shown in Fig. 20.16 has the following data :*

$$R_s = 50\ k\Omega,\ r_d = 100\ k\Omega \text{ and } g_m = 0.005 \text{ mho}.$$

*A 50 Ω ammeter with a full scale deflection current of 1 mA is used. If the input voltage is 1 V, what value of calibration resistance will produce full scale current ?*

**Solution.** From Eqn. 20.13, the current through the meter is :

$$i = \frac{v_1}{2/(g_m) + R_m}$$

This expression, however, does not include the resistance of the calibrating resistor, $R_{cal}$ and is modified as :

$$i = \frac{v_1}{2/(g_m) + R_m + R_{cal}}$$

$$\because \quad R_{cal} = \frac{v_i}{i} - \frac{2}{g_m} - R_m$$

$$= \frac{1}{1 \times 10^{-3}} - \frac{2}{0.005} - 50 = 350\ \Omega.$$

## 20.7 D.C. VOLTMETER WITH DIRECT COUPLED AMPLIFIER

The d.c. electronic voltmeters consist of an ordinary d.c. meter movement (PMMC instrument) preceded by a d.c. amplifier of one or more stages. When a very high input resistance is required, it is convenient to use an FET at the input stage. The output of the FET can usually be directly coupled to the input of a BJT.

Direct coupled amplifiers are normally used in low priced d.c. voltmeter. Figure 20.17 shows an FET input direct coupled d.c. amplifier. Bipolar transistor $Q_2$ along with resistors forms a balanced bridge circuit. FET $Q_1$ serves as a source follower and is used to provide impedance transformation between the input the base of $Q_2$.

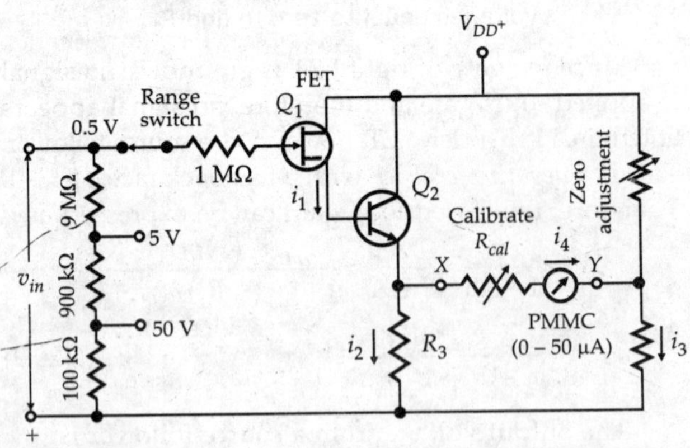


**Fig. 20.17** Basic d.c. voltmeter circuit with FET input.

The bias on $Q_2$ is such that $i_2 = i_3$, when the input voltage $V_{in} = 0$. Under that condition $v_x = v_y$, so that no current flows through the meter movement *i.e.*, $i_4 = 0$. The bias on $Q_2$ is controlled by input voltage $V_{in}$. Thus when an unknown input voltage $V_{in}$ is applied, the bias on $Q_2$ increases, which causes $v_x$ to increase.

Since $v_x$ becomes greater than $v_y$, current $i_4$ is no longer zero. The magnitude of this current, hence the deflection of the meter is proportional to $V_{in}$.

The value of $V_{in}$ that causes maximum meter deflection is the basic range of the instrument. This is generally the lowest range on the Range switch in non-amplified models. Higher ranges can be obtained by using an input attenuator, and lower ranges can be obtained by a pre-amplifier.

The input attenuator in Fig. 20.17 is a calibrated front panel control in the form of a resistance voltage divider. The full-scale voltage appears across the divider so that the voltage at each tap is a progressively lower fraction of the full input voltage.

Bridge balance is obtained by adjusting the zero set potentiometer when $V_{in}$ is zero. Full-scale calibration is obtained by adjusting the potentiometer marked calibration in series with the ammeter.

The ***advantages*** of this meter are :

(*i*) It decreases the amount of power drawn from the circuit under test by increasing the input impedance using an amplifier with unity gain.

(*ii*) The source follower drives an emitter follower. This combination is capable of thousand-fold or more increase in impedance while maintaining a voltage gain of nearly one.

(*iii*) The input impedance of this meter is 10 MΩ, which requires a power of 0.025 μW for a 0.5 V deflection as compared to 25 μW for an unamplified meter, thereby giving at increased sensitivity of 100 times.

A block diagram of a meter used for measurement of small voltage and currents is shown in Fig. 20.18. The input voltage is amplified and applied to a meter. If the amplifier has a gain of 10, the sensitivity of the measurement is increased by a like amount. A d.c.-coupled amplifier, that is, an amplifier with no coupling capacitors and having a well-controlled dc gain, is used to provide the necessary amplification. An amplifier capable of a fixed dc gain of 10 is not difficult to construct and to keep stable. A simple op-amp plus the required feedback components will do a suitable job for this application.

Direct current gains of much more than 10 are required to use a standard d'Arsonval meter movement to measure very small currents and voltages such as microvolt and nanoampere. To amplify nanoampere to drive a milliampere meter requires a gain of $10^6$. In theory, this requires an op-amp and two resistors, and a simple circuit. However, when gains this large are desired, all the defects of an operational amplifier become significant. Offset current, offset voltage, and bias currents become so troublesome that it is practically impossible to achieve acceptable performance with a standard op-amp (See Chapter 26). Many of these defects can be reduced or eliminated by the use of trim adjustments accessible from the front panel in a similar fashion as the calibrate and Zero functions discussed above. However, temperature- and time-induced drifts would soon render the amplifier unusable, and the adjustment would have to be repeated. Direct-coupled amplifiers that have been optimized for low-temperature drift and low offset and bias currents are called *instrumentation amplifiers* and are manufactured by semiconductor suppliers.

20

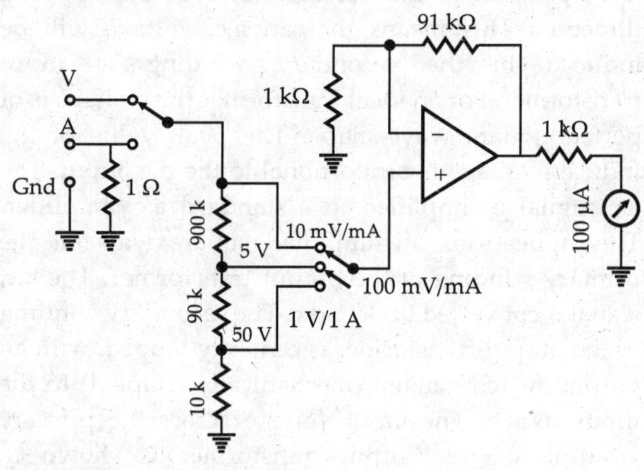


**Fig. 20.18** Amplified voltage and current meter.

## 20.8 CHOPPER STABILIZED AMPLIFIERS

One technique for amplifying direct currents and relatively low-frequency alternating currents is the *Chopper stabilized amplifier*. This circuit illiminates the effects of d.c. offset currents and the drift of other d.c. parameters by using an a.c. coupled amplifier for necessary gain. The technique as used in Fig. 20.19, is to convert the input signal to an a.c. signal and, after, high gain amplification, reconstruct the d.c. from the amplified a.c. signal.

Simple a.c. amplifiers may be used to amplify a d.c. input through use of an additional circuit component known as **chopper.** The approach is used to build a d.c. amplifier wherein the d.c. is first converted to an equivalent a.c. signal that is amplified by a standard a.c. amplifier. The a.c. signal is finally converted back to a d.c. signal.

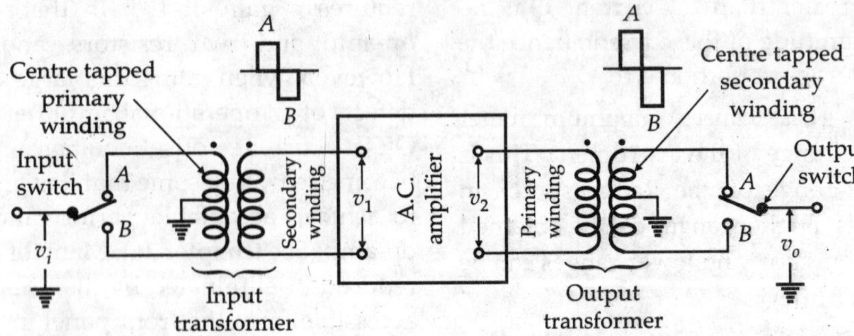


**Fig. 20.19** Chopper type d.c. amplifier.

Figure 20.19 shows a chopper type d.c. amplifier, $v_i$ is the input d.c. voltage. This voltage is alternately connected to terminals $A$ and $B$. When the switch is in position $A$, the direction of flow of current in the primary winding is in one direction. When the switch is in position $B$, the current flows in the opposite direction. This means that an a.c. voltage will be induced in the secondary winding of input transformer. For an ideal transformer this voltage is of perfect square wave-shape. The peak value of the induced voltage is proportional to the d.c. input. The a.c. signal is amplified in a standard a.c. amplifier. This appears as an amplified square wave at the primary winding of the output transformer. The a.c. signal is converted back to d.c. The secondary winding of the output transformer is centrally tapped, with an output switch ganged (mechanically coupled) to the input switch. The input voltage across the primary winding of the output transformer is shown in Fig. 20.20($a$). The output of the secondary winding is shown in Fig. 20.20($b$).

**Fig. 20.20** Output of chopper amplifier.

The amplifer of Fig. 20.19 is referred to as a **chopper amplifer** since the input d.c. voltage is literally chopped to produce an a.c. signal. The chopping action may be accomplished by either mechanical or electronic means. A vibrating reed is used when mechanical method is employed.

The synchronized chopping restores the d.c. value of the signal amplified by the a.c. gain of amplifier. Because, the amplifier does not provide any d.c. gain, the effects of d.c. offset voltages and currents are eliminated.

Enormous gains can be achieved in this fashion, and the chopper-stabilized amplifier can provide gains of more than $10^6$ with excellent d.c. stability. All this does not come without problems. First, when dealing with very small currents and voltages, unusual problems can occur. One significant problem is with the chopper. This device must be specially made to avoid generating voltages from thermocouple effects. When two dissimilar metals are joined, depending on the temperature, small voltages can be generated. The chopper is specially made to reduce these thermally generated voltages.

The electromechanical chopper, being a mechanical device, has a relatively short life span when compared to other electronic devices. Various types of all-electronic choppers have been devised to replace the venerable mechanical chopper. The most important characteristic of the chopper is that it must not inject any current into the circuit being chopped, especially for the input chopper. Bipolar transistors, light-activated devices, and field effect transistors have been used for choppers with the MOS field effect

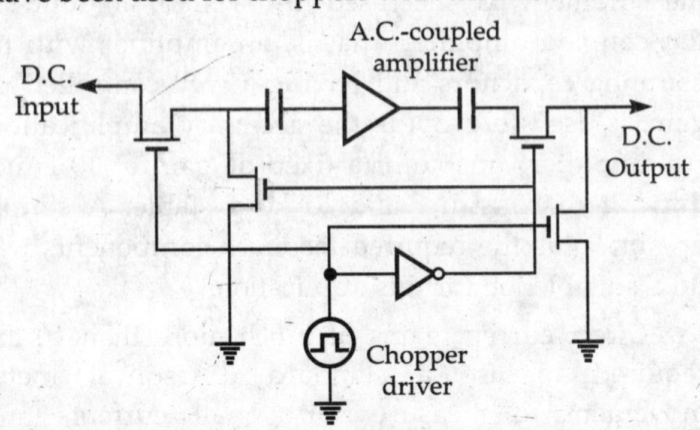


**Fig. 20.21** An electronic chopper circuit using Field effect transistors.

transistor being the most successful. Because the MOS transistor has no junction as a source of leakage current, very little current is transmitted from the chopping signal to the input. Figure 20.21 shows a series-shunt chopper using two MOS field effect transistors. The chopping signal is fed to the inverter, which drives the two chopper FETs, one on each half of the chopping cycle.

The input impedance of the chopper-stabilized amplifier is very high for direct current. Looking into the chopper-stabilized amplifier, the series chopper switches the input to the ac-coupled amplifier every half-cycle ; however, because the amplifier is a.c. coupled, it appears as an infinite resistance to direct current. The series chopper switch is always open before the shunt switch is closed, and thus there is no path to ground.

### 20.8.1 Chopper Amplifier Type Voltmeter Using Photodiodes

A chopper amplifier is normally used for the first stage of amplification in very sensitive instruments of a few μV range. In such an amplifier the d.c. voltage is chopped to a low frequency, 200 to 300 Hz a.c. It is passed through a blocking capacitor, amplified and then passed through another blocking capacitor in order to remove d.c. drift or offset from the amplified signal. By suitable selection of the blocking capacitors it is possible to manufacture a sensitive d.c. voltmeter which is virtually drift free, has high d.c. input impedance and has high degree of accuracy.

Figure 20.22 shows the block diagram of chopper type amplifier used in sensitive voltmeters. The arrangement employs photodiodes as non-mechanical choppers for modulation *i.e.*, conversion from d.c. to a.c. and demodulation *i.e.*, conversion from a.c. to back to d.c. A photoconductor has a low resistance of the order to a few hundred ohm when it is illuminated by an incandescent or a neon lamp. Its resistance sharply increases to several megohms, when it is not illuminated. An oscillator drives two neon lamps into illumination on alternate half cycles of oscillation as shown in Fig. 20.22.

Each neon lamp illuminates one photo-diode in the input circuit of the amplifier and one in the output circuit. The two photodiodes in the input circuit form a series shunt half wave modulator or chopper. Together they act like a switch across the input to the amplifier, alternately opening and closing at a rate determined by the frequency of the neon oscillator.

20

The input signal to the amplifier is a square wave voltage whose amplitude is proportional to the input voltage level and its frequency is equal to the frequency of oscillator. The a.c. amplifier gives an output square wave with an amplified magnitude. The photodiodes in the amplifier output circuit, operating in antisynchronism with the input chopper, produce a demodulating action giving a d.c. output voltage. This d.c. signal is passed through a low pass filter to remove any residual a.c. components. This amplified drift free d.c. output is then applied to a PMMC movement for measurement. It should be

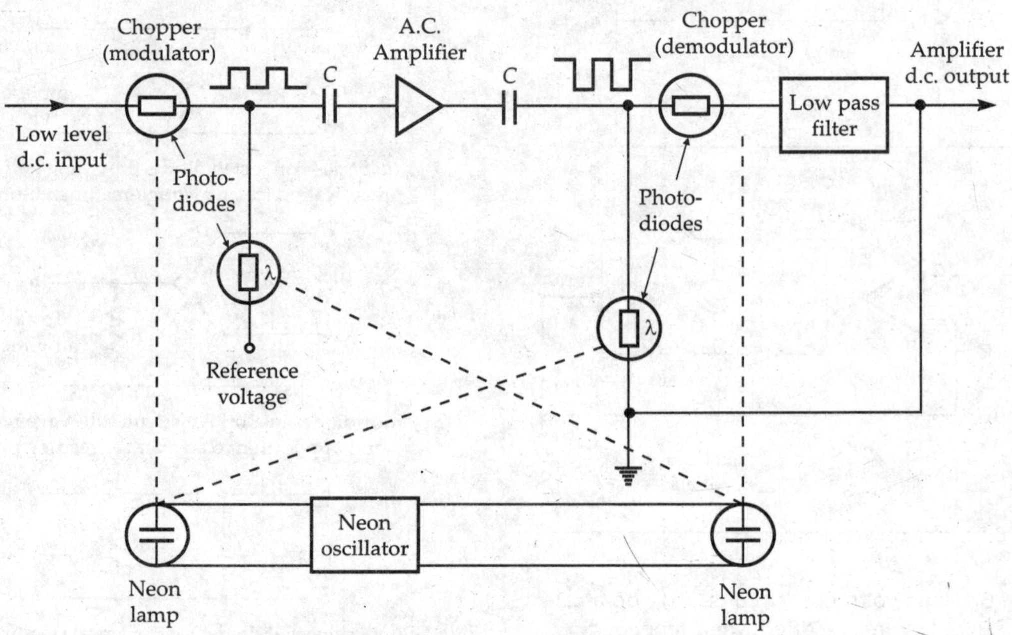


**Fig. 20.22** Photoconductive chopper amplifier.

noted that the amplifier is not stabilized by the chopper photodiodes. The chopper only eliminates the necessity for a high gain d.c. amplifier with its inherent drift and stability problems.

The input impedance of chopper amplifier d.c. voltmeter is normally of the order of 10 MΩ or higher, except on very low input ranges. In order to eliminate errors caused by high source impedances, an arrangement for nulling is included in the meter circuit. A control facility provided on the front panel of the instrument permits to null the input voltage with a bucking voltage. When the bucking voltage is equal to the input voltage, a null is indicated and the meter exhibits an infinite impedance and therefore the loading effects are totally eliminated. The input voltage is then removed, and the bucking voltage which is equal to the input voltage is indicated by the meter.

A commercially available instrument using a photo chopper amplifier has an input impedance of 100 MΩ, a resolution of 0.1 μV and input ranges from 3 μV full scale to 1 kV full scale with an accuracy of ± 2% of full scale deflection.

## 20.9 ELECTRONIC VOLTMETERS USING RECTIFIERS

The d.c. electronic voltmeters discussed previously may be used to measure a.c. voltages by first detecting the alternating voltage, as shown in Fig. 20.23. In some situations, rectification takes place before amplification, in which case a simple diode circuit precedes the amplifier and meter as in Fig. 20.23. This amplifier ideally requires zero-drift characteristics and unity voltage gain, a d.c. meter movement with adequate sensitivity.

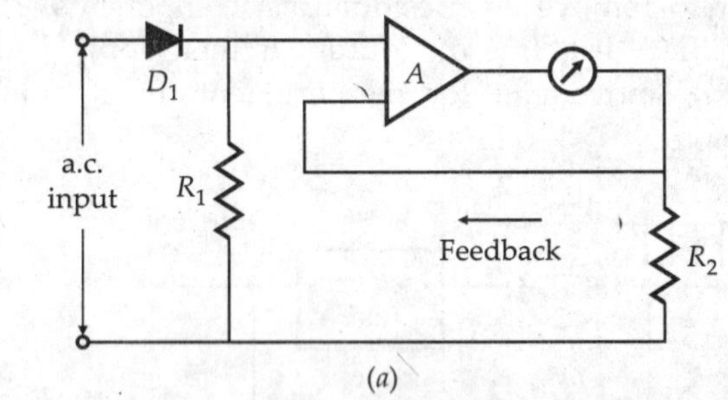

(a)

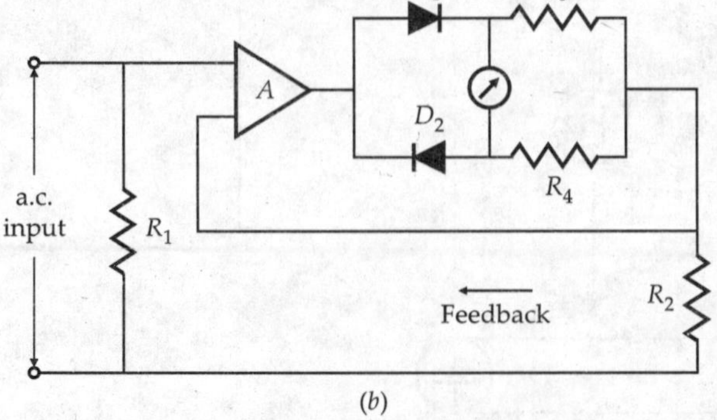

(b)

**Fig. 20.23** Basic ac voltmeter circuits. (*a*) The a.c. signal is first rectified, then amplified. (*b*) The a.c. signal is first amplified, then rectified.

In another method, rectification takes place after amplification as shown in Fig. 20.23(*b*). This method generally uses a high open-loop gain and large negative feedback to overcome the non-linearity of the rectifier diodes.

Alternating-current voltmeters that use half-wave or full-wave rectification are usually of the average-responding type, with the meter scale calibrated in terms of the rms value of a waveform instead of the average value. Thus, most meters are calibrated in terms of both rms and peak values. Since so many waveforms encountered in electronics are sinusoidal, these methods are satisfactory and much less expensive than a true rms-reading voltmeter. Non-sinusoidal waveforms, however, will cause this type of meter to read high or low, depending on the form factor of the waveform. The form factor is the ratio of the rms value to the average value of this waveform. It can be expressed as $K_f = V_{rms} / V_{av}$.

The main advantage of the a.c. voltmeter is that using negative feedback greatly reduces the response time. However, there is the disadvantage of reduced sensitivity, unless compensated by corresponding larger open-loop gains that are normally required.

A few basic rectifier circuits are shown in Fig. 20.24. The series connected diode as shown in Fig. 20.24(*a*) gives half wave rectification, and the

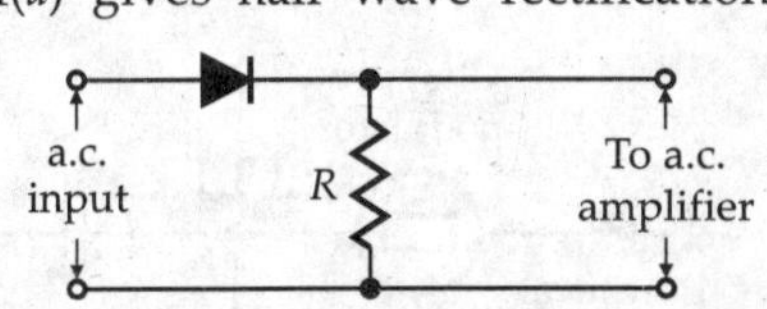

(*a*) Series-connected diode, providing half-wave rectification for an average-reading voltmeter.

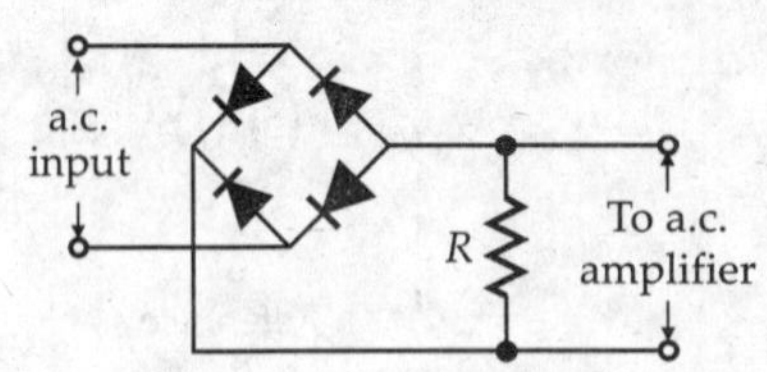

(*b*) Four diodes in a bridge circuit full-wave rectification and application to an average-reading meter.

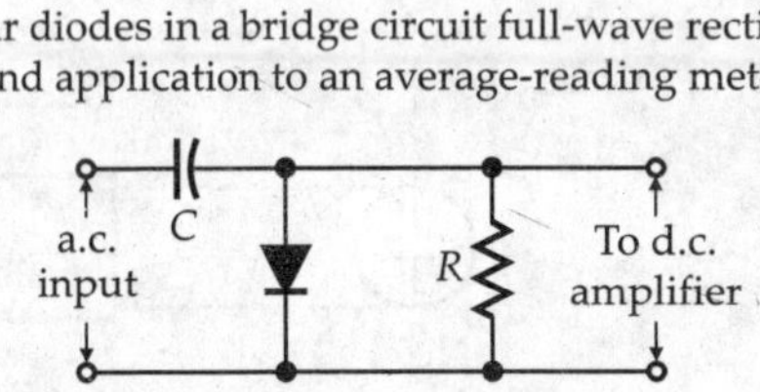

(*c*) Shunt-connected diode used in a peak-reading voltmeter

**Fig. 20.24** Rectifier circuits used in ac voltmeters.

average value of the half wave is developed across the resistor and is applied to the input terminals of the amplifier. The bridge circuit of Fig. 20.24(*b*) obtains full wave rectification, where the average value of the sine wave is applied to the amplifier and the meter circuit. In some case, there may be a requirement to measure the peak value of a waveform, instead of average value, the circuit of Fig. 20.24(*c*) may be used. In the peak form of voltmeter, the rectifier diode charges the small capacitor to the peak value of applied voltage and therefore the meter becomes **"peak reading"**. In most cases the meter scale is calibrated in terms of both rms and peak values of sinusoidal input waveform.

The waveform form factor for full wave rectification is 1.11. Therefore, when an average reading voltmeter has scale markings corresponding to rms value of sinusoidal wave, these markings are actually corrected by a factor 1.11 from the true (average) value of the applied voltage.

Nonsinusoidal waveforms, when applied to this meter, will call this meter to read either high or low, depending upon the waveform of the input supply. These are illustrated in the following examples.

***Example 20.4*** *A symmetrical square-wave voltage (Fig. 20.25) is applied to an average reading a.c. voltmeter with scale calibrated in terms of rms value of a sinusoidal wave. Calculate the error.*

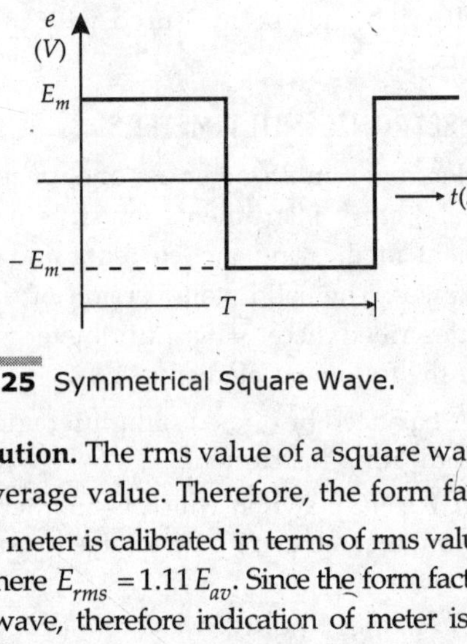


**Fig. 20.25** Symmetrical Square Wave.

**Solution.** The rms value of a square wave is 1 and so its average value. Therefore, the form factor is 1.

The meter is calibrated in terms of rms value of a sine wave where $E_{rms} = 1.11\,E_{av}$. Since the form factor is 1 for a square wave, therefore indication of meter is high by a factor $(1.11-1)/1 \times 100 = 11\%$. The error is also 11% high.

***Example 20.5*** *A sawtooth voltage has a peak value of 50 V and a time period of 3.0 s as shown in Fig. 20.26. Calculate the error when measuring this voltage with an average reading voltmeter calibrated in terms of rms value of a sinusoidal wave.*

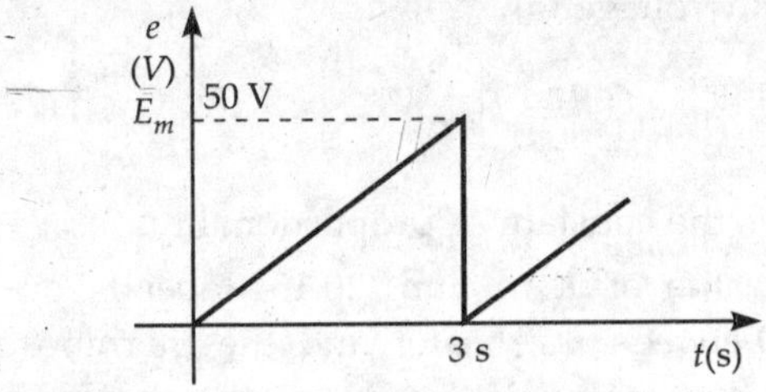


**Fig. 20.26** Sawtooth wave.

**Solution.**

$$E_{rms} = \sqrt{\frac{1}{T}\int_0^T e^2\,dt} = \sqrt{\frac{1}{3}\int_0^3 (50\,t)^2\,dt} = 50\sqrt{3}\text{ V}.$$

$$E_{av} = \frac{1}{T}\int_0^T e\,dt = \frac{1}{3}\int_0^3 50\,dt = 75\text{ V}.$$

$$\text{Form factor,}\quad K_f = \frac{E_{rms}}{E_{av}} = \frac{50\sqrt{3}}{75} = 1.155.$$

The ratio of the two form factors is $= \dfrac{1.1}{1.155} = 0.961.$

Therefore, the meter reads low by a factor of 0.961.

$$\therefore\ \text{Percentage error} = \frac{0.961-1}{1} = -3.9\,\%.$$

## 20.10 TRUE RMS READING VOLTMETERS

Complex waveforms are most accurately measured with a **true rms reading voltmeter**. This instrument indicates the rms value of any waveform (such as a sine wave, square wave, or sawtooth wave) by using an rms detector that responds directly to the heating value of the input signal.

To measure the rms value of an arbitrary waveform, we may feed an input signal to a heating element in close proximity to a **thermocouple**, as shown in Fig. 20.27.

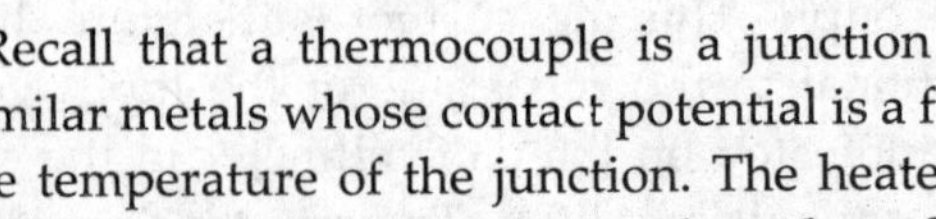


**Fig. 20.27** An rms detector using a thermocouple.

Recall that a thermocouple is a junction of two dissimilar metals whose contact potential is a function of the temperature of the junction. The heater raises the temperature of the thermocouple and produces an output voltage that is proportional to the power delivered to the heater.

$$P = \frac{V_{rms}^2}{R_{heater}} \qquad \ldots(20.14)$$

20

Output voltage

$$V_0 = f(P) = f\left(\frac{V_{rms}^2}{R_{heater}}\right) = KV_{rms}^2 \quad ...(20.15)$$

where $K$ is the constant of proportionality.

The value of $K$ in Eqn. 20.15 depends on the distance between the heater and the thermocouple and on the materials used in the heater and thermocouple. The difficulty with this method is that the thermocouple often displays a nonlinear characteristic. This problem is overcome in some instruments by placing two thermocouples in the same environment as shown in Fig. 20.28. The nonlinear characteristic of the input-measuring thermocouple is cancelled by similar nonlinear effects of the balancing thermocouple in the feedback circuit.

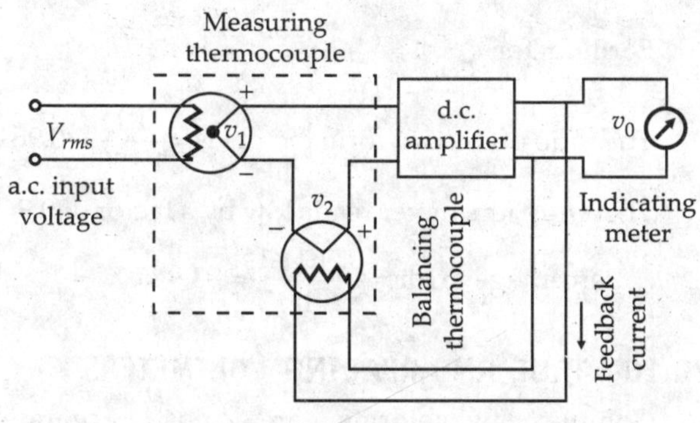


**Fig. 20.28** Block diagram of true rms-reading voltmeter. The measuring and balancing thermocouples are located in the same thermal environment.

The two thermocouples form part of a balanced bridge applied to the input circuit of a dc amplifier. The a.c. input voltage is applied to the heater of the measuring thermocouple. An output voltage $v_1$ is produced that aspects the balance of the bridge. This unbalanced voltage is amplified by the dc amplifier and fed back to the heater of the balancing thermocouple. When the output of both thermocouples is equal, bridge balance will be re-established. Thus, the dc feedback current is equal to the ac current in the input thermocouple. This d.c. current is therefore directly proportional to the rms value of the input voltage and is indicated on the d.c. voltmeter in the output circuit of the dc amplifier. To verify this we observe that :

$$v_0 = A(v_1 - v_2) \quad ...(20.16)$$

where $A$ is the voltage gain of the d.c. amplifier. Rearranging Eqn. 20.16, we get,

$$v_1 - v_2 = \frac{v_0}{A} \approx 0 \quad ...(20.17)$$

when $A$ is a very large number for a high-gain amplifier. Therefore,

$$v_1 \approx v_2 \quad ...(20.18)$$

From Eqn. 20.18, it is clear that,

$$KV_{rms}^2 = Kv_0^2$$

or

$$V_{rms} = v_0 \quad ...(20.19)$$

The last result tells us that the voltage measured by the d.c. voltmeter is approximately equal to the rms value of the input signal. Hence, we have an rms voltmeter. The true rms value is measured independently of the waveform of the a.c. signal ; thus, the waveform of the input signal is immaterial. If the ac input voltage is very small, an a.c. amplifier may be used to increase the signal level before applying it to the input thermocouple. Sensitivities in the millivolt region are possible with such an arrangement.

A typical laboratory-type rms-responding voltmeter provides accurate rms readings of complex waveforms having a *crest factor* (ratio of peak value to rms value) of 10/1. At 10 percent of full-scale meter deflection, where there is less change of amplifier saturation, waveforms with crest factors as high as 100/1 could be accommodated, Voltages throughout a range of 100 μV to 300 V within a frequency range of 10 Hz to 10 MHz may be measured with most good instruments.

## 20.11 ELECTRONIC MULTIMETERS

An electronic multimeter is one of the most versatile laboratory instruments which is capable of measurement of d.c. and a.c. currents and voltages and resistances. The solid state version of the multimeter is described here. This multimeter generally consists of the following elements :

(*i*) A balanced bridge d.c. amplifier and PMMC indicating instrument.

(*ii*) A **RANGE** switch which is an attenuator to limit the value of input voltage to the desired value.

(*iii*) A rectifier to convert a.c. to a proportionate d.c. value.

(*iv*) An internal battery and additional circuitry for the measurement of resistance.

(*v*) A **FUNCTION** Switch to select various measurement functions of the instrument like current, voltage and resistance.

In addition, the instrument has a built-in power supply for operation on power mains, and many cases one or more batteries for operation as a portable test instrument.

Figure 20.29 shows the schematic diagram of a balanced bridge d.c. amplifier using two FETs which are reasonably well matched for current gain to ensure thermal stability of the circuit. The two FETs form the upper arms of the bridge circuit. Source resistors $R_1$ and $R_2$, together with **ZERO** adjustment resistor $R_3$, form the lower bridge arms. The PMMC instrument is connected between the source terminals of FETs, representing two opposite corners of the bridge.

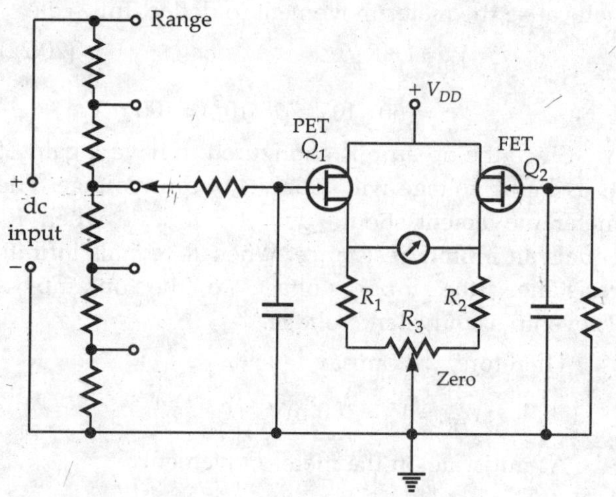


**Fig. 20.29** Balanced d.c. amplifier with input attenuator and indicating meter.

In case no input signal is applied to the gate terminals of FETs, they (gate terminals) are at ground potential. Ideally, no current should flow through the meter. However, on account of some mismatch between the FETs and difference in values of resistors a current does flow. This current is reduced to zero by the **ZERO** adjustment resistor $R_3$. The bridge thus becomes balanced.

When a positive voltage is applied to the gate of input transistor $Q_1$, its drain current increases which causes the voltage at the source terminal to rise. The resulting unbalance between the $Q_1$ and $Q_2$ source voltages is indicated by the meter movement, whose scale is calibrated to agree with the magnitude of the applied input voltage.

The maximum voltage that can be applied to the gate of $Q_1$ is determined by the operating range of FET and is usually of the order of a few volt. The range of input voltages can easily be extended by an input attenuator or **RANGE** switch, as shown in Fig. 20.30. The unknown d.c. input voltage is applied through a large resistor in the probe body to a resistive voltage divider. Thus, with the RANGE switch in the 3-V position as shown, the voltage at the gate of the input FET is developed across 8 MΩ of the total resistance of 11.3 MΩ and the circuit is so arranged that the meter deflects full scale with 3 V applied to the tip of the probe. With the **RANGE** switch in the 12-V position, the gate voltage is developed across 2 MΩ of the total divider resistance of 11.3 MΩ an input voltage of 12 V is required to cause the same full scale meter deflection.

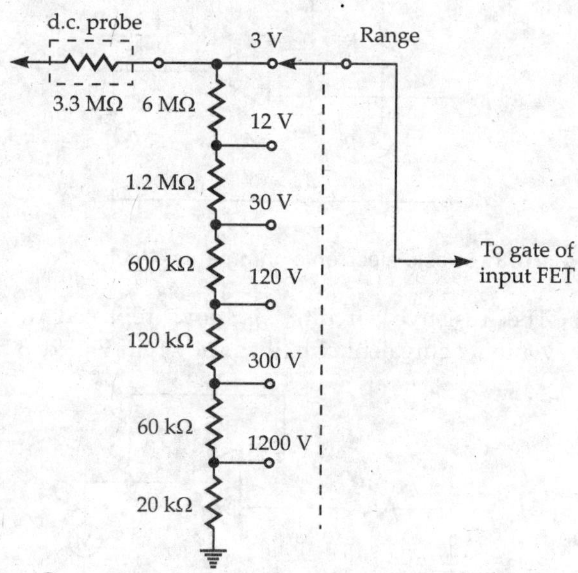


**Fig. 20.30** Typical input voltage attenuator for a multimeter. The range switch on the front panel of the multimeter allows selection of the desired voltage range.

### 20.11.1 Electronic Ohmmeter

When the function switch of the multimeter is placed in the **OHM** position, a resistor is connected in series with an internal battery, and the meter simply measures the voltage drop across the unknown resistor.

A basic electronic ohmmeter is shown in Fig. 20.31. As can be seen, this circuit uses an operational amplifier (op-amp) configured as a noninverting amplifier with a gain of one. The very high input impedance seen looking into the noninverting input of the op-amp, effectively isolates the meter movement from the resistive circuitry of the ohmmeter. The terminals to which a resistor of unknown value should be connected are identified as $x$ and $x'$. The circuit is best analyzed by use of Thevenin's theorem to find an equivalent circuit for the resistive network as seen at points $A$ and $B$. We must either disconnect the op-amp or treat it as an infinite impedance, which we can justifiably do since it does represent a very high impedance. Thevenin's equivalent voltage is computed as,

$$V_{TH} = V\frac{R_2}{R_1 + R_2} \qquad ...(20.20)$$

and Thevenin's equivalent resistance is computed as,

$$R_{TH} = \frac{R_1 R_2}{R_1 + R_2} \qquad ...(20.21)$$

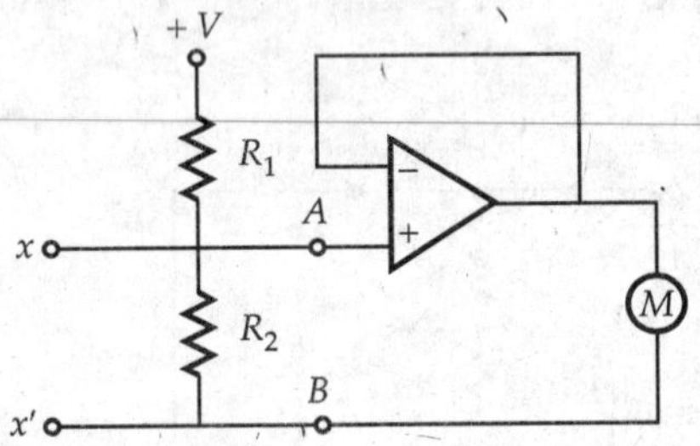


**Fig. 20.31** Basic electronic ohmmeter circuit.

The resistive circuitry is now replaced with Thevenin's equivalent circuit as shown in Fig. 20.32.

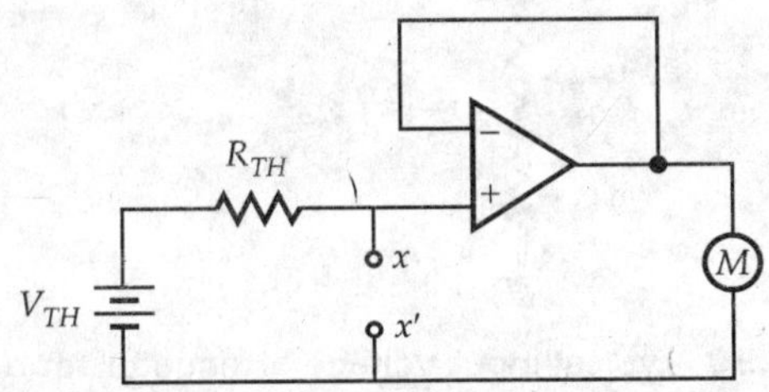


**Fig. 20.32** Circuit of Fig. 20.31 with input Thevenized.

By observation we can see that if points $x$ and $x'$ are shorted together, which represents a measurement of zero ohm, the input voltage to the op-amp is 0 V. Therefore, the meter movement should show an indication of 0 Ω. In addition, if the test points are open indicating an infinite resistance, the meter movement should show full-scale deflection ; if there is an unknown resistance, $R_x$, that is equal in value to $R_{TH}$, the meter movement should show exactly half-scale deflection.

The best way to show these relationship quantitatively is to start with a meter movement with known paremeters and work back to the input as shown in the following example.

***Example 20.6*** *Assume a meter movement with 50 μA full-scale deflection current and 2 kΩ internal resistance and solve for $R_1$, $R_2$ and $V$ for the circuit in Fig. 20.33.*

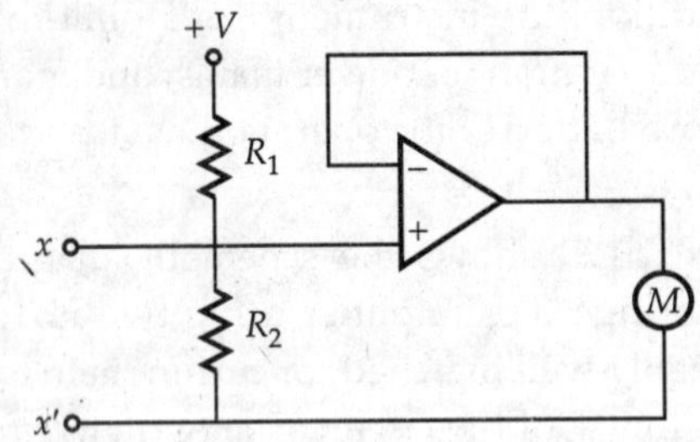


**Fig. 20.33** Circuit for Example 20.6.

**Solution.** Given the full-scale deflection current and internal, resistance of the meter movement, we can solve for the output voltage of the op-amp that will cause the meter movement to deflect full scale

$$V_0 = I_{fs} R_m \qquad ...(20.22)$$

$$= (50 \times 10^{-6} \times 2 \times 10^3) = 100 \text{ mV}.$$

Since the op-amp is configured to have a gain of 1, its input voltage will equal its output voltage. The meter movement should deflect full scale when $R_x$ equals an infinite resistance. When $R_x$ equals infinite resistance, the input voltage to the op-amp is Thevenin's equivalent voltage.

Therefore, we can say,

$$V_{TH} = V_0 = 100 \text{ mV}$$

At midscale on the meter movement :

$$i_m = 0.5\, i_{fs} = 25\ \mu\text{A}$$

or

$$V_0 = (25 \times 10^{-6} \times 2 \times 10^3) = 50 \text{ mV}.$$

Since $V_0 = \frac{1}{2} V_{TH}$ we can see that $R_x = R_{TH}$. Although $R_{TH}$, we can be any parallel combination of $R_1$, and $R_2$, we will make them equal in value. Their values will determine the midscale marking on the ohmmeter scale. Suppose we want the midscale reading to be 100 Ω, which means the pointer should deflect to midscale when $R_x = 100\ \Omega$. $R_{TH}$ also equals 100 Ω, and both $R_1$ and $R_2$ equal 200 Ω if $V_{TH} = 100$ mV, then we can solve for $V$ by use of the voltage divider equation.

$$V_{TH} = V\frac{R_2}{R_1 + R_2}$$

Solving for $V$ yields

$$V = V_{TH}\frac{R_1 + R_2}{R_2}$$

or

$$= (100 \times 10^{-3})\left(\frac{200}{100}\right) = 200 \text{ mV}$$

We can construct a multiple-range electronic ohmmeter by incorporating a switching arrangement such as the one shown in Fig. 20.34. If both $R_1$ and $R_2$ equal 20 Ω, then

$$R_{TH} = R_1 R_2 / (R_1 + R_2) = 10\ \Omega$$

An unknown resistor of 10 Ω will cause half-scale deflection. Therefore, mid-scale is marked as 10 Ω.

If $R_1'$ and $R_2'$ equal 200 Ω, then,

$$R_{TH} = R_1' R_2' / (R_1' + R_2') = 100\ \Omega$$

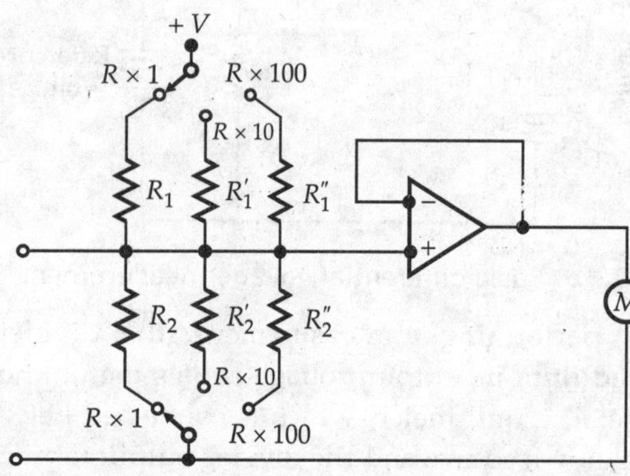


**Fig. 20.34** Multiple-range electronic ohmmeter.

An unknown resistor of 100 Ω will cause half-scale deflection on the $R \times 10$ range, which means the unknown resistor equals a reading of 10 times 10 or 100 Ω.

If both $R_1''$ and $R_2''$ equal 2 kΩ, then

$$R_{TH} = R_1'' R_2'' / (R_1'' + R_2'') = 1\ \text{k}\Omega.$$

An unknown resistor of 1 kΩ will cause half-scale deflection on the $R \times 100$ range, which now means the unknown resistor equals a reading of $10 \times 100$ or 1 kΩ.

## 20.12 CONSIDERATIONS IN SELECTING AN ANALOG VOLTMETER

The selection most appropriate instrument for a particular voltage measurement depends on the performance required in a given situation. Some important considerations in selecting a voltmeter are given below :

1. ***Input impedance.*** In order to avoid loading effects, the input resistance or impedance of the voltmeter should be at least an order of magnitude higher than the impedance of the circuit under measurement. For example, when a voltmeter with a 10 MΩ input resistance is used to measure the voltage across a 100 kΩ resistor, the circuit is hardly disturbed and the loading effect of the meter on the circuit is negligible. The same meter placed across a 10 MΩ resistor, however, seriously loads the circuit and causes an error in measurement of approximately 50 percent.

The input impedance of the voltmeter is a function of the inevitable shunt capacitance across the input terminals. The loading effect of the meter is particularly noticeable at the higher frequencies, when the input shunt capacitance greatly reduces the input impedance.

In some applications, a passive voltage-divider probe can be used to reduce the input capacitance at the point of measurement at the sacrifice of perhaps 20 db of sensitivity. With such a probe, measurements can be easily made at random points without disturbing the circuit under test.

2. ***Voltage ranges.*** The voltage ranges on the meter scale may be in the 1-3-10 sequence with 10 db of separation, or in the 1-5-5-15 sequence, or in a single scale calibrated in db. In any case, the scale divisions should be compatible with the accuracy of the instrument. For example, a linear meter with a 1 percent full-scale accuracy should have 100 divisions on the 1.0-V scale so that 1 percent can be easily resolved. An instrument with an accuracy of 1 percent or less should also have a mirror-backed scale to reduce parallax and to improve accuracy.

3. ***Decibels.*** Use of the decibel scale can be very effective in measurements that cover a *wide range* of voltages. A measurement of this kind is found, for example, in the frequency response curve of an amplifier or a filter, where the output voltage is measured as a function of the frequency of the applied input voltage. Almost all voltmeters with db scales are calibrated in db m, referenced to some particular impedance. The 0-db m reference for a 600-Ω system is 0.7746 V; for a 50-Ω system it is 0.2236 V. In many applications only a 0-db reference is needed. In this ease, 0 dbv (relative to 1 V) can be used for any impedance system.

4. ***Sensitivity versus bandwidth.*** Noise is a function of bandwidth. A voltmeter with a broad bandwidth will pick up and generate more noise than one operating over a narrow range of frequencies. In general, an instrument with a bandwidth of 10 Hz to 10 MHz has a sensitivity of 1 mV. A voltmeter whose bandwidth extends only to 5 MHz could have a sensitivity of 100 μV.

5. ***Battery operation.*** For field work, a voltmeter powered by an internal battery is essential. If an area contains troublesome groundloops, a battery-powered instrument is preferred over a mains-powered voltmeter to remove the ground paths.

20

6. ***A.C. current measurements.*** Current measurements can be made by a sensitive a.c. voltmeter and a series resistance. In the usual case, however, an a.c. *current probe* is used which enables the operator to measure an ac current without disturbing the circuit under test. The current probe simply clips around the wire carrying the unknown current and in effect makes the wire the one-turn primary of a transformer formed by a ferrite core and a many-turn secondary within the current-probe body. The signal induced in the secondary winding is amplified and the output voltage of the amplifier is applied to a suitable ac voltmeter for measurement. Normally, the amplifier is designed so that 1 mA in the wire being measured produces 1 mV at the amplifier output. The current is then read directly on the voltmeter using the same scale as for voltage measurements.

In summarizing the preceding considerations, the following general guidelines can be stated :

(*a*) For measurements involving d.c. applications, select the meter with the broadest capability meeting the circuit's requirements.

(*b*) For a.c. measurements involving sine waves with only modest amounts of distortion (< 10 percent), the average-responding voltmeter provides the best accuracy and most sensitivity per unit investment.

(*c*) For high-frequency measurements (> 10 MHz), the peak-responding voltmeter with a diode-probe input is the most economical choice. Peak-responding circuits are acceptable if the inaccuracies caused by distortion in the input waveform can be tolerated.

(*d*) For measurements where it is important to determine the effective power of waveform which depart from the true sinusoidal form, the rms-responding voltmeter is the appropriate choice.

## 20.13 DIFFERENTIAL VOLTMETER

One of the most accurate methods of measuring an unknown voltage is the *differential voltmeter* technique, where the voltmeter is used to indicate the difference between the unknown voltage and a known voltage. The principle of operation of the differential voltmeter is similar to that of the potentiometers and, for this reason, the instrument is sometimes called a *potentiometric* voltmeter.

The classic differential voltage measurement is shown in basic form in the circuit of Fig. 20.35. In this circuit, the null meter is connected between the unknown voltage source and the output terminals of a precision voltage divider, and hence indicates the difference between the two. This voltage divider is connected across a *reference* voltage source and can be adjusted to provide accurately known fractions of the reference voltage.

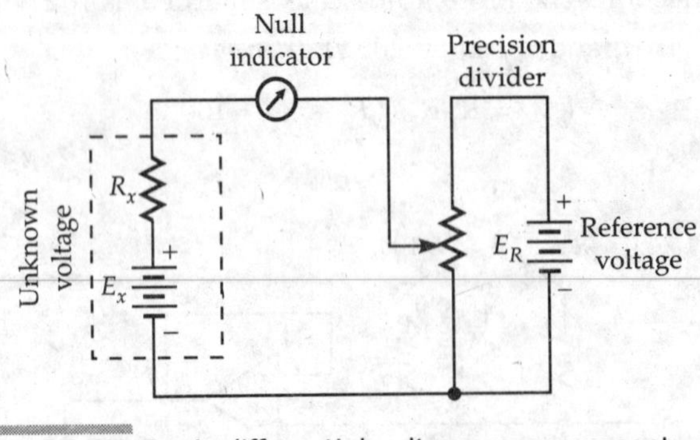


**Fig. 20.35** Basic differential voltage measurement.

In performing a measurement, the divider is adjusted until its output voltage equals the unknown voltage. The null meter, which is connected between the unknown source and the divider-output terminals, indicates zero volt when the two voltages are equal. In this *null condition* neither the source nor the reference supplies current to the meter, and the differential voltmeter therefore presents an infinite impedance to the source under test. Note that the null meter serves to indicate only the residual differential between the known voltage and the unknown voltage. To detect small differences in unbalance potentials, a sensitive meter movement is required ; accuracy of the meter is of secondary importance since the meter is not used to indicate the absolute value of the unknown voltage.

The reference source usually consists of a low-voltage d.c. standard, such as a 1-V d.c. laboratory reference standard or a low-voltage Zener-controlled precision supply.

When measuring high voltages, a high-voltage reference supply can be used. In the usual case, however, a voltage divider is placed across the unknown source to reduce the voltage to a sufficiently low value for direct comparison against the usual low-voltage dc standard. The main drawback of this system is that a differential voltmeter with input divider has a relatively low input resistance, especially for unknown voltages much higher than the reference standard. This low input resistance is undesirable because of its loading effects. A differential voltmeter offers an input resistance approaching infinity only at the null condition and then *only* if an input divider is not used.

The a.c. *differential voltmeter* is a modification of the d.c. instrument and incorporates a precision *rectifier* circuit. The unknown a.c. voltage is applied to the rectifier for conversion to a d.c. voltage equivalent to the average value of the a.c. The resulting d.c. is then applied to the potentiometric voltmeter in the usual manner. The simplified block diagram of an a.c. differential voltmeter shown in Fig. 20.36 is self-explanatory.

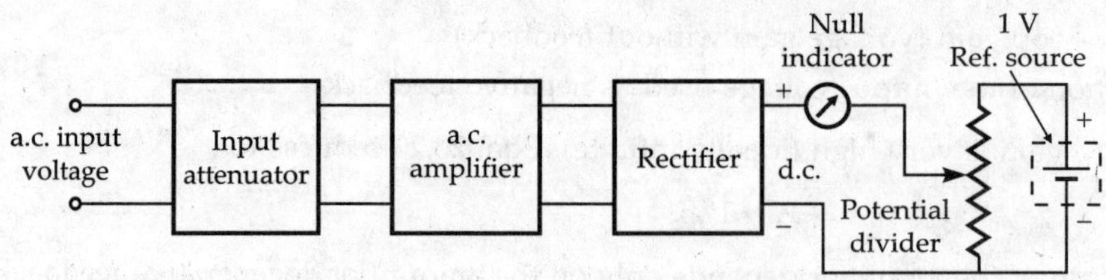


**Fig. 20.36** Simplified block diagram of an a.c. differential voltmeter.

The differential voltmeter requires a reference source to make a measurement and a meter circuit to detect an unbalance between the unknown and the known voltage. Some manufacturers combine the various elements in a multi-function laboratory instrument called a d.c. standard differential voltmeter. This instrument has three modes of operation : as a *d.c. voltage standard,* as a *d.c. differential voltmeter* and as a *d.c. voltmeter*. Function switching allows many of the same basic circuits to be used in each mode of operation.

20

The block diagram of Fig. 20.37 illustrates the *standard* mode of operation, where the instrument generates precision output voltages from 0 V to 1,000 V as a reference source for many laboratory applications. A temperature-controlled reference supply generates a very stable + 1 V d.c., which is applied to a decimal divider network. The divider ratios are controlled by a set of front panel switches allowing the reference supply to be adjusted in 1-μV steps from 0 V to 1 V. This reference output voltage is applied to a high-gain d.c. amplifier with degenerative voltage feedback to obtain precisely controlled gain characteristics.

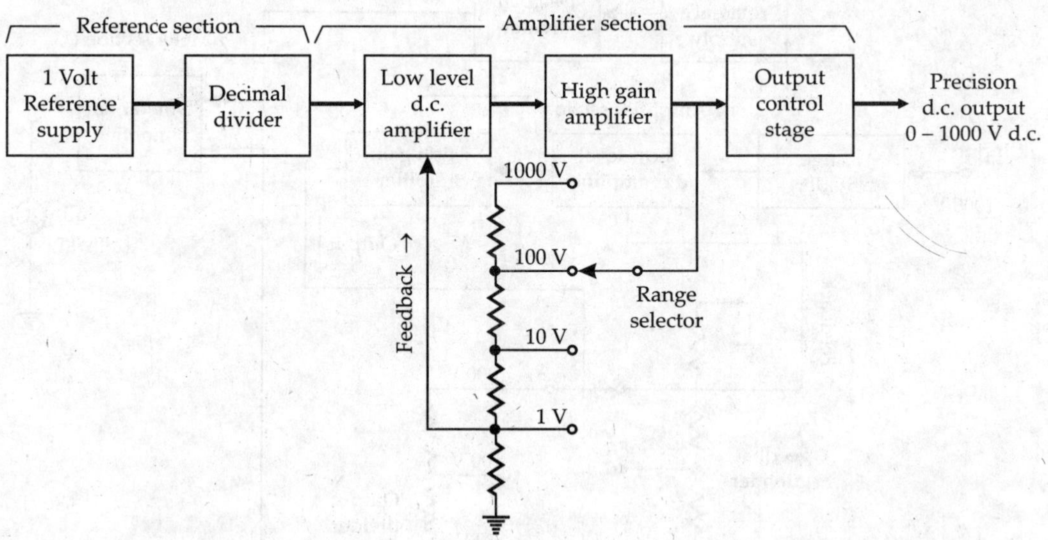


**Fig. 20.37** Block diagram of the d.c. standard/differential voltmeter in the *standard* mode of operation. The reference section in conjunction with the d.c. amplifier section provides precision d.c. output voltages from 0 V to 1,000 V.

The d.c. amplifier, consisting of several stages in cascade, provides an open loop gain of $10^3$ or higher.

The feedback network monitors the actual output voltage and feeds a controlled fraction of the output back to the amplifier input.

The closed-loop gain of the feedback amplifier can be expressed by the relationship

$$A' = \frac{A}{1+\beta A} \qquad ...(20.23)$$

where $A'$ = closed-loop gain (voltage gain with feedback)

$A$ = open-loop gain (voltage gain without feedback)

$\beta$ = fraction of the output voltage used as negative feedback

If the open-loop gain is very high (ideally infinite), Eqn. 20.23 reduces to :

$$A' = 1/\beta \qquad ...(20.24)$$

showing that the gain of the amplifier depends only on the amount of degenerative feedback. The accuracy of the closed-loop gain therefore depends only on the accuracy of the voltage divider which determines β. The feedback divider, shown in the block diagram of Fig. 20.37, is made from stable, wire-wound precision resistors, enabling the amplifier to have precisely controlled closed-loop gain characteristics. The output terminals of the instrument in the *standard* mode of operation provide the following precision voltage ranges :

| | |
|---|---|
| 0 – 1 V in 1 μV steps | (1-V range) |
| 0 – 10 V in 10 μV steps | (10-V range) |
| 0 – 100 V in 100 μV steps | (100-V range) |
| 0 – 1,000 V in 1 μV steps | (1,000-V range) |

In the second mode of operation, the instrument is connected as a *differential voltmeter*, using the same building blocks as before, together with a meter circuit. This is illustrated in the block diagram of Fig. 20.38. The unknown d.c. voltage is applied to the input terminals of the amplifier section. One half of the divider network

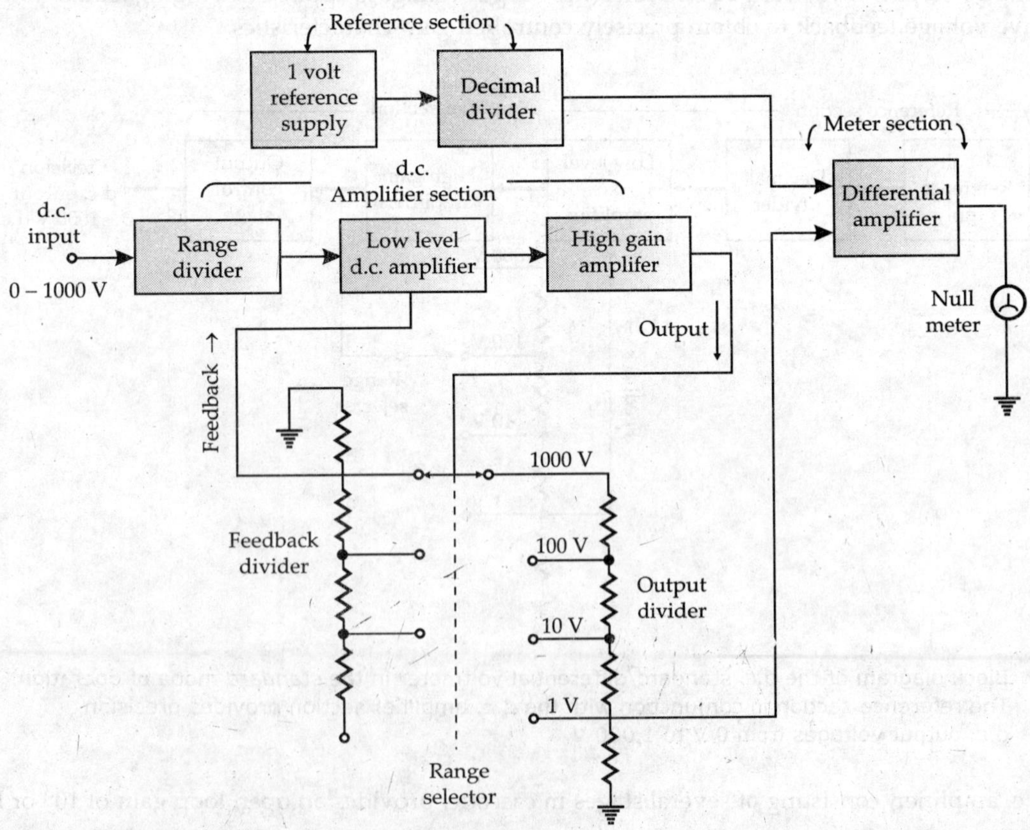


**Fig. 20.38** Block diagram of the d.c. standard/differential voltmeter in the *differential* mode of operation. The meter section indicates the voltage balance between the reference section and the d.c. amplifier section.

feeds a fraction of the output voltage back to the input stage and thereby controls the closed-loop gain of the amplifier. The second section of the divider network applies a fraction of the output voltage to the differential input of the meter amplifier.

The meter circuit measures the difference between the feedback voltage and the reference voltage, and indicates a null when the two voltages are equal.

The range selector on the front panel of the instrument controls both the feedback voltage and the voltage that is applied in opposition to the reference divider output in such a way that the 1-V capability of the reference supply is never exceeded.

In the third mode of operation, the instrument is connected as a *voltmeter* and the d.c. amplifier acts as a buffer stage to provide high-input impedance to the unknown voltage source. The input voltage is amplified, and the d.c. output voltage is applied directly to the meter circuit. The meter circuit incorporates a feedback-controlled amplifier and allows selection of its sensitivity by adjustment of the feedback loop through a front panel control, marked *sensitivity*. This feature provides for extreme sensitivity of the meter circuit, often on the order of 1 μV full-scale deflection. Meaningful measurements at the very high sensitivities, however, are difficult to make because of the problems of noise generation and pickup.

An a.c.-to-d.c. converter can be incorporated in the instrument to provide the capability of a.c. voltage measurement by potentiometric methods.



## 20.14 VECTOR VOLTMETER

A *vector voltmeter* measures the amplitude of a signal at two points in a circuit and simultaneously measures the phase difference between the voltage waveforms at these two points. This instrument can be used in a wide variety of applications, especially in situations where other methods are very difficult or time-consuming. The vector voltmeter is useful in VHF applications and can be used successfully in such measurements as

(*i*) Insertion Losses
(*ii*) Complex impedance of mixers
(*iii*) *S* parameters of transistors
(*iv*) Radio frequency distortion
(*v*) Amplitude modulation index
(*vi*) Amplifier gain and phase shift
(*vii*) Filter transfer functions
(*viii*) Two port network parameters.

The vector voltmeter basically converts two RF signals of the same fundamental frequency (from 1 MHz to GHz) to two IF signals with 20-kHz fundamental frequencies. These IF signals have the same amplitudes, waveforms, and phase relationships as the original RF signals. Consequently, the fundamental components of the IF signals have the same amplitude and phase relationships as the fundamental components of the RF signals. These fundamental components are filtered from the IF signals and are measured by a voltmeter and a phase meter.

The block diagram of Fig. 20.39 shows that the instrument consists of five major sections as follows : two RF-to-IF converters, an automatic phase control section, a phase meter circuit, and a voltmeter circuit. The RF-to-IF converters and the phase control section produce two 20-kHz sine waves with the same amplitudes and the same phase relationship as the fundamental components of the RF signals applied to channels *A* and *B*. The phase meter section continuously monitors these two 20-kHz sine waves and indicates the phase angle between them. The voltmeter section can be switched to channel *A* or channel *B* to provide a meter display of the amplitude.

Each RF-to-IF converter consists of a sampler and a tuned amplifier. The sampler produces a 20-kHz replica of the RF input waveform, and the tuned amplifier extracts the 20-kHz fundamental component from this waveform replica. *Sampling* is a time-stretching process, with which a high-frequency repetitive signal is

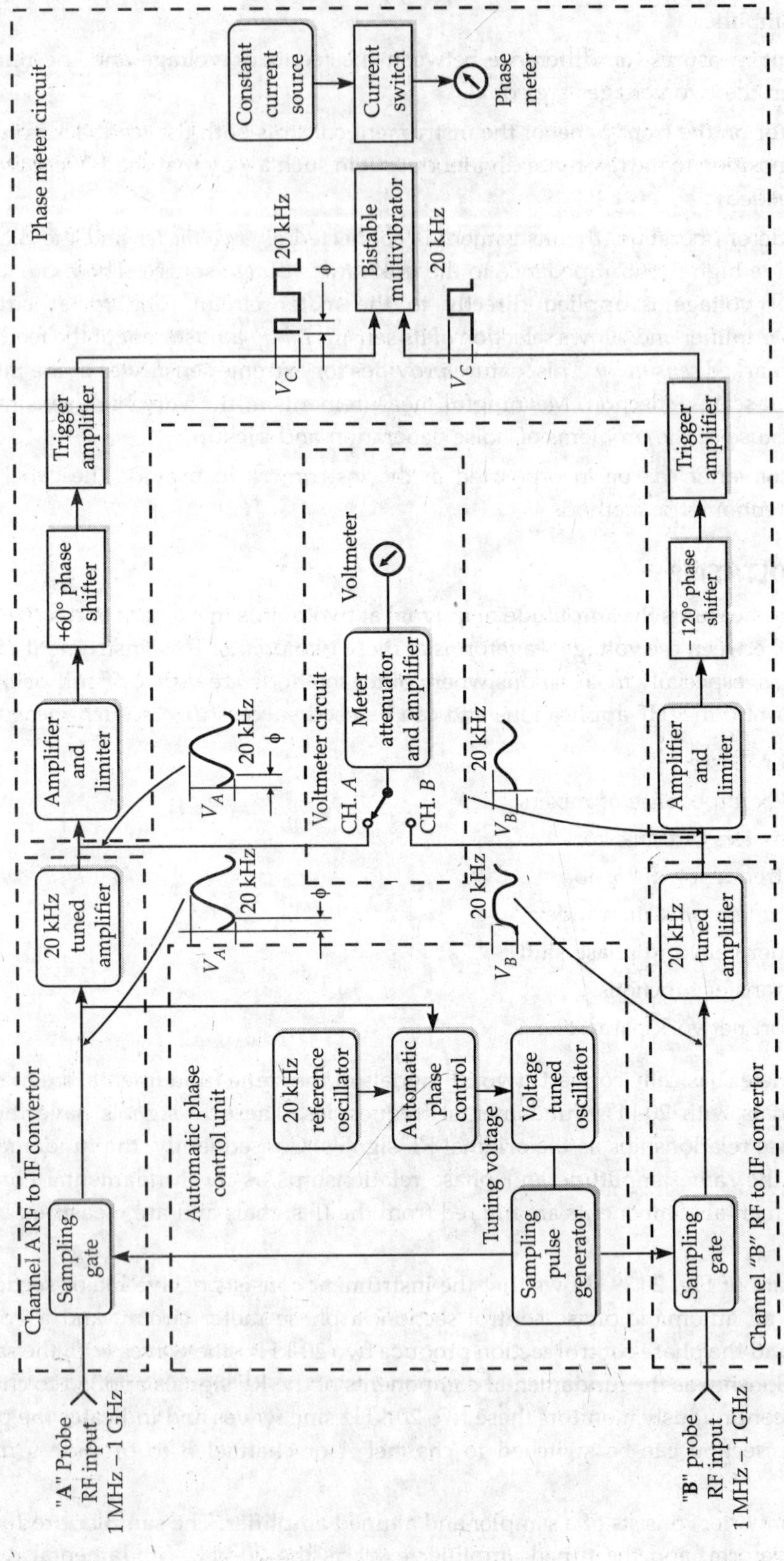


**Fig. 20.39** Block diagram of the vector voltmeter.

duplicated at a much lower frequency. The process is illustrated in the diagram of Fig. 20.40. An electronic switch is connected between the RF input waveform and a storage capacitor. Each time the switch is momentarily closed, the capacitor is charged to the instantaneous value of the input voltage and holds this until the next switch closure. With appropriate timing, samples are taken at progressively later points on the RF waveform. Provided that the RF waveform is repetitive, the samples reconstruct the original waveform at a much lower frequency. Each input channel has a sampler consisting of a sampling gate and a storage capacitor. The sampling gates are controlled by pulses from the same pulse generator. Samples are taken in each channel at exactly the same instant, and the phase relationship of the input signals is therefore preserved in the IF signals.

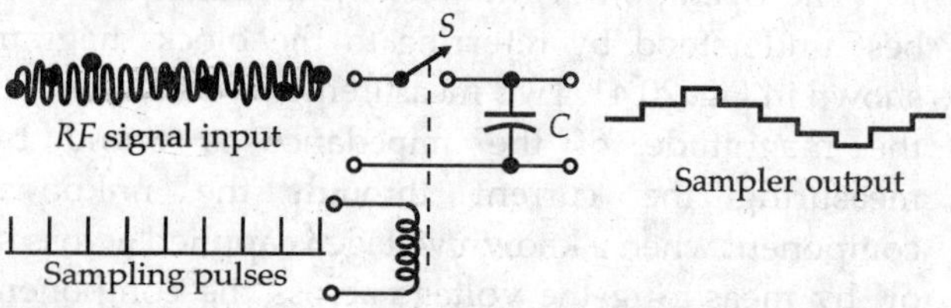


**Fig. 20.40** Simplified diagram of a sampling circuit.

The phase control unit is a rather sophisticated circuit that generates the sampling pulses for both RF-to-IF converters and automatically controls the pulse rate to produce 20-kHz IF signals. The sampling pulse rate is controlled by a voltage-tuned oscillator (VTO) for which the tuning voltage is supplied by the automatic phase control section. This section locks the IF signal of channel *A* to a 20-kHz reference oscillator. To get initial locking, the phase control section applies a ramp voltage to the VTO. This ramp voltage sweeps the sampling rate until channel A IF is 20 kHz and in phase with the reference oscillator. Then the sweep stops and channel, A IF is held in phase with the reference oscillator.

The tuned amplifier passes only the 20-kHz fundamental component of the IF signal of each channel. The output of each tuned amplifier then consists of a signal that has retained its original phase relationship with respect to the signal in the other channel and also its correct amplitude relationship. The two filtered IF signals can be connected to the voltmeter circuit by a front panel switch, marked *channel A* and *channel B*. The voltmeter circuit contains an input attenuator to provide the appropriate meter range. This attenuator is also a front panel control, marked *amplitude range*. The meter amplifier consists of a stable fixed-gain feedback amplifier, followed by a rectifier and a filter section. The rectified signal is applied to a dc voltmeter.

To determine the *phase difference* between the two IF signals, the tuned amplifiers are followed by the phase meter circuit. Each channel is first amplified and then limited, resulting in square-wave signals at the inputs to the IF phase-shifting circuits. The circuit in channel *A* shifts the phase of the square-wave signal by + 60° ; the circuit in channel *B* shifts the phase of its signal by – 120°. Both phase shifts are accomplished by a combination of capacitive networks and inverting and non-inverting amplifiers whose vector-sum outputs provide the desired phase shift. The outputs of the phase-shift circuits are amplified and clipped, producing square waveforms, and applied to the trigger amplifiers. These circuits convert the square-wave input signals to positive spikes with very fast rise times. The bistable multivibrator is triggered by pulses from both channels. Channel *A* is connected to the *set* input of the MV ; channel *B* is connected to the *rest* input of the MV. If the initial phase shift between the RF signals at the probes was 0°, the trigger, pulses into the multivibrator are 180° out of phase owing to the action of the phase-shift circuits. The MV then produces a square-wave output voltage which is symmetrical about zero. Any phase shift at the RF probes carries through the entire system and varies the trigger pulses from their 180° relationship, producing-an asymmetrical waveform.

The (asymmetrical) square wave controls the current switch, which is a transistor switched into conduction by the negative portion of the square wave. The switch connects the constant current supply to the *phase meter*. At 0° phase shift at the RF input, the switch is turned off and on for equal amounts of time and the current supply is adjusted to cause the meter to read 0° or centre scale. Any RF phase shift results in an asymmetrical waveform and allows either more or less current to the phase meter, depending on whether the phase shift caused the negative half-cycle of the square wave to be larger or smaller. An input phase shift of 180° would cause the square wave to collapse into either a positive or a negative dc voltage and the switch would then allow no current or maximum current to the phase meter. These maximum deviations from the centre reading of 0° are marked on the meter face as + 180° and – 180°. The *phase range* can be selected by a front panel switch that places a shunt across the phase meter and changes its sensitivity.

The instrument contains a power supply section, which is not shown on the block diagram of Fig. 20.39. The power supply generates all the necessary supply voltages for the various sections of the instrument.

20

## 20.15 VECTOR-IMPEDANCE METER

The measurement of impedance is concerned with both the magnitude (Z) and the phase angle (θ) of a component. In many applications it is not only necessary to determine the magnitude of the impedance, but it may also be important to find the ratio of the reactance to the resistance and whether the reactance is inductive or capacitive.

The measurement of voltage and current is usually sufficient to determine the magnitude of the impedance at frequencies lower than 100 MHz. The phase difference between the voltage waveform and the current waveform indicates whether the component is inductive or capacitive. If the phase angle can be determined, for example, by using a CRO displaying a Lissajous pattern, the reactance can be determined. The properties of a component should be determined at several different frequencies if the component has to be fully specified. Therefore, many measurements are required.

These measurements become highly elaborate and time-consuming especially at high frequencies and therefore many steps are required to obtain the desired information.

With the development of instruments, such as the **vector impedance meter,** it has been possible to simplify impedance measurements over a wide frequency range. The vector impedance meter makes it possible to obtain **sweep frequency plots** of impedance and phase angle *versus* frequency, thereby providing complete information within the frequency band of interest.

A typical vector impedance meter makes simultaneous measurements of impedance and phase angle over a frequency range of 5 Hz to 500 kHz. The unknown component is simply connected across the input terminals of the instrument, the desired frequency is selected by adjusting the controls provided on the front panel. Two meters on the front panel indicate the magnitude of the impedance and the phase angle.

The operation of the vector-impedance meter is best understood by referring to the block diagram shown in Fig. 20.41. Two measurements are made : (1) the magnitude of the impedance is found by measuring the current through the unknown component when a known voltage is applied across it, or, by measuring the voltage across the component

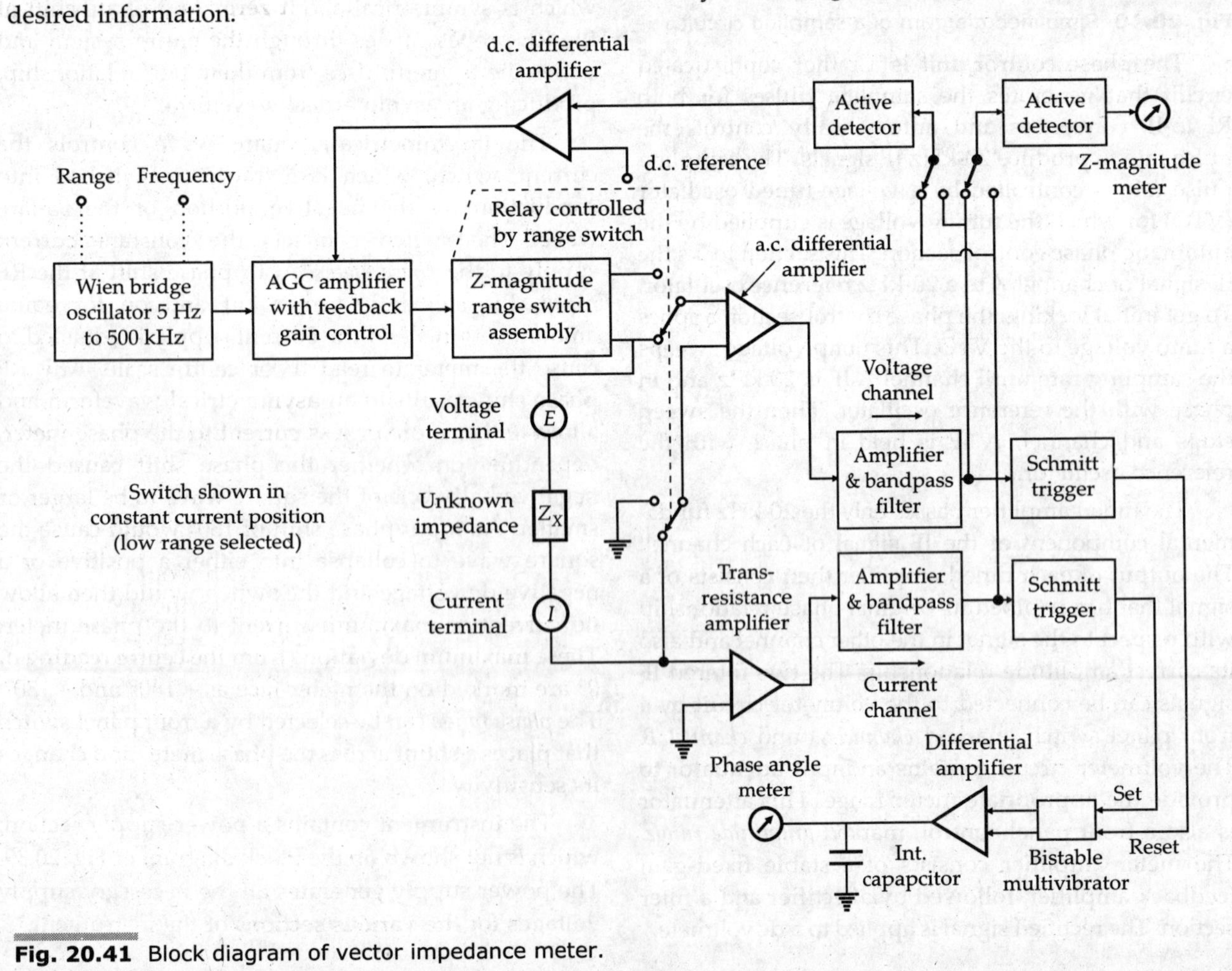


**Fig. 20.41** Block diagram of vector impedance meter.

when a known current is passed through it ; (2) the phase angle is determined by finding the phase difference between the voltage across the component and the current through the component.

Referring to the block diagram shown in Fig. 20.41, the meter contains a **signal source** (a Wien bridge oscillator) with front panel controls for selecting the frequency range and continuous adjustment of the frequency selected. The output of the Wien bridge oscillator is fed to an **AGC amplifier** which permits accurate gain adjustment of the oscillator output by means of its feedback voltage. This gain adjustment is obtained by an internal control actuated by the setting of the **impedance range switch**, to which the AGC amplifier output is connected. The impedance range switch is a precision attenuator network which controls the oscillator output voltage and at the same time determines the manner in which the unknown component is to be connected into the circuitry which follows the range switch.

There are two modes in which the impedance switch can operate the instrument. They are : (*i*) **constant-current mode** and (*ii*) **constant-voltage mode**. The three lower ranges ( × 1, × 10 and × 100) operate in the constant-current mode and the four higher ranges (× 1 k, × 10 k × 100 k, and × 1 M) operate in the constant-voltage mode.

In the constant-current mode, the unknown impedance is connected across the input of the a.c. differential amplifier. The current supplied to the unknown impedance depends on the setting of the impedance range switch. The current is held constant by the action of the trans-resistance amplifier ($R_T$) which converts the current through the unknown impedance to a voltage output equal to the current times its feedback resistance. The trans-resistance amplifier is an operational amplifier where the output voltage is proportional to the input current. The output of the $R_T$ amplifier is fed to a detector circuit and compared to a d.c. reference voltage. The resulting control voltage regulates the gain of the AGC amplifier and hence the voltage applied to the impedance range switch. The output of the ac differential amplifier is applied to an amplifier and filter section which contains high and low band filters which are changed with the frequency range in order to limit the amplifier bandwidth. The output of the bandpass filter is connected, when selected, to a detector which drives the **Z-magnitude meter**. Since the current through the unknown impedance is kept constant by the $R_T$ amplifier, the Z-magnitude meter, which measures the voltage across the unknown, impedance deflects in proportion to the magnitude of the unknown impedance and is therefore calibrated accordingly.

When the instrument is operating in the constant voltage mode, the two inputs to the differential amplifier are switched. The terminal which was connected to the input of the trans-resistance amplifier in the constant-current mode is now grounded. The other input of the differential amplifier which was connected to the voltage terminal of the unknown component, is now connected to a point on the Z-magnitude range switch which is held at a constant potential. The voltage terminal of the unknown is connected to this same point of constant potential or, depending on the setting of the Z–magnitude range switch, to a decimal fraction of voltage. In any case, the voltage across the unknown is held constant. The current through the unknown impedance is applied to the trans-resistance amplifier which again produces an output voltage proportional to its input current.

The roles of the a.c. differential amplifier and the trans-resistance amplifier are now reversed. The voltage output of the $R_T$ amplifier is applied to the detector and then to the Z-magnitude meter. The output voltage of the differential amplifier controls the gain of the AGC amplifier in the same manner that the $R_T$ amplifier did in the constant-current mode.

***Phase-angle measurements.*** These measurements are carried out simultaneously. The outputs of both the voltage channel and the current channel are amplified and each output is connected to a Schmitt trigger circuit. The Schmitt trigger circuits produce a positive-going spike every time that the input sine wave goes through a zero crossing. These positive spikes are applied to a **binary phase detector circuit**. A bistable multivibrator, a differential amplifier, and an integrating capacitor combined together to act as phase detector, a differential amplifier, and an integrating capacitor. The positive-going pulse from the constant current channel **sets** the multivibrator, and the pulse from the constant voltage channel resets the multivibrator. The "set" time of the multivibrator is therefore determined by the zero crossings of the voltage waveform and the current waveform. The "set" and "reset' outputs of the multivibrator are applied to the differential amplifier, which applies the difference voltage to the integrating capacitor. The capacitor voltage is directly proportional to the zero-crossing time interval and is applied to the phase-angle meter which then indicates the phase difference, between the voltage and current waveforms in degrees.

## 20.16 A.C. VOLTAGE MEASUREMENTS

Electronic a.c. voltmeters are essentially indicated to d.c. voltmeters except that a.c. signal may be rectified before it can be applied to the PMMC meter movement, when the a.c. voltage is rectified before amplification a d.c. amplifier is needed. While when rectification takes place after amplification, an a.c. amplifier is used.

## 20.17 CURRENT MEASUREMENTS USING ELECTRONIC INSTRUMENTS

1. ***D.C. current measurements.*** Electronic voltmeters are frequently constructed to act as multipurpose instruments so that they can be used to measure current as well as voltage. The unknown current is made to flow through a known standard resistance. The voltage drop across this resistance is proportional to the current and is measured by a VTVM or a TVM. The scale of the meter is calibrated in terms of current.

2. ***A.C. current measurements.*** When alternating current is to be measured a rectifier is used to change the alternating current to a corresponding direct current, which is then measured by a VTVM or TVM as explained above.

Another method employs an a.c. current probe which enables the a.c. current to be measured without disturbing the circuit under test. The a.c. current probe clips around the wire carrying the current and in effect makes the wire a one turn primary of a current transformer. The C.T. has a ferrite core and the secondary consists of a large number of turns. The voltage induced in the secondary winding is amplified and the amplifier's output can be measured by any suitable a.c. voltmeter. Normally the amplifier is designed so that 1 mA current in the wire being measured produces 1 mV at the amplifier output. The current is then read directly on the voltmeter, using the same scale as for voltage measurements.

## 20.18 MEASUREMENT OF POWER AT AUDIO FREQUENCIES

We frequently need to measure the output power of certain electron devices such as amplifiers and oscillators. The frequencies involved in such devices are generally **audio frequencies** (upto approximately 15 kHz) or radio frequencies (above 15 kHz). The dynamometer type of wattmeters are not suitable for measurement of power at high frequencies since they are designed to operate on power frequencies.

There are a number of methods for measurement of power at high frequencies. The simplest method, perhaps, is to connect a dummy load to the output terminals of the device under test. Then either the voltage drop across the dummy load or the current through the dummy load is measured using instruments like rectifier or thermocouple types (these instruments are used because they can work at higher frequencies). Knowing the resistance, $R$, of the dummy load we can calculate the power input.

Power output $P = E^2 / R = I^2 R$

where $E$ is the voltage across the dummy load while $I$ is the current through it.

The basic circuit for measurement of power output of an audio frequency amplifier or oscillator is shown in Fig. 20.42.

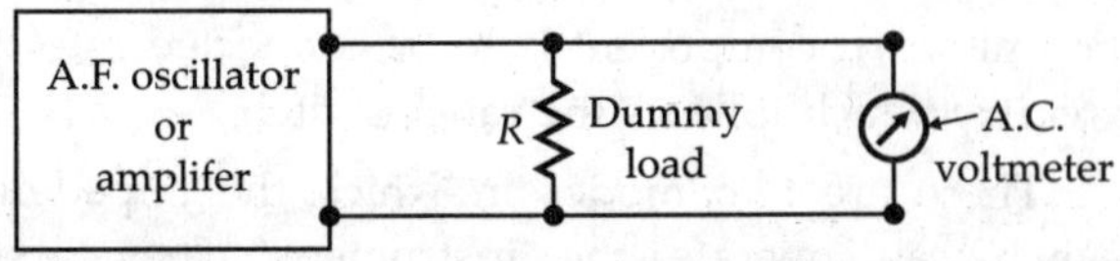


**Fig. 20.42** Basic circuit for measurement power output at audio frequencies.

A non-inductive type resistance $R$ forms the dummy load and is connected across the output terminals of the device under test. The value of resistance $R$ should be approximately equal to that of the output impedance of the device. Also the resistance value should be high enough to dissipate safely the power output. The a.c. voltmeter is a high resistance rectifier type of instrument. If this voltmeter is used with a fixed value of $R$ (resistance of dummy load) its scale can be directly calibrated in terms of power output at audio frequencies.

**Decibels (db).** The a.c. voltmeters used for measurement of power output are provided with **decibel** (abbreviated as db) scales. The scales provide measurement of power levels in decibels under certain conditions. Measurements of power output in terms of decibels is quite customary in audio frequency ranges.

The decibel is a logarithmic unit to express power ratios. If the powers being compared are $P_1$ and $P_2$, then,

$$\text{decibels db} = 10 \log_{10} P_2 / P_1 \quad ...(20.25)$$

The sign associated with decibels indicates which power is greater. A positive sign means that $P_2$ is greater than $P_1$ while a negative sign means that $P_2$ is less than $P_1$.

The advantages of a db scale lie in its logarithmic nature. This permits the use of a very wide scale range of power to be expressed in terms of db without running into inconveniently large numbers, while at the same time permitting small ratios to be conveniently expressed. For example, 0 db represents a ratio of 1 while 1 db represents an approximate ratio of 1.25 and 60 db represents a ratio of 1000,000.

In our case, the power is measured in terms of a voltage and a resistance or a current and a resistance. If the resistance value is the same, we can express decibels as :

$$db = 10\log_{10}\frac{E_2^2/R}{E_1^2/R} = 20\log_{10}\frac{E_2}{E_1} \quad ...(20.26)$$

$$= 10\log_{10}\frac{I_2^2 R}{I_1^2 R} = 20\log_{10}\frac{I_2}{I_1} \quad ...(20.27)$$

It is clear from Eqn. 20.26, the scale of a voltmeter can be expressed in terms of db.

There are several different reference levels used for decibel scale. One type establishes the zero reference level as 1 mW power dissipated in a 600 Ω resistance. Now $P = E^2 / R$.

$$\therefore \quad E = \sqrt{1\times 10^{-3}\times 600} = 0.7746 \text{ V}.$$

Since 0 db means the two powers $P_1$ and $P_2$ should be equal, the 0 db point corresponds to 0.7746 V on the voltmeter scale.

The second method places the 0 db reference as 6 mW is 500 Ω. This method places the 0 db point at 1.732 V on the voltmeter scale.

Meters using these methods will read directly in decibel when they are across the load resistance for which their 0 db reference is established, and the range switch is in 0 db position. For all other range-switch settings algebraic sum of the scale reading and the range switch indication must be taken. Power levels above 0 db are termed positive and those below 0 db are negative.

Decibels may be converted into watt by the following relationship which follows from Eqn. 20.25.

$$\text{Power output } P_2 = P_1 \text{ antilog}_{10} \text{ db/10 watt} \quad ...(20.28)$$

## 20.19 DIODE SENSOR BASED INSTRUMENTS

Radio frequency (RF) is essentially a.c. voltage, except that the frequencies involved are much greater than that which are encountered in power distribution, audio frequency amplifiers, or control systems. Radio frequencies extend well into the GHz region, where it is difficult to amplify and great care must be taken because normal components are often useless.

Radio-frequency voltage is measured by rectifying the alternating voltage and amplifying the resulting dc output. Because of the difficulty in amplifying the RF signal itself, the RF voltage is first rectified and the dc output is amplified.

The diodes used to rectify the RF waveform are not like the rectifiers used in a conventional ac meter, discussed in Chapter 9. The diodes used to rectify the RF signal are either Schottky barrier or point contact diodes. Conventional junction diodes with small geometries can be used for lower frequencies, but most detector diodes are not PN junction diodes. There are two significant problems with diodes used for RF rectification. First, most diodes have excessive capacitance for high-frequency RF rectification, and second, most diodes have excessive reverse recovery time.

When diodes are operated at low forward-biased potentials, the rectified output does not equal the peak of the input. This means that for rather low amplitude RF voltage the resulting dc output is even lower, and a chopper-stabilized amplifier or other amplifier stabilized for dc drifts is required. Figure 20.43 shows a block diagram of a sensitive RF millivoltmeter. The actual RF rectifier or detector is usually mounted on a probe so that measurements can be made with the least amount of interconnecting RF cable, as even the losses of coaxial cable can cause significant errors at very high frequencies. The detected output is in the very low millivolt region, and often even lower, and is amplified via a chopper-stabilized amplifier, digitized and displayed on a digital readout.

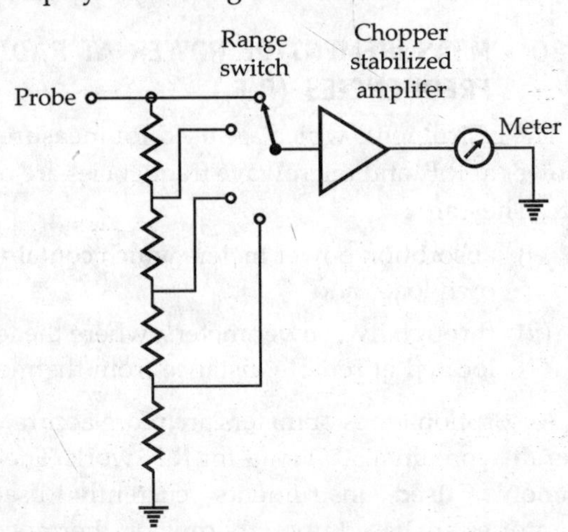


**Fig. 20.43** Block diagram of the RF millivoltmeter.

20

The type of measurement made by the RF millivoltmeter depends on the type of probe used. Voltage measurements are made with a probe similar to the one shown in Fig. 20.44(*a*). Voltage measurements are made with a relatively high impedance, but some capacitance is inescapable. This probe would be used within circuits where the impedances vary and the circuit cannot be isolated and terminated externally.

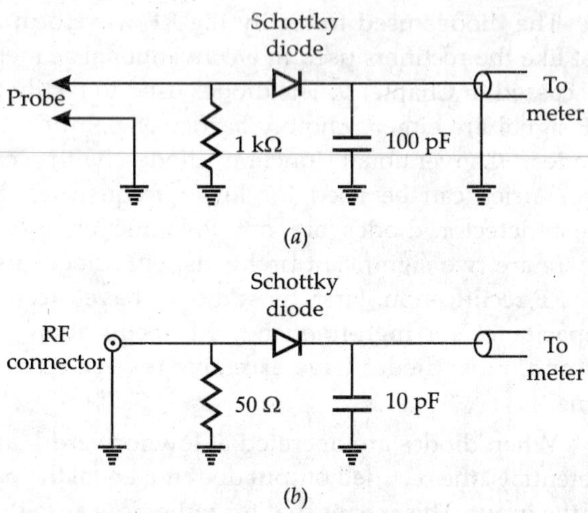


**Fig. 20.44** Two different RF probes for use with an RF millivoltmeter.

Many high-frequency circuits can be disconnected and terminated, usually in 50 Ω, externally, and the probe of Fig. 20.44(*b*) is used. This probe is more a power measuring probe rather than a voltage probe and can be used to measure powers to the nanowatt region. This power measurement is not a true rms measurement, and care must be taken in interpreting measurements, especially when the signal being measured has modulation applied.

## 20.20 MEASUREMENT OF POWER AT RADIO FREQUENCIES (R.F.)

The instruments which are used for measurement of power at R.F. and microwave frequencies are of two types. They are :

(*i*) absorbtion power meters which contain their own load, and

(*ii*) through live power meters, where the load is located at remote distance from the meter.

Absorption types of meters are more accurate and generally contain a 50 Ω load for R.F. work. The most commonly used instruments currently used in laboratories are based upon thermistor, thermocouple and diode sensors.

A power sensor should accept all the r.f. power input, since the meter connected to the sensor measures the power which has been accepted. However, due to mismatch between impedances of the r.f. source and sensor, some of the power is reflected back, and not dissipated in the sensor. If $P_i$ denotes the input or incident power, $P_r$ the reflected power, and $P_d$ the power dissipated in the sensor, then Eqns. 20.29 and 20.30 may be derived.

$$P_i = P_r + P_d \quad \text{...(20.29)}$$

$$P_r = \rho_r^2 P_i \quad \text{...(20.30)}$$

where $\rho_r$ is called the *reflection coefficient* of the sensor. Ideally the reflection coefficient should be zero, although it is acceptable to have a value of 0.05 *i.e.*, 5% reflection.

R.F. power may also be dissipated in parts of the instrument other than the sensing element. Examples of such losses include dissipation in the conducting walls of the power sensor, leakage into instrumentation, and power radiated into space. These losses are not measured by the power meter, and should be minimised. The amount of loss is given by the *effective efficiency* ($\eta_e$) of the sensor, which should ideally be 1.00 (100%).

The *calibration factor* ($K_b$) of the sensor connects together the reflection coefficient and effective efficiency, and is given by,

$$K_b = \eta_e(1-\rho_\rho^2) \quad \text{...(20.31)}$$

A calibration factor of 0.90 (90%) means that the meter will read 10% below the value of the incident power $P_i$. Power meters usually have a dial, to allow them to be set to the correct value of $K_b$ for the sensor being used, and so compensate for errors. The values of $K_b$, at different frequencies are usually specified by the sensor manufacturer.

Instruments must be carefully set up for high frequency power measurements, to minimise losses. Figure 20.45 shows a typical set up. The R.F. source is a signal generator, or a sweeper which is used below 100 mW (or 20 db) only. The input attenuator is required for low power work, below 1 mW (or 0 db), and sometimes in the medium power range 1 mW-10 W (or 0 + 40 db) to allow the input level to be set at a safe value for the device under test. The circulator is used to channel reflections back to the load, and prevent them reaching the source. Circulators are generally used at high power levels, above 10 W, to protect the source.

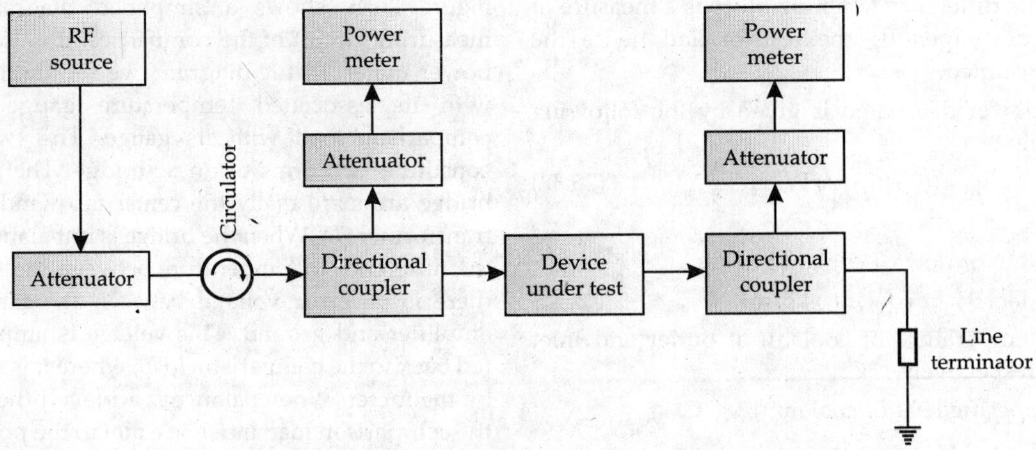


**Fig. 20.45** Test set up to measure power at high frequencies.

Directional couplers monitor the input and output to the device under test. At low power levels the indicator may be placed in series with the load, but for high power it is placed off the coupler. At high microwave powers the power level at each component in the test set up is also very important, and must be limited to the safe value for that device.

## 20.20.1 Voltmeter Based Instruments

For frequencies below about 100 MHz, the voltage across a standard resistor $R_s$ may be measured and the power calculated by using the relationship $P = V_s^2 / R_s$. This method is essentially used at higher frequencies and is incorporated in a single instrument called the "*absorbtion type power meter*". The absorbtion type meter is nothing more than a resistor, which is designed to maintain constant resistance over the range of frequencies of interest and a high frequency. Voltmeter which is calibrated in terms of power. Figure 20.46 shows the circuit of an absorbtion type voltmeter.

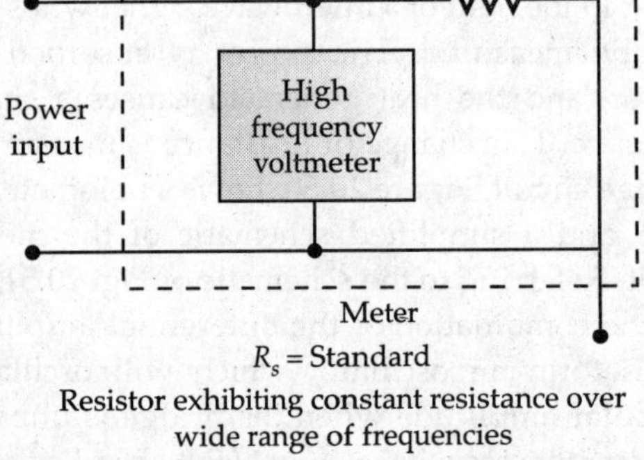


**Fig. 20.46** Circuit of an Absorbtion meter.

Standard resistors are available which exhibit constant resistance for frequencies ranging from d.c. to 4 GHz. Although, the value of resistance is fairly constant over the desired range, the resistance versus frequency characteristics of standard resistors are available for any correction to be applied. This type of meter is normally used for frequencies below 500 MHz. For measurement of power, at higher frequencies well into the range of microwave, other types of absorbtion types of meters are used.

The two other types of absorbtion meters which are extensively used in the band of frequencies ranging from 500 MHz to 40 GHz are the *calorimeter* and *bolometer* power meters.

## 20.20.2 Calorimeter Power Meter

Calorimeters are used for measurement of high power in a standard laboratory rather than in industry. The basic principle of operation of the calorimeter-type power meter is quite simple. A resistor is heated by the signal to be measured. This resistor is kept in a completely enclosed, well insulated bath as shown in Fig. 20.47. The insulation of the walls of the bath are such that very little heat is allowed to radiate from the bath. If the bath temperature is measured before and after the signal is

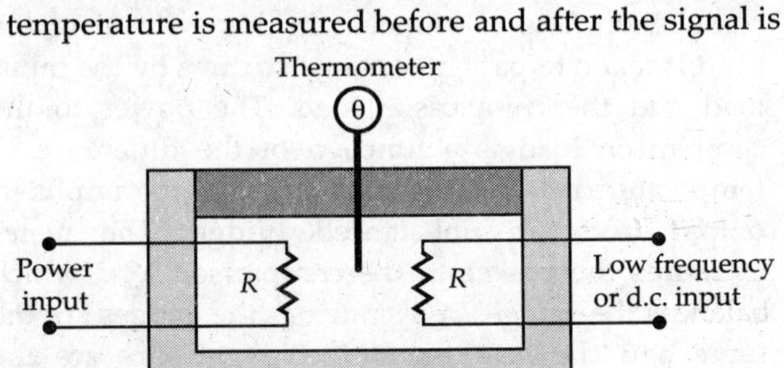


**Fig. 20.47** Basic calorimeter power meter.

20

applied, the difference in temperature is a measure of the heat developed by the resistor and hence the power consumed.

The power dissipated is given by the following relationship :

$$P = (\theta_0 - \theta_i) r d s \text{ W} \qquad ...(20.32)$$

where

$r$ = rate of flow of coolant, $m^3/s$,

$d$ = density of coolant ; $kg/m^3$,

$\theta_0, \theta_i$ = temperature of coolant, at outlet and inlet respectively ; °C

$s$ = specific heat of coolant, J/kg°C.

The need to know the exact mass and specific heat of the calorimeter bath material, coupled with the extreme difficulty of obtaining complete thermal isolation from the environment, restricts the usage of the basic calorimeter power meter. Some of these difficulties are avoided in an instrument called the substitution calorimeter. In this instrument the signal is applied to the terminating resistors in the bath, and the equilibrium, temperature of the bath is recorded. Now the signal is removed and dc or low-frequency ac is applied until the same equilibrium temperature is reached. The power can easily be measured with an accurate voltmeter. Figure 20.48 shown the basic substitution power meter.

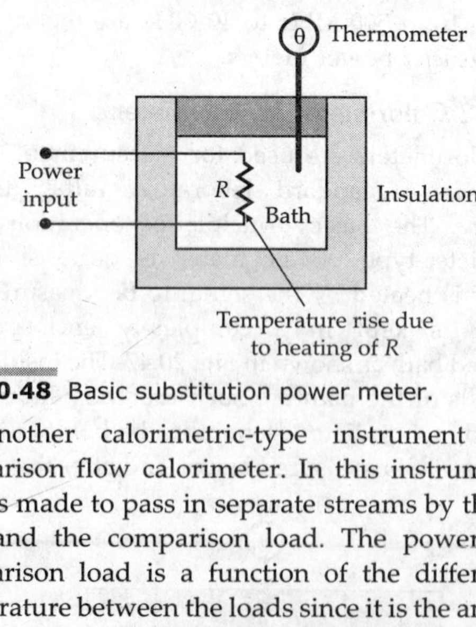


**Fig. 20.48** Basic substitution power meter.

Another calorimetric-type instrument is the comparison flow calorimeter. In this instrument the fluid is made to pass in separate streams by the input load and the comparison load. The power to the comparison load is a function of the difference in temperature between the loads since it is the amplified output from an unbalanced bridge. The meter measures the power to the comparison load which balances the bridge. In as much as the gauges are the same and the heat transfer characteristics are the same, the power in each load is the same ; therefore the meter can be calibrated in input power.

Figure 20.49 shows a simplified diagram of the measuring circuit of the comparison flow calorimeter power meter. In the diagram, we see the input load with its associated temperature gauge and the comparison load with its gauge. The two gauges constitute two arms of an ac bridge. The remaining bridge arms are really the center tap winding of the transformer ($T$). When the bridge is unbalanced due to the difference in temperature between the two loads, there is an error voltage between the input of the amplifier and ground. This voltage is amplified and fed back to the comparison load, where it is monitored by the meter. When balance is achieved the power in the comparison load must be equal to the power in the input load, so that the meter which is calibrated in power units reads input power.

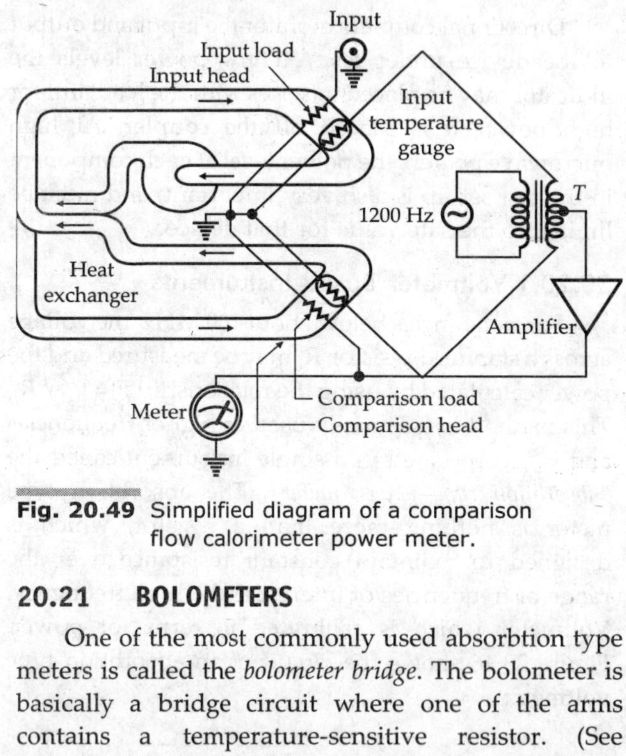


**Fig. 20.49** Simplified diagram of a comparison flow calorimeter power meter.

## 20.21 BOLOMETERS

One of the most commonly used absorbtion type meters is called the *bolometer bridge*. The bolometer is basically a bridge circuit where one of the arms contains a temperature-sensitive resistor. (See Fig. 20.50). The temperature-sensitive resistor is placed in the field of a microwave signal whose power is to be measured. The power is absorbed by the resistor, and the heat generated causes a change in resistance. This change of resistance is measured with a bridge circuit. Figure 20.50 shows a bolometer power meter and a simplified schematic of the measuring circuit. Referring to the schematic of Fig. 20.51, we see that the combination of the differential amplifier and bridge form an oscillator which will oscillate at a particular amplitude when the bridge is unbalanced. The sensitive resistive element ($R_T$) absorbs the power and heats itself until the bridge is very nearly balanced. The d.c. bias is then adjusted until exact balance is achieved. Now the element is placed in the microwave field. The element absorbs the power and is heated, which causes a change in its resistance, thereby unbalancing the bridge. The imbalance is in a direction opposite to that of the initial cold resistance. This action causes the output from the oscillator to decrease so as to affect a balance. The Electronic Voltmeter circuit measures the amount of this power decrease from the oscillator and displays the measurement as a power increase which is that supplied by the microwave field.

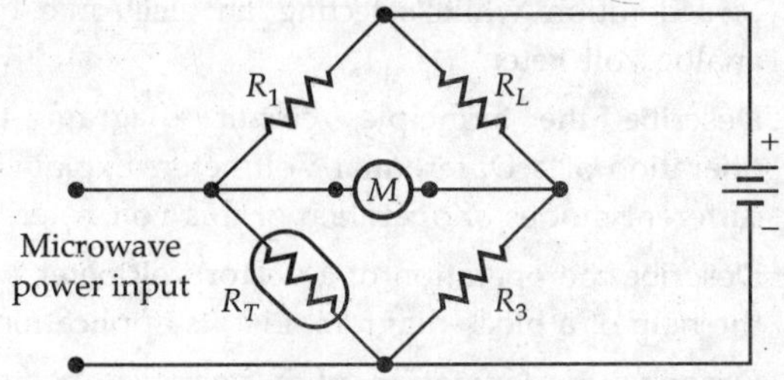


**Fig. 20.50** Basic bolometer bridge circuit.

Bolometer bridges use two types of elements : barretters and thermistors. Barretters are metallic wires or film in which the temperature coefficient of resistance is positive ; that is, the higher the temperature, the higher the resistance. Thermistors are resistors made of semiconductor material which has a negative temperature resistance coefficient ; this means that the resistance of the element decreases as the temperature increases. In general barretters are very delicate, whereas thermistors are much more rugged. The bolometer is customarily used for power

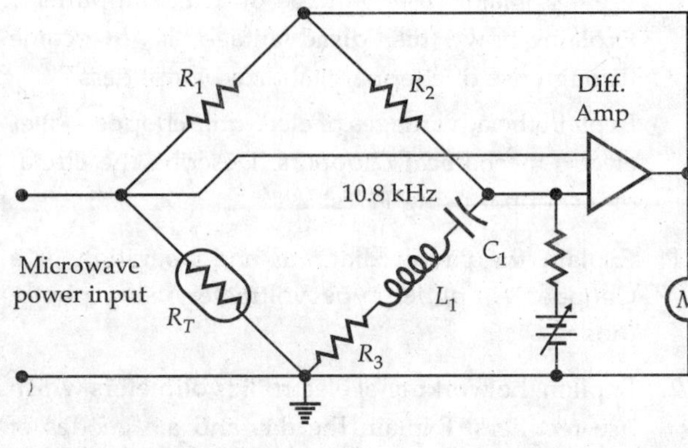


**Fig. 20.51** Simplified diagram of the measurement circuit of a modern bolometer type power meter.

measurements in the range 0.01 – 10 mW. In the range of 10 mW to 10 W a comparison-type calorimeter is quite suitable. Above 10 W, power measurements can be made by using attenuators and a low-power bolometer.



# Review Questions

1. Explain the advantages of Electronic voltmeters over conventional type voltmeters are regards
   (*i*) detection of low level signals
   (*ii*) power consumption (*iii*) loading effects and
   (*iv*) frequency range.
2. Explain the construction and working of an average reading VTVM. Describe its advantages and disadvantages.
   Explain how this meter can be converted to a rms reading voltmeter.
3. Explain the construction and working of the following types of peak reading VTVMs :
   (*i*) Series type
   (*ii*) compensated shunt type.
4. Explain the working of a differential amplifier using two FETs. Derive the expression for the output voltage.
5. Describe the circuit diagram and operation of an Electronic voltmeter using a difference amplifier. Derive its equivalent circuit and find the expression for the current flowing through the meter.
6. Describe the circuit diagram and operation of an Electronic voltmeter using a difference amplifier. Explain the functions of zero setting and calibration resistors. Describe how the range of the voltmeter can be extended.
7. Explain the circuit diagram and operation of a Source Follower Electronic Voltmeter. Derive its equivalent circuit and find the expression for current through the meter.
8. Explain the working of a Source Follower Electronic voltmeter. Describe how the range of this voltmeter can be extended. Explain the use of zero adjustment and calibration resistors.

9. Describe the circuit diagram and operation of a d.c. voltmeter using a direct coupled amplifier. Explain its advantages.

   Describe, with the help of a circuit diagram, how small values of voltages and currents can be measured with it.

10. What is major disadvantage of a d.c. amplifier ? Explain how this disadvantage is overcome through use of chopper stabilized amplifiers.

    Explain the advantages of electronic choppers over electro-mechanical choppers. Describe the circuit of a chopper using FETs.

11. Explain the circuit diagram and operation of a Chopper Amplifier type voltmeter using photo-diodes.

12. Explain the working of electronic voltmeters which use rectifiers. Explain the d.c. and a.c. modes of operation and describe why negative feedback is used in them.

13. Explain the circuit diagram of following types of electronic voltmeters :

    (*i*) voltmeters using a series connected diode

    (*ii*) voltmeters using a full wave bridge rectifier

    (*iii*) peak reading voltmeters using shunt connected diode.

    Explain, with examples, the errors caused in case in the input voltages are non sinusoidal.

14. Describe the circuit diagram and operation of a true rms reading voltmeter using thermocouples. Explain how these voltmeters are free from waveform errors.

15. Explain, with the help of a block diagram, the various parts of an electronic multimeter.

16. Explain the ohmmeter part of an electronic multimeter. Describe how $R\times1$, $R\times10$ and $R\times100$ ranges are obtained in this instrument.

17. Describe the various factors which are taken into consideration while selecting an electronic type analog voltmeter.

18. Describe the principle, circuit diagram and operation of a Differential Voltmeter. Explain the different modes of operation of this voltmeter.

19. Describe the operation of a vector voltmeter with the help of a block diagram. List its applications.

20. Describe the operation of a vector impedance meter with the help of a block diagram. Explain the function of each block. Describe how phase angle measurements are carried out with it.

21. Describe the methods of measurement of power at audio frequencies.

22. Describe the methods of measurement of voltage and power at radio frequencies.

# Unsolved Problems

1. What is the lowest full scale voltage that could be displayed with a 100 μA meter movement with an internal resistance of 150 Ω ? What is the sensitivity of this meter and its power consumption ? Is there any way this meter could be used to construct a lower full scale voltage reading ?
   [**Ans.** 15 mV, 10 kΩ/V, 1.5 μW]

2. A 25 mA full scale current meter with an internal resistance of 100 Ω is available for constructing an a.c. voltmeter with a voltage range of 200 V rms. The meter uses a bridge configuration for the rectifier of the instrument. If each diode has a forward resistance of 500 Ω and infinite reverse resistance, calculate the value of the series resistance to limit the current to the rated value at rated voltage. [**Ans.** 6100 Ω]

3. An electronic voltmeter is shown in Fig. 20.9. Assume the following values : $R_D = 15\text{k}\Omega$, a.c. drain resistance = 100 kΩ and trans-conductance = 0.003 mho. If the meter has a resistance of 1800 Ω and a full scale current of 5 mA, what voltage must be applied to the gate of left hand FET to produce full scale current if the gate of right handed FET is grounded ? [**Ans.** 3.57 V]

4. The electronic voltmeter of Example 3 is to have full scale current for $V_1 = 5$ V, calculate the value of the resistance to be connected in series with the ammeter. [**Ans.** 11.25 kΩ]

5. An electronic voltmeter shown in Fig. 20.12 reads 1 V at full scale. If a 300 V range is to be added to the potential divider, show the new values for various sections of the potential divider if the total resistance of the divider is 10 MΩ.
   [**Ans.** 6.67 MΩ, 2.33 MΩ, 0.67 MΩ, 0.23 MΩ, 66.6 kΩ., 33.3 kΩ]

6. Figure 20.13 shows a Source Follower Voltmeter. Find the relationship between ammeter current and input voltage if a.c. drain resistance, $r_d$ = 100 kΩ, trans-conductance, $g_m$ = 0.003 mho, $R_s$ = 15 kΩ and meter resistance $R_m$ = 1800. **[Ans.** $i = 84 \times 10^{-6}\ V_1$]

7. Figure 20.13 shows a Source Follower Voltmeter with $r_d$ = 10 kΩ, $g_m$ = 0.003 mho, $R_s$ = 15 kΩ and $R_m$ = 1800 Ω. How much current through the meter for 1 V input. Find the value of resistance to be added in series with the ammeter to limit to 0.1 mA for 1 V input. **[Ans.** 0.389 mA, 7030 Ω]

8. A practical version of a source follower electronic voltmeter is shown in Fig. 20.16. The data given is : $R_s$ = 40 kΩ, $r_d$ = 200 kΩ, $g_m$ = 0.004 mho. An ammeter having an internal resistance 1000 Ω and a full scale current of 50 mA is used. If the input signal is a 1 V rms sine wave, what value of calibration resistor produces full scale deflection ? **[Ans.** 18.5 kΩ]

9. A half wave rectified voltage is applied to an average reading voltmeter with a scale calibrated in terms of the rms value of a sine wave. Calculate the error in meter indication. **[Ans.** –50%]

10. What would a true rms reading voltmeter indicate if a pulse waveform of 5 V peak and a duty cycle of 25% is applied to it ? **[Ans.** 2.5 V]

# Objective Type Questions

20

***Tick (√) the most appropriate answer*** :

1. A PMMC meter has an internal resistance 200 Ω and the current required for its full scale deflection is 50 μA. The meter is capable of measuring, on its own, a maximum voltage of :
   (*a*) 5 mV (*b*) 10 mV
   (*c*) 5 μV
   (*d*) 10 μV.

2. A PMMC meter has an internal resistance of 500 Ω and the current required for its full scale deflection is 100 μA. The power consumed by the meter is :
   (*a*) 5 μW
   (*b*) 5 mW
   (*c*) 5 W
   (*d*) 0.5 μW.

3. An average reading VTVM uses one diode connected in series with an external resistance. A high value of series resistance is used so that the instrument should have :
   (*a*) a high input impedance
   (*b*) a linear *v* - *i* characteristics
   (*c*) low power consumption
   (*d*) all the above.

4. In a compensated shunt type peak reading VTVM a bucking battery with a series adjustable resistance is used for :
   (*a*) providing power for the circuit
   (*b*) zero setting
   (*c*) the purpose of amplifying low level signals
   (*d*) all the above.

5. In a differential amplifier using two FETs, a resistance has to be used to reduce the meter current to zero even when no voltage is applied to the circuit. This is necessitated on account of :
   (*a*) mismatch between characteristics of the FETs
   (*b*) difference between tolerance values of resistors used in the circuit even though they are marked nominally equal.
   (*c*) variations in the operating voltage of the circuit.
   (*d*) both (*a*) and (*b*).

6. The resistances, $R_D$, connected in the bridge circuit of a difference amplifier type of Electronic Voltmeter are each 10 kΩ. The value of $R_D$ is much smaller than the value a.c. drain resistance, $r_d$, of FETs. The trans-conductance of FETs is 0.005 mho. The value of current through the meter when an input voltage of 1 V is applied to the gate of left FET with the gate of right hand FET grounded, and the meter resistance almost negligible is :
   (*a*) 5 mA (*b*) 2.5 mA
   (*c*) 10 mA
   (*d*) 1 mA.

7. In a source follower type electronic voltmeter, the trans-conductance of FETs is 0.005 mho and the resistance of the meter is 100 Ω. The approximate value of current through the meter, when the input voltage is 1 V, is :
   (*a*) 2.5 mA (*b*) 5 mA
   (*c*) 2 mA
   (*d*) 1 mA.

8. The power consumption of a d.c. voltmeter using a direct coupled amplifier when measuring a voltage 0.5 V is of the order of :
(a) a few watt (b) a few milliwatt
(c) a few microwatt (d) a few nanowatt.

9. A chopper stabilized a.c. amplifier may use :
(a) an electro-mechanical chopper
(b) bipolar transistors as choppers
(c) MOS field effect transistors as choppers
(d) light activated devices as choppers
(e) all the above.

10. Electronic voltmeters which use rectifiers employ negative feedback. This is done
(a) to increase the overall gain
(b) to improve stability
(c) to overcome nonlinearity of diodes
(d) none of the above.

11. A voltage having symmetrical square waveshape is applied to an average reading voltmeter with scale calibrated in terms of rms value of a sinusoidal wave has an error of :
(a) – 3.9% (b) + 3.9%
(c) – 11% (d) + 11%.

12. A true rms reading voltmeter uses two thermocouples in order
(a) to increase sensitivity
(b) that the second thermocouple cancels out the no-linear effects of the first thermocouple
(c) to prevent drift in the d.c. amplifier
(d) all the above.

13. In an electronic ohmmeter, an OP-AMP is used as a :
(a) Summer (b) Multiplier
(c) Buffer amplifier (d) Integrator.

14. A voltmeter with a broad band width has :
(a) low noise level and high sensitivity
(b) high noise level and high sensitivity
(c) high noise level and low sensitivity
(d) low noise level and low sensitivity.

15. In circuits of RF voltmeters
(a) vacuum tube diodes are used
(b) conventional PN junction doides are used
(c) Schottky barrier or point contact type diodes are used
(d) all the above.

## Answers

1. (b) 2. (a) 3. (d) 4. (b) 5. (d) 6. (b) 7. (c) 8. (d) 9. (e) 10. (c)
11. (d) 12. (b) 13. (c) 14. (c) 15. (c)

# 21
# Cathode Ray Oscilloscope (CRO)

## 21.1 INTRODUCTION

The cathode ray oscilloscope (CRO) is a very useful and versatile laboratory instrument used for display, measurement and analysis of waveforms and other phenomena in electrical and electronic circuits. CROs are in fact very fast *X-Y* plotters, displaying an input signal versus another signal or versus time. The "stylus" of this "plotter" is a luminous spot which moves over the display area in response to an input voltage. The luminous spot is produced by a beam of electrons striking a fluorescent screen. The extremely low inertia effects associated with a beam of electrons enables such a beam to be used for following the changes in instantaneous values of rapidly varying voltages. The extremely low inertia of electrons as compared to metallic conductors used in an electro-mechanical Duddell's oscilloscope enables a CRO to be used at frequencies much above the highest on which the Duddell's oscilloscope can be used.

The normal form of a CRO uses a horizontal input voltage which is an internally generated ramp voltage called **"Time Base"**. This horizontal voltage moves the luminous spot periodically in a horizontal direction from left to right over the display area or screen. The vertical input to the CRO is the voltage under investigation. The vertical input voltage moves the luminous spot up and down in accordance with the instantaneous value of the voltage. The luminous spot thus traces the waveform of the input voltage with respect to time. When the input voltage repeats itself at a fast rate, the trace (display) on the screen appears stationary on the screen. The CRO thus provides a means of visualizing time varying voltages. As such, the CRO has become a universal tool in all kinds of electrical and electronic investigations.

Cathode Ray Oscilloscopes operate on voltages. However, it is possible to convert current, strain, acceleration, pressure and other physical quantities into voltages with the help of transducers and thus to present visual representations of a wide variety of dynamic phenomena on CROs.

Cathode Ray Oscilloscopes are also used to investigate waveforms, transient phenomena, and other time varying quantities from a very low frequency range to the radio frequencies.

Oscilloscopes have been evolved continuously, and they are now available which can measure frequencies upto 1 GHz, and observer events as small as 20 Hz in duration.

Many additional features are available with some oscilloscopes and these include built in digital multi-meters and counters. The oscilloscopes are progressively getting smarter, and many are microprocessor controlled. They have the ability to calculate several features, such as rise time or pulse width of the measured waveform, and to display these values along with the display of waveforms. They are easier to use and internal routines often act as a guide for the user, and display a warning if there is any error in setting.

Many oscilloscopes are now available with IEEE 488 bus capabilities, so that they can be used as a part of measurement test bed, with the instrument's controls set at a remote location, and the readings digitised and retrieved for the purposes of recording and analysis.

Although, most oscilloscopes are monochromatic, colour oscilloscopes are finding increasing applications in computers and in television.

The applications of oscilloscopes have been enhanced on account of many recent developments. Most present day oscilloscopes are capable of accepting two or more inputs displaying them simultaneously. This may be achieved through using a split beam or by using a multiple beam tube. Sampling oscilloscopes are used for high speed applications.

These oscilloscopes employ time sampling and through their use it is possible to measure signals of about 20 GHz. They can only detect a repetitive waveform, and work on the principle of taking a sample once every cycle, over several cycles, each sample point being shifted from the previous point. The complete picture of the waveform is stored, and can be displayed as a stationary signal.

Storage oscilloscopes can be used for capturing transient signals, and then display them for periods which may vary from a few minutes to several years. An analog oscilloscope uses a modified form of a conventional cathode ray tube to store the trace. The digital storage oscilloscope first converts the analog signal to a digital form and stores it in digital memory. The signal can then be recalled for display as and when required.

## 21.2 CATHODE RAY TUBE (CRT)

A cathode ray oscilloscope consists of a cathode ray tube (CRT), which is the heart of the tube, and some additional-circuitry to operate the CRT. The main parts of a CRT are :

(*i*) Electron gun assembly,
(*ii*) Deflection plate assembly,
(*iii*) Fluorescent screen,
(*iv*) Glass envelope,
(*v*) Base, through which connections are made to various parts.

The main parts of a CRT are shown in Fig. 21.1. Before going into details of working of various parts of a CRT, a summary of functions of the different parts is given below :

The "**Electron gun assembly**" produces a sharply focused beam of electrons which are accelerated to high velocity. This focused beam of electrons strikes the fluorescent screen with sufficient energy to cause a luminous spot on the screen.

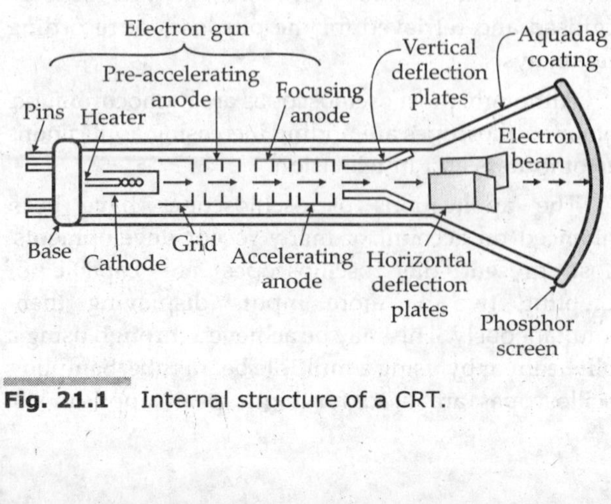


**Fig. 21.1** Internal structure of a CRT.

After leaving the electron gun, the electron beam passes through two pairs of "**Electrostatic deflection plates**". Voltages applied to these plates deflect the beam. Voltages applied to one pair of plates move the beam vertically up and down and the voltages applied to the other pair of plates move the beam horizontally form one side to another. These two movements *i.e.*, horizontal and vertical are independent of each other and thus the beam may be positioned anywhere on the screen.

The working parts of a CRT are enclosed in an evacuated glass envelope so that the emitted electrons are able to move about freely from one end of the tube to the other.

## 21.3 ELECTRON GUN

The source of focused and accelerated electron beam is the electron gun. The electron gun, which emits electrons and forms them into a beam consists of a heater, a cathode, a grid, a pre-accelerating anode, a focusing anode and an accelerating anode.

In smaller CRTs, connections to the various electrodes are brought out through pins in the base or the tube as shown in Fig. 21.1. Large and medium sized high performance tubes operate at very high voltages, and these leads are usually brought out through the sides of the glass envelope.

Electrons are emitted from the indirectly heated cathode. A layer of barium and strontium oxide is deposited on the end of the cathode-which is a cylinder – to obtain high emission of electrons, at moderate temperatures. The typical values of current and voltage required by an indirectedly heated cathode are 600 mA at 6.3 V. High efficiency systems use 300 mA at 6.3 V. The special low power designs use 140 mA at 1.5 V. These electrons pass through a small hole in the "**control grid**". This control grid is usually a nickel cylinder, with a centrally located hole, co-axial with the CRT axis. This is usually a metal cup of low permeability steel, about 15 mm in diameter and 15 mm long. An aperture of about 0.25 mm is drilled in the cap of the grid for the electrons to flow through. The intensity of electron beam depends upon the number of electrons emitted from the cathode. The grid with its negative bias controls the number of electrons emitted from the cathode and hence the **intensity** is controlled by the grid.

The electrons, emitted from the cathode and passing through the hole in the control grid are accelerated by the high positive potential which is applied to the "**pre-accelerating**" and "**accelerating anodes**".

The electron beam is focused by the "**focusing anode**". The accelerating and focusing anodes are cylindrical in form, with small openings located in the centre of each electrode, coaxial with the tube axis. After leaving the focusing anodes, the electron beam passes through the vertical and horizontal deflection plates and then goes on to the fluorescent screen.

The pre-accelerating anode and the accelerating anode are connected to a common positive high voltage of about 1500 V. The focusing anode is connected to a lower adjustable voltage of 500 V.

There are *two* methods of focusing an electron beam :

(*i*) Electrostatic focusing and

(*ii*) Electromagnetic focusing.

The CRO uses electrostatic method of focusing as compared to a TV picture tube which employees electromagnetic focusing.

## 21.4 ELECTROSTATIC FOCUSING

Figure 21.2 shows an electron at rest placed in an electric field produced two parallel plates. Force on the electron is,

$$F = -e\varepsilon \text{ newton} \qquad ...(21.1)$$

where $\varepsilon$ = electric field intensity ; V/m,

and $e$ = charge of electron = $1.602 \times 10^{-19}$ C.

The minus sign indicates that the force acts in the opposite direction to that of the field. The above discussion is valid only if the electron is situated in a field of uniform intensity. In practice, however, the field is not uniform. The lateral repulsion of the electric field lines causes a spreading of space between the lines, resulting in curved field lines at the ends. Thus the field intensity will be less at the ends. Figure 21.2 also shows equipotential surfaces, indicated by solid lines. Since the force is in a direction opposite the field and the equipotential surfaces are perpendicular to the field, the force on an electron is in a direction normal to the equipotential surfaces.

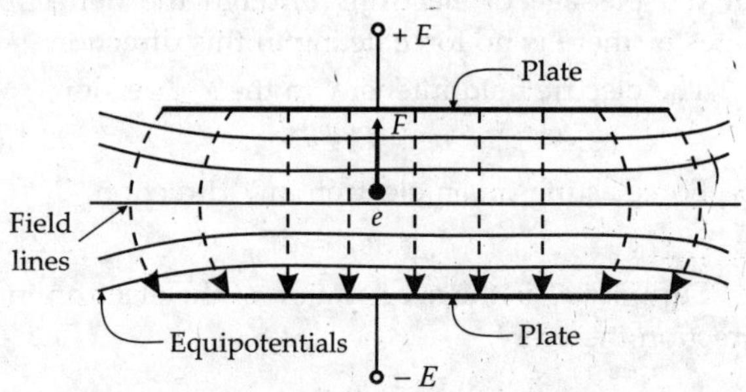


**Fig. 21.2** Electric field between parallel plates.

Figure 21.3 shows two concentric cylinders with a potential applied between them. Lateral repulsion again causes the spreading of the flux lines producing a field as shown. The equipotential surfaces are shown as solid lines. It is clear from the diagram that the

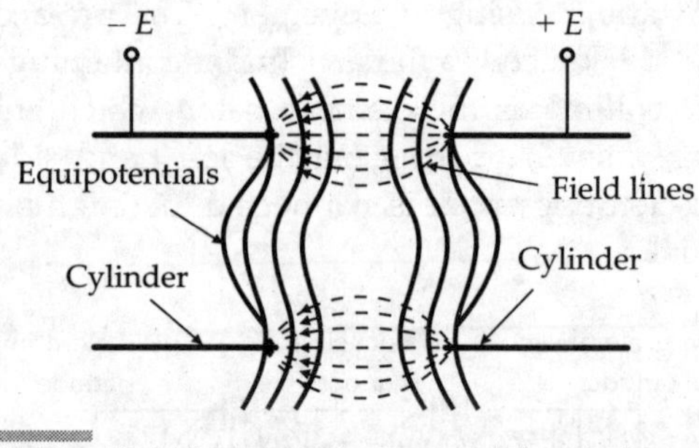


**Fig. 21.3** Field between two co-axial cylinders.

equipotential surfaces are curved. Let us consider the region on the two sides of an equipotential surface $S$ as shown in Fig. 21.4. The potential on the left side of the surface is $-V$ and on the right side is $+V$. Let an electron moving in a direction AB enter the area to the left of $S$. This electron experiences a force which is normal to the surface S and is thus accelerated.

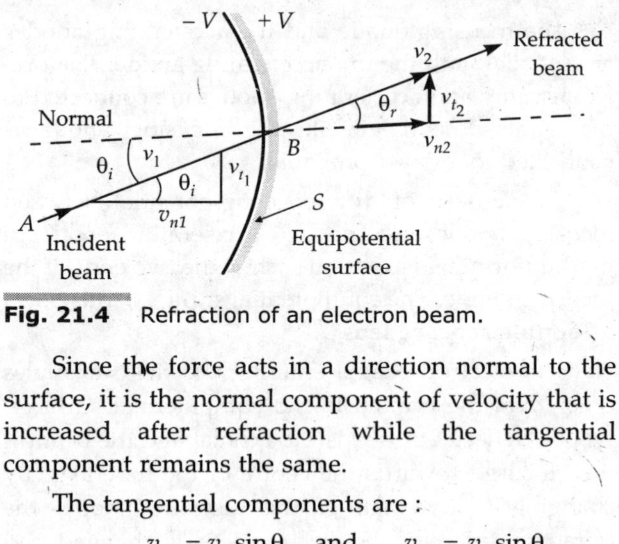


**Fig. 21.4** Refraction of an electron beam.

Since the force acts in a direction normal to the surface, it is the normal component of velocity that is increased after refraction while the tangential component remains the same.

The tangential components are :

$$v_{t1} = v_1 \sin\theta_i \quad \text{and} \quad v_{t2} = v_2 \sin\theta$$

Now $\quad v_{t1} = v_{t2} \quad$ or $v_1 \sin\theta_i = v_2 \sin\theta_r$

$$\therefore \quad \frac{\sin\theta_i}{\sin\theta_r} = \frac{v_2}{v_1} \qquad ...(21.2)$$

where

$v_1$ = initial velocity of electrons,

$v_2$ = velocity of electrons after leaving surface $S$,

$\theta_i$ = angle of incidence, and

$\theta_r$ = angle of refraction.

Equation 21.2 is identical to the expression relating the refraction of a light beam in geometrical optics.

The refraction of an electron beam follows as the bending of a light beam at a refracting surface such as an optical lens. For this reason the focusing system in a CRT is known as **electron lens.**

Figure 21.5 shows the functional diagram of an electrostatic focusing arrangement. The pre-accelerating anode, which is a metal cylinder containing many baffles, collimates the electron beam which enters it through a small opening on the left hand side. The pre-accelerating anode is connected to a high positive potential.

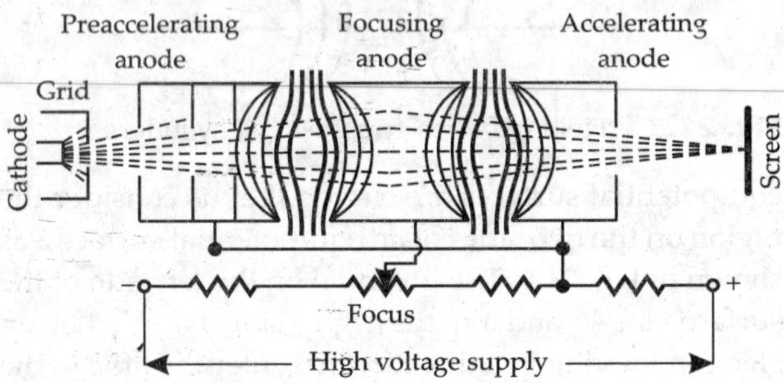


**Fig. 21.5** Electrostatic focusing arrangement.

The focusing anode and the accelerating anodes are co-axial with the pre-accelerating anode. The pre-accelerating and accelerating anodes are connected to the same potential while the focusing anode is connected to a lower potential.

On account of difference of potential between focusing anode and the two accelerating anodes a non-uniform field exists on each of the two ends of the focusing anode. The equipotential surfaces, thus, form a **"double concave lens"**.

The electron beams entering the field at angles other than the normal to the equipotential surfaces, will be deflected towards the normal and the beam is thus focused towards the centre of the tube axis. By changing the voltage of the focusing anode, the refractive index of the electron lens is changed and therefore the focal point of the beam can be changed. The change in voltage is brought about by changing the setting of a potentiometer. This control is brought to the front panel of CRO and is marked **Focus.**

## 21.5 ELECTROSTATIC DEFLECTION

Figure 21.6 shows a general arrangement for electrostatic deflection. There are two parallel plates with a potential applied between. These plates produce a uniform electrostatic field in the Y direction. Thus any electron entering the field will experience a force in the Y direction and will be accelerated in that direction. There is no force either in X direction or Z direction and hence there will be no acceleration of electrons in these directions.

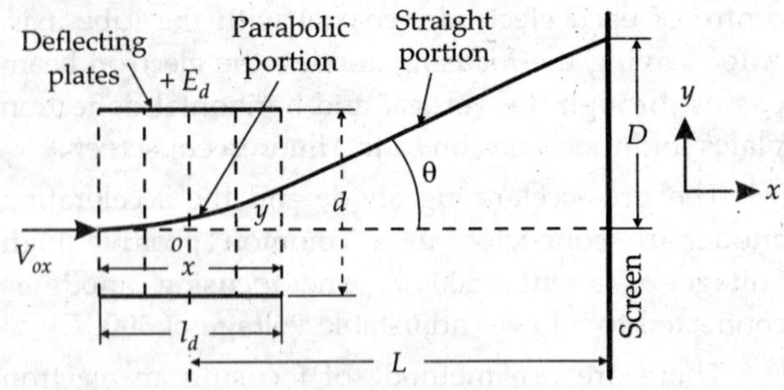


**Fig. 21.6** Electrostatic deflection.

Let

$E_0$ = voltage of pre-accelerating anode ; V,
$e$ = charge of an electron ; C,
$m$ = mass of electron ; kg,
$v_{ox}$ = velocity of electron when entering the field of deflecting plates ; m/s,
$E_d$ = potential between deflecting plates ; V,
$d$ = distance between deflecting plates ; m,
$l_d$ = length of deflecting plates ; m,
$L$ = distance between screen and the centre of the deflecting plates ; m, and
$D$ = deflection of electron beam on the screen in Y direction ; m.

The loss of potential energy (P.E.) when the electron moves from cathode to accelerating anode ;

$$\text{P.E.} = e\,E_a \qquad ...(21.3)$$

The gain in kinetic energy (K.E.) by an electron

$$\text{K.E.} = (1/2)mv_{ox}^2 \qquad ...(21.4)$$

(The mass of an electron is $m = 9.109 \times 10^{-31}$ kg)

Equating the two energies, we have :

$$v_{ox} = (2\,e\,E_a / m)^{1/2} \qquad ...(21.5)$$

This is the velocity of the electron in the X direction when it enters the deflecting plates. The velocity in the X direction remains the same throughout the passage of electrons through the deflecting plates as there is no force acting in this direction.

The electric field intensity in the Y direction,

$$\varepsilon_y = E_d / d \qquad ...(21.6)$$

Force acting on an electron in Y direction

$$F_y = e\varepsilon_y = e\,E_d / d \qquad ...(21.7)$$

Suppose $a_y$ is the acceleration of the electron in Y direction therefore,

$$F_y = ma_y$$

or

$$a_y = e\varepsilon_y / m \qquad ...(21.8)$$

As there is no initial velocity in the $Y$ direction the displacement $y$ at any instant $t$ in the $Y$ direction is :

$$y = \frac{1}{2} a_y t^2 = \frac{1}{2} \frac{e\varepsilon_y}{m} t^2 \quad \text{...(21.9)}$$

As the velocity in $X$ direction is constant, the displacement in $X$ direction is given by :

$$x = v_{ox}\, t \quad \text{...(21.10)}$$

$$\therefore \quad t = x / v_{ox} \quad \text{...(21.11)}$$

Substituting the above value of $t$ in Eqn. 21.9, we have :

$$y = \frac{1}{2} \frac{e\varepsilon_y}{m v_{ox}^2} x^2 \quad \text{...(21.12)}$$

This is the equation of a parabola.

The slope at any point $(x,y)$ is

$$\frac{dy}{dx} = \frac{e\varepsilon_y}{m v_{ox}^2} \quad \text{...(21.13)}$$

Putting $x = l_d$ in Eqn. 21.13, we get the value of $\tan\theta$.

$$\tan\theta = \frac{e\varepsilon_y}{m v_{ox}^2} l_d = \frac{e E_d l_d}{m d v_{ox}^2} \quad \text{...(21.14)}$$

After leaving the deflection plates, the electrons travel in a straight line. The straight line of travel of electrons is tangent to the parabola at $x = l_d$ and this tangent intersects the X-axis at point $O'$. The location of this point is given by :

$$x = \frac{y}{\tan\theta} = \frac{e\varepsilon_y l_d^2}{2 m v_{ox}^2} \Bigg/ \frac{e\varepsilon_y}{m v_{ox}^2} l_d = \frac{l_d}{2}$$

The apparent origin is thus at the centre of deflection plates. The deflection $D$ on the screen is given by :

$$D = L\tan\theta = \frac{L e E_d l_d}{m d v_{ox}^2} \quad \text{...(21.15)}$$

Substituting the value $v_{ox}^2 = \frac{2 e E_a}{m}$ in Eqn. 21.15, we get

$$D = \frac{L e E_d l_d}{m d} \cdot \frac{m}{2 e E_a} = \frac{L l_d E_d}{2 d E_a} \quad \text{...(21.16)}$$

From Eqn. 21.16, we conclude :

(*i*) For a given accelerating voltage $E_a$, and for particular dimensions of CRT, the deflection of the electron beam is directly proportional to the deflecting voltage. This means that the CRT may be used as a linear indicating device.

(*ii*) The discussion above assumes that $E_d$ is a fixed d.c. voltage. The deflection voltage is usually a time varying quantity and the image on the screen thus follows the variations of the deflection voltage in a linear manner.

(*iii*) The deflection is independent of the *e/m* ratio. In a cathode ray tube, in addition to the electrons many types of negative ions such as oxygen, carbon, chlorine etc. are present. With electrostatic deflection system, because deflection is independent of *e/m*, the ions travel with the electrons and are not concentrated at one point. Hence cathode ray tube with electrostatic deflection system does not produce an ion burn.

***Deflection sensitivity.*** The deflection sensitivity of a CRT is defined as the deflection of the screen per unit deflection voltage.

$\therefore$ Deflection sensitivity

$$S = \frac{D}{E_d} = \frac{L l_d}{2 d E_a} \text{ m/V} \quad \text{...(21.17)}$$

The **Deflection Factor** of a CRT is defined as the reciprocal of sensitivity.

$\therefore$ Deflection factor

$$G = \frac{1}{S} = \frac{2 d E_a}{L l_d} \text{ V/m} \quad \text{...(21.18)}$$

It is clear from Eqn. 21.17, that the sensitivity can be increased by decreasing the value of accelerating voltage $E_a$. But this has a disadvantage as the luminosity of the spot is decreased with decrease in $E_a$. On the other hand a high value of $E_a$, produces a highly accelerated beam and thus produces a bright spot.

However, a high accelerating voltage ($E_a$) requires a high deflection potential ($E_d$) for a given deflection. Also highly accelerated beam is more difficult to deflect and is sometimes called **hard beam.** Typical values of sensitivities are 0.1 mm/V to 1.0 mm/V corresponding to deflection factors of 10 V/mm and 1 V/mm.

## 21.6 POST DEFLECTION ACCELERATION OF ELECTRON BEAM

After electrons pass beyond the deflection plates, they may or may not experience additional acceleration. This depends primarily upon the maximum frequencies to be applied to CRT. In an electrostatic system of deflection, Eqn. 21.17 indicates that a low value of accelerating voltage, $E_a$, should be used. This voltage is usually kept below 4 kV. Although this voltage gives good sensitivity but

21

reduces brightness, which can be seriously impaired at high frequencies. In general, mono-accelerator tubes such as shown in Fig. 21.1 may be used, if the maximum frequency to be displayed is below 10 MHz. However, if signals of frequencies higher than 10 MHz are to displayed, *post deflection acceleration* is necessary to increase the brightness of the trace which otherwise would be dim.

When measurements are to be made above a frequency of 10 MHz, *post-accelerators* or *post deflection acceleration tubes* (PDA) are used. These have a large beam accelerating bias after the deflection plates so as to increase the beam energy and give a bright display on the screen.

Several types of PDA structures can be used. Figure 21.7(*a*) shows a spiral accelerator in which a high resistance narrow spiral of graphite is painted over a considerable length of the inside of the envelope funnel. The spiral is connected to the aluminium film on the phosphor, if present. A voltage of about 10 kV is applied to this spiral, and the effect is to accelerate the beam, but also to bend it slightly towards the X-axis. This decreases the deflection sensitivity, which can be compensated for by increasing the length of the tube. Alternatively spherical mesh may be inserted into the helix tube, as in Fig. 21.7(*b*). This shapes the accelera- ting field and prevents it from affecting the original beam deflection, so giving the same sensitivity as a mono-accelerator tube of equal length.

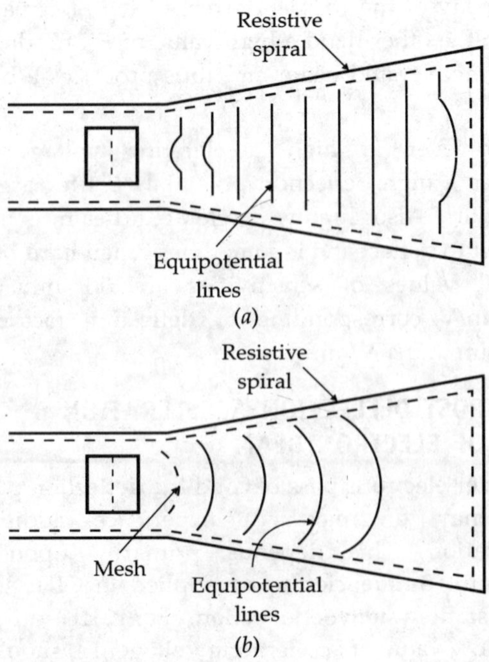


**Fig. 21.7** CRT post accelerators : (*a*) spiral (*b*) mesh.

Both the systems illustrated in Fig. 21.7 are limited in scan angle to between 35° and 40°. An alternative PDA system, called a high expansion mesh tube, uses a spherical shaped mesh without a helix tube. The mesh is shaped so as to expand the deflection of the beam, and with this system beam deflections up to 90° be obtained. However the spot size is also increased in proportion to the increase in sensitivity.

## 21.7 EFFECT OF BEAM TRANSIT TIME AND FREQUENCY LIMITATIONS

In the derivation of Eqn. 21.16, the plate voltage $E_d$ is assumed constant during the motion of the electrons through the deflecting field. If the voltage applied to the vertical deflecting plates changes during the transit time of the electrons through the horizontal plates, the deflection sensitivity gets decreased.

$$\text{Transit time } t_1 = \frac{l}{v_{ox}} \quad \text{...(21.19)}$$

The transit time imposes a limitation of the upper frequency limit. An upper limiting frequency is defined as that frequency at which the transit time is equal to one quarter of the period of the voltage applied to vertical plates.

∵ Upper limiting frequency,

$$f_c = \frac{1}{4t_1} = \frac{v_{ox}}{4l} \quad \text{...(21.20)}$$

The frequency range of the oscilloscopes can be increased by subdividing the deflecting plates in a number of sections in the path of the electron beam. The voltage being measured is applied to the vertical plates through an iterative network, whose propagation time corresponds to the velocity of electrons, thereby the voltage applied to the vertical plates is made to synchronize with the velocity of the beam. The use of this technique allows the cathode ray oscilloscope to be used upto frequencies of 500 MHz and above.

## 21.8 DEFLECTION PLATES

Referring to Fig. 20.1, we observe that the electron beam, after leaving the electron gun, passes through two pairs of deflection plates. One pair of plates is mounted horizontally and produces an electric field in the vertical plane. This pair produces a vertical deflection and is thus called **Vertical Deflection Plates** or **Y Plates**. The other pair of plates is mounted vertically and produces a horizontal deflection. This pair of plates is called **Horizontal Deflection Plates** or **X Plates**. The plates are flared so as to allow the beam to pass through them without striking the plates.

## 21.9 SCREEN FOR CRTs

The front of the CRT is called the face plate. It is flat for screen sizes up to about 100 mm × 100 mm, and is slightly curved for larger displays. The face plate is formed by pressing molten glass in a mould and then annealing it. Some CRTs have a face plate which is made entirely from fibre optics, which has special characteristics. It is also feasible to have a small fibre optic region which gives direct contact exposure of photographic film, if required, without needing a lens system.

The inside surface of the face plate is coated with phosphor. This consists of very pure inorganic crystalline phosphor crystals, about 2-3 microns in diameter, to which traces of other elements, called activators, have been added. *Activators* in current use are metals such as silver, manganese, copper and chromium, and they affect the characteristics of the phosphor, such as its luminous efficiency, spectral emission and persistence.

Many types of phosphors are used for CRTs and the characteristics of a few of them are given in Table 20.1. A phosphor converts electrical energy to light energy. When an electron beam strikes phosphor crystals it raises. their energy level. This is known as *cathodoluminescence*. Light is emitted during phosphor excitation and this is called fluorescence. When the electron beam is switched off the phosphor crystals return to their initial state, and release a quantum of light energy. This is called *phosphorescence* or *persistence*.

The writing speed of a phosphor is measured by its fluorescence rise time, which is the time from the beginning of excitation to reach 90% of the maximum emission state, and by the decay time, which is the time to fall from the maximum state to the 10% level. Writing speed is determined by the phosphor type, crystal size, impurity content, and the manufacturing process. A phosphor must be refreshed by electrons, before the end of its decay time, in order to give a flicker free display. Short persistence phosphors require more frequent refreshes, whereas long persistence phosphors may result in characters fading slowly, giving a trace after glow or trails. Long persistence phosphors are also more prone to permanent discolourations, and loss of luminous efficiency, at high beam current or fixed data use. Most modern oscilloscopes use short persistence phosphors (*e.g.*, P1, P2, P11, P31) because of their fast refresh capability, whilst the slower speed signals seen in medical applications need longer persistence phosphors (P7 and P39). For use in radars, very long persistence is needed (P19, P26, P33). Long persistance is even needed in storage type oscilloscopes.

The electron beam on striking the phosphor gives off light and heat. Excessive beam current on one spot of the screen, over a long time, can degrade the light output from the phosphor, and can burn a hole in it. Phosphors can be classified as having low burn resistance (P19, P26, P33), medium burn resistance (P1, P2, P4, P7, P11) and high burn resistance (P15, P31).

The luminance of the phosphor is a measure of its brightness. It is determined by the luminance efficiency of the phosphor, and by the beam energy, which is a product of the beam current density, the accelerating potential and the writing time. Therefore one can have high light output, at low beam accelerating potential provided the sweep speed is low, but for fast single-shot transient recording, a high beam current and accelerating voltage, and phosphor with high luminance, are required.

Each phosphor has a unique colour, the human eye response peaks at 555 nm, in the yellow-green region, so phosphors which gives light in these regions (P2, P31, P39) would appear to be brighter than others such as P11, P16, P19. A camera film, however, is sensitive to ultraviolet or blue, so for high speed photography phosphors P11 or Pl6 may be better. The spectral emission from a phosphor is broadband, *i.e.*, a range of wavelengths go to determine its colour. The gradation of colour (hue) and the chromatic purity with respect to white (saturation) can be precisely specified by $X$ and $Y$ co-ordinates on a chromaticity diagram.

A thin film of metal, such as aluminium (0.1 – 0.15 μm) is usually deposited on the non-viewing side the phosphor.

This has *three* effects :

(*i*) The metal layer acts as a heatsink and reduces the danger of phosphor burn.

(*ii*) The light scatter from the phosphor is reduced, and the aluminium reflects light going back into the tube towards the viewer, so increasing the brightness.

(*iii*) Electrons which strike the phosphor release secondary electrons. These secondary electrons are usually collected by the aquadag. Under certain conditions the electrons may remain on the screen, and cause a negative voltage build up. This would decrease the accelerating voltage of the beam. This charge build up is prevented by the aluminium layer, which is connected electrically to the aquadag layer.

**Table 21.1** *Phosphor Characteristics*

| Type | Colour | | Persistence (to 10% level) | Luminance* | Writing speed* | Intended use |
|---|---|---|---|---|---|---|
| | Fluorescent | Phosphorescence | | | | |
| P1 | yellow-green | yellow-green | 1 – 100 ms | 4 | 7 | oscilloscope, radar |
| P2 | yellow-green | yellow-green | 1 – 100 ms | 2 | 4 | oscilloscopes |
| P4 | white | white | 10 μs – 1 ms | 3 | 3 | monochrome television radar, medical |
| P7 | blue | yellow | 10 μs – 1 ms(b)<br>100 ms – 1 s(y) | 4 | 2 | radar, medical |
| P11 | blue | blue | 10 μs – 1 ms | 5 | 1 | photographic |
| P15 | ultra-violet | green | < 1 μs (uv)<br>1 – 10 μs (g) | 9 | 8 | TV (flying spot scanner), |
| P16 | ultra-violet | ultra-violet | < 1 μs | 10 | 8 | TV (flying spot scanner), photographic |
| P18 | white | white | 10 μs – 100 ms | – | – | projection TV |
| P19 | orange | orange | > 1 s | 5 | 10 | radar |
| P22 | white (red, blue, green) | white (red, blue, green) | 10 μs – 1 ms | – | – | tricolour TV |
| P26 | orange | orange | > 1 s | 8 | 10 | radar |
| P28 | yellow-green | yellow-green | 100 ms – 1 s | 3 | 5 | radar, medical |
| P31 | green | green | 10 μs – 1 ms | 1 | 3 | oscilloscopes, bright TV |
| P33 | orange | orange | > 1 s | 6 | 9 | radar |
| P39 | yellow-green | yellow-green | 100 ms – 1 s | 3 | 6 | radar, computer graphics |
| P40 | blue | yellow-green | 10 μs 1 ms(b)<br>100 ms – 1s(yg) | – | – | low repetition rate P16 |
| P44 | yellow-green | yellow-green | 1 – 100 ms | 3 | – | bistable storage |
| P45 | white | white | 1 – 100 ms | 4 | – | monochrome TV display |

* Highest luminance or fastest writing speed.

The electron beam needs more energy to penetrate the aluminium layer in front of the phosphor. Therefore at low acceleration potentials the aluminised phosphor screen has lower efficiency than a standard screen, but the efficiency increases rapidly at higher potentials.

The ideal phosphor for a laboratory oscilloscope CRT would be one which gives off a colour to which the eye has maximum response ; has a high burn resistance, to prevent accidental damage ; has a high luminance and writing speed; and has short persistence to avoid multiple images when recording fast displays. A typical choice of phosphor is P31.

## 21.10 GRATICULE

The graticule is a grid of lines that serves as a scale when making time and amplitude measurements.

There are *three* types of graticules.

(*i*) *External graticule.* This is scribed on a Plexiglas acrylic plastic and fitted to screen. This graticule can be easily changed, to make different types of measurements and its position can be adjusted to align it with the trace on CRT. External graticules suffer from parallax errors as they are not in the plane of the phosphor.

(*ii*) *Internal graticule.* This is deposited on the internal surface of the CRT face plate, and is therefore on the same surface as the phosphor. Although there is now no parallax, an internal graticule cannot be changed, and also needs some method of electrical trace alignment. The graticule is also difficult to illuminate for photography, unless special illumination is provided in the oscilloscope.

(*iii*) *Porjected graticule.* It is available with some cameras and provides flexibility which can include additional features such as legends on the graticule.

The face plate is sometimes tinted neutral-grey to reduce ambient light interference, or external filters can be used to reduce glare. Black wire mesh filters increase the contrast ratio and restrict the viewing angle, so reflections from oblique light sources are minimised. Polarised filters can also be used in high glare applications. Light filters are sometimes utilised to affect selective colour components of the phosphor. For example, an amber filter on a P7 phosphor would eliminate the short persistence blue component, and increase the longer persistence yellow component ; and a blue filter would do the reverse.

The graticule may be etched or silk-screened on a plastic CRT face plate or it may be chemically deposited on the face of CRT along with phosphor. If the graticule is put on the plastic face plate, it should be positioned on the inside surface, that is, the surface that will be in contact with the face of CRT. This places the display produced by the electron beam and the graticule on the same plane, thereby eliminating the errors on account of parallax. Parallax errors occur when the trace and the graticule are in different planes and the observer's eye is shifted from the direct line of sight as shown in Fig. 21.8.

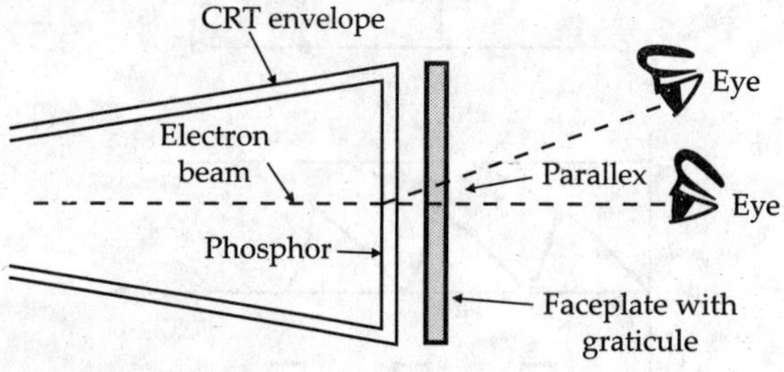


**Fig. 21.8** Illustration of parallax error.

The graticules are generally laid out in an 8 × 10 pattern as shown in Fig. 21.9 with 8 major divisions along the vertical axis and 10 major divisions along the horizontal axis. The tick marks on the centre graticule represent minor divisions or subdivisions.

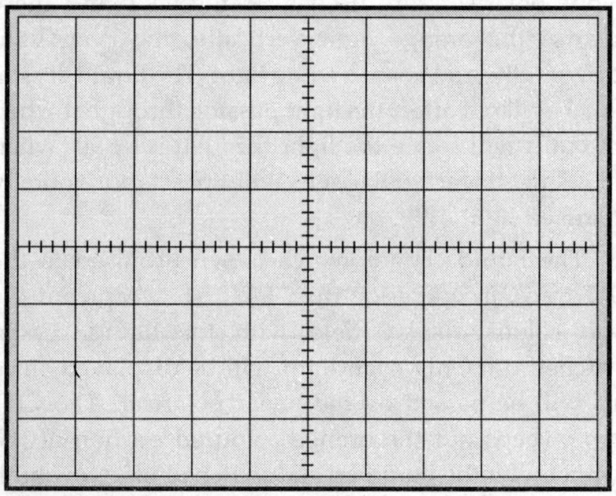

**Fig. 21.9** The CRT graticule.

## 21.11 AQUADAG

The bombarding electrons, striking the screen, release secondary emission electrons. These secondary electrons are collected by an aqueous solution of graphite called '*Aquadag*' which is connected to the second anode, collection of secondary electrons is necessary to keep the CRT screen in a state of electrical equilibrium.

## 21.12 COLOUR CRT DISPLAYS

Although colour CRTs, such as shadow mask tubes, are common in television and computer displays, they are not used for oscilloscopes. One of the reasons is that they are relatively expensive, and colour does not give many advantages when viewing single or multiple traces which are separated from each other. However, limited colour can have advantages, such as in medical oscilloscopes, and an instrument is available for this application. It uses a standard CRT, and a liquid crystal shutter to produce a three colour display, as shown in Fig. 21.10.

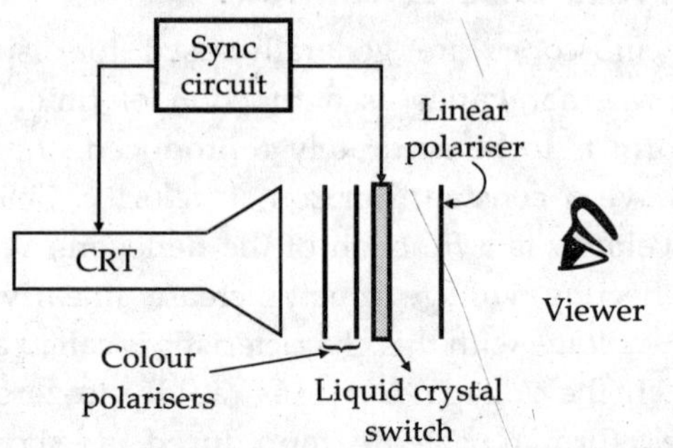


**Fig. 21.10** Exploded view of a liquid crystal switch used with a CRT.

21

The CRT has a multi-component phosphor, which gives peak outputs of oranges and cyan light. Two colour polarisers are used alter the CRT and these polarise the orange light vertically, the cyan light horizontally, and absorb white light. The liquid crystal switch will not affect the light passing through it when on, but it will rotate the light through it by 90° when off. The linear polariser will pass horizontally polarised light only.

The liquid crystal switch is synchronised to the CRT trace. In operation the CRT first sweeps out the cyan colour display field with the liquid crystal switched on. Only cyan light will be displayed since this will be passed by the linear polariser. The CRT then sweeps out the orange coloured portion of the display with the liquid crystal switched off. Now both the cyan and orange lights from the colour polarisers are rotated, so that the orange is horizontally polarised and is passed by the linear polariser. If any part of the display is to be yellow then it is traced by the CRT, when the liquid crystal switch is both on and off, so the cyan and orange colours combine to give yellow.

The liquid crystal shutter oscilloscope has a good contrast ratio since the shutter attenuates the non lighted areas giving a black background. However the brightness is low since the light is emitted 50% of the time for any one colour, due to the duty cycle, and the polarisers also attenuate about half the light going through them. Overall, the brightness is less than 20% of that of a CRT without a liquid crystal shutter. The most serious problem with the liquid crystal shutter system is that it is limited in switching speed. By making the liquid crystal switch narrow, and the rotational distance of the molecules small, the switching time can be reduced to about 0.5 ms. Even then the bandwidth of the oscilloscope is limited to under 100 kHz.

## 21.13 TIME BASE GENERATORS

Oscilloscopes are generally used to display a waveform that varies as a function of time. If the waveform is to be accurately reproduced, the beam must have a constant horizontal velocity. Since the beam velocity is a function of the deflecting voltage, the deflecting voltage must increase linearly with time. A voltage with this characteristic is called a *ramp voltage*. If the voltage decreases rapidly to zero with the waveform repeatedly reproduced, as shown in Fig. 21.11, the pattern is generally called a sawtooth waveform.

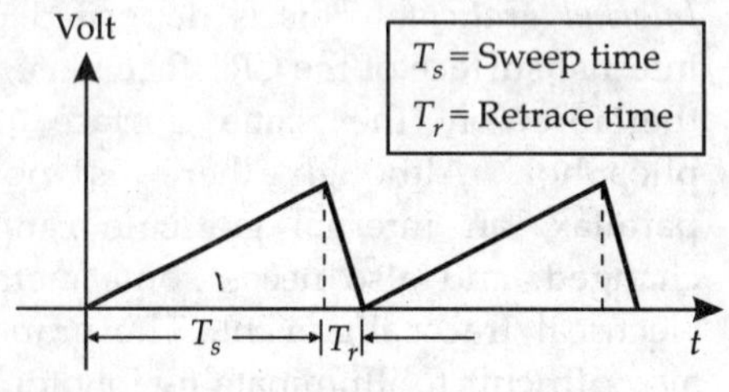


**Fig. 21.11** Typical sawtooth waveform applied to the horizontal deflection plates.

During the **sweep time**, $T_s$, the beam moves from left to right across the CRT screen. The beam is deflected to the right by the increasing amplitude of the ramp voltage and the fact that the positive voltage attracts the negative electrons. During the **retrace time**, or **flyback time**. $T_r$, the beam returns quickly to the left side of the screen. The control gird is generally "gated off", which blanks out the beam during retrace time and prevents an undesirable retrace pattern from appearing on the screen.

Since signals of many different frequencies will be observed with the oscilloscope, the sweep rate must be adjustable. We can change the sweep rate in steps by switching different capacitors into the circuit. The front panel control for this adjustment is marked time/div or sec/div. The sweep rate can be adjusted in minor ways by making the resistor, *R*, in Fig. 21.12(*a*) a variable resistor.

The circuit shown in Fig. 21.12(*a*) is a simple sweep circuit in which the capacitor *C* charges

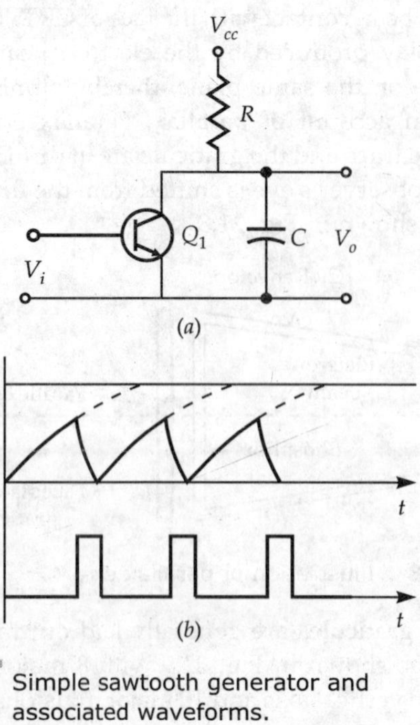


**Fig. 21.12** Simple sawtooth generator and associated waveforms.

through the resistor $R$. The capacitor discharges periodically through the transistor, $Q_1$, which causes the waveform shown in Fig. 21.12(*b*) to appear across the capacitor. The signal, $V_i$, which must be applied to the base of the transistor to turn it "on" for short time intervals is also shown in Fig. 21.12(*b*). When the transistor is turned completely "on," it presents a low-resistance discharge path through which the capacitor discharges quickly.

If the transistor is not turned "on" the capacitor will charge exponentially to the supply voltage $V_{cc}$ according to the equation,

$$V_o = V_{cc}[1 - \exp(-t/RC)] \qquad ...(21.21)$$

where $v_o$ = instantaneous voltage across the capacitor at time $t$ ; V,

$V_{cc}$ = supply voltage ; V,

$t$ = time of interest ; s,

$R$ = value of series resistor ; Ω,

$C$ = value of capacitor ; F.

***Example 21.1*** *A trigger pulse is applied to the sweep generator in Fig. 21.13 every 10 ms. Compute the amplitude of the voltage, $V_o$, across the capacitor when the trigger pulse is applied.*

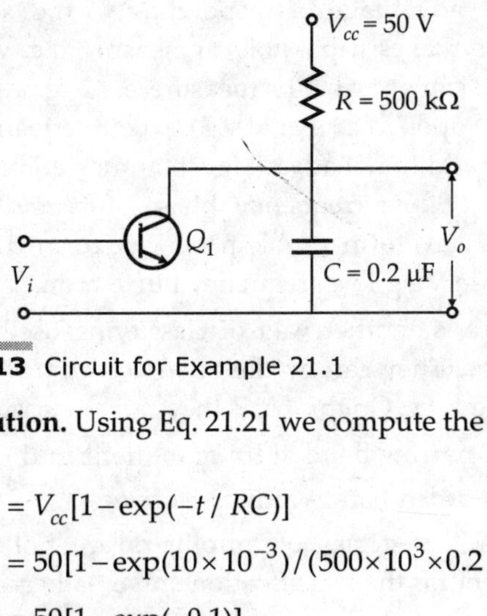


**Fig. 21.13** Circuit for Example 21.1

**Solution.** Using Eq. 21.21 we compute the voltage as

$$V_o = V_{cc}[1 - \exp(-t/RC)]$$
$$= 50[1 - \exp(10\times 10^{-3})/(500\times 10^{3}\times 0.2\times 10^{-6})]$$
$$= 50[1 - \exp(-0.1)]$$
$$= 50(1 - 0.905) = 4.76\ V.$$

Since $V_o$ is less than 10% of $V_{cc}$, the charge curve should still be quite linear at this point.

When a trigger pulse $V_i$ is applied to the transistor $Q_1$ in Fig. 21.12 (*a*), the capacitor C discharges through the resistance presented by the "on" transistor according to the equation

$$V_o' = V_o[\exp(-t_1/R_1C)] \qquad ...(21.22)$$

where

$V_o'$ = voltage across capacitor C at time $t_r$ during retrace ; V,

$V_o$ = amplitude of the voltage across capacitor C at the start of discharge ; V,

C = value of capacitor C ; F,

$t_1$ = time of interest after start of retrace ; s,

$R_1$ = resistance from collector to emitter of the saturated transistor ; Ω.

***Example 21.2*** *An input pulse, $V_i$, of 5 ns duration is applied to the circuit in Fig 21.13 at the instant $V_o$, reaches 4.76 V. What is the voltage across the capacitor after 50 μs if the saturated transistor presents a resistance of 0.2 kΩ to the circuit ?*

**Solution.** Using Eqn. 21.22, we find the voltage as :

$$V_o' = V_o e^{-t_1/R_1C}$$
$$= 4.76\exp[-50\times 10^{-6}/(0.2\times 10^{3}\times 0.2\times 10^{-6})]$$
$$= 1.36\ V$$

The relationship between the current charging a capacitor and the voltage across the capacitor is given as

$$V_o = \frac{I}{C}\int_0^t i_c\,dt \qquad ...(21.23)$$

If the current is constant, the solution of 21.23 is,

$$V_o = It/C \qquad ...(21.24)$$

Substituting $V_{cc}/R$ for $I$, we have,

$$V_o = \frac{V_{cc}}{RC}t \qquad ...(21.25)$$

Equation 21.15 is the equation of a straight line with a slope of $V_{cc}/RC$. The voltage $V_o$ would be a "linear ramp" if it followed this equation. Figure 21.14 compares the graph of the exponential function of Eqn. 21.21 to the graph of the linear function of Eqn. 21.25.

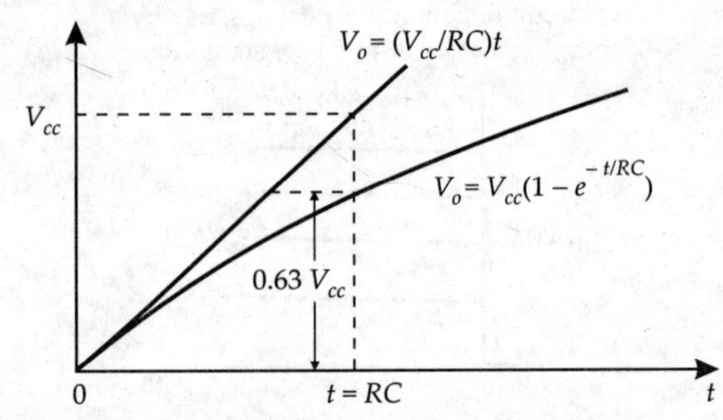


**Fig. 21.14** Graph comparing a capacitor charged by a constant-current source to one charged by a constant-voltage source.

21

As can be seen by comparing the linear and exponential curves, the exponential function is fairly linear during the first 10% or 15% of the graph. Some inexpensive oscilloscopes use this type RC, of circuit as a sweep generator with reasonably satisfactory results. However, linearity can be improved by using one of several other types of circuits as a sweep generator. The circuits most commonly used are :

1. Some form of constant-current generator to charge a capacitor at a constant rate.
2. A bootstrap circuit in which the current is kept constant by maintaining a constant voltage across a charging resistor.
3. An operational amplifier in a circuit known by one of the following names: integrating amplifier, Miller integrator, Miller sweep generator.

The circuit shown in Fig. 21.15(*a*) utilizes an ordinary bipolar junction transistor, $Q_1$, connected in the common-base configuration to charge capacitor *C* at a constant rate. If the resistor, *R* has a large value of resistance, the emitter current will be nearly constant as will the collector current. Since the collector current charges the capacitor *C*, it will charge at a nearly constant rate. Typical common-base output characteristic curves are shown in Fig. 21.15(*b*). Note that the curves are very nearly horizontal in the active region

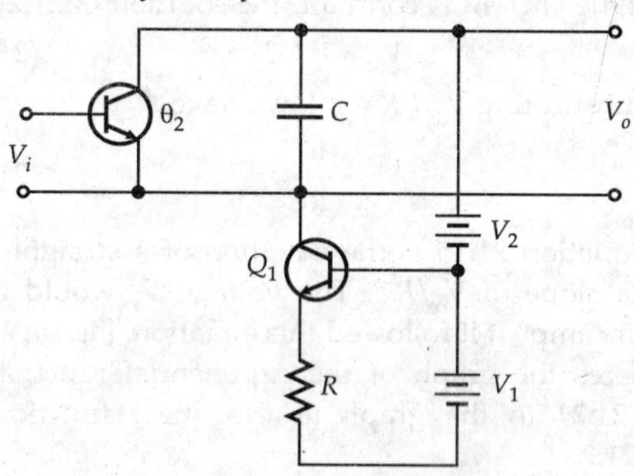


(*a*) Circuit to generate a ramp voltage by charging a capacitor at a constant rate.

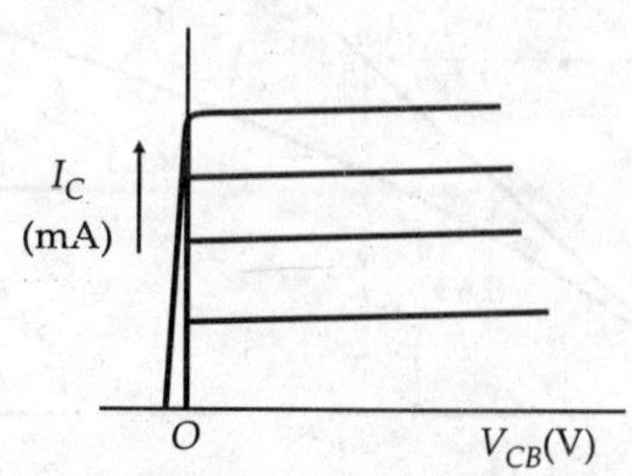


(*b*) The characteristic curves for transistor $Q_1$.

**Fig. 21.15**

of the transistor. The collector-to-base voltage $V_{CB}$ can change considerably without any appreciable effect on the collector current.

In order to have a repetitive sweep voltage, the capacitor must be discharged after it has reached its maximum voltage. This can be done by connecting transistor $Q_2$ across the capacitor as shown. At specific time intervals, a voltage pulse applied to the base of $Q_2$ turns the transistor "on" providing a discharge path for the capacitor.

## 21.14 OSCILLOSCOPE AMPLIFIERS

The purpose of an oscilloscope is to provide a faithful representation applied to its input terminals, considerable attention has to be paid to the designing of these amplifiers for faithful representation of input signals. The oscillographic amplifiers can be classified into *two* major categories

(*i*) a.c. coupled amplifiers, and

(*ii*) d.c. coupled amplifiers.

Low cost oscilloscopes use a.c. coupled amplifiers A.C. amplifiers are commonly used in oscilloscopes used for laboratory purposes.

The d.c. coupled amplifiers are expensive, they offer the advantage of responding to d.c. voltages. Their use makes it possible to measure d.c. voltages, pure a.c. signals, and to measure a.c. signals superimposed upon d.c. signals. D.c. coupled amplifiers have the additional advantage, that they eliminate the problem of low frequency phase shift and accompanying waveform problem of waveform distortion when observing low frequency pulse train.

There is another way of classifying oscilloscopic amplifiers. They can be classified according to bandwidth they use. On this basis they can be classified as :

(*i*) narrow bandwidth amplifiers, and

(*ii*) broad bandwidth amplifiers.

These categories of amplifiers are not sharply divided, nor is the classification universally accepted.

If the frequency band does not extend upto the television colour sub-carrier frequency of 3.58 MHz, the amplifier can be classified as narrow band. However, if the frequency response curve of the amplifier is flat beyond 3.58 MHz, the amplifier is classified as a broad band amplifier. Most general purpose, laboratory quality oscilloscopes respond fully to frequencies in excess of 5 MHz. Therefore, the responsible amplifier, which is the vertical amplifier, is broadband.

### 21.14.1 Vertical Amplifier

The *vertical amplifier* is the principal factor in determining the sensitivity and bandwidth of an oscilloscope. In general, greater sensitivity, which is expressed in terms of V/cm of vertical deflection at the midband frequency, is obtained at the expense of bandwidth. As a rule of thumb, an amplifier at some specific cost may be obtained with a certain gain-bandwidth product, which is the product of the voltage gain of the amplifier and its bandwidth. Voltage gain may be sacrificed in favour of greater bandwidth, or vice versa, without significantly affecting the cost of the amplifier. However, if the gain-bandwidth product increases, the cost of the amplifier will increase.

The gain of the vertical amplifier determines the smallest signal that the oscilloscope can satisfactorily reproduce on the CRT screen. The sensitivity of an oscilloscope is directly proportional to gain of the vertical amplifier that is, as gain increases sensitivity increases, which allows us to observe smaller-amplitude signals.

The vertical sensitivity is a measure of how much the electron beam will be deflected for a specified input signal. The CRT screen is covered with a plastic grid pattern called a graticule. The spacing between the grid lines is generally 10 mm. However, vertical sensitivity is generally expressed in volt per division. On the front panel of the oscilloscope shown in Fig. 21.28, one can see a knob attached to a rotary switch labeled gain. The gain could be designated V/Div. The rotary switch is electrically connected to the input attenuator network, which will be discussed in subsequent paragraphs. The setting of the rotary switch indicates what amplitude signal is required to deflect the beam vertically by one division.

The vertical sensitivity of an oscilloscope is the smallest deflection factor that can be selected with the rotary switch. As an example, if the most sensitive position on the volt/division rotary switch is 5 mV/division, then the vertical sensitivity of the oscilloscope is 5 mV/division.

The *bandwidth* of an oscilloscope determines the range of frequencies that can be accurately reproduced on the CRT screen. The greater the bandwidth, the wider is the range of frequencies that can be observed with the instrument. Ideally, the gain of the broadband amplifier should be constant from direct current to near the upper limit of the range of frequencies that can be observed with the oscilloscope.

A typical curve for a broadband amplifier of gain versus frequency, called a frequency-response curve, is shown in Fig. 21.16.

The bandwidth of an oscilloscope is the range of frequencies over which the gain of the vertical amplifier is within 3 db of the midband frequency gain. The upper limit of the bandwidth is the frequency, $f_2$, at which the gain has decreased by 3 db. The bandwidth can be increased by using feedback in designing the amplifier. However, this reduces the gain, which is why the gain-bandwidth product for an amplifier is constant.

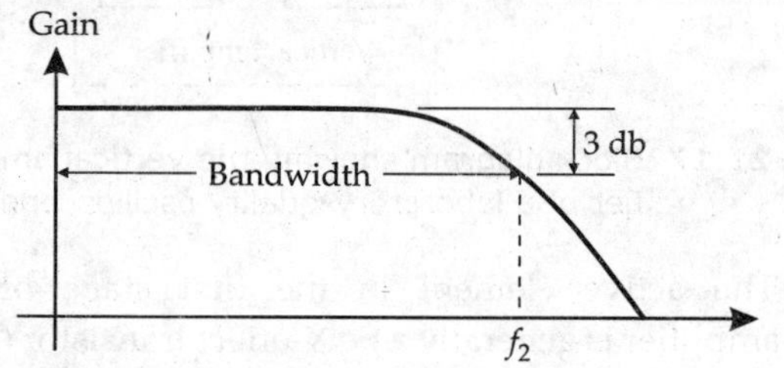


**Fig. 21.16** Typical frequency response curve for a broadband amplifier.

One of the specifications generally included on the specifications sheet for the vertical amplifier of a broadband oscilloscope is the *rise time*. The rise time of a pulse is defined as the time required for the edge to rise from 10% to 90% of its maximum amplitude. When an oscilloscope is used to observe a pulse or a square wave, the rise time of the instrument must be faster than the rise time of the pulse or square wave, otherwise the observed signal will not be accurately reproduced. Although the exact relationship between bandwidth and the rise time varies slightly with amplifier design, an approximate relationship is given as

$$t_r \times BW = 0.35 \qquad ...(21.26)$$

where

$t_r$ = rise time measured in second,

BW = bandwidth in hertz.

**Example 21.3** *If the bandwidth of an oscilloscope is given as direct current to 10 MHz, what is the fastest rise time a sine wave can have to be accurately reproduced by the instrument?*

**Solution.** Using Eq. 21.26, we can compute the maximum rise time as,

$$t_r = \frac{0.35}{10 \times 10^6} = 35 \text{ ns.}$$

The vertical amplifier of a laboratory-quality oscilloscope frequently consists of the two major circuit blocks shown in Fig. 21.17, a pre-amplifier and the main vertical amplifier or mainframe amplifier.

21

The pre-amplifier is sometimes a separate, interchangeable plug-in unit that can quickly and easily be plugged into the mainframe of the oscilloscope. If different plug-in units are available for specific measurement applications, the measurement capabilities of an oscilloscope can be broadened considerably at a reasonable cost.

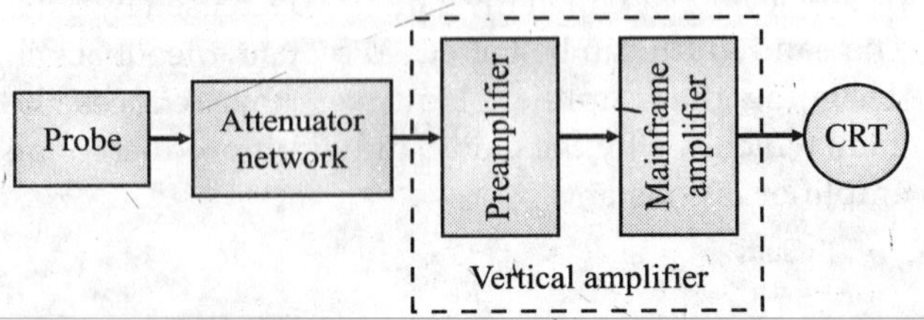


**Fig. 21.17** Block diagram showing the vertical amplifier of a laboratory-quality oscilloscope.

The active element in the first stage of the pre-amplifier is generally a field-effect transistor (FET) which provides a high input impedance for the oscilloscope. The final stage of the mainframe amplifier is generally a push-pull amplifier which provides voltages of equal amplitude but out of phase by 180° to achieve balanced deflection.

### 21.14.2 Horizontal Amplifiers

The *horizontal amplifier* basically serves *two* purposes.

1. When the oscilloscope is being used in the ordinary mode of operation to display a signal applied to the vertical input, the horizontal amplifier will amplify the sweep generator output.
2. When the oscilloscope is being used in the X-Y mode, the signal applied to the horizontal input terminal will be amplified by the horizontal amplifier.

When the oscilloscope is being used in its ordinary mode of operation, the gain and bandwidth requirements for the horizontal amplifier are not as stringent as those for the vertical amplifier. Although the vertical amplifier must be able to reproduce faithfully low-amplitude, high-frequency signals with fast rise times, the horizontal amplifier is required to provide only a faithful reproduction of the sweep signal, which has a relatively high amplitude and a slow rise time.

As with the vertical amplifier, the final stage of the horizontal amplifier is a push-pull amplifier, as shown in Fig. 21.18. The attenuator network given in the figure reduces, by voltage division, the amplitude of the horizontal input signal to a level equal to the sensitivity of the horizontal amplifier.

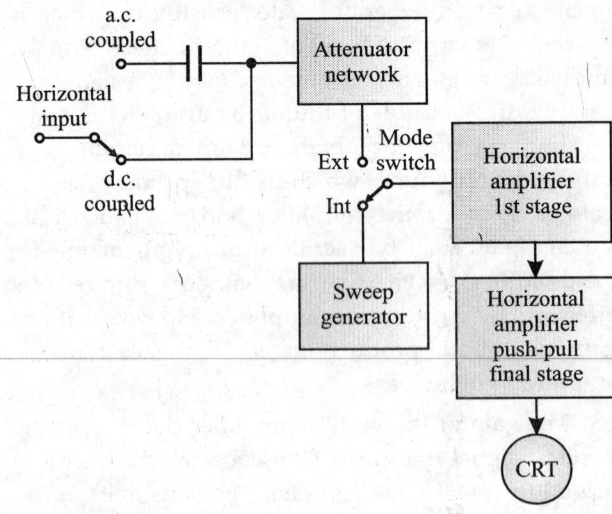


**Fig. 21.18** Block diagram of the horizontal amplifier of an oscilloscope.

## 21.15 VERTICAL INPUT AND SWEEP GENERATOR SIGNAL SYNCHRONIZATION

Most waveforms that we will have occasion to observe with an oscilloscope will be changing at a rate much faster than the eye can follow, perhaps many million times per second. If we are to be able to observe such rapid changes, the beam must retrace the same pattern repeatedly. If the pattern is retraced in such a manner that the pattern always occupies the same location on the screen, the eye will see a stationary display. The beam will retrace the same pattern at a rapid rate if the vertical input signal and the sweep generator signal are synchronized ; which means that the frequency of vertical input signal must be equal to or an exact multiple of, the sweep generator signal frequency, as shown in Fig. 21.19. If

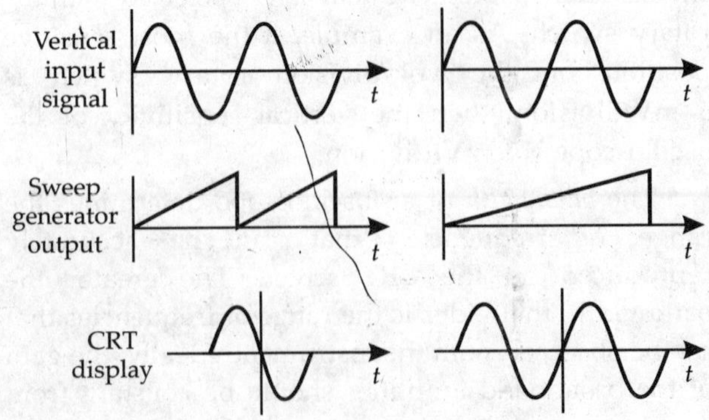


**Fig. 21.19** Synchronized waveforms and CRT display.

the vertical input frequency is not exactly equal to or an exact multiple of, the sawtooth frequency, the waveforms will not be synchronized and the display "walks" across the screen. If the pattern moves toward the right, the frequency of the sawtooth waveform is too high. Movement of the pattern towards the left indicates that the frequency of the sawtooth is too low.

Synchronization of the waveforms can be accomplished in two different ways.

(*i*) ***Free running sweep.*** In very basic oscilloscopes the sweep generator is continuously changing and discharging a capacitor. One ramp voltage is followed immediately by another; hence, the sawtooth pattern appears. A sweep generator operating in this manner is said to be *"free running"*. In order to present a stationary display on the screen, the sweep generator signal must be forced to run in synchronization with the vertical input signal. In basic oscilloscopes this is accomplished by carefully adjusting the sweep frequency to a value very close to the exact frequency of the vertical input signal, or a sub-multiple of this frequency. With both signals at the same frequency, an internal synchronizing pulse will lock the sweep generator into the vertical input signal. This method of synchronization has some serious limitations when an attempt is made to observe low-amplitude signals. However, the most serious limitation is probably the inability of the instrument to maintain synchronization when the amplitude or frequency of the vertical signal is not constant, such as with voice or music signals.

(*ii*) ***Triggered sweep.*** The above limitations are overcome by incorporating a *trigger circuit* into the oscilloscope as shown in Fig. 21.20. The trigger circuit may receive an input from one of three sources depending on the setting of the trigger selector switch. The input signal may come from an external source when the trigger selector switch is set to EXT, from a low-amplitude ac voltage at line frequency when the switch is set to LINE, or from the vertical amplifier when the switch is set to INT. When set for internal triggering (INT), the trigger circuit receives its input from the vertical amplifier. When the vertical input signal that is being amplified by the vertical amplifier reaches a certain level, the trigger circuit provides a pulse to the sweep generator, thereby ensuring that the sweep generator output is synchronized with the signal that triggers it.

One type of circuit frequently used in the "trigger circuit" block of Fig. 21.20 is called a Schmitt trigger or a voltage-level detector. Basically, the Schmitt trigger compares an input voltage, in this case from the vertical amplifier, with a voltage at a point in the circuit. When the input voltage exceeds the voltage to which it is being compared, the circuit changes states, which means the output voltage goes to a high state. The point at which this occurs is called the *upper trigger point* (UTP). When the input voltage drops below a certain level called the *lower trigger paint* (LTP), the Schmitt trigger output returns to its original level, and thus the output of the circuit is a square wave. Unless this square wave is of very short duration, it will not be suitable to trigger the sweep generator directly. It is common practice to apply the square wave to an *RC* circuit called a differentiator whose output is a short-duration spike suitable for triggering the sweep generator.

21

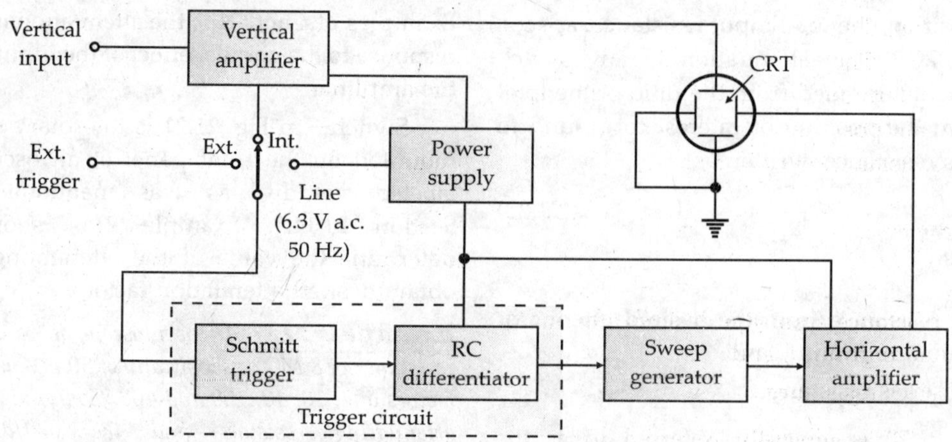


**Fig. 21.20** Block diagram of an oscilloscope with triggered sweep.

## 21.16 ATTENUATORS

The voltage at the input terminal of the vertical amplifier, which causes the beam to be deflected off the CRT screen, is quite low in amplitude. So that high-amplitude signals may be displayed, an **attenuator** network is placed between the vertical input terminal and the input terminal of the vertical amplifier. The term *attenuate* means to *"reduce in size"*. The purpose of the attenuator is to reduce the amplitude of the vertical input signal before applying it to the vertical amplifier. The most basic atenuator is a simple resistive voltage divider such as the one shown in Fig. 21.21. With this circuit, the input voltage $V_i$ will be attenuated by a factor of 10,000 with the switch $S$ set to

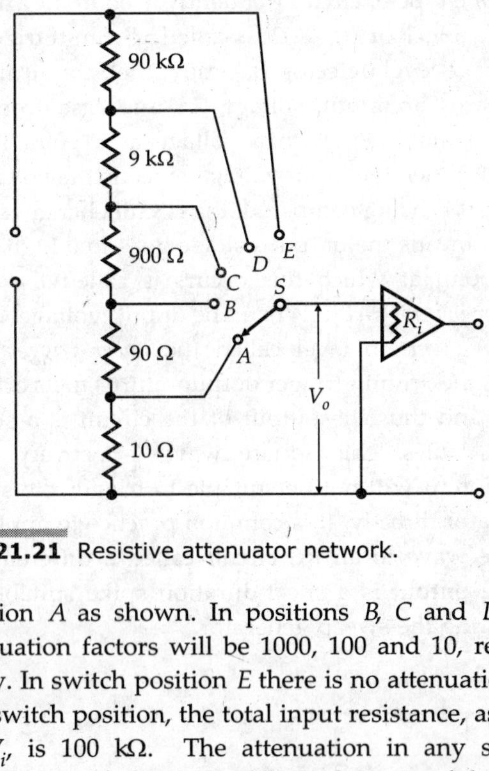


**Fig. 21.21** Resistive attenuator network.

position $A$ as shown. In positions $B$, $C$ and $D$, the attenuation factors will be 1000, 100 and 10, respectively. In switch position $E$ there is no attenuation. In any switch position, the total input resistance, as seen by $V_i$, is 100 kΩ. The attenuation in any switch position can be determined from the ratio of the total resistance from the position of interest to ground to the total series resistance, written as :

$$\frac{V_o}{V_i} = \frac{R}{R_t} \qquad \text{...(21.22)}$$

where

$R$ = total resistance from the desired attenuator terminal to ground, and

$R_t$ = total series resistance.

Equation 21.22 is generally referred to as the voltage divider equation. The attenuation factor is the reciprocal of the voltage divider ratio.

**Example 21.4** *If the switch in Fig. 21.21 is set of position D, what is the attenuation factor ?*

**Solution.** From Eqn. 21.21, we obtain the voltage divider ratio as :

$$\frac{V_o}{V_i} = \frac{R}{R_t}$$

$$= \frac{9\times10^3 + 900 + 90 + 10}{100\times10^3} = 0.1$$

Therefore, the attenuation factor = 1/0.1 = 10.

The resistance values shown in Fig. 21.21 will provide the desired attenuation only if the input resistance $R_i$ of the vertical amplifier is much greater than the attenuator resistance values. If the amplifier input resistance is not much larger than the attenuator resistance with which it is in parallel, appreciable error in the attenuation factor will be noted since the attenuator is "loaded down" by the amplifier.

**Example 21.5** *If the switch in Fig. 21.21 is in position D, as in Example 21.4, and the input resistance of the amplifier $R_i$ is 100 kΩ, what is the attenuation factor ?*

**Solution.** Since the amplifier input resistance is in parallel with the attenuator resistance, this parallel combination must be used in the voltage divider equation. The parallel resistance is

$$R_p = \frac{R_i R}{R_i + R}$$

$$= \frac{100\times10^3 \times 10\times10^3}{100\times10^3 + 10\times10^3} = 9.09\ \text{k}\Omega$$

Therefore,

$$\frac{V_o}{V_i} = \frac{9.09}{100} = 0.0909$$

The attenuation factor is 1/0.0909 = 11.

As can be seen by comparing the results of Examples 21.4 and 21.5, the attenuation is increased in response to the parallel effect of the input resistance of the amplifier.

Switch $S$ in Fig. 21.21 is the rotary switch that is mounted on the front panel of an oscilloscope and labeled volt/div. as was mentioned briefly in Section 21.14.1. Example 21.6, shows how to determine the value of the attenuating resistors to obtain desired attenuation factors.

**Example. 21.6** *An oscilloscope is to have an input resistance of 8 MΩ, a sensitivity of 50 mV, and attenuation factors of 4, 10, 40, 100 and 400. Compute the value of the attenuating resistors and the volt/division value corresponding to each attenuation factor. Assume $R_i$ of the vertical amplifier is sufficiently high to be ignored.*

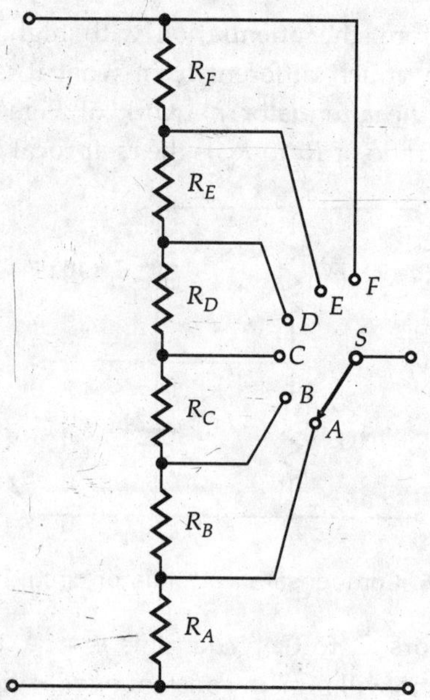


**Fig. 21.22** Circuit for Example 21.6.

**Solution.** The attenuating network required is shown in Fig. 21.22. When switch *S* is set to position *A*, as indicated, maximum voltage can be applied across the network because of the voltage divider action. Therefore, at this position the attenuation factor will be 400. The attenuation factor equals the reciprocal of the voltage divider ratio. For position *A* then

$$\text{attenuation factor} = 400 = \frac{1}{V_o / V_i} \text{ or } 400 = \frac{V_i}{V_o}$$

Since $V_o$ equals the value of the sensitivity (50 mV/div) because $V_o$ is applied directly to the amplifier in each switch position, we can solve for $V_i$ as :

$$V_i = 400\,V_o$$
$$= 400\ (50\ \text{mV/div}) = 20\ \text{V/div at position } A.$$

From this result we see that we can multiply the attenuation factor by the sensitivity to obtain the value of the volts/division at each switch setting.

Position *B* : $V_i = 100$ (50 mV/div) = 5 V/div.

Position *C* : $V_i = 40$ (50 mV/div) = 2 V/div.

Position *D* : $V_i = 10$ (50 mV/div) = 0.5 V/div.

Position *E* : $V_i = 4$ (50 mV/div) = 0.2 V/div.

Position *F* : There is no attenuation or the attenuation factor is 1 ; therefore,

$$V_i = 1\ (50\ \text{mV/div}) = 50\ \text{mV/div}.$$

Solving for the attenuating resistors, we obtain the following :

$$R_A = R_T(V_o / V_i) = (8\,\text{M}\Omega)\ (30\ \text{mV}/20\ \text{V}) = 20\ \text{k}\Omega$$
$$R_A + R_B = R_T(V_o / V_i) = (8\ \text{M}\Omega)\ (50\ \text{mV}/5\ \text{V}) = 80\ \text{k}\Omega$$
$$R_B = (R_A + R_B) - R_A = 80\ \text{k}\Omega - 20\ \text{k}\Omega = 60\ \text{k}\Omega$$

$$R_A + R_B + R_C = R_T(V_o / V_i)$$
$$= (8\ \text{M}\Omega)\ (50\ \text{mV}/2\ \text{V}) = 200\ \text{k}\Omega$$
$$R_C = (R_A + R_B + R_C) - (R_A + R_B)$$
$$= 200\ \text{k}\Omega - 80\ \text{k}\Omega = 120\ \text{k}\Omega$$
$$R_A + R_B + R_C + R_D = R_T(V_o / V_i)$$
$$= (8\ \text{M}\Omega)\ (50\ \text{mV}/0.5\ \text{V})$$
$$= 800\ \text{k}\Omega$$
$$R_D = (R_A + R_B + R_C + R_D) - (R_A + R_B + R_C)$$
$$= 800\ \text{k}\Omega - 200\ \text{k}\Omega = 600\ \text{k}\Omega$$
$$R_A + R_B + R_C + R_D + R_E = V_o / V_i$$
$$= (8\,\text{M}\Omega)(50\,\text{mV} / 0.2\ \text{V})$$
$$= 2\ \text{M}\Omega$$
$$R_E = (R_A + R_B + R_C + R_D + R_E) - (R_A + R_B + R_C + R_D)$$
$$= 2\ \text{M}\Omega - 800\,\text{k}\Omega = 1.2\ \text{M}\Omega$$
$$R_F = R_T - (R_A + R_B + R_C + R_D + R_E)$$
$$= 8\ \text{M}\Omega - 2\ \text{M}\Omega = 6\ \text{M}\Omega.$$

The complete attenuator network is shown in Fig. 21.23.

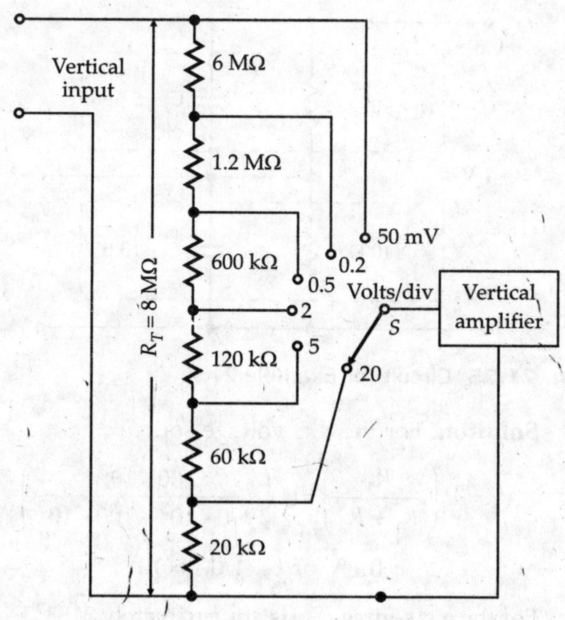


**Fig. 21.23** Complete attenuator network for Example 21.6

If we were measuring dc voltages exclusively, our discussion of voltage divider attenuators might end here. However, oscilloscopes are used primarily to display ac waveforms. This introduces the problem of the stray capacitance of the attenuator network as

21

shown in Fig. 21.24. For a.c. signals, the attenuation factor is actually related to the ratio of the impedance given by the voltage divider ratio as :

$$\frac{\mathbf{V}_o}{\mathbf{V}_i} = \frac{\mathbf{Z}_2}{\mathbf{Z}_1 + \mathbf{Z}_2} \quad ...(21.23)$$

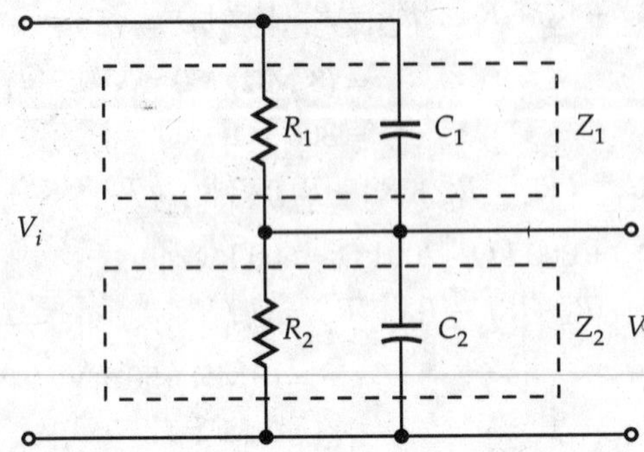


**Fig. 21.24** Attenuator network with associated stray capacitance.

If the stray capacitance values are not correctly related to the resistance values, the voltage divider ratio will not have the same value at all frequencies.

***Example 21.7*** *Compare the output voltage of the voltage divider attenuator shown in Fig. 21.25 for a d.c. voltage and a 10 MHz a.c. signal.*

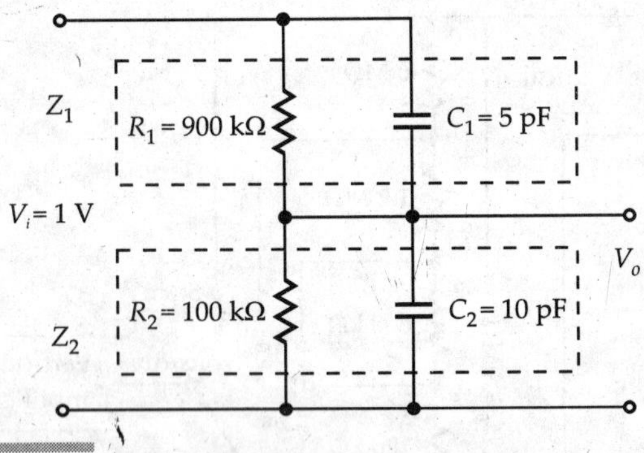


**Fig. 21.25** Circuit for Example 21.7.

**Solution.** For the d.c. voltage source

$$V_o = V_i\left(\frac{R_2}{R_1 + R_2}\right) = (1)\left(\frac{100 \times 10^3}{900 \times 10^3 + 100 \times 10^3}\right)$$

$$= 0.1 \text{ V or } k = 1/0.1 = 10$$

For the a.c. source, $X_{c1}$ is approximately 3183 Ω and $X_{c2}$ is approximately 1592 Ω. Since these values are much less than the value of the resistances in parallel with them, we can approximate the output voltage as :

$$V_o = (1)\left(\frac{X_{c2}}{X_{c1} + X_{c2}}\right) = (1)\left(\frac{1592}{3183 + 1592}\right)$$

$$= 0.33 \text{ V or } k = 1/0.33 \simeq 3.$$

Therefore, the attenuation with high frequency a.c. is very much different from what it is with d.c. Consider the attenuator network of Fig. 21.26. The resistance ratio of $R_2$ to $R_1$ is the reciprocal of the ratio

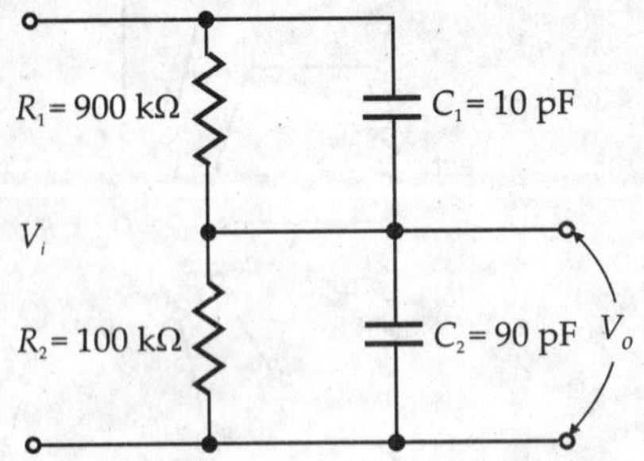


**Fig. 21.26** Compensated attenuator circuit.

of capacitors $C_2$ to $C_1$. When this is true, the voltage divider ratio will remain constant over a wide range of frequencies. Moreover, for a frequency-compensated voltage divider, the following relationships apply :

$$R_2 = \frac{R_1}{k - 1} \quad ...(21.24)$$

and $$C_2 = C_1(k - 1) \quad ...(21.25)$$

where $k$ is the voltage divider ratio, $V_i / V_o$.

with these values, $X_{c1} = 1592\ \Omega$ and $X_{c2} = 177$.

$$\therefore \quad V_o = (1)\left(\frac{177}{1592 + 177}\right) = 0.1 \text{ V}$$

which is the same as with d.c.

## 21.17 BASIC CRO CIRCUITS

Figure 21.27 shows a block diagram of a typical oscilloscope. In the following few pages, we describe the basic circuitry of cathode my oscilloscope.

**1. *Vertical (Y) deflection system.*** The signals to be examined are usually applied to the vertical or Y deflection plates through an input attenuator and a number of amplifier stages. Vertical amplifier is required because the signals are not strong enough to produce measurable deflection on the CRT screen. The amplifier response must be wide enough to pass faithfully the entire band of frequencies to be measured.

When high voltage signals are to be examined, they must be attenuated to bring them within the range of vertical amplifiers. The vertical amplifier output is also applied to the synchronizing amplifier through the synchronizer selector switch in the internal position. This permits the horizontal sweep circuit to be triggered by the signal being investigated.

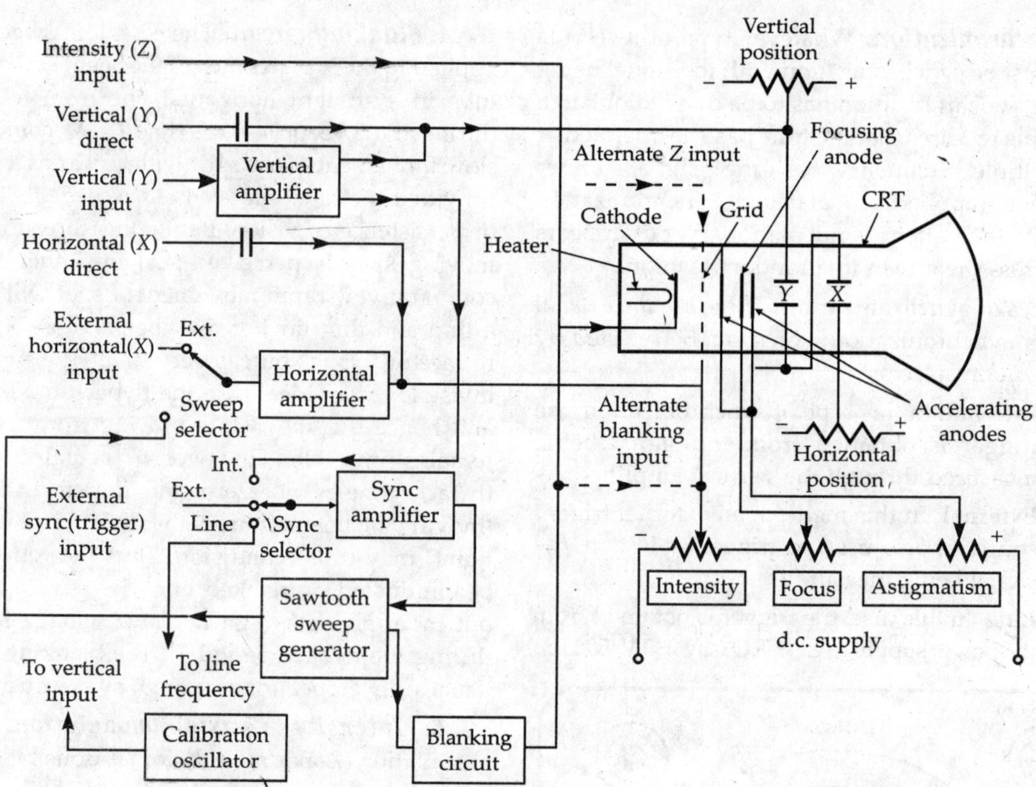


**Fig. 21.27** Simplified block diagram of a general purpose oscilloscope.

2. ***Horizontal (X) deflection system.*** The horizontal (X) deflection plates are fed by a sweep voltage that provides a time base. The horizontal plates are supplied through an amplifier, but they can be fed directly when voltages are of sufficient magnitude. When external signals are to be applied to the horizontal deflection system, they can also be fed through the horizontal amplifier, *via* the sweep selector switch in the external position. When the sweep selector switch is in the internal position, the horizontal amplifier receives an input from the saw tooth sweep generator which is triggered by the synchronizing amplifier.

**Types of sweeps.** There are *four* basic types of sweeps :

(*i*) *Free running or recurrent sweep.* In the free running or recurrent sweep, the sawthooth waveform is repetitive. A new sweep is started immediately after the previous sweep is terminated and the circuit is not initiated by any external signal.

(*ii*) *Triggered sweep.* A waveform to be observed on the CRO may not be periodic but may perhaps occur at irregular intervals. In this case it is desirable that the sweep circuit remain inoperative and the sweep be initiated by the waveform under examination. In some cases the waveform may be periodic, but it may be that the interesting part of the waveform is of a very short duration compared to the period of the waveform. Under such cases a triggered sweep is used.

In triggered sweep or single sweep, the spot is swept once across the screen in response to a trigger signal. The triggered sweep is used for examination of transients or one time signals and the waveform is photographed for record. The trigger can be obtained from the signal under investigation or by an external source.

(*iii*) *Driven sweep.* In most cases, a driven sweep is used where the sweep is recurrent but triggered by the signal under test.

(*iv*) *Non saw tooth sweep.* For some applications like comparison of two frequencies or for finding phase shift between two voltages, non sawtooth sweep voltages are utilized for the sweep circuit. Sweep frequencies vary with the type of oscilloscope. A laboratory oscilloscope may have sweep frequencies up to several MHz ; a simple oscilloscope for audio work has an upper limit of 100 kHz. Most TV services require a sweep voltage frequency up to 1 MHz.

21

**3. *Synchronization.*** Whatever type of sweep is used, it must be synchronization with the signal being measured. Synchronization has to be done to obtain a pattern. This requires that the time base be operated at a submultiple frequency of the signal under measurement (applied to $Y$ plates). If synchronization is not done, the pattern is not stationary, but appears to drift across the screen in a random fashion.

**Sources of synchronization.** There are *three* usual sources for synchronization which can be selected by synchronization selector :

(*i*) **Internal**. In this type of synchronization, the trigger is obtained from the signal being measured through the vertical amplifier.

(*ii*) **External**. In this method, an external trigger source is also used to trigger or initiate the signal being measured

(*iii*) **Line**. In this case, the trigger is obtained from the power supply to the CRO (say 230 V, 50 Hz).

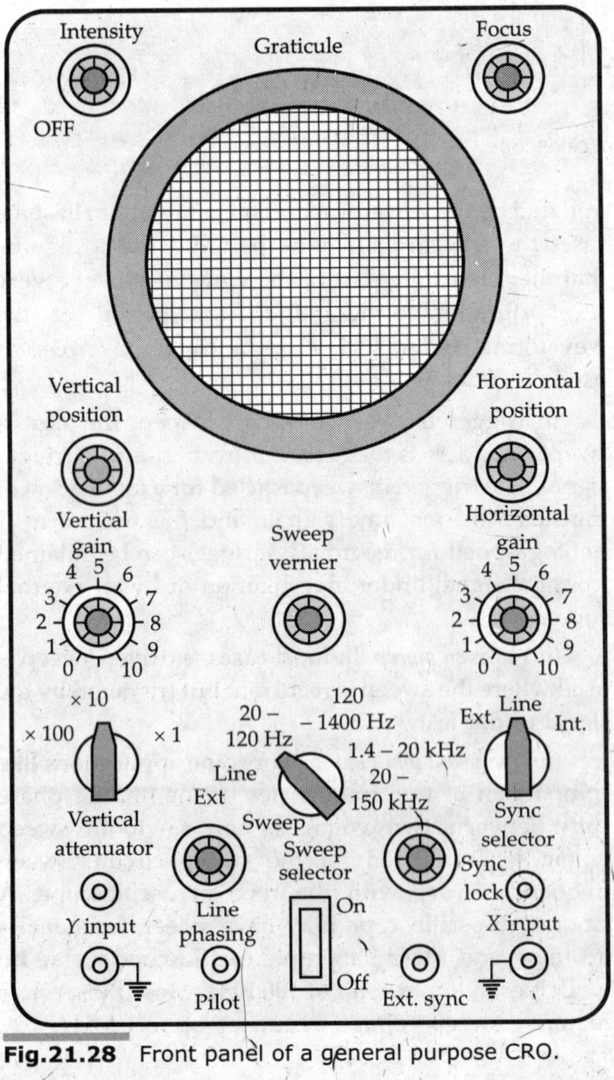


**Fig.21.28** Front panel of a general purpose CRO.

**4. *Blanking circuit.*** The sawtooth sweep voltage applied to the X plates moves the beam across the CRT tube in a straight horizontal line from left to right during the sweep or trace time $T_s$. At comparatively slow movement of the spot will appear as a solid line, provided the rate of movement exceeds the threshold of persistence of vision. Below this threshold limit, a moving spot is perceived. On the other hand, the comparatively rapid movement of spot will appear as a thin and dim line, or may be invisible. Thus if the retrace or flyback time is very small, the spot remains invisible. In an ideal case the flyback time, $T_r$ is zero and hence the spot while moving from right to left remains invisible. However in actual practice the flyback time is not zero and therefore the retrace (moving of beam from right to left *i.e.*, its starting point) may cause confusion. Thus the retrace should be eliminated or blanked out. The retrace is blanked out by applying a high negative voltage to the grid during the flyback period $T_r$. The blanking voltage is usually developed (or triggered) by sweep generator.

**5. *Intensity (Z-axis) modulation.*** Intensity modulation (Z-axis modulation) is done by inserting a signal between the ground and the cathode (or control grid). Z-axis modulation is applied during normally visible portion of the trace.

The Z-axis modulation can be used for brightening the display. Periodic positive pulses are applied to the grid (alternatively negative pulses are applied to cathode) to brighten the beam during its sweep period. These periodically brightened spots may be used as markers for time calibration of the main waveform.

**6. *Positioning controls.*** It is necessary to provide some means of positioning the trace on the screen. The positioning of the trace is done by applying small independent, internal d.c. voltages to the deflecting plates and control can be exercised by varying the voltage with help of potentiometers.

**7. *Focus control.*** As mentioned earlier, the focusing electrode acts like a lens whose focal length can be changed. This change can be brought about by changing the potential of the focusing anode.

**8. *Intensity control.*** The intensity of the beam is varied by the Intensity control potentiometer which changes the grid potential with respect to cathode. The grid potential determines the amount of electrons leaving the cathode and thus controls the intensity of the beam.

**9. *Calibration circuit.*** Laboratory oscilloscopes normally have an internally generated and stabilized voltage of known amplitude which is used for calibration purposes. Usually the calibrating voltage has a square waveform.

10. ***Astigmatism.*** In most modern oscilloscopes there is an additional focusing control marked **Astigmatism.** This is used to correct an effect which exactly is analogous to astigmatism in optical lenses. To focus the spot correctly, it is necessary to stop it near the centre of the screen by switching off the time base and adjusting the X and Y positioning controls. The spot is then made as sharp as possible by successive adjustment of focus and astigmatism controls.

## 21.18 OBSERVATION OF WAVEFORM ON CRO

In order to observe waveform on a CRO, the waveform of voltage under test is applied to Y plates and a voltage obtained from a sawtooth generator is applied to X plates. Let us assume that the sawtooth waveform has an idealized waveshape.

When simultaneously with the horizontal sawtooth (ramp) voltage, an input voltage is applied to vertical deflection (Y) plates, the beam is under the influence of *two* forces : (*i*) one in the horizontal direction moving the beam at a linear rate from left to right, and (*ii*) second in the vertical direction moving the beam up and down. Since the deflection is proportional to the voltage applied to the deflection plates, the horizontal movement is proportional to the voltage applied to X plates at any instant and since the ramp voltage is linear it traces a straight line on the CRT screen. The vertical deflection is proportional to the voltage applied to the Y plates at any instant and thus the beam moves up and down according to the

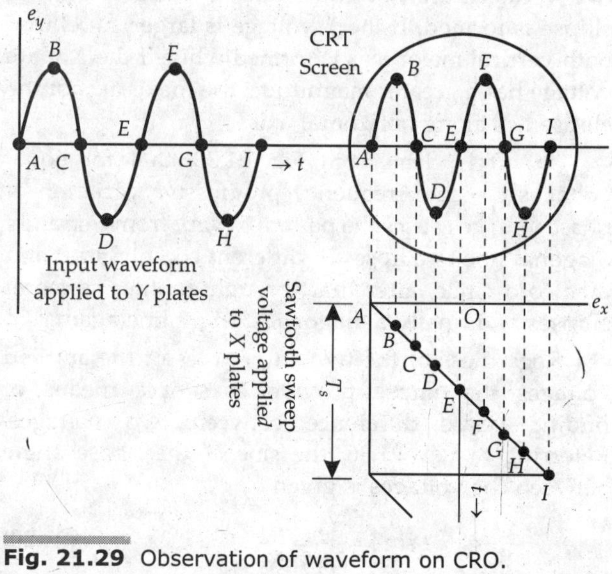


**Fig. 21.29** Observation of waveform on CRO.

magnitude and polarity of the input voltage. Figure 21.29 shows the waveform displayed on a CRT tube due to an input sinusoidal voltage.

At the end of one sweep cycle, the sweep voltage abruptly drops down and the spot is immediately transferred to its original position. The process is then repeated again, with the result, that a stationary image is seen on the screen.

For the case shown the frequency of the input voltage is twice that of sawtooth (sweep) voltage. To observe more than one cycle of the input voltage, the sweep voltage frequency has to be a submultiple of the input voltage frequency.

## 21.19 MEASUREMENT OF VOLTAGES & CURRENTS

The expression for electrostatic deflection, given in Eqn. 21.16, shows that the deflection is proportional to the deflection-plate voltage. Thus the cathode-ray tube will measure voltage. It is usual to calibrate the tube under the given operating conditions by observing the deflection produced by a known voltage. Direct voltages may be obtained from the static deflection of the spot, alternating voltages from the length of the line produced when the voltage is applied to Y plates while no voltage is applied to X plates. The length of this line corresponds to the peak-to-peak voltage. When dealing with sinusoidal voltages, the rms value is given by dividing the peak-to-peak voltage by $2\sqrt{2}$.

Laboratory oscillographs frequently incorporate voltage-measurement facilities by including constant-gain amplifiers and calibrated shift controls. The Y-shift control is adjusted so that positive peak of the test voltage coincides with some datum line on the screen ; the shift control is then operated until the negative peak coincides with the datum. The movement of the control is arranged to read directly the peak-to-peak voltage. The value of a current can be obtained by measuring the voltage drop across a known resistance connected in the circuit.

## 21.20 MEASUREMENT OF PHASE & FREQUENCY (LISSAJOUS PATTERNS)

It is interesting to consider the characteristics of patterns that appear on the screen of a CRT when sinusoidal voltages are simultaneously applied to horizontal and vertical plates. These patterns are called **'Lissajous Patterns'.**

When two sinusoidal voltages of equal frequency which are in phase with each other are applied to the

21

horizontal and vertical deflection plates, the pattern appearing on the screen is a straight line as is clear from Fig. 21.30.

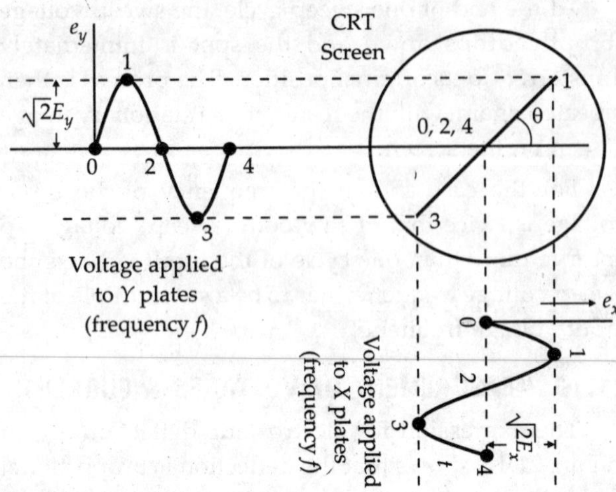


**Fig. 21.30** Lissajous pattern with equal frequency voltages and zero phase shift.

Thus when two equal voltages of equal frequency but with 90° phase displacement are applied to a CRO, the trace on the screen is a circle. This is shown in Fig. 21.31.

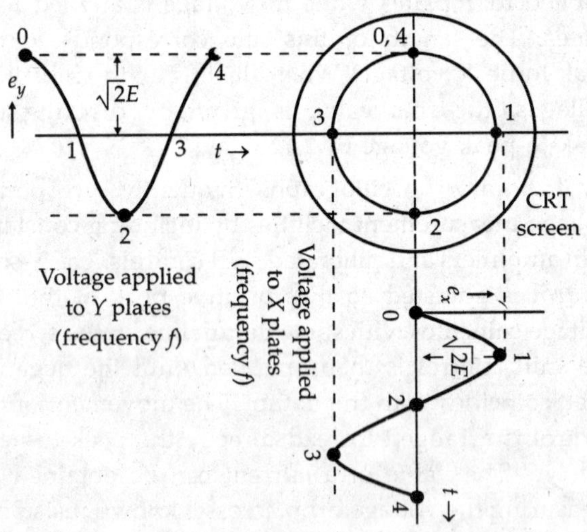


**Fig. 21.31** Lissajous pattern with equal voltages of equal frequency and a phase shift of 90°.

When two equal voltages of equal frequency but with a phase shift $\phi$ (not equal to 0° or 90°) are applied to a CRO we obtain an ellipse as shown in Fig. 21.32. An ellipse is also obtained when unequal voltages of same frequency are applied to the CRO.

A number of conclusions can be drawn from the above discussions. When two sinusoidal voltages of same frequency are applied :

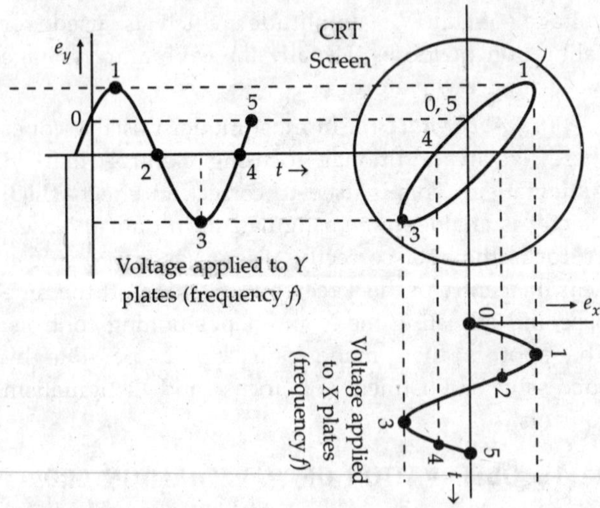


**Fig 21.32** Lissajous pattern with two equal voltages of same frequency and phase shift of $\phi$.

(*i*) A straight line results when the two voltages are equal and are either in phase with each other or 180° out of phase with each other. The angle formed with the horizontal is 45° when the magnitudes of voltages are equal. An increase in the vertical deflection voltage causes the line to have an angle greater than 45° with the horizontal. On the other hand a greater horizontal voltages makes the angle less than 45° with the horizontal.

(*ii*) Two sinusoidal waveforms of the same frequency produce a Lissajous pattern, which may be a straight line, a circle or an ellipse depending upon the phase and magnitude of the voltages.

A circle can be formed only when the magnitude of the two singals are equal and the phase difference between them is either 90° or 270°. However, if the two voltages are not equal and/or out of phase an ellipse is formed. If the $Y$ voltage is larger, an ellipse with vertical major axis is formed while if the $X$ plate voltage has a greater magnitude, the major axis of the ellipse lies along horizontal axis.

(*iii*) It is clear from Fig. 21.33, that for equal voltages of same frequency progressive variation of phase voltage causes the pattern to vary from a straight diagonal line to ellipses of different eccentricities and then to a circle, after that through another series of ellipses and finally a diagonal straight line again.

Regardless of the two amplitudes of the applied voltages the ellipse provides a simple means of finding phase difference between two voltages. Referring to Fig. 21.36, the sine of the phase angle between the voltages is given by :

$$\sin\phi = \frac{Y_1}{Y_2} = \frac{X_1}{X_2} \qquad ...(21.20)$$

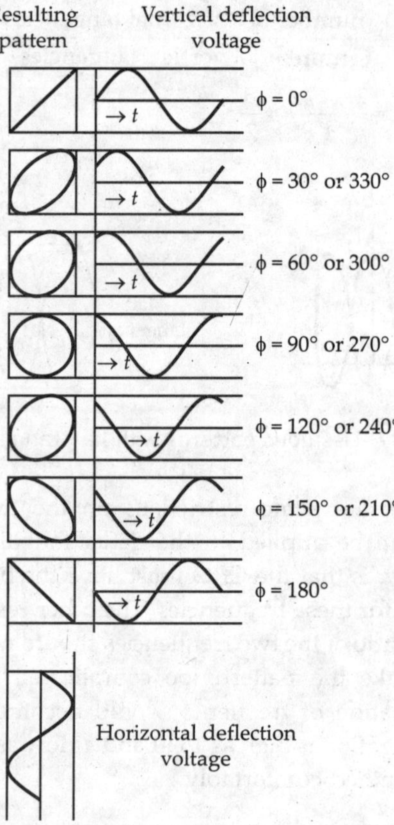


**Fig. 21.33** Lissajous pattern with different phase shifts.

For convenience, the gains of the vertical and horizontal amplifiers are adjusted so that the ellipse fits exactly into a square marked by the lines on the graticule.

If the major axis of the ellipse lies in the first and third quadrants (*i.e.,* its slope is positive) as in Fig. 21.34 (*a*) the phase angle is either between 0° to 90° or between 270° to 360°. When the major axis of ellipse lies in second and fourth quadrants *i.e.* when its slope is negative as in Fig. 21.34 (*b*), the phase angle is either between 90° and 180° or between 180° and 270°.

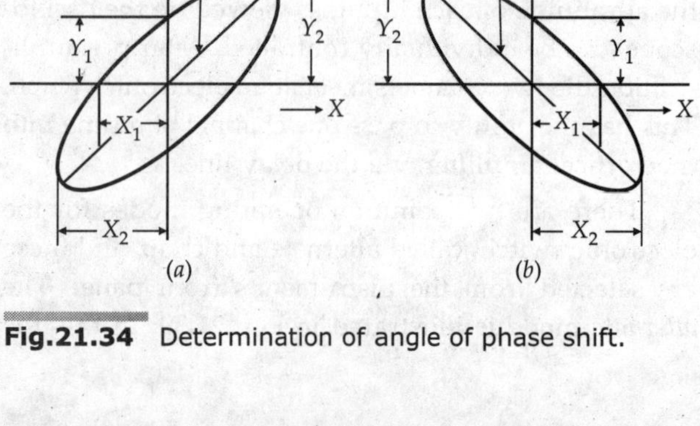


**Fig.21.34** Determination of angle of phase shift.

**Frequency measurements.** Lissajous patterns may be used for accurate measurement of frequency. The signal, whose frequency is to be measured, is applied to the *Y* plates. An accurately calibrated standard variable frequency source is used to supply voltage to the *X* plates, with the internal sweep generator switched off. The standard frequency is adjusted unit the pattern appears as a circle or an ellipse, indicating that both signals are of the same frequency. Where it is not possible to adjust the standard signal frequency to the exact frequency of the unknown signal, the standard is adjusted to multiple or a submutliple of the frequency of the unknown source so that the pattern appears stationary.

Let us consider an example. Suppose sine waves are applied to *X* and *Y* plates as shown in Fig. 21.35. Let the frequency of wave applied to *Y* plates is twice that of the voltage applied to *X* plates. This means that the CRT spot travels two complete cycles in the vertical direction against one in the horizontal direction.

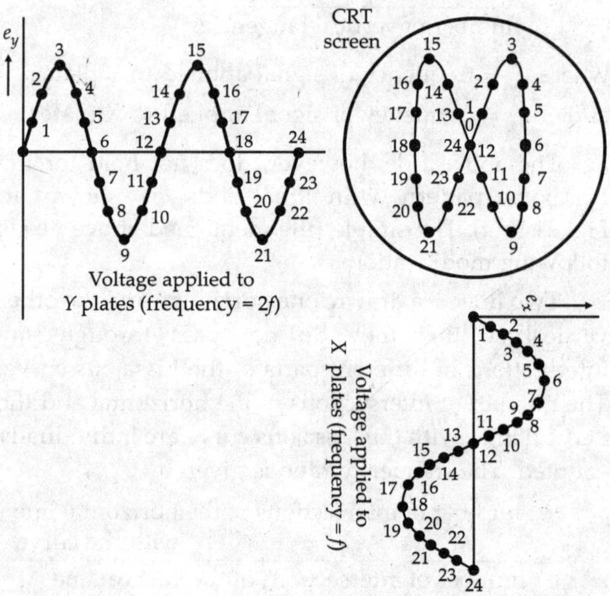


**Fig. 21.35** Lissajous pattern with frequency ratio 2:1.

The two waves start at the same instant. Lissajous pattern may be constructed in the usual way and a 8 shaped pattern with two loops is obtained. If the two waves do not start at the same instant we get different patterns for the same frequency ratio. The Lissajous patterns for other frequency ratios can be similarly drawn. Some of the these patterns are shown in Fig. 21.36.

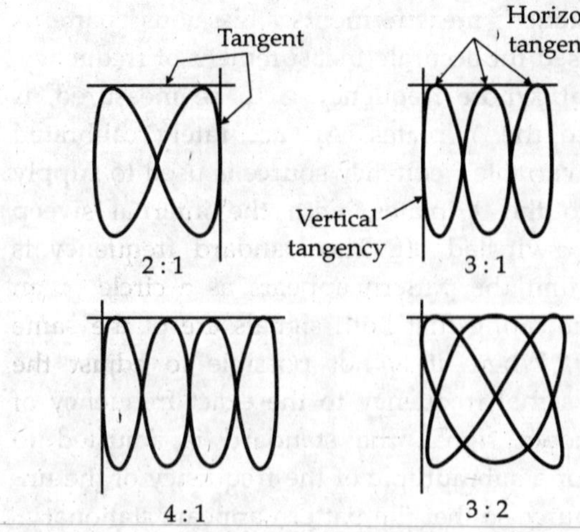


**Fig. 21.36** Lissajous patterns with different frequency ratios.

It can be shown that for all the above cases, the ratio of the two frequencies is :

$$\frac{f_y}{f_x} = \frac{\text{number of times tangent touches top or bottom}}{\text{number of times tangent touches either side}}$$

$$= \frac{\text{number of horizontal tangencies}}{\text{number of vertical tangencies}}$$

where $f_y$ = frequency of signal applied to $Y$ plates,

and $f_x$ = frequency of signal applied to $X$ plates

The above rule, however, does not hold for the Lissajous pattern with free ends as shown in Fig. 21.37($a$). The simple rule mentioned above needs following modifications :

Two lines are drawn, one horizontal and the other vertical so that they do not pass through any intersections of different parts of the Lissajous curve. The number of intersections of the horizontal and the vertical lines with the Lissajous curve are individually counted. The frequency ratio is given by :

$$\frac{f_y}{f_x} = \frac{\text{number of intersections of the horizontal line with the curve}}{\text{number of intersections of the vertical line with the curve}}$$

The applications of this rule to Fig. 21.37 gives a frequency ratio $\frac{f_y}{f_x} = \frac{5}{2}$.

The modified rule is applicable in all cases whether the Lissajous pattern is open or closed.

The ratio of frequencies when open ended Lissajous patterns are obtained can also be found by treating the open ends as half tangencies as shown in Fig. 21.37($b$). The pattern of Fig. 21.37($a$) is reproduced

$$\therefore \frac{f_y}{f_x} = \frac{\text{number of horizontal tangencies}}{\text{number of vertical tangencies}}$$

$$= \frac{2 + 1/2}{1} = \frac{5}{2}$$

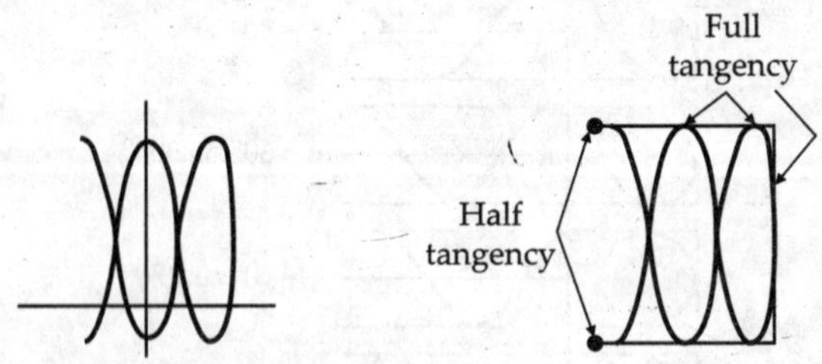


**Fig. 21.37** Lissajous patterns with half tangencies.

There are some restrictions on the frequencies which can be applied to the deflection plates. One obviously, is that the CRO must have the bandwidth required for these frequencies. The other restriction is that the ratio of the two frequencies should not be such as to make the pattern too complicated otherwise determination of frequency would become difficult. As a rule ratios as high as 10 : 1 and as low as 10 : 9 can be determined comfortably.

## 21.21 MULTI-INPUT OSCILLOSCOPES

Oscilloscopes can have multiple input and display facilities. Two inputs is the most common, although four and eight inputs are available for special applications. There are two primary types : *single beam* which can then be converted into several traces, and *dual beam,* which may also subsequently be converted into a further number of traces. Two input oscilloscopes are described in this section, under the headings of dual trace and dual beam, although the principles are applicable to any number of inputs.

### 21.21.1 Dual Trace Oscilloscopes

Figure 21.38 illustrates the construction of a typical dual trace oscilloscope. There are two separate vertical input channels, $A$ and $B$, and these use separate attenuator and preamplifier stages. Therefore the amplitude of each input, as viewed on the oscillo-scope, can be individually controlled. After preampli-fication the two channels meet at an electronic switch. This has the ability to pass one channel at a time into the vertical amplifier, via the delay line.

There are two common operating modes for the electronic switch, called alternate and chop, and these are selected from the instrument's front panel. The *alternate* mode is illustrated in Fig. 21.39. In this the

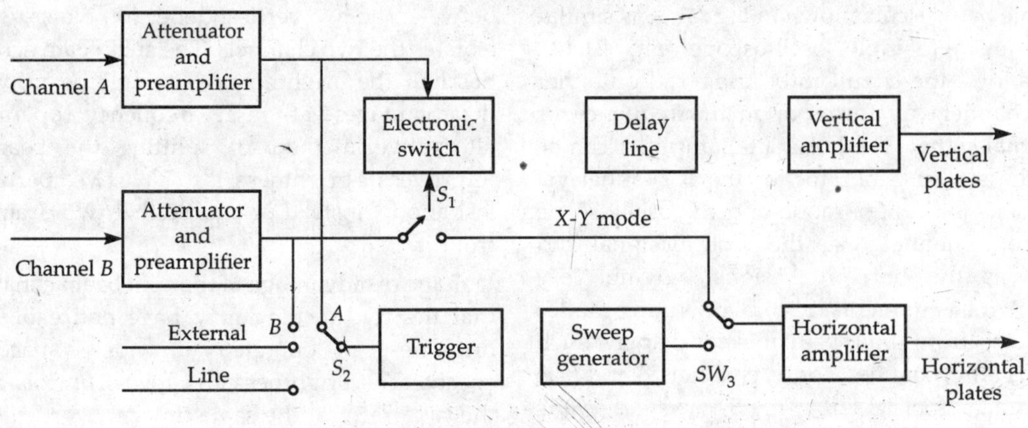


**Fig. 21.38** A dual trace oscilloscope block diagram.

electronic switch alternates between channels $A$ and $B$, letting each through for one cycle of the horizontal sweep. The display is blanked during the flyback and hold-off periods, as in a conventional oscilloscope. Provided the sweep speed is much greater than the decay time of the CRT phosphor, the screen will show a stable display of both the waveform at channels $A$ and $B$. The alternate mode cannot be used for displaying very low frequency signals.

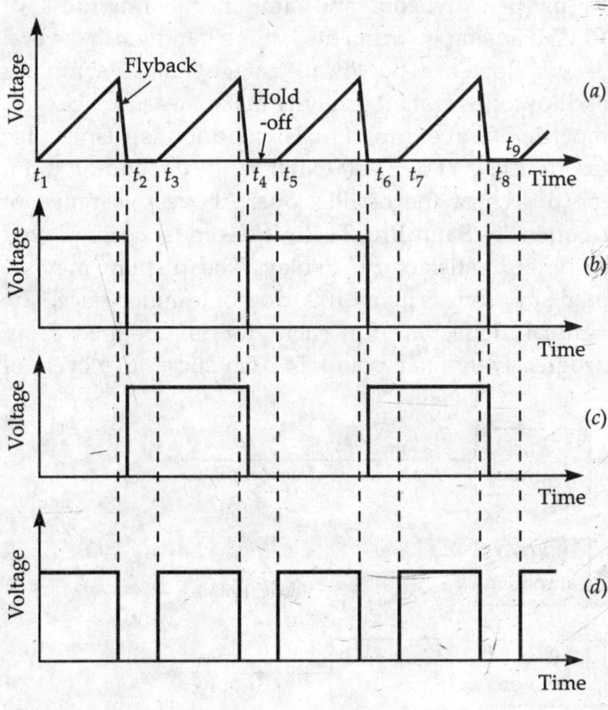


**Fig. 21.39** Waveforms for a dual channel oscilloscope operating in alternate mode ; (*a*) horizontal sweep voltage, (*b*) voltage to channel *A*. (*c*) voltage to channel *B*, (*d*) grid control voltage.

The *chopped* operating mode of the electronic switch is shown in Fig. 21.40. In this mode the electronic switch free runs at a high frequency of the order of 100 kHz to 500 kHz. The result is that small segments from channels $A$ and $B$ are connected alternately to the vertical amplifier, and displayed on the screen. Provided the chopping rate is much faster than the horizontal sweep rate, the display will show a continuous line for each channel. If the sweep rate approaches the chopping rate then the individual segments will be visible, and the alternate mode should now be used.

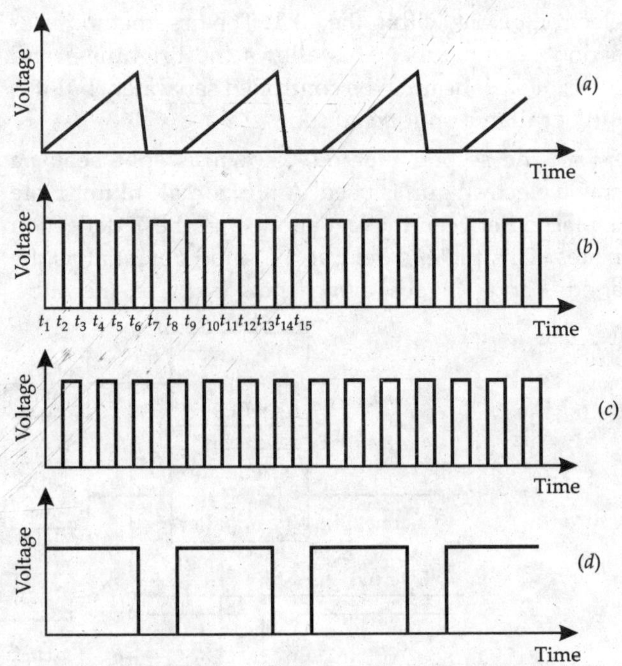


**Fig. 21.40** Waveforms of a dual channel oscilloscope operating in chopped mode. (*a*) Horizontal sweep voltage (*b*) Voltage to channel *A*, (*c*) *Voltage* to channel *B*,(*d*) grid control voltage.

The time base circuit shown in Fig. 21.38 is similar to that of a single input oscilloscope (Fig. 21.12). Switch $S_2$ allows the circuit to be triggered on either the $A$ or $B$ channel waveforms, or on line frequency, or on an external signal. The horizontal amplifier can be fed from the sweep generator, or the $B$ channel via switch $S_1$. This is the $X$-$Y$ mode and the oscilloscope operates from channel $A$ as the vertical signal and channel $B$ as the horizontal signal, giving very accurate $X$-$Y$ measurements. Several operating modes can be selected from the front panel for display, such as channel $A$ only, channel $B$ only, channels $A$ and $B$ as two traces, and signals $A + B$, $A - B$, $B - A$ or $-(A + B)$ as a single trace.

### 21.21.2 Dual Beam Oscilloscope

The dual trace oscilloscope cannot capture two fast transient events, as it cannot switch quickly enough between traces. The dual beam oscilloscope has two separate electron beams, and therefore two completely separate vertical channels, as in Fig. 21.39. The two channels may have a common time base system, as in Fig. 21.38, or they may have independent time base circuits, as in Fig. 21.41. An independent time base allows different sweep rates for the two channels but increases the size and weight of the oscilloscope.

Two methods are used for generating the two electron beams within the CRT. The first method uses a double gun tube. This allows the brightness and focus of each beam to be controlled separately but it is bulkier than a split beam tube.

In the second method, known as split beam, a single electron gun is used. A horizontal splitter plate is placed between the last anode and the $Y$ deflection plates. This plate is held at the same potential as the anode, and it goes along the length of the tube, between the two vertical deflection plates, It therefore isolates the two channels. The split beam arrangement has half the brightness of a single beam, which has disadvantages at high frequency operation. An alternative method of splitting the beam, which improves its brightness, is to have two apertures in the last anode, instead of one, so that two beams emerge from it.

The disadvantage of the split beam construction is that the two display may have noticeably different brightness, if operated at widely spaced sweep speeds. The brightness and focus controls also affect the two traces at the same time.

## 21.22 SAMPLING OSCILLOSCOPES

A sampling oscilloscope is used to examine very fast signals. It is similar in principle to the use of stroboscopic light to look at fast mechanical motion. Samples are taken at different portions of the waveform, over successive cycles, and then the total picture is stretched, amplified by relatively low bandwidth amplifiers, and displayed as continuous wave on the screen. The advantage of a sampling oscilloscope is that it can measure very high speed events, which require sweep speeds of the order of 10 ps per division, and amplifier bandwidths of 15 GHz, using instruments having bandwidth several orders lower. The disadvantage of a sampling oscilloscope is that it can only make measurements on repetitive waveform signals, continuous display for frequencies in the 50–300 MHz range depending upon the design of the oscilloscope. Above this range of frequencies **Sampling Techniques** must be employed to obtain satisfactory display. The display may be made up from as many 1000 dots of luminescence. The vertical deflection for each dot is obtained from progressively later points in each successive cycle of

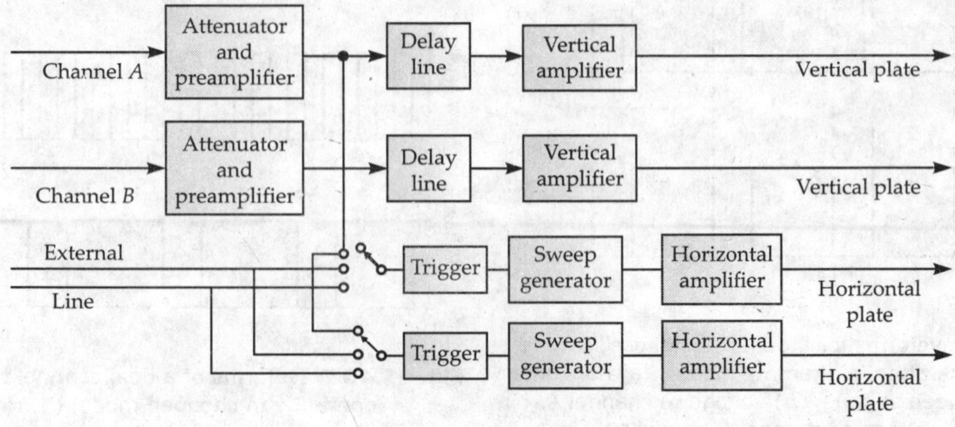


**Fig. 21.41** Dual beam oscilloscope with independent time bases.

the input waveform as shown in Fig. 21.42. The horizontal deflection of the electron beam is obtained by application of a staircase waveform to X deflection plates.

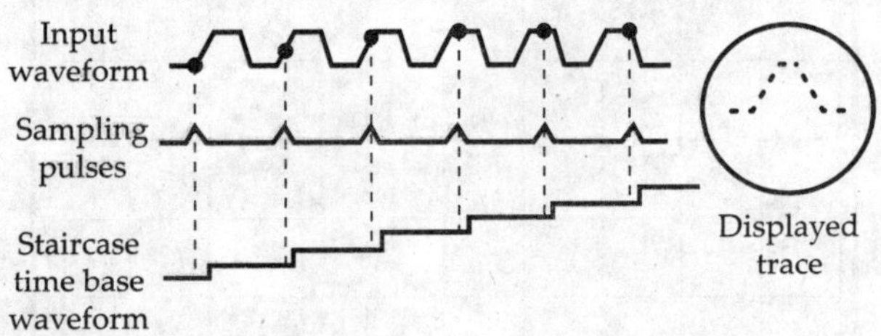


**Fig. 21.42** Principle of sampling oscilloscopes.

The sampling oscilloscope is able to respond and store rapid bits of information and present them in a continuous display. It is this ability that enables the sampling oscilloscope to side-step the usual limitations in conventional high frequency oscilloscopes which have limited sensitivity and bandwidth and small display size. The sampling techniques immediately transform the input signal into lower frequency. domain, where conventional low frequency circuitry is then capable of producing a highly effective display.

The sampling oscilloscopes can be used beyond 50 MHz into the UHF range around 500 MHz and beyond up to 10 GHz. However, it should be understood that sampling techniques cannot be used for the display of transient waveforms.

Figure 21.43 shows a block schematic for a typical sampling oscilloscope. The input signal is delayed, and then sampled by a diode gate. The sampled signal is saved on a capacitor store, then amplified and fed to the vertical plates. Unity feedback is used from the amplifier to the sampling diode gate. This ensures that the voltage on the capacitor store is only increased by the incremental value of the input voltage change, between each sample.

The staircase is reset after a certain number of steps, typically 100 to 1000, and it then starts again. Therefore up to about 1000 points are used to create the waveform on the screen.

The staircase waveform also feeds the horizontal plates of the CRT, and it is used to move the spot across the screen in a series of rapid movements. Figure 21.42 shows the screen display.

The sample frequency used in sampling oscilloscopes can be as low as one hundredth of the signal frequency, so a signal frequency of 1 GHz needs an amplifier bandwidth of only 10 MHz.

***Delayed sweep.*** Many laboratory-quality oscilloscopes include a **delayed-sweep** feature. This feature increases the versality of the instrument by making it possible to magnify a selected portion of the undelayed sweep, measure waveform jitter or rise time, and check pulse-time modulation, as well as many other applications.

Delayed sweep is a technique that adds a precise amount of time between the trigger point and the beginning of the scope sweep. When the scope is being used in the delayed-sweep mode, the start of the delayed sweep can range from a few microseconds to perhaps 10 seconds or more. The delayed sweep operation allows the instrument user to view a small segment of a waveform-for example, an oscillation or ringing that occurs during a small portion. of a lower-frequency waveform.

21

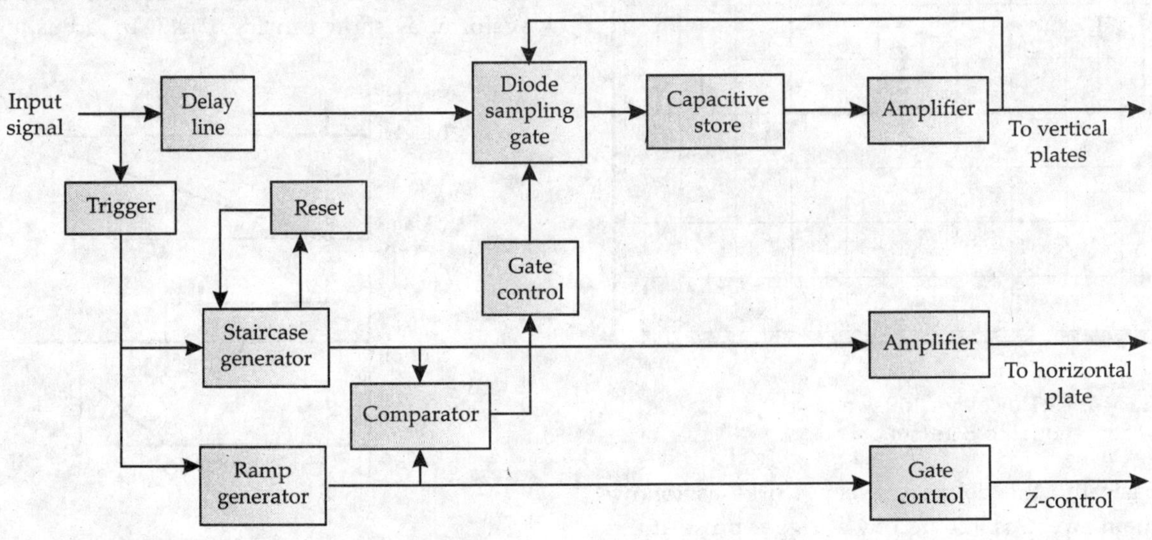


**Fig. 21.43** Sampling oscilloscope block diagram.

Sometimes the delayed-sweep feature is used for convenience to allow triggering at some other point than at the leading or trailing edge. However, in some situations a measurement is possible only if the delayed-sweep feature is used. For example, suppose the part of waveform that is to be measured is too far from the only available trigger point to permit a stable display on CRT screen. The problem can be solved by using delayed sweep to trigger at the only available trigger point and then starting the sweep at the point of interest.

Although there are a few exceptions, delayed sweep is normally a feature of dual-time base oscilloscopes which have two completely separate sweep generators. One sweep functions as a main sweep, and the other serves as the delayed sweep. The main sweep is initiated by a trigger pulse at the leading edge of pulse 1 shown in Fig. 21.44. Suppose we wish to observe in detail a portion of the waveform near its trailing edge by expanding the waveform and using a higher sweep speed-for example, a sweep speed of 0.1 μs/cm. As can be seen in Fig. 21.43, the portion of the waveform of interest to us is completely off the screen to the right when the sweep speed is set to 0.1 μs/cm. To observe the portion of the waveform that is of interest at the higher sweep speed, we must use the delayed-sweep feature and reset the sweep speed to 5 μs/cm.

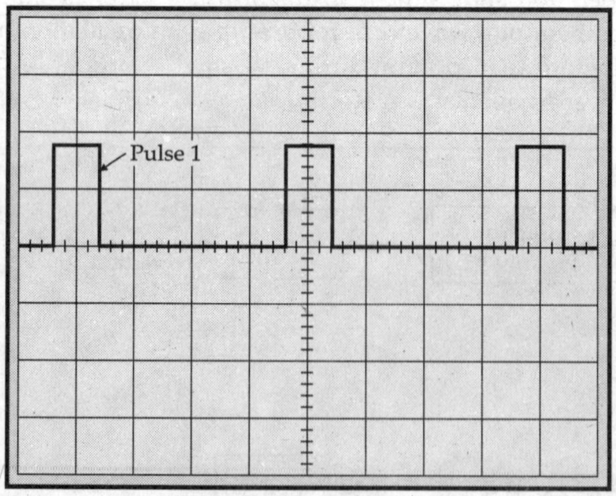


**Fig. 21.44** CRT display with sweep speed equal to 5 μs/cm.

Basically, the delayed sweep works as follows. The main sweep is initiated by a trigger pulse at time $t_0$, shown in Fig. 21.46. This time corresponds to the

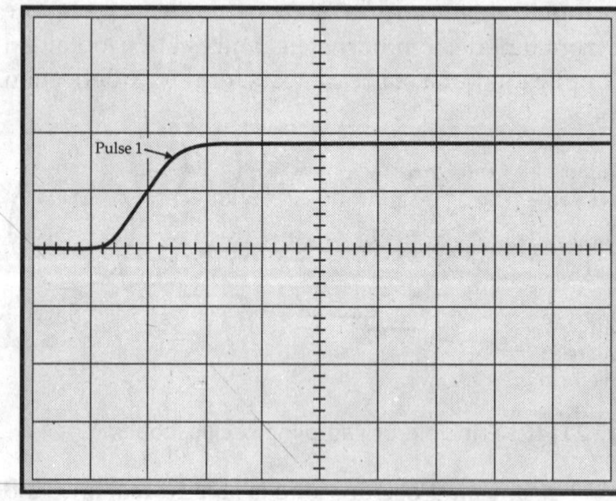


**Fig. 21.45** CRT display with sweep speed equal to 0.1 μs/cm.

leading edge of pulse 1 in Fig. 21.45. The main sweep generator ramp, which is applied to a comparator along with a voltage from a *delay control* circuit, increases linearly until it trips the comparator at time $t_1$. When the comparator changes states, the delayed-sweep generator is triggered, which intensifies a portion of the original display. Adjusting the *delay time* should intensify the portion of the waveform that is of interest. The exact front panel adjustments that are necessary to display the expanded portion of the waveform on the CRT screen depend on the oscilloscope being used. Some oscilloscopes have an alternate *sweep separation* control so that two waveforms are displayed: the original waveform, with the portion of the waveform of interest intensified, and a second expanded waveform, as shown in Fig. 21.47. By adjusting the

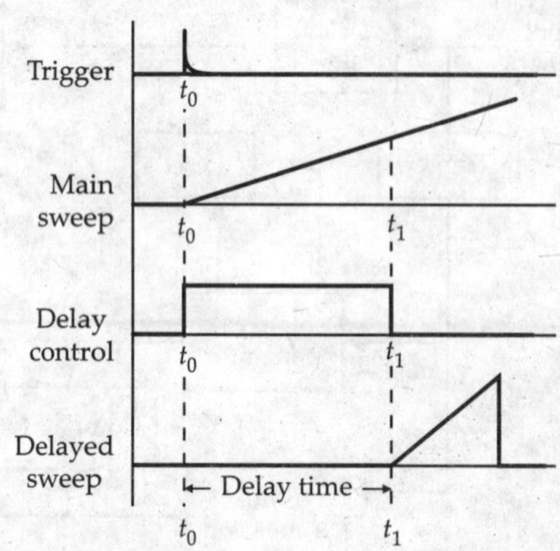


**Fig. 21.46** Delayed-sweep triggering waveforms.

horizontal mode to channel *B*, we can now increase the sweep time so that we can view the portion of the waveform that is of interest.

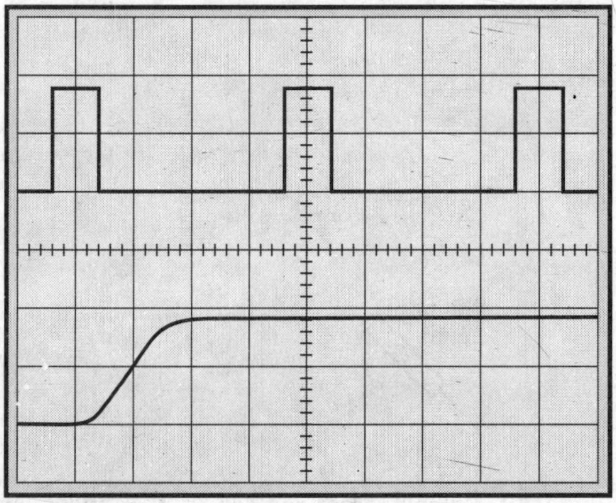

**Fig. 21.47** Delayed-sweep measurement.

## 21.23 ANALOG STORAGE OSCILLOSCOPES

Storage oscilloscopes, capable of retaining the image on the screen for longer than that possible with conventional high persistence phosphors, have many applications. Examples of these are the capture and storage of transients, and the steady display of a very low frequency signal. Two techniques are used to store signals in an oscilloscope, and these are called analog and digital storage. Analog storage is capable of higher speeds, but is less versatile than digital storage. This section describes the principles of analog storage, and the two methods most frequently used, called variable persistence and bistable storage. Digital storage oscilloscopes are covered in Art. 21.24.

### 21.23.1 Principle of Secondary Emission

Both the variable persistence and bistable storage oscilloscopes depend, for their operation, on the principle of secondary emission. Figure 21.48 shows a simple electrode arrangement to illustrate this principle. When a beam of electrons from the electron gun strikes the target it emits secondary electrons, which are gathered by the collector. The collector is at a positive potential of $V_2$, the target is at a potential of $V_1$, which can be varied, and the electron gun is at ground potential.

If $I_p$ is the value of the current in the primary electron beam, coming from the electron gun, and $I_s$ is the electron current emitted from the target and collected by the collector, then the ratio $I_s / I_p$ is called the secondary emission ratio. The value of this ratio depends on the primary electron velocity and intensity, and on the chemical composition of the target.

Figure 21.49 shows the variation of the secondary emission ratio with target voltage, for the arrangement of Fig. 21.48. When the target voltage is much greater than the collector voltage, all secondary electrons emitted from the target are attracted back to it. Therefore the collector current, or secondary current, is zero, and so also is the secondary emission ratio. The operating point is now well to the right of $V_2$ in Fig. 21.49. If the target voltage is slightly negative, as at point *F*, then all the electrons from the gun are deflected on-to the collector, before they reach the target. Therefore although there is no secondary emission the collector current equals the beam current, and the secondary emission ratio is unity. This point is known as the lower stable point. As the target voltage increases from this point electrons are attracted away from the collector, but they do not have enough energy to release secondary electrons from the target. Therefore the secondary emission ratio falls to a minimum at *A*. Beyond the minimum point secondary emission from the target starts to occur, and these electrons are accumulated by the collector, so increasing the secondary emission ratio.

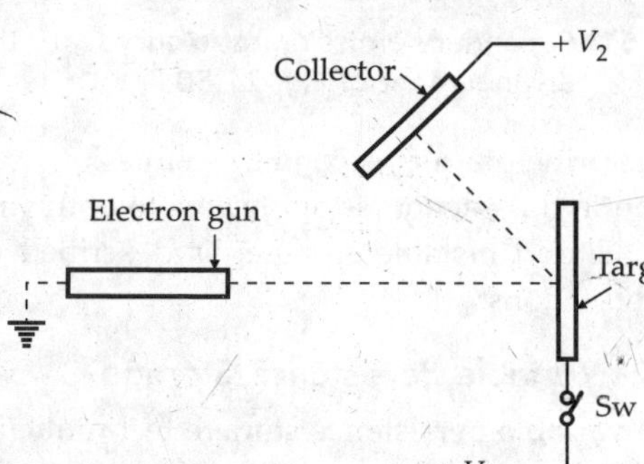


**Fig. 21.48** Simple electrode arrangement to illustrate secondary emission.

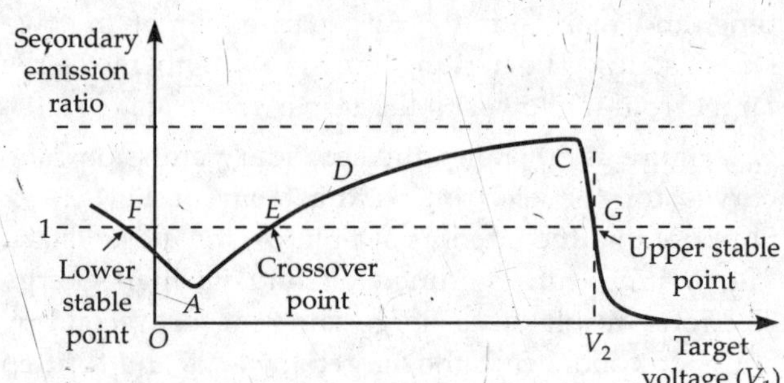


**Fig. 21.49** Variation of secondary emission ratio for the electrode arrangement shown in Fig. 21.48.

21

The secondary emission ratio increases through the crossover point $E$ until it reaches a peak at $C$. Beyond this point secondary electrons emitted from the target are attracted back, in numbers greater than those which reach the collector, so that the secondary emission ratio decreases sharply. The curve reaches the upper stable point at $G$, where the primary and secondary currents are equals, and then decreases to zero.

The lower and upper stable points represent the erased and written conditions of the CRT screen, and in the absence of a target voltage the target can remain in one of these two stages only. Suppose for instance that the target is in a state to the left of the crossover point $E$ and switch SW is opened to remove the target supply voltage. Because the secondary emission ratio is less than one, fewer electrons leave the target than arrive at its surface, so the target is driven progressively negative, until point $F$ is reached. Now the number of electrons reaching the target equals the number which are emitted from it. This represents a stable erased state. If the target was to the right of the crossover point when switch SW is opened, then, since the secondary emission ratio is greater than one, more electrons leave the target than arrive at its surface. The target is therefore driven positive unitl it reaches the upper stable point $G$ which represents the written condition of the screen. The target will stay at the upper or lower stable point so long as the electron gun is on. At the crossover point the target is unstable, and it will move up or down on the curve depending on the direction in which it is sent by noise.

Figure 21.50 shows an arrangement using a segmented multiple target, with a collector mesh in front through which the electrons can pass. Two guns are used, a high energy writing gun, which emits a narrow beam of electrons, and a flood gun which covers the target area with a continuous stream of low energy electrons. The flood gun remains on all the time, and maintains the target in either of the two stable states. The writing gun is used to move the target from the lower to the upper stable state.

Figure 21.50 shows the secondary emission ratio curves for the electron arrangement of Fig. 21.50. Suppose that the target is at point $A$, the erased state. The writing gun is truned on and its high energy electrons result in a large amount of secondary emission, and drive the target towards the written state. Up to point $C$ the flood gun opposes the writing gun, but thereafter it aids it ; so the writing gun can be swtiched off and the target will reach the written state at $D$. To erase the target the collector mesh is

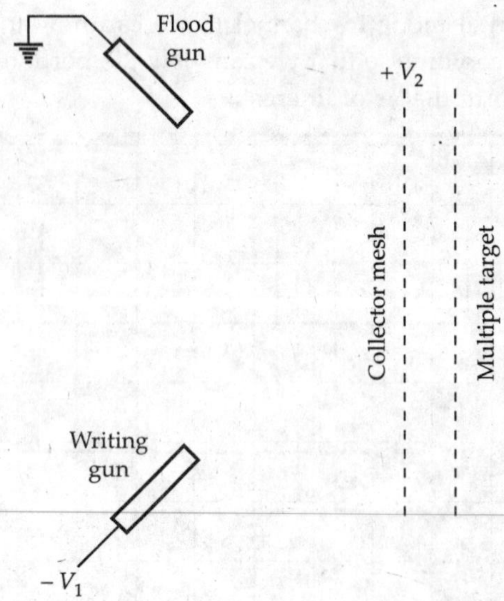


**Fig. 21.50** Multiple target and flood gun arrangement.

momentarily taken to a negative voltage. This repels secondary emission back into the target, moving its potential negative due to the electrons reaching it from the flood gun. When the target reaches point $A$ no more chanrging occurs. The collector mesh is now returned to its original voltage of $+V_2$. This is done gradually to prevent the target from being accidentally driven beyond the crossover point, due to capacitive coupling, and so reaching a written state again.

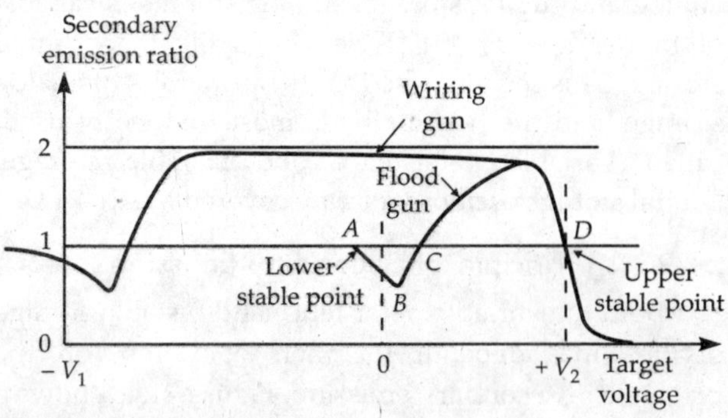


**Fig. 21.51** Secondary emission ratio curves for the arrangement of Fig. 21.50.

The principle of secondary emission storage described in this section, is applicable to both variable persistence and bistable storage, as described in the next two sections.

### 21.23.2 Variable Persistence Storage

The variable persistence storage technique is also known as halftone storage or mesh storage. Figure 21.52 illustrates the construction of a CRT using this storage technique. There are two screens, a storage mesh which retains the image traced on it by the writing gun, and the phosphor screen, which is very similar to that used in a conventional CRT.

Dielectric material, consisting of a thin layer of material such as magnesium fluoride, is deposited on the storage mesh, and this acts as the stroage target. The writing gun is at a high negative voltage, the flood gun at a few volts negative, the collector mesh at about 100 V positive, and the storage mesh at ground potential or a few volts negative. The collimator consists of a conductive coating on the inside surface of the CRT. It is biased so as to distribute the flood gun electrons evenly over the surface of the target, and causes the electrons to be perpendicular to the storage mesh.

When the writing gun is aimed at the storage target it causes areas where it strikes to be charged to a positive potential, due to secondary emission as in Fig. 21.51. These areas are maintained at their upper stable point, even after the writing gun is switched off, due to the action of the flood gun. Electrons from the flood gun also pass through those areas which are positively charged, causing the phosphor beyond to glow, and displaying the original signal traced by the writing gun.

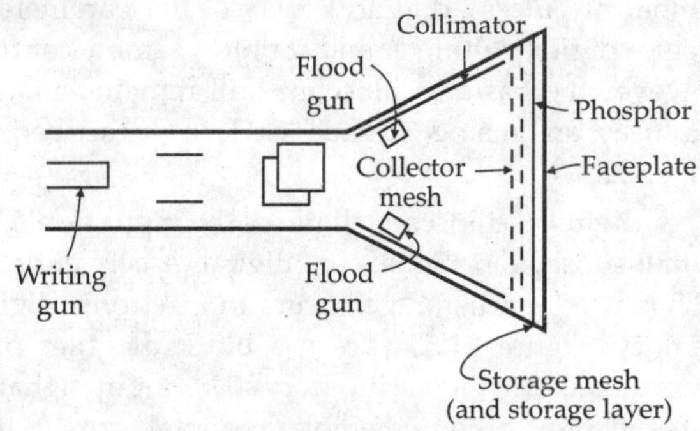


**Fig.21.52** Construction of a variable persistence storage tube.

The pattern stored on the storage target lasts for about one hour, but it can display a bright image for about one minute. The stored pattern fades due to electrons from the flood gun charging other parts of the storage surface, giving an impression that the whole screen has been written. This is known as fading positive. To erase an image which has been stored the storage mesh is momentarily raised to the same positive potential as the collector mesh.

The variable persistence storage tube is capable of producing halftone or gray scale images. As the charge on the storage surface is increased, beyond a critical threshold value, more electrons pass through it onto the display phsophor. The charge on different parts of the storage screen can be varied either by control of the time spent by the beam on different areas, or by varying the beam current over different areas. Usually four levels can be readily obtained, for example a mesh voltage of – 10 V represents an off state, – 5 V and – 2.5 V represent intermediate states, and 0 V represents full brightness.

Variable persistence tubes enable the persistence or storage time of the image to be varied from a few milliseconds to several hours. It can be done in the CRT shown in Fig. 21.52 by applying a repetitive negative pulse, of – 4 V to – 11 V, to the storage mesh, so as to discharge the positively charged areas. The control over the discharge time is usually obtained by operating the pulses at a constant frequency of about 1 kHz, but varying the pulse width ; the narrowest pulse gives the largest persistence as it takes longer to discharge the storage layer.

Variable persistence storage finds many applications, such as the storage of an entire waveform of a slow moving signal, which then fades before the next trace is written. It can also be used to store several traces before the first one fades, so as to see how the signal changes with time.

### 21.23.3 Bistable Storage Oscilloscope

The bistable storage tube is between two and ten times slower than a comparable variable persistence tube. However it is capable of much longer storage times, measured in hours rather than in minutes as for variable persistence. The bistable tube is also capable of operating in a split screen mode, where half the screen has storage capability and the other half is a conventional phosphor tube.

Figure 21.53 shows the construction of a bistable storage tube. Unlike the variable persistence tube the same phosphor screen is used for both storage and display. The screen consists of P1 phosphor, doped to have good secondary emission characteristics and deposited on a conductive backplane made from a transparent metal film. The phosphor layer consists of a thin coating of scattered particles, so as to give a discontinuous surface. This stops the boundary migration of stored charge. The thin phosphor coating also has a short life since it suffers from light output reduction with time.

The conductive film is held at a low positive potential, so as to attract a cloud of low energy electrons from the flood gun. These electrons have insufficient energy to penetrate the phosphor, and are gathered by the collimator. When the write gun is switched on, its high energy electrons result in secondary emission from areas traced on the screen, moving these to the upper stable point. The trace is therefore at a high positive potential, and this is maintained due to the low leakage of the phosphor. The low energy electrons from the flood gun are now attracted to the positive areas of the screen, and go through the phosphor to reach the metal film at the back. In passing through the phosphor they cause it to glow, displaying the area traced out by the writing gun. The screen can be erased by setting the metal film to a negative voltage, repelling the electrons back into the storage area, and returning the phosphor to the low stable point.

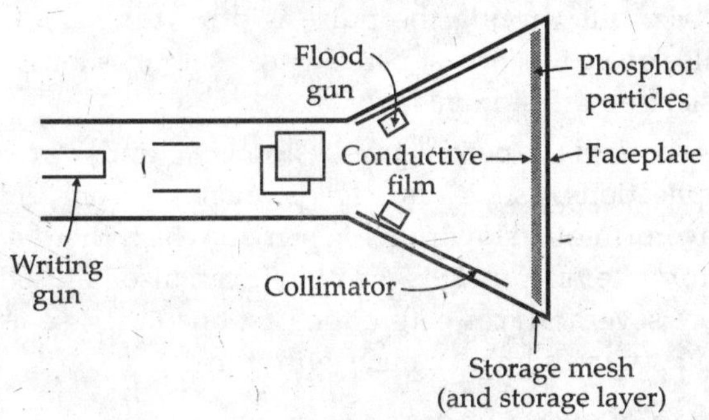


**Fig. 21.53** Construction of a bistable storage tube.

### 21.23.4 Fast Storage Oscilloscopes

Two modifications are made to the conventional storage CRTs, to enhance the speed with which they can capture transient information. These methods are known as transfer storage and expansion storage.

*Transfer storage* uses an intermediate mesh target, called the fast mesh, which has been optimized for speed. The waveform is initially written onto this mesh and it is then transferred, at lower speed, onto a second mesh which has been optimized for storage time. The second mesh can be bistable or variable persistence.

*Expansion storage,* shown in Fig. 21.54 uses an intermediate mesh which is about one fifth the size of the viewing screen. This mesh can be written on very quickly, and it is placed between the collector and accelerator meshes. The image on the storage mesh is magnified by a static electron lens system, to give a full size display on the viewing. screen. This system is capable of writing speeds of 2000 cm/μs, and can capture signals at frequencies of 300 MHz.

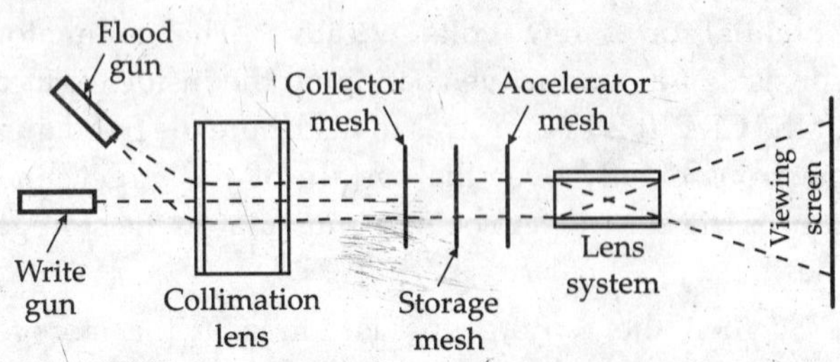


**Fig. 21.54** Expansion storage tube construction.

## 21.24 DIGITAL STORAGE OSCILLOSCOPES

### 21.24.1 Principle of Operation

The availability of electronic circuitry at low cost has enabled many digital features to be added to analog oscilloscopes. Examples of these are generation of a trigger after an elapsed time or after a count of a number of pulses ; digital display of the parameters; integral digital voltmeter and counter : remote control. However the basic oscilloscope still remains analog, and uses an analog storage CRT, as described in Art. 21.2.

A digital oscilloscope digitises the input signal, so that all subsequent signals are digital. A conventional CRT is used, and storage occurs in electronic digital memory. Figure 21.55 shows a block diagram of a basic digital storage oscilloscope. The input signal is digitised and stored in memory in digital form. In this state it is capable of being analysed to produce a variety of different information. To view the display on the CRT the data from memory is reconstructed in analog form.

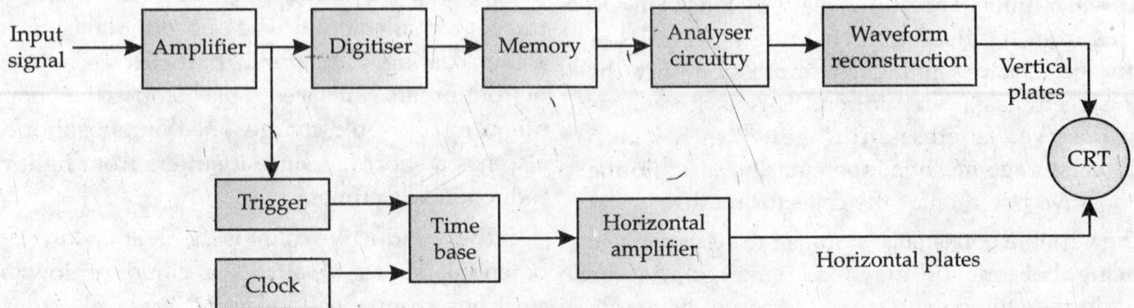


**Fig. 21.55** Block diagram of a basic digital storage oscilloscope.

Digitising occurs by taking a sample of the input waveform at periodic intervals. In order to ensure that no information is lost, sampling theory states that the sampling rate must be at least twice as fast as the highest frequency in the input signal. If this is not done then aliasing will result, as shown in Fig. 21.56. This requirement for a high sampling rate means that the digitiser, which is an analog to digital converter, must have a fast conversion rate. This usually requires expensive flash analog to digital converters, whose resolution decreases as the sampling rate is increased. It is for this reason that the bandwidth and resolution of a digital oscilloscope is usually limited by its analog to digital converter.

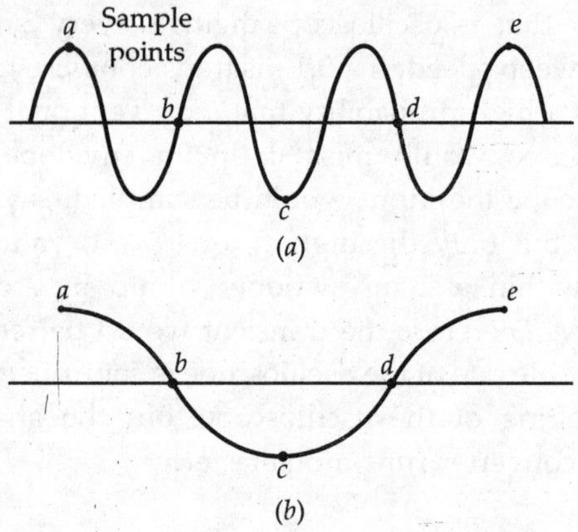


**Fig. 21.56** Effect of a low sampling frequency : (*a*) input signal, (*b*) aliased signal.

One method of overcoming the need for a high performance converter is to use an analog store, as in Fig. 21.57. The input signals are sampled, and these are stored in an analog shift register. They can then be read out at a much slower rate to the analogue to digital converter, and the results stored in a digital store. This method allows operation at up to 100 mega samples per second, and has the advantage that a low cost analog to digital converter can be used, whose resolution does not decrease as the sampling rate is changed. The disadvantage is that the oscilloscope cannot accept data during the digitising period, so it has a blind spot. At low sweep speed operations it is usual to switch out the analog memory, feeding the analog to digital converter in real time.

Many different input channels are used with digital storage oscilloscopes. However if all these channels share a common store, through a multiplexer, then the memory available to each channel is reduced. Oscilloscopes with up to 40 channels are commercially obtainable, with a storage capability of 25000 dots. Several oscilloscopes also have floppy disc storage capability to allow non volatile storage of waveforms, which can later be recalled into the oscilloscope and manipulated.

### 21.24.2 Waveform Reconstruction

Although the input signal may be sampled at greater than twice the highest signal frequency, aliasing can still result when the output is present as a series of dots, corresponding to the sampled values. This is illustrated in Fig. 21.58(*a*), where the user's mind connects together the dots which are physically closest to each other, rather than those which are closest on the time scale.

In the illustration of Fig. 21.58(*a*) it is difficult to visualise the final waveform, and oscilloscopes generally have the facility to interpolate between the dots, if required by the user. Two techniques are used,

21

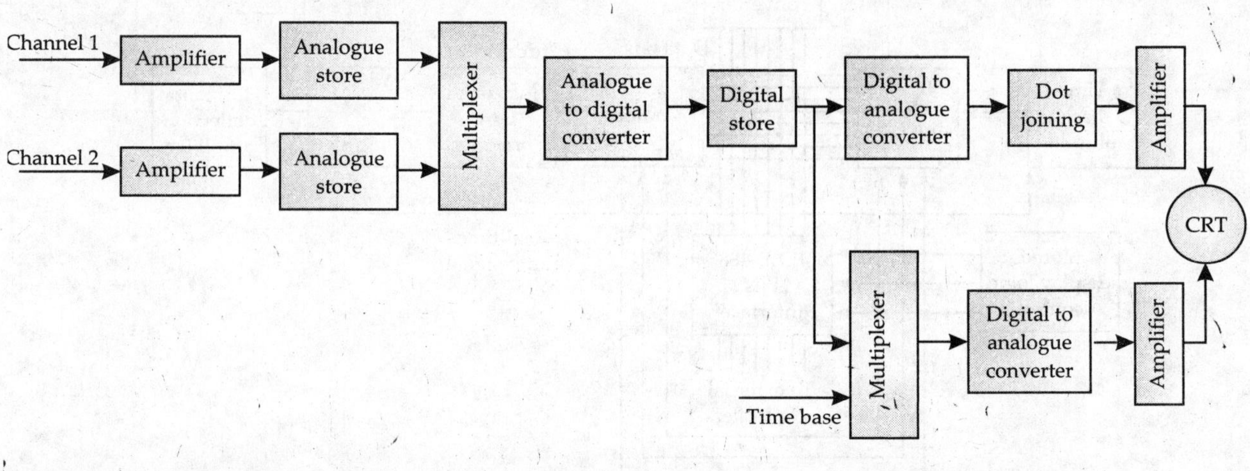


**Fig. 21.57** A digital oscilloscope which uses analogue storage to eliminate the need for a very fast analogue to digital converter.

linear interpolation and sinusoidal interpolation. In *linear interpolation*, shown in Fig. 21.58(*b*) a straight line is used to connect the dots together. This works well on a pulsed or square waveform, but not on a sinusoidal wave, Fig. 21.58(*c*) shows that *sinusoidal interpolation* gives a much better fit for sine waves, although it is not suitable for pulse or square waves.

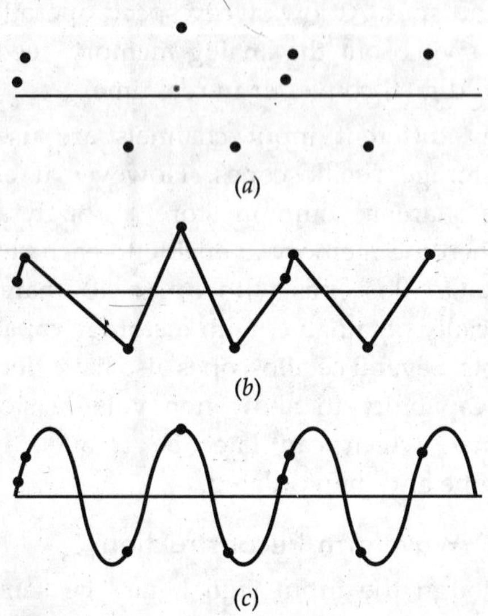


**Fig. 21.58** Illustration of interpolation methods : (*a*) without interpolation, (*b*) Linear interpolation. (*c*) sinusoidal interpolation.

Another practical diagram of a digital oscilloscope is shown in Fig. 21.59.

Another problem with the sampling technique used in digital oscilloscopes is that it can miss short term transient, or 'glitches', which occur in between the sample points. To overcome this problem *envelope mode oscilloscopes* may be used. These have special logic circuitry which causes the sample and digitising circuitry to run at a high speed, independent of the setting of the display time. At each sample the value is compared with the previous stored sample, and the higher (or lower) value is stored. This is continued for the screen interval, so that for that interval the highest and lowest points are always stored. For example, suppose that an oscilloscope digitises every 2 ms, at a given sweep speed. If a 0.1 ms transient were to occur there is a high probability that a conventional digital oscilloscope would miss it. In an envelope mode oscilloscope the input would be sampled, say, every 200 ns, but only the highest, or lowest, values that occur within a 2 ms window would be stored in memory. Therefore the transient would be recorded. The sample rate of the oscilloscope is controlled by the time setting of the oscilloscope, but the analog to digital converter runs much faster.

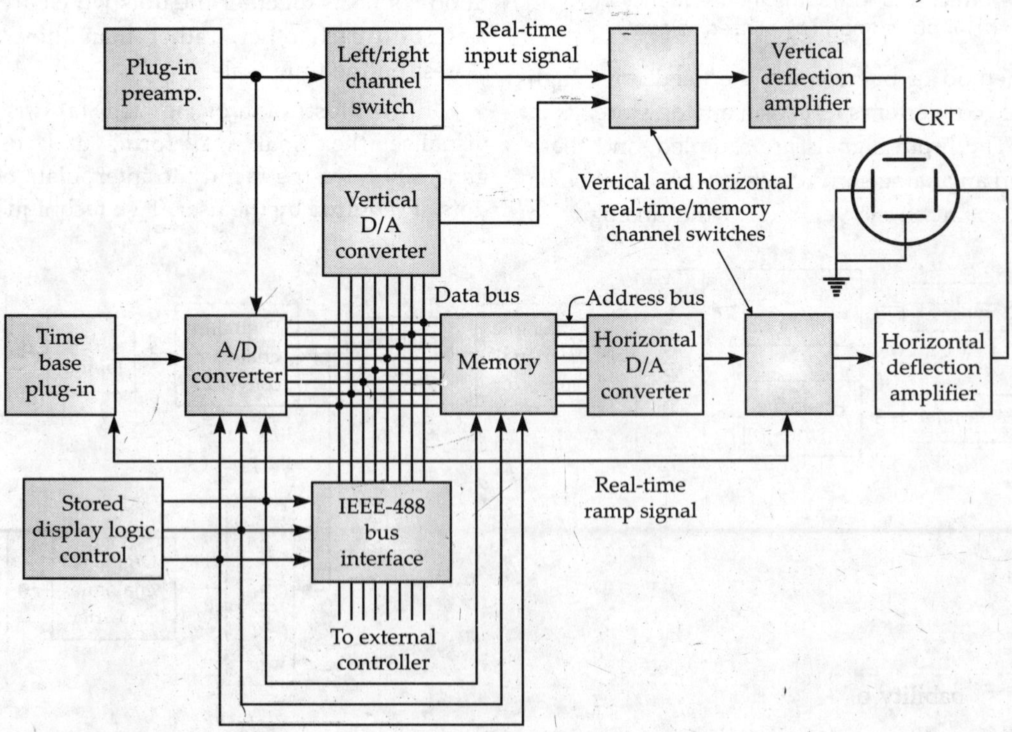


**Fig. 21.59** Block diagram of a digital storage oscilloscope.

## 21.25 COMPARISON BETWEEN ANALOG AND DIGITAL STORAGE OSCILLOSCOPES

The advantage of the analog storage oscilloscope is that it has a higher bandwidth and writing speed than a digital oscilloscope, being capable of operating speeds of about 15 GHz. The digital oscilloscope is primarily limited in speed by the digitising capability of the analog to digital converter. Aliasing effects also limit the useful storage bandwidth (usb) of the oscilloscope to a value given by the ratio.

$$\text{usb} = \frac{\text{maximum digitising rate}}{\text{constant } C} \quad \text{...(21.26)}$$

The value of constant $C$ is dependent on the interpolation method used between the dots. For a dot display $C$ should be about 25, to give an eligible display: for straight line interpolation it should be about 10, and for sinusoidal interpolation $C$ should be about 2.5.

The digital storage oscilloscope has a CRT which is much cheaper than an analog storage oscilloscope, making replacements more economical. The digital oscilloscope is also capable of an infinite storage time, using its digital memory. Furthermore it can operate with a constant CRT refresh time, so giving a bright image even at very fast signal speeds. The digital storage oscilloscope is not, however, capable of functioning in a variable persistence storage mode.

The time base in a digital oscilloscope is generated by a crystal clock, so that it is more accurate and stable than an analog oscilloscope, where the time base is generated by a ramp circuit. The analog to digital converter used in a digital oscilloscope also gives it higher resolution than an analog oscilloscope. For example a twelve bit digitser can resovle one part in 4096. A conventional analog oscilloscope typically resolves to about 1 mm on the screen, that is to about one part in 50, equivalent to 6 bit resolution.

Digital storage oscilloscopes are also capable of operating in a look back mode, as described for waveform recorders. An analog oscilloscope collects data after it has been triggered. A digital storage oscilloscope is always collecting data, and the trigger tells it when to stop. The oscilloscope can stop immediately on trigger, so that all the stored information is pretrigger, or it can stop some time after it has received the trigger. If the delay is longer than the storage capability of the oscilloscope, then all the storage capability of the oscilloscope, then all the stored information is post trigger, as for an analog oscilloscope. The digital oscilloscope is also able to operate in a baby sitting mode. When the scope is triggered it prints out the stored results onto a hard copy recorder (or disc storage), and then re-arms itself ready for another reading.

## 21.26 ACCESSORIES OF CATHODE RAY OSCILLOSCOPES

The cathode-ray oscilloscope is one of the most useful instruments in the electronic industry. The usefulness of the oscilloscope is further extended by provision of accessories or auxiliary equipment. Some of the accessories are described below.

### 21.26.1 Calibrators

Many oscilloscopes have a built-in reference source of voltage which has usually a square waveform with a frequency of 1 kHz. The voltage may have a single magnitude or may be of several selectable magnitudes. Provision of this facility enables the oscilloscope amplifiers and time base to be calibrated and checked for accuracy.

### 21.26.2 Probes

The probe performs the very important function of connecting the test circuit to the oscilloscope without altering, loading, or otherwise disturbing the test circuit.

The probes are of *three* different types :

(*i*) direct reading probe,

(*ii*) circuit isolation probe, and

(*iii*) detector probe.

**1. *Direct reading probe*.** This probe is simplest of all the probes and uses a shielded co-axial cable. It avoids stray pick-ups which may create problems when low level signals are being measured. It is, usually used for low frequency or low impedance circuits. However, in using the shielded probe, the shunt capacitance of the probe and cable is added to the input impedance and capacity of the scope and acts to lower the response of the oscilloscope to high impedance and high frequency circuits.

External high impedance probes are used to increase the input resistance and reduce the effective input capacitance of an oscilloscope. A resistor and capacitor combination can be added to an oscilloscope as shown in Fig. 21.60, in effect, moving the input terminals from the front panel of the instrument to the end of the probe, supposing, it is intended to attenuate the signal by a factor of 10. By using Eqns. 21.24 and 21.25,

21

where $k = 10$, the values of $R_1$ and $C_1$ can be computed as under

$$R_1 = R_2(k-1) = (1 \times 10^6)(10-1) = 9\,M\Omega$$

$$C_1 = C_2/(k-1) = 30 \times 10^{-12}/(10-1) = 3.33\,pF$$

New input impedance $R_i$ is the total resistance,

where $R_i = R_1 + R_2 = 10\,M\Omega$

and $C_i = \dfrac{C_1 C_2}{C_1 + C_2} = 3\,pF$

The input resistance has been increased by a factor of 10 while, at the same time, the input capacitance has been decreased by a factor of 10. The combination of $R_1$ and $C_1$ is called ×10 (times 10) probe.

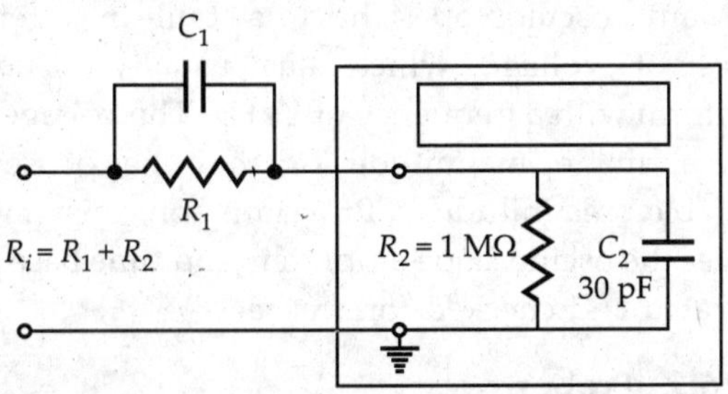


**Fig. 21.60** External High Impedance Probe.

Capacitor $C_1$, is usually adjustable to compensate for differences in input capacitance between oscilloscopes. If the probe capacitance is adjusted to the wrong value, the oscilloscope will exhibit a factor frequency response. The adjustment of probe is usually checked by displaying a square wave on the CRT screen. If the probe is not properly compensated, the display of square waveform will be adversely affected as shown in Fig. 21.61. If the value of $C_1$ is too

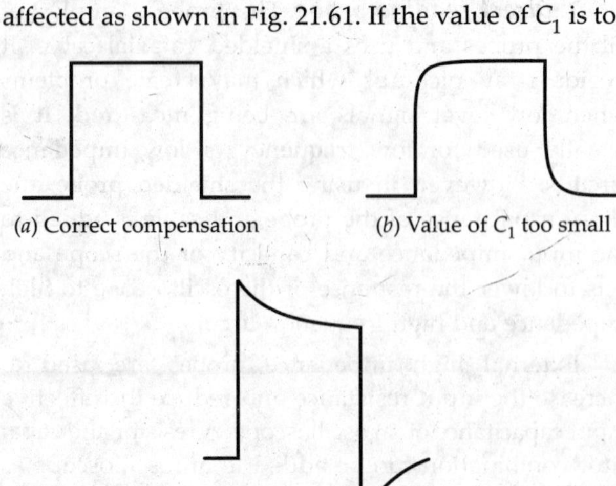

(*a*) Correct compensation (*b*) Value of $C_1$ too small

(*c*) Value of $C_1$ too large.

**Fig. 21.61** Effect of probe capacitance compensation on display of square waveform.

small, the leading edge of the square wave is rounded off but if value of $C_1$ is too large, the leading edge of square wave overshoots.

**2.** ***Isolation probe.*** Isolation probe is used in order to avoid the undesirable circuit loading effects of the shielded probe. The isolation of the probe, which is used along with a capacitive voltage divider, decreases the input capacitance and increases the input resistance of the oscilloscope. This way the loading effects are drastically reduced.

**3.** ***Detector probe.*** When analyzing the response to modulated signals used in communication equipment like AM, FM and TV receivers, the detector probe functions to separate the low frequency modulation component from the high frequency carrier. The amplitude of the modulator carrier (which is proportional to the response of the receiver to the much high frequency carrier signal) is displayed on the oscilloscope by rectifying and bypassing action. This permits an oscilloscope capable of audio-frequency response to perform signal tracing tests on communication signals in the range of hundreds of MHz, a range, which is beyond the capabilities of all oscilloscopes except the highly specialized ones.

### 21.26.3 Cameras

The best method of permanent records of oscilloscope traces for analysis is the use for photography. Special cameras are used for the purpose and they are of the following *two* types :

(*i*) **35 mm film.** The 35 mm unit is usually prefocused and bolted over the tube face. Some cameras have a fixed lens aperture, the exposure is simply controlled by hand operation of flapper shutter. More sophisticated arrangements use exposure and aperture control. Panchromatic and special blue sensitive films are commonly used because they are faster and hence are better suited for recording of high speed transients. Both these films give a "negative" record from which positive prints having white traces on a black background are made. Another type of 35 mm film gives a photograph which has a black trace on a white background. These films are, however, difficult to enlarges.

(*ii*) **Polaroid film.** This photographic technique approaches instant photography. A permanent record is obtained within 10-20 s. However, this process is much more expensive than the 35 mm photographic technique. But the relative cost may be reduced by using the facility of some of the cameras to be moved relative to the screen, making it possible to place several traces on a single print. The camera is again prefocused, but is normally provided with aperture and exposure controls. The photographic record is obtained on Polaroid film which is a white trace on a black background. Since in many situations, it is the time which is much more important, the Polaroid photography is normally used under such circumstances.

### 21.26.4 Electronic Switch

The electronic switch is an extremely useful accessory for displaying two signals simultaneously on single trace oscilloscopes. The electronic switch is essentially a square wave generator. The two signals which are to be displayed are each connected to a separate grid of a dual-section amplifier tube, operating with cutoff bias. The square wave is applied in such a manner that it alternately unblanks each tube section, thereby presenting each signal alternately to the vertical amplifier of the oscilloscope. When the switching rate, which is determined by the square wave frequency, is high enough each signal appears continuous even though it is actually interrupted periodically. The switching rates may be as high as 100 kHz.

**Example 21.8** *Calculate the maximum velocity of the beam of electrons in a CRT having a cathode anode voltage of 800 V. Assume that the electrons to leave the cathode with zero velocity. Charge of electron = $1.6 \times 10^{-19}$ C and mass of electron = $9.1 \times 10^{-31}$ kg.*

**Solution.** From Eqn. 21.5, velocity of electron is :

$$v_{ox} = \sqrt{\frac{2eE_a}{m}}$$

$$= \sqrt{\frac{2 \times 1.6 \times 10^{-19}}{9.1 \times 10^{-31}}}$$

$$= 16.8 \times 10^3 \text{ m/s.}$$

**Example 21.9** *An electrically deflected CRT has a final anode voltage of 2000 V and parallel deflecting plates 1.5 cm long and 5 mm apart. If the screen is 50 cm from the centre of deflecting plates, find (a) beam speed, (b) the deflection sensitivity of the tube, ana (c) the deflection factor of the tube.*

**Solution.** Velocity of beam

$$v_{ox} = \sqrt{\frac{2 \times 1.6 \times 10^{-19}}{9.1 \times 10^{-31}} \times 2000}$$

$$= 26.5 \text{ m/s.}$$

From Eqn. 21.17, deflection sensitivity,

$$S = \frac{Ll_d}{2dE_a} = \frac{0.5 \times 1.5 \times 10^{-2}}{2 \times 5 \times 10^{-3} \times 2000} \text{ m/V} = 0.375 \text{ mm/V.}$$

$$\text{Deflection factor, } G = \frac{1}{S} = \frac{1}{0.375} = 2.66 \text{ V/mm.}$$

**Example 21.10** *A CRT has an anode voltage of 2000 V and parallel deflecting plates 2 cm long and 5 mm apart. The screen is 30 cm from the centre of the plates. Find the input voltage required to deflect the beam through 3 cm. The input voltage is applied to the deflecting plates through amplifiers having an overall gain of 100.*

**Solution.** From Eqn. 21.16, deflection

$$D = \frac{Ll_d E_d}{2dE_a}$$

∴ Voltage applied to deflecting plates

$$E_d = \frac{2dE_a D}{Ll_d} = \frac{2 \times 5 \times 10^{-3} \times 2000 \times 3 \times 10^{-2}}{0.3 \times 2 \times 10^{-2}}$$

$$= 100 \text{ V.}$$

∴ Input voltage required for a deflection of 3 cm

$$= \frac{E_d}{\text{gain}} = \frac{100}{100} = 1 \text{ V.}$$

**Example 21.11** *Calculate the velocity of the electron beam in an oscilloscope if the voltage applied to its vertical deflection plates is 2000 V. Also calculate the cutoff frequency if the maximum transit time is 1/4 of a cycle. The length of horizontal plates is 50 mm.*

**Solution.** Velocity of electron beam

$$v_{ox} = \sqrt{\frac{2 \times 16 \times 10^{-19}}{9.1 \times 10^{-31}} \times 2000} = 26.5 \times 10^6 \text{ m/s.}$$

Cutoff frequency

$$f_c = \frac{v_{ox}}{4l} = \frac{26.5 \times 10^6}{4 \times 50 \times 10^{-3}}$$

$$= 132.5 \times 10^6 \text{ Hz} = 132.5 \text{ MHz.}$$

**Example 21.12** *Voltage $E_1$ is applied to the horizontal input and voltage $E_2$ is applied to the vertical input of a CRO. $E_1$ and $E_2$ have the same frequency. The trace is an ellipse. The slope of the major axis is positive. The maximum vertical value is 2.5 divisions and the point where the ellipse crosses the vertical axis is 1.25 divisions. The ellipse is symmetrical about the horizontal and vertical axis. Determine the possible phase angles of $E_2$ with respect to $E_1$.*

**Solution.** Referring to Fig. 21.34(*a*),

$$\sin\phi = \frac{Y_1}{Y_2} = \frac{1.25}{2.50} = 0.5 \qquad \therefore \quad \phi = 30°$$

Thus possible phases are 30° or 330°.

21

***Example 21.13*** *The sketches shown in Fig. 21.62 display Lissajous patterns for cases where voltages of same frequency out of different phase are connected to Y and X plates of the oscilloscope. Find the phase difference in each case. The spot generating the patterns moves in a clockwise direction. Calculate the angles if the spot generating the patterns moves in the anticlockwise direction.*

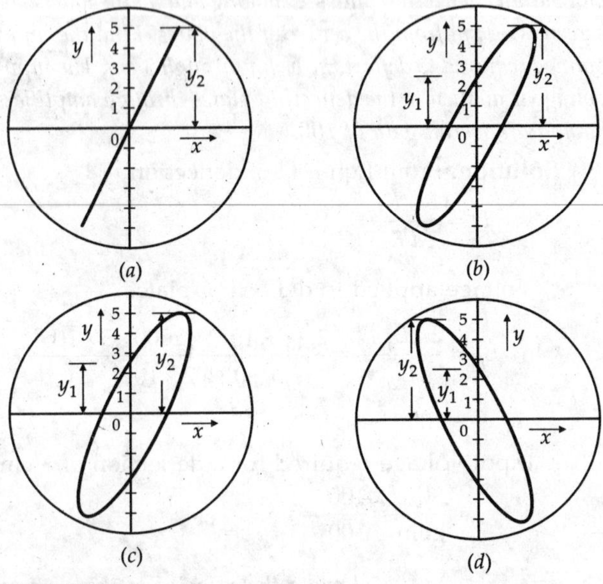


**Fig. 21.62** Diagrams of Example 21.13.

**Solution.** The spot generating the patterns moves in clockwise direction.

(*a*) $\sin\phi = \frac{Y_1}{Y_2} = \frac{0}{5} = 0$

$\therefore \phi = 0°$

(as the Lissajous pattern is in 1st and 3rd quadrants).

(*b*) $\sin\phi = \frac{2.5}{5} = 0.5$

$\therefore \phi = 30°$

(as the major axis of the pattern is in 1st & 3rd quadrants).

(*c*) $\sin\phi = \frac{3.5}{5} = 0.7$

$\therefore \phi = 45°$

(as the major axis of the pattern is in 1st & 3rd quadrants).

(*d*) $\sin\phi = \frac{2.5}{5} = 0.5$

$\therefore \phi = 180° - 30° = 150°$

(as the major axis is in 2nd and 4th quadrants).

If the spot generating the patterns moved in the counter clockwise direction, the angles would be :

(*a*) 180°, (*b*) –30°,

(*c*) –45°, (*d*) 180 + 30 = 210°.

# Review Questions

1. Describe an overview of applications of a CRO.
2. Draw the block diagram of a general purpose CRO and explain the functions of the following controls
   (*i*) intensity
   (*ii*) focus
   (*iii*) horizontal and vertical positioning
   (*iv*) synchronization.
3. Describe the different parts of a CRT.
4. Derive an expression for vertical deflection of an electron beam in a CRT.
5. Describe the different types of phosphor materials used in a CRO and list their applications.
6. Explain the different types of graticules used in a CRO. Describe their advantages and disadvantages.
7. Explain the functioning of a time base generator in a CRO.
8. What are the different types of amplifiers used for CROs ? Describe the basis on which they are classified.
9. Describe in details the vertical amplifier used in a CRO.
10. Describe the phenomenon of synchronization of vertical input signal to its sweep generator.
11. Describe the different types of sweeps used in a CRO. Explain their spheres of application.
12. Describe the function of attenuators in CROs. Explain how are they designed with particular reference to frequency compensation.
13. Describe the following :
    (*i*) Sources of Synchronization.
    (*ii*) Blanking circuit.
    (*iii*) Z-axis modulation.
    (*iv*) Astigmatism control.
14. Describe the following types of oscilloscopes
    (*i*) Dual trace type (*ii*) Dual beam type.
15. Describe in details the construction and working of an analog type storage oscilloscope. Explain the principle of secondary emission.

16. Explain the following features of an analog type storage oscilloscope :
 (*i*) bistable persistence storage
 (*ii*) bistable storage
 (*iii*) fast storage.
17. Describe the principle of working and circuit diagram of a digital oscilloscope.
18. Discuss the advantages and disadvantages of analog and digital type of oscilloscopes.
19. Discuss in details the delayed sweep.
20. Describe how the following measurements can be made with the use of a CRO :
 (*i*) frequency
 (*ii*) phase angle.

# Unsolved Problems

1. The deflection sensitivity of an oscilloscope is 35 V/cm. If the distance from the deflection plates to the CRT screen is 16 cm, the length of the deflection plates is 2.5 cm, and the distance between the deflection plates is 1.2 cm. What is the acceleration anode voltage ? **[Ans.** 583 V]
2. What is the amplitude of the voltage $V_0$ in Fig. 21.63 after 20 ms ? **[Ans.** 22 V]

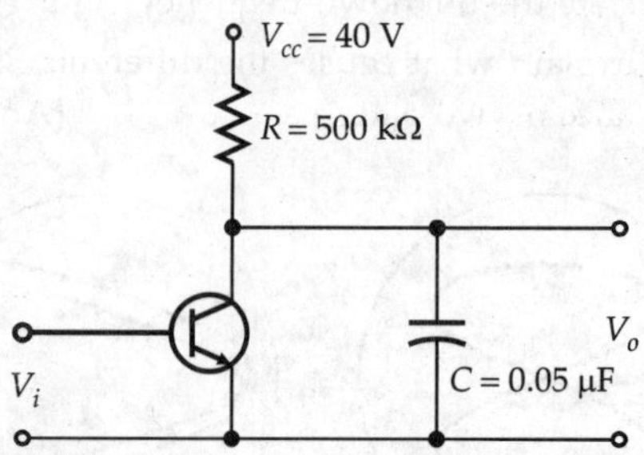


**Fig. 21.63** Circuit for Problem 2.

3. Identify each of the items indicated in the drawing in Fig. 21.64. **[Ans.** 22 V]

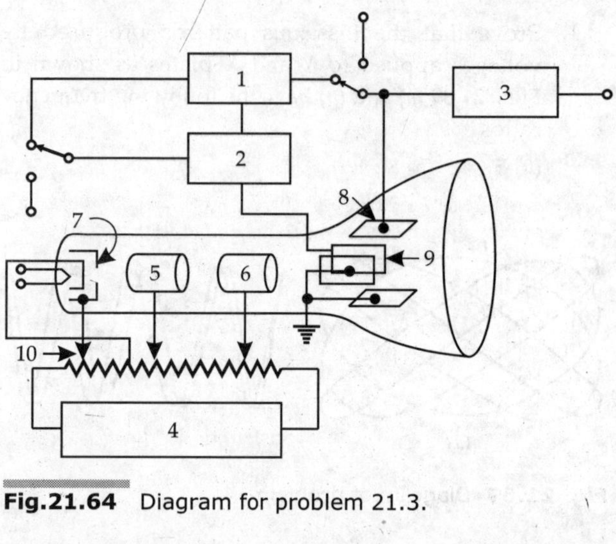


**Fig.21.64** Diagram for problem 21.3.

4. Compute the value of resistors in the attenuator network shown in Fig. 21.65 to provide the volts/division values indicated for each position.

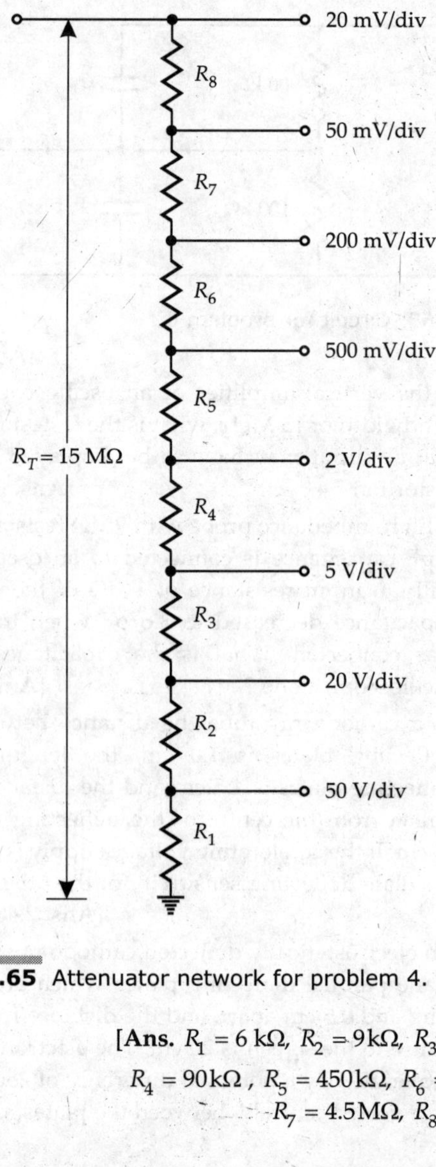


**Fig. 21.65** Attenuator network for problem 4.

**[Ans.** $R_1 = 6$ kΩ, $R_2 = 9$kΩ, $R_3 = 45$kΩ, $R_4 = 90$kΩ, $R_5 = 450$kΩ, $R_6 = 900$kΩ, $R_7 = 4.5$MΩ, $R_8 = 9$MΩ]

21

5. The waveform shown in Fig. 21.66 is observed on a CRT screen. If the time/div switch is set 10 μs and the volt/div switch is set to 200 mV. Determine the frequency and peak-to-peak amplitude of the signal. [**Ans.** 33.33 kHz, 600 mV]

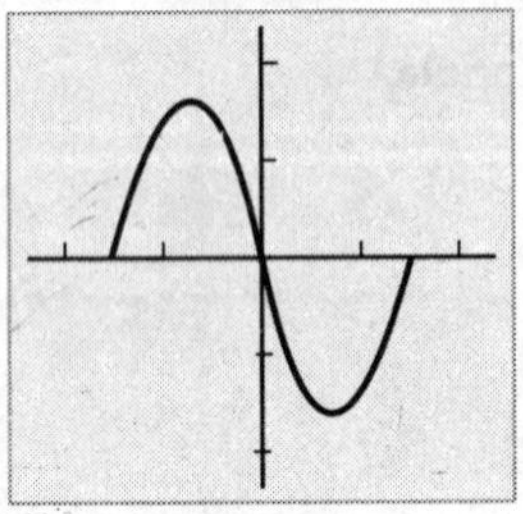

**Fig.21.66** CRT display for problem 5.

6. What value should $C_1$ have for $V_0$ to be equal to $0.1V_i$ in the circuit shown in Fig. 21.67 ?

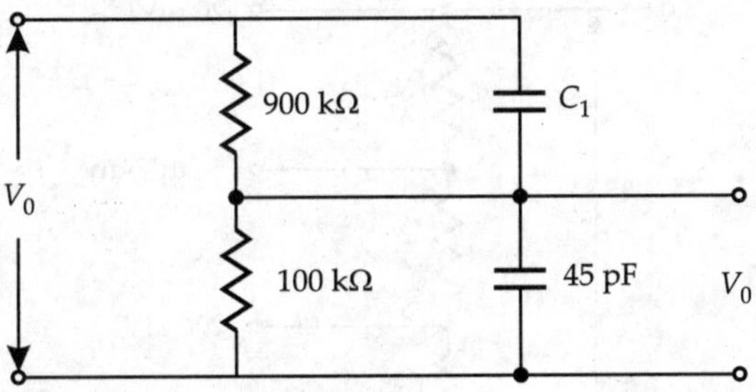


**Fig. 21.67** Circuit for problem 6.

[**Ans.** 5 pF]

7. If the vertical amplifier of an oscilloscope has a bandwidth of 15 MHz, what is the fastest rise time that an input may have to be displayed without distortion ? [**Ans.** 23.33 ns]

8. A high-impedance probe with 9 MΩ resistance and 4 pF capacitance is connected to an oscilloscope with an input resistance of 1 MΩ. If the effective capacitance decreased to 3.6 pF when the probe was connected. What is the capacitance of the oscilloscope alone ? [**Ans.** 36 pF]

9. In a cathode ray tube the distance between the deflecting plates is 1.0 cm, the length of the deflecting plates is 4.5 cm and the distance of the screen from the centre of the deflecting plates is 33 cm. If the accelerating voltage supply is 300 volt, calculate deflecting sensitivity of the tube. [**Ans.** 2.48 mm/V]

10. An electrostatically deflected cathode ray tube has plane parallel deflecting plates which are 2.5 cm long and 0.5 cm apart, and the distance from their centre to the screen is 20 cm. The electron beam is accelerated by a potential difference of 2500 V and is projected centrally between the plates. Calculate the deflecting voltage required to cause the beam to strike a deflecting voltage and find the corresponding deflection of the screen. [**Ans.** 200 V ; 4 cm]

11. Voltage $E_1$ is applied to the horizontal input and $E_2$ to the vertical input of a CRO. $E_1$ and $E_2$ have same frequency. The trace on the screen is an ellipse. The slope of major axis is negative. The maximum vertical value is 3 divisions and the point where the ellipse crosses the vertical axis is 2.6 divisions. The ellipse is symmetrical about horizontal and vertical axis. Determine the possible phase angle of $E_2$ with respect to $E_1$. [**Ans.** 120° or 210°]

12. A Lissajous pattern on an oscilloscope is stationary and has 5 horizontal tangencies and 2 vertical tangencies. The frequency of horizontal input is 1000 Hz. Determine the frequency of vertical input. [**Ans.** 2500 Hz]

13. In the oscilloscope pattern shown in Fig. 21.68 the signal connected to the vertical plates has the same frequency in parts (*a*) and (*b*). If the input to the horizontal plates is a 50 Hz voltage :
   (*a*) state the unknown frequency, and
   (*b*) explain what causes the difference in appearance of two patterns. [**Ans.** 25 Hz].

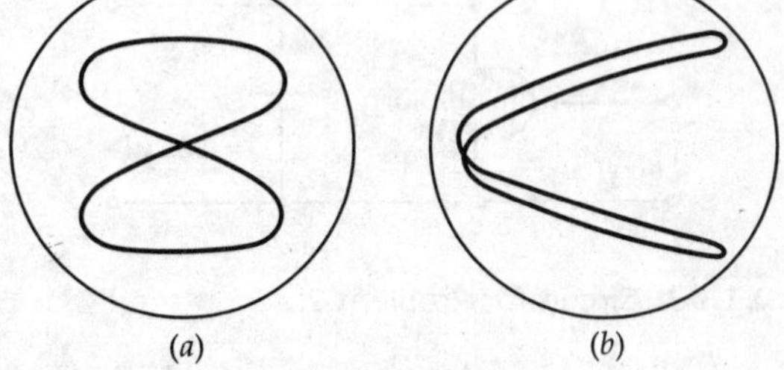


**Fig. 21.68** Diagram of problem 13.

14. Prove that the Lissajous patterns produced by voltages applied to *Y* and *X* plates as shown in Figs. 21.69 (*a*) and (*b*) have the following frequency ratios :
   (*a*) 5 : 4 (*b*) 8 : 1.

**Fig. 21.69** Diagram of problem 14.

**15.** Find the frequency of the vertical plates if the frequency applied to horizontal plate is 50 Hz for the patterns shown in Figs. 21.70 (*a*) and (*b*).
[**Ans.** (*a*) 25 Hz, (*b*) 100 Hz]

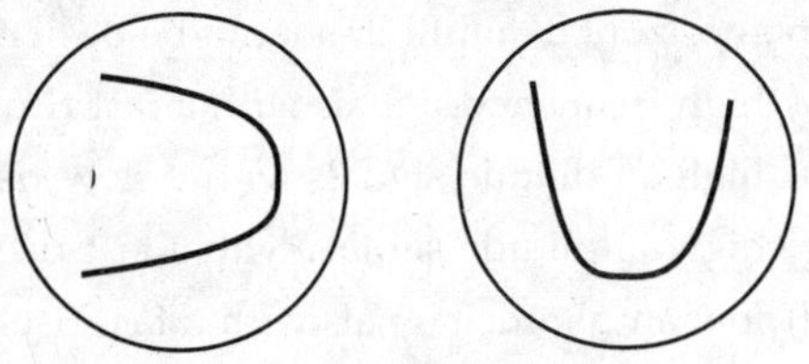


**Fig. 21.70** Diagram of problem 15.

**16.** A sinusoidal input is applied to the vertical input of an oscilloscope starting at $t = 0$. The following Lissajous patterns are obtained when a sinusoidal input is applied to the horizontal terminals. Estimate the phase shift between vertical and horizontal inputs. [**Ans.** 0°, 90°, 180°, 45°, 60° leading]

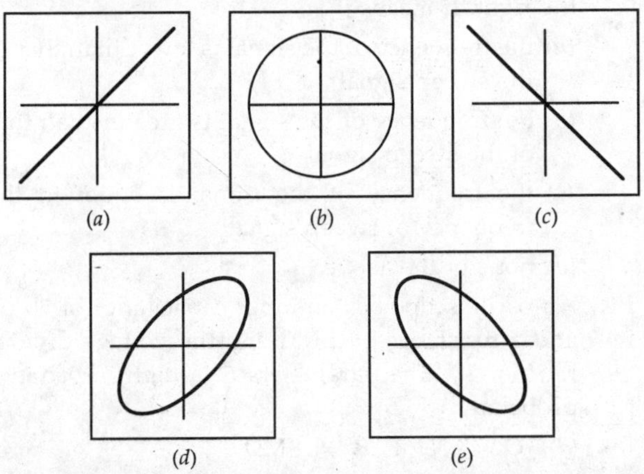


**Fig. 21.71** Lissajous figures of problem 16.

**17.** If the horizontal frequency is 51 Hz, determine the vertical frequency for the Lissajous pattern shown in Fig. 21.72

**Fig. 21.72** Lissajous figure for problem 17.

**18.** Calculate the ratio of vertical to horizontal frequencies for an oscilloscope which displays the following Lissajous figures shown in Fig. 21.73. The input in the case of (*d*) is applied to Z-axis.

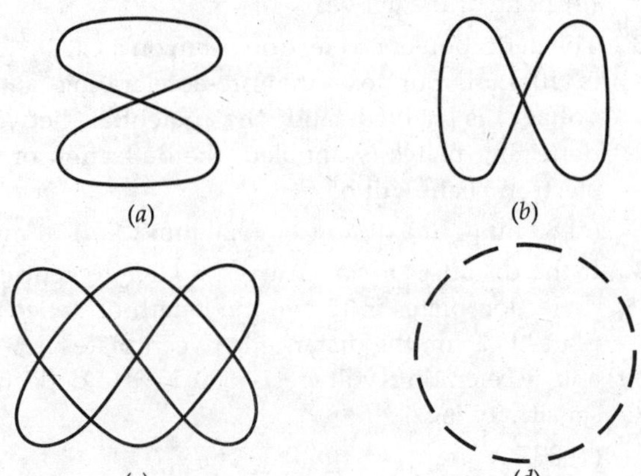


**Fig. 21.73** Lissajous Figures of problem 18.

[**Ans.** *fx* / *fy*: (*a*) 2 : 1 (*b*) 1 : 2 (*c*) 2 : 3 (*d*) 1 : 8]

**19.** Determine the phase shift between two sine waves, which is indicated by the pattern shown in Fig. 21.74. [**Ans.** 53°]

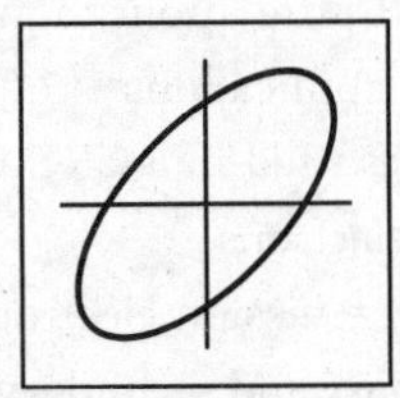

**Fig. 21.74** CRT display for problem 19.

**20.** A sampling oscilloscope is being used to observe a 400 MHz sine wave. A sampling pulse occurs every 3 ns. Draw five cycles of the 400-MHz signal and place a dot at the sampled point on each of the five cycles.

21

# Objective Type Questions

***Tick (√) the most appropriate answer :***

1. The source of emission of electrons in a CRT is :
   (*a*) *PN* function diode
   (*b*) a barium and strontium oxide coated cathode
   (*c*) accelerating anodes
   (*d*) post-accelerating anodes.

**2.** In a CRT the focusing anode is located
(*a*) between pre-accelerating and accelerating anodes
(*b*) after accelerating anode
(*c*) before pre-accelerating anode
(*d*) none of the above.

**3.** The deflection of an electron beam on a CRT screen is 10 mm. Suppose the pre-accelerating anode voltage is halved and the potential between deflecting plates is doubled, the deflection of the electron beam will be :
(*a*) 80 mm (*b*) 40 mm (*c*) 20 mm (*d*) 10 mm.

**4.** If the distance of screen from a CRT to centre of deflection plates is 15 cm, the length of deflection plates is 2 cm, the distance between plates is 1 cm and accelerating voltage is 500 V, the deflection sensitivity is :
(*a*) 33.2 V/cm (*b*) 0.03 cm/V
(*c*) 60.4 V/cm (*d*) 0.015 cm/V.

**5.** Post acceleration is needed in a CRO if the frequency of the signal is :
(*a*) less than 1 MHz (*b*) more than 1 MHz
(*c*) more than 10 MHz (*d*) more than 10 Hz.

**6.** $P_1$ phsophor material is used for display in CRTs for :
(*a*) photographic applications
(*b*) general purpose applications
(*c*) television applications (*d*) all the above.

**7.** An aquadag is used in a CRO to collect :
(*a*) primary electrons
(*b*) secondary emission electrons
(*c*) both primary and secondary emission electrons
(*d*) none of the above.

**8.** A thin aluminium film is usually deposited on the non viewing side of the phosphor because,
(*a*) it acts as a heatsink and prevents phosphor burn
(*b*) the light scatter from the phosphor is reduced
(*c*) it does not allow the screen to be negatively charged
(*d*) all the above.

**9.** During the retrace time, the electrons forming the horizontal beam
(*a*) move from left to right on the screen
(*b*) move from right to the left on the screen
(*c*) move from bottom to top of screen
(*d*) move from top to bottom of screen.

**10.** The bandwidth of a CRO is from 0 - 20 MHz. The fastest rise time a sine wave can have to be accurately reproduced by the instrument is
(*a*) 35 ns (*b*) 35 μs (*c*) 17.5 ns (*d*) 0.175 μs.

**11.** The horizontal amplifier should be designed for
(*a*) high frequency signals with a fast rise time
(*b*) high amplitude signals with a slow rise time
(*c*) high amplitude signals with a fast rise time
(*d*) low amplitude signals with a fast rise time.

**12.** A vertical amplifier for a CRO can be designed for :
(*a*) only a high gain (*b*) only a broad bandwidth
(*c*) a constant gain times bandwidth product
(*d*) all the above.

**13.** In an oscilloscope, when the unknown signal applied to the vertical plates is being synchronized with the sweep signal applied to horizontal plates, the pattern seen on the CRO screen moves towards the right. It means that,
(*a*) the frequency of the signal is lower than that of the sweep signal
(*b*) the frequency of the signal is greater than that of the sweep signal
(*c*) the frequency of the signal is equal to the frequency of sweep signal
(*d*) none of the above.

**14.** An oscilloscope has an input capacitance of 50 pF and a resistance of 2 MΩ and the voltage divider ratio of 10. The parameters of a high impedance probe are
(*a*) $C_1 = 5.55$ pF, $R_1 = 18$MΩ
(*b*) $C_1 = 3.33$ pF, $R_1 = 9$MΩ
(*c*) $C_1 = 11.1$ pF, $R_1 = 18$MΩ
(*d*) $C_1 = 5.55$ pF, $R_1 = 9$MΩ

**15.** In a digital storage oscilloscope, the input signals are
(*a*) directly applied to the oscilloscope
(*b*) multiplexed, converted to digital form and stored and applied to oscilloscope
(*c*) multiplexed, converted to digital form stored, converted to analog form and applied to oscilloscope
(*d*) applied to amplifier, stored as analog signals, multiplexed, converted to digital form, stored in digital form, converted to analog form, and applied to CRO through an amplifier.

## Answers

**1.** (*b*) **2.** (*a*) **3.** (*b*) **4.** (*b*) **5.** (*c*) **6.** (*b*) **7.** (*b*) **8.** (*d*) **9.** (*b*) **10.** (*c*)
**11.** (*b*) **12.** (*c*) **13.** (*a*) **14.** (*a*) **15.** (*d*)

# 22

# Instruments for Generation of Waveforms

## 22.1 SIGNAL SOURCES

Signal sources provide a variety of waveforms for testing of electronic circuits at low power levels. The waveforms generated by several different kinds of instruments, which range in complexity from simple fixed frequency sine-wave oscillators to highly sophisticated instruments such as those used for testing complex communication circuits. Some of the applications of signal generators include measuring the frequency response of amplifiers and the alignment of radio receivers. There are various types of signal generators, but the following types of characteristics are common to all types :

(*i*) the frequency of the signal should be well known and stable,

(*ii*) the amplitude of the output should be controllable from very small to relatively large values,

(*iii*) the output signal should be free from distortion.

There are many variations of the above requirements, especially for specialized signal generators such as function gerenators, pulse generators and pulse frequency generators.

Sine-wave generators, both in the audio and radio frequency ranges are called **Oscillators**. Although, the terminology is not universal, the term **Oscillator** is generally used for an instrument that provides only a sinusoidal output signal. The term **Function Generator** is applied to an instrument that provides several output waveforms, including sinewave, square wave, triangular wave and pulse trains as well as *amplitude modulation* of the output signal.

Although we speak of oscillators as '*generators*' of signals, it should be emphasized that no energy is created, it is simply converted from a *dc* source into *ac* energy at some specified frequency.

## 22.2 OSCILLATORS

An oscillator is one of the most basic and useful of the electrical and electronic measuring instruments. Oscillators provide a convenient source of power of test voltage for practically measurements. An oscillator provides a sine-wave signal output. Depending upon the oscillator design or application, terms such as oscillator, test oscillator, signal generator or function generator are used. **An oscillator** is the basic element of all signal sources and generates a sine-wave signal of known frequency and amplitude. This instrument covers a frequency range from a few Hz to GHz. A simple oscillator has two basic blocks (*i*) an oscillator and (*ii*) an attenuator. The oscillator's performance depends upon the functioning of these two blocks. The accuracy, stability and freedom from distortion depends upon the design of oscillator, while the amplitude accuracy depends upon the design of attenuator.

### 22.2.1 Classification of Oscillators

Oscillators may be classified in several ways. Here they are classified on *two* bases : (*i*) the design principle used, and (*ii*) the frequency range over which they are used.

The classification based on design principle categorises the oscillators as : (*i*) Feedback oscillators and (*ii*) Negative resistance oscillators.

The classification based on frequency band is :

| | | |
|---|---|---|
| (*i*) | Audio frequency (AF) oscillators | 20 Hz - 20 kHz. |
| (*ii*) | Radio frequency (RF) oscillators | 20 kHz - 30 MHz. |
| (*iii*) | Video frequency (VF) oscillators | d.c. - 5 MHz. |
| (*iv*) | High frequency (HF) oscillators | 1.5 MHz - 30 MHz. |
| (*v*) | Very high frequency (VHF) oscillators | 30 MHz - 300 MHz. |

**Feedback oscillators** are more commonly used. They use an active device such as an amplifier whose output is feedback in phase to its input (positive feedback) to cause regenerative action resulting in oscillations.

**Negative resistance oscillators** use an active device with an inherent negative resistance characteristics which provide its own regeneration and hence oscillations.

## 22.3 FEEDBACK OSCILLATORS

A basic feedback circuit is shown in Fig. 22.1.

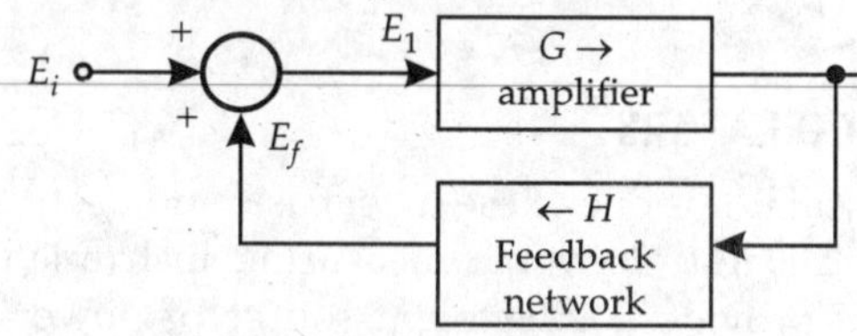


**Fig. 22.1** Basic feedback circuit

Voltage gain of amplifier

$$\mathbf{G} = \frac{\mathbf{E}_0}{\mathbf{E}_1} \qquad \text{...(22.1)}$$

where $\mathbf{E}_1$ and $\mathbf{E}_0$ are phasors and **G** is a complex function of frequency. A portion $\mathbf{E}_f$ of output voltage $\mathbf{E}_0$ is feedback into the input.

Feedback factor $$\mathbf{H} = \frac{\mathbf{E}_f}{\mathbf{E}_0} \qquad \text{...(22.2)}$$

where **H** is a complex quantity.

Now amplifier input voltage $\mathbf{E}_1$ is the sum of the signal $\mathbf{E}_i$ and feedback voltage $\mathbf{E}_f$.

$$\therefore \quad \mathbf{E}_1 = \mathbf{E}_i + \mathbf{E}_f \qquad \text{...(22.3)}$$

Amplifier output voltage

$$\mathbf{E}_0 = \mathbf{G}\mathbf{E}_1 = \mathbf{G}(\mathbf{E}_i + \mathbf{E}_f) = \mathbf{G}(\mathbf{E}_i + \mathbf{H}\mathbf{E}_0)$$

or $$\mathbf{E}_0 = \frac{\mathbf{G}\mathbf{E}_i}{1 - \mathbf{G}\mathbf{H}} \qquad \text{...(22.4)}$$

$\therefore$ Gain with feedback

$$\mathbf{G}_f = \frac{\mathbf{E}_0}{\mathbf{E}_i} = \frac{\mathbf{G}}{1 - \mathbf{G}\mathbf{H}} \qquad \text{...(22.5)}$$

Equation 22.5 indicates that if the complex quantity called *loop gain* **GH** is positive, as it approaches unity the gain increases without limit. Hence, if the value of $\mathbf{GH} = 1 + j0$, the overall gain is infinite. Physically interpreted, this situation means that the signal input voltage $\mathbf{E}_i$ can be reduced to zero without affecting the output voltage. Thus the amplifier can provide an output without an input and such a self-excited source is termed as a **Feedback Oscillator**.

In general the operating conditions for regenerative action or oscillations of feedback oscillators are :

1. The frequency of a sinusoidal feedback oscillator is determined by the condition that the loop phase shift is zero. Another way of explaining it would be that the phase shift between the amplifier input signal and the signal which is returned through amplifier and the feedback circuit must be equal to zero.
2. The product of voltage of amplifier, **G**, and the feedback factor **H** must be equal to or greater than 1.

These two conditions for sustained oscillations are called **Barkhausen Criteria**.

It is not necessary to supply an input signal to initiate oscillators. A noise voltage or a transient is sufficient to start oscillations. It is usually desirable for the loop gain to be greater than unity to start oscillations. However, if the loop gain is too large, the output waveform will not be sinusoidal as the circuit will operate under conditions of saturation in this case.

## 22.4 TYPES OF FEEDBACK OSCILLATORS

Feedback oscillators are classified into *three* categories :

(*i*) *RC* (resistance-capacitance) oscillators,

(*ii*) *LC* (Inductance-capacitance) oscillators, and

(*iii*) Crystal oscillators.

### 22.4.1 RC Oscillators

*RC* oscillators may be one of the *two* following forms :

(*i*) Wien's bridge oscillator, and

(*ii*) Phase shift oscillator.

#### 1. Wien's Bridge Oscillator

The Wien's bridge *RC* oscillator has become the standard circuit for variable frequency test circuits in the audio frequency range. This oscillator is simple and stable in operation. Also it is far less cumbersome than the *LC* oscillator.

A Wien's bridge oscillator is an audio oscillator is useful for testing equipment that operates in the **audio-frequency** range. Such instruments always produce a sine-wave signal, variable in both amplitude and frequency and usually provide a square-wave output as well. The maximum amplitude of the output waveform is typically on the order of

25 V whereas the range of frequencies covers at least the audio-frequency range from 20 Hz to 20 kHz. The most common output impedances for audio oscillators are 75 Ω and 600 Ω.

The two most common audio oscillator circuits are the **Wien's bridge oscillator** and the **phase-shift oscillator** both of which employ *RC* feedback networks. The Wien's bridge offers some very attractive features, including a straightforward design, a relatively pure sine-wave output and a very stable frequency.

The **Wien's bridge oscillator** is essentially a feedback amplifier in which the Wien's bridge serves as the phase-shift network. The Wien's bridge is an a.c. bridge the balance of which is achieved at one particular frequency. The basic Wien's bridge oscillator is shown in Fig. 22.2.

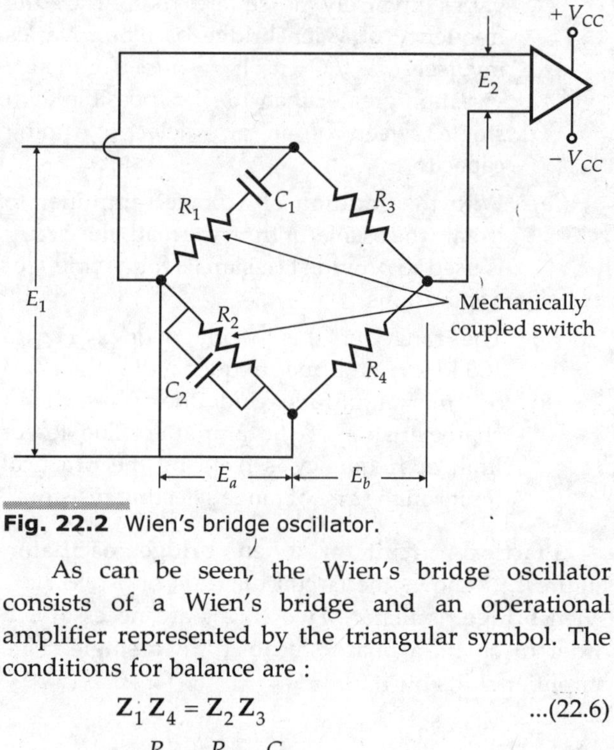


**Fig. 22.2** Wien's bridge oscillator.

As can be seen, the Wien's bridge oscillator consists of a Wien's bridge and an operational amplifier represented by the triangular symbol. The conditions for balance are :

$$\mathbf{Z}_1 \mathbf{Z}_4 = \mathbf{Z}_2 \mathbf{Z}_3 \quad ...(22.6)$$

$$\frac{R_3}{R_4} = \frac{R_1}{R_1} + \frac{C_2}{C_2} \quad ...(22.7)$$

and
$$f_0 = \frac{1}{2\pi\sqrt{R_1 R_2 C_1 C_2}} \quad ...(22.8)$$

If $C_1 = C_2 = C$ and $R_1 = R_2 = R$, Eqns. 22.7 and 22.8 simplify to

$$\frac{R_3}{R_4} = 2 \quad ...(22.9)$$

$$f_0 = \frac{1}{2\pi RC}\,\text{Hz} \quad ...(22.10)$$

The values of $R_1$ and $R_2$ are kept equal through a mechanically operated switch.

The first condition for oscillations requires that the phase shift between $E_1$ and $E_2$ to be zero. This can only happen when the bridge is balanced and as per the analysis given in Art. 16.8.1, the resonant frequency $f = \frac{1}{2\pi RC}$ as shown in Eqn. 22.10.

Equation 22.10 indicates that bridge resonance is independent of the value of other two bridge arms *i.e.* $R_3$ and $R_4$.

The second condition for oscillations requires that the product of amplifier gain **G**, and the feedback factor **H** should be equal or greater than 1. As the bridge is to be used as a feedback network for the oscillator, voltage $\mathbf{E}_2$ should not be equal to zero in magnitude but the phase shift between $\mathbf{E}_1$ and $\mathbf{E}_2$ should be zero.

This means that the value of **H** should not be equal to zero.

The impedances $\mathbf{Z}_1$ and $\mathbf{Z}_2$ have the same phase angle at resonance since $R_3$ and $R_4$ are pure resistance. At resonance, therefore :

$$\mathbf{Z}_1 = R - \frac{j}{\omega C} = (1-j)R$$

and
$$\mathbf{Z}_2 = \frac{1}{(1/R) + j\omega C} = (1-j)\frac{R}{2}$$

Thus
$$\mathbf{E}_a = \frac{\mathbf{Z}_2}{\mathbf{Z}_1 + \mathbf{Z}_2} = \frac{1}{3}\mathbf{E}_1 \quad ...(22.11)$$

and
$$\mathbf{E}_b = \frac{R_4}{R_3 + R_4}\mathbf{E}_1 \quad ...(22.12)$$

(The above relationships indicate that both $\mathbf{E}_a$ and $\mathbf{E}_b$ are in phase with $\mathbf{E}_1$ at resonant frequency.)

At null point : $\mathbf{E}_2 = \mathbf{E}_a - \mathbf{E}_b = 0.$

Thus if $R_4/(R_3 + R_4) = \frac{1}{3}$ or $R_3 = 2R_4$, the output voltage $E_2$ will be equal to zero. We do not want to make $E_2$ equal to zero and therefore, the ratio, $R_4(R_3 + R_4)$ should be smaller than $\frac{1}{3}$.

Let
$$\frac{\mathbf{E}_b}{\mathbf{E}_1} = \frac{R_4}{R_3 + R_4} = \frac{1}{3} - \frac{1}{\alpha} \text{ where } \alpha > 3.$$

Then
$$\mathbf{H} = \frac{\mathbf{E}_2}{\mathbf{E}_1} = \frac{\mathbf{E}_a - \mathbf{E}_b}{\mathbf{E}_1} = \frac{\mathbf{E}_a}{\mathbf{E}_1} - \left(\frac{1}{3} - \frac{1}{\alpha}\right) \quad ...(22.13)$$

At resonant frequency $E_a / E_1 = \frac{1}{3}$ and hence $\mathbf{H} = 1/\alpha.$

22

Thus if the gain, G, of the amplifier is made equal to α, the condition $GH = 1$ is satisfied.

It is clear from above that :

(i) The frequency of oscillations is equal to the resonant frequency $f_o$.

(ii) At any frequency other than resonant frequency $\mathbf{E}_a$ is not in phase with $\mathbf{E}_1$ and therefore $\mathbf{E}_2 = \mathbf{E}_a - \mathbf{E}_b$ is not in phase with $\mathbf{E}_1$.

Thus the condition for oscillations that $\mathbf{E}_2$ be in phase $\mathbf{E}_1$ is satisfied only at resonant frequency.

(iii) The relationship **GH** = 1 is satisfied only at resonant frequency.

Capacitors, C, are variable air capacitors mounted on the same shaft. Continuous variation of frequency is accomplished by varying the value of C. Resistors R are provided with switches, so that different frequency ranges may be selected.

In order that the amplitude of oscillations be held steady over a wide range of frequencies, a negative feedback is necessary. This negative feedback is provided by resistors $R_1$ and $R_2$. Resistance $R_2$ is often a temperature sensitive resistor with a positive resistance temperature co-efficient. $R_2$ is usually an incandescent lamp operated at a temperature lower than its illumination level. Since the resistance of the lamp changes with amplitude of driving signal, thus changing the voltage division of the bridge arms accordingly. Thus, as the amplitude of the oscillations increases, the resistance $R_2$ increases. This reduces the negative feedback which reduces the gain of the amplifier and thus amplitude of oscillations is restored to normal.

The above statement may be elaborated further as under :

The amplitude of oscillations depends upon how much the value of **GH** is greater than unity. If the feedback factor **H** is constant, the amplitude is determined by the value of gain, **G** of the amplifier. If **G** increases, the amplitude increases till the increase is limited by the non-linear characteristics of transistors. If the output of the amplifier tends to increase, the increased current through $R_2$ increases its resistance. This reduces the value of **H** and would then tend to keep the value of product **GH** constant. Thus the amplitude is regulated.

A vacuum tube Wien bridge oscillator depends upon a temperature sensitive element $R_2$. This allows the operation of the amplifier within its linear region of its transfer characteristics. Thus we get a low distortion sinusoidal output. However, when using transistors, a peak detector circuit is used to provide negative feedback signal proportional to the output voltage. This is because the current drawn by an incandescent lamp is too much of drain on the power supplies and thus is incompatible with the transistors and battery power sources. Also the use of an incandescent lamp causes a thermal lag.

### *Advantages*

(i) The Wien bridge oscillator provides a stable low distortion sinusoidal output voltage over a wide range of frequency.

(ii) The frequency range can be selected by using simple decade resistance boxes.

(iii) The frequency of oscillations in an RC network is $f_0 = 1/2\pi RC$. Thus unlike L-C networks where the resonant frequency varies inversely as square root of C, the frequency of Wien bridge oscillator varies inversely as C itself. Thus frequency variation greater than 10 : 1 is possible with single sweep of an air dielectric tuning capacitor.

(iv) With the addition of a power amplifier to isolate the oscillator from the load, the circuit is used to provide test signals for a variety of applications.

(v) The range of this oscillator is 2 Hz to 100 kHz. The upper frequency limit is fixed by the amplitude and the phase shift characteristics of the amplifier. The lower limit of frequency is fixed by the practical limits on the size of range selecting resistors.

**Practical circuit of Wien bridge oscillator.** Figure 22.3 shows the circuit diagram of a two state Wien bridge oscillator. Two stages are necessary in order to obtain a phase shift of 0° as a single stage amplifier shifts by 180°.

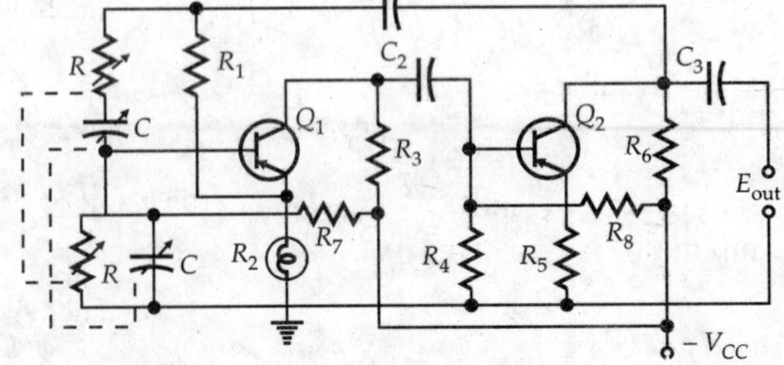


**Fig. 22.3** Practical Wien bridge circuit.

The *four* arms of the Wien bridge are :

(*i*) $R$ and $C$ in series

(*ii*) $R$ and $C$ in parallel

(*iii*) resistor $R$ and

(*iv*) resistor $R_2$.

$Q_1$ and $Q_2$ are two transistors and the positive feedback is applied from the collector of $Q_2$ through the coupling capacitor $C_1$ to top of the bridge circuit.

**Example 22.1** *Determine the frequency of oscillations shown in a Wien's bridge circuit of Fig. 22.4 ; where*

$$R_1 = R_2 = 6 \text{ k}\Omega, \quad C_1 = C_2 = C = 3 \text{ nF},$$
$$R_3 = 12 \text{ k}\Omega, \quad R_4 = 6 \text{ k}\Omega$$

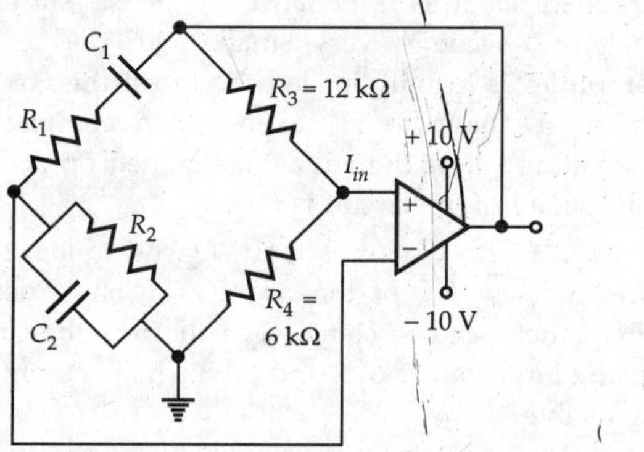


**Fig. 22.4** Wien's bridge oscillator for Example 22.1.

**Solution.** The conditions of a wien bridge are fulfilled with :

$$R_3 / R_4 = 12/6 = 2.$$

With $R_1 = R_2 = R$ and $C_1 = C_2 = C$, the output frequency is,

$$f_0 = \frac{1}{2\pi RC} = \frac{1}{2\pi \times 6 \times 10^3 \times 3 \times 10^{-9}} = 8.84 \text{ kHz}$$

**Example 22.2** *Design a Wien's bridge oscillator around the following specifications, $f = 15$ kHz, $V_{CC} = \pm 10$ V, $I_{in} = 1\mu A$, $I_{R4} = 100\ I_{in}$.*

**Solution.** $I_{R4} = 100\ I_{in} = 100 \times 1\ \mu A = 100\ \mu A.$

If the peak value of the sinusoidal waveform is 90% of $V_{CC}$, we have :

$$R_3 + R_4 = \frac{0.9\ V_{CC}}{100 \times 10^{-6}} = \frac{0.9 \times 10}{100 \times 10^{-6}} = 90 \text{ k}\Omega.$$

For balance in a Wien's bridge, $R_3 = 2\ R_4$.

$\therefore\ 2\ R_4 + R_4 = 90$ kΩ or $R_4 = 30$ kΩ.

and $\quad R_3 = 60$ kΩ.

It is generally convenient to have

$$R_1 = R_2 = R_4 = 30 \text{ k}\Omega$$

or $\quad R = 30$ kΩ.

Now $\quad f_0 = \dfrac{1}{2\pi RC}$

$\therefore$ Capacitance

$$C = \frac{1}{2\pi R f_0}$$
$$= \frac{1}{2\pi \times 30 \times 10^3 \times 15 \times 10^{-3}} = 354 \text{ kHz}$$

## 2. Phase Shift Oscillator

When an audio-frequency (AF) oscillator is to be used over comparatively a limited frequency range, or on a specified frequency, simplified *RC* networks are usually employed. For example, Fig. 22.5 shows a simple **"phase shift oscillator"**, which can be used for generation of sinusoidal voltages upto a frequency range of several hundred kHz. The circuit consists of a single transistor amplifier stage, followed by three cascaded arrangements of resistor *R* and capacitor *C* to provide the necessary positive feedback.

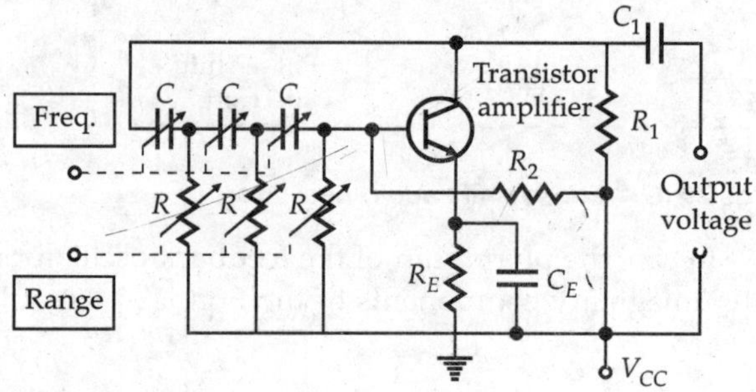


**Fig. 22.5** Phase shift oscillator.

The single transistor amplifier shifts the phase of any voltage by 180° which appears at its base. The *RC* network is necessary to provide an additional phase shift of exactly 180°. This additional phase shift is obtained from the *RC* network only at a specific frequency. Hence, at a frequency where the *RC* network provides exact 180° phase shift, the total phase shift from the base of transistor, around the circuit, back to the base is exactly 0°. Thus the feedback voltage is in phase with the voltage input to the amplifier at that frequency. Thus, if the amplification of the transistor is sufficiently large, the circuit will act as an oscillator at that frequency.

The phase shift network for the phase oscillator is shown in Fig. 22.6. It consists of equal valued capacitors and resistors connected in cascade. Each stage provides a phase shift 60° with the total phase shift being 180°.

The phase shift oscillator is analysed by ignoring any loading effects produced by the amplifier.

The expression for feedback factor is :

$$\mathbf{H} = \frac{\mathbf{V}_i}{\mathbf{V}_o} = \frac{1}{1 - \frac{5}{(\omega RC)^2} + j\left[\frac{1}{(\omega CR)^3} - \frac{6}{\omega CR}\right]} \quad ...(22.14)$$

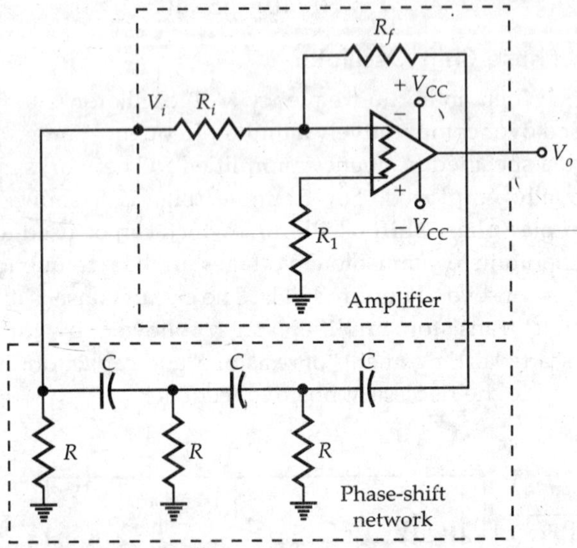


**Fig. 22.6** Phase shift audio oscillator.

In case the phase shift of the feedback oscillator is 180°, imaginary components of the Eqn. 22.14 should be zero.

$$\therefore \quad \frac{1}{(\omega RC)^3} - \frac{6}{\omega CR} = 0 \quad ...(22.15)$$

This requires that the frequency of oscillations is,

$$f_0 = \frac{\omega}{2\pi} = \frac{1}{2\sqrt{6}\pi RC} \text{ Hz} \quad ...(22.16)$$

and

$$\omega_0 = \frac{1}{\sqrt{6}\,RC} \text{ rad/s}^{-1} \quad ...(22.17)$$

Substituting the value of $\omega_0$ in Eqn. 22.14, we have,

$$\mathbf{H} = \frac{\mathbf{V}_i}{\mathbf{V}_o} = \frac{1}{1 - 5/(1/6) + j(6\sqrt{6} - 6\sqrt{6})} \quad ...(22.18)$$

or

$$\mathbf{H} = \frac{\mathbf{V}_i}{\mathbf{V}_o} = \frac{1}{1 - 5 \times 6} = -\frac{1}{29} \quad ...(22.19)$$

$$\therefore \quad \mathbf{V}_o = -29\, V_1 \quad ...(22.20)$$

which means that the gain of the amplifier must be at least 29 if the circuit is to sustain oscillations.

**Example 22.3** *Determine the frequency of oscillations of a phase shift oscillator with a three section feedback network consisting of 13 Ω resistors and 100 μF capacitors.*

**Solution.** From Eqn. 22.16, frequency of oscillations is :

$$f_0 = \frac{1}{2\sqrt{6}\pi RC} = \frac{1}{2\sqrt{6}\pi \times 13 \times 100 \times 10^{-6}} = 50 \text{ Hz}$$

### Advantages and Disadvantages

(*i*) The circuit of a phase-shift oscillator is much simpler than the Wien bridge circuit. This is because this circuit dispenses with the negative feedback and the stabilization arrangements.

(*ii*) The phase shift oscillator may be used over a wide frequency range, from a few Hz to several kHz. The upper frequency range is limited since at high frequencies, the impedance of phase shifting *RC* network becomes very small, with the result the amplifier is heavily loaded. To offset this, corrective measures have to be taken which complicate the circuit and thus the advantage gained on account of simplicity of the circuit is lost.

(*iii*) The circuit is particularly suited for low frequencies, say of the order of 1 Hz, since these frequencies can be obtained easily from the circuit by using large values of *R* and *C* which are commercially available.

(*iv*) The phase shift oscillator has decidedly an advantage over *L-C* tuned oscillators in the low frequency range as in this frequency range the large inductors required are cumbersome and impracticable.

(*v*) The frequency stability is not as good as that of Wien bridge oscillator. The distortion is greater and changing of frequency is in convenient as each capacitor is to be individually adjusted.

### 22.4.2 L.C. Oscillators and R.F. Oscillators

The first oscillators were the L.C. variety. These oscillators are particularly suited for the radio frequency (R.F.) range. There are many types of (L.C.) oscillators and they use a conventional type of L.C. tank (R.F.) circuit. Their principle of operation is simple and almost similar for any of the many variations of the basic circuit. The opeation of the basic circuit is as follows :

An L.C. oscillator uses a tank circuit which comprises of an inductor *L* in parallel with a capacitor *C*. The parallel combination of *L* and *C* is excited into oscillations and the output voltage of the tank circuit is amplified by either a transistor or a vacuum tube amplifier. A fraction of the amplified voltage is fed back into the tank circuit by either an inductive or a capacitive coupling. This is done in order to compensate for the power loss in the tank circuit. This **"regenerative feedback"** results in a constant amplitude output voltage at the resonant frequency of the tank circuit. The resonant frequency of the tank circuit is

$$f_0 = 1/2\pi\sqrt{LC}.$$

The Barkhausen criteria must be satisfied to produce sustained oscillations.

It is evident from the expression for $f_o$, that these circuits are particularly suited for radio frequencies since the value of inductance required to produce oscillations is quite small at these frequencies. The *L-C* oscillators can operate at very high frequencies, up to several hundred MHz.

**Armostrong Oscillator**

Figure 22.7 shows the basic circuit of an Armstrong oscillator. Inductor, $L_1$, and capacitor, $C_1$, form the tank circuit of the oscillator. Coil $L_1$ is inductively coupled to a coil $L_2$. When the plate supply voltage $E_{bb}$ is first connected, current flows through the plate circuit. This current flows through coil $L_1$ and since $L_2$ is inductively coupled to coil $L_1$, a voltage is induced in the later. This voltage is in such a direction that grid of the tube is driven positive. This increases the plate current at a faster rate and the induced voltage in coil $L_1$, increases further. Thus a high positive voltage is built up across the tank circuit. Capacitor $C_1$, charges with a positive polarity on its top plate. During this process, since the grid is positive the grid current charges the capacitor $C_g$ to the peak value with a polarity as shown. As the plate current approaches its saturation point the rate of rise of plate current

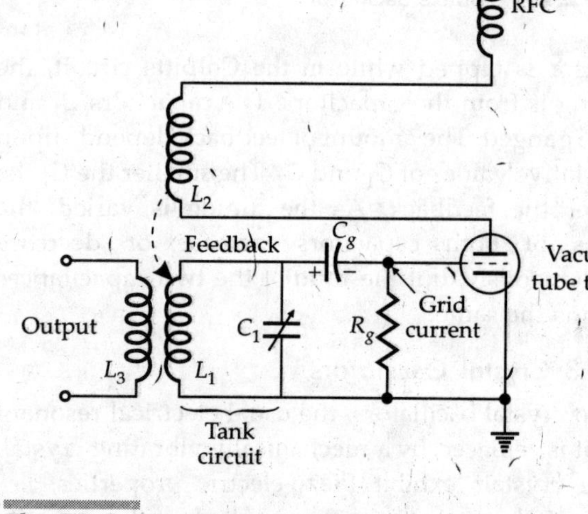


**Fig. 22.7** Armstrong oscillator circuit.

decreases and therefore the induced voltage decreases. Capacitor $C_g$, therefore, discharges through resistance $R_g$. This makes the grid negative and therefore a chain reaction is set off.

With the grid becoming negative, the plate current starts decreasing and as a result the magnetic field of $L_2$ is wiped out. This induces a negative voltage in coil $L_1$. The capacitor, $C_g$ is still discharging and this drives the tube to a cutoff region and therefore the plate current becomes zero. The negative induced voltage across $L_1$ causes $C_1$ to discharge and then charge up again to the peak value of the negative induced voltage.

The tube is cut-off and $C_1$ is charged with a negative polarity at its top plate. During the next half cycle, the tank circuit takes out the tube from its cut-off region. $C_1$ starts to discharge through $L_1$, and the grid is raised in potential till the tube starts conducting again. As soon as the tube conducts, the energy is transferred from the plate circuit to the tank circuit. Capacitor, $C_1$ charges up again to the peak value of the positive voltage. The entire cycle repeats itself as described earlier. The output is taken from coil $L_3$ which is inductively coupted to coil $L_1$.

**Hartley Oscillator**

Figure 22.8 shows the transistorized version of a Hartley oscillator. This circuit uses only one coil instead of two coils as Armstrong circuit. The coil has a tap which corresponds to the common a.c. ground of the Armstrong circuit. The variable (tuning) capacitor $C_1$ is now in parallel with the entire coil $L_1 + L_2$. The output is now available through an *RC* coupling instead of an inductive coupling as in Armstrong circuit.

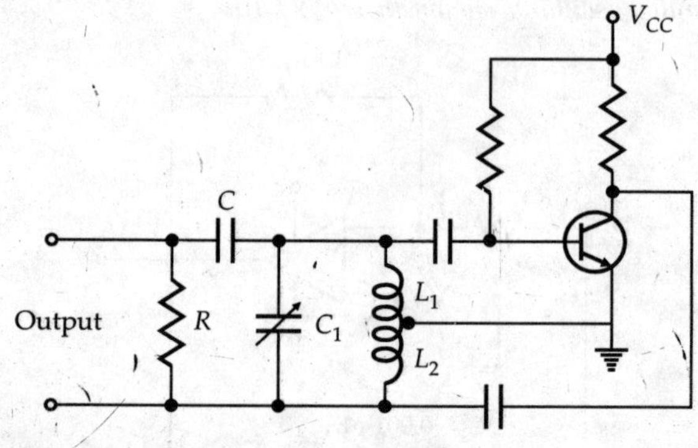


**Fig. 22.8** Hartley oscillator circuit.

Phase shift network consists of a tapped conductor consisting of sections $L_1$ and $L_2$ and an justable capacitor $C_1$ to vary the frequency of oscillations.

The feedback factor is $\mathbf{H} = -L_1 / L_2$ ...(22.21)

22

The negative sign means that there must be a 180° phase shift across the amplifier. This is accomplished by connecting the amplifier in the inverting configuration as shown in Fig. 22.9.

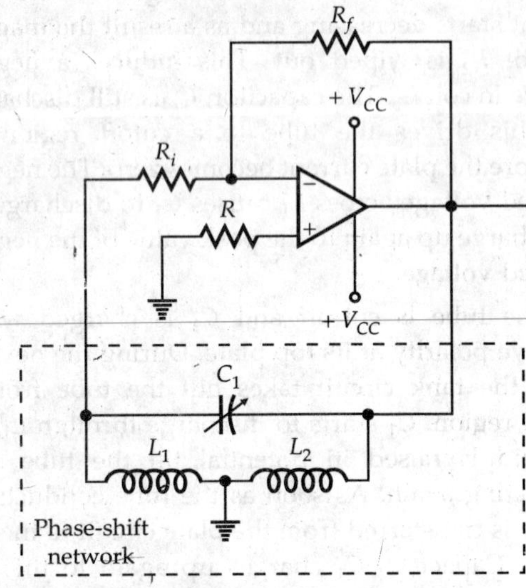


**Fig. 22.9** Basic Hartley oscillator.

The circuit must satisfy the Barkhausen criteria for sustained oscillations for which $\mathbf{GH} \geq 1$ or

$$\mathbf{GH} \geq 1/H \quad \text{...(22.22)}$$

or

$$\mathbf{G} \geq -L_2/L_1 \quad \text{...(22.23)}$$

Equation 22.23 states that the gain of the amplifier must be greater than or equal to $L_2/L_1$. Using the equation for the gain of an inverting amplifier

$$A = -R_f' / R_i \quad \text{...(22.24)}$$

***Example 22.4*** *Determine the frequency of oscillations and the minimum value of $R_f$ to sustain oscillations in a Hanley oscillator shown in Fig. 22.10.*

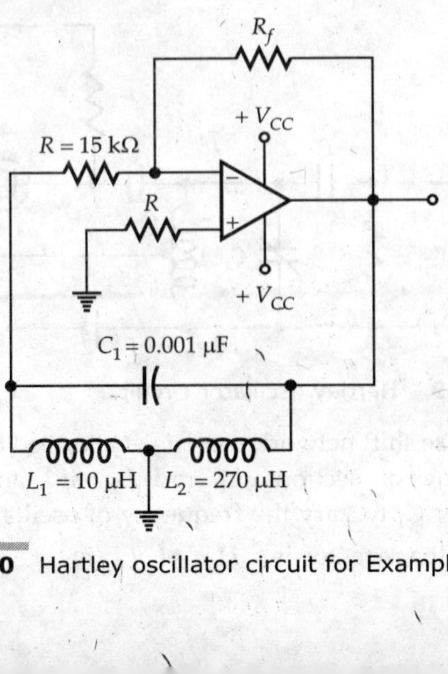


**Fig.22.10** Hartley oscillator circuit for Example 22.4.

**Solution.** Frequency of oscillations

$$f_0 = \frac{1}{2\pi\sqrt{(L_1 + L_2)C_1}} = \frac{1}{2\pi\sqrt{280 \times 10^{-6} \times 0.001}}$$

$$= 300 \text{ Hz}.$$

The minimum gain of the amplifier is computed using Eqn. 22.23 as,

$$G_{min} = -\frac{L_2}{L_1} = \frac{270}{10} = -27$$

Using Eqn. 22.24 the value of the feedback resistor $R_f$ is

$$R_f = G_{min} R_i = 27 \times 15 = 405 \text{ k}\Omega$$

**Colpitts Oscillator**

The Colpitts circuit in its transistorized form is shown in Fig. 22.11. It is, in fact, another form of basic Armstrong circuit. The tank circuit consists of a fixed inductor $L_1$, and two variable capacitors $C_1$ and $C_2$ in series. This circuit differs from the Hartley's circuit only in the form of tapping. In the Hartley's circuit, the

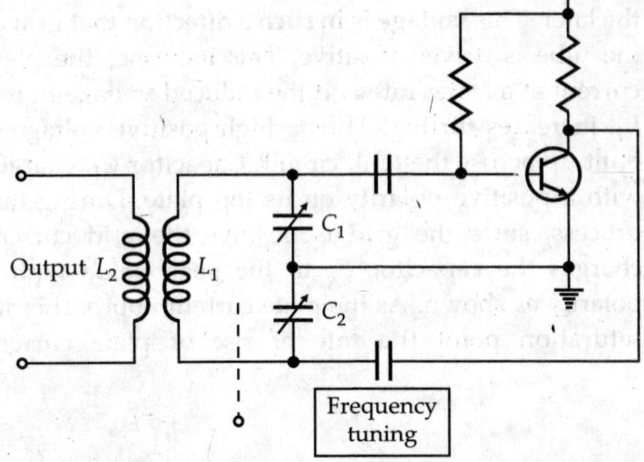


**Fig. 22.11** Colpitts oscillator.

inductor is tapped while in the Colpitts circuit, the tapping is from the capacitors. The capacitors $C_1$ and $C_2$ are ganged. The amount of feedback depends upon the relative values of $C_1$ and $C_2$. The smaller the $C_1$ the greater the feedback. As the tuning is varied, the values of both capacitors increase or decrease simultaneously, but the ratio of the two capacitances remains the same.

### 22.4.3 Crystal Oscillators

In crystal oscillators, the usual electrical resonant circuit is replaced by a mechanically vibrating crystal. These crystals exhibit Piezo-electric properties, *i.e.*, when mechanical stresses are applied on the opposite faces of the crystal, electrical charges appear at some other faces and *vice-versa*. Thus when an alternating voltage is applied to appropriate faces, mechanical vibrations are produced at some other faces. Therefore if the frequency of alternating voltage is equal to the frequency of mechanical vibrations, the vibrations will be intense.

Piezo-electric properties are exhibited by a number of natural crystal substances like quartz, Rochelle salt, etc. Quartz is the normally used material for crystal oscillators because of its frequency control in oscillators which comes from its permanence, low temperature co-efficient and high mechanical $Q$. Quartz is preferred to Rochelle salt on account of the above-mentioned qualities even though Quartz has a lesser Piezo-electric effect.

The electrical equivalent of a crystal is shown in Fig. 22.12(*b*). The circuit has *two* resonant frequencies

(*i*) Series resonant frequency $f_1 = 1/2\pi\sqrt{LC}$

(*ii*) Parallel resonant frequency

$$f_2 = \frac{1}{2\pi}\sqrt{\frac{1+C/C'}{LC}}$$

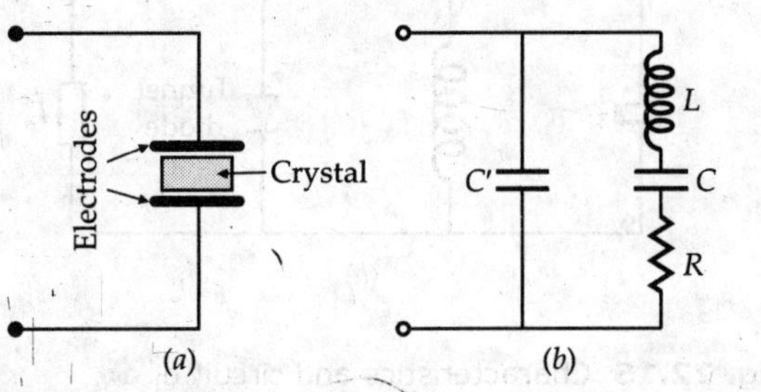


**Fig. 22.12** Crystal and its equivalent electric circuit.

It is apparent that $f_2$ is greater than $f_1$. But the value of ratio $C/C'$ is very small and hence the two frequencies $f_1$ and $f_2$ are close to each other.

Because of the close analogy between a crystal and an $LC$ network, the crystal oscillator designs are similar to those of $LC$ oscillators. Figure 22.13 shows the circuit of a Pierce oscillator. This circuit is similar to that of the Colpitts oscillator.

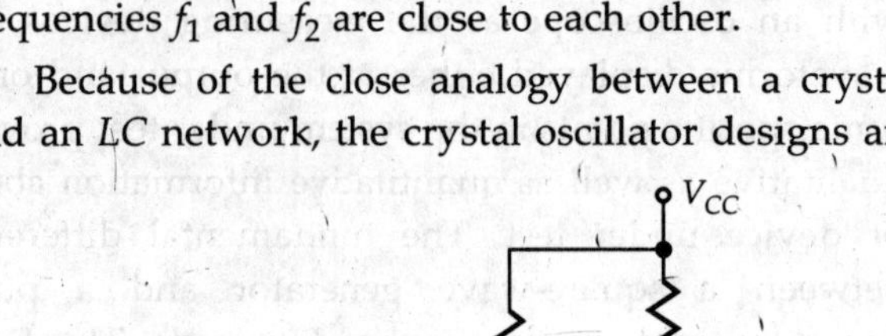


**Fig. 22.13** Crystal oscillator circuit.

### *Advantages and Disadvantages*

(1) The frequency stability of the oscillations obtained from a crystal vary and that too over a long period of time.

(2) The frequency of oscillations generated by other RF oscillators using electrical resonating (tank) circuits is effected considerably by changes in supply voltage, load and temperature. Crystal oscillators do not face these problems and thus their frequency stability is high. Hence these oscillators are the standard means of maintaining frequency.

(3) The vibrating crystal structure has a high value of $Q$. Values in excess of 20,000 are always to be expected and values over 1,000,000 have been obtained under special circumstances.

(4) Since the ratio of capacitance $C/C'$ is very small, the coupling between crystal resonator and external circuit is inherently small.

(5) The crystal oscillators have a very limited tuning range (or not at all). The crystal oscillators are used for frequencies greater than 100 kHz. Their use is mainly at radio frequencies (RF).

## 22.5 BEAT FREQUENCY OSCILLATOR (B.F.O.)

Before the advent of the Wien bridge oscillator, the beat frequency oscillator was the most commonly oscillator to obtain variable frequency output in the audio frequency (AF) range. Now the use of B.F.O. has decreased. It is mostly used when it is desired to cover a very large frequency range with a single dial rotation (without the need for switching bands).

A beat frequency oscillator circuit is shown in Fig. 22.14 in block diagram form. One of the oscillators generates a fixed frequency and the other a variable frequency. The outputs of these oscillators are fed to a mixer. The frequency of output of the mixer is the difference of two frequencies. Thus if the fixed

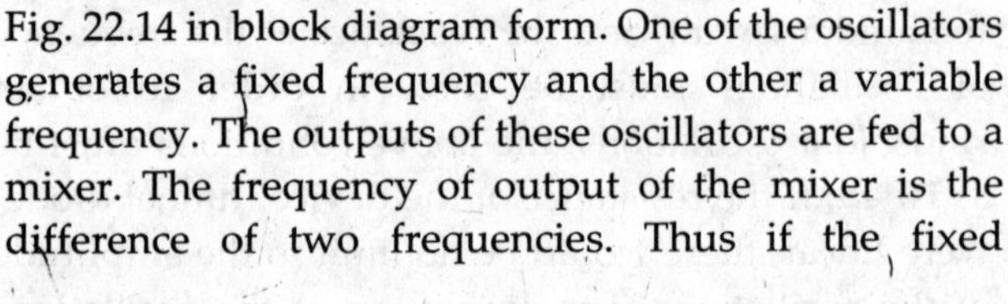


**Fig. 22.14** Beat frequency oscillator.

frequency oscillator generates a frequency of 100 kHz and the variable frequency oscillator has a frequency range of 100 kHz to 120 kHz, the output frequency varies between 0 to 20 kHz. Therefore, frequency dial is marked 0 to 20 kHz (*i.e.,* the difference of the two frequencies). This then is an audio frequency (AF) output. This output is amplified by an AF amplifier.

It is obvious that circuit of a B.F.O. is much more complicated than that of a Wien bridge oscillator and hence B.F.O. is becoming obsolete.

## 22.6 NEGATIVE RESISTANCE OSCILLATORS

Many negative resistance elements are available for use in negative-resistance oscillators. These are tetrodes, tunnel diodes, neon tubes, and unijunction transistors. There are *two* types of negative resistance oscillators used in practice.

(*i*) ***Dynatron.*** Its action is based upon the negative resistance region of the characteristics of a tetrode coupled with an *LC* tank circuit.

(*ii*) ***Tunnel diode oscillator.*** This is solid state equivalent of the dynatron with additional advantages of very high speed operation. Therefore this oscillator is very useful in the very high frequency (VHF) and ultra high frequency (UHF) ranges.

Figure 22.15(*a*) shows the characteristics of a tunnel diode. The characteristics are stable if driven by a voltage source. The voltage current response can be used to cause or sustain oscillations. The load lines as shown are unstable ; that is there are two stable intersections and one unstable intersection. If the tunnel diode is connected to an impedance which may be represented by $Z = R + jX$, we find that the criteria for instability is : $R + R_D = 0$. Thus the device resistance, $R_D$, must cancel out the resistance of the remainder circuit, and the reactance at the frequency of oscillations must be zero. In order to obtain sinusoidal oscillations, the above conditions must be satisfied. If the resistance of the device (tunnel diode) were linear, there would be no limit to the amplitude of the oscillations. The amplitude of the oscillations can be controlled, if the positive resistance line is almost tangent to the negative resistance line. The negative resistance will then have an average value over one complete cycle which is equal to positive resistance.

The tunnel diode, which has a negative resistance, is very useful as it may operate at very high frequencies. It has been difficult to obtain high frequency operation of transistors because of the difficulty of making their base regions and because of relatively slow motion of minority carriers in semi-conductors. Frequencies of the order of 100 – 500 MHz are the upper limit. The tunnel diode operates because of majority carriers and hence is thus very fast. It is possible to make tunnel diode oscillators which operate in the microwave frequency region. A simple circuit shown in Fig. 22.15 can be made to oscillate if the correct bias current is applied to the tunnel diode. The tunnel diode circuits may use negative feedback for bias stability, just as feedback oscillators do.

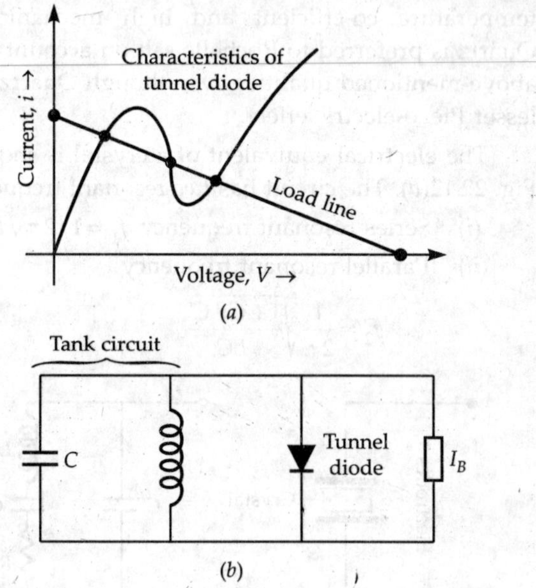


**Fig. 22.15** Characteristics and circuit of a tunnel diode oscillator.

## 22.7 SQUARE-WAVE AND PULSE GENERATORS

Square-wave and pulse generators are often used with an oscilloscope as the measuring device. The waveforms displayed either at the output and or at some specific points in the system under test provide qualitative as well as quantitative information about or device under test. The fundamental difference between a square-wave generator and a pulse generator is dependent upon *duty cycle.* The Duty Cycle is defined as the ratio of average value of the pule over one cycle and the peak value are inversely related to their time duration. Since the average value and peak value are inversely related to their time duration, the duty cycle can be defined in terms of *pulse width* and the *period* or *pulse repetition time.*

$$\text{Duty cycle} = \frac{\text{pulse width}}{\text{period}} \qquad ...(22.25)$$

A square wave voltage is shown in Fig. 22.16(*a*)

$$\therefore \text{ Duty cycle for a square wave} = \frac{T/2}{T} = 0.5$$

or according to another definition

$$= \frac{\text{average value}}{\text{peak value}}$$

$$= \frac{1/2 \text{ peak value}}{\text{peak value}} = 0.5$$

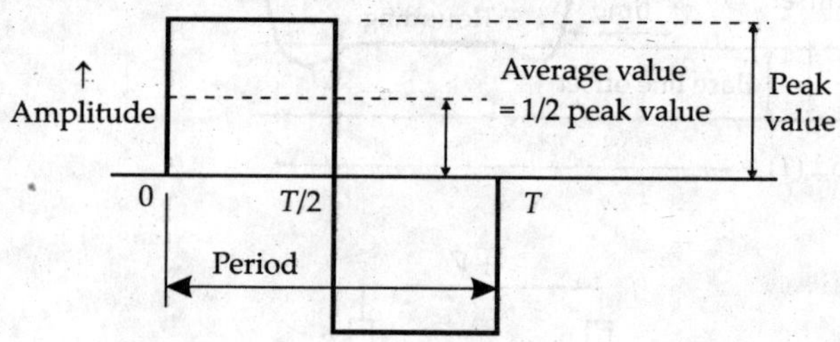


(*a*) Square wave

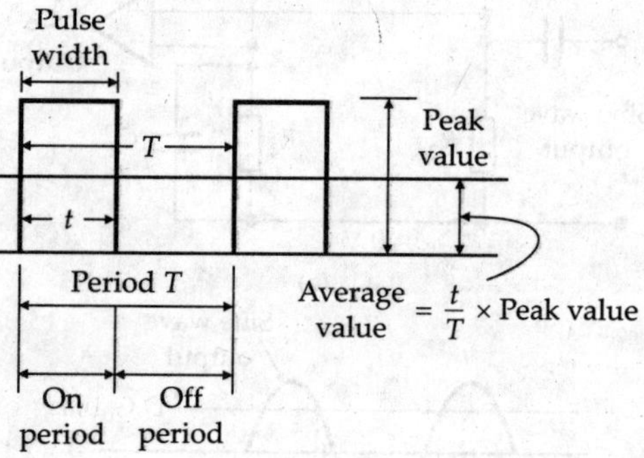


(*b*) Pulse wave

**Fig. 22.16** Square and Pulse waveforms.

Thus *square wave generators* produce an output voltage with equal *on* and *off* periods, their duty cycle is 0.5 or 50%, as the frequency of oscillations is varied. This is on account of the fact that whatever may be the frequency, the positive and negative half cycles always extend over half of the total period.

For a pulse the on and off periods can be varied.

$$\text{Duty factor for a pulse wave} = \frac{t}{T}$$

According to the other definition.

Duty cycle for a pulse wave

$$= \frac{\text{average value}}{\text{peak value}} = \frac{t/T \times \text{peak value}}{\text{peak value}} = t/T \quad ...(22.26)$$

The duty cycle of a pulse generator may vary (normally between 50% to 95%) depending upon the duration of the *on* period. The very short duration pulses give a low duty cycle. The pulse generator generally can supply more power during its *on* period than a corresponding square wave generator can, short duration pulses reduce power dessipation in the component under test. For instance, measurements of transistor gain can be made with pulses having short duration which does not cause overheating of junctions.

Square wave generators are used whenever low frequency characteristics of a system are being investigated. These applications include testing of audio-systems. Square wave pulses are also preferred to short duration pulses if the transient response of a system requires a comparatively high settling time.

### 22.7.1 Pulse Characteristics and Terminology

The characteristics of a pulse are shown in Fig. 22.17. The base *line* is referred to the d.c. level and is the line at which the pulse starts and finishes. The shift of this line from zero volts, or the expected value, is called the *base line offset*. The *amplitude* of the pulse is measured from the base line to the steady state pulse value.

**1. *Pulse rise and fall time.*** The pulse *rise time* is the time needed for the pulse to go from 10% to 90% of its amplitude, and the *fall time* is the time for the trailing edge to go from 90% to 10%. These times are also called leading edge and trailing edge *transition times*.

**2. *Linearity.*** The *linearity* of the pulse is the *deviation* of an edge, from the straight line drawn through the 10% and 90% points, expressed as a per cent of the pulse amplitude.

**3. *Pulse preshoot.*** The pulse *preshoot* is the deviation prior to reaching the base line at the start of the pulse. The *overshoot* is the maximum height immediately following the leading edge.

**4. *Ringing.*** It is the positive and negative peak distortion, excluding overshoot.

**5. *Settling time.*** It is the period needed for pulse ringing to be within a specified percentage of the pulse amplitude, measured from the 90% point of the leading edge.

**6. *Pulse droop or sag.*** It is the fall in pulse amplitude with time. Pulse *rounding* is the curvature of the pulse at the leading and trailing edges.

**7. *Pulse width.*** The *width* of the pulse is measured, in units of time, between the 50% points on the leading and trailing edges. The pulse *period* is the time between equal points on the waveform.

**8. *Pulse repetition rate.*** The pulse *repetition rate* is a measure of how frequently a pulse occurs. It is equal to the reciprocal of the pulse period, and is measured in units of frequency.

22

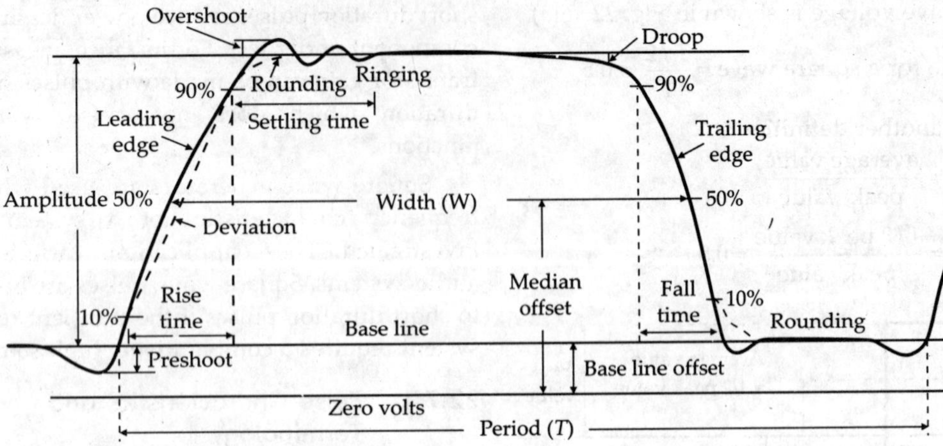


**Fig. 22.17** Illustration of critical pulse parameters.

**9. *Duty cycle.*** The duty cycle of the pulse is the ratio of its width to its period, usually expressed as a percentage.

**10. *Pulse jitter.*** *Pulse jitter* is a measure of short term instability of one event with respect to another, for example, instability in the starting time, or the pulse width, or the pulse amplitude. It is usually expressed as a percentage of the main parameter.

## 22.8 PULSE AND SQUARE WAVE CIRCUITS

The pulse and square wave generating circuits are of *two* types :

(*i*) *Passive circuits.* These circuits shape an input waveform, usually a sine wave to give a pulse output.

(*ii*) *Active circuits.* These circuits include a blocking oscillator or a multivibrator.

The circuits are briefly described as below :

### 22.8.1 Pulse Shaping Circuit

Figure 22.18 illustrates a circuit in which a sine-wave, from an oscillator, is converted into a pulse train. The amplifier is saturated between two limits, by the sine wave, and the sharpness of the pulse depends on the magnitude of the sine wave relative to the d.c., bias, and the gain of the amplifier. This circuit can also be used to 'clean up' a pulse input having overshoots, undershoots, rings and so on.

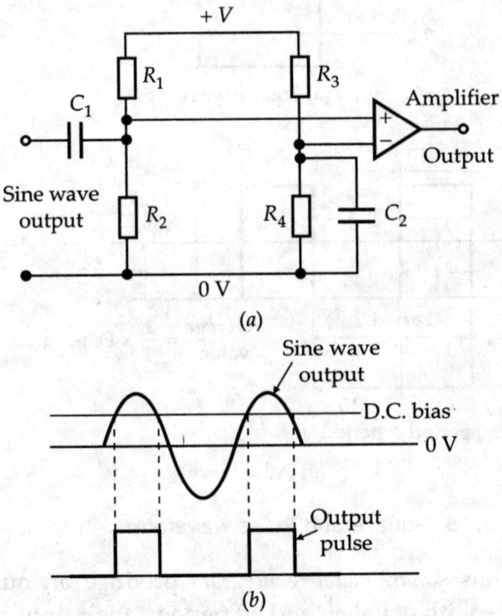


**Fig. 22.18** Generating a pulse from a sine wave : (*a*) circuit diagram, (*b*) waveforms.

### 22.8.2 Blocking Oscillator

The circuit, shown in Fig. 22.19 operates by driving transistor $Q_1$ hard into and out of saturation, so that the output is a series of pulses. The transistor gives a 180° phase shift, and the transformer provides a further 180° shift, so the circuit free runs and maintains oscillations. $R_1$ adjusts the frequency of the oscillator, which is given by Eqn. 22.27, where $R$ is the total resistance in series with capacitor $C$ and the transformer has a step down ratio of 1 : $N$

$$f = \frac{N+1}{RC} \qquad ...(22.27)$$

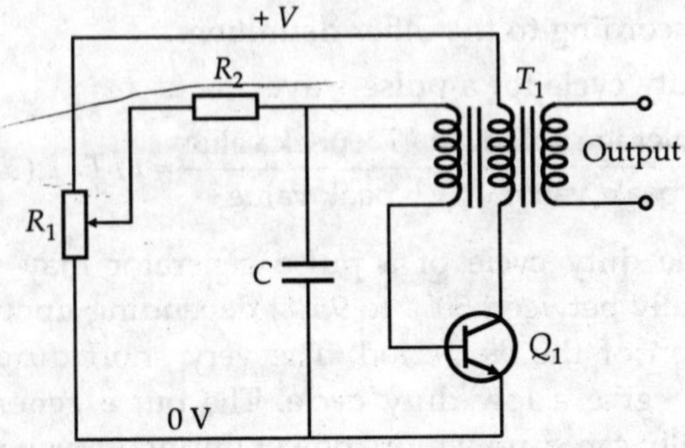


**Fig. 22.19** A blocking oscillator pulse generator.

The width of the pulse is determined by the characteristics of the transformer, mainly the self inductance and self capacitance of the primary winding.

### 22.8.3 Multivibrators

There are *three* types of multivibrators : bistables or flip flops, which provide an output pulse for every two input trigger signals; monostables which give a timed pulse out for a trigger input ; and astable or free running multivibrators, which give a string of pulses. The period and pulse widths of these circuits can be readily adjusted by timing resistors and capacitors. They are usually made as integrated circuits, and are widely used in industry.

**Astable Multivibrator**

The *astable* or *free running multivibrator* is widely used for the generation of pulses. It can be made to produce either square wave or pulses depending on the choice of circuit components. A typical free-running multivibrator is shown in Fig. 22.20.

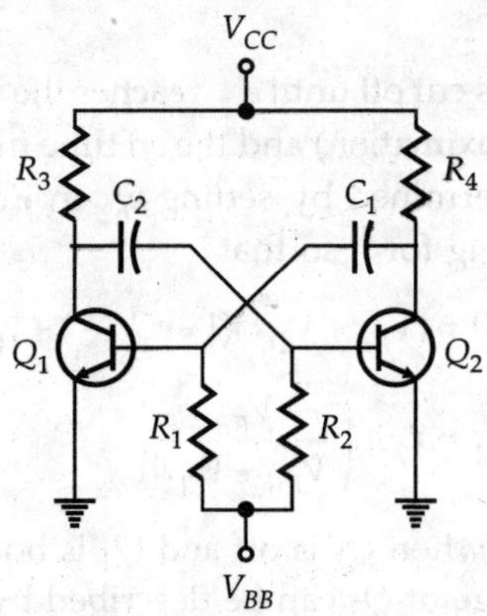


**Fig. 22.20** Astable or free-running multivibrator.

Essentially, the circuit consists of a two-stage *RC*-coupled amplifier, with the output of the second stage ($Q_2$) coupled back to the input of the first stage ($Q_1$) via capacitor $C_1$. Similarly, the output of $Q_1$ is coupled via $C_2$ to the input of $Q_2$. Since the coupling between the two transistors is taken from the collectors, the circuit is known as a *collector-coupled* astable multivibrator.

The usual qualitative analysis of the circuit proceeds as follows : When the power is first applied to the circuit, both transistors start conducting. Because of small differences in their operating characteristics, one of the transistors will conduct slightly more than the other. This starts a series of events. Assume that $Q_1$ initially conducts more than $Q_2$. This means that the collector voltage of $Q_1$ ($e_{c1}$) drops more rapidly than the collector voltage of $Q_2$ ($e_{c2}$). The decrease in $e_{x1}$ is applied to $R_2$ $C_2$ network, because the charge on $C_2$ cannot change instantaneously, the full negative-going change appears across $R_2$. This decreases the forward bias on $Q_2$ which in turn decreases the collector current of $Q_2$ ($i_{c2}$), and the collector voltage of $Q_2$ rises. This rise in $Q_2$ collector voltage is applied via the $R_1C_1$, network to the base of $Q_1$, increasing its forward bias. $Q_1$ therefore conducts even more heavily and its collector voltage drops still more rapidly. This negative-going change is coupled to the base of $Q_2$, further decreasing its collector current. The entire process is cumulative until $Q_2$ is entirely cut off and $Q_1$ conducts heavily (*bottoms*).

With $Q_2$ cut off, its collector voltage practically equals the supply voltage, $V_{CC}$, and capacitor $C_1$ charges rapidly to $V_{CC}$ through the low-resistance path from emitter to base of the conducting transistor $Q_1$. When the circuit action turns $Q_1$ fully on, its collector potential drops to approximately 0 V and since the charge on $C_2$ cannot change instantaneously, the base of $Q_2$ is at least $- V_{CC}$ potential, driving $Q_2$ deep into cutoff.

The switching action now begins. $C_2$ begins to discharge exponentially through $R_2$. When the charge on $C_2$ reaches 0 V, $C_2$ attempts to charge up to the value of $+ V_{BB}$, the base supply voltage. But this action immediately places a forward bias on $Q_2$ and this transistor starts to conduct. As soon as $Q_2$ starts conducting, its collector current causes a decrease in collector voltage $e_{c2}$. This negative-going change is coupled to the base of $Q_1$ which starts to conduct less, *i.e.*, it comes out of saturation. This cumulative action repeats until $Q_1$ finally cuts off and $Q_2$ conducts heavily. At this instant, the collector voltage of $Q_1$ reaches its maximum value of $V_{CC}$. Capacitor $C_2$ charges to the full value of $V_{CC}$, and a full cycle of operation has been completed.

The waveforms appearing at the base and the collector of each transistor are the result of a *symmetrical* or *balanced* operation : The time constants $R_1C_1$ and $R_2C_2$, the transistors themselves, and the supply voltages are all identical. The conducting and non-conducting periods are therefore of almost the same duration. The waveforms for each of the two transistors are given in the waveform diagram of Fig. 22.21.

Assume that at time $t = 1$, transistor $Q_1$ is fully turned on and transistor $Q_2$ is cut off. This makes the

22

collector voltage $e_{c1}$ of $Q_1$ a minimum (practically 0 V) and the collector voltage $e_{c2}$ of $Q_2$ a maximum ($V_{CC}$). Capacitor $C_1$ is charging through the emitter-to-base resistance of $Q_1$ toward the supply voltage $V_{CC}$ and reaches its full charge rapidly (low emitter-to-base resistance). Since $e_{c1}$ is 0 V, capacitor $C_2$ begins to charge exponentially through $R_2$ toward the base supply voltage $V_{BB}$ with a time constant equal to $R_2C_2$.

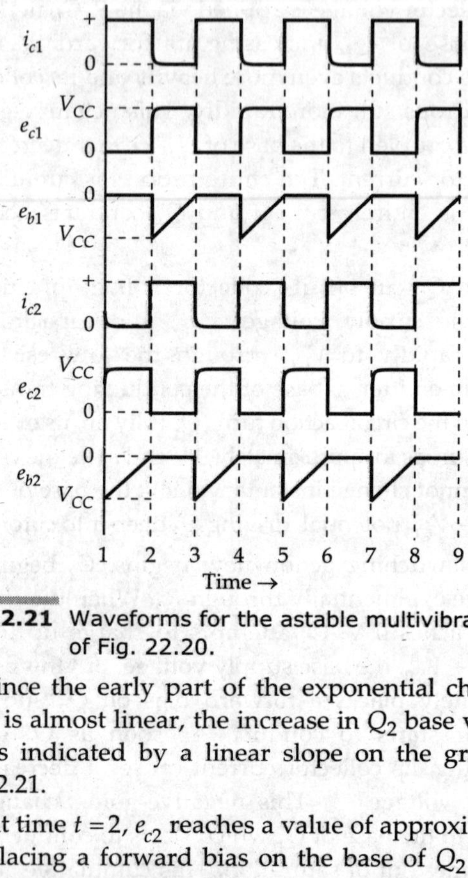


**Fig. 22.21** Waveforms for the astable multivibrator of Fig. 22.20.

Since the early part of the exponential charging curve is almost linear, the increase in $Q_2$ base voltage ($e_{b2}$) is indicated by a linear slope on the graph of Fig. 22.21.

At time $t = 2$, $e_{c2}$ reaches a value of approximately 0 V placing a forward bias on the base of $Q_2$ which then starts to conduct. Within a very short time, the collector current of $Q_2$ reaches its maximum and collector voltage $e_{c2}$ drops to 0 V. When $Q_2$ starts to draw current, the base of $Q_1$ becomes negative and $Q_1$ is quickly driven into cutoff. Its collector voltage, $e_{c1}$, reaches the $V_{CC}$ value and collector current $i_{c1}$ becomes zero. Within a very small fraction of the total $Q_2$ conducting time, capacitor $C_2$ is hilly charged to $V_{CC}$ through the low-resistance emitter-to-base path of $Q_2$.

Between times $t = 2$ and $t = 3$, transistor $Q_1$ is cut off, and its collector current and voltage remain constant. Similarly, the collector voltage and current for $Q_2$ remain constant. Only capacitor $C_1$ is charging and the base voltage $e_{b1}$ of $Q_1$ is rising exponentially toward $V_{BB}$. At time $t = 3$, the base voltage of $Q_1$ exceeds the cutoff value (approximately 0 V) and $Q_1$ starts conducting again. Obviously, one complete cycle of operation, from time $t = 1$ to time $t = 3$, depends on the time required for the base voltage of the cutoff transistor to reach the forward bias value. This time depends on two things : the magnitude of the reverse bias ($-V_{CC}$) and the time constant of the capacitor charging circuit involved, namely, $R_1C_1$, or $R_2C_2$.

The *analytical* evaluation of circuit operation proceeds as follows :

During its non-conducting period, the collector voltage of $Q_1$ equals

$$e_{c1} = V_{CC}(1 - e^{-1/\tau_3}) \quad \text{...(22.28)}$$

where $\tau_3 = R_3C_2$.

When $Q_1$ switches on, its collector voltage is at ground potential and the base voltage of $Q_2$ becomes $-V_{CC}$ with respect to ground. The subsequent rise in $Q_2$ base voltage, through the $R_2C_2$ charging circuit, is described by

$$e_{b2} = (V_{BB} + V_{CC})(1 - e^{-t/\tau_2}) - V_{CC} \quad \text{...(22.29)}$$

where $\tau_2 = R_2C_2$.

$Q_2$ remains cut off until $e_{b2}$ reaches the value of 0 V (by good approximation) and the *off* time interval $T_2$ of $Q_2$ can be determined by setting $e_{b2}$ in Eqn. 22.29 to zero and solving for $t$, so that

$$0 = (V'_{BB} + V_{CC})(1 - e^{-t/\tau_2}) - V_{CC} \quad \text{...(22.30)}$$

and $$T_2 = \tau_2 \ln\left(\frac{V_{BB}}{V_{BB} + V_{CC}}\right) \quad \text{...(22.31)}$$

Similarly, when $Q_2$ is off and $Q_1$ is bottomed, the collector voltage of $Q_2$ can be described by

$$e_{c2} = V_{CC}(1 - e^{-t/\tau_4}) \quad \text{...(22.32)}$$

where $\tau_4 = R_4C_1$.

When now $Q_2$ is switched on, its collector voltage drops to 0 V and the base voltage of $Q_1$ is given by

$$e_{b1} = (V_{BB} + V_{CC})(1 - e^{-t/\tau_1}) \quad \text{...(22.33)}$$

where $\tau_1 = R_1C_1$.

Solving for the *off* time interval $T_1$ of $Q_1$ by setting $e_{b1}$ in Eqn. 22.33 to zero, we obtain.

$$0 = (V_{BB} + V_{CC})(1 - e^{-t/\tau_1}) - V_{CC} \quad \text{...(22.34)}$$

and $$T_1 = \tau_1 \ln\left(\frac{V_{BB}}{V_{BB} + V_{CC}}\right) \quad \text{...(22.35)}$$

The total period of oscillation is given by,

$$T = T_1 + T_2 \quad \text{...(22.36)}$$

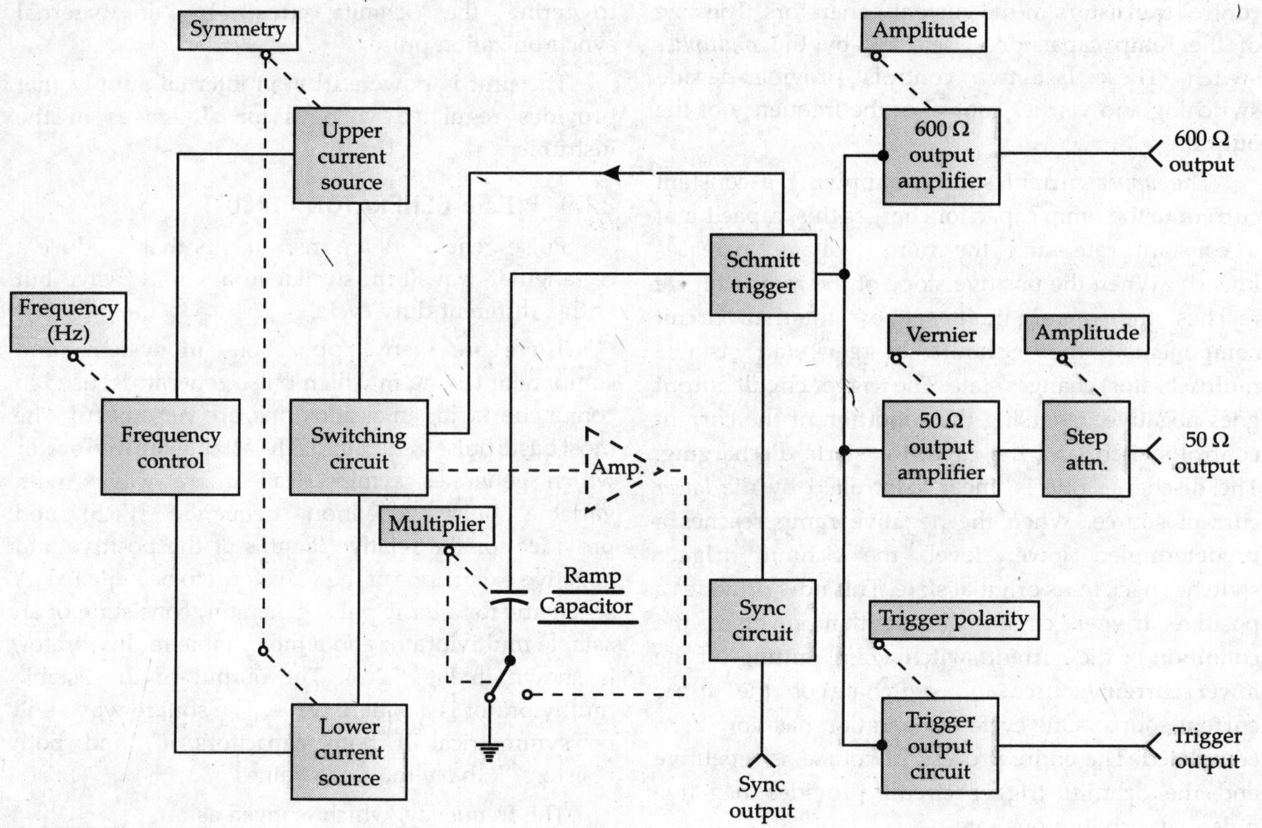


**Fig. 22.22** Block diagram of a pulse generator.

In the case of *symmetrical* operation, when time constants $R_1C_1$ and $R_2C_2$ are equal, the waveform is a symmetrical square wave. By making time constant $R_1C_1$ larger than time constant $R_2C_2$, the output waveform becomes a pulse train because the *off* time of $Q_1$ will be larger than the *off* time of $Q_2$.

### 22.8.4 Laboratory Square-wave and Pulse Generator

The block diagram of a typical general-purpose generator providing negative pulses of variable frequency, duty cycle, and amplitude is given in Fig. 22.22. The frequency range of the instrument is covered in seven decade steps from 1 Hz to 10 MHz, with a linearly calibrated dial for continuous adjustment on all ranges. The duty cycle can be varied from 25 per cent to 75 per cent. Two independent outputs are available: a 50Ω source that supplies pulses with rise-and fall times of 5 ns at 5 V peak amplitude, and a 600 Ω source that supplies pulses with rise-and fall times of 70 ns at 30 V peak amplitude. The instrument can be operated as a free-running generator or it can be synchronized with external signals. Trigger output pulses for synchronization of external circuits are also available.

The basic generating loop, which is redrawn for greater clarity in Fig. 22.23 consists of two current sources, the ramp capacitor, the Schmitt trigger circuit, and the current-switching circuit (indicated by a

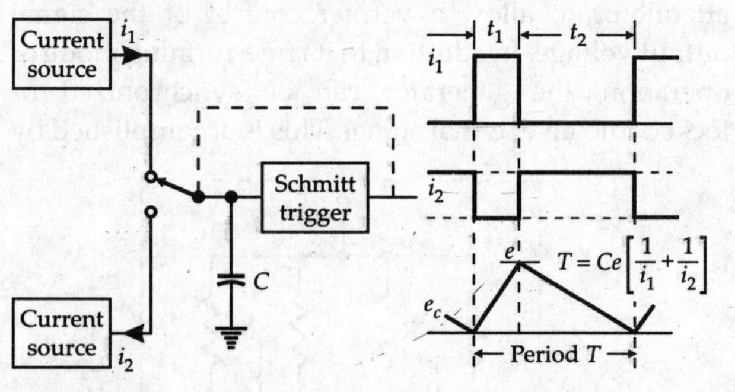


**Fig. 22.23** Simplified current source operation.

simple switch). The two current sources provide a constant current for charging and discharging the ramp capacitor. The ratio of these two currents is determined by the setting of the *symmetry* control which then determines the duty cycle of the output waveform. The *frequency* dial controls the sum of the two currents from the current sources by applying appropriate control voltages to the bases of the current control transistors in the current generators. The size of the ramp capacitor is selected by the *multiplier* switch. These last two controls provide decade switching and vernier control of the frequency of the output.

The *upper* current source, supplying a constant current to the ramp capacitor charges this capacitor at a constant rate, and the ramp voltage increases linearly. When the positive slope of the ramp voltage reaches the upper limit set by internal circuit components, the Schmitt trigger (a bistable multivibrator) changes state. The trigger circuit output goes negative, reversing the condition of the current control switch, and the capacitor starts discharging. The discharge rate is linear, controlled by the *lower* current source. When the negative ramp reaches a predetermined lower level, the Schmitt trigger switches back to its original state. This now provides a positive trigger circuit output that reverses the condition of the current switch again, cutting off the lower current source and switching on the upper current source. One cycle of operation has now been completed. The entire process, of course, is repetitive and the Schmitt trigger circuit provides negative pulses at a continuous rate.

The output of the Schmitt circuit is passed to the trigger output circuit and to the 50 Ω and 600 Ω amplifiers. The trigger output circuit differentiates the square-wave output from the Schmitt trigger, inverts the resulting pulse, and provides a positive triggering pulse. The 50 Ω amplifier is provided with an output attenuator to allow a vernier control of the signal output voltage. In addition to its free-running mode of operation, the generator can be synchronized or locked into an external signal. This is accomplished by triggering the Schmitt circuit by an external synchronization pulse.

The unit is powered by an internal supply that provides regulated voltages for all stages of the instrument.

## 22.9 PULSE GENERATOR CIRCUIT

Pulse generators are instruments that produce a rectangular waveform similar to a square wave but with a different duty cycle.

There are many applications in systems and component testing in which pulse generators, used in conjunction with an oscilloscope, are very useful. The most basic pulse generator is the astable **multivibrator** which generates symmetrical square waves with variable pulse repetition frequency (PRF) and provides for the relative lengths of the positive and negative excursions of the waveform to be adjusted. A schematic for a basic pulse generator, consisting of an astable multivibrator and a monostable multivibrator, is shown in Fig. 22.24. The output of the astable multivibrator is a square wave. The square wave will be symmetrical if both capacitors, C, and both resistors, *R*, have the same values.

The frequency, which is given as

$$f = \frac{0.693}{RC} \qquad \ldots(22.37)$$

can be changed, while maintaining the symmetry of the waveform, by changing both resistors or both capacitors. The duty cycle of the square wave can be varied by changing the value of one capacitor or one resistor. Common practice would be to use several switch-selectable capacitors to provide different ranges of frequency and to provide for vernier control by making one of the resistors variable.

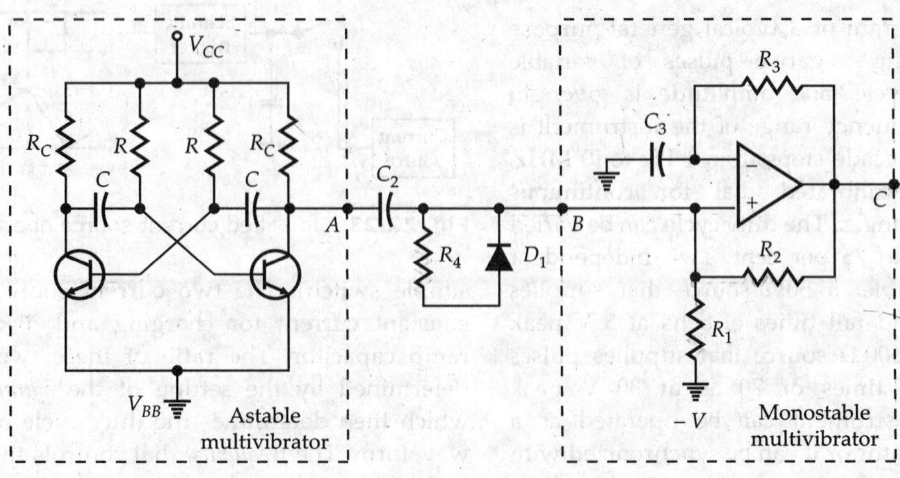


**Fig. 22.24** Basic pulse generator circuit.

The purpose of the "one-short," or monostable multivibrator, is to provide a short-duration output pulse each time a pulse is applied to its input. The monostable multivibrator has one stable state and one unstable state. The circuit operates in its stable state until an input pulse drives it into its unstable state, thus producing an output pulse. The duration of the output pulse is computed as

$$t = R_3C_3 \ln\left(\frac{R_1 + R_2}{R_3}\right) \qquad \ldots(22.38)$$

If resistors $R_1$ and $R_2$ are equal in value, as they frequently are, then the expression simplifies to

$$t = 0.693\, R_3C_3 \qquad \ldots(22.39)$$

The circuit $C_2R_4$ is called a *differentiator.* Its purpose is to provide a short-duration "spike" to trigger the monostable multivibrator. The purpose of diode $D_1$ is to bypass to ground the negative spike from the differentiator. The waveforms at points *A, B,* and C are shown in Fig. 22.25.

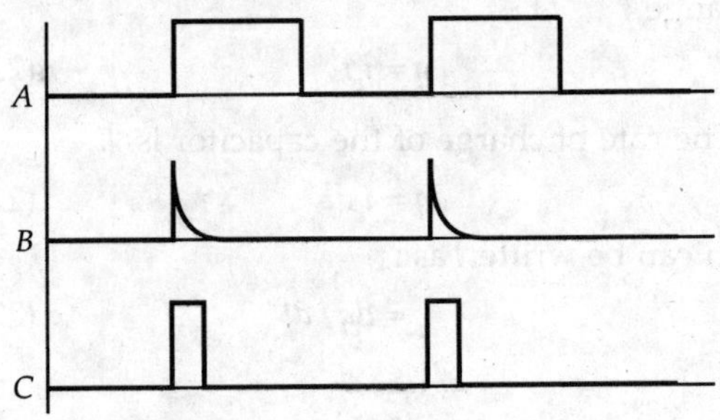


**Fig. 22.25** Waveforms for the pulse generator in Fig. 22.24.

***Example 22.5*** *Determine the frequency, pulse width, and duty cycle of the circuit shown in Fig. 22.25.*

**Solution.** The frequency of the square wave generated by the astable multivibrator is computed as

$$f = \frac{0.693}{RC}$$

$$= \frac{0.893}{75 \times 10^3 \times 0.01 \times 10^{-6}} = 924 \text{ Hz}$$

The pulse width is computed as

$$t = 0.69\, R_3C_3$$

$$= 0.69 \times 50 \times 10^3 \times 0.005 \times 10^{-6}$$

$$= 0.173 \text{ ms}$$

To compute the duty cycle, we must first compute the pulse period which is

$$T = \frac{1}{f} = \frac{1}{924\,\text{Hz}} = 1.08 \text{ ms}$$

We can now compute the duty cycle as

$$\text{Duty cycle} = \frac{\text{Pulse width}}{\text{Pulse period}} \times 100\%$$

$$= \frac{0.173 \times 10^{-3}}{1.08 \times 10^{-3}}$$

$$= 16\%$$

Some of the typical specifications of a laboratory-quality pulse generator given below :

- Pulse repetition rate : 0.3 Hz to 20 MHz, five ranges.
- Pulse width : 25 nsec to 3 sec in five ranges.
- Maximum duty cycle : greater than 80% from 0.3 Hz to 1 MHz; greater than 50% form 1 to 10 MHz.
- Output impedance: 50 Ω.
- Amplitude continuously variable from 0.03 to 10 V.



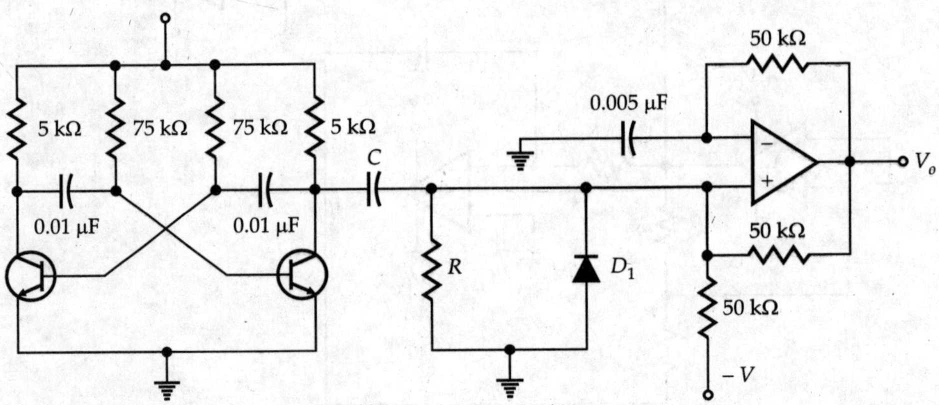


**Fig. 22.25** Pulse generator circuit for Example 22.6.

## 22.10 TRIANGULAR WAVESHAPE GENERATOR

The circuit diagram of a triangular waveshape generator is shown in Fig. 22.26. At the first stage, $A_1$ which is a voltage comparator generates a square wave output.

The output of $A_1$ is driven to saturation; therefore, the square wave is either at $+V_{CC}$ or $-V_{CC}$. The second stage, $A_2$, is an integrator which generates a triangular output. The square wave is applied to a square-to-sine wave converter that filters out the odd harmonics making up the square wave while passing on only the fundamental sine wave.

The operation of the circuit can be analyzed by starring at the output of the comparator, which is at either $+V_{CC}$ or $-V_{CC}$. Consider $V_{01}$, to be at $-V_{CC}$. The voltage $V_{01}$ will remain at $-V_{CC}$ until the voltage at the inverting input of $A_1$ exceeds the voltage at the noninverting input, which in this case is at zero volt. The noninverting input voltage, $V_x$, is due, in part, to the voltage $V_{01}$, and, in part, to the voltage $V_{02}$, according to the expression

$$V_x = -V_{CC}\frac{R_1}{R_1+R_2} + V_{02}\frac{R_2}{R_1+R_2} \quad \text{...(22.40)}$$

the output $V_{01}$ changes states when $V_x = 0$; therefore, we can say

$$0 = -V_{CC}\frac{R_1}{R_1+R_2} + V_{02}\frac{R_2}{R_1+R_2} \quad \text{...(22.41)}$$

which simplifies to

$$V_{02}R_2 = V_{CC}R_1 \quad \text{...(22.42)}$$

From Eqn. 22.42 we can determine the maximum amplitude of the triangular output, $V_{02}$, which is expressed as :

$$V_{02} = V_{CC}\frac{R_1}{R_2} \quad \text{...(22.43)}$$

When the output voltage $V_{02}$ reaches the amplitude given by Eqn. 22.43, the output of the comparator changes states and the triangular wave begins to decrease linearly. Since the output is symmetrical about 0 V. Equation 22.43 also expresses the minimum value of $V_{02}$ at which switching occurs. The wave-forms at $V_x$, $V_{01}$, and $V_{02}$ are shown in Fig. 22.27 for the situation in which $R_1 = R_2$.

The frequency of the circuit is controlled by the RC time constant of the integrator. To obtain an expression for the frequency, we begin with the expression relating capacitor current, charge and time of change :

$$q = i_c t \quad \text{...(22.44)}$$

The rate of charge of the capacitor is

$$dq = i_c dt \quad \text{...(22.45)}$$

which can be written as :

$$i_c = dq / dt \quad \text{...(22.46)}$$

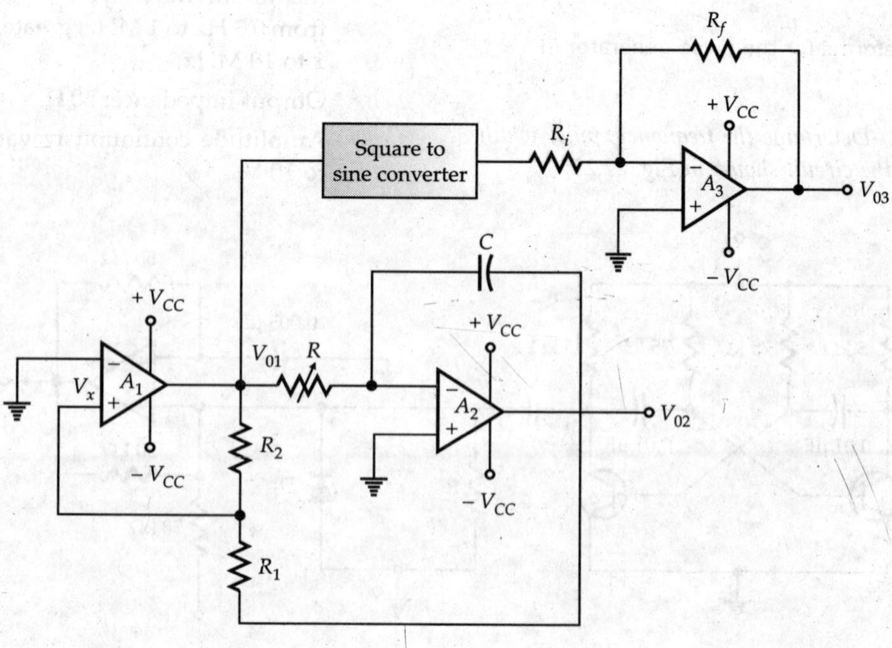


**Fig. 22.26** Circuit for a basic function generator.

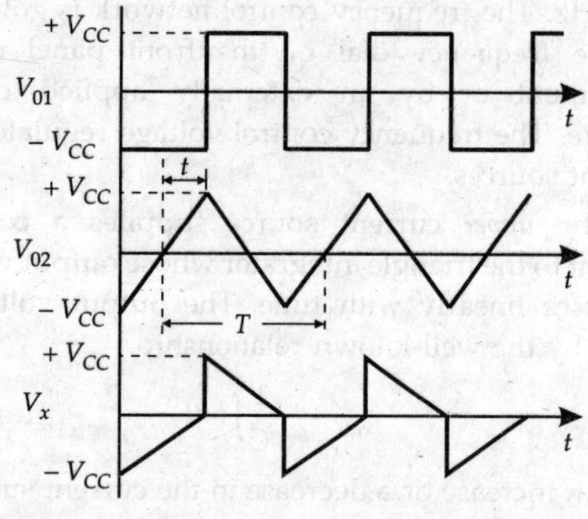


**Fig. 22.27** Waveforms for the function generator of Example 22.6.

As the capacitor charges, the relationship between charge, capacitance, and voltage across the capacitor plates is

$$q = CV_{02} \qquad \text{...(22.47)}$$

Substituting Eqn. 22.47 into Eqn. 22.46 yields

$$i_c = C\frac{d(V_{02})}{dt} \qquad \text{...(22.48)}$$

Since the input resistance of the operational amplifier is very high, the current through resistor $R$ is approximately equal to the charging current of the capacitor. Therefore, we can write

$$i_R = C\frac{d(V_{02})}{dt} \qquad \text{...(22.49)}$$

In addition, since the voltage gain of the operational amplifier is very high, the voltage at the input to the amplifier is very nearly zero.

Therefore,

$$i_R = \frac{V_{01} - 0}{R} \qquad \text{...(22.50)}$$

Substituting Eqn. 22.50 into Eqn. 22.49, we obtain,

$$d(V_{02}) = \frac{1}{RC}V_{01}\,dt \qquad \text{...(22.51)}$$

Integrating both sides of Eqn. 22.51, we obtain

$$V_{02} = \frac{1}{RC}\int V_{01}\,dt = \frac{V_{01}}{RC(t)} \qquad \text{...(22.52)}$$

Substituting Eqn. 22.43 into Eqn. 22.52, gives

$$V_{CC}\frac{R_1}{R_2} = \frac{V_{01}}{RC}t \qquad \text{...(22.53)}$$

Since $V_{01} = V'_{CC}$, Eqn. 22.53 simplifies to :

$$t = RC\frac{R_1}{R_2} \qquad \text{...(22.54)}$$

The development of Eqn. 22.54 began with Eqn. 22.44, which allows us to compute the charge on a capacitor after a period of time $t$. Equation 22.44 is valid only if the initial charge and, therefore, the initial voltages on the capacitor are zero. Therefore, the time $t$ in Eqn. 22.54 is the time for the capacitor to charge from 0 V until switching occurs, which is at one-fourth cycle as shown in Fig. 22.27. Since $t = T/4$, Eqn. 22.54 becomes

$$T = 4RC\left(\frac{R_1}{R_2}\right) \qquad \text{...(22.55)}$$

The frequency, which is the reciprocal of the period, is now expressed as

$$f = \frac{1}{4RC}\left(\frac{R_2}{R_1}\right) \qquad \text{...(22.56)}$$

***Example 22.6*** *Compute the frequency and the peak amplitude of the triangular output of the circuit shown in Fig. 22.28.*

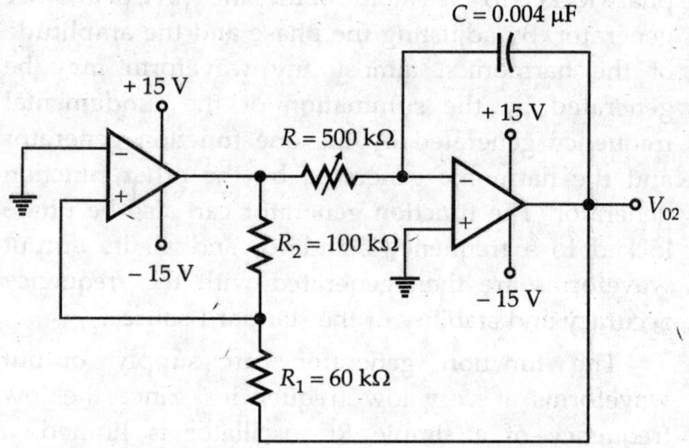


**Fig. 22.28** Function generator for Example 22.7.

The output frequency can be computed using Eqn. 22.56 as :

$$f = \frac{1}{4RC}\left(\frac{R_2}{R_1}\right)$$

$$= \frac{1}{4(500\times 10^3)(0.004\times 10^{-6})} \times \left(\frac{100\times 10^3}{60\times 10^3}\right)$$

$$= 208 \text{ Hz}$$

The amplitude of the triangular waveform can be computed from Eqn. 22.43 as

$$V_{02} = V_{CC}\frac{R_1}{R_2} = 15\times\frac{60\times 10^3}{100\times 10^3}$$

$$= 9 \text{ V}.$$

## 22.11 FUNCTION GENERATOR

A function generator is a versatile instrument that delivers a choice of different waveforms whose frequencies are adjustable over a wide range. The most common output waveforms are the sine, triangular, square, and sawtooth waves. The frequencies of these waveforms may be adjusted from a fraction of a hertz to several hundred kilohertz.

The various outputs of the generator may be available at the same time. For instance, by providing a square wave for linearity measurements in an audio system, a simultaneous sawtooth output may be used to drive the horizontal deflection amplifier of an oscilloscope providing a visual display of the measurement results. The capability of the function generator to *phase lock* to an external signal source is another useful feature. One function generator may be used to phase lock a second function generator, and the two output signals can be displaced in phase by an adjustable amount. In addition, one generator may be phase locked to a harmonic of the sine wave of another generator. By adjusting the phase and the amplitude of the harmonics, almost any waveform may be generated by the summation of the fundamental frequency generated by the one function generator and the harmonic generated by the other function generator. The function generator can also be phase locked to a frequency standard, and all its output waveforms are then generated with the frequency accuracy and stability of the standard source.

The function generator can supply output waveforms at very low frequencies. Since the low frequency of a simple *RC* oscillator is limited, a different approach is used in the function generator of Fig. 22.29. This instrument delivers sine, triangular, and square waves with a frequency range of 0.01 Hz to 100 kHz. The frequency control network is governed by the frequency dial on the front panel of the instrument or by an externally applied control voltage. The frequency control voltage regulates two current sources.

The *upper* current source supplies a constant current to the triangle integrator whose output voltage increases linearly with time. The output voltage is given by the well-known relationship.

$$e_{out} = -\frac{1}{C}\int i\,dt \qquad \ldots(22.57)$$

An increase or a decrease in the current supplied by the upper current source increases or decreases the slope of the output voltage. The voltage comparator multivibrator changes state at a predetermined level on the positive slope of the integrator's output voltage. This change of state cuts off the upper current supply to the integrator and switches on the lower current supply.

The *lower* current source supplies a reverse current to the integrator so that its output decreases linearly with time. When the output voltage reaches a predetermined level on the negative slope of the output waveform.

## 22.12 SIGNAL GENERATING INSTRUMENTS

The previous sections of this chapter have considered the circuits which go to produce the sine, pulse and square waveforms used in signal generating equipment. In the remainder of this chapter the instruments are introduced. These are categorised as signal generators, swept frequency generators, synthesizers, pulse generators; and function generators, although the dividing lines between these groups are often blurred.

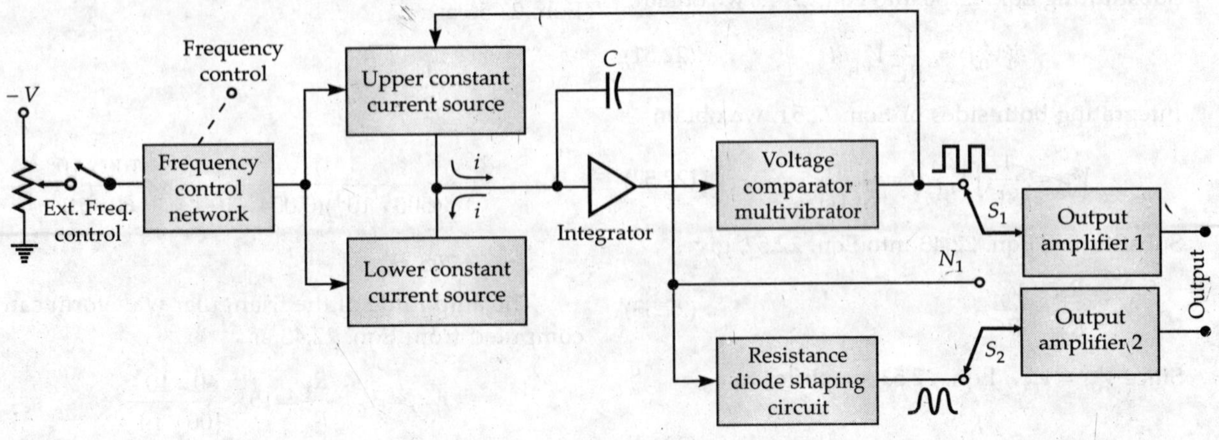


**Fig. 22.29** Basic element of a function generator.

### 22.12.1 Signal Generators

These instruments usually produce a fixed frequency sine wave, whose output can be frequency or amplitude modulated by another signal. The instruments cover a frequency range of 0.001 Hz–50 GHz, but not from the same device. Figure 22.30 shows a signal generator circuit. Frequency modulation is achieved by varying the voltage across a variable capacitance diode in the tuning circuit of the oscillator. This gives a system with low output distortion, for modulation depths below 1% of the carrier frequency. Above this modulation level the waveform applied to the tuning diode needs to be deliberately distorted, in order to compensate for its non-linear characteristics. During frequency modulation manual or automatic methods may be used to keep the amplitude of the output constant.

Amplitude modulation is most conveniently done by varying the supply voltage to the oscillator. This method is, however only suitable for small modulation depths, up to about 50%. It also gives phase modulation due to the effect on the components used within the oscillator circuit. Feedback can be used to reduce output distortion, as shown in Fig. 22.30, by detecting the output to obtain the modulation envelope, comparing this with the amplitude modulation input, and then amplifying and feeding back the difference as the modulation signal. This technique is known as envelope feedback.

The output amplifier shown in Fig. 22.30 provides the required signal level, and buffers the oscillator from changes in the load impedance. The attenuator is needed to give low level output signals.

The heterodyne principle is sometimes used, as in Fig. 22.31, to give a continuously variable, wide frequency range output from a single instrument. The signal quality is good, but the frequency stability is poor, especially at low frequencies. This is because the output is a difference frequency ($f_2 - f_1$) so a slight change in one of the frequencies, when they are close to each other, can give a wide drift in the output. The output also has a considerable amount of noise and spurious signals.

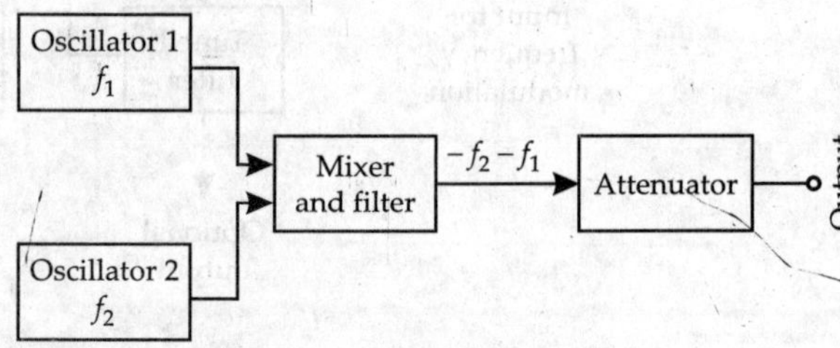


**Fig. 22.31** A heterodyne oscillator.

Multiplier and divider techniques can be used to extend the frequency range available from a single instrument. In the multiplier generator, shown in Fig. 22.32, the output from the fixed frequency oscillator is fed through a series of tuned multipliers, each having a non-linear amplifier which produces harmonics. The output from each stage is fed to a tuned filter which selects the high frequency output. Frequency modulation may be applied to the master oscillator, and amplitude modulation is achieved by varying the d.c. supply to the last multiplier stage. The disadvantage of the multiplier generator is that it produces a large amount of spurious signal around the desired frequency.

The divider signal generator uses a high frequency master oscillator, whose output is divided down by a series of electronic stages. The outputs from the divider stages are square waves, which need to be filtered to produce sine waveforms. Frequency modulation is applied to the main oscillator, if required, and amplitude modulation is obtained by diode modulators in the output amplifier. The divider generator does not produce any spurious signal at frequencies below the required frequency.

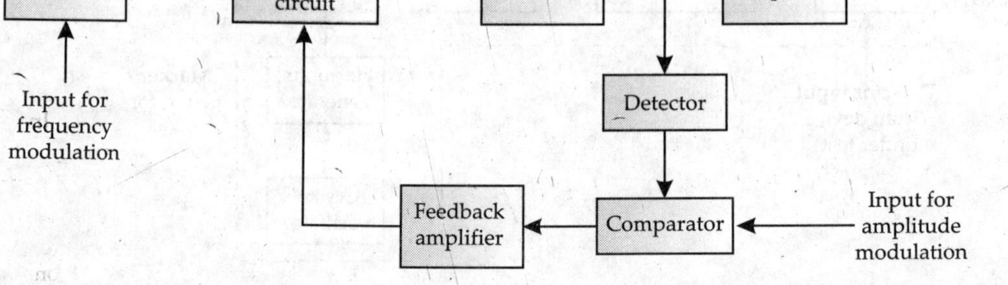


**Fig. 22.30** A signal generator using envelope feedback for amplitude modulation.

22

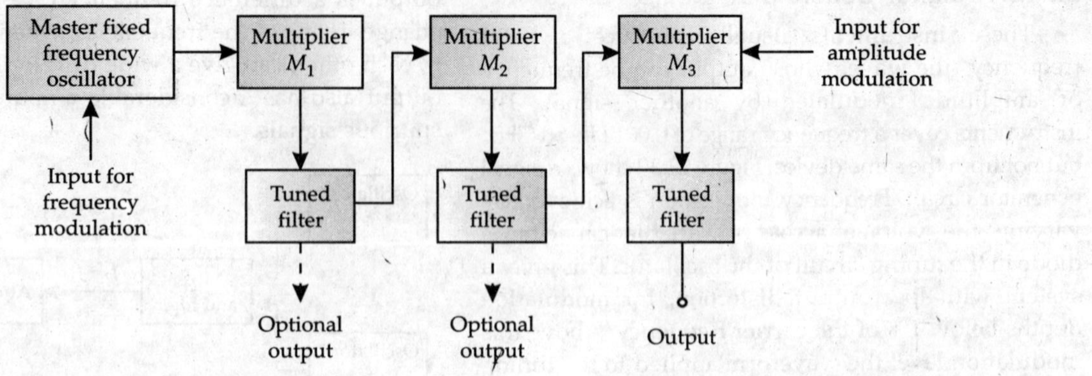


**Fig. 22.32** A multiplier frequency generator.

## 22.12.2 Swept Frequency Generator

A swept frequency generator, or sweeper, is a special type of signal generator in which the output frequency is cyclically swept through a range of frequencies. The instrument may have a display built into it, to show the variation of amplitude with frequency, or it may provide signals for use with an external oscilloscope.

Figure 22.33 shows the block diagram of a typical sweep generator. The time base is usually adjustable, to give output sweep times in the range from 10 ms to greater than 100 s. It is also often possible to control the sweep manually from the front panel of the instrument. The time base is frequently a triangular or sawtooth waveform.

Two modes are used to set the swept frequency range :

(*i*) *The stop start or* ($f_2 - f_1$) *mode*. In this mode the stop and start frequencies are set from the front panel, and the instrument sweeps between these limits. This mode is used for wide sweep widths.

(*ii*) *Delta frequency mode* ($\Delta f$). In this mode the centre frequency, and the maximum excursion about this frequency, are set from the front panel. It is used for narrow sweep widths.

The frequency range of the swept frequency generator usually extends over three bands, 0.001 Hz – 100 kHz (low frequency to audio), 100 kHz–1500 MHz

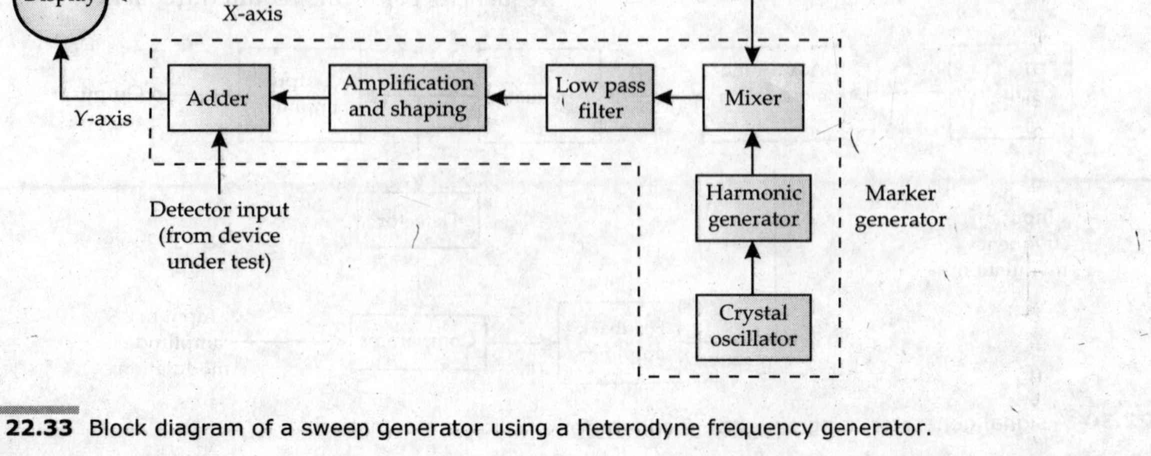


**Fig. 22.33** Block diagram of a sweep generator using a heterodyne frequency generator.

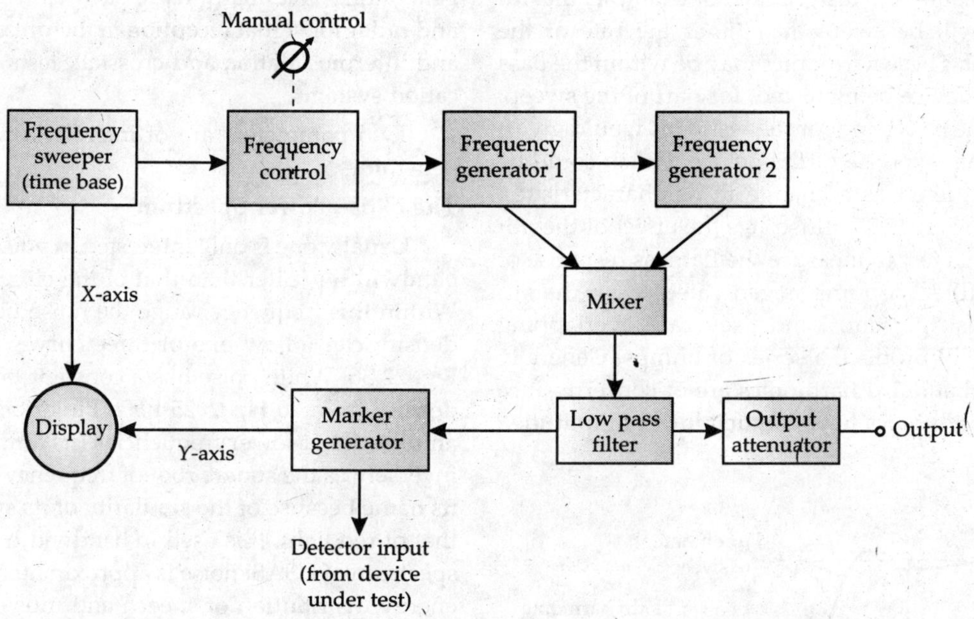


**Fig. 22.34** Block diagram of a sweep generator using a heterodyne frequency generator.

(r.f. range), and 1–200 GHz (microwave range). Three approaches may be used to cover a wide band of swept frequencies from a single instrument.

(*i*) Manually switching between different frequency oscillators. The problem occurs when the frequency range needed overlaps two bands.

(*ii*) Stacked switching. In this the bands are automatically selected by electronic switches, so that one can sweep the whole instrument range as one continuous band.

(*iii*) Heterodyne control. In this method, illustrated in Fig. 22.34, two high frequency signals are mixed to give a lower difference frequency output, as one continuous band. If, for example, generator 1 can be varied over a frequency range 3-5 GHz (*i.e.*, a change of 2 : 3), and generator 2 has a fixed frequency output of 3 GHz. then the output can vary from 1 to 2000 MHz (*i.e.*, a change of 2000 : 1). Since the tuning range needed for this wide output variation small, the instrument has good linearity. However, a non-linear mixer will produce many other frequencies in addition to the required one.

Output level control is used in a swept frequency generator to keep the output amplitude to the-value set on the front dial. Usually an r.f. detector monitors the r.f. amplitude of the output. This is compared with a signal corresponding to the required amplitude, and the error is fed back to an electronic attenuator circuit, to keep the output constant.

The graticule on the display is not accurate enough to be able to interpret the frequency being displayed, and some swept frequency generators have the facility for putting a series of accurately known frequency markers onto the screen, to act as a reference. One method of generating the marker frequencies, as shown in Fig. 22.33, is to pass the output from a stable crystal oscillator through a harmonic generator, which gives a series of narrow pulses, spaced at harmonic intervals. These are mixed with a sample of the r.f. output from the sweeper. The mixer will produce a series of low difference frequency output bursts, called birdies, as the sweeper frequency approaches and passes through each harmonic frequency. The birdies are shaped and amplified, and then combined with the signal received back from the device under test (via a detector), before being fed to the display. This gives a composite picture of the frequency response characteristic, of the device being tested, and the frequency calibration markers. An alternative method, of generating a marker, is to pass a single fixed frequency signal to the mixer, to give a spot of accurately known frequency on the screen.

### 22.12.3 Sweeper Errors

Several errors occur in swept frequency generator measurements, due to imperfections in the output of the generator. Any harmonics produced in the frequency source will be swept along with the fundamental, but at a faster rate. For example, the $n$th harmonics will be swept at $n$ times the rate of the fundamental. These harmonics may be within the pass band of the device being tested, for part of the sweep, and out of the pass band for the rest. This would give a change in the response of the device, which could be falsely interpreted as a change in its characteristics. For example Fig. 22.35, illustrates the effect of the $n$th harmonic on a test to measure the flatness response of a low pass filter. A bump is generated when the $n$th harmonic is present, and several overlapping harmonics will produce a series of bumps. Generally only the 2nd and 3rd harmonics are of concern, since the higher harmonics have a much lower amplitude.

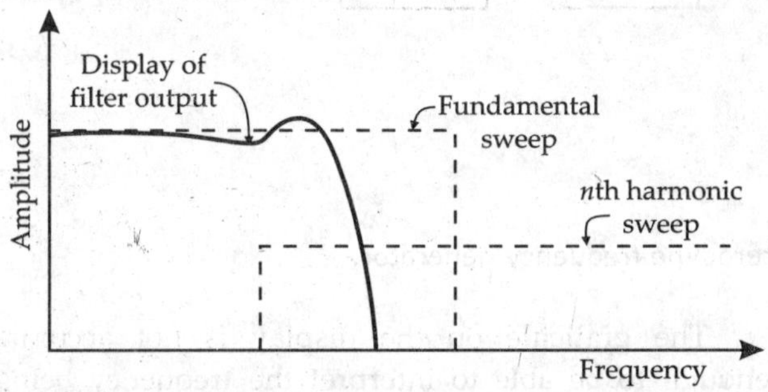


**Fig. 22.35** Effect of sweep generator harmonics on a low pass filter response measurement.

The accuracy of the output, and the change in its amplitude with frequency, is another important consideration. The amplitude varies with the sweep rate, since the amplitude feedback correction circuits have less time to work at faster sweep rate. The effect of this error can be minimised by sweeping at the lowest rate acceptable for the device being measured, and by making readings with respect to the markers, since these are also affected by a change in amplitude.

The output from a sweeper is not linear, which means that the spacing between markers is unequal, so that one cannot easily estimate a value at a frequency between two markers. The error is usually very small, but if an exact reading is required then the sweeper is set to its manual mode. Its frequency can then be changed until the desired frequency appears on the display, and this can be read with an external accurate frequency meter.

### 22.12.4 Random Noise Generators

A random noise generator produces a signal output whose instantaneous amplitude varies at random, and has no periodic frequency component. Generators are available which cover frequency bands from low audio to microwave. Random noise generators have many uses, such as testing of radio and radar for signal reception in the presence of noise, and intermodulation and crosstalk tests in communication systems.

Two parameters are of interest in random noise generators :

**The Noise-Power Spectrum**

Usually one is only interested in noise which has a bandwidth greater than that of the circuit under test. Within this frequency range the noise power spectral density can follow one of three curves, as shown in Fig. 22.36. White noise has a constant power spectral density from 20 Hz to 25 kHz. Pink noise has higher amplitudes at lower frequencies (its amplitude varies inversely as the square root of frequency) and is given its name because of the similarity of its spectrum with that of red light. It is used in bandwidth analysis. The spectrum of USASI noise is approximately equal to the energy distribution of speech and music frequencies, and it is used for testing audio systems.

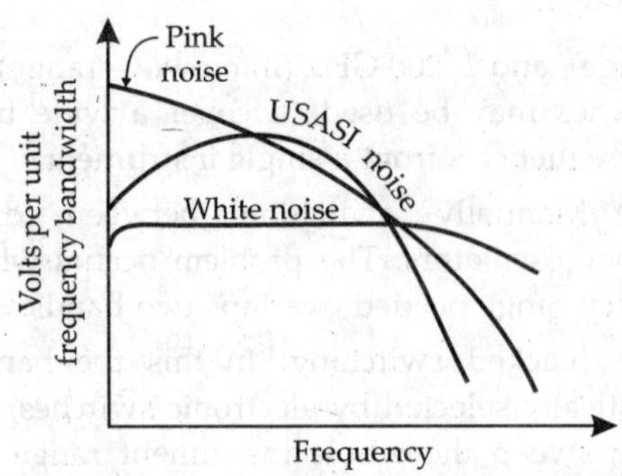


**Fig. 22.36** Noise spectrum.

**Probability Distribution Function (PDF)**

The power spectrum of a noise signal generator gives its frequency content, but does not characterise its waveshape; this is done by its p.d.f. The p.d.f. is the statistical calculation of the portion of time which the signal spends at the various amplitudes. Usually, for naturally occurring random signals, this function follows a Gaussian curve.

Random noise can be generated by using a gas discharge tube or a zener diode generator. A semiconductor noise diode gives an output frequency in the band from 50 kHz to 200 kHz. This can be amplified and modulated to produce an audio frequency band. The signal is then passed through a filter to get one of the spectra shown in Fig. 22.36.

The disadvantage of the random noise generation method is that sometimes the total output is subject to long term variations, giving an unpredictable power spectrum. An alternative method is to generate a pseudo-random noise, such as by clocking a long shift register having several feedback loops. The noise pattern will now be repeatable, but within a cycle the signal behaves as truly random noise. The output from the shift register is fed through a low pass filter to give the pseudo-random analogue noise signal.

## 22.13 FREQUENCY SYNTHESIZER

Frequency generators are of *two* types :

(*i*) Free running, in which the output can be tuned continuously, over a frequency range, by mechanical or electronic methods. These are the types described so far, and they have good overall performance, although their frequency, accuracy and stability are relatively poor.

(*ii*) Synthesizer, which has its output derived from a fixed frequency, highly stable oscillator, and covers the range in a series of steps.

Synthesizers are of two types, direct and indirect. The *direct synthesizer* uses a stable crystal oscillator, followed by a series of harmonic multipliers and mixers, to provide the range of different output frequencies. The *indirect frequency synthesizer* uses a phase locked loop to give an output which is a fraction of that of a stable crystal oscillator. The synthesizer technique provides a generator with excellent frequency accuracy and stability, but careful design is needed to overcome the problem of spurious signals in the output.

Figure 22.37 shows the basic indirect frequency synthesizer. The output from the phase detector is a signal proportional to $f_1 \pm f_2$. The low pass filter selects the signal, which is then amplified and fed to the voltage controlled oscillator (VCO). The output from the VCO is fed to a programmable divider, and then back to the phase detector. The system drives $f_2$ towards $f_1$ until they are equal, and at this stage there is still a phase difference between the two frequencies, which produces sufficient d.c. signal to keep $f_2$ locked to $f_1$. The output frequency is $Nf_1$ and this can be changed by varying $N$, usually through a keyboard from the front panel of the instrument.

The switching speed, between frequency setting, of the frequency generator depends on the time needed for $f_2$ to lock to $f_1$ and on the setting time. For large frequency differences the lock time can be long. It is minimised by setting the VCO close to its final value, usually under microprocessor control, and then letting the phase lock loop pull it to its exact value. The setting time is the time, after lock, for the output to reach within a given range of the final output setting: it is typically a few milliseconds.

The problem, with the basic system shown in Fig. 22.37 is that for a wide frequency range, with small incremental steps, a low reference frequency, and a VCO which can cover a wide range, are required.

The programmable divider also needs a large divider ratio, and this produces noise within the loop bandwidth. A multiloop frequency synthesizer, as shown in Fig. 22.38, overcomes these disadvantages. The low frequency synthesizer has a small range, and can operate with small steps, whilst the high frequency synthesizer operates over a wide range with wide steps. Combining these two signals, in a mixer, gives a system which can work over a wide frequency range, with small steps.

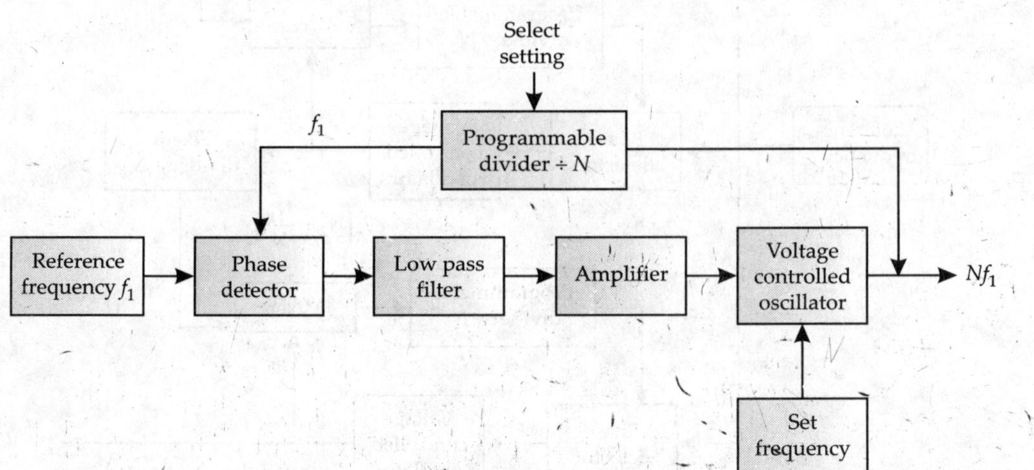


**Fig. 22.37** An indirect frequency synthesizer

22

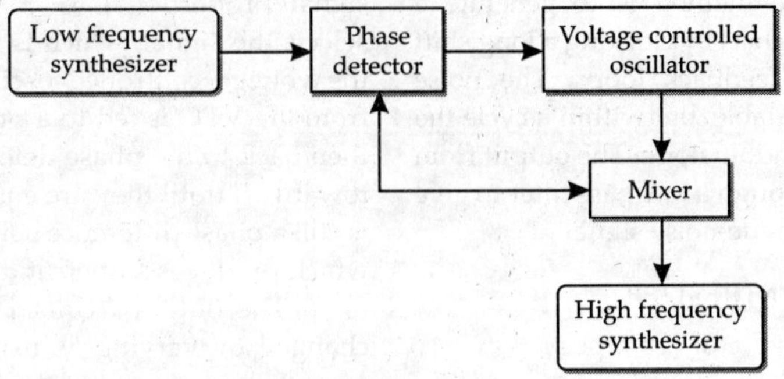


**Fig. 22.38** An indirect multiloop frequency synthesizer.

## 22.14 PROGRAMMABLE DECADE FREQUENCY SYNTHESIER

Figure 22.39 shows an alternative system, which is called a *decade synthesier* as the fixed dividers have a ratio of 10. The range of ratios needed from each programmable divider is much less. The fixed divider frequency synthesier uses a high reference frequency allowing a wide loop bandwidth, and thereby giving a faster lock time and better noise performance commercial synthesiers are available which cover frequency ranges from 10 kHz to 3000 MHz and from 10 MHz to 30 GHz and with modulation capabilities for use a signal generator from the same instrument, and by their flexibility. The three basic types of outputs produced are sine, square and triangular wave, over a frequency range of 1–50 MHz, although some function generators also have the capability of producing pulses, the performance of the output from a function generator is usually lower than that from a dedicated instrument.

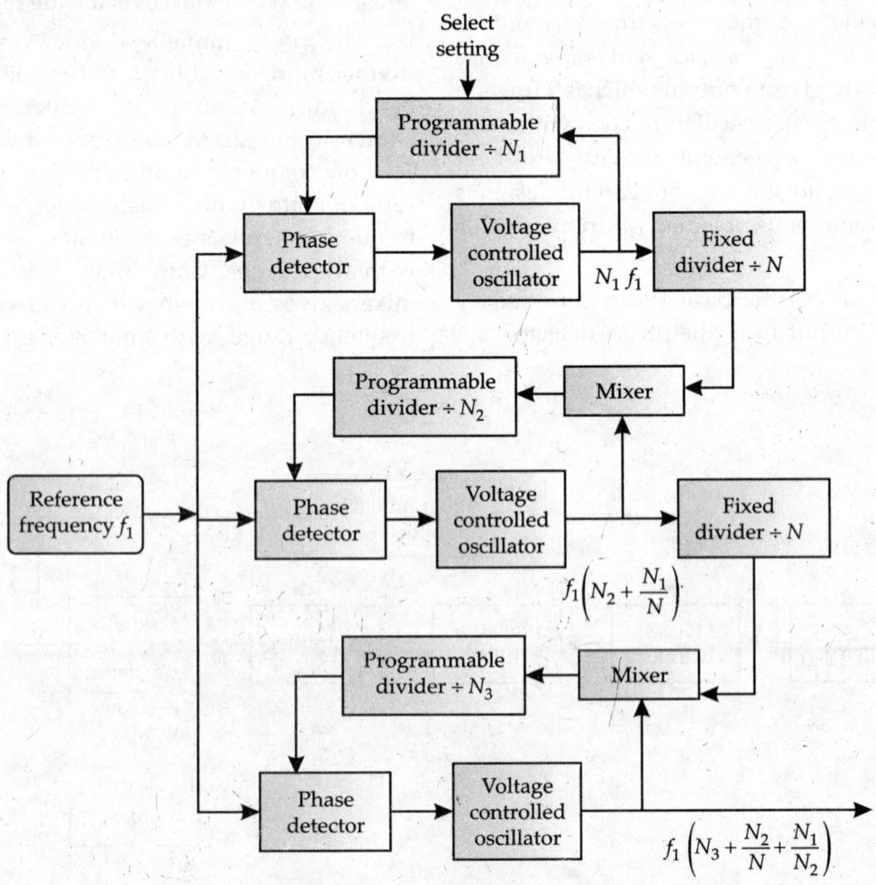


**Fig. 22.39** A fixed divider (decadic) indirect frequency synthesizer.

## Review Questions

1. What are signal sources? What are the desirable characteristics of a signal? Give an overview of different signal sources used.
2. Describe the meaning of an oscillator. Describe the classification of oscillators
   (*i*) depending upon the principle they operate
   (*ii*) depending upon the frequency range used by them.
3. Describe the Barkhausen criteria for production of oscillations. Describe how the Barkhausen criteria is met within Wien's bridge oscillator.
4. Describe the principle of working of a Wien's bridge oscillator. Derive the conditions for its balance.
5. Describe the practical circuit for a Wien's bridge oscillator and list its advantages and disadvantages.
6. Explain the circuit of a R.C. phase shift oscillators. Describe how Barkhausen criteria is satisfied in this oscillator. Give its advantages and disadvantages.
7. Prove that in a R.C. shift oscillator the gain of the amplifier should be at least 29 to sustain oscillations.
8. Describe the circuit of a Hartley oscillator for production of radio frequencies. Derive the condition for production of oscillations.
9. Explain the circuit of a Colpitts' oscillator for production of radio frequencies. Compare its operation with that of Hartley's oscillator.
10. Describe the working of a crystal oscillator. List its advantages and disadvantages.
11. Explain the principle of working of a negative resistance oscillator and describe the frequency range over which it is used.
12. Define the duty cycle for a pulse wave. Describe the characteristics and terminology associated with pulse waves.
13. Distinguish between the active and passive circuits used for production pulse and square waveshapes. Describe a passive pulse shaping circuit.
14. Describe in details the circuit and working of an astable multivibrator.
15. Describe the circuit diagram and working of a laboratory type square wave and pulse generator. Explain the working of a current source.
16. Explain with a neat circuit diagram how a triangular waveshape is produced using a square wave.
17. Describe the circuit of a square wave generator which generates square, triangular and sine waveshapes.
18. Describe the following :
    (*i*) A signal generator using envelop feedback for amplitude modulation.
    (*ii*) A heterodyne oscillator.
19. Describe the working of a sweep frequency generator. What are the sweeper errors ?
20. What is a frequency synthesizer? Describe its circuit details.

22

## Unsolved Problems

1. Determine the gain with feedback $A$ for the amplifier in Fig. 22.40, for sustained oscillations.
   **[Ans.19.96]**

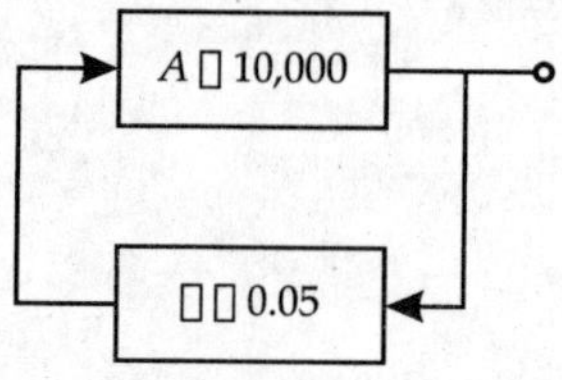


**Fig. 22.40** Circuit for Problem 1.

2. Determine the values of $C_1$ and $C_2$, that will cause the circuit in Fig. 22.41 to oscillate at 12 kHz. Assume $C_1$ □ $C_2$
   **[Ans. 1.32 nF]**

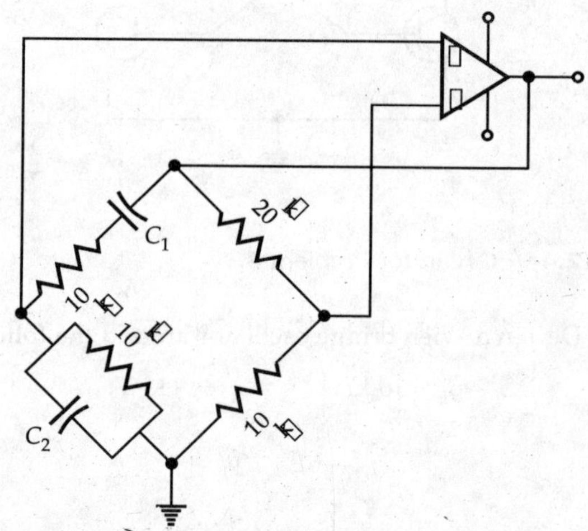


**Fig. 22.41** Circuit for Problem 2.

3. Determine the frequency of oscillations of a *RC* phase shift oscillator with a three section feedback network consisting of 20 Ω resistors and 1.5 nF capacitors. **[Ans.** 2.16 MHz]
4. Determine the value of $L_2$ in the circuit shown in Fig. 22.42 if the frequency of oscillations is to be 100 kHz. **[Ans.** 153 μH]
5. Determine the minimum value of $R_1$ in Fig. 22.42 to sustain oscillations if $L_2$ = 125 μH. **[Ans.** 83.3 kΩ]

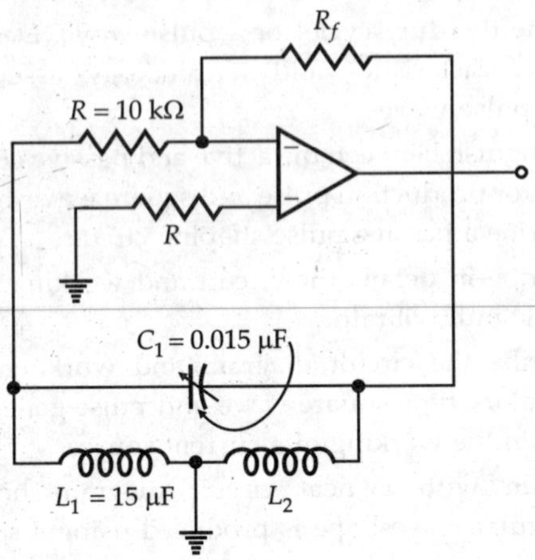


**Fig. 22.42** Circuit for Problem 4.

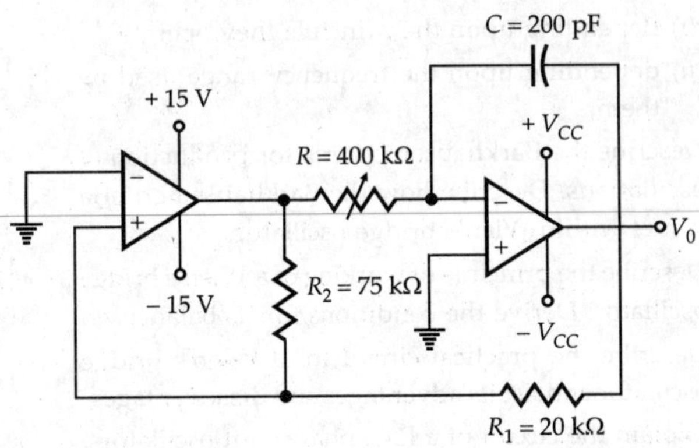


**Fig. 22.43** Circuit for Problem 6.

6. Determine the frequency of the triangular output of the circuit shown in Fig. 22.43. **[Ans.** 11.72 kHz]
7. Determine the peak to peak amplitude of the output signal for the circuit shown in Fig. 22.43. **[Ans.** 8 V]
8. Determine the frequency and duty cycle of the astable multivibrator output in the circuit of Fig. 22.44
9. Determine the pulse width of the output pulse $V_0$ in the circuit of Fig. 22.44 **[Ans.** 69 μs]

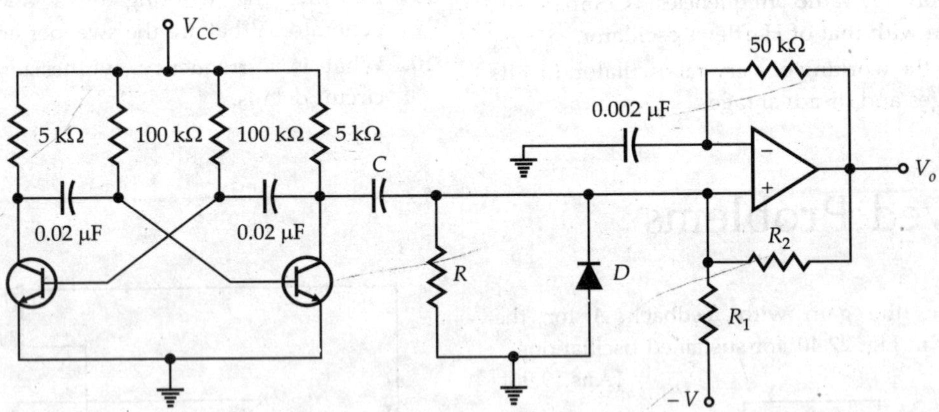


**Fig. 22.44** Circuit for Problem 8.

10. Design a Wien bridge oscillator around the following specifications :

$f_0$ = 10 kHz    $V_{CC}$ = 115 V,    $I_{in}$ = 1 μA    $I_{R4}$ + 100 $I_{in}$

(Refer to Fig. 22.4)

**[Ans.** $R_3$ = 90 kΩ, $R_4$ = 45 kΩ, $c$ = 35.4 pk]

# Objective Type Questions

**Tick (✓) the most appropriate answer**

1. The desirable characteristics of a signal source are :
   (*a*) the frequency of the signal should be well known and stable
   (*b*) the amplitude of a signal source should be controllable from very small to relatively large values
   (*c*) the output signal should be free from distortion
   (*d*) all the above.
2. In signal generators
   (*a*) energy is created
   (*b*) energy is generated
   (*c*) energy is converted from a simple d.c. source into a.c. energy at some specific frequency
   (*d*) all the above.
3. A video frequency oscillator has a frequency range of :
   (*a*) 20 Hz to 20 kHz (*b*) 20 kHz to 30 MHz
   (*c*) d.c. to 5 MHz (*d*) 2 MHz to 30 MHz.
4. Feedback oscillators have a closed loop gain of
   (*a*) $\frac{\mathbf{G}}{1-\mathbf{GH}}$ (*b*) $\frac{\mathbf{G}}{1+\mathbf{GH}}$
   (*c*) $\frac{\mathbf{G}}{1\pm\mathbf{GH}}$ (*d*) $\frac{\mathbf{H}}{1+\mathbf{GH}}$
   where **G** and **H** are respectively the gains of forward and feedback paths.
5. A Wien's bridge oscillator circuit is shown in Fig. 22.45.

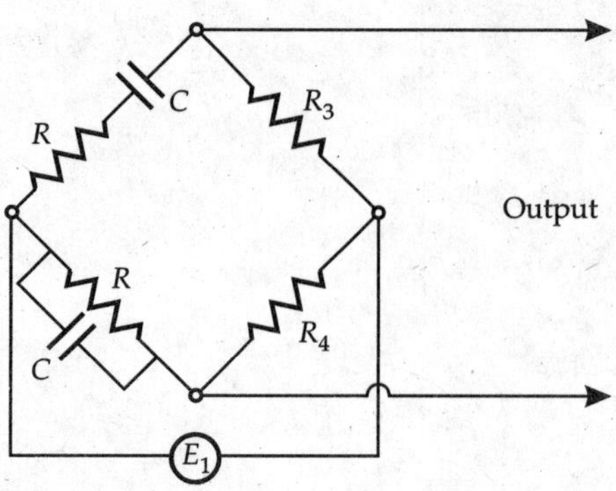


**Fig. 22.45** Fig. of Q. 5.

The values of components

$R = 6\ k\Omega,\quad C = 3\ nF,\quad R_3 = 6\ k\Omega,\quad R_4 = 12\ k\Omega.$

The frequency of oscillations is :
   (*a*) 8.84 kHz
   (*b*) 4.42 kHz
   (*c*) 8.84 MHz
   (*d*) No oscillations occur as Barkhausen criteria is not satisfied.
6. In an *RC* phase shift oscillator having three stages which have equivalued resistors and capacitors, the amplifier, in order to satisfy, Barkhausen criterion should have a minimum gain of
   (*a*) 6
   (*b*) 29
   (*c*) 31
   (*d*) none of the above.
7. In a Hartley oscillator, the inductor is divided into two parts. The inductor which is on the input of the amplifier has a value of 20 μH while that connected to the output side of the amplifier has an inductance of 300 μH. If the resistance connected to the input of the amplifier is 15 kΩ, what should be the value of resistance in the feedback path of the amplifier ?
   (*a*) 300 kΩ
   (*b*) 225 kΩ
   (*c*) 1 kΩ
   (*d*) 15 kΩ.
8. The pulse rise time is defined as the time taken by the pulse
   (*a*) to go from 10% to 90% of its amplitude
   (*b*) to go from 0% to 100% of its amplitude
   (*c*) to go from 0% to 90% of its amplitude
   (*d*) to go from 10% to 100% of its amplitude.
9. The duty cycle of a pulse shown in Fig. 22.46 is

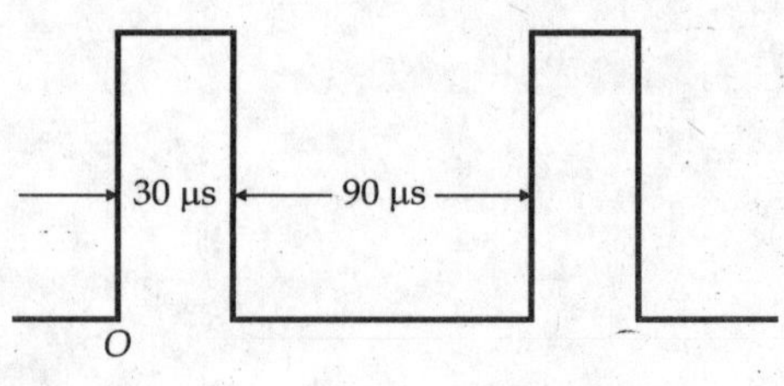


**Fig. 22.46** Fig. of Q. 9.

   (*a*) 33% (*b*) 25%
   (*c*) 300% (*d*) 75%.

22

**10.** An astable multivibrator

(*a*) gives one output pulse for every two input pulses
(*b*) gives a timed output pulse for a trigger input
(*c*) gives a train of output pulses for a trigger input
(*d*) gives four output pulses for a single input pulse.

**11.** An astable multivibrator uses a resistance of 100 kΩ and capacitance 0.01 μF. The frequency of square waves generated by it is

(*a*) 924 Hz (*b*) .593 Hz
(*c*) 693 Hz (*d*) 110 Hz

**12.** A triangular waveshape is obtained :

(*a*) by integrating a square wave
(*b*) by differentiating a in wave
(*c*) by differentiating a square wave
(*d*) by integrating a sine wave

**13.** In a sweep frequency generator, two oscillators, one with frequency range of 3 GHz to 5 GHz is heterodyned with a second oscillator having a fisced frequency output of 3 GHz, the output frequency varies from

(*a*) 0 to 2 GHz
(*b*) 6 GHz to 8 GHz
(*c*) 0 to 3 GHz
(*d*) 0 to 5 GHz.

**14.** A random noise generator produces a signal

(*a*) whose amplitude varies randomly
(*b*) which has no periodic frequency
(*c*) has an unpredictable power spectrum
(*d*) all the above.

**15.** A voltage controlled oscillator ($V_{CO}$) is an instrument

(*a*) whose frequency is dependent upon the amplitude of input signal
(*b*) whose frequency is independent of the amplitude of the input signal
(*c*) whose frequency is dependent upon the frequency of the input signal
(*d*) none of the above.

## Answers

**1.** (*d*) **2.** (*c*) **3.** (*c*) **4.** (*a*) **5.** (*d*) **6.** (*b*) **7.** (*b*) **8.** (*a*) **9.** (*b*) **10.** (*c*)
**11.** (*c*) **12.** (*a*) **13.** (*a*) **14.** (*a*) **15.** (*a*)

# 23
# Signal Analysers

## 23.1 INTRODUCTION

The analysis of electrical signals is used in many applications. The different instruments which are used for signal analysis are wave analysers, distortion analysers, spectrum analysers, audio analysers and modulation analysers.

All signal analysis instruments measure the basic frequency properties of a signal, but they use different techniques to do so. A spectrum analyser sweeps the signal frequency band and displays a plot of amplitude versus frequency. It has an operating range of about 0.02 Hz–250 GHz. A wave analyser is a voltmeter which can be accurately tuned to measure the amplitude of a single frequency, within a band of about 10 Hz–40 MHz.

Distortion analysers operate over a range of 5 Hz–1 MHz and give a measure of the energy present in a signal outside a specified frequency band. They therefore tune out the fundamental signal and give an indication of the harmonics. An audio analyser is similar to a distortion analyser but can measure additional functions, such as noise. Modulation analysers tune to the required signal and recover the whole amplitude modulated (a.m.), frequency modulated (f.m.) or phase modulation (p.m.) envelope for display or analysis.

## 23.2 WAVE ANALYSERS

It is well known that any periodic waveform can be represented as the sum of a d.c. component and a series of sinusoidal harmonics. Analysis of a waveform consists of determination of the values of amplitude, frequency, and sometime phase angle of the harmonic components. Graphical and mathematical methods may be used for the purpose but these methods are quite laborious. The analysis of a complex waveform can be done by electrical means using a band pass filter network to single out the various harmonic components. Networks of these types pass a narrow band of frequency and provide a high degree of attenuation to all other frequencies.

**A wave analyser**, in fact, is an instrument designed to measure relative amplitudes of single frequency components in a complex waveform. Basically, the instrument acts as a frequency selective voltmeter which is tuned to the frequency of one signal while rejecting all other signal components. The desired frequency is selected by a frequency calibrated dial to the point of maximum amplitude. The amplitude is indicated either by a suitable voltmeter or a CRO.

There are *two* types of wave analysers, depending upon the frequency ranges used,

(*i*) Frequency Selective wave analyser, and

(*ii*) Heterodyne wave analyser.

### 23.2.1 Frequency Selective Wave Analyser

This wave analyser is used for measurements in the audio frequency range (*i.e.*, from 20 Hz to 20 kHz). The analyser has a filter section with a very narrow passband which can be tuned to the frequency of interest. The block diagram of this wave analyser is shown in Fig. 23.1(*a*).

The waveform to be analysed in terms of its separate frequency components is applied to an input attenuator that is set by the *meter range* switch on the front panel. A driver amplifier feeds the attenuated waveform to a high-$Q$ active filter. The filter consists of a cascaded arrangement of $RC$ resonant sections and filter amplifiers. The passband of the total filter section is covered in decade steps over the entire audio range by switching capacitors in the $RC$ sections. Close-tolerance polystyrene capacitors are

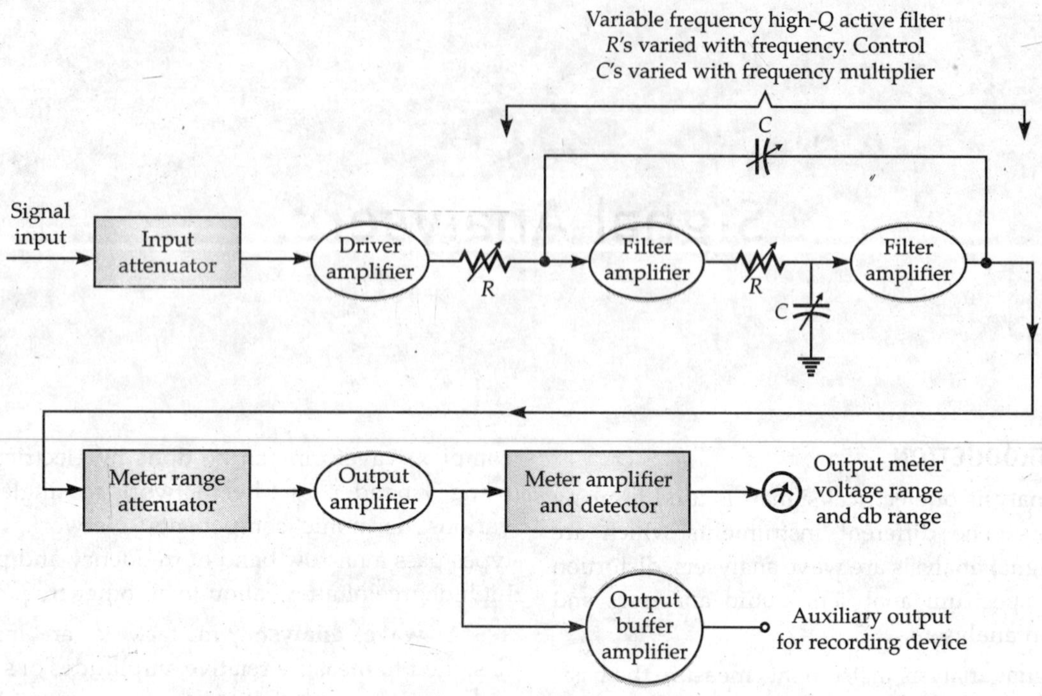


(a) Block diagram

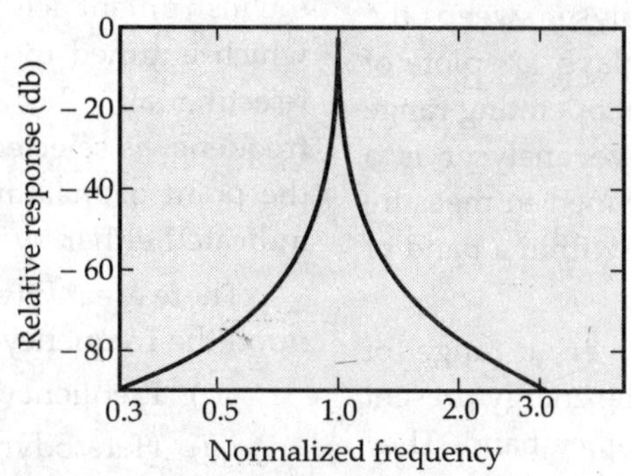


(b) Attenuation characteristics

**Fig. 23.1** Audio-range wave analyser and Characteristics of the active filter.

generally used for selecting the frequency ranges. Precision potentiometers are used to tune the filter to any desired frequency within the selection pass-band.

A final amplifier stage supplies the selected signal to the meter circuit and to an untuned buffer amplifier. The buffer amplifier can be used to drive a recorder or an electronic counter. The meter is driven by an average-type detector and usually has several voltage ranges as well as a decibel scale.

The bandwidth of the instrument is very narrow, typically about 1 per cent of the selected frequency. Figure 23.1(*b*) shows a typical attenuation curve of a wave analyzer.

## 23.2.2 Heterodyne Wave Analyser

For frequency measurements in the megahertz range, a heterodyne wave analyser construction is used, as shown in Fig. 23.2. The input signal is fed through an attenuator and amplifier before being mixed with a local oscillator. The frequency of this oscillator is adjusted to give a fixed frequency output which is in the passband of the i.f. amplifier. This signal is then mixed with a second crystal controlled oscillator, whose frequency is such that the output from the mixer is centred on zero frequency. The subsequent active filter has a controllable bandwidth, and passes the selected component of the frequency to the indicating meter.

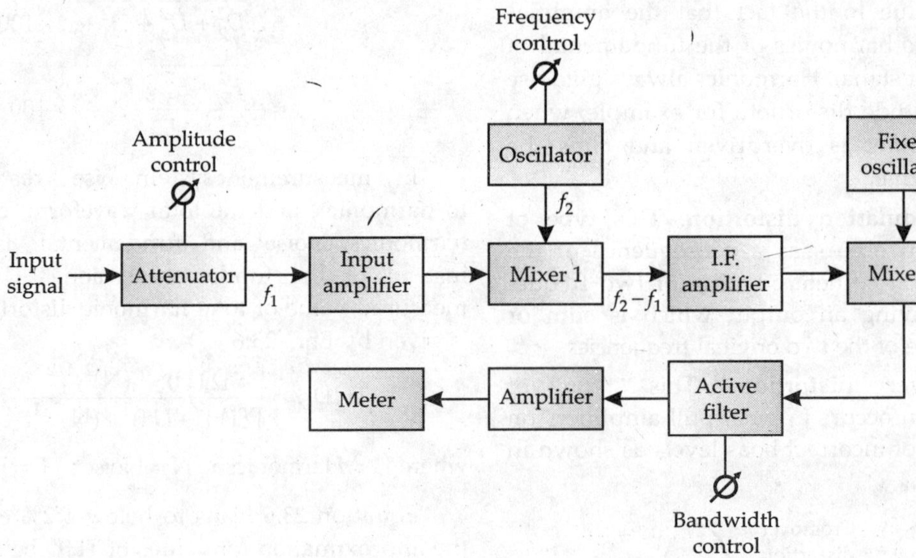


**Fig. 23.2** Block diagram of a heterodyne wave analyser.

Good frequency stability in a wave analyser is obtained by using frequency synthesisers, which have high accuracy and resolution, or by automatic frequency control (AFC). In an AFC system the local oscillator locks to the signal, and so eliminates the drift between them.

### 23.2.3 Applications of Wave Analysers

Wave analysers have very important applications in the following fields :

(*i*) Electrical measurements
(*ii*) Sound measurements and
(*iii*) Vibration measurements

The wave analysers are applied industrially in the field of reduction of sound and vibrations generated by rotating electrical machines and apparatus. The source of noise and vibrations is first identified by wave analysers before it can be reduced or eliminated A fine spectrum anylsis with a wave analyser shows various discrete frequencies and resonances that can be related to the motion of machines. Once, these sources of sound and vibrations are detected with the help of wave analysers, ways and means can be found to eliminate them.

## 23.3 HARMONIC DISTORTION ANALYSERS

The application of a sinusoidal input signal to an electronic device, such as an amplifier should result in generation of a sinusoidal output waveform. Generally, however, the output waveform is not an exact replica of the input waveform because of various types of distortions that may occur.

Distortion may be a result of the inherent non-linear characteristics of different components used in an electronic circuit. Nonlinear behaviour of circuit elements introduces harmonics in the output waveform and the resultant distortion is often referred to as *harmonic distortion* (HD).

### Types of Distortion

Distortion is caused by many devices and components which form an electronic circuit. In this section the different types of distortions caused by amplifiers are considered. The various types of distortions which occur are :

(*i*) **Frequency distortion.** This type of distortion occurs because the amplification factor of the amplifier is different for different frequencies.

(*ii*) **Phase distortion.** This distortion occurs on account of the energy storage elements in the system which cause the output signal to be displaced in phase with the input signal. If signals of all frequencies are displaced by the same amount, the phase shift distortion would not be noticed. However, in actual practice, signals at different frequencies are shifted in phase by different angles and therefore the phase shift distortion becomes noticeable.

23

(*iii*) **Amplitude distortion.** Harmonic distortion occurs due to the fact that the amplifier generates harmonics of the fundamental of the input signal. Harmonics always give rise to amplitude distortion, for example, when an amplifier is overdriven and clips the input signal.

(*iv*) **Intermodulation distortion.** This type of distortion occurs as a consequence of the interaction or heterodyning of two frequencies, giving an output which is sum or difference of the two original frequencies.

(*v*) **Cross-over distortion.** This type of distortion occurs in push-pull amplifiers on account of incorrect boas levels as shown in Fig. 23.3.

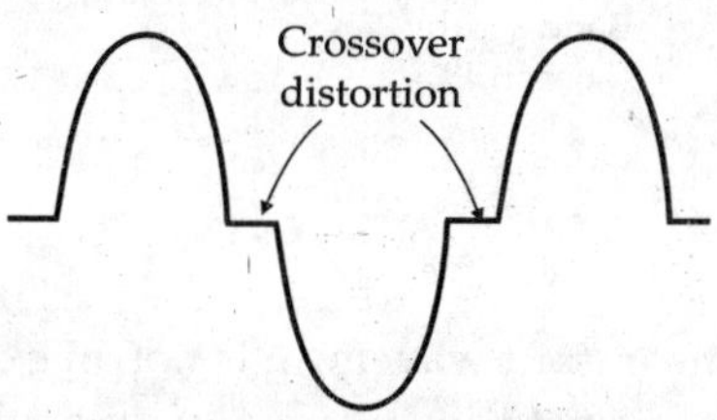


**Fig. 23.3** Illustration of crossover distortion.

## 23.4 TOTAL HARMONIC DISTORTION

A non-linear system produces harmonics of an input sine wave, the harmonics consisting of sine waves with frequencies which are multiples of the fundamental of the input signal. Total harmonic distortion (THD) is measured in terms of the harmonic content of the wave, as given by Eqn. 23.1.

$$\text{THD} = \frac{[\Sigma(\text{Harmonics})^2]^{1/2}}{\text{Fundamental}} \quad ...(23.1)$$

A measure of distortion represented by a particular harmonic is simply the ratio of amplitude of harmonic to that of the fundamental Harmonic distortion. (HD) is then represented by,

$$D_2 = \frac{E_2}{E_1},\ D_3 = \frac{E_3}{E_1},\ D_4 = \frac{E_4}{E_1}$$

When $D_n\ (n = 2, 3, 4, ...)$ distortion of $n$th harmonic and $E_n$ represents the amplitude of $n$th harmonic.

$E_1$ is the amplitude of fundamental.

Total harmonic distortion (THD) is defined as :

$$\text{THD} = \sqrt{D_2^2 + D_3^2 + D_4^2 + ...} \quad ...(23.2)$$

$$= \frac{\sqrt{E_2^2 + E_3^2 + E_4^2 + ...}}{E_1} \quad ...(23.3)$$

Percentage harmonic distortion

$$= \sqrt{D_2^2 + D_3^2 + D_4^2 + ...} \times 100 \quad ...(23.4)$$

$$= \frac{\sqrt{E_2^2 + E_3^2 + E_4^2 + ...}}{E_1} \times 100 \quad ...(23.5)$$

In a measurement system noise is read in addition to harmonics, and the total waveform, consisting of harmonics, noise and fundamental, is measured instead of the fundamental alone. Therefore the measured value of total harmonic distortion ($\text{THD}_M$) is given by Eqn. 23.6.

$$\text{THD}_M = \frac{\{\Sigma(\text{H})^2 + (\text{N})^2\}^{1/2}}{\{\Sigma(\text{F})^2 + (\text{H})^2 + (\text{N})^2\}^{1/2}} \quad ...(23.6)$$

where H = Harmonics ; N = Noise ; F = Fundamental

Equation 23.6 leads to below 1/2% error, due to the approximation for values of THD below 10%.

Figure 23.4 shows a harmonic distortion analyser which is used to measure THD. The signal source has very low distortion and this can be checked by reading its output distortion by connecting directly into the analyser. The signal from the source is fed into the amplifer under test. This generates harmonics and the original fundamental frequency. The fundamental frequency is removed by the notch filter.

In the manual system of Fig. 23.4(*a*) the switch Sw is first placed in position 1 and the total content of fundamental and harmonics ($E_T$) is measured. Then

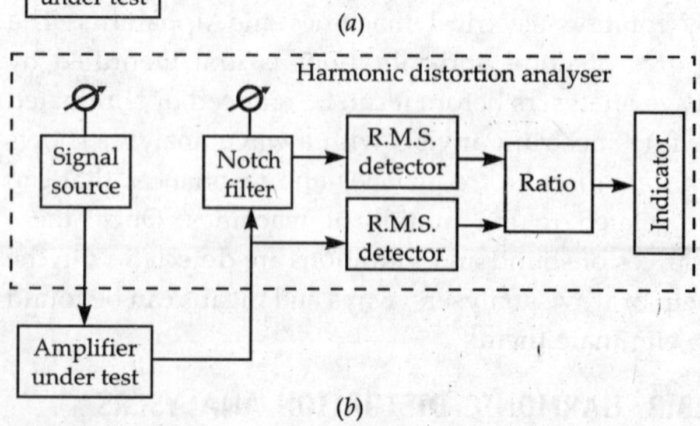


**Fig. 23.4** Simplified block diagrams of fundamental suppression harmonic distortion analysers ; (*a*) manual reading, (*b*) ratio reading.

the switch is moved to position 2 to measure just the harmonics $E_H$. The value of THD is then found from Eqn. 23.6 as in Eqn. 23.7.

$$\text{THD} = \frac{E_H}{E_T} \times 100 \qquad \text{...(23.7)}$$

The meter can be calibrated by putting the switch in position 1 and adjusting the reading for full scale deflection. With the switch in position 2 the meter reading is now proportional to THD. Figure 23.4(*b*) shows an alternative arrangement, where the values of $E_T$ and $E_H$ are read simultaneously and their ratio calculated and displayed as THD on the indicator. For good accuracy the notch filter must have excellent rejection and high pass characteristics. It should attenuate the fundamental by 100 db or more and the harmonics by less than 1 db. The filter also needs to be tuned accurately to the fundamental of the signal source. This is difficult to achieve manually and most distortion analysers do this automatically. A common form of notch filter is a Wien bridge. This balances at one frequency only and at this frequency the output voltage at the bridge null detector is minimum.

## 23.5 INTERMODULATION DISTORTION

When a high frequency ($f_1$) signal and a low frequency ($f_2$) signal are mixed in a linear circuit the output contains the two frequencies only. When mixed in a non-linear circuit, for example, an amplifier having distortion, modulation occurs. The output now contains the original frequencies ($f_1$ and $f_2$) and also the sum and difference frequencies ($f_1 + f_2$) and ($f_1 - f_2$) in addition to several harmonics and their sum and difference frequencies.

Figure 23.5 shows an instrument for measuring intermodulation distortion. The SMPTE (Society of Motion Picture and Television Engineers) defines the two signal source ($f_1$ and $f_2$) frequencies as 7 kHz and 60 Hz, having amplitudes in the ratio 1 : 4 respectively. The lower frequency has a much higher amplitude since a low degree of non-linearity is expected. In Europe the values of $f_1$ and $f_2$ are defined by DIN 45403 as 8 kHz and 250 Hz again having an, amplitude ratio of 1 : 4.

The output from the amplifier under test is fed into a high pass filter, giving a waveform at point *A* which consists of the high frequency carrier and is low frequency modulation as in Fig. 23.6. The waveform can be observed on an oscilloscope and intermodulation distortion (IMD) calculated from (Eqns. 23.8 or 23.9)

$$\text{IMD} = \frac{\text{Amplitude of modulation}(A_M)}{\text{Amplitude of carrier}(A_C)} \times 100 \qquad \text{...(23.8)}$$

$$= \frac{Q - P}{P} \times 100 \qquad \text{...(23.9)}$$

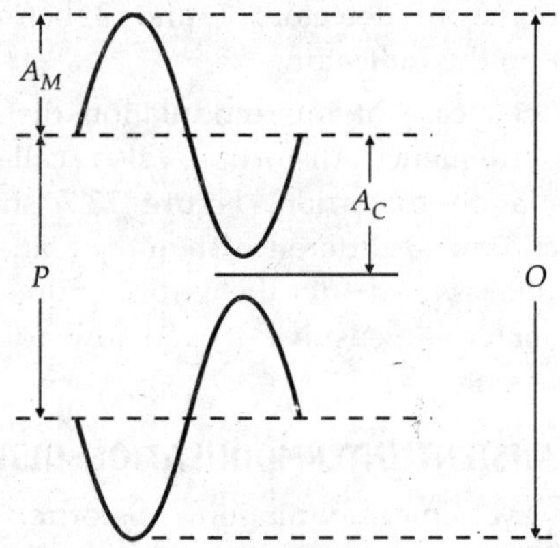


**Fig. 23.6** Waveform output from an intermodulation distortion meter.

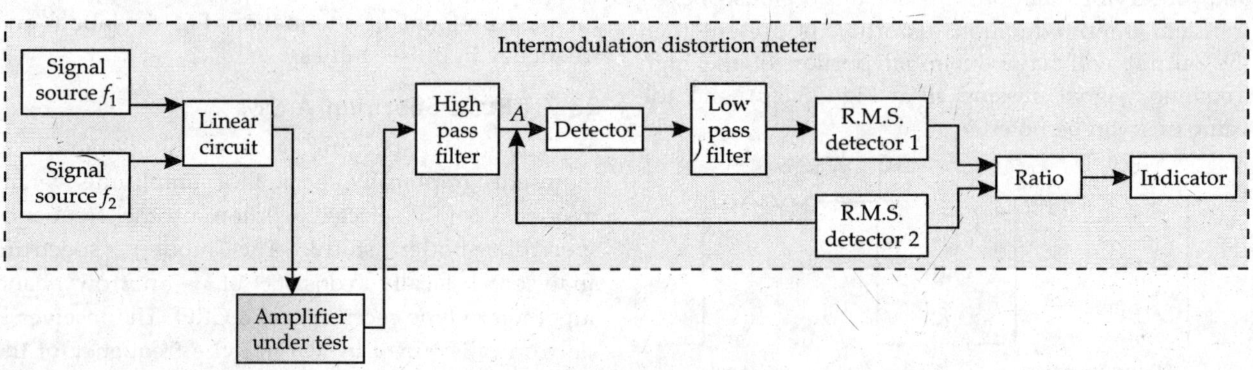


**Fig. 23.5** Simplified block diagram of an intermodulation distortion meter.

23

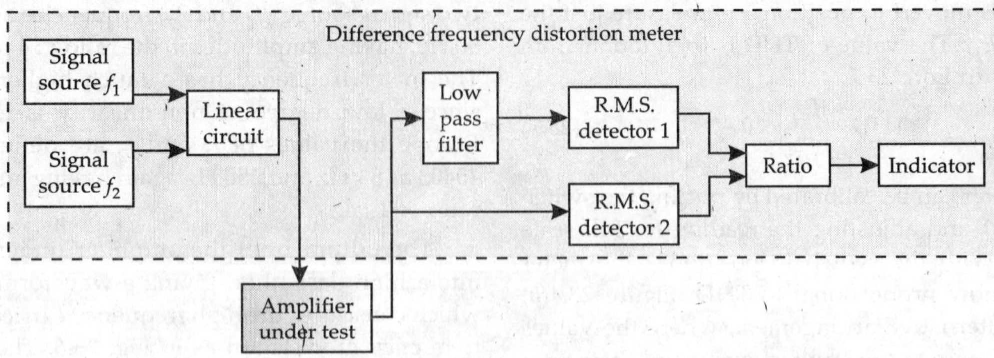


**Fig. 23.7**

The analyser shown in Fig. 23.5 is capable of calculating IMD directly and displaying its value on an indicator. The r.m.s. value of the waveform at A is measured by r.m.s detector 2. The waveform at *A* is also demodulated by a detector and the high frequency components removed by the low pass filter, leaving the low frequency distortion components. This is measured by r.m.s. detector 1 and the ratio is taken of the outputs from detectors 1 and 2 before being displayed on the indicator.

A special case of intermodulation distortion is difference frequency distortion, also called CCIF intermodulation distortion. Figure 23.7 shows the cons- truction of a difference frequency analyser. A low pass filter is used after the amplifier under test so that the meter is sensitive to the low (difference) frequencies only.

## 23.6 TRANSIENT INTERMODULATION DISTORTION

Transient intermodulation distortion occurs because an amplifier is not able to respond rapidly to changing inputs. It can be tested by applying a square wave signal with a superimposed high frequency sine wave, to the input of the amplifier, as in Fig. 23.8(*a*) and observing the output on an oscilloscope. If transient intermodulation distortion is present then the output will have an initial portion of the high frequency signal missing as at Fig. 23.8(*b*), and the value of *T* can be noted.

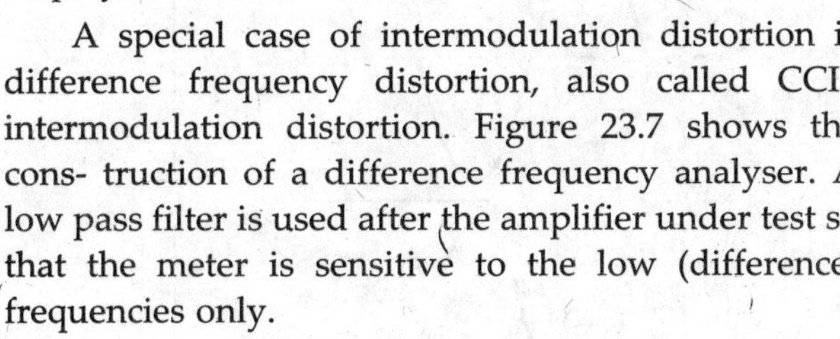


**Fig. 23.8** Waveforms to measure transient intermodulation distortion.

## 23.7 SPECTRUM ANALYSERS

Spectrum analysis is defined as the study of energy distribution across the frequency spectrum of a given electrical signal. The study gives valuable information about bandwidth, effects of different types of modulation and spurious signal generation. The knowledge of the above quantities and phenomena are useful in the design and testing of radio frequency (RF) and pulse circuitry.

The spectrum analysis is divided into two major categories on account of instrumentation limitations and capabilities. They are :

(*i*) Audio frequency (AF) analysis, and

(*ii*) Radio frequency (RF) spectrum analysis.

The RF spectrum analysis covers a frequency range of 10 MHz to 40 GHz, and hence is more important, because it includes the vast majority of communication, navigation, radar, and industrial instrumentation frequency bands.

The spectrum analysers are sophisticated instruments which are capable of portraying graphically the amplitude as a function of frequency in a portion of RF spectrum. These instruments find wide applications for measurement of attenuation, FM deviation, and frequency in pulse studies.

### 23.7.1 Basic Spectrum Analyser

The basic spectrum analyser is designed to represent graphically, a plot of amplitude versus frequency of a selected portion of the frequency spectrum under study. The modern spectrum analyser basically consists of a narrow band superheterodyne receiver and a CRO. The receiver is electronically tuned by varying the frequency of the local oscillator. A simplifed block diagram of a swept frequency spectrum analyser is shown in Fig. 23.9.

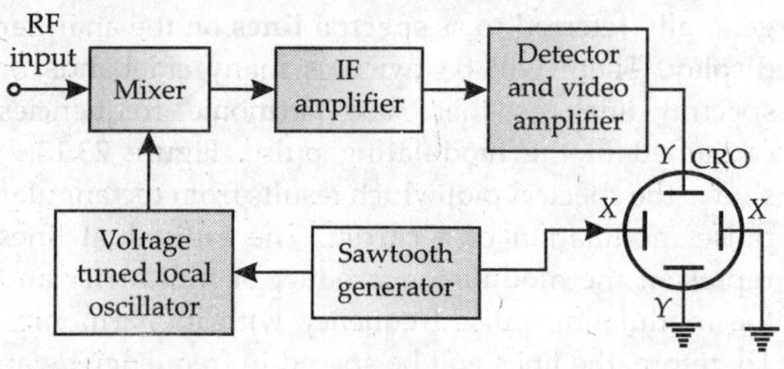


**Fig. 23.9** Basic swept receiver spectrum analyser.

The circuit incorporates a sawtooth generator which supplies a ramp voltage to the frequency control element of the voltage tuned local oscillator. The local oscillator then sweeps through its frequency band at a recurring linear rate. The same sawtooth voltage is simultaneously applied to the horizontal plates of the CRO. The RF signal to be tested is applied to the input of the mixer stage.

The sawtooth generator makes the local oscillator sweep through its frequency band to beat with the input signal to produce the desired intermediate frequency (IF). An IF component is produced only when the corresponding component is present in the RF input signal. The resulting IF signals are amplified, and then detected. After that they are applied to the vertical deflection plates of the CRO, thereby producing a display of amplitude versus frequency on the screen.

## 23.7.2 Spectral Displays

Normally a CRO is used for display of electrical signals with respect to time, with the X-axis of CRO calibrated to read directly the rise time, pulse width, and the repetition rate. Such measurements are said to be in **Time Domain**. However, in the case of spectrum analysers, the signals are broken down into their individual frequency components and displayed along X-axis of the CRO which is calibrated in terms of frequency. Therefore, the signal amplitude is displayed versus frequency. These measurements are then said to be in the **Frequency Domain**.

A three dimensional representation of a fundamental frequency, $f_1$ and its second harmonic, $f_2 - 2f_1$, is shown in Fig. 23.10. The two signals and the sum instantaneous values, as displayed on a CRO screen are shown in Fig. 23.10(*b*). The two signals are shown in Fig. 23.10 (*c*) on amplitude-frequency plane and are portrayed on the CRO as two components of the composite signal, as the window of the spectrum analyser sweeps across the frequency range of the signal.

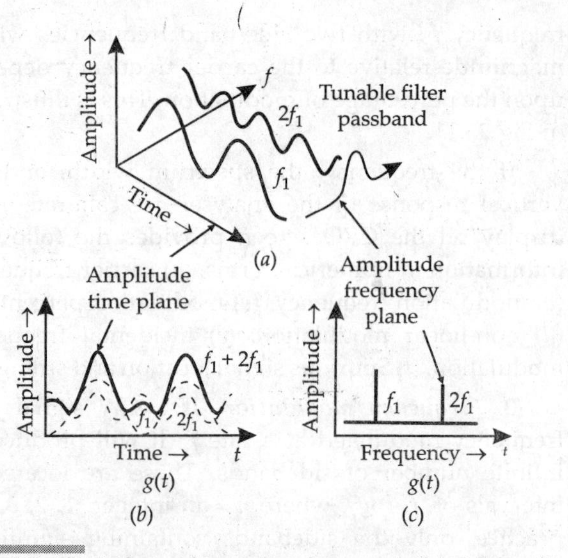


**Fig. 23.10** A three-dimensional presentation of amplitude, frequency, and time (*a*) The addition of a fundamental and its second harmonic. (*b*) View seen in the Time-Amplitude plane. On an oscilloscope, only the composite $f_1 + 2f_1$ would be seen. (*c*) View seen in the Frequency-Amplitude plane. The components of the composite signal are clearly seen.

## 23.7.3 Spectra of Different Signals

Let us consider some of the commonly used signals in order to illustrate the spectra which are displayed on the CRO when they are applied to the spectrum Analyser.

**1. *Continuous wave (CW) signals.*** When a continuous wave (CW) input signal is slowly swept through by the spectrum analyser's local oscillator, the response displayed on the screen is a plot of the IF amplifier pass band. Since the CW signal has energy at only one frequency and therefore the display on the screen is a single spike. This occurs in case the total RF sweep width or **Spectrum Width** is wide as compared to the IF band width in the analyser.

**2. *Amplitude modulation.*** When a continuous wave signal of frequency, $f_c$, is amplitude modulated by an input signal of frequency, $f_s$, two side-band frequencies of $f_c + f_s$ and $f_c - f_s$ are produced. The display on the spectrum analyser screen is a signal of

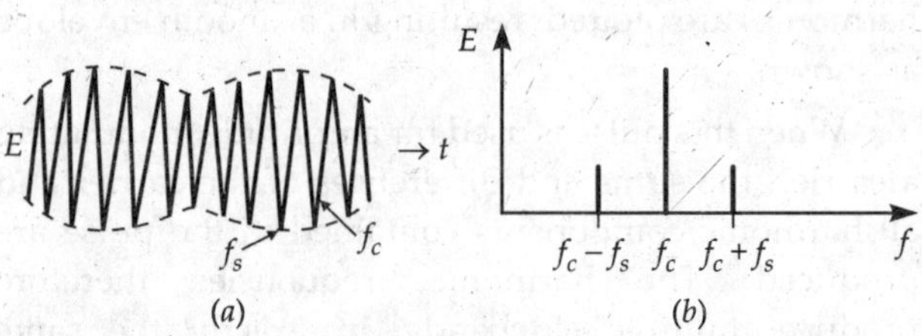


**Fig. 23.11** A single-tone amplitude modulation. (*a*) Time-amplitude plot. (*b*) Frequency-amplitude plot.

frequency $f_c$, with two side-band frequencies whose magnitude relative to the carrier frequency depends upon the percentage of modulation. This is illustrated in Fig 23.11.

If the frequency, the spectrum width, and the vertical response of the analyser are calibrated, the display on the CRO screen provides the following information in numerical terms, (*a*) carrier frequency, (*b*) modulation frequency, (*c*) modulation percentage, (*d*) non-linear modulation, (*e*) incidental frequency modulation, (*f*) Spurious signal location and strength.

**3. *Frequency modulation.*** If a CW signal $f_c$ is frequency modulated at a rate $f_r$, it will produce an infinite number of sidebands. These are located at intervals of $f_c + nf_r$, where $n$ = an integer, 1, 2, 3... In practice, only the sidebands containing significant power are usually considered. A frequency modulated display is shown in Fig. 23.12.

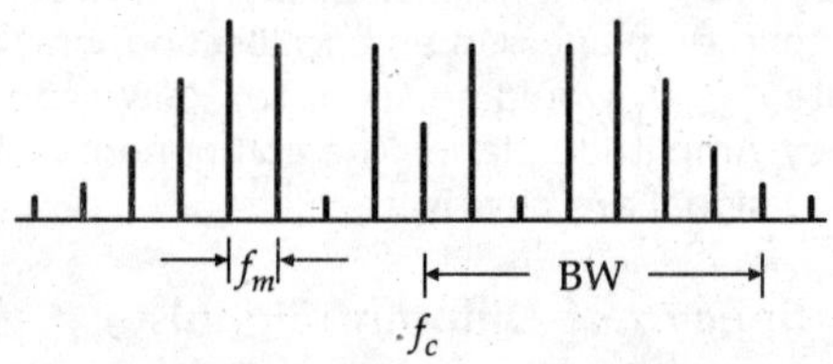


**Fig. 23.12** Amplitude spectrum of single-tone frequency modulation.

**4. *Pulse modulation.*** An idealized rectangular waveform, with zero rise time and no overshoot is shown in Fig 23.13(*a*). This pulse is shown in the time domain, but when its frequency spectrum is to be analyzed, it must be broken down into its individual frequency components. This is shown in Fig. 23.13(*b*), where a constant voltage, a fundamental frequency, and its third harmonic are added algebraically to form a wave which eventually becomes a square wave as more odd harmonics are added in phase with the fundamental. The pulse becomes perfectly rectangular if an infinitely large number of odd harmonics are added. A spectral plot, in the frequency domain, would have the form given in Fig. 23.13(*c*), where the amplitude and phases of an infinite number of harmonics are plotted, resulting in a smooth envelope as shown.

When this pulse is used for amplitude modulating a carrier, the sums and differences of the carrier and all harmonic components contained in the pulse are produced. The harmonic frequencies therefore produce multiple sidebands, in exactly the same manner that the modulating signal in amplitude modulation does. These multiple sidebands are generally referred to as **spectral lines** on the analyser display. There will be twice as many sidebands or spectral lines as there are harmonic frequencies contained in the modulating pulse. Figure 23.13(*d*) shows the spectral plot which results from rectangular pulse modulation of a carrier. The individual lines represent the modulation product of the carrier and the modulating pulse frequency with its harmonics. Therefore, the lines will be spaced in frequency by an

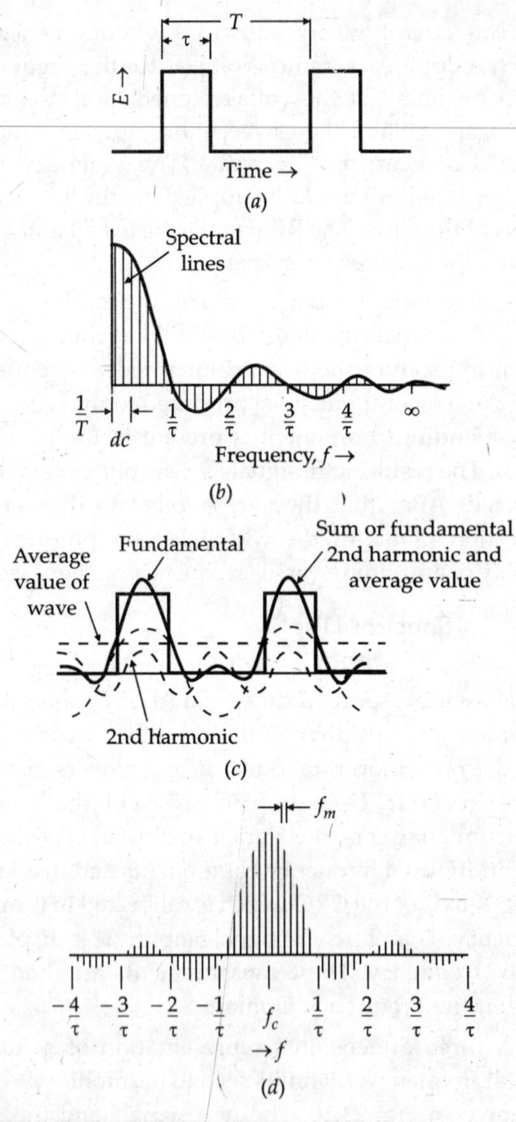


**Fig. 23.13** Pulse modulation : (*a*) Periodic rectangular pulse train. (*b*) Addition of a fundamental cosine wave and its harmonics to form rectangular pulses. (*c*) Spectrum of a perfectly rectangular pulse. Amplitudes and phases of an infinite number of harmonics are plotted resulting in smooth envelope as shown (*d*) Resultant spectrum of a carrier amplitude, modulated with a rectangular pulse.

amount equal to the pulse repetition rate of the original pulse waveform. The **main lobe** in the centre and the **side lobes** are shown as groups of spectral lines extending above and below the base line. For a perfectly, rectangular pulse, the number of side lobes is infinite. The main lobe contains the carrier frequency, represented by the longest line in the centre.

There are a number of other useful applications of this instrument. A few typical examples are cited :

1. Pulse-width and repetition rate measurements.
2. Tuning a parametric amplifier.
3. FM deviation measurement.
4. RF interference testing.
5. Antenna pattern measurements.

## 23.8 STANDING WAVE RATIO

When a transmission line is terminated by an impedance different from its characteristic impedance some of the forward signal wave is reflected back. This reflected wave mixes with the forward wave and the resultant amplitude at any point is the algebraic sum of the amplitudes of the two waves, as in Fig. 23.14.

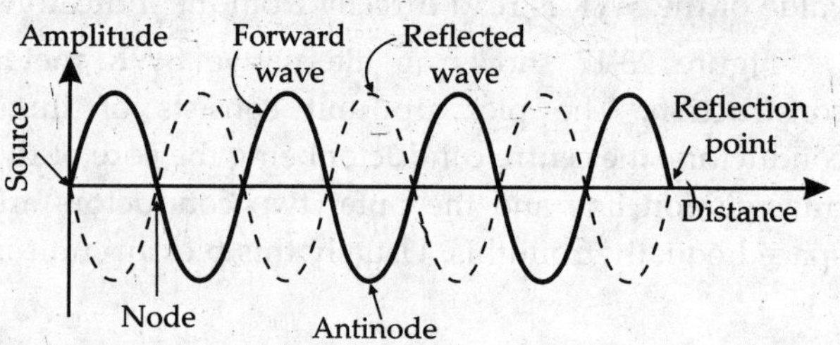


**Fig. 23.14** Illustration of standing wave.

The nodes and antinodes do not move relative to the transmission line *i.e.*, they are stationary and the waves are called standing waves.

An important consideration in transmission line and antenna design is the standing wave ratio (SWR) and this is illustrated using Fig. 23.15 which represents the standing wave on a line. If the line is open circuited or short circuited (load impedance being infinite or zero respectively), then all the forward wave is totally reflected back to the source and $V_1$ is zero. For an open circuit line the antinodes are located at half wavelengths along the line, starting back from the open circuit point. For a short circuited line the nodes are located at half wavelengths, starting back from the short circuit point. Therefore although open and short circuited lines give identical standing wave patterns, they are shifted relative to the positions of the nodes and antinodes.

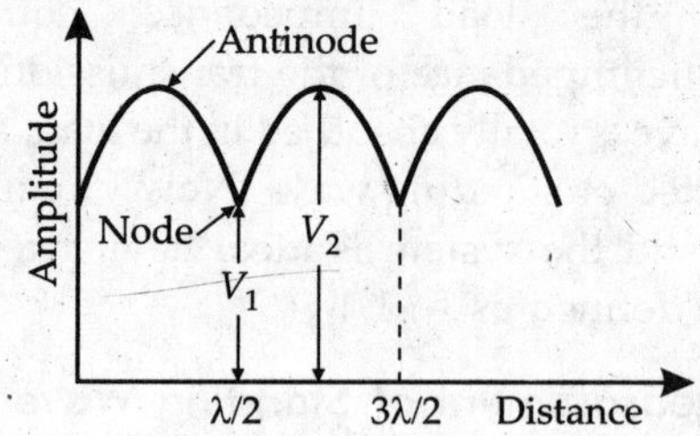


**Fig. 23.15** Illustration of standing wave ratio (SWR = $V_2$ / $V_1$)

In most systems some of the forward wave will be absorbed by the load, for example, in the case of an antenna some of the forward energy is radiated into space. Therefore the reflected wave has a lower amplitude than the forward wave, resulting in a standing wave similar to that in Fig. 23.15. The standing wave ratio (SWR) is then given by the ratio of the maximum and minimum amplitudes of this wave. Both voltage and current ratios can be used and these are referred to as VSWR and ISWR respectively and are given by Eqns 23.10 and 23.11.

$$\text{VSWR} = \frac{V_2}{V_1} \quad \text{...(23.10)}$$

$$\text{ISWR} = \frac{I_2}{I_1} \quad \text{...(23.11)}$$

The standing wave ratio is also given by (Eqns. 23.12 and 23.13) using impedance, power, or forward and reflected voltage as the parameters.

$$\text{SWR} = \frac{Z_C}{Z_L} \text{ (for } Z_C > Z_L) \quad \text{...(23.12)}$$

or

$$\text{SWR} = \frac{Z_L}{Z_C} \text{ (for } Z_L > Z_C) \quad \text{...(23.13)}$$

where $Z_C$ is the characteristic impedance of the transmission line and $Z_L$ is the characteristic impedance of the load.

$$\text{SWR} = \frac{1+(P_R / P_F)^{1/2}}{1-(P_R / P_F)^{1/2}} \quad \text{...(23.14)}$$

where $P_F$ is the forward power from the source to the load and $P_R$ is the power reflected back from the load.

$$\text{SWR} = \frac{1+V_R / V_F}{1-V_R / V_F} \quad \text{...(23.15)}$$

where $V_F$ is the voltage component of the forward wave, and $V_R$ is the voltage component of the reflected wave.

23

When the load impedance equals the characteristic impedance of the transmission line, the forward wave is totally absorbed in the load, and there is no reflected or standing wave. Now $V_1$ equals $V_2$ in Fig. 23.15 and the system is ideal having a standing wave ratio defined as 1 : 1.

### 23.8.1 Measurement of Standing Wave Ratio

The standing wave ratio (SWR) is one of the most fundamental parameters as it checks for mismatch between the transmitter, transmission line and antenna. The SWR varies with frequency, since the impedance of the source, line and load are unlikely to change by the same amount with frequency ; so the SWR should always be measured at several frequencies.

Any of the equations (23.10 to 23.15) can be used to find SWR and meters are available which are based on all of these. Most low cost instruments use voltage or current as their measurement parameter. These instruments give correct readings only when the load is not too reactive that is the SWR ratio being measured is low. The meters must also be connected to take readings at the load feed point or at half wavelength distances along the transmission line, back from the load. This usually means that the transmission line needs to have a length which is an integral value of its electrical length. The electrical length for half wavelength and quarter wavelength transmission lines is respectively given by Eqns. 23.16 and 23.17

$$E_L = 150\,F / f \qquad ...(23.16)$$

$$= 75\,F / f \qquad ...(23.17)$$

where $f$ = frequency of antenna and

$F$ = velocity factor whose value varies from 0.67 to 0.97 depending upon the dielectric used.

SWR meters based on power measurements give results which are independent of length. A true RF wattmeter is used fitted with a directional coupler so as to measure the forward and reflected power separately and the SWR is calculated from Eqn. 23.14.

The radiation resistance of an antenna can also be measured using a resistance bridge and the SWR found from (Eqn. 23.12) or (23.13). Figure 23.16 shows a Wheatstone bridge arrangement in which resistors $R_B$, $R_A$ and $R_X$ form three of the bridge arms, and the radiation resistance of the antenna forms the fourth arm. The supply to the bridge is a low power R.F signal generator, which provides the excitation. The value of resistor $R_X$ is chosen depending on the antenna type. *i.e.*, a 50 ohm or a 75 ohm system. When making measurements switch Sw is first set to

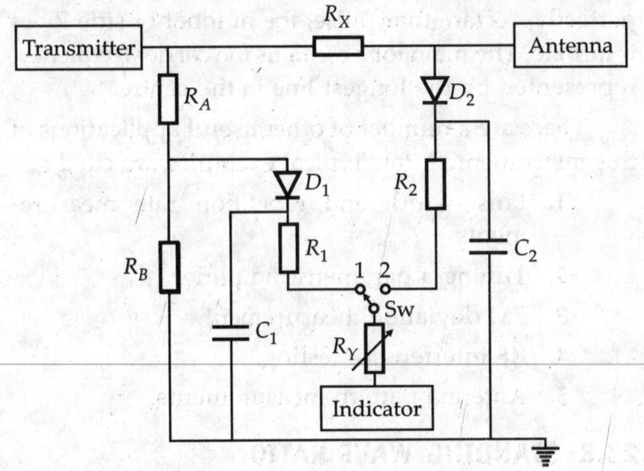


**Fig. 23.16** An SWR meter arrangement based on a Wheatstone bridge.

position 1 and the value of $R_Y$ is adjusted to give full scale defection on the indicator. Having calibrated the instrument switch SW is moved to position 2, and the value of the SWR is read directly from the indicator.

Figure 23.17 shows an alternative SWR meter construction. The pick up unit consists of three conductors, the centre conductor being the core of the transmission line and the outer two conductors are spaced equally from this. Usually this pick up unit is

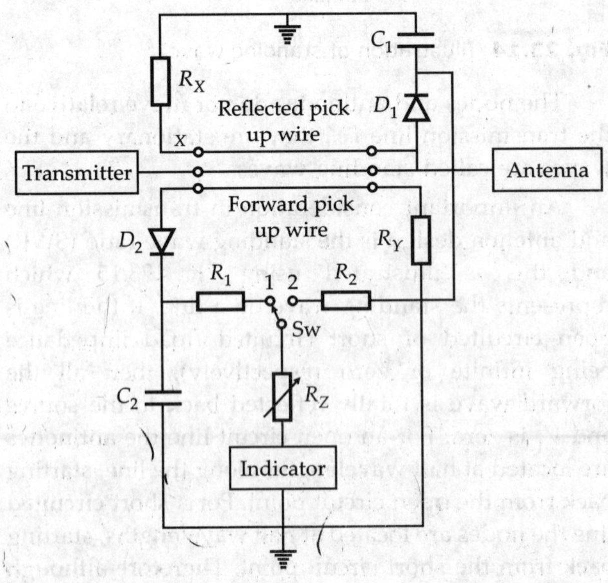


**Fig. 23.17** An SWR meter arrangement based on forward and reflected wave measurements.

made as a pattern on a printed circuit board. Resistors $R_X$ and $R_Y$ are selected to have values equal to the characteristic impedance of the line *i.e.*, 50 ohms or 75 ohms. In operation any R.F. signal on one of the pick up wires will be rectified by diodes $D_1$ or $D_2$ smoothed by capacitors $C_1$ or $C_2$ and then applied as d.c. to the indicator. The forward signal from the transmitter to the antenna picked up by the reflected wire, is absorbed by resistor $R_X$ so no forward voltage appears on the pick up line after point X. However this signal remains on the forward pick up wire, and is applied to the indicator. Similarly, the reflected signal is picked up by the reflected pick up wire, but is absorbed in resistor $R_Y$ of the forward pick up circuit.

When making measurements with the meter shown in Fig. 23.18 the switch is set to position 1 and $R_Z$ is adjusted to give a reading at a set calibration point. The switch is then moved to position 2 and the SWR read from the indicator.

A good antenna has a standing wave ratio which varies between 1 : 1 and 1 : 1.2. Poor antennas have ratios greater than 1 : 1.5 and most SWR meters do not give a reading beyond 1 : 3.

Instruments are usually multipurpose for example, combining a standing wave ratio meter with a field strength meter and power output meter. Figure 23.18 shows a block diagram of an impedance noise bridge, which is a low cost instrument for measuring several parameters on antennas and transmission lines ; The noise source produces a wide spectrum of pseudogaussian noise. This is amplified and then goes through an isolator before being coupled to three output. The detector gives an indication of the received signal and shows a high noise level at all frequencies which are separate from the resonant frequency of the device under test (usually an antenna or a transmission line). The electrical length of a transmission line differs from its physical length due to the velocity factor F. The noise bridge can be used to find the correct physical length which gives an electrical length equal to half a wavelength. The transmission line should initially be cut to a length slightly greater then that given by (Eqn. 23.16) of (23.17). $R_X$ is set to it low value, and one end of the cable under test is connected to the test socket, whilst test is connected to the test socket whilst the other end is short circuited. The detector is set to the-frequency at which the cable is to be half a wavelength in length and the frequency is slowly decreased until null is reached. Now the cable is gradually trimmed in stages. After each trim its end is again short circuited and it is re-tested until eventually the null point occurs at the required frequency.

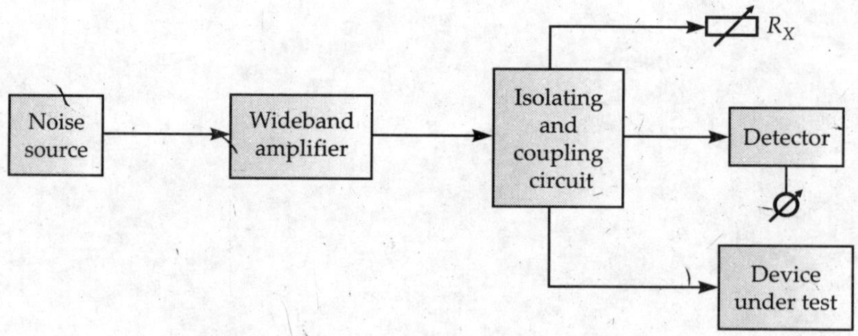


**Fig. 23.18** Simplified block diagram of an impedance noise bridge.

# Review Questions

1. Describe the circuits and working of wave analysers used for audio frequency and megahertz ranges.
2. Describe the engineering applications of wave analysers.

3. Explain the different types of distortions caused by amplifiers.
4. Explain the term "total harmonic distortion". Describe the functioning of a total harmonic distortion meter.
5. What is Intermodulation Distortion ? Describe the working of an intermodulation distortion meter with the help of a block diagram.
6. Describe the working of a difference frequency distortion analysers with the help of a block diagram.
7. Describe the basic circuit of a spectrum analyser. Explain how the spectra of the following is displayed :
   (*i*) continuous wave signals
   (*ii*) amplitude modulated signals
   (*iii*) frequency modulated signals
   (*iv*) pulse modulated signal.
8. Explain the following terms
   (*i*) SWR
   (*ii*) VSWR
   (*iii*) ISWR
9. Explain the term Electrical length for transmission lines. How is this affected by the operating frequency and the velocity factor ?
10. Describe an arrangement for measurement of Standing wave ratio.

# 24

# High Frequency Measurements

## 24.1 INTRODUCTION

In this chapter methods for measurement of inductance, capacitance and resistance in the frequency range of 0.1 to 100 MHz are described. No doubt, some of the methods described earlier *i.e.*, bridge methods can be used in this frequency range but these methods give erroneous results on account of the fact that these methods do not take into consideration the special effects which are inherent in any measurement at high frequencies. Some of these special effects which are predominant at high frequencies are :

(*i*) high reactance due to stray series inductance and its variation with frequency ;

(*ii*) low reactance due to stray shunt capacitance and its variation with frequency ;

(*iii*) increase of effective resistance with frequency on account of skin effects ;

(*iv*) increase of effective resistance on account of proximity effects ;

(*v*) increase of effective resistance due to production of eddy currents ;

(*vi*) variation of dielectric loss of capacitors with frequency.

These effects make the high frequency measurements complex. If bridge circuits are used at high frequencies, elaborate shielding arrangements have to be adopted to nullify the effects of factors listed above.

The normal approach for measurements at high frequencies is to adopt such methods which are as simple as possible so that the effects of stray electric and magnetic fields are easy to remove. Since errors of the order of 1 to 5 percent are acceptable in measurements in this frequency range, the methods used involve use of very simple apparatus which may include :

(*i*) a set of inductors from 1.0 μH to 10 mH.

(*ii*) a variable air capacitor of maximum capacitance 1000 pF,

(*iii*) a decade resistance box of resistance upto 100 Ω,

(*iv*) a thermo electric ammeter or an electronic voltmeter (EVM) in order to indicate resistance conditions.

## 24.2 RESONANCE METHODS

Most of the methods used for high frequency measurements employ a series *RLC* resonant circuit as shown in Fig. 24.1. The emf for the circuit is obtained by coupling the coil *L* with the output coil of an oscillator. A loose coupling is desirable in order that the changes in conditions of the *RLC* circuit do not effect the oscillator circuit. A tight coupling on the other hand, will affect the oscillator.

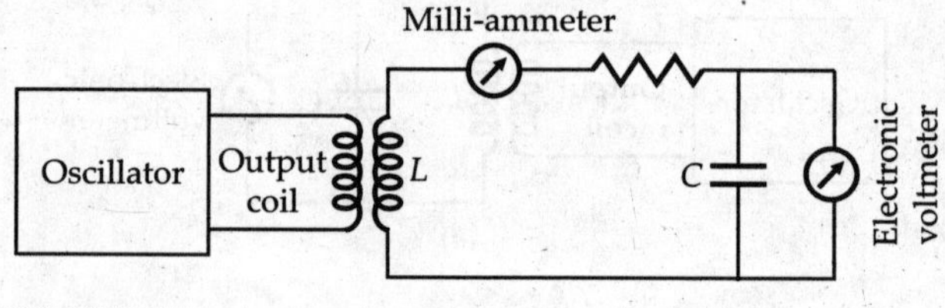


**Fig. 24.1** *RLC* resonant circuit.

Resonance is obtained either by variation of frequency of the oscillator or by variation of the capacitance and keeping the frequency constant. The resonance conditions are indicated by the thermal milli-ammeter connected in the circuit. (A thermal milli-ammeter is necessary as it can be worked at high frequencies.) Under resonance conditions the milli-ammeter indicates maximum reading. However, in practice it is convenient to use an electronic voltmeter connected across the capacitor C to indicate the resonance conditions since the ammeter introduces an extra series resistance in the circuit which may be undesirable under many circumstances. The resonance is indicated when the voltmeter reads maximum. However, it is interesting to note that the circuit conditions which give maximum current are not the same as those which show maximum voltage across capacitor C, but, for high Q circuits the difference between the two conditions is negligible. Hence, in general, it is assumed that when the voltmeter connected across C shows maximum reading, the circuit is having maximum current *i.e.*, the circuit is working under resonance conditions.

Another assumption, which is made for resonance methods, is that the resonance curve is symmetrical about its maximum ordinate. This condition is valid only when :

(*i*) the curve is highly peaked *i.e.*, highly selective which is the case when the circuit has a high Q factor ;

(*ii*) the readings are taken in the vicinity of the resonance point. To satisfy the condition

(*iii*) measurements should be limited to values of current greater than 50% of the maximum value of current.

## 24.3 MEASUREMENT OF INDUCTANCE

Figure 24.2(*a*) shows a simple circuit for measurement of inductance, L, of a coil. This coil under measurement is loosely coupled to the output coil of a variable frequency oscillator. The dial of the oscillator is calibrated in terms of frequency. A

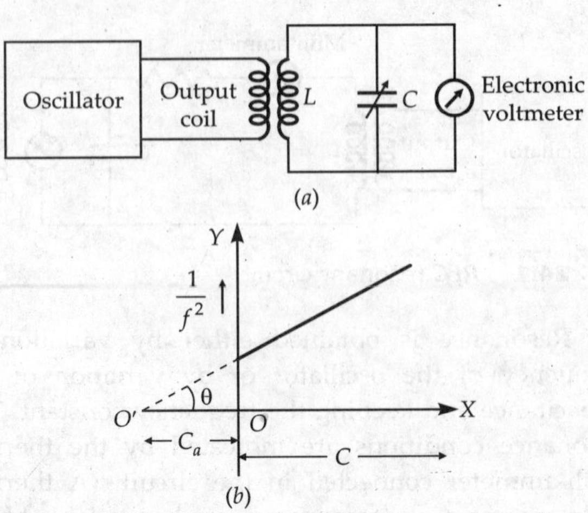


**Fig. 24.2** Measurement of inductance.

variably capacitor, C, is used for obtaining resonance conditions which are indicated by an electronic voltmeter connected across it

$$\text{At resonance} \quad LC = 1/\omega^2 = 1/4\pi f^2 \qquad \text{...(24.1)}$$

Thus knowing the value of frequency, $f$ and capacitance, C, the value of inductance can be calculated. The above method apparently looks very simple but the value of inductance obtained this way is rarely accurate. The reasons are discussed below :

The above method does not take into account the additional capacitances which are introduced into the circuit. Also errors are caused due to the fact that the dial setting of capacitor, C, may not be accurately known. To offset the influence of above effects and to get improved accuracy a number of readings are taken by varying the frequency, $f$.

Let $C_a$ be the value of additional capacitance. Therefore, for resonance at any frequency

$$L(C + C_a) = 1/4\pi^2 f^2 \qquad \text{...(24.2)}$$

$$1/f^2 = 4\pi^2 L(C + C_a) = 4\pi^2 LC + 4\pi^2 LC_a \qquad \text{...(24.3)}$$

From Eqn. 24.3, it is clear that if a curve is plotted between $1/f^2$ and C, we get a straight line as shown in Fig. 24.2 (*b*). The line produced back cuts the $x$-axis at O′. The slope of the line is $\tan\theta = 4\pi^2 L$

$$\therefore \quad L = \tan\theta / 4\pi^2 \qquad \text{...(24.4)}$$

Thus knowing the value of slope of the line, the value of inductance, L, may be obtained.

The value of additional capacitance, $C_{a'}$ is equal to the intercept OO′. Now,

$$C_a = C_d + C_v \qquad \text{...(24.5)}$$

where $C_d$ = self-capacitance of the coil,

$C_v$ = capacitance of voltmeter.

Equation 24.5 is true only when the calibration of capacitance C includes the effects of its screen capacitances. Under condition, the method affords a means of measurement of self-capacitance of coils.

In the above analysis we have assumed that the value of inductance, L, remains constant. However, this is not strictly true as the effective value of inductance changes with frequency. Thus this method gives erroneous results for coils whose effective self-inductance varies considerably over the range of frequency used. It is clear from Eqn. 24.3 that we get a straight line relationship between C and $1/f^2$ only if L remains constant over the range of frequency used. Thus if the value of self-inductance, L, changes with frequency we do not get a straight-line but a curve.

Therefore, in order to determine the value of L at a particular frequency, a tangent is drawn to the curve at a value the frequency where the value of inductance is required. The value of inductance is determined from the slope of this curve using Eqn. 24.4.

***Example 24.1*** *A capacitance of 250 pF produces resonance with a coil at a frequency of* $(2/\pi)\times 10^6$ *Hz, while at the second harmonic of this frequency resonance is produced by a capacitance of 50 pF. Calculate :*

(*a*) *the self-capacitance, and*

(*b*) *the inductance of the coil.*

*Neglect the effect voltmeter capacitance and also other stray capacitances.*

**Solution.** Suppose

C = capacitance of variable capacitor,

$C_d$ = Self-capacitance of coil, and

L = inductance of coil.

At resonance,

$$L = 1/4\pi^2 f^2 (C + C_d)$$

$$= \frac{1}{4\pi^2 \times (2/\pi)^2 \times 10^{12}(250 + C_d)\times 10^{-12}}$$

$$= \frac{1}{16(250 + C_d)} \qquad ...(i)$$

and also

$$L = \frac{1}{4\pi^2 \times (4/\pi)^2 \times 10^{12}(50 + C_d)\times 10^{-12}}$$

$$= \frac{1}{64(50 + C_d)} \qquad ...(ii)$$

From (*i*) and (*ii*), $16(250 + C_d) = 64(50 + C_d)$

or self-capacitance $C_d = 16.67$ pF.

From (*i*),

inductance, $L = \dfrac{1}{16(250 + 16.67)}$ H = 234 μH.

***Example 24.2*** *A coil of unknown inductance and self-capacitance is connected in series with a standard variable capacitor C. An electronic voltmeter is connected across C. A variable frequency oscillator is loosely coupled to the coil and the circuit is adjusted to resonance for each different oscillator frequency f by adjusting C, as in the following table :*

| *f, kHz* | 350 | 400 | 450 | 500 | 600 |
|---|---|---|---|---|---|
| *C, pF* | 132 | 98 | 73 | 55 | 31 |

*Determine :*

(*a*) *the inductance, and*

(*b*) *the self-capacitance of the coil.*

*The input capacitance of the electronic voltmeter is 4.5 pF and the lead capacitance is 1.5 pF.*

**Solution.** The method has been explained in Art. 24.3. A curve is drawn between $1/f^2$ and C. The points for the graph are :

| *C, pF* | *f, kHz* | $1/f^2$ |
|---|---|---|
| 132 | 350 | $8.16\times 10^{-12}$ |
| 98 | 400 | $6.25\times 10^{-12}$ |
| 73 | 450 | $4.93\times 10^{-12}$ |
| 55 | 500 | $4.0\times 10^{-12}$ |
| 31 | 600 | $2.78\times 10^{-12}$ |

The graph is plotted in Fig. 24.3.

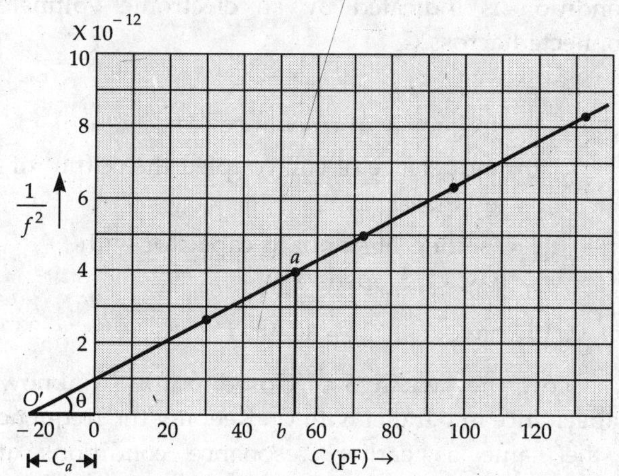


**Fig. 24.3** Plot of $1/f^2$ v/s C for Example 24.2.

From the graph, additional capacitance,

$C_a = 20$ pF.

Capacitance of voltmeter and leads

$= 4.5 + 1.5 = 6.0$ pF.

∴ Self-capacitance of coil

$C_d = 20 - 6 = 14$ pF.

Slope of line

$$\tan\theta = \frac{ab}{O'b} = \frac{4\times 10^{-12}}{75\times 10^{-12}} = \frac{4}{75}$$

From Eqn. 24.4, inductance of coil

$$L = \frac{\tan\theta}{4\pi^2} = \frac{4}{75}\times\frac{1}{4\pi^2}\text{ H} = 1.35\text{ mH}$$

## 24.4 MEASUREMENT OF CAPACITANCE

Substitution method is normally used for measurement of capacitance at high frequencies. Figure 24.4 shows a simple circuit for measurement of capacitance. C is the unknown capacitance connected in parallel with a variable standard capacitance $C_s$.

The standard capacitance is of the same order as the unknown capacitance.

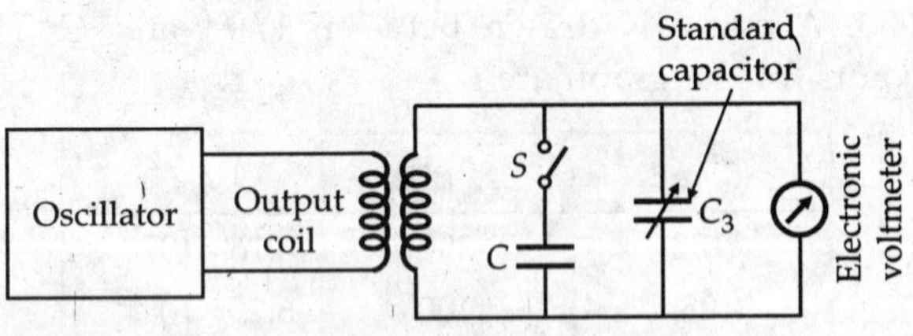


**Fig. 24.4** Measurement of capacitance by substitution method.

With switch $S$ open, resonance is obtained by varying the standard capacitance $C_s$. The resonance condition is indicated by an electronic voltmeter connected across $C_s$.

Let

$f$ = frequency at resonance ; Hz,

$L$ = inductance of coil coupled the output of oscillator ; H,

$C_{s1}$ = setting of standard capacitor with switch $S$ open ; F.

$\therefore$ We have $C_{s1} = 1/4\pi^2 f^2 L$ ...(24.6)

Now the switch $S$ is closed putting unknown capacitance in parallel with $C_s$. Keeping the frequency $f$ the same as earlier, resonance conditions are obtained by varying $C_s$.

Let $C_{s2}$ = setting of standard capacitor with switch closed.

$\therefore$ We get $C + C_{s2} = 1/4\pi^2 f^2 L$ ...(24.7)

From Eqns. 24.6 and 24.7, value of unknown capacitance is :

$$C = C_{s1} - C_{s2}.$$

Thus the value of unknown capacitance is directly known from the difference of the two settings of the standard capacitor.

## 24.5 MEASUREMENT OF EFFECTIVE RESISTANCE

There are *two* methods of measurement of effective resistance of a series *RLC* resonant circuit.

(*i*) Resistance variation method, and

(*ii*) Reactance variation method.

## 24.6 RESISTANCE VARIATION METHOD

The effective series resistance of a resonant circuit is measured by adding extra resistance from outside.

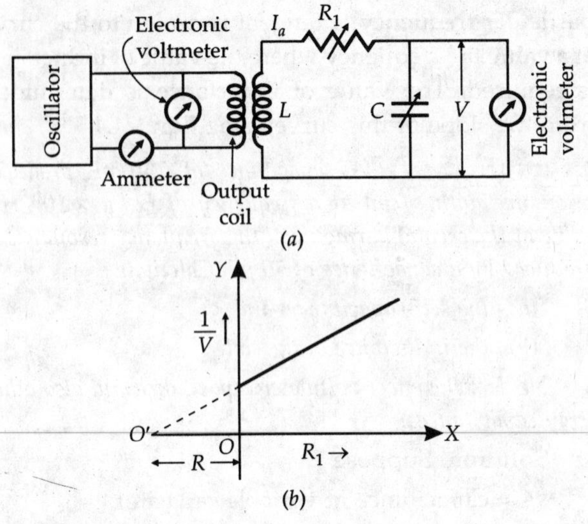


**Fig. 24.5** Measurement of effective resistance by resistance variation method.

Let

$E$ = induced emf in the resonant circuit,

$I_0$ = current in the resonant circuit,

$R$ = effective resistance of resonant circuit excluding resistance $R_1$,

$R_1$ = additional resistance inserted in the resonant circuit,

$L$ = inductance of coil,

$C$ = capacitance,

$V$ = reading of voltmeter connected across $C$,

and $f_0$ = resonant frequency.

At resonance $I_0 = \dfrac{E}{R + R_1}$ ...(24.9)

$\therefore$ Voltage $V = \dfrac{I_0}{2\pi f_0 C} = \dfrac{E}{(R + R_1) 2\pi f_0 C}$ ...(24.10)

or $R + R_1 = E/2\pi f_0 CV$ ...(24.11)

If the values of $E$, $f_0$ and $C$ are kept constant, we get

$R + R_1 = K/V$ ...(24.12)

where $K$ = a constant $= E/2\pi f_0 C$ ...(24.13)

Thus, under the above conditions if a curve is plotted between $1/V$ and $R_1$, we get a straight line as shown in Fig. 24.5(*b*).

The effective resistance is given by the intercept $OO'$ provided that the loss resistance of capacitor and the resistance of the leads is negligible.

From the above analysis we conclude :

(*i*) We can determine the value of effective resistance of a resonant circuit by inserting an external variable resistance $R_1$. The circuit is set into resonance by varying the value of $C$. This value of capacitance $C$ is kept constant afterwards. We go on increasing the value of $R_1$. This does not disturb the resonance conditions and its effect is only to decrease the value of resonance current $I_0$ and to decrease the value of voltage $V$ across the capacitor $C$. Thus an increase in the value of $R_1$ decreases $V$ and increases the value of $1/V$.

(*ii*) The analysis is valid only if the value of induced emf $E$ in the resonant circuit is constant. Also the resonant frequency $f_0$ is constant. This requires that the coupling between the oscillator circuit and the resonant circuit be very loose otherwise changes in the resonant circuit would affect the oscillator circuit. In order to ensure that the value of induced emf and the resonant frequency remain constant for all the readings, the oscillator should possess a high degree of frequency and amplitude stability.

(*iii*) We have explained above that the value of induced emf $E$ in the resonant circuit should remain constant. This can be checked by connecting a thermal ammeter in series with the output coil of the oscillator or by connecting an electronic voltmeter across the output coil of the oscillator.

## 24.7 REACTANCE VARIATION METHOD

The effective resistance of a resonant circuit can be measured by variation of reactance of the circuit. Figure 24.6 shows the circuit for such a measurement.

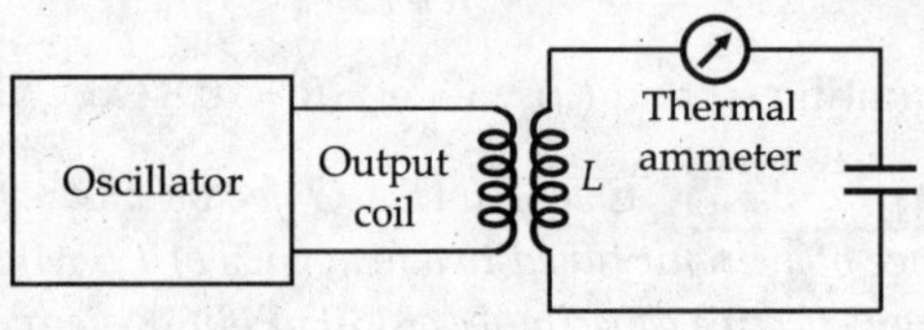


**Fig. 24.6** Measurement of effective resistance by reactance variation method.

In this circuit a thermal ammeter is used to measure the value of current flowing. The circuit reactance is varied by either varying the value of capacitance $C$ or the frequency $f$. The reactance is adjusted till the ammeter reads maximum value. This happens when the circuit is under resonance conditions.

The maximum value of current :

$$I_0 = E / R \qquad \text{...(24.14)}$$

where $E$ = emf induced, and

$R$ = effective value of resistance.

The reactance is then adjusted so that the current

$$I = I_0 / \sqrt{2}$$

$$\therefore \quad I = \frac{E}{\sqrt{R^2 + X^2}} = \frac{I_0}{\sqrt{2}} = \frac{E}{\sqrt{2}\,R}$$

Hence,

$$\sqrt{R^2 + X^2} = \sqrt{2}\,R \text{ or } X = \pm R \qquad \text{...(24.15)}$$

There are, therefore, two values of reactance $X$ which give a current, $I = I_0 / \sqrt{2}$ .

The difference of these two values or reactance gives twice the value or effective resistance *i.e.*, $2R$.

As mentioned above there are *two* ways of varying the reactance :

(*i*) variation of capacitance, $C$, and

(*ii*) variation of frequency $f$

The two methods are discussed below :

**1. *Variation of capacitance.*** In this method, the value of frequency $f$ is kept constant and the capacitance is varied. Let the value of capacitance which gives the maximum current $I_0$ be $C_0$ and the two values of capacitance which give a current of $I_0 / \sqrt{2}$ be $C_1$ and $C_2$.

Therefore,

$$2R = 1/\omega C_2 - 1/\omega C_1$$

or

$$R = \frac{C_1 - C_2}{2\omega C_1 C_2} = \frac{C_1 - C_2}{4\pi f C_1 C_2} \qquad \text{...(24.16)}$$

Now if the circuit loss is small and the resonance is peaked and symmetrical,

$$C_1 C_2 \approx C_0^2 \qquad \text{...(24.17)}$$

$$\therefore \quad R \approx \frac{C_1 - C_2}{2\omega C_0^2} \approx \frac{C_1 - C_2}{4\pi f C_0^2} \qquad \text{...(24.18)}$$

Also

$$Q = \omega L / R = 1/\omega C_0 R \qquad \text{...(24.19)}$$

$$\approx \frac{2C_0}{C_1 - C_2} \qquad \text{...(24.20)}$$

The accuracy of this method depends upon the accuracy with which the difference $C_1 - C_2$ can be known. For accurate determination of this difference a micrometer capacitor is put in parallel with the main capacitor. From Eqn. 24.20 it is clear that the circuit $Q$ can be determined in terms of two values of capacitance.

**2. *Variation of frequency.*** In this method the capacitance $C$ is kept constant and the frequency is varied. Let $\omega_0$ be the angular frequency to give maximum current $I_0$ and $\omega_1$ and $\omega_2$ be the two frequencies which give a value of current

$$I = I_0 / \sqrt{2}.$$

$$2R = (\omega_1 L - 1/\omega_1 C) - (\omega_2 L - 1/\omega_2 C)$$
$$= (\omega_1 - \omega_2)L + 1/C(1/\omega_2 - 1/\omega_1)$$
$$= (\omega_1 - \omega_2)L + \omega_0^2 L$$
$$= (\omega_1 - \omega_2)/(\omega_1\omega_2) \text{ as } \omega_0^2 L = 1/C.$$

Now for a peaked and symmetrical resonance curve :

$$\omega_0^2 = \omega_1\omega_2 \quad ...(24.21)$$

$$\therefore \quad 2R \approx (\omega_1 - \omega_2)L + \omega_0^2 L \frac{\omega_1 - \omega_2}{\omega_0} \approx 2(\omega_1 - \omega_2)L$$

or $$R \approx (\omega_1 - \omega_2)L \quad ...(24.22)$$

$$\approx \frac{\omega_1 - \omega_2}{\omega_0^2 C} \quad ...(24.23)$$

and $$Q = \frac{\omega_0 L}{R} = \frac{\omega_0}{\omega_1 - \omega_2} \quad ...(24.24)$$

The frequency variation method has a disadvantage because it is quite difficult to keep the value of induced emf $E$ constant with change of frequency. But the error introduced on this account is small and can be allowed for.

**Example 24.3** *A coil of inductance L is tuned by a capacitance C to resonate at 1 MHz. A standard capacitor of 230 pF is tuned to resonate with the same coil at 2 MHz. If the resonance is indicated by a VTVM of capacitance 8 pF in parallel with the circuit, calculate the values of L and C.*

**Solution.** We have

$$C + C_v = 1/4\pi^2 L(1\times 10^6)^2$$

or $$(C+8)\times 10^{-12} = 1/(4\pi^2 L \times 10^{12}) \quad ...(i)$$

and $$(C_s + C_v)\times 10^{-12} = 1/4\pi^2 L(2\times 10^6)^2$$

or $$(230+8)\times 10^{-12} = 1/(16\pi^2 L\times 10^{12}) \quad ...(ii)$$

where $C_v$ = capacitance of voltmeter
$= 8\times 10^{-12}$ F

and $C_s$ = capacitance of standard capacitor
$= 230\times 10^{-12}$ F

From relation (*ii*), we get,
Inductance

$$L = \frac{1}{238\times 16\pi^2} \text{H} = 26.6\ \mu\text{H}$$

From relation (*ii*), capacitance

$$C = \frac{1}{4\pi^2 \times 26.6} - 8 = 944 \text{pF}.$$

**Example 24.4** *A constant rms voltage of frequency 3 MHz is applied to a coil of inductance L and resistance R, a calibrated variable capacitor C, and a thermojunction all in series. The uncalibrated ammeter is found to show the same scale reading I for the two values of C – namely 37 and 45 pF, one being below and the other above the capacitance $C_0$ needed to produce maximum current.*

*The capacitor is now adjusted to a value $C_0$, and an added series resistance of 75 Ω is found to reduce the current to previous value of I. Determine values of L and R.*

**Solution.** Let the two values of capacitance which produce the same current be $C_1$ and $C_2$.

$$\therefore \quad C_1 = 45 \text{pF and } C_2 = 37 \text{pF}$$

Let the resonance curve be peaked and symmetrical and therefore capacitance required for resonance is :

$$C_0 = \frac{C_1 + C_2}{2} = \frac{45+37}{2} = 41 \text{pF}.$$

At resonance, inductance

$$L = \frac{1}{4\pi^2 f^2 C_0} = \frac{1}{4\pi^2 \times (3\times 10^6)^2 \times 41\times 10^{-12}} \text{H}$$
$$= 69.2\ \mu\text{H}.$$

When the capacitance in the circuit is $C_1$, the value of current

$$I = \frac{E}{\sqrt{R^2 + (\omega L - 1/\omega C_1)^2}}$$

$$= \frac{E}{\sqrt{R^2 + \left(2\pi\times 3\times 10^6 \times 69.2\times 10^{-6} - \frac{1}{2\pi\times 3\times 10^6 \times 45\times 10^{12}}\right)^2}} \quad ...(i)$$

At resonance, $$I = \frac{E}{R+75} \quad ...(ii)$$

Equating (*i*) and (*ii*) we get, $R = 70.8\ \Omega$.

**Example 24.5** *Calculate the Q factor and effective resistance of a circuit tuned to a frequency of 1.5 MHz and having an effective capacitance of 150 pF. In this circuit the current falls to 70.7% of its resonant value when the frequency of an emf of constant magnitude injected in series with the circuit deviates from the resonant frequency by 5 kHz.*

**Solution.** The analysis of this problem has been done in Art. 24.5.

From Eqn. 24.24,

$$Q \text{ factor} = \frac{\omega_0}{\omega_1 - \omega_2} = \frac{f_0}{f_1 - f_2}.$$

Resonant frequency

$$f_0 = 1.5 \text{ MHz} = 1.5\times 10^6 \text{ Hz}$$

The two frequencies at which the current is 70.7% of or $1/\sqrt{2}$ time the current at resonant frequency are :

$$f_1 = (1.5\times10^6 + 5\times10^3)\text{ Hz}$$

and
$$f_2 = (1.5\times10^6 - 5\times10^3)\text{ Hz}$$

$$\therefore \quad f_1 - f_2 = 10\times10^3 \text{ Hz}$$

Hence,
$$Q = \frac{1.5\times10^6}{10\times10^3} = 150$$

From Eqn. 24.23, resistance

$$R = \frac{\omega_1-\omega_2}{\omega_0^2 C} = \frac{f_1 - f_2}{2\pi f_0^2 C}$$

$$= \frac{10\times10^3}{2\pi(1.5\times10^6)^2\times150\times10^{-12}} = 4.7\Omega$$

## 24.8 *T* NETWORKS

*T* networks are very useful for measurement of inductance, capacitance, resistance and frequency in the high frequency range. These networks employ null techniques.

There are *two* types of *T* networks used :

(*i*) Parallel *T* networks, and

(*ii*) Bridge *T* networks.

## 24.9 PARALLEL *T* NETWORK

Figure 24.7(*a*) shows a parallel *T* network with source null detector. In fact there are two *T*s formed by impedances $Z_1$, $Z_2$, $Z_3$, and $Z_4$, $Z_5$, $Z_6$ in parallel. The network can be transformed into another form shown in Fig. 24.7(*b*) by star delta transformation. A null is indicated by the detector if the admittance of $Z_A$ and $Z_D$ in parallel is equal to zero. The other impedances *i.e.*, $Z_B$, $Z_C$, $Z_E$, $Z_F$ do not affect the null conditions.

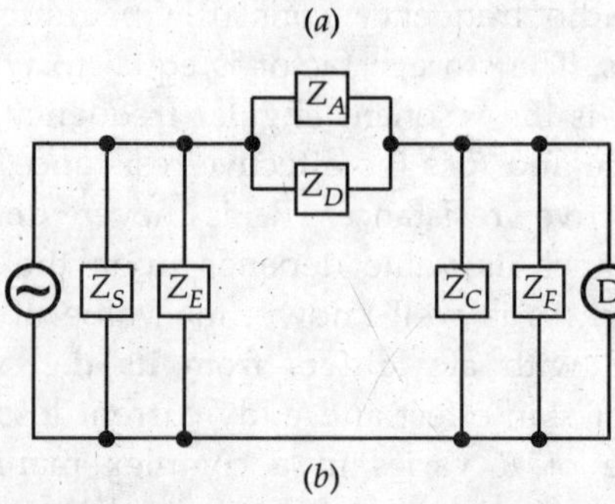


**Fig. 24.7** Parallel *T* network.

Therefore at null point :

$$\frac{1}{Z_A} + \frac{1}{Z_D} = 0 \quad \text{or} \quad Z_A + Z_D = 0$$

or
$$Z_A = -Z_D \qquad ...(24.25)$$

But
$$Z_A = Z_1 + Z_2 + Z_1Z_2/Z_3$$

and
$$Z_D = Z_4 + Z_5 + Z_4Z_5/Z_6$$

Therefore, at null point,

$$Z_1 + Z_2 + Z_1Z_2/Z_3 = -Z_4 - Z_5 - Z_4Z_5/Z_6 \qquad ...(24.26)$$

A common parallel *T* network for measurement of frequency is shown in Fig. 24.8. Comparing Fig. 24.8 with circuit of Fig. 24.7(*a*).

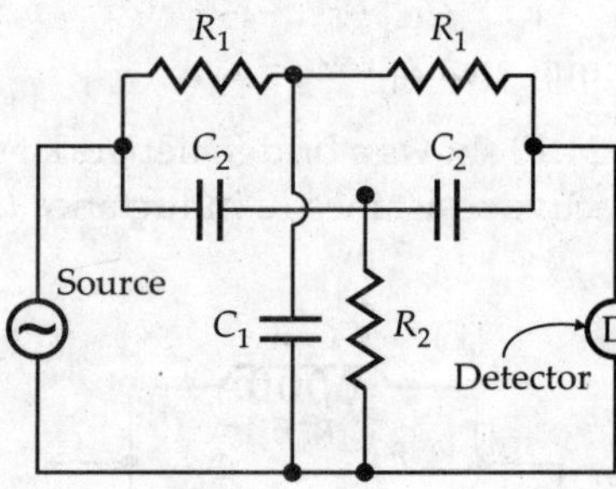


**Fig. 24.8** Parallel *T* network for measurement of frequency.

We have $Z_1 = R_1$, $Z_2 = R_1$, $Z_3 = -j/\omega C_1$, $Z_4 = -j/\omega C_2$, $Z_5 = -j/\omega C_2$ and $Z_6 = R_2$.

Applying Eqn. 24.26 for null conditions, we have,

$$R_1 + R_2 + \frac{R_1^2}{-j/\omega C_1} = \frac{j}{\omega C_2} + \frac{j}{\omega C_2} + \frac{1}{\omega^2 C_2^2 R_2}$$

or
$$2R_1 + jR_1^2\omega C_1 = \frac{2j}{\omega C_2} + \frac{1}{\omega^2 C_2^2 R_2}$$

Equating real and imaginary terms we get,

$$2R_1 = 1/\omega^2 C_2^2 R_2 \quad \text{or} \quad \omega^2 = 1/2R_1R_2C_2^2 \qquad ...(24.27)$$

and
$$R_1^2\omega C_1 = 2/\omega C_2 \quad \text{or} \quad \omega^2 = 2/R_1^2 C_1 C_2 \qquad ...(24.28)$$

This circuit (Fig. 24.8) therefore can be used to measure frequency. From Eqns. 24.27 and 24.28, we have,

$$1/2R_1R_2C_2^2 = 2/R_1^2C_1C_2$$

This condition is fulfilled if

$$R_1 = 4R_2 \quad \text{and} \quad C_1 = C_2.$$

24

## 24.10 BRIDGE *T* NETWORK

Bridge *T* network is a simplification of parallel *T* network. If the impedance $Z_3$ is infinite and the combination of impedances $Z_1$, and $Z_2$ is made into a single impedance $Z_1$, it becomes a bridge '*T*' network. Figure 24.9 shows a bridge *T* network.

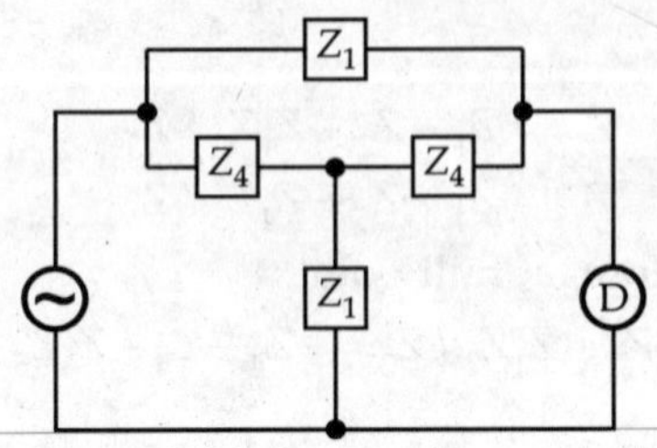


**Fig. 24.9** Bridge Network.

From Eqn. 24.26 for a bridge *T* network

$$Z_1' = -Z_4 - Z_5 - Z_4 Z_5 / Z_6 \qquad ...(24.29)$$

as $Z_3$ is infinite and $Z_1 + Z_2 = Z_1'$.

Figure 24.10 shows a bridge network for measurement of effective resistance *R*, inductance *L* and *Q* of a coil.

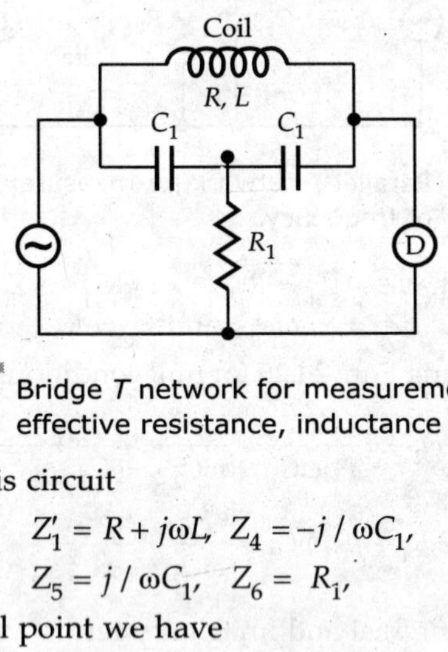


**Fig. 24.10** Bridge *T* network for measurement of effective resistance, inductance and *Q*.

For this circuit

$$Z_1' = R + j\omega L,\ Z_4 = -j / \omega C_1,$$
$$Z_5 = j / \omega C_1,\ \ Z_6 = R_1,$$

At null point we have

$$Z_1' = -Z_4 - Z_5 - Z_4 Z_5 / Z_6 \text{ (Eqn. 24.29)}$$

$$R + j\omega L = \frac{j}{\omega C_1} + \frac{j}{\omega C_1} + \frac{1}{\omega^2 C_1^2 R_1}$$

or $\quad R + \frac{2j}{\omega C_1} + \frac{1}{\omega^2 C_1^2 R_1}$

Equating real imaginary terms, we get,

$$R = 1 / \omega^2 C_1^2 R_1 \qquad ...(24.30)$$

and

$$L = 2 / \omega^2 C_1 \qquad ...(24.31)$$

$$Q = \frac{\omega L}{R} = \frac{2\omega}{\omega^2 C_1} \times \omega^2 C_1^2 R_1 = 2\omega C_1 R_1 \qquad ...(24.32)$$

The *T* networks have an advantage over ordinary bridge circuits. It is clear that these networks have a common point between source and detector. This point may be earthed and thus the screening of whole of the circuit becomes easy.

**Example 24.6** *Two variable capacitors $C_1$, mechanically coupled so that their capacitances are always equal, are connected in series between terminals A and B to form the series arms of a bridge 'T' network. Their junction is connected through a variable resistor $R_1$ to the terminal C. A coil of unknown inductance L and resistance R is connected across terminals A and B. The input voltage is applied between A and C, and a detector is connected between B and C. Under what conditions will the detector show an output voltage approaching zero ? Evaluate R and L if this balance condition is obtained at a frequency of 5 MHz with $C_1 = 45$ pF and $R_1 = 4.2\ \Omega$.*

**Solution.** The network is similar to the one shown in Fig. 24.10. The analysis of the network is carried out in Art. 24.10.

Equations 24.30 and 24.31 give conditions for no output voltage as :

$$R = \frac{1}{\omega^2 C_1^2 R_1}$$

and $\quad L = \frac{2}{\omega^2 C_1}$

$$R = \frac{1}{4\pi^2 \times (5\times10^6)^2 \times (45\times10^{-12})^2 \times 4.5}\ \Omega$$
$$= 0.115\ \text{M}\Omega$$

$$L = \frac{2}{4\pi^2 \times (5\times10^6)^2 \times 45\times10^{-12}}$$
$$= 45\ \mu\text{H}.$$

## 24.11 *Q* METER

The determination of the storage factor *Q* is one of the most widely used means in the laboratory for testing radio frequency coils (RFC), inductors and capacitors. The storage factor is equal to $Q = \omega_0 L / R$ where $\omega_0$ is the resonant angular frequency, *L* is the inductance and *R* is the effective resistance of a coil. The effective resistance, *R*, is never determined directly since its value depends upon the value of frequency. As is well known, the value of effective resistance with a.c. differs from its d.c. value on account of skin effect and eddy current losses. Since the value of *R* varies in a complex manner with frequency, it is indirectly determined by measuring the value of *Q*.

The $Q$ meter is an instrument which is designed to measure the value of $Q$ directly and as such is very useful in measuring the characteristics of coils and capacitors.

**Principle of working.** The principle of working of this useful laboratory instrument is based upon the well-known characteristics of a resonant series $R$, $L$, $C$ circuit. Figure 24.11(*a*) shows a coil of resistance $R$ and inductance $L$ in series with a capacitor $C$.

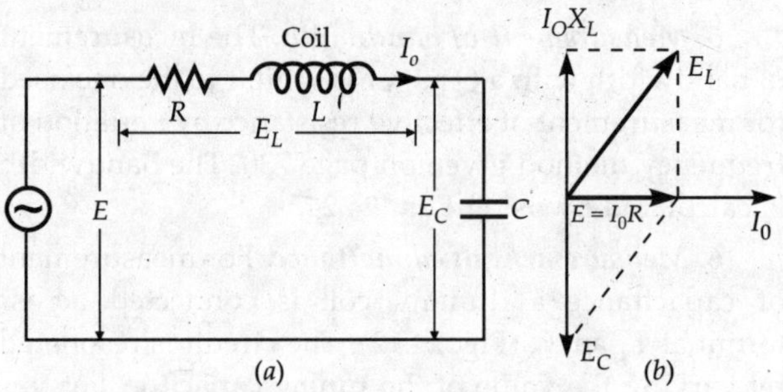


**Fig. 24.11** A resonant *RLC* circuit and its phasor diagram.

At resonant frequency $f_0$, we have,

$$X_C = X_L$$

where capacitive reactance $X_C = 1/2\pi f_0 C = 1/\omega_0 C$

and inductive reactance $X_L = 2\pi f_0 L = \omega_0 L$,

resonant frequency $$f_0 = \frac{1}{2\pi\sqrt{LC}}$$

and current at resonance $$I_0 = \frac{E}{R}$$

The phasor diagram is shown in Fig. 24.11(*b*).

Voltage across capacitor, $E_C = I_0 X_C = I_0 X_L = I_0 \omega_0 L$

Input voltage $E = I_0 R$

$$\therefore \quad \frac{E_C}{E} = \frac{I_0 \omega_0 L}{I_0 R} = \frac{\omega_0 L}{R} = Q$$

or $$E_0 = QE \qquad \text{...(24.32)}$$

Thus the input voltage $E$ is magnified $Q$ times.

If the input voltage $E$ is kept constant, the voltage appearing across the capacitor is $Q$ times $E$ and a voltmeter connected across the capacitor can be calibrated to read the value of $Q$ directly.

## Practical Circuit

A practical $Q$ meter circuit is shown in Fig. 24.12. It consists of a self-contained variable frequency RF oscillator. This oscillator delivers current to a low value shunt resistance $R_{sh}$. This low resistance is of the order of a hundredths of an ohm ; a typical value may be 0.02 Ω. Through this resistance a small value of voltage $E$ is injected into the resonant circuit. This voltage is measured by a thermocouple voltmeter. Since the value of shunt resistance is very low, it introduces almost no resistance into the oscillatory circuit and therefore represents a voltage source of magnitude $E$, with a very small internal resistance. A calibrated standard variable capacitor $C$ is used for resonating the circuit. An electronic voltmeter is connected across this capacitor. The coil under test is connected to terminals $T_1$ and $T_2$.

## Application

**1. *Measurement of Q.*** The circuit for measurement of $Q$ is shown in Fig. 24.12. The oscillator is set to the desired frequency and then the tuning capacitor is adjusted for maximum value of $E_0$. As discussed earlier under these conditions $Q = E_0 / E$ and if the voltage is kept constant, the voltmeter connected across the capacitor may be calibrated to read the value of $Q$ directly. This measured value of $Q$ is commonly regarded as the $Q$ of the coil under test. However, there is an error. The measured value of $Q$ is the $Q$ of whole the circuit and not of the coil. There are errors caused on account of the shunt resistance and also due to the distributed capacitance of the circuit.

24

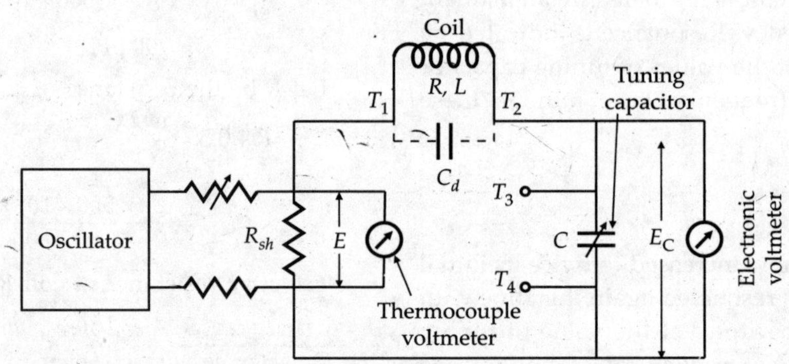


**Figure 24.12** Circuit of a $Q$ meter.

**Correction for shunt resistance.** Measured value,

$$Q_{meas} = \frac{\omega_0 L}{R + R_{sh}}$$

True value $\quad Q_{true} = \omega_0 L / R$

$$\therefore \quad Q_{true} = Q_{meas}\left(1 + \frac{R_{sh}}{R}\right) \quad ...(24.33)$$

Thus the measured value of $Q$ is smaller than the true value. Now if coils of high resistance (low $Q$ coils) are being measured, the difference between the two values may be negligible. But when measurements are done on low resistance (high $Q$) coils, the error caused on this account may be serious.

**Correction for distributed capacitance.** The distributed capacitance or self-capacitance of measuring circuit modifies the true value of $Q$. The measured value of $Q$ is less than the true value by a factor that depends upon the value of the distributed (self) capacitance and the tuning capacitor. It can be shown that

$$Q_{true} = Q_{meas}\left(1 + \frac{C_d}{C}\right) \quad ...(24.34)$$

where $C_d$ = distributed or self-capacitance, and
$C$ = tuning capacitance.

A method for measurement of distributed (self) capacitance is given later.

2. ***Measurement of inductance.*** The value of inductance is given by $L = 1/4\pi^2 f_0^2 C$.

The values $f_0$ and $C$ are known and therefore the value of inductance may be calculated.

3. ***Measurement of Effective resistance.*** The value of the effective resistance may be computed from the relation $R = \omega_0 L / Q_{true}$.

4. ***Measurement of self-capacitance.*** The self-capacitance is measured by making two measurements at different frequencies. The capacitor is set to a high value, and the circuit is resonated by adjustment of the oscillator frequency. Resonance is indicated by the circuit $Q$ meter. Let the values of tuning capacitor be $C_1$ and that of frequency be $f_1$ under these conditions. Therefore,

$$f_1 = \frac{1}{2\pi\sqrt{L(C_1 + C_d)}}$$

The frequency is now increased to twice its initial value and the circuit is resonated again this time with the help of tuning capacitor. Let the value of tuning capacitor be $C_2$ and that of frequency be $f_2$ under these conditions.

Thus,

$$f_2 = \frac{1}{2\pi\sqrt{L(C_2 + C_d)}}$$

Now $\quad f_2 = 2 f_1$

$$\therefore \quad \frac{1}{2\pi\sqrt{L(C_2 + C_d)}} = 2 \times \frac{1}{2\pi\sqrt{L(C_1 + C_d)}}$$

or distributed capacitance $\quad C_d = \dfrac{C_1 - 4C_2}{3} \quad ...(24.35)$

5. ***Measurement of bandwidth.*** The measurement of bandwidth with a $Q$ meter is similar to the one used for measurement of effective resistance by variation of frequency method given on page 730. The bandwidth is calculated by using Eqn. 24.24.

6. ***Measurement of capacitance.*** For measurement of capacitance a dummy coil is connected across terminal $T_1$ an $T_2$ (Fig. 24.12). The circuit is resonated by varying the value of the tuning capacitor. Let this value be $C_1$.

The capacitor under test $C_T$, is now connected across terminals $T_3$ and $T_4$. This puts the test capacitor in parallel with tuning capacitor. The circuit is resonated again by varying the value of tuning capacitor. Let this value be $C_2$. Therefore value of capacitance under test is $C_T = C_1 - C_2$.

This method is similar to the one given in Art. 24.4 on page 727.

**Example 24.7** *A circuit consisting of a coil, a resistance an a variable capacitor connected in series is tuned to resonance using a Q meter. If the frequency is 500 kHz, the resistance 0.5 Ω and the variable capacitor set to 350 pF. Calculate the effective inductance and resistance of the coil, if the Q meter indicates 90.*

**Solution.** At resonance ;

Inductance $\quad L = \dfrac{1}{4\pi^2 f_0^2 C}$

$$= \frac{1}{4\pi^2 \times (500 \times 10^3)^2 \times 350 \times 10^{-12}} H$$

$$= 290\ \mu H.$$

Let $R$ be the resistance of coil.

$$\therefore \quad R + 0.5 = \frac{\omega_0 L}{Q}$$

$$= \frac{2\pi \times 500 \times 10^3 \times 290 \times 10^{-6}}{90} = 10.25\ \Omega$$

Effective resistance of coil $R = 10.25 - 0.5 = 9.75\ \Omega$

**Example 24.8** *A coil of resistance 10 Ω is connected in the Q meter circuit as shown in Fig. 24.12. Resonance occurs at a frequency of 1 MHz with the tuning capacitor* *set at 65 pF. Calculate the percentage error introduced in the calculated value of Q if a resistance of 0.02 Ω is used across the oscillator circuit.*

**Solution.** True value of Q is :

$$Q_{true} = \frac{1}{\omega CR}$$

$$= \frac{1}{(2\pi\times 10^6)\times(45\times 10^{-12})\times 10} = 245$$

Measured value of

$$Q_{meas} = \frac{1}{\omega C(R+0.02)}$$

$$= \frac{1}{(2\pi\times 10^6)\times(45\times 10^{-12})\times 10.02}$$

$$= 244.5$$

∴ %age error

$$= \frac{245-244.5}{245}\times 100 = 0.2\% \text{ high}$$

**Example 24.9** *The self-capacitance of a coil is measured by a Q meter. The circuit is set into resonance azt 2 MHz and the tuning capacitor is as value of 460 pF. The frequency is now adjusted to 4 MHz and resonance conditions are obtained the tuning capacitor at 100 pF. Calculate the value of se1f-capacitance of the coil.*

**Solution.**

The problem has already been analysed in previous page.

From Eqn. 24.35, self-capacitance of coil

$$C_d = \frac{C_1 - 4C_2}{3}$$

$$= \frac{400-4\times 100}{3} = 20 \text{ pF}$$

**Example 24.10** *Tests using a Q meter, on a radio tuning coil to find its self-capacitance gave the following results :*

1. *With a radio coil connected normally, the resonance was obtained at 1 MHz with the tuning capacitor set at 80 pF.*
2. *With a standard inductor connected in place of the radio coil, resonance was obtained at 3 MHz and this condition was not altered when the radio coil was connected in parallel with the standard inductor. Calculate the self-capacitance of the radio coil.*

**Solution.** Let

$C_1$ = capacitance of tuning capacitor for 1st setting.

$\omega_1$ = angular frequency for 1st setting.

∴ Inductance of coil

$$L = 1/\omega_1^2(C_1 + C_d) \quad \text{...(i)}$$

In this second setting; the reason for the resonance conditions to remain unchanged is that a very high impedance has been placed in parallel, this being the resonant circuit of the radio coil and its self capacitance. In other words, the radio coil in parallel with its self capacitance is resonant at 3 MHz and thus it may be written

$$L = 1/\omega_2^2 C_d \quad \text{...(ii)}$$

From (*i*) and (*ii*) we have, self capacitance of coil is,

$$C_d = \frac{C_1\omega_1^2}{\omega_2^2-\omega_1^2} = C_1\left(\frac{f_1^2}{f_2^2-f_1^2}\right)$$

$$= 80\times\frac{(1)^2}{(3)^2-(1)^2} = 10 \text{ pF}$$

24

# Review Questions

1. Discuss the problems associated with measurement of resistance, inductance and capacitance at high frequencies.
2. Describe the resonance method for measurement of inductance at high frequencies. Discuss the assumptions made. How is the effect of additional capacitances in the tuning capacitor taken care of ?
3. Describe the method for measurement of capacitance at high frequencies using resonance method.
4. Discuss the following two methods of measurement of effective resistance of a coil at high frequencies :
   (*i*) Resistance variation method.
   (*ii*) Reactance variation method.
5. Describe the following methods of measuring effective resistance of a coil using resonant methods :
   (*i*) Variation of capacitance
   (*ii*) Variation of frequency.

6. Explain how frequency can be measured using a parallel $T$ network. Derive the expression for the frequency in terms of values of components used in a parallel $T$ network.
7. Describe how $Q$ factor of a coil can be measured using a Bridge $T$ network. Derive the expression for $Q$.
8. Describe the circuit and working of a $Q$ meter. Describe its applications.
9. Describe how corrections for shunt resistance and distributed capacitance are applied when measuring $Q$ factor of a coil with a $Q$ meter.
10. Describe the measurement of following using a $Q$ meter :
    (*i*) $Q$ factor (*ii*) inductance
    (*iii*) effective resistance (*iv*) Self capacitance
    (*v*) Bandwidth (*vi*) capacitance

# Unsolved Problems

1. A capacitor of capacitance 245 pF produces resonance at an angular frequency of $5 \times 10^6$ rad/s, while a capacitor of capacitance 50 pF produces resonance with the second harmonic of this frequency. Calculate :
   (*a*) the inductance, and
   (*b*) the self capacitance of the coil.
   **[Ans.** 154 μH, 15 pF]
2. With a suitable standard inductor connected to a Q-meter, resonance is obtained with an angular frequency ω with capacitor setting $C_1$; the indicated $Q$ factor is $Q_1$. The unknown impedance is connected in series with the standard inductor and resonance is re-established by resetting the capacitor to $C_2$, the corresponding $Q$ factor being $Q_2$. Calculate :
   (*a*) the resistive, and
   (*b*) reactive components of the unknown impedance when
   $C_1 = 208$ pF, $Q_1 = 170$,
   $C_2 = 185.5$ pF and $Q_2 = 52$.
   The frequency is 159 kHz. **[Ans.** 35 Ω, 584 Ω]
3. A coil of unknown impedance in series with a capacitor of 224 pF and an ammeter of negligible impedance was connected to a variable frequency of constant voltage and negligible impedance. The frequency was adjusted both above and below resonant frequency till the reading of ammeter was reduced to 70.7% of its value at resonance. This occurred at frequencies of 876 and 892 kHz. Determine effective resistance, inductance and $Q$ of the coil. **[Ans.** 14.5 Ω, 0.145 mH, 55.3]
4. A test coil is connected across a calibrated loss free variable capacitor having in parallel with it a VIVM of input capacitance 3 pF. A constant current, of varying frequency is passed through an injector coil loosely coupled to the test coil, and at each frequency the capacitor is adjusted to produce a maximum reading on the voltmeter.
   The readings are as follows :

| *Frequency kHz* | *Capacitance pF* | *Maximum voltage, V* |
|---|---|---|
| 200 | 345 | 4 |
| 300 | 145 | – |
| 400 | 75 | – |
| 500 | 43 | – |
| 600 | 25 | 16 |

   Determine :
   (*a*) the inductance of the test coil,
   (*b*) its self capacitance, and
   (*c*) the percentage increase in its resistance as the frequency varies from 100 kHz to 600 kHz.
   **[Ans.** 1.732 mH, 12 pF, 125%]
5. The effective resistance of a coil at a frequency of 0.2 MHz is measured by the reactance variation method. The coil is connected in series with a calibrated variable capacitor and a 10 Ω resistor. A constant emf at 0.2 MHz is injected into the circuit. An electronic voltmeter connected across the resistor is used to indicate resonance. The capacitor settings to reduce the resonant current by one half are 150 pF and 170 pF. Calculate the effective resistance of the coil. **[Ans.** 170 Ω]
6. A coil of unknown impedance in series with a capacitor of capacitance 0.02 μF and an ammeter of negligible impedance was connected to a variable frequency source having constant voltage. The frequency was adjusted both above and below resonance until the reading of the ammeter was reduced to 0.8 of the value at resonance. This occurred at frequencies of 0.97 and 1.01 MHz. Find the inductance and resistance of the coil.
   **[Ans.** 1.29 μH ; 0.424 Ω]

7. A test on a radio tuning coil, using $Q$ meter, gave the following results :
   (*a*) With a radio coil connected normally, resonance was obtained at a frequency of 1 MHz with the $Q$ meter capacitor set at 480 pF.
   (*b*) With standard inductor connected in place of the radio coil, the resonance was obtained at 5 MHz and this condition was not altered when the coil was connected in parallel with the standard inductor. Calculate the self capacitance of radio coil. [**Ans.** 20 pF]

8. A $Q$ meter gave the following readings during a certain test on a coil :

$$Q = 250,\ C = 150 \text{ pF},\ f = 750 \text{ kHz}.$$

A VTVM was then connected across the standard capacitor and resonance re-established. The corresponding readings now become

$$Q = 150,\ C = 150 \text{ pF},\ f = 750 \text{ kHz}.$$

Calculate for the VTVM :
   (*a*) the input resistance, and
   (*b*) the input capacitance. [**Ans.** 496 kΩ, 10 pF]

9. A coil was tested using a $Q$ meter and following information was obtained. At a frequency of 3 MHz, the value of capacitance to give maximum voltage across the capacitor was 251 pF while at 6 MHz frequency it was 50 pF. Calculate the self capacitance of the coil. [**Ans.** 17 pF]

10. A coil with a resistance of 0.1 Ω is connected in a $Q$ meter circuit as shown in Fig. 24.12. Resonance occurs when the oscillator frequency is 40 MHz and the tuning capacitor is set at 135 pF. Calculate the percentage error introduced in the calculated value of $Q$ by a 0.02 Ω shunt resistance connected across the oscillator circuit. Comment upon me result. [**Ans.** 17%]

# Objective Type Questions

***Tick (✓) the most appropriate answer :***

1. The effective reactance of an inductive coil,
   (*a*) increase because of stray capacitances as the frequency increases
   (*b*) decrease because of stray capacitances as the frequency increases
   (*c*) remains the same irrespective of the increase in frequency even if stray capacitances are present
   (*d*) none of the above.

2. The effective reactance of an inductive coil
   (*a*) increases due to stray series inductances as frequency increases
   (*b*) decreases due to stray series inductances as frequency increases
   (*c*) remains the same irrespective of the increase in frequency even if stray series inductances, are present
   (*d*) none of the above.

3. The effective resistance of a coil at high frequencies is more than its d.c. resistance on account of
   (*a*) skin effects
   (*b*) proximity effects
   (*c*) eddy current losses
   (*d*) all the above.

4. A series $RLC$ circuit is used for measurement of inductance of a coil using resonant method. Resonance conditions are produced by varying the capacitance. A thermoelectric type ammeter is used in series with the circuit. Also an electronic voltmeter is connected across the tuning capacitor
   (*a*) The conditions of resonance are more reliably indicated when the ammeter reads maximum
   (*b*) The conditions of resonance are more reliably indicated when the voltmeter connected across the capacitor reads maximum
   (*c*) The conditions of resonance are the same when both ammeter and voltmeter read maximum
   (*d*) The conditions of resonance are indicated when both ammeter and voltmeter read minimum.

5. If the $Q$ factor of a coil which is $\omega_0 L/R$ is measured by varying the frequency
   (*a*) the plot between $Q$ and frequency is linear
   (*b*) the value of $Q$ initially decreases with increase of frequency and afterwards increases with increase of frequency
   (*c*) the value of $Q$ initially increases with increase of frequency and afterwards decreases with increase of frequency
   (*d*) the $Q$ factor remains constant irrespective of the value of frequency.

24

6. The $Q$ factor of a coil at the resonant frequency 1.5 MHz of an $RLC$ series circuit is 150. The bandwidth is :

(a) 225 MHz (b) 1.06 MHz
(c) 10 kHz (d) none of the above.

7. The resonant frequency of an $RLC$ series circuit is 1.5 MHz with the tuning capacitor set at 150 pF. The bandwidth is 10 kHz. The effective resistance of the circuit is :

(a) 29.5 Ω (b) 14.75 Ω
(c) 9.4 Ω (d) 4.7 Ω

8. In a $Q$ meter, distributed capacitance of coil is measured by changing the capacitance of the tuning capacitor. The values of tuning capacitor are $C_1$ and $C_2$ for resonant frequencies $f_1$ and $2f_1$ respectively. The value of distributed capacitance is :

(a) $\frac{C_1 - C_2}{2}$ (b) $\frac{C_1 - 2C_2}{3}$
(c) $\frac{C_1 - 4C_2}{3}$ (d) $\frac{C_1 - 3C_2}{2}$.

9. The true value of $Q$ of a coil is 245 and the measured value is 244.5. The ratio of distributed capacitance to self capacitance of the coil is

(a) $2.04 \times 10^{-3}$ (b) 1.002
(c) 0.997 (d) none of the above.

10. In a $Q$ meter, the value of shunt resistance connected across the oscillator is typically of the order of

(a) Ω (b) mΩ
(c) μΩ (d) kΩ.

## Answers

1. (b) 2. (a) 3. (d) 4. (b) 5. (c) 6. (c) 7. (d) 8. (c) 9. (a) 10. (b)

# 25

# Primary Sensing Elements and Transducers

## 25.1 INTRODUCTION

The measurand in an instrumentation system makes its first contact with a *Primary Detection Element* or an *Input Device*. There is a multiplicity and variety of measurands to be measured. These include process variables like temperature, pressure and flow rate which are widely employed in process and production plants. The measurands also include electrical quantities like current, voltage, resistance, inductance, capacitance, frequency, phase angle, power and magnetic quantities like flux, flux density, reluctance etc.

All these quantities require a primary detection element and/or a transducer to be converted into another analogous format which is acceptable by the later stages of the measurement system.

The measurand or the input signal is called an *information* for the measurement system, The information may be in the form of a physical phenomenon or it may be an electrical signal. The process of detection and conversion of the information into an acceptable form requires energy. This energy may be extracted from the measurand, but in that case it will not be represented in its faithful form as it would be subjected to loading errors. In order that a measurand is represented in its faithful form undistorted, no energy should be extracted from it during the process of conversion *i.e.*, it should not be subjected to any kind of loading effects. In fact, efforts should be made to supply energy required for conversion from outside sources so that the measurand is not distorted during the process of conversion in order that it be faithfully reproduced in its analogous form. The ideal conversion is where absolutely no energy is extracted from the measurand during the process of conversion and all the energy that is required for conversion is supplied from outside, so that the measurand is not distorted and the analogous output of the detector is a faithful representation of the measurand.

## 25.2 MECHANICAL DEVICES AS PRIMARY DETECTORS

In order to extract information from mechanical systems, only mechanical displacement or velocity can be used, and therefore the importance of mechanical sensing elements is obvious. Some of the commonly used mechanical sensing elements are springs – which convert a force or a torque into a displacement ; a diaphragm, a capsule, bellows or Bourdon tube – which convert pressure into a displacement, a bimetallic strip converts temperature into a displacement ; a mass damper system is used for measurement of acceleration, velocity and displacement. Some input devices may involve more than one mechanical conversion for example, fluid flow measurements may involve conversion of fluid rate into pressure differential using an orifice, venturi tube or pitot tube and then in turn this pressure is converted into displacement for purposes of measurement.

There are a number of mechanical quantities which are to be measured. Some of these quantities are listed in Table 25.1 along with their modes of operation for the purposes of measurement.

These mechanical quantities include force, pressure, displacement, flow rate, temperature and the list is perhaps unending.

**Table 25.1** *Mechanical quantities and their Modes of Operation*

| | *Type* | *Operation* |
|---|---|---|
| A. | Contacting spindle, pin or finger | Displacement to displacement |
| B. | **Elastic Member** | |
| | 1. Prooving ring | Force to displacement. |
| | 2. Bourdon tube | Pressure to displacement. |
| | 3. Bellows | Pressure to displacement. |
| | 4. Diaphragm | Pressure to displacement. |
| | 5. Spring | Force to displacement. |
| C. | **Mass** | |
| | 1. Seismic mass | Forcing function to displacement. |
| | 2. Pendulum scale | Force to displacement. |
| | 3. Manometer | Pressure to displacement. |
| D. | **Thermal** | |
| | 1. Thermocouple | Temperature to electric current. |
| | 2. Bimaterial | Temperature to displacement. |
| | 3. Temp-stik | Temperature to phase. |
| E. | **Hydropneumatic** | |
| | 1. *Static* | |
| | (*a*) Float | Fluid level to displacement. |
| | (*b*) Hydrometer | Specific gravity to displacement. |
| | 2. *Dynamic* | |
| | (*a*) Orifice | Velocity to pressure. |
| | (*b*) Venturi | Velocity to pressure. |
| | (*c*) Pitot tube | Velocity to pressure. |
| | (*d*) Vanes | Velocity to force. |
| | (*e*) Turbines | Linear to angular velocity. |

The initial concept of converting an applied force into a displacement is basic to many types of primary sensing elements. The mechanical elements which are used to convert the applied force into displacement are usually elastic members. There are many types of these elastic members. They can be classified into three categories, although some elastic members may fall into a combination of these categories.

The *three* categories are :

(*i*) Direct tension or compression type :

(*ii*) Bending type ;

(*iii*) Torsion type.

## 25.3 MECHANICAL SPRING DEVICES

Most mechanical-input measuring systems employ mechanical springs of one form or another. The displacements are usually small and engineering approximations for small displacements or deflections are valid. Various common types of springs are shown in Fig. 25.1. These range from cantilever, helical and spiral springs to torsion bars, prooving (proof) rings and spring flexture pivots.

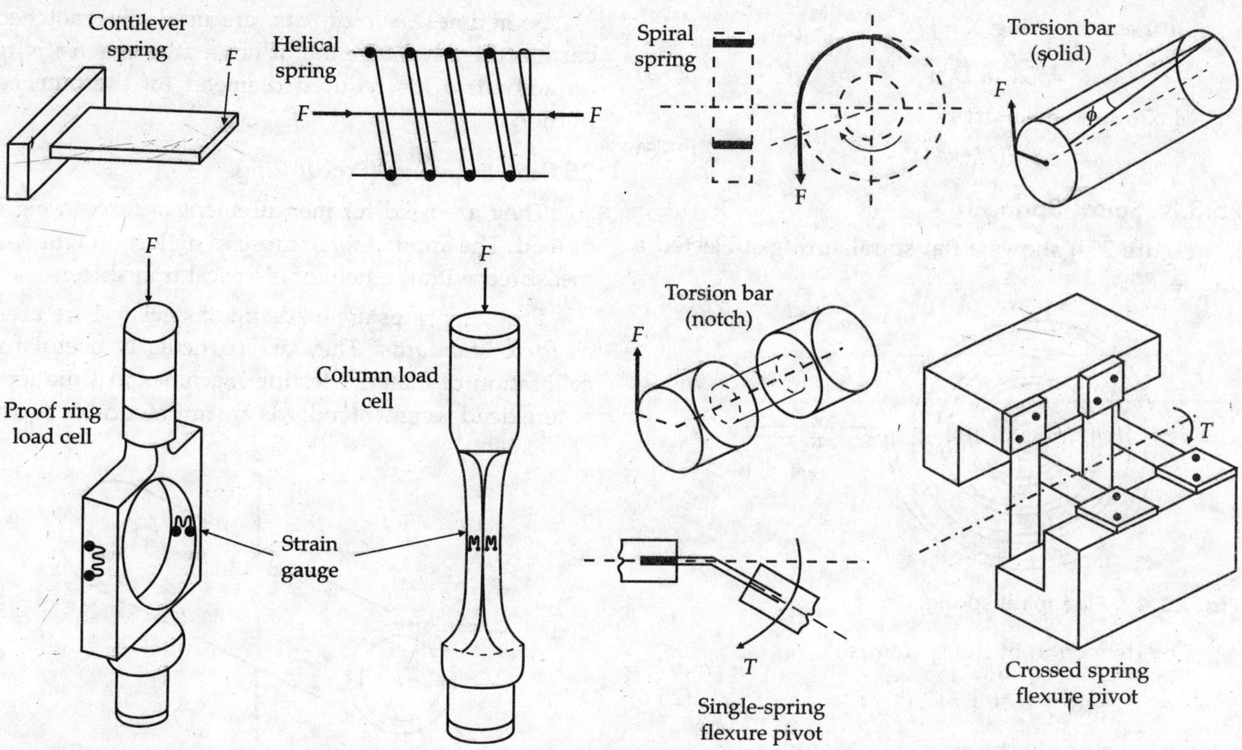


**Fig. 25.1** Spring elements used for sensing force ($F$) or torque ($T$).

### 25.3.1 Cantilever

A cantilever is shown in Fig. 25.2 which is subjected to a force at its free end.

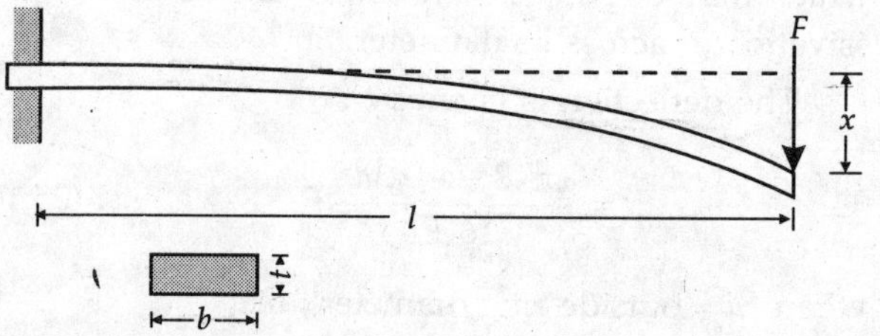


**Fig. 25.2** Cantilever.

Deflection at the free end

$$x = Fl^3 / 3\,EI \qquad \text{...(25.1)}$$

where

$F$ = applied force ; N,

$l$ = length of cantilever, m ;

$E$ = modulus of elasticity ; $N/m^2$

$I$ = moment of inertia $= \left(\frac{1}{12}\right) bt^3$

$b$ = width of cantilever ; m

$t$ = thickness of cantilever ; m

Stiffness of cantilever

$$K = F/x = 3\,EI/l^3 \qquad \text{...(25.2)}$$

### 25.3.2 Helical Spring

Figure 25.3 shows a close coiled helical spring subjected to a compressive force $F$.

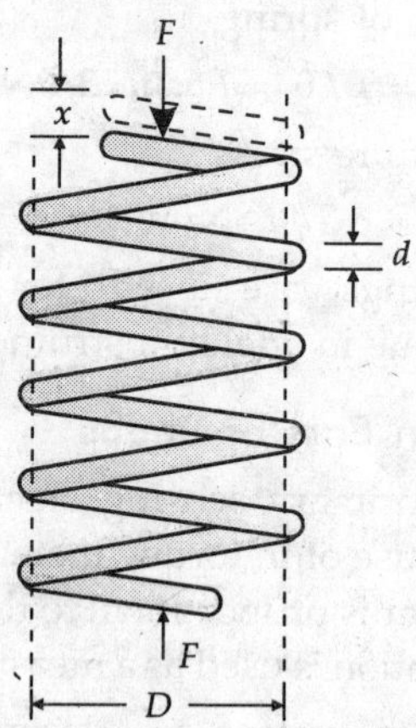


**Fig. 25.3** Closed coiled helical spring.

Displacement of spring :

$$x = \frac{8\,FD^3 n}{Gd^4} \qquad \text{...(25.3)}$$

where $F$ = applied force ; N,

$D$ = mean diameter of coiled spring ; m,

$d$ = diameter of spring wire ; m,

$n$ = number of wires,

$G$ = shear modulus ; $N/m^2$.

25

Stiffness of spring $K = F/x$

$$= Gd^4/8\,D^3 n \quad ...(25.4)$$

Maximum shear stress,

$$\tau = 8\,FD/(\pi d^3) \quad ...(25.5)$$

### 25.3.3 Spiral Spring

Figure 25.4 shows a flat spiral spring subjected a torque $T$.

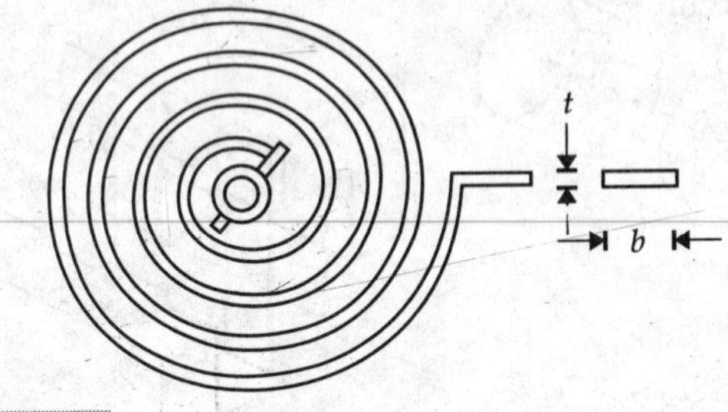


**Fig. 25.4** Flat spiral spring.

The deflection of the spring is :

$$\theta = E\,bt^3T / 12\,l, \text{ rad} \quad ...(25.6)$$

where $E$ = modulus of elasticity, N / m$^2$,

$b$ = width of spring ; m,

$t$ = thickness of spring ; m,

$l$ = length of spring ; m,

$T$ = torque ; Nm

∴ Stiffness of spring

$$K = T/\theta = Ebt^3/12\,l \text{ Nm/rad} \quad ...(25.7)$$

Maximum fibre stress

$$S_{max} = 6\,T/bt^2 \text{ N/m}^2 \quad ...(25.8)$$

Spiral springs are used for production of controlling torque in analog instruments.

### 25.3.4 Torsion Bars or Shafts

These are primary sensing elements for torque. They are made use of in torque meters. The deflection or twist of the bar is proportional to the applied torque and the deformation is used as a measure of the torque.

Some torque meters are designed so that the angular displacement due to twisting of the bar is measured with the help of a displacement transducers. In others, the strain in the surface of the bar, which is proportional to the torque, is measured with the help of strain gauges. The shear strain is a measure of the torque.

Angle of twist $\theta = 16T/\pi G d^3$ rad, ...(25.9)

where $T$ = applied torque ; Nm,

$G$ = shear modulus ; N/m$^2$,

$d$ = diameter of bar ; m

Sometimes, *notched bars*, are used. The notched bar has the advantage that it has a greater sensitivity on account of its reduced diameter (of the notched portion).

### 25.3.5 Prooving (Proof) Rings

They are used for measurement of force, weight or load. The applied force causes a deflection which is measured with the help of electrical transducers.

Prooving rings are made up of steel and are used as force standards. They are particularly useful for calibration of material testing machines in situations where dead weight standards are impracticable to use

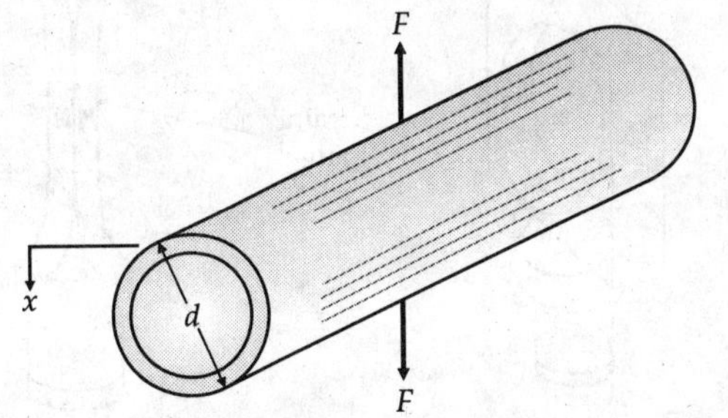


**Fig. 25.5** Prooving ring.

on account of their bulk. A *prooving ring* is a circular ring of rectangular cross-section as shown in Fig. 25.5 which may be subjected to either tensile or compressive forces across its diameter.

The deflection is given by :

$$x = \frac{(\pi/2 - 4/\pi)d^3}{16\,EI}F \quad ...(25.10)$$

where $d$ = outside ring diameter ; m.

The other terms have been explained earlier.

The common practice for measurement of displacement is to attach a displacement transducer between the top and bottom of the prooving ring. When the force is applied, the relative displacement can be measured. An LVDT is normally used for measurement of deflection which is of the order of 1 mm or so. Another method is to use strain gauges for measurement of strain caused by the applied force. The strain, then can be used to compute the applied force.

### 25.3.6 Load Cells

Load cells utilise an elastic member as the primary transducer and strain gauges as secondary transducers as shown in Fig. 25.6.

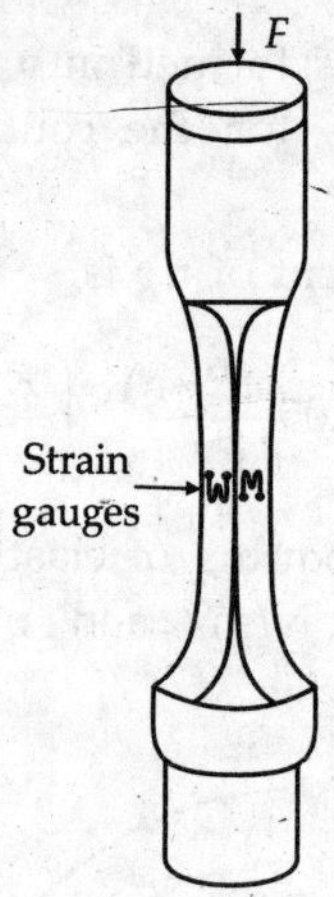


**Fig. 25.6** Load cells.

### 25.3.7 Spring Flexture Pivots

Figure 25.1 shows two different types of arrangements for flexture pivots, one for the single spring flexture pivot and the second for crossed spring flexture pivot. The crossed-spring flexture pivot is widely used in measurement work for the following reasons :

(*i*) it is practically frictionless

(*ii*) the pivot sensitivity *i.e.,* the angular deflection per unit applied torque is virtually constant for angular relations less than 15°.

## 25.4 PRESSURE SENSITIVE PRIMARY DEVICES

Most pressure measuring devices use elastic members for sensing pressure at the primary stage. These elastic members are of many types and convert the pressure into mechanical displacement which is later converted into an electrical form using a secondary transducer. These devices are many a time known as **force summing devices**. Figure 25.7 shows some of the commonly *pressure sensitive primary devices*.

The principle of working of these devices is explained as : the fluid whose pressure is to be measured is made to press the pressure sensitive element and since the element is an elastic member, it deflects causing a mechanical displacement. The displacement is proportional to the pressure applied. The displacement is then measured with the help of electrical transducers. The output of the electrical transducers is proportional to the displacement and hence to the applied input pressure.

The commonly used pressure sensitive devices are described below :

### 25.4.1 Bourdon Tubes

The Bourdon tubes are designed in various forms. These are :

(*i*) C type (*ii*) Spiral

(*iii*) Twisted tube and (*iv*) Helical.

The Bourdon tubes are made out of an elliptically flattened tube bent in such a way as to produce the shapes shown in Fig. 25.7. One end of the tube is sealed or closed. The other end is open for the fluid to enter. When the fluid whose pressure is to be measured enters the tube, the tube tends to straighten



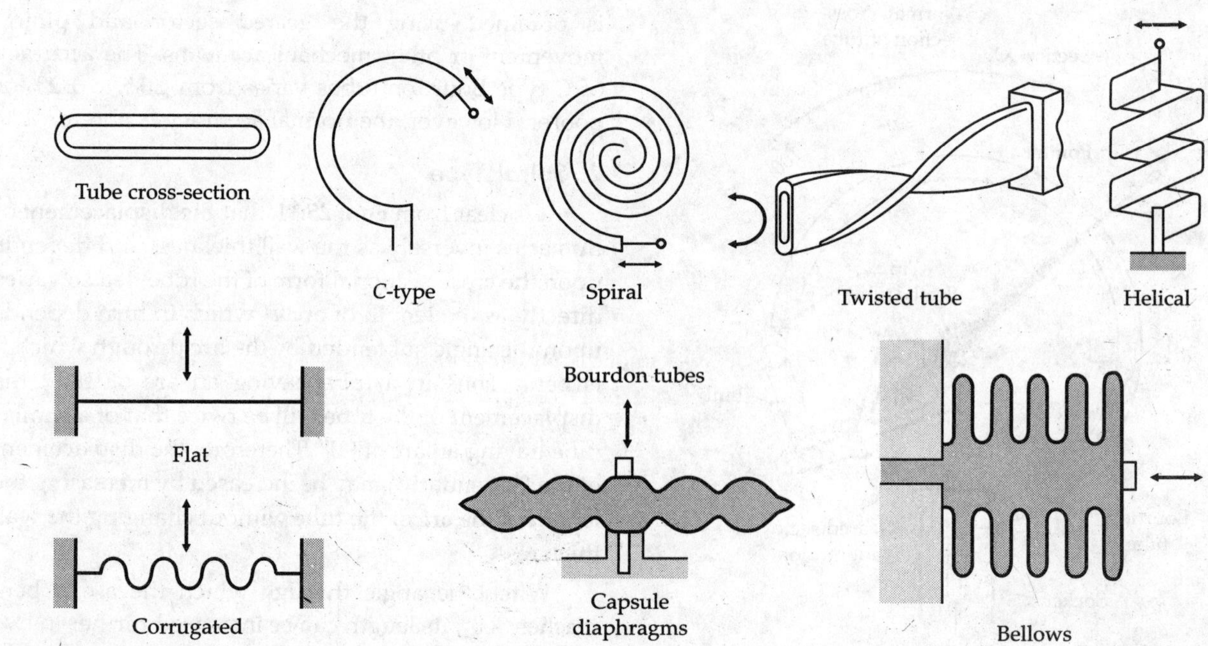


**Fig. 25.7** Pressure sensitive primary devices.

out on account of the pressure applied. This causes a movement of the free end (closed end) and the displacement of this end is amplified through mechanical linkages. The amplified displacement of the free end may be used to move a pointer on a scale calibrated in terms of pressure or may be applied to a electrical displacement transducer whose output may be calibrated in terms of the pressure applied.

Bourdon tube elements have several distinct advantages and these include low cost, simple construction, high pressure range, good accuracy except at low pressures, and improved designs at the pressure for maximum safety. Their greatest advantage is that they are easily adapted for designs for obtaining electrical outputs.

The disadvantages of Bourdon tubes are : their low spring gradient which their use limits for precision measurements upto a pressure of 3 MN/m$^2$, are susceptible to shocks and vibrations and are subject to hysteresis.

The Bourdon tube has the following forms :

## 1. C Type

The C type of Bourdon element is most commonly used for local indication but it is also used for pressure transmission and control applications. The tube which is oval in section is formed into an arc of 250° and hence the name C for the configuration which is shown in Fig. 25.8.

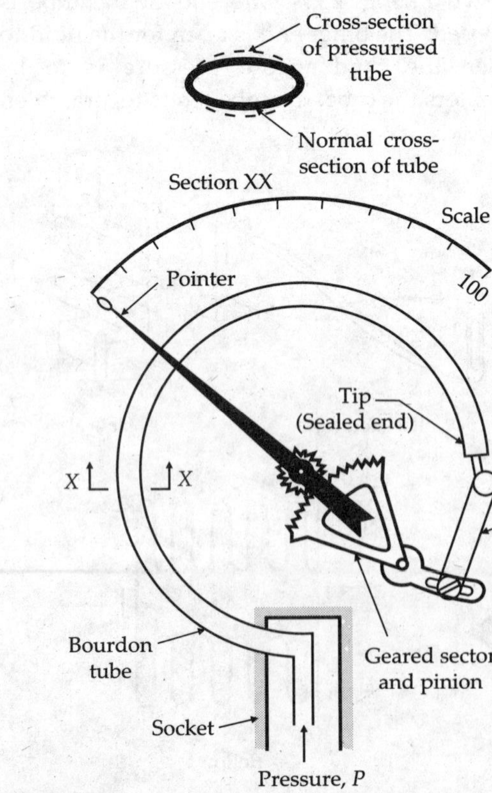


**Fig. 25.8** C type Bourdon tube.

The displacement of tip is,

$$\Delta a = 0.05 \frac{aP}{E}\left(\frac{r}{t}\right)^{0.2}\left(\frac{x}{y}\right)^{0.33}\left(\frac{x}{t}\right)^{3} \quad \text{...(25.11)}$$

where $E$ is the modulus of elasticity ; N/m$^2$, $P$, the applied pressure, N/m$^2$ and other terms are as depicted in Fig. 25.9.

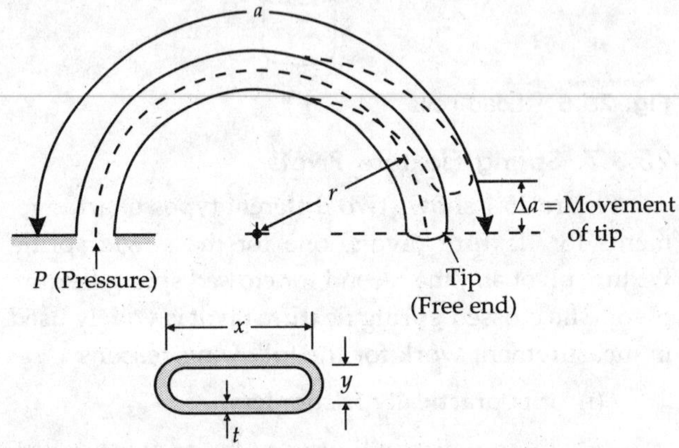


**Fig. 25.9** Geometry of C type Bourdon tube.

Thus relationship between the displacement of tip and the applied pressure is non-linear *i.e.*, each pressure increment does not produce a corresponding tip movement. However, a linear relationship between the deflection of the pointer and the applied pressure is obtained using the geared sector and pinion movement or other mechanical means. The accuracy of C type Bourdon tubes varies from ±0.5 to ± 2% or poorer. However, the normal accuracy is about ± 1%.

## 2. Spiral Type

It is clear from Eqn. 25.11, that the displacement of tip varies inversely as the wall thickness and depends upon the cross-sectional form of the tube. It also varies directly as the length of arc '*a*' which in turn depends upon the angle subtended by the arc through which it is bent. Thus in a tube having an arc of 180°, the displacement of the tube will be twice that of a similar tube having an arc of 90°. Therefore, the displacement of the free end (tip) may be increased by increasing the length of the arc of the tube without changing the wali thickness.

When the angle through which the arc is bent reaches 360°, its length can be increased further in two ways : the tube can be made in the form of a *spiral*, or it can be made in the form of a *helix*. An increased displacement of the free end can be obtained by increasing the number of turns in the spiral or helix thereby avoiding the need for further magnification. It has been mentioned earlier that magnification is obtained through the use of geared sector and pinion is used in C type Bourdon tube. But while using spiral or helix type Bourdon tubes there is no need for further amplification and hence the geared sector and pinion arrangement is not used in their case. The absence of geared sector and pinion arrangement eliminates the backlash which tends to occur when they (geared sector and pinion) become worn owing to continued use.

A spiral type Bourdon tube is shown in Fig. 25.10. Spiral tubes are made by winding several turns of the tube with its flattened cross-section in the form of a spiral.

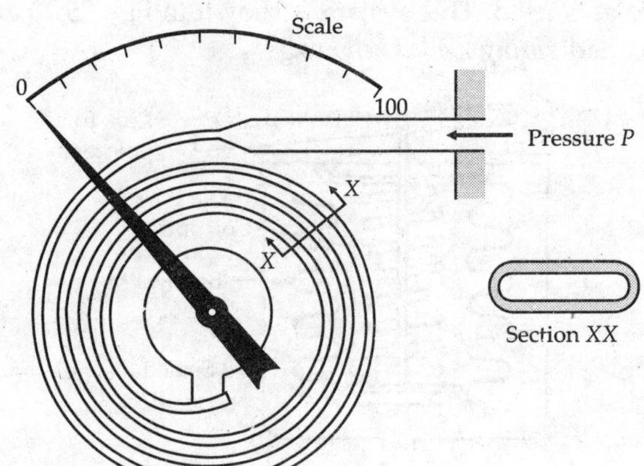


**Fig. 25.10** Spiral type Bourdon tube.

When the pressure to be measured is applied to the spiral, it tends to uncoil producing a relatively long movement of the tip whose displacement can be used for indication or transmission.

The accuracy of spiral tube elements is higher than that of C type elements on account of absence of magnifying elements and is typically about ± 0.5%.

### 3. Helical Type

A helical type Bourdon tube is shown in Fig. 25.11. As explained earlier, a helical bourdon element is similar to the spiral element, except it is wound in the form of a helix. The displacement of the tip of a helical element is larger than that of spiral element. Usually a central shaft is installed within a helical element and the pointer is driven from this shaft by connecting links. This system transmits only the circular motion of the tip to the pointer which is directly proportional to the changes in pressure.

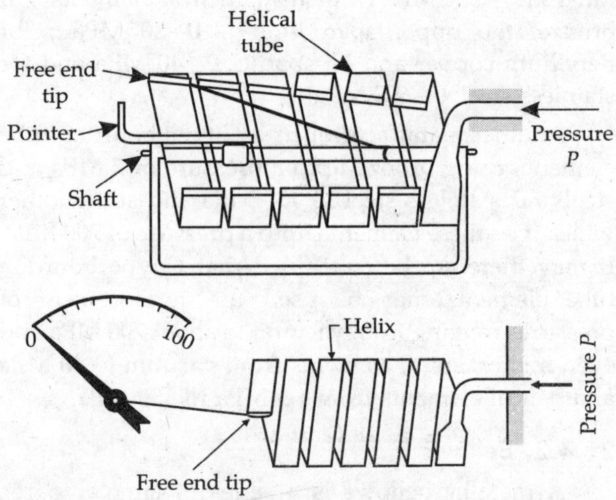


**Fig. 25.11** Helical type Bourdon tube.

The advantages of helix elements include the high over-range capabilities which may be in the ratio as high as 10 : 1, its stability in fluctuating pressure applications, and its adaptability for high pressure service.

The number of coils employed in helix elements depends upon the pressure to be measured. Helix type of pressure elements use as few as three coils while elements used for measurement for high pressures may have as many as 16 coils or even more. The accuracies obtainable from helical elements may vary from ± 0.5% to ± 1% of span.

25

Bourdon tubes are made of different materials which include brass, alloy steel, stainless steel, bronze, phosphor bronze, beryllium copper, *K* Monel, Monel and Ni-span C. The choice of type of material to be used depends upon its elastic characteristics suitable for the pressure range, process medium, temperature and corrosive resistance to the media to be used.

Phosphor bronze is used in low pressure applications where the atmosphere is non-corrosive while in applications where corrosion and/or high pressure is a problem, stainless steel or Monel are used.

Pressure gauges using bourdon tube elements are made with ranges from 760 mm of mercury to 700 MPa or higher for special applications with the minimum span being about 70 kPa. In fact, there is quite an overlap of ranges at which the various bourdon tube elements are designed to operate and these ranges differ from one manufacturers, to another. One manufacturer, for example, lists spiral bourdon tube elements from 0 – 75 to 0 – 1500 kPa when bronze, beryllium copper and Ni-span C are used as the materials for the element. The same manufacturer uses the aforesaid materials for pressure ranges 1.5 – 550 MPa while using helical elements. For bronze, the upper span limit is 0 – 30 MPa ; for beryllium copper and Ni-span C, 0 – 40 MPa and for stainless steel 0 – 600 MPa.

Another manufacturer uses C type bourdon tube elements using bronze upto a pressure of 7 MPa and steel and stainless steel to 150 MPa and still another which uses these elements upto a pressure of 700 MPa. It may therefore be concluded that C-type bourdon tube elements may be used for measurement of pressures ranging from vacuum to about 700 MPa and even higher, spiral elements from vacuum to 30 MPa and helical elements from 1.5 MPa to 550 MPa.

### 25.4.2 Bellows

A metallic bellows is a series of circular parts, resembling the folds in an accordian as shown in Fig. 25.12. These parts are formed or joined in such a manner that they are expanded or contracted axially by changes in pressure. The metals used in the construction of bellows, must be thin enough to be flexible, ductile enough for reasonably easy fabrication, and have a high resistance to fatigue failure. Materials commonly used are brass, bronze, beryllium copper, alloys of nickel and copper, steel and Monel. Most of the bellows used in pressure gauges are seamless and are made from drawn tubing by hydraulic or other methods of rapid forming. These methods produce uniform walls that give a higher life expectancy. Other methods such as soldering and welding of annular sections, rolling, spinning and turning from solid stock may also be used for manufacture of bellows.

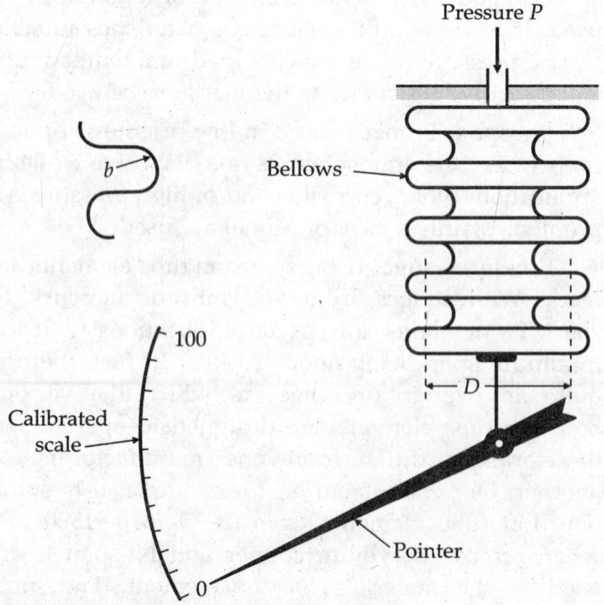


**Fig. 25.12** Bellows element.

The displacement of bellows element is given by,

$$d = \frac{0.453\ Pb\ n\ D^2 \sqrt{1-\nu^2}}{Et^3} \quad \text{...(25.12)}$$

where $P$ = pressure, N/m$^2$ ;
$b$ = radius of each corrugation, m ;
$n$ = number of semi-circular corrugations ;
$t$ = thickness of wall, m ;
$D$ = mean diameter, m ;
$E$ = modulus of elasticity, N/m$^2$ ;
and $\nu$ = Poisson's ratio.

Normally a Bellows has the ability to move over a greater distance than required in a pressure application and, therefore, to give it maximum life and to have better accuracy, its movement is generally opposed by a calibrated spring so that only a part of the maximum stroke is used. This system is shown in Fig. 25.13 and is called *spring loaded bellows.*

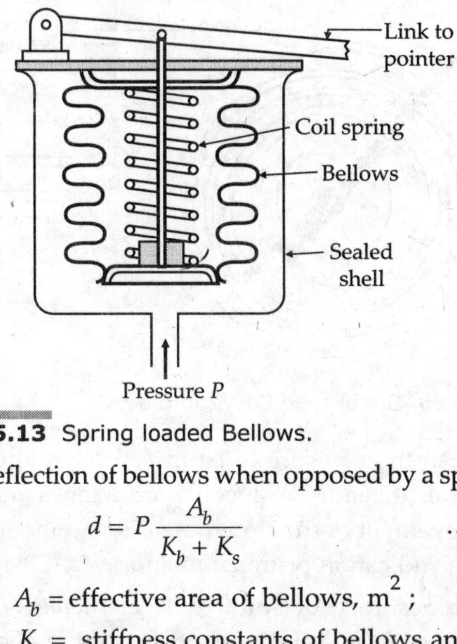


**Fig. 25.13** Spring loaded Bellows.

Deflection of bellows when opposed by a spring is,

$$d = P\frac{A_b}{K_b + K_s} \quad \text{...(25.13)}$$

where $A_b$ = effective area of bellows, m$^2$ ;
$K_b$, $K_s$ = stiffness constants of bellows and spring respectively, N/m$^2$,

$$\therefore \text{ Pressure } \quad P = d\frac{(K_p + K_s)}{A_b} \quad \text{...(25.14)}$$

If the bellows assembly operates an electric switch or some other mechanism, we have,

$$P = \frac{F + d_s(K_b + K_s)}{A_b} \quad \text{...(25.15)}$$

where $F$ = force required to operate switch or mechanism, N ;
$d_s$ = deflection required to operate switch or mechanism, m

There are three main configurations in which bellows elements are used and these are for measurement of absolute, gauge and differential pressures. These configurations may be illustrated with the help of Fig. 25.14.

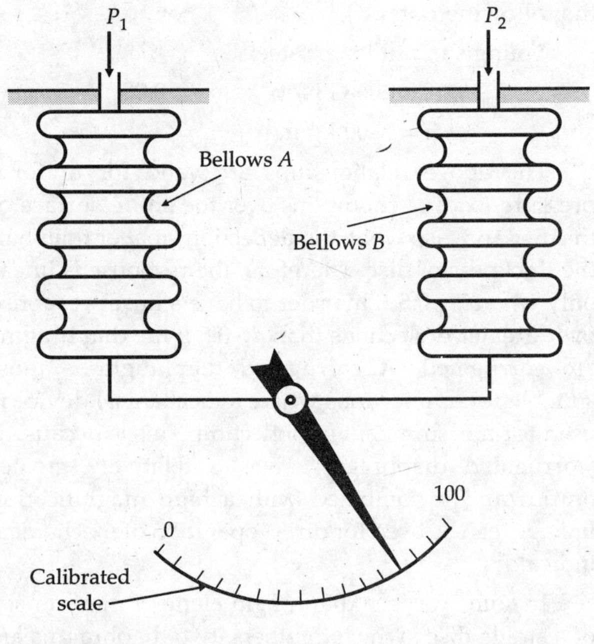


**Fig. 25.14** Measurement of pressure with bellows.

There are two bellows, $A$ and $B$. The pressure applied to bellows $A$ is $P_1$ and that to bellows $B$ is $P_2$. In case it is desired to measure the absolute pressure, bellows $B$ is evacuated and the resultant pressure, $P_1$ is the absolute pressure. When measurement of gauge pressure is desired, bellows $B$ is opened up to atmosphere with pressure $P_2$ is equal to the atmospheric pressure and therefore the reading of the gauge is the gauge pressure. The measurement of **differential pressure** $P = P_1 - P_2$ is done when pressure $P_1$ is applied to bellows $A$ and pressure $P_2$ is applied to bellows $B$.

The advantages of bellows include their simple and rugged construction, moderate price, their usefulness for measurement of low, medium and high pressures, and their applicability for use in measurement of absolute, gauge and differential pressures. Bellows elements, like many other elements have been greatly improved over the past few years. The reduction of drift and hysteresis allows their use in functions requiring ± 0.5% of full span accuracy. They deliver relatively high forces and are well adapted to vacuum and low pressure measurements.

The disadvantages of bellows are that are not suited for dynamic measurements on account of their greater mass and longer relative movement. Also they need temperature compensating device to avoid errors resulting from changes in ambient temperature.

### 25.4.3 Diaphragms

The operating principle of diaphragm elements is similar to that of the bellows. The pressure to be measured is applied to the diaphragm, causing it to deflect, the deflection being proportional to the applied pressure. The movement of the diaphragm depends on its thickness and diameter. The movement is small and hence a diaphragm element does not require any springs as is the case in bellows.

The movement of a diaphragm is a convenient way of sensing a pressure. The unknown pressure is applied to one side of a diaphragm. The edge of the diaphragm is rigidly fixed and causes a deflection on account of the applied pressure. The displacement of the diaphragm may be measured to determine the value of applied pressure, $P$.

The diaphragm, which is a very thin membrane under pressure and therefore under radial tension, may use capacitive or inductive transducers, to produce an electrical output which is proportional to the output of the transducers. The use of diaphragms using thin membranes is limited to measurement of low level pressure fluctuations. Since membranes can withstand only limited values of force (produced by pressure), diaphragms using thin membranes are suitable for resistance and piezoelectric transducers which require greater displacements than are produced by membrane type of diaphragms. Therefore, in place of membranes, it is usual to employ thin circular plates which are either clamped, around their circumference between two solid rings, or are machined from a solid piece of metal. In order to maintain linear relationship between deflection of plates and applied pressure, the deflection of the centre of plate should be less than half its thickness or $d_m \leq 0.5t$ which are essentially flexible discs.

The diaphragms, are of *two* types :

(*i*) Flat type and

(*ii*) Corrugated type.

These are shown in Fig. 25.15.

Corrugated diaphragms, on account of their increased surface area, give greater deflections, typically 2 per cent of the diaphragm diameter.

25

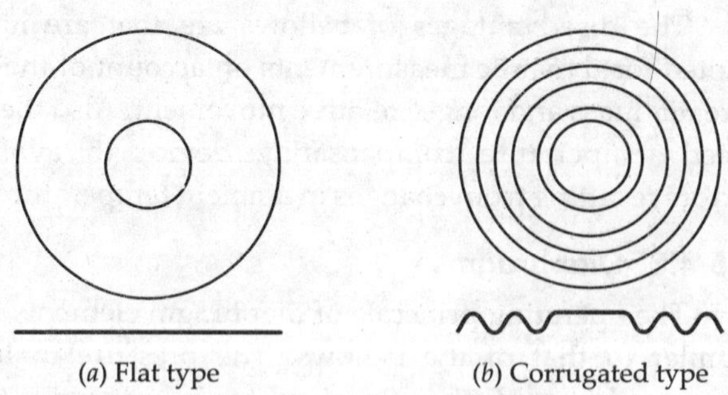


**Fig. 25.15** Single diaphragm elements.

A flat diaphragm is shown in Fig. 25.16.

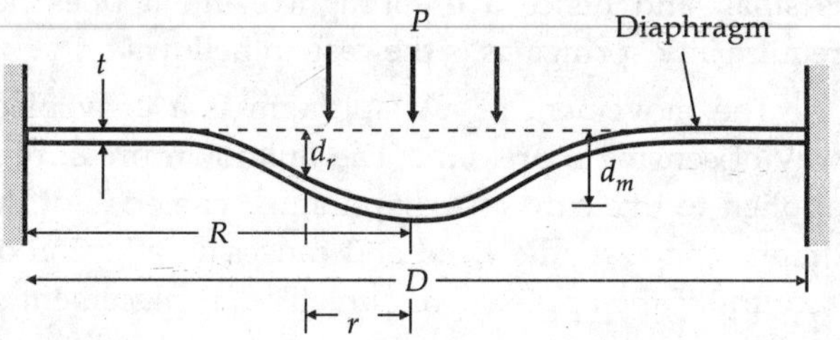


**Fig. 25.16** Flat diaphragm.

$$\text{Pressure } P = \frac{256\ Et^3 d_m}{3(1-\nu^2)\,D^4}\ \text{N/m}^2 \qquad \ldots(25.16)$$

where $E$ = Young's modulus ; N/m$^2$,

$t$ = thickness of diaphragm ; m,

$D$ = diameter of diaphragm ; m,

$R$ = radius of diaphragm ; m,

$\nu$ = Poisson's ratio,

and $d_m$ = deflection at the centre of the diaphragm ; m

The above relationship between pressure, $P$ and the deflection at the centre, $d_m$ is linear. But linearity holds good as long as $d_m \le 0.5\ t$ and not otherwise.

The deflection at the centre is :

$$d_m = \frac{3P(1-\nu^2)}{16\ Et^3} D^4$$

$$= \frac{3P(1-\nu^2)}{16\ Et^3} R^4 \qquad \ldots(25.17)$$

The deflection at a radius $r$ from the centre is :

$$d_r = \frac{3P(1-\nu^2)}{16\ Et^3}(R^2 - r^2)^2 \qquad \ldots(25.18)$$

Maximum stress at the circumference

$$S_m = \frac{3\ D^2 P}{16 t^2}\ \text{N/m}^2 \qquad \ldots(25.19)$$

Lowest natural frequency (for air or gas medium)

$$\omega_n = \frac{20t}{D^2}\sqrt{\frac{E}{3\rho\,(1-\nu^2)}}\ \text{rad/s} \qquad \ldots(25.20)$$

where $\rho$ = density of diaphragm material, kg/m$^3$

Design equations for the flat diaphragm may use the following data :

Young's modulus for steel

$E = 206$ G N/m$^2$ ; $\nu = 0.28$

and $\rho = 7800$ kg/m$^3$

The above relationships are valid for uniform pressure loading conditions over the entire surface of the disc and also when the deflection smaller than half the thickness of disc. Therefore, the response is linear only when $d \le 0.5\ t$. In order to have a linear response over greater deflections than $d > 0.5\ t$, the diaphragms are corrugated. A corrugated diaphragm is most suitable for applications where a mechanical device is used for measurement of deflection. This is because it (corrugated diaphragm) gives a relatively greater output and if combined with a high magnification linkage, can be used for direct operation of mechanical indicators.

In some cases, a diaphragm element may consist of a single disc ; which in others, two diaphragms are bonded together at their circumference by soldering or pressure welding to form a **capsule**. A diaphragm element may consist of one *capsule* or two or more *capsules* connected together with each capsule deflecting on the application of pressure. The total deflection is the sum of the deflections of individual capsules. Figure 25.17 shows as diaphragm element consisting of three capsules. In this assembly, the individual capsule is connected axially with the next one and is allowed to expand without any restraints. The element may be provided with stops for over-range and under-range protection.

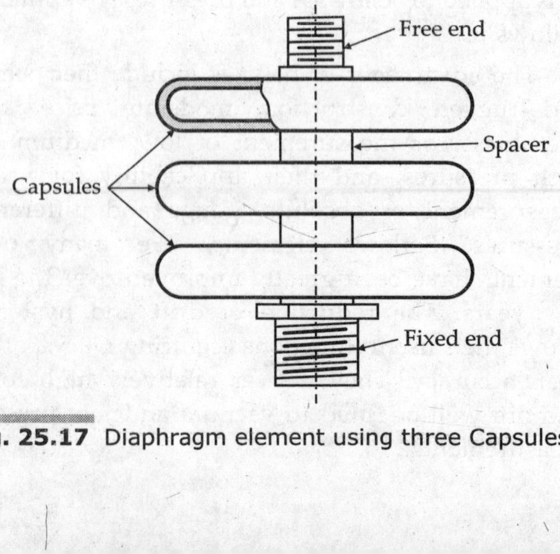


**Fig. 25.17** Diaphragm element using three Capsules.

Two different arrangements are used for capsular elements, the convex and the nested type as shown in Fig. 25.18.

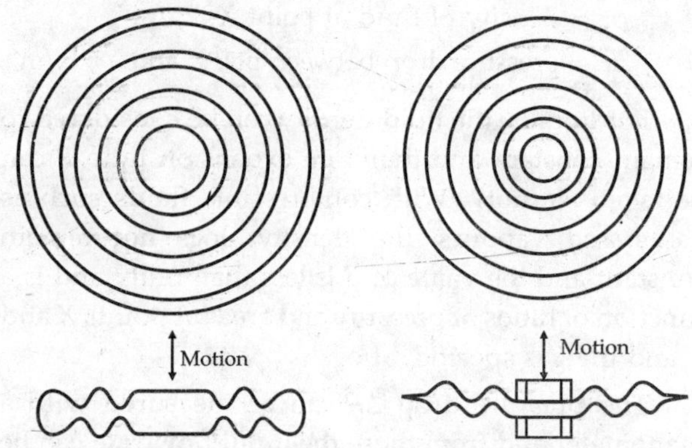


**Fig. 25.18** Types of capsules.

Materials used for diaphragms include phosphor bronze, stainless steel, beryllium copper, Ni-span C, Inconel, Monel and nickel. Non-metallic materials are used for some applications. Buna N rubber, nylon and Teflon are used in environments that corrode metals.

The diaphragm elements find extensive use in applications where measurement of low pressures including vacuum is involved. The ranges are 0 – 50 $N/m^2$ to 0 – 200 $kN/m^2$. Accuracies range from ± 0.5% to ± 1.25% of full span.

## 25.5 FLOW RATE SENSING ELEMENTS

### 25.5.1 Introduction

The measurement of flow is very important in a large number of industries. Flow measurements account for a high percentage of the process variables measured in the chemical industries. Flow is the most important variable in plant operation. The plant material balances, quality control and continuous processes will be impossible without flow measurements.

Flow of material in a process or system can be measured by a variety of methods depending upon the material and its condition, the type of flow, the volume and the mass of the quantity flowing, the range and rangeability, the pressure and temperature and the accuracy required. Many accurate and reliable methods are available for the measurement of flow. Some of these methods are applicable only to liquids, some only to gases and vapours and some to both. Fluids, may be of different types, like they may be clear or opaque, clean or dirty, wet or dry, erosive and corrosive. Fluid streams may be multiphase, vapour, liquid or slurries. The flow may be turbulent or laminar. Viscosities may vary. Pressures may vary from near vacuum to many times atmospheric, and the temperatures may vary from cryogenic to hundreds of degrees of celsius.

The characteristics and properties which need to be considered are viscosity, density, compressibility, temperature and pressure.

In most applications, the knowledge of type of flow, whether laminar and turbulent is very important. All liquids have a force of friction, called viscosity; which tends to inhibit the formation of turbulent eddies and vortices, and maintain streamline flow. Viscosity enters into determination of flow constants, particularly the discharge co-efficient C which is a function of *Reynold's number*.

The Reynold's number is given by :

$$Re = vd\,\rho / \eta \qquad \text{...(25.21)}$$

where $v$ = velocity of flow ; m/s,

$d$ = diameter of pipe throat or orifice ; m,

$\rho$ = density of fluid ; $kg/m^3$

and $\eta$ = viscosity ; $Ns/m^2$.

The Reynold's number is dimentionless, and a useful criterion by which to compare flows in geometrically similar installations but with different flow conditions. It provides information regarding conditions where streamline flow ceases and turbulent flow begins. It has been found by experiments that the flow is *streamline or laminar* for Reynold's number less than 2000 and turbulence sets in increasingly as Reynold's number increases beyond 2000. Flow becomes completely *turbulent* for Reynold's number greater than $10^5$.

Flow measurements are concerned with the following :

(*i*) velocity of flow ; m/s,
(*ii*) volumetric flow rate ; $m^3/s$,
(*iii*) mass flow rate ; kg/s,
(*iv*) total volumetric flow ; $m^3$, over a set period of time,
(*v*) total mass flow ; kg, over a set period of time.

### 25.5.2 Types of Flow Rate Meters

The flow rate meters can be classified into *two* categories :

(*i*) Inferential meters
(*ii*) Absolute or positive displacement meters.

In *inferential meters*, the flow is not directly measured but is inferred from measurements of other quantities which are related. Inferential methods of measurement are based upon differential pressure or head obtained from orifice plate, a venturi tube, a pitot tube or a weir. Flow rate can also be inferred from turbine and electromagnetic flow meters.

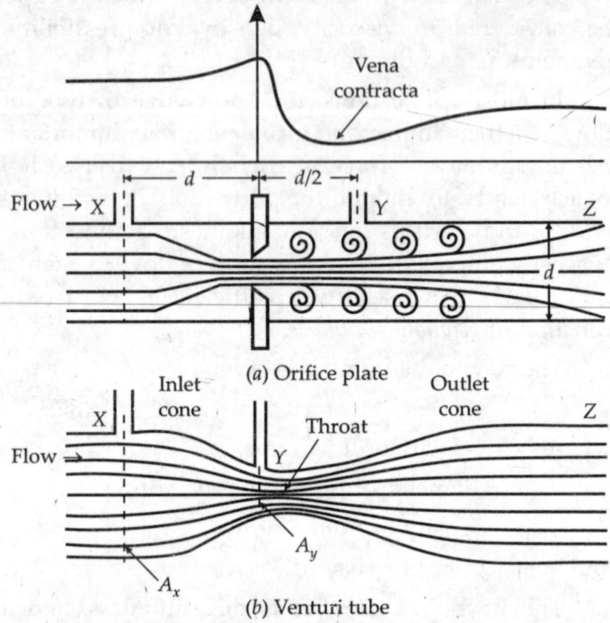


**Fig. 25.19** Flow meters based upon producing pressure drop using constrictions.

The purpose of this section is to briefly consider the flow rate measuring devices with pressure related techniques (inferential in nature) which are widely used. The two basic techniques used are :

(*i*) a pressure drop is produced by means of a constriction,

(*ii*) the kinetic energy of the flowing fluid is transformed into potential energy in the form of a static head.

Two different forms of producing constriction are shown in Figs. 25.19(*a*) and 25.19(*b*). Figure 25.19(*a*) shows a circular orifice plate while Fig. 25.19(*b*) shows a venturi tube.

The fundamental equation for any flow rate sensing element which changes the local velocity by using a constriction or reduction in area of flow is given by

$$Q = CEBA_y \sqrt{2\Delta P / \rho_x} \quad \ldots(25.22)$$

where $Q$ = flow rate, $m^3/s$,

$C$ = co-efficient of discharge

$$= \frac{\text{actual flow rate}}{\text{theoretical flow rate}}$$

$$E = \sqrt{1-(A_y / A_x)^2} = \text{velocity approach factor} \quad \ldots(25.23)$$

$B$ = expansion ratio,

$A_y$ = area at point $Y$ ; $m^2$,

$A_x$ = area at point $X$ ; $m^2$,

$\rho_x$ = density of fluid at point $X$, $kg/m^3$,

$\Delta P$ = pressure drop between pts. $X$ and $Y$ ; $N/m^2$

For liquids, the fluid density can be considered to remain constant and therefore expansion ratio $B$ can be taken as unity. With compressible fluids such as gases and vapours, the density does not remain constant and the value of $B$ is less than unity and is a function of ratios of pressure and areas at points $X$ and $Y$ and the gas specific ratio.

The pressure drop $\Delta P$ can be measured with a manometer and from there the input flow rate can be inferred.

Orifice plates, venturi tubes and nozzles provide a simple means of inducing a pressure drop. The flow rate is proportional to the square root of the pressure drop and hence there is a non-linear relationship between output (pressure drop) and input (flow rate).

The discharge co-efficient for an orifice plate is about 0.6 for Reynold's numbers above $10^4$ and the permanent pressure loss is 60 – 70 percent of differential pressure $\Delta P$. Therefore, an orifice plate which is easy and cheap to install, may cause six or seven times the pumping power as compared to a venturi tube which has a discharge co-efficient of 0.96 for Reynold's numbers above $10^4$ and less tendency to trap suspended material in the fluid being metered. However, venturi tubes occupy greater space and are costlier.

*Pitot tubes* are one of the most common forms of flow rate measuring devices which work on the principle of converting kinetic energy of the fluid being metered into a static head. Pitot tubes are essentially laboratory type of instruments are not normally used for industrial applications. Figure 25.20

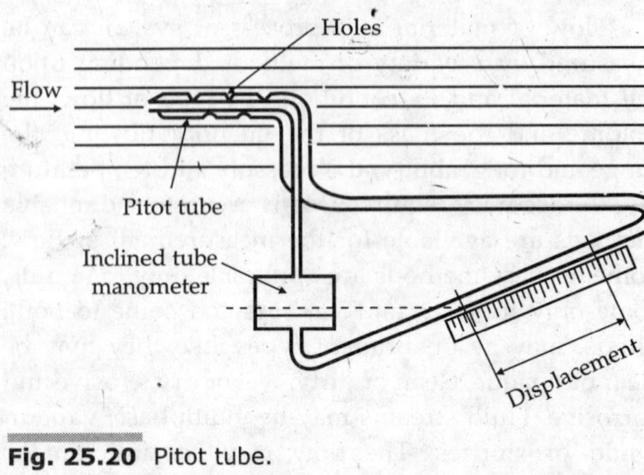


**Fig. 25.20** Pitot tube.

shows a pitot tube installed in a pipeline for measurement of flow-rate. The '*tube*' actually consists of two concentric tubes. The open end of the inner tube faces the incoming flow. The outer tube has a closed end, but has a number of holes in its walls. Both the tubes contain the same fluid as is flowing in the main pipeline. The pressure in the outer tube is the static pressure in the pipeline. The total pressure in the inner tube, however, is greater than this static pressure. The additional pressure on the inner tube is on account of the impact of the fluid stream on the stationary inner tube.

Application of Bernoullis' theorem shows that the velocity at this point in the cross-section is :

$$v = \sqrt{\left(\frac{2\,(P - P_0)}{\rho}\right)} \quad \ldots(25.24)$$

where $P_0$ = static pressure ; $N/m^2$,

$P$ = pressure at the entrance of the tube ; $N/m^2$,

and $\rho$ = density of liquid ; $kg/m^3$

Thus, the velocity at a particular point may be determined from the pressure differential generated by the pitot tube, which is measured by an inclined tube manometer.

The velocity of a fluid across a section of pipe is not uniform but varies from zero at the pipe surface to a maximum usually at the centre of the pipe. In order to find the mean velocity it is necessary to make a traverse of the pipe *i.e.*, readings are taken by placing the Pitot tube at various positions at different points along the cross-section of the pipe. For Reynold's numbers above $10^5$, the ratio of mean velocity to the velocity at the centre of pipe is 0.82. Once this ratio is determined exactly for a given situation, the Pitot tube can be placed at the centre of pipe to sense the velocity. The mean velocity is,

$$v_{(mean)} = C_v \sqrt{\frac{28}{w}(P - P_0)}$$

$$= C_v \sqrt{\frac{2}{\rho}(P - P_0)} \quad \ldots(25.25)$$

where $C_v$ is the *velocity correction factor* and is the ratio of mean velocity to the velocity at the centre.

The advantages of Pitot tubes are that they produce a negligible loss of head when inserted in the pipe and their cost is much smaller than that of venturi meters or orifice meters.

The major disadvantage of Pitot tubes is that they are laboratory type instruments as for their operation they require mean velocity to be determined by making transverse measurements across the diameter of pipe and hence they cannot be used for industrial applications which require an instant readout. The other disadvantages are that they produce a low differential head on account of which their sensitivity is low, and they require high flow velocities of about 15 m/s to produce measurable heads. The small opening of Pitot tubes may get clogged if fluids containing solid particles are used which may give rise to wrong results.

*Weirs and flumes* are commonly used devices for open channel measurements. A detailed description of open channel measurements is beyond the scope of this text. The purpose here is to acquaint the reader with this type of measurements.

### (*i*) Weirs

A weir is essentially a dam with a notched opening at the top through which liquid flows as shown in Fig. 25.21.

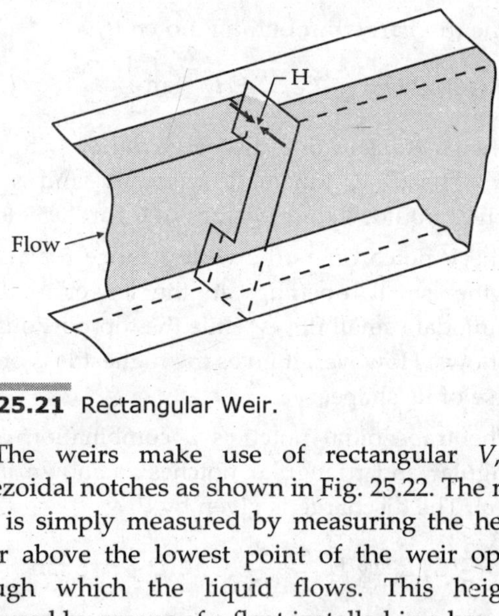


**Fig. 25.21** Rectangular Weir.

The weirs make use of rectangular *V*, and trapezoidal notches as shown in Fig. 25.22. The rate of flow is simply measured by measuring the head of water above the lowest point of the weir opening through which the liquid flows. This height is measured by means of a float installed in a box called stilling well, which is a part of the total structure. The float is so placed that it is not disturbed by the velocity of flow or by turbulence of the stream.

The discharge through a rectangular notch is,

$$Q = \frac{2}{3} C_d L \sqrt{2g}\, H^{1.5} \quad \ldots(25.26)$$

The co-efficient of discharge, $C_d$, varies with the dimensions, of the weir and is usually less than 0.7. The rectangular notch is the most commonly used notch on account of its simplicity, ease of construction and accuracy.

25

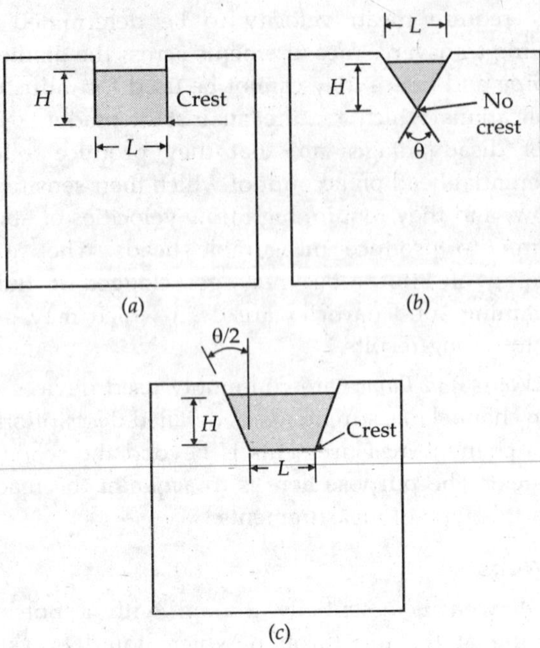


Fig. 25.22 Types of Notches.

The discharge through a $V$ notch is,

$$Q = \frac{8}{15} C_d \sqrt{2g}\, H^{2.5} \tan\frac{\theta}{2} \qquad \ldots(25.27)$$

The co-efficient of a discharge varies with notch angle $\theta$, head $H$, kinematic viscosity and surface tension of liquid. Typical values of $C_d$ are 0.58 to 0.7.

The $V$ notch offers the widest range for a single size, the small opening at the lower end can accommodate small flows while the top portion is for large flows. However, it gives the highest loss of head because of its shape.

The trapezoidal notch is a combination of the rectangular and triangular notches as shown in Fig. 25.22($c$). The discharge is given by :

$$Q = \frac{2}{3} C_d L\sqrt{2g}\, H^{1.5} + \frac{8}{15} C_d L\sqrt{2g}\, H^{2.5} \tan\frac{\theta}{2} \qquad \ldots(25.28)$$

Applications of weirs are primarily in water works and waste and sewage systems. Measurements can be made from a few litres per minute to millions of litres per day.

### (ii) Flumes

Figure 25.23 shows a venturi flume. It is an open channel whose sides are arranged sharply to form a throat, followed by a fairly long expansion section. Stilling wells are provided for measuring the head at both entrance and throat as shown in Fig. 25.24.

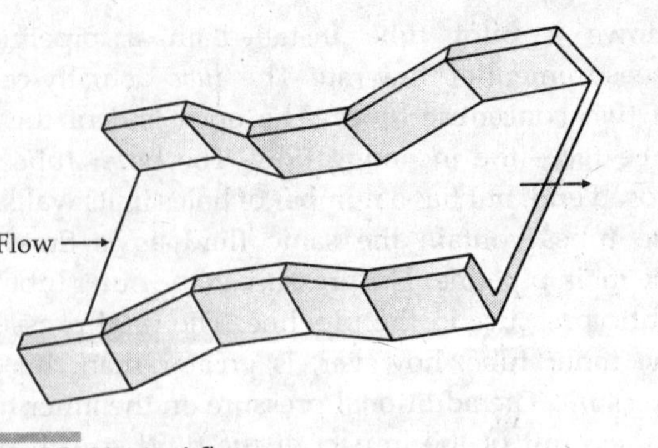


Fig. 25.23 Venturi flume.

The discharge can be calculated as in the case of a venturi meter. The actual discharge is,

$$Q_{(actual)} = \frac{CA_2}{\sqrt{1+(A_2/A_1)^2}}\sqrt{2gh} \qquad \ldots(25.29)$$

$$= CM\sqrt{2gh} \qquad \ldots(25.30)$$

where $h = h_1 - h_2$

$M = A_2 / \sqrt{1+(A_2/A_2)^2}$

$C$ = venturi flume co-efficient (0.95 to 1)

The maximum value of flow occurs when

$$h_2 = \left(\frac{2}{3}\right) h.$$

∴ Maximum discharge is given by,

$$Q_{max} = 1.7\ b_2\ H^{1.5} \qquad \ldots(25.31)$$

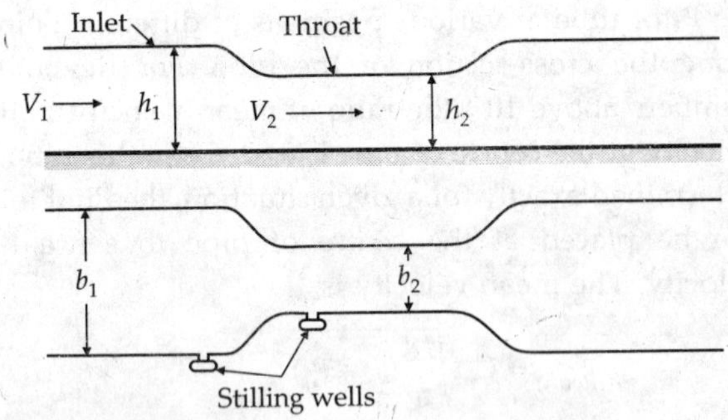


Fig. 25.24 Elevation and plan of flume.

Flumes are capable of better accuracies than weirs because the loss of head in improved versions of flumes is about one fourth of that of weir.

Weirs and flumes have not previously found application in process industries, but their use is increasing with the increasing demand for water pollution control. These measurements are a must when large volumes of water are handled in waste treatment facilities.

***Example 25.1*** *A C type bourdon tube made of Monel metal is show in Fig. 25.9. The dimensions of the tube are:*

$$r = 36.5\ mm,\quad x = 16\ mm,\quad y = 3\ mm,\quad t = 0.35\ mm$$

*Calculate the displacement of the free end if a pressure of 1500 kPa is applied. The modulus of elasticity for Monel metal is 180 GN/m².*

**Solution.** From Eqn. 25.11, the displacement of the tip is given by,

$$\Delta a = 0.05 \frac{aP}{E}\left(\frac{r}{t}\right)^{0.2}\left(\frac{x}{y}\right)^{0.33}\left(\frac{x}{t}\right)^{3}$$

$$= 0.05 \times \frac{\pi \times 36.5 \times 10^{-3} \times 1500 \times 10^{3}}{180 \times 10^{9}} \times \left(\frac{36.5}{0.35}\right)^{0.2} \times \left(\frac{16}{3}\right)^{0.2} \times \left(\frac{16}{0.35}\right)^{3}$$

$$= 0.02\ m = 2\ mm$$

***Example 25.2*** *A differential bellows arrangement for measuring absolute pressure is shown in Fig. 25.25. It uses two bellows each of natural length 50 mm, effective area = 1500 mm², and stiffness = 0.5 N/mm. Bellows A is evacuated and a contains a spring of stiffness 3 N/mm. Find the required natural length of the spring if the bellows are to be equally compressed to a length 40 mm when a pressure of 100 kN/m² absolute is applied to B. Also find the displacement of the output point C for a change of 10 kN/m² in applied pressure.*

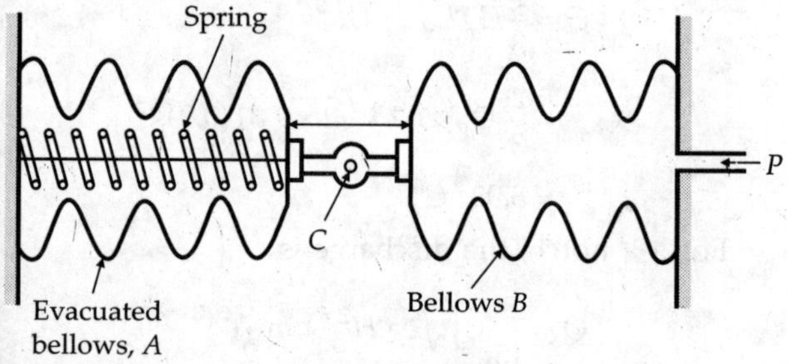


**Fig. 25.25** Bellows arrangement for measurement of absolute pressure.

**Solution.**

When a pressure of 100 kN/m² is used.

Force applied

$$F = PA = 100 \times 10^{3} \times 1500 \times 10^{-6} = 150\ N$$

$\therefore$ Compression of spring = 150/3 = 50 mm

Thus natural length of spring = 50 + 40 = 90 mm

When a pressure of 10 kN/m² is applied, the force is,

$$F = 10 \times 10^{3} \times 1500 \times 10^{-6} = 15\ N$$

Total stiffness of spring plus stiffness of bellows

$$= 3 + 2 \times 0.5 = 4\ N/mm$$

$\therefore$ Displacement of point C = 15/4 = 3.75 mm

***Example 25.3*** *A flat circular diaphragm of mild steel has a diameter of 15 mm. For mild steel, Young's modulus = 200 GN/m² and Poisson's ratio ν = 0.28. Find the thickness of the diaphragm if the maximum stress is not to exceed 300 MN/m² when the applied pressure is 300 kN/m². Find the deflection at the centre for a pressure kN/m².*

*Also calculate the natural frequency of the diaphragm if the density of mild steel is 7800 kg/m³.*

**Solution.** From Eqn. 25.19, maximum stress,

$$S_m = \frac{3D^2P}{16\,t^2}$$

$\therefore$ Thickness of diaphragm

$$S = \sqrt{\frac{3D^2P}{16\,S_m}} = \sqrt{\frac{3 \times (15 \times 10^{-3})^2 \times 30 \times 10^{3}}{16 \times 300 \times 10^{6}}}$$

$$= 0.205 \times 10^{-3}\ m = 0.0205\ mm$$

From Eqn. 25.17, the deflection at the centre for a pressure of 150 kN/m² is,

$$d_m = \frac{3P(1-\nu^2)}{256\,Et^2}$$

$$D^4 = \frac{3 \times 150 \times 10^{3}(1-0.28^2) \times (15 \times 10^{-3})^4}{256 \times 200 \times 10^{9} \times (0.205 \times 10^{-3})^3}$$

$$= 47.6 \times 10^{-6}\ m = 0.0476\ mm$$

The natural frequency is given by Eqn. 25.20,

$$\omega_n = \frac{20\,t}{D^2}\sqrt{\frac{E}{3\rho(1-\nu^2)}}$$

$$= \frac{20 \times 0.205 \times 10^{-3}}{(15 \times 10^{-3})^2}\sqrt{\frac{200 \times 10^{9}}{3 \times 7800\,(1-0.28^2)}}$$

$$= 51950\ rad/s = 8.3\ kHz$$

***Example 25.4*** *A mild steel torsion bar of 30 mm diameter is used for measurement of a torque of 100 Nm. Calculate the angle of twist if the shear modulus of mild steel is 80 GN/m².*

**Solution.** From Eqn. 25.9,

Angle of twist

$$\theta = \frac{16T}{\pi G d^3} = \frac{16 \times 100}{\pi \times 80 \times 10^{9} \times (30 \times 10^{-3})^3}$$

$$= 0.236 \times 10^{-3}\ rad = 0.81'$$

***Example 25.5*** *A venturi tube of throat diameter 60 mm is placed in a water pipe of diameter 100 mm to measure the volumetric flow. The volumetric flow rate through the tube is 0.08 m³/s and the water has a density of 10³ kg/m³ and viscosity of 10⁻³ Ns/m².*

*(a) Determine the Reynold's number for these conditions.*

25

(b) *The co-efficient of discharge is 0.99, determine the upstream-to-throat differential pressure.*

(c) *The differential pressure calculated in (b) is applied across an unstressed flat diaphragm. The diaphragm is made of mild steel for which Young's modulus is 206 GN/m$^2$ and Poisson's ratio = 0.28, and has a diameter of 10 mm and a thickness of 0.2 mm. Determine the deflection of the diaphragm.*

**Solution.**

(*a*) Area of throat

$$= (\pi/4)\times(60\times10^{-3})^2 = 2827\times10^{-6}\,\text{m}^2$$

Velocity through throat

$$v = \frac{0.08}{2827\times10^{-6}} = 28.3\text{ m/s}$$

From Eqn. 25.21, Reynold's number is,

$$\text{Re} = \frac{vdP}{\eta} = \frac{28.3\times60\times10^{-3}\times10^{-3}}{10^{-3}} = 1.7\times10^6$$

(*b*) From 25.23, velocity approach factor,

$$E = \sqrt{1-(A_x/A_y)^2} = \sqrt{1-(60/100)^2} = 1.25$$

From Eqn. 25.22, flow rate,

$$Q = CEB\,A_y\sqrt{2\Delta P/\rho_x}$$

The value of expansion factor *B* for liquids is 1

$$\therefore\ Q = 0.99\times1.25\times1\times\frac{\pi}{4}(60\times10^{-3})^2\sqrt{\frac{2\Delta P}{1\times10^3}} = 0.08$$

∴ Differential pressure

$$\Delta P = 261\text{ kN/m}^2$$

(*c*) The deflection at the centre of the diaphragm is given by Eqn. 25.17

$$d_m = \frac{3P(1-v^2)D^4}{256\,Et^3}$$

$$= \frac{3\times261\times10^3(1-0.28^2)\times(10\times10^{-3})^4}{256\times206\times10^9\times(0.2\times10^{-3})^3}$$

$$= 17.1\times10^{-6}\text{ m} = 17.1\ \mu\text{m}$$

**Example 25.6** *A Pitot tube is used for measurement of velocity of flow of water having a density of 1000 kg/m$^3$.*

(a) *Determine the velocity of flow at the head of the pitot tube if it produces a differential pressure of 10 kN/m$^2$ between its two outlets.*

(b) *The same differential pressure is obtained in air at an altitude where the density of air is 0.65 kg/m$^3$. Determine the velocity of air flow.*

**Solution.** (*a*) For a Pitot tube the velocity,

$$= \sqrt{\frac{2(P-P_0)}{\rho}} \qquad \text{(See. Eqn. 25.24)}$$

$$= \sqrt{\frac{2\times10\times10^3}{1000}} = 4.47\text{ m/s}$$

(*b*) Velocity of air

$$= \sqrt{\frac{2\times10\times10^3}{0.65}} = 175.4\text{ m/s}$$

A correction to the result obtained above has to be applied for taking into account the Mach number. The Mach number is the ratio of the velocity of an object moving through a gas, to the velocity of sound in air. The velocity of sound in air at atmospheric pressure is about 330 m/s, and therefore the Mach number for the air velocity being measured is 175.4/330 ≃ 05. At this subsonic Mach number, a small correction of about 3 per cent is applied. Therefore, the correct velocity of air is $(1-0.03)\times175.4 = 170$ m/s.

**Example 25.7** *Water is flowing with a depth of 0.9 m over a rectangular notch 1.2 m wide. Find the depth of flow over a right angled notch if the same were to pass over it. Assume that both the notches have the same co-efficient of discharge.*

**Solution.** For rectangular notch, the discharge is,

$$Q = \frac{2}{3}C_d\,L\sqrt{2g}\,H^{1.5}$$

$$= \frac{2}{3}C_d\times1.2\times\sqrt{2\times9.81}(0.9)^{1.5}$$

$$= 3.02\ C_d$$

For a *V* notch, the discharge is,

$$Q = \frac{8}{15}C_d\sqrt{2g}\,H^{2.5}\tan\frac{\theta}{2}$$

or $\quad 3.02\,C_d = \frac{8}{15}C_d\sqrt{2\times9.81}\,H^{2.5}\tan45°$

or $\quad H = 1.1$ m

**Example 25.8** *A right angled notch has a co-efficient of discharge of 0.6. If head above the still is 0.5 ± 0.01 m, calculate the discharge.*

**Solution.** For a right angled *V* notch, discharge is,

$$Q = \frac{8}{15}C_d\sqrt{2g}\,H^{2.5}$$

$$= \frac{8}{15}\times0.6\sqrt{2\times9.81}\times0.5^{2.5}$$

$$= 0.2506\text{ m}^3/\text{s}$$

Now $\frac{dQ}{dH} = \frac{8}{15} C_d \sqrt{2g} \times 2.5 \times H^{1.5}$

or $\frac{dQ}{Q} = 2.5 \frac{dH}{H}$

or $dQ = \pm 2.5 \times 0.2506 \times 0.01 / 0.5$

$= \pm 0.0153 \text{ m}^3/\text{s}$

Hence, the discharge can be written as 0.2506 ± 0.0153 $m^3$/s.

## 25.6 TRANSDUCERS

An electronic instrumentation system consists of a number of components to perform a measurement and record its results. As explained earlier a generalized measurement system consists of *three* major components.

(*i*) an input device,

(*ii*) a signal conditioning or processing device, and

(*iii*) an output device.

The input device receives the measurand or the quantity under measurement and delivers a proportional or analogous electrical signal to the signal conditioning device. Here the signal is amplified, attenuated, filtered, modulated, or otherwise modified in format acceptable to the output device.

The input quantity for most instrumentation systems is a "*non-electrical quantity*". In order to use electrical methods and techniques for measurement, manipulation or control, the non-electrical quantity is generally converted into an electrical form by a device called a "*transducer*". We can define a **transducer** as a device which, when actuated transforms energy from one form to another.

The broad definition of a **transducer** includes, for example, devices which convert *mechanical force* into an electrical signal. These devices form a very large and important group of transducers commonly used in industrial instrumentation area. The instrumentation engineers and technologists are therefore primarily concerned with this area of instrumentation. Many other *physical* parameters such as heat, intensity of light, flow rate, liquid level, humidity and pH value may also be converted into electrical form by means of transducers. These transducers provide an output signal when stimulated by a mechanical or a non-mechanical input : a photoconductor converts light intensity into change of resistance, a thermocouple converts heat energy into electrical voltage, a force produces a change of resistance in a strain gauge, an acceleration produces a voltage in a piezo-electrical crystal and so on. In all cases, however, the electrical output is measured by *standard* methods, giving the magnitude of the input quantity in terms of an analogous output.

## 25.7 ELECTRIC TRANSDUCERS

The art of electrical measurements has been chiefly used for measurement of electrical quantities but its value in making measurements of non-electrical quantities in this new era of automation is rapidly growing. In order to measure non-electrical quantities a detector is used which usually converts the physical quantity into a displacement. This displacement actuates an **electric transducer**, which acting as a secondary transducer, gives an output that is electrical in nature. The electrical quantity so produced is measured by standard methods used for electrical measurements. The result (electrical output) gives the magnitude of the physical quantity or condition being measured.

The electrical signal may be a current or a voltage or a frequency and production of these signals is based upon electrical effects which may be resistive, capacitive, inductive etc. in nature.

The first stage of a measurement system may simply be called a **transducer stage** instead of **detector transducer stage** by redefining a **transducer**. A **transducer**, in general form, may be defined as a device which converts energy from one form to another. However, this definition has to be restricted, many a time especially in the field of **electrical instrumentation**. Keeping this restriction in view, a **transducer** may be defined as a device which converts a physical quantity or a physical condition into an electrical signal. Another name for a transducer is **pick up.**

25

### 25.7.1 Advantages of Electrical Transducers

There are a number of transducers which transform a variety of physical quantities and phenomena into electrical signals. The reasons for transforming a physical phenomenon into electrical form are numerous. The advantages of converting physical quantities into analogous electrical quantities are :

(*i*) Electrical amplification and attenuation can be done easily and that too with static devices.

(*ii*) The mass-inertia effects are minimised. In fact, when dealing with electrical or electronic signals, the inertia effects are due to electrons which have negligible mass. In many situations, we do not come across mass or inertia problems at all.

(*iii*) The effects of friction are minimised.

(*iv*) The electrical or electronic systems can be controlled with a very small power level.

(*v*) The electrical output can be easily used, transmitted and processed for the purpose of measurement.

(*vi*) Telemetry is used in almost all sophisticated measurement systems. The entire aerospace research and development is based upon telemetry and remote control. The ever-enlarging field of radio monitoring in space research has left us with no alternative but to resort to electronic means. This completely eliminates the data transmission through mechanical means and hence electrical and electronic principles have to be employed for these conditions. The remote indication or recording is an essential part of modern day instrumentation technology.

(*vii*) There has been an explosive development in the field of electronic components and devices. This development is on account of the fact that electronic devices are very amenable to miniaturization. Components which are compact, have always an, advantage. **Miniaturization** on account of use of **ICs** (integrated Circuits) has completely revolutionised the field of instrumentation.

In short, it can be said that the reasons for transforming a physical phenomenon into electrical form is that the electrical output can be easily used, transmitted and processed for the purposes of measurement. Modern digital computers make the use of these transducers absolutely essential.

When the definition of **transducer** is confined to a device that covers the entire **detector transducer stage,** wherein the transducer converts a non-electrical quantity into an analogous electrical signal, the transducer may be thought of consisting of two important and closely related parts. These *two* parts are :

- Sensing Element, and
- Transduction Element.

In addition, there may be many other auxiliary parts, such as amplifiers and other signal processing equipment, power supplies, calibrating and reference sources, and mechanical mounting features.

**1. *Sensing or Detector Element.*** A detector or a sensing element is that part of a transducer which responds to a physical phenomenon or a change in a physical phenomenon. The response of the sensing element must be closely related to the physical phenomenon.

**2. *Transduction element.*** A transduction element transforms the output of a sensing element to an electrical output. The **transduction element,** in a way, acts as a **secondary transducer.**

## 25.8 CLASSIFICATION OF TRANSDUCERS

The transducers can be classified

(*i*) on the basis of transduction form used,

(*ii*) as primary and secondary transducers,

(*iii*) as passive and active transducers,

(*iv*) as analog and digital transducers, and

(*v*) as transducers and inverse transducers.

### 25.8.1 Classification based upon Principle of Transduction

The transducers can be classified on the basis of principle of transduction as resistive, inductive, capacitive etc. depending upon how they convert the input quantity into resistance, inductance or capacitance respectively. They can be classified as piezoelectric, thermoelectric, magnetorestrictive, electrokinetic and optical. Table 25.2 lists the classification of transducers based upon principle of transduction.

### 25.8.2 Primary and Secondary Transducers

Let us consider the case of a Bourdon's tube as shown in Fig. 25.26. The Bourdon tube acting as a primary detector senses the pressure and converts the pressure into a displacement of its free end. The displacement of the free end moves the core of a linear variable differential transformer, (L.V.D.T.) which produces an output voltage which is proportional to the movement of the core, which is proportional to the displacement of the free end which in turn is proportional to the pressure. Thus, there are two stages of transduction, firstly the pressure is converted into a displacement by Bourdon tube then the displacement, is converted into an analogous voltage by L.V.D.T. The Bourdon tube is called a **"Primary Transducer"** while the L.V.D.T. is called a **"Secondary Transducer".**

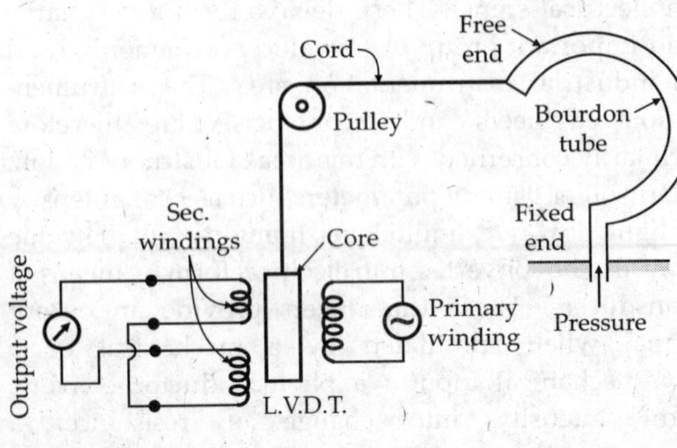


**Fig. 25.26** Measurement of pressure using Bourdon tube an L.V.D.T. (Example of primary and Secondary transducers).

**Table 25.2** *Types of Electrical Transducers*

| *Electrical parameter and class of transducer* | *Principle of Operation* | *Typical applications* |
|---|---|---|
| | *Passive transducers (externally powered)* | |
| **Resistance** | | |
| Potentiometer device | Positioning of the slider by an external force varies the resistance in a potentiometer or a bridge circuit. | Pressure, displacement. |
| Resistance strain gauge | Resistance of a wire or semiconductor is changed by elongation or compression due to externally applied stress. | Force, torque, displacement. |
| Pirani gauge or hot wire meter | Resistance of heating element is varied by convection cooling of a stream of gas. | Gas flow, gas pressure. |
| Resistance thermometer | Resistance of pure metal wire with a large positive temperature co-efficient of resistance varies with temperature. | Temperature, radiant heat |
| Thermistor | Resistance of certain metal oxides with negative temperature coefficient of resistance varies with temperature. | Temperature, flow |
| Resistance hygrometer | Resistance of a conductive strip changes with moisture content. | Relative humidity. |
| Photoconductive cell | Resistance of the cell as a circuit element varies with incident light. | Photosensitive relay. |
| **Capacitance** | | |
| Variable capacitance pressure gauge | Distance between two parallel plates is varied by an externally applied force. | Displacement, pressure. |
| Capacitor microphone | Sound pressure varies the capacitance between a fixed plate and a movable diaphragm. | Speech, music, noise. |
| Dielectric gauge | Variation in capacitance by changes in the dielectric or dielectric constant. | Liquid level, thickness. |
| **Inductance** | | |
| Magnetic circuit transducer | Self-inductance or mutual inductance of a.c. excited coil is varied by changes in the magnetic circuit. | Pressure, displacement. |
| Reluctance pick up | Reluctance of the magnetic circuits is varied by changing the position of the iron core of coil. | Pressure, displacement, vibrations, position. |
| Differential transformer | The differential voltage of two secondary windings of a transformer is varied by positioning the magnetic core through an externally applied force. | Pressure, force, displacement, position |
| Eddy current gauge | Inductance of a coil is varied by the proximity of an eddy current plate. | Displacement, thickness. |
| Magnetostriction gauge | Magnetic properties are varied by pressure and stress. | Force, pressure, sound. |
| **Voltage and Current** | | |
| Hall effect pickup | A potential difference is generated across a semiconductor plate (germanium) when magnetic flux interacts with an applied current. | Magnetic flux, current, power. |
| Ionization chamber | Electron flow induced by ionization of gas due to radio-active radiation. | Particle counting, radiation. |
| Photoemissive cell | Electron emission due to incident radiation upon photoemissive surface. | Light and radiation. |
| Photomultiplier tube | Secondary electron emission due to incident radiation on photosensitive cathode. | Light and radiation, photosensitive relays. |
| | *Self-generating transducers (no external power)* | |
| Thermocouple and thermopile | An emf is generated across the junction of two dissimilar metals or semiconductors when that junction is heated. | Temperature, heat flow, radiation. |
| Moving coil generator | Motion of a coil in a magnetic field generates a voltage. | Velocity, vibrations. |
| Piezoelectric pickup | An emf is generated when an external force is applied to certain crystalline materials, such as quartz. | Sound, vibrations, acceleration, pressure changes. |
| Photovoltaic | A voltage is generated in a semiconductor junction device when radiant energy stimulates the cell. | Light meter, solar cell. |

Let us take another example which is the case of measurement of a compressive force with the help of a load cell in conjunction with strain gauges as shown in Fig. 25.27. (Load cells and strain gauges are explained later in this chapter.)

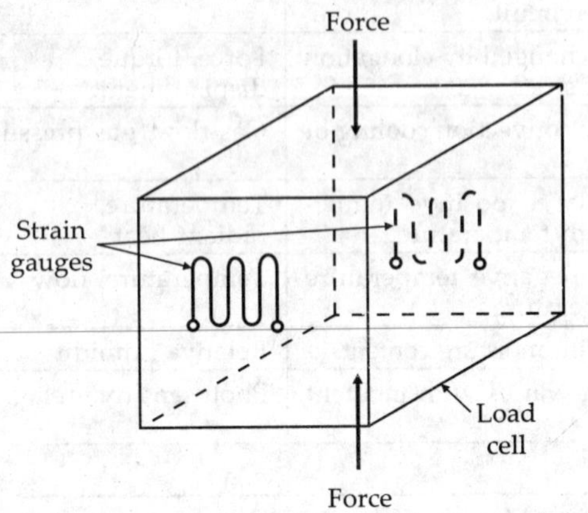


**Fig. 25.27** Force measurement with load cell and strain gauges.

The load cell is a short column or a strut with resistance wire strain gauges bonded to it. The measurand, in this case, is a force and is applied to the column thereby producing strain. The force is first detected by the column and is converted into strain which is a mechanical displacement. The higher the force, the higher is the strain and thus the input signal (force) is converted into an analogous output (strain). This strain changes the resistance of the strain gauges. Thus we have an output which is a change in the value of resistance *i.e.*, electrical in form. Hence, in this case, it takes two processes to convert an input into an analogous output. The first process involves conversion of force into mechanical displacement which is done by the column, while the second process involves conversion of mechanical displacement into change of resistance which is done by strain gauges.

Thus we see that the force is *detected* by the column in the first stage and hence it is called a **Primary Transducer**. The output signal from the primary transducer is converted subsequently into a usable output by the strain gauges and therefore they are known as **Secondary Transducers.**

In most of measurement systems, there is a suitable working combination wherein a **Mechanical device** acts as a **primary detector transducer** and the **electrical device** acts as the **secondary transducer** with **mechanical displacement** serving as the **intermediate signal**

### 25.8.3 Passive and Active Transducers

Transducers may be classified according to whether they are *passive* or *active*.

**1.** ***Passive Transducers.*** Passive transducers derive the power required for transduction from an auxiliary power source. They also derive part of the power required for conversion from the physical quantity under measurement. They are also known as *"externally powered transducers"*. Typical examples of passive transducers are resistive, inductive and capacitive transducers.

A typical example of a passive transducer is a 'POT' which is used for measurement of displacement. A 'POT' is a resistive transducer powered by a source voltage $e_i$ as shown in Fig. 25.28. This 'POT' is used for measurement of linear displacement $x_i$.

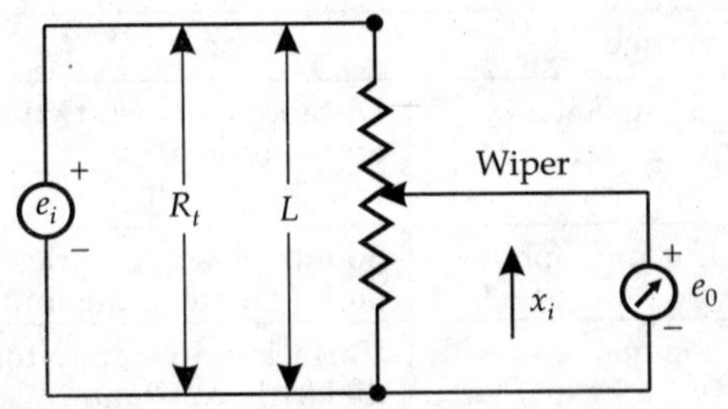


**Fig. 25.28** Linear potentiometer (POT), a passive transducer.

Suppose $L$ is the total length of potentiometer whose total resistance $R_t$. The input displacement is $x_i$.

∴ Output voltage

$$e_0 = \frac{x_i}{L}e_i \quad \text{or} \quad x_i = \left(\frac{e_0}{e_i}\right)L$$

In the absence of external power, the transducer cannot work and it hence is called a *passive transducer.*

**2.** ***Active Transducers.*** Active transducers are those which do not require an auxiliary power source to produce their output. They are also known as *self generating type* since they develop their own voltage or current output. The energy required for production of output signal is obtained from the physical quantity being measured.

Velocity, temperature, light intensity and force can be transduced with the help of active transducers. These transducers include tachogenerators, thermocouples, photovoltaic cells and piezoelectric crystals. Consider the case of a piezoelectric crystal used for measurement of acceleration as shown in Fig. 25.29 The crystal is sandwiched between two metallic electrodes, and the entire sandwich is fastened to a base which may be the floor of a rocket. A fixed mass is placed on the top of the sandwich.

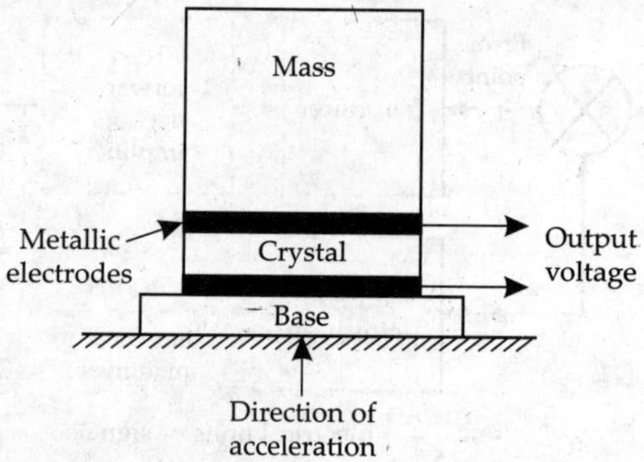


**Fig. 25.29** Piezo-electric crystral measuring acceleration – an active transducer.

The property of the piezo-electric crystals is that when a force is applied to them, they produce an output voltage. The mass exerts a certain force on account of acceleration on the crystal due to which a voltage is generated. The acceleration is applied to the base, due to which the mass produces a force. The mass being fixed, the force is proportional to acceleration. The voltage output is proportional force and hence is proportional to acceleration (the mass being fixed).

It should be noted from above that this transducer called *"accelerometer"* which converts acceleration into electrical voltage does not need any auxiliary power source to convert a physical phenomenon (acceleration in this case) to an electrical output (voltage in this case) and therefore is an *active transducer*.

### 25.8.4 Analog and Digital Transducers

The transducers can be classified on the basis of the output which may be a continuous function of time or the output may be in discrete steps.

**1.** ***Analog Transducers***. These transducers convert the input quantity into an analog output which is a continuous function of time. Thus a strain gauge, an L.V.D.T., a thermocouple or a thermistor may be called as *"Analog Transducers"* as they give an output which is a continuous function of time.

**2.** ***Digital Transducers.*** These transducers convert the input quantity into an electrical output which is in the form of pulses.

As the binary system uses only two symbols 0 and 1 it can be easily represented by opaque and transparent areas on a *glass scale* or non-conducting and conducting areas on a *metal scale.* A scale constructed to show the linear position on a movable object and having five digits is shown in Fig. 25.30. The complete binary number denoting position is obtained by scanning the pattern across the scale at a stationary index mark.

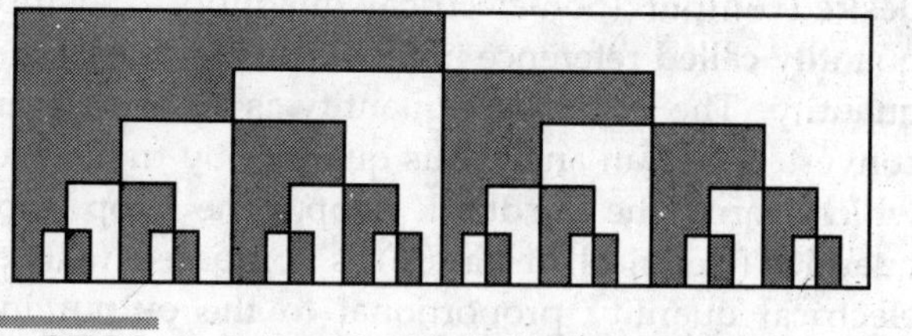

**Fig. 25.30** 5-digit scale for digital indication of linear position of a movable object.

Glass scales can be read optically by means of a light source, an optical system and photocells. Metal scales are scanned by brushes making electrical contact with individual tracks.

The resolution depends upon the digits comprising the binary number and is $1/2n$ of full scale where $n$ is the number of digits.

### 25.8.5 Transducers and Inverse Transducers

There is a strong association of control with measurement. The basic requirement for control of physical quantities such as position, speed, temperature, pressure and flow rate in ah industrial planet is the ability to measure these quantities. The control action is only possible if the physical quantity can be measured. For example, in a position control system called *"servomechanism"*, it is desired to control the position of a shaft. This requires an accurate method for measurement of the shaft position in order that its position be accurately controlled. Further, if it is desired that the shaft be accelerated in a controlled manner, then the position measuring device must be able to measure shaft position for rapid changes *i.e.*, the device must have a fast dynamic response.

Figure 25.31 shows the block diagram of closed loop control system. The controlled (output) quantity is usually a non electrical quantity. The control action is through an input quantity that corresponds to the

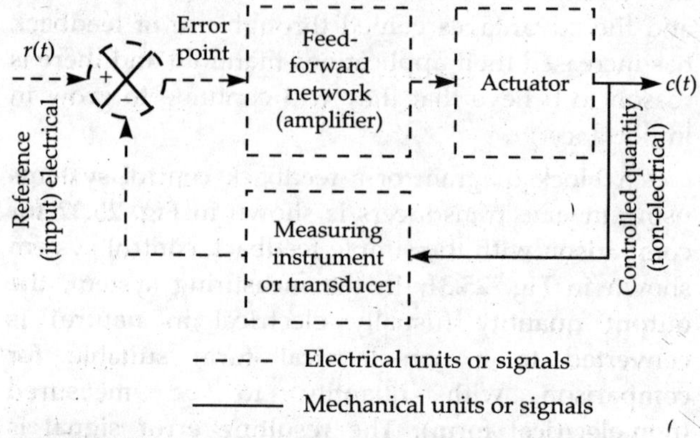


**Fig. 25.31** Feedback control system for control of non-electrical quantities.

25

desired output (non-electrical quantity). The input quantity called reference input is usually an electrical quantity. The controlled quantity is measured and converted into an analogous quantity by transducers which form the feedback loop. The loop input quantity (electrical in nature) is compared with the electrical quantity proportional to the output in a *comparator*. In case, the two are not equal an error signal is produced. This error signal is amplified and applied to an *actuator* in the forward path, which corrects the output quantity till the output quantity reaches the desired level.

**1. *Transducers.*** A transducer can be broadly defined as a device which converts a non-electrical quantity into an electrical quantity.

**2. *Inverse Transducers.*** An *inverse transducer* is defined as a device which converts an electrical quantity into a non-electrical quantity. It is a precision actuator which has an electrical input and a low power non-electrical output. A piezoelectric crystal acts as an inverse transducer because when a voltage is applied across its surfaces, it changes its dimensions causing a mechanical displacement.

A current carrying coil moving in a magnetic field is also an inverse transducer because current carried by it is converted into a force which causes translational or rotational displacement. Many data indicating and recording devices are inverse transducers. An analog ammeter or voltmeter converts current into mechanical displacement. However, such devices which include instruments like indicating instruments, pen recorders, oscilloscopes that convert the electrical signals to a mechanical movement are placed at the output stage (data presentation stage) are called *output transducers.*

The most useful application of inverse transducers is in feedback measuring systems. The development of transducers and inverse transducers, and the advantages gained through use of feedback has increased their applications manifold and there is reason to believe that they will continue to grow in importance.

A block diagram of a feedback control systems using inverse transducers is shown in Fig. 25.32 for comparison with the simple feedback control system shown in Fig. 25.31. In the measuring system, the output quantity (usually electrical in nature) is converted to a non-electrical form suitable for comparison with quantity to be measured (non-electrical form). The resulting error signal is usually transduced into electrical form and amplified to give output indication.

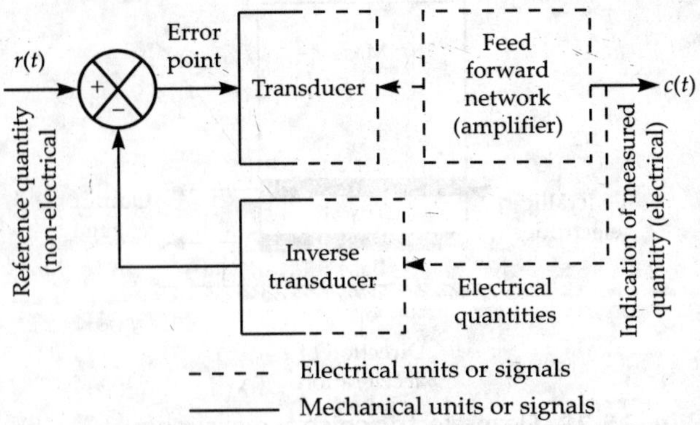


**Fig. 25.32** Feedback control system for control of electrical quantities using inverse transducer.

It is evident that the operation of the feedback is similar for both control and measuring systems as regards accuracy and stability. The essential differences are in power level of the systems and the performance requirements of transducing elements. The control systems handle relatively high power levels through the actuator to directly determine the output quantity. Therefore, the actuator may have poor regulation without affecting the control action. The transducer and noise at the amplifier input largely determine the accuracy of control. On the other hand, the feedback measuring system involves use of fairly low power devices at the output. These devices include the indicator and inverse transducer. However, the inverse transducer essentially determines the characteristics of the system. The noise connected with the transducer and amplifier input stage is very important.

## 25.9 CHARACTERISTICS AND CHOICE OF TRANSDUCERS

When choosing a transducer for any application the *input, transfer* and *output characteristics* have to be taken into account.

### 25.9.1 Input Characteristics

**1. *Type of Input and Operating Range.*** The foremost consideration for the choice of a transducer is the input quantity it is going to measure and its operating range. The type of input, which can be any physical quantity, is generally determined in advance. A physical quantity may be measured through use of a number of transducers. However, the choice of a particular transducer that is selected for the purpose, depends upon the useful range of input quantity over which the transducer can be used. The useful operating range of the transducer may be a decisive factor in selection of a transducer for a particular application. The upper limit is decided by the transducer capabilities while the lower limit of range is normally determined by the transducer error or by the unavoidable noise originating in the transducer. In fact the transducer should maintain a good resolution throughout its operating range.

**2. *Loading Effects.*** Ideally a transducer should have no loading effect on the input quantity being measured. In theory, it is impossible, although in practice steps may be taken to reduce the loading effects to negligible proportions. The magnitude of the loading effects can be expressed in terms of force, power or energy extracted from the quantity under measurement for working of the transducers. Therefore, the transducer, that is selected for a particular application should ideally extract no force, power or energy from the quantity under measurement in order that the latter is measured accurately.

## 25.9.2 Transfer Characteristics

The transfer characteristics of transducers require attention of *three* separate elements, *viz.*,

- transfer function
- error, and
- response of transducer to environmental influences.

### 25.9.2A Transfer Function

The *transfer function* of a transducer defines a relationship between the input quantity and the output. The transfer function is,

$$q_o = f(q_i) \qquad \ldots(25.32)$$

where $q_o$ and $q_i$ are respectively output and input of the transducer.

The *sensitivity* of a transducer is defined as the differential quotient,

$$S = \frac{dq_o}{dq_i} \qquad \ldots(25.33)$$

In general, the sensitivity of transducers is not constant but is dependent upon the quantity $q_i$.

However, in some cases the relationship between the output $q_o$ and input $q_i$, is linear. In that case the sensitivity remains constant over the entire range of the transducer and is defined as,

$$S' = \frac{dq_o}{dq_i} = \frac{dq_o}{dq_i} \qquad \ldots(25.34)$$

The *scale factor* is defined as the inverse of sensitivity and is therefore,

$$\text{scale factor} = \frac{1}{S} = \frac{dq_o}{dq_i} \qquad \ldots(25.35)$$

Some manufacturers specify the scale factor and still call it sensitivity.

### 25.9.2B Error

The errors in transducers occur because they do not follow, in many situations the input-output relationship given by $q_o = f(q_i)$. Any departure from the above relationship results in errors. For example, the output on account of input, $q_i$ has to be $q_o$ but in practice an output $q_o$ is obtained, then the error of the instrument is,

$$\varepsilon = q_o' - q_o \qquad \ldots(25.36)$$

The error, $\varepsilon$ can be expressed in terms of either input or the output quantity.

The error analysis has already been given in Chapter 3. The error can be split into three components, which are,

- scale error
- dynamic error, and
- error on account of noise and drift.

**Scale Error**

The scale error comprises of four different types of errors :

(*i*) ***Zero Error.*** In this case the output deviates from the correct value by a constant factor over the entire range of the transducer. This is shown in Fig. 25.33.

25

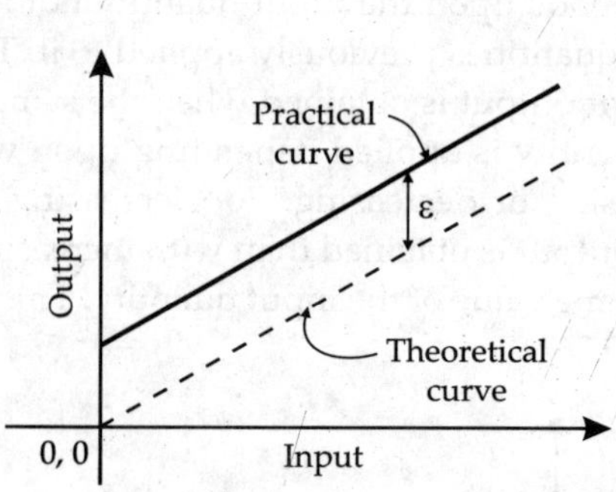


**Fig. 25.33** Transducer zero error.

(*ii*) ***Sensitivity Error.*** Sensitivity error occurs where the observed output deviates from the correct value by a constant value. Suppose the correct output is $q_o$, the output would be $Kq_o$, over the entire range of the transducer, where $K$ is a constant. This is shown in Fig. 25.34.

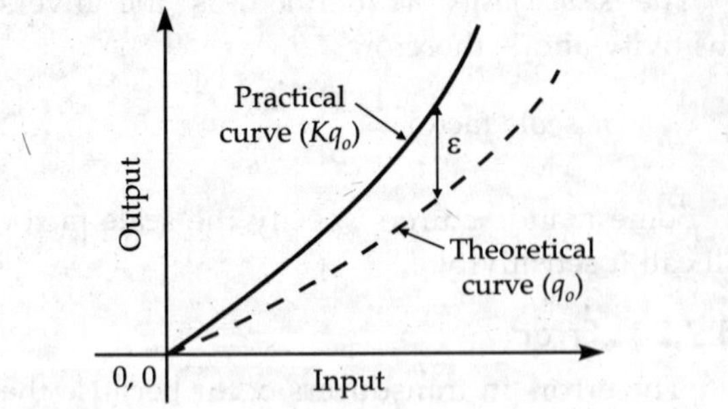


**Fig. 25.34** Transducer sensitivity error.

(*iii*) ***Non-conformity.*** This pertains to a case in which the experimentally obtained transfer function deviates from the theoretical transfer function for almost every input. This is shown in Fig. 25.35.

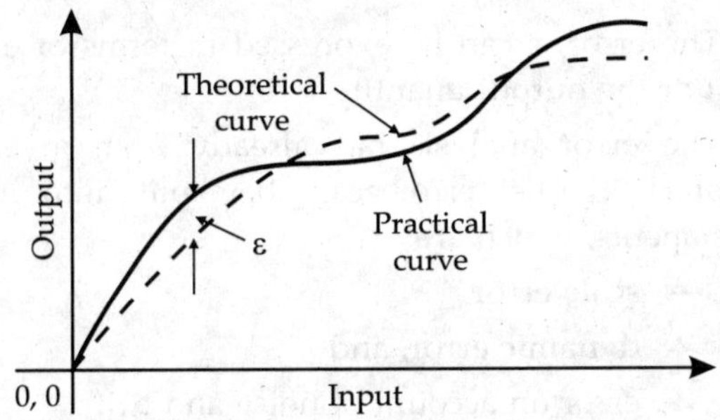


**Fig. 25.35** Transducer Non-conformity.

In the special case of a theoretical linear relationship between input and output quantities, this error is called *non-linearity* or *non-linear distortion*.

(*iv*) ***Hysteresis.*** All transducers are subject to the effects of hysteresis. The output of a transducer not only depends upon the input quantity but also upon in input quantities previously applied to it. Therefore, a different output is obtained when the same value of input quantity is applied depending upon whether it is increasing or decreasing. For decreasing values, a greater output is obtained than with increasing values for the same value of the input quantity. This is shown in Fig. 25.36.

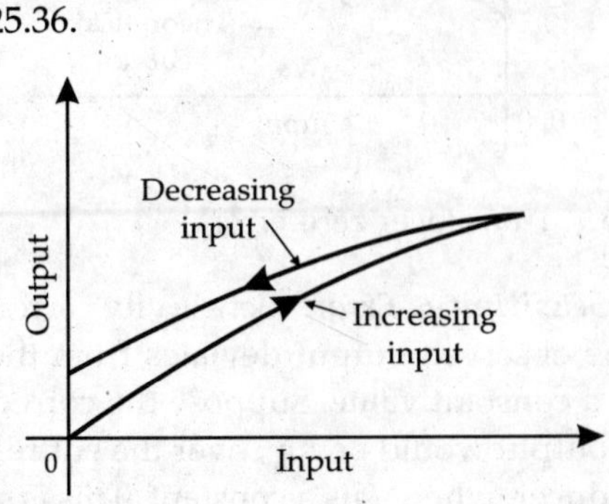


**Fig. 25.36** Transducer hysteresis.

## Dynamic Error

Dynamic errors occur only when the input quantity is varying with time. This is on account of the fact the systems contain energy storage elements and due to this the output cannot follow the input exactly but with a time lag.

Let us consider an *R-C* series circuit to which a step input of magnitude *E* is applied, the voltage across the capacitor after a time, *t* of application of step input is,

$$e_c = E[1 - \exp(-t/\tau)]$$

where $\tau$ = time constant = *RC*

The response is shown in Fig. 25.37.

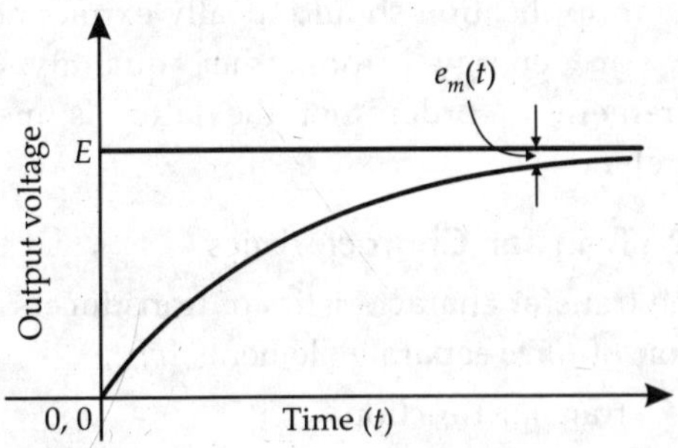


**Fig. 25.37** Response of *RC* circuit to a step input.

The dynamic or measurement error is,

$$e_m(t) = E - e_c = \exp(-t/\tau)$$

The *dynamic or measurement error* can be made small by having a small time constant. It should be understood that the dynamic error is a function of time. As the time after application of the input increases, the dynamic error reduces. The steady state error,

$$\underset{t\to\infty}{e_{ss} = e_m(t)} = \underset{t\to\infty}{e^{-t/\tau}} = 0$$

in this particular case.

## Errors due to Noise and Drift

Noise and drift signals originating from the transducers vary with time and are superimposed on the output signal. The difference between noise and drift is that *noise* consists of a signal of random amplitude and random frequency whilst *drift* is a slow change with time. The magnitude of the noise and drift is normally independent of the magnitude of the input signal.

### Errors due to Change of Frequency (Frequency Response)

*Frequency response* and the high frequency cut off are the two specifications that describe the response of a transducer to a variable frequency sine wave input applied to it. For a reasonably, linear transducer, a sine wave input yields a sine wave output. As the frequency of the sine wave input is increased, the transducer is required to respond more and more quickly. Ultimately, beyond a particular frequency the transducer can no larger respond as rapidly as its sinusoidal input is changing. So the output of the transducer becomes smaller and also the phase shift between the input and output increases. Thus as the frequency increases the output of the transducer falls. This *roll off* of amplitude of output with an increase in

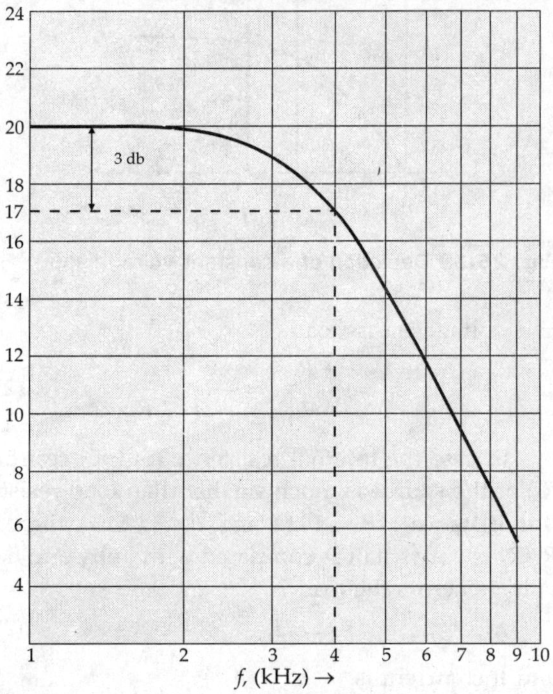


**Fig. 25.38** Frequency response plot.

input frequency is the *frequency response*. The frequency response of a transducer is shown in Fig. 25.38. The vertical axis is the gain of the transducer. This may be simply the output to input ratio. However, the gain is specified in terms of decibels.

$$\text{Gain in db} = 20 \log (\text{output/input}) = 20 \log (q_o/q_i) \quad ...(25.37)$$

The horizontal axis represents the frequency, and is plotted on a logarithmic scale in order to cover a wide range of frequencies.

The high frequency, cutoff ($f_h$) is the frequency at which the transducers gain has fallen to 0.707 of the value it had at its stable low frequency. If the vertical axis is scaled in db, $f_h$ occurs when the gain has decreased by –3 db. Figure 25.38 shows that the low frequency gain is 20 db but drops of 17 db (20 – 3 = 17 db) at a frequency of 4 kHz. Therefore, the high frequency cutoff takes place at 4 kHz in this case.

The high frequency cutoff can be related to the rise time by the following relationship :

$$t_r = 0.35/fh \quad ...(25.38)$$

Therefore, even if the transient response specifications are not given, the speed of a transducer can be determined from its high frequency cutoff.

### 25.9.3 Transducer Response

The response of the transducer to environmental influences is of a great importance. This is often given insufficient attention when choosing the best transducer for a particular measurement. This gives rise to results that are not as accurate as expected, or, worse, results that are accepted as more accurate than they actually are. The performance of the transducer is fully defined by its transfer function and errors, provided that the transducer is in constant environments and not subject to any disturbances like stray electromagnetic and electrostatic fields, mechanical shocks and vibrations temperature changes, pressure and humidity changes, changes in supply voltage and improper mechanical mountings. If transducers are subjected to the above environmental disturbances, which they are, precautions are taken, so that changes in transfer function and resulting errors therefrom do not occur.

25

Therefore, the transducer selected must be guarded against the interfering and modifying inputs.

### 25.9.4 Output Characteristics

The three conditions in the output characteristics which should be considered are,

- Type of electrical output,
- Output impedance, and
- Useful ouput range.

### Type of Electrical Output

The types of output which may be available from the transducers may be a voltage, current, impedance or a time function of these amplitudes. These output quantities may or may not be acceptable to the latter stages of the instrumentation system. They may have to be manipulated *i.e.*, their magnitudes changed or they may have to be changed in their format by signal conditioning equipment so as to make them drive the subsequent stages of instrumentation system.

## Output Impedance

The output impedance, $\mathbf{Z_o}$, of a transducer determines to the extent the subsequent stages of instrumentation is loaded. Ideally, the value of output impedance should be zero if no loading effects are there on the subsequent stage. However, the output impedance, $\mathbf{Z_o}$, cannot made equal to zero and therefore, its value should be kept as low as possible to minimise the loading effects.

The output impedance determines the amount of power that can be transferred to the succeeding stages of the instrumentation system for a given output signal level. If the output impedance is low compared to the forward impedance of the system, the transducer has the characteristics of a *constant voltage source* (provided the output of the transducer is a voltage), while in case the forward impedance is high as compared with the output impedance of transducer, it behaves as *constant current source.*

This can be explained as under :

An ideal constant voltage source, furnishes at its output terminals, a voltage that is independent of the load connected to the source or of the current drawn from the source. The voltage may or may not vary in time. In fact, it must vary in time, if it represents a signal which is true of transducers since their output is dependent upon the input signals which vary with time.

If a constant voltage source having a voltage of $E$ is [Fig. 25.39(*a*)] is applied across a resistance $R$, the current is $E/R$ as shown in Fig. 25.39(*c*). A constant voltage source in a strict sense does not exist ; the output voltage will always vary with the current drawn from the source.

This is because a practical voltage source has an internal resistance (output resistance), $R_o$, on account of which the voltage across the load resistance, $E_L$, decreases with increase of current or decrease of load resistance $R_L$.

Figure 25.39(*d*) shows a practical voltage source which consists of a voltage source $E$ in series with an internal resistance $R_o$, and a load resistance $R_L$.

The current in this circuit is given by,

$$I = \frac{E}{R_o + R_L} \quad \text{...(25.39)}$$

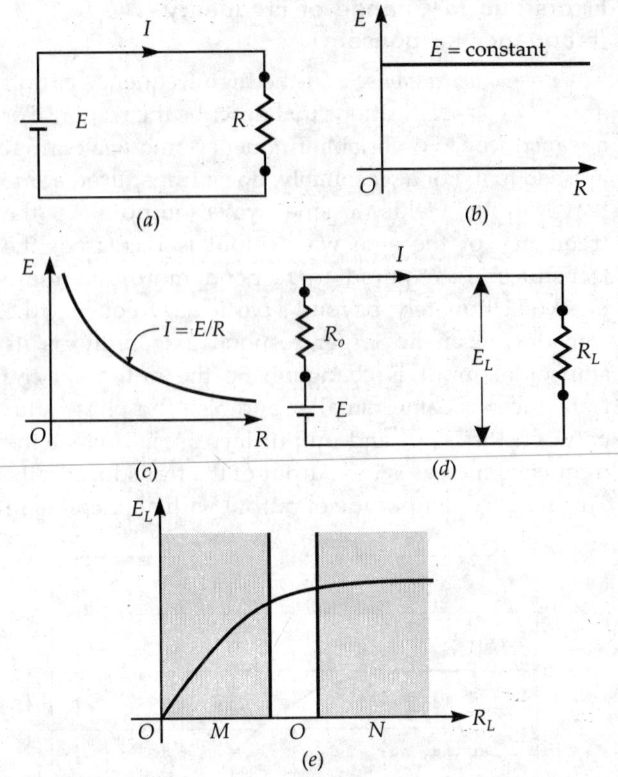


**Fig. 25.39** Depiction of a constant voltage source.

and voltage across load

$$E_L = \frac{E R_L}{R_o + R_L} = \frac{E}{1 + R_o / R_L} \quad \text{...(25.40)}$$

In case the internal resistance (output resistance) $R_o$, of the source is much smaller than load resistance (for instance : $R_o = 1\ \text{k}\Omega$ and $R_L = 1\ \text{M}\Omega$) then ratio $R_o/R_L$ is very small as compared with unity and hence, voltage across the load is,

$$E_L = E \quad \text{...(25.41)}$$

and the current is,

$$I = E / R_L \quad \text{...(25.42)}$$

Thus the voltage across the load is constant and the current is independent of the internal resistance $R_o$ of the source. Thus the source behaves as a constant voltage source as indicated in region $N$ of Fig. 25.39(*e*).

An ideal current source furnishes a current that is independent of the load resistance. It remains constant even if the load resistance varies. The current produced by a constant current source may or may not vary with time.

If a constant current passes through a resistor $R$, as shown in Fig. 25.40(*a*), a potential difference occurs between the terminals which is given by $E = IR$. In a strict sense, as in the case of constant voltage source, a constant current source does not exist. A practical current source contains a voltage source $E$ which causes the current. A constant current source has an internal resistance $R_o$, which is connected in parallel with the current source, and is very large as compared with the load resistance $R$. The current in such a system is determined primarily by magnitudes of $E$ and $R_L$.

The current through load resistance is,

$$I_L = \frac{IR_o}{R_o + R_L} = \frac{I}{1 + R_L / R_o} \quad ...(25.43)$$

If the internal resistance $R_o$ is very large as compared to the load resistance (for instance $R_o = 1\ k\Omega$ and $R_L = 1\ \Omega$) the ratio $R_L/R_o$ is very small so compared with $I$ and therefore, load current,

$$I_L = I \quad ...(25.44)$$

Therefore, the source behaves as a constant current source as depicted in region $M$ of Fig. 25.40($e$). There is a region $O$, between regions $M$ and $N$ where the source neither behaves as a constant voltage source or a constant current source.

Therefore, we conclude, if the output impedance of the transducer is low as compared with that of load impedance of the successive stages, it has the characteristics of a constant voltage source. On the other hand, if the output impedance of the transducer is much higher than that of the succeeding stages, it exhibits characteristics of a constant current source.

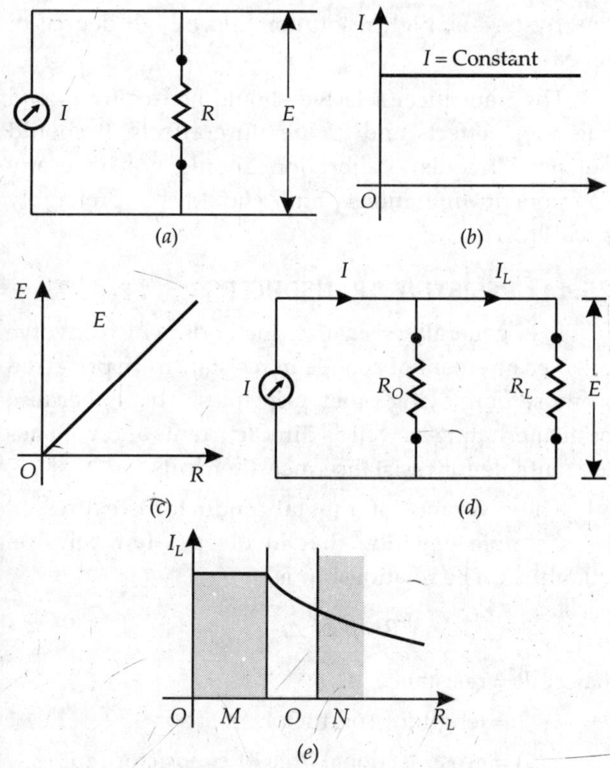


**Fig. 25.40** Depiction of constant current source.

When the output impedance of the transducer is equal to that of the following stages of instrumentation system, *matching* takes place and maximum power is transferred from the transducer to the succeeding stages. However, it must be understood that in case maximum power transfer takes place, when the output resistance of transducer, is equal to the resistance of the succeeding stages, the efficiency is only 50%. Also, when dealing with a.c. systems, the impedance matching takes place when the impedance of the load is $R_L - jX_L$ in case the output impedance is $R_L + jX_L$.

### Useful Output Range

The output range of a transducer is limited at the lower end by noise signals which may shroud the desired input signal. The upper limit is set by the maximum useful input level. The output range can be increased, in some cases, by the inclusion of amplifier in the transducer. However, the inclusion of an amplifier also increases the noise level and therefore in such situations the amplifier may not be of any use at all.

## 25.10 SUMMARY OF FACTORS INFLUENCING THE CHOICE OF TRANSDUCERS

25

In the following parts of this chapter, an attempt has been made to give the details and applications of the transducers for measurement of different physical quantities. It will be realised that there are many ways for measurement of a physical quantity, and in many cases, there is no best way. The transducers and the methods used may depend upon the instrumentation already available and also on the experience of the user.

Unfortunately most transducers are not sensitive to just one quantity. If measurements are to be made under conditions where there is likelihood of two or more input quantities influencing the transducer, it is desirable to select a transducer which is sensitive to the desirable quantity and insensitive to the unwanted quantity. If this is not possible, ways and means should be found to eliminate or compensate for the effects of the unwanted input quantity. The following is the summary of the factors influencing the choice of a transducer for measurement of a physical quantity.

**1. *Operating Principle.*** The transducers are many a times selected on the basis of operating principle used by them. The operating principles used may be resistive, inductive, capacitive, optoelectronic, piezo-electric etc.

**2. *Sensitivity.*** The transducer must be sensitive enough to produce detectable output.

**3. *Operating Range.*** The transducer should maintain the range requirements and have a good resolution over its entire range. The rating of the transducer should be sufficient so that it does not breakdown while working in its specified operating range.

**4. *Accuracy.*** High degree of accuracy is assured if the transducer does not require frequent calibration and has a small value for repeatability. It may be emphasised that in most industrial applications, repeatability is of considerably more importance than absolute accuracy.

**5. *Cross sensitivity.*** *Cross sensitivity* is a further factor to be taken into account when measuring mechanical quantities. There are situations where the actual quantity is being measured is in one plane and the transducer is subjected to variations in another plane. More than one promising transducer design has had to be abandoned because the sensitivity to variations of the measured quantity in a plane perpendicular to the required plane has been such as to give completely erroneous results when the transducer has been used in practice.

**6. *Errors.*** The transducer should maintain the expected input-output relationship as described by its transfer function so as to avoid errors.

**7. *Transient and Frequency Response.*** The transducer should meet the desired time domain specifications like peak overshoot, rise time, settling time and small dynamic error. It should ideally have a flat frequency response curve. In practice, however, there will be cutoff frequencies and higher cut off frequency should be high in order to have a wide bandwidth.

**8. *Loading Effects.*** The transducer should have a high input impedance and a low output impedance to avoid loading effects.

**9. *Environmental Compatibility.*** It should be assured that the transducer selected to work under specified environmental conditions maintains its input-output relationship and does not break down. For example, the transducer should remain operable under its temperature range. It should be able to work in corrosive environments (if the application so requires), should be able to withstand pressures and shocks and other interactions to which it is subjected to.

**10. *Insensitivity to Unwanted Signals.*** The transducer should be minimally sensitive to unwanted signals and highly sensitive to desired signals.

**11. *Usage and Ruggedness.*** The ruggedness both of mechanical and electrical intensities of transducer versus its size and weight must be considered while selecting a suitable transducer.

**12. *Electrical aspects.*** The electrical aspects that need consideration while selecting a transducer include the length and type of cable required.

Attention also must be paid to signal to noise ratio in case the transducer is to be used in conjunction with amplifiers. Frequency response limitations must also be taken into account.

**13. *Stability and Reliability.*** The transducer should exhibit a high degree of stability to be operative during its operation and storage life.

Reliability should be assured in case of failure of transducer in order that the functioning of the instrumentation system continues uninterrupted.

**14. *Static Characteristics.*** Apart from low static error, the transducers should have a low non-linearity, low hysteresis, high resolution and a high degree of repeatability.

The transducer selected should be free from load alignment effects and temperature effects. It should not need frequent calibration, should not have any component limitations, and should be preferably small in size.

## 25.11 RESISTIVE TRANSDUCERS

It is generally seen that methods which involve the measurement of change in resistance are preferred to those employing other principles. This is because both alternating as well as direct currents and voltages are suitable for resistance measurements.

The resistance of a metal conductor is expressed by a simple equation that involves a few physical quantities. The relationship is

$$R = \rho L / A,$$

where $R$ = resistance ; $\Omega$,

$L$ = length of conductor ; m,

$A$ = cross-sectional area of conductor ; $m^2$,

and $\rho$ = resistivity of conductor material ; $\Omega$-m

Any method of varying one of the quantities involved in the above relationship can be the design basis of an electrical **resistive transducer**. There are a number of ways in which resistance can be changed by a physical phenomenon. The translational and rotational potentiometers which work on the basis of change in the value of resistance with change in length of the conductor can be used for measurement of translational or rotary displacements. **Strain gauges** work on the principle that the resistance of a conductor or a semi-conductor changes when strained. This property can be used for measurement of displacement, force and pressure. The resistivity of materials changes with change of temperature thus causing a change of resistance. This property may be used for measurement of **temperature**. Thus electrical resistance transducers have a wide field of application.

## 25.12 POTENTIOMETERS

Basically a resistance potentiometer, or simply a **POT**, (a resistive potentiometer used for the purposes of voltage division is called a **POT**) consists of a resistive element provided with a sliding contact. This sliding contact is called a **wiper**. The motion of the sliding contact may be translatory or rotational. A linear pot and a rotary pot are shown in Figs. 25.41(*a*) and (*b*) respectively.

Some **POTS** use the combination of the two motions, *i.e.*, translational as well as rotational. These POTS have their resistive element in the form of a helix and therefore, they are called **helipots**.

The translational resistive elements are straight devices and have a stroke of 2 mm to 0.5 m. The rotational devices are circular in shape and are used for measurement of angular displacement. They may have a full scale angular displacement as small as 10°. A full single turn potentiometer may provide accurate measurements upto 357°. Multiturn potentiometers may measure upto 3500° of rotation through use of **helipots**.

The helical resistive elements are multiturn rotational devices which can be used for measurement of either translational or rotary motion. The resistive element of the POT may be excited by either d.c. or a.c. voltage. The POT is a **passive transducer** since it requires an external power source for its operation.

The resistive body of potentiometer may be wire wound. A very thin, 0.01 mm diameter of platinum or nickel alloy is carefully wound on an insulated former. The resistance elements are also made up from cermet, hot moulded carbon, carbon film and thin metal.

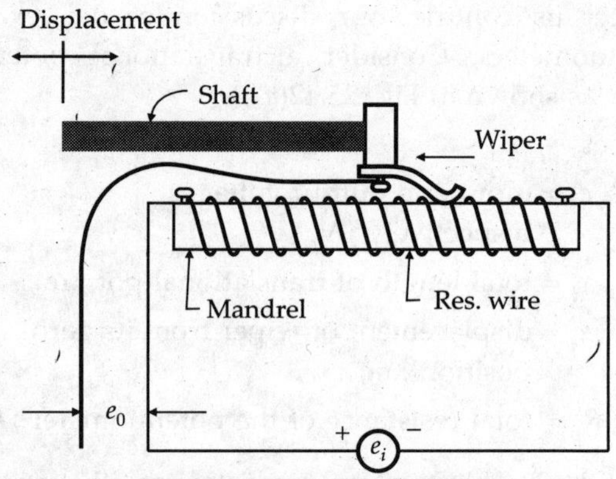


(*a*) Linear (translational) POT

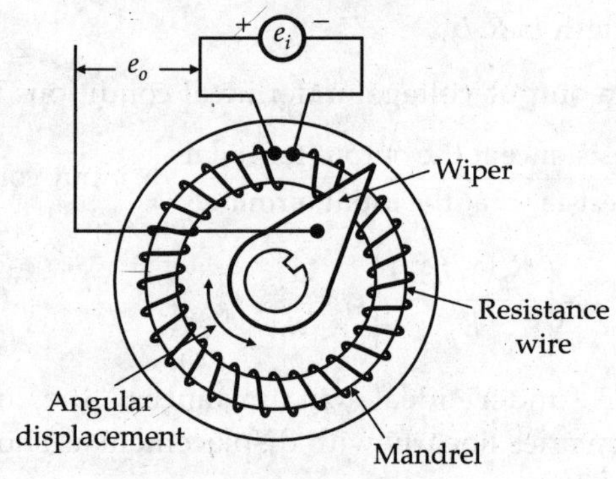


(*b*) Rotary POT

**Fig. 25.41** Resistive potentiometers (POTs)

Figure 25.42 shows the diagrams for translational, single turn rotational, and multiturn helix potentiometers.

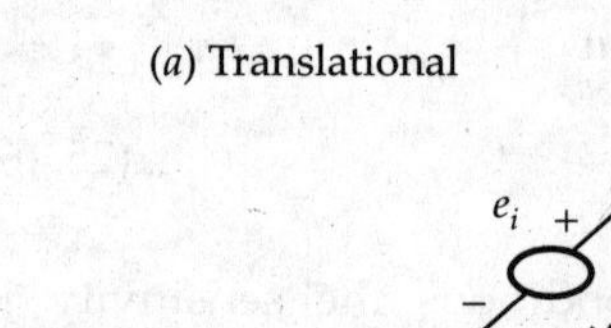


(*a*) Translational

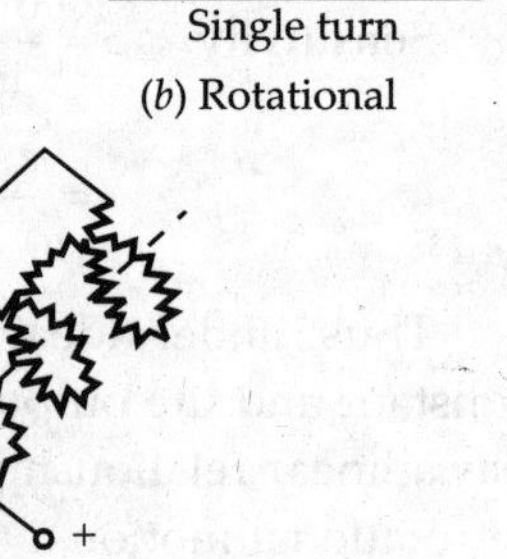


(*b*) Rotational

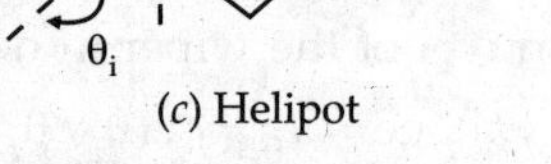


(*c*) Helipot

**Fig. 25.42** Diagrams for translational, rotational and helipots.

Let us confine, our discussion of d.c. excited potentiometers. Consider a translational potentiometer as shown in Fig. 25.42($a$).

Let,

$e_i$ & $e_0$ = input and output voltages respectively ; V,

$x_t$ = total length of translational pot ; m,

$x_i$ = displacement of wiper from its zero position ; m,

$R_p$ = total resistance of the potentiometer ; Ω

If the distribution of the resistance with respect to translational movement is linear, the resistance per unit length is $R_p/x_t$.

The output voltage under ideal conditions is :

$$e_0 = \left(\frac{\text{resistance at the output terminals}}{\text{resistance at the input terminals}}\right) \times \text{input voltage}$$

$$= \left[\frac{R_p(x_i/x_t)}{R_p}\right] e_i = \frac{x_i}{x_t} \times e_i \qquad \text{...(25.45)}$$

The under ideal circumstances, the output voltage varies linearly with displacement as shown in Fig. 25.43($a$).

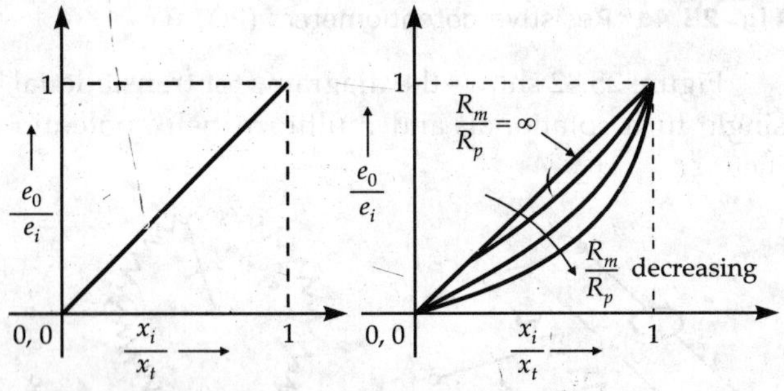


**Fig. 25.43** Characteristics of potentiometers.

$$\text{Sensitivity} \quad S = \frac{\text{output}}{\text{input}}$$

$$= \frac{e_0}{x_i} = \frac{e_i}{x_t} \qquad \text{...(25.46)}$$

Thus under ideal conditions the sensitivity is constant and the output is faithfully reproduced and has a linear relationship with input. The same is true of rotational motion.

Let $\theta_i$ = input angular displacement in degrees, and $\theta_t$ = total travel of the wiper in degrees.

$\therefore$ Output voltage $\quad e_0 = e_i . (\theta_i / \theta_t) \qquad$ ...(25.47)

This is true of single turn potentiometers only.

The circuits shown in Fig. 25.42 are called **potentiometer dividers** since they produce an output voltage which is a fraction of the input voltage. Thus the input voltage is "divided". The potential divider is a device for dividing the potential in a ratio determined by the position of the sliding contact.

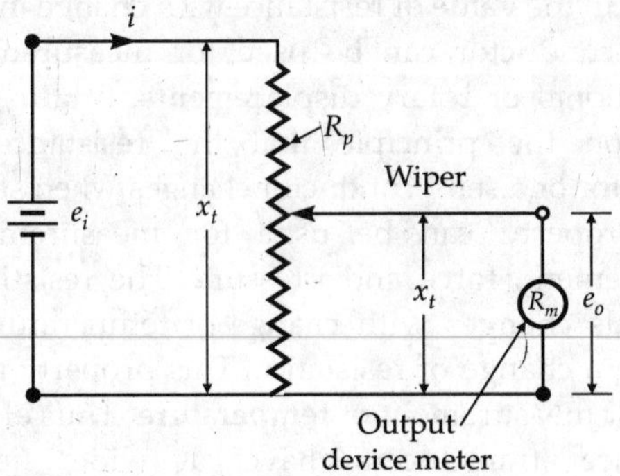


**Fig. 25.44** Loaded potentiometer.

Equations 25.46 and 25.47 are based upon the assumption that the distribution of resistance with respect to linear or angular displacement is uniform and the resistance of the voltage measuring device (*i.e.* output device) is infinite.

However, in practice, the output terminals of the **pot** are connected to a device whose impedance is finite. Thus, when an electrical instrument, which forms a load for the **pot** and is connected across the output terminals, the indicated voltage is less than that given by Eqn. 25.46. The error, which is referred to as a **loading error** is caused by the input resistance of the output device.

Let us consider the case of a translational potentiometer as shown in Fig. 25.44. Let the resistance of a meter or a recorder monitoring the output be $R_m$.

As explained earlier if the resistance across the output terminals is infinite, we get a linear relationship between the output and the input voltage.

$$e_0 = (x_i / x_t)\, e_i = Ke_i \qquad \text{...(25.48)}$$

where $\quad K = x_i/x_t$

However, under actual conditions the resistance, $R_m$, is not infinite. This causes a non-linear relationship between the output and input voltages.

### 25.12.1 Loading Effect

The resistance of the parallel combination of load resistance and the portion of the resistance of the potentiometer is :

$$\frac{(x_i/x_t) R_p R_m}{(x_i/x_t) R_p + R_m} = \frac{K R_p R_m}{K R_p + R_m} \qquad \text{...(25.49)}$$

The total resistance seen by the source is :

$$R = R_p(1-K) + \frac{KR_pR_m}{KR_p + R_m}$$

$$= \frac{KR_p^2(1-K) + R_pR_m}{KR_p + R_m} \quad ...(25.50)$$

$\therefore$ Current $i = \frac{e_i}{R}$

$$= \frac{e_i(KR_p + R_m)}{KR_p^2(1-K) + R_pR_m}$$

The output voltage under load conditions is :

$$e_0 = i\frac{KR_pR_m}{KR_p + R_m}$$

$$= \frac{e_i(KR_p + R_m)}{KR_p^2(1-K) + R_pR_m} \times \frac{KR_pR_m}{(KR_p + R_m)}$$

$$= \frac{e_iK}{K(1-K)(R_p/R_m)+1} \quad ...(25.51)$$

The ratio of output voltage to input voltage under load conditions is :

$$\frac{e_0}{e_i} = \frac{K}{K(1-K)(R_p/R_m)+1} \quad ...(25.52)$$

Equation 25.52 shows that there exists a non-linear relationship between output voltage $e_0$ and input displacement $x_i$ since $K = x_i/x_t$. In case $R_m = \infty$, $e_0/e_i = K$.

It is evident from Eqn. 25.52 that as the ratio of $R_m/R_p$ decreases, the non-linearity goes on increasing. This is shown in Fig. 25.43(*b*). Thus, in order to keep linearity, the value of $R_m/R_p$ should be as large as possible. However, when we have to measure the output voltage with a given meter, the resistance of the potentiometer, $R_p$, should be as small as possible.

$\therefore$ Error = output voltage under load − output voltage under no load

$$= \frac{e_iK}{[K(1-K)(R_p/R_m)+1]} - e_iK$$

$$= -e_i\left[\frac{K^2(K-1)}{K(1-K)+R_m/R_p}\right] \quad ...(25.53)$$

Based upon full-scale output, this relationship may be written as :

$$\%\ \varepsilon = -\left[\frac{K^2(K-1)}{K(1-K)+(R_m/R_p)}\right] \times 100 \quad ...(25.54)$$

Except for the two end points where $K = 0$ *i.e.*, $x_i = 0$ and $K = 1$ where $x_i = x_t$ the error is always negative. Figure 25.45 shows a plot of the variation in error with the slider position for different ratios of the load (output device or meter) resistance to the potentiometer resistance.

The error as indicated in Fig. 25.45 is actually **negative**. Examine Fig. 25.45, the maximum error is about 12 per cent of full scale if $R_m/R_p = 1$. This error drops down to about 1.5 per cent when $R_m/R_p = 10$. For values of $R_m/R_p > 10$, the position of maximum error occurs in the vicinity of $x_i/x_t = 0.67$.

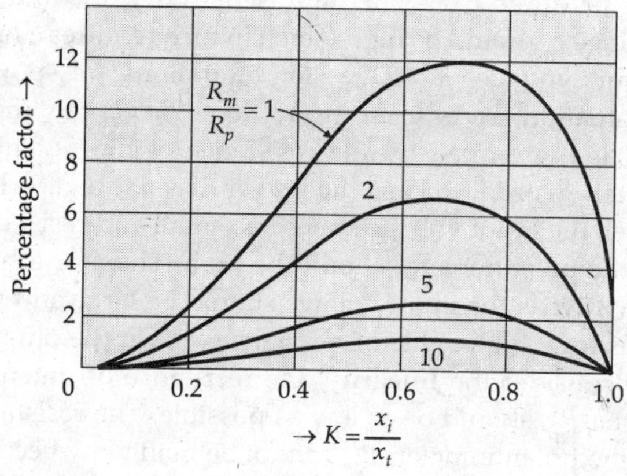


**Fig. 25.45** Variation of error due to loading effect of a potentiometer.

Maximum percentage error

$$\varepsilon_{max} = 15 \times (R_p/R_m) \quad ...(25.55)$$

It should be understood that the error in POTS is on account of the *non-linearity effect* produced by the output device used for measurement whose input resistance, $R_m$, is finite. If meter resistance were infinite there would be no linearity effects and the output voltage will be a linear function of input displacement $x_i$ as shown in Fig. 25.43 and consequently, there would no error because of the absence of non-linearity.

### 25.12.2 Power Rating of Potentiometers

The potentiometers are designed with a definite power rating which is related directly to their heat dissipating capacity. The manufacturer normally designs a series of potentiometers of single turn with a diameter of 50 mm with a wide range of ohmic values ranging from 100 Ω to 10 kΩ in steps of 100 Ω. These potentiometers are essentially of the same size and of the same mechanical configuration. They have the same heat transfer capabilities. Their rating is typically

25

5 W at an ambient temperature of 21°C. This limits their input excitation voltage. Since power $P = e_i^2 / R_p$, the maximum input excitation voltage that can be used is :

$$(e_i)_{max} = \sqrt{PR_p} \text{ volt} \qquad \ldots(25.56)$$

### 25.12.3 Linearity and Sensitivity

It has been explained earlier that in order to achieve a good linearity, the resistance of potentiometer $R_p$, should be as low as possible when using a meter for reading the output voltage which has a fixed value of input resistance $R_m$.

In order to get a high **sensitivity** the output voltage $e_0$ should be high which in turn requires a high input voltage, $e_i$. Due to limitations of power dissipation as is clear from Eqn. 25.56, the input voltage is limited by the resistance of the potentiometer. In order to keep the power dissipation at a low level, the input voltage should be small and resistance of the potentiometer should be high. Thus for a high **sensitivity**, the input voltage should be large and this calls for a high value of resistance $R_p$. On the other, if we consider the **linearity**, the resistance of potentiometer $R_p$, should be as low as possible. The resistance of the potentiometer, $R_p$, cannot be made low because if we do so the power dissipation goes up with the result that we have to make the input voltage small to keep the power dissipation to the acceptable level. This results in lower **sensitivity**.

Thus **linearity** and **sensitivity** are therefore two conflicting requirements. If $R_p$ is made small, the **linearity** improves, but a low value of $R_p$, requires a lower input voltage $e_i$ in order to keep down the power dissipation and a low value of $e_i$ results in a lower value of output voltage $e_0$ resulting in lower **sensitivity**. Thus the choice of potentiometer resistance, $R_p$, has to be made considering both the **linearity** and **sensitivity** and a compromise between the two conflicting requirements has to be struck.

The maximum available sensitivity of potentiometer varies considerably from type to type and also with size in a given type. The sensitivity can be calculated from the data supplied by the manufacturers which includes maximum allowable voltage, current or power and the maximum stroke. The shorter stroke devices have generally a higher sensitivity. Extreme values are of the order of 15 V/degree for short stroke sector type rotational potentiometers and 12 V/mm for short stroke (6.25 mm) translational potentiometers. These, in fact, are maximum achievable values and usual values of sensitivity are 10 to 100 times smaller.

The linearity can be improved by using a resistance $R_{in}$ (made equal to the meter resistance $R_m$) as shown in Fig. 25.46 without using a meter of higher resistance. The input-output characteristics with and without $R_{in}$ in the circuit are shown in Fig. 25.46(*b*).

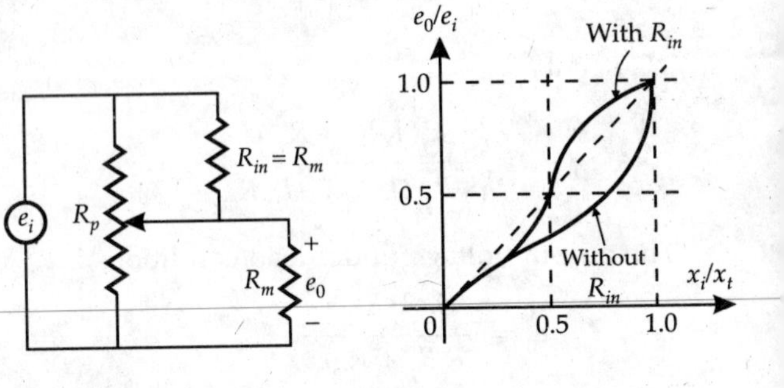


**Fig. 25.46** Linearization of POT characteristics, through use of an external resistance $R_{in} = R_m$.

### 25.12.4 Construction of Potentiometers

The resolution of the potentiometers influences the construction of their resistance elements. Normally, the resistive element is a single wire of conducing material which gives a continuous stepless variation of resistance as the wiper travels over it. Such potentiometers are available but their length (in the case of translational potentiometers) and diameter (in the case of rotational potentiometers) restrict their use on account of space considerations.

The resolution of the potentiometers is dependent upon the construction of the resistive element and in order to get high values of resistance in small space **wire wound** potentiometers are used extensively. The resistance wire is wound on a mandrel or a card for translational displacement as shown in Fig. 25.47(*a*). For the measurement of rotational motion these mandrels or cards are formed into a circle or a helix.

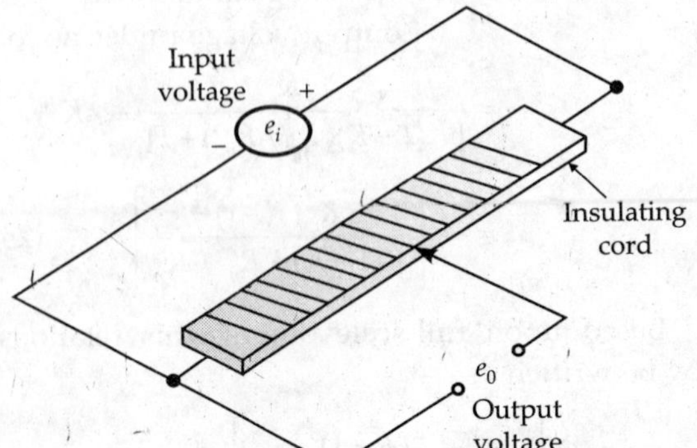


(*a*) Linear potentiometer.

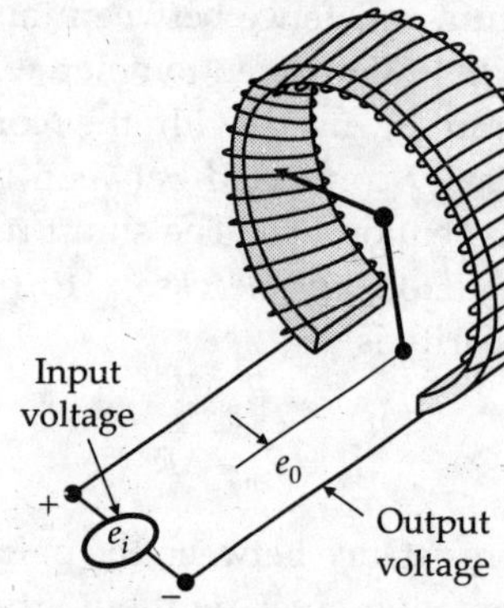


(b) Circular (Rotational) potentiometer.

**Fig. 25.47** Wire wound potentiometers.

This is shown in Fig. 25.47(*b*). If wire wound type of construction is adopted, the variation of resistance is not a linear continuous change but is in small steps as the sliding contact (wiper) moves from one turn to another. This is shown in Fig. 25.48.

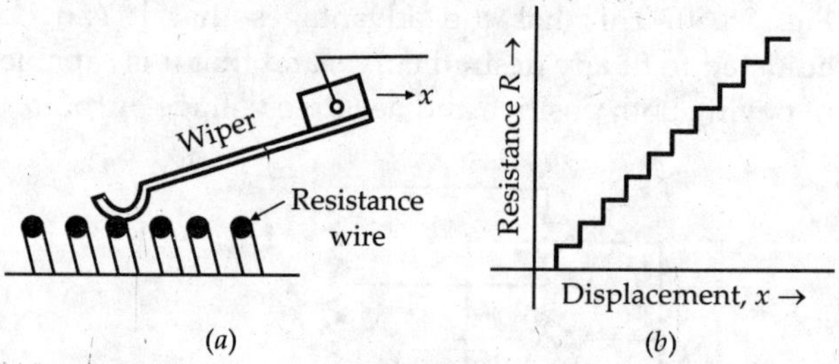


**Fig. 25.48** Translational potentiometer and its characteristics.

Since the variation in resistance is in steps the resolution is limited. For instance, a translational potentiometer has about 500 turns of a resistance wire on a card of 25 mm in length and for this device the resolution is limited to 25/500 = 0.05 mm = 50 μm.

The actual practical limit is 20 and 40 turns per mm. Thus for translational devices the resolution is limited to 25 – 50μm.

For rotational devices, the best angular resolution

$$= \frac{3 \text{ to } 6}{D} \text{ degrees} \qquad \ldots(25.57)$$

where $D$ = diameter of the potentiometer ; mm

In order to get higher resolution, thin wires which have a high resistance have to be put close to each other and they can be closely wound on account of their small diameter. Thus the resolution and total resistance are interdependent.

In case a fine resolution and high resistance are required a carbon film or a conductive-plastic resistance elements are used. Carbon film resistive elements have a resolution of 12.5 nm.

### 25.12.5 Helipots

The resolution can be increased by using multi-turn potentiometers. These are called **helipots**. The resistance element is in the form of a helix and the wiper travels along a "lead screw". The number of turns is still limited to 20 to 40 turns per mm but an increase in resolution can be obtained by using a gearing arrangement between the shaft whose motion is to be measured and the potentiometer shaft. As an example, one rotation of the measured shaft can cause 10 rotations of the potentiometer shaft. This increases the resolution of the measured shaft motion by 10 times. Multiturn potentiometers are available to about 60 turns.

Similar magnifying or amplifying techniques can be used for translational motion as well.

## 25.13 NON-LINEAR POTENTIOMETERS FUNCTION GENERATORS

In many instrumentation problems, non-linear conditions arise. These may include the effect of magnetic saturation, the non-linearity of the voltage produced by a thermocouple as a function of temperature, the air speed expressed in Mach number as a function of wind-tunnel pressures, trigonometric functions, etc. One simple method of solving these non-linear functions is by the use of non-linear potentiometers. A standard self-balancing potentiometer is used to produce a displacement proportional to the input voltage $e_i$ is shown in Fig. 25.49. This displacement also operates a non-linear resistance which produces an output voltage $e_0$ which is the desired function of the input voltage $e_i$. This operation is illustrated in the following example. Let the input voltage $e_i$ vary from 0 to 10 V. The desired function of this voltage is as plotted in Fig. 25.49(*b*).

$$e_0 = 0.3\, e_i + 0.5\, e_i^2 \qquad \ldots(25.58)$$

The maximum output voltage

$$e_{0(\max)} = 0.3 \times 10 + 0.5 \times (10)^2 = 53 \text{ V}.$$

A displacement proportional $e_i$ is obtained using an amplifier and a servomotor as shown in Fig. 25.49(*a*). A good approximation of the required resistance of the non-linear potentiometer is obtained by shunting a linear potentiometer at 10 equally spaced taps with the correct resistances. The desired voltage above ground is computed from Eqn. 25.58 and are listed in Table 25.3. The current is assumed to be 1 mA. The resistance $R$ is the resistance from each tap to ground, and $\Delta R$ is the resistance between adjacent taps. These resistances are computed and

25

listed in Table 25.3. The circuit of the non-linear potentiometer with resistances between each tap is shown in Fig. 25.49.

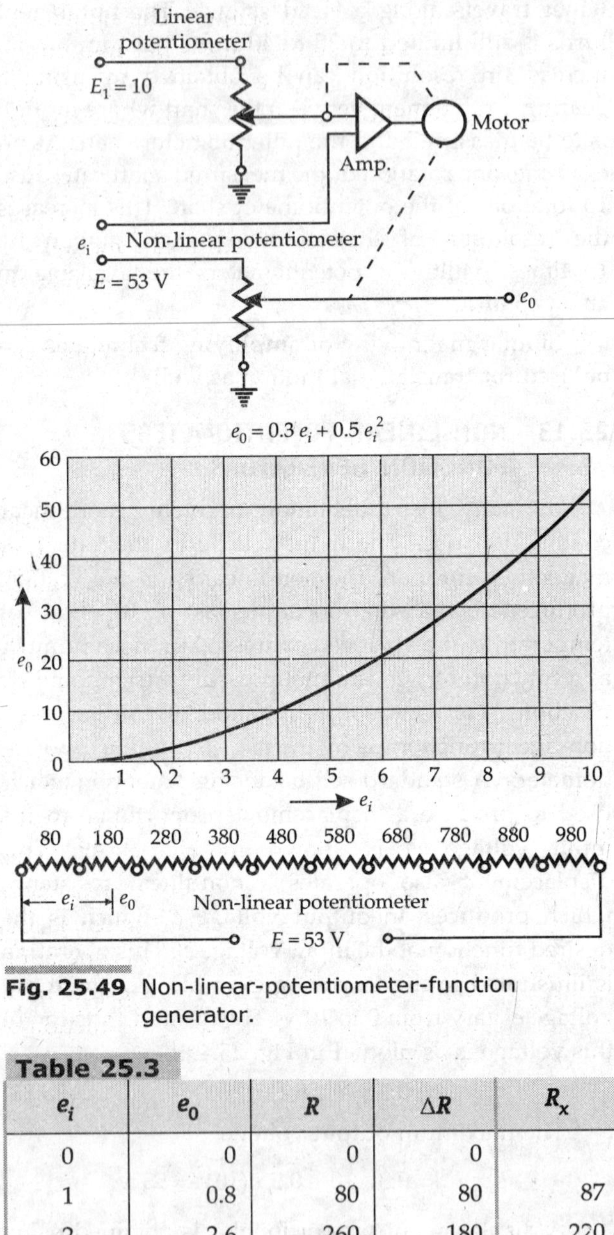


**Fig. 25.49** Non-linear-potentiometer-function generator.

**Table 25.3**

| $e_i$ | $e_0$ | $R$ | $\Delta R$ | $R_x$ |
|---|---|---|---|---|
| 0 | 0 | 0 | 0 | |
| 1 | 0.8 | 80 | 80 | 87 |
| 2 | 2.6 | 260 | 180 | 220 |
| 3 | 5.4 | 540 | 280 | 319 |
| 4 | 9.2 | 920 | 380 | 613 |
| 5 | 14.0 | 1,400 | 480 | 923 |
| 6 | 19.8 | 1,980 | 580 | 1,380 |
| 7 | 26.6 | 2,660 | 680 | 2,124 |
| 8 | 34.4 | 3,440 | 780 | 3,540 |
| 9 | 43.2 | 4,320 | 880 | 7,330 |
| 10 | 53.0 | 5,300 | 980 | 49,000 |

The maximum resistance between taps is 980 Ω. Then a 10,000 Ω, 10-tap potentiometer with 1,000 Ω between taps can be used with the correct shunt resistance externally connected between the taps for the non-linear potentiometer. The shunt resistance $R_x$ between each pair of taps necessary to produce the desired non-linearity is,

$$R_x = \frac{R_p \Delta R}{R_p - \Delta R} \qquad \ldots(25.59)$$

where $R_p$ is the resistance between the pair of adjacent taps of the potentiometer without the shunting resistors, and $\Delta R$ is the desired resistance.

These values are given in Table 25.3. The difference between the desired function and the approximation obtained with this 10-tap potentiometer is the difference between the curve in Fig. 25.49 and the straight lines connecting the points on the curve.

An adjustable tapped potentiometer is shown in Fig. 25.50. This has the advantages that it can be adjusted to fit any desired curve, and that it is capable of having both positive and negative voltage outputs.

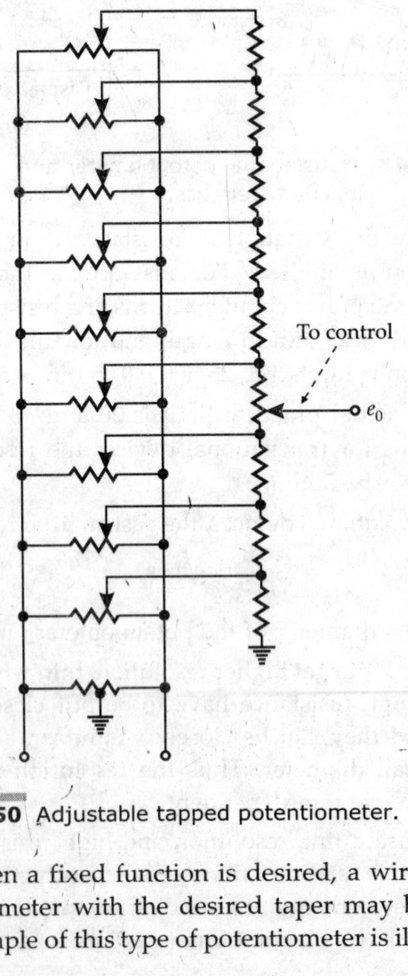


**Fig. 25.50** Adjustable tapped potentiometer.

When a fixed function is desired, a wire-wound potentiometer with the desired taper may be made. An example of this type of potentiometer is illustrated

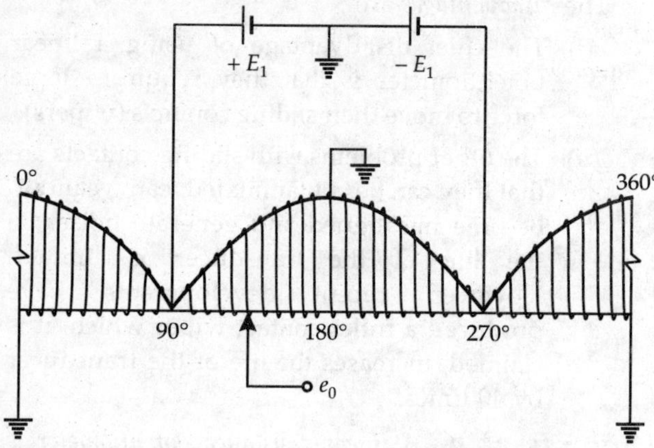


**Fig. 25.51** Sine-cosine, tapered potentiometer.

in Fig. 25.51, where the taper is shown for a continuous rotating potentiometer with a sine-cosine output function. The potentiometer shaft is rotated at the desired angular velocity with a servomotor when a sine-wave input voltage is desired, or it may be operated with a self-balancing potentiometer circuit, as shown in Fig 25.49 when the desired output is,

$$e_0 = a \sin e_i$$

or when the axis is shifted 90°.

$$e_0 = b \cos e_i$$

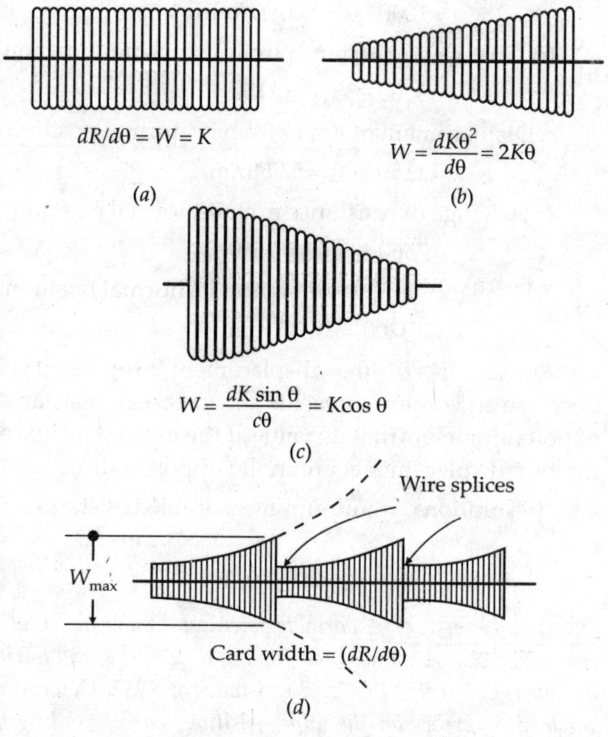


**Fig. 25.52** Generation of non-linear functions with potentiometers.

It will be remembered that servo circuits are comparatively slow and are inadequate for fast analog computers, yet these circuits are very accurate, usually ± 0.5 percent or better.

If the slope of the body on which the resistive wire is wound is kept accurately uniform, there will be a linear relationship between the output voltage and the position of the wiper (input displacement). However, if the shape of the body varies, it will cause a prescribed non-linear transfer function. This is shown in Fig. 25.52.

There are many standard functions available, or custom position-to-voltage relationship can be produced. This non-linear relationship may be used to compensate for a non-linearity in the quantity being measured or to introduce a special function into the control and instrumentation scheme without having to resort to special mathematics electronics or a computer.

## 25.14 MATERIALS USED FOR POTENTIOMETERS

The materials used for POTS may be classified as wire wound and non-wire wound.

### 1. Wire Wound Potentiometers

These are platinum, nickel chromium, nickel copper, or some other precious resistance elements. Wire wound potentiometers carry relatively large currents at high temperatures. Their resistance temperature co-efficient is usually small, is of the order of $20 \times 10^{-6}/°C$ or less. Their resolution is about 0.025 – 0.05 mm and is limited by the number of turns that can be accommodated on the card.



It should be noted that the interwinding capacitance between turns, and between windings and shaft, housing etc. limits the use of wire wound potentiometers to low frequencies. The response is limited to about 5 Hz.

### 2. Non-Wire Potentiometers

*These are also called continuous potentiometers.* The non-wire wound potentiometers provide improved resolution and life. This is because the resolution is no longer limited by the number of turns that can be wrapped onto a body. The wiper moves across a smooth surface (not bouncing from turn to turn), wear, bounce and the resulting failures are decreased. The maximum speed that a wire wound potentiometer is turned is about 300 rpm. About that, the noise created as the wiper bounces from turn to turn becomes significant. A continuous potentiometer may be turned at a speed of 2000 rpm. However, they are more sensitive to temperature changes, have a higher wiper contact resistance, which is variable and can carry only moderate currents.

The materials used for non-wire wound (or continuous) potentiometers are :

(*i*) ***Cermet.*** Cermet uses precious metal particles fused into ceramic base. These fused metal particles act as resistance elements. The advantages of using Cermet are large power ratings at high temperatures, low cost and moderate temperature coefficients of the order $100 \times 10^{-6}$ to $200 \times 10^{-6}/°C$. Cermet is very useful for a.c. applications.

(*ii*) ***Hot moulded carbon.*** The resistance element is fabricated by moulding together a mixture of carbon and a thermosetting plastic binder. Hot moulded carbon units are useful for a.c. applications.

(*iii*) ***Carbon film.*** A thin film of carbon deposited on a non-conductive base forms the resistance element. The advantage of carbon film potentiometers is their low cost. Temperature coefficients are upto $1000 \times 10^{-6}/°C$.

(*iv*) ***Thin metal film.*** A very thin, vapour deposited layer of metal on glass or ceramic base is used as a resistance element. The advantages of this potentiometer are its excellent resistance to changes in environments and use on a.c. The cost is also moderate.

## 25.15 ADVANTAGES AND DISADVANTAGES OF RESISTANCE POTENTIOMETERS

Resistance potentiometers have the following major *advantages* :

(*i*) They are inexpensive.

(*ii*) They are simple to operate and very useful for applications where the requirements are not particularly severe.

(*iii*) They are very useful for measurement of large amplitudes of displacement.

(*iv*) Their electrical efficiency is very high and they provide sufficient output to permit control operations without further amplification.

(*v*) It should be understood that while the frequency response of wire wound potentiometers is limited, the other types of potentiometers are free from this problem.

(*vi*) In wire wound potentiometers the resolution is limited while in Cermet and metal film potentiometers, the resolution is infinite.

The *disadvantages* are :

(*i*) The chief disadvantage of using a linear potentiometer is that they require a large force to move their sliding contacts (wipers).

(*ii*) The other problems with sliding contacts are that they can be contaminated, can wear out, become misaligned and generate noise. So the life of the transducer is limited. However, recent developments have produced a roller contact wiper which, it is claimed, increases the life of the transducer by 40 times.

**Example 25.9** *A linear resistance potentiometer is 50 mm long and is uniformly wound with a wire having a resistance of 10,000 Ω. Under normal conditions, the slider is at the centre of the potentiometer. Find the linear displacement when the resistance of the potentiometer as measured by a Wheatstone bridge for two cases is : (i) 3850 Ω, (ii) 7560 Ω. Are the two displacements in the same direction ?*

*If it is possible to measure a minimum value of 10 Ω resistance with the above arrangement, find the resolution of the potentiometer in mm.*

**Solution.** The resistance of the potentiometer at its normal position

$$= 1000/2 = 5000\ \Omega$$

Resistance of potentiometer per unit length

$$= 1000/50 = 200\ \Omega/\text{mm}$$

(*i*) Change of resistance from its normal position

$$= 5000 - 3850 = 1150\ \Omega$$

∴ Displacement of wiper from its normal position

$$= 1150/200 = 5.75 \text{ mm}$$

(*ii*) Change of resistance from its normal position

$$= 7560 - 5000 = 2560\ \Omega$$

∴ Displacement of wiper from its normal position

$$2560/200 = 12.80 \text{ mm}$$

Since, one of the displacements represents a decrease and other represents an increase in resistance of potentiometer from its value at the normal position, the two displacements are in the opposite direction.

$$\text{Resolution} = \text{minimum measurable resistance} \times \text{mm}/\Omega$$

$$= 10 \times 1/200 = 0.05 \text{ mm}$$

**Example 25.10** *A variable potential divider has a total resistance of 2 kΩ and is fed from a 10 V d.c. supply. The output is connected to a load resistance of 5 kΩ. Determine the loading errors for the wiper positions corresponding to $K = x_i/x_t = 0, 0.25, 0.5, 0.75$ and 1.0. Use the results to plot a rough graph of error versus $x_i/x_t$.*

**Solution.** We have $R_p = 2\ k\Omega$ and $R_m = 5\ k\Omega$

From Eqn. 25.53, error is,

$$-e_i\left[\frac{K^2(K-1)}{K(1-K)+R_m/R_p}\right]$$

Substituting the values, the results are tabulated below :

| K | Error, V |
|---|---|
| 0 | 0 |
| 0.25 | −0.174 |
| 0.5 | − 0.454 |
| 0.75 | − 0.524 |
| 1 | 0 |

The graph between error versus $K = x_i/x_t$ is plotted in Fig. 25.53. (The errors are no doubt – ve but they are shown as +ve in Fig. 25.53.)

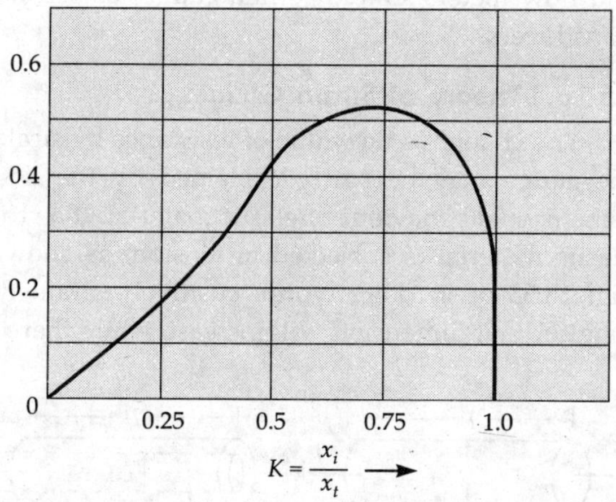


**Fig. 25.53** Graph of results of Example 25.10.

***Example 25.11*** *A resistive potential divider $R_1R_2$ with a resistance of 5000 Ω and a shaft stroke of 125 mm is used in the arrangement shown in Fig. 25.54. Potentiometer $R_3R_4$ has a resistance of 5000 Ω and $e_i = 5.0$ V. The initial position to be used as reference point is such that $R_1 = R_2$ i.e., the wiper is at midstroke. At the start of the test*

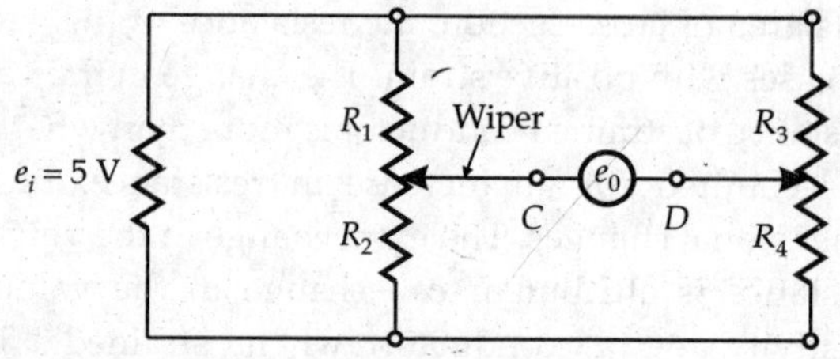


**Fig. 25.54** Diagram of Example 25.11.

*potentiometer $R_3R_4$ is adjusted so that the bridge is balanced and $e_0 = 0$. Assuming that the displacement being measured will move a maximum distance of 12.5 mm towards A, calculate the value of $e_0$.*

**Solution.** Total length of $AB = 125$ mm ; Midpoint of $AB = 62.5$ mm from either $A$ or $B$. If the wiper moves 12.5 inward towards $A$ from midstroke, its distance from $B$ becomes $62.5 + 12.5 = 75$ mm.

$$\therefore \quad R_2 = (75/125) \times 5000 = 3000\ \Omega$$

Hence, output voltage

$$e_0 = \left(\frac{R_2}{R_1+R_2} - \frac{R_4}{R_3+R_4}\right)e_i$$

$$= \left[\frac{3000}{5000} - \frac{2500}{5000}\right] \times 5 = 0.5\ \text{V}$$

***Example 25.12*** *The output of a potentiometer is to be read by a recorder of 10 kΩ input resistance. The non-linearity must be held to 1 percent. A family of potentiometers having a thermal rating of 5 W and resistances ranging from 100 Ω to 10 kΩ in steps of 100 Ω are available. Choose from the family of potentiometers, a potentiometer that has the greatest possible sensitivity and which meets the non-linearity requirements. Find the maximum excitation voltage permissible with this potentiometer. What is the sensitivity if the potentiometer is a single turn (360° unit) ?*

**Solution.** From Eqn. 25.55, maximum possible percentage linearity $= 15\ R_p/R_m$. This is to be limited to 1 percent $15\ R_p/R_m = 1$. Therefore, the maximum value of resistance of potentiometer is,

$$R_p = \frac{R_m}{15} = \frac{10{,}000}{15} = 666.7\ \Omega$$

Thus we are left with choice of potentiometers having resistances of 100 Ω, 200 Ω, 300 Ω, 400 Ω, 500 Ω and 600 Ω.

The potentiometer with the highest value of resistance has the highest sensitivity. Therefore, a potentiometer with $R_p =$ 600 Ω is selected. With a power dissipation of 5 W, the maximum value of excitation voltage is,

$$e_{i(\max)} = \sqrt{PR_p} = \sqrt{5 \times 600} = 54.8\ \text{V}$$

The sensitivity of potentiometer when it is a single turn unit $S = 54.8/360 = 0.152$ V/degree.

***Example 25.13*** *A helipot is provided with 400 turns/mm. The gearing arrangement is such that the motion of the main shaft causes 5 revolutions of potentiometer shaft. Calculate the resolution of the potentiometer.*

25

**Solution.** The resolution of potentiometer without the gearing arrangement = 1/400 mm = 25 μm. With the gearing arrangement which causes 5 revolutions of the potentiometer shaft with one revolution of main shaft = 25/5 = 5 μm.

**Example 25.14** *It is necessary to measure the position of an object. It moves 0.8 m. Its position must be known within 1 mm. Part of the mechanism which moves the object is a shift in a shaft that rotates 250° when the object is moved from one extreme to the other. A control potentiometer has been found which is rated at 300° full scale movement. It has one thousand turns of wire. Is the potentiometer suitable for the application ?*

**Solution.** The shaft provides a

$$\frac{250°}{0.8} = 312.5°/\text{m or } 0.3125°/\text{mm conversion}$$

A resolution of 1 mm at the object translates into

$$1 \text{ mm} \times 0.3125° = 0.3125°$$

Required resolution for the potentiometer.

The potentiometer actually has a resolution

$$\frac{300}{1000} = 0.300°$$

∴ The resolution of the potentiometer is higher than that required for the application and hence the potentiometer is very much suited for the application.

**Example 25.15** *A control potentiometer is rated as : resistance = 150 Ω, power rating = 1 W. De-rate the potentiometer by 10 mW/°C above 65°C, thermal resistance = 30° C/W.*

*Can the potentiometer be used with a 10 V supply at 80°C ambient temperature ?*

**Solution.** Power dissipated by potentiometer

$$P = (10)^2/150 = 0.667 \text{ W} = 667 \text{ mW}$$

The actual temperature of the potentiometer is the sum of the ambient temperature and temperature rise produced due to self heating.

$$\theta_{pot} = 80° + (667) \times 30 \times 10^{-3} = 100°\text{C}$$

The allowable power dissipation must be decreased by 10 mW for each °C above 65°C

Dissipation allowed

$$= 1.0 - 10 \times 10^{-3}(100 - 35)$$

$$= 0.65 \text{ W} = 650 \text{ mW}$$

A first glance, it would appear that the potentiometer is well below the rated 1 W power dissipated. This could have been true if the temperature of the potentiometer were 65°C. However, the temperature of the potentiometer is 100°C and therefore de-rating in power dissipation has to done on account of increase in temperature. The de-rated power dissipation is 650 mW while the potentiometer is dissipating 667 mW and hence it will fail.

## 25.16 STRAIN GAUGES

If a metal conductor is stretched or compressed, its resistance changes on account of the fact that both length and diameter of conductor change. Also there is a change in the value of resistivity of the conductor when it is strained and this property is called **piezoresistive effect**. Therefore, resistance strain gauges are also known as **piezoresistive gauges**. The strain gauges are used for measurement of strain and associated stress in experimental stress analysis. Secondly, many other detectors and transducers, notably the load cells, torque meters, diaphragm type pressure gauges, temperature sensors, accelerometers and flow meters, employ strain gauges as secondary transducers.

### 25.16.1 Theory of Strain Gauges

The change in the value of resistance by straining the gauge may be partly explained by the normal dimensional behaviour of elastic material. If a strip of elastic material is subjected to tension, as shown in Fig. 25.55 or in other words positively strained, its longitudinal dimension will increase while there will

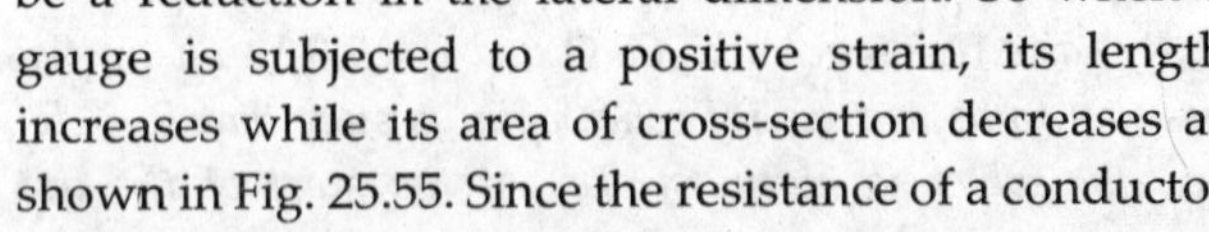


**Fig. 25.55** Change in dimensions of a strain gauge element when subjected to a tensile force.

be a reduction in the lateral dimension. So when a gauge is subjected to a positive strain, its length increases while its area of cross-section decreases as shown in Fig. 25.55. Since the resistance of a conductor is proportional to its length and inversely proportional to its area of cross-section, the resistance of the gauge increases with positive strain. The change in the value resistance of strained conductor is more than what can be accounted for an increase in resistance due to dimensional changes. The extra change in the value of resistance is attributed to a change in the value of resistivity of a conductor when strained. This property, as described earlier, is known as **peizoresistive effect.**

Let us consider a strain gauge made of circular wire. The wire has the dimensions : length = $L$, area = $A$, diameter = $D$ before being strained. The material of the wire has a resistivity $\rho$.

$\therefore$ Resistance of unstrained gauge $R = \rho L/A$.

Let a tensile stress $s$ be applied to the wire. This produces a positive strain causing the length to increase and area to decrease as shown in Fig. 25.55. Thus when the wire is strained there are changes in its dimensions. Let $\Delta L$ = change in length, $\Delta A$ = change in area, $\Delta D$ = change in diameter and $\Delta R$ = change in resistance.

In order to find how $\Delta R$ depends upon the material physical quantities, the expression for $R$ is differentiated with respect to stress $s$.

Thus we get :

$$\frac{dR}{ds} = \frac{\rho}{A}\frac{\partial L}{\partial s} - \frac{\rho L}{A^2}\frac{\rho A}{\partial s} + \frac{L}{A}\frac{\partial \rho}{\partial s} \quad \ldots(25.59)$$

Dividing Eqn. 25.59 throughout by resistance $R = \rho L/A$, we have

$$\frac{1}{R}\frac{dR}{ds} = \frac{1}{L}\frac{\partial L}{\partial s} - \frac{1}{A}\frac{\partial A}{\partial s} + \frac{1}{\rho}\frac{\partial \rho}{\partial s} \quad \ldots(25.60)$$

It is evident from Eqn. 25.60, that the per unit change in resistance is due to :

(*i*) per unit change in length = $\Delta L/L$,

(*ii*) per unit change in area = $\Delta A/A$, and

(*iii*) per unit change in resistivity = $\Delta \rho/\rho$

$$\text{Area} \quad A = \frac{\pi}{4}D^2 \quad \therefore \quad \frac{\partial A}{\partial s} = 2.\frac{\pi}{4}D.\frac{\partial D}{ds} \quad \ldots(25.61)$$

or
$$\frac{1}{A}\frac{dA}{ds} = \frac{(2\pi/4)D}{(\pi/4)D^2}\frac{\partial D}{\partial s} = \frac{2}{D}\frac{\partial D}{\partial s} \quad \ldots(25.62)$$

$\therefore$ Equation 25.60 can be written as :

$$\frac{1}{R}\frac{dR}{ds} = \frac{1}{L}\frac{\partial L}{\partial s} - \frac{2}{D}\frac{\partial D}{\partial s} + \frac{1}{\rho}\frac{\partial \rho}{\partial s} \quad \ldots(25.63)$$

Now, Poisson's ratio

$$\nu = \frac{\text{lateral strain}}{\text{longitudinal strain}} = -\frac{\partial D/D}{\partial L/L} \quad \ldots(25.64)$$

or
$$\partial D/D = -\nu \times \partial L/L$$

$$\therefore \quad \frac{1}{R}\frac{dR}{ds} = \frac{1}{L}\frac{\partial L}{\partial s} + \nu\frac{2}{L}\frac{\partial L}{\partial s} + \frac{1}{\rho}\frac{\partial \rho}{\partial s} \quad \ldots(25.65)$$

For small variations, the above relationship can be written as :

$$\therefore \frac{\Delta R}{R} = \frac{\Delta L}{L} + 2\nu\frac{\Delta L}{L} + \frac{\Delta \rho}{\rho} \quad \ldots(25.66)$$

The gauge factor is defined as the ratio of per unit change in resistance to per unit change in length.

$$\text{Gauge factor} \quad G_f = \frac{\Delta R/R}{\Delta L/L} \quad \ldots(25.67)$$

or
$$\frac{\Delta R}{R} = G_f\frac{\Delta L}{L} = G_f \times \varepsilon \quad \ldots(25.68)$$

where $\varepsilon = \text{strain} = \frac{\Delta L}{L}$

The gauge factor can be written as :

$$= 1 + 2\nu + \frac{\Delta \rho/\rho}{\varepsilon} \quad \ldots(25.69)$$

$$= \underset{\substack{\text{Resistance}\\\text{change due to}\\\text{change of length}}}{1} + \underset{\substack{\text{Resistance}\\\text{change due to}\\\text{change in area}}}{2\nu} + \underset{\substack{\text{Resistance}\\\text{change due to}\\\text{piezoresistive effect}}}{\frac{\Delta\rho/\rho}{\varepsilon}}$$

$$G_f = \frac{\Delta R/R}{\Delta L/L} = 1 + 2\nu + \frac{\Delta \rho/\rho}{\Delta L/L}$$

The strain is usually expressed in terms of microstrain. 1 microstrain = 1 μm / m.

If the change in the value of resistivity of a material when strained is neglected, the gauge factor is :

$$G_f = 1 + 2\nu \quad \ldots(25.70)$$

Equation 25.70 is valid only when **Piezoresistive Effect** *i.e.*, change in resistivity due to strain is almost negligible.

The Poisson's ratio for all metals is between 0 and 0.5. This gives a gauge factor of approximately, 2. The common value for Poisson's ratio for wires is 0.3. This gives a value of 1.6 for wire wound strain gauges.

Table 25.4 gives the value of gauge factors for the various materials.

**Table 25.4** *Gauge Factors*

| *Material* | *Gauge* | *Material* | *Gauge Factor* |
|---|---|---|---|
| Nickel | − 12.1 | Platinum | + 4.8 |
| Manganin | + 0.47 | Carbon | + 20 |
| Nichrome | + 2.0 | Doped Crystals | 100 − 5000 |
| Constantan | + 2.1 | | |
| Soft iron | + 4.2 | | |

**Example 25.16** *A resistance wire strain gauge uses a soft iron wire of small diameter. The gauge factor is + 4.2. Neglecting the piezoresistive effects, calculate the Poisson's ratio.*

25

**Solution.** The gauge factor is given by Eqn. 25.69,

$$G_f = 1 + 2\nu + \frac{\Delta\rho/\rho}{\varepsilon}$$

If piezoresistive effect is neglected, the gauge factor is given by Eqn. 25.70 as :

$$G_f = 1 + 2\nu$$

$$\therefore \text{ Poisson's ratio } \nu = \frac{G_f - 1}{2} = \frac{4.2 - 1}{2} = 1.6$$

***Example 25.17*** *A compressive force is applied to a structural member. The strain is 5 micro-strain. Two separate strain gauges are attached to the structural member, one is a nickel wire strain gauge having a gauge factor of – 12.1 and the other is nichrome wire strain gauge having a gauge factor of 2. Calculate the value of resistance of the gauges after they are strained. The resistance of strain gauges before being strained is 120 Ω.*

**Solution.** According to our convention, the tensile strain taken as positive while the compressive strain is taken as negative.

Therefore, strain $\varepsilon = -5 \times 10^{-6}$

(1 micro strain = 1 μm / m)

Now $\Delta R / R = G_f \varepsilon$ (See Eqn. 25.68)

∴ Change in value of resistance of nickel wire strain gauge :

$$\Delta R = G_f \varepsilon \times R$$
$$= (-12.1) \times (-5 \times 10^{-6}) \times 120$$
$$= 7.26 \times 10^{-3}\,\Omega = 7.26 \text{ m}\Omega$$

Thus there is an increase of 7.26 mΩ in the value of resistance.

For nichrome, the change in the value of resistance is :

$$\Delta R = (2) \times (-5 \times 10^{-6}) \times 120$$
$$= 1.2 \times 10^{-3}\,\Omega = -1.2 \text{ m}\Omega$$

Thus, with compressive strain, the value of resistance gauge shows a decrease of 1.2 mΩ in the value of resistance.

## 25.17 TYPES OF STRAIN GAUGES

The following are the major types of strain gauges :

- Unbonded metal strain gauges
- Bonded metal wire strain gauges
- Bonded metal foil strain gauges
- Vacuum deposited thin metal film strain gauges
- Sputter deposited thin metal strain gauges
- Bonded semiconductor strain gauges
- Diffused metal strain gauges.

Strain gauges are broadly used for two major types of applications and they are :

(*i*) Experimental stress analysis of machines and structures, and

(*ii*) Construction of force, torque, pressure, flow and acceleration transducers.

### 25.17.1 Unbonded Metal Strain Gauges

An unbonded metal strain gauge is shown in Fig. 25.56. This gauge consists of a wire stretched between two points in an insulating medium such as air. The wires may be made of various copper nickel, chrome nickel or nickel iron alloys. They are about 0.003 mm in diameter, have a gauge factor of 2 to 4 and sustain a force of 2 mN. The length of wire is 25 mm or less.

In Fig. 25.56, the flexture element is connected via a rod to a diaphragm which is used for sensing of pressure. The wires are tensioned to avoid buckling when they experience a compressive force.

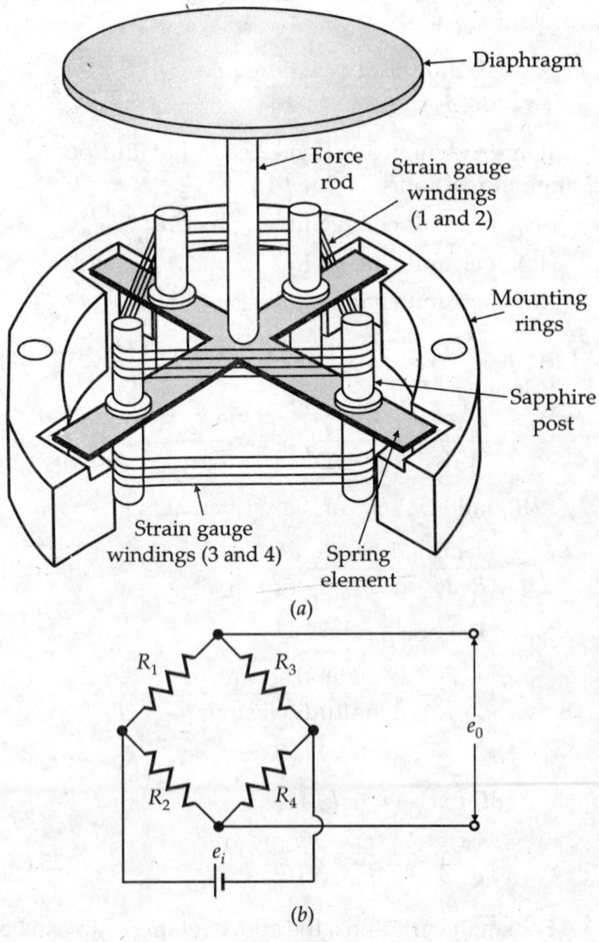


**Fig. 25.56** Set-up of a unbonded strain gauge and measurement with a Wheatstone bridge.

The unbonded metal wire gauges, used almost exclusively in transducer applications, employ preloaded resistance wires connected in a wheatstone bridge as shown in Fig. 25.56($b$). At initial preload, the strains and resistances of the four arms are nominally equal, with the result the output voltage of the bridge, $e_0 = 0$. Application of pressure produces a small displacement which is about 0.004 mm (full scale), the displacement increases tension in two wires and decreases it in the other two, thereby increase the resistance of two wires which are in tension and decreasing the resistance of the remaining two wires. This causes an unbalance of the bridge producing an output voltage which is proportional to the input displacement and hence to the applied pressure.

Electric resistance of each arm is 120 Ω to 1000 Ω, the input voltage to the bridge is 5 to 10 V, and the full scale output of the bridge is typically about 20 mV to 50 mV.

### 25.17.2 Bonded Wire Strain Gauges

The bonded metal-wire strain gauges are used for both stress analysis and for construction of transducers.

A resistance wire strain gauge consists of a grid of fine resistance wire of about 0.025 mm in diameter or less. The grid is cemented to carrier (base) which may be a thin sheet of paper, a thin sheet of bakelite or a sheet of teflon. The wire is covered on top with a thin sheet of material so as to prevent it from any mechanical damage. The spreading of wire permits a uniform distribution of stress over the grid. The carrier is bonded with an adhesive material to the specimen under study. This permits a good transfer of strain from carrier to grid of wires. The wires cannot buckle as they are embedded in a matrix of cement and hence faithfully follow both the tensile and compressive strains of the specimen. Since, the materials and the wire sizes used for bonded wire strain gauges are the same as used for unbonded wire strain gauges, the gauge factors and resistances for both are comparable.

The size of the strain gauges varies with application. They can be as small as 3 mm by 3 cm square. Usually they are larger but seldom more than 25 mm long and 12.5 mm wide.

For excellent and reproducible results, it is desirable that the resistance wire strain gauges should have the following characteristics :

($i$) The strain gauge should have a high value of gauge factor $G_f$. A high value of gauge factor indicates a large change in resistance for a particular strain resulting in high sensitivity.

($ii$) The resistance of the strain gauge should be as high as possible since this minimizes the effects of undesirable variations of resistance in the measurement

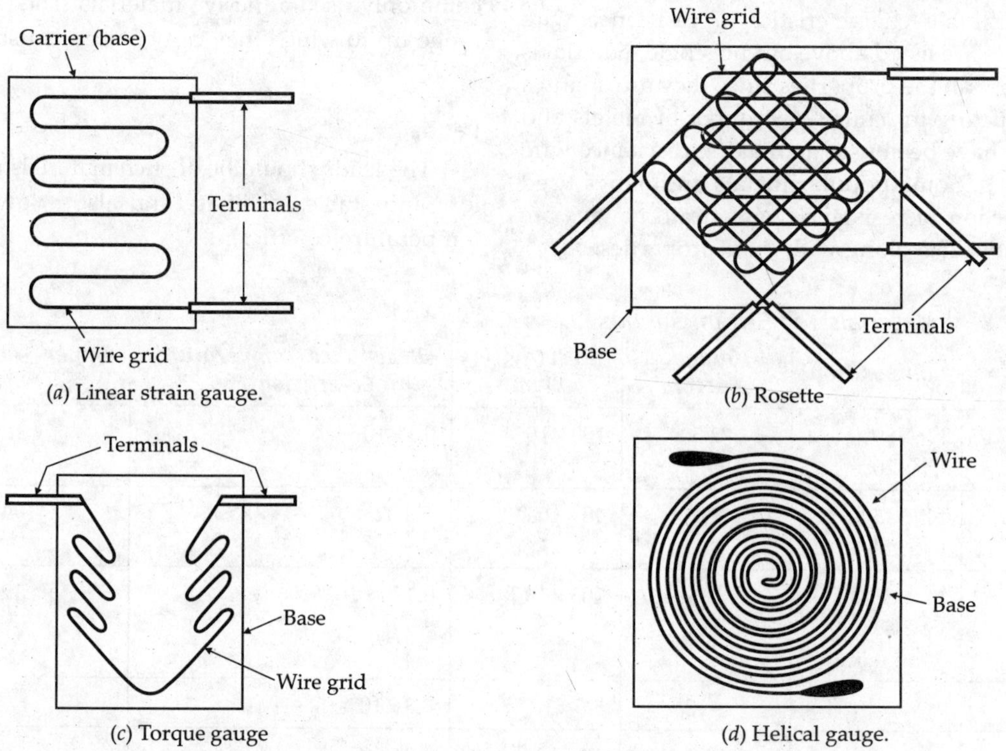


**Fig. 25.57** Resistance wire strain gauge.

circuit. Although undesirable, but necessary, are the resistance of connecting leads and terminals, etc. Typical resistances of strain gauges are 120 Ω, 350 Ω and 1000 Ω. Although a high resistance value of strain gauges is desirable from the point of view of swamping out the effects of variations of resistance in other parts of the bridge circuit in which they are invariably used but it results in lower sensitivity.

Thus in order to get high sensitivity higher bridge voltages have to be used. The bridge voltage is limited by the maximum current carrying capacity of the wires which is typically 30 mA.

(*iii*) The strain gauges should have a low resistance temperature co-efficient. This is essential to minimise errors on account of temperature variations which affect the accuracy of measurements.

(*iv*) The strain gauge should not have any hysteresis effects in its response.

(*v*) In order to maintain constancy of calibration over the entire range of the strain gauge, it should have linear characteristics *i.e.*, the variations in resistance should be a linear function of the strain.

(*vi*) The strain gauges are frequently used for dynamic measurements and hence their frequency response should be good. The linearity should be maintained within accuracy limits over the entire frequency range.

The desirable characteristics of resistance wire strain gauges are listed above but no single material is able to satisfy all the properties since they are, many a times, conflicting in nature. A number of metals and their alloys have been used in making resistance wire strain gauges. Some popular metals and alloys used for construction of resistance wire strain gauges are listed in Table 25.5 along with their properties.

### Base (Carrier) Materials

Several types of base or carrier materials are used to support the wires. Impregnated paper is used for room temperature applications. The range of some of the other materials is :

Epoxy – 200°C to 150°C, Bakelite cellulose or fibre glass materials – upto 200°C for continuous operation and 300°C for limited operation.

### Adhesives

The adhesives act as bonding materials. Like other bonding operations, successful strain gauge bonding depends upon careful surface preparation and use of the correct bonding agent. In order that the strain be faithfully transferred on to the strain gauge, the bond has to be formed between the surface to be strained and the plastic backing material on which the gauge is mounted. It is important that the adhesive should be suited to this backing ; numerous materials are used for backings which require different adhesive materials. It is usually desirable that the adhesive material should be of quick drying type and also be insensitive to moisture in order to have good adherence.

Ethylcellucose cement, nitrocellucose cement, bakelite cement and epoxy cement are some of the commonly used adhesive materials. The temperature range up to which they can be used is usually below 175°C.

### Leads

The leads should be of such materials which have low and stable resistivity and also a low resistance temperature co-efficient.

**Table 25.5** *Materials for Strain Gauges*

| *Material* | *Composition* | *Gauge Factor* | *Resistivity* Ω *m* | *Resistance Temperature Co-efficient/°C* | *Upper Temperature °C* |
|---|---|---|---|---|---|
| Nichrome | Ni : 80%<br>Cr : 20% | 2.5 | $100 \times 10^{-8}$ | $0.1 \times 10^{-3}$ | 1200 |
| Constantan | Ni : 45%<br>Cu : 55% | 2.1 | $48 \times 10^{-8}$ | $\pm 0.02 \times 10^{-3}$ | 400 |
| Isoelastic | Ni : 36%<br>Cr : 8%<br>Mo : 0.5% | 3.6 | $105 \times 10^{-8}$ | $0.175 \times 10^{-3}$ | 1200 |
| Nickel | – | –12 | $6.5 \times 10^{-8}$ | $6.8 \times 10^{-3}$ | – |
| Platinum | – | 4.8 | $10 \times 10^{-8}$ | $4.0 \times 10^{-3}$ | – |

### 25.17.3 Bonded Metal Foil Strain Gauges

This class of strain gauges is only an extension of the bonded metal wire strain gauges. The bonded metal wire strain gauges have been completely superseded by bonded metal foil strain gauges.

Metal foil strain gauges use identical or similar materials to wire strain gauges and are used today for most general-purpose stress analysis applications and for many transducers. The metals and alloys used for foil type strain gauges are shown in Table 25.5. A typical bonded metal foil strain gauge is shown in Fig. 25.58.

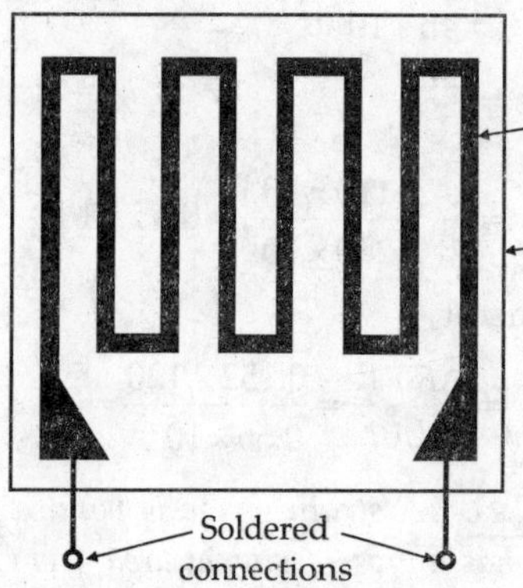


**Fig. 25.58** Bonded metal foil gauge.

Foil type gauges have a much greater heat dissipation capacity as compared with wire wound strain gauges on account of their greater surface area for the same volume. For this reason, they can be used for higher operating temperature range. Also the large surface area leads to better bonding.

The sensing elements of foil gauges are formed from sheets less than 0.005 mm thick by photo-etching processes, which allow greater flexibility with regard to shape. In Fig. 25.59, for example, the three linear grid gauges are designed with fat end turns. *This local increase in area reduces the transverse sensitivity which is a spurious input since the gauge is designed to measure the strain component along the length of grid elements.*

For foil type strain gauges, the manufacturing process also easily provides convenient soldering tabs, which are integral to the sensing grid, on all four gauges as shown in Fig. 25.59.

Foil type of gauges are employed for both stress analysis and as well as for construction of transducers. Foil type of gauges are mounted on a flexible insulating *carrier film* about 0.025 mm thick which is made of polymide, glass phenolic etc. Typical gauge resistances are 120, 350 and 1000 Ω with the allowable gauge current of 5 to 40 mA which is determined by the heat dissipation capabilities of the gauge. The gauge factors typically range from 2 to 4.

Minimum practical gauge size is constrained by the manufacturing limitations and handling attachment problems. The smallest gauge sizes are about 0.38 mm long. Foil type of gauges can be applied to curved surfaces ; the minimum safe bend radius can be as small as 1.5 mm in some strain gauges.

The maximum measurable strain varies from 0.5 to 4 percent. However, special *postyield* gauge devices allow measurement upto 0.1. Fatigue life of gauges varies with conditions. However, 10 million cycles at ± 1500 microstrain can be applied to foil gauges without causing failure.

Many different *adhesives* have been developed for fastening strain gauges to specimens. Gauges and fastening methods are available to cover temperature ranges from – 269°C to 816°C.

It is interesting to carry out simple calculations to find out what effect an applied stress has on a metal

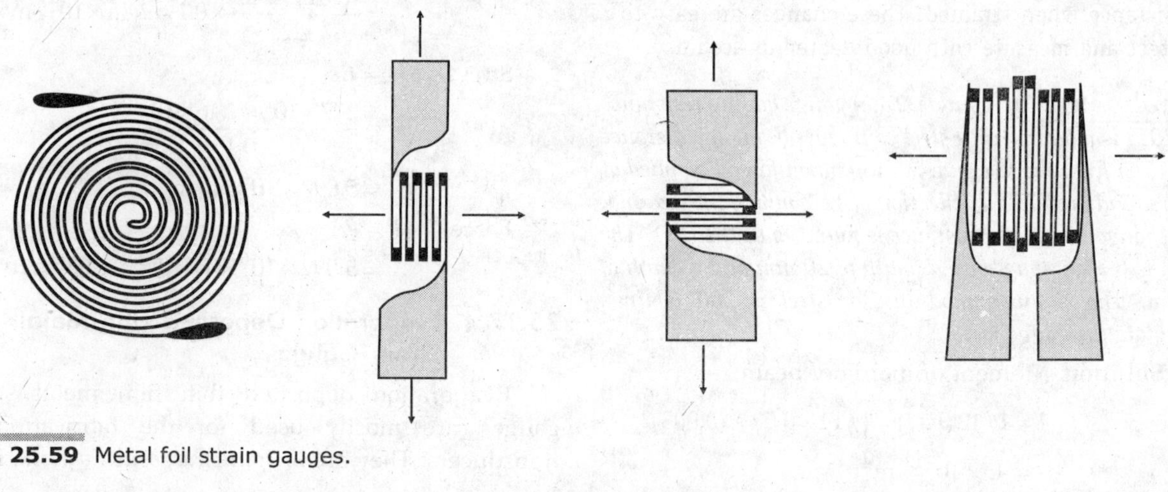

**Fig. 25.59** Metal foil strain gauges.

strain gauge. Hooke's law gives a relationship between stress and strain for linear stress-strain curve (*i.e.*, for elastic limits) in terms of modulus of elasticity of the material under stress. Hooke's law may be written as :

$$\text{Strain} \quad \varepsilon = s / E \qquad \ldots(25.71)$$

where $G$, $s$ and $E$ are respectively the strain, stress and modulus of elasticity. The units for stress and modulus of elasticity are $N/m^2$.

The change in the value of resistance is quite small as is clear from the following example.

**Example 25.18** *A resistance, wire strain gauge with a gauge factor of 2 is bonded to a steel structural member subjected to a stress of 100 $MN/m^2$. The modulus of elasticity of steel is 200 $GN/m^2$. Calculate the percentage change in the value of the gauge resistance due to the applied stress. Comment upon the results.*

**Solution.**

$$\text{Strain} \quad \varepsilon = \frac{s}{E} = \frac{100 \times 10^6}{200 \times 10^9}$$

$$= 500 \times 10^{-6} \text{ (500 microstrain)}$$

We have, $$\frac{\Delta R}{R} = G_f \varepsilon = 2 \times 500 \times 10^{-6} = 0.001 = 0.1\%$$

∴ The change in resistance is only 0.1%.

**Comments.** The above example illustrates that a very heavy stress of 100 $MN/m^2$ results in resistance change of only 0.1 per cent, which is by all means a very small change. This may present difficulties in measurement. Lower stresses produce still lower changes in resistance which may not be perceptible at all or the methods required to detect these changes may have to be highly accurate. To overcome this difficulty we must use strain gauges which have a high gauge factor which produce large changes in resistance when strained. These changes are easy to detect and measure with good degree of accuracy.

**Example 25.19** *A single strain gauge having resistance of 120 Ω is mounted on a steel cantilever beam at a distance of 0.15 m from the free end. An unknown force F applied at the free end produces a deflection of 12.7 mm of the free end. The change in gauge resistance is found to be 0.152 Ω. The beam is 0.25 m long with a width of 20 mm and a depth of 3 mm. The Young's modulus for steel is 200 $GN/m^2$. Calculate the gauge factor.*

**Solution.** Moment of inertia of beam,

$$I = 1/12\,(bd^3) = 1/12 \times 0.02 \times (0.003)^3$$

$$= 45 \times 10^{-12} \text{ m}^4$$

$$\text{Deflection} \quad x = \frac{Fl^3}{3EI}$$

$$\therefore \text{ Force } F = \frac{3EIx}{l^3}$$

$$= \frac{3 \times 200 \times 10^9 \times 45 \times 10^{-12} \times 12.7 \times 10^{-3}}{(0.25)^3}$$

$$= 22 \text{ N}$$

Bending moment at 0.15 m from free end

$$M = Fx = 22 \times 0.15 = 3.3 \text{ Nm}$$

Stress at 0.15 m from free end

$$s = \frac{M}{I} \cdot \frac{t}{2} = \frac{3.3}{45 \times 10^{-12}} \times \frac{0.003}{2} = 110 \text{ MN/m}^2$$

$$\text{Strain} \quad \varepsilon = \frac{\Delta L}{L}$$

$$= \frac{s}{E} = \frac{110 \times 10^6}{200 \times 10^9} = 0.55 \times 10^{-3}$$

∴ Gauge factor

$$= \frac{\Delta R / R}{\Delta L / L} = \frac{0.152 \times 120}{0.55 \times 10^{-3}} = 23$$

**Example 25.20** *A strain gauge is bonded to a beam 0.1 m long and has a cross-sectional area 4 $cm^2$. Young's modulus for steel is 207 $GN/m^2$. The strain gauge has an unstrained resistance of 240 Ω and a gauge factor of 2.2. When a load is applied, the resistance of gauge changes by 0.013 Ω. Calculate the change in length of the steel beam and the amount of force applied to the beam.*

**Solution.** We have gauge factor

$$G_f = \frac{\Delta R / R}{\Delta L / L}$$

and therefore change in length

$$\Delta L = \frac{\Delta R / R . L}{G_f}$$

$$= \frac{(0.013/240)}{2.2} \times 0.1 = 2.46 \times 10^{-6} \text{ m}$$

$$\text{Stress} \quad s = E\varepsilon$$

$$= \frac{207 \times 10^9 \times 2.46 \times 10^{-6}}{0.1}$$

$$= 5.092 \times 10^6 \text{ N/m}^2$$

$$\text{Force} \quad F = sA$$

$$= 5.092 \times 10^6 \times 4 \times 10^{-4} = 2.037 \times 10^3 \text{ N}$$

### 25.17.4 Evaporation Deposited Thin Metal Strain Gauges

Evaporation deposited thin film metal strain gauges are mostly used for the fabrication of transducers. They are of *sputter deposited variety*. Both processes begin with a suitable elastic metal element. The elastic metal element converts the physical quantity into a strain. To cite an example of a pressure transducer, a thin, circular metal diaphragm is formed. Both the evaporation and sputtering processes form all the strain gauge elements directly on the strain surface, they are not separately attached as in the case of bonded strain gauges.

In the evaporation process, the diaphragm is placed in a vacuum chamber with some insulating material. Heat is applied until the insulating material vaporises and then condenses, forming a thin dielectric film on the diaphragm. Suitably shaped templates are placed over the diaphragm, and the evaporation and condensation processes are repeated with the metallic gauge material, forming the desired strain gauge pattern on top of the insulating substrate.

In the *sputtering process*, a thin dielectric layer is deposited in vacuum over the entire diaphragm surface. The detailed mechanism of deposition is, however, entirely different from the evaporation method. The complete layer of metallic gauge is sputtered on the top of the dielectric material without using any substrate. The diaphragms are now removed from the vacuum chamber, and microimaging techniques using photomasking materials are used to form the gauge pattern. The diaphragms are then returned to the vacuum chamber. Sputter etching techniques are used to remove all unmasked metal layer, leaving behind the desired gauge pattern.

Resistance and gauge factors of film gauges are identical to those of foil gauges. Since no organic cementing materials are used, thin film gauges exhibit a better time and temperature stability.

### 25.17.5 Semi-conductor Strain Gauges

It has been explained above in order to have a high sensitivity, a high value of gauge factor is desirable. A high gauge factor means a relatively higher change in resistance which can be easily measured with a good degree of accuracy.

Semiconductor strain gauges are used where a very high gauge factor and a small envelope are required. The resistance of the semi-conductors changes with change in applied strain. Unlike in the case of metallic gauges where the change in resistance is mainly due to change in dimensions when strained, the semi-conductor strain gauge depend for their action upon **piezo-resistive effect** *i.e.*, the change in the value of the resistance due to change in resistivity.

Semi-conducting materials such as silicon and germanium are used as resistive materials for semi-conductor strain gauges. A typical strain gauge consists of a strain sensitive crystal material and leads that are sandwiched in a protective matrix. The production of these gauges employs conventional semi-conductor technology using semi-conducting wafers or filaments which have a thickness of 0.05 mm and bonding them on a suitable insulating substrates, such as teflon. Gold leads are generally employed for making the contacts. Some of the typical semi-conductor strain gauges are shown in Fig. 25.60. These strain gauges can be fabricated along with integrated circuit (IC) operational amplifiers which can act as pressure sensitive transducers.

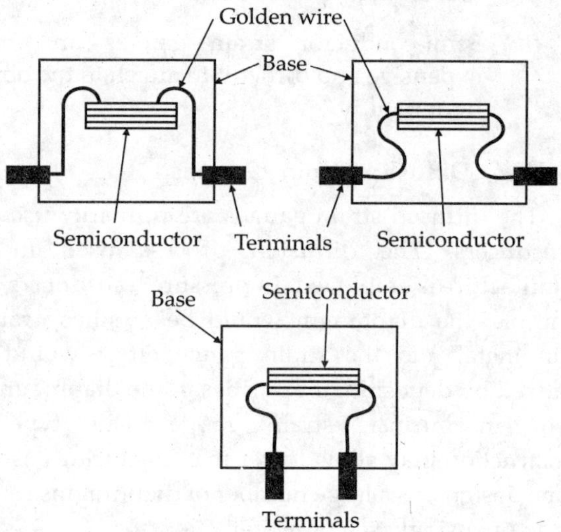


**Fig. 25.60** Semi-conductor strain gauge.

***Advantages***

(*i*) Semi-conductor strain gauges have the advantage that they have a high gauge factor of about ± 130. This allows measurement of very small strains of the order of 0.01 microstrain.

(*ii*) Hysteresis characteristics of semi-conductor strain gauges are excellent. Some units maintain it to less than 0.05%.

(*iii*) Fatigue life is in excess of $10 \times 10^6$ operations and the frequency response is upto $10^{12}$ Hz.

(*iv*) Semi-conductor strain gauges can be very small ranging in length from 0.7 to 7 mm. They are very useful for measurement of local strains.

***Disadvantages***

(*i*) The major and serious disadvantage of semi-conductor strain gauges is that they are very sensitive to changes in temperature.

(*ii*) Linearity of the semi-conductor strain gauge is poor. The equation for the fractional change in resistance is :

$$\Delta R / R = As + B\,\varepsilon^{-4}\,\text{m}$$

where *A* and *B* are constants.

This gauge is rather non-linear at comparatively high strain levels. The gauge factor varies with strain. For example, if the gauge factor is 130 at 0.2 percent strain, then it is 112 at 0.4 percent strain. The characteristics can be made linear by proper doping.

(*iii*) Semi-conductor strain gauges are more expensive and difficult to attach to the object under study.

### 25.17.6 Diffused Strain Gauges

The diffused strain gauges are primarily used in transducers. The diffusion process used in IC manufacture is employed. In pressure transducers, for example, the diaphragm would be of silicon rather than metal and the strain gauge effect would be realized by depositing impurities in the diaphragm to form an intrinsic strain gauge. This type of construction may allow lower manufacturing costs in some designs, as a large number of diaphragms can be made on a single silicon wafer.

## 25.18 ROSETTES

In addition to single element strain gauges, a combination of strain gauges called *"Rosettes"* are available in many combinations for specific stress analysis or transducer applications.

In practical problems, an element may be subjected to stresses in any direction and hence it is not possible to locate the direction of principal stress. Therefore, it is not possible to orient the strain, gauges along the direction of principal stress. Hence there is a necessity to evolve a strain gauge measurement system which measures the values of principal strains and stresses without actually knowing their directions. The solution to the problem lies in using three strain gauges to form a unit called a *Rosette*. Some forms of Rosettes are shown in Fig. 25.61.

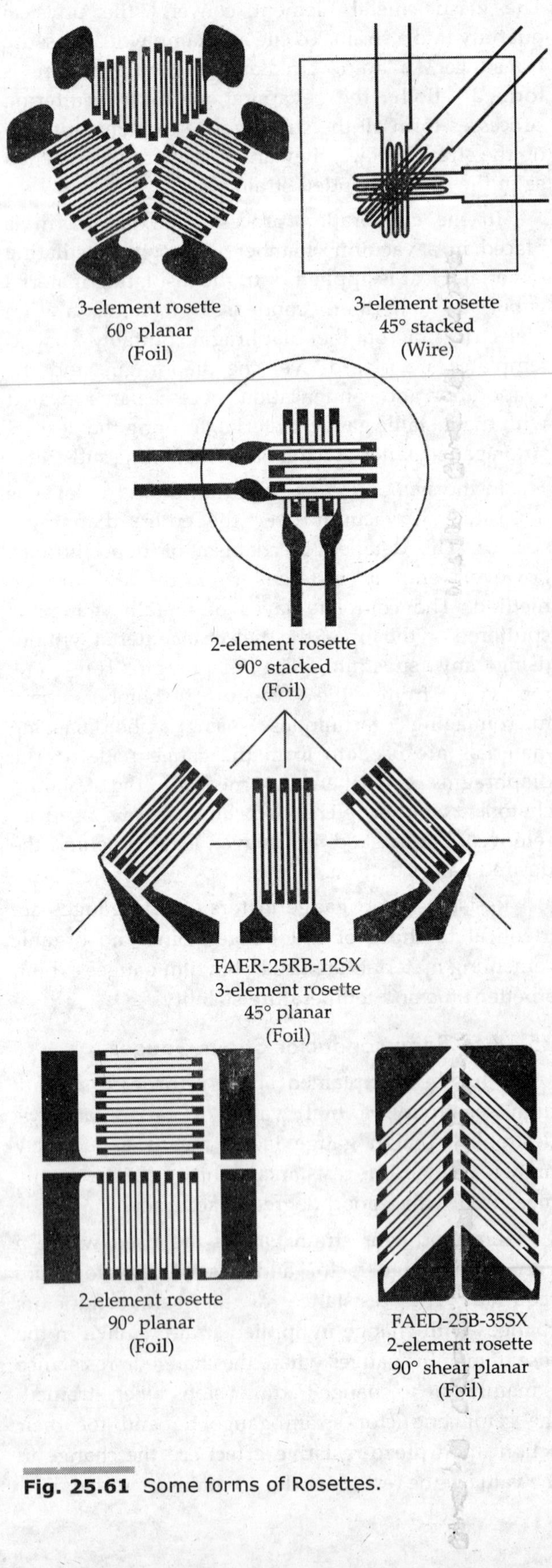


**Fig. 25.61** Some forms of Rosettes.

## 25.19 RESISTANCE THERMOMETERS

The resistance of a conductor changes when its temperature is changed. This property is utilized for measurement of temperature.

The variation of resistance $R$ with temperature $T$(°K) can be represented by the following relationship for most of the metals as :

$$R = R_0(1+\alpha_1 T+\alpha_2 T^2+...+\alpha_n T^n+...) \quad ...(25.72)$$

where $R_0$ = resistance at temperature $T = 0$

and $\alpha_1, \alpha_2, \alpha_3, ...\alpha_n$ are constants.

The resistance thermometer uses the change in electrical resistance of conductor to determine the temperature.

The resistivity of metals showed a marked dependence on temperature was discovered by Sir Humphry Davy. A few years later Sir William Siemens preferred the use of platinum as an element in the resistance thermometer. His choice proved most propitious, as platinum is used to this day as the primary element in all high accuracy resistance thermometers. In fact, the platinum resistance temperature detector (PRTD) is used today as an interpolation standard from oxygen point (– 182.96°C) to antimony point (630.74°)

*Platinum* is especially suited for this purpose, as it can withstand high temperatures while maintaining excellent stability. As a noble metal, it shows limited susceptibility to contamination. All metals produce a positive change in resistance with temperature. This, of course, is the main function of an RTD. The system error is minimized when the nominal value of RTD is large. This implies a metal with a high value of resistivity should be used for RTDs. The lower is the resistivity of the metal, the more material we will have to use.

The requirements of a conductor material to be used in RTDs are :

(*i*) The change in resistance of material per unit change in temperature should be as large as possible.

(*ii*) The material should have a high value of resistivity so that minimum volume of material is used for the construction of RTD.

(*iii*) The resistance of materials should have a continuous and stable relationship with temperature.

**Gold** and **Silver** are rarely used for construction of RTDs on account of their low resistivities. **Tungsten** has relatively a high resistivity, but is reserved for high temperature applications as it is extremely brittle and difficult to work. **Copper** is used occasionally as an RTD element. Its low resistivity forces the element to be longer than the platinum element, but its low linearity and low cost make it an economical alternative. Its upper limit of temperature is about 120°C.

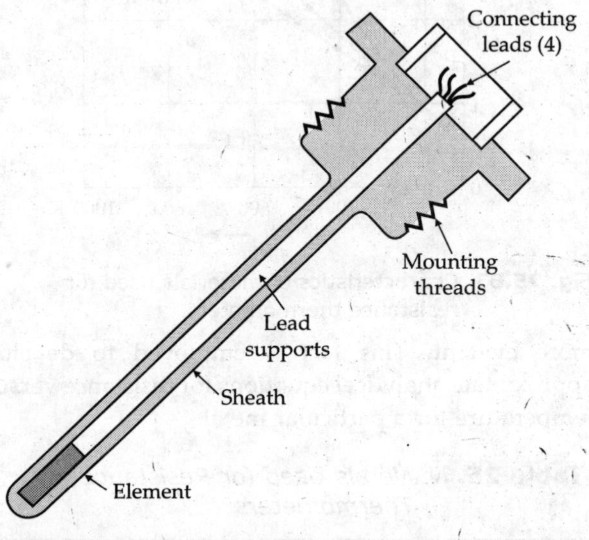


**Fig. 25.62** Industrial platinum resistance thermometer.

The most common RTDs are made of either platinum, nickel or nickel alloys. The economical **nickel wires** are used over a limited temperature range. They are quite non-linear and tend to drift with time. For measurement integrity, **platinum** is the obvious choice.

The common values of resistance for a platinum RTD range from 10 Ω for the bird cage model to several thousands ohm for the film RTD. The single most common value is 100 Ω at 0°C with a resistance temperature co-efficient of 0.00385/°C. The more chemically pure platinum wire has a resistance temperature co-efficient of 0.00392/°C.

The construction of an industrial type of platinum RTD is shown in Fig. 25.62.

The characteristics of various materials used for resistance thermometers are plotted in Fig. 25.63.

Metals commonly used for resistance thermometers are listed in Table 25.6 along with their salient properties.

An examination of the resistance versus temperature curves of Fig. 25.63, shows that the curves are nearly linear. If fact, when only short temperature spans are considered, the linearity is

25

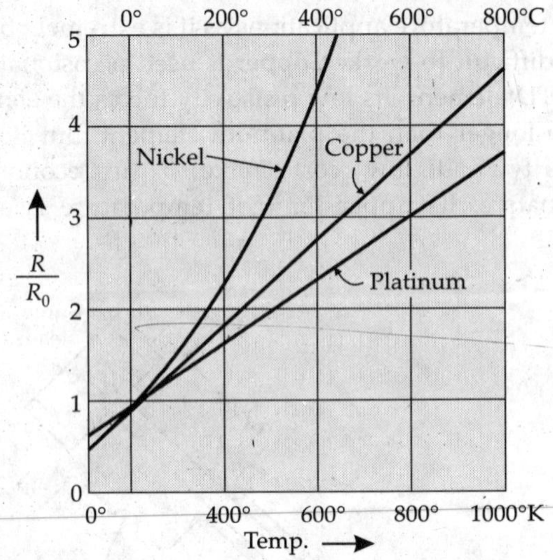


**Fig. 25.63** Characteristics of materials used for resistance thermometers.

more evident. This fact is employed to develop approximate analytical equations for resistance versus temperature for a particular metal.

**Table 25.6** *Metals Used for Resistance Thermometers*

| Metal | Resistance temperature Co-efficient /°C | Temperature range °C | | Melting point °C |
|---|---|---|---|---|
| | | Min | Max | |
| Platinum | 0.39 | –260 | 110 | 1773 |
| Copper | 0.39 | 0 | 180 | 1083 |
| Nickel | 0.62 | –220 | 300 | 1435 |
| Tungsten | 0.45 | –200 | 1000 | 3370 |

### 25.19.1 Linear Approximation

A linear approximation means that we may develop an equation for a straight line which approximates the resistance versus temperature curve over a specified span. Figure 25.64 shows a curve of variation of resistance $R$ with temperature $\theta$°C.

Here a straight line has been drawn between the points of the curve which represent $\theta_1$°C and $\theta_2$°C with $\theta_0$°C representing the mid point temperature. The equation of this straight line is the *linear approximation* of the curve from $\theta_1$°C to $\theta_2$°C. The equation of the straight line is written as :

$$R_\theta = R_{\theta_0}(1 + a_{\theta_0}\,\Delta\theta) \text{ with } \theta_1 < \theta_0 < \theta_2 \qquad \ldots(25.73)$$

where $R_\theta$ = approximate resistance at $\theta$°C; Ω,

$R_{\theta_0}$ = approximate resistance at $\theta_0$°C; Ω,

$\Delta\theta = \theta - \theta_0$ = change in temperature ; °C,

$\alpha_{\theta_0}$ = resistance temperature co-efficient at temperature $\theta_0$°C; /°C

The reason for using $\alpha_{\theta_0}$ as the fractional slope of resistance-temperature curve is that this same constant be used for conductors having the same material with different dimensions. The value of $\alpha_{\theta_0}$ can be found from the value of resistance and temperatures as shown in 25.64. In general,

$$\alpha_{\theta_0} = \frac{1}{R_{\theta_0}} \times (\text{slope at } \theta_0)$$

$$= \frac{1}{R_{\theta_0}}\left[\frac{R_{\theta_2} - R_{\theta_1}}{\theta_2 - \theta_1}\right] \qquad \ldots(25.74)$$

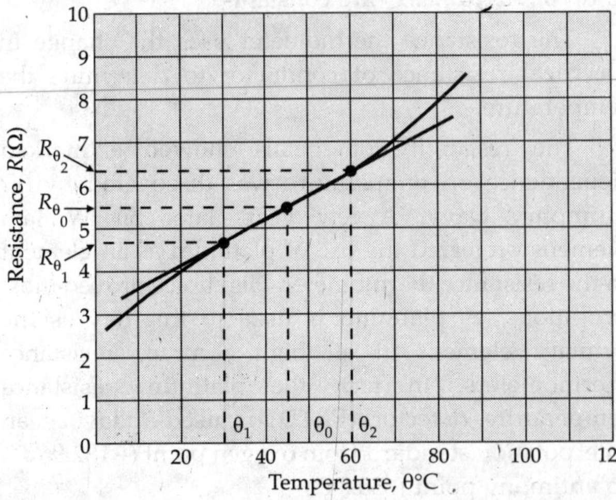


**Fig. 25.64** A straight line representing an approximate relationship between resistance $R$ and temperature $\theta$.

***Example 25.21*** *Find the linear approximation between 30°C to 60°C using the resistance-temperature curve as shown in Fig. 25.64.*

**Solution.** We have

$$\theta_1 = 30°\text{C} \quad \text{and} \quad \theta_2 = 60°\text{C}$$

Mean temperature

$$\theta_0 = (30 + 60)/2 = 45°\text{C}$$

From Fig. 25.64, we have :

Resistance at 30°C,

$R_{\theta_1} = 4.8\ \Omega$, resistance at 60°C,

$R_{\theta_2} = 6.2\ \Omega$ and resistance at 45°C

$R_{\theta_0} = 5.5\ \Omega$

From Eqn. 25.74,

$$\alpha_{\theta_0} = \frac{1}{R_{\theta_0}} \times \frac{R_{\theta_2} - R_{\theta_1}}{\theta_2 - \theta_1}$$

$$= \frac{1}{5.5} \times \frac{6.2 \times 4.8}{60 - 30} = 0.0085°\text{C}.$$

Hence, the linear approximation is :

$$R_\theta = 5.5[1 + 0.0085(\theta - 45)]\ \Omega$$

### 25.19.2 Quadratic Approximation

A quadratic approximation of the resistance temperature curve is more accurate representation of the curve over a limited range of temperature. The quadratic approximation relationship includes both a linear term as in Eqn. 25.73 and an addition has a term which varies as the square of the difference in temperature. The approximation is written as :

$$R_\theta = R_{\theta_0}[1+\alpha_1\Delta\theta+\alpha_2(\Delta\theta)^2] \quad ...(25.75)$$

where

$\alpha_1$ = linear fractional change in resistance ; /°C

and $\alpha_2$ = quadratic fractional change in resistance, /(°C)$^2$

Values of $\alpha_1$ and $\alpha_2$ are found from tables and graphs as indicated in the example given below by using values of resistance and temperature at three different points. Two equations are formed and values of $\alpha_1$ and $\alpha_2$ are calculated from these.

**Example 25.22** *Use the following values of resistance versus temperature for an RTD to find the linear and quadratic approximations of resistance between 100°C and 130°C about a mean temperature of 115°C.*

| ***Temperature, °C*** | ***Resistance Ω*** |
|---|---|
| 90 | 562.66 |
| 95 | 568.03 |
| 100 | 573.40 |
| 105 | 578.77 |
| 110 | 584.13 |
| 115 | 589.48 |
| 120 | 594.84 |
| 125 | 600.18 |
| 130 | 605.52 |

**Solution. 1. Linear Expansion** : Given :

$\theta_1 = 100°C,\ \theta_2 = 130°C,\ \theta_0 = 115°C$

$R_{\theta_1} = 573.40\ \Omega,\ R_{\theta_2} = 605.52\ \Omega,\ R_{\theta_0} = 589.48\ \Omega$

From Eqn. 25.74,

$$\alpha_{\theta_1} = \alpha_1 = \frac{1}{R_{\theta_0}} \times \frac{R_{\theta_2} - R_{\theta_1}}{\theta_2 - \theta_1}$$

$$= \frac{1}{598.48} \times \frac{605.52 - 573.40}{130 - 100}$$

$$= 0.00182\ /°C$$

∴ The linear approximation is :

$$R_\theta = 598.48[1 + 0.00182(\theta - 115)]\ \Omega$$

**2. Quadratic Expansion :** The resistance $R_\theta$ at any temperature θ°C is given by :

$$R_\theta = R_{\theta_0}[1+\alpha_1\Delta\theta+\alpha_2(\Delta\theta)^2]$$

We can find the quadratic terms, by forming two equations using two points about the mean temperature $\theta_0 = 115°C$

We have, $R_{\theta_0} = 598.48\ \Omega$ at $\theta_0 = 115°C$.

Now using 100°C and 130°C as the two points, we have :

$$573.40 = 589.48[1+\alpha_1(100-115)+\alpha_2(100-115)^2]$$

$$605.52 = 589.48[1+\alpha_1(130-115)+\alpha_2(130-115)^2]$$

From above, we have,

$$\alpha_1 = 1.823 \times 10^{-3}/°C$$

and $$\alpha_2 = -0.22 \times 10^{-6}/(°C)^2$$

Hence,

$$R_\theta = 589.48[1 + 1.823 \times 10^{-3}(\theta - 115) - 0.22 \times 10^{-6}(\theta - 115)^2]\ \Omega$$

**Example 25.23** *A platinum thermometer has a resistance of 100 Ω at 25°C.*

(*a*) *Find its resistance at 65°C if the platinum has a resistance temperature co-efficient of 0.00392/°C.*

(*b*) *If the thermometer has a resistance of 150 Ω, calculate the temperature.*

**Solution.** (*a*) Using the linear approximation, the resistance at any temperature θ°C, is

$$R_\theta = R_{\theta_0}[1+\alpha_{\theta_0}\Delta\theta]$$

∴ Resistance at 65°C is,

$$R_{65} = 100[1+0.00392(65-25)]$$

$$= 115.68\ \Omega$$

(*b*) Suppose θ is the unknown temperature,

$$150 = 100[1+0.00392(\theta-25)]$$

or $$\theta = 152.55°C$$

**Example 25.24** *A copper resistor at 20°C is to used to indicate the temperature of bearings of a machine. What resistance should not be exceeded if the maximum bearing temperature is not to exceed 150°C ? The resistance temperature co-efficient of copper is 0.00393/°C at 20°C.*

**Solution.** The value of resistance in case the temperature is not to exceed 150°C can be calculated as under.

$$R_{150} = 10[1+0.00392(150-20)] = 15.11\ \Omega$$

**Example 25.25** *A temperature alarm unit with a time constant of 120 s is subjected to a sudden rise of temperature of 50°C because of fire. If an increase of 30°C is required to actuate the alarm, what will be the delay in sudden temperature increase ?*

25

**Solution.** Assume the thermometer be a first order system, the variation of indicated temperature $\theta$, to a step input temperature $\theta_0$ is,

$$\theta = \theta_0[1 - \exp(-t/\tau)]$$

$$30 = 50[1 - \exp(-t/120)]$$

or $$t = 110 \text{ s}$$

The alarm would be delayed by 110 s.

## 25.20 THERMISTORS

Thermistor is a contraction of a term **"thermal resistors"**. Thermistors are generally composed of semi-conductor materials. Although positive temperature co-efficient of units (which exhibit an increase in the value of resistance with increase in temperature) are available, most thermistors have a negative coefficient of temperature resistance *i.e.* their resistance decreases with increase of temperature. The negative temperature coefficient of resistance can be as large as several percent per degree celsius. This allows the thermistor circuits to detect very small changes in temperature which could not be observed with an RTD or a thermocouple. In some cases the resistance of thermistor at room temperature may decrease as much as 5 percent for each 1°C rise in temperature. This high sensitivity to temperature changes makes thermistors extremely useful for precision temperature measurements control and compensation.

Thermistors are widely used in applications which involve measurements in the range of – 60°C to 15°C. The resistance of thermistors ranges from 0.5 Ω to 0.75 MΩ. Thermistor is a highly sensitive device. The price to be paid off for the high sensitivity is in terms of linearity. The thermistor exhibits a highly non-linear characteristic of resistance versus temperature.

### 25.20.1 Construction of Thermistors

Thermistors are composed of sintered mixture of metallic oxides such as manganese, nickel, cobalt, copper, iron and uranium. They are available in variety of sizes and shapes. The thermistors may be in the form of beads, rods and discs. Some of the commercial forms are shown in Fig. 25.65.

A thermistor in the form of a bead is smallest in size and the bead may have a diameter of 0.015 mm to 1.25 mm. Beads may be sealed in the tips of solid glass rods to form probes which may be easier to mount than the beads. Glass probes have a diameter of about 2.5 mm and a length which varies from 6 mm to 50 mm. Discs are made by pressing material under high pressure into cylindrical flat shapes with diameters ranging from 2.5 mm to 25 mm.

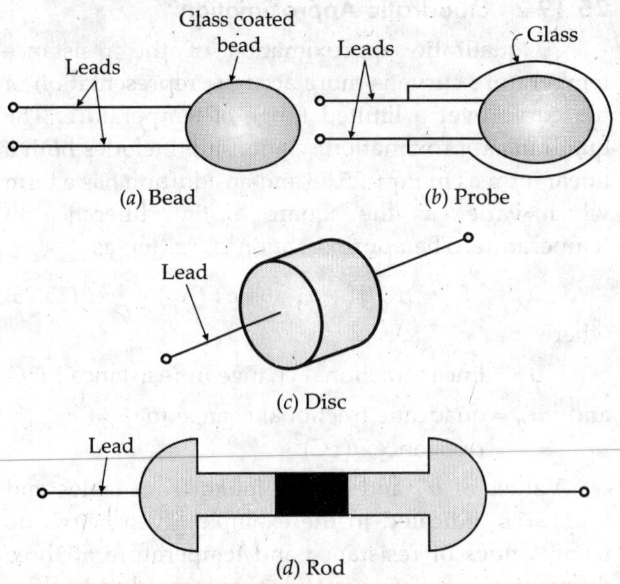


**Fig. 25.65** Different forms of construction of thermistors.

### 25.20.2 Resistance-Temperature Characteristics of Thermistors

The mathematical expression for the relationship between the resistance of a thermistor and absolute temperature of thermistor is :

$$R_{T1} = R_{T2} \exp\left[\beta\left(\frac{1}{T_1} - \frac{1}{T_2}\right)\right] \quad \text{...(25.76)}$$

where $R_{T1}$ = resistance of the thermistor at absolute temperature $T_1$ ; °K,

$R_{T2}$ = resistance of the thermistor at absolute temperature $T_2$ ; °K,

and $\beta$ = a constant depending upon the material of thermistor, typically 3500 to 4500 °K

The resistance temperature characteristics of a typical thermistor are given in Fig. 25.66. The resistance temperature characteristics of Fig. 25.66 shows that a thermistor has a very high negative temperature co-efficient of resistance, making it an ideal **temperature transducer**. Figure 25.66 also shows the resistance temperature characteristics of platinum which is a commonly used material for resistance thermometers. Let us compare the characteristics of the two materials. Between – 100°C and 400°C, the thermistor changes its resistivity from $10^5$ to $10^{-2}$ Ωm, a factor of $10^7$, while platinum changes its resistivity by a factor of about 10 within the same temperature range. This explains the high sensitivity of thermistors for measurement of temperature.

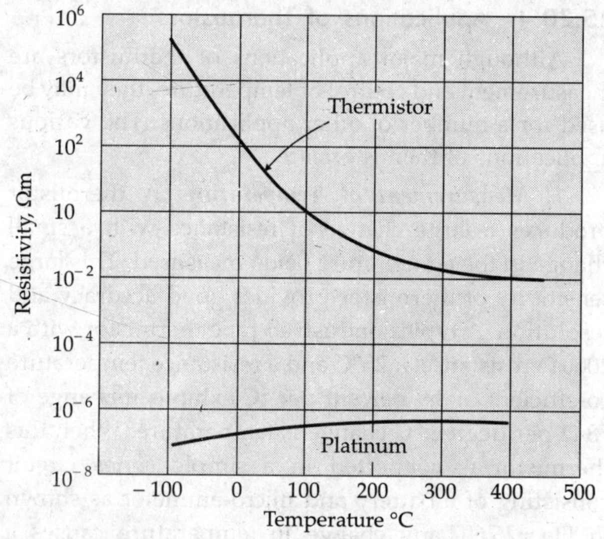


**Fig. 25.66** Resistance-temperature characteristics of a typical thermistor and platinum.

The characteristics of thermistors are no doubt non-linear but a linear approximation of the resistance-temperature curve can be obtained over a small range of temperature. Thus, for a limited range of temperature, the resistance of a thermistor varies as given by Eqn. 25.77.

$$R_\theta = R_{\theta_0}[1 + \alpha_{\theta_0}\Delta\theta] \quad ...(25.77)$$

A thermistor exhibits a negative resistance temperature co-efficient which is typically about 0.05/°C.

An individual thermistor curve can be closely approximated through the **Steinhart-Hart equation :**

$$\frac{1}{T} = A + B\log_e R + C(\log_e R)^3 \quad ...(25.78)$$

where $T$ = temperature ; °K,

$R$ = resistance of thermistor ; Ω,

$A, B, C$ = curve fitting constants.

$A$, $B$ and $C$ are found by selecting three data points on the published data curve and solving the three simultaneous equations. When the data points are chosen to span no more than 100°C within the nominal centre of thermistors temperature range, this equation approaches a remarkable ±0.2°C curve fit.

A simpler equation is :

$$T = \frac{B}{\log_e R - A} - C \quad ...(25.79)$$

where $A$, $B$ and $C$ are found by selecting three $(R, T)$ data points and solving three resultant simultaneous equations. Equation 25.79 must be applied over a narrower temperature range in order to approach the accuracy achieved by Steinhart-Hart equation. Another relationship that can be conveniently used for resistance-temperature curve of thermistors is :

$$R_T = aR_0\exp(b/T) \quad ...(25.80)$$

where, $R_T, R_0$ = resistance of thermistor at temperature $T$°K and ice point respectively.

### 25.20.3 Voltage-Current and Current-Time Characteristics of Thermistors

Three important characteristics of thermistor make them extremely useful in measurement and control applications. These are :

- the *resistance-temperature* characteristics.
- the *voltage-current* characteristics.
- the *current-time* characteristics.

**1.** ***Resistance-Temperature Characteristics.*** The resistance-temperature characteristics have already been described in Art. 25.20.2. The other two characteristics are described below.

**2.** ***Voltage-Current Characteristics.*** These characteristics are shown in Fig. 25.67. Figure 25.67 shows that the voltage drop across a thermistor increases with increasing current until is reaches a peak value beyond which the voltage drop decreases as the current increases. In this portion of the curve, the thermistor exhibits a *negative resistance* characteristic. If a very small voltage is applied to the thermistor, the resulting small current does not produce sufficient heat to raise the temperature of the thermistor above ambient. Under this condition, Ohm's law is followed and the current is proportional to the applied voltage. Larger currents, at larger

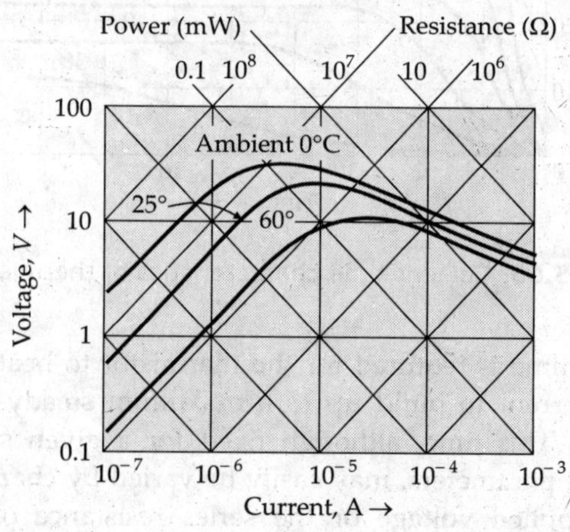


**Fig. 25.67** Voltage-current characteristics of thermistors.

25

applied voltages, produce enough heat to raise the thermistor temperature above the ambient temperature and its resistance then decreases. As a result, more current is then drawn and the resistance decreases further. The current continues to increase until the heat dissipation of the thermistor equals the power supplied to it. Therefore, under any fixed ambient conditions, the resistance of a thermistor is largely a function of the power being dissipated within itself, provided that there is enough power available to raise its temperature above ambient. Under such operating conditions, the temperature of the thermistor may rise 100°C or 200°C and its resistance may drop to one-thousandth of its value at low current.

This characteristic of *self-heat* provides an entirely new field of uses for the thermistor. In the self-heat state, the thermistor is sensitive to anything that changes the rate at which heat is conduced away from it. It can so be used to measure *flow, pressure, liquid level, composition of gases*, etc. If, on the other hand, the rate of heat removal is fixed, then the thermistor is sensitive to power input and can be used for *voltage* or *power-level* control.

**3. *Current Time Characteristics.*** The current-time characteristics shown in Fig. 25.68 indicate the time delay to reach maximum current as a function of the applied voltage. When the heating effect just described occurs in a thermistor network, a certain

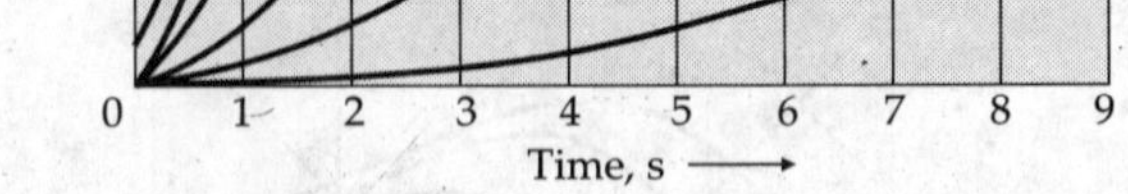


**Fig. 25.68** Current-time characteristics of thermistors.

finite time is required for the thermistor to heat and the current to build up to a maximum steady-state value. This time, although fixed for a given set of circuit parameters, may easily be varied by changing the applied voltage or the series resistance of the circuit. This time-current effect provides a simple and accurate means of achieving time delays from milliseconds to many minutes.

## 25.20.4 Applications of Thermistors

Although major applications of thermistors are measurement and control of temperature, they may be used for a number of other applications. The various applications of themistors are :

**1. *Measurement of Temperature.*** A thermistor produces a large change of resistance with a small change in the temperature being measured. This large sensitivity of thermistor provides good accuracy and resolution. A typical industrial-type thermistor with a 2000 Ω resistance at 25°C and a resistance temperature co-efficient of 3.9 percent per °C exhibits a change of 78 Ω per degree °C change is temperature. When this thermistor is connected in a simple series circuit consisting of a battery and micro-ammeter as shown in Fig. 25.69, any change in temperature causes a change in the resistance of thermistor and corresponding change in the circuit current. The micro-ammeter may be directly calibrated in terms of temperature. The micro-ammeter may be able to give a resolution of 0.1°C.

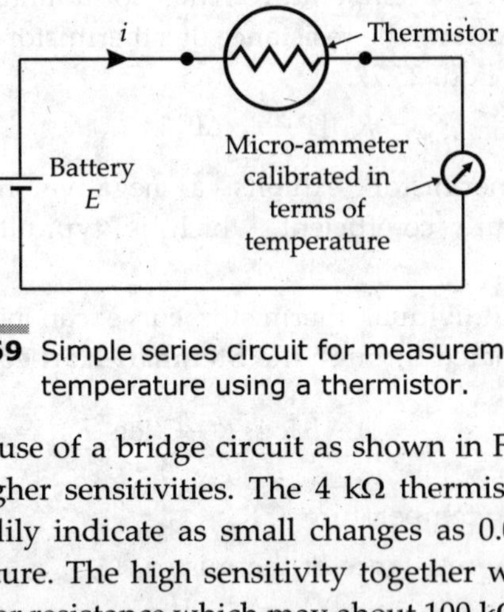


**Fig. 25.69** Simple series circuit for measurement of temperature using a thermistor.

The use of a bridge circuit as shown in Fig. 25.70 gives higher sensitivities. The 4 kΩ thermistor used will readily indicate as small changes as 0.005°C in temperature. The high sensitivity together with high thermistor resistance which may about 100 kΩ, makes the thermistor ideal for remote measurement or control, as the changes in contact or transmission line resistances due to change in the ambient temperature have almost a negligible effect on the accuracy of measurement or control. For example a 150 m long transmission line made of copper when subjected to change of 25°C will affect the accuracy of measurement or control by approximately 0.05°C.

**2. *Control of Temperature.*** A simple temperature control circuit may be constructed by replacing the micro-ammeter shown in the typical thermistor temperature control circuit of 25.70 with a relay. This is shown in the typical thermistor temperature control

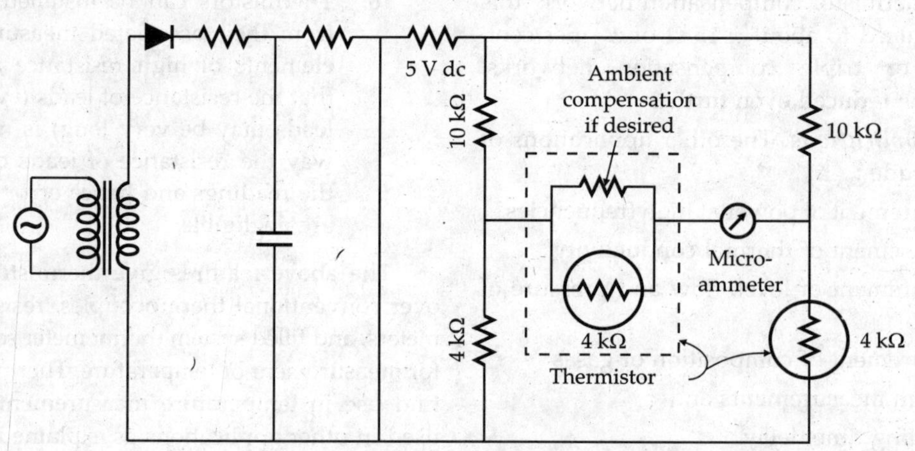


**Fig. 25.70** Measurement of temperature using a thermistor and a bridge circuit for getting higher sensitivities.

circuit of Fig. 25.71. It uses a 4 kΩ thermistor connected in an a.c. excited bridge. The unbalance voltage is fed to an a.c. amplifier whose output excites a relay coil. The relay contacts are used to control the current in the circuit which generates the heat. These circuits can be controlled to a precision of 0.00005°C.

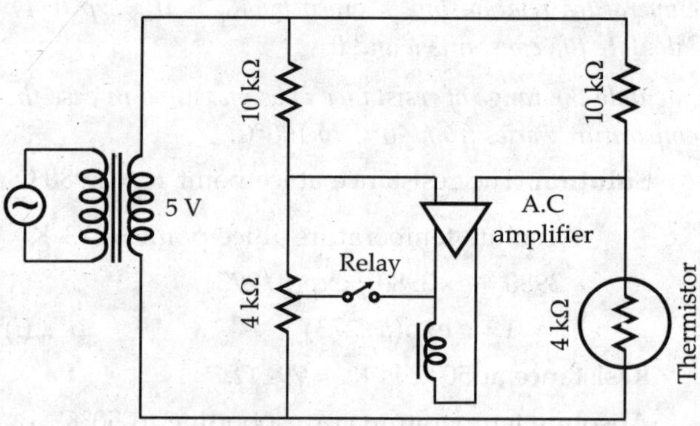


**Fig. 25.71** Typical thermistor temperature control circuit.

Thermistor control systems are inherently sensitive, stable and fast acting and require relatively simple circuitry. The voltage output of the standard bridge circuit at 25°C is about 18 mV/°C using a 4 kΩ thermistor in the configuration of Fig. 25.70.

3. ***Temperature Compensation.*** Because thermistors have a negative temperature coefficient of resistance-opposite to the positive coefficient of most electrical conductors and semiconductors—they are widely used to compensate for the effects of temperature on both component and circuit performance. Disk-type thermistors are used for this purpose where the maximum temperature does not exceed 125°C. A properly selected thermistor, mounted against or near a circuit element, such as a copper meter coil, and experiencing the same ambient temperature changes, can be connected in such a way that the total circuit resistance is constant over a wide range of temperatures. This is shown in the curves of Fig. 25.72, which illustrates the effect of a compensation network.

The compensator consists of a thermistor, shunted by a resistor. The negative temperature coefficient of this combination equals the positive coefficient of the copper coil. The coil resistance of 5,000 Ω at 25°C, varies from approximately 4,500 Ω at 0°C to 5,700 Ω at 60°C, representing a change of about ± 12 per cent.

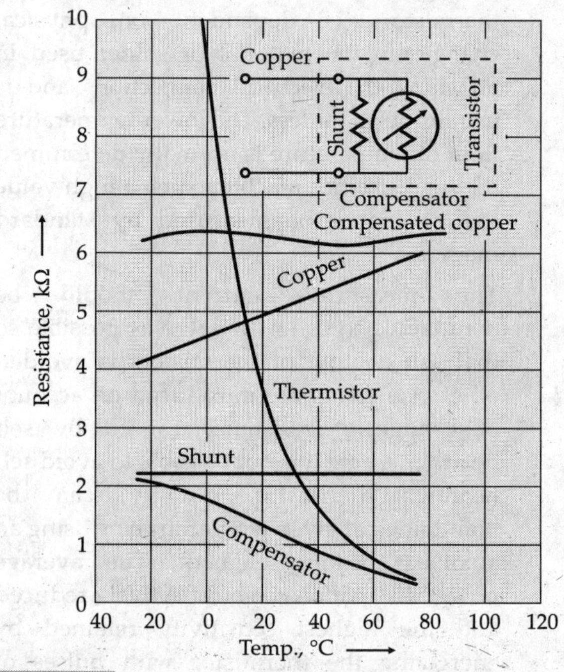


**Fig. 25.72** Temperature compensation of a copper conductor by means of a thermistor network.

25

With a single thermistor compensation network, this variation is reduced to about $\pm$ 15 $\Omega$ or $\pm\frac{1}{4}$ per cent. With double or triple compensation networks, variations can be reduced even further.

**4. *Other Applications.*** The other applications of thermistors include :

(*i*) measurement of power at high frequencies
(*ii*) measurement of thermal conductivity
(*iii*) measurement of level, flow and pressure of liquids
(*iv*) measurement of composition of gases
(*v*) vacuum measurements and
(*vi*) providing time delay.

### 25.20.5 Salient Features of Thermistors

1. Thermistors are compact, rugged and inexpensive.
2. Thermistors when properly aged, have good stability.
3. The response time of thermistors can vary from a fraction of a second to minutes, depending on the size of the detecting mass and thermal capacity of the thermistor. It varies inversely with the dissipation factor. The power dissipation factor varies with the degree of thermal isolation of the thermistor element.
4. The upper operating limit of temperature for thermistors is dependent on physical changes in the material or solder used in attaching the electrical connections and is usually 400°C or less. The lower temperature limit of temperature is normally determined by the resistance reaching such a high value that it cannot be measured by standard methods.
5. The measuring current should be maintained to as low a value as possible so that self heating of thermistors is avoided otherwise errors are introduced on account of change of resistance caused by self heating. Where it is not possible to avoid self heating, thermistor stability can be maintained at given temperature by using an auxiliary heating element. The average power dissipation can be effectively reduced and the highest sensitivity retained by energizing the thermistor with pulses of measuring power.
6. Thermistors can be installed at a distance from their associated measuring circuits if elements of high resistance are used such that the resistance of leads (even though the leads may be very long) is negligible. This way the resistance of leads does not affect the readings and hence errors on this count are negligible.

The above features give thermistors advantages over conventional thermocouples, resistance thermometers, and filled system thermometer sensing elements for measurement of temperature. Thermistors not only find use in temperature measurements but are also used in other applications as explained in 25.20.4.

***Example 25.26*** *A thermistor has a resistance temperature co-efficient of –5% over a temperature range of 25°C to 50°C. If the resistance of the thermistor is 100 Ω at 25°C, what is the resistance at 35°C ?*

**Solution.** Resistance at a temperature of 35°C is :

$$R_{35} = 100[1-0.05(35-25)] = 50°\Omega$$

***Example 25.27*** *A thermistor has a resistance of 3980 Ω at the ice point (0°C) and 794 Ω at 50°C. The resistance-temperature relationship is given by $R_T = aR_0 \exp(b/T)$. Calculate the constants a and b.*

*Calculate the range of resistance to be measured in case the temperature varies from 40°C to 100°C.*

**Solution.** The resistance at ice point $R_0 = 3980\ \Omega$

Absolute temperature at ice point = 273°K

$$\therefore \quad 3980 = a \times 3980 \times \exp(b/273)$$

$$\text{or} \quad 1 = a\exp(b/273) \qquad ...(i)$$

Resistance at 50°C is $R_T = 794\ \Omega$.

Absolute temperature corresponding to 50°C

$$T = 273 + 50° = 323\text{ K°}.$$

$$\text{Hence, } 794 = a \times 3980 \exp(b/323) \qquad ...(ii)$$

$$= 3980 \exp(b/323)$$

Solving (*i*) and (*ii*), we have,

$$a = 30 \times 10^{-6} \text{ and } b = 2843$$

Absolute temperature at 40°C

$$= 273 + 40 = 313\text{ K°}.$$

$\therefore$ Resistance at 40°C

$$= 30 \times 10^{-6} \times 3980 \times \exp(2843/313)$$

Absolute temperature at 100°C

$$= 273 + 100 = 373\ °\text{K}$$

$\therefore$ Resistance at 100°C

$$= 30 \times 10^{-6} \times 3980 \times \exp(2843/373)$$

Thus the range of resistance is 244 Ω to 1051 Ω.

**Example 25.28** *An automatic temperature control arrangement for an electric oven is shown in Fig. 25.73. The thermistor has a resistance of 2000 Ω at 70°C. A potentiometer is used for obtaining balance conditions. The temperature of oven changes and therefore the rheostat has to be set at 1800 Ω to obtain balanced conditions. Using the linear approximation for the resistance temperature curve obtain the change in the value of temperature of oven. The resistance temperature coefficient can be assumed as – 0.05/°C.*

**Solution.** Using linear approximation, we have :

$$1800 = 2000[1 - 0.05\,(\theta - 70)] \text{ or } \theta = 72\,°C$$

∴ Change in temperature of oven = 2°C

**Example 25.29** *A thermistor has a resistance of 10 kΩ at 25°C. The resistance temperature co-efficient is – 0.05/°C. A Wien's bridge oscillator uses two identical thermistors in the frequency determining part of the bridge. The value of capacitance used in the bridge is 500 pF. Calculate the value of frequency of oscillations for*

*(i) 20°C, (ii) 25°C, and (iii) 30°C.*

*The frequency of oscillations is*

$$f = 1/2\pi RC \text{ Hz}$$

*where R and C are resistance and capacitance respectively.*

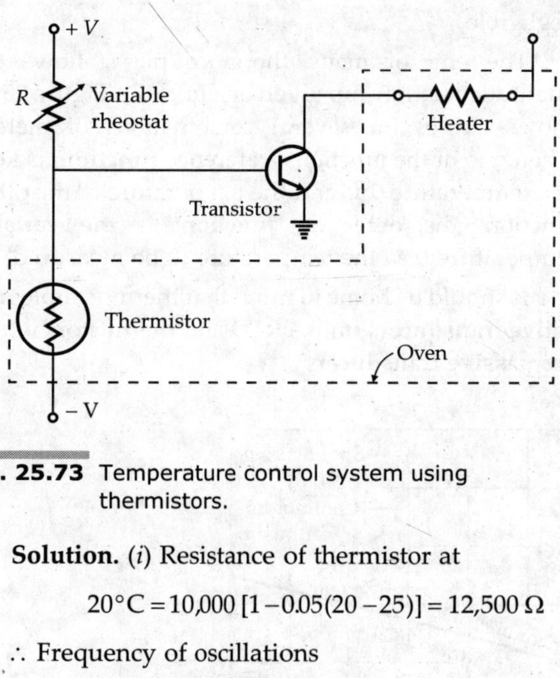


**Fig. 25.73** Temperature control system using thermistors.

**Solution.** (*i*) Resistance of thermistor at

$$20°C = 10{,}000\,[1 - 0.05(20 - 25)] = 12{,}500\ \Omega$$

∴ Frequency of oscillations

$$f = \frac{1}{2\pi RC}$$

$$= \frac{1}{2\pi \times 12500 \times 500 \times 10^{-12}} \text{ Hz}$$

$$= 25.46 \text{ kHz}$$

(*ii*) Resistance of thermistor at 25°C = 10,000 Ω

∴ Frequency of oscillations

$$= \frac{1}{2\pi \times 7500 \times 500 \times 10^{-12}} \text{ Hz}$$

$$= 31.83 \text{ kHz}$$

(*iii*) Resistance of thermistor at 30°C

$$= 10{,}000[1 - 0.05(30 - 25)] = 7500\ \Omega$$

∴ Frequency of oscillations

$$= \frac{1}{2\pi \times 5700 \times 500 \times 10^{-12}} \text{ Hz}$$

$$= 42.44 \text{ kHz}$$

## 25.21 THERMOCOUPLES

The operation of thermocouples has been explained in Chapter 9.

The emf produced in a thermocouple circuit is given by :

$$E = a(\Delta\theta) + b(\Delta\theta)^2$$

(Eqn. 9.71 on page 278)

where Δθ = difference in temperature between the hot thermocouple junction and the reference junction of the thermocouple ; °C,

and *a*, *b* = constants.

*a* is usually very large as compared with *b* and therefore emf thermocouple is $E \simeq a\,(\Delta\theta)$ or $\Delta\theta \simeq E/a$.

In a thermocouple temperature measuring circuit, the emf set-up is measured by sending a current through a moving coil instrument, the deflection being directly proportional to the emf. Since emf is a function of temperature difference Δθ, the instrument can be calibrated to read the temperature. The emf may also be measured by a potentiometer.

Figure 25.74 shows a typical circuit of an iron constantan thermocouple.

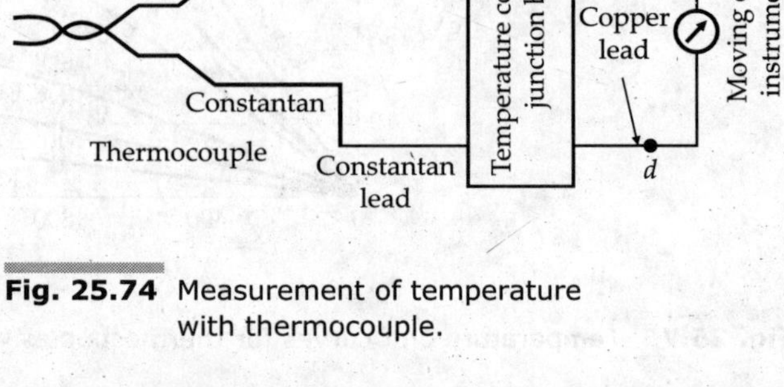


**Fig. 25.74** Measurement of temperature with thermocouple.

25

**Table 25.7** *Industrial Thermocouples*

| | ***Base-metal couples*** | | | | ***Rare metal couples*** | | |
|---|---|---|---|---|---|---|---|
| Positive Wire | Copper | Iron | Chromel 90% Cr. 10% Ni | Chromel | Platinum 90% rhodium 10% | Tungsten 95% rhenium 5% | Rhodium iridium |
| Negative Wire | Constantan 40% Ni, 60% Cu approx. | Constantan | Alumel 94% Ni, 2% Al, + Si and Mn | Constantan | Platinum | Tungsten 72% rhenium 26% | iridium |
| Temp. range (°C) | -- 250 to + 400 | – 200 to + 850 | – 200 to + 1100 | – 200 to + 850 | 0 to + 1400 | 0 to + 2600 | 0 to + 2100 |
| Spot max. (°C) | 500 | 1100 | 1300 | 1100 | 1650 | | |
| Characteristics | Resists oxidising and reducing atmospheres up to 350°C. Requires protection from acid fumes. | Low cost. Corrodes in the presence of moisture oxygen, and sulphur-bearing gases. Suitable for reducing atmospheres. | Resistant to oxidising but not to reducing atmosphere. Susceptible to attack by carbon-bearing gases sulphur, and cyanide fumes. | Suitable for oxidising but not for reducing atmospheres, carbon-bearing gases and cyanide fumes. High emf. | Low emf. Good resistance to oxidising atmospheres, poor with reducing atmospheres, Calibration is affected by metallic vapours and contact with metallic oxides. | For use in non-oxidising atmospheres only. The 5% rhenium arm is brittle at room temperatures. | Similar to platinum rhodium-platinum |

Since the thermo-electric emf depends upon the difference in temperature between the hot junction and the reference junction, the temperature of the latter should remain absolutely constant in order that the calibration holds good and there are no errors on account of change in ambient temperature. The temperature of the reference junction is controlled for this purpose. The reference junction temperature is usually 0°C.

Thermocouples are used for measurement of temperature upto 1400°C. The common types of thermocouples are given in Table 25.7 together with useful temperature range. It should be mentioned here that the combination of metals be so chosen that a rise in temperature should always produce a linear rise in emf *i.e.*, the value of '*b*' (Eqn. 9.71) should be negligible.

The emf of many thermocouples follows the quadratic relationship given by Eqn. 9.71. Figure 25.75 shows curves for several combinations of metals, when one of the junctions, **reference junction**, is kept at a temperature 0°C and the temperature of the other junction, the **detecting junction**, is the variable temperature (*i.e.*, the temperature to be measured).

It should be home in mind that thermocouples are **active transducers** unlike RTD and thermistors which are **passive transducers.**

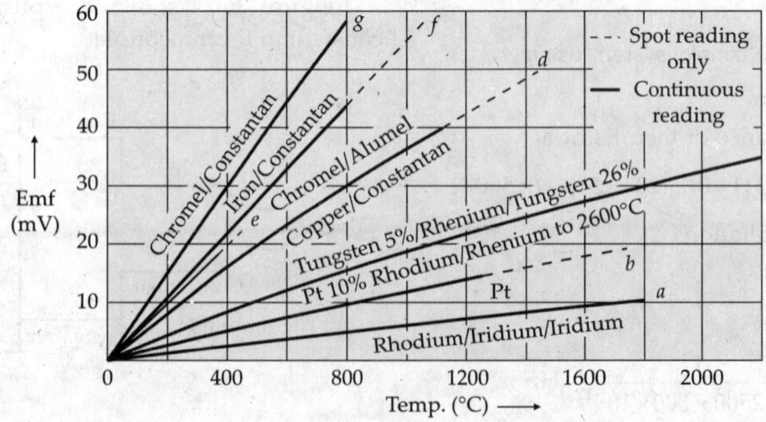


**Fig. 25.75** Temperature emf curves for thermocouples with reference junction at 0°C.

### 25.21.1 Construction of Thermocouples

In industrial applications the choice of materials used to make up a thermocouple depends upon the temperature range to be measured, the kind of atmosphere to which the material will be exposed, to output emf and its stability, mechanical strength, and the accuracy required in measurements. Thermocouple materials are divided into *two* categories : (*i*) Rare metal types using platinum, rhodium, etc. and (*ii*) Base metal type.

Several combinations of dissimilar metals make good thermocouples for industrial use. These combinations apart from having linear response and high sensitivity, should be physically strong to withstand high temperatures, rapid temperature changes, and the effect of corrosive and reducing atmospheres. Based on years of experience in application of thermocouples, industry has standardized a few wire combinations.

A pair of two dissimilar metals that are in physical contact with each other form a thermocouple. These metals may be twisted, screwed, peened, clamped or welded together. The most commonly used method for fabricating is to weld the metals together.

Thermocouples do not use bare conductors except in applications where atmospheric conditions permit their use. These conditions obtain when temperatures to be measured are low and the atmosphere is non-corrosive. Industrial thermocouples employ protective sheathing surrounding the junction and a portion of the extension leads. The leads and the junction are internally insulated from the sheath, using various potting compounds, ceramic beads or oxides. The type of insulation used depends upon the process being monitored. The different types of protective sheaths used for thermocouples are shown in Fig. 25.76.

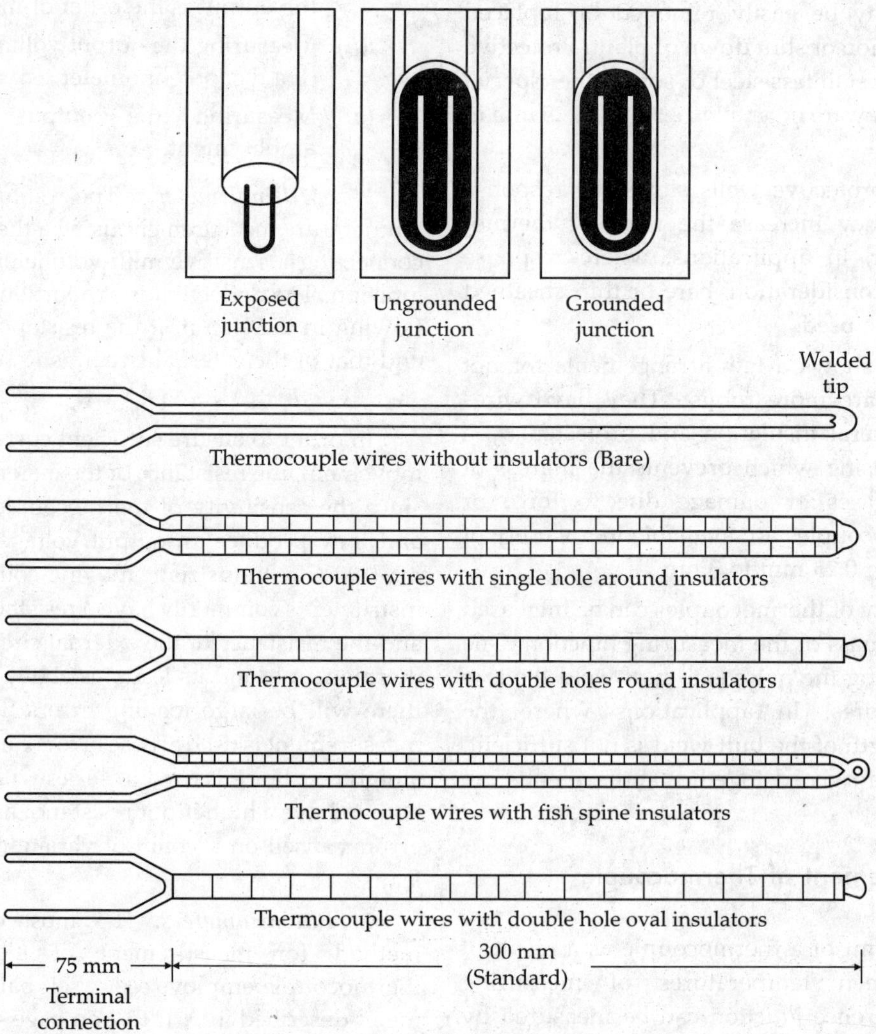


**Fig. 25.76** Protective sheaths for thermocouples.

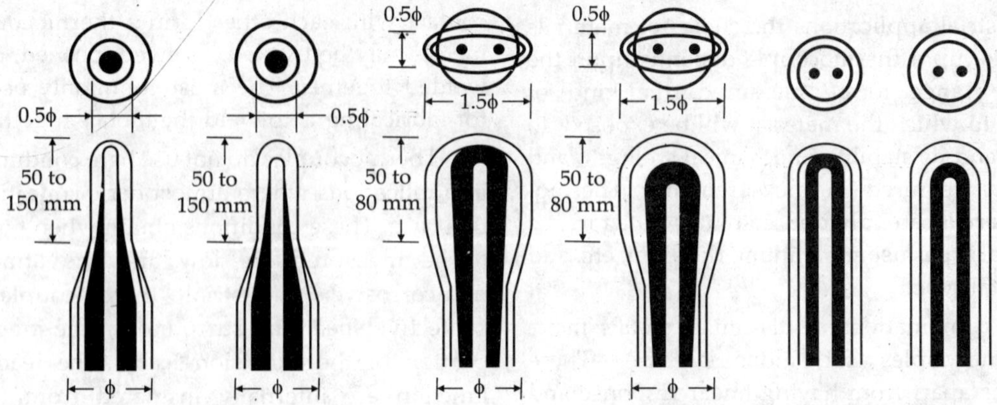


**Fig. 25.77** Construction of hot junctions.

Thermocouples are normally not installed in pipelines, vessels and other pieces of equipment directly. They are usually placed inside protective wells so that may be easily removed or replaced without interruption or shut down of plant. Protective wells are made of stainless steel or some other special alloy material. They are normally 12.5 mm to 25 mm in diameter.

The use of protective wells slows the response appreciably as they increase the mass of thermocouple. However, in applications, where response time is primary consideration, bare or thin sheathed thermocouples are used.

Figure 25.77 shows a few arrangements for hot junctions which are more robust. They have wires insulated by mineral insulation and protected by a stainless outer casing which prevents the ingress of damaging hot gases or damage due to force or pressure. Thermocouples are available in a variety of size, ranging from 0.25 mm to 3 mm.

The sensitivity of thermocouples can be increased by reducing the mass of the measuring junction. One method of reducing the mass is to butt weld the two thermocouple wires. In applications where the mechanical strength of the butt weld is not sufficient, the two wires are twisted together and the ends are welded.

### 25.21.2 Measurement of Thermocouple Output

The output emf of a thermocouple as a result of difference between temperatures of measuring junction and reference junction can be measured by the following methods :

(*i*) Measuring the output voltage directly with a permanent magnet moving coil (PMMC) instrument. A millivoltmeter is used since the output is the order of mV.

(*ii*) Measuring the output voltage with the help of a d.c. potentiometer.

(*iii*) Measuring the output voltage after amplifying it.

**1. *Millivoltmeter.*** The simplest type of temperature measurement using a thermocouple is by connecting a sensitive milli-voltmeter across the cold junction. The deflection is proportional to the current flowing in the circuit. If the resistance of meter is $R_m$ and that of the external circuit is $R_e$, the current is

$$i = E(R_m + R_e)$$

In order to ensure sufficient current to deflect the movement, the resistance of the meter should be small since the sensitivity of thermocouples is quite small and they produce an output voltage which is a few mV/100°C. Industrial moving-coil thermocouple instruments commonly have a resistance of about 50 Ω and the resistance of the external circuit (resistance of thermocouple and its leads) is about 10 — 20 Ω. Thus there will be large loading errors if this method of measurement is used. These errors can be eliminated if a high resistance is used in series with the instrument both to have a high input resistance and to swamp out errors caused on account of variation in resistance of leads.

**2. *Potentiometers.*** The most commonly used method for measurement of temperature with thermocouple employs a d.c. self balancing potentiometer described in Art. 15.2 on page 466. There are no loading errors as at balance no current is drawn from the thermocouple whose emf is being measured. The resistance variation problems are also absent when potentiometers are used.

Many types of automatic potentiometers have been developed both for automatic recording of temperature on chart recorders and for automatic process control.

3. ***Amplifiers with Output Devices.*** The use of transistors and integrated circuit methods has led to the development of compact and cheap d.c. amplifiers. The amplifiers amplify the output of the thermocouple and feed to indicating or recording devices. These devices may be either analog or digital type. In this case also, the loading errors and errors on account variation of resistance are absent. Therefore there are no circuit problems. But a disadvantage of these methods is that they require an auxiliary source of power while direct deflection method do not require any separate source.

### 25.21.3 Compensating Circuit

When temperatures near ambient are to be measured with a thermocouple and it is inconvenient to use a fixed reference junction and therefore compensating circuits must be employed in the measuring system. An arrangement for automatic compensating is shown in Fig. 25.78. A temperature sensitive bridge is included in the thermocouple circuit, such that variations in the ambient temperature are compensated for by the changes in the resistance $R_c$ and the compensating junction.

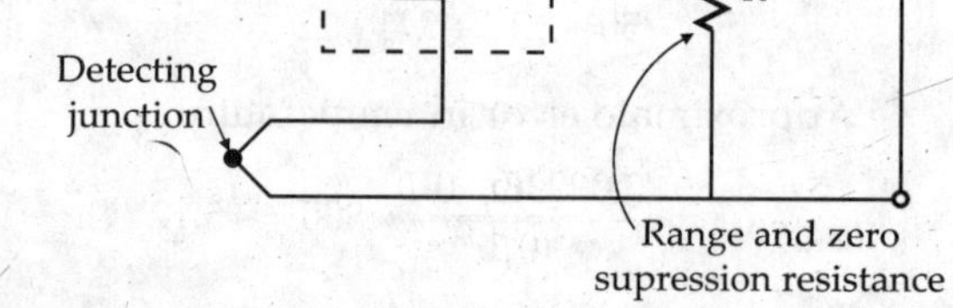


**Fig. 25.78** Thermocouple with bridge type compensation.

Multichannel thermocouple systems sometimes use a heated, thermostatically controlled reference junction as shown in Fig. 25.79. By using a regulated temperature control of the reference junction, an accuracy of ± 3% can be obtained.

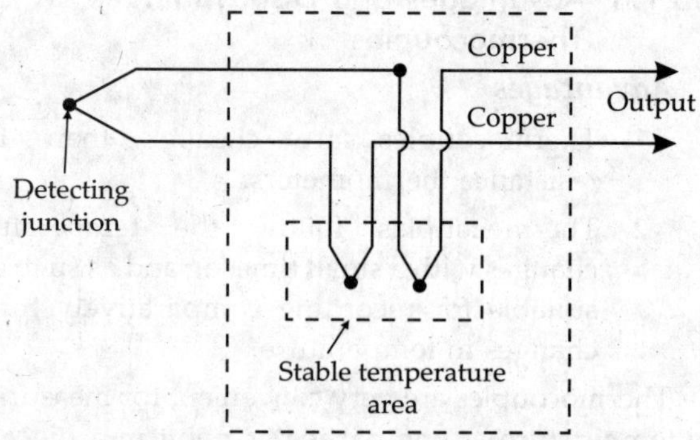


**Fig. 25.79** Thermocouple reference junction.

### 25.21.4 Reference Junction Compensation

A factor which is important in the use of thermocouple is the requirement of a known reference temperature of the reference junction. This is because when the reference junction is not held at 0°C, the observed value must be corrected by adding to it a voltage that has resulted from a temperature difference equal to the amount by which the reference junction is above 0°C. (This is because the thermocouples are calibrated with temperature of reference junction as 0°C.)

Now $E_T = E_t + E_0$ where $E_T$ is the total emf at temperature $T$, $E_t$, is the emf on account of temperature difference between detecting (hot) and the reference junction and $E_0$ is the emf due to temperature of the reference junction being above 0°C. Since, there exists a non-linear relationship between the emf and the temperature, it is important that temperatures are determined by the above process rather than converting an emf to temperature and then adding it to ambient temperature.

### 25.21.5 Lead Compensation

In many applications it is desirable to place the reference junction at a point far removed from the measurement junction. The connecting wires from the thermocouple head to the meter are, therefore, very long and are usually not at the same temperature throughout their length. This causes errors, which can be avoided by using connecting wires made of the same material as the thermocouple wires. The implementation of this arrangement may not be possible in many cases due to cost and other consideration. Under these circumstances, materials are chosen such that the relationship between emf and temperature is the same or almost the same as that for thermocouple wires. These wires then called **Compensating Leads.**

25

### 25.21.6 Advantages and Disadvantages of Thermocouples

***Advantages***

1. Thermocouples are cheaper than the resistance thermometers.
2. Thermocouples follow the temperature changes with a small time lag and as such are suitable for recording comparatively rapid changes in temperature.

Thermocouples are very convenient for measuring the temperature at one particular point in a piece of apparatus.

***Disadvantages***

1. They have a lower accuracy and hence they cannot be used for precision work.
2. To ensure long life of thermocouples in their operating environments, they should be protected in an open or closed-end metal protecting tube or well. To prevent contamination of the thermo-couple, when precious metals like platinum or its alloys are being used, the protecting tube has to be made chemically inert and vacuum tight.
3. The thermocouple is placed remote from measuring devices. Connections are thus made by means of wires called extension wires. Maximum accuracy of measurement is assured only when compensating wires are of the same material as the thermocouple wires. The circuitry is, thus, very complex.

**Example 25.30** *Calculate the thermoelectric sensitivity of a device using bismuth and tellurium as the dissimilar metals. Estimate the maximum output voltage for a 100°C temperature difference at room temperature using one junction. The sensitivity of bismuth is –72 μV/°C and that of tellurium is 500 μV/°C.*

**Solution.** Sensitivity of thermocouple

$$= 500 - (-72) = 572\ \mu V/°C$$

∴ Voltage output for a temperature difference of

$$100°C = 572 \times 10^{-6} \times 100 = 57.2\ mV$$

**Example 25.31** *A chromel-alumel hot junction is connected to a potentiometer. This is at a temperature of 20°C and gives a reading of 27.07 mV. Determine the measured temperature assuming the thermocouple to conform to the values given in Fig. 25.75 which are based on reference junction at 0°C. The emf corresponding to a temperature of 20°C is 0.8 mV. Consult curves given in Fig. 25.74.*

**Solution.** The measured emf is the algebraic sum of emfs at the hot and reference junctions.

∴ Required emf is

$$E_T = E_t + E_0 = 27.07 + 0.8$$
$$= 27.87\ mV$$

Corresponding to this emf the temperature is 620°C.

**Example 25.32** *A thermocouple circuit uses a chromel-alumel thermocouple which gives an emf of 33.3 V when measuring a temperature of 800°C with reference temperature 0°C. The resistance of the meter coil, $R_m$ is 50 Ω and a current of 0.1 mA gives full scale deflection. The resistance of junctions and leads, $R_e$, is 12 Ω. Calculate :*

(*a*) *resistance of the series resistance if a temperature of 800°C is to give full scale deflection ;*

(*b*) *the approximate error due to rise of 1 Ω in $R_e$ ;*

(*c*) *the approximate error due to a rise of 10°C in the copper coil of the meter.*

*The resistance temperature co-efficient of coil is 0.00426/°C.*

**Solution.** (*a*)

$$\text{Emf } E = i(R_m + R_s + R_e)$$

or $$33.3 \times 10^{-3} = 0.1 \times 10^{-3}(50 + R_s + 12)$$

or series resistance $R_s = 271\ \Omega$

(*b*) Current in the circuit with increased resistance

$$= \frac{33.3 \times 10^{-3}}{(50 + 271 + 1 + 12)} = 0.0997\ mA$$

∴ Approximate error in temperature

$$= \frac{0.0997 - 0.1}{0.1} \times 800 = -2.4°C$$

(*c*) Change in resistance of coil with a temperature increase of $10°C = 50 \times 0.00426 \times 10 = 2.13\ \Omega$

Current in the circuit with increase resistance of coil

$$= \frac{33.33 \times 10^{-3}}{50 + 2.13 + 271 + 12} A = 0.09936\ mA$$

∴ Approximate error in temperature

$$= \frac{0.09936 - 0.1}{0.1} \times 800 = -5.12°C$$

**Example 25.33** *The simple potentiometer circuit of Fig. 25.80 is to work from a platinum/platinum rhodium 10 per cent thermocouple and have a measuring range of 900°C – 1200°C.*

*The scale readings are to be correct for a reference junction temperature of 20°C. The slide wire resistance is 2.5 Ω and the circuit is standardized to give 1.08 V between A and B.*

*Find the values of resistance $R_1$ and $R_2$. Data from thermocouple tables for platinum/platinum- rhodium 10 per cent with 0°C reference junction temperature is :*

| | |
|---|---|
| *emf for* | $20°C = 0.112$ *mV* |
| *emf for* | $900°C = 8.446$ *mV* |
| *emf for* | $1200°C = 1.946$ *mV* |

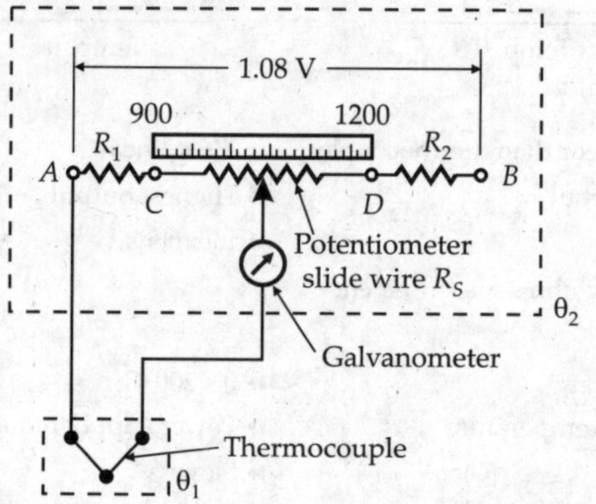


**Fig. 25.80** Diagram of Example 25.33.

**Solution.** The resultant emfs at the detecting junction temperature of $\theta_1$, and reference junction temperature of $\theta_2 = 20°C$, are

at 900°C, $E_1 = 8.446 - 0.112 = 8.334$ mA

and at 1200°C $E_2 = 11.946 - 0.112 = 11.834$ mV

With the same standardising current, the emfs are :

$$E_{ac} = \frac{1.08}{R_1 + 2.5 + R_2} \times R_1 = 8.334 \times 10^{-3} \quad ...(i)$$

and
$$E_{ad} = \frac{1.08}{R_1 + 2.5 + R_2} \times (R_1 + 2.5) = 11.834 \times 10^{-3} \quad ...(ii)$$

Dividing (*ii*) by (*i*), we have

$$\frac{R_1 + 2.5}{R_1} = \frac{11.834}{8.334} \quad \text{or} \quad R_1 = 5.95\ \Omega$$

Substituting the value of $R_1$ in (*i*), we get

$$R_2 = 762.6\ \Omega$$

## 25.22 INTEGRATED CIRCUIT TEMPERATURE TRANSDUCERS

Each of the three temperature transducers described earlier, *i.e.*, RTD, thermistors and thermocouples have some significant limitations. For example, thermocouples have a low output signal which varies non-linearly with temperature. Also they need some form of reference compensation with regards to ice point, RTDs are more linear than thermocouples but the changes in their resistance is very small even for large change in input temperatures *i.e.*, they have low sensitivity. The thermistors have a large change in resistance with temperature and thus have a high sensitivity. However, they exhibit a highly non-linear resistance-temperature characteristics. Their typical properties are shown in 25.81.

For each of these transducers, electronic compensation circuits have to be used in order to overcome their shortcomings. Also, additional circuitry may be needed to increase their voltage or current outputs. Usually, this additional electronics circuitry takes the form of monolithic integrated circuits. Thus, it requires to combine the temperature sensing element with signal conditioning electronics to produce single monolithic integrated circuit package.

Three integrated circuit (IC) packages are described in this section. They are :

(*i*) LM 335 – it provides an output of 10 mV/°K

(*ii*) LM 34 – it provides an output of 10 mV/°F

(*iii*) AD 592 – it provides a current output of 1 μA/°K

**1. *LM 335 Series.*** The LM 335 is a temperature sensitive zener diode, which when reverse biased into its breakdown region, gives an output of :

$$V_z = \frac{10}{T} \theta \text{ mV} \quad ...(25.81)$$

where $T$ = temperature ; °K

$\theta$ = temperature ; °C.

The size of degree kelvin and degree celsius are the same, and therefore there is 273° offset.

Hence, Eqn. 25.81 can be written as :

$$V_z = 2.73 \times 10^3 + 10\,\theta \text{ mV} \quad ...(25.82)$$

The three temperature ranges available are listed in Table 25.8. Notice that none of these ranges come closer to the temperature range of thermocouples or RTDs. This is inherent in the silicon material of integrated circuit.

**Table 25.8** *LM/135/235/2335 Temperature Ranges*

| *Device* | *Range (°C)* | *Use* |
|---|---|---|
| LM 135 | –55 to + 150 | Defence |
| LM 235 | –45 to + 125 | Industrial |
| LM 335 | –45 to + 100 | Commercial |

25

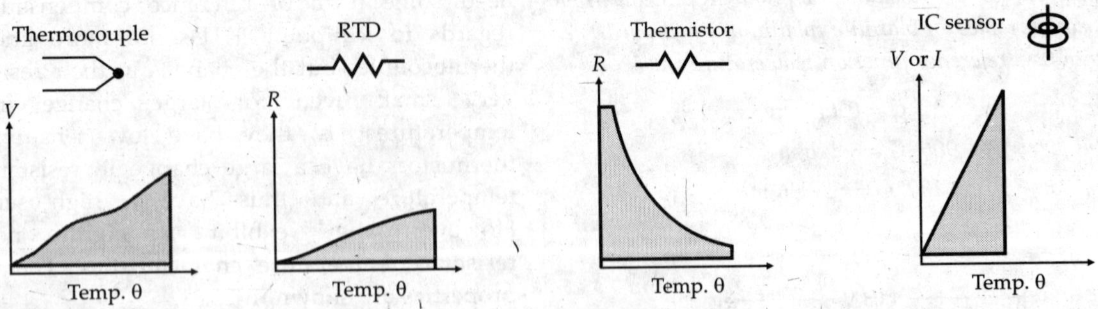


| | Thermocouple | RTD | Thermistor | IC sensor |
|---|---|---|---|---|
| **Advantages** | ▸▸ Self-powered | ▸▸ Wide variety | ▸▸ More linear than thermocouple | ▸▸ Most linear |
| | ▸▸ Simple | ▸▸ Wide temperature range | ▸▸ High output | ▸▸ Highest output |
| | ▸▸ Rugged | ▸▸ Most stable | ▸▸ Fast | ▸▸ Inexpensive |
| | ▸▸ Inexpensive | ▸▸ Most accurate | ▸▸ Two-wire ohms measurement | |
| **Disadvantages** | ▸▸ Non-linear | ▸▸ Expensive | ▸▸ Non-linear | ▸▸ 0 < 200°C |
| | ▸▸ Low voltage | ▸▸ Current source required | ▸▸ Limited temperature | ▸▸ Power supply required |
| | ▸▸ Reference required | ▸▸ Small ΔR | ▸▸ Fragile | ▸▸ Slow |
| | ▸▸ Lest stable | ▸▸ Low absolute resistance | ▸▸ Current source required | ▸▸ Self-heating |
| | | ▸▸ Self-heating | ▸▸ Self-heating | ▸▸ Limited configuration |

**Fig. 25.81** Properties of temperature transducers.

A simple circuit using LM 335 is shown in Fig. 25.82. Since LM 335 is a zener diode, a reverse bias voltage is applied to produce zener current. The current should be limited to :

$$400\ \mu A < I_z < 5\ mA$$

where $I_z$ is the zener current.

The testing at the manufacturer's end is done at $I_z$ = 1 mA, which is a reasonable choice. At higher currents, LM 335 will heat itself on account of power ($I_z\ V_z$) which it has to dissipate. At current levels, below 1 mA, the accuracy decreases.

A simple temperature transducer circuit using LM 335 is shown in 25.82.

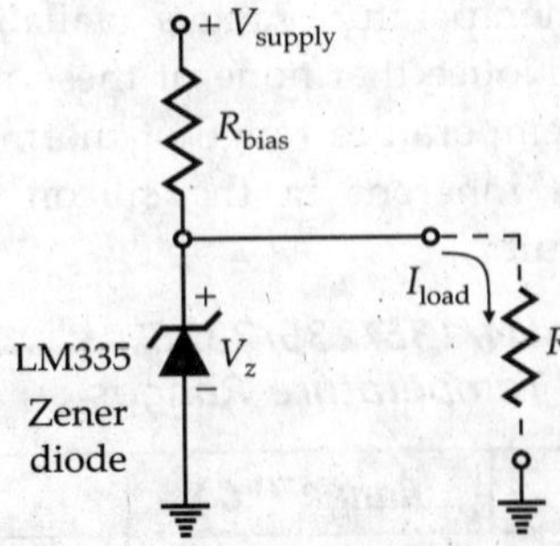


**Fig. 25.82** Simple temperature transducer using LM 335.

In order to determine the proper size of series limiting resistor as shown in Fig. 25.82, the voltage across the zener diode at nominal temperature is first calculated. Let this voltage be $V_{nominal.}$

$$R_{bias} = \frac{V_{supply} - V_{nominal}}{10^{-3}}\Omega \qquad ...(25.83)$$

$$(1\ mA = 10^{-3}\ A)$$

For linear results, it is important that the load current be small compared to the minimum current that will actually flow through the zener diode. This requires that the following inequalities should be followed :

$$I_{load} \leq I_{z\ min} \qquad ...(25.84)$$

$$\frac{V_{max\,T}}{R_L} \leq \frac{V_{supply} - V_{max\,T}}{R_{bias}} \qquad ...(25.85)$$

The rated non-linearity is ± 1°C . However, it will be desirable to adjust the transducer so that maximum accuracy is achieved over the desired range.

The 2.73 V offset voltage (See Eqn. 25.82) which accrues on account of conversion from Kelvin scale to Celsius scale is nuisance. Ideally, 0°C should produce 0 V d.c.

The circuit of Fig. 25.83 provides opportunity for two point calibration capability. Initially, the potentiometer is adjusted to give $-2.73\,V(-2.73\times 10^3\ mV)$ at its wiper and the 10 kΩ potentiometer is set at its midrange. The temperature sensor is placed at its lowest temperature. The zero potentiometer is adjusted to remove *half* of the error. Then the temperature sensor is placed at the maximum temperature and the 10 kΩ potentiometer is adjusted to remove *half* of the upper point error.

Alternately, the temperature sensor is placed at its two extreme ends and half of the error is adjusted each time. The zero adjustment at the bottom and 10 kΩ adjustment at the top.

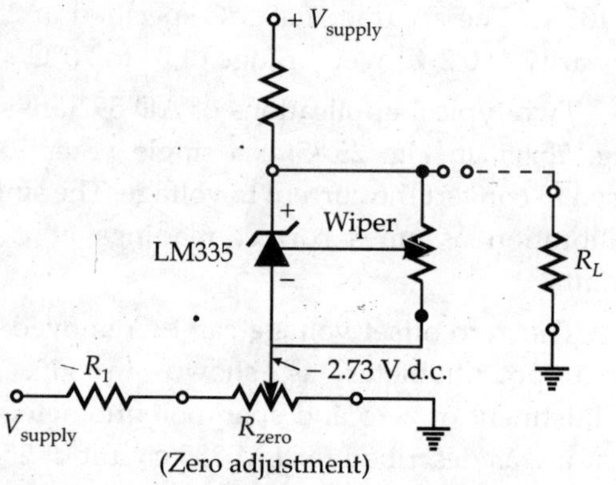


**Fig. 25.83** Two point calibrations for LM 335 temperature sensor.

The temperature sensors should be allowed enough time to stabilize to their new temperature before making any adjustments. The value of $R_{zero}$ zero should be small as compared to $R_{bias}$ in order that the adjustment of $R_{zero}$, does not affect the zener current $I_Z$.

**Example 25.34** *Design a circuit using a LM 335 to cover a range of – 10 V to + 50 V with a nominal temperature of 20°C. Assume a supply voltage of 15 V. Also, calculate the minimum recommended loud resistance. The output of zener diode is,*

$$V_z = 2.73 + 10 \times 10^{-3}\,\theta\,\text{V}$$

*The zener current is limited to 1 mA.*

**Solution.** Using the circuit of Fig. 25.83, at the nominal temperature,

$$V_z = 2.73 + (10 \times 10^{-3}) \times 20 = 2.93\text{ V}$$

But since the anode of zener is at –2.73 V the output voltage is,

$$V_{out} = V_z + V_{offset}$$

$$= 2.93 - 2.73 = 0.2\text{ V} = 200\text{ mV}$$

$$R_{bias} = \frac{5 - 0.2}{1 \times 10^{-3}} = 4.8\text{ k}\Omega$$

To minimize the effects of adjusting $R_{zero}$ on $I_{bias}$, select $R_{zero} << R_{bias}$.

Let $R_{zero} = 500\ \Omega$.

To check load resistance, solve the inequality for $R_L$

$$R_L >> \frac{V_{max\ T}\, R_{bias}}{V_{supply} - V_{max\ T}}$$

In order to find out $V_{max\,T'}$

$$V_z = 2.73 + 10 \times 10^{-3} \times 50 = 3.23\text{ V}$$

$$V_{max\,T} = 3.23 - 2.73 = 0.5\text{ V}$$

$$R_L >> \frac{0.5 \times 4.7 \times 10^3}{5 - 0.5} = 522\ \Omega$$

**2. *LM 34 Series.*** The LM 34 series are precision integrated circuit sensors, whose output voltage is linearly proportional to Fahrenheit temperature (it provides an output of 10 mV/°F). The LM 34 thus hits advantage over linear temperature sensors calibrated in terms of degrees Kelvin, as the user is not required to subtract a large constant voltage from its output to obtain convenient Fahrenheit scaling. The LM 34 does not require any external calibration or trimming to provide typical accuracies of ±0.5°F at room temperature or ±1.5 °F over its full scale temperature range of –50 to 300°F. It has a low cost, low output impedance, linear output, and precise inherent calibration which makes its interfacing to readout or control circuitry especially easy.

It draws a very small current, of the order, of 70 μA from its supply and therefore its self heating is small typically about 0.2°F. The LM 34 can be applied easily in the same way as other IC temperature sensors. It can be glued or cemented to a surface and its temperature will be within ±0.02°F of the surface temperature.

Several circuits for using the LM 34 are shown in Fig. 25.84. The simplest, Fig. 25.84 (*a*) requires no extra components. It gives an output which varies between 50 mV to 3 V d.c. indicating temperatures ranging from +5°F to 300°F. In case temperatures below 0°F arc required, the IC must be provided with a negative source of power.

Often, it is necessary to locate the temperature sensor more than a few centimetres from the support electronics. For this, there are two additional requirements. Only two leads should be run to the sensor. The signal returned by the sensor should be current and not voltage. This eliminates the effects or series resistance in the wiring. The circuits shown in Figs. 25.84(*c*) and 25.84(*d*) meet these requirements.

In Fig. 25.84(*d*) it is necessary to ground the sensor.

25

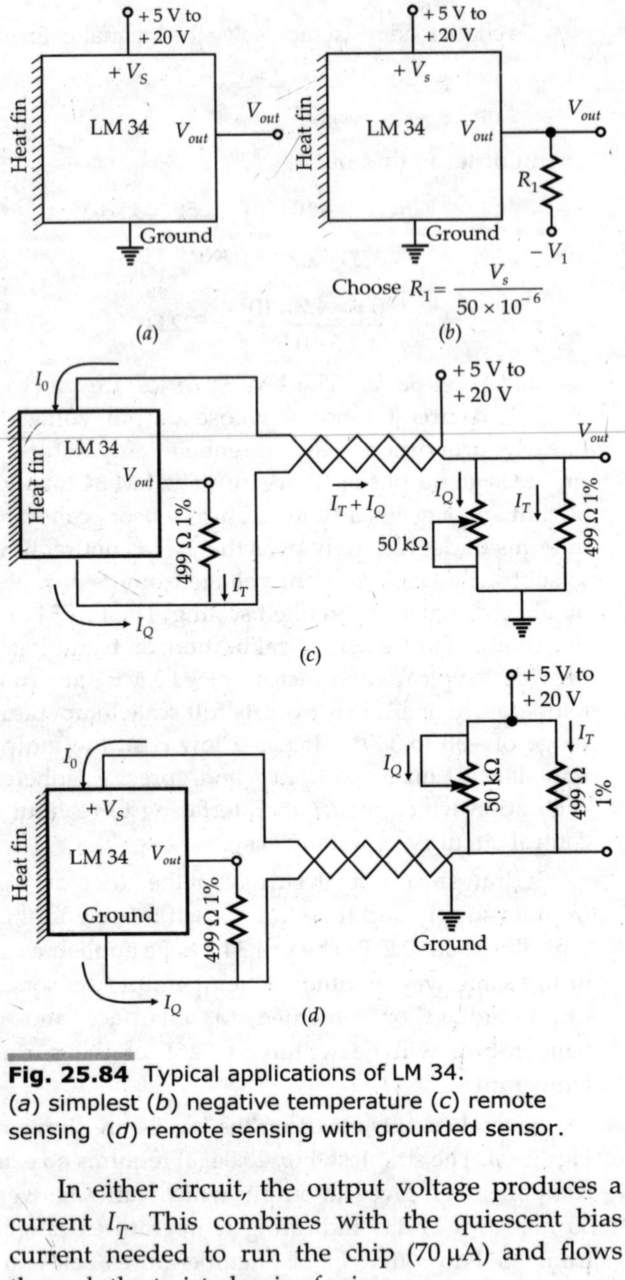


**Fig. 25.84** Typical applications of LM 34.
(*a*) simplest (*b*) negative temperature (*c*) remote sensing (*d*) remote sensing with grounded sensor.

In either circuit, the output voltage produces a current $I_T$. This combines with the quiescent bias current needed to run the chip (70 μA) and flows through the twisted pair of wires.

$$I_T = V_{out} / 499 \qquad \text{...(25.86)}$$

At the processing electronics the bias current is split off and flows through 50 kΩ potentiometer. So $I_T$ passes through 499 Ω resistor to recover the original voltage. It should be noticed that when the remote sensor is grounded, the output voltage must be taken differentially between + *V* and the signal returned the sensor.

**3. *AD 592 Series.*** In case the signal is to be transmitted over a large distance, AD 592 is a better choice as its output is current and a current signal is not affected by the resistance of the wiring. The current output of AD 592 is,

$$I_{out} = 1 \times 10^{-6}\, T \qquad \text{...(25.87)}$$

where $T$ = temperature, °K

or $$I_{out} = 273 \times 10^{-6} + 1 \times 10^{-6}\, \theta \text{ A} \qquad \text{...(25.88)}$$

where θ = temperature, °C

AD 592 has a low cost. Its range is –25°C to + 105°C. The accuracy is 0.5°C specified at 25°C. The linearity is 0.2°C over a range of 0° to 70°C.

Two typical applications of AD 592 are shown in Fig. 25.85. In Fig. 25.85(*a*) a single potentiometer is used to convert the current to voltage. The single point calibration assumes correct readings at calibration point.

The zero offset voltage can be removed with the zero potentiometer as shown in Fig. 25.85(*b*). Adjustment of zero and span potentiometers is done just it was described for LM 335 circuit of 25.83.

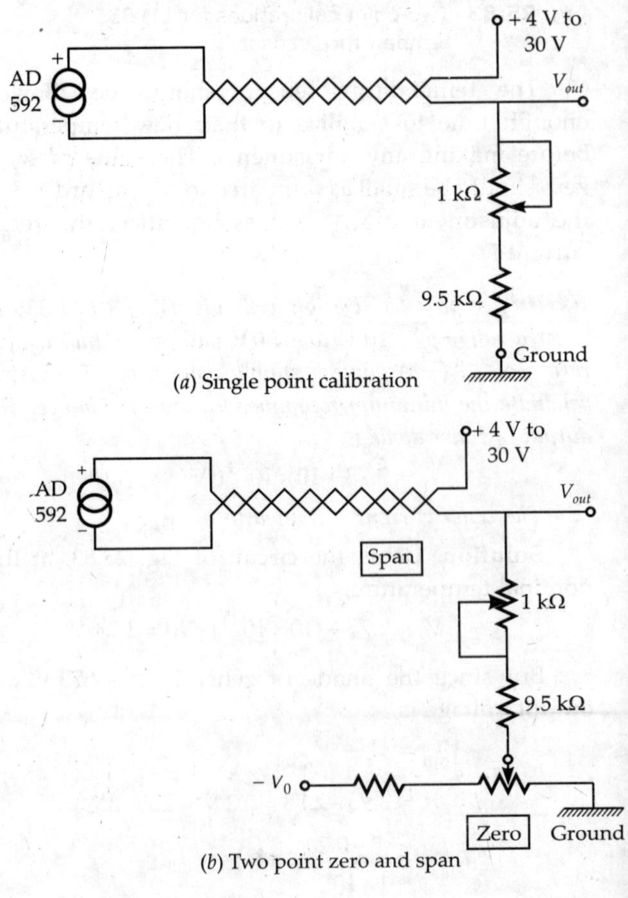


**Fig. 25.85** Temperature sensing and conversion with AD 592.

## 25.23 VARIABLE INDUCTANCE TRANSDUCERS

The variable inductance transducers work, generally, upon one of the following three principles :

(*i*) change of self inductance,

(*ii*) change of mutual inductance, and

(*iii*) production of eddy currents,

### 1. Transducers working on principle of change of Self Inductance

The self inductance of a coil $L = N^2 / R$

where $N$ = number of turns, and

$R$ = reluctance of the magnetic circuit.

The reluctance of the magnetic circuit $R = l / \mu A$

$$\therefore \text{ Inductance } \quad L = N^2 \mu (A / l) = N^2 \mu G \quad \text{...(25.89)}$$

where $\mu$ = effective permeability of the medium in and around the coil ; H/m

$G = A / l$ = geometric form factor

$A$ = area of cross-section of coil ; $m^2$

and $l$ = length of coil ; m

It is clear from Eqn. 25.89 that the variation in inductance may be caused by :

(*i*) change in number of turns, $N$,

(*ii*) change in geometric configurations, $G$, and

(*iii*) change in permeability, $\mu$.

Inductive transducers are mainly used for measurement of displacement. The displacement to be measured is arranged to cause variation of any of three variables in Eqn. 25.89 and thus alter the self inductance $L$ by $\Delta L$.

The different types of inductive transducers for measurement of translational and rotary displacements are shown in Fig. 25.86.

***Differential Output.*** Normally the change in self inductance $\Delta L$ is adequate for detection for subsequent stages of instrumentation system. However, if the succeeding instrumentation responds to $\Delta L$, rather than to $L + \Delta L$ the sensitivity and accuracy will be much higher. The transducer can be designed to provide two outputs one of which is an increase of self inductance and the other is a decrease in self inductance. The succeeding stages of instrumentation system measure the difference between these outputs *i.e.* $2\Delta L$. This is known as the differential output. The advantages of differential outputs are :

(*i*) The sensitivity and accuracy are increased.

(*ii*) The output is less affected by external magnetic fields.

(*iii*) The effective variations due to temperature changes are reduced.

(*iv*) The effects of changes in supply voltage and frequency are reduced.

The differential arrangement consists of a coil which is divided into two parts. In response to a physical signal, which is normally a displacement, the inductance of one part increases from $L$ to $L + \Delta L$ while that of the other part decreases from $L$ to $L - \Delta L$. The change is measured as the difference of the two resulting in an output of $2\Delta L$ instead $\Delta L$ when only a single winding is used. The differential arrangements are shown in Fig. 25.86.

### 2. Transducers working on principle of change of Mutual Inductance

An inductive transducer working on the principle variation of mutual inductance uses multiple coils. The mutual inductance between two coils is

$$M = K\sqrt{L_1 L_2}.$$

where $L_1$ and $L_2$ = self-inductances of two coils,

and $K$ = co-efficient of coupling.

Thus mutual inductance between the coils can be varied by variation of self-inductances or the co-efficient of coupling. However, the mutual inductance can be converted into a self-inductance by connecting the coils in series. The self-inductance of such an arrangement varies between $L_1 + L_2 - 2M$ to $L_1 + L_2 + 2M$ with one of the coils being stationary while the other is movable. The self-inductance of each coil is constant but the mutual inductance changes depending upon the displacement of the movable coil.

The different arrangements of measurement of translational and rotary displacements are shown in Fig. 25.86.

In the differential arrangement, the fixed coil is divided into two parts. The movement of the movable coil increases the mutual inductance of one part by $\Delta M$ and decreases that of the other by $\Delta M$.

Air or iron cored coils can be used for inductive transducers. Both have their own advantages and disadvantages.

***Air Cored Coils.*** Air cored coil transducers can be operated at a higher carrier frequency because of absence of eddy current losses in air cores. The inductance of air cored coils is independent of the current carried by the coil as the permeability of air is constant and does not depend upon the current carried by the coil. Hence air cored coil transducers can be used for measurement of displacement variations occurring at fairly high frequencies.

25

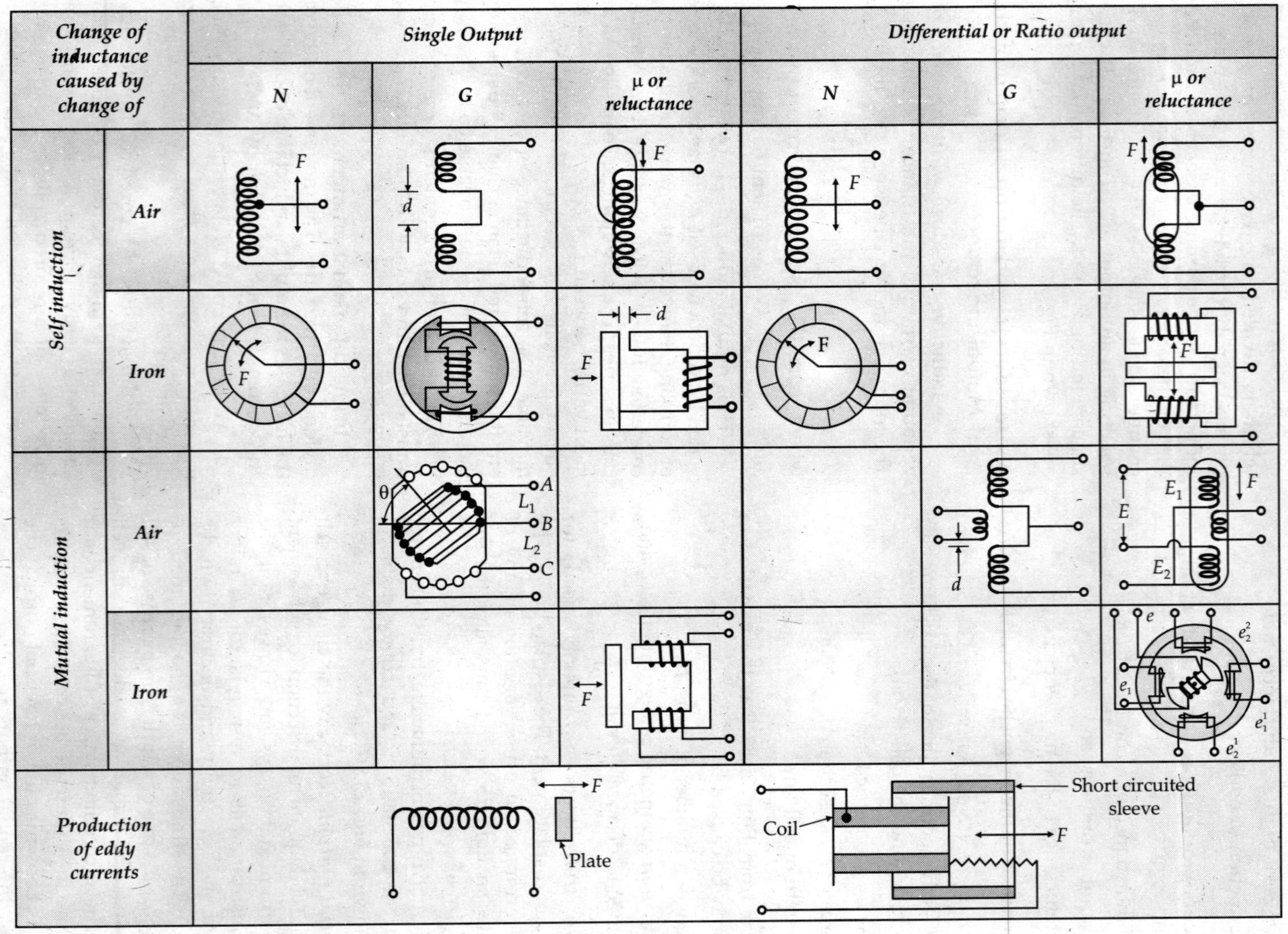


**Fig. 25.86** Variable Inductance Transducers.

***Iron Cored Coils.*** The greatest disadvantage of iron cored coils transducers is that their inductance is not constant but depends upon the value of the current carried by the coil. Also, at high frequencies, the eddy current loss tends to be high and therefore iron cored coil transducers cannot be used beyond a particular frequency. The frequency of supply voltage should not exceed 20 kHz for iron core transducers to keep the core losses to acceptable values. Hence, for accurate measurements the frequency of the input displacement should not exceed 2 kHz.

The advantages of iron cored coil transducers are :

(*i*) Their size is much smaller than that air cored transducers on account of high permeability of iron cores.

(*ii*) Iron cored transducers are less likely to cause external magnetic fields because their magnetic field is confined to the iron core of the transducer on account of high permeability and are less affected by stray magnetic fields on account of the high magnetic field produced by them.

Most iron cored transducers are of the variable reluctance type where the length of air gap in the magnetic circuit is varied. In most applications the reluctance of magnetic circuit is primarily that of air gap.

### 3. Transducers working on principle of production of Eddy Currents

These inductive transducers work on the principle that if a conducting plate is placed near a coil carrying alternating current, eddy currents are, produced in the conducting plate. The conducting plate acts as a short-circuited secondary winding of a transformer. The eddy currents flowing in the plate produce a magnetic field of their own which acts against the magnetic field produced by the coil. This results in reduction of flux and thus the inductance of the coil is reduced. The nearer is the plate to the coil, the higher are the eddy currents and thus higher is the reduction in the inductance of the coil. Thus the inductance of the coil alters with variation of distance between the plate and the coil.

A number of arrangements are possible and two arrangements are shown in Fig. 25.86. The plate may be at right angle to the axis of the coil. The displacement of the plate causes a change in the inductance of the coil. In the other arrangement a conducting sleeve runs in parallel and coaxially over a coil. If the short-circuited sleeve is away from the coil, the inductance of the coil is high while if the sleeve is covering the coil, its inductance is low. The change in inductance is a measure of displacement.

**Example 25.35** *In a variable reluctance type proximity inductive transducer shown in Fig. 25.87 the coil has an inductance of 2 mH when the target made of ferromagnetic material is 1 mm away from the core. Calculate the value of inductance when a displacement of 0.02 mm is applied to the target in a direction moving it towards the core. Show that the change in inductance is linearly proportional to the displacement. Neglect the reluctance of the iron parts.*

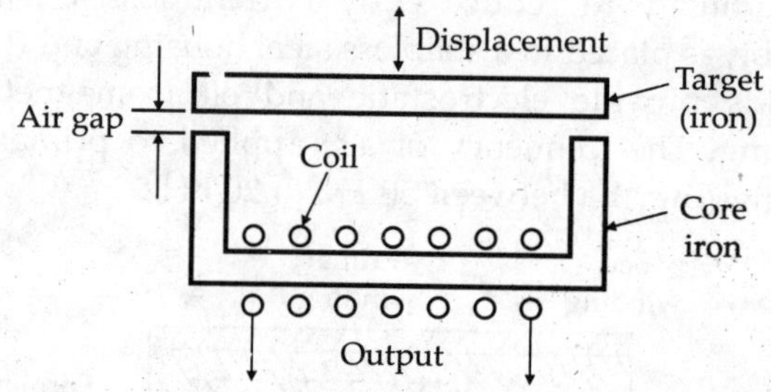


**Fig. 25.87** Variable reluctance inductive transducer.

**Solution.** Inductance with air gap length of 1.00 mm,

$$L = 2 \text{ mH}$$

Length of air gap when a displacement of 0.05 mm is applied,

$$= 1.00 - 0.02 = -0.98 \text{ mm}$$

Now inductance is inversely proportional to the length air gap as the reluctance of flux paths through iron are neglected. Since the gap length decreases, the inductance increases by $\Delta L$

$$\therefore \quad L + \Delta L = 2 \times \frac{1}{0.98} = 2.04 \text{ mH}$$

Change in inductance $\quad \Delta L = 0.04$ mH

The ratio of change in inductance to the original inductance is 0.04/2 = 0.02, and ratio of displacement to original gap length is also 0.02. Hence the change in inductance is linearly proportional to the displacement. This linear relationship, however, is true of only very small values of displacements.

## 25.24 LINEAR VARIABLE DIFFERENTIAL TRANSFORMER (LVDT)

The most widely used inductive transducer to translate the linear motion into electrical signals is the *linear variable differential transformer* (LVDT). The basic construction of LVDT is shown in Fig. 25.88. The transformer consists of a single primary winding $P$ and two secondary windings $S_1$ and $S_2$ wound on a cylindrical former. The secondary windings have equal number of turns and are identically placed on either side of the primary winding. The primary winding is connected to an alternating current source. A movable soft iron core is placed inside the former. The displacement to be measured is applied to the arm attached to the soft iron core. In practice the core is made of high permeability, nickel iron which is hydrogen annealed. This gives low harmonics, low null voltage and a high sensitivity. This is slotted longitudinally to reduce eddy current losses. The assembly is placed in a stainless steel housing and the end lids provide electrostatic and electromagnetic shielding. The frequency of a.c. applied to primary windings may be between 50 Hz to 20 kHz.

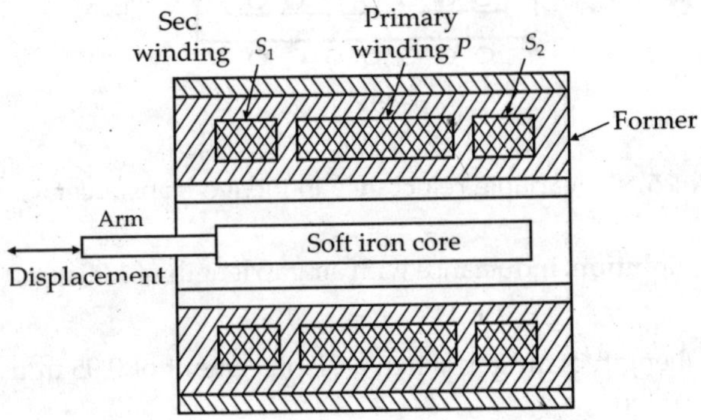


**Fig. 25.88** Linear variable differential transformer (L.V.D.T.).

Since the primary winding is excited by an alternating current source, it produces an alternating magnetic field which in turn induces alternating current voltages in the two secondary windings.

The output voltage of secondary. $S_1$, is $E_{s1}$ and that of secondary, $S_2$, is $E_{s2}$. In order to convert the outputs from $S_1$ and $S_2$ into a single voltage signal, the two secondaries $S_1$ and $S_2$ are connected in series opposition as shown in Fig. 25.89(*b*). Thus the output voltage of the transducer is the difference of the two voltages.

Differential output voltage,

$$\mathbf{E_0 = E_{s1} - E_{s2}} \quad \text{...(25.90)}$$

When the core is at its normal (**NULL**) position, the flux linking with both the secondary windings is equal and hence equal emfs are induced in them. Thus at null position : $E_{s1} - E_{s2}$. Since the output voltage of the transducer is the difference of the two voltages, the output voltage $E_0$ is zero at null position.

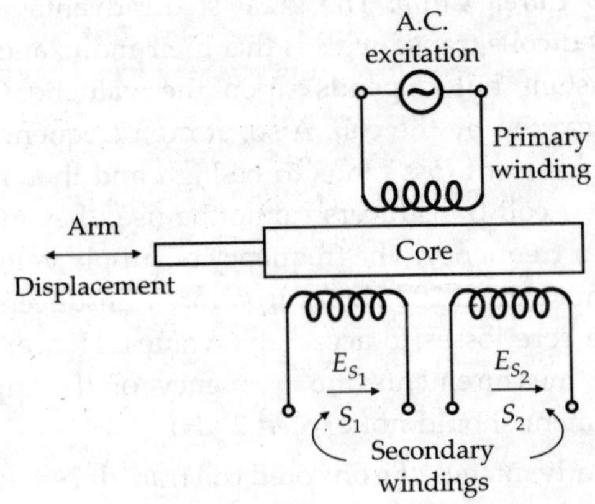


(*a*)

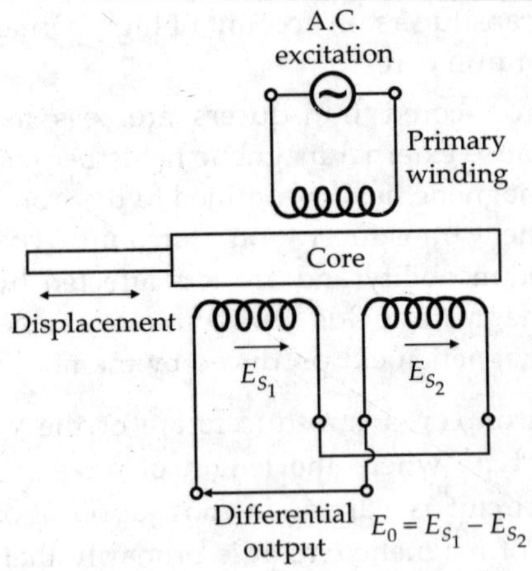


(*b*)

**Fig. 25.89** Circuits of an LVDT.

Now if the core is moved to the left of the **NULL** position, more flux links with winding $S_1$ and less with winding $S_2$. Accordingly output voltage $E_{s1}$, of the secondary winding $S_1$, is more than $E_{s2}$, the output voltage of secondary winding $S_2$. The magnitude of output voltage is, thus, $\mathbf{E_0 = E_{s1} - E_{s2}}$ and the output voltage is in phase with say the primary voltage. Similarly, if the core is moved to the right of the null position, the flux linking with winding $S_2$ becomes larger than that linking with winding $S_1$. This results in $E_{s2}$ becoming larger than $E_{s1}$. The output voltage in this case is $\mathbf{E_0 = E_{s1} - E_{s2}}$ and is 180° out of phase with the primary voltage. Therefore, the two differential voltages are 180° out of phase with each other.

The amount of voltage change in either secondary winding is proportional to the amount of movement of the core. Hence, we have an indication of amount of linear motion. By noting which output voltage is increasing or decreasing, we can determine the direction of motion. In other words, any physical displacement of the core causes the voltage of one secondary winding to increase while simultaneously reducing the voltage in the other secondary winding. The difference of the two voltages appears across the output terminals of the transducer and gives a measure of the physical position of core and hence the displacement.

As the core is moved in one direction from the null position, the differential voltage *i.e.,* the difference of the two secondary voltages, will increase while maintaining an in-phase relationship with the voltage from the input source. In the other direction from the null position, the differential voltage will also increase, but will be 180° out of phase with the voltage from the source. By comparing the magnitude and phase of the output (differential) voltage with that of the source, the amount and direction of the movement of the core and hence of displacement may be determined.

The amount of output voltage may be measured to determine the displacement. The output signal may also be applied to a recorder or to a controller that can restore the moving system to its normal position.

The output voltage of an LVDT is a linear function of core displacement within a limited range of motion, say, about 5 mm from the null position. Figure 25.90 shows the variation of output voltage against displacement for various positions of core. The curve is practically linear for small displacements (upto about 5 mm as mentioned above). Beyond this range of displacement, the curve starts to deviate from a straight line.

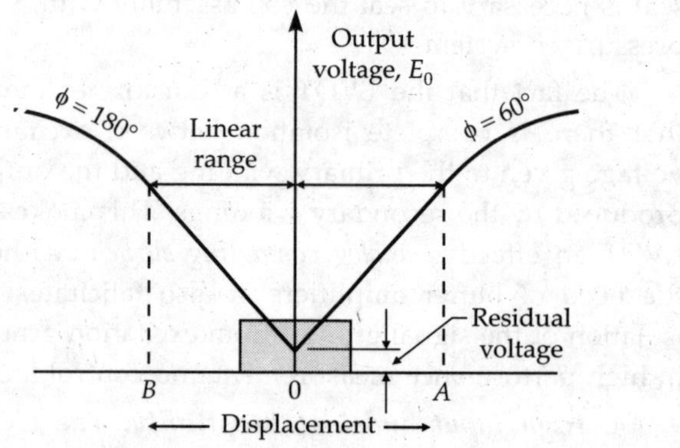


**Fig. 25.90** Variation output voltage with linear displacement for an LVDT.

Figure 25.90 shows the variation of output voltage versus displacement for various positions of core. The curve is practically linear for a limited range of displacement from the null position. Beyond this range of displacement the curve starts to deviate from a straight line.

Figure 25.91 shows the core of an **LVDT** at three different positions. In Fig. 25.91(*b*) the core is at null position, it is symmetrical with respect to both the secondary windings. This is called the *null* position. At this position $\mathbf{E}_{s1} - \mathbf{E}_{s2}$ hence the output voltage $\mathbf{E}_0 = 0$.

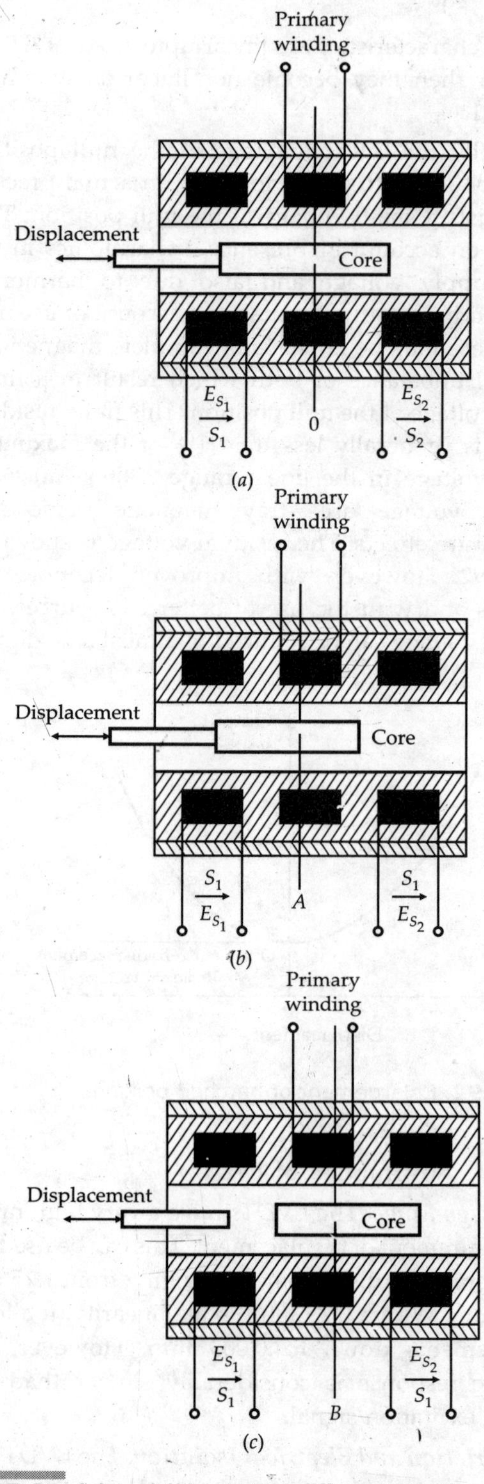


**Fig. 25.91** (*a*) and (*b*) LVDT at different positions, (*c*) Core of LVDT at different positions.

25

When the core is moved to the left as in Fig. 25.91(*a*) and is at *A*, $E_{s1}$ is greater than $E_{s2}$ and therefore phase angle $\phi_l = 0$. When the core is moved to the right towards *B* shown in Fig. 25.91(*c*) $E_{s2}$ is greater than $E_{s1}$ and hence the output voltage is negative or a phase angle of 180°.

The characteristics are linear upto $0 - A$ and $0 - B$ but after then they become non-linear as shown in Fig. 25.91.

Ideally the output voltage at the null position should be equal to zero. However, in actual practice there exists a small voltage at the null position. This may be on account of presence of harmonics in the input supply voltage and also due to harmonics produced in the output voltage on account of use iron core. There may be either an incomplete magnetic or electrical unbalance or both which result in a finite output voltage at the null position. This finite residual voltage is generally less than 1% of the maximum output voltage in the linear range. Other causes of residual voltage are stray magnetic fields and temperature effects. The residual voltage is shown in Fig. 25.92. However, with improved technological methods and with the use of better a.c. sources, the residual voltage can be reduced to almost a negligible value.

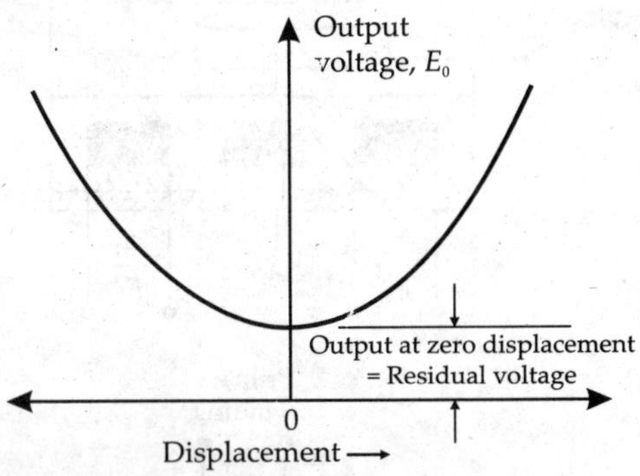


**Fig. 25.92** Enlargement of hatched portion of Fig. 25.90.

### 25.24.1 Advantages of LVDT

1. *High range*. The LVDTs have a very high range for measurement of displacement. This can be used for measurement of displacements ranging from 1.25 mm to 250 mm. With a 0.25% full scale linearity, it allows measurements down to 0.003 mm. However, the dynamic response is considerably slower than the 2.5 kHz excitation signal.

2. *Friction and Electrical Isolation*. The LVDT has many commendable features that make it useful for a wide variety of applications. Some of these features are unique to the LVDT and are not available in any other transducers. The features arise from the basic fact that LVDT is an electrical transformer with a separable non-contacting core.

Ordinarily, there is no physical contact between the movable core and coil structure which means that the LVDT is a frictionless device. This permits its use in critical measurements that cannot tolerate the addition of low mass core but cannot tolerate friction loading. The absence of friction between coil and core of an LVDT means that there is no wear out. This gives an LVDT essentially infinite mechanical life. This is of paramount importance in applications such as the fatigue-life testing of materials and structures. The infinite mechanical life is also important in high reliability mechanisms and systems found in aircrafts, missiles, space vehicles and critical industrial equipment.

The frictionless operation of the LVDT combined with the induction principle by which the LVDT function is truly infinite resolution. This means that LVDT can respond to even minute motion of the core and produce an output.

3. ***Immunity front External Effects***. The separation between LVDT core and LVDT coils permits the isolation of media such as pressurized, corrosive, or caustic fluids from the coil assembly by a non-magnetic barrier interposed between the core and inside of the coil. It also makes the hermetic sealing of the coil assembly possible and eliminates the need for a dynamic seal on the moving member. Only a static seal is necessary to seal the coil assembly within the pressurized system.

The fact that the LVDT is a transformer means that there is complete isolation between excitation voltage given to the primary winding and the output produced by the secondary windings. This makes an LVDT an effective *analog computing element* without the need of buffer amplifiers. It also falicitates the isolation of the signal ground from excitation ground in high performance measurement and control loops.

4. ***High input and high sensitivity***. The LVDT gives a high output and many a times there is no need for amplification. The transducer possesses a high sensitivity which is typically about 40 V/min.

5. ***Ruggedness***. These transducers can usually tolerate high degree of shock and vibrations especially when the core is spring loaded without any adverse effects. They are simple in construction and by virtue of their being small and light in weight, they are stable and easy to align and maintain.

6. ***Low Hysteresis.*** LVDTs show a low hysteresis and hence repeatability is excellent under all conditions.

7. ***Low Power Consumption.*** Most of LVDTs consume power which is less than 1 W.

### 25.24.2 Disadvantages of LVDTs

**1.** Relatively large displacements are required for appreciable differential output.

**2.** They are sensitive to stray magnetic fields but shielding is possible. This is done by providing magnetic shields with longitudinal slots.

**3.** Many a times, the transducer performance is affected by vibrations.

**4.** The receiving instrument must be selected to operate on a.c. signals or a demodulator network must be used if a d.c. output is required.

**5.** The dynamic response is limited mechanically by the mass of the core and electrically by the frequency of applied voltage. The frequency of the carrier should be at least ten times the highest frequency component to be measured.

**6.** Temperature affects the performance of the transducer. But when temperature effects are expected to influence the performance, manganin wire may be used instead of copper wire. Unfortunately manganin has a high resistivity and therefore its use results in loss of sensitivity. The sensitivity with manganin coils may be as low as 1/5 of that with copper coils. Temperature also causes phase shifting effects which may be minimized by using a capacitor across one of the secondary windings.

### 25.24.3 Uses of LVDTs

The uses of LVDTs are numerous. Some of the major applications of LVDTs are listed below :

**1.** The LVDT can be used in all applications where displacements ranging from fraction of a mm to a few cm have to be measured. The LVDT acting as a primary transducer converts the displacement direct into an electrical output proportional to displacement. This is a fundamental conversion *i.e.*, the mechanical variable (displacement in this case) is directly converted into an analogous signal (voltage in this case) in one stage only. In contrast, the electrical strain gauge requires the assistance of some form of a sensing element to act as primary transducer to convert the mechanical displacement into strain which in turn is converted into an electrical signal by the strain gauge acting as a secondary transducer. Therefore, two stages of signal conversion are involved in strain gauges while there is only one in case of LVDT when displacement is being measured.

**2.** Acting as a secondary transducer it can be used as a device to measure force, weight and pressure etc. The force measurement can be done by using a load cell as the primary transducer while fluid pressure can be measured by using Bourdon tube which acts as primary transducer. The force or the pressure is convened into a voltage. In these applications the high sensitivity of LVDTs is a major attraction.

Some of the typical applications of LVDTs are shown in Fig. 25.93. In Fig. 25.93(*a*) four LVDTs are used for measurement of weight or pressure exerted by liquid in a tank. They (LVTDs) are excited in parallel to increase the sensitivity.

Figure 25.93(*b*) shows two LVDTs which are used for measurement and control of thickness of a metal sheet being rolled. When the thickness equals the desired value, the two LVDTs are balanced out.

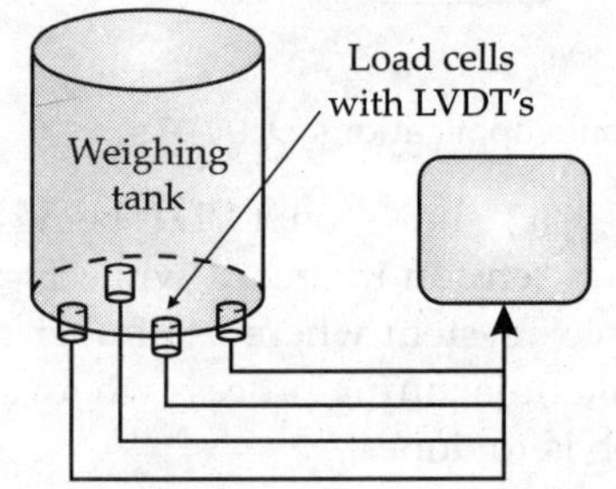

(*a*)

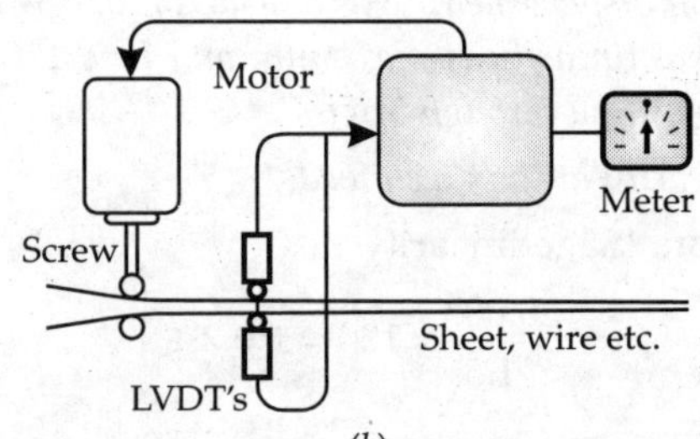

(*b*)

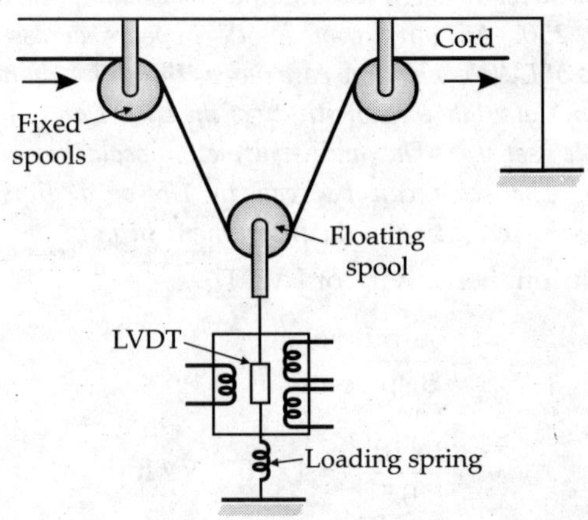

(*c*)

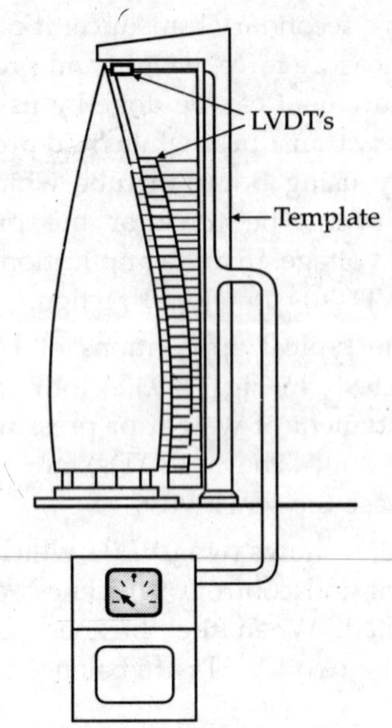


(d)

**Fig. 25.93** Sonic applications of LVDTs.

Figure 25.93(*c*) shows an LVDT being used for measurement of tension in a cord while Fig. 25.93(*d*) shows a complex system where a number of LVDTs are used in a manufacturing process where a complex machining gob is to done.

***Example 25.36*** *The output voltage of a LVDT is 1.5 V at maximum displacement. At a load of 0.5 MΩ, the deviation from linearity is maximum and it is ± 0.003 V from a straight line through origin.*

*Find the linearity at the given load.*

**Solution.** %age linearity

$$= \pm\frac{0.003}{1.5}\times 100 = \pm 0.2\%$$

***Example 25.37*** *The output of an LVDT is connected to a 5 V voltmeter through an amplifier whose amplification factor is 250. An output of 2 mV appears across the terminals of LVDT when the core moves through a distance of 0.5 mm. Calculate the sensitivity of the LVDT and that of the whole set-up. The milli-voltmeter scale has 100 divisions. The scale can be read to 1/5 of a division. Calculate the resolution of the instrument in mm.*

**Solution.** Sensitivity of LVDT

$$= \frac{\text{output voltage}}{\text{displacement}}$$

$$= \frac{2\times 10^{-3}}{0.5} = 4\times 10^{-3}\ \text{V/mm}$$

$$= 4\ \text{mV/mm}$$

Sensitivity of instrument

$$= \text{amplification factor} \times \text{sensitivity of LVDT}$$

$$= 4\times 10^{-3}\times 250 = 1\ \text{V/mm} = 1000\ \text{mV/mm}$$

1 scale division = 5/100 V = 50 mV

Minimum voltage that can be read on the voltmeter

$$= (1/5)\times 50 = 1\ \text{mV}$$

∴ Resolution of instrument

$$= 1\times(1/1000) = 1\times 10^{-3}\ \text{mm}$$

***Example 25.38*** *A steel cantilever is 0.25 m long, 20 mm wide and 4 mm thick.*

(*a*) *Calculate the value of deflection at the free end for the cantilever when a force of 25 N is applied at this end. The modulus of elasticity for steel is 200 GN/m*$^2$.

(*b*) *An LVDT with a sensitivity of 0.5 V/mm is used. The voltage is read on a 10 V voltmeter having 100 divisions. Two tenths of a division can be read with certainty.*

(*c*) *Calculate the minimum and maximum value of force that can be measured with this arrangement.*

**Solution.** (*a*) Moment of area of cantilever

$$I = \frac{1}{12}bt^3$$

$$= \frac{1}{12}\times(0.02)\times(0.004)^3$$

$$= 0.107\times 10^{-9}\ \text{m}^4$$

Deflection $x = \dfrac{Fl^3}{3EI}$

$$= \frac{25\times(0.25)^3}{3\times 200\times 10^9\times 0.107\times 10^{-9}}$$

$$= 6.08\times 10^{-3}\ \text{m} = 6.08\ \text{mm}$$

(*b*) Deflection per unit force

$$\frac{x}{F} = \frac{6.08}{25}$$

$$= 0.2432\ \text{mm/N}$$

Overall sensitivity of measurement system :

$$= 0.2432\ \frac{\text{mm}}{\text{N}}\times 0.5\frac{\text{V}}{\text{mm}}$$

$$= 0.1216\ \text{V/N}$$

1 scale division = (10/1000) = 0.1 V

Since two tenths of a scale division can be read with certainty, resolution = (2 / 10) × 0.1 = 0.02 V

(*c*) Minimum force that can be measured

= 0.02/0.1216 = 0.1645 N

Maximum force that can be measured

= (10/0.1216) = 82.2 N

## 25.25 ROTARY VARIABLE DIFFERENTIAL TRANSFORMER (RVDT)

A variation of linear variable differential transformer (LVDT) may be used to sense angular displacement. This is the **Rotary Variable Differential Transformer (RVDT)**. The circuit of a RVDT is shown in Fig. 25.94. It is similar to the LVDT except that its core is cam shaped and may be rotated between the windings by means of a shaft.

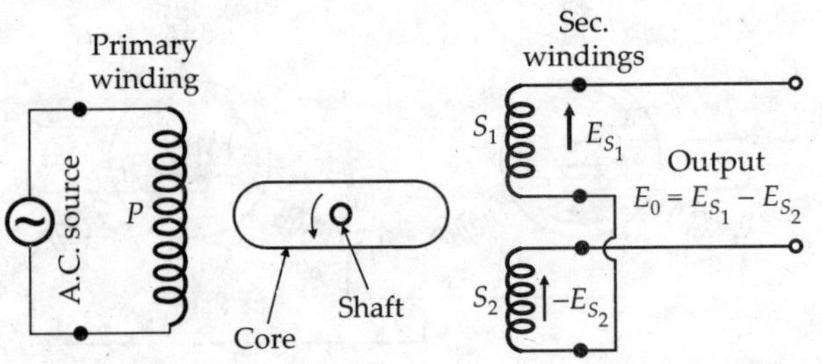


**Fig. 25.94** Rotary variable differential transformer (RVDT).

The operation of a RVDT is similar to that of an LVDT. At the Primary **null** position of the core, the output voltage of secondary windings $S_1$ and $S_2$ are equal and in opposition. Therefore, the net output is zero. Any angular displacement from the null position win result in a differential voltage output. The greater this angular displacement, the greater will be the differential output. Hence the response of the transducer is linear.

Clockwise rotation produces an increasing voltage of a secondary winding of one phase while counter clock-wise rotation produces an increasing voltage of opposite phase. Hence, the amount of angular displacement and its direction may be ascertained from the magnitude and phase of the output voltage of the transducer.

## 25.26 SYNCHROS

A *synchro* is an *electromagnetic transducer* which is commonly used to convert the angular position of a shaft into an electric signal. Although the name *"Synchro"* is universally used in the instrumentation field, trade names such as Selsyns, Microsyns and Autosyns are used for these instruments.

There are *two* types of synchro systems :

(*i*) control or error detecting type and

(*ii*) torque transmission type.

Torque transmission type of systems are used only to drive very light loads, such as pointers. Ironically, torque transmission systems have very little output torque. When large torques and high accuracies are needed, control type synchros are used.

### 25.26.1 Control Type Synchro Systems

These systems are used as error detectors in positional control systems.

The classical synchro system consists of *two* units :

(*a*) a synchro transmitter and

(*b*) a synchro receiver.

The basic synchro unit is called a *"Synchro Transmitter"*. Its construction is similar to that of a three phase alternator. The stator is made up of laminated steel in order to reduce core losses. It is slotted to accommodate a balanced three phase winding, with the axis of each stator winding displaced by 120° in space. The three stator coils are identical and are $Y$ (star) connected. The rotor has a salient pole construction, is dumb-bell shaped and is wound with a concentric coil. An a.c. voltage is applied to die rotor through two slip rings. The constructional features and schematic diagram are shown in Figs. 25.95 and 25.96 respectively.

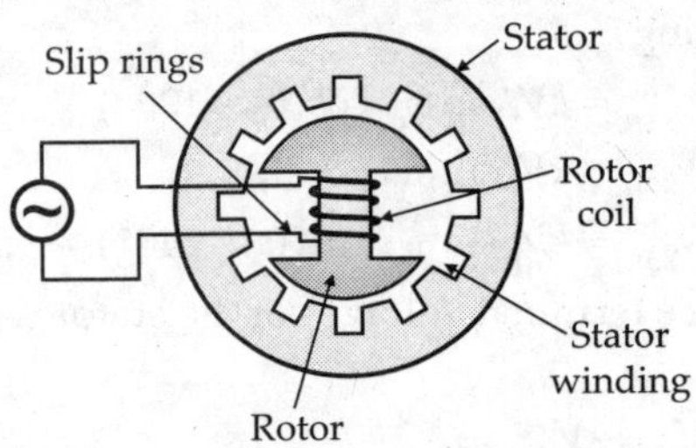


**Fig. 25.95** Constructional features of synchro transmitter.

Let an a.c. voltage be applied to the rotor of a synchro-transmitter as shown in Fig. 25.96. The voltage is :

$$v_r = \sqrt{2}\, V_r \sin \omega_c t$$

where $V_r$ = r.m.s. value of rotor voltage, V,

$\omega_c$ = carrier frequency ; rad/s

This voltage produces a magnetising current and consequently an alternating sinusoidally varying flux along its axis and distributed nearly sinusoidal in space. Voltages are induced in the three stator windings. As the air gap flux is sinusoidally distributed in space, the flux linking with any stator windings is proportional to the cosine of the angle between the rotor and the stator winding axes and so is the voltage induced in each stator winding. It should understood, that since the voltages in each

25

stator winding are produced by transformer action, the three stator voltages may have different values (as they link with different values of flux) but they are in time phase with each other. Thus, the rotor as the primary winding of a single phase transformer and the three stator windings act as secondary windings.

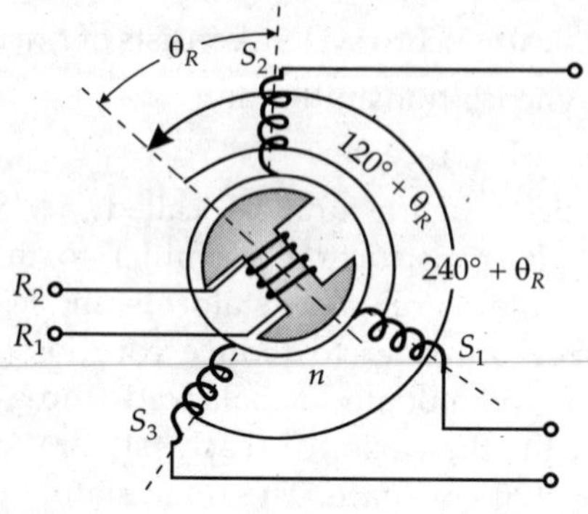


**Fig. 25.96** Schematic diagram of synchro transmitter.

Let $V_{s1n}$, $V_{s2n}$ and $V_{s3n}$ respectively be the voltages induced in the stator windings $S_1$, $S_2$ and $S_3$ with respect to neutral $n$. Then, for the rotor position of the synchro transmitter as shown in Fig. 25.96, where the rotor axis makes an angle $\theta_R$ with respect to axis of stator winding $S_2$.

$$V_{s1n} = kV_r \sin\omega_c t \cos(\theta_R + 120°) \quad \text{...(25.91)}$$

$$V_{s2n} = kV_r \sin\omega_c t \cos\theta_R \quad \text{...(25.92)}$$

$$V_{s3n} = kV_r \sin\omega_c t \cos(\theta_R + 240°) \quad \text{...(25.93)}$$

The three terminal voltages of the stator windings are :

$$V_{s1s2} = V_{s1n} - V_{s2n}$$

$$= \sqrt{3}\, kV_r \sin(\theta_R + 240°) \sin\omega_c t \quad \text{...(25.94)}$$

$$V_{s2s3} = V_{s2n} - V_{s3n}$$

$$= \sqrt{3}\, kV_r \sin(\theta_R + 120°) \sin\omega_c t \quad \text{...(25.95)}$$

$$V_{s3s1} = V_{s3n} - V_{s1n}$$

$$= \sqrt{3}\, kV_r \sin\theta_R \sin\omega_c t \quad \text{...(25.96)}$$

The variation of the scator terminal voltages with angle $\theta_R$ is shown in Fig. 25.97.

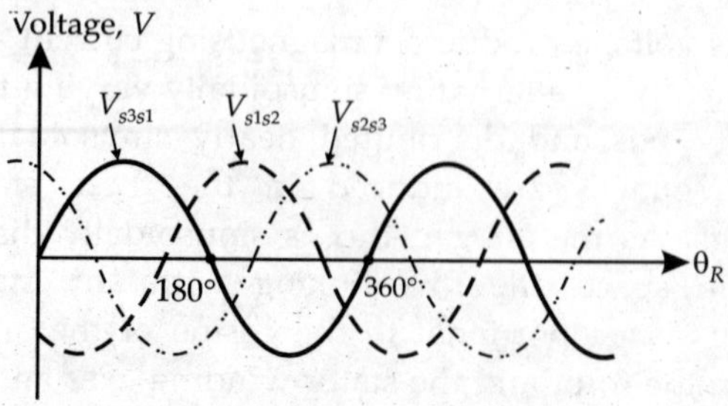


**Fig. 25.97** Variation of stator terminal voltages with $\theta_R$.

When $\theta_R = 0$, from Eqn. 25.92 it is seen that maximum voltage is induced in the stator winding $S_2$, while it follows from Eqn. 25.96 that the terminal voltage $V_{s3s1}$, is Zero. This position of the rotor is defined *Electrical zero* of the transmitter and is used as reference for specifying angular position of the rotor (See Fig. 25.98).

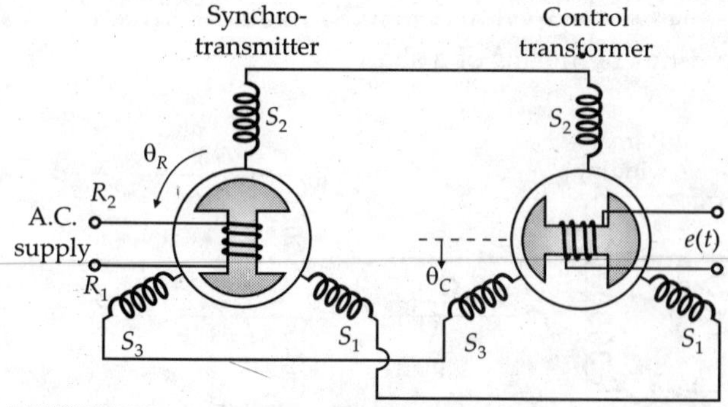


**Fig. 25.98** Syncho error detector.

The output of the synchro transmitter is applied to the stator windings of a *Synchro Control Transformer*. The synchro control transformer is similar in construction to a synchro transmitter except that the rotor of a control transformer is not of salient pole type but is cylindrical in construction. The use of cylindrical construction has an advantage because whatever may be the positron of the rotor, the length of air gap remains constant, and therefore irrespective of the position of the rotor, the input impedance of the control transformer remains the same.

The *synchro-transmitter-control transformer pair* acts as an error detector. The arrangement for *error-detector* is shown in Fig. 25.98. Circulating currents of the same phase are of different magnitudes flow through the stator windings transmitter and control transformer. This results in establishment of an identical flux pattern in the air gap of control transformer. The control transformer flux axis is in the same position as the flux axis in the transmitter. The voltage induced in the rotor of the control transformer is proportional to the cosine of the angle between axes of the two rotors. This voltage is given by :

$$e(t) = k'V_r \cos\phi \sin\omega_c t \quad \text{...(25.97)}$$

where $\phi$ = angular displacement between axes of rotor of synchro transmitter and rotor of synchro control transformer.

when $\phi = 90°$, *i.e.*, the axis of rotor of synchro-transformer is at right angles to that of axis of rotor transmitter, the voltage induced in the control transformer rotor is,

$$e(t) = k'V_r \cos(90°) \sin\omega_c t = 0$$

This position of the rotor of control transformer is known as *Electrical Zero.*

In Fig. 25.98, the rotors of transmitter and the control transformer are respectively shown in their *zero positions.*

Let the rotor of transmitter rotate through an angle $\theta_R$ in the direction indicated and let the rotor of control transformer rotate by an angle $\theta_c$ in the same direction resulting in a net angular separation of $\phi = (90° - \theta_R + \theta_C)$ between the two rotors.

From Eqn. 25.97, the voltage at the rotor terminals of synchro transformer is :

$$e(t) = k'V_r \cos(90° - \theta_R + \theta_C) \sin \omega_c t$$
$$= k'V_r \sin(\theta_R - \theta_C) \sin \omega_c t \quad \ldots(25.98)$$

For small angular displacements between the two rotor positions.

$$\sin(\theta_R - \theta_C) \simeq \theta_R - \theta_C \quad \ldots(25.99)$$

From Eqn. 25.98 for small angular displacements between the two rotor positions

$$e(t) = k'V_r \sin(\theta_R - \theta_C) \sin \omega_c t \quad \ldots(25.100)$$

The synchro transmitter – control transformer pair thus acts as an error detector giving a voltage signal at the rotor terminals of control transformer which is proportional to the difference between the transmitter and control transformer shaft positions.

The error signal is fed to an amplifier which in turn feeds a servomotor as shown in Fig. 25.99. The servomotor through gears moves the rotor of the control transformer thus giving a closed loop positional control system.

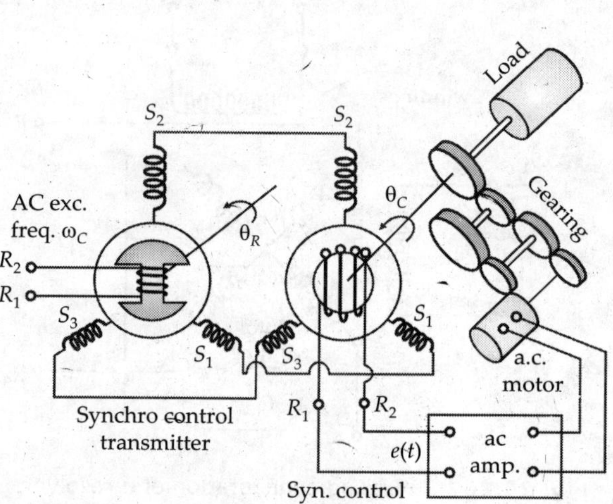


**Fig. 25.99** Positional control system.

Equation 25.100 is represented graphically in Fig. 25.100 for an arbitrary time variation of $(\theta_R - \theta_C)$.

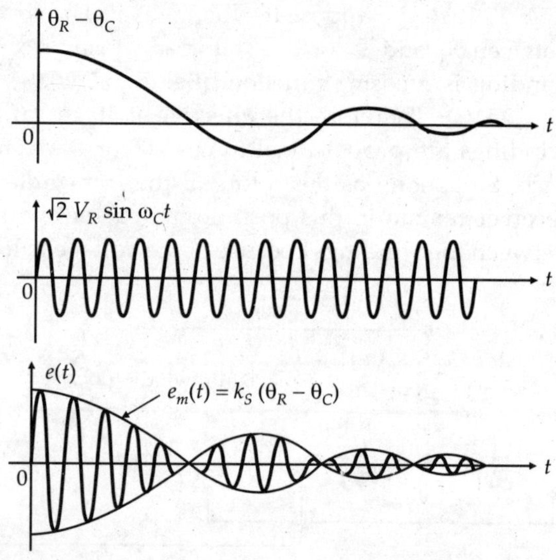


**Fig. 25.100** Typical wave forms of synchro error detector.

We see from this diagram that the output of synchro error detector is a modulated signal, the modulating wave has the information regarding the lack of correspondence between the two rotor positions and the carrier wave is the a.c. input to the rotor of the synchro transmitter. This type of modulation is known as *suppressed-carrier modulation.* From Eqn. 25.100 the modulating signal representing the discrepancy between the two shaft positions is,

$$e_m(t) = K_s(\theta_R - \theta_C) \quad \ldots(25.101)$$

where $K_s$ is known as *sensitivity* of error detector and has the units of volt per degree of angular difference of the shafts of the synchro pair.

As pointed out earlier the rotor of the control transformer is made cylindrical in shape so that the air gap is practically uniform. This is essential for a control transformer, since its rotor terminals are usually connected to an amplifier therefore the change in the rotor output impedance with rotation of the shaft must be minimised. Another distinguishing feature is that the stator winding of the control transformer has a higher impedance per phase. This feature allows several control transformers to be fed from a single transmitter.

### 25.26.2 Synchros as Torque Transmitters

In instrumentation systems ; synchros are normally used in the torque transmission mode. In this mode, the synchro transmitter and receiver are connected as shown in Fig. 25.101.

Initially winding $S_2$ of the stator of transmitter is positioned for maximum coupling with rotor winding

25

[Fig. 25.101(*a*)]. Suppose its voltage is $V$. The coupling between $S_1$ and $S_2$ of the stator and primary (rotor) winding is a cosine function. (See Eqns. 25.91, 25.92 and 25.93). Therefore the effective voltages in these windings are proportional to cos 60° or they are $V/2$ each. So as long as the rotors of the transmitter and receiver remain in this position, no current will flow between the windings because of the voltage balance.

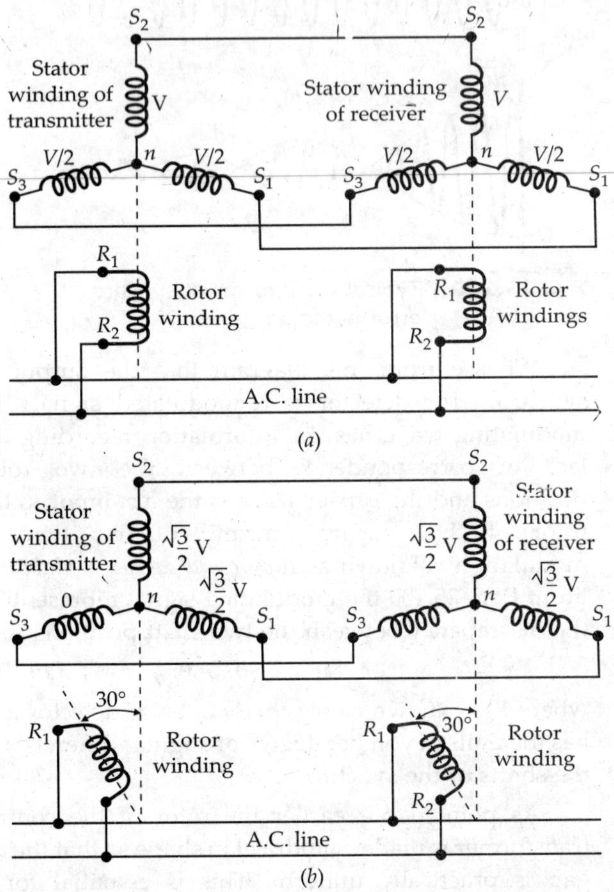


**Fig. 25.101** Torque transmission using synchros.

When the rotor of the transmitter is moved to a new position, the voltage balance is disturbed. Assume that the rotor of the transmitter is moved through 30° [See Fig. 25.101(*b*)], the stator winding voltages of the transmitter will be changed to 0, $\sqrt{3}/2$ V and $\sqrt{3}/2$ V respectively. Thus there is a voltage imbalance between the stator windings of the transmitter and the receiver. This voltage imbalance between the windings causes currents to flow between the windings producing a torque that tends to rotate the rotor of the receiver to a position where the voltage balance is again restored. This balance is restored only if the receiver turns through the same angle as the transmitter and also the direction of rotation is the same as that of the transmitter *i.e.*, 30° in this case. Thus a synchro can be used to determine the magnitude and the direction of an angular displacement.

## 25.27 RESOLVERS

Resolvers are used for conversion of angular position of a shaft into cartesian co-ordinates. The output of the transducer is in the form of two signals, one proportional to the sine of the angle and the other proportional to cosine of the angle.

A *resolver* is a very precise electromagnetic device comprising of two stator and two rotor windings. The construction of a resolver is similar to that of a two phase, two pole wound rotor induction motor. The two stator windings are identical and are housed in a magnetic structure, with the axis of two windings 90° to each other. Similarly, the two rotor windings are placed in a magnetic structure and are mutually perpendicular to each other.

Stator windings are supplied with an alternating voltage that produces an alternating magnetic flux which induces voltages in the two rotor windings. The output voltage of the rotor windings is proportional to the stator voltage and the coupling between stator and rotor windings. The way in which the windings are placed, the rotor output voltages are proportional to the sine and cosine of the rotor angle.

Referring to Fig. 25.102, which shows the winding configuration of a resolver, when one of the stator windings say $S_1S_3$ is excited by an a.c. source, with the

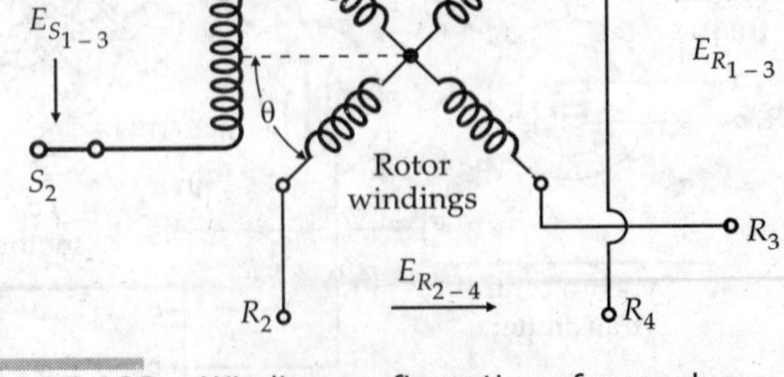


**Fig. 25.102** Winding configuration of a resolver.

other stator winding $S_2S_4$ short circuited, the following output voltages are obtained from the rotor.

$$E_{R1-3} = E_{S1-3} \cos\theta \quad \text{...(25.102)}$$

$$E_{R2-4} = -E_{S1-3} \sin\theta \quad \text{...(25.103)}$$

When the two stator windings are excited, the outputs are as under :

$$E_{R1-3} = E_{S1-3} \cos\theta + E_{S2-4} \sin\theta \quad ...(25.104)$$

$$E_{R2-4} = E_{S2-4} \cos\theta - E_{S1-3} \sin\theta \quad ...(25.105)$$

When two rotor windings are excited, we get the following outputs from the stator windings :

$$E_{S1-3} = E_{R1-3} \cos\theta - E_{R2-4} \sin\theta \quad ...(25.106)$$

and $$E_{S2-4} = E_{R2-4} \cos\theta + E_{R1-3} \sin\theta \quad ...(25.107)$$

For all above equations $\theta$ = angular displacement of rotor.

### 25.27.1 Classification of Resolvers

Resolvers are classified in *two* groups :

- Computing resolvers and
- Synchro resolvers.

*Computing Resolvers* are used for generating sine, cosine, and tangent functions as well as for solving geometric relationships.

*Synchro Resolvers* are used for data transmission. They perform the same functions as synchro transmitters, receivers and control transformers but with a better accuracy.

### 25.27.3 Applications of Resolvers

The applications of resolvers are :

(*i*) Vector resolution,

(*ii*) Vector composition,

(*iii*) Vector angle and component resolution,

(*iv*) Pulse amplitude control and pulse resolution,

(*v*) Phase shifting.

## 25.28 CAPACITIVE TRANSDUCERS

The principle of operation of capacitive transducers is based upon the familiar equation for capacitance of a parallel plate capacitor.

$$\text{Capacitance} \quad C = \varepsilon A/d$$

$$= \varepsilon_r \varepsilon_0 A/d \quad ...(25.108)$$

where $A$ = overlapping area of plates ; $m^2$,

$d$ = distance between two plates ; m,

$\varepsilon = \varepsilon_r \varepsilon_0$ = permittivity of medium ; F/m,

$\varepsilon_r$ = relative permittivity,

$\varepsilon_0$ = permittivity of free space ; $8.85 \times 10^{-12}$ F/m

A parallel plate capacitors is shown in Fig. 25.103.

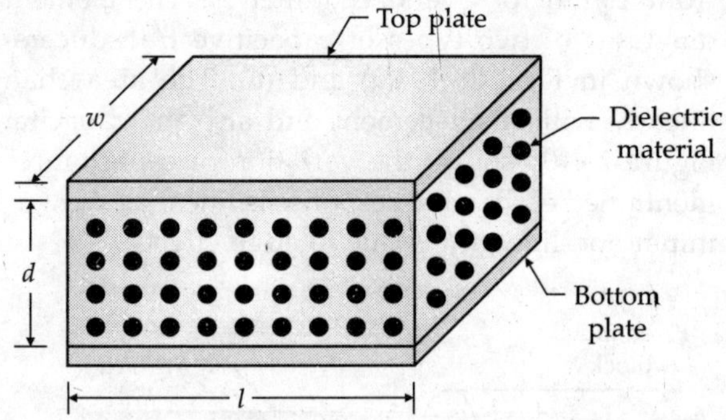


**Fig. 25.103** Schematic diagram of a parallel plate capacitive transducer.

The capacitive transducer works on the principle of change of capacitance which may be caused by :

(*i*) Change in overlapping area $A$,

(*ii*) Change in the distance $d$ between the plates, and

(*iii*) Change in dielectric constant.

These changes are caused by physical variables like displacement, force and pressure in most of the cases. The change in capacitance may be caused by change in dielectric constant as is the case in measurement of liquid or gas levels.

The capacitance may be measured with bridge circuits. The output impedance of a capacitive transducer is :

$$X_c = 1/2\pi f C,$$

where $C$ = capacitance and

$f$ = frequency of excitation in Hz

In general, the output impedance of a capacitive transducer is high. This fact calls for a careful design of the output circuitry.

The capacitive transducers are commonly used for measurement of linear displacement. These transducers use the following effects :

(*i*) Change in capacitance due to change in overlapping area of plates, and

(*ii*) Change in capacitance due to change in distance between the two plates.

### 25.28.1 Transducers Using Change in Area of Plates

Examining the equation for capacitance (Eqn. 25.108), it is found that the capacitance is directly proportional to the area, $A$ of the plates. Thus the capacitance changes linearly with change in area of plates. Hence this type of capacitive transducer is useful for measurement of moderate to large displacements say from 1 mm to several centimetres. The elementary diagrams of two types of capacitive transducers are shown in Figs. 25.104(*a*) and (*b*). The area changes linearly with displacement and also the capacitance. Figure 25.105 shows the variation of capacitance. As mentioned earlier, the response is linear as shown, the initial non-linearity is due to edge effects.

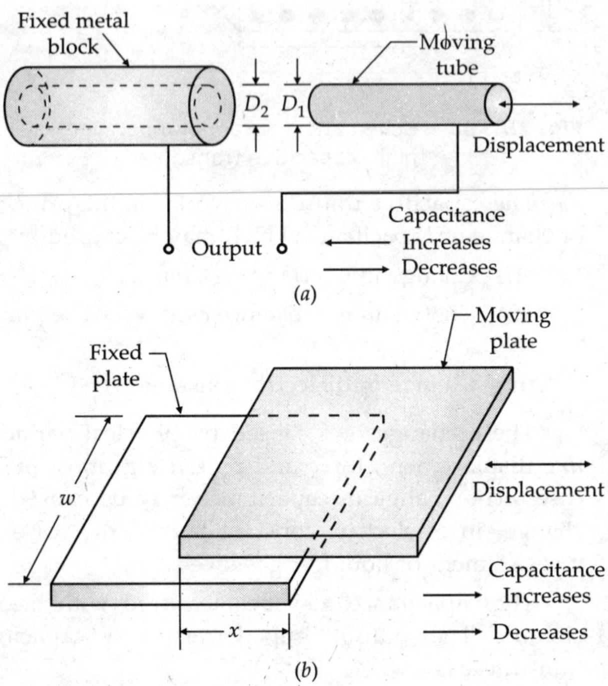


**Fig. 25.104** Capacitive transducers working on the principle of change of capacitance with change of area.

For a parallel plate capacitor, the capacitance is :

$$C = \frac{\varepsilon A}{d} = \frac{\varepsilon x w}{d} \text{ F} \qquad \ldots(25.109)$$

where $x$ = length of overlapping part of plates ; m,

and $w$ = width of overlapping part of plates ; m

$$\text{Sensitivity} \quad S = \frac{\partial C}{\partial x} = \varepsilon \frac{w}{d} \text{ F/m} \qquad \ldots(25.110)$$

The sensitivity is constant and therefore there is linear relationship between capacitance and displacement.

Sensitivity for a fractional change in capacitance

$$= S' = \frac{\partial C}{C \partial x} = \frac{1}{x} \qquad \ldots(25.111)$$

This type of a capacitive transducer is suitable for measurement of linear displacements ranging from 1 mm to 10 mm. The accuracy is as high as 0.005%.

For a cylindrical capacitor the capacitance is :

$$C = \frac{2\pi\varepsilon x}{\log_e(D_2 / D_1)} \text{ F} \qquad \ldots(25.112)$$

where $x$ = length of overlapping part of cylinders ; m,

$D_2$ = inner diameter of outer cylindrical electrode ; m

and $D_1$ = outer diameter of inner cylindrical electrode ; m

Sensitivity

$$S = \frac{\partial C}{\partial x}$$

$$= \frac{2\pi\varepsilon}{\log_e(D_2 / D_1)} \text{ F/m} \qquad \ldots(25.113)$$

Therefore, the sensitivity is constant and the relationship between capacitance and displacement is linear as shown in Fig. 25.105.

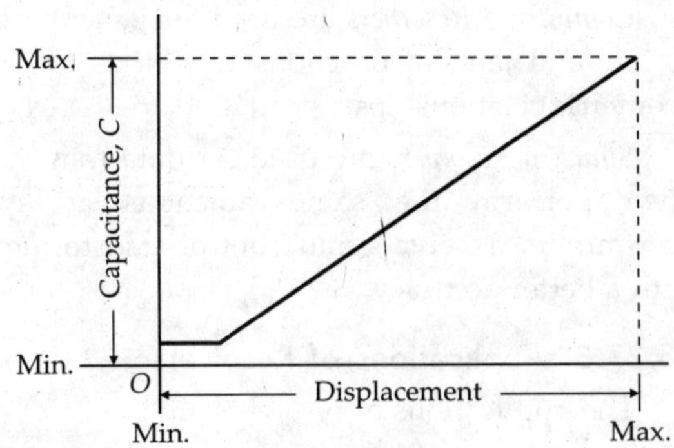


**Fig. 25.105** Capacitance displacement curve of capacitive transducer (working on principle of change of plate area caused by change in displacement).

The principle of change of capacitance with change in area can be employed for measurement of angular displacement. Figure 25.106(*a*) shows a two-plate capacitor. One plate is fixed and the other is movable. The angular displacement to be measured is applied to movable plate. The angular displacement changes the effective area between the plates and thus changes the capacitance. The capacitance is maximum when the two plates completely overlap each other *i.e.*, when $\theta = 180°$.

∴ Maximum value of capacitance

$$C_{max} = \frac{\varepsilon A}{d} = \frac{\pi \varepsilon r^2}{2d} \qquad \ldots(25.114)$$

Capacitance at angle θ is,

$$C = \frac{\varepsilon \theta r^2}{2d} \qquad \ldots(25.115)$$

where θ = angular displacement in radian.

Sensitivity $$S = \frac{\partial C}{\partial \theta} = \frac{\varepsilon r^2}{2d} \qquad \ldots(25.116)$$

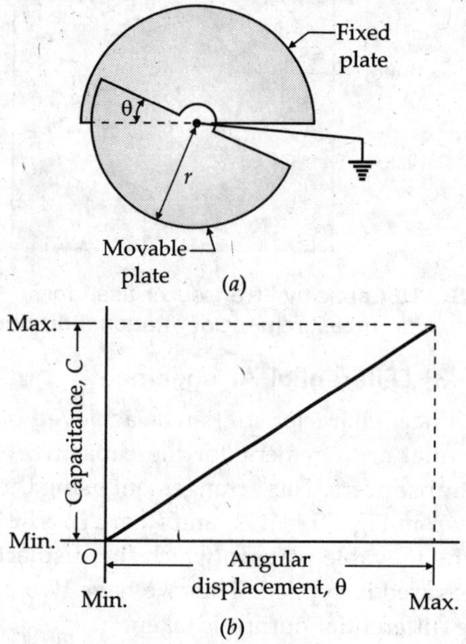


**Fig. 25.106** Capacitive transducer for measurement of angular displacement.

Therefore, the variation of capacitance with angular displacement is linear. This is shown in Fig. 25.106(*b*). It should be understood that the above, mentioned capacitive transducer can be used for a maximum angular displacement of 180°.

## 25.28.2 Transducers Using Change in Distance between Plates

Figure 25.107(*a*) shows the basic form of a capacitive transducer utilizing the effect of change of capacitance with change in distance between the two plates. One is a fixed plate and the displacement to be measured is applied to the other plate which is movable. Since, the capacitance, *C*, varies inversely as the distance *d*; between the plates the response of this transducer is not linear and as shown in Fig. 25.107(*b*). Thus, this transducer is useful only for measurement of extremely small displacements.

Sensitivity $$S = \frac{\partial C}{\partial x} = -\frac{\varepsilon A}{x^2} \qquad \ldots(25.117)$$

From Eqn. 25.105, it is clear that the sensitivity of this type of transducer is not constant but varies over the range of the transducer. Thus, as explained earlier this transducer exhibits non-linear characteristics.

The relationship between variation of capacitance, *C* with variation of distance between plates, *x*,

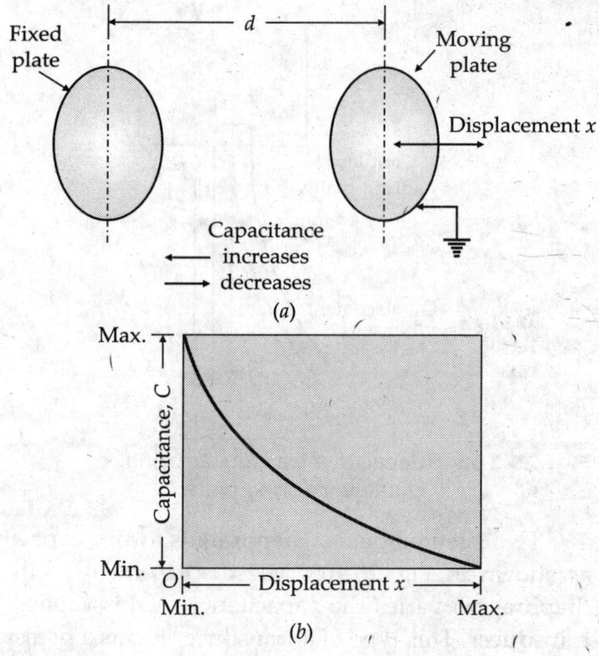


**Fig. 25.107** Capacitive transducer using the principle of change of capacitance with change of distance between plates.

is hyperbolic and is only approximately linear over a small range of displacement. The linearity can be closely approximated by use of a piece of dielectric material like mica having a high dielectric constant. In this type of transducer, a thin piece of mica thinner than the minimum gap distance is inserted between the plates.

25

Theoretically, the sensitivity of the transducer can be increased to any desirable value by making the distance between the plates extremely small. But a practical limit is reached when the electric field strength in the air gap exceeds the breakdown voltage. The breakdown limit in air at atmospheric pressure is about 3 kV/mm.

Figure 25.108 shows an arrangement for measurement of linear displacement. The displacement when applied to the cantilever type spring plate moves it towards the second plate decreasing the distance. This increases the capacitance of the capacitor. It is clear that the capacitance of air dielectric capacitor does not vary linearly with change in distance between the plates and therefore, this arrangement is fundamentally non-linear. However, linearity can be closely approximated by keeping the change in the distance small or by having a medium of high dielectric constant in the space between the two plates.

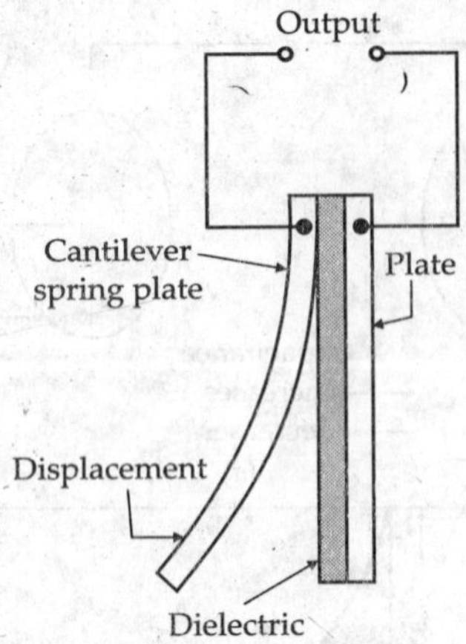


**Fig. 25.108** Capacitive transducer using cantilever spring plate

Two silvered quartz diaphragms form a capacitor as shown in Fig. 25.109. The displacement of these diaphragms varies the capacitance of this capacitive transducer. This type of a transducer forms a pressure gauge which works very satisfactorily. If the pressure is reduced to perfect vacuum between the diaphragms, the transducer measures the absolute pressure of the surrounding medium. Quartz is used because it has a very low mechanical hysteresis. The disadvantage of this type of transducer is that it has high temperature co-efficient and this must be taken into consideration when making precision type measurements.

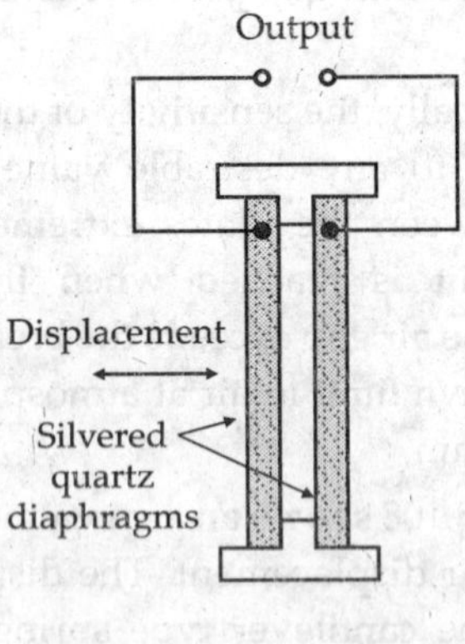


**Fig. 25.109** Capacitive transducer using Quartz diaphragms.

Rotational displacement can be measured with an arrangement shown in Fig. 25.110. As the rotor plates of the capacitor are displaced in the counter clockwise direction the capacitance increases. The change in the capacitance is a measure of the angular displacement. This capacitive transducer can be effectively used for measurement of torque.

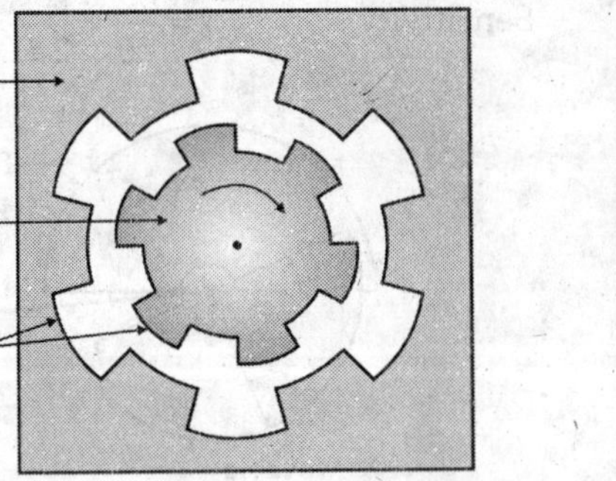


**Fig. 25.110** Capacitive transducer used for measurement of rotational displacement.

### 25.28.3 Differential Arrangement

A linear characteristic can be achieved by using a differential arrangement for the capacitive displacement transducers. This arrangement using three plates is shown in Fig. 25.111. $P_1$ and $P_2$ are fixed plates and $M$ is the movable plate to which the displacement to be measured is applied. Thus we have two capacitors whose differential output is taken.

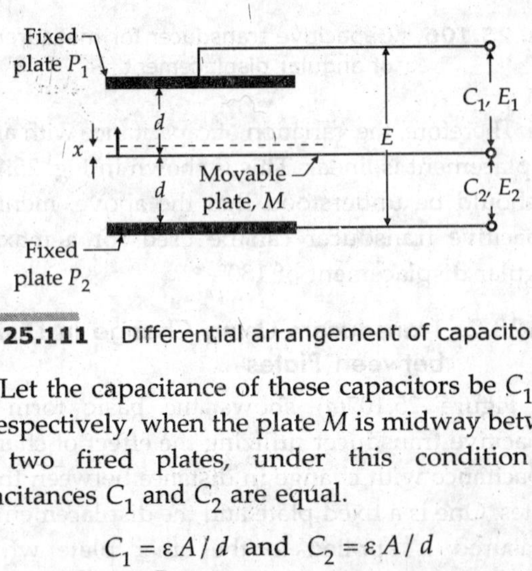


**Fig. 25.111** Differential arrangement of capacitors.

Let the capacitance of these capacitors be $C_1$ and $C_2$ respectively, when the plate $M$ is midway between the two fired plates, under this condition the capacitances $C_1$ and $C_2$ are equal.

$$C_1 = \varepsilon A/d \text{ and } C_2 = \varepsilon A/d$$

An alternating current voltage $E$ is applied across plates $P_1$ and $P_2$ and the difference of the voltages across the two capacitances is measured. When the movable plate is midway between the two fixed plates $C_1 = C_2$ and therefore $E_1 = E_2 = E/2$.

Voltage across $C_1$ is,

$$E_1 = EC_2/(C_1 + C_2) = E/2$$

and voltage across $C_2$ is

$$E_2 = EC_1/(C_1 + C_2) = E/2$$

$\therefore$ Differential output when the movable plate is midway $\Delta E = E_1 - E_2 = 0$.

Let the movable plate be moved up due to displacement $x$. Therefore the values $C_1$ and $C_2$ become different resulting in a differential voltage output.

Now $\quad C_1 = \dfrac{\varepsilon A}{d-x}$

and $\quad C_2 = \dfrac{\varepsilon A}{d+x}$

$\therefore \quad E_1 = \dfrac{C_2 E}{C_1 + C_2}$

$$= \frac{\varepsilon A/(d+x)}{\varepsilon A/(d-x)+\varepsilon A/(d+x)} E = \frac{d-x}{2d} E$$

and $\quad E_2 = \dfrac{C_1 E}{C_1 + C_2}$

$$= \frac{\varepsilon A/(d-x)}{\varepsilon A/(d-x)+\varepsilon A/(d+x)} E = \frac{d+x}{2d} E$$

$\therefore$ Differential output voltage

$$= \Delta E = E_2 - E_1$$

$$= \frac{d+x}{2d} E - \frac{d-x}{2d} E$$

$$= \frac{x}{d} E \qquad \text{...(25.118)}$$

Therefore the output voltage varies linearly as the displacement $x$.

Sensitivity $\quad S = \dfrac{\Delta E}{x} = \dfrac{E}{d} \qquad$ ...(25.119)

The differential method can be used for displacements of $10^{-8}$ mm to 10 mm with an accuracy of 0.1%.

### 25.28.4 Variation of Dielectric Constant for Measurement of Displacement

The third principle used in capacitive transducers is the variation of capacitance due to change in dielectric constant. Figure 25.112 shows a capacitive transducer for measurement of linear displacement working on the above mentioned principle. It has a dielectric of relative permittivity $\varepsilon_r$.

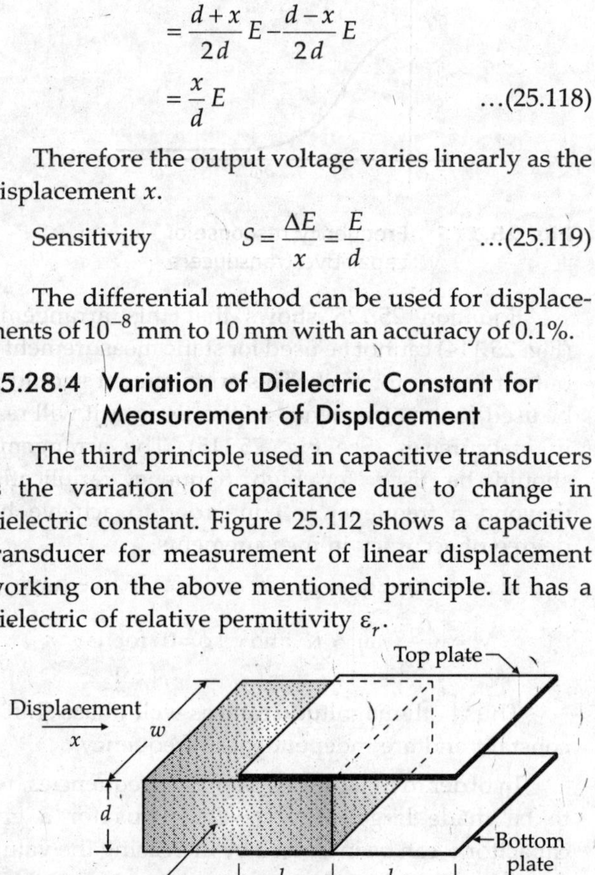


**Fig. 25.112** Capacitive transducer using principle of change in dielectric constant for measurement of displacement.

Initial capacitance of transducer

$$= C = \varepsilon_0 \frac{wl_1}{d} + \varepsilon_0 \varepsilon_r \frac{wl_2}{d}$$

$$= \varepsilon_0 \frac{w}{d}[l_1 + \varepsilon_r l_2] \qquad \text{...(25.120)}$$

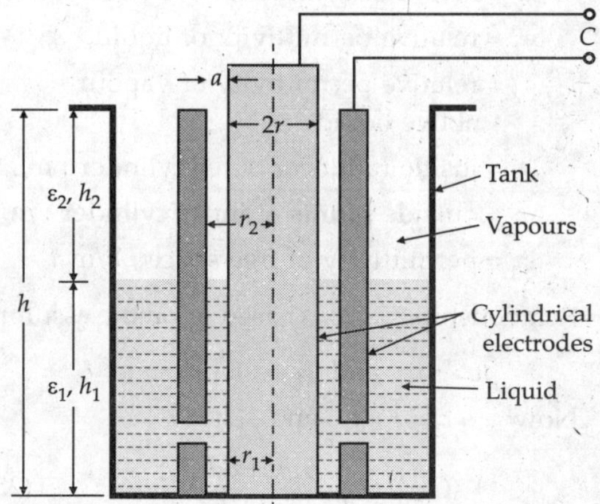


**Fig. 25.113** Capacitive transducer for measurement of level of a non-conducting liquid.

Let the dielectric be moved through a distance $x$ in the direction indicated. The capacitance changes from $C$ to $C + \Delta C$.

$$\therefore \quad C + \Delta C = \varepsilon_0 \frac{w}{d}(l_1 - x) + \varepsilon_0 \varepsilon_r \frac{w}{d}(l_2 + x)$$

$$= \varepsilon_0 \frac{w}{d}[l_1 - x + \varepsilon_r (l_2 + x)]$$

$$= \varepsilon_0 \frac{w}{d}(l_1 + \varepsilon_r l_2) + \varepsilon_0 \frac{wx}{d}(\varepsilon_r - 1)$$

$$= C + \varepsilon_0 \frac{wx}{d}(\varepsilon_r - 1) \qquad \text{...(25.121)}$$

$\therefore$ Change in capacitance

$$\Delta C = \varepsilon_0 \frac{wx}{d}(\varepsilon_r - 1) \qquad \text{...(25.122)}$$

Hence the change in capacitance is proportional to displacement.

### 25.28.5 Variation of Dielectric Constant for Measurement of Liquid Level

Capacitive transducers using the principle of change of capacitance with change of dielectric are normally used for measurement of liquid levels. Figure 25.113 shows a capacitive transducer used for measurement of level of a non-conducting liquid.

The electrodes are two concentric cylinders and the non-conducting liquid acts as the dielectric. At the lower end of the outer cylinder there are holes which allow passage of liquid. In case these holes are small, they provide mechanical damping of the surface variation.

The value of capacitance for this capacitor is,

$$C = 2\pi\varepsilon_0 \frac{\varepsilon_1 h_1 + \varepsilon_2 h_2}{\log_e(r_2/r_1)} \quad \ldots(25.123)$$

where $h_1$ = height of liquid ; m,
$h_2$ = height of cylinder above liquid ; m,
$\varepsilon_1$ = relative permittivity of liquid,
$\varepsilon_2$ = relative permittivity of vapour above liquid,
$r_2$ = inside radius of outer cylinder ; m,
$r_1$ = outside radius of inner cylinder ; m,
$\varepsilon_0$ = permittivity of free space ; F/m

Relationship 25.123 is based upon the assumption

$$h >> r_2 \text{ and } r_2 >> r_2 - r_1 >> a$$

Now $r_2 = r + a$ and $r_1 = r$

$$\therefore \quad C = 2\pi\varepsilon_0 \frac{\varepsilon_1 h_1 + \varepsilon_2 h_2}{\log_e(1 + a/r)} \quad \ldots(25.124)$$

## 25.28.6 Frequency Response of Capacitive Transducers

Figure 25.114 shows a simple circuit for capacitive transducers using change in capacitance due to change in distance between plates. Let the distance between the plates be $x_0$ when they are stationary. Under this condition no current flows and output voltage is $e_0 = E_b$.

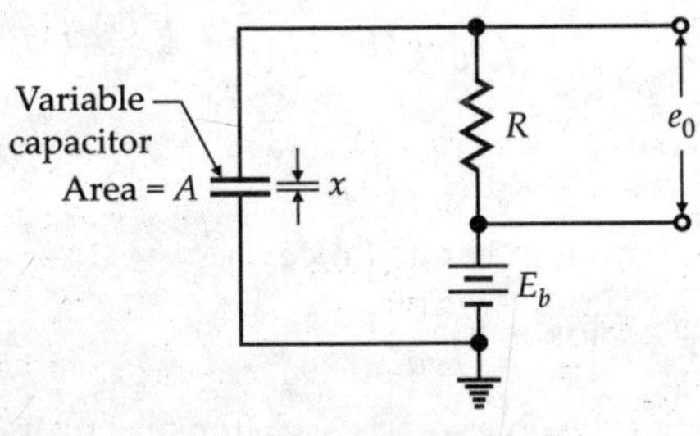


**Fig. 25.114** Capacitive transducer.

Thus, then a relative displacement $x_i$ from the $x_0$ position, a voltage $e_0$ is produced.

The following analysis is done for small displacements with $x_1/d < 0.1$.

$$\text{Transfer function} = \frac{E_0(s)}{X_i(s)} = \frac{K\tau s}{1 + \tau s}$$

where $K = \frac{E_b}{x_0}$ V/m,

$\tau$ = time constant = $RC$

Sinusoidal transfer function

$$\frac{E_0(j\omega)}{X_1(\omega)} = \frac{j\omega K\tau}{1 + j\omega\tau}$$

The amplitude ratio is thus :

$$M = \left|\frac{E_0}{KX_i}(j\omega)\right| = \frac{\omega\tau}{\sqrt{1+\omega^2\tau^2}} = \frac{1}{\sqrt{1+(1/\omega\tau)^2}} \quad \ldots(25.125)$$

Phase shift

$$\phi = \pi/2 - \tan^{-1}\omega\tau \text{ rad.} \quad \ldots(25.126)$$

The frequency response curves are shown in Fig. 25.115.

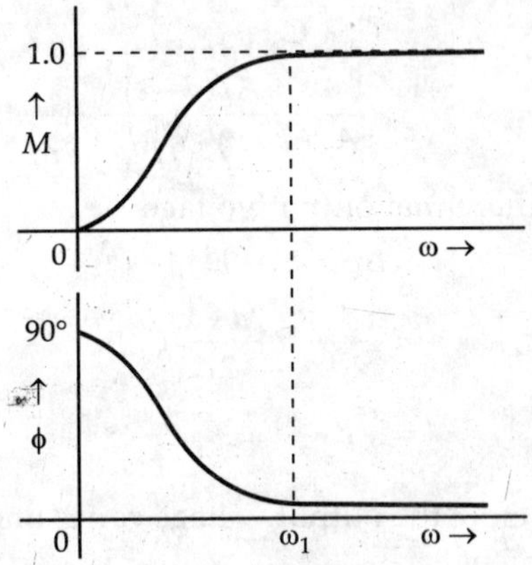


**Fig. 25.115** Frequency response of capacitive transducers.

Equation 25.125 shows that this arrangement (Fig. 25.114) cannot be used for static measurements as output is zero for $\omega = 0$. This arrangement should not be used for low frequency applications as it will result in inaccuracies (See Fig. 25.115). The arrangement should be used for high frequency applications (beyond a frequency $\omega_1$) in order to achieve high degree of accuracy in measurements.

Examining Eqns. 25.125 and 25.126, we have

$$M = \left|\frac{E_0}{X_i}(j\omega)\right| \approx K \text{ and } \phi = 0 \text{ for } \omega\tau >> 1$$

Thus both magnitude ratio as well phase shift are constant and are independent of frequency.

In order to make $\omega\tau >> 1$ for low frequencies, $\tau$ has to be made large. Now $\tau = RC$. Thus for a given capacitor $\tau$ can be increased by increasing the value of resistance $R$. This resistance is of the order of 1 MΩ or more. In order to prevent loading of capacitive transducers, readout or measuring device should have an impedance of at least 10 MΩ (or more) such as provided by FETs.

### 25.28.7 Advantages of Capacitive Transducers

The major advantages of capacitive transducers are :

(*i*) They require extremely small forces to operate them and hence are very useful for use in small systems.

(*ii*) They are extremely sensitive.

(*iii*) They have a good frequency response. This response is as high as 50 kHz and hence they are very useful for dynamic studies.

(*iv*) They have a high input impedance and therefore the loading effects are minimum.

(*v*) A resolution of the order of $2.5 \times 10^{-3}$ mm can be obtained with these transducers.

(*vi*) The capacitive transducers can be used for applications where stray magnetic fields render the inductive transducers useless.

(*vii*) The force requirements of capacitive transducers are very small and therefore they require small power to operate them.

### 25.28.8 Disadvantages of Capacitive Transducers

The principal disadvantages of capacitive transducers are :

(*i*) The metallic parts of the capacitive transducers must be insulated from each other. In order to reduce the effects of stray capacitances, the frames must be earthed.

(*ii*) The capacitive transducers show non-linear behaviour many a times on account of edge effects. Therefore guard rings must be used to eliminate this effect. Guard rings are also a must in order to eliminate the effect of stray electric fields, especially when the transducers have a low value of capacitance of the order of pF.

(*iii*) The output impedance of capacitive transducers tends to be high on account to their small capacitance value. This leads to loading effects. The output impedance depends upon the frequency of the signal used to determine the value of capacitance. For capacitances lying between 10–500 pF, the frequencies used are such that they give an output impedance in the range of 1 kΩ to 10 MΩ. This high value of output impedance means that the insulation resistance must be kept high, to avoid the shunting of the capacitance unduly and reducing the sensitivity. Both the above conditions are difficult to meet in practice especially when there is humidity.

(*iv*) The cable connecting the transducer to the measuring point is also a source of error. The cable may be source of loading resulting loss of sensitivity. Also loading makes the low frequency response poor.

Let us take a capacitive transducer of capacitance C. This capacitor is used with its polarizing source to produce an emf $E_0$ at no load. Let a load impedance $Z_L$ be connected across it. The output value of voltage with load connected is :

$$E_L = \frac{E_0}{1 + Z_0/Z_L} \quad \text{(See Eqn. 2.37 page 35)}$$

and impedance of transducer is,

$$Z_0 = 1/j\omega C$$

$$\therefore \quad E_L = \frac{E_0}{1 + 1/j\omega C Z_L} \quad \ldots(25.127)$$

In order to have good response $\omega C Z_L$ should be as large as possible. This means that for a particular value of C and $Z_L$, ω should be large. Thus capacitive transducers should be used for dynamic measurements at high frequencies.

In order to have good response at low frequencies the value of the capacitance should be large so that output impedance $Z_0$ of the transducer is high. This thereby makes $E_L$ nearly equal to $E_0$.

The capacitive transducers are connected to the second stage of the instrumentation system through cables. The cables are a source of loading. Consider a capacitive transducer as shown in Fig. 25.116.

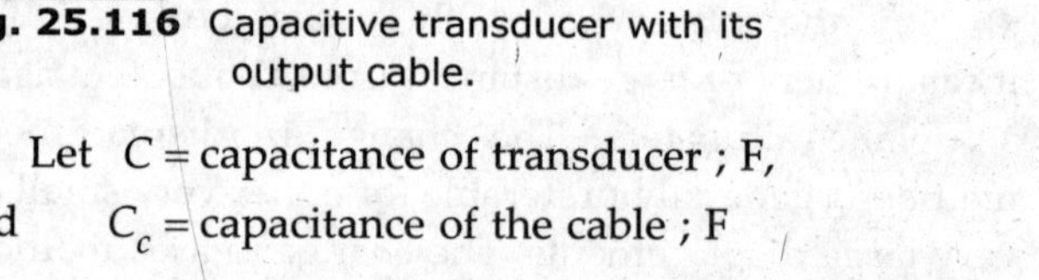


**Fig. 25.116** Capacitive transducer with its output cable.

Let $C$ = capacitance of transducer ; F,

and $C_c$ = capacitance of the cable ; F

If we neglect the leakage resistance, the output impedance of transducer is $Z_0 = 1/j\omega C$.

Impedance of load (taking cable as the load) is : $Z_L = 1/j\omega C_c$

∴ Voltage at the second stage

$$E_L = \frac{E_0}{1 + j\omega C_c / j\omega C} = \frac{C}{C + C_c} E_0 \quad \ldots(25.128)$$

Thus in order that the loading effects be minimised the capacitance of the cable should be as small as possible as compared to capacitance of transducer or $C_c << C$. This requires short lengths of cables to have small loading errors. Consider the polarizing voltage to be short-circuited it is then seen that C and $C_c$ are connected in parallel.

∴ Minimum resonant frequency

$$f_{min} = \frac{1}{2\pi R_l/(C+C_c)} \text{Hz} \quad \text{...(25.129)}$$

where $R_l$ = leakage resistance.

If the cable capacitance $C_c$ is high, the frequency $f_{min}$ decreases. Hence cable must have small capacitance to improve low frequency response.

(*v*) The capacitance of a capacitive transducer may be changed on account of presence of extraneous matter like dust particles and moisture.

(*vi*) The capacitive transducers are temperature sensitive and therefore any change in temperature adversely affects their performance.

(*vii*) The instrumentation circuitry used with these transducers is very complex.

### 25.28.9 Uses of Capacitive Transducers

(*i*) Capacitive transducers can be used for measurement of both linear and angular displacements. The capacitive transducers are highly sensitive and can be used for measurement of extremely small displacements down to the order of molecular dimensions *i.e.*, $0.1 \times 10^{-6}$ mm.

This is on account of the fact that small capacitance changes produced on account of small displacements can be measured. In practice it is possible to detect capacitance changes of the order of 1 aF* and that too with a good degree of accuracy.

On the other hand, they can be used for measurement of large distances upto about 30 m as in aeroplane altimeters. The change in displacement method is generally preferable for either very small or very large displacements. The change in area method is used for measurement of displacements ranging from 10 mm to 100 mm.

(*ii*) Capacitive transducers can be used for the measurement of force and pressure. The force and pressure to be measured are first converted to displacement which causes a change of capacitance.

(*iii*) Capacitive transducers can also be used directly as pressure transducers in all those cases where the dielectric constant of a medium changes with pressure. For instance, the dielectric constant of Benzene changes by 0.5 per cent in the range of pressure of 1 to 1000 times the atmospheric pressure. Similarly, the dielectric constant of air at 19°C changes from 1.0006 to 1.0548 in the same pressure range. Variation of dielectric constant with pressure also occurs in solids, primarily in crystals like MgO, LiF, NaCl, KCl and KBr.

(*iv*) Capacitive transducers are used for measurement of humidity in gases since the dielectric constant of gases changes with change in humidity thereby producing a change in capacitance. The change in capacitance is quite small but is detectable. An idea of change of dielectric constant can be had from the example of air. The dry air at 45°C has a dielectric constant of 1.000247 and that of air saturated with water is 1.000593 at the same temperature. These measurements are carried out with microwave techniques and the frequency is of the order of 10 GHz.

(*v*) Capacitive transducers are commonly used in conjunction with mechanical modifiers for measurement of volume, density, liquid level, weight etc.

**Example 25.39** *Figure 25.117 shows a capacitive transducer using five plates. The dimensions of each plate are 25 × 25 mm and the distance between plates is 0.25 mm. This arrangement is to be used for measurement of displacement by observing the change in capacitance with distance x. Calculate the sensitivity of the device. Assume that the plates are separated by air. The permittivity of air is* $8.85 \times 10^{-12}$ *F/m.*

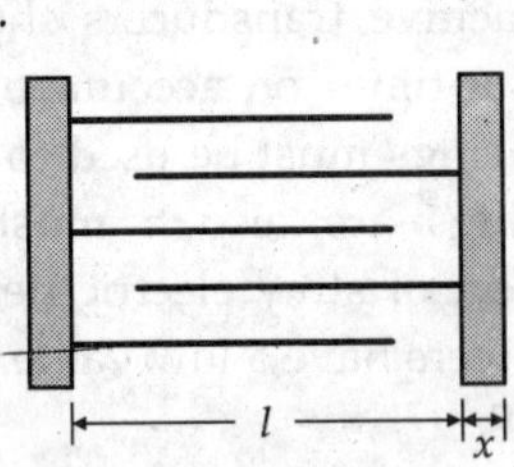


**Fig. 25.11.7** Capacitive transducer for measurement of linear displacement.

**Solution.** The five plate transducer forms a combination of four capacitors connected in parallel. If the movable plate is moved through a distance $x$ on the right side,

Capacitance of each capacitor

$$C' = \frac{\varepsilon_0(l-x)w}{d}$$

where $w$ = width of each plate.

* 1 aF = $10^{-18}$ F.

Capacitance of transducer

$$C = 4C' = \frac{4\varepsilon_0(l-x)w}{d}$$

Sensitivity of transducer

$$\frac{\partial C}{\partial x} = \frac{-4\varepsilon_0 w}{d}$$

$$= -\frac{4 \times 8.85 \times 10^{-12} \times 25 \times 10^{-3}}{0.25 \times 10^{-3}}$$

$$= -3540 \text{ pF/m}$$

$$= 3.54 \text{ pF/mm (Disregarding the sign).}$$

**Example 25.40** *A capacitive transducer uses two quartz diaphragms of area 750 mm$^2$ separated by a distance of 3.5 mm. A pressure of 900 kN/m$^2$ when applied to the top diaphragm produces a deflection of 0.6 mm. The capacitance is 370 pF when no pressure is applied to the diaphragms. Find the value of capacitance after the application of a pressure of 900 kN/m$^2$.*

**Solution.** Suppose $C_1$ and $C_2$ are respectively the values of capacitance before and after application of pressure. Let $d_1$ and $d_2$ be the values of distance between the diaphragms for the corresponding pressure conditions.

$$C_1 = \varepsilon A / d_1 \quad \text{and} \quad C_2 = \varepsilon A / d_2$$

or $$\frac{C_2}{C_1} = \frac{d_1}{d_2}$$

$\therefore$ $$C_2 = C_1 \times \frac{d_1}{d_2}$$

But $d_1 = 3.5$ mm and $d_2 = 3.5 - 0.6 = 2.9$ mm

$\therefore$ Value of capacitance after application of pressure

$$C_2 = 370 \times 3.5/2.9 = 446.5 \text{ pF}$$

**Example 25.41** *A pressure measuring instrument uses a capacitive transducer having a spacing of 4 mm between its diaphragms. A pressure of 600 kN/m$^2$ produces an average deflection of 0.3 mm of the diaphragm of the transducer. The transducer which has a capacitance of 300 pF before application of pressure and is connected in an oscillator circuit having a frequency of 100 kHz. Determine the change frequency of the oscillator after the pressure is applied to the transducer.*

**Solution.** The subscript 1 is used for values before application of pressure and 2 is used for values after application of pressure.

Resonant frequency

$$f_0 = 1/2\pi\sqrt{LC}$$

The inductance is constant but the capacitance is variable.

$$\therefore f_{01} = 1/2\pi\sqrt{LC_1} \quad \text{and} \quad f_{02} = 1/2\pi\sqrt{LC_2}$$

Thus we have,

$$f_{02}/f_{01} = \sqrt{C_1/C_2}$$

But $C_1 = \varepsilon A / d_1$

and $C_2 = \varepsilon A / d_2$

$\therefore$ $$\frac{f_{02}}{f_{01}} = \sqrt{\frac{C_1}{C_2}} = \sqrt{\frac{d_2}{d_1}}$$

The frequency of the oscillator after the application of pressure is :

$$f_{02} = \sqrt{\frac{d_2}{d_1}} \times f_{01}$$

$$= \sqrt{\frac{-0.3 + 4.0}{4.0}} \times 100 = 96.2 \text{ kHz}$$

$\therefore$ Change in frequency of oscillator

$$= 100 - 96.2 = 3.8 \text{ kHz.}$$

**Example 25.42** *A capacitive transducer is made up of two concentric cylindrical electrodes. The outer diameter of the inner cylindrical electrode is 3 mm and the dielectric medium is air. The inner diameter of the outer electrode is 3.1 mm. Calculate the dielectric stress when a voltage of 100 V is applied across the electrodes. Is it within safe limits ? The length of electrodes is 20 mm. Calculate the change in capacitance if the inner electrode is moved through a distance of 2 mm. The breakdown strength of air is 3 kV/mm.*

**Solution.** Length of air gap between the two electrodes $(3.1 - 3)/2 = 0.05$ mm

$\therefore$ Dielectric stress

$$= 100/0.05 = 2000 \text{ V/mm} = 2 \text{ kV/mm}$$

The breakdown strength of air is 3 kV/mm and hence the dielectric is safe.

Capacitance of the transducer

$$C = \frac{2\pi\varepsilon l}{\log_e(D_2/D_1)}$$

$$= \frac{2\pi \times 8.85 \times 10^{-12} \times 20 \times 10^{-3}}{\log_e(3.1/3)} \text{F} = 33.9 \text{ pF}$$

The moving in electrode is shifted through a distance of 2 mm.

$\therefore$ $$l = 20 - 2 = 18 \text{ mm}$$

$$= 18 \times 10^{-3} \text{ m}$$

New value of capacitance

$$= \frac{2\pi \times 8.85 \times 10^{-12} \times 18 \times 10^{-3}}{\log_e(3.1/3)} \text{F} = 33.9 \text{ pF}$$

$\therefore$ Change in value of capacitance

$$\Delta C = 33.9 - 30.5 = 3.4 \text{ pF}$$

**Example 25.43** *A capacitive transducer circuit used for measurement of linear displacement. The transducer is a parallel plate air capacitor wherein the capacitance can be changed by changing the distance between the plates. This transducer is to be used for dynamic measurements. Suppose a flat frequency response with an amplitude ratio within 5% is required down to a frequency range of 20 Hz. What is the minimum allowable value of time constant ? Calculate the phase shift at this frequency. The area of plates is 300 mm² and the distance between plates is 0.125 mm. Calculate the value of series resistance R. What is the amplitude ratio at 5 Hz with the above time constant ?*

*Calculate the high frequency voltage sensitivity of the transducer if the battery voltage is 100 V.*

**Solution.** For a flat response within 5 percent, the amplitude ratio is $M = 1 - 0.05 = 0.95$

From Eqn. 25.125

$$M = \frac{1}{\sqrt{1+(1/\omega\tau)^2}}$$

or
$$0.95 = \frac{1}{\sqrt{1+(1/2\times\pi\times 20\times\tau)^2}}$$

or Minimum time constant

$$\tau = 24.2\times 10^{-3}\text{ s} = 24.2\text{ ms}$$

From Eqn. 25.126, phase shift $\phi = \pi/2 - \tan^{-1}\omega\tau = \pi/2 - \tan^{-1}(2\pi\times 20\times 24.2\times 10^{-3}) = 18.2°$

Capacitance

$$C = \frac{8.85\times 10^{-12}\times 300\times 10^{-6}}{0.125\times 10^{-3}}$$

$$= 21.24\times 10^{-12}\text{ F.}$$

∴ Series resistance

$$R = \frac{\tau}{C} = \frac{24.2\times 10^{-3}}{21.24\times 10^{-12}}$$

$$= 1.14\times 10^{9}\,\Omega = 1140\,\Omega$$

Amplitude ratio at 5 Hz is :

$$M = \frac{1}{\sqrt{1+(1/2\times\pi\times 5\times 24.2\times 10^{-3})^2}}$$

$$= 0.605$$

Thus the output is reduced nearly to 60% of its high frequency value. Therefore this transducer arrangement cannot be used for low frequency applications.

High frequency sensitivity of the transducer

$$K = \frac{E_b}{x} = \frac{100}{0.125\times 10^{-3}}$$

$$= 800\times 10^{3}\text{ V/m} = 800\text{ kV/m}$$

**Example 25.44** *A parallel plate capacitive transducer uses plates of area 500 mm² which are separated by a distance 0.2 mm. Calculate the value of capacitance when the dielectric is air having a permittivity of 8.85 × 10⁻¹² F / m.*

(*i*) *Calculate the change in capacitance if a linear displacement reduces the distance between the plates to 0.18 mm. Also calculate the ratio of per unit change of capacitance to per unit change of displacement.*

(*ii*) *Suppose a mica sheet 0.01 mm thick is inserted in the gap. Calculate the value of original capacitance and change in capacitance for the same displacement. Also calculate the ratio of per unit change in capacitance to per unit change in displacement. The dielectric constant of mica is 8.*

**Solution.** Initial capacitance

$$C = \frac{\varepsilon_0 A}{d}$$

$$= \frac{8.85\times 10^{-12}\times 500\times 10^{-6}}{0.2\times 10^{-3}}\text{ F}$$

$$= 22.125\text{ pF}$$

(*i*) Change in displacement

$$\Delta d = 0.2 - 0.18 = 0.02\text{ mm}$$

Capacitance after application of displacement,

$$C + \Delta C = \frac{8.85\times 10^{-12}\times 500\times 10^{-6}}{0.18\times 10^{-3}}\text{ F}$$

$$= 24.583\text{ pF}$$

Change in capacitance

$$\Delta C = 24.583 - 22.125$$

$$= 2.458\text{ pF.}$$

Ratio
$$= \frac{\Delta C/C}{\Delta d/d}$$

$$= \frac{(2.458/22.125)}{(0.02/0.2)} = 1.111$$

(*ii*) Initially the displacement between the plates is 0.2 mm. Since the thickness of mica is 0.01 mm, the length of air gap between the plates $= 0.2 - 0.01 = 0.19$ mm.

Initial capacitance of transducer,

$$C = \frac{\varepsilon_0 A}{d_1/\varepsilon_1 + d_2/\varepsilon_2}$$

$$= \frac{8.85\times 10^{-12}\times 500\times 10^{-6}}{(0.19/1 + 0.01/8)\times 10^{-3}}\text{ F}$$

$$= 23.137\text{ pF}$$

When a displacement of 0.02 mm is applied, the length of air gap is reduced to $(0.19 - 0.02) = 0.17$ mm.

∴ Capacitance with displacement applied,

$$= \frac{8.85\times10^{-12}\times500\times10^{-6}}{(0.17/1+0.01/8)\times10^{-3}}\text{F} = 25.839\,\text{pF}.$$

Change in capacitance

$\Delta C = 25.839 - 23.137 = 2.702$ pF

Ratio

$$\frac{\Delta C/C}{\Delta d/d} = \frac{(2.702/25.839)}{(0.02/0.2)} = 1.046$$

**Example 25.45** *A variable capacitance angular velocity pickup is shown in Fig. 25.118(a). Prove that the current, i, flowing in the circuit in which the transducer is connected across a battery of constant d.c. voltage $E_b$ is directly proportional to the angular velocity $d\theta/dt$. Since voltage signals are readily manipulated, how might the current signal be transduced to a proportional voltage ? The area of capacitor is $A = K\theta$.*

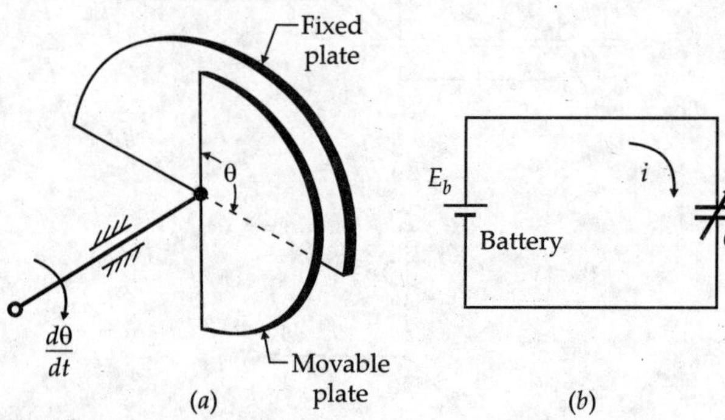


**Fig. 25.118** Capacitive transducer for measurement of angular velocity.

**Solution.** The battery voltage, $E_b$, is constant while the capacitance, C, is variable and its value depends upon the angular displacement. The overlapping area of plates is proportional to angular displacement, θ. Let $A = K\theta$, where $K$ = a constant.

Suppose the charge is $q$ and the capacitance is C.

∴ Current $\quad i = \frac{dq}{dt} = \frac{d(E_b C)}{dt} = E_b\frac{dC}{dt}$.

Now capacitance

$$C = \varepsilon A/d \text{ and}$$

∴ Current $\quad i = E_b\frac{\varepsilon K}{d}\left(\frac{d\theta}{dt}\right)$

Now $E_b$, $d$ and $K$ are constants.

∴ $\quad i = K'\frac{d\theta}{dt}$

where $K'$ = a constant.

Hence current $i$ is proportional to $d\theta/dt$ or the angular velocity.

A circuit for conversion of current to voltage signals is shown in Fig. 25.119.

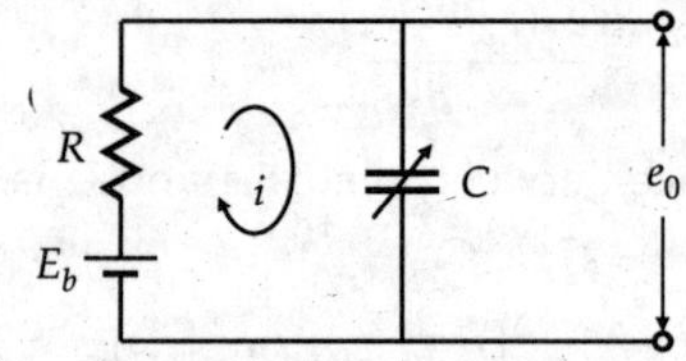


**Fig. 25.119** Voltage output from a capacitive transducer.

**Example 25.46** *A push-pull non-contact capacitive transducer is shown in Fig. 25.120(a). It consists of four parallel plates separated by air. Plates A, C and D are fixed and plate B can be moved. Plate B has a thickness t and is at a distance d from plates on either side. Plates B, C, D are all of length l, while plate A has length 2l. All plates have a width w. The gap between plates C and D can be considered as negligible. Neglecting the end effects, derive expressions for capacitances $C_{AC}$ and $C_{AD}$ for movement of the midpoint of plate B between $x = \pm l/2$. $x = 0$ is the position of symmetry.*

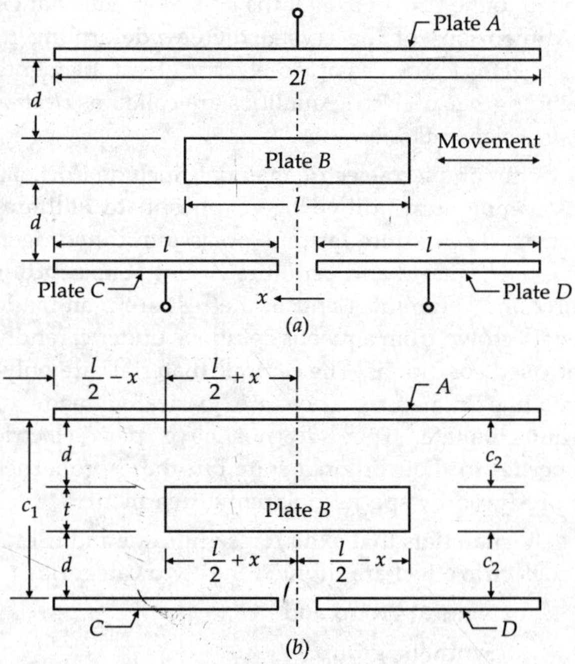


**Fig. 25.120** Push-pull capacitive transducer.

**Solution.** Figure 25.120(*b*) shows a simplified arrangement where the central plate *B* is moved to the left by a distance *x*.

The capacitance between plates *A* and C is combination of capacitance $C_1$ in parallel with a series combination of two capacitances each of value $C_2$, where :

$$C_1 = \frac{\varepsilon_0(l/2 - x)w}{2d + 1}$$

and $$C_2 = \frac{\varepsilon_0(l/2 + x)\,w}{d}$$

Capacitance between plates $A$ and $C$ is,

$$C_{AC} = C_1 + C_2$$

$$= \varepsilon_0 w\left[\frac{2\,ld + lt/2 + tx}{2d\,(2d + t)}\right]$$

Similarly, capacitance between plates $A$ and $D$,

$$C_{AD} = \varepsilon_0 w\left[\frac{2\,ld + lt/2 - tx}{2d\,(2d + t)}\right]$$

## 25.29 PIEZO-ELECTRIC TRANSDUCERS

A piezo-electric material is one in which an electric potential appears across certain surfaces of a crystal if the dimensions of the crystal are changed by the application of a mechanical force. This potential is produced by the displacement of charges. The effect is reversible, *i.e.*, conversely, if a varying potential is applied to the proper axis of the crystal, it will change the dimensions of the crystal thereby deforming it. This effect is known as **piezo-electric effect**. Elements exhibiting piezo-electric qualities are called as *electro-resistive* elements.

Common piezo-electric materials include Rochelle salts, ammonium dihydrogen phosphate, lithium sulphate, dipotassium tartrate, potassium dihydrogen phosphate, quartz and ceramics $A$ and $B$. Except for quartz and ceramics $A$ and $B$, the rest are man-made crystals grown from aqueous solutions under carefully controlled conditions. The ceramic materials are poly-crystalline in nature. They are, basically, made of barium titanate. They do not have piezo-electric properties in their original state but these properties are produced by special polarizing treatment.

The materials that exhibit a significant and useful piezoelectric effect are divided into two categories :

- Natural group and
- Synthetic group.

Quartz and Rochelle salt belong to *natural group* while materials like lithium sulphate, ethylene diamine tartrate belong to the *synthetic group*.

The piezo-electric effect can be made to respond to (or cause) mechanical deformations of the material in many different modes. The modes can be : thickness expansion, transverse expansion, thickness shear and face shear. The mode of motion affected depends on the shape of the body relative to the crystal axis and location of the electrodes. A piezo-electric element used for converting mechanical motion to electrical signals may be thought as charge generator and a capacitor. Mechanical deformation generates a charge and this charge appears as a voltage across the electrodes. The voltage is $E = Q/C$.

The piezo-electric effect is direction sensitive. A tensile force produces a voltage of one polarity while a compressive force produces a voltage of opposite polarity.

A piezo-electric crystal is shown in Fig. 25.121.

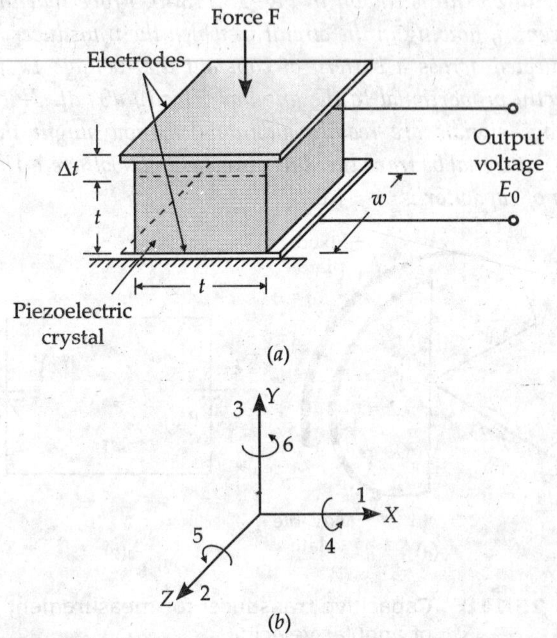


**Fig. 25.121** (*a*) Piezo-electric crystal used for measurement of force, (*b*) Axis numbering system for the crystal.

The magnitude and polarity of the induced surface charges are proportional to the magnitude and direction of the applied force $F$. The polarity of induced charges depends upon the direction of applied force.

Charge $Q = d \times F$ coulomb ...(25.130)

where $d$ = charge sensitivity of the crystal ;
$C/N$ : (it is constant for a given crystal)

and $F$ = applied force, $N$

The force $F$ causes a change in thickness of the crystal.

$$\therefore \quad F = \frac{AE}{t}\Delta t \text{ newton} \quad ...(25.131)$$

where $A$ = area of crystal ; $m^2$

$t$ = thickness of crystal ; m, and

$E$ = Young's modulus $N/m^2$

Young's modulus

$$E = \frac{\text{stress}}{\text{strain}} = (F/A)\cdot\frac{1}{\Delta t/t}$$

$$= \frac{Ft}{A\Delta t}\ \text{N/m}^2 \qquad \ldots(25.132)$$

Area $A = wl$

where $w$ = width of crystal : m, and

$l$ = length of crystal ; m

∴ From Eqns. 25.130 and 25.131, we have, charge :

$$Q = dAE(\Delta t/t) \qquad \ldots(25.133)$$

The charge at the electrodes gives rise to an output voltage $E_0$

Voltage $E_0 = Q/C_p$ ...(25.134)

where $C_p$ = capacitance between electrodes ; F

Capacitance between electrodes

$$C_p = \varepsilon_r\varepsilon_0 A/t \qquad \ldots(25.135)$$

From Eqns. 25.130, 25.134, 25.135,

$$E_0 = \frac{Q}{C_p} = \frac{dF}{\varepsilon_r\varepsilon_0 A/t} = \frac{dt}{\varepsilon_r\varepsilon_0}\frac{F}{A} \qquad \ldots(25.136)$$

But $F/A = P$ = pressure or stress in N/M$^2$

$$\therefore \quad E_0 = \frac{d}{\varepsilon_r\varepsilon_0}tP \qquad \ldots(25.137)$$

$$= g\,t\,P \qquad \ldots(25.138)$$

where $g = d/\varepsilon_r\varepsilon_0$ ...(25.139)

'$g$' is the voltage sensitivity of the crystal. This is constant for a given crystal cut. Its units are Vm/N.

Now $g = \frac{E_0}{tP} = \frac{E_0/t}{P}$ ...(25.140)

But $E_0/t$ = electric field strength, V/m

Let $\varepsilon = E_0/t$ = electric field

$$\therefore \quad g = \frac{\text{electric field}}{\text{stress}} = \frac{\varepsilon}{P} \qquad \ldots(25.141)$$

Now $E_0/t$ is the electric field intensity in the crystal and $P$ is the pressure or the applied stress to the crystal. Therefore, crystal voltage sensitivity, $g$, can be defined as the ratio of electric field intensity to pressure (or stress). Now $E_0/t = \varepsilon$ is the electric field intensity in the crystal and $P$ is the pressure or the stress applied to the crystal. Therefore, crystal voltage sensitivity, $g$, can be defined as the ratio of the electric field intensity to pressure (or stress). The units of $g$ are Vm/N.

From Eqn. 25.139,

charge sensitivity $d = \varepsilon_r\varepsilon_0\, g$ C/N ...(25.142)

The values of crystal voltage sensitivity, permittivity and charge sensitivity for barium titatate and quartz are shown in Table 25.8.

**Table 25.9** *Properties of Barium Titanate and Quartz*

| Material | Voltage Sensitivity g Vm/N | Permittivity, ε F / m | Charge Sensitivity, d pC/N |
|---|---|---|---|
| Barium titanate | $12 \times 10^{-3}$ | $12.5 \times 10^{-9}$ | 150 |
| Quartz | $50 \times 10^{-3}$ | $40.6 \times 10^{-12}$ | 2 |

Sometimes it is desired to express the output voltage or charge in terms of deflection rather than in terms of either stress or force. This is because it is really the deformation that causes the charge generation. Thus we must know the modulus of elasticity of the material for this purpose. The values of modulus of elasticity are :

Barium titanate : $12 \times 10^{10}$ N/m,

Quartz : $8.6 \times 10^6$ N/m$^2$

It has been stated earlier that the piezo-electric effect is direction sensitive. The main characteristics of piezo-electric motion to voltage transducers can be illustrated by considering only one common mode of deformation *i.e.*, thickness expansion. For this mode the physical arrangement is shown in Fig. 25.121(*a*). Various double-subscripted physical constants are used to describe numerically the phenomena occurring. The convention is that first subscript refers to the direction of the electrical effect and the second to that of the mechanical effect. The axis numbering system is given in Fig. 25.121(*b*).

The two main families of constants *i.e.*, the '$d$' constants and '$g$' constants are considered. For barium titanate the commonly used constants are $d_{33}$ and $g_{33}$

$$g_{33} = \frac{\text{field produced in direction 3}}{\text{stress applied in direction 3}}$$

$$= \frac{E_0/t}{F/A} \qquad \ldots(25.143)$$

Voltage output

$$E_0 = g_{33} \times \frac{F}{A} \times t = g_{33}\,t\,P \qquad \ldots(25.144)$$

Thus if $g$ is known for a particular material, the voltage output per unit stress can be calculated by knowing the value of $t$.

25

In order to relate the applied force to the generated charge, the $d$ constants are used. One of the $d$ constants can be defined as :

$$d_{33} = \frac{\text{charge generated in direction 3}}{\text{force applied in direction 3}}$$

$$= \frac{Q}{F} \qquad \text{...(25.145)}$$

Actually $d_{33}$ can be calculated from $g_{33}$ if the relative permittivity of the material is known. From Eqn. 25.139, we get

$$d_{33} = \varepsilon_r \varepsilon_0 \, g_{33} \qquad \text{...(25.146)}$$

When dealing with quartz, subscripts 11 are used because in quartz the thickness – expansion mode is along the crystallographic axis conventionally called axis 1.

The piezo-electric transducer is cut from a larger crystal in the direction of any of the electrical or mechanical axes perpendicular to the optical or crystal axis. The values of $d$ and $g$ are not necessarily the same but are dependent upon the axis of cut.

### 25.29.1 Modes of Operation of Piezo-electric Crystals

The piezo-electric crystals are used in many modes. These modes are :

(i) Thickness shear, (ii) Face shear,

(iii) Thickness expansion and

(iv) Transverse expansion.

These modes are shown in Fig. 25.122.

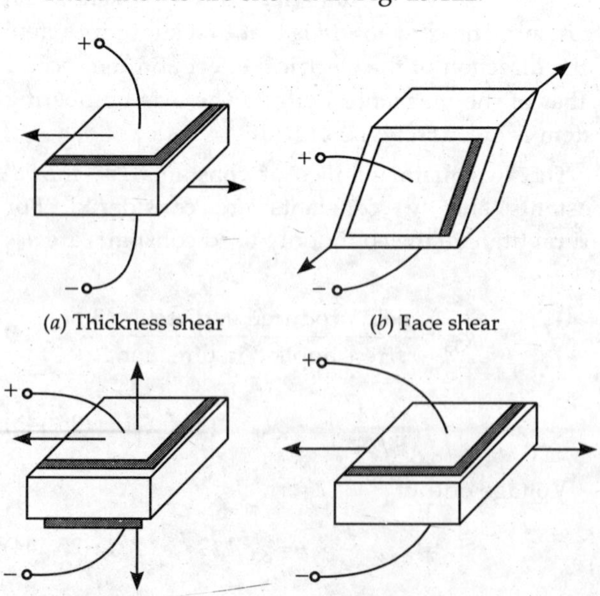


(a) Thickness shear (b) Face shear

(c) Thickness expansion (d) Transverse expansion

**Fig. 25.122** Modes of operation of piezo-electric crystals.

By cementing two crystals together so that their electrical axes are perpendicular, '**benders**' or **twisters** can be produced (See Fig. 25.123). This means that a bending motion applied to a bender produces an output voltage. Similarly, a twisting motion applied to a twister produces an output.

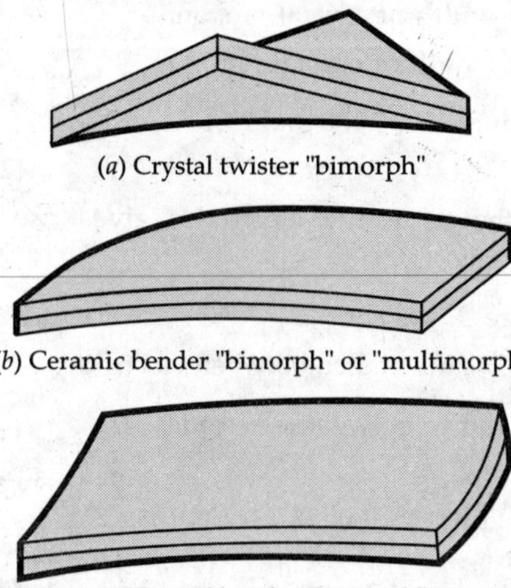

(a) Crystal twister "bimorph"

(b) Ceramic bender "bimorph" or "multimorph".

(c) Crystal bender "bimorph".

**Fig. 25.123** Curvature of "twister" and "bender" piezo-electric transducers when voltage is applied.

Piezo-electric crystals can be used in another mode for force measurement. A crystal controlled electronic oscillator uses a thin plate of quartz. The natural frequency of mechanical oscillation of the plate determines the frequency of electrical oscillation.

### 25.29.2 Properties of Piezo-electric Crystals

The desirable properties of piezo-electric materials are stability, high output insensitivity to temperature and humidity and the ability to be formed into most desirable shape.

Quartz is the most stable piezo-electric material. However, its output is quite small. On the other hand, Rochelle salt provides the highest output but it can be worked over a limited humidity range and has to be protected against moisture. The highest temperature is limited to 45°C.

Barium titanate has the advantage that it can be formed into a variety of shapes and sizes since it is polycrystalline. It has also a higher dielectric constant. Natural crystals possess the advantages that they have higher mechanical and thermal stability, can withstand higher stresses, have low leakage (their volume resistivity is about $10^{16} \Omega$ - m) and have a good frequency response. The synthetic materials, in general, have a higher voltage sensitivity.

***Example 25.47*** *A quartz piezo-electric crystal having a thickness of 2 mm and voltage sensitivity of 0.055 V - m/ N is subjected to a pressure of 1.5 MN/m². Calculate the voltage output. If the permittivity of quartz is 40.6 × 10⁻¹² F / m, calculate its charge sensitivity.*

**Solution.** From Eqn. 25.138, voltage output

$$E_0 = gt\,P = 0.055 \times 2 \times 10^{-3} \times 10^6 \times 1.5 \times 10^6$$
$$= 165 \text{ V}$$

Charge sensitivity

$$d = \varepsilon_0 \varepsilon_r\, g \qquad \text{(See Eqn. 25.139)}$$
$$= 40.6 \times 10^{-12} \times 0.055$$
$$= 2.23 \times 10^{-12} \text{ C/N} = 2.23 \text{ pC/N}$$

***Example 25.48*** *A piezo-electric crystal having dimensions of 5 mm × 5 mm × 1.5 mm and a voltage sensitivity of 0.055 V - m/ N is used for force measurement. Calculate the force if the voltage developed is 100 V.*

**Solution.** From Eqn. 25.138, the applied pressure is

$$P = \frac{E_0}{g\,t} = \frac{100}{0.055 \times 1.5 \times 10^{-3}} \text{ N/m}^2$$
$$= 1.2 \text{ MN/m}^2$$

Force $\quad F = PA$

$$= 12 \times 10^6 \times 5 \times 5 \times 10^{-6} = 30 \text{ N}$$

***Example 25.49*** *A barium titanate pickup has the dimensions of 5 mm × 5 mm × 1.25 mm. The force acting on it is 5 N. The charge sensitivity of barium titanate is 150 pC/N and its permittivity is 12.5 × 10⁻⁹ F / m. If the Modulus of elasticity of barium titanate is 12 × 10⁶ N / m², calculate the strain. Also calculate the charge and the capacitance.*

**Solution.** Area of plates

$$A = 5 \times 5 \times 10^{-6} = 25 \times 10^{-6} \text{ m}^2$$

Pressure $P = 5/(25 \times 10^{-6})$ N/m²

$$= 0.2 \text{ MN/m}^2$$

Voltage sensitivity

$$g = \frac{d}{\varepsilon_0 \varepsilon_r} = \frac{150 \times 10^{-12}}{12.5 \times 10^{-9}}$$
$$= 12 \times 10^{-3} \text{ Vm/N}$$

Voltage generated

$$E_0 = g\,t\,P$$
$$= 12 \times 10^{-3} \times 1.25 \times 10^{-3} \times 0.2 \times 10^6 = 3 \text{ V}$$

Strain $\quad \varepsilon = \Delta t$

$$= \frac{\text{stress}}{\text{Young's modulus}}$$
$$= \frac{0.2 \times 10^6}{12 \times 10^6} = 0.0167.$$

Charge $\quad Q = dF$

$$= 150 \times 10^{-12} \times 5 \text{ C} = 750 \text{ pC}$$

Capacitance

$$C_p = \frac{Q}{E_0} = \frac{750 \times 10^{-12}}{3} \text{ F} = 250 \text{ pF}$$

### 25.29.3 Equivalent Circuit of Piezo-electric Transducers

The basic equivalent circuit of a piezo-electric transducer is shown in Fig. 25.124(*a*).

The source is a charge generator. The value of the charge is $Q = dF$.

The charge generated is across the capacitance, $C_p$, of the crystal and its leakage resistance $R_p$.

The charge generator can be replaced by an equivalent voltage source having a voltage of

$$E_0 = \frac{Q}{C_p} = \frac{dF}{C_p} \qquad \ldots(25.147)$$

in series with a capacitance, $C_p$, and resistance, $R_p$, as shown in Fig. 25.124(*b*).

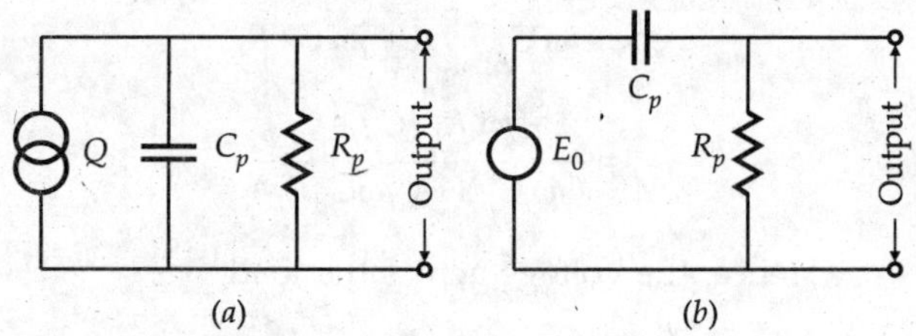


**Fig. 25.124** Equivalent circuits of piezo-electric transducers.

### 25.29.4 Loading Effects and Frequency Response

Let the transducer be loaded by a capacitance $C_L$ and a resistance $R_L$. The capacitance $C_L$ is the combination of the capacitance of the load, the capacitance of the cable and the stray capacitance. The diagram showing the load connected to a piezo-electric transducer is given in Fig. 25.125(*a*).

The value of leakage resistance, $R_p$, of the crystal is very large. It is of the order of $0.1 \times 10^{12}\ \Omega$. The value of load resistance, $R_L$, is considerably smaller than $R_p$, and hence the equivalent circuit of the Piezo-electric crystal under load conditions is as shown in Fig. 25.125(*b*) in which the leakage resistance, $R_p$, of the crystal has been dropped.

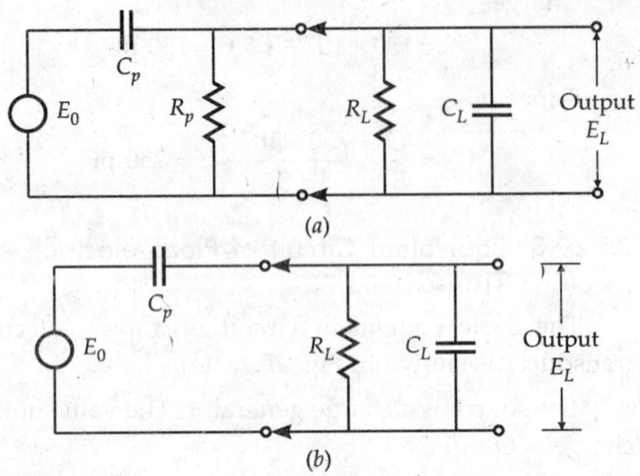


**Fig. 25.125** Circuit of a Piezo-electric crystal under conditions of load.

The voltage output of the transducer under no load conditions, is therefore $E_0$. Under conditions of load, impedance of load

$$Z_L = \frac{R_L}{1 + j\omega C_L R_L}$$

Total impedance of circuit

$$Z_t = \frac{1}{j\omega C_p} + \frac{R_L}{1 + j\omega C_L R_L}$$

$$= \frac{1 + j\omega R_L (C_p + C_L)}{(j\omega C_p)(1 + j\omega C_p R_L)}$$

Hence, the voltage across the load,

$$E_L = \frac{Z_L}{Z_t} \times E_L$$

$$= \frac{R_L}{(1 + j\omega C_L R_L)} \times \frac{(j\omega C_p)(1 + j\omega C_p R_L)}{1 + j\omega R_L (C_p + C_L)} E_0$$

$$= \left[\frac{j\omega C_p R_L}{1 + j\omega (C_p + C_L) R_L}\right] E_0$$

The magnitude of voltage across the load is :

$$E_L = \left[\frac{\omega C_p R_L}{\sqrt{1 + \omega^2 (C_p + C_L)^2 R_L^2}}\right] E_0 \quad \text{...(25.148)}$$

$$\therefore E_L = \frac{dF}{C_p}\left[\frac{\omega C_p R_L}{\sqrt{1 + \omega^2 (C_p + C_L)^2 R_L^2}}\right] \text{ as } E_0 = \frac{dF}{C_p}$$

$$= dF\left[\frac{\omega R_L}{\sqrt{1 + \omega^2 (C_p + C_L)^2 R_L^2}}\right] \quad \text{...(25.149)}$$

At medium and high frequencies :

$$\omega^2 (C_p + R_L)^2 R_L^2 >> 1$$

$$\therefore \quad E_L = \frac{E_0 C_p}{(C_p + C_L)} \quad \text{...(25.150)}$$

Thus at medium and high frequencies, the response is independent of frequency, but is dependent upon $C_L$ *i.e.*, the capacitance of load circuit. In practice this transducer is nearly always coupled to *virtual Earth* point of an amplifier which has a feedback capacitor, this arrangement being known as *charge amplifier*.

From Eqn. 25.148, it is clear that under steady state conditions *i.e.*, when $\omega = 0$, the transducer does not provide any output. As far as the maximum frequency limit is concerned, it is imposed by the mechanical resonance of the piezo-electric crystal and its associated mountings.

The piezo-electric transducers are mainly used for measurement of displacement. They can be used for measurement of force, pressure or acceleration. These quantities when measured with piezo-electric transducers are first converted into displacement and the displacement is subsequently applied to these transducers to produce an output voltage. Hence the conversion of displacement into voltage by piezo-electric crystals is considered here.

For the purpose of analysis it is necessary to consider the transducer, the connecting cable and the amplifier as a unit. The impedance of the transducer is very high and hence an amplifier with a high input impedance has to be used in order to avoid loading errors.

Figure 25.126(*a*) shows the complete set-up for measurement of displacement.

Charge produced

$$q = K_q x_i \text{ coulomb} \quad \text{...(25.151)}$$

where $K_q$ = sensitivity ; C/m,

and $x_i$ = displacement ; m

Figure 25.126(*b*) shows the equivalent circuit of the measurement set up.

$R_p$ = leakage resistance of transducer ; Ω,

$C_p$ = capacitance of transducer ; F

$C_c$ = capacitance of cable ; F,

$C_A$ = capacitance of amplifier ; F

$R_A$ = resistance of amplifier ; Ω

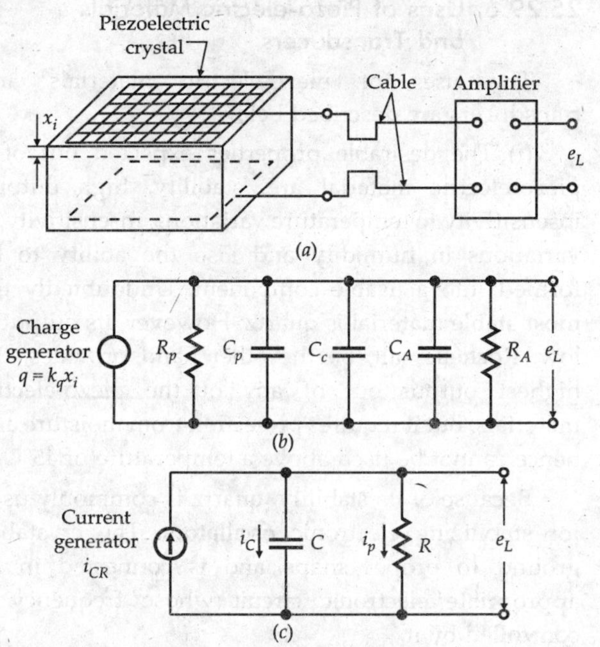


**Fig. 25.126** Set-up of a piezo-electric transducer and its equivalent circuit.

The charge generator is converted into a constant current generator as shown in Fig. 25.126(*c*). The capacitance connected across the current generator is C, where :

$$C = C_p + C_c + C_A \quad \text{...(25.152)}$$

Resistance $$R = \frac{R_A R_p}{R_A + R_p} \quad \text{...(25.153)}$$

Since the leakage resistance of transducer is very large (of the order of $0.1 \times 10^{12}\ \Omega$) and therefore,

$$R = R_A$$

Converting the charge generator into a current generator

$$i_{CR} = \frac{dq}{dt} = K_q\left(\frac{d\,x_i}{dt}\right) \quad \text{...(25.154)}$$

where $i_{CR}$ is the current of the constant current generator.

Now $\quad i_{CR} = i_C + i_R$

$\therefore$ Output voltage at load

$$e_L = e_c = \frac{1}{C}\int i_C\, dt = \frac{1}{C}\int (i_{CR} - i_R)\, dt$$

or $$\frac{d(e_L)}{dt} = \frac{1}{C}(i_{CR} - i_R)$$

or $$C\frac{d(e_L)}{dt} = i_{CR} - i_R$$

$$= K_q\frac{d\,(x_i)}{dt} - \frac{e_L}{R}$$

or $$RC\frac{d(e_L)}{dt} + e_L = K_q R\frac{d(x_i)}{dt} \quad \text{...(25.155)}$$

$$\tau\frac{d(e_L)}{dt} + e_L = K\tau\frac{d(x_i)}{dt} \quad \text{...(25.156)}$$

where $\quad K = \text{sensitivity} = K_q / C\ \text{V/m} \quad$ ...(25.157)

Taking Laplace transform, we get :

$$(\tau s + 1)\, E_L(s) = K\tau s\, X_i(s)$$

$\therefore$ Transfer function

$$\frac{E_L(s)}{X_i(s)} = \frac{K\tau s}{1 + \tau s} \quad \text{...(25.158)}$$

Sinusoidal transfer function $\dfrac{E_L}{X_i}(j\omega) = \dfrac{j\omega K\tau}{1 + j\omega\tau}$

The amplitude ratio is :

$$M = \left|\frac{E_L}{KX_i}(j\omega)\right| = \frac{\omega\tau}{\sqrt{1 + (\omega^2\tau^2)}}$$

$$= \frac{1}{\sqrt{1 + (1/\omega\tau)^2}} \quad \text{...(25.159)}$$

The phase shift $\phi = \pi/2 - \tan^{-1}\omega\tau$ rad ...(25.160)

At high frequencies $\omega >> 1 \quad \therefore \quad M = 1$ and $\phi = 0$

Hence high frequency sensitivity is :

$$K = e_L / x_i \quad \text{...(25.161)}$$

Thus the frequency response of piezo-electric transducers is the same as that of capacitive transducers. This is shown in Fig. 25.125.

From Eqn. 25.159, we see that steady state response piezo-electric transducers to a constant displacement $x_i$ is zero. Therefore, piezo-electric transducer cannot be used for measurement of static displacements.

### 25.29.5 Impulse Response of Piezo-electric Crystals

Let a displacement $x_i$ be applied to the transducer :

where $\quad x_i = A, \ 0 \le t < T$

and $\quad x_i = 0, \ T < t < \infty$

The function is shown in Fig. 25.127(*a*).

Therefore the charge is suddenly increased to $K_q A$ and the crystal voltage rises to $e_L$ $A/C$. Thus the initial condition is $e_L = K_q\ A/C$ at $t = 0^+$.

From Eqns. 25.156 and 25.157, we have :

$$\tau\frac{d(e_L)}{dt} + e_L = K\tau\, d\frac{(x_i)}{dt}$$

$$= \frac{K_q}{C}\tau\frac{d(x_i)}{dt} \quad \text{...(25.162)}$$

25

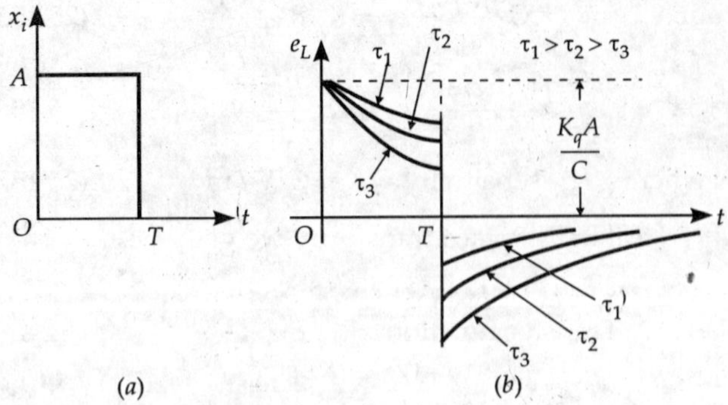


**Fig. 25.127** Pulse response of piezo-electric transducers.

Solving the differential equation with initial condition, we get,

$$e_L = (K_q \ A/C) \cdot [\exp(-t/\tau)] \quad \ldots(25.163)$$

For $T < t < \infty$, the value of $x_i = 0$ and hence the differential equation is :

$$\tau \frac{d(e_L)}{dt} + e_L = 0 \quad \ldots(25.164)$$

At $t = T^-$, Eqn. 25.163 still holds good and therefore $e_L = (K_q \ A/C)[\exp(-t/\tau)]$

Value of $e_L$ at $t = T$ is :

$$e_L = (K_q \ A/C) \exp(-T/\tau) \quad \ldots(25.165)$$

However, at $t = T$, $x_i$ suddenly decreases by a value $A$. This causes a sudden decrease in charge by an amount $K_q \ A$ resulting in a sudden change of $e_L$ by an amount $K_q \ A/C$ from its value at $t = T^-$.

Hence at $t = T^+$, the value of $e_L$ is :

$$e_L = (K_q \ A/C)[\exp(-T/\tau)] - K_q \ A/C$$

$$= K_q \ A/C[\exp(-T/\tau) - 1] \quad \ldots(25.166)$$

This becomes the initial condition for Eqn. 25.163. The solution is :

$$e_L = \frac{K_q \ A}{C}[\exp(-T/\tau - 1)] \cdot \exp[-(t-T)/\tau] \quad \ldots(25.167)$$

For $T < t < \infty$

Figure 25.127(*b*) shows the impulse response with three different time constants. In order to have faithful reproduction of input displacement $x_i$, the value of time constant $\tau$ should be large. If the decay and the under shoot is to be kept within say, 5 per cent of true value, $\tau$ should be at least 20 $T$.

## 25.29.6. Uses of Piezo-electric Materials and Transducers

The uses of piezo-electric materials and transducers are described below :

(*i*) The desirable properties expected out of a piezo-electric material are, stability, high output, insensitivity to temperature variations, insensitivity to variations in humidity and also the ability to be formed into a usable component. Undoubtedly the most stable material is quartz. However, its output is low. Rochelle salt, on the other hand provides, the highest output out of any of the piezo-electric materials. But it requires protection from moisture and hence cannot be used above a temperature of 45°C.

Because of its stability, quartz is commonly used for stabilizing electronic oscillators. The crystal is ground to proper shape and is connected in an appropriate electronic circuit whose frequency is controlled by it.

(*ii*) The use of piezo-electric transducer elements is confined primarily to dynamic measurements. The voltage developed by application of strain is not held under static conditions. Hence, the elements are primarily used in the measurement of such quantities as surface roughness and in accelerometers and vibration pickups.

(*iii*) Ultrasonic generator elements also use barium titanate, a piezo-electric material. Such elements are used in industrial cleansing apparatus and also in underwater detection system known as sonar.

**Example 25.50** *A piezo-electric quartz crystal has an area of 100 mm$^2$ and is 1 mm thick. It is held between two metal electrodes and is used for measurement of changes in force across the crystal. Young's modulus for the material is 90 GN/m$^2$, the charge sensitivity is 2 pC/N, the relative permittivity is 5.*

*The resistivity is $10^{12}$ Ω m. A 20 pF capacitance and a resistance of 100 MΩ are connected in parallel across the electrodes. If a force $F_i = 0.01 \sin 1000\,t$ newton is applied, find*

(*a*) *peak to peak voltage swing across the electrodes under open circuit and under load conditions and*

(*b*) *the maximum change in crystal thickness. [$F_i$ is the instantaneous value of force F]. The permittivity of free space is 8.85 pF/m.*

**Solution.** The rms value of voltage under open circuit

$$E_0 = \frac{dt}{\varepsilon_r \varepsilon_0} \times \frac{F}{A} \quad \text{(See Eqn. 25.136)}$$

∴ Maximum value of voltage under open circuit

$$E_{0\,max} = \sqrt{2}\,\frac{dt}{\varepsilon_r \varepsilon_0} \times \frac{F}{A}$$

$$= \frac{dt}{\varepsilon_r \varepsilon_0} \times \frac{F_{max}}{A}$$

as maximum value of force $F_{max} = \sqrt{2}\,F$.

Hence, peak to peak value of voltage under open circuit :

$$E_{0\,(peak\ to\ peak)} = 2\,E_{0\,max} = 2\,\frac{dt}{\varepsilon_r \varepsilon_0} \times \frac{F_{max}}{A}$$

$$= 2 \times \frac{2 \times 10^{-12} \times 1 \times 10^{-3}}{5 \times 8.85 \times 10^{-12}} \times \frac{0.01}{100 \times 10^{-6}}$$

$$= 9.03 \times 10^{-3}\ \text{V} = 9.3\ \text{mV}$$

Leakage resistance of crystal

$$R_p = \frac{\rho t}{A} = 10^{12} \times \frac{1 \times 10^{-3}}{100 \times 10^{-6}} = 10 \times 10^{12}\ \Omega$$

Load resistance $R_L = 100\ \text{M}\Omega = 100 \times 10^6\ \Omega$

Hence the resistance of the crystal is very large as compared with resistance of load *i.e.*, $R_p >> R_L$ and therefore we can neglect $R_p$.

Capacitance of crystal

$$C_p = \frac{\varepsilon_r \varepsilon_0 A}{d}$$

$$= \frac{5 \times 8.85 \times 10^{-12} \times 100 \times 10^{-6}}{1 \times 10^{-3}}$$

$$= 4.43 \times 10^{-12}\ \text{F}$$

The peak to peak output voltage under load conditions is :

$$E_{o\,(peak\ to\ peak)} = \frac{2d\,F_{max}\,t}{\varepsilon_r \varepsilon_0 A}\left[\frac{\omega C_p R_L}{\sqrt{1 + \omega^2 (C_p + C_L)^2 R_L^2}}\right]$$

$$E_{o\,(peak\ to\ peak)} = \frac{2 \times 2 \times 10^{-12} \times 0.01}{5 \times 8.85 \times 10^{-12} \times 100 \times 10^{-6}}$$

$$\left[\frac{1000 \times 4.43 \times 10^{-12} \times 100 \times 10^{6}}{\sqrt{1 + (1000)^2 \times (4.43 + 20)^2 \times (10^{-12}) \times (100 \times 10^6)^2}}\right]$$

$$= 1.58\ \text{mV}$$

This means that the output signal is attenuated to a great extent due to loading.

From Eqn. 25.132, the maximum value of change in thickness is :

$$\Delta t = \frac{F_{max}\,t}{AF}$$

$$= \frac{0.01 \times 1 \times 10^{-3}}{100 \times 10^{-6} \times 90 \times 10^{9}}$$

$$= 1.1 \times 10^{-12}\ \text{m}$$

$$\therefore \Delta t_{(peak\ to\ peak)} = 2 \times 1.1 \times 10^{-12} = 2.2 \times 10^{-12}\ \text{m}$$

**Example 25.51** *In a piezo-electric transducer a flat frequency response within 5% is required. Find the value of minimum frequency in terms of time constant for which it can be used. If the time constant of the transducer is 1.5 ms, find the value of minimum frequency. Find the phase shift at this frequency.*

**Solution.** In order to obtain a flat response within 5%, the amplitude ratio should be :

$$M = 1 - 0.05 = 0.95$$

Now $\quad M = \dfrac{1}{\sqrt{1 + (1/\omega\tau)^2}}$ or 0.95

$$= \frac{1}{\sqrt{1 + (1/\omega\tau)^2}} \text{ or } \omega t = 3.04$$

∴ Minimum value of frequency

$$\omega = \frac{3.04}{\tau} = \frac{3.04}{1.5 \times 10^{-3}}$$

$$= 2.02 \times 10^3\ \text{rad/s.}$$

Phase shift $\quad \phi = \pi/2 - \tan^{-1} \omega\tau$ rad.

$$= 90° - \tan^{-1} 3.04 = 90° - 71.8° = 18.2°$$

**Example 25.52** *A piezo-electric transducer has a capacitance of 1000 pF and a charge sensitivity of 40 × $10^{-3}$ C/m. The connecting cable has a capacitance of 300 pF while the oscilloscope used for readout has a readout input resistance of 1 MΩ with a parallel capacitance of 50 pF.*

*(a) What is the sensitivity (V/m) of the transducer alone ?*

*(b) What is the high frequency sensitivity (V/m) of the entire measuring system ?*

*(c) What is the lowest frequency that can be measured with 5 per cent amplitude error by the entire system ?*

*(d) What is the value of an external shunt capacitance that can be connected in order to extend the range of 5 pet cent error down to 10 Hz ?*

*(e) With external capacitance calculated in (d) connected in the circuit, what is the system high frequency sensitivity ?*

**Solution.** *(a)* Charge sensitivity of transducer

$$K_q = 40 \times 10^{-3}\ \text{C/m}$$

Capacitance of transducer

$$C_p = 100 \times 10^{-12}\ F$$

$\therefore$ Sensitivity of transducer

$$K = \frac{K_q}{C_p} = \frac{40 \times 10^{-3}}{1000 \times 10^{-12}} = 40 \times 10^6\ \text{V/m}.$$

(*b*) The total capacitance of the measuring circuit

$$C = C_p + C_c + C_A$$
$$= 1000 + 300 + 50 = 1350\ \text{pF}$$

$\therefore$ High frequency sensitivity of entire system

$$= \frac{K_q}{C} = \frac{40 \times 10^{-3}}{1350 \times 10^{-12}}$$
$$= 29.63 \times 10^6\ \text{V/m}$$

(*c*) Time constant of system

$$\tau = RC$$
$$= 1 \times 10^6 \times 1350 \times 10^{-12}\ \text{s} = 1.35\ \text{ms}$$

The lowest frequency that can be used for 5 percent magnitude error is found as under :

$$0.95 = \frac{1}{\sqrt{1 + (1/\omega\tau)^2}} \quad \text{or} \quad \omega\tau = 3.04$$

or
$$\omega = \frac{3.04}{1.35 \times 10^{-3}} = 22.54\ \text{rad/s}$$

$\therefore$ Minimum frequency that can be used with 5% error $f = 22.54/(2\pi) = 358.7$ Hz

(*d*) It is now required that a frequency of 10 Hz be measured with 5% error.

$\therefore$ Time constant

$$\tau = \frac{3.04}{\omega} = \frac{3.04}{2\pi \times 10} = 48.38 \times 10^{-3}\ \text{s}$$

Total capacitance

$$C = \frac{48.38 \times 10^{-3}}{1 \times 10^6}\ \text{F} = 48.380\ \text{pF}$$

$\therefore$ External capacitance required

$$C_e = 48380 - 1350 = 47030\ \text{pF}$$

(*e*) High frequency sensitivity with external capacitance

$$= \frac{40 \times 10^{-3}}{48380 \times 10^{-12}}\ \text{V/m} = 827\ \text{kV/m}$$

**Example 25.53** *A piezo-electric transducer has a capacitance of 2000 pF and a charge sensitivity of $100 \times 10^{-12}$ C/N. The resistance of transducer is $10^6$ MΩ and the impedance of the measuring system consists of a capacitance of 500 pF in parallel with a resistance of 1 MΩ.*

*Find the response if the applied force is :*

*$F = 0.1$ N for $0 < t < 2$ ms,*

*and $F = 0$ N for $2\ \text{ms} < t < \infty$*

*Find the value of voltage just before and just after the impulse is terminated. Also find the voltage after 10 ms of application of the pulse.*

**Solution.** The leakage resistance of the transducer is very large as compared with the resistance of the measuring system $R = 1\ \text{M}\Omega = 10^6\ \Omega$

The capacitance of the system

$$C = C_p + C_c + C_A$$
$$= 2000 + 500 = 2500\ \text{pF}$$

Time constant of circuit

$$\tau = RC = 10^6 \times 2500 \times 10^{-12}$$
$$= 2.5 \times 10^{-3}\ \text{s} = 2.5\ \text{ms}$$

Voltage across the load

$$e_L = \frac{dF}{C}\exp(-t/\tau)$$
$$= \frac{100 \times 10^{-12}}{2500 \times 10^{-12}} \times 0.1 \times \exp(-t/2.5 \times 10^{-3})$$
$$= 4 \times 10^{-3}\exp(-400t) \quad \text{for} \quad 0 < t < 2\ \text{ms}$$

At $t = 2$ ms,

$$e_L = 4 \times 10^{-3}\exp(-400 \times 2 \times 10^{-3})$$
$$= 1.8 \times 10^{-3}\ \text{V} = 1.8\ \text{V}$$

This is voltage at just before $t = 2$ ms

The voltage output just after 2 ms :

$$e_L = \frac{dF}{C}[\exp(-t/T) - 1]$$
$$= \frac{100 \times 10^{-12} \times 0.1}{2500 \times 10^{-12}}\left[\exp\left(\frac{-2 \times 10^{-3}}{2.5 \times 10^{-9}}\right)^{-1}\right]$$
$$= 4 \times 10^{-3} \times (0.55)$$
$$= -2.2 \times 10^{-3}\ \text{V} = -2.2\ \text{mV}$$

The voltage output beyond 2 ms is given by :

$$e_L = \frac{dF}{C}[\exp(-T/\tau - 1)][\exp-(t-T)/\tau]$$
$$= \frac{100 \times 10^{-12} \times 0.1}{2500 \times 10^{-12}}\exp\left[\frac{-2 \times 10^{-3}}{2.5 \times 10^{-3}} - 1\right]$$
$$\exp[-(t - 2 \times 10^{-3})/10^{-3}]$$
$$= -2.2 \times 10^{-3}\exp[-400\,t - 0.8]$$

for $2\ \text{ms} < t < \infty$.

Voltage after 10 ms of application of pulse :

$$= -2.2 \times 10^{-3}\exp[-400 \times 10 \times 10^{-3} - 0.8]$$
$$= -0.09 \times 10^{-3}\ \text{V} = -0.09\ \text{mV}$$

***Example 25.54*** *A pulse is applied to a piezo-electric transducer for a time T. Prove that in order to keep the undershoot to a value within 5%, the value of time constant should be approximately 20 T.*

**Solution.** The value of voltage at $t = T$ is given by :

$$e_L = \frac{K_q A}{C} \exp(-T/\tau)$$

The voltage reduces by a factor $\exp(-T/\tau)$ from its value at $t = 0$.

For 5% undershoot, the value of voltage at $t = T$ should be 0.95 of its initial value

$$\therefore \quad 0.95 = \exp(-T/\tau)$$

or $\quad T/\tau = 0.0513$ or $\tau = 19.5\ T$

Therefore in order to keep the undershoot to 5% within its initial value $\tau \approx 20\ T$.

## 25.30 HALL EFFECT TRANSDUCERS

The principle of *Hall Effect Transducers* has already been explained in chapter 7 on page 179. This is being re-introduced for continuity.

The principle of working of a Hall Effect Transducer is that if a strip of conducting material carries a current in the presence of a transverse magnetic field as shown in Fig. 25.128, a difference of potential is produced between the opposite edges of the conductor. The magnitude of the voltage depends upon the current, the strength of magnetic field and the property of the conductor called *Hall Effect*. The Hall effect is present in metals and semiconductors in varying amounts, depending upon the densities and mobilities of carriers.

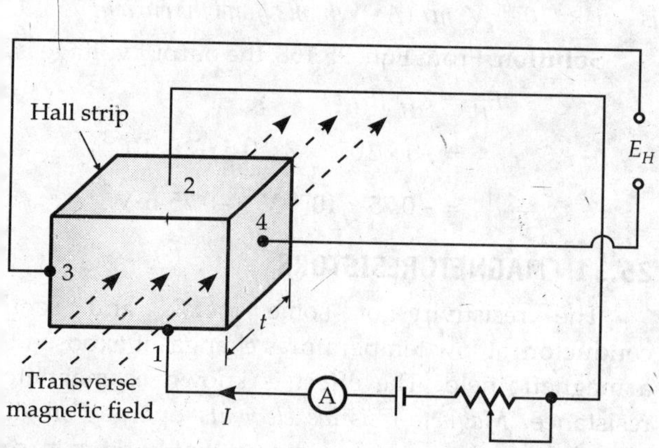


**Fig. 25.128** Hall Effect Element.

Let us consider Fig. 25.128, current is passed through leads 1 and 2 of the strip. The output leads connected to edges 3 and 4 are at the same potential when there is no transverse magnetic field passing through the strip. When a transverse magnetic field passes through the strip, an output voltage appears across the output leads. This voltage is proportional to the current and the field strength.

The output voltage is,

$$E_H = K_H\ IB/t \qquad \text{...(25.168)}$$

where $K_H$ = Hall effect coefficient ; $\dfrac{\text{V - m}}{\text{A - Wb m}^{-2}}$

$t$ = thickness of strip ; m,

and $I$ and $B$ are respectively the current in ampere and flux density in Wb/m$^2$.

Thus the voltage produced may be used for measurement of either the current $I$ or the magnetic field strength $B$.

Typical values of Hall Effect Co-efficient of different materials are given in Table 25.9.

**Table 25.10** Hall Effect Co-efficients

| *Material* | *Field strength* $Wb/m^2$ | *Temperature* °C | $K_H$ $\dfrac{V\text{-}m}{A\text{-}Wb\text{-}m^2}$ |
|---|---|---|---|
| As | 0.4 to 0.8 | 20 | $4.52 \times 10^{-9}$ |
| C | 0.4 to 1.8 | Room | $-11.73 \times 10^{-9}$ |
| Bi | 0.113 | 20 | $-1 \times 10^{-6}$ |
| Cu | 0.8 to 2.2 | 20 | $-52 \times 10^{-12}$ |
| Fe | 1.7 | 22 | $1.1 \times 10^{-9}$ |
| n - Ge | 0.001 – 0.8 | 25 | $-8 \times 10^{-3}$ |
| Si | 2 | 23 | $4.1 \times 10^{-6}$ |
| Sn | 0.4 | Room | $-2 \times 10^{-12}$ |
| Te | 0.3 – 0.9 | 20 | $53 \times 10^{-6}$ |

The Hall effect emf is very small in conductors and is difficult to measure. But in some semiconductors, such as germanium, the emf produced is sufficiently large to be measured by a sensitive moving coil instrument.

### 25.30.1 Applications of Hall Effect Transducer

The following are some of the applications of the *Hall Effect Transducer*.

**1. *Magnetic to Electric Transducer.*** The Hall Effect Transducer can be used as a Magnetic to Electric transducer. A semiconductor plate is inserted into the magnetic field to be measured. The magnetic lines of force are perpendicular to the semiconducting plate. The transducer gives an output voltage which is proportional to $B$, the magnetic field density. The

25

system has the advantage of requiring a very small space in the direction of the magnetic field and therefore, the Hall element can be inserted in narrow gaps for magnetic measurements in air spaces. Also, the element gives out a continuous electric signal in direct response to the magnetic field strength.

Its disadvantages are a high sensitivity to temperature variations and to the fact that the Hall co-efficient may vary from plate to plate, thereby necessitating individual calibration in each case.

**2. *Measurement of Displacement.*** The Hall effect element can be used for the measurement of the location or displacement of a structural element *i.e.*, it can serve as an indirect acting position displacement or proximity transducer in cases where a change of geometry of a magnetic structure causes a change of magnetic field strength. An example is shown in Fig. 25.129 which shows a ferro-magnetic structure having a permanent magnet. The Hall effect transducer is located-in the gap, adjacent to the permanent magnet. The field strength produced by the permanent magnet in the gap, where the Hall effect element is located, is varied by changing the position of a ferro-magnetic plate. The voltage output of the Hall effect transducer is proportional to the field strength in the gap which is a function of the position of the ferromagnetic plate from the structure *i.e.*, the displacement. The method permits measurements of displacement down to 0.025 mm.

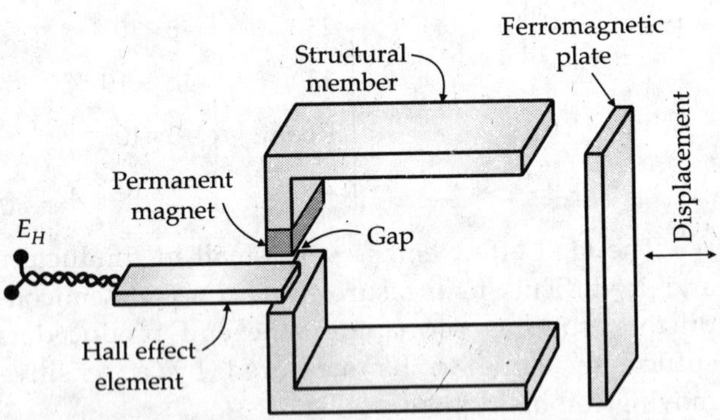


**Fig. 25.129** Measurement of Displacement using Hall effect transducer.

**3. *Measurement of Current.*** An interesting application of Hall effect transducer is shown in Fig. 25.130. The device serves to measure current in a conductor without the need for interrupting the circuit and without making electrical connection between the conductor circuit and the meter. (This is similar in action to the clamp on ammeter described on page 323.) A current (d.c. or a.c.) passes through the conductor and sets up a magnetic field surrounding the conductor. This magnetic field is proportional to the current. A Hall effect transducer is placed in a slotted ferro-magnetic tube which acts as a magnetic concentrator. The voltage produced at the output terminals is proportional to the magnetic field strength and hence is proportional to the current flowing in the conductor. The system can be used for measurement of currents from less than a mA to thousands of ampere. At high current levels, the magnetic concentrator can be omitted since the magnetic field is fairly strong in the vicinity of the Hall effect element and thus can cause appreciable output voltages which can be easily detected.

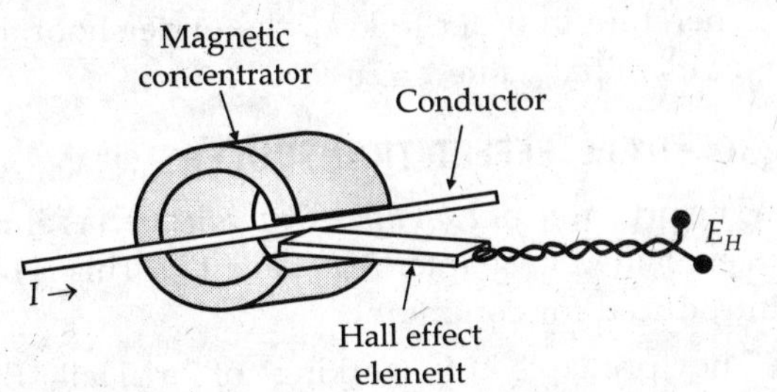


**Fig. 25.130** Measurement of current using Hall effect transducer.

**4. *Measurement of Power.*** The method for measurement of power using a Hall effect transducer has been explained in Art. 11.7 on page 365.

***Example 25.55*** *A Hall effect transducer is used for the measurement of a magnetic field of 0.5 Wb/m$^2$. The 2 mm thick slab is made of Bismuth for which the Hall's co-efficient is $-1 \times 10^{-6}$ V m/(A - Wb m$^{-2}$) and the current is 3 A.*

**Solution.** From Eqn. 25.168, the output voltage is,

$$E_H = K_H \; IB/t$$

$$= -1 \times 10^{-6} \times 3 \times 0.5/(2 \times 10^{-3})$$

$$= -0.75 \times 10^{-3} \text{ V} = -0.75 \text{ mV}$$

## 25.31 MAGNETORESISTORS

The resistivity of some metals and semi-conductors at low temperatures changes if exposed to a magnetic field. The effect is known as magneto-resistance. *Magneto-resistive elements* operate on the law of electrodynamics which says that Lorentz forces act upon mobile charge carriers in a magnetic field, causing electrons to move in an indirect route, thereby lengthening the current path and increasing the resistance of material. The amount of deflection of the electrons depends upon electron mobility. It is the highest in a semiconductor such as indium antimonide which has a mobility of 7.8 $m^2$/Vs as compared to $50 \times 10^{-4} m^2 / Vs$ for a metal. Magneto-resistors are made from indium antimonide or indium arsenide which have a mobility of 2.4 $m^2$/Vs.

Magneto-resistors are sensitive to the total magnetic field and not to its rate of change. Figure 25.131 shows the magneto-resistive characteristics. $H_0$ is the effective anistropy field in the material and is sum of the demagnization anistropy field and the anisotropy field of elemental deposition. With no field applied the domain, magnetization is along elemental length at 0. As the field increases the resistance increases,

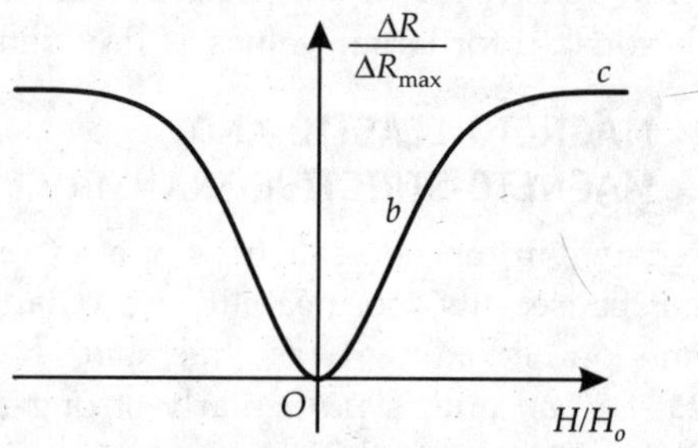


**Fig. 25.131** Basic magneto-resistive characteristics.

until at $b$ the elements are rotated by 45° to the elemental length. Any further increase in field leads to saturation at C. The magneto-resistive element may be operated at $O$, or at $b$ by means of an external bias magnet. Operation at $b$ gives linear characteristics.

The magneto-resistive effect is increased as the ratio of length ($l$) to width ($w$) of the material decreases. This is shown in Fig. 25.132.

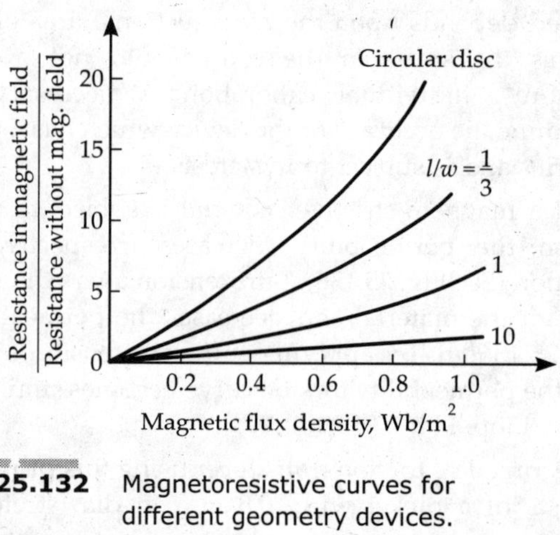


**Fig. 25.132** Magnetoresistive curves for different geometry devices.

Magneto-resistors are built by depositing a film of about 25 μm of indium antimonide/nickel antimonide into a 0.1 mm thick substrate. The film is in the form of a meander and the value of resistance at zero magnetic field can be varied by changing the dimensions of this meander and the number of loops. The film is insulated from the substrate, which may be made from a magnetic base coated with a thin insulating film of a non-magnetic material, such as ceramic or plastic.

The value of resistance depends on the direction of the magnetic field, as indicated in Fig. 25.133. The maximum change in resistance occurs when the nickel antimonide crystals are parallel to each other and the magnetic and electric fields are at right angles to each other. The temperature response of the magneto-resistor depends on the material doping. Generally resistance decreases with temperature and the drop in resistance is greater after the application of the magnetic field, so that the ratio of resistance in a magnetic field to resistance without a field decreases with temperature.

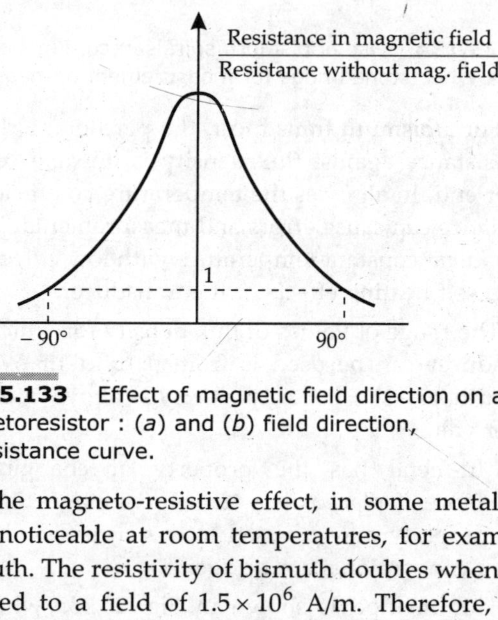


**Fig. 25.133** Effect of magnetic field direction on a magnetoresistor : (*a*) and (*b*) field direction, (*c*) resistance curve.

25

The magneto-resistive effect, in some metals, is even noticeable at room temperatures, for example, Bismuth. The resistivity of bismuth doubles when it is exposed to a field of $1.5 \times 10^6$ A/m. Therefore, this metal could be used for measurement of flux density.

The bismuth transducer consists of a thin wire wound in a flat bipolar spiral 0.5 mm to 30 mm in diameter cemented between two sheets of mica.

Figure 25.134 shows an arrangement of using a bismuth transducer for measurement of flux density. A bismuth conductor has the property of experiencing a change in electrical resistance if the conductor is placed in a transverse magnetic field. This effect has been used by several experimenters. One or two bismuth spirals are placed in some sort of thin mounting that can be inserted in the magnetic field, as for example, between armature and poles of a d.c. motor. The spiral can be made one arm of a Wheatstone bridge (or two spirals may be used in opposite arms as shown in Fig. 25.134). By calibration the resistance can be related to field strength. One difficulty is a rather low sensitivity to weak fields. Another and more serious, trouble is the effect of temperature on the resistance of bismuth, so that large errors can result unless the temperature is carefully controlled or the effect compensated.

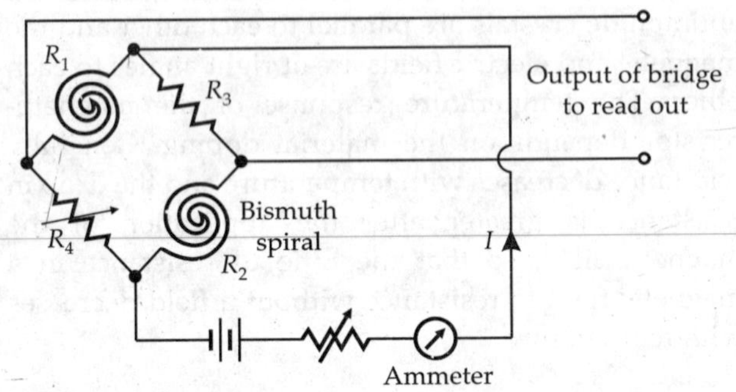


**Fig. 25.134** Use of bismuth spirals in conjunction with a Wheatstone bridge for measurement of magnetic.

For a bismuth transducer, the percentage change of resistance against flux density is linear except at lower end. In all cases the temperature co-efficient of resistance can cause errors and measurements should be made at constant temperature with low currents to avoid self heating effects of the transducer.

The range of values of flux density for which this transducer can be used is from 0.01 to 1.5 Wb/m$^2$. Bismuth gives a linearity of $\pm 1\%$ above a critical lower value.

Mu-metal has the property of changing its impedance if placed in a transverse magnetic field. This property may be used to measure flux density with a conventional a.c. bridge.

This method is shown in Fig. 25.135 who two mu-metal wires are placed in the unknown magnetic field. The impedance of the wires is a function of the strength of magnetic field and is measured with an a.c. bridge.

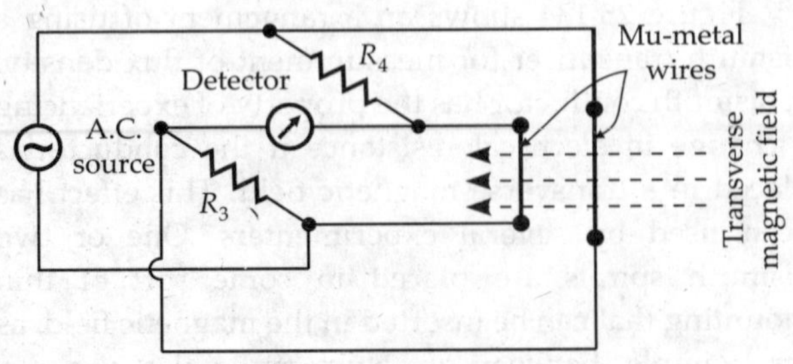


**Fig. 25.135** Measurement of flux density with Mu-metal wires.

This transducer produces an acceptable emf for the input to an instrumentation system and is capable of measuring field strength of the order of $10^{-7}$ Wb/m$^2$.

More modern transducers use either indium antimonide or indium arsenide which are more sensitive than either bismuth or mu-metal. The magneto-resistive effect is more marked in indium antimonide but indium arsenide is usually used for transducers because the room temperature resistivity of indium arsenide is 100 times greater than that of indium antimonide.

The change in resistance is proportional to square of the flux density for small values but deviates from this relationship for larger values of flux densities.

## 25.32 MAGNETO-ELASTIC AND MAGNETO-STRICTIVE TRANSDUCERS

A change in mechanical stress of a ferromagnetic material causes its permeability to change. This phenomenon is known as the *Villari effect* or *magneto-striction* and is particularly strong in Nickel Iron alloys. The permeability changes in response to dimensional changes, such as compression, tension or torque. This magnetic-elastic effect can cause the induction of voltages and hence can be used in mechanic-electric transducers.

A transducer utilising the Villari effect basically consists of a coil wound on a core of magneto-strictive material to which a force, causing a stress in the core, is applied. The stress causes a change in the permeability of the core and hence the inductance.

The range of force over which this transducer can be used depends upon the cross-sectional area of the core as the stress in the core should not exceed 70 M/m$^2$. Individual calibration is necessary to determine the accuracy of the device which has a poor linearity and is subject to hysteresis.

If a magneto-strictive material is subject to pure torsion the permeability increases irrespective of direction (See Fig. 25.136). Pure tension alters the $B$- $H$ curve of the material and decreases the permeability [Fig. 25.136(*b*)]. By applying sufficient tension to the wire the permeability torsion curve becomes similar to a $B$- $H$ loop, Fig. 25.136(*c*).

A practical torsion transducer using this principle consists of a nickel wire, 0.5 mm in dia, stretched between the poles of a magnet and having a small stylus rigidly fixed to the mid point. Before attaching the wire between the poles, the wire is twisted. This results in additional torsional stress. Two pick-up coils are connected in series are wound around the two halves of the wire.

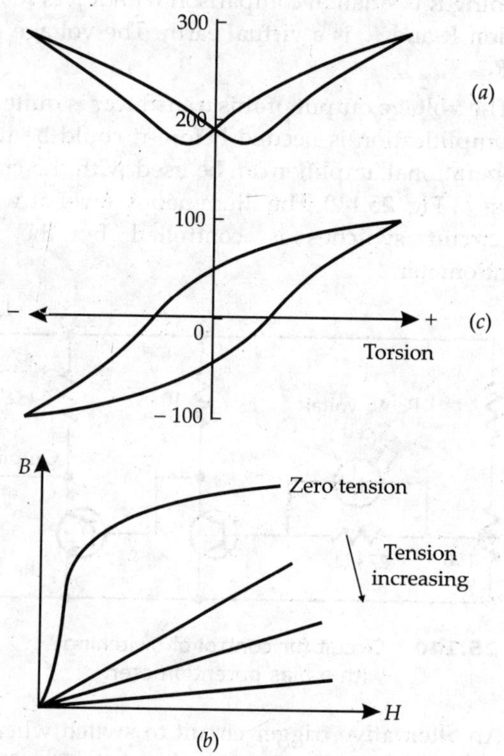


**Fig. 25.136** (*a*) Effect of pure torsion on the permeability of nickel. (*b*) Effect of pure tension on the *B-H* curve of nickel. (*c*) Effect of torsion and simultaneous strong tension on the permeability of nickel.

As the displacement of stylus increases the stress on one side of the torsion wire and decreases it by an equal amount on the other half. This produces an increase of flux in one half and a decrease in other half. These changes in the magnetic flux produce voltages in the two pick up coils as the movement takes place. The resultant voltage is a function of the stress applied. The frequency response of these transducers is flat over a range of 150 Hz to 15 kHz.

Figure 25.137 illustrates the functioning of a *magneto-elastic accelerometer*. Acceleration of the mass *M* causes

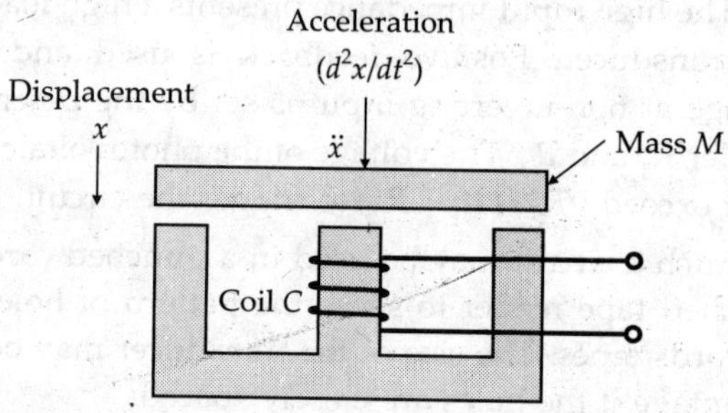


**Fig. 25.137** Schematic diagram of Magneto-elastic transducer.

a force and hence produces a stress in the laminated core thereby changing its permeability. The change in permeability produces a voltage in coil C surrounding the core. Since, the force and consequently the stress are proportional to the acceleration, that is force

$$F = Md^2 x / dt^2 = M\ddot{x}$$

and since the voltage is proportional to *dF/dt*, the output is proportional to third derivative of displacement ($\dddot{x}$) of the mass *M*.

### Magnetostrictive Effect

When a ferromagnetic magnetic material is subjected to changing magnetisation, its permeability changes with change in magnetisation (on account of non-linear *B-H* characteristics of material) thereby causing a change in the dimensions of the material. This effect is *magnetostriction* and is very much pronounced in nickel iron alloys, although in these materials, the change in dimensions with change in magnetisation is quite small.

If a rod is premagnetised by a d.c. field produced by a coil $C_1$ as shown in Fig. 25.138 and an a.c. field produced by another coil $C_2$ is superposed, it will

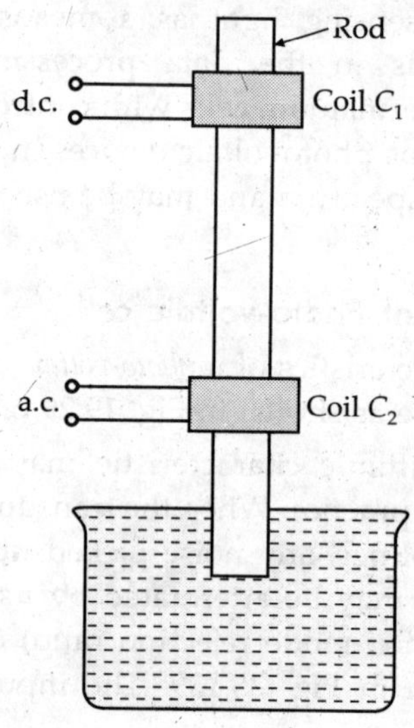


**Fig. 25.138** Magneto-strictive transducer for generation of ultrasound.

periodically change the length of rod on account of change in magnetization. The effect will be maximum at resonance *i.e.*, if the frequency coincides with the lengthwise oscillation frequency of the rod. This frequency is in general, at the high end of the audio frequency range or in the *ultrasonic range*.

25

Arrangements of the type shown in Fig. 25.138 can be used as powerful generators for ultrasonic oscillations which may cause effects like :

(*i*) cavitation in liquids

(*ii*) breaking up of particles such bacterial cells and

(*iii*) formation of stable emulsions of oil and water or even mercury and water. The same system can also be used as a tuned magneto-elastic transmitter for ultrasound.

## 25.33 OPTO-ELECTRONIC TRANSDUCERS

The opto-electronic transducers include photo voltaic cell, semiconductor photodiode and photo-transistors. These transducers have already been explained in Chapter 19. However, some additional information is being given here.

### 25.33.1 Photo-voltalic cell

The *"photo-voltaic cell"*, or *"solar cell"* as it is sometimes called, produces an electrical current when connected to a load. Both silicon (Si) and selenium types are used.

Photo-voltaic cells may be used in a number of applications. Multiple-unit silicon voltaic devices may be used for sensing light as a means of reading punched cards in the data processing industry. Gold-doped germanium cells with controlled spectral responses act as photovoltaic devices in the infra-red region of the spectrum and may be used as infra-red detectors.

### Applications of Photo-voltaic cell

The characteristics of a *photo-voltaic cell* are logarithmic in nature as shown in Fig. 19.29 on page 588.

A logarithmic characteristic may have useful application in practice. When the transducer is remote from the amplifier, any noise picked up or the connecting leads may be attenuated by using the high CMRR (common-mode rejection ratio) of the amplifier as shown in Fig. 25.139. The input impedance

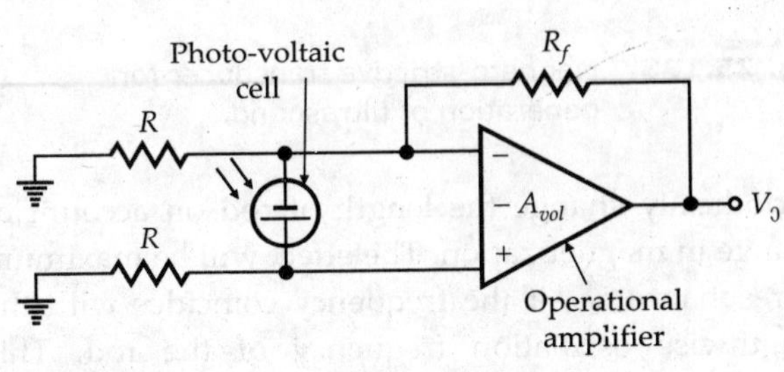


**Fig. 25.139** Amplifier having a high value of CMRR.

assuming $R$ is small in comparison with $R_f$, is $R$ since junction $R$ and $R_f$ is a virtual earth. The voltage gain is $R/R$.

The voltage output of this transducer is quite low and amplification is needed before it could be used. An operational amplifier can be used with the circuit shown in Fig. 25.140. The illumination level at which the circuit switches is controlled by the bias potentiometer.

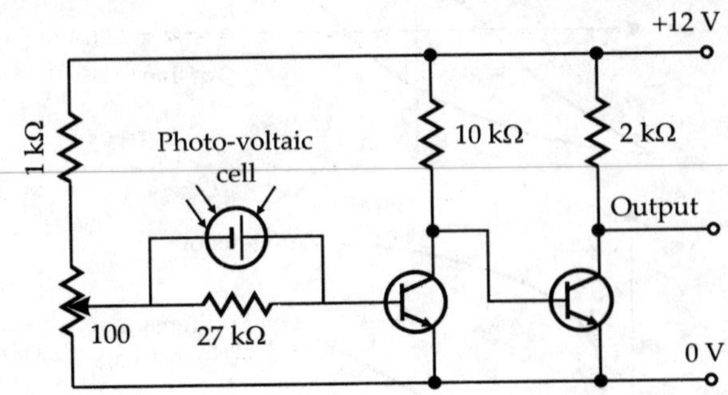


**Fig. 25.140** Circuit for control of switching with a bias potentiometer.

An alternative trigger circuit to switch when the light exceeds a certain value is shown in Fig. 25.141. This circuit uses an operational amplifier.

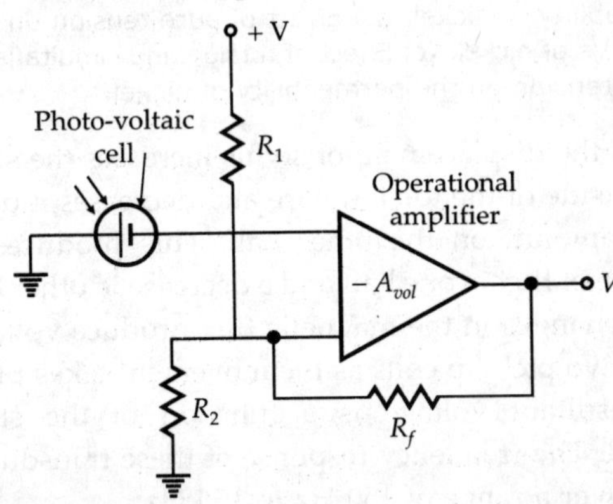


**Fig. 25.141** Trigger circuit used for switching when illumination exceeds a predetermined level.

The high input impedance presents a high load to the transducer. Positive feedback is used and the voltage at non-inverting input is set by the potential divider $R_1$ and $R_2$. The voltage of the photovoltaic cell must exceed $VR_2/(R_1 + R_2)$ to trigger the circuit.

Such a circuit may be used in a punched card or punched tape reader to sense the pattern of holes in the cards/tapes. The size of the transducer may be an advantage if the holes are closely spaced.

The extension of the above method may be used to add the outputs of several photocells together. Such

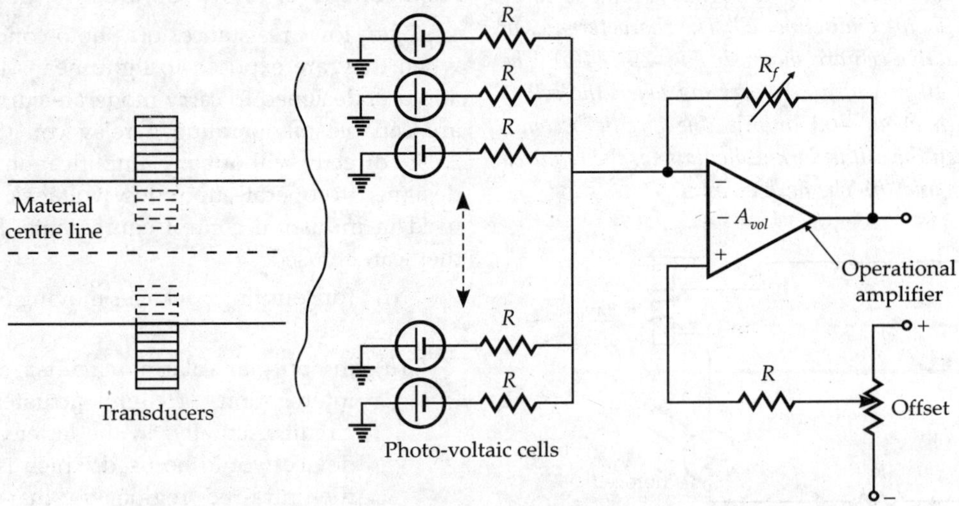


**Fig. 25.142** Array of Photo-voltaic cells to determine the width of material passing.

an application may occur while measuring the width of a material passing over rollers if the mean position of the materials is subject to variations. Two rows of photo-voltaic cells are arranged so that both edges of material partially obscure (cut off) a row. The total output is summed to give an output proportional to the width of the material. This is shown in Fig. 25.142.

The accuracy and resolution is determined by the physical size of transducers, which should be small to obtain good results.

The off-set control is adjusted to give zero output voltage with the transducers in the dark. The gain of the amplifier is adjusted by varying the value of feedback resistor $R_f$ so that a full scale voltage of 10 V is obtained with all the transducers moderately illuminated (of course equally).

Such an array of transducers can be especially manufactured, sometimes on a single chip of a semi-conducting material. They have excellent thermal tracking between individual cells, closely matched outputs and accurate pitching dimensions. They can be encapsulated in special mounts to suit the specifications of the customers.

The other areas of applications of photovoltaic cell which have been described earlier are their use

(*i*) in photographic exposure meters and

(*ii*) in space crafts to supply electric energy.

They are also being increasing used as an unconventional source of electric power in remote areas.

## 25.33.2 Photoconductive Cell

Another Photo-electric effect that has proved very useful is the *photo-conductive effect,* which is used in photo-conductive cells or *photo-cells.* In this type of device, the electrical resistance of the material varies with the amount of light energy striking it.

A typical control circuit utilizing a photo-conductive cell is illustrated in Fig. 25.143. The potentiometer is used to make adjustments to compensate for manufacturing tolerances in photocell sensitivity and relay operating sensitivity. When the photo-cell has the appropriate light incident upon it, its resistance is low and the current through the relay is consequently high to operate the relay. When the light is interrupted or shut off partially or completely, the resistance of the photocell increases thereby reducing the current through the relay. The current may drop down to a level where it will not be able to operate the relay thereby de-energizing the relay.

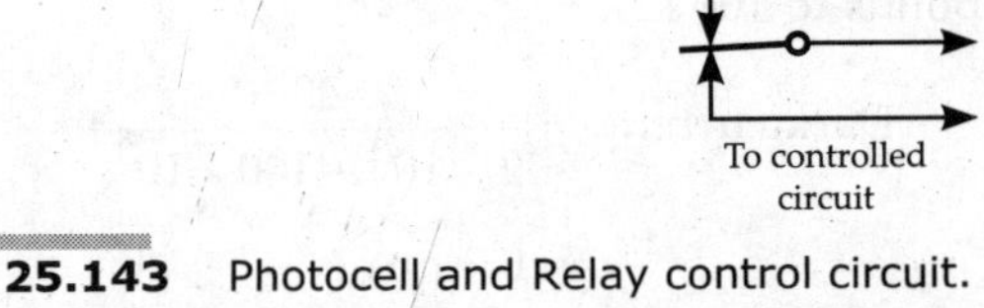


**Fig. 25.143** Photocell and Relay control circuit.

25

***Example 25.56*** *The relay of Fig. 25.144(a) is to be controlled by a photo-conductive cell. The characteristics of the photo-conductive cell are shown in Fig. 25.144(b). The circuit delivers 10 mA at a 30 V setting when the cell is illuminated with about 400 lm/m$^2$. The circuit becomes deenergised when the cell is dark. Calculate (a) the required series resistance and (b) The dark current.*

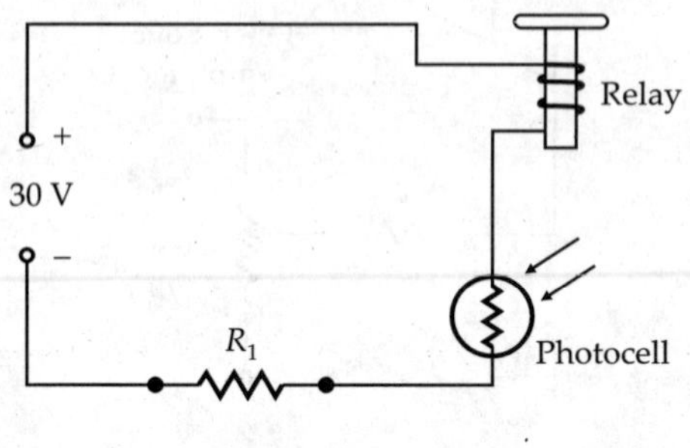


**Fig. 25.144** (*a*) Relay control by a Photo-conductive cell.

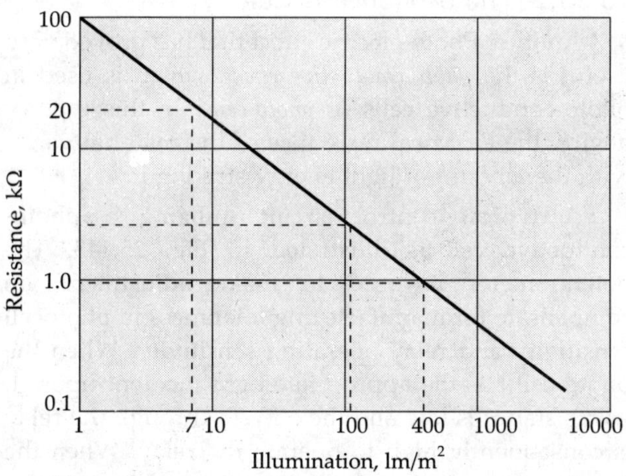


**Fig. 25.144** (*b*) Photocell illumination characteristics.

**Solution.** (*a*) Referring to Fig. 25.144(*b*), the resistance of cell corresponding to an illumination level of 400 lm/m$^2$ $\approx$ 1 kΩ.

Hence from Fig. 25.144(*a*), External resistance

$$R_1 = (30/10 \times 10^{-3}) - 1000$$

$$= 2{,}000\ \Omega = 2\ \text{k}\Omega$$

(*b*) By extrapolating the resistance versus illumination curve [Fig. 25.144 (*b*)], dark resistance corresponds to 100 kΩ.

$$\therefore \quad \text{Dark current} = \frac{30}{2 \times 10^3 + 100 \times 10^3}$$

$$\approx 0.3\ \text{mA}.$$

## Applications of Photo-Conductive Cells

The low resistance of photo-conductive cells when they are exposed to light means that they can and are, designed to carry moderate currents, such as are capable of operating a relay coil (See Example 25.56) directly without any amplification. They can be designed to operate upon low voltages and are thus used in industrial control equipments. For example, they can be used

(*i*) for counting packages moving on a conveyor belt

(*ii*) in burglar alarm circuits, wherein the interception of light actuates an alarm circuit. Actually, in the latter application a device would be used which is sensitive in the infra-red region, so that the burglar would not see the beam of light.

Another application makes use of their light dependant resistance property. If a photo-conductive cell is inserted as a component in another circuit, light can be made to control a parameter of that circuit. To cite a few examples, the insertion of a photo-conductive cell may alter the bias of a transistor or change the gain of an amplifier. Such circuits are used in automatic brightness compensation of T.V. receivers, so that when the room light is switched on, the brightness of the screen is automatically adjusted. The volume of an audio amplifier can be controlled by light–this is particularly useful if several channels have to be controlled together. The device may also be used as a *light controlled switch.*

Figure 25.145 shows a light measuring circuit consisting of a balanced bridge for measurement of light alternation at Photo-conductive cell $R_2$. This is a form of a differential circuit. The two photo-conductive cells are similar and they keep the bridge under conditions of balance whatever may be the environmental conditions. This reduces the effects of power supply variations, light history and temperature drifts of photo-conductive cells and the variations of light emitted by the lamp, as both halves of the bridge experience similar effects and the bridge remains balanced in case initially $R_1R_4 = R_2R_3$.

If a transluscent object is inserted in the path of light falling on photo-conductive cell $R_2$, the light to it is partially cut off *i.e.*, light is attenuated. This changes the resistance of photo-conductive cell $R_2$, thereby rendering the bridge unbalanced. The output voltage of the bridge, on account of the unbalance is amplified and measured with the help of an appropriate detector. The output of the detector is a measure of the *light attenuation.*

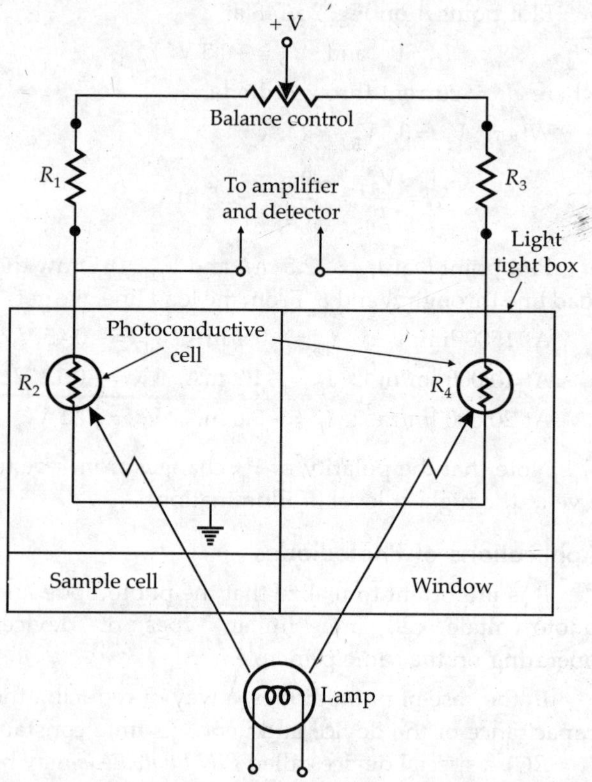


**Fig. 25.145** Measurement of Light Attenuation.

***Example 25.57*** *The bridge circuit of Fig. 25.146 is used for measurement of attenuation of light. $R_1$ and $R_3$ are resistors having a value of 1 kΩ each. $R_2$ and $R_4$ are two photo-conductive cells whose values of resistance depend upon the light incident upon them. The bridge input voltage is 10 V d.c. and the output of the bridge is measured with a detector which can be considered to have an infinite input impedance.*

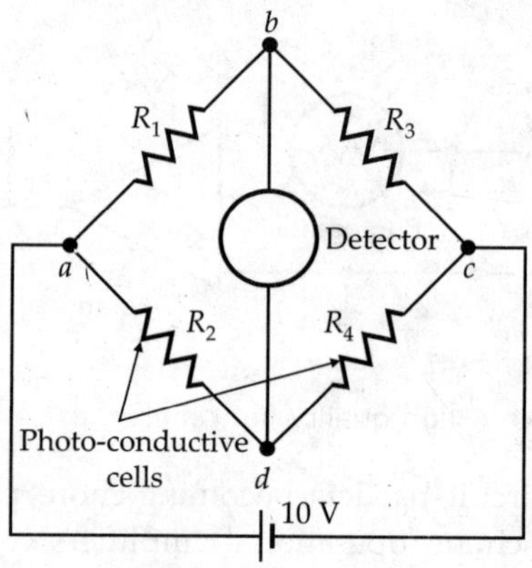


**Fig. 25.146** Bridge circuit of Example 25.57.

*Initially the intensity of light incident on both $R_2$ and $R_4$ is 400 lm/m$^2$ with the result both $R_2$ and $R_4$ have a resistance value of 1 kΩ each [Refer to Fig. 25.144(b)]. The light falling on $R_2$ is now attenuated to 100 lm/m$^2$ and therefore its resistance increases to 2 kΩ. The intensity of light falling on $R_4$ remains the same. Find the output voltage of the bridge.*

**Solution.** Initially each of the four resistances forming the bridge is 1 kΩ each with the result that the bridge is balanced and its output voltage is zero.

When the light falling on cell $R_2$ is attenuated, its resistance increases to 2 kΩ, with the other three resistances remaining 1 kΩ each and the bridge becomes unbalanced and an output voltage appears across terminals *b* and *d*. This voltage is the open circuit voltage, $E_{bd}$ between terminals *b* and *d* since the detector has an infinite impedance. This voltage can be found by redrawing the bridge circuit as shown in Fig. 25.147.

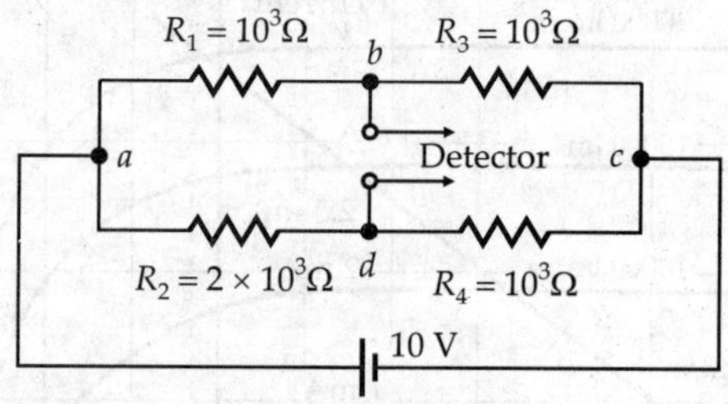


**Fig. 25.147** Diagram of Example 25.57.

Potential of point

$$b = 10 - \frac{10}{10^3 + 10^3} \times 10^3 = 5 \text{ V}$$

Potential of point

$$d = 10 - \frac{10}{2 \times 10^3 + 10^3} \times 2 \times 10^3$$

$$= 3.33 \text{ V}$$

∴ Output voltage of the bridge when the light falling on cell $R_2$ is attenuated is

$$E_{bd} = 5 - 3.33 = 1.67 \text{ V}$$

### 25.33.3 Semi-conductor Photo-diode

A reverse biased semi-conductor diode passes only a small leakage current (a fraction of 1 μA in typical silicon diodes) if the junction is not exposed to light. Under illumination, however, the current rises almost in direct proportion to the light intensity. Thus, the *photodiode* can be used in applications similar to those in a photo-conductive cell.

When the device operates with a reverse voltage applied, it functions as a photo-conductive device. When operating without the reverse voltage, it functions as a photo-voltaic device. It is also possible to arrange for a photodiode to change from photo-conductive mode to photo-voltaic mode. However, the photodiode has very important advantages over the photo-conductive cell. These advantages are its response, time is much faster, so that it may be used in applications in which light fluctuations occur at high frequencies.

The photo-conductive cell is useful only at very low frequencies.

Figure 25.148 shows the symbol and typical illumination characteristics for a silicon photodiode.

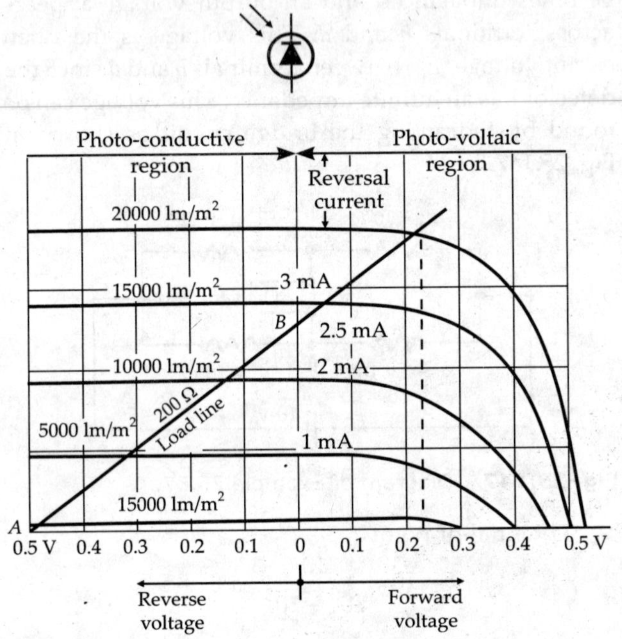


**Fig. 25.148** The symbol and typical illumination characteristics for the silicon photodiode.

***Example 25.58*** *A photodiode with illumination characteristics is shown in Fig. 25.148. It is connected in series with a 200 Ω resistance and a 0.5 V supply. The supply polarity is reverse biases the device. Draw the d.c. load line for the circuit and determine diode currents and voltages at 1500, 10,000 and 20,000 lm/m² illumination.*

**Solution.** A circuit is shown in Fig. 25.149 when

$$I_D = 0, \quad V_{R1} = I_D R_1 = 0,$$

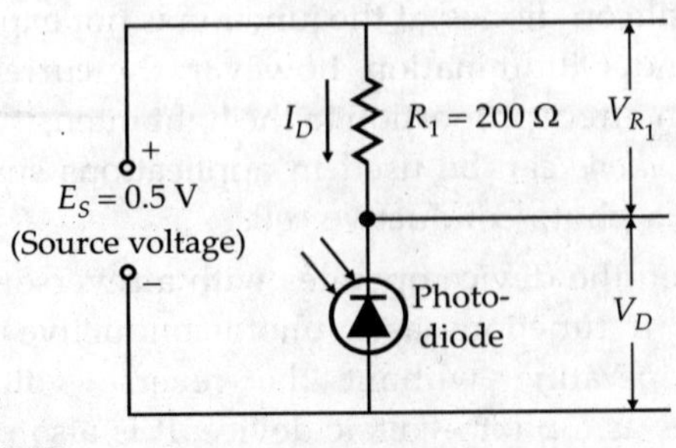


**Fig. 25.149** Photodiode with load resistance.

Plot point $A$ on Fig. 25.148 at

$$I_D = 0 \quad \text{and} \quad V_D = -0.5 \text{ V}$$

where $I_D$ = current through diode.

When $V_D = 0, \ V_{R1} = E_s,$

$$\therefore \quad I_D = \frac{V_{R1}}{R_1} = \frac{-0.5}{200} = -2.5 \text{ mA}$$

Plot point $B$ at $I_D = -2.5$ mA and $V_D = 0$. Draw the load line through $A$ and $B$. From the load line, we get

At 1500 lm/m² : $I_D = -0.2$ mA, $V_D = -0.45$ V

At 10,000 lm/m² : $I_D = -1.9$ mA, $V_D = -0.12$ V

At 20,000 lm/m² : $I_D = -3.6$ mA, $V_D = 0.22$ V.

Note that the polarity of $V_D$ changes from –ve to +ve at the highest level of illumination.

## Applications of Photodiodes

It is important to realize that the photodiode and photo-voltaic cell are similar types of devices operating on the same principle.

In the case of photodiodes, a way of reducing the capacitance of the device and hence its time constant ($\tau = RC$), a special device called *PIN Photodiode* may be used. An $I$ or intrinsic layer is introduced between $P$ and $N$ ends. This effectively increases the width of depletion layer thus reducing the capacitance. Thus faster response and low noise are obtained.

Typical values for a photodiode are :

Junction capacitance = 10 pF,

Reverse resistance = 50 MΩ and

Forward resistance = 100 Ω

The following circuit (Fig. 25.150) is used for the photodiode.

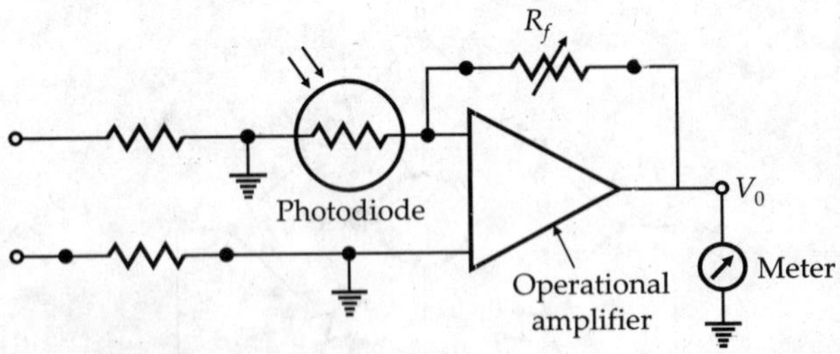


**Fig. 25.150** High quality light meter.

This circuit used is becoming more popular as integrated circuit operational amplifiers come down in price. It has the disadvantage of requiring a dual power supply but has a linear output voltage capable of driving the meter directly. Thus it is used in high quality light meters for photometry and studio photographic work. Some of the applications of photo-voltaic cell can be applied to a photodiode. Because of their fast response time, photodiodes are used in film sound track readers. The sound track is printed on the film as a variable area. As it moves through the protector, it is examined through a fine (<0.1 mm slit) and the light falling on the photodiode is varied. The highest mode that can be reproduced is thus dependant upon the response of the photodiode as well as the film speed, the width of slit and the definition of photographic emulsion. This is shown diagrammatically in Fig. 25.151.

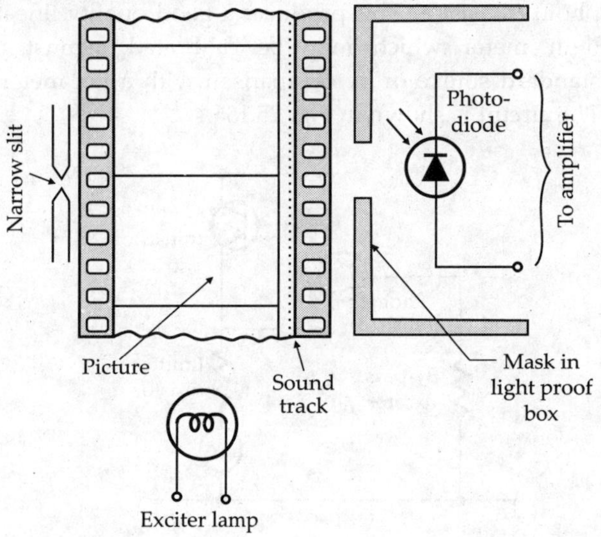


**Fig. 25.151** Sound track reader.

Photodiodes can be used as detectors of modulated light in optical communication systems.

The photodiodes can be used in switching circuits as they have a fast response time. A switching circuit using a photodiode is shown in Fig. 25.152. The transistor is normally 'ON' due to the bias resistor. When the photodiode is illuminated, the base current is reduced turning the transistor 'OFF'.

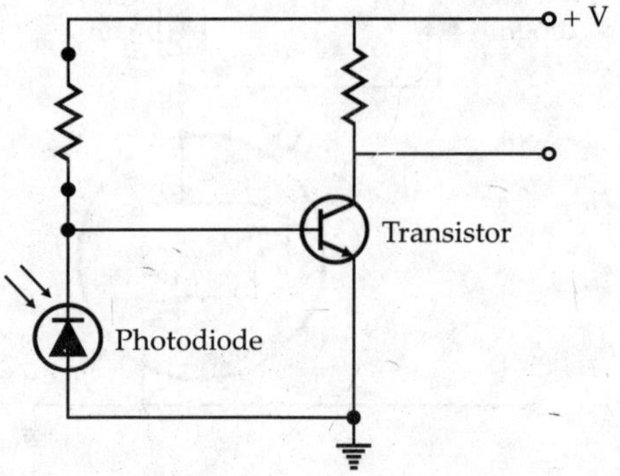


**Fig. 25.152** A switching circuit using a photodiode.

### 25.33.4 Phototransistors

The sensitivity of a photodiode can be increased by as large a factor 100 by addition of a junction which makes it a N-P-N device *i.e.,* a *photo-transistor*. A simple representation of the construction is shown in Fig. 25.153, which also shows the circuit representation of a photo-transistor.

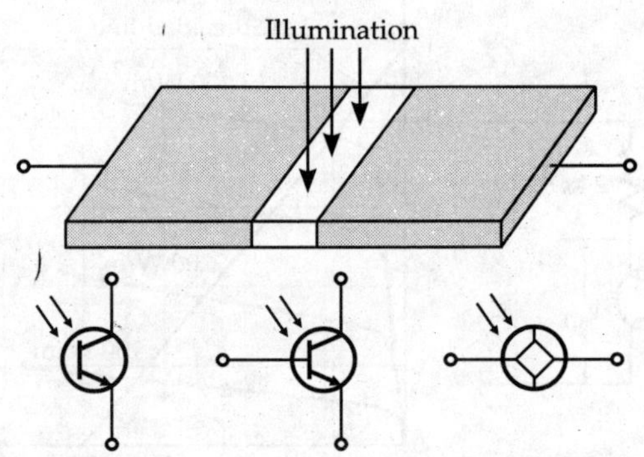


**Fig. 25.153** Photo-transistor and its circuit representation.

Illumination of the central region causes the release of electron hole pairs here. This lowers the barrier potential across both junctions, causing an increase in the flow of electrons from left hand region into the central region and on to the right hand region.

For a given amount of illumination on a very small area, the *photo-transistor* provides a much larger output current than that is available from a photodiode and therefore photo-transistor is much more sensitive than a photodiode. The typical characteristics of a photo-transistor are shown in Fig. 25.154.

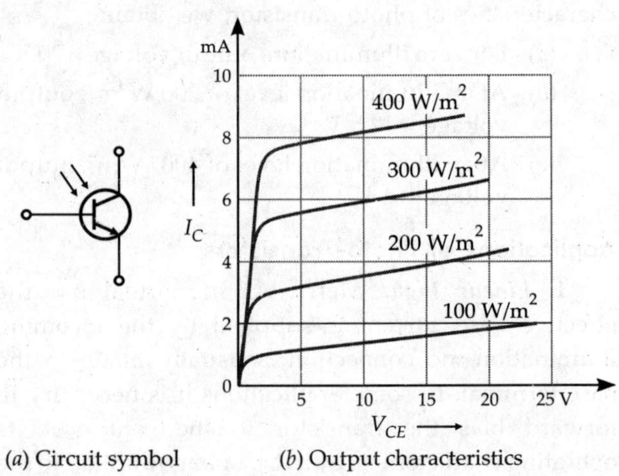


**Fig. 5.154** The circuit symbol and output characteristics for the phototransistor.

25

**Example 25.59** *A photo-transistor of circuit shown in Fig. 25.155 (a) with characteristics shown in Fig. 25.155 (b) has a supply voltage of 20 V and a collector load resistance of 2 kΩ. Determine the output voltage when the illumination level is :*

*(a) Zero (b) 200 W/m² and (c) 400 W/m².*

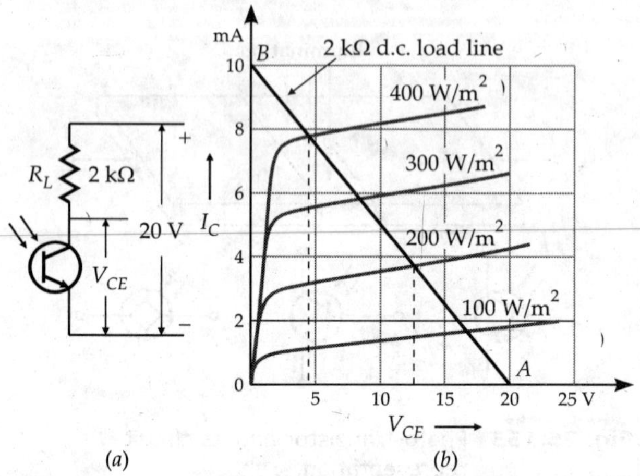


**Fig. 25.155** The phototransistor circuit and d.c. load line of Example 25.59.

**Solution.** The circuit and load line are shown in Fig. 25.155. The load line for a 2 kΩ resistance is drawn with the following considerations :

When $I_C = 0$, $V_{CE} = 20$ V (Point *A*)

When $I_C = 10$ mA,

$$V_{CE} = 20 - 10 \times 10^{-3} \times 2 \times 10^{3}$$
$$= 0 \text{ V (Point } B)$$

From the intersection of the load line and the characteristics of photo-transistor, we obtain :

(*a*) For zero illumination, output voltage = 20 V

(*b*) At an illumination level of 200 W/m², output voltage = 12.5 V

(*c*) At an illumination level of 400 W/m², output voltage = 4 V

## Applications of Photo-Transistors

**1. *Linear Light Meter.*** As in normal use the effective base current is supplied by the incoming illumination, no connection is usually made to the base terminal. In some applications it is necessary to forward bias the transistor to the centre of its operating characteristic by a d.c. voltage applied to the base terminal. Thermal stabilisation techniques may be employed, or an electrical threshold control is possible.

Another use of the base terminal is at extremely low light levels. Here the dark current is important and can be reduced by connecting a high resistance between base and emitter. The resistance acts as a by-pass and greatly reduces the dark current through the photo-transistor.

Although a meter in the collector circuit will indicate illumination, the circuit is sensitive to variations. If another transistor is used to amplify the transducer current, feedback is available across the emitter resistor for stabilising the gain of the photo-transistor. This produces a good quality linear light meter which may be calibrated against a standard source or by comparison with a lux meter. The circuit is shown in Fig. 25.156.

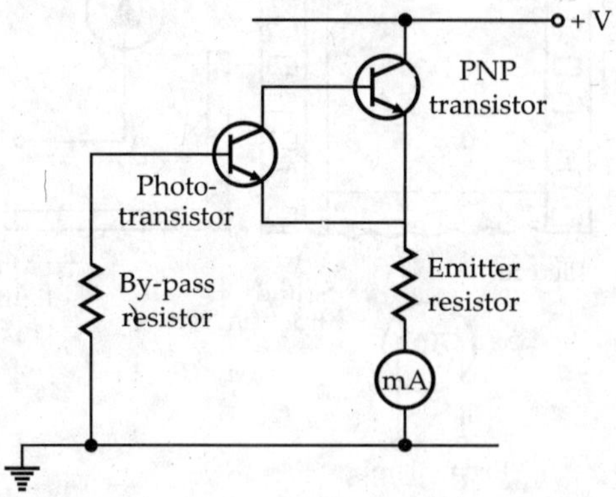


**Fig. 25.156** Linear Light Meter.

**2. *Operation of Relay.*** A device for further increasing the sensitivity is the photo-Darlington transistor, shown diagrammatically in Fig. 25.157. Here the photo-transistor has a second transistor connected as a Darlington pair in the same case. The

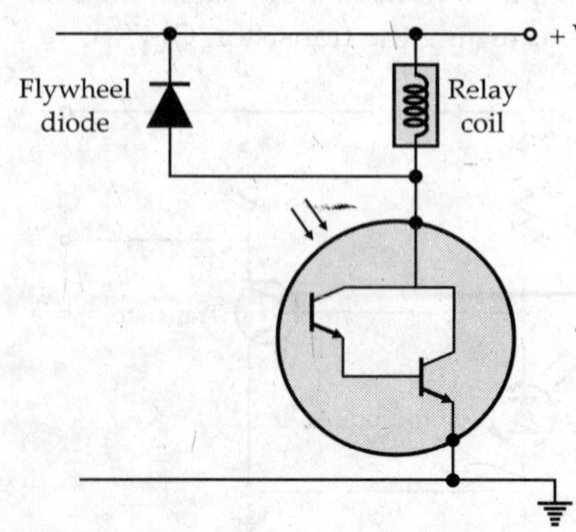


**Fig. 25.157** Application of a photo-Darlington transistor.

two transistors can be matched and the photo-current is multiplied by the combined $h_{FE}$ of both transistors. This gives enough current to operate a larger relay directly as shown in the diagram.

It should be evident that although it has disadvantages for analog work, the photo-transistor will make an excellent switching device for digital applications. Here load resistance is not so much a consideration as we are only concerned whether the transistor is 'on' or 'off'.

**3. *Modulated Light.*** If the light directed on the photo-transistor is coded and the circuit of Fig. 25.158 responds only to that code, stray light can be ignored. In practice an oscillator operating at a few kilo-hertz drives a light-emitting diode. The photo-transistor is followed by a tuned amplifier such that it responds only to the frequency of operation. Stray light thus produces no output. If the beam is interrupted all output ceases and the appropriate safety device can be operated.

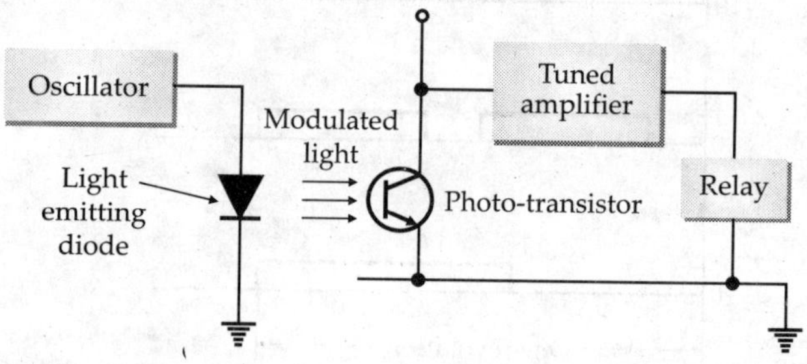


**Fig. 25.158** Use of modulated light.

Modulated light is also used in communication systems, for example, to transmit speech down a light beam. Such a system has the advantages of secrecy and freedom from interference but the transmitter and receiver must be in line of sight and the range is limited to a few hundred metres.

Photo-transistors can also be used with logic circuits, either via an amplifier or trigger circuit, or directly with TTL logic gates as shown in Fig. 25.159. These are used in tape readers for computers. Another application is to sense an object between the doors of an automatic lift and this information is entered into the logic circuits controlling the lift together with information about the position of the lift and which button is pressed.

With the circuit shown, a controlled amount of hysteresis in the trigger circuit is provided by the resistor $R$. Two gates are necessary to achieve the correct phase relationship.

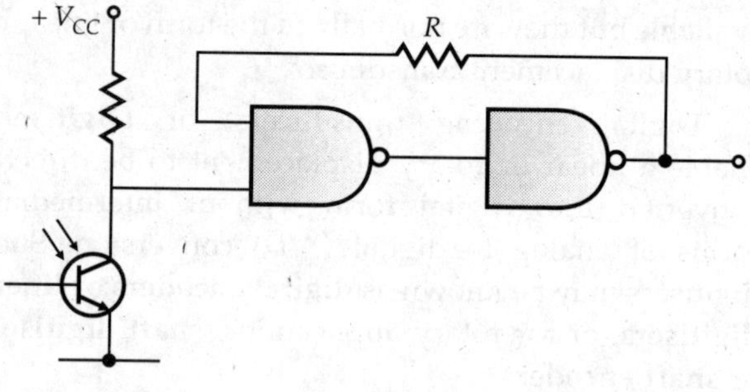


**Fig. 25.159** Use of photo-transistor with logic.

**4. *Use in Shaft Encoders.*** Photo-transistors are used in shaft encoders which translate angular position into a digital code. This is done by shining lamps through a coded *perspex disc* on to photo-transistors. The Gray code is normally used which is then translated into binary code by a logic network and used as the input to a digital control system.

**5. *Use as Switching Devices for Digital Circuits.*** It is evident that photo-transistors have disadvantages as regards analog circuits, but they can be used as excellent switching devices for digital applications. In these applications, the load resistance is not of much consequence as we are concerned only to know whether the transistor is "ON" or "OFF". They are useful in these applications because they have small rise and fall times. Photo-transistors can be used in a wide range of applications which include,

(*i*) computer punch card/tape readers
(*ii*) counting of objects
(*iii*) digital machine control
(*iv*) digital process control
(*v*) pattern recognition

## 25.34 DIGITAL ENCODING TRANSDUCERS

The transducers often communicate with digital computers and therefore transducers which have a digital output are preferable as they are convenient to use since they can be directly interfaced with a digital computer. However, few transducers exist which can provide a *direct digital output*. In most of the situations in measurement systems, *we come across transducers* which provide only an analog output. Therefore, with such transducers we have to use an analog to digital (*A/D*) converters to realize the digital data which could be handled by digital computers.

25

Digital transducers are called **Encoders**. They are available but they are normally in the form of linear or rotary displacement transducers.

Digital encoding transducers, or **Digitisers,** enable a linear or rotary displacement to be directly converted into digital form without intermediate forms of analog to digital (*A/D*) conversion. Such digitisers may be known as **digital encoders** or **linear digitisers,** or for rotary applications, **shaft digitisers** or **shaft encoders.**

A *digitiser* is perhaps the most elementary form of analog to digital (*A/D*) converter because it converts a continuous displacement an analog quantity to be defined incrementally in some binary or decimal code. There are several techniques used for achieving this conversion, each with its own advantages and limitations, which include cost, simplicity of associated circuitry and reliability etc.

## 25.35 CLASSIFICATION OF ENCODERS

The **Digital Displacement transducers** can be classified into three major categories :

- tachometer transducers
- incremental transducers
- absolute transducers.

### 25.35.1 Tachometer Encoders

A *tachometer encoder* has only a single output signal which consists of a pulse for each increment of displacement. This is shown in Fig. 25.160. If the motion were always in one direction, a digital counter could accumulate these pulses to determine the displacement from a known starting point. However, any motion in the opposite direction, would also produce identical pulses, which would produce errors. Therefore, this digital transducer, is usually used for measurement of speed, rather than for displacement, and in situations where the rotation never reverses.

### 25.35.2 Incremental Encoders

The problems caused by reverse motion in the case tachometer encoder are solved by using air *incremental encoder*. The incremental encoder uses at least two (and sometimes a third) signal generating elements. The two tracks (the tachometer encoder uses only one track), in the case incremental encoder are mechanically shifted by $\frac{1}{4}$ cycle relative to each other. This allows detection of motion which signal rises first. Thus an up down pulse counter can be used to subtract pulses whenever the motion reverses. A third output, which produces one pulse per revolution at a distinct point, is sometimes provided for zero reference.

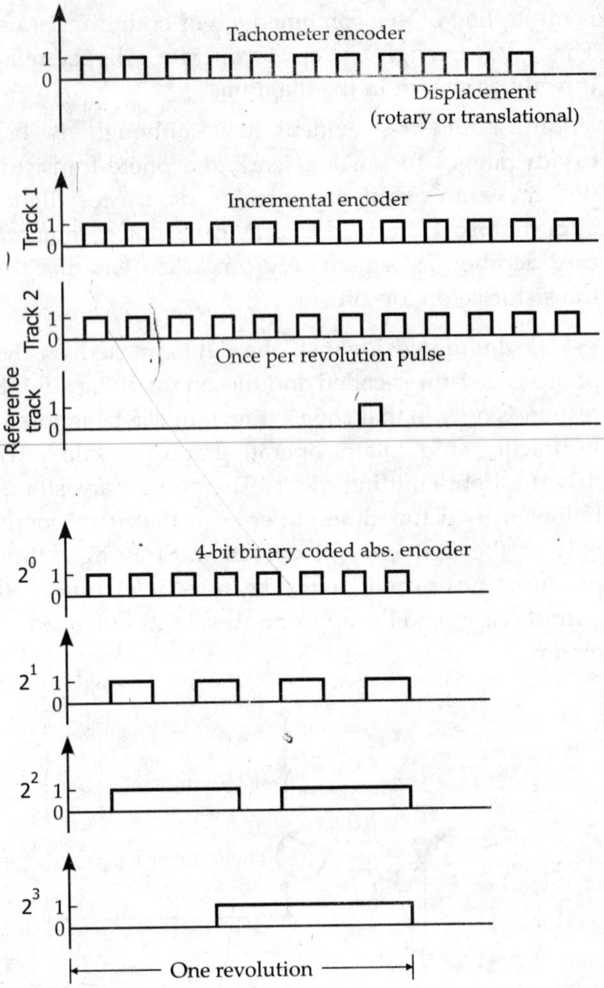


**Fig. 25.160** Classification of Encoders.

An *incremental encoder* has the advantage of being able to rotate through as many revolutions as the application requires. However, any false pulses resulting from electric noise will cause errors that persist even when the noise disappears. The failure of system power also causes total information about the position data which cannot be retrieved even after re-application of power.

### 25.35.3 Absolute Encoders

These are generally limited to measurement of a single revolution. They use multiple tracks and outputs, which are read out in parallel to produce a binary representation of the angular shaft input position. Since, there is a one-to-one correspondence between binary output, position data are recovered when power is restorted after an outage. The transient electric noise causes only transient measurement errors.

## 25.36 CONSTRUCTION OF ENCODERS

Encoders of all three types mentioned above, can be constructed as

- Contacting type and
- Non-contacting type.

### 25.36.1 Contacting or Brush Type (Resistive Encoders)

The shaded areas are made of conducting material and the unshaded areas are made of non-conducting materials. Brushes are placed on the transducer which act as sliding contacts. The circuit of sliding contacts, which come in contact with conducting areas are completed and the circuits of others which make contacts with insulated area are not completed. Thus the encoder gives out a digital read out which is an indication of position and hence the encoder determines the displacement.

In Fig. 25.161, the read-out lamps are only shown for an explanatory function. The voltages on the four lamps-lines could be sent to a digital computer directly. If a visual read-out were desired, these four voltages would be applied to a binary-to-decimal conversion module and then read out decimally on a display.

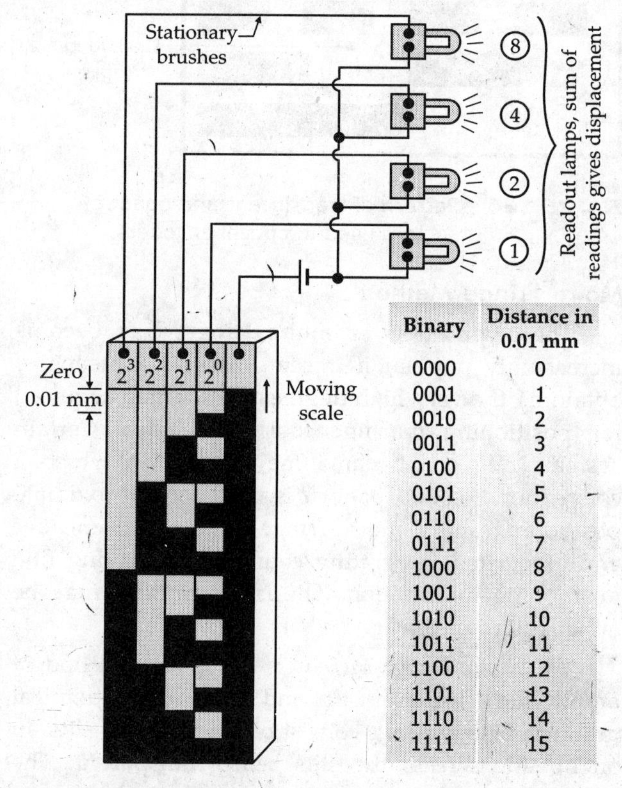


| Binary | Distance in 0.01 mm |
|---|---|
| 0000 | 0 |
| 0001 | 1 |
| 0010 | 2 |
| 0011 | 3 |
| 0100 | 4 |
| 0101 | 5 |
| 0110 | 6 |
| 0111 | 7 |
| 1000 | 8 |
| 1001 | 9 |
| 1010 | 10 |
| 1011 | 11 |
| 1100 | 12 |
| 1101 | 13 |
| 1110 | 14 |
| 1111 | 15 |

**Fig. 25.161** Translational Encoder.

While the code pattern of Fig. 25.161 is most convenient for explaining how motion is represented in the familiar natural binary system, many commercial encoders use a different code patterns such as Gray code in order to avoid errors resulting from small misalignments which occur in practical applications. For example (Fig. 25.162), at the midpoint of binary scale, if the shaded area of $2^2$ bit were displaced slightly to the left, instead of going

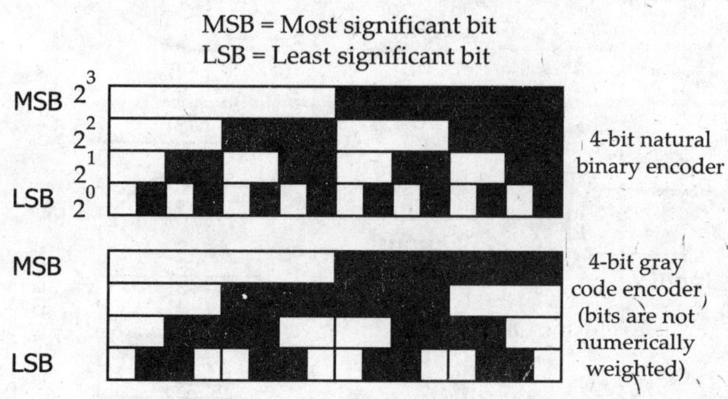


**Fig. 25.162** Linear displacement transducers using Natural binary and Gray codes.

from 0111(7) to 1000(8), the count would go from 0111(7) to 0011(3). The Gray code shown in the same diagram does not suffer from this problem since only one bit changes at each transition in the case of Gray code. Since the Gray code output may not be compatible with the read-out device, conversion from Gray code to natural binary system (or vice versa) may be necessary. This is accomplished by using standard logic gates as shown in Fig. 25.163.

Gray code is used for overcoming the inherent disadvantages of a natural binary code. In a natural binary code, a change of decimal number by one may result in a change of number of bits. In the example quoted above, the transition from natural number to 7 to 8, would need a change in all four bits as number 7 is represented by 0111 and number 8 by 1000. Thus, an error of one bit in a large digital number, in practice, will result in large error when the binary number is converted to decimal number. In the case of digital transducers, an increment of this order, would result in large errors when the result is reflected back into the decimal system.

In case of Gray code, in which case the change in one value to the next is by one bit only. For example, the change in value from 7 to 8 (in decimal system) is 0100 (corresponding to 7) to 1100 (corresponding to a value of 8). Thus there is a change in one bit only. Therefore, the use of Gray system greatly reduces the errors.

25

| Input 1 | Input 2 | Output |
|---|---|---|
| 0 | 0 | 0 |
| 0 | 1 | 1 |
| 1 | 0 | 1 |
| 1 | 1 | 0 |

Exclusive-OR Truth table

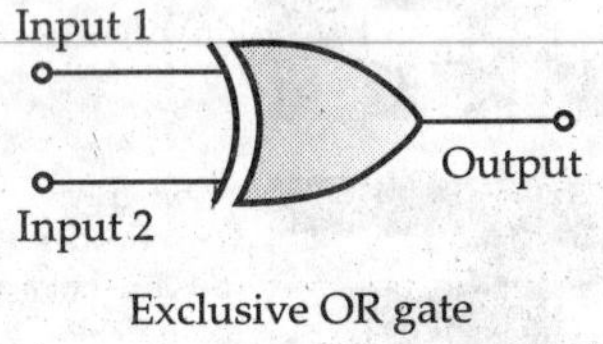


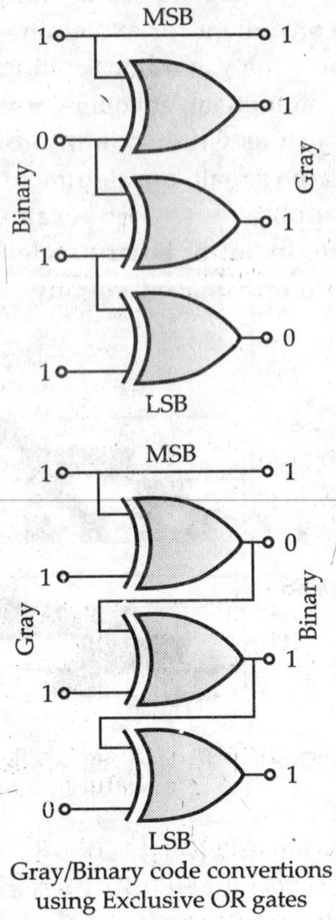


**Fig. 25.163** Natural and Gray binary codes.

## Advantages and Disadvantages

The *advantages* of this type of encoder are :

(*i*) It is relatively inexpensive.

(*ii*) It can be made to any degree of desired accuracy provided that the sector is made large enough to accommodate the required number of rows for binary numbers, and are quite adequate for slowly moving systems.

The *disadvantages* are :

(*i*) The major problem with the use of these encoders is on account of wear of contactors and maintenance of the contacts.

(*ii*) There is often an ambiguity of one digit in the least significant binary digit.

The resolution of digital transducers depends upon the number of digits comprising the binary number. The resolution is $1/2^n$ of full scale where $n$ is the number of digits.

The range is up to several metres with an accuracy of 1 part in 20,000 of full scale. The accuracy obviously depends on the accuracy with which the scale is drawn.

## 25.36.2 Optical Displacement Transducers

Optical transducers follow the following two types of principles for their operation :

(*i*) they consist of a system of coded tracks consisting of transparent and opaque sections and associated lamps and photocells to detect the corresponding switching sequence,

(*ii*) they rely on the use of Moire's fringe techniques, capable of much higher resolution when used for incremental measurement.

A sector (Fig. 25.164) may be designed with a pattern of opaque and translucent areas. A photo-electric sensor and a light source is placed on the two sides of the sector. The displacement is applied to the sector and therefore changes the amount of light falling on the photo-electric sensor. The pattern of illuminated sensor then carries the information as to the location of the sector. Figure 25.164 shows a possible sector of pattern of opaque and translucent areas. The number of levels in the encoder determines the accuracy with which device operates.

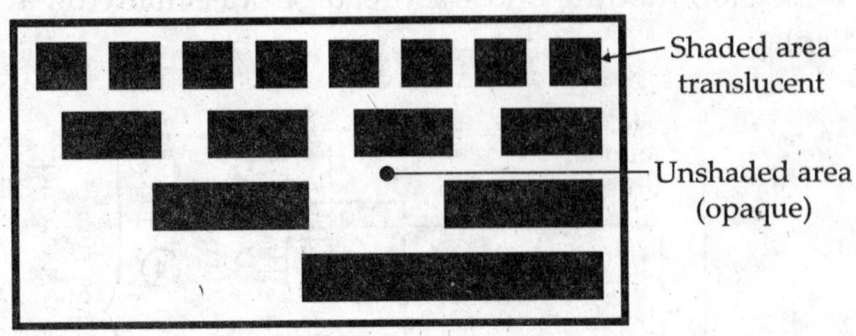


**Fig. 25.164** Pattern of translucent and opaque sectors used in optical encoders.

### Moire Fringe Method

The application of moire fringes has become increasingly popular in recent years as a means of obtaining the very high degree of resolution required for positioning machine tool cross-slides. Certain systems rely on a coarse/fine solution in which a conventional coded pattern is used for approximate positioning and a moire fringe incremental encoder and associated integrating counter for deriving fine positioning information. The latter encoder may be either rotary or linear.

*The Moire Fringe Method* is an optical method of amplifying displacement and uses two identical gratings. The gratings consist of a number of slits on an opaque screen, the slits being transparent. The pitch of the gratings is quite small, and may be between 0.005 mm to 0.05 mm.

Figure 25.165 shows two gratings of the same pitch, mounted face-to-face, with the ruling inclined at angle θ to each other. As shown in figure, a set of dark bands, called Moire fringes, is obtained, with fringe spacing very large as compared to pitch of the grating.

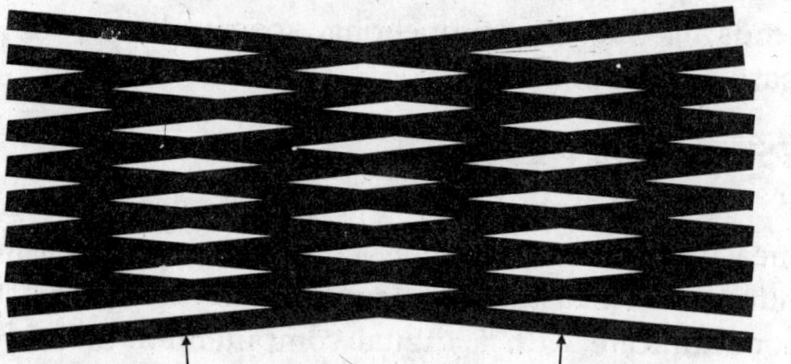


**Fig. 25.165** Moire Fringes.

A movement of the grating in a direction perpendicular to the gratings would move the fringes in a direction perpendicular to the fringes by a larger amount. The movement of the fringes can be measured and from the same, movement of the grating can be calculated or alternatively, the number of fringes passing a given point can be counted, using a photo-electric transducer. This method of displacement measurement is used for measuring the movement of work in machine tools with an accuracy of ± 0.001 percent over a large range.

The essential element of a moire fringe measuring system, be it linear or rotary, is a length of transparent material engraved transversely or in the case of a rotary transducer, radially with a precisely known number of lines per unit length or angle rotation. When two similarly engraved sections are superimposed at a slight angle, a beam of light projected through the twin layer produces a dark area; caused by the angular intersection of the individual lines.

Travel of one section in the direction of its long axis produces a movement of this dark area (fringe) at right angles to the movement; travel in the opposite direction produces a reversal of the fringe movement.

One complete movement of the fringe across the field represents a distance of one line division, thus a very small movement is translated into an easily readable quantity resulting in a very accurate system for measurement–particularly when the engravings are finely spaced. Linear resolutions of 2.5 μm to an accuracy of ±1.2 μm in 250 mm travel are obtainable. Defects due to possible minor local blemishes are integrated out because the fringe signal is derived from the interference pattern of a considerable number of lines. The use of Moire fringe technique reproduces a typical fringe pattern and resulting waveforms on the outputs of four photo-cells evenly spaced across the track is shown in Fig. 25.166.

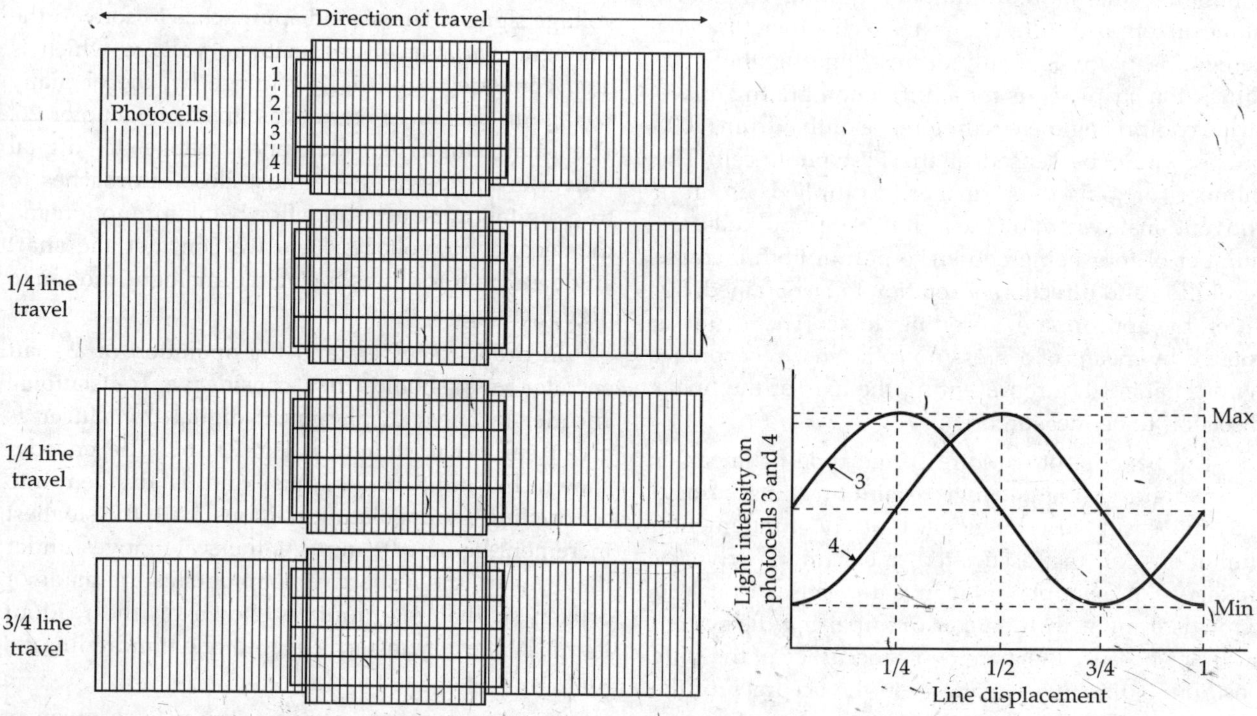


**Fig. 25.166** Moire fringe measuring technique.

A simple form of rotary encoder based on moire fringes is illustrated schematically in Fig. 25.167. It comprises a drum, lamp, collimating lens, prism and photocell assembly. The drum consists of a metal strip 0.05 mm thick and photo-etched to leave webs comprising the grating, and formed as the periphery between two end cheeks. Up to 500 lines may be obtained for a drum size up to 75 mm diameter. In operation the light from the lamp is collimated by the lens and projected through the drum, via a prism. This enables a lens of reasonable focal length to be accommodated by doubling back the light beam, across the body of the encoder using a prism. The form of construction of the drum and optical system has the advantage over the more conventional approach, the latter having one fixed and one moving section, in that

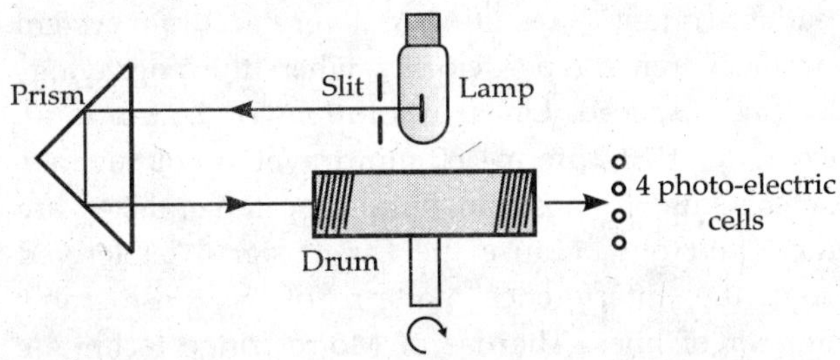


**Fig. 25.167** Drum type Moire fringe optical encoder.

the gratings on opposite sides of the drum, while simulating the more traditional grating assembly, move in opposite directions to each other, thereby generating twice as many Moire fringes as there are diffraction strips. Thus for a diffraction grating of 500 strips comprising the circumference of the drum, 1000 pulses would be sensed at the first photo-cell. The number of pulses is further multiplied as in a conventional system by the number of photo cells, the number of four being chosen to permit both increase resolution and direction of rotation to be obtained, the latter by appropriate decoding logic. The drum is rotated by means of a pressure roller in direct contact with a flat surface representing the axis of the linear direction to be measured.

The benefits of a Moire fringe type of linear or rotary digitiser (namely high resolution and accuracy) may be easily lost by neglect of the fundamental limitations of the associated mechanical drive or linkage. For example, attaching a digitiser to a lead screw is fraught with danger of impairing the system accuracy because the incremental accuracy of the slide position is directly dependant on the accuracy of the lead screw, and due allowance must be made for wear of the screw and associated nuts. The advent of the recirculating ball type of lead screw has offered one practical solution to this problem however, and marked a major step forward in pursuit of the 2.5 μm goal of machine tool users. The linear type of Moire fringe digitiser possesses the advantage that it can be used to measure incremental motion directly but tends to require engineering specifically to fit a particular application.

## 25.37 SHAFT ENCODER

A shaft encoder is a digital device used for measurement of angular position. An increasing number of measuring instruments are being-used to communicate with a digital computer for measurement and control applications. The amount of raw data generated by large scale test programmes is so great that automated computer reduction of these data in meaningful form becomes a necessity. In addition, the feedback control systems for complex processes are becoming increasingly dependent on digital computers for part or complete action of computers for generation of control action. Therefore, there is a necessity of measuring devices that form a basic part of these systems whose output is compatible with digital nature of the computer. In practice, there are two forms of generation of digital signals. The first form converts the analogue variable to a shaft rotation (or translation in linear measurements) and then using many types of **shaft angle encoder** to generate digital voltage signals. The second approach is to convert the analogue variable to an analogue voltage which is then converted to a digital voltage by one of many voltage to digital converters as given in Chapter 26. The shaft angle encoder and voltage to digital converter are perhaps the most close approaches to true digital transducers, the first being for motion and the second for voltage. Here we consider the **shaft angle encoder** for transducing the analogue motion to digital voltage.

In order to understand the operation of a shaft encoder, we should first consider a translational encoder (a linear displacement digital transducer as given earlier) is shown in Fig. 25.161. The encoder shown has four tracks (bits) and is divided into conducting and insulating portions, with the smallest increment being 0.01 mm. As the scale moves under the brushes, the respective lamp circuits are made or broken so that the numbers shown on the readout lamps is at every instant equal to one hundredth of a mm.

For angular displacements, the pattern given in Fig. 25.164 is simply deformed so that the length of the scale becomes the circumference of a circle on a flat disc. The brushes are then placed along a radial line on the disc. Figure 25.168 shows the disc.

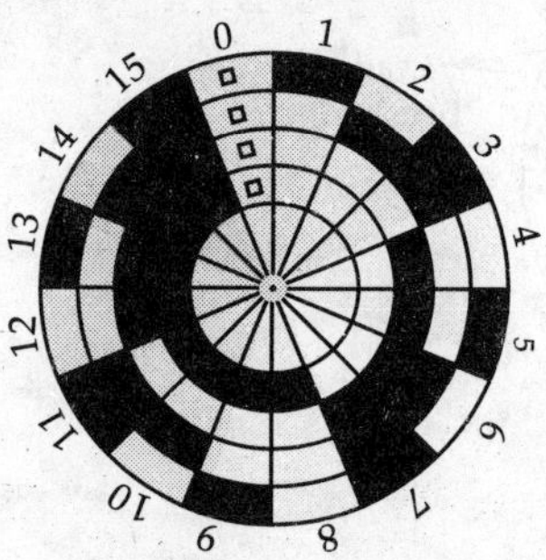


**Fig. 25.168** A rotary shaft encoder rising four track.

The disc is divided into concentric circular tracks and each of these is then divided into segments in a manner depending upon the code being used. For pure binary code, the inner track is halved, the next quartered and the next divided into 8 parts and so on. Each track has twice as many segments as the adjacent one near the centre. The detection method determines the treatment of the disc. The alternate segments on each track are made transparent and opaque, if transmitted light and photo-cells are used. If the segments are made reflecting and non-reflecting, reflected light and photo-cells are used. Electrical methods are used for detection in case the segments are made conducting and non-conducting alternatively.

Thus if the scale is scanned radially a binary 0 or 1 is obtained from each track depending upon the angular position of the disc. Each position of the disc, provided the resolution angle is exceeded between positions, gives a unique number comprising the same number of digits as there are tracks on the disc.

The accuracy depends upon the number of tracks and if there are $n$ tracks, the accuracy obtained is : $\frac{360°}{2^n}$ and is easy to obtain ± 10 s of arc accuracy.

Such discs are manufactured with diameters ranging from 50 mm to 250 mm. The sequence and order of indications represent the position of shaft in coded form.

### 25.37.1 Coding Limitations

Irrespective of the form of the transducer the predominant problem associated with the absolute position encoder is that of providing a noise free, stable and unambiguous output. Assuming that practical limitations preclude obvious solutions such as stepping the shaft by means of an intermittent motion mechanism and electrically inhibiting the output during movement, two main techniques are adopted, depending on the form of encoder. These are the use of 'Gray codes' and 'Double-brush systems'. The former are preferred for optical encoders which tend to be inherently free from noise, while combinations of both techniques together with digital-complement systems, have been used for brush-type encoders.

The problem of obtaining an unambiguous digital output, irrespective of shaft position may be appreciated from a study of a four bit pure binary brush type encoder illustrated in Figure 25.169. It is formed from a number of coded concentric tracks in which live segments are separated by means of insulating segments and connected to a collector ring for applying an energising voltage via a contacting brush. Four coding brushes, (one per track) are shown, to enable a binary output to be detected from the appropriate patterns on the discs.

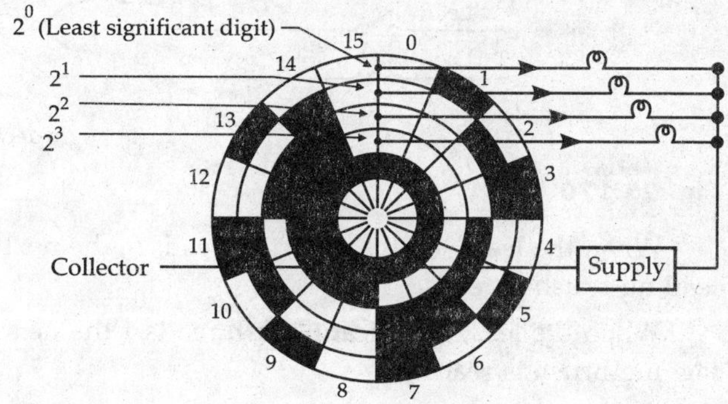


**Fig. 25.169** Binary coded encoder disc.

The example shown is impractical, because if the shaft comes to rest at a transition between say, 0 and 15, mechanical positioning limitations preclude any certainty that all four brushes were reading a binary 1 on a conducting segment, or a binary '0' on an adjacent insulator.

### 25.37.2 Brush Type Encoders

Continuing with the simple brush type encoder, let us look at application of a double brush system to 'V scanning' as shown in Fig. 25.170($a$). The same binary pattern is used as shown in Fig. 25.169 but with two sets of brushes disposed symmetrically as shown about the least significant ($2^0$) brush. The brushes are spaced so that each pair is nominally equal to the digit length of the previous lesser significant track. The brushes are connected by a logic network as shown in Fig. 25.170($b$) and the output is transmitted according to the following rules of operation.

25

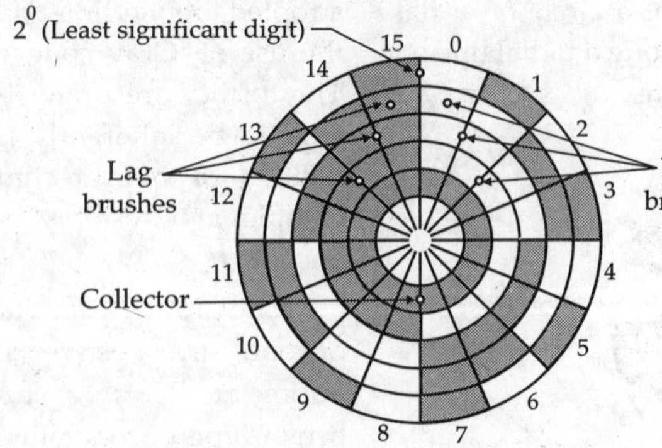


(*a*) Binary coded encoder disc.

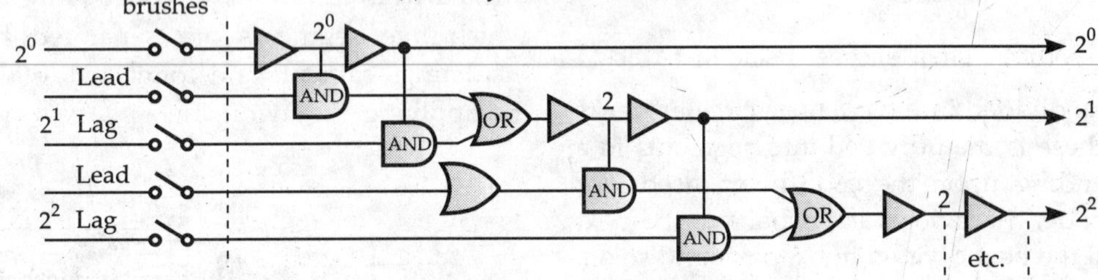


(*b*) Logic decoding for *V*-scan.

**Fig. 25.170**

When the least significant brush reads 0, the next leading brush is read.

When the least significant brush reads 1 the next lagging brush is read.

Thus for the position shown in Fig. 25.170, the least significant brush is reading '0' and the leading brush is read to also give a '0', and so on for successive brushes. Thus although the least significant brush is on a point of transition the total encoder output is stable on all '0"s' or '1"s without ambiguity. An alternative to '*V*' scan is '*U*' scan which is similar except that the brushes are spaced one least significant digit apart as shown in Fig. 25.171(*a*) with associated logic in Fig. 25.171(*b*) which has a faster readout speed.

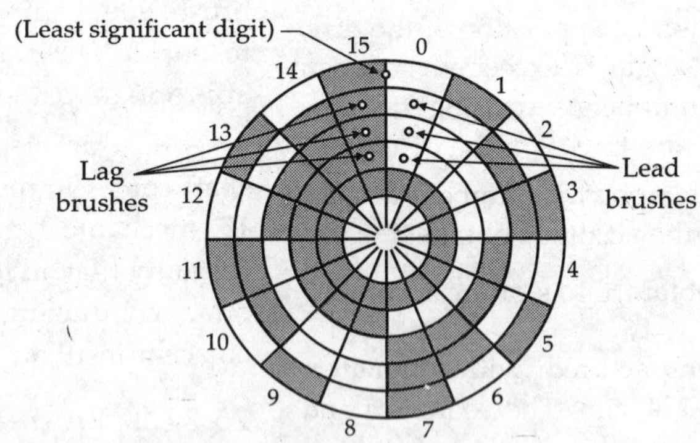


(*a*) Binary coded disc with *U*-scanning.

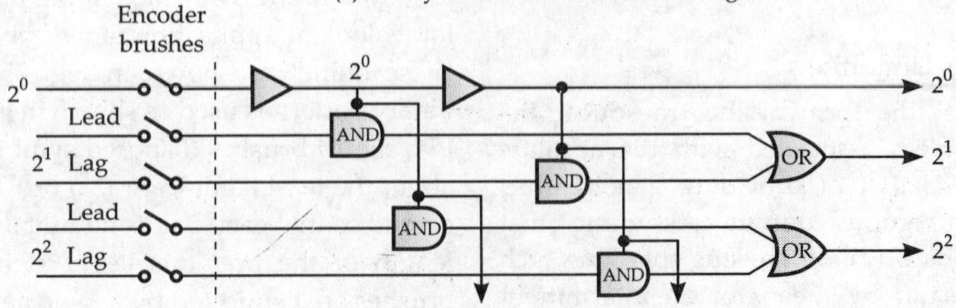


(*b*) Logic decoding for *U*-scan.

**Fig. 25.171**

### 25.37.3 Codes and Decoding

Possibly the simplest way of overcoming the risk of ambiguity is the use of a cyclic code in which only one digit is changed at a time.

An objective of Gray's innovation was to reduce coding errors in pulse code transmission systems to provide a code and corresponding encoder in which the coding error is never greater than one counting unit. Another purpose was to simplify and to reduce the cost of manufacturing encoders. An example of a popular Gray coded disc is shown in Fig. 25.172(*a*) from which it can be seen that for any transition point *e.g.*, 0 to 15 the readout encounters a change in one brush position only. The code requires conversion to pure binary and involves the decode logic shown in Fig. 25.172(*b*) based on exclusive OR gates. Starting

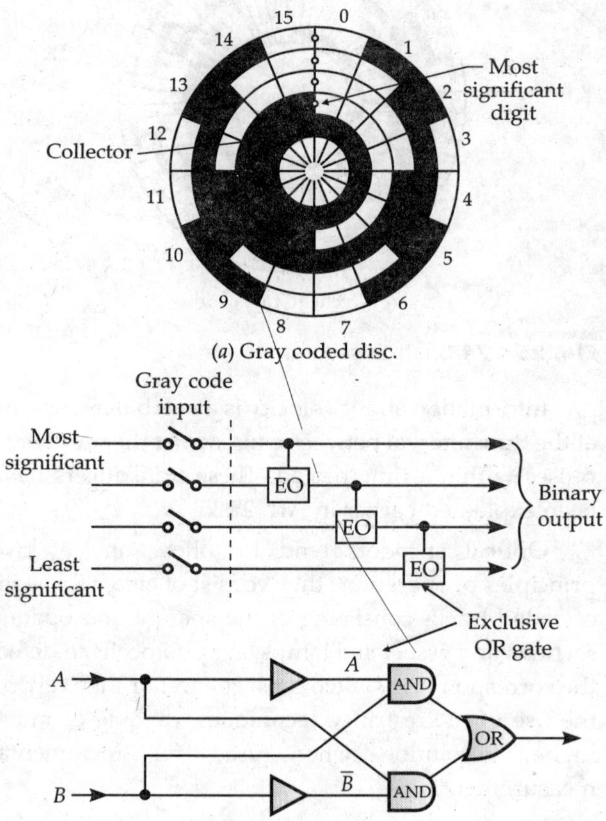


(*a*) Gray coded disc.

(*b*) Gray to binary decoder.

**Fig. 25.172**

with the most significant digit this time, the state of each brush is either confirmed or reversed depending on whether the sum of the previous more significant '1' is even or odd, respectively. Thus in Fig. 25.172(*b*) the output of each gate is '1' when only one input is present.

In general, the construction of brush-type digitisers has evolved over some years with the result that reliable performance can be claimed. Early attempts to produce the coded tracks were based on conventional printed circuit technology but these proved unsatisfactory and sophisticated techniques based on photo engraving are now applied to encode the required pattern in gold-plated copper interspersed with insulated melamine sectors ; the latter chosen for wear properties that match the conducting sections. Precious metal brushes are shaped to conform to particular stiffness criteria so that vibrations are not induced and 'bounce' problems are minimised.

In general, it is not usual to exceed three decade (12 bit B C D) on one disc, and if greater resolution is required two or more discs may be geared together.

The alternative magnetic or optical forms of encoders are generally conceded to be capable of higher resolution, greater accuracy and to be less prone to wear than the brush type, because wiping contacts are eliminated. They tend to be more expensive when applied to a system however, because the signal output is several orders lower than that obtainable from a brush-type digitiser. A brush-type digitiser can switch voltage levels reliably up to 12 V d.c. and may consequently be used for subsequent processing or to switch indicator tubes directly, whereas the magnetic or optical encoder requires associated electronics to raise the signal level and shape the output to a useful form for subsequent processing.



### 25.37.4 Optical Encoders

There are two basic generic styles of optical encoders :

- incremental
- absolute.

The *incremental encoder* provides a pulse each time the shaft rotates through a predefined distance while an absolute encoder provides a "whole world" output with a unique code pattern representing each position.

An eight bit absolute optical shaft encoder is shown in Fig. 25.173. The output code is derived from independent tracks on the encoder disc corresponding to individual photo detectors. The output from these detectors would then be high or low (1 or 0) depending upon the code disc pattern for that particular position.

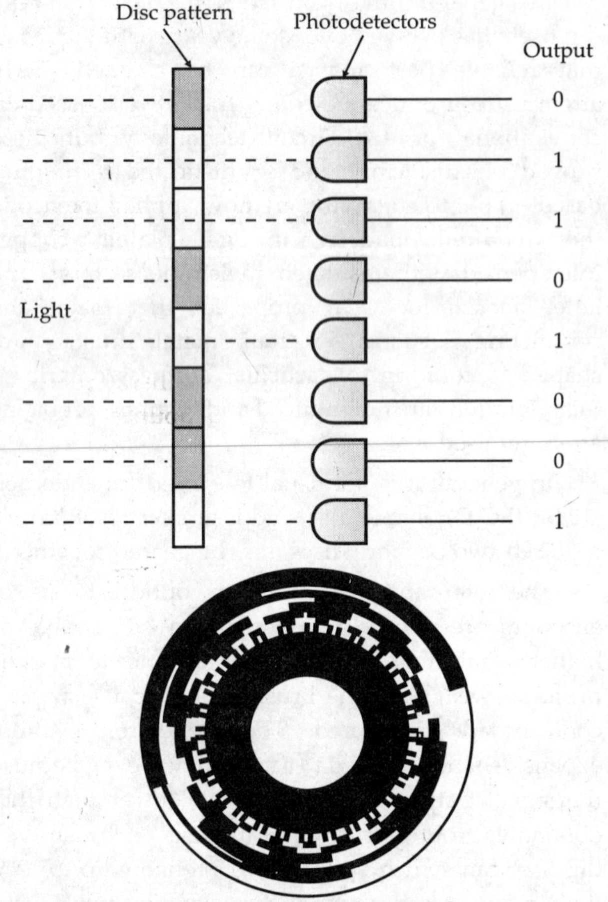


Fig. 25.173 An 8 bit absolute optical shaft encoder.

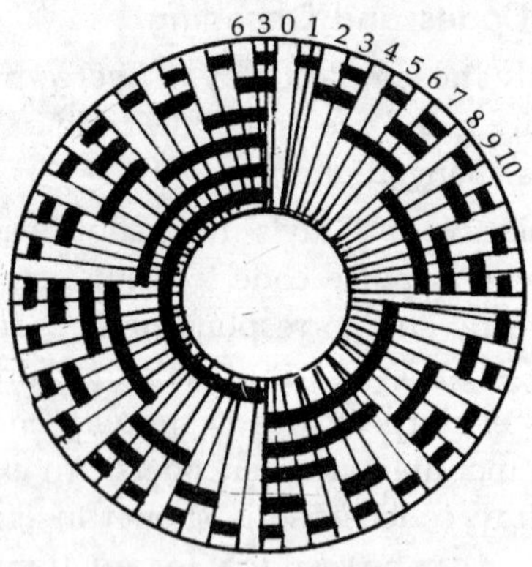


(a) Straight binary system.

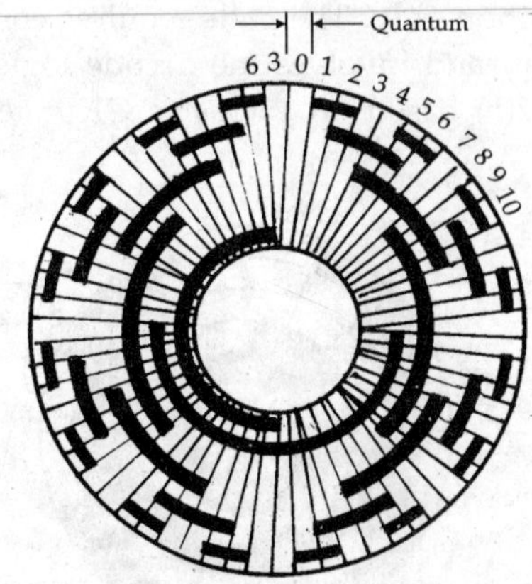


(b) Cyclic (Gray) code.

**Fig. 25.174** Shaft Encoders.

***Absolute Encoders*** are used in applications where a device is inactive for long periods of time or moves at slow rates, such as for flood control, telescopes, cranes etc.

Absolute encoders use the natural binary code or Gray code and BCD in addition to many other codes.

The **incremental encoder** creates a series of equally spaced signals corresponding to the mechanical increment required. For example, if we divide a shaft rotation into 100 parts, an encoder would be selected to supply 1000 square wave cycles per revolution. By using a counter to count these cycles, we can find out how much the shaft has rotated. Suppose the count is 100, then the shaft has rotated through $360 \times (100/1000) = 360°$.

The simpler type of incremental encoder is the **tachometer encoder**. Its output waveform and code track on the disc are shown in Fig. 25.174. A tachometer encoder is sometimes called a **single track incremental encoder** because it has only one output and cannot detect direction. The output is usually a square wave.

Information about velocity is available by looking at the time interval between pulses or at the number of pulses within a time period. These transducers have been explained earlier in Art. 25.35.

Optical encoders tend to follow one of two principles of operation ; they consist of either a system of coded tracks consisting of transparent and opaque sections and associated lamps and photocells to detect the corresponding switching sequence, or they rely on the use of moire fringe techniques, capable of much higher resolution when used for incremental measurement.

The absolute digitiser comprises an assembly consisting of a Gray-coded pattern photographically reproduced on a glass disc mounted on the input shaft. The code consists of ten annular tracks each with a pattern of opaque and transparent sections.

The code reading system employs a filament lamp and collimating lens from which light passes through the disc and a narrow radial slit, to be detected by ten photovoltaic cells. Depending on the angular position of the shaft, certain cells receive light from the transparent portions of the disc and enable the outputs from all ten cells to reproduce the shaft position directly in parallel-Gray-coded form. The output, which is noise free, is suitable for amplification and subsequent processing for use in digital servo systems, computers, data logging and visual displays.

Figure 25.174 shows the shaft encoders using straight binary and Gray codes.

## 25.38 AUTOSYN

It is a high resolution *incremental encoder* based upon the electromagnetic coupling between a fixed scale provided with an a.c. excited serpentive conductor, produced by printed circuit techniques, and a similar but smaller sensing winding which travels over a scale as show in Fig. 25.175. When the alignment is as shown in Fig. 25.175(*b*), the output is at its positive maximum. A displacement of $s/2$ results in minimum output while a movement of $s$ gives negative maximum output. Further, a displacement of

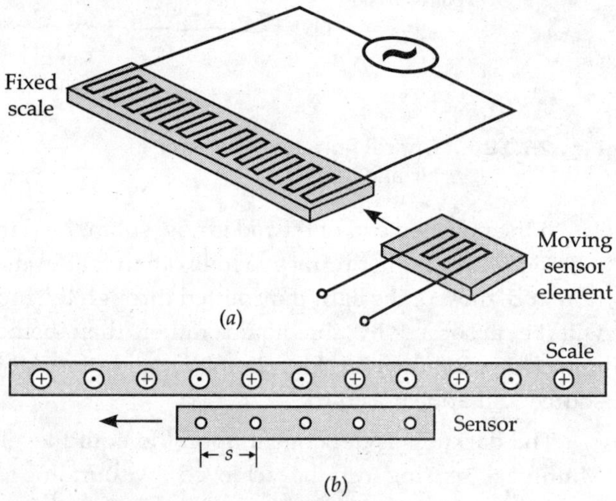


**Fig. 25.175** Inductosyn transducer.

$3\,s/2$ gives minimum again, and a travel of $2s$ gives a positive maximum. The output variation over $2s$ cycle length is essentially *consinusoidal*. A *coarse* digital output is obtained by counting the cycles of spacing $2s$, while fine resolution is obtained by electronically digitizing the analog voltage variation within each cycle. As in other incremental encoders, to detect *direction* of motion, the sensor element includes a second winding displaced $s/2$ from the first, providing a sinusoidal signal. With both a sine and cosine output available, the device behaves essentially as a resolver and can use resolver-to-digital type of electronics similar to that discussed earlier under synchros. Inductosyns are available in both translational (best accuracy ± 0.001 mm) and rotary (best accuracy ± 0.5 second of arc) forms. The standard spacings $s$ available are 1.25, 0.625 and 2 mm.

## 25.39 PHOTO OPTIC TRANSDUCERS

The *fibre optic transducers* can be classified into two categories as explained before in Art. 19.20.5 on page 599. This classification places the optic transducers as :

(*i*) intensity or intrinsic transducers and

(*ii*) interferometric or extrinsic transducers.

The following paragraphs describe how physical parameters can be measured with fibre optic transducers.

### 25.39.1 Temperature Transducers

Optical fibres can be used for measurement of temperature. This is because the temperature induces changes in refractive index of the fibre. The core and claddings of the fibres have different indices of refraction because of the difference in their composition. The refractive index is a function of both composition and temperature, and it changes at different rates in response to a temperature change. A higher temperature change causes the critical angle of the fibre to change slightly, which in turn changes the amount of light lost through leakage in the cladding as shown in Fig. 25.176. Total internal reflection occurs at angles of incidence greater than the critical angle. At angles of incidence less than the critical angle, some light is lost due to refraction.

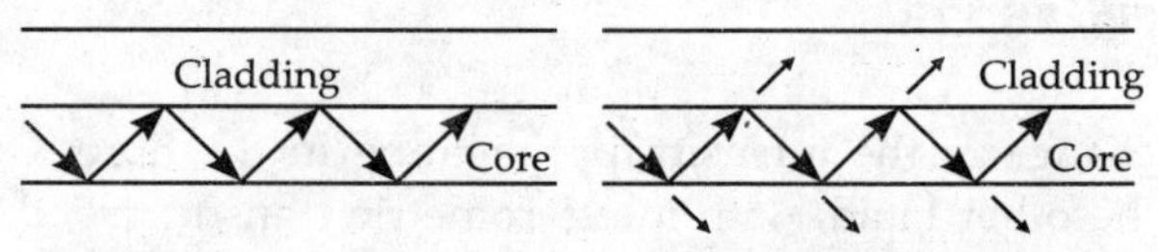


(*a*) Total internal reflection (*b*) Light leakage caused by changes in the refractive incidence of core and cladding

**Fig. 25.176** Fibre optic temperature transducer.

### 25.39.2 Measurement of Sound

The measurement of sound level can be done with either

(*i*) an intensity fibre optic transducer or with

(*ii*) an interferometric type transducer. In this case we have chosen an interferometric transducer as shown in Fig. 25.177.

Part of the beam passes through the reference coil, which is enclosed in stable environments and the remaining passes through the sensing coil. Sound

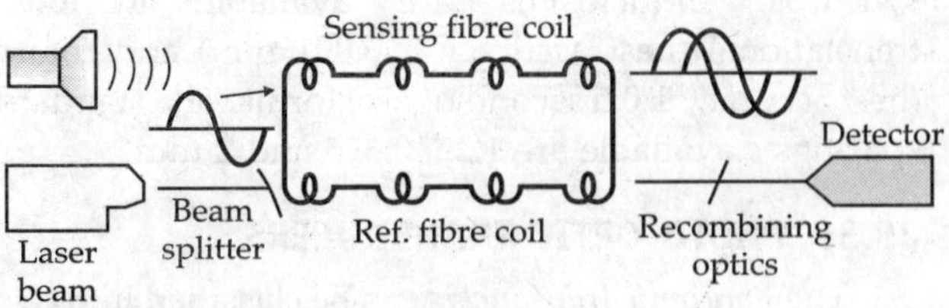


**Fig. 25.177** Interferometer-type optical fibre sound transducer.

waves distort the sensing fibre, which causes its optical length to change, thus altering the relative phases of the light in each fibre. The sound level is a function of the phase shift, which will be seen in the interference pattern when the two beams are recombined.

Figure 25.178 shows in details the operation of Intensity and interferometric transducers. With intensity senors, the measured variable causes a

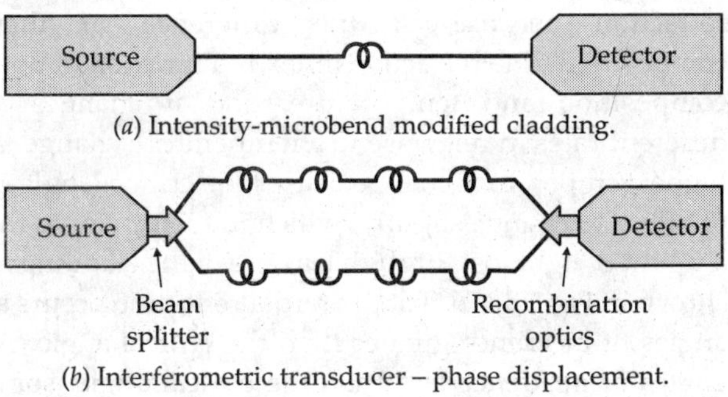


**Fig. 25.178**

change in the intensity propagating in the fibre. On the other hand, with interferometric transducers, the measured variable causes a phase change of the light propagating in the fibre. This beam is then compared with the reference beam as shown in Fig. 25.179 and the phase displacement owing to the sensed beam causes a change in intensity of interference pattern shown in Fig. 25.179.

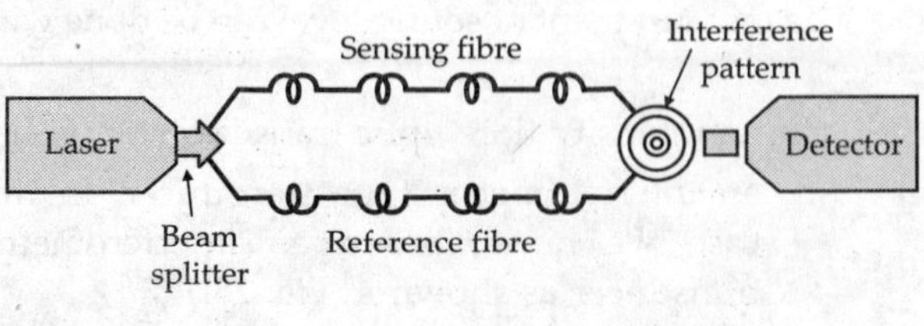


**Fig. 25.179** Change in intensity of the interference pattern caused by phase displacement.

### 25.35.8 Measurement of Level

Measurement of level is another application of fibre optic sensors.

The principle of sensing liquid level using fibre optic technology is straight forward. If one end of the cable or rod is cut and polished properly to form a prism and the prism is in air, light transmitted from the other end is reflected back with negligible loss. Under these conditions, the end near the light source will appear bright as shown in Fig. 25.180(*a*).

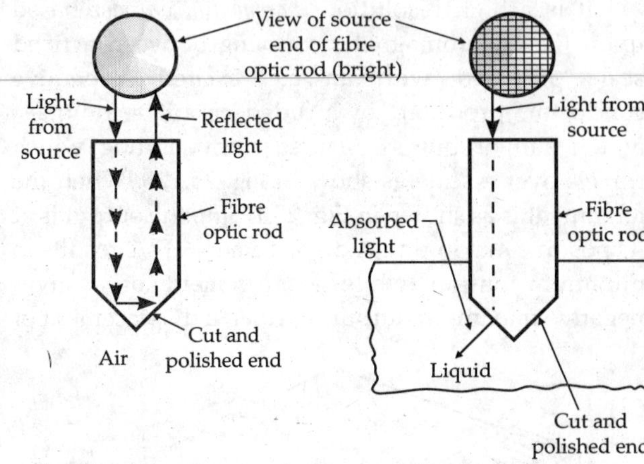


**Fig. 25.180** Optical fibre level detector in air and liquid.

If the polished end of the rod is now submerged in a liquid with a higher refractive index than that of the fibre rod, most of the light transmitted through the rod will be absorbed by the liquid rather than being reflected as shown in Fig. 25.180(*b*) and the light source will appear as dark.

The darkness is dependent upon the liquid level. Multilevel sensing can be achieved by bunding a number of fibre optic rods of different lengths as shown in Fig. 25.181.

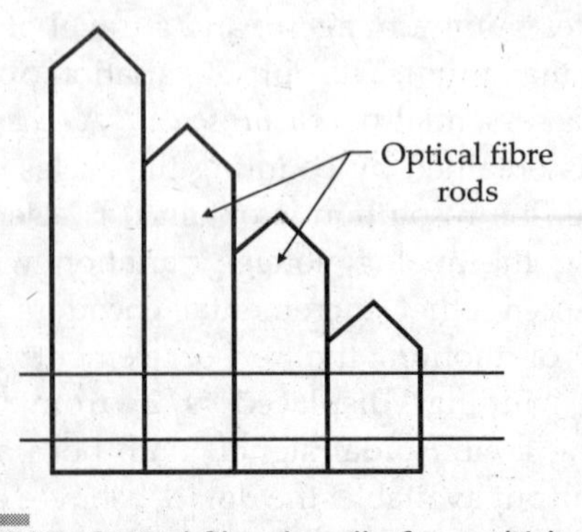


**Fig. 25.181** Optical fibre bundle for multi-level sensing.

# Review Questions

1. Describe with the help of neat diagrams, the mechanical devices used as primary detectors. Derive their output in terms of their physical dimensions.
2. Derive for a metallic diaphragm, expressions for deflection, maximum stress and natural frequency in terms of the physical dimensions, when it is subjected to a pressure on one side.
3. Describe with neat sketches the following types of primary detecting elements :
   (*i*) Bourdon tubes
   (*ii*) Bellows and
   (*iii*) Diaphragms.
4. Describe the methods for measurement of flow using (*i*) venturimeters and (*ii*) orifice plates. Derive the expressions for flow rate in each case.
5. Explain how a Pitot tube can be used for measurement of flow. Describe how the velocity profile of a liquid flowing in a pipe is taken care of in the measurements when using Pitot tubes.
6. Describe the different types of weirs and flumes are used for measurement of flow in open channels. Derive the related expressions.
7. Differentiate between the following using suitable examples :
   (*i*) control systems using feedback
   (*ii*) measurement systems using feedback.
8. Define the following :
   (*i*) Transducer
   (*ii*) Inverse transducer and
   (*iii*) Output transducer
   Give suitable examples.
9. Differentiate between the following citing suitable examples :
   (*i*) Transducers and Inverse transducers
   (*ii*) Active and passive transducers
   (*iii*) Primary and secondary transducers
   (*iv*) Analog and digital transducers
   (*v*) Output and inverse transducers
10. Describe the different criteria for selection of transducers for a particular application.
11. For a transducer, describe the following :
    (*i*) Input characteristics
    (*ii*) Transfer characteristics
    (*iii*) Output characteristics.
12. For a transducer, describe the following :
    (*i*) transfer function (*ii*) error.
13. Explain the following types of errors for a transducer :
    (*i*) Scale errors
    (*ii*) Dynamic errors and
    (*iii*) Noise and drift errors.
14. Define the following with reference to transducers :
    (*i*) Zero error,
    (*ii*) Sensitivity error,
    (*iii*) Non-conformity error,
    (*iv*) Hysteresis error,
    (*v*) Dynamic error, and
    (*vi*) Cross sensitivity.
15. Describe the constructional details of a resistance potential divider and derive the expression for its output voltage when connected across a meter of finite impedance.
16. Derive the expression for error for a resistance potential divider when connected across a load of finite resistance. Draw typical curves to shown the variation of error with input displacement for different values of load resistance. Derive the condition for maximum error.
17. Explain that sensitivity land linearity are two conflicting requirements in a resistance potential divider.
18. Explain the construction of wire wound strain gauges and derive the expression for the gauge factor.
19. (*a*) Describe the construction of foil type strain gauges and explain their advantages over wire wound strain gauges.
    (*b*) Explain the construction of semi-conductor strain gauges and explain their advantages and disadvantages.
20. Explain with diagrams, the bonded and unbonded types of strain gauges.
    For bonded strain gauges, describe the materials used for base, and adhesive materials and also the materials used for leads.
21. Describe the working and construction of resistance thermometers. Describe the materials used for RTDs, along with their properties. Sketch their typical characteristics.
22. What are thermistors ? Explain their different forms of construction. Draw their resistivity versus

temperature characteristics and show that they have a very high value of sensitivity as compared with that of metal resistance thermometers. Describe their applications.

**23.** Draw and describe the following for thermistors :
(*i*) Resistance-temperature characteristics
(*ii*) Voltage-current characteristics, and
(*iii*) Current-time characteristics.

**24.** Describe the construction, theory and working of thermocouples. Describe the different types of compensations used and also the methods of measurement of their output voltage.

**25.** Describe the method for measurement of temperature with use of
(*i*) RTDs (*ii*) Thermistors, and
(*iii*) IC sensors.
Describe their advantages and limitations.

**26.** Give an overview of the inductive transducers explaining their principle of operation like variation of number of turns, geometric configuration and permeability. Draw neat sketches to show the above effects.

**27.** How is a differential output taken from an inductive transducer ? Explain the advantages when inductive transducers are used in push-pull configuration.

**28.** Explain the construction and principle of working of a linear voltage differential transformer (L.V.D.T). Explain how the magnitude and direction of the displacement of core of an L.V.D.T. detected ? Why is the frequency of excitation of primary winding kept very high as compared to the frequency of the signal being detected ?

**29.** Explain the different principles of working of capacitive transducers.

**30.** Explain how by using a differential arrangement a capacitive transducer which works on the principle of variation of capacitance with displacement between two plates the response can be made linear.

**31.** Describe the properties of materials used for piezo-electric transducers. Derive expressions for voltage and charge sensitivities.

**32.** Describe the different modes of operation of piezoelectric transducers. Define and sketch binders and twisters. Explain the applications of piezo-electric transducers.

**33.** Draw the equivalent circuit of piezo-electric transducers. Derive the expression for magnitude of voltage across the load by making simplifying assumptions. Prove that for medium and high frequencies the magnitude of the voltage across the load is independent of frequency.

**34.** Derive the expressions for frequency response characteristics of piezo-electric transducers.

**35.** Derive the expression for impulse response of piezo-electric transducers. Sketch the response curves.

**36.** Describe the construction. principle of working and applications of Hall Effect transducers.

**37.** Explain the constructional features of
(*i*) synchro-transmitters and
(*ii*) synchro transformers.
Explain how the above pair can be used as an error detector in a servomechanism.

**38.** Derive the expressions for voltages in the three phases of a synchro-transmitter. Derive also the equations for voltages between line and the neutral point and explain the concept of 'zero' in a synchro-transmitter and a synchro-transformer.

**39.** Explain how a synehro-transmitter and receiver pair can be used as transducers in the torque transmission mode.

**40.** Explain the construction, working and applications of resolvers.

**41.** Explain the principle of working, constructional details and applications of light dependent resistors.

**42.** Explain the principle of working, constructional details and applications of photo-diodes. Draw the characteristics.

**43.** Explain the principle of working constructional details and applications of photo-transistors. Explain why are they preferred for digital applications. Draw their characteristics.

**44.** Explain the principle of working and constructional details of a photo-voltaic cell. Explain, why is it very useful for space applications.

**45.** Define a digital transducer and explain its applications for measurement of linear and angular displacements.

**46.** Describe the following :
(*i*) Tachometer Encoder (*ii*) Incremental Encoder
(*iii*) Absolute Encoder.

**47.** Describe in details the contact type of encoders and explain why Gray code is used in them.

**48.** Describe the following types of shaft encoders :
(*i*) resistive type (*ii*) optical type.

**49.** Describe the method of using Moire Fringes for measurement of linear and rotational displacements in encoders. Describe its advantages and limitations.

**50.** Describe how fibre optic transducers are used for :
(*i*) measurement of temperature
(*ii*) measurement of sound level
(*iii*) measurement of liquid level.

# Unsolved Problems

1. A steel diaphragm is used for pressure measurement. It is 50 mm in diameter and is to be designed to measure a maximum pressure of 1.5 $MN/m^2$. The modulus of elasticity of steel is 200 $GN/m^2$ and Poisson's ratio is 0.3. Calculate the thickness of diaphragm in order that the maximum deflection is not more than 1/3 of its thickness. Calculate the natural frequency of diaphragm. The density of steel is 7800 $kg/m^3$. **[Ans.** 1.1 mm, 26,600 Hz]

2. A torsion bar is made of mild steel and has a diameter of 25 mm. The shear modulus of mild steel is 80 $GN/m^2$. Calculate the value of shear sensitivity. Suppose a torque of $10 \times 10^3$ Nm is applied to the bar, calculate the value of shear strain angle. If the bar is notched to 15 mm diameter, what is the value of shear sensitivity and the shear angle for the above torque ?
**[Ans.** $4 \times 10^{-6}$ rad/Nm, $18.8 \times 10^{-6}$ rad/Nm, 0.188 rad]

3. A vertical venturimeter has its inlet and throat diameters as 20 mm and 100 mm respectively. A differential mercury manometer connected to the inlet and throat points gives a reading of 250 mm. Determine the rate of flow of water if the co-efficient of discharge is 0.98.
**[Ans.** 0.0628 $m^3/s$]

4. A venturi tube of throat diameter 60 mm is placed in a water pipe of diameter 100 mm to measure volumetric flow. The volumetric flow rate through the tube is $80 \times 10^{-3}\ m^3/s$. Water has a density of $10^3\ kg/m^2$ and a viscosity of $10^{-3}\ Ns/m^2$.
   (*a*) Determine Reynold's number.
   (*b*) Given that the co-efficient of discharge is 0.99, determine the upstream and throat differential pressure.
   (*c*) The differential pressure is applied at the centre of a mild steel diaphram having the following data :
   diameter = 10 mm, thickness = 0.2 mm
   modulus of elasticity = 206 $GN/m^2$ and Poisson's ratio = 0.28.

   Calculate the deflection of the centre of diaphragm.
   **[Ans.** $1698 \times 10^3$, 356.1 $kN/m^2$, $23.3 \times 10^{-6}$ mm]

5. An orifice plate is situated in a water pipe of 50 mm diameter and has an orifice of 30 mm diameter. For a Reynold's number of 1,00,000 the discharge co-efficient is 0.61. Calculate this nominal exit velocity $V_2$ at the orifice to give this flow value and also the pressure differential. The value of density is 1000 $kg/m^3$ and the kinematic viscosity is $10^{-6}\ m^2/s$. **[Ans.** 3.33 m/s, 130 $kN/m^2$]

6. The density of air in a duct is 1.2 $kg/m^3$ and its maximum velocity is 20 m/s. The duct is 100 mm in diameter and the velocity is to be measured using a 80 mm diameter orifice and a differential manometer. If the co-efficient of discharge is assumed to be 0.6, calculate the necessary range of the manometer in mm of $H_2O$. **[Ans.** 100 mm of $H_2O$]

7. A Pitot tube is to be used to measure the speed of an aeroplane. The pressure differential is measured by a U tube manometer using water. The density of air is 1.22 $kg/m^3$. If the reading of the manometer is 80 mm of water and the velocity correction factor is 0.98, calculate the velocity of the aeroplane. Assume that air is incompressible. **[Ans.** 35.15 m/s]

8. A Pitot tube with a thermometer attached is used to measure the velocity of air in a duct. A differential pressure of 100 mm of water is recorded. The temperature is 27°C. The static pressure is 10 $kN/m^2$. The barometer reading is 760 mm of mercury. The gas constant is 287 J/kg and the specific gravity of mercury is 13.6. The velocity correction factor is 0.98. Calculate the mean velocity of air. **[Ans.** 38.2 m/s]

9. A submarine moves horizontally in sea and has its axis 15 m below the surface of water. A Pitot tube properly placed in front of submarine and along its axis, is connected to the two limbs of a U tube containing mercury. The difference of mercury level in the two limbs is 170 mm. Find the speed of submarine. Assume that sea water is 1.26 times as dense as compared to normal water. The specific gravity of mercury is 13.6. **[Ans.** 6.4 m/s]

10. Water flows over a rectangular notch 1 m wide over a depth of 0.15 m and afterwards passes through a triangular right angled notch. Find the depth of water through the latter notch. The co-efficient for discharge are 0.62 and 0.59 respectively for rectangular and triangular notches. **[Ans.** 0.357 m]

11. A 5000 Ω voltage dividing potentiometer feeds a load of 8000 Ω. Calculate the percentage error based upon full scale for slider positions at 0, 0.2, 0.4, 0.6, 0.8 and 1.0 p.u. of the total travel. Plot the graph between p.u. travel and percentage error.
**[Ans.** Percentage error : 0, 1.8. 5.2. 7.8. 7.2. 0]

12. A potentiometer has a resistance of 5000 Ω and is rated at 3 W. What is the maximum allowable excitation voltage ? Calculate the value of the sensitivity and resolution if the length of potentiometer is 0.1 m and there are 200 turns. Also calculate the percentage leading error at 0.67 of the travel if a meter of 5000 Ω is connected across the pot. **[Ans.** 122 V, 12.2 V/mm, 0.61 V, – 12.1%]

13. A voltage dividing potentiometer is used to measure an angular displacement. The angle of displacement is 60° and total angle of travel of potentiometer is 355°. Calculate the voltage output on open circuit if the potentiometer is excited by a 60 V source. Calculate the actual value of the output voltage at this setting if a voltmeter of 1 MΩ resistance is connected across the output. The resistance of the potentiometer is 1 kΩ. The turns are uniformly distributed. Calculate the percentage error. **[Ans.** 10.14 V, 10.1386 V, – 0.014%]

14. A resistive displacement transducer with a shaft stroke of 25 mm is applied to the circuit shown in Fig. 25.182. The applied voltage is 10 V. What is the displacement indicated for each of the voltage readings 3.0 V, 5.0 V, 8.0 V ? Assume that the resistance of output device $R_m$ is infinite.

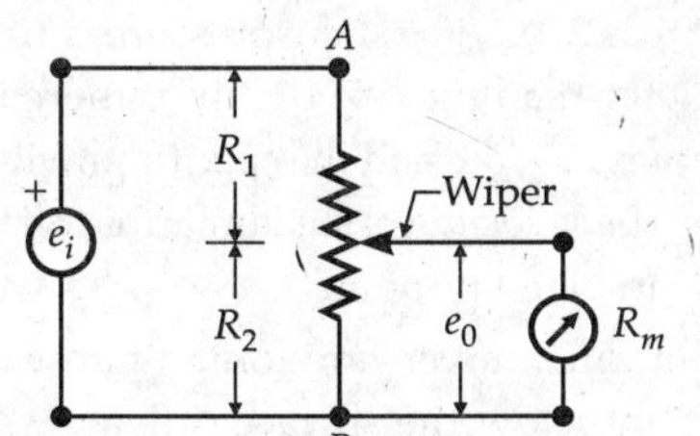


**Fig. 25.182** Resistance potentiometer.

**[Ans.** 7.5 mm, 12.5 mm, 20 mm]

15. The circuit shown in Fig. 25.182 is used by a resistive displacement transducer. The input voltage is 5 V. If the resistance of potentiometer ($R_1 + R_2$) is 1000 Ω and the wiper is set so that $R_2$ = 300 Ω, calculate the value of output voltage. Assume the resistance, $R_m$ to be infinity. **[Ans.** 1.5 V]

16. A resistive displacement transducer with a shaft stroke of 50 mm is used in the circuit shown in Fig. 25.182. The total resistance of potentiometer $R_1 + R_2$ = 400 Ω and the input voltage is 4 V. The wiper is 37.5 mm from *B* (*a*). What is the value of $R_2$ ? (*b*) What is the value of output voltage ? Assume the value of $R_m$ to be infinity. **[Ans.** 300 Ω, 3 V]

17. In a resistive potentiometer the maximum percentage error is 15 $R_p / R_m$ where $R_p$ and $R_m$ are respectively the resistance of potentiometer the load respectively. A position measurement may have a maximum non-linearity of 0.5 percent when driving a load of 10 kΩ resistance. Find out the maximum value of resistance of potentiometer.

**[Ans.** 333 Ω]

18. A resistive potentiometer is rated as 200 Ω, 1 W (derate at 10 mW/°C above 60°C) 30°C/W thermal resistance. It is used with 10 V supply at 70°C ambient temperature. Calculate,

(*i*) the power dissipated by potentiometer

(*ii*) actual temperature of potentiometer

(*iii*) power dissipation allowed.

Can the potentiometer be used safely under the above conditions ?

**[Ans.** 500 mW, 85°C, 750 mW, yes]

19. Find the strain that results from a tensile force of 1000 N applied to a 10 m long aluminium bar having cross-sectional area of $4 \times 10^{-4}$ m$^2$. The modulus of elasticity of aluminium is 69 GN/m$^2$.

**[Ans.** 38.23 microstrain]

20. A resistance wire strain gauge having a nominal resistance of 350 Ω is subjected to a strain of 500 microstrain. Find the change in the value of resistance neglecting the piezo-resistive effect.

**[Ans.** 0.35 Ω]

21. Compare the resistance change produced by a strain of 150 microstrain if a strain gauge of nominal resistance of 120 Ω is used (*i*) when made of wire resistance having a gauge factor of 2.13 and (*ii*) when made of a semiconducting material having a gauge factor of 151.

**[Ans.** (*i*) 0.03834 Ω, (*ii*) 2.718 Ω]

22. A strain gauge has a gauge factor of 4. If the strain gauge is attached to a metal bar that stretches from 0.25 m to 0.255 m when strained, what is the percentage change in resistance ? If the unstrained value of gauge is 120 Ω, what is the resistance value of gauge after application of strain ?

**[Ans.** 8%, 129.6 Ω]

23. A metallic strain gauge has a resistance of 120 Ω and a gauge factor of 2. It is installed on an aluminium structure which has a yield point stress of 0.2 GN/m$^2$ and Young's modulus of 68.7 GN/m$^2$, determine the change in resistance of the gauge that would be caused by loading the material to yield point. **[Ans.** 0.7 Ω]

24. A semiconductor gauge having a resistance of 1000 Ω and a gauge factor of – 133 is subjected to a compressive strain of 500 microstrain. Calculate the new value of resistance of the gauge.

**[Ans.** 1066.5 Ω]

**25.** A strain gauge having a gauge factor of 4 is used for testing a machine. If the gauge resistance is 10 Ω, and the strain is $20 \times 10^{-6}$, how much will the resistance of strain gauge change ?

**[Ans.** $8 \times 10^{-3}$ Ω]

**26.** A strain gauge has a resistance of 120 Ω unstrained and the gauge factor is – 12. What is the resistance value if the strain is 1%. **[Ans.** 105.6 Ω]

**27.** A strain gauge with a gauge factor of 4 has a resistance of 500 Ω. It is to be used in a test in which the strain to be measured may be as low as $5 \times 10^{-6}$. What will the change in resistance of gauge be ?

**[Ans.** $10 \times 10^{-3}$ Ω]

**28.** A strain gauge is bonded to a steel beam 0.25 m long and has a cross-sectional area of $0.4 \times 10^{-3}$ m$^2$. Young's modulus of elasticity for steel is 207 GN/m$^2$. The strain gauge has a unstrained resistance of 240 Ω and a gauge factor of 2.20. When the load is applied, the gauge's resistance changes by 0.013 Ω. Calculate the change in length of the steel beam and the amount of force applied to the beam. **[Ans.** $2.46 \times 10^{-6}$ m, $2.037 \times 10^3$ N]

**29.** A platinum resistance thermometer has a resistance of 150 Ω at 0°C. What is the resistance when the temperature is 200°C ? When the thermometer has a resistance of 400 Ω, what is the value of the temperature ? The resistance temperature co-efficient of plantinum is 0.0039/°C.

**[Ans.** 170.25 Ω, 427°C]

**30.** The following table gives the variation of resistance with temperature for a RTD.

| *Temperature* °C | *Resistence* Ω |
|---|---|
| 15 | 106.06 |
| 18 | 167.14 |
| 21 | 108.22 |
| 24 | 109.3 |
| 26.5 | 110.38 |
| 29.5 | 111.46 |
| 33 | 112.75 |

Find the linear and quadratic approximation of the above resistance temperature curve for temperature variations between 15°C and 33°C about a mean temperature of 24°C.

**[Ans.** $R_\theta = 109.3\,[1 + 3.41 \times 10^{-3}\,(\theta - 24)]$,
$R_\theta = 109.3\,[1 + 3.55 \times 10^{-9}\,(\theta - 24.3) - 0.66 \times 10^{-6}\,(0 - 24)^2]$]

**31.** The resistance temperature relationship for a thermistor is given by :

$$R_{T2} = R_{T1} \exp \beta\,(1/T_2 - 1/T_1)$$

If the value of β is 4000 °K and the resistance of thermistor is 200 kΩ at – 100°C, find the value of resistance at 400°C. Find the ratio of two resistances. Find the ratio of two resistances for platinum over the same temperature range. Platinum has a resistance temperature co-efficient of 0.0039/°C. **[Ans.** $6.93 \times 10^{-3}$ Ω, $28.8 \times 10^6$, 2.95]

**32.** A thermistor may be assumed to have a linear temperature resistance over a limited temperature range. The variation resistance is – 05°C rise of temperature. The thermistor has a resistance of 1000 Ω at 20°C. Calculate the value of its resistance at 25°C. Supposing this thermistor is used in series with a copper coil, what is the value of the resistance of copper coil if the resistance at 20°C and 25°C is the same for the series connected circuit comprising of coil and the thermistor ? The resistance temperature co-efficient of copper may be assumed as 0.004°C. **[Ans.** 750 Ω, 12.5 kΩ]

**33.** An LVDT has an output of 6 V rms when the displacement is $0.4 \times 10^{-3}$ mm. Determine the sensitivity of this instrument in V/mm. A 10 V voltmeter with 100 scale divisions is used to read the output. Two-tenths of a division can be estimated with ease. Determine the resolution of the voltmeter.

The above arrangement is used in a pressure transducer for measuring the deflection of a diaphragm. The diaphragm is deflected through $0.5 \times 10^{-3}$ mm by a pressure of 1000 N/m$^2$. Determine the sensitivity and resolution of this instrument.

**[Ans.** 15000 V/rmn, 0.02 V, 7.5 m V/N-m$^{-2}$, 2.67 Nm]

**34.** An LVDT with a secondary voltage of 5 V has a range of ± 25 mm.

(*a*) Find the output voltage when the core is – 18.75 mm for the centre

(*b*) Plot the output voltage versus core position for a core movement going from + 18.75 mm to – 10 mm.

**[Ans.** (*a*) – 3.75 V (*b*) 3.75 V to – 2 V linearly]

**35.** An LVDT with a secondary voltage of 5 V has a range of ± 25 mm.

(*a*) Find the core movement from centre if the output voltage is – 3 V

(*b*) Plot the core positions versus output voltages varying from + 3 V to – 4.5 V.

**[Ans.** (*a*) – 15 mm (*b*) 15 mm to – 22.5 mm linearly]

25

36. A parallel plate capacitive transducer has plates of 600 $mm^2$ area which are separated by air by a distance of 0.2 mm. The resistance of the transducer is $20 \times 10^6\ \Omega$. Calculate the time constant of the transducer and find the attenuation of the output at 1000 Hz. The permittivity of air is $8.85 \times 10^{-12}$ F/m. **[Ans.** 0.531 ms, $4.2^{-2}$]

37. Two plates of parallel plate capacitive transducer are 30 mm apart and the space is filled with two different dielectric materials, one material is 1 cm thick with a dielectric constant of 5 and the other material is 20 mm thick with a dielectric constant 10. If the capacitive transducer were to be made up of a single dielectric material ? What is the dielectric constant of that material ? **[Ans.** 7.5]

38. A parallel plate capacitive transducer uses air as the dielectric medium for measurement of displacement as shown in Fig. 25.114. The plates are 300 $mm^2$ in area and are separated by a distance of 1.25 mm. Find the time constant an 1 series resistance if the output is attenuated by 5% by a frequency of 30 Hz. The permittivity of air is $8.85 \times 10^{-12}$ F/m. **[Ans.** 16.1 ms, 7580 MΩ]

39. In a variable capacitance transducer the diaphragms are 20 mm in diameter and 4 mm apart. If a pressure produces an average deflection of 0.25 mm, calculate the value of capacitance after the application of force. The capacitance before application of force is 400 pF. **[Ans.** 427 pF]

40. A displacement capacitive transducer uses a differential arrangement with two outer plates which are fixed and a central plate which is movable. The distance between fixed and movable plates is 5 mm when no displacement is applied. A voltage of 1000 V rms is applied across the fixed plates. Find the differential output voltage if a displacement of 0.01 mm is applied to the central plate. Find also the sensitivity of the transducer: **[Ans.** 2 V ; 200 V/mm]

41. For vibration measurement a d.c. voltage of 90 V is applied across the series combination of a parallel plate air capacitor and a C.R.O. The parallel plate air capacitor has plates of 625 $mm^2$ area displaced by 0.25 mm. The oscilloscope may be considered as a resistance of 10 MΩ in parallel with a capacitance of 10 pF. The sensitivity of the oscilloscope is 1 mV/cm. Determine the displacement sensitivity and the frequency for which a flat response within 5% can be obtained. **[Ans.** $3.6 \times 10^6$ ; 22 kHz]

42. Air exhibits a breakdown strength of about 3 kV/mm. What limitation does this impose on the ultimate sensitivity of a variable displacement capacitive transducer ?

43. The voltage sensitivity for barium titanate is $12 \times 10^{-2}$ Vm / N Vm/N while that of quartz is $50 \times 10^{-2}$ Vm / N. The permittivity of barium titanate is $12.5 \times 10^{-9}$ F / m and that of quartz is $40.6 \times 10^{-12}$ F / m. Calculate the charge sensitivities and their ratio. If the thickness of crystal in each case is 2.5 mm, calculate the sensitivities in each case and their ratio.
**[Ans.** 150 p C/N, 2 pC/N, 75 ; $30 \times 10^{-6}$ V/N-$m^{-2}$, $125 \times 10^{-6}$ V/N-$m^{-2}$, 0.24]

44. A quartz crystal has the dimensions of 2 mm × 2 mm × 1 mm. Quartz has the following properties charge sensitivity = 21C / N. Young's modulus = $8.6 \times 10^{10}$ N/$m^2$.
Permittivity = $40.6 \times 10^{-12}$ F / m.
Calculate the value of force, charge and voltage if the crystal is subjected to a strain of $10 \times 10^{-6}$ m / m. **[Ans.** 34 N, 6.88 pC, 43 V]

45. A barium titanate crystal has dimensions of 5 mm × 5 mm × 1.5 mm. The Young's modulus of barium titanate is $12 \times 10^6$ N / $m^2$, its charge sensitivity is 150 pC/N and permittivity $12.5 \times 10^{-9}$ F/m. A capacitance of 10 pF in parallel with 100 MΩ resistance is connected across the crystal. Calculate the rms value of voltage under open circuit and load conditions when a force of 0.0142 sin 100 $t$ newton is applied to the crystal. Calculate the rms value of deflection. The resistance of the crystal may be neglected.
**[Ans.** 72 mV, 62.4 mV, 0.05 mm]

46. A flat frequency response within 10% is required from a piezo-electric crystal. Find the value of minimum frequency for which it can be used if the time constant is 2 ms. Find also the phase shift. **[Ans.** 164 Hz, 26°]

47. A piezo electric transducer has an input
$$x_i = A_t \quad 0 < t < T, \quad x_i = 0 \quad T < t < \infty.$$
Solve the differential equation to find the value of output voltage. For $= t = T$, find the error. Calculate this error by using the following approximation.
$$\exp - T/\tau = 1 - \frac{T}{\tau} + \frac{1}{2}\left(\frac{T}{\tau}\right)^2 + \ldots$$
Explain the error as a percentage of the ideal value of output voltage. What must be the value of $T/\tau$ if the error is to be 5 percent ? For this value of $T/\tau$ evaluate the error caused by truncating the series.

48. The hot junction of a chromel alumel thermocouple is connected to a potentiometer terminals are at

24°C. The potentiometer whose terminals are at 24°C reads 25.76 mV. What is the temperature of the thermocouple junction ?

The calibration chart of the thermocouple is :

| Temperature °C | Voltage, mV |
|---|---|
| 20 | 0.8 |
| 24 | 0.95 |
| 28 | 1.12 |
| ... | ... |
| 480 | 26.25 |
| 488 | 26.72 |
| 493 | 26.04 |

[**Ans.** 488°C]

**49.** A thermopile arrangement of the copper constantan thermocouple consists of three junction pairs and has the reference junction at 200°C. If the output voltage is 3.3 mV, determine the temperature of the detecting junction.

The calibration chart for a thermocouple junction is :

| Temperature °C | Voltage, mV |
|---|---|
| 100 | 4.22 |
| 200 | 9.23 |
| 250 | 11.95 |

[**Ans.** 171.95°C]

**50.** A PMMC instrument has a resistance of 5 Ω and is calibrated to read correctly when the resistance of the external circuit is 2 Ω. Calculate the percentage error if the resistance of the circuit is changed to 4 Ω. What would have been the error if the PMMC instrument had a resistance of 500 Ω ?

[**Ans.** 22.2%. 0.397%]

**51.** The emf of an iron constantan thermocouple is to be measured by the potentiometer shown in Fig. 25.80. A potential difference of 1.0186 V is applied over points *A* and *B* and is standardized using a Weston cell. The current through the resistor is to be 2 mA, and the range of temperature measurement is to be from 250°C to 850°C. Calculate the value of resistances $R_1$, $R_2$ and $R_s$, for an ambient temperature of 20°C. The emf of the iron constantan couple with reference junction at 0°C is 1.019 mV at 20°, 13.553 mV at 250°C and 48.716 mV at 850°C.

[**Ans.** $R_1 = 6.27\ \Omega$, $R_2 = 485.5\ \Omega$, $R_s = 17.58\ \Omega$]

**52.** A germanium crystal having dimensions of $5 \times 5 \times 2$ mm is used for measurement of flux density. Hall effect co-efficient is $-8 \times 10^{-13}$ V - m / A Wb m$^{-2}$. If the current density in the crystal is 0.3 A/mm$^2$ and the voltage developed is $-0.36$ V, find the value of flux density.

[**Ans.** 0.012 Wb/m$^2$]

**53.** The circuit of Fig. 25.144(*a*) uses a supply voltage of 20 V d.c. A current of 5 mA flows when the cell is illuminated with 400 lm/m$^2$ and de-energized when the cell is dark. Find (*a*) the value of $R_1$ and (*b*) dark current. The characteristics of the photocell are shown in Fig. 25.144(*b*).

[**Ans.** (*a*) 3 kΩ (*b*) 0.194 mA]

**54.** The circuit of a photodiode is shown in Fig. 25.149 and its characteristics curves with load line are shown in Fig. 25.148. At an illumination level of 5000 lm/m$^2$, find

(*a*) the diode current

(*b*) voltage drop across resistance $R_1$ and

(*c*) diode voltage drop.

[**Ans.** (*a*) 0.9 mA, (*b*) 0.18 V, (*c*) 0.32 V]

**55.** The circuit and characteristic curves of a phototransistor are shown in Fig. 25.155. At an illumination level of 300 W/m$^2$, find (*a*) $V_{CE}$, (*b*) $V_{RL}$ (*c*) $I_C$.

[**Ans.** (*a*) 8.5 V (*b*) 11.5 V (*c*) 5.6 mA]

25

# Objective Type Questions

***Tick (√) the most appropriate answer :***

1. A transducer converts
   (*a*) mechanical energy into electrical energy
   (*b*) mechanical displacement into electrical signal
   (*c*) one form of energy into another form of energy
   (*d*) electrical energy into mechanical form.

2. An Inverse transducer converts
   (*a*) electrical energy to any other form of energy
   (*b*) electrical energy to light energy
   (*c*) mechanical displacement into electrical signal
   (*d*) electrical energy to mechanical form.

3. One of the following can act as an Inverse transducer
   (*a*) electrical resistance potentiometer
   (*b*) L.V.D.T.
   (*c*) capacitive transducer
   (*d*) piezo electric crystals.

4. One of the following is an active transducer
   (*a*) strain gauge (*b*) selsyn
   (*c*) photovoltaic cell (*d*) photo-emissive cell.

5. A strip chart recorder is
   (*a*) an active transducer (*b*) an inverse transducer
   (*c*) an output transducer (*d*) (*b*) and (*c*).

6. In a transducer, the experimentally obtained transfer function is different from the theoretical transfer function, the errors result from this difference are called
   (*a*) Zero errors (*b*) Sensitivity errors
   (*c*) Non-conformity errors
   (*d*) Dynamic errors.

7. In a transducer, the observed output deviates from the correct value by a constant factor the resulting error is called
   (*a*) Zero Error (*b*) Sensitivity Error
   (*c*) Non-conformity Error
   (*d*) Hysteresis Error.

8. A transducer has an output impedance of 1 kΩ and the load resistance is 1 MΩ, the transducer behaves as
   (*a*) a constant current source
   (*b*) a constant voltage source
   (*c*) a constant power source
   (*d*) none of the above.

9. If a transducer has an output impedance of 1 Ω and a load resistance of 1 kΩ, it behaves as
   (*a*) a constant current source
   (*b*) a constant voltage source
   (*c*) a constant impedance source
   (*d*) none of the above.

10. The lower limit of useful working range of a transducer is determined by
   (*a*) minimum useful input level
   (*b*) by transducer error and noise
   (*c*) cross sensitivity
   (*d*) dynamic response.

11. While selecting a transducer for a particular application
   (*a*) only the input characteristics should be considered
   (*b*) only the output characteristics should be considered
   (*c*) only the transfer characteristics should be considered
   (*d*) input, output and transfer characteristics should be considered.

12. A cantilever is used as a primary sensing element for measurement of force. The measured deflection 0.5 mm for a particular set up. Suppose each of the dimensions *i.e.*, length, width and thickness of the cantilever of the cantilever are doubled with the force and material of cantilever remaining the same, the deflection is
   (*a*) 1 mm (*b*) 2 mm
   (*c*) 0.251 mm (*d*) 0.125 mm.

13. A closed coil spring has a stiffness of $20 \times 10^{-6}$ N/m. Suppose the diameter of the wire is doubled, its outer diameter is made two times and its number of turns are made 2 times, the stiffness of the new spring will be
   (*a*) $20 \times 10^{-6}$ N/m (*b*) $80 \times 10^{-6}$ N/m
   (*c*) $10 \times 10^{-6}$ N/m (*d*) $160 \times 10^{-6}$ N/m.

14. A diaphragm has a natural frequency of 30 kHz. If both its diameter and thickness are halved, the natural frequency is
   (*a*) 15 kHz (*b*) 240 kHz
   (*c*) 60 kHz (*d*) 120 kHz.

15. A pressure of 256 kN/m$^2$ acting on a diaphragm produces a deflection of 0.2 mm at the centre. What pressure would produce the same deflection if the diameter is 2 times the earlier one and the thickness is $\frac{1}{2}$ of the earlier one
   (*a*) 2 kN/m$^2$ (*b*) 128 kN/m$^2$
   (*c*) 64 kN/m$^2$ (*d*) 0.5 kN/m$^2$.

16. A pressure applied to a bellows element produces a linear displacement of 2 mm. Suppose the number of bellows elements is made 3 times and thickness of the bellows element is halved, what would be the displacement of the element for the same applied pressure ?
   (*a*) 24 mm (*b*) 12 mm
   (*c*) 48 mm (*d*) 3 mm.

17. A torsion bar of 30 mm diameter produces a twist of 1 minute when torque of 200 Nm is applied. Supposing the diameter of the torsion bar is reduced to 15 mm, what would be the angle of twist for the same torque ?
   (*a*) 4 minute (*b*) 8 minute
   (*c*) 2 minute (*d*) 0.125 minute.

18. A Reynold's number of 1000 indicates
(*a*) turbulent flow (*b*) laminar flow
(*c*) a flow which can either be turbulent or laminar
(*d*) none of the above.

19. In a venturi meter the flow is 0.15 $m^3/s$ when the differential pressure is 30 $kN/m^2$. What is the value of flow when the differential pressure is 60 $kN/m^2$.
(*a*) 0.212 $m^3/s$ (*b*) 0.106 $m^3/s$
(*c*) 0.3 $m^3/s$ (*d*) 0.075 $m^3/s$.

20. Pitot tubes are useful
(*a*) for industrial applications
(*b*) for laboratory experimental purposes
(*c*) for both industrial and laboratory applications
(*d*) none of the above.

21. Water is flowing with a depth of 0.9 m over a rectangular notch 1.2 m wide. Find the depth of flow over a right angled notch if the same were to pass over it. Assume that both the notches have the same co-efficient of discharge
(*a*) 0.91 n (*b*) 1.1 m
(*c*) 2.2 m (*d*) 0.45 m.

22. A resistance potentiometer is a
(*a*) first order instrument
(*b*) zero order instrument
(*c*) second order instrument
(*d*) none of the above.

23. In a resistance potentiometer, the non-linearity
(*a*) increases with increase of load to potentiometer resistance
(*b*) decreases with increase of load to potentiometer resistance
(*c*) is not dependent upon load to potentiometer resistance
(*d*) none of the above.

24. In a resistance potentiometer high value of resistance of POT leads to
(*a*) high value of sensitivity
(*b*) low value of sensitivity
(*c*) low value of non-linearity
(*d*) low value of error.

25. A 100 Ω resistive potentiometer is used with an input supply voltage of 10 V. The temperature of the POT is of the thermal resistance 30°C/W and the ambient temperature is 40°C
(*a*) 60°C (*b*) 80°C
(*c*) 70°C (*d*) 100°C.

26. A helipot is used for measurement of linear displacement of 0.1 m. It uses a 10 turn potentiometer (3000 of movement full scale) with a total of 5000 turns of wire. Its resolution is
(*a*) 0.1 mm (*b*) 0.2 mm
(*c*) 0.01 mm (*d*) 0.02 mm.

27. In wire wound strain gauges, the change in resistance on application of strain is mainly due to
(*a*) change in length of wire
(*b*) change in diameter of wire
(*c*) change in both length and diameter of wire
(*d*) change in resistivity.

28. Unbonded strain gauges are
(*a*) exclusively used for transducer applications
(*b*) exclusively used for stress analysis
(*c*) commonly used for both transducer applications as well as for stress analysis
(*d*) none of the above.

29. The gauge factor is defined as
(*a*) $\frac{\Delta L/L}{\Delta R/R}$ (*b*) $\frac{\Delta R/R}{\Delta L/L}$
(*c*) $\frac{\Delta R/R}{\Delta D/D}$ (*d*) $\frac{\Delta R/R}{\Delta \rho/\rho}$

where *L*, *D*, ρ and *R* are respectively the length, diameter, resistivity and resistance of strain gauge.

30. For wire wound strain gauges the approximate value of gauge factor is
(*a*) $1 + 2\Delta\rho/\rho$ (*b*) $1 + 2\Delta R/R$
(*c*) $1 + 2\nu$ (*d*) $1 + 2\Delta D/D$.

where ρ, *R*, ν and *D* are respectively resistivity, resistance, poisson's ratio and diameter.

31. Metal foil type of strain gauges are superior to wire wound metal strain gauges because
(*a*) they have a higher heat dissipation capacity
(*b*) they have a larger surface area which makes the bonding easier
(*c*) they can be fabricated economically using techniques which are similar to those used for fabrication of microelectronic circuitry
(*d*) all the above.

32. Metal foil gauges use fat end turns in order to
(*a*) reduce transverse sensitivity
(*b*) reduce longitudinal sensitivity
(*c*) reduce cross sensitivity
(*d*) all the above.

33. In semi-conductor strain gauges, the change in resistance on application of strain
(*a*) is much higher than that in the case of metal strain gauges and this is mainly due to change in dimensions

25

(*b*) is much higher than that in the case of metal strain gauges and this is mainly due to change in resistivity
(*c*) is lower than that in the case of metal strain gauges
(*d*) none of the above.

**34.** In semi-conductor strain gauges, when tensile strain is applied
(*a*) resistance increases in N-type of materials
(*b*) resistance increases in P-type of materials
(*c*) resistance increases in both P and N type materials
(*d*) resistance decreases in both P and N type materials.

**35.** The RTDs use the principle of change of resistance with temperature. The properties of a conductor material to be used as an element of an RTD should possess the following properties
(*a*) the change in resistance per unit change in temperature should be as small as possible
(*b*) the resistance of the materials should not have a continuous and stable relationship with temperature
(*c*) the change of resistance with temperature should not be a linear function
(*d*) none of the above.

**36.** Platinum is the commonly used metal for RTDs. This is because
(*a*) platinum has a constant value of resistance temperature co-efficient of 0.004°C for a temperature range between 0 to 100°C
(*b*) the resistivity of platinum tends to increase less rapidly at higher temperatures
(*c*) platinum is available in pure form for commercial applications and has a stability over higher ranges of temperature
(*d*) all the above.

**37.** The resistance of a thermometer is 5 Ω at 30°C and 6.5 Ω at 60°C. Using linear approximation, the value of resistance temperature co-efficient at 45°C
(*a*) 0.009/°C (*b*) 0.0087/°C
(*c*) 0.0085/°C (*d*) 0.01/°C.

**38.** The sensitivity of thermistors as compared with sensitivity of platinum resistance temperature detector over a temperature range of −100°C to 400°C to changes in temperature is
(*a*) 100 times (*b*) $10^6$ times
(*c*) $10^7$ times
(*d*) 10° times.

**39.** A thermistor exhibits
(*a*) only a negative change of resistance with increase in temperature
(*b*) only a positive change of resistance with increase in temperature
(*c*) can exhibit either a negative or positive change of resistance with increase of temperature depending upon the type of material used
(*d*) none of the above.

**40.** The resistance value of a thermistor is 5000 Ω at 25°C and its resistance temperature co-efficient is 0.04/°C. A measurement with a lead resistance of 10 Ω will cause an error of
(*a*) 0.02°C (*b*) 0.05°C
(*c*) 0.1°C (*d*) none of the above.

**41.** The temperature transducers exhibit non-linear behaviour. The order in which they exhibit non-linearity (highest to lowest) is
(*a*) Thermocouples, RTDs. thermistors
(*b*) Thermistors, thermocouples, RTDs
(*c*) RTDs, thermocouples, thermistors
(*d*) Thermistors, RTDs, thermocouples.

**42.** Three types of temperature transducers are compared as regards their sensitivity. The order in which they exhibit their sensitivities (highest to lowest) is
(*a*) Thermistors, RTDs, thermocouples
(*b*) Thermocouples. RTDs, thermistors
(*c*) TDs, thermistors, thermocouples
(*d*) RTDs, thermocouples, thermistors.

**43.** Thermocouples are
(*a*) passive transducers
(*b*) active transducers
(*c*) both active and passive transducers
(*d*) output transducers.

**44.** Thermocouples
(*a*) are most commonly used temperature transducers
(*b*) require reference junction compensation
(*c*) have a low output voltage level
(*d*) all the above.

**45.** A thermocouple produces a voltage of 50 mV. Its internal resistance is 50 Ω. The resistance of its leads is 10 Ω. Its output is read by a PMMC meter having an internal resistance of 120 Ω. The output voltage indicated is
(*a*) 33.3 mV (*b*) 25 mV
(*c*) 40 mV (*d*) 10 mV.

**46.** Air cored Inductive transducers are suitable for use
(*a*) at lower frequencies
(*b*) at higher frequencies
(*c*) at equal frequencies
(*d*) as are employed for iron cored transducers.

**47.** The size of air cored transducers as compared with their iron cored counterparts is
(*a*) smaller (*b*) bigger
(*c*) same (*d*) all the above.

**48.** An iron cored transducer is exited with a frequency of 20 kHz. The maximum value of frequency of displacement to be measured with this transducer should not exceed
(*a*) 0.5 kHz (*b*) 1 kHz
(*c*) 1.5 kHz (*d*) 2 kHz.

**49.** Inductive transducers are used in differential configuration because the output is not influenced by
(*a*) external magnetic fields
(*b*) temperature changes
(*c*) variations of supply voltage and frequency
(*d*) all the above.

**50.** Two inductive transducers working on the principle of change of self inductance $L$, are connected in a push pull arrangement. If the change of inductance of transducers is $\Delta L$, the change of inductance exhibited at the output terminals is
(*a*) $\Delta L$ (*b*) $2\Delta L$
(*c*) $\pm 2\Delta L$ (*d*) zero.

**51.** In an L.V.D.T., the two secondary windings are connected in differential to obtain
(*a*) higher output voltage
(*b*) an output voltage which is phase sensitive *i.e.*, the output voltage has a phase which can lead us to a conclusion whether the displacement of the core took place from right to left or from left to right
(*c*) in order to establish the null or the reference point for the displacement of the core
(*d*) both (*b*) and (*c*).

**52.** In an L.V.D.T., the core is made up of a
(*a*) non-magnetic material
(*b*) a solid ferro-electric material
(*c*) high permeability, nickel-iron hydrogen annealed material in order to produce low harmonics, low null voltage, and high sensitivity. The core is slotted to reduce eddy current losses
(*d*) all the above.

**53.** An L.V.D.T.
(*a*) exhibits linear characteristics upto a displacement of ± 5 mm
(*b*) has a linearity of 0.05%
(*c*) has an infinite resolution and a high sensitivity which is of the order of 40 V/mm
(*d*) all the above.

**54.** An L.V.D.T.
(*a*) has very good frequency response as the mass of the core to which the input signal is applied is fairly large
(*b*) is immune to stray magnetic fields
(*c*) has a zero voltage at the null point even if the supply voltage contains and harmonics and even if the two secondary windings are not exactly similar and even asymmetrically placed with respect to primary winding
(*d*) none of the above.

**55.** There are two modes in which synchro transmitter –transmitter pairs are used *i.e.*, (*i*) error detection mode (*ii*) torque transmission mode. The torque in the torque transmission mode
(*a*) is greater than that in the error transmission mode
(*b*) is smaller than that in the error transmission made
(*c*) the torque is the same in both the modes
(*d*) none of the above.

**56.** The rotor of a synchro-transmitter
(*a*) uses a salient pole type of construction
(*b*) uses a cylindrical pole type of construction
(*c*) may use either salient pole or cylindrical type of construction
(*d*) none of the above.

**57.** The electrical zero of synchro-transmitter is when the voltage induced
(*a*) in coil $S_1$, of stator is maximum
(*b*) in coil $S_3$, of stator is maximum
(*c*) in coil $S_2$, of stator is maximum
(*d*) in coil $S_2$, of stator is maximum and the voltage across terminals $S_3 S_1$ is zero.

**58.** The electrical zero of a synchro-transformer is
(*a*) displaced by an angle 180° from the zero of a synchro-transmitter
(*b*) displaced by an angle 90° from the zero of a synchro-transmitter
(*c*) displaced by an angle 0° from the zero of the transmitter
(*d*) none of the above.

59. In a synchro-transmitter, the voltages induced in the three windings of the stator are
(a) displaced 120° in time phase
(b) are in time phase with each other
(c) displaced 90° in time phase
(d) none of the above.

60. When a synchro-transmitter synchro control transformer pair is used as an error detector the output voltage obtained from the rotor terminals of the synchro-transformer is
(a) sinusoidal
(b) suppressed carrier modulated
(c) half wave rectified
(d) none of the above.

61. The rotor of a synchro-transformer is made cylindrical in construction
(a) in order that air gap remains uniform
(b) there is no change of reactance of the rotor winding with change of rotor position
(c) there should not be any change in the rotor winding impedance with position of rotor since the rotor winding is connected to an amplifier
(d) all the above.

62. A synchro-transmitter transformer pair is used as error detector. The sensitivity of the pair is 100 V/rad. The rotor of the transformer turns through an angle of 0.5 rad while the transformer rotor turns through an angle of 0.45 rad. The value of error signal representing the two shaft positions is :
(a) 5 V (b) 99.5 V (c) 90 V (d) 10 V.

63. How many windings does a simple resolver have ?
(a) one stator and two rotor windings
(b) one rotor and two stator windings
(c) two stator and two rotor windings
(d) a one stator and one rotor winding.

64. In Fig. 25.102, a simple resolver is shown. The stator windings are excited with voltages $E_{S1-3}$ and $E_{S2-4}$. Find the voltages across rotor windings $E_{R-13}$ and $E_{R2-4}$ when an angle θ
(a) $E_{R1-3} = E_{S1-3}\sin\theta + E_{S2-4}\cos\theta$
$E_{R2-4} = E_{S1-3}\sin\theta - E_{S2-4}\cos\theta$
(b) $E_{R1-3} = E_{S1-3}\cos\theta - E_{S2-4}\sin\theta$
$E_{R2-4} = E_{S1-3}\cos\theta + E_{S2-4}\sin\theta$
(c) $E_{R1-3} = E_{S1-3}\cos\theta + E_{S2-4}\sin\theta$
$E_{R2-4} = E_{S1}\cos\theta - E_{S24}\sin\theta$
(d) $E_{R1-3} = E_{S1-3}\cos\theta + E_{S2-4}\sin\theta$
$E_{R2-4} = E_{S1-3}\cos\theta + E_{S2-4}\sin\theta$.

65. Capacitive transducers are normally used for
(a) static measurements
(b) dynamic measurements
(c) both static and dynamic measurements
(d) transient measurements.

66. The dynamic characteristics of capacitive transducers are similar to those of
(a) low pass filters (b) high pass filters
(c) notch filters (d) band stop filters.

67. A capacitive transducer working on the principle of change of capacitance with change of displacement, exhibits non-linear characterics. The response of these transducers can be made linear by using
(a) differential arrangement
(b) use of an OPAMP
(c) working them over a small displacement range
(d) all the above.

68. Capacitive transducers can be used for measurement of liquid level. The principle of operation used in this case is
(a) change of capacitance with change of distance between plates
(b) change of area of plates
(c) change of dielectric strength
(d) none of the above.

69. What is the order of minimum displacement that can be measured with capacitive transducers ?
(a) 1 cm (b) 1 mm
(c) 1 μm (d) $1 \times 10^{-12}$ m.

70. The capacitive transducers have a high output impedance and therefore,
(a) to reduce loading effects long lengths of cable should be used
(b) to increase the value of resonant frequency long lengths of cables should be used
(c) long cable lengths should be used in conjunction with capacitive transducers in order to improve their frequency response
(d) none of the above.

71. The voltage sensitivities of Barium titanate and quartz are respectively $12 \times 10^{-3}$ Vm/N and $50 \times 10^{-3}$ Vm/N. Their respective permittivities are $12.5 \times 10^{-9}$ F/m and $40.6 \times 10^{-12}$ F/m. What are their charge sensitivities ?
(a) $1.04 \times 10^{6}$ C/N, $1.23 \times 10^{9}$ C/N
(b) 150 pC/N, 2 pC/N
(c) 20 pC/N, 50 pC/N
(d) none of the above.

72. Ceramic materials are used for piezo-electric transducers
(a) they are polycrystalline in nature
(b) they are basically made of barium titanate
(c) they do not have piezo-electric properties in their original state but these properties are produced by special polarizing treatment
(d) all the above.

73. Quartz and Rochelle salt belong to
(a) natural group of piezo-electric materials
(b) synthetic group of piezo-electric materials
(c) can belong to natural or synthetic group of piezo-electric materials provided properly polarized
(d) all the above.

74. Piezo-electric transducers are
(a) passive transducers (b) active transducers
(c) inverse transducers (d) (b) and (c).

75. A barium titanate crystal has a thickness of 2 mm. Its voltage sensitivity is $12 \times 10^{-3}$ Vm/N. It is subjected to a pressure of 0.5 MN/$m^2$. Calculate the voltage generated
(a) 3 V (b) 6 V
(c) 12 V (d) 5 V.

76. A pieze-electrical transducer has an output voltage of 3 V at no load conditions. It has a capacitance 250 pF. It is connected to load capacitance of 125 pF. Find the voltage across the load at high frequencies.
(a) 1 V (b) 2 V
(c) 9 V
(d) cannot be determined with the data given.

77. Magneto-elastic transducers work on the principle of
(a) change of dimensions with change of applied stress
(b) change of permeability with change in stress
(c) change of co-ercive force with change of stress
(d) none of the above.

78. Magnetostriction is the effect produced when a magnetic material is subjected to change of magnetization
(a) the change of magnetization results in change of permeability
(b) the change in magnetization results in change of dimensions
(c) the change in magnetization results in change in temperature
(d) none of the above.

79. Tuned magneto-elastic transducers can be used for generation of
(a) radio waves (b) electric current
(c) ultrasound waves (d) all the above.

80. A Hall effect transducer can be used for measurement of
(a) power (b) current
(c) displacement (d) all the above.

81. A Hall effect transducer is used for measurement of a magnetic field of 1.5 Wb/$m^2$ with a copper transducer for which the Hall effect co-efficient is $-52 \times 10^{-12}$ V-m / A-Wb $m^{-2}$. The thickness of the element is 2 mm and the current passing is 5 A, find the voltage generated
(a) – 0.195 μV (b) 0.195 μV
(c) –390 mV (d) $-390 \times 10^{-9}$ V.

82. A photo-voltaic cell is connected across a very large value of resistance, its sensitivity at higher illumination levels is typical
(a) infinity (b) zero
(c) some finite value which is very large
(d) some finite value which is very small and nears zero.

83. Which of the following photo-electric devices is most suitable for digital applications ?
(a) photo-emissive cell (b) photo-diode
(c) photo-transistor (d) photo-voltaic cell.

84. Which of the photo-electric transducers is used for production of electric energy by converting solar energy
(a) photo-emissive cell (b) photo-diode
(c) photo-transistor (d) (b) and (c).

85. Which of the following optical transducers is an active transducer ?
(a) photo-emissive cell (b) photo-diode
(c) photo-transistor (d) photo-voltaic cell.

86. A cadmium sulphide cell has the highest response for a light of wavelength of nearly
(a) 300 nm (b) 800 nm
(c) 500 nm (d) 600 nm.

87. A photo-diode functions as
(a) photo-conductive device while working with reverse voltage
(b) photo-voltaic device while working without a reverse voltage
(c) photo-voltaic devices while working with or without reverse voltage
(d) none of the above.

**88.** A photo-conductive cell is useful for
(*a*) high frequency applications
(*b*) medium frequency applications
(*c*) low frequency applications
(*d*) all the above.

**89.** The photo-diode as compared to a photo-transistor has
(*a*) faster switching time (*b*) lower sensitivity
(*c*) higher size for the same value of output current
(*d*) all the above.

**90.** For measurement of temperature
(*a*) only intensity type fibre optic transducers can be used
(*b*) only interferometric type fibre-optic transducers can be used
(*c*) both intrinsic and interferometric type of transducers can be used
(*d*) none of the above.

**91.** When measuring sound level-with a interferometric type transducer
(*a*) a laser beam is used with a beam splitter
(*b*) measured variable causes a phase shift in the light propagating through optic fibre
(*c*) the phase shift is a measure of sound level
(*d*) all the above.

**92.** While measuring liquid level with fibre optic transducers, the fibre rod or cable is cut in the form of
(*a*) sphere (*b*) rectangle
(*c*) prism (*d*) none of the above.

**93.** A linear displacement transducer (digital) normally uses
(*a*) straight binary code (*b*) BCD
(*c*) Gray code (*d*) Hexadecimal code.

**94.** A tachometer encoder has
(*a*) one output (*b*) two outputs
(*c*) three outputs
(*d*) all the above.

**95.** A tachometer encoder can be used for measurement of displacement
(*a*) in both the directions
(*b*) in one direction only
(*c*) its output pulses are counted only for the forward direction
(*d*) its output pulses are counted for only the reverse direction.

**96.** Incremental counters use
(*a*) one channel (*b*) two channels
(*c*) two channels and sometimes three
(*d*) none of the above.

**97.** Absolute Encoders are used for
(*a*) normally one revolution
(*b*) continuous speed
(*c*) continuous speed in both the directions
(*d*) none of the above.

**98.** Moire Fringes are used for measurement of rotary displacement along with
(*a*) only contact type encoders
(*b*) only optical encoders
(*c*) both contact type and optical encoders
(*d*) none of the above.

**99.** The resolution of a 8 bit shift encoder is
(*a*) $360°/2^7$ (*b*) $360°/2^9$
(*c*) $360°/2^8$ (*d*) none of the above.

**100.** Absolute encoders are used where
(*a*) fast varying signals are not used
(*b*) when position data is to be recovered even alter a temporary power out age
(*c*) transient noise can be tolerated as it causes only a transient disturbance
(*d*) all the above.

## *Answers*

| | | | | | | | | | |
|---|---|---|---|---|---|---|---|---|---|
| **1.** (*c*) | **2.** (*a*) | **3.** (*d*) | **4.** (*c*) | **5.** (*d*) | **6.** (*c*) | **7.** (*b*) | **8.** (*b*) | **9.** (*a*) | **10.** (*b*) |
| **11.** (*d*) | **12.** (*c*) | **13.** (*a*) | **14.** (*c*) | **15.** (*a*) | **16.** (*c*) | **17.** (*b*) | **18.** (*b*) | **19.** (*a*) | **20.** (*b*) |
| **21.** (*b*) | **22.** (*b*) | **23.** (*b*) | **24.** (*a*) | **25.** (*c*) | **26.** (*d*) | **27.** (*c*) | **28.** (*a*) | **29.** (*b*) | **30.** (*c*) |
| **31.** (*d*) | **32.** (*a*) | **33.** (*b*) | **34.** (*b*) | **35.** (*d*) | **36.** (*d*) | **37.** (*b*) | **38.** (*b*) | **39.** (*c*) | **40.** (*b*) |
| **41.** (*b*) | **42.** (*a*) | **43.** (*b*) | **44.** (*d*) | **45.** (*a*) | **46.** (*b*) | **47.** (*b*) | **48.** (*d*) | **49.** (*d*) | **50.** (*c*) |
| **51.** (*d*) | **52.** (*c*) | **53.** (*d*) | **54.** (*d*) | **55.** (*b*) | **56.** (*a*) | **57.** (*d*) | **58.** (*b*) | **59.** (*b*) | **60.** (*b*) |
| **61.** (*d*) | **62.** (*a*) | **63.** (*c*) | **64.** (*c*) | **65.** (*b*) | **66.** (*b*) | **67.** (*d*) | **68.** (*c*) | **69.** (*c*) | **70.** (*d*) |
| **71.** (*b*) | **72.** (*d*) | **73.** (*a*) | **74.** (*d*) | **75.** (*c*) | **76.** (*b*) | **77.** (*b*) | **78.** (*b*) | **79.** (*c*) | **80.** (*d*) |
| **81.** (*a*) | **82.** (*d*) | **83.** (*d*) | **84.** (*d*) | **85.** (*d*) | **86.** (*c*) | **87.** (*a*) | **88.** (*c*) | **89.** (*d*) | **90.** (*c*) |
| **91.** (*d*) | **92.** (*c*) | **93.** (*c*) | **94.** (*a*) | **95.** (*b*) | **96.** (*c*) | **97.** (*a*) | **98.** (*b*) | **99.** (*c*) | **100.** (*d*) |

# 26

# Signal Conditioning

## 26.1 SIGNAL CONDITIONING

The measurand, which is basically a physical quantity as is detected by the first stage of the instrumentation or measurement system. The first stage, with which we have become familiar, is the *"Detector Transducer stage"*. The quantity is detected and is transduced into an electrical form in most of the cases. The output of the first stage has to be modified before it becomes usable and satisfactory to drive the signal presentation stage which is the third and the last stage of a measurement system. The last stage of the measurement system may consist of indicating, recording, displaying, data processing elements or may consist of control elements.

In this chapter, methods used for modifying the transduced signal into a usable format for the final stage of the measurement system are described.

Measurement of dynamic physical quantities requires faithful representation of their analog or digital output obtained from the intermediate stage *i.e., signal conditioning stage* and this places a severe strain on the signal conditioning equipment. The signal conditioning equipment may be required to do linear processes like amplification, attenuation integration, differentiation, addition and subtraction. They are also required to do non-linear processes like modulation, demodulation, sampling, filtering, clipping and clamping, squaring, linearizing or multiplication by another function etc. These tasks are by no means simple. They require ingenuity in proper selection of components and the selection of most faithful methods of reproduction of output signals for the final data presentation stage.

The **signal conditioning or data acquisition equipment** in many a situation is an excitation and amplification system for **passive transducers**. It may be an **amplification system** for active transducers. In both the applications, the transducer output is brought up to a sufficient level to make it useful for conversion, processing indicating and recording. Excitation is needed for passive transducers because these transducers do not generate their own voltage or current. Therefore, passive transducers like strain gauges, potentiometers, resistance thermometers, inductive and capacitive transducers require excitation from external sources. The active transducers like technogenerators, thermocouples, inductive pick-ups and piezo-electric crystals, on the other hand, do not require an external source of excitation since they produce their own electrical output on account of application of physical quantities. But these signals usually have a low voltage level, and hence, need amplification.

The excitation sources may be an alternating or a d.c. voltage source. The d.c. system is comparatively simple as shown in Fig. 26.1.

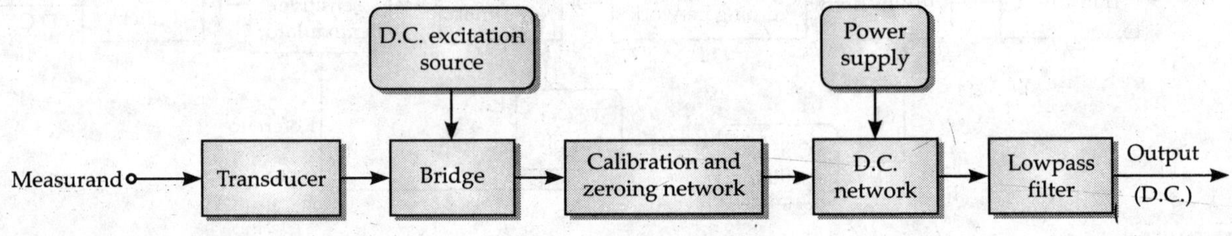


**Fig. 26.1** D.C. Signal conditioning system.

The resistance transducers like strain gauges, constitute one arm or more than one arm of a wheatstone bridge which is excited by an isolated d.c. source. The bridge can be balanced by a potentiometer and can also be calibrated for unbalanced conditions.

The desirable characteristics of a d.c. amplifier are :

(*i*) It may need balanced differential inputs giving a high **Common mode rejection ratio** (CMRR). This is elaborated later on in this chapter.

(*ii*) It should have an extremely good thermal and long term stability.

The advantages of a d.c. amplifier are that :

(*i*) It is easy to calibrate at low frequencies.

(*ii*) It is able to recover from an overload condition unlike its a.c. counterpart.

But the greatest disadvantage of a d.c. amplifier is that it suffers from the problem of drift. Thus low frequency spurious signals come out as data information. For this reason special low drift d.c. amplifiers are used. The d.c. amplifier is followed by a lowpass filter which is used to eliminate high frequency components or noise from the data signal.

In order to overcome the problems that are encountered in d.c. systems, a.c. systems are used. In a.c. systems, the **Carrier-type a.c. Signal Conditioning systems** are used as shown in Fig. 26.2.

The transducers used are the variable resistance or variable inductance transducers. They are employed between carrier frequencies of 50 Hz to 200 kHz. The carrier frequencies are much higher, they are at least 5 to 10 times the signal frequencies.

Transducer parameter variations amplitude modulate the carrier frequencies at the bridge output and waveform is amplified and demodulated. The demodulation is **Phase Sensitive** so that the polarity of d.c. output indicates the direction of the parameter change in the bridge output.

In a carrier system amplifier drift and spurious signals are not of much importance unless they modulate the carrier. However, it is more difficult to achieve a stable carrier oscillator than a comparable d.c. stabilized source. In carrier systems, it is easy to obtain very high rejection of mains frequency pick up. Active filters can be used to reject this frequency and prevent overloading of a.c. amplifier. The phase-sensitive, demodulators filter out carrier frequency components of the data signal.

D.C. systems are generally used for common resistance transducers such as potentiometers and resistance strain gauges. A.C. systems have to be used for variable reactance transducers and for systems where signals have to be transmitted via long cables to connect the transducers to the signal conditioning equipment.

After the physical quantities like temperature, pressure, strain, acceleration etc. have been transduced into their analogous electrical form and amplified to sufficient current or voltage levels (say 1 V to 10 V), they are further processed by electronic circuits. In some applications, the signal does not need any further processing and the amplified signal may be directly applied to indicating or recording or control instruments. But many applications involve further processing of signals which involve linear and non-linear operations as mentioned earlier.

The signal may be applied to **sample and hold (S/H) circuit** as shown in Fig. 26.3. This may be fed to an **analog multiplexer** and **analog to digital (A/D) converter**. If the signal is in digital form it may be applied to a variety of digital systems like a digital computer, digital controller, digital data logger or a digital data transmitter.

The sample and hold units shown in Fig. 26.3 sample the different inputs at a specified time and then hold the voltage levels at their output while analog multiplexer performs the **time division multiplexing (TDM)** operation between different

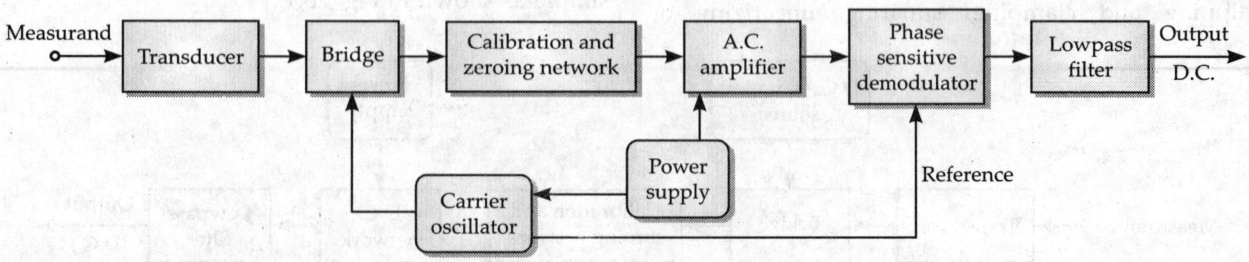


**Fig. 26.2** A.C. Signal conditioning system

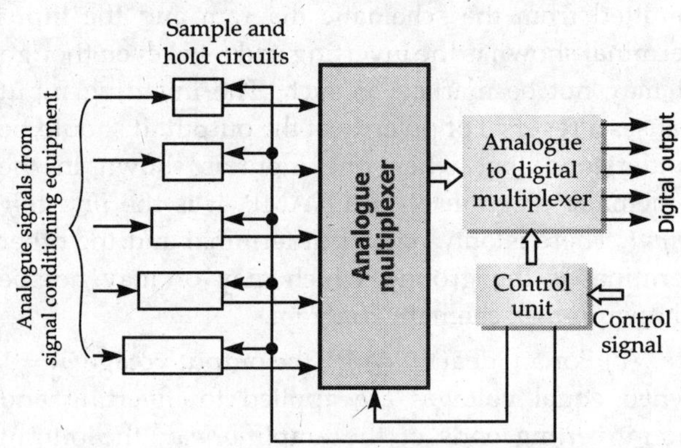


**Fig. 26.3** Data acquisition and conversion system.

data inputs. Time Division Multiplexing means that each input channel is sequentially connected to the multiplexer for a certain specified time. (The input signals are not applied to the multiplexer continuously but are connected in turn to the multiplexer thereby sharing time). The timing of the various input channels is controlled by a **control unit**. This unit controls the sample and hold (S/H) circuits, the multiplexer and the analog to digital (A/D) converter. The control unit may be controller itself.

In case time division multiplexing is not used, the **frequency division multiplexer (FDM)** may be used. In this case, the multiple data analog inputs can remain in analog form and are transmitted all at the same time, using frequency division multiplexing (FDM). The voltage input from the signal conditioning equipment is converted into frequency. Thus any change in voltage input of the measurand produces a corresponding change in frequency.

Earlier circuits comprising of discrete electronic components were used where impedance transformations, amplification, and other signal conditioning were required. The requirement to produce designs from discrete components has given way to easier and more reliable methods of signal conditioning which use integrated circuits (ICs). Many special circuits and general purpose amplifiers are now contained in IC packages producing a quick solution to signal conditioning problems, together with small size, low power consumption and, low cost.

## 26.2 OPERATIONAL AMPLIFIER

The operational amplifier (OPAMP) manufactured with integrated circuit technology contains transistors, diodes, resistors and capacitors, is an extremely versatile device that does countless jobs in many electronic circuits. We limit our discussions regarding OPAMP to an external description and to those applications, where it serves as a dependant source for important electronic circuits as well as a device for performing such important functions as isolation, inversion, addition, subtraction, multiplication and division. It can also be used for performing mathematical operations like integration and differentiation which together with its summing capabilities puts it in a position to model differential equations of physical systems.

One of the most popular OPAMPs is the 741 type, which has been produced by several manufacturers for many years. Also several improved versions have been developed having different internal circuitry to the original 741 OPAMPs.

The 741 is a plastic encapsulated device with *dual in line* (DIL) rows of pins as shown in Fig. 26.4(*a*). Such devices are said to be directly *pin compatible.* The pin connections are shown in Fig. 26.4(*b*). The 741 is an eight pin device. The operational amplifier has two types of connections :

(*i*) *Inverting and*

(*ii*) *Non-inverting.*

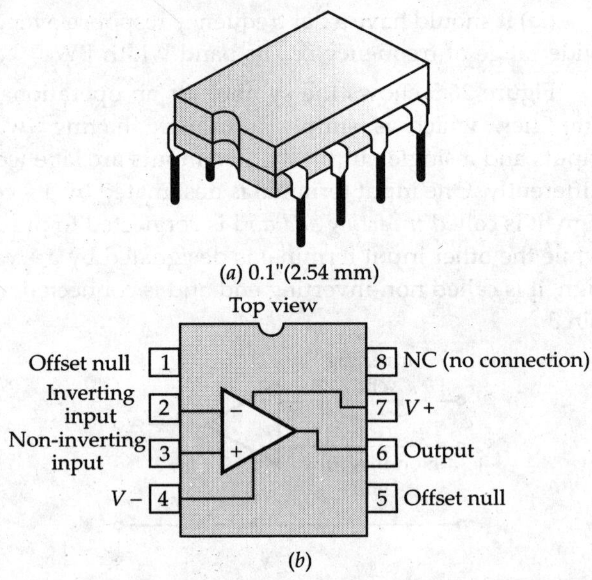


**Fig. 26.4** Integrated circuit 741 OPAMP (*a*) DIL package (*b*) Pin connections.

When a voltage is applied to *inverting input* pin 2, the output voltage has a sign which is opposite to that of the input. In case, the input voltage is applied to pin 3, which is designated as *non-inverting*, the output voltage has the same sign as the input.

The operational amplifier is shown by a triangular symbol with inverting and non-inverting inputs being connected at pins 2 and 3 respectively.

The OPAMP output appears at pin 6. Several other pins are labelled in the diagram. The most important of these are 7 and 4, where positive and negative constant supply voltages, $V^+$ and $V^-$ are connected to supply energy to the integrated circuit. Usually this supply voltage is fixed at ±15 V with respect to ground reference. The OPAMP does not work without the supply voltage.

Two other pins (1 and 5) are provided on 741 and are labelled *offset null*. The offset effects are one of the several types of non-ideal behaviour of OPAMPs and are explained later in this chapter.

### 26.2.1 Ideal Operational Amplifier

The properties of an ideal operational amplifier are :

(*i*) It should have an infinite input impedance *i.e.*, $Z_i = \infty$.

(*ii*) It should have a zero output impedance *i.e.*, $Z_0 = 0$.

(*iii*) It should have an infinite open loop gain *i.e.*, $A_{vol} = -\infty$.

(*iv*) It should have a flat frequency response over a wide range of frequency *i.e.*, its band-width BW = ∞.

Figure 26.5 shows the symbol for an operational amplifier, which is simply a triangle having two inputs and a single output the two inputs are labelled differently. One input terminal is designated by a – ve sign, it is called *inverting end* and is connected to pin 2 while the other input terminal is designated by a + ve sign, it is called non-inverting end and is connected to pin 3.

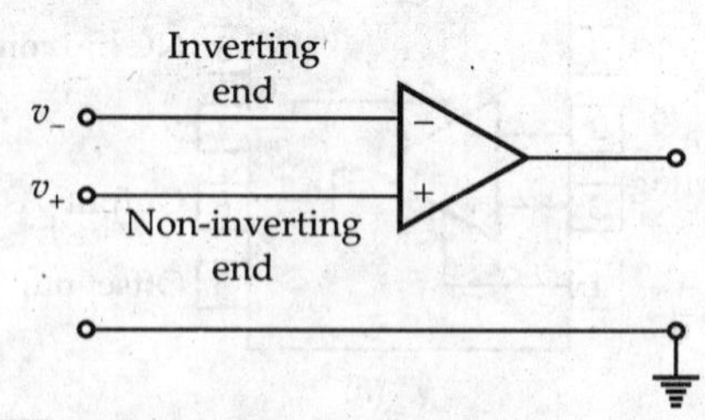


**Fig. 26.5** Symbol for an OPAMP.

A negative voltage ($v_-$) applied to inverting end appears as a +ve voltage $v_0$ at pin 6 while a positive voltage ($v_+$) applied to the non-inverting end appears a +ve voltage at pin 6.

Frequently, the non-inverting end is not used and some operational amplifiers do not have this terminal for use. When non-inverting end is not available, it is omitted from the schematic diagram and the input terminal shown is the inverting end (–ve) even though it may not be marked as such. The inverting input causes a reversal of polarity at the output. It should be understood that when one input is shown in the schematic diagram of an OPAMP, it is the *inverting input*. There is only one input terminal and the other terminal is the ground which may or may not be shown in the schematic diagram.

(*v*) For an ideal OPAMP, the output votage $v_0 = 0$, when equal voltages are applied to inverting and non-inverting ends of the amplifier *i.e.*, the output voltage is zero *i.e.*, $v_0 = 0$ when ($v_-$ input) – ($v_+$ input) = 0 Where $v_-$ and $v_+$ are the voltages applied to inverting and non-inverting ends of OPAMP.

The OPAMP can accommodate both positive and negative input signals simultaneously or one at a time. The schematic diagram showing the input signals and +ve and –ve power supplies is shown in Fig. 26.6(*a*). The signal $v_+$ is applied to pin 3, which produces a non-inverting voltage output. That is, the output voltage appearing at pin 6 has the same polarity as that of the input signal. When a signal $v_-$ is applied to pin 2, it gets inverted and we obtain a signal output of opposite polarity at pin 6. In the interest of simplicity the symbol most commonly used for an OPAMP is shown in Fig. 26.6(*b*). The design of the OPAMP is such that the output voltage is related to two input voltages as,

$$v_0 = A_{vol}(v_+ - v_-) = A_{vol}\, v_d \qquad \ldots(26.1)$$

where $v_d$ = net input differential voltage = $v_+ - v_-$

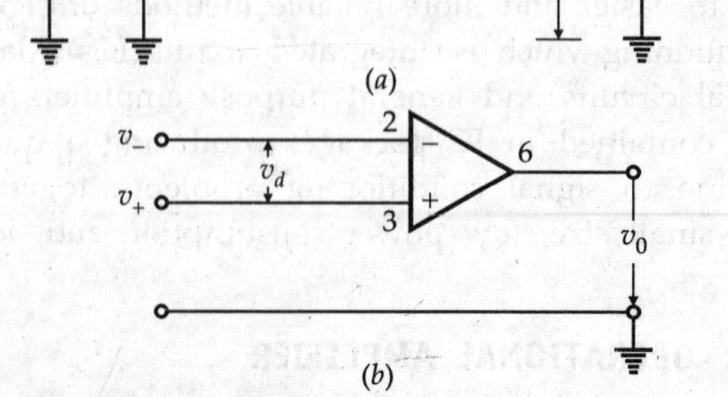


**Fig. 26.6** Schematic diagram of an OPAMP. (*a*) Model showing supply voltages and signal inputs with connection to ground reference. (*b*) Commonly used simplified symbol for modeling an OPAMP.

Three notable features characterise the OPAMP model shown in Fig. 26.6. In the interest of stressing a simplified strategy for analysis, the *idealised versions* of these features are specified.

These are :

(*i*) Input impedance $\mathbf{Z}_i$ should be infinite. The actual value may be of the order of several megohms, but this is tantamount to infinity.

(*ii*) Output impedance $\mathbf{Z}_0$ should ideally be zero. In actual amplifiers, the output resistance is below 100 Ω, but for the level of the load resistors generally used across the output terminals of the OPAMP in applications, this value can be considered as negligible.

(*iii*) Open loop gain $A_{vol}$ which is the ratio of $v_0$ to $v_d$ is considered to be infinity.

Although we know this gain to be finite and in the vicinity of 1,00,000, this figure is high enough so that the practical results obtained with finite gain are the same as if the OPAMP was treated with an infinite gain.

$v_d$ is the net input differential voltage. $A_{vol}$ is the open loop gain which is about 1,00,000. This means that the net differential voltage is increased at the output terminals (pin 6) by this factor. Clearly, then if overloads are to be avoided, it is necessary to restrict $v_d$ to those values that will keep $v_0$ atleast below the magnitude of supply voltage.

Accordingly, for the case where

$A_{vol} = 100{,}000$, it follows that $v_d$ must be limited to,

$$v_d = 15/100{,}000 = 150\ \mu\text{V}$$

where 15 V is the supply voltage, which is obviously a small quantity. Fortunately, there are techniques available for increasing this limit to more practical values. These are considered as below.

One of the characteristics of OPAMP is that saturation of the output voltage sets in rapidly once the differential input voltage reaches a threshold value such as 150 μV as calculated above. This feature is illustrated in Fig. 26.7.

Figure 26.7 shows the transfer characteristics of an OPAMP. Here $v_{TH}$ represents the threshold value of $v_d$ beyond which saturation sets in.

The saturation value is approximately taken to be supply voltage $V_{CC}$ for convenience. Actually the threshold voltage is less than $V_{CC}$. The slope of the curve represents the open loop gain $A_{vol}$.

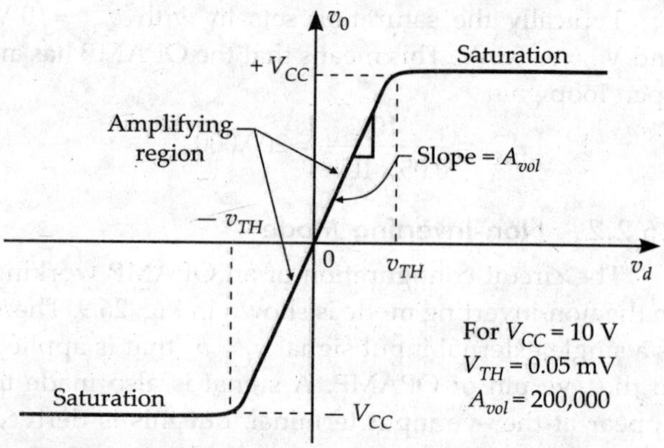


**Fig. 26.7** Transfer characteristics of an OPAMP.

The transfer characteristics indicate that the amplifier region for the OPAMP in which $v_0 = A_{vol}\, v_d$ is restricted to an output voltage range of $v_{0\,min} \le v_0 \le v_{0\,max}$. In this region, each stage of OPAMP is restricted to amplifier operation. If the signal level drives the output stage (or any stage) into saturation or cut off, the output voltage of OPAMP gets fixed as shown in Fig. 26.8 (*b*) outside the amplifier region. Furthermore, *positive saturation* of OPAMP occurs when $v_0 = v_{0\,max}$ and *negative saturation* occurs when $v_0 = v_{0\,min}$.

The restriction on output voltage of the OPAMP indicates that the input voltage difference of voltage ($v_d$) of OPAMP has similar restrictions. In fact, the range of $v_d$ that provides amplifier operation is easily determined if we realize that the slope of transfer characteristics is $A_{vol}$. Thus the range for amplifier region is given by :

$$\frac{v_{0\,min}}{A_{vol}} \le v_d \le \frac{v_{0\,max}}{A_{vol}}$$

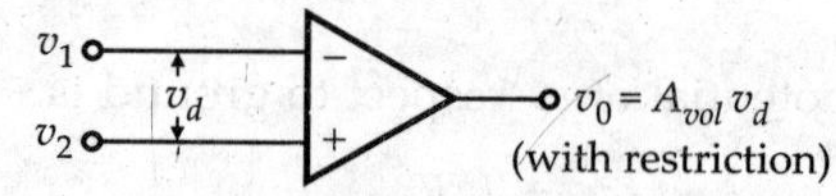


(*a*) Circuit symbol

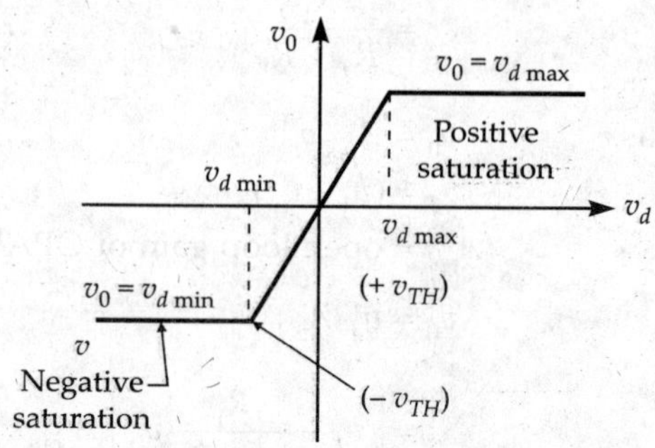


(*b*) Voltage transfer characteristics.

**Fig. 26.8** The OPAMP.

26

Typically the saturation sets in with $v_{CC} = 10$ V and $V_d = 0.05$ mV. This means that the OPAMP has an open loop gain

$$A_{vol} = \frac{10}{0.05 \times 10^{-3}} = 200{,}000.$$

### 26.2.2 Non-Inverting Mode

The circuit configuration of an OPAMP working in the non-inverting mode is shown in Fig. 26.9. There is a single external input signal $v_1 = v_+$ that is applied to the +ve pin of OPAMP. A signal is also made to appear at the –ve input terminal, but this is derived from resistors $R_1$ and $R_f$.

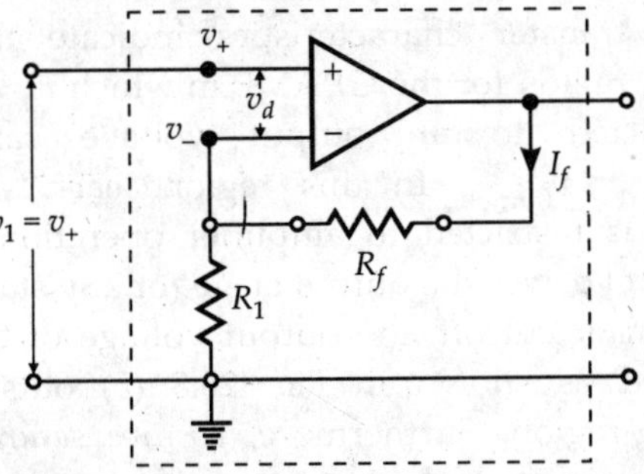


**Fig. 26.9** OPAMP used in non-inverting mode.

In order to determine the value of $v_-$, we must note that when an output voltage $v_0$ appears at the output terminals OPAMP, we have, from ohm's law,

$$v_0 = I_f(R_f + R_1) \quad \text{...(26.2)}$$

The connection of junction point of $R_f$ and $R_1$ to the negative input pin causes no effect in the circuit involving $R_f$ and $R_1$ because of the assumption of infinite resistance for OPAMP. For this very reason current produced by input voltage $v_1 = v_+$ which is $I_p$ is zero.

The potential of $v_-$ respect to ground is,

$$v_- = I_f R_1 \quad \text{...(26.3)}$$

$$= \frac{R_1}{R_f + R_1} v_0 \quad \text{...(26.4)}$$

Thus, $$v_0 = A_{vol} v_d \quad \text{...(25.5)}$$

where $$v_d = (v_+ - v_-)$$

and $A_{vol}$ = open loop gain of OPAMP.

$$\therefore \quad v_0 = v_d A_{vol}$$

$$= \left(v_1 - \frac{R_1}{R_1 + R_f} v_0\right) A_{vol} \quad \text{...(26.6)}$$

Dividing by $A_{vol}$ on both sides of Eqn. 26.6, we get,

$$\left(v_1 - \frac{R_1}{R_1 + R_f} v_0\right) = \frac{v_0}{A_{vol}} \quad \text{...(26.7)}$$

Next we introduce characteristic information about OPAMP. The output voltage $v_0$ is a finite quantity and is often less than 15 V. The open loop gain $A_{vol}$ is very large and may be taken as infinite.

$$\text{i.e.} \quad v_1 - \frac{R_1}{R_1 + R_f} v_0 = 0$$

or $$v_0 = \frac{R_1 + R_f}{R_1} v_1 \quad \text{...(26.8)}$$

Hence, the closed loop gain of the amplifier is,

$$A = \frac{v_0}{v_1} = \frac{R_1 + R_f}{R_1} = \left(1 + \frac{R_f}{R_1}\right) \quad \text{...(26.9)}$$

Equation 26.9 makes it obvious that the magnitude of the output voltage can be controlled by properly selecting the values for resistors $R_f$ and $R_1$

In case, impedances are used, closed loop gain is,

$$A = \left(1 + \frac{Z_f}{Z_1}\right) \quad \text{...(26.10)}$$

where $Z_f$ and $Z_1$ are the feedback and input resistances respectively.

### 26.2.3 Voltage Follower (Isolation) Mode

An interesting result is obtained when the feedback resistor $R_f = 0$ as shown in Fig. 26.10.

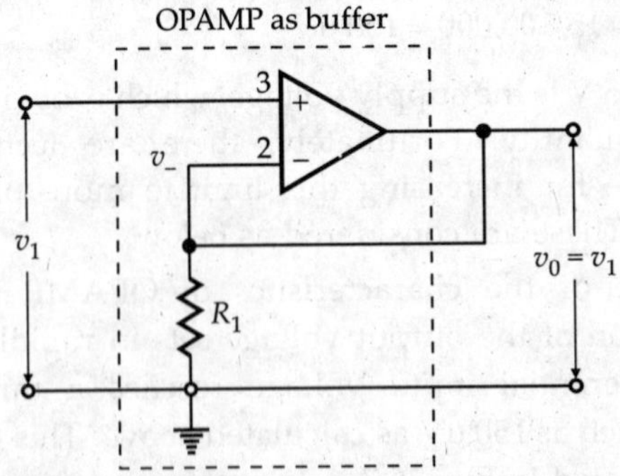


**Fig. 26.10** Voltage follower.

For such a circuit we have,

$$v_0 = v_1 \quad \text{...(26.11)}$$

In other words, the output of the OPAMP exactly tracts the input voltage in sign and magnitude. This is the reason why this circuit is called a *voltage follower*.

The result of Eqn. 26.11 can be found from direct application of Eqn. 26.1. Observe in Fig. 26.10 that by physical connection $v_-$ is equal to $v_0$. Therefore, from Eqn. 26.1,

$$v_0 = A_{vol}\, v_d = A_{vol}\,(v_1 - v_0) \quad \text{...(26.12)}$$

$$\therefore \quad \frac{v_0}{v_1} = \frac{A_{vol}}{1 + A_{vol}} = 1 \quad \text{...(26.13)}$$

as $\quad A_{vol} >> 1$

The direct connection of the output voltage to the non-inverting terminal of the OPAMP represents the case when 100% *negative feedback* of the output to the input. In such cases the output quantity is always related to the input quantity by the Eqn. 26.13. Of course when the open loop gain $A_{vol}$ is allowed to assume large values compared to unity the corresponding closed loop gain becomes approximately unity. However, a unity gain is not the only feature of the voltage follower. A more detailed analysis of the equivalent circuit of the OPAMP reveals that the resistance looking into the input terminals (*i.e.*, pin 3 and ground) takes on the value of $A_{vol}\, R_i$. It will be recalled that when OPAMP is used in open loop mode (*i.e.*, without feedback connection from the output) $R_i$ has already a value of 1 or 2 MΩ. Hence for a typical value of $A_{vol} = 10^5$, it follows that the input resistance for voltage follower is $10^5$ MΩ or more. This virtually means that the OPAMP has no effect on the input signal even if the latter is characterised by a high internal resistance.

Detailed analysis of the equivalent circuit of an OPAMP reveals that the output resistance, $R_0$, looking into the output terminals, which is just under 100 Ω, is reduced to $R_0 / A_{vol}$ when 100% feedback, as in the case of voltage follower, is used. Consequently, the output impedance of the OPAMP has a value of 0.001 Ω. This means that on the output side, the OPAMP connected in a voltage follower configuration behaves like a voltage source *i.e.*, a source having a negligible output (internal) resistance.

The *voltage follower,* therefore, possesses the following important characteristics :

(*i*) it has large input resistance (impedance)

(*ii*) it has a unity gain

(*iii*) it has a small value of output resistance (impedance).

With high input impedance and a low output impedance this OPAMP when connected in the circuit will have negligible loading effects. Therefore, this circuit can serve as an ideal **Buffer** (or isolation circuit). In this way, it can be made to prevent the disturbance of one part of a circuit affecting another can be prevented.

### 26.2.4 Inverting Mode

Figure 26.11 shows a circuit wherein an OPAMP is depicted in an inverting mode. The inververting terminal (pin 3) is tied to the ground reference and therefore $v_+ = 0$. Insertion of this fact into the basic equation 26.1 yields the following result :

$$v_0 = A_{vol}(v_+ - v_-) = -A_{vol}\, v_- \quad \text{...(26.14)}$$

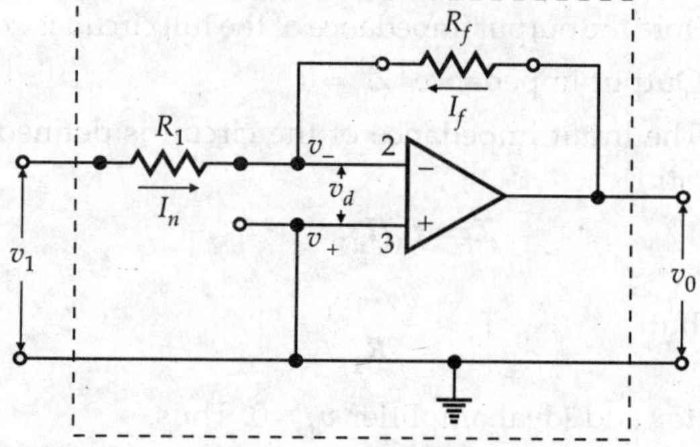


**Fig. 26.11** OPAMP working in inverting mode.

This expression may now be used to derive information about the magnitude of $v_-$ in this particular application of OPAMP. Typical values of $v_0$ in these applications is determined by value of supply Voltage $V_{CC}$. Thus for a $V_{CC} = 15$ V, and $A_{vol} = 1{,}00{,}000$ the maximum values to be expected for $v_-$ for normal operation is therefore, 15/1,00,000 $v$ or 15 μV. This quantity is so small that it can be accepted as being at ground potential. Under these circumstances it is interesting to note that the resistance looking into the circuit is $R_1$. Therefore, when the OPAMP is used in inverting mode, the input resistance is given by :

$$R_i = R_1$$

Note that the input resistance for the inverting mode is considerably less than the very high value encountered in the non-inverting mode of the OPAMP.

Applying Kirchhoff's current law at 2 in Fig. 26.11,

$$I_n + I_f = 0$$

$$\frac{v_1}{R_1} + \frac{v_0}{R_f} = 0$$

or

$$v_0 = -\frac{R_f}{R_1} v_1 \quad \text{...(26.15)}$$

26

Closed loop gain is,

$$A = \frac{v_0}{v_1} = -\frac{R_f}{R_1} \qquad \text{...(26.16)}$$

In case impedances are used in place of resistances, closed loop gain is,

$$A = -\frac{Z_f}{Z_1} v_1 \qquad \text{...(26.17)}$$

It should be remembered that there is simple relationship between $v_0$ and $I_n$ ; $v_0 = -I_n R_f$ as $(I_f = -I_n)$. The input and output impedances of the inverting amplifier (Fig. 26.11) circuit are of interest. The output impedance of an ideal OPAMP is zero and therefore the output impedance of the full circuit is zero.

Output impedance $\mathbf{Z_0} = 0$

The input impedance of the circuit is defined by the ratio

$$Z_i = v_1 / I_n .$$

But $$I_n = \frac{v_d - v_1}{R_1}$$

and for and ideal amplifier $v_d = 0$. Thus,

$$I_n = v_1 / R_1$$

and so input impedance, $\mathbf{Z_i = R_1}$.

### 26.2.5 Non-Ideal Operational Amplifiers

The characteristics of an actual operational amplifier differ from those of an ideal OPAMP as follows :

(*i*) Input resistance, $R_i = 10^4$ to $10^9\,\Omega$.

(*ii*) Output resistance, $R_0 = 10\,\Omega - 5\,\text{k}\Omega$.

(*iii*) Open loop gain, $A_{vol} = 10^4$ to $10^9$.

(*iv*) Limited Bandwidth.

The effect of gain of the practical amplifiers not being infinitely large can be found as under. For non-ideal amplifiers $v_d \neq 0$.

Now, $$I_n + I_f = 0$$

Therefore, $$\frac{v_1 - v_d}{R_1} + \frac{v_1 - v_d}{R_f} = 0$$

and $$v_d = -v_0 / A_{vol}$$

The above relationship can easily be re-arranged to give an overall gain as,

$$A = \frac{v_0}{v_1} = -\frac{R_f}{R_1} \cdot \frac{A_{vol}\dfrac{R_1}{R_1 + R_f}}{1 + A_{vol}\dfrac{R_1}{R_1 + R_f}} \qquad \text{...(26.18)}$$

The term $\dfrac{R_1}{R_1 + R_f}$ is the feedback factor β which is the ratio of output voltage feedback to input.

$$\text{Feedback factor} \quad \beta = \frac{R_1}{R_1 + R_f} \qquad \text{...(26.19)}$$

Equation 26.18 can be written as,

$$A = -\frac{R_f}{R_1}\left[\frac{1}{1 + (1 / A_{vol}\beta)}\right] \qquad \text{...(26.20)}$$

The term $A_{vol}\,\beta$ is referred to as *loop gain* and is extensively used in determining how the non-ideal amplifier operates.

$$\text{Loop gain} = A_{vol}\beta \qquad \text{...(26.21)}$$

If the loop gain $(A_{vol}\,\beta)$ is much greater than 1, the voltage gain for an inverting amplifier reduces to $(R_f / R_1)$ as shown by Eqn. 26.20.

This is the same as given by Eqn. 26.16 which is used for idealised amplifier.

While feedback reduces the gain of OPAMP by a factor equal to loop gain $A_{vol}\,\beta$, there is a corresponding improvement in all other amplifier specifications.

An amplifier's stability, input resistance, bandwidth, and output resistance are improved as the loop gain increases.

For an OPAMP with finite open loop gain,

$$R_{if} = R_i(1 + A_{vol}\beta) \qquad \text{...(26.22)}$$

$$BW_f = BW(1 + A_{vol}\beta) \qquad \text{...(26.23)}$$

$$R_{0f} = \frac{R_0}{1 + A_{vol}\beta} \qquad \text{...(26.24)}$$

where $R_{if}$, $BW_f$ and $R_{0f}$ are respectively the input resistance, bandwidth and output resistance with feedback.

Now examine what effect the parameters of a practical OPAMP have on the input and output impedances of an inverting amplifier. For the output impedance it is necessary to derive an expression for the self impedance $Z_0$ of the Thevenin Equivalent circuit which represents the output of the inverting voltage amplifier. A voltage generator having a voltage of $A_{vol}\,V_1$ is connected in series with $Z_0$, provides an inverted voltage signal as shown in Fig. 26.12.

By replacing the input signal by a short circuit $V_1 = 0$, the generator voltage $A_{vol}\,V_1$ disappears, thus leaving $Z_0$ on its own in the Thevenin equivalent circuit. Thus by connecting a hypothetical voltage source $V$ at the output terminals, and determining the current $i$ that flows, the output impedance $Z_0 = V.i$ is obtained.

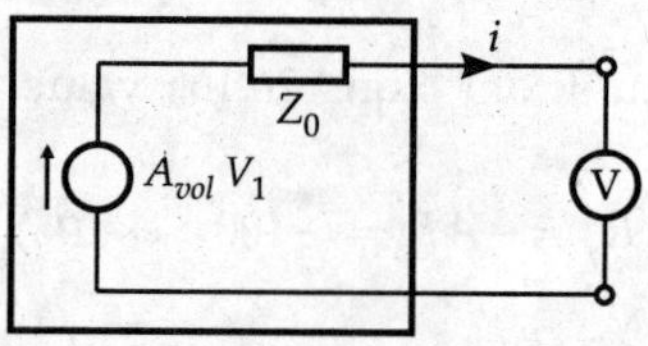


**Fig. 26.12** OPAMP circuit with a hypothetical voltage **V**.

Applying these constraints to the inverting amplifier gives the circuit of Fig. 26.13. The OPAMP model is based on a voltage controlled source with a gain of $-A_{vol}$. The OPAMP output impedance $r_{oa}$, has been added to the model to determine its effect on the circuit impedance $Z_0$. The OPAMP input impedance $r_{ia}$ is included for completeness.

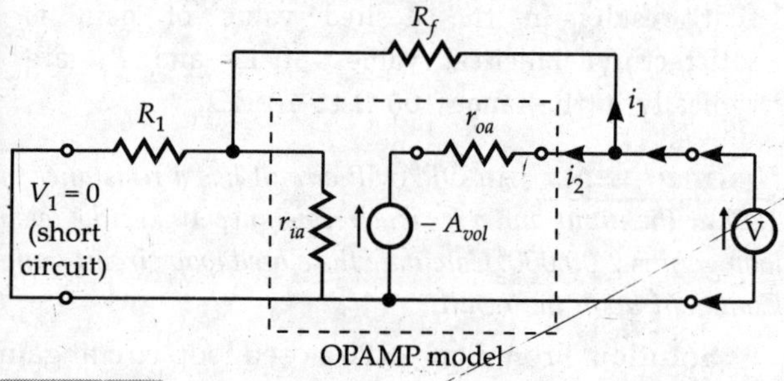


**Fig. 26.13** Inverting voltage amplifier incorporating OPAMP model.

The first step is to notice that current $i$, produced by a hypothetical voltage $v$ is split into two parts,

$$i = i_1 + i_2 \quad ...(26.25)$$

current $i_1$, flows through $R_f$ and through $R_1$ and $r_{ia}$ which are effectively connected in parallel. Thus,

$$i_1 = \frac{V}{R_f + \dfrac{R_f\, r_{ia}}{R_f + r_{ia}}} \quad ...(26.26)$$

current $i_2$ flows through $r_{oa}$

$$i_2 = \frac{R - (-A_{vol}\, v_d)}{r_{oa}} = \frac{V + A_{vol}\, v_d}{r_{oa}} \quad ...(26.27)$$

The voltage $v_d$ has been introduced into the expression. This has to be eliminated because the output impedance $Z_0 = V/i$ involves only the voltage $V$. This is achieved by noting that in Fig. 26.13, $v_d$ is related to $V$ by the potential divider formed, by $\dfrac{R_1 r_{ia}}{R_1 + r_{ia}}$ and $R_2$ and thus,

$$v_d = \frac{R_1 r_{ia} / (R_1 + r_{ia})}{R_f + (R_1 r_{ia}) / (R_1 + r_{ia})} V) \quad ...(26.28)$$

Substituting for $v_d$ in Eqn. 26.27 and then substituting Eqns. 26.27 and 26.25 into Eqn. 26.25.

$$i = \frac{V}{R_f + \dfrac{r_{ia}\, R_1}{R_1 + r_{ia}}} + \frac{V}{r_{oa}} + \frac{R_1 r_{ia}}{R_f + \dfrac{r_{ia}\, R_1}{R_1 + r_{ia}}} V$$

$\therefore$ Output impedance

$$Z_0 = \frac{V}{i} = \frac{r_{oa}}{1 + \dfrac{r_{oa}}{R_f + \dfrac{R_1 r_{ia}}{R_1 + r_{ia}}} + A_{vol}\dfrac{\dfrac{R_1 r_{ia}}{R_1 + r_{ia}}}{R_f + \dfrac{R_1 r_{ia}}{R_1 + r_{ia}}}} \quad ...(26.29)$$

In order to find the input impedance, the equivalent circuit of an OPAMP is redrawn as shown in Fig. 26.14.

$$Z_i = \frac{v_1}{i_1} = \frac{i_1 R_1 + v_d}{i_1} = R_1 + Z_i' \quad ...(26.30)$$

It remains to find $Z_i' = v_d / i_1$

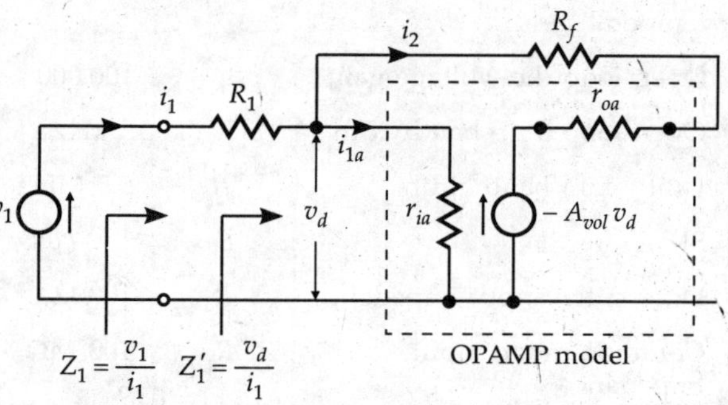


**Fig. 26.14** Input impedance of inverting voltage amplifier.

The current $i_1$ divides itself into two parts *i.e.*, $i_2$ and $i_{1a}$,

$$i_1 = i_2 + i_{1a} \quad ...(26.31)$$

The current $i_2$ flows through $R_2$ and $r_{oa}$ therefore, $i_2$ equals the voltage difference across this series combination divided by $(R_2 + r_{oa})$

$$i_2 = \frac{v_d - (-A_{vol}\, v_d)}{R_2 + r_{oa}} = v_d \frac{1 + A_{vol}}{R_f + r_{oa}} \quad ...(26.32)$$

Also, $\qquad i_{1a} = v_d / r_{ia} \quad ...(26.33)$

Substituting both Eqns. 26.32 and 26.33 in 26.31 gives

$$i_1 = v_d \left( \frac{1}{r_{ia}} + \frac{1 + A_{vol}}{R_f + r_{oa}} \right)$$

26

$$Z_1' = \frac{v_d}{i} = \frac{1}{\frac{1}{r_{ia}} + \frac{1 + A_{vol}}{R_f + r_{oa}}} \quad \ldots(26.34)$$

Total circuit impedance $\mathbf{Z}_i = R_1 + \mathbf{Z}_i$

$$= R_1 + \frac{1}{\frac{1}{r_{ia}} + \frac{1 + A_{vol}}{R_f + r_{oa}}} \quad \ldots(26.35)$$

Table 26.1 shows the typical values of parameters of an IC operational amplifier.

**Table 26.1** *Typical Parameters IC Operational Amplifier*

| *Parameter* | *Symbol* | *Value* |
|---|---|---|
| Input offset voltage | $V_{os}$ | 2 mV |
| Input offset current | $I_{os}$ | 10 nA |
| Input bias current | $I_B$ | 100 nA |
| Input offset temperature drift | – | 5 μV/°C |
| Input offset current temperature drift | – | 10 nA/°C |
| Open loop d.c. voltage gain | $A_{vol}$ | 100,000 |
| Open loop 3 db bandwidth | $BW_{vol}$ | 5 kHz |
| Unity gain bandwidth | $ft$ | 0.5 MHz |
| Slew rate | $S_r$ | 1 V / μs |
| Differential input impedance | $R_i$ | 100 kΩ |
| Common mode input impedance | $R_c$ | 100 MΩ |
| Output impedance | $R_o$ | 50 Ω |
| Common Mode Rejection Ratio | *CMRR* | 80 db |

The operational amplifiers utilize standard resistors whose resistance values are given in Table 26.2. These values apply to the first decade ranging from 1 to 9.1 Ω. The standard tolerance values for these resistances is ±5%. However, all resistance values in the first row are available as ±10% resistances. All entries in the third row are available with ±10% and 20±% tolerances.

**Table 26.2** *Standard Resistance Values*

| | | | | | |
|---|---|---|---|---|---|
| 1.0 | 1.5 | 2.2 | 3.3 | 4.7 | 6.8 |
| 1.1 | 1.8 | 2.4 | 3.6 | 5.1 | 7.5 |
| 1.2 | 1.8 | 2.7 | 3.9 | 5.6 | 8.2 |
| 1.3 | 2.0 | 3.0 | 4.3 | 6.2 | 9.1 |

**Example 26.1** *Design an OPAMP voltage amplifier to have an input resistance of 0.1 kΩ and a gain of – 100. Assume an infinite closed loop gain. Comment upon the result.*

**Solution.** From Eqn. 26.16, value of feedback resistance

$$R_f = -A R_1 = (-100 \times 1 \times 10^3) = 100 \text{ k}\Omega$$

**Comments**

In the above worked example, the requirement for input impedance (resistance in this case) to be a particular value results in unique set of design values for $R_1$ and $R_f$. In many practical cases the signal source has negligibly small self impedance, as occurs for example, when the signal emanates from the output of a previous OPAMP circuit, so that the choice of $R_1$ is not critical. In these cases any reasonable pair of values for $R_1$ and $R_f$ with a ratio that results in the desired value of gain is satisfactory. Practical values of $R_1$ and $R_f$ are typically in the range 100 Ω to 10 MΩ.

**Example 26.2** *An OPAMP circuit has a resistance of 1kΩ at the input and a feedback resistance 10 kΩ the open loop gain is 2,00,000. Calculate the closed loop circuit gain. Comment upon the result.*

**Solution.** From Eqn. 26.20 closed loop circuit gain with finite open loop gains

$$A = \frac{-R_f}{R_1}\left[\frac{1}{1 + (1/A_{vol})(R_1 + R_f)/R_1}\right]$$

$$= \frac{-10}{1}\left[\frac{1}{1 + (1/2{,}00{,}000)(1 + 10)/1}\right] = -9.9995$$

Open loop gain $A_{vol} = (-10/1) = -10.$

**Comments**

In this example, the closed loop gain is hardly affected by the finite gain of the OPAMP. Because the open loop gain of OPAMP is 2,00,000 which is very much larger than the desired circuit gain of – 10. Using the negative feedback terminology, the feedback is very strong (high open loop gain) so closed loop gain is almost independent of $A_{vol}$.

**Example 26.3** *An OPAMP which works in the non-inverting mode has the following specifications :*

| | |
|---|---|
| *Open loop gain* | = 200,000 |
| *Resistance at input end* $R_1$ | = 1 kΩ |
| *Resistance in the feedback loop* $R_f$ | = 10 kΩ |
| *Output resistance* $r_{oa}$ | = 75 Ω |
| *Input resistance* $r_{oi}$ | = 2 MΩ |

*Calculate the output and input resistances of the amplifier considering it as non-ideal one. Comment upon the results.*

**Solution.** From Eqn. 26.29, the output impedance of an OPAMP with a finite gain is,

$$\mathbf{Z}_0 = \frac{r_{oa}}{1 + \dfrac{r_{oa}}{R_f + \dfrac{R_1 \times r}{R_1 + r_{ia}}} + \dfrac{A_{vol}\dfrac{R_1 r_{ia}}{R_1 + r_{ia}}}{R_f + \dfrac{R_1 r_{ia}}{R_1 + r_{ia}}}}$$

Substituting the values given $\mathbf{Z}_0 = 0.0041\ \Omega$.

This is extremely a low value and is close to zero which is the output impedance of an ideal OPAMP.

The input impedance of OPAMP with non-ideal characteristics is given by Eqn. 26.35.

Input impedance $$\mathbf{Z}_i = R_1 + \frac{1}{\dfrac{1}{r_{ia}} + \dfrac{1 + A_{vol}}{R_f + r_{oa}}}$$

Substituting the different values, we get

$$\mathbf{Z}_i = 1000.5\ \Omega$$

**Comments**

For an ideal amplifier the output impedance $Z_0 = 0$ and for non-ideal amplifier we get a value of 0.0041 Ω which is extremely a low value close to zero. The input impedance of an ideal amplifier is $V_1 / R_1 = 1000\ \Omega$ and the value obtained with non-ideal amplifier is 1000.5 which differs by a very small amount for that obtained for an ideal amplifier.

The conclusion which can be drawn from the foregoing analysis is that provided the open loop gain of the OPAMP is very high, the circuit behaviour is very close indeed to that obtained if the OPAMP were assumed to be ideal. This fact is generally true for all OPAMPs with negative feedback. Together with low cost OPAMPs, it provides the reason for the extensive use of these circuits. This is because the circuit performance is defined almost entirely by the circuit components surrounding the OPAMP and is accurately defined if these components have low tolerances.

**Example 26.4** *In the OPAMP non-inverting circuit shown in Fig. 26.9, the OPAMP is shown to have an open loop gain of $10^5$ and the resistor values $R_1 = 1\ k\Omega$ and $R_f = 39\ k\Omega$*

(a) *Find the actual value of voltage gain.*

(b) *Find the ideal value of gain treating the open loop gain as finite.*

(c) *Calculate the percent error caused by working with ideal gain.*

**Solution.** From Eqn. 26.4 the potential of negative terminal of OPAMP is

$$v_- = \frac{R_1}{R_1 + R_f} v_0 \quad \ldots(i)$$

The differential input to the OPAMP is given by,

$$v_d = v_+ - v_- = v_1 - \frac{R_1}{R_1 + R_f} v_0 \quad \ldots(ii)$$

From (*i*) and (*ii*), the closed loop gain is,

$$A = \frac{v_0}{v_i} = \frac{A_{vol}}{1 + \dfrac{A_{vol} R_1}{R_1 + R_f}} = \frac{10^5}{1 + \dfrac{10^5 \times 1}{1 + 39}}$$

$$= 39.98.$$

(*b*) For ideal case, closed loop gain

$$A = \frac{v_0}{v_1} = 1 + \frac{R_f}{R_1} = 1 + \frac{39}{1} = 40$$

(*c*) The percent error of the ideal value compared to actual value is

$$\%\ \text{error} = \frac{40 - 39.98}{40} \times 100 = 0.05\%$$

which clearly is a small error. This helps to explain why the ideal characteristics are used in the analysis of circuits where OPAMP is used as a component.

## 26.3 OPERATIONAL AMPLIFIER CHARACTERISTICS

There are certain characteristics of operational amplifiers which must be considered for design and applications. In addition to open loop gain, input and output impedances and bandwidth, following are the important characteristics of operational amplifiers.

When an OPAMP is used as a d.c. amplifier, it is important that for zero d.c. input voltage, the output is zero. In actual practice, there will be an output d.c. voltage, called the *offset voltage* for a zero input voltage. This offset voltage, which is a serious limitation in the accuracy of a d.c. amplifier is caused by a mismatch at the differential amplifier input stage.

The *three* parameters, which cause the output offset voltage as specified by OPAMP manufacturers are the following :

- Input offset voltage ($V_{os}$)
- Input offset current ($I_{os}$) and
- Input bias current ($I_B$).

### 1. Input offset voltage ($V_{os}$)

A mismatch in the base to emitter voltage of the input differential transistor amplifiers is the cause of a d.c. offset voltage at the output. Manufacturers specify the input offset voltage ($V_{os}$) as required input to the OPAMP such that its output voltage is zero.

When an OPAMP is used as an inverting or non-inverting amplifier it will have a d.c. output offset voltage due to $V_{os}$, given by,

$$v_{o(\text{offset})} = V_{os}(1 + R_f / R_1) \quad \ldots(26.36)$$

with $\quad 1 + (R_f / R_1) = (R_1 + R_f)/ R_1 = \beta \quad \ldots(26.37)$

$$v_{o(\text{offset})} = V_{os} / \beta \quad \ldots(26.38)$$

## 2. Input Bias Current ($I_B$)

For an operational amplifier to operate properly, it is necessary to supply a d.c. bias current at the input differential amplifier stage. This current is of the order of pA to nA. The average of the two bias currents, with reference to Fig. 26.15.

Bias current $\quad I_B = \dfrac{I_{B1} + I_{B2}}{2} \quad \ldots(26.39)$

Input offset current

$$I_{os} = I_{B1} - I_{B2} \quad \ldots(26.40)$$

Input bias current

$$I_B = \frac{I_{B1} + I_{B2}}{2}$$

Input offset current

$$I_{os} = I_{B1} - I_{B2}$$

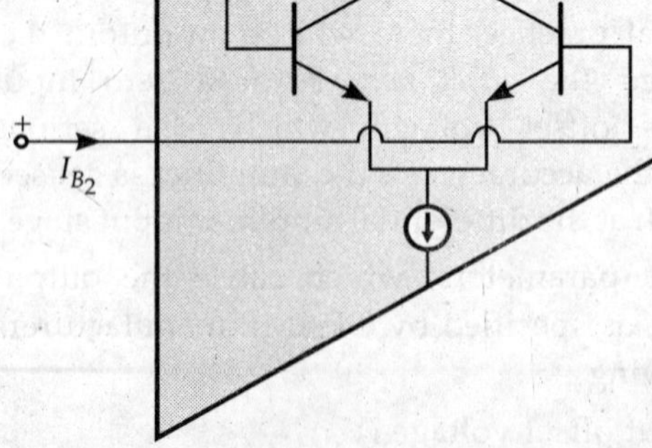


**Fig. 26.15** Input stage of an OPAMP consisting of a differential amplifier.

## 3. Input offset current

The two input transistors forming an OPAMP are not perfectly matched and a different bias current is required at each input. The difference between the two bias currents is referred to as input offset current ($I_{os}$).

Input offset current $\quad I_{os} = I_{B1} - I_{B2} \quad \ldots(26.40)$

With the OPAMP operating as an inverting and non-inverting amplifier as shown in Fig. 26.16(*a*) and 26.16(*b*) respectively, there will be an offset d.c. voltage due to the input bias currents.

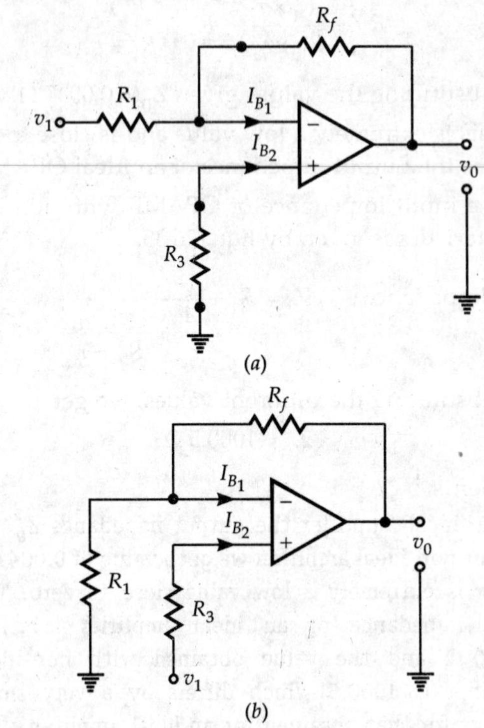


**Fig. 26.16** (*a*) Inventing configuration (*b*) Non-inverting configuration.

This offset voltage is given by :

$$v_{o(\text{offset})} = I_{B1}R_f - I_{B2}R_3(1 + R_f / R_1) \quad \ldots(26.41)$$

Now, feedback factor

$$\beta = \frac{R_1}{R_1 + R_f}$$

or $\quad 1/\beta = 1 + R_f / R_1 \quad \ldots(26.42)$

$$\therefore \quad v_{o(\text{offset})} = I_{B1}R_f - \frac{I_{B2}R_3}{\beta} \quad \ldots(26.43)$$

The output offset voltage is minimized if

$$R_3 = R_1 R_f / (R_1 + R_f)$$

with this condition,

$$v_{o(\text{offset})} = (I_{B1} - I_{B2})R_f$$

Now $\quad I_{os} = I_{B1} - I_{B2}$

$$\therefore \quad v_{o(\text{offset})} = I_{os} R_f \quad \ldots(26.44)$$

For the case where $R_3 = 0$, and since $I_B \simeq I_{B1} \simeq I_{B2}$

$$v_{o(\text{offset})} = I_B R_f$$

where $$I_B = \frac{I_{B1} + I_{B2}}{2}$$

A typical OPAMP operating as an inverting or non-inverting amplifier will have an output offset voltage due to input offset voltage ($V_{os}$), input bias current $I_B$ and input offset current, $I_{os}$, operating simultaneously. With all the three conditions operating, the output offset voltage is given by,

$$v_{o(\text{offset})} = \frac{V_{os}}{\beta} + I_{B1} R_f - \frac{I_{B2} R_3}{\beta}$$

For the special case where

$$R_3 = R_1 R_f / (R_1 + R_f)$$

$$v_{o(\text{offset})} = \frac{V_{os}}{\beta} + I_{os} R_f \qquad \ldots(26.45)$$

For $$R_3 = 0$$

$$v_{o(\text{offset})} = \frac{V_{os}}{\beta} + I_B R_f \qquad \ldots(26.46)$$

**Example 26.5** *An OPAMP has the following input parameters : $V_{os} = 6$ mV, $I_{os} = 200$ nA and $I_B = 1.5\ \mu$ A. The amplifier has a gain of 10, with $R_f = 100$ kΩ, $R_1 = 10$ kΩ and $R_3 = 9$ kΩ. Referring to Fig. 26.16.*

(i) *determine the output offset voltage if the amplifier is to be used in its inverting configuration.*

(ii) *what is the output offset voltage if $R_3 = 0$ ?*

**Solution.** We have,

$$\frac{R_f R_1}{R_f + R_1} = \frac{100\times10^3 \times 10\times10^3}{100\times10^3 + 10\times10^3} = 9000\ \Omega = 9\ \text{k}\Omega.$$

$$\beta = \frac{R_1}{R_1 + R_f} = \frac{10\times10^3}{10\times10^3 + 100\times10^3} = \frac{1}{11}$$

(i) $$v_{o(\text{offset})} = \frac{V_{os}}{\beta} + I_{os} R_f \qquad \text{(See Eqn. 26.45)}$$

$$= 6\times10^{-3}(11) + 200\times10^{-9}(100\times10^3)$$

$$= 86 \text{ mV}$$

(ii) With $R_3 = 0$

$$v_{o(\text{offset})} = \frac{V_{os}}{\beta} + I_B R_f \qquad \text{(See Eqn. 26.46)}$$

$$= 6\times10^{-3}\times11 + 1.5\times10^{-6}(100\times1000)$$

$$= 216 \text{ mV}.$$

## 4. OPAMP Frequency Response

Although the OPAMP has a very high differential gain, this very much depends upon the operating frequency. At low and moderate frequencies the gain usually begins to fall off. The frequency response of 741 OPAMP is shown in Fig. 26.17. The gain of OPAMP has a high value, typically 2,00,000 or 106 db at zero frequency. But as the signal frequency increases to 10 Hz, a capacitor inside the OPAMP causes the gain to fall off. This is the *dominant high frequency effect*. The fall in gain continues at the rate of 20 db/decade for many decades of frequency until the next high frequency comes into play at 1 MHz. After this frequency, the fall in gain accelerates with increase in frequency.

Although the first high frequency effect occurs at a low frequency of 10 Hz, the usable signal bandwidth of most circuits using OPAMP is considerably wider than this.

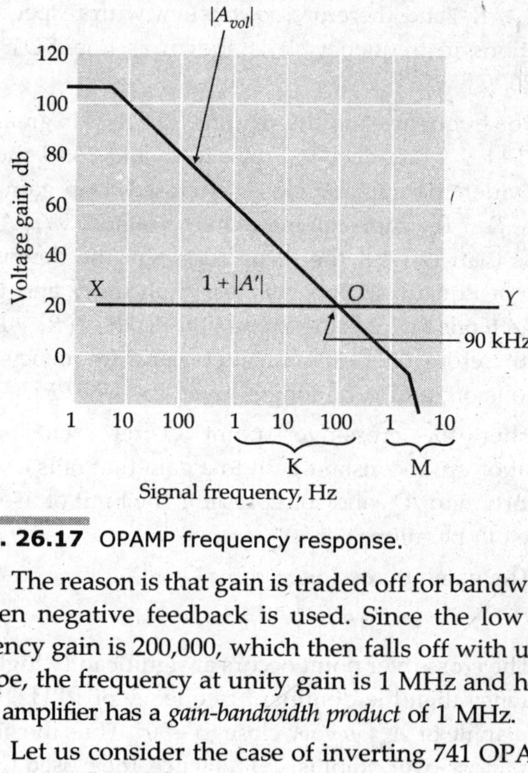


**Fig. 26.17** OPAMP frequency response.

The reason is that gain is traded off for bandwidth when negative feedback is used. Since the low frequency gain is 200,000, which then falls off with unity slope, the frequency at unity gain is 1 MHz and hence the amplifier has a *gain-bandwidth product* of 1 MHz.

Let us consider the case of inverting 741 OPAMP having a closed loop gain of − 10. It is desired to estimate the closed loop bandwidths of the circuit.

From Eqn. 26.18, the closed loop gain is given as

$$A = -\frac{R_f}{R_1} \cdot \frac{A_{vol}\dfrac{R_1}{R_1 + R_f}}{1 + A_{vol}\dfrac{R_1}{R_1 + R_f}} \qquad \ldots(26.47)$$

and $$A \approx -R_f / R_1 = A' \qquad \ldots(26.48)$$

provided following condition is satisfied

$$|A_{vol}| \frac{R_1}{R_1 + R_f} >> 1 \qquad \ldots(26.49)$$

26

The fall in $|A_{vol}|$ as signal frequencies are increased means that eventually the above condition is not satisfied. The closed loop gain $A$ then ceases to be well defined by the resistor ratio $R_f / R_1$ and this marks the limit of useful frequency range of the circuit.

Re-arranging Eqn. 26.49 gives,

$$|A_{vol}| >> 1 + R_f / R_1 = 1 + |A'| \quad \ldots(26.50)$$

The left hand side of this inequality is the differential gain $|A_{vol}|$ and its variation is given in Fig. 26.17 by the manufacturer. The right hand side of the inequality is defined by the desired closed loop gain $A'$ and so, $1 + |A'| = 1 + 10 = 11$. Expressed in db this is 20.8. This quantity is given by the resistor ratio $(1 + R_f / R_1)$ and therefore, is constant with respect to variations in frequency. It is shown as line $XOY$ in Fig. 26.17.

For frequencies in the region $XO$, $|A_{vol}|$ is greater than $1 + |A'|$, reference to Eqns. 26.47 and 26.49 show that wider the gap the closer is closed loop gain to ratio $-R_f / R_1$. For frequencies in the region $OY$, $|A_{vol}|$ is less than $1 + |A'|$, the second term in the denominator of Eqn. 26.47 becomes less than unity and the gain. $A$ tends towards the expression $A_{vol} R_f / (R_1 + R_f)$ and therefore falls off with increasing frequency as $A_{vol}$ falls off, and is no longer given by $-R_f / R_1$.

Thus the cross-over point $O$ represents the transition from constant gain to a gain that falls off in proportion to $A_{vol}$ and so represents the limit of useful range of operation.

The gain at cross-over point $O$ is ;

$$|A_{vol}| = 1 + |A'| = 11.$$

The cross-over point occurs at significantly higher frequency than the dominant frequency of 10 Hz, the phase angle of $A_{vol}$ is very close to $-90°$. Thus the gain $A_{vol}$ at cross-over point is $-j\,11$. Hence, the closed loop gain at this point is

$$A = -(10) \cdot \frac{j11 \times \frac{1}{11}}{1 + j\,11 \times \frac{1}{11}} = \frac{10}{1-j}$$

and

$$|A| = \frac{10}{\sqrt{(-1)^2 + (-1)^2}} = \frac{10}{\sqrt{2}}$$

Therefore, the closed loop gain at cross over point is reduced by a factor of $\sqrt{2}$. Hence, point $O$ corresponds to upper half power frequency point of the closed loop amplifier. In Fig. 26.16, this point is located at approximately 90 kHz. The OPAMP gain remains high down to a frequency of 0 Hz and therefore 0 Hz is the lower half power point. Therefore, the closed loop bandwidth is,

$$BW_f = 90 \text{ kHz}.$$

### 5. Slew Rate and Full Power Bandwidth

If a differential input is applied to an OPAMP, and this input changes (steps) suddenly from zero volt level to some constant voltage, the output voltage changes rapidly but not instantaneously at a rate called **Slew Rate**. It is found that the rate of change of output voltage is not dependent upon the magnitude of the input voltage step. Hence, slew rate represents the maximum rate of change that is available at the OPAMP output. The *Slew Rate* is measured in terms of V/s.

The reason why a slew rate limitation exists in an OPAMP is due to presence of capacitances in the path of the input signal, across which the voltage cannot be changed instantaneously.

The slew rate for a typical 741 OPAMP is 5,00,000 V/s. This is quite a large value and so for convenience slew rate is expressed in V / μs. Thus the slew rate for a typical 741 OPAMP is 0.5 V / μs.

In order to see the slew rate limiting effects let us consider the example of an inverting voltage amplifier when it is driven to produce a sine wave output. Figure 26.18 shows the output voltage for three different signal frequencies.

For the low frequency $f_1$ the output waveform the slope (rate of change of voltage with time) at no instant exceeds the slew rate and therefore, there is no slew rate limiting problem. At point $A$, the frequency is increased to $f_2$. The slope of the waveform is low at this point but increases with time until at point $B$ the slope of the sine wave is equal to the slew rate. Slew rate limiting then occurs.

The actual output waveform must follow a straight line path which deviates from the desired sinusoidal path. This path is shown by a broken line. This condition continues until point $C$, where the desired sine wave coincides with the slew rate limiting line. At this point the control is regained by the OPAMP. This phenomenon occurs in a similar way on the following downward slope, and so on for every half cycle as long as the frequency is $f_2$.

At point $D$ the frequency is increased to $f_3$. Slew rate limiting again occurs at point $E$ and continues, to

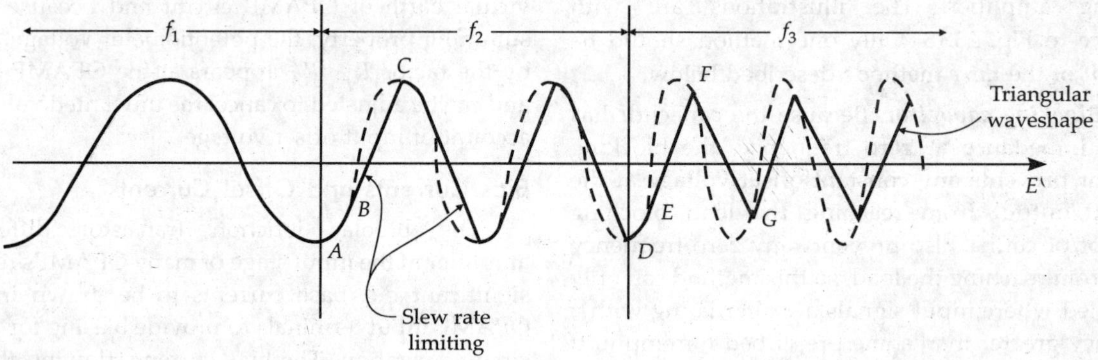


**Fig. 26.18** Progressive effects of slew rate limiting on sinusoidal output of an OPAMP circuit.

point $F$. However, at point $F$ the sine wave has a slope which exceeds slew rate and so slew rate-limiting immediately occurs thus taking the OPAMP along a straight line to $G$.

The effect of slew rate limiting, therefore is to take pieces out of the desired output waveform. This is not tolerated in many applications.

For any OPAMP circuit the following simple rule applies. The circuit is free from slew rate effects provided the slope of the waveform is always less in magnitude than the slew rate.

In many cases, the desired output signal is sinusoidal and a useful parameter for an OPAMP is the **Full-Power Bandwidth,** which is defined as the greatest frequency of a full wave sinusoidal voltage that can be the output from an OPAMP without any slew rate limitations.

The maximum output voltage, $V_{om}$, from an OPAMP is controlled by the choice of power supply voltage. If the signal tries to exceed $\pm V_{om}$, then amplitude clipping of the signal occurs. Therefore, the maximum amplitude sine wave at the output of OPAMP is,

$$v_o = V_{om} \sin (2\pi ft + \phi)$$

The slope of the output waveform is,

$$\frac{dv_o}{dt} = (2\pi f)\, V_{om} \cos(2\pi ft + \phi)$$

The greatest amplitude of slope occurs where $\cos(2\pi ft + \phi) = \pm 1$

Thus,

$$\left|\frac{dv_o}{dt}\right|_{max} = 2\pi f\, V_{om}\, |\pm 1| = 2\pi f\, V_{om} \quad \text{...(26.51)}$$

At full power bandwidth frequency $f = f_{FPBW}$, the slope of $v_o$ does not exceed the slew rate (Sr) but at one or more points equals $Sr$.

Therefore,

$$Sr = \left|\frac{dv_o}{dt}\right|_{max} \text{ at frequency } f_{FPBW} = 2\pi f_{FPBW}\, V_{om} \quad \text{...(26·52)}$$

Hence, the full power bandwidth is,

$$f_{FPBW} = \frac{1}{2\pi}\frac{Sr}{V_{om}} \quad \text{...(26.53)}$$

## 6. Unity Gain Frequency

In many cases, specifications include the frequency response including unity gain frequency. This is the frequency at which the open loop gain of the amplifier becomes unity (0 db). The low frequency gain is about 200,000 and falls to unity at about 1 MHz as shown in Fig. 26.18 the amplifier is said have a 1 MHz gain bandwidth product.

## 7. Common Mode Rejection Ratio (CMRR)

This term has been explained later in this chapter.

## 26.4 REMEDIES FOR OFFSET VOLTAGE EFFECTS

Although in negative feedback OPAMP circuit the unwanted output offset voltage is usually quite small, this voltage cannot be always neglected. This depends upon the application. If, for example, high precision is required, even a small value of offset voltage may cause the circuit output to deviate from the ideal output voltage by an unacceptable value. Difficulties can also arise if the input signal to the circuit is small as compared with $V_{os}$ and large amplification is required. Large amplification means use of large resistor ratios, which can result in large offset voltages. This is because, the offset voltages are also amplified by the same ratio as the desired input signal.

When the output offset voltages cannot be neglected, the following techniques can be used in an

26

inverting amplifier. The illustrations are with reference to Fig. 26.18. Only one method should be used out of the *three* methods described below.

1. ***Blocking capacitor.*** Because the capacitor has infinite impedance at zero frequency, the blocking capacitor prevents any constant offset voltage at the OPAMP output, from reaching the load blocking capacitor of course also prevents any zero frequency signal from reaching the load, so this method can only be applied where input signals are alternating with a frequency greater than some prescribed minimum. If the input signals contain zero frequency components, then capacitor C should not be inserted and therefore this method should not be used.

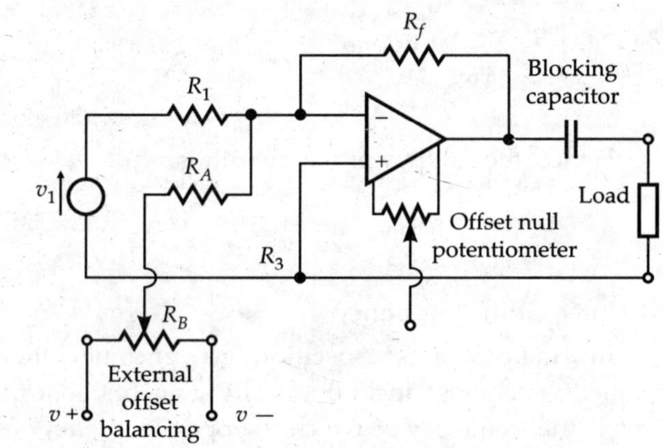


**Fig. 26.19** Methods used removal of offset voltages.

2. ***Offset null potentiometer.*** Many OPAMPS are provided with arrangements for balancing out (*i.e.*, nulling) the offset voltage. In the 741 OPAMP, a 10 kΩ potentiometer is connected between pins 1 and 5 and the sliding contact of the potentiometer is connected to negative end of supply, *V*. The itself can be used. The imbalances inside the OPAMP are corrected by adjusting the potentiometer control unit, until with $v_1 = 0$, and output $v_1 = 0$ is observed.

3. ***External offset balancing.*** The idea behind this method is to infect a voltage or current into the circuit to produce a component of output voltage which is equal and opposite to the unwanted offset component, thus canceling the latter. A variety of methods exist for doing this adjustment. The main requirement for any chosen method is that it should not interfere with the normal signal amplifying properties of the OPAMP circuit. The arrangement shown on resistor potentiometer $R_B$ in Fig. 26.19 achieves this. The potentiometer $R_B$ is connected across the supplies $V_+$ and $V_-$ and allows the voltage applied to $R_A$ to be adjusted. The other end of $R_A$ is connected to the virtual earth of OPAMP circuit and because of the summing property, the potentiometer voltage, scaled by the factor $R_f / R_A$ appears at the OPAMP output and can be adjusted to cancel the unwanted voltage on account of input offset voltage.

### Bias Currents and Offset Currents

The bipolar junction transistor differential amplifier at the input stage of many OPAMPs requires significant d.c. base currents to be drawn into the OPAMP input terminals to provide biasing for proper circuit operation. The bias currents flowing through resistances into the inverting end ($I_{B1}$) and non-inverting end ($I_{B2}$) of an OPAMP generate unwanted d.c. voltages. The general purpose 741 OPAMP has a junction transistor input stage and the maximum bias currents are specified at 0.5 μA at 25°C. Where lower bias currents are needed, FETs are used for the input stage for which typical bias current is 10 pA.

The unwanted effects of equal bias currents can be reduced by using

$$R_3 = \frac{R_1 R_f}{R_1 + R_f} \qquad \text{...(26.54)}$$

in Fig. 26.16.

However, if the two bias currents $I_{B1}$ and $I_{B2}$ are not equal on account of mismatch between the transistors, an offset current $I_{os} = I_{B1} - I_{B2}$ flows. In order to compensate for the offset current,

$$\frac{R_1 R_f}{R_1 + R_f} I_{B1} = R_3 I_{B2} \qquad \text{...(26.55)}$$

## 26.5 OPAMPS CIRCUITS USED IN INSTRUMENTATION

Some of the commonly used operational amplifier circuits are described below :

### 26.5.1 Inverter

The circuit of an operational amplifier used as an inverter is shown in Fig. 26.20. The feedback resistance $R_f$ is made equal to the resistance, $R_1$, connected to the inverting end of the amplifier.

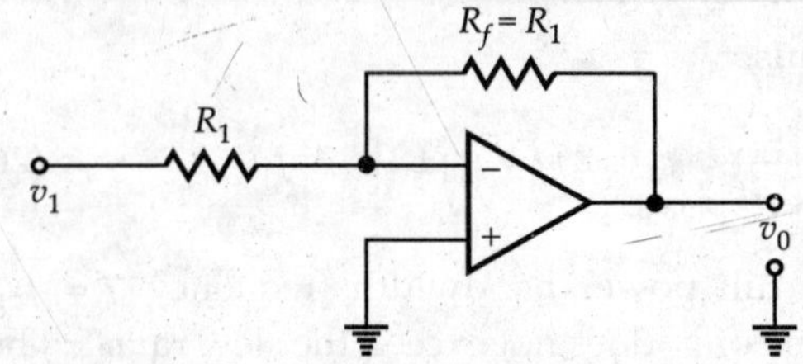


**Fig. 26.20** Operational amplifier as an inverter.

From Eqn. 26.18, output voltage,

$$v_o = -\frac{R_f}{R_1}v_1 = -v_1 \text{ (since } R_f = R_1) \quad ...(26.56)$$

It is clear that the output voltage is 180° out of phase with the input voltage.

### 26.5.2 Adder

The circuit that performs the addition of signals with amplification (if desired) is shown in Fig. 26.21 using the superposition theorem, we have,

$$\text{Output voltage, } v_o = -\left(\frac{R_f}{R_1}v_1 + \frac{R_f}{R_2}v_2 + \frac{R_f}{R_3}v_3\right) \quad ...(26.57)$$

Note that the input signals can be added together with different weighing factors, if desired.

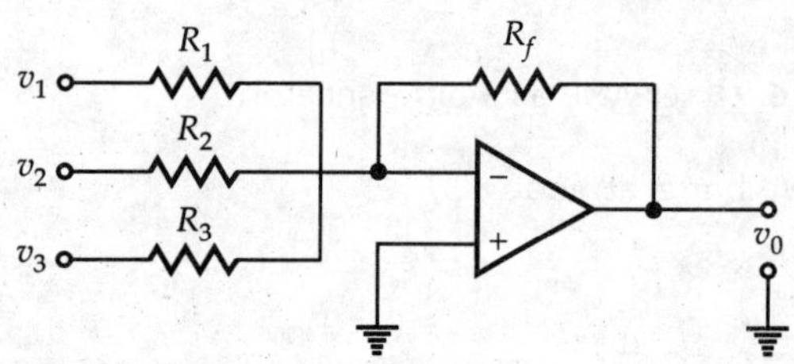


**Fig. 26.21** Operational amplifier as an adder.

However, if all the resistors are chosen to be equal, the circuit acts as a pure adder and adds the input voltages together at the output *i.e.*, in case $R_1 = R_2 = R_3 = R_f$, the circuit acts as a pure adder and output voltage is,

$$v_o = -(v_1 + v_2 + v_3)$$

*i.e.*, sum of the individual input voltages. The inversion that occurs cannot be avoided.

### 26.5.3 Subtractor

The operational amplifier circuit used for subtraction of two input signals is shown in Fig. 26.22. The output of the second operational amplifier is,

$$v_o = -\left(-v_1\frac{R_f}{R_1}\cdot\frac{R_f}{R_3} + v_2\frac{R_f}{R_2}\right) \quad ...(26.58)$$

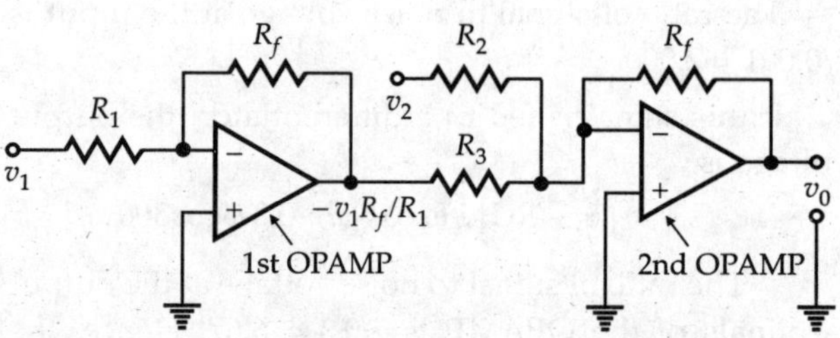


**Fig. 26.22** Operational amplifier as a subtractor.

In case, $R_f = R_1 = R_2 = R_3$, the circuit acts as a pure subtractor, with the output being :

$$v_o = v_1 - v_2$$

Another circuit which uses only one operational amplifier for the purposes of subtraction is shown in Fig. 26.23.

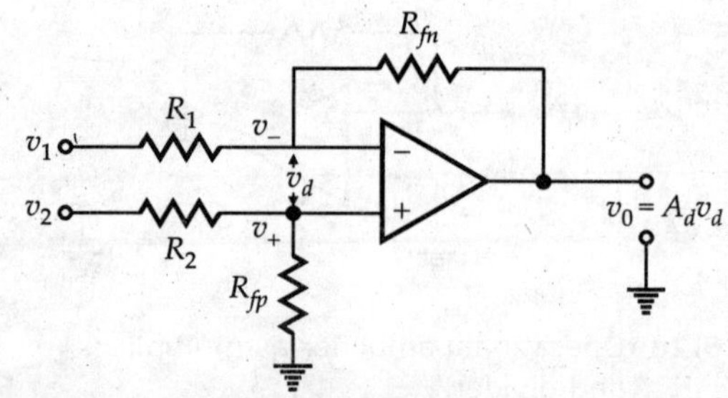


**Fig. 26.23** Operational amplifier as a subtractor.

This circuit performs the subtraction of one signal from another with amplification if desired. In order to show that the voltage is proportional to difference in input voltages, we use superposition theorem since the circuit behaves linearly.

The output voltage due to $v_1$ alone (*i.e.*, with $v_2 = 0$) is simply that of an inverting amplifier, and is equal to,

$$v_{o1} = -(R_{fn} / R_1)v_1 \quad ...(26.59)$$

The contribution to output voltage $v_0$ from $v_2$ (with $v_1 = 0$) is obtained by first realising that the voltage at the output terminals is (for large $R_i$),

$$v_+ = \left(\frac{R_{fp}}{R_{fp} + R_2}\right)v_2 \quad ...(26.60)$$

The output voltage can be written as,

$$v_{o2} = [(R_{fn} + R_1)/R_1]v_+$$

and substituting the value of $v_+$ from Eqn. 26.60, we obtain,

$$v_{o2} = \left(\frac{R_{fn} + R_1}{R_1}\right)\left(\frac{R_{fp}}{R_{fp} + R_2}\right)v_2 \quad ...(26.61)$$

However, if $R_{fn} = R_{fp} = R_f$ and $R_1 = R_2 = R$

$$v_{o2} = \left(\frac{R_f + R}{R}\right)\left(\frac{R_f}{R_f + R}\right)v_2 = \left(\frac{R_f}{R}\right)v_2 \quad ...(26.62)$$

Eqn. 26.59 becomes

$$v_{o1} = -(R_f / R)\,v_1 \quad ...(26.63)$$

Hence by using superposition by adding Eqn. 6.62 to Eqn. 26.63 gives,

$$v_0 = v_{o1} + v_{o2} = (R_f / R)(v_2 - v_1) \quad ...(26.64)$$

Now if $R_f = R$, the circuit will perform subtraction without amplification.

### 26.5.4 Multiplier and Divider

The output voltage of an OPAMP in the inverting mode is $v_0 = -(R_f / R_1)v_1$. In case $R_f > R_1$, the circuit of Fig. 26.24 acts as a multiplier and in case $R_f < R_1$, it acts as a divider.

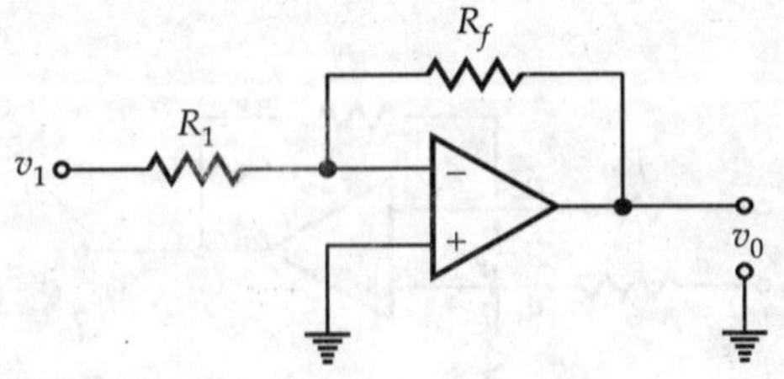


**Fig. 26.24** Operational amplifier as multiplier and divider.

Suppose $R_f = 100\ \text{M}\Omega$ and $R_1 = 10\ \text{M}\Omega$

$$v_0 = -(100/10)\ V_1 = -10v_1$$

Thus the input gets multiplied by a factor 10.

In case $R_f = 10\ \text{M}\Omega$ and $R_1 = 100\ \text{M}\Omega$

$$v_0 = -(10/100)\ v_1 = -v_1/10$$

Therefore, the input gets divided by a factor of 10.

By properly, choosing the values of $R_f$ and $R_1$, the multiplier and the divider circuits can be designed for multiplication and division by any number.

### 26.5.5 Integrator

A circuit in which the output voltage is proportional to the integral of the input voltage is shown in Fig. 26.25.

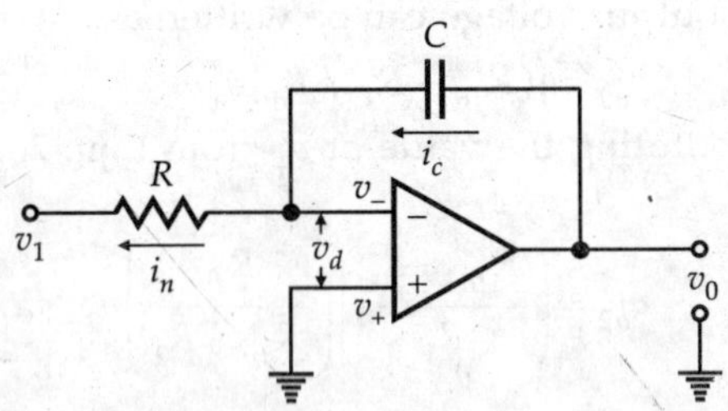


**Fig. 26.25** Operational amplifier as an integrator.

To show that the circuit shown in Fig. 26.25, acts as integrator, we use *KCL* at node $v_-$,

$$i_n = i_c$$

or
$$\frac{v_- - v_1}{R} = C\frac{d}{dt}(v_0 - v_-)$$

For infinite differential gains, $v_- = 0$

$$\therefore \quad -\frac{v_1}{R} = C\frac{d}{dt}v_0$$

By integration, we get,

$$v_0 = -\frac{1}{RC}\int v_1\, dt \qquad \text{...(26.65)}$$

The convenient values $R$ and $C$ are in the MΩ and μF range respectively.

### 26.5.6 Differentiator

The differentiating amplifier circuit is obtained by interchanging the position of resistance $R$ and capacitor C as shown in Fig. 26.26.

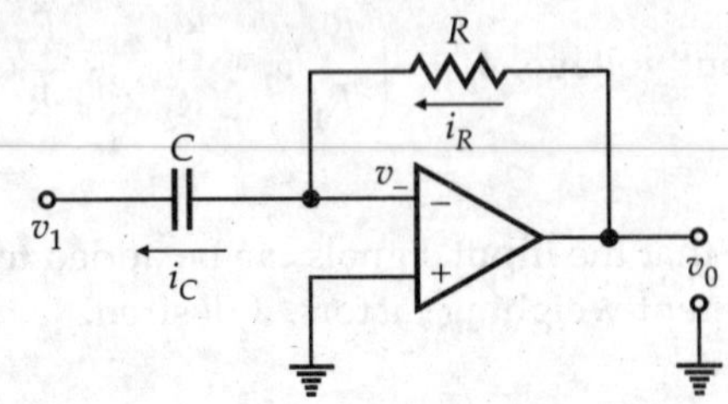


**Fig. 26.26** OPAMP as a differentiator.

We have, at node $v_-$,

$$i_c = i_R$$

$$C\frac{d}{dt}(v_- - v_1) = \frac{v_0 - v_-}{R}$$

Now $\quad v_- = 0$

$$\therefore \quad -C\frac{d}{dt}(v_1) = \frac{v_0}{R}$$

or output voltage,

$$v_0 = -RC\frac{d}{dt}(v_1) \qquad \text{...(26.66)}$$

Therefore, the output voltage is equal to the differentiated input voltage.

The operational amplifiers are normally not used as differentiators as they tend to decrease the signal to noise (S/N) ratio. This is illustrated by the following example.

Suppose the input signal is,

$$v_1 = 2\sin\omega t + 0.001\sin 300\,\omega t$$

The ratio of signal to noise voltage at the input is 2/0.001 = 2,000.

If this signal is fed to a differentiator, the output voltage is,

$$v_0 = -RC\,[2\,\omega\cos\omega t + 0.3\,\omega\cos 300\,\omega t]$$

∴ The ratio of signal to noise voltage at the output terminals of the OPAMP is 2/0.3 = 6.67. Hence, the noise level at the output increases immensely.

## 26.5.7 Active Rectifiers

In the circuits using diodes for the purposes of rectification, it is observed that the output voltage of the diode half wave rectifier circuits exhibit amplitude reduction and shortening of pulse width such that an exact half wave corresponding to the input is not reproduced.

Figures 26.27 and 26.28 show two possibilities for active half wave rectifiers which use OPAMPs without distortion. Also shown are the corresponding voltage transfer characteristics. The operation of these circuits is understood by considering the conditions necessary to open and short the diode. The circuit of Fig. 26.27 provides voltage amplification whereas that of Fig. 26.28 does not.

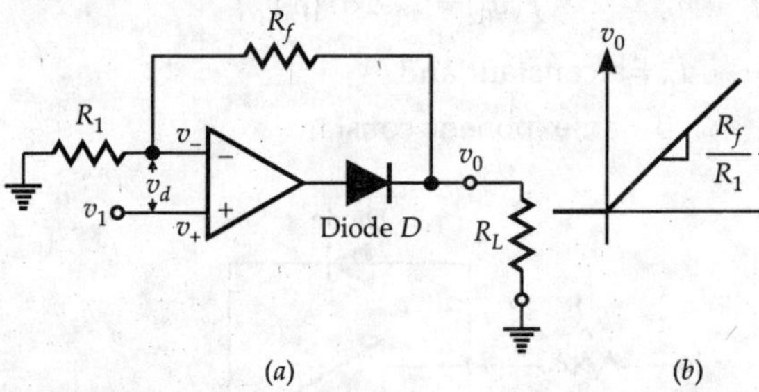


**Fig. 26.27** Active Half-wave non-inverting amplifier (*a*) Circuit (*b*) Transfer characteristics.

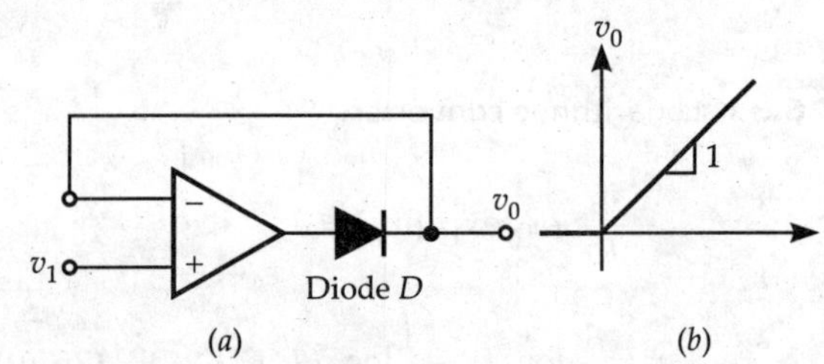


**Fig. 26.28** Half-wave voltage follower (*a*) Circuit (*b*) Transfer characteristics.

In order to demonstrate the transfer characteristics shown, consider Fig. 26.27. When $v_1 < 0$, the output voltage of OPAMP ($A_{vol}\, v_d$) is negative (since the input is connected to non-inverting end). Therefore, the diode is reverse biased and open. Therefore, the output $v_0$ is zero and this portion ($v_1 < 0$) of the transfer characteristics shown in Fig. 26.27 is verified.

For $v_1 > 0$, the output voltage of the OPAMP ($A_{vol}\, v_d$) is positive, and the diode is forward biased.

Thus the forward voltage is 0.7 V (which represents the voltage drop across the diode under forward conducting conditions). Therefore, we can write,

$$v_0 = A_{vol} v_d - 0.7 \qquad \ldots(26.67)$$

However, $v_d = v_+ - v$

where $v_+ = v_1$

and $v_- = \frac{R_1}{R_f + R_1} v_0$

Substitution into Eqn. 26.67 yields,

$$v_0 = A_{vol}\left(v_1 - \frac{R_1}{R_f + R_1}\right) - 0.7$$

Re-arranging this expression and solving for $v_0$ gives,

$$v_0 = \frac{A_{vol} v_1}{A_{vol}\left(\frac{R_1}{R_f + R_2}\right) + 1} - \frac{0.7}{A_{vol}\left(\frac{R_1}{R_f + R_1}\right) + 1}$$

Hence, for $A_{vol} >> (R_f + R_1)/R_1$ we neglect the 1 in the denominator in each term to obtain,

$$v_0 = \left(\frac{R_f + R_1}{R_1}\right) v_1 - \left(\frac{R_f + R_1}{R_1 A_{vol}}\right) \times 0.7 \qquad \ldots(26.68)$$

Notice that the last term in Eqn. 26.68 is negligible because $A_{vol}$ is very large. Thus,

$$v_0 = \left(\frac{R_f}{R_1} + 1\right) v_1 \qquad \ldots(26.69)$$

The above expression verifies the region of Fig. 26.27 in which $v_1 > 0$. Furthermore, we have accounted for diode forward voltage drop of 0.7 V. However, it has absolutely no effect on output voltage $v_0$ because of large voltage amplification of the OPAMP. Figure 26.28 is also verified since in this case $R_f = 0$ and $R_1 = \infty$.

### Active Full Wave Rectifiers

Full wave rectification can be achieved by using two half wave rectifiers in tandom. There are many circuits available for use as full wave rectifiers and one of them is shown in Fig. 26.29.

To describe the operation of this circuit, we use the results of active half wave rectifiers as obtained above. Note that the upper and lower OPAMP circuits of Fig. 26.29 are set up so that they will each provide unity gain. Also, the top portion of the circuit (containing OPAMP 2) is on when the bottom portion (containing OPAMP 1) is off and vice versa.

For positive $v_1$, OPAMP 2 produces an output voltage $v_0 = v_1$. Meanwhile in the bottom portion of, the circuit, diode $D_1$ is open and zero current flows through equal resistors ($R_f$) since the voltage at each end is $v_1$.

26

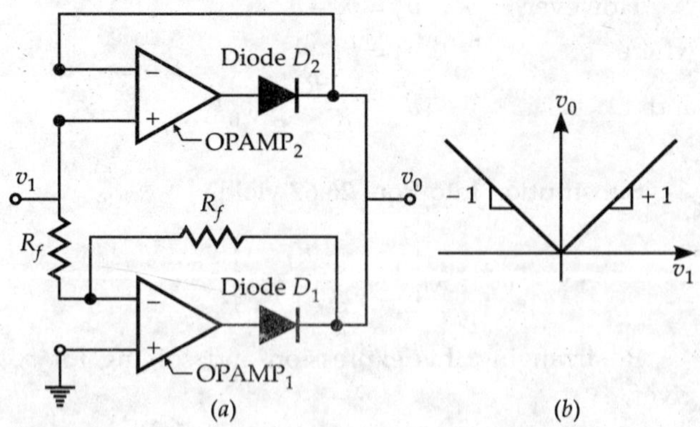

Fig. 26.29 Active Full-wave rectifier (a) Circuit (b) Transfer characteristics.

For negative $v_1$, the bottom portion of the circuit of Fig. 26.29 provides a voltage

$$v_0 = -v_1 = +|v_1|.$$

Diode $D_2$ is forced open and OPAMP 2 is in negative saturation ($v_{d2} = -2\,|v_1|$). Also with negative $v_1$ applied to the positive terminal of OPAMP 2, the output voltage of OPAMP 2 is large and negative which keeps diode $D_2$ open. Thus the full wave transfer characteristic of Fig. 26.29(b) is verified.

### 26.5.8 Comparators

Comparators are similar to OPAMPs except that $A_{vol}$ is made larger for these integrated circuits by including positive feedback in the internal circuitry. The same circuit symbol as that of OPAMP is used to represent the comparator applications with external positive feedback. The basic comparator is shown in Fig. 26.30(a) and the corresponding characteristic is

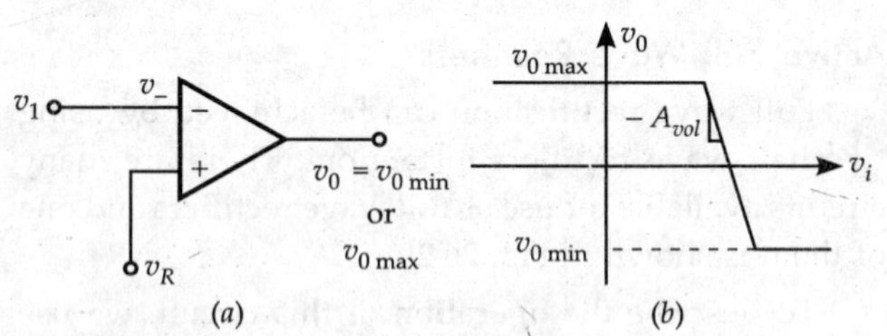

Fig. 26.30 The comparator (a) Basic Circuit (b) Transfer characteristics.

shown in Fig. 26.30(b). The extremely high voltage amplification (on account of large $A_{vol}$) leads directly to a transfer characteristic that displays an abrupt change in $v_0$ when $v_2 = v_R$. The output voltage essentially provides digital operation in that only two possible values exist, $V_{0\,min}$ and $V_{0\,max}$, since $A_{vol}$ is very large. These values are provided when $v_1 < v_R$ and $v_1 > v_R$ respectively. Note that one possibility for the reference voltage is $V_R = 0$.

In this basic comparator circuit, we see that there is no external feedback and the reference voltage $V_R$ is fixed.

### 26.5.9 Logarithmic Converter

Another application of the single ended amplifier is a logarithmic converter which is shown in Fig. 26.31. A silicon diode is placed in the feedback loop. The current of the diode depends upon the output voltage. For Fig. 26.31 we get,

$$\frac{v_1}{R_1} + f(v_0) = 0$$

where $f(v_0)$ = variation of the current with voltage and is exponential in the case of a diode

$$f(v_0) = F_0 \exp[\beta v_0]$$

where $F_0$ = a constant and

$\beta$ = an exponent constant.

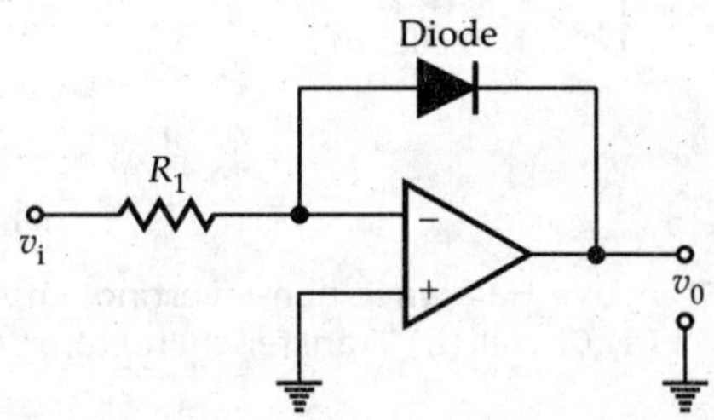

Fig. 26.31 Logarithmic converter.

$$\therefore \quad -\frac{v_1}{R_1} + F_0 \exp(\beta v_0) = 0$$

or
$$v_0 = \frac{1}{\beta}\log_e v_1 - \frac{1}{\beta}\log_e(F_0 R_1) \quad \ldots(26.70)$$

Thus the output $v_0$ is a logarithmic function of input voltage $v_1$.

### 26.5.10 Attenuated Feedback

In many applications use of low and medium value resistors are used for the attenuation of feedback as shown in Fig. 26.32. This has the advantage of using resistors which can be easily manufactured within reasonable limits of accuracy while retaining a high degree of gain. Let us consider an example. Suppose a closed loop gain of 1000 and an input impedance of 10 kΩ is desired. If the circuit of a multiplier as shown in Fig. 26.24 is used, the feedback resistor $R_f$ should have a value of $1000 \times 10 \times 10^3 = 10$ MΩ. It is difficult to obtain resistors of such high value with a tolerance limit of ±5%. Therefore, the use of high value resistors causes errors.

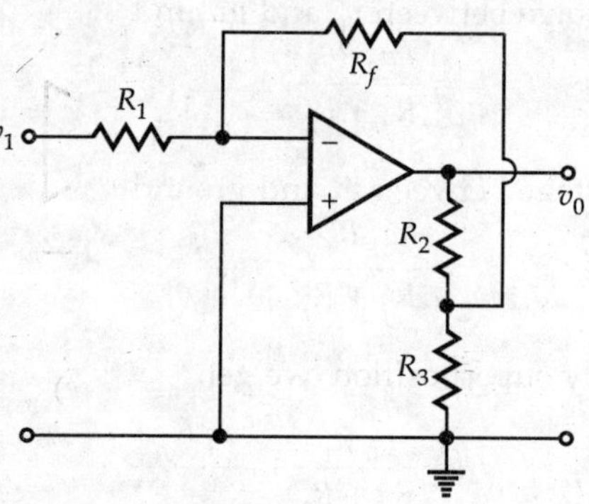


**Fig. 26.32** Operational amplifier with attenuated feedback.

The alternative is to use attenuated feedback which requires the use of low value resistors that can be manufactured with much lower tolerance limits thereby reducing the guarantee errors. The closed loop gain of circuit of Fig. 26.32 which uses an attenuated feedback is :

$$A \approx \frac{v_0}{v_1} = \frac{R_f}{R_2}\left(\frac{R_2 + R_3}{R_3}\right) \text{ with } R_1 << R_f \quad ...(26.71)$$

Now with,

$$R_1 = 10 \text{ k}\Omega, \quad R_2 = 1 \text{ k}\Omega \quad \text{and} \quad R_f = 100 \text{ k}\Omega,$$

the value of $R_3$ works out to be only 10 Ω.

The values of resistance used are much lower especially the feedback resistance which is $R_f = 100$ kΩ as against 10 MΩ in the earlier circuit. These values of resistors have much lower tolerance limits are much more stable.

### 26.5.11 Current-to-Voltage and Voltage-to-Current Converter

A *current to voltage converter* is a circuit the provides an output voltage that is directly proportional to input current. Figure 26.33 shows this type of converter with an OPAMP connected in the inverting configuration and $R_1 = 0$.

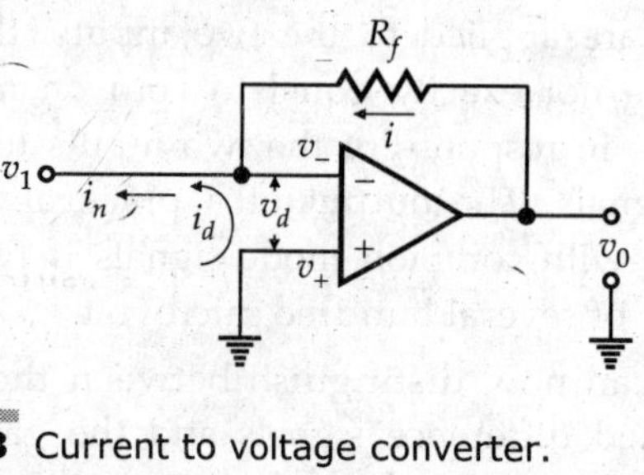


**Fig. 26.33** Current to voltage converter.

Since the OPAMP input current $i_d$ is essentially zero, the input current directly flows through $R_f$.

Also, since $v_-$ terminal is a virtual ground, the output voltage is

$$v_0 = i_n R_f \quad ...(26.72)$$

Thus the desired result is obtained.

Note that the input impedance of current to voltage converter of Fig. 26.33 is approximately zero since $v_-$ terminal is virtual ground. The input impedance is $R_f / A_{vol}$. The output impedance is quite small since $R_1 = 0$.

The circuit diagram of a voltage-to-current converter is shown in Fig. 26.34. Since $v_d$ is quite small, the input voltage $v_1$ is applied directly across $R_1$ and hence $i_{R1} = v_1 / R_1$. However this current must be supplied by the output of OPAMP. Therefore,

$$i_L = i_{R1} = \frac{v_1}{R_1} \quad ...(26.73)$$

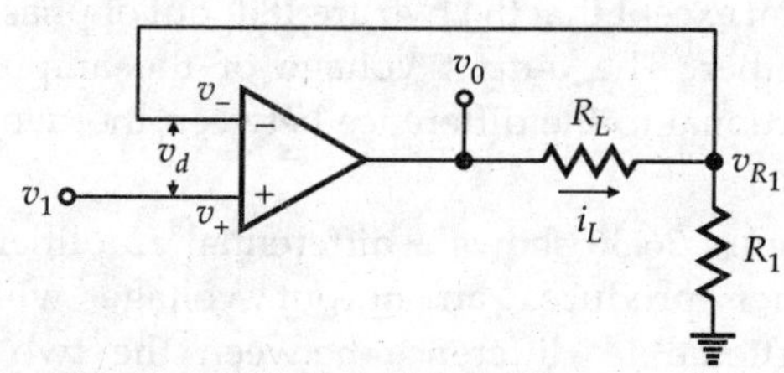


**Fig. 26.34** Voltage to current converter.

Thus, the circuit acts as voltage to current converter. It is clear from Eqn. 26.73 that the current through $R_L$ is independent of resistance $R_L$. Hence, this circuit can be used as constant current source controlled by voltage. Note that the input impedance is quite large.

### 26.5.12 Buffer Amplifier

The circuit of a unity gain buffer amplifier also called a **voltage follower** is shown in Fig. 26.35.

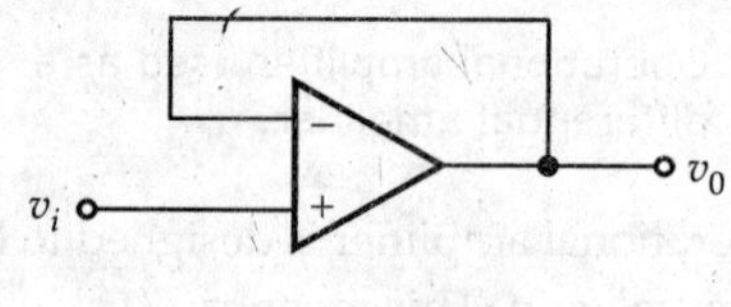


**Fig. 26.35** Unity gain buffer amplifier (voltage follower).

This circuit has a unity gain and a very high input impedance. The input impedance is essentially the input impedance of the operational amplifier itself which may be greater than 100 MΩ. The output voltage is equal to the input voltage. In fact, the output voltage tracks the input voltage from plus to negative saturation levels. Current output is limited to short circuit current of the operational amplifier and the output impedance is small, typically less than 100 MΩ. Many manufacturers market operational amplifiers with internally provide feedback. These are used as buffer amplifiers and have a very high input impedance.

The buffer amplifier is essentially an impedance transformer which converts a voltage at high impedance to the same voltage at low impedance. The use of unity gain buffer amplifiers greatly reduces the loading effects in measurement systems.

## 26.6 DIFFERENTIAL AMPLIFIER

The operational amplifier which is very important in an instrumentation system is the *differential amplifier*. In its basic form it has two inputs and two outputs. The signal available to the two outputs are identical except that the two are 180° out of phase with each other. The output voltage of the amplifier is proportional to the difference between the two input voltages.

Figure 26.36 shows a differential amplifier. This amplifier produces an output voltage which is proportional to difference between the two input voltages.

$$v_0 = A_d(v_+ - v_-) \quad ...(26.74)$$

where $A_d$ = differential gain

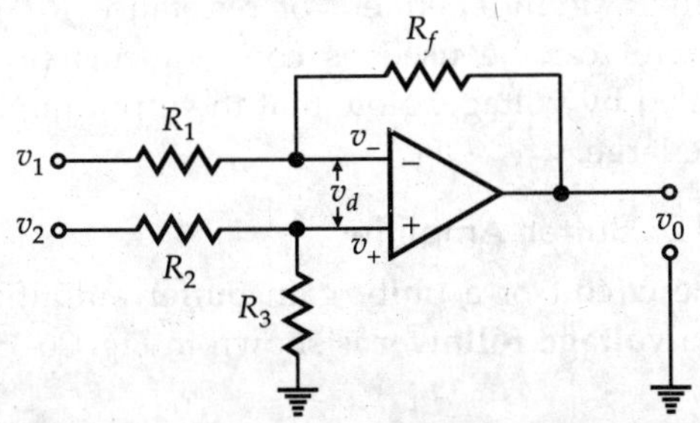


**Fig. 26.36** Operational amplifier used as a differential amplifier.

The operational amplifier is designed to have a very high differential gain. This means $v_0/(v_+ - v_-) = 10^5$ or more. Also this high gain implies that $v_+ \approx v_-$. Since the output voltage $v_0$ has a limited amplitude and therefore, the term, $v_0/(v_+ - v_-)$ is very large. The differential voltage $v_d = (v_+ - v_-)$ is very small and hence we can safely assume $v_+ = v_-$.

The input impedance of a differential amplifier is quite large. For negative feedback, the path must be from output to $v_-$. Now $v_0$ is shifted by 180° with respect to $v_-$ and 0° with respect to $v_+$.

Voltage between $v_0$ and ground

$$= \frac{v_0 R_1}{R_f + R_1} \quad ...(26.75)$$

Voltage between $v_0$ and ground

$$= \frac{v_1 R_f}{R_f + R_1} \quad ...(26.76)$$

∴ By superposition, we get,

$$v_- = \frac{v_0 R_1}{R_f + R_1} + \frac{v_1 R_f}{R_f + R_1} \quad ...(26.77)$$

Now $\quad v_+ = \dfrac{v_2 R_3}{R_2 + R_3} \quad ...(26.78)$

Since $v_- \approx v_+$, we have,

$$\frac{v_0 R_1}{R_f + R_1} + \frac{v_1 R_f}{R_f + R_1} = \frac{v_2 R_3}{R_2 + R_3}$$

or $\quad v_0 = v_2 \dfrac{R_3(R_f + R_1)}{R_1(R_2 + R_3)} - v_1 \dfrac{R_f}{R_1} \quad ...(26.79)$

Putting

$$R_f / R_1 = R_3 / R_2 = A_d, \text{ we get,}$$

$$v_0 = v_2 A_d - v_1 = A_d(v_2 - v_1) \quad ...(26.80)$$

$A_d$ is called **Difference Mode Gain**. The signal $v_d = (v_+ - v_-)$ is called the **Difference Mode Signal** or simply **Difference Signal**,

$$A_d = v_0 / (v_+ - v_-) = v_0 / v_d \quad ...(26.81)$$

### 26.6.1 Common Mode Signals

The output voltage of a differential amplifier is proportional to the difference between the input voltages. Thus if there is no difference between the two input voltages *i.e.*, when they are equal, the output voltage is zero. Equal inputs are known as **common mode signals** because the input signal is common to both inputs.

However, in actual practice when equal input voltages are applied to the two inputs, the output voltage is not exactly equal to zero on account of difference in response of the two inputs to common mode signals. The output of a practical difference amplifier with common mode signals is typically of the order of several hundred microvolt.

We can now distinguish between the gain for the desired difference signals and the gain for, the common mode signals referring to the former as the **difference mode gain** and the latter as the **common mode gain**.

Common mode gain

$$A_c = v_0 / v_c \qquad ...(26.82)$$

where, $v_c$ = common mode input signal.

In fact $v_c$ is the value of the input signal that is common to both inputs, and $v_0$ is the output signal resulting from the common mode input signal.

**Common mode rejection ratio (CMRR).** The common mode rejection ratio (*CMRR*) is defined as :

$$\text{CMRR} = A_d / A_c \qquad ...(26.83)$$

The common mode rejection ratio (CMRR) is frequently expressed in db.

$$\therefore \qquad \text{CMRR} = 20 \log_{10} (A_d / A_c) \qquad ...(26.84)$$

In an ideal amplifier, this ratio should be infinity but in practice, it is not, and has a finite value. CMRR is a measure of ratio of the desired signal to the undesired signal. The larger the CMRR, the better is the amplifier.

### 26.6.2 Advantages of Differential Amplifiers

The advantages of differential amplifiers are :

#### 1. Noise Immunity

The differential amplifier can be used in situations where the operation of single ended operational amplifiers is impossible on account of interference from pick up and ground potential differences. When a difference amplifier is used, ideally it responds only to the difference signal between input terminals and ignores common mode signals like the noise pick-up and ground voltages that appear in phase on both signal lines.

The differential amplifiers are extensively used in equipment such as electronic voltmeters and oscilloscopes. Usually the first stage of such an instrument is known as **Differential Input**, or **Balanced Input** or **Push-pull Input**.

The advantage of this form of construction is that the two input terminals are located physically close to each other. Consequently noise pick-up (such as hum from power lines or electrical equipment, "static" from flourescent lights etc.) tends to be equal on the two terminals. Also, the wires can be twisted together to equate the pick-up on both the terminals. Therefore the noise and hum pick-up will be a common mode signal and so the response of a differential amplifier to this is negligible on account of high value of CMRR.

On the other hand, the desired signal can be arranged to be equal and 180° out of phase at the two input terminals also, so that the difference signal $v_d = v_+ - v_-$ is additive. Thus a 50 mV signal is made + 50 mV at one terminals and – 50 mV at the other terminal, with the result that the difference mode input signal is $v_+ - v_- = 100$ mV. When a signal source provides this type of output signal, it is known as a **"balanced output"** or **"push-pull"** source. Fig. 26.37 shows an arrangement when the output voltage of a Wheatstone bridge is measured with the help of an instrument with an amplifier in-between. The signal source (Wheatstone bridge) and instrument are both

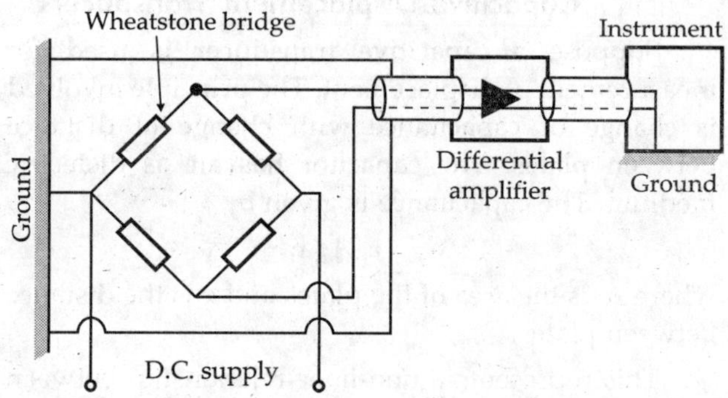


**Fig. 26.37** Use of a differential amplifier, when both source and instrument have an earth point.

earthed. Therefore, a common mode signal is applied to the amplifier and naturally it requires the use of a differential amplifier so as to eliminate spurious signals on account of earth potentials.

#### 2. Drift Immunity

One of the major problems in amplifier design is the change in the value and of voltage levels due to aging and variations in temperature .These changes take place slowly and are therefore, called **Drift**. The problem of drift may assume very serious proportions in electronic instrumentation systems as even an extremely small change may affect the results drastically.

The differential amplifier has inherent capabilities of eliminating problem of drift. The two inputs and two outputs that characterise the amplifier are the result of the fact that it is made up of two identical amplifiers connected side by side. Thus any changes, due to aging or variations in temperature which occur on one side are duplicated on the other side. Therefore, these effects may be considered as two identical signals, for in other words a common mode signal. Since, a differential amplifier has at high value of CMRR, the net effect of the drift is almost negligible. In case the two sides of the amplifier are made absolutely identical the effect on account of aging and temperature changes will affect the two sides equally and hence there will be no output, on account of these changes. Thus, the drift problem is completely eliminated.

On account of features described above the differential amplifier construction is used for the early stages of oscilloscope and electronic voltmeter amplifiers, where low drift is extremely important.

### 26.6.3 Use of Operational Amplifier with Capacitive Displacement Transducers

Suppose a capacitive transducer is used for measurement of displacement. The principle involved is change of capacitance with change of distance between plates. The capacitor has air as dielectric medium. The capacitance is given by :

$$C_x = \varepsilon_0 A / x$$

where $A$ is the area of the plates and $x$ is the distance between plates.

This represents a non-liner relationship between output *i.e.,* output capacitance and the input displacement, $x$. A linear relationship between output and input can be obtained, by using an operational amplifier as shown in Fig. 26.38.

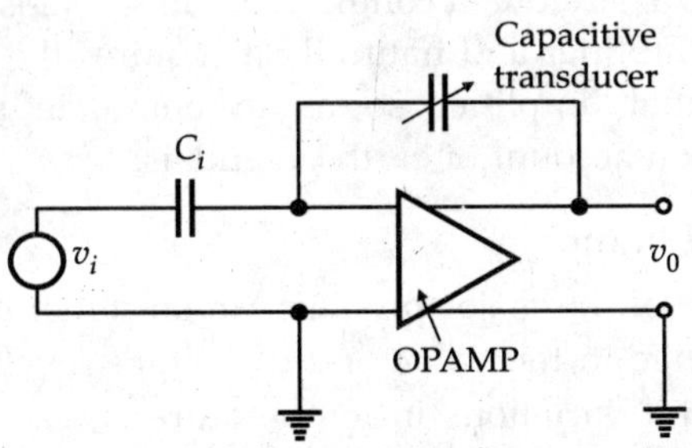


**Fig. 26.38** Operational amplifier used in conjunction with capacitive transducers.

This is done by connecting the operational amplifier in the feedback loop. Therefore, we have :

$$\frac{v_0}{v_i} = -\frac{Z_f}{Z_i} = -\frac{1}{j\omega C_x} \times j\omega C_i = -\frac{C_i}{\varepsilon_0 A / x}$$

$$\therefore \text{ Output voltage } v_0 = \left(-\frac{C_i v_i}{\varepsilon_0 A}\right) x = Kx \quad ...(26.85)$$

$$\text{where sensitivity} \quad K = \left(-\frac{C_i}{\varepsilon_0 A}\right) \text{V/m.} \quad ...(26.86)$$

Thus a linear relationship is obtained between output and input. In commercial instruments the input voltage $v_i$ is a 50 kHz sine wave of fixed amplitude. The output $v_0$ is also a 50 kHz sine wave which is connected to rectifiers to obtain a d.c. output that is read by d.c. instruments. This system is used for measurement of static displacements. For dynamic displacements, the signal $v_0$ is a modulated wave and is fed to demodulators and low pass filters to recover the true output voltage

### 26.6.4 Charge Amplifiers

Piezo-electric transducers are extensively used in accelerometers, pressure pick-ups, and load cells. The output of a piezo-electric transducers is charge. The charge is converted to voltage. This has led to development of a special kind of an amplifier, called a **charge amplifier**. The charge amplifier offers some advantages over the usual voltage amplifiers in these applications.

A charge amplifier can be described by considering it as an operational amplifier with capacitive feedback. A block diagram of the charge amplifier is shown in Fig. 26.39.

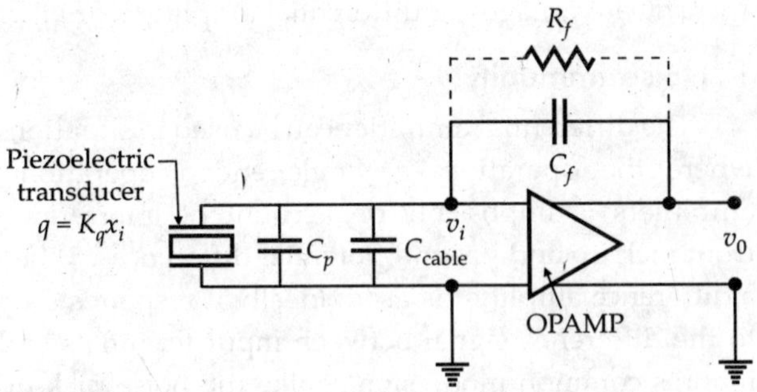


**Fig. 26.39** Charge amplifier.

$$\text{Now,} \quad \frac{v_0}{v_i} = -\frac{C_p + C_{cable}}{C_f} \quad ...(26.87)$$

where $C_p$ = capacitance of crystal ; F

$C_{cable}$ = capacitance of cable ; F

Let $C = C_p + C_{cable}$

$$\frac{v_0}{v_i} = -\frac{C}{C_f} \quad ...(26.88)$$

But $v_i = q / C$ where $q$ is the charge generated.

$$\therefore \quad v_0 = -\frac{C}{C_f} v_i = -\frac{C}{C_f} \times \frac{q}{C} = -\frac{q}{C_f} \quad ...(26.89)$$

Let $K_q$ = charge sensitivity ; C/m

and $x_i$ = input displacement ; m

$$\therefore \quad v_0 = -\frac{q}{C_f} = -\frac{K_q x_i}{C_f} \quad ...(26.90)$$

Equation 26.90 indicates that the output voltage $v_0$ is linearly related to input displacement $x_i$. It also indicates that the output changes instantaneously with input without loss in the steady state response. But unfortunately in practice, it is not so. This is because the operational amplifier does not have an infinite input resistance as is assumed, but has a finite input resistance though very high. The input resistance and the leakage resistance of $C_f$, exhibit a steady charging of $C_f$ by the leakage current till the amplifier is saturated. To overcome this problem, a resistance $R_f$ is connected across capacitor $C_f$ in the feedback path. This prevents the small leakage current to charge the capacitor heavily. The connection of $R_f$ is shown by a dotted line. The output-input relationship is now :

$$\frac{E_0(s)}{X_i(s)} = \frac{K\tau s}{1+\tau s} \quad ...(26.91)$$

where $K$ = voltage sensitivity $= k_q / C_f$ ; V/m,

and $\tau$ = time constant $= R_f C_f$ : s.

The amplitude ratio is

$$M = \frac{1}{K}\left|\left(\frac{v_0(j\omega)}{x_i(j\omega)}\right)\right| = \frac{1}{\sqrt{1+1/\omega^2\tau^2}} \quad ...(26.92)$$

Therefore, the frequency response of a charge amplifier is similar to that of a piezo-electric transducer. The amplifier exhibits attenuation of output at low frequencies. This response is shown in Fig. 26.40.

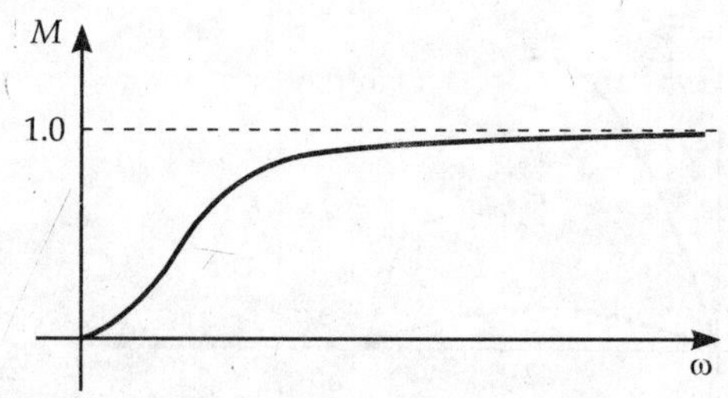


**Fig. 26.40** Frequency response of a charge amplifier.

The advantages of a charge amplifier are :

- The output voltage is proportional to the charge produced by the piezo-electric transducer.
- The amount of charge present is not affected by the cable capacitance. Since the amplifier detects charge rather than voltage, the system is completely independent of shunt capacitances and the changes in their value. This is important because of the fact that the output is not influenced by cable capacitance and hence large cable lengths can be used without any distortion of the input signal. Also the system temperature characteristics are dependent only on the charge versus temperature characteristics of the transducer and are not affected by changes in the capacitance of transducer or the cable.
- The sensitivity, K, as well as the time constant are independent of the capacitance of the crystal and also that of connecting cables. These advantages are not available in ordinary voltage amplifiers.

There are certain disadvantages of charge amplifiers which may be present in certain applications. These disadvantages may be :

- The signal to noise ratio (S/N) tends to be small.
- The natural frequency of the transducer is reduced due to loss of stiffness caused by what amounts to a short circuit across the crystal.

In practical piezo-electric transducers the value of $C_f$ varies from 10 to 1,00,000 pF. The value of $R_f$ used is between $10^{10}$ and $10^{14}$ Ω. The high value of time constant results in almost a d.c. response.

In addition to above applications operational amplifiers are used in Sample Hold Circuits voltage to current converters, current to voltage converters and digital to analog (D/A) converters.

26

**Example 26.6** *A 741 OPAMP has its input terminals short circuited. Estimate the maximum output voltage which might be expected at the OPAMP output if the offset voltage is 5 mV and the open loop gain of the amplifier is 2,00,000. The saturation limit of the OPAMP is ± 10 V.*

**Solution.** The offset voltage may get applied to either inverting or non-inverting end of the amplifier as shown in Fig. 26.41.

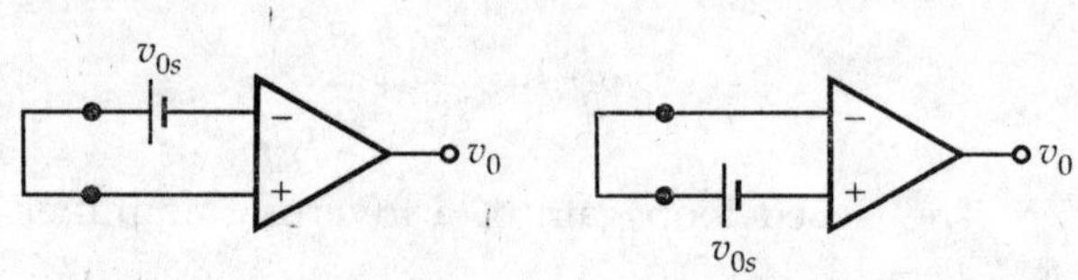


**Fig. 26.41** Offset voltage applied to OPAMP (Diagram of Example 26.6).

The output voltage

$$v_0 = A_{vol}\, v_{os}$$
$$= 2,00,000 \times 5 \times 10^{-3} = 1000 \text{ V}$$

∴ Output voltage due to input offset voltage may be either –1000 V or +1,000 V. However, the OPAMP gets saturated at ± 10 V. Therefore zero voltage applied to the OPAMP produces an output of ± 10 V.

**Example 26.7** *Derive a formula for the output voltage of a non-inverting amplifier circuit owing to voltage offset effects. Estimate the maximum output voltage due to an input voltage offset if a 741 OPAMP is used and the circuit is designed to have a gain of 10. The offset voltage is 5 mV. Assume the only OPAMP non-ideality present is offset voltage $V_{os}$. Is the output saturated on account of offset voltage if the saturation level is ± 10 V ?*

**Solution.** The circuit of an inverting OPAMP with offset voltage $v_{os}$ is shown in Fig. 26.42.

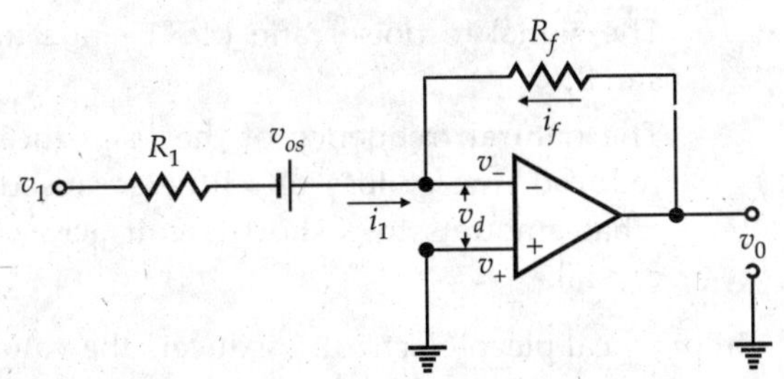


**Fig. 26.42** Inverting OPAMP with offset voltage $v_{os}$.

We have $$i_1 = \frac{v_1 + v_{os}}{R_1}$$

and $$i_f = \frac{v_{os} + v_{os}}{R_f}$$

Now $$i_1 + i_f = 0$$

$$\frac{v_1 + v_{os}}{R_1} + \frac{v_o + v_{os}}{R_f} = 0$$

or output voltage $$v_o = -v_1\frac{R_f}{R_1} - v_{os}\left(1 + \frac{R_f}{R_1}\right)$$

Therefore, the output voltage on account of input offset voltage $v_{os}$ is,

$$v_0 = -v_{os}\left(1 + \frac{R_f}{R_1}\right)$$

The closed loop gain of a inverting amplifier is $R_f / R_1 = 10$

∴ Output voltage due to offset voltage is

$$v_o = -5 \times 10^{-3}(1 + 10) = 55 \times 10^{-3}\text{ V} = 55\text{ mV}.$$

For a real circuit the actual output voltage owing to offset lies in the range – 55 mV to + 55 mV.

The output is not saturated as the saturation voltage is 10 V.

**Example 26.8** *The slew rate of an operational amplifier is 20 V/μs. The OPAMP has saturation voltages of ± 10 V, calculate the maximum switching time from one saturation region to the other.*

**Solution.** Difference in voltage of saturation levels of OPAMP = 10 – (–10) = 20 V.

∴ Switching time

$$t = \frac{\text{Voltage difference}}{\text{slew rate}} = \frac{20}{20} = 1\ \mu\text{s}$$

**Example 26.9** *Sketch the output voltage for the circuit shown in Fig. 26.43 with $v_1$ = 5 sin ωt, $R_f$ = 10 kΩ and $R_1$ = 1 kΩ.*

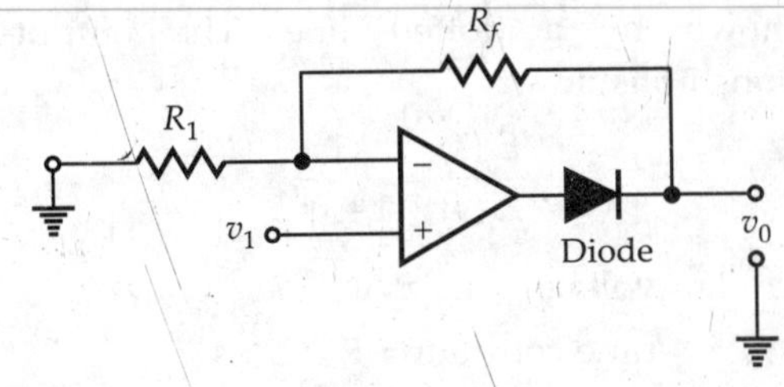


**Fig. 26.43** Diagram of a half wave rectifier of Example 26.9.

**Solution.** The circuit shown in Fig. 26.43 is a half wave circuit. The sine wave is amplified by a factor $(R_f / R_1 + 1) = (10/1 + 1) = 11$

The input and output voltages are shown in Fig. 26.44.

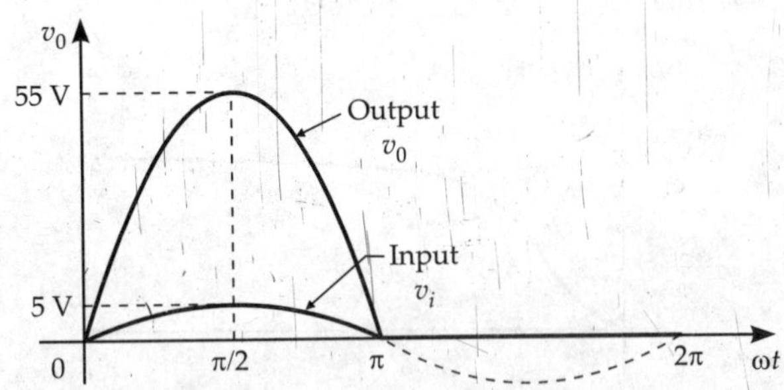


**Fig. 26.44** Half wave rectification (Example 26.9).

**Example 26.10** *For the OPAMP circuit shown below, calculate the output voltage 500 kΩ.*

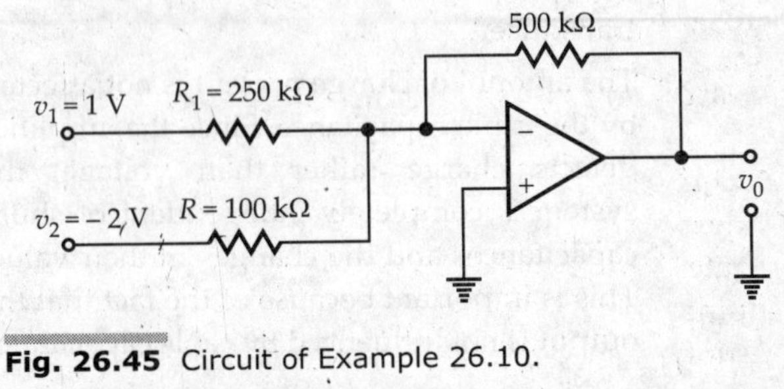


**Fig. 26.45** Circuit of Example 26.10.

**Solution.** This is simply an adder circuit for which the output voltage is

$$v_0 = -\left(\frac{R_f}{R_1}v_1 + v_2\frac{R_f}{R_2}\right)$$

$$= -\left(\frac{500}{250}\times 1 - 2\times\frac{500}{100}\right) = 8\text{ V}$$

***Example 26.11*** *(a) An inverting OPAMP has a resistance of 100 kΩ in its feeds back path and a resistance of 1 kΩ at its input terminals. What is the gain ? (b) An operational amplifier is to be built with its input resistance 10 kΩ. What should be the value of feedback resistance if it acts as a multiplier with a factor of 10 ?*

**Solution.** (*a*) Gain

$$A = \frac{R_f}{R_1} = \frac{100\times 10^3}{1\times 10^3} = 100$$

(*b*) For a multiplier

$$\frac{v_0}{v_1} = \frac{R_f}{R_1}$$

or $$10 = \frac{R_f}{10} \quad \text{or} \quad R_f = 100\text{ k}\Omega$$

***Example 26.12*** *Three signals $v_1$, $v_2$ and $v_3$ are to be weighed in the ratio of 1, 0.5 and 0.333, multiplied by – 10 and added by an operational amplifier. If the feedback resistance of the operational amplifier is 10 kΩ, what are the values of input resistances $R_1$, $R_2$ and $R_3$ ? For 1 V on each input, the output is to be sum of – 10 V, – 5 V and – 3.33 V.*

**Solution**. Output voltage

$$v_0 = -\left[\frac{R_f}{R_1}v_1 + \frac{R_f}{R_2}v_2 + \frac{R_f}{R_3}v_3\right]$$

Now $$\frac{R_f}{R_1} = 10 \quad \therefore \quad R_1 = \frac{10}{10} = 1\text{ k}\Omega$$

$$\frac{R_f}{0.5\,R_2} = 10 \quad \therefore \quad R_2 = \frac{10}{0.5\times 10} = 2\text{ k}\Omega$$

$$\frac{R_f}{0.333\,R_3} = 10 \quad \therefore \quad R_3 = \frac{10}{0.333\times 10} = 3\text{ k}\Omega$$

***Example 26.13*** *An operational amplifier is used as an integrator to produce a ramp signal of – 10 V/ms. Design the circuit for this.*

**Solution.** A ramp output signal is obtained by integrating a constant voltage as shown in Fig. 26.46.

$$v_0 = -\frac{1}{RC}\int v_1\,dt = -\frac{t}{RC}v_1$$

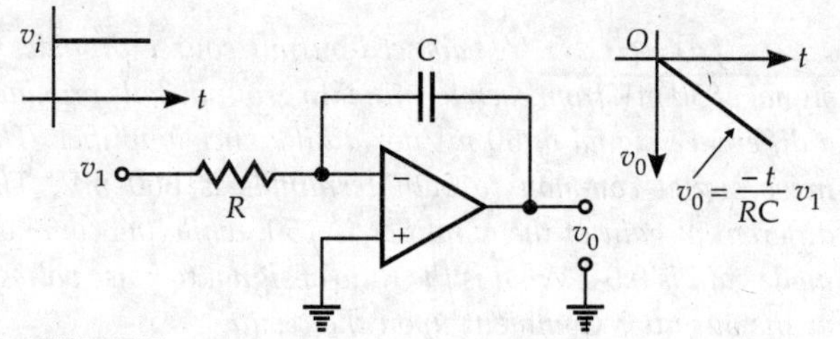


**Fig. 26.46** Generation of a ramp voltage.

Let $\quad v_1 = 10\text{ V}$

$\therefore \quad v_0 = (-10/RC)t$

Thus for production of a ramp of $v_0/t = -10/RC = -10\text{ V/ms}$.

$$RC = 1\text{ ms}$$

$$C = 1\,\mu\text{F} \quad \text{and} \quad R = 1\text{k}\Omega,$$

$$RC = 1\times 10^3 \times 1\times 10^{-6} = 10^{-3}\text{ s} = 1\text{ ms}$$

The ramp generator requires a constant voltage source of amplitude 10 V, an input resistance $R = 1\text{ k}\Omega$ and a capacitance $C = 1\,\mu\text{F}$ in the feedback path.

***Example 26.14*** *(a) A differential amplifier has two inputs of 5 mV and 3 mV. The output is 300 mV. What is the amplifier gain ? (b) If now inputs of 153 mV and 155 mV are applied, what is output voltage ?*

**Solution.** (*a*) Difference mode gain

$$A_d = \frac{v_0}{v_2 - v_1} = \frac{300\times 10^{-3}}{(5-3)\times 10^{-3}} = 150.$$

(*b*) Output voltage

$$e_0 = A_d(v_1 - v_2) = 150(155 - 153) = 300\text{ mV}$$

Note that the inputs in the second case are much larger than what they are in the first case. Same output voltage is obtained since the difference of the two input voltages is the same *i.e.*, 2 mV.

***Example 26.15*** *A differential amplifier with a common-mode input of 500 mV and a difference-mode input of 30 mV has an output of 5 mV due to the common mode input and 3 V due to difference-mode input. Find the difference-mode and the common mode gains. Find also the common-mode rejection ratio.*

**Solution.** Difference-mode gain

$$A_d = \frac{v_0}{v_d} = \frac{3}{30\times 10^{-3}} = 100$$

Common-mode gain

$$A_c = \frac{v_0}{v_c} = \frac{5\times 10^{-3}}{500\times 10^{-3}} = 0.01$$

$\therefore$ Common-mode rejection ratio CMRR

$$= \frac{A_d}{A_c} = \frac{100}{0.01} = 10{,}000 = 80\text{ db}$$

26

**Example 26.16** *A balanced-output source provides a signal of 30 mV from each terminal to ground. This provides a difference signal of 60 mV for a difference amplifier. The noise signal common to both terminals is 600 mV. The differential gain of the amplifier is 150, while the common mode gain is 0.04. What is the ratio of signal to noise voltage at the output ? Comment upon the result.*

**Solution.** The balanced source provides an output of 30 mV from each terminal to ground.

Thus $v_2 = 30 \text{ mV}, \quad v_1 = -30 \text{ mV}$

$\therefore$ Difference-mode input

$$v_d = 30 - (-30) = 60 \text{ mV}$$

Signal output voltage

$$v_{os} = A_d\, v_d = 150 \times 60 \times 10^{-3} = 9 \text{ V}$$

The noise at the input terminals has a magnitude of 600 mV.

$\therefore$ Noise output voltage

$$v_{oN} = A_c\, v_c = 0.04 \times 600 \times 10^{-3} \text{ V} = 24 \text{ mV}.$$

$\therefore$ Signal to noise voltage ratio

$$= \frac{v_{os}}{v_{oN}} = \frac{9}{24 \times 10^{-3}} = 375$$

Common mode rejection ratio CMRR

$$= \frac{A_d}{A_c} = \frac{150}{0.04} = 3750$$

**Comments**

We find that the noise signal is 10 times as great as the desirable signal at the input but at the output, the noise signal is 1/375 of the desired signal. Therefore, the noise has been reduced by a factor $10 \times 375 = 3750$, which is equal to (CMRR) by the use of a difference amplifier. Therefore, difference amplifiers are very useful where elimination of common-mode undesirable signals is concerned.

**Example 26.17** *A capacitive displacement transducer has plates of area 200 mm$^2$. The distance between plates is variable. The dielectric medium is air. This capacitor is used in the feedback loop of an operational amplifier. The capacitor is connected at the inverting input end of the operational amplifier has a capacitance of 10 pF. Calculate the sensitivity and also calculate the voltage output if the displacement is 1 μm.*

**Solution.** The sensitivity is given by Eqn. 26.86.

$$\text{Sensitivity } K = -\left(\frac{C_i}{\varepsilon_0 A} v_i\right)$$

$$= -\left(\frac{100 \times 10^{-12}}{8.85 \times 10^{-12} \times 200 \times 10^{-6}} \times 10\right)$$

$$= 0.057 \times 10^6 \text{ V/mm}$$

Voltage output corresponding to a displacement of 1 μm is :

$$v_0 = 0.057 \times 10^6 \times 10^{-6} = 0.057 \text{ V} = 57 \text{ mV}$$

**Example 26.18** *Ceramic type capacitive transducers have capacitance in the range of 10 to 1000 pF and shunt resistance of $10^8$ to $10^{10}$ Ω. Find the maximum and minimum time constants. Calculate the values of minimum and maximum frequencies for which they can be used with an attenuation of ± 5% of the input.*

**Solution.** Maximum time constant

$$= 10^{10} \times 1000 \times 10^{-12} = 10 \text{ s}$$

Minimum time constant

$$= 10^8 \times 10 \times 10^{-12} = 10^{-3} \text{ s}$$

Let us find out the lowest frequency for which these transducers can be used for a response within ±5%.

Amplitude ratio $0.95 = \dfrac{1}{\sqrt{1 + (1/2\pi f \times 10)^2}}$ or

minimum allowable frequency $f = 0.048$ Hz.

For a time constant $\tau = 10^{-3}$, the minimum frequency can similarly be found.

or $0.95 = \dfrac{1}{\sqrt{1 + (1/2\pi f \times 10^{-3})^2}}$ $\quad\therefore\quad f = 484$ Hz

## 26.7 INSTRUMENTATION AMPLIFIERS

The instrumentation amplifier is a dedicated differential amplifier with extremely high input impedance. Its gain can be precisely set by a single internal or external resistor. The high common mode rejection makes this amplifier very useful in recovering small signals buried in large common-mode offsets and noise.

The instrumentation amplifier (*IA*) is not a general building block, like the OPAMP. The instrumentation amplifier (*IA*) is a *closed loop device* with carefully set gain. The OPAMP itself is an open loop device with some very large (but variable) gain. This allows the instrumentation amplifier (*IA*) to be optimized for its role as signal conditioner of low level (often dc.), signals in large amounts of noise. The OPAMP in contrast, can be used to build a wide variety of circuits but does not make as good as a difference amplifier as does an Instrumentation Amplifier (*IA*). Instrumentation amplifiers consist of two stages. The first stage offers very high input impedance to both input signals and allows to set the gain with a single resistor. The second stage is a differential amplifier with the output, negative feedback, and ground connections all brought out.

The input stage is shown in Fig. 26.47. It consists, of two carefully matched OPAMPS. Each input $v_1$ and $v_2$ is applied to the non-inverting input terminal of its OPAMP. This OPAMP configured as a voltage follower, produces the instrumentation amplifier's very high input impedance. The outputs of the OPAMPs are connected together through a string of resistors. The two resistors $R$ are internal to integrated circuit while $R_g$ is the gain setting resistor. It may be internal or connected externally.

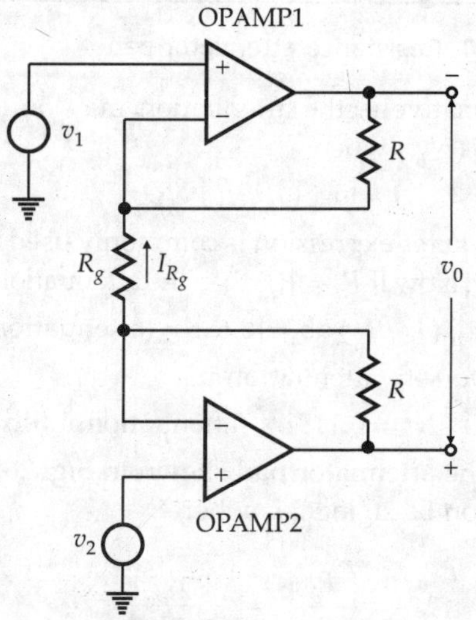


**Fig. 26.47** Instrumentation amplifier input stage.

The output voltage is taken between outputs of two OPAMPs.

$$v_0 = (v_2 - v_1)\left(1 + \frac{2R}{R_g}\right) \quad ...(26.93)$$

Equation 26.93 can be derived first noting that there can be no difference in potential between the inverting and non-inverting inputs of OPAMPS. This means that $v_1$ is at the top of $R_g$ and $v_2$ is at the bottom of $R_g$. That is, voltage across $R_g$ is

$$v_{R_g} = v_2 - v_1 = I_g\, R_g$$

$$I_{R_g} = \frac{v_2 - v_1}{R_g} \quad ...(26.94)$$

This current must flow through all the three resistors because none of the current can flow into OPAMP's input. So the output voltage is,

$$v_0 = I_{R_g} \times (2R + R_g)$$

$$= \frac{v_2 - v_1}{R_g}(2R + R_g)$$

$$= (v_2 - v_1)\left(1 + \frac{2R}{R_g}\right) \quad ...(26.94)$$

Equation 26.95, it is clear that decrease in the value of $R_g$ will increase the output voltage $v_0$. Therefore, to increase the value of gain, the value of $R_g$ has to be decreased.

The second stage of the instrumentation amplifier is a unity gain differential amplifier. The complete diagram of an instrumentation amplifier (including the first and second stage) is shown in Fig. 26.48. Three terminals are brought out. The sense terminal gives an access to the feedback loop. The reference terminal allows to establish d.c. reference *i.e.*, ground potential of the output. For normal operation the sence terminal is directly connected to output (completing the feedback loop) and tie the reference to ground. This configures the output of a standard differential amplifier as shown in Fig. 26.48.

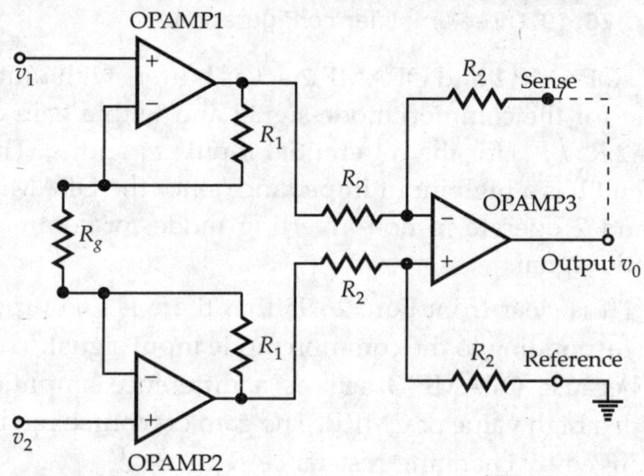


**Fig. 26.48** Instrumentation amplifier full schematic.

### 26.7.1 Three Amplifier Configuration

This circuit is being described to show the elimination of a common mode signal. The circuit of Fig. 26.49 is used for this purpose. This circuit uses three operational amplifiers and hence it is called "three amplifier configuration". $v_1$ and $v_2$ are the desired signals and $v_{cm}$ is the common mode signal.

For simplicity it is assumed $R_2 = R_3 = R_4 = R_5$

Therefore, we can write,

$$v_3 = \left(1 + \frac{R_1}{R_g}\right) v_1 - \left(\frac{R_1}{R_g}\right) v_2 + v_{cm}$$

$$v_4 = \left(1 + \frac{R_1}{R_g}\right) v_2 - \left(\frac{R_1}{R_g}\right) v_1 + v_{cm}$$

and $$v_0 = v_4 - v_3$$

$\therefore$ The output voltage is

$$v_0 = \left(1 + \frac{2R_1}{R_g}\right)(v_2 - v_1) \quad ...(29.96)$$

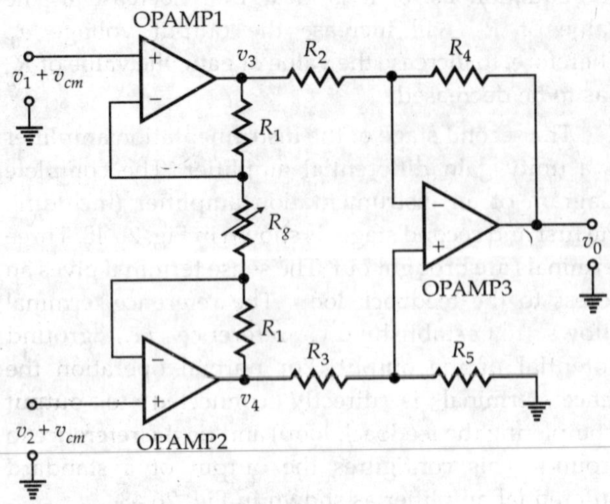


**Fig. 26.49** Three amplifier configuration.

OPAMP 1 and OPAMP 2 act as buffers with unity gain for the common mode signal and with a gain of $(1+2R_g/R_1)$ for the differential inputs $v_1$ and $v_2$. The circuit has a high input impedance since the OPAMPs 1 and 2 operate in non-inverting mode for common mode signal.

It is clear from Eqn. 26.95 that there is no output corresponding to the common mode input signal. This is because OPAMP 3 acts as a difference amplifier with a high value of CMRR. The gain off course can be changed by changing resistance $R_g$.

## 26.8 ATTENUATORS

The term attenuator is often loosely used to denote a device which reduces the voltage and/or power conducted between the circuits connected to its input and output terminals. Such an interpretation of the term may be applied to all the potential dividers, shunts, current and voltage transformers. A purer interpretation restricts the use of the term to apply only to those dividers which are constructed so that in addition to providing the reduction in voltage or power, the impedance of the divider is matched to the input and output circuits, and for a multiratio attenuator these impedances are constant irrespective of ratio setting.

### 26.8.1 Resistance Attenuator

Figure 26.50 shows a simple form of resistance attenuation pad for which $R_S = R_i$, $R_L = R_0$ and $V_i$ is the input voltage and $V_0$ the output voltage.

It should be noted that the attenuation $(A)$ is normally quoted as a power ratio in terms of decibels, that is :

$$A = 10\log_{10}(P_S/P_L) \quad ...(26.97)$$

$$A = 20\log_{10}(V_i/V_0)\ \text{db} \quad (\text{provided } R_S = R_L) \quad ...(26.98)$$

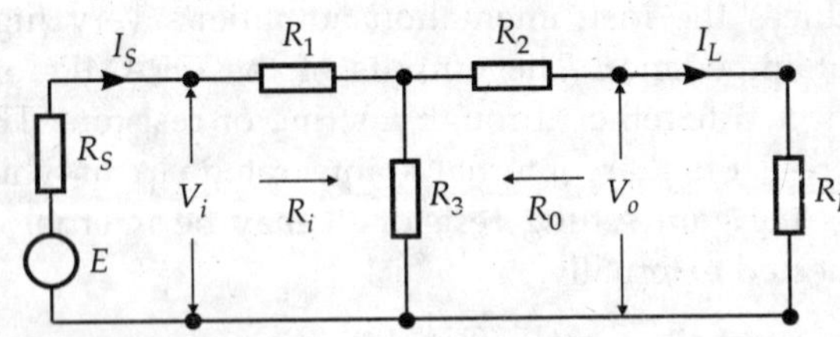


**Fig. 26.50** Resistance attenuator pad.

Alternatively, the attenuation may be expressed in nepers ($N_p$) when

$$A = \log_e(I_S/I_L)\,N_p \quad ...(26.99)$$

Thus latter expression is commonly used in theoretical work. Now if $R_s = R_L$, then the attenuation in db is,

$A = 20\log_{10}(I_S/I_L)$ db = [8.686 × (attenuation in $N_p$)]

Conversely attenuation in

$N_p = 0.1151 \times$ (attenuation in db)

For the attenuator pad shown in Fig. 26.50 let the attenuation be $20\log_{10} k$, where

$$k = \left(\frac{P_S}{P_L}\right)^{\frac{1}{2}} = \left(\frac{I_S^2 . R_S}{I_L^2 R_L}\right)^{\frac{1}{2}} \quad ...(26.100)$$

Now $$R_i = R_S = R_1 + \frac{R_3(R_2+R_L)}{R_2+R_3+R_L} \quad ...(26.101)$$

and $$R_c = R_L = R_2 + \frac{R_3(R_1+R_S)}{R_1+R_3+R_S} \quad (26.102)$$

also $$\frac{I_S}{I_L} = \frac{R_2+R_L+R_3}{R_3} \quad ...(26.103)$$

From equations 26.100 to 26.102 it can be shown that,

$$R_1 = R_S\left\{\frac{k^2+1}{k^2-1}\right\} - 2(R_S R_L)^{\frac{1}{2}}\left\{\frac{k}{k^2-1}\right\} \quad ...(26.104)$$

$$R_2 = R_L\left\{\frac{k^2+1}{k^2-1}\right\} - 2(R_S R_L)^{\frac{1}{2}}\left\{\frac{k}{k^2-1}\right\} \quad ...(26.105)$$

$$R_3 = 2(R_S R_L)^{\frac{1}{2}}\left\{\frac{k}{k^2-1}\right\} \quad ...(26.106)$$

### 26.8.2 Symmetrical *T* Attenuator

Referring to Fig. 26.51. If the load and source impedance are equal the *T* attenuator will become symmetrical and

$$R_1 = R_2 = R_L\left(\frac{k-1}{k+1}\right) \quad ...(26.107)$$

and $$R_3 = \frac{2R_L k}{k^2-1} \quad ...(26.108)$$

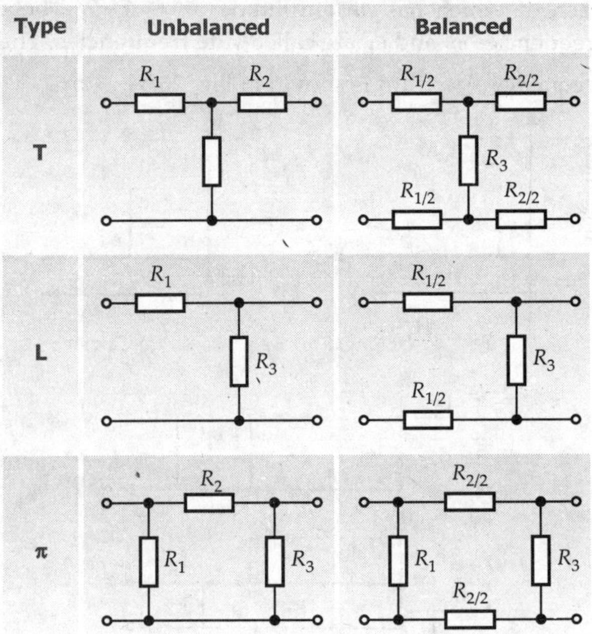


**Fig.26.51** Balanced and unbalanced attenuators.

### 26.8.3 L-type Attenuator

If the attenuator/pad is required only for matching purposes it will be arranged to have minimum attenuation and under these conditions $R_2 = 0$ resulting in an $L$ type attenuator (26.51) for which,

$$R_1 = [R_S(R_S - R_L)]^{\frac{1}{2}} \quad ...(26.109)$$

and

$$R_3 = \left(\frac{R_S R_L}{R_S - R_L}\right) \quad ...(26.110)$$

### 26.8.4 π Attenuator

Another common form of attenuator is the π type (Fig. 26.51) which, when used for matching, requires

$$R_1 = R_S \left\{ \frac{k^2 - 1}{k^2 - 2k\left(\frac{R_S}{R_L}\right)^{\frac{1}{2}} + 1} \right\} \quad ...(26.111)$$

$$R_2 = \frac{(R_S R_L)^{\frac{1}{2}}}{2} \left\{ \frac{k^2 - 1}{k} \right\} \quad ...(26.112)$$

$$R_3 = R_L \left\{ \frac{k^2 - 1}{k^2 - 2k\left(\frac{R_S}{R_L}\right)^{\frac{1}{2}} + 1} \right\} \quad ...(26.113)$$

and for attenuation alone, that is when $R_S = R_L$

$$R_1 = R_3 = R\left[\frac{k+1}{k-1}\right] \quad ...(26.114)$$

and

$$R_2 = R\left[\frac{k^2+1}{2k}\right] \quad ...(26.115)$$

The above attenuator have all been unbalanced and in some applications it is desirable that the legs of the circuit should be balanced. This is simply arranged by dividing the series arm in half and placing the equal halves in each leg (See Fig. 26.51).

Resistance attenuators of this form may be used from d.c. to several hundreds of MHz. For higher frequencies wave guide devices are used.

## 26.9 AMPLITUDE MODULATION

The term **Modulation** means "to change" or "to modify".

The interest in amplitude modulation signals in measurement systems is on account of *two* reasons :

1. Physical data that are to be measured, processed and interpreted are, many a times, amplitude modulated.
2. When carrying out dynamic measurements, the measurement system are intentionally designed to introduce amplitude modulation on account of various advantages.

There are, invariably, two parts to any method of modulation :

(*i*) **Carrier.** This is the medium that carries the information or intelligence, and

(*ii*) **Signal.** The signal is the information or intelligence to be carried.

In **amplitude modulation (AM)**, the signal or the information modulates the amplitude of the carrier. Qualitatively, amplitude modulation may be defined as the modification of the amplitude of the carrier in proportion to the amplitude of the modulating signals. In general, the signal that modulates the amplitude of a carrier wave may be of any form, it may be a single sine wave, a periodic function, a random wave or a transient phenomenon. Similarly, the carrier may also be of different forms *i.e.*, it may be a sine wave or a square wave etc.

In order to understand the concept of amplitude modulation, the simplest form is a sinusoidal signal modulating the amplitude of a sinusoidal carrier wave. The modulation process, fundamentally, is the multiplication of the signal carrying the information by a carrier wave of constant frequency and amplitude as shown in Fig. 26.52.

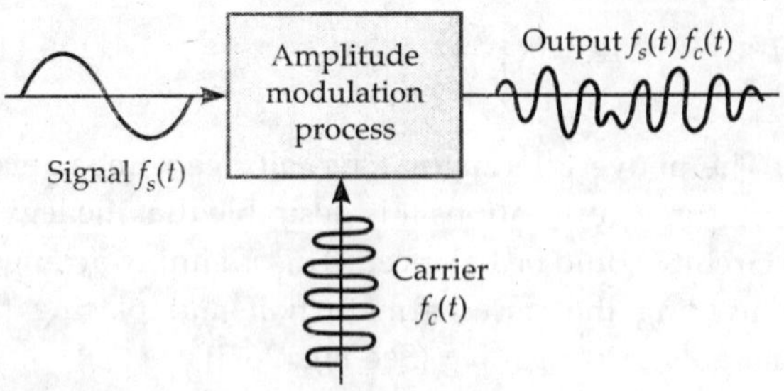


**Fig. 26.52** Amplitude modulation of a sinusoidal signal.

Let $e_s = A_s \sin \omega_s t$ and $e_c = A_c \sin \omega_c t$.

where $A_s, A_c$ = amplitude of signal and carrier respectively,

$\omega_s, \omega_c$ = angular frequency of signal and carrier respectively.

∴ Output

$$e_{cm} = (A_s \sin \omega_s t)(A_c \sin \omega_c t)$$

$$= \frac{A_s A_c}{2}[\cos(\omega_c - \omega_s)t - \cos(\omega_c + \omega_s)t] \quad ...(26.116)$$

The carrier frequency, $\omega_c$, is usually considerably greater than the signal frequency, $\omega_s$. The carrier signal and the modulated output waveforms are shown in Fig. 26.53.

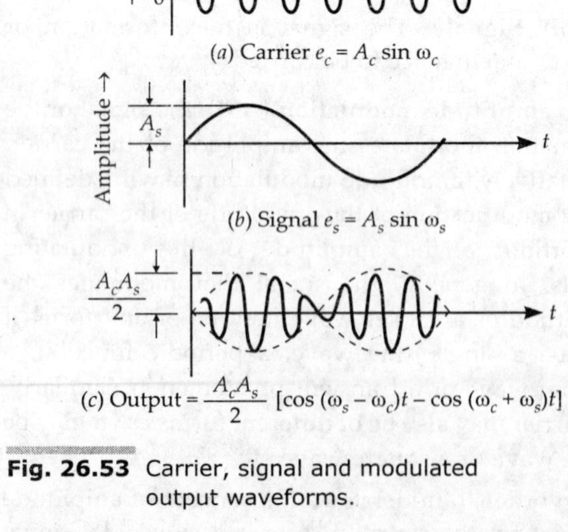


**Fig. 26.53** Carrier, signal and modulated output waveforms.

From Eqn. 26.116, it is clear that the output represents a signal composed of two frequencies :

$$\omega_1 = \omega_c - \omega_s \quad \text{and} \quad \omega_2 = \omega_c + \omega_s$$

each of which has an amplitude of $A_c A_s / 2$. These frequencies $\omega_1$ and $\omega_2$ are called **side frequencies**. The frequency spectrum is shown in Fig. 26.54.

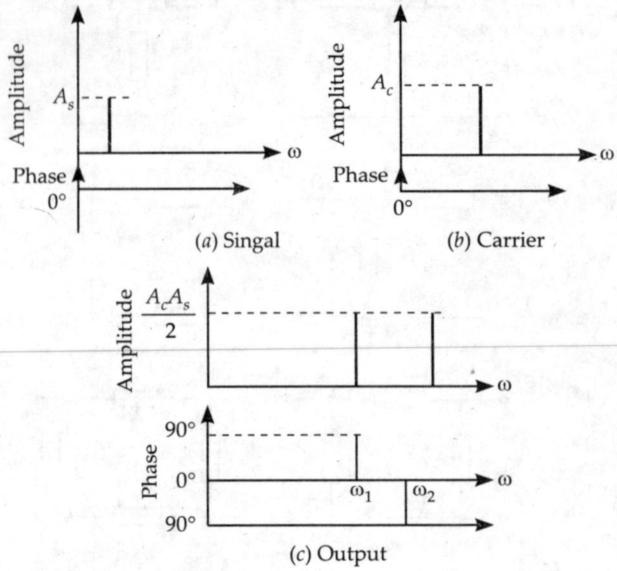


**Fig. 26.54** Frequency spectrum of amplitude modulated signals.

The concept of amplitude modulation is used in **carrier amplifiers**. The output signals coming from transducers like strain gauges etc are extremely small and therefore high gain amplifiers are used to produce perceptible outputs. A high gain a.c. amplifier is much easier to build than a d.c. amplifier as latter suffers from the disadvantage that they have inherent problem of drift. On the other hand, an a.c. amplifier does not amplify static and slow varying dynamic inputs on account of their low frequency cutoff. This is shown in Fig. 26.52. Thus, as such a.c. amplifiers are unsuitable for static and slow varying inputs. This problem can be overcome by the use of amplitude modulations techniques.

Supposing strain is to be measured with the help of a strain gauge. This strain gauge forms an arm of a Wheatstone bridge. The bridge is excited by a carrier voltage of 5 V at 3000 Hz as shown in Fig. 26.55. When a static strain is applied to the strain gauge, its resistance changes and an output varying at a frequency of 3000 Hz appears at the output terminals of the bridge.

This output can be easily amplified by an a.c. bridge. Supposing now we are interested in measurement of dynamic strains varying at a frequency of up to 20 Hz. The output of a bridge (and the input to the amplifier) will be an amplitude modulated wave having a frequency spectrum :

$$\omega_1 = 3000 - 20 = 2980 \text{ Hz to } \omega_2 = 3000 + 20 = 3020 \text{ Hz}$$

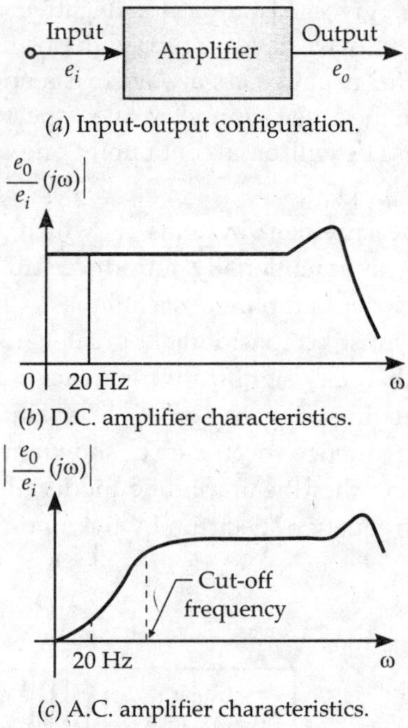


**Fig. 26.55** Characteristics of d.c. and a.c. amplifiers.

This range of frequencies can be easily handled by an a.c. amplifier and are reproduced faithfully with a constant gain. This shifting of the information frequencies from one part of the frequency range to another forms the basis of many useful applications of amplitude modulation.

There is an additional advantage of using amplitude modulation. Supposing the wires leading from bridge to the amplifier circuit in Fig. 26.56 are subjected to a 50 Hz stray magnetic field produced by the surrounding wiring system and a.c. machines. This produces a 50 Hz noise or hum which is superimposed upon the desired signals. If d.c. amplifiers are used, this 50 Hz noise could easily be greater than the desired output signals. However with the use of carrier system, this noise may easily be removed by simply designing the a.c. amplifier that does not respond to 50 Hz signal at all *i.e.*, the amplification factor is almost zero for 50 Hz input as shown in Fig. 26.57. Since the desired band of frequencies is between 2980 and 3020 Hz, making the low frequency cutoff of the amplifier greater than 50 Hz is quite easy.

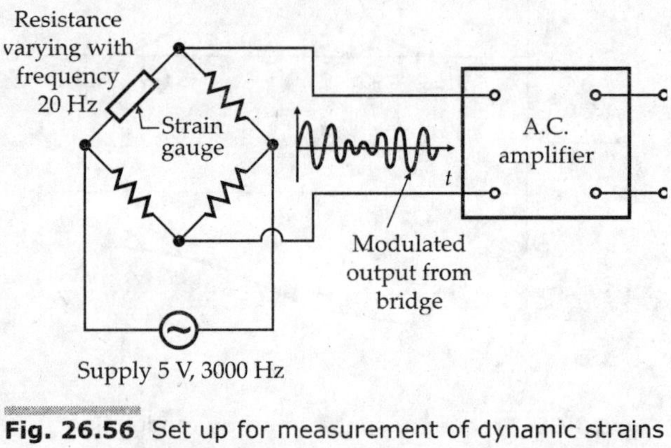


**Fig. 26.56** Set up for measurement of dynamic strains and frequency response of an a.c. amplifier.

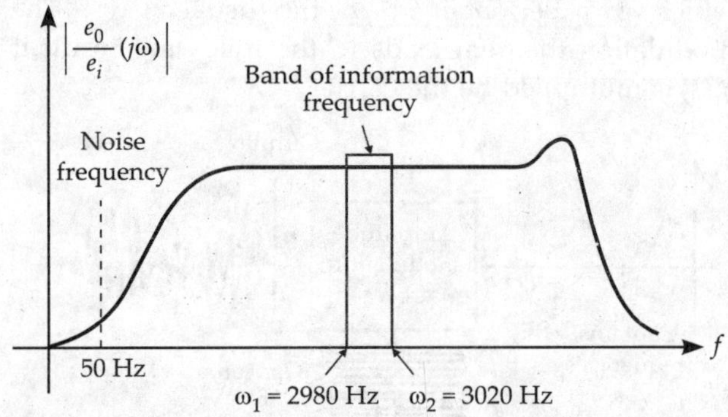


**Fig. 26.57** Rejection of noise through amplitude modulation.

The signals encountered in measurement systems are not always pure sinusoidal signals. They may be periodic waves of any shape. Let the modulating signal be a periodic function $f_i(t)$.

The modulating periodic signal $f_i(t)$ may be expanded by Fourier series to get an output,

$$= \left[ f_{i,av} + \frac{1}{L}\left( \sum_{n=1}^{\infty} a_n \cos \frac{n\pi t}{L} + \sum_{n=1}^{\infty} b_n \sin \frac{n\pi t}{L} \right) \right] A_c \sin \omega_c t \quad \text{...(26.117)}$$

which may be written as :

Output

$$= A_0 A_c \sin \omega_c t + (A_1 A_c \cos \omega_1 t \sin \omega_c t + A_2 A_c \cos \omega_2 t \sin \omega_c t + \ldots) + (B_1 A_c \sin \omega_c t \sin \omega_c t + B_2 A_c \sin \omega_2 t \sin \omega_c t + \ldots) \quad \text{...(26.118)}$$

$$= A_0 A_c \sin \omega_c t + C_1 \{\sin [(\omega_c + \omega_1)t - \alpha_1] + \sin[(\omega_c - \omega_1)t + \alpha_1]\} \quad \text{...(26.119)}$$

$$\text{where } C_1 = \frac{A_c}{2}\sqrt{A_1^2 + B_1^2} \quad \text{...(26.120)}$$

$$\text{and} \quad \alpha_1 = \tan^{-1}\frac{B_1}{A_1} \quad \text{...(26.121)}$$

26

We see that the spectrum of the output signal is a discrete spectrum containing the frequencies $\omega_c$, $\omega_c \pm \omega_1$, $\omega_c \pm \omega_2$, $\omega_c \pm \omega_3$, etc. That is, each frequency component of the modulating signal produces one pair of side frequencies (see Fig. 26.58).

If the modulating signal is a transient, the spectrum of the modulator output may be determined with the help of the modulation theorem for Fourier (or Laplace) transforms. If the modulating signal is a transient $f_i(t)$, it will have a Fourier transform $F_i(i\omega)$, which cart be obtained in the usual ways. The modulation theorem leads to the following result if $f_i(t)$ is multiplied by the carrier.

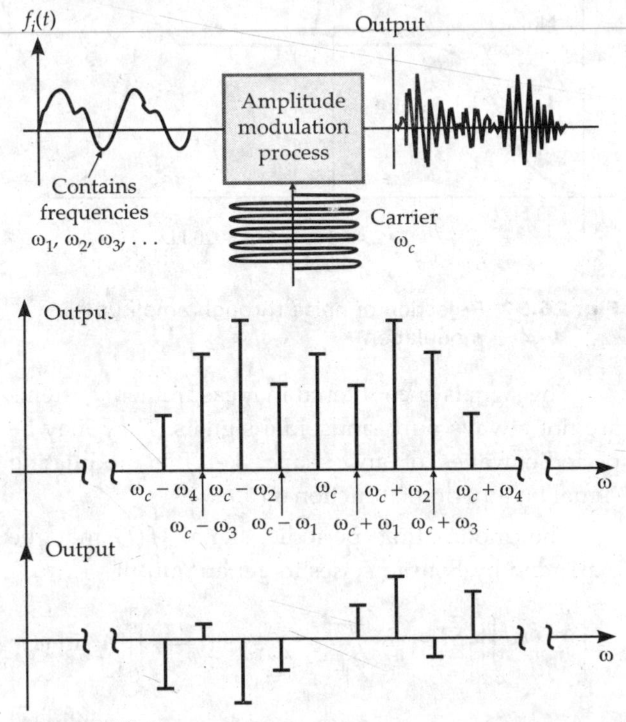


**Fig. 26.58** Frequency spectrum when modulating signal is periodic.

$A_c \sin \omega_c t$ to produce the modulated output :

Fourier transform of modulated output

$$\triangleq |Q_i(i\omega)| \angle Q_i(i\omega) \qquad \ldots(26.122)$$

where $$|Q_i(i\omega)| = \frac{A_c}{2} \text{ magnitude } \{F_i[i(\omega - \omega_c)]\} \qquad \ldots(26.123)$$

$$\angle Q_i(i\omega) = \text{angle } \{F_i[i(\omega - \omega_c)]\} \qquad \ldots(26.124)$$

and $0 \leq \omega < \infty$. Note the argument (independent variable) of the $F_i$ function is now $i(\omega - \omega_c)$, so that if we want to find $Q_i(i4{,}000)$ and $\omega_c = 3{,}000$, we must evaluate $F_i(i1{,}000)$. Also note that, to get $Q_i(i\omega)$ for $0 < \omega < \infty$, we must know $F_i(i\omega)$ for *negative* $\omega$'s, since any $\omega < \omega_c$ gives $F_i[i(\omega - \omega_c)]$ a negative argument. While generally we have worked with positive $\omega$'s, the transform for negative $\omega$'s always exists and is easily found from the given symmetry rules, spectrum given by Eqn. 26.119 will be a continuous one, and if the modulated.

For measurement systems in which amplitude modulation is intentionally introduced to allow the use of carrier amplifier techniques, the carrier frequency must be considerably greater (usually 5 to 10 times) than any significant frequencies present in the modulating signal. For such a situation, the pertinent frequency spectra are as shown in Fig. 26.59. We see again that the amplitude modulation process shifts the frequency spectrum by the amount $\omega_c$.

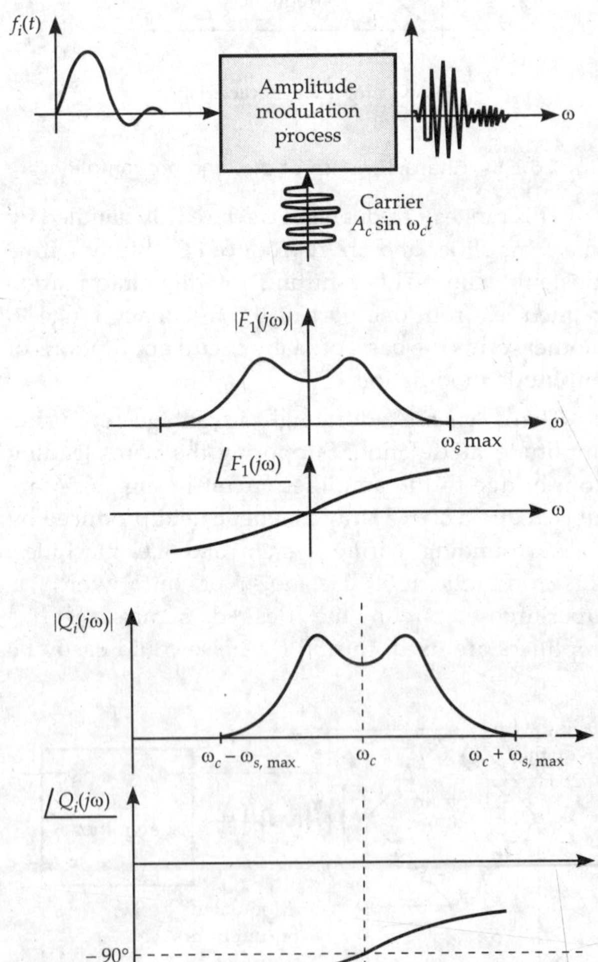


**Fig. 26.59** Frequency spectrum when modulating signal is transient.

## 26.10 AMPLITUDE DEMODULATION

When amplitude modulation is intentionally introduced in a system for data handling, it is used as an intermediate step. The amplitude modulated signal cannot be used to drive the final read-out system as such because it is not suitable for purpose. In fact, the original form of the signal, which is the output of the transducers, should be applied to the final stage (*i.e.*, the data presentation stage) of the instrumentation system. The process, required for recovery of original signal from the modulated waveform is called **demodulation** or **detection**. The detection or demodulation of an amplitude modulated signal requires rectification of the modulated signal followed by elimination of carrier frequency by proper filtering.

The demodulation may be :

(*i*) half wave, (*ii*) full wave,

and

(*i*) phase sensitive,

(*ii*) non-phase sensitive as shown in Fig. 26.60.

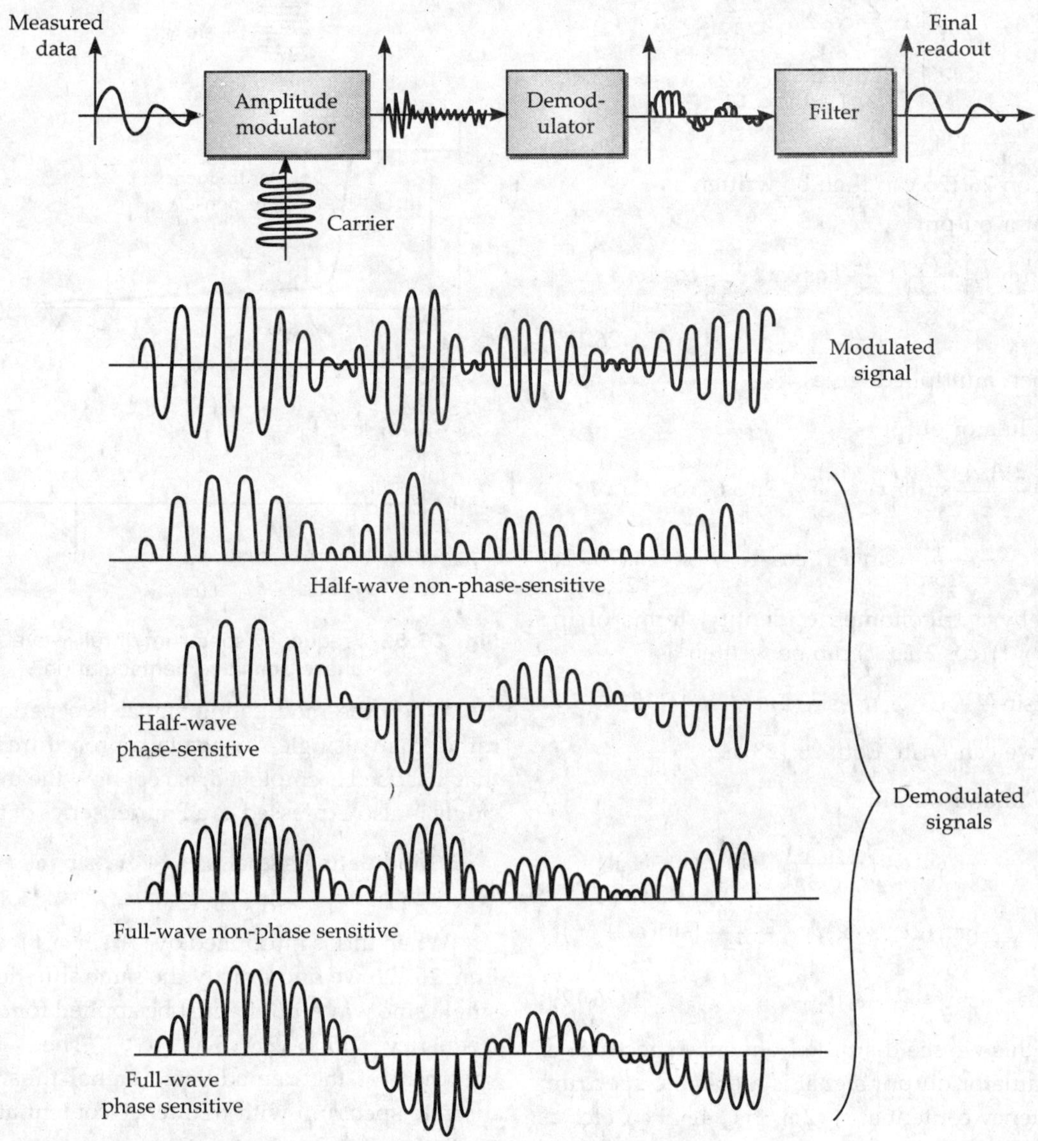


**Fig. 26.60** Types of modulation.

Here we treat the form giving the best reproduction of the original data, full-wave phase-sensitive demodulation and consider only the process, not the hardware for accomplishing it. Again, it is necessary to consider whether the form of the original signal was a single sine wave, periodic function, or transient. For a single sine wave $A_s \sin\omega_s t$ which is modulating a carrier $A_c \sin\omega_c t$, the expression for the full-wave phase-sensitive demodulated signal is

Demodulator output

$$= (A_s \sin\omega_s t)[A_c \sin\omega_c t] \quad ...(26.125)$$

as seen from Fig. 26.61(*a*). Now $|A_c \sin\omega_c t|$ is a periodic function and may be expanded in a Fourier series. The results are :

$$|A_c \sin\omega_c t| = \frac{2A_c}{\pi}\left(1 - \frac{2}{3}\cos 2\omega_c t - \frac{2}{15}\cos 4\omega_c t + ... + \frac{2}{1-4n^2}\cos 2n\omega_c t + ...\right)$$

$$n = 3, 4, 5 \quad ...(26.126)$$

Equation 26.125 can then be written as

Demodulator output

$$= (A_s \sin\omega_s t)\left[\frac{2A_c}{\pi}\left(1 - \frac{2}{3}\cos 2\omega_c t - \frac{2}{15}\cos 4\omega_c t + ...\right)\right] \quad ...(26.127)$$

which, when multiplied, gives

Demodulator output

$$= \frac{2A_cA_s}{\pi}\sin\omega_s t - \frac{4A_cA_s}{3\pi}\sin\omega_s t\cos 2\omega_c t - \frac{4A_cA_s}{15\pi}\sin\omega_s t\cos 4\omega_c t + ... \quad ...(26.128)$$

Now, by a trigonometric identity, terms of the form $(\sin\omega_s t)(\cos 2n\omega_c t)$ can be written as

$$[\sin(2n\omega_c + \omega_s)t - \sin(2n\omega_c - \omega_s)t]/2$$

Thus we can write Eqn. 26.128 as :

Demodulator output

$$= \frac{2A_cA_s}{\pi}\sin\omega_s t - \frac{2A_cA_s}{3\pi}[\sin(2\omega_c + \omega_s)t] - [\sin(2\omega_c - \omega_s)t] - \frac{2A_cA_s}{15\pi}[\sin(\omega_c + \omega_s)t - \sin(4\omega_c - \omega_s)t] + ... \quad ...(26.129)$$

From this we see than the frequency spectrum of the demodulator output signal is a discrete spectrum with frequency content at $\omega_s, 2\omega_c \pm \omega_s, 4\omega_c \pm \omega_s$, etc., as shown in Fig. 26.61(*b*). This signal were an input to a system of known frequency response (such as the filter of Fig. 26.60. If the output of the filter is to look like the original data, the filter must be designed to *reject* frequencies $2\omega_c \pm \omega_s, 4\omega_c \pm \omega_s$, etc., while *passing* with a minimum of distortion the signal frequency $\omega_s$. The design of such a low-pass filter is made simpler if the pass band and the rejection band are more widely separated. This is the basis of our earlier statement that carrier frequencies usually are chosen to be 5 to 10 times the highest expected signal frequency.

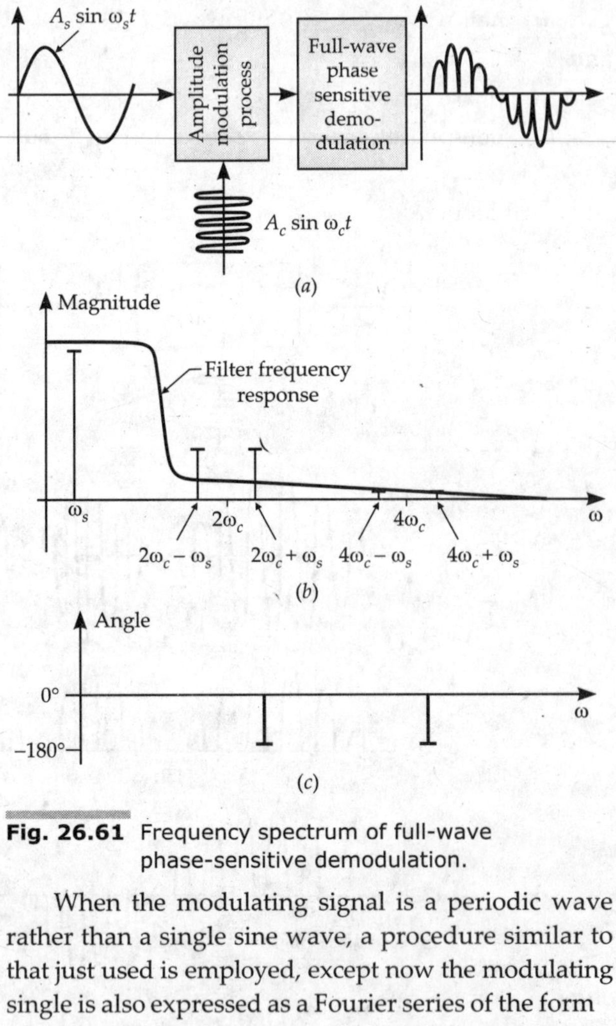


**Fig. 26.61** Frequency spectrum of full-wave phase-sensitive demodulation.

When the modulating signal is a periodic wave rather than a single sine wave, a procedure similar to that just used is employed, except now the modulating single is also expressed as a Fourier series of the form

$$\text{Modulating signal} = A_{s0} + A_{s1}\sin(\omega_s t + \alpha_1) + A_{s2}\sin(2\omega_s t + \alpha_2) + ... \quad ...(26.130)$$

When this is multiplied by $|A_c \sin\omega_c t|$ as given by Eqn. 26.126 we find exactly the same situation as for a single sine wave, but it must be applied for *each* signal frequency $(0, \omega_s, 2\omega_s, 3\omega_s$, etc.). The frequency spectrum of the demodulated signal thus will be a discrete spectrum with frequency content at $\omega = 0$, $(\omega_s, 2\omega_c, 2\omega_c \pm \omega_s, 4\omega_c \pm \omega_s, 6\omega_c, 6\omega_c \pm \omega_s$, etc.), $(2\omega_s, 4\omega_c \pm 2\omega_s, 6\omega_c \pm 2\omega_s)$ etc.

Figure 26.62 illustrates these concepts.

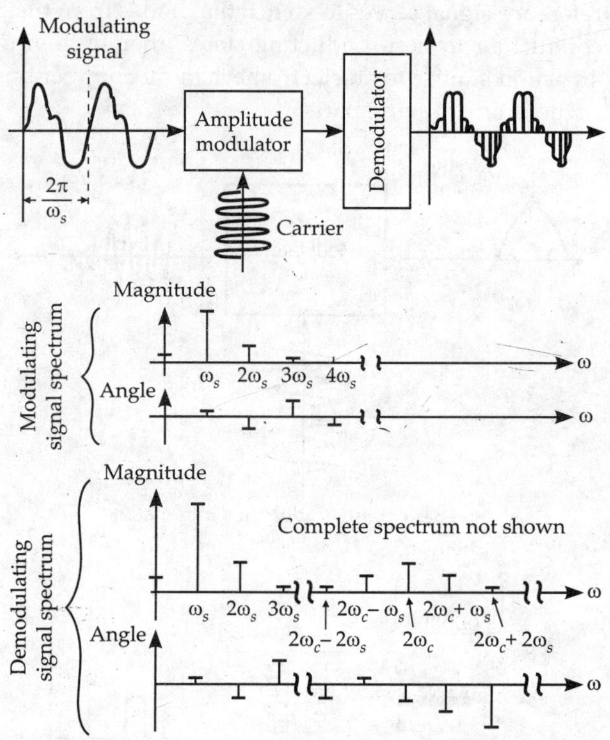


**Fig. 26.62** Demodulation spectrum for periodic input.

When the modulating signal is a transient $f_i(t)$, the demodulated signal will be $f_i(t)|\, A_c \sin \omega_c t\,|$, which can be written as

Demodulated signal

$$= f_i(t)\left[\frac{2A_c}{\pi}\left(1-\frac{2}{3}\cos 2\omega_c t - \frac{2}{15}\cos 4\omega_c t + \ldots\right)\right] \quad \ldots(26.131)$$

or, after multiplying,

Demodulated signal

$$= \frac{2A_c}{\pi} f_i(t) - \frac{4A_c}{3\pi} f_i(t)\cos 2\omega_c t - \frac{4A_c}{15\pi} f_i(t)\cos 4\omega_c t + \ldots \quad \ldots(26.132)$$

Application of the modulation theorem to each of the modulated terms of Eqn. 26.129 leads to the result.

Fourier transform of demodulated signal

$$\triangleq |Q_i(i\omega)|\,\angle Q_i(i\omega)$$

where $|Q_i(i\omega)| = \frac{2A_c}{\pi}|F_i(i\omega)| + \frac{2A_c}{3\pi}|F_i[i(\omega-2\omega_c)]|$

$$+ \frac{2A_c}{15\pi}|F_i[i(\omega-4\omega_c)]| + \ldots \quad \ldots(26.133)$$

and $0 \le \omega \le \infty$. The expression for $\angle Q_i(i\omega)$ is not easily given since in Fourier transforming Eqn. 26.132 (each term may be treated separately since superposition is allowed), we find $Q_i(i\omega)$ as a *sum* of complex numbers rather than a product. To find the overall angle of a sum of complex numbers, we must express *each* number as $a_i + ib_i$ and then add the $a$'s and $b$'s to get the overall $a + ib$. The angle is then $\tan^{-1}(b/a)$. The angle due to transforming $f_i(t)$ is $\angle F_i(i\omega)$, the angle due to transforming $-f_i(t)\cos 2\omega_c t$ is $-180° + \angle F_i[i(\omega-2\omega_c)]$, the angle due to transforming $-f_i(t)\cos 4\omega_c t$ is $-180° + \angle F_i[i(\omega-4\omega_c)]$, etc. These results allow calculation of $\angle Q_i(i\omega)$. The frequency spectrum of such a signal is shown in Fig. 26.63. Since it is a continuous spectrum, it is assumed that the $F_i(i\omega)$ is practically zero for $\omega > \omega_{s,\max}$, and that $\omega_c > \omega_{s,\max}$. For such a situation, the phase-angle calculation is greatly simplified, since the individual terms of Eqn. 26.133 do not co-exist over any frequency range. That is, for $0 < \omega < \omega_{s,\max}$, only the first term is nonzero ; for $(2\omega_c - \omega_{s,\max}) < \omega < (2\omega_c + \omega_{s,\max})$, only the second term is nonzero ; etc, Thus the overall phase angle is determined by one term only within each of the specified frequency bands.

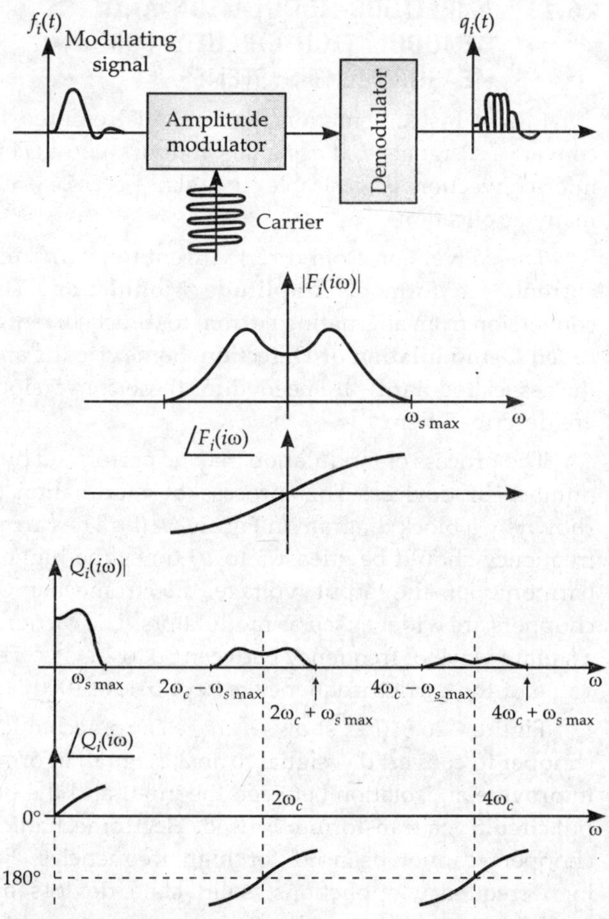


**Fig. 26.63** Demodulation spectrum for transient input.

26

This concludes our treatment of amplitude-modulated and demodulated signals. The methods developed can be readily applied to the other common variations such as half-wave demodulation, non-phase-sensitive demodulation, non-sinusoidal- carrier waveforms (square wave, for instance), etc. Non-phase-sensitive demodulation cannot detect a sign change in the modulating signal. Half-wave systems shift the side frequencies of demodulated signals to $\omega_c \pm \omega_s$ rather than $2\omega_c \pm \omega_s$, thus making the filtering problem more difficult. They also have less amplitude than full-wave systems. Non-sinusoidal carriers may be useful in reducing heating effects in resistive transducers such as strain gages. The carrier waveform is such as to give a high ratio of peak value to rms (effective heating) value. The output signal is related to peak value while the power dissipated in the strain gage is related to rms value ; thus a high peak/rms ratio increases output for a given allowable heating level. A carrier in the form of a train of high, narrow pulses satisfies this sort of criterion.

## 26.11 AMPLITUDE MODULATION AND DEMODULATION CIRCUITS FOR MEASUREMENT SYSTEMS

In any Instrumentation System, it is required to convert a.c. signals to d.c. signals and vice-versa. This inter-convection is desirable and also necessary for many applications.

The conversion from direct current to alternating current is a form of **Amplitude Modulation**. The conversion from alternating current to direct current is called **Demodulation** or **Detection**. Some circuits and the associated hardware needed for these conversions are described here.

The process of modulation may be performed by a number of devices. This process of modulation is shown by a block diagram in Fig. 26.64 (*a*). The carrier frequency should be atleast 5 to 10 times the highest harmonic of the input voltage. Electromechanical choppers are widely used as modulators. The carrier is a square wave of frequency between 50 to 400 Hz. The value of the carrier frequency is kept below 1000 Hz.

Figure 26.64(*b*) shows an electromechanical chopper to convert d.c. signal to an a.c signal. In order to provide an isolation between the input and the out put circuitry, a transformer is used. Electromechanical choppers cannot be used for high frequencies. For high frequency applictions, solid state devices are used for chopping. A solid state chopper as used for high frequency applications as shown in Fig. 26.64(*c*).

This diagram shows a ring type diode modulator. The reference signal serves to switch the diodes from their conducting to non-conducting states, thereby giving an action similar to an electromechanical chopper but without any moving parts.

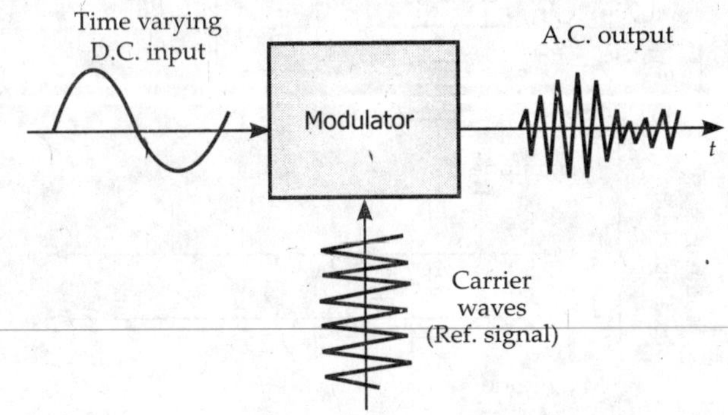


(*a*) Basic configuration for a modulator.

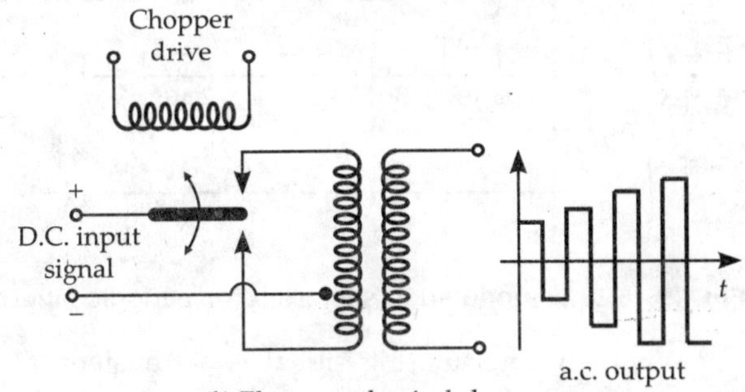


(*b*) Electromechanical chopper

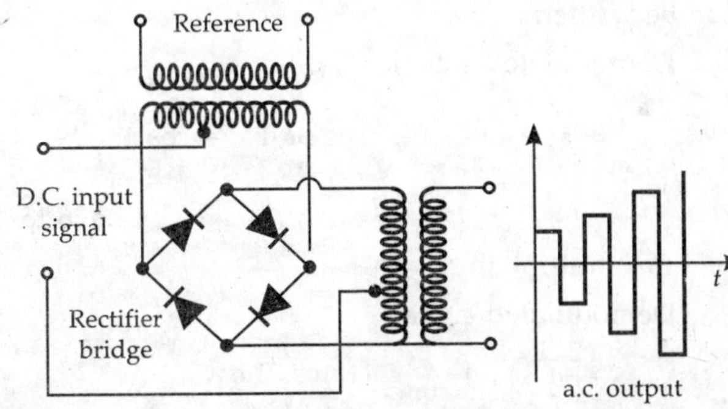


(*c*) Semi-conductor modulator.

**Fig. 26.64** Amplitude Modulation.

### 26.11.1 Input Modulator

A typical input modulator is shown in Fig. 26.65. The battery represents the input d.c. signal. If the d.c. input is zero both diodes are driven by the oscillator signal alone. Since the two diodes are identical and primary winding of the transformer is center tapped, there will be no output. During the positive half cycle of the oscillator input, both diodes are turned on. The currents $i_1$ and $i_2$ are equal, and since these currents flow in the opposite direction, and therefore there will not be any input to the transformer and hence no output. During the negative half cycle, both the diodes are non-conducting and hence there is no output. Therefore this modulator is called a Balanced Modulator.

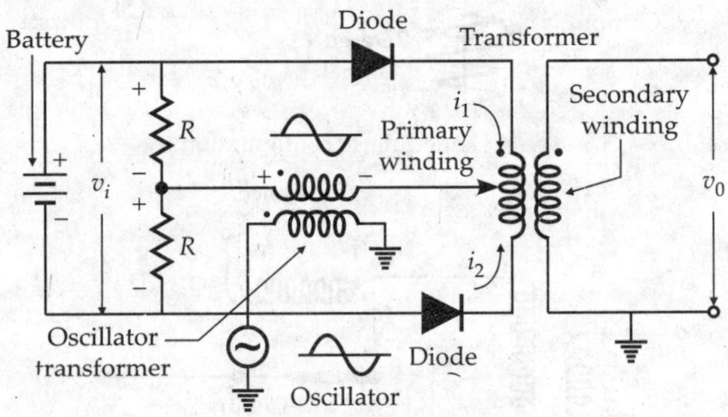


**Fig. 26.65** Input modulator.

The waveforms for the currents under zero input signal condition are shown in Fig. 26.66(*a*). The two currents $i_1$ and $i_2$ are half wave rectified sinusoids with equal maximum values. Suppose now a d.c. signal is applied to the input. A positive d.c. voltage as shown in Fig. 26.65 will produce plus – minus d.c. across resistors *R*.

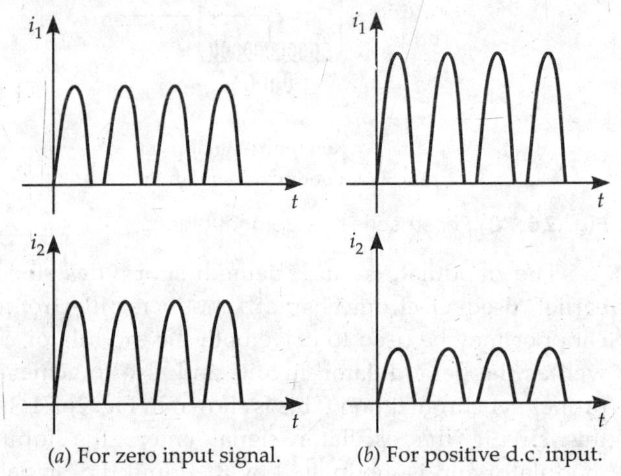


**Fig. 26.66** Input modulator currents.

During the positive half cycle of the oscillator signals plus – minus d.c. voltages appears across the oscillator secondary. In upper loop, the d.c. voltage across the resistor *R* aids the oscillator signal, while, on the other hand, the lower loop voltages are in opposition. On account of this reason the upper diode conducts heavily than the lower diode. This means that current $i_1$ is greater than $i_2$ as shown in Fig. 26.66(*b*). On account of difference between $i_1$ and $i_2$, a net flux is produced by the primary of transformer giving rise to an output voltage in the secondary winding. The magnitude of the output voltage is proportional to the difference between currents $i_1$ and $i_2$ and hence to the d.c. input signal. The output wave form is shown in Fig. 26.67(*a*).

In case the polarity of the battery voltage in Fig. 26.65 (which represents the input signal) is reversed, the lower diode conducts more heavily than the upper one. The resulting output voltage waveform is shown in Fig. 26.67(*b*). Thus a negative going output is produced.

In case, low frequency input signals are applied, an output waveform shown in Fig. 26.67(*c*) is obtained. The output signals of modulator may be amplified prior to demodulation.

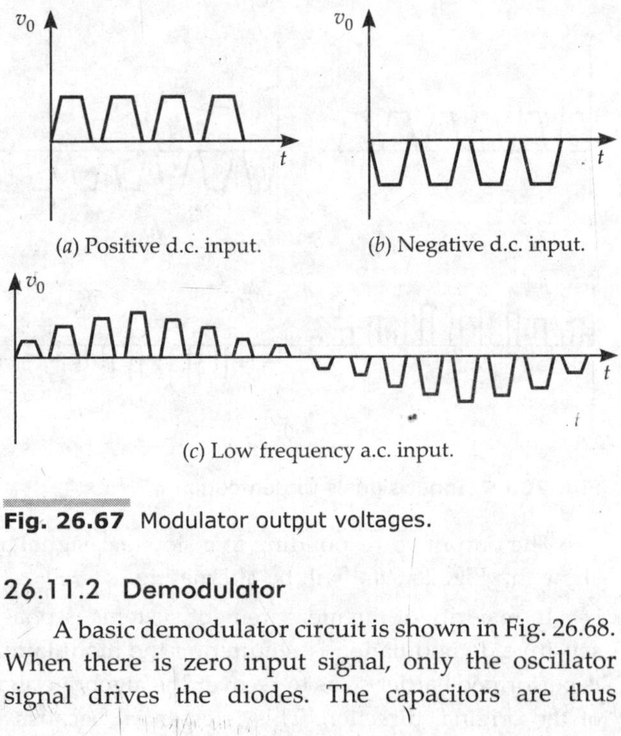


**Fig. 26.67** Modulator output voltages.

## 26.11.2 Demodulator

A basic demodulator circuit is shown in Fig. 26.68. When there is zero input signal, only the oscillator signal drives the diodes. The capacitors are thus

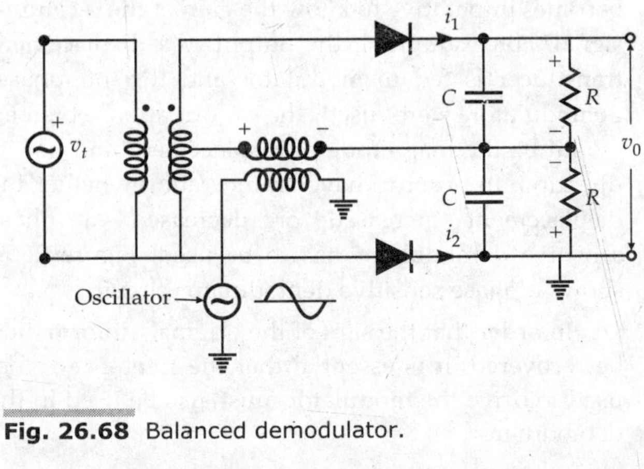


**Fig. 26.68** Balanced demodulator.

charged to the peak value of the oscillator signal. Since the voltages across the capacitors are opposite in polarity, the output voltage is zero. Thus for zero input voltage there is zero output.

Suppose now the input signal is a positive going voltage as shown in Fig. 26.69(*a*). This input voltage signal will induce voltages across the secondary with the polarities indicated. The first half cycle of input signal is in phase with oscillator signal. Due to this the upper loop is aiding and the lower loop is opposing. This means that the voltage across the upper capacitor is greater than that across the lower one. The output voltage is the difference between the capacitor voltages and hence will be positive. If the input to the demodulator is a negative signal, the output will be negative. as shown in Fig. 26.69(*b*).

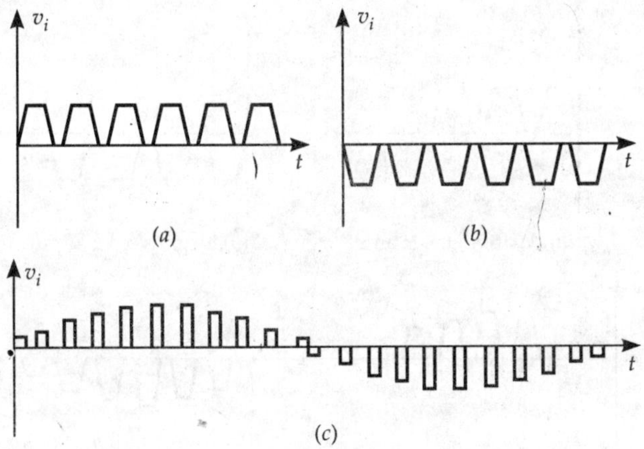


**Fig. 26.69** Input signals to demodulator.

The output corresponding to a slow a.c. signal as shown in Fig. 26.69(*c*) will be alternating.

In majority of the measurement systems, a phase sensitive demodulation is required, if the modulation is performed earlier so as to recover the algebraic sign of the original direction. This is essential because it becomes imperative to know the gain of the d.c. input signal. For example, if the output of a displacement transducer is fed to modulator and if a nonphase demodulator were used the information obtained would be the magnitude of displacement and not its direction. In order to have an indication whether the displacement increased or decreased, a phase sensitive demodulator has to be used. Figure 26.70 shows a phase sensitive demodulator circuit.

In order that the sign of the original information be recovered, it is essential that the reference signal used to drive the modulator must also be used in the demodulator.

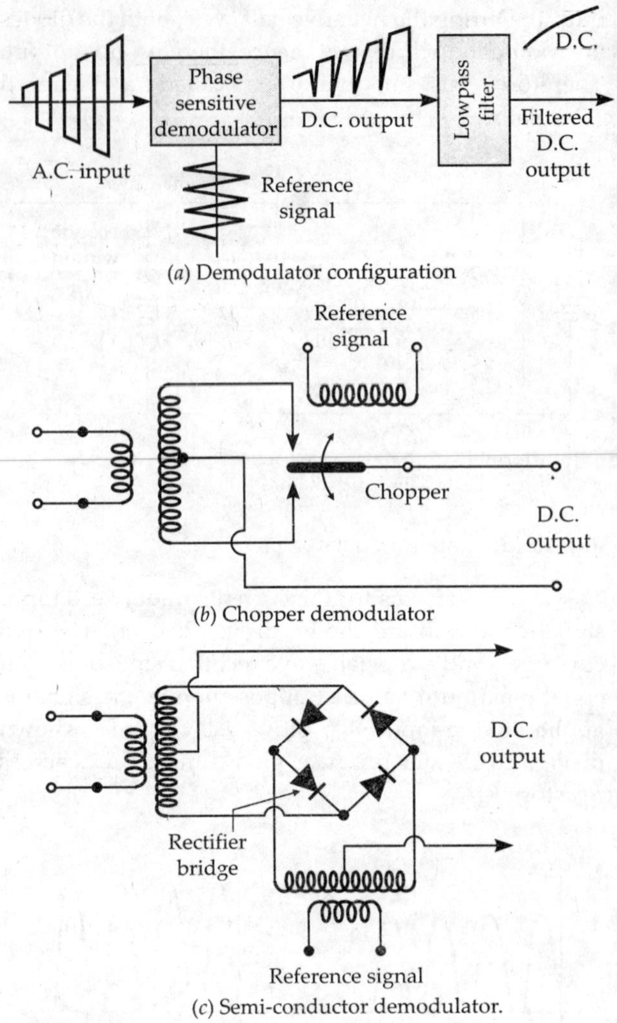


**Fig. 26.70** Phase sensitive demodulation.

The modulators and demodulators described earlier used electromechanical choppers. Electronic chopper may be used to drive both the modulator as well as the demodulator circuits and also to achieve proper synchronization. This is shown in Fig. 26.71. In this circuit, the oscillator signal enters the input modulator and is modulated by the input d.c. signal. The output of the modulator is a signal of the same frequency as that of the oscillator. The peak value of this signal is proportional to the magnitude of the d.c. input. In case, there is no d.c. input, the output is zero. The modulator is called balanced because there is no output in the absence of a d.c. signal. The modulator input is amplified by an a.c. amplifier. This output is demodulated to recover the original d.c. signal.

The electronic system shown in Fig. 26.71 can also be used for low frequency input signals. In this case, the oscillator signal is modulated by the low frequency input signal. The output of modulator is amplified and then demodulated to recover the signal of original frequency. However, in order to eliminate the noise from the output a low-pass filter has to be introduced at the output of the demodulator.

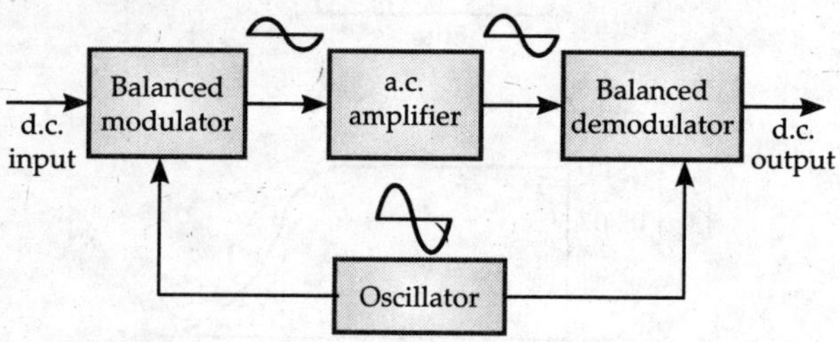


**Fig.26.71** Electronic chopper type d.c. amplifier.

## 26.12 FILTERS

The signal originating from a transducer is fed to the signal conditioning equipment. In order to measure the output signal of the transducer originating on account of variations of a physical change, it is desirable that the output signal be reproduced faithfully. For faithful reproduction of signal it becomes necessary to eliminate any kind of spurious or unwanted signals which may get introduced into the system either at the transduction stage or at the signal conditioning stage. The filters are thus designed to pass the signals of wanted frequencies and to reject the signals of unwanted frequencies which may be unwanted harmonics and noise. The harmonics or noise may be due to some form of distortion.

## 26.13 TYPES OF FILTERS

Filters may be of any physical form. They may be electrical, mechanical, pneumatic, hydraulic, acoustical etc. The most commonly used filters are electrical in form. The basic electrical filters are of two forms as regards the components constituting them. They are (*i*) Passive filters and (*ii*) Active filters.

**Passive Filters** only use passive circuit elements like resistors, capacitors and inductors while **Active Filters** use active elements like operational amplifiers in addition to passive elements like resistance, inductance and capacitance. The study of other filters may be carried out by analogies with electrical filters.

Both *passive* and *active filters* may be classified further as :

- Low pass filters,
- High pass filters,
- Band pass filters and
- Band stop filters.

Let us consider a basic configuration of an electrical filter as shown in Fig. 26.72. The source is sinusoidal of variable frequency. The filter circuit may be so designed that some frequencies are passed from input to output of the filter with very little attenuation while other frequencies are greatly attenuated.

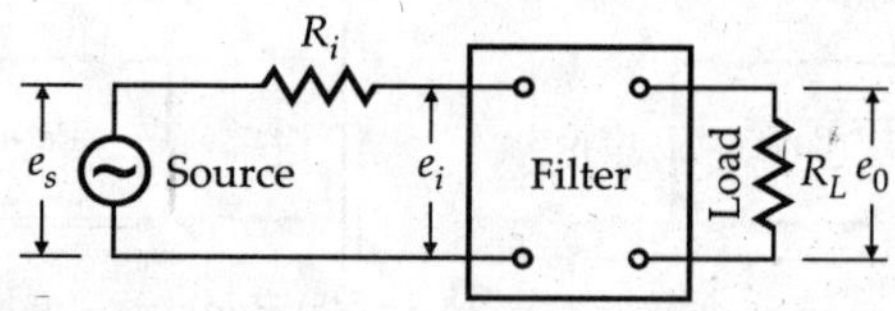


**Fig. 26.72** Basic configuration of an electrical filter.

The respones of various filters are shown in Fig. 26.73. These are ideal responses but cannot be achieved in actual practive.

The ideal response of a **Low Pass Filter** is shown in Fig. 26.73(*a*). The voltage gain is :

$$A = \left|\frac{e_0}{e_i}\right|$$

This gain is constant over a frequency range starting from 0 to a cut off frequency $\omega_c$. The output of any signal having a frequency greater than mc, will be rejected altogether *i.e.*, there will not be any output voltage corresponding to signals having frequency greater than $\omega_{c+}$ whatever their input magnitude be. The signals having a frequency range of 0 to $\omega_c$, will be faithfully reproduced with a constant gain of A. The output corresponding to signals above frequency $\omega_c$ is equal to zero *i.e.*, they are rejected.

The characteristics of a **High Pass** filter are shown in Fig. 26.73(*b*). The high pass filter has a gain of zero starting from zero frequency up to cut off frequency $\omega_c$. Above the cut off frequency, the gain is constant and is equal to *A*. Thus signals of any frequency beyond $\omega_c$ will be faithfully reproduced with a constant gain A while signals having frequency below $\omega_c$ will not be reproduced at all *i.e.*, rejected.

The **Band Pass** filter shown in Fig. 26.73(*c*), will faithfully reproduce signals falling within the range $\omega_{c1}$ and $\omega_{c2}$ faithfully while signals of frequency between 0 to $\omega_{c1}$ and of frequency greater than $\omega_{c2}$ will be rejected. There will be an output corresponding to signals having frequencies between $\omega_{c1}$ and $\omega_{c2}$, and there will not be any output for signals having frequencies below $\omega_{c1}$ and above $\omega_{c2}$. Thus this filter passes a band of frequencies. Figure 26.73(*d*) represents the characteristics of a **Band Stop** filter.

This filter rejects a particular band of frequencies from $\omega_{c1}$ and $\omega_{c2}$ while passing the signals of other frequencies (with a constant gain A) starting from 0 to $\omega_{c1}$ and $\omega_{c2}$ onwards. This is also called a **Notch Filter**.

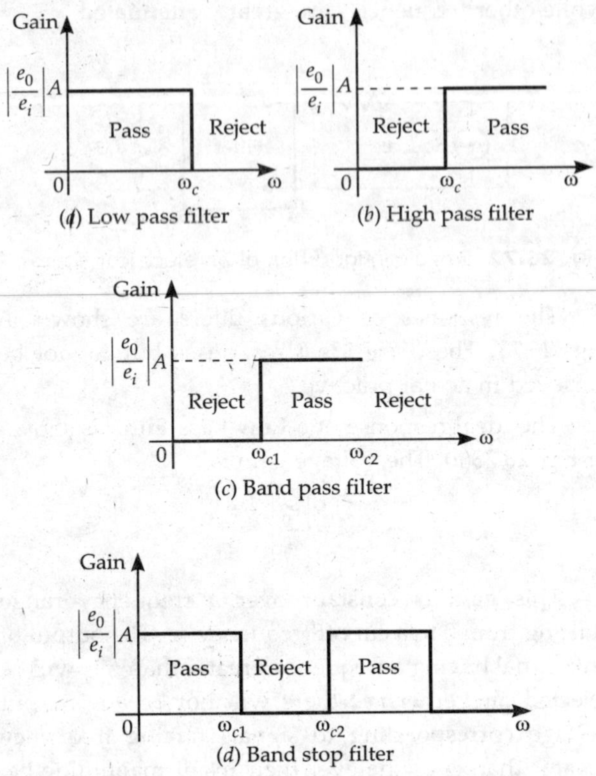


**Fig. 26.73** Ideal characteristics of filters.

The filters discussed above have idea characteristics *i.e.,* they have as harp cut off but in actual practice, the characteristics as shown in Fig. 26.73 are not practically realizable. Therefore, we discuss below the characteristics of filters which can be practically realized.

## 26.14 PASSIVE FILTERS

### 26.14.1 Low Pass Filters

An electric low pass filter is shown in Fig. 26.74(*a*). It is *R*-*C* network. At low frequencies, the capacitive reactance is very high and therefore the capacitor circuit can be considered as an open circuit. Under these conditions $e_0 = e_i$ or the voltage again is equal to unity. At very high frequencies, the capacitive reactance is very low and therefore the output voltage $e_0$ is very small as compared with the input voltage $e_i$. Thus the gain is low and drops off gradually as the frequency is increased. This is shown in the characteristics of a low pass in Fig. 26.74(*b*).

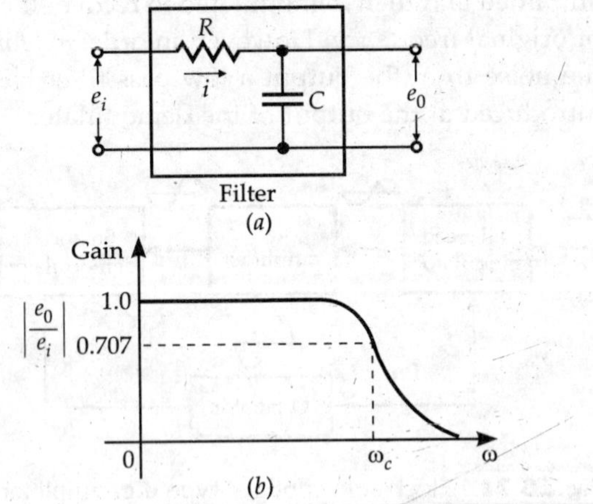


**Fig. 26.74** Low pass *R-C* filter and its characteristics.

The transfer function is

$$\frac{E_0}{E_i}(s) = \frac{1/sC}{R + 1/sC} = \frac{1}{1 + sRC}$$

$$= \frac{1}{1 + s\tau} \quad ...(26.134)$$

The sinusoidal transfer function of a low pass filter is :

$$\frac{e_0}{e_i}(j\omega) = \frac{1}{1 + j\omega RC} = \frac{1}{1 + j\omega\tau} \quad ...(26.135)$$

$$\therefore \text{ Gain } \quad A = \left|\frac{e_0}{e_i}(j\omega)\right| = \frac{1}{\sqrt{1 + (\omega RC)^2}}$$

$$= \frac{1}{\sqrt{1 + (\omega\tau)^2}} \quad ...(26.136)$$

The gain drops to 0.707 at cut off frequency $\omega_c$.

$$0.707 = \frac{1}{\sqrt{1 + (\omega_c RC)^2}}$$

or cut off frequency

$$\omega_c = \frac{1}{RC} = \frac{1}{\tau} \text{ rad/s} \quad ...(26.137)$$

The cut off frequency in Hz is :

$$f_c = \frac{1}{2\pi RC} = \frac{1}{2\pi\tau} \quad ...(26.138)$$

Let us have Bode plot of low pass *RC* filter. Consider the sinusoidal transfer function of Eqn. 26.135.

When $|j\omega\tau| << 1$, the gain is unity and the gain in terms of db is zero. The corner frequency is

$$\omega = 1/\tau = 1/RC.$$

For $|j\omega\tau| >> 1$, the gain $1/\omega\tau$ (*i.e.,* $-20 \log \omega\tau$ in terms of db).

∴ For frequencies greater than $1/\tau$, the asymptote has a slope of – 20 db/decade or – 6 db/octave. This is shown in Fig. 26.75.

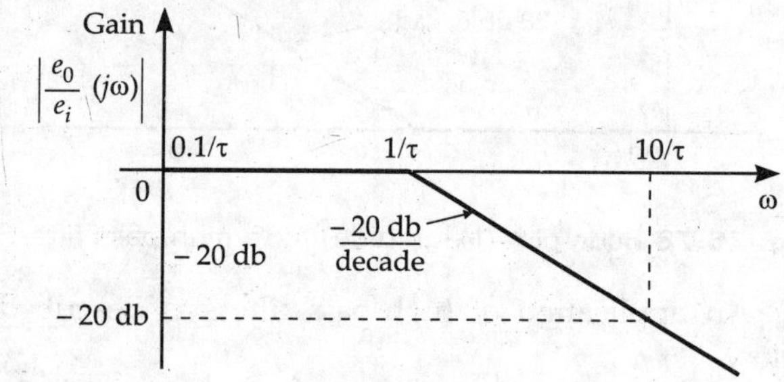


**Fig. 26.75** Bode plot (Magnitude) for a low-pass filter.

Let us cite examples of use of a low pass filters in measurement systems. Supposing a dynamic strain varying at a frequency of 1 Hz is being measured by a strain gauge bridge-shown in Fig. 26.76(*a*). There is an ample possibility that the output measurement circuit may pick up signals from 50 Hz power lines, thereby distorting the output from the strain gauge bridge. Therefore, it is essential that the 50 Hz signal be eliminated so that there is no spurious signal applied to the measurement circuit.

Figure 26.76(*b*) shows the input signal to the measurement system which contains the desired signal arising out on account of the change in strain (at a frequency of 1 Hz) plus the 50 Hz power line

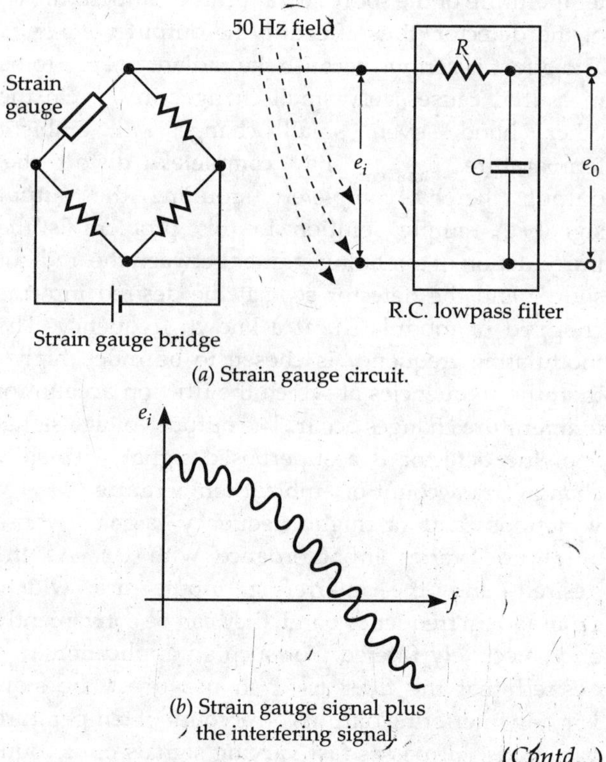


(*Contd...*)

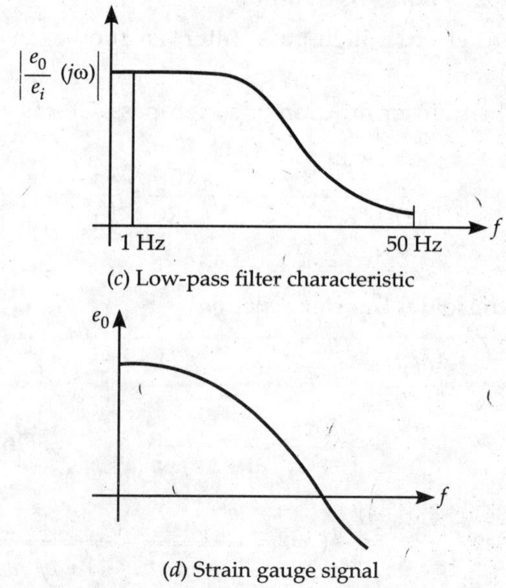


**Fig. 26.76** Output filtering of strain gauge bridge signal by a low-pass filter.

interfering signal. A low-pass *RC* filter can be designed to have a characteristics as shown in Fig. 26.76(*c*) so that the desired signal (1 Hz in this case) is reproduced faithfully, while the high frequency signal of 50 Hz is greatly attenuated as shown in Fig. 26.76(*d*).

**Example 26.19** *A simple RC low-pass filter is to be so designed that the output voltage be attenuated by 3 db at 50 Hz. Calculate the time constant and suitable values of R and C.*

**Solution.** The output voltage is attenuated by 3 db.

$$\therefore \quad 10 \log \frac{E_0}{E_i} = -3 \text{ or } \frac{E_0}{E_i} = 0.501.$$

From Eqn. 26.136, for a low pass filter,

$$\frac{E_0}{E_i} = \frac{1}{\sqrt{1+\omega^2\tau^2}}$$

$$= \frac{1}{\sqrt{1+(2\pi/50)^2\tau^2}} = 0.501$$

or time constant

$$\tau = 5.5 \times 10^{-3} \text{ s.}$$

Now $\quad \tau = RC$

Take $\quad R = 10 \text{ k}\Omega$

$$\therefore \quad C = \frac{5.5\times10^{-3}}{10\times10^3} \text{ F}$$

$$\doteq 0.55 \ \mu\text{F}$$

## 26.14.2 High Pass Filter

An electric high-pass filter is shown in Fig. 26.77(*a*).

The transfer function of a high-pass filter is given by :

$$\frac{E_0(s)}{E_i(s)} = \frac{R}{R+1/sC} = \frac{sRC}{1+sRC} = \frac{\tau s}{1+\tau s} \quad ...(26.139)$$

Sinusoidal transfer function

$$\frac{e_0}{e_i}(j\omega) = \frac{j\omega RC}{1+j\omega RC}$$

$$= \frac{j\omega\tau}{1+j\omega\tau} = \frac{1}{1+1/j\omega\tau} \quad ...(26.140)$$

Gain $$A = \left|\frac{e_0}{e_i}(j\omega)\right| = \frac{\omega\tau}{\sqrt{1+\omega^2\tau^2}} = \frac{1}{\sqrt{1+1/\omega^2\tau^2}} \quad ...(26.141)$$

∴ When the frequency is low the gain is small and when the frequency is large, the gain approaches unity. Thus this circuit passes high frequencies while rejecting the low frequencies. The response of a high pass filter is shown in Fig. 26.77(*b*) when $\omega\tau << 1$ or $\omega << 1/\tau$, gain is 0 (in db). Therefore, the low frequencies are greatly attenuated. The corner frequency is $\omega = 1/\tau$ when $\omega\tau >> 1$ or $\omega >> 1/\tau$, the gain is $A = \omega\tau$.

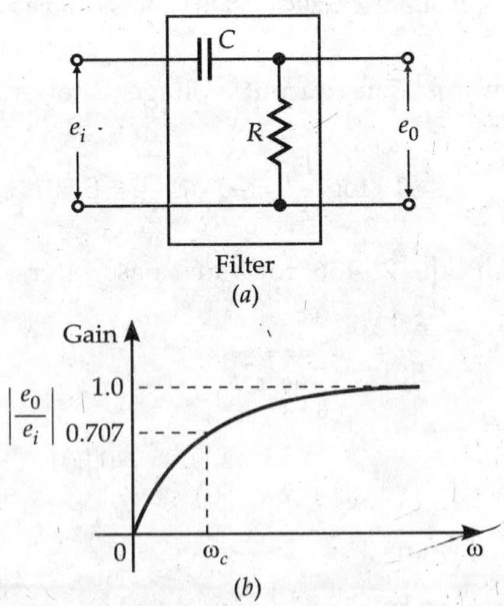


**Fig. 26.77** A high-pass filter and its response.

The asymptote has thus a slope of 20 db/decade or 6 db/decade as shown in Fig. 26 78.

Cutoff frequency is :

$$f_c = 1/2\pi RC \quad ...(26.142)$$

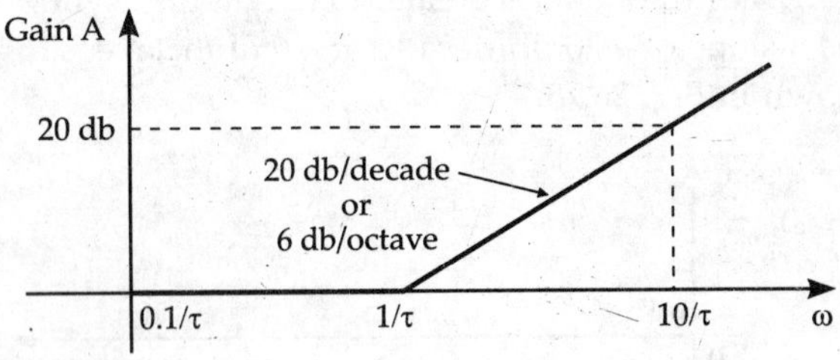


**Fig. 26.78** Bode plot (Magnitude) for a high pass filter.

An application of high-pass filters is described below :

A chopped radiometer is shown in Fig. 26.79 in a simplified form. The objective of this device is to sense the temperature, $\theta_s$, of some object in terms of the infra red radiant energy it emits. The emitted energy is focussed on a detector, which responds to this type of radiation. This produces a change in $\theta_d$, the temperature of the detector. Since the detector is sensitive to change in temperature, it will produce a voltage $e_i$, which is amplified to $e_0$ by an amplifier. Thus if $\theta_s$ changes the temperature of the detector $\theta_d$ changes, with the result that $e_i$, changes and consequently the output voltage $e_0$ changes. An elementary set-up of this transducer is shown in Fig. 26.79.

The major problem with these devices is that the ambient temperature $\theta_{ambient}$ as well as the temperature of the source $\theta_s$, affect the temperature $\theta_d$ of the detector, thus affecting the output voltage $e_0$. The effect is serious because the radiant energy to be measured causes very small changes in $\theta_d$. On the other hand, even small changes in ambient temperature, $\theta_{ambient}$ may completely distort the, output, thereby completely srouding the output signal. A simple solution to this problem is the introduction of a rotating shutter between the radiant source and the detector so that the desired input is chopped or modulated, at a known frequency. The modulating frequency is chosen to be much higher than the frequencies at which the drift on account of temperature changes occur. The output voltage signal $e_i$ of the detector is a superposed signal with slow changes on account of ambient temperature ($\theta_{ambient}$ variations and a high frequency signal whose amplitude varies in accordance with $\theta_s$). As the desired and the interfering inputs are widely separated in frequency band, they can be conveniently and selectively filtered. For such an application, it is desired that the filter used should reject the slow varying interfering outputs on account of temperature variations but passes fast varying signals on account of variations in $\theta_s$. This calls for a **high-pass filter**. In fact such applications require the use of an a.c. amplifier. Since a.c. amplification, even otherwise needed, the use of an a.c. amplifier with low frequency cut-off solves both the problems *i.e.*, the problem of attenuating a low frequency signal and amplification of a high frequency a.c. signal without distortion. A wide frequency band is necessary and this is conveniently provided by the arrangement shown in Fig. 26.79.

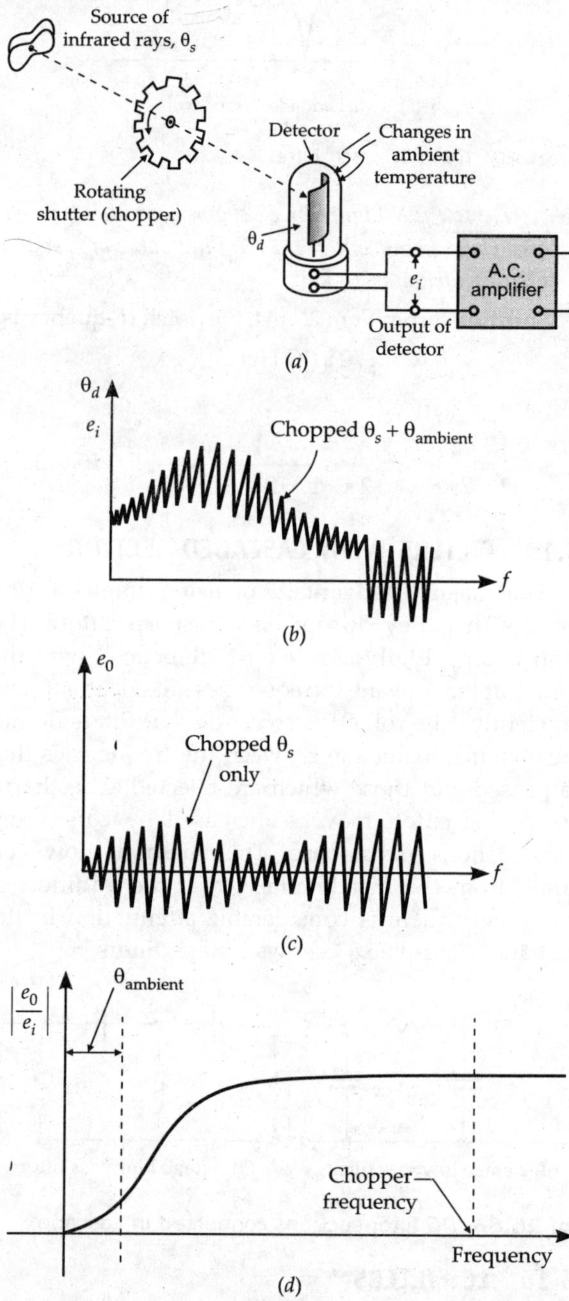


**Fig. 26.79** A high-pass filter for measurement infra-red radiant energy.

### 26.14.3 Band Pass Filters

A simple band-pass filter can be constructed by cascading a low-pass and high-pass filter as shown in Fig. 26.80(*a*).

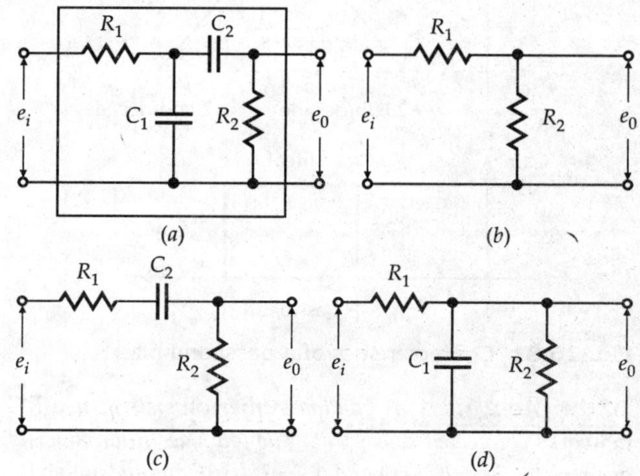


**Fig. 26.80** *RC* passband filter.

There is a passband, with a lower cutoff frequency, $f_{C1}$, and an upper frequency cutoff at $f_{C2}$. In the passband, the circuit behaves like a voltage divider network shown in Fig. 26.80(*b*). At frequencies below the passband, it behaves like a high-pass band filter as shown in Fig. 26.80(*c*). (Since capacitor $C_1$ is short circuited and capacitor $C_2$ behaves as an open circuit). Above passband frequencies, the capacitor $C_2$ behaves as a short circuited element and the circuit behaves like a lowpass filter as shown in Fig. 26.80(*d*).

Thus if $R_2 > 10R_1$, and $f_{C2} > 10 f_{C1}$.

The lower cutoff frequency is

$$f_{C1} = 1/2\pi R_2 C_2$$

and the upper cutoff frequency is

$$f_{C2} = 1/2\pi R_1 C_1$$

The low-pass and high-pass filters are connected in cascade, the transfer function is :

$$\frac{E_0(s)}{E_i(s)} = \frac{1}{(1+s\tau_1)} \times \frac{\tau_2 s}{(1+s\tau_2)} \quad ...(26.142)$$

with $\tau_1 = R_1C_1$ and $\tau_2 = R_2C_2$ with $\tau_2 > \tau_1$.

The characteristics of a bandpass filter are shown in Fig. 26.81.

In this passband, the circuit behaves almost as a resistive network and therefore the gain in the passband is :

$$A = \left|\frac{e_0}{e_i}\right| = \frac{R_2}{R_1 + R_2} \quad ...(26.143)$$

26

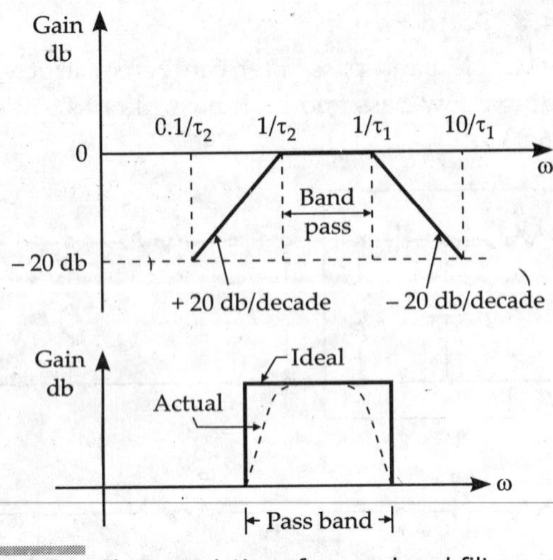


**Fig. 26.81** Characteristics of a passband filter.

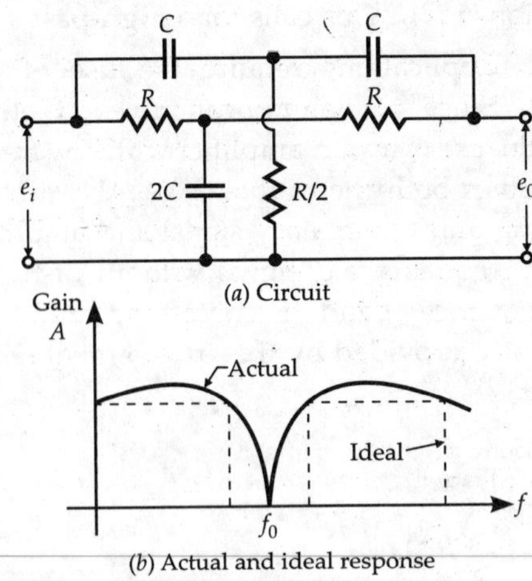


**Fig. 26.82** Twin *T* Notch Filter.

***Example 26.20*** *A bandpass filter consists of two RC networks connected in cascade. The low pass filter consists of a resistor $R_1 = 10$ kΩ and $C_1 = 100$ pF and the high pass filter consists of $R_2 = 1$ MΩ and $C_2 = 0.01$ μF. Find the lower and the upper cutoff frequencies and the passband gain.*

**Solution.** Passband gain

$$A = \frac{R_2}{R_1 + R_2} = \frac{1 \times 10^6}{1 \times 10^3 + 1 \times 10^6} = 0.99$$

Lower cutoff frequency

$$f_{C1} = \frac{1}{2\pi R_2 C_2} = \frac{1}{1 \times 10^6 \times (0.01) \times 10^{-6}} = 15.9 \text{ Hz.}$$

Upper cutoff frequency

$$f_{C1} = \frac{1}{2\pi R_1 C_1} = \frac{1}{2\pi \times 10 \times 10^3 \times 100 \times 10^{-12}} \text{Hz} = 159 \text{ kHz.}$$

### 26.14.4 Band Rejection (band stop) Filters

A simple *RC* band rejection (band stop) filters use either a Wien Bridge circuit of a Twin '*T*' circuit.

#### Twin *T* Band Stop filter

Figure 26.82 shows a Twin T band stop filter. At very low and very high frequencies, the gain is almost unity. Somewhere in between there is a frequency where the gain becomes zero.

The output at frequency, $f_0$, is equal to zero. This type of filter is often called a **notch filter** because it completely rejects a particular frequency and attenuates a range of frequencies greatly while passing the frequencies below and above the notch. Frequency $f_0$ is called **notch frequency.**

Figure 26.82(*b*) shows the actual and ideal response of a notch filter.

Notch frequency $f_0 = 1/2\pi RC$ ...(26.144)

***Example 26.21*** *Figure 26.82 shows a parallel twin 'T' network. The resistance R = 1 MΩ. Find value of C so that the notch frequency is 10 kHz.*

**Solution.** From Eqn. 26.144, the notch frequency is

$$f_0 = 1/2\pi RC \text{ Hz.}$$

∴ Capacitance C

$$= \frac{1}{2\pi R f_0} = \frac{1}{2\pi \times 1 \times 10^6 \times 10 \times 10^3} \text{F} = 15.9 \text{ pF}$$

## 26.15 FILTERS WITH CASCADED SECTIONS

The major disadvantage of using single section filters is that they do not have a sharp cutoff. The cutoff is only 20 db/decade (or 6 db/octave) with the result that the unwanted frequencies are not attenuated sufficiently. The roll off is so gentle that filters do not give a sharp distinction between the frequencies that are passed and those which are rejected. The sharpness of the cutoff may be increased by connecting more sections in cascade. This method, however, suffers from the disadvantage, that the addition of more stages there is considerable attenuation in the passband. Figure 26.83 shows 2-stage filters.

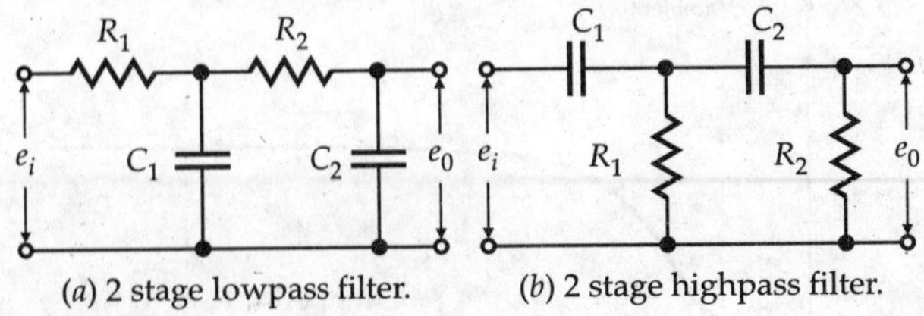


**Fig. 26.83** *RC* filter sections connected in cascade.

## 26.16 *LC* FILTERS

In order to have sharper cutoff and also prevent any attenuation in passband, *LC*, filters are used.

Figure 26.84 shows some basic configurations of **Low Pass LC Filters**. At low frequencies, the inductance presents a short while the capacitance is open circuited with output almost equal to input. At high frequencies the inductance presents a high series impedance while the capacitance is short circuited. This gives a zero output. Therefore the circuits are basically low pass filters.

Characteristic resistance

$$R_0 = \sqrt{L/C} \qquad ...(26.145)$$

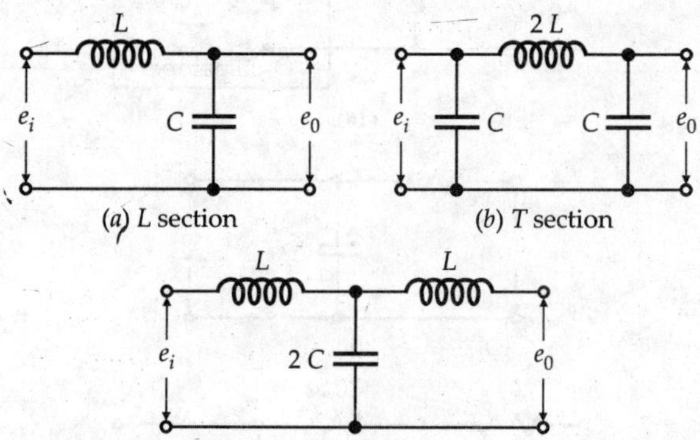


**Fig. 26.84** Basic configuration of *LC* lowpass filters.

The source and the load with which, LC filters used must have the same resistance values as $R_0$.

$$\text{Cutoff frequency } f_c = 1/2\pi\sqrt{L/C} \qquad ...(26.146)$$

$$= 1/2\pi R_0 C \qquad ...(26.147)$$

Attenuation in the stopband

$$= 8.7\, n\alpha \text{ db} \qquad ...(26.148)$$

where $\alpha = 2\cosh^{-1}(f/f_c)$ and

$n$ = number of stages

Figure 26.85 shows different configurations of **highpass LC filters.**

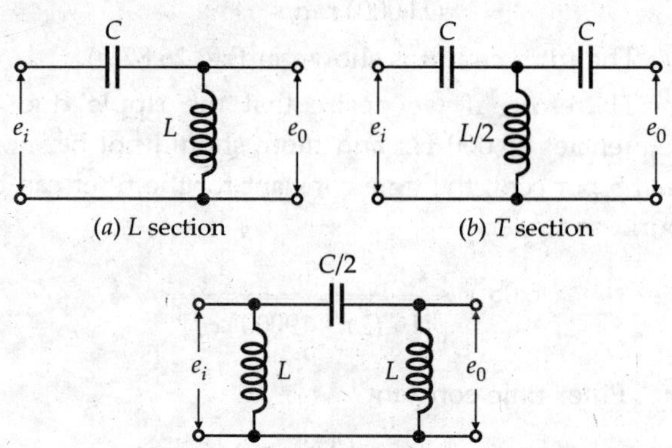


**Fig. 26.85** Basic configurations of Highpass LC Filters.

At low frequencies the capacitance presents an open circuit and inductance presents a short and therefore the output is almost zero. At higher frequencies, the capacitor is almost shorted while inductance presents an open circuit with the result that the output is almost equal to input.

$$\alpha = 2\cosh^{-1}(f_c/f) \text{ for high pass filters.} \qquad ...(26.149)$$

***Example 26.22*** *A basic low pass T section filter uses an inductance of 1000 mH and a capacitance of 5000 pF. Find the size of the load and the source resistances that should be used with this filter. Find also the cutoff frequency.*

**Solution.** The values of L and C used in Eqn. 26.146 refer to the values of the basic L-section filter. The T-section is made up of two back to back L-sections. Therefore shunt capacitance of the T-section is 2C. Hence, 2 C = 5000 pF or C = 2500 pF.

Characteristics resistance

$$R_0 = \sqrt{L/C} = \sqrt{\frac{1000\times10^{-3}}{2500\times10^{-12}}}\ \Omega = 20\ \text{k}\Omega$$

Cutoff frequency

$$f_c = \frac{1}{2\pi\sqrt{LC}} = \frac{1}{2\pi\sqrt{1000\times10^{-3}\times2500\times10^{-12}}} = 3180\ \text{Hz}$$

The source and the load resistances should also be equal to 20 kΩ.

***Example 26.23*** *Design a low pass LC filter with a cutoff frequency of 500 Hz with a load of resistance 16 Ω. Find out the attenuation at 1000 Hz (an octave above cutoff frequency).*

**Solution.** The characteristic resistance should be equal to load resistance

$$\therefore \quad R_0 = 16\ \Omega$$

$$\text{Cutoff frequency} \quad f_c = \frac{R_0}{2\pi L}$$

∴ Inductance

$$L = \frac{R_0}{2\pi f_c} = \frac{16}{2\pi\times500} = 5.1\times10^{-3}\ \text{H} = 5.1\ \text{mH}$$

Capacitance

$$C = \frac{1}{2\pi R_0 f_c} = \frac{1}{2\pi\times16\times500} \approx 20\times10^{-6}\ \text{F} = 20\ \mu\text{F}$$

A single stage $LC$ low pass filter is shown in Fig. 26.86($a$).

$$\alpha = 2\cosh^{-1}(f/f_c) = 2\cosh^{-1}(1000/500) = 2.64$$

Attenuation at a 1000 Hz = 8.7 $n$

$$\alpha = 8.7 \times 1 \times 2.64 = 23 \text{ db.}$$

This $LC$ filter gives a much shaper cutoff of 23 db/octave instead of 6 db/octave given by a simple $RC$ filter. If a more sharper cutoff is desired more sections can be added in cascade.

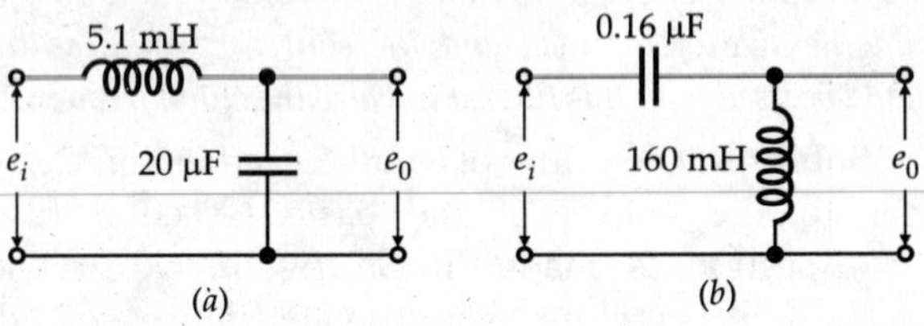


**Fig. 26.86** *LC* filters.

***Example 26.24*** *Design a high-pass LC filter with a load resistance of 1000 Ω and a cutoff frequency of 1000 Hz. What is the attenuation at one octave below the cutoff frequency i.e., at 500 Hz. If there are two stages what will be attenuation at 500 Hz ?*

**Solution.** The characteristics resistance should be equal to the load resistance.

$\therefore \quad R_0 = 1000\ \Omega$

Inductance $\quad L = \dfrac{R_0}{2\pi f_c}$

$$= \frac{1000}{2\pi \times 1000} = 0.16 \text{ H} = 160 \text{ mH}$$

Capacitance, $\quad C = \dfrac{1}{2\pi R_0 f_c}$

$$= \frac{1}{2\pi \times 1000 \times 1000} \text{F} = 0.16\ \mu\text{F}$$

A single section $LC$ high pass filter is shown in Fig. 26.86($b$).

For a high-pass filter

$$\alpha = 2\cosh^{-1}(f_c/f) = \cosh^{-1}(1000/500) = 2.64$$

$\therefore$ For a single stage high pass filter :

Attenuation at 500 Hz is = 8.7 × 1 × 2.64 = 23 db

In case, there are two stages connected in cascade, the attenuation at 500 Hz is :

$$= 8.7 \times 2 \times 2.64 = 46 \text{ db}$$

Thus the low frequencies are considerably attenuated.

***Example 26.25*** *The primary winding of an LVDT is excited by a 10,000 Hz source. The input displacement to the core has a highest frequency content of 1000 Hz. The output of the LVDT is connected to a phase sensitive demodulator which in turn is connected to an RC lowpass filter. Find the time constant of the filter in case the ripple due to the demodulated signal is limited to 5 percent. Is the arrangement satisfactory ? If not, suggest a better arrangement for the filter circuit.*

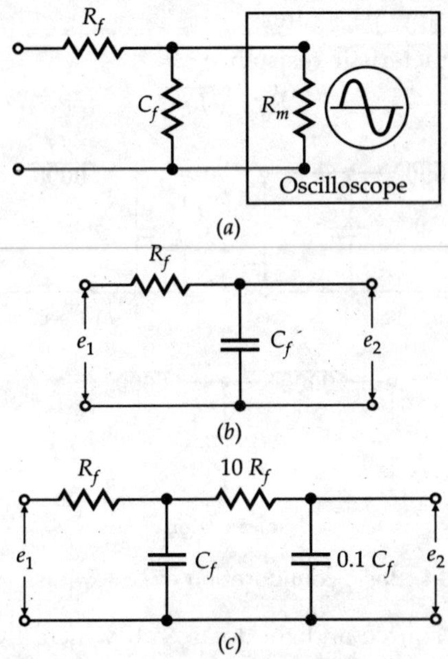


**Fig. 26.87** Diagram of Example 26.23.

**Solution.** The excitation frequency of the LVDT 10,000 Hz and that of the signal is 1000 Hz. Therefore,

$$\omega_c = 2\pi \times 10{,}000 \text{ rad/s ;}$$

and $\quad \omega_s = 2\pi \times 1000$ rad/s

The frequency bands produced by the demodulation process are :

$$2\omega_c - \omega_s = 2\pi(2 \times 10{,}000 - 1000) = 2\pi(19{,}000) \text{ rad/s.}$$

and $\quad 2\omega_c + \omega_s = 2\pi(2 \times 10{,}000 + 1000)$

$$= 2\pi(21{,}000) \text{ rad/s}$$

The filter circuit is shown in Fig. 26.87($b$).

Therefore, if we desire that the ripple due to frequencies 19,000 Hz and more should not be more than 5 per cent, the time constant for the filter can be found as under :

$$0.05 = \frac{1}{\sqrt{[1 + (2\pi \times 19000)\tau_f]^2}}$$

or $\quad$ Filter time constant

$$\tau_1 = 0.167 \times 10^{-3} \text{ s} = 0.167 \text{ ms}$$

The amplitude ratio at the highest motion frequency of 1000 Hz is given by :

$$M = \left|\frac{e_0}{e_i}(j\omega)\right| = \frac{1}{\sqrt{1+\omega^2\tau_f^2}}$$

$$= \frac{1}{\sqrt{1+(2\pi\times 1000\times 0.167\times 10^{-3})^2}} = 0.69$$

Phase shift

$$\phi = \tan^{-1}\omega\tau_f$$
$$= -\tan^{-1}(2\pi\times 1000\times 0.167\times 10^{-3})$$
$$= -46.4°$$

Thus this filter has an amplitude ratio of 0.69 and a phase shift of −46.4°. This will distort the high frequency portion of the input displacement $x_i$ considerably. Therefore, this filter is not suited for this particular application. It has a cut-off rate of −20 db/decade. A filter of sharper cut-off rate is desired.

A more suitable proposition would be to use a double *RC* filter as shown in Fig. 26.87(*c*). For this filter two *RC* filters are connected in cascade with same time constant.

The frequency response is :

$$\frac{e_2}{e_1}(j\omega) = \frac{1}{(1+j\omega R_f C_f)^2}$$

This filter gives a cut-off rate of − 40 db/decade.

∴ The ratio of output voltage to input voltage is :

$$M = \frac{1}{1+\omega^2 R_f^2 C_f^2} = \frac{1}{1+\omega^2\tau_f^2}$$

$$\therefore \quad 0.05 = \frac{1}{1+(2\pi\times 19{,}000\times\tau_f)^2}$$

or $$\tau_f = 36.5\times 10^{-6}\ \text{s} = 36.5\ \mu\text{s}$$

The amplitude ratio for a frequency of 1000 Hz is :

$$M = \frac{1}{\sqrt{1+(2\pi\times 1000\times 36.5\times 10^{-6})^2}} = 0.95$$

Phase angle

$$\phi = -2\tan^{-1}$$
$$\omega\tau_f = -2\tan^{-1}(2\pi\times 1000\times 36.5\times 10^{-6})$$
$$= -26°.$$

## 26.17 ACTIVE FILTERS

In general impedances are used in the inverting amplifier shown in Fig. 26.88. The basic relationship in case of inverting amplifiers is :

$$\frac{V_0}{V_1} = \frac{Z_f}{Z_1}$$

The above relationship can be used to obtain desired filter sections. Since, the voltage can also be amplified these circuits are used as *active filters*.

Figure 26.88 shows a low pass filter. The output voltage of this filter is given by

$$V_0 = \frac{-R_{fH}\times 1/j\omega C_{fH}}{R_{1H}(R_{fH}+1/gj\omega C_{fH})}V_1$$

$$= -\frac{R_{fH}/R_{1H}}{gj\omega C_{fH}R_{fH}+1}V_1$$

where the subscript *H* refers to the characteristic high frequency

$$\omega_H = \omega_{c1} = 1/R_{fH}C_{fH} \qquad \text{...(26.150)}$$

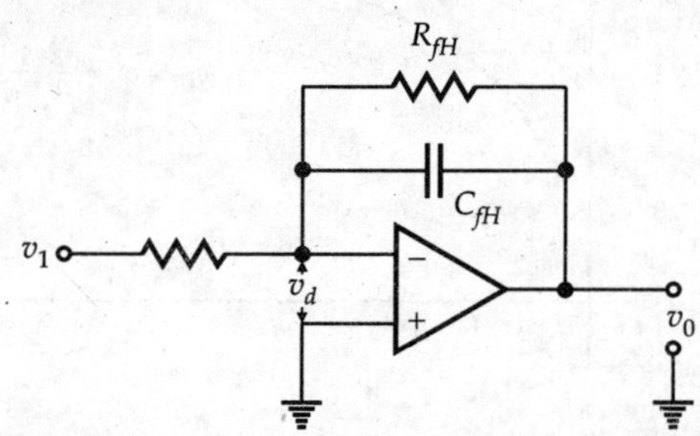


**Fig. 26.88** Circuit of a lowpass active filter.

The characteristics of an active low pass filter are shown in Fig. 26.89.

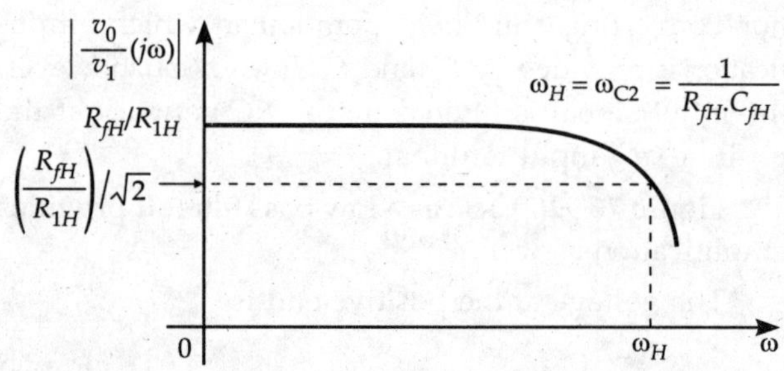


**Fig. 26.89** Characteristics of a low pass active filter.

The diagram of a high pass active filter is shown in Fig. 26.90.

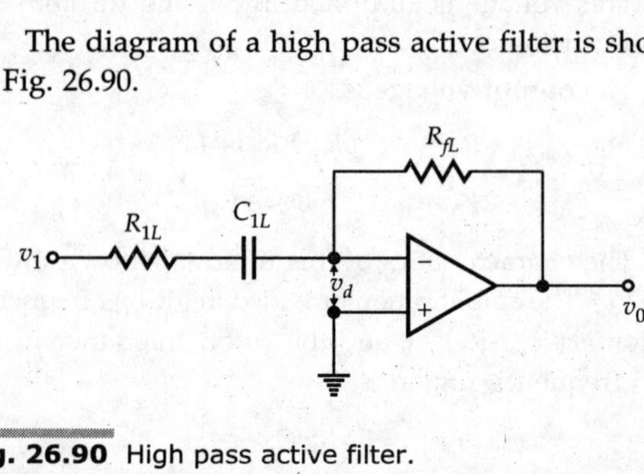


**Fig. 26.90** High pass active filter.

For a high pass active filter shown in Fig. 26.90.

$$V_0 = \frac{-R_{fL}}{R_{1L} + 1/j\omega C_{1L}} V_1 = -\frac{j\omega C_{1L} R_{fL}}{1 + j\omega C_{1L} R_{1L}} V_1 \qquad ...(26.151)$$

where the subscript $L$ refers to characteristic low frequency,

$$\omega_L = \omega_{c1} = 1/R_{1L} C_{1L} \qquad ...(26.152)$$

The characteristics of a high pass active filter are shown in Fig. 26.91.

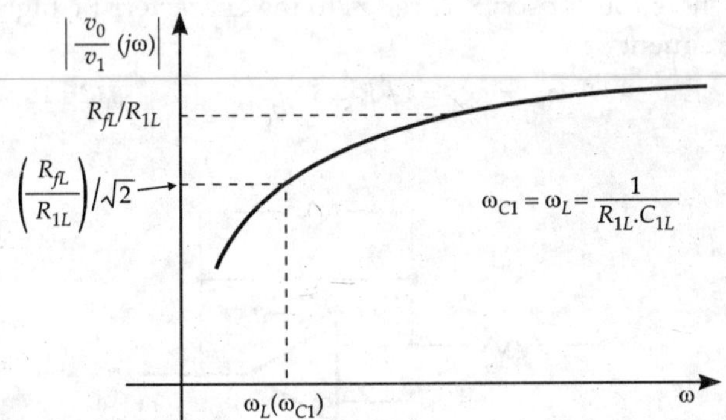


**Fig. 26.91** Characteristics of a high pass active filter.

Additional practical configurations used for low pass and high pass filters are shown in Figs. 26.92($a$) and ($c$). These circuit configurations explicitly carry out the non-converting amplifier operation in which amplification is provided by $R_f$ and $R_1$. In addition, however, the input signal is filtered by the $RC$ network at the positive ($v_+$) input terminal.

Figure 26.92($a$) acts as a low pass filter. It provides amplification as well.

The voltage at the positive end is

$$V_+ = \frac{1}{1 + j\omega R_H C_H} V_1 \qquad ...(26.153)$$

This voltage is amplified in passing through the non-inverting configuration.

The output voltage is,

$$V_0 = \left(1 + \frac{R_f}{R_1}\right) V_+ = \frac{(R_f/R_1) + 1}{1 + j\omega C_H R_H} V_1 \qquad ...(26.154)$$

The characteristics of this filter are shown in Fig. 26.92($b$). The amplification provided in the low frequency region is $(R_f/R_1) + 1$, and the cutoff frequency in the high frequency region is

$$\omega_{c2} = \omega_H = 1/R_H C_H$$

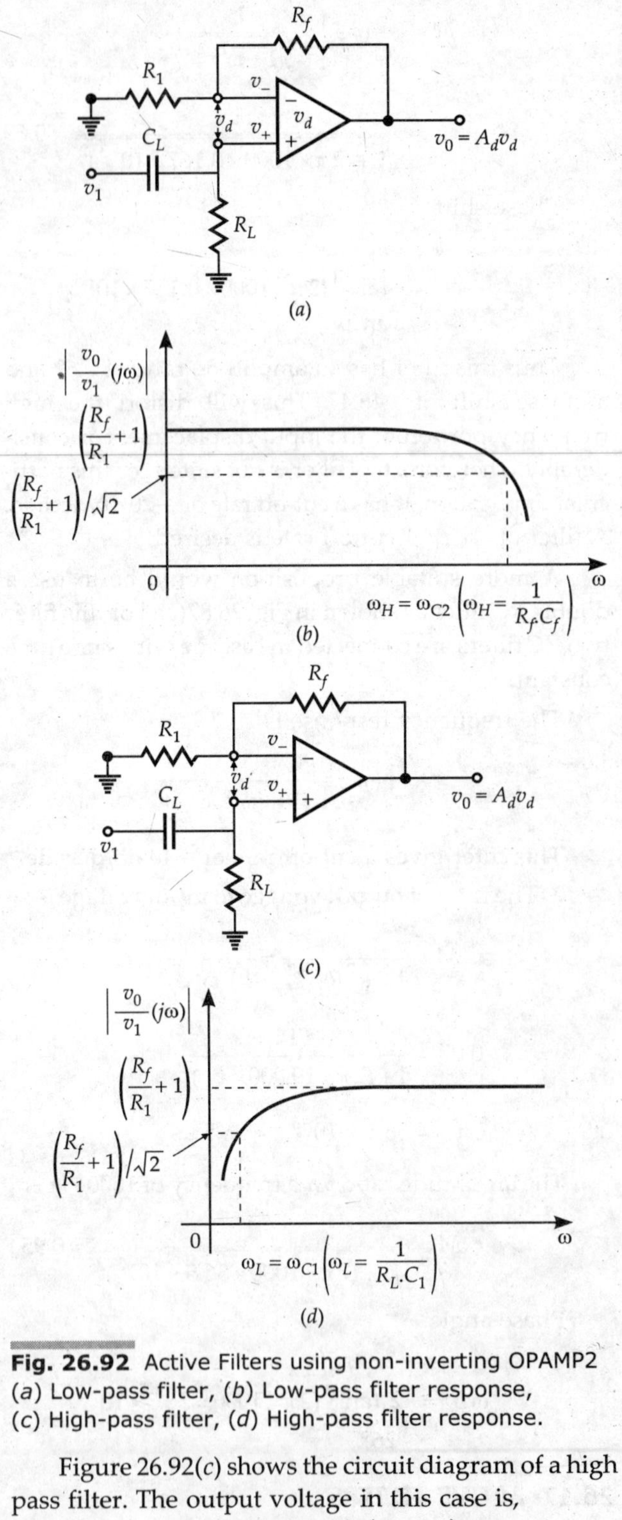


**Fig. 26.92** Active Filters using non-inverting OPAMP2 ($a$) Low-pass filter, ($b$) Low-pass filter response, (c) High-pass filter, ($d$) High-pass filter response.

Figure 26.92($c$) shows the circuit diagram of a high pass filter. The output voltage in this case is,

$$V_0 = \left(\frac{j\omega R_L C_L}{1 + j\omega R_L C_L}\right)\left(\frac{R_f}{R_1} + 1\right) V_1 \qquad ...(26.155)$$

The characteristics of this filter are shown in Fig. 26.92($d$). The lower cutoff frequency in this case is,

$$\omega_{C1} = \omega_L = 1/R_L C_L$$

By combining the actions of the low and high pass filters, a *band pass active filter* is obtained as shown in Fig. 26.93. The output voltage is given by the product of results obtained in Eqns. 26.154 and 26.155. Output voltage

$$V_0 = \frac{(R_{f1}/R_1)+1}{1+j\omega R_H C_H}\cdot\frac{j\omega R_L C_L}{1+j\omega R_L C_L}\left(\frac{R_{f2}}{R_2}+1\right)V_1 \quad ...(26.156)$$

The variation of magnitude $\left|\frac{v_0}{v_1}(j\omega)\right|$ is shown in Fig. 26.93(*b*).

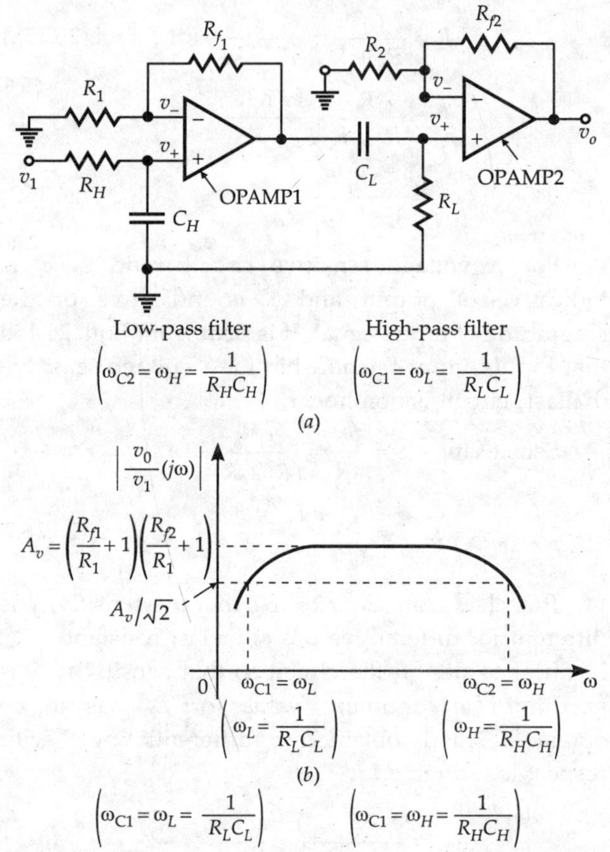


**Fig. 26.93** Active band pass filter and its characteristics.

**Example 26.26** *Calculate the upper and lower cutoff frequencies and voltage amplification between these frequencies for a band pass active filter shown in Fig 26.93 which has :*

$$R_1 = R_2 = 10\text{ k}\Omega,\ R_{f1} = R_{f2} = 100\text{ k}\Omega,$$
$$R_L = R_H = 10\text{k}\Omega,\quad C_L = 1\,\mu\text{F and } C_H = 1\text{pF}.$$

**Solution.** From Eqn. 26.156, voltage amplification

$$\frac{V_0}{V_1} = \frac{(R_{f1}/R_1)+1}{1+j\omega R_H C_H}\cdot\frac{j\omega R_L C_L}{1+j\omega R_L C_L}\left(\frac{R_{f2}}{R_2}+1\right)$$

Now $\quad R_1 = R_2 = R,\ R_{f1} = R_{f2} = R_f$

$$\therefore \quad \frac{V_0}{V_1} = \frac{(R_f/R)+1}{1+j\omega R_H C_H}\cdot\frac{j\omega R_L C_L}{1+j\omega R_L C_L}\left(\frac{R_f}{R}+1\right)$$

Therefore, at midpoint between frequencies, ($\omega_{C1} = \omega_L$ and $\omega_{C2} = \omega_H$)

$$\frac{V_0}{V_1} = \left(\frac{R_{f1}}{R_1}+1\right)\left(\frac{R_{f2}}{R_2}+1\right)$$

as the response is independent of frequency

$$= \left(\frac{R_f}{R}+1\right)\left(\frac{R_f}{R}+1\right)$$

$$= \left(\frac{100\times10^3}{10\times10^3}+1\right)\left(\frac{100\times10^3}{10\times10^3}+1\right) = 121$$

The lower cutoff frequency

$$\omega_{C1} = \omega_L = \frac{1}{R_L C_L}$$

$$= \frac{1}{10\times10^3\times1\times10^{-6}} = 100\text{ rad/s}$$

The upper cutoff frequency

$$\omega_{C2} = \omega_H = \frac{1}{R_H C_H}$$

$$= \frac{1}{10\times10^3\times1\times10^{-12}} = 10^8\text{ rad/s}$$

## 26.18 INPUT CIRCUITS

The following types of input circuits are used for signal conditioning of electrical transducers. Though the list is not complete but it includes most of the important items. The input circuits used are :

(*i*) Simple current sensitive circuits,
(*ii*) Ballast circuits,
(*iii*) Voltage dividing circuits,
(*iv*) Voltage balancing potentiometer circuits,
(*v*) Bridge circuits, and
(*vi*) Resonant circuits.

## 26.19 CURRENT SENSITIVE CIRCUITS

Figure 26.94 shows a simple current sensitive circuit. This input circuit uses the flow of current in through a passive resistance transducer as an indication of the value of the resistance. The transducer may employ any one of the various forms of variable resistance elements. The resistance of transducer changes when there is a change in the physical quantity which is being measured, thereby, causing a change in current. This current is sensed and the physical quantity is therefore measured or recorded. The measurement or the recording device may be calibrated to read the physical quantity directly.

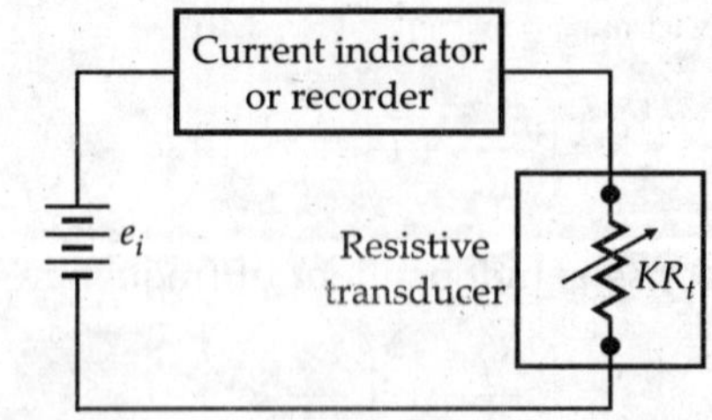


**Fig. 26.94** Current sensitive circuit.

The output current depends upon the input voltage $e_i$ and therefore in order to maintain the calibration of the current sensitive circuit, the value of $e_i$ should be kept constant.

## 26.20 BALLAST CIRCUIT

A ballast circuit is only a simple variation of the current sensitive circuit. In this case at voltage sensitive device is connected across the transducer in place of a series connected current sensitive device. The circuit is also called a voltage sensitive circuit.

Figures 26.95(*a*) and (*b*) show a Ballast circuit. The circuit includes a series resistance, $R_b$ which is called **ballast resistance.**

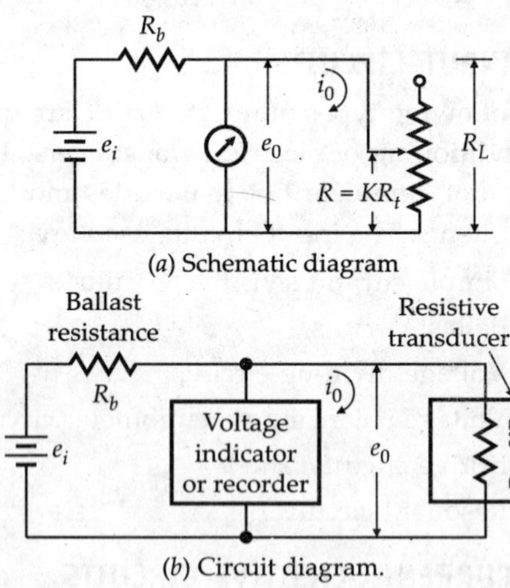


**Fig. 26.95** Ballast circuit (voltage sensitive circuit).

A ballast resistance is a must in a voltage sensitive circuits since in the absence of the ballast resistance, $R_b$, the voltage device will always record the same voltage *i.e.*, $e_i$ irrespective of the transducer resistance.

It is assumed that the voltage measuring or recording device has an infinite resistance so that it does not draw any current.

Let $R_t$ = maximum resistance of transducer.

$R = KR_t$ = resistance of transducer when measuring a physical quantity.

The value of $K$ may vary from 0 to 1 and is a function of the magnitude of the input signal.

$$\text{Output current} \quad i_0 = \frac{e_i}{R_b + R} = \frac{e_i}{R_b + KR_t} \quad \text{...(26.157)}$$

Output voltage

$$e_0 = i_0$$

$$R = \frac{e_i R}{R_b + R} = \frac{e_i}{R_b + KR_t} \times KR_t \quad \text{...(26.158)}$$

$$\therefore \quad \frac{e_0}{e_i} = \frac{R}{R_b + R}\,\frac{KR_t}{R_b + KR_t}$$

$$= \frac{KR_t / R_b}{1 + KR_b / R_t} \quad \text{...(26.159)}$$

For a voltage sensitive circuit ratio $e_0 / e_i$ is indicative of output and $K$ is indicative of the magnitude of input signal. It is clear from Eqn. 26.150 that input-output relationship for a voltage sensitive (ballast) circuit is non-linear.

$$\text{Sensitivity} \quad S = \frac{de_0}{dR} = \frac{e_i R_b}{(R_b + R)^2}$$

$$= \frac{e_i R_b}{(R_b + KR_t)^2} \quad \text{...(26.160)}$$

It is clear from Eqn. 26.160, that the sensitivity is different for different values of ballast resistance, $R_b$. In order to design the circuit so that sensitivity $S$ is maximum an optimum value for $R_b$ has to be obtained. This is obtained by differentiating $S$ with respect to variable $R_b$.

$$\frac{dS}{dR_b} = \frac{e_i(R - R_b)}{(R_b + R)^3} = \frac{e_i(KR_t - R_b)}{(R_b + KR_t)^3}$$

The maximizing condition is :

$$\frac{dS}{dR_b} = \frac{e_i(KR_t - R_b)}{(R_b + KR_t)^3}$$

$$= 0$$

$\therefore$ We get, $R_b = KR_t = R$ for maximum sensitivity.

The value of $R_b$, $= R$ must be based upon a compromise since $R_b$, a constant, cannot always have the value of $R = KR_t$, a variable. But since the resistance of the transducer $KR_t$ is variable, we may select the value of $R_b$ only for the range of $R = KR_t$ where the sensitivity is to be maximum.

The advantage of the voltage sensitive circuit over a current, sensitive circuit is that a voltage measurement is easier to perform than a current measurement.

The disadvantages of a voltage sensitive circuit are :

- Changes in the input voltage $e_i$, results in greater change in output than similar changes in $K$ (which are on account of change in input signal). Therefore for constancy of calibration, the input voltage $e_i$, should be kept absolutely constant. This requires a voltage regulated source.
- The input-output relationships are non-linear.

**Example 26.27** *A ballast circuit is to be used for a thermistor which has a nominal resistance of 100 Ω.*

*(a) What should be the value of the ballast resistor $R_b$, for maximum sensitivity at the nominal transducer resistance ?*

*(b) If the ballast resistance as calculated in (a) is used along with an input of 10 V, what output voltage will be obtained when the transducer has resistance equal to its nominal resistance ?*

*(c) If the conditions of (b) remain, except the transducer resistance is reduced by 5 percent what output voltage will be indicated ?*

*(d) What are the sensitivities for (b) and (c) ?*

**Solution.** (*a*) Nominal value of transducer resistance $R_t = 100\ \Omega$.

At nominal resistance of transducer $K = 1$.

For maximum sensitivity : ballast resistance $R_b = KR_t = 1 \times 100 = 100\ \Omega$.

(*b*) From Eqn. 26.159, output voltage with $R_b = 100\ \Omega$ :

$$e_0 = \frac{KR_t / R_b}{1 + KR_t / R_b} e_i = \frac{1 \times 100 / 100}{1 + (1 \times 100 / 100)} \times 10 = 5 \text{ V}.$$

(*c*) If the resistance of the transducer decreases by 5%, we have $K = 0.95$.

$\therefore$ Output voltage

$$= \frac{0.95 \times 100 / 100}{1 + (0.95 \times 100 / 100)} \times 10 = 4.87 \text{ V}$$

(*d*) From Eqn 26.160, sensitivity

$$S = \frac{e_i R_b}{(R_b + KR_t)^2}$$

For (*c*), $K = 1$, and the sensitivity,

$$S = \frac{10 \times 100}{(100 + 1 \times 100)^2} = 0.025 \text{ V/}\Omega$$

For (*d*), $K = 0.95$ and the sensitivity,

$$S = \frac{10 \times 100}{(100 + 0.95 \times 100)^2} = 0.0263 \text{ V/}\Omega$$

**Example 26.28** *A simple ballast circuit is used to measure the output of a pressure pickup. The circuit is designed so that the internal resistance is six times the total transducer resistance. A source of 100 V is used to energize the circuit. Calculate the voltage output at 25, 50, 60 and 80 percent full load on the transducer.*

**Solution.** Let the transducer resistance be $R_t$ at full load. Therefore, the ballast resistance $R_b = 6\,R_t$.

$$\text{Output voltage} \quad e_0 = \frac{KR_t / R_b}{1 + (KR_t / R_b)} e_i$$

Now $R_b = 6\,R_t$ and $e_i = 100$ V

$$\therefore \quad e_0 = \frac{K \times 1/6}{1 + (1/6)K} \times 100$$

$$= \frac{16.7\,K}{1 + 0.167\,K}$$

$\therefore$ Voltage output at 25 percent load

$$= \frac{16.7 \times 0.25}{1 + 0.167 \times 0.25} = 4 \text{ V}$$

Similarly, voltage output at 50% load is 7.7 V, at 60% load 9.11 V and 80% load 11.78 V.

## 26.21 BRIDGE CIRCUITS

Bridge circuits of various types are used in Instrumentation systems for the measurement of resistance, inductance and capacitance. The bridges may be either of the two types *viz* :

- d.c. type, and
- a.c. type

In case the output of resistive type of transducers is to be measured, either d.c. or a.c. bridges may be used. But when measuring inductance or capacitance which may be on account of output of inductive or capacitive transducers, a.c. bridges are used.

## 26.22 WHEATSTONE BRIDGE

It is most commonly used d.c. bridge for measurement of resistance. This bridge is used for measurement of small resistance changes that occur in passive resistive transducers like strain gauges, thermistors and resistance thermometers. Detailed description of a Wheatstone bridge has already been given in chapter 14. (See Art. 14.2.3 on page 424)

Figure 26.96 shows the basic circuit of a Wheatstone bridge. It consists of four resistive arms

26

with a source of emf (a battery) and a meter which acts as a detector. The detector is usually a current sensitive galvanometer. Measurements may be carried out either by balancing the bridge or by determining the magnitude of unbalance.

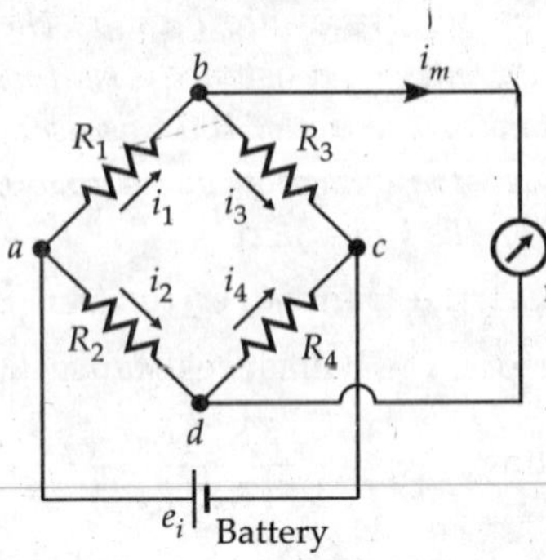


**Fig. 26.96** Wheatstone bridge.

This means that there are two ways in which a Wheatstone bridge can be used. Thus there are two types of Wheatstone bridge. They are : (*i*) Null type bridge, and (*ii*) Deflection type bridge.

### 26.22.1 Null Type Bridge

When using this type of measurement, adjustments are made in various arms of the bridge so that the voltage across the detector is zero and hence no current flows through it. When no current flows through detector the bridge is said to be **balanced.**

Under conditions of balance :

$$R_1 = R_2(R_3 / R_4) \qquad ...(26.161)$$

Resistance $R_1$ represents the resistance of a resistive transducer whose value depends upon the physical variable being measured. The ratio of resistors $R_3$ and $R_4$, is fixed for a particular measurement. The bridge is balanced by varying the value of resistor $R_2$. Thus if three resistances are known the fourth may be determined from Eqn. 26.161.

In instrumentation work, it is the change $\Delta R_1$ in the transducer resistance $R_1$ which is to be found. The change unbalances the bridge and therefore resistor $R_2$ has to be adjusted by an amount $\Delta R_2$ to restore balance.

The rebalanced conditions give :

$$\begin{aligned} R_1 + \Delta R_1 &= (R_2 + \Delta R_2)(R_3 / R_4) \\ &= R_2(R_3 / R_4) + \Delta R_2(R_3 / R_4) \\ &= R_1 + \Delta R_2(R_3 / R_4) \end{aligned}$$

$$\therefore \quad \Delta R_2 = \Delta R_1(R_4 / R_3) \qquad ...(26.162)$$

Some applications such as temperature measuring systems (where the resistance, $R_1$ of the transducer changes on account of temperature), an automatic bridge balancing control system may be used as shown in Fig. 26.97.

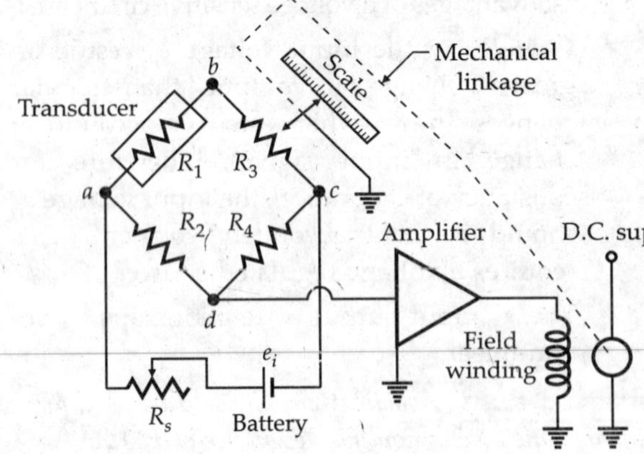


**Fig. 26.97** Self (automatic) balancing Wheatstone bridge.

When there is change in resistance $R_1$, it produces a voltage output which is amplified and applied as an error signal to the field winding of a d.c. motor. The motor is coupled to a moving contact. This movable contact is actuated so as to reduce the unbalance and hence the error voltage. When the bridge is balanced, there is no error voltage and therefore there is no voltage across the field winding of the motor. The motor stops and thus the slider comes to rest. The read out scale may be calibrated in resistance change values or in terms of temperature or any other physical variable being measured. In this case resistances $R_2$ and $R_4$ are fixed and $R_3$ is made variable.

### 26.22.2 Deflection Type Bridge

The null type Wheatstone bridge is accurate but the problem with this bridge is that the balancing, even if done automatically, is not instantaneous. Therefore this bridge is unsuitable for dynamic applications where the changes in resistance are rapid. For measurement of rapidly changing a input signals, the *Deflection* type Bridge is used. When the input changes, the resistance $R_1$ changes producing an unbalance causing a voltage to appear across the meter. The deflection of meter is indicative of the value of resistance and the scale of the meter may be calibrated to read the value resistance directly.

For static inputs, an ordinary galvanometer may be used. However for dynamic inputs, output signal may be displayed by a cathode ray oscilloscope or may be recorded by a recorder.

The deflection type bridge circuit is provided with a zero setting arrangement as shown in Fig. 26.98. The series resistance $R_s$, is used to change the bridge sensitivity.

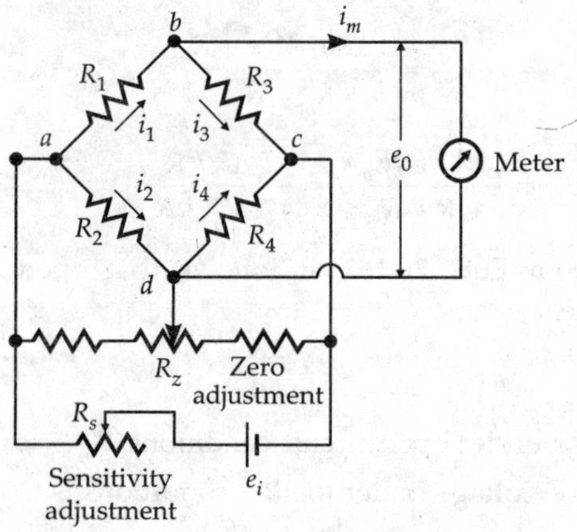


**Fig. 26.98** Deflection Type Wheatstone bridge provided with zero adjustment and sensitivity adjustment arrangements.

When a deflection type bridge is used, the bridge output on account of the unbalance may be connected either to a high input impedance device or to a low input impedance device. If the output of the bridge is connected directly to a low impedance devices like a current galvanometer or a PMMC instrument, a large current flows through the meter. In this case, the bridge is called a *Current Sensitive Bridge.*

In most of the applications of deflection type bridge, the bridge output is fed to an amplifier which has a high input impedance and therefore the output current $i_m$0. This would also be the case if the bridge output is connected to a CRO or an electronic voltmeter (EVM) a digital voltmeter. This bridge thus used is a *Voltage Sensitive Bridge.*

### 26.22.3 Voltage Sensitive Bridge

For analysis of this bridge, we refer to Fig. 26.96. Let us assume that the input impedance of the meter is infinite and therefore $i_m = 0$

Hence $i_1 = i_3$ and $i_2 = i_4$, output voltage $e_0$ = voltage across terminals $b$ and $d = i_1 R_1 - i_2 R_2$

But $$i_1 = \frac{e_i}{R_1 + R_3} \text{ and } i_2 = \frac{e_i}{R_2 + R_4}$$

$$\therefore \quad e_0 = \left[\frac{R_1}{R_1 + R_3} - \frac{R_2}{R_2 + R_4}\right] e_i$$

$$= \left[\frac{R_1 R_4 - R_2 R_3}{(R_1 + R_3)(R_2 + R_4)}\right] e_i \qquad \text{...(26.163)}$$

Suppose now $R_1$ changes by an amount $\Delta R_1$. This causes a change $\Delta e_0$ in the output voltage.

Thus :

$$e_0 + \Delta e_0 = \left[\frac{(R_1 + \Delta R_1)\, R_4 - R_2 R_3}{(R_1 + \Delta R_1 + R_3)(R_2 + R_4)}\right] e_i$$

$$= \left[\frac{1 + \Delta R_1 / R_1 - R_2 R_3 / R_1 R_4}{\{1 + (\Delta R_1 / R_1) + R_3 / R_1\}\{1 + R_2 / R_4\}}\right] \qquad \text{...(26.164)}$$

In order to simplify the relationship, let us assume that initially all the resistance comprising the bridge are equal *i.e.*, $R_1 = R_2 = R_3 = R_4 = R$.

Under these conditions

$$e_0 = 0$$

and $$\Delta e_0 = \left[\frac{(\Delta R / R)}{4 + 2(\Delta R / R)}\right] e_i \qquad \text{...(26.165)}$$

It is clear from Eqn. 26.165 that the input-output relationship *i.e.*, relationship between $\Delta R$ and $\Delta e_0$ is non-linear. However, if the change in resistance is very- small as compared to initial resistance, then we have, $2(\Delta R / R) << 4$.

$$\therefore \quad \Delta e_0 = \frac{(\Delta R / R)}{4} e_i \qquad \text{...(26.166)}$$

Hence, bridge sensitivity

$$\frac{\Delta e_0}{\Delta R_1} = \frac{e_i}{4R} \qquad \text{...(26.167)}$$

For such cases, the input-out relationship is linear. The major disadvantage of a voltage sensitive deflection bridge as compared to voltage sensitive null bridge is that the calibration of the former is dependent upon the value of supply voltage $e_i$ as is evident from Eqns. 26.166 and 26.167. Therefore for constancy of calibration, the input voltage should be absolutely constant.

### 26.22.4 Current Sensitive Bridge

When using galvanometers at the output terminals, the resistance of the meter cannot be neglected as galvanometers draw an appreciable current. Under such circumstances, where in the measuring device has a low impedance, the deflection is on account of the meter current $i_m$, which cannot be assumed to be zero, the bridge is said to be a *current sensitive bridge.*

In order to find the input-output relationship and the bridge sensitivity for this general case, it is necessary to convert the bridge circuit of Fig. 26.96, to a Thevenin generator looking into the output terminals $b$ and $d$.

Open circuit voltage of the Thevenin generator

$$e_0 = \frac{R_1R_4 - R_2R_3}{(R_1+R_3)(R_2+R_4)}e_i$$

The internal resistance of the Thevenin generator is found from looking into terminals $b$ and $d$ and shorting terminals $a$ and $c$ as shown in Fig. 26.99($a$). The internal resistance of the Thevenin generator looking into the terminals $b$ and $d$ and is :

$$R_0 = \frac{R_1R_3}{R_1+R_3} + \frac{R_2R_4}{R_2+R_4}$$

$$= \frac{R_1R_3(R_2+R_4)+R_2R_4(R_1+R_3)}{(R_1+R_3)(R_2+R_4)} \quad ...(26.168)$$

The equivalent Thevenin generator is shown in Fig. 26.99($b$).

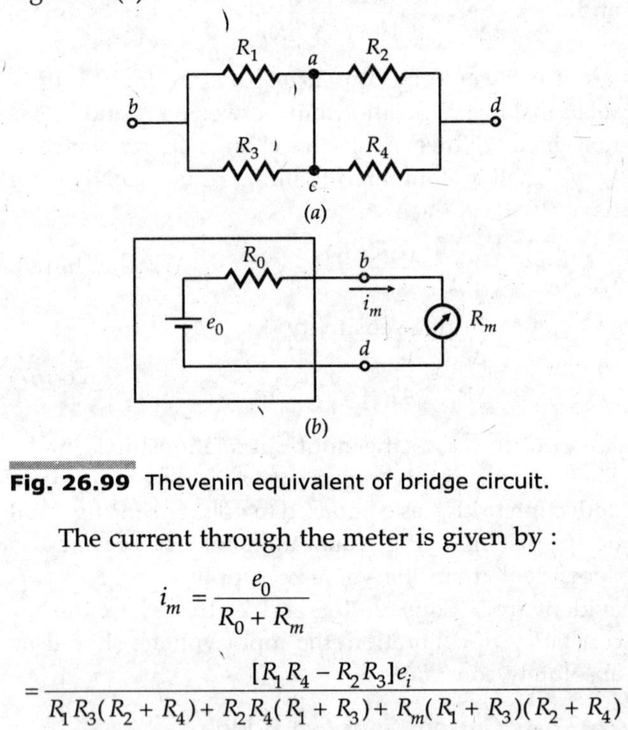

**Fig. 26.99** Thevenin equivalent of bridge circuit.

The current through the meter is given by :

$$i_m = \frac{e_0}{R_0+R_m}$$

$$= \frac{[R_1R_4 - R_2R_3]e_i}{R_1R_3(R_2+R_4)+R_2R_4(R_1+R_3)+R_m(R_1+R_3)(R_2+R_4)} \quad ...(26.169)$$

Suppose initially, all the resistances are equal *i.e.*,

$$R_1 = R_2 = R_3 = R_4 = R$$

Let there be a change $\Delta R$ in $R_1$,

Meter current,

$$i_m = \frac{[(R+\Delta R)R - R^2]e_i}{(R+\Delta R)R(2R) + R^2(R+\Delta R+R) + R_m(\Delta R+R+R)(2R)}$$

$$= \frac{(\Delta R/R^2)e_i}{2(1+\Delta R/R)+(2+\Delta R/R)+2R_m(\Delta R/R^2+2/R)} \quad ...(26.170)$$

when $R >> \Delta R$, we have,

$$i_m = \frac{(\Delta R/R^2)e_i}{4(1+R_m/R)} \quad ...(26.171)$$

Voltage output under load conditions is :

$$e_L = i_m R_m = \left[\frac{(\Delta R/R^2)e_i}{1+R_m/R}\right]\frac{R_m}{4}$$

$$= \frac{(\Delta R/R)R_m e_i}{4(R+R_m)} = \frac{\Delta R/R}{4(1+R/R_m)}e_i \quad ...(26.172)$$

From Eqn. 26.162, it follows that open circuit voltage

$$e_0 = \frac{\Delta R}{4R}e_i \quad ...(26.173)$$

because under open circuit conditions $R_m = \infty$.

The voltage under loaded conditions is :

$$e_L = \frac{e_0 R_m}{R_0 + R_m}$$

∴ Ratio of voltage under loaded and no load

$$\frac{e_L}{e_0} = \frac{1}{1+R_0/R_m} \quad ...(26.174)$$

**Example 26.29** *A resistance bridge has the configuration shown in Fig. 26.96, in which $R_1$ = 120.4 Ω $R_2$ = 119.0 Ω and $R_3$ = 119.7 Ω (a). What resistance must $R_4$ have for resistance balance ? (b) If $R_4$ has a value of 121.2 Ω, and if the input voltage is 12 V d.c. what is the output voltage of the bridge, assuming it to be voltage sensitive bridge ?*

**Solution.** ($a$) Value of $R_4$ for balance

$$= (R_2 \times R_3 / R_1) = 119 \times (119.7/120.4)$$

$$= 118.3\ \Omega.$$

($b$) With $R_4$ = 121.2 Ω, the bridge is unbalanced. Therefore from Eqn. 26.163, the output voltage on open circuit is :

$$e_0 = \left[\frac{R_1R_4 - R_2R_3}{(R_1+R_3)(R_2+R_4)}\right]e_i$$

$$= \left[\frac{120.4\times121.2 - 119\times119.7}{(120.4+119.7)(119+121.2)}\right]\times 12$$

$$= 0.0724\ \text{V} = 72.4\ \text{mV}.$$

**Example 26.30** *A simple equal arm voltage sensitive bridge is initially in balance (null). Three of the arms consist of ordinary resistors, while the fourth is a thermistor. Each arm has a nominal resistance of 10,000 Ω, and the bridge is energised with a 6 V d.c. source (a). If the temperature change causes a + 5% change in thermistor resistance, what* *output voltage will be indicated from the bridge ? (b) If a –5% change in resistance is caused, what will be the bridge output ?*

**Solution.** (*a*) Let the thermistor form resistance $R_1$, of bridge circuit shown in Fig. 26.96.

Initially $R_1 = R_2 = R_3 = R_4 = R = 10{,}000$

Change in resistance

$$\Delta R / R = +5\% = +0.05$$

Output voltage of equal arm bridge

$$e_0 = \left[\frac{(\Delta R / R)}{4 + 2(\Delta R / R)}\right] e_i$$

(See Eqn. 26.165)

$$= \left[\frac{0.05}{4 + 3 \times 0.05}\right] \times 6 = 0.0732 \text{ V}$$

$$= 73.2 \text{ mV}.$$

(*b*) Change in resistance $\Delta R / R = -5\% = -0.05$.

$$\therefore \text{ Output voltage, } e_0 = \left[\frac{-0.05}{4 - 2 \times 0.05}\right] \times 6$$

$$= -0.077 = -77 \text{ mV}$$

**Example 26.31** *It is known that a certain transducer has a resistance of approximately 800 Ω. This transducer is connected as resistor $R_1$, in the Wheatstone bridge shown in Fig. 26.96. The other three arms of the bridge have resistances of exactly 800 Ω. A 4 V battery with negligible internal resistance is used in the circuit. The galvanometer resistance is 100 Ω and the galvanometer current is 0.8 μA. Calculate the resistance of the unknown resistor.*

**Solution.** The resistances are :

$$R_2 = R_3 = R_4 = 800 \,\Omega$$

and meter resistance $R_m = 100\,\Omega$.

Galvanometer current

$$i_m = 0.8 \,\mu\text{A} = 0.8 \times 10^{-6} \text{ A}$$

Battery voltage $e_i = 4$ V.

Using the approximate formula given in Eqn. 26.171, we have,

$$i_m = \frac{(\Delta R / R^2) e_i}{4(1 + R_m / R)}$$

or $$0.8 \times 10^{-6} = \frac{\Delta R \times 4}{(800)^2 [4 + (100/800)]}$$

or $$\Delta R = 0.528 \,\Omega$$

$$\therefore \quad R_1 = R + \Delta R = 800 + 0.528 = 800.528 \,\Omega$$

**Example 26.32** *In a Wheatstone bridge three out of four resistors have a value of 1 kΩ each, and the fourth resistor equals 1010 Ω. If the battery voltage is 100 V, what is the approximate value of the open circuit voltage ? If the output of the bridge is connected to a 4 kΩ resistor, how much current would flow through the resistor ?*

**Solution.** This problem can conveniently be solved by Thevenin theorem.

Open circuit voltage

$$e_0 = \left(\frac{1010}{1010 + 1000} - \frac{1000}{1000 + 1000}\right) \times 100 = 0.249 \text{ V}$$

Internal resistance of the bridge looking into its output terminals,

$$R_0 = \frac{1010 \times 1000}{1010 + 1000} + \frac{1000 \times 1000}{1000 + 1000}$$

$$= 1002.5 \,\Omega$$

Meter current

$$I_m = \frac{E_0}{R_0 + R_m} = \frac{0.249}{1002.5 + 4000} = 50 \,\mu\text{A}$$

**Example 26.33** *A Wheatstone bridge is used for measuring the value of change of resistance of a strain gauge. All the arms of the bridge including the strain gauge have a resistance of 100 Ω. The maximum allowable power dissipation from the strain gauge is 250 mW. Find out the value of maximum permissible current through strain gauge and the maximum allowable value of bridge supply voltage. Suppose a source of 20 V is available, find the value of the series resistance $R_s$ to limit the input voltage of the bridge to the permissible level. What is the power dissipation in the series resistor ?*

**Solution.** Suppose $e_i$ is the input voltage. The value of resistance of each arm is $R = 100\,\Omega$

Maximum power dissipation from the strain gauge

$$P = 250 \text{ mW} = 250 \times 10^{-3} \text{ W}$$

Now $$P = i^2 R$$

where $i$ = maximum permissible current.

$$\therefore \quad i = \sqrt{P / R}$$

$$= \sqrt{250 \times 10^{-3} / 100}$$

$$= 0.05 \text{ A} = 50 \text{ mA}.$$

Maximum value of input voltage for the bridge

$$e_i = 2iR = 2 \times 50 \times 10^{-3} \times 100 = 10 \text{ V}.$$

The supply voltage is 20 V and the maximum allowable voltage is 10 V. Therefore a series resistance $R_s$ has to be inserted to reduce the voltage to 10 V.

Since all the arms of the bridge have equal resistance value of 100 Ω each, the resistance of the bridge looking into output terminals is $R = 100\,\Omega$. As

26

the bridge and the series resistance have equal voltage drops (10 V) across them, the value of series resistance $R_s$, is also 100 Ω.

∴ Power dissipation in series resistance

$$= 10^2 / 100 = 1 \text{ W}$$

**Example 26.34** *A thermistor of resistance 1 kΩ temperature co-efficient of resistance 4.5 percent/°C and an internal temperature rise of 0.2°C/mW around 27°C is included in a d.c. bridge with three fixed resistors each of value of 1 kΩ. A d.c. amplifier of high input impedance is connected across the bridge output and an indicating device connected to the amplifier output.*

*Calculate the maximum voltage sensitivity of the bridge if the internal temperature rise is not to exceed 0.1°C.*

**Solution.** Power dissipation in the thermistor due to a change 0.1°C, $P = 0.1/0.2 = 0.5$ mW.

Supposing $e_{im}$ is the maximum allowable input voltage across the bridge. Now each arm of the bridge has a resistance $R = 1$ kΩ.

∴ Voltage across each arm $= e_{im}/2$

Power dissipation in thermistor

$$P = \left(\frac{e_{im}}{2}\right)^2 \times \frac{1}{R}$$

∴ Maximum allowable input voltage

$$e_{im} = 2\sqrt{PR}$$

$$= 2\sqrt{0.5 \times 10^{-3} \times 10^3} = 1.41 \text{ V}.$$

Change in resistance of thermometer

$$\Delta R_1 = \alpha \Delta\theta R$$

$$= (4.5/100) \times 0.1 \times 1000 = 4.5\ \Omega$$

Bridge output voltage, when all the arms have equal initial resistance

$$= \Delta_{e_0} = \frac{\Delta R_1 / R}{4} e_i$$

∴ Maximum output voltage

$$e_{om} = \frac{4.5/1000}{4} \times 1.41$$

$$= 1.59 \times 10^{-3} \text{ V} = 1.59 \text{ mV}.$$

This voltage change is caused by temperature input of 0.1°C.

∴ Maximum bridge sensitivity

$$S = \frac{\Delta e_0}{\Delta\theta} = \frac{1.59}{0.1} = 15.9 \text{ mV/°C}.$$

## 26.23 A.C. BRIDGES

Alternating current bridge methods are of outstanding importance in electrical measurements. The theory of a.c. bridges has been explained in chapter 16. A.C. bridges are widely used for variable inductance and capacitance bridges. This section deals with special a.c. bridges used for instrumentation purposes.

### 26.23.1 A.C. Bridges using Push-Pull Transducers

The a.c. bridges are suited for accepting inputs push-pull transducers, *i.e.*, transducers which accept input from transducers working on differential principle. When a transducer works on a push-pull principle it produces a more linear relationship with the bridge output and the measured quantity than is produced by a single element transducer. Another advantage of the push-pull transducers is that when they are connected in the adjacent arms of a bridge, any equal changes such as due to temperature in the two parts of the transducer act against each other and hence are cancelled out.

### 26.23.2 A.C. Bridges with Push-Pull Inductive Transducers

A push-pull inductance transducer consists of two parts each having an initial inductance value $L$. If an input signal is applied to this transducer the inductance of part increases by $\Delta L$ while that of other part reduces by $\Delta L$. Thus the inductance of one part becomes $L + \Delta L$ while that of other $L - \Delta L$

Figure 26.100(*a*) shows an arrangement of connecting an inductive push-pull transducer in an a.c bridge. The bridge is initially balanced but when a change occurs in the inductance of the transducer on account of application of an input signal, it becomes unbalanced. For transducers using high $Q$ coils have $\omega L >> R$ and therefore the changes in resistance of transducer coils is insignificant in comparison with change of inductance.

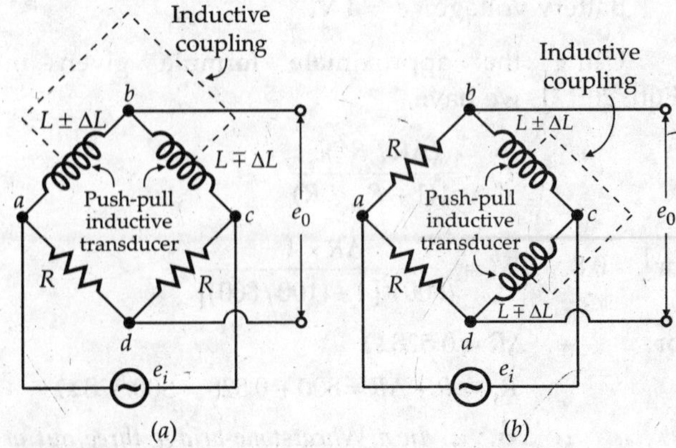


**Fig. 26.100** Push-Pull inductive transducers connected in a.c bridges.

Hence,

$$\text{Output voltage } e_0 = \frac{1}{2}\frac{\Delta L}{L} e_i \quad ...(26.175)$$

Another arrangement is shown in Fig. 26.100(*b*). In case the $Q$ factor is unity *i.e.*, $\omega L = R$.

$$\text{Output voltage } e_0 = (\Delta L / L)\, e_i \quad ...(26.176)$$

Hence, the sensitivity of this bridge is twice that of the one shown in Fig. 26.100(*a*).

### 26.23.3 Inductive Transducer Blumlein Bridge

Blumlein bridge, as shown in Fig. 26.101(*a*) uses two closely coupled inductive ratio arms and other two aims consist of one part each of a push-pull inductive transducer of a high $Q$ value. The equivalent T network of the closely coupled arms is shown in Fig. 26.101(*b*).

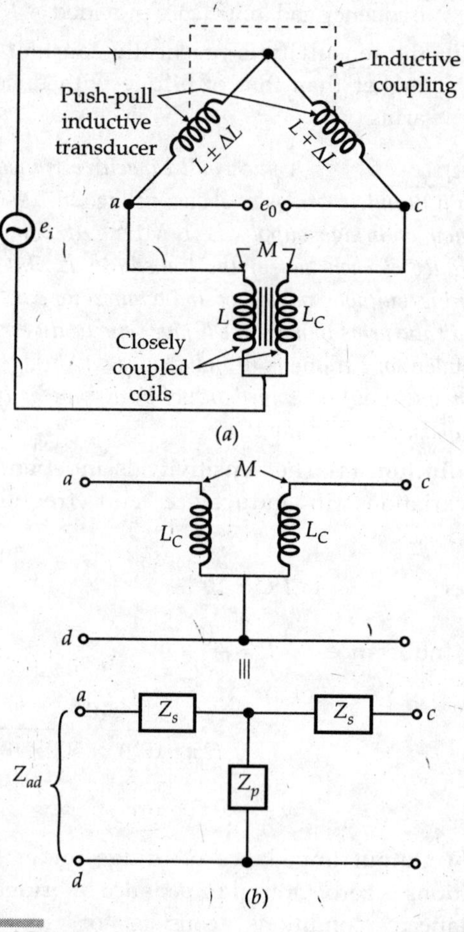


**Fig. 26.101** Inductive transducer Blumlein bridge.

For networks to be equivalent,

$$Z_{ad} = Z_S + Z_P$$
$$= j\omega M$$

The coupling factor between the closely coupled coils is $K = \pm M / L_c$ and for perfect coupling $K = \pm 1$.

At the bridge balance, the currents in the two coils are equal and are in the same direction and for a particular winding sense and therefore $K = +1$.

Therefore under balance conditions, impedance across ratio arms is :

$$2\mathbf{Z}_s = 2j\omega(L_C - M) \text{ and with } K = 1 \text{ we get } 2\mathbf{Z}_s = 0$$

Thus it is evident that there is no voltage drop across the ratio arms and any stray or cable capacitance which is in parallel with the ratio arms becomes ineffective. Thus shielding and earthing problems are greatly simplified. This is the biggest advantage of Blumlein bridge.

When the bridge is not balanced, the change in the currents on the two sides of the bridge can be considered as a circulating current around the bridge. Thus for this condition the coupling factor $K$ changes sign and is $K = -1$ and therefore $2\mathbf{Z}_s = 4j\omega L_c$.

In order to find the sensitivity of the bridge, the parallel impedance of $T$ network $\mathbf{Z}_P (= -j\omega L_c)$ may be considered to be connected in series with bridge impedance $\mathbf{Z}_B$ with bridge supply voltage across them.

Bridge impedance

$$\mathbf{Z}_B = \frac{j\omega L(1 + 2L_c/L + \Delta L/L)(1 + 2L_c/L - \Delta L/L)}{2(1 + 2L_c/L)} \quad ...(26.177)$$

Neglecting $\mathbf{Z}_P$, the open circuit output voltage

$$e_0' = \frac{\Delta L}{L} \cdot \frac{4L_c/L}{(1 + 2L_c/L + \Delta L/L)(1 + 2L_c/L - \Delta L/L)} e_i \quad ...(26.178)$$

with $\mathbf{Z}_P$ in series with $\mathbf{Z}_B$, we have, open circuit voltage

$$e_0 = \frac{\mathbf{Z}_B}{\mathbf{Z}_B + \mathbf{Z}_P} e_0' = \frac{\Delta L}{L} \cdot \frac{4L_c/L}{(1 + 2L_c/L)} e_i \quad ...(26.179)$$

Equation 26.179 is based upon the assumption that $(\Delta L / L)^2 << 1$.

Bridge sensitivity factor

$$= \frac{e_0}{e_i(\Delta L / L)} = \frac{4L_c/L}{(1 + 2L_c/L)} \quad ...(26.180)$$

Figure 26.102 shows the variation of bridge sensitivity factor with variation in ratio $L_c/L$ for a bridge with tightly coupled ratio aims ($K = -1$) and with uncoupled ratio aims ($K = 0$). In both the cases the bridge sensitivity is independent of frequency. It is observed from Fig. 26.102, that the bridge with tightly couped ratio has a higher sensitivity throughout and when $L_c >> L$, the sensitivity factor is 2 and is essentially independent of ratio arm inductance variations.

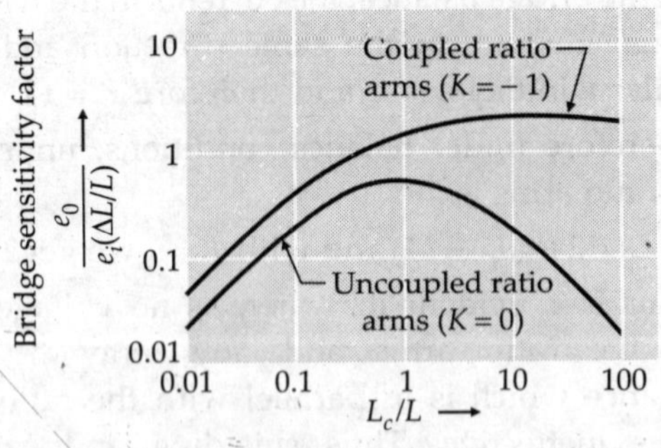


**Fig. 26.102** Variation of bridge sensitivity with $L_c$ / L.

## 26.23.4 Capacitive Transducer Blumlein Bridge

Blumlein bridge using push-pull capacitive transducers is shown in Fig. 26.103(*a*). Analysing this bridge in the same way as was done for inductive Blumlein bridge, it can be shown that output voltage is

$$e_0 = \frac{\Delta C}{C} \cdot \frac{4\omega^2 L_c C}{(2\omega^2 LC - 1)} e_i \quad ...(26.181)$$

∴ Bridge sensitivity factor

$$= \frac{e_0}{e_i(\Delta C / C)} = \frac{4\omega^2 L_c C}{(2\omega^2 L_c C - 1)} \quad ...(26.182)$$

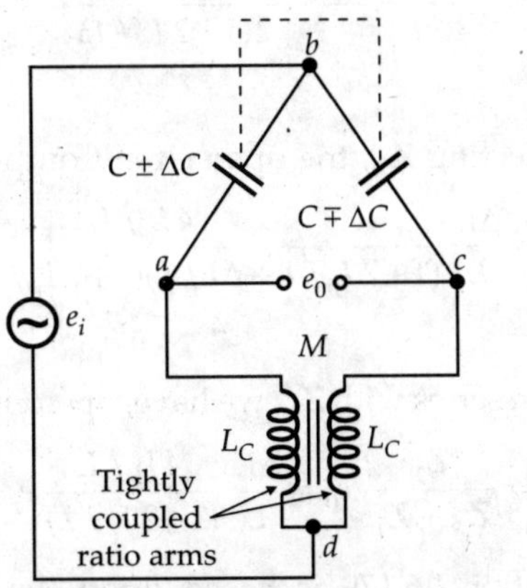


(*a*) Bridge

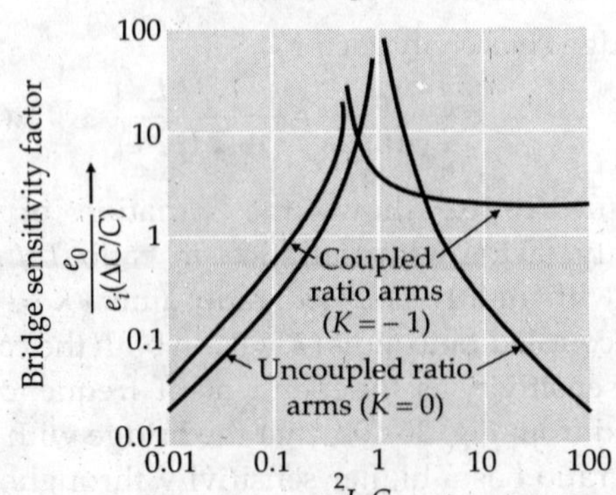


(*b*) Variation of bridge sensitivity factor with $\omega^2 L_C C$

**Fig. 26.103** Capacitive transducer Blumlein bridge.

The variation of bridge sensitivity factor with $\omega^2 LC$ is shown in Fig. 26.99(*b*). Resonance occurs when $2\omega^2 L_c C = 1$, and above resonance, when $\omega^2 L_c C >> 1$, the sensitivity factor is 2 and is independent of the variations in bridge frequency and ratio arm inductance.

The variation of sensitivity factor for an uncoupled bridge is also shown. It is noted that the uncoupled arm bridge has no horizontal characteristics and there is no region where sensitivity is independent of frequency or inductance.

Thus the advantages of a capacitive transducer bridge with tightly coupled arms are :

(*i*) there are no noise and pickup problems,

(*ii*) the bridge is insensitive to variations in frequency and inductance provided $\omega^2 L_c C >> 1$,

(*iii*) the sensitivity is practically constant and is higher than that of bridge with uncoupled arms.

***Example 26.35*** *A push-pull capacitive transducer is used in a Blumlein bridge as shown in Fig. 26.100 (a). The frequency of bridge supply is 1 MHz. At balance, the capacitance of each half of the transducer is 500 pF. (a) Determine suitable values of inductance for the tightly coupled ratio arms to make the bridge sensitivity essentially independent of variations in inductance and bridge supply frequency, (b) output impedance of bridge under balanced and unbalanced conditions.*

**Solution.** (*a*) The sensitivity is independent of the variations in inductance and frequency if $\omega^2 L_c C >> 1$.

Let $\omega^2 L_c C = 10$

∴ Inductance $L_c = \frac{10}{\omega^2 C}$

$$= \frac{10}{(2\pi \times 10^6)^2 \times 500 \times 10^{-12}} \text{H}$$

$$= 0.506 \text{ mH}$$

(*b*) Output impedance of bridge under balance conditions is zero. Output impedance of bridge under unbalanced conditions consists of impedances $4j\omega L_c$ and $2/(j\omega C)$ in parallel. However, $4j\omega L_c$ is very large as compared with $2/\omega C$ and hence the output impedance is approximately equal to $2/\omega C$.

$$\text{Output impedance} = \frac{2}{2 \times 1 \times 10^6 \times 500 \times 10^{-12}}$$

$$= 2000\ \Omega = 2\ \text{k}\Omega$$

## 26.24 ANALOG/DIGITAL/ANALOG CONVERSION TECHNIQUES

The most naturally occurring phenomena are analog in nature. Analog quantities are continuous functions with time and most transducers give an analog output.

With the introduction of digital computer, the data taken in a physical system has to be converted into a digital form. When system engineers made their first attempts at applying computational power, speed and data handling capability of digital computers to automatic control and instrumentation systems, they were faced with a tremendous task of making the digital computer to work with analog devices.

The data to be fed to digital devices, normally appears in analog form. For example, a temperature difference would be represented by the voltage output of a thermocouple, the strain of a mechanical member would be represented the voltage resulting from an unbalanced strain gauge resistance bridge. Therefore the need arises for a device that converts analog information into digital form. Therefore, **analog to digital (A/D) conversion devices** are used in measurement and instrumentation systems.

It should be borne in mind that digital quantities are discrete and vary in equal steps. Each digital number is a fixed sum of equal steps which is defined in that number.

### 26.24.1 Resolution and Quantization

In A/D converters the original analog signal has essentially an infinite resolution as the signal is continuous. The digital representation of this signal would, of course, reduce, this resolution as digital quantities are discrete and vary in equal steps. Each digital number is a fixed sum of equal steps which is defined by that number.

In-order to convert an analog quantity to a digital number involves *quantization*. The characteristics of a Quantizer are shown in Fig. 26.104.

This quantizer is used for conversion of analog voltages of 0 – 7 V. Now in binary code, decimal number 7 = binary 111. Therefore the digital number has 3 bits.

The analog input is shown in Fig. 26.104(*a*) on the horizontal axis and the discrete output voltage level on the vertical axis. The basic operation of A/D conversion involves *two* processes, *viz.*, (*i*) *quantization* and (*ii*) *coding*.

The decision levels of the A/D convertor are at .05, 1.5, 2.5. 3.5, etc. Thus the design levels are spaced 1 V apart. Analog values between two decision levels cannot be coded. In fact, analog values between $1 \pm 0.5$ are read as digital value 001. The distance between decision levels is $Q$, the quantization size. An $n$ bit A/D converter has $2^n$ discrete levels with resolution $= 1/2^n$ or 1 part in $2^n$ and analog decision levels = $2^n - 1$.

Therefore, for an $n$ bet converter

$$\text{Discrete levels} = 2^n \quad \text{...(26.183)}$$

$$\text{Resolution} = 1/2^n \quad \text{...(26.184)}$$

$$\text{and decision levels} = 2^n - 1 \quad \text{...(26.185)}$$

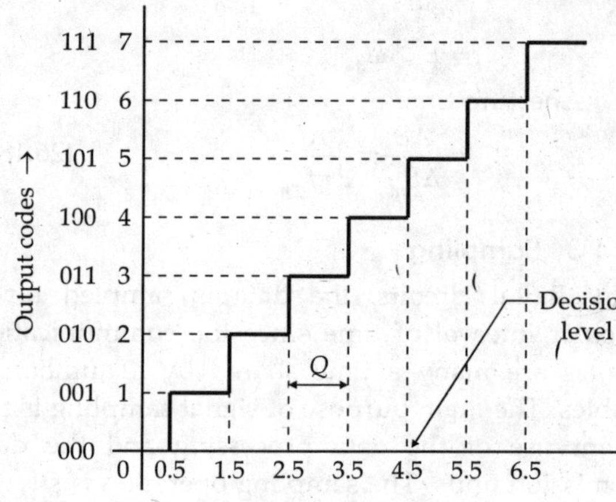

(*a*) Analog level (*Q*)

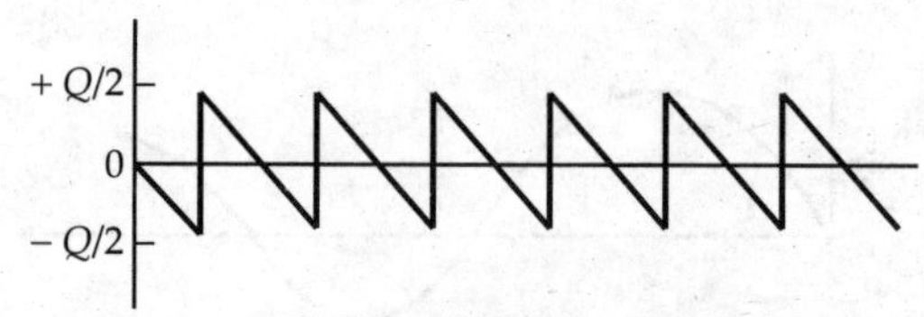

(*b*) Quantization error.

**Fig. 26.104** Characteristics of a Quantizer.

If the input to the quantizer is moved through its full range and subtracted from the discrete output levels, the error signal is shown in Fig. 26.104(*b*). This is called *quantization error*. The quantization error is dependent on the number of quantization levels. The quantization levels are an index of the resolution of the instrument. The quantization error is of a sawtooth waveform as shown in Fig. 26.104(*b*). The peak to peak value is $Q$. The output of a quantizer can be considered as a noise signal with an rms value of

$$E_q = Q/2\sqrt{3} \quad \text{...(26.186)}$$

which is the *quantization error*.

### 26.24.2 Aperture Time

In order to perform the operation of quantizing and coding a signal, an A/D converter requires an *aperture time.* A sample-hold S/H system is used in order to avoid the use of very fast and expensive A/D converters. The use of a sample-hold circuit is useful because it provides a very small time for taking a very rapid sample of the signal and then holding its value till it is converted. For a sinusoidal signal to be converted into a digital form, the maximum rate of change occurs at the instant where the voltage goes through zero. The amplitude change of voltage at this instant is :

$$\Delta E = \left[\frac{d}{dt}(E_m \sin \omega t)\right]_{t=0} \times t_a$$

$$= E_m \, \omega t_a$$

∴ Aperture time

$$t_a = \frac{\Delta E}{\Delta E_m} = \frac{\Delta E}{2\pi f E_m} \quad ...(26.187)$$

### 26.24.3 Sampling

In digital circuits, the data is sampled for a particular interval of time since the communication channels are many a time shared by a number of variables. The main purpose of signal sampling is the efficient use of the data processing and the data transmission units. The sampling operation is shown in Fig. 26.105. This shows that an analog signal and a train of periodic sampling signal whose 'on' time is extremely short as compared with the total period of the signal.

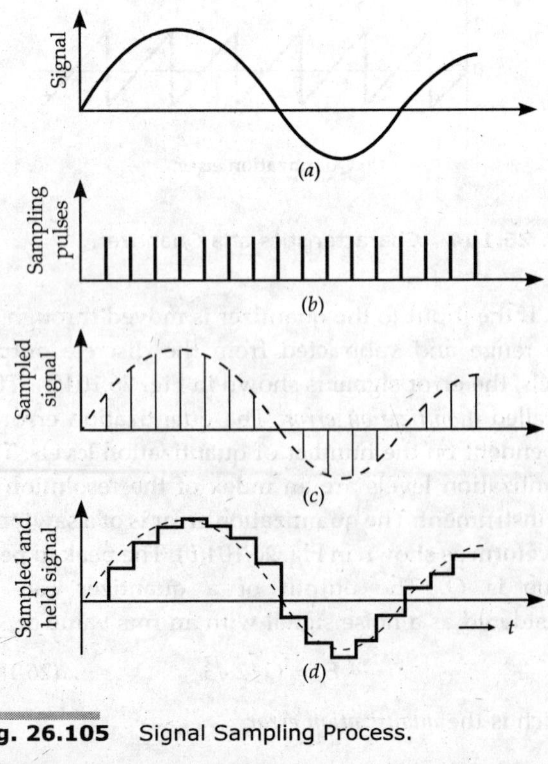


**Fig. 26.105** Signal Sampling Process.

The result of the sampling process is identical to multiplying the analog signal by a train of pulses of unit magnitude. The resultant modulated signal is shown in Fig. 26.105(*c*). This preserves the amplitude of the analog signal in the modulation envelope of the pulses.

Now the question is how long a signal be sampled so as not to lose its identity. The sampling theorem states that if the highest frequency content in the input signal is $f_h$ Hz, then input signal can be recovered without distortion if it is sampled at a rate of atleast $2 f_h$ samples per second, This is called the *Nyquist rate.* However, in practice it is necessary to sample at least $5 f_h$ samples per second in order to reduce the effects of noise and non-sinusoidal filters. The sample and hold circuit acts as a low pass filter with a cut off frequency $f_C = f_s/2$ where $f_s$ = sampling frequency.

***Example 26.36*** *(a) An analog transducer with a 0 – 10 V input is able to distinguish a change of 10 mV in its input signal. Calculate its resolution. (b) Calculate the number of bits of an A/D converter so that the digital output has almost the same resolution as the transducer. The A/D converter uses a binary code. Calculate also (c) the quantization error, and (d) the number of decision levels. Comment upon the results.*

**Solution.** (*a*) Resolution of analog instruments

$$= \frac{10 \text{ mV}}{10 \text{ V}}$$

$$= \frac{10 \times 10^{-3}}{10} = \frac{1}{1000}$$

$$= 1 \text{ part in } 1000$$

(*b*) Let the A/D converter have *n* bits to give a resolution of 1 part in 1000.

From Eqn. 26.184,

$$\text{Resolution} \quad \frac{1}{2^n} = \frac{1}{1000}$$

or number of bits $n = 10$

as $2^{10} = 1024$. (Nearly equal to 1000)

(*c*) Quantization interval

$$Q = \frac{10}{2^n}$$

$$= \frac{10}{1024} \text{ V} \approx 1 \text{ mV} \quad \text{(See Eqn. 26.183)}$$

Quantization error or noise

$$E_q = \frac{Q}{2\sqrt{3}}$$

$$= \frac{1\times 10^{-3}}{2\times\sqrt{3}}$$

$$= 0.29\times 10^{-3}\text{ V} = 0.29\text{ mV}.$$

(See Eqn. 26.186)

(*d*) The number of decision levels $= 2^n - 1$

$= 1024 - 1 = 1023.$ (See Eqn. 26.185)

**Comments**

A resolution of 10 bits means that a 10 V signal can be resolved down to (10/ 1024) V, or approximately 0.01 (1 mV) per bit. As the analog signal starts increasing from zero volt, the A/D converter will not have an output value until the analog signal equals or exceeds 0.01 V. After the A/D converter recognises this minimum signal and produces the correct output, that output will not change again until the output analog signal reaches a value of 0.02 V. Therefore, this 10 bit A/D converter with an analog input of 10 V cannot resolve, or recognize, changes less than 0.01 V (1 mV) in magnitude. It is therefore said to **quantize** the input, or chop it into chunks (quanta) of 0.01 V (1 mV) in size. Each 0.01 V (1 mV) chunk is considered to be a **dead band.**

## 26.25 GENERAL CONSIDERATIONS OF A/D AND D/A CONVERSION

A/D and D/A converters relate analog quantities to digital quantities and vice versa through an appropriate code. As mentioned earlier, the most commonly used code is the **binary code**, in which a number is represented by :

$$N = d_n\times 2^n + d_{n-1}\times 2^{n-1} + \ldots + d_2\times 2^2 + d_1^2\times 2^1 + d_0\times 2^0 \quad \ldots(26.188)$$

where the co-efficients $d_n, d_{n-1}, \ldots d_2, d_1, d_0$ assume the values of either 0 or 1. The leftmost bit is $d_n 2^n$ ($d_n$ may be 1 or 0). This is called the *Most significant bit* (MSB) and that it has as weight (maximum value) which is twice as large as the next most significant bit (next bit between MSB and binary point). This is evident as $2^n$ has twice large a weight as $2^{n-1}$. Each binary bit has a weight which is twice the weight of the next less significant bit. The *least significant bit* (*LSB*) is on the extreme right of the binary code.

Let us apply the binary weighing scheme to a d.c. voltage. Irrespective of the number of bits in the binary word to be converted, the MSB has to have twice the maximum weight, maximum value, importance or voltage capability as the next less significant bit. This means that if MSB can cause the output voltage to assume a value equal to one half of its maximum possible magnitude, then the next less significant bit must be able to cause the output voltage to change by one half of that value (MSB's value). Since the MSB changes the output by one half, therefore, the next less significant bit should be able to change the output by one fourth of the maximum possible.

Thus for an *n* bit converter :

$$\text{range of MSB} = \frac{1}{2}\times \text{range of converter} \quad \ldots(26.189)$$

$$\text{and range of LSB} = \frac{1}{2^n}\times \text{range of converter} \quad \ldots(26.190)$$

The full scale output voltage of an *n* bit converter is given by :

$$E_0 = E_R\times(d_{n-1}\times 2^{-1} + d_{n-2}\times 10^{-2} + \ldots + d_1 \times 2^{-(n-1)} + d_0\times 2^{-n}) \quad \ldots(26.191)$$

$E_R$ can be any convenient reference analog value.

Thus there is a difference between $E_0$ and $E_R$ and hence the error. The difference $E_0 - E_R$ decreases as the number of bits, $n_1$ increases.

Resolution = analog value of

$$\text{LSB} = \frac{\text{full scale}}{2^n} \quad \ldots(26.192)$$

Resolution often specified in terms of percentage of full scale. The resolution of a converter can be conveniently related to its *Dynamic Range* expressed in db.

$$\text{Dynamic range} = 6\times(\text{number of bits})\text{ db} \quad \ldots(26.193)$$

***Example 26.37*** *A 5 bit converter is used for a d.c. voltage range of 0 – 10 V. Find the weight of MSB and LSB. Also the exact range of the converter and the error. Find the error if a 10 bit converter is used.*

**Solution.**

$$\text{Range of MSB} = \frac{1}{2}\times\text{ range of converter} = \frac{1}{2}\times 10$$

$$= 5\text{ V.} \quad \text{(See Eqn. 26.189)}$$

$$\text{Range of LSB} = \frac{1}{2^n}\times\text{range of the converter}$$

$$= \frac{1}{2^5}\times 10 = 0.3125\text{ V}$$

(See Eqn. 26.190)

26

From Eqn. 26.191, the exact range of converter

$$E_0 = E_R(d_{n-1}\times 2^{-1} + d_{n-2}\times 2^{-2} + ... + d_1 \times 2^{-n-1} + d_0 \times 2^{-n})$$

$$= 10(1\times 2^{-1} + 1\times 2^{-2} + 1\times 2^{-3} + 1\times 2^{-4} + 1\times 2^{-5})$$

$$= 9.6875 \text{ V}$$

$$\text{Error} = 10 - 9.6875$$

$$= 0.3125 \text{ V} = \frac{0.3125}{10}\times 100 = 3.125\%$$

The exact range of the converter when 10 bits are used is :

$$= 10(2^{-1} + 2^{-2} + 2^{-3} + 2^{-4} + 2^{-5} + 2^{-6} + 2^{-7} + 2^{-8} + 2^{-9} + 2^{-10})$$

$$= 9.99 \text{ V}.$$

$$\text{Error} = 10 - 9.99 = 0.01 = \frac{0.01}{10}\times 100 = 0.1\%$$

Thus if a large number of bits are used, the error reduces considerably. But the use of a converter with a large number of bits results in higher cost of the converter itself and also of the system where it is used. Also, a higher number of bits adds to the complexity of the system.

**Example 26.38** *A control valve has a linear variation of opening as the input voltage varies from 0 to 10 V. A microcomputer outputs an 8 bit output word to control valve opening using an 8 bit D/A converter to generate the value voltage. (a) Find the reference voltage required to obtain full value opening (10 V), (b) Find the percentage of valve opening for all bit change in the input word.*

**Solution.** (*a*) With 8 bit A/D converter with all the bits being in 1 state, the output is 10 V.

$$\therefore 10 = E_R\left[\frac{1}{2}+\frac{1}{4}+\frac{1}{8}+\frac{1}{16}+\frac{1}{32}+\frac{1}{64}+\frac{1}{128}+\frac{1}{256}\right]$$

or reference voltage $E_R = 10.0039$ V.

(*b*) Change in voltage due to change in LSB

$$= 2^{-n} E_R = 2^{-8}\times 10$$

$$= 0.0392 \text{ V} = 39.2 \text{ mV}$$

$$\therefore \text{Percentage change} = \frac{0.0392}{10}\times 100 = 0.392\%$$

**Example 26.39** *An analog voltage signal whose highest significant frequency is 1 kHz is to be digitally coded with a resolution of 0.01 per cert covering a voltage range of 0 – 10 V. Determine : (i) minimum number of bits in the digital code, (ii) analog value of LSB, (iii) rms value of the quantization error, (iv) minimum sampling rate, (v) aperture time required for the A/D converter and (vi) dynamic range of converter in db.*

**Solution.**

(*i*) Resolution $= 0.01\% = \frac{0.01}{100} = \frac{1}{10000} = \frac{1}{2^n}$

$\therefore$ Minimum number of bits $n = 14$

as $2^{14} = 16384$ (we cannot choose $n = 13$ as $2^{13} = 8192$ which is less than 10000).

(*ii*) Analog value of LSB

$$= \frac{1}{2^n}\times 10 = \frac{1}{16384}\times 10 \text{ V} = 610.4 \text{ }\mu\text{V}$$

(*iii*) Rms value of quantization error

$$E_q = \frac{Q}{2\sqrt{3}}$$

$$= \frac{10}{(16384)\times 2\sqrt{3}} \text{ V} = 176 \text{ }\mu\text{V}.$$

(*iv*) In order that the signal may not be lost, the sampling frequency should be at least 5 times the highest frequency in the signal.

$\therefore$ Minimum sampling frequency

$$f_s = 5 f_h = 5\times 1 = 5 \text{ kHz}$$

(*v*) From Eqn. 26.187, aperture time

$$t_a = \frac{\Delta E}{2\pi f E_m}$$

but $\Delta E / E_m = 1/16384$

$$\therefore \quad t_a = \frac{1}{2\pi\times 1\times 10^3}\times \frac{1}{16384} \text{ s}$$

$$= 9.71 \text{ ms}$$

(*vi*) Dynamic range of converter $= 6\times$ number of bits

$$= 6\times 14 = 84 \text{ db} \quad \text{(See Eqn. 26.193)}$$

## 26.26 DIGITAL-TO-ANALOG (D/A) CONVERSION

It would seem logical to commence discussion of A/D and D/A conversion techniques with a presentation of A/D converters. In practice, however, some popular A/D converters techniques use D/A converters for functioning. Therefore, we start with discussion on D/A converters first. The D/A converters are, no doubt, used as the basis of some A/D converters but they have a large important uses in their own right. Among these uses are computer driven CRT displays, digital generation of analog waveforms, and digital control of automatic process control systems.

The voltage output of an $n$ bit converter is

$$E_0 = (d_{n-1}2^{-1} + d_{n-2}2^{-2} + ... d_1 2^{-(n-1)} + d_0 2^{-n}) E_R \quad ...(26.194)$$

where co-efficients $d_n$ represent the binary word and $d_n$, is either 1 or 0 if the $n$th bit is 1 or 0. The MSB corresponds to $d_{n-1}$ and its weight is $E_R/2$ while the LSB corresponds to $d_0$ and its weight is $E_R/2^n$.

Equation 26.194 can be written as

$$E_0 = (d_{n-1}2^{n-1} + d_{n-2}2^{n-2} + ... + d_1 2^1 + d_0 2^0)\frac{E_R}{2^n}$$

$$= (d_{n-1}2^{n-1} + d_{n-2}2^{n-2} + ... + d_1 2 + d_0)\frac{E_R}{32} \quad ...(26.195)$$

Let us consider a 5 bit converter. For this, $n = 5$ and we have :

$$E_0 = (16d_4 + 8d_3 + 4d_2 + 2d_1 + d_0)\frac{E_R}{32}$$

Let us consider binary number 10000. Its decimal equivalent is 16.

For this $d_4 = 1$ and $d_3 = d_2 = d_1 = d_0 = 0$.

$$\therefore \quad E_0 = (16 + 0 + 0 + 0 + 0)\frac{E_R}{32} = 16\frac{E_R}{32}$$

Let us consider another binary number 11000. Its decimal equivalent is 24.

For this number

$$d_4 = d_3 = 1 \text{ and } d_2 = d_1 = d_0 = 0$$

$$\therefore \quad E_0 = (16 + 8 + 0 + 0 + 0)\frac{E_R}{32} = 24\frac{E_R}{32}$$

Similarly, for binary number 11001, the decimal equivalent is 25 and the output is :

$$\therefore \quad E_0 = (16 + 8 + 0 + 0 + 1)\frac{E_R}{32} = 25\frac{E_R}{32}$$

If all the 5 bits are 1, the decimal equivalent is 31 and the output is :

$$\therefore \quad E_0 = (16 + 8 + 4 + 2 + 1)\frac{E_R}{32} = 31\frac{E_R}{32}$$

The above analysis shows that the analog output voltage is proportional to the digital input.

## 26.27 D/A CONVERTER

Figure 26.106 shows a **binary weighted resistance D/A converter**. This converter consists of the following major components :

(*i*) ***Switches.*** There are $n$ switches, one for each bit. These are : $S_{n-1}, S_{n-2}, ... S_1, S_0$.

(*ii*) ***Resistances.*** There is a resistive network of $n$ resistors. One resistor is connected in the input circuit of each bit. The resistances are weighted *i.e.*, if the resistance in the circuit for MSB is $R$, the resistance in the circuit of next significant bit is $2R$ and so on. The resistance in the LSB, circuit is $2^{n-2}R$.

(*iii*) ***Reference source.*** The voltage of the reference source is ER and is stable.

(*iv*) ***Current to voltage converter.*** It is an operational amplifier that adds up the currents flowing in the resistive network to develop an output analog voltage signal which is proportional to the digital input signal.

Suppose switch $S_{n-1}$ is connected to positive end of the reference source and rest of the switches are earthed. Under the conditions only MSB is 1 and rest are 0. Therefore output voltage $E_0$, should be equal to $E_R/2$ for proper digital to analog conversion.

$$\frac{E_0}{E_R} = \frac{R_f}{R}, \text{ but } \frac{E_0}{E_R} = \frac{1}{2} \quad \text{or} \quad \frac{R_f}{R} = \frac{1}{2}$$

Resistance in the feedback circuit of OPAMP is :

$$R_f = R/2$$

Now suppose for a binary input MSB = 1 and also the next significant figure is also 1 with the rest equal to 0.

Under those conditions

$$E_0 = \left(\frac{1}{2} + \frac{1}{2^2}\right)E_R = 0.75\,E_R$$

Suppose $R_1$ is the resistance in the circuit of the bit next MSB. Input resistance for OPAMP

$$R_i = \frac{R R_1}{R + R_1}$$

$$E_0 = \frac{R_f}{R_i}E_R = 0.75\,E_R$$

$$\text{or} \quad \frac{R_f}{R_i} = 0.75 \quad \text{or} \quad \frac{R}{2R_i} = 0.75$$

$$\text{or} \quad R_i = \frac{R}{1.5} \quad \text{but} \quad R_i = \frac{R R_1}{R + R_1}$$

$$\therefore \quad \frac{R R_1}{R + R_1} = \frac{R}{1.5} \quad \text{or} \quad R_1 = 2R.$$

Suppose all the bits in a $n$ bit converter are 0 except for LSB which is 1.

Suppose $R_n$ is the resistance in the LSB circuit.

Output voltage for LSB, $E_0 = E_2/2^n$

Therefore, for proper digital to analog conversion

$$E_0 = \frac{R_f}{R_i}E_R$$

26

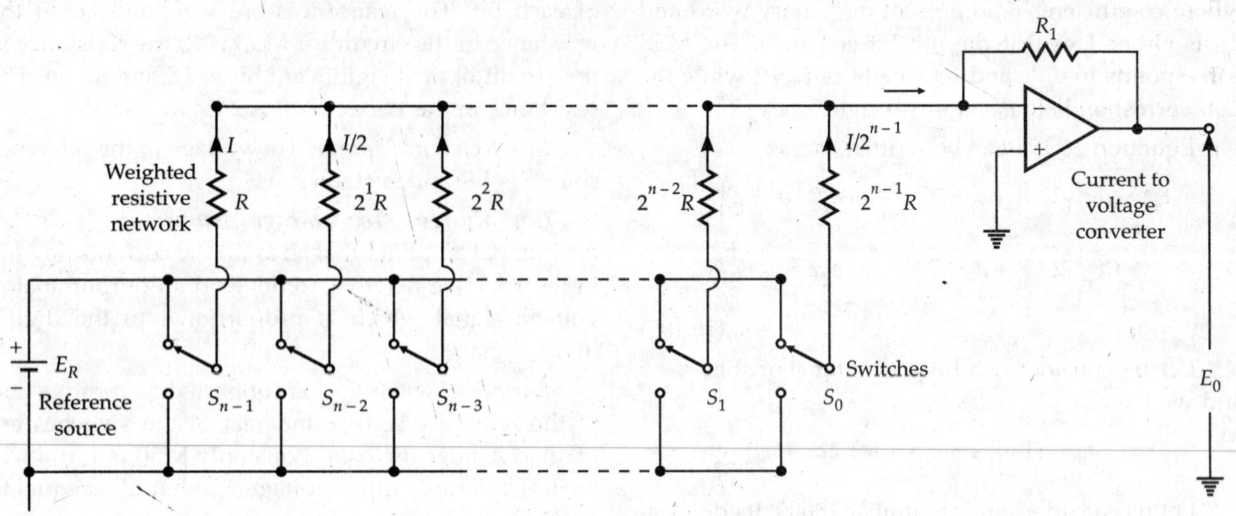


**Fig. 26.106** Binary weighted resistance D/A converter.

But $\quad E_0 = \frac{E_R}{2^n}$, $R_f = \frac{R}{2}$ and $R_i = R_n$

$$\therefore \quad \frac{E_R}{2^n} = \frac{R}{2} \times \frac{1}{R_n} \times E_R$$

Hence, $R_n = 2^{n-1} R$

It is evidently clear from above that if proper digital to analog conversion is to be carried out, the resistive network should be weight as shown in Fig. 26.106.

Output current

$$I_0 = \left( I + \frac{I}{2} + \frac{I}{4} + ... + \frac{I}{2^{n-2}} + \frac{I}{2^{n-1}} \right)$$

$$= I\left( 1 + \frac{1}{2} + \frac{1}{4} + ... + \frac{1}{2^{n-2}} + \frac{1}{2^{n-1}} \right)$$

$$= \frac{E_R}{R}\left( 1 + \frac{1}{2} + \frac{1}{4} + ... + \frac{1}{2^{n-2}} + \frac{1}{2^{n-1}} \right) \quad ...(26.196)$$

This is true only if all the bits are 1.

The general expression is

$$I_0 = \frac{E_R}{R}\left[ d_{n-1} + \frac{d_{n-2}}{2} + \frac{d_{n-3}}{4} + ... + \frac{d_1}{2^{n-2}} + \frac{d_0}{2^{n-1}} \right] \quad ...(26.197)$$

The output current is maximum when all the bits are 1.

Maximum output current

$$I_{max} = \frac{E_R}{2^{n-2} R}(2^n - 1) \quad ...(26.198)$$

In practical D/A circuits, the resistive network, called a DAC, (digital to analog conversion module), is connected to a flip flop register which holds the digital number. As the divider is simply a passive network, the digital input voltage (the on and off levels) determines the output voltage. Because digital voltage levels are usually not as precise as required in an analog system, level amplifiers may by placed between the flip flop register and the divider network. These amplifiers switch the inputs to the divider network between ground and a reference voltage supplied by a precision reference source. The analog output voltage then falls between these two levels.

A practical digital to analog converter is shown in Fig. 26.107. The basic components are

(*i*) Flip flop (FF) register.

(*ii*) DAC modules which include level amplifiers and reference supplies.

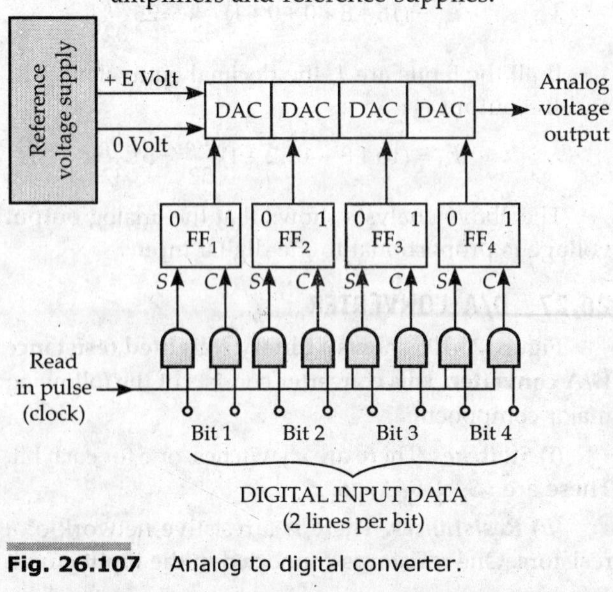


**Fig. 26.107** Analog to digital converter.

The digital signals are dropped into the register by a read in pulse (usually a clock pulse). These pulse are automatically converted by the DAC network to analog voltage.

**Example 26.40** *A D/A converter has 6 bits and a reference voltage of 10 V. Calculate the minimum value of R such that the maximum value of output current does not exceed 10 mA. Find also the smallest quantized value of output current.*

**Solution.**

From Eqn. 26.198, the minimum value of

$$R = \frac{E_R \times (2^n - 1)}{2^{n-2}} \times \frac{1}{I_{max}}$$

$$= \frac{10 \times (2^6 - 1)}{2^5} \times \frac{1}{10 \times 10^{-3}} = 1969\ \Omega$$

$$\approx 2\ k\Omega.$$

Current with LSB

$$= \frac{E_R}{2^{n-1} R} = \frac{10}{2^5 \times 2000}$$

$$= 156 \times 10^{-6}\ A = 156\ \mu A$$

**Example 26.41** *Consider a 6 bit D/A converter with a resistance of 320 kΩ in LSB position. The converter is designed with weighted resistive network. The reference voltage is 10 V. The output of the resistive network is connected to an OPAMP with a feedback resistance of 5 kΩ. What is the output voltage for a binary input of 111.010 ?*

**Solution.** Output current

$$I_0 = \frac{E_R}{R}\left[d_{n-1} + \frac{d_{n-2}}{2} + \ldots + \frac{d_1}{2^{n-2}} + \frac{d_0}{2^{n-1}}\right]$$

Now $n = 6$

∴ Resistance in LSB

$$= 2^{n-1} R = 2^5 R = 320\ k\Omega$$

or $R = 10\ k\Omega$

**Hence,** output current

$$\frac{10}{10 \times 10^3}\left[1 \times 1 + 1 \times \frac{1}{2} + 1 \times \frac{1}{4} + 0 \times \frac{1}{8} + 1 \times \frac{1}{16} + 0 \times \frac{1}{32}\right] A$$

$$= 1.8125 = mA$$

Therefore, output voltage

$$E_0 = I_0 R_f$$

$$= -1.8125 \times 10^{-3} \times 5 \times 10^3 = -9\ V$$

## 26.28 ANALOG TO DIGITAL (A/D) CONVERSION TECHNIQUES

Analog to digital (A/D) conversion is slightly more complex than digital-to analog (D/A) conversion. A number of methods may be used for this purpose. The following four methods are most commonly used :

(*i*) successive approximation method (potentiometric type)

(*ii*) voltage to-time conversion method (ramp type)

(*iii*) voltage to frequency conversion method, (integration type)

(*iv*) dual slope integration method, and

(*v*) re-circulating remainder system.

### 26.28.1 Potentiometric A/D Converter

The potentiometric A/D converter is probably the most widely used in general practice on account of its high resolution and high speed. The potentiometric type A/D converter works on the principle of successive approximation. This converter compares the analog input to a Decade Assembly Counter (DAC) reference voltage which is repeatedly divided in half. The process is shown in Fig. 26.108, where a four digit binary number 1000, representing the full reference voltage $E_r$ is divided in half (binary number 100), to corresponding voltage $\frac{1}{2} E_r$. A comparison of this reference voltage $\frac{1}{2} E_r$, and the analog voltage is made. If the result of this comparison shows that this first approximation is too small (*i.e.*, $\frac{1}{2} E_r$, is smaller

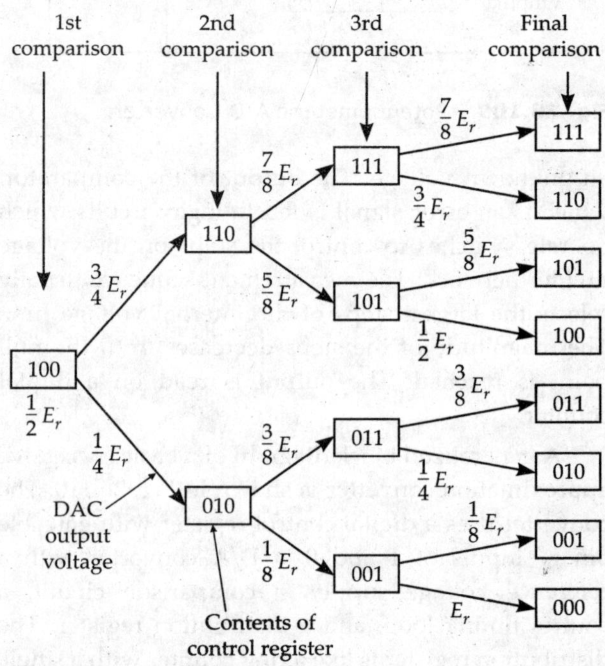


**Fig. 26.108** Operation of successive approximation A/D converter.

than the analog input), then the next comparison is made against $\frac{3}{4}E_r$ (binary number 110). If the comparison showed that the first approximation was too large $C\frac{1}{2}E_r$, larger than the analog input), then the next comparison will be made against $C\frac{1}{4}E_r$, binary number 010). After four successive approximations the digital number is resolved. A six digit number will be resolved in six successive approximations.

The potentiometric A/D converter circuit block diagram is shown in Fig. 26.109. It contains a voltage divider network with curse and fine steps. The voltage divider network is connected through electronic switches to a voltage comparator which compares an internal voltage provided by a reference source with

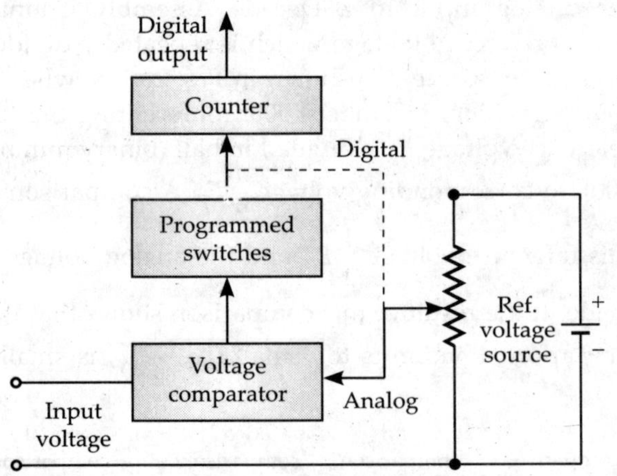


**Fig. 26.109** Potentiometeric A/D Converter.

an unknown voltage. The output of the comparator, which is an error signal is fed to logic circuits which operate switches to control the steps on the voltage divider network. The measurement sequence usually selects the largest steps of the internal voltage first. The magnitude of the steps decreases until the null point is reached. The output is read on a digital counter.

A generalized block diagram of a basic successive approximation converter is shown in Fig. 26.110. The converter uses a digital control register with gateable binary inputs of 1 and 0, a D/A converter with a reference voltage supply, a comparison circuit, a control timing loop, and a distribution register. The distribution register is like a ring counter with a single 1 circulating in it to determine which step is taking place.

At the start of conversion cycle, both the control register and the distribution register are set with a 1 in the MSB and a 0 in all bits of less significance. Thus the distribution register shows that the cycle has started and the process is in its first phase. The control register shows 1000, and this causes an output voltage at the D/A converter section of one half of reference supply. At the same time, a pulse enters the time delay circuitry. By the time that the D/A converter and the comparator have settled, this delayed pulse is gated with the comparator output.

When the next MSB is set in control register by the action of the timing circuit, the MSB remains in the 1 state or it is reset to 0 depending upon the comparator output. The single 1 in the distribution register is shifted to the next position and keeps track of the comparison made.

The procedure repeats itself following the diagram of Fig. 26.110 until the final approximation has teen corrected and the distribution register indicates the end of the conversion.

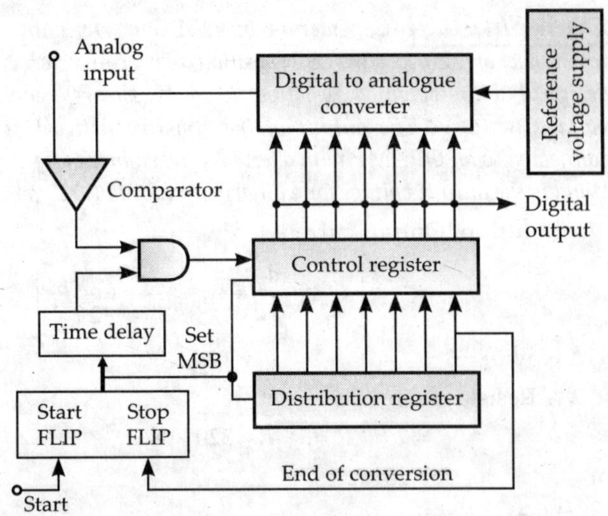


**Fig. 26.110** Simplified block diagram of the successive approximation A/D converter.

For a successive approximation converter, the digital output corresponds to some value that the analog input had during the conversion. Thus the aperture time is equal to the conversion time. The aperture time can be reduced by using **Sample Hold (S/H)** circuit.

**Example 26.42** *Find the successive approximation A/D output for a 4 bit converter to a 8.217 V input if the reference is 5 V.*

**Solution.**

(*i*) Set $d_3 = 1$

$\therefore$ Output $= 5/2^1 = 2.5$ V

Now $3.217 > 2.5$ $\therefore$ Set $d_3 = 1$

(*ii*) Set $d_2 = 1$

$\therefore$ Output $= 2.5 + \frac{5}{2^2} = 3.75$

Now $3.217 < 3.75$ $\therefore$ Set $d_2 = 0$

(*iii*) Set $d_1 = 1$

$\therefore$ Output $= 2.5 + \frac{5}{2^3} = 3.125$

Now $3.125 < 3.217$

$\therefore$ Set $d_1 = 1$

(*iv*) Set $d_0 = 1$

$\therefore$ Output $= 3.125 + \frac{5}{2^4} = 3.4375$

Now $3.4375 > 3.217$ $\therefore$ Set $d_0 = 0$

Thus output of A/D converter is 1010.

## 26.28.2 Voltage to Time A/D Converter (Ramp Type)

This type of A/D converter utilizes digital counter techniques to measure the time required for a voltage ramp to rise from zero to the input voltage level. The time, however, may be for the ramp to drop down from the input voltage level to zero. Thus, this converter may be used when the ramp is either positive going or negative going.

A well defined, internally generated ramp voltage, is fed to two voltage comparators as shown in Fig. 26.111. The ramp voltage may have either a negative or a positive going slope. Let us consider a ramp voltage with a positive going slope. When the input voltage is equal to the ground voltage, the ground comparator emits a start signal to a gate which opens and permits the passage of pulses of a clock (1 MHz) from a crystal oscillator to a digital counter. When the ramp voltage is equal to the level of the input voltage of the second comparator it produces a signal to close the gate and thus further count is prevented.

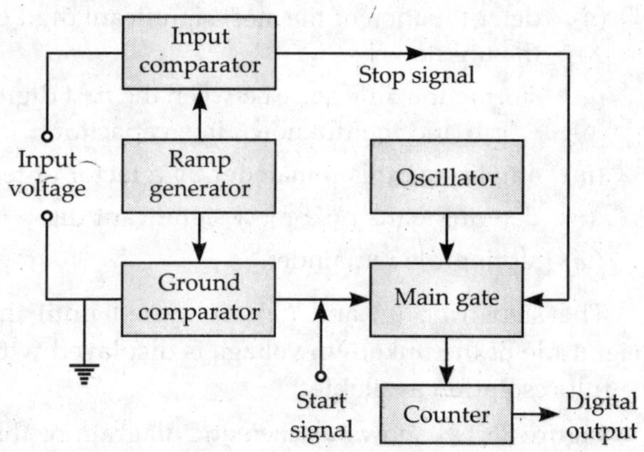


**Fig. 26.111** Voltage to time A/D converter.

The count, with suitable scaling, may be made equal to magnitude of the difference in the two input voltage levels of the input voltage.

## 26.28.3 Voltage-to-Frequency Converter (Integrating Type)

A block diagram representation of a voltage to frequency converter is shown in Fig. 26.112. The analog input is applied to an integrator. The integrator produces a ramp signal whose slope is proportional to the input voltage signal level. When this ramp signal reaches a preset threshold level, a trigger pulse is produced. Also a current pulse is produced which discharges the capacitor of the integrator, after which a new ramp is initiated.

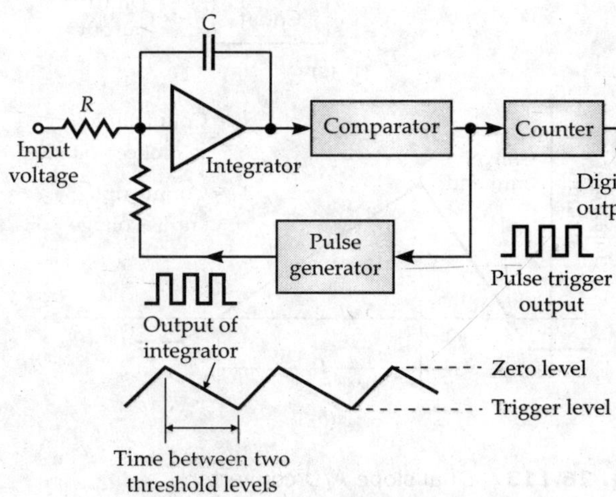


**Fig. 26.112** Voltage to frequency A/D converter (integrating type).

The time between successive threshold level crossings is inversely proportional to the slope of the ramp. Since the slope of the ramp is proportional to the input analog voltage, hence the frequency of output pulses from the comparator is directly proportional to the input voltage. The output frequency may be measured with the help of a digital frequency counter. Frequency counters are explained in chapter 28.

## 26.28.4 Dual Slope Integration A/D Converter

The block diagram of a dual slope integration A/D converter is shown in Fig. 26.113($a$). The reference voltage and the input analog voltage are sequentially connected to the integrator with the help of a switch. The reference voltage and the input analog voltage must be of opposite polarity. The input voltage is integrated for a fixed input sample time. The integrated value is then discharged at a fixed rate and the time to do this is measured by a counter.

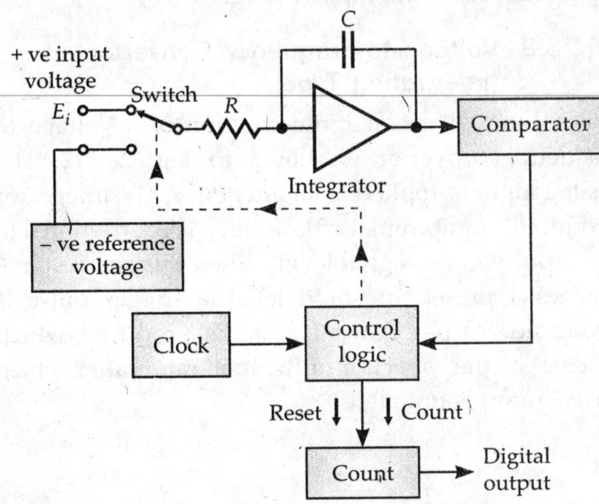


($a$)

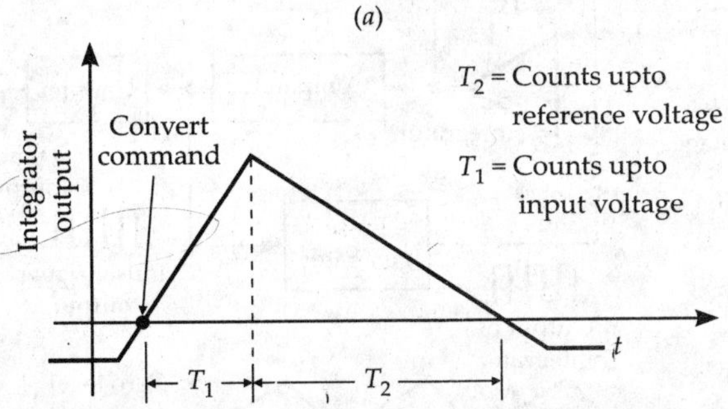


($b$)

**Fig. 26.113** Dual slope A/D converter.

Two types of commands are given by the logic control to the counter and they are : ($i$) Reset and ($ii$) Convert.

When a convert command is received by the counter, it automatically resets to all zeros and the switch connects the input voltage to the integrator. The output from the comparator is designed such that at this time it will permit the counter to count up for a output from the integrator will be steadily increasing in value.

On the next count after the converter has counted all the way up to all is ($i.e.$, the next count will cause it to go all zeros and start over again), the switch changes position. The input voltage is dis-connected and the switch connects the reference voltage to the integrator. Therefore the integrator now integrates the opposite polarity voltage, which causes the output to decrease towards zero voltage. Meanwhile, the counter is counting up from zero again.

When the output of the integrator goes to zero, it causes the comparator to switch its output, thereby stopping the counter via the control logic. The binary number in the counter at this time is proportional to the amount of time that it took the integrator to integrate down from its starting point to zero. Therefore, the binary count is proportional to input voltage. Figure 26.113($b$) illustrates the output from the integrator during a conversion cycle. The difference in the slopes of the curve is due to difference in voltage level of input voltage $E_i$ and reference voltage $E_r$.

Suppose

$T_1$ = Counts when the input voltage is applied,

$T_2$ = Counts when the reference voltage is applied.

Input voltage is,

$$E_i = (T_2 / T_1) \times E_r \qquad ...(26.199)$$

## 26.28.5 Recirculating Remainder System

The counters found in the methods of conversion so far described have been fairly conventional, that is a complete counting circuit had been used for each decade of display. In the recirculating remainder system only one counting circuit is used, the output from this being routed to each display digit in turn. This system, which reduces the number of component parts of the instrument, is made possible by a sequence of operations as follows :

($i$) sample the unknown voltage

($ii$) determination of the most significant digit of the unknown.

($iii$) storing the difference between the first digits worth and the unknown in a capacitor

($iv$) amplifying this remainder by a factor of ten

($v$) determine the next most significant digit

($vi$) storing the remainder.

The steps ($iv$), ($v$) and ($vi$) are cycled until the magnitude of the unknown voltage is displayed with the full resolution available.

Figure 26.114 shows a schematic diagram of this type of d.v.m. which is comparatively new, makes greater utilisation of component parts, and has a

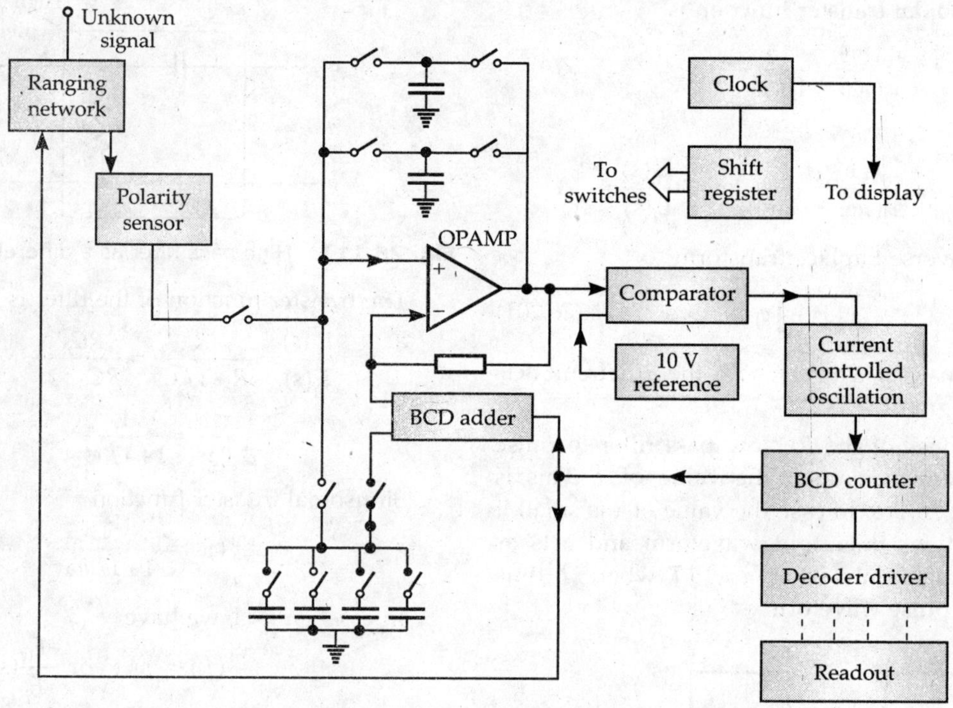


**Fig. 26.114** Recirculating remainder system.

strobed display (thereby consuming less power) facilitating battery operation. Since the same capacitors are used to store all the digits, each is determined with the same precision. The sampling time is small (4 ms) and the conversion fast, giving a reading rate of 40 per second.

The accuracy limiting factors are :

(*i*) stability of the operational amplifiers which can b made of the order of a few p.p.m.,

(*ii*) the reference voltage, which is derived from a reference amplifier – a combined zener diode and transistor *i.e.,* having zero temperature coefficient over normal ambient temperatures,

(*iii*) leakage and dielectric storage of the remainder storage capacitor, which is much less of a problem in this conversion technique than in the dual slope method since here only the remainder is stored.

## 26.29 INTEGRATION AND DIFFERENTIATION USING R.C. CIRCUITS

In many measurement systems it becomes necessary to obtain integrals and/or derivatives of signals with respect to time. It is easy to integrate a signal as integration leads to alternation of noise at the output terminals and hence this process improves the signal to noise (S/N) ratio. However, differentiation is much harder to carry out since differentiation leads to decrease in signal to noise (S/N) ratio. The integration and differentiation may be carried out by R.C. networks without the use of operational amplifiers as described below.



### 26.29.1 Low Pass R.C. Filter as an Integrator

A low pass filter shown in Fig. 26.115 acts as an integrator.

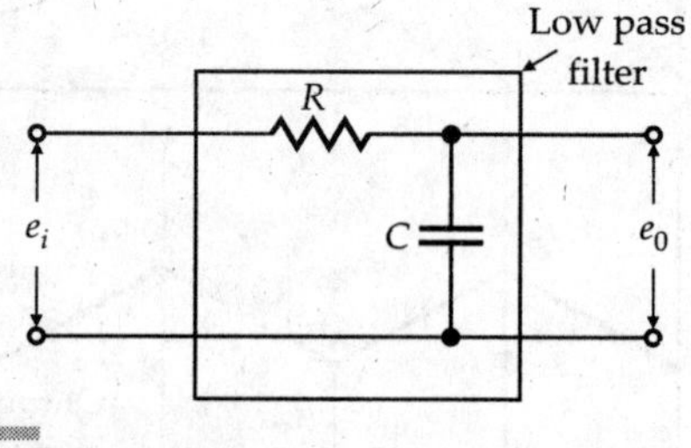


**Fig. 26.115** Low pass R.C. filter as an integrator.

The transfer function of the filter is

$$\frac{E_0(s)}{E_i(s)} = \frac{1/Cs}{R + 1/Cs}$$

$$= \frac{1}{1 + RCs} = \frac{1}{1 + \tau s} \qquad ...(26.200)$$

where $\tau = RC$ = time constant

The sinusoidal transfer function is

$$\frac{e_0(j\omega)}{e_i(j\omega)} = \frac{1}{1 + j\omega\tau}$$

Now if $\omega\tau >> 1$,

$$\frac{e_0(j\omega)}{e_i(j\omega)} = \frac{1}{j\omega\tau} \quad \text{or} \quad \frac{E_0(s)}{E_i(s)} = \frac{1}{s\tau}.$$

Taking inverse Laplace transform

$$e_0 = \frac{1}{\tau}\int e_i\,dt. \qquad ...(26.201)$$

This means that when $\omega\tau >> 1$, the $RC$ circuit acts as a pure integrator.

The response of the $RC$ low pass filter to pulse wave input depends upon the value of $\tau$. This is shown in Fig. 26.116. In case the value of $\tau$ is small it causes distortions in output waveform and acts as perfect integrator only when $\tau > 15\,T$ where $T$ time period of the pulse waveform.

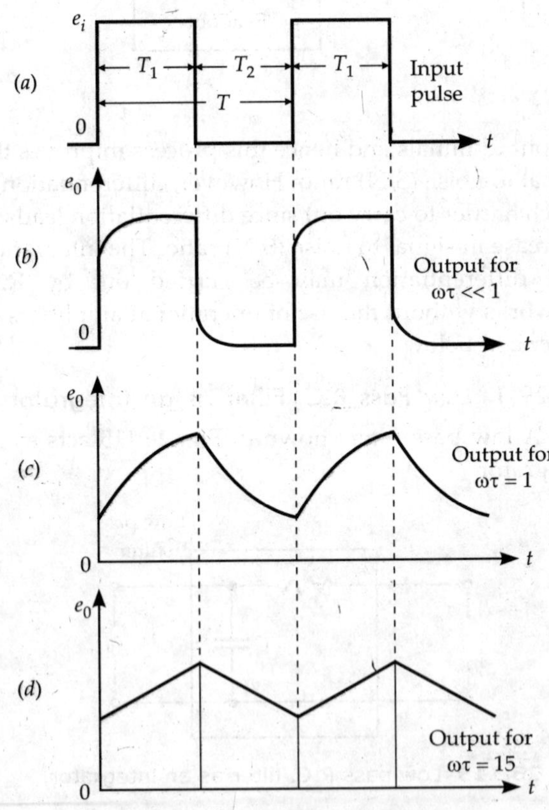


**Fig. 26.116** Output waveforms for a integrator for a periodic pulse input for different values of time constant $\tau$.

## 26.29.2 High pass *RC* Filter as Differentiator

A high pass filter which acts as a differentiator is shown in Fig. 26.117.

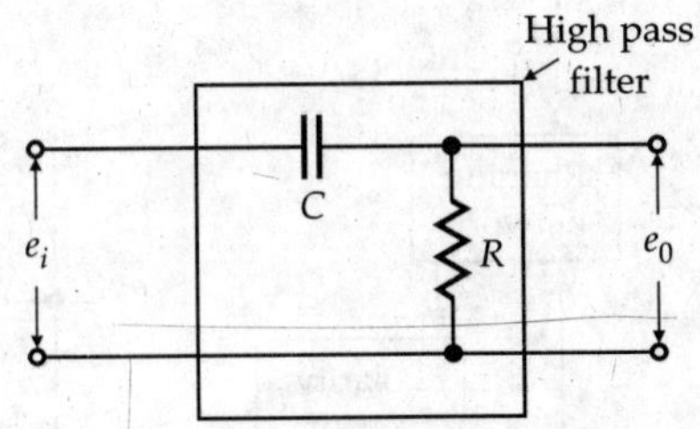


**Fig. 26.117** High pass filter as a differentiator.

The transfer function of the filter is

$$\frac{E_0(s)}{E_i(s)} = \frac{R}{R + 1/Cs} = \frac{RCs}{RCs + 1}$$

$$\tau s = \frac{\tau s}{1 + \tau s} = \frac{1}{1 + 1/\tau s}$$

Sinusoidal transfer function

$$\frac{e_0}{e_i}(j\omega) = \frac{1}{1 + 1/j\omega\tau}$$

In case $\omega\tau << 1$, we have

$$\frac{e_0}{e_i}(j\omega) = j\omega\,\tau \quad \text{or} \quad \frac{E_0}{E_i}(s) = s\tau$$

Taking inverse Laplace transform

$$e_0 = \tau\frac{d\,e_i}{dt} \qquad ...(26.202)$$

Thus the circuit acts as a perfect differentiator for $\omega\tau << 1$.

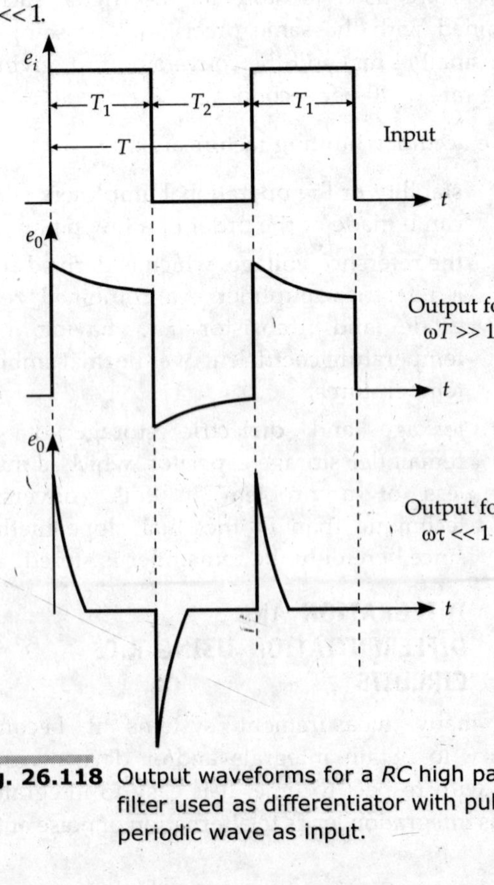


**Fig. 26.118** Output waveforms for a *RC* high pass filter used as differentiator with pulse periodic wave as input.

The input waveforms applied to an $RC$ high pass filter are periodic pulse waves as shown in Fig. 26.118. The nature of the output waveforms depend upon the time constant. In case the time constant is large, perfect differentiation does not take place and we have distorted waveform as shown in Fig. 26.118($b$). The distortion less differentiation is shown in Fig. 26.118($c$) when $\omega\tau << 1$.

**Example 26.43** *A triangular input voltage of peak value 20 V is applied to a high pass filter having $C = 1\,\mu F$ and $R = 1\ k\Omega$. The time period of the wave is 400 ms, calculate the output d.c. voltage.*

**Solution.** The wave rises to its peak value in 1/4 of its cycle. $\therefore\ T = 400/4 = 100$ ms

Charging current

$$i = C(dv/dt) = 1\times10^{-6} \times (20/100\times10^{-3}) = 0.2\times10^{-3}\ \text{A}.$$

The output voltage is a square wave of amplitude

$$e_0 = iR = 0.2\times10^{-3}\times1\times10^{3} = 0.2\ \text{V}$$

[The high pass filter acts as a differentiator. The input is a triangular wave $e_i = kt$ and the differential is $e_0 = k$].

## 26.30 CLIPPING CIRCUITS

These are diode-operated circuits or networks that change their $EI$ characteristics in response to the level of the applied signal. A diode can be considered a voltage-controlled switch and can be used as such in analog circuits. A number of such circuits are described below.

Figures 26.119 ($a$) and ($b$) show the symbol and the $EI$ characteristics of a diode when it is considered a voltage-controlled switch. Details about the voltage current characteristics such as a curvature in the vicinity of the zero point can be neglected. When the anode is positive with respect to the cathode, the diode conducts with negligible forward resistance ; when the anode is negative with respect to the cathode, the diode acts as an insulator or as an open switch.

An example of the application of a diode switch is shown in Figs. 26.119($c$) and ($d$). When the input voltage $E_i$ is negative, the diode is forward-biased and behaves as a short circuit ; therefore, there will be no voltage drop across it that is, $E_0 = 0$. When the input voltage is positive, the diode behaves as an open circuit, and the total input voltage appears at the output, that is, $E_0 = E_i$. The resistance $R$ is inserted to limit the current through the diode when the input voltage is negative.

The point $P$ at which the slope changes can be varied by the insertion of a bias voltage $E_B$ in series with the diode as illustrated in Figs. 26.119 ($e$) and ($f$). The voltage applied to the diode $D$ is composed of $E_B$ and the applied input voltage $E_i$. A current $I$ will pass through the diode as long as $E_i$ in the polarity shown, is smaller than $E_B$ (or is negative). As long as a current passes through the diode; it can be considered a closed switch so that the output voltage is $E_0 = E_B$. If the input voltage $E_i$ exceeds the bias voltage $E_B$, the diode switch opens, and the output voltage is $E_0 = E_i$. Reversal of the polarity of $E_B$ has the effect that the point $P$ comes to lie in the third quadrant as indicated by point $P'$ and the dotted lines.

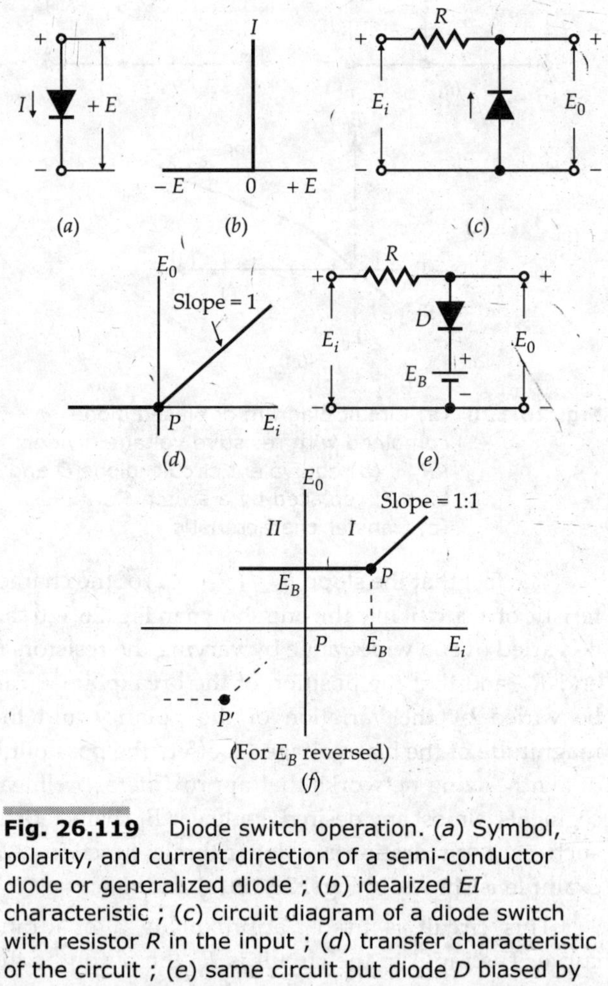


**Fig. 26.119** Diode switch operation. ($a$) Symbol, polarity, and current direction of a semi-conductor diode or generalized diode ; ($b$) idealized $EI$ characteristic ; ($c$) circuit diagram of a diode switch with resistor $R$ in the input ; ($d$) transfer characteristic of the circuit ; ($e$) same circuit but diode $D$ biased by a source $E_B$ ; ($f$) transfer characteristics, solid line for polarity of $E_B$ as shown ; dotted line for polarity of $E_B$ reversed.

The slope of that part of the transfer characteristic for which the diode switch is closed can be varied by the insertion of a resistor $R_D$ in series with the diode. Figure 26.120 shows a circuit of this type. As long as the input voltage $E_i$ is smaller than the bias voltage $E_B$, the diode acts as an open circuit, and the output voltage $E_0$ is equal to the input voltage $E$. If the input voltage, with the polarity as shown in Fig. 26.120 exceeds the bias voltage $E_B$, the circuit acts as the voltage divider illustrated in Fig. 26.120. The transfer function has a breakpoint $P$ at $E_i = E$. The slope of the characteristic for $E_i < E_B$ is one, and for $E_i > E_B$, the slope is $R_D/(R + R_D)$ as shown in Fig. 26.120.

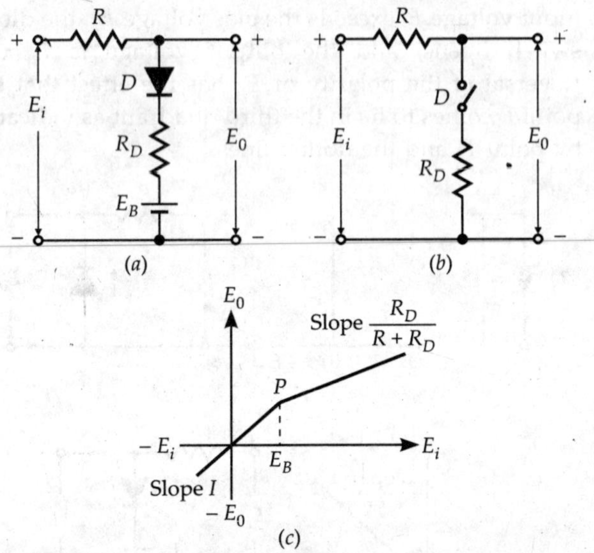


**Fig. 26.120** (a) Circuit diagram of biased diode combined with resistive voltage divider $RR_D$ ; (b) equivalent circuit, diode $D$ and bias $E_B$ replaced by a switch $S$ ; (c) transfer characteristic.

The fact that the slope $R_D/(R + R_D)$ of the characteristic of a circuit like the one shown in Fig. 26.120 can be varied over a wide range by varying the resistors $R$ and $R_D$ and that the position of the breakpoint $P$ can be varied by the variation of the polarity and the magnitude of the bias voltage $E_B$ offers the possibility of synthesizing networks that approximate, by linear segments, almost any desired nonlinear $E_1$ characteristic such as a quadratic or a logarithmic function. An example is shown in Figs. 26.121(*a*) and (*b*).

This circuit is used approximate a quadratic curve. The input to the circuit is $E_i$, the output is the current $I$, and the relationship which is to be approximated is $I = kE_i^2$ for the current range $0 < 1 < 9$ mA.

When $E_i$ is negative (polarity shown in Fig. 26.121), there will be no current in the circuit as all three diodes are open. When $E_i$ becomes positive, diode $D_0$ closes and a current $I = E_i / R_0$ will flow. The slope of the $IE$ curve, therefore, becomes $G_0 + 1/R_0$. When $E_1 < E_i < E_2$, both $D_0$ and $D_1$ are conducting, and the slope of the curve becomes $G_A = G_0 + 1/R_1$. When $E_i > E_2$, all three diodes are conducting ; all three resistors are connected in parallel ; and the resulting slope of the IE curve is the sum of the three conductances, $G = G_0 + G_A + G_B$.

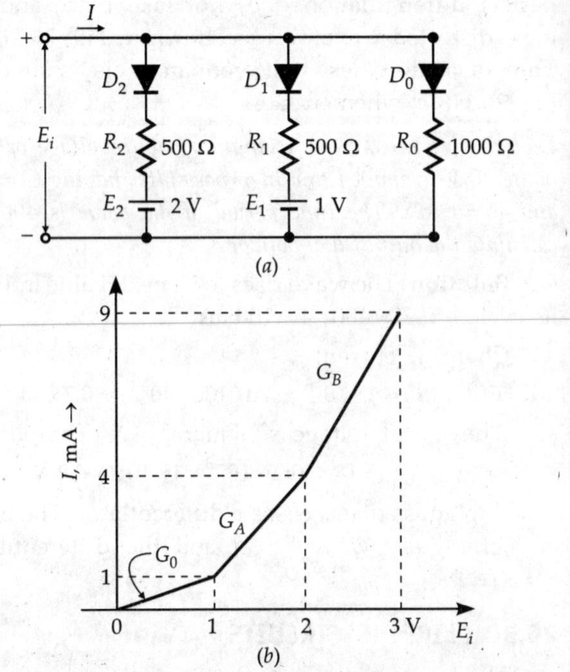


**Fig. 26.121** (a) Diagram of a circuit which causes curved *EI* characteristic by synthesis of linear segments; (b) approximately quadratic characteristic formed from three segments.

The diode switch circuit shown in Fig. 26.121(*a*) contains two basic diode switch circuits of the type shown in Fig. 26.116 ; the two diodes and their bias sources are in opposite directions. If an input voltage $E_i$ is applied with the polarity as shown in Fig. 26.121 and if $E_i < E_1$, both diodes are open, and the output voltage $E_0$ is equal to the input voltage $E_i$. For $E_i > E_1$, the diode $D_1$ becomes a short circuit, and the out, at voltage $E_0$ remains at the value of the bias voltage $E_1$. For an applied input voltage of opposite polarity an analogous process will occur at the diode $D_2$. If $E_1$ and $E_2$ are alike, a transfer characteristic will result like that shown in Fig. 26.122 (*b*). If the applied voltage $E_i$ alternates periodically [Fig. 26.122 (*c*)] and varies with respect to amplitude and frequency, then the output voltage closely resembles a square wave of constant amplitude $E_1$ and of the same fundamental frequency as the input voltage, provided that the excursion of the input voltage is large compared to the bias voltage $E_1$. This clipper circuit can be used to separate any information contained in the frequency from that contained in the amplitude of the applied voltage.

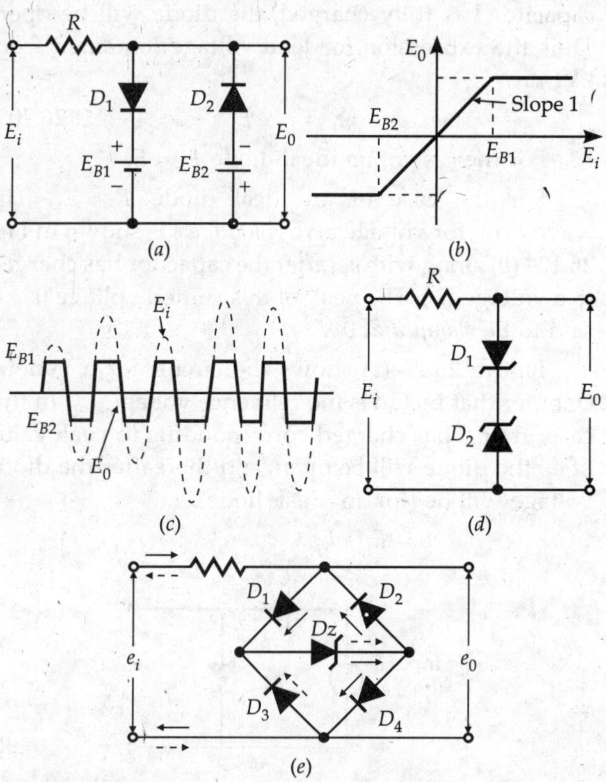


**Fig. 26.122** Clipper or slicer. (*a*) Circuit diagram for a symmetrical clipper ; (*b*) *EI* characteristic ; (*c*) clipper action on a sinusoidal input $E_i$, causing approximately square wave output $E_0$ ; (*d*) the same clipping effect as shown in (*c*) can be obtained with two zener diodes connected back to back ; (*e*) circuit diagram for symmetrical clipping or slicing.

The same result can be obtained by a circuit like that shown in Fig. 26.122(*d*). The circuit contains two zener diodes connected back to back and a series resistor *R*. If the input $e_i$ is positive, diode $D_2$ is conducting, but $D_1$ is open until the zener voltage breakpoint is reached. At that point, both diodes are conducting and the output voltage $e_0$ can rise no farther. The same condition exists for the opposite polarity, so that the output $e_0$ approaches a square-wave function like that shown in Fig. 26.122 (*c*). If symmetry of the positive and of the negative half-waves is essential, and if a matched pair of zener diodes is not available, the modified bridge circuit shown in Fig. 26.122(*c*) may offer some advantages.

As is evident from the arrows signifying the flow of current in the positive (solid arrows) and in the negative (dotted arrows), half-cycles, the current flows through the zener diode $D_2$ in each half-cycle in direction. Therefore the clipping points in both half-cycles are identical, provided that the characteristics of the diodes $D_1$ to $D_4$ are reasonably alike.

## Diode Limiter

The basic *diode limiter circuit* is shown in Fig. 26.123(*a*). The voltage transfer function is displayed in Fig. 26.123(*b*). In this characteristics the output voltage, $e_0$, is plotted against the input voltage $e_i$ and is obtained by varying $e_i$ through all possible values and determining the corresponding values of $e_0$. Accounting for the forward voltage drop $V_0$, we observe that when $e_i < (E_R + V_0)$, the diode is reverse biased and no current flows with result diode *D* is open circuited and $e_i = e_0$. When $e_i > (E_R + V_0)$, the diode is forward biased and conducts *i.e.*, it is short circuited, thereby the output voltage $e_0 = e_i$. Since the output voltage is limited to a maximum value $(E_R + V_0)$ the circuit is called a *diode limiter*.

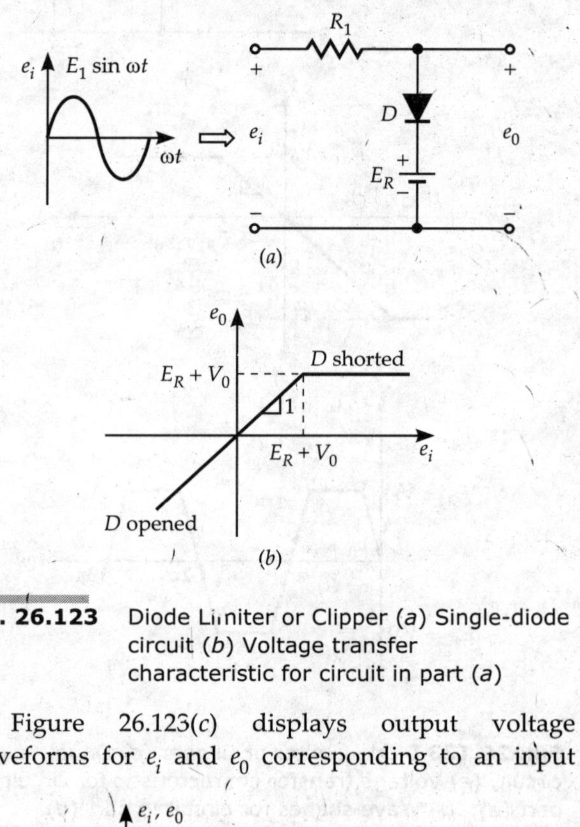


**Fig. 26.123** Diode Limiter or Clipper (*a*) Single-diode circuit (*b*) Voltage transfer characteristic for circuit in part (*a*)

Figure 26.123(*c*) displays output voltage waveforms for $e_i$ and $e_0$ corresponding to an input

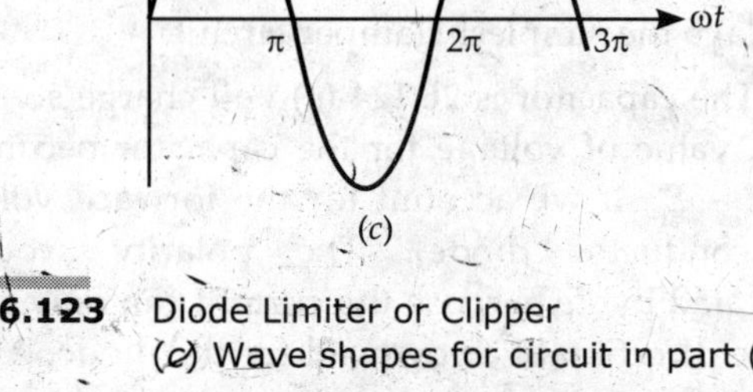


**Fig. 26.123** Diode Limiter or Clipper (*c*) Wave shapes for circuit in part (*a*)

26

sinusoidal voltage $e_{s1} = E_{s1} \sin \omega t$ and $E_{s1} > E_R$. Note that the output voltage wave shape is clipped off $e_1 > (E_R + V_0)$. Hence this circuit is called a *diode clipper*.

Figure 26.123(*d*) shows a diode limiter that limits the negative peak value of $e_0$ as well as the positive peak, values. The voltage transfer characteristic of Fig. 26.123(*e*) is obtained, by considering the state of each diode for all values of $e_i$. Voltage waveforms for this case are shown in Fig. 26.123(*f*).

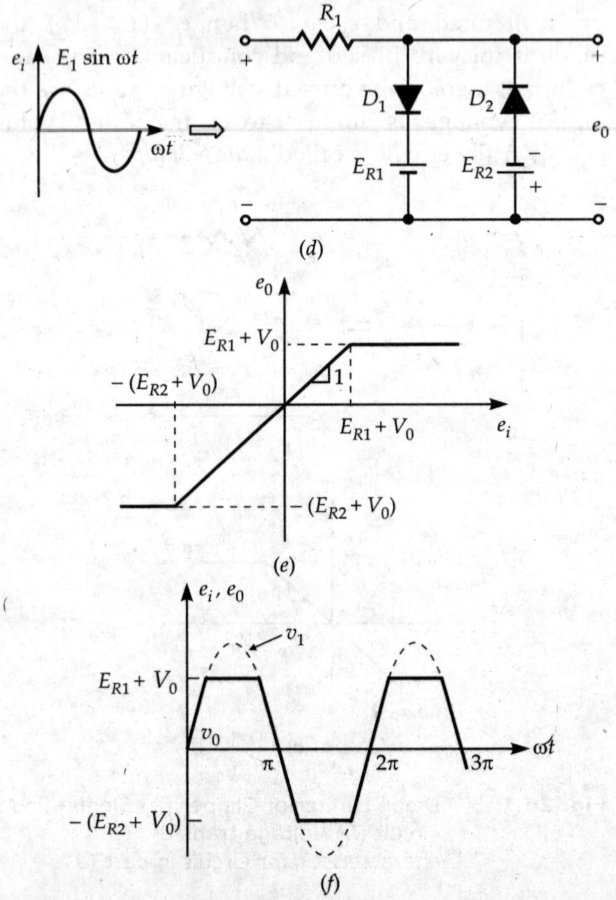


**Fig. 26.123** Diode Limiter or Clipper (*d*) Double-diode circuit, (*e*) Voltage transfer characteristic for circuit in part (*d*), (*f*) Wave shapes for circuit in part (*d*).

## Clamper Circuit

The *diode clamper* is a waveshaping circuit that shifts the signal voltage to a desired level. Fig. 26.124(*a*) displays the simplest clamper circuit.

The capacitor is 26.124 (*a*) will charge so that the peak value of voltage for the capacitor becomes $E_1$ (or $E_1 - E_0$, if we account for the forward voltage of semiconductor diode). The polarity would be indicated by $Ec$ because the current can only conduct in the direction of arrowhead for the diode. After the capacitor has fully charged, the diode will be open. Thus the expression for load voltage (obtained from kVL) is,

$$e_L = e_1 - E_c \qquad ...(26.203)$$

For the case of an ideal diode $E_c = E_1$

For the case of an ideal diode $E_c = E_1$. The waveform, for voltage across load, $e_L$, is shown in Fig. 26.124 (*b*) along with $e_1$ (after the capacitor has charged to a voltage $E_1$). The peak of this shifted voltage ($e_L$) is said to be *clamped* at 0 V.

Figure 26.124(*c*) shows the circuit for a general clamper that includes the reference voltage $E_R$. In this case, after C has charged corresponding to peak value of $e_1$, the diode will be open, and thereafter the diode voltage will be (for an ideal diode).

$$E = E_1 + E_R.$$

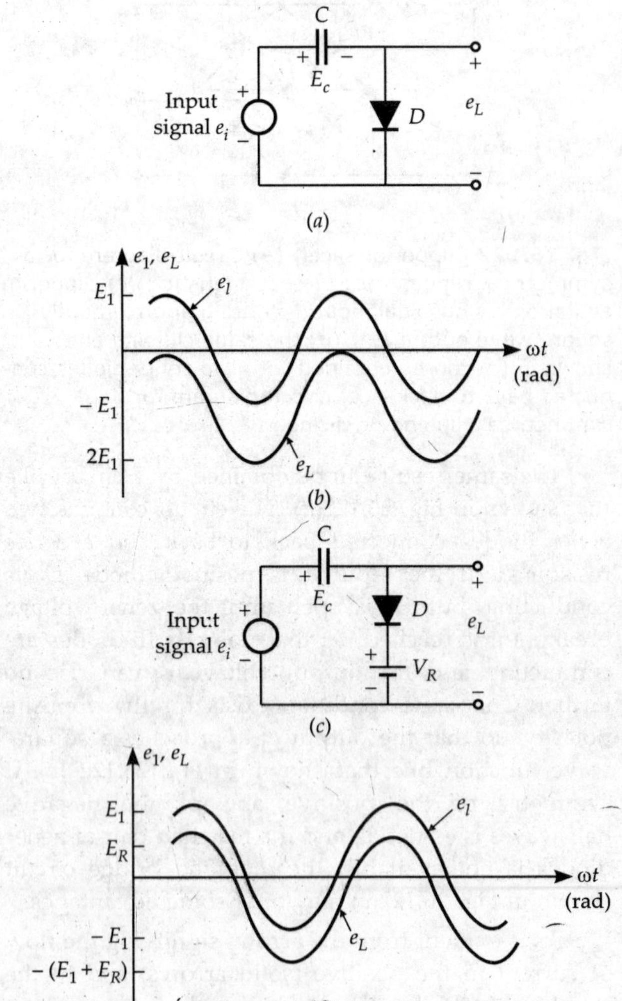


**Fig. 26.124** Diode Clamper ; (*a*) Basic circuit, (*b*) Waveshapes of $v_s$ and $v_L$ for basic circuit, (*c*) General circuit, (*d*) Waveshapes of $v_s$ and $v_L$ for general circuit.

From kVL, voltage $e_L$ is given by

$$e_L = e_1 - E_c$$

or substituting for $E_c$

$$e_L = e_1 - E_1 + E_R$$

This waveshape is the signal voltage $e_1$ clamped at reference voltage $E_R$ as shown in Fig. 26.124(*d*). Note that $V_R$ can be either positive or negative.

In case the diode is considered to be non-ideal its forward resistance $R_D$ and its forward voltage drop $V_0$ have to be considered and compensated for. It should be mentioned that the signal voltage $e_1$ need not be sinusoidal. The capacitor in the clamper circuit will change to the peak value of the input signal voltage, and the current will continue to operate in a similar manner. For the general clamper, the output voltage would consist of non-sinusoidal signal voltage shifted by the voltage $E_R$.

**Example 26.44** *A sinusoidal voltage is shown in Fig. 26.125 (a). It has a peak value of 50 V. It is applied to a circuit containing an ideal diode. Sketch the waveform of $e_0$.*

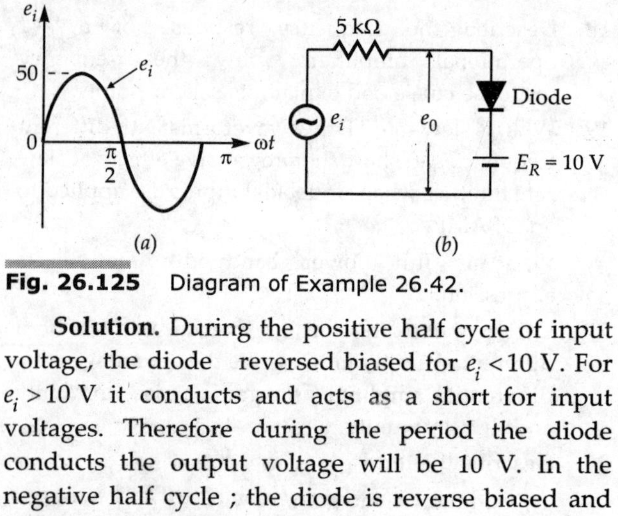


**Fig. 26.125** Diagram of Example 26.42.

**Solution.** During the positive half cycle of input voltage, the diode reversed biased for $e_i < 10$ V. For $e_i > 10$ V it conducts and acts as a short for input voltages. Therefore during the period the diode conducts the output voltage will be 10 V. In the negative half cycle ; the diode is reverse biased and hence the output voltage will follow the input voltage.

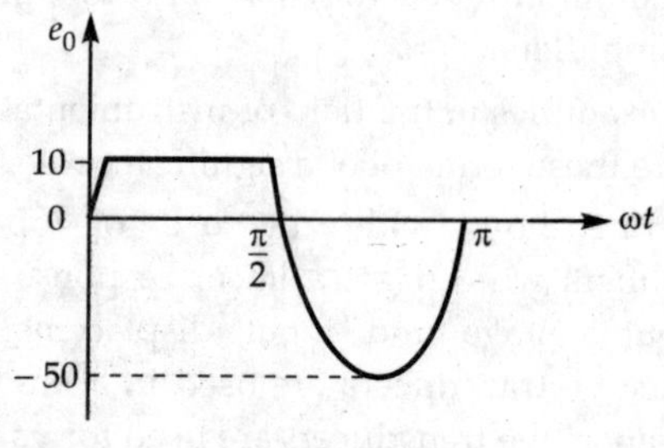


**Fig. 26.126** Output voltage of Example 26.42

**Example 26.45** *Sketch the output waveform for the clipping circuit shown in Fig. 26.127.*

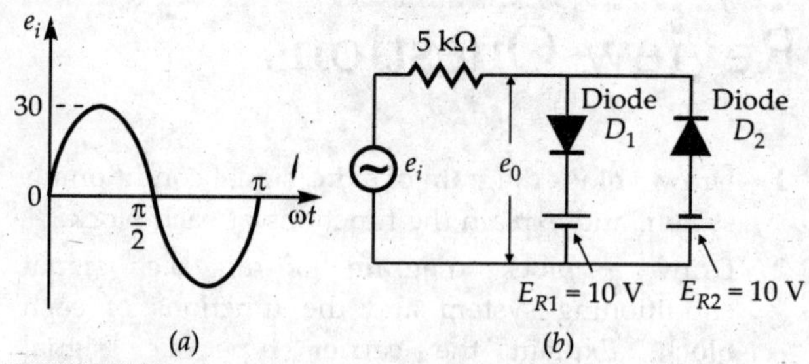


**Fig. 26.127** Diagram of Example 26.43.

**Solution.** It is clear that diode $D_1$ and its 10 V battery $E_{R1}$ act as clipper during the positive half cycle with clipping level at +10 V. Similarly, during the negative half cycle diode $D_2$ and its 10 V battery as clipper with clipping level at – 10 V.

The output voltage waveform is shown in Fig. 26.128.

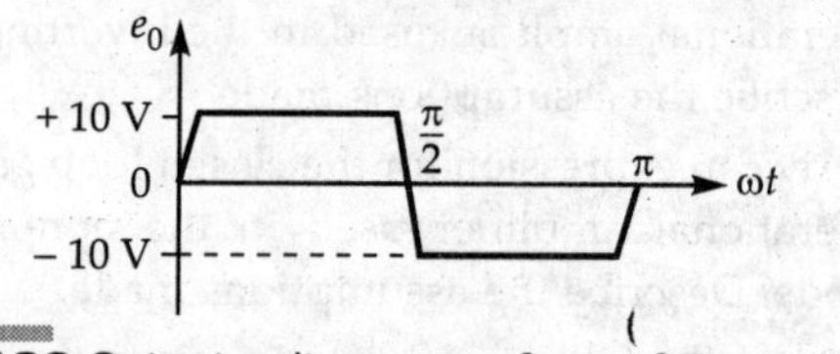


**Fig. 26.128** Output voltage waveform of Example 26.43.

**Example 26.46** *Consider the general clamper circuit shown in Fig. 26.129(a).*

*Let $V_R = 1$ V and assume that C has charged to its final value at $t = 0$. Determine the sketch of load voltage for $t > 0$ for the triangular input signal shown in Fig. 26.129(b).*

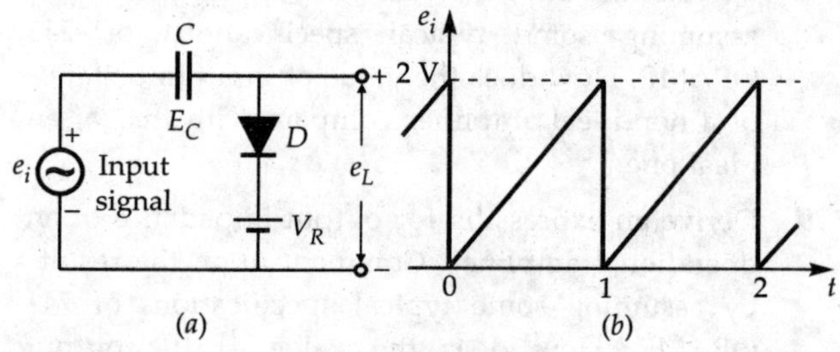


**Fig. 26.129** Diagram of Example 26.44.

**Solution.** The capacitor charges to $E_c = E_1 - V_p = 1$ V during the time $t < 0$. Thus for $t > 0$, the load voltage is

$$e_L = e_1 - E_c = e_1 - 1$$

The wave shape is sketched in Fig. 26.130.

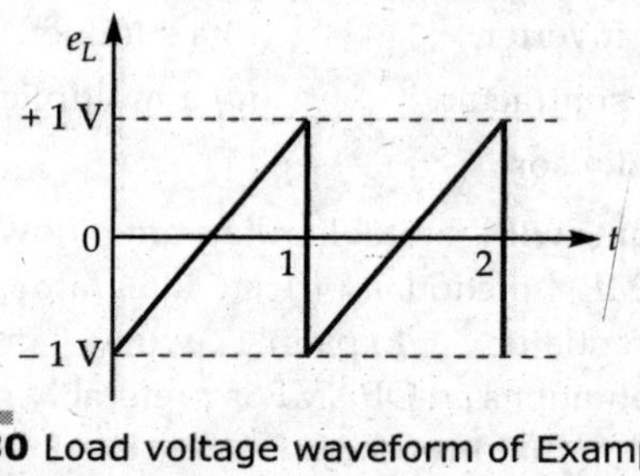


**Fig. 26.130** Load voltage waveform of Example 26.44.

26

# Review Questions

1. Draw a block diagram of a d.c. signal conditioning system and explain the functions of each block.
2. Draw a block diagram of an a.c. signal conditioning system and the functions of each block. Explain the carrier type a.c. signal conditioning system and its advantages.
3. Explain the characteristics and an ideal operational amplifier. Explain the functions of its two input terminals.
4. Draw the pin diagram of an IC used as an operational amplifier. Explain the functions of the different pin connections.
5. Derive an expression for the closed loop gain of an operational amplifier used in the inverting mode. Describe the assumptions made.
6. Derive an expression for the closed loop gain of an operational amplifier used in the non-inverting mode. Describe the assumptions made.
7. Explain the following terms in connection with an operational amplifier :
   (*i*) Input offset voltage (*ii*) Input offset current
   (*iii*) Bias current (*iv*) Slew rate and
   (*v*) Unity gain frequency.
8. Derive the expression for input impedance of an operational amplifier. Comment upon the results assuming some typical specifications of 741 OPAMP. How does the value of input impedance of a non-ideal amplifier compare with that of an ideal one ?
9. Derive an expression for output impedance of an operational amplifier. Comment upon the results by assuming some typical specifications of 741 OPAMP. How does the value of the output impedance of a non-ideal OPAMP compare with that of an ideal one ?
10. Describe the methods for elimination of effects of input offset voltages in OPAMPs.
11. Explain with circuit diagrams, how does an OPAMP act as :
    (*i*) an inverter, (*ii*) a summer,
    (*iii*) a subtractor, (*iv*) a multiplier and
    (*v*) a divisor.
12. Explain with suitable diagrams how does an OPAMP function as (*i*) an integrator and (*ii*) a differentiator. Explain why in analog computations an OPAMP is preferably not used as a differentiator.
13. Explain the functioning of an OPAMP as a cathode follower.
14. Describe with circuit diagrams how an OPAMP act as
    (*i*) precision half wave rectifier and
    (*ii*) precision full wave rectifier.
15. Explain with suitable diagrams the following OPAMP circuits,
    (*i*) attenuated feedback
    (*ii*) logarithmic converter
    (*iii*) limiter.
16. Derive expressions to show the effect of feedback on
    (*i*) input impedance,
    (*ii*) output impedance and
    (*iii*) bandwidth of operational amplifiers.
17. Explain the functioning of a buffer amplifier.
18. Describe the frequency response of a 741 operational amplifier. Draw the frequency response curve and explain its salient points.
19. What is slew rate ? Draw waveforms with different frequencies to show the progressive effects of slew rate limiting when sinusoidal inputs are applied to an OPAMP.
    What is full power bandwidth Derive its expression.
20. Derive the expression for output voltage of a differential amplifier. Describe the advantages of a differential amplifier as regards noise immunity and drift immunity.
21. Describe the terms
    (*i*) Common mode gain,
    (*ii*) Differential mode gain and
    (*iii*) Common mode rejection ratio for differential amplifiers.
    Cite examples in the field of instrumentation where these terms play a significant role.
22. Describe how with the use of operational amplifiers we can get a linear response between output voltage and input displacement when capacitive transducers are used in a mode where the capacitive transducers are used for variation of capacitance with distance between the plates.
23. Describe the functioning of a charge amplifier using an operational amplifier with piezo-electric transducer used at the input end for measurement

of displacement. Derive the expression for transfer function and draw its a curve showing its frequency response.

Explain the advantages and disadvantages of a charge amplifier.

**24.** Explain the term "Common mode rejection ratio" CMRR and explain its significance in the field of instrumentation.

**25.** Describe the circuit of a 3 amplifier configuration of an Instrumentation amplifier. This amplifier is divided into two stages. Describe how the first stage offers a very high impedance to both the input signals and the gain to be set and the second stage acts as a difference amplifier.

Derive the expression for output voltage in terms of the two input voltages.

**26.** What is amplitude modulation ? Why is it used in the field of instrumentation ? Cite examples.

**27.** Derive the expression for output voltage when a sinusoidal signal is amplitude modulated using a sinusoidal carrier frequency. Draw the frequency spectrum (both for amplitude and phase angle) of the modulated output wave.

**28.** Derive the expression for output voltage when a non-sinusoidal periodic input signal is amplitude modulated using a sinusoidal carrier wave. Draw the frequency spectrum.

**29.** Draw the frequency spectrum of output when a transient input wave is amplitude modulated using a sinusoidal carrier.

**30.** Why is amplitude demodulation necessary when the input is amplitude modulated in an Instrumentation system ? A non-sinusoidal input signal is amplitude modulated with a sinusoidal carrier wave. Draw the waveforms of the modulated wave and also of demodulated waves in case the following types of demodulations are used :

(*i*) Half wave-non phase sensitive.
(*ii*) Half wave-phase sensitive.
(*iii*) Full wave-non phase sensitive.
(*iv*) Full wave-phase sensitive.

**31.** Derive the expression for the full wave demodulated signal if the input signal is $A_s \sin \omega_s t$ and the carrier signal is $A_c \sin \omega_c t$.

Describe the frequency spectrum of output wave.

**32.** Describe the different de-modulator circuits used.

**33.** Describe the following types of passive *RC* filters

(*i*) low pass, (*ii*) high pass.
(*iii*) band pass and (*iv*) band stop.

Draw their ideal and practical characteristics.

**34.** Describe a low pass passive *RC* filter. Derive the expression for its transfer function. Draw its characteristics and Bode diagram. Derive the expression for cut off frequency.

**35.** Describe a high pass passive *RC* filter. Derive the expression for its transfer function. Draw its characteristics and also the Bode diagram. Derive the expression for cutoff frequency.

**36.** Describe the use of passive *RC* filter low pass and high pass filters in Instrumentation. Cite an example for each case illustrating your answer.

**37.** Derive the expression for transfer function of a band pass *RC* filter. Draw its characteristics and also the Bode diagram.

**38.** Describe the different types of low pass and high pass *LC* filters. Derive the expression for cutoff frequency in each case.

**39.** What are active filters ? Describe a low pass active filter and derive the expression for the cutoff frequency. Draw its frequency characteristics.

**40.** Derive the expression for output voltage of an active high pass filter. Derive the expression for cutoff frequency and draw the frequency response curve.

**41.** Describe how a band stop active filter circuit can be obtained by combining an active low pass filter and an active high pass filter. Derive the expressions for output voltage, and lower and upper cutoff frequencies. Draw its frequency response curve.

**42.** Describe a ballast circuit. Find the condition that maximizes its sensitivity. Describe why it is impossible to obtain this optimum sensitivity when using a resistive transducer whose resistance changes in response to changing inputs ?

**43.** Derive the expression for bridge sensitivity in the case of a voltage sensitive Wheatstone bridge having equal arms. Start deriving the expression by first assuming that the arms have unequal resistances and finally assume that the arms have equal resistances.

**44.** Derive the expression for current in the meter connected across the output terminals of a Wheatstone bridge when it is slightly unbalanced. Find the ratio of output voltage when the meter is not connected to the output voltage, when the meter is connected, when an equi-arm Wheatstone bridge is slightly unbalanced.

**45.** Find the expressions for bridge sensitivity factor Blumlein bridges having.

(*a*) inductors in push pull configuration
(*b*) capacitors in push pull configuration.

**46.** Describe the circuit of an $RC$ low pass filter which can be used as an integrator. Derive the transfer function. Draw the waveforms of the output voltage for different time constant when the input is a square wave.

**47.** Describe the circuit of an $RC$ high pass filter which can be used as a differentiator. Derive the transfer function. Draw the waveforms of the output voltage for different values of time constant when the input is a square waveform.

**48.** Describe one method of digital to analog (D/A) conversion.

**49.** Describe the following terms used in conjunction with digital to analog (D/A) conversion

(*i*) Discrete levels, (*ii*) Resolution,
(*iii*) Quantization error (*iv*) Decision levels,
(*v*) Aperture time.

**50.** Describe the following in conjunction with digital to analog conversion :

(*i*) Range of converter,
(*ii*) MSB,
(*iii*) LSB.

Derive the expression for an $n$ bit A/D converter.

**51.** Describe at least two methods of analog to digital (A/D) conversion.

**52.** Describe in details the successive approximation method of analog to digital (A/D) conversion.

**53.** Describe in details the dual slope method of analog to digital (A/D) conversion.

**54.** What are diode clipping circuits ? Describe how they are used as limiters.

**55.** Describe a clamping circuit and derive the relevant relationships.

# Unsolved Questions

**1.** An OPAMP with an open loop gain of $10^5$ impedance = ∞ and output impedance, $R_0 = 0$ and with $V_{max} = -V_{0\,min} = 10$ V. If a sinusoidal voltage with 1 V amplitude is applied directly across the input terminals of the OPAMP, determine the corresponding voltage, $v_0$. Also determine the length of time during which the OPAMP is operating, in the amplifier region if the period of sinusoid is 10 s. [**Ans.** Square wave voltage with peaks at ± 10 V, $t = 0.037$ s]

**2.** The supply voltage to an OPAMP is 10 V and its differential (open loop) gain is $10^5$. In case overloads have to be avoided, the output voltage should not exceed the supply voltage. Calculate the limiting value of differential voltage $v_d = v_+ - v_-$. [**Ans.** 10 μV]

**3.** Design a non-inverting amplifier circuit that is capable of providing a voltage gain of 10. Assume an ideal operational amplifier. Use standard values of resistors as shown in Table 26.2. [**Ans.** $R_f = 27$ kΩ, $R_1 = 3$ kΩ]

**4.** Design an inverting amplifier circuit that is capable of providing a voltage gain of 10. Assume an ideal OPAMP. Use standard values of resistance as given in Table 26.2. [**Ans.** $R_f = 27$ kΩ, $R_1 = 10$ kΩ]

**5.** The OPAMP circuitry of Fig. 26.131 is used to provide amplification of an input voltage of 0.1 V.

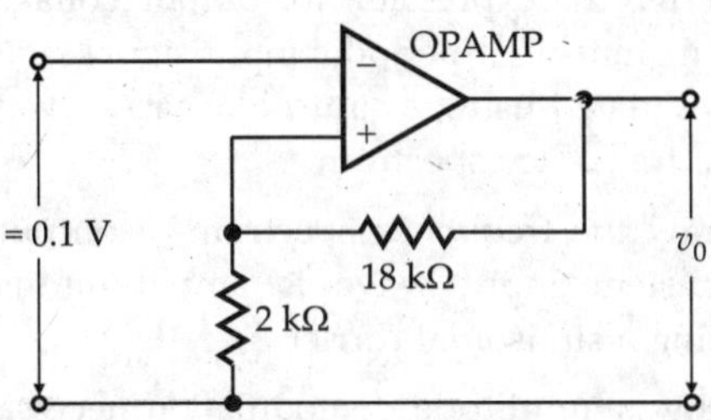


**Fig. 26.131** Diagram of Example 5.

(*a*) If $A_{vol} = 10^5$, find the actual value of output voltage

(*b*) Calculate the value of differential input to the OPAMP. [**Ans.** 1 V, 10 μV]

**6.** Determine the input impedance, output impedance and the voltage amplification for an inverting amplifier with $R_1 = 1$ MΩ, $R_f = 10$ MΩ. The parameters of the OPAMP are $r_{ia} = 100$ kΩ, $r_{oa} = 100\,\Omega$ and the open loop gain is $10^5$. [**Ans.** 10, 1 MΩ, 0.11 mΩ]

**7.** Determine the input impedance, the output impedance and the voltage amplification for a non- inverting amplifier with $R_1 = 100$ kΩ, $R_f = 100$MΩ.

The operational amplifier parameters are :

$$r_{ia} = 100\text{ k}\Omega,\ r_{oa} = 100\,\Omega$$

and the open loop gain is $10^5$. [**Ans.** 101, $10^5$ kΩ, 0.2 mΩ]

8. A physical process is described by equation

$$z = 2x - 3y$$

where $x$ and $y$ are process variables that may be represented. by voltages $v_1$ and $v_2$ which are well within the threshold value of OPAMP. Design an OPAMP circuit that makes quantity $z$ available as an output voltage $v_0$. Confine the design to the use of standard resistors as given in Table 26.2. Assume that $x$ and $y$ originate as positive quantities.

**Hint.** Figure 26.132 shows the solution of this problem.

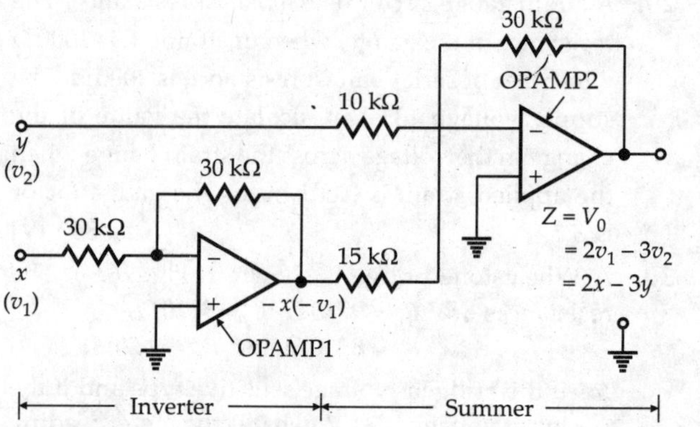


**Fig 26.132** Diagram of Example 8 (OPAMP solution of $z = 2x - 3y$ where 2 is the generated output voltage $v_0$.

9. For the OPAMP circuit shown in Fig. 26.133, calculate the output voltage. Assume the amplifier to be an ideal one. **[Ans.** 10 V]

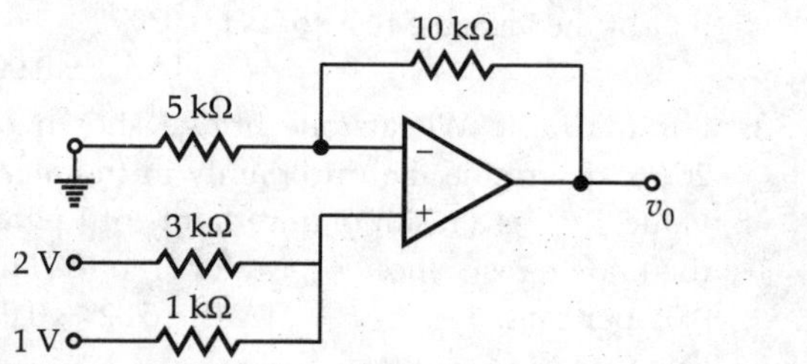


**Fig.26.133** Diagram of Problem 9.

10. For an OPAMP shown in Fig. 26.134, calculate the value of output voltage. **[Ans.** $\frac{37}{3}$ V.]

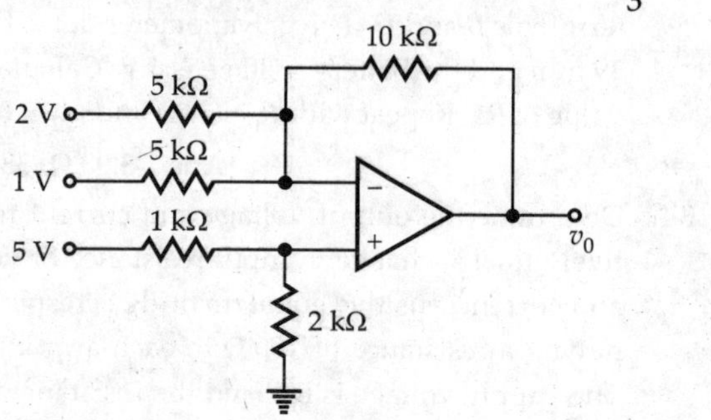


**Fig. 26.134** Diagram of problem10.

11. Design an OPAMP circuit that provides an output voltage $v_0 = -5v_1 + 6v_2$ for inputs $v_1$ and $v_2$. Use resistors in the 1 to 10 kΩ range and only one amplifier. **[Ans.** 1 kΩ and 5 kΩ]

12. A circuit of an ideal OPAMP is shown in Fig. 26.135. Determine the function of this circuit for the following cases :

(*a*) $Z_1 = 1\,\text{k}\Omega$ and $Z_f = 100\,\text{k}\Omega$

(*b*) $Z_1 = 1/j\omega C$ and $Z_f = R_f$

(*c*) $Z_1 = j\omega L$ and $Z_f = R_f$

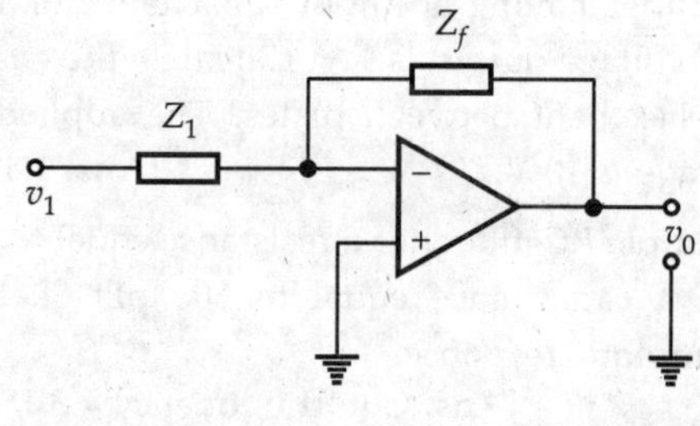


**Fig. 26.135** Diagram of Problem 12.

**[Ans.** (*a*) amplifier, (*b*) differentiator, (*c*) integrator]

13. A differential amplifier has two inputs + 100 mV and + 250 mV connected to its two terminals. The output voltage is +1.5 V. What is the gain ? **[Ans.** 10]

14. A differential amplifier with a difference mode gain of 20 has + 100 mV applied to terminal 1 and – 250 mV applied to terminal 2. What is the output signal ? **[Ans.** 7 V]

15. A differential amplifier has an input $e_1$ = +150 mV and $e_2$ = +150 mV. If this amplifier was an ideal one, what would its output ? Suppose the amplifier is not ideal but instead, has an output of 1.5 mV. What is the value of its common mode gain ? **[Ans.** zero, 0.01]

16. A differential amplifier has a difference mode gain of 1000 and common mode gain of 0.05. What is the value of the common mode rejection ratio ? **[Ans.** 20,000]

17. A balanced output source provides a signal of 50 mV from each terminal to ground. There is a hum pickup of 50 mV (common mode signal) on each terminal. A differential amplifier of differential gain of 200 and a CMRR of 25000 is used to amplify the signal.

(*a*) What is the ratio of signal to hum at the amplifier input ?

(*b*) What is the ratio of signal to hum at the output ? **[Ans.** (*a*) 1, (*b*) $8 \times 10^{-3}$]

26

**18.** Three signals are to be added together without changing their values. The feedback resistance is 1kΩ, what values should $R_1$, $R_2$, $R_3$ be ? [**Ans.** 1 kΩ each]

**19.** The input signal to an operational amplifier acting as a differentiator is : $e = 10\sin 20t + 0.1 \sin 377$ t. What is the signal to noise ratio before and after differentiation ? [**Ans.** 100, 5.3]

**20.** A capacitive transducer uses air as dielectric medium. It has plates of 250 mm$^2$ area and is connected in the feedback of an operational amplifier having an input capacitance of 150 pF. The voltage output is 1 V. Calculate the change in displacement between plates. The applied input voltage is 10 V. [**Ans.** 1.475 μm]

**21.** A simple *RC* filter has a resistance equal to 100 kΩ and a capacitance equal to 500 pF. Sketch the asymptotic response.
[**Ans.** Cutoff frequency = 3.18 kHz :
The slope at 31.8 kHz = – 20 db]

**22.** A low pass filter intended to have its 3 db point at 50 kHz is to be added at the output of an amplifier. The amplifier's internal resistance is 5000 Ω. What size capacitor should be connected in parallel with the amplifier's output terminals ?

**23.** Design a high-pass filter with a characteristic resistance of 50 Ω and a cutoff frequency of 20 MHz, and an attenuation of 80 db at 5 MHz.
[**Ans.** Inductance = 0.398 μH,
capacitance C = 159 pF,
*L* sections = 3 ; User *T* filter]

**24.** Using Fig. 26.80 the following values apply :

$R_1 = 500\,\Omega$, $R_2 = 100\,\text{k}\Omega$,

$C_1 = 680$ and $C_2 = 0.1\,\mu\text{F}$.

Find the approximate passband gain and cutoff frequencies. [**Ans.** 0.995, 15.9 Hz, 468 Hz]

**25.** Find the notch frequency of the twin *T* network of Fig. 26.82 for *R* = 50 kΩ and *C* = 2000 pF. [**Ans.** 1.59 kHz]

**26.** A notch filter is to be built to prevent an indesired signal at 2 MHz from entering an electronic system. If an inductance of 1 mH is used, what value of the capacitance will be required ? [**Ans.** 6.33 pF]

**27.** A metallic strain gauge made of aluminium has a nominal resistance of 120 Ω. It is connected in series with a ballast resistor of 120 Ω in a voltage sensitive circuit. The input voltage to the circuit is 30 V. Calculate the change in the output voltage across the gauge when the stress of 100 MN/m$^2$ is applied to it. The modulus of elasticity of aluminium is 70 GN/m$^2$. The gauge factor is 1.6. [**Ans.** 17.1 mV]

**28.** The output of a resistive transducer with a total resistance of 150 Ω is to be measured with a voltage sensitive circuit. The sensitivity is to be maximum at the midpoint of the transducer. Calculate the sensitivity at the 25 and 75 percent positions if the voltage of the input source is 100 V. [**Ans.** 0.592 V/Ω, 0.213 V/Ω]

**29.** A strain gauge circuit uses a ballast resistance. The resistance of the gauge, when unstrained, is 1000 Ω while that of series ballast resistance is 3000 Ω. The supply voltage 135 V. Calculate the value of the change in the voltage across the strain gauge when the applied strain is 0.001 m/m. The gauge factor is 3.5. [**Ans.** 88 mV]

**30.** A Wheatstone bridge is shown in Fig. 26.96. The resistances are $R_1 = 9725\,\Omega$, $R_2 = 8820\,\Omega$,
$R_3 = 8550\,\Omega$ and $R_4 = 9876\,\Omega$

(*a*) If the bridge is voltage sensitive type and if the input voltage is 24 V, what is the meter reading connected at the output terminals ?

(*b*) If $R_3$ is variable, what value should it have for null balance ? [**Ans.** (a) 1.45 V, (b) 10888 Ω]

**31.** Calculate the possible error in the measurement of a resistance of 1 Ω using a null type Wheatstone bridge if the accuracy of bridge ratio is ± 0.05% and that of the variable resistor is t 0.03%. [**Ans.** ± 0.0008 Ω]

**32.** For the basic Wheatstone bridge shown in Fig. 26.96, determine the uncertainty in the measured value of $R_1$ as a result of uncertainty of 1 percent in the known resistance. Repeat for an uncertainty of 0.05 per cent. [**Ans.** 1.732%, 0.0366%]

**33.** The bridge of Fig. 26.96 has

$R_1 = 6000\Omega$, $R_2 = 600\,\Omega$, $R_3 = 340\,\Omega$.

A galvanometer with a resistance of 70 Ω and a sensitivity of 0.04 μA/mn is connected between terminals *B* and *D*. The galvanometer deflection is 39 mm and the battery voltage is 4 V. Calculate the value of R4. Repeat with $R_1 = 60\,\Omega$ and $R_2 = 600\,\Omega$. [**Ans.** 33.93 Ω, 33.98 Ω]

**34.** Determine the output voltage and current respectively for 1% change in input resistance of voltage and current sensitive equal rm bridges respectively having a resistance of 100 Ω in each arm initially. The supply voltage is 6 V and the resistance of the galvanometer is 200 Ω. [**Ans.** 15 mV, 50 μA]

**35.** A current sensitive bridge is shown in Fig. 26.96. The initial value of the transducer resistance $R_1 = R$ and the other resistors have a value of $R_2 = 10\,R$, $R_3 = R$ and $R_4 = 10\,R$. The output of the bridge is connected to the galvanometer of an ultraviolet recorder having an input resistance of 0.00 Ω. Determine the sensitivity of the bridge in ampere per percentage change in resistance. The initial value of the transducer resistance is 500 Ω and the bridge supply voltage is 4 V.

[**Ans.** 5.96 μA% change of $R$]

**36.** A thermistor of resistance 1000 Ω at 50°C temperature is connected in a bridge circuit having all arms of resistance 1000 Ω. The supply voltage to the bridge is 40 V. The resistance of the thermistor increases by 5 Ω for a decrease of temperature of 1°C. Find the temperature if the open circuit voltage of the bridge is 50 mV. [**Ans.** 49°C]

**37.** An equal arm bridge has R = 10 kΩ. The supply voltage to bridge is 30 V. An amplifier with a gain of 26 db is connected across the output. Find the change in output voltage of the amplifier if the resistance change in one of the arms is 50 Ω.

[**Ans.** 0.45 V.

**Hint :** $e_0 = \dfrac{(30\times 50)/10000}{4\left(1+\dfrac{10{,}000}{15{,}000}\right)} = 0.0225$ V.

$\text{Gain} = \text{Antilog}\,\dfrac{26}{20} = 20 \therefore \text{output} = 20\times 0.0225 = 0.45$ V]

**38.** A good quality analog transducer having an input of 0−8 V is able to distinguish a change of 1 mV in its input signal. Calculate the number of bits the A/D converter should have in order to achieve a resolution not lower than that of the analog device.

[**Ans.** 13 bits]

**39.** An analog voltage signal whose highest significant frequency is 500 Hz is to be coded with a resolution of 0.1 percent approximately covering a range of 0−10 V.

Determine the :

(*i*) number of bits in the digital code,

(*ii*) analog value of LSB,

(*iii*) r.m.s. value of quantization error,

(*iv*) minimum sampling rate so that signal is lost,

(*v*) aperture time required by A/D converter and

(*vi*) dynamic range of converter in db.

[**Ans.** 10 bits, 9.77 mV, 2.8 mV, 2.5 kHz, 31.1 μs , 60 db]

**40.** A 8 bit D/A converter has a reference voltage of 12 V. It uses a weighted resistive network. Find the minimum value of resistance $R$ to be connected in the MSB input circuit such that the maximum output current does not exceed 10 mA. Also find the smallest quantized value of current.

[**Ans.** 2.4 kΩ, 39.1 μA]

26

# Objective Type Questions

***Tick (√) the most appropriate answer.***

**1.** Excitation and amplification systems are needed for :

(*a*) for active transducers only

(*b*) for passive transducers only

(*c*) for both active and passive transducers

(*d*) for both passive and output transducers.

**2.** A dc. amplifier

(*a*) needs to have a balanced differential inputs with a high common mode rejection ratio (*CMRR*) to give very good thermal and long term stability

(*b*) easy to calibrate at low frequencies and has ability to recover from overload conditions

(*c*) is prove to drift and low frequency spurious signals come out as data information

(*d*) is followed by a low pass filter to eliminate high frequency components including noise from the data signal

(*e*) all the above.

**3.** When using a.c. signal conditioning system for capacitive transducers, the carrier frequencies,

(*a*) range between 50 Hz and 20 kHz

(*b*) should be of the order of 0.5 MHz

(*c*) should be of the order of 20 MHz

(*d*) none of the above.

**4.** An a.c. signal conditioning system is normally used for,

(*a*) resistive transducers like strain gauges

(*b*) inductive and capacitive transducers

(*c*) piezoelectric transducers

(*d*) all of the above.

5. In a carrier system, drift and spurious signals are important,
   (a) because they modulate the carrier
   (b) because they do not modulate the carrier
   (c) because it is easier to achieve a stable carrier than a stabilized d.c. source .
   (d) none the above.
6. When using a d.c. signal conditioning system, with a carrier of 3 kHz, the data frequency should be limited to :
   (a) 1 kHz (b) 5 Hz
   (c) 600 Hz (d) 2 MHz
7. An operational amplifier has an open loop gain of 200,000 Ω. Its output exhibits saturation at 10 V. The thereshold differential voltage of the amplifier is :
   (a) 25 μV (b) 50 μV
   (c) 0.5 mV (d) 10 V
8. An ideal OPAMP has a gain of – 100. The input is connected to inverting end and the input resistance is 1 kΩ. The feedback resistance is,
   (a) 100 kΩ (b) 10 Ω
   (c) 100 Ω (d) 1000 kΩ
9. An ideal OPAMP has a gain of – 100. The input is connected to the inverting end. The value of input resistance is 1 kΩ. The value of feedback resistance is
   (a) 101 kΩ (b) 99 kΩ
   (c) 9.9 kΩ (d) 1.01 kΩ
10. The properties of an ideal OPAMP are,
   (a) it should have zero input impedance
   (b) it should have an infinite output impedance
   (c) it should have a zero open loop gain
   (d) none of the above.
11. The closed loop gain of an OPAMP is dependent upon whether the OPAMP is used,
   (a) in inverting mode
   (b) in non-inverting mode
   (c) is independent of the fact whether the input is connected to inverting or the non-inverting terminal
   (d) is dependent upon the fact whether the input is connected to inverting or non-inverting terminal.
12. The differential (open loop) gain of an OPAMP is 20,000. The input resistance is 2 kΩ and the feedback resistor has a value of 200 kΩ. The closed loop gain of the amplifier is :
   (a) – 9.9995 (b) – 9.9955
   (c) – 10 (d) +10
13. A buffer amplifier has gain of
   (a) infinity (b) zero
   (c) unity
   (d) dependant upon the circuit parameters.
14. An operational amplifier has a bandwidth of 100 kHz. Negative feedback is introduced with a loop gain of 50. The bandwidth with feedback is
   (a) 2 kHz (b) 49 kHz
   (c) 5.1 MHz (d) 1.96 kHz.
15. The order of input resistance of 741 OPAMP is,
   (a) $10\,\Omega$ to $10^3\,\Omega$ (b) $10^3\,\Omega$ to $10^6\,\Omega$
   (c) $10^3\,\Omega$ to $10^9\,\Omega$
   (d) $10\times10^3\,\Omega$ to $10^9\,\Omega$
16. The order of output resistance of 741 OPAMP is,
   (a) $0.1\,\Omega$ to $10\,\Omega$ (b) $10\,\Omega$ to $10^5$ kΩ
   (c) $10\times10^3\,\Omega$ to $10^9\,\Omega$ (d) $10^3\,\Omega$ to $10^6\,\Omega$.
17. The gain of a 741 OPAMP falls off at a low frequency of,
   (a) 10 kHz (b) 10 Hz
   (c) 100 Hz (d) 1000 Hz
18. An OPAMP having an open loop gain of $A_{vol}$ is designed with a feed factor β where $\beta = \dfrac{R_1}{R_1 + R_f}$ is the input resistance connected to non-inverting end and R is the resistance in the feedback path, the closed loop gain is :
   (a) $-\dfrac{R_f}{R_1}\left[\dfrac{1}{1+(1/A_{vol}\beta)}\right]$
   (b) $\dfrac{R_f}{R_1}\left[\dfrac{1}{1+(1+A_{vol}\beta)}\right]$
   (c) $\dfrac{R_f}{R_1}\left[\dfrac{1}{1+(1-A_{vol}\beta)}\right]$
   (d) $-\dfrac{R_f}{R_1}\left[\dfrac{1}{1+(1+A_{vol}\beta)}\right]$.
19. An OPAMP with an open loop gain is designed with a loop gain ($A_{vol}$ β) where
   $$\beta = \frac{R_1}{R_1 + R_f} \text{ with}$$
   $R_1$ = resistance connected to non-inverting end, and
   $R_f$ = resistance connected in the feedback path, the input resistance of the amplifier ($R_{in\,f}$) with feedback is :
   (a) $R_{in\,f} = R_{in}(1 + A_{vol}\beta)$

(*b*) $R_{in\,f} = R_{in}(1 - A_{vol}\beta)$
(*c*) $R_{in\,f} = R_{in}(1+\beta)$
(*d*) $R_{in\,f} = R_{in}(1-\beta)$.

where $R_{in}$ = input resistance of amplifier without feedback.

**20.** An OPAMP with an open loop gain is designed with a loop gain $A_{vol}\,\beta$ where

$$\beta = \frac{R_1}{R_1 + R_f} \text{ with}$$

$R_1$ = resistance connected to non-inverting end, and

$R_f$ = resistance connected in the feedback path,

the output resistance of the amplifier, $R_{of}$, with feedback is

(*a*) $R_{of} = \dfrac{R_0}{1 - A_{vol}\beta}$ (*b*) $R_{of} = \dfrac{R_0}{1-\beta}$

(*c*) $R_{of} = \dfrac{R_0}{1+\beta}$ (*d*) $R_{of} = \dfrac{R_0}{1 + A_{vol}\beta}$.

where $R_o$ = output resistance of amplifier without feedback.

**21.** Offset voltages in OPAMPs are produced because of :
(*a*) variations in the input voltage applied to the amplifier
(*b*) mismatch between the two differential amplifiers which form the IC of the OPAMP chip
(*c*) mismatch between the input signals applied to the OPAMP.

**22.** The typical value of input offset voltage of a 741 OPAMP is :
(*a*) 2 μV (*b*) 12 mV
(*c*) 2 V (*d*) 0.2 V

**23.** The typical value of input offset current of a 741 OPAMP is :
(*a*) 10 nA (*b*) 10 mA
(*c*) 10 pA (*d*) 100 mA

**24.** A 741 OPAMP has,
(*a*) 6 pins (*b*) 7 pins
(*c*) 8 pins (*d*) 9 pins

**25.** A 741 OPAMP has an open loop gain 200,000. The input offset voltage is 2 mV. If the input terminals are shorted, the output voltage is :
(*a*) 0 V (*b*) ∞
(*c*) 400 V (*d*) ± 400 V.

**26.** In order to eliminate the output voltage on account of input offset voltage,
(*a*) A nulling potentiometer is connected between pins 1 and 5 and potentiometer is adjusted to give zero output voltage. No voltage input is applied
(*b*) A nulling potentiometer is connected between pins 1 and 5 and the sliding contact of the POT is connected to pin 4 (V—of supply). The POT is adjusted to give zero, No input voltage is applied
(*c*) A blocking capacitor is used at the input of non inverting end
(*d*) An external offsetting potentiometer is used whose sliding contact is connected to the output terminal. This potentiometer is adjusted to give zero output voltage with no input voltage applied

**27.** The output of the following circuit is :

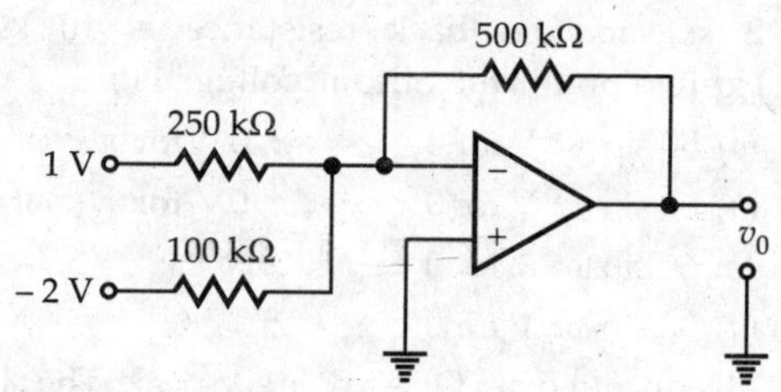


**Fig. 26.136** Circuit of Problem 27.

(*a*) – 8 V (*b*) + 8 V
(*c*) + 12 V (*d*) – 12 V

**28.** The output of the following circuit is :

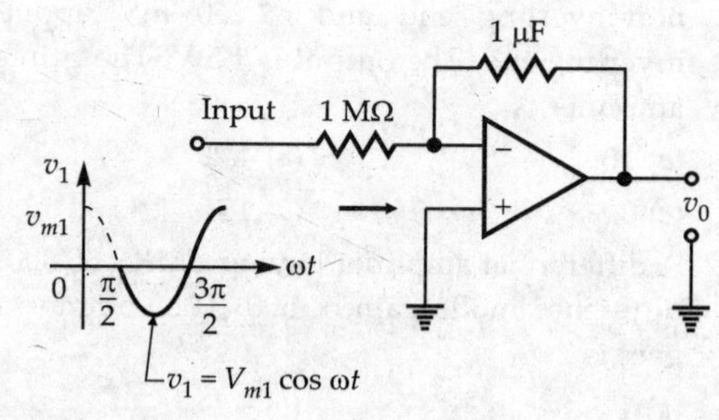


**Fig. 26.137** Circuit of Problem 28.

(*a*) $V_{m_1}\cos \omega t$

(*b*) $-\dfrac{V_{m_1}}{\omega}\cos \omega t$

(*c*) $-\dfrac{V_{m_1}}{\omega}\sin \omega t$

(*d*) $\dfrac{V_{m_1}}{\omega}\sin \omega t$

**29.** The output of the following circuit is :

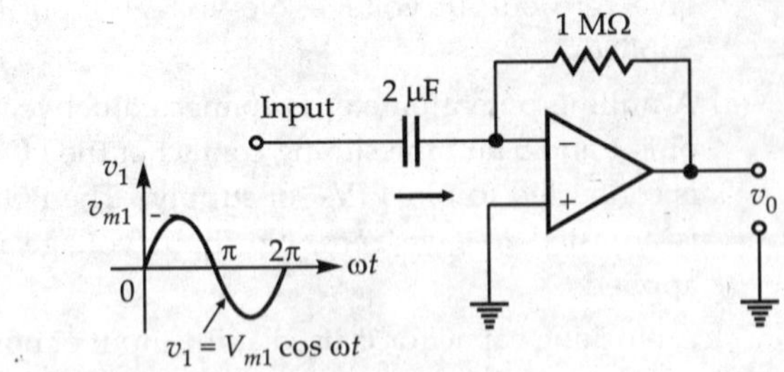


**Fig. 26.138** Circuit of Problem 29.

(a) $-2\omega V m_1 \cos \omega t$ (b) $\frac{\omega}{2} V m_1 \cos \omega t$

(c) $-2\omega V m_1 \sin \omega t$ (d) $+2\omega V m_1 \cos \omega t$

**30.** An active half wave voltage follower circuit has an input of 10 sin $\omega t$ V. The input resistance has value 2 kΩ and feedback resistance is 10 kΩ. The expression for the output voltage is

(a) 50 sin $\omega t$ V for $0 < \omega t < \pi$, 0V for $\pi < \omega t < 2\pi$

(b) 60 sin $\omega t$ V for $0 < \omega t < \pi$, 0V for $\pi < \omega t < 2\pi$

(c) 60 sin $\omega t$ V for $0 < \omega t < 2\pi$

(d) 50 sin $\omega t$ V for $0 < \omega t < 2\pi$.

**31.** The output of an OPAMP is 2 V peak. The slew rate is 5 V/μ. The input sinusoidal which can be reproduced with no distortion has the maximum frequency of :

(a) 398 Hz (b) 1592 kHz

(c) $125 \times 10^3$ Hz (d) 398 kHz

**32.** A differential amplifier has + 100 mV applied to non-inverting end and + 250 mV applied to inverting end. The output is 1.5 V. The gain of the amplifier is :

(a) 10 (b) 4.29

(c) 0.6 (d) 15

**33.** A differential amplifier having *CMRR* 50,000 has a difference mode gain of 500. The common mode gain is :

(a) 1 (b) 0.1

(c) 0.01 (d) 0.001

**34.** A differential amplifier has a comm mode gain of 0.02. It has + 200 mV signals applied to each of its inputs. The amplitude of output signal is :

(a) 0 V (b) 8 mV

(c) 4 mV (d) none of the above .

**35.** The output from a transducer has a frequency of 50 Hz. It modulates (amplitude modulation) a carrier having a frequency of 1 kHz. The output waveshape contains frequencies which are :

(a) 1000 Hz, 1050 Hz (b) 1000 Hz, 950 Hz,

(c) 950 Hz and 1050 Hz (d) 2000 Hz ± 100 Hz

**36.** The output from a transducer is a non sinusoidal wave having frequencies $\omega_1, \omega_2, \omega_3, \omega_4$. The carrier frequency is $\omega_c$. The frequency spectrum of output wave contains frequencies

(a) $\omega_c \pm \omega_1$

(b) $\omega_c \pm \omega_2$

(c) $\omega_c \pm \omega_1, \omega_c \pm \omega_2, \omega_c \pm \omega_3$

(d) $\omega_c \pm \omega_1, \omega_c \pm \omega_2, \omega_c \pm \omega_{c3}, \omega_c \pm \omega_4$

**37.** A low pass filter has a time constant of τ. Its gain at a frequency ω is,

(a) $\sqrt{1+(\omega\tau)^2}$ (b) $\frac{\omega\tau}{\sqrt{1+\omega^2\tau^2}}$

(c) $\frac{1}{\sqrt{1+(\omega\tau)^2}}$ (d) none of the above.

**38.** A high pass filter has a time constant of τ. Its gain at a frequency ω is,

(a) $\sqrt{1+(\omega\tau)^2}$ (b) $\frac{\omega\tau}{\sqrt{1+\omega^2\tau^2}}$

(c) $\frac{1}{\sqrt{1+(\omega\tau)^2}}$ (d) none of the above.

**39.** The following figure shows some Bode plots. Recognize the Bode plot for a Bandpass filter.

(a)

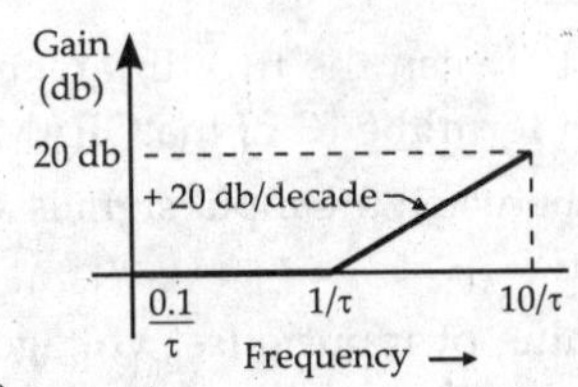


(b)

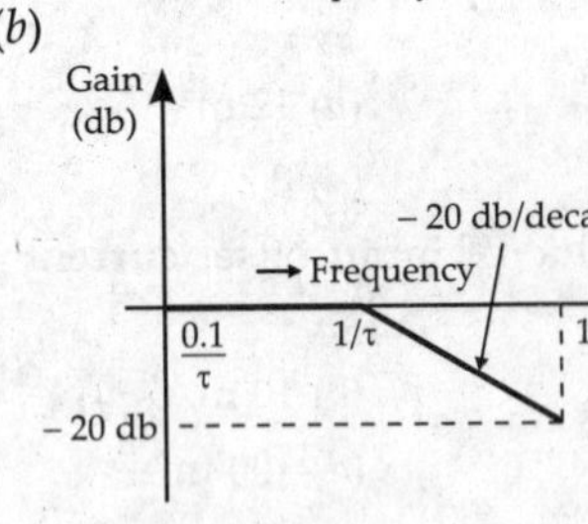


(c)

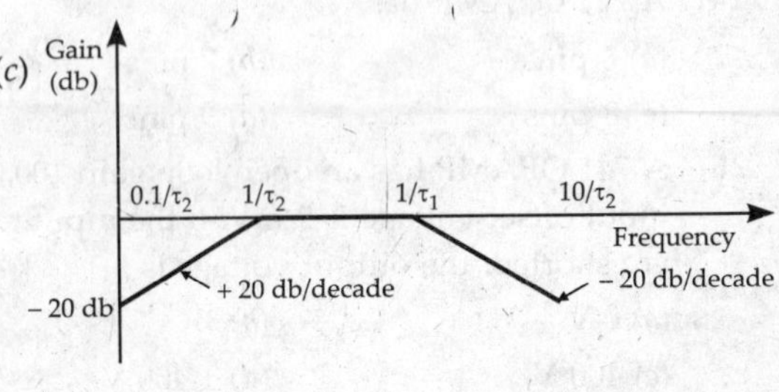


(d) none of the above.

**40.** The input circuit of variable, resistance transducer consists of a ballast resistance, in series with the transducer resistance $R_t$. The resistance of the transducer for a particular input is $kR_t$, where $k$ is a constant. If the input circuit is supplied by an input voltage $e_i$, the sensitivity of the transducer is

(a) $\dfrac{e_i\, R_b}{(R_b + kR_t)}$ (b) $\dfrac{e_i\, R_b}{(R_b + kR_t)^2}$

(c) $\dfrac{e_i\, R_b}{(R_b + kR_t)^3}$ (d) $\dfrac{e_i\, R_b\, kT_t}{(R_b + kR_t)^3}$

**41.** In a voltage sensitive wheatstone, having each arm having a resistance, R, the resistance of one of the arms is changed to $R + \Delta R$ where $\Delta R << R$. The wheatstone bridge is supplied with an input voltage of $e_i$. The output voltage on account of unbalance is,

(a) $\left(\dfrac{\Delta R / R}{2 + \Delta R / R}\right) e_i$ (b) $\left(\dfrac{\Delta R / R}{4 + \Delta R / R}\right) e_i$

(c) $\left(\dfrac{2\Delta R / R}{4 + \Delta R / R}\right) e_i$ (d) $\left(\dfrac{\Delta R / R}{4 + 2\Delta R / R}\right) e_i$

**42.** The bridge sensitivity factor for a Blumlein bridge using two inductive transducers in push pull arrangement is,

(a) $\dfrac{2\, L_c / L}{(1 + L_c / L)}$ (b) $\dfrac{4\, L_c / L}{(1 + 2\, L_c / L)}$

(c) $\dfrac{4\, L_c / L}{(1 + L_c / L)}$ (d) $\dfrac{2\, L_c / L}{(1 + 2\, L_c / L)}$

where $L$ = nominal value of inductance of inductive transducer and $L_c$ = self inductance of one of the tightly coupled ratio arms.

**43.** In a Blumlein bridge using two capacitive transducers in push pull configuration, the bridge sensitivity factor above the resonance frequency for a tightly coupled bridge is,

(a) 2 and is independent of the variations of the bridge frequency and ratio arm inductance

(b) 2 and is dependent upon the variations of the bridge frequency and ratio arm inductance

(c) practically constant and is lower than that of bridge with uncoupled arms

(d) variable and is dependent upon the frequency even though the ratio arms are tightly coupled.

**44.** A low pass RC filter acts as a pure integrator when

(a) $\omega\tau = 1$ (b) $\omega\tau << 1$

(c) $\omega\tau >> 1$ (d) $\omega\tau = 0$

where $\omega$ = applied frequency and $\tau$ time constant of RC circuit.

**45.** A high pass RC filter acts as a pure differentiator when :

(a) $\omega\tau = 1$ (b) $\omega\tau << 1$

(c) $\omega\tau >> 1$ (d) $\omega\tau = 0$

where $\omega$ = applied frequency and $\tau$ = time constant of RC circuit.

**46.** An 8 bit converter is used for a d.c. range of 0 – 10V. Find the weight of LSB

(a) 39 mV (b) 78 mV

(c) 39.2 mV (d) none of the above.

**47.** A 5 bit D/A converter has a current output. The digital input is 10100 and the output current corresponding to this is 10 mA. What will be output current for digital input of 11101 ?

(a) 10.5 mA (b) 14.5 mA

(c) 29 mA (d) none of the above.

**48.** A successive approximation A/D converter has a resolution 20 mV. What will be its digital output for an analog input of 2.17 V ?

(a) 01101100 (b) 01101 101

(c) 01101011 (d) none of the above.

**49.** A analog transducer has range 0 – 10 V. Calculate its of an converter if-the resolution is 5 mA.

(a) 10 (b) 9

(c) 11 (d) none of the above.

**50.** The 0 – 10V A / D converter has to have a, resolution of 0.025 percent. Find the r.m.s. value of quantization error.

(a) 176 μV (b) 705 μV

(c) 352 μV (d) 1410 μV.

26

## Answers

| | | | | | | | | | |
|---|---|---|---|---|---|---|---|---|---|
| 1. (d) | 2. (e) | 3. (b) | 4. (b) | 5. (a) | 6. (c) | 7. (b) | 8. (a) | 9. (b) | 10. (d) |
| 11. (d) | 12. (a) | 13. (c) | 14. (c) | 15. (d) | 16. (b) | 17. (b) | 18. (a) | 19. (a) | 20. (d) |
| 21. (b) | 22. (b) | 23. (a) | 24. (c) | 25. (d) | 26. (b) | 27. (b) | 28. (c) | 29. (a) | 30. (a) |
| 31. (d) | 32. (a) | 33. (c) | 34. (c) | 35. (c) | 36. (d) | 37. (c) | 38. (b) | 39. (c) | 40. (b) |
| 41. (d) | 42. (b) | 43. (a) | 44. (c) | 45. (b) | 46. (a) | 47. (b) | 48. (a) | 49. (c) | 50. (b) |

# 27

# Data Transmission and Telemetry

## 27.1 INTRODUCTION

In modern measurement systems, the various components comprising the system are usually located at a distance from each other. It, therefore, becomes necessary to transmit the data or an information between them through some form of communication channels.

The terms *data transmission* and *telemetry* refer to the process by which information regarding the quantity being measured, may be using a transducer and signal conditioning equipment, is transferred to a remote location, perhaps to be processed, recorded and displayed. **Telemetry** is the technology which enables a user to collect data from several measurement points at inaccessible or inconvenient locations, transmit that data to a convenient location, and present the several individual measurements in a usable form.

## 27.2 METHODS OF DATA TRANSMISSION

The transmission of a measured variable to a remote point is an important function in Instrumentation systems because of the size and complexity of modern industrial plants. The most common variables encountered in industrial plants are temperature, pressure and flow. Most measuring devices for these variables such as mercury thermometers, pressure gauges or flow rate meters would require fluid-line connections of great length from the place of measurement to the place of data recording or display. This will result into excessive *measurement lags*. Hence, there is a need for fast transmission of data.

The methods employed for data transmission depend upon the variable and the distance over which it has to be transmitted. The following methods may be used for data transmission :

(*i*) Hydraulic transmission,

(*ii*) Pneumatic transmission, and

(*iii*) Electrical and Electronic transmission.

The electrical and electronic methods of data transmission are extensively used in Instrumentation and measurement systems. In this text, data transmission by electrical and electronic means is only discussed.

## 27.3 GENERAL TELEMETRY SYSTEM

Telemetry may be defined as measurement at a distance. A general telemetering system is shown in Fig. 27.1. The primary detector and the end device of the telemetering system have the same position and functional roles as in a generalized measurement system. However, there are three system elements in the intermediate stage which are peculiar to a telemetering system, they are :

(*i*) telemeter transmitter, (*ii*) telemeter channel, (*iii*) telemeter receiver.

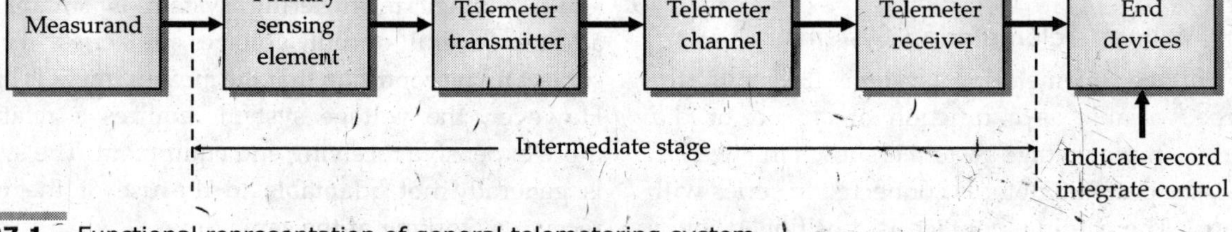


**Fig. 27.1** Functional representation of general telemetering system.

The function of the *telemeter transmitter* is to convert the output of a primary sensing element into an electrical signal and to transmit it over a telemetering channel. This signal is in electrical format and is received by a receiver placed at a remote location. This signal is converted into a usable form by the *receiver* and is indicated or recorded by an end device, which is graduated in terms of the measurand. The end device may be a control element which may be used for the control of the input quantity (measurand), through a feedback loop to produce desired output.

## 27.4 TYPES OF TELEMETRY SYSTEMS

*Two* types of telemetering systems are used :

- Land line telemetry,
- R.F. (radio frequency) telemetry.

## 27.5 LAND LINE TELEMETERING SYSTEM

A land line telemetring system requires a telemeter channel which is a physical link between the telemeter transmitter and receiver. This physical link may be a cable, a specially laid out wire, existing telephone and telegraph cables or a power line carrier. The land line telemetering is, in fact, a direct transmission of information through cables and transmission lines. The direct transmission via cables employs current, voltage, frequency, position or impulses to convey the information. Current, voltage and position type systems can be used only for short distances while for long distance telemetry, pulse and frequency types of systems are used. The information may be in the form of analog or digital signals. While current, voltage, position, frequency and pulse types of signals can be used for analog telemetry, only pulse signals can be used for digital telemetry.

The land line telemetry systems can be classified as :

(*i*) voltage telemetering systems,

(*ii*) current telemetering systems, and

(*iii*) position telemetering systems.

The frequency and pulse telemetering systems are described later along with R.F. (radio frequency) telemetering systems.

### 27.5.1 Voltage Telemetering Systems

A voltage telemetering system transmits the measured variable as a function of an a.c. or d.c. voltage. A simple voltage system is shown in Fig. 27.2. A slide wire potentiometer is connected in series with a battery. The sliding contact is positioned by a pressure sensitive bourdon tube. The telemetering channel consists of a pair of wires connected to a voltage measuring device such as a null balance d.c. potentiometer indicator or recorder. As the measured pressure changes, the bourdon tube actuates the sliding contact thereby changes the voltage. The d.c. null balance potentiometer measures the voltage and positions a pointer on a scale calibrated in terms of the pressure being measured.

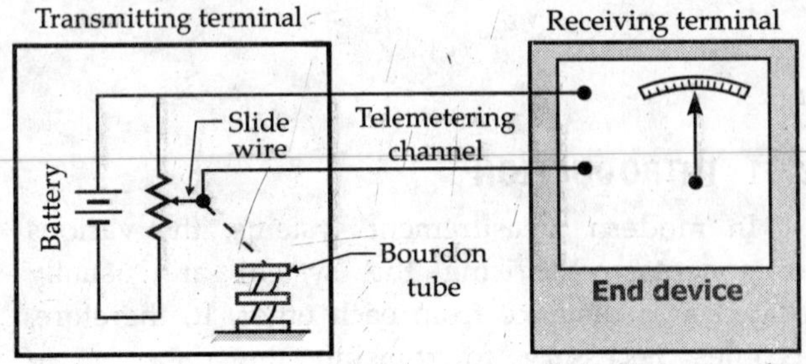


**Fig. 27.2** Voltage telemetering system.

The use of a null balance d.c. potentiometer reduces the current carried by the telemetering (or communication) channel to minimum, where the resistance is negligible.

Most of the systems use primary elements which produce a voltage signal. These elements include thermocouples tachometers and differential transformers. The application of voltage systems in industrial plants is limited to distances up to about 300 metres. Sell-balancing potentiometers are the usual receivers for such systems. The deflection type of indicators may also be used if they are calibrated for the line resistance involved.

Voltage telemetering system requires high quality circuits than a current system. The signal to noise (S/N) ratio must be comparatively high. Since, the power level is small in voltage telemetering systems, the transmission channel must be protected from sources of interference which are of the same order as that of the signal.

A voltage telemetering system is suitable for adding several output voltages in series. This is subject to the condition that the measurement is linear. However, the voltage system requires a relatively more expensive receiving end equipment. The system is generally not adaptable to the use of the many receivers working at the same time.

### 27.5.2 Current Telemetering System

The earlier version of a current telemetering system is shown in Fig. 27.3. It has a slide wire potentiometer connected in series with a battery. The sliding contact of the potentiometer is positioned by a pressure sensitive bourdon tube. The telemetering channel consists of a pair of wires connected to a current measuring device.

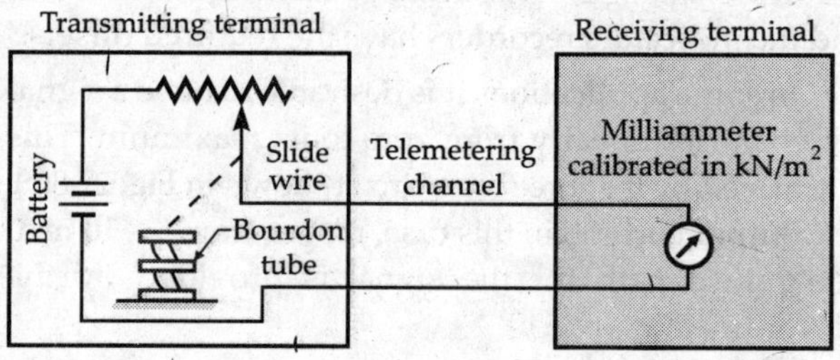


**Fig. 27.3** Basic current telemetering system.

As the pressure, which is the measurand, changes the bourdon tube moves and changes the position of the sliding contact on the slide wire, thereby changing the current in the circuit. This current is measured with the help of a milliammeter whose scale is graduated terms of pressure.

The commonly used current telemetering systems are motion and force balance types which are improved forms of the basic current telemetering system described above.

In a **Motion Balance System** the slide wire is replaced by a position detector like an LVDT as shown in Fig. 27.4. A capacitive transducer may also be used.

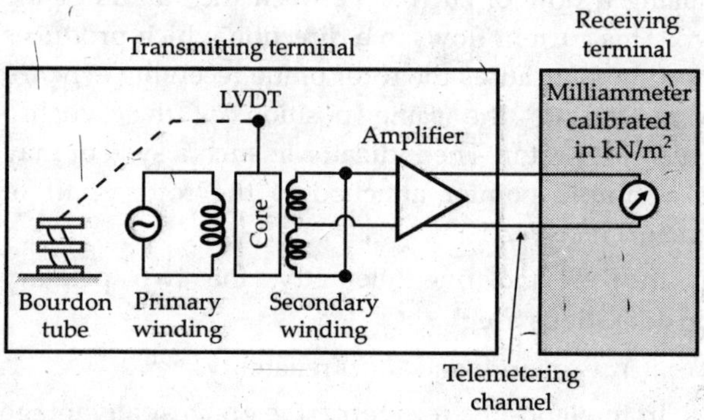


**Fig. 27.4** Motion balance current telemetering system.

The pressure acting on the bourdon tube causes a displacement which moves the core of the LVDT, thereby producing a voltage output which is amplified and rectified. This voltage produces a d.c. current of the order of 20 mA in the telemetering channel and is measured by a d.c. milli-ammeter. The scale of the d.c. milli-ammeter is directly calibrated in terms of pressure being measured.

A **Force Balance System** is shown in Fig. 27.5. In this system, a part of the current output is feedback to oppose the motion of the input variable. The system is operated by the bourdon tube which rotates the feedback force coil which in turn changes the flux linkages between the primary and the secondary coils. The change in flux linkages varies the amplitude of the amplifier. The output signal is connected to the feedback force coil which produces a force opposing the bourdon tube input.

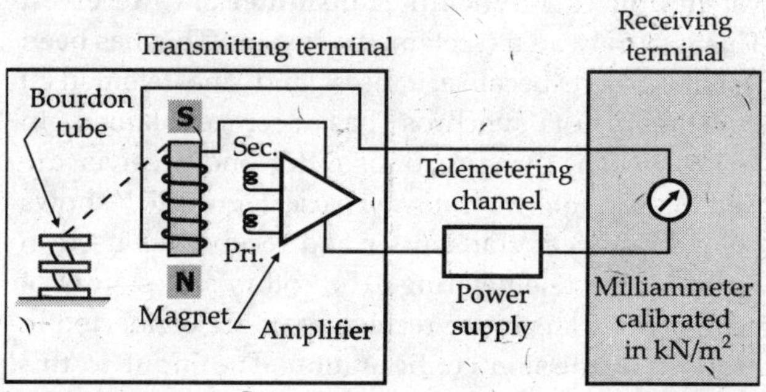


**Fig. 27.5** Force balance current telemetering system.

A force balance system **in**creases the accuracy as smaller motions are required which result in better linearity.

### 27.5.3 Position Telemetering System

A position telemetering system transmits and reproduces the measured variable by positioning variable resistors or other electrical components in a bridge circuit form so as to produce proportional changes at both the transmitter and the receiver ends. This is known as **Bridge type System.**

Figure 27.6 shows two potentiometers, one at transmitting end and the other at the receiving end.

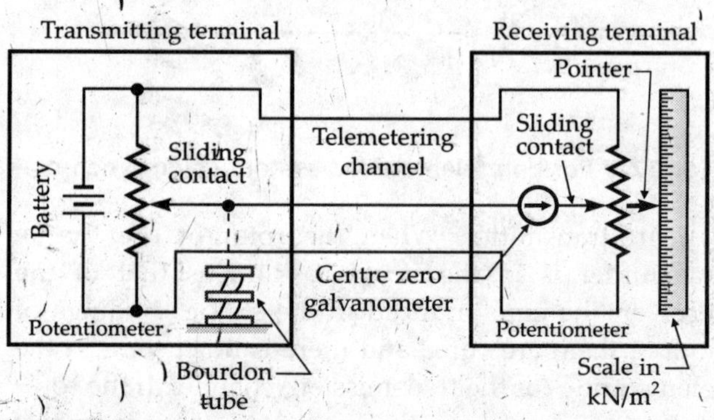


**Fig. 27.6** Position telemetering system using a bridge type configuration.

27

The two potentiometers are energized by a common power supply. The sliding contact at the transmitting end is positioned by the bourdon tube as pressure is applied to the latter. If the sliding contact at the receiving end is positioned until the centre zero galvanometer indicates Zero, the position of the contact will assume the same position as the contact at the transmitter. The receiving contact moves the pointer which indicates on the scale the pressure which is being measured (the scale is directly calibrated in terms of pressure). The principle involved is the same as that of a Wheatstone bridge.

Another most commonly used position telemetry system utilizes a **Synchro Transmitter** and **Receiver**. (This is purely an a.c. telemetry system. This has been described here because it uses land line telemetry.) The theory of synchros has been explained in Art. 25.26. The synchro transmitter and receiver are used in the *torque transmission mode*. Figure 27.7 shows a pair of synchro transmitter and receiver as used in position type telemetering. The rotors and stators of the synchro transmitter-receiver pair are connected in torque transmission configuration. The input to this type of system is angular position of the

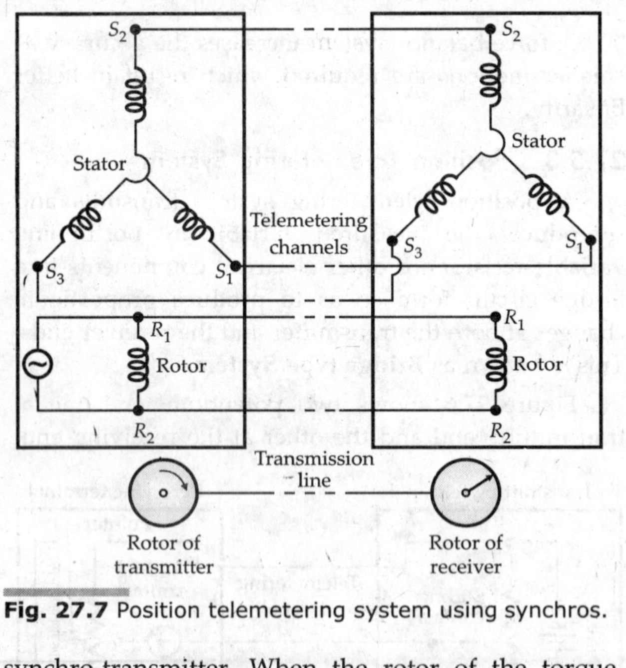


**Fig. 27.7** Position telemetering system using synchros.

synchro-transmitter. When the rotor of the torque transmitter is in the same position as that of the receiver, the emfs induced in the stator windings of both of them are equal and there is no current in the telemetering (or the transmission) channel. If the rotor of the transmitting synchro is rotated, there is an emf unbalance in the stator windings of synchro transmitter and the receiver, resulting in a current flow in the telemetering channel. This current flowing in the stator of receiver produces a torque on the receiver rotor till it occupies the same position as the rotor of the synchro transmitter.

The indicator in such a system may be a simple pointer attached to rotor shaft of the receiver synchro for the end devices. The telemetering channel includes a connecting line and its associated terminal equipment.

The *standing current* must be maintained constant and the indicators recorders have the required offsets.

In some applications it is desirable to have a signal which varies linearly from zero to its maximum. This is achieved by the three wire circuit shown in Fig. 27.8(*b*). The output current, in this case, ranges from 0 – 10 mA. The return path for the signal is provided by the power lines.

The *voltage type of land line telemetry system* employs a voltage balance method. The detector generates a voltage which is a function of the measurand. This voltage is transmitted to the receiving end of the line and the series current is reduced to zero by an opposing voltage, which thus becomes a measure of the quantity under measurement.

The most commonly used land line *position telemetry system* is, the one that employs a synchro-transmitter and receiver pair as shown in Fig 27.7. Here the synchro-transmitter and receiver pair are used in torque transmission mode as explained in Art. 25.26.2.

Briefly, we can state that if the rotor of the transmitting synchro is turned, so that it is no longer in the same position as the rotor of the receiving synchro, the terminal emfs of the two machines differ causing a flow of current between the stators of the two. This current flows in a direction which produces a torque that causes the rotor of the receiving synchro to move into the same position as the synchro transmitter rotor. The indicator in such a system may be a simple pointer attached to the rotor shaft of synchro receiver.

In the land line telemetry, the two primary considerations are :

(*i*) economy and (*ii*) reliability.

In the land-line telemetry, the greatest advantage is the simplicity of the set-up as space required and weight are generally of no consideration. However, the disadvantages are :

(*i*) all the transmission link distortions are directly introduced into the system. This is particularly true of induction effects of mains frequency from nearby cables.

(*ii*) the effects of thermo-electric emfs may be substantial in case d.c. telemetry systems are used.

(*iii*) the frequency response is limited.

(*iv*) signal multiplexing tends to be impracticable.

### 27.5.4 Land Line Telemetry Feed-back System

The current and voltage types of land line telemetry systems use feed back systems. The current type land line telemetry system employs a *torque-balance* method which is shown in Fig. 27.8. In this method, the d.c. in wires connecting a measurement transmitter to receiving devices is used to apply a torque to balance a torque generated by the measurand.

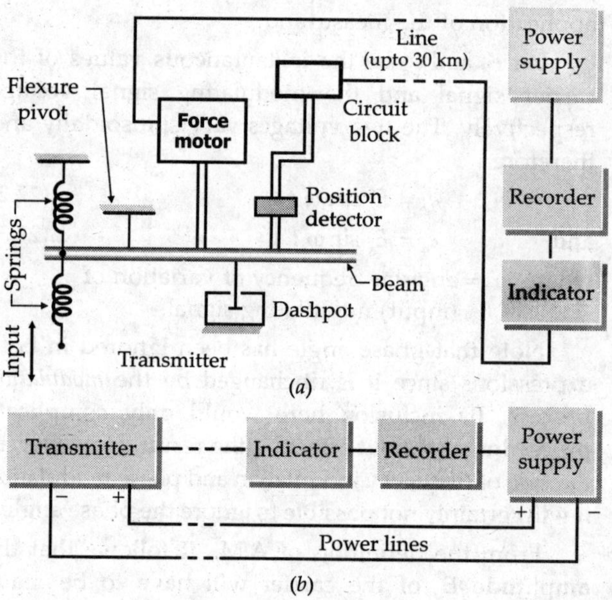


**Fig. 27.8** Torque balance Telemetry
(*a*) Two wire transmitter,
(*b*) Three wire transmitter.

Let the receiving device be an ammeter, which is used to apply to a torque to a balance a torque produced by the measurand. The current in the connected circuit or line is a known function of the measurand. A torque transmitter is shown in Fig. 27.8(*a*). An input movement on a spring, due to a signal from measurand, produces a torque on a beam with a flexure pivot. Any movement of the beam is detected by a position detector and a current is produced in the force motor (which acts as an inverse transducer) to produce the balancing torque. Thus there is only a small movement of the beam and the output d.c. on account of this movement is linearly related to input movement. This movement may be caused by the closed end of a C type Bourdon tube on account of the pressure (measurand) applied.

The d.c. signal can be transmitted via a two or a three wire system over distances that are determined by the d.c. resistance of the line. For two wire transmission, the maximum allowable resistance may be 500 Ω while it is 2000 Ω in the case of three wire transmission. If the resistance values are kept within these limits, the output is not affected by the resistance of lines or by the number of indicators and recorders used.

The two wire transmitter shown in Fig. 27.8 (*a*), is an instrument with a *live zero* current output. It is connected in series with its indicators and recorders which are located at the receiving end. The current carried by wires can be considered to have two components namely :

(*i*) a constant *standing current* of approximately 4 mA to drive the transmitter circuits. The power associated with this current is about 100 mW, and

(*ii*) a variable current (a component of signal current) which is 0 – 16 mA that represents the value of the measurand.

In some applications it is desirable to have a signal which varies linearly from zero to a maximum. This is achieved by the 3 wire circuit of Fig. 27.8(*b*). The output current of this transmitter ranges from 0 to 10 mA. The return path for the signal is in the power supply lines.

## 27.6 RADIO FREQUENCY (R.F.) TELEMETRY

The telemetry, earlier on, has been defined, as a technology that enables the user to collect data from several measurement points at inaccessible or inconvenient locations. This is very true of applications which require Radio Frequency (R.F.) telemetry, as in such applications, there is no physical link between the transmitting and receiving stations. The link between the transmission station (where the actual measurements are being carried out) and the receiving station (where the measurand is measured, recorded and information used for control purposes) can only be established through radio links. For example, an aircraft on a test flight has many parameters which change during the flight *i.e.*, fuel flow, amount of fuel, engine performance, vibrations of critical parts, performance of avionic systems, temperature of various components. The test pilot can monitor some of these, of course, by observing the readings of the meters on the panel, but time permits reading those instruments which are critical to the safety of the flight and even those are not sufficiently accurate for detailed flight analysis. An alternative would be taken along a team of specialist engineers involved in the design, production of the air-craft and analysis of the data produced and to suggest the corrective actions. This solution is just not practicable. Another, alternative would be to put a sophisticated computer on the flight for analysis of data. This again is not a satisfactory solution. The best way is to locate the team of engineers on a land based station and the test data be transmitted through radio links from the air-craft to the land station. The team of engineers and the computer stationed on land are in a much better environments to analyse the data and suggest the corrective measures.

The rocket or unmanned space craft presents more obvious need for a radio link based telemetry. The vehicle in this case is too small to carry even one person, much less the entire team of engineers and also a computer. Here, R.F. telemetry monitors all information which enables the team of engineers to evaluate performance of the test vehicle with the help of computer, while the flight is in progress.

R.F. telemetry is usually more suitable if the data is to be transmitted over distances greater than 1 km. Certain parts of the radio-frequency spectrum have been allocated for telemetry, and microwave links above about 4 MHz. Radio waves, at these frequencies tend to travel in straight lines, requiring repeater stations with disc like antennas on high buildings and towers every 30 to 60 km.

Radio links with airborne instrumentated flight vehicles often use Pulse Duration Modulation (PDM) – Frequency Modulated (FM) systems, with pulse duration varying from a minimum of 100 μs (representing zero level, signal) to a maximum of about 700 μs (representing maximum level signal).

## 27.7 MODULATION METHODS

The modulation methods used for transmission in R.F. (radio frequency) airborne telemetry are also applicable to land line transmission.

The expression for a carrier wave is :

$$e_c = A_c \sin (2\pi f_c t + \theta) \quad ...(27.1)$$

where $A_c$ = amplitude of carrier,

$f_c$ = frequency of carrier ; Hz,

$\theta$ = relative phase shift of carrier, rad.

A signal can be described by its (*i*) amplitude, (*ii*) frequency and (*iii*) phase shift. Accordingly, *three* methods of modulation are used and they are :

(*i*) amplitude modulation,

(*ii*) frequency modulation and

(*iii*) phase modulation.

### 27.7.1 Amplitude Modulation

In amplitude modulation, the amplitude of a *carrier signal* is varied by a *modulating voltage signal* whose frequency is much lower than that of the carrier. Formally, AM is defined as system of modulation in which the amplitude of the carrier is proportional to the instantaneous amplitude of the modulating signal. In case of instrumentation systems, the modulating signal is the output voltage of a transducer which is generated on account of application of the measurand.

Let $e_c$ and $e_s$ be the instantaneous values of the carrier signal and the modulating signal voltage respectively. The two voltages vary sinusoidally and therefore :

$$e_c = E_c \sin \omega_c t \quad ...(27.2)$$

and

$$e_s = E_s \sin \omega_s t \quad ...(27.3)$$

where $\omega_s$ = angular frequency of variation of (input) modulating signal.

Note that phase angle has been ignored in both expressions since it is unchanged by the *modulation* process. Its inclusion here would only complicate proceedings without affecting the result. However, in the case of frequency modulation and phase modulation it will certainly not possible to ignore the phase angle.

From the definition of A.M., it follows that the amplitude $E_c$ of the carrier will have to be made proportional to instantaneous modulating voltage $E_s \sin \omega_s t$ when the carrier is amplitude modulated.

From Eqns. 27.2 and 27.3, the amplitude of amplitude modulated voltage is,

$$A = E_c + E_s \sin \omega_s t = E_c + mE_c \sin \omega_s t$$

$$= E_c(1 + m \sin \omega_s t) \quad ...(27.4)$$

where

$$m = E_s / E_c \quad ...(27.5)$$

$m$ is called *modulation index* and its value lies between 0 and 1.

The instantaneous voltage of the resulting amplitude modulated wave is,

$$e_{cm} = A \sin \theta = A \sin \omega_c t = E_c (1 + m \sin \omega_s t) \sin \omega_c t$$

$$= E_c \sin \omega_c t + \frac{m E_c}{2} \cos (\omega_c - \omega_s)t - \frac{m E_c}{2} \cos (\omega_c + \omega_s)t \quad ...(27.6)$$

Equation 27.6 contains *three* terms

(*i*) $E_c \sin \omega_c t = E_c \sin 2\pi f_c t$

▲ *This represents the original carrier wave*

(*ii*) $\frac{mE_c}{2} \cos(\omega_c - \omega_s)t = \frac{mE_c}{2} \cdot \cos 2\pi (f_c - f_s)t$

▲ *This represents the lower side band frequency (LSB)*

(*iii*) $\frac{mE_c}{2} \cos(\omega_c - \omega_s t) = \frac{mE_c}{2} \cos 2\pi (f_c + f_s)t$

▲ *This represents the upper side band frequency (USB).*

Therefore, we can write,

$$\text{Lower band frequency (LSB)} = f_c - f_s \quad ...(27.7)$$

$$\text{Upper band frequency (USB)} = f_c + f_s \quad ...(27.8)$$

the frequency of AM. wave contains three discrete frequencies as shown in Fig. 27.9.

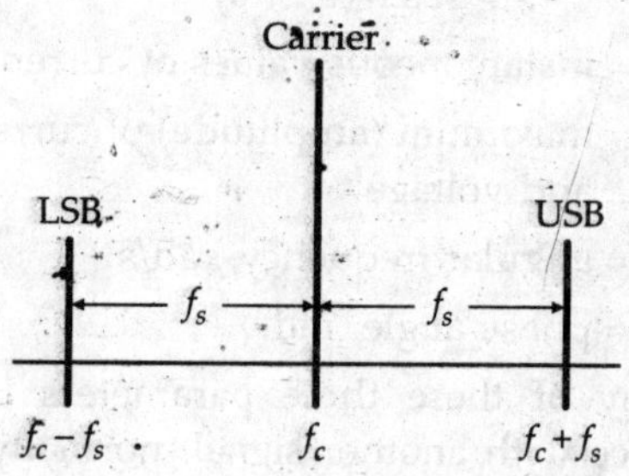


**Fig. 27.9** Frequency spectrum of an amplitude modulated (A.M.) wave.

The central frequency (carrier frequency $f_c$) has the highest amplitude, and the other two frequencies which are disposed symmetrically about it *i.e.,* LSB frequency $= (f_c - f_s)$ and USB frequency $= (f_c + f_s)$ have amplitudes which are equal to each other, but have an amplitude which can never exceed half the carrier amplitude for modulation index $m < 1$.

$$\therefore \text{Bandwidth BW} = (f_c + f_s) - (f_c - f_s) = 2 f_s \quad ...(27.9)$$

The amplitude modulated wave for one cycle of modulating wave is shown in Fig. 27.10.

The maximum amplitude or *top envelope* of the A.M. wave given by :

$$A = E_c + E_s \sin \omega_s t \quad ...(27.11)$$

The maximum negative amplitude or *bottom envelope* is,

$$-A = -(E_c + E_s \sin \omega_s t) \quad ...(27.12)$$

The modulated wave extends between these two limiting envelopes and has repetition rate equal to the unmodulated carrier frequency.

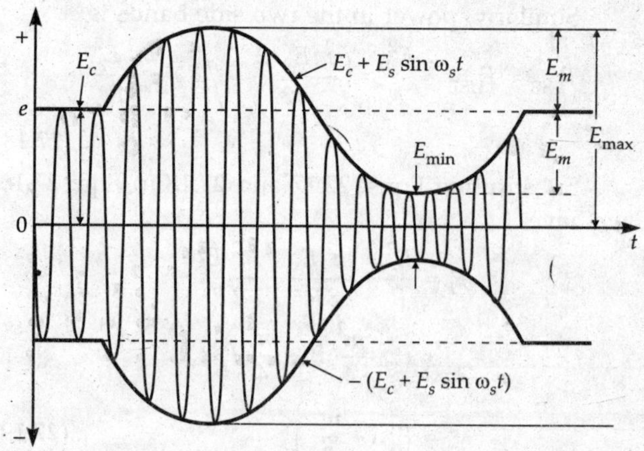


**Fig. 27.10** Amplitude modulated wave.

It will be recalled that $E_s = mE_c$ and therefore it is possible to calculate the index (or depth) of modulation from the waveform of Fig. 27.10.

$$\text{Now,} \quad E_m = \frac{E_{max} - E_{min}}{2} \quad ...(27.13)$$

$$\text{and} \quad E_c = E_{max} - E_{min} = E_{max} - \frac{E_{max} - E_{min}}{2}$$

$$= \frac{E_{max} + E_{min}}{2} \quad ...(27.14)$$

From Eqns. 27.13 and 27.14, we have, modulation index

$$m = \frac{E_s}{E_c} = \frac{E_{max} - E_{min}}{E_{max} + E_{min}} \quad ...(27.15)$$

It has been shown that the carrier component of the modulated wave has the same amplitude as the unmodulated carrier. However, the modulated wave, of course, contains two side bands in addition to carrier. Therefore, it is obvious, that the modulated wave contains more power than the carrier had before its modulation took place. Moreover, since the amplitude of side bands depends upon the modulation index, it is logical that the total power in the modulated wave will depend on the modulation index also.

The total power in modulated wave

$$P_t = \frac{E_{carr}^2}{R} + \frac{E_{LSB}^2}{R} + \frac{E_{USB}^2}{R} \quad ...(27.16)$$

Where all three voltages are *rms* values and R of the resistance in which the power is dissipated.

The first term of Eqn. 27.16 is the power in the unmodulated carrier and is given by :

$$P_c = \frac{E_{carr}^2}{R} = \frac{(E_c/\sqrt{2})^2}{R} = \frac{E_0^2}{R} \quad ...(27.17)$$

Similarly, power in the two side bands is,

$$P_{LSB} = P_{USB} = \frac{E_{sB}^2}{R} = \left(\frac{mE_c/2}{\sqrt{2}}\right)^2 \Big/ R = \frac{m^2 E_c^2}{8R} \quad ...(27.18)$$

Substituting Eqns. 27.17 and 27.18 in Eqn. 27.16, we have,

$$P_t = \frac{E_c^2}{R} + \frac{m^2 E_c^2}{8R} + \frac{m^2 E_c^2}{8R}$$

$$= P_c + \frac{m^2}{4} P_c + \frac{m^2}{4} P_c$$

$$= P_c\left(1 + \frac{m^2}{2}\right) \quad ...(27.19)$$

$$\therefore \quad \frac{P_t}{P_c} = 1 + \frac{m^2}{2} \quad ...(27.20)$$

The maximum power in the A.M. wave is :

when $\quad m = 1$

or $\quad P_t = 1.5\ P_c. \quad ...(27.21)$

***Example 27.1*** *A 1000 kHz carrier is simultaneously modulated with 300 Hz, 800 Hz and 2 kHz audio sine waves. What will be frequencies present in the output ?*

**Solution.** In addition to the carrier frequency of 1000 kHz, the other frequencies present would be the side band frequencies on account of the modulating signals. There are three modulating signals and consequently there are three LSB frequencies and three USB frequencies in the output. These are :

$$f_c \pm f_{s_1} = 1000 \pm 0.3 = 999.7,\ 1000.3 \text{ kHz}$$

$$f_c \pm f_{s_2} = 1000 \pm 0.8 = 999.2,\ 1000.8 \text{ kHz}$$

$$f_c \pm f_{s_3} = 1000 \pm 2 = 998,\ 1002 \text{ kHz}.$$

***Example 27.2*** *The tuned circuit in a simple A.M. transmitter uses a 50 μH inductance and 1 nF capacitance. If the oscillator output is modulated by radio frequencies up to 10 kHz, what is the frequency range occupied by side bands ?*

**Solution.** Carrier frequency

$$f_c = \frac{1}{2\pi\sqrt{LC}} = \frac{1}{2\pi\sqrt{50\times10^{-6}\times1\times10^{-9}}}$$

$$= 712 \times 10^3 \text{ Hz}.$$

Since the highest modulating frequency is 10 kHz, the lower side band frequency

$$= f_c - f_s = 712 - 10 = 702 \text{ kHz}$$

and the upper side band frequency

$$= f_c + f_s = 712 + 10 = 722 \text{ kHz}.$$

∴ The frequency range occupied by side bands is 702 to 722 kHz.

***Example 27.3*** *A broadcast A.M. transmitter radiates 50 kW of carrier power. What will be the radiation power at 85 percent of modulation ?*

**Solution.** From Eqn. 27.19, total power is,

$$P_t = P_c\left(1 + \frac{m^2}{2}\right)$$

Now $P_c$ = 50 kW and $m$ = 0.85, therefore, total power,

$$P_t = 50\left[1 + \frac{(0.85)^2}{2}\right] = 68.1 \text{ kW}.$$

### 27.7.2 Frequency Modulation

Frequency modulation is a system in which the amplitude of the modulated carrier is kept constant, while its frequency is varied by the modulating signal. The general equation of an unmodulated wave, or carrier, can be written as :

$$x = A \sin(\omega t + \theta) \quad ...(27.22)$$

where $x$ = instantaneous values of current or voltage

$A$ = maximum (amplitude) of current and voltage

$\omega$ = angular frequency, rad/s

$\theta$ = phase angle, rad.

If any of these three parameters is varied in accordance with another signal, normally of a lower frequency, then the second signal is called *modulating signal*, and the first is *modulated* by the second. Amplitude modulation, already discussed, is achieved when amplitude $A$ is varied, alteration of phase angle $f$ will yield, phase modulation. Finally, if the frequency of the carrier is made to vary, frequency modulated waves are obtained.

For simplicity, it is again assumed that the modulating signal is sinusoidal. This signal has two important parameters which must be represented by the modulation process, namely its amplitude and frequency. It is assumed that the phase relations of a complex modulation signal will be preserved. By the definition of frequency modulation, the amount by which the carrier frequency is varied from its unmodulated value is called the *derivation*, is made *proportional to the instantaneous value of the modulating voltage*. The rate at which this frequency variation, takes place is naturally equal to the modulating frequency.

The situation is illustrated in Fig. 27.11 which shows the modulating voltage and the resulting frequency modulated wave. Figure 27.11 also shows the frequency variation with time of the modulating voltage. The result of using that modulating voltage to produce AM is also shown for comparison. As an example of FM, all signals having the same amplitude will deviate the carrier frequency by the same amount, say 45 kHz, no matter what their frequencies. Similarly, all signals of the same frequency, say 2 kHz, will deviate the carrier at the same rate of 2000 per second, no matter their amplitudes. *The amplitude of the frequency modulated wave remains constant at all times.* This is, in fact, the greatest single advantage of F.M.

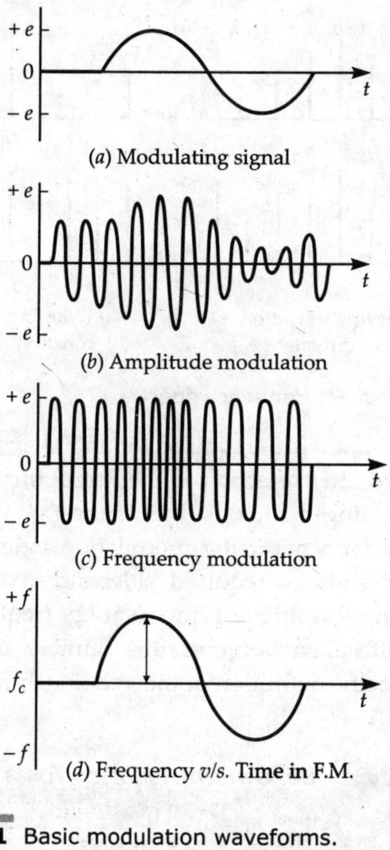


**Fig. 27.11** Basic modulation waveforms.

From Fig. 27.11(*d*) it is seen that the instantaneous frequency *f*, of the modulated frequency wave is,

$$f = f_c(1 + kE_s \cos \omega_s t) \qquad ...(27.23)$$

where, $f_c$ = unmodulated (or average) carrier frequency ; Hz,

$E_s \cos \omega_s t$ = instantaneous value of modulating voltage ; V.

The maximum deviation from this particular signal occurs when the cosine term has its maximum value *i.e.*, ± 1. Under these conditions, the instantaneous frequency is,

$$f = f_c(1 \pm k E_s) \qquad ...(27.24)$$

So that the maximum deviation is,

$$\delta = k\, E_s f_c \qquad ...(27.25)$$

The instantaneous amplitude of F.M. signal is of the form

$$e_{f_m} = A \sin[F(\omega_c, \omega_s)] = A \sin\theta \qquad ...(27.26)$$

where $F(\omega_c, \omega_s)$ is a function of the carrier and the modulating frequencies.

Figure 27.12 shows, θ is the angle traced out by phasor *A* in time *t* = *t*. If *A* were rotating at a constant velocity, say *p*, angle θ = *pt*. However the velocity *p* is not constant, but is variable and is governed by angular frequency $\omega = 2\pi f$ as given by Eqn. 27.22, *i.e.*,

$$\omega = \omega_c(1 + k\, E_s \cos \omega_s t). \qquad ...(27.27)$$

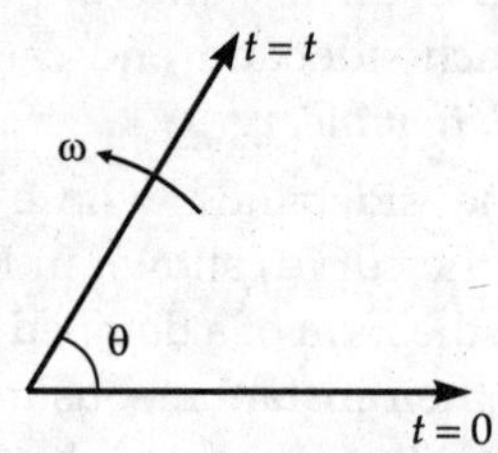


**Fig. 27.12** Frequency modulated signals.

Therefore,

$$\theta = \int \omega\, dt = \int \omega_c (1 + k\, E_s \cos \omega_s t)\, dt$$

$$= \omega_c\left[t + \frac{k\, V_s \sin \omega_s t}{\omega_s}\right] = \omega_c t + \frac{k\, E_s\, \omega_c \sin \omega_s t}{\omega_s}$$

$$\omega_c t + \frac{k\, E_s\, f_c \sin \omega_s t}{f_s}$$

$$= \omega_c t + \frac{\delta}{f_s} \sin \omega_s t \qquad ...(27.28)$$

Equation 27.28 is substituted in Eqn. 27.26 to obtain,

$$e_{f_m} = A \sin\left[\omega_c t + \frac{\delta}{f_s} \sin \omega_s t\right] \qquad ...(27.29)$$

The modulation index for F.M. is defined as,

$$m_f = \frac{\text{maximum frequency deviation}}{\text{modulating frequency}} = \frac{\delta}{f_s} \qquad ...(27.30)$$

Substituting Eqn. 27.30 in Eqn. 27.29, we obtain

$$e_{f_m} = A \sin(\omega_c t + m_f \sin \omega_s t) \qquad ...(27.31)$$

It should be noted that as the modulating frequency, $f_s$, decreases and the modulating voltage amplitude δ remains constant, the modulation index $m_f$ decreases. This is the basis for distinguishing frequency modulation from phase modulation.

27

The description of the F.M. wave is considerably more complex than the equivalent A.M. wave. The AM consisted of a constant strength carrier with an addition of an upper and lower side band displaced from the carrier by the frequency of modulating signal. FM on the other hand, contains many sidebands. Each successive side band is displaced from the next adjacent band by frequency of modulating signal. The number of sidebands and the *signal strength* of each sideband is determined by the *modulation index.*

Mathematically, the sidebands are defined by set of tables or curves known as Bessel functions shown in Fig. 27.13. These functions are plotted relative to the amplitude of the carrier and the trace of relative amplitude of each sideband and carrier for different values of *modulation index.*

Each of the sidebands whose amplitudes are determined by the curves shown in Fig. 27.13 present a wave shape in the form of a damped oscillation, which is a function of their instantaneous modulation index. The carrier signal represented by $J_0$ actually disappears at specific high values of $m_f$. At a particular value of $m_f$, the energy will be distributed over the many sidebands in a manner indicated in the chart.

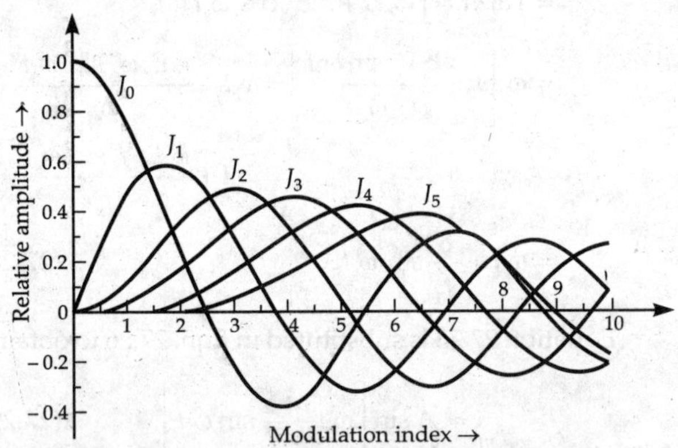


**Fig. 27.13** Bessel functions.

F.M. also differs from A.M. with regards to variation in total transmitted power with modulation. A.M. increases total power transmitted power with a modulation up to a maximum increase of 50%. During the process, the carrier power remains constant and the 50% increase appears as sideband power. In the F.M. wave, the total power remains constant, but the distribution of energy over the sidebands varies with modulation.

The modulation index ($m_f$) for F.M. varies directly with the deviation δ and inversely with amplitude modulating signal $f_m$. Therefore, $m_f$ can be varied by holding the frequency constant and bands that are frequency spaced by the same amount, but the amplitudes (any extent) of the sidebands differ. This is shown in Fig. 27.14(*a*). Another, possibility for $m_f$ variation is to keep the deviation constant and vary the frequency of the modulating signal. This results in sidebands that are spaced differently as shown in Fig. 27.14(*b*). In practice both the amplitude and frequency of the modulating signal are varying and the result is a complex FM waveshape described mathematically by Bessel functions.

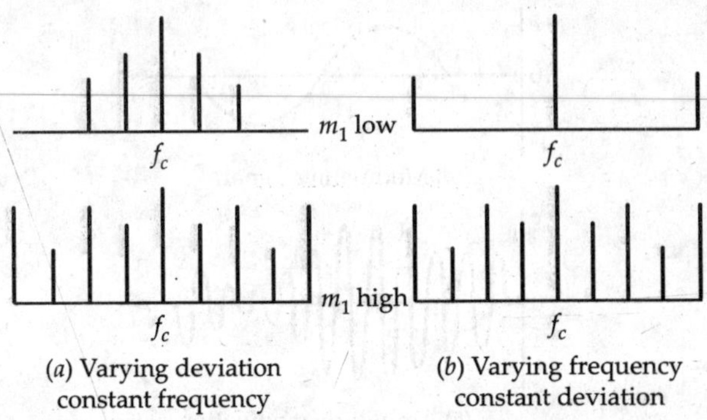


**Fig. 27.14** FM frequency spectra.

The bandwidth required for an FM system can be calculated after referring to the Bessel function tables and deciding which is the highest $J$ term to be included for a particular modulation index. A rough approximation of required sideband is the sum of deviation plus highest modulating frequency. The bandwidth then becomes this number doubled. A more specific definition of this relationship is listed in Table 27.1.

**Table 27.1** *Bandwidth Required for a FM Signal*

| Modulation Index ($m_f$) | Number of Significant Sidebands | Bandwidth: As Multiple of Modulating Frequency | Bandwidth: As Multiple of Frequency Deviation |
|---|---|---|---|
| 0.1 | 2 | 2 | 20 |
| 0.5 | 4 | 4 | 8 |
| 1 | 6 | 6 | 6 |
| 2 | 8 | 8 | 4 |
| 5 | 16 | 16 | 3.2 |
| 10 | 28 | 28 | 2.8 |

FM broadcast standards specify a maximum deviation of 75 kHz and a maximum modulating frequency of 15 kHz. This equates to an $m_f$ of 75/15 or 5.

The frequency components of an F.M. signal extend to infinity on either side of the carrier frequency. However, their amplitudes fall off rapidly outside a certain frequency range. According to Carson's rule, the approximate value of bandwidth of an F.M. signal is :

Bandwidth $\quad BW = 2\,(\delta + f_h)$ Hz $\quad$ ...(27.32)

where $\delta$ = maximum deviation in the carrier from its unmodulated frequency

$f_h$ = highest frequency in the message signal.

Frequency Modulation requires a greater bandwidth than amplitude modulation. It is used for both land line telemetering and radio frequency telemetering.

**Example 27.4** *In an F.M. system, when the audio frequency (A.F.) is 500 Hz and the A.F. voltage is 2.4 V, the deviation is 4.8 kHz. If the A.F. is now increased to 7.2 V, what is the new deviation ? If the A.F. voltage is raised to 10 V while its frequency is decreased 20 Hz, what is the deviation ? Find the modulation index in each case.*

**Solution.** The maximum deviation is given by Eqn. 27.25 as,

$$\delta = K\,E_s\,f_c.$$

Now $f_c$ is constant, and therefore,

$$\delta \propto E_s \quad \text{or} \quad \delta = K'\,E_s$$

For $\delta = 4.8$ kHz and $E_s = 2.4$ V,

$K' = 4.8/2.4 = 2$ kHz/V.

$E_s = 7.2$ V

$\delta = K'\,E_s = 2 \times 7.2 = 14.4$ kHz.

For $E_s = 10$ V

$\delta = 2 \times 10 = 20$ kHz

The modulation indices for different signal frequencies are :

$$m_{f_1} = \frac{\delta_1}{f_{s_1}} = \frac{4.8}{500 \times 10^{-3}} = 9.6$$

$$m_{f_2} = \frac{\delta_2}{f_{s_1}} = \frac{14.4}{500 \times 10^{-3}} = 28.8$$

$$m_{f_3} = \frac{\delta_3}{f_{s_1}} = \frac{20}{200 \times 10^{-3}} = 100.$$

**Example 27.5** *Find the carrier and modulating frequencies, the modulation index and the maximum deviation of the FM wave represented by :*

$$e_{f_m} = 12 \sin(6 \times 10^8 t + 5 \sin 1200\ t)\ V$$

*What power will this F.M. wave dissipate in a 10 Ω resistor ?*

**Solution.** The given F.M. wave is,

$$e_{f_m} = 12 \sin(6 \times 10^8 t + 5 \sin 1250\ t)\ \text{V}.$$

The expression for the instantaneous value of voltage of a F.M. wave is given by Eqn. 27.31 as,

$$A \sin[\omega_c t + m_f \sin \omega_s t]$$

Comparing the *two* expressions :

▲ Angular frequency of carrier

$\omega_c = 6 \times 10^8$ rad/s

or carrier frequency

$$f_c = \frac{6 \times 10^8}{2\pi} = 95.5 \text{ MHz}$$

▲ Angular frequency of signal

$\omega_s = 1250$ rad/s

or frequency of modulating signal

$$f_s = \frac{1250}{2\pi} = 199 \text{ Hz}$$

The modulation index $mf = 5$

∴ Maximum deviation

$\delta = m_f f_s = 5 \times 199 = 995$ Hz.

r.m.s. value of voltage of F.M. wave $= 12/\sqrt{2}$ V.

∴ Power dissipated by the wave in a 10Ω resistor,

$$P = \frac{(12/\sqrt{2})^2}{10} = 7.2 \text{ W}$$

**Example 27.6** *What is the bandwidth required for an F.M. signal in which the modulating frequency is 2 kHz and the maximum deviation is 10 kHz ?*

**Solution.** The modulation index $m_f = \delta / f_s = 5$.

From Table 27.1, the bandwidth as multiple of frequency of modulating signal is 16.

∴ Bandwidth $\quad BW = 16\,f_s = 16 \times 2 = 32$ kHz

### 27.7.3 Phase Modulation

Strictly speaking, there are *two* types of continuous wave modulation :

(*i*) amplitude modulation and

(*ii*) angle modulation

Angle modulation, may be subdivided into two types :

(*i*) Frequency modulation (F.M.) and

(*ii*) Phase modulation (P.M.).

Thus, the frequency modulation and phase modulation systems are closely related. It is possible to obtain frequency modulation from phase modulation. It must be stressed, however, that phase modulation as such is not used in practical analog telemetering systems.

27

If the phase angle $\theta$ in the equation $e = A \sin(\omega_c t + \theta)$ is varied so that its magnitude is proportional to the instantaneous amplitude of modulating voltage, the resulting wave is phase modulated. The expression for a P.M. wave is :

$$e_{pm} = A \sin(\omega_c t + \theta_m \sin \omega_s t) \quad ...(27.32)$$

where $\theta_m$ = maximum value of phase change introduced by this particular modulating voltage, and is proportional to the maximum amplitude of this modulation. For the sake of uniformity, Eqn. 27.32 is rewritten as,

$$e_{pm} = A \sin(\omega_c t + m_p \sin \omega_s t) \quad ...(27.33)$$

where $m_p = \theta_m$ ...(27.34)

= modulation index for phase modulation

Comparison of Eqns. 27.31 and 27.34 shows them to be identical except for the different definitions of modulation index.

***Example 27.7*** *A P.M. transmitter operating at a carrier of 100 MHz with a carrier voltage of 8 V. The modulating signal has an amplitude of 3 V and a frequency of 6 kHz resulting in a deviation of 60 kHz. Write the voltage equations for the following conditions :*

(*i*) *Original values,*

(*ii*) *Audio amplitude increased by 4 V,*

(*iii*) *Audio frequency increased to 8 kHz, and*

(*iv*) *Audio changed to 2 V and 3 kHz.*

**Solution.** (*i*) Angular frequency of carrier

$$\omega_c = 2\pi f_c$$
$$= 2\pi \times 100 \times 10^6 = 0.6283 \times 10^9 \text{ rad/s}$$

Angular modulating frequency

$$\omega_s = 2\pi \times 6 \times 10^3 = 37.7 \times 10^3 \text{ rad/s}$$
$$m_p = m_f = \delta / f_s = 60/6 = 10$$

$\therefore$ Expression for P.M. modulated signal is,

$$e_{pm} = A \sin(\omega_c t + m_p \sin \omega_s t)$$
$$= 8 \sin(0.6283 \times 10^9 t + 10 \sin 37.7 \times 10^3 t) \text{ V}$$

(*ii*) For signal voltage increased to 4 V,

$$m_p = \frac{4}{3} \times 10 = 13.33$$

$$\therefore \quad e_{pm} = 8 \sin(0.6283 \times 10^9 t + 13.33 \sin 37.7 \times 10^3 t) \text{ V}.$$

(*iii*) Audio frequency does not change $m_p$.

However,

$$f_s = 2\pi \times 8 \times 10^3 = 50.27 \times 10^3 \text{ rad/s}$$

$$\therefore \quad e_{pm} = 8 \sin(0.6283 \times 10^9 t + 10 \sin 50.27 \times 10^3 t) \text{ V}$$

(*iv*) An audio amplitude of 2 V changes $m_p$ to $\frac{2}{3} \times 10 = 6.67$

An audio change to 3 kHz results in,

$$\omega_s = 2\pi \times 3 \times 10^3 = 18.85 \times 10^3 \text{ rad/s}$$

$$\therefore \quad e_{pm} = 8 \sin(0.6283 \times 10^9 t + 6.67 \sin 18.85 \times 10^3 t) \text{ V}.$$

### 27.7.4 Frequency Modulation (F.M.) Telemetry Systems

One of the earliest techniques for mixing (multiplexing) data channels in a telemetry system is the frequency modulation (F.M.). This still accounts for about 20% of the applications in the telemetry field.

An F.M. system which is used for R.F. telemetering applications is shown in Fig. 27.15. The output of each transducer/signal conditioner modulates the frequency of a voltage-controlled subcarrier oscillator. Several oscillators, each operating in a dedicated part of the frequency spectrum, are mixed for radio transmission. This can be linked to several F.M. broadcast stations in a city. Each is assigned a unique place in the frequency spectrum, and each can be modulated within the assigned band without interfering with others.

At the receiving station a F.M. modulator, which is usually referred to as a *discriminator* is tune to the frequency of sub-carrier and has a bandwidth equal to that of the modulated sub-carrier.

As the measurement value, changes at the source, the discriminator output voltages changes accordingly.

Consider an example, a signal conditioner has a maximum amplitude of ± 2.5 V. The VCO is operating at a central frequency, of 400 Hz and will deviate ± 7.5% form central frequency if a signal ± 2.5 V is applied to it. Therefore, the output of VCO is a signal varying 400 (1 ± 7.5/100) between 370 Hz and 430 Hz. Suppose this channel is monitoring full level 300 *l*. It tank, the signal level may put out – 2.5 V when the tank is empty and + 2.5 V when the tank is full. This will drive the VCO frequency to 370 Hz for the empty tank, and 430 Hz for full tank and the corresponding values in between (50 *l* causes a 1 Hz change in VCO frequency). The discriminator senses frequency variations and converts each frequency into the appropriate voltage. Thus calibrating the – 2.5 V to + 2.5 V output of the discriminator into a 0 *l* to 3000 *l* display, one can get a good indication of liquid level, in tank which may be thousands of metres or thousands of km away from the receiving station through radio link.

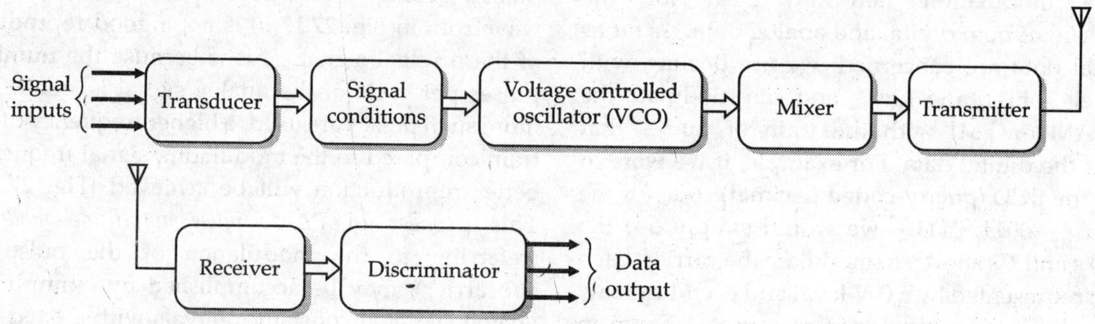


**Fig.27.15** Basic Frequency Modulation (F.M.) system.

### 27.7.5 Comparison between Frequency and Amplitude Modulation

Frequency and amplitude modulations are compared on a different basis from that for FM and PM. These are both practical systems, quite different from each other, and so the performance and characteristics of the two systems will be compared. To begin with, frequency modulation has the following advantages :

**1.** The amplitude of the frequency-modulated wave is constant. It is thus independent of the modulation depth, whereas in AM modulation depth governs the transmitted power. All the transmitted power in FM is useful, whereas in AM most of it is in the transmitted carrier, which does not indicate any modulation changes.

**2.** FM receivers can be fitted with amplitude limiters to remove the amplitude variations caused by noise, this makes FM reception a good deal more immune to noise than AM reception.

**3.** It is possible to reduce noise still further by increasing the deviation. This is a feature which AM does not have, since it is not possible to exceed 100 percent modulation without causing severe distortion.

**4.** Commercial FM broadcasts began in 1940, decades after their AM counterparts. Consequently, they have a number of advantages due to better planning and other circumstances. The following are the most important ones :

(*a*) Standard* frequency allocations provide a guard band between commercial FM stations, so that there is less adjacent-channel interference than in AM.

(*b*) FM broadcasts operate in the upper VHF and UHF frequency ranges, at which there happens to be less noise than in the MF and HF ranges occupied by AM broadcasts ;

(*c*) At the FM broadcast frequencies, the *space wave* is used for propagation, so that the radius of operation is limited to slightly more than line of sight. It is thus possible to operate several independent transmitters on the same frequency with considerably less interference than would be possible with AM.

The advantages are not all one-sided, or there would be no AM transmissions left. The following are some of the disadvantages of FM :

**1.** A much wider channel is required by FM up to 10 times as large as that needed by AM. This is the most significant disadvantage of FM.

**2.** FM transmitting and receiving equipment tends to be more complex, particularly for modulation and demodulation.

**3.** Since reception is limited to line of sight, the area of reception for FM is much, smaller than for AM. This may be an advantage for cochannel allocations, but it is a disadvantage for FM mobile communications over a wide area. Note that this is due not to so much to in intrinsic properties of FM but rather to the frequencies employed for its transmission.

## 27.8 PULSE MODULATION (PM)

*Pulse modulation* (*PM*) is, in reality, not a different, type of modulation. It refers, rather, to the *type* of modulating *signal* involved, a pulse. We may, then have pulse amplitude modulation or pulse frequency modulation. The important question here is how the information is represented by the pulse or pulses. For example, we might consider the amplitude of the pulse as the important parameter, or its width, or its location in a series of pulses, and so on. We consequently have a great variety of pulse modulation techniques. All will use either AM or FM as the mode of transmission.

* Allocated worldwide by the International Radio Consultative Committee (CCIR) of the I.T.U.

Pulse modulation may be used for the transmission of both digital and analog data. As far as the digital data are concerned, we are dealing with logic levels – high and low – and can modulate the carrier (AM or FM) with the train of pulses that represent the digital data. For example, if we were to transmit the BCD (binary-coded decimal) code for the number 37 – 0011, 0111 – we would simply use the train of 0's and 1's shown to modulate the carrier. Each 0 may be represented by a 0-V level and each 1 by, say, a 5-V level. The resulting train of pulses shown in Fig. 27.16 with a concurrent clock train to identify the position of the 1's and 0's. The order of the pulses in the train shown is important. First, we transmit the LSD (least significant digit), 7, then later in time the MSD, 3. The order of bits for each digit is MSB (most significant bit) first in time. Note that even though all pulses have the full 5-V amplitude, digital levels are usually permitted to vary over a wide range without affecting system performance. For example, the logic "1" level may vary between 2.4 and 5.5 V.

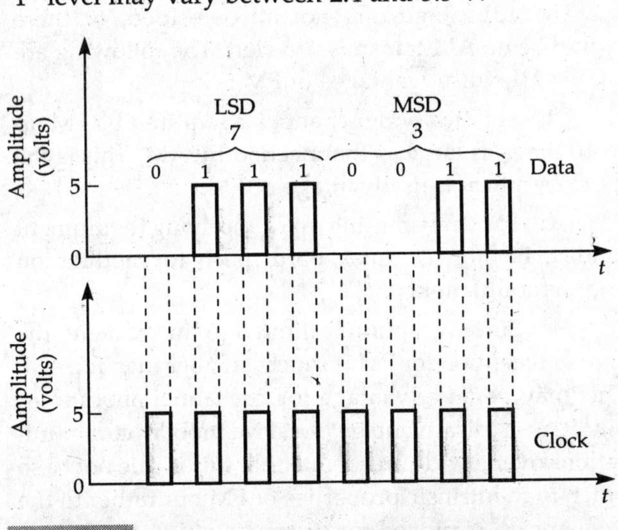


**Fig. 27.16** Pulse train representing the number 37 in BCD (least significant digit first).

To utilize pulse techniques for transmission of analog data, we must first convert the analog data into pulse format. This conversion is also referred to as modulation, since the analog data are used to modulate – modify – a train of pulses or a pulse subcarrier. Figure 27.17(*a*) shows a sine wave amplitude- modulating a train of pulses. The amplitude of each pulse in the modulated train depends on the instantaneous amplitude of the analog signal. The sine wave may be reconstructed from the modulated pulse train by simple filtering.

Figure 27.17(*b*) shows graphically the process of reconstructing the original signal by connecting the pulses, using straight lines. The reconstructed wavefrom in Fig. 27.17(*b*) is not a good reproduction of the original signal. This is because the number of pulses per cycle of the analog signal is small. When more such pulses are used, a higher frequency of pulse train compared to the modulating signal frequency, a better reproduction will be achieved [Fig. 27.17(*c*)]. This process of PAM, *pulse amplitude modulation* (referring to the modulation of the pulse train subcarrier), may be accomplished by sampling the analog signal at constant intervals with a fixed-width sampling pulse. The samples are pulses whose amplitude is that of the original analog signal at the sampling instant. The sampling rate (samples per second) must be selected to be at least twice the highest analog signal frequency. Usually, for better reproducibility the sampling frequency is about 5 times that of the highest modulating frequency.

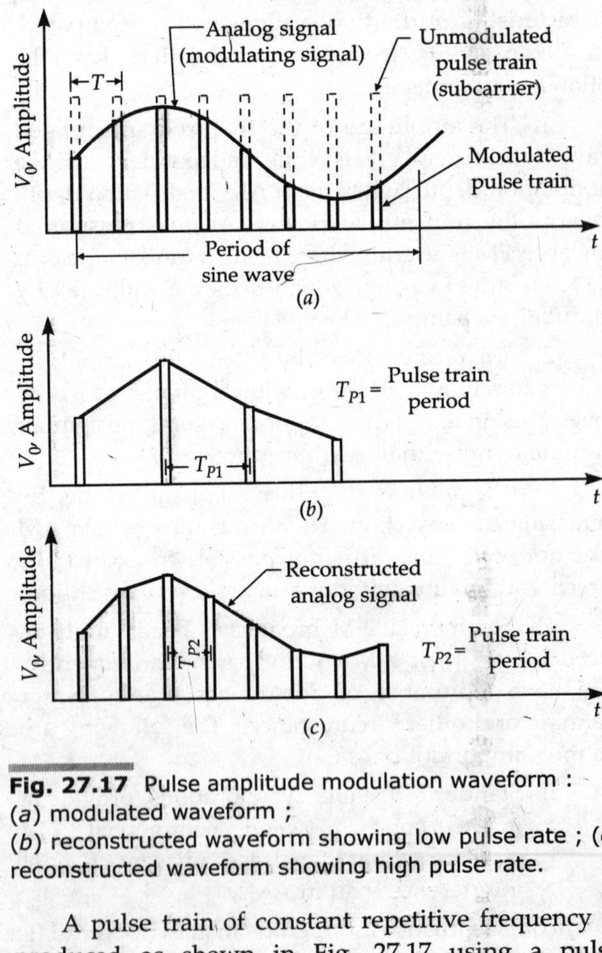


**Fig. 27.17** Pulse amplitude modulation waveform : (*a*) modulated waveform ; (*b*) reconstructed waveform showing low pulse rate ; (*c*) reconstructed waveform showing high pulse rate.

A pulse train of constant repetitive frequency is produced as shown in Fig. 27.17 using a pulse generator.

$$\text{Repetitive frequency } f = \frac{1}{T} \text{ ; Hz} \qquad ...(27.35)$$

where $T$ = pulse period ; s.

In fact, the carrier is modulated by pulses whose height varies proportional with the unknown signal. Therefore, amplitude of the unknown pulses $V_0$ varies proportionately with the unknown quantity $x$ or

$$V_0 = K_A x \quad ...(27.36)$$

where $K_A$ = a constant.

Pulse amplitude modulation is similar in principle to amplitude modulation process. Errors occur on account of damping of pulse amplitude. The pulse amplitude modulation (PAM) assumes special significance in time division multiplexing systems.

PAM is only *one* type of pulse modulation. Some other types are :

- PDM – pulse duration modulation (or pulse width modulation – PWM)
- PFM – pulse frequency modulation
- PCM – pulse code modulation.

*Pulse duration modulation* converts the sampled voltage levels into a series of pulses whose width (duration) is directly proportional to the amplitude of the sampled voltage (see Fig. 27.18). Note that the amplitude of the pulses is fixed ; only the pulse width varies in accordance with the modulating signal. The *sample interval* — the interval between samples — is also fixed.

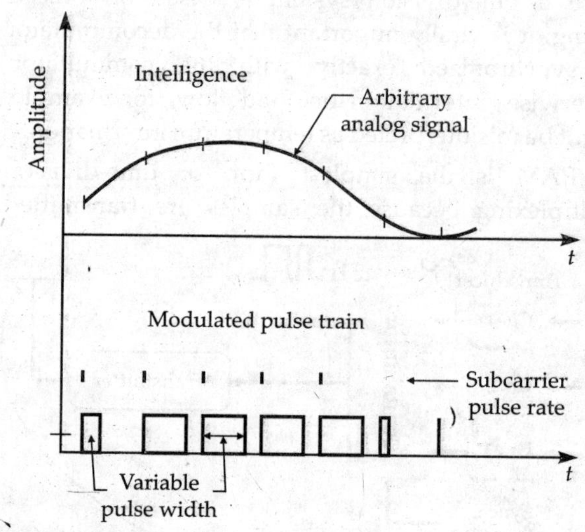


**Fig. 27.18** Pulse duration modulation.

In pulse duration modulation system, the magnitude of all the pulses remains constant but the duration of pulses varies. Pulse duration systems may consist of electromechanical or electronic devices. The input may be a mechanical displacement or an electric analog signal like voltage or current. The primary measuring device may be a low torque device such as a flow meter or a pressure gauge.

The unknown quantity, $x$ is converted into a rectangular pulse voltage $V_0$ having a duration $t_x$, where

$$t_x = K_d x \quad ...(27.37)$$

where $K_d$ = a constant

*Pulse frequency modulation* converts the sampled voltages into a train of pulses whose instantaneous frequency, or repetition rate, is directly related to the sampled voltage magnitude. Here again the amplitude of the pulses is fixed ; only its frequency changes. This is essentially the same as frequency-modulating a carrier consisting of a train of pulses instead of a sine wave, as in FM.

The frequency of the pulse train is :

$$f = K_f x \quad ...(27.38)$$

where $K_f$ = a constant.

The time period $T = 1/f$, and is therefore inversely proportional to the unknown quantity, $x$.

Sum of pulses $$N = \int f\,dt = K_f \int x\,dt \quad ...(27.39)$$

The pulse frequency method is shown in Fig. 27.19. From Eqn. 27.39 it is clear that sum of the pulses is a measure of time integral of quantity being measured.

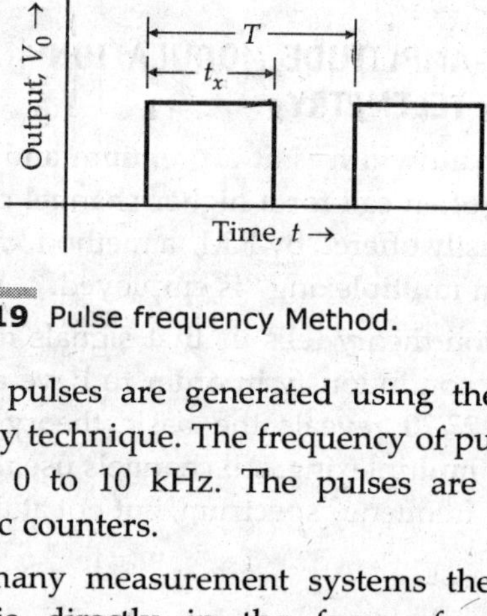


**Fig. 27.19** Pulse frequency Method.

The pulses are generated using the voltage to frequency technique. The frequency of pulse train lies between 0 to 10 kHz. The pulses are counted by electronic counters.

In many measurement systems the transducer output is directly in the form of pulses. Such transducers include armatures, turbine type flow meters for liquids and gases which are provided with electromagnetic or photo-electric, pick ups to provide pulse outputs for the telemetering systems.

*Pulse code modulation* converts the sampled voltage into a coded message. For example, if the sampled voltage ($E_s$) is 5.5 V, it may be represented by the binary number 101.10 = 5.5 by the use of an A/D converter. The coded message 101.10 represents the particular sample voltage $E_s$. A similar code, in this case a binary code, is associated with every sample.

The string of these coded messages is a set of numbers representing successive samples. The code may be any code ; binary 6-bit, as above or binary *N*-bit, or BCD, and so forth (see Fig. 27.16).

The above modulation schemes are only representative of a large range of techniques used. It is worth repeating that the modulation discussed here refers to subcarrier modulation that is, modulation of train of pulses which are then used in an AM or FM modulation system. We wind up with two successive modulations. *First,* the intelligence modulates a train of pulses. This first modulation may be PAM, PDM, PFM, PCM, or any other type. *Second,* the intelligence-carrying subcarrier modulates the sine wave carrier.

Pulse frequency modulation of a sine wave carrier results in $\Delta\omega_c$ (frequency deviation from the carrier frequency $\omega_c$) having discrete noncontinuous values. For example, the PM of logic levels "0" and "1" (0 V, 5 V) yields two frequencies – one $\omega_c$ (for logic "0") and the, other $\omega_c + \Delta\omega_c$ (for logic "1"). In essence, we simply shift the carrier frequency from $\omega_c + \Delta\omega_c$, to represent logic "1". This type of FM is also called FSK, *frequency-shift keying,* and is commonly used in telegraphy or other digital communications. The PLL (phase-locked loop) may be used to recover the logic levels from the FSK carrier.

## 27.9 PULSE-AMPLITUDE MODULATION (PAM) TELEMETRY

Because bandwidth is at a premium and system requirements often call for a higher channel capacity than can be easily offered by F.M, a method known as "time-division multiplexing" is employed.

Information theory tells us that signals need not be monitored continuously in order to have accurate data. Figure 27.20 reveals the basic theory behind time-division multiplexing. All channels use the same portion of the frequency spectrum, but not at the same time. The signal in each channel is sampled in sequence by a commutator, and the amplitude of each is an indication of the instantaneous data value at that point.

When all channels have been sampled, the sequence starts over at the first channel. Thus, samples, from a particular channel are interleaved in time with samples from all the other channels, and the amplitude of each is modulated by its data input.

Since no channel is monitored continuously in a time-division system, the sampling must be rapid enough that the signal amplitude in any channel does not change too much between sampling intervals. Practical telemetry systems use high sampling rates to preserve all the information in the original signal. The sampling rate in a typical telemetry system is about five times the highest frequency component in the sampled signal. For example, if the highest frequency component in a particular channel is 40 Hz, the channel is sampled about 200 times per second. If there are eight such channels in a system, the commutator must take at least 1600 samples per second.

At the receiving end of the system, a decommutator operating at exactly the same frequency as the commutator distributes the parts of the multiplexed signal to the proper output channels. Since a time-division system is based on precise timing, it is vitally important that the decommutator be synchronized exactly with the commutator. Otherwise, information on fluid flow, for example, might be misinterpreted as temperature information.

PAM is the simplest form of time-division multiplexing because the samples are transmitted

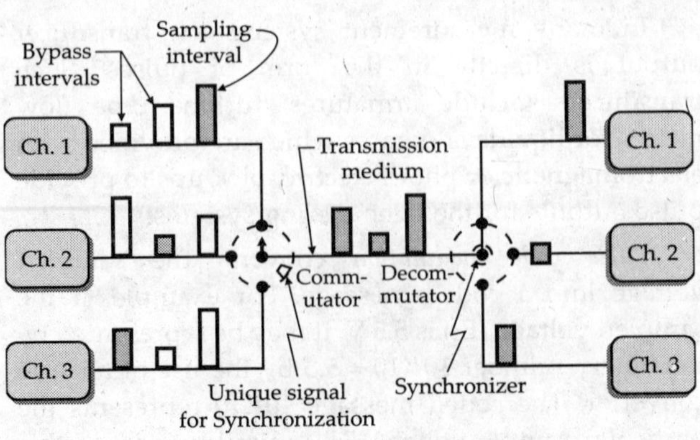


**Fig. 27.20** Basic time-division multiplexing system.

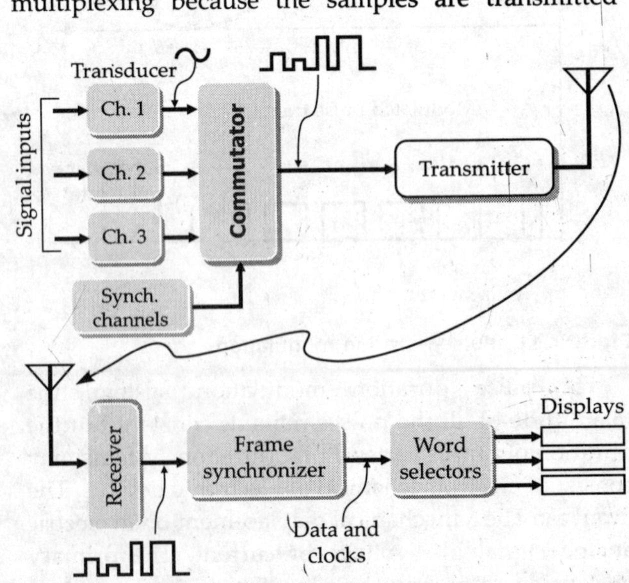


**Fig. 27.21** Basic PAM System.

without being modified. As stated previously, it is vitally important in a time-division system that the decommutator be synchronized exactly with the commutator. Synchronization channels must be introduced as part of the commutation scheme as shown in, Fig. 27.21. On the receiving end, no data is decommutated until these channels have been properly recognized and "time-tagged," at which time the decommutation process can begin.

## 27.10 PULSE-CODE MODULATION (PCM) TELEMETRY

At first glance, a pulse-code modulation (PCM) telemetry system (Fig. 27.22) may appear to be like a PAM system. Here, as before, we use "time-division multiplexing" to sample all the measurement points for a test sequentially with a commutator. However, a closer look shows another element in the transmitting system, an "encoder," has been added. This device accepts each PAM sample in turn, converts the amplitude of that sample into a binary number, and shifts the bits of that number out serially. The encoder may convert a zero-amplitude pulse into the binary number "0", the full-scale pulse into the binary number "1023", and pulses between those extremes into appropriate binary numbers so that each number is almost exactly proportional to the instantaneous amplitude of its measurement point.

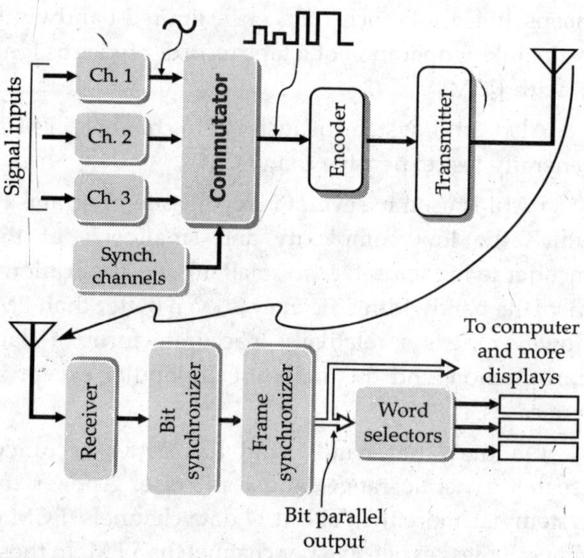


**Fig. 27.22** Basic PCM System.

The receiving station must synchronize on the serial data stream, identify the sequence of bits which make up each binary number, and convert those bits sequences or "words" into computer data, analog values, or other useful outputs.

Since the system makes binarily weighted "codes" of the measurement data, we call it a pulse-code modulation (PCM) system.

In PCM Modulation, the data to be measured are quantized and transmitted serially as data words in pulse code. This is the most efficient of the pulse coding systems. It consists of sending analog information by transforming into a series binary digits as shown in Fig. 27.23. The analog signal is sampled at regular intervals and then each sample value is converted into a coded form (The conversion of analog signals into digital signals is called **Coding** or **Encoding**). The process of coding introduces errors.

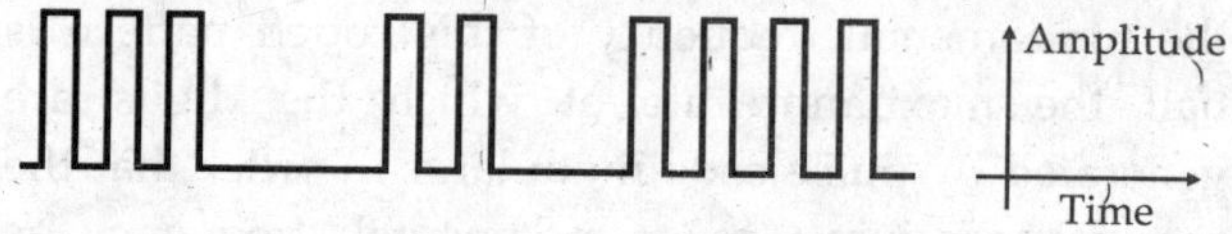


**Fig. 27.23** Pulse code modulation (PCM).

This is illustrated by the following example:

***Example 27.8*** *In a PCM system, a 5 bit encoder is used. Suppose each level represents 1 V, what is the range of encoder ?*

*Find out the quantization error on account of PCM if a voltage of 27.39 V is involved.*

**Solution.** Number of quantization levels

$$= 2^n = 2^5 = 32.$$

Range of encoded signal

$$= (2^n - 1) \times 1 = (2^5 - 1) \times 1$$

$$= 31 \text{ V}.$$

Therefore, the range is 0 – 31 V with each step representing 1 V. Since the decision level is taken at 1 V, the voltage involved is transmitted as 27 V in binary digits as : 11011

as $2^4 + 2^3 + 0 + 2^1 + 2^0 = 27$ V.

$\therefore$ Error on account of modulation

$$= 27.39 - 27 = 0.39 \text{ V}.$$

From the above example it is clear that modulation introduces an additional error in the message signal. This noise may be called as quantization noise. In order to reduce quantization noise it is necessary to a use higher bit encoder.

The bandwidth of PCM signal depends upon the bandwidth of the original signal. It also depends upon the number of bits used to represent the analog signal.

The receiver decodes the signal (**Decoding** means conversion of digital number into an analog quantity). Thus, it is essential for the receiver to distinguish between '1' and '0' levels.

The bandwidth of the original signal extends from 0 to the highest frequency $f_h$ Hz. The maximum sampling rate is $2f_h$ Hz. Thus if an $n$ digit code is used, the minimum bandwidth required as :

Minimum bandwidth

$$= \frac{1}{2} \times 2\, f_h \times n = nf_h \text{ Hz}$$

This is because the coded pattern containing alternate 1's and 0's contains the highest frequency components and the bandwidth required to transmit the fundamental frequency of this coded pattern is half the maximum rate at which the digits are generated by pulse code modulator in order that the receiver receiving the fundamental frequency is adequate to be able to distinguish between 1's and 0's. Therefore, if an '$n$' digit code is used the bandwidth is increased $n$ times.

The number of digits in a digital transmission can be arbitrarily increased by suitable choice of the quantizing interval. A binary number of $n$ digits can transmit a maximum of $2^n$ quantum levels of the same weightage and hence enables a relative error of $2^{-n}$ to be achieved corresponding to a transmission error of ±1 in the least significant figure. However, no additional advantage is achieved by increasing the accuracy of the transmission much larger than the errors of the pickups of the transducers. Generally 8 to 12 digit-binary numbers are used in measurement systems. These correspond to 256 to 4096 quantization levels.

The PCM system often employs a 7 digit code, giving $2^7 = 128$ quantum levels, and the signal is transmitted using FM, the least significant figure sent first. Apart from the quantizing noise, no other noise is introduced during transmission. Land line repeater stations can easily generate *'pure pulse'* or *no pulse* waveforms onto succeeding repeater stations about 2 km away. The basis of decoder output is simply a capacitor ($C$) and resistor ($R$) in parallel, the time constant ($RC$) being such that the voltage across the circuit decreases to one half of its initial value in the time between successive pulses received. In twice this time the voltage would have fallen to one quarter, and in three times this interval of time to one eighth. In fact, the time constant provides just the correct weighing for digits if the least significant digit is received first. For a 5 digit coded sample 11001 assuming that all 1's have an amplitude of 16 V, the final voltage would be

$$16\left[\frac{1}{16} + \frac{0}{8} + \frac{0}{4} + \frac{1}{2} + 1\right] = 25 \text{ V}.$$

## 27.11 COMPARISON : FM, PAM AND PCM

In summary, telemetry systems for general use employ one of *three* methods of data multiplexing :

- FM,
- PAM or
- PCM.

There is not a clear-cut choice between FM, PAM, and PCM (otherwise everyone would drop two and concentrate on the third !).

However, for each application there is likely to be a good reason for choosing one to the exclusion of the other two.

FM has one clear-cut advantage, it is about twice as efficient as PCM in translation of input data bandwidth into multiplex bandwidth ; for example, it enables a user to put ten 1000 Hz data measurements into a multiplex with a top frequency of about 125 kHz. While the same measurements encoded into PCM would occupy a spectrum of about 250 kHz. This means twice as much data on a limited-bandwidth radio link or operation of a tape recorder twice as long as with PCM.

Also, in a system of just a few channels, FM is generally less expensive than PCM.

PAM is used in several Navy missile programs in which the low complexity and small size of the encoder lend themselves to small missile applications. Also, the bandwidth efficiency is even better than FM. However, it is a relatively inaccurate form of data transmission, and this makes it unpopular except in rare cases.

On the other hand, PCM has better accuracy, greater dynamic range, and less noise. And, if the system has more than about 14 data channels, PCM is generally less expensive per channel than FM. In those cases in which the eventual destination of data is a digital computer, there are many advantages of using PCM multiplexing, transmission and storage.

The preceding comparisons are presented in tabular form in Table 27.2.

**Table 27.2** *Comparison of FM, PAM, PCM*

| | *FM* | *PAM* | *PCM* |
|---|---|---|---|
| Efficiency in use of radio or tape recorder bandwidth | medium | best | worst |
| Cost of a small transmitting system | lowest | low | highest |
| Size | smallest | small | largest |
| Cost of a large transmitting system | highest | lowest | medium |
| Size | largest | smallest | large |
| Cost of a small receiving system | lowest | higher | highest |
| Cost of a large receiving system | highest | high | high |
| Accuracy | poor | poor | excellent |
| Percent of use (approximate) | 20 | 15 | 65 |

## 27.12 COMPARISON OF FM, PAM AND PCM

Since PCM and FM account for about 85% of the telemetry use at this time, this ILM concentrates on those two methods from this point.

## 27.13 TRANSMISSION CHANNELS AND MEDIA

The most extensively used transmission media are cables and electromagnetic radiation radio links. However, optical, ultrasonic and magnetic induction data links are also used for many applications. Land-line telemetry utilizes cables or wires to transmit data from industrial process plants and power generation stations to central monitoring and control units.

When the data is to be transmitted over lengths greater than 1 km, radio links are usually more suitable. Certain parts of the radio frequency spectrum are allocated for telemetry. The microwave links are used above about 4 MHz. Radio waves at these frequencies tend to travel in a straight line, requiring repeater stations every 30 – 60 km away. Radio links with airborne instrumented-flight vehicles usually use PDM-FM system.

Short range non-contact data links of approximate lengths up to 50 m, such as those used force rotating machinery and power lines generally use either FM radio, optical or ultrasonic methods. Optical links using lasers and pulse transmission are in the developmental stages. Magnetic induction finds its use in biomedical applications.

The common communication channels used for telemetering are :

(*i*) wire line channels
(*ii*) microwave channels,
(*iii*) radio channels, and
(*iv*) power-line carrier channel.

The major difference between the channels is the range of frequencies that each channel transmits.

### 27.13.1 Wire Line Channels

For land-line telemetry wire line channels are used. The wire line is the simplest type of channel for transmission of information. The wire lines may be : (*a*) privately owned or (*b*) leased from Posts and Telegraphs Department. In case of leased type of lines from *P* and *T* department the channels may be telegraph, teletype or telephone lines.

The lines may be aerial lines or may be cables. The line telemetering equipment is subject to the following hazards :

(*i*) direct of induced voltages due to lightning,
(*ii*) voltages due to unintentional contact with adjacent high voltage power circuits, and
(*iii*) voltage induced electromagnetically or electrostatically by high voltage power circuits that run parallel to the data transmission lines.

The lines and equipment must be protected against the above mentioned undesirable effects.

Cables have advantages over the aerial lines in that :

(*i*) being unexposed to lightning are more reliable,
(*ii*) they have high insulation, and
(*iii*) their parameters remain the same irrespective of the changes in the atmospheric conditions.

For land-line telemetry an enormous variety of connectors are employed. They range from, a pair of open wires, a pair of twisted wires, wires of a telephone cable, to co-axial cables. Non co-axial cables

27

can be used for bandwidths upto 10 MHz over a length of a few hundred meters. The co-axial cables have a much wider range.

Cable installations are expensive and also are difficult to move once installed, but they are more efficient and reliable.

### 27.13.2 Radio Channels

Where longer distances are involved, or where measurements have to be made in a missile or moving vehicle, it may be necessary to use radio frequency (RF) techniques to transmit information.

There are *two* major reasons for resorting to radio channels for telemetry in the case of spacecrafts, missiles or aircraft rather than record the data on board. They are :

(*i*) It is many a times difficult and in some situations impossible to recover records from flights, and

(*ii*) The limitations of space, weight and operating conditions can be met more easily by telemetering the information to ground stations.

In radio frequency telemetry, multiplexing is used, as it is not possible to have many channels on account of antenna complications and due to desired saving of space, weight and power.

### 27.13.3 Microwave Channels

Microwave is a special case of radio transmission. The necessity for a large number of communication channels in industries like power and gas transmission and the overcrowding of spectrum in the radio frequency range has led to the increased use of microwave radio transmission. Several bands in the range of 890 MHz to 30 GHz have been allocated for microwave transmission.

The microwave frequencies are similar to television and radar in the sense that they are line-of-sight beamed directly from one antenna to other antenna and thus their use depends upon the type of terrain. The line-of-sight distance is small because of the curvature of the earth. So principle of repeated relaying is used for increasing the range of transmission with the help of intermediate receiving transmitting stations set up every 50-80 km depending upon the terrain.

The basic cost of the transmitter and the receiver is high, but cost of additional channels is small. Hence the cost of microwave equipment becomes competitive in case of multichannel transmission. The equipment is rugged and reliable and as mentioned above can provide a number of channels.

### 27.13.4 Power Line Carrier Channels

The use of power line carriers has been prevalent for a long time to transmit information between the central generating station and the local substations. They utilize the power line equipment for this purpose. An a.c. carrier is coupled to the a.c. voltage transmission line and the information to be telemetered is superposed upon the transmitted power.

## 27.14 RADIO LINK

The radio link in the radio telemetry system forms the communications path between an inconvenient or inaccessible data source and the user's laboratory.

The radio link consists of the radio transmitter, transmitting antenna (one or more), receiving antenna (possibly auto-tracking) ; radio receiver (one or two) ; and possibly also some or all of these : premodulation filter, power splitter, pre-amplifier and diversity combiner, interconnected as shown in Fig. 27.24.

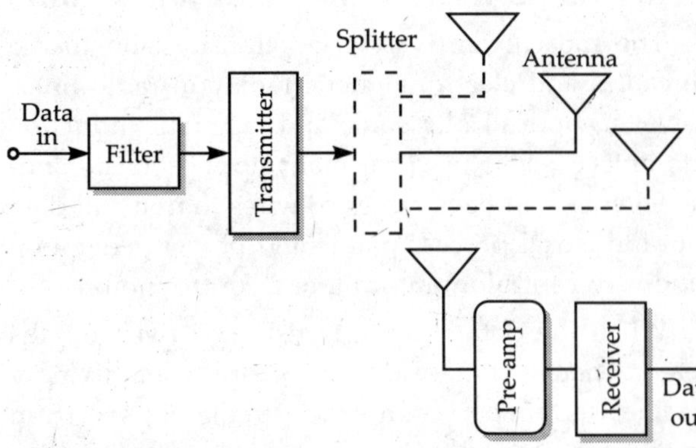


**Fig. 27.24** Radio Link.

U.S. government test ranges use the *"L"* band (1435-1540 MHz) for manned vehicle tests, and the *"S"* band (2200—2300 MHz) for unmanned vehicles. Equipment is readily available from manufacturers for these bands. Industrial telemetry is authorized in the *"P"* band, around 225 MHz and equipment is available in that band also.

### 27.14.1 Transmitter Modulation Data

The typical telemetry transmitted is frequency modulated (FM) by data from the multiplexer. Frequency stability is extremely good in the quiet state (typically within 0.003% of the assigned location, even with temperature and power supply variations, and in the presence of vibration and shock). Modulation is typically ±500 kHz per volt rms input and output power is whatever the user requires (typically 1 to 20 W).

### A. Transmitter Modulation, PCM Data

In the typical PCM system, in order to get maximum data transmission per unit of transmission bandwidth, non-return-to-zero (NRZ) code is output from the encoder.

This output is a square wave as shown in the top part of Fig. 27.25.

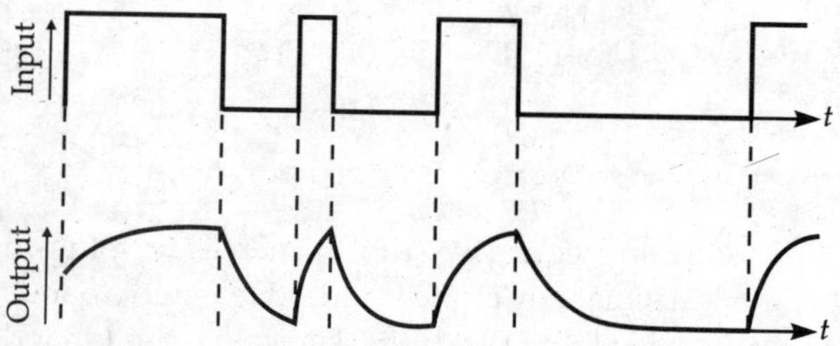


**Fig. 27.25** Premodulation Filtering.

If this square wave signal were applied directly to modulate the transmitter, the high-frequency components of the sharp edges would cause unacceptable spurious radiations from system. In order to avoid these radiations (which would disturb other users of the nearby spectrum), a premodulation filter is used on the NRZ signal. This low-pass filter has a Bessel (linear-phase) characteristic, six poles. The filter output does not contain sharp edges and looks like the bottom part of Fig. 27.25.

The nominal cutoff frequency of the filter is generally 0.5 of bit rate.

Typically, the transmitter carrier is deviated ± 0.35 of bit rate ; therefore, the deviation bandwidth is from + 0.35 bit rate to – 0.35 bit rate, or 0.7 bit rate total.

Modulation voltage comes from the encoder or premodulation filter and is adjustable at that point. The typical transmitter modulation sensitivity is ± 500 kHz per volt rms. A properly filtered 1-megabit-per-second NRZ data stream should deviate the transmitter ± 350 kHz ; therefore the modulation voltage in this case should be 350/500 of one volt rms, or 0.7 volt rms. This is approximately 2.0 volt peak-to-peak.

### B. Transmitter Modulation, FM Data

The amplitude of a multiplex of FM subcarriers is difficult to measure. Since several frequencies are present, some of the subcarriers are going positive together at any given instant while others are going negative together — their relationship to each other is changing constantly. No meter can give an adequate measure of the thing that really matters, namely, what will this multiplex do to the instantaneous deviation of an FM transmitter ? The user has two choices : *first*, observe the multiplex on an oscilloscope and adjust the amplitude until the peak-to-peak excursion reaches the amount necessary to swing the transmitter carrier ± 500 kHz (or whatever is the recommended nominal range) ; or *second*, observe the transmitted carrier on a spectrum analyzer, and adjust for ± 500 kHz (or nominal range).

### 27.14.2 Transmitting Antenna

Generally the telemetry transmitter is located in a moving vehicle. This means that antenna location and orientation are critical; all the frequencies used for telemetry are "line-of-sight" frequencies (they are functional only when there is a clear line of sight from the transmitting antenna to the receiving antenna), and the transmitting antenna must be placed where it can send a signal to the receiving station regardless of the orientation of the vehicle. The choice of an antenna type is quite simple then — it must be an "omnidirectional" (all-directions) antenna, placed on a vehicle surface which always has a line-of-sight path to the receiving station.

Often, because of the difficulty in selecting a vehicle mounting site for assured operation, a "power splitter" is used to drive two (or even three) antennas on the moving vehicle.

### 27.14.3 Receiving Antenna

Several general types of receiving antennas are available to the telemetry user. The choice is dependent on several factors – gain, directivity, beam width and price considerations, primarily. The trade-offs should be studied carefully by the system designer to determine the optimum type for a given set of conditions. General types are :

(*i*) Omnidirectional antenna, with essentially equal gain in all directions of azimuth. This antenna is generally not good enough for a telemetry receiving site because of the low gain.

(*ii*) Three-element array, fixed in azimuth, with a little higher gain than the omni, and a fairly wide beam.

(*iii*) Horn, with still higher gain and a wide beam.

(*iv*) Single-axis (azimuth-only) tracking antenna, often the optimum unit for aircraft flight tests. The antenna may be :

- Multi-element array, or Disc type.
- Dual-axis (azimuth and elevation) tracking antenna, narrow beam, high gain, ideal for missile tests.

The antenna is available as :

- Mono-pulse tracker, or Conical-scan tracker.

27

**Table 27.3** *Antenna Comparison*

| | Omni-directional | Three-element fixed | Horn | Steerable array or single-axis tracter: 12-element array | Steerable array or single-axis tracter: 6-foot dish | Two-axis tracker: Monopulse or Conscan |
|---|---|---|---|---|---|---|
| Gain, db (type) | 4-5 db | 8-10 db | 15 db | 16 db | 25 db | > 30 db |
| Beam width (type) | 360° | 27° | 45° | 16° Az<br>40° El | 8° Az<br>40° El | 5° |
| Price, normalized to single-axis tracking system | 0.01 – 0.02 | 0.02 | 0.05 | 1.0 | 1.0 – 1.3 | 2.0 – 3.0 |

It is not necessary for the prospective user to understand the intricacies of all these types in order to choose the optimum antenna for a system. Certain key features will define everything which is important for making at least a tentative selection. These features are shown in Table 27.3.

Antenna patterns for each of the basic types are depicted in Fig. 27.25 ; gain is shown on a scale, as is relative beam width.

### 27.14.4 Receiver

A high-quality telemetry receiver is shown by the functional diagram, Fig. 27.26, which is typical of any high-quality telemetry rceiver.

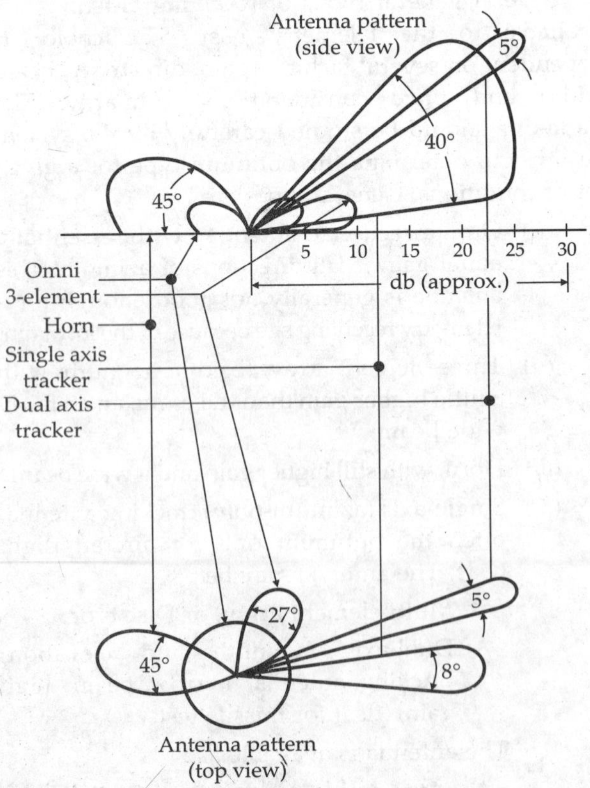


**Fig. 27.26** Antenna Patterns.

The radio-frequency (RF) turner is a plug-in module, such that any of the standard telemetry bands can be received by the unit. Typically, an *L*-band (1435—1540 MHz) tuner is used for flight tests, and *S*-band (2200—2300 MHz) tuner is used for missile tests, and other tuners are used for other tests as dictated by test range or other government regulations.

The tuner receives frequency modulated (FM) signals from the antenna and provides a 50 MHz or 160 MHz output.

The *first* intermediate frequency (IF) amplifier is optional and may not be needed in relatively quiet areas. However, it is always necessary if a spectrum display unit is included in the receiver.

A *second* local oscillator (L.O.) feeds a second mixer, the output of which is a second IF of 10 MHz.

Plug-in second IF amplifiers are available for several different bandwidths, and this is the opportunity to optimize the receiver for rejection of unwanted signals and amplification of those containing telemetry data. The modules of interest range from 1.0 MHz bandwidth to 6.0 MHz bandwidth.

What IF bandpass should be used ? The selection is not simple ; on the one hand, you want it wide to insure that your carrier will not drift outside the passband, but on the other hand, you want it narrow in order to shut out noise and improve data quality. A good compromise is to select a bandpass equal to :

- PCM : Twice the peak carrier deviation plus twice the premodulation filter cutoff.
- FM : Twice the peak carrier deviation plus twice the highest subcarrier frequency.

Several bandbass filters should be kept on hand at a receiving station, in order to optimize the station for each day's operation.

For frequency modulation (FM) carriers, a plug-in FM detector retrieves the telemetry "video" signal. Again, a user has a selection of bandwidth characteristics.

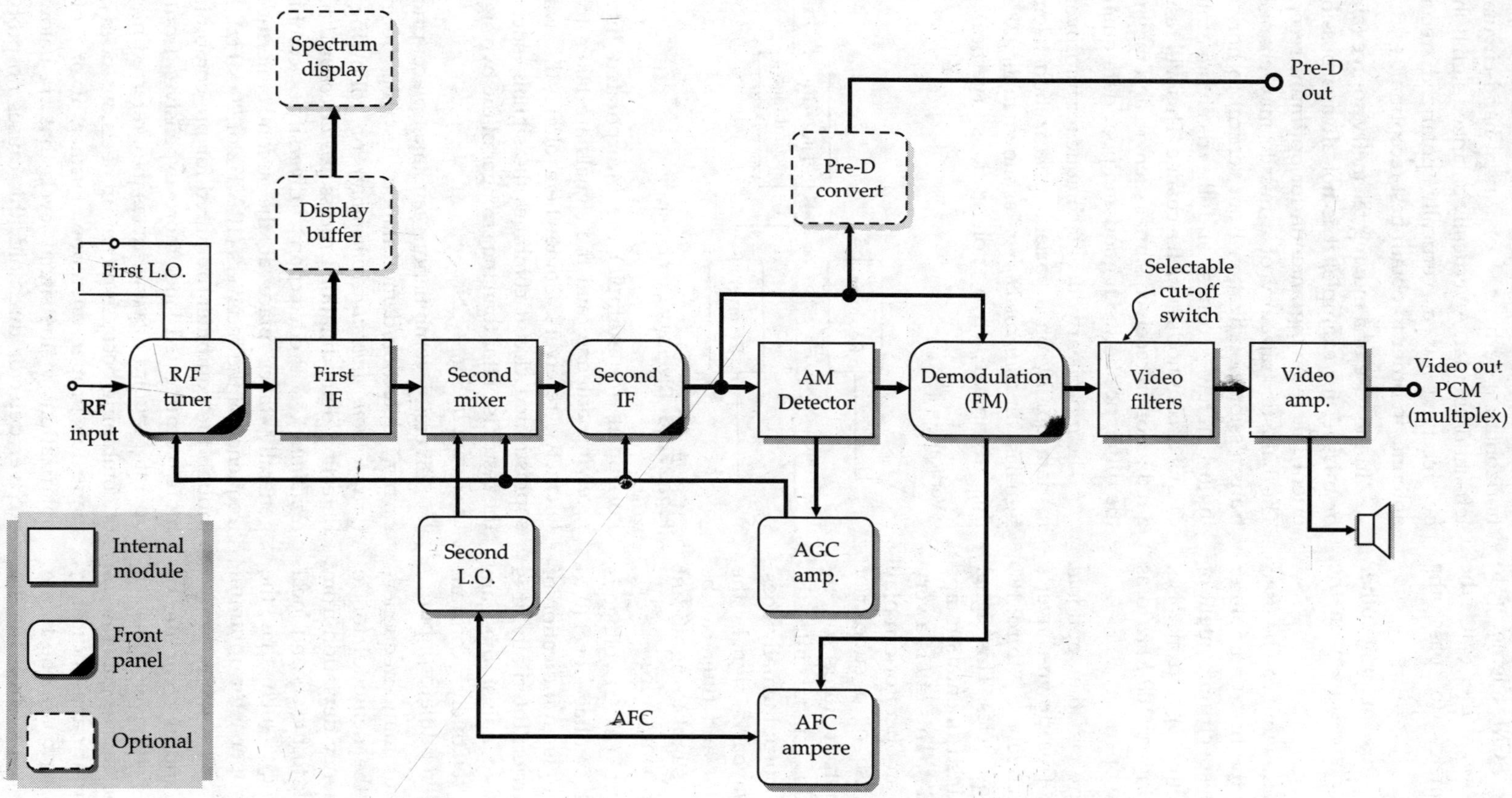


**Fig. 27.27** Receiver Block Diagram.

Within the video filter module is a switch-selectable low-pass output filter which rolls off at one of several points. The nominal cutoff of the four-pole filter is selected for PCM telemetry to approximate the premodulation filter at the transmitter, or for FM telemetry to pass the highest-frequency subcarrier.

Several other features of this or any other well-designed telemetry receiver are shown on the block diagram. For example :

(*i*) A plug-in specturm display unit can be used with a receiver, to operate from the first-IF amplifier signal. This is able to display the spectrum content of the carrier and the surrounding spectrum also. Edge-to-edge display can be as low as 100 kHz to as high as 6 MHz as selected by the operator.

(*ii*) Another useful feature is the optional plug-in predetection converter. This down-converts the RF carrier to what is effectively a "third IF" centered at either of several frequencies. This signal can be recorded on a wideband instrumentation tape recorder before detection ("pre-d"). This offers a very distinct advantage in telemetry ; an operator can play back the pre-d recording by use of a subcarrier discriminator and low-pass output filter, or by up-converting and inserting this signal into the receiver where the second mixer output usually goes. On demodulation of this playback signal, the characteristics of the demodulator and filter can be optimized to "squeeze" a slight-to-significant improvement in signal-to-noise performance.

(*iii*) An automatic gain control (AGC) voltage is generated at the demodulator. This voltage level (0 to 8 volts, either positive or negative) is an indication of the amplitude of the received carrier. Internally, it is fed back to front-end modules to adjust the gain automatically, such that a relatively constant amplitude is maintained at the demodulator input.

(*iv*) Another voltage, automatic-frequency-control (AFC), is generated in the demodulator to represent the carrier center frequency deviation from optimum. The voltage is fed back to the second local oscillator to adjust its frequency, such that the demodulator input is centered properly for optimum operation.

(*v*) To notify the operator and/or computer of loss of carrier signal, the receiver has a carrier-operated relay. Loss of the carrier generates a logic change which operates a front-panel indicator, and this signal could be used as a bit in the frame synchronizer status word or as an interrupt for the computer.

### 27.14.5 Diversity Combining

Telemetry radio signals are radiated from the source with a distinct "polarization" —vertical / horizontal, perhaps, or right-hand-circular / left-hand-circular. A vertically polarized signal should be received by a vertically polarized receiving antenna, in order to obtain best reception.

In the case of a moving telemetry source (aircraft or missile, for example) it is not always possible to orient the receiving antenna for optimum reception. The ballistic trajectory of a missile may change the transmission polarization from vertical to horizontal in just a few seconds, and the receiving antenna cannot accommodate such a change. This will result in a reduction in received signal strength and may make the difference between good and poor data quality.

The most nearly ideal technique for improvement in data quality on a variable-polarization telemetry signal is "*diversity combining,*" as shown in Fig. 27.28. In this technique, two polarizations (vertical and

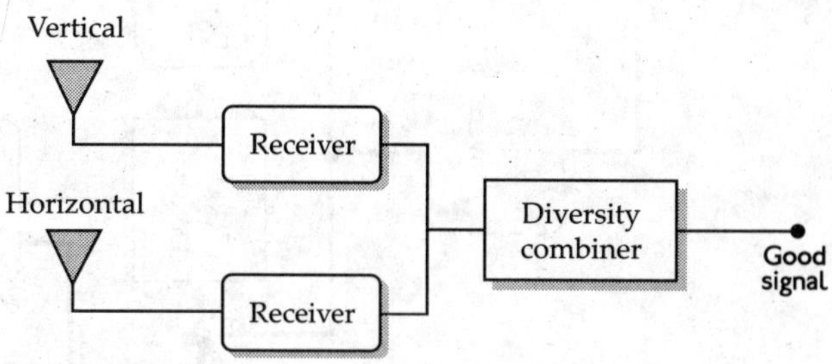


**Fig. 27.28** Diversity Combiner.

horizontal, for example) are incorporated into the receiving antenna, and the signals are taken to two telemetry receivers. Somewhere along the way, a sophisticated circuit combines the signals, such that the result is better than the stronger of the two signals, as shown in Fig. 27.29.

Diversity combination can take place prior to carrier detection/demodulation in the receivers (*predetection diversity combining*) or following the detection/demodulation process (*postdetection diversity combining*). Predetection diversity combining generally takes place at the second intermediate frequency of the receivers (10 MHz or 20 MHz). Even though the equipment required for pre-combining is more complicated and more expensive, from the operator's point of view this is the preferred method of combining because no set-up decisions need to be made for proper operation. Further, if one of the polarizations yields signal levels that are below the receiver detector threshold, this causes no problems since the combiner is located ahead of the detector.

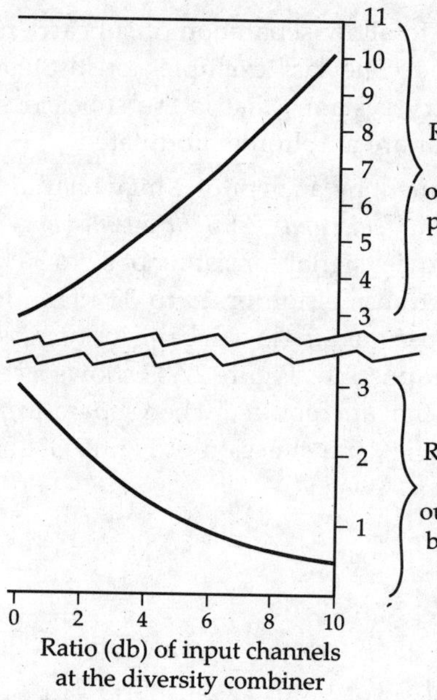


**Fig. 27.29** Diversity Combiner Performance.

The diversity combiner in a receiver system is controlled by the automatic gain controls (AGC) from the two receivers. These voltage levels give continually tracking indications of the relative signal levels in the two receivers.

### 27.14.6 Propagation Losses

The formula for calculation of free-space attenuation of radio waves is :

$$\text{Path Loss (Db)} = 37 + 20\log_{10} D \text{ (miles)} + 20\log_{10} \mathbf{f(MHz)} \quad \ldots(27.40)$$

**Multipath** is a term indicating occasional reception of a strong reflected signal in support of and later in opposition to the real signal, the result being peaks and valleys in signal strength.

## 27.15 TELEMETRY SCIENCE

*Telemetry* is the science involved in performing a measurement at a remote location and transmitting the data to a central location for processing and/or storage. Three basic parts comprise the typical telemetry system (Fig. 27.30) : (1) the data source, which is usually a transducer that converts the measured parameter into an electrical signal ; (2) the mode of transmission of the data ; (3) the receiving station and the recovery of the transmitted data. This section will concentrate on various modes of transmission and touch only briefly on the other two areas. When discussing telemetering techniques, our attention will focus on methods that provide the most efficient use of the transmission media.

The following example demonstrates what is meant by efficient use of the transmission media. Consider a 100 MHz carrier that is amplitude modulated (Fig. 27.31). Assume that the permissible transmission bandwidth is ± 5 kHz. The intelligence, which modulates the carrier, has a bandwidth of 1 kHz. From our knowledge of AM, we realize that the modulated carrier will have a bandwidth of 100 MHz ± 1 kHz.

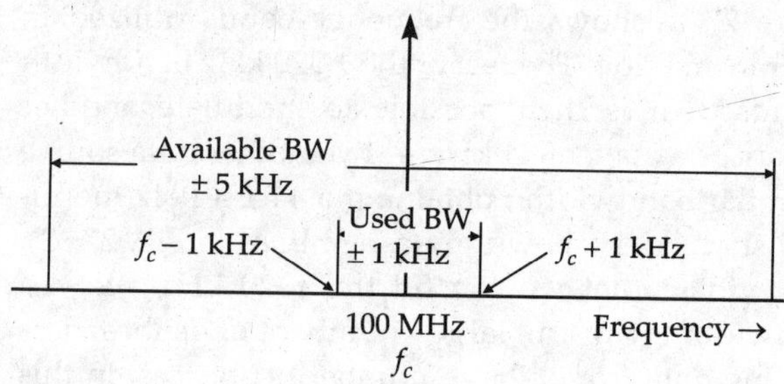


**Fig. 27.31** 100 MHz carrier with AM $\omega_m = 0 - 1$ kHz, $\omega_p$ = permissible BW, ± 5 kHz.

This is bandwidth necessary to transmit a 1 kHz data band. Clearly, since we are permitted to have a ± 5 kHz modulation bandwidth out of which the transmitted intelligence uses only ± 1 kHz, there is substantial waste. Theoretically speaking, it should be possible to use the ± 5 kHz bandwidth to transmit 5 times the 1 kHz data proposed. Conceivably, we might be able to transmit five 1 kHz, data channels via

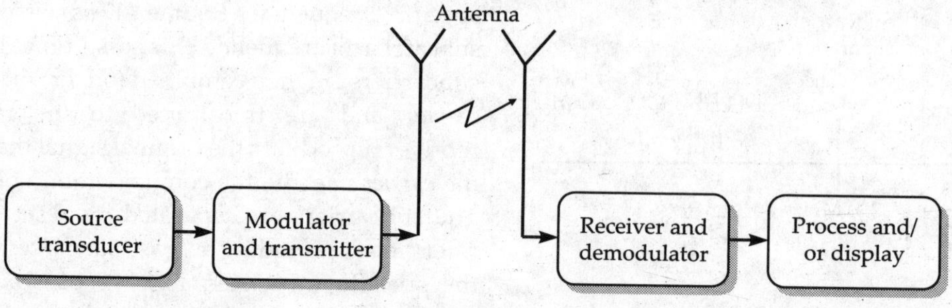


**Fig. 27.30** Typical telemetry system.

the single 100 MHz modulated carrier. A number of methods have been developed to accomplish this increase in transmission efficiency. The two most common are frequency division, or frequency multiplexing, and time division or time multiplexing.

### 27.15.1 Frequency Division (Frequency Multiplexing)

A typical telemetry carrier frequency of 230.0 MHz may be used with a bandwidth or ± 320 kHz. This implies that when used in amplitude modulation (AM), the intelligence that can be transmitted without distortion may have a bandwidth of 320 kHz. Most telemetry applications, however, involve intelligence (data) of much narrower bandwidth. For demostration purposes let us assume a data bandwdith of 4 kHz. Instead of directly modulating the 230.0 MHz with the data, we can first modulate a subcarrier of, say, 32 kHz. This subcarrier modulation will yield a signal of 32 ± 4 kHz (assuming AM). The modulated subcarrier may now modulate the 230-MHz carrier. Figure 27.32 shows the frequency bands utilized in this transmission. The rest of the ± 320 kHz bandwidth is unused. It is then possible to modulate another subcarrier — say, of 44 kHz — by another data source of similar bandwidth ; obtaining a 44 ± 4-kHz modulated subcarrier (shown in dashed lines in Fig. 27.32). It is evident that we can fill the ±320 kHz allowed bandwidth with a large number of subcarriers carrying data from a large number of sources. In this example the intelligence bandwidth has been arbitrarily limited to 4 kHz. Note that an unused band (here of 4 kHz) separates the modulated subcarriers : 32 ± 4 kHz yields 28 to 36 kHz, and 44 ± 4 kHz yields 40 to 48 kHz [4 kHz between the highest frequency of the lower subcarrier (36 kHz) and the lowest frequency of the higher subcarrier (40 kHz)]. This separation is necessary to avoid inter channel

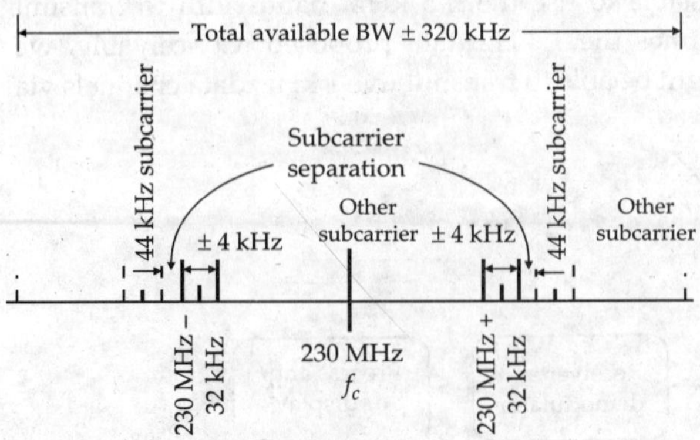


**Fig. 27.32** AM subcarriers : carrier = 230 MHz, subcarrier BW = ± kHz.

interference and to allow separation of subcarriers at the receiving end. The last example constitutes an AM/AM telemetry system ; that is the subcarrier as well as the carrier are amplitude-modulated.

An improvement by a factor of 2 may result from the use of ***single*** *sideband (SSB) transmission,* the modulated subcarrier signal consisting of 32 to 36 kHz upper sideband transmission or 28 to 32 kHz lower sideband transmission. Single sideband transmission is used whenever possible. Figure 27.33 shows a set of frequency filters that are required when separating the *double sideband (DSB)* transmission of Fig. 27.33. Here

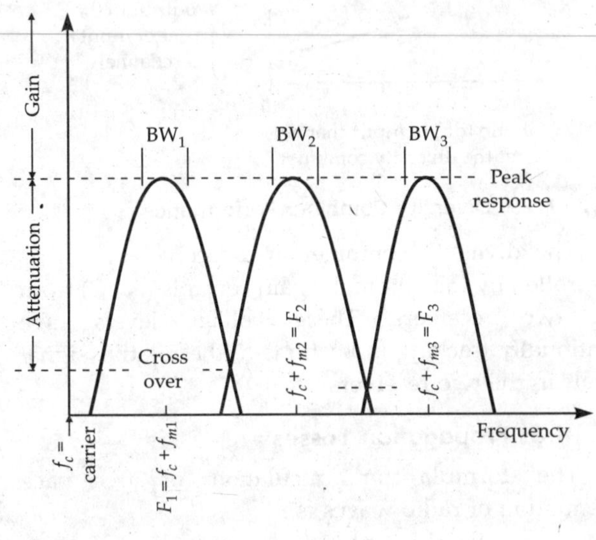


**Fig. 27.33** Filtering for frequency division.

$F_1$, $F_2$ and $F_3$ are three subcarriers, and $BW_1$, $BW_2$ and $BW_3$ are three subcarrier bandwidths (± 4 kHz = 8 kHz in this example). The crossover point would be well below the peak response — say, 80 db down in order to assure minimum internal interference. It should be stressed that $BW_1$, $BW_2$ and $BW_3$ do not have to be the same. These are selected to fit data requirements. Figure 27.34 gives a simplified block diagram of a three-channel frequency-division communication system. Each data source modulates a subcarrier of a specific frequency. Source 1 is associated with subcarrier of frequency $F_1$ and so on. All modulated subcarriers are then combined (mixer) to modulate the carrier and are transmitted to the receiver. The receiver reproduces the original signal that modulated the *carrier*-namely, the conglomeration of subcarriers. The subcarriers are separated by a set of frequency filters, each providing a "pass band" compatible with the specific subcarrier. Filter 1 passes a frequency band around the center frequency of carrier 1 ($F_1$) (see Fig. 27.33). The output of each filter consists of an amplitude-modulated subcarrier where the modulating signal corresponds to the respective data sources. Note that two detectors are used. The first "detects" or reproduces the carrier modulations, while the second reproduces the subcarrier modulation. This is to be expected, since the system consists of two successive amplitude modulations (AM/AM). We now have a communication system where each data source is allotted a specific frequency slot. To provide DSB reception, each filter must provide a pass band for two frequency ranges, corresponding to the upper and lower sidebands. The two sidebands of a specific channel are then combined to yield the channel output.

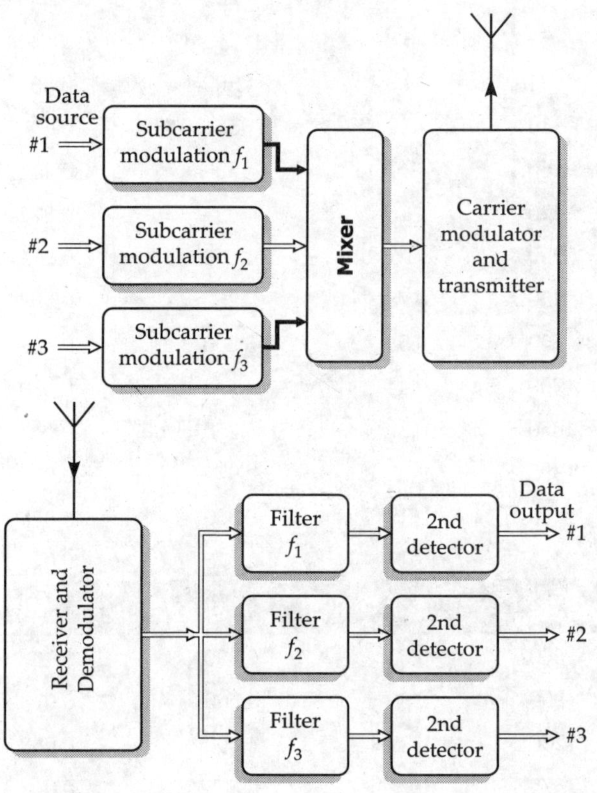


**Fig. 27.34**

The system discussed so far is an AM/AM system. Other modulation schemes such as AM/FM or FM/FM are frequently used in telemetry.

The use of the RF spectrum for telemetry and other applications is regulated by various government agencies. The Inter-Range Instrumentation Group (IRIG) has issued a set of *Telemetry Standards* revised May 1973 by Telemetry Group Inter-Range Instrumentation Group, Range Commanders Council, Doc. 10673, which specifies all pertinent details of telemetry applications. We shall discuss here some of these standards as they are related to an FM/FM telemetry system.

Twenty-one data channels, with subcarrier center frequencies ranging from 400 Hz to 165 kHz, have been designated for telemetry use. Figure 27.35 gives the details of the 21 subcarriers. As noted, all have a ± 7.5% frequency deviation assuming a modulation index of 5, we can obtain a recommended intelligence bandwidth. For example, 7.5% of channel 1,400 Hz, yields 30 Hz.

Then,

$$\text{Intelligence} \quad BW = \frac{\text{Frequency deviation}}{\text{Modulation index}}$$

$$= \frac{30}{5} = 6 \text{ Hz}.$$

The maximum frequency response shown in Fig. 27.35 is based on a modulating index of 1 (30 Hz in ms above example). The rise times given are related to the bandwidth via $T_r = 0.35/BW$ (where $T_r$ is in msec and BW is in kHz), so that channel 1 has a 30-Hz maximum response and 0.35/0.03 = 11.5 ms minimum $T_r$. Nominal $T_r$ is based on a modulation index of 5. Clearly, if a wider bandwidth is required for specific data, a larger frequency deviation, ± 15%, may be required (assuming no change in the modulation index). A ± 15 % frequency deviation may be used in conjunction with the last eight bands as shown in Fig. 27.35. Note that it is not necessary to use all eight bands with the ± 15% deviation (actually, it is not possible to use all eight bands). For example, we may use channel A (instead of 14) at ± 15 % deviation and then use channels 16 through 21 at ± 7.5% (excluding channel 15—the channel adjacent to *A*), or use channels *C* and *E* at ± 15% (in place of channels 16 and 18, excluding channels 17, *B*, *D* and *F*).

The data given in Fig. 27.35 are based on a modulating index of 1, maximum response and 5, nominal frequency response. With a "quiet" environment and good reception, an index of 1 may be used. Normal transmission conditions usually dictate the use of a modulation index of 5. It should be clear that the total cumulative bandwidth of all subcarriers must be less than the bandwidth of the carrier. The main carrier bandwidth must allow not only for the subcarrier bandwidths but also for subcarrier separation. Typically, the carrier bandwidth for FM/FM applications is ± 320 kHz in the assigned carrier frequency range of 225–260 MHz. Other frequency ranges with varying bandwidth allocations

| Channel | Centre Frequencies (Hz) | Lower Deviation Limit[a] (Hz) | Upper Deviation Limit[a] (Hz) | Nominal Frequency Response (Hz) | Nominal Rise Time (ms) | Maximum Frequency Response[b] (Hz)[a] | Minimum Rise Time[b] (ms) |
|---|---|---|---|---|---|---|---|
| ± 7.5 % Channels | | | | | | | |
| 1 | 400 | 370 | 430 | 6 | 58 | 30 | 11.7 |
| 2 | 560 | 518 | 602 | 8 | 42 | 42 | 8.33 |
| 3 | 730 | 675 | 785 | 11 | 32 | 55 | 6.40 |
| 4 | 960 | 886 | 1,032 | 14 | 42 | 72 | 4.86 |
| 5 | 1,300 | 1,202 | 1,398 | 20 | 18 | 98 | 3.60 |
| 6 | 1,700 | 1,572 | 1,828 | 25 | 14 | 128 | 2.74 |
| 7 | 2,300 | 2,127 | 2,473 | 35 | 10 | 173 | 2.03 |
| 8 | 3,000 | 2,775 | 3,225 | 45 | 7.8 | 225 | 1.56 |
| 9 | 3,900 | 3,607 | 4,193 | 59 | 6.0 | 293 | 1.20 |
| 10 | 5,400 | 4,995 | 5,805 | 81 | 4.3 | 405 | 0.864 |
| 11 | 7,350 | 6,799 | 7,901 | 110 | 3.2 | 551 | 0.635 |
| 12 | 10,500 | 9,712 | 11,288 | 160 | 2.2 | 788 | 0.444 |
| 13 | 14,500 | 13,412 | 15,588 | 220 | 1.6 | 1,088 | 0.322 |
| 14 | 22,000 | 20,350 | 23,650 | 330 | 1.1 | 1,650 | 0.212 |
| 15 | 30,000 | 27,750 | 32,250 | 450 | 0.78 | 2,250 | 0.156 |
| 16 | 40,000 | 37,000 | 43,000 | 600 | 0.58 | 3,000 | 0.117 |
| 17 | 52,500 | 48,562 | 56,438 | 790 | 0.44 | 3,938 | 0.089 |
| 18 | 70,000 | 64,750 | 75,250 | 1050 | 0.33 | 5,250 | 0.067 |
| 19 | 93,000 | 86,025 | 99,975 | 1395 | 0.25 | 6,975 | 0.050 |
| 20 | 124,000 | 114,700 | 133,300 | 1860 | 0.19 | 9,300 | 0.038 |
| 21 | 165,000 | 152,624 | 177,375 | 2475 | 0.14 | 12,375 | 0.020 |
| ± 15 % Channels | | | | | | | |
| A | 22,000 | 18,700 | 25,300 | 660 | 0.53 | 3,330 | 0.106 |
| B | 30,000 | 25,500 | 34,500 | 900 | 0.39 | 4,500 | 0.078 |
| C | 40,000 | 34,000 | 46,000 | 1200 | 0.29 | 6,000 | 0.058 |
| D | 52,500 | 44,625 | 60,375 | 1575 | 0.22 | 7,875 | 0.044 |
| E | 70,000 | 59,500 | 80,500 | 2100 | 0.17 | 10,500 | 0.033 |
| F | 93,000 | 79,050 | 106,950 | 2790 | 0.13 | 13,950 | 0.025 |
| G | 124,000 | 105,400 | 142,600 | 3720 | 0.09 | 18,600 | 0.018 |
| H | 165,000 | 140,250 | 189,750 | 4950 | 0.07 | 24,750 | 0.014 |

[a]Rounded off to nearest Hz.
[b]The indicated data for maximum frequency response and minimum rise time are based upon the maximum theoretical response that can be obtained in a bandwidth between the upper and lower frequency limits specified for the channels.
[c]Channels *A* through *H* may be used by omitting adjacent lettered and numbered channels. Channels 13 and *A* may be used together with some increase in adjacent channel interference.

**Fig. 27.35** Proportional bandwidth subcarrier channels. (From IRIG, *Telemetry Standards*)

| Transmitter | |
|---|---|
| Frequency tolerance Power | 0.01% of specified centre frequency (see list below).<br>100 W max. Power used should be no more than necessary for reliable telemetry transmission. |
| Bandwidth (modulated) | Full-power $f_c \pm 320$ MHz. 40 db down from carrier power for the ranges $f_c + 320$ kHz to $f_c + 500$ kHz and $f_c - 320$ kHz to $f_c - 500$ kHz. |
| **Receiver** | |
| Local oscillator tolerance | ± 0.001% of assigned frequency (cumulatively for all local oscillators in the receiver). |
| Bandwidth | ± 600 kHz (Go-db points). Usable bandwidth is substantially less. (See transmitter specifications). |
| Assigned centre frequencies (MHz) | 226.7, 230.4, 231.9, 232.9, 235.0, 237.0, 239.4, 240.2, 244.3, 245.3, 246.3, 248.6, 250.7, 253.8, 256.2, 258.3, 259.7. |

**Fig. 27.36** FM/FM transmitter and receiver specifications for the 225-260 MHz range. (*From IRIG Telemetry standards*)

have been specified by the IRIG *Telemetry Standards*. For example, the band 1435 – 1535 MHz has been designated for government and non-government telemetry use, primarily for flight test applications (manned aircraft —1435 – 1485 MHz ; unmanned aircraft —1485 – 1535 MHz). The band 2200 – 2300 MHz has been allocated for use in other space research applications such as launch vehicles, missiles, research rockets and space vehicles. The IRIG standards give complete specifications for the carriers and subcarriers including frequency stability, transmitting power etc. (see Fig. 27.36).

The subcarrier frequency deviations, as specified in Fig. 27.35 (±7.5% or ±15 %), are proportional to the centre frequency ; that is, if the centre frequency is higher, the frequency deviation is proportionally higher. This *frequency multiplexing* (or *frequency division*) scheme is referred to as having a *proportional bandwidth* format. This implies that only high subcarrier frequencies are suitable for wide-band intelligence. Another format, the *constant bandwidth*, assigns a fixed frequency deviation to all subcarrier channels. For example, 21 channels between 16 and 176 kHz may have maximum frequency deviations of ± 2 kHz (centre frequencies of 16 kHz, 24 kHz, etc.) or ± 4 kHz (32 kHz, 48 kHz, 64 kHz, etc.) or ± 8 kHz (32 kHz, 64 kHz, 96 kHz, etc.). Here the intelligence bandwidth, assuming a modulation index of 5, will be 400 Hz, 800 Hz, and 1600 Hz for the three respective frequency deviations ± 2 kHz, ± 4 kHz and ± 8 kHz. Once a specific frequency deviation is selected, the intelligence bandwidth is fixed for all subcarriers.

### 27.15.2 Time Division (Time Multiplexing)

*Time multiplexing* (sometimes called *time sharing* or *time division*) involves the transmission of data samples rather than continuous data transmission typical of frequency division. When telemetering slow-changing, low bandwidth data (*e.g.*, temperature : it is enough to transmit temperature data at a slow rate — say, once every 10 seconds), it would be extremely wasteful if these temperature measurements occupied a full RF communication link. To increase efficiency, the communication link should be used for other measurements during the "dead" interval between temperature samples. We see that efficient use of the communication system may be achieved by time-sharing the communication channel among a number of measured parameters, each being transmitted at a rate compatible with the rate of change of the parameters. This time sharing simply assigns time slots for specific measurands. During the 10 s interval, in our example, a number of divergent pieces of data may be transmitted. The various measurands are transmitted via a single transmit-receive link, one after the other, each in its own time slot. The receiving station must be capable of unscrambling the flow of data and producing data lines corresponding to the original measurands. It is necessary to keep track of time, or identify each time slot, so that the receiving end will recognize each data source. Figure 27.37 shows the time assignments and a block diagram of a typical time-shared telemetry system.

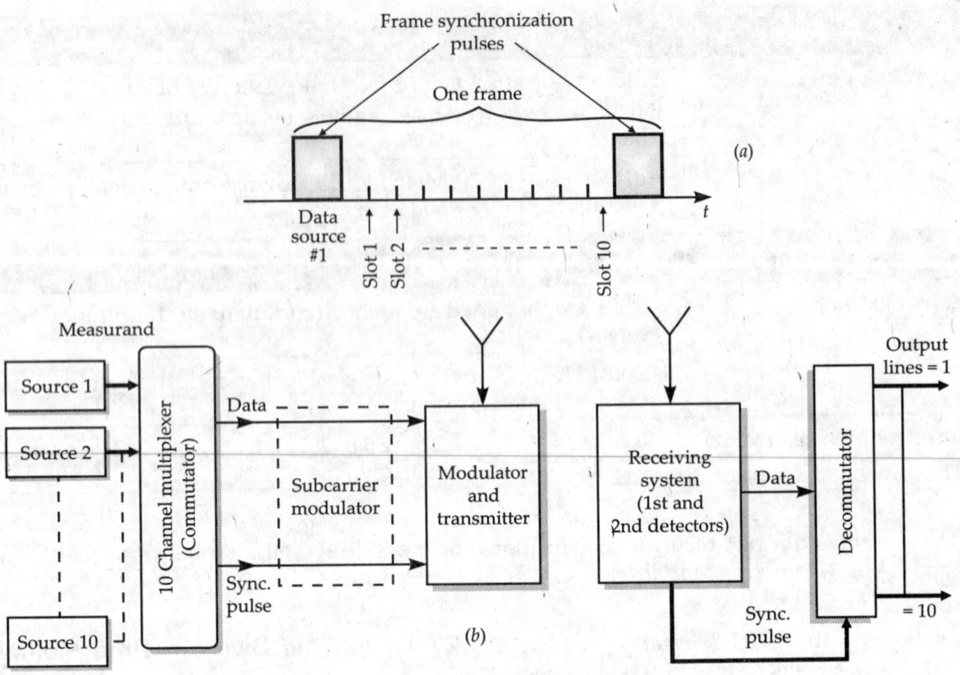


**Fig. 27.37** Time division – multiplexing : (*a*) time slot assignments (10 channels) ; (*b*) simplified system block diagram.

A common method of identifying each time slot is by measuring its position with respect to a distinct synchronization pulse that occurs at the beginning of a transmission cycle-the beginning of a "frame". A more detailed block diagram of the commutator and decommutator is shown in Fig. 27.38(*a*). The commutator multiplexes the many input lines, the many data sources, into one transmission channel. The counter identifies each time slot and hence each data source. For example, line 5, the fifth data line in the multiplexing scheme, is connected to the communication channel during the time the counter is in position 5 or count 5. Figure 27.38(*b*) shows a simplified switching arrangement of commutation and decommutation.

When the commutator switch is in position 1 so is the decommutator switch. Hence, data from line 1 are being transmitted and received. The two switches step in synchronism. The "frame sync" pulse provides positive synchronization, assuring that the commutator and decommutator start at the same position-upon occurrence of the frame sync pulse. Note that the equipment used in the commutator and decommutator is identical, the difference being the direction of data flow.

Since the commutation and decommutation is controlled by a fixed clock frequency, the stepping rate is fixed and the duration of each time slot is constant. This may be advantageous in cases where the different data sources require greatly varying bandwidths. In order to appreciate the relation between bandwidth and commutation rate, a brief discussion of data sampling is required.

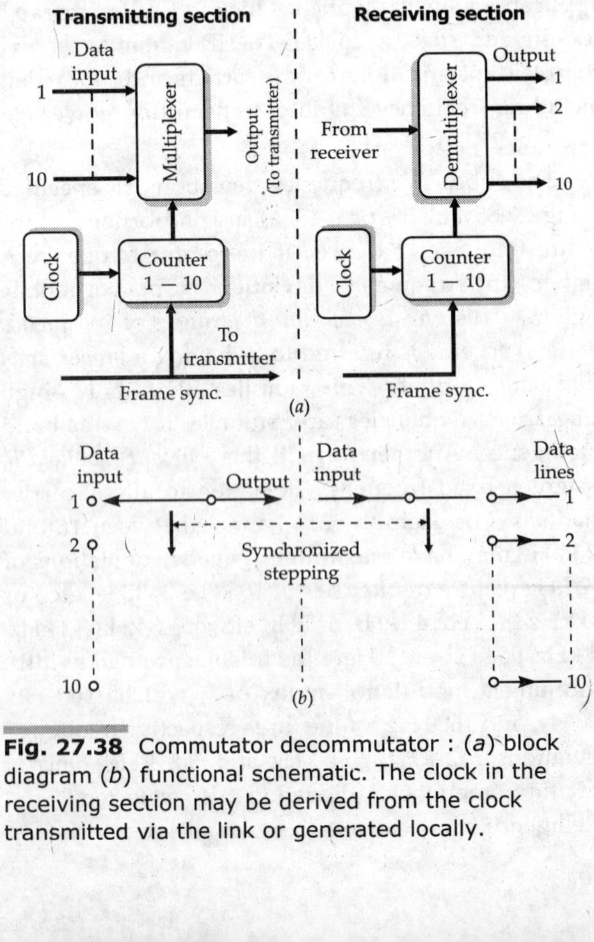


**Fig. 27.38** Commutator decommutator : (*a*) block diagram (*b*) functional schematic. The clock in the receiving section may be derived from the clock transmitted via the link or generated locally.

As we have previously noted, a sine wave can be reproduced from a set of voltage samples. To reproduce a 1 kHz sine wave with good fidelity (distortion of less than 1%) we require at least 5 samples for each cycle of the intelligence. The 1 kHz signal must then be sampled at a rate of 5000 samples/s, producing 5 samples per cycle of the data. If we propose to commutate 10 data sources, each requiring a sampling rate of 5000 samples/s (each having an intelligence bandwidth of 1 kHz), we need a commutation rate of 10 × 5000 samples/s = 50,000 samples/s. The commutator must switch from source to source at a 50 kHz rate (20 μsec) so that each source will be visited once every 10 steps, once every 200 μs – *i.e.*, at a 5 kHz rate. The frame rate, the number of frames per second, will be 5000 frames/sec. The commutation rate will be frame rate times the number of data sources in the system, or frame rate times the number of samples per frame (5000 × 10 = 50,000 sample/s). The communication link must be capable of transmitting the high data rate, 50,000 samples/s, without appreciable degradation. This would imply the need for a communication system with a bandwidth capability of well over 50,000 Hz, corresponding to the rate of 50,000 samples/s.

The samples of the different data sources are shown in Fig. 27.37(*b*) to modulate the carrier directly. While this direct modulation is possible, the samples are often used to modulate a subcarrier which then modulates the carrier, as shown in broken lines in Fig. 27.37(*b*). The samples are transmitted on one of the subcarriers of a frequency-multiplexed system. This permits the flexibility of combining both multiplexing methods in one communication link. The samples themselves are pulse amplitude modulation (PAM) : that is, the data is pulse-amplitude-modulated, resulting in pulses whose amplitude represents the data. Since these PAM signals modulate the subcarrier (say, FM) which then modulates the carrier (say, FM), we obtain a PAM/FM/FM system.

We may now continue to analyze an example intended to demonstrate the effect of sampling on the bandwidth efficiency of the system.

Consider a carrier of 100 MHz which is modulated (FM) by a subcarrier of 70 kHz center frequency. The intelligence is carried by an FM modulation of the 70 kHz subcarrier. We thus have an FM/FM communication channel. To conform to standards we have to restrict the frequency deviation of the subcarrier to ± 15%. With a modulation index of 5 this would mean an intelligence bandwidth capability of about 2100 Hz, far below the 50,000 Hz bandwidth required for the proposed multiplexed system. If the number of samples per frame were reduced to 1—that is, a single data source-a commutation rate of 5 kHz would be required, still beyond the 2100 Hz available in the 70 kHz subcarrier. Note that with a single data source no multiplexing is necessary, and hence continuous transmission is possible (no sampling). This would permit an intelligence bandwidth of 2100 Hz, double the required for a single source (1 kHz in our previous example). This deterioration in bandwidth utilization-with sampling we need a bandwidth of 5 kHz, without sampling only 1 kHz—is due to the sampling involved. When we use 5 samples for each cycle of the data we require more than 5 times the bandwidth to transmit this data. While there is inefficiency in the utilization of the bandwidth available, there is an advantage in using a single subcarrier for a large number of data sources. Time division, which utilizes sampling techniques, is used mainly in applications with low bandwidth requirements. High-frequency data may be transmitted by the use of discrete sampling. In this method the duration of each sample is much *longer* than the period of the intelligence ; it contains 5 or more cycles, of the intelligence. This simply means that the data is unaffected by the sampling. It also means that we transmit the data in "bursts," one source at a time. We must be assured of no loss of data during the time the transmission is cut off for a particular source.

The mode of transmission above was assumed to be FM/FM. Hence, the frequency of the subcarrier (the frequency deviation) during a particular time slot represents the magnitude of the measurand being sampled at that time. Each time slot contains a signal whose frequency deviation from the subcarrier center frequency is produced via FM by the data source. The width of these time slots is fixed and identifiable with reference to the synchronization pulse. The sync pulse consists of a maximum frequency deviation and of a duration which is double the normal time slot. The extra width is necessary to distinguish the sync pulse from the data slots.

The characteristics of the transmission system are standardized and controlled by various governmental or international bodies (depending on the specific application : satellite telemetry—international agreement, industrial telemetry—government control, etc.). For example, frame rate must be kept constant to it ± 5% (long-term stability) ; the length of a frame is limited to 128 channel time slots ; and so forth (IRIG, *Telemetry Standards*). Note that at high subcarrier frequencies the available bandwidth is greater ; hence, the commutation rate may be higher.

For improved efficiency it is sometimes useful not to have the same sampling rate for different sources. The wide-band data source must be sampled more often than the low-frequency one. This is easily accomplished by simple changes in the commutator and decommutator interconnection. For example, if we interconnect stations 1 and 5 in a 10-point commutator (multiplexer), the data source connected to stations 1 and 5 will be sampled twice within a single frame and hence at double the rate. It is also possible to subcommutate-that is, to select one or-more time slots during which an additional set of data sources is multiplexed. The sample time of the main frame becomes a subframe for the subcommutator. These techniques permit extreme flexibility in adapting the system to a wide range of bandwidth requirements.

### 27.15.3 The Complete Telemetry Package

We have so far described separate telemetry techniques. To round off the discussion let us present a telemetry system which utilizes all the various telemetry methods. This is not to imply that such a complex system is the typical telemetry system ; it allows us, however, to interrelate the various techniques.

Figure 27.39(*a*) and (*b*) show transmitting and receiving telemetry stations. As shown, the system consists of a mixture of techniques and handles 39

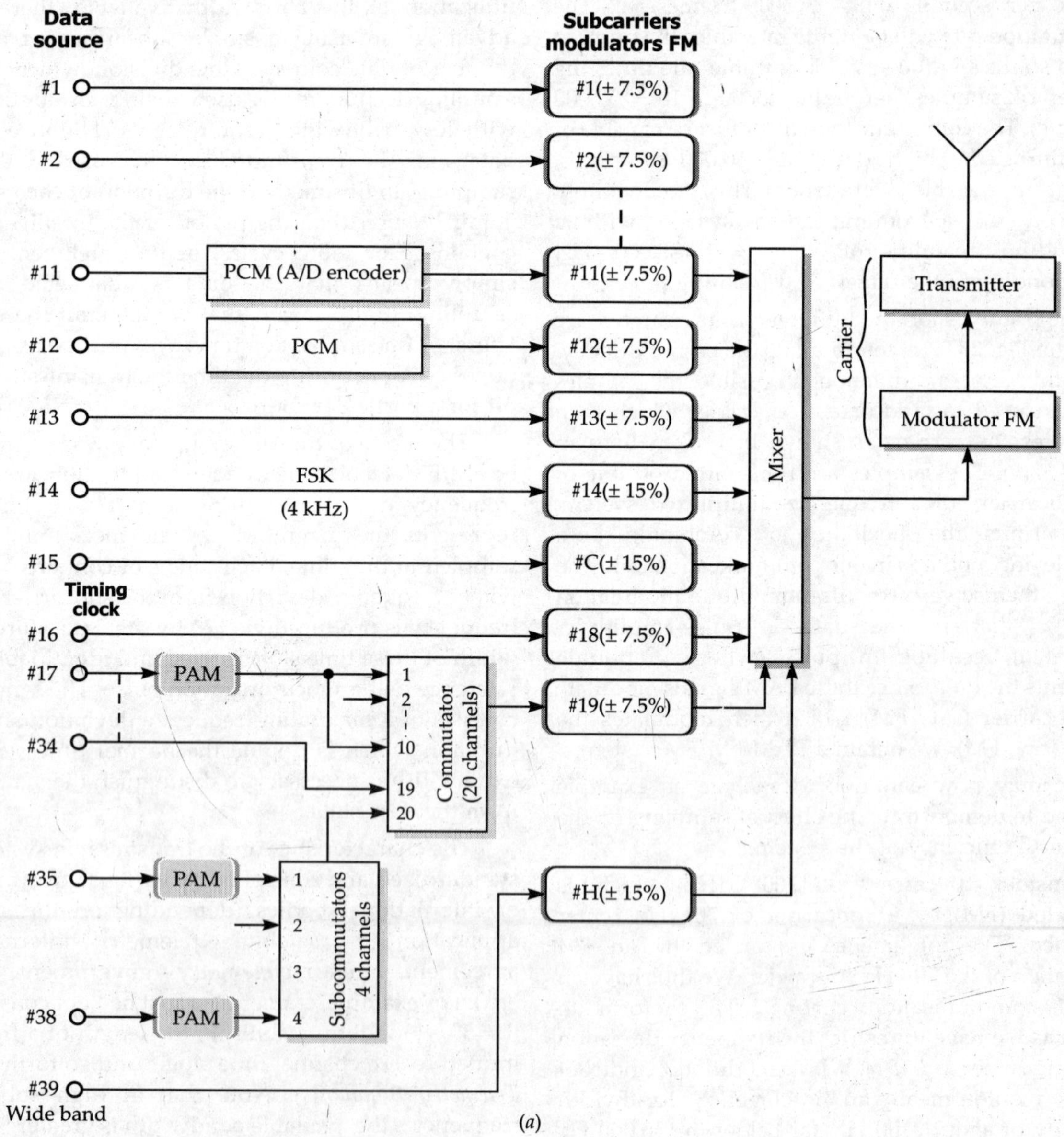


**Fig. 27.39** (*a*) Telemetry transmitting system.

different lines of data. The 18 subcarriers shown in Fig. 27.39(*a*) (see frequency and deviation data in Fig. 27.35) provide continuous transmission. Subcarrier 19 is utilized in conjunction with a commutator and subcommutator. Subcarrier 19 (93 kHz) has a bandwidth of 1395 Hz (nominal), a relatively large bandwidth, and good rise time (0.25 ms). This means that the commutated sources may not consist, of signals with a rise time shorter than 0.25 ms. In reality, the standards require that the sampling rate be no faster than the nominal rise time (0.25 ms here). We may now assume a sampling rate of 1 kHz, 1 sample/ms, or 20 ms for a single frame. This will permit a subcommutation rate of 4 steps ms or 0.25 ms per sample (data lines 35 through 38). Note that data line 17 is connected to two points on the commutator (data line 34), hence, it is sampled twice per frame. Data line 17 can be a signal that needs sampling every – 10 ms, while all others are sampled only once every 20 ms. The 4 subcommtated data lines are sampled at a 0.25 ms rate (each sample may last 0.25 ms), once every 20 ms, the same as the other commutated lines. Subcarrier C is shown to carry a *frequency-shift-keyed* (FSK) signal of 4 kHz. This 4-kHz signal may be the

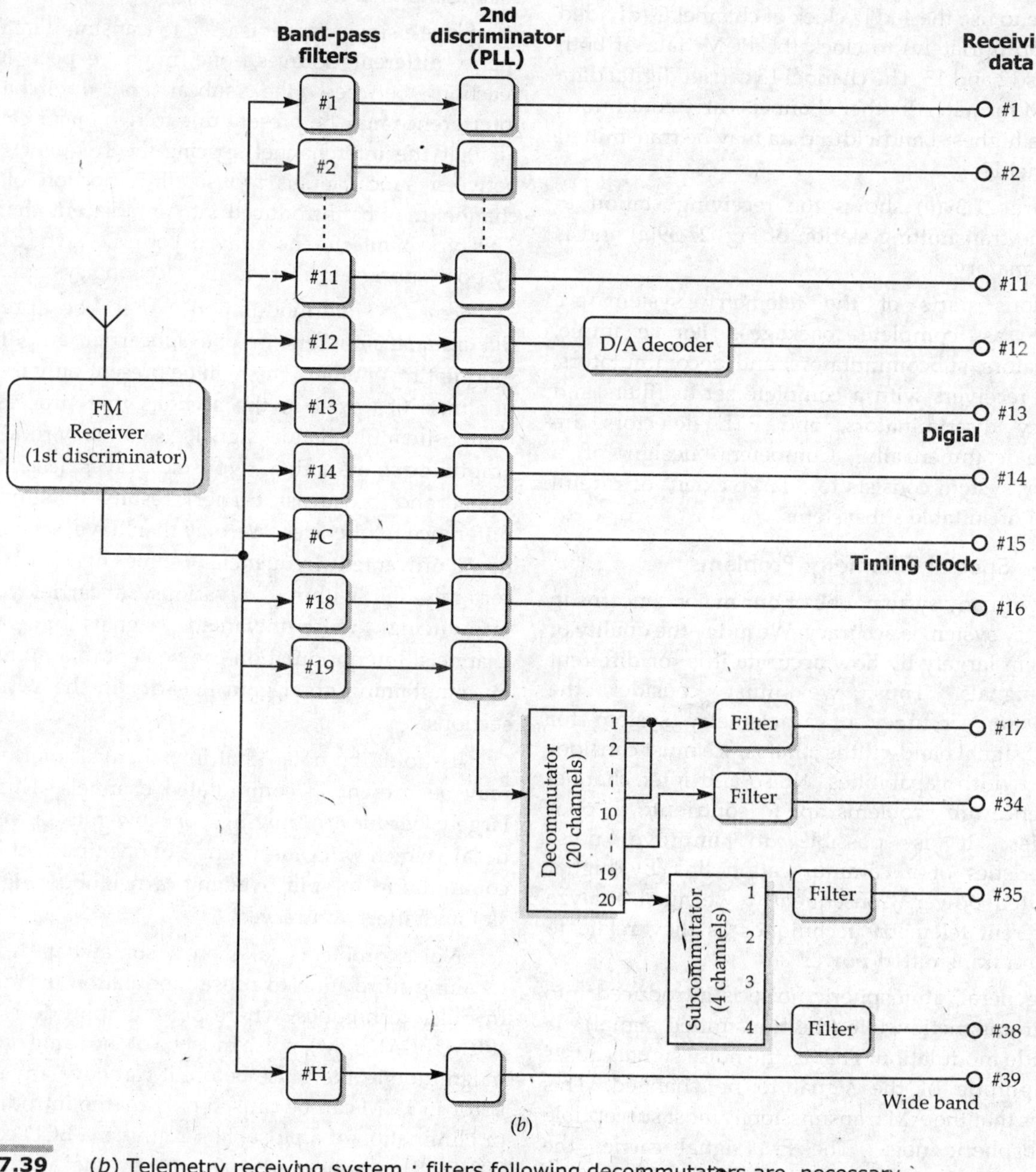


**Fig. 27.39** (*b*) Telemetry receiving system ; filters following decommutators are necessary to reconstruct analog data from PAM samples.

clock controlling the commutating section (1 kHz is divided down from 4 kHz) an the decommutator operation, providing extra synchronization between commutator and decommutator (this not essential, since the sync pulse in the frame sequence provides synchronization). It should be noted that 20 commutated lines are shown in an assumed 20 ms period, not allowing for the sync pulse. This is done only to preserve simplicity.

Data lines 1 through 11 are of low-bandwidth data. Line 12 consists of a PCM signal obtained by digitizing and analog signal (A/D conversion). It is advisable to use the 4 kHz clock of channel C (divided down appropriately) to clock the PCM data of both data lines 12 and 13. The channel 13 carrier digital data is in PCM format. All other channels carry continuous data. The highest-bandwidth data may be transmitted on channel H.

Figure 27.39(*b*) shows the receiving station to match the transmitting station of Fig. 27.39(*a*) and is self-explanatory.

Various parts of the telemetry system are available as complete packages. For example, commutators, subcommutators, and decommutators, FM-AM receivers with a complete set of filters and frequency discriminators, and PLL detectors are available commercially. Competent design of a telemetry system consists to a large extent of careful selection of suitable subsystems.

### 27.15.4 Special Telemetry Problems

As with any system, one of our major concerns in a telemetry system is accuracy. We judge the quality of the system largely by how accurate it is for different input signals. Thus, we must consider the reproduction accuracy of a telemetry system for different signal bandwidths ; that is, we must consider the bandwidth capabilities. Noise and inter channel interference are problems apt to contribute to poor accuracies. It is possible to improve noise characteristics of a communication link by higher transmitted power. Consequently, we must analyze the different telemetry techniques as they relate to maximum transmitted power.

In general, atmospheric noise is introduced into an electromagnetic wave (the transmitted signal) via amplitude modulation. That is, the noise signal causes the amplitude of the signal to be changed. This indicates that the AM transmission is most susceptible to atmospheric noise. The FM signal carries the intelligence in frequency and not amplitude variations ; hence, an amplitude change can be eliminated by the "limiter" in the receiver. The limiter is designed to equalize the FM signal amplitude. It maintains a constant FM signal amplitude and reduces all AM components. The FM technique is usually employed at high carrier frequencies (100 MHz and above) and has a much larger bandwidth than the AM. The high carrier frequency used makes the FM system more compact and more efficient in terms of power. It also permits better electromagnetic propagation, further improving the noise characteristics of FM. Since most telemetry systems involve subcarrier operation, we must consider the effects of the multiple subcarriers as they relate to noise and other spurious signals.

Since a single carrier is used to transmit data from many different sources, one might expect inter-reaction between the subcarriers. Interchannel interference may be present due to two major causes. *First,* if the interchannel spacing-the frequency gap between subcarriers-is too small, a portion of one channel may be introduced into an adjacent channel. This type of interference may, of course, also be caused by poor subcarrier filters in the receiving system.

*Second,* "intermodulation" may be present. *Intermodulation* refers to one subcarrier amplitude-modulating another. This will be present only if some nonlinerrities exist in the sections that process the composite multichannel signal. As you may recall, the amplitude modulation two sine waves (*e.g.*, voice signal and broadcast carrier) results in sum and difference frequencies. We may then have a variety of new, unwanted frequencies, some of which will certainly fall within the various subcarrier bands, introducing noise-unwanted signals-into these channels. Intermodulation can be kept to a minimum by maintaining good gain linearity in the pertinent sections.

It should be noted that interchannel interference may be present in commutated channels. This will largely be due to "ringing" or slow rate of voltage decay, which will cause signals of one time slot in the commutator' to spill over into a neighbouring time slot and affect accuracy.

Noise problems take on a somewhat different meaning in relation to pulse modulation techniques. In pulse techniques, where pulse amplitude is fixed (PCM, PDM, PFM), noise must be of the same order of magnitude as the pulse before it can have any effect. Error in PCM can be caused only by the introduction or elimination of a pulse. For example, a BCD code of 0001 = 1 may become 1001 = 9 by the action of a spurious pulse. The magnitude of the error may be enormous ; however, a substantial noise signal is required to introduce such an error. In practice, PCM is highly immune to noise. Similarly, PDM and PFM have-high noise immunity, PAM, where the pulse *amplitude* is the parameter representing the intelligence, is much more susceptible to noise interference.

### 27.15.5 Telemetry Hardware and Applications

Figure 27.40 gives a block diagram of a PCM telemetry package. The subsystem shown contains only the input and pulse-processing section. This permits modular construction of telemetry systems with a number of similar units (as many as needed for a particular application) in conjunction with a communication link. It is important to note that a unit such as this may be used not only for wireless transmission. The digital data may be frequency-shift-keyed into a *voice line,* a line capable of carrying voice data-that is, intelligence with about a 3000-Hz bandwidth.

Figure 27.40 is for the most part self-explanatory. We have a signal conditioner to amplify and condition the transducer signals. This is usually necessary since most transducer signals are of millivolt magnitude. An analog-processing section includes an analog multiplexer with submultiplexer or subcommutator, a sample-hold circuit, and an A/D converter. The digital data are entered via a parallel-to-serial converter, since most digital data are in parallel format, and via digital multiplex. This means that a number of analog and digital data sources are commutated and formatted to produce the stream of PCM data. The analog-digital word-selector (right side of diagram) controls the sequence of data commutations and feeds the encoder, which is designed to produce the specific levels and codes suitable for either line or radio link transmission. (This complete subsystem may be used in conjunction with a single subcarrier.) A sync word and frame identification generator is shown. The sync word provides frame synchronization. In PCM it is common to use a complete word with a specific code, which hopefully occurs only once in a frame (at sync time). This sync word performs the function of the frame sync pulse. The timing control for the subsystem is provided by a precision oscillator-pulse generator, with a number of frequency dividers and various control logic circuits. The subsystem shown is capable of handling word length of 1 to 16 bits and frame length of 1 to 32 words, and the number of subframes may be between 2 and 32. The rate at which the various circuits operate (*e.g.*, bit rate, frame rate) is controlled by the main control unit. Provision has been made for a wide range of bit and frame rates.

A special class of telemetry hardware, the *modem,* is manufactured by a large number of companies. The modem, an acronym for *modulation-demodulation,* handles the modulation and demodulation of telemetry signals. In recent years with the expansion of digital technology, digital modems have come into being. These handle only digital data, similar to the subsystem of Fig. 27.40. The modulation and demodulation techniques used vary from unit to unit. To accommodate high-speed serial data, the modem converts the high-speed serial word into a number of parallel low-speed bit streams. These parallel stream are used in a frequency-multiplexed system. For example, a rate of 12,00 bps (bits per second) is obtained by a 16-channel frequency-multiplexing of a voice line bandwidth of 375 to 3025 Hz. Each of the 16 channels carries FSK data of a 75 bps rate for of transmission rate of $75 \times 16 = 1200$ bps. The assigned channels are 170 Hz apart, from 425 Hz to 2975 Hz. The FSK signal consists of a tone shift of 85 Hz centered around the assigned channel frequency ($f_c$). This actually provides three recognizable levels $f_c$ + 42.5 Hz, $f_c$ – 42.5 Hz and $f_c$. The-last represents no data. Hence, both the mark and the space (or logic "1" and logic "0") are distinct and separate from the channel frequency.

The receiving-end filters demodulate the 16 FSK channels and recombine them to produce the original serial "stream."

Some modems use PSK (phase-shift keying) instead of FSK. This technique shifts the phase of the tone in the frequency-multiplexed voice channel with respect to a reference signal. Typically, a phase shift of 45°, 135°, 225° and 315° represents 2 bits (two sets of logic, "0" and "1"). The phase shift is then measured or detected at the receiving end and a corresponding logic level produced.

The modems discussed above use a narrow-band transmission channel. The techniques used are clearly suitable for wide-band channel transmission. The great advantage of wide-band transmission is the very high data rates that may be obtained and hence the elimination of the serial-to-parallel conversion. These wide-band transmissions usually operate in conjunction with microwave transmission where noise effects are less deleterious. A bandwidth of, say, 48 kHz may permit a total data rate of 48 kbps (kilo bits per second). Rites of up to 3.8 Mbps (Mega bits per second) are theoretically possible.

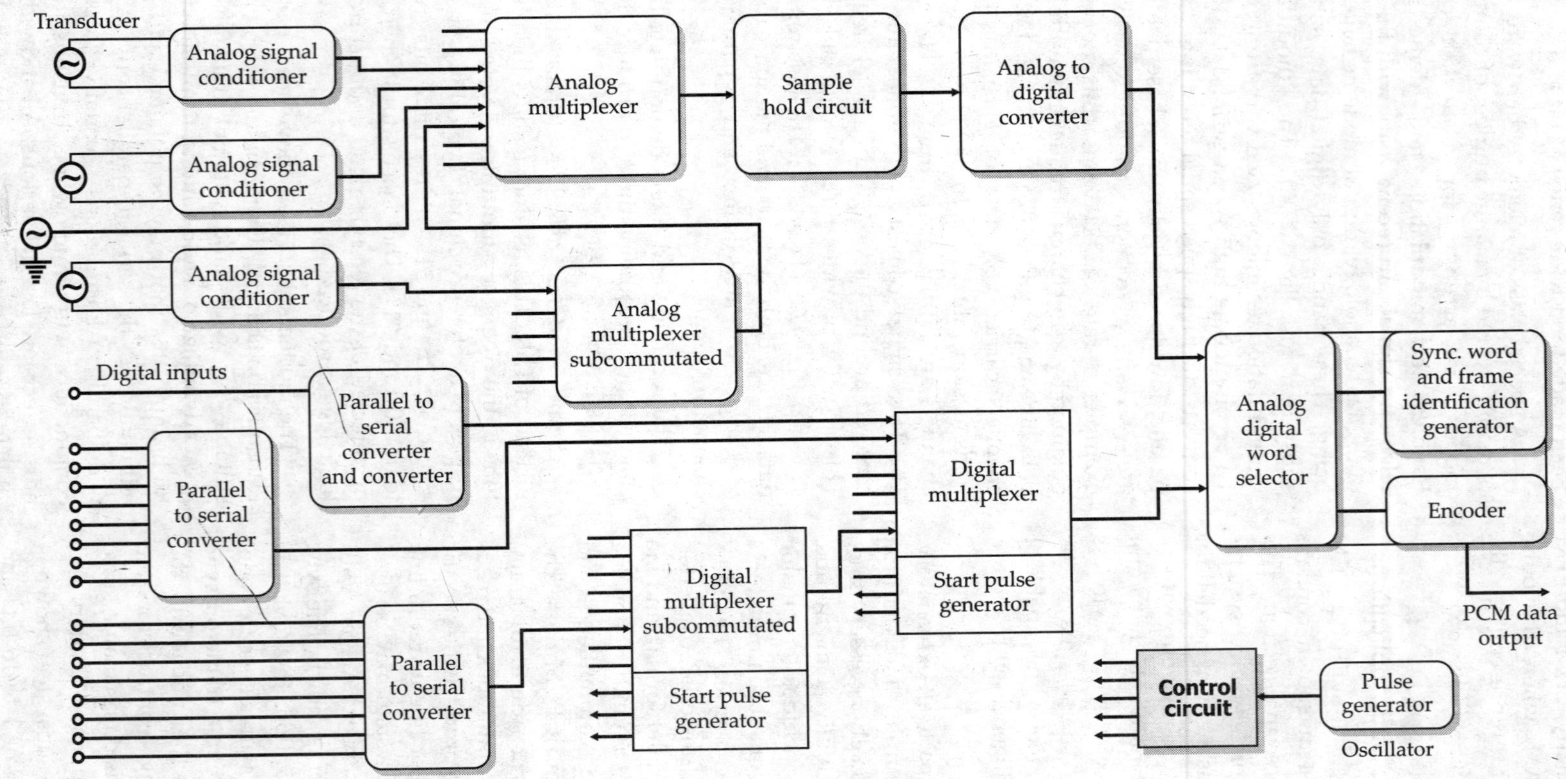


**Fig. 27.40** Block diagram of a typical PCM telemetry system.

## 27.16 BANDWIDTH AND NOISE RESTRICTIONS

A consideration of fundamental importance in the measurement of a quantity is the signal-to-noise ratio at the output indicator or recorder. Noise or interference introduced in the measurement system is superimposed on the signal and obscures or alters the information. There are many potential sources of noise in a system and they may well be significant if signals have a low level. Noise stems mainly from electric and magnetic fields, and thermal and acoustic sources. Examples range from d.c. drift and ground currents, a.c. pickup via electrostatic and electromagnetic coupling, mains frequency hum and supply ripple, switching and line transients, to mechanical contact and vibration, thermal e.m.f.s, thermal and semiconductor noise. Atmospheric and stellar noise may cause interference in some kinds of telemetry signals. Other examples of unwanted signals are 'quantization noise' from analogue to-digital converters, aliasing signals in connection with samplers and uncompensated 'wow and flutter' noise from magnetic tape recorders.

If 0.1 per cent accuracy of transmitted information is to be maintained, interference must be suppressed such that the signal-to-noise ratio is better than 60 db. With careful equipment design, the non-random noise sources can usually be avoided. Random noise is characterized by randomness of amplitude and frequency distribution. Most random noise in instrument system arises as a result of charge motion within resistors, semiconductors, or electronic valves in the signal channels. Thermal, or Johnson, noise is due to thermal agitation of the current carriers in the conductor or semiconductor. The r.m.s. noise voltage $V_{rms}$ developed in a resistor, is stated as

$$V_{rms} = \sqrt{(4\,kTRf)}$$

where $T$ is temperature (K), $k$ is Boltzmann's constant ($1.38 \times 10^{-23}$ J/K), $R$ is the resistance of the noise source (Ω) and $f$ is the frequency bandwidth of interest (Hz). The generation of thermal noise is uniform over the entire frequency spectrum and such noise is called 'white noise' ; thus the response bandwidth of the instrument system, irrespective of its centre frequency, determines noise amplitude appearing at the output. At room temperature thermal noise power amounts to approximately $1.6 \times 10^{-20}$ W/Hz.

Composition and deposited carbon resistors generate noise proportional to applied voltage and is typically between 0.1 and 10 r.m.s. μV per applied volt per frequency decade. In active devices, such as transistors and electronic valves; noise is generated by thermal agitation as in resistors, by the random generation of charge carriers (called shot noise), and by a 'flicker' effect due to recombination of carriers and electrode surface effects. Shot noise has a uniform frequency distribution, but its magnitude depends on the operating current in the active device. Flicker noise power is concentrated at low frequencies, having a distribution approximately proportional to the inverse of the center frequency of the passband ($1/f$), down to very low frequency. This noise component predominates in low-frequency and d.c. amplifiers. In amplifiers using low-noise transistors, however, the $1/f$ noise component loses its importance above about 1 kHz.

Random noise is generally of importance only in the low-level part of the signal channel. Such noise arising in the higher level parts of the channel is usually negligible compared to the amplified noise from the input. It is evident that an increase in the bandwidth of an instrument system increases the noise amplitude, but it may also increase the signal amplitude by passing more frequency components in the signal. Such an increase in bandwidth may in some cases result in an improvement in received signal-to-noise ratio, while in other cases there is a reduction in received signal-to-noise ratio.

The noise figure ($F$) for a device or system gives a measure of the noise generation in the device or system, and is defined, for impedance-matched input and output, as the ratio of the signal-to-noise power ratio at the input to that at the output. It is usually expressed in decibels, that is $F_{db} = 10 \log_{10} F$. For example, a typical low-noise preamplifier employing a low-noise field-effect transistor input stage has a 1 db or better noise figure over the frequency range 1 Hz to 500 kHz with source resistances from 5 kΩ to 26 MΩ. The noise power density at the amplifier output, for shorted input, would be about 5 $nV_{rms} / \sqrt{Hz}$ above 1 kHz.

The *bandwidth* and *rise-time* of a system are closely related. The rate at which the output voltage of an electrical system rises when an input voltage is suddenly applied is commonly expressed in terms of the time it takes the output voltage to rise from 10 to 90 per cent of its full amplitude. This rise-time is inversely proportional to the bandwidth of the amplifier, and when the overshoots are less than 5 per cent it is given with good accuracy by the relation

$$\text{rise-time (in seconds)} = 0.35/\text{BW} \quad ...(27.41)$$

where bandwidth BW in hertz is defined as the frequency band for which the overall response of the system does not fall below 70.7 per cent (3 db) of the mid-frequency response. Rise-times add together as the square root of the sum of the squares of the rise-times of the individual components involved. For example, a step-function test signal with a given rise-time injected into a single time-constant amplifier having the same rise-time gives a signal output with a rise-time which is 1-4 times the rise-time of the input test signal.

Measured information is transmitted by some quantity assuming different values in succession. In all forms of communication something must change its value or level. There sue two very important and fundamental limits to the quantity of information which can be transmitted in a given time.

*Firstly*, there is a limit to the number of different, levels which may be distinguished : noise is responsible for this limit.

*Secondly*, a finite time is needed for the quantity to change from one level to another : mechanical inertia or capacitive and inductive circuit elements are responsible for this limit, and bandwidth and rise-time express the limit.

It is possible to quantify the amount of information that can be transmitted by a single signal which may be at any one of $N$ levels. The usual unit of information is the binary digit or *bit*. A single answer which must be either 'yes' or 'no' provides one bit of information. This is strictly true only if the answers 'yes' and 'no' are equally likely. If there are four possible levels, 0, 1, 2 and 3, then the level intended could be fixed by the answers to two 'yes' or 'no' type questions, such as 'Is it greater than 1 ?' 'Yes'. 'Is it number 2 ?' 'No'. Therefore the required level must be number 3. Similarly, to fix which of 8 the possible levels is intended requires 3 bits, 16 levels need 4 bits and 32 levels need 5 bits.

Now $4 = 2^2, 8 = 2^3, 16 = 2^4$ and $32 = 2^5$.

The number of bits of information contained in the knowledge that a signal is at a particular level, when it could equally likely be at any one of $N$ possible levels, is $\log_2 N$ bits. To apply this to an electrical system, we will assume that the input signal power is $P_s$ and that the input noise power is $P_n$, so that the total input power is $P_s + P_n$. Voltage levels differing by less than the noise voltage cannot easily be distinguished and so, since voltage is proportional to the square root of power, the greatest number of distinguishable levels is $\sqrt{[(P_s + P_n)/P_n]}$. A single signal transmitted over a system in which the signal-to-noise power ratio is $P_s / P_n$ is therefore capable of containing up to

$$\log_2 \sqrt{\left(\frac{P_s + P_n}{P_n}\right)} = \frac{1}{2} \log_2 \left[1 + \frac{P_s}{P_n}\right] \quad ...(27.42)$$

bits of information.

If a system can respond to all frequencies up to $f$, then the rise-time is approximately one half of the period of the sine wave of this frequency, that is ; the rise-time is $1/(2f)$. Figure 27.41 shows a dotted sinewave of this highest frequency to which the system responds, and the thick line shows that this is equivalent to a rise-time of approximately one half of the complete period. Since the time required to change from one level to another is $1/(2f)$, there are $2f$ changes in level effected in one second when the bandwidth is $f$ Hz. This is not really a proof, and there are the difficulties of how we define rise-time and bandwidth. However, the final result we shall obtain is true and is capable of rigorous proof.

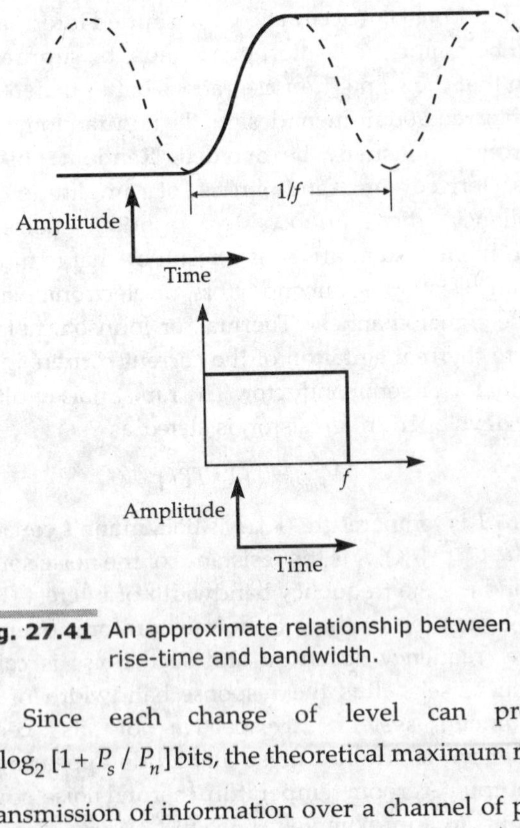


**Fig. 27.41** An approximate relationship between rise-time and bandwidth.

Since each change of level can provide $\frac{1}{2} \log_2 [1 + P_s / P_n]$ bits, the theoretical maximum rate of transmission of information over a channel of power signal-to-noise ratio $P_s / P_n$ and bandwidth $f$ Hz is,

$$2f \times \frac{1}{2} \log_2 \left[1 + \frac{P_s}{P_n}\right] = f \log_2 \left[1 + \frac{P_s}{P_n}\right] \text{ bits/s} \quad ...(27.43)$$

Although it has been realized for a long time that a high rate of information requires a wide bandwidth and a good signal-to-noise ratio, this very important equation, due to C.E. Shannon (1948), provides a quantitative relationship. If the signal-to-noise voltage ratio is $V_s / V_n$, then the data-transmission rate (C) is given by the equation

$$C = 2 f \log_2 [1 + V_s / V_n] \text{ bits/s} \quad ...(27.44)$$

The number of bits of information transmitted in time $\tau(H)$ will be given by the expression

$$H = 2 f\tau \log_2 [1 + V_s / V_n] \text{ bits} \quad ...(27.45)$$

In an ideal system it is possible to 'trade' bandwidth for signal-to-noise ratio. If a data-transmission system is split into two transmission channels, (*a*) a primary information channel (before modulation and after detection), and (*b*) a carrier transmission channel, then ideally information in channels (*a*) and (*b*) is the same.

That is

$$\left[1 + \frac{V_s}{V_n}\right] = \left[1 + \frac{V'_s}{V'_n}\right]^{f'/f} \quad ...(27.46)$$

where $f$ and $V_s / V_n$, refers to channel (*b*), and $f'$ and $V'_s / V'_n$ refers to channel (*a*) ; $f' / f$ is called the bandwidth expansion factor. Thus, increasing the bandwidth of the carrier transmission channel sometimes allows an increase in the detected signal-to-noise ratio.

**Example 27.9** *A data signal having frequency component from d.c. to 1 kHz is to be transmitted using a modulation method. Determine the minimum carrier channel bandwidth using the following methods of modulation :*

(*a*) *amplitude modulation.*

(*b*) *frequency modulation with maximum deviation in carrier frequency being 1.5 kHz.*

(*c*) *pulse code modulation using an 8 digit code.*

**Solution.** (*a*) For amplitude modulation, the minimum bandwidth of carrier channel is twice of the data

∴ Minimum width of carrier channel

$$= 2 \times 1 = 2 \text{ kHz.}$$

(*b*) For frequency modulation, the minimum bandwidth of carrier channel is given by Carson's rule. (Eqn. 27.32)

Minimum bandwidth of carrier channel

$$= 2\,(\delta + f_h) = 2\,(1.5 + 1) = 5 \text{ kHz.}$$

(*c*) For pulse code modulation (PCM), the minimum bandwidth of carrier channels is given by Eqn. 27.9. Minimum bandwidth of carrier channels.

$$= n\, f_h = 8 \times 1 = 8 \text{ kHz}$$

**Example 27.10** *An F.M. telemetry system uses a 370 - 430 Hz voltage controlled oscillator to carry the fuel level of 3000 l tank, where 370 Hz represents an empty tank and 430 Hz represents full tank. What level does 403 Hz represent ?*

**Solution.** Change in frequency from empty tank to full tank

$$= 430 - 370 = 60 \text{ Hz}$$

At 403 Hz, change in frequency from empty tank

$$= 403 - 370 = 33 \text{ Hz}$$

∴ A frequency of 403 Hz represents a fuel level

$$= 33 \times \frac{3000}{60} = 1650 \text{ l}$$

**Example 27.11** *A 5 channel d.c. to 50 Hz telemetry system uses PAM and PCM systems. What is the lowest practical sampling rate (excluding the frame synchronization sequence) for good quality data ?*

**Solution.** In order to recover good quality data, the sampling rate should be at least 5 times the data frequency if there is one channel. For five channels, the sampling rate should be

$$5 \times 5 \times 50 = 1250 \text{ sample per second.}$$

**Example 27.12** *It is necessary to transmit measurement data from one point to another using a 7 digit code in a pulse code modulation system.*

(*a*) *Given that the transmission channel bandwidth is 100 kHz and the signal to noise voltage ratio at the modulation detector is 7 : 1, determine the maximum possible data transmission rate of the channel.*

(*b*) *If the data are obtained from various transducers and in all cases contain frequency components upto 1 kHz, determine the theoretical maximum possible number of channels that can be accommodated without loss of information using time division multiplexing.*

**Solution.** (*a*) Maximum possible data transmission rate is :

$$H = 2\, f \log_2 (1 + V_s / V_n)$$
$$= 2 \times 10^3 \log_2 [1 + 7]$$
$$= 6 \times 10^5 \text{ bits per second.}$$

(*b*) The minimum sampling rate per channel

$$= 2\, f_h = 2 \times 10^3 \text{ samples per second}$$

Now 7 bits are required for each sample, so the total possible number of maximum samples per second

$$= \frac{6 \times 10^5}{7} = 85714.$$

$$\therefore \text{ Maximum number of channels} = \frac{85714}{2000} = 42$$

(This ignores time required for synchronization)

***Example 27.13*** *You have a PCM system operating at a data rate of 100 k bits/s. The output code is NRZ - L. Select a frequency for the premodulation signal.*

**Solution.** The nominal cut-off frequency of the filter is generally 0.5 of bit rate.

$\therefore$ Cut off frequency = 0.5 × 100 kHz = 50 kHz

***Example 27.14*** *In a A.M. system the carrier has a frequency of 1 MHz. The permissible bandwidth is ± 5 kHz. The intelligence which modulates the carrier, has a bandwidth of 1 kHz. What is the number of transmission channels for efficient use of transmission media ?*

**Solution.** From our knowledge of AM., we realise that the modulated carrier will have a bandwidth of 100 MHz ± 1 kHz. This is the bandwidth necessary to transmit 1 kHz data band.

It is clear that we are permited to have a ± 5 kHz modulation bandwidth out of which the intelligence uses only ± 1 kHz. This leaves the transmission bandwidth unoccupied and results in waste. Therefore, for efficient utilization of transmission media, it should be possible to use the ± 5 kHz bandwidth to transmit 5 times the 1 kHz data. Therefore we may be able to transmit five 1 kHz data via 100 MHz modulated carrier. Therefore, we can have 5 channels each transmitting a 1 kHz data.

***Example 27.15*** *Twenty one data channels, with subscarrier center frequencies ranging from 400 Hz to 165 kHz have been designated for telemetry use. All have a frequency deviation of ± 7.5%, assuming a modulation index of 5, obtain intelligence bandwidth of channel 1 for a centre frequency of 165 kHz. Also find the rise time. Assume that frequency division has been used.*

**Solution.** Frequency deviation of channel 1

$$= \frac{7.5}{100} \times 165 \times 10^3 = 12375 \text{ Hz}$$

Bandwidth of intelligence

$$= \frac{\text{frequency deviation}}{\text{modulation index}}$$

$$= \frac{12375}{5} = 2475 \text{ Hz.}$$

$$\text{Rise time } T_r = \frac{0.35}{2475 \times 10^{-3}} = 0.14 \text{ ms.}$$

# Review Questions

1. Define the term 'Telemetry' and explain why it is necessary to use it in an Instrumentation system.
2. Describe the different methods of data transmission. Explain the block diagram of a general telemetry system.
3. What are the different types of telemetering systems ? Explain the land-line telemetering system and describe its advantages.
4. Explain the land-line telemetering system. Describe the torque balance telemetering system.
5. Explain the land-line telemetering system using a synchro transmitter-receiver pair used in torque transmission mode.
6. Describe the method of amplitude modulation as applied to telemetry. Derive the expressions for modulation index, the power conveyed by the carrier and the power transmitted by upper and lower frequency bands. Find the condition for maximum power in the A.M. wave.
7. Define amplitude modulation and modulation index. Use a sketch of a sinusoidally modulated A.M. wavefrom to help explain the definition.
8. Explain why is it essential to use Radio Frequency (R.F.) telemetry. Describe it with some relevant examples.
9. Describe the frequency modulation systems. What is the significance of Bessel functions in FM ?
10. Explain the purpose of the device called a "multiplexer" in the telemetry system. Also explain "demultiplexer".
11. Describe the different methods used for pulse modulation in telemetry.
12. Describe the salient features of A.M.) and F.M. telemetry and compare and contrast them.
13. Describe in details the Pulse Amplitude Modulation system as used for telemetry.
14. Describe the Pulse Code Modulation telemetry system.

15. Describe the comparison between FM, PAM and PCM telemetering systems.
16. Describe in details a radio line system used for telemetry.
17. Explain in details the term "Diversity combining".
18. Define "Time Division Multiplexing" and "Frequency Division Multiplexing" as applied to telemetry.
19. Describe the different types of channels used for telemetry. Explain their advantages and disadvantages.
20. Describe the Noise and Bandwidth restrictions in a telemetry system.
21. What type of system (FM, PAM or PCM) would you use in the following cases and why ?
    (*a*) Ten data channels with cost of the entire system being the most important factor.
    (*b*) 200 data channels, with accuracy of atmost importance.
    (*c*) 50 data channels, with the transmitting system demolished at the end of each test. Thus the system must be a low cost device.
22. Automobiles are tested at a test range. The receiving site is 5 km away from the end of the range, so that a 40° angle of coverage will pickup signal anywhere on the range. The transmitter can deliver as much power as is needed. What type of antenna should be used ? Explain.
23. Carriers of two polarizations are received at equal levels (0 db difference). What is the improvement (db) over either carrier when a diversity combiner is used ?
24. Explain the term 'Diversity Combining' as used in radio links.
25. Describe (*a*) PAM, (*b*) PDM, (*c*) PFM, (*d*) PCM.
26. A 100 MHz carrier is used in an FM system transmitting PAM data. Give a block diagram of the system and explain.
27. Use a PLL to demodulate an FSK carrier. Sketch a block diagram for explanation.
28. Explain the following terms :
    (*a*) Commutator, (*b*) Subcommutation,
    (*c*) Commutation frame,
    (*d*) Synchronization for time division,
    (*e*) Intermodulation.
29. Describe the problems of interchannel interference in both time sharing and frequency division systems.
30. What is a modem and how does it operate ?

# Unsolved Problems

1. A broadcast AM transmitter radiates 50 kW of power at carrier frequency. What will be the radiated power at 85% modulation ? **[Ans.** 6.81 kW]
2. When the modulation percentage is 75, an A.M. transmitter produces 10 kW. How much of this is carrier power ? What would be the power saving if the carrier and one of the side bands were suppressed before the transmission took place ? **[Ans.** 7.81 kW, 89%]
3. A 360 W carrier is simultaneously is modulated by two audio waves with modulating percentages of 55 and 65 respectively. What is the total side band power radiated ? **[Ans.** 130 W]
4. In an amplitude modulated waveform

   $$E_{max} = 18 \text{ V and } E_{min} = 6 \text{ V},$$

   determine
   (*i*) index of modulation and
   (*ii*) percentage modulation. **[Ans.** 0.5, 50 %]
5. An A.M. broadcast station is transmitting at its assigned frequency of 880 kHz. The carrier is modulated by a 3.5 kHz sine wave. Determine the transmitted frequencies. **[Ans.** 880 kHz, 876.5 kHz, 883.5 kHz]
6. Assume an FM system has a frequency deviation of 10 kHz when the modulating signal has an amplitude of 4 V and a frequency of 2 kHz. Determine the frequency deviation if the audio signal is,
   (*a*) increased to 6 V
   (*b*) changed to 5 kHz
   (*c*) increased to 10 V and a frequency of 8 kHz.
   **[Ans.** (*a*) 15 kHz, (*b*) 10 kHz, (*c*) 25 kHz]
7. For an F.M. signal represented by the equation : $e = 10 \sin(8 \times 10^8\, t + 7 \sin 6 \times 10^4 t)$ V.
   Determine,
   (*a*) carrier frequency,
   (*b*) modulating frequency,
   (*c*) modulation index,
   (*d*) maximum deviation.
   **[Ans.** (*a*) 127.3 MHz, (*b*) 9.549 kHz, (*c*) 7, (*d*) 66.84 kHz].

27

8. An F.M. telemetry system uses a 370—430 Hz voltage controlled oscillator to carry a fuel level of 3000 lt tank, where 370 Hz represents an empty tank and 430 Hz represents a full tank. What does a frequency of (*i*) 392 Hz, (*ii*) 429 Hz represent ?
[**Ans.** (*i*) 1100 lt, (*ii*) 2950 lt]

9. In a PCM system, a 7 bit encoder is used. Suppose each level represents a voltage of 1 V. What is the range of the encoder ? Find out the quantization error on account of PCM if a voltage of 126.80 involved. [**Ans.** 127 V, 0.8 V]

10. In order to reproduce a 1 kHz sine wave, the sampling rate should be atleast 5 times the intelligence. It is proposed to commutate 10 such data channels. What is the switching rate of commutator if F.M. is used ?
[**Ans.** 50,000 samples per second]

11. Twenty one data channels, with subcarrier centre frequencies ranging from 400 Hz to 165 kHz have been designated for telemetry use. All have a frequency deviation of ± 7.5%. Assuming a modulation index, of 5, obtain intelligence bandwidth of channel 1 for a center frequency of 70 kHz. Also find the rise time. Assume that frequency division has been used.
[**Ans.** 1050 Hz, 0.33 ms]

12. An F.M. system is modulating the transmitter ± 500 kHz with a 1 kHz – 100 kHz multiplex. What IF bandpass would you select (round off to nearest 100 kHz) ? [**Ans.** 1200 kHz]

13. In time division multiplexing, in order to reproduce a 2 kHz sine wave with good fidelity (distortion less than 1% ), we require atleast 5 samples for each cycle of intelligence. What is the sampling rate ? It is proposed to commutate 10 data sources, what is the commutation rate ? What is the time elapsed between switching from one source to another ? What is the frame rate ? Calculate the commutation rate.
[**Ans.** 10,000 samples-per second, 100,000 samples per second 10 μs, 10,000 frames/s, 10,000 samples/s]

14. A carrier of 100 MHz is modulated by FM by a subcarrier of 70 kHz centre frequency. The intelligence is carried by an FM modulation of 70 kHz subcarrier. The frequency deviation is ± 15% and the modulation index is 5. What is the intelligence bandwidth capability ?
[**Ans.** 2100 Hz]

15. A modem converts a high speed serial word into a number of parallel low speed bit streams. Each of the 16 channels carries FSK data 75 bits per second. What is the transmission rate ? The voice line bandwidth available is 375 to 3025 Hz. The assigned channels are 170 Hz apart. What is frequency range occupied by them ? The FSK signal consists of a tone shift of 90 Hz centered around the assigned channel frequency $f_c$. What are the three recognizable frequency levels ?
[**Ans.** 1200 bps, 425 Hz to 2975 Hz, $f_c + 45$ Hz, $f_c - 45$ Hz, $f_c$ Hz]

## Objective Type Questions

1. The tuned circuit of the oscillator in a simple A.M. transmitter uses a 100 μH inductor and 0.5 nF capacitor. If the oscillator output is modulated by audio frequencies upto 20 kHz, what is the range occupied by side frequencies ?
(*a*) 702 to 722 kHz (*b*) 702 to 732 kHz
(*c*) 692 to 732 kHz, (*d*) 682 to 722 kHz.

2. An A.M. broadcast radio transmitter radiates 10 kW when the modulation percentage is 60. What is the power of the carrier ?
(*a*) 7.69 kW (*b*) 8.47 kW
(*c*) 9.17 kW (*d*) 5 kW

3. In an amplitude modulated system, the total power radiated 600 W. The power of the carrier is 400 W. What is modulation index ?
(*a*) 1
(*b*) 0.5
(*c*) 0.75
(*d*) none of the above.

4. In an A.M. system, the modulated output voltage has $E_{min} = 15$ V, the modulation index is :
(*a*) 2 (*b*) 1.5
(*c*) 1 (*d*) 0.5

5. For an F.M. wave represented by the voltage equation

$$e = 8 \sin (6 \times 10^8 t + 5 \sin 4 \times 10^4 t) \text{ V}$$

the carrier and modulating frequencies are :
(*a*) 127.3 MHz, 9.5 kHz
(*b*) 95.5 MHz, 6.37 kHz
(*c*) 190 MHz, 6.37 kHz
(*d*) 95.5 MHz, 12.74 kHz

6. For an F.M. wave represented by the voltage equation

$$e = 8 \sin (6 \times 10^8 t + 5 \sin 4 \times 10^4 t) \text{ V}$$

the modulation index and the maximum deviation are :

(a) 5, 31.85 kHz (b) 40, 31.85 kHz
(c) 5, 95.5 MHz (d) 5, 12.74 kHz

7. The F.M. telemetry as compared with A.M. telemetry requires a channel that is
(a) equal to that of A.M. telemetry
(b) smaller than what is required for A.M. telemetry
(c) 100 times of that required for A.M. telemetry
(d) 10 times of that required for AM telemetry.

8. In F.M. systems operate at
(a) VHF and UHF (b) only VHF
(c) only MF (d) MF and HF.

9. F.M. systems as compared to A.M. systems
(a) are equally effected by noise
(b) are less effected by noise
(c) are more effected by noise
(d) are highly effected by noise sense they operate at VHF and UHF.

10. A typical telemetry carrier frequency 230 MHz may be used with a bandwidth of ± 320 kHz using A.M. A data with a bandwidth of 4 kHz has to be transmitted. Subcarriers having a frequency 32 kHz and 44 kHz are modulated. The frequency bands utilized are
(a) 28 kHz to 40 kHz
(b) 28 kHz to 36 kHz, 40 kHz to 48 kHz
(c) 28 kHz to 48 kHz
(d) 230 KHz ± 324 kHz

11. Twenty-one data channels, with subcarrier center frequencies ranging from 400 Hz to 165 kHz have been designated for telemetry use. All have a frequency deviation of ± 7.5%. If the index of modulation is 5, the intelligence bandwidth of channel 1 for a center frequency of 3000 Hz is,
(a) 6 Hz
(b) 20 Hz
(c) 60 Hz
(d) 45 Hz
Assume the frequency multiplexing has been used.

12. In frequency multiplexing systems, if the subcarrier frequency deviations are proportional to the centre frequency, the scheme is referred to having :
(a) a proportional bandwidth format
(b) a constant bandwidth format
(c) a variable bandwidth format
(d) a consistent bandwidth format.

13. Time division multiplexing requires
(a) constant data transmission
(b) transmission of data samples
(c) transmission of data at random
(d) transmission of data of only one measurand.

14. Time division multiplexing is used when
(a) the data to be transmitted is slow changing.
(b) the data to be transmitted has a small bandwidth
(c) the data be transmitted is slow changing and has a low bandwidth
(d) the data to transmitted is fast changing and has a high bandwidth.

15. Modem is an acronym of
(a) modulation
(b) demodulation
(c) modulation and demodulation
(d) all the above.

27

## Answers

1. (c) 2. (a) 3. (a) 4. (d) 5. (b) 6. (a) 7. (d) 8. (a) 9. (b) 10. (b)
11. (d) 12. (a) 13. (b) 14. (c) 15. (c)

# 28 Display Devices and Recorders

## 28.1 INTRODUCTION

The last stage of a measurement system is the data presentation stage which consists of **data presentation elements.** This stage consists of display devices and recorders. In order that the results of a measurement system are meaningful, they must be displayed for instant observation or for storage for observation at a later stage. The first device is called a "**display device**" and the second, a "**recorder**". The data presentation devices may be called as **output devices**. These output devices may not only be used as display devices indicating the output, but may be used as control devices. The control of the process is done through **inverse transducers.** The control aspect is not within the scope of this book and hence it is not discussed here and so are the output devices connected with such an operation. In fact, in this chapter, we are concerned with devices which represent the output of a measurement system. The devices discussed in this chapter pertain to electrical and electronic devices which display and record the output data of a measurement system.

The choice between the display devices and the recorders depends upon :

(*i*) The expected use of the output and

(*ii*) The information content of the output.

The first factor concerns whether output is meant for human observation or the output is to be stored or the output is going to be an input to a digital computer. The second factor is influenced by whether a single value is desired or the output is needed as a function of time and also by the frequency content of the output.

**1. *Single number output devices.*** These devices are designed to indicate the value of some particular quantity under conditions such that the value to be measured can be regarded as time invariant over the time interval during which the measurement is made. Thus a single number will represent the measurement. The time for which the single number represents a value is very short and hence readings may have to be taken after certain intervals of time. The examples of this class of devices are indicating instruments and digital display units.

**2. *Time domain output devices.*** If the values of the quantity are to be taken as a function of time, the indicating instruments or the digital display units are no longer satisfactory except for applications where the output varies at a very slow rate. For fast changing outputs *i.e.,* where signal waveform or shape is the desired information, general purpose cathode ray tube (CRT) has to be used. In case information is to be stored over a short period of time say-up to an hour a storage type of CRT is used. However, where a permanent record of the variation of the output with time is to be kept cathode ray tube photographs, direct writing recorders, magnetic tape recorders etc. are used.

**3. *Machine interpretable output.*** The output devices described above provide an output that requires interpretation by human beings. In modern applications it is necessary that the output data should be in such a form that it can be "*read*" by a machine. These machine interpretable "*outputs*" can be had from

- magnetic tapes,
- punched paper tapes,
- punched cards and
- teletype writer.

The "**machine interpretable output**" can either be in analog form or in digital form. A signal recorded on a magnetic tape in analog form may be useful in activating output devices like CRT, a recorder or an indicating instrument.

On the other hand, a signal recorded in digital format on a magnetic tape, punched paper tape or punched cards a floppy disc or a hard disc may be used as an input for output devices like digital display units, a printer or a digital computer.

## 28.2 ELECTRICAL INDICATING INSTRUMENTS

The electrical indicating instruments are used extensively for measurement of current, voltage, resistance and power. These instruments can be broadly classified as :

- Analog instruments and
- Digital instruments.

The analog indicating instruments have already been discussed in this text. The digital indicating instruments are discussed in this chapter along with analog and digital recorders.

## 28.3 DIGITAL INSTRUMENTS

The analog instruments display the quantity to be measured in terms of deflection of a pointer *i.e.*, an analog displacement or an angle corresponding to the electrical quantity. The **digital instruments** indicate the value of the measurand in the form of a **decimal number**. The digital meters work on the principle of quantization. The analog quantity to be measured is first subdivided or quantized into a number of small intervals upto many decimal places. The objective of the digital instrument is then to determine in which portion of the subdivision the measurand can thus be identified as an integral multiple of the smallest unit called the quantum, chosen for subdivision. The measuring procedure thus reduces to one of counting the number of quanta present in the measurand.

The reading accuracy can be arbitrarily increased by increasing the number of decimal places *i.e.*, by increasing the quantizing levels (Art. 26.4.1). The advantages of digital instruments are given below.

### 28.3.1 Advantages of Digital Instruments

(*i*) The digital instruments indicate the readings directly in decimal numbers and therefore errors on account of human factors like errors due to parallax and approximation are eliminated.

(*ii*) The readings may be carried to any number of significant figures by merely positioning the decimal point.

(*iii*) Since the output of digital instruments is in digital form and therefore the output may be directly fed into memory devices like tape recorders, printers, floppy discs, hard discs and digital computers etc., for storage and for future computations.

(*iv*) The power requirements of digital instruments are considerably smaller.

The chief advantages of analog instruments are that they are cheap and simple and accordingly for ordinary purposes these (analog) instruments will not be completely displaced by digital instruments. However, where cost considerations and complexity of digital instruments are not of much consequence, digital instruments are certainly preferred over their analog counterparts. (At present digital instruments are costlier than the corresponding analog instruments but with the developments in modern techniques, the cost gap will be narrowed down.) Also there are some applications where only digital instruments can be used.

### 28.3.2 Digital versus Analog Instruments

When it is necessary to decide between digital and analog instruments, the choice depends upon many factors. Some of these factors are :

**1. *Accuracy.*** The best analog instruments are rated usually within ± 0.1 percent of full scale. Digital instruments can be made to much greater accuracies.

**2. *Reaction to environment.*** Analog meter movements are relatively simple and will operate under a wide range of environments. Digital instruments are relatively complex and consist of large number of parts which individually will react to change in temperature and humidity. However, the advantage of digital instruments is that they can be made without any moving part thus removing the errors which are caused on account of movements.

**3. *Resolution.*** This is sometimes referred to as *readability* below which differences can no longer be differentiated. In analog instruments the limit is one part in several hundreds. Digital instruments can be made with a resolution of one part in several thousands.

**4. *Power requirements.*** Digital instruments draw only negligible power whereas the analog instruments may load the circuit under measurement and thus indicate an erroneous reading. Digital instruments have input impedances of the order of 10 MΩ or even higher.

**5. *Cost and portability.*** Analog instruments are extremely portable and usually do not require an outside source of supply for measurements. Analog instruments are low in cost and can be moved from one location to another with ease. On the other hand, digital instruments are not easily portable and require an external source of power. However, on account of modern developments in integrated circuit technology digital instruments can be made extremely portable and low in cost.

6. ***Range and polarity.*** Most digital instruments are essentially d.c. instruments which measure upto 100 V and upto 1000 V by means of the range attenuator. A.C. instruments use an a.c. to d.c. converter. Many digital instruments incorporate automatic polarity and range indication (**Automatic polarity selection** and **Auto-ranging** facilities) which reduce operator training, measurement error and possible damage through overloads.

7. ***Freedom from observational errors.*** The digital instruments are free from observational errors like parallax and approximation errors. They directly indicate the quantity being measured in decimal form with the help of readout and display devices.

Analog instruments usually have a scale which is cramped at the lower end and therefore give considerable observational errors. Also when using analog instruments inexperienced personnel often tend to select a scale that does not correspond to the setting of the range switch thereby introducing large operating errors.

Before the digital meters are described it is essential that the reader be familiarized with the working of electronic counters and the digital display devices.

## 28.4 ELECTRONIC COUNTERS

Electronic counters are capable of making many measurements involving frequency, time, phase angle, radiation events and totalizing electric events.

The electronic counter normally employs a frequency divider circuit known as a **scaler**. A scaler produces a single pulse for every set of number of input pulses. For example, a 2 : 1 scaler produces one output pulse for 2 input pulses. A scaler is essentially a frequency divisor. The basis of counters is frequency division.

This is done by a 2 : 1 scaler called a **bistable multivibrator** or a **Flip-flop (FF) circuit**.

### 28.4.1 RS Flip-Flop

A flip-flop is a bistable multivibrator which has an output which is either a high or a low voltage. The low voltage output is called 0 and the high voltage is called 1. Thus a flip-flop has two stable states, either 0 or 1. The output stays in its original state, either 0 or 1 unless the circuit is driven by an input called **trigger**. On application of trigger, the output changes its state. Thus if a trigger is applied when output is in 1 state, it changes to 0 state while if the output is in 0 state, it changes to 1 state. It should be understood that in a bistable multivibrator the output continues to be in its original state indefinitely unless triggered. Therefore it can be said that a Flip-flop has a **memory**. Figure 28.1 shows an R-S flip-flop.

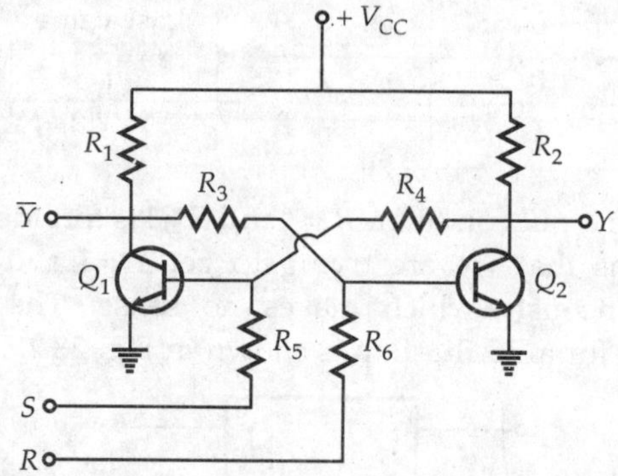


**Fig. 28.1** RS Flip-Flop (Bistable Multivibrator).

There are two similar transistors $Q_1$ and $Q_2$. They are NPN type. The collector of each transistor is cross-coupled to the base of the transistor. The cross-coupling between collector and base results in a positive feedback.

If $Q_1$ is conducting there is a large voltage drop in resistance $R_1$ and therefore $Q_1$ is **saturated** and voltage $V_{CE}$ at $Y$ is only about 0.1 V. This voltage is applied to the base of transistor $Q_2$ and is not sufficient to turn on the transistor $Q_2$ which requires a voltage of $V_{BE} = 0.7$ V (assuming that both $Q_1$ and $Q_2$ are silicon transistors). Thus transistor $Q_2$ is **cutoff**. Under these conditions output $\overline{Y} = 0$ and $Y = 1$.

Similarly, if by some means we can get transistor $Q_2$ to conduct there will be large voltage drop across resistor $R_2$ and therefore the transistor $Q_2$ gets saturated. The potential at $Y$ becomes 0.1 V forcing the transistor $Q_1$ to cutoff. Hence under these conditions $\overline{Y} = 1$ and $Y = 0$.

Hence there are two stable states for the flip-flop and these are :

(*i*) $\overline{Y} = 1$ and $Y = 0$ (*ii*) $\overline{Y} = 0$ and $Y = 1$

In order to control the flip-flop, trigger inputs are applied. If a positive going pulse is applied to the $S$ (**set**) input terminal, then $Q_1$ conducts heavily (gets saturated) and $Q_2$ is cut off. Similarly, when a positive going pulse is applied to $R$ (**reset**) input terminal $Q_2$ gets saturated and $Q_1$ is cut off.

28

Thus application of a positive going pulse to $S$ input terminal is called **Setting** the flip-flop and result in a binary output where : $Y = 1$ and $\overline{Y} = 0$.

Application of a positive going pulse to the $R$ input terminal is called **Resetting** of the flip-flop and results in a binary output where $Y = 0$ and $\overline{Y} = 1$.

The truth table for an RS flip-flop is :

| ***R*** | ***S*** | ***Y*** |
|---|---|---|
| 0 | 0 | Last value |
| 0 | 1 | 1 |
| 1 | 0 | 0 |

The input condition $R = 1$ and $S = 1$ is forbidden as it means that we are trying to get $Y = 1$ and $Y = 0$ simultaneously which makes no sense. The logic symbol for a RS flip-flop is shown in Fig. 28.2.

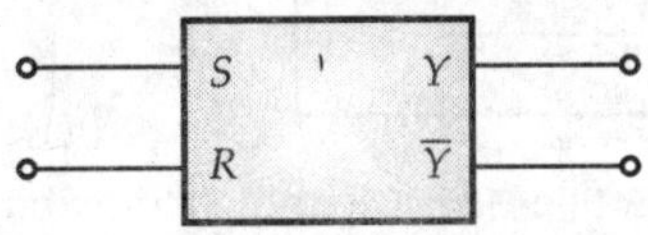


**Fig. 28.2** Logic symbol for RS flip-flop.

### 28.4.2 T and RST Flip-Flops

Figure 28.3 shows a **Toggle (T)** flip-flop. A train of narrow pulses is applied to the $T$ input terminal. This circuit uses two **AND** gates and a RS flip-flop. Every time a pulse is applied, the output of the flip-flop changes state.

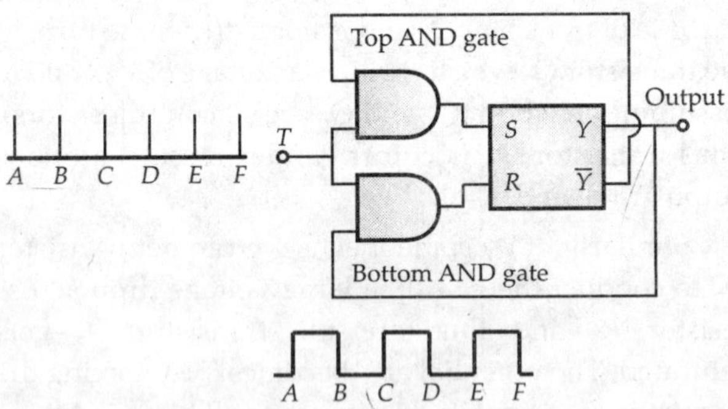


**Fig. 28.3** *T* flip-flop.

Suppose $Y = 0$ and $\overline{Y} = 1$ just before the pulse $A$ is applied to terminal $T$. On application of this pulse, the inputs to the top AND gate are 1 and 1 and therefore the top gate is **enabled** (opened).

A pulse is applied to terminal $S$. On the other hand, the two inputs to bottom AND gate are 1 and 0 and is thus **disabled** (closed) and no input is applied to terminal $R$.

Hence, the flip-flop is set or $Y = 1$ and $\overline{Y} = 0$

When the next pulse $B$ is applied to terminal $T$, the input to the top gate is 1 and 0 (since $\overline{Y} = 0$) and hence it is **disabled.** Therefore no pulse is applied to terminal $S$. On the other hand, the bottom AND gate is **enabled** as the two inputs to it are 1 and 1 as $Y = 1$ and therefore a pulse is applied to terminal $R$. This resets the flip-flop resulting in $Y = 0$ and $\overline{Y} = 1$.

Thus each incoming positive pulse is alternately steered into the set and reset inputs, and hence the flip-flop toggles *i.e.,* alternately sets and resets producing 1 or 0 states at the output. Therefore two input pulses produce one output pulse. This means that the frequency of the output is half of that of the input. Thus a $T$ flip-flop acts as a **frequency divisor** which divides the input frequency by two.

We have till now considered flip-flops which respond to positive going pulses. However, many a designs are possible for a $T$ flip-flop. For example, it is possible to build a $T$ flip-flop that responds to a negative going pulse. Hence, onwards we will discuss $T$ flip-flops which respond to **negative edge** of a **square wave.**

Figure 28.4 shows a $T$ flip-flop. The flip-flop changes its state only on the negative edge of the input signal.

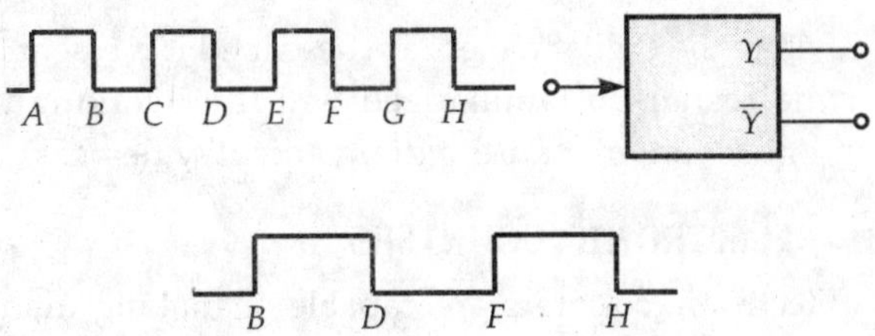


**Fig. 28.4** *T* flip-flop.

A RST flip-flop combines an RS flip-flop and a T flip-flop. It can set, reset or toggle. The most important flip-flop is the one which sets or resets to positive going $R$ and $S$ inputs and toggles for negative going inputs.

The symbol for an RST flip-flop is shown in Fig. 28.5.

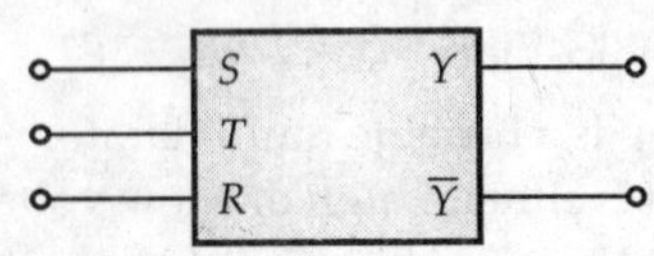


**Fig. 28.5** RST flip-flop.

Thus, summarising :

| Input | Sets/ Resets/ Toggles | Final State |
|---|---|---|
| +ve going applied to $S$ | Sets | $Y=1,\ \overline{Y}=0$ |
| +ve going applied to $R$ | Resets | $Y=0,\ \overline{Y}=1$ |
| –ve going applied to $T$ | Toggles | switches from 1 to 0 and 0 to 1 |

### 28.4.3 Decade Counter

Four flip-flops can be in cascade to act as counters for the input pulses. A decade counter uses four RST flip-flops $A$, $B$, $C$, $D$. Figure 28.6 shows four RST flip-flops in cascade.

A square wave is input to the counters. This square wave is called a **"clock"**. The output from flip-flop $A$ drives the flip-flop $B$. The output of flip-flop $B$ drives flip-flop $C$ which in turn drives flip-flop $D$.

To start the operation, all the flip-flops are reset and therefore the states of the four RS flip-flops are :

$$D\ C\ B\ A = 0\,0\,0\,0 \text{ and } \overline{D}\ \overline{C}\ \overline{B}\ \overline{A} = 1\,1\,1\,1$$

When the first clock pulse is applied to $T$ input terminal of RS flip-flop $A$, it changes its state on the negative going edge of the pulse. Thus at the end of the 1st pulse the state of the flip-flop is :

$$D\ C\ B\ A = 0\,0\,0\,1 \text{ and } \overline{D}\ \overline{C}\ \overline{B}\ \overline{A} = 1\,1\,1\,0$$

The two inputs to the AND gate are $A$ and $\overline{D}$ and both are 1 state. Therefore the AND gate is **enabled** (opened). A positive going pulse is applied to the $T$ terminal of $B$ flip-flop and therefore it does not change its state. It remains at 0 state and so do $C$ and $D$ flip-flops. On the application of 2nd pulse, $A$ goes from 1 to 0. Since the AND gate is open a negative going pulse is applied to $B$ flip-flop. This changes the state of flip-flop $B$. It changes from 0 to 1. Therefore a positive going pulse is applied to flip-flop $C$ and it remains in its original state and so does flip-flop $D$.

The states at the end of 2nd pulse are :

$$D\ C\ B\ A = 0\,0\,1\,0 \text{ and } \overline{D}\ \overline{C}\ \overline{B}\ \overline{A} = 1\,1\,0\,1$$

This **disables** (closes) the AND gate since $A=0$ and $\overline{D}=1$

On application of the third pulse, flip-flop $A$ changes its state from 0 to 1. Since the AND gate is closed no pulses are applied to flip-flops $B$, $C$ and $D$ and their states remain unchanged.

Thus at the end of 3rd pulse, the states are :

$$D\ C\ B\ A = 0\,0\,1\,1 \text{ and } \overline{D}\ \overline{C}\ \overline{B}\ \overline{A} = 1\,1\,0\,0$$

Since the inputs to AND gate $A$ and $\overline{D}$ are both 1, it is **enabled** or the gate is open now. On the application of the 4th pulse, $A$, goes from 1 to 0 and therefore a negative going pulse is applied to $B$. Flip-flop $B$ changes its state and it goes from 1 to 0. Thus a negative going pulse is applied to flip-flop $C$ and its state changes from 0 to 1. The change of state of flip-flop $C$ is applied to flip-flop $D$. Since flip-flop $C$ changes its state from 0 to 1, a positive going pulse is applied to flip-flop $D$ and therefore it does not change its state. At the end of 4th pulse, the states of flip-flops are :

$$D\ C\ B\ A = 0\,1\,0\,0 \text{ and } \overline{D}\ \overline{C}\ \overline{B}\ \overline{A} = 1\,0\,1\,1$$

The AND gate is **disabled** (closed). On the application of 5th pulse, flip-flop $A$ changes its state from 0 to 1. Since the AND gate is closed no pulses is applied to flip-flops $A$, $B$, $C$, $D$ and they therefore remain in their original state.

Thus at the end of 5th pulse the states of the flip-flops are :

$$D\ C\ B\ A = 0\,1\,0\,1 \text{ and } \overline{D}\ \overline{C}\ \overline{B}\ \overline{A} = 1\,0\,1\,0$$

This means that the AND gate is **enabled** (open) again on the application of 6th pulse Flip-flop $A$ changes its state from 1 to 0. Since the AND gate is open its output is applied to flip-flop $B$ which does not change its state and so do flip-flops $C$ and $D$. Therefore at the end of 6th pulse, the states of flip-flops are :

$$D\ C\ B\ A = 0\,1\,1\,0 \text{ and } \overline{D}\ \overline{C}\ \overline{B}\ \overline{A} = 1\,0\,0\,1$$

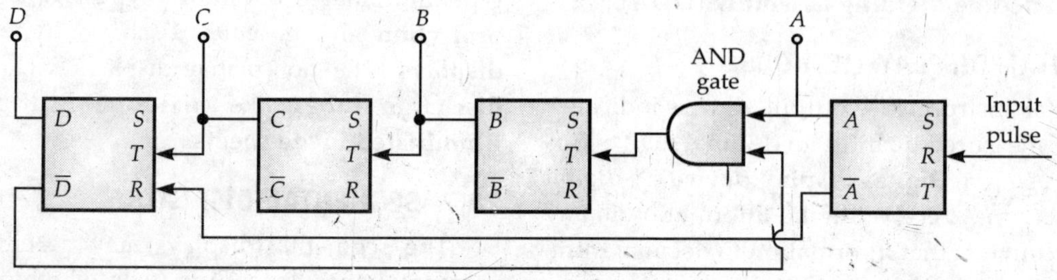


**Fig. 28.6** A decade counter.

The AND gate is closed as $A=0$ and $\overline{D}=1$

On the application of 7th pulse, flip-flop $A$ changes its state from 0 to 1. As the AND gate is closed and therefore flip-flops $B$, $C$ and $D$ remain in their original state. Hence, at the end of the 7th the pulse states of flip-flops are :

$$D\ C\ B\ A = 0\ 1\ 1\ 1 \quad \text{and} \quad \overline{D}\ \overline{C}\ \overline{B}\ \overline{A} = 1\ 0\ 0\ 0$$

The AND gate is open now as $A=1$ and $\overline{D}=1$. On application of 8th pulse, flip-flop $A$ goes from 1 to 0. This applies a negative going pulse to flip-flop $B$ which goes from 1 to 0. The output form flip-flop $B$ is applied to flip-flop C which also receives a negative going pulse and therefore it goes from 1 to 0. So a negative going pulse is applied to flip-flop $D$ changing its state from 0 to 1. The states at the end of 8th pulse are :

$$D\ C\ B\ A = 1\ 0\ 0\ 0 \quad \text{and} \quad \overline{D}\ \overline{C}\ \overline{B}\ \overline{A} = 0\ 1\ 1\ 1$$

Since the inputs to the AND gate are $A=0$ and $\overline{D}=0$, it is closed.

On the application of 9th pulse, flip-flop $A$ changes its state from 0 to 1. Since the AND gate is closed ; the other three flip-flops $D$, $C$, $B$ remain in their original state. Hence at the end of 9th pulse, the states are :

$$D\ C\ B\ A = 1\ 0\ 0\ 1 \quad \text{and} \quad \overline{D}\ \overline{C}\ \overline{B}\ \overline{A} = 0\ 1\ 1\ 0$$

The AND gate is closed as $A=0$ and $\overline{D}=0$

On application of 10th pulse, $A$ changes its state from 1 to 0. Since the gate is closed no pulse is applied to flip-flops $B$ or $C$ and they remain in their original state. $A$ changes from 0 to 1. This positive pulse is applied to $R$ terminal of flip-flop $D$ which resets itself. It was in 1 state and therefore comes to 0 state.

Thus at the end of 10th pulse, the states are :

$$D\ C\ B\ A = 0\ 0\ 0\ 0 \quad \text{and} \quad \overline{D}\ \overline{C}\ \overline{B}\ \overline{A} = 1\ 1\ 1\ 1$$

The counting now can start all over again.

Thus we get one output pulse for ten input pulses.

Hence a decade counter acts as a decade frequency divider since it divides the frequency by 10 and therefore it is called a **decade dividing assembly (DDA)**.

## 28.5 DIGITAL DISPLAY METHODS

In digital instruments, output device indicates, the value of measured quantity in decimal digits. This is done by using a **Digital display device**. A digital display device may receive digital information in any form but it converts that information to decimal form. Thus the display device indicates the value directly in decimal digits. The number of digits correspond to the significant figures needed to represent the value. The basic element in a digital display device is the display for a single digit because a multiple digit display is nothing else but a group of single digit displays. Figure 28.7 shows a multiple digit display consisting of 4 single digit displays.

**Fig. 28.7** Multiple digit display.

A single digit display is capable of indicating the numbers from 0 to 9. There is also usually provision for a decimal point between each of the numerals. One of these is selected and activated in accordance with the range selection controls of the instrument. Some instruments have automatic range selection, commonly called **autoranging**. The input to the digit display is a code indicating the particular number to be displayed, or the excitation of one of the ten inputs designating the number to be displayed.

A typical binary code on four input lines is given in Chapter 30. The digit displayed will depend on which combination of excitation or non-excitation is present, 0 indicates non-excitation while 1 indicates excitation. It should be understood that the signal should be decoded and decoding circuits are a part of the display units.

## 28.6 DIGITAL DISPLAY UNITS

There are many ways of classifying digital display units. One of the methods of classifying them is based upon the format used. The display can be **planar** *i.e.,* the entire read-out characters and in the same plane or **non-planar** where the characters are displayed in different planes.

The planar displays may be illuminated **segmental type**, illuminated **dot matrix type**, display using **rear projection** and **gaseous discharge** type **segmental displays**. The non-planar displays include **gaseous discharge tubes** like **nixies** and displays using **illuminated lucite sheets.**

## 28.7 SEGMENTAL DISPLAYS

The segmental displays may be either 7 or 14 segmental ones depending upon whether numeric or alphanumeric displays are required.

### 28.7.1 Seven Segmental Display

Figure 28.8 shows a seven segmental display. This is used for numeric display. It consists of seven segments *a, b, c, d, e, f* and *g*. A segmental display forms the digit to be displayed by illuminating proper segments from the group.

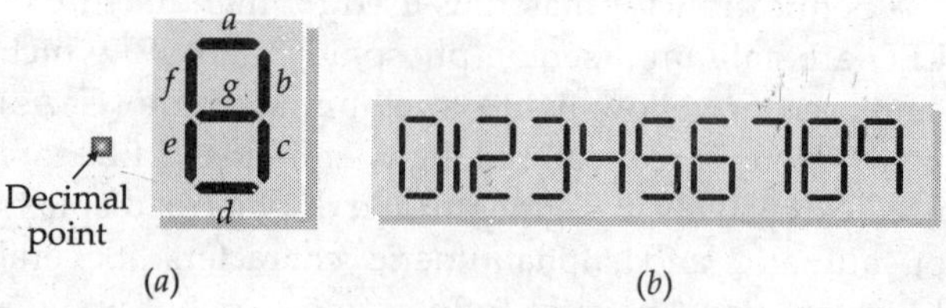


**Fig. 28.8** Seven segmental display.

By illuminating the proper combination of these 7 segments numbers 0 through 9 can be displayed. The display is incandescent and operates on low voltages (12.48 V) and requires about 10 to 50 mA current when using **LEDs** (Light emitting diodes). **LCDs** (Liquid crystal diodes) are also used for segmental displays.

### 28.7.2 Fourteen Segmental Display

For display of **Alphanumeric Characters** (both numerals as well as alphabets) a 14 segmental display unit shown in Fig. 28.9 is used.

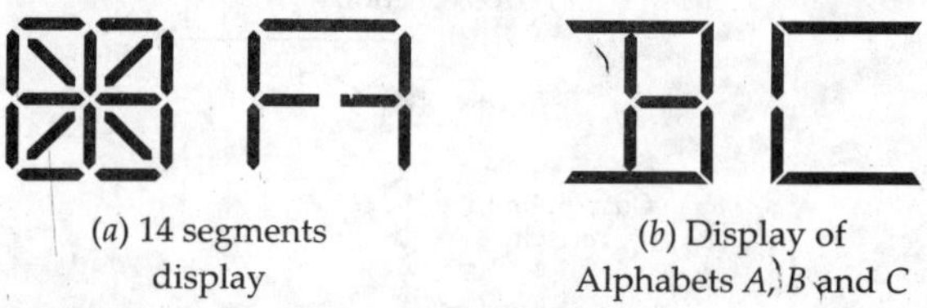


**Fig. 28.9** Fourteen segmental display and formation of alphabets.

The diagram shows how alphabets *A, B* and *C* are displayed by illumination of proper segments.

## 28.8 DOT MATRICES

Dot matrices may be used for display of numeric and alpha numeric characters.

### 28.8.1 A 3 × 5 Dot Matrix

A 3 × 5 dot matrix shown in Fig. 28.10, may be used for display of numeric characters.

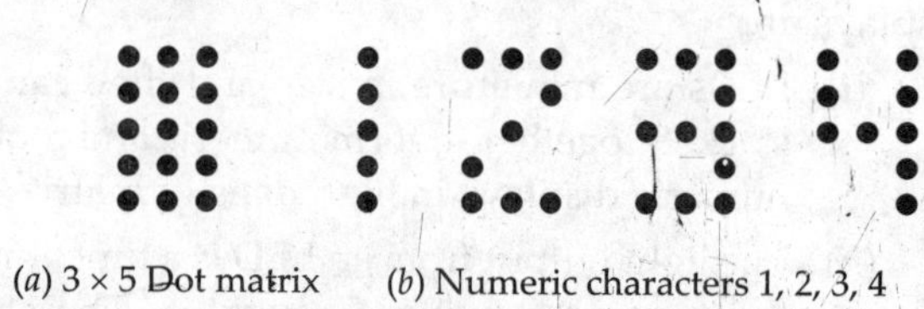


**Fig. 28.10** A 3 × 5 Dot Matrix and formation of numeric characters.

### 28.8.2 Dot Matrix Utilizing 27 Dots

Another system using 27 dots is shown in Fig. 28.11. This system displays the numeric characters as shown in the diagram.

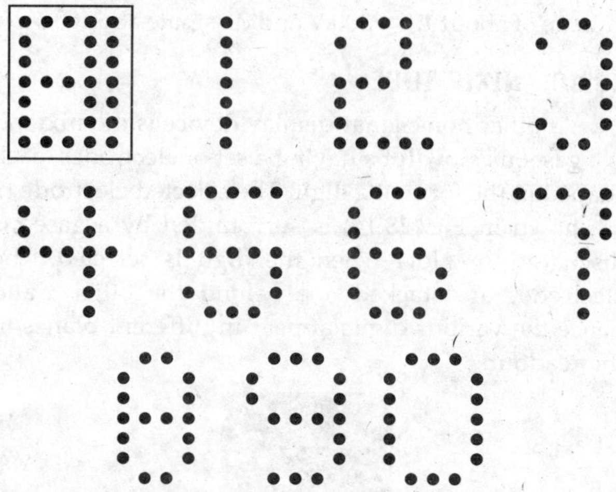

**Fig. 28.11** Digital display unit with grid illuminated dots.

The dots may be square or round with 0.4 mm side or diameter, LEDs and LCDs are used for display (illumination) of dots.

### 28.8.3 A 5 × 7 Dot Matrix

For display of Alphanumeric characters a 5 × 7 dot matrix is used as shown in Fig. 28.12.

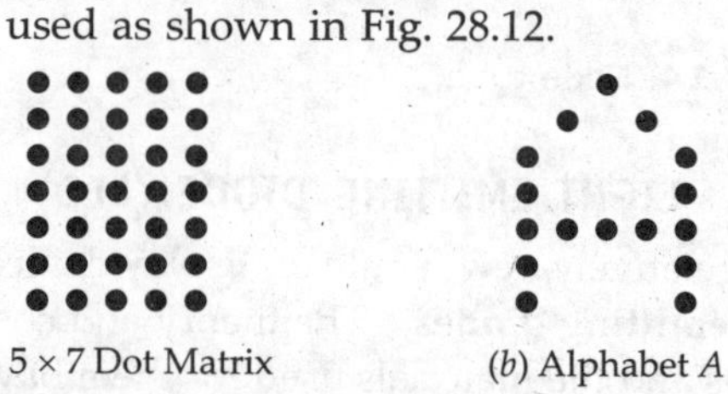


**Fig. 28.12** A 5 × 7 Dot matrix and alphabet A.

## 28.9 REAR PROJECTION DISPLAY

A cutout view of a typical **rear projection display** is shown in Fig. 28.13. Each of the 12 incandescent lamps when energized by the input signal illuminates

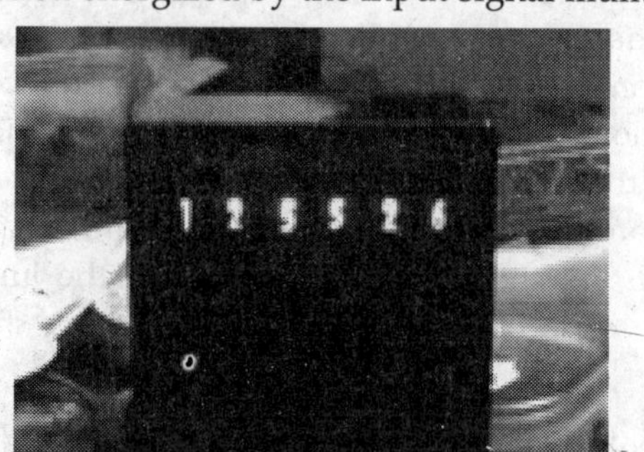


**Fig. 28.13** Rear projection display.

a different part of the filmstrip. The lens system projects the illuminated part of the film onto a viewing screen. The rear projection is slower than **NIXIE** (discussed later). It requires a lower voltage (6.3–28 V) and dissipates between 1 to 3 W. Nixie tubes use a voltage of about 170 – 300 V and dissipate about 0.5 W.

## 28.10 NIXIE TUBE

Another non-planar display device is the **nixie.** It is a gaseous glow tube having a set of electrodes, each shaped in the form of a digit. The selected electrode (3 as shown in Fig. 28.14) is surrounded by a gaseous discharge, or glow when the digit is selected. The electrodes are stacked one behind the other ; and hence the various digits appear in different planes in the readout.

**Fig. 28.14** Nixie tube.

## 28.11 LIGHT EMITTING DIODE (LED)

A relatively new family of display devices utilizes "**light emitting diodes**". The theory and construction of LEDs and the materials used for them have already been described on pages 580 to 582. The LED is perhaps the most important of the display devices available today for use in Instrumentation systems. The **LED** is a *PN* junction device which emits light when a current passes through it in the forward direction.

Charge carrier recombination occurs at a PN junction as electrons cross from *N* side and recombines with holes on the *P* side. When recombination takes place, the charge carriers give up energy in the form of heat and light. If the semiconducting material is translucent the light is emitted, and the junction is source of light. This is the light emitting diode *i.e.*, **LED**.

Figure 28.15(*a*) shows a cross-sectional view of a typical LED charge carrier recombinations takes place in the *P* type material. Therefore, the *P* region becomes the surface of the devices. For maximum light emission, a metal film anode is deposited around the edge of the *P* type material. The cathode connection for the device is usually a gold film at the bottom of the *N* type region. This helps in reflecting the light to the surface.

Semiconductor materials used for manufacture of LED are gallium arsenide phosphide (GaAsP) which emits red or yellow light of gallium arsenide (GaAs) which gives green or red light emission. LEDs are used extensively in segmental and dot matrix displays of numeric and alphanumeric characters. Several LEDs are used in series to form one segment while a single LED may be used to form a decimal point. LEDs are available in many colours like green, yellow, amber and red.

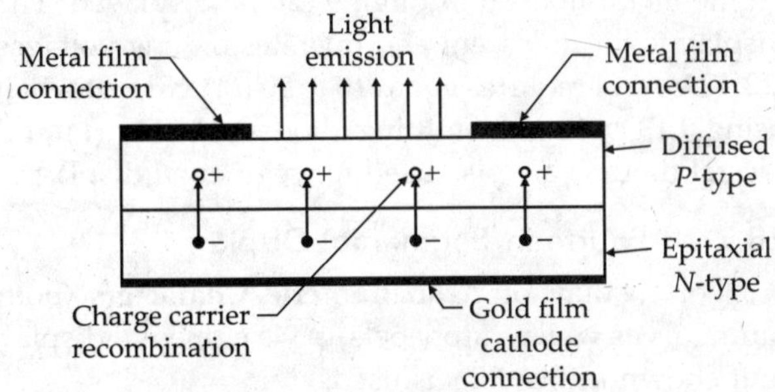


(*a*) Cross-section

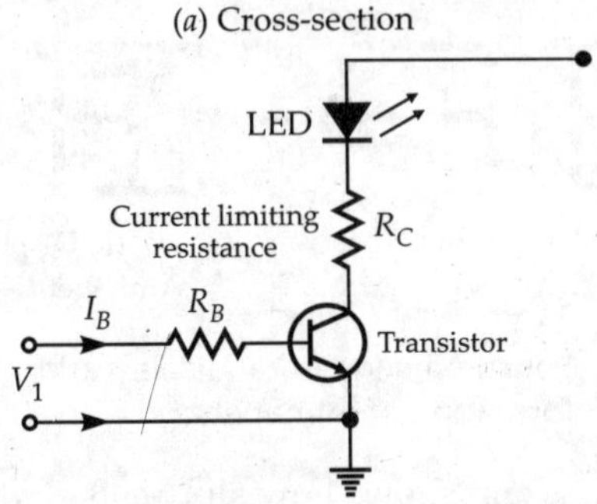


(*b*) LED controlled by a transistor switch

**Fig. 28.15** LEDs.

A simple transistor can be used for OFF/ON of an LED as shown in Fig. 28.15(*b*). When the transistor is driven into saturation by base current $I_B$, it conducts heavily (switch is closed and the LED emits light). The LED current is limited by a resistance $R_C$.

The major advantages of LEDs in electronic displays are :

(*i*) LEDs are miniature in size and they can be stacked together to form numeric and alpha-numeric displays in high density matrix.

(*ii*) The light output from an LED is a function of the current flowing through it. Therefore intensity of light emitted from LEDs can be smoothly controlled.

(*iii*) LEDs have a high efficiency as emitters of electromagnetic radiation. They require moderate power for their operation. A typical voltage drop of 1.2 V and a current of 20 mA is required for full brightness. Therefore, LEDs are useful where miniaturization with low d.c. power are important.

(*iv*) LEDs are available which emit light in different colours like red, green, yellow and amber.

(*v*) The switching time (both on and off) is less than 1 ns and therefore they are very useful where dynamic operation of large number of arrays is involved.

(*vi*) LEDs are manufactured with the same type of technology as is used for transistors and ICs and therefore they are economical and have a high degree of reliability.

(*vii*) LEDs are rugged and can therefore withstand shocks and vibrations. They can be operated over a wide range of temperature say 0 – 70 °C.

The disadvantage of LEDs as compared with LCDs is their high power requirement. Also LEDs are not suited for large area displays, primarily because of their high cost. For large displays, devices using gas filled plasma are used.

## 28.12 LIQUID CRYSTAL DIODES (LCD)

Liquid crystal cell displays (LCDs) are used in similar applications where LEDs are used. These applications are display of numeric and alphanumeric characters in dot matrix and segmental displays.

The LCDs are of two types :

- Dynamic scattering type, and
- Field effect type.

The construction of a dynamic scattering liquid crystal cell is shown in Fig. 28.16. The liquid crystal material may be one of the several organic compounds which exhibit optical properties of a crystal though they remain in liquid form. Liquid crystal is layered between glass sheets with transparent electrodes deposited on the inside faces.

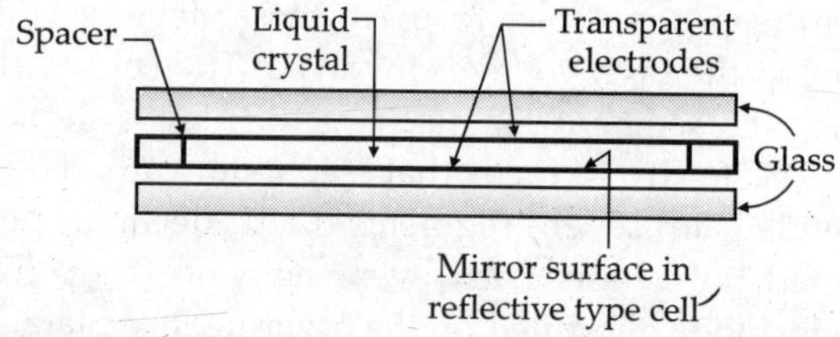


**Fig. 28.16** Liquid crystal diode cell (LCD).

When a potential is applied across the cell, charge carriers flowing through the liquid disrupt the molecular alignment and produce turbulence. When the liquid is not activated, it is transparent. When the liquid is activated the molecular turbulence causes light to be scattered in all directions and the cell appears to be bright. The phenomenon is called **dynamic scattering**.

The construction of a **field effect** liquid crystal display is similar to that of the dynamic scattering type, with the exception that two thin polarizing optical filters are placed at the inside of each glass sheet. The liquid crystal material in the field effect cell is also of different type from that employed in the dynamic scattering cell. The material used is twisted nematic type and actually twists the light passing through the cell when the latter is not energized. This allows the light to pass through the optical filters and the cell appears bright. When the cell is energised, no twisting of light takes place and the cell appears dull.

Liquid crystal cells are of two types :

- Transmittive type, and
- Reflective type.

In the **transmittive type** cell, both glass sheets are transparent, so that light from a rear source is scattered in the forward direction when the cell is activated.

The **reflective type** cell has a reflecting surface on one side of glass sheets. The incident light on the front surface of the cell is dynamically scattered by an activated cell. Both types of cells appear quite bright when activated even under ambient light conditions.

The liquid crystals are light reflectors or transmitters and therefore they consume small amounts of energy (unlike light generators). Considering the case of seven segmental display, the current is about 25 μA for dynamic scattering cells and 300 μA for field effect cells. Unlike LEDs which can work on d.c. the LCDs require a.c. voltage supply. A typical voltage supply to dynamic scattering LCD is 30 V peak to peak with 50 Hz.

The **advantages** of LCDs are :

(*i*) They have a low power consumption. A seven segmental display requires about 140 μW (20 μW/segment). This is great advantage over LEDs which require about 40 mW per numeral.

(*ii*) They have a low cost.

28

The chief **disadvantages** of LCDs are :

(*i*) LCDs are very slow devices. The turn on and the turn off times are quite large. The turn on time is typically of the order a few milliseconds while the turn-off is ten milliseconds.

(*ii*) When used on d.c. their life span is quite small. Therefore they are used with a.c. supplies having a frequency less than 500 Hz.

(*iii*) They occupy a large area.

On account of the reasons described above LCDs are normally used for seven segmental displays.

## 28.13 NIXIE TUBE

The basic construction of a digital indicator tube is shown in Fig. 28.17. It is a cold cathode glow discharge tube, which is popularly known as **Nixie** which is the trade mark of M/s Burrough's Corporation, U.S.A. The display works on the principle that when a gas breaks down, a glow discharge is produced. A gauze electrode with a positive voltage supply functions as an anode, and there are 10 separate wire cathodes,

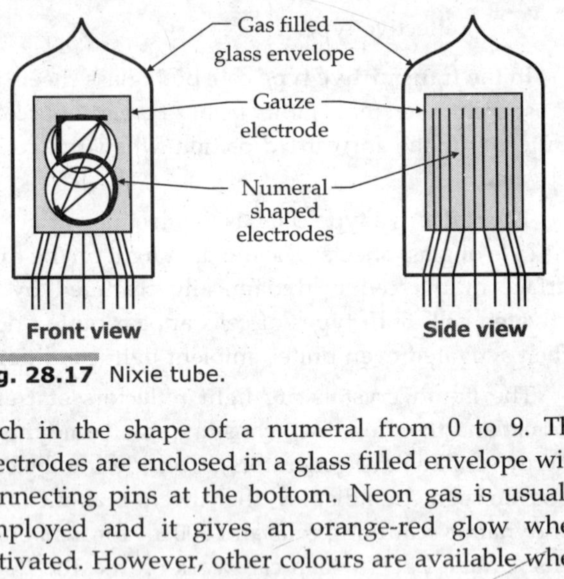


**Fig. 28.17** Nixie tube.

each in the shape of a numeral from 0 to 9. The electrodes are enclosed in a glass filled envelope with connecting pins at the bottom. Neon gas is usually employed and it gives an orange-red glow when activated. However, other colours are available when different gases are used.

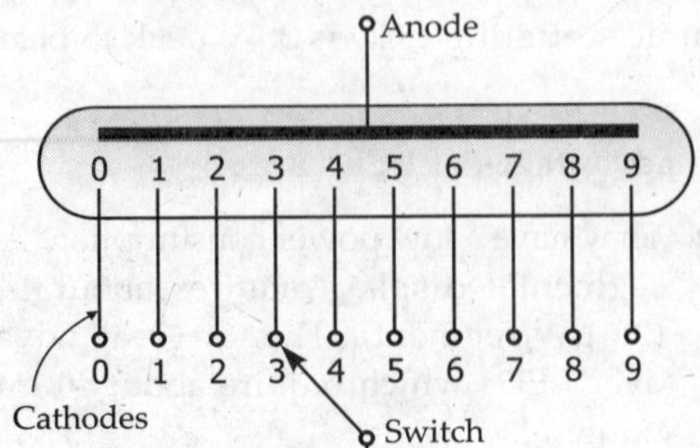


**Fig. 28.18** Schematic symbol of Nixie.

There is one anode and 10 cathodes. After a negative voltage is applied to the selected cathode, a simple gas discharge diode is formed which lights the selected digit.

A transistor gate is usually employed at each cathode so that the desired numeral can be switched on.

The circuitry driving the nixie tubes is simpler than that for seven segment displays. However, high voltages (150 – 220 V) are required to produce glow discharge. The current required is of the order of 1 – 5 mA. The Nixie tubes are bulkier in size than the seven segmental displays.

**Modern nixies.** Modem Nixies vary in construction. The original Nixie had 10 cathodes representing numerals 0 – 9 but additional cathodes displaying decimal symbol and + and – signs are being used.

Special Nixies (Burroughs B 7971, shown in Fig. 28.19) have 15 cathodes constituting 15 segments that be used to produce numeric as well as alphanumeric characters.

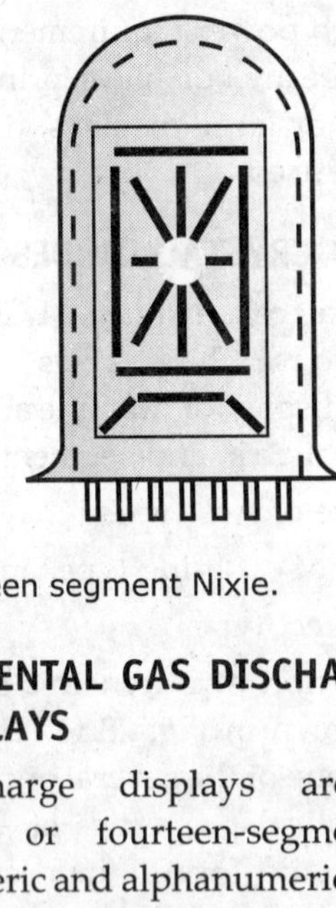

**Fig. 28.19** Fifteen segment Nixie.

## 28.14 SEGMENTAL GAS DISCHARGE DISPLAYS

Gas discharge displays are available in seven-segment or fourteen-segment format for display of numeric and alphanumeric characters. They can be built up to operate on the gaseous discharge glow principle as Nixies do. Integrated circuits (ICs) have been developed to drive these devices and to develop the high voltages involved. The construction of a seven segment gas discharge display is shown in Fig. 28.20. It is seen that separate cathodes are provided in seven segments (and decimal point) formed on a base. Each even segment group has a single anode deposited on the covering face plate. The gas is contained in the space between anodes and cathodes. The connections are provided with the help of pins for all electrodes at the rear. A keep alive cathode is also enclosed with each group of segments.

A 50 μA current maintained through the keep alive cathode provides a source of ions. This improves the switching speed of the display.

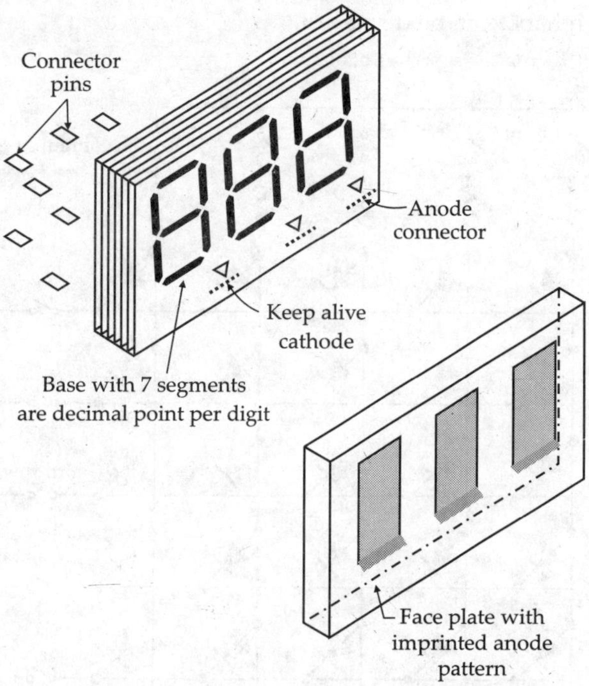


**Fig. 28.20** Seven segmental display using gaseous discharge.

The most serious disadvantage of the gas discharge displays is the high voltage required to operate them. This voltage is between 150 V to 220 V. High voltage transistors must be employed as switches for the cathodes. In fact, a separate high voltage supply is required for this purpose. However, a major advantage that offsets the disadvantages given above is that we can have bright displays with tube currents as low as 200 μA. Therefore the drain on power supply is extremely small.

## 28.15 DECADE COUNTING ASSEMBLIES (DCAs)

A single numeric display unit can display numbers from 0 to 9. This is called a Decade Counting Assembly (DCA).

Modern electronic counters use triggering units along with decade counting assemblies and display devices to totalize and display the count.

A typical decimal indicating system uses 10 characters to display numbers 0 to 9. Ten trigger circuits are needed for each decade counting assembly (DCA). A pulse from first trigger actuates circuit of 1 to indicate the first event counted. The second event triggers the second circuit, displaying 2 and feeds a pulse back to the first circuit, for it to turn off. This sequence continues upto a count of 9, and the decade recycles. However, in practice, a single decade is not used but a number of DCAs connected in cascade, are used as shown in Fig. 28.21.

Figure 28.21 shows 5 decade counting assemblies connected in cascade. They can count upto 99999. When the first unit counts upto 9, it returns to 0 on the 10th count and activates 1 of the 2nd (tens decade) with DCA indicating a count of 10.

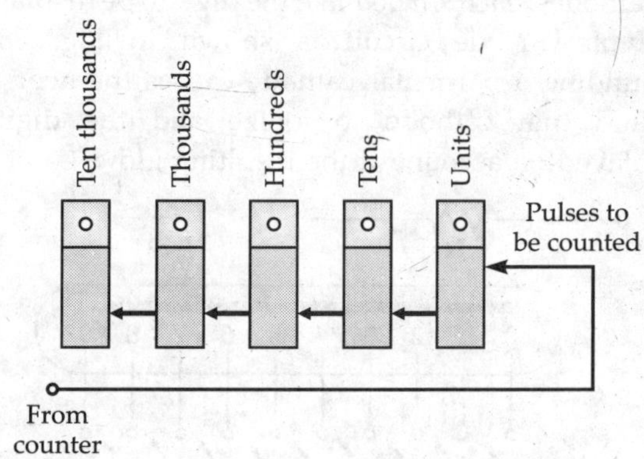


**Fig. 28.21** Decade counting assemblies (DCAs) with their readout.

When the tens decade recycles, it activates 1 of the 3rd (hundreds decade) DCA. As each decade recycles and activates the higher DCA.

Thus, this readout does not require any interpolation and the number is clearly displayed.

## 28.16 DISPLAY SYSTEMS

Electronic displays serve as output units of instrumentation systems. The digital data resulting from a measurement application is binary coded decimal (BCD) form. Thus in display applications, where machine communicates with a human being some sort of **Decoding** is required. If the language of the machine is BCD, we have to decode (translate) it into a decimal code that can be easily understood by the human operator. In many cases, decoding binary into octal might facilitate man machine communications. The decoder converts the binary code, such as BCD, into a non binary code that can be understood by man, such as decimal or 7-segment code. The decimal code has 10 different symbols while the 7 segment code has 7 separate lines used in activating 7 segment decimal displays.

We have discussed earlier that there are three types of display formats commonly used viz. segmental (7 segment), dot matrix (5×7 matrix) and decimal indicator (Nixie tube). Thus, each of these formats require the use of an appropriate code converter between the BCD data input and the output display. The code converter ( for operating Nixie tube has simply to be a 10 position switch which operates in accordance with the input code and therefore it is simply a decoder. However, in the case of segmental and dot matrix displays, a suitable decoder-encoder combination is used.

In the case of a Nixie tube there is one anode and 10 cathodes, each shaped like the digit to be displayed. A typical Nixie circuit is shown in Fig. 28.22. Grounding a particular cathode causes the neon gas around that cathode to ionize and the digit is diaplayed on account of the resulting glow.

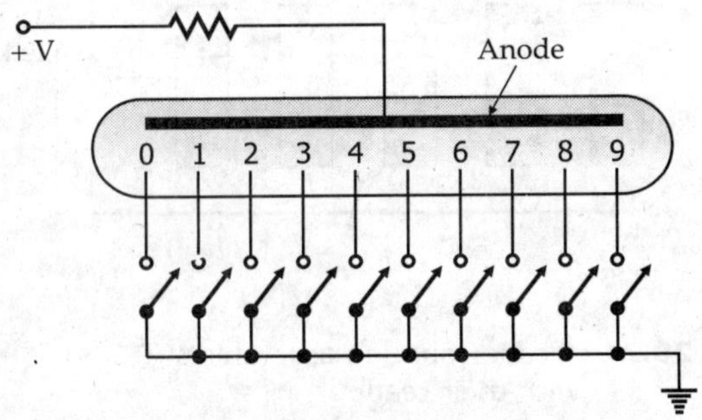


**Fig. 28.22** Display system of Nixie tube.

In the case of segmental displays, one or more than one light sources may be involved in the display of a particular character. Supposing digit 0 is to be displayed by a 7 segment display as shown in Fig. 28.23. This requires that segments *a, b, c, d, e* and *f* should be lit up. Similarly, for digit 2, segments *a, b, d, e* and *g* should be lit up. Thus by properly choosing the segments any number from 0 to 9 can be displayed.

A common supply voltage drives the anodes of the LEDs and when a switch closes, the corresponding LED is forward biased and emits light.

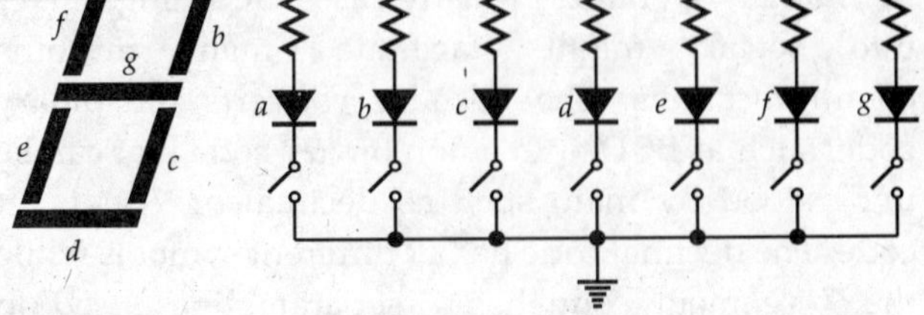


**Fig. 28.23** Seven segment readout and circuit.

Figure 28.24 shows a typical 5×7 dot matrix using LEDs. In order to light an LED in this matrix, a voltage has to be applied to its anode and its cathode has to be grounded. In Fig. 28.24, the circled LED lights up because a voltage is applied to 3rd row and 5th column. By applying voltage to more than one column and grounding more than one row, a decimal digit or alphabet can be displayed.

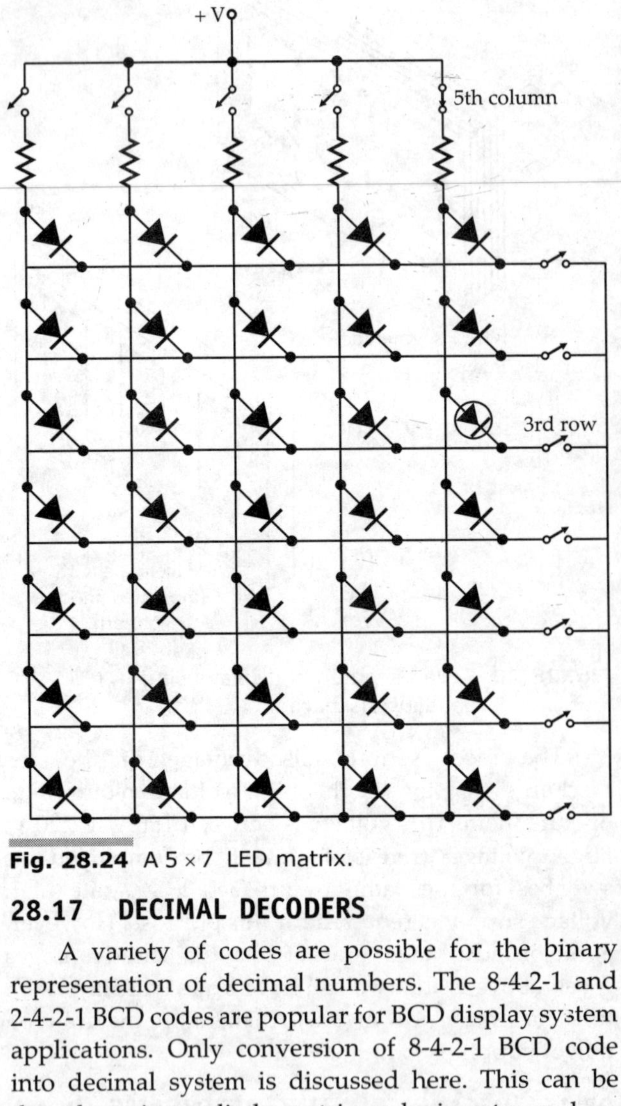


**Fig. 28.24** A 5×7 LED matrix.

## 28.17 DECIMAL DECODERS

A variety of codes are possible for the binary representation of decimal numbers. The 8-4-2-1 and 2-4-2-1 BCD codes are popular for BCD display system applications. Only conversion of 8-4-2-1 BCD code into decimal system is discussed here. This can be done by using a diode matrix or logic gates such as TTL (Transistor logic).

### 28.17.1 Diode Matrix

One number system may be converted into another number system with the help of a diode matrix circuit. For example, BCD number system can be converted into a decimal readout by using the diode matrix circuit of Fig. 28.25. Switches 1, 2, 4 and 8 provide input to the matrix and the decimal indicators 0 to 9 are the output of the matrix. In order to energize

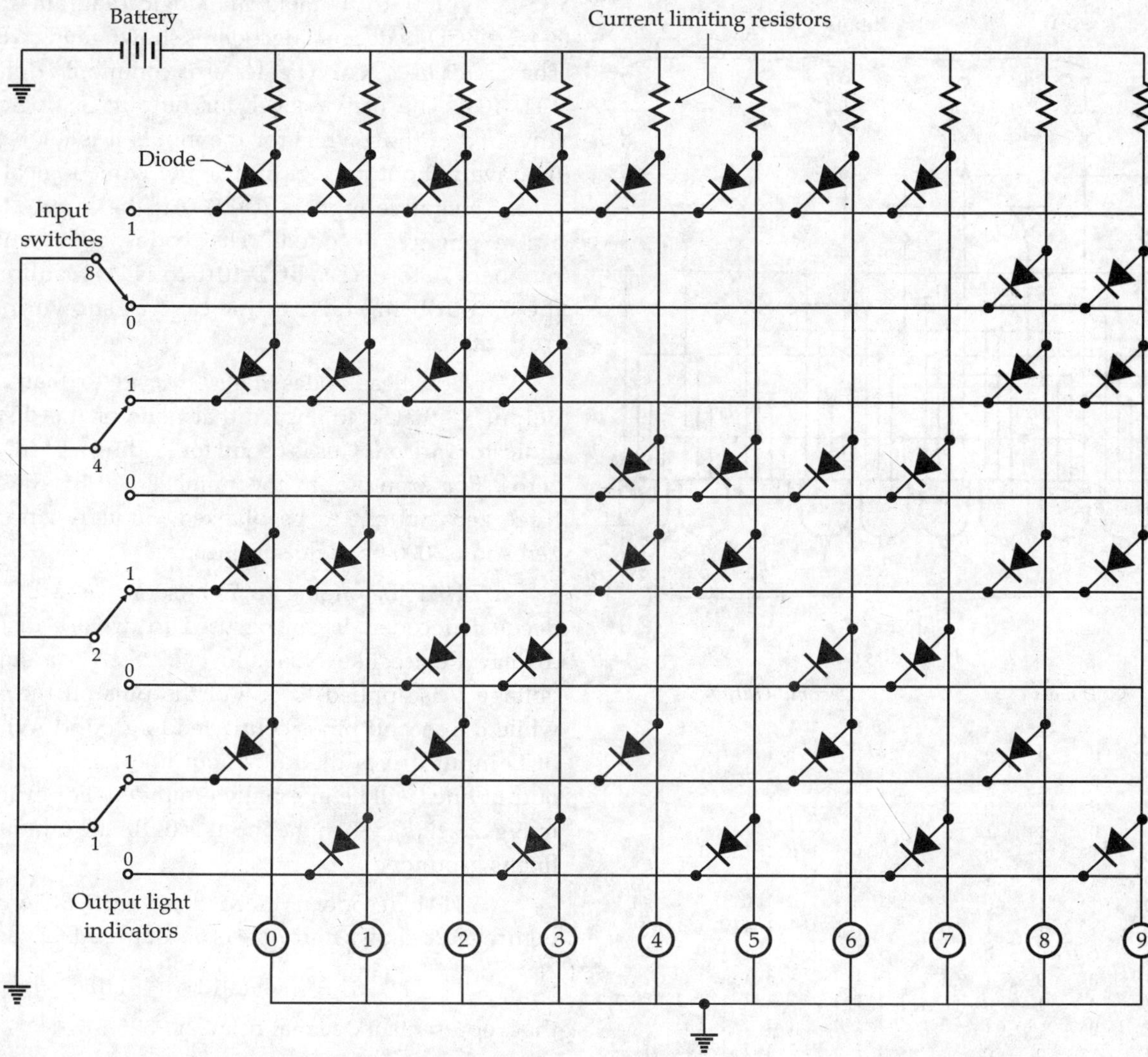


**Fig. 28.25** Diode matrix circuit for conversion of BCD system to decimal readout.



an output decimal indicator, the connection from the battery to the indicator must not be connected to ground through a diode as under these conditions the diode will short the indicator.

Figure 28.25 shows how number 7 is indicated. For number 7 to be indicated corresponding to a BCD input, switch 8 should be at its 0 position, switch 4 should be at its 1 position, switch 2 should be at its 1 position, and switch 1 should be at its 1 position. Under these conditions :

Switch 8 shorts indicators 8 and 9,

Switch 4 shorts indicators 0, 1, 2 ,3, 8 and 9,

Switch 2 shorts indicators 0, 1, 4, 5, 8 and 9,

Switch 8 shorts indicators 0, 2, 4, 6 and 8.

Thus all the indicators with the exception of 7 are short-circuited. Thus only light indicator 7 will be lit. Thus the readout is 7.

This way any number can be indicated. The number 0 is indicated when all the input switches are in 0 position.

This system may be used for Nixie tubes.

### 28.17.2 TTL Logic

The 7400 series, a line of circuits introduced in 1964, has become the most widely used group of bipolar ICs. Various decoders are available (See Texas Instruments Inc., 7441 through 7448 in the TTL 7400 series). Electrical characteristics such as input loading, output drive capability and switching characteristics do not vary within specific family of devices. The variations mostly involve drive capabilities tailored to suit particular loads *i.e.,* Nixie tubes, 7 segmental displays, LED displays etc.

A typical truth table and logic diagram for a 8-4-2-1 BCD to decimal decoder is shown in Fig. 28.26. The circuit uses NAND gates as is commonly done in TTL ICs. The active level for outputs is logic '0' (inverted output) while for the inputs it is logic 1. In the inverted output logical 0 is true while logical 1 is false. (Negative logic is used). All the input states are explicitly decoded. The codes representing numbers 10 to 15 (*i.e.*, BCD 1010 to 1111) result in all the output being false or logical 1 as shown in the truth table.

Any of these codes may be used to blank the output — that is to ensure that none of the display indicators are on. This is useful for leading and trailing zeros. For example, in the number 000486 the first three zeros ought to be blanked. Similarly, trailing zeros in 6.3000 have no significance.

A 74141 (belonging to 7400 series) is a BCD to decimal decoder. It can be used to drive a discrete display device like Nixie. In Fig. 28.27(*a*) a supply voltage *V* is applied to one of the pins on the right while the second pin is connected to ground. When a BCD input drives the four input lines, one of the 10 output lines will be internally grounded. For example in Fig. 28.27(*b*), when the input is 0110, the 6th output line is grounded.

A 74141 decoder is needed for each BCD digit. Figure 28.28 shows four 74141s decoding a BCD digit :

1001 0101 0010 0110

that represents a decimal number 9526.

Since each 74141 functions as a BCD-to-decimal converter, the thousands decoder grounds 9th output, line, the hundreds decoder grounds 5th output line, the tens decoder grounds the 2nd output line while the units decoder grounds the 6th output line.

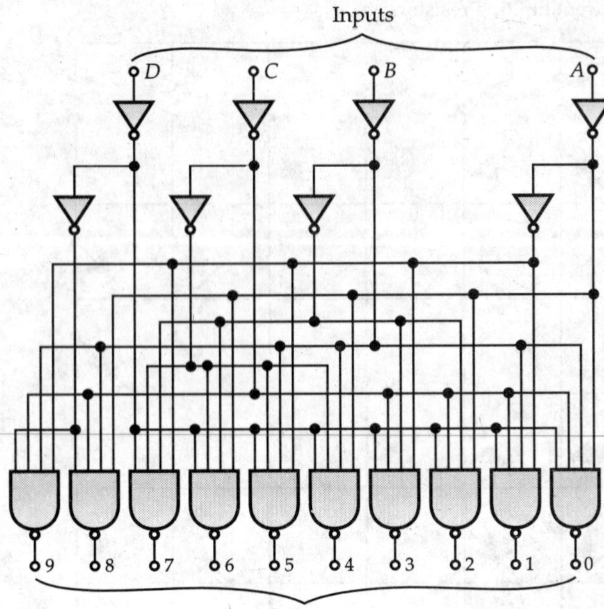

(*a*) Logic diagram

| BCD Input | | | | Decimal Output | | | | | | | | | |
|---|---|---|---|---|---|---|---|---|---|---|---|---|---|
| *D* | *C* | *B* | *A* | 0 | 1 | 2 | 3 | 4 | 5 | 6 | 7 | 8 | 9 |
| 0 | 0 | 0 | 0 | 0 | 1 | 1 | 1 | 1 | 1 | 1 | 1 | 1 | 1 |
| 0 | 0 | 0 | 1 | 1 | 0 | 1 | 1 | 1 | 1 | 1 | 1 | 1 | 1 |
| 0 | 0 | 1 | 0 | 1 | 1 | 0 | 1 | 1 | 1 | 1 | 1 | 1 | 1 |
| 0 | 0 | 1 | 1 | 1 | 1 | 1 | 0 | 1 | 1 | 1 | 1 | 1 | 1 |
| 0 | 1 | 0 | 0 | 1 | 1 | 1 | 1 | 0 | 1 | 1 | 1 | 1 | 1 |
| 0 | 1 | 0 | 1 | 1 | 1 | 1 | 1 | 1 | 0 | 1 | 1 | 1 | 1 |
| 0 | 1 | 1 | 0 | 1 | 1 | 1 | 1 | 1 | 1 | 0 | 1 | 1 | 1 |
| 0 | 1 | 1 | 1 | 1 | 1 | 1 | 1 | 1 | 1 | 1 | 0 | 1 | 1 |
| 1 | 0 | 0 | 0 | 1 | 1 | 1 | 1 | 1 | 1 | 1 | 1 | 0 | 1 |
| 1 | 0 | 0 | 1 | 1 | 1 | 1 | 1 | 1 | 1 | 1 | 1 | 1 | 0 |
| 1 | 0 | 1 | 0 | 1 | 1 | 1 | 1 | 1 | 1 | 1 | 1 | 1 | 1 |
| 1 | 0 | 1 | 1 | 1 | 1 | 1 | 1 | 1 | 1 | 1 | 1 | 1 | 1 |
| 1 | 1 | 0 | 0 | 1 | 1 | 1 | 1 | 1 | 1 | 1 | 1 | 1 | 1 |
| 1 | 1 | 0 | 1 | 1 | 1 | 1 | 1 | 1 | 1 | 1 | 1 | 1 | 1 |
| 1 | 1 | 1 | 0 | 1 | 1 | 1 | 1 | 1 | 1 | 1 | 1 | 1 | 1 |
| 1 | 1 | 1 | 1 | 1 | 1 | 1 | 1 | 1 | 1 | 1 | 1 | 1 | 1 |

(*b*) Truth table

**Fig. 28.26** BCD-to-decimal decoder.

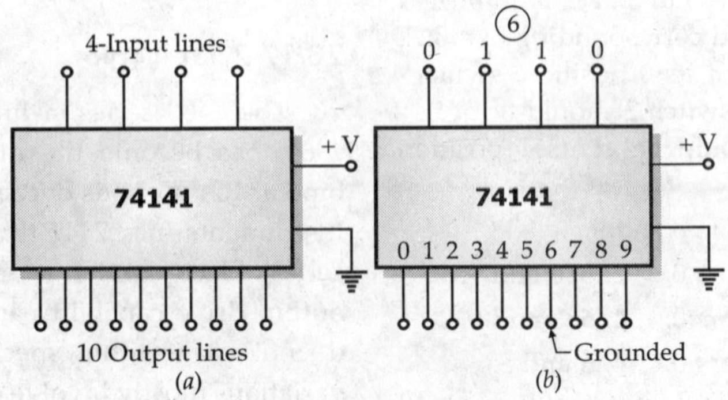


**Fig. 28.27** 74141 BCD-to-decimal decoder.

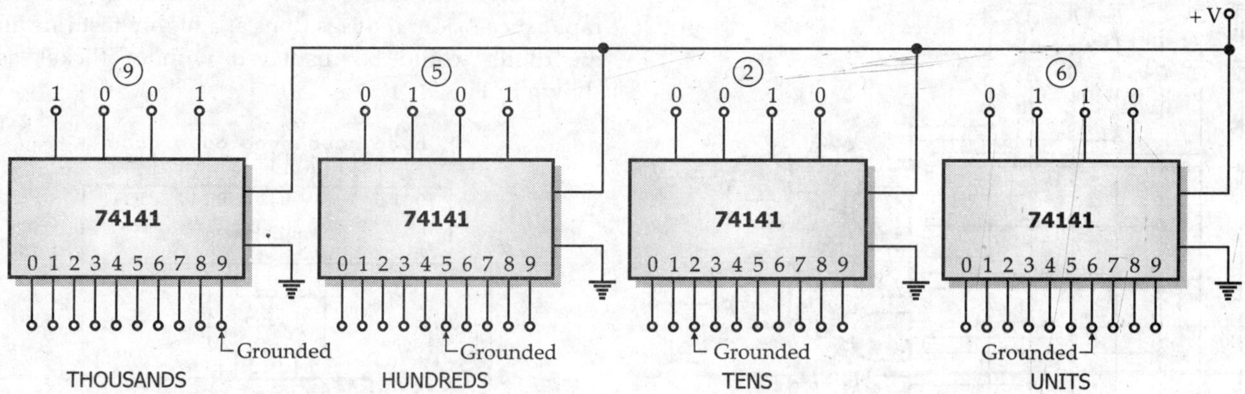


**Fig. 28.28** Decoding four BCD digits.

## 28.18 BCD TO 7-SEGMENT CONVERTER

A BCD-to-7 segment converter is required to convert a four bit BCD input to control the outputs for driving a 7 segment numeric characters such as those which use LEDs and LCDs.

The conversion of 8-4·2-1 BCD code to 7 segment code is given in Table 28.1 (Refer to Fig. 28.23)

The output (segment select) code is logic 1 type *i.e.,* positive logic used. However, inverted type (the complement of code with active 1's) may also be used where it is logic 0. The inverted 7 segment logic is obtained by interchanging 0s and 1s.

In addition to 10 BCD inputs 0000 through 1001 representing decimal numbers 0 through 9, selected symbols may be used 6 additional inputs *i.e.,* 1010 to 1111. Since 16 characters of 7 bit output code are handled by this code converter, it is also called a 16×7 converter. The converter using NAND gates which are ideally suited for TTL IC configuration is shown in Fig. 28.29. Since all the 16 states from 0000 to 1111 are acceptable, there is 'false data rejection'.

The 7447 is a BCD-to-7 segment converter. It has four inputs for the BCD and 7 outputs for driving 7 segmental displays. For a given BCD input it grounds the appropriate 7 segment LED display. For example if the BCD input is 1001 (9), it grounds the *a, b, c, f* and *g* output lines. These grounds then activate segments *a, b, c, f* and *g* to display digit 9 as shown in Fig. 28.30.

The display of a 5 digit decimal number say 48572 will require one 7447 and a seven segment display for each of the five digits. Thus five 7447s and five 7 segment displays will be needed. With this arrangement all the five digits are excited simultaneously.

**Table 28.1** *BCD to 7-Segment Code Conversion*

| Decimal Number | BCD Code | | | | 7 segment Output | | | | | | |
|---|---|---|---|---|---|---|---|---|---|---|---|
| | D | C | B | A | a | b | c | d | e | f | g |
| 0 | 0 | 0 | 0 | 0 | 1 | 1 | 1 | 1 | 1 | 1 | 0 |
| 1 | 0 | 0 | 0 | 1 | 0 | 1 | 1 | 0 | 0 | 0 | 0 |
| 2 | 0 | 0 | 1 | 0 | 1 | 1 | 0 | 1 | 1 | 0 | 1 |
| 3 | 0 | 0 | 1 | 1 | 1 | 1 | 1 | 1 | 0 | 0 | 1 |
| 4 | 0 | 1 | 0 | 0 | 0 | 1 | 1 | 0 | 0 | 1 | 1 |
| 5 | 0 | 1 | 0 | 1 | 1 | 0 | 1 | 1 | 0 | 1 | 1 |
| 6 | 0 | 1 | 1 | 0 | 0 | 0 | 1 | 1 | 1 | 1 | 1 |
| 7 | 0 | 1 | 1 | 1 | 1 | 1 | 1 | 0 | 0 | 0 | 0 |
| 8 | 1 | 0 | 0 | 0 | 1 | 1 | 1 | 1 | 1 | 1 | 1 |
| 9 | 1 | 0 | 0 | 1 | 1 | 0 | 1 | 0 | 0 | 1 | 1 |

28

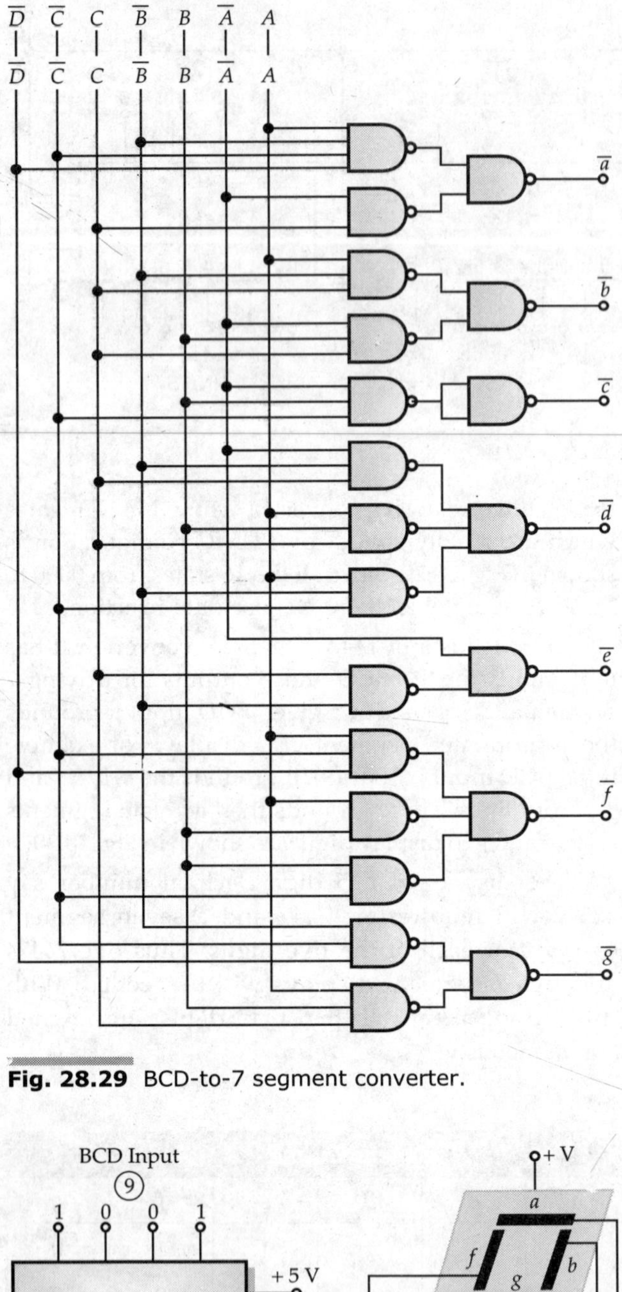


**Fig. 28.29** BCD-to-7 segment converter.

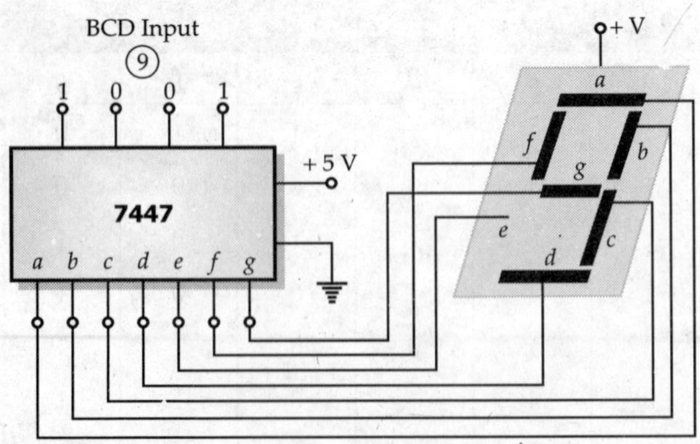


**Fig. 28.30** 7447 BCD-to-7 segment decoder.

However, multiplexing can be used to reduce the number of decoders, wiring and the displays. In the arrangement using multiplexing, instead of activating all the digits simultaneously, we can display them in rapid succession. If this is done at a highly fast rate all the digits would be displayed without flicker as shown in Fig 28.31.

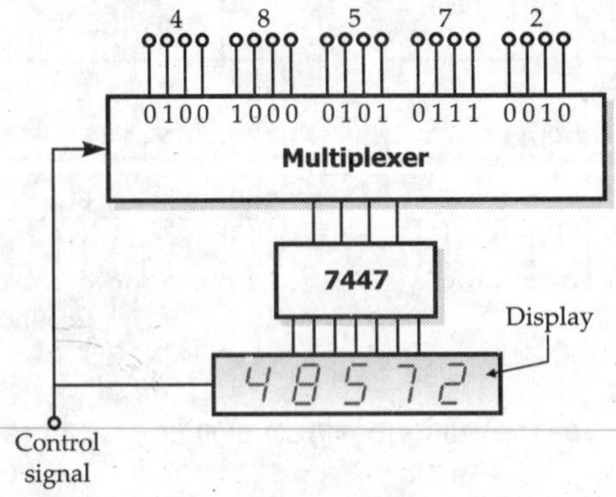


**Fig. 28.31** Multiplexing and digital display.

## 28.19 BCD-TO-DOT MATRIX CONVERTER

A 3×5 dot matrix is required for display of numeric characters while the display of alphanumeric characters requires a 5×7 dot matrix. A 4 bit BCD input is required for conversion to appropriate dot selection code which will activate the dots to display the character. Considering 16 characters are to be accommodated in the input code a converter of 16×15 bits capacity is required for 3×5 dot matrix while the converter capacity is 16×35 bits for a 5×7 dot matrix.

A six bit code is required for display of alphanumeric characters as there are 26 alphabets, 10 decimal number and 28 special characters totaling 64. Thus the capacity of alphanumeric code converter is 64×35 bits. These code conversions are done through special building blocks called read-only memory **(ROM)**.

A 5×7 LED dot matrix is shown in Fig. 28.24. All the LEDs required for display of a character are not illuminated simultaneously. Usually they are activated a row at a time. For example, the first row of LEDs is first grounded and voltage is applied to the appropriate columns. After displaying for a short interval, the second row of LEDs is grounded and voltages applied to the appropriate columns. The third and subsequent rows are grounded successively with voltages applied to required columns. This process is repeated at a very fast rate, so that the character is displayed without flicker.

The circuits driving the dot matrix converter scan through the rows and apply voltages to appropriate columns at the same time. Therefore the circuitry is complex. A ring counter scans the rows, while a read-only memory (ROM) applies voltages to columns.

## 28.20 RESOLUTION IN DIGITAL METERS

The number of digit positions (or simply, the number of digits) used in a digital meter determines the resolution. Thus a three digit display on a digital voltmeter. (**DVM**) for 0-1 V range will be able to indicate values from zero to 999 mV, with smallest increment or resolution of 1 mV.

In practice a fourth digit usually capable of indicating either 0 or 1 only, is placed to the left of active digits. This permits going above 999 to 1999 to give overlap between ranges for convenience. This is called **over-ranging**. This type of display is known as a **$3\frac{1}{2}$ digit display**. Figure 28.32 shows an auto ranging portable digital voltmeter. The resolution of a digital meter, however, is determined by the number of **active** or **full** digits used.

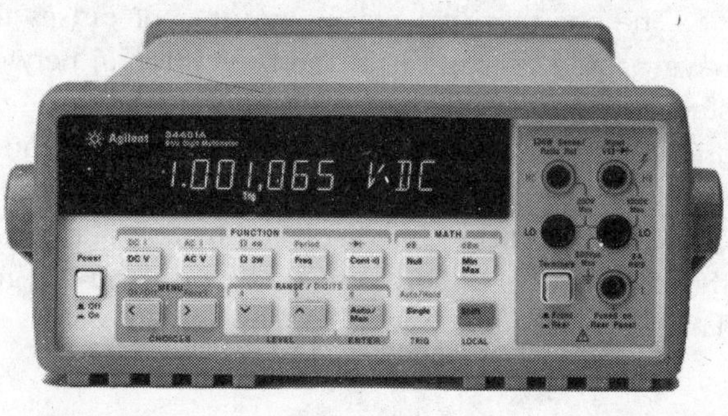


**Fig. 28.32** Digital multimeter.

If $n$ = number of full digits. Resolution = $1/10^n$ ...(28.1)

Therefore, a 4 digit display has a resolution of $1/10^4$ = 0.0001 or 0.01 percent. This resolution is also referred to as one part in 10,000.

## 28.21 SENSITIVITY OF DIGITAL METERS

Sensitivity is the smallest change in input which a digital meter is able to detect. Therefore it is lowest voltage range's full scale value multiplied by the meter's resolution.

Sensitivity $S = (f.s.)_{min} \times R$ ...(28.2)

where $(f.s.)_{min}$ = lowest full scale value of meter and

$R$ = resolution expressed as decimal.

## 28.22 ACCURACY SPECIFICATION FOR DIGITAL METERS

Accuracy is usually expressed as a percentage of the reading plus a percentage of full scale. Often the percentage of full scale part of the statement is given in terms of digits. When this is done, the digits referred to are the least significant digits (**LSD**).

**Example 28.1** *(a) What is the resolution of a $3\frac{1}{2}$ digits display ? (b) Find the resolution of a $3\frac{1}{2}$ digit meter in case its range is 1 V. (Some manufacturers may specify the range as 2 V since a $3\frac{1}{2}$ digit meter can count upto 1999 mV) (c) Find the resolution of the meter for 10 V range.*

**Solution.** (*a*) Number of full digits in case of $3\frac{1}{2}$ digits display = 3.

$\therefore$ Resolution $= 1/10^n = 1/10^3 = 0.001$ or 0.1%.

Hence the meter cannot distinguish between values that differ from each other by less than 0.001 to full scale.

(*b*) Full scale reading = 1.000

$\therefore$ Resolution on 1 V range $= 1 \times 0.001 = 0.001$ V

Hence, on 1 V range it will not be able to distinguish between readings which differ by less than 0.001 V.

(*c*) Full scale reading = 10.00

$\therefore$ Resolution on 10 V range $= 10 \times 0.001 = 0.01$ V.

Hence, on 10 V scale it will not be able to distinguish between readings which differ by less than 0.01 V.

**Example 28.2** *A $4\frac{1}{2}$ digit voltmeter is used for voltage measurements. (a) Find its resolution. (b) How would 12.98 V be displayed on 10 V range ? (c) How would 0.6973 be displayed on 1 V range ? (d) How would 0.6973 be displayed on 10 V range ?*

**Solution.** (*a*) Number of full digits on a $4\frac{1}{2}$ digit display = 4

$\therefore$ Resolution $= 1/10^4 = 0.0001$ or 0.01%

(*b*) There are 5 digit places in a $4\frac{1}{2}$ digit display. Therefore, 12.98 V will be displayed as 12.980 on its 10 V scale.

(*c*) Resolution on 1 V range $= 1 \times 0.0001 = 0.0001$ V

Therefore, on 1 V range, any reading can be displayed to 4th decimal place. Hence, 0.6973 V will be displayed as 0.6973 on 1 V range.

(*d*) Resolution on 10 V range $= 10 \times 0.0001$ V $= 0.001$

Therefore, on 10 V range, readings can be displayed only upto 3rd place of decimal. Hence, 0.6973 V will be displayed as 00.697 on 10 V scale. The digit 3 in the fourth decimal place will be lost. Using the appropriate range *i.e.*, 1 V, the digit 3 can be retained in the readout.

**Example 28.3** *A $3\frac{1}{2}$ digit DVM has an accuracy specification of ± 0.5 percent of reading ± 1 digit. (a) What is the possible error in volt, when the instrument is reading 5.00 V on the 10 V range ? (b) What is the possible error in volt, when reading 0.10 V on the 10 V range ? (c) What percentage of the reading is the possible error in the case of (c) ?*

28

**Solution.** (*a*) 0.5 percent of reading

$= 0.005 \times 5.00 = 0.025$ V.

The display for 5.00 V read on 10 V scale of $3\frac{1}{2}$ digit meter is 05.00 as there are four digit positions.

The digit in the LSD has a value 0.01 V.

$\therefore$ Total possible error $= 0.025 + 0.01 = 0.035$ V

(*b*) When reading 0.10 V on 10 V range, we have,

$\pm$ 5% of reading $= 0.005 \times 0.10 = 0.0005$ V

$\pm$ 1 digit $= \pm$ 0.01 V

$\therefore$ Total possible error $= \pm$ 0.0105 V

(*c*) Error as a percent of reading $= \dfrac{0.0105}{0.100} = 10.5\ \%$

This is a large error and demonstrates the inherent danger of taking readings low on a scale.

## 28.23 DIGITAL FREQUENCY METER

The signal whose frequency is to be measured is converted into a train of pulses, one pulse for each cycle of the signal. Then the number of pulses appearing in a definite interval of time is counted by means of an electronic counter. Since the pulses represent the cycles of unknown signal, the number appearing on the counter is a direct indication of frequency of the unknown signal. Since the electronic counters are extremely fast, the frequency of high frequency signals may be known.

### 28.23.1 Basic Circuit

The block diagram of the basic circuit of a digital frequency meter is shown in Fig 28.33. The unknown frequency signal is fed to a **Schmitt trigger**.

The signal may be amplified before being applied to Schmitt trigger. In a **Schmitt trigger**, the signal is converted into a square wave with very fast rise and fall times, then differentiated and clipped. As a result, the output from a Schmitt trigger is a train of pulses, one pulse, for each cycle of the signal.

The output pulses from the Schmitt trigger are fed to **start stop gate**. When this gate opens (**start**), the input pulses pass through this gate and are fed to an electronic counter which starts registering the input pulses. When the gate is closed (**stop**), the input of pulses to counter ceases and it stops counting.

The counter displays the number of pulses that have passed through it in the time interval between start and stop. If this interval is known, the pulse rate and hence the frequency of the input signal can be known. Suppose $f$ is the frequency of unknown signal, $N$ the number of counts displayed by counter and $t$ is the time interval between start and stop of gate. Therefore, frequency of unknown signal $f = N/t$.

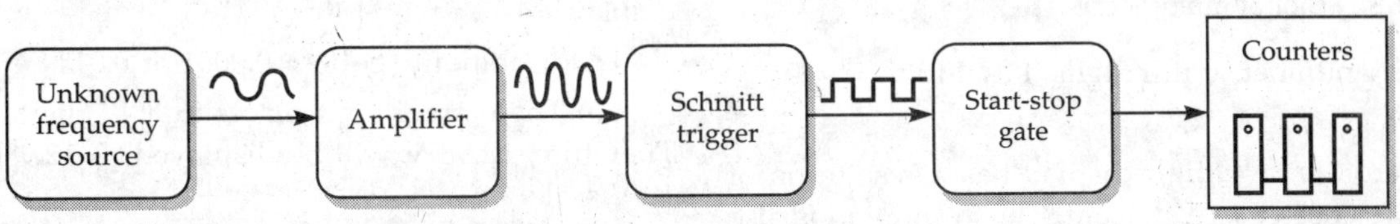


**Fig. 28.33** Basic circuit of a digital frequency meter.

### 28.23.2 Time Base

It is abundantly clear that in order to know the value of frequency of input signal, the time interval between start and stop of gate must be accurately known. This time interval known as **time base** can be determined by the circuit given in Fig. 28.34. The time-base consists of a fixed frequency crystal oscillator. This oscillator, which is known as **clock oscillator** must be very accurate. In order to insure its accuracy, the crystal is enclosed in a constant temperature oven. The output of this constant frequency oscillator is fed to a Schmitt trigger which converts the input to an output consisting of a train of pulses at a rate equal to the frequency of the clock oscillator. The train of pulses then passes through a series of frequency divider decade assemblies connected in cascade. Each decade divider consists of a decade counter as explained in Art. 28.4.3 on page 1009 and divides the frequency by 10. Connections are taken from the output of each decade in the series chain, and, by means of a selector switch, any output may be selected.

In the block diagram of Fig. 28.34, the clock oscillator frequency is 1 MHz or $10^6$ Hz. Thus the output of Schmitt trigger is $10^6$ pulses per second. At the $\times$ 1 tap of the switch there are $10^6$ pulses per second, and thus the time interval between two consecutive pulses is $10^{-6}$ second or 1 μs.

At $\times 10^{-1}$ tap, the pulses, having gone through decade divider 1, are reduced by a factor 10, and now

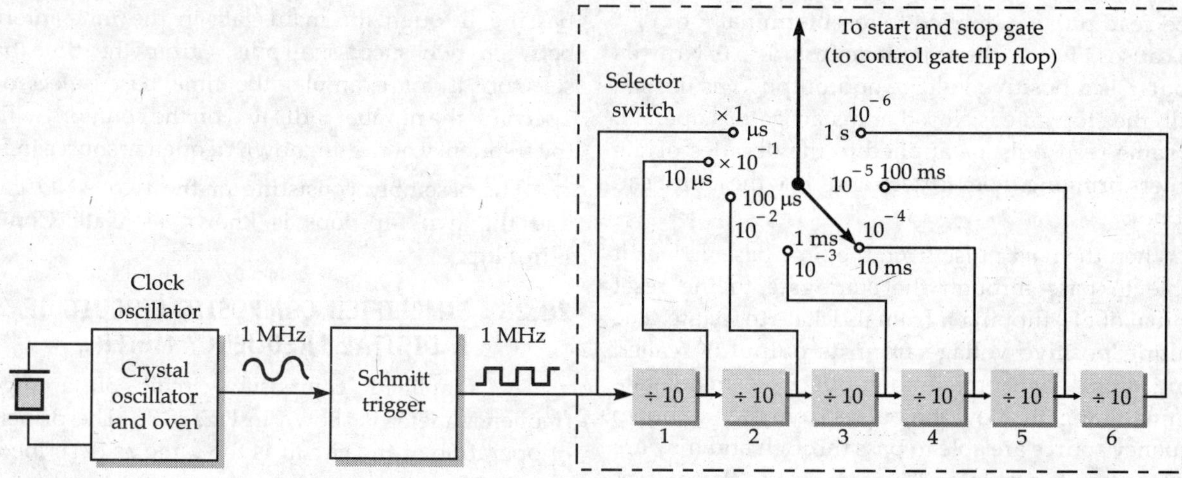


**Fig. 28.34** Time base selector.

there are $10^5$ pulses per second. Therefore the time interval between them is 10 μs. Similarly, there are $10^4$ pulses per second at tap × $10^{-2}$ and the time interval is 100 μs ; $10^3$ pulses at tap × $10^{-3}$ and the time interval is 1 ms ; 100 pulses per second at tap × $10^{-4}$ and the time interval 10 ms, 10 pulses per second at tap × $10^{-5}$ and time interval 100 ms ; one pulse per second at tap × $10^{-6}$ and the time interval is 1 second.

This time interval between the pulses is the time base and it can be selected by means of the selector switch.

### 28.23.3 Start and Stop Gate

The flip-flops (FF) act as start and stop gates.

## 28.24 CIRCUIT FOR MEASUREMENT OF FREQUENCY

The complete circuit for measurement of frequency is shown in Fig. 28.35. The positive pulses from the unknown frequency source called counted signal are arriving at input *A* of the main gate and positive pulses from time base selector are arriving at input *B* of the start gate and input *B* of stop gate. Initially, flip-flop $FF_1$ is its 1 state. The resulting positive voltage from output *Y*, applied to input *A* of stop gate opens this gate. The zero voltage from output *Y* of flip-flop $FF_1$, applied to input *A* of the start gate closes that gate.

As the stop gate is open, the positive pulses from the time base, can get through to set input terminal *S* of flip-flop $FF_2$ and keep it in state 1. The resulting zero output voltage from *Y* of the flip flop is applied to terminal *B* of main gate keeping it closed. Hence no pulses from the unknown frequency source can pass through the main gate to register on the counter.

28

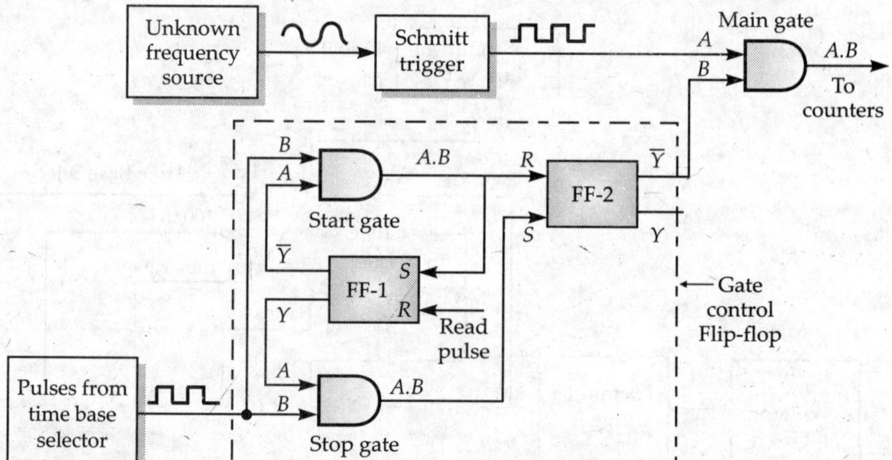


**Fig. 28.35** Circuit for measurement of frequency.

In order to start the operation, a positive pulse, called **read pulse** is applied to reset terminal $R$ of $FF_1$. This causes $FF_1$ to reverse its state from 1 to 0. Now its output $Y$ is a positive voltage and output $Y$ is 0. As a result, the stop gate is closed and start gate is opened. The same read pulse is applied to the decades of the counters bringing them to zero and thus the count can start now.

When the next pulse from the time base arrives it is able to pass through the start gate to the reset terminal of $FF_2$ flipping it from its l state to 0 state. The resulting positive voltage from its output $Y$ (called gating signal) is applied to input $B$ of main gate opening that gate. Now the pulses from the unknown frequency source are able to pass through and they are registered on the counter. The same pulse that passes through the start gate is applied so the set input $S$ of $FF_1$ changing its state from 0 to 1. This results in closing of start gate and the stop gate is opened. However, since the main gate is still open, pulses from the unknown frequency source continue to get through to the counter.

The next pulse from the time base selector passes through the open stop gate to the set input terminal $S$ of $FF_2$ changing it back to its 1 state. Its output from terminal $Y$ becomes zero and so the main gate is closed and the counting stops.

Thus the counter registers the number of pulses passing through the main gate in the time interval between two successive pulses from the time base selector. If for example, the time base selected is 1 second, the number indicated on the counters will be the frequency of the unknown frequency source in Hz.

The assembly consisting or the two AND gates and the two flip-flops is known as **Gate Control Flip-Flop**.

## 28.25 SIMPLIFIED COMPOSITE CIRCUIT OF A DIGITAL FREQUENCY METER

A simplified composite circuit of a digital frequency meter is shown in Fig. 28.36. The principle of operation of the circuit is the same as described in Art. 28.24.

There are two signals to be traced :

(*i*) ***Input signal*** – (or the counted signal). The signal whose frequency is to be measured.

(*ii*) ***Gating signal*** – This determines the length of time during which the counters (which consist of decade counter assemblies) are allowed to totalize the pulses separated by the period of the original input signal.

The input signal is amplified and is applied to a Schmitt trigger where it is converted to a train of pulses separated by the period of the original input signal.

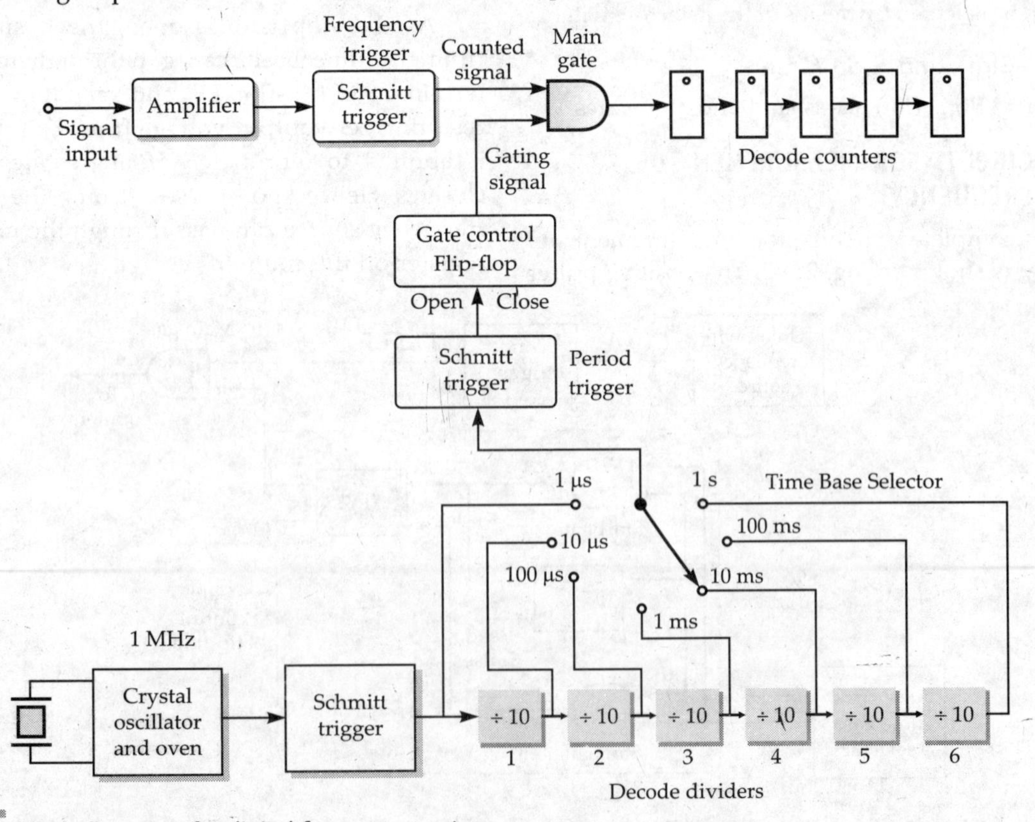


**Fig. 28.36** Block diagram of a digital frequency meter.

The oscillation frequency is 1 MHz. Therefore, the time base output is shaped by a Schmitt trigger into positive pulses, 1 μs apart. These pulses are applied to 6 decade divider assemblies (DDAs). A selector switch allows the time interval to be selected from 1 μs to 1 s.

The first output pulse from the time base selector switch-passes through the Schmitt trigger to the gate control flip-flop. The gate control flip-flop assumes a state such that an enable signal is applied to the main gate. The main gate being an **AND** gate, the input signal pulses are allowed to enter the DCAs where they are totalized and displayed. This process continues till a second pulse arrives at the control gate flip-flop from the DDAs (Decade dividing assemblies). The control gate reverses its state which removes the enabling signal from the main gate and no more pulses are allowed to get to counting assemblies since the main gate closes. Thus the number of pulses which have passed during a specific time are counted and displayed on the DCAs. The frequency can be read directly in Hz, kHz or, MHz in case the time base selector moves the decimal point in the display area.

## 28.26 HIGH FREQUENCY MEASUREMENTS

The direct count range of digital frequency meters extends from d.c. to a few hundred MHz. The limitation is on account of the counters used in conjunction with the digital frequency meters. The counters can work upto a limited rate of counting and not beyond it and high frequency measurements demand a very high rate of counting.

This range (*i.e.*, a few hundred MHz) covers only a small portion of the frequency spectrum in use now-a-days. So techniques other than direct counting are used to extend the range of digital frequency meters to above 40 GHz. The frequency of the input signal is reduced before it is applied to a digital counter. This is done by special techniques.

Some of the techniques in use are :

**1. *Prescaling.*** The high frequency signal is divided by some numbers — 2, 4, 6, 8, etc. by using high speed divider circuits to get it within the frequency range of the digital frequency meter.

**2. *Heterodyne converter.*** The high frequency signal is reduced in frequency to a range within that of the meter by using heterodyne techniques.

**3. *Transfer oscillator.*** A harmonic of a tunable low frequency continuous wave oscillator is zero beat with the unknown high frequency signal. The low frequency oscillator frequency is measured. This measured frequency is then multiplied by an integer which equal to the ratio of the two frequencies in order to determine the value of the unknown high frequency.

**4. *Automatic divider.*** The high frequency signal is reduced by some factor such as 100 : 1 or 1000 : 1 using automatically tuned circuits which generate an output frequency equal to 1/100 or 1/1000 of the input frequency to get it within the range of the digital frequency meter.

## 28.27 PERIOD MEASUREMENT

Sometimes it is desirable and necessary to measure the period of an input signal rather than its frequency.

This is specially true when measuring the low frequencies because measurement of frequencies in the low frequency range using frequency mode of operation gives low accuracy. To get good accuracy at low frequencies, we should make measurement of period to know the unknown frequency rather than make direct frequency measurements. Now period of unknown frequency, $T = 1/f$. Thus measurement of period can be done directly by interchanging the two input signals to the main gate.

The circuit for measurement of frequency shown in Fig. 28.36 can be used for measurement of time period but the counted signal and the gating signal are reversed. Figure 28.37 shows the circuit for measurement of time period.

The gating signal is derived from the unknown input signal which now controls the opening and closing of the main gate. In the diagram (Fig. 28.37), the time base is set at 10 μs (frequency 0.1.MHz). The number of pulses which occur during one period of the unknown signal are counted and displayed by the decade counting assemblies.

The only drawback in using period measurements is that to get accuracy at low frequencies, the operator must take the reciprocal of the answer displayed by the display of the counters if he wants to know the input frequency.

For example, when measuring the period of a 60 Hz frequency, the electronic counter might display 16.6673 ms.

$$\therefore \text{ Frequency } f = \frac{1}{16.6673 \times 10^{-3}} = 59.9977 \text{ Hz}$$

The accuracy of the period measurement and hence of frequency can be greatly increased by using the **multiple period** average mode of operation. In this mode, the main gate is held open for more than one period of the unknown signal.

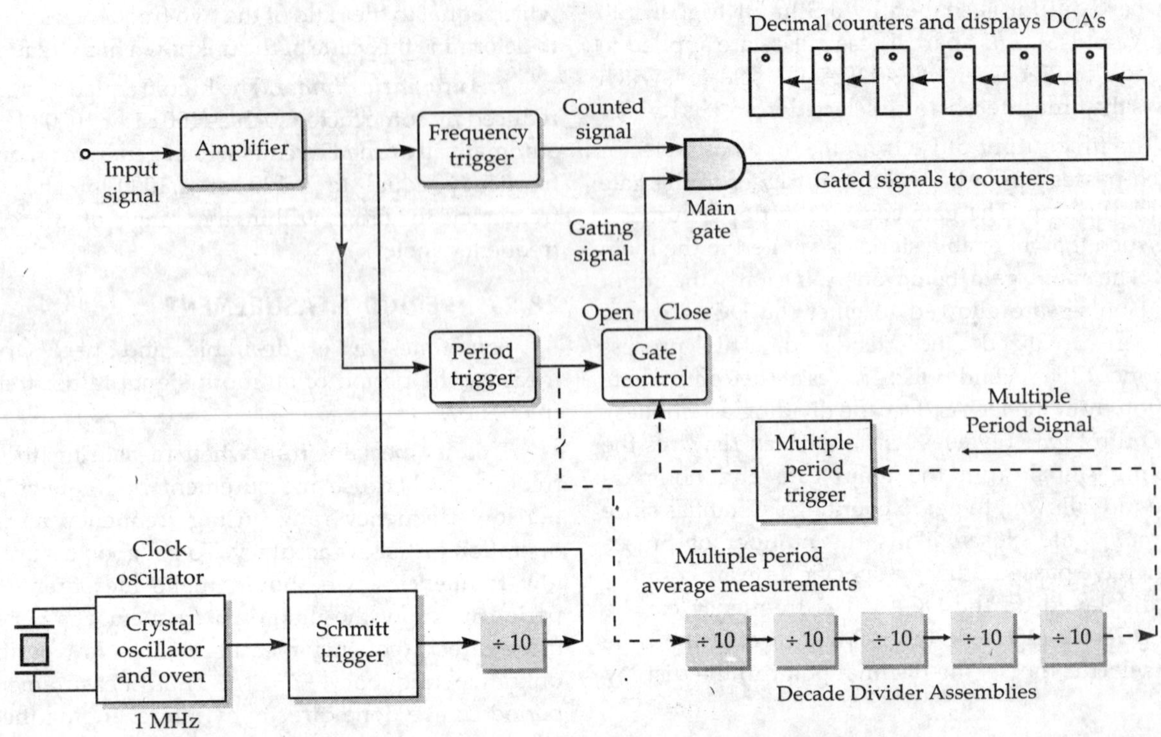


**Fig. 28.37** Block diagram of single and multiple period (average) measurement mode of operation.

This is done by passing the unknown signal through one or more decade divider assemblies (DDAs) so that the period is extended by a factor of 10, 100 or more.

Hence the digital display on the counters will show more digits of information thus increasing the accuracy. However, the decimal point location and measurement units are usually changed each time an additional decade divider is added so that the display is always in terms of the period of 1 cycle of the input signal, even though the measurement may have lasted for 10 or 100 or more cycles.

Figure 28.37 shows the multiple period average mode of operation by the dashed portion of the block diagram. In this diagram, 5 more decade dividing assemblies (DDAs) have been added so that the gate now remains open for an interval of $10^5$ times than it did with that only one DDA.

## 28.28 RATIO AND MULTIPLE RATIO MEASUREMENTS

The ratio measurement involves the measurement of ratio of two frequencies. The measurement in effect is a period measurement. The lower frequency is used as a gating signal while the higher frequency is the counted signal. Thus the lower frequency takes the place of the time base. The block diagram for ratio and multiple ratio measurements is shown in Fig. 28.38.

The number of cycles of the high frequency signal $f_1$ which occur during the period of lower frequency signal $f_2$ are counted and are displayed by DCAs.

In the multiple ratio measurements, the periods of the lower frequency signal is extended by a factor of 10, 100, etc. by using DDAs.

## 28.29 TIME INTERVAL MEASUREMENTS

In this measurement, the electronic counter used as an output readout device performs the function of an **electronic stop watch** *i.e.,* it measures the time interval.

The block diagram for time interval measurement is shown in Fig. 28.39. There are two parallel input channels where one channel supplies the **enabling pulse** for the main gate and the other a **disenabling pulse**.

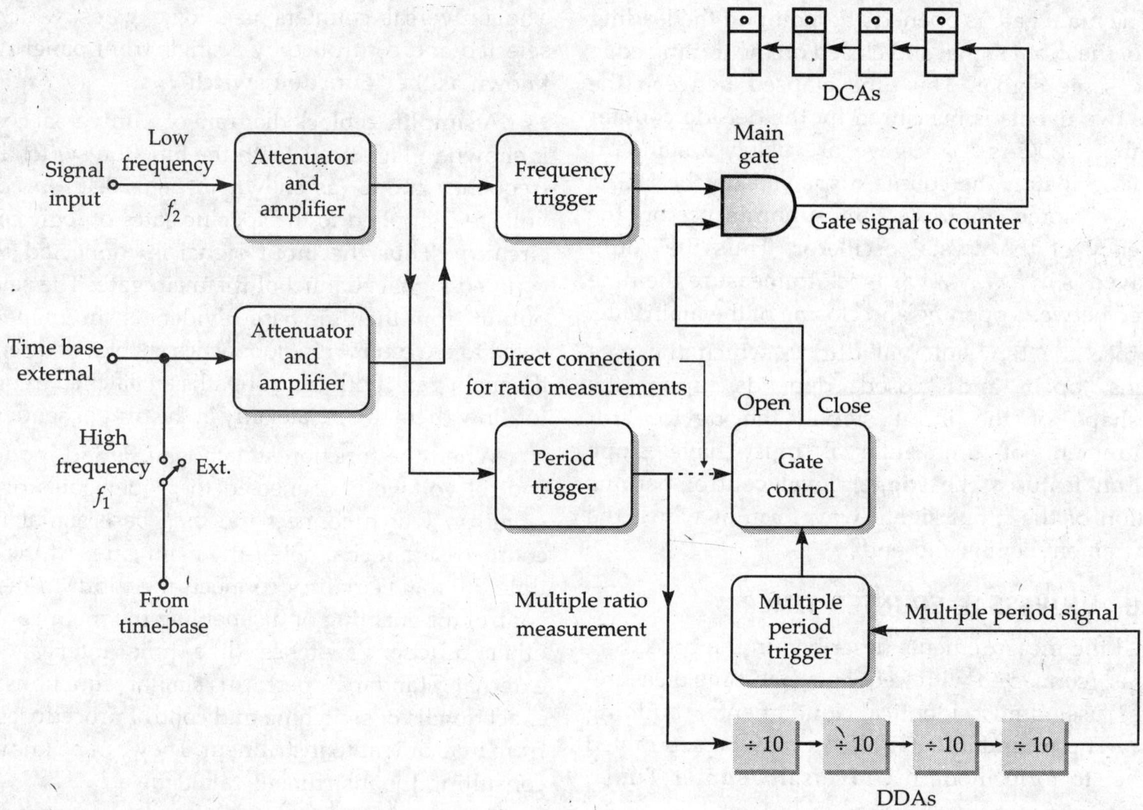


**Fig. 28.38** Ratio and multiple ratio mode of operation.

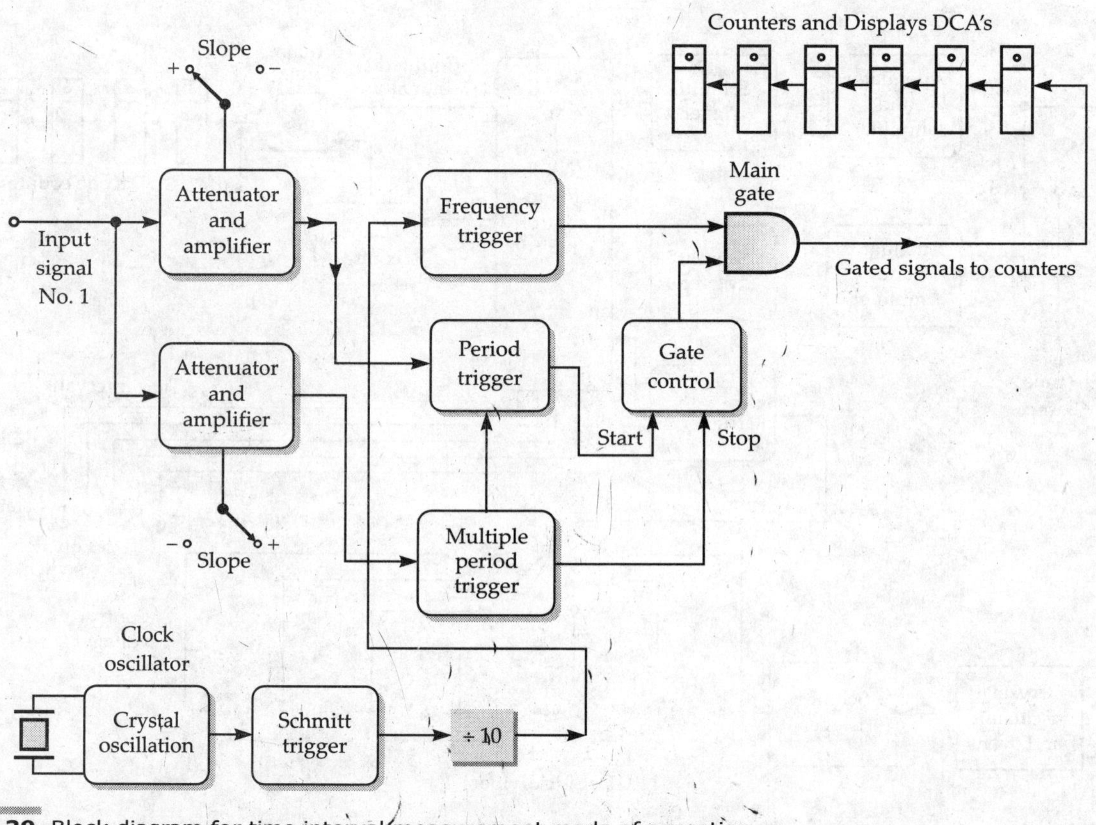


**Fig. 28.39** Block diagram for time interval measurement mode of operation.

The main gate is opened at a point on the leading edge of the input signal and closed on the trailing edge of the same signal. The time elapsed between the above two events is measured by the decade counter assemblies (DCAs). Since we are using a standard crystal oscillator, the counts displayed on the DCAs must be some multiple on submultiple of the frequency of the crystal oscillator. Thus the count displayed is in a way a yardstick to measure the time elapsed between opening and closing of the main gate.

Since the time interval during which the gate remains open and closed depends upon the waveshape of the input signal, the devices for measurement of time interval must have **slope selection** feature. The **trigger level** control permits selection of the input signal waveform at which the measurement begins and ends.

## 28.30 UNIVERSAL COUNTER TIMER

All the measurements described in the preceding sections, (connected with frequency and time measurements) use almost identical fundamental building blocks. These fundamental blocks may be assembled together to form modern **Universal Counter Timer.** The universal counters use logic gates which are selected and controlled by a single front panel switch known as the **"Function switch".**

A simplified block diagram of a universal counter is shown in Fig. 28.40. With the function switch in the frequency mode (as shown in Fig. 28.40), a control voltage is applied to the specific gates of logic control circuitry. Thus the input signal is connected to the counted-signal channel of the main gate. The selected output from the time base dividers is simultaneously gated to the control flip-flop, which enables or disenables the main gate. Both control paths are latched internally to allow them to operate only in the proper sequence.

When the function switch is in Period mode, the control voltage is connected to proper gates of logic circuitry, which connect the time base signal to the counted-signal channel of the main gate. At the same time the logic circuitry connects the input to the gate control for enabling or disenabling the main gate. The other function switches (like time interval, ratio, external standard) perform similar functions. The exact details of switching and control procedures vary from instrument to instrument. They can be known by consulting the instrument catalogues.

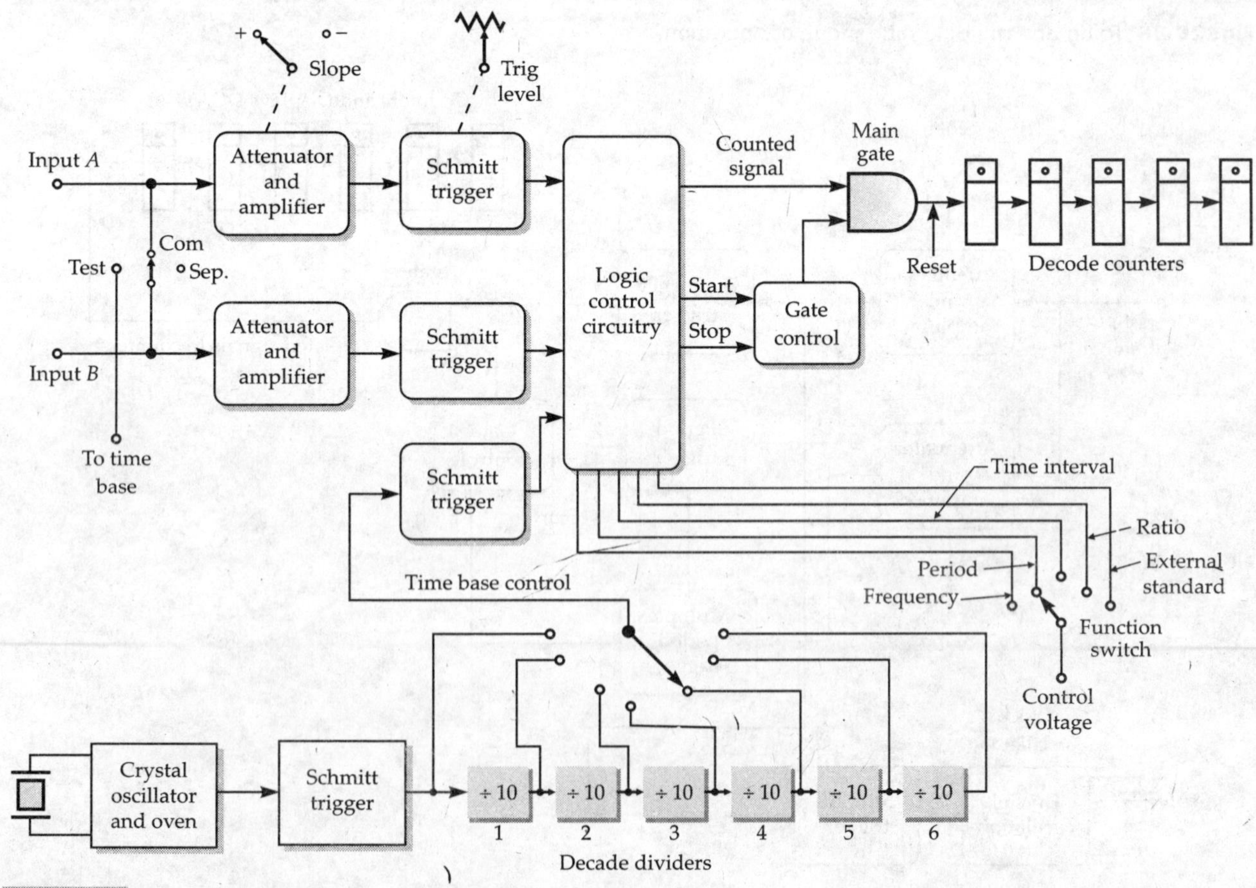


**Fig. 28.40** Block diagram of a universal counter-time.

**Example 28.4** *Gating periods of 1 ms, 10 ms, 100 ms, 1 s and 10 s are provided on a digital counter-time-frequency meter having a 3 digit display. A gating period of 10 ms is selected to measure an unknown-frequency and a reading of 034 is obtained. What is the likely value of frequency ? What steps be taken (a) to check the validity of result ; (b) to obtain a more accurate result ?*

**Solution.** (*a*)

$$\text{Frequency } f = \frac{N}{t} = \frac{034}{10 \times 10^{-3}} = 3400 \text{ Hz} = 3.4 \text{ kHz}$$

(*b*) In order to check the validity of the result, we should use a lower gating time *i.e.*, 1 ms. In case the frequency is between 3000 to 3499 Hz, the readings will be :

$$3000 \times 1 \times 10^{-3} = 3 \text{ and } 3499 \times 1 \times 10^{-3} = 3.499$$

Since the meter has a 3 digit display it can show the readings as 0.003 in both the cases.

(*c*) In order to get better results (better resolution) we must use a higher gating time *i.e.*, 100 ms.

Suppose the frequency is 3420 Hz instead of 3400 Hz. The reading of the meter will be :

$$3420 \times 100 \times 10^{-3} = 342$$

No useful purpose will be served if the gating time is increased to 1 s or 10 s. Suppose the frequency is 3424 Hz and the gating time is kept as 1 s. The reading is $3424 \times 1 = 3424$

But since the meter has only 3 digits, the meter will show an overflow. Same is the case for the gating time of 10 s.

**Example 28.5** *A digital timer with eight digit readout is stated to have an accuracy of 0.005 percent of reading, ± 1 in the final digit. Readout is in s, ms and μs. Assuming that the instrument meets its specifications, what are the maximum likely errors when the reading is : (a) 05 00 0000 μs (b) 00 000 500 s ? (c) What is the maximum nominal accuracy in time units with which reading (b) could be made with this instrument ?*

**Solution.** (*a*) The reading is 05 000 000 μs or

$$\text{Reading} = 5000\,000 \ \mu\text{s} = 5 \times 10^6 \ \mu\text{s}$$

$$0.005 \text{ percent of reading} = \pm \frac{0.005}{100} \times 5 \times 10^6$$

$$= \pm 250 \ \mu\text{s}$$

The digit in LSD has a value of 1 μs

∴ Maximum error = ±250 ± 1 = ± 251 μs

(*b*) The reading is 00 000 500 s or Reading = 500 s

$$0.005 \text{ percent of reading} = \pm \frac{0.005}{100} \times 500 = \pm 0.025 \text{ s}$$

The digit in LSD has now a value of 1 s

∴ Maximum error = ± 0.025 ±1 = ± 1.025 s

(*c*) Maximum accuracy means minimum error.

Minimum error is obtained when the time is read on the μs readout.

$$500 \text{ s} = 500 \times 10^6 \ \mu\text{s} = 500\,000\,000 \ \mu\text{s}$$

This readout requires 9 digit positions and hence when the meter is put on the readout will show an overflow as the meter has 8 digit display. Hence a reading of 500 s cannot betaken the μs range.

Let us try the ms readout.

$$500 \text{ s} = 500 \times 10^3 \text{ ms} = 500\,000 \text{ ms}$$

∴ The ms readout will display the reading as 00500 000

0.005 percent of reading

$$= \pm \frac{0.005 \times 500 \times 10^3}{100} = \pm 25 \text{ ms}$$

The LSD has a value of 1 ms.

∴ Maximum possible accuracy with which a reading of 500 s can be read by this meter is ±25 ± 1 = ±26 ms.

## 28.31 DIGITAL VOLTMETERs (DMVs)

A digital voltmeter (DVM) displays the value of a.c. or d.c. voltage being measured directly as discrete numerals in the decimal number system. Numerical readout of DVMs is advantageous since it eliminates observational errors committed by operators. The errors on account of parallax and approximations are entirely eliminated. The use of digital voltmeters increases the speed with which readings can be taken. Also the output of digital voltmeters can be fed to memory devices for storage and future computations.

A digital voltmeter is a versatile and accurate voltmeter which has many laboratory applications. On account of developments in the integrated circuit (IC) technology, it has been possible to reduce the size, power requirements and cost of digital voltmeters. In fact, for the same accuracy, a digital voltmeter now is less costly than its analog counterpart. The decrease in size of DVMs on account of use of ICs, the portability of the instruments has increased.

### 28.31.1 Types of DVMs

The increasing popularity of DVMs has brought forth a wide number of types employing different circuits. The various types of DVMs in general use are :

(*i*) Ramp type DVM,

(*ii*) Integrating type DVM,

(*iii*) Potentiometric type DVM,

(*iv*) Successive approximation type DVM, and

(*v*) Continuous balance type DVM

The circuits described here do not represent those of any specific make of a digital voltmeter. These circuits are being described merely to explain the voltage measuring principles on which these instruments operate.

In every case, the basic function that is performed, is an analog to digital (A/D) conversion. For example, a voltage value may be changed to a proportional time interval, which starts and stops a clock oscillator.

In turn the oscillator output is applied to an electronic counter which is provided with a readout in terms of voltage values. The A/D conversion techniques have already been explained in Art. 26.28.

## 28.32 RAMP TYPE DIGITAL VOLTMETER

The operating principle of a ramp type digital voltmeter is to measure the time that a linear ramp voltage takes to change from level of input voltage to zero voltage (or *vice versa*). This time interval is measured with an electronic time interval counter and the count is displayed as a number of digits on electronic indicating tubes of the output readout of the voltmeter.

The conversion of a voltage value of a time interval is shown in the timing diagram of Fig. 28.41.

At the start of measurement a ramp voltage is initiated. A negative going ramp shown in Fig. 28.41 but a positive going ramp may also be used. The ramp voltage value is continuously compared with the voltage being measured (unknown voltage). At the instant the value of ramp voltage is equal to that of unknown voltage a coincidence circuit, called an input comparator, generates a pulse which opens a gate (See Fig. 28.42). The ramp voltage continues to decrease till it reaches ground level (zero voltage). At this instant another comparator called ground comparator generates a pulse and closes the gate.

The time elapsed between opening and closing of the gate is $t$ as indicated in Fig. 28.41. During this time interval pulses from a clock pulse generator pass through the gate and are counted and displayed.

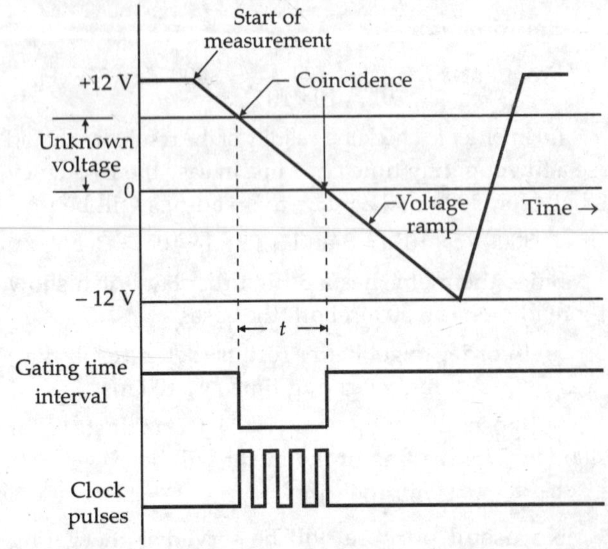


**Fig. 28.41** Timing diagram showing voltage to time conversion.

The decimal number as indicated by the readout is a measure of the value of input voltage.

The sample rate multivibrator determines the rate at which the measurement cycles are initiated. The sample rate circuit provides an initiating pulse for the ramp generator to start its next ramp voltage. At the same time it sends a pulse to the counters which sets all of them to 0. This momentarily removes the digital display of the readout.

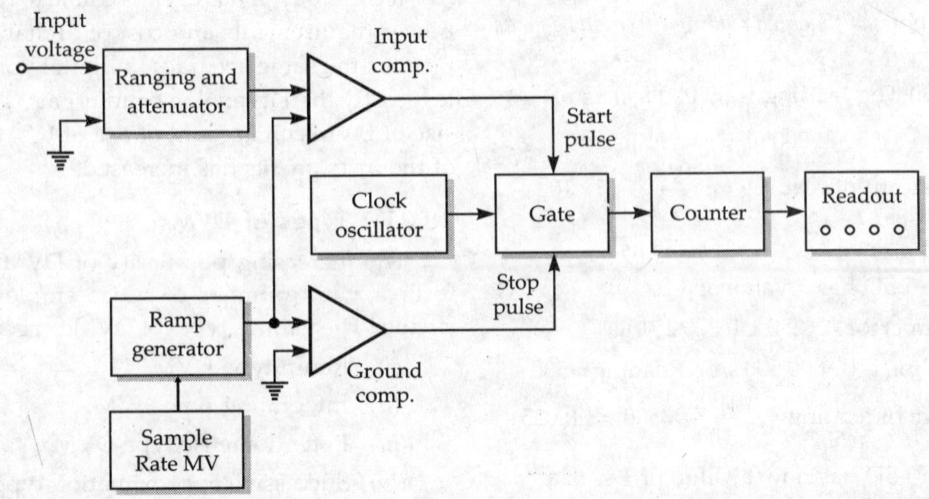


**Fig. 28.42** Block diagram of a ramp DVM.

## 28.33 INTEGRATING TYPE DIGITAL VOLTMETER

This voltmeter measures the true average value of the input voltage over a fixed measuring period. In contrast the ramp type DVM samples the voltage at the end of the measuring period. This voltmeter employs an integration technique which uses a voltage to frequency conversion. The voltage to frequency (V/F) converter functions as a feedback control system which governs the rate of pulse generation in proportion to the magnitude of input voltage.

Actually when we employ the voltage to frequency conversion techniques, a train of pulses, whose frequency depends upon the voltage being measured, is generated. Then the number of pulses appearing in a definite interval of time is counted. Since the frequency of these pulses is a function of unknown voltage, the number of pulses counted in that period of time is an indication of the input (unknown) voltage.

The heart of this technique is the **operational amplifier** acting as an **Integrator.**

Output voltage of integrator $E_0 = -E_i \frac{1}{RC} t$

Thus if a constant input voltage $E_i$ is applied, an output voltage $E_0$ is produced which rises at a uniform rate and has a polarity opposite to that input voltage. In other words, it is clear from the above relationship, that for a constant input voltage the integrator produces a ram output voltage of opposite polarity.

Let us examine Fig. 28.43. Here the graphs showing relationship between input voltages of three different values and their respective output voltages are shown. It is clear that the polarity of the output voltage is opposite to that of input voltage. Not only that, the greater the input voltage the sharper is the rate of rise, or slope, of output voltage.

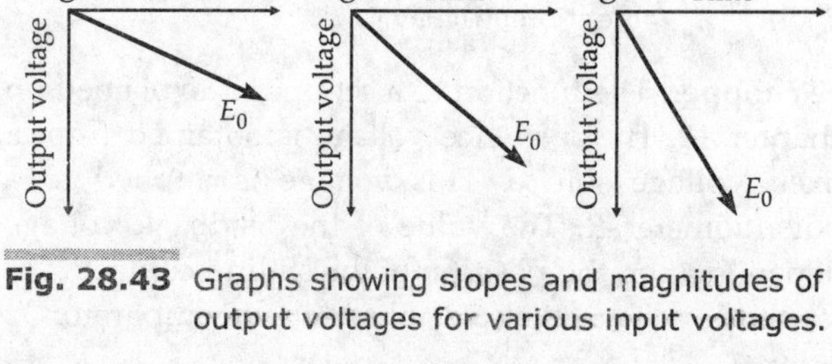


**Fig. 28.43** Graphs showing slopes and magnitudes of output voltages for various input voltages.

The basic block diagram of a typical integrating type of DVM is shown in Fig. 28.44. The unknown voltage ($E_i$) is applied to the input of the integrator, and the output voltage ($E_0$) starts to rise. The slope of output voltage ($E_0$) is determined by the value of input voltage ($E_i$). This voltage is fed a **level detector,** and when $E_0$ reaches a certain reference level, the detector sends a pulse to the pulse generator gate.

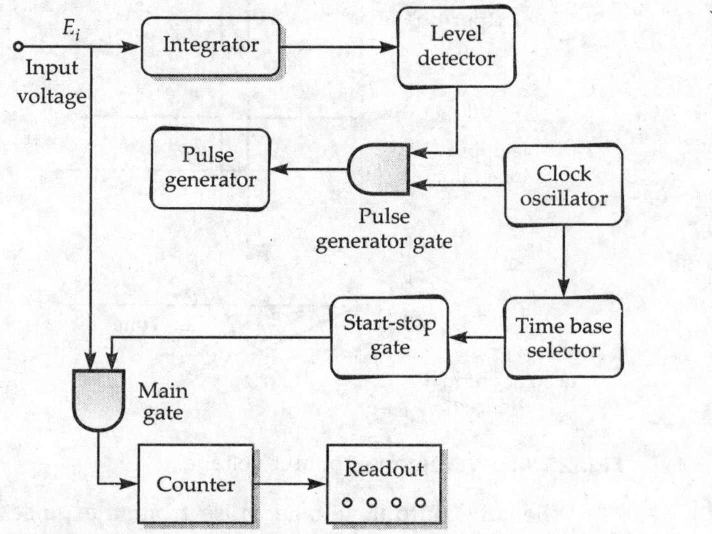


**Fig. 28.44** Block diagram of integrating type DVM.

The level detector is a device similar to a voltage comparator. The output voltage from integrator ($E_0$) is compared with the fixed voltage of an internal reference source, and, when $E_0$ reaches that level, the detector produces an output pulse. It is evident that greater the value of input voltage $E_i$ the sharper will be the slope of output voltage $E_0$, and quicker the output voltage $E_0$ will reach its reference level.

The output pulse of the level detector opens the pulse generation gate, permitting pulses from a fixed frequency **clock oscillator** to pass through **pulse generator.** This generator is a device such as a Schmitt trigger, that produces an output pulse of fixed amplitude and width for every pulse it receives.

This output pulse, whose polarity is opposite to that of $E_i$ and has a greater amplitude, is feedback to the input of the integrator. Hence, the net input to the integrator is now of a reversed polarity (opposite to that of $E_i$ as is clear from Fig. 28.45). As a result of this reversed input, the output $E_0$ drops back to its original level. Since $E_0$ is now below the reference level of the level detector, there is no output from the detector to the pulse generator gate and the gate gets closed. Thus no more pulses from the clock oscillator can pass through to trigger the pulse generator.

28

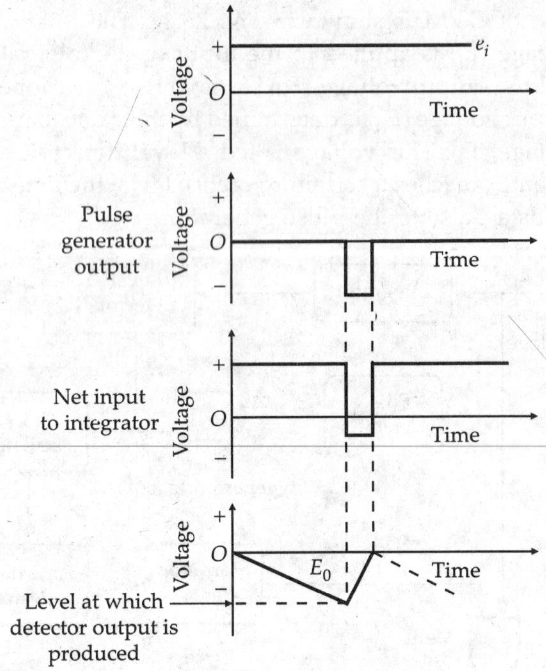


Fig. 28.45 Waveform of output voltage.

When the output voltage pulse from the pulse generator has passed, $E_i$ is restored to its original value and $E_0$ starts its rise again. When it reaches the level of reference voltage again, the pulse generator gate is opened. The pulse generator is triggered by a pulse from the clock generator and the entire cycle is repeated again. Thus, the waveform of $E_0$ is a sawtooth wave whose rise time is dependent upon the value of input voltage $E_i$ and the fall time is determined by the width of the output pulse from the pulse generator.

Thus the frequency of the sawtooth wave ($E_0$) is a function of the value of $E_i$, the voltage being measured. Since one pulse from the pulse generator is produced for each cycle of the sawtooth wave, the number of pulses produced in a given time interval and hence the frequency of sawtooth wave is an indication of the value of voltage being measured.

The frequency of sawtooth wave may be measured by counting the number of pulses in a given interval of time.

Pulses from the clock oscillator are applied to a time base selector. The first pulse passes through start-stop gate, producing an output which is applied to the main gate, thus opening the gate. As a result of this, the same output pulses from the pulse generator (that are applied to integrator) also pass through the main gate.

The next pulse from the time base closes the start-stop gate and also the main gate. Thus no more pulse generator pulses can pass through. Hence the counters and their associated readout indicate the number of pulses that have passed during at known interval of time. This count is an indication of the voltage being measured. In order to make the counter read directly in terms of voltage, the amplitude and width of pulse generator pulses can be suitably adjusted.

## 28.34 POTENTIOMETRIC TYPE DIGITAL VOLTMETER

A potentiometric type of DVM, employs voltage comparison technique. In this DVM the unknown voltage is compared with a reference voltage whose value is fixed by the setting of the calibrated potentiometer. The potentiometer setting is changed to obtain balance (*i.e.*, null conditions). When null conditions are obtained the value of the unknown voltage, is indicated by the dial setting of the potentiometer.

In potentiometric type DVMs, the balance is not obtained manually but is arrived at automatically. Thus, this DVM is in fact a self-balancing potentiometer. The potentiometric DVM is provided with a readout which displays the voltage being measured.

The block diagram of basic circuit of a potentiometric DVM is shown in Fig. 28.46. The unknown voltage is filtered and attenuated to suitable level. This input voltage is applied to a **comparator** (also known as **error detector**). This error detector may

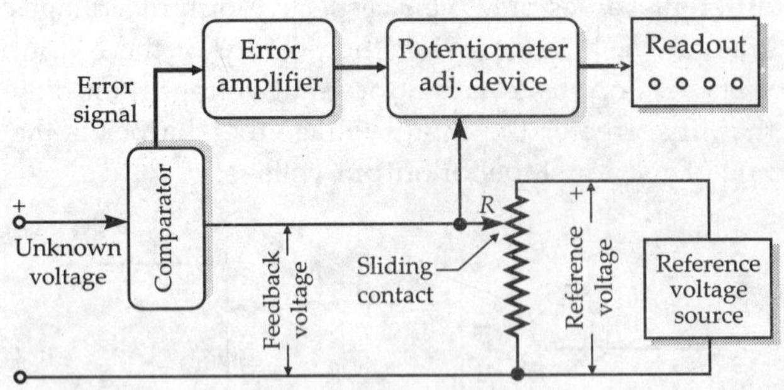


Fig. 28.46 Basic block diagram of a potentiometric DVM.

be **chopper**. The function of a chopper is explained on chapter 12. The reference voltage is obtained from a fixed voltage source. This voltage is applied to a potentiometer $R$. The value of the feedback voltage depends upon the position of the sliding contact. The feedback voltage is also applied to the comparator.

The unknown voltage and the feedback voltages are compared in the comparator. The output voltage of the comparator is the difference of the above two voltages. The difference of voltage is called the **error signal.** The error signal is amplified and is fed to a potentiometer adjustment device which moves the sliding contact of the potentiometer. This magnitude by which the sliding contact moves depends upon the magnitude of the error signal. The direction of movement of slider depends upon whether the feedback voltage is larger or the input voltage is larger. The sliding contact moves to such a place where the feedback voltage equals the unknown voltage. In that case, there wil1 not be any error voltage and hence there will be no input to the device adjusting the position of the sliding, contact and therefore it (sliding contact) will come to rest.

The position of the potentiometer adjustment device at this point is indicated in numerical form on the digital readout device associated with it. Since the position at which no voltage appears at potentiometer adjustment device is the one where the unknown voltage equals the feedback voltage, the reading of readout device indicates the value of unknown voltage.

The potentiometer adjustment device *i.e.,* the device which moves the sliding contact is a 2 phase servomotor. The automatic action of adjustment of sliding contact with the help of a servomotor has already been explained on chapter 12.

The reference voltage source must be extremely stable and generally consists of a standard cell or a Zener diode sources.

## Practical potentiometric DVM

A practical potentiometric DVM does not consist of a single resistor *R* but consists of a number of decade resistors connected in the Kelvin Varley slide fashion.

Figure 28.47 shows a practical potentiometric DVM circuit. A stepping switch operates the variable

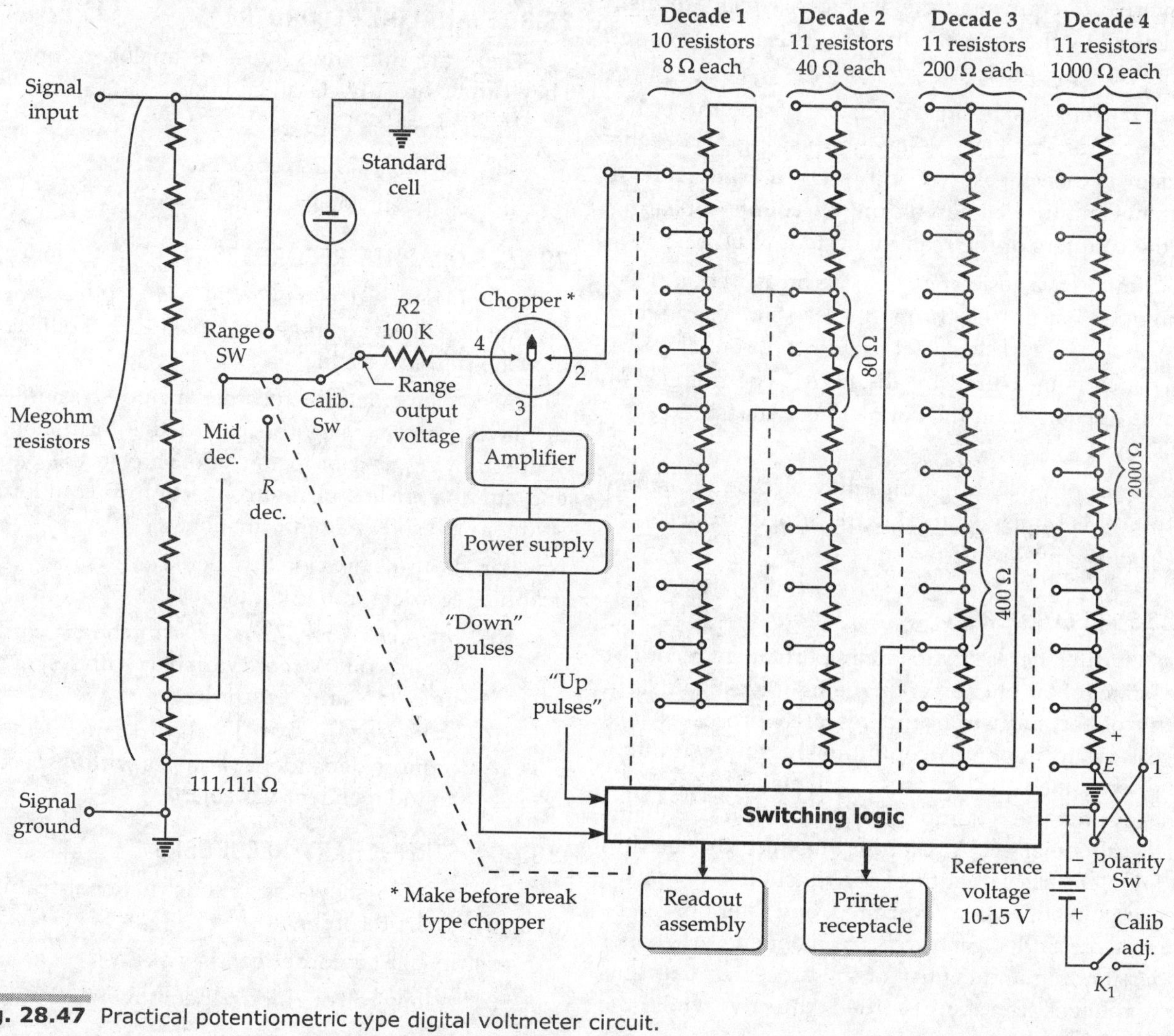


**Fig. 28.47** Practical potentiometric type digital voltmeter circuit.

voltage divider, which is composed of wire wound resistors and divides the reference voltage into a number of precisely equal parts in a decimal number system. This circuit has four resistance decades and therefore converts the input (unknown) analog voltage to a 4-digit decimal number. There are four stepping switches (which may be called decade switches). An error amplifier compares the feedback voltage with the input signal and commands stepping switches to move until they come to a position where the two voltages are within the limits of resolution of the instrument. At this point the stepping switch motion stops and the readout indicates the value of the input (unknown) voltage.

## 28.35 OTHER TYPES OF MEASUREMENTS WITH A DVM

The digital voltmeter is essentially, a d.c. voltmeter. If a.c. voltages are to be measured, they are first converted to equivalent d.c. voltages, which are then measured by the digital voltmeter. This is done by an a.c. to d.c. converter.

If current is to be measured, it is passed through a standard resistor. The voltage appearing across the resistor is then measured with a DVM. The value of current can thus be known from the counts appearing on the counters of the readout system of DVM.

Similarly, a resistance is to be measured, a fixed current is first passed through a resistor. The voltage drop, across the resistor, which is proportional to the resistance is measured by the DVM. The number of counts appearing on the counters is a measure of the resistance.

Other physical quantities may also be measured with DVMs by using suitable transducers to feed the input of the DVM.

## 28.36 RECORDERS

It is often necessary to have a permanent record or the state of a phenomenon being investigated. In many of the industrial and research processes it is necessary to monitor continuously the condition, state, or value of the process variables such as flow, force, pressure, temperature, current, voltage, electrical power etc. A recorder thus records electrical and non-electrical quantities as a function of time. This record may be written or printed, and later on, can be examined and analyzed to obtain a better understanding and control of the processes. Currents and voltages can be recorded directly while the non-electrical quantities are recorded indirectly by first converting them to equivalent currents or voltages with the help of sensors or transducers.

The ever increasing emphasis on automation, continuously recording instruments are finding many applications in industry.

## 28.37 RECORDING REQUIREMENTS

One of the important considerations in an instrumentation system is the method by which the data acquired is recorded. The recording method should be consistent with the type of system. If we are dealing with a wholly analog system, then **analog recording techniques** should be used. While on the other hand, if the system has a digital output, **digital recording devices** are used.

Thus there are two types of recording devices :

- Analog recorders, and
- Digital recorders.

## 28.38 ANALOG RECORDERS

There are numerous types of analog recorders. They can be broadly classified into :

(*i*) Graphic recorders
(*ii*) Oscillographic recorders
(*iii*) Magnetic tape recorders.

## 28.39 GRAPHIC RECORDERS

Graphic recorders generally are devices which display and store a pen-and-ink record of the history of some physical event.

Basic elements of a recorder include a chart for displaying and storing the recorded information, a stylus moving in a proper relationship to the paper and suitable means of interconnection to couple the stylus to the source of information.

For the purposes of this book, we classify the graphic recorders into two categories :

(*i*) ***Strip chart recorders.*** A strip chart recorder records one or more variables with respect to time. It is an X-t recorder.

(*ii*) ***X-Y recorders.*** An X-Y recorder records one or more dependent variables with respect to an independent variable.

## 28.40 STRIP CHART RECORDERS

Figure 24.48 shows basic constructional features of a strip chart recorder.

A strip chart recorder consists of :

(*i*) A long roll of graph paper moving vertically.

(*ii*) A system for driving the paper at some selected speed. A speed selector switch is generally provided. Chart speeds of 1 – 100 mm/s are usually used.

(*iii*) A stylus for making marks on the moving graph paper. The stylus moves horizontally in proportional to the quantity being recorded.

(*iv*) A stylus driving system which moves the stylus in a nearly exact replica or analog of the quantity being recorded.

A range selector switch is used so that input to the recorder drive system is within the acceptable level.

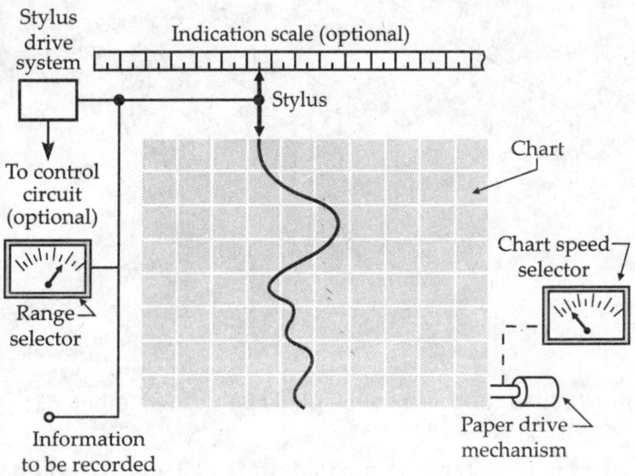


**Fig. 28.48** Strip chart recorder.

Most recorders use a pointer attached to the stylus. This pointer moves over a calibrated scale thus showing the instantaneous value of the quantity being recorded. An external control circuit for the stylus may be used.

## A. Paper Drive Systems

The paper drive system should move the paper at a uniform speed. A spring wound mechanism may be used but in most of the recorders a synchronous motor is used for driving the paper.

## B. Marking Mechanisms

There are many types of mechanisms used for making marks on the paper. The most commonly used ones are :

**1. *Marking with Ink filled stylus.*** The stylus is filled with ink by gravity or capillary actions. This requires that the pointer shall support an ink reservoir and a pen, or contain a capillary connection between the pen and a pen reservoir as shown in Fig. 28.49. In general, red ink is used but other colours are available and in instrumentation display a colour code can be adopted. The stylus, moving over the paper with preprinted scales, traces the variations of the input signal. This method is most commonly employed as ordinary paper can be used and therefore the cost is low. Other advantages are that with this system, operation over a very wide range of recording speeds is possible and also there is little friction between the stylus tip and the paper. The disadvantage of this method is that ink splatters at high speeds, batches at low speeds and clogs when the stylus is at rest. The frequency limit of recorders incorporating this method of writing is only a few Hz.

**2. *Marking with heated stylus.*** Some recorders use a heated stylus which writes on a special paper. This method overcomes the difficulties encountered in ink writing systems.

The heated stylus melts a thin, white wax like coating on a black paper base. Since the paper required is a special one, the cost is high. This method cannot be used for recording certain processes which produce heat which indirectly affect the recordings. But this method is quite reliable and offers high contrast traces. Sophisticated recorders using papers with waxed surfaces and special pens, have a frequency response upto 40 Hz are available.

**3. *Chopper bar.*** If a chart made from a pressure sensitive paper is used a simple recording process is possible. A V-shaped pointer is passed under a chopper bar which presses the pen into the paper once per second (or any other selected interval) thus making a series of marks on the special paper. In fact this system is not purely continuous and hence is suitable for recording some slowly varying quantities, for example, those which have a variation of 1 cycle per hour. This type of marking has the advantage of a straight line horizontal scale without the use of complex linkage arrangement.

**4. *Electric stylus marking.*** This method employs a paper with a special coating which is sensitive to current. When current is conducted from the stylus to the paper, a trace appears on the paper. It is clear that the electric stylus marking method has a wide range of marking speeds, has low stylus friction and a long stylus life. The disadvantage is that the cost of paper is very high.

**5. *Electrostatic stylus.*** This method uses a stylus which produces a high voltage discharge thereby producing a permanent trace on an electrosensitive paper. This arrangement has been incorporated in a recorder having a 50 mm wide chart nine voltage ranges from 10 mV/mm to 5 V/mm ; eight chart speeds from 300 mm/s to 10 mm/min and a frequency response of 60 Hz at maximum amplitude of 1 db.

**6.** ***Optical marking method.*** This method uses a beam of light to write on a photosensitive paper. Thus this method allows higher frequencies to be recorded and permits a relatively large chart speed with good resolution. The disadvantages are that the paper cost is very high. Secondly the writing process is a photographic one, the paper must be developed before a record is available and hence this method is not suitable for processes where instantaneous monitoring is to be done.

## C. Tracing Systems

There are two types of tracing systems used for producing graphic representations.

**1.** ***Curvilinear system.*** In the curvilinear system, the stylus is mounted on a central pivot and moves through an arc which allows a full-width chart marking. If the stylus makes a full range recording, the line drawn across the chart will be curved and the time intervals will be along this curved segments. This type of system is used on many records, with PMMC galvanometers actuating the stylus filled with ink as shown in Fig. 28.50. The disadvantage of this method of tracing is that the charts are difficult to analyze because of curved time base lines.

**2.** ***Rectilinear system.*** Figure 28.51 shows the rectilinear system of tracing. It is noticed that a line of constant time is perpendicular to the time axis and therefore this system produces a straight line across the width of the chart. Here the stylus is actuated by a drive cord over pulleys to produce the forward and reverse motion as determined by the drive mechanism. The stylus may be actuated by a self-balancing potentiometer system, a photoelectric deflection system, a photoelectric potentiometer system, or a bridge balance system. This system is usually used with thermal or electric writing.

## 28.41 TYPES OF STRIP CHART RECORDERS

The different types of strip chart recorders are :

1. **Galvanometer type.** This type of strip chart recorder operates on the deflection principle. The deflection is produced by a galvanometer which produces a torque on account of a current passing through its coil. This current is proportional to the quantity being measured.
2. **Null type.** This type of recorder operates on comparison basis.

## 28.42 GALVANOMETER TYPE RECORDERS

These recorders use a d'Arsonval galvanometer described in Chapter 8. The pointer is equipped with a recording pen mechanism (stylus). A cut-away view of the moving coil element is shown in Fig. 28.49. As the current flows through the coil, it deflects. The greater the amplitude of the incoming signal (which is proportional to the quantity being measured), the greater is the deflection. When the pointer comes to rest on account of controlling torque exerted by

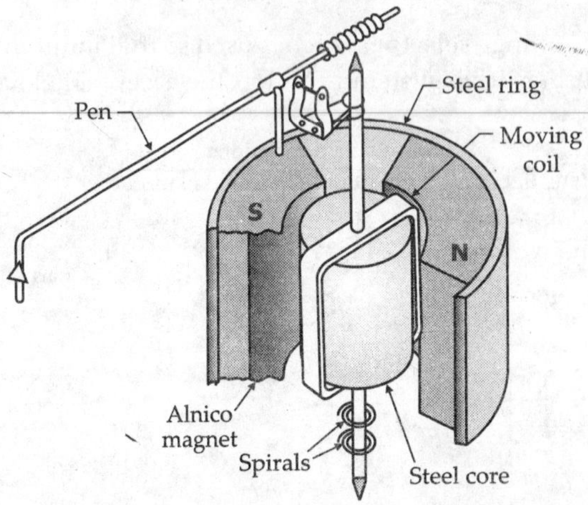


**Fig. 28.49** Cut-away view of moving coil element.

springs, the stylus also comes to rest. Thus, the value of the quantity is recorded. The design of the d'Arsonval galvanometer used in recorders is different from the one described in Chapter 8. This is necessitated by the fact that in recorders, the movement of the instrument requires an appreciable torque. To obtain this torque the d'Arsonval movement consists of a large moving coil situated in a strong magnetic field.

It should be understood that the instrument must be critically damped or nearly critically damped so that there is no significant overshoot. But this results in slow response, the response time being 0.75 to 1.5 s.

Thus this type of recorder is not useful for recording fast variations in either current or voltage or power. This records only the average values and hence it should be designed for these. A galvanometer type recorder is shown in Fig. 28.49. It is a modified version of the PMMC instrument. The modification is done in order that the chart may be driven at a constant speed by a clockwork mechanism or an electric motor. The type of chart used depends upon the form of movement. The recorder shown in Fig. 28.49 uses a chart having a curvilinear system of tracing.

This system is used because it allows the direct use of simple moving coil movement. However, some instruments employ additional linkage system in the system which allow the use of rectilinear system of tracing as shown in Fig. 28.50.

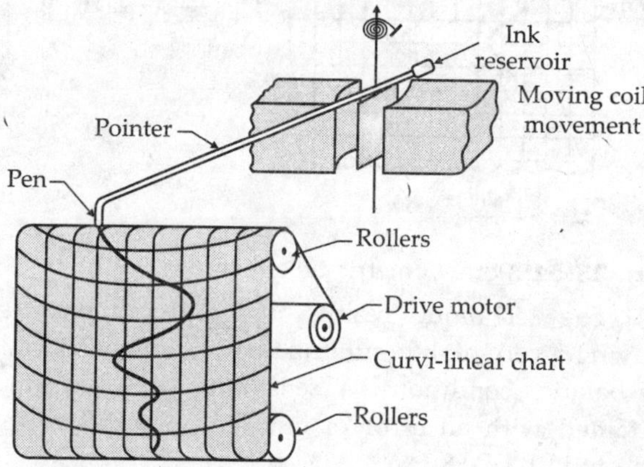


**Fig. 28.50** Components of a galvanometer type recorder.

The recorders can work on ranges from a few mA to several mA or from a few mV to several mV.

This moving galvanometer type recorder is comparatively inexpensive instrument having a narrow bandwidth of 0 to 10 Hz. It has a sensitivity of 0.4 mV/mm or from a chart of 100 mm width a full scale deflection of 40 mV is obtained. For measurement of smaller voltage linear amplifiers are used.

In these recorders the bearings must be substantially larger than those used in indicator instruments, because of the large mass of coil and stylus. On account of large mass of the moving mechanism, friction becomes a factor which cannot be ignored.

The analysis of the galvanometer movement in Chapter 8 neglected the sliding friction which is also called **Coulomb Friction**. This friction is constant and when we consider it, the equation of motion of coil becomes :

$$J\frac{d^2\theta}{dt^2} + D\frac{d\theta}{dt} + K\theta + F = Gi \qquad \ldots(28.3)$$

where $F$ = torque due to Coulomb friction, N-m

Note that Eqn. 28.3 is a modified form of Eqn. 8.10 given on page 199.

The solution of Eqn. 28.3, under steady state condition is :

$$K\theta + F = Gi \quad \text{or} \quad \theta = \frac{Gi}{K} - \frac{F}{K} \qquad \ldots(28.4)$$

Therefore the deflection of the stylus is proportional to the product of constant $G/K$ and the current $i$, less a constant $F/K$. The term $F/K$ is the error term which produces a dead zone (see Art. 2.22 page 33) Therefore care must be taken to reduce coulomb friction to a minimum and to produce a large driving torque so that $F/K$ is very small as compared to $Gi/K$. Since $G = NBld$ (See Eqn. 8.5, page 198) therefore a large coil with many turns produces a large value of G resulting in large deflecting torque. (The deflecting torque being $Gi$). Also by increasing the flux density, $B$, the value of G can be increased. The value or flux density can be increased by reducing the air gap. These recorders give an accuracy of 2 percent or even better.

***Example 28.6*** *The coil of a recording ammeter is 65 mm long and 25 mm wide. The rated current of the coil is 10mA. The flux density in the air gap is* $4.6 \times 10^{-3}$ *Wb/m*$^2$. *The damping constant is* $8 \times 10^{-3}$ *N-m/rad-s*$^{-1}$. *The moment of inertia is* $8 \times 10^{-3}$ *kg-m*$^2$. *The spring constant is* $16 \times 10^{-3}$ *N-m/rad. The coulomb friction is* $0.2 \times 10^{-6}$ *N-m.*

(*a*) *Determine if the meter is underdamped, critically damped or over-damped.*

(*b*) *Determine the number of turns on the coil to produce a deflection of 100° at rated current.*

(*c*) *Determine the current required to overcome coulomb friction.*

**Solution.** (*a*) Damping constant

$$D = 8 \times 10^{-3} \text{ N-m/rad}^{-1} \text{ s}$$

$$\therefore \quad D^2 = (8 \times 10^{-3})^2 = 64 \times 10^{-6},$$

$$4JK = 4 \times 10^{-3} \times 16 \times 10^{-3} = 512 \times 10^{-6}.$$

Hence, $D^2 < 4JK$ and therefore the instrument is underdamped.

(*b*) Deflection $\theta = 100° = 100 \times \pi / 180 = 1.8$ rad

From Eqn. 28.4, Displacement constant

$$G = \frac{K\theta + F}{i} = \frac{16 \times 10^{-3} \times 1.8 + 0.2 \times 10^{-6}}{10 \times 10^{-3}}$$

$$= 2.88 \text{ Nm/A}$$

Number of turns

$$N = \frac{G}{Bld}$$

$$= \frac{2.88}{4.6 \times 10^{-3} \times 65 \times 10^{-3} \times 25 \times 10^{-3}} = 387000$$

(*c*) Current required to overcome friction

$$= \frac{F}{G}$$

$$= \frac{0.2 \times 10^{-6}}{2.88} \text{ A} = 0.07 \ \mu\text{A}$$

28

## 28.43 NULL TYPE RECORDERS

Many recorders operate on the principle whereby a change in its input, produced by the signal from the sensor or transducer (which is used to convert a non-electrical quantity to an equivalent electric signal), upsets the balance of the measuring circuit of the recorder. As a result of this unbalance, an **Error Signal** is produced that operates some device which restores balance or brings the system to **Null conditions**. The amount of movement of this balance restoring device then, is an indication of the magnitude of the error signal, and the direction of the movement is an indication of the direction of the quantity being measured has deviated from normal.

The signal from the transducer may take any of the several forms. It may be a voltage (a.c. or dc.), a current (a.c. or d.c.) or it may be a value of resistance, inductance or capacitance. The recorder, therefore, must be of a type able to accept the form of the input signal.

There are a number of null type recorders. They are :

- Potentiometric recorders
- Bridge recorders and
- LVDT recorders.

The principle of operation for all these recorders is the same *i.e.*, to obtain null condition and hence only potentiometric recorders are described.

## 28.44 POTENTIOMETRIC RECORDERS

The basic disadvantage of a galvanometer type of recorder is that it has a low input impedance and a limited sensitivity. A simple method of overcoming the input impedance problem is to use an amplifier between the input terminals and the display or indicating instrument. However, this technique, while producing a high input impedance (so as to reduce loading effects) and improved sensitivity, results in an instrument having low accuracy. The accuracy can be improved if the input signal is compared with a reference voltage by using a potentiometer circuit (See Fig. 28.51).

The error signal which is the difference between the input signal and the potentiometer voltage, is amplified and is used to energize the field coil of a d.c. motor. A wiper is mechanically connected to the armature of the d.c. motor. This wiper moves over the potentiometer in the appropriate direction to reduce the magnitude of the error signal and to obtain balance. The wiper comes to rest when the unknown signal voltage is balanced against the voltage of the potentiometer.

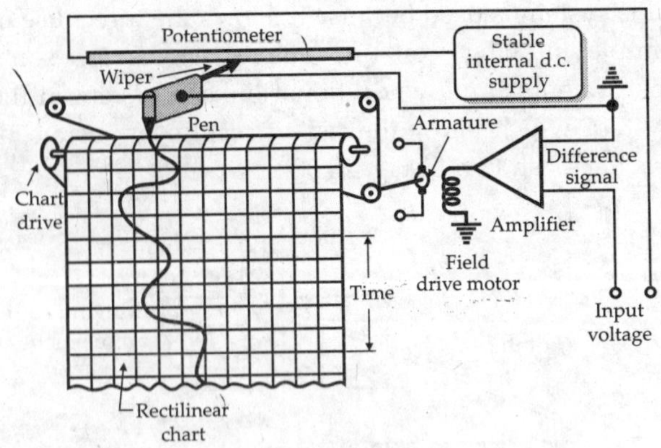


**Fig. 28.51** Potentiometric type recorder.

The technique results in graphical recorders having a very high input impedance, which is infinity at balance conditions. A sensitivity of 4 VB/mm is attained with an error of less than ± 0.25% with a bandwidth of 0.8 Hz.

The chart drive for most potentiometric recorders is obtained from a motor synchronized to power line frequency. Different speeds may be obtained by using a gear train that uses different gear ratios.

The most common application of potentiometric recorders is for recording and control of process temperatures. This has already been described under the heading of "**Self Balancing Potentiometers**" in Art. 15.2 page 466.

The scale of this recorder is calibrated in terms of temperatures. The chart for such a temperature recorder may be rectilinear or circular. A circular scale is shown in Fig. 28.52.

The circular chart has concentric circles ruled out. In addition it has equally spaced arcs extending from the centre to the rim of chart. These are called time arcs. The time markings are placed along the circumference of the rim. The values of the measurand lie along the arcs. These values depend upon the linkage geometry of the instrument.

## 28.45 SINGLE POINT RECORDERS

Analog recorders are available in several types. These recording instruments can be entirely mechanical entirely electrical, or a combination of both. There are two sets of units shown. One set indicates the value of the measured variable, for example, pressure. The second set indicates time. Therefore me graphic record shows variations of pressure with time.

Instruments that record changes of only one measured variable are called **single point recorders**. The trace on such instruments is usually in the form of single continuous curve.

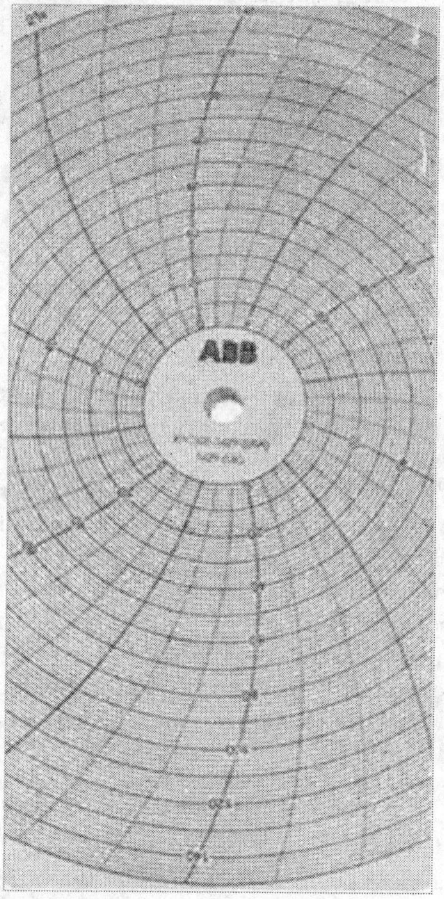


**Fig. 28.52** Circular chart.

## 28.46 MULTIPOINT RECORDERS

Rectangular strip chart recorders do not usually incorporate control features but many models are available in which one recorder may be used for recording several inputs. In process industry it becomes necessary to record simultaneously variables like temperature, pressure, flow rate, liquid level, etc. This feature may be performed either by having several (Maximum 4) pens which overlap each other and record the inputs simultaneously, or by replacing the pen by print wheel geared to a selector switch so that when a particular input is connected to the potentiometer balance circuit, a point plus its identifying character is printed on the chart. This form of recorder is called a **Multipoint Recorder** and may have as many as 24 inputs, with traces displayed in 6 colours.

## 28.47 X-Y RECORDERS

A strip chart recorder records the variations of a quantity with respect to time while a X-Y Recorder is an instrument which gives a graphic record of the relationship between two variables.

In strip chart recorders, usually self-balancing potentiometers are used. These self-balancing potentiometers plot the emf as a function of time. In **X-Y Recorders**, an emf is plotted as a function of another emf. This is done by having one self-balancing potentiometer control the position of the rolls (*i.e.*, the paper) while another self-balancing potentiometer controls the position of the recording pen (stylus).

In some *X-Y* recorders, one self-balancing potentiometer circuit moves a recording pen (stylus) in the *X* direction while another self-balancing potentiometer circuit moves the recording pen (stylus) in the *Y* direction at right angles to the X direction, while the paper remains stationary.

There are many variations of *X-Y* recorders. The emf, used for operation of *X-Y* recorders, may not necessarily measure only voltages. The measured emf may be the output of a transducer that may measure displacement force, pressure, strain, light intensity or any other physical quantity. Thus with the help of *X-Y* recorders and appropriate transducers, a physical quantity may be plotted against another physical quantity.

Hence an **X-Y Recorder** consists of a pair of servo-systems, driving a recording pen in two axes through a proper sliding pen and moving arm arrangement, with reference to a stationary paper chart. Attenuators are used to bring the input signals to the levels acceptable by the recorder.

Figure 28.53 shows a block diagram of a typical *X-Y* recorder. A signal enters each of the two channels. The signals are attenuated to the inherent full scale range of the recorder (often 0.5 mV), the signal then passes to a balance circuit where it is compared with an internal reference voltage. The error signal (*i.e.*, the difference between the input signal voltage and the reference voltage is fed to a chopper which converts d.c. signal to an a.c. signal. The signal is then amplified in order to actuate a servometer which is used to balance the system and hold it in balance as the value of the quantity being recorded changes.

The action described above takes place in both axes simultaneously. Thus we get a record of one variable with respect to another.

An *X-Y* recorder may have a sensitivity of 10 μV/mm, a slewing speed of 1.5 m/s and a frequency response about 6 Hz for both the axes. The chart size is 250 × 180 mm. The accuracy of *X-Y* recorder is about ± 0.3%.

28

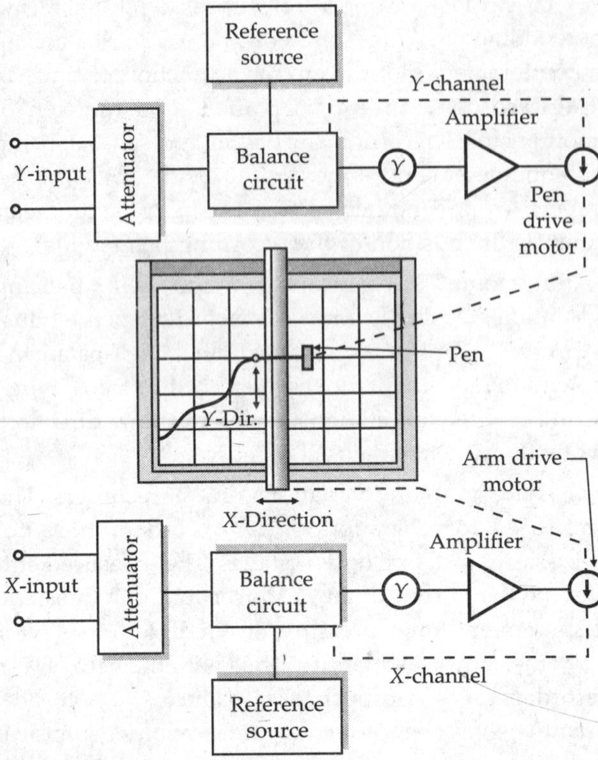


**Fig. 28.53** *X-Y* Recorder.

The use of *X-Y* recorders in laboratories greatly simplifies and expedites many measurements and tests. A few examples are being given below :

1. Speed torque characteristics of motors.
2. Lift Drag wind tunnel tests.
3. Plotting of characteristics of vacuum tubes, zener diodes rectifiers and transistors, etc.
4. Regulation curves of power supplies.
5. Plotting stress-strain curves, hysteresis curves and vibrations amplitude against swept frequency.
6. Electrical characteristics of materials such as resistance versus and temperature plotting the output from electronic calculators and computers.

## 28.48 ULTRAVIOLET RECORDERS

These are basically electro-mechanical oscillographic recorders and are modified version of the Duddel's oscillographs.

The recorder consists of a number of galvanometer (moving coil) elements mounted in a single magnet block as shown in Fig. 28.54. This is unlike the Duddel's multichannel oscillograph where a separate magnet assembly is used for every galvanometer element and hence there is great reduction in size and cost. The galvanometer uses a source of ultraviolet light in place of white light. A paper sensitive to ultraviolet light is used for producing a trace for the purpose of recording. The u.v. light is projected on the paper with the help of mirrors attached to the moving coils.

### 28.48.1 Principle of Operation

When a current is passed through the moving (galvanometer) coil, it deflects under the influence of the magnetic field of the permanent magnet. The u.v. light falling on the mirrors is deflected and is projected on to the u.v. light sensitive paper through a lens and mirror system. The paper is driven past the moving light spot and thus a trace of variation of current with respect to time is produced. In many u.v. recorders,

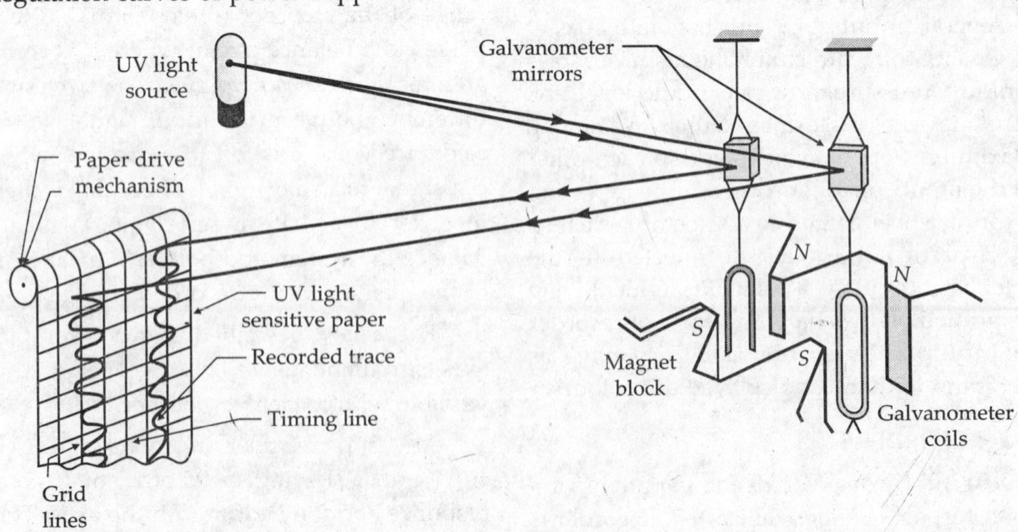


**Fig. 28.54** Ultraviolet (UV) recorder.

arrangements are provided to select a suitable paper driving speed out of as many as 12 different speeds. Some galvanometers incorporate an arrangement for controlling the speed of the paper with the help of an externally applied voltage. The u.v. sensitive paper may be processed by photodeveloped, permanized or photocopied.

The recorder, in addition to the traces of the input currents, may have the following additional traces :

(*i*) ***Grid lines.*** These lines are along the length of the paper and may be obtained by shining the u.v. light on the paper through a comb.

(*ii*) ***Timing lines.*** These lines are along the width of the paper and are obtained from a vapour tube actuated from an external source or an internal source of known frequency.

(*iii*) ***Trace identification.*** The u.v. recorders are multi-channel recorders and many recorders have 25 channels. Since, each channel may produce a 100 mm wide peak to peak trace on a 30 cm wide paper; there will be considerable overlapping of traces produced by different channels. Therefore, it is essential to provide an identification mark for each trace so as to avoid confusion. A simplified identification process is to interrupt each trace momentarily in turn and to coincide this interruption with a numeral marked on the side of the record by passing u.v. light through cutouts of the numeral.

### 28.48.2 Recorder Galvanometers

The recorder galvanometers are Pencil galvanometers. The construction and transient performance of these has been explained in Art. 8.18 on pages 227 – 230.

### 28.48.3 Applications

The u.v. recorder may be used for d.c. and a.c. signals having a fundamental frequency up to 400 to 500 Hz. The frequency range depends on the recorder being used and paper driving speeds. The recording of high frequency inputs is possible if recorders with high paper speeds of about 10 m/s are available.

Typical applications of u.v. recorders are in recording

(*i*) output of transducers,

(*ii*) control system performance and

(*iii*) regulation transients of generators.

These recorders are also used for recording the magnitude of low frequency signals which cannot be measured with analog (pointer) type instruments.

## 28.49 MAGNETIC TAPE RECORDERS

It is frequently desirable and in many cases necessary, to record data in such a way that they can be retrieved or reproduced in electrical form again. The most common and most useful way of achieving this is through the use of magnetic tape recording.

The recorders described till now are basically low-frequency recorders but magnetic tape recorders have response characteristics which enable them to be used at higher frequencies. Therefore magnetic tape recorders are extensively used in Instrumentation systems.

### 28.49.1 Advantages of Magnetic Tape Recorders

(*i*) Magnetic tape recorders have a wide frequency range from d.c. to several MHz.

(*ii*) They have a wide dynamic range which exceeds 50 db. This permits the linear recording from full scale signal level to approximately 0.3% of full scale.

(*iii*) They have a low distortion.

(*iv*) The magnitude of the electrical input signal is stored in magnetic memory and this signal can be reproduced whenever desired. The reproduced signal can be analyzed by automatic data reduction methods.

(*v*) The recorded signal is immediately available, with no time lost in processing. The recorded signal can be played back, or, reproduced as many times as desired without loss of signal.

(*vi*) When the information has been processed, the tape can be erased and reused to record a new set of data.

(*vii*) Magnetic tape recording permits multi-channel recording. A tape facilitates the continuous record of a number of signals, which may have a wide range of frequency, to be made simultaneously. This is a great advantage especially when recording transient and "once only" signals.

(*viii*) The use of magnetic tape recorders provides a convenient method of changing the time base. Data may be recorded at very fast speeds (1.52 or 3.05 m/s) and played back at speeds (4.76 or 2.38 cm/s) slow enough to be recorded with low frequency recorders like graphic recorders.

28

## 28.49.2 Basic Components of a Tape Recorder

A magnetic tape recorder consists of the following basic components :

1. ***Recording heat.*** This device responds to an electrical signal in such a manner that a magnetic pattern is created in a magnetisable medium. The construction of a magnetic head as shown in Fig. 28.55 is similar to that of a transformer having a toroidal core with a coil. A fine air gap of length 5 – 15 μm is shunted by the passing magnetic tape. A current in the

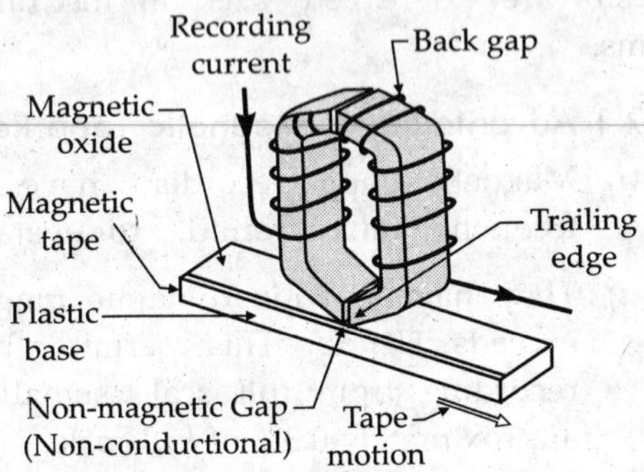


**Fig. 28.55** Magnetic tape recording head.

coil causes a flux of the same shape to bridge the air gap and hence to pass through the magnetic tape; thereby magnetizing the iron oxide particles as they pass the gap. The state of magnetization of the oxide as it leaves the gap is retained, thus the actual recording takes place at the trailing edge of the gap. Any signal recorded on the tape appears as a magnetic pattern dispersed in space along the tape, similar to the original coil current variation with time.

2. ***Magnetic tape.*** Magnetic tape is composed of a coating of fine magnetic iron oxide particles ($Fe_2O_3$) on a plastic ribbon. A typical tape is 12.7 mm wide and 25.4 μm thick. The magnetic particles conform to the magnetic pattern induced in them and retain it.

3. ***Reproducing head.*** The reproducing head detects the magnetic pattern stored in them and converts it back to original electrical signal. The reproducing is similar in appearance to that of a recording head.

4. ***Tape transport mechanism.*** This mechanism moves the tape along the recording or the reproducing heads at a constant speed. The tape mechanism must be capable of handling the tape dining various modes of operation without straining, distorting or wearing the tape. This requires that the mechanism must use arrangements to guide the tape past the magnetic heads with great precision, maintain proper tension and obtain sufficient tape to magnetic head contact. Arrangements for fast winding and reversing are also provided. A simple arrangement is shown in Fig. 28.56. A capstan and pinch roller are used to drive the tape. Some models use a closed loop drive.

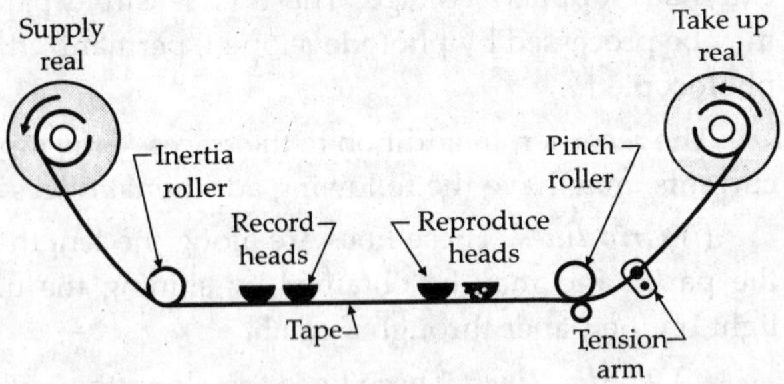


**Fig. 28.56** Tape transport mechanism.

5. ***Conditioning devices.*** These devices consist of amplifiers and filters required for modifying the signal to a format that can be properly recorded on a tape.

## 28.49.3 Principle of Tape Recorders

When a magnetic tape is passed through a recording head, any signal recorded on the tape appears as magnetic pattern dispersed in space along the tape, similar to the original coil current variation with time.

The same tape when passed through a reproduce or playback head produces variations in the reluctance of the winding thereby inducing a voltage in the winding dependent upon the direction of magnetization and its magnitude on the magnetic tape. The induced voltage is proportional to rate of change of flux linkages. Therefore the emf induced in the winding of reproducing head is proportional to rate of change of the level of magnetization on the tape, *i.e.*, $e_{rep}$ α $N\, d\phi/dt$, where $N$ is the number of turns of the winding put on the reproduce head. Since the voltage in the reproduce head is proportional to $d\phi/dt$, the reproduce head acts as a differentiator.

For example, let the original signal be $A \sin \omega t$. The current in the record head winding and the flux produced will be proportional to this voltage.

$$\therefore \quad \phi = K_1 A \sin \omega t, \quad \text{where } K_1 = \text{constant}$$

Assuming that the tape retains this flux pattern and regenerates it in the reproduce head, the voltage induced in the reproduce head winding is :

$$e_{rep} = N\frac{d\phi}{dt} = K_1 NA\,\omega \cos \omega t = K_2 A\,\omega \cos \omega t \qquad \ldots(28.42)$$

Thus the output signal from the reproduce head is a derivative of the input signal. The magnitude of the output signal is not only proportional to the flux recorded on the tape but also the frequency of the recorded signal.

It is clear from Eqn. 28.42 that the output doubles for every octave rise of frequency of the recording signal, and therefore the reproduced head output is subject to a 6 db octave rise (see Fig. 28.57). Since, the recording is done at a constant current level for all frequencies, and therefore, in order to have an overall flat frequency response, *i.e.,* a high fidelity the reproduce head characteristics and the characteristics

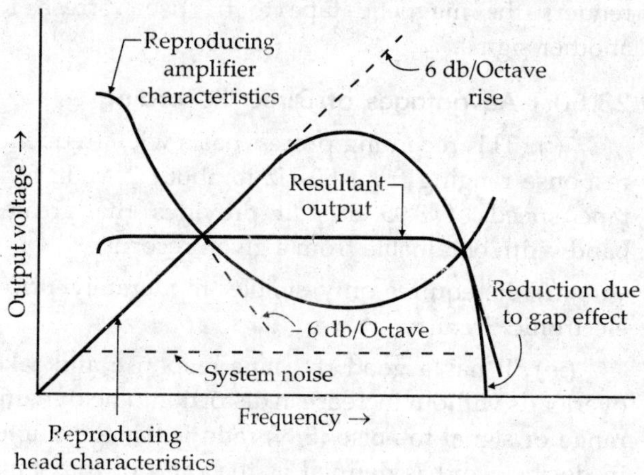


**Fig. 28.57** Reproduce head and reproduce amplifier characteristic and equalization.

of the amplifier connected in the reproduce head circuit must be complementary, *i.e.,* the amplifier must have a response of - 6 db octave as shown in Fig. 28.57. This process of compensation is known as **Equalization.**

### 28.49.4 Methods of Recording

There are three methods of magnetic tape recording which are used for instrumentation purposes. They are :

(*i*) Direct recording,

(*ii*) Frequency Modulation (FM) recording and

(*iii*) Pulse Duration Modulation (PM) recording.

Frequency modulation (FM) recorders are generally used for instrumentation purposes. The PDM recording is used in instrumentation systems for special applications where a large number of slowly changing variables have to be recorded simultaneously.

## 28.50 DIRECT RECORDING

It is the simplest method of recording and usually requires one tape track for each channel. The signal to be recorded is amplified mixed with a high frequency bias and fed directly to the recording head as a varying electric current. The reason for introducing bias is explained below :

The magnetic pattern impressed upon the tape is directly proportional to the flux density produced at the air gap. The relationship between current in the winding and the flux density in the gap is non-linear.

This non-linearity produces a distortion in the signal applied to the tape as shown in Fig. 28.58 where input signal is sinusoidal while the magnetization produced on the tape is non-sinusoidal. In order to avoid distortion, a high frequency bias of constant amplitude is applied with the signal input to the recorder. Both the amplitude and frequency of the bias is larger than the maximum amplitude and highest frequency contained in the input signal. The bias frequency is usually four times the highest signal frequency. The bias current is usually between 5 to 30 times the signal current and the exact value depends the tape and head characteristics. The combining of the bias and the input signal is accomplished by a linear mixing process (and is not an amplitude modulation process).

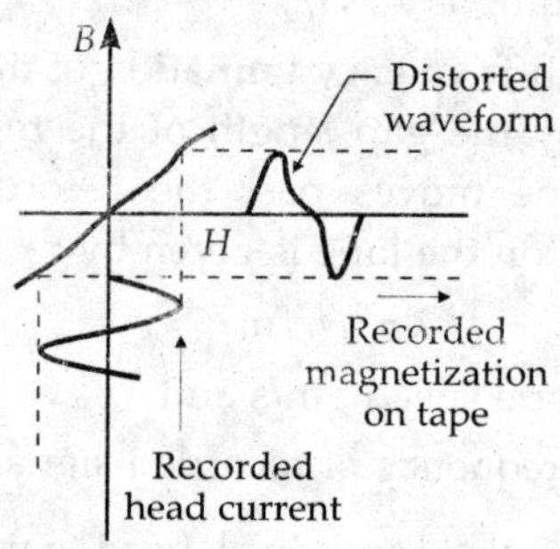


**Fig. 28.58** Distortion due to non-linear *B-H* magnetization curve of magnetic tape.

The combined input and bias signal is shown in Fig. 28.58. This is the signal which is applied to the record head. When the magnitude of the input signal is limited so that the peak values of input plus bias signal lie on the straight portion of BH curve of tape material, the waveform recorded as shown in Fig. 28.59. When the tape is passed through the reproduce head, this very waveform is reproduced. This waveform is conditioned by filtering out the high frequency component, thus reproducing the waveform of the original signal. The output signal may be amplified to give a suitable magnitude.

Figure 28.57 shows that, as the recorded frequency approaches zero, the output drastically falls down to zero (since the output voltage of the reproduce head is proportional to frequency). Thus direct recording

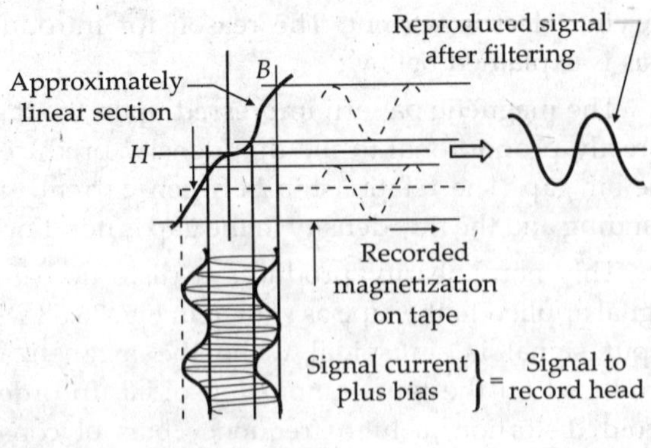


**Fig. 28.59** Representation of direct recording.

process cannot be used for d.c. (as the reproduce head voltage is zero for d.c. signals). Direct recording cannot function below a certain frequency. As the frequency decreases, the output of the reproduce head decreases. A limit is reached when the reproduce head voltage is equal to the voltage due to the inherent noise in the system. If the system is used below this frequency, the signal will be completely shrouded by the noise signal. The lower operating frequency for direct recording is 50 Hz.

The high-frequency limitation of direct recording is limited by the gap length of the reproduce head. When a tape moves past the record air gap, the wavelength on the tape is given by :

$$\lambda = v / fm \qquad \text{...(28.43)}$$

where $v$ = tape speed ; m/s and

$f$ = frequency of recorded signal ; Hz

Now, if the reproduce head gap has a width which is significant proportion of $\lambda$, then the reproduce circuits cannot respond to small changes in the signal. Therefore, it follows that the gap length in the reproduce head must be small compared with the wavelength of the highest frequency to be detected. Thus, as the speed of the tape is increased, the wavelength at any given frequency is increased, thereby increasing the frequency response of the system. If a gap length of 6.4 μm is used at a speed of 1524 mm/s, a response of 100 kHz is typical. The smallest gap length is 1.3 μm and with this gap length the maximum response is about 1 MHz for a tape speed of 3048 mm/s.

In fact the high frequency response of the direct recording system is limited by several factors, the most important of which is the gap length of the reproducing head. The output voltage of the reproducing head increases with frequency, simply because $d\phi/dt$ increases. This is true upto the point where the length of gap equals one-half the recorded wavelength. Beyond this point the output voltage decreases rapidly and drops to zero when the gap lengths equals the recorded wavelength. At this point there is no magnetic gradient spanned and therefore the output voltage is zero. This phenomenon is called gap effect and is most serious restriction on the high frequency response of the tape recorder. (See Fig. 28.57).

In the **erase head**, a signal of high frequency and level sweeps the magnetic tape thereby completely wiping out the information contained therein. This renders the magnetic tape to be used afresh for another signal.

### 28.50.1 Advantages of Direct Recording

(*i*) This recording process has a wide frequency response ranging from 50 Hz to about 2 MHz for a tape speed of 3.05 m/s. It provides the greatest bandwidth obtainable from a given recorder.

(*ii*) It requires only simple, moderately priced electronic circuitry.

(*iii*) It has a good dynamic response and takes overloads without increase in distortion. The dynamic range or signal-to-noise (S/N) ratio is usually quoted in decibels and is defined as the ratio of maximum signal to the minimum signal which can be recorded at a certain amount of total harmonic distortion (THD). In general the instrumentation recorders have a S/N ratio of 22 db to 30 db at 1 per cent THD and a frequency down to 400 Hz.

(*iv*) It is used to record signals where information is contained in the relation between frequency and amplitude, such as spectrum analysis of noise.

(*v*) It can be used for recording voice and in multiplexing a number of channels of information into one channel of tape recording.

### 28.50.2 Disadvantages of Direct Recording

(*i*) The direct recording process is characterised by some amplitude instability caused by random surface inhomogenities in tape coating. Some portions of the tape may not be perfectly recorded owing to dirt or poor manufacture, and this is called' **"Drop Out"**. At long wavelengths (low frequencies) the amplitude variations caused by in homogenities may be only a few percent ; however for frequencies near the upper bandwidth limit, for a given tape speed, amplitude variations can exceed 10% and momentary decreases of over 50% due to drop outs may occur. This also shows one of the difficulties associated with magnetic digital tape recording.

(*ii*) The direct recording is used only when maximum bandwidth is required and when variations in amplitude are acceptable.

(*iii*) The direct recording, no doubt, can be used for Instrumentation purposes but it is mainly used for recording of speech and music. In sound recordings, the ear averages the amplitude variation errors. While, the audio tape recorder utilizes the direct recording process, it is seldom satisfactory to use an audio recorder for Instrumentation purposes. The audio tape recorder is designed to take advantage of the rather peculiar spectral energy characteristics of speech and music, whereas the requirements of Instrumentation recorder are uniform response over its entire range.

## 28.51 FREQUENCY MODULATED (FM) RECORDING

Frequency modulation (FM) as the name implies, uses the variation of frequency to carry the desired information instead of varying the amplitude.

The frequency modulation overcomes some of the basic limitations of the direct recording process at the cost of high frequency response.

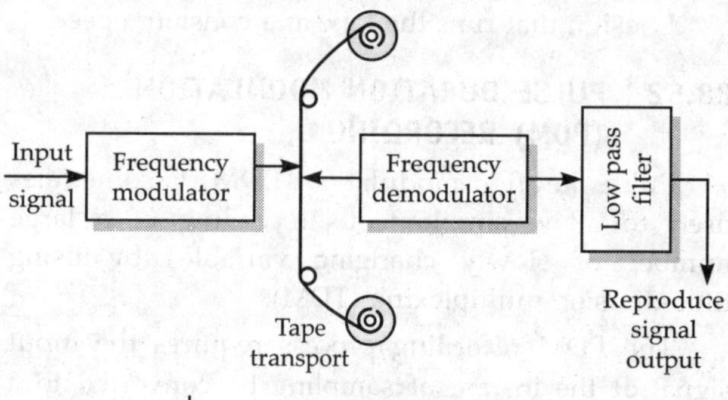


**Fig. 28.60** Frequency modulation (FM) recording system.

This system is shown in Fig. 28.60. In this system a carrier oscillator frequency, $f_c$, (called centre frequency) is modulated by the level of the input signal. When the input is zero, the modulation contains oscillation with a centre frequency $f_c$. Application of a positive input voltage deviates the carrier frequency by a specified percentage in one direction, whereas application of a negative voltage deviates the frequency by an appropriate percentage in the opposite direction. The output of the modulation which is fed to the tape is a signal of constant frequency for d.c. inputs while for a.c. inputs the signal is of varying frequency. The variation of frequency is directly proportional to the amplitude of input signal. On playback the output of the reproduce head is modulated and fed through a filter which removes the carrier and other unwanted frequencies produced due to modulation process.

The operation of an FM modulator can easily be checked by applying a known input voltage and measuring the output frequency with an electronic counter. This signal is applied to the tape with no further conditioning, as the signal is independent of the amplitude. The FM demodulator converts the difference between centre frequency and the frequency on the tape to a voltage proportional to the difference in the frequencies. This system can thus record frequencies from d.c. to several thousand Hz. Residual carrier signals and any out of band noise are removed by a low pass filter.

There are two important factors in FM recording :

### 1. Percentage deviation

It is defined as the carrier deviation to centre frequency.

$\therefore$ Percentage deviation or modulation index

$$M = \frac{\Delta f}{f_c} \times 100 \qquad \ldots(28.44)$$

where $\Delta f$ = carrier deviation from centre frequency,

and $f_c$ = centre or carrier frequency.

In a typical FM recording system ± 40% deviation of the carrier frequency corresponds to plus and minus full scale of input signal. To illustrate this, let the centre frequency be 50,000 Hz. Let an input signal of 40 percent deviation be applied to the modulator. Hence from Eqn. 28.44.

Carrier deviation from centre frequency,

$$\Delta f = \frac{m f_c}{100}$$

$$= \frac{40 \times 50{,}000}{100} = 20{,}000 \text{ Hz}$$

Thus a 1 volt input will produce an output frequency of : $f_0 + \Delta f$ = 50,000 + 20,000 = 70,000 Hz from the modulator.

Now let us apply d.c. to the same modulator. In this case $m=0$ and therefore $\Delta f = 0$. Hence the output of the modulator has a frequency of $f_0 + \Delta f = 50{,}000$ Hz.

**2. Deviation ratio**

It is defined as the ratio of carrier deviation from centre frequency to signal or modulating frequency.

$$\therefore \text{ Deviation factor } \delta = \frac{\Delta f}{f_m} \qquad \ldots(28.45)$$

where $f_m$ = data signal (frequency of the input signal to the modulator).

A system with a high deviation ratio generally has a low noise figure. However, $\Delta f$ is limited by recorder bandwidth and $f_m$ must be kept high to accommodate all the data signals.

### 28.51.1 Advantages of FM Recording

(*i*) FM recording is useful primarily when the d.c. component of the input signal is to be preserved or when the amplitude variations of the direct recording process cannot be tolerated.

(*ii*) This system has wide frequency range and can record from d.c. voltages to several kHz.

(*iii*) It is free from dropout effects.

(*iv*) It is independent of amplitude variations and accurately reproduces the waveform of the input signal.

(*v*) It is used extensively for recording the voltages from the force, pressure and acceleration transducers.

(*vi*) It is extremely useful for multiplexing in instrumentation systems.

### 28.51.2 Disadvantages of FM Recording

(*i*) FM recording is extremely sensitive to tape speed fluctuations (flutter). This is because, in either the recording or reproducing mode, tape speed variations introduce unwanted apparent modulation of the carrier (noise), reducing the dynamic range of the system. With a low percentage deviation, such as 7.5% in FM telemetry subcarrier systems, the effect of flutter becomes pronounced. For example, a 1 percent deviation caused by flutter in a system having 7.5 percent deviation corresponds to $1/75 \times 100 = 13.3$ percent noise. The same flutter imposed on a 40 percent deviation system causes only $1/40 \times 100 = 2.5$ percent noise signal. Thus systems with a higher percentage deviation are generally less influenced by tape speed variations.

(*ii*) This circuitry of an FM recording system is more complicated than that of a direct recording system. This complexity of circuitry is on account of separate modulation systems.

(*iii*) The FM system has a limited high frequency response or about 80 kHz.

(*iv*) It requires a high tape speed.

(*v*) It requires a high quality of tape transport and speed control and, therefore, expensive than the direct recording system.

To sum up, a well designed FM system gives a reasonably good amplitude accuracy or the order of 1% against 5% given by direct recording, d.c. response and d.c. linearity and low distortion.

Frequency response for a given tape speed is greatly reduced, and the recording and reproducing circuits are much more complex than those of the direct recording system. In order to keep the flutter figure low the tape transport mechanism must be of good design that runs the tape at a constant speed.

## 28.52 PULSE DURATION MODULATION (PDM) RECORDING

Pulse duration modulation (PDM) is sometimes used to allow simultaneous recording of a large number of slowly changing variables by using time-division multiplexing (TDM).

The PDM recording process requires the input signal at the instant of sampling be converted to a pulse, the duration of which is proportional to the amplitude of the signal at that instant. As an example, for recording a sine wave, it is sampled and recorded at uniformly spaced discrete intervals instead of continuously recording the instantaneous values. The original sine wave can be reconstructed on playback by passing the discrete readings through an appropriate filter.

A schematic diagram of PDM recording system employing TDM for a sine wave signal is shown in Fig. 28.61.

The system makes 900 samples through 90 scanner (multiplexer) positions. The number of channels per track are 86 provided that the input frequencies are less than 1.5 Hz. The rest of the four scanner positions are used for zero and maximum voltage levels and for synchronizing purposes.

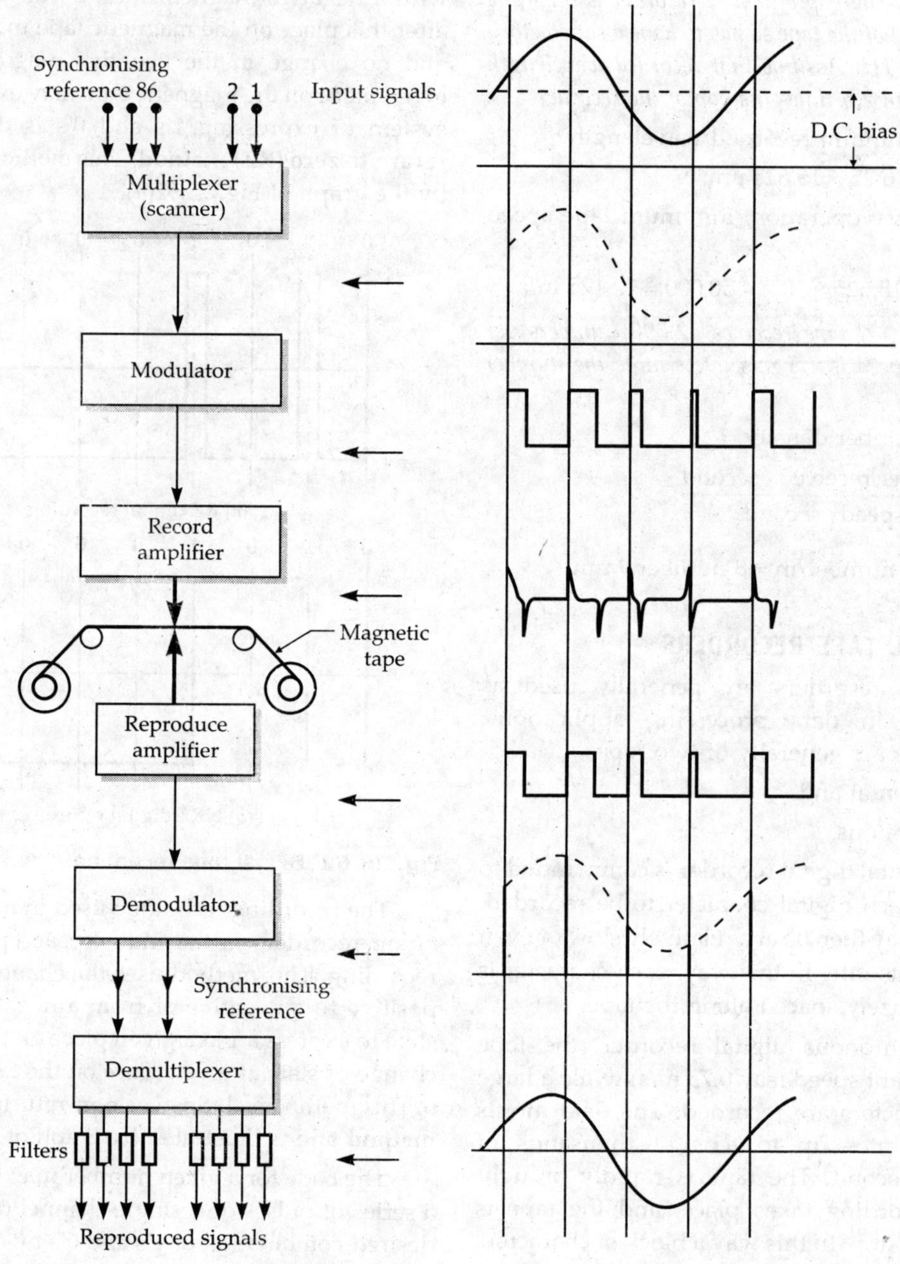


**Fig. 28.61** Schematic diagram of Pulse Duration Modulation (PDM) recording system.

## 28.52.1 Advantages of PDM Recording

The major advantages of a PDM recording system are :

(*i*) It has the ability to simultaneously record information from a large number of channels.

(*ii*) It has a high accuracy due to the fact that it can be self-calibrated.

(*iii*) It has a high S/N ratio.

## 28.52.2 Disadvantages of PDM Recording

The disadvantages of a PDM recording system are :

(*i*) It has a limited frequency response.

(*ii*) It has a highly complex electronic circuitry and therefore the reliability of such systems is low.

(*iii*) It is used only for special applications such as flight recorders, where a large number of slowly changing variables are involved.

***Example 28.7*** *The gap of a tape recorder is 6.25 μm. Determine the speed of the tape so has to have a satisfactory response at 50,000 Hz. Assume that recorded wavelength must be greater than 2.5 times the gap of the recorder.*

**Solution.** Minimum recorded wavelength

$$\lambda = 2.5 \times 6.25 = 15.625 \ \mu m.$$

For satisfactory operation, minimum tape speed is :

$$\eta = \lambda f = 15.625 \times 10^{-6} \times 50{,}000 = 7.8125 \text{ m/s}$$

***Example 28.8*** *A tape receives 12,000 numbers per second. The tape speed is 1.5 m/s. Determine the number density of the tape.*

**Solution.** Number density

$$= \frac{\text{number received / second}}{\text{speed / second}}$$

$$= \frac{12000}{1.5} \text{ number/m} = 8 \text{ numbers/mm.}$$

## 28.53 DIGITAL TAPE RECORDERS

Digital tape recorders are generally used as storage devices in data processing applications. Digital tape units are generally of two types :

- Incremental and
- Synchronous.

The incremental digital recorder is commanded to step ahead for each digital character to be recorded. The input data may then be at a relatively slow or even at a discontinuous rate. In this way, each character is equally and precisely spaced along the tape.

In the synchronous digital recorder, the tape moves at a constant speed (say 0.75 m/s) while a large number of characters are recorded. The data inputs are at precise rates up to tens of thousands of characters per second. The tape is rapidly brought upto speed, recording takes place, and the tape is brought to a fast stop. In this way a block of characters (a record) is written with each character spaced equally along the tape. Blocks of data are usually separated from each other by an erased area on the tape called a "record gap". The synchronous unit starts and stops the tape for each block of data to be recorded.

Characters are represented on magnetic tape by combination of 1-bits in the appropriate tracks across the tape width.

Instrumentation systems and computers use binary codes made up of combination of two symbols : 0 and 1. As far as digital tapes are concerned a positive magnetized space at a given place on a magnetic tape with a negative magnetization on the tape before or after this place on the magnetic tape may express a 1, and no change in the negative magnetization at a given place on the magnetic tape may express a 0. This system of expressing 1's and 0's is defined as the **return to zero (RZ) method**. This method is illustrated by the graph of Fig. 28.62(*a*).

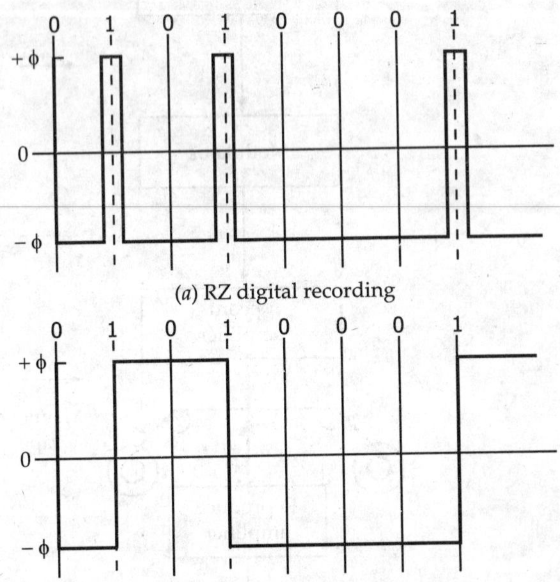


(*a*) RZ digital recording

(*b*) NRZ digital recording

**Fig. 28.62** Digital tape recording.

The recording technique used in most instrumentation recorders is the industry accepted IBM NRZ recording. This method uses the change from either a positive to a negative or from a negative to positive state to express a 1 at a given place on the tape, and no change of state at given place on the tape to express a 0. This method is defined as **non-return-to zero (NRZ) method** and is illustrated by graph of Fig. 28.62(*b*).

The code for a given number may be expressed as a series of pulses on a single channel representing the desired combination of 1's and 0's by either the RZ or NRZ method. This is a serial system of writing numbers. A faster and more common system uses a multichannel tape and inscribes the common number at a given place on the tape. This requires a recording channel for each of the digits, 1 or 0, required to express the number. This is a parallel system of writing numbers.

The process of recording numbers in a digital coded form is very simple. It requires only an amplifier is not recorder and another amplifier in the reproducer to condition the signal. The "on" and "off" conditions are neither a function of frequency nor of amplitude. Yet there are a few problems in the digital recording.

The major problem is the dropout errors or loss of pulses and spurious (extraneous) pulses. The loss of pulses becomes more critical as the rate of pulses decreases. This is expressed in pulses per mm of tape. Many recorders use 8 pulses/mm and 2.4 m/s (240 mm/s) speed, but some tape recorders use 20 pulses/mm and a speed of a 4.8 m/s (480 mm/s). Packing densities greater than 40 pulses/mm have been obtained.

As a check on the dropout errors, some systems record the information twice. Thus it is called use of **redundancy** (because the second recording would be redundant if the system were perfect).

A second method uses the logic of the number code to indicate, on a separate channel of the tape, if one of the 1's is missing. If an even number of 1's is missing, no error is recorded. This is known as **parity check**. Systems of parity checking have been developed which check not only the words written across the tape but also that of the numbers recorded in each of the tape channels. In some equipment the parity checks have been used to insert the missing information due to dropouts on the magnetic tape, as well as to indicate the errors that have been made. Thus this system is very effective in detecting and reducing tape dropout errors.

In parallel recording of digital information the recording head and the reproducing head should be at right angles to the tape. If the tape becomes skewed, the signals cannot be read at the same time and hence errors will be introduced. Thus an accurate alignment of tape is necessary to avoid this error.

The present day digital recording techniques two codes with which we will deal most are non-return-to-zero-level (NRZ-L) and bi-phase level (Bi-Phase-L). They are defined below, and some of their relative merits are shown in Table 28.1.

## NRZ-L code

This has been explained earlier. Further information is being given in regards to telemetry. This is a popular code for radio transmission of telemetry data ; it offers no more than one logic-level transition per bit period, and thus it does not have excessively high frequency components. However, long strings of ones or zeros may occur in normal sequences ; these introduce low-frequency or DC components which are not compatible with "direct" tape recording. The spectrum of an NRZ-L data sequence with some of these long strings of bits without logic transitions looks like Fig. 28.63.

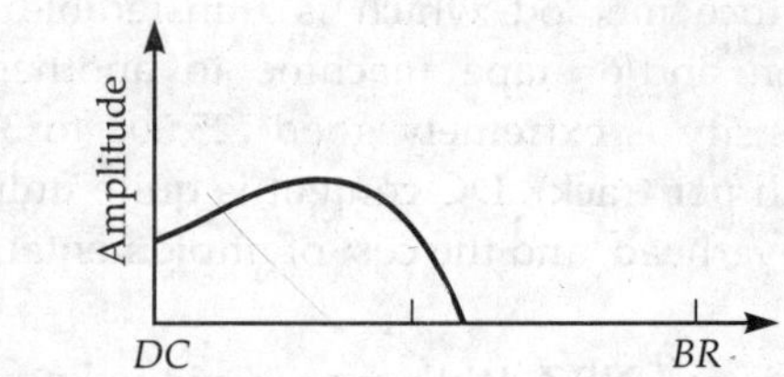


**Fig. 28.63** NRZ-L Spectrum.

## Bi-Phase-L code

This has at least one logic level transition per bit period and often has two transitions. Therefore, the tape must run faster to record the inherently higher frequencies. However, the guaranteed transitions ensure no DC components. This spectrum looks like Fig. 28.64.

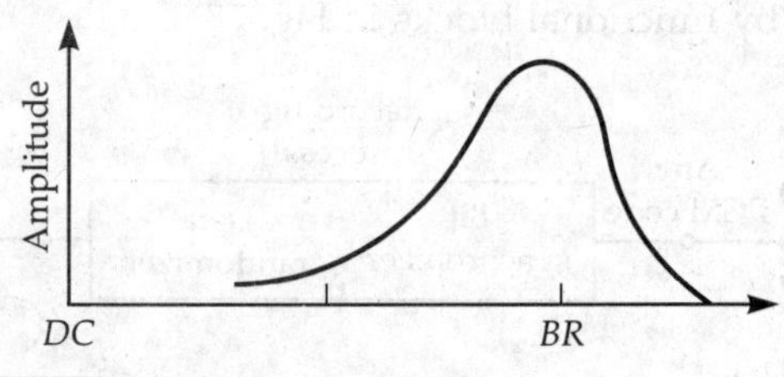


**Fig. 28.64** Bi-Phase Spectrum.

**Table 28.1** *NRZ and Bi-phase-L codes*

| Code Name | Descriptions | Transitions | DC Component |
|---|---|---|---|
| NRZ-L | One = Positive Level<br>Zero = Negative Level<br>Easy to generate and decode ; not self-clocking | Only when bits change ; highest rate is equal to data rate | Can approach 100% for long runs |
| Biϕ-L | One = Positive Transition<br>Zero = Negative<br>Transition (with additional Transitions at bit-cell beginning where required to satisfy above.) Requires twice as much reproduce bandwidth as NRZ for the same data. | Highest rate twice that of NRZ codes, for same data | Can never be significant |

28

Either of these techniques can be used for serial digital recording. However, the user may be prohibited from using "direct record" electronics if the DC component is excessive ; may be required to run the tape faster than desired to handle the data transition rates ; or may incur excessive overhead or other disadvantages with the selected techniques.

**Serial digital recording using randomizing technique.** A technique which offers near-optimum performance is known as "randomized NRZ" or "serial high-density digital recording." This technique was adopted by the IRIG Committee in its 106-80 standard for tape recording, thereby defining a record/produce method which is transferable from one location and/or tape machine to another. Bit packing density is extremely good (25,000 to 33,000 bits per inch per track), DC content is quite unlikely, here is no overhead, and the cost of implementation is reasonable.

**Randomized NRZ.** With the technique known as pseudo-random data modification, or simply "randomizing," normal NRZ data is modified in accordance with a pseudo-random pattern such that transitions are guaranteed, and the reverse process takes place on data playback to "derandomize" the data and recover the original NRZ data. The process does not add extraneous bits, nor does it impose any difficulty in bit decisions.

Operation of a randomizer/derandomizer is shown by functional blocks in Fig. 28.65.

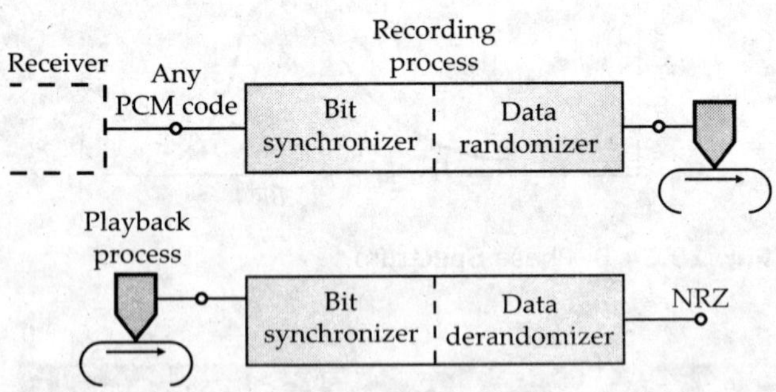


**Fig. 28.65** Randomizer/Derandomizer.

The data randomizer and derandomizer functions are performed by pseudo-random data generators, connected as shown by the hypothetical diagrams in Figs. 28.66 and 28.67.

The randomizer shift register feedback modifies incoming NRZ data in a known manner, guaranteeing frequent data transitions without the insertion of extraneous "overhead" bits. Such randomized NRZ ("RNRZ") is suitable for tape recording.

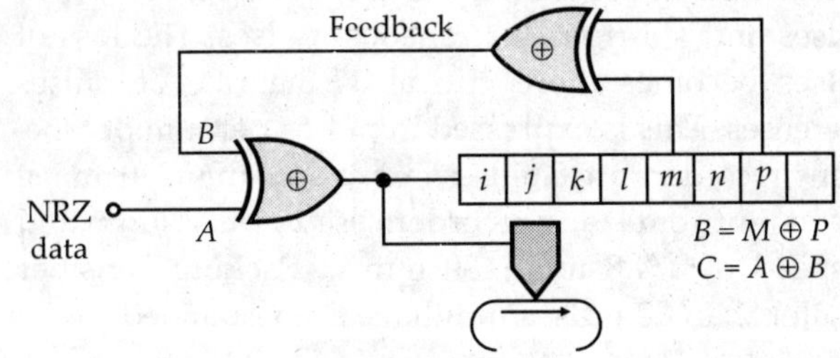


**Fig. 28.66** Record Process.

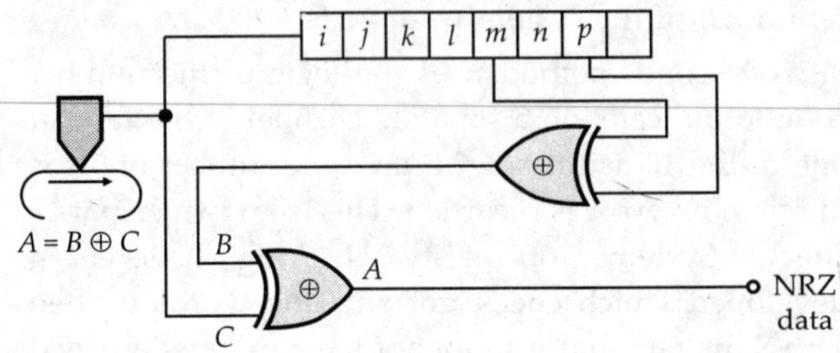


**Fig. 28.67** Playback Process.

On playback, an identical shift register with the same feedback points derandomizes the RNRZ, and the output NRZ is identical to that which is input to the randomizer.

IRIG standard 106-86 defines the algorithm for derandomized NRZ by specifying that the shift register be 15 bits long and that the feedback points be from the fourteenth and fifteenth stages. Tapes made from an IRIG-standard randomizer can be replayed on any IRIG-standard derandomizer.

Frequency content of NRZ compares to typical RNRZ, as shown in Fig. 28.68.

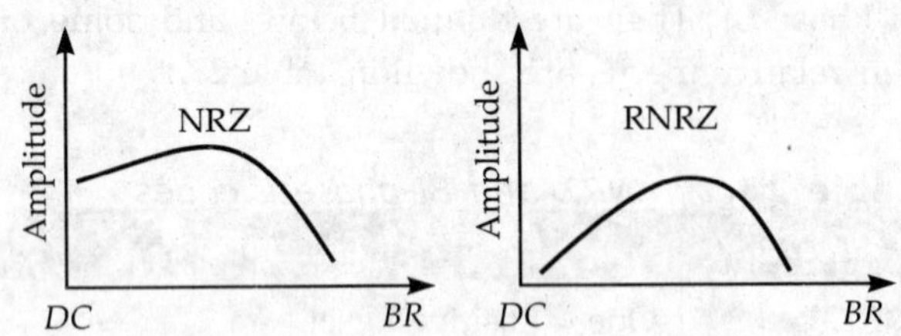


**Fig. 28.68** NRZ and RNRZ Spectrum.

The implementation as shown involves use of a PCM bit synchronizer and the direct record/reproduce electronics of a tap recorder. Packing densities of up to 25,000 bits per inch per track with normal wideband heads and electronics (or up to 50,000 bits per inch per track with "double-frequency" heads) are practical with this implementation.

### Serial High-Density Digital Recording

A form of randomized NRZ is called serial high-density digital recording, or serial HDDR, or serial HDR. In this case, however, randomization / derandomization takes place in tape record/reproduce electronics rather than in a separate PCM bit synchronizer. The recording process is accomplished without a bias signal, and the tape is driven to near-saturation by a properly adjusted HDDR record electronics module (in place of the direct or FM record electronics module).

On playback, the HDDR reproduce electronics module uses high-boost playback equalizers to optimize the occurrence of each data transition.

Both the randomizer and derandomizer are identical to those shown previously, standardized by the IRIG committee. The special optimized tape electronics, however, enable a user to 13,200 handle upto bits per cm per track with normal wide-band heads (or up to 26,400 bits per cm per track with "double frequency" heads).

# Review Questions

**1.** Explain the advantages of digital indicating instruments over their analog counterparts.

**2.** Explain the working of a *RS* flip-flop circuit.

**3.** Explain the working of decade counter.

**4.** Give an overview of different digital display devices.

**5.** Explain the theory land working of an LED. Describe the advantages of LEDs.

**6.** Explain the theory and working of LCDs. Describe the difference between light scattering and field effect types of LCDs. Also explain the advantages of LCDs.

**7.** Describe the principle and working of Nixie tubes.

**8.** Explain the functioning of a 5×7 LED matrix display.

**9.** Describe the functioning of a diode matrix for conversion of BCD system to decimal readout.

**10.** Draw the logic diagram and draw up the truth table for a BCD-to decimal decoder using TTL logic with 7441 chip.

**11.** Draw the logic diagram and draw up the truth table for conversion of 8-4-2-1 BCD code to 7 segmental code using 7447 chip.

**12.** Explain the following terms as applied to digital displays :

(*i*) Resolution

(*ii*) Difference between $3\frac{1}{2}$ digit and 4 digit displays

(*iii*) Sensitivity of digital meters

(*iv*) Accuracy specifications of digital meters.

**13.** Draw and explain the circuit of a digital frequency meter. What are the different methods used for high frequency determination ?

**14.** Describe with the help of suitable circuit diagrams, how the following types of measurements are carried out using a digital frequency meter

(*i*) single and multiple period measurements

(*ii*) time interval measurements

(*iii*) multiple ratio measurements.

**15.** Describe with the help of a circuit diagram, the working of a universal time counter. Explain the various functions it is capable of performing.

**16.** Explain the functioning of a ramp type digital voltmeter.

**17.** Describe the working of an integrating type digital voltmeter.

**18.** Explain the functioning of a potentiometric type digital voltmeter.

**19.** Explain the functioning of a basic type of strip chart recorder. Explain the different types of marking mechanisms used in it.

**20.** Describe the working of a galvanometric type strip chart recorder. What are the different types of tracing systems used in it ? Explain with the help of suitable diagrams.

**21.** Explain, with the help of a diagram, the working of a null type recorder. Distinguish between single point and multipoint recorders.

**22.** What is an *X*-*Y* recorder ? How do you distinguish it from a *X*-*t* or a *Y*-*t* recorder ?

Explain, with suitable circuit diagram, the working of an *X*-*Y* recorder. Describe its applications.

**23.** Explain the working of an ultra-violet recorder. Describe its applications.

24. Describe the basic components of a magnetic tape recorder used for instrumentation applications using direct recording techniques. Describe its advantages and disadvantages.
25. Describe how equalization is carried out in a magnetic tape recorder using direct recording techniques.
26. Explain the F.M. method of magnetic tape recording and explain its advantages and disadvantages.
27. Describe the method of pulse duration modulation (PDM) as used in magnetic tape recorders. Explain its advantages and disadvantages.
28. Describe the different methods used for digital tape recording. Explain its advantages and disadvantages.
29. Distinguish between RZ and NRZ techniques of digital tape recording.
30. Explain the randomized NRZ technique of digital tape recording.

# Unsolved Problems

1. A $3\frac{1}{2}$ digit voltmeter is used for measuring voltage.
   (*a*) Find the resolution of the instrument.
   (*b*) How would n voltage of 14.53 V be displayed on 10 V scale ?
   (*c*) How would a reading of 14.53 V be displaced on 100 V scale ?
   **[Ans.** (*a*) 0.001, (*b*) 15.43, (*c*) 014.5]
2. The lowest range on a $4\frac{1}{2}$ digit DVM is 10 mV full scale. What is sensitivity of this meter ? **[Ans.** 1 μV]
3. A certain $3\frac{1}{2}$ digit DVM as an accuracy specifications of ± 0.5 per cent of reading ± 2 digits.
   (*a*) What is the possible error, in volt, when the instrument is reading 5.00 V on its 10 V range ?
   (*b*) What is the possible error in volt, when reading 0.10 on the 10 V range ?
   (*c*) What percentage of the reading is the possible error in the case of (*b*) ? Comment upon the result.
   **[Ans.** (*a*) ± 0.45 V, (*b*) ± 0.0205 V, (*c*) 20.5%]
4. Gating periods of 1 ms, 10 ms, 100 ms, 1 s, 10 s are provided on a digital counter-timer-frequency meter having a 4 digit display. When a gating period is used a reading of 0095 is obtained.
   (*a*) Find the likely frequency.
   (*b*) What steps should be taken to obtain most accurate result ?
   (*c*) Suppose the frequency is 2.745 kHz, find the display for various gating times.
   **[Ans.** (*a*) 9500 Hz, (*b*) select a gating time of 1 s, (*c*) 0027, 0274, 2745, overflow]
5. The gap of a reproducing head in a tape recorder is 6.4 μm. Determine the minimum speed of the tape for satisfactory response at 50 kHz. Assume the wavelength on the tape is greater than 2.5 times the gap length. **[Ans.** 0.8 m/s]

# Objective Type Questions

1. If an information is required to be stored over a short interval of time
   (*a*) a single number/devices should be used
   (*b*) a storage type oscilloscope should be used
   (*c*) a CRO with photographic equipment should be used
   (*d*) a direct writing recorder or a magnetic tape recorder should be used.
2. The machine interpretable outputs from an analog transducer can be had from
   (*a*) magnetic tapes
   (*b*) punched cards and tapes
   (*c*) teletypewriter
   (*d*) all the above.
3. Digital instruments have input impedance of the order of
   (*a*) Ω (*b*) kΩ
   (*c*) MΩ (*d*) mΩ

**4.** If four flip-flops are connected in cascade, they serve as

(*a*) 2 : 1 scalar (*b*) 4 : 1 scalar
(*c*) 8 : 1 scalar (*d*) 16 : 1 scalar.

**5.** The LEDs, or their display require

(*a*) a voltage of 1.2 V and a current of 20 mA
(*b*) a voltage of 25 V and a current of 20 mA
(*c*) a voltage of 1.2 V and a current of 100 mA
(*d*) a voltage of 25 V and a current of 100 mA.

**6.** The switching time of LEDs is of the order of

(*a*) 1 s (*b*) 1 ms
(*c*) 1 μs (*d*) 1 ns

**7.** LEDs emit light

(*a*) only in red colour
(*b*) only in yellow colour
(*c*) only in green colour
(*d*) in red, green, yellow and amber colour.

**8.** An LCD requires a power of

(*a*) 20 W (*b*) 20 mW
(*c*) 20 μW (*d*) 20 nW

to light up one segment of a seven segmented display system.

**9.** The turn on and turn off times of a LCD are of the order of

(*a*) 1 s (*b*) 1 ms
(*c*) 10 ms (*d*) 10 ns

**10.** The power requirement of an LED is

(*a*) 40 mW per numeral
(*b*) 40 μW per numeral
(*c*) 10 W per numeral
(*d*) 10 μW per numeral.

**11.** A Nixie tube requires

(*a*) 10 cathodes
(*b*) 14 cathodes
(*c*) 15 cathodes
(*d*) 20 cathodes.

**12.** Nixie tubes require

(*a*) 150 V – 220 V
(*b*) 100 V – 150 V
(*c*) 150 mV – 220 mV
(*d*) 100 mV – 150 mV

to produce glow discharge.

**13.** In a digital frequency meter, the Schmitt trigger is used for

(*a*) sinusoidal waveforms into rectangular pulses
(*b*) scaling of sinusoidal waveforms
(*c*) providing time base
(*d*) none of the above.

**14.** Pulses of a frequency of 1 MHz are applied to the time base selector of a digital frequency meter which consists of 6 frequency dividers, each dividing the incoming frequency by a factor of 10. What is the time base setting at the output of 4th frequency divisor ?

(*a*) 1 ms (*b*)10 ms
(*c*) 100 ms (*d*) 1 ms

**15.** Period measurement is done in frequency meters for achieving high accuracy in the case of

(*a*) high frequencies
(*b*) medium frequencies
(*c*) d.c.
(*d*) low frequency.

**16.** In a digital frequency meter, the multi-period average mode of operation, the main gate is held open for more than one period of the unknown signal

(*a*) so that the period is extended in order that counters show more digits thereby an increased accuracy is obtained
(*b*) so that resolution is increased
(*c*) so that range of the instrument is extended
(*d*) so that errors on account of dead time are eliminated.

**17.** A digital voltmeter uses an A/D converter which needs a start pulse, uses an analog comparator and has a relatively fixed conversion time independent of the applied voltage.

The A/D converter is

(*a*) successive approximation converter
(*b*) digital ramp converter
(*c*) divial slope converter
(*d*) all the above.

**18.** X-*Y* recorders

(*a*) record one quantity with respect to another quantity
(*b*) record one quantity on *X*-axis with respect to time on *Y*-axis
(*c*) record one quantity on *Y* axis with respect to time on *X*-axis.

**19.** In a direct magnetic tape recording system, the reproducing head and amplifier characteristics are

(*a*) – 6 db/octave, + 6 db/octave
(*b*) – 6 db/octave, + 20 db/decade
(*c*) + 6 db/octave, + – 6 db/octave
(*d*) – 20 db/decade, + 20 db/decade.

28

**20.** The advantages of F.M. magnetic tape recording are

(*a*) it can record from d.c. to several kHz

(*b*) it is free from dropout effects

(*c*) it is independent of amplitude variations and accurately reproduces the waveform of input signal

(*d*) all the above.

## Answers

| | | | | | | | | | |
|---|---|---|---|---|---|---|---|---|---|
| **1.** (*b*) | **2.** (*d*) | **3.** (*c*) | **4.** (*d*) | **5.** (*a*) | **6.** (*d*) | **7.** (*d*) | **8.** (*c*) | **9.** (*c*) | **10.** (*a*) |
| **11.** (*c*) | **12.** (*a*) | **13.** (*a*) | **14.** (*b*) | **15.** (*d*) | **16.** (*a*) | **17.** (*a*) | **18.** (*a*) | **19.** (*c*) | **20.** (*d*) |

# 29

# Measurement of Non-Electric Quantities

## 29.1 LINEAR DISPLACEMENT TRANSDUCERS

Most of the transducers which are used for measurement of linear displacement have already been described in great details in Chapter 25. These transducers are :

- Resistive potentiometers,
- Strain gauges,
- Variable inductance transducers,
- Linear variable differential transformers (LVDTs),
- Capacitive transducers,
- Piezo-electric transducers,
- Hall effect transducers,
- Digital transducers.

## 29.2 MEASUREMENT OF ROTARY DISPLACEMENT

The transducers required for measurement of rotary displacement have been described in detail in Chapter 25. These transducers are:

- Resistive potentiometers,
- Variable inductance transducers,
- Rotary variable differential transformers (RVDTs),
- Variable reluctance transducers,
- Synchros,
- Capacitive transducers,
- Shaft Encoders.

## 29.3 STRAIN GAUGES AND MEASUREMENT OF STRAIN

### 29.3.1 Strain Gauge Circuits

The theory and construction of strain gauges have already been explained in Art. 25.16 on page 776.

The gauge factor of a strain gauge is given by Eqn. 25.69, page 777.

Gauge factor

$$G_f = \frac{\Delta R / R}{\varepsilon} = 1 + 2v + \frac{\Delta\rho/\rho}{\varepsilon} \qquad ...(29.1)$$

The various symbols have already been explained.

When we consider the sensitivity of a metallic strain gauge, we find that it is extremely versatile and reliable.

Typical values of gauge factor and resistance for commonly used strain gauges are :

$$G_f = 2 \text{ and } R = 120\,\Omega$$

Strain gauges are used for measurements of strains as low as 1 microstrain and therefore the corresponding change in resistance is :

$$\Delta R = G_f\,\varepsilon\,R = 2 \times 10^{-6} \times 120$$
$$= 10^{-6}\,\Omega = 0.00024\,\Omega$$

This small change in resistance which is 0.002% of original resistance has to be measured. It is quite resistance evident that in order to measure a change in resistance of this low order will require the use of extremely, sensitive and sophisticated instrumentation.

Strain gauges utilize circuits like voltage measuring potentiometer circuits, ballast circuits and Wheatstone bridges which have already been described. The purpose here is to review them.

## 29.4 BALLAST CIRCUIT

The ballast circuit has been described in Art. 26.20 on page 924. Figure 29.1 shows a ballast circuit using a strain gauge.

If the $R_g$ is the resistance of strain gauge the output voltage when the gauge is not strained is,

$$e_0 = \frac{R_g}{R_b + R_g}\,e_i \qquad ...(29.2)$$

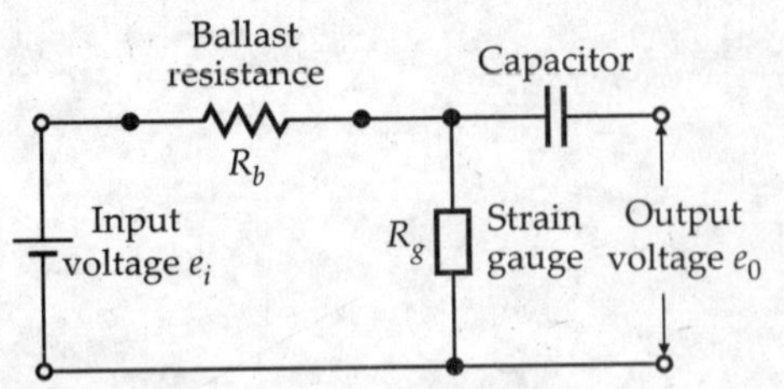


**Fig. 29.1** Strain gauge using ballast circuit.

∴ Change in output voltage when the gauge is strained :

$$d_{e0} = \frac{R_b}{(R_b + R_g)^2} dR_g \cdot e_i = \frac{R_b R_g}{(R_b + R_g)^2} \frac{dR_g}{R_g} e_i$$

$$= \frac{R_b R_g}{(R_b + R_g)^2} G_f \varepsilon e_i \qquad ...(29.3)$$

∴ Change in output voltage when the gauge is strained is directly proportional to the strain.

Maximum sensitivity is obtained when the ballast resistance is equal to the strained resistance of the gauge or $R_b = R_g$. For this arrangement, change in output voltage when the gauge is strained

$$\Delta e_0 = (G_f \varepsilon / 4) e_i \qquad ...(29.4)$$

Some of the inherent limitations in the ballast circuit can be explained by the following example. Consider a gauge of resistance of 120 Ω and gauge factor of 2. Let the output voltage be 8 V. The voltage across the strain gauge when it is not strained is $e_0 = e_i / 2 - 4$ V

When the strain gauge is subjected to strain, the change in output voltage is,

$$\Delta e_0 = 2 \times (8/4)\,\varepsilon = 4\varepsilon$$

Now if a strain equal to 1 microstrain is to be measured the change in output voltage is,

$$\Delta e_0 = 4 \times 1 \times 10^{-6} = 4 \times 10^{-6} = 4\ \mu\text{V}$$

Thus the indicator connected to output terminals to provide indication of strain must sense a change of voltage of 4 μV in a range/of 4 V *i.e.,* a change 0.0001%. This is a stringent requirement which practically eliminates the use of ballast circuits for static strain measurements. However, the circuit is used for dynamic strain measurements where static strain components are ignored. Thus a capacitor is connected in series in one of the leads as shown in Fig. 29.1. It acts as a high pass filter. It does not allow d.c. components of output voltage on account of static strain to go to output terminals. Only a.c. output voltage components due to dynamic strains are allowed to pass.

**Example 29.1** *A strain gauge having a resistance 100Ω and gauge factor of 2 is connected in series with a ballast resistance of 100 Ω across a 12 V, supply. Calculate the difference between the output voltage with no stress applied and a stress of 140 MN/m². The modulus of elasticity is 200 GN/m².*

**Solution.** When no stress is applied voltage across the gauge ; output voltage is,

$$e_0 = \frac{R_g}{R_b + R_g} \cdot e_i = \frac{100}{100 + 100} \times 12 = 6\text{ V}$$

$$\text{Strain} = s/E = (140 \times 10^6 / 200 \times 10^9) = 0.7 \times 10^{-3}$$

Change in resistance of gauge

$$\Delta R_g = G_f \varepsilon R_g = 2 \times 0.7 \times 10^{-3} \times 100 = 0.014\ \Omega$$

∴ Voltage across the gauge when stress is applied :

$$= \frac{100 + 0.014}{2100 + 100 + 0.014} \times 12$$

$$= 6.00042\text{ V}$$

Change in voltage

$$= 6.00042 - 6 = 0.00042\text{ V} = 0.42\text{ mV}$$

**Example 29.2** *A strain gauge having a resistance of 120 Ω and a gauge factor of 2 is subjected to a strain of : (60 +10 sin 314t) × 10⁻⁶.*

*(a) Find the expression for the change in output voltage when the strain gauge is connected in ballast circuit having a ballast resistance of 120 Ω and an input voltage of 6 V.*

*(b) If a capacitor is connected in one of the output leads and a true rms reading voltmeter of infinite impedance is connected across the output terminals, find the reading of the voltmeter.*

**Solution.** (*a*) Change in output voltage

$$\Delta e_0 = \frac{G_f}{4} \varepsilon e_i$$

$$= \frac{2}{4} \times 6 \times [20 + 10 \sin 314\, t] \times 10^{-6}\text{ V}$$

$$= [60 + 30 \sin 314\, t] \times 10^{-6}\text{ V}$$

$$= [60 + 30 \sin 314\, t]\ \mu\text{V}$$

(*b*) When a capacitor is connected in series with one of the leads, the d.c. component is blocked and only the a.c. component is allowed to pass.

∴ Signal applied to a true rms reading voltmeter

$$= 30 \sin 314\, t\ \mu\text{V}$$

∴ Reading of voltmeter

= rms value of output voltage

$$= 30/\sqrt{2} = 21.2\ \mu\text{V}$$

## 29.5 WHEATSTONE BRIDGES

The Wheatstone bridge is of two types :

- Null type and
- Deflection type.

### 29.5.1 Null Type Wheatstone Bridge

The null type Wheatstone bridge is shown in Fig. 26.92, page 922. When the bridge is unbalanced on account of change in resistance $\Delta R_1$ of arm 1, the resistance required to balance it,

$$\Delta R_2 = \Delta R_1 (R_3 / R_4)$$

Suppose arm 1 is composed of a strain gauge having a resistance $R_g$ and initially with

$$R_2 = R_3 = R_4 = R_g$$

$$\therefore \quad \Delta R_2 = \Delta R_g$$

where $\Delta R_g$ = change in the value of gauge resistance when strained

But $\quad \Delta R_g = G_f \varepsilon R_g$

and $\quad \Delta R_2 = G_f \varepsilon R_g$

Hence change in the value of resistance $R_2$ is a direct measure of strain.

### 29.5.2 Deflection Type Wheatstone Bridges

There are two types of deflection type Wheatstone bridges as described in Art. 26.22.2, on page 926. They are :

(*i*) Voltage sensitive bridge, and

(*ii*) Current sensitive bridge.

***(i) Voltage sensitive bridge.*** The output of a voltage sensitive bridges when the resistance of arm 1 is changed is given by ;

$$\Delta e_0 = \left[\frac{\Delta R_1 / R}{4 + 2(\Delta R_1 / R)}\right] e_i$$

(See Eqn. 26.165, page 927)

This relationship pertains to a condition wherein the values of resistance of all the arms are initially equal, *i.e.*, $R_1 = R_2 = R_3 = R_4 = R$ and the unbalance is created on account of change of resistance equal to $\Delta R_1$ in $R_1$.

Suppose now the strain gauge is in arm 1 and unstrained resistance of gauge is equal to $R_g$. The bridge is balanced initially with :

$$R_g = R_1 = R_2 = R_3 = R_4$$

When the strain gauge is subjected to strain, its value changes by $\Delta R_g$.

$\therefore$ Output voltage

$$\Delta e_0 = \left[\frac{\Delta R_g / R_g}{4 + 2(\Delta R_g / R_g)}\right] e_i \quad \text{...(29.5)}$$

However, the change in the value of resistance is quite small and therefore : $2\Delta R_g / R_g << 4$

$$\text{Hence, } \Delta e_0 = \frac{\Delta R_g / R_g}{4} = \frac{G_f \varepsilon}{4} e_i \quad \text{...(29.6)}$$

Therefore the change in output voltage is linearly proportional to strain. This bridge is called a **Quarter Bridge** since only one of the four arms contains an **Active Gauge.**

***(ii) Current sensitive bridge.*** The analysis of this bridge has been done in Art. 26.22.4, page 927 by using Thevenin theorem. The meter current under unbalance conditions is directly proportional to the strain.

**Example 29.3** *A Wheatstone bridge is shown in Fig. 26.96 on page 926 with $R_3 = R_4 = 100\,\Omega$. The galvanometer resistance is $50\,\Omega$. The strain gauge is connected in arm 1 and has an unstrained resistance of $120\,\Omega$. The value of $R_2$ is adjusted for balance under zero strain conditions. The gauge factor is 2. Calculate the galvanometer current when a strain of $400\,\mu m / m$ is applied. Take the battery voltage as 4 V. Calculate the voltage output when a meter of infinite impedance is connected across output terminals.*

**Solution.** Suppose the strain forms arm 1 of the bridge. The value of $R_2$ for initial balance is $120\,\Omega$.

Resistance of gauge when strained

$$R_1 = 120(1 + 2 \times 400 \times 10^{-6}) = 120.096\,\Omega$$

This causes unbalance resulting in the appearance of an output voltage. The value of Thevenin source voltage is,

$$e_0 = \left[\frac{R_1}{R_1 + R_3} - \frac{R_2}{R_2 + R_4}\right] e_i$$

$$= \left[\frac{120.096}{120.096 + 100} - \frac{120}{120 + 100}\right] \times 4 = 793\,\mu V$$

The Thevenin source resistance can be found by short-circuiting the battery terminals and finding the resistance looking into output terminals. Thevenin source resistance is :

$$R_0 = \frac{R_1 R_3}{R_1 + R_3} + \frac{R_3 R_4}{R_2 + R_4}$$

$$= \frac{120 \times 100}{120 + 100} + \frac{120 \times 100}{120 + 100} = 109\,\Omega.$$

$\therefore$ Meter current

$$I_m = \frac{e_0}{R_0 + R_m} = \frac{793 \times 10^{-6}}{109 + 50} \simeq 5 \times 10^{-6}\,A = 5\,\mu A$$

29

## 29.6 GAUGE SENSITIVITY

Figure 29.2 shows a single strain gauge mounted on a cantilever beam. When a force is applied to the

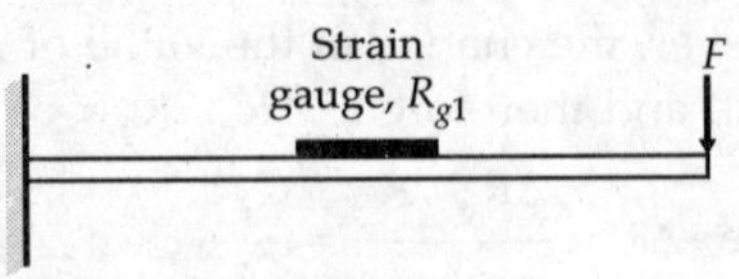


**Fig. 29.2** Force acting on a cantilever and measured by using single strain gauge.

beam, the gauge is subjected to a tensile stress. Therefore, the resistance of the strain gauge increases. This increase in resistance is measured using a Wheatstone bridge as shown in 29.3.

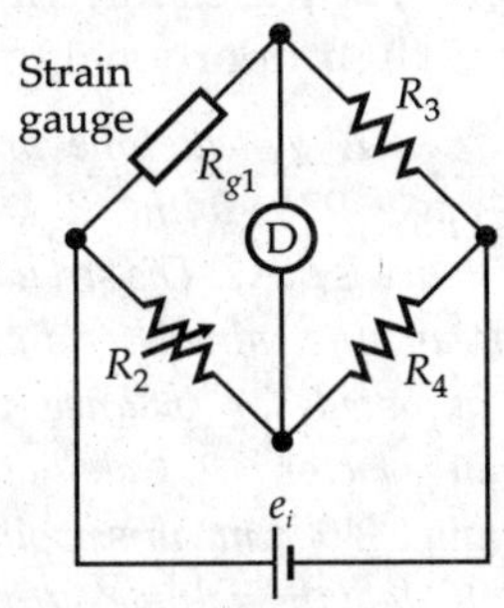


**Fig. 29.3** Strain gauge bridge for arrangement of Fig. 29.2.

The bridge is balanced under unstrained conditions, but becomes unbalanced when the gauge is strained.

The change in resistance of gauge

$$\Delta R_g = \Delta R_2 (R_3 / R_4)$$

where $\Delta R_2$ = change in $R_2$, to rebalance the bridge after the strain is applied

Now $\Delta R_2 = (R_4 / R_3) R_{g1}$

Output of bridge = $K \Delta R_2$

where $K$ = scale factor of $\Delta R_2$ readout or indicator.

Gauge sensitivity

$$S_g = \frac{K\Delta R_2}{\varepsilon} = \frac{K\Delta R_2}{\Delta R_{g1} / R_{g1}} G_f \quad \text{...(29.7)}$$

$$= K \frac{R_4}{R_3} \Delta R_{g1} \frac{G_f}{\Delta R_{g1} / R_{g1}}$$

$$= K \frac{R_4}{R_3} R_{g1} G_f \quad \text{...(29.8)}$$

When $R_4 = R_3$ and $R_{g1} = R_g$

Gauge sensitivity

$$S_g = K R_g G_f \quad \text{...(29.9)}$$

Now if $R_{g1} = R_2 = R_3 = R_4 = R$

The output voltage from the bridge is

$$\Delta e_0 = \frac{\Delta R / R}{4} e_i = \frac{G_f \varepsilon}{4} e_i \quad \text{...(29.10)}$$

## 29.7 TEMPERATURE COMPENSATION

The resistive type strain gauges are sensitive to temperature. Therefore, it becomes necessary to account for variations in the strain gauge resistance which occur on account of temperature changes. If changes in resistance due to temperature variations are not accounted for or not compensated, the results would be in error. This is because the resistance of the strain gauge changes both with strain as well as with temperature. Temperature changes act as spurious inputs and cause change in the resistance in two ways :

(*i*) the resistance of the wire grid of strain gauge changes with change in temperature, and

(*ii*) a change in resistance occurs due to differential expansion exsiting between the gauge and the test surface to which the strain gauge is bonded.

A strain may be caused because of the differential co-efficients of expansion of strain gauge and the surface to which it is bonded to even when there is no load on the test surface.

Temperature effects may be handled by :

(*i*) compensation or cancellation, and

(*ii*) evaluation as a part of the data reduction problem.

The first method is extensively used for both metallic as well as semi-conductor strain gauges while the second method is used for semi-conductor gauges only.

## 29.8 TEMPERATURE COMPENSATION AND CANCELLATION TECHNIQUES

Compensation may be provided by :

(*i*) use of adjacent arm balancing or compensating gauge,

(*ii*) self-compensation, and

(*iii*) use of special external control circuitry.

### 29.8.1 Adjacent Arm Compensating Gauge

#### 1. Use of Dummy Gauge

One of the ways in which temperature error can be eliminated by using adjacent arm compensating gauge is to use a **dummy gauge** in the adjacent arm.

This arrangement is shown in Fig. 29.4. Gauge 1 is installed on the lest specimen (called active gauge) while gauge 3, called a dummy gauge is installed on a like piece of material and is not subjected to any strain. The gauges installed on the test piece and the dummy gauge are at the same temperature. A gauge is called **dummy gauge** in case it is not subjected to any strain. **Active gauge** is one which is subjected to strain. Initially when the bridge is balanced,

$$\frac{R_1}{R_3} = \frac{R_2}{R_4}$$

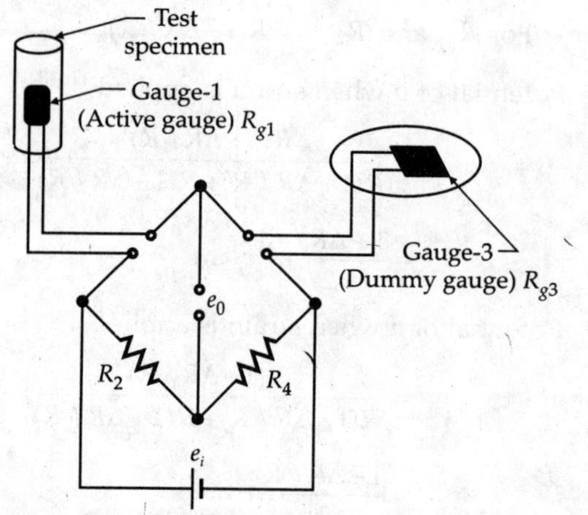


**Fig. 29.4** Use of dummy for temperature compensation.

Supposing a change in temperature occurs, the resistances $R_1$ and $R_3$ change by an amount $\Delta R_1$ and $\Delta R_3$ respectively.

Hence for balance,

$$\frac{R_1 + \Delta R_1}{R_3 + \Delta R_3} = \frac{R_2}{R_4}$$

or $$\frac{R_4}{R_3}(R_1 + \Delta R_1) = (R_3 + \Delta R_3)$$

or $$\frac{R_4}{R_2} R_1 + \frac{R_4}{R_2} \Delta R_1 = R_3 + \Delta R_3$$

But $$\frac{R_4}{R_3} R_1 = R_3$$

$\therefore$ $$\frac{R_4}{R_2} \Delta R_1 = \Delta R_3$$

Suppose $R_4 = R_2$.

This requires that $\Delta R_1 = \Delta R_3$

It means that for the bridge to remain insensitive to variations in temperature the gauges $R_1$ and $R_3$ should have their resistances change by equal amount when subjected to variation in temperature. Therefore the active gauge $R_1$ and the dummy gauge $R_3$ should be identical.

The use of dummy gauge for temperature compensation is simple and effective and should be employed whenever possible.

## 2. Use of Two Active Gauges in Adjacent Arms

In certain applications, where equal and opposite strains are known to exist, it is possible to attach two similar gauges in such a way that one gauge experiences a positive strain and the other a negative strain. Thus instead of having an arrangement wherein one gauge acts as the active gauge and the other as the dummy gauge, we have now an arrangement wherein both the gauges are active gauges.

Figure 29.5 shows the two gauges mounted on a cantilever. The gauge $R_{g1}$ is on top of the cantilever and hence experiences tension or a positive strain. The $R_{g3}$ is at the bottom surface of the cantilever and hence experiences a compression or a negative strain.

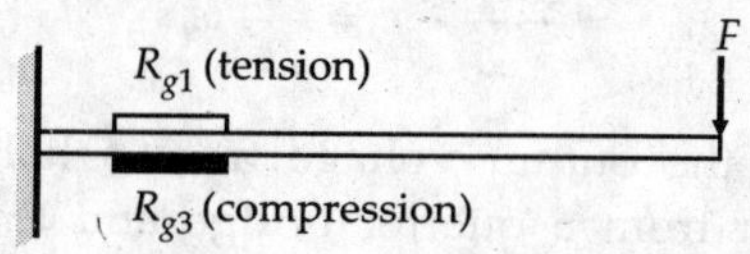


**Fig. 29.5** Two gauges used for measurement of strain.

The bridge arrangement for the two gauges is shown in Fig. 29.6. There are two active gauges in the 4 arm bridge and hence it is called a **Half Bridge.**

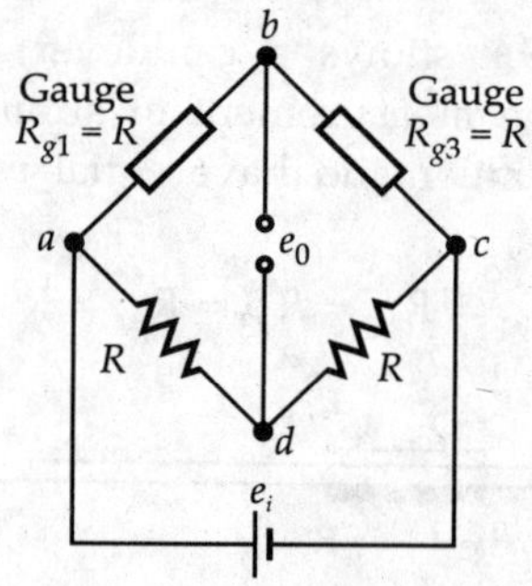


**Fig. 29.6** Adjacent arm compensation using two active gauges.

The temperature effects are cancelled out by having $R_2 = R_3$ and using two identical gauges in the opposite arms of the bridge.

Suppose $R_{g1} = R_{g3} = R_2 = R_4 = R$

When no strain is applied both points $b$ and $d$ are at the same potential, $e_i/2$ and the value of output voltage $e_0 = 0$.

When the arrangement shown in Fig. 29.5 is subjected to strain, the resistance of gauge 1 increases and that of gauges $R_{g3}$ decreases.

Resistance of gauge $R_{g1}$, when strained is

$$R(1+\Delta R/R)$$

Resistance of gauge $R_{g3}$ when strained is

$$R(1-\Delta R/R)$$

Now $\quad R_2 = R_4 = R$

$\therefore$ Potential of point $d = e_i/2$

$\therefore$ Potential of point $b$

$$= \frac{R(1+\Delta R/R)}{R(1+\Delta R/R)+R(1-\Delta R/R)} \times e_i$$

$$= \frac{1+\Delta R/R}{2} e_i$$

$\therefore$ Change in output voltage when strain applied is,

$$\Delta e_0 = \frac{1+\Delta R/R}{2} e_i - \frac{e_i}{2}$$

$$= \frac{\Delta R/R}{2} e_i = \frac{G_f \varepsilon}{2} e_i \quad \text{...(29.11)}$$

Thus the output voltage from a half bridge is twice that from a quarter bridge and therefore the sensitivity is doubled. In addition, the temperature effects are cancelled. The gauge, sensitivity of a half bridge is,

$$S_g = 2\, KR_g G_f \quad \text{...(29.12)}$$

### 3. Use of Four Active Gauges

Figure 29.7 shows a cantilever using 4 strain gauges for the measurement of strain. All the four gauges are similar and have equal resistance when strained *i.e.,*

$$R_{g1} = R_{g2} = R_{g3} = R_{g4} = R$$

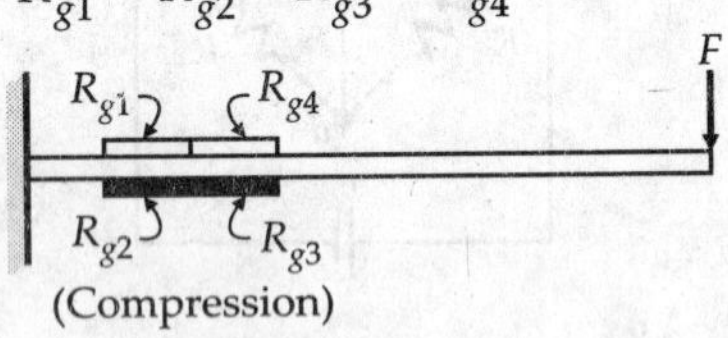


**Fig. 29.7** Use of four strain gauges for measurement of strain.

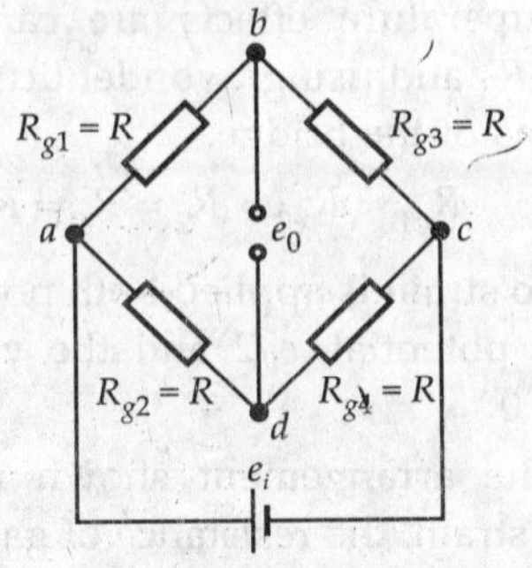


**Fig. 29.8** Bridge circuit for measurement of strain four using active gauges.

These gauges are connected in the arms of a Wheatstone bridge as shown in Fig. 29.8. Since the bridge has 4 active gauges with one gauge in each of the four arms, it is called a **Full Bridge.**

When no strain is applied the potential of points $b$ and $d$ are both equal to $e_i/2$ and hence the output voltage $e_0 = 0$.

When strained, the resistance of various gauges are :

For $R_{g1}$ and $R_{g4}$ : $\quad R(1+\Delta R/R)$

and $\quad$ For $R_{g2}$ and $R_{g3}$ : $\quad R(1-\Delta R/R)$

Potential of $b$ when strain is applied

$$= \frac{R(1+\Delta R/R)}{R(1+\Delta R/R)+R(1-\Delta R/R)} \times e_i$$

$$= \frac{1+\Delta R/R}{2} e_i$$

Potential of $d$ when strain is applied

$$= \frac{R(1-\Delta R/R)}{R(1-\Delta R/R)+R(1+\Delta R/R)} \times e_i$$

$$= \frac{1-\Delta R/R}{2} e_i$$

Therefore change in output voltage

$$\Delta e_0 = \frac{1+\Delta R/R}{2} e_i - \frac{1-\Delta R/R}{2} e_i$$

$$= (\Delta R/R)$$

$$= G_f \varepsilon\, e_i \quad \text{...(29.13)}$$

Four active-active-arm bridges are extensively used when strain gauges are used as secondary transducers to give maximum sensitivity combined with full temperature compensation. The effect of increasing the number of active gauges is the same if a low impedance detector is used.

The gauge sensitivity of a full bridge

$$S_g = 4\, KR_g G_f \quad \text{...(29.14)}$$

### 4. Poisson's Method

This method of temperature compensation utilises two active gauges $R_{g1}$ and $R_{g3}$ which are bonded at right angles to the structural member as shown in Fig. 29.9(*a*).

Measuring gauge is in tension, but gauge $R_{g3}$ reduces in length depending upon the Poisson's ratio of the material. Poisson's gauge is connected, in the adjacent arm of the bridge as shown in Fig. 29.9(*b*). The advantage of Poisson's gauge is that in addition to temperature compensation, it acts as an active gauge. The bridge sensitivity is increased by a factor $(1+\nu)$ accordingly.

Change in output voltage when strain is applied

$$\Delta e_0 = (1+\nu)\frac{G_f\,\varepsilon}{4}\,e_i \qquad ...(29.15)$$

***Practical strain bridge.*** A strain gauge bridge may be rigged up by the user or may be purchased as a 'package' which is built up as a unit and only requires connections to the active strain gauge and to a source of power supply. The 'package' consists of an amplifier and also other signal conditioning equipment for amplification and monitoring of the output of the strain gauge bridge.

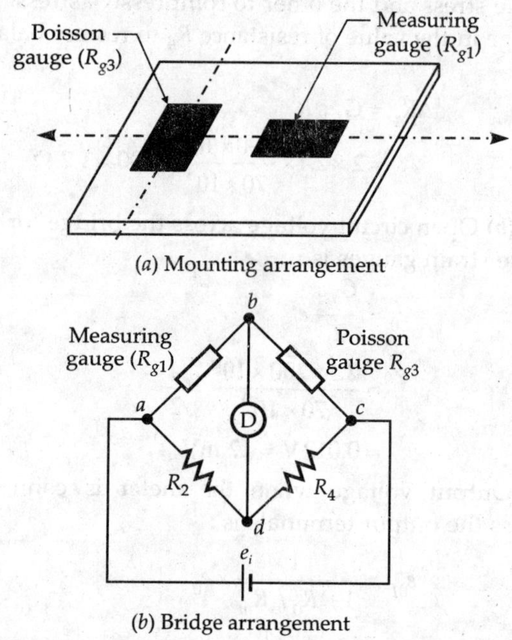


**Fig. 29.9** Use of Poisson's gauge.

**Example 29.4** *A strain gauge bridge is shown in Fig. 29.10.*

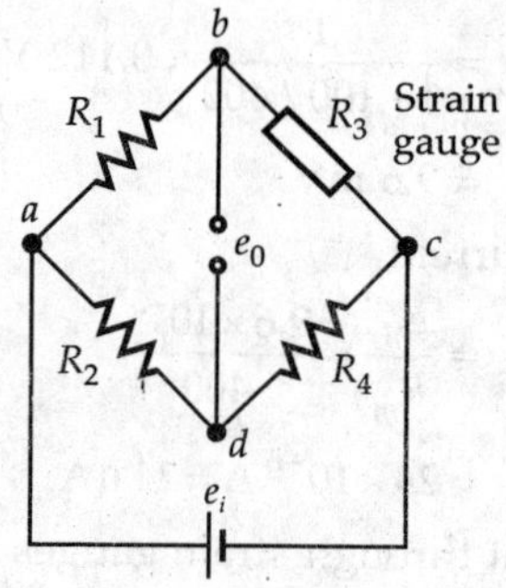


**Fig. 29.10** Bridge of Example 29.4.

*It has two fixed resistors $R_1$ and $R_2$ of 120 Ω each. The gauge is represented by resistance $R_3$. The variable resistance $R_4$ is 120.00 Ω at zero strain and has value of 120.63 Ω with strain ε. The gauge factor is 2.04. Determine the strain in the beam at the point where the strain gauge is attached.*

**Solution.** Change in value of resistance when strain is applied :

$$\Delta R_4 = 120.63 - 120 = 0.63\ \Omega.$$

∴ Change in resistance of gauge when strained :

$$\Delta R_3 = (R_1 / R_2)(\Delta R_4)$$

$$= \frac{120}{120} \times 0.63 = 0.63\ \Omega$$

Now $\quad \dfrac{\Delta R_3}{R_3} = G_f\,\varepsilon$

$$\therefore \text{ Strain } \varepsilon = \frac{\Delta R_3}{R_3} \times \frac{1}{G_f}$$

$$= \frac{0.63}{120} + \frac{1}{2.04} = 2.57 \times 10^{-3}\ \text{m}$$

**Example 29.5** *A strain gauge bridge comprises of two fixed resistors each of value 120 Ω, one active gauge, and an unstrained temperature compensation gauge. The two gauges are of unstrained resistance 120 Ω and gauge factor 2.2. Find the bridge output voltage (with output open circuited) for a supply voltage of 3 V, when the active gauge is subjected to 600 micro strain.*

**Solution.** The bridge has equal arms when there is no strain. It has only one active gauge. Therefore, output voltage is :

$$\Delta e_0 = \frac{G_f\,\varepsilon}{2}\,e_i = \frac{2.2 \times 600 \times 10^{-6} \times 3}{4}$$

$$= 0.99 \times 10^{-3}\ \text{V} = 0.99\ \text{mV}$$

**Example 29.6** *A single strain gauge is mounted to measure the axial strain in a simple tensile member. If the recorded strain is 380 microstrain, what is the axial stress :*

*(i) If the member is of steel ?*

*(ii) If the member is of aluminium ?*

*The modulus of elasticity of steel is 200 GN/m² and that of aluminium is 70 GN/m².*

**Solution.** Stress $s = \varepsilon E$

where $\varepsilon$ = strain and $E$ = modulus of elasticity.

(*i*) Stress in steel

$$s = 380 \times 10^{-6} \times 200 \times 10^9$$

$$= 76 \times 10^6\ \text{N/m}^2 = 76\ \text{MN/m}^2$$

(*ii*) Stress in aluminium

$$s = 380 \times 10^{-6} \times 70 \times 10^9$$

$$= 26.6 \times 10^6\ \text{N/m}^2 = 26.6\ \text{MN/m}^2$$

29

***Example 29.7*** *A single electrical resistance strain gauge of resistance 120 Ω and having a gauge factor of 2 is bonded to steel having an elastic limit stress of 400 MN/m² and modulus of elasticity 200 GN/m². Calculate the change in resistance,*

(*a*) *due to a change in stress equal to 1/10 of the elastic range ; and*

(*b*) *due to a change of temperature of 20°C, if the material is advance alloy.*

*The resistance temperature co-efficient of advance alloy is* $20 \times 10^{-6}$ */°C.*

(*c*) *Calculate the strain due to differential expansion of the gauge metal and steel if, co-efficient of linear expansion of steel is* $12 \times 10^{-6}$ */°C and that of advance alloy is* $16 \times 10^{-6}$ */°C.*

*Calculate also the change in resistance. Comment upon the results.*

**Solution.** (*a*) Change of stress

$$s = 400 \times 10^{-6} \times 1/10$$
$$= 40 \times 10^{6}\,\text{N/m}^2 = 40\ \text{MN/m}^2$$

Change of strain

$$\varepsilon = \frac{s}{E} = \frac{40 \times 10^6}{200 \times 10^9} = 200 \times 10^{-6}$$

∴ Change in resistance

$$\Delta R = G_f\,\varepsilon\,R$$
$$= 2 \times 200 \times 10^{-6} \times 120 = 0.048\ \Omega$$

(*b*) Change in resistance due to change in temperature :

$$= \alpha \Delta\theta R = 20 \times 10^{-6} \times (20) \times 120$$
$$= 0.048\ \Omega$$

The strain due to differential expansion of the gauge metal and steel is :

$$\varepsilon = (12 - 16) \times 10^{-6} \times 20$$
$$= -80 \times 10^{-6} = -80\ \text{microstrain}$$

$$\therefore \quad \Delta R = G_f\,\varepsilon\,R = 2 \times 80 \times 10^{-6} \times 120$$
$$= -0.019\ \Omega$$

**Comments.** The change of resistance due to change in temperature is of the same order as that due change in resistance caused by application of strain. If the temperature effects are not compensated for, they will lead to extremely large errors in the measurement of strain. The measurement in fact does not have any meaning in that case.

It is seen that change due to differential expansion partly offsets the change due to co-efficient of resistivity in the example. The effect may become additive instead of subtractive in certain combinations of gauge metal and the metal used for mounting.

***Example 29.8*** *Two electrical strain gauges are bonded to a Duralumin cantilever (Fig. 29.5) and connected into a bridge circuit as shown in Fig. 29.6. Each gauge has a resistance of 100 Ω and a gauge factor of 2.1. The input voltage is 4 V. The stress is 200 M/N², calculate*

(*a*) *the change in the value of resistances* $R_4$ *to restore balance,*

(*b*) *the output voltage if the detector has a resistance of 10 kΩ,*

(*c*) *the current through the detector if its resistance is 400 Ω,*

(*d*) *current through the strain gauges.*

*The modulus or elasticity for Duralumin is 70 GN/m².*

**Solution.** (*a*) One of the gauges is subjected to tensile stress and the other to compressive stress. The change in the value of resistance $R_4$ to restore balance is :

$$\Delta R_4 = G_f\,\varepsilon\,R$$
$$= 2 \times 2.1 \times \frac{200 \times 10^6}{70 \times 10^8} \times 100 = 1.2\ \Omega$$

(*b*) Open circuit voltage across the bridge for two active strain gauges is :

$$e_0 = \frac{G_f \varepsilon}{2} e_i$$
$$= \frac{2.1 \times 200 \times 10^6}{70 \times 10^9} = \frac{4}{2}$$
$$= 0.012\,\text{V} = 12\ \text{mV}$$

Output voltage when the meter is connected across the output terminals is :

$$e_{0L} = \frac{1}{1 + R_0 / R_m} \times e_0$$
$$= \frac{1}{1 + 100/10 \times 10^3} \times 0.012\ \text{V}$$
$$= 11.9\ \text{mV}$$

(*c*) When a meter of 400 Ω is connected across the bridge output terminals, the output voltage is :

$$e_{0L} = \frac{1}{1 + 100/400} \times 0.112\ \text{V}$$
$$= 9.6\ \text{mV}$$

∴ Meter current

$$I_m = \frac{e_{0L}}{R_m} = \frac{9.6 \times 10^{-3}}{400}$$
$$= 24 \times 10^{-6}\ \text{A} = 24\ \mu\text{A}$$

(*d*) Current through strain gauges

$$= \frac{e_i}{2R} = \frac{4}{2 \times 100} = 0.02\ \text{A} = 20\ \text{mA}$$

**Example 29.9** *In order to measure strain in a cantilever beam, a single strain gauge of resistance 1 kΩ, and gauge factor 2 and a temperature co-efficient $10 \times 10^{-6}/°C$ is mounted on the beam and connected in one arm of a bridge circuit. The other three arms of the bridge have a resistance of 100 Ω each. The bridge detector resistance is 100 Ω and its sensitivity is 10 mm/μA.*

*(a) Calculate the detector deflection for 0.1 percent strain.*

*(b) Calculate the change in effective strain indicated when the room temperature increases by 10°C.*

**Solution.** (*a*) Change in strain $\varepsilon = 0.1\% = 0.001$

Change in resistance

$$\Delta R = 0.001 \times 2 \times 1000 = 2\,\Omega$$

∴ Resistance of gauge

$$= 1000 + 2 = 1002\,\Omega$$

Open circuit voltage of Thevenin generator :

$$e_0 = \left(\frac{1002}{1000+1002} - \frac{100}{100+100}\right) \times 10\text{ V}$$

$$= 5\text{ mV}$$

Resistance of Thevenin generator

$$R_0 = \frac{1002 \times 1000}{2002} + \frac{100 \times 100}{200}$$

$$= 550.5\,\Omega$$

∴ Current through detector

$$I_m = \frac{5 \times 10^{-3}}{550.5 + 100}\text{ A}$$

$$= 7.7\,\mu\text{A}$$

and deflection of galvanometer $= 7.7 \times 10 = 77$ mm

(*b*) Change in resistance due to change in temperature $\Delta R = 10 \times 10^{-6} \times 1000 \times 10 = 0.1\,\Omega$

The equivalent strain

$$\varepsilon = \frac{\Delta R}{G_f R} = \frac{0.1}{2 \times 1000}$$

$$= 50 \times 10^{-6} = 50\text{ microstrain.}$$

**Example 29.10** *A strain gauge is used for measurement of dynamic strain as shown in Fig. 29.10. The strain gauge has a resistance of 1 kΩ when unstrained. A resistance of 1 kΩ is connected in the arm adjacent to strain gauge. The resistances of other two arms are 10 kΩ each.*

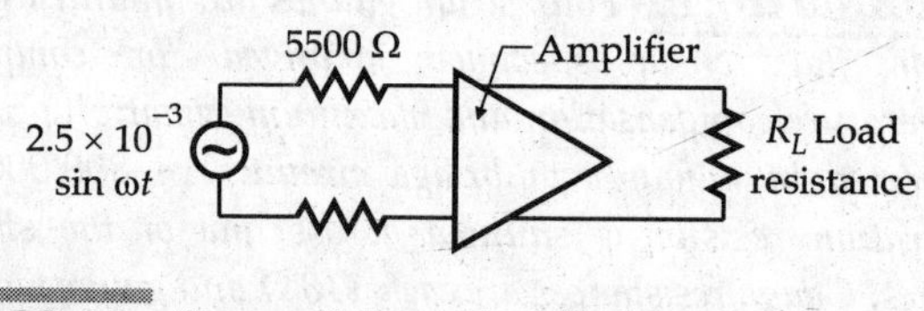


**Fig.29.11** Figure of Example 29.10.

*The strain gauge is bonded to structural member and is subjected to a sinusoidal time varying strain. As a result of this strain, the resistance value of strain gauge between extremes of 999 Ω and 1001 Ω. The bridge output is connected to an amplifier. Determine the equivalent circuit. The supply voltages is 10 V.*

**Solution.** Open circuit voltage of bridge under maximum tension :

$$= \left[\frac{1001}{1001+1000} + \frac{10{,}000}{10{,}000+10{,}000}\right] \times 10\text{ V}$$

$$= 2.5\text{ mV}$$

Open circuit voltage under maximum compression :

$$= \left[\frac{999}{999+1000} + \frac{10{,}000}{10{,}000+10{,}000}\right] \times 10\text{V}$$

$$= -2.5\text{ mV}$$

Therefore the output voltage alternates with an amplitude of 2.5 mV.

Thevenin source resistance is,

$$R_0 = \frac{1000 \times 1000}{2000} + \frac{10{,}000 \times 10{,}000}{20{,}000}$$

$$= 5500\,\Omega.$$

The equivalent circuit is shown in Fig. 29.11.

NOTE For More Solved and Unsolved Examples on Strain Gauges, see pages 777, 778, 782, 862.

## 29.9 STRAIN GAUGE CALIBRATION

A theoretical objection to the use of strain gauges is that, in most applications, it is impossible to check the accuracy of the readings obtained from them. Once the strain gauge is bonded in its measuring position, it cannot be removed or transferred and subjected to a known strain for calibration. Strain gauges are often used in applications where no other form of strain measurement is possible. Since no check on performance can be made, the value of the gauge factor as specified by the manufacturer has to be relied upon. However, this leads to errors because the value of gauge factor is specified with certain tolerance limits and hence each gauge should be individually calibrated.

Simple strain gauge bridges usually employ readout systems having scales marked in arbitrary units. The basic motive of practical calibration of strain gauges is to establish a relationship between arbitrary units and the measured strain. In fact, calibration of strain gauges involves making necessary adjustments to obtain correct scale reading.

29

The technique employed is to produce a known change in the resistance of one of the bridge arms to simulate a particular value of strain. The value of the change in resistance required for this purpose may be calculated if the unstrained resistance of the gauge, and the gauge factor, are known.

The procedure for calibration of strain gauges, involves a known change in the value of the resistance of the strain gauge by shunting it with a high value of resistance and then measuring the output corresponding to the known change of resistance. This is shown in Fig. 29.12. A shunt resistance $R_{sh}$, is used, the resistance of the gauge being $R_g$.

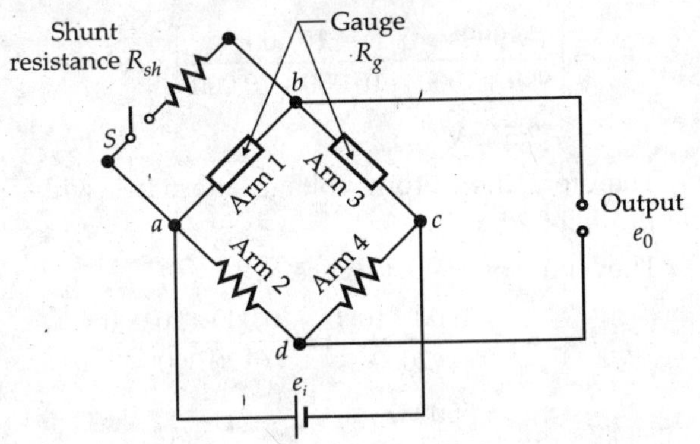


**Fig. 29.12** Calibration of strain gauges.

Resistance of arm 1 of bridge before switch $S$ is closed $= R_g$

Resistance of arm 1 of bridge after switch $S$ is closed

= Resistance of parallel combination of $R_g$ and $R_{sh}$

$$= \frac{R_g R_{sh}}{R_g + R_{sh}}$$

∴ The change in resistance of arm 1 resulting from introduction of shunt resistance $R_{sh}$ is :

$$\Delta R_g = R_g - \frac{R_g R_{sh}}{(R_g + R_{sh})}$$

$$= \frac{R_g^2}{(R_g + R_{sh})} \quad ...(29.16)$$

Equivalent strain

$$\varepsilon_e = \frac{1}{G_f} \cdot \frac{\Delta R_g}{R_g} = \frac{1}{G_f} \cdot \frac{1}{R_g} \cdot \frac{R_g^2}{(R_g + R_{sh})}$$

$$= \frac{R_g}{G_f(R_g + R_{sh})} \quad ...(29.17)$$

Shunt resistance

$$R_{sh} = \frac{R_g(1 - G_f \varepsilon_e)}{G_f \varepsilon_e} \quad ...(29.18)$$

In practice $G_f \varepsilon_e << 1$, and therefore the term $G_f \varepsilon_e$ in the numerator of Eqn. 29.18 can be neglected without introducing any significant error.

$$R_{sh} \simeq \frac{R_g}{G_f \varepsilon_e} \quad ...(29.19)$$

The value of effective strain $\varepsilon_e$ may be determined, exactly from Eqn. 29.17 or 29.18 or nearly exactly from Eqn. 29.19. If a linear relationship between the readout (output indicator system) and strain is assumed, only one value of $R_{sh}$ is required for the calibration. This value of $R_{sh}$ should be so chosen that the readout (output indicator) gives its full scale deflection.

The calibration of bridges using more than one active strain gauge may be done in the same manner. Suppose there are $n$ active strain gauge elements in the bridge, then the strain represented by a given value of $R_{sh}$ is $1/n$ (approximately) of the strain given by Eqn. 29.17.

∴ Effective strain

$$\varepsilon_e = \frac{R_g}{nG_f(R_g + R_{sh})} \quad ...(29.20)$$

**Example 29.11** *A strain gauge bridge comprises of two fixed 120 Ω resistors, and one is an active gauge and one unstrained temperature compensating gauge. The two gauges have an unstrained resistance value of 120 Ω each. The gauge factor is 2.2. Find the strain which would be represented by 100 kΩ calibration resistance shunting one arm of bridge.*

**Solution.** From Eqn. 29.17, the effective strain represented by the arrangement is :

$$\varepsilon_e = \frac{R_g}{G_f(R_g + R_{sh})}$$

$$= \frac{120}{2.2(120 + 100 \times 10^3)}$$

$$= 0.545 \times 10^{-3} = 545 \text{ microstrain}$$

**Example 29.12** *Four strain gauges are mounted on a simple flat tensile specimen arranged for complete temperature compensation and maximum sensitivity when connected in a four arm bridge circuit. An 800,000 Ω calibration resistor is shunted across one of the strain gauges. Gauge resistances are each 118 Ω and gauge factors are 1.22.*

*If the strain gauge readout is 150 divisions when the calibration switch is closed and 240 divisions when the load is applied, what is the axial strain in the specimen ?*

**Solution.** The bridge has four active elements.

∴ Effective strain

$$\varepsilon_e = \frac{R_g}{n\,G_f(R_g + R_{sh})}$$

$$= \frac{188}{4 \times 1.22\,(188 + 800{,}000)} = 48.1 \times 10^{-6}$$

When the calibration switch is closed, the readout gives a reading of 150 divisions.

∴ 1 division

$$= \frac{48.1}{150} \times 10^{-6} = 0.321 \times 10^{-6}$$

When the strain gauges' arrangement is loaded, the strain is :

$$\varepsilon = 0.321 \times 10^{-6} \times 240 = 77 \times 10^{-6} = 77 \text{ microstrain}$$

## 29.10 LOAD CELLS

Load cells utilise an elastic member as the primary transducer and strain gauges as secondary transducers as explained in Art. 25.3.6, page 742. Strain gauges may be attached to any elastic member (as shown in Fig. 29.13) on which there exists a, suitable plane area to accommodate them. This arrangement may then be used to measure loads applied to deform or deflect the member, provided that the resultant strain is large enough to produce detectible outputs. When the **strain gauge-elastic member** combination is used for weighing it is called a **load cell.**

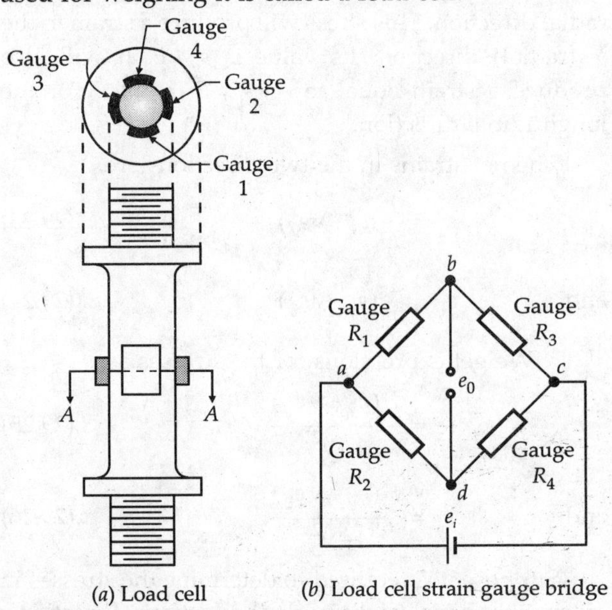


**Fig. 29.13** Tension cômpression resistance strain gauge cell.

### 29.10.1 Tensile-Compressive Cell

Figure 29.13 shows a tensile-compressive cell which is a cylinder. This arrangement uses four strain gauges each mounted at 90° to each other. The bridge circuit is also shown. Two of strain gauges experience tensile stress while the other two are subjected to compressive stress.

The output of a bridge with equal arms and using two strain gauges mounted 90° to each other is :

$$\Delta e_0 = (1+\nu)\left[\frac{\Delta R_1 / R}{4 + 2(\Delta R_1 / R)}\right] e_i \quad \text{...(29.21)}$$

In the present case, there are two sets of gauges mounted 90°C to each other, with one set experiencing tensile stress while the other a compressive stress.

∴ Output voltage of Wheatstone bridge :

$$\Delta e_0 = 2\,(1+\nu)\left[\frac{\Delta R_1 / R}{4 + 2(\Delta R_1 / R)}\right] e_i \quad \text{...(29.22)}$$

The above relationship is clear from the following explanation :

In the case of a cylinder, an axial compressive load causes a negative strain in the vertical gauges, and a positive strain in the circumferential gauges. The two strains are not equal in this case. These two strains. are related to each other by a factor ν, called the Poisson's ratio. Since the value of the strain is of no interest, the objective is to produce a force transducer. This force transducer is calibrated empirically. In case all the gauges are similar, temperature compensation is obtained as all the gauges contribute equally to unbalance the bridge. 29

Compression cells of this type have been used with a capacity of 300 MN.

Unbonded strain gauge transducers are frequently used for force measurement, particularly if high sensitivity is required. Highest sensitivity is obtained when the gauge itself performs the function of elastic member, *i.e.*, all the input force is used fort producing strain in the gauge.

**Example 29.13** *A load cell consists of a solid cylinder of steel 40 mm in diameter with four strain gauges bonded to it and connected into the four arms of a voltage sensitive bridge. The gauges are mounted to have Poisson's arrangement.*

*If the gauges are each of 100 Ω resistance and the gauge factor, 2.1, the bridge excitation voltage 6 V, determine the sensitivity of the cell in V/kN. Modulus of elasticity for steel is 200 GN/m² and the Poisson's ratio is 0.29.*

**Solution.** Suppose a load of 1 kN is applied to the load cell.

Stress $s = \dfrac{1 \times 10^3}{(\pi / 4)(40 \times 10^{-3})^2}$

$= 0.796 \times 10^6 \, N / m^2$

Strain $\varepsilon = \dfrac{s}{E}$

$= \dfrac{0.796 \times 10^6}{200 \times 10^9} = 3.98 \times 10^{-6}$

$\therefore \quad \dfrac{\Delta R}{R} = \varepsilon \, G_f$

$= 3.98 \times 10^{-6} \times 2.1 = 8.36 \times 10^{-6}$

From Eqn. 29.22, the voltage output of the bridge is :

$$\Delta e_0 = 2\,(1+\nu)\left[\frac{\Delta R_1 / R}{4 + 2(\Delta R_1 / R)}\right] e_i$$

$$= 2(1+0.29)\left[\frac{8.36 \times 10^{-6}}{4 + 2 \times 8.36 \times 10^{-6}}\right] \times 6$$

$$= 0.324 \times 10^{-6} \, V = 0.324 \, \mu V$$

Hence, the sensitivity is 0.324 μV / kN.

## 29.11 STRESS STRAIN RELATIONSHIPS

In this section cases of uniaxial and biaxial strain measurements are analyzed.

**1. *Simple Uni-axial stresses.*** This case arises where a simple tensile or compressive loading is applied to a member. The stress may be calculated from the observed strain using the modulus of elasticity, $E$, of the material used provided the operation is within the elastic limits.

The axial stress is, $s = \varepsilon E$

**2. *Bi-axial stresses.*** Many a times strain gauges are used in applications wherein the test piece is subjected to stress in more than one direction. If the test piece is on a free surface, as is usually the case, the condition is called as *bi-axial*. A typical example of this case exists on the outer surface or shell, of a cylindrical vessel. The *Hoop Stresses* acting circumferentially (in the radial direction) tend to open up to seam in the longitudinal direction. Thus there are longitudinal stresses which produce stresses in the radial direction. This is shown in Fig. 29.14.

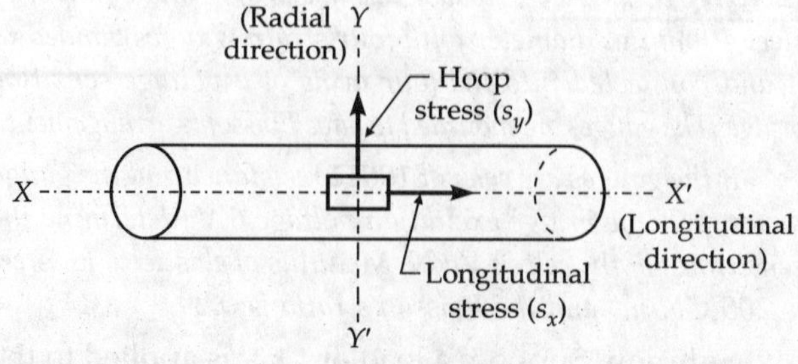


**Fig. 29.14** A free surface subjected to bi-axial stresses.

Let us consider a general element as shown in Fig. 29.14 that is subjected to bi-axial tresses. Suppose the stresses $s_x$ and $s_y$ are applied at the same time. This is shown in Fig. 29.15.

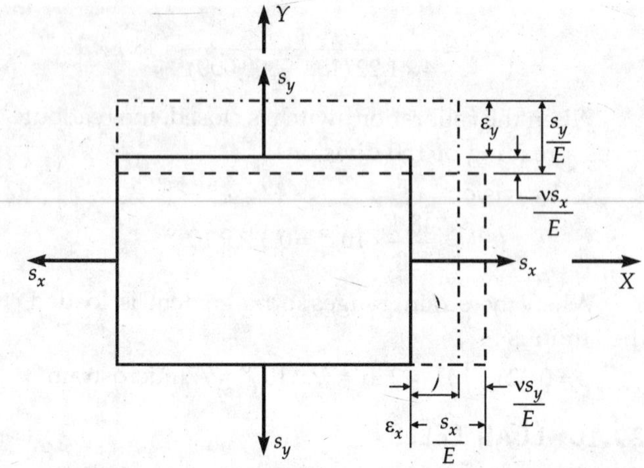


**Fig. 29.15** An element subjected to bi-axial stresses.

Suppose a stress $s_x$ is applied in the longitudinal direction. This will produce a strain $\varepsilon_x = s_x / E$ in the longitudinal or *X-direction.*

At the same time, because of the Poisson's ratio effect, there will be a stress produced in the radial direction or along *Y-axis. This produces a strain equal to* $-\nu s_x / E$.

Now suppose that a stress $s_y$ is applied in the radial direction. This stress will produce a strain in the $Y$ (radial) direction. Its value is $s_y / E$. It will also produce a strain equal to $-\nu\, s_y / E$ in the $X$ or the longitudinal direction.

The net strains in the two directions are,

$$\varepsilon_x = \frac{1}{E}(s_x - \nu s_y) \qquad ...(29.23)$$

and

$$\varepsilon_y = \frac{1}{E}(s_y - \nu s_x) \qquad ...(29.24)$$

$\therefore$ We get expressions for the stresses as,

$$s_x = \frac{E\,(\varepsilon_x + \nu \varepsilon_y)}{1 - \nu^2} \qquad ...(29.25)$$

and

$$s_y = \frac{E\,(\varepsilon_y + \nu \varepsilon_x)}{1 - \nu^2} \qquad ...(29.26)$$

Suppose it is required to determine the stresses in given orthogonal (right angled) directions. Equations. 29.25 and 29.26 show that in order to determine these stresses, it is necessary to know the modulus of elasticity *E*, Poisson's ratio ν and also the strains $\varepsilon_x$ and $\varepsilon_y$ in the *X* and *Y* directions. Even if the stress is desired to be known in one direction it becomes necessary to measure strains in both the directions.

The problem becomes more complicated if a third stress acts in the third orthogonal direction. Let the stress in the *Z* direction be $s_z$. The stress strain relationships are,

$$\varepsilon_x = \frac{1}{E}[s_x - \nu(s_y + s_z)] \quad ...(29.27)$$

$$\varepsilon_y = \frac{1}{E}[s_y - \nu(s_z + s_x)] \quad ...(29.28)$$

and
$$\varepsilon_z = \frac{1}{E}[s_z - \nu(s_x + s_y)] \quad ...(29.29)$$

**Example 29.14** *Two strain gauges attached to the surface of a cylindrical pressure vessel, one in axial and one in circumferential direction, gave the strain values of 0.00018 and 0.00072 respectively. Calculate the hoop and the longitudinal stress values if the cylinder is of steel having a, modulus of elasticity of 200 GN/m² and Poisson's ratio = 0.29.*

**Solution.** Strain in the longitudinal direction

$$\varepsilon_x = 0.00018$$

strain in the circumferential direction

$$\varepsilon_y = 0.00072$$

From Eqn. 29.26, the hoop stress is,

$$s_y = \frac{E(\varepsilon_y + \nu\varepsilon_x)}{1-\nu^2}$$

$$= \frac{200\times10^9(0.00072 + 0.29\times0.00018)}{[1-(0.29)^2]}\ \text{N/m}^2$$

$$= 168.6\ \text{MN/m}^2$$

From Eqn. 29.25, the longitudinal stress is,

$$s_x = \frac{E(\varepsilon_x + \nu\varepsilon_y)}{1-\nu^2}$$

$$= \frac{200\times10^9(0.00018 + 0.29\times0.00072)}{[1-(0.29)^2]}\ \text{N/m}^2$$

$$= 84.9\ \text{MN/m}^2$$

**Example 29.15** *A simple tension member having an area of 125 mm² is subjected to a load of 2700 kg. Strains of 1620 and – 455 microstrain are measured in the axial and tranverse directions respectively. Determine the value of Young's modulus and Poisson's ratio.*

**Solution.** Axial stress

$$s_x = \frac{2700\times9.81}{125\times10^{-6}} = 212\times10^6\ \text{N/m}^2$$

Transverse stress $s_y = 0$

Axial strain

$$\varepsilon_x = \frac{1}{E}(s_x - \nu s_y) = \frac{1}{E}s_x$$

or Modulus of elasticity

$$E = \frac{s_x}{\varepsilon_x}$$

$$= \frac{212\times10^6}{1620\times10^{-6}}$$

$$= 131\times10^9 = 131\ \text{GN/m}^2$$

Equation 29.24 gives the value of strain, the transverse direction

$$\varepsilon_y = \frac{1}{E}(s_y - \nu s_x)$$

$$= -\nu\frac{s_x}{E} \text{ as } s_y = 0$$

$$\therefore \quad -455\times10^{-6} = -\nu\times\frac{212\times10^6}{131\times10^9}$$

or Possion's ratio $\nu = 0.281$

## 29.12 STRAIN GAUGE ROSETTES

The installation of a strain gauge on a bar specimen shown in Fig. 29.16 is a useful application of the gauge, but it is restricted. The strain that is measured in such a situation is a principal strain since we assumed that the bar is operating under only a tensile load.

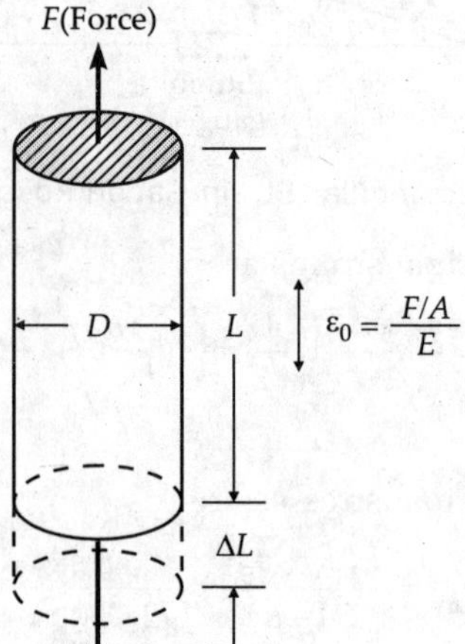


**Fig. 29.16** Simple bar under axial strain.

The axial strain is,

$$\varepsilon_a = \frac{F/A}{E} = \frac{s_a}{E} = \frac{\Delta L}{L}$$

where $\varepsilon_a$ and $s_a$ are respectively the axial strain and stress respectively.

29

The strain that is measured in situations as shown in Fig. 29.16, is a principal strain since we have assumed that the bar is operating under only a tensile load. Obviously, a more general measurement problem will involve strains in more than one direction, and the orientation of the principal stress axis will remain unknown. In a practical problem, an element may be subjected to stresses in any direction and hence it will not be possible to locate the direction of principal stress. Therefore, as a consequence, it will not be possible to orient the strain gauge along the direction of principal stress. Hence, there is a necessity to evolve a strain gauge measurement system which measures the values of principal strains and stresses, without actually knowing their directions. The solution to the problem lies in using three strain gauges *oriented in different directions* to form a unit is called a **Rosette**.

Consider the rosette shown in Fig. 29.17. The three strain gauges forming a rosette are oriented as shown, and the three strains measured by these gauges are $\varepsilon_1$, $\varepsilon_2$ and $\varepsilon_3$.

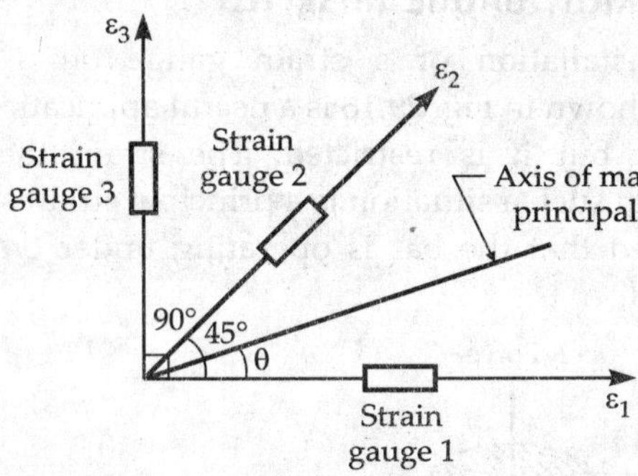


**Fig. 29.17** Rectangular Strain Gauge Rosette.

The principal strains are :

$$\varepsilon_{max}, \varepsilon_{min} = \frac{\varepsilon_1 + \varepsilon_3}{2} \pm \frac{1}{\sqrt{2}} \pm \left[(\varepsilon_1 - \varepsilon_2)^2 + (\varepsilon_2 - \varepsilon_3)^2\right]^{1/2} \quad ...(29.30)$$

The principal stresses are :

$$s_{max}, s_{min} = \frac{E(\varepsilon_1 + \varepsilon_3)}{2(1-\mu)} \pm \frac{E}{\sqrt{2}(1+\mu)} \pm \left[(\varepsilon_1 - \varepsilon_2)^2 + (\varepsilon_2 - \varepsilon_3)^2\right]^{1/2} \quad ...(29.31)$$

The maximum shear stress,

$$\tau_{max} = \frac{E}{\sqrt{2}(1+\mu)}\left[(\varepsilon_1 - \varepsilon_2)^2 + (\varepsilon_2 - \varepsilon_3)^2\right]^{1/2} \quad ...(29.32)$$

The principal stress axis is located with the angle θ is

$$\tan 2\theta = \frac{2\varepsilon_2 - \varepsilon_1 - \varepsilon_3}{\varepsilon_1 - \varepsilon_3} \quad ...(29.33)$$

This is the axis at which the maximum stress, $s_{max}$, occurs.

A problem arises with the determination of the quadrant for θ since there will be two values obtained from the solution of Eqn. 29.33. The angle will lie in the first quadrant $0 < \theta < \pi/2$, if,

$$\varepsilon_2 > \frac{\varepsilon_1 + \varepsilon_3}{2} \quad ...(29.34)$$

Another type of strain-gauge rosette in common use is the **Delta Rosette** shown in Fig. 29.18. The principal strains in this case are,

$$\varepsilon_{max}, \varepsilon_{min} = \frac{\varepsilon_1 + \varepsilon_2 + \varepsilon_3}{3} \pm \frac{\sqrt{2}}{3}\left[(\varepsilon_1 - \varepsilon_2)^2 + (\varepsilon_2 - \varepsilon_3)^2 + (\varepsilon_3 - \varepsilon_1)^2\right]^{1/2} \quad ...(29.35)$$

The principal stresses are,

$$s_{max}, s_{min} = \frac{E(\varepsilon_1 + \varepsilon_2 + \varepsilon_3)}{3(1-\mu)} \pm \frac{\sqrt{2}\,E}{3(1+\mu)}\left[(\varepsilon_1 - \varepsilon_2)^2 + (\varepsilon_2 - \varepsilon_3)^2 + (\varepsilon_3 - \varepsilon_1)^2\right]^{1/2} \quad ...(29.36)$$

The maximum shear stress is,

$$\tau_{max} = \frac{\sqrt{2}\,E}{3(1+\mu)}\left[(\varepsilon_1 - \varepsilon_2)^2 + (\varepsilon_2 - \varepsilon_3)^2 + (\varepsilon_3 - \varepsilon_1)^2\right]^{1/2} \quad ...(29.37)$$

The principal stress axis is located according to :

$$\tan 2\theta = \frac{\sqrt{3}(\varepsilon_3 - \varepsilon_2)}{2\varepsilon_1 - \varepsilon_3 - \varepsilon_3} \quad ...(29.38)$$

Angle θ will be in the first quadrant when $\varepsilon_3 > \varepsilon_2$ and in the second quadrant $\varepsilon_2 > \varepsilon_3$.

It is worthwhile to mention that the resistance strain gauges may be sensitive to transverse as well as to axial strains. The resistance change produced by a transverse strain, however, is usually less than 2 to 3 percent of the change produced by axial strain. For this reason, it may be neglected in many applications. If the transverse strain is to be considered, the above rosette formulae may be modified accordingly.

Rosettes are of many types. Figure 25.57 on page 779 shows different types of strain gauges and rosettes.

***Example 29.16*** *A rectangular rosette as shown in Fig. 29.18 is maintained on a steel plate having a modulus of elasticity $E = 200$ GN/m$^2$ and the Poisson's ratio is 0.3. The three strains measured are :*

$\varepsilon_1 = +500\ \mu$ *strain*, $\varepsilon_2 = +400\ \mu$ *strain*, $\varepsilon_3 = -100\ \mu$ *strain*

*Calculate the principal strains and stresses and the maximum shear stress. Locate the axis of principal stress.*

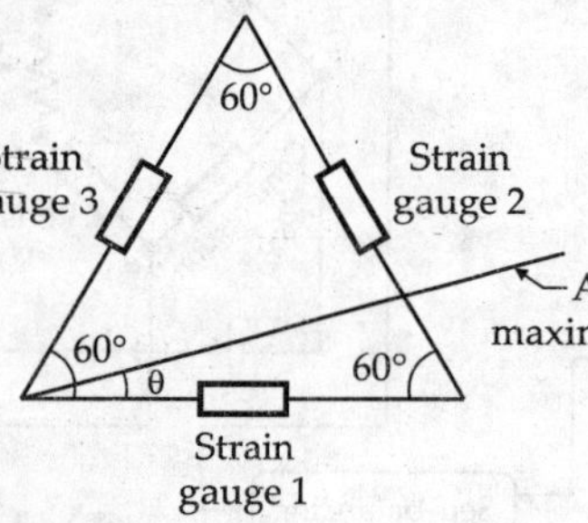


**Fig. 29.18** Delta Strain gauge rosette.

**Solution.** As an intermediate step, we calculate the quantities

$$A = \frac{\varepsilon_3 + \varepsilon_1}{2} = \frac{-100 + 500}{2} = 200\ \mu \text{ strain}$$

$$B = [(\varepsilon_1 - \varepsilon_2)^2 + (\varepsilon_2 - \varepsilon_3)^2]^{1/2}$$

$$= [(500 - 400)^2 + (400 + 100)^2]^{1/2} \times 10^{-6}$$

$$= 510\ \mu \text{ strain}$$

Then,

$$\varepsilon_{max} = A + \frac{B}{\sqrt{2}} = 200 + \frac{510}{\sqrt{2}} = 561\ \mu \text{ strain}$$

$$\varepsilon_{min} = A - \frac{B}{\sqrt{2}} = 200 - \frac{510}{\sqrt{2}} = -161\ \mu \text{ strain}$$

Maximum stress

$$s_{max} = \frac{EA}{1-\mu} + \frac{EB}{\sqrt{2}\ (1+\mu)}$$

$$= \frac{200 \times 10^9 \times 200 \times 10^{-6}}{1 - 0.3} + \frac{200 \times 10^9 \times 510 \times 10^{-6}}{\sqrt{2}\ (1 + 0.3)}$$

$$= 57.1 \times 10^6 + 55.4 \times 10^6 = 112.5\ \text{MN/m}^2$$

Minimum stress

$$s_{min} = 57.1 \times 10^6 - 55.4 \times 10^6$$

$$= 1.7\ \text{MN/m}^2$$

Maximum shear stress

$$\tau_{max} = \frac{EB}{\sqrt{2}\ (1+\mu)} = 55.4\ \text{MN/m}^2$$

Now, we have,

$$\tan 2\theta = \frac{2\,\varepsilon_2 - \varepsilon_1 - \varepsilon_3}{\varepsilon_1 - \varepsilon_3}$$

$$= \frac{2 \times 400 - 500 - (-100)}{500 - (-100)}$$

$\therefore \quad 2\theta = 33.7°$ or $213.7°$

or $\quad \theta = 16.8°$ or $106.8°$

We choose the first quadrant angle $\theta = 16.8°$ in accordance with Eqn. 29.34.

**Example 29.17** *The maximum and minimum strains in a rectangular rosette are given by*

$$\varepsilon_{max}, \varepsilon_{min} = \frac{\varepsilon_1 + \varepsilon_3}{2} \pm \frac{1}{\sqrt{2}}\left[(\varepsilon_1 - \varepsilon_2)^2 + (\varepsilon_2 - \varepsilon_3)^2\right]^{1/2}$$

*and then is located at an angle,*

$$\tan 2\theta = \frac{2\,\varepsilon_2 - \varepsilon_1 - \varepsilon_3}{\varepsilon_1 - \varepsilon_3}$$

*Show the* $\varepsilon_{max} = \varepsilon_1$ *and* $\varepsilon_{min} = \varepsilon_3$ *when* $\theta = 0$

**Solution.** When $\theta = 0$, $\tan 2\theta = 0$

and hence,

$$\frac{2\,\varepsilon_2 - \varepsilon_1 - \varepsilon_3}{\varepsilon_1 - \varepsilon_3} = 0$$

or $\quad \varepsilon_2 = \dfrac{\varepsilon_1 + \varepsilon_3}{2}$

$$\therefore \quad \varepsilon_{max} = \frac{\varepsilon_1 + \varepsilon_3}{2} + \frac{1}{\sqrt{2}}\left[(\varepsilon_1 - \varepsilon_2)^2 + (\varepsilon_2 - \varepsilon_3)^2\right]^{1/2}$$

Now $\quad \varepsilon_2 = \dfrac{\varepsilon_1 + \varepsilon_3}{2}$

Also $\quad 2\,\varepsilon_2 = \varepsilon_1 - \varepsilon_3$

or $\quad \varepsilon_1 - \varepsilon_2 = \varepsilon_2 + \varepsilon_3$

$\therefore$ Therefore maximum strain

$$\varepsilon_{max} = \frac{\varepsilon_1 + \varepsilon_3}{2} + \frac{1}{\sqrt{2}}\left[(\varepsilon_1 - \varepsilon_2)^2 + (\varepsilon_2 + \varepsilon_3)^2\right]^{1/2}$$

$$= \varepsilon_2 + \frac{1}{\sqrt{2}}\left[(\varepsilon_1 - \varepsilon_2)^2 + (\varepsilon_1 - \varepsilon_2)^2\right]^{1/2}$$

$$= \varepsilon_2 + \frac{\sqrt{2}}{\sqrt{2}}(\varepsilon_1 - \varepsilon_2) = \varepsilon_1$$

The value of the minimum strain can be obtained in the same way

$$\varepsilon_{min} = \varepsilon_3.$$

## 29.13 STRAIN GAUGE CIRCUITRY

There are many strain gauge instrumentation circuits available commercially. The type instrumentation system depends upon the following considerations :

(*i*) whether static or dynamic strains have to be measured,

(*ii*) whether an analog or a digital readout is desired,

(*iii*) whether output is to be displayed on a recorder for instantaneous reading or for permanent record,

(*iv*) whether a single channel is required or multichannel measurement is necessary.

Depending upon the requirement a suitable instrumentation may be selected.

The most commonly used circuit for strain measurements is a Wheatstone bridge with d.c. excitation. The strain may be measured with the help of a voltmeter connected across the output terminals. The scale of the voltmeter may be calibrated in terms of strain. The voltmeter may be called a **"strain indicator"**. The voltmeter may be a d.c. analog meter and digital voltmeter. Since the output is low, it may have to be amplified before being fed to the voltmeter. This requires the use of a d.c. amplifier which has its inherent problems.

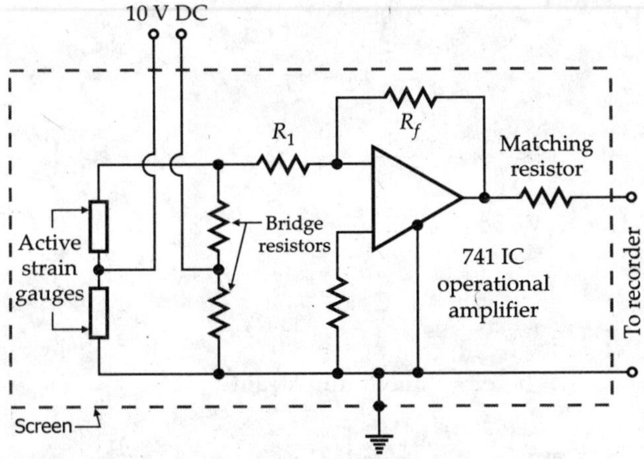


**Fig. 29.19** Strain gauge bridge circuit with IC operational amplifier.

It has been found that by using IC operational amplifiers placed in the proximity of a strain gauge bridge using two active gauges, inexpensive load cells of various sensitivities may be constructed with sufficient output to actuate galvanometric recorders (ultraviolet recorders are normally used). This arrangement is shown in Fig. 29.19. However, this arrangement involves screening and earthing to ensure the removal of power line (50 Hz) pickup.

Figure 29.20 shows a highly sophisticated system which employs four active strain gauge connected in a bridge circuit. The galvanometer of the usual bridge circuit is replaced by a self-balancing potentiometer. A low resistance slide wire is connected between $R_{g3}$ and $R_{g4}$ to adjust the bridge so that the self-balancing potentiometer reads zero when no strain is applied.

The output device of a self-balancing potentiometer could be an indicator, a digital voltmeter, a strip chart recorder for recording purposes. The output may also be indicated or recorded by numerous combinations of the above readout systems.

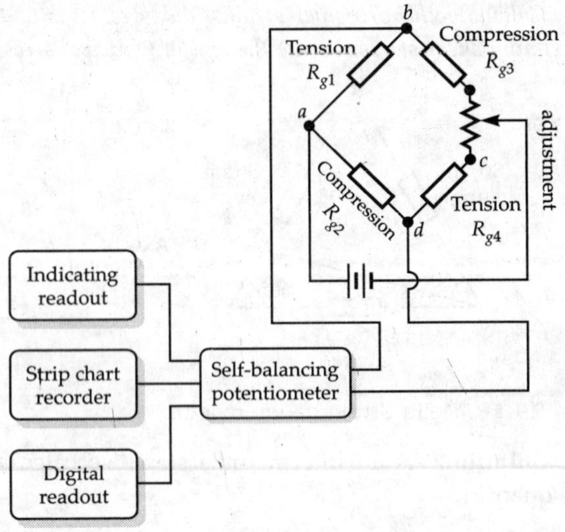


**Fig. 29.20** Strain gauge measurement using different types of readout devices.

Commercial strain gauge bridges normally use an a.c. power supply and subsequently demodulate and amplify the output so that greater sensitivity could be achieved. The normal excitation frequency is 1000 Hz, providing frequency response up to 100 Hz. Since the normal frequency response is about 10% of the carrier frequency, problems requiring better response will necessitate higher carrier frequencies. Instruments utilizing 20 kHz excitation frequency are commercially available (Fig. 29.21). They usually have their own recording oscillograph with provision for viewing the waveform on an external oscilloscope display.

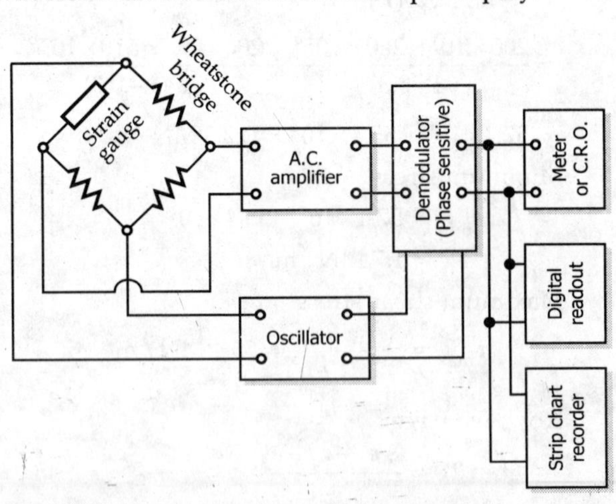


**Fig. 29.21** Strain gauge circuitry for dynamic measurement and recording.

The output may be fed to a strip chart recorder to record the variations. A galvanometer type recorder provided with mirror arrangement is used with moving photographic paper or film for permanent and reproducible records.

### Multipoint Measurements

In many applications of strain gauges, it is desirable to measure strain at more than one point. This necessitates the use of a number of strain gauges which may be arranged so that either they have a bridge circuit and an a display device each. Instead an arrangement where in only one bridge circuit and a single display device are used which are connected to each strain gauge circuit in turn through a switch. The latter method is very much less expensive. However, this method has some disadvantages like introduction of errors due to switch contact resistance, and different zero settings for each strain gauge circuit.

The strain gauge circuits may use one, two or four active strain gauges and suitable arrangements must be developed whereby each of these situations may be incorporated in a multipoint measuring system.

Figure 29.22(*a*) shows a system where there is only one active strain gauge in each circuit while Fig. 29.22(*b*) shows an arrangement for four active gauges in each circuit.

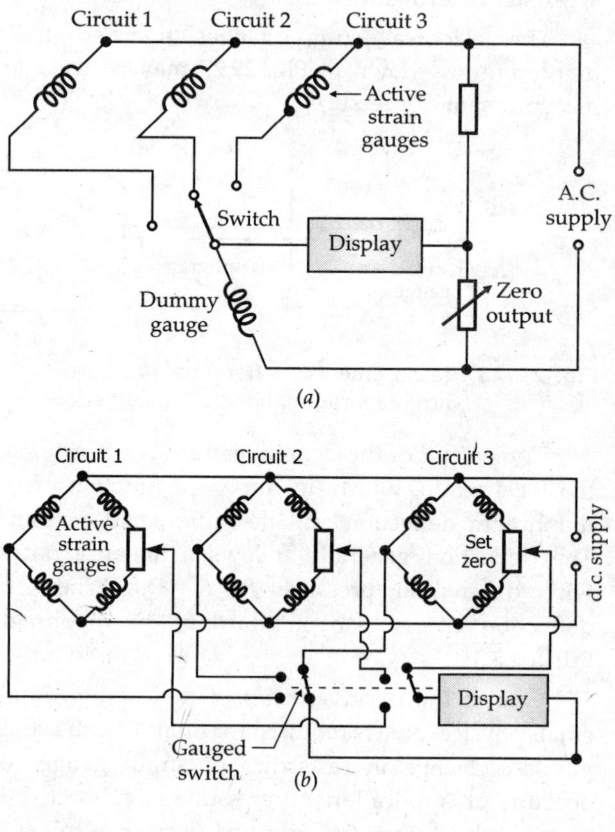


**Fig. 29.22** Multipoint arrangements.
(*a*) Single active gauge arrangement ;
(*b*) four active gauge arrangement.

## 29.14 USES OF STRAIN GAUGES

1. Strain gauges are used extensively for analysing the dynamic strains in complex structures such as the stress and the strain in bridges, automobiles, roads etc. where the output of the bridge can be recorded with an oscillograph calibrated to read the instantaneous strain or stress.
2. Resistance strain gauges are very useful for measurement of tension, torque, force, stresses in structures. They are particularly useful where a minimum disturbance by the presence of transducer is required as in wind tunnel measurements.
3. Strain gauges are used in load cells and proving rings to measure force by the strain produced. The proving ring is usually made of steel. The strain produced in the load ring is measured with a strain gauge. The strain is a measure of the force applied to the ring.

## 29.15 MEASUREMENT OF PRESSURE

Pressure measurements are one of the most important measurements made in industry. The number of instruments used are by far greater in number than the instruments used for any other type of measurement.

The principles used in the measurement of pressure are also, applied in the measurement of temperature, flow, and liquid levels. Thus it is essential to have complete knowledge of working principles of these instruments.

Pressure is represented as force per unit area. As such it may be considered as a type of stress since stress is also defined as force per unit area. In this section, the term 'pressure' refers to the force per unit area exerted by a fluid on a containing wall. Therefore, the discussion of pressure measurement is limited to fluid systems. The forces that arise as a result of strains in solids are designated as stresses and have been discussed earlier.

## 29.16 TYPES OF PRESSURE MEASUREMENT DEVICES

In industrial applications the pressure is usually measured by means of indicating gauges or recorders. These instruments may be mechanical, electro-mechanical, electrical or electronic in operation.

(*i*) ***Mechanical Instruments.*** These instruments may be classified into two groups. The first group includes those instruments in which the pressure measurement is made by balancing an unknown force with a known force. Instruments using this principle include manometers and ring and belt type of gauges. The second group includes those instruments which employ the balancing of unknown force through a force produced on or known area due to stress in an elastic member. This group employs quantitative deformation of an elastic member for measurement of pressure. The elastic members used are Bourdor tubes, bellows and diaphragms.

(*ii*) ***Electro-Mechanical Instruments.*** These instruments usually employ a mechanical means for detecting the pressure and electrical means for indicating or recording the detected pressure.

(*iii*) ***Electronic Instruments.*** Electronic pressure measuring instruments normally depend on some physical change that can be detected and indicated or recorded electronically.

In this text, the discussion is limited only to Electro-mechanical and Electronic instruments.

## 29.17 MEASUREMENT OF PRESSURE USING ELECTRICAL TRANSDUCERS AS SECONDARY TRANSDUCERS

The measurement of force or pressure can be done by converting the applied force or pressure into a displacement by elastic elements which act as primary transducers. This displacement, which is a function of pressure may be measured by transducers which act as secondary transducers. The output of the secondary transducers is a function of displacement, which in turn is a function of pressure. Mechanical methods have to be used to convert the applied force or pressure into displacement. These devices are called **Force Summing Devices.**

They possess elasticity, when deformed, the stresses established in the summing devices establish equilibrium with the applied pressure. The choice and design of the type of summing element used depends on the magnitude of force or pressure to be measured.

The most commonly used summing devices are :

- Flat or corrugated diaphragms.
- Bellows.
- Circular or twisted Bourdon tube.
- Straight tube.
- Single or double mass cantilever suspension.
- Pivot torque.

Examples of these force summing devices are shown in Fig. 25.4 on page 742. For description of force summing devices like Bourdon tube, bellows and diagphrams see pages 742 – 749. Pressure transducers generally use one of the first four types of force summing members. While mass cantilever suspension and pivot torque types are found in accelerometers and velocity transducers.

**Secondary transducers.** The displacement created by the action of the force summing member is converted into a change of some electrical parameter. The force summing member actuates a transducer which converts the displacement into an output of electrical format. The various transducers used are of the following types :

- Resistive,
- Inductive,
- Differential transformers,
- Capacitive,
- Photo-electric,
- Piezo-electric,
- Ionization, and
- Oscillation.

### 1. Resistive Transducers

The electrical strain gauges attached to a diaphragm as shown in Fig. 29.23 may be used for measurement of pressure.

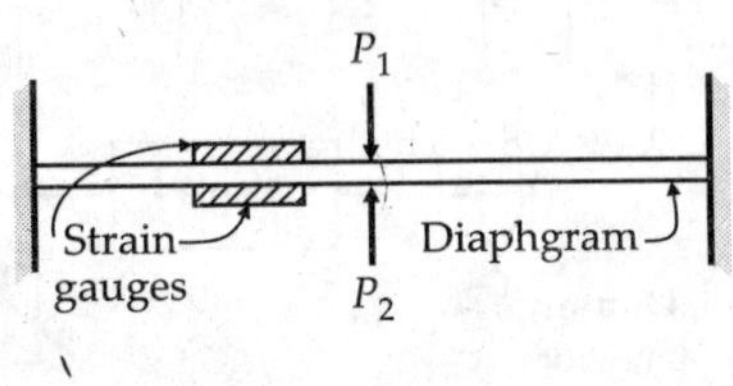


**Fig. 29.23** Measurement of differential pressure with diaphragm and strain gauges.

The output of these strain gauges is a function of the local strain, which, in turn, is a function of the diaphragm deflection and the differential pressure. The deflection generally follows a linear variation with differential pressure $P = P_2 - P_1$. (When the deflection is less than one-third of the diaphragm thickness.)

One of the disadvantages of the method is the small physical area is required for mounting the strain gauges. Change in resistance of strain gauges on account of application of pressure is calibrated in terms of the differential pressure. Gauges of this type are made in sizes having a lower range of : 100 kN/m$^2$ to 3 MN/m$^2$ to an upper range of 100 kN/m$^2$ to 100 MN/m$^2$.

## 2. Inductive Transducers

Inductive transducers have been successfully used as secondary transducers along with a diaphragm for measurement of pressure. Figure 29.24 shows an arrangement which uses two coils ; an upper and a lower coil which form the two arms of an a.c. bridge. The coils have equal number of turns. The other two arms of the bridge are formed by two equal resistances each of value $R$.

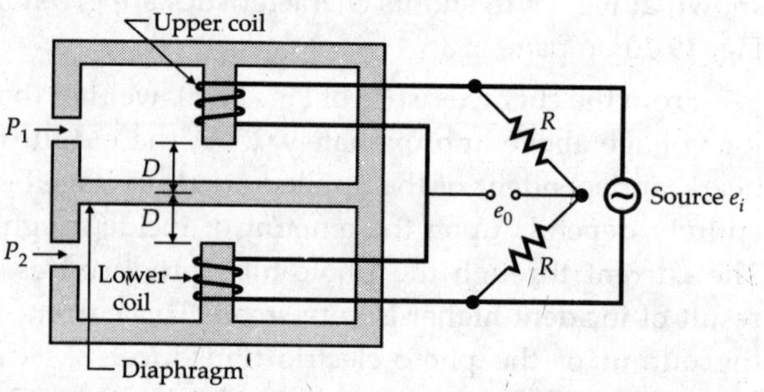


**Fig. 29.24** Measurement of pressure with diaphragm element and inductive transducer.

The diaphragm is symmetrically placed with respect to the coils and so when $P_1 = P_2$, the reluctances of the paths of magnetic flux for both the coils are equal and hence the inductances of the coils are equal.

Now initial self-inductance $= N^2 / \mathbf{R}_0$

where $N$ = number of turns,

and $\mathbf{R}_0$ = initial reluctance of flux path

Under this condition the bridge is balanced and the output, $e_0$, of the bridge is zero.

Suppose $P_2$ is greater than $P_1$ and therefore the differential pressure $P = P_2 - P_1$, deflects the diaphragm upwards through a distance $d$. For small displacements of diaphragm, the reluctance of the flux path of the upper coil is $\mathbf{R}_1 = \mathbf{R}_0 + K(D-d)$ and that of the lower coil is $\mathbf{R}_2 = \mathbf{R}_0 + K(D+d)$.

Hence, the inductance of the upper coil

$$L_1 = N^2 / \mathbf{R}_1 = N^2 / [\mathbf{R}_0 + K(D-d)]$$

and that of lower coil is

$$L_2 = N^2 / \mathbf{R}_2 = N^2 / [\mathbf{R}_0 + K(D+d)].$$

The bridge becomes unbalanced and the approximate value of output voltage is given by :

$$e_0 = \left[\frac{1}{2} - \frac{L_2}{L_1 + L_2}\right] e_i$$

$$= \left\{\frac{1}{2} - \frac{N^2 / [\mathbf{R}_0 + K(D+d)]}{N^2 / [\mathbf{R}_0 + K(D-d)] + N^2 / [\mathbf{R}_0 + K(D-d)]}\right\} \times e_i$$

$$= \frac{Kd \times e_i}{2(\mathbf{R}_0 + KD)} \quad \text{...(29.39)}$$

Since $K$, $\mathbf{R}_0$, $D$ and $e_i$ are constant, the output voltage is directly proportional to displacement $d$, of the diaphragm. Displacement $d$, is directly proportional to differential pressure $P = P_2 - P_1$. Hence the output voltage $e_0$ may be calibrated in terms of the differential pressure $P$.

It is seen that there exists a linear relationship between output voltage $e_0$ and the differential pressure for small values of deflection $d$.

It is possible to determine whether $P_2 > P_1$ or $P_1 > P_2$ with reference to the phase of output voltage, $e_0$, with respect to source voltage $e_i$.

Standard laboratory equipment, such as an electronic voltmeter, an oscilloscope, as well as recorders may be used for display and recording of the output. The gauge may be used for pressures ranging from $0 - 5\ \text{kN/m}^2$ to $0 - 100\ \text{kN/m}^2$.

## 3. LVDT

The LVDT is used as a secondary transducer for measurement of pressure with Bellows or Bourdon tube acting as the transducer *i.e.*, as a force summing device. The two arrangements are shown in Figs. 29.25 and 29.26. The pressure is converted into displacement which is sensed by the LVDT and transduced into a voltage.

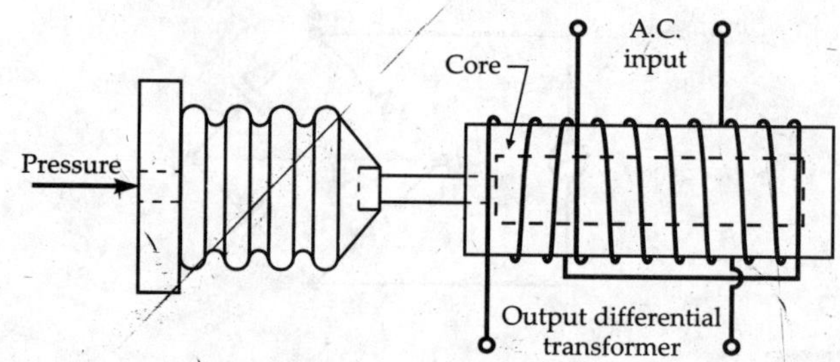


**Fig. 29.25** Measurement of pressure with Bellows and LVDT.

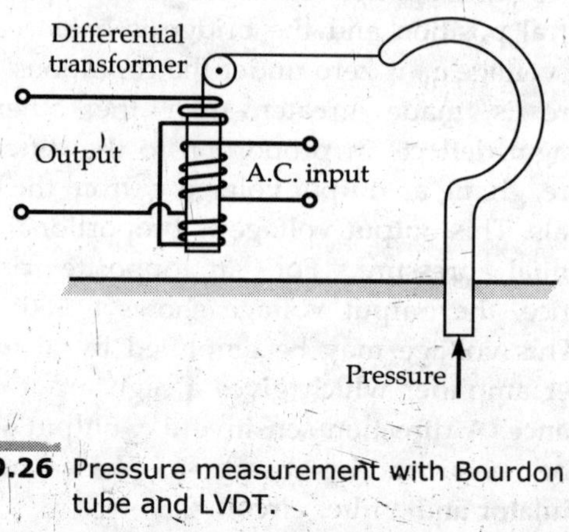


**Fig. 29.26** Pressure measurement with Bourdon tube and LVDT.

29

## 4. Capacitive Transducers

Capacitive transducers are used for measurement of pressure by converting the pressure into a displacement. The displacement is sensed by a capacitive transducer using a differential arrangement as explained in Art. 25.28.3 on page 818. This arrangement gives a linear relationship between output voltage and displacement (and hence pressure) as shown in Eqn. 25.111.

The use of a three terminal variable differential circuit capacitor is shown in Fig. 29.27. Spherical depressions of a depth of about 0.025 mm are ground into the glass discs. These depressions are coated with gold to form the two fixed plates of the differential capacitor. A thin stainless steel diaphragm is clamped between the discs, acts as the movable plate. With equal pressures applied (*i.e.*, $P_1 = P_2$) the diaphragm is in neutral position and the bridge is balanced. The output voltage $e_0$, is zero under the conditions. If one pressure is made greater than the other, the diaphragm deflects in proportion to the differential pressure, giving an output voltage, $e_0$ from the bridge terminals. This output voltage is proportional to the differential pressure. For an opposite pressure difference, the output voltage shows a 180° phase shift. This voltage may be amplified by an emitter follower amplifier which gives a high input output impedance. A direction sensitive d.c. output voltage may be obtained by using a phase sensitive demodulator and a filter circuit.

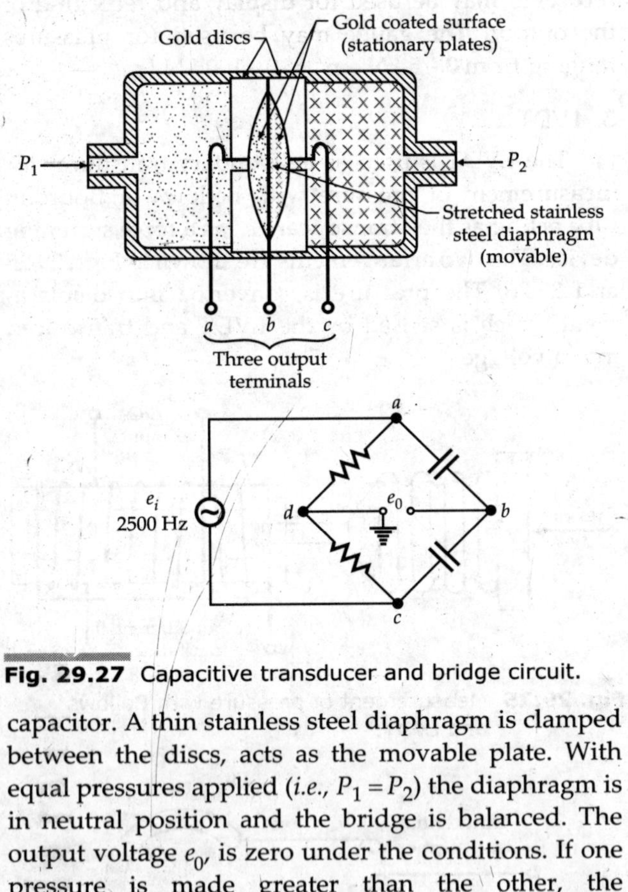


**Fig. 29.27** Capacitive transducer and bridge circuit.

The use of capacitive transducers is not common because of low sensitivity. Also capacitive transducers require high carrier frequencies (typically 2500 Hz) for dynamic pressure measurements.

## 5. Photoelectric Transducers

The photo-electric transducer makes use of the properties of a photo-emissive cell or photo-tube. A photo tube is described in details on page 583. It is shown at Fig. 19.18 and its characteristics are given in Fig. 19.20 on page 583.

From the characteristics of Fig. 19.20, we find that for voltage above approximately 20 V, the output is nearly independent of the applied anode voltage but entirely depends upon the amount of incident light. The current through the photo-tube produced as a result of incident higher is very small. This current is the output of the photo-electric transducer. As the current is small (of the order of a few μA), it must be amplified to provide a usable output.

The photo-electric transducer of Fig. 19.28 uses a photo-tube and a light source separated by a small window, whose aperture is controlled by the force summing member of the pressure transducer.

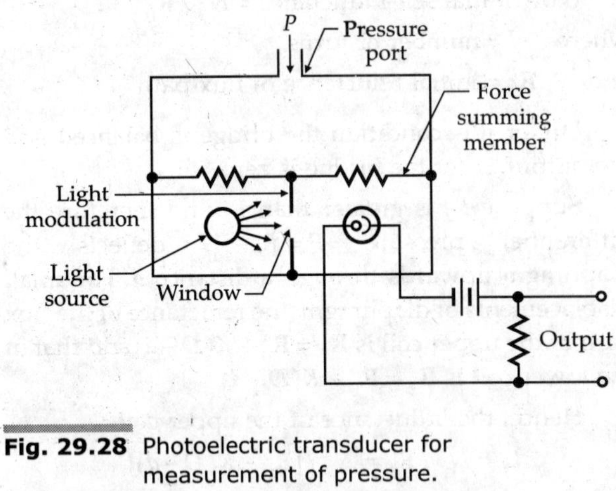


**Fig. 29.28** Photoelectric transducer for measurement of pressure.

The displacement of the force summing member modulates quantity of incident light falling on the photo-tube (since the applied pressure or force changes the position of the force summing member which in turn changes the position of the window thus causing a change in incident light). According to curves given in Fig. 19.20 page 583 a change in light intensity varies the photo-emissive properties at a rate approximately linear with displacement. This transducer can use either a stable source of light or an a.c. modulated light.

***Advantages*** :

(*i*) It has a high efficiency.

(*ii*) It can be used for both static and dynamic conditions but it does not respond to high frequency light variations.

***Disadvantages*** :

(*i*) It has a poor long term stability.

(*ii*) It requires a large displacement of the force summing member in order to produce a detectable output.

### 6. Piezoelectric Transducers

Piezoelectric crystals produce an emf when they are deformed. The force or displacement or pressure to be measured is applied to the crystal. The pressure is applied to the crystal through a force summing member. This causes a deformation which produces an emf that is a function of the deformation. This output emf may be measured to know the value of applied force and hence the pressure.

***Advantages*** :

(*i*) This transducer needs no external power and is therefore self-generating (active types).

(*ii*) It has a very good high-frequency response.

***Disadvantages*** :

(*i*) The principal disadvantage is that this transducer cannot measure static pressure.

(*ii*) The output of the transducer is affected by changes in temperature. Therefore temperature compensating devices have to be used.

### 7. Oscillation Transducers

These transducers use a force summing member to change the capacitance, *C*, or inductance, *L*, of an *LC* oscillator circuit. Figure 29.29 shows the basic elements of *LC* transistor oscillator whose output frequency is affected by a change in the inductance of a coil. The change in inductance is caused by the force summing member acting upon an inductive device.

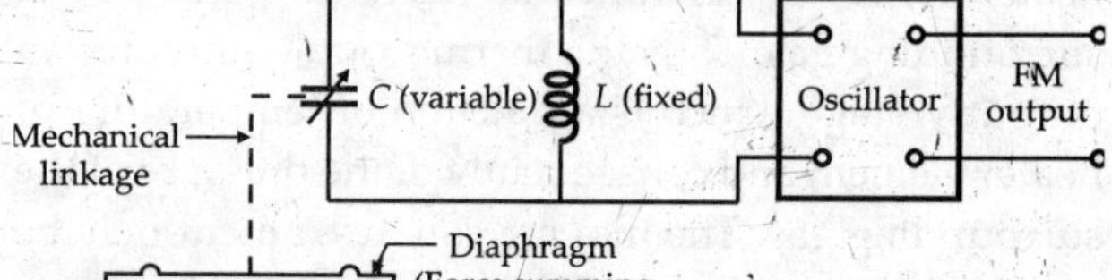


**Fig. 29.29** Basic elements of an oscillation transducer.

The output of oscillator is a modulated output and can be demodulated and calibrated in terms of pressure or force applied.

***Advantages*** :

(*i*) This transducer measures both dynamic and static phenomena.

(*ii*) This transducer is very useful for telemetry applications.

***Disadvantages*** :

(*i*) This transducer has a very hunted temperature range.

(*ii*) It has poor thermal stability.

(*iii*) It has a low accuracy and therefore is used only in low accuracy applications.

***Example 29.18*** *A piezoelectric pressure transducer gives an output of 1 mV per 100 kN/m$^2$. It feeds into an amplifier which may be considered as having negligible output impedance and a voltage gain of 25. The amplifier output is taken to an oscilloscope having calibrated vertical sensitivities of 1, 5, 20, 100 and 500 mV/mm. The viewing area of the oscilloscope screen is 100 mm × 100 mm. The transducer is used to measure a pressure fluctuating at a frequency of 50 Hz between upper and lower level of about 700 kN/m$^2$ and 100 kN/m$^2$ respectively. Assuming that the trace seen on the screen is steady and the peak to peak amplitude may be estimated correctly within ± 1 mm, what is*

*(a) optimum setting of the vertical sensitivity, and*

*(b) the likely maximum reading error in a measurement of the peak amplitude of wave in kN/m$^2$.*

**Solution.** (*a*) The two extreme pressures of 700 kN/m$^2$ and 100 kN/m$^2$ give the outputs as

$$700 \times (1/100) \times 25 = 175 \text{ mV}$$

and $$100 \times (1/100) \times 25 = 50 \text{ mV}$$

The corresponding peak to peak voltages are :

$$2 \times 175 = 350 \text{ mV}$$

and $$2 \times 25 = 50 \text{ mV}$$

Suppose we use the sensitivity of 1 mV/mm

∴ Deflection on the screen corresponding to maximum pressure $= 350/1 = 350$ mm.

Since the length of the screen is 100 mm, the waveform is out of range and hence the 1 mV/mm should not be used. Let us try 5 mV/mm setting.

The peak to peak deflection $= 350/5 = 70$ mm

This is within the range of the screen of the scope.

If we use 20, 100 or 500 mV/mm, no doubt, the pattern will be visible on the screen but the sensitivity will be lower as the deflections are lower. Hence the setting 5 mV/mm is the optimum setting.

(*b*) A ± 1 mm error corresponds to an input peak to peak pressure of

$$\pm\left[1\,\text{mm} \times \frac{5\,\text{mV}}{\text{mm}} \times \frac{1}{25} \times \frac{100\,\text{kN/m}^2}{1\,\text{mV}}\right] = \pm 20\ \text{kN/m}^2$$

## 29.18 MEASUREMENT OF LOW PRESSURE (VACUUM GAUGES)

These pressure gauges are used primarily for measuring pressure below atmospheric pressure, which is often referred to as vacuum. The science of low pressure measurement is rather a specialized field which requires considerable care on the part of the experimentalist. The wide range of pressures to be measured under the general heading of vacuum measurement makes the problem an extremely extensive one. This range extends from the normal atmospheric pressure of 760 mm of mercury column down to $10^{-8}$ mm of mercury column.

A common unit of low pressure is the **micron,** which is one-millionth of a meter (0.001 mm) of mercury column. Very low pressure may be defined as any pressure below 1 mm of mercury and an ultra low pressure is a pressure is less than a millimicron ($10^{-3}$ micron) or $10^{-6}$ mm of mercury. Torr is 1 mm of mercury column and therefore a millimicron is 1 micro torr.

There are two basic methods for measurement of low pressure :

**1. *Direct methods.*** The direct methods of measurement involve measurement of a displacement as a result of application of the pressure.

**2. *Indirect or inferential methods.*** These methods involve the measurement of pressure through measurement of certain other properties which depend upon the pressure to be measured. The pressure controlled properties which may be utilized for measurement of pressure are volume and thermal conductivity etc.

Pressure is measured by direct methods using spiral Bourdon tubes, flat and corrugated diaphragms, capsules and various forms of manometers. These devices have been discussed earlier. They need not be discussed further except to state that these devices are useful for measurement of pressures down to about 10 mm of mercury column. For measurement of pressures below this value, indirect methods are used.

Indirect or inferential methods for pressure measurement are used which involve a property that is pressure dependent and is examined for the measurement of pressure. The property selected must be sensitive to change in pressure in the ranges required.

### 29.18.1 Thermocouple Vacuum Gauge

This gauge operates on the principle that at low pressures the thermal conductivity of a gas is a function of pressure. A thermocouple vacuum gauge consists of a heater element (heated a temperature of 50° to 400°C by a known constant current) having a thermocouple in contact with its centre shown in Fig. 29.30.

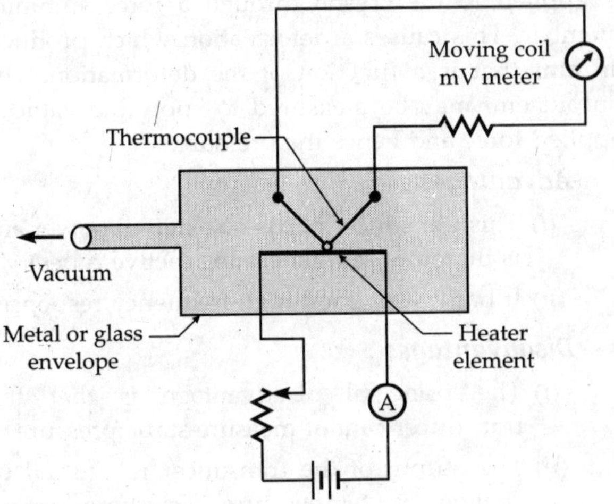


**Fig. 29.30** Thermocouple vacuum gauge.

The heater element and thermocouple are enclosed in a glass or metal envelope which is valued into the vacuum system. The heater element is heated by a constant current and its temperature depends upon the amount of heat which is lost to the surroundings by conduction and convection. At pressures below $10^{-3}$ mm Hg, the temperature of the heater wire is a function of the pressure of surrounding gas. Thus the thermocouple provides an output voltage which is a function of temperature of heater element and consequently of the pressure of the surrounding gas. The moving coil instrument may be directly calibrated to read the pressure.

The advantages of the thermocouple gauges are that they are inexpensive. However, they are subject to burnout if exposed at rugged and atmospheric pressure when hot and the current is flowing.

### 29.18.2 Pirani Gauges

The operation of a Pirani gauge depends on variation of the thermal conductivity of a gas with pressure. For pressures down to about 1 mm Hg the thermal conductivity is independent of pressure, but below this an approximately linear relationship exists between pressure and the thermal conductivity. At very low pressures the amount of heat conducted becomes very small and the method cannot be used.

The thermal conductivity of the gas is measured by detecting the amount of heat lost from an electrically heated wire placed in the gas. Heat is conducted from wire by conduction through the gas and greater the thermal conductivity of the gas, the lower will be the temperature of the heater wire. Now since electrical resistance varies with temperature, the resistance of the heater wire is a measure of the pressure. A Wheatstone bridge as shown in Fig. 29.31

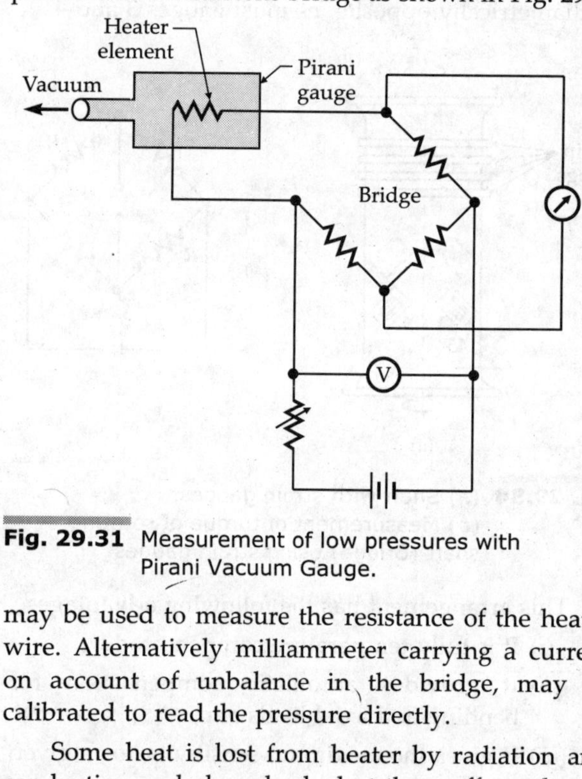


**Fig. 29.31** Measurement of low pressures with Pirani Vacuum Gauge.

may be used to measure the resistance of the heater wire. Alternatively milliammeter carrying a current on account of unbalance in the bridge, may be calibrated to read the pressure directly.

Some heat is lost from heater by radiation and conduction and along leads, but these effects do not depend on pressure or on the presence of gas. The compensation fo: this effect may be carried out by introducing a similar Pirani element in an opposite arm of bridge. This second element is enclosed in a sealed container evacuated to a very low pressure.

Pirani gauge is useful for pressures ranging from $10^{-1}$ to $10^{-3}$ mm of Hg. Pirani gauges are rugged, inexpensive and usually more accurate than thermocouple gauges. However, they must be individually calibrated and checked frequently.

### 29.18.3 Ionization Type Vacuum

**Gauge**. Ionization is the process of removing and electron from an atom producing a free electron and a positively changed ion. Ionization may be produced by the collision of a high speed electron from the atom. Figure 29.32 shows the essential features of an ionization type gauge. Electrons are emitted from heated cathode. using a filament and are accelerated

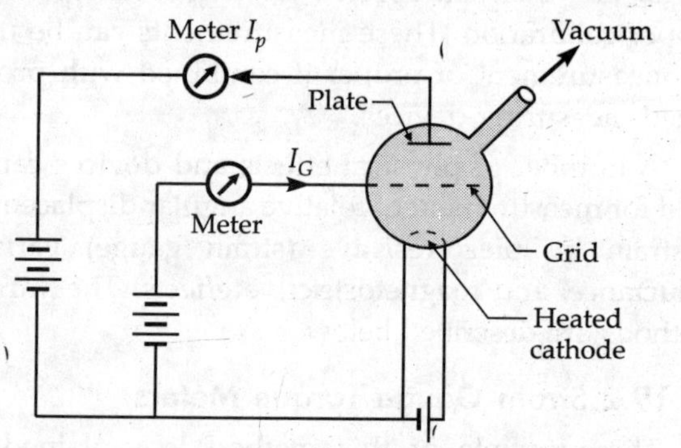


**Fig. 29.32** Ionization type vacuum gauge.

towards the grid, which is positively charged. Some of the electrons are captured by the grid, producing grid current $I_G$. Electrons having high kinetic energy pass through and cause ionization of gas atoms.

The positive ions so produced are attracted to plate (which is at negative potential) and a current $I_p$ is produced in the plate circuit.

It is found that the pressure of gas is proportional to ratio of plate to grid current.

$$P = \frac{1}{S}\frac{I_p}{I_G} \qquad \text{...(29.40)}$$

where $S$ = constant of proportionality.

$S$ is called the sensitivity of the gauge. A typical value for nitrogen is $S = 20$ torr$^{-1}$. However, the exact value must be determined by calibration of the particular gauge since sensitivity $S$ is a function of the geometry of the tube and the gas filled in it. Pressures that can be measured by ionization gauge range from $10^{-3}$ to $10^{-8}$ mm of Hg.

The disadvantage of these gauges is that the filament can burn out quickly if it is heated before the pressure is at a low value. For this reason, it is usually necessary to have a Pirani or thermocouple gauge in the system to ensure a low enough vacuum. Also an automatic cutout circuit is used to protect the ionization tube in case of a system leak to break.

These gauges must be carefully installed – in particular, absorbed gases must be eliminated. Voltages required are in the range 50 to 300 V.

## 29.19 MEASUREMENT OF TORQUE

### 29.19.1 Torque Transducers

Dynamic measurement of torque transmitted by a rotating shaft is based upon the angular displacement or twist in the shaft in a calibrated length of torque tube attached to the shaft. The strain is sensed by transducers and is measured. The strain measurements are then interpreted in terms of torque by proper calibration. These measurements can be used for measurement of proper if combined with proper speed measuring devices.

A number of physical effects and devices can be used for measurement of relative angular displacement – strain by piezo-resistive (strain gauge) variable inductance, and magnetostrictive effects. The various methods are described below.

### 29.19.2 Sϊrain Gauge Torque Meters

The principle of this method is explained by Fig. 29.33. Two strain gauges are mounted on a shaft at an angle 45° to each other. The torque is given by

$$T = \frac{\pi G(R^4 - r^4)}{2L}\theta \text{Nm} \quad ...(29.41)$$

where $G$ = modulus of rigidity ; N / m$^2$,

$R$ = outer radius of shaft ; m,

$r$ = inner radius of shaft ; m

$L$ = length of shaft ; m,

and $\theta$ = angular deflection of shaft ; rad

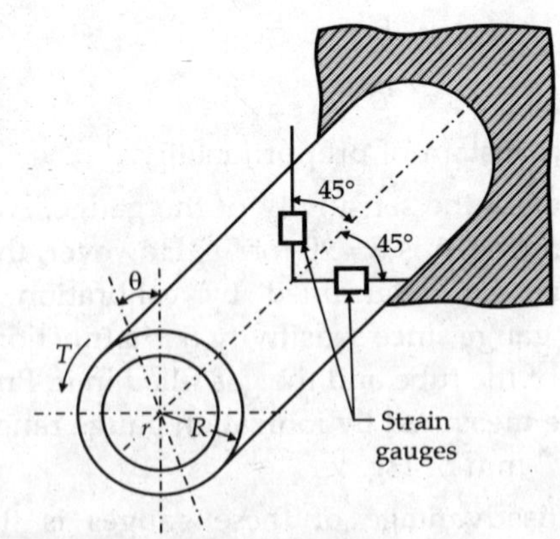


**Fig. 29.33** Measurement of stress in a hollow shaft.

The strain gauges attached at 45° to the axis of the shaft as shown will indicate strains of

$$\varepsilon_{45^\circ} = \pm \frac{TR}{\pi G(R^4 - r^4)} \quad ...(29.42)$$

A strain may be measured by electrical means to indicate the torque. Multiple strain gauges may be installed and connected in a bridge circuit configuration so that any deformation due to axial or traverse loads is cancelled out in the final readout.

The strain in the shaft may be measured by means of strain gauges attached to its surface. The gauges should be so mounted that they give maximum sensitivity to the strains produced by torsion. The theory of two dimensional stress systems shows that, for a shaft subjected to pure torsion, the gauges will be strained in the directions of their major axis if they are mounted at 45° to the axis of the shaft. The normal method is to mount a complete strain gauge bridge on the shaft. The strain bridge configuration generally used for measurement of torque is shown in Fig. 29.34(*a*). In this arrangement two strain gauges are subjected to tensile stresses while the other two experience compressive stresses. The gauges must be precisely at 45° with the shaft axis. Gauges 1 and 2 must be diametrically opposite, as must gauges 3 and 4.

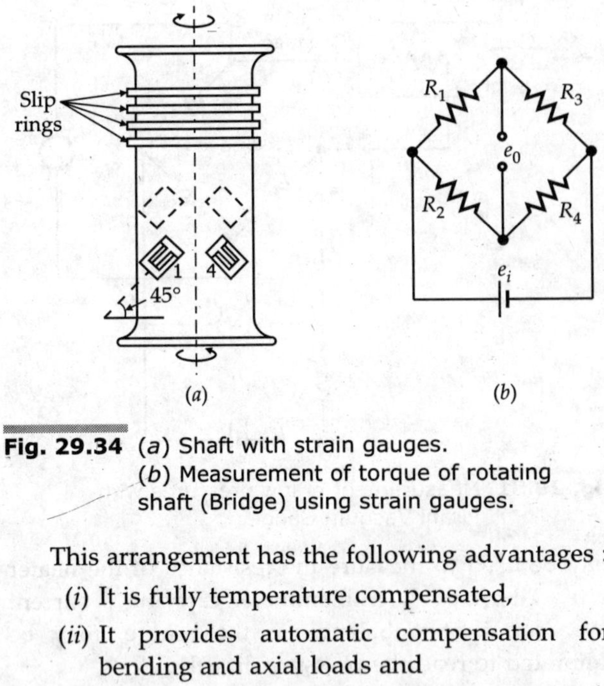


**Fig. 29.34** (*a*) Shaft with strain gauges. (*b*) Measurement of torque of rotating shaft (Bridge) using strain gauges.

This arrangement has the following advantages :

(*i*) It is fully temperature compensated,

(*ii*) It provides automatic compensation for bending and axial loads and

(*iii*) It gives the maximum sensitivity for a given torque.

The main difficulties associated with the use of this arrangement is the connection of the bridge to its power source and display arrangement. **Slip rings** are used for this purpose. These are conducting rings attached to the shaft, but insulated from it, with one of the slip rings connected to each of the bridge terminals. Rubbing contact is made between the rings and the stationary brushes, which are connected to the input and output equipment.

### 29.19.3 Inductive Torque Transducer

Figure 29.35(*a*) shows the use of an inductance transducer. Flange *A* carries a coil and flange *B*, an iron core. This core moves in and out of the coil according to relative displacement of the two flanges. Therefore inductance of the coil is altered on account of relative displacement. The coil is used as an arm of an a.c. bridge. The output of the a.c. bridge depends upon the inductance of the coil which in turn depends upon the position of core and thus on the displacement. Since the displacement is dependent upon the torque and hence the bridge output can be directly calibrated to read the torque.

A scheme which gives higher sensitivity and better linearity uses four inductive transducers with the coils connected as four arms of an a.c. bridge. The arrangement is shown in Fig. 29.35(*b*).

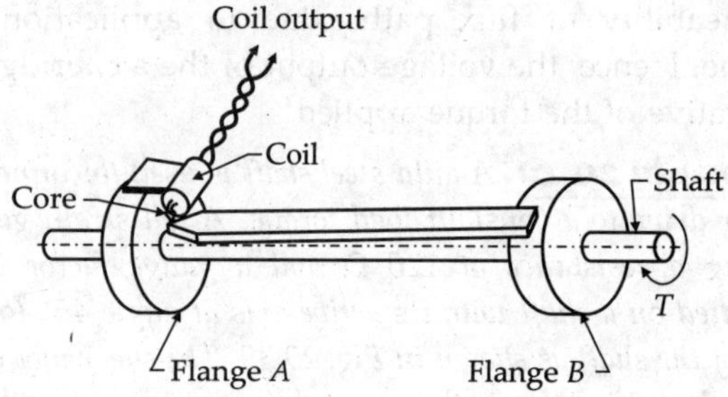

(*a*) Inductive torque transducer

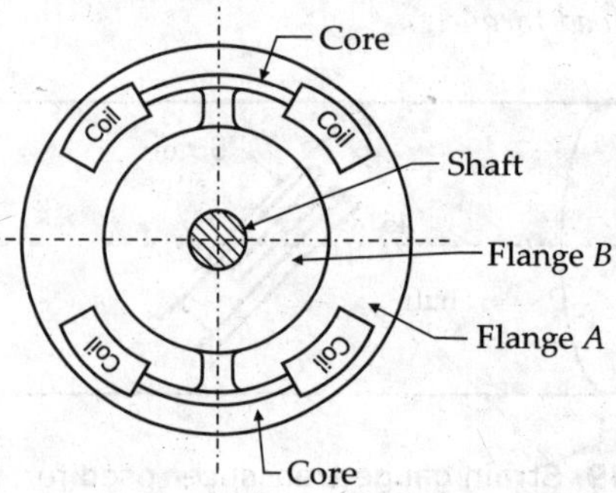

(*b*) Arrangement using four inductive transducers

**Fig. 29.35** Arrangement for torque transducers.

This arrangement is such that a torque applied to the shaft in a given direction moves the cores in such a way that the inductance of two coils increases, while the inductance of the other two coils decreases. (This arrangement is similar to use of four strain gauges, two in tension and two in compression.) Thus the sensitivity of the bridge is increased four times as compared to the sensitivity obtained with a bridge using only one inductive transducer.

### 29.19.4 Digital Methods

Digital timing techniques can also be used for determination of relative displacement between two flanges *A* and *B*. Suppose the flanges are made in the form of single toothed wheels as shown in Fig. 29.36. The teeth produce voltage pulses in inductive pickups *C* and *D* respectively. When no torque is applied to the shaft, the teeth are perfectly aligned and hence the voltage pulses are produced in *C* and *D* simultaneously. If the two pulses are compared by an electronic timer, the time difference shown will be zero. When a torque is applied to the shaft, there is a relative displacement between the two flanges. This produces a phase shift between the pulses generated in the inductive pickups C and *D*. When these pulses are compared with the help of an electronic timer, it will show a time interval between the two pulses. This time interval is proportional to the relative displacement of the two flanges which in turn is proportional to torque. Therefore, the ratio of this interval to the time taken for one complete revolution gives the relative displacement as a fraction of a revolution.

The advantage of digital methods is that they eliminate errors arising from the use of slip rings. There is no leakage of the signal and also there is no noise problem.

There are other ways of obtaining output signals from digital pickup. Figure 29.37 shows multi-toothed wheels which replace the single toothed wheels of Fig. 29.36. The pickups are either magnetic or photoelectric. In case multiple toothed wheels are replaced by single toothed wheels, the output of the pickoffs is perfectly sinusoidal. The two outputs are exactly in phase if the two wheels are correctly aligned.

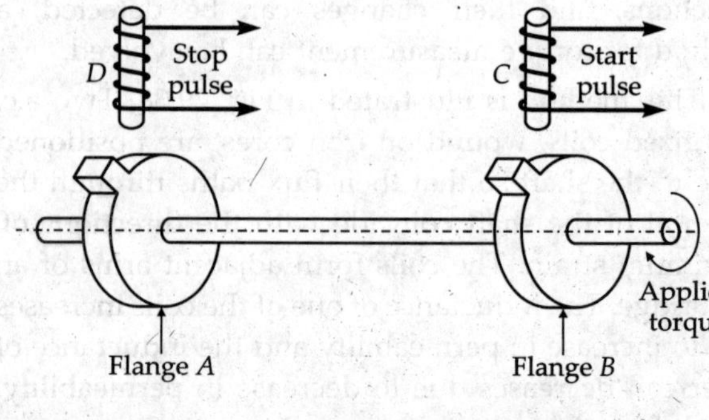


**Fig. 29.36** Torque transducer using single toothed flanges and inductive pickups.

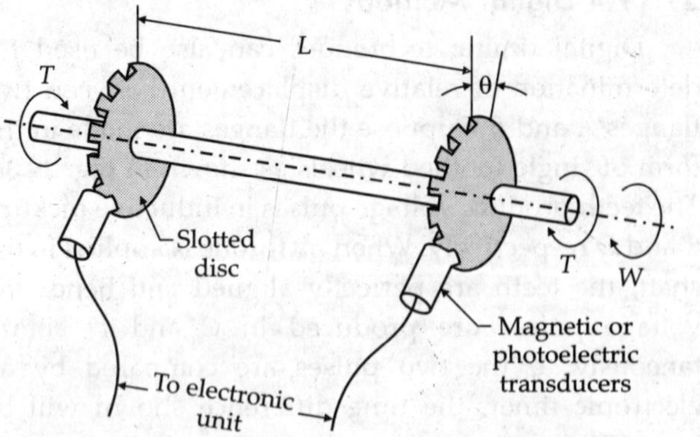


**Fig. 29.37** Torque measurements of rotating shafts using slotted discs and pickups.

The output voltage progressively becomes out of phase as the torque increases since an increase in torque results in relative displacement of the two flanges. The phase difference may be measured with the help of a cathode ray oscilloscope. The phase difference may also be measured by converting it to suitable analog or digital signals.

### 29.19.5 Magneto-Strictive Transducers

The action of **Magneto-strictive** transducers depends upon the change which occurs in, the permeability of magnetic materials when they are subjected to strain. The permeability decreases wit positive strain and increases with negative strain.

The torque transducers described earlier use an attachment with the shaft whose torque is to be measured resulting in substantial increase in the length of shaft. Magneto-strictive transducers (see pages 838) can operate without any attachments or modifications to the shaft.

It is well known that a shaft subjected to pure torsion, the highest positive and negativestrains are experienced in the two directions at 45° to the shaft axis. Thus, if flux paths can be established in these two directions, and their changes can be detected, a method for torque measurement can be evolved.

This method is illustrated in Fig. 29.38. Two a.c. energized coils, wound on iron cores, are positioned close to the shaft so that their flux paths through the material of the shaft coincide with the directions of maximum strain. The coils form adjacent arms of an a.c. bridge. The inductance of one of the coils increases due to increase in permeability and the inductance of other coil decreases due to decrease in permeability. The bridge is initially balanced when there is no torque applied and the two coils have equal indutance(or reactance). But when the torque is

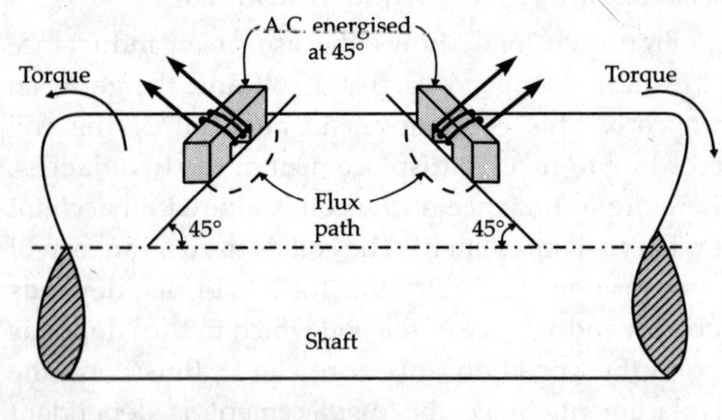


**Fig. 29.38** Magneto-strictive transducer for measurement of torque.

applied the bridge becomes unbalanced on account of increase in inductance (or reactance) of one coil and decrease in inductance (or reactance) of the other coil. This is due to differential change in inductance (or reactance) of the two coils, caused by change in permeability of flux paths due to application of torque. Hence, the voltage output of the a.c. bridge is indicative of the torque applied.

**Example 29.19** *A mild steel shaft is used to connect a motor drive to a constant load torque. A foil strain gauge having a resistance of 120 Ω and a gauge factor 2 is mounted on a shaft with its active axis at angle 45° to the axis of the shaft as shown in Fig. 29.39. The shear modulus of steel is 80 GN/m², the shaft radius is 15 mm and the change in strain gauge resistance due to the load is 0.24 Ω. Find the load torque.*

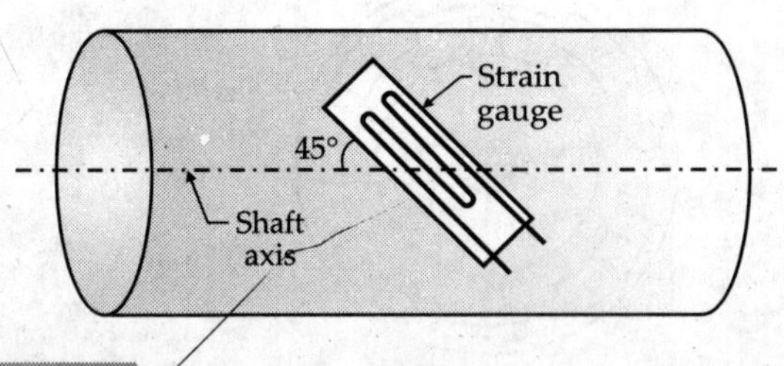


**Fig. 29.39** Strain gauge transducer used for measurement of torque (Example 29.19).

**Solution.** Angle of shear $\theta = \dfrac{2T}{\pi G r^3}$

where $G$ is the Shaft Shear modulus, $r$ is the radius of the shaft and $T$ is the applied torque.

An area of the shaft surface, originally square with the sides of unit length, is deformed by strain to a parallelogram. The original length of the diagonal is $\sqrt{2}$. If the angle of shear, θ, is small, the length of the diagonal of the parallelogram is longer than the diagonal of the square. The difference in lengths is $\theta/\sqrt{2}$.

Therefore the longitudinal strain is :

$$\varepsilon = \frac{\Delta L}{L} = \frac{\theta/\sqrt{2}}{\sqrt{2}} = \frac{\theta}{2}$$

But $\frac{\Delta R}{R} = G_f \varepsilon = 2 \times \frac{\theta}{2} = \theta$

or $\theta = \frac{\Delta R}{R} = \frac{0.24}{120} = 2 \times 10^{-3}$ rad

$\therefore$ Torque $T = \frac{\pi G r^3}{2} \theta$

$$= \frac{\pi \times 80 \times 10^9 \times (15 \times 10^{-3})^3}{2} \times 2 \times 10^{-3}$$

$= 848$ Nm

**Example 29.20** *A shaft is to transmit power up to 44 kW at a constant speed of 25 rps and it is proposed that the torque be sensed by a pair of torque strain gauges bonded to a specially machine portion of the shaft. The gauges are to be connected pushpull in an equiarmed voltage sensitive bridge, the output of which is to be calibrated in power units. If the maximum strain value of the gauges is 0.0015, their resistance 120 Ω and gauge factor 2.1, calculate :*

*(a) the diameter of steel shaft to which they should be bonded if its modulus of elasticity is 200 × $10^9$ N/m$^2$,*

*(b) the output voltage at full power if the excitation voltage of the bridge is 6 V,*

*(c) the sensitivity of the bridge is in V/kW.*

**Solution.**

(a) Angular speed $\omega = 2\pi \times 25 = 157$ rad/s

Torque $T = \frac{P}{\omega} = \frac{44000}{157} = 280.25$ Nm

Maximum value of strain $\varepsilon_{max} = 0.0015$

$\therefore$ Maximum value of direct stress

$s_{max} = 0.0015 \times 200 \times 10^9 \text{ N/m}^2 = 300 \text{ MN/m}^2$

For pure torsion, the maximum shear stress, τ has the same value.

Using relationship, $\frac{T}{I_p} = \frac{\tau}{d/2}$

we can find the diameter.

Polar moment of inertia $I_p = \frac{\pi}{32} d^4$

$$\therefore \quad \frac{280.25}{(\pi/32)d^4} = \frac{300 \times 10^6}{d/2}$$

or diameter of shaft $d = 0.0168$ m $= 16.8$ mm.

(b) The output voltage of an equal armed bridge using two active elements is

$$e_0 = \left(\frac{\Delta R/R}{2}\right) e_i$$

Now $\frac{\Delta R}{R} = \varepsilon G_f$

$= 0.0015 \times 2.1 = 3.15 \times 10^{-3}$

$\therefore$ Output voltage

$$e_0 = \frac{3.15 \times 10^{-3}}{2} \times 6 \text{ V} = 9.45 \text{ mV}$$

(c) Overall sensitivity of bridge

$$S = \frac{9.45 \times 10^{-3}}{44} \text{ V/kW} = 215 \text{ μV/kW}$$

**Example 29.21** *A digital timer is used to determine the torque in a rotating shaft using single toothed flages and inductive pickups. Static calibration shows that the flanges twist by an angle of one degree for an applied torque of 1000 Nm. In a test with the shaft rotating at 500 rpm, the torque calculated from the timer readings is 1200 Nm. What is the maximum probable error, of final digit on the timer display represents units of $10^{-5}$ s, and accuracy of instrument is specified is.0.05 per cent of the reading ± 1 in the final digit.*

**Solution.** 1 degree = 1000 Nm

The shaft is rotating at a speed of 500 rpm

$\therefore$ Angular speed

$$= \frac{360 \times 500}{60} = 3000 \text{ degrees/second}$$

or 1 s = 3000 degrees = 3000 × 1000

= 30,00,000 Nm

$10^{-5}$ s $= 10^{-5} \times 30{,}00{,}000 = 30$ Nm

The error is thus :

$$\frac{0.05}{100} \times 1200 = 0.6 \text{ Nm} = \frac{+\,30 \text{ Nm}}{30.6 \text{ Nm}} \text{ Total}$$

Hence percentage error $= \frac{30.6}{1200} \times 100 = 2.55\%$

## 29.20 MEASUREMENT OF LINEAR VELOCITY

The most commonly used transducer for measurement of linear velocity is the **electro-magnetic transducer**. This transducer utilises the voltage produced in a coil on account of change in flux linkages resulting from change in reluctance.

In general, the output voltage output from a coil is given by

$$e_0 = \frac{d\phi}{dt} = \frac{N}{\mathbf{R}}\frac{di}{dt} - \frac{Ni}{\mathbf{R}^2}\frac{d\mathbf{R}}{dt}$$

Supposing $i$ is constant $\therefore e_0 = \frac{Ni}{\mathbf{R}^2} \cdot \frac{d\mathbf{R}}{dt}$ ...(29.43)

Hence output voltage : $e_0 \propto \frac{d\mathbf{R}}{dt}$ ...(29.44)

Equation 29.44 is derived on the basis of following explanation.

If the average value of reluctance **R** is considerably greater than the variations in **R**, the quantity $Ni/R^2$ is approximately constant. Therefore the rate of change of flux is directly proportional to the rate of change of reluctance. The change in flux produces an output voltages and hence the output voltage is directly proportional to the rate of change of reluctance. The reluctance varies directly as the length of air gap and therefore the output voltage is directly proportional to rate of change of air gap and hence to the velocity. (This condition is only satisfied when the length of air gap is small.) Thus this type of transducer is velocity sensitive.

The electro-magnetic transducers are classified into two categories :

(*i*) Moving magnet type, and
(*ii*) Moving coil type.

## 29.20.1 Moving Magnet Type

The constant mmf (polarizing field) $Ni$, can be provided by a solenoid of $N$ turns and carrying a constant current $i$. But the electromagnetic transducers use a permanent magnet which provides a constant polarizing field.

Figure 29.40 shows a moving magnet type of transducer. The sensing element is a rod that is rigidly coupled to the device whose velocity is being measured. This rod is a permanent magnet. There is a coil surrounding the permanent magnet. The motion of the magnetindtices a voltage in the coil and the amplitude of the voltage is directly proportional to the velocity. For a coil placed in a magnetic field, the voltage induced in the coil is directly proportional to the velocity.

The polarity of the output voltage determines the direction of motion. The sensitivity of this transducer is stated in terms of mV/mm-s.

***Advantages*** **:**

(*i*) The maintenance requirements of these transducers are negligible, because there are no mechanical surfaces or contacts.
(*ii*) The output voltage is linearly proportional to velocity.
(*iii*) These transducers can be used as event markers which are robust and inexpensive to manufacture.

***Disadvantages*** **:**

(*i*) The performance of these transducers is adversely affected by stray magnetic fields. These fields cause noise.
(*ii*) The frequency response is usually limited and is stated.
(*iii*) These transducers are not very useful for measurement of vibrations because their calibration deteriorates as contact with steel tools etc. leads to progressive demagnetization.

There are two other forms of electromagnetic transducers which are described below :

## 29.20.2 Moving Coil type Velocity Transducer

The moving coil type velocity transducer shown in Fig. 29.41 operates essentially through the action of a coil moving in a magnetic field. A voltage is generated in the coil which is proportional to the velocity of the

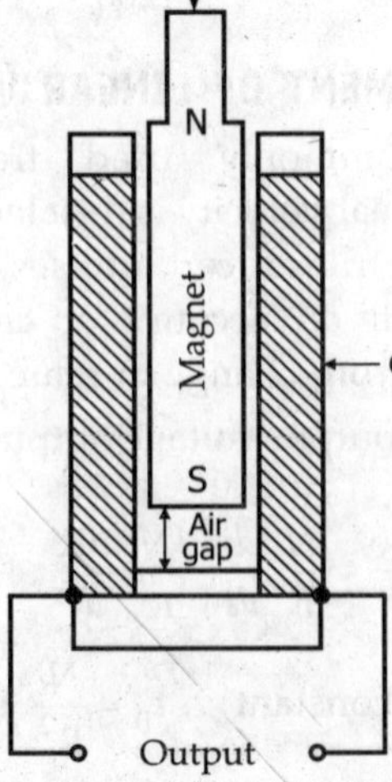


**Fig. 29.40** Moving magnet type transducers.

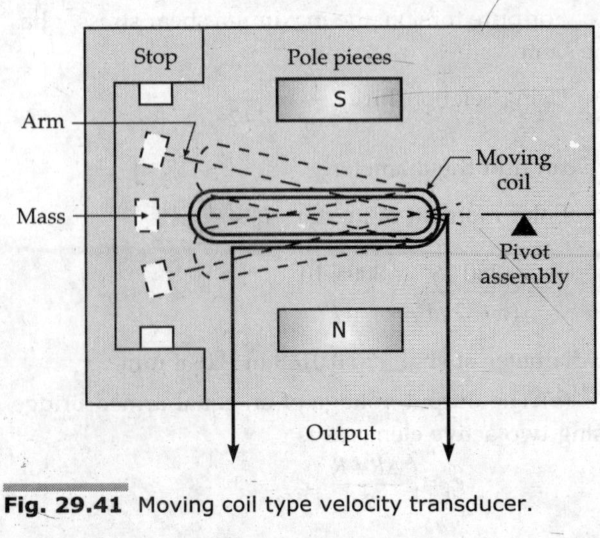


**Fig. 29.41** Moving coil type velocity transducer.

coil. (Note that in this transducer, the functions of permanent magnet and the coil are reversed from their functions as used in transducer of Fig. 29.47.)

The advantages of this transducer are :

(*i*) This is a more satisfactory arrangement as the system now forms a closed magnetic circuit with a constant air gap.

(*ii*) The whole device is contained in an antimagnetic case which reduces the effects of stray magnetic fields.

Another type of velocity transducers uses a pivoted arm on which a coil is mounted. There is a mass attached at the end of the arm. The velocity to be measured is applied to the arm and therefore the coil moves in the field of a permanent magnet.

A voltage is generated on account of motion of the coil in the magnetic field. The output voltage is proportional to the velocity. This type of transducer is generally used for measurement of velocities developed in a linear sinusoidal or random manner. Damping is obtained electrically, thus assuring high stability under varying tempemture conditions.

### 29.20.3 Seismic Tape Velocity Transducer

In this type of construction a permanent magnet is supported between two springs and fitted with low friction bearing rings as shown in Fig. 29.42. If this transducer is attached to an object which is vibrating at a frequency that is greater than the low natural

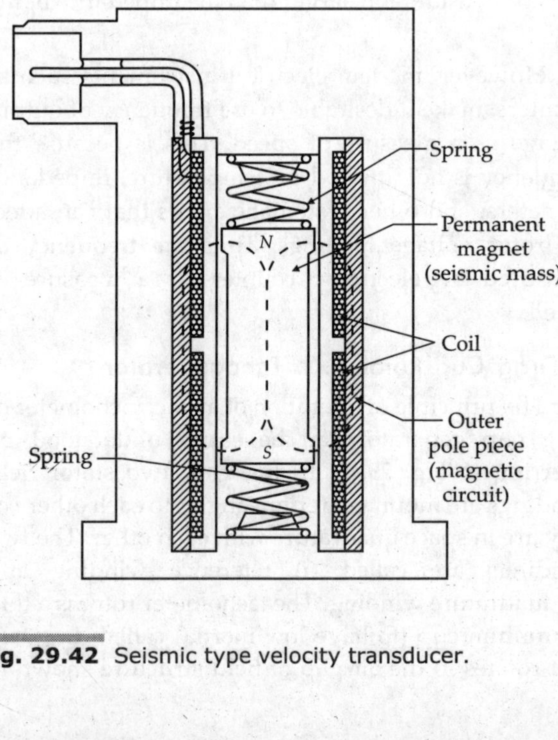


**Fig. 29.42** Seismic type velocity transducer.

frequency of the transducer, the magnet appears to be stationary in space. The case and coil oscillate around the magnet resulting in an output voltage which is proportional to the amplitude of the vibration. The frequency of output is equal to the frequency of the vibration.

Both moving coil and seismic transducers have the disadvantage that the connections must be made to the moving member.

## 29.21 MEASUREMENT OF ANGULAR VELOCITY

In many cases the only way to measure linear velocity is to convert it into angular velocity. For example, a speedometer uses the wheel rotational speed as a measure of the linear road speed. The disadvantage with measurement of linear velocity arises because a fixed reference must be used and if the moving object has to travel large distances, the detection becomes impossible. Hence angular velocity transducers are used.

The measurement of angular speed may be made with tachometers which may be either mechanical or electrical type.

## 29.22 ELECTRICAL TACHOMETERS

The electrical tachometers are preferred over mechanical tachometers for all applications because these tachometers offer all the advantages associated with electrical transducers. The various types of electrical tachometers are discussed below :

### 29.22.1 Electromagnetic Tachometer Generators

There are two types of electromagnetic tachometer generators called **tachogenerators.**

These are :

- D.C. tachometer generators, and
- A.C. tachometer generators.

### 1. D.C. Tachometer Generators

D.C. tachometer generators consist of a small armature which is coupled to the machine whose speed is to be measured. This armature revolves in the field of a permanent magnet. The emf generated is proportional to the product of flux and speed. Since the flux of the permanent magnet is constant, the voltage generated is proportional to speed. The polarity of output voltage indicates the direction of rotation. This emf is measured with the help of a moving coil voltmeter having a uniform scale and calibrated directly in terms of speed.

Figure 29.43 shows a d.c. tachometer generator. A series resistance is used in the circuit for the purpose of limiting the current from the generator in the event of a short circuit on the output side.

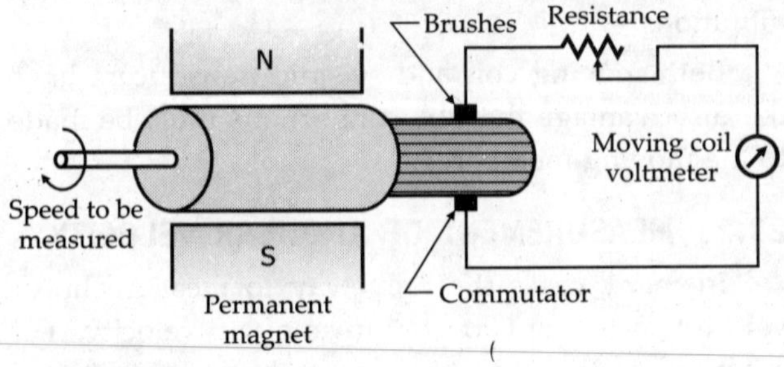


**Fig. 29.43** D.C. Tachometer Generator.

***Advantages*** :

The chief advantages of this tachometer are :

(*i*) The direction of rotation is directly indicated by the polarity of the output voltage.

(*ii*) The output voltage is typically 10 mV/rpm and can be measured with conventional type d.c. voltmeters.

***Disadvantages*** :

The disadvantages are :

(*i*) Brushes on small tachometer generators often produce maintenance problems, as their contact resistance may vary and produce appreciable error. Thus the commutator and the brushes require periodic maintenance.

(*ii*) The input resistance of meter should be very high as compared with output resistance of generator. This is required to limit the current to small value. If the armature current is large, the field of the permanent magnet is distorted giving rise to non-linearity.

## 2. A.C. Tachometer Generators

In order to overcome some of the difficulties mentioned above, a.c. tachometer generators are used. The tachometer generator has rotating magnet which may be either a permanent magnet or an electro-magnet. The coil is wound on the stator and therefore the problems associated with commutator (as in d.c. tachometers) are absent.

The rotation of the magnet causes an emf to be induced in the stator coil. The amplitude and frequency of this emf are both proportional to the speed of rotation. Thus either amplitude or frequency of induced voltage may be used as a measure of rotational speed.

When amplitude of induced voltage is used as a measure of speed, the circuit of Fig. 29.44 is used. The output voltage of a.c. tachometer generator is rectified and is measured with a permanent magnet moving coil instrument.

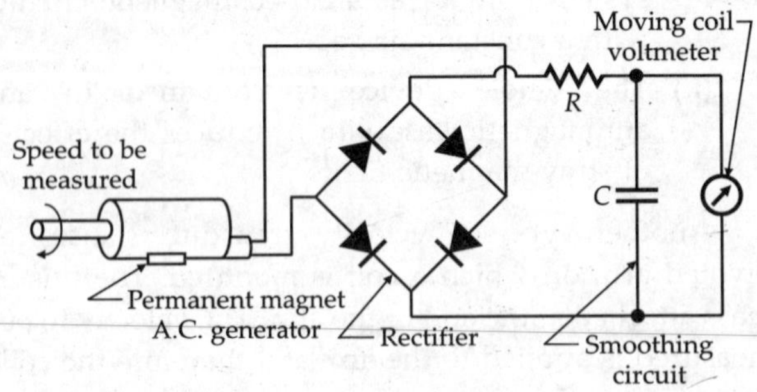


**Fig. 29.44** A.C. Tachometer Generator.

**Limitations :**

(*i*) The difficulty with this system is that at low speed the frequency of output voltage is low and hence it is very difficult to smooth out the ripples in the output voltage wave shape and hence a.c. tachometer generators are designed to have a large number of poles so that the frequency of output voltage is high even at low speeds.

(*ii*) High speeds also present a problem. At high Reference speeds, the frequency increases and therefore, the impedance of the coils of tacho-generator increases. If good linearity is to be maintained the input impedance of the display device must be considerably larger than the impedance of the coils.

However, modem electrical instruments such as counters, make it desirable to use frequency of output voltage as a measure of speed. This is because the frequency is not affected by temperature, impedance of meters, and other systematic errors that can affect the rotor voltage readings. Thus the frequency as measured by electronic counters is a measure of speed.

## 3. Drag Cup Rotor A.C. Tahogenerator

The principle of operation of an a.c. tachometer or drag cup generator can be easily understood by referring to Fig. 29.45. In this case two stator field windings are mounted at right angles to each other *i.e.*, they are in space quadrature with each other. The two windings are called (*i*) reference winding and (*ii*) quadrature winding. The tachometer rotor is a thin aluminium cup (to have low inertia) called drag cup that rotates in the air gap of field structure shown in

Fig. 29.52. This cup, which has a very low inertia and made of a high conductive material (aluminium) proves to be a uniformly short circuited conductor.

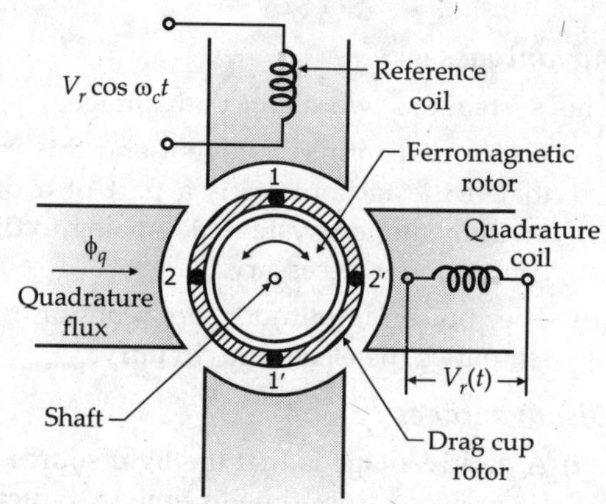


**Fig. 29.45** A.C. Tachometer Generator.

An alternating current voltage is applied to the reference winding and the output winding is taken from the quadrature winding.

Let the voltage applied to reference coil

$$= V_r \cos \omega_c t$$

where $\omega_c$ =carrier frequency, rad/s

This produces a reference flux $\Phi_r \sin \omega_c t$ lagging the reference voltage by an angle 90° if the resistance and reactance coil is considered to be negligible.

Let the rotor speed be

$$\omega(t)\,\theta°(t) = \theta_m \cos a\,\omega_c t$$

The rotor is replaced by two imaginary coils 1, 1' and 2, 2'

Speed (motional) emf in coil 1, 1'

$$= k_1 (\phi_r \sin \omega_c t)(\theta'_m \cos a\,\omega_c t)$$

Assuming the rotor reactance to be negligible, the current in coil 1, 1' will be proportional to the induced emf. This will cause a quadrature flux whose magnitude is,

$$\phi_q = K_2\,\theta'_m \cos a\,\omega_c t \sin \omega_c t$$

Because of the quadrature flux, a transformer emf will be induced in the quadrature coil. This emf is

$$e_q = K_3\,\theta'_m \frac{d}{dt}[\cos a\,\omega_c t \sin \omega_c t]$$

$$= K_3\,\theta'_m[-a\,\omega_c \sin a\,\omega_c t \sin \omega_c t + \omega_c \cos a\,\omega_c t \cos \omega_c t]$$

If the carrier frequency is very large $a << 1$, the sinusoidally varying speed $\theta(t)$ is very much less than the carrier frequency $\omega_c$,

$$e_q \simeq K_t\,\theta'_m \cos a\,\omega_c t \cos \omega_c t$$

$$e_q = K_t\,\theta'(t) \cos \omega_c t = v_t(t) \cos \omega_c t$$

or $v_t(t) = K_t\,\theta'(t)$

The emf induced in the quadrature coil is thus proportional to rotor speed and is phase with the voltage applied to rotor speed and is phase with the voltage applied to the reference coil. The modulating signal (emf) available at the terminals of the reference coil can be expressed as

$$v_t(t) = K_t\,\theta'(t) = K_t\,\omega(t) \quad ...(29.45)$$

where $K_t$ = tachometer constant.

Hence, the voltage output available at the terminals of quadrature coil is a function of the speed.

The rotation of the rotor causes an induced voltage in the sensing winding and this voltage is proportional to the instantaneous value of speed if the excitation frequency is very large as compared with speed.

***Advantages :***

(*i*) The drag cup type of a.c. tachogenerator described above is in fact a two phase squirrel cage induction motor. One of the windings is excited by carrier and the other winding develops an output voltage proportional to the speed. With the rotor stationary, the output voltage is zero. The rotation of rotor in one direction produces an output voltage at the terminals of the sensing winding. The frequency of the a.c. output voltage is the same as that of the excitation frequency with its amplitude proportional to the speed. The output voltage is in phase with the input excitation voltage. Reversal of the rotation produces an output voltage which is proportional to the speed but is 180° out of phase with the input voltage. This can be detected with a phase sensitive demodulator. Thus the arrangement affords a discrimination in the direction of the applied speed.

(*ii*) A linear relationship between the output voltage and speed is obtained if the carrier frequency is 5 to 10 times larger than the frequency of the speed signal. For this purpose the excitation winding is supplied with a frequency of 400 Hz.

(*iii*) The drag cup type tachometers are rugged and are in-expensive. Also, they require little maintenance.

(*iv*) In some situations these tachometers are very useful since they give a ripple free output.

***Disadvantages :***

(*i*) The output voltage is proportional to the product of speed and input voltage. Thus, in order that the calibration of the tachometer holds good, the input voltage should be maintained absolutely constant.

(*ii*) These tachometers are hard to calibrate.

(*iii*) At high speed there exists a non-linear relationship between output voltage and the input speed. Hence corrections should be applied or otherwise higher carrier frequencies should be used. It is no doubt possible to excite the tachometer excitation winding with frequencies higher than 400 Hz if necessary, but some performance characteristics will change their value in the process.

## 29.23 DIGITAL METHODS

The electromechanical methods for measurement of angular velocity are satisfactory up to about a speed of 10,000 rpm. Higher speed measurements are possible with digital pickups which work in conjunction with digital-frequency meters.

The biggest advantage of digital methods is that no direct physical contact is required with the shaft whose speed is to be measured. Therefore, no load is imposed upon the shaft by measuring device.

The digital pickups are of two types :

- photo-electric type, and
- inductive type.

### 29.23.1 Photoelectric Tachometer

This method of measuring speed of rotation consists of mounting an opaque disc on the rotating shaft as is shown in Fig. 29.46.

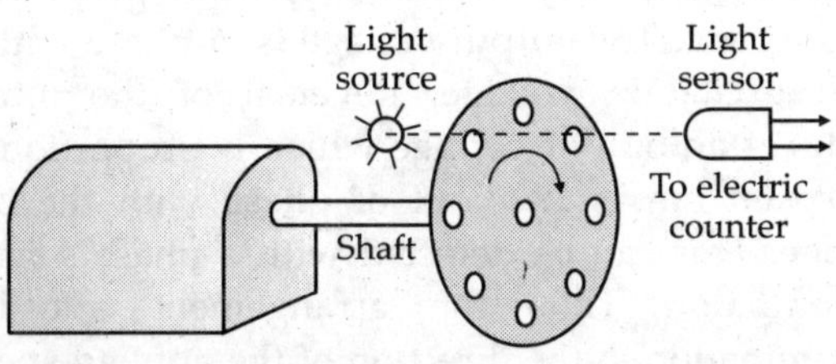


**Fig. 29.46** Photoelectric tachometer.

The disc has a number of equidistant holes on its periphery. At one side of the disc a light source is fixed and at the other side of the disc, and on line with the light source, a light sensor such as a photo tube or some photosensitive semi-conducting device is placed. When the opaque portion of the disc is between the light source and the light sensor, the latter is unilluminated and produces no output. But when a hole appears between the two, the light falling upon the sensor produces an output pulse.

The frequency at which these pulses are produced depends upon the number of holes in the disc and its speed of rotation. Since the number of holes is fixed, the pulse rate is a function of speed of rotation. The pulse rate can be measured by an electronic counter which can be directly calibrated in terms of speed in rpm.

***Advantages :***

This system has two distinct advantages :

(*i*) The output format is digital and this means that if the tachometer is a part of a digital instrumentation system, no analog to digital conversion is necessary.

(*ii*) The pulse amplitudes are constant. This simplifies the electronic circuitry.

***Disadvantages :***

(*i*) A disadvantage is that the light source must be replaced from time to time. A typical life time for light source is 50,000 hours.

(*ii*) The accuracy of this method depends principally on the error represented by one pulse. The digital meters measure frequency by counting the number of input pulses which occur in short period of time called **gating period.** If this period is too small serious errors maybe caused. The gating period should therefore, be chosen to give a sufficiently large count. In general, all the digits on the digital display should be utilized.

The factors which the user can control to minimise the errors are :

(*i*) gating period, and

(*ii*) number of pulses generated per revolution.

### 29.23.2 Toothed Rotor Variable Reluctance Tachometer

This tachometer generator consists of a metallic toothed rotor mounted on the shaft whose speed is to be measured. This is shown in Fig. 29.47. A magnetic pickup is placed near the toothed rotor.

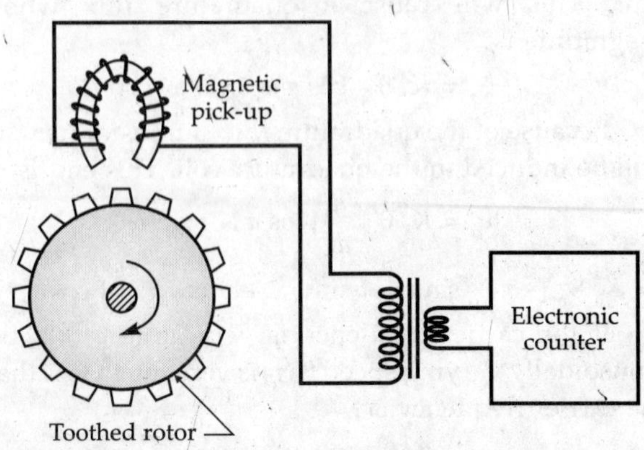


**Fig. 29.47** Toothed rotor tachometer generator.

The magnetic pick up consists of a housing containing a small permanent magnet with a coil wound round it. When the rotor rotates, the reluctance of the air gap between pickup and the toothed rotor changes giving rise to an induced e.m.f. in the pickup coil. This output is in the form of pulses, with a variety of wave shapes.

The frequency of the pulses of induced voltage will depend upon the number of teeth of the rotor and its speed of rotation. Since the number of teeth is known, the speed of rotation can be determined by measuring the frequency of pulses with an electronic counter. Suppose the rotor has $T$ teeth, the speed of rotation is $n$ rps and number of pulses per second is $P$.

$\therefore$ Number of pulses per revolution $= T$

Hence speed

$$n = \frac{\text{pulses per second}}{\text{number of teeth}} = \frac{P}{T} \text{ rps}$$

$$= \frac{P}{T} \times 60 \text{ rpm} \qquad \text{...(29.46)}$$

A typical rotor has 60 teeth. Thus if the countre counts the pulses in one second, the counter will directly display the speed in rpm.

We have mentioned above that the pulses have a variety of wave shapes. This is immaterial, as this tachometer is always connected to an electronic counter, whose requirement is merely that the amplitude be great enough to trigger a count.

Variable reluctance tachometer probably is the most common type of tachometer in use today.

***Advantages* :**

The advantages of this tachometer are :

(*i*) It is simple and rugged in construction.

(*ii*) It is maintenance free.

(*iii*) It is easy to calibrate. This has been illustrated earlier, if the rotor has 60 teeth and the pulses are counted by counter in one second, the count displayed by the counter gives the speed directly in r.p.m.

(*iv*) The information from this device can be easily transmitted.

**Example 29.21** *A variable reluctance type tachometer has 60 rotor teeth. The counter records 3600 counts per second. Determine the speed in rpm.*

**Solution.**

$$\text{Speed} = \frac{\text{pulses per second}}{\text{number of teeth}}$$

$$= \frac{3600}{60} \text{ rps} = 3600 \text{ rpm.}$$

**Example 29.22** *An inductive pickoff operating from a 120 tooth wheel is used with a digital frequency meter to measure the speed of rotation of the shaft on which the wheel is mounted. The gating period is set to $10^4$ μs, and a reading of 0030 is obtained on the four digit display. What shaft speed does this represent in r.p.s. ? If the available gating periods are $10^2$, $10^3$, $10^4$, $10^5$, $10^6$, $10^7$ μs respectively, what would be the optimum setting of gating period for making this measurement ?*

**Solution.** $\therefore$ Number of pulses per second

$$= \frac{\text{reading of digital meter}}{\text{gating period}}$$

$$= \frac{0030}{10^4 \times 10^{-6}} = 3000$$

$$\therefore \quad \text{Speed} = \frac{\text{number of pulses per second}}{\text{number of teeth}}$$

$$= \frac{3000}{120} = 25 \text{ rps.}$$

The setting of the gating period for this measurement is $10^4$ μs and this gives a reading of 0030 on a four digit display. It is seen that out of four digit place only two are utilised. For optimum setting of gating period, all the four digits places should be utilized. This required that the gating time be increased by a factor $10^2$.

$\therefore$ Optimum gating time setting

$$= 10^4 \times 10^2 = 10^6 \ \mu\text{s}$$

## 29.24 STROBOSCOPE AND STROBOSCOPIC METHODS

The stroboscope is a simple, portable manually operated device which may be used for measurement of periodic or rotary motions. Basically, the instrument is a source of variable frequency flashing brilliant light, the flashing frequency being set by the operator. The circuit used is based upon variable frequency oscillator which controls the flashing frequency. The speed is measured by adjusting the frequency so that the moving objects are visible only at specific intervals of time.

The method of use of the stroboscope depends upon imperfect dynamic response of the human eye. If a strong light is caused to flash on a moving object which, at the time each flash occurs, occupies a given position, the object will appear to be stationary. Therefore the method is useful for only those types of motions which occur regularly after a fixed interval of time, such as oscillation or rotation.

The stroboscope consists of a source of flashing light whose frequency can be varied and controlled. This source is called a **strobotron**.

### 29.24.1 Strobotron

It is hot cathode gaseous discharge tube as shown in Fig. 29.48. It has a cathode, an anode and two grids, one is called the inner grid and other as the outer grid.

If the potential of the outer grid is increased or that of inner grid is decreased beyond a certain limit, the conduction starts. Once the conduction starts, it can be stopped only by removing the anode potential.

The flashing of light is started by a signal from an oscillator or a multivibrator. When the tube is ionized and starts giving out flashes the capacitor C is discharged and a heavy current drawn by it reduces the anode potential due to large voltage drop in resistance $R_1$. The ionization stops and so does the flashing of light. In the mean time the capacitor is recharged and waits for the next pulse to start the flashing again. The tube has a maximum flashing rate of 300 per second with effective discharge time of 10 to 100 μs. The discharge current may be as high as 300 A.

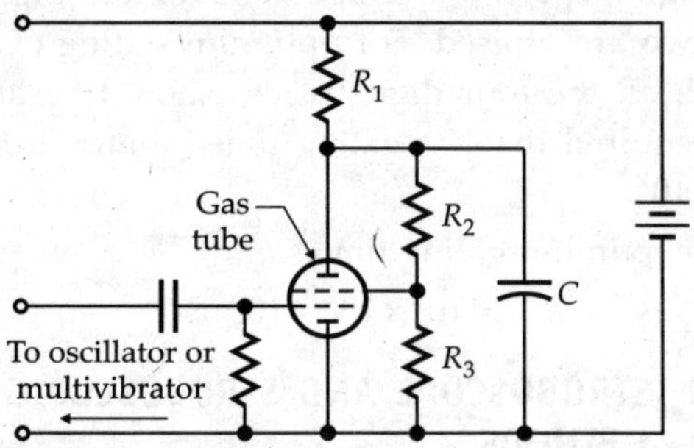


**Fig. 29.48** Strobotron.

### 29.24.2 Shaft Speed Measurements

A distinctive mark is made on the shaft or on a disc attached to the shaft as shown in Fig. 29.49. A stroboscope is made to flash light directly on the mark. The flashing frequency is adjusted until the mark appears stationary. Under these conditions, the speed is equal to the flashing frequency provided that the approximate speed of the shaft is known in advance and the flashing frequency is not allowed to depart too much away from this value.

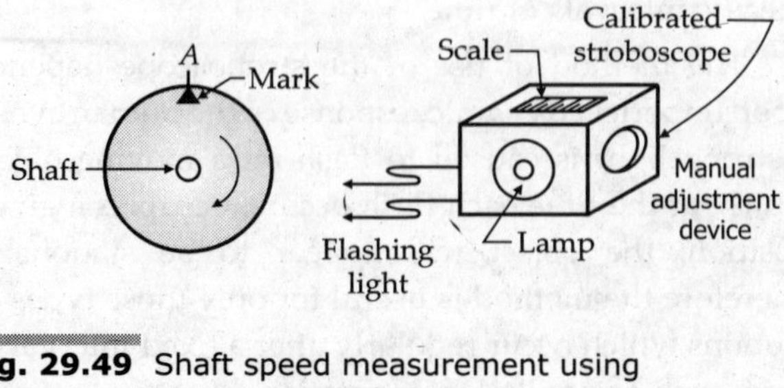


**Fig. 29.49** Shaft speed measurement using stroboscope.

The scale of the stroboscope is calibrated in terms of speed which can be directly read off.

If the two conditions outlined above are not met or if there are several identical marks on the shaft like spokes of a wheel or jaws of a chuck, serious errors in the measurement may arise.

Consider first the case where there is only one mark as shown in Fig. 29.49. If the mark is at $A$ every time a flash occurs, a stationary image appears. This is the case when $n = f$ where $n$ = speed of shaft, rps and $f$ = flashing frequency, Hz.

However, if $n = 2f, 3f, 4f, ...$ etc. *i.e.,* and if speed of shaft is any whole multiple of flashing frequency the mark would appear as stationary. Therefore, the mere fact that a stationary image is obtained does not guarantee the shaft speed with certainty.

It should be noted that this problem does not arise if $n < f$. Under these circumstances if $n$ is a sub-multiple of $f$, **Multiple images** are obtained as shown in Figs. 29.50(*a*) and (*b*).

The reason for formation of these multiple images may be visualized considering the case where $n = f/2$ as shown in Fig. 29.50(*a*). Suppose a flash occurs when the mark is at $A$. With $n = f/2$, the next flash will occur after half a revolution *i.e.,* when the mark is at $B$, and the next when the mark is at $A$ and so on. Thus although the mark will be repeated at $A$ when the flashes occur, it will also repeatedly be at $B$ at the time of alternate flashes. Thus for $n = f/2$, two stationary marks will be seen.

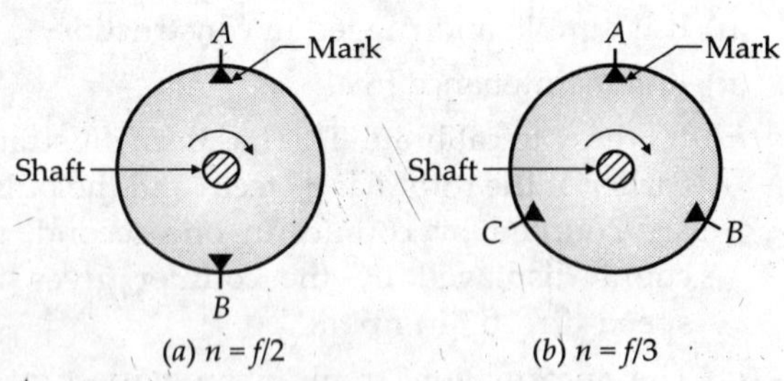


**Fig. 29.50** Formation of multiple images.

Figure 29.50(*b*) shows the case where $n = f/3$, where in a stationary mark is seen at three different places $A$, $B$ and $C$. Thus the argument can be extended for cases $n = f/3, f/4, f/5...$ etc. where 3, 4, 5.....etc. stationary marks are seen for one mark on shaft.

Let us now consider the case, where there are multiple marks on the shaft. Figure 29.51 shows a shaft with three identical, equi- spaced marks on the shaft. It is evident by considering the position of the marks as successive flashes occur, it can be seen that not only will the stationary pattern be obtained when $n = f, 2f, 3f$, etc. but also for certain values of $n$ which are smaller than $f$, namely $n = f/3, 2f/3, 4f/3, 5f/3$, etc. This leads to the conclusion that there is a great possibility of error when using stroboscopic methods.

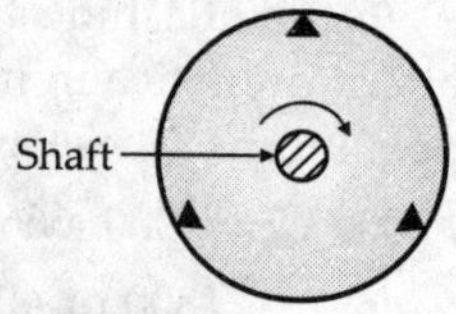


**Fig. 29.51** Stroboscope patterns with (multiple) three identical marks.

From the above discussion, it is clear that in order to avoid gross mistakes in measurements :

(*i*) it is safest to work with a single mark and to find the highest flashing frequency at which a true image is seen, and

(*ii*) as a check that the correct value of speed, $n$ has been found, the frequency may be doubled which should produce a double image.

A method for finding speed of a multiple marked shaft is given below :

Figure 29.52 shows a disc having 6 marks. This disc is attached to the shaft of the rotating machine and is illuminated by a series of successive flashes by a stroboscope. If the speed is such that each mark of star moves forward a distance of one point pitch ($p$) during the interval between successive flashes the pattern appears to be stationary. This speed is called **primary speed.** The pattern will also appear to be stationary if the speed or rotation is exactly twice, three times of any multiple of this speed.

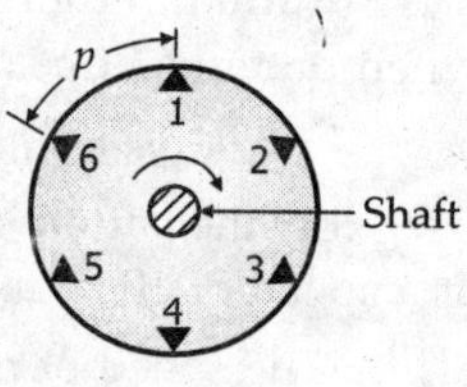


**Fig. 29.52** Pattern having six marks.

When the disc appears to be stationary, the speed of rotating machine is given by

$$n = f / m \qquad ...(29.47)$$

where $f$ = number of flashes per second ;

and $m$ = number of point marks.

The stroboscopic dial may be directly calibrated to read the speed.

The disc appears stationary for all speeds which are a multiple of primary speed. Therefore, to avoid confusion an approximate value of actual speed should either be known or measured by other means.

The value of the exact speed may be found as under : Single line image is obtained by adjusting the stroboscope at its highest flashing frequency. The flashing rate is gradually reduced and the flashing frequencies are noted for all single line images. If single line images are obtained at $m$ different flashing rates $f_1, f_2, ..., f_n$, the shaft speed is given by

$$n = \frac{f_m\, f_1 (m-1)}{(f_m - f_1)} \qquad ...(29.48)$$

where $f_m$ = highest flashing frequency,

$f_1$ = lowest flashing frequency,

$m$ = number of flashing frequencies.

(These flashing frequencies refer to the frequencies at which single line images are obtained.)

If the shaft (and alongside the disc) rotates at a speed slightly higher than the primary speed, the pattern appears to rotate slowly forward. On the other hand, if the speed is slightly less than the, primary speed and the pattern apparently moves in the reverse direction to that of the direction of rotation of shaft. This is very useful for measurement of slip speed in induction motors.

### 29.24.3 Advantages and Disadvantages of Stroboscopic Methods

***Advantages*** :

(*i*) This method imposes no load on the shaft.

(*ii*) It requires no special attachments with the shaft.

(*iii*) This method is particularly useful where it is inconvenient or impossible to make contact with the shaft.

(*iv*) It is very convenient to use a stroboscope for spot checks on machinery speeds and for laboratory work.

***Disadvantages*** :

(*i*) The circuit of the variable frequency oscillator cannot be stabilized to give a fixed frequency. Therefore, this method is less accurate than the methods utilizing digital meters.

(*ii*) The stroboscope cannot be used in surroundings where the ambient light is above a certain level. The stroboscope requires subdued lighting conditions for efficient operation.

Commercial models of stroboscopes are usually available in the speed range of 600 to 20,000 rpm.

29

**Example 29.23** *The speed of a shaft rotating at 2880 rpm is measured using a stroboscope. The stroboscope dial is slowly turned from setting of 4320 rpm to 1400 rpm corresponding to flash rate of 96 to 24 per second (a) Indicate the speed settings which give single, double, and triple steady images. (b) What is the observation when the flashing rate is 50 per second ?*

**Solution.** Speed of shaft $n = 2880/60 = 48$ rps

(*a*) If the flashing frequency is 48 per second the shaft completes one revolution during the interval between successive flashes and therefore we get a stationary image. If the flashing frequency is 24 per second, the shaft completes two revolutions during the interval between successive flashes, and hence again we get a stationary single image.

Therefore, for getting one stationary image the flashing rates should be either 24 or 48 per second. The other flashing frequencies which give stationary single line images lie outside the range of the instrument.

Double images are obtained if the shaft turns through, $\frac{1}{2}$, $1\frac{1}{2}$ revolutions during successive flashes.

$\therefore$ Flashing frequencies for double images are

$$48 \times 2 = 96 \quad \text{or} \quad 48 \times (2/3) = 32,$$

*i.e.,* 32 and 96 flashes per second.

Similarly, for triple images, the flashing frequencies are :

$48 \times 3 = 144$ (not within range), $48 \times (3/2) = 72$,
$48 \times (3/4) = 36$, and $48 \times (3/5) = 28.8$ flashes/second

(*b*) Rotation of shaft in $1/50$ s $= (1/50) \times 48 = 48/50$ revolution.

$\therefore$ The image falls by $(-48/50) = 2/50$ revolution within a time of 1/50 s or by 2 rps.

**Example 29.24** *A stroboscope is directed at a rotating disc having five equispaced radial lines on it. The highest flashing frequency at which a true pattern is observed is 2000 flashes minute. Give two other flashing frequencies which would produce (a) a 5 line pattern and (b) a 10 line pattern.*

**Solution.** (*a*) The highest flashing frequency to give a 5 line pattern is 2000 per minute.

The other flashing frequencies which give a 5 line pattern are a submultiple or the highest frequency.

These are :

$2000/2 = 1000$, $2000/3 = 666.7$, $2000/4 = 500$,
$2000/5 = 400$ or 1000, 6.667, 500, 400 flashes/minute.

(*b*) The flashing frequencies which give a 10 line pattern are twice in magnitude of those which give 5 line patterns. These are :

$2000 \times 2 = 4000$. $666.7 \times 2 = 1333$, $400 \times 2$
$= 800$ or 4000, 1333, 800 flashes/ minute

(Flashing frequencies of 2000 and 1000/minute are not included in the above list since they give 5 line patterns.)

**Example 29.25** *While measuring speed of a steam turbine with stroboscope single line images were observed for stroboscope setting of 3000, 4000 and 5230 rpm. Calculate the speed of the turbine.*

**Solution**. The speed is given by Eqn. 29.47 :

$$n = \frac{f_m\, f_1\,(m-1)}{(f_m - f_1)}$$

Now $\quad f_m = 5250$ rpm, $f_1 = 3000$ rpm and $m = 3$

$\therefore$ Speed of steam turbine

$$= \frac{5250 \times 3000\,(3-1)}{(5250-3000)} = 14000 \text{ rmp}$$

**Example 29.26** *A disc mounted on the shaft of a machine has 12 pattern points. The number of flashes projected on the disc by a stroboscope is 6000 in a minute.*

(*a*) *Find the speed of the machine if the disc appears stationary and has single image of 12 points.*

(*b*) *If the disc appears to move forward in the direction of rotation at 10 rpm, find the speed of the disc.*

**Solution.** From Eqn. 29.46, speed of machine when patterns appear stationary,

$$n = f/m = 6000/12 = 500 \text{ rpm}$$

Now the patterns move in the direction of rotation at a speed of 10 rpm.

$\therefore$ The speed of machine in this case is :

$$n = 500 + 10 = 510 \text{ rpm}$$

## 29.25 MEASUREMENT OF VIBRATIONS

The need for making measurement of vibrations has arisen mainly because of the growth of environmental testing. Specifications, many a times, require that the equipment should withstand stated levels of vibrations. This can be done quantitatively only through vibration measurement. Vibration measurements are frequently carried out on rotating and reciprocating machinery for analysis, design and trouble-shooting purposes.

Much knowledge has been gained in the recent years and computer solutions of various vibration problems have been developed. However, many a times to make actual measurements of vibration characteristics by tests, during development, either on the machine itself or on its prototype due the fact that it is impossible to build a perfect mathematical model with all its intracies and complexities.

Vibration monitoring is carried out on such important machines as power station turbines and generators to give an early warning of impending conditions which may develop and lead to complete failure and destruction of the equipment.

### 29.25.1 Nature of Vibrations

Most vibrations are sinusoidal displacement of the vibrating member about its mean position. A vibration of this nature may be defined by its amplitude and frequency.

For a sinusoidal vibration, the displacement is given by :

$$x = x_m \sin \omega t$$

where $x_m$ = amplitude,

$\omega$ = angular frequency ; rad/s,

Velocity $v = \dot{x} = x_m \omega \cos \omega t$

and maximum velocity $v_0 = x_m \omega$.

Acceleration $a = \ddot{x} = -x_m \omega^2 \sin \omega t$

and maximum acceleration $a_0 = -x_m \omega^2$

A sinusoidal vibration can, therefore, be defined by specifying its frequency, plus its amplitude or maximum velocity or maximum acceleration.

### 29.25.2 Quantities Involved in Vibration Measurements

The quantities required to be measured in a vibrating system are displacement, velocity and acceleration. The peak values of the quantities should be known. Frequencies are measured and modes of vibration at particular frequency values may be observed.

Displacement, velocity and acceleration are related to each other. If one of the three variables concerned (displacement, velocity or acceleration) is measured, it is possible to determine the other two by integration or differentiation using electronic devices.

Measurements of vibration may therefore, be made with a transducer sensitive to amplitude (displacement), velocity or acceleration. The output of this transducer, after necessary signal conditioning is fed to the display which indicates or records the original measured variable or another variable derived from it. Since the transducer output is pulsating the frequency may be measured by a digital frequency meter at a suitable stage of signal conditioning process.

## 29.26 SEISMIC TRANSDUCERS

A schematic diagram of a seismic transducer is shown in Fig. 29.53. It is called a **seismic accelerometer** also. The mass is connected through a parallel spring and damper arrangement to a housing frame. The housing frame is connected to the source of vibrations whose characteristics are to be measured.

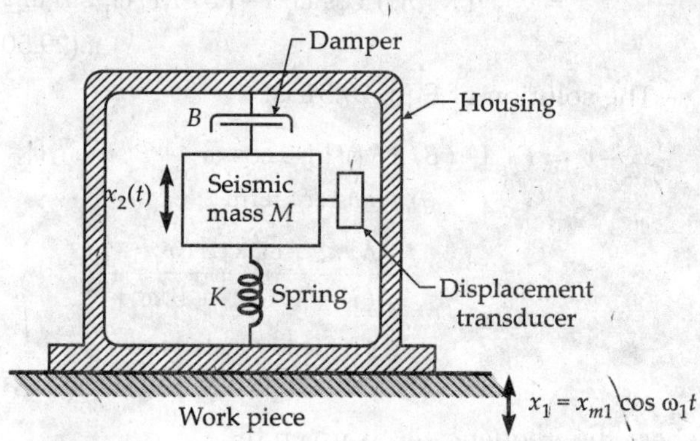


**Fig. 29.53** Schematic diagram of a seismic transducer.

The mass has the tendency to remain fixed in its spatial position so that the vibrational motion is registered as a relative displacement between mass and housing frame. This displacement is sensed and indicated by an appropriate transducer.

The seismic transducer may be used in two different modes :

- Displacement mode, and
- Acceleration mode.

The mode to be selected depends upon the proper selection of mass, spring and damper combinations. In general, a large mass and a soft spring are suited for displacement mode measurements, while a relatively small mass and a stiff spring are used for acceleration mode measurements.

Figure 29.54 shows the diagrammatic representation of the system to be analyzed.

The input is : $x_1 = x_{x1} \cos \omega_1 t$.

Equation for motion can be written as

$$M\ddot{x}_2 + B\dot{x}^2 + Kx_2 = B\dot{x}_1 + Kx_1 \qquad ...(29.49)$$

We are interested in the relative displacement $x_0 = (x_2 - x_1)$ in terms of the above impressed motion.

29

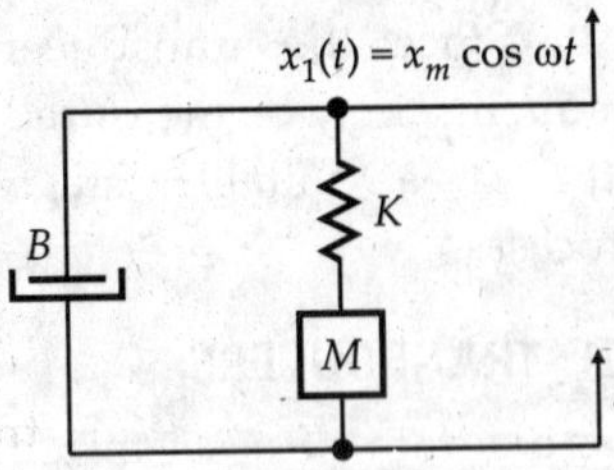


**Fig. 29.54** Mechanical system for seismic transducer.

Therefore we can write the equation of motion as :

$$\ddot{x}_2 + (B/M)\,\dot{x}_2 + (K/M)\,x_2 = x_{m1}\,[(K/M)\cos\omega_1 t - (B/M)\,\omega_1 \sin\omega_1 t] \quad ...(29.50)$$

The solution of Eqn. 29.51 is,

$$x_2 - x_1 = E_{xp}\,[-(B/2M)t]\,[C\cos\omega_d t + B\sin\omega_d t]$$

(*i*) Transient term

$$+\frac{M\,x_{m1}\,\omega_1^2 \cos(\omega_1 t + \phi)}{[(K - M\,\omega_1^2)^2 + B^2\omega_1^2]^{1/2}}$$

(*ii*) Steady state term

...(29.51)

where the frequency $\omega_d$ is given by

$$\omega_d = \left[\frac{k}{M} - \left(\frac{B}{2M}\right)^2\right]^{1/2} \text{rad/s} \quad ...(29.52)$$

and the phase angle by

$$\phi = \tan^{-1}\frac{B\omega_1}{K - M\omega_1^2}\ \text{rad} \quad ...(29.53)$$

*C* and *D* are constants of integration which can be determined from the initial or boundary conditions.

Note that Eqn. 29.52 is composed of (*i*) a transient term involving an exponential function and (*ii*) a steady state term. This means that after the initial transient has died out a steady state harmonic motion is established in accordance with the second term. The frequency of this steady state term is the same as that of the impressed motion. The steady term is,

$$x_2 - x_1 = \frac{M\,x_{m1}\,\omega_1^2 \cos(\omega_1 t + \phi)}{[(K - M\omega_1^2)^2 + B^2\,\omega_1^2]^{1/2}} \quad ...(29.54)$$

Amplitude of steady state term is,

$$(x_2 - x_1)_m = \frac{M\,x_{m1}\,\omega_1^2}{[(K - M\omega_1^2)^2 + B^2\,\omega_1^2]^{1/2}} \quad ...(29.55)$$

Dividing throughout by *M*, we have

$$(x_2 - x_1)_m = \frac{x_{m1}\,\omega_1^2}{[(K/M - \omega_1^2)^2 + (B/M)^2\,\omega_1^2]^{1/2}} \quad ...(29.56)$$

But, natural frequency

$$\omega_n = \sqrt{K/M}\ \text{rad/s} \quad ...(29.57)$$

and damping ratio

$$\xi = B/2\sqrt{KM} \quad ...(28.58)$$

Hence Eqn. 29.56 can be written as

$$(x_2 - x_1)_m = \frac{x_{m1}\,(\omega_1/\omega_n)^2}{\{[1 - (\omega_1/\omega_n)^2]^2 + [2\,\xi(\omega_1/\omega_n)]^2\}^{1/2}} \quad ...(29.59)$$

$$= \frac{x_{m1}\,u^2}{[(1-u^2)^2 + (2\,\xi u)^2]^{1/2}} \quad ...(29.60)$$

where $u$ = normalized frequency = $\omega_1/\omega_n$ ...(29.61)

The phase angle may be written as,

$$\phi = \tan^{-1}\left(\frac{2\,\xi u}{1-u^2}\right) \text{rad} \quad ...(29.62)$$

A plot of Eqn. 29.60 is given in Fig. 29.55. It is clear from the diagram that the output amplitude is almost equal to the input when $\zeta = 0.7$ and $\omega_1/\omega_n > 2$.

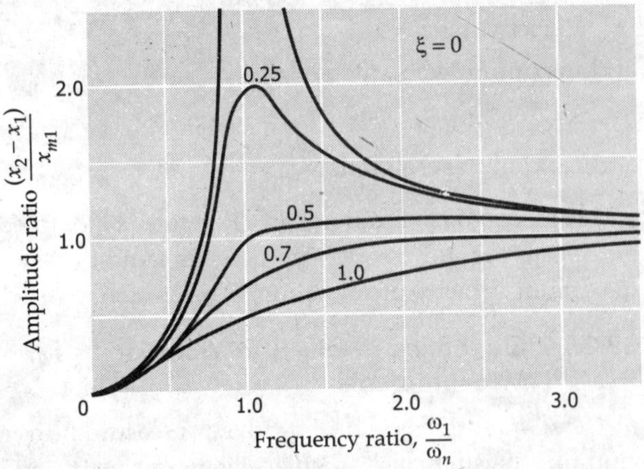


**Fig. 29.55** Dynamic response of a Seismic transducer.

For low values of damping ratio, $\zeta$ the amplitude may become quite large. The output becomes essentially a linear function of input at high frequency ratios. Thus a seismic-vibration transducer should be used for measurement of displacement amplitude at frequencies substantially higher than its natural frequency. Now the natural frequency is $\omega_n = \sqrt{K/M}$. Therefore, this type of transducer (which requires small natural frequencies) should use weak springs (which have low value of *K*) mid heavy mass, *M*.

The acceleration amplitude of the input vibration is :

$$a_m = \ddot{x}_1 = \omega_1^2\,x_{m1} \quad ...(29.53)$$

We may thus use the measured output of the instrument as an indication of acceleration and call their instrument as **accelerometer**. However, there are problems associated with this application. Figure 29.62 shows the plot of $(x_2 - x_1)_m \ \omega_n^2 / a_m$ versus $\omega_1 / \omega_n$.

It is clear from Fig. 29.56 that the frequency response is unsatisfactory for frequency ratios greater than 0.4 as the curves tend to be non-linear. Thus if the device is to be used for acceleration measurements, the input frequency should be much lower than the natural frequency of the accelerometer. In order that the instrument be used for higher frequencies, it should be designed to have a high natural frequency. This requires a stiff spring (high value of $K$) and a small mass $M$.

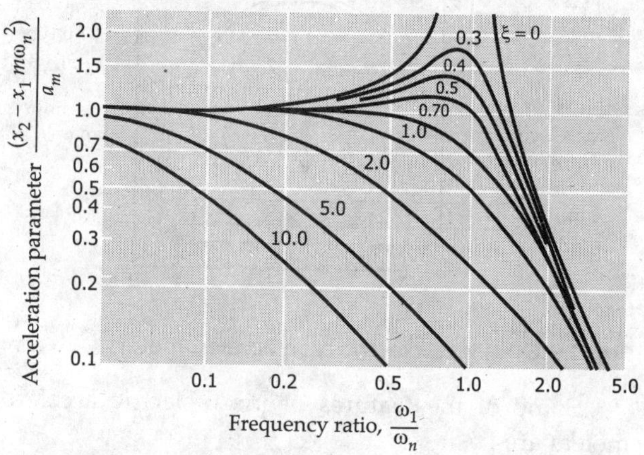


**Fig. 29.56** Frequency response of a Seismic transducer.

The most important transducer for vibration, shock and general purpose absolute motion is the **accelerometer**. The instrument is commercially available in a wide variety of types and ranges in order to meet diverse application requirements.

## 29.27 TYPES OF ACCELEROMETERS

The variety of accelerometers used results from different applications with varied requirement of range of natural frequency and damping. The specification sheet for an accelerometer gives the natural frequency, damping ratio, and a scale factor which relates output with the acceleration input.

### 29.27.1 Potentiometric Type Accelerometer

This is the most simple type of accelerometer is shown in Fig. 29.57. The seismic mass is attached to the wiper arm or resistance potentiometer. The relative motion of the mass with respect to the transducer frame is sensed either as a change in resistance or as a change in voltage output (if the potentiometer is used as potential divider). The damping may be provided by filling the housing of the accelerometer completely with a viscous fluid or it may be provided by a dashpot. Proper damping is necessary because it increases the range of frequencies over which the transducer may be used.

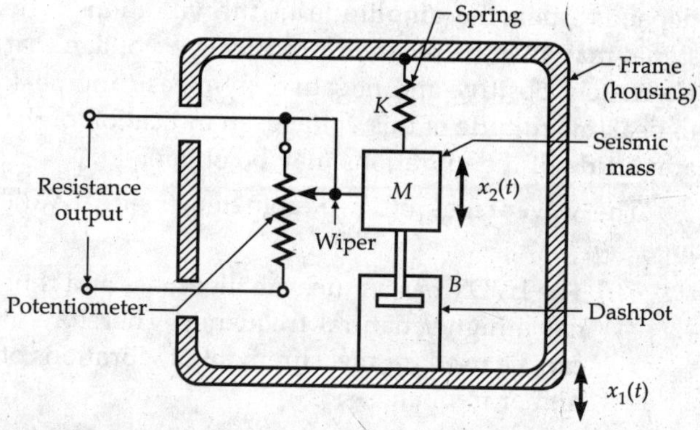


**Fig. 29.57** Potentiometric Accelerometer.

The major drawbacks of a seismic accelerometer using resistance potential divider are its limited resolution and a rather low natural frequency. This frequency is generally lower than 100 Hz and hence its application is limited to input frequencies lower than 50 Hz. Hence, this instrument is useful only for low frequency applications.

The instrument also gives errors on account of its sliding contacts.

### 29.27.2 LVDT Accelerometers

Figure 29.58 shows a seismic accelerometer using a linear voltage differential transformer (LVDT). The core of the LVDT acts as the mass and two flexible reeds, attached at each end of the rods of the core, provide the necessary spring action. The reeds are attached to a housing, which is subject to vibrations.

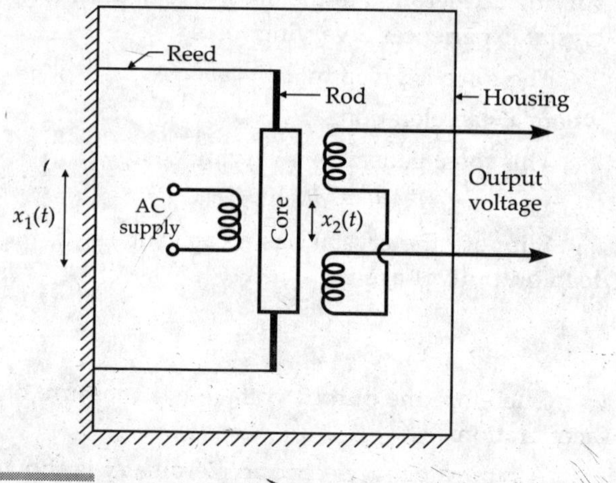


**Fig. 29.58** Seismic accelerometer using LVDT.

The above arrangement is necessary in order that the core of the LVDT is maintained at its null position. As the sensor moves up and down on account of vibrations, the LVDT secondaries give an a.c. output voltage, first of one phase and then, alternately of the opposite phase. The magnitude of this output signal depends upon the amplitude of the vibrations. The signal may then be rectified producing a voltage that alternately positive and negative. By measuring peak to peak magnitude of this voltage, an indication of the amplitude of the vibrations may be obtained.

The advantages of accelerometers using LVDT are :

(*i*) The LVDT has a much smaller mass and thus has a higher natural frequency. Therefore, it can, be used for measurement of vibrations of higher frequencies.

(*ii*) The LVDT offers a lower resistance to the motion than the potentiometer and is capable of a much better resolution.

(*iii*) It is a contactless device and is therefore free from problems which arise on account of moving contacts.

The LVDT accelerometers are used for steady state and low frequency vibration measurements.

### 29.27.3 Piezo-Electric Accelerometers

Piezo-electric transducers have been explained in details in Art. 25.29, page 826.

When a force $F$ is applied to a piezo-electric crystal it develops a charge $Q = dF$ coulomb (See Eqn. 25.130) where $d$ = charge sensitivity of crystal ; C/N.

Through incorporation of a mass, $m$, in direct contact with the crystal, we have the essential components of an acceleration transducer. By applying a varying acceleration to the mass-crystal assembly, the crystal experiences a varying force.

The force is given by $F = m \times a$

where $a$ = acceleration.

This force generates a varying charge :

$$Q = dF = dMa$$

Suppose the crystal has a capacitance, C, the no load output voltage is

$$e_0 = \frac{Q}{C} = \frac{dF}{C} = d\,\frac{Ma}{C} \qquad \text{...(29.63)}$$

Therefore the output voltage is a measure of the acceleration.

A typical piezo-electric accelerometer is shown in Fig. 29.59. The piezo-electric crystal is spring loaded with seismic mass in contact with the crystal. When subjected to an acceleration, the seismic mass stresses the crystal to a force $F = Ma$, resulting in a voltage generated across the crystal. This force, generates an output voltage which is proportional to the acceleration.

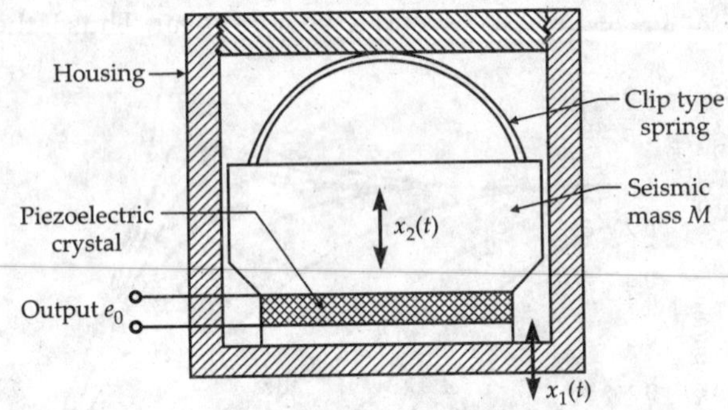


**Fig. 29.59** Piezo-electric type accelerometer.

Some of the features of piezo-electric accelerometers are :

(*i*) The instrument is quite small in size and has a small weight (Typically 0.025 kg).

(*ii*) The natural frequency is very high. It may be as high as 100 kHz and therefore the accelerometer is useful for high frequency applications. They can be used for any vibration and shock applications.

The primary elements of importance in shock measurements are that the device should have a natural frequency which is greater than 1 kHz and a range typically greater than 500 g (*i.e.*, $g = 9.81$ m/s$^2$). The only accelerometer that can usually satisfy these requirements is the piezo-electric type.

(*iii*) The crystal is a source with a high output impedance and in order to avoid loading effect, a voltage monitoring source of a high input impedance should be used.

Electrical impedance matching between transducer and readout circuitry is usually a critical matter requiring a very careful design consideration.

(*iv*) These accelerometers are useful for high input frequencies and their response is poor at low frequencies. Therefore, they should not be used for applications where the input frequency is lower than 10 Hz.

**Example 29.28** *An accelerometer has a seismic mass of 0.05 kg and a spring constant of $3 \times 10^3$ N/m. Maximum mass displacement is ± 0.02 m (before the mass hits the stop). Calculate (a) maximum measurable acceleration and (b) natural frequency.*

**Solution.** (*a*) Natural frequency

$$\omega_n = \sqrt{K/M} = \sqrt{3\times10^3/0.05} = 245 \text{ rad/s.}$$

Maximum acceleration

$$a_m = \omega^2 x_m = (245)^2 \times (0.02) = 1200 \text{ m/s}^2$$

(*b*) Natural frequency $f = 245/2\pi = 20$ Hz.

***Example 29.29*** *An accelerometer has a damping ratio of 0.7. Calculate the value of frequency ratio so that ratio of steady relative displacement to amplitude of input displacement is 0.99 (i.e., the error is 1.0%).*

**Solution.** From Eqn. 29.60, we have ratio of amplitudes of output and input

$$\frac{(x_2-x_1)_0}{x_{m1}} = \frac{u^2}{\sqrt{(1-u^2)^2+(2\xi u)^2}}$$

$$= \frac{u^2}{\sqrt{(1-u^2)^2+(2\times0.7u)^2}} = 0.99$$

$\therefore$ $u$ = normalized frequency $= \omega_1/\omega_n = 2.45$

***Example 29.30*** *A seismic instrument has a natural frequency of 4 Hz and a damping ratio of 0.66. If the system is excited by a frequency 6 Hz, determine the error due to the proximity of excited frequency with natural frequency of the instrument.*

**Solution.** The ratio of output displacement to input displacement is given by

$$\frac{(x_2-x_1)_m}{x_{m1}} = \frac{u^2}{\sqrt{(1-u^2)^2+(2\xi u)^2}} \quad \text{(See Eqn. 29.60)}$$

Now normalized frequency $u = 6/4 = 1.5$

$$\therefore \frac{(x_2-x_1)_m}{x_{m1}} = \frac{(1.5)^2}{\sqrt{(1-1.5^2)^2+(2\times0.66\times1.5)^2]^{1/2}}} = 0.961$$

$\therefore$ Error $= (0.961-1)\times100 = -3.9\%$

***Example 29.31*** *An LVDT is used in an accelerometer to measure seismic mass displacements. The LVDT and signal conditioning outputs are 0.31 mV/mm with a ± 20 mm core displacement. The spring constant is 240 N/m and the core mass is 0.05 kg. Find (a) relation between acceleration in m/s² and the output voltage, (b) natural frequency, and (c) maximum acceleration measurable.*

**Solution.** (*a*) The amplitude of displacement is proportional to the amplitude of vibration.

$\therefore$ Sensitivity of accelerometer

$$= \frac{\text{m/s}^2}{\text{mm}} \times \frac{\text{mm}}{\text{mV}} = \frac{96}{20} \times \frac{1}{0.31} = 15.5 \text{ ms}^{-2}/\text{mV}$$

(*b*) Natural frequency

$$\omega_n = \sqrt{K/M} = \sqrt{240/0.05} = 69.3 \text{ rad/s}$$

Maximum measurable acceleration

$$= \omega^2 A = (69.3)^2 \times 20\times10^{-3} = 96 \text{ m/s}^2$$

***Example 29.32*** *An accelerometer gives an output of 14 mV per g where = 9.81 ms⁻². Design signal conditioning that provides :*

(*a*) *a velocity signal scaled at 0.25 V/ms⁻¹ and*

(*b*) *determine the gain of the system and feedback resistance ratio.*

**Solution.** Sensitivity $= \dfrac{14\text{ mV}}{9.81\text{ ms}^{-2}} = 1.43\dfrac{\text{mV}}{\text{ms}^{-2}}$

This is an acceleration signal to be converted to a velocity signal, hence an integrator is used.

The circuit is shown in Fig. 29.60.

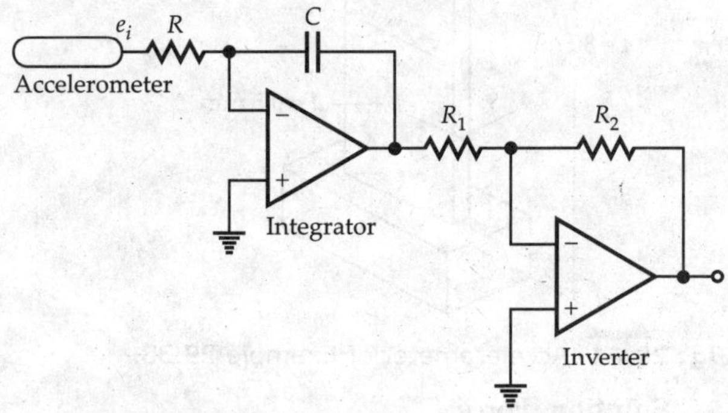


**Fig. 29.60** An integrator and an inverter used for obtaining velocity information for accelerometer (Example 29.32).

Output of the integrator $= -\dfrac{1}{RC}\int e_i\,dt$

Take $RC = 1$ s

$\therefore$ Output of integrator is :

$$\frac{1.43\text{ mV}}{\text{ms}^{-2}} \times \frac{1}{1\text{ s}} = -1.43 \text{ mV/ms}^{-1}$$

$$\text{Required gain} = \frac{0.25\text{ V/ms}^{-1}}{1.43\times10^{-3}\text{ V/ms}^{-1}} = 174.8$$

This can be obtained by using an inverter with

$$R_2 = 174.8 \text{ k}\Omega \quad \text{and} \quad R_1 = 1 \text{ k}\Omega$$

***Example 29.33*** *An accelerometer shown in Fig. 29.61 consists of mass weight 0.5 N attached to a spring cantilever. The spring is made up of a material of modulus of elasticity 200 GN/m² and is of uniform rectangular cross-section. The output is obtained from a strain gauge bridge comprising four active strain gauges of gauge factor 2. Two strain gauges are mounted on each side of the cantilever. The gauges are connected in a bridge circuit with input voltage of 3 V d.c. Find the bridge output voltage at open circuit for a horizontal acceleration of 4 g where g = 9.81 ms⁻². The stress is given by s = 6 Fl / bd² N / m².*

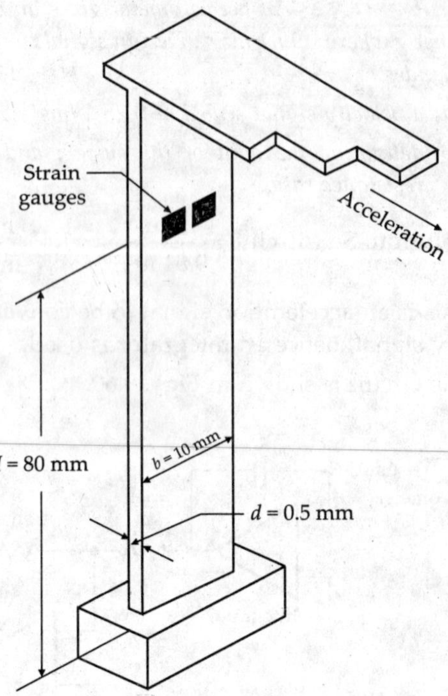


**Fig. 29.61** Accelerometer of Example 29.33.

**Solution.** Force $F$

$$= m \times a = \frac{0.5}{g} \times 4g = 2 \text{ N}$$

Stress $s$

$$= 6\ Fl / bd^2$$

$$= 6 \times 2 \times 80 \times 10^{-3} / (10 \times 10^{-3}) \times (0.5 \times 10^{-3})^2$$

$$= 384 \times 10^6 \text{ N/m}^2 = 383 \text{ MN/m}^2$$

Strain $\varepsilon$

$$= \frac{s}{E} = \frac{384 \times 10^6}{(200 \times 10^9)} = 1.92 \times 10^{-3}$$

Relative change in resistance of gauge

$$= \Delta R / R = G_f \varepsilon$$

$$= 2 \times 1.92 \times 10^{-3} = 3.84 \times 10^{-3}\ \Omega$$

Open circuit voltage of a bridge using 4 active strain gauges is :

$$e_0 = \Delta R / R\ e_i = 3.84 \times 10^{-3} \times 3 \text{ V} = 11.52 \text{ mV}$$

## 29.28 MEASUREMENT OF TEMPERATURE

The following devices are used for the measurement of temperature :

- Resistance thermometers,
- Semiconductor thermometers,
- Thermistors,
- Thermocouples, and
- Bimetallic Thermometers,
- Radiation Pyrometers.

### 29.28.1 Electrical Resistance Thermometer

The principle of electrical resistance thermometers has already been explained in Art. 25.19, page 975. The resistance of metals changes with changes with change in temperature. The resistance thermometer uses the change in electrical resistance of a metallic conductor to determine the temperature. The requirements of a conductor material to be used in resistance thermometers are :

(*i*) the change in resistance of the material per °C should be as large as possible, and

(*ii*) the resistance of the material should have a continuous and stable relationship with temperature.

The resistance temperature characteristics of nickel, copper and platinum are shown in Fig. 25.63, page 786.

### 29.28.2 Platinum Resistance Thermometer

The most commonly used material for metallic resistance thermometer is platinum. The resistance temperature characteristics of pure platinum are very well defined and show a high degree of capability. Therefore, these resistance thermometer elements are used over a wide range to establish the Internal temperature scale.

Resistance typè temperature bulbs use sensing elements in the form of wires or foil. The films deposited on insulating surfaces are also used for temperature sensing. In the wire type, the arrangement is commonly a helical coil wound as a double wire to avoid inductive effects. The laboratory type resistance thermometers have the temperature sensing element wound on a cross mica former and enclosed in a Pyrex tube as shown in Fig. 29.62(*a*). The tube may be evacuated or filled with in gas to protect the platinum.

The industrial type of thermometer is shown in Fig. 29.62(*b*) the former being of grooved ceramic and the wire being protected by a glass coating or by a stainless steel tube. The element is normally sealed in glass when used for temperatures upto 150 °C and ceramic for use in temperatures up to 850 °C. This sealing has the two-fold advantage of providing structural strength and protection from chemical attack. Resistance thermometers are sometimes used above 850 °C but they have a reduced life.

Resistance elements are also available as thin etched grids of metal foil similar in shape to foil type strain gauges. They are constructed of platinum and may be bonded to a plastic backing for attachment to a surface. They may be open-faced or coated and have a fast response compared with bulb type thermometers. Thin film sensors have an extremely fast response and they are extensively used in ares- space industry.

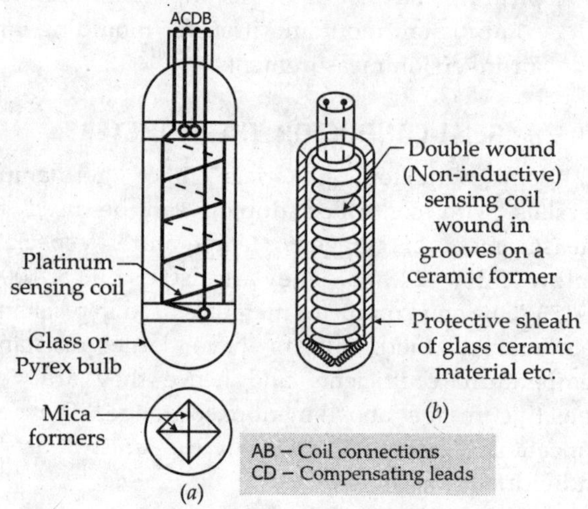


**Fig. 29.62** Resistance thermometer.

### 29.28.3 Measurement of Resistance of Thermometers

The measurement of change of resistance of thermometer due to temperature changes is measured by Wheatstone bridge. Ordinary Wheatstone bridge is not used for measurement of changes in resistance as it has many disadvantages as explained below :

(*i*) The contact resistance of the adjustable standard resistor may be large enough to produce an error when measuring the change in resistance of thermometer.

(*ii*) The leads from the thermometer to the bridge may introduce an error due to change of their resistance produced by temperature changes.

(*iii*) The current through the thermometer produces a heating effect equal to the product of the current squared and the resistance of thermometer.

Slight modification of the Wheatstone bridge, such as a double slide wire bridge, eliminate most of these problems.

#### 1. Three Lead Method

A double slide wire bridge is shown in Fig. 29.63. It has two slide wire resistors $S_1$ and $S_2$ which are tied together so that the fraction of $S_1$ in series with the resistance $R_2$ is equal to the fraction of $S_2$ in series with resistance $R_3$. This fraction is defined as $f$.

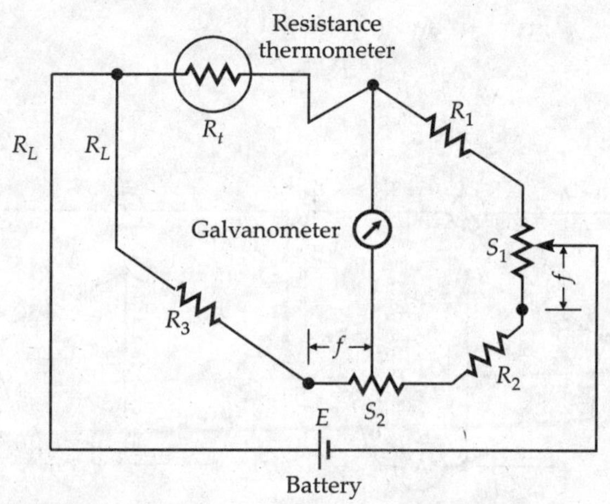


**Fig. 29.63** Measurement of temperature with double slide wire bridge and resistance thermometer.

There are three equal leads from the thermometers to the bridge. The resistance of each lead is $R_L$

$\therefore$ At balance we have :

$$\frac{R_t + R_L}{fS_2 + R_3 + R_L} = \frac{R_1 + S_1 - fS_1}{fS_1 + R_2 + S_2 - fS_2}$$

If the right hand side of the above expression is unity, the resistance of thermometer element is

$$R_t = fS_1 + R_3$$

This condition is obtained when :

$$R_1 + S_1 - fS_1 = fS_1 + R_2 + S_2 - fS_2$$

This is an identity when :

$$R_1 - R_2 = S_1 = 0.5\ S_2$$

The bridge is designed using these values.

Since there are three leads coming out of the resistance thermometers, the method is known as **three lead method.**

#### 2. Four Lead Method

In applications, where the highest degree of accuracy is required, the **four lead method** is used. Such a system is used with a platinum resistance thermometer employed as a laboratory standard for calibration purposes. In this method, two circuit arrangements are used. They are shown in Figs. 29.64(*a*) and (*b*). In fact, both the arrangements are required for measurement purposes. First a measurement is made using circuit of Fig. 29.72(*a*) and then a second reading is taken by using circuit of Fig. 29.72(*b*). The average of the two readings is taken to give the correct result.

For circuit of Fig. 29.64(*a*), we have :

$$R_a + C = R_t + T \qquad ...(i)$$

For circuit of Fig. 29.64(*b*), we have :

$$R_b + T = R_t + C \qquad ...(ii)$$

∴ From (*i*) and (*ii*) we get :

$$R_t = \frac{R_a + R_b}{2}$$

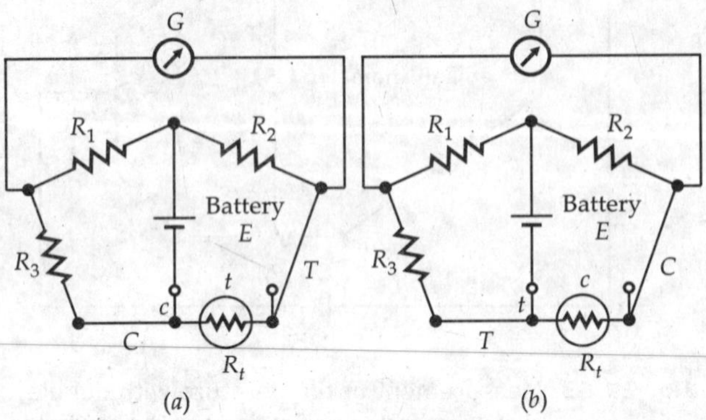


**Fig. 29.64** Four lead method for connection of resistance thermometer.

It is evident that this method should be used only when accuracy of the highest order is desired since the method is both time consuming and inconvenient.

It should be noted that two separate leads are connected directly to each end of the resistance winding.

These are leads C and *c* on one side and *T* and *t* on the other side.

### Salient Features of Resistance Wire Thermometers

- Resistance wire thermometers have a higher degree of accuracy. Precision laboratory instruments may be calibrated for measuring temperatures to within ± 0.01 °C. Industrial instruments called **"resistance temperature detectors (RTD)"** can be calibrated to detect the actual temperatures to within ± 0.25 °C up to 120 °C, and ± 0.5 °C from 120°C to 550 °C.
- Resistance thermometers are normally designed for fast response, as well as accuracy to provide close control of processes in which narrow ranges or small temperature spans must be maintained.
- Each type of resistance thermometer is interchangeable in a process without compensation or relibration, because each type is calibrated with reference to a standard resistance-temperature curve. Thus if a unit is damaged it can be easily replaced.
- Typical industrial applications of resistance thermometers for measuring temperature are cooling processes, heating ovens, drying ovens, kilns, process vessels, baths, quenches, refining, controlled cold storage plants, steam and power generation condensates, steam exhausts, pickling and plating plants, injection moulding, compression moulding, transfer moulding, and air inversion measurements.

## 29.28 SEMI-CONDUCTOR THERMOMETERS

Semi-conductor materials like germanium crystals with controlled doping can be used for measurement of cyragenic temperatures especially below 25 K ( – 248 °C). They are also usable at 100 K (–173 °C). Semi-conductor materials used in resistance temperature thermometers have large resistance temperature coefficients and hence they are very sensitive in the above-mentioned range. They are repeatable within 0.01 °C, but the units must be individually calibrated.

Silicon crystals are now being used in the range – 48 °C to 252 °C. In this range, their resistance increases with increasing temperature, and the relationship is sufficiently linear. These thermometers must be individually calibrated.

Carbon resistors like those commonly used in electronic circuits are also used as temperature sensitive elements.

The ***advantages*** are :

(*i*) They are readily available and are inexpensive.

(*ii*) They are rugged and sensitive.

(*iii*) Their sizes are small. Common sizes vary from 0.25 to 1 watt power and 10 to 500 Ω resistance.

(*iv*) They have a good frequency response characteristics.

Their ***disadvantages*** are :

(*i*) They are not so reproducible as germanium.

(*ii*) They have a large negative resistance temperature co-efficient below – 213 °C.

## 29.30 THERMISTORS

The principle of working and constructional details of thermistors have already been explained in Art. 25.20 on page 788.

## 29.31 THERMOCOUPLES

The thermocouple is one of the simplest and most commonly used methods of measuring process temperatures. The operation of a thermocouple is based upon Seebeck effect. In 1821, Seebeck discovered that when heat is applied to junction (hot junction) of two dissimilar metals, an emf is generated which can be measured at the other junction (cold junction). The two dissimilar metals form an electric circuit, and a current flows as a result of the generated emf as shown in Fig. 29.65. This current will continue to flow as long as $T_1 > T_2$. Metal $B$ is described as – ve with respect to a metal $A$ if current flows into it at the cold junction.

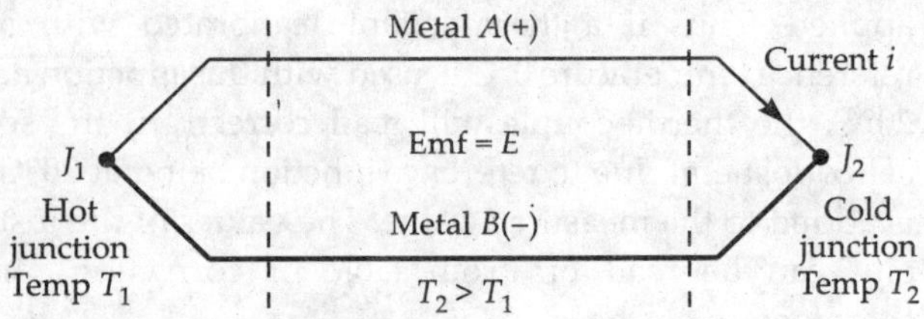


**Fig. 29.65** Basic circuit of thermocouple.

The emf produced is function of the difference in temperature of hot and cold junctions and is given by :

$$E = a\Delta\theta \qquad ...(29.64)$$

(See Art. 25.21, page 793)

where $\Delta\theta$ = difference between temperatures of hot and cold junctions.

### 29.31.1 Thermo-electric Laws

Several thermo-electric phenomena have been discovered which are called thermo-electric laws.

These are :

**1.** The application of heat to a single homogeneous metal is in itself not capable of producing or sustaining an electric current.

**2.** A thermo-electric emf is produced when the junctions of two dissimilar homogeneous metals are kept at different temperatures. This emf is not affected by temperature gradients along the conductors shown in Fig. 29.66.

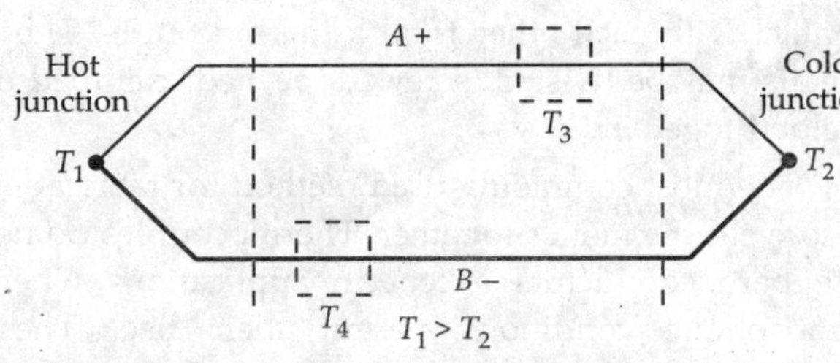


**Fig. 29.66** No emf is generated by temperature gradients in homogeneous conductors.

**3.** In a circuit consisting of two dissimilar homogeneous metals having the junctions at different temperatures, the emf developed will not be affected when a third homogeneous metal is made a part of the circuit, provided the temperatures of its two junctions are the same as shown in Fig. 29.67. rms is called *Law of Intermediate Metals.*

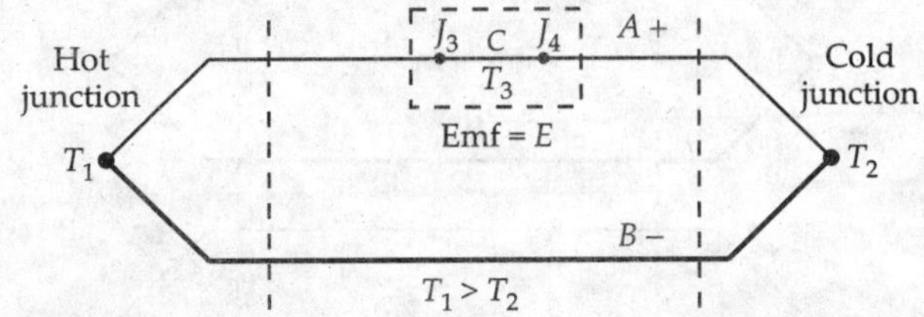


**Fig. 29.67** Introduction of a third metal does not affect the emf provided its two junctions are at the same temperature.

In practice, some means of measuring emf produced by a thermocouple have to be used. This usually requires the introduction of a third metal into the circuit. The law of intermediate metals states that the net emf in the circuit remains unaltered if a third metal is of the third metal are at the same temperature. The third metal may be introduced anywhere in series with the circuit without affecting the net emf provided junctions $J_3$ and $J_4$ are kept at the same temperature.

The applications of this law are :

(*i*) The law makes it possible to use extension wires of a metal different from the metals used for thermocouple. For example, the extension wires used for Platinum/Platinum rhodium thermocouple may be made of copper which is an inexpensive material.

(*ii*) The law enables a measuring instrument to be introduced into the circuit without affecting the emf generated by the thermocouple. (This is subject to the condition that the meter has an infinite impedance and therefore there are no loading effects.)

(*iii*) Another important consequence of this law is that the wires forming the junction can be soldered or brazed together without altering the performance of the junction.

(*iv*) The thermal emf of any two homogeneous metals with respect to another is the algebraic sum of their individual emfs with respect to a third homogeneous metal. Figure 29.68 shows that emf of metal $A$ with respect to $B$ is $E_{AB}$ and that of $B$ with respect to $C$ is $E_{BC}$ and therefore the emf of $A$ with respect to $C$ is $E_{AC} = E_{AB} + E_{BC}$.

29

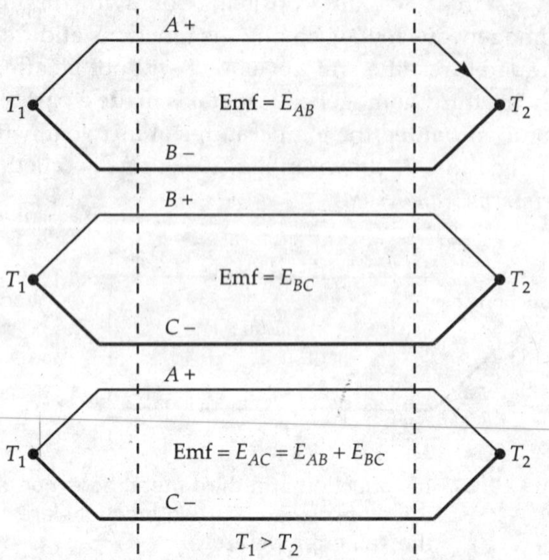


**Fig. 29.68** Thermal emfs of two metals with respect to one another is the algebraic sum of their individual emfs with respect to a third metal.

**4.** The thermal emf produced when a circuit of two homogeneous metals exists between a first temperature and a second and thermal emf produced when the same circuit exists between the second temperature and a third are algebraically equal to the thermal emf produced when the circuit exists between first and third temperatures. This is called *Law of Intermediate Temperatures.*

Figure 29.69 illustrates this law. The emf generated with junction temperatures $T_1$ and $T_3$ is equal to the sum of emfs produced by two similar thermocouples, one operating with junction temperature $T_1$ and $T_2$ and the other with $T_2$ and $T_3$ where $T_2$ lies between $T_1$ and $T_3$.

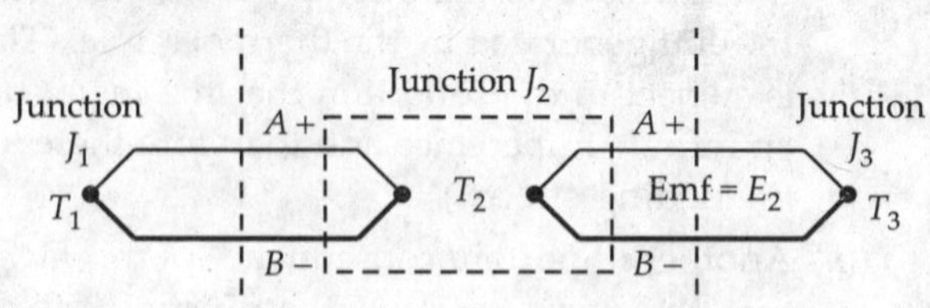


**Fig. 29.69** The algebraic sum of thermal emfs produced between junctions $J_1$ and $J_2$ and between $J_2$ and $J_3$ is equal to emf $E_3$ produced in a similar circuit between junctions $J_1$ and $J_3$.

Suppose the emf between junctions $J_1$ and $J_2$ is $E_1$ and that between junctions $J_2$ and $J_3$ is $E_2$, the resultant emf between junctions $J_1$ and $J_3$ is $E_3 = E_1 = E_2$.

This law is used while making corrections to the thermocouple readings in case reference (cold) junction temperature is different from one for which the thermocouple was calibrated. The calibration curves and charts are available for a reference junction temperature of 0°C. In case the temperature of reference junction is different from 0°C, the calibration curves no longer hold good and corrections have to be applied. Thus if a thermocouple calibrated with a reference temperature 0°C is used with this junction at 20°C, the thermocouple will read correctly if the so called lost emf due to reference junction being at 20°C is added to the measured value. The values of the lost emf can be read off from table of correction for thermocouple used.

**5.** The algebraic sum of the emfs produced in a circuit containing two or more thermocouples all at the satire temperature is zero.

**6.** The net emf of a circuit containing two thermocouples is unaffected by the addition of more thermocouples at the same temperature as any of the others as shown in Fig. 29.70.

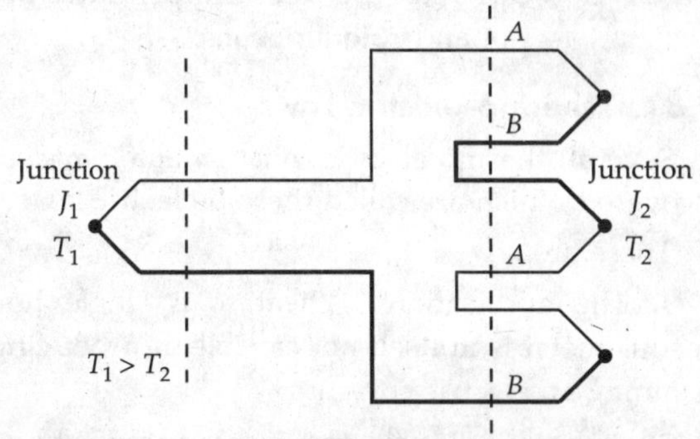


**Fig. 29.70** Additional thermocouples at junctions $J_1$ and $J_2$ not affect the net emf.

## 29.32 CONSTRUCTION OF THERMOCOUPLES

A pair of two dissimilar metals that are in physical contact with each other form a thermocouple. These metals may be twisted, screwed, peened, clamped or welded together.

The most commonly used method for fabricating is to weld the metals together. Thermocouples do not use bare conductors except in applications where atmospheric conditions permit their use. These conditions obtain when temperatures to be measured are low and the atmosphere is non-corrosive. Industrial thermocouples employ protective sheathing surrounding the junction and a portion of the extension leads. The leads and the junction are internally insulated from the sheath, using various potting compounds, ceramic beads or oxides. The type of insulation used depends upon the process being monitored. The different types of protective sheaths are shown in Fig. 29.71.

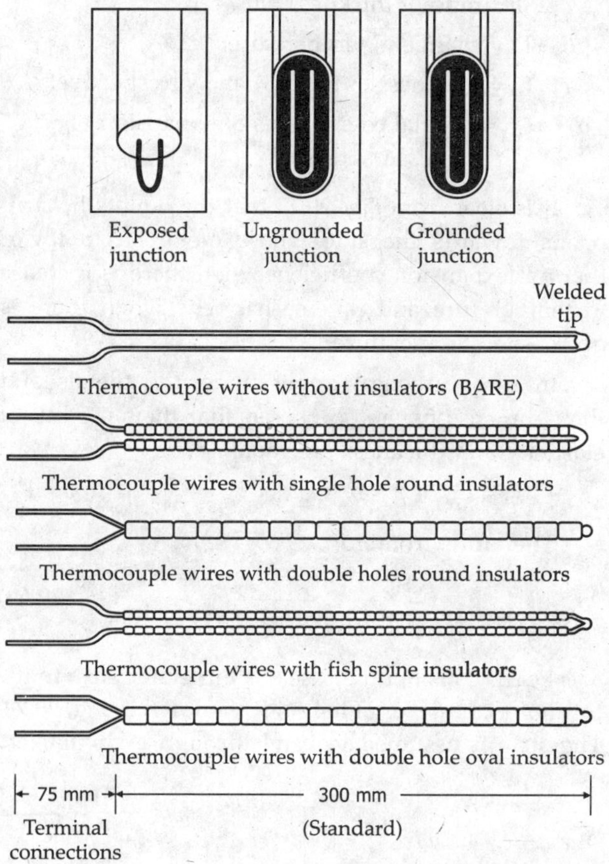


**Fig. 29.71** Protective sheaths for thermocouples.

Thermocouples are normally not installed in pipelines vessels or other pieces of equipment directly. They are usually placed inside protective wells so that they may be easily removed or replaced without interruption or shut down of the plant. Protecting wells are made of stainless steel and some other special alloy materials. They are normally 12.5 mm to 25 mm in diameter.

The use of protective wells slows down the response appreciably as they increase the mass of thermocouples. However, in applications, where response time is the primary consideration, bare or thin sheathed thermocouples are used.

Read about the rest of thermocouples from pages 793 to 798.

## 29.33 QUARTZ CRYSTAL THERMOMETER

A quartz crystal possesses a property that its resonant frequency changes with the change in temperature. This property is made use of in **quartz crystal thermometer** for measurement of temperature. This is a novel and extremely accurate method for measurement of temperature.

If the quartz crystal is given a proper angle cut there exists a linear relationship between resonant frequency and the temperature. The resonant frequency is, therefore, a measure of the temperature. The frequency is measured with the help of digital frequency meters and is displayed.

The advantage of this thermometer is that sensitivities of the order of 0.001 °C are obtained. Also the thermometer is insensitive to stray pickups in the connecting cables since the output is are frequency which is measured by digital techniques.

## 29.34 BIMETALLIC THERMOMETERS

Bimetallic thermometers are extensively used in process industries for local temperature measurements. These thermometers use two fundamental principles :

(*i*) all metals expand or contract with change in temperature and,

(*ii*) the temperature co-efficient of expansion is not the same for all metals and therefore their rates of expansion or contraction are different. The difference in thermal expansion rates is used to produce deflections proportional to temperature changes.

A bimetallic thermometer consists of a bimetallic strip which is by bonding together two thin strips of two different metals such that they cannot move relative to each other. Since all metals try to change their physical dimensions at different rates when subjected to same change in temperature, these two metallic strips change their lengths at different rates. The differential change of expansion of two metals results in bending of the bimetallic strip with change in temperature.

Figure 29.72 shows a bimetallic strip in the form of a straight cantilever beam. With one end fixed, the temperature changes cause the free end to deflect.

The range over which a linear relationship exists between deflection and temperature depends upon the combination of metals used for the bimetallic strip. The deflection of the free end is directly proportional to the temperature change and square of the length of strip, and inversely proportional to the thickness throughout the linear portion of deflection-temperature characteristics.

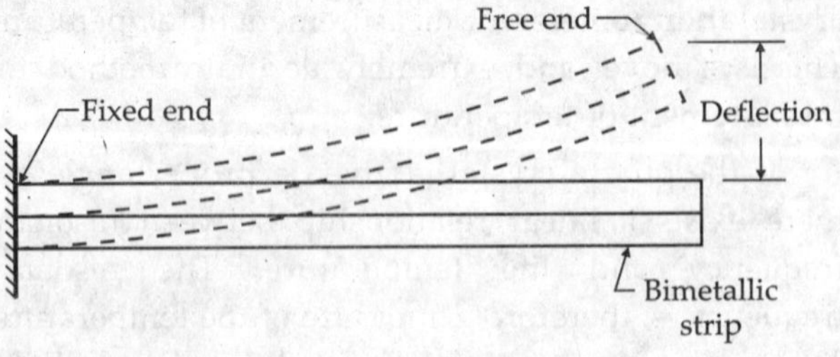


Fig. 29.72 Bimetallic strip fixed at one end.

Figure 29.73 shows a bimetallic strip made up of two metals $A$ and $B$, having different thermal expansion coefficients, bonded together at a temperature $T_1$. A change in temperature, $(T_2 - T_1)$ causes a differential expansion of the strip and if the motion is unstrained,

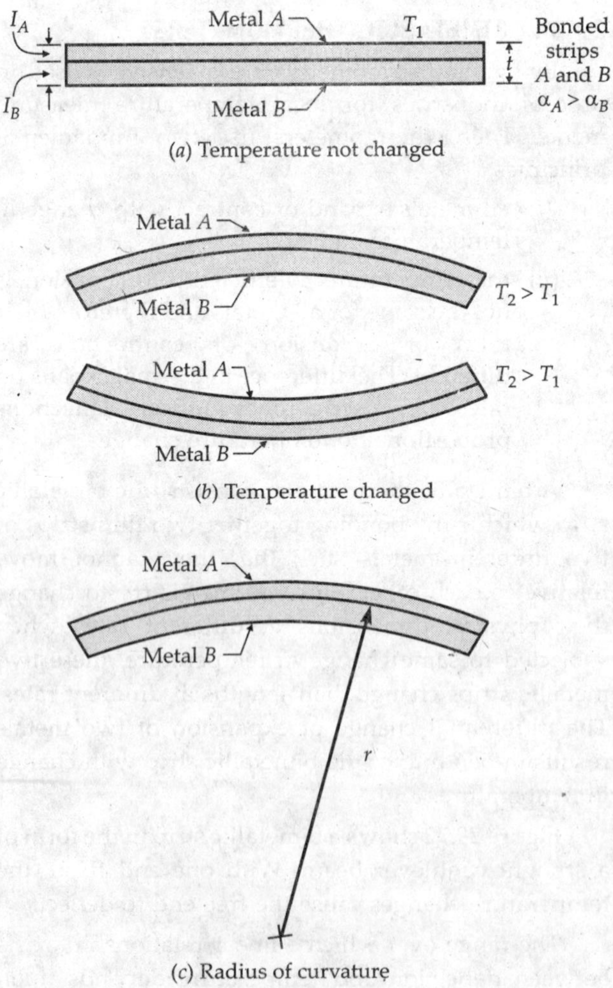


Fig. 29.73 Bonded bimetallic strip (unstrained) and its deflection when subjected to change on temperature.

the strip deflects into a uniform circular arc. The radius of arc is given by :

$$r = \frac{t[3(1+m)^2 + (1+mn)(m^2 + 1/mn)]}{6(\alpha_A - \alpha_B)(T_2 - T_1)(1+m)^2} \quad ...(29.65)$$

where, $t$ = total thickness of strip,

$n$ = ratio of moduli of elasticity = $E_B / E_A$,

$m$ = ratio of thickness = $t_B / t_A$,

$T_2 - T_1$ = change in temperature,

$t_A, t_B$ = thickness of metal $A$ and $B$ respectively,

$\alpha_A, \alpha_B$ = thermal co-efficients of expansion of metals $A$ and $B$

It is clear from Fig. 29.73 that the bimetallic strip bends towards the side whose metal has a lower thermal expansion coefficient when there is increase in temperature and reverse happens when there is decrease in temperature.

In most practical applications, the metals and their dimensions are so chosen that their moduli of elasticity and thickness are equal *i.e.*,

$$E_B = E_A \text{ or } n = 1 \quad \text{and} \quad t_A = t_A = t_B \text{ or } m = 1$$

Therefore, from Eqn. 29.65, we have,

$$r = \frac{2t}{3(\alpha_A - \alpha_B)(T_2 - T_1)} \quad ...(29.66)$$

Let us consider the case of a bimetallic strip in the form of a cantilever of length $L$ as shown in Fig. 29.74. The strip is assumed to bend through a circular arc

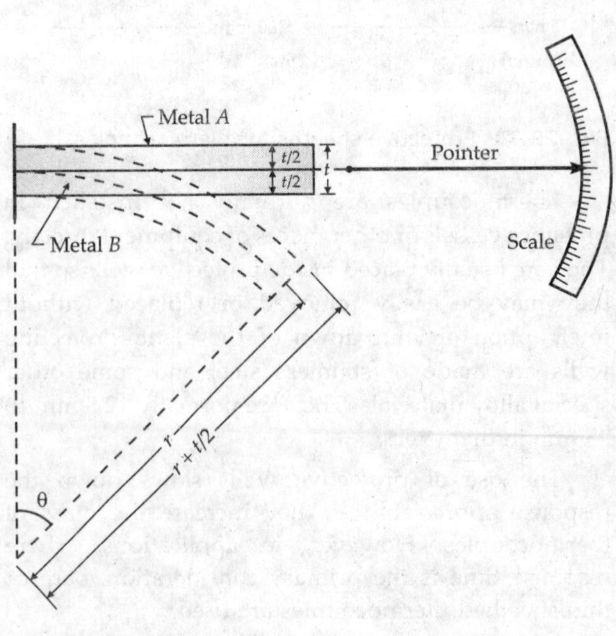


Fig. 29.74 Deflection of Bimetallic Strain.

when subjected to a change in temperature. The thickness of each metal forming the strip is $t/2$. Therefore,

$$\frac{r+t/2}{r} = \frac{\text{expanded length of strip } A}{\text{expanded length of strip } B}$$

$$= \frac{L[1+\alpha_A(T_2-T_1)]}{L[1+\alpha_B(T_2-T_1)]}$$

or
$$r = \frac{t}{2}\left[\frac{1+\alpha_B(T_2-T_1)}{(\alpha_A-\alpha_B)(T_2-T_1)}\right] \quad ...(29.67)$$

In case one of the metals has a very low thermal expansion co-efficient say $B$, we have $\alpha_B \approx 0$.

$$\therefore \quad r = \frac{t}{2\alpha_A(T_2-T_1)} \quad ...(29.68)$$

From the above relationship it is clear that if one end of the bimetallic strip is fixed, the deflection of the free end is a direct indication of the temperature of strip. The displacement of the free end can be converted into an electrical signal through use of secondary transducers like variable resistance, inductance and capacitance transducers.

The bimetallic strip described above is unsuitable for use in industrial applications because the deflection is small and hence the sensitivity is low. The sensitivity can be increased by increasing the length of strip. In order to keep the size of the thermometer within manageable limits, it is formed in the shape of either a spiral or a helix. This arrangement permits a very long length of strip to be compressed into a fairly small space.

A bimetallic strip using a spiral is shown in Figure 29.75(*a*). The curvature of strip varies with temperature thereby causing the pointer to deflect. Figure 29.75(*b*) shows a bimetallic strip wound in the form of a helix. One end of the helix is fastened to the causing of the bulb and the other end is connected to the pointer. The pointer sweeps over a circular dial to indicate the temperature.

The spiral strip is often used in ambient temperature measurement devices and air conditioning thermostats. The helical strip is used for most process applications because of its ability to be manufactured into a small diameter sheath while maintaining ruggedness.

In making bimetallic thermometers metals are chosen which have widely different thermal expansion co-efficient. Invar (an alloy of nickel and iron) is the mostly commonly used low expansion material. Its thermal expansion co-efficient remains stable over a wide temperature range. Nickel Iron alloys with chromium and manganese added are often used for thermal expansion material.

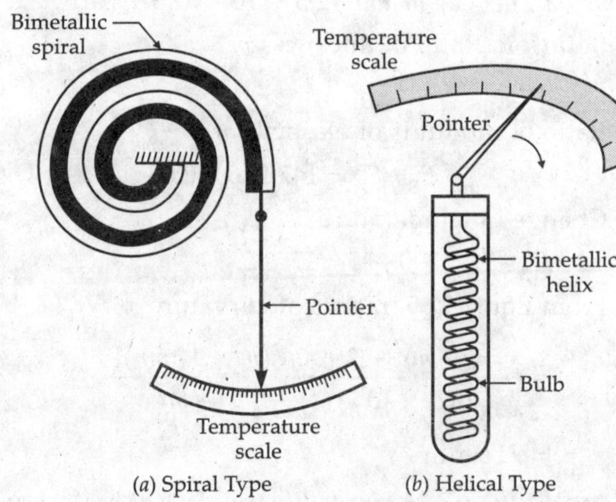


**Fig. 29.75** Industrial type bimetallic thermometers.

Bimetallic thermometers, like other industrial temperature measuring devices, are usually mounted in wells to provide protection against wear and corrosion. However, the use of protective wells increases the response time.

The advantages of bimetallic thermometers are that they are simple, robust and inexpensive. Their accuracy ranges from ± 0.5% for laboratory type of about ± 2% for process type instruments. These thermometers, in general can withstand about 50% over-range in temperature.

Bimetallic thermometers are not recommended for use at temperatures above 400°C for continuous duty of above 550°C for intermittent duty. All metals have physical limitations and are subject to permanent warp distortion. This means that metals do not return to their normal condition and therefore, temperatures indicated are not correct. Bimetallic thermometers are used for measurement of temperature between – 40°C and 550°C and are expected to measure with an accuracy of ± 1% when used below 400°C for continuous service.

Bimetallic thermometers are used in refineries, oil burners, tire vulcanizers, hot solder tanks, hot wire heaters, tempering tanks, and imperegnating tanks.

***Example 29.33*** *A bimetallic thermometer is made up of strips of a nickel chromium alloy and Invar bonded together at 25°C. Each strip has a thickness of 1 mm and a length of 50 mm. Calculate the radius of curvature produced* *when the strip is unstrained and is subjected to a temperature of 200°C. For nickel chrome alloy and Invar the moduli of elasticity and co-efficients of expansion are :*

*216 GN/m², 147 GN/m² and 12.5 × 10⁻⁶°C, 1.7 × 10⁻⁶°C.*

**Solution.** Ratio of thickness $m$

$$= t_A / t_B = 1$$

Ratio of modulus of elasticity $n$

$$= E_B / E_A = 147/216 = 0.68$$

Change in temperature

$$T_2 - T_1 = 200 - 25 = 175°C$$

From Eqn. 29.65, radius of curvature

$$r = t\frac{[3(1+m)^2 + (1+mn)(m^2 + 1/mn)]}{6(\alpha_A - \alpha_B)(T_2 - T_1)(1+m)^2}$$

$$= 711 \text{ mm}$$

***Example 29.34*** *A bimetallic strip element has one end fixed and other free, with the length of cantilever being 40 mm. The thickness of each metal is 1 mm, and the element is initially straight at 20°C. Calculate the movement of the free end in a perpendicular direction from the initial line when the temperature is raised to 180°C. One of the metals is Invar with a negligible thermal expansion co-efficient while the second is a nickel chrome alloy with an expansion co-efficient of $12.5 \times 10^{-6}$/°C.*

**Solution.** Thickness of bimetallic element $t = 2 \times 1 = 2$ mm. The co-efficient of expansion of one of the metals is negligible and hence the radius of curvature is given by Eqn. 29.84.

$$r = \frac{t}{2(T_2 - T_1)\alpha_A} = \frac{2}{2(180 - 0) \times 12.5 \times 10^{-6}}$$

$$= 500 \text{ mm}$$

Referring to Fig. 29.76

$$\theta = 40/500 = 0.08 \text{ rad} = 4.58°$$

∴ Vertical displacement

$$y = r(1 - \cos\theta) = 1.6 \text{ mm}$$

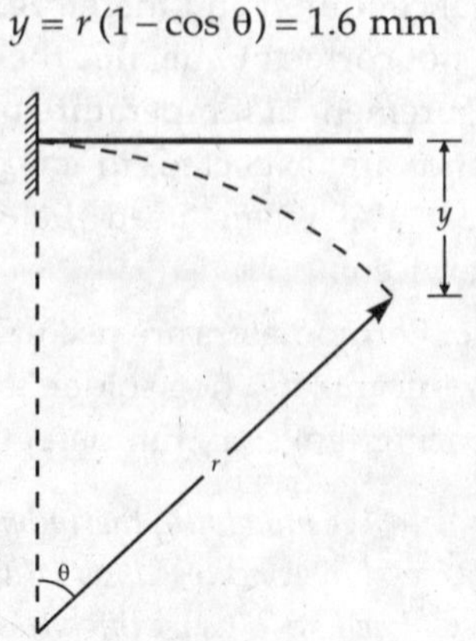


**Fig. 29.76** Diagram of Example 29.34.

## 29.35 RADIATION PYROMETERS

When temperatures being measured are high and physical contact with the process to be measured is impracticable or impossible, use is made of thermal radiation methods or optical pyrometers are used. These pyrometers are used under conditions where corrosive vapours or liquids would destroy thermocouples, resistance thermometers and thermistors if made to come in contact with the measured medium. These pyrometers find applications for temperatures which are above the range of thermocouples and also for rapidly moving objects.

Radiation pyrometry measures the radiant heat emitted or reflected by a hot object. Thermal radiation is electromagnetic radiation emitted as a result of temperature. Thermal radiation lies in the wavelength region from about 0.1 to 100 μm. Practical radiation pyrometers are sensitive to a limited band of radiant energy although the theory indicates that they should be sensitive to entire spectrum of energy radiated by the object. The operation of thermal radiation pyrometers is based upon blackbody concepts. The total thermal radiation emitted by a blackbody is :

$$q_b = \sigma T^4 \text{ W/m}^2 \quad ...(29.69)$$

where $\sigma$ = Stefan Boltzmann constant

$= 57.2 \times 10^{-9}$ W/m²–°K

and $T$ = absolute temperature ; °K

In practical applications for thermal pyrometers, the transfer of thermal energy takes place at temperatures above absolute zero. Prevost's theory of exchange states that for two black bodies in sight, each will radiate energy to the other, and

$$q_b = \sigma(T_1^4 - T_2^4) \text{ W/m}^2 \quad ...(29.70)$$

and $T_1$ and $T_2$ are absolute temperatures with $T_1 > T_2$

If $T_1$ is much higher than $T_2$, it is safe to preserve that the radiation is proportional to $T_1^4$, as, the term $T_2^4$ becomes insignificant. For example, suppose that the hot body temperature is 600°C and the instrument is at room temperature of say 30°C then $T_2^4$ is only about 1.5 per cent of $T_1^4$ and hence can be neglected.

In general a rough black surface radiates more heat than a smooth bright surface. This effect is called emissivity and is expressed as :

$$\varepsilon = q / q_b$$

$q$ = heat radiated by a gray body, W/m² ...(29.71)

$$= \varepsilon\,\sigma T^4 \text{ W/m}^2 \quad ...(29.72)$$

The value of emissivity from 0 to 1.

The energy is radiated over a wide range of frequencies of the electromagnetic spectrum. The distribution for any particular wavelength, $\lambda$ is given by Planck's radiation law,

$$q_{b\lambda} = \frac{C_1 \lambda^{-5}}{(e^{C_2/\lambda T} - 1)} \quad ...(29.73)$$

where $C_1$ and $C_2$ are constants.

Energy distribution curves for perfect black bodies are drawn in Fig. 29.77 for three different temperatures. The small band of visible radiation is indicated.

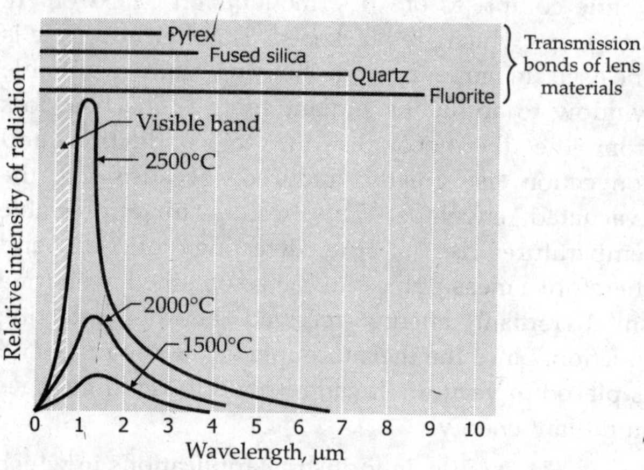


**Fig. 29.77** Distribution of constant for black bodies.

If a vertical line is drawn at a particular frequency value, it represents a particular intensity at each temperature. If a vertical band representing a range of frequencies is drawn, the energy radiated at a particular temperature is given by area in the band under that temperature curve. The values given in Fig. 29.85 are for perfect black bodies. These values should be multiplied by emissivity in order to get results for actual surfaces. Actual surfaces frequently exhibit highly variable emissivities over the wavelength spectrum. For the purpose of analysis, the actual surface is approximated as a gray body. A gray body is one which has constant emissivity for all wavelengths.

It is apparent that the intensity of radiation varies appreciably with wavelength. Also it is observed that the point of maximum radiant intensity shifts to the shorter wavelengths as the temperature increases. The point of maximum radiant intensity is given by Wien's displacement law :

$$\lambda_m T = 2900 \quad ...(29.74)$$

where $\lambda_m$ = wavelength at which maximum intensity occurs ; μm

This is a common phenomenon observed in the change of colour of a body being heated. A metal gradually heated changes its colour from red, which has a long wavelength to yellow and white as the intensity of radiation increases at the shorter wave lengths of the visible spectrum.

## 29.36 PRINCIPLE USED FOR RADIATION TEMPERATURE MEASURING DEVICES

There are two principles used for construction of radiation temperature measuring devices

**1. *Total Radiation Pyrometery.*** In this case the total radiant energy from a heated body is measured. This energy is represented by the area under the curves of Fig. 29.77 as explained earlier and is given by Stefan-Boltzmann law. The radiation pyrometer then, is intended to receive maximum amount of radiant energy at widest range of wavelengths possible.

**2. *Selective (or Partial) Radiation Pyrometery.*** In this case, we may measure the spectral radiant intensity of the radiated energy from the heated body at a given wavelength. For example, if a vertical line is drawn on Fig. 29.77, the variation of intensity with temperature for given wavelength can be found. The optical pyrometer uses this principle.

Actually it is difficult to build a practical pyrometer that is responsive to radiation of all wavelengths or to spectral radiation of one wavelength. Therefore practical radiation pyrometers respond to a wide band of radiation of approximately 0.1 to 8 μm width in the visible and infrared band of radiation spectrum.

29

## 29.37 BLACK-BODY CONDITIONS

The realization of good black-body conditions under actual working conditions is not difficult ; as otherwise radiation methods of temperature measurement would never have become practical. One of the foremost reasons for this is that the radiation from many types of surfaces and certain cavities is almost indistinguishable from black-body radiation.

Where a hot body is totally enclosed by walls at the same temperature, then both walls and body radiate and absorb heat at the same rate. If a small hole is made in the container. this area will behave as a perfect blackbody, since rays leaving the enclosure would have been reflected many a times.

This principle applies very closely to a furnace with a small hole through which a radiation pyrometer may be sighted either on the body as shown in Fig. 29.78 or on the ceiling of the furnace.

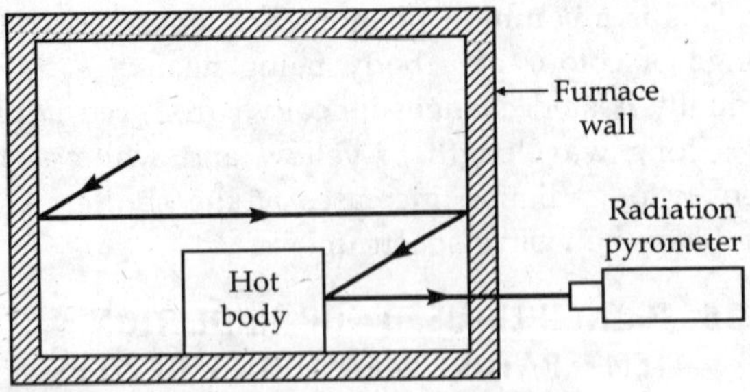


**Fig. 29.78** Heated body inside a furnace.

This can be explained as under :

The walls and the heated body are at the same steady temperature. The body absorbs a part of radiation from the walls and reflects the rest. That is,

$$\alpha + \rho + \tau = 1 \qquad ...(29.75)$$

where $\alpha, \rho, \tau$ are respectively the absorption, reflection and transmission factors.

The transmission factor for most solids is zero and, therefore.,

$$\tau = 0 \qquad ...(29.77)$$

or

$$\alpha + \rho = 1$$

Kirchhoff's identity gives an additional relationship :

$$\varepsilon = \alpha \qquad ...(29.78)$$

When the heated body is viewed from outside, the reflected radiant energy is seen. Since $\varepsilon = \alpha$ for black-body radiation, the heated body must radiate the same amount of energy that it absorbs otherwise temperature equilibrium will not be established. Therefore, when heated body is viewed from outside, the absorbed part of energy is received as well as the reflected part. Since these two parts constitute the total black-body radiation, then the heated body inside the walls having the same temperature produces black-body radiation. Thus, the heated body is considered to be a black-body.

Where the temperature of an object in the open is measured, due regard must be made to the emissivity or the surface. In case the body cannot be approximated as black-body, its temperature may be measured by determining the total radiation energy emitted from the body and the calculating the temperature.

Now, $\quad q = \varepsilon\sigma T^4 \quad$ or $\quad T = (q/\varepsilon\sigma)^{1/4}$

and therefore in order to determine temperature value of emissivity $\varepsilon$ must be known.

The apparent black-body temperature is the value calculated with $\varepsilon = 1$, or

apparent temperature $T_a = (q/\sigma)^{1/4} = \varepsilon^{-1/4} T$ ...(29.78)

If the apparent temperature is taken as measured value, the error in temperature due to non-black-body conditions is :

$$\text{Error} = \frac{T - T_a}{T} = \frac{1 - T_a}{T} = 1 - \varepsilon^{1/4} \qquad ...(29.79)$$

## 29.38 RADIATION RECEIVING ELEMENTS

The purpose of a radiation temperature measuring device is to convert the radiant energy into a suitable form for indication of temperature. For this purpose the following devices are used.

**1. *Vacuum Thermocouple.*** A vacuum thermocouple comprises of a thermo-junction of extremely thin strips which are blackened. The thermocouple is enclosed in an evacuated housing with a suitable window to admit the radiant energy. The heat loss from the thermocouple due to conduction and convection is greatly reduced because of the evacuated envelope. This results in appreciable temperature rise of the detecting junction and therefore a measurable emf can be obtained even for a small radiant energy received at the detecting junction. Since the thermocouple has a low mass and is placed in vacuum, it can respond to rapid changes in radiant energy.

**2. *Thermopile.*** In industrial applications in which a vacuum thermocouple is needed for its speed of response, it is current practice to assemble several thermocouples in series to form a thermopile. A thermopile consists of a group of very small thermocouples as shown in Fig. 29.79 so connected in series that their emfs are additive. This gives an increased sensitivity. The tiny thermocouples junctions, about pin point size, are flattened and blackened so that they absorb all the radiant energy reaching them.

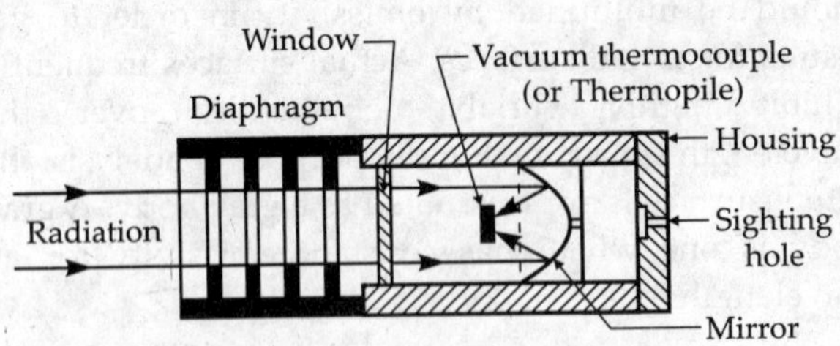


**Fig. 29.79** Mirror type radiation reeeiving device.

Pyrometers incorporating thermocouple or thermopiles do not require compensating cable between the pyrometery and the display devices. The sensitive element is required to measure its own temperature change due to radiant energy received relative to the temperature of the immediate surroundings *i.e.*, the body of the instrument. This is achieved by locating the reference junction or junctions within the body of the instrument, in a position shielded from direct radiation from the heated body.

Varying ambient temperatures, which may affect the thermopile, are compensated for by a nickel resistance spool. This spool provides a variable shunt across the emf produced. As the ambient temperature increases or decreases, the corresponding resistance changes of the nickel coil vary the emf output of the head resulting in accurate compensation over the entire range of the instrument.

A thermopile has longer speed of response but normally speech of response of less than several seconds are not important in most industrial applications. Typical response time is 2 s and higher.

3. ***Bolometer.*** A bolometer is a thermal device that changes its electrical resistance with temperatures. The resistance of the bolometer changes in response to the thermal radiation focused on it. A bolometer is made of thin ribbon or platinum or nickel depending upon the response required. The change in resistance is measured by a Wheatstone bridge. In fact two thin strips are used which form two arms of Wheatstone bridge. One strip is exposed to the radiation. The other strip is shielded and compensates for any change in the ambient temperature. Absorption of radiant thermal energy by the exposed strip results in increase of resistance of the strip which is measured by the bridge which is calibrated in terms of temperature.

Bolometers are expensive to construct and are less rugged than other detectors but have a fast response.

4. ***Photo-electric Transducers.*** Photo-electric transducers are discussed in details in Chapter 19. Photo-electric transducers used for detection of radiant energy are photo-emissive cells, photo-conductive cells and photovoltaic cells. The output of these cells varies with the amount of radiant energy incident on them. In general the photo-electric transducers are sensitive to given portions of the spectrum and therefore they are used with partial radiation and optical radiation pyrometers.

These detectors are very rugged and also have a very fast response.

## 29.39 TOTAL RADIATION PYROMETERS

The total radiation pyrometer receives virtually all the radiation from a particular area of hot body and focusess it on a sensitive temperature transducer like thermocouple, thermopile, bolometer etc. The term **total radiation** includes both the visible (light) and invisible (infra-red) radiations.

The total radiation pyrometer consists of a radiation-receiving element and a measuring device to indicate the temperature directly.

The mirror type radiation receiver is shown in Fig. 29.79. Here diaphragm unit along with a mirror is used to focus the radiation on a radiant energy sensing transducer. The lens to transducer (a vacuum thermocouple or a thermopile) distance is adjustable for proper focus. The mirror arrangement has an advantage that since there is no lens, the absorption and reflection effects are absent.

The mirror type of radiation receiver shown in Fig. 29.79 is so arranged that the image of the front diaphragm is focused on the thermocouple by the mirror. Therefore, the temperature measurements are independent of the distance of the target if the field of view is filled.

Presence of any absorbing media between target and the transducer reduces the radiation received therefore the pyrometer reads low. Substances like smoke, dirt and gases absorb radiation and therefore cause negative static errors. On the other hand, presence of heat sources like hot gases, high temperature particles and flame cause the meter to read high.

Due to the fourth power law ($q \alpha T^4$) the characteristics of total radiation pyrometer are non-linear and the device exhibit poor sensitivity in the lower temperature ranges. For example, any output voltage 2 mV for temperature change of 0°C to 500°C is typical. Inspite of this lower sensitivity, total radiation pyrometers are used for low temperature applications instead of partial radiation pyrometers which give better results in applications involving measurement of high temperatures. Total radiation pyrometers cannot be used for temperatures lower than 600°C since at lower temperatures error may be introduced by the fact that the temperature of the pyrometer itself may not be negligible as compared with that of hot body. Therefore total radiation pyrometers are used for a temperature range of 1200°C to 3500°C.

The output from a total radiation pyrometer whether applified or not, is usually taken to a PMMC instrument, or to a self-balancing potentiometer. Digital displays are also being increasingly used. The output may be fed to a recorder for recording or to a controller for control purposes.

## 29.40 INFRA-RED PYROMETERS

Infra-red pyrometers are **partial** or **selective radiation pyrometers**. Infra-red energy is invisible to the human eye, but can be felt. There is a proportional increase in infra-red energy as the temperature of the surface radiating body increases. Above temperatures of approximately 550°C, a surface starts to radiate visible light energy and simultaneously there is a proportional increase in the infra-red energy. This proportional increase in infra-red energy with surface energy makes infra-red pyrometry possible by combining a suitable detector, electronic circuitry, and means of indication and/or control. The infra-red spectrum ranges from 0.22 μm to 17 μm and the commonly used portion is 2 to 7 μm. Some manufacturers use only 5 to 7 μm portion of the range for the purposes of pyrometry.

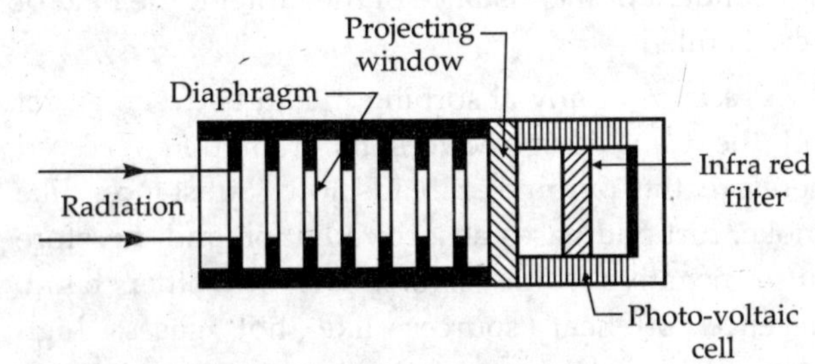


**Fig. 29.80** Infra-red pyrometer.

Infra-red principles using vacuum thermo-couples, thermopiles and bolometers have been successfully employed in infra-red spectrometers as well as in total radiation pyrometers.

Various types of photo-electric transducers are most commonly used for Infra-red transducers. The most successful transducer used for industrial applications is photo-voltaic cell. The cell used in radiation pyrometers, in fact, responds to wavelengths in infra-red region, rather than to visible light, and may be used to measure temperatures down to above 400°C.

An outstanding feature of the pyrometers based on photo-voltaic cells is their high speed of response. The time to reach 98% of full response to a step input are of the order of 1′ ms, as compared with several seconds for a total radiation pyrometer.

The infra-red radiation is focused on to a photo-voltaic cell and therefore it is necessary to insure that the cell does not become overheated. A radiant energy receiver is shown in Fig. 29.80. The cone of radiation passing ultimately to the photo-cell is defined by the area of the first diaphragm. The protective window is made of thin glass and serves to protect the cell and filter from physical damage. The filter is used in the range of 1000 °C to 1200°C in order to reduce the infra-red radiation passed to the photo-cell. This helps in preventing the photo-cell from getting overheated.

All infra-red systems depend upon the transmission of the infra-red radiant energy being emitted by a heated body to a detector in the measuring system through surroundings. There is no direct contact with the surface whose temperature is being measured. The sensor head is focused on the object whose temperature is being measured and/or controlled. The infrared energy falling on the detector either changes the detector resistance in proportion to temperature, as in the case of a thermistor, or generates an emf in the detector, such as a thermopile. The change in resistance or generated emf is then indicated on a meter or a digital display or is used to operate a controller for the process.

## 29.41 OPTICAL PYROMETERS

The radiations from a heated body at high temperatures fall within the visible region of the electromagnetic spectrum. For a given wavelength in the visible region the energy radiated is greater at higher temperatures. Within the visible region a given wavelength has a fixed colour and the energy of radiation is interpreted as intensity or brightness. Therefore if we measure brightness of the light of a given colour emitted by a hot source, we can have an indication of temperature. This is the principle on which an optical pyrometer works.

In an optical pyrometer the wavelength of radiation accepted is restricted by means of a colour filter and the brightness is measured by comparison with a standard lamp.

### Disappearing Filament Optical Pyrometer

The most common type of optical pyrometer is the *"Disappearing Filament Pyrometer"*. The schematic diagram of this pyrometer is shown in Fig. 29.81.

An image of the radiating source is produced by a lens and made to coincide with the filament of an electric lamp. The current through the lamp filament is made variable so that lamp intensity can be adjusted. The filament is viewed through an eye piece and a filter. The current through filament is adjusted until the filament and the image are of equal brightness.

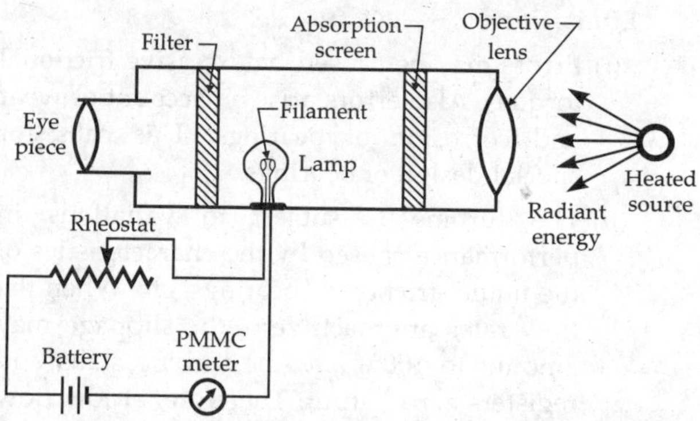


**Fig. 29.81** Disappearing filament type optical pyrometer.

When the brightness of image produced by the source and brightness produced by the filament are equal, the outline of the filament disappears as shown in Fig. 29.82(*c*).

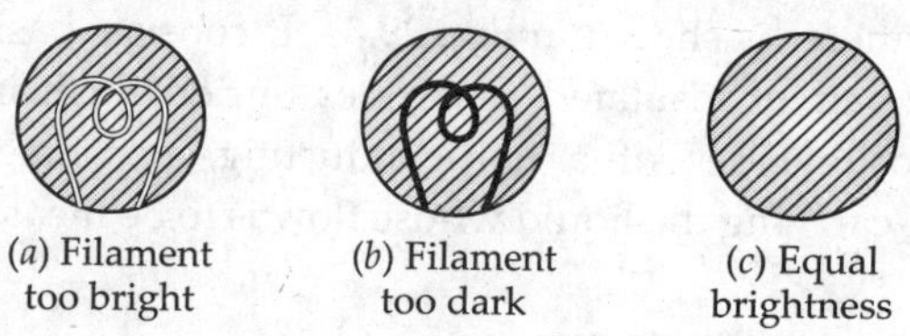


**Fig. 29.82** Field view of disappearing-filament pyrometer.

However, if the temperature of the filament is higher than that required for the equality of brightness, it (filament) becomes too bright as shown in Fig. 29.82(*a*). On the other hand, if the temperature of filament is lower, it becomes dark as shown in Fig. 29.82(*b*).

Since the intensity of light of any wavelength depends upon the temperature of the radiating body and the temperature of filament depends upon the current flowing through the lamp, the instrument may be directly calibrated in terms of the filament current. However the filament current depends upon the resistance of the filament,. therefore modern pyrometers are calibrated in terms of resistance of filament circuit.

The range of temperature which can be measured by instruments of this type is governed by the maximum allowable temperature of the lamp which is of the order of 1400°C. The range can be extended by using an absorption type screen near objective lens. Thus as known fraction of radiant energy enters the pyrometer for comparison. The range can be extended to 3000°C by adopting this technique.

The optical pyrometer is widely used for accurate measurement of temperature of furnaces, molten metals and other heated materials.

**Example 29.36** *The lens of an optical pyrometer is clouded so that the transmission factor is 0.8. The instrument indicates a temperature of 1480°C. What is the true temperature ?*

**Solution.** Apparent absolute temperature indicated

$$T_t = 1480 + 273 = 1753\ °K$$

Using Eqn. 29.78 (which can be used for the purposes of calculation of actual temperature by using emissivity in place of transmission factor) true absolute temperature

$$T = \varepsilon^{-1/4}\, T_a = (0.8)^{-1/4} \times 1753 = 1853.7\ °K$$

$$\text{True temperature} = 1853.7 - 273 = 1580.7\ °C$$

**Example 29.37** *The emitted radiant energy from a piece of metal is measured and the temperature is found to be 1065°C assuming a surface emissivity of 0.82. It was later found out that the true emissivity is 0.75. Calculate the error in temperature measurement.*

**Solution.** Absolute temperature with emissivity of 0.82

$$= 1065 + 273 = 1338\ °K$$

∴ Apparent absolute temperature is,

$$T_a = (0.82)^{-1/4}(1338) = 1273°K$$

Actual absolute temperature when the emissivity is 0.75,

$$T_a = (0.75)^{-1/4}\,(1273) = 1368\ °K$$

$$\text{Actual temperature} = 1368 - 273 = 1095\ °C$$

Hence error in temperature measure

$$= 1065 - 1095 = -30\,°C$$

## 29.42 MEASUREMENT OF FLOW

There are a number of devices for sensing the rate of fluid flow. They mainly operate on the principle of placing an obstruction in the path of fluid causing a change in fluid pressure which is dependent upon the rate of flow. Thus by measuring the difference in pressure before and after the obstruction by means of a differential pressure sensor, the rate of flow may be determined. They are, in fact, mechanical means of measuring rate of liquid flow. The examples of these flow meters are : Orifice plate, venturi tube and Rotameter. Some of these methods have been explained in Chapter 25. These are indirect means of measuring fluid flow. However, there are some direct methods of measuring fluid flow which are described below. These methods employ electrical means directly.

### 29.42.1 Turbine Meters

There have been extremely rapid developments in turbine meters in recent years, partly because of the

29

advances in electronics technology and the case with which the output may be used to indicate rate of flow or be integrated to give total flow. An exploded view of the meter is shown in Fig. 29.83.

Turbine flow meters are volumetric flow meters and are available in wide ranges. The output is usually in the form of a digital electrical signal whose frequency is directly proportional to flow rate and whose total count is proportional to the total quantity, as each pulse represents a discrete volume.

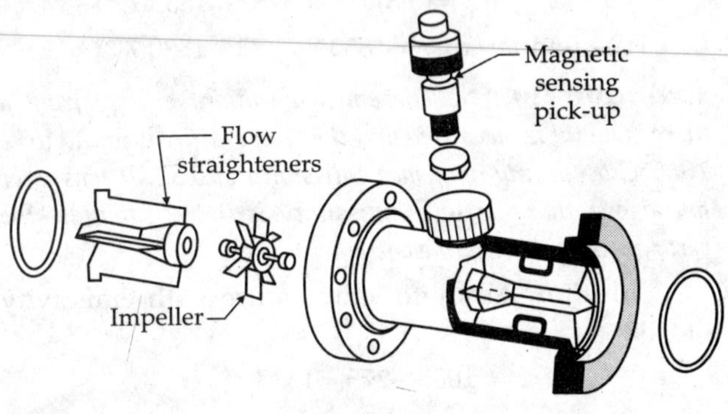


**Fig. 29.83** Turbine flow meter.

Figure 29.83 shows a magnetic pickup type of turbine flow meter. A feature of this turbine meter is a hydraulically supported turbine rotor. A permanent magnet sealed inside the rotor body is polarized at 90° to the axis of the rotation. As the rotor rotates so does the magnet and therefore rotating magnetic field is produced. This produces an a.c. voltage pulse in the pick-up coil located external to the meter housing. The frequency of this voltage is directly proportional to the rate of flow. The pulse can be totalized by a counter to give the value of total flow over a particular interval of time.

**Advantages :**

(*i*) The output is in electrical digital form which lends itself admirably to line or radio telemetry for recording or control at a distant point.

(*ii*) Passage of rotor past the pick up coil produces an emf which can be converted to a d.c. analog voltage by a D/A converter for indication on a conventional d.c. instrument or recorded on a potentiometric recorder.

(*iii*) The pulse output may be summed on a simple counter giving totalized flow without loss of accuracy. The error is approximately ± 0.5%.

**Limitations :**

(*i*) Errors may be caused by excessive frictional torques. Also errors arise on account of wear and corrosion of bearings. This calls for special design of bearings.

(*ii*) The turbine is subject to variations in performance caused by the characteristics of the liquid, friction or bearings, etc. When the flow rates approach zero, the slippage may amount to 100%, *i.e.*, the turbine stalls and registers zero output. Therefore, at low flow rates, there is a possibility of large errors.

### 29.42.2 Electromagnetic Flow Meters

Electromagnetic flow meters are particularly suitable for the flow measurements of slurries, sludge and any electrically conducting liquid.

A schematic diagram of an electromagnetic flowmeter is shown in Fig. 29.84. It consists basically of a pair of insulated electrodes buried flush in the opposite sides of a non-conducting, non-magnetic pipe carrying the liquid whose flow is to be measured.

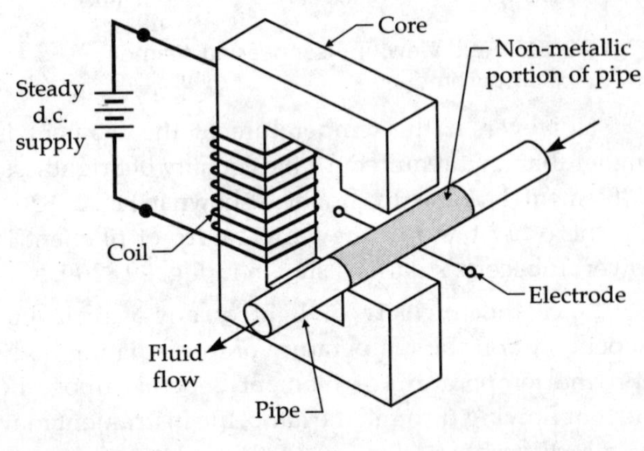


**Fig. 29.84** Electromagnetic flow meter.

The pipe is surrounded by an electromagnet which produces a magnetic field. The arrangement is analogous to a conductor moving across a magnetic field. Therefore, voltage is induced across the electrodes. This voltage is given by :

$$E = Blv \text{ volt}$$

where $B$ = flux × density ; Wb / m$^2$,

$l$ = length of conductor = diameter of pipe ; m,

and $v$ = velocity of conductor (flow) ; m/s

Thus, assuming a constant magnetic field, the magnitude of the voltage appearing across the electrodes will be directly proportional to velocity.

Non-conducting pipe has to be used as the output voltage gets short circuited if metallic pipes are used. This is true when liquids of low conductivity are being measured. But when liquids of high conductivity are measured the short circuiting has no effect. Stainless steel pipes can then be used. The voltages produced are small specially at low flow rates. Therefore, the meter relies greatly on a high gain amplifier to convert the induced voltage into a usable form.

***Advantages*** :

(*i*) The electromagnetic flow meters may be manufactured to measure flow in pipes of any size provided powerful magnetic field can be produced.

(*ii*) The major advantage from a fluid handling point of view is that with the electro-magnetic flow meter there is no obstruction to flow that may cause pressure drops.

(*iii*) The output (voltage) is linearly rated to the input (flow rate).

(*iv*) The output is unaffected by changes is characteristics of liquid such as viscosity, pressure, and temperature.

***Limitations*** :

(*i*) The operating costs are high particularly if heavy slurries are handled.

(*ii*) The conductivity of the liquid being metered should not be less than $10\,\mu\,\Omega/\text{m}$. As a general rule it will be found that most aqueous solutions are adequately conductive while majority of hydrocarbon solutions are not sufficiently conductive.

## Comparison of D.C. and A.C. Excitation

The magnetic field used in electromagnetic now meters can either be d.c. or a.c. giving rise to d.c. or an a.c. output signal respectively. The d.c. is used in limited applications due to many reasons :

(*i*) When d.c. excitation is used for materials of very low conductivity and flowing at slow speeds, the output emf is too small to be easily read off. This has to be amplified for display and recording purposes. The d.c. amplifiers have many inherent problems especially at low levels. High amplification can be more easily cheaply and more reliably done with a.c. than with d.c. Therefore these applications require the use of a.c. excitation of electromagnet to produce a.c. emfs which can be easily amplified by a.c. amplifiers and then fed to the display units.

(*ii*) Many hydrogen bearing or aqueous solutions exhibit polarization effects when the excitation is d.c. The positive ions migrate to the negative electrode and disassociate, forming an insulating pocket of geseous hydrogen. There is no such action when a.c. is used.

(*iii*) Another phenomenon which happens with use of d.c. is that the d.c. field may distort the fluid-velocity profile by magnetohydrodynamic (MHD) action. An a.c. field (50 Hz) has little effect on velocity profiles because fluid inertia and friction forces at 50 Hz are sufficient to prevent any large fluid motions.

(*iv*) Since the output of electromagnetic flow meters is quite small (a few mV), interfering voltage inputs due to thermocouple type of effects and galvanic action of dissimilar metals used in meter construction may be of the same order as the signal. Since the spurious interfering inputs are generally drifts of very low frequency, the 50 Hz a.c. system can use high pass filters to eliminate them.

(*v*) While a.c. systems predominate, d.c. type of systems have been used for flow measurements of liquid metals like mercury. Here, no polarization problem exists. Also, an insulating pipe linear or a non-metallic pipe is not needed since the conductivity of the liquid metal is very good relative to an ordinary metal pipe. (Usually stainless steel pipes are used.) This means that the metal pipe is not very effective as a short circuit for the voltage induced in the flowing liquid metal. When metallic pipes are used as with d.c. excitation no special electrodes are necessary. The output voltage is tapped off the metal pipe itself at the points of maximum potential difference.



## 29.42.3 Hot Wire Anemometers

The hot wire anemometer is a device that is most often used in research applications to study varying flow conditions.

When a fluid flows over a heated surface, heat is transferred from the surface and therefore its (surface's) temperature reduces. The rate of reduction of temperature is related to flow rate.

In a hot wire anemometer, heat is supplied electrically to a fine wire placed in the flow stream. The temperature of wire is determined by measuring its resistance with a Wheatstone bridge. One method involves adjusting the current through the wire so that the temperature remains constant, and measuring the heating current. In this way the bridge remains always balanced. The current is measured by finding the voltage drop across a standard resistor connected in series with the heating wire (See Fig. 29.85). The voltage drop is found by using a potentiometer.

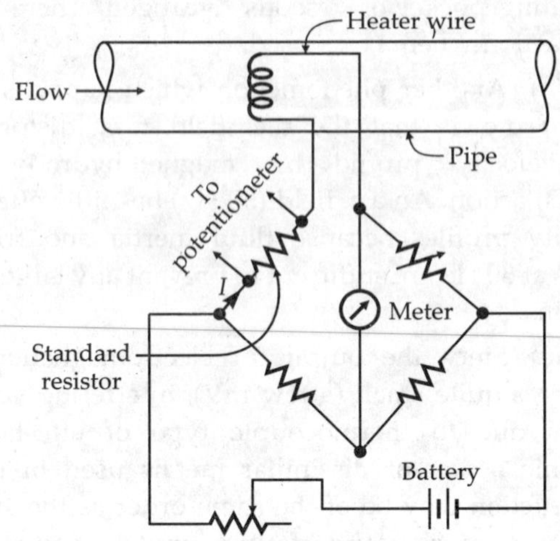


**Fig. 29.85** Measurement of rate of fluid flow using a hot wire anemometer.

It can be shown that the loss of heat from the heated wire :

$$= a(v\rho + b)^{1/2} \text{ J/s}$$

where $v$ = velocity of heat flow,

$\rho$ = density of fluid,

and $a$ and $b$ are constants.

Constants $a$ and $b$ depend upon dimensions and physical properties of wire and fluid. The values of these constants are found by calibrating the instrument against a static pitot tube.

Now supposing a current $I$ flows through the wire having a resistance $R$. Therefore under equilibrium conditions,

heat generated = heat lost

or $$I^2 R = a(v\rho + b)^{1/2}$$

or $$v = \frac{(I^4 R^2 / a^2 - b)}{\rho}$$

Thus if the resistance and the temperature of the wire are kept constant, the rate of fluid flow can be measured by measuring current, $I$, through the heater wire.

The transient conditions may be studied with an oscilloscope. Time constants of the order of 0.1 ms are obtained by using 0.0025 mm diameter platinum and tungsten wires.

### 29.42.4 Flow Meter using Thermistors

Flow can be measured by using two thermistors connected in two separate arms of a bridge circuit as shown in Fig. 29.86. One thermistor is sealed in a cavity in a brass block and the other thermistor is mounted in a small pipe. When air flows through the pipe, the temperature of the thermistor (place inside the pipe) decreases because of conduction of heat. This thermistor is thus cooled, and its resistance increases which unbalances the bridge causing a current to flow through the meter. The amount of cooling is proportional to the rate of flow of air and the meter may be calibrated to read directly the rate of flow in the pipe.

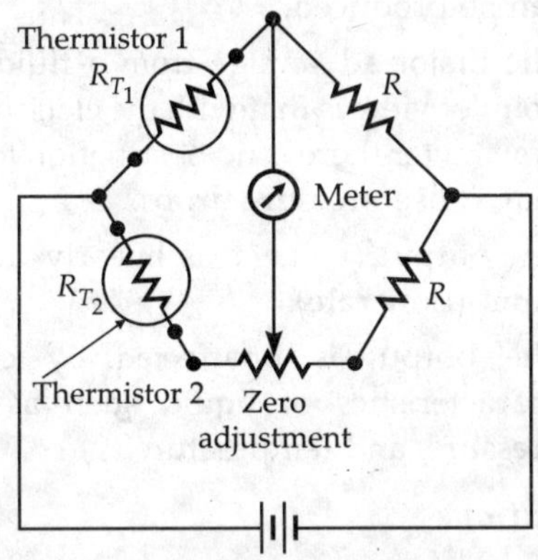


**Fig. 29.86** Flow measurements using a bridge circuit having two thermistors.

This instrument can be designed to measure flows as low as $15 \times 10^{-12} \text{m}^3/\text{s}$. The instrument can measure flow rates over a range of 100,000 : 1 by switching the resistance in series with the output meter.

When the sensing thermistor of the flow meter is placed in free air, the instrument becomes an anemometer which measures velocity of air. It can be calibrated to measure velocities of air from light breeze to supersonic speeds.

***Example 29.37*** *An electromagnetic flow meter is used to measure the average flow rate of an effluent in a pipe of 50 mm diameter. The velocity profile is symmetrical and can be assumed uniform. The flux density in the liquid has a peak value of 0.1 Wb/m², the output from the flow meter electrodes is taken to an amplifier of gain 1000 and impedance between the electrodes is 250 kΩ. The input impedance of the meter is 250 MΩ.*

*(a) Determine the effluent average velocity when the peak to peak voltage at the amplifier output is 0.2 V.*

*(b) Given that the effluent conductivity decreases by 20 percent with the same flow rate, determine the percentage change in reading at the amplifier output.*

**Solution.** (*a*) Peak to peak voltage at amplifier output = 0.2 V

Peak voltage at amplifier output = 0.2/2 = 0.2 V

This is the voltage under loaded conditions.

From Eqn. 2.37, page 35, we have :

Voltage under loaded conditions

$$E_L = \frac{E_0}{1+Z_0 Z_L}$$

or peak open circuit voltage

$$E_0 = E_L\,(1+Z_0/Z_L)$$

$$= 0.1\,[1+(250\times 10^3)/2.5\times 10^6] = 0.11 \text{ V}$$

The peak open circuit voltage at the amplifier output terminal is :

$$E_0 = B\,l\,v \times \text{gain}$$

∴ Average flow rate

$$v = \frac{0.11}{0.1\times 50\times 10^{-3}\times 10^3} = 0.0002 \text{ m/s}$$

(*b*) The effluent conductivity decreases by 20 per cent and therefore its impedance increases to :

$$\mathbf{Z}_0 = 1.2\times 250\times 10^3 = 300\times 10^3\ \Omega = 300\ \text{k}\Omega$$

Therefore, for the same flow rate, the peak to peak of voltage under loaded conditions

$$= \frac{2\times 0.11}{[1\times(300\times 10^3)/(2.5\times 10^6)]} = 0.1964 \text{ V}$$

Percentage decrease in voltage

$$= \frac{0.2-0.1964}{0.2}\times 100 = 1.78\%$$

## 29.42.5 Ultrasonic Flow Transducer

Basically an ultrasonic transducer for flow rate consists of two piezoelectric crystals in the liquid or gas separated by a distance. One of the crystals' acts as a transmitter and the other as a receiver.

The transmitter emits an ultrasonic pulse which is received at the receiver a time $\Delta t$ later. The transit time in the direction of flow is,

$$\Delta t_1 = \frac{d}{c+v} \qquad \text{...(29.80)}$$

where $d$ = distance between transmitter and receiver ; m,

$c$ = velocity of sound propagation in medium ; m/s,

$v$ = linear velocity of flow ; m/s.

When the signal is travelling in the opposite direction against the flow

$$\Delta t_2 = \frac{d}{c-v} \qquad \text{...(29.81)}$$

Similarly, a sinusoidal signal of frequency $f$ Hz travelling in the flow direction has a phase shift of :

$$\Delta\phi_1 = \frac{2\pi f d}{c+v} \text{ rad} \qquad \text{...(29.82)}$$

and that travelling against the direction of flow has a phase shift of :

$$\Delta\phi_2 = \frac{2\pi f d}{c-v} \text{ rad} \qquad \text{...(29.83)}$$

Velocity can, therefore, be determined by either measuring the transit time or the phase shift.

Figure 29.87 shows a system which can be used external to the pipe carrying the liquid. $T$ and $R$ are respectively transmitting and receiving crystals. They are either pressed to the exterior of pipe or are immersed in the liquid so that the signal is transmitted through the liquid.

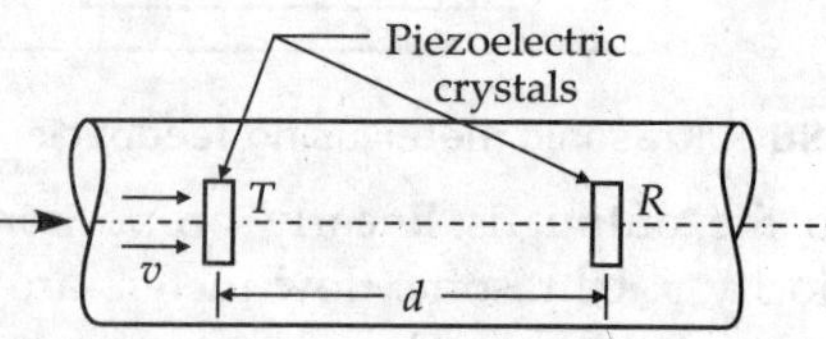


**Fig. 29.87** Ultrasonic method for measurement of flow.

The oscillator provides a sinusoid signal of about 100 kHz to crystal $T$ whereas crystal $R$ acts as the receiver. The functions of $T$ and $R$ are reversed periodically by a commutating switch. The difference in transit times is,

$$\Delta t = \Delta t_2 - \Delta t_1 = \frac{2dv}{c^2-v^2} \qquad \text{...(29.84)}$$

This is measured by a phase sensitive detector driven synchronously with the commutator. Usually $c >> v$.

$$\therefore \qquad \Delta t \approx \frac{2dv}{c^2} \qquad \text{...(29.85)}$$

Hence, time $\Delta t$ is linearly proportional to flow velocity $v$. This system, though gives a linear relationship, is subject to an error on account of uncertainty of the value of $c$.

A system based upon frequency may be evolved. Figure 29.88 shows two self-excited oscillating systems created because of using the received pulses to trigger the transmitted pulses in feed back arrangement. The pulse repetition frequency in the forward propagating loop is $1/\Delta t_1$ while that in the backward loop is $1/\Delta t_2$.

$$\Delta t_1 = \frac{d}{(c + v\cos\theta)} \quad \text{or} \quad f_1 = \frac{(c + v\cos\theta)}{d}$$

$$\Delta t_2 = \frac{d}{(c + v\cos\theta)} \quad \text{or} \quad f_2 = \frac{(c - v\cos\theta)}{d}$$

Hence difference in frequency,

$$\Delta f = f_1 - f_2 = \frac{2v\cos\theta}{d} \qquad ...(29.86)$$

Equation 29.86 shows that the output is independent of $c$, and is, therefore, not subject to errors on account of errors involved in the value of $c$.

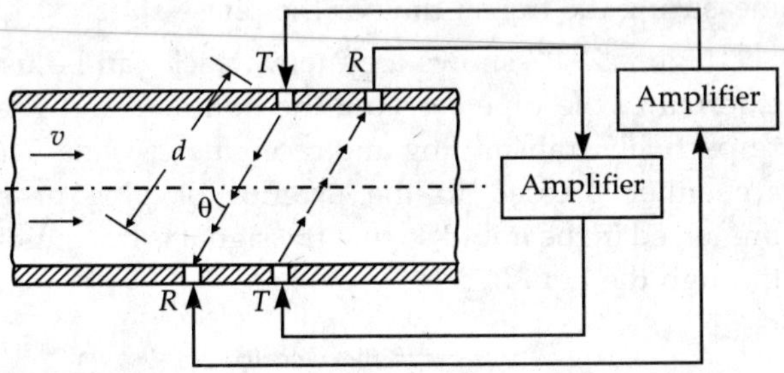


**Fig. 29.88** Ultrasonic meter using feedback.

Also the output is linearly proportional to the flow velocity $v$. Ultrasonic flow meters are presently used for mainly liquids. The advantages are that there is no obstruction to the flow, they are insensitive to variations in viscosity, density and temperature, there are no moving parts, there exists a linear relationship between input and output, their dynamic response is excellent and they lend themselves to bidirectional flow.

However the complexity and relatively high cost limit the use of this instrument for individual applications.

The range is 0 – 80 km/h for air with an accuracy of ± 0.5%.

## 29.43 MEASUREMENT OF LIQUID LEVEL

### Electrical Methods

The direct conversion to liquid level position to electrical signal is used in many instances. The measurement is generally done by two conversions, so that the liquid level is determined indirectly. The first conversion usually is liquid level to a displacement through a float in a liquid or a spring loaded plate in contact with the surface in the case of granular solids. This displacement is then converted into an electrical signal by a secondary transducer connected to float or plate. There are, however, many applications where this is not possible and hence other methods like optical or acoustic means or gamma rays are used.

The electric transducers used for level measurements are :

- Resistive,
- Inductive, and
- Capacitive.

### 29.42.1 Resistive Method

This method uses mercury as a conductor as shown in Fig. 29.89. A number of contact rods are placed at various liquid levels. As head $h$ increases, the rising level of mercury above the datum, shorts successive resistors $R$ and increases the value of $h$ directly.

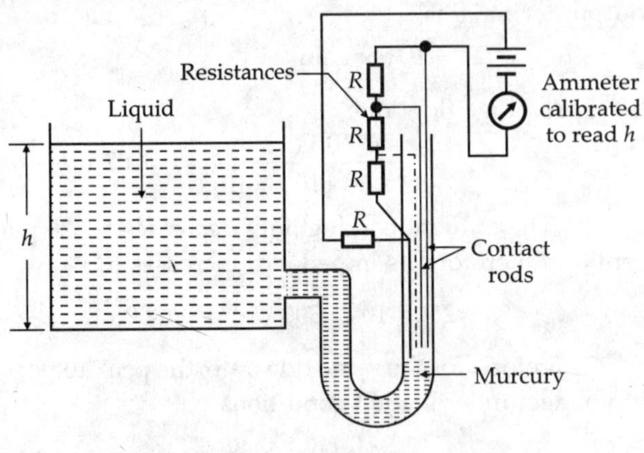


**Fig. 29.89** Measurement of level of liquids by resistive method.

***Advantages :***

(*i*) Where there is a need for a fairly continuous record of the level, more and more contact rods can be added, with separate signal outputs for each contact rod. The signal can be used for indication and also for initiating some control action to actuate valves or pumps as well as warning lights and alarms depending upon the level.

(*ii*) The system uses low voltage to eliminate danger to the operators and to prevent arcing at the contact points.

(*iii*) The signal can be transmitted to any desired point.

(*iv*) These transducers can be used in pressurised containers without packing glands or shafts.

(*v*) The unit is simple to calibrate since the distance between the levels of contact rods can be accurately measured and the indicated value may be checked for each measured value.

***Disadvantages*** :

(*i*) Due to arcing at the contact points, it is not safe to use this transducer in explosive atmosphere.

(*ii*) In order to have a stepless indication of the liquid level, an extremely large number of contact rods are needed.

(*iii*) The contact rods are corroded by corrosive liquids. In addition, the electric charges promote corrosion.

(*iv*) These systems also present difficulties when there is saturated vapour above the liquid phase.

(*v*) Any changes in the conductivity of the liquid causes serious errors.

### 29.43.2 Inductive Methods

The inductive level transducers are mainly used for measurement of level of conductive liquids employing **variable permeability method.** The arrangement shown in Fig. 29.90 uses two coils $L_1$ and $L_2$ wound around a steel tube containing the liquid. The coils are connected in series through a resistance and the circuit is energised by an alternating current source. The inductance of each coil is initially equal say about 250 μH. One coil say $L_1$ acts as the search coil. It can be set at a predetermined level. The inductance of the search coil changes rapidly as the conducting liquid moves into the plane of the coil. The method works well because the tape material is weakly magnetic and the liquid metal is a conductor which allows eddy currents to flow in it. The relationship between the output voltage and the liquid level is essentially non-linear.

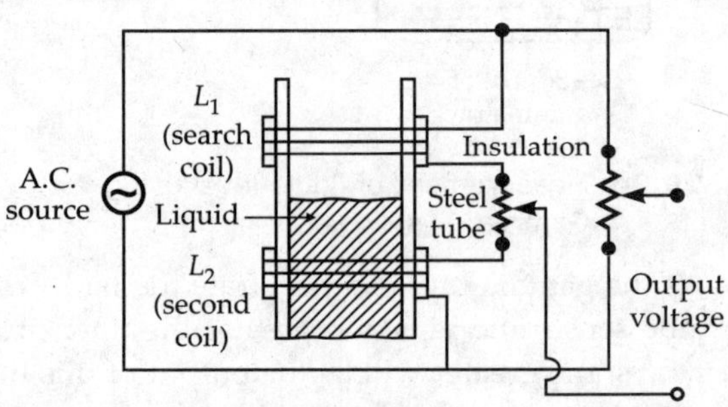


**Fig. 29.90** Measurement of liquid level with variable permeability method.

Another method uses the loading of secondary winding of a transformer. This method is applicable to good electrical conducting materials like mercury. This is shown in Fig. 29.91.

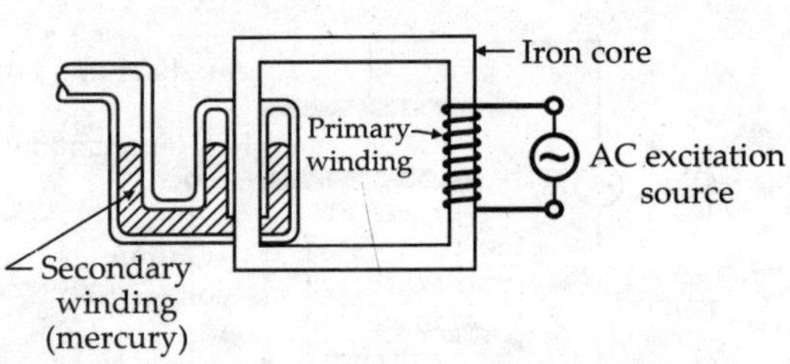


**Fig. 29.91** Liquid level determination by loading of secondary winding.

A coil is wound round one core of a two limbed transformer. The mercury column surrounding the iron core forms the secondary of the transformer. The resistance of the secondary winding depends upon the height of the mercury column and therefore the power consumption as monitored on the primary side is indicative of the liquid level.

### 29.43.3 Capacitive Methods

#### 1. Variable Area Method

The variable area capacitive transducer is used for measurement of levels of both solids and liquids. The electrical conducting container containing the materials is used as one connection point of the transducer. The other point is a metal rod completely covered by insulating material inside the container. The insulating material acts as the dielectric medium and the capacitance varies linearly with the height of the material.

The relationship is given by :

$$C = \frac{2\pi\varepsilon h}{\log_e (d_2 / d_1)} \text{ F} \qquad \text{...(29.87)}$$

where $\varepsilon$ = permittivity of the insulator : F/m

$d_1$ = diameter of the metal rod ; m,

$h$ = height of material ; m,

and $d_2$ = external diameter of the insulator ; m

The container should be earthed to avoid any danger of electric shock to the personnel and to prevent any errors due to external metallic objects.

#### 2. Capacitive Voltage Divider Method

In case, the conductivity of a liquid is high, its surface can be used as one electrode of the capacitor. The other electrode is a fixed reference plate parallel to the surface of the liquid. A system incorporating these features is shown in Fig. 29.92. This uses an auxiliary electrode $P_1$ placed at a fixed distance above the reference electrode $P_2$. The two electrodes $P_1$ and $P_2$ are electrically insulated from each other. An a.c. voltage is applied between the liquid and the electrode $P_1$.

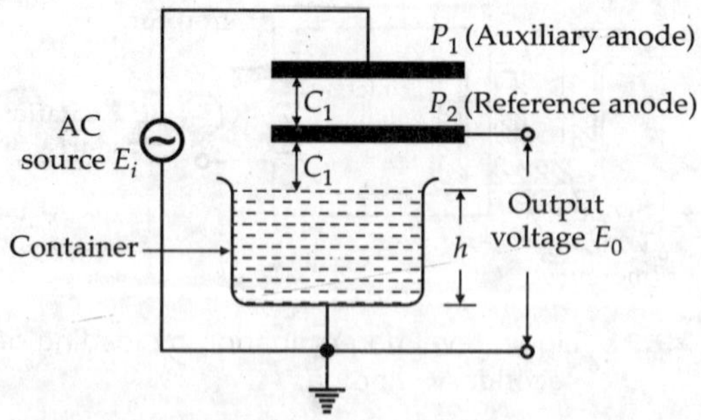


**Fig. 29.92** Capacitive voltage divider method for measurement of level of high conductivity liquid.

Potential of electrode $P_2$ with respect to earth

$$E_0 = E_i \frac{C_1}{C_1 + C_2} \text{ V}$$

Capacitance $C_2$ is inversely proportional to the distance between the liquid surface and $P_2$. Thus the output voltage decreases with rise of liquid level and therefore the relationship between them is non-linear.

### 3. Variable Dielectric Constant Method

If the liquid is non-conducting it can be used as a dielectric in a capacitor. Article 25.28.5 on page 819 explains the measurement of liquid level with variation of dielectric constant.

The arrangements for measurement of liquid level for non-conducting liquids are explained below :

Figure 29.93(*a*) shows an insulated metal electrode firmly fixed near and parallel to the metal wall of the tank. If the liquid is non-conductive, the electrode and the tank wall form the plates of a parallel plate capacitor with the liquid in between them acting as the dielectric. If the liquid is conductive the rod and the liquid form the plates of the capacitor, and the insulation between them is the dielectric.

The capacitance of this capacitor depends, among other factors, upon the height of the dielectric between the plates. The greater the height, the greater the capacitance. The lesser the height, the smaller is the capacitance. Thus, the capacitance is proportional to the height of the liquid in the tank.

Where the tank is not of metal, two parallel insulated rods (electrodes) kept at a fixed distance apart are used as shown in Fig. 29.93(*b*). The two rods act as two plates of a parallel plate capacitor. The higher the liquid level, the greater is the capacitance.

The capacitance in the above cases may be measured and this measured capacitance is an indication of liquid levels.

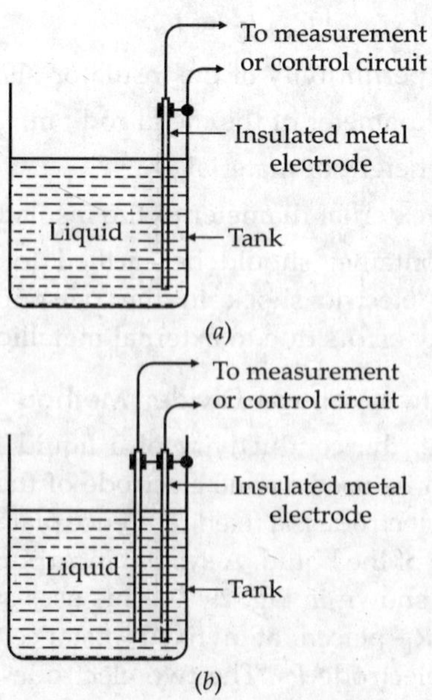


**Fig. 29.93** Measurement of liquid level using capacitive transducers.

### 29.43.4 Measurement of Liquid Level with Gamma Rays

Gamma rays are being increasingly used for detection of liquid levels. A source of gamma rays is placed at the bottom of the tank. A sensor of gamma rays like a Geiger Müller tube is placed outside the tank near the top. (See Fig. 29.94.)

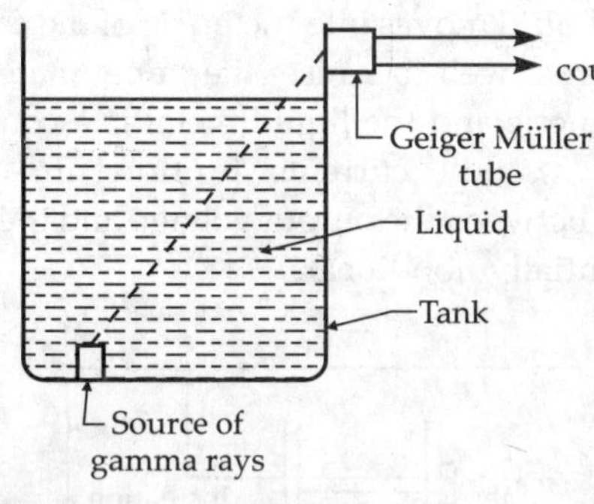


**Fig. 29.94** Measurement of liquid level using Gamma rays.

As the gamma rays can penetrate the tank walls, the tube senses the rays and greater the intensity of other rays, the greater will be the output of this tube. Now the intensity of the rays will depend upon the liquid level. Maximum radiation will reach the Geiger Müller tube if the tank is empty. But if there is some liquid in it, some of the rays will be absorbed by the liquid and the radiations reaching the tube will be small, and hence its output is small. The higher the level of the liquid, the greater is the absorption and hence lesser will be the output of the Geiger Müller tube. Thus the output of the Geiger Müller tube is inversely proportional to the liquid level. The output of the Geiger Müller tube is in the form of pulses which may be counted by a counter. Thus the counter may be directly calibrated in terms of the liquid level.

### 29.43.5 Ultrasonic Method

An ultrasonic transmitter receiver can be mounted on the top of for measurement of level of either solids or liquids as shown in Fig. 29.95. The beam is projected downwards by the transmitter and

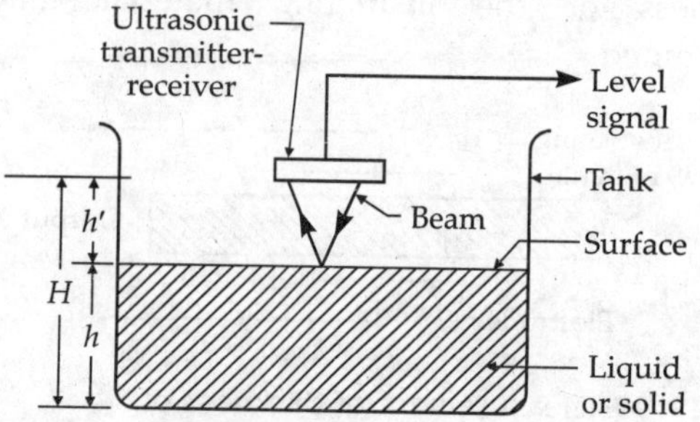


**Fig. 29.95** Measurement of level by ultrasonic method.

is reflected back by the surface of the solid or liquid contained in the tank. The beam is received by the receiver. The time taken by the beam is a measure of the distance travelled by the beam. Therefore, the time '$t$' between transmitting and receiving a pressure pulse is proportional to the distance '$h'$' between the ultrasonic set and surface of the contents of the tank.

or $\quad t \propto h' \propto (H-h)$

Since distance $H$ between ultrasonic set and the bottom of the tank is fixed, time '$t$' is measure of level '$h$'.

### 29.43.6 Measurement of Liquid Level using Float

A float operated voltage potential divider is shown in Fig. 29.96. As one liquid level rises in the

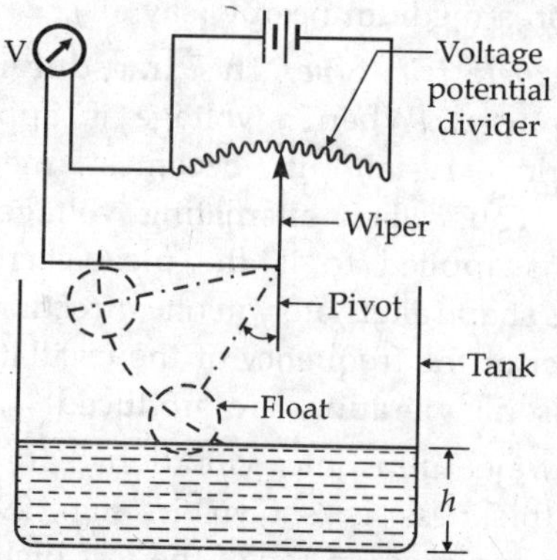


**Fig 29.96** Float operated voltage potential divider for liquid level measurements.

tank, the float, which is generally a hollow ball, is raised. Its arm causes the wiper to move over the potential divider whose output terminals are connected to a voltameter. As a float rises, a greater part of the potential divider is included in the output circuit giving an increased output voltage.

∴ The output voltage $V$ is proportional to the liquid level $h$.

The output terminals from the potential divider may also be taken to a remote location for display and control.

## 29.44 MEASUREMENT OF THICKNESS

The electrical methods of measuring thickness are :

- Resistive,
- Inductive,
- Capacitive,
- Ultrasonic and
- Nuclear.

### 29.44.1 Inductive Methods

Inductive methods of thickness measurement use : (*i*) reluctance variation transducers, and (*ii*) eddy current transducers. Figure 29.97(*a*) shows the measurement of thickness of a magnetic material using a reluctance variation transducer. This transducer has a *U*-shaped magnetic core over which a coil is wound. The test piece completes the magnetic circuit. Inductance of the coil depends upon the reluctance of the magnetic circuit. The reluctance, in turns depends upon the thickness of the test piece. Higher the thickness, lower is the reluctance and higher is the inductance. Therefore, the inductance of the coil provides a measure of the thickness of the magnetic test piece.

29

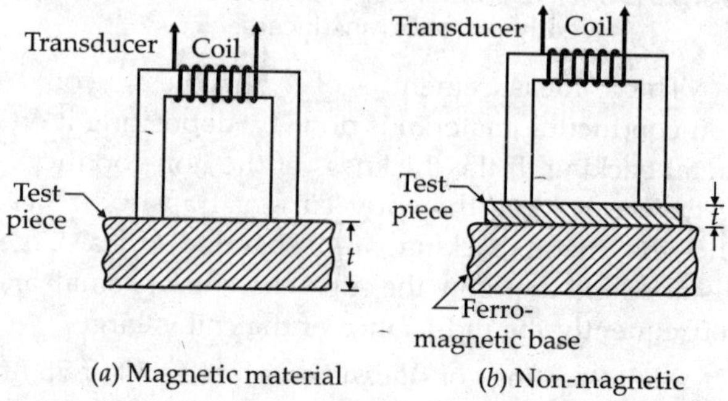


**Fig. 29.97** Measurement of thickness using reluctance variation transducer.

The thickness of non-magnetic materials can also be measured by the reluctance variation transducer. The non-magnetic material is deposited or placed on a magnetic base as shown in Fig. 29.97(*b*). The ferromagnetic base is so thick that the reluctance of the magnetic circuit is determined by the distance between the ends of the *U* core and the base plate, *i.e.*, by the thickness of the test piece. An increase in thickness causes an increase in reluctance and a decrease in inductance. Thus by measuring inductance, the thickness can be known.

The measurement of thickness of conducting and non-conducting materials using eddy current transducers is shown in Figs. 29.98(*a*) and (*b*). The eddy current transducer used for measurement of thickness of non-magnetic but conducting materials consists of a coil wound on an insulated core. The coil is excited by an alternating current supply. The alternating magnetic field produces eddy currents in the test piece or its backing. The magnetic field produced by eddy currents opposes the magnetic field of the coil and therefore the inductance of the coil is reduced. This system is suitable for measurement of thickness of both magnetic and non-magnetic materials, as well as non-conducting coatings like paints on a metal backing. In the case of conducting material as shown in Fig. 29.98(*a*), the higher the thickness of test piece, higher will be the eddy currents and lower would be the inductance of the coil.

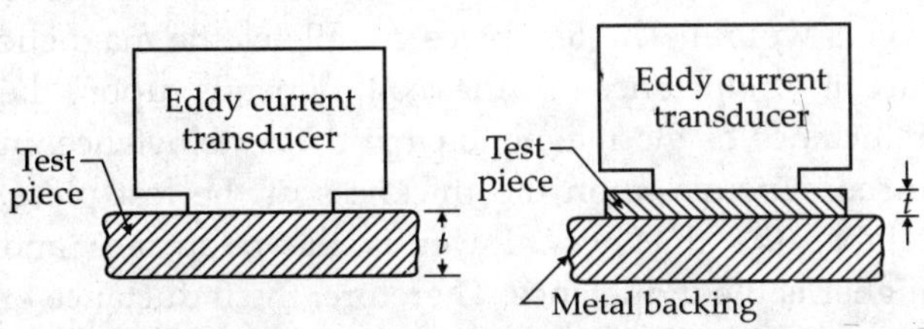


(*a*) Non-magnetic conducting material (*b*) Non-magnetic non-conducting material

**Fig. 29.98** Measurement of thickness using eddy current transducers.

The measurement of thickness of a non-conducting material is done by depositing it on a metal backing. If the thickness of the non-conducting test piece is large the eddy current transducer head and the metal backing are separated by a larger distance and therefore the eddy currents are small and consequently the inductance of the coil is large.

Metal coatings of thickness as low as 50 or 75 nm and aluminium films down to a thickness of 125 – 500 pm can be measured with the help of eddy current transducers.

## 29.44.2 Capacitive Method

If the material is being tested is an insulator, capacitive method using an arrangement shown in Fig. 29.99 may be used. Two metal electrodes are placed on the two sides of the insulating material being tested. This arrangement forms a parallel plate capacitor, the two electrodes acting as the two plates with the insulating material acting as the dielectric. The capacitance, naturally depends upon the thickness of the insulating material under test. Thus by measuring the capacitance of this arrangement, the thickness of the insulating material may be determined.

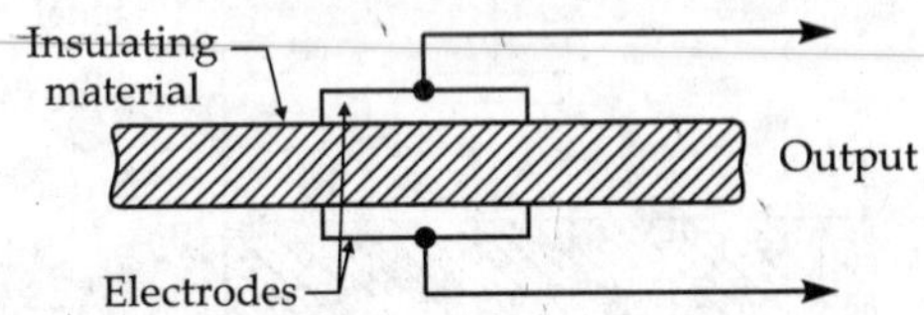


**Fig. 29.99** Measurement of thickness of insulating materials.

## 29.44.3 Measurement of Thickness using Ultrasonic Vibrations

Ultrasonic vibrations are mechanical vibrations employing a gas, liquid or solid as medium, and whose frequencies are above the audio frequency range. These vibrations are produced by a transducer which converts the electrical output of an oscillator to ultrasonic vibrations of corresponding frequencies.

There are two types of such transducers :

(*i*) Magnetostrictive type and

(*ii*) Piezoelectric type.

**1. *Magnetostrictive type.*** This transducer consists of magnetic rod placed in coil. The coil is excited by an oscillator. The alternating field produces deformations in the magnetic rod, making it longer and shorter alternatively. This results in corresponding vibrations produced in a medium nearby (say air).

**2. *Piezoelectric type.*** This transducer is more commonly used. When a voltage is applied to a piezoelectric crystal, it becomes mechanically deformed. Thus when alternating voltage from an oscillator is applied to it, the piezoelectric crystal changes its shape alternately in one direction and then in the other at the frequency of the oscillator. In this way, ultrasonic vibrations are produced.

The way ultrasonic vibrations are used to measure thickness is shown in Fig. 29.100. The transducer is placed on top of the test piece, and the ultrasonic vibrations pass through it. When the frequency of the oscillator is varied and standing waves are set up of at certain frequencies. The value of these frequencies depend upon the thickness of the test piece. The setting up of standing waves is indicated by a sudden sharp rise in power absorbed from the transducer. Thickness $t = c/2f$ where $c$ is the velocity of sound in the test piece, and $f$ is the lowest frequency at which standing waves occur.

Ultrasonic method can be used for measurement of thickness of steel ranging from 3 mm to 30 cm.

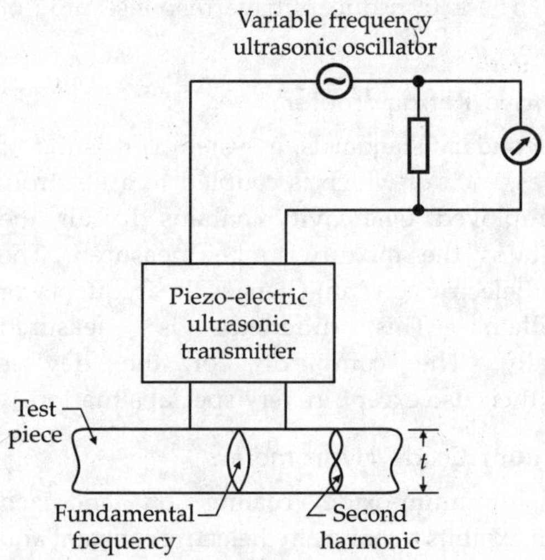


**Fig. 29.100** Measurement of thickness by ultrasonic transducer.

### 29.44.4 Nuclear Radiation Methods

Nuclear radiation methods use either alpha, beta, gamma or X-rays as radiation obtained from radioactive sources. The radiation source is shielded all around except in the direction of the detector. A radiation detector such as a Geiger Müller tube ionization chamber or a scintillation counter explained in Arts. 29.52 and 29.53 are used for measuring the amount of radiation reaching the detector.

A method using X-rays for measurement of thickness is shown in Fig. 29.101. The X-ray source radiates to two detecting cells through a sheet of known thickness and through a similar sheet whose thickness is to be measured. The detecting cells are electron multiplier cells modified to respond to X-rays only. When the thickness of two sheets *i.e.*, the thickness of sheet of known thickness and the thickness of the sheet under test are equal, the amplifier input is zero and hence the indicating instrument reads zero. (The indicating instrument is a centre zero instrument.) The indicating instrument may be directly calibrated in terms of the relative thicknesses of the two sheets.

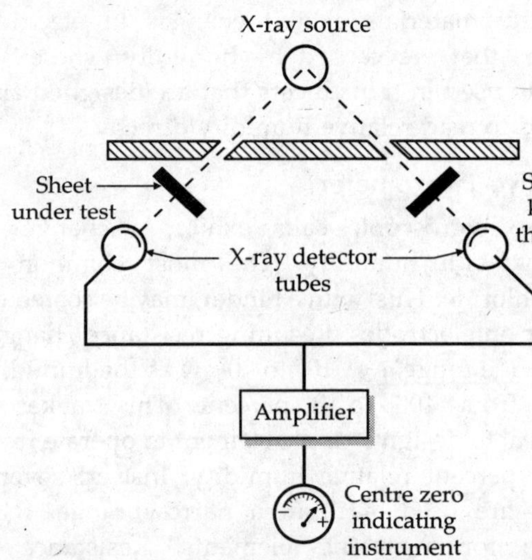


**Fig. 29.101** Measurement of thickness using X-ray methods.

The advantages of using X-rays system are :

(*i*) It measures thickness without contact with the material under test.

(*ii*) This method is well-suited to measurement of thickness of sheet in rapid motion like in rolling etc.

Gamma and X-rays are normally used for heavy metals and thick test pieces. Beta rays are used for thin metal sheets or foils, paper, rubber and plastics. Alpha-rays are used for very thin foils whose thickness is of the order of mm.

Beta, Gamma and X-rays may be used in the examination of hidden flows in castings. In this case the penetration power of different X-rays may be used advantageously.

29

## 29.45 MEASUREMENT OF HUMIDITY

**Humidity.** Humidity is the measure of water vapour present in a gas. It is usually measured as absolute humidity, relative humidity or dew point temperature.

**Absolute humidity.** It is the mass of water vapour present per unit volume.

**Relative humidity.** It is the ratio of water vapour pressure actually present to water vapour pressure required for saturation at a given temperature. The ratio is expressed in percent. Relative humidity (RH) is always dependent upon temperature.

### 29.45.1 Hygrometers

A hygrometer measures the value of humidity directly. Generally, the output of a Hygrometer is used to indicate relative humidity.

Several materials exhibit changes in electrical properties that are caused by humidity. These are frequently used in transducers that are designed and calibrated to read relative humidity directly.

### 1. Resistive Hygrometer

Some hygroscopic salts exhibit a change in resistivity with humidity. The most common is lithium chloride. This, with a binder, may be coated on a wire or on electrodes. Resulting resistance changes cover a wide range, *e.g.*, $10^4$ to $10^9$ $\Omega$ as the humidity changes from 100 to 0 percent. This makes it impractical to design a single element to operate from 1 to 100 percent relative humidity. Instead several elements are used, each in a narrow range, with provision for switching elements. Resistance is measured either with a Wheatstone bridge or by a combination of current and voltage measurements.

Most of these must not be exposed to conditions of 100 percent humidity as the resulting condensation may damage the device. Either they must be operated in a constant temperature environment or temperature corrections must be made. These are accurate to within ± 2.5 percent or ± 1.5 percent in some cases. Response times are typically of the order of a few seconds. These are currently the most common electronic hygrometers.

A typical resistive hygrometer is shown in Fig. 29.102. It shows a mixture of lithium chloride and carbon which acts as conducting film. This is put on an insulating substrate between metal electrodes.

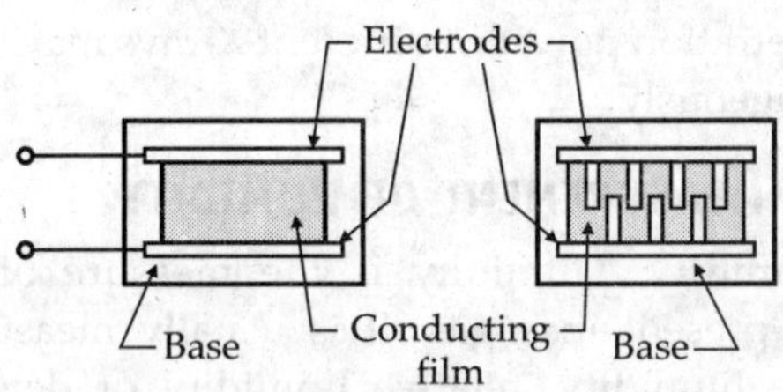


**Fig. 29.102** Resistive hygrometer.

The resistance of the element changes when it is exposed to variations in humidity. The higher the relative humidity, the more moisture the lithium chloride will absorb, and lower will be its resistance.

The resistance of the sensing unit is a measure of the relative humidity. Resistance should be measured by applying a.c. to the Wheatstone bridge. D.C. voltage is not applied because it tends to breakdown the lithium chloride to its lithium and chlorine atoms. The current flow is a measure of the resistance and hence of the relative humidity.

### 2. Capacitive Hygrometer

Some hygroscopic materials exhibit a change in dielectric constant with humidity changes. In addition, the presence of water vapour in air changes the dielectric constant of the mixture. In either case, the changes are small, and the change in capacitance is usually measured by including it as the frequency-determining element in an oscillator, heterodyning this signal with a beat frequency oscillator, and measuring the resulting difference frequency. These transducers have response time of about 1 s.

### 3. Microwave Refractometer

For secondary standards, a system consisting of two cavities, each of which is coupled to a klystron, may be employed. One cavity contains dry air, the other contains the mixture to be measured. The change in dielectric constant changes the frequency of one oscillator. This difference is measured electronically. The complexity of the devices precludes their use except in very special situation.

### 4. Aluminium Oxide Hygrometers

The aluminium-oxide coating on anodized aluminium exhibits a change in dielectric constant and a change in resistivity with a change in humidity. A hygrometer that uses this is built by using the base aluminium as one electrode and depositing a very thin electrode — usually gold — on the opposite side of the oxide as shown in Fig. 29.103.

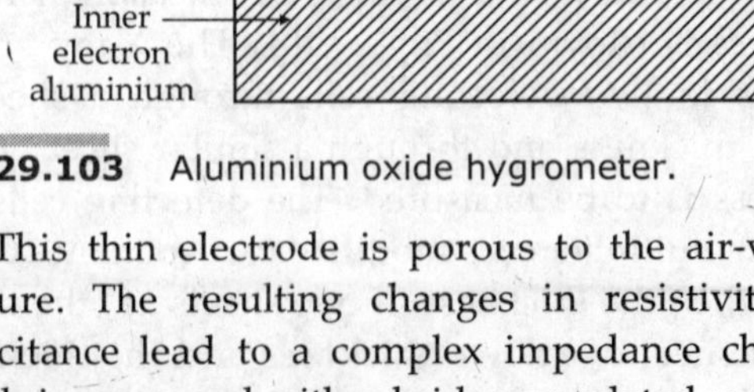


**Fig. 29.103** Aluminium oxide hygrometer.

This thin electrode is porous to the air-vapour mixture. The resulting changes in resistivity and capacitance lead to a complex impedance changes, which is measured with a bridge or related methods. Errors are less than ± 3 percent and response times are about 10 s. The changes are large, but often only one sensing element is needed for the entire range to be covered. This hygrometer is rapidly becoming a very important component in electronic systems.

### 5. Crystal Hygrometer

Some crystals are hygroscopic, and others may be coated with a hygroscopic material. The coating materials are hygroscopic polymers. The crystals are used as frequency determination elements in electronic oscillators. Frequency shifts with humidity as the mass of the crystal changes with amount of water absorbed by coating are measured electronically. These transducers are useful if a telemetry system is needed because the frequency range can be chosen as a standard telemetry frequency.

A crystal hygrometer using quartz is shown in Fig. 29.104.

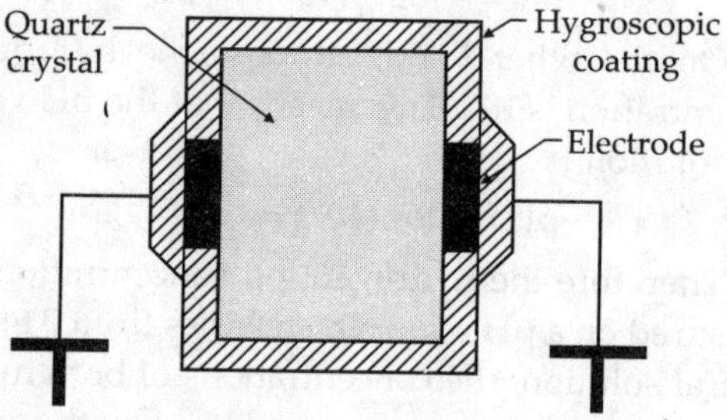


**Fig. 29.104** Crystal hygrometer.

## 29.46 MEASUREMENTS OF SOUND USING MICROPHONES

### 29.46.1 Microphones

A microphone is a special type of pressure transducer converting acoustic energy — pressure waves — into electrical energy for measurement. Invariably the sensing element is a diaphragm. A variety of transduction elements are available. The range of pressure to be measured is normally small compared with ambient (atmospheric) pressure, even for the most intense sounds. Desirable characteristics include linearity over a wide range of amplitudes and frequencies. Directional patterns must be considered. Sound pressure levels are commonly expressed in db.

The commonly used pressure reference being $20 \times 10^{-5}$ N/m$^2$.

This reference corresponds to the lowest audible sound in a normal human ear. A pressure of 90 db or 0.632 N/m$^2$ is normally the maximum sustained level human beings can tolerate (threshold of pain).

The sensitivity of a microphone is a measure of the voltage developed per unit sound pressure. It is given by :

$$\text{Sensitivity (db)} = 20 \log_{10} (\text{voltage output}) + 74 \text{ db (applied sound pressure in db)}$$

Thus a sensitivity of – 20 db means that a voltage of 0.1 V will be developed by a sound signal of pressure level 74 db which is the reference condition of sensitivity.

### 29.46.2 Types of Microphones

#### 1. Carbon Microphones

The moving diaphragm is mounted so as to compress together grains of carbon, thus changing the resistance in a circuit. The device is externally powered from a constant-voltage source and the resulting current is thus a function of the sound pressure. The most common application of these is in telephones. Their frequency range is limited to a maximum of about 5 kHz. The output impedance is low, and the sensitivity is high, about 400 db. These find little application in the field of sound measurements.

#### 2. Capacitive Microphones

The moving diaphragm is used as one plate of the capacitor. A constant charge is maintained on the plates from a 200 to 500 V supply. The changes in capacitance cause a change in the capacitor voltage. An emitter follower is usually located close to the microphone, reducing the inherently high output impedance and reducing the distortion caused by noise. These are expensive, but accurate, instruments. They are used for precision measurements, and as standards. Their frequency range is up to 50 kHz and are reasonably sensitive. A typical sensitivity is –50 db. The high voltage supply, however, is often inconvenient to provide.

#### 3. Dynamic Microphones

These microphones use the electromagnetic transduction principle. A dynamic microphone is shown in Fig. 29.105. An a.c. voltage is generated

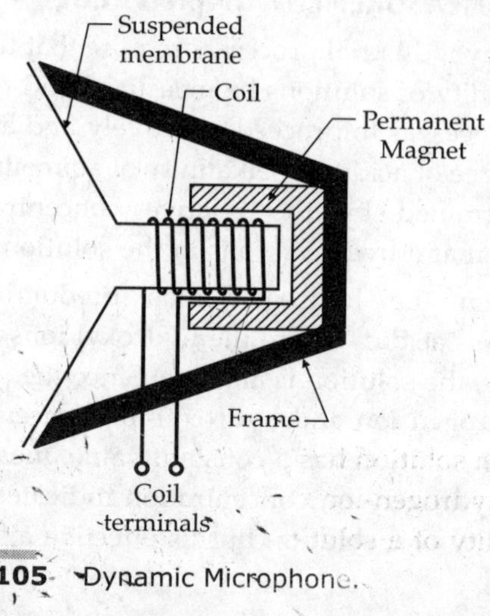


**Fig. 29.105** Dynamic Microphone.

when a coil is moved in a magnetic field. A membrane is mechanically linked to a moving coil surrounded by a permanent magnet. The membrane and the coil are suspended so that they can move back and forth. The sound pressure waves force the membrane towards the magnet, moving the coil across the magnetic field. The voltage developed across the coil terminals is a direct result of the motion of the coil, which clearly depends upon the force of pressure waves. They are self-generating and have low sensitivity, typically – 80 db and a low output impedance. Their frequency response is limited to about 20 kHz. These find field application but are seldom used in precision measurements.

### 4. Inductive Microphones

The diaphragm causes changes if the magnetic properties of certain materials. In turn, these changes appear as changes in dimensions, which are sensed by displacement transducers. These have low output impedance, low sensitivity (– 100 db), and operate in the ultrasonic region. They find application in underwater transducers.

### 5. Piezoelectric Microphones

These self-generating microphones have been developed for use in a wide variety of situations. The problems of piezoelectric devices in general apply. These problems include high output impedance, temperature dependence, and vibration sensitivity. Microphones built with this transduction element have a frequency response extending to the ultrasonic region, above 100 kHz, and they have a low sensitivity – 50 to – 100 db. They also are very linear over a wide range of amplitudes and are widely used in sound measurement systems.

## 29.47 MEASUREMENT OF pH VALUE

Many industrial processes require that the acidity or alkalinity of solutions be measured and controlled if the process is to proceed effectively and efficiently. The degree of acidity or alkalinity of aqueous solution is determined by the relative concentrations of hydrogen and hydroxyl ions in the solution.

When the hydrogen ions predominate, the solution is acidic. Where the hydroxyl ions are in the majority, the solution is alkaline. Since the product of the hydrogen ion and hydroxyl-ion concentration in any such solution has a constant value, measurement of the hydrogen-ion concentration indicates not only the acidity of a solution but its effective alkalinity as well.

The dissociation constant is the product of hydrogen ($H^+$) ions and hydroxyl $(OH)^-$ ions and this product is always equal to $10^{-14}$.

Hydrogen ion concentration is measured on a scale called pH scale.

**pH value** of a solution is defined as the negative logarithm of the hydrogen ion concentration or

$$pH = -\log_{10}(H^+)$$

This scale ranges from 0 to 14.

For example, suppose the hydrogen ion concentration is $10^{-14}$. Therefore, hydroxyl ion concentration is $10^0$ as the product of two has to be $10^{-14}$. The pH value for such a solution is :

$$pH = -\log_{10}(10^{-14}) = 14.$$

On the other hand, suppose the hydrogen ion concentration is $10^0$. This means that the pH value for this solution is :

$$pH = -\log(10^0) = 0$$

Therefore the hydrogen ion concentration can be measured on a pH scale which varies from 0 to 14. In a neutral solution, the concentrations of both hydrogen and hydroxyl ions are equal *i.e.*, both are $10^{-7}$. Therefore, for a neutral solution :

$$pH = -\log(10^{-7}) = 7$$

Thus a neutral solution like pure water has a pH value of 7.

Suppose the solution is acidic. Therefore the hydrogen ion concentration is more. Suppose it is $10^{-5}$, therefore the hydroxyl ion concentration is $10^{-9}$.

The pH value for this solution is :

$$pH = -\log_{10}(H^+) = -\log_{10}(10^{-5}) = 5$$

Thus for acidic solutions the pH value lies between 0 to 7. In case of alkaline solution the pH value is between 7 to 14.

For alkaline solutions the hydroxyl ion concentration is more. Supposing hydroxyl ion concentration is $10^{-2}$ and hence the hydrogen ion concentration is $10^{-12}$. The pH value for this solution is :

$$pH = -\log_{10}(10^{-12}) = 12$$

Thus we conclude that hydrogen ion concentration is measured on a scale (the pH scale) that ranges from 0 to 14. On this scale, neutrality is 7.0, the value obtained when the hydrogen and hydroxyl-ion are equal and balance each other, as is the case with pure water. As the solution becomes more acidic, its pH scale reading decreases below 7. As the solution becomes more alkaline, its pH scale reading increases above 7.

The measurement of pH value is obtained by immersing a pair of electrodes into the solution to be measured and measuring the voltage developed across them. The action is somewhat similar to that of a voltaic cell where a pair of dissimilar electrodes are immersed into an electrolyte.

In the pH cell, one of the electrodes, called the **reference electrode**, is at a constant potential regardless of the pH value of the solution under test. The potential of the other electrodes, called the **measuring electrode**, is determined by the pH value of the solution. Thus the potential difference between the two electrodes depends upon the pH value of the solution.

The reference electrode [See Fig. 29.106(*a*)] is made of glass and consists of an inner assembly containing a solution of calomel (mercury chloride) and mercury. This assembly is surrounded by a larger glass tube, and the space between the two contains an accurate solution of potassium chloride (KCl). A tiny opening in the bottom of the electrode permits the potassium chloride to diffuse very slowly into the solution under test. In this way, electrical contact is made between this solution and the calomel solution of the electrode.

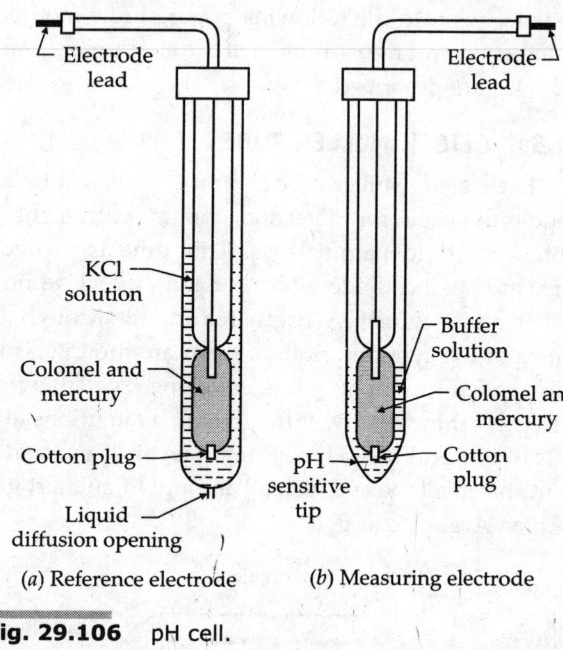


**Fig. 29.106** pH cell.

The measuring electrode (also known as the *glass electrode*) is somewhat similar to the reference electrode [See Fig. 29.106(*b*)]. However, the mercury calomel element is surrounded by a buffer solution of constant pH. The bottom of the outer tube has no opening. Instead, it tapers down to a tip made of thin glass of special composition. At this tip a potential difference is developed between the buffer solution and the solution under test because of the difference in the pH value of the two solutions. Since the pH value of the buffer solution is constant, the net potential of this electrode is a function of the pH value of the solution being tested.

Both electrodes are mounted to form an assembly that is inserted into the solution. The difference of potential between the electrodes, which is a function of the pH value of the solution under test, may be measured by a voltmeter calibrated, units of pH value or, after amplification it may be employed to actuate some control device.

## 29.48 MEASUREMENT OF THERMAL-CONDUCTIVITY (GAS ANALYZER)

When the chemical purity of a gas is to be measured, the thermal conductivity gas analyzer is used. The operation of this type of analyzer is based on the vast difference in heat conductivity among gases. Figure 29.107 shows the basic circuit for a thermal conductivity gas analyzer.

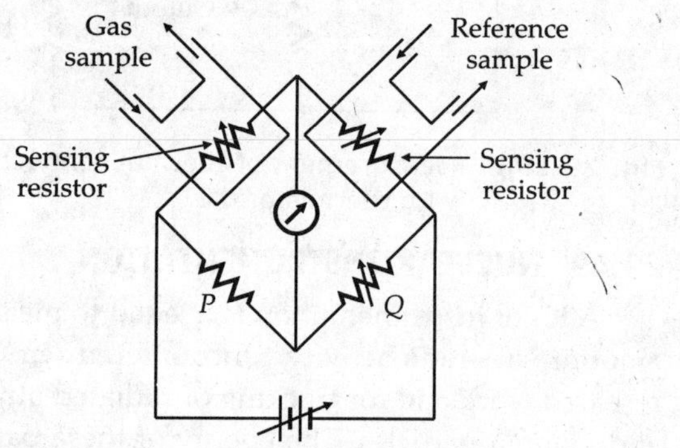


**Fig. 29.107** Basic thermal-conductivity gas analyzer circuit.

Note that a balanced bridge circuit is formed with two sensing resistors acting as balanced legs of the bridge. One sensing resistor is surrounded by the gas to be analysed, whereas the other resistor is surrounded by a reference gas (such as oxygen, pure atmosphere etc.). Where practical, both gases are maintained at the same pressure, water content, and so forth.

The bridge is first balanced by exposing both resistors to the same gas. Current flows through both sensing resistors, as well as through resistors *P* and *Q*. Resistor *Q* is adjusted for a "balance" or "zero set" condition on the meter. Then the resistors are exposed to the reference sample and gas sample.

If the gas sample contains elements having a different thermal conductivity than the reference sample, the bridge will be unbalanced. In some cases, the meter reads out in terms of the thermal-conductivity, whereas in other analyzers, the indication is on a *go-no go* or *good-bad* basis.

### Measurement of Thermal Conductivity using Thermistors

For this application two thermistors are connected in the adjacent arms of a Wheatstone bridge as shown in Fig. 29.108. The bridge supply is high enough to raise the thermistors above ambient temperature, typically to about 150°C. One thermistor is mounted in a static area to provide temperature compensation while the other is placed in the medium to be measured.

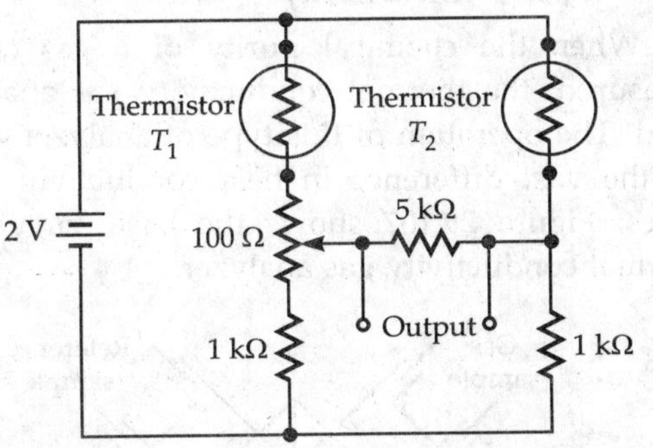


**Fig. 29.108** Measurement of thermal conductivity with thermistors.

## 29.49 NUCLEAR INSTRUMENTATION

A lot of instrumentation is needed to measure the amount of radioactivity encountered in nuclear research work and for tracking of radiated alpha, beta and gamma particles. The energy of these particles is absorbed by other molecules, producing chemical changes, ionization, or other reactions that produce visual evidence of the presence of other particles. These processes are used to detect, count and trace the movement of such particles.

## 29.50 TYPES OF RADIATIONS

Radioactive materials emit nuclear radiations which mainly consist of four parts :

**1. *Alpha-particles.*** The alpha-particle is a helium nucleus consisting of two protons and one neutron. It has a mass of $62 \times 10^{-27}$ kg and carries a positive charge of $3.2 \times 10^{-9}$ C. The alpha-particle is the heaviest and the slowest, travelling at speeds of from 1/10 to 1/100 of the velocity of light. Since it has the greatest mass and slowest speed, it is least penetrating. It penetrates about 0.02 mm of aluminium foil.

**2. *Beta-particles.*** The beta-particle is an electron with a mass of $9.03 \times 10^{-31}$ kg and a charge of $1.6 \times 10^{-19}$ C. Since beta-particle is much lighter than alpha-particle and has a speed approximately equal to that of light, it can penetrate much farther than alpha-particle – about 100 times the penetration of alpha-particle.

**3. *Gamma-particles.*** Gamma-particles or gamma rays, are an electromagnetic radiation with a wavelength of approximately 0.03 to 3 angstrom. Gamma-particles travel at the speed of light, and because of their shorter wavelengths, have very great penetrating power. Gamma rays can penetrate about 20 cm through a mass of lead.

**4. *Neutrons.*** The mass of the neutron is intermediate between that of alpha and beta-particles, having the same mass as a proton. The speeds of neutrons may be high or low depending upon the source from which they are emitted. Because neutron has no charge it is very difficult to detect. However, its penetrating power is small.

The above mentioned properties and other properties of these particles must be considered in the methods and processes for their detection and measurement. In the following pages, a few methods of detection and measurement of the above mentioned particles are described.

## 29.51 GEIGER MÜLLER TUBE

The Geiger Müller tube or simply Geiger tube are frequently used for detection and measurement of alpha, beta and gamma-rays. This tube is a diode, consisting of a cathode which is a long metal cylinder and an anode which is a fine wire running through the centre of the cylinder. Both of them are mounted in a thin walled, an tight, glass envelope, sealed by an extremely thin window through which radiations may pass at one end (See Fig. 29.109). The air is evacuated from the envelope and a small amount of an inert gas, such as Argon, is added.

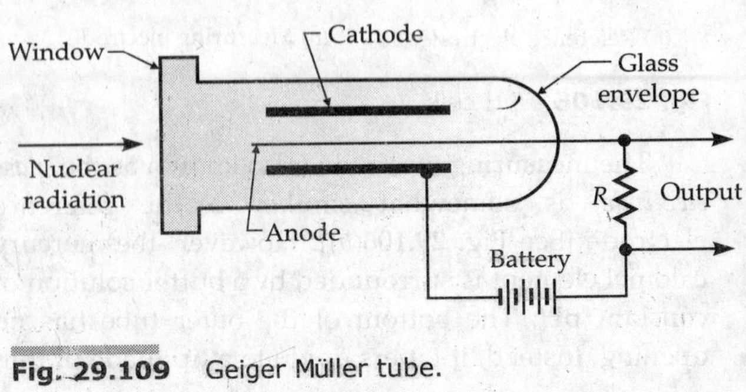


**Fig. 29.109** Geiger Müller tube.

The voltage of the battery is kept below the ionization potential of the gas. Now if radiation penetrates through the window and enters the envelope it would ionize some of the gas atoms. The resulting negative ions go towards the anode and positive ions towards the cathode. In their passage, the ions collide with some of the gas atoms, causing them to be ionized in turn. This process continues till whole of the gas atoms are ionized. In fact complete ionization takes place in no time. A pulse of current thus flows through the tube. This current flows through the resistance, *R*, which is connected in the anode circuit. The resulting voltage drop across *R* is the output voltage.

Once ionization is established, it is expected to continue indefinitely. However, since resistor *R*, which has a high resistance, is connected in series with the anode and the battery, when the gas ionizes a heavy current flows through resistance causing a large voltage drop. This voltage drop is sufficient to reduce the anode voltage below the ionization potential and therefore ionization ceases.

As long as the gas is not ionized no current flows through the resistance *R* and hence there is no output. Thus when the gas deionizes, the current ceases to flow through resistance *R* and so there is no output.

Now if radiation again enters the tube, the ionization starts and a current again flows through resistance *R* giving an output voltage. This way, a series of alpha or beta-particles or bursts of gamma rays, cause a series of current pulses to pass through the anode circuit of the tube and hence through resistance *R*. The output pulses from resistance *R* may be amplified and registered by a counting device. By counting the number of pulses, we can know the number of particles entering the Geiger Müller tube in a particular interval of time. The number counted gives the intensity of radiation. The pulses may be stored in some cases and the total count may be calibrated directly in terms of radio-activity.

The counting rate of this tube is seldom greater than $10^3$ counts per second.

## 29.52 IONIZATION CHAMBER

This is quite similar to the Geiger Müller tube as shown in Fig. 29.110. The ionization chamber is a metal cylinder (the outer electrode) sealed at one end by a window. The other end of the cylinder is closed, and a metal rod (the centre electrode) protrudes outside the cylinder on this end.

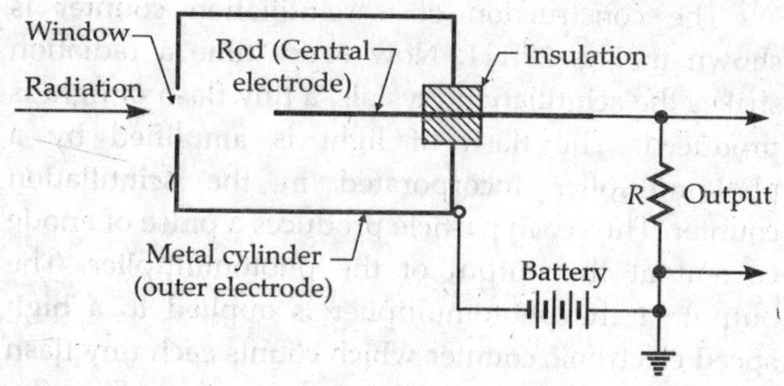


**Fig. 29.110** Ionization chamber.

The central electrode is grounded through resistance *R*. Resistor *R* has a very high resistance. A large positive potential is applied to the outer electrode. The chamber contains a gas which may be air, carbon dioxide, nitrogen, argon or methane.

The action of this ionization chamber is similar to that of Geiger Müller tube.

Let us consider that the chamber is filled with air. When a nuclear radiation enters the chamber through the window, some of the atoms of the air are ionized, the resulting ions travel towards their respective electrodes *i.e.*, positive ions towards the central electrode and the negative ions towards the outer electrode. In the process of rushing electrodes, these ions collide with other atoms of air causing ionization of these atoms. The action is instantaneous.

As the positive ions collect on the central electrode (which is at ground potential), the d.c. potential of this electrode is increased. This increased potential, which is the output signal of the chamber, may then be amplified and measured. Since the increase in this potential is a function of the number of radiation particles that have entered the chamber, the intensity of the nuclear radiation may be determined. The main use of this chamber, is to detect and measure alpha-particles.

## 29.53 SCINTILLATION COUNTERS

Certain crystals such as zinc sulphide, sodium iodide, anthracene and naphthalene etc., produce a brief flash of light each time they are struck by an alpha or beta or gamma-particles. The production of a flash of light by striking the crystals mentioned above, with alpha or beta or gamma rays is called **Scintillation**.

This principle is employed in a nuclear radiation sensor called **"Scintillation Counter"**.

The construction of a scintillation counter is shown in Fig. 29.111. Now every time a radiation strikes the scintillation crystals, a tiny flash of light is produced. The flash of light is amplified by a photomultiplier incorporated in the scintillation counter. Thus each particle produces a pulse of anode current at the output of the photomultiplier. The output of the photomultiplier is applied to a high speed electronic counter which counts each tiny flash generated by the scintillization crystals. Thus by counting the number of pulses in a given period of time, the intensity of radiation may be detected.

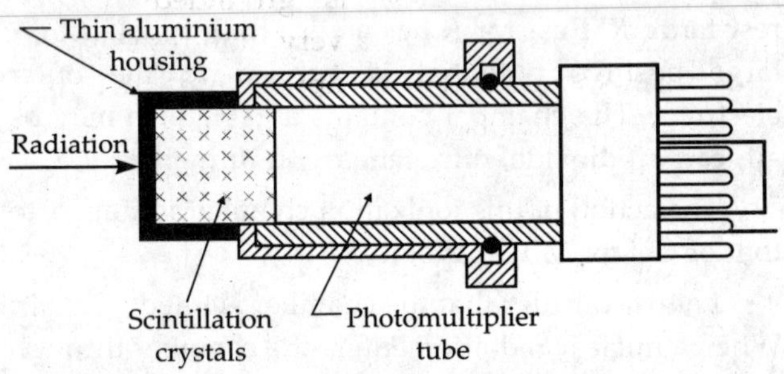


**Fig. 29.111** Scintillation counter.

The advantages of a scintillation counter are :

(*i*) This is more sensitive than the Geiger Müller tube and hence can detect lower levels of radiation.

(*ii*) The crystals used in this device produce a flash of light when struck by X-rays. Hence this transducer may be used to detect X-rays as well.

(*iii*) Counting rate is very fast. Pulses of as short a duration as 2 *ns* may easily be recorded.

## 29.54 SOLID STATE IONIZATION SENSING ELEMENTS

A system for counting the number of particles or photons whose energy level exceeds about 1,000 electron volt is built by using extrinsic semi-conductors mounted so that the radiation ionizes an atom in the crystal, and a pulse appears in the output circuitry. Operation is normally at liquid nitrogen temperatures. Counting rates as high as $10^6$ per second are possible.

## 29.55 DETECTION OF NEUTRONS

The neutrons do not carry any charge and hence are very difficult to detect. Thus we have to adopt indirect methods to detect them. Neutrons detecting methods require an arrangement where alpha or beta-ray particles are produced by neutrons and these particles detected by the methods described above.

An indirect method of detection of neutrons is described below :

A neutron is made to strike the atom of a certain isotope of uranium. The uranium atom splits into two or more parts on account of fission. This releases, in the process, nuclear radiation in the form of alpha, beta and gamma-rays, which may be detected by methods described above. Thus, by measuring the intensity of the radiations, the intensity of neutron beam producing them maybe detected.

Only slow moving neutrons will produce fission of uranium atoms. Where fast moving neutrons are to be detected, a layer of paraffin is put on top of the uranium layer of detector. The paraffin converts the fast moving neutrons to the slow moving neutrons required for fission.

# Review Questions

1. A strain gauge having an unstrained value of $R_g = R$ is cemented to a cantilever [Fig. 29.112(*a*)] and is connected in bridge circuit where $R_3 = R_4 = R_3 = R_g = R$ [Fig. 29.112(*b*)]. The input voltage to the bridge is $e_i$.
Calculate the output voltage when the strain is ε and the gauge factor for the strain gauge is $G_f$.
Calculate also the gauge sensitivity.

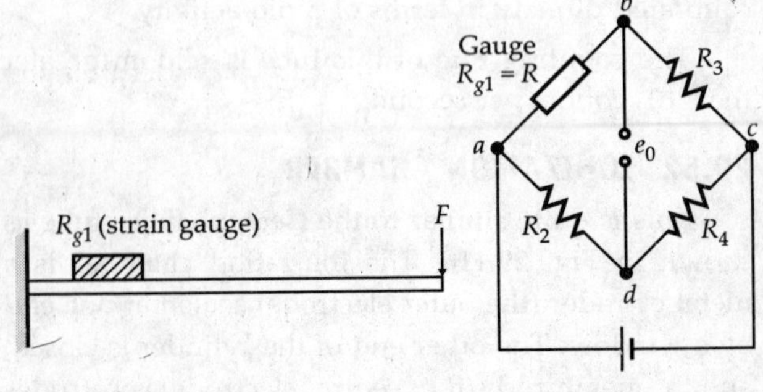


**Fig. 29.112** Diagram of Example 29.1.

Two strain gauges, each having an instrained value of $R_g = R$ are cemented to a cantilever as shown in Fig. 29.113(*a*). They are connected in a bridge circuit as shown in Fig. 29.113(*b*). Calculate the output voltage when they are subjected to a strain $\varepsilon$. They have a gauge factor of $G_f$ and the input voltage to the bridge is $e_i$ and $R_3 = R_4 = R$.

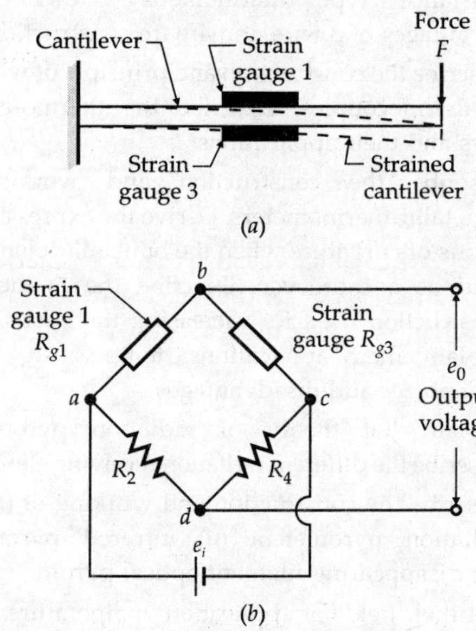


**Fig. 29.113** Diagram of Example 29.2.

3. Four strain gauges each having a strained resistance $R_{g1} = R_{g2} = R_{g3} = R_{g4} = R$ are cemented to a cantilever. They are subjected to a strain $\varepsilon$. They are connected in a bridge circuit as shown in Fig. 29.114. Calculate the output voltage and also the gauge sensitivity.

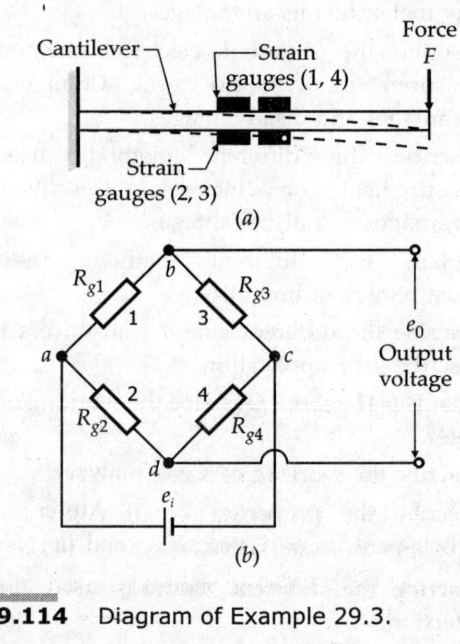


**Fig. 29.114** Diagram of Example 29.3.

4. Why are dummy gauges used ? In what way do they affect the output of a strain gauge bridge ?
5. Derive the expression of output from the following types of strain gauge bridges. Assume that the bridge is equiarmed :
   (*i*) quarter bridge, (*ii*) half bridge, (*iii*) full bridge.
6. Describe the different methods used for compensation and cancellation the effects of temperature changes which affect the resistance elements used in strain gauge bridges. Assume that all the four arms of the strain gauge have the same value of resistance.
7. Explain the method of calibration of strain gauges.
8. Derive the expressions for stress in an element subjected to bi-axial stress.
9. In a half bridge, the two strain gauges are connected in Poisson's configuration. The unstrained resistances of the two gauges $R_{g1}$ and $R_{g2}$ are equal as shown in Fig. 29.115 to the resistances connected in the rest of two arms *i.e.*, $R_{g1} = R_{g3} = R_{g2} = R_{g4} = R$. Calculate the bridge sensitivity.

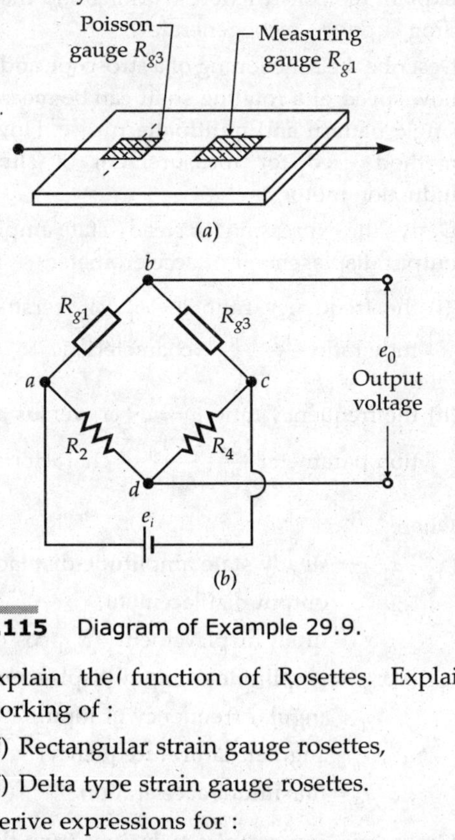


**Fig. 29.115** Diagram of Example 29.9.

10. Explain the function of Rosettes. Explain the working of :
    (*i*) Rectangular strain gauge rosettes,
    (*ii*) Delta type strain gauge rosettes.
    Derive expressions for :
    (*i*) minimum and maximum stresses and strains,
    (*ii*) angle of principal stress,
    (*iii*) shear stress,
    in each case.

29

11. Describe the method of measurement of differential pressure using an inductive transducer.
12. Describe the methods of measurement of pressure using
    (*i*) Capacitive transducers,
    (*ii*) Photo-electric transducers.
13. Describe the following methods used for measurement of low pressures using :
    (*i*) Thermocouple vacuum gauge,
    (*ii*) Pirani gauge,
    (*iii*) Ionization type vacuum gauge.
14. Explain the working of strain gauge type of torque transducers. Explain its advantages and disadvantages.
15. Explain the working of strain gauge type of torque transducers. Explain its advantages and disadvantages.
16. Explain the following methods of measurement of torque :
    (*i*) digital methods, (*ii*) magnetostrictive methods.
17. Explain the construction, working and theory of a drag cup type tachogenerator.
18. Describe the functioning of a stroscope and explain how speed of a rotating shaft can be measured by single pattern and multipattern disc. How is this method used for measurement of slip of an induction motor ?
19. Derive the expression for steady state amplitude of output displacement of accelerometer.
    (*i*) the frequency ratio $u = \omega_1 / \omega_n$ versus amplitude ratio $\frac{(x_2 - x_1)_m}{x_{m1}}$ characteristics.
    (*ii*) the frequency ratio $u = \omega_1 / \omega_n$ versus acceleration parameter $\frac{(x_2 - x_1)_m \omega_n^2}{a_m}$ characteristics.

    where

    $(x_2 - x_1)_m$ = steady state amplitude displacement
    $x_2$ = output displacement,
    $x_1$ = input displacement = $x_{m1} \cos \omega_1 t$,
    $x_{m1}$ = amplitude of input displacement,
    $\omega_1$ = angular frequency of input,
    $\omega_n$ = angular natural frequency,
    $a_0$ = maximum acceleration

    Draw some important conclusions from the above curves with regards to operation of accelerometers working in,
    (*i*) displacement mode and
    (*ii*) acceleration mode.
20. Describe the construction and working of the following types of accelerometers :
    (*i*) Potentiometric type, (*ii*) LVDT type,
    (*iii*) Piezoelectric type.
    Describe their advantages and disadvantages.
21. Describe the construction and working of a resistance type thermometer. Describe the advantages of 3 wire and 4 wire constructions.
22. Describe the construction and principle of working of thermocouples. Describe the thermo-electric laws and their applications.
23. Describe the construction and working of Bimetallic thermometers. Derive the expression for radius of curvature when the bimetallic element is used as a cantilever. Describe the methods of construction used for increasing their sensitivity. Explain their applications, temperature range, advantages and disadvantages.
24. Explain the theory of radiation pyrometers. Describe the different radiation receiving elements.
25. Describe the construction and working of (*i*) total radiation pyrometers (*ii*) infrared pyrometers (*iii*) disappearing filament optical pyrometers. List their fields of application, temperature range, advantages and disadvantages.
26. Describe the construction, working and theory of an electromagnetic type flow meter. Compare the operations of this meter when it is excited by
    (*i*) d.c. and (*ii*) a.c.
27. Describe the working of an nemometer when used for measurement of flow of fluids.
28. Describe the working and theory of an ultrasonic flow meter. List its advantages.
29. Describe the different electrical methods for measurement of liquid level. Compare their advantages and disadvantages.
30. Describe the different methods used for measurement of thickness. Describe their advantages and disadvantages.
31. Explain the different methods used for measurement of humidity.
32. Describe the different sound transducers in use. Describe their applications.
33. What is p-H value ? Describe the working of a p-H meter.
34. Describe the working of a gas analyzer.
35. Describe the properties of (*i*) Alpha-particles, (*ii*) Beta-particles, (*iii*) Gamma-rays and, (*iv*) Neutrons.
    Describe the different methods used for their detection.

# Unsolved Problems

1. A strain gauge having an unstrained resistance of 350 Ω and a gauge factor of 2 is connected in series with a ballast resistance across a 10 V supply. The ballast resistance is designed to give maximum sensitivity. The gauge is subjected to a dynamic strain of (10 + 20 sin 314 *t*) microstrain.
   (*a*) Find the expression for the change in output voltage on account of strain.
   (*b*) If a capacitor is connected in one output leads and if the voltage is read by a true rms reading voltmeter of infinite impedance, find the output voltage.

   [**Ans.** (*a*) [50 + 100 sin 314 *t*] μV, (*b*) 70.7 μV]

2. The resistance of a strain gauge is $R_g$ = 120 Ω and its gauge factor is 2. It is connected in a current sensitive Wheatstone bridge in which all resistances are 120 Ω. The input voltage is 4 V. Calculate the detector current in μA for 1 microstrain. The resistance of the galvanometer is 100 Ω.

   Calculate the voltage output if 1 microstrain is applied to the gauge and the voltmeter has an infinite input impedance. [**Ans.** 9.1 nA, 2 μV]

3. A strain bridge uses a strain gauge of 100 Ω resistance. Under no strain condition all the bridge arms have equal resistance. The resistance $R_1$ has to be changed to a value 100.52 Ω (Figs. 29.2 and 29.3) to obtain balance when the gauge is subjected to strain. The gauge factor is 2. Find the value of strain.

   [**Ans.** 0.0026]

4. A strain gauge having a gauge factor 2 is connected in a bridge circuit having an excitation voltage of 8 V. The resistances are all equal. It is subjected to a strain of 0.006. If this output is to represent 2/3 of full scale deflection of a recorder, what should be the gain of the amplifier. The full scale input voltage of recorder is 1 V.

   [**Ans.** 28]

5. A single strain gauge of gauge factor of 2 is used in a Wheatstone bridge where each resistance is 120 Ω. If the bridge exciting voltage is 8 cos 1000 *t* and if the gauge is subject to sinusoidally varying strain of 300 μm / m peak value and a frequency of 10 Hz
   (*i*) find the equation for the output voltage as observed on a CRO (assuming infinite impedance for the CRO,
   (*ii*) reading of rectifier type voltmeter having an infinite impedance and using full wave rectification. Assume that the scale is calibrated in terms of average value of voltage.

   [**Ans.** (*i*) 1.2 cos 1000 *t* sin 20 π *t* mV, (*ii*) 0.76 mV)

6. A bridge circuit has two fixed resistors and two strain gauges all of which have a value of 120 Ω. The gauge factor is 2.04 and strain applied to twin strain gauges, one in tension and the other in compression is $165 \times 10^{-6}$. If the battery current is 50 mA, determine the
   (*i*) voltage output of the bridge,
   (*ii*) the sensitivity in volt per unit strain.

   If the galvanometer connected to output terminals reads 100 μV per scale division and if 1/10th of a division can be read with confidence, determine the resolution.

   [**Ans.** (*i*) 0.202 V, (*ii*) 1.225 mV / μs, 0.008 μs]

7. In order to measure strain in a cantilever, two gauges are used one on top of the cantilever and the other at the bottom of the cantilever. The resistance of strain gauges is 1 kΩ each. The other two resistances in the circuit have resistance of 100 Ω each. The bridge detector resistance is 100 Ω and it sensitivity is 10 m / μA. Calculate the deflection of the galvanometer for 0.1% strain. The bridge supply voltage is 10 V. [**Ans.** 154 mm]

8. A gauge is made of a material having a resistance temperature co-efficient of $12 \times 10^{-6}$/°C. It has a resistance of 120 Ω and a gauge factor of 2. It is connected in a bridge circuit having resistances of 120 Ω each. The bridge is balanced at ambient temperature. Supposing there is a change of 20°C in the temperature of the gauge.
   (*a*) Find the output voltage of the bridge if the input voltage is 10 V.
   (*b*) What is the equivalent strain represented by change in temperature ?

   [**Ans.** (*a*) 0.55 mV, (*b*) 120 μs]

9. Calculate the fractional change of resistance of strain gauge made of nichrome V material
   (*a*) due to a stress of 200 MN/m$^2$ in the metal to which it is bonded,
   (*b*) for a temperature change of 100°C when bonded to steel,
   (*c*) when bonded to Duralumin.

29

The modulus of elasticity for steel is 200 $GN/m^2$ while that for Duralumin is 70 $GN/m^2$. The gauge factor is 2.5. The coefficients for linear expansion are :

Steel : $12 \times 10^{-6}$/°C, Duralumin : $23 \times 10^{-6}$/°C, Nichrome V : $11.9 \times 10^{-6}$/°C.

[**Ans.** $2.5 \times 10^{-3}$, $0.025 \times 10^{-3}$, $7.15 \times 10^{-3}$, $2.7 \times 10^{-3}$]

**10.** A Wheatstone bridge is shown in Fig. 29.6. If $R_1 = R_2 = R_3 = R_4$ at balance and if $R_1$ and $R_2$ are two strain gauges subjected to equal strains but of opposite nature such that $\Delta R_1 = -\Delta R_3$ while $\Delta R_2 = \Delta R_4 = 0$, prove the voltage $e_0$ of the bridge is linear function $\Delta R_1$, no matter how large $\Delta R_1$ gets.

**11.** A gauge has a resistance of 120 Ω and a gauge factor of 2.1. It is shunted by a resistance of 100 kΩ. What equivalent strain will be indicated when the gauge is shunted ? [**Ans.** 570 microstrain]

**12.** A gauge of resistance 120 Ω and a gauge factor of 2.1 is to provide an equivalent strain of 100 microstrains. Find the value of shunt resistance. [**Ans.** 571, 880 Ω]

**13.** For an electrical strain gauge of resistance 100 Ω and a gauge factor of 2, determine the value of shunt re. istance to calibrate for a strain of

(*a*) 0.001, (*b*) 0.0005, (*c*) 0.0001

[**Ans.** (*a*) 499900 Ω, (*b*) 99900 Ω, (*c*) 49900 Ω]

**14.** Calculate the hoop and longitudinal stresses in a cylindrical pressure vessel. The modulus of elasticity of the material is 100 $GN/m^2$. The hoop and the longitudinal strains are $425 \times 10^{-6}$ and $15 \times 10^{-9}$ respectively. The Poisson's ratio is 0.28. [**Ans.** 49.6 $MN/m^2$, 25.4 $MN/m^2$]

**15.** A rectangular rosette is mounted on a steel plate having modulus of elasticity $E$ = 200 $GN/m^2$ and Poisson's ratio = 0.3. The three strains measured are :

$$\varepsilon_1 = 72 \times 10^{-6}, \quad \varepsilon_2 = 120 \times 10^{-6} \text{ and } \quad \varepsilon_3 = 248 \times 10^{-6}$$

Calculate the principal strains and stresses, the maximum shear stress and the orientation angle for principal axes. [**Ans.** $256.5 \times 10^{-6}$, $63.5 \times 10^{-6}$, 60.6 $MN/m^2$, 102.3°]

**16.** A transducer is intended to measure forces in the range of 5 N to 40 N. This transducer consists of four strain gauges mounted on a cantilever as shown in Fig. 29.116. The dimensions of cantilever are $l$ = 50 mm and $b$ = 10 mm.

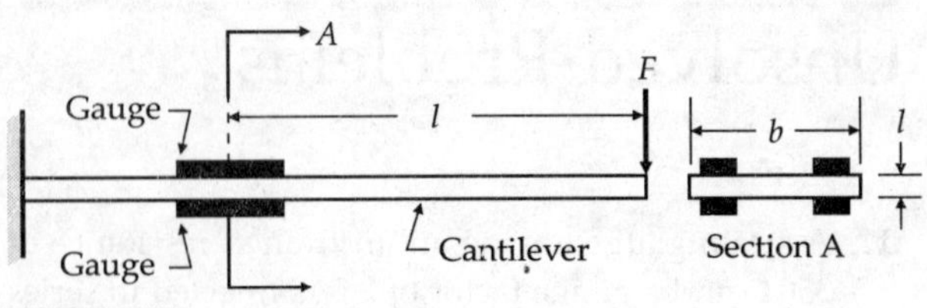

**Fig. 29.116** Diagram of Problem 29.16.

Find the required thickness, $t$, of the cantilever if the stress on section A-A is to be limited to 75 $MN/m^2$. The stress, $s$, on the cantilever is given bys = $Fl / 6\,bt^2$.

If the gauge factor is 2, the bridge supply voltage 4 V, and the modulus of elasticity 200 $GN/m^2$, find the minimum unbalance voltage which will have to be found for the force range specified. Assume that the detector has an infinite impedance. [**Ans.** 375 mV]

**17.** A point on a vibrating body has a simple harmonic motion. Determine at frequency of 10 Hz and 10 kHz,

(*a*) the amplitude of acceleration if its displacement amplitude is 1 mm,

(*b*) its displacement amplitude if the acceleration amplitude is 1 g.

[**Ans.** (*a*) 0.4 g, 400,000 g, (*b*) 2.5 m, 0.005 μm].

**18.** An inductive pickup operating from a 60 tooth wheel is used with a digital frequency meter to determine the speed of rotation of a shaft on which the wheel is mounted. The gating period is set to $10^4$ μs and a reading of 0.024 is obtained on the 4 digit display. What shaft speed does this represent in r.p.s. ? If the available gating periods are $10^2$, $10^3$, $10^4$, $10^6$, $10^7$ and 10 μs respectively, what would be the optimum setting of gating period for making this measurement ? [**Ans.** 40 r.p.m., $10^7$ μs]

**19.** In order to measure a speed higher than flashing rate of stroboscope the flashing rate was gradually reduced from the highest noting all the speeds at which single image is observed of a mark on the rotating object. These speeds are 20,000, 15,000, 12,000 and 10,000 rpm. Calculate the actual speed. [**Ans.** 60,000 rpm]

**20.** A stroboscope projects 6000 flashes per minute on a disc with 10 patterns mounted on the shaft of a machine.

(*a*) Find the speed of machine if the disc appears stationary and has a single image of points.

(*b*) What would be the two possible shaft speeds if the 10 points appear to be revolving once in 12 s ?

[**Ans.** (*a*) 600 rpm, (*b*) 605, 595 rpm]

**21.** For the purposes of checking the dial calibration of a stroboscope the dial was set to 1000 rpm and flashes were allowed to fall on the shaft of a synchronous motor running perfectly at 1000 rpm.

The single image took 15 s to complete a rotation and it was seen moving in the direction of rotation of the shaft :

(*a*) What should have been the dial reading instead ?

(*b*) If the image had been moving opposite to shaft rotation, what should have been the dial reading ? **[Ans.** (*a*) 996 rpm, (*b*) 1004 rpm]

**22.** Calculate the frequency ratio for which the error in acceleration is 1% if the damping factor is 0.7. **[Ans.** 0.306]

**23.** A seismic accelerometer sensing displacement has an undamped frequency of 20 Hz and a damping ratio 0.7. Calculate

(*a*) its damped frequency

(*b*) the amplitude ratio and phase angle between the motion of the seismic mass and the applied vibration if the later is a sinusoidal displacement at a frequency of (*i*) 30 Hz and(*ii*) 1 kHz.

**[Ans.** (*a*) 14 Hz, (*b*) 0.92, 120° 46′, 10.178°.24]

**24.** A piezoelectric accelerometer has a transfer function of 61 mV/g and a natural frequency of 4500 Hz. In a vibration test at 110 Hz, a reading of 3.6 V peak is obtained. Find the vibration peak displacement. **[Ans.** 1.21 mm]

**25.** An accelerometer shown in Fig. 29.33 consists of a mass 50 g attached to a spring cantilever. The spring is made from mild steel having an elastic modulus of 206 GN/m$^2$ and is of uniform rectangular cross-section. The output is obtained from a strain gauge bridge and amplifier. Two active gauges are used, one is each side of cantilever. The unstrained resistance of gauges is 200 Ω each and the resistance of other two arms are 10 kΩ each. Determine the horizontal acceleration if the amplifier output is 1 V. The output of amplifier is given by

$$e_0 = 30\,(R_2 / R_1)\, G_f \varepsilon$$

The gauge factor is 2.

The dimensions of the frame are : $l = 90$ mm, $b = 10$ mm, $d = 0.5$ mm.

**26.** An accelerometer for shock is designed as shown in Fig. 29.117. Find the relation between strain gauge resistance change and shock. The mass is 0.02 kg. The force rod has a modulus if elasticity of 0.1 GN/m$^2$. The cross-sectional area is $0.2 \times 10^{-3}$ m$^2$. The strain gauge has an unstrained resistance of 120 Ω and a gauge rector of 2.03.

**[Ans.** $2.39 \times 10^{-3}$ Ω/g]

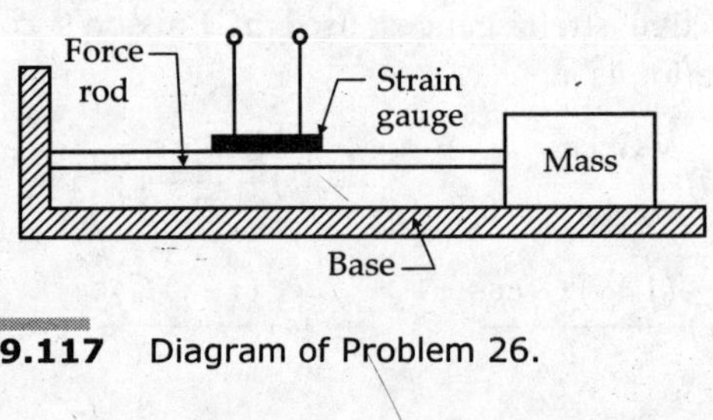


**Fig. 29.117** Diagram of Problem 26.

**27.** A resistance wire thermometer with a fundamental interval of 10 Ω is connected by two copper wires into a Wheatstone bridge, with no compensating leads. Each of the connecting wires has a resistance of 4 Ω at the ambient temperature at the time of calibration. If the range of the bridge indicator is 0 – 250 °C, find the approximate error, as a percentage of f.s.d. which will be caused by a change of 10°C in the temperature of leads. The resistance temperature co-efficient of copper is 0.0044/°C. **[Ans.** 1.4%]

**28.** The hot junction of a chromel alumel thermocouple is connected to a potentiometer terminals is at 24°C. The potentiometer whose terminals are 24°C reads 25.76 mV. What is the temperature of the thermocouple junction ? The calibration chart of the thermocouple is :

| *Temperature °C* | *Voltage, mV* |
|---|---|
| 20 | 0.8 |
| 24 | 0.95 |
| 28 | 1.12 |
| ...... | ...... |
| 480 | 26.25 |
| 488 | 26.72 |
| 493 | 26.04 |

**[Ans.** 488°C]

**29.** A thermopile arrangement of the copper constantan thermocouple consists of free junction parts and has the reference junction at 200°C. If the output voltage is 3.3 mV, determine the temperature of the detecting junction. The calibration chart for a thermocouple junction is :

| *Temperature °C* | *Voltage, mV* |
|---|---|
| 100 | 4.22 |
| 200 | 9.23 |
| 250 | 11.95 |

**[Ans.** 171.95°C]

**30.** A PMMC instrument has a resistance of 5 Ω and is calibrated to read correctly when the resistance of the external circuit is 2 Ω. Calculate the percentage

29

error if the resistance of the circuit is changed to 4 Ω. What would have been the error if the PMMC instrument had a resistance of 500 Ω ?

[**Ans.** 22.2%, 0.397%]

**31.** The emf of an iron-constantan thermocouple is to be measured by the potentiometer shown in Fig. 29.69. A potential difference of 1.0186 V is applied over points *AB* and is standardized using a Weston cell. The current through the resistors is to be 2 mA, and the range of temperature measurement is to be from 250°C to 850°C. Calculate the value of resistance $R_1$, $R_2$ and $R_3$ for an ambient temperature of 20°C. The emf of the iron constantan couple with reference junction at 0°C is :

emf at 20 °C = 1.019 mV,

emf at 250 °C = 13.553 mV,

emf at 850 °C = 48.716 mV

[**Ans.** $R_1$ = 6.27 Ω, $R_2$ = 485.5 Ω, $R_s$ = 17.58 Ω]

**32.** In a radiation pyrometer, the energy emitted from a piece of metal is measured. A surface emissivity of 0.82 is assumed and the temperature is determined as 1000 K. However, it is later found that the emissivity is 0.76. Find the error in the temperature determination. [**Ans.** 19°C]

**33.** The liquid level in a tank is to be measured using the upthrust from a float shown in Fig. 29.118. The upthrust deflects a cantilever on which a float is mounted. The deflection is measured by a voltage potential divider and a voltmeter. The sensitivity of the transducer is 10 V/mn. If the maximum height *h* of the water in the tank is 2 m and cross-section area of the float is 0.2 m$^2$, calculate the required stiffness of the cantilever in N/mm if the meter range is 50 V.

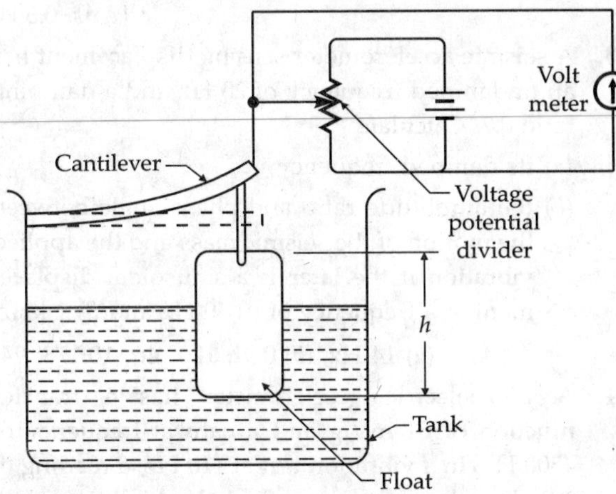


**Fig. 29.118** Diagram of Problem 33.

# Objective Type Questions

1. In a ballast circuit, the resistance of ballast and the strain gauge are 100 Ω each. The input voltage is 10 V. The gauge factor of the strain gauge is 2 and a strain of $10^{-6}$ is applied, the output voltage of the ballast circuit is :
   (*a*) 0.25 μV (*b*) 1 μV
   (*c*) 0.5 μV (*d*) 2 μV
2. When measuring strain, ballast circuits use a capacitor to act as high pass filter. This is done when :
   (*a*) static strains are being measured
   (*b*) dynamic strains are being measured
   (*c*) both static and dynamic strains are being measured
   (*d*) none of the above.
3. Two strain gauges are used to measure strain in a cantilever. One gauge is mounted on top of the cantilever and the other is placed at the bottom. The two strain gauges form two arms of a voltage sensitive Wheatstone bridge. This bridge configuration is called :
   (*a*) a quarter bridge (*b*) a half bridge
   (*c*) a full bridge (*d*) a null bridge.
4. Dummy strain gauges are used for :
   (*a*) compensation of temperature changes
   (*b*) increasing the sensitivity of bridge in which they are included
   (*c*) compensating for different expansion
   (*d*) calibration of strain gauges.
5. The output of a strain gauge bridge using two active strain gauges used in Poisson's configuration is
   (*a*) $\frac{\nu G_f \varepsilon e_i}{4}$ (*b*) $\frac{\nu G_f \varepsilon e_i}{2}$
   (*c*) $\frac{(1+\nu) G_f \varepsilon e_i}{2}$ (*d*) $\frac{(1+\nu) G_f \varepsilon e_i}{4}$

where $v$ = Poission's ratio,
$G_f$ = gauge factor,
$\varepsilon$ = strain
and $e_i$ = excitation voltage of bridge.

6. Hoop stresses act :
(a) in radial direction
(b) in axial direction
(c) in both radial and axial directions
(d) none of the above.

7. Strain gauge rosettes are used
(a) when the direction of principal stress is known
(b) when the direction of principal stress is not known
(c) when the direction of hoop stress is not known
(d) when the direction of longitudinal stress is not known.

8. For a rectangular rosette shown in Fig. 29.17, $\varepsilon_1 = 20 \times 10^{-6}$, $\varepsilon_2 = -40 \times 10^{-6}$ and $\varepsilon_3 = 100 \times 10^{-6}$, the angle of principal stress is
(a) 34.1° (b) 68.2°
(c) 248.2° (d) 124.1°

9. In a thermocouple pressure gauge, the temperature of heater element is a function of pressure for pressure ranges
(a) above atmosphere
(b) below 1 mm of Hg
(c) below $10^{-3}$ mm of Hg
(d) below 1 μm of Hg.

10. Pirani gauge is useful for measurement of pressure
(a) between $10^{-1}$ to $10^{-3}$
(b) between $10^{-1}$ to $10^{-6}$ mm of Hg
(c) $10^{-1}$ to $10^{-9}$ of Hg
(d) between $10^{-1}$ to $10^{-12}$ mm of Hg.

11. In ionization types of vacuum gauges, the pressure is $P$ is
(a) $P \propto I_G / I_P$
(b) $\propto I_P / I_G$
(c) $\propto I_K / I_G$
(d) $I_P / I_K$
where $I_G$ = grid current, $I_P$ = plate current and $K_K$ = cathode current.

12. In strain gauge torque transducers, the strain gauges should be mounted at
(a) 0° to the shaft axis
(b) at 45° degrees to shaft axis
(c) 90° to the shaft axis
(d) 60° to shaft axis.

13. In d.c. tachogenerators used for measurement of speed of a shaft, frequent calibration has to done because
(a) the contacts wear off
(b) the strength of permanent magnet decrease with age
(c) the armature current produces heating effects
(d) all the above.

14. In a drag cup type a.c. tachogenerator, the output voltage is :
(a) sinusoidal
(b) in the form of pulses
(c) modulated waveform
(d) constant d.c. because rectifiers are used.

15. A toothed type tachogenerator has 60 teeth. A magnetic pick up is used in conjunction with it. The number of pulses generated per second in the magnetic pick up is :
(a) 3000 (b) 1500
(c) 1800 (d) 1200
if the speed of the shaft to which the toothed wheel connected is 25 rps.

16. An induction motor is rotating at a speed of 1470 rpm. A disc marked with one star is attached to it. A stroboscope flashes light on it at a frequency of 12.5 flashes per second :
(a) it is observed that the star mark moves at a speed of 30 rpm in the direction of rotation
(b) it is observed that the star mark moves at a speed of 30 rpm against the direction of rotation
(c) it is observed that the star mark remains stationary
(d) it is observed the star mark moves against the direction of rotation at a speed of 60 rpm.

17. When seismic transducers are used in the acceleration made they should be designed with
(a) stiff springs and small mass
(b) heavy mass and weak springs
(c) weak springs and small mass.

18. When accelerometers operate in the displacement mode the ratio of forcing frequency to natural frequency should be
(a) below 1 (b) below 2
(c) above 2 (d) above 200

19. Piezoelectric accelerometers meters :
(a) should not be used for high frequencies above 100 Hz
(b) should be used for low frequencies

29

(c) should use a monitoring source of low input impedance

(d) have a low natural frequency.

**20.** Law of intermediate metals in thermocouples allows them to :

(a) use reference junction compensation

(b) use meters for measurement without disturbing the circuit conditions

(c) use extension wires of materials other than the one used for making thermocouples

(d) both (b) and (c).

**21.** A thermocouple :

(a) has a low time constant when it is bare

(b) has a low time constant if it is provided with a sheath

(c) has the same time constant whether it is bare or is provided with sheath

(d) none of the above.

**22.** A bimetallic thermometer is made of two metals $A$ and $B$ as shown in Fig. 29.119. The thermal co-efficients of expansion are :

$$\alpha_A = 20 \times 10^{-6}/°C$$

and $$\alpha_B = 10 \times 10^{-6}/°C.$$

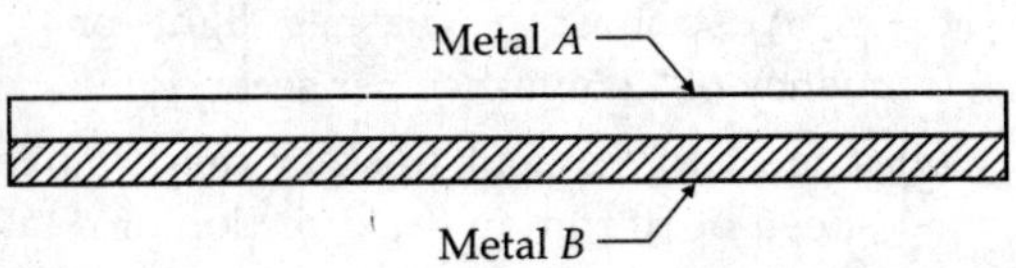


**Fig. 29.119** Diagram of question 22.

The temperature of the metal strips when build is $\theta_1 = 10°C$. Suppose the thermometer is raised to a temperature of $\theta_2 = 30°$. The thermometer will bend as :

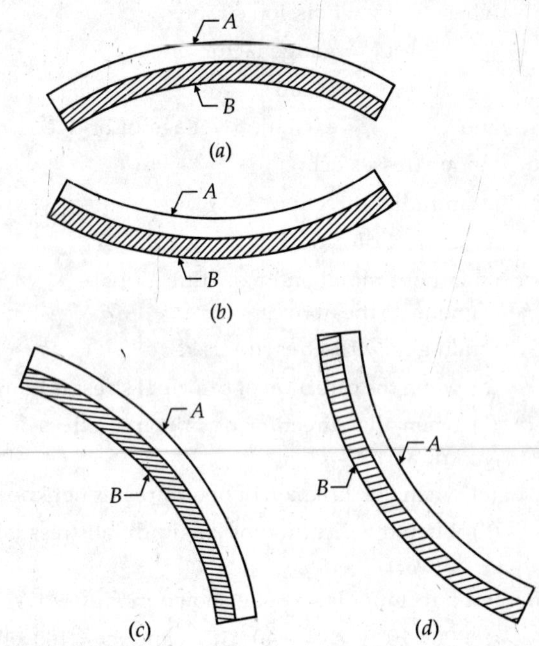


**Fig. 29.120**

**23.** If the temperature of a radiating body is 1500°C, the wavelength at which the maximum radiant energy occurs :

(a) 2.8 μm (b) 1.63 μm

(c) 1 μm (d) 20.8 μm

**24.** Radiation pyrometers are used in the temperature range of

(a) 0 – 500°C (b) 500 – 1000°C

(c) – 250 to 500°C (d) 1200 – 2500°C

**25.** In nuclear instrumentation which are the heaviest and slow moving particles ?

(a) beta-particles (b) gamma-rays

(c) alpha-particles (d) neutrons.

## Answers

| | | | | | | | | | |
|---|---|---|---|---|---|---|---|---|---|
| 1. (c) | 2. (b) | 3. (b) | 4. (a) | 5. (d) | 6. (a) | 7. (b) | 8. (d) | 9. (c) | 10. (a) |
| 11. (b) | 12. (b) | 13. (b) | 14. (c) | 15. (a) | 16. (b) | 17. (a) | 18. (c) | 19. (b) | 20. (d) |
| 21. (a) | 22. (a) | 23. (b) | 24. (d) | 25. (c) | | | | | |

# 30 Microprocessors

## 30.1 INTRODUCTION

The basic idea behind introduction of this chapter is to introduce to the users of this text the applications of digital systems and microprocessors used for measurement systems and control.

The major reasons for digital (and microprocessor control) are :

(*i*) Stability and accuracy of control,
(*ii*) Lower cost per function,
(*iii*) Flexibility,
(*iv*) Greater reliability and equipment life,
(*v*) Human factors favouring digital interface.

### 30.1.1 Digital Precision

A typical analog controller of high quality has a drift rate of approximately 0.1% per hour, or specifically 5 mV in a 0 – 5 V signal. Noise in an industrial environment adds at least another 10 mV. The total error is thus $5 + 10 = 15$ mV which is 0.3% precision in a signal of 5 V. This is true of even the frequently calibrated analog instruments. If we express process quantities by discrete numbers (as in a digital system and microprocessors), there is no limitation on accuracy. Suppose we can discriminate one unit in four decimal numbers say 1 in 9999, the precision is nearly 0.01%. Furthermore, this precision can be increased by adding more numbers.

Therefore, digital systems and microprocessors are capable of much better precision in Instrumentation and control applications as compared to their analog counterparts.

### 30.1.2 Versatility and Cost per Junction

A modem solid state electronic analog controller has about 20 active components like semiconductor diodes and transistors required for its operation. These chips can now be printed on a simple small silicon chip, less than 0.25 mm square, with many tens of thousands of components. Thus, the equivalent of 1000 or more controllers could conceivably placed on a single chip costing less than Rs. 300. Therefore, cost per controller function is vastly lowered.

The most general and versatile circuit that can be placed on a single chip is the digital *microprocessor*. The microprocessor is versatile because it can be programmed to perform an almost unlimited computing tasks. Programming means that the circuitry follows a sequence of coded orders that is carried in another component called *memory*. The processor responds to each of these coded orders as if it were reconnected to perform a new task. These tasks can be carried out in microseconds and therefore they can be switched and alternated so rapidly which gives the effect of many devices working in parallel. The microprocessor jumps rapidly between one task to another task, changing its function with every jump. This versatility and multiple functionality stem from the digital nature of microprocessor that permits it to be controlled by rapidly changing program. The versatility is enhanced further by letting the results of a sequence of steps alter the program sequence that follows. Thus the microprocessor control can even adapt itself to events that occur in the process.

The microprocessor provides the functions of many special devices like controller at relatively low cost, through its fast speed and time sharing ability. In addition, a microprocessor is capable of providing all sorts of functions not usually associated with controllers, such as binary logic, sequencing and numerical calculations.

### 30.1.3 Reliability and Cost

It has been described above that the digital integrated circuits are more precise and stable than analog circuits, have lower cost per function, and are versatile *i.e.,* they can perform multi-functions. The lower cost of chips is on account of accommodation of thousands of circuits on a small single chip, which can be mass produced.

The power requirement for driving the circuits is very small which require lower voltages for operation and hence there is very small power is wasted. Thus there is less heat to be dissipated. This results in lower electric and thermal stresses which lead to greater life and hence greater reliability. Another factor which leads to vastly improved reliability is the absence of soldered connections in integrated circuits.

### 30.1.4 Interface with Humans and Computers

Numbers are the natural output of Microprocessors. Versatile microprocessors can perform many operations on the numbers before presenting them to humans, such as converting them to decimal form or to engineering units, highlighting limits by flashing or changing colour.

### 30.1.5 System Control

Many separate Control and Instrumentation devices must be co-ordinated to accomplish the so called high level tasks such as the overall optimization of a process unit or entire plant. Today, high level control coordination means a supervising computer, given hierarchal authority over individual process loop controls. This requires a central control room in which plant variables are displayed by digital CRT monitors. Since the supervisory or display computers are digital, they can best communicate with lower level digital microprocessor controllers since both "speak the same language". This results in precise control and large reduction in errors.

## 30.2 NUMBER SYSTEMS

In the history of mankind different number systems were developed. Today the decimal system is universally used. However, it is neither the only number system nor it is the simplest to use in many applications. In recent years, primarily due to the development of the digital computers and the digital instruments, the binary system has become increasingly popular.

## 30.3 DECIMAL NUMBER SYSTEM

The **decimal number system** uses 10 Arabic Indian symbols or digits. The digits are : 0, 1, 2, 3, 4, 5, 6, 7, 8 and 9 in order of ascending value. In any number system the number of different digits that may be employed is called **radix**. Thus, since the decimal system employs 10 digits its radix is 10. Now, in a decimal system the highest digit is 9. Numbers larger than 9 are constructed by assigning values or **weights** to the position of the symbol with respect to the decimal point.

Each position in a decimal number has a weight assigned to it which is ten times the value of the next position to the left of the decimal point. For example, in the decimal number 47, the positional value of digit 4 is ten times the positional value of digit 7. Every positional weight is a multiple of ten and can be expressed as 10 raised to some power. Starting at the decimal point the positional weight of the ones position is $10^0 = 1$. The next position to the left is the tens' position which carries a weight of $10^1 = 10$. The next position to the left, the hundreds' position, carries a positional weight of $10^2 = 100$. This progression of increasing exponents can be carried as far as left to the decimal point as desired. As mentioned earlier each digit has a **weight** or **digit value**. Standing by itself, the digit 4 has a value of four units. However, when the digit is associated with other digits to form a multidigit number, it has a positional **weight** in addition to its digit value. The positional weight of the digit at the extreme right of the number is unity or one. This is called the **Least Significant Digit** or simple **LSD**. The digit occupying the extreme left position has the greatest positional weight and is called **Most Significant Digit (MSD)**.

We have seen that in a decimal system, the positional weights increase tenfold each time we move by one position to the left. Thus these positional weights may be represented in ascending order of powers of 10.

We have considered whole number integers till now. When we write a **decimal fraction** (*i.e.,* a value less than unity), we place a dot, called **radix point** (which is called a **decimal point** in the decimal system) between the integers and the fraction to separate two sets of digits. Then each digit of the fraction has a proportional weight that is one tenth the weight of the position to its left. Since positional weights, change one-tenth time every time we move from left to right by one position and thus a similar progression can be carried to the right of the decimal point but here the exponents are negative. The first position to the right of decimal point has a weight of $10^{-1} (= 1/10)$, the second has a weight of $10^{-2} (= 1/100)$ and so on.

A decimal system can he represented as in Fig. 30.1. The digit placed in each box indicates how many multiples of the indicated power of ten are part of the total quantity represented by that number.

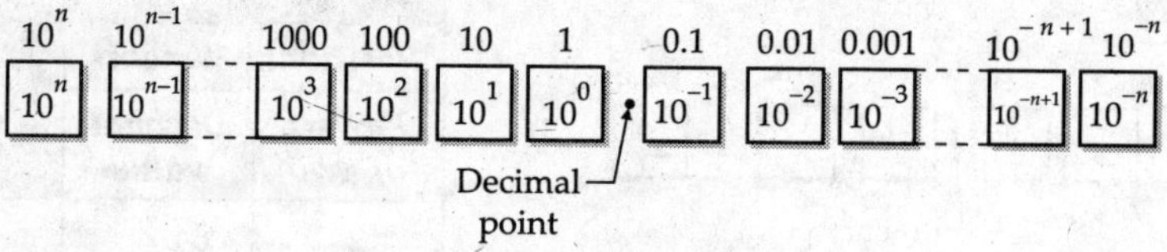


**Fig. 30.1** Diagrammatic representation of a Decimal System.

Consider the decimal number of 5642.37. We can write the value of each digit as represented by its relative position to the decimal point.

$$5642.37 = 5\times10^3 + 6\times10^2 + 4\times10^1 + 2\times10^0 + 3\times10^{-1} + 7\times10^{-2}$$

$$= 5000 + 600 + 40 + 2 + 0.3 + 0.07 = 5642.37$$

## 30.4 A GENERAL NUMBER SYSTEM

Any number can be used as a radix of a number system as long as the positional weights are adjusted accordingly. A positive digital number may be expressed as :

$$N = d_n\, d_{n-1} \dots d_3\, d_2\, d_1\, d_0\, d_{-1}\, d_{-2} \dots \qquad \dots(30.1)$$

where $d_n, d_{n-1}, \dots, d_3, d_2, d_1, d_0, d_{-1}, d_{-2}$, etc, are individual digit symbols.

Magnitude

$$N = d_n R^n + d_{n-1} R^{n-1} + \dots + d_2R^2 + d_1R^1 + d_0R^0 + d_{-1}R^{-1} + d_{-2}R^{-2} + \dots \qquad \dots(30.2)$$

where $R$ = radix of the system. The radix of a number system refers to the number of basic symbols used.

Let us take an example from the decimal system. In a decimal system, there are 10 basic symbols 0, 1, 2, 3, 4, 5, 6, 7, 8, 9 and therefore, $R = 10$. Consider a number 46753.28.

Now $N = 46753.28$. We have $d_4 = 4$, $d_3 = 6$, $d_2 = 7$, $d_1 = 5$, $d_0 = 3$, $d_{-1} = 2$ and $d_{-2} = 8$

Applying Eqn. 30.2, we get magnitude of number as :

$$N = d_4R^4 + d_3R^3 + d_2R^2 + d_1R^1 + d_0R^0 + d_{-1}R^{-1} + d_{-2}R^{-2}$$

$$= 4\times10^4 + 6\times10^3 + 7\times10^2 + 5\times10^1 + 3\times10^0 + 2\times10^{-1} + 8\times10^{-2}$$

$$= 40{,}000 + 6{,}000 + 700 + 50 + 3 + 0.2 + 0.08 = 46753.28$$

This agrees with our earlier result.

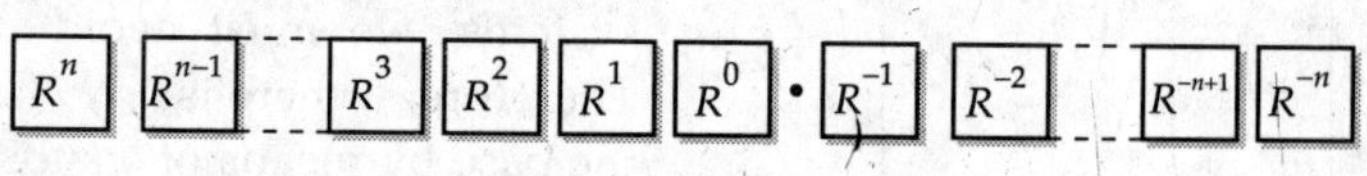


**Fig. 30.2** Skeleton of a general number system.

The diagrammatic representation of a general number system with radix $R$ is given in Fig. 30.2.

There are some other number systems in use. A binary system has a radix of 2 while an **octal** system has a radix of 8. The **duodecimal** system has a radix of 12 while **hexadecimal** system has a radix of 16. Except for binary system, the use of other system is limited.

## 30.5 BINARY NUMBER SYSTEM

A binary number system has a **radix** $R = 2$ and uses only two symbols **0** and **1**. These symbols represent an off-on system. Off state is represented by symbol **0** and On state by **1** and thus the binary system is very well suited to switching circuits. The numbers 0 and 1 are known as bits. In a binary system, the positional weights increase from right to left as in decimal system, only this time the increase is in ascending powers of 2. The skeleton of binary system is represented as in Fig. 30.3.

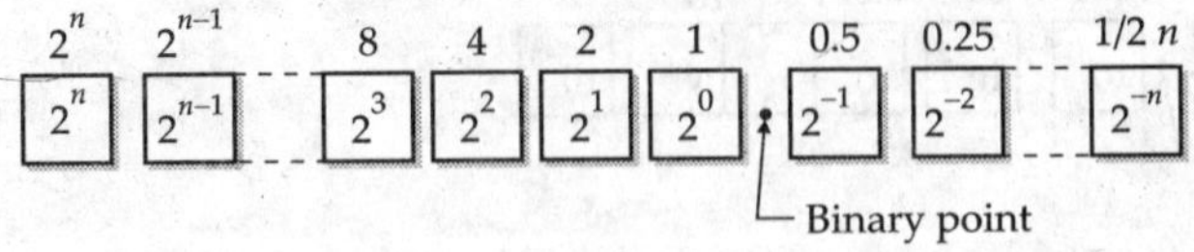


**Fig. 30.3** Skeleton of binary system.

Thus the position weight of the first digit on the **right** of binary point is $2^0 = 1$ ; the second position has a weight of $2^1 = 2$ ; the third has a weight of $2^2 = 4$ and so on. Similarly, the positional weights of digits on the **right** of the binary point are :

1st digit $= 2^{-1} = 1/2 = 0.5,$

2nd digit $= 2^{-2} = 1/2^2 = 0.025$

3rd digit $= 2^{-3} = 1/2^3 = 0.0125...$

*n*th digit $= 2^{-n} = 1/2^n$

Decimal number $1 = 1 \times 2^0 = 1$ in binary,

Decimal number $2 = 1 \times 2^1 + 0 \times 2^0 = 10$ in binary system.

Decimal number $3 = 1 \times 2^1 + 1 \times 2^1 = 11$ in binary system

The conversion of decimal numbers to binary is given in Table 30.1.

**Table 30.1** Conversion of Decimal Numbers to Binary Numbers

| Decimal | Binary |
|---|---|
| 0 | 0 |
| 1 | 1 |
| 2 | 10 |
| 3 | 11 |
| 4 | 100 |
| 5 | 101 |
| 6 | 110 |
| 7 | 111 |
| 8 | 1000 |
| 9 | 1001 |

Table 30.2 is a partial table showing the positional weights of the digits of the binary system.

**Table 30.2** Positional Weights of Digits of a Binary System

| Digit Positions | | | |
|---|---|---|---|
| Positional Weights | | Digit values | |
| Powers of two | Decimal value | | |
| $2^6$ | 64 | 0 | 1 |
| $2^5$ | 32 | 0 | 1 |
| $2^4$ | 16 | 0 | 1 |
| $2^3$ | 8 | 0 | 1 |
| $2^2$ | 4 | 0 | 1 |
| $2^1$ | 2 | 0 | 1 |
| $2^0$ | 1 | 0 | 1 |
| $2^{-1}$ | 0.5 | 0 | 1 |
| $2^{-2}$ | 0.25 | 0 | 1 |
| $2^{-3}$ | 0.125 | 0 | 1 |
| $2^{-4}$ | 0.0625 | 0 | 1 |

It should be understood that the prefixes before the powers of 2 in a binary system have to be either of the two bits *viz.* 1 or 0.

For example, we can write :

Decimal number

$$28 = 3 \times 2^3 + 1 \times 2^2 + 0 \times 2^1 + 0 \times 2^0 = 3100$$

Now it involves an extra digit 3 which is not included in the binary system. Therefore we write decimal number 28 in binary system as :

Decimal number

$$28 = 1 \times 2^4 + 1 \times 2^3 + 1 \times 2^2 + 1 \times 2^2 + 0 \times 2^1 = 1110 \text{ (in binary)}.$$

Since we share symbols between different number systems, we must avoid confusion between them. Therefore, if there is any chance of mixup, we denote the base by means of a subscript or spell it out as

| | |
|---|---|
| Decimal number | $237 = 23_{10}$ |
| Binary number | $101 = 101_2$ |
| Octal number | $456 = 456_8$ |

It should also be understood that the binary number $1110_2$ is read as one one one zero and not one thousand one hundred and ten.

### 30.5.1 Binary to Decimal Conversion

Binary quantities can be converted into their decimal equivalents. This can be electronically converted but we must examine the techniques behind them. This conversion is usually done by :

- ▲ Direct method, and
- ▲ Double dabble method.

The direct method is only described.

***Direct method.*** This method employs Eqn. 30.2. Let us illustrate that with an example.

***Example 30.1*** *Convert binary number 11001 into its decimal equivalent.*

**Solution.** For binary number 11001,

we have $d_4 = 1,\ d_3 = 1,\ d_2 = 0,\ d_1 = 0$ and $d_0 = 1.$

Magnitude of

$$N = d_4R^4 + d_3R^3 + d_2R^2 + d_1R^1 + d_0R^0$$
$$= 1\times2^4 + 1\times2^3 + 0\times2^2 + 0\times2^1 + 1\times2^0$$
$$= 16 + 8 + 0 + 0 + 1 = 25$$

$\therefore \quad 11001_2 = 25_{10}$

The conversion of binary mixed numbers and factors is illustrated by the following example.

***Example 30.2*** *Convert binary mixed number 100.110 into its decimal equivalent.*

**Solution.** Binary number = 100.110

$\therefore \quad d_2 = 1,\ d_1 = 0,\ d_0 = 0,\ d_{-1} = 0,\ d_{-2} = 1,\ d_{-3} = 0$

$\therefore$ Magnitude of binary number

$$= 1\times2^2 + 0\times2^1 + 0\times2^0 + 1\times2^{-1} + 1\times2^{-2} + 0\times2^{-2}$$

$\therefore \quad 100110_2 = 4 + 0 + 0 + 0.5 + 0.25 + 0 = 4.75_{10}$

### 30.5.2 Decimal to Binary Conversion

The binary to decimal conversion becomes a necessity at the output of terminals of a binary device such as a digital computer.

***(i) Direct method.*** In this method the largest power of 2 is subtracted from the decimal number. The next step is to again subtract the largest power of 2 from the remainder. This process is repeated successively till no remainder is left.

***(ii) Double Dabble system.*** This system of converting decimal into binary system is reverse of that of converting a binary system into a decimal system. The various steps involved are :

(*a*) Divide the decimal number by 2. Record the remainder.

(*b*) Divide the first quotient by 2. Record the remainder.

(*c*) Divide the next quotient by 2. Record the remainder.

(*d*) Repeat the process till the final quotient is 0.

(*e*) The remainders starting with the last one make up the equivalent binary number.

***Example 30.3*** *Convert the decimal number 93 into its binary equivalent by double dabble method.*

**Solution.**

| *Number* | *Quotient* | *Remainder* |
|---|---|---|
| 93 ÷ 2 | 46 | 1 [MSB] |
| 46 ÷ 2 | 23 | 0 |
| 23 ÷ 2 | 11 | 1 |
| 11 ÷ 2 | 5 | 1 |
| 5 ÷ 2 | 2 | 1 |
| 2 ÷ 2 | 1 | 0 |
| 1 ÷ 2 | 0 | 1 [LSB] |

$\therefore$ Reading the remainder column from bottom to top, we get :

Decimal $93_{10}$ = Binary $1011101_2$.

#### Conversion of Decimal Fractions to Binary

The method is best illustrated by the following example :

***Example 30.4*** *Convert decimal fraction 0.6372 into its binary equivalent.*

**Solution.**

| *Number* | *Product* | *Integer* |
|---|---|---|
| 0.6372 × 2 = | 1.2744 | 1 [MSB] |
| 0.2744 × 2 = | 0.5488 | 0 |
| 0.5488 × 2 = | 1.0976 | 1 |
| 0.0976 × 2 = | 0.1952 | 0 |
| 0.1952 × 2 = | 0.3904 | 0 |
| 0.3904 × 2 = | 0.7808 | 0 |
| 0.7808 × 2 = | 1.5616 | 1 |
| 0.5616 × 2 = | 1.1232 | 1 |

and so on.

$\therefore$ Decimal 0.6372 = Binary 0.10100011

### 30.5.3 Arithmetic Processes in a Binary System

The bases of arithmetic processes of addition, subtraction, multiplication and division in the binary system are the same as those used in the decimal system, except that only two digit symbols *i.e.,* 0 and 1 are used.

30

**1. *Binary addition.*** There are *four* types of additions in a binary system.

(*i*) $0+0=0$, (*ii*) $0+1=1$

(*iii*) $1+0=1$ (*iv*) $1+1=0$ with a carry of 1.

**Example 30.5** *Find the addition of the two binary quantities : 10011 and 1111.*

**Solution.**

| | | | | | | |
|---|---|---|---|---|---|---|
| | | 1 | 1 | 1 | 1 | **Carry** |
| | 1 | 0 | 0 | 1 | 1 | *First number* |
| + | | 1 | 1 | 1 | 1 | *Second number* |
| 1 | 0 | 0 | 0 | 1 | 0 | *Total* |

Now $10011_2 = 19_{10}$ and $1111_2 = 15_{10}$

and $100010_2 = 34_{10}$.

Therefore, addition of binary numbers 10011 and 1111 leads to 100010 which is equivalent to decimal number 34.

**Example 30.6** *Find the sum of binary numbers 101.101 and 110.100.*

**Solution.** Arranging the numbers in columns and allowing space for carries,

| | | | | | | | |
|---|---|---|---|---|---|---|---|
| | 1 | 1 | 1 | | | | **Carry** |
| | 1 | 0 | 1 | .1 | 0 | 1 | *First number* |
| + | 1 | 1 | 0 | .1 | 0 | 0 | *Second number* |
| 1 | 1 | 0 | 0 | .0 | 0 | 1 | *Total* |

**2. *Binary subtraction.*** The simple rules of binary subtraction are :

(*i*) $0-0=0$ (*ii*) $1-1=0$

(*iii*) $1-0=1$ (*iv*) $0-1=1$ with a borrow of 1.

**Example 30.7** *Subtract 110 from 11010. Both are binary numbers.*

**Solution.**

| | | | | | | |
|---|---|---|---|---|---|---|
| | | | 0 | 1 | 0 | **Changes in the minuend after borrowing** |
| | 1 | 1 | 0 | 1 | 0 | *Original minuend* |
| – | | | 1 | 1 | 0 | *Subtrahend* |
| | 1 | 0 | 1 | 0 | 0 | *Remainder* |

Let us check the result

$$11010_2 = 26_{10} \quad \text{and} \quad 110_2 = 6_{10}$$

$$11010_2 - 110_2 = 10100_2 = 20_{10}$$

which is $26 - 6 = 20$

## 30.6 MULTIPLICATION OF BINARY NUMBERS

The multiplication of binary numbers is a simple affair since both multiplicand and the multiplier contain only digits 0 and 1. The rules are :

(*i*) $0\times1=0$ (*ii*) $1\times0=0$

(*iii*) $1\times1=1$

However, when the partial products are added, rules for binary addition must be adhered to.

**Example 30.8** *Multiply binary numbers 1101 and 11011. Check the answer by converting numbers into equivalent decimal numbers.*

**Solution.**

```
          1 1 0 1     multiplicand
      × 1 1 0 1 1     multiplier
      -----------
          1 1 0 1
        1 1 0 1
      0 0 0 0
    1 1 0 1
  1 1 0 1
  -----------------
  1 0 1 0 1 1 1 1 1   final product
```

Now Binary number $1101_2$

$$= 1\times2^3 + 1\times2^2 + 0\times2^1 + 1\times2^0 = 13_{10}$$

and $$11011_2 = 1\times2^4 + 1\times2^3 + 0\times2^2 + 1\times2^1 + 1\times2^0 = 27_{10}$$

Product $13\times27 = 351$ (decimal).

Binary number

$$101011111_2 = 1\times2^8 + 0\times2^7 + 1\times2^6 + 0\times2^5 + 1\times2^4 + 1\times2^3 + 1\times2^2 + 1\times2^1 + 1\times2^0 = 351_{10}$$

This confirms with the answer obtained above.

## 30.7 DIVISION OF BINARY NUMBERS

The general rules of division in decimal system are applicable to binary system. In this context it may be emphasised that we must obey the rules of binary multiplication and binary subtraction.

**Example 30.9** *Divide binary number 110 by binary number 11.*

**Solution**

```
     10      quotient
  11)110     dividend
     11
     ---
     000
     000
     ---
       0     remainder
```

$\therefore$ Quotient $= 10$ (binary),

and Remainder $= 0$ (binary),

Now Binary $110_2 = 1\times2^2 + 1\times2^1 + 1\times2^0$

$= 6_{10}$. (decimal)

and Binary $11_2 = 1\times2^1 + 1\times2^0 = 3_{10}$ (decimal).

Now $6 + 3 = 2.$

In binary system 10 means a value of 2.

Thus the answer confirms to the solution obtained above.

***Example 30.10*** *Find the quotient of 10011 divided by 101.*

**Solution.**

$$
\begin{array}{r l}
11 & \longleftarrow \text{quotient} \\
10\overline{)10011} & \longleftarrow \text{dividend} \\
\underline{101} & \\
1001 & \\
\underline{101} & \\
100 &
\end{array}
$$

The answer in binary is $11\dfrac{100}{101}$

Now Binary $11_2 = 3_{10}$ (decimal),

Binary $100_2 = 4_{10}$ (decimal),

and Binary $101_2 = 5_{10}$ (decimal).

The answer in decimal system is $3\dfrac{4}{5}$

This confirms to the decimal system as verified below.

Binary $110011_2$ = decimal $19_{10}$

Binary $101_2$ = decimal $5_{10}$

$$
\begin{array}{r}
3 \\
5\overline{)19} \\
\underline{15} \\
4
\end{array}
$$

$\therefore$ The answer in decimal system is $3\dfrac{4}{5}$

## 30.8 OCTAL NUMBER SYSTEM

The octal number system has a radix of eight (8) and uses the symbols 0, 1, 2, 3, 4, 5, 6 and 7. The positional value of each digit in an octal number system follows the skeleton number of Fig. 30.4.

| $8^4$ | $8^3$ | $8^2$ | $8^0$ | $8^1$ | $8^{-1}$ | $8^{-2}$ | $8^{-3}$ | $8^{-4}$ | $8^{-5}$ |
|---|---|---|---|---|---|---|---|---|---|

Octal point

**Fig. 30.4** Octal System.

The powers of 8, associated with each position, increase to the left of the octal point decrease to the right of the octal point. This is shown in Table 30.3.

Table 30.3 gives the equivalence of decimal and octal systems.

**Table 30.3** Equivalence of Decimal and Octal System

| *Decimal Number* | *Octal Number* |
|---|---|
| 1 | 1 |
| 2 | 2 |
| 3 | 3 |
| 4 | 4 |
| 5 | 5 |
| 6 | 6 |
| 7 | 7 |
| 8 | 10 |
| 9 | 11 |
| 10 | 12 |
| — | — |
| — | — |
| 15 | 17 |
| 16 | 20 |

### 30.8.1 Octal to Decimal Conversion

Octal numbers are easily converted to their decimal equivalents by using the technique of positional value of the digits. The technique is identical to that converting binary numbers to their decimal equivalents.

***Example 30.11*** *Convert an octal number 372 to its decimal equivalent.*

**Solution.** The octal number is rearranged with the corresponding powers of 8 for each bit.

$$372_8 = 3\times(8)^1 + 7\times(8) + 2\times(8)^0$$

$$= 3\times64 + 7\times8 + 2\times1$$

$$= 256_{10}.$$

Thus octal number 372 is equivalent to binary number 256.

***Example 30.12*** *Convert octal number 24.6 into its decimal equivalent.*

**Solution.**

$$24.6_8 = 2\times(8)^1 + 4\times(8)^0 + 6+(8)^{-1}$$

$$= 16 + 4 + 0.75 = 20.75_{10}.$$

### 30.8.2 Decimal to Octal Conversion

The best method for converting a decimal number into its octal equivalent is the *double dabble* method as was used for conversion of decimal numbers into binary numbers. The steps to be followed are similar with the exception that the radix in the case of binary system is 2 while it is 8 in the octal system.

The following steps are required :

1. Divide the decimal integer by 8 and record the remainder in the second column.
2. Divide the quotient obtained in (1) by 8. Record the remainder again.
3. Repeat the process until the quotient is zero.
4. The octal equivalent is given by the column of remainders written from bottom to top.

**Example 30.13** *Convert the decimal number 266 into its octal equivalent.*

**Solution.** A simple table is set up and the method outlined above is followed :

| Number | | Quotient | Remainder |
|---|---|---|---|
| 266 ÷ 8 | = | 33 | 2 |
| 33 ÷ 8 | = | 4 | 1 |
| 4 ÷ 8 | = | 0 | 4 |

Reading the remainder column from bottom to top yield,

$$266_{10} = 412_8$$

### 30.8.3 Octal to Binary Conversion

The primary advantage of the octal number system is the ease with which conversion can be made between binary and octal numbers. The conversion from octal to binary is performed by converting each digit to its 3 bit binary equivalent. The eight digits are converted as shown in Table 30.4.

**Table 30.4** Octal to Binary Conversion

| Octal Digit | Binary Equivalent |
|---|---|
| 0 | 0 0 0 |
| 1 | 0 0 1 |
| 2 | 0 1 0 |
| 3 | 0 1 1 |
| 4 | 1 0 0 |
| 5 | 1 0 1 |
| 6 | 1 1 0 |
| 7 | 1 1 1 |

Using these conversions any octal number is converted to binary by individually converting each digit.

**Example 30.14** *Convert octal digit 472 to its binary equivalent.*

**Solution.**

Octal digit 4 → 100 in binary

Octal digit 7 → 111 in binary

Octal digit 2 → 010 in binary

∴ Octal digit $472_8 = 100111010_2$ (Binary).

### 30.8.4 Conversion from Binary to Octal System

Since octal system finds applications in digital computers, where binary numbers are often recorded in octal notation, conversions from the binary to octal system and from octal to binary system must be understood. Also, a large decimal number is most easily translated into its binary equivalent by first converting it into octal notation.

To convert a binary number into octal number the following rules must apply :

1. Starting at binary point, group the binary numbers into three digit clusters, both left and right to the binary point.
2. In order to complete the groups of three digits when forming the clusters, zeros must be placed at the extreme right and left ends of the original binary number.
3. Read out each group of three binary numbers. The combined readout is the equivalent octal equivalent of the binary number.

**Example 30.15** *Convert the binary number 11010110.1001 into its octal number.*

**Solution.** The original binary number is separated into group-of-3 clusters, adding zeros if necessary

011 010 110 . 100 100

Reading the binary value of each cluster gives :

011 010 110 100. 100 = 326.44

3 2 6 4 4

$$\therefore \quad 11010110.1001_2 = 326.44_8.$$

### 30.8.5 Usefulness of Octal System

The case with which conversion can be made between octal and binary systems makes the octal systems attractive as a *shorthand* means of expressing large binary numbers. In computer work, binary numbers with upto 64 bits are not uncommon. These binary numbers, do not always represent a numerical quantity but are often some kinds of codes which convey non-numerical information.

In computers, the binary numbers represent :

1. exact numerical data
2. numbers corresponding to a location (address) in memory
3. an instruction code
4. a code representing alphabetic and non-numerical characters
5. a group of bits representing the status of devices internal and external to computer.

When dealing with a large quantity of binary numbers of many bits it is convenient and more efficient to write octal numbers rather than binary numbers. Octal system is used for convenience of operators.

## 30.9 HEXADECIMAL NUMBER SYSTEM

The hexadecimal system uses base 16. Thus, it has 16 possible digit symbols. It uses the digits 0 through 9 plus the letters *A, B, C, D, E* and *F* as the 16 digit symbols. Table 30.5 shows the relationships among hexadecimal, decimal and binary. Note that each hexadecimal digit represents a group of four binary digits. It is important to remember that hex (abbreviation for hexadecimal) digits *A* through *F* are equivalent to the decimal values 10 through 15.

**Table 30.5**

| *Hexadecimal* | *Decimal* | *Binary* |
|---|---|---|
| 0 | 0 | 0000 |
| 1 | 1 | 0001 |
| 2 | 2 | 0010 |
| 3 | 3 | 0011 |
| 4 | 4 | 0100 |
| 5 | 5 | 0101 |
| 6 | 6 | 0110 |
| 7 | 7 | 0111 |
| 8 | 8 | 1000 |
| 9 | 9 | 1001 |
| A | 10 | 1010 |
| B | 11 | 1011 |
| C | 12 | 1100 |
| D | 13 | 1101 |
| E | 14 | 1110 |
| F | 15 | 1111 |

### 30.9.1 Hex-to-Decimal Conversion

A hex number can be converted to its decimal equivalent by using the fact that each hex digit position has a weight that is a power of 16. The LSD has a weight of $16^0 = 1$, the next higher digit has a weight of $16^1 = 16$, the next higher digit has a weight of $16^2 = 256$, and so on. The conversion process is demonstrated in the examples below :

$$365_{16} = 3\times16^2 + 5\times16^1 + 6\times16^0$$
$$= 768 + 80 + 6$$
$$= 854_{10}$$

$$2\text{AF}_{16} = 2\times16^2 + 10\times16^1 + 15\times16^0$$
$$= 512 + 160 + 15$$
$$= 687_{10}$$

Note that in the second example the value 10 was substituted for *A* and the value 15 for *F* in the conversion to decimal.

For practice, verify that $1\text{BC}2_{16}$ is equal to $7106_{10}$.

### 30.9.2 Decimal-to-Hex Conversion

Recall that we did decimal-to-binary conversion using repeated division by, 2 and decimal-to-octal using repeated division by 8. Likewise, decimal-to-hex conversion can be done using repeated division by 16. The following two examples will illustrate.

**Example 30.16** *Convert the decimal number 423 into hex.*

**Solution.**

| *Number* | *Quotient* | *Remainder* |
|---|---|---|
| 423 ÷ 16 | 26 | 7 |
| 26 ÷ 16 | 1 | 10 |
| 1 ÷ 16 | 0 | 1 |

$$\therefore \quad 423_{10} = 1\text{A}7_{16}$$

as 10 = A in hex.

**Example 30.17** *Convert decimal number 214 to hexadecimal.*

**Solution.**

| *Number* | *Quotient* | *Remainder* |
|---|---|---|
| 214 ÷ 16 | 13 | 6 |
| 13 ÷ 16 | 0 | 13 |

$$\therefore \quad 214_{10} = \text{D}6_{10}$$

as 13 = D in hexadecimal system.

Note that the remainders greater than 9 are represented by letters from *A* through *F*.

30

### 30.9.3 Hexadecimal to Binary Conversion

Like the octal system, the hexadecimal system is used primarily as *shorthand* method for representing binary numbers. It is relatively simple matter to convert a hex number to binary. Each hex digit is converted to its 4 bit binary equivalent.

***Example 30.18*** *Convert $9F2_{16}$ to its binary equivalent.*

**Solution.**

| 9 | F | 2 |
|---|---|---|
| ↓ | ↓ | ↓ |
| 1001 | 1111 | 0010 |

Therefore the binary equivalent is 100111110010.

### 30.9.4 Binary to Hex Conversion

Conversion from binary to hex is just the reverse process explained above. The binary number is grouped into clusters of four, bits, and each is converted to its hex digit. Zeros are added, as needed, to complete a 4 bit group.

***Example 30.19*** *Convert binary number 1110100110 into a hexadecimal number.*

**Solution.**

| 0011 | 1010 | 0110 |
|---|---|---|
| ↓ | ↓ | ↓ |
| 3 | A | 6 |

Zeros are added to the group on the left side.

$\therefore \quad 1110100110_2 = 3A6_{16}.$

***Example 30.20*** *Convert decimal 378 to a 16 bit binary number by first converting to hexadecimal.*

**Solution.**

$378 \div 16 = 23 +$ remainder of 10

$23 \div 16 = 1 +$ remainder of 7

$1 \div 16 = 0 +$ remainder of 1

Thus $378_{10} = 7A_{16}$

This hex value can be easily convened to binary.

| 1 | 7 | A |
|---|---|---|
| ↓ | ↓ | ↓ |
| 0001 | 0111 | 1010 |

Finally, we can express $378_{10}$ as a 16 bit by adding four leading zeros.

$$378_{10} = 0000\ 0001\ 0111\ 1010_2$$

***Example 30.21*** *Convert $B2F_{16}$ to octal.*

**Solution.** It is easiest to first convert to binary and then to octal.

$$
\begin{aligned}
B2F_{16} &= 1011 \quad 0010 \quad 1111 && [\textit{convert to Binary}]\\
&= 101 \quad 100 \quad 101 \quad 111 && [\textit{group bits by } 3_8]\\
&= 5 \quad 4 \quad 5 \quad 7_8 && [\textit{convert to octal}]\\
&= 5457_8
\end{aligned}
$$

## 30.10 BINARY CODED DECIMAL SYSTEMS (BCD)

The decimal numbers have been used for centuries and their use would continue. But the digital systems require the use of binary numbers. The compromise between the two systems can be had by using **binary coded decimals (BCD)**. These codes combine the features of both decimal and binary numbers where a decimal read-out of a binary is required, it is sometimes convenient to use a modified version of binary code in which the length of the code sequence is restricted to 10 digits before it is repeated. In fact binary coded decimal (BCD) notation is a special way of expressing decimal quantities in binary form. In BCD notation, each digit of the original decimal number is expressed directly as a four digit binary group. The number of binary groups is then equal to the number of digits in the original decimal number.

The main binary codes used are :

- 4 bit binary codes,
- Excess-3 code,
- Parity code and
- Gray code.

### 30.10.1 The 8-4-2-1 Code

The most commonly used BCD code is the so called 8-4-2-1 code. This is a four bit code which does not conform to Eqns. 30.1 and 30.2 but is a composite system using binary numbers from 0 to 1001 to represent decimal numbers from 0 to 9. The 8-4-2-1 code expresses each decimal digit by its four-bit binary equivalent. We cite here a few examples of equivalents between decimal and BCD (8421) systems.

| Decimal | BCD (8421) | | | |
|---|---|---|---|---|
| 0 | 0000 | 0000 | 0000 | 0100 |
| 19 | 0000 | 0000 | 0001 | 1001 |
| 100 | 0000 | 0001 | 0000 | 0000 |
| 5000 | 0101 | 0000 | 0000 | 0000 |

The first group of four digits on the right in the BCD numbers are the binary equivalent numbers of the first digit on the right in the decimal number. The other groups of 4 digits from right to left are binary equivalent of decimal digits of tens, hundreds, thousands respectively.

Table 30.6 gives more information about the 8-4-2-1 code. It is seen that each decimal digit is changed to its equivalent four bit groups. It should be noted that 1001 is the largest four-bit group in the 8-4-2-1 code.

**Table 30.6** Decimal Numbers and Their Equivalents in BCD (8421)

| ***Decimal*** | ***BCD (8-4-2-1)*** | | | |
|---|---|---|---|---|
| 0 | 0000 | 0000 | 0000 | 0000 |
| 1 | 0000 | 0000 | 0000 | 0001 |
| 2 | 0000 | 0000 | 0000 | 0010 |
| 3 | 0000 | 0000 | 0000 | 0011 |
| 4 | 0000 | 0000 | 0000 | 0100 |
| 5 | 0000 | 0000 | 0000 | 0101 |
| 6 | 0000 | 0000 | 0000 | 0110 |
| 7 | 0000 | 0000 | 0000 | 0111 |
| 8 | 0000 | 0000 | 0000 | 1000 |
| 9 | 0000 | 0000 | 0000 | 1001 |
| 10 | 0000 | 0000 | 0001 | 0000 |
| 11 | 0000 | 0000 | 0001 | 0001 |
| 12 | 0000 | 0000 | 0001 | 0010 |
| 13 | 0000 | 0000 | 0001 | 0011 |
| 14 | 0000 | 0000 | 0001 | 0100 |
| 15 | 0000 | 0000 | 0001 | 0101 |
| 98 | 0000 | 0000 | 1001 | 1000 |
| 99 | 0000 | 0000 | 1001 | 1001 |
| 100 | 0000 | 0001 | 0000 | 0000 |
| 101 | 0000 | 0001 | 0000 | 0001 |
| 109 | 0000 | 0001 | 0000 | 1001 |
| 579 | 0000 | 0101 | 0111 | 1001 |
| 4728 | 0100 | 0111 | 0010 | 1000 |

Therefore, only 10 of the 16 possible four-bit groups are used. The 8-4-2-1 code does not use the numbers 1010, 1011, 1100, 1101, 1110 and 1111. An error will occur if any of these numbers are fed into a digital computing machine. These numbers, in fact, are forbidden for a 8-4-2-1 BCD.

The 8421 code is identical to binary upto the decimal number 9. Above 9, the 8-4-2-1 code differs from the binary number code. For example, the binary number for 15 is 1111, but the 8421 number for 15 is 0001, 0101. Taking another example, the decimal number 24 is 11000 in binary but it becomes 0010, 0100 in the 8-4-2-1 code. Thus, above decimal number 9, every binary number differs from the corresponding 8-4-2-1 number.

The 8-4-2-1 code is identical to binary through the decimal number 9. Because of this, it is called 8-4-2-1 code. The weights in the groups are 8, 4, 2, 1 reading from left to right.

However, above 9, the 8421 BCD code differs from the binary number code. For example, the binary number for 14 is 1110, but the same number as used in BCD (8421) is 0001, 0100.

The first group of four digits on the right in the BCD numbers are the binary equivalent numbers equivalent of the first digit on the right in the decimal number. The other groups of 4 digits from right to left are the binary equivalents of decimal digits of tens, hundreds and thousands respectively. This system is extensively used to keep the numbers in a form that can easily be read in the decimal a system and easily interpreted by the observer and recorded in decimal form.

**Example 30.22** *Express 6932 in the 8-4-2-1 code.*

**Solution.** Each digit of the decimal number is expressed in the pure binary form as :

$$6 = 0\times2^3 + 1\times2^2 + 1\times2^1 + 0\times2^0 = 0110$$
$$9 = 1\times2^3 + 0\times2^2 + 0\times2^1 + 1\times2^0 = 1001$$
$$3 = 0\times2^3 + 0\times2^2 + 1\times2^1 + 1\times2^0 = 0011$$
$$2 = 0\times2^3 + 0\times2^2 + 1\times2^1 + 0\times2^0 = 0010$$

The binary groups are moved together to form the BCD notation and

Decimal 6932 = 0110 1001 0011 0010

**Example 30.23** *Convert the following BCD 8421 numbers into Binary system :*

(*i*) 1000 0101 0110 0011

(*ii*) 0011 0101 0001 1001 0111

**Solution.** Divide the BCD (8421) numbers into groups of four and decode them.

(*i*) 1000 0101 0110 0011
↓ ↓ ↓ ↓
8 5 6 3

Therefore, BCD 1000 0101 0110 0011 is equivalent to $8563_{10}$

(*ii*) 0011 0101 0001 1001 0111

↓ ↓ ↓ ↓ ↓

3 5 1 9 7

∴ BCD 0011 0101 0001 1001 0111

$= 35197_{10}$

The main advantage of the 8-4-2-1 code is the case of converting to and from decimal numbers. We need only remember the binary numbers 0 to 9 because we encode only one decimal digit a time. A disadvantage, however, of the 8-4-2-1 code is that the rules for binary addition do not apply to the entire 8421 number, but only to the individual four bit groups.

**Example 30.24** *Add decimal numbers 12 and 9 in binary and BCD (8-4-2-1) system. Comment upon the result.*

**Solution.** Decimal 12 = Binary 1100

Decimal 9 = Binary 1001

Decimal-addition = 12 + 9 = 21

Binary addition

```
  1 1 0 0
  1 0 0 1
1 0 1 0 1
```

In BCD (8-4-2-1)

Decimal 12 = 0000 0000 0001 0010

Decimal 19 = 0000 0000 0000 1001

BCD addition = 0000 0000 0001 1011

The result is 0001 1011. This is untenable as it is not possible to decode 1011 as it does not exist in the 8-4-2-1 code. The largest 8421 group is 1-0-0-1 (9). Therefore addition of numbers in 8-4-2-1 code is not so simple as for binary numbers.

### 30.10.2 Other BCD Codes

The 8-4-2-1 code is one of the many codes referred to as binary coded decimal systems (BCD). A large number of such codes can be evolved. Since four bit BCD codes use only 10 of 16 possible states, we can evolve (16! / 6!) possible codes, which are approximately 30 billion. The 8-4-2-1 code is a mixed base code. It is a binary within each group of four bits, but at the same time it is decimal from group to group. Since the 8-4-2-1 BCD code is the most natural type of BCD code, it is frequently referred to as BCD without any reference to it. Thus, without any reference to the contrary, we refer the BCD code as a 8-4-2-1 code.

Many other four bit codes exist like Aiken 2-4-2-1 and Johnson code. Table 30.7 show some commonly used BCD codes.

**Table 30.7** Four Bit BCD Codes

| *Decimal* | *4221* | *2421* | *Excess 3* | *5211* |
|---|---|---|---|---|
| 0 | 0000 | 0000 | 0011 | 0000 |
| 1 | 0001 | 0001 | 0100 | 0001 |
| 2 | 0010 | 0010 | 0101 | 0011 |
| 3 | 0011 | 0011 | 0110 | 0101 |
| 4 | 1000 | 0100 | 0111 | 0111 |
| 5 | 1001 | 0101 | 1000 | 1000 |
| 6 | 1010 | 0110 | 1001 | 1001 |
| 7 | 1011 | 0111 | 1010 | 1011 |
| 8 | 1110 | 1110 | 1011 | 1101 |
| 9 | 1111 | 1111 | 1100 | 1111 |

**Table 30.8**

| *Decimal* | *Binary Code* | *Gray Code* | *Decimal* | *Binary Code* | *Gray Code* |
|---|---|---|---|---|---|
| 0 | 0000 | 0000 | 8 | 1000 | 1100 |
| 1 | 0001 | 0001 | 9 | 1001 | 1101 |
| 2 | 0010 | 0011 | 10 | 1010 | 1111 |
| 3 | 0011 | 0010 | 11 | 1011 | 1110 |
| 4 | 0100 | 0110 | 12 | 1100 | 1010 |
| 5 | 0101 | 0111 | 13 | 1101 | 1011 |
| 6 | 0110 | 0101 | 14 | 1110 | 1001 |
| 7 | 0111 | 0100 | 15 | 1111 | 1000 |

## 30.11 GRAY CODE

The *Gray code* belongs to a class of codes called *minimum-change codes,* in which only *one* bit in the code group changes when going from one step to the next. The Gray code is an *unweighted* code, meaning that the bit positions in the code groups do not have any specific weight assigned to them. Because of this, the Gray code is not suited for arithmetic operations but finds application in input/output devices and some types of analog-to-digital converters.

Table 30.8 shows the Gray-code representation for the decimal numbers 0 through 15, together with the straight binary code. If we examine the Gray-code groups for each decimal number, it can be seen that in going from *any* one decimal number to the next, only *one* bit of the Gray code changes. For example, in going from 3 to 4, the Gray code changes from 0010 to 0110, with only the second bit from the left changing. Going from 14 to 15 the Gray-code bits change from 1001 to 1000, with only the last bit changing. This is the principal characteristic of the Gray code. Compare this with the binary code, where anywhere from one to all of the bits change in going from one step to the next.

The Gray code is often used in situations where other codes, such as binary, might produce erroneous or ambiguous results during those transitions in which more than one bit of the code is changing. For instance, using binary code and going from 0111 to 1000 requires that all four bits change simultaneously. Depending on the device or circuit that is generating the bits, there may be a significant difference in the transition times of the different bits. If so, the transition from 0111 to 1000 could produce one or more intermediate states. For example, if the most significant bit changes faster than the rest, following transitions will occur :

| | | | | | |
|---|---|---|---|---|---|
| 0 | 1 | 1 | 1 | — | *Decimal 7* |
| ↓ | | | | | |
| 1 | 1 | 1 | 1 | — | *Erroneous code* |
| | ↓ | ↓ | ↓ | | |
| 1 | 0 | 0 | 0 | — | *Decimal 8* |

The occurrence of 1111 is only momentary but it could conceivably produce erroneous operation of the elements that are being controlled by the bits. Obviously, using the Gray code would eliminate this problem, since only one bit change occurs per transition and no "race" between bits can occur.

## 30.12 ALPHANUMERIC CODES

In addition to numerical data, a computer must be able to handle non-numerical information. In other words, a computer should recognize codes that represent letters of the alphabet, punctuation marks, and other special characters as well as numbers. These codes are called *alphanumeric codes*. A complete alphanumeric code would include the 26 lowercase letters, 26 uppercase letters, 10 numeric digits, 7 punctuation marks, and anywhere from 20 to 40 other characters, such as +, /, #, %, *, and so on. We can say that an alphanumeric code represents all of the various characters and functions that are found on a standard typewriter (or computer) keyboard.

### ASCII CODE

The most widely used alphanumeric code, the American Standard Code for Information Interchange (ASCII), is used in most miorocomputers and minicomputers, and in many mainframes.

The ASCII code (pronounced "askee") is a 7-bit code, and so it has $2^7 = 128$ possible code groups. This is thus enough to represent all of the standard keyboard characters as well as control functions such as the (RETURN) and (LINEFEED) functions. Table 30.9 (on next page) shows a partial listing of the ASCII code. In addition to the binary code group for each character, the table gives the octal and hexadecimal equivalents.

**Example 30.25** *The following is a message encoded in ASCII code. What is the message ?*

*1001000 1000101 1001100 1010000*

**Solution.** Convert each 7-bit code to its hex equivalent. The results are :

48 45 4C 50

Now locate these hex values in Table 30.9 given on next page and determine the character represented by each. The results

H E L P

The ASCII code is used for the transfer of alphanumeric information between a computer and input/output devices such as video terminals or printers. A computer also uses it internally to store the information that an operator types in at the computer's keyboard.

The following example illustrates this.

**Example 30.26** *An operator is typing in a BASIC program at the keyboard of a certain microcomputer. The computer converts each keystroke into its ASCII code and stores the code in memory. Determine the codes that will be entered into memory when the operator types in the following BASIC statement.*

GOTO 25

**Solution.** Locate each character (including the space) in Table 30.9 and record its ASCII code.

| | |
|---|---|
| G | 1000111 |
| O | 1001111 |
| T | 1010100 |
| O | 1001111 |
| (space) | 0100000 |
| 2 | 0110010 |
| 5 | 0110101 |

30

**Table 30.9** Partial Listing of ASCII Code

| Character | 7-Bit ASCII | Octal | Hex |
|---|---|---|---|
| A | 100 0001 | 101 | 41 |
| B | 100 0010 | 102 | 42 |
| C | 100 0011 | 103 | 43 |
| D | 100 0100 | 104 | 44 |
| E | 100 0101 | 105 | 45 |
| F | 100 0110 | 106 | 46 |
| G | 100 0111 | 107 | 47 |
| H | 100 1000 | 110 | 48 |
| I | 100 1001 | 111 | 49 |
| J | 100 1010 | 112 | 4A |
| K | 100 1011 | 113 | 4B |
| L | 100 1100 | 114 | 4C |
| M | 100 1101 | 115 | 4D |
| N | 100 1110 | 116 | 4E |
| O | 100 1111 | 117 | 4F |
| P | 101 0000 | 120 | 50 |
| Q | 101 0001 | 121 | 51 |
| R | 101 0010 | 122 | 52 |
| S | 101 0011 | 123 | 53 |
| T | 101 0100 | 124 | 54 |
| U | 101 0101 | 125 | 55 |
| V | 101 0110 | 126 | 56 |
| W | 101 0111 | 127 | 57 |
| X | 101 1000 | 130 | 58 |
| Y | 101 1001 | 131 | 59 |
| Z | 101 1010 | 132 | 5A |

| Character | 7-Bit ASCII | Octal | Hex |
|---|---|---|---|
| 0 | 011 0000 | 060 | 30 |
| 1 | 011 0001 | 061 | 31 |
| 2 | 011 0010 | 062 | 32 |
| 3 | 011 0011 | 063 | 33 |
| 4 | 011 0100 | 064 | 34 |
| 5 | 011 0101 | 065 | 35 |
| 6 | 011 0110 | 066 | 36 |
| 7 | 011 0111 | 067 | 37 |
| 8 | 011 1000 | 070 | 38 |
| 9 | 011 1001 | 071 | 39 |
| blank | 010 0000 | 040 | 20 |
| . | 010 1110 | 056 | 2E |
| ( | 010 1000 | 050 | 28 |
| + | 010 1011 | 053 | 2B |
| $ | 010 0100 | 044 | 24 |
| * | 010 1010 | 052 | 2A |
| ) | 010 1001 | 051 | 29 |
| – | 010 1101 | 055 | 2D |
| / | 010 1111 | 057 | 2F |
| , | 010 1100 | 054 | 2C |
| = | 011 1101 | 075 | 3D |
| (RETURN) | 000 1101 | 015 | 0D |
| (LINEFEED) | 000 1010 | 012 | 0A |

## 30.13 PARITY METHOD FOR ERROR DETECTION

The movement of binary data and codes from one location to another is the most frequent operation performed in digital systems. Here are some examples :

- ▲ Transmission of digitized voice over a microwave link
- ▲ The storage and retrieval of data from external memory devices such as magnetic tape and disk
- ▲ The transmission of information from a computer to a remote user terminal or another computer.

Whenever information is transmitted from one device (the transmitter) to another device (the receiver), there is a possibility that errors can occur such that the receiver does not receive the identical information that was sent by the transmitter. The major cause of any transmission errors is **electrical noise**, which consists of spurious fluctuations in voltage or current that are present in all electronic systems to varying degrees. Figure 2.1 on page 21 is a simple illustration of a type of transmission error.

The transmitter sends a relative noise-free serial digital signal over a signal line to a receiver. However, by the time the signal reaches the receiver, it contains a certain degree of noise superimposed on the original signal. Occasionally, the noise is large enough in amplitude that it will alter the logic level of the signal as it does at point *x*. When this occurs, the receiver may incorrectly interpret that bit as a logic 1, which is not what the transmitter has sent.

Most modern digital equipment is designed to be relatively error-free, and the probability of errors such as that shown in Fig. 30.5 is very low. However we

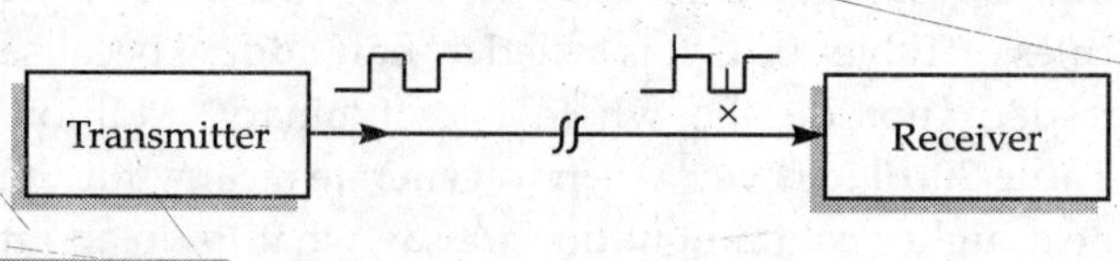


**Fig. 30.5** Example of noise causing an error in the transmission of digital data.

must realize that digital systems often transmit thousands, even millions of bits per second, so that even a very low rate of occurrence of errors can produce an occasional error that might prove to be bothersome, if not disastrous. For this reason, many digital systems employ some method for detection (and sometimes correction) of errors. One of the simplest and most widely used schemes for error detection is the **parity method.**

### Parity Bit

A *parity bit* is an extra bit that is attached to a code group that is being transferred from one location to another. The parity bit is made either 0 or 1, depending on the number of 1s that are contained in the code group. Two different methods are used.

In the *even-parity* method, the value of the parity bit is chosen so that the total number of 1s in the code group (including the parity bit) is an *even* number. For example, suppose that the group is 1000011.

This is the ASCII character C. The code group has *three* 1s. Therefore, we will add a parity bit of 1 to make the total number of 1s an even number. The *new* code group, including the parity bit, thus becomes

**1** 1 0 0 0 0 1 1
↑______ added parity bit*

If the code group contains an even number of 1s to begin with, the parity bit is given a value of 0. For example, if the code group were 1000001 (the ASCII code for "*A*"), the assigned parity bit would be 0, so that the new code, including the parity bit, would be 01000001.

The *odd-parity* method is used in exactly the same way except that the parity bit is chosen so the total number of 1s (including the parity bit) is an *odd* number. For example, for the code group 1000001, the assigned parity bit would be a 1. For the code group 1000011, the parity bit would be a 0.

Regardless of whether even parity or odd parity is used, the parity bit becomes an actual part of the code word. For example, adding a parity bit to the 7-bit ASCII code produces an 8-bit code. Thus, the parity bit is treated just like any other bit in the code.

The parity bit issued to detect any *single-bit* errors that occur during the transmission of a code from one location to another (*e.g.*, from a computer to a video terminal). For example, suppose the character "*A*" is being transmitted and *odd* parity is being used. The transmitted code would be :

**1** 1 0 0 0 0 0 1

When the receiver circuit receives this code, it will check that the code contains an odd number of 1s

* The parity bit can be placed at either end of code group but is usually placed to the left of MSB.

(including the parity bit). If so, the receiver will assume that the code has been correctly received. Now, suppose that because of some noise or malfunction the receiver actually receives the following code :

**1** 1000000

The receiver will find that this code has an *even* number of 1s. This tells the receiver that there must be an error in the code, since presumably the transmitter and receiver have agreed to use odd parity. There is no way, however, that the receiver can tell which bit is in error, since it does not know what the code is supposed to be.

It should be apparent that this parity method would not work if *two* bits were in error, because two errors would not change the "oddness" or "evenness" of the number of 1s in the code. In practice, the parity method is used only in situations where the probability of a single error is very low and the probability of double errors is essentially zero.

When the parity method is being used, the transmitter and receiver must have agreement, in advance, as to whether odd or even parity is being used. There is no advantage of one over the other, although even parity seems to be used more often. The transmitter must attach an appropriate parity bit to each unit of information that it transmits. For example, if the transmitter is sending ASCII-coded data, it will attach a parity bit to each 7-bit ASCII code group. When the receiver examines the data that it has received from the transmitter, it checks each code group to see that the total number of 1s (including the parity bit) is consistent with the agreed-upon type of parity. This is often called *checking the parity* of the data. In the event that it detects an error, the receiver may send a message to the transmitter asking it to retransmit the last set of data. The exact procedure that is followed when an error is detected will depend on the particular system design.

**Example 30.27** *A transmitter is sending ASCII-coded data to a receiver with an even-parity bit. Show the actual codes when the transmitter is sending the message "HELLO".*

**Solution.** First look up the ASCII codes for each character in the message. Then for each code, count the number of 1s. If it is an even number, attach a 0 as the MSB. If it is an odd number, attach a 1. Thus the resulting 8-bit codes will all have an even number of 1s (including parity).

*attached even-parity bits*

| | | | | | | | | |
|---|---|---|---|---|---|---|---|---|
| H — | 0 | 1 | 0 | 0 | 1 | 0 | 0 | 0 |
| E — | 1 | 1 | 0 | 0 | 0 | 1 | 0 | 1 |
| L — | 1 | 1 | 0 | 0 | 1 | 1 | 0 | 0 |
| L — | 1 | 1 | 0 | 0 | 1 | 1 | 0 | 0 |
| O — | 1 | 1 | 0 | 0 | 1 | 1 | 1 | 1 |

## 30.14 BINARY ADDITION

The rules of binary addition have been explained earlier. However, to understand the design of a binary adder in a microprocessor, let us state the rules explicity in Table 30.10.

**Table 30.10** Rules of Binary Addition in Microprocessors

| *Augend* | *Addend* | *Sum* | *Carry* |
|---|---|---|---|
| 0 + | 0 = | 0 + | 0 |
| 0 + | 1 = | 1 + | 0 |
| 1 + | 0 = | 1 + | 0 |
| 1 + | 1 = | 0 + | 1 |

Table 30.10 gives the rules for adding 2 bits in a single column of bits. The *carry* is the bit resulting from 1 + 1 addition that should be *carried* over into the next higher order column of the bits. The only obvious instruction is what to do when a carry bit results from addition in the last and highest order column. Let us put this aside for the moment and consider only a single column.

A "black box" to add only one column is called a 1-bit adder or *half adder* and has the input-output functions shown in Fig. 30.6. Like any other black box, we need not to know what is inside it but only that inputs result in the desired outputs as specified by the rules of Table 30.10. It is called a "**half adder**" because it does not do the whole job of binary addition. **Table 30.10** tells you when to generate a carry for the next higher column but doesn't say what to do about an *incoming carry* from a lower-order column of bits. After learning how the half adder works, we will take up this question.

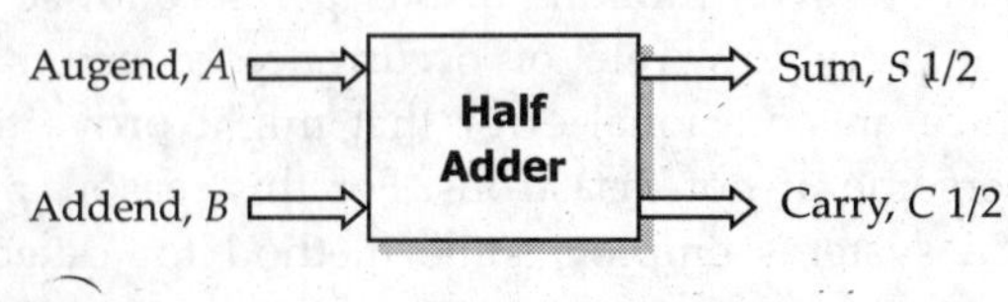


**Fig. 30.6** Half Adder "Black Box".

The information of Fig. 30.6 can be put in a slightly different form called a *truth table.* A truth table specifies all the inputs and outputs of a logical box or system such as the half adder. The inputs $A$ and $B$, of Table 30.11 are the same augend and addend as shown in Fig. 30.6 and the outputs are the sum $S$ 1/2 and carry $C$ 1/2 of the half adder.

**Table 30.11** Truth Table for Half Adder

| *Inputs* | | *Outputs* | |
|---|---|---|---|
| **A** | **B** | **S 1/2** | **C 1/2** |
| 0 | 0 | 0 | 0 |
| 1 | 0 | 1 | 0 |
| 0 | 1 | 1 | 0 |
| 1 | 1 | 0 | 1 |

Note that the regular patterns of the inputs $A$ and $B$ cover all possible combinations of 1 and 0 for the inputs. The outputs follow the specified rules of binary addition. Any combination of devices that results in the correct outputs as a function of the inputs will, by definition, be a half adder.

A truth table need not be restricted to two inputs or outputs. Any binary function can be specified by a truth table. If there are three input variables instead of only two there will be 2 to the third power or 8 possible combinations that must be specified ; four inputs will require 2 to the fourth power or 16, and so on.

***Example 30.28*** *Write down in orderly fashion all combinations of three binary input values and find the output of a half adder.*

**Solution.** The results are tabulated below.

| *Inputs* | | | *Outputs* | |
|---|---|---|---|---|
| **A** | **B** | **C** | **S 1/2** | **C 1/2** |
| 0 | 0 | 0 | 0 | 0 |
| 1 | 0 | 0 | 1 | 0 |
| 0 | 1 | 0 | 1 | 0 |
| 0 | 0 | 1 | 1 | 0 |
| 1 | 1 | 0 | 0 | 1 |
| 0 | 1 | 1 | 0 | 1 |
| 1 | 0 | 1 | 0 | 1 |
| 1 | 1 | 1 | 1 | 1 |

### 30.14.1 Binary Logic

A binary logic device like a half adder is constructed of combinations of elementary logical functions, each of which can be considered as a black box with inputs and output(s) as shown in Fig. 30.7. Because there are four possible combinations of the two inputs of Fig. 30.7, there are 2 to the fourth power or 16 possible logic functions corresponding to 16 possible output columns. Some of these are trivial : for example, the output can always be 0 regardless of input or it can always be 1. (Perhaps these would correspond to a shorted switch or to an open circuit in an electrical "black box".) But the others are meaningful elementary functions that can be used to construct all other types of logic functions, including the adders and arithmetic units of microprocessors.

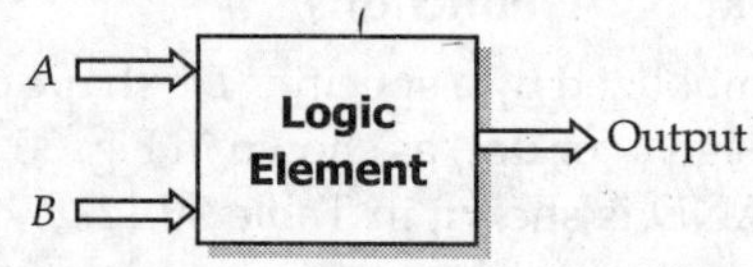


**Fig. 30.7** General Logic Element.

***Example 30.29*** *Construct a table showing all of the possible outputs columns corresponding to elementary logical functions of two binary input variables. You should find 16 of these, including the "all 0" and "all 1" functions just mentioned.*

**Solution.**

**AND Functions**

| *Inputs* | | *Outputs* |
|---|---|---|
| **A** | **B** | **C** |
| 0 | 0 | 0 |
| 1 | 0 | 0 |
| 0 | 1 | 0 |
| 1 | 1 | 1 |

**NOT Functions**

| *Inputs* | | *Outputs* |
|---|---|---|
| **A** | **B** | **C** |
| 0 | 0 | 1 |
| 1 | 0 | 0 |
| 0 | 1 | 0 |
| 1 | 1 | 0 |

**OR Functions**

| *Inputs* | | *Outputs* |
|---|---|---|
| **A** | **B** | **C** |
| 0 | 0 | 0 |
| 1 | 0 | 1 |
| 0 | 1 | 1 |
| 1 | 1 | 1 |

**Half Adder Functions**

| *Inputs* | | *Outputs* | |
|---|---|---|---|
| **A** | **B** | **S 1/2** | **S1/2** |
| 0 | 0 | 0 | 0 |
| 1 | 0 | 1 | 0 |
| 0 | 1 | 1 | 0 |
| 1 | 1 | 0 | 1 |

It can be shown that a small number, two or three, of these logical functions can be chosen as basic and can be used to build up any other possible elementary function, or even more complex logical devices, such as the half adder. The elements regarded as basic are somewhat a matter of choice, but one intuitively obvious set is that consisting of the AND and OR, together with the one-variable NOT function. Any possible logical result can be constructed from these three functions. Each of them can be described by a unique truth table and can be symbolized by a specific variation of Fig. 30.7.

## 30.14.2 AND, OR, NOT Functions

The AND is symbolized by a specific "D" shape or by a multiplication sign (× or dot) as shown in Fig. 30.8. The truth table of AND is shown in Table 30.12.

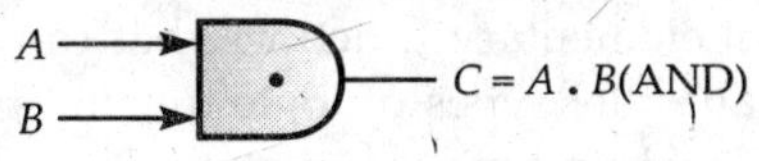


**Fig. 30.8** AND Function Symbol.

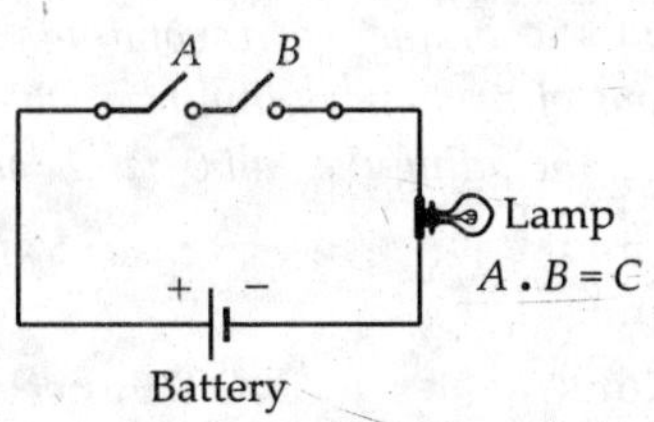


**Fig. 30.9** AND Function Circuit Implementation.

**Table 30.12** Truth Table for AND Gate

| Inputs | | Outputs |
|---|---|---|
| A | B | C |
| 0 | 0 | 0 |
| 1 | 0 | 0 |
| 0 | 1 | 0 |
| 1 | 1 | 1 |

Literally, the AND function is TRUE (meaning equal to 1) when both $A$ and $B$ are true (1). It is FALSE (0) when either or both inputs are zero (0).

One can imagine a hardware implementation of an AND function constructed by connecting two single-pole switches in series to a lamp and a battery, as shown in Fig. 30.9. If the two switches represent the $A$ and $B$ functions and are TRUE when closed, and if the lamp represents the output $C$ and is TRUE when lit, then this circuit follows exactly the rules of the truth table, Table 30.12, and is therefore the representation of an AND.

The OR is represented by a dished-in "D" and the symbol "+", as shown in Fig. 30.10. Its definition is the truth table, Table 30.13.

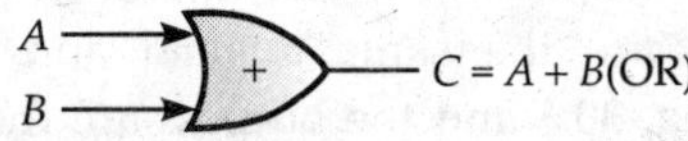


**Fig. 30.10** OR Function Symbol.

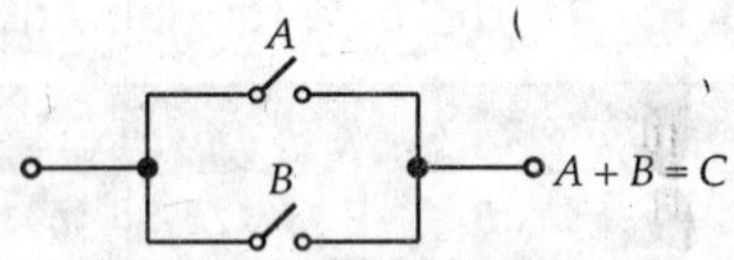


**Fig. 30.11** OR Function Circuit Implementation.

**Table 30.13** Truth Table for OR Gate

| Inputs | | Outputs |
|---|---|---|
| A | B | C |
| 0 | 0 | 0 |
| 1 | 0 | 1 |
| 0 | 1 | 1 |
| 1 | 1 | 1 |

As implied by its name, the OR is TRUE (equal to 1) when either the input $A$ or the input $B$ or both are TRUE (1). A circuit representing an OR can be made of two parallel switches, Fig. 30.10, that are substituted for the two series switches of Fig. 30.9. Closure of either switch or both will light the lamp.

Note that if we change the definition of TRUE to mean "lamp is not lit," that the OR circuit will behave like an AND while the AND becomes an OR. This demonstrates that it is important to define exactly what is meant by TRUE, FALSE, 1 and 0 when analyzing logical circuits or devices. Practical integrated circuits (ICs) embodying logic elements such as AND and OR typically employ a high-voltage state such as 5 volt to represent 1 and a low voltage (less than 1 volt) to represent 0. This arrangement is termed as *positive logic*. If the opposite convention is adopted, so that the TRUE state is low voltage or 0, it is called *negative logic*. Thus, we can change a positive logic OR to a negative logic AND using the same hardware, merely by adopting a different rule. Of course, the same rules must apply to all the logical elements that are connected together in any way.

A special logic element is the NOT, since it is the simplest and has only one input and one output. The output is merely the inverse of the input, whatever that may be, 0 or 1. Thus, the NOT has the truth table shown in Table 30.14.

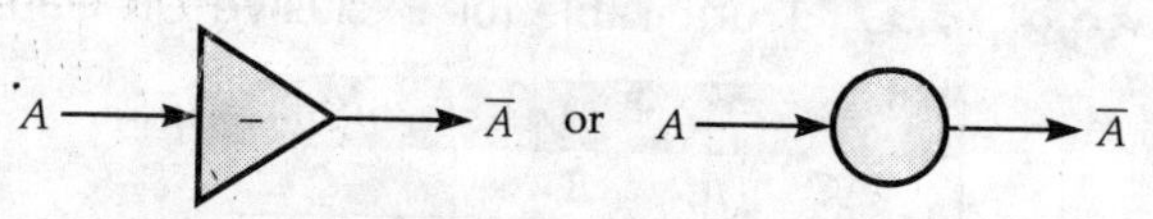


**Fig. 30.12** NOT Function Symbol.

**Table 30.14** Truth Table for NOT Gate

| *Input* | *Output* |
|---|---|
| 0 | 1 |
| 1 | 0 |

The NOT function is symbolized by a small circle or by a triangle representing a one-stage (inverting) amplifier as shown in Fig. 30.12.

### 30.14.3 NAND, NOR Functions

All of the 16 possible functions of two inputs you find in Example 30.29 can be assembled from some combination of AND, NOT and OR elements. For example, an AND followed by a NOT (see Fig. 30.13) can be defined by the following truth table.

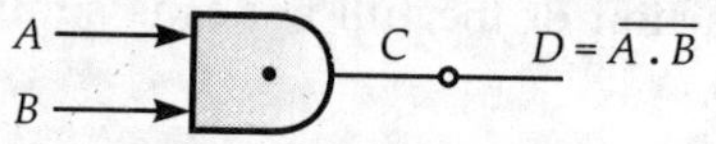


**Fig. 30.13** NOT-AND Logic.

**Table 30.15** Truth Table for NAND Gate

| *Inputs* | | *AND OUT* | *NOT-AND* |
|---|---|---|---|
| **A** | **B** | **C** | **D = (NAND) $\overline{A.B}$** |
| 0 | 0 | 0 | 1 |
| 1 | 0 | 0 | 1 |
| 0 | 1 | 0 | 1 |
| 1 | 1 | 1 | 0 |

The result is called "NOT-AND," abbreviated "NAND." If the first element had been an OR instead of AND we would have constructed a "NOT-OR" or "NOR" device (See Fig. 30.14 and Fig. 30.15 for symbols of these devices). The NAND and NOR themselves can be considered basic elements ; just as the NOT, OR and AND can be used to construct all other binary logic properties, so can the two elements NAND and NOR. Multiple NAND and NOR elements are very commonly implemented in MSI (medium-scale integration) IC's because of this versatility. However, you should note that it usually requires more NAND and NOR elements to construct some logical function if we restrict ourselves to these two, than if we permit the use of AND, OR and NOT elements as well.

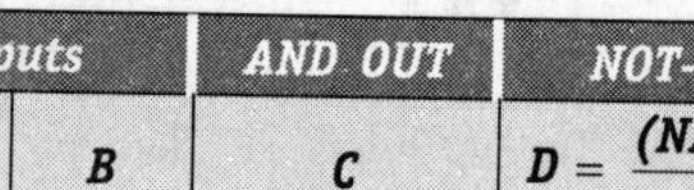


**Fig. 30.14** NAND Symbol.

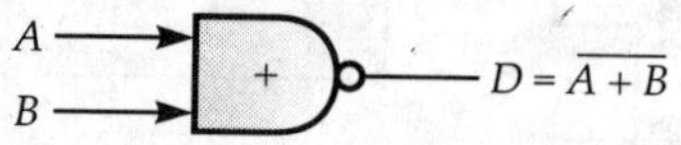


**Fig. 30.15** NOR Symbol.

It should be noted that $\overline{A.B}$ and $\overline{A+B}$ are respectively complements of $A.B$ and $A + B$.

**Table 30.16** Truth Table for NOR Gate

| *Inputs* | | *Outputs* |
|---|---|---|
| **A** | **B** | **D** |
| 0 | 0 | 1 |
| 1 | 0 | 0 |
| 0 | 1 | 0 |
| 1 | 1 | 0 |

***Example 30.30*** *Three elements, two ANDs and OR, plus two NOTs, are connected as in Fig. 30.16. Develop a truth table for this function. (The result will be an important function used in the half adder, as we shall see in the next section.)*

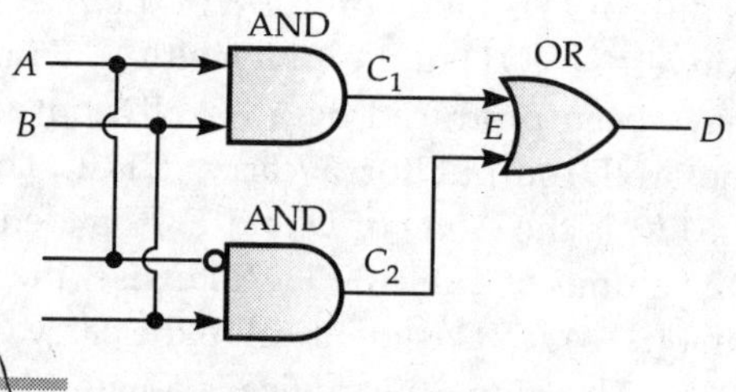


**Fig. 30.16** Diagram of Example 30.30.

**Solution.** This is given in the following Table :

| A | B | $C_1 = A\,.\,\overline{B}$ | $C_2 = \overline{A}\,.\,B$ | $D = C_1 + C_2$ |
|---|---|---|---|---|
| 0 | 0 | 0 | 0 | 0 |
| 1 | 0 | 1 | 0 | 1 |
| 0 | 1 | 0 | 1 | 1 |
| 1 | 1 | 0 | 1 | 1 |

### 30.14.4 Half Adder and X-OR

We have now put together the background to analyze and implement the half adder, Table 30.11, as we intended. Actually, there are many different logic combinations that will result in this function. Figure 30.17 is one that is economically composed of four elements, two ANDs, an OR and a NOT.

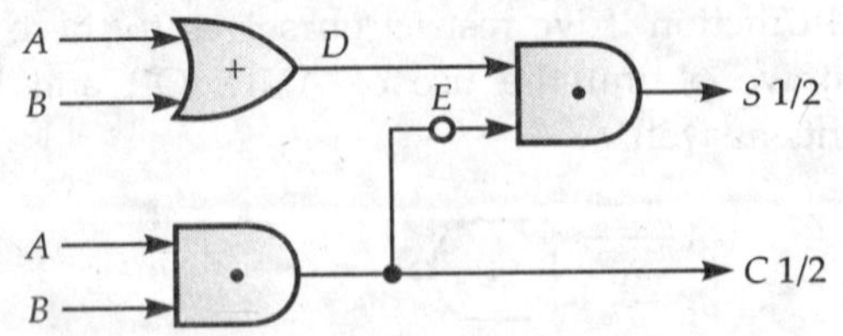


**Fig. 30.17** Half Adder Logic Implementation.

To prove that Fig. 30.17 is indeed a half adder, all we need to do to construct its truth table and compare the outputs with Table 30.11. These details are worked out in Table 30.16, which you should study carefully to be sure you understand what is happening at each step.

**Table 30.16** Proof of Half Adder Logic

| A | B | D (OR) | E (NOT AND) | S 1/2 | C 1/2 |
|---|---|---|---|---|---|
| 0 | 0 | 0 | 1 | 0 | 0 |
| 1 | 0 | 1 | 1 | 1 | 0 |
| 0 | 1 | 1 | 1 | 1 | 0 |
| 1 | 1 | 1 | 0 | 0 | 1 |

The output $D$ is merely the OR of the $A$, $B$ inputs. $E$ is the inverse (NOT) of the AND with $A$, $B$ inputs. $E$ could have been produced by a NAND, but we also need the AND output for a carry, $C1/2$. The sum output, $S1/2$ is the AND of $D$ and $E$. It is identical to the $S1/2$ column of Table 30.11. Likewise, the output of the first ($A$, $B$) AND is identical to the carry $C1/2$ of Table 30.11. Therefore, the device is a half adder.

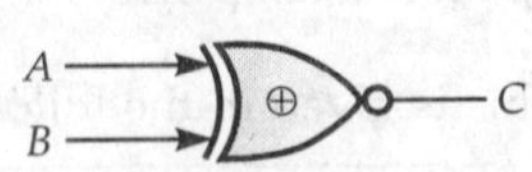


**Fig. 30.18** Exclusive-OR (X-OR) Symbol.

The $S1/2$ output is one of those you have obtained in Example 30.28 ; it is itself an elementary logical function. It is known as the Exclusive-OR or X-OR. The symbols for the X-OR are shown in Fig. 30.17. Its truth table can be copied from the $A$, $B$, and $S1/2$ columns of Table 30.16. Note that it is identical to the result of Example 30.30, Fig. 30.16. Therefore, an X-OR is equal to the OR of two AND functions, as in Fig. 30.16.

**Table 30.17** Truth Table for Exclusive OR Gate

| Inputs | | Output |
|---|---|---|
| A | B | C |
| 0 | 0 | 0 |
| 1 | 0 | 1 |
| 0 | 1 | 1 |
| 1 | 1 | 0 |

### Full adder

The addition problems for a column of bits is only partially solved by the half adder. The full adder not only must generate a carry bit for the next higher column but also must know what to do with carry bit coming in from the lower column. Thus the full column adder has three inputs — the augend and addended as in the half adder and carry bit from the next lower stage. The outputs, as before, are the sum $S$ and the carry $C_1$. If at least two of the inputs are l's, a carry is generated. Thus the full adder can be cascaded for any number of steps to provide simultaneous addition of multiple columns of bits or words such as byte. The action of the full adder is summarized in Table 30.18.

**Table 30.18** Truth Table for Full Adder

| Inputs | | | Outputs | |
|---|---|---|---|---|
| A | B | C | S | $C_1$ |
| 0 | 0 | 0 | 0 | 0 |
| 1 | 0 | 0 | 1 | 0 |
| 0 | 1 | 0 | 1 | 0 |
| 1 | 1 | 0 | 0 | 1 |
| 0 | 0 | 1 | 1 | 0 |
| 1 | 0 | 1 | 0 | 1 |
| 0 | 1 | 1 | 0 | 1 |
| 1 | 1 | 1 | 1 | 1 |

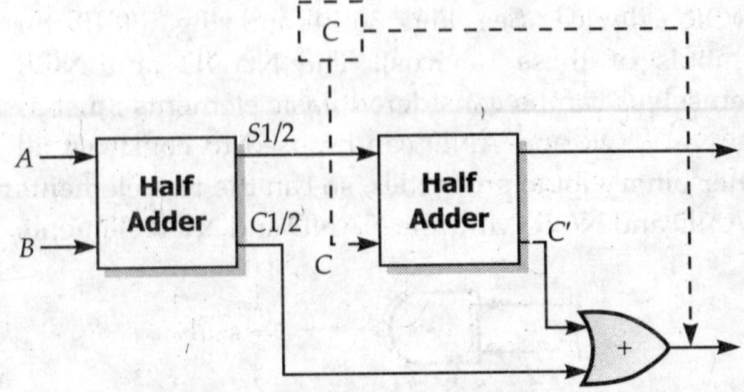


**Fig. 30.19** Full Adder Logic Implementation.

The dashed box $C$ in Fig. 30.18 shows how it is possible to use a single adder circuit in time serial fashion to add multiple columns of bits. The inputs $(A, B)$ for each column, beginning with the lowest order, are presented to the adder in sequence. The outgoing carry for each column is saved in a temporary 1-bit memory (the dashed box), and is presented in an incoming carry when the next higher order $A$, $B$ pair is entered. This *ripple method* is seldom used in contemporary microprocessors because of the additional processing time to perform the entire multibit add. However, as we will show in a subsequent example, it is similar to the principal of multibyte precision arithmetic.

## 30.15 BOOLEAN LOGIC EQUATIONS

The design of logic circuits can lead to considerable complications. Therefore, it is advisable to simplify the logic equations before using Them. Logic equations have to be manipulated to simplify them using certain rules. Some of the rules for manipulating logic expressions are :

| No. | Expression | | Comments |
|---|---|---|---|
| 1. | $A + 1 = 1$ | ...(30.3) | Whether $A$ is 0 or 1, the output is 1. |
| 2. | $A + 0 = A$ | ...(30.4) | When $A = 1$, the output is 1, when $A = 0$ the output is 0. |
| 3. | $A.0 = 0$ | ...(30.5) | Whether $A = 0$ or 1, the output is 0. |
| 4. | $A.1 = A$ | ...(30.6) | When $A = 1$, the output is 1 and when $A = 0$, the output is 0. |
| 5. | $A + A = A$ | ...(30.7) | When $A = 1$, the output is 1 and when $A = 0$, the output is 0. |
| 6. | $A.A = A$ | ...(30.8) | When $A = 1$, the output is 1 and when $A = 0$, the output is 0. |
| 7. | $A.\overline{A} = 0$ | ...(30.9) | When $A = 1$, $A = 0$ and therefore the output is 0. |
| 8. | $A + \overline{A} = 1$ | ...(30.10) | When $A = 1$, $\overline{A} = 0$ and therefore the output is 1.<br>When $A = 0$, $\overline{A} = 1$ and therefore the output is 1. |
| 9. | $A + B = B + A$<br>$A . B = B . A$ | ...(30.11)<br>...(30.12) | Commutative law. |
| 10. | $(A + B) + C = A + (B + C)$<br>$(A . B) . C = A . (B . C)$ | ...(30.13)<br>...(30.14) | Associative law. |
| 11. | $AB + AC = A.(B + C)$ | ...(30.15) | Distributive law. |
| 12. | $A + BC = (A + B)(A + C)$ | ...(30.16) | |

## 30.16 DEMORGAN'S THEOREMS

There are *two* DeMorgan's Theorems :

1. $\overline{A + B} = \overline{A}.\overline{B}$ ...(30.17)
2. $\overline{A.B} = \overline{A} + \overline{B}$ ...(30.18)

▲ The first equation states that the complement of a sum equals the product of the complements.

▲ The second equation states that the complement of product equals the sum of the complements.

Let us prove the first equation.

**Case 1.** When

$A = 0,\ B = 0,\ A + B = 0 \quad \therefore \quad \overline{A + B} = \overline{0} = 1$

$\overline{A} = 1 \quad \overline{B} = 1 \quad \therefore \quad \overline{A}.\overline{B} = 1$

**Case 2.** When

$A = 1,\ B = 0,\ A + B = 1 \quad \therefore \quad \overline{A + B} = \overline{1} = 0$

$\overline{A} = 0 \quad \overline{B} = 1 \quad \therefore \quad \overline{A}.\overline{B} = 0$

**Case 3.** When

$A = 0,\ B = 1,\ A + B = 1 \quad \therefore \quad \overline{A + B} = \overline{1} = 0$

$\overline{A} = 1 \quad \overline{B} = 0 \quad \therefore \quad \overline{A}.\overline{B} = 0$

**Case 4.** When

$A = 1,\ B = 1,\ A + B = 1 \quad \therefore \quad \overline{A + B} = \overline{1} = 0$

$\overline{A} = 0 \quad \overline{B} = 0 \quad \therefore \quad \overline{A}.\overline{B} = 0$

30

Therefore for all possible cases

$$\overline{A+B} = \overline{A}.\overline{B}$$

The truth table is shown below :

**Table 30.19**

| $A$ | $B$ | $\overline{A+B}$ | $\overline{A}.\overline{B}$ |
|---|---|---|---|
| 0 | 0 | 1 | 1 |
| 1 | 0 | 0 | 0 |
| 0 | 1 | 0 | 0 |
| 1 | 1 | 0 | 0 |

Let us have the circuit meaning of the first theorem. $\overline{A+B}$ represents a logic circuit in which a **NOT** gate follows an **OR** gate as shown in Fig. 30.20($a$).

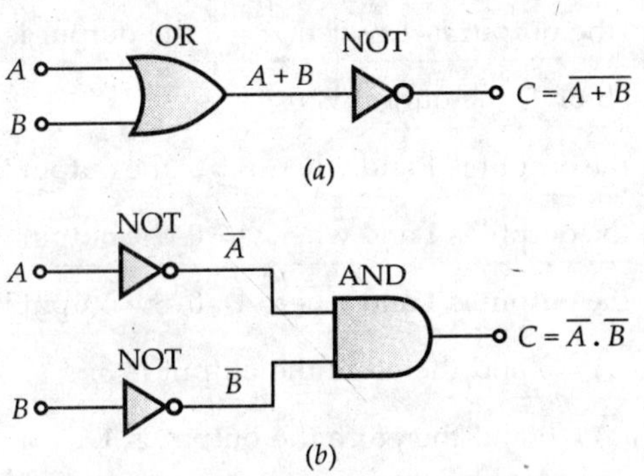


**Fig. 30.20** Circuits representing DeMorgan's first theorem.

Expression $\overline{A}.\overline{B}$ is described by two NOT gates whose outputs serve as an input for an AND gate as shown in Fig. 30.20($b$). DeMorgan's theorem states that the logic circuits are interchangeable.

In Fig. 30.20($a$), a NOT gate follows an OR gate, we call this particular combination as NOT-OR or simply **NOR**.

The second DeMorgan's theorem is

$$\overline{A.B} = \overline{A}+\overline{B}$$

This can also be proved in a manner similar to the one followed for the first theorem. The truth table for this is shown in Table 30.20.

**Table 30.20**

| $A$ | $B$ | $\overline{A.B}$ | $\overline{A}+\overline{B}$ |
|---|---|---|---|
| 0 | 0 | 1 | 1 |
| 1 | 0 | 1 | 1 |
| 0 | 1 | 1 | 1 |
| 1 | 1 | 0 | 0 |

The circuit representation $\overline{A.B}$ is shown in Fig. 30.21($a$) and for $\overline{A}+\overline{B}$ in 30.21($b$). Since the two expressions are equal the logic circuits are interchangeable.

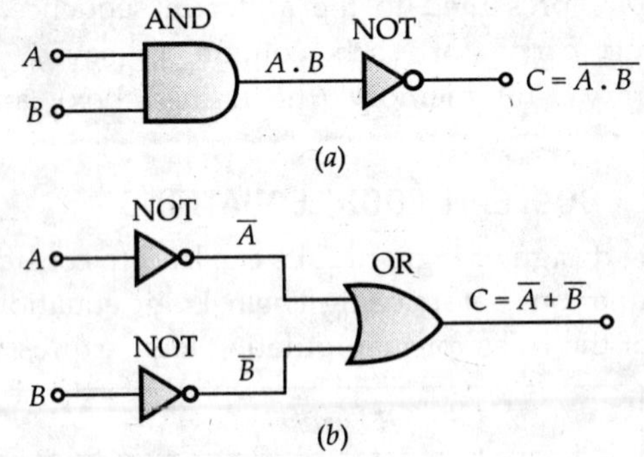


**Fig. 30.21** Circuit representation for DeMorgan's second theorem.

In Fig. 30.21($a$), a NOT gate follows an AND gate. So this combination is NOT-AND gate or simply **NAND** gate.

Some additional useful theorems are :

**3.** $(\overline{\overline{A}}) = A$ ...(30.19)

**4.** $(A+\overline{B})\,B = A.B$ ...(30.20)

**5.** $A.\overline{B}+B = A+B$ ...(30.21)

***Example 30.31*** *Simplify the following Boolean expressions :*

($i$) $\overline{A}A+B$ ($ii$) $\overline{A}A.B$

($iii$) $A+A+B$ ($iv$) $(A+\overline{A}).B$

**Solution.**

($i$) $\overline{A}A+B$

Now $\overline{A}A = 0$ as when $A=0$, $\overline{A}=1$ and when $A=0$, $A=1$

$\therefore \quad \overline{A}A+B = 0+B = B$

($ii$) From above we have seen

$\overline{A}A = 0 \quad \therefore \quad \overline{A}A.B = 0.B = 0$

($iii$) $A+A+B = A(1+A)+B$

But $(1+A) = 1$, when $A=0$

$= 1$, when $A=1$

$\therefore \quad (1+A) = 1$ (This is given in Eqn. 30.3)

$\therefore A+A+B = A+B$

($iv$) $(A+\overline{A}).B$

$A+\overline{A} = 1$, since when $A=0$, $\overline{A}=1$

$= 1$, since when $A=1$, $\overline{A}=0$.

$\therefore (A+\overline{A}).B = 1.B = B.$

*Example 30.32* *Prove that :*

(*i*) $AB + A\overline{B} = A$

(*ii*) $(A+B)(A+\overline{B}) = A$

(*iii*) $D = \overline{A}(BA + C)$ where $D = \overline{A}C$

**Solution.** (*i*) $AB + A\overline{B} = A(B+\overline{B})$

But $(B+\overline{B}) = 1$

$\therefore \; A(B+\overline{B}) = A.1 = A$

Hence $\quad AB + A\overline{B} = A$

(*ii*) $(A+B)(A+\overline{B}) = A$

Expanding the above expression, we get,

$$A.A + A.\overline{B} + B.A + B.\overline{B} = A.A + A.B + A.\overline{B} + B.\overline{B}$$
$$= A + A.B + A.\overline{B} + 0 \text{ as } A.A = A \text{ and } B.\overline{B} = 0$$
$$= A + A.B + A.\overline{B}$$
$$= A + A(B + \overline{B})$$
$$= A + A \text{ as } B + \overline{B} = 1$$
$$= A$$

(*iii*) $D = \overline{A}(BA + C)$

Expanding the R.H.S., we get, $BA\overline{A} + \overline{A}C$

But $\quad A\overline{A} = 0,$

$\therefore \quad BA\overline{A} = 0$

Hence, R.H.S. $= \overline{A}C$

But $\quad D = \overline{A}C$

*Example 30.33* *Write the Boolean expression for the output C of Fig. 30.22(a) verify the results with a truth table.*

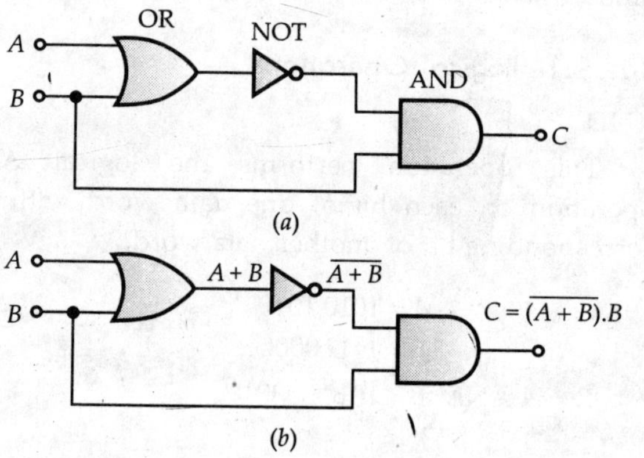


**Fig. 30.22** Diagram of Example 30.33.

The solution can be had with the help of Fig. 30.22(*b*).

The output is thus $C = \overline{(A+B)}.B$

From DeMorgan's Theorem,

$$\overline{A+B} = \overline{A}.\overline{B}$$

$$\therefore \quad C = \overline{(A+B)}.B$$
$$= \overline{A}.\overline{B}.B = 0 \text{ as } \overline{B}.B = 0$$

This can be verified as by a truth table as under

| **A** | **B** | **A + B** | $\overline{A+B}$ | $\overline{B}$ | **C = $\overline{(A+B)}$.B** |
|---|---|---|---|---|---|
| 0 | 0 | 0 | 1 | 0 | 1.0 = 0 |
| 1 | 0 | 1 | 0 | 1 | 0.1 = 0 |
| 0 | 1 | 1 | 0 | 0 | 0.0 = 0 |
| 1 | 1 | 1 | 0 | 1 | 0.1 = 0 |

## 30.17 MICROPROCESSORS

**Number systems.** Information in computers is stored as patterns of zeros and ones. End zero or one is called a **BIT** and a group of eight bits handled as a single unit is called a **BYTE**. Any group of bits may be referred to as a **word**, though the term word is used to describe the number of bits normally operated at one time by the computer. The word length of the microprocessors mentioned in this text is 8 bits or one byte. The different number systems used in digital systems have already been described.

## 30.18 POSITIVE AND NEGATIVE INTEGERS

The following conventions are used for indicating the sign and magnitude of a binary integer :

**1.** ***One's complement.*** The most common significant bit is 0 for positive numbers, and rest of the bits indicate the magnitude. Negative numbers have each bit of positive number *"flipped"* $(1 \to 0, 0 \to 1)$ including the most significant bit

*For example,*

Decimal number $\quad +7_{10}$ is flipped to $0111_2$

and $\quad -7_{10}$ is flipped to $1000_2$

This system is occasionally used. It has separate codes for + 0 and – 0.

**2.** ***Twos complement.*** This is the usual system for indicating negative numbers in Microprocessors. A way to find the two's complement of a number is to find the one's complement and then add 1.

*For example,*

$$+7_{10} = 0111_2 \text{ and } -7_{10} = 1001_2$$

The two's complement of a number may be found in other ways. For an $n$ bit number $x_1$ the two's complement of $x = 2^n - x$. Alternatively, the two's complement of a number may be found by examining each bit in turn, starting with *LSB*, taking no action until the first 1 is found. This is left unaltered and all the more significant bits are inverted. Note that the two's complement operation reverses the sign of positive or negative numbers correctly. If the operation is performed twice, the original number is recovered. Addition and subtraction circuits for negative numbers are simple.

Another view of the two's complement code may be obtained by considering the place value of the digits in the code. In the natural binary code, the most significant bit in an $n$-bit number has the place value of $2^{n-1}$. In the two's complement code, the most significant bit has the place value of $-2^{n-1}$ and all the other bits have the same place values as they have in the natural binary code. Thus for an 8-bit two's complement code, the place values are :

$$-128 \quad +64 \quad +32 \quad +16 \quad +8 \quad +4 \quad +2 \quad +1$$

The system employed in a particular microprocessor to indicate a negative number can be found by examining the bit pattern when a negative number is known to be formed, for example when +1 is subtracted from zero.

### Decimal Numbers

One decimal digit can be stored in four bits of memory, so an 8-bit byte can hold two decimal digits. Microprocessors usually have special provision for dealing with decimal arithmetic, so the user is able to perform addition and subtraction on decimal numbers, and obtain a correctly coded decimal answer. The special provision may be a decimal adjust instruction, which corrects false codes, or a switch under program control to a decimal mode of operation.

The details of decimal arithmetic in microprocessors are not well standardized, and demand careful attention before practical use.

### Floating Point Numbers

The number $2.998 \times 10^8$ may also be expressed as 299.800,000, these being decimal floating point and decimal fixed point integer versions of the same number.

The same number may be expressed in binary form, and the example shows 32-bit floating point and signed, fixed point formats (Fig. 30.23).

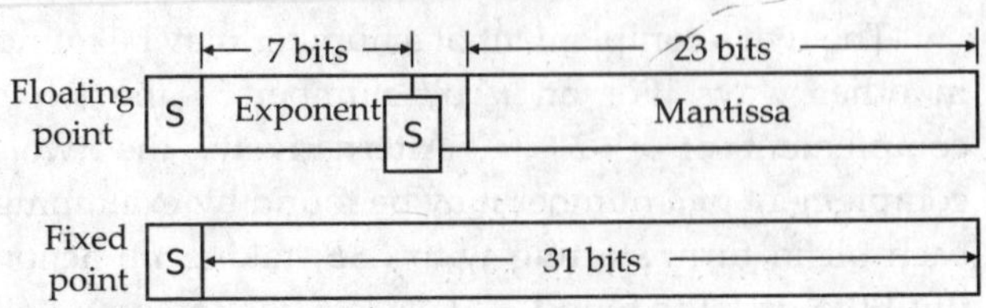


**Fig. 30.23** Floating and fixed point formats.

In the floating point form, the sign of the mantissa can occupy the bit position that would have been given to the most significant bit of the mantissa, as this is always adjusted to be 1, so need not be stored. Floating point numbers in this 32-bit notation can range from $2^{127}$ to $2^{-128}$, whereas the largest 32-bit binary integer is $(2^{32} - 1)$. If floating point numbers are to be used, a floating point package in software or hardware will convert and manipulate the numbers, offering convenience and avoidance of overflow at the cost of speed. When using a floating point package, it is seldom necessary to follow its inner workings in detail, as there will be routines included in the package for conversion of fixed point numbers to and from floating point form.

## 30.19 LOGICAL OPERATIONS AND FLAGS

Most computer instructions operate on a complete word of data, affecting every bit in the word according to the same rules. This section describes how the operations found in microprocessors affect data. It does not discuss the internal hardware of the microprocessor which is needed to carry out the operations. Any microprocessor will have single instructions for executing most of the instructions discussed in this section. Those operations which cannot be executed using a single instruction can be implemented using a combination of other instructions, as manufacturers are careful to arrange that 'any microprocessor can compute anything–given time'.

### 30.19.1 Logical Operations

**AND**

This operation performs the logical AND operation on each bit of one data word with the corresponding bit of another data word.

$$A = 1010\,1010$$
$$B = 1111\,0000$$
$$R = A \text{ AND } B = 1010\,0000$$

**Note.** A AND A gives A again.

**OR**

This operation performs the logical inclusive – *OR* operation on each bit of one data word with the corresponding bit of another data word.

$$A = 1010\,1010$$
$$B = 1111\,0000$$
$$R = A \text{ OR } B = 1111\,1010$$

**Note.** A OR A = A.

**EXOR ≡ XOR**

This operation performs the logical exclusive OR function (either/or, but not both) on each bit of one data word with the corresponding bit of another data word.

$$A = 1010\,1010$$
$$B = 1111\,0000$$
$$R = A \text{ EXOR } B = 0101\,1010$$

**Note.** A EXOR A gives all zeros.

**$\overline{\text{EXOR}}$**

The exclusive NOR is not usually included in the instruction set. It operates on two data words bit-by-bit, as in the example.

$$A = 1010\,1010$$
$$B = 1111\,0000$$
$$R = A\,\overline{\text{EXOR}}\,B = 1010\,0101$$

**Note.** A $\overline{\text{EXOR}}$ A gives all ones.

**COMPLEMENT**

This instruction operates on each bit of a single word, changing each bit from 0 to 1 or from 1 to 0.

$$A = 1010\,1010$$
$$\text{COMPLEMENT}(A) = 0101\,0101$$

**Note.** This is the one's complement not the two's complement.

**CLEAR**

This operation fills all the bits of a word with zeros. It is not usually included in an instruction set, as there are other ways of obtaining this result, such as subtracting the word from itself. Instructions to clear single bit registers are often available.

**SET**

This operation fills all the bits of a word with ones. It is usually implemented using another instruction, such as moving a word full of ones to the register.

## 30.19.2 Flags

A flag is a single-bit signal held in a flip-flop to indicate that an event has occurred. Flags are used to indicate certain properties of the result of computer operations. Other names for flags are condition codes or status bits.

Only some of the instructions in a microprocessor affect flags ; others leave them unaltered. There are two ways a programmer can be trapped into error by flags — either by forgetting that an instruction sets or resets a flag, or by assuming that an instruction controls a flag when in fact the instruction has no effect on the flag.

The main use of flags is to control program flow between alternative paths. All microprocessors have a sufficient set of flags, so that performance and preference are not determined by the flag set. There are differences in the detail of the flags used in different microprocessors, so the action of flags must be studied with care. The names and symbols given to a flag by a manufacturer may be different from those used here.

**ZERO flag (Z)**

This shows that the result of a computer operation is to leave only zero in a register. When the result is zero, the Z flag = 1. When the result is non-zero, the Z flag = 0.

**SIGN flag (S)**

This indicates the sign of the contents of a register, assuming the contents to be an integer coded using the two's complement convention for negative numbers. Its value is therefore the value of the most significant bit in the register. When the number is positive or zero, $S = 0$. When the number is negative, $S = 1$.

**CARRY flag (CY)**

The primary use of this flag is to indicate an arithmetic carry out of the most significant bit of the register. The carry flag takes the value of the carry out of the register. Where logical instructions affect a carry flag, they always reset it to zero, as logical operations do not cause any carry from bit to bit. The same flag is used to indicate that a borrow was needed in a subtraction operation. Some microprocessors set $CY$ to 1 to indicate a borrow ; others reset $CY$ to 0 for the same purpose.

A second distinct use of the $CY$ flag is to hold the bit shifted out of a register in a SHIFT or ROTATE operation.

**OVERFLOW flag (V)**

This indicates that overflow has occurred while adding two signed numbers in the two's complement code, so that the result is too large to be held in the register.

```
 + 93  =        0 1 0 1   1 1 0 1
       +      +
 + 84  =          0 1 0 1   0 1 0 0
                0 1 0 1 1   1 0 0. ← carry bits
 + 177 (V = 1)1011 0001 = −79
```

In the example, the correct result (+177) is too large to be held in an 8-bit register, and the 8-bit result would be incorrectly interpreted as a negative number (–79). The $V$ flag is formed from ((carry into Most

30

Significant Bit) EXOR (carry out of Most Significant Bit)). In the example, the two relevant carry bits are underlined. The $V$ flag will also signal overflow when two negative numbers are added, or as a result of subtraction. The $V$ flag not only detects overflow, but permits correction of the false 8-bit result. Not all microprocessors have a $V$ flag.

**PARITY flag (*P*)**

The parity flag indicates whether the number of ones in a word is even or odd. If $P$ is one when the number of bits is even, the flag is an even parity flag.

Not all microprocessors have a $P$ flag.

## 30.19.3 Arithmetic Operations

**1. *ADD.*** This operation adds the contents of two registers, leaving the total in one of them, and leaving the contents of the other register unchanged. In the microprocessor considered in this texts the two's complement convention is used for representing negative numbers, and the ADD instruction gives the correct total for the addition of positive or negative numbers. The ADD instruction will affect the $Z$, $S$, $CY$, $V$ and $P$ flags.

| | | |
|---|---|---|
| + 84 | 0 1 0 1 | 0 1 0 0 |
| ADD | | |
| + 30 | 0 0 0 1 | 1 1 1 0 |
| +114 | 0 1 1 1 | 0 0 1 0 |

| $Z = 0$ | $S = 0$ | $CY = 0$ | $V = 0$ | $P = 0$ |
|---|---|---|---|---|
| no zero | positive | no carry | no overflow | even parity |

**2. *Add with carry.*** This operation adds the contents of two registers and the value of the carry flag. It is used for multibyte addition where a carry must propagate from byte to byte. If the $CY$ flag is reset to zero, the ADC instruction has the same effect as ADD. Some microprocessors have both ADD and ADC instructions ; others have only one of the two.

**3. *SUB subtract.*** This operation subtracts the contents of one register from the contents of another, leaving the answer in one of the registers, while leaving the contents of the other register unchanged. The $CY$ flag is affected if a borrow is required in the most significant bit. It is necessary to check the microprocessor manual for details of the borrow, as there is no uniform convention.

**4. *SBB subtract with borrow.*** This operation performs the operations described under SUB, and also subtracts the borrow as indicated by the $CY$ flag. It is used for subtraction of multibyte numbers. As with SUB there is no uniform convention for the treatment of the $CY$ flag by this operation. Some microprocessors do not have both SUB and SBB instruction.

**5. *CMP compare.*** This operation sets flags in exactly the same way as SUB, but does not change the content of either of the registers being compared. It is (used when two bit patterns are to be compared, either for identity, as given by $Z = 1$, or for relative magnitude, and when it is wished to preserve both of the bit patterns.

**6. *INC increment.*** This operation adds one to the contents of the register. Some extended instruction sets allow INC($n$), when the register is incremented by $n$. There is no general agreement about which flags are controlled by this instruction.

**7. *DEC decrement.*** This operation subtracts one from the contents of a register. It is often used in reducing a number to zero to count the number of times a loop is still to be traversed. It will obey rules equivalent to those for INC for a particular microprocessor.

**8. *MULT multiply.*** Eight-bit microprocessors usually do not have an instruction for multiplying two numbers, and rely on a sequence of instructions in a subroutine or an external specialist arithmetic processor. Sixteen-bit microprocessors are likely to include a multiply instruction which takes two $n$-bit unsigned integers and gives a $2n$-bit result. The value of $n$ may be 8 or 16. Longer sequences of operations are required for dealing with signed numbers and with floating point numbers.

**9. *DIV divide.*** The comments on the multiply instruction also apply to DIV, though the dividend must be of length $2n$, and the divisor, quotient, and remainder of length $n$.

**10. *SHIFT or ROTATE.*** A shift-left instruction moves each bit in a register one place to the left. A shift-right instruction moves each bit one place to the right. Strictly this instruction is neither an arithmetic nor logical instruction.

The value of the bit to be shifted into the empty place may be decided in several ways. The bit shifted out of the register may also be treated in several ways.

| New bit for empty position is : | Bit shifted out of register is : |
|---|---|
| the bit shifted out of the register, | sent to the other end of the register, |
| the CY flag, | |
| always 1, | saved in the CY flag, |
| always 0, | lost. |
| the same as bit ($n-1$) to extend the sign. | |

A microprocessor will have one or more shift-left and shift-right instructions, and the details of these must be studied before use. Notice that adding a number to itself or multiplying it by two is equivalent to a left shift with zero entering the least significant bit.

## 30.20 MICROCOMPUTER ARCHITECTURE

The term architecture as applied to computers is used to describe the function of each of the main units of the computer, the way each block is related to other blocks, and the ways signals are passed from one unit to another. Thus two computers which differed only in the size of the memory would not be said to have different architectures, but the use of a data bus or an index register in one computer and not in another would constitute a clear difference of architecture.

The main parts of a simple microcomputer are shown in Fig. 30.24. Each of the three main blocks can be varied in power independently, so the division is along natural lines. The CPU drives the rest of the system, and carries out the operations on data. Memory may usefully be divided into fast program storage, fast data storage, and slower back-up memory of a much greater capacity than the fast memory. As computers have developed the physical forms of memory have changed repeatedly, but the functions have persisted.

Input and output devices vary widely according to the needs of the system. Keyboards, visual display units (VDUs), and printers are needed where the source and destination of the computer's data flow is a human user. Where the microcomputer is part of an instrument or a larger control system, then output and input may pass through transducers where conversion to and from analogue form is required. If the signals are already in digital form, then parallel and serial ports form a simple boundary between the computer and the outside world.

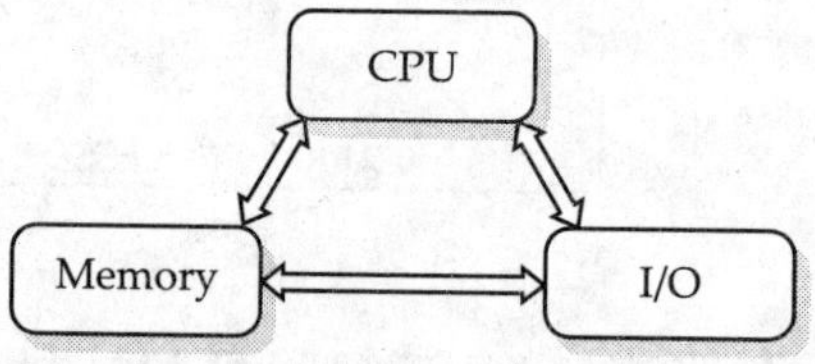


**Fig. 30.24** There are three main functional blocks in a simple computer. These are the central processing unit (CPU) which includes control and an arithmetic and logic unit ; memory for program and data ; and input and output devices for communication or control.

Several examples of computers which fit the structure of Fig. 30.24 are shown in Fig. 30.25. Where a microcomputer forms part of an instrument the structure of Fig. 30.25(*a*) is suitable. The CPU is chosen so that it has a sufficient rate of computation, the required rate being set by the frequency with which the instrument must respond to data, and the complexity of computation.

Figure 30.25(*b*) shows the basis of a personal computer which has access to a magnetic disc store. Depending on the form of the disc, $10^5$ to $10^7$ bytes of memory are accessible with access times set by the rotation frequency of the disc, and the time taken to select a track on the disc.

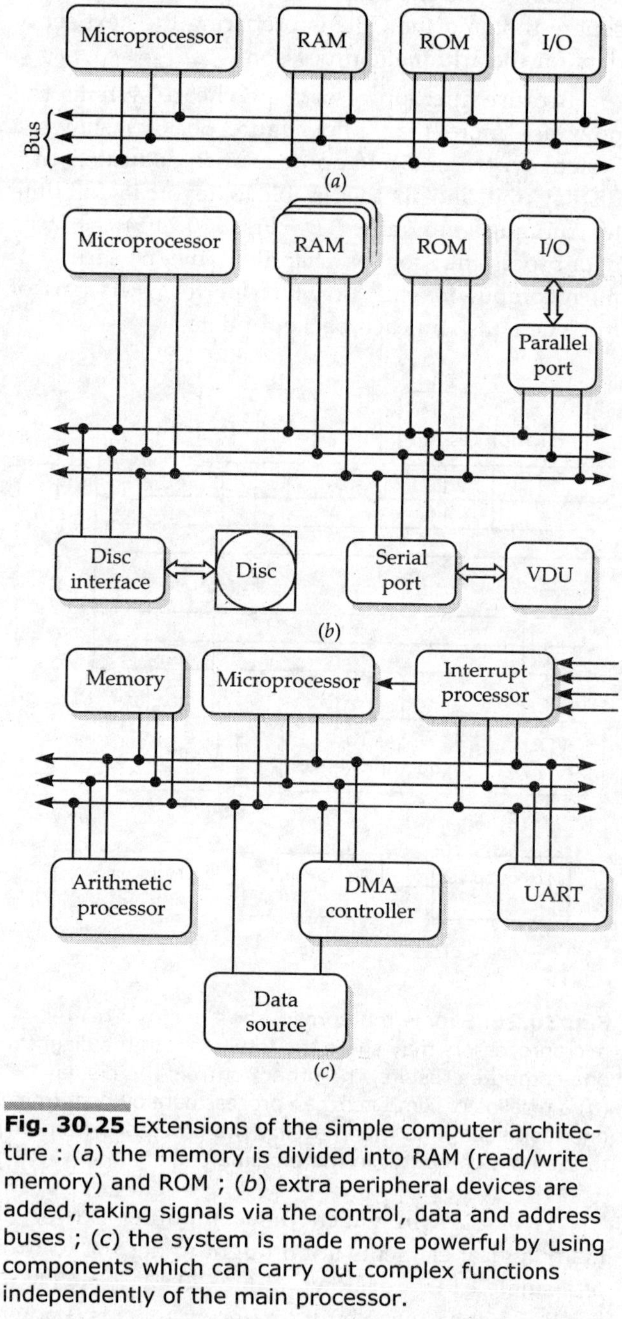


**Fig. 30.25** Extensions of the simple computer architecture : (*a*) the memory is divided into RAM (read/write memory) and ROM ; (*b*) extra peripheral devices are added, taking signals via the control, data and address buses ; (*c*) the system is made more powerful by using components which can carry out complex functions independently of the main processor.

Figure 30.25(*c*) shows a more powerful computer, where each of the blocks is a complex device specializing in one kind of function. Some manufacturers produce sets of devices for this purpose which are designed to be mutually compatible. The system formed from such a set of devices has two advantages : each specialist device can execute its function more rapidly than a general purpose device is able to, and several tasks can be handled simultaneously. Thus the interrupt controller can be establishing whether an interrupting device has sufficient priority to deserve attention, while the arithmetic processor is finding a square root and the CPU is preparing the next set of data for the arithmetic processor.

Because microprocessors are cheap, systems can now use more than one. Figure 30.26(*a*) shows a system where one microprocessor is dedicated to a display unit, and another is able to devote its full time to running programs. In Fig. 30.24(*b*) a large data-processing job is handled by a series of microcomputers, each of which performs its part of the whole task on each packet of data.

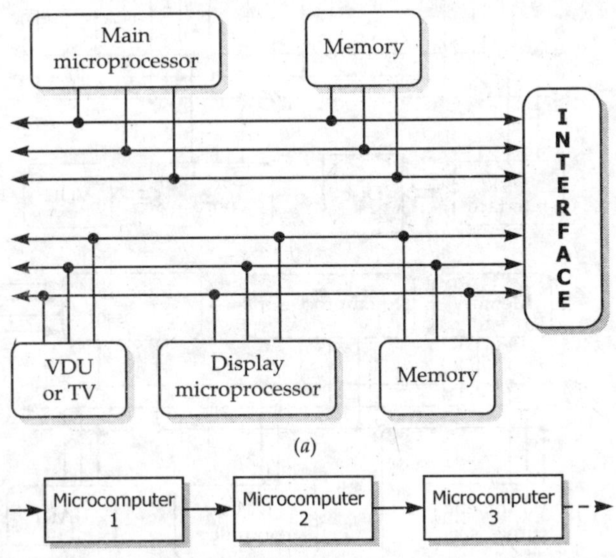


**Fig. 30.26.** Simple multiprocessor systems : (*a*) two microprocessors may share the total task. In the diagram, one computes results, the other controls the display. (*b*) a pipeline of computers can process data by performing one task each, and then passing the processed item on to the next computer in the pipeline.

Two ways in which the computing power of many logical elements are brought under the control of a single CPU are shown in Fig. 30.27. The CPU in Fig. 30.27(*a*) has been built up of components (known as bit-slices) which each deal with a few bits in the word being processed. The number of bits in the word is thus under the designer's control, and the individual bit-slice components can have high performance. The price to be paid for the high performance and flexible structure is a much larger design effort. The array processor in Fig. 30.27(*b*) includes a large number of arithmetic units which operate under the control of a single master controller. All arithmetic units carry out the same operation at any one time, and very high total rates of computation are possible on suitable problems. Array processors have been mentioned here to show that computer systems can be much more elaborate than the simple systems of Fig. 30.24.

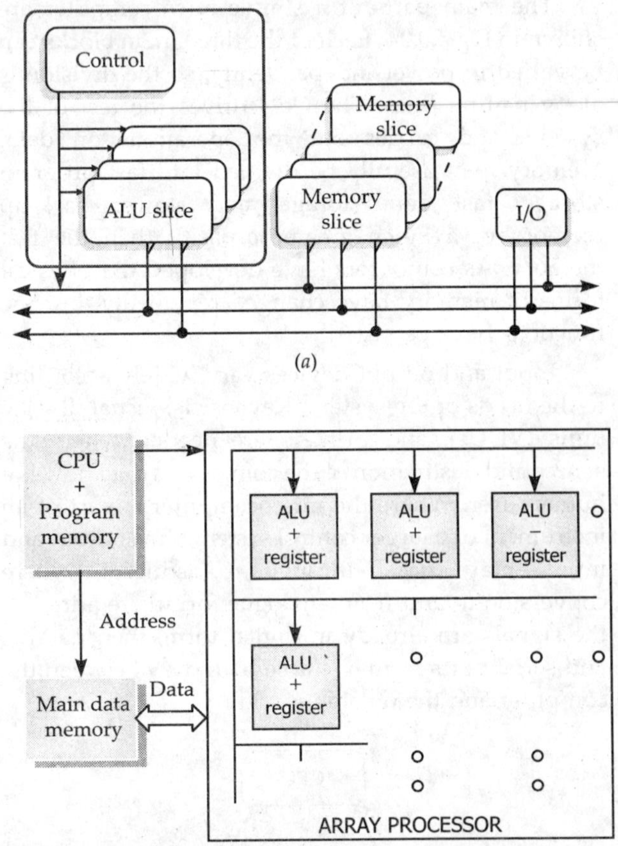


**Fig. 30.27.** Alternative architectures : (*a*) a bit-slice design where the CPU is built up from several integrated circuits ; (*b*) an array processor, where the same operation is performed simultaneously on an array of data words.

## 30.21 BUS STRUCTURES

A system of electrical buses has turned out to be the effective way of connecting the parts of a computer. In the same way as a motor-bus connects several points along the route where passengers may get on or off, so a computer bus connects several parts of a computer system where signals may be transmitted or received.

One possible form of transmitting and receiving circuit suitable for attaching to a bus is shown in Fig. 30.28. The tristate output circuit allows the single bus line to carry signals in both directions.

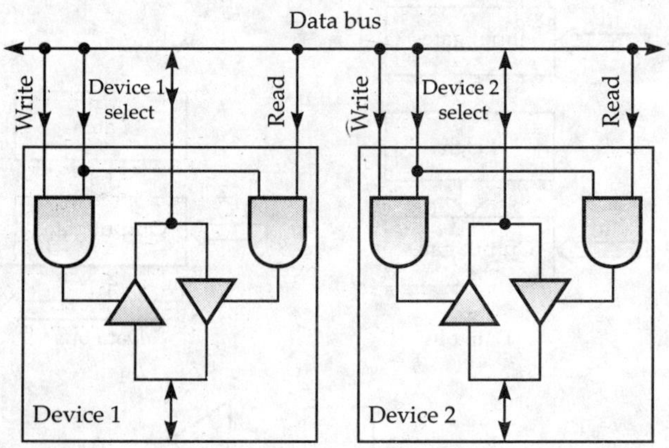


**Fig. 30.28** Circuits to allow two devices to be connected to a data bus, and to transmit or to receive data over it.

Frequently a bus handles many bits in parallel, and then the control signals are the same for each bit. A conventional grouping of signals on buses is into data, address and control. This is a functional distinction, as a number may be stored as data, and then used later as an address. Data buses are commonly 8 or 16 bits wide in microcomputers. Bus structures have proved useful both inside computers and external to them, an internal bus will have many lines in parallel and can accept high speed signals. External buses are more likely to have fewer lines, and to transfer data at lower rates. To provide sufficient flexibility in their operation, the same line will serve different functions at different times, transmitting first control information, then addresses, then data.

A complete system will have several interlinked bus structures. Figure 30.29 shows such a system, where the internal bus is of concern only to the designer of the microprocessor chip. The system bus must be understood by a computer engineer who needs to interconnect the microprocessor to other components in order to assemble a microcomputer.

The external bus is shown connected through a port, which retains data under the control of the program for as long as is required. The signals on the external bus are therefore relatively simple to work with, and a detailed knowledge of the operation of a computer is not needed for the design of interface circuits. What is needed is reasonable skill in programming the computer and in conventional electronic circuit design.

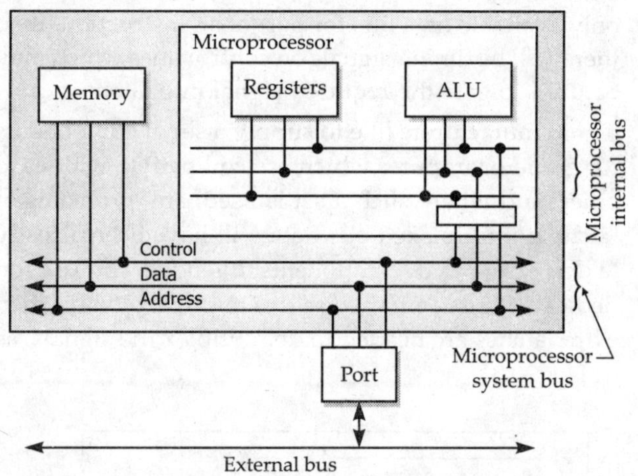


**Fig. 30.29.** Three levels of bus in a microprocessor-based microcomputer system. The internal buses have more lines, are shorter and have more precise timing requirements.

The *three* main groups of signal that travel over microcomputer system buses they are control, address and data. The control signals require a dedicated set of lines for their exclusive use, but the need to keep the total number of pins on an integrated circuit to a minimum has led to several bus systems where the same line is used first for one signal, and then for another. Figure 30.30 shows three ways that have been used. If each signal can have a line dedicated to it, then

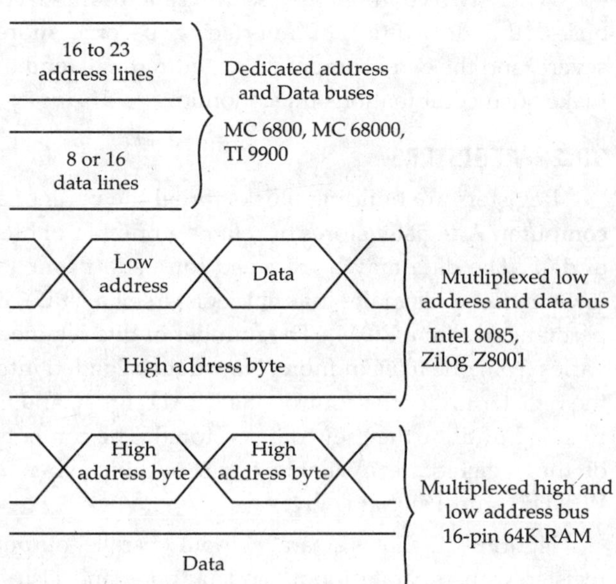


**Fig. 30.30** Three data and address bus variations found in microcomputer components.

30

system design is straightforward. Where a signal is only available on a line for part of an instruction, then there will be timing signals on control lines which may be used to gate the required signal into latches.

Manufacturers like to supply a set of components for a microcomputer which are compatible with each other, and where such a set is used, any problems of latching multiplexed bus data will have been already taken care of. If components intended for use on dedicated buses are used with a multiplexed bus, then extra latches are needed to demultiplex the signals, as in Fig. 30.31.

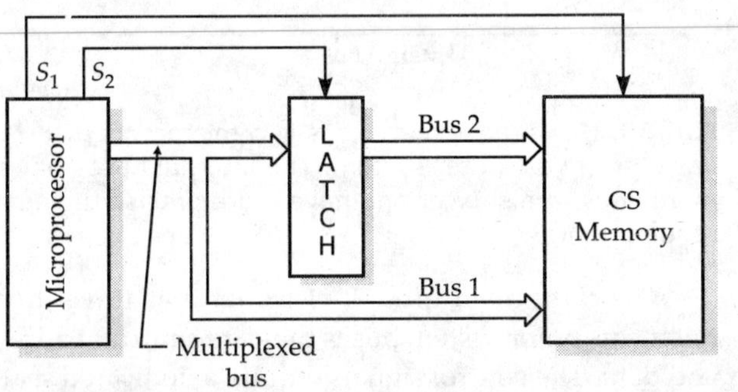


**Fig. 30.31** The use of a latch to demultiplex the signals on a bus. Only one latch is needed, as the memory is assumed to have a suitable input to which the $S_1$ timing signal may be connected. $S_1$ and $S_2$ indicate that the signals sent over BUS1 and BUS2 respectively are valid.

Where two components use different multiplexed buses, the difficulties of interfacing become more severe, and the extra components required will tend to make such combinations uneconomic.

## 30.22 REGISTERS

Registers are building blocks for all the parts of a computer. A register stores or retains a number of bits of data. The data may be retained for a short time (a few nanoseconds in the internal registers of a CPU) or permanently (in a ROM). The number of bits retained varies from one to 64 or more. New data is loaded into the register under the control of a READ signal, and is retained until further new data is stored. The contents of the register are available at its outputs when a WRITE control signal is given.

Figure 30.32 shows a parallel-input / parallel-output register with separate inputs and outputs, and also a modified version of this with a single set of lines acting as either input or output. The J-K flip-flop and tristate-gate in Fig. 30.32(*c*) are one way of making a single bit register.

On an *n*-bit register, the READ and WRITE signals are applied in parallel to all the bits at once.

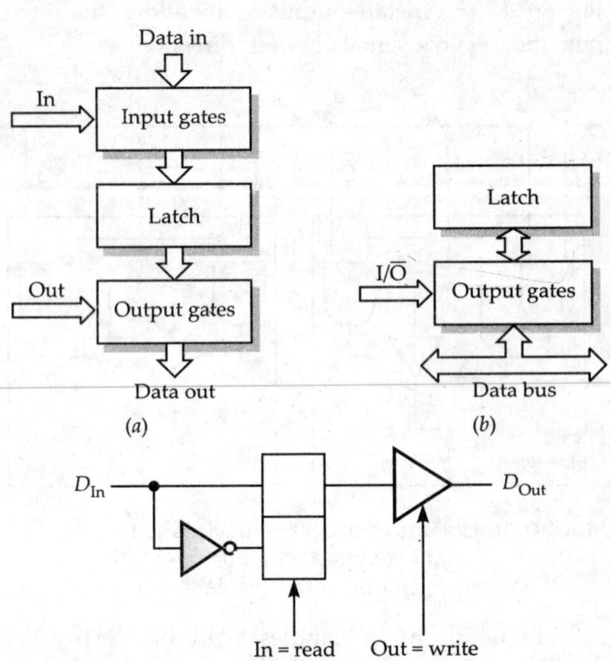


**Fig. 30.32** Registers. (*a*) a parallel-input parallel-output. (*b*) a register which can read or write from a data bus. (*c*) a possible gate circuit for a one-bit register.

A memory is a set of registers, with circuits which allow any one of the registers to be selected for reading or writing.

Figure 30.33 shows registers arranged to act as an addressable memory.

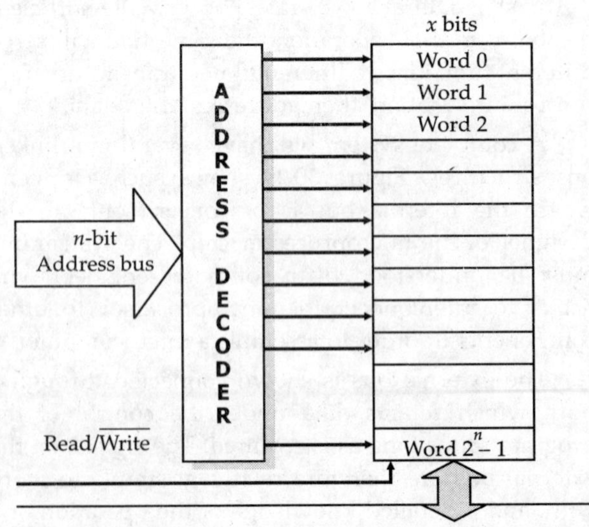


**Fig. 30.33** A read/write Random Access Memory shown as a selectable set of registers.

Each register has a unique binary address in the range 0 to ($2^n - 1$), where $n$ is the number of bits in the address supplied to the memory. Thus if the address $0003_{16}$ is present on the address bus, then depending on the state of the memory read/write control line, either the content of register 3 is read via the data bus, or new data is stored in register 3.

In order that the memory size of different systems may be compared, it is normal to measure the size of a memory in bytes of eight bits rather than words whose length may not be the same in the two systems to be compared. The unit of memory size is $2^{10}$ (=1024) which is given the symbol K. Most 8-bit microprocessors have a 16-bit address bus and can address $2^{16}$ or 65 536 words of memory, and if each word is one byte, the memory size is 64K.

Sixteen-bit microprocessors arrange to address several megabytes of memory, using 20 to 24 address lines.

## 30.23 SPECIAL REGISTERS

1. ***Accumulator.*** The standard arithmetic or logical operation takes the contents of two registers, combines them, and leaves the result in a register which is called the accumulator. The accumulator is usually one of the two registers which hold the initial data (Fig. 30.34), so that there is a feedback loop from output back to input. A temporary store for data (as in a master-slave J-K flip-flop) is therefore needed to avoid an endless cycle where a new input changes the output which changes the input which in its turn causes a further change in the output. This extra data storage register will be present, but will not be visible to a user.

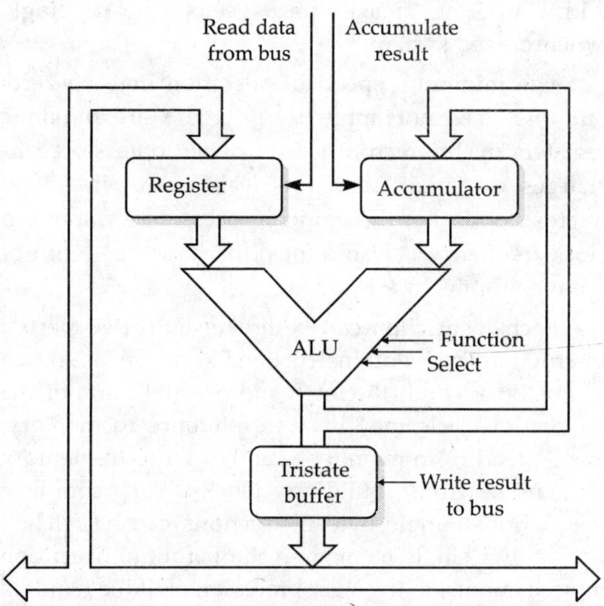


**Fig. 30.34** An arithmetic and logic unit (ALU) working with an accumulator and a data bus.

Accumulators will operate on many bits in parallel. Eight and 16 bit accumulators are common in microprocessor design but the existence of a single-bit microcontroller, and of very wide accumulators in minicomputers should be noted.

One accumulator is necessary. The 6800 family of microprocessors provide two accumulators, and some systems make the accumulator function available at many registers, either in the microprocessor, or in memory.

2. ***Index registers.*** An index register holds a number which when added to numbers held elsewhere forms an address. There are several ways that this can be done, and these are described earlier. Many, though not all, microprocessors have an index register. One of the operations frequently needed on the contents of an index register is increment (or decrement), and this is likely to be available in the instruction set.

3. ***Stack pointer.*** A stack pointer allows the user to control the use of a stack directly and in detail. The operation of a stack is shown in Fig. 30.35. The function of a stack is to provide a simple way of saving temporary data or addresses until they are needed again. The normal sequence of use is 'last in, first out'. It is convenient though not universal to use normal RAM for the stack, and the stack pointer is loaded with the address of the starting location of the stack. Each time new data is placed on the stack (PUSHED), the stack pointer is automatically updated. The reverse process is termed POP or PULL.

Points to check in a specific application are whether the stack builds using successively higher or lower addresses, and whether the change of address occurs before or after each use. A further point is the range of addresses that the stack can occupy – the full 64K may be possible, or alternatively only a part of this. Most microprocessors have at least one stack pointer. If a microprocessor has no stack pointer nor the corresponding instructions it is possible to obtain the same result with a series of program instructions. It is a valuable exercise to design PUSH and POP sequences.

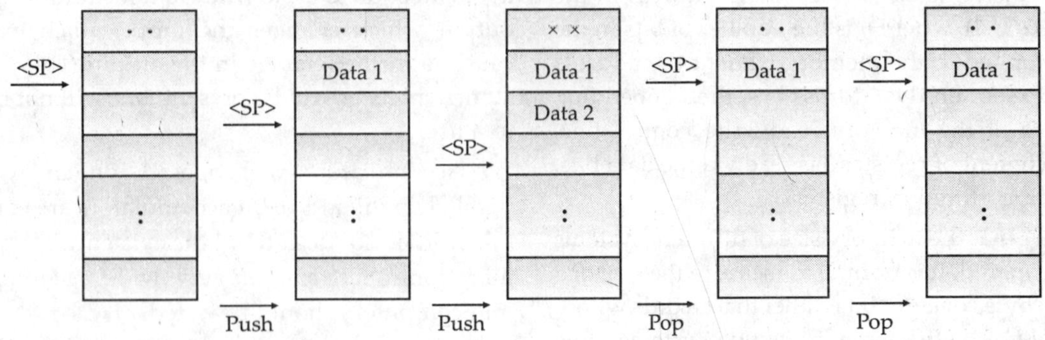


**Fig. 30.35** A stack in RAM with a stack pointer SP. Two PUSH and two POP operations are shown. This stack builds from high to low addresses, and the stack pointer points to the next unused location.

**4. *Program counter or instruction counter.*** This holds the address of the next instruction to be executed. After each instruction it is automatically incremented by the right amount to point to the next instruction in the program, its contents can be controlled explicitly by the programmer using direct or indirect jump instructions. During program writing the contents of the program counter are not usually in the forefront of the programmer's mind, but during attempts to debug a faulty program a knowledge of the contents of the program counter is of great value. To start the microcomputer running in a predictable way, the contents of the program counter are forced to a fixed value by the RESET signal. This value is not the same for all microprocessors. Similarly, some microprocessors respond to interrupt signals by entering a predetermined address into the program counter, and then continuing to execute the instructions found at that address and the subsequent ones.

**5. *Instruction register.*** This register holds the binary pattern which is the code for the instruction currently being executed. The programmer's concern is normally with the sequence of instructions in memory rather than with the contents of the instruction register, though a knowledge of the contents is useful during debugging operations.

## 30.24 MACHINE CYCLES

Each program instruction in a microprocessor requires a sequence of actions for its complete implementation. The number of actions varies from instruction to instruction, and from one microprocessor to another. We are concerned here with the effects that are detectable outside the microprocessor. The internal sequence is of concern to the microprocessor designer.

The observed effects are timed by a clock signal, and internal logic in the microprocessor controls the sequence. Each clock pulse puts the microprocessor into a new state, and the number of states required to complete an instruction is one way of measuring the length of an instruction. The clock is often controlled by a quartz crystal oscillator, so that a microcomputer with this kind of clock signal runs with quartz crystal precision timing.

The maximum frequency that a microprocessor can accept is stated by the manufacturer. It is possible to run a microprocessor at less than its maximum speed by using a lower frequency clock signal. The reason for doing this might be to permit the use of slightly slower (and cheaper) memory or peripheral devices, or to use a single clock for microprocessors with differing maximum speeds in a single synchronized system.

The minimum speed of operation may be zero, but this does not apply where there are dynamic registers in the microprocessor, when data stored as charges on small capacitors leaks away in a few microseconds. For these microprocessors, a minimum clock frequency is also quoted, perhaps one-tenth of the maximum clock frequency.

Each instruction can be broken into two parts ; Instruction Fetch and Instruction Execution, as shown in Fig. 30.36. Both the fetch and execute parts of the instruction cycle may involve reference to memory, and read-from-memory and write-to-memory sequences from building blocks for complete instructions. Single byte instructions can be fetched with only a single memory-read operation. Execution may be internal to the microprocessor, or, if the required address has been stored in the microprocessor registers already, the execution of the instruction may include memory reference cycles.

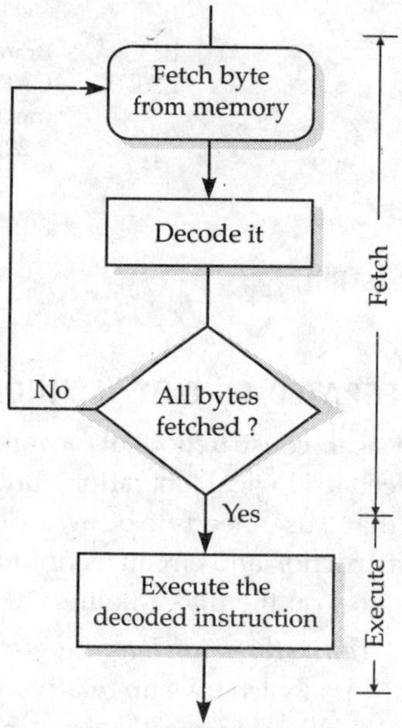


**Fig. 30.36** The fetch and execute sequence for a computer instruction. In a multibyte instruction bytes after the first may not need decoding if they only provide an address or data for the instruction to use.

When the first byte of a multibyte instruction has been transferred to the CPU, it is decoded and the full number of bytes to be transferred is noted. Instruction fetch cycles can then be repeated until the full instruction is in the CPU and it can then be executed.

Three byte instructions are often met, where the second and third bytes contain a 16-bit address. Even longer instructions may be met in 16-bit microprocessors.

## 30.25 ADDRESSING MODES

The way in which a computer instruction refers to memory is known as an addressing mode.

A sufficient variety of addressing modes is of value to a user as functions may then be programmed with few instructions. However, making provision for a wide range of addressing modes adds to the size and cost of a microprocessor chip, and may reduce the simplicity and speed of the device.

The notion of a target location may be introduced here. The target location of an instruction is the location which it will finally refer to. The process of calculating the target address may be simple as in direct addressing or more complex as in indirect or indexed addressing.

**1. *Direct addressing or absolute addressing.*** The address of a memory location is part of the instruction.

In 8-bit microcomputers, the address needs two bytes more after the op-code to refer to a 16-bit address, though some instructions use a shorter form with two bytes in total referring to an 8-bit address.

*Example* :

| | | | |
|---|---|---|---|
| 8085 code | 3A | LDA $1039_{16}$ | Load the accumulator with the contents of memory location $1039_{16}$ |
| | 39 | | |
| | 10 | | |

Where the instruction is stored in ROM, the address is fixed and cannot be changed during the program to refer to a succession of locations, as would be needed when examining a table. The target location might very well be in RAM, and hold different numbers successively dining a program run.

**2. *Immediate addressing.*** An instruction of this type refers to the location which immediately follows it. The address is not included in the instruction, but the data appears as part of the full instruction.

*Example* :

| | | | |
|---|---|---|---|
| 8085 code | 04 | MVI, B $0F_{16}$ | Move the hexa-decimal number 0F into the B register. |
| | 3F | | |

This addressing mode will provide constants in fixed places economically.

**3. *Indirect addressing.*** The instruction does not contain the address of the target memory location. Instead it contains an address where the address of the target location may be found. The intermediate location may be in normal memory, or in a CPU register.

The indirect idea could be repeated to further stages, but this does not occur in the microprocessors referred to in this text.

*Example* :

Indirect addressing *via* memory.

| | | | |
|---|---|---|---|
| 6502 code | 6C | JMP (4031) | Jump to the address contained in memory location $4031_{16}$ |
| | 31 | | |
| | 40 | | |

*Example* :

Indirect addressing *via* a CPU register.

**8085 code.** In the 8085 instruction set, there are instructions which take the contents of the H, L register pair as an address and are thus examples of indirect addressing via a register.

30

*Example :*

| H | L | 7E | MOV M, A | Move the contents of the A register into address ($2047_{16}$) pointed by H.L. |
|---|---|---|---|---|
| 20 | 47 | | | |

The value of an indirect addressing mode is that the same instruction can be made to refer to many different locations without any need to modify the instruction itself. All that is needed is to change the contents of the register (in memory or the CPU) which points to the final address.

**4. *Indexed addressing.*** When indexed addressing is used the address of the target location is the total of two numbers. One of the numbers is part of the instruction; and the other is held in a special register. Most microprocessors have one index register, some none, and some more than one. The address held in the instruction may be known as the base address. If the base address is that of the first entry in a table of data, each item in the table can be referred to by incrementing the contents of the index register.

*Example :*

| 6502 code | BD | LDA 2345, X | Load the accumulator with the contents of the location whose address is (2345 + (X)) |
|---|---|---|---|
| | 45 | | |
| | 23 | | |

*X* is an 8-bit index register. If *X* holds 05, then the target address is $234A_{16}$.

Indexed addressing is found to be valuable, and extra features may be added to it. There can be provision for incrementing or decrementing the index register as part of some instructions. Another variation is to combine indexing with indirect memory reference. The final result depends which happens first, the indexing or taking the address from the intermediate memory location. Both the possibilities occur in the 6502 and must be carefully distinguished.

**5. *Relative addressing.*** The name 'relative addressing' is short for 'relative to the value in the program counter'. The address is calculated by adding a number held in the instruction to the value of the program counter. The programmer thinks in terms of a number of memory locations earlier or later than that of the current instruction. A point of detail to be observed is whether the program counter holds the value of the present instruction or the next one. In microprocessors relative addressing is often found for jump instructions, and rarely for other forms of memory reference.

*Example :*

| 6502 code | 90 | BCC LABEL | Branch on no CARRY to (next address + 22) |
|---|---|---|---|
| | 22 | | |
| Address | | | |
| Address +22 LABEL : | | | |

## 30.26 INTEGRATED CIRCUIT TECHNOLOGY

The physical construction of a microprocessor does not affect its logical operation, but does have some effects that a user needs to be aware of. The main classes of construction and circuit techniques are listed below with notes on the major points.

**1. *TTL transistor-transistor logic.*** TTL uses bipolar transistors which take up relatively large areas of silicon, and are thus unsuitable for complex circuits with many gates. Logic levels are about 2.5 V and 0.7 V, with power supplies of 0 and 5.0 V. Speeds are high, but power consumption is also high. Output drive current from a TTL gate is high, which is good, but the current required at the input, particularly that required to pull an input to a logical low level is also high, and may be too high for some MOS circuits to supply. A common specification on a logic output is that it can drive one standard TTL gate. TTL components are available from many manufacturers, but are all coded 74xx. Subdivisions of TTL are Low power TTL 74Lxx, and Schottky diode (high speed) TTL, 74Sxx.

**2. *P-MOS.*** *p*-channel metal-oxide semiconductor. P-MOS uses MOS transistors on an *n*-type silicon substrate as switching elements and loads. The conducting path in the transistors is formed by inverting the *n*-type silicon to *p*-type by application of a positive gate voltage, thereby justifying the name. P-MOS was at first easier to manufacture reliably than N-MOS, but it was slower as the current carriers in P-MOS (holes) have a lower mobility than the current carriers in N-MOS (electrons). MOS devices take less silicon area than bipolar transistors, so allow more components on a chip of a given size.

**3. *N-MOS.*** *n*-channel metal-oxide-semiconductor. N-MOS offers about three times the operating speed of P-MOS, if the technologies are otherwise equivalent. Both P-MOS and N-MOS have moderate power requirements. They require only small input currents and can supply moderate output currents.

**4. *C-MOS.* (Complementary MOS)**. C-MOS uses a *p*-channel and an *n*-channel device in series in every logic circuit. One of the two is ON and the other OFF when the channel is in either of its two states, so there is never a high steady current. As a result C-MOS uses very little power-below a microwatt per gate rather than milliwatts. Speed is lower than N-MOS or TTL, but can be varied by changing power supply voltage-higher voltage gives higher speed.

C-MOS circuits do not interface to TTL circuits directly. In some cases an interface is built in to the C-MOS device, to provide TTL compatible outputs. In other cases the use of an interface chip is needed to convert voltage levels and provide drive currents.

**5. *ECL Emitter-coupled logic***

**6. *$I^2L$ Current (I) injection logic***

Both of these are high speed, high power circuit techniques which use bipolar transistors. Where they are employed, there will be TTL interface circuits on the chip, so a microcomputer designer does not have a need to consider their properties explicitly.

## 30.27 MICROPROCESSORS

Points to look for in assessing a microprocessor are :

- The word length.
- The instruction set.
- The addressing modes available.
- The flags and how they can be set and used.
- The number of registers available for use in the CPU, and whether they are special purpose or general purpose.
- The size of the memory that can be addressed.
- The time taken for a typical instruction to be executed.
- The provision for interrupts, both in the hardware and the instruction set.
- The provision made for a stack in memory.
- The power supplies required.

In this chapter, only Intel 8085 microprocessor and its applications are described.

## 30.28 INTEL 8085

Intel introduced the first microprocessor, and have continued to develop their designs. The Intel 8085 is an 8-bit microprocessor, which means that it handles eight bits of data simultaneously. It uses a 16-bit address for memory. The 8085 was designed to use the same set of binary instructions as the earlier 8085, with two instructions (RIM and SIM) added.

Features which distinguish the 8085 from other 8-bit microprocessors are :

1. It has several general purpose registers in the CPU.
2. It has few addressing modes.
3. The less significant byte of the address and the data are multiplexed on to the same eight bus lines.
4. There are separate instructions and addressing methods for input/output and for memory.
5. There are five interrupt lines and two serial input/output lines.

The 8085 has been widely used. Apart from its technical qualities, its early availability and the existence of suitable development systems contributed to its success.

### 30.28.1 8085 Inputs and Outputs and Reserved Memory Locations

The 40 input and output lines of the Intel 8085 are shown in Fig. 30.37(*a*) and grouped functionally in Fig. 30.37(*b*).

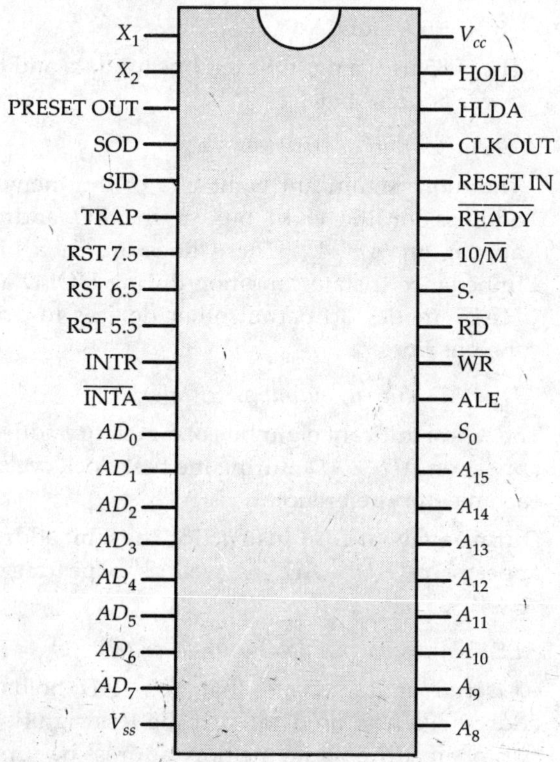


**Fig. 30.37.** (*a*) Physical pin-out.

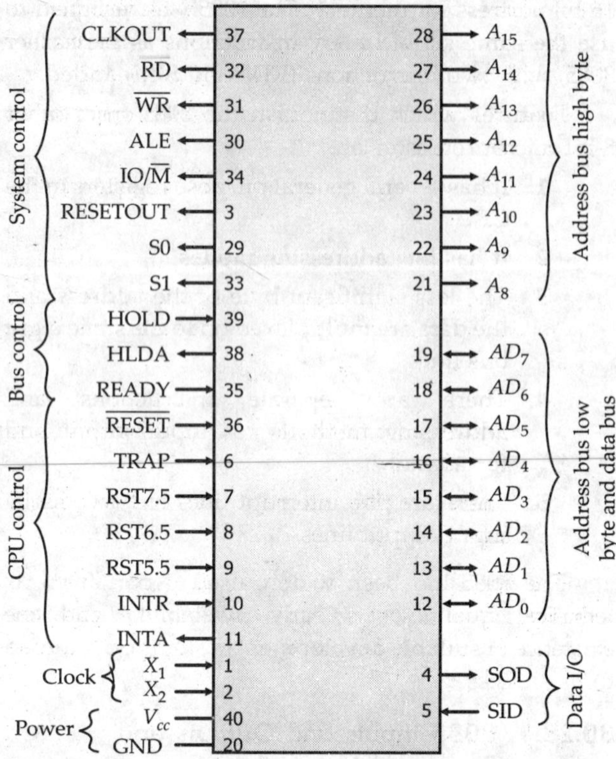


**Fig. 30.37** (*b*) Functional pin-out : the CPU and system control use the pins set free by the use of a multiplexed data and address bus.

▲ *Power*

+5.0 C ± 5% and 0 V

The 8085 uses a multiplexed bus for data and the lower address byte.

▲ $A_8 - A_{15}$ *(output or tristate)*

The more significant eight bits of the memory address, or the eight bits of the I/O address appear on $A_8 - A_{15}$. These lines are in a high impedance tristate condition during HOLD and HALT modes, to permit other devices to drive the bus lines.

▲ $AD_0 - AD_7$ *(input, output or tristate)*

The less significant eight bits of a memory address appear on $AD_0 - AD_7$ during the first clock cycle of each memory reference.

During an IN or OUT instruction the 8-bit address appears on $AD_0 - AD_7$ as well as appearing on $A_8 - A_{15}$.

▲ *ALE Address Latch Enable (output or tristate)*

ALE true (high) indicates that $AD_0 - AD_7$ holds an address, so it is used to latch the less significant part of an address into memory address buffers of bus demultiplexers.

▲ $S_0$ *and* $S_1$ *(outputs)*

These show the status of the data bus.

| Status | $S_0$ | $S_1$ |
|---|---|---|
| Halt | 0 | 0 |
| Write data | 0 | 1 |
| Read data | 1 | 0 |
| Fetch Instruction | 1 | 1 |

▲ $\overline{AD}$ *Active low Read (output or tristate)*

A low level on $\overline{RD}$ is the signal for the addressed device to place data on the $AD_0 - AD_7$ lines. The $\overline{RD}$ line is placed in a tristate high impedance condition during HOLD and HALT states, so that other devices such as a DMA controller can take command of the data transfer to memory.

▲ $\overline{WR}$ *Active low WRITE (output or tristate)*

A low level on $\overline{WR}$ is the signal for the addressed device to take data from $AD_0 - AD_7$. *It is tristated in the same way as* $\overline{RD}$.

▲ *READY (input)*

This input allows the 8085 to accept a signal from a device on the address or data buses to wait for the device to be ready to proceed with data transfer. If READY is low, WAIT states are inserted in the instruction cycle until READY goes high. Because this input line is provided, many peripheral devices can be used with the 8085.

▲ *HOLD (input)*

The HOLD input accepts a request from an external device to take control of the data and address buses. The 8085 CPU completes the current instruction, sets HLDA high, and relinquishes control of the data and address buses, and $\overline{RD}$, $\overline{WR}$ and IO/$\overline{M}$ until HOLD goes low. This input permits systems with more than one bus controller to be constructed.

▲ *HLDA Hold Acknowledge (output)*

HLDA is normally low, but goes high one clock cycle before the 8085 hands over control of the buses to another device.

▲ *TRAP (input)*

This is the highest priority, non-maskable interrupt. It is recognized on the penultimate clock cycle of the instruction.

▲ *RST 5.5, RST 6.5, RST 7.5 Restart 5.5, 6.5, 7.5 (inputs)*

These restart interrupt inputs force transfer of program control to locations that are 8 × (the number in the name of the interrupt). Thus RST 5.5 causes the program counter to hold $44_{10}$.

▲ *INTR Interrupt (input)*

INTR is the lowest priority interrupt, recognized in the same clock cycle as TRAP.

▲ $\overline{\text{INTA}}$ *Active low Interrupt Acknowledge (output)*

Interrupt acknowledge is output during the first instruction cycle after a signal on the INTR input line is accepted. The timing is the same as that of a RD signal. The $\overline{\text{INTA}}$ signal is used to activate circuits that put a Restart or CALL instruction on the buses, to divert program control to an interrupt handling sequence.

▲ $\overline{\textit{RESET IN}}$ *active low Reset (input)*

$\overline{\text{RESET IN}}$ forces the program counter to $0000_{16}$, resets the Interrupt Enable and HLDA bits, and resets the SOD output line. The CPU stays in the RESET condition as long as $\overline{\text{RESET IN}}$ is low.

▲ $\overline{\textit{RESET OUT}}$ *Active low (output)*

This provides a version of $\overline{\text{RESET IN}}$ which is synchronized to the system clock, and which can be used as a reset signal for other parts of the microcomputer.

▲ *IO /* $\overline{M}$ *Input/Output (when high) or Memory (when low) (output or tristate)*

When IO/$\overline{\text{M}}$ is low memory is being accessed. When IO/$\overline{\text{M}}$ is high, an IN or OUT instruction is being processed. The IO/$\overline{\text{M}}$ line is in a high-impedance state during HOLD and HALT modes.

▲ *SID Serial Input Data (input)*

Data on SID are loaded into the most significant bit of the *A* register by the RIM instruction.

▲ *SOD Serial Output Data (output)*

Data are supplied to SOD from the interrupt mask (bit 7). The data are loaded into the interrupt mask by the SIM instruction from bit 7 of the A register.

▲ $X_1$*,* $X_2$ *(inputs) and CLK (output)*

$X_1$ and $X_2$ are connection points for an *R/C* or crystal timing circuit. $X_1$ can be used as an input from an external clock. CLK is an output from the internal divide-by-two circuits which supply the CPU clock signal. A 6 MHz crystal connected at $X_1$ and $X_2$ will cause CLK to be a 3 MHz square wave.

## Reserved Memory Locations

The memory locations which are reserved for specific functions are listed below. They are all locations whose address is forced into the program counter by some action.

| **Address** | ***Cause of forced jump*** |
|---|---|
| $0000_{16}$ | RESET or RST 0 |
| $0008_{16}$ | RST 1 instruction |
| $0010_{16}$ | RST 2 instruction |
| $0018_{16}$ | RST 3 instruction |
| $0020_{16}$ | RST 4 instruction |
| $0024_{16}$ | TRAP interrupt input |
| $0028_{16}$ | RST 5 instruction |
| $002C_{16}$ | RST 5.5 interrupt input |
| $0030_{16}$ | RST 6 instruction |
| $0034_{16}$ | RST 6.5 interrupt input |
| $0038_{16}$ | RST 7 instruction |
| $003C_{16}$ | RST 7.5 interrupt input |

The only essential location is $0000_{16}$ for the power-on reset. Notice that there are no reserved locations near $FFFF_{16}$. The concept of a page of 256 words is not used in 8085-based systems, and the stack may be in any range of memory.

## 30.28.2 The Instruction Cycle

The timing of the 8085 instructions is governed by a single-phase clock whose frequency may be set by an external crystal or by a waveform injected from an external source.

There is an internal divide-by-two circuit so that a 6 MHz crystal produces a 3 MHz clock waveform.

The number of clock cycles taken by an 8085 instruction varies from 4 to 18. The simplest instructions are those that refer only to one or two registers in the 8085, and do not require a reference to any other component once the instruction has been fetched from memory. Examples of instructions that take four clock cycles are ADD *B* – add the content of the *B* register to the accumulator ; or MOV C, *E* – move the content of the *E* register to the C register. Each reference to memory is known as a machine cycle, and adds three clock cycles (usually) to the instruction, so that MVI C, $FF_{16}$ will take seven clock cycles. The instruction in hexadecimal form is *OE FF*, occupying two bytes, and therefore needing two memory references to bring it to the CPU. The final execution of the instruction is internal to the 8085 and can therefore be fitted into the last part of the second memory read cycle.

30

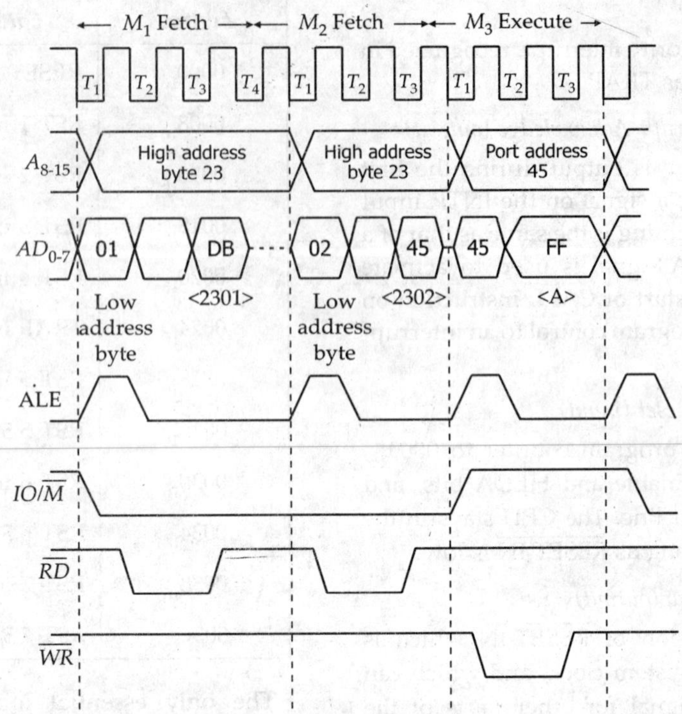


**Fig. 30.38** The timing of the instruction OUT 45, with the hexadecimal form DB 45 held in memory locations $230_{16}$ and $2302_{16}$. Each T cycle takes two periods of the oscillator crystal.

The timing diagram for a ten clock cycle instruction is shown in Fig. 30.38. The instruction has two memory-read machine cycles followed by a machine cycle where $\overline{WR}$ is low and IO/$\overline{M}$ is high, implying writing to an output port. There is only one instruction with this pattern, namely 'OUT'. The first seven clock cycles read the code of OUT and the port address, and the last three clock cycles send the content of the accumulator to the I/O port.

Comments on the separate signals in Fig. 30.38 are :

- **CLK.** This is the output from the CLK on the 8085. It is a square wave of half the frequency of the crystal or other input at $X_1$ and $X_2$ on the 8085. The falling edge of CLK is generally used as a timing reference.
- $A_8 - A_{15}$. These bus lines are not multiplexed, and retain the high order address for three or four clock cycles. These are the signals which are supplied to chip-select inputs on memory chips.
- $AD_{0-7}$. Timing for transfer of data to and from this bus is critical, as it handles both data and addresses.

  The addresses are latched when ALE falls, and the data are latched when $\overline{RD}$ or $\overline{WR}$ rises.
- **ALE Address latch enable.** This signal must be connected either to the ALE input in a chip designed to operate with the 8085, or to a latch which demultiplexes the $AD_{0-7}$ bus. It goes high each time an address appears on the $AD_{0-7}$ bus.
- **$\overline{RD}$ or $\overline{WR}$.** These are mutually exclusive, and control the signal flow to and from the 8085.
- **$S_0$, $S_1$. STATUS.** These indicate whether a read, write or instruction fetch is occurring. It is possible to build a system without using these signals. They could be compared to $\overline{RD}$ or $\overline{WR}$ to provide an indication of faulty operation, as combinations such as $S_0 = 1$, $S_1 = 0$, $\overline{WR} = 0$ should not occur.

Addition of extra clock states to the timing of an instruction is possible by using the READY input on the 8085.

After each memory address is placed on the address bus, the 8085 checks READY. If it is low, then WAIT states are executed, where all the internal processes of the microprocessor are halted. When READY goes high again, the sequence of operations to execute the operation resumes.

The 8085 signal ALE occurs at every memory reference, so it can be taken as the signal for a slow memory device to pull READY low.

## 30.28.3 8085 Registers

The resources available to the designers of the Intel 8080 and 8085 have been concentrated on the register set and the instructions which refer to registers. Effective use of the 8085 requires that the registers be used to their full capability.

The registers in the 8085 which are of concern to the user are shown in Fig. 30.39.

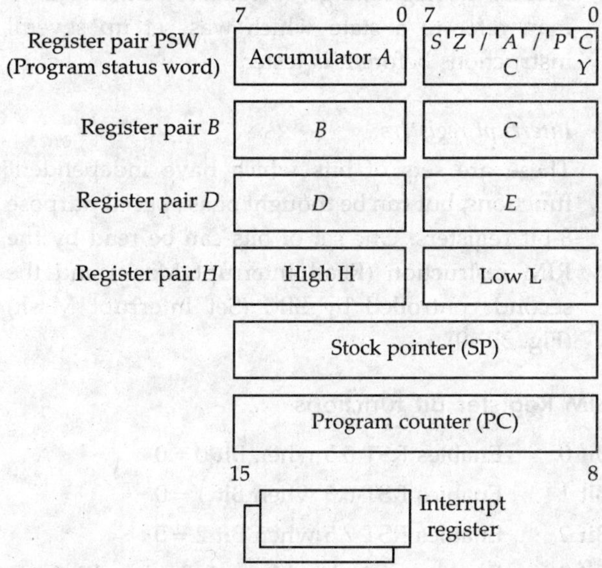


**Fig. 30.39** The 8085 registers that are available to a user.

▲ *A register*

The '*A*' register or accumulator is a standard 8-bit accumulator. The results of nearly all arithmetic and logical operations on data are stored in the *A* register, and there are more instructions which refer to this register than to any other. It is the only register for which there are shift instructions. For transfer of data to and from the stack (PUSH and POP), it is coupled with the flag register, the combination being known as the program status word (PSW). Note that the flags do not always reflect the latest content of the *A* register.

▲ *B, C, D and E registers*

These are either scratch-pad or address-holding registers. They can be used as four 8-bit registers, or as two 16-bit registers, *B*, *C* and *D*, *E* where *B* and *D* are the more significant bytes.

In the scratch-pad function they are used to hold data for a short time. A common use is to count the number of times a loop is traversed. The register pairs can hold addresses for movement of data to and from memory, the address being readily changed during the program, as in Problem 5.

▲ *H, L registers*

The *H* and *L* 8-bit registers are usually used as a 16-bit register pair to hold addresses which can be used by data movement, logical and arithmetic instructions. There is provision for transfer of addresses to the stack, the program counter and the *D*, *E* register pair. The *H*, *L* register pair may also act as an accumulator for 16-bit addition, so its use has to be carefully planned.

▲ *Stack pointer (SP)*

The 16-bit stack pointer can refer to a stack in memory anywhere in the range of a 16-bit address. The stack builds from high memory towards low memory, and the stack pointer contains the address of the most recently filled location on the stack. The contents of the stack pointer may be found during a program by using the DAD SP instruction. The stack is used automatically to hold return addresses from subroutines and interrupts, and explicitly as a safe temporary store for data.

▲ *Program counter*

The 16-bit program counter can be controlled directly by the PCHL or JMP instructions, and its contents are stored on the stack by subroutine call or restart instructions. Normally there is no need to consider the value held in the program counter during program construction, but during testing the program counter contents are a valuable diagnostic guide. An input on the RESET pin of the 8085 forces the program counter to zero, so every system must have a program sequence starting from $0000_{16}$.

▲ *Index register*

There is no index register. The functions of a normal index register are implemented by computing an address and holding it in one of the register pairs (*B*, *C*), (*D*, *E*), or (*H*, *L*).

▲ *Flag register and flags*

The flag register (*F*) contains five single-bit flags, and three unused bits. (Although an unused bit may appear to be affected systematically by the microprocessor, this could be altered in any future example of the same device, so may not be relied on. The same consideration applies to the ten unused op-codes.)

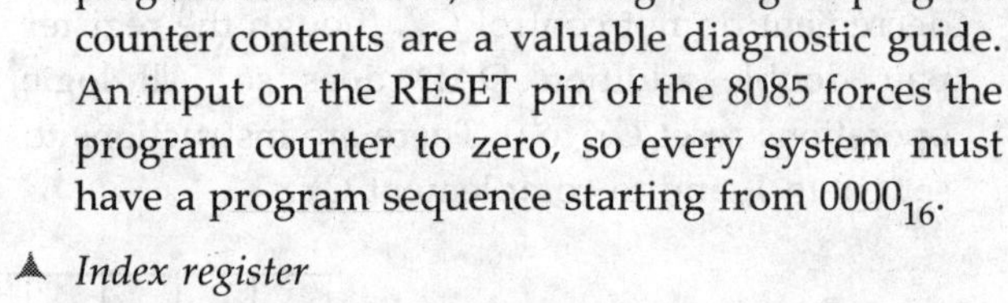

The flags are affected by some instructions and not by others, so that the list of instructions should be consulted for the full details. A nearly complete summary follows :

***Sign* (S).** *This is conventional.* $S = 1$ indicates that the last arithmetic or logical operation on the accumulator, or an increment or decrement or any register on memory location resulted in a negative number. $S = 0$ indicates a positive number or zero. The same instructions control $S$, $Z$, $AC$ and $P$.

***Zero* (Z).** *This is conventional.* $Z = 1$ indicates zero and $Z = 0$ indicates non-zero. The same instructions control $Z$ and $S$.

***Auxiliary carry* (AC).** This indicates a carry form bit 3 to bit 4 in any operation noted in the 'sign' description. Its use is to assist BCD arithmetic operations, when two BCD digits occupy one byte.

***Parity* (P).** This is conventional odd parity. $P = 1$ indicates an odd number of ones in the byte. $P = 0$ indicates an even number of ones. The same instructions control $P$ and $S$.

***Carry* (CY).** This indicates a carry out of the most significant bit of a register. Increment and decrement do not control $CY$, though the register pair double addition DAD does so. All logic operations reset $CY$ to 0. There are instructions to set $CY$ to 1, and to complement $CY$.

The subtraction operations (SUB, SBB and CMP) affect the carry flag, treating it as $\overline{\text{BORROW}}$. SBB also treats CY as equivalent to a $\overline{\text{BORROW}}$ input in computing a result.

There are conditional jump, call and return instructions which test for 0 or 1 on $S$, $Z$, $P$ or $CY$. It is worth-noting that many instructions do not modify the flag settings, so that a conditional jump may refer to a state which was set up several instructions before it is tested.

▲ *Interrupt registers*

These are sets of bits which have independent functions, but can be thought of as special purpose 8-bit registers. One set of bits can be read by the RIM instruction (Read Interrupt Mask) and the second controlled by SIM (Set Interrupt Mask) (Fig. 30.40).

**SIM Register Bit functions**

| | |
|---|---|
| Bit 0 | Enables RST 5.5 when Bit 0 = 0 |
| Bit 1 | Enables RST 6.5 when Bit 1 = 0 |
| Bit 2 | Enables RST 7.5 when Bit 2 = 0 |
| Bit 3 | Permits control of bits 0, 1, 2 when Bit 3 = 1 |
| Bit 4 | Resets Interrupt 7.5 flag to zero |
| Bit 5 | not used |
| Bit 6 | Permits control of bit 8 when Bit 6 = 1 |
| Bit 7 | Outputs to the SOD line |

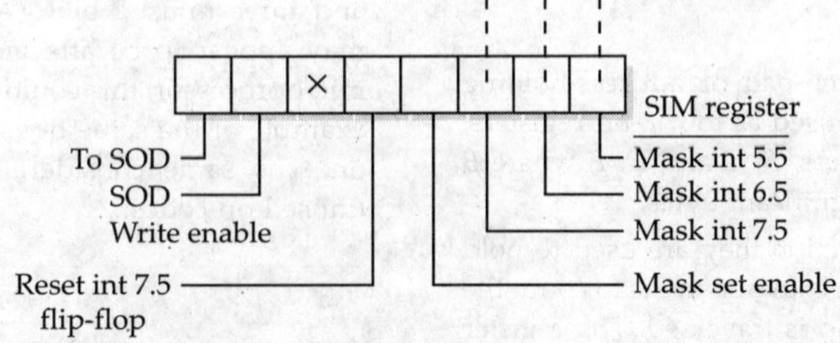


**Fig. 30.40** The two aspects of the interrupt register in the 8085. The SIM register is set by the instruction SIM. The two enable bits allow SOD and the interrupt masks to be separately altered. The RIM register can be read by the RIM instruction.

SIM carries out three distinct functions, that of controlling the value of the serial output line, the RST 7.5 interrupt flag, and the masks for RST 7.5, 6.5 and 5.5. Each function can be carried out without checking the existing state of the other bits by using the two enable bits to mask off the parts of the register which it is not wished to disturb. Thus to set SOD to zero the accumulator must be filled with 0.1 x 0.0 xxx before using SIM.

### 8085 Addressing Modes

The main addressing modes used in the 8085 are noted briefly here.

***DIRECT***

The second and third bytes hold the address of the target location.

*Examples :*

*LDA* $1234_{16}$ *JMP* $4567_{16}$

***IMMEDIATE***

The second or second and third bytes hold the data.

*Examples :*

*MVI B,* $OF_{16}$

***DIRECT via a register pair.***

H, L is often used to hold the address of a target location. *B, C* or *D, E* are used in two instructions each in this way.

*Examples :*

*MOV A,M LDAXB*

### 30.28.4 The 8085 Instruction Set

Each instruction is explained here. A list of all the hexadecimal op-codes is given on pages 1495, 1496 and 1497. Two points are worth noting : in the mnemonic form of the instruction, the destination is mentioned first, and then the source, so that movement should be read as 'into *xx* from *yy*' ; and in the binary or hexadecimal form of an instruction, addresses are always presented low byte first.

In the notes that follow :

- data 8 means one byte of data
- data 16 means two bytes of data
- addr 8 means one byte of address for I/O
- addr 16 means two bytes of address.

### Data Movement

**MOV $r_1$, $r_2$** Move data into $r_1$ from $r_2$.

Move the content of register $r_2$ into register $r_1$, where $r_1$ and $r_2$ may be *A, B, C, D, E, H* or *L*.

| No flag affected, | 1 byte | 4 clock cycles |
|---|---|---|

*Examples :*

**MOV M, *r*** Move into memory from register.

**MOV *r*, M** Move into register from memory

In each case *r* may be *A, B, C, D, E, H* or *L*, and the address of the memory location is held in *H* and *L*.

| No flag affected | 1 byte | 7 clock cycles |
|---|---|---|

**MVI *r*, data 8** Move immediate to register *r*.

Move the content of the second byte of the instruction to register, where *r* is *A, B, C, D, E, H* or *L*.

| No flag affected | 2 bytes | 7 clock cycles |
|---|---|---|

$36_{16}$ **MVI M, data 8** Move immediate to memory.

Move the content of the second byte of the instruction into memory. The address of the memory location is held in *H* and *L*.

| No flag affected | 2 bytes | 10 clock cycles |
|---|---|---|

**LXI register pair, data 16** Load extended immediate.

Load the next byte to *C, E, L* or stack pointer low byte, and the following byte to *B, D, H* or stack pointer high byte.

| No flag affected | 3 bytes | 10 clock cycles |
|---|---|---|

*Example :*

| | | |
|---|---|---|
| 31 | LXI SP, 20CO | Put 20 CO in the stack pointer |
| CO | | |
| 20 | | |

$3A_{16}$ **LDA addr 16** Load the accumulator.

$32_{16}$ **STA addr 16** Store the accumulator.

Transfer data between accumulator and the memory location whose address is given in the next two bytes.

| No flag affected | 3 bytes | 13 clock cycles |
|---|---|---|

*Example :*

| | | |
|---|---|---|
| 3A | LDA 5540 | Load register *A* with the content of memory location 5540 |
| 40 | | |
| 55 | | |

$OA_{16}$ **LDAX B** Load *A* indirect via *B, C*

$1A_{16}$ **LDAX D** Load *A* indirect via *D, E*

$02_{16}$ **STAX B** Store *A* indirect via *B*, *C*

$12_{16}$ **STAX D** Store *A* indirect via *D*, *E*

Transfer data between accumulator and the memory location whose address is held in the *B* and *C* register pair, or in the *D* and *E* register pair.

| No flag affected | 1 byte | 7 clock cycles |
|---|---|---|

$2A_{16}$ **LHLD addr 16** Load *H* and *L* direct.

$22_{16}$ **SHLD addr 16** Store *H* and *L* direct.

Transfer two bytes of data between *L* and the memory location whose address is in the second and third bytes of the instruction, and between *H* and the succeeding memory location.

| No flag affected | 3 bytes | 16 clock cycles |
|---|---|---|

*Example* :

| | | |
|---|---|---|
| 2*A* | LHLD 4130 | Move content of 4130 to L, and the content of 4131 to *H*. |
| 30 | | |
| 41 | | |

$EB_{16}$ **XCHG** Exchange *H* and *L* with *D* and *E*.

| No flag affected | 1 byte | 4 clock cycles |
|---|---|---|

## Arithmetic and Logic

**ADD *r* or *M*** Add register *r* or Add memory.

Add the content of register *r* or the content of a memory location to the accumulator, leaving the total in the accumulator. Register *r* may be *A*, *B*, *C*, *D*, *E*, *H* or *L*. The address of *M* is held in *H*, *L*. ADD *A* is allowed, and gives a shift to the left.

| All flags affected | 1 byte | 4 clock cycles for ADD *r*<br>7 clock cycles for ADD *M* |
|---|---|---|

*Example* :

$80_{16}$ ADD *B* Add the contents of the *B* register to the accumulator.

**ADC *r* or *M*** Add with carry register *r* or memory.

Add the content of register *r* or the content of memory to the accumulator, and then add the value of the *CY* flag. *r* may be *A*, *B*, *C*, *D*, *E*, *H* or *L* and the address of *M* is held in *H* and *L*.

| All flags affected | 1 byte | 4 clock cycles for ADD *r*<br>7 clock cycles for ADD *M* |
|---|---|---|

*Example* :

ADC *E* Add content of *E* and *CY* flag to accumulator.

**SUB *r* or *M*** Subtract-register *r* or memory.

Subtract the content of register *r* or a memory location from the accumulator, leaving the result in the accumulator. Register *r* may be *A*, *B*, *C*, *D*, *E*, *H* or *L*. The address of *M* is held in *H* and *L*.

If the subtraction causes a borrow at the most significant bit, then the *CY* flag is set to 1.

| All flags affected | 1 byte | 4 clock cycles for SUB *r*<br>7 clock cycles for SUB *M* |
|---|---|---|

**CMP *r* or *M*** Compare register *r* or memory with accumulator.

Subtract the content of register *r* or of a memory location from the accumulator and set the flags accordingly, but do not change the content of the accumulator. Register *r* may be *A*, *B*, *C*, *D*, *E*, *H* or *L*. The address of *M* is held in *H* and *L*.

| All flags affected | 1 byte | 4 clock cycles for CMP *r*<br>7 clock cycles for CMP *M* |
|---|---|---|

| **For unsigned inters 225 to 0** | **Flag test is** |
|---|---|
| Acc. Greater | *CY* = 0 |
| Equal | Z = 0 |
| Acc. Less | *CY* = 1 |

**SBB *r* or *M*** Subtract register or memory with borrow.

The action is that of SUB *r* or *M*, but the existing value of the CY flag is also subtracted from the accumulator.

| All flags affected | 1 byte | 4 clock cycles for SBB *r*<br>7 clock cycles for SBB *M* |
|---|---|---|

*Example* :

9*D* SBB *L* Subtract *L* from *A* with borrow

If *L* holds $03_{16}$, *A* holds $16_{16}$ and *CY* = 1 before the instruction, then after it, *L* still holds $03_{16}$, *A* holds $12_{16}$ and *CY* = 0.

**AND *r* or *M*** (AND register *r*) or (AND memory) with accumulator.

The contents of register *r* or a memory location are ANDed bit-by-bit with the contents of the accumulator, the result being left in the accumulator. Register *r* may be *A*, *B*, *C*, *D*, *E*, *H* or *L*. The address of *M* is held in *H* and *L*.

| All flags affected ($CY \rightarrow 0, AC \rightarrow 0$) | 1 byte | 4 clock cycles for AND *r*<br>7 clock cycles for AND *M* |
|---|---|---|

**ORA** ***r*** Or register *r* with accumulator,

**ORA** ***M*** Or memory with accumulator.

The content of register *r* or the content of a memory location are ORed bit-by-bit with the content of the accumulator, the result being left in the accumulator. Register *r* may be *A, B, C, D, E, H* or *L*. The address of *M* is held in *H* or *L*.

| All flags affected (CY → 0, AC → 0) | 1 byte | 4 clock cycles for ORA *r*<br>7 clock cycles for ORA *M* |
|---|---|---|

***XRA r*** **or** ***M*** Exclusive-OR register *r* or memory with accumulator.

The content of register *r* or the content of a memory location is exclusive. ORed bit-by-bit with the content of the accumulator, the result being left in the accumulator. Register *r* may be *A, B, C, D, E, H* or *L*. The address of *M* is held in *H* and *L*.

| All flags affected (CY → 0, AC → 0) | 1 byte | 4 clock cycles for XRA *r*<br>7 clock cycles for XRA *M* |
|---|---|---|

$C6_{16}$ **ADI data 8 Add immediate.**

Add the content of the second byte of the instruction to the accumulator, leaving the result in the accumulator.

| All flags affected | 2 bytes | 7 clock cycles |
|---|---|---|

$CE_{16}$ **ACI data 8** Add immediate with carry.

Add the content of the second byte of the instruction and the value of CY to the accumulator.

| All flags affected | 2 bytes | 7 clock cycles |
|---|---|---|

$D6_{16}$ **SUI data 8** Subtract immediate.

Subtract the content of the second byte of the instruction from the accumulator, leaving the result in the accumulator.

| All flags affected | 2 bytes | 7 clock cycles |
|---|---|---|

$DE_{16}$ **SBI data 8** Subtract immediate with borrow.

Subtract the content of the second byte of the instruction and the value of the CY flag from the accumulator, leaving the result in the accumulator.

| All flags affected | 2 bytes | 7 clock cycles |
|---|---|---|

$E6_{16}$ **ANI data 8** AND immediate.

The content of the second byte of the instruction are ANDed bit-by-bit with the content of the accumulator, leaving the result in the accumulator.

| All flags affected | 2 bytes | 7 clock cycles |
|---|---|---|

$EE_{16}$ **XRI data 8** Exclusive-Or immediate.

The contents of the second byte of the instruction are exclusive-ORed bit-by-bit with the contents of the accumulator, leaving the result in the accumulator.

| All flags affected | 2 bytes | 7 clock cycles |
|---|---|---|

$F6_{16}$ **ORI data 8** OR immediate.

The contents of the second byte of the instruction are ORed bit-by-bit with the contents of the accumulator, leaving the result in the accumulator.

| All flags affected | 2 bytes | 7 clock cycles |
|---|---|---|

$FE_{16}$ **CPI data 8** Compare immediate.

Subtract the content of the second byte of the instruction from the content of the accumulator and set flags accordingly, leaving the accumulator contents unchanged.

| All flags affected | 2 bytes | 7 clock cycles |
|---|---|---|

**DAD** ***rp*** Double add register pair to *H, L*.

Add the contents of the register pair *rp H* and *L* acting as a 16 bit-accumulator. Register pair *rp* may be (*B, C*), (*D, E*), (*H, L*) or (Stack Pointer). Note that DAD H acts as a left shift on (*H, L*).

A carry out of *L* is added to *H*. A carry out of *H* sets *CY* to 1.

| CY flag affected | 1 byte | 10 clock cycles |
|---|---|---|

$27_{16}$ **DAA** Decimal adjust accumulator.

This instruction can follow ADD or ADC operating on two b.c.d. coded bytes to put the result in correct binary coded decimal. It does not convert binary numbers to binary coded decimal form.

| All flags affected | 1 byte | 4 clock cycles |
|---|---|---|

**INR** ***r*** **or** ***M*** Increment register or memory.

Add one to the content of register *r* or memory. Register *r* may be *A, B, C, D, E, H* or *L*. The address of the memory location is held in *H* and *L*.

| Z, S, P and AC flags affected (but not the CY flag) | 1 byte | 4 clock cycles for INR *r*<br>10 clock cycles for INR *M* |
|---|---|---|

**DCR, or** ***M*** Decrement register or memory.

Subtract one from the content of register *r* or memory. Register *r* may be *A, B, C, D, E, H* or *L*.

The address of the memory location is held in *H* and *L*.

| Z, S, P and AC flags affected (but not the CY flag) | 1 byte | 4 clock cycles for INR *r*<br>10 clock cycles for INR *M* |
|---|---|---|

$03_{16}$ **INX** ***B*** Increment register pair *B, C*
$13_{16}$ **INX** ***D*** Increment register pair *D, E*
$23_{16}$ **INX** ***H*** Increment register pair *H, L*
$33_{16}$ **INX SP** Increment the 16-bit stack pointer

Add one to the content of the register pair. Any carry out of the low register is a carry into the high register, so the operation acts on a 16-bit register.

| No flag affected (compare with INR) | 1 byte | 6 clock cycles |
|---|---|---|

$OB_{16}$ **DCX** ***B*** Decrement register pair *B, C*
$IB_{16}$ **DCX** ***D*** Decrement register pair *D, E*
$2B_{16}$ **DCX** ***H*** Decrement register pair *H, L*
$3B_{16}$ **DCX SP** Decrement the 16-bit stack pointer.

Subtract one from the 16-bit register pair.

| No flag affected (compare with DCR) | 1 byte | 6 clock cycles |
|---|---|---|

$07_{16}$ **RLC** Rotate left.

Rotate the contents of the accumulator left as in the diagram.

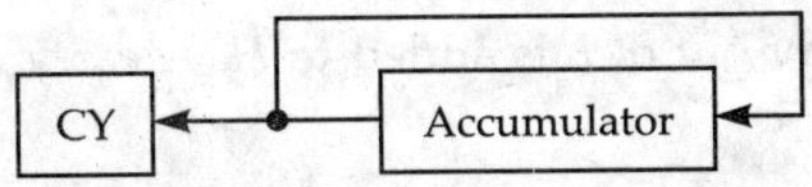


| CY flag affected | 1 byte | 4 clock cycles |
|---|---|---|

$0F_{16}$ **RRC** Rotate right.

Rotate the contents of the accumulator right as in the diagram.

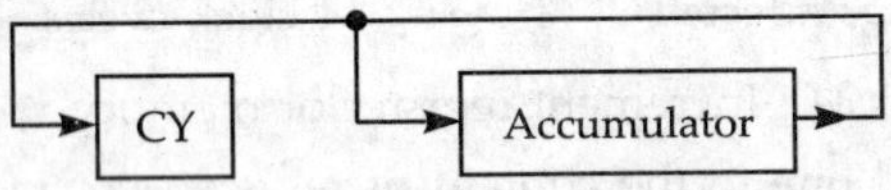


| CY flag affected | 1 byte | 4 clock cycles |
|---|---|---|

$17_{16}$ **RAL** Rotate left through carry.

Rotate the contents of the accumulator and carry left. Notice that the mnemonic is apparently not a good description of the operation for this and the next instruction.

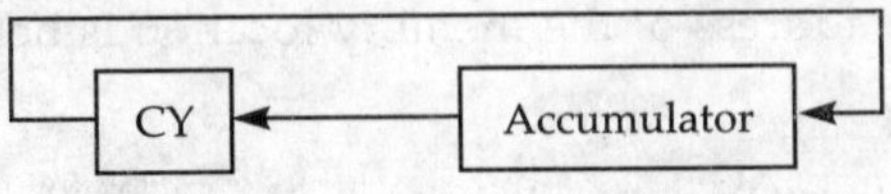


| CY flag affected | 1 byte | 4 clock cycles |
|---|---|---|

$1F_{16}$ **RAR** Rotate right through carry.

Rotate the contents of the accumulator and carry to the right.

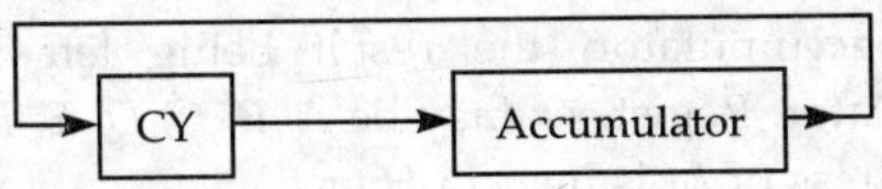


| CY flag affected | 1 byte | 4 clock cycles |
|---|---|---|

$2F_{16}$ **CMA** Complement accumulator.

Change all the zeros in the accumulator to ones and all the ones to zeros.

| No flag affected | 1 byte | 4 clock cycles |
|---|---|---|

$3F_{16}$ **CMC** Complement carry.

Complement the carry flag. No other changes occur.

| CY flag affected | 1 byte | 4 clock cycles |
|---|---|---|

$37_{16}$ **STC** Set carry.
Set CY to 1.

| CY flag affected | 1 byte | 4 clock cycles |
|---|---|---|

## Instructions which alter the program counter

$C3_{16}$ **JMP** addr 16 Jump

The program jumps to the address in bytes 2 and 3 of the instruction.

| No flag affected | 3 bytes | 10 clock cycles |
|---|---|---|

*Example* :

| | | |
|---|---|---|
| C3 | JMP 4410 | Jump to location $4410_{16}$ for the next instruction. |
| 10 | | |
| 44 | | |

$CD_{16}$ **CALL** addr 16 Call subroutine.

The address of the next instruction is stored on the stack, with the higher byte at the higher stack address. The content of the stack pointer is decremented by 2. This prepares for the later return from the subroutine. The program now jumps to the address held in bytes 2 and 3 of the instruction.

| No flag affected | 3 bytes | 18 clock cycles |
|---|---|---|

$C9_{16}$ **RET** Return from subroutine.

The address of the next instruction is taken from the stack which is then incremented by two. The lower byte of the address is taken from the lower stack location.

| No flag affected | 1 byte | 10 clock cycles |
|---|---|---|

## Condition Codes

There are eight conditions that may be tested before a conditional JUMP, CALL or RETURN is executed. They test the *Z*, *S*, *P* and CY flags for 1 or for 0. In the table below, X stands for *J* (MP), *C* (ALL) or *R* (ET).

X on not zero *XNZ* (Z = 0)
X on no Carry *XNC* (C = 0)
X on zero *XZ* (Z = 1)
X on parity odd *XPO* (*P* = 0)
X on parity even *XPE* (*P* = 1)
X on carry *XC* (CY = 1)
X on plus *XP* (*S* = 0)
X on minus *XM* (S = 1)

***J*** (*condition*) Jump if condition holds.

Jump as specified for the unconditional jump if the condition of the relevant flag holds. Otherwise go on in normal program sequence.

| No flag affected | 3 bytes | 10 clock cycles if the condition holds<br>7 clock cycles if the condition does not hold |
|---|---|---|

C (condition) Call subroutine if condition holds.

Call subroutine as described for the unconditional call instruction if the condition of the relevant flag holds. Otherwise go on in normal program sequence.

| No flag affected | 3 bytes | 18 clock cycles if the condition holds<br>9 clock cycles if the condition does not hold |
|---|---|---|

***R*** (*condition*) Return from subroutine if condition holds.

Return from a subroutine as described for the unconditional return if the condition of the relevant flag holds. Otherwise go on the normal program sequence.

| No flag affected | 1 byte | 12 clock cycles if the condition holds<br>6 clock cycles if the condition fails |
|---|---|---|

**RST *n*** Restart 0 to 7.

Save the program counter on the stack and then Jump unconditionally to the address which is eight times '*n*' where *n* may be 0 to 7.

| No flag affected | 1 byte | 12 clock cycles |
|---|---|---|

*Example*

DF RST3 Jump to location $24_{10}$

$E9_{16}$ **PCHL**
Jump indirect (*H* and *L* to program counter).
Jump to the address held in *H* and *L*.

| No flag affected | 1 byte | 6 clock cycles |
|---|---|---|

## Stack Operations

$D5_{16}$

| | | |
|---|---|---|
| $C5_{16}$ | **PUSH B** | Push *B*, *C* on to stack |
| $D5_{16}$ | **PUSH D** | Push *D*, *E* on to stack |
| $E5_{16}$ | **PUSH H** | Push *H*, *L* on to stack |
| $F5_{16}$ | **PUSH PSW** | Push *A* and flags on to stack = Push program status word. |

The contents of the register pair are moved to the stack in memory as pointed to by the stack pointer. At the end of the operation the stack pointer holds the address of the contents of *C*, *E*, *L* or the flag register.

| No flag affected | 1 byte | 12 clock cycles |
|---|---|---|

| | | |
|---|---|---|
| $R1_{16}$ | **POP B** | Pop *B*, *C* off stack |
| $D1_{16}$ | **POP D** | Pop *D*, *E* off stack |
| $E1_{16}$ | **POP H** | Pop *H*, *L* off stack |
| $F1_{16}$ | **POP PSW** | Pop *A* and flags off stack = Pop program status word off stack. |

The contents of the memory location pointed to by the stack pointer are moved to *C*, *E*, *L* or the flags. The contents of the next higher memory location are moved to *B*, *D*, *H* or the accumulator.

The stack pointer is then incremented by two.

| No flag affected for POP *B*, *D*, *H* | 1 byte | 10 clock cycles |
|---|---|---|

All flags affected for POP PSW

$E3_{16}$ **XTHL** Exchange top of stack with *H* and *L*.

*L* is exchanged with the location pointed to by the stack pointer. *H* is exchanged with the next higher memory location.

The stack pointer remains unaltered.

| No flag affected | 1 byte | 16 clock cycles |
|---|---|---|

**SPHL** Move to stack pointer from *H* and *L*.

| No flag affected | 1 byte | 6 clock cycles |
|---|---|---|

## Input/Output and Machine Control

$D3_{16}$ **IN** addr 8 Input from port.

Move data from the port whose address is held in the second byte of the instruction to the accumulator.

| No flag affected | 2 bytes | 10 clock cycles |
|---|---|---|

*Example*

| D3 | IN $44_{16}$ | Read from port address $44_{16}$ to accumulator |
|---|---|---|
| 44 | | |

$DB_{16}$ **OUT addr 8** Output to port.

Move data from the accumulator to the port whose address is held in the second byte of the instruction.

| No flag affected | 2 bytes | 10 clock cycles |
|---|---|---|

$FB_{16}$ **EI** Enable interrupts.

Enables interrupts, but not until the following instruction is complete :

| No flag affected | 1 byte | 4 clock cycles |
|---|---|---|

$F3_{16}$ **DI** Disable interrupts.

Disables interrupts from the start of the DI instruction.

| No flag affected | 1 byte | 4 clock cycles |
|---|---|---|

$76_{16}$ **HLT** Halt

Program execution stops until a RESET or interrupt signal occurs.

| No flag affected | 1 byte | 5 clock cycles |
|---|---|---|

$00_{16}$ **NOP** No operation.

This instruction causes the program counter to be incremented, but has no other effects.

| No flag affected | 1 byte | 4 clock cycles |
|---|---|---|

$20_{16}$ **RIM** Read interrupt mask

The contents of the RIM register as detailed in Section 2.4 are copied to the accumulator. The manufacturer calls this register the interrupt mask, but it is necessary to distinguish the two registers that are referred to by RIM and SIM, as only three of the bits are common.

| No flag affected | 1 byte | 4 clock cycles |
|---|---|---|

$30_{16}$ **SIM**

The contents of the accumulator are copied to the SIM mask as detailed in Section 2.4.

| No flag affected | 1 byte | 4 clock cycles |
|---|---|---|

## Instruction set for 8085 microprocessor

| *Hex* | *Mnemonic* | *Flags* | *Hex* | *Mnemonic* | *Flags* | *Hex* | *Mnemonic* | *Flags* |
|---|---|---|---|---|---|---|---|---|
| *Move data REG ← REG* | | | *Add register to A* | | | *AND register with A* | | |
| 7F | MOV A,A | none | 87 | ADD A | all | A7 | ANA A | all |
| 78 | MOV A,B | none | 80 | ADD B | all | A0 | ANA B | all |
| 79 | MOV A,C | none | 81 | ADD C | all | A1 | ANA C | all |
| 7A | MOV A,D | none | 82 | ADD D | (4) all | A2 | ANA D | (4) all |
| 7B | MOV A,E | none | 83 | ADD E | all | A3 | ANA E | all |
| 7C | MOV A,H | none | 84 | ADD H | all | A4 | ANA H | all |
| 7D | MOV A,L | none | 85 | ADD L | all | A5 | ANA L | all |
| 47 | MOV B,A | none | | | | | | |

| Code | Instruction | Flags | Code | Instruction | Flags | Code | Instruction | Flags |
|---|---|---|---|---|---|---|---|---|
| 40 | MOV B,B | none | | **ADD to A with CARRY** | | | **EXOP register with A** | |
| 41 | MOV B,C | none | 8F | ADC A | all | AF | XRA A | all |
| 42 | MOV B,D | none | 88 | ADC B | all | A8 | XRA B | all |
| 43 | MOV B,E | none | 89 | ADC C | all | A9 | XRA C | all |
| 44 | MOV B,H | none | 8A | ADC D (4) | all | AA | XRA D (4) | all |
| 45 | MOV B,L | none | 8B | ADC E | all | AB | XRA E | all |
| 4F | MOV C,A | none | 8C | ADC H | all | AC | XRA H | all |
| 48 | MOV C,B | none | 8D | ADC L | all | AD | XRA L | all |
| 49 | MOV C,C | none | | | | | | |
| 4A | MOV C,D | none | | **Subtract from A reg** | | | **OR register with A** | |
| 4B | MOV C,E | none | 97 | SUB A | all | B7 | ORA A | all |
| 4C | MOV C,H | none | 90 | SUB B | all | B0 | ORA B | all |
| 4D | MOV C,L | none | 91 | SUB C | all | B1 | ORA C | all |
| 57 | MOV D,A | none | 92 | SUB D (4) | all | B2 | ORA D (4) | all |
| 50 | MOV D,B (4) | none | 93 | SUB E | all | B3 | ORA E | all |
| 51 | MOV D,C | none | 94 | SUB H | all | B4 | ORA H | all |
| 52 | MOV D,D | none | 95 | SUB L | all | B5 | ORA L | all |
| 53 | MOV D,E | none | | | | | | |
| 54 | MOV D,H | none | | **Subtract register** | | | **Use 1 byte of memory** | |
| 55 | MOV D,L | none | | **and CARRY from A** | | | **Address in H and L** | |
| 5F | MOV E,A | none | 9F | SBB A | all | 36 | MVI M, d8 (10) | none |
| 58 | MOV E,B | none | 98 | SBB B | all | 7E | MOV A,M | none |
| 59 | MOV E,C | none | 99 | SBB C | all | 46 | MOV B,M | none |
| 5A | MOV E,D | none | 9A | SBB D (4) | all | 4E | MOV C,M | none |
| 5B | MOV E,E | none | 9B | SBB E | all | 56 | MOV D,M | none |
| 5C | MOV E,H | none | 9C | SBB H | all | 5E | MOV E,M | none |
| 5D | MOV E,L | none | 9D | SBB L | all | 66 | MOV H,M | none |
| 67 | MOV H,A | none | | | | 6E | MOV L,M | none |
| 60 | MOV H,B | none | | **Increment register** | | 77 | MOV M,A 7 | none |
| 61 | MOV H,C | none | 3C | INR A | Z S P AC | 70 | MOV M,B | none |
| 62 | MOV H,D | none | 04 | INR B | Z S P AC | 71 | MOV M,C | none |
| 63 | MOV H,E | none | 0C | INR C (4) | Z S P AC | 72 | MOV M,D | none |
| 64 | MOV H,H | none | 14 | INR D | Z S P AC | 73 | MOV M,E | none |
| 65 | MOV H,L | none | 1C | INR E | Z S P AC | 74 | MOV M,H | none |
| 6F | MOV L,A | none | 24 | INR H | Z S P AC | 75 | MOV M,L | none |
| 68 | MOV L,B | none | 2C | INR L | Z S P AC | 86 | ADD M | all |
| 69 | MOV L,C | none | | | | 8E | ADC M | all |
| 6A | MOV L,D | none | | **Decrement register** | | 96 | SUB M | all |
| 6B | MOV L,E | none | 3D | DCR A | Z S P AC | 9E | SBB M | all |
| 6C | MOV L,H | none | 05 | DCR B | Z S P AC | 34 | INR M | Z S P AC |
| 6D | MOV L,L | none | 0D | DCR C (4) | Z S P AC | 35 | DCR M (10) | Z S P AC |
| | | | 15 | DCR D | Z S P AC | A6 | ANA M | all |
| | **Compare register with A** | | 1D | DCR E | Z S P AC | AE | XRA M (7) | all |
| BF | CMP A | all | 25 | DCR H | Z S P AC | B6 | ORA M | all |
| B8 | CMP B | all | 2D | DCR L | Z S P AC | BE | CPM M | all |
| B9 | CMP C | all | | | | | | |
| BA | CMP D (4) | all | | **Machine control** | | | **FLAGS** | Z = Zero |
| BB | CMP E | all | 20 | RIM | 30 SIM (4) | | | S = Sign |
| BC | CMP H | all | 00 | NOP (4) | 76 HLT (5) | | | P = Parity |
| BD | CMP L | all | F3 | DI | FB EI (4) | | | CY = Carry |
| | | | | | | | | AC = Auxillary carry |

(*n*) means *n* clock cycles.

30

| Hex. | Mnemonic | Description | Flags |
|---|---|---|---|
| | **JUMP** | | |
| C3 | JMP, adr | Jump. (10) | none |
| C2 | JNZ, adr | Jump. on no zero | — |
| CA | JZ, adr | Jump. on zero | — |
| D2 | JNC, adr | Jump. on no carry | — |
| DA | JC, adr | Jump. on carry (10/7) | — |
| E2 | JPO, adr | Jump. on parity odd | — |
| EA | JPE, adr | Jump. on parity even | — |
| F2 | JP, adr | Jump. on positive | — |
| FA | JM, adr | Jump. on minus | — |
| E9 | PCHL | H, L → PC = indirect jump (6) | — |
| | **CALL SUBROUTINE PC ← STACK** | | |
| CD | CALL, adr | Call subroutine | none |
| C4 | CNZ, adr | Call on no zero | — |
| CC | CZ, adr | Call on zero | — |
| D4 | CNC, adr | Call on no carry | — |
| DC | CC, adr | Call on carry | (18/9) |
| E4 | CPO, adr | CAll on parity odd | — |
| EC | CPE, adr | Call on parity even | — |
| EC | CP, adr | Call on positive | — |
| FC | CM, adr | Call on minus | — |
| | **RETURN FROM SUBROUTINE** | | |
| C9 | RET | Return from subrtn. (10) | none |
| C0 | RNZ | Return on no zero | — |
| C8 | RZ | Return on zero | — |
| D0 | RNC | Return on no carry | — |
| D8 | RC | Return on carry | (12/6) |
| E0 | RPO | Return on parity odd | — |
| E8 | RPE | Return on parity even | — |
| F0 | RP | Return on positive | — |
| F8 | RM | Return on minus | — |
| | **RESTART AND SAVE PC (12)** | | |
| C7 | RST | 0 none CF RST 1 | none |
| D7 | RST | 2 none DF RST 3 | none |
| E7 | RST | 4 none EF RST 5 | none |
| F7 | RST | 6 none FF RST 7 | none |
| | **ROTATE A REGISTER** | | |
| 07 | RLC | Rotate A left (4) | CY |
| 0F | RRC | Rotate A right | CY |
| 17 | RAL | Rotate A left through CY | CY |
| 1F | RAR | Rotate A right through CY | CY |
| | **COMBINE A RIGISTER AND NEXT BYTE** | | |
| C6 | ADI,d8 | ADD byte to A | all |
| CE | ACI,d8 | Add with CARRY | all |
| D6 | SUI,d8 | Subtract from A | all |
| DE | SBI,d8 | Subtract with CY | all |
| E6 | ANI,d8 | AND with A (7) | all |
| EE | XRI,d8 | EXOR with A | all |
| F6 | ORI,d8 | OR with A | all |
| FE | CPI,d8 | Compare with A | all |
| | **SPECIALS** | | |
| 27 | DAA | Decimal adjust A | all |
| 2F | CMA | Complement A (4) | none |
| 37 | STC | Set CARRY flag | CY |
| 3F | CMC | Complement CY | CY |

| Hex. | Mnemonic | Description | Flags |
|---|---|---|---|
| | **MOVE IMMEDIATED TO REGISTER** | | |
| 3E | MVI | A,d8 Next byte ← A | none |
| 06 | MVI | B,d8 Next byte ← B | none |
| 0E | MVI | C,d8 Next byte ← C | none |
| 16 | MVI | D,d8 Next byte ← D | none (7) |
| 1E | MVI | E,d8 Next byte ← E | none |
| 26 | MVI | H,d8 Next byte ← H | none |
| 2E | MVI | L,d8 Next byte ← L | none |
| | **LOAD REGISTER PAIR IMMEDIATE** | | |
| 01 | LXI | B,d16 Next bytes ← C, B | none |
| 11 | LXI | D,d16 Next bytes ← E, D | none (10) |
| 21 | LXI | H,d16 Next bytes ← L, H | none |
| 31 | LXI | Sp,d16 Next bytes ← SP | none |
| | **LOAD/STORE A/MEMORY** | | |
| 3A | LDA, adr | Load A from mem. | none (13) |
| 32 | STA, adr | Store A in mem. | none |
| 0A | LDAX B | M → A Adr in B, C | none |
| 02 | STAX B | A → m Adr in B,C | none (7) |
| 1A | LDAX D | M → A Adr in D, E | none |
| 12 | STAX D | A → M Adr in D, E | none |
| | **REGISTER PAIR MOVES** | | |
| 2A | LHLD, adr | M → L M + 1 → H | none (16) |
| 22 | SHLD, adr | L → m H → M + H | none (16) |
| EB | XCHG | H → D L → E | none (4) |
| E3 | XTHL | stack top → H,L | none (16) |
| F9 | SPHL | H,L → SP | none (6) |
| | **STACK USE** | | |
| C5 | PUSH B | B, C → stack | none |
| D5 | PUSH D | D, E → Stack | none (12) |
| E5 | PUSH H | H, L → Stack | none |
| F5 | PUSH PSW | A, Flags → stack | none |
| C1 | POP B | stack → B, C | none |
| D1 | POP D | stack → D, E | none (10) |
| E1 | POP H | stack → H, L | none |
| F1 | POP PSW | stack → A, flags | all |
| | **DOUBLE ADD TO H AND L** | | |
| 09 | DAD | B Add B , C to H, L | CY |
| 19 | DAD | D Add D, E to H, L | CY (10) |
| 29 | DAD | H Add H, L to H, L | CY |
| 39 | DAD | SP Add SP to H, L | CY |
| | **INCREMENT/MECREMENT REGISTER PAIR** | | |
| 03 | INX | B Increment B, C | none |
| 13 | INX | D Increment D, E | none |
| 23 | INX | H Increment H, L | none |
| 33 | INX | SP Increment SP | none (6) |
| 0B | DCX | B Decrement B, C | none |
| 1B | DCX | DE Decrement D, E | none |
| 2B | DCX | H Decrement H, L | none |
| 3B | DCX | SP Decrement SP | none |
| | **INPUT/OUTPUT** | | |
| D3 | OUT,d8 | Output A to port | none |
| DB | OUT,d8 | Output A to port | none |

*Note* d8 means 8 bits of data
d16 means 16 bits of data
adr mean 2 bytes of address

## 30.29 EXAMPLE PROGRAMS

The five examples perform the same tasks in each of the four microprocessors that are considered in detail. The programs here illustrate the use of the instruction set and registers in the 8085. There may be shorter or faster ways of achieving the same result.

**Problem 1** *Move an 8-bit byte from one memory location to another. The address of the source location is held in memory locations $1000_{16}$ and $1001_{16}$ and the address of the destination is held in memory locations $1008_{16}$ and $1009_{16}$.*

| Address | Before | After |
|---|---|---|
| 1000 | 12 | 12 |
| 1001 | 34 | 34 |
| | ⋮ | ⋮ |
| 1008 | AB | AB |
| 1009 | CD | CD |
| | | |
| 3412 | EF | EF |
| | | |
| CDAB | 00 | EF |

**Solution.**

| *Address* | *Hex.* | *Mnemonic* | *Comments* |
|---|---|---|---|
| 2000 | 2A 00 10 | LHLD $1000_{16}$ | Put source address in *H*, *L* |
| 2003 | 7E | MOV A, M | Move the byte to *A*. |
| 2004 | 2A 08 10 | LHLD $1008_{16}$ | Put destination address in *H*, *L*. |
| 2007 | 77 | MOV *M*, *A* | Move the byte to destination. |
| 2008 | 76 | HLT | |

**Problem 2** *Add two n-digit binary-coded-decimal numbers X and Y, where n is even and less than + 128. The number n is held in location $1000_{16}$ and the least significant two digits of X are held in $1001_{16}$. The first digits of Y follow immediately after the last digits of X. The result should replace X, with any final carry being held in the carry flag.*

*Example :*

```
      12 22 32
     +29 00 81
 (0)  41 23 13
```

| Address | Before | | After | |
|---|---|---|---|---|
| $1000_{16}$ | 06 | *n* | 06 | |
| | 32 | | 13 | |
| | 22 | X | 23 | Result |
| | 12 | | 41 | |
| | 81 | | 81 | |
| | 00 | Y | 00 | |
| | 29 | | 29 | |
| | *CY* = ? | | *CY* = 0 | |

30

**Solution.** The indirect-via-register addressing mode is exploited. The number of bytes is first found, and the address of X is then placed in *D*, *E* and of *Y* in *H*, *L*. The bytes are added via the accumulator, corrected to b.c.d form and the result stored at once.

| Address | Hex. Label | Mnemonic | Comments |
|---|---|---|---|
| 2000 | 11 00 10 | LXID $1000_{16}$ | Point to number of digits. |
| 2003 | 1A | LDAX D | Put *n* in accumulator. |
| 2004 | 37 | STC | Clear carry flag. |
| 2005 | 3F | CMC | |
| 2006 | 17 | RAL | Divide by 2 to give number of bytes. |
| 2007 | 47 | MOV B, A | *B* holds number of bytes. |
| 2008 | 13 | INXD | *D, E* now points to low byte of X. |
| 2009 | 6F | MOV L, A | |
| 200A | 26 00 | MVI H, $00_{16}$ | |
| 200C | 19 | DAD D | *H, L* now points to low byte of *Y*. |
| 200D | A7 | ANA A | Clear carry flag (another method). |
| 200E | 1A NEXBYT | LDAX D | Put byte of X in accumulator. |
| 200F | 8E | ADC M | Add byte of Y to X with any carry. |
| 2010 | 27 | DAA | Correct the result to b.c.d form |
| 2011 | 12 | STAX D | Put the result away. |
| 2012 | 13 | INX H | |
| 2013 | 23 | INX D | Point to next bytes. |
| 2014 | 05 | DCR B | Check for end of numbers. |
| 2015 | C2 0E 20 | JNZ NEXBYT | Continue if there is more to do. |
| 2018 | 76 | HLT | |

**Problem 3** *Form the logical function* $f = (P.Q) + R$, *and leave* $FF_{16}$ *in the accumulator if* $f = 1$ *and* $00_{16}$ *in the accumulator if* $f = 0$. *P, Q and R are the three least significant bits of the accumulator contents at the start of the program.*

*Example 1 :*

Before

| x | x | x | x | x | 0 | 0 | 1 |
|---|---|---|---|---|---|---|---|

P Q R $f = 1$

After

| 1 | 1 | 1 | 1 | 1 | 1 | 1 | 1 |
|---|---|---|---|---|---|---|---|

*Example 2 :*

Before

| x | x | x | x | x | 1 | 0 | 0 |
|---|---|---|---|---|---|---|---|

P Q R $f = 0$

After

| 0 | 0 | 0 | 0 | 0 | 0 | 0 | 0 |
|---|---|---|---|---|---|---|---|

**Solution.** We use the fact that $f = 0$ only when the accumulator hods $00_{16}$, $02_{16}$ or $04_{16}$, as can be seen by constructing the Karnaugh map for *f*. This seems to be a simpler way to implement the required function than selecting bits and using OR and AND instructions.

| Address | Hex | Label | Mnemonic | Comments |
|---|---|---|---|---|
| 2000 | E6 07 | | ANI $07_{16}$ | Mask off all except the three |
| 2002 | FE 00 | | CPI $00_{16}$ | relevant bits. |
| 2004 | CA 16 20 | | JZ BLANK | |
| 2007 | FE 02 | | CPI $02_{16}$ | Compare accumulator contents |
| 2009 | CA 16 20 | | JZ BLANK | possible value of $\overline{f}$. |
| 200C | FE 04 | | CPI $04_{16}$ | |
| 200E | CA 16 20 | | JZ BLANK | |
| 2011 | 3E FF | | MVI A $FF_{16}$ | $\overline{f}$ not matched, therefore $\overline{f} = 1$. |
| 2013 | C3 18 20 | | JMP END | |
| 2016 | 3E 00 | BLANK | MVI A $00_{16}$ | Action when $f = 0$. |
| 2018 | 76 | END | HLT | |

**Problem 4** *Shift a number in an 8-bit memory location with address $1000_{16}$, n places to the left. Introduce zeros to the empty places at the right. The number n is held in binary form in the accumulator, which should be clear at the end of the program.*

*Example :*

| | Before | After |
|---|---|---|
| Accumulator | 0 0 0 0 0 0 1 1 | 0 0 0 0 0 0 0 0 |
| Memory $1000_{16}$ | 1 0 1 1 1 1 0 1 | 1 1 1 0 1 0 0 0 |

**Solution.** Use *B* as a counter, and *A* to shift the word.

| Address | Hex. | Label | Mnemonic | Comments |
|---|---|---|---|---|
| 2000 | 47 | | MOV B, A | Save count of shifts |
| 2001 | 3A 00 10 | | LDA $1000_{16}$ | Move word so that it can be shifted |
| 2004 | A7 | LOOP : | ANA A | Force Carry to 0 so that zeros are shifted in. |
| 2005 | 17 | | RAL | Rotate the word. |
| 2006 | 05 | | DCR B | Decrement count of shifts. |
| 2007 | C2 04 20 | | JNZ LOOP | Repeat if not finished. |
| 200A | 32 00 10 | | STA $1000_{16}$ | Put the result away. |
| 200D | 97 | | SUB A | Clear *A* as required. |

**Problem 5** *Attempt to match a string held in memory to one of a set of strings. The new string starts at NEWSTRING, and the last character is (ASCII Ret). A list of the strings starts at LISTSTR, and contains the addresses of the first byte of each string. Each address occupies two bytes. The list ends with $0000_{16}$.*

*Each string in the list ends with (ASCII Ret). If a match is found at the Nth word in the list, leave N in the accumulator. If no match is found, leave $00_{16}$ in the accumulator. The addresses in LISTSTR are stored with the more convenient byte (in this case the less significant byte) at the lower address.*

*Example :*

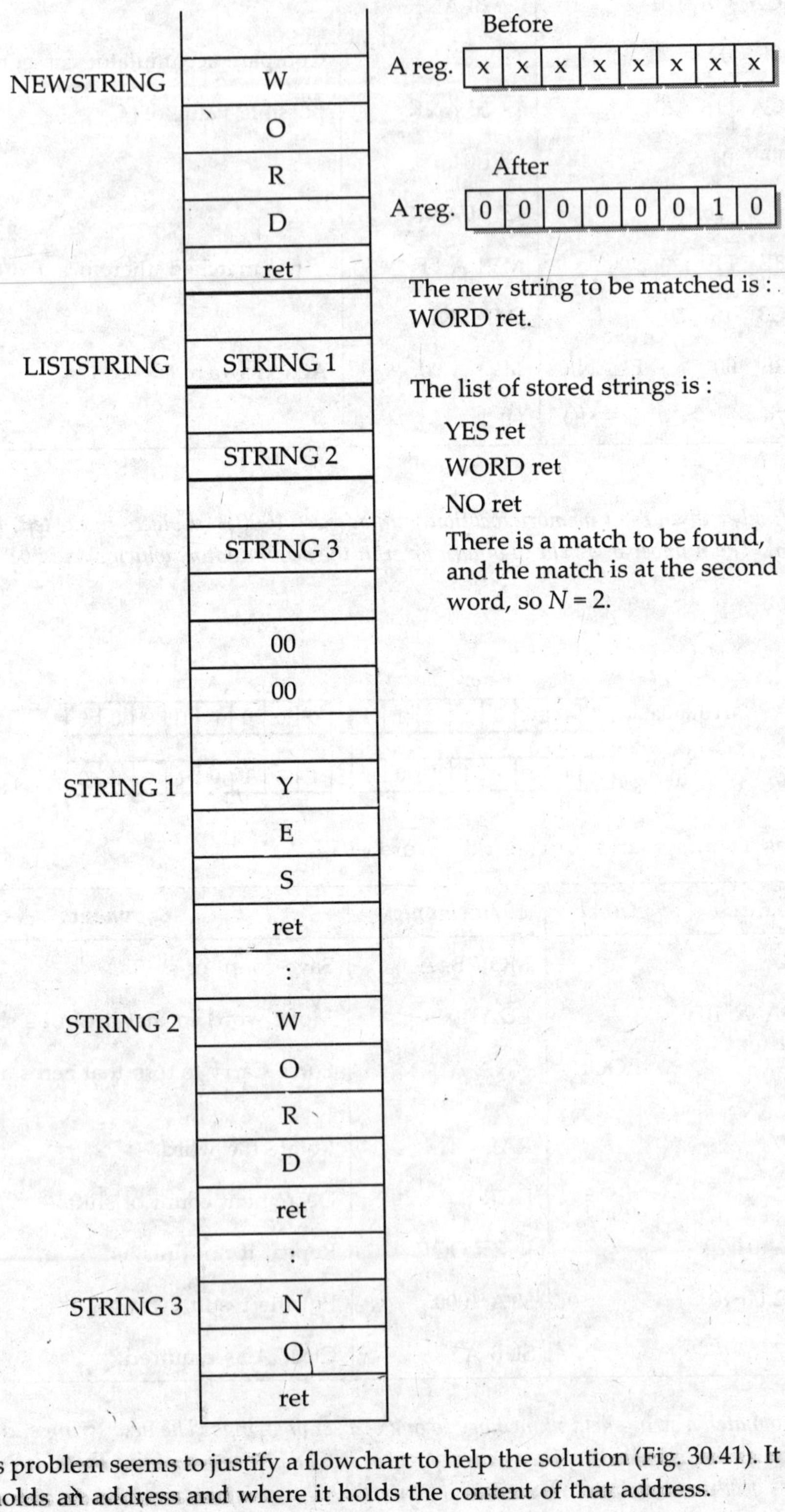


**Solution.** This problem seems to justify a flowchart to help the solution (Fig. 30.41). It is necessary to be clear where a register holds an address and where it holds the content of that address.

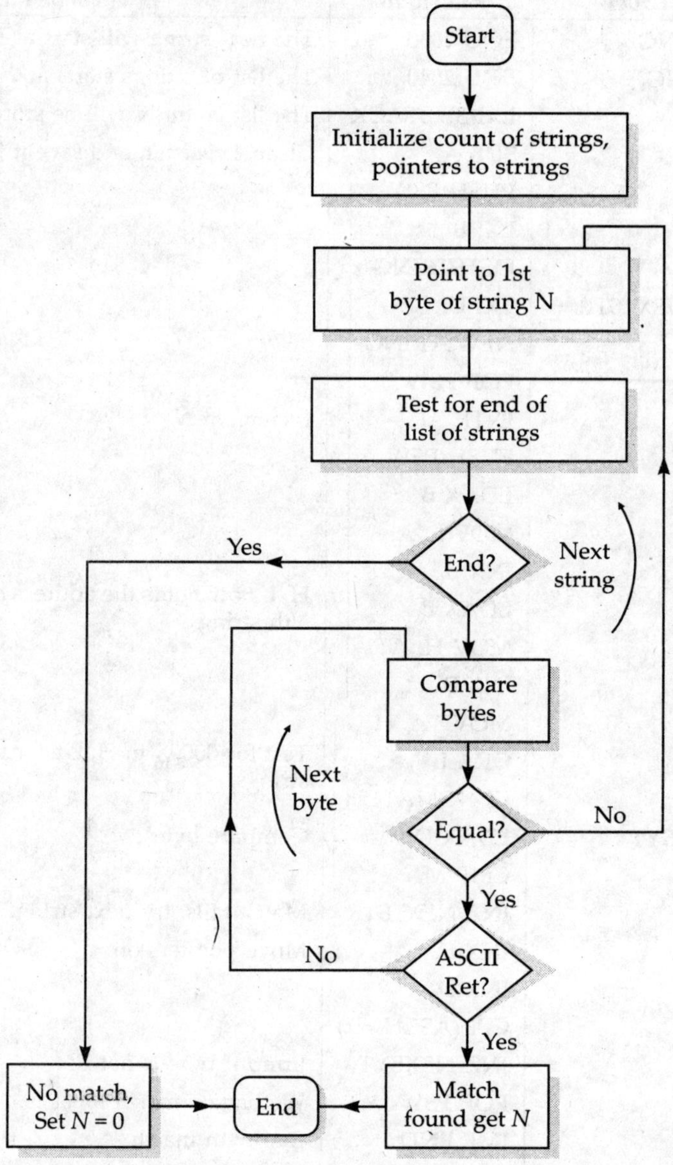


**Fig. 30.41** Flowchart for the string matching Problem 5.

The use made of register is :

| | |
|---|---|
| *A* | no permanent use |
| *B, C* | points to the list of strings |
| *D, E* | holds the address of a byte of the new string |
| *H, L* | holds the address of a byte of the *N*th string in the list |

The inner loop of the program (the seven instructions from NEXB YT) compares two strings. The outer loop (starting from NEXTSTRING) prepares the registers to read the strings.

To make the program explicit, NEWSTRING and LISTSTRING have been given values using EQU pseudo-instructions.

30

| Address | Hex. Label | Mnemonic | Comments |
|---|---|---|---|
| | NEWSTRING | EQU $2030_{16}$ | The new string will start at Location $2030_{16}$. |
| | LISTSTRING | EQU $2040_{16}$ | The list of strings starts at $2040_{16}$. |
| 2000 | 31 60 20 | LXI SP STACK | The list count N will be stored on the stack. |
| 2003 | 97 | SUB A | Clear *A* register, and save it to initialize list count *N*. |
| 2004 | F5 | PUSH PSW | |
| 2005 | 01 40 20 | LXI B LISTSTRING | |
| 2008 | 02 30 20 NEXTSTR | LXI D NEWSTRING | |
| 200B | F1 | POP PSW | |
| 200C | 3C | INTR A | Increment *N*. |
| 200D | F5 | PUSH PSW | |
| 200E | 0A | LDAX B | |
| 200F | 6F | MOV L, A | |
| 2010 | 03 | INX B | |
| 2011 | 0A | LDAX B | H, L now holds the address of the first byte of the *N*th string. |
| 2012 | 67 | MOV H, A | |
| 2013 | 03 | INX B | |
| 2014 | 7C | MOVE A, H | |
| 2015 | B5 | ORA L | Test for $0000_{16}$ in H, L as marker for end of list of strings. |
| 2016 | CA 26 20 | JZ NOMATCH | |
| 2019 | 1A NEXTBYT | LDAX D | Compare bytes. |
| 201A | BE | CMP M | |
| 201B | C2 08 20 | JNZ NEXTSTR | Match fails, try next string. |
| 201E | 23 | INX H | Move pointers on. |
| 201F | 13 | INX D | |
| 2020 | FEOD | CPI (ASCH ret) | |
| 2022 | C2 19 20 | JNZ NEXBYT | Jump if byte is not Ret. |
| 2025 | F1 | POP PSW | We have a match ! Get *N*. |
| 2026 | C3 2B 20 | JMP END | Exit with match. |
| 2029 | F1 NOMATCH | POP PSW | Restore state of stack. |
| 202A | 97 | SUB A | Clear *A* as required for no match. |
| 202B | 76 END | HLT | |

## 30.30 INTERRUPTS, INPUT/OUTPUT AND PERIPHERAL DEVICES

### Interrupts

The provision of interrupts on the 8085 is generous, but requires the mastery of many details for full comprehension. There are five interrupt inputs with fixed relative priority. They are, highest priority first, TRAP, RST 7.5, RST 6.5, RST 5.5 and INT. The four highest priority interrupts force the program counter to take a value fixed by the 8085 design. The existing contents of the program counter are saved on the stack for use when the computer returns to its main task. There are four memory words reserved for the interrupt process following each of these addresses, and it is expected that a JMP instruction would be placed there, so that program control is transferred by two stages from a main program to an interrupt service routine.

Each of the four interrupts discussed is thus suitable for dealing with a single interrupting device without the need for any further steps to identify the interrupting device.

The fifth interrupt INTR ('interrupt') operates in a different way; which was used in the earlier Intel 8080. When the 8085 recognizes a true signal on the INT line, it suspends normal operation, and sends out a pulse on the INTA output line. The interrupting device must be able to respond to this signal by putting the code for CALL or RST 0 — 7 on the data bus. The 8085 decodes this as an instruction, and acts accordingly, asking for two bytes of address if it received a CALL There is thus provision for a number of devices to share the INT line, as each can use CALL with a different address. This is then a true vectored interrupt.

It is often convenient to be able to prevent interrupts during part of a program, and to permit them during others. This is possible with the 8085 because all the interrupts except TRAP can be enabled or disabled by setting bits in the interrupt mask. There are two stages in setting up the system to allow an interrupt. First, the interrupt mask must be loaded with a suitable pattern of bits. Also the interrupt must be enabled by the EI instruction. The following sequence enables RST 5.5 and RST 7.5 and INTR, but does not enable RST 6.5. TRAP is always enabled.

| | |
|---|---|
| Restart | hardware input to define initial conditions |
| EI | INTR enabled after next instruction |
| MVI A<br>0000 $1010_2$ | sets up bit pattern for interrupt mask |
| SIM | RST 5.5 and RST 7.5 now enabled |
| : | |

When an interrupt occurs, the existing status of the system can be saved by storing eight register contents on the stack with four push instructions. At the end of the interrupt service routine the status is restored by four pop instructions, interrupt enable and return to take the address from the stack which was placed there by the interrupt mechanism.

General interrupt service routine

```
PUSB  PSW
PUSB  B        Save status
PUSB  D
PUSB  H
   :
functional body of interrupt service routine
   ;
POP   H
POP   D        Restore status
POP   B
POP   PSW
EI             Enable interrupts after the
               RET instruction.
RET
```

In summary, there must be three pieces of program for any interrupt — an enabling routine in the main program, a JMP to a service routine at the location required by the interrupt, and the service routine itself, which may be located anywhere in the address space.

Where a large number of distinct interrupting devices must be serviced, then a more complex system must be built up. This may employ external logic to establish the priority of competing interrupts and additional program sequences to poll the interrupting devices.

## Supporting Components

There are some components which are designed to interface directly with the 8085. Examples are the 8155/56 RAM, the 8355 ROM, and the 8755 EPROM. Other components assist in interfacing the 8085 bus structure to components which are not designed to be used with the 8085, but which are cheap, compact or can do things that cannot be done otherwise.

**8185** (a 1024 × 8-bit static RAM for use with the 8085 microprocessor)

This is a straightforward RAM, allowing 1024 words to be stored and retrieved, and offering no other functions. The address on the multiplexed $AD_{0-7}$ bus is latched in when ALE goes low, and data is supplied to the $AD_{0-7}$ bus without any other components being required. Figure 30.42 the 8185 package. Notice that there are two chip enable ($\overline{CE_1}$, $CE_2$) pins as well as a chip select. Figure 30.43 shows a functional description of the 8185, and also how it may be connected to the 8085 microprocessor. The pins $AD_0 - AD_7$, $A_8$, $A_9$, $\overline{RD}$, $\overline{WR}$ and ALE must be connected exactly as shown, but the remaining pins $CE_1$, $CE_2$, and $\overline{CS}$ act in combination to select a specific 8185, and alternative connections are possible.

ALE has overall control over the sequence of events, as when it goes low the lowest 8 bits of the address are latched into the 8185 within 80 ns, and valid data

30

appears on the same $AD_{0-7}$ lines after 300 ns, assuming that the chip is fully enabled and a READ cycle has been demanded.

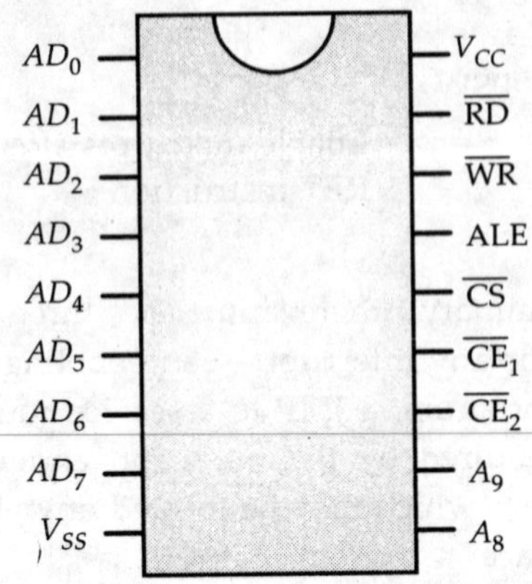


| | |
|---|---|
| $AD_0 - AD_7$ | Address/data |
| $A_8 - A_9$ | Address |
| $\overline{CS}$ | Chip select |
| $\overline{CE_1}$ | Chip enable |
| $\overline{CE_2}$ | Chip enable |
| ALE | Address latch enable |
| $\overline{RD}$ | Read |
| $\overline{WR}$ | Write |

**Fig. 30.42** Pin functions for the 8185 1 K × 8 static RAM.

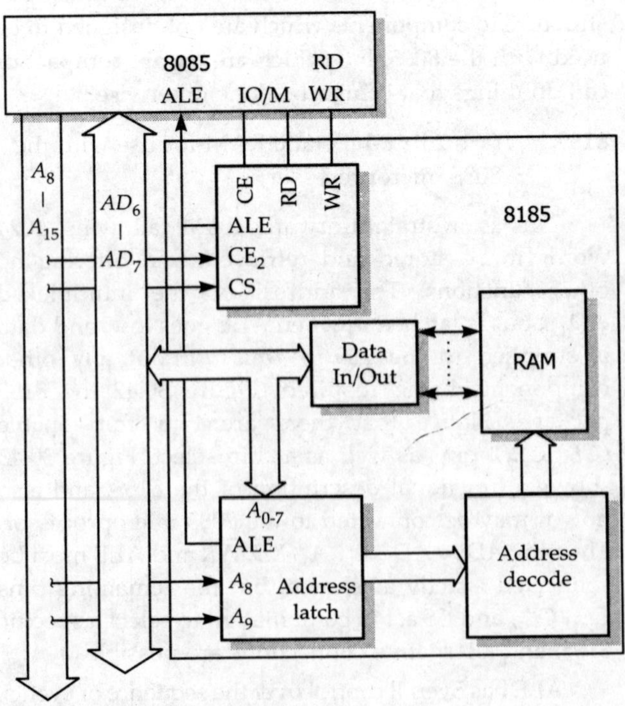


**Fig. 30.43** The block diagram for the 8185 1 K × 8, 8085-compatible RAM.

Notice that the IO/ $\overline{M}$ (input/output or memory) signal from the 8085 has been used as a chip select. One effect of this is to reduce power consumption to a standby level during input/output cycles.

**8755** (EPROM with I/O, 2048 × 8-bit EPROM for use with the 8085)

The 8755 is an electrically programmable (1 ← 0) and UV erasable (0 or 1 ← 1) ROM in a 40-pin D.I.L. package.

The block diagram is shown in Fig. 30.44.

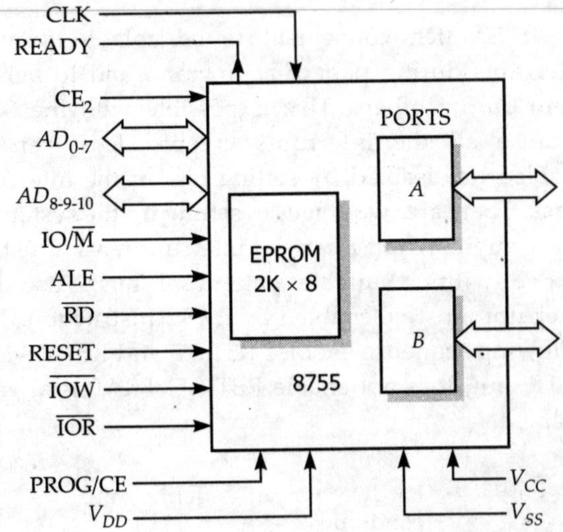


**Fig. 30.44** The 8755 device, which includes EPROM and input/output ports.

## Memory Section

The lowest 8 address lines are part of the common address/data bus, and the designer of a microcomputer system has no choice of how they are connected. Similarly, address lines $A_8$, $A_9$, and $A_{10}$ will be taken from the least significant lines of the 8085 address bus. The location of the chip on the system memory map is determined by the signals selected by the designer to drive $\overline{CE_1}$, $CE_2$ and IO/$\overline{M}$. Table 30.21 shows two possible sets of connections, and the resulting addresses for memory and the input/output ports.

For the memory mapped I/O, IO/ $\overline{M}$ is selected by an address line ($A_{13}$) and I/O access is via memory reference instructions such as LDA or MOV *M*, *B*. The input/output ports thus appear to the programmer as memory locations and must have addresses that are not duplicated by any other memory device.

**Table 30.21** 8755 Addressing

| | 8085 outputs | 8755 inputs | Value to select memory | Value to select I/O |
|---|---|---|---|---|
| **Memory mapped I/O** | $A_{11} \rightarrow$ | $\overline{CE}_1$ | 0 | 0 |
| Memory address : | | | | |
| $4096 - 6143_{10} = 1000 - 17FF$ | $A_{12} \rightarrow$ | $CE_2$ | 1 | 1 |
| I/O address : | | | | |
| $12388 - 12391_{10} = 3000 - 3003_{16}$ | $A_{13} \rightarrow$ | $IO/\overline{M}$ | 0 | 1 |
| **Standard I/O** | $A_{11} \rightarrow$ | $\overline{CE}_1$ | 0 | 0 |
| Memory address : | | | | |
| $4096 - 6143_{10} = 1000 - 17FF_{16}$ | $A_{12} \rightarrow$ | $CE_2$ | 1 | 1 |
| I/O address : | | | | |
| $08 - 12_{10} = 08 - 0B_{16}$ | $IO/\overline{M} \rightarrow$ | $IO/\overline{M}$ | 0 | 1 |

For the standard I/O, the device is being used in the way the pin names suggest, with the I/O function being selected when the IO/ $\overline{M}$ signal is high. In use, the address and the signals on $\overline{CE}_1$, $CE_2$ and IO/$\overline{M}$ are latched in on the falling edge of ALE, and if the set of signals demands a read from memory, the contents of the ROM location that has been addressed appear on the AD bus.

## 8755 Input/Output Section

There are two 8-bit ports (*A* and *B*) on each 8755, and each line in each port may be programmed to act as an input or output.

Before use, the function of a port should be defined by writing to its Data Direction Register (DDR), with an 8-bit word, a '0' defining a line as an input, and a '1' as an output. The DDRs and ports are referred to by a programmer using the same instructions, but with the DDR having an address equal to (port address + 2).

**Table 30.22** Port and DDR Address on 8755

| $A_1$ | $A_0$ | |
|---|---|---|
| 0 | 0 | Port A |
| 0 | 1 | Port B |
| 1 | 0 | DDR A |
| 1 | 1 | DDR B |

If the chip is using memory mapped I/O as in Table 30.21, then a set of instructions to output the contents of location 2000 via port *A* could be as shown below.

| Hex | Memory mapped I/O instructions | |
|---|---|---|
| | **Mnemonic** | **Comments** |
| 3E FF | MVI FF H | Put all 1s in register *A* |
| 32 02 30 | STA 3002 | Make port *A* to be output |
| 30 00 20 | LDA 2000 | Load register *A* with contents of 2000 (= input data). |
| 30 00 30 | STA 3000 | Put data into output port *A* |

The data remains valid on the output lines from the 8755 until it is over-written, or a RESET occurs.

| ***Standard I/O instructions*** |
|---|
| The form of the IN and OUT instructions is,<br>IN = DB ; OUT = D3<br>address xx address xx |

The effect of these instructions is to transfer data between register *A* and the addressed port or DDR. The address appears on the $AD_{0-7}$ bus and is *repeated* on the $A_{9-15}$ bus. The five most significant bits are used for selecting the right chip, and the remaining three ($A_2$, $A_1$ and $A_0$) are available for defining the port or DDR. On the 8755, only $A_1$ and $A_0$ are used in this way, and $A_2$ need not be defined, but in the 8155. RAM all 3 lines are effective.

A program sequence follows to define port *A* as an 8-bit output, and send the contents of location 2000 to it, for an 8755 connected as standard I/O. The addresses of DDRA and PORT *A* in binary form are,

| | | |
|---|---|---|
| 0 0 0 0 1 | 0 1 0 = 0A H | DDRA |
| 0 0 0 0 1 | 0 0 0 = 80 H | PORT A |
| $A_{15} \cdots A_{11}$ | $A_2 \cdots A_0$ | |

| ***Hex*** | ***Mnemonic*** | ***Comments*** |
|---|---|---|
| 3E FF | MVI FF H | Put all 1s in register A |
| D3 0A | OUT 0A | Program port A as output on chip selected by 00001 |
| 3A 00 20 | LDA 2000 | Put data into register A from 2000 |
| D3 08 | OUT 80 | Output data to selected port A |

**8355 (*ROM with I/O 2048 × 8 bit ROM*)**

The 8355 is a mask-programmed ROM for use with the 8085 microprocessor, which is fully compatible with the 8755 EPROM.

**8155/8156 (*256 × 8 bits static RAM for use with 8085 microprocessor*)**

The 8155 and 8156 differ only in that the 8155 is enabled when CE is low, while the 8156 is enabled when CE is high.

The decoding of the multiplexed AD bus is performed by the chip so it is another of the family of chips which may be connected directly to the 8085. There are three input/output ports which may be configured by program instructions to act as normal inputs or outputs, or for one port to act as a handshake, controlling data transfer on the other two.

There is a 14-bit counter which can count external pulses. The counter may be preset under program control so that simple counting or timing does not need to occupy the full microcomputer system.

**8112 (*8-bit latch with 3 state output buffer*)**

This device has many uses. One worth special mention is that of latching the low order address bits $A_{0-7}$, so that memory devices without such a facility may be used with the 8085.

**8205 (*1 out of 8 binary decoder*)**

On the 8205 there are three address lines which select eight output lines. One use for the 8205 is to generate chip-select signals for up to eight memory devices.

Fuller details of the devices discussed in this section and of other devices may be found in the *MCS-85 User's Manual*, published by Intel Corporation.

## 30.31 9085-BASED MICROCOMPUTER DESIGNS

Two design routes are possible for microcomputers based on the 8085. Either the special components may be used which are directly compatible with the 8085 signals and bus structure, or the 8085 signals and bus may be adapted to suit the needs of standard components. An example of each kind of design is included in this section.

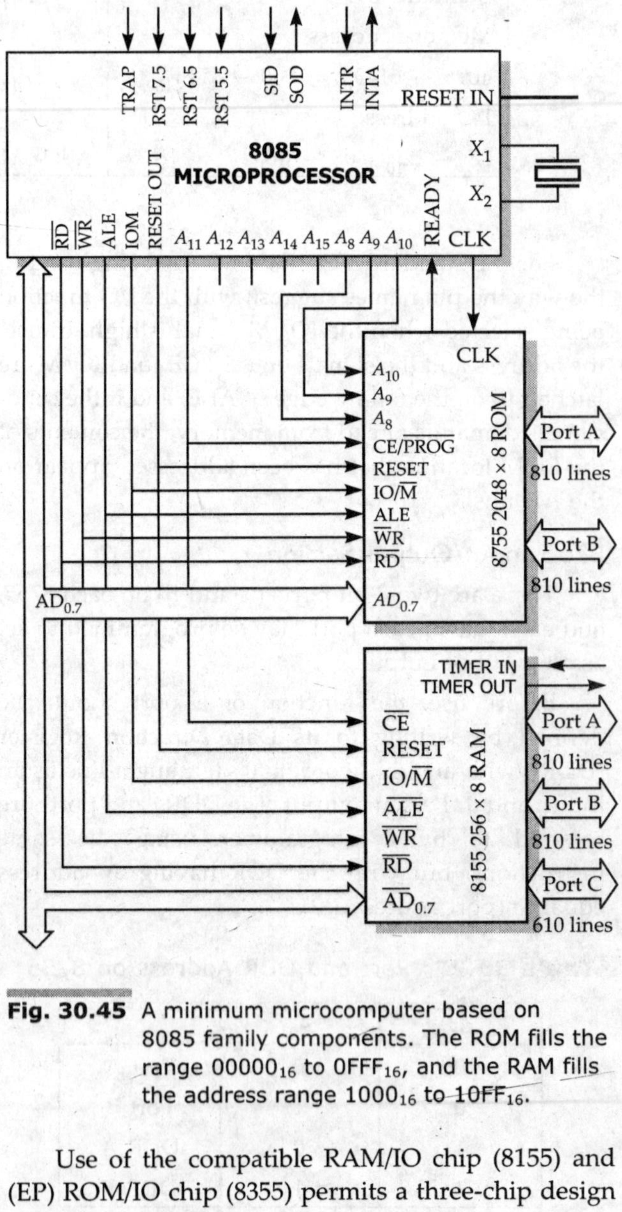


**Fig. 30.45** A minimum microcomputer based on 8085 family components. The ROM fills the range $00000_{16}$ to $0FFF_{16}$, and the RAM fills the address range $1000_{16}$ to $10FF_{16}$.

Use of the compatible RAM/IO chip (8155) and (EP) ROM/IO chip (8355) permits a three-chip design (Fig. 30.35) which includes :

| | |
|---|---|
| 2048 | 8-bit words of ROM for program storage |
| 256 | 8-bit words of RAM for stack and storage of temporary results |

| | |
|---|---|
| 4× | 8-bit ports, programmable as input or output |
| 1× | 6-bit port, programmable as input or output or handshake |
| 1 | 14-bit counter, controllable and readable from the 8085 |
| 1 | serial input line |
| 1 | serial output line |
| 5 | interrupt lines |
| 46 | separate I/O lines |

The three integrated circuits, a clock crystal and the pull-up resistors that are required occupy a circuit board about half the size of one of the pages of this book.

The low 8 bits of the address put out by the 8085 are latched by the two memory components on the ALE signal.

The $IO/\overline{M}$ input on each memory component is connected to the $IO/\overline{M}$ output of the 8085. This causes the ports to be addressed only when the $IO/\overline{M}$ signal is high, which only happens on IN and OUT instructions. The ports are thus addressed in the I/O address space, not the memory address space.

The 8355 is selected by supplying + 5 V, to the *CE* input and 0 to the $PROG/\overline{CE}$ input. In Fig. 30.44, this has been done by connecting *CE* to a permanent + 5 V and by connecting $PROG/\overline{CE}$ to *A*11. Thus the 8355 responds to address $0000_{16}$, when *A*11 is zero. This is required so that the system can start operating after each RESET, which forces $0000_{16}$ on to the program counter.

The chip enable input of the 8155 is connected to *A*11 on the 8055, so the lowest range of addresses which refer to the 8155 are $1000_{16}$ to $10FF_{16}$.

The ports on the 8155 and 8355 are selected when $IO/\overline{M}$ is high and a suitable address is applied to *CE* or $\overline{CE}$. For the circuit in Fig. 30.46, the port address will be

| | | |
|---|---|---|
| 8355 ROM | Port A | $00_{16}$ |
| | Port B | $01_{16}$ |
| | Data Direction Register *A* | $02_{16}$ |
| | Data Direction Register *B* | $03_{16}$ |
| 8155 RAM | Command/Status Register | $80_{16}$ |
| | Port A | $81_{16}$ |
| | Port B | $82_{16}$ |
| | Port C | $83_{16}$ |
| | Timer low byte | $84_{16}$ |
| | Timer high byte | $85_{16}$ |

## The Expanded System

The 8085 generates a 16-bit address which can distinguish 65354 different locations. To make a large memory using standard memory components there are three tasks that need doing. These are :

1. Convert the multiplexed data and address buses to separate 8-bit data and 16-bit address buses. This necessity is peculiar to the 8085 design.
2. Provide a greater current source capability to drive a large number of components simultaneously. An expanded microcomputer based on any microprocessor will require this provision of extra drive.
3. Decode combinations of the more significant address lines to generate mutually exclusive chip selection signals, once again, this feature is needed in all microcomputers.

All the three functions are included in Fig. 30.46. The less significant addresses are captured and fed on

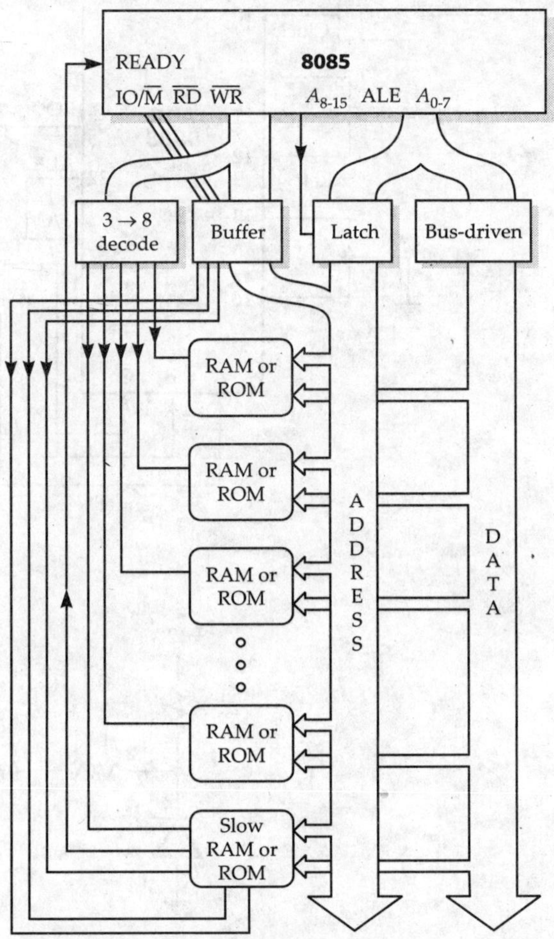


**Fig. 30.46** An expanded 8085 system, with maximum memory capability.

30

to an address bus by the 8212 latch, which takes ALE as the signal to latch its inputs. It also acts as a bus driver. Data on $AD_{0-7}$ passes through the bi-directional buffer, which provides sufficient source current to supply all the components that are connected to the address bus.

The signals needed to control the bi-directional buffer are $\overline{RD}$ and $\overline{WR}$ from the 8085. The upper byte of the address is split into two pans. The less significant bits combine with $A_{0-7}$ and are connected to the address inputs of the components. The more significant bits are the logical inputs to decoders which provide outputs to select one memory block at a time. The arrangement in Fig. 30.47 has the three most significant address bits providing inputs to a 3-to-8 line decoder, so that eight blocks of memory may be distinguished. The remaining five bits of the address only need a one-way buffer to provide sufficient drive current. More distinct chip-enable signals can be generated by the use of further 8205 decoders.

The structure of the data and address signals is now suitable for standard random access memory components (both RAM and ROM). Figure 30.44 shows such a system. Notice that there has been much less provision for I/O lines in this large system than in the smaller system in Fig. 30.47.

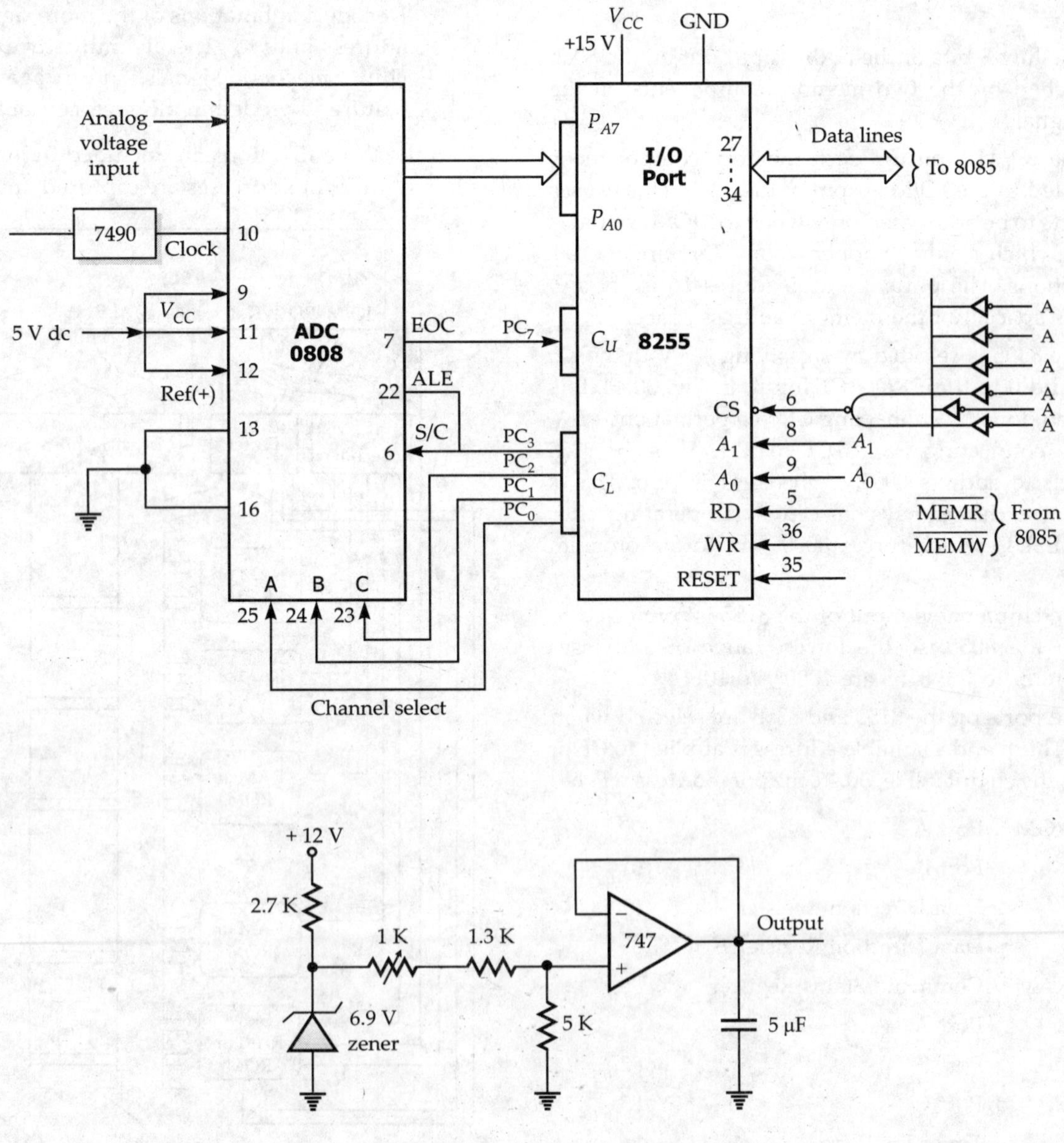


**Fig. 30.47** (*a*) Interfacing of ADC 0808 with Intel 8085. (*b*) Circuit for $V_{ref}$.

## 30.32 MICROPROCESSOR BASED SYSTEMS

Microprocessor-based systems are suitable for dedicated applications in industries, instrumentation etc. and being very small and compact can form a part of the equipment to be controlled. For the measurement and control of physical quantities transducers are used to convert physical quantities into electrical voltage. The electrical voltages are proportional to physical quantities like speed, pressure, displacement, temperature etc. The electrical voltage is an analog quantity. Before it is applied to the microprocessor, it is converted into digital form by A/D converter. If more than one quantity is to be processed multiplexers are used. If a.c. voltage is used, sample and hold circuits are used to keep the desired instantaneous voltage constant during conversion. Digital output obtained is fed to the microprocessor through the interfacing device. In a microprocessor based system the designer has to select suitable input/output devices to interface them to the microprocessor. The selected input/output devices should be compatible with microprocessor. If a particular device is not compatible additional electronic circuits have to be designed through which the device may be interfaced to the CPU.

### 30.32.1 Measurement of Voltage

An A/D converter is used for the measurement of voltage. The voltage is applied to the input terminal of the A/D converter. The output of the A/D converter is digital quantity equivalent to the input voltage. The output terminals of the A/D converter are connected to the I/O port. The microprocessor reads I/O port and the voltage in digital form is transferred to the accumulator. The digital voltage obtained can be calibrated either by computation or using look-up table. Figure 30.47(*a*) shows interfacing of ADC 0808 to inter 8085 microprocessor. An analog input is connected to IN3. 5V d.c. applied to pin 12 *i.e.*, Ref ( + ) should be accurate. A circuit shown in Fig. 30.47 (*b*) can be used to obtain 5 V d.c. ($V_{ref}$).

The assembly language program is as follows :

| Label | Mnemonics | Operands | | Comments |
|---|---|---|---|---|
| | MVI | A, Cntr word | ; | Initialize I/O ports of 8255 |
| | OUT | CR | | |
| | MVI | A, 03 | ; | Switch on multiplexer Channel IN3 |
| | OUT | Port C | | |
| | MVI | A, OB | ; | Start of conversion pulse |
| | OUT | Port C | | |
| | MVI | A, 03 | | |
| | OUT | Port C | | |
| READ : | IN | Port C | ; | Read of end of conversion signal |
| | RAL | | | |
| | JNC | READ | ; | Is conversion over ? No. jump to READ |
| | IN | PORT A | ; | Read digital output of A/D converter |
| | STA | RESULT | ; | Store the result at memory location RESULT |
| | HLT | | | |

#### Port addresses

$\overline{CS}$, $A_0$ AND $A_1$ are device select signals. $\overline{CS}$ is connected to decoded address lines $A_0$ and $A_1$ respectively. The $\overline{CS}$ signal is master chip select and $A_0$ and $A_1$ specify one of the I/O ports or the control register (*CR*) as given below :

| $\overline{CS}$ | $A_1$ | $A_0$ | Selected |
|---|---|---|---|
| 0 | 0 | 0 | PORT A |
| 0 | 0 | 1 | PORT B |
| 0 | 1 | 0 | PORT C |
| 0 | 1 | 1 | CR (Control Register) |

30

The $\overline{CS}$ line goes low when $A_0$ to $A_7$ are at logic 0. When these signals are combined with $A_0$ and $A_1$, the port address ranges from 00 H to 03 H as shown in Fig. 30.47(*a*).

**Control word**

Control word (Contr word) specifies an I/O function for each port. Control word is always written into control register to initialize the ports.

| $D_7$ | $D_6$ | $D_5$ | $D_4$ | $D_3$ | $D_2$ | $D_1$ | $D_0$ |
|---|---|---|---|---|---|---|---|
| 1 | 0 | 0 | 1 | 1 | 0 | 0 | 0 = 98H |
| I/O Function | Port A in Mode O | | Port A as Input Port (data) | Port Cu as Input Port (EOC) | Port B in Mode O | Port B = Output | Port $C_L$ Output (SOC and to select the channel) |

The logic to switch on IN3 is as follows :

| C | B | A | Address lines |
|---|---|---|---|
| $PC_2$ | $PC_1$ | $PC_0$ | Port pints |
| 0 | 1 | 1 | = 03 H |

To send the start of conversion (SOC) pulse first $PC_3$ (of Port C) is made high and then low without disturbing the multiplexer channel IN3. The logic is as follows :

| $PC_7$ $PC_6$ $PC_5$ $PC_4$ | $PC_3$ | $PC_2$ | $PC_1$ | $PC_0$ | |
|---|---|---|---|---|---|
| Upper C port | 1 | 0 | 1 | 1 | = OB.H. |
| | for SOC | for IN3 | | | |

Again to give low signal :

| $PC_7$ $PC_6$ $PC_5$ $PC_4$ | $PC_3$ | $PC_2$ | $PC_1$ | $PC_0$ | |
|---|---|---|---|---|---|
| Upper C port | 0 | 0 | 1 | 1 | = 03 H |
| | | for SOC | | | for IN3 |

### 30.30.2 Interfacing of Stepper Motor

General inter-connections for a stepper motor are shown in Fig 30.48. A 12 V a.c. supply is used to energise the poles. Pulses sent by the microprocessor switch on rated voltage to the windings of the desired poles. A delay subroutine is used in the program. After energizing one set of pole windings some delay is provided then the power supply is switched on to the other set of pole windings. This delay time governs the speed of the motor.

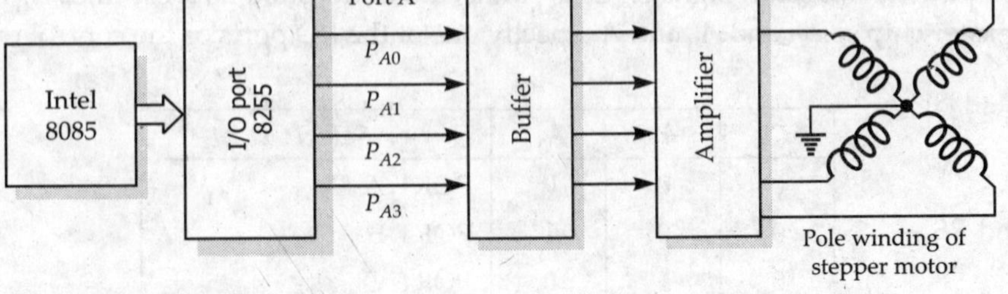


**Fig. 30.48** Interfacing of stepper motor.

The program is given below to rotate the motor in the clockwise direction through 90°.

```
Label       Mnemonics   Comments
            MVI         E, COUNT     ;   load no. of steps of motor
            MVI         A, CW        ;   load control word
            OUT         CR           ;   out to control register
LOOP# 2:    MVI         D, 04 H      ;   NO. OF POLES
            LXI         H, CODE      ;   set pointer to get the pole status word
LOOP# 3:    MOV         A, M
            OUT         PORT A
            CALL        DELAY        ;   Introduce required delay
            INX         H
            DCR         D
            JNZ         LOOP # 3
            DCR         E
            JNZ         LOOP # 2
            HLT
CODE        DB          FAH          ;   sequence for clockwise motor rotation
            DB          F 6 H
            DB          F 5 H
            DB          F 9 H
```

## 30.33 BUS INTERFACE STANDARDS

The micro-computer is a bus-oriented system where by subsystems or peripherals are interconnected through the bus architecture. The design approach should be such that systems are modular, expandable and multipurpose. The user should be able to select a peripheral from any manufacturer and plug it into the system. Similarly, the user should be able to print out programs as well as collect data from various instruments. To design microcomputers with such features, a common understanding of equipment specifications among manufacturers is needed. This is known as defining standards. In the field of electronics, these standards are generally defined by professional organizations such as IEEE (Institute of Electrical and Electronics Engineers) ; EIA (Electronic Industries Association) or the standards are forced upon the industry by common practice. The need for expandability and modularity gave rise to various bus standards as listed in Table. The bus GPIB and RS 232 described here.

| Buses | I/O Mode | Description |
|---|---|---|
| 1. S – 100 (IEEE — 696) | Parallel | To interconnect various components within the micro-computer. It has 100 signals. |
| 2. STD Bus | Parallel | This is a bus competing with S – 100. It has 56 signals. |
| 3. Multibus (IEEE — 796) | Parallel | Interface between modules with the master/slave type communication. It has 86 signals. |
| 4. GPIB (IEEE – 488) | Parallel | Interface between micro-computer and measurement equipment such as a voltmeter. It has 24 signals. |
| 5. RS — 232 C | Serial | Interface between the micro-computer and serial peripherals. |
| 6. RS — 422 and RS — 423 | Serial | High speed serial communication for distances longer than 20 meter. |
| 7. Current loop | Serial | |

30

### 30.33.1 GPIB Interface Bus (IEEE 488)

This bus was developed by Hewlett-Packard. It facilitates interfacing of programmable instruments. Later it was accepted as the IEEE 488 interface standard. The bus standard is also known as the General Purpose Interface Bus (GPIB) or the Hewlett-Packard Interface Bus (HP-IB).

Some of the features of the bus are given below :

1. Data transfer among the interconnected devices is digital.
2. Fifteen devices may be connected to one continuous bus.
3. Total transmission path is limited to twenty metre or two metre per device.
4. Data rate on any signal line is limited to 1 M-byte/s.

#### Bus Signals

The bus has 24 signals ; eight bi-directional data lines, five general bus management lines, three handshake lines and eight grounds. Figure 30.49 shows four types of devices that can be connected to the bus. These devices are classified as follows :

**1. *Listener*** is a device capable of receiving data when addressed. Fourteen devices can listen at a time.

**2. *Talker*** is a device capable of transmitting data when addressed. Only one device can be active at a time.

**3. *Listener/Talker*** is device that can receive as well as transmit data over the interface. A programmable digital voltmeter (DVM) is a listener/talker device.

**4. *Controller*** is a device that controls signals and specifies which device can talk and which device can listen.

#### RS-232 C

In serial I/O, data can be transmitted either as current or as voltage. Typically, 20 mA or 60 mA current loops are used in teletype equipment. When data are transmitted as voltage, the commonly used standard is known as RS-232 C. It is defined in reference to Data Terminal Equipment (DTE) and Data Communication Equipment (DCE)-terminal and modem as shown in Fig. 30.50(*a*).

Figure 30.50(*b*) shows the RS-232 C 25-pin connector and its signals.

The signals are divided into 4 groups :

1. data signals.
2. control signals.
3. timing signals.
4. grounds.

For data lines, the voltage level from +3 V to +15 V is defined as logic 0 and from –3 V to –15 V as logic 1. This is negative true logic. Because of incompatibility with TTL logic, voltage translators called line drivers and line receivers are required to interface TTL logic with the RS-232 signals. The minimum interface requires three lines : pins 2, 3 and 7 as shown in Fig. 30.50(*a*). These line are defined in relation to the DTE, the terminal transmits on 2 and receives on pin 3. On the other hand, the DCE transmits an pin 3 and receives on pin 2. Typically, data transmission with a handshake requires eight lines. Specific function on handshake lines differ in different peripherals and therefore, should be referred to in the manufacturer's manuals.

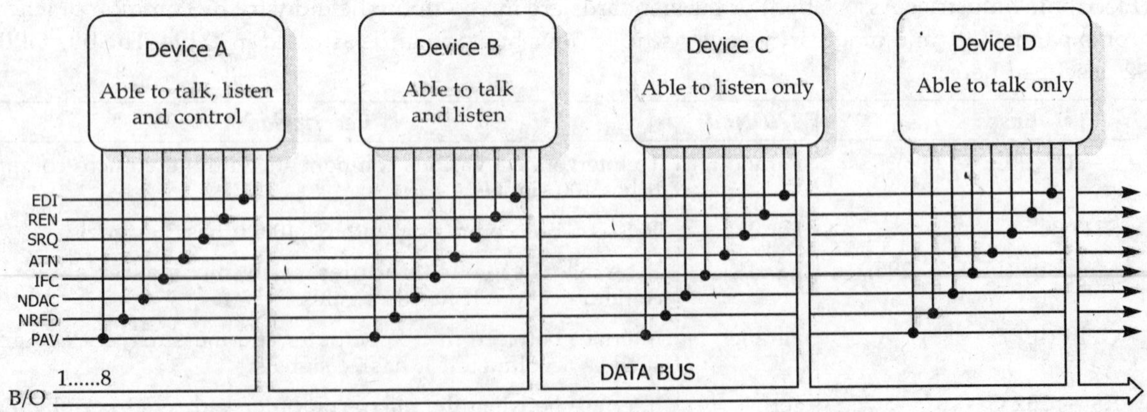


EOI, REN, SRQ, ATN, IFC signals are for General Bus Management
NDAC, NRFD and DAV are handshake lines.

**Fig. 30.49** GPIB Bus Interface.

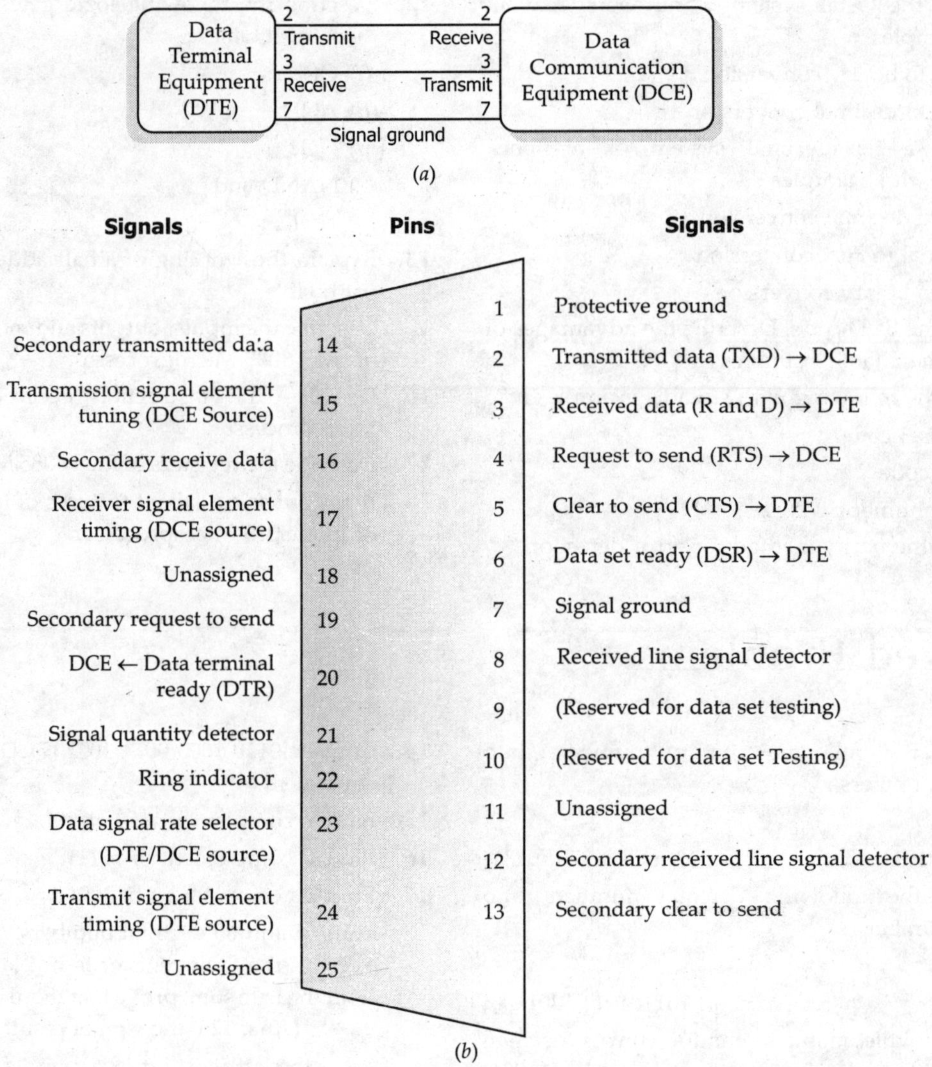


**Figure 30.50** (*a*) Minimum configuration of control signals between DTE and DCE.
(*b*) RS-232 C Signal Definitions and Pin assignments.

# Review Questions

1. Compare the analog and digital modes of control as regards the following :
   (*i*) stability and accuracy of control
   (*ii*) cost per control function
   (*iii*) flexibility
   (*iv*) reliability and
   (*v*) human factors.
2. Describe the general structure of a number system.
3. Describe the binary number system. Draw tables for conversion of decimal numbers from 0 to 9 to binary system.
4. Describe a method of binary to decimal conversion. Cite an example.
5. Describe the two methods used for conversion of decimal to binary numbers. Cite examples.
6. Describe with examples the following arithmetic processes in a binary system :
   (*i*) Addition
   (*ii*) Subtraction
   (*iii*) Multiplication and
   (*iv*) Division.

7. Describe the Octal system of numbers. Explain with examples
   (*i*) Octal to binary conversion
   (*ii*) Octal to decimal conversion.
8. Describe a hexadecimal system of numbers. Describe with examples :
   (*i*) Hex to decimal conversion
   (*ii*) Decimal to Hex conversion
   (iii) Hex to binary conversion.
9. Explain the BCD code. Describe the advantages of BCD with suitable examples.
10. Explain the following with suitable examples :
    (*i*) Excess-3 code
    (*ii*) Gray code
    (*iii*) Alphanumeric codes.
11. Describe the parity method for error detection.
12. Describe the following logic gates and draw up their truth tables :
    (*i*) OR
    (*ii*) AND
    (*iii*) NOT
    (*iv*) NAND and
    (*v*) NOR.
13. Explain the working of a half adder. Draw up its truth table.
15. Describe the inputs, outputs and memory locations of Intel 8085 Microprocessor.
16. Describe the instruction cycle of Intel 8085 Microprocessor.
17. Describe the registers of Intel 8085 Microprocessor.
18. Bring out some salient features of instructional set of Intel 8085 Microprocessor.

# Unsolved Problems

1. Convert the following binary numbers into decimal numbers :
   (*a*) 11011
   (*b*) 10110101 **[Ans.** (*a*) 27 (*b*) 181]
2. Convert the following decimal numbers into binary numbers :
   (*a*) 45,
   (*b*) 76 **[Ans.** (*a*) 101101 (*b*) 1001100]
3. Use the double dabble method to convert decimal number 25 into binary. **[Ans.** 11001]
4. Use the double dabble method to convert decimal number 37 into binary. **[Ans.** 100101]
5. Convert octal number 372 into decimal number. **[Ans.** 250]
6. Convert octal number 24.6 into decimal number. **[Ans.** 20.75]
7. Convert decimal number 266 into an octal number. **[Ans.** 412]
8. Convert octal number 5413 into binary number. **[Ans.** 101100011001]
9. Convert binary number 100111010 into an octal number. **[Ans.** 326]
10. Convert decimal number 423 to hex. **[Ans.** 1A7]
11. Convert decimal number 214 to hex. **[Ans.** D6]
12. Convert hex number BA6 into a binary number. **[Ans.** 101110100110]
13. Convert the binary number 101011111 into hex. **[Ans.** 15F]
14. Convert 0110 1000 0011 1001 (BCD) to its decimal equivalent. **[Ans.** 6839]
15. Find 1′ complement of 100110. **[Ans.** 011001]
16. Find 1′ complement of 11111. **[Ans.** 00000]
17. Find 2′s complement of 10011. **[Ans.** 01101]
18. Add 65 and 60 decimal numbers. Why is the sign +ve ? If the 8-bit register is not larger enough to hold the true sum, prove that there is no overflow.
    **[Ans.** 125, the sign of result is +ve because the left most bit is zero, no overflow occurs since the sum is less than 127]
19. Add 65 and 65 decimal numbers. Comment upon the result.
    **[Ans.** The sum is 130 which is greater than 127 therefore, there is an overflow. In most microprocessors, an overflow flag is set when this condition occurs. The flag is a bit in a special register called status register]
20. Convert the following decimal numbers into binary coded system (BCD-8-4-2-1) :
    (*i*) 8 **[Ans.** 0000 0000 0000 1000]
    (*ii*) 17 **[Ans.** C000 0000 0001 0111]
    (*iii*) 800 **[Ans.** 0000 1000 0000 0000]
    (*iv*) 9999 **[Ans.** 1001 1001 1001 1001]
21. Convert the following binary numbers into Gray code :
    (*i*) 10110 (*ii*) 10011101001 (*iii*) 10101101.
    **[Ans.** (*i*) 1101 (*ii*) 110100110101 (*iii*) 11111011001010]

**22.** Convert the following Gray numbers into binary numbers :

(*i*) 101110101 (*ii*) 11100100011

**[Ans.** (*i*) 110100110 (*ii*) 10111000010]

**23.** A system has three inputs : *A*, *B* and *C*. Show the truth table if its output is *Y*

$$Y = (A + B)(A + CC)$$

**[Ans.** *ABCY* are: 0000, 0010, 0111,1001, 1011, 1101, 111]

**24.** Work out the value for $C = \overline{B} + \overline{A}B$ for four possible conditions. **[Ans.** *ABC* are 000, 011, 101, 110]

**25.** Show that $\overline{ABCD} = \overline{A}.\overline{B}.\overline{C}.\overline{D}$.

**26.** Reduce $C = (A + B)(A + \overline{B})(\overline{A} + B)$ by using Boolean Algebra.

**27.** What is the Boolean expression for the output of the system shown in Fig. 30.51 ?

**[Ans.** $C = \overline{A + B}.B$]

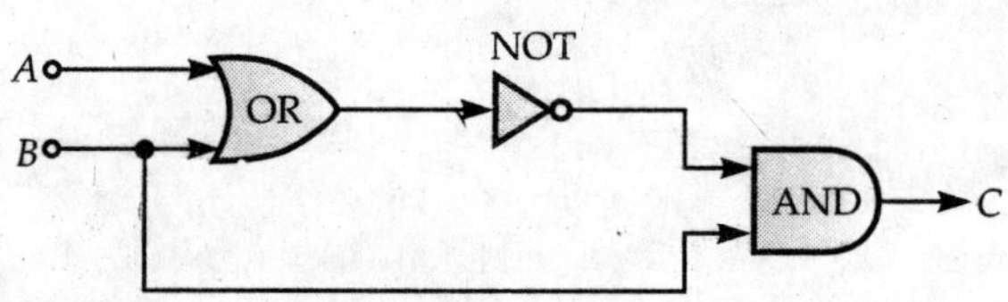


**Fig. 30.51** Diagram of Problem 30.27.

**28.** What is the Boolean expression for the output of circuit shown in Fig. 30.52. **[Ans.** $\overline{A} + B$]

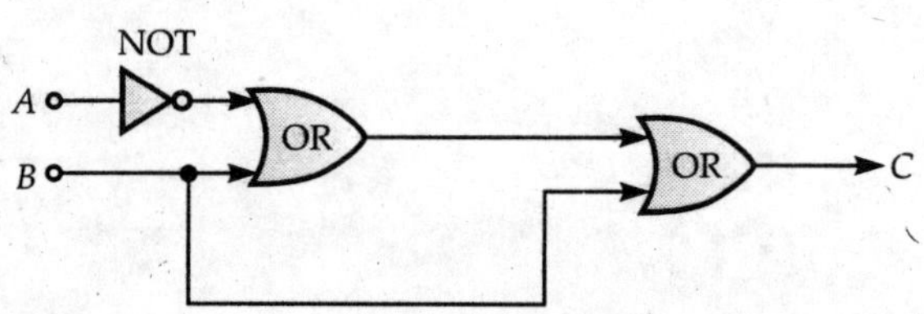


**Fig. 30.51** Diagram of Problem 30.28.

**29.** Write an assembly language program to add the contents of memory locations 2420 H and 2421 H. The result should be stored in memory location 2422 H and 2423 H.

**30.** Write a program starting from memory location 2400 H to subtract digital number 48 stored in memory location 2421H from decimal number 99 stored in memory location 2420 H. The answer should be in decimal numbers and stored in memory location 2422 H.

**[Ans.** 2422 H = 51]

**31.** A string of *n* numbers is stored in memory location starting from 2500 H. The length of the string (*i.e.*, *n*) is stored in memory location 2601 H. Write a program to arrange the string in the ascending order starting with location 2700 H.

**32.** Calculate the control word under Mode 0 for the following port configurations of PPI 8255 chip :

Port *A* — Input port

Port *B* — Output port

Port *C* (upper) — Input port

Port *C* (lower) — Output port. **[Ans.** 98 H]

**33.** A 3 MHz system clock is connected to the TIMER IN input of an 8156 chip. Write a program that divides this clock by a factor of 2500 and produces a train of pulses from the TIMER OUT pin.

**34.** Execute the following program according to Intel 8085 instructions :

```
L×1    H, 2000 H
MOV    B, M
       HLT
```

**35.** Execute the following program according to Intel 8085 instructions :

```
L×1    H, 2400 H
MVI    M, 08
       HLT
```

30

# 31
# Data Acquisition Systems

## 31.1 INSTRUMENTATION SYSTEMS

An instrumentation system is an aggregation or assembly of devices united by some form of regular interaction of an interdependence. It is a group of diverse units or devices so combined by nature or by an art to form an integral whole, and to function, operate or move in unison, and often in obedience to some form of control.

Data acquisition systems via analog signals are used in communications, electronic and medical applications. Conversion to digitized systems is widely used today because complex circuits are low cost, accurate, simple to implement. In addition, there is rapid growth of microprocessors and microcomputers to perform difficult measurement and control functions.

The demands for more and faster information from physical systems and devices used for space and nuclear research and development, and for industrial automation, have produced many new and effective instrumentation systems for data acquisition and reduction.

Data acquisition systems are used to measure and record analog signals in basically *two* different ways :

(*i*) Signals which originate from direct measurement of electrical quantities. These signals may be d.c. or a.c. voltages, frequency, or resistance, etc.

(*ii*) Signals which originate from use of transducers.

## 31.2 TYPES OF INSTRUMENTATION SYSTEMS

The instrumentation systems can be classified into *two* distinct categories :

▲ Analog Systems ▲ Digtial Systems

1. ***Analog systems.*** These systems deal with information in analog form. An analog signal may be defined as continuous function, such as a plot of voltage versus time or displacement versus force.

2. ***Digital systems.*** A digital quantity may consist of a number of discrete or discontinuous pulses whose time relationship contains information about the magnitude and the nature of the quantity under measurement.

## 31.3 COMPONENTS OF AN ANALOG DATA ACQUISITION SYSTEM

An analog data acquisition system typically consists of some or all of the following elements :

1. ***Transducers.*** Transducers have been described in details in Chapter 25. It is desirable that an emf, obtained from the transducer proportional to the quantity being measured, is used as an input to the data acquisition system. Therefore, transducers such as thermocouples, strain gauge bridges, piezo-electric devices and photosensitive devices are used.

As an exception to this usual function of transducers, some sensors produce a frequency which can be counted with an electronic counter to obtain the integral of the measured quantity or it can modulate a telemetring transmitter, and can be reduced to a voltage with the help of a discriminator.

2. ***Signal conditioning equipment.*** Signal conditioning equipment includes any equipment that assists in transforming the output of transducer to the desired magnitude or form required by the next stage of the data acquisition system. It also produces the required conditions in the transducers so that they work properly.

Signal conditioners may include devices for amplifying, refining, or selecting certain positions of these signals.

Examples of signal conditioning equipment include known constant voltage sources for strain gauge bridges, zero bridge balance devices for strain gauge circuits, temperature control devices for thermocouple junctions, voltage amplifiers and servo-systems.

3. ***Multiplexer.*** Multiplexing is the process of sharing a single channel with more than one output. Thus a **multiplexer** accepts multiple analog inputs and connects them sequentially to one measuring input.

Multiplexing is a means of using the same transmission channel for transmitting more than one quantity. Multiplexing becomes necessary in measurement systems when the distance between transmitting and receiving point is large and many quantities are to be transmitted. If a separate channel is used for each quantity, the cost of installation, maintenance, and periodic replacement becomes prohibitively large and therefore a single channel is used which is shared by the various quantities.

Multiplexing is commonly accomplished by either time or frequency sharing of the transmission channel between the individual quantities.

4. ***Calibrating equipment.*** Before each test there is a pre-calibration, and often after each test there is a post-calibration. This usually consists of a millivolt calibration of all input circuits and shunt calibration of all bridge-type transducer circuits.

5. ***Integrating equipment.*** It is often desirable to know the integral or summation of a quantity. There are several ways of determining the time integral of a quantity. An analog integrating circuit can be used for a qualitative test. It has the possible danger of becoming overloaded and also its accuracy is low. Therefore, digital techniques are normally used for integration purposes.

Another method of effective summation, utilizes the emf from the conditioning equipment to be applied to a saturable core reactor in the resonant circuit of an oscillator. The change in reactance is proportional to the applied emf and produces a change in the frequency. This frequency is beat with a frequency equal to frequency of the oscillator when the input is zero. The beat frequency, which is directly proportional to the input voltage of the oscillator, is amplified and counted with a counter.

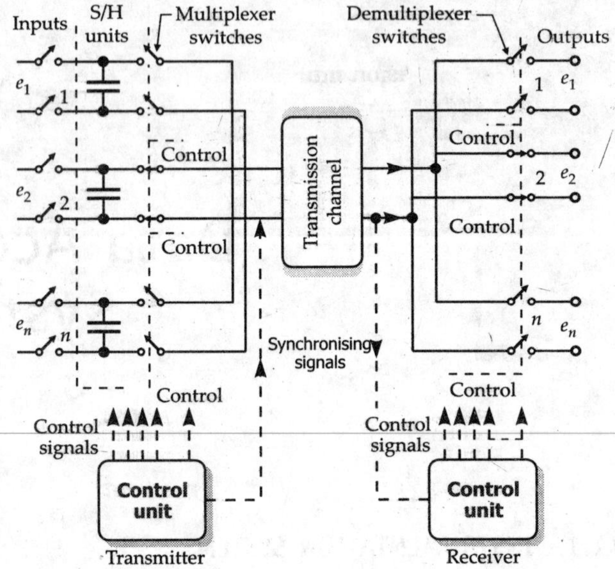


**Fig. 31.1** Time Division Multiplexing.

6. ***Visual display devices.*** Visual display devices are required for continuous monitoring of the input signals. These devices include panel mounted meters, numerical displays, single or multichannel CROs and storage type CROs, etc.

7. ***Analog recorders.*** The methods of recording data in analog form have been discussed in details in Chapter 28. Analog recorders include strip chart recorders, Duddell's oscillograph, magnetic tape recorders and cathode ray oscilloscopes with photographic equipment.

8. ***Analog computers.*** The function of a data acquisition system is not only to record data acquired by the transducers and the sensors but also to reduce this data to the desired form. An analog computer may be used as a data reduction device. The output voltage of an analog computer can either be recorded in analog form or be converted to a digital form for recording and further computations.

It should be noted that analog computations are usually less accurate than digital computations, yet it is generally more economical to compute data that are acceptable by analog methods using analog computers.

9. ***High speed cameras and TV equipment.*** In many industrial processes, engine testing, and aerodynamic testing it is not possible for the test operator to have a view of the equipment being tested. Therefore, closed circuit TV is used to enable the operator to make visual observations of the test. Also high speed cameras are employed to obtain a complete visual record of the process for further analysis.

### 31.3.1 Types of Multiplexing Systems

There are *two* methods of multiplexing :

(*i*) Time division multiplexing (TDM), and

(*ii*) Frequency division multiplexing (FDM).

#### 1. Time Division Multiplexing

In time division multiplexing the information from different measuring points is transmitted serially one after another on the same communication channel. The measurements are sampled sequentially and transmitted until all the quantities have been sampled. After this the cycle is again repeated. The time to complete transmission of all the measurements may be sufficiently small so that one quantity cannot change considerably during the interval between sampling. The order of transmission is controlled either cyclically by means of pulse generators or by command signals issued corresponding to the freely programmable requirements of the measurement system.

Multiplexers are used for switching of the measuring channels to the relevant transmission channel as shown in Fig. 31.2. The diagram shows a simple example of a system used for multiplexing of analog signals. A number of measuring points which monitor physical quantities like temperature, pressure, etc. are used as transmitters. They are shown as 1, 2, 3 ,...... *n*. Each point is sequentially connected to its corresponding receiver. The system connects transmitter 1 to receiver 1 for a specified interval of time. Then transmitter 2 is connected to receiver 2 and so on.

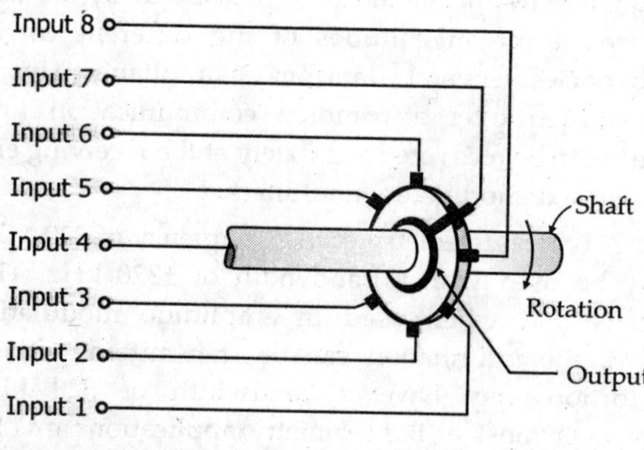


**Fig. 31.2** A time division multiplexing system using a machine commutator.

In order to ensure that the rotary scanning switches do not get out of step because of power failure in the transmission channel, the two rotary switches must be synchronized. This is done by additional circuitry that stops the transmitter switch on the synchronization point until the receiver switch also is at its synchronization point.

***Time division multiplexing*** involves the transmission of data samples rather than continuous data transmission which is typical of frequency division. The time division multiplexing is especially useful when telemetering slow-changing, low bandwidth data. An example of this is telemetering temperature. It is enough to transmit temperature data at a slow rate-say at an interval of 10 s. It will be extremely wasteful if the temperature measurements occupied a full radio frequency (RF) communication link. In order to increase efficiency, the communication link should be used for telemetering other process variables during the 'dead' time (*i.e.*, the time between two consecutive temperature measurements). Thus the efficient use of the communication system may be achieved by time sharing the communication channel among a number of measured parameters each being transmitted at a rate compatible with the rate of change of parameter. This time sharing simply assigns time slots for specific measurands. For example, in our case, during the time interval of 10 s a number of different data may be transmitted thereby utilizing the communication link fully. The different measurands are transmitted via a single link, one after another in a sequence, during a specified time slot. The receiving station must be capable of unscrambling the flow of incoming data and producing data lines corresponding to original measurands. It is essential to keep track of time, or identify each time slot, so that the receiving end recognises each data source.

***Time division equipment.*** An example of a time-division multiplexing system is shown in Fig. 31.2. It is a commutator with only one bar which successively connects the output of the device to one of the eight inputs. If the output of the commutator were connected to an instrumentation system, the system would read each input for a little less than 1/8 of 1 revolution. Thus the instrumentation system shares the time of each revolution with the eight inputs. This system is limited to the revolutions per minute at which good low resistance contact on the commutator can be maintained. The system can be designed for any number of inputs.

A block diagram of an electronic time-division multiplexing system is shown in Fig. 31.3. There are eight input circuits which are successively connected to the output circuit through a diode bridge. When junction '$a$' is positive with respect to junction $b$, the diodes in the bridge have infinite resistance, and the circuit between junctions $c$ and $d$ is effectively an open circuit. When junction $b$ is positive with respect to point $a$, however, the diodes have very low resistance and the bridge becomes a balanced bridge circuit with very low resistance arms, and the potentials of the junctions $c$ and $d$ are the same, then the voltage at $c$ is equal to the input voltage at $d$, or the circuit from $c$ to $d$ is a closed circuit.

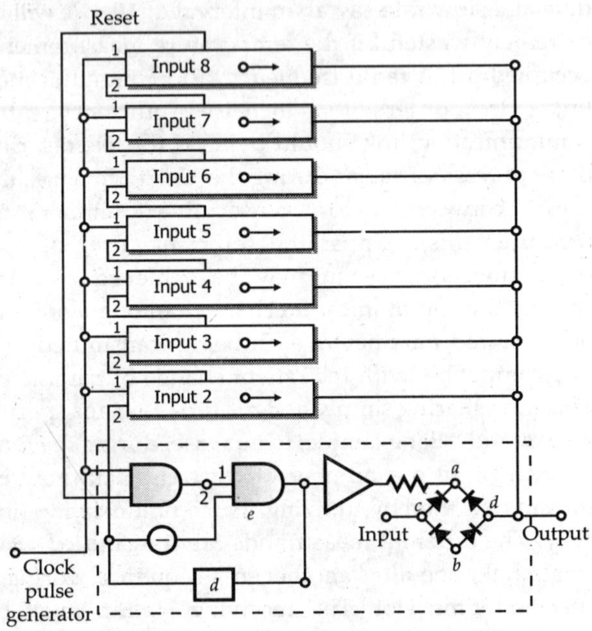


**Fig. 31.3** An electronic time division multiplexing system.

Now the switching circuit that controls the potential at points $a$ and $b$ is operated by the coincident application of a clock pulse and a reset signal at the first and second inputs, respectively, of the first AND circuit of the control system.

The clock pulse is of very short duration and has a magnitude of +15 V. During the remainder or the time, it is –15 V. The reset signal is generated by the operation of the preceding switching circuit. This reset signal is applied to the next switching circuit before the clock pulse is generated and it continues for a good part of the cycle after the clock pulse. It is +15 V during this part of the cycle and remains at –15 V during the rest of the cycle. These two pulses produce a short +15 V pulse at the output of the first AND circuit which is applied to the first input of the second AND circuit in this control system. The second input to the second AND circuit was already +15 V. These two inputs produce +15 V at the output junction $e$ of the second AND circuit which is fed back to the first input of the second AND circuit. This locks the second AND circuit in the 1 position and keeps the output of the second AND circuit at junction e at +15 V as long as the second input to the second AND circuit is +15 V. The second input to the second AND circuit is a feedback circuit from the junction $e$ through a time-delay circuit and a NOT circuit. The output of the time-delay circuit is also applied to the reset of the next switch to be operated. The +15 V pulse from junction $e$ will arrive at the reset of the next switch to be operated just before the next clock pulse, and the switch will operate at the arrival of the next clock pulse. At the same time the voltage through the NOT circuit will shut off the switch that has been in operation, and the potential of point $e$ will become –15 V and will remain at this value until the two inputs to the first AND circuit of this switch are both +15 V.

When junction $e$ is +15 V, junction $b$ is +15 V and junction $a$, which is supplied from the junction $e$ through an inverter, is –15 V. This makes the voltage at $d$ equal to the voltage at $c$, and the input voltage 1 is applied to the output of the switch. When the junction $e$ is –15 V, the reverse hence reset the circuit. The switches used are electronic switches and are activated by digital logic levels.

## 2. Frequency Division Multiplexing (FDM)

Several information can be simultaneously transmitted using different carrier frequencies and employing modulation techniques described earlier in Chapter 27. Each of these methods makes use of a specific frequency band for the modulation of the frequency by the signal to be measured. By suitable choice of the magnitudes of the different carrier frequencies, several data may simultaneously be transmitted over a common communication link. These can be recovered separately at the receiving end by using demodulators and filters.

A typical telemetry carrier frequency of 230 MHz may be used with a bandwidth of ±320 kHz. This means that when used in amplitude modulation (AM), the information can be transmitted without distortion may have a bandwidth of 320 kHz. However most of the telemetry applications involve information (data) of much narrower bandwidth. For the purposes of illustration let us assume a data bandwidth of 4 kHz. Instead of directly modulating the 230 MHz carrier with the data, we can first modulate a subcarrier of say, 32 kHz. Assuming amplitude modulation (AM), this carrier will give a signal of 32 ± 4 kHz. The modulated subcarrier may now be used to modulate the 230 MHz carrier. Figure 27.31 on page 985 shows the frequency bands utilized in this transmission. The rest of the ±320 kHz band is unused.

It is then possible to modulate another sub carrier-say of 44 kHz — by another data source of similar bandwidth. Thus a 44 ± 4 kHz modulated subcarrier is obtained which is shown by dotted lines in Fig. 31.4. It is evident that we can fill the 320 kHz allowed bandwidth for a large number of subcarriers carrying data from a large number of sources. In this

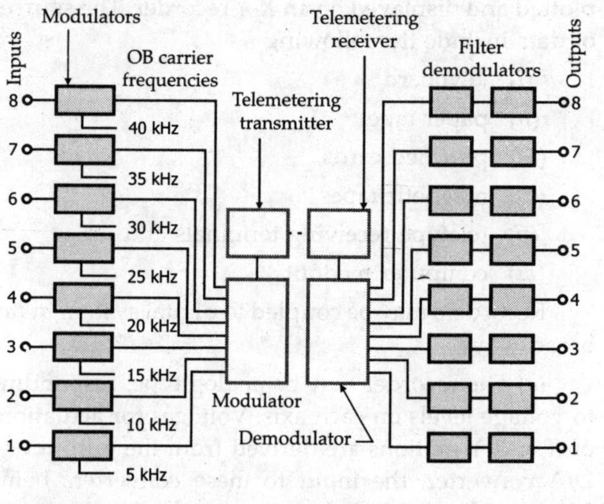


**Fig. 31.4** Block diagram of an AM frequency-division multiplexing system.

example, the bandwidth of data has been taken as 4 kHz arbitrarily. It should be noted that an unused band of 4 kHz separates the modulated subcarriers. This is because 32 ± 4 kHz gives 28 to 36 kHz, and 44 ± 4 kHz gives 40 to 48 kHz. Thus there is 4 kHz separation between the highest frequency of lower subcarrier (36 kHz) and lowest frequency of the higher subcarrier (40 kHz). This separation is necessary to avoid interchannel interference and to allow separation of subcarriers at the receiving end. An improvement by a factor of 2 is obtained by using the **Single Sideband (SSB) Transmission.** The modulated subcarrier signal consists 32 to 36 kHz upper sideband transmission or 28 to 32 kHz lower sideband transmission. Single sideband transmission is used wherever possible.

A block diagram of an AM frequency division multiplexing system is shown in Fig. 31.4. Each input modulates an assigned subcarrier at comparatively low frequencies (5 kHz to 40 kHz) in this case. These subcarrier frequencies are combined in a mixer. This combination modulates the main high frequency carrier. The modulated carrier is transmitted to a receiver where the main high frequency signal is demodulated. This process reproduces a combination of the subcarrier frequencies. Each of the sub-carrier frequencies is separated from the other subcarrier frequencies with the help of bandpass filters. These subcarrier frequencies are then demodulated to recover the original data. The AM frequency division multiplexing system are well adapted to analog signals, as all inputs and output voltages are continuous whereas TDM systems have discontinuous outputs and are well suited to digital information especially when sample hold (S/H) circuits are used.

The AM-FDM system permits a number of data to be transmitted over a common channel. Similarly an FM-FDM can be used to place data on one recording channel of a magnetic tape recorder.

Examples of data usually sent by FDM systems are vibration, acceleration and dynamic strain.

## 31.4 DIGITAL DATA ACQUISITION SYSTEM

A generalized diagram of a digital data acquisition system is shown in Fig. 31.5.

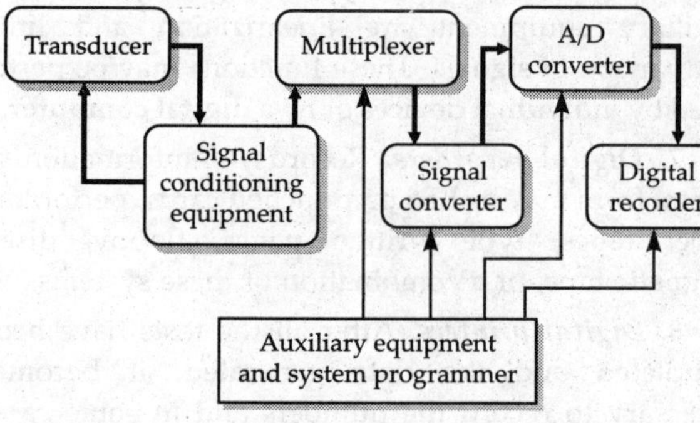


**Fig. 31.5** Digital data acquisition system.

A digital data acquisition system may include some or all of the components shown in Fig. 31.5. The essential functional operations of a digital data acquisition system are :

(*a*) handling of analog signals,
(*b*) making the measurement,
(*c*) converting the data to digital form and handling it, and
(*d*) internal programming and control.

The various components and their functions are described below :

**1. *Transducers.*** They convert a physical quantity into an electrical signal which is acceptable by the data acquisition system.

31

2. ***Signal conditioning equipment.*** Signal conditioning has already been described in details in Chapter 26.

3. ***Multiplexer.*** Multiplexing is the process of sharing a single channel with more than one input. Thus a **multiplexer** accepts multiple analog inputs and connects them sequentially to one measuring instrument. Another name for a multiplexer is "**scanner**".

4. ***Signal converter.*** A signal converter translates the analog signal to a form acceptable by the analog to digital (A/D) converter. An example of the signal converter is an amplifier for amplifying the low-level signal voltages produced by transducers.

5. ***Analog to digital converter (A/D converter).*** An A/D converter converts the analog voltage to its equivalent digital form. The output of the A/D converter may be fed to digital display devices for visual display or may be fed to digital recorders for recording. It may be fed to a digital computer for data reduction and further processing.

6. ***Auxiliary equipment.*** This contains devices for system programming functions and digital data processing. Some of the typical functions done by auxiliary equipment are linearization and limit comparison of signals. These functions may be performed by individual devices or by a **digital computer**.

7. ***Digital recorders.*** Records of information in digital form may be had on punched cards, perforated paper tapes, type written pages, floppy, discs, magnetic tape, or a combination of these systems.

8. ***Digital printers.*** After all the tests have been completed and the data generated, it becomes necessary to record the numbers and in some cases reduce the data to a more meaningful form. A digital printer can be specified to interface with an electronic instrumentation system in order to perform this work, and thus provide a high quality hard copy for records and minimizing the labour of the operating staff.

## 31.5 USES OF DATA ACQUISITION SYSTEMS

Data acquisition systems are being used in ever increasing, large and wide fields in a variety of industrial and scientific areas, including the aerospace, biomedical and telemetry industries. The type of data acquisition system to be used depends upon the application and the-intended use of recorded input data.

Analog data acquisition systems are used when wide frequency width is required or when lower accuracies can be tolerated.

Digital data acquisition systems are used when the physical quantity being monitored has a narrow bandwidth (*i.e.,* when the quantity varies slowly). Digital systems are also used when high accuracy and low per channel cost is required.

Digital data acquisition systems are in general, more complex than analog systems, both in terms of instrumentation involved and the volume and complexity of the data they can handle.

## 31.6 USE OF RECORDERS IN DIGITAL SYSTEMS

Data originating from digital sources may be plotted and displayed on an *X-Y* recorder. The sources of data include the following :

(*i*) keyboard,
(*ii*) paper tape,
(*iii*) punched cards,
(*iv*) magnetic tapes,
(*v*) teletype receiving terminals and
(*vi*) computer readout.

Recorders may be coupled to digital system in *two* broad ways :

(*a*) The recorder may be analog type, responding to voltage levels on each axis. Voltages for actuations of X and *Y* motions are derived from the output of a D/A converter, the input to these converters being furnished by the digital system involved.

(*b*) The recorder is designed solely for digital systems and is fitted with step-by-step driver systems. The data enters, the recorders at a bit rate suitable for the stepping speed of the digital recording system.

1. ***Systems using analog recorders.*** Figure 31.6 shows an outline of a circuit used to produce analog recorder signals from digital inputs. The controls shown are closed by keyboard keys. But in the case of tape or card reader, the controls are closed by either mechanical or solid state relays.

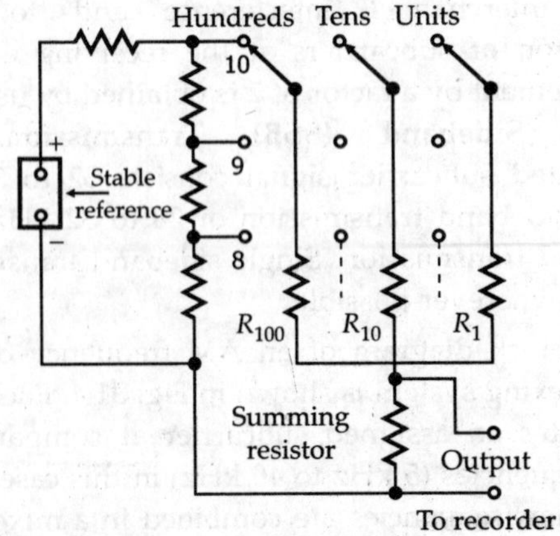


**Fig. 31.6** Keyboard circuit diagram.

Figure 31.7 shows in block form, the main elements of digital systems feeding an X-Y recorder. In each case the digital information must be conditioned to put it into analog voltage form.

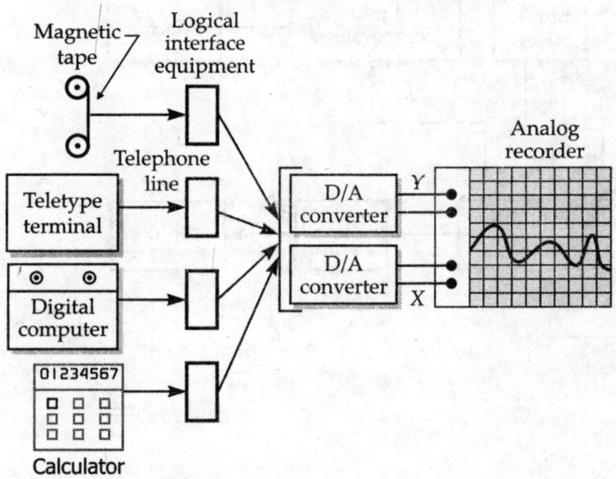


**Fig. 31.7** Digital recording using analog recorder.

2. ***Systems using digital recorders.*** Figure 31.8 shows the general set up of systems employing digital recorders. The computer is programmed to produce information broken down into X and Y coordinates form and is operated at a rate acceptable to the recorder.

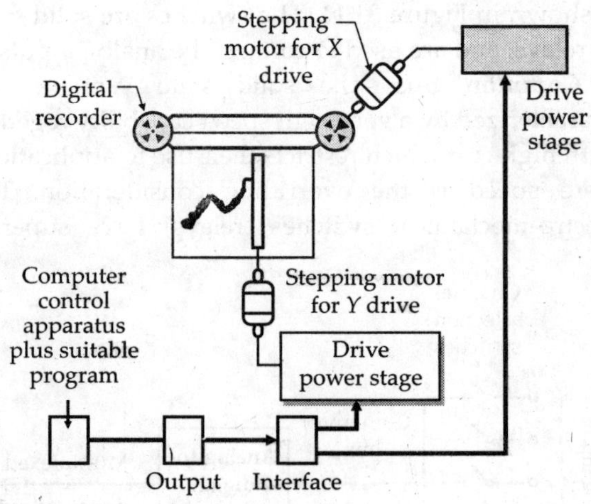


**Fig. 31.8** Digital recording using digital recorder.

The recorder axes are driven by powerful and positive stepping motors giving an incremental movement to the recording system for each small step of data entered.

Successful application of these digital systems requires a thorough understanding of computer technology and programming.

## 31.7 DIGITAL RECORDING SYSTEMS

The most important feature of the digital systems is their ability to drive a printer, floppy discs or a paper punch.

This enables the data recording system to be built around a digital instrument. A schematic diagram for accomplishing this job is shown in Fig. 31.9. A number of analog signals are multiplexed and fed to a digital converter, which is a digital voltmeter (DVM). The output of the DVM is fed to a printer or paper punch through a drive unit.

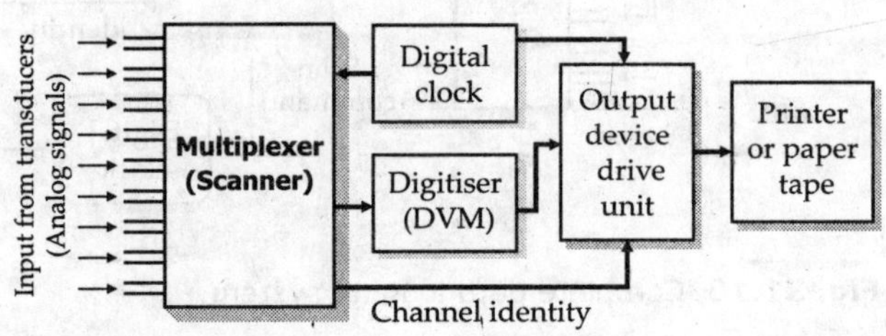


**Fig. 31.9** Block diagram or a digital data recording system.

A digital clock sends a command pulse to initiate the scanning (multiplexing) of the analog inputs from transducers at some preselected time intervals (which may be 1 second, 1 minute or 1 hour). The output device is supplied with a signal from a clock simultaneously so that the time at the start of multiplexing is recorded. The magnitude of the signal of each channel is identified in the record with recording the channel number through a channel identification adjacent to the magnitude of the signal.

The arrangement shown in Fig. 31.10 shows the important components of a **data logger** using a digital voltmeter (DVM). The data logger will assume greater importance if the different measurands like temperature, pressure, etc. are displayed in terms of their proper units instead of mere voltages in order to have proper interpretation of the data acquired from different sources. This requires pressure be displayed in terms of $kN/m^2$ and temperature be displayed in °C and so on. This desirable condition may be obtained by making some additions to the basic data logging system. For example, the compensation for ambient temperature changes may be made from the output of thermocouples (used for monitoring temperature) and the input to the digitiser modified so that the output may be directly printed as temperature in °C. Similarly, the outputs from other transducers should be converted so that the data logger output is displayed in terms of units which are appropriate to

31

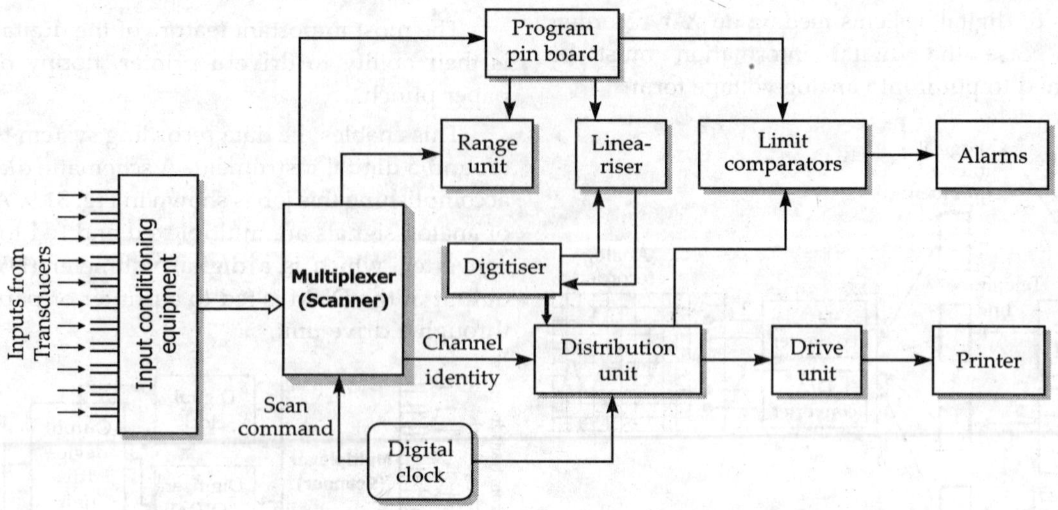


**Fig. 31.10** Complete data logging system.

the measurand of each channel. The arrangement of a modified data logger which is a comprehensive unit is shown in Fig. 31.10 so as to incorporate the requirements listed above.

The functions of various parts of a complete data logging system are explained below :

### 31.7.1 Input Conditioning Equipment

The output of a transducer which is an input to the data logging system may either an a.c. or d.c. signal. If the input signal is a.c. a rectifying circuit must be inserted between the transducer and the data logger. This is because the input to a data logger which uses a DVM as a **digitiser** should normally be in the range of 0 – 50 mV d.c. However, the lower and upper limits at which the multiplexer works are 10 μV and 100 V respectively. Therefore, voltages lower than 10μV are amplified which voltages higher than 100 V are attenuated, in case the signals are in the form of current, they must be converted to voltage drops (by passing the current through a known standard resistance) while pulse or variable frequency signals are converted to voltage signals by appropriate circuits.

### 31.7.2 Digitliser

The digitliser used in the data logging system is a digital voltmeter (DVM). The digital voltmeters have been discussed in Chapter 28. These DVMs use counters which use a complete counting circuit for each decade of display.

### 31.7.3 Multiplexer

The multiplexer of a data logger is a time controlled switch which selects and connects the inputs in a predetermined order and for a preselected period.

In an analog-to-digital conversion, it is convenient to multiplex the analog inputs rather than the digital outputs. One system of analog-to-digital multiplexing is shown in Figure 31.11. The switches are solid state or relays, and are used to connect the analog inputs to a **Common Bus**. The solid state switches are characterized by a very high speed capability together with high cost which restricts their use to applications were speed is the overriding consideration. The electro-mechanical switches (relays) have superior

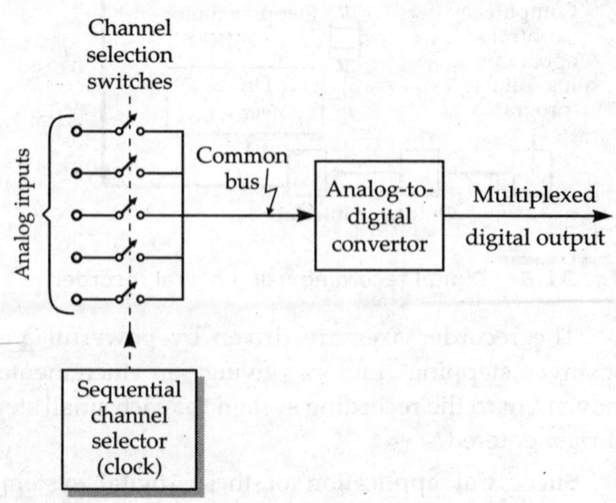


**Fig. 31.11** Multiplexed A/D Conversion.

switching characteristics, and are capable of switching speeds up to 100 channels per second. The switching of both types is normally time controlled by the digital clock. The common bus is connected to a single analog to digital converter which is used for all the channels. The analog inputs are switched sequentially to the bus by channel selector control circuitry. In case, simultaneous time samples are required, a sample and hold (S/H) circuit may be used before each multiplexer switch. In this manner, all the channels are sampled simultaneously and then switched to the A/D converter sequentially.

### 31.7.4 Program Pinboard

This is used for routing the signals to each channel via proper paths. Program pinboards are not used in all data loggers but are very useful in reducing the restrictions on the input signals thereby reducing the amount of scaling required to be performed on each channel at the input conditioning stage.

### 31.7.5 Lineariser

The inputs must be converted (scaled) to their appropriate form before being displayed. Many transducers produce an output voltage which is linearly proportional to the magnitude of the measurand and therefore for these, only a suitable voltage dividing network is needed to reduce the voltage to the compatible level of the data logger. However, the output characteristics, of some transducers like thermocouples, are non-linear. Therefore the output of these transducers is routed through an appropriately programmed circuit which does the linearisation of the signal (produces and output linearly proportional to the signal) and also converts the signal to the original form (*i.e.*, converts the output of a thermocouple from voltage to temperature).

### 31.7.6 Digital Clock

The digital clock is synchronised to the power supply or to an internal crystal oscillator. This clock governs the timing throughout the system by applying triggering pulses to different parts of the system thereby initiating the sequences of events for recording each signal. The digital clock is also used for supplying signals in order that the time of the day or the period elapsed may be recorded.

The principle of working of digital clocks is explained below :

The sequence of events is invariably controlled in digital systems. Therefore, it becomes essential to display and/or record time digital form. This function is performed by a digital clock. The time base for a digital clock is frequently derived from a power line frequency but for better accuracy a crystal oscillator or an atomic resonator can be used. Figure 31.12 shows the block diagram of a digital clock which uses the power line frequency of 50 Hz for deriving its time base.

This clock is capable of displaying hours, minutes, and seconds on a 24 hour basis.

Instead of deriving 50 pulses per second from the power line, which may have additional spurious signals, the power line may be used to drive a synchronous motor which is coupled to a toothed rotor tachogenerator that produces 3000 or more pulses per second giving system command pulses and a resolution of 20 ms.

### 31.7.7 Limit Detectors

A series of limit detectors are provided in order to protect the different components of the automatic system against overloads. The limit detectors compare the magnitude of the signal of the selected signal with a preset safe voltage level. In case the level of the signal is in excess of the preset value, the protective devices are activated to switch off the system or alternatively alarm circuits are energised. The overloads may also be recorded by a special recorder say in red ink on a recording paper.

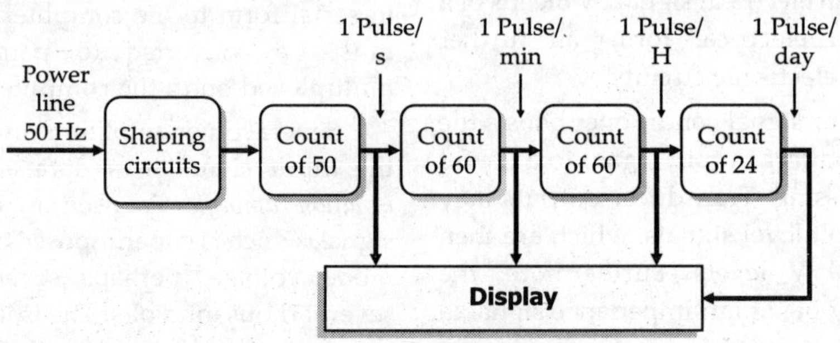


**Fig. 31.12** Schematic diagram of a power line synchronized digital clock.

31

### 31.7.8 Output Devices

A permanent record may be obtained from devices like a line printer, an electronic type writer, a digital magnetic tape recorder, floppy discs or a storage type digital oscilloscope.

## 31.8 MODERN DIGITAL DATA ACQUISITION SYSTEM

The electronic devices that perform the interfacing function between the analog and digital world are the analog-to-digital converters abbreviated as A/D converters or ADC, and digital to analog converters abbreviated as D/A converters or DAC. They have been described in Chapter 28.

Some of the specific applications in which data converters are used include data telemetry systems, pulse coded communications, automatic test systems, computer display systems, value signal processing systems, data logging systems, and sampled-data control systems. In addition, every laboratory digital multimeter or digital panel meter contains an A/D converter.

Besides A/D and D/A converters, data-acquisition systems may employ one or more of the following circuit functions :

1. Transducers
2. Amplifiers
3. Filters
4. Nonlinear analog functions
5. Analog multiplexers
6. Sample-holds.

The interconnection of these components is shown in the diagram of the data-acquisition portion of a computerized feedback control system in Fig. 31.13.

The input to the system is a *physical parameter*, such as temperature, pressure, flow, acceleration, or position, which are analog quantities. The parameter is first converted into an electrical signal by means of a transducer ; once in electrical form, all further processing is done by electronic circuits.

Next an amplifier or signal conditioner boosts the amplitude of the transducer output signal to a useful level for further processing. Transducer outputs may be microvolt or millivolt level signals, which are then amplified to 1 to 10 V levels. Furthermore, the transducer output may be a high impedance signal, a differential signal with common-mode noise, a current output, a signal superimposed on a high voltage, or a combination of these. The amplifier, in order to convert such signals into a high level voltage, may be one of several specialized types.

The amplifier is frequently followed by a low-pass *active filter*, which reduces high-frequency signal components, unwanted electrical interference noise, or electronic noise from the signal. The amplifier is sometimes also followed by a special *nonlinear analog function* circuit that performs a nonlinear operation on the high-level signal. Such operations include squaring, multiplication, division, rms conversion, log conversion, or linearization.

The processed analog signal next goes to an *analog multiplexer*, which sequentially switches between a number of different analog input channels. Each input is in turn connected to the input of the multiplexer for a specified period of time by the multiplexer switch. During this connection time a *sample hold* circuit acquires the signal voltage and then holds its value while an *analog-to-digital converter* converts the value into digital form. The resultant digital word goes to a computer data bus or to the input of a digital circuit.

Thus the analog multiplexer, together with the sample-hold, time shares the A/D converter with a number of analog input channels. The timing and control of the complete data-acquisition system is done by a digital circuit called a *programmer-sequencer*, which in turn is under control of the computer. In some cases the computer itself may control the entire data acquisition system.

While this is perhaps the most commonly used data acquisition system configuration, there are alternatives. Instead of multiplexing high-level signals, low-level multiplexing is sometimes used with the amplifier following the multiplexer. In such cases only one amplifier is required, but its gain may have to be changed from one channel to the next during multiplexing. Another method is to amplify and convert the signal into digital form at the transducer location and send the digital information in serial form to the computer. Here the digital data must be converted to parallel form and then multiplexed on to the computer data bus.

Several other amplifiers are useful in conditioning the input signal in a data acquisition system. An *isolation amplifier* is used to amplify a differential signal, which is superimposed on a very high common mode voltage, perhaps several hundred or even several thousand volts. The isolation amplifier has the characteristics of an instrumentation amplifier with a very high common-mode input voltage capability.

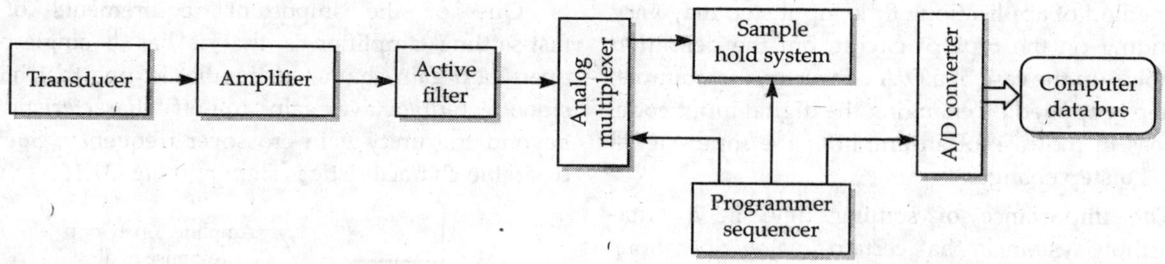


**Fig. 31.13** Digital Data Acquisition system.

Another special amplifier, the *chopper stabilized amplifier*, is used to accurately amplify microvolt-level signals to the required amplitude. This amplifier employs a special switching stabilizer that gives extremely low input offset voltage drift. Another useful device, the *electrometer amplifier*, has ultralow input bias currents, generally less than 1 picoampere (pA) and is used to convert extremely small signal currents into a high-level voltage.

### 31.8.1 Filters

A *low-pass filter* frequently follows the signal-processing amplifier to reduce signal noise. Low-pass filters are used for the following reasons : to reduce man-made electrical interference noise, to reduce electronic noise, and to limit the bandwidth of the analog signal to less than half the sampling frequency in order to eliminate frequency folding. When used for the last reason, the filter is called a *presampling filter* or *intialiasing filter*.

Man-made electrical noise is generally periodic, as for example in power line interference, and is sometimes reduced by means of a special filter such as a *notch filter*. Electronic noise, on the other hand, is random noise with noise power proportional to bandwidth and is present in transducer resistances, circuit resistances, and in amplifiers themselves. It is reduced by limiting the bandwidth of the system to the minimum required to pass desired signal components.

No filter does a perfect job of eliminating noise or other undesirable frequency components, and therefore the choice of a filter is always a compromise. Ideal filters, frequently used as analysis examples, have flat passband response with infinite attenuation at the cutoff frequency, but are mathematical filters only and not physically realizable.

In practice, the systems engineer has a choice of cutoff frequency and attenuation rate. The attenuation rate and resultant phase response depend on the particular filter characteristic and the number of poles in the filter functions. Some of the more popular filter characteristics include Butter-worth, Chebychev, Bessel, and elliptic. In making this choice, the effect of overshoot and non-uniform phase delay must be carefully considered. Figure 31.14 illustrates some practical low-pass filter response characteristics.

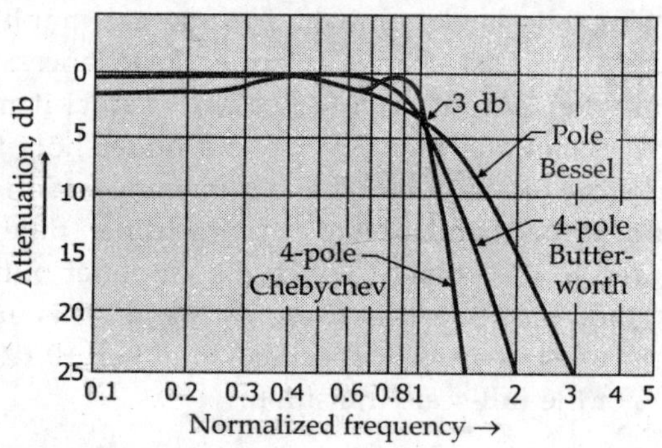


**Fig. 31.14** Some practical low-pass filter characteristics.

Passive *RLC* filters are seldom used in signal-processing applications today owing chiefly to the undesirable characteristics of inductors. Active filters are generally used now since they permit the filter characteristics to be accurately set by precision, 'stable resistors and capacitors. Inductors, with their undesirable saturation and temperature drift characteristics, are thereby eliminated. Also, because active filters use operational amplifiers, the problems of insertion loss and output loading are eliminated.

### 31.8.2 Settling Time

A parameter that is specified frequently in data acquisition and distribution systems is *settling time.* The term *settling time* originates in control theory, but it is now commonly applied to amplifiers, multiplexers, and D/A converters.

Settling time is defined as the time elapsed from the application of a full-scale step input to a circuit to the time when the output has entered and remained within a specified error band around its final value.

The method of application of the input step may vary depending on the type of circuit, but the definition still holds. In the case of a D/A converter, for example, the step is applied by changing the digital input code, whereas in the case of an amplifier the input signal itself is a step change.

The importance of settling time in a data-acquisition system is that certain analog operations must be performed in sequence, and one operation may have to be accurately settled before the next operation can be initiated. Thus a buffer amplifier preceding an A/D converter must have accurately settled before the conversion can be initiated.

Settling time for an amplifier is illustrated in Fig. 31.15. After application of a full-scale step input, there is a small delay time, following which the amplifier output slew's or changes at its maximum rate. *Slew rate* is determined by internal amplifier currents that must change internal capacitances. As the amplifier output approaches final value, it may first overshoot and then reverse and undershoot this value before finally entering and remaining within the specific error band. Note that settling time is measured to the point at which the amplifier output *enters* and *remains* within the error band. This error band in most devices is specified to either ±0.1% or ±0.01% of the full-scale transition.

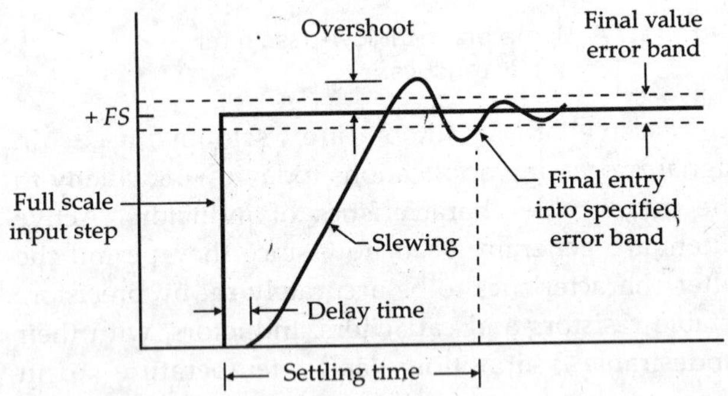


**Fig. 31.15** Amplifier settling time.

## 31.8.3 Amplifier Characteristics

Settling time, unfortunately, is not readily predictable from other amplifier parameters such as bandwidth, slew rate, or overload recovery time, although it depends on all of these. It is also dependent on the shape of the amplifier open-loop gain characteristics, its input and output capacitance, and the dielectric absorption of any internal capacitances. An amplifier must be specifically designed for optimized settling time, and settling time is a parameter that must be determined by testing.

One of the important requirements of a fast-settling amplifier is that it has a single-pole open-loop gain characteristic, that is, one that has a smooth 6-db/octave gain roll-off characteristic to beyond the unity gain crossover frequency. Such a desirable characteristic is shown in Fig. 31.16.

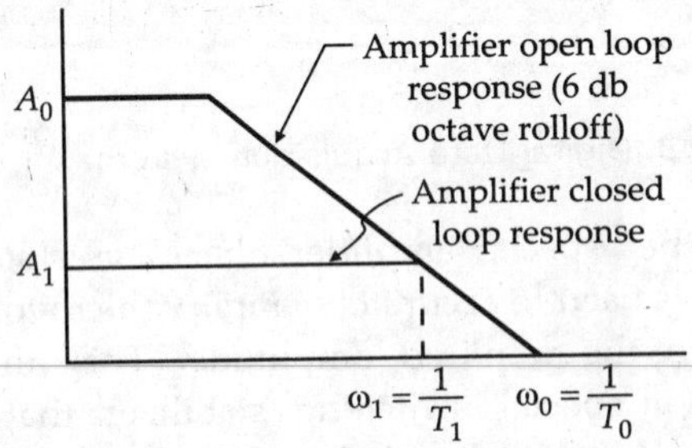


**Fig. 31.16** Amplifier single-pole loop gain characteristic.

It is important to note that an amplifier with a single pole response can never settle faster than the time indicated by the number of closed-loop time constants to the given accuracy. Figure 31.17 shows the output error as a function of the number of time constants τ, where

$$\tau = \frac{1}{2\pi f} \qquad ...(31.1)$$

and $f$ is the closed-loop 3-db bandwidth of the amplifier.

Actual settling time for a good-quality amplifier may be significantly longer than that indicated by the number of closed-loop time constants due to slew-rate limitation and overload recovery time. For example, an amplifier with a closed-loop bandwidth of 1 MHz has a time constant of 160 ns which indicates a settling time of 1.44 μs (9 time constants) to 0.01% of final value. If the slew rate of this amplifier is 1V / μs, it will take more than 10 μs to settle to 0.01% for a 10 V change.

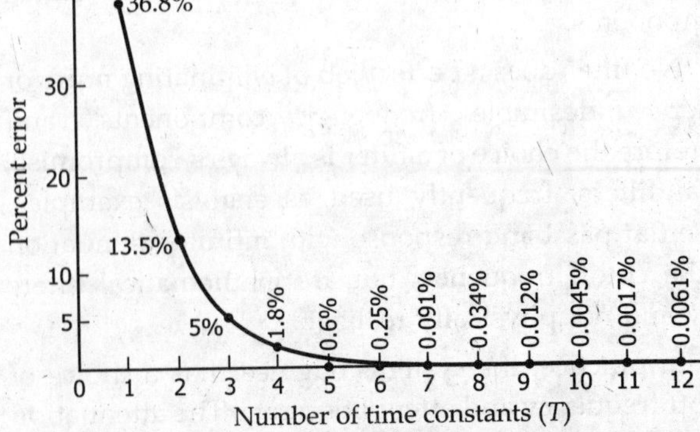


**Fig. 31.17** Output settling error as a function of number of time constants.

If the amplifier has a nonuniform gain roll-off characteristic, then its settling time may have one of two undesirable qualities. First, the output may reach the vicinity of the error band quickly, but then take a long time to actually enter it. Second, it may overshoot the error band and then oscillate back and forth through it before entering and remaining inside it.

Modern fast-settling operational amplifiers come in many different types, including modular, hybrid, and monolithic amplifiers. Such amplifiers have settling times of 0.1% or 0.01% of 2 μs down to 100 ns and are useful in data acquisition and conversion applications. Figure 31.17 an example of an ultrafast settling operational amplifier of the hybrid type.

## 31.9 ANALOG MULTIPLEXER OPERATION

Analog multiplexers are the circuits that time share an A/D converter among a number of different analog channels. Since the A/D converter in many cases is the most expensive component in a data acquisition system, multiplexing analog inputs to the A/D is an economical approach. Usually, the analog multiplexer operates into a sample-hold circuit that holds the required analog voltage long enough for A/D conversion.

As shown in Fig. 31.18 an analog multiplexer consists of an array of parallel electronic switches connected to a common output line. Only one switch is turned on at a time. Popular switch configurations include 4, 8 and 16 channels, which are connected in single (single-ended) or dual (differential) configurations.

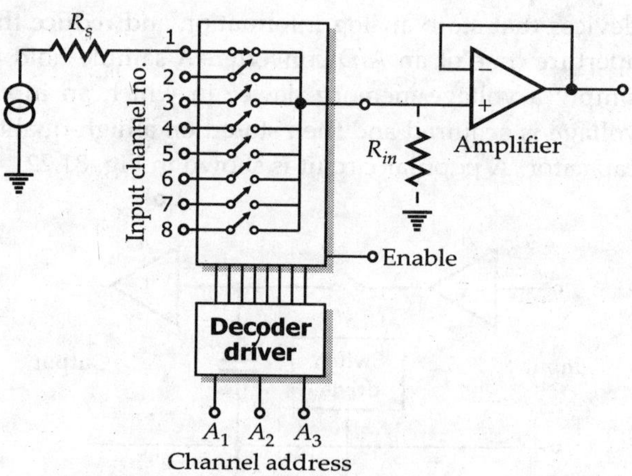


| $A_1$ | $A_2$ | $A_3$ | EN | ON CHAN |
|---|---|---|---|---|
| X | X | X | X | None |
| 0 | 0 | 0 | 1 | 1 |
| 0 | 0 | 1 | 1 | 2 |
| 0 | 1 | 0 | 1 | 3 |
| 0 | 1 | 1 | 1 | 4 |
| 1 | 0 | 0 | 1 | 5 |
| 1 | 0 | 1 | 1 | 6 |
| 1 | 1 | 0 | 1 | 7 |
| 1 | 1 | 1 | 1 | 8 |

**Fig. 31.18** Analog multiplexer circuit.

The multiplexer also contains a decoder-driver circuit that decodes a binary input word and turns on the appropriate switch. This circuit interfaces with standard transistor-transistor-logic (TTL) inputs and drives the multiplexer switches with the proper control voltages. For the 8-channel analog multiplexer shown, a one-of-eight decoder circuit is used.

Most analog multiplexers today employ the CMOS switch circuit shown in Fig. 31.19. A CMOS driver controls the gates of parallel-connected *P*-channel and *N*-channel MOSFETs. Both switches turn on together, with the parallel connection giving relatively uniform on-resistance over the required analog input voltage range. The resulting on-resistance may vary from about 50 Ω to 2 kΩ, depending on the multiplexer ; this resistance increases with temperature.

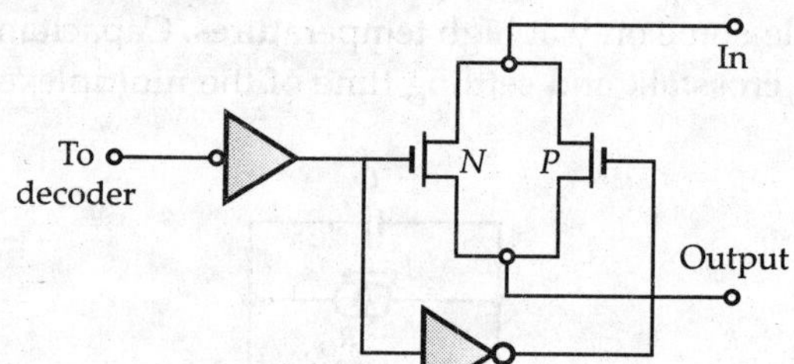


**Fig. 31.19** CMOS analog switch circuit.

### 31.9.1 Analog Multiplexer Characteristics

Because of the series resistance, it is common practice to operate an analog multiplexer into a very high load resistance such as the input of a unity gain buffer amplifier. The load impedance must be large compared with the switch on-resistance and any series source resistance in order to maintain high transfer accuracy. *Transfer error* is the input-to-output error of the multiplexer with the source and load connected ; error is expressed as a percent of input voltage.

Transfer errors of 0.1% to 0.01% or less are required in must data acquisition systems. This is readily achieved by using operational amplifier buffers with typical input impedances from $10^8$ to $19^{12}$ Ω. Many sample-hold circuits also have very high input impedances.

Another important characteristic of analog multiplexers is *break-before-make* switching. There is a small time delay between disconnection from the previous channel and connection to the next channel, which assures that two adjacent input channels are never instantaneously connected together.

31

*Settling time* is another important specification for analog multiplexers ; it has the same definition previously given for amplifiers except that it is measured from the time the channel is switched on.

*Throughput rate* is the highest rate at which a multiplexer can switch from channel to channel with the output settling to its specified accuracy.

*Cross talk* is the ratio of output voltage to input voltage with all channels connected in parallel and off : it is generally expressed as an input-to-output attenuation ratio in decibels.

As shown in the representative equivalent circuit of Fig. 31.20, analog multiplexer switches have a number of leakage currents and capacitances associated with their operation. These parameters are specified on data sheets and must be considered in the operation of the devices. Leakage currents, generally in picoampere at room temperature, become trouble-some only at high temperatures. Capacitances affect crosstalk and settling time of the multiplexer.

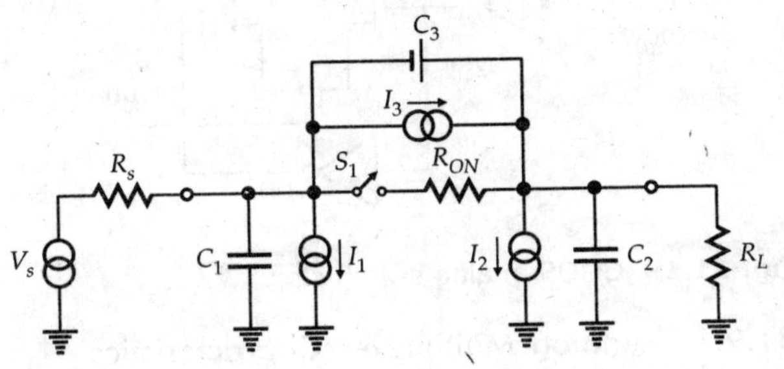


**Fig. 31.20** Equivalent circuit of analog multiplexer switch.

### 31.9.2 Analog Multiplexer Applications

Analog multiplexers are employed in two basic types of operation : high level and low level. In *high-level multiplexing,* the most popular type, the analog signal is amplified to the 1 to 10 V range ahead of the multiplexer. This has the advantage of reducing the effects of noise on the signal during the remaining analog processing. In *low-level multiplexing,* the signal is amplified after multiplexing ; therefore, great care must be exercised in handling the low-level signal up to the multiplexer. Low-level multiplexers generally use two-wire differential switches in order to minimize noise pickup. Reed relays, because of essentially zero series resistance and the absence of switching spikes, are frequently employed in low-level multiplexing systems. They are also useful for high common-mode voltages.

A useful specialized analog multiplexer is the *flying-capacitor* type. This circuit, shown as a single channel in Fig. 31.21 has differential inputs and is particularly useful with high common-mode voltages. The capacitor connects first to the differential analog input, charging up to the input voltage, and is then switched to the differential output, which goes to a high-input-impedance instrumentation amplifier. The differential signal is therefore transferred to the amplifier input without the common-mode voltage and is then further processed up to A/D conversion.

To realize large numbers of multiplexed channels, you can connect analog multiplexers in parallel using the enable input to control each device. This is called *single-level multiplexing.* You can also connect the output of several multiplexers to the inputs of another to expand the number of channels ; this method is *double-level multiplexing.*

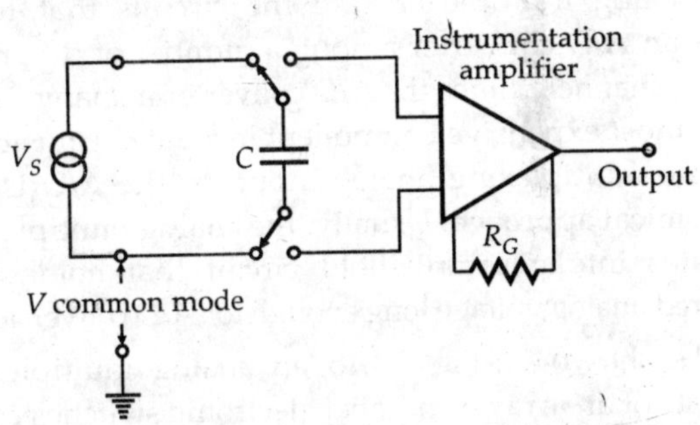


**Fig. 31.21** Flying-capacitor multiplexer switch.

## 31.10 OPERATION OF SAMPLE-HOLD CIRCUITS

Sample-hold circuits, discussed earlier, are the devices that store analog information and reduce the aperture time of an A/D converter. A sample hold is simply a voltage-memory device in which an input voltage is acquired and then stored on a high-quality capacitor. A popular circuit is shown in Fig. 31.22.

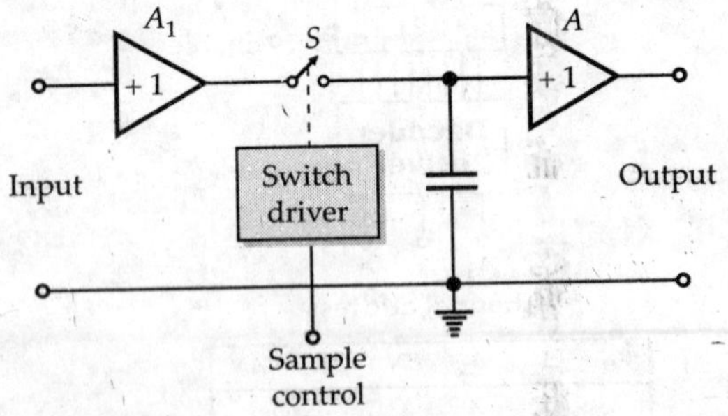


**Fig. 31.22** Popular sample-hold circuit.

$A_1$ is an input buffer amplifier with a high input impedance so that the source, which may be an analog multiplexer, is not loaded. The output of $A_1$ must be capable of driving the hold capacitor with stability and enough drive current to charge it rapidly. $S_1$ is an electronic switch, generally an FET, which is rapidly switched on or off by a driver circuit that interfaces with TTL inputs.

C is a capacitor with low leakage and low dielectric absorption characteristics ; it is a polystyrene, polycarbonate, polypropylene, or Teflon type. In the case of hybrid sample-holds, the MOS-type capacitor is frequently used.

$A_2$ is the output amplifier that buffers the voltage on the hold capacitor. It must therefore have extremely low input bias current, and for this reason an FET input amplifier is required.

There are two modes of operation for a sample-hold *sample* (or tracking) *mode,* when the switch is closed ; and *hold mode,* when the switch is open. Sample-holds are usually operated in one of two basic ways. The device can continuously track the input signal and be switched into the hold mode only at certain specified times, spending most of the time in tracking mode. This is the case for a sample-hold employed as a deglitcher at the output of a D/A converter, for example.

Alternatively, the device can stay in the hold mode most of the time and go to the sample mode just to acquire a new input signal level. This is the case for a sample-hold used in a data-acquisition system following the multiplexer.

### 31.10.1 Sample-hold as a Data-Recovery Filter

A common application for sample-hold circuits is *data-recovery,* or *signal-reconstruction, filters.* The problem is to reconstruct a train of analog samples into the original signal ; when used as a recovery filter, the sample-hold is known as a *zero-order hold.* It is a useful filter because it fills in the space between samples, providing data smoothing.

As with other filter circuits, the gain and phase components of the transfer function are of interest. By an analysis based on the impulse response of a sample hold and use of the Laplaee transform, the transfer function is found to be :

$$G_0(f) = \frac{1}{f_s}\left[\frac{\sin \pi(f/f_s)}{\pi(f/f_s)}\right]\exp(-j\pi f/f_s) \quad \ldots(31.2)$$

where $f_s$ is the sampling frequency.

This function contains the familiar $(\sin x)/x$ term plus a phase term, both of which are plotted in Fig. 31.23.

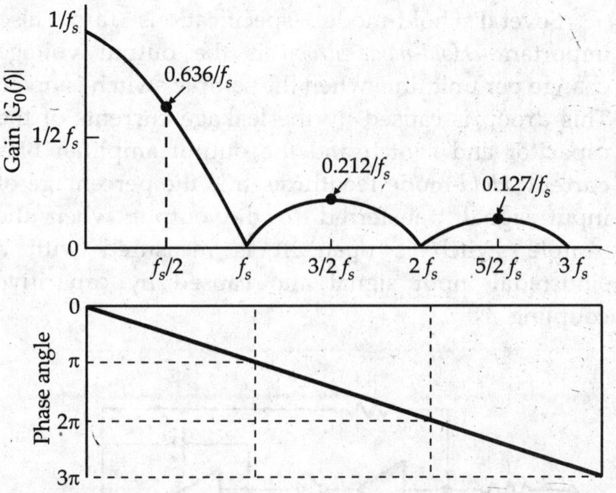


**Fig. 31.23** Gain and phase components of zero-order hold transfer function.

The sample-hold is therefore a low-pass filter with a cutoff frequency slightly less than $f_s/2$ and a linear phase response that results in a constant delay time of $T/2$, where $T$ is the time between samples. Notice that the gain function also has significant response lobes beyond $f_s$. For this reason, a sample-hold reconstruction filter is frequently followed by another conventional low-pass filter.

### 31.10.2 Other Sample-Hold Circuits

In addition to the basic circuit of Fig. 31.24, several other sample-hold circuit configurations are frequently used. Figure 31.25 shows two such circuits, which are closed-loop circuits, as contrasted with the open-loop circuit of Fig. 31.24. Figure 31.25(*a*) uses an operational integrator and another amplifier to make a fast, accurate inverting sample-hold. A buffer amplifier is sometimes added in front of this circuit to give high input impedance. Figure 31.25(*b*) shows a high input impedance noninverting sample-hold circuit.

The circuit in Fig. 31.22, although generally not as accurate as those in Fig. 31.24 can be used with a diode-bridge switch to realize ultrafast.

### 31.10.3 Sample-Hold Characteristics

A number of parameters are important in characterizing sample-hold performance. Probably most important of these is *acquisition time.* The definition is similar to that of settling time for an amplifier. It is the time required, after the sample-command is given, for the hold capacitor to change to a full-scale voltage charge and remain within a specified error band around final value.

Several hold-mode specifications are also important. *Hold-mode droop* is the output voltage change per unit time when the sample switch is open. This droop is caused by the leakage currents of the capacitor and switch and the output amplifier bias current. *Hold*-mode feedthrough is the percentage of input signal transferred to the output when the sample switch is open. It is measured with a sinusoidal input signal and caused by capacitive coupling.

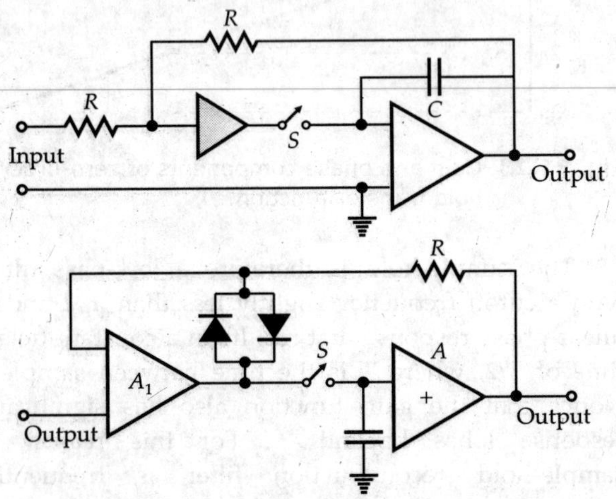


**Fig. 31.24** Closed loop Sample-hold circuits.

The most critical phase of sample-hold operation is the transition from the sample mode to the hold mode. Several important parameters characterize this transition. *Sample-to-hold offset* (or step) *error* is the change in output voltage from the sample mode to the hold mode, with a constant input voltage. It is caused by the switch transferring change onto the hold capacitor as it turns off.

*Aperture delay* is the time elapsed from the hold command when the switch actually opens ; it is generally much less than a microsecond. *Aperture uncertainty* (or *aperture jitter*) is the time variation, from sample to sample, of the aperture delay. It is the limit on how precise is the point in time of opening the switch. Aperture uncertainty is the time used to determine the aperture error due to rate of change of the input signal. Several of these specifications are illustrated in Fig. 31.25.

Sample-hold circuits are simple in concept, but generally difficult to fully understand and apply. Their operation is full of subtleties, and they must therefore be carefully selected and then tested in a given application.

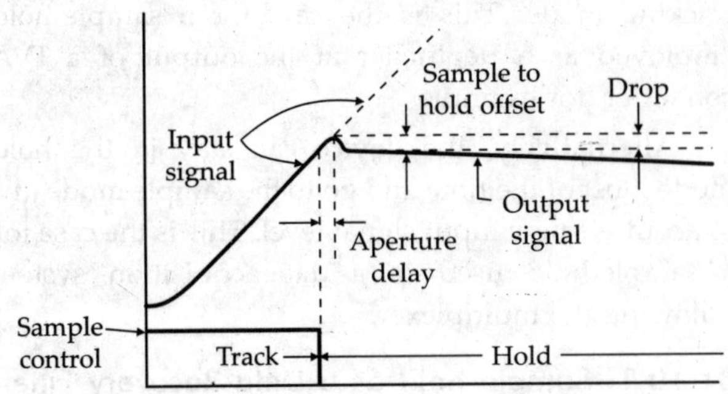


**Fig. 31.25** Some sample-hold characteristics.

# 32

# Modern Sensors

## 32.1 INTRODUCTION

In last chapters we have discussed on micro probes fitted with membrane and electrodes for measurement purposes. Modern day technological advances ensure analysis of almost every type of Biosecretion like enzymes, blood glucose, other biocatalysts, crude cell extract, viable plant cells etc. with the help of properly designed biprobe or nanoprobe. Measurement of nonviable microbial cells or micro organisms or viable microbial cells is also possible with these types of biosensors.

## 32.2 TYPES OF MODERN SENSORS

The different types of biological materials are possible starting from enzyme, antibody organelle, hormones, nucleic acid or a whole cell which may be immobilized or mobilized. Immobilized enzymes are prepared by soluble enzyme protein into a solid form of catalyst by different methods. The immobilization technique is mainly developed due to reuse, or continuous use of the enzyme. The biosensor, with its tip coated with immobilized enzyme and fitted with a ion selective semi permeable membrane, is with intimate contact of the bio-material. The biological signal produced is converted into electrical signal. The process electrode used, or detection by oxygen electrode or sensing by photocell or a thermistor. The bio-substances to be analysed pass through the membrane and interact with immobilized bio-material at the tip of the probe and transformed into heat or gas (oxygen) or layer of electrons or ions like Ammonium ions. The product passes through another membrane above the immobilized bio-tip to the transducer *i.e.*, gas (oxygen) detector or ion detector or thermistor for heat detection or capacitive detection for electrons or ions, and milivolt order electrical signal is generated. Also, there is colour change at the bio tip which can be analysed by colorimetric detection. This signal is amplified and processed and the read out is recorded.

### 32.2.1 Nanotechnology

Referring to Nanoscience, which is the study of the fundamental principle of molecule or atoms whose diameter is in between 1–100 nanometers, the structure is referred to a nanostructure. Nano technology is used to analyse the nanostructure in which biostructure and micro organism is included. For example, photorefractive polymers have highly unusual structure, with mobile electronic charges like metal. The mobile charges can be excited with sunlight or electric field so that the new position of these changed particles forms a code which is read by shinning it with different colour of light. This code is a characteristic of the analytical properties of the polymer. This type of nanoscale analysers or sensors are called nano-sensors.

Nanosensors have wide range of applications ranging from water, temperature, light, sound, electricity, molecules, biological secretions, *i.e.*, enzyme, catalyst etc. and micro-organisms, toxins and DNA. These sensors use the principle of molecules recognition. Dr. Hupp made some elegant molecules that are identified as molecule metal required and designed to recognise particular target molecules called as analytes. Another major direction of work is nanoscale biostructure. These structure are designed in the nanoscale, can affect a biological process or interact with biological entity. As the biological entities are generally nanostructures, biomedical applications and investigation constitute a major par' is the Nanoscience. One of the developments in Nasnoscience is Graetzel cell, which is a dye molecule used to capture energy from sun light. The excited molecule jumps to higher energy state. In the high energy state the molecule separates charge by passing of electron from the dye molecule to a nanoparticle of white crystal of titanium dioxide. The positively charged dye molecule and the negatively charged titanium dioxide (nanoparticle) are allowed to recombine chemically with the solar energy converted to a current passing in the external circuit. So this is an example of nanoscale solar energy capture and transformation to electricity and torque. This electricity is used to produce light in the light emitting diode. This is an instance of sensing of electro-magnetic radiation.

Now-a-days, there is a break through in transducer technology with the invention of porous silicon technology. In porous silicon the active surface to volume ratio is very high. Measurement, starting from pressure, temperature, strain, distance, acceleration and electrical voltage, current and true power, can be measured with porous silicon sensors. The design and fabrication of porous silicon wafer has been machined mechanically, hence, they are grouped as Micro-machined devices or MEMS. The main advantage of MEMS is the on-chip fabrication of detector, controller or display or it is called System-ON-Chip (SOC). The other advantage is the customised IC designed (ASICs) which are fabricated for particular industry specific use.

### 32.2.2 Biocatalyst

Biocatalysts are special type of bio-product which accelerates reactions happening in Animal body. The most prominent biocatalysts are enzymes. There are other types of biocatalyst like cell extract. Plant and animal cell, viable microbial cells or micro organism and nonviable microbe cell.

### 32.2.3 Enzymes

Commonly used animal enzymes are lipases, rennets, triprin etc. and plant enzymes are papain, analyses, soyabean lipoxy guase. Enzymes are widely used in leather plant for softing of leather and detergent manufacturing. Enzymes are found in *three* methods

1. Due to natural secretion from cell *i.e.*, extracellular secretion,
2. intracellular enzymes which remain in the cell, and
3. microbial enzyme which is prepared by the way of fermentation in presence of microorganism like bacteria virus or toxin. The production of primary and secondary metabolides is caused in presence of various enzymes.

Intracellular enzymes, due to its presence in the cell, need the cell wall to be destroyed during its preparation. The cell wall is destroyed in presence of high electric field or in some biochemical processes like homogenizer or bread mill. Whereas the extracellular enzymes is prepared with the help of microorganism like cellulose, polymethylgalacturonase etc. After the extraction of proper enzymes, its purification has been done and other elements of cell and different acids forming the cell has been eliminated.

The enzyme prepared due to fermentation process in presence of microorganism are more economical for large scale enzyme production and its amount depends on the size of farmenter. The process of preparation of microbial enzymes is of less cost, easily extraction is possible and easy to purify. For growth of intracellular and extracellular enzyme specific environmental condition *i.e.*, temperature pressure and humidity and pH should be reached and control of the ambient condition is necessary. But microbial enzyme produced by the fermentation with microorganism is prepared in varying environmental conditions and a wide variety of enzymes can be made by microbial fermentation.

The microbial enzymes can use Bacteria as microorganism (Bacillus crews) to reach to a enzyme named penicillinease. Similarly, the residual enzyme urate oxides is the product when asperyilus flavus fungi (microorganism) is used. These microbial enzymes are used to get fructose from glucose, widely used in food industry and detergent industry.

#### 32.2.3A Properties of Enzyme

The main properties of enzyme are given below :

- For two closely related, species the enzyme protein macromolecule, they differ inspite of some similar molecular structure.
- The extracellular unlike intracellular enzymes are influenced by environmental conditions *i.e.*, temperature, pressure, pH, humidity etc. in which they are prepared and their stability, activity differs with the ambient condition of microbial growth.
- A particular microorganism may not decompose a organic matter whereas it is possible that a community of different microorganism can cause secretion of a particular enzyme. It is also some specific enzyme, that can cause the secretion of a

particular enzyme in higher amount and utilize the substrate is a more economical way.

- The same enzyme secretion from different sources show different responses in presence of the same activator of inhibitor.

### 32.2.3B Production of Enzymes

The different methods used in the preparation of enzyme are

1. Isolate microorganism or development of strain
2. Culture medium formation and inoculum
3. Sterilization and inoculation of medium
4. Purification of enzyme
5. Immobilization.

### 32.2.4 Isolation of Microorganism

Production of enzyme passes through this process of isolation of microorganism because the fermentation process occurs in a short time for production of high amount of enzyme rather than other notabilities and to develop a low cost culture medium. The suitable microorganism for an enzyme, is used to form a cultured medium at particular temperature and pH of the medium and in this optimised culture mediums improvement of strains is developed. The development of strains in micro-organism is catered by the use of mutagenic chemical in the cultured medium with exposure to ultraviolet light.

The other constituents of culture medium are sources of carbon, nitrogen, amino acids, growth promoters trace element and a little salt. Proper pH value and temperature is maintained although the fermentation process in the cultured medium.

The culture medium is sterilized continuously before fermentation. After proper sterilization sufficient amount of inoculum is done so that fermentation starts. The inoculum, for enzyme production, floats on the upper surface of cultured medium and this is called surface culture technique. Also, there is submerged culture medium, so that it is less prone to infection and better yield of enzyme. The growth of enzyme is optimised with proper temperature, pH and oxygen flow into the fermentation process.

Enzymes are purified by preparation of concentrated solution by vacuum evaporation technique with low temperature. It is also done by ultra filtration to remove stresses of microorganism. The addition of preservative like calcium salt, protein starch, sugar as stabilizer is an important process.

Drying and packaging of enzyme is also a process in enzyme production.

### 32.2.5 Immobilization

Immobilization is a process in which enzyme has been separated from the bulk phase by some means. Polymer has been created from the enzyme particularly for reuse of the enzyme product. The bulk phase contains substrate, effector or inhibitor molecules in a dispersed way and by close monitoring, the enzyme as a polymer has been separated.

The pure enzyme may not be economical until they are prepared for reuse. The immobilized enzyme is reused, not only that, they are useful for continuous use and needs less labour incentive.

The different ways for immobilization are

1. Absorption
2. Covalent bonding
3. Entrapment
4. Encapsulation
5. Cross linking.

The process of immobilization is the absorption of loosely bound enzyme molecules on a polymer carrier surface. Absorption is a static process enzyme is immobilized on polymer carrier by stirring a container containing enzyme and polymer. Besides **absorption**, there may be other processes like covalent **bonding** or cross-linking of enzyme with polymer **molecule** (Fig. 32.1).

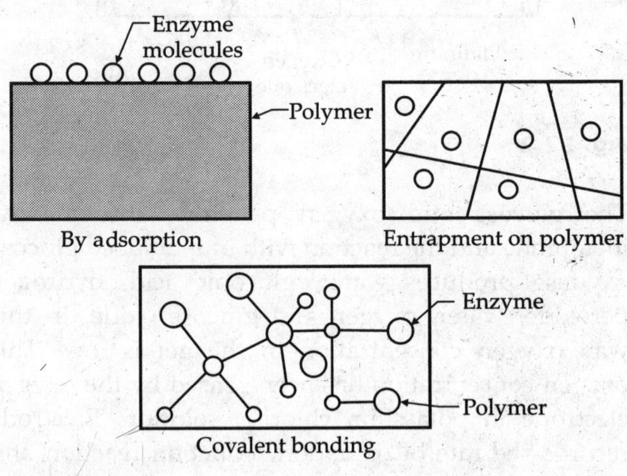


**Fig. 32.1**

### 32.2.6 Biosensor

The biological product when comes to intimate contact to a biological sensor, with immobilized layer, the sensor transduces biological signal to electrical sensor. These immobilized layer sensor which analyse biological product like enzyme, hormone, antibody, nucleic acid, protein, organelle is called the biosensor. There may be any type of electrode like ion-sensitive electrode, oxygen electrode, thermistor or photocell. The immobilized biological material in contact with the electrode polymer tip is fitted with a semi permeable membrane through which the biological material to be analysed, can pass and interact with immobilized bio-material on the sensor tip. This interaction may produce gas, heat, electrons, oxygen, hydrogens ions and other ions. The product passes through other membrane with specific permeability of the product to the sensor which either analyse the ionic activity or by thermistor to measure temperature or oxygen ion concentration to produce an electrical signal which is recorded.

### 32.2.7 Glucose Electrode

The glucose electrode shown in Fig. 32.2, is separated from immobilized glucose oxydase solution with a semipermeable membrane of cellulose acetate with specific permeability for glucose ion and oxygen.

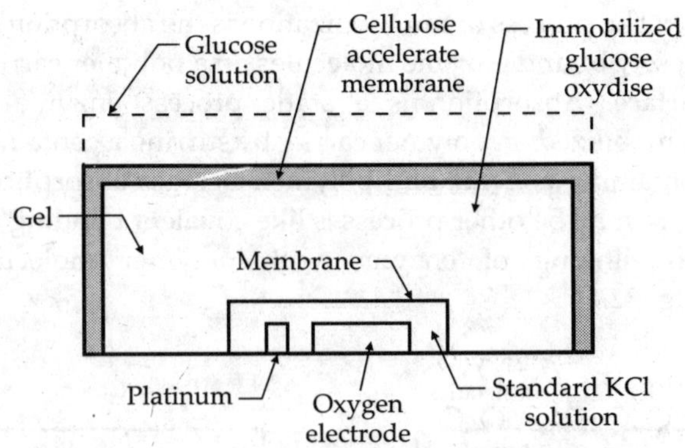


**Fig. 32.2**

The glucose and oxygen penetrate through the membrane and the reaction with immobilized glucose oxydase produces water, gluconic acid, hydrogen peroxide, water, oxygen and glucose oxide. In this way oxygen concentration in the gel is low. This oxygen concentration has been sensed by the oxygen electrode in potassium chloride solution. Electrode records the rate of oxidation reduction reaction and the rate of decrease in oxygen level is directly proportional to the concentration of glucose in the solution. The range of glucose concentration in which this biosensor is applicable is $10^{-1} - 10^{-5}$ mol $dm^{-5}$, with a response time of 1.5 minutes.

Biosensors are of different types in view of the type of electrode used. Firstly, is electro-chemical biosensors, Field effect transistor or light emitting diodes are present which transfer the charge layer produced by reaction between bio-material and its specific enzyme to current or optics. As field effect transistor provides high input impedance and low output impedance it is used to pick the low e.m.f. produced, and amplify it. The second from of biosensor is of amperometric type which analyse reaction between bio material with corresponding enzyme so that electron has been generated directly or through a mediator.

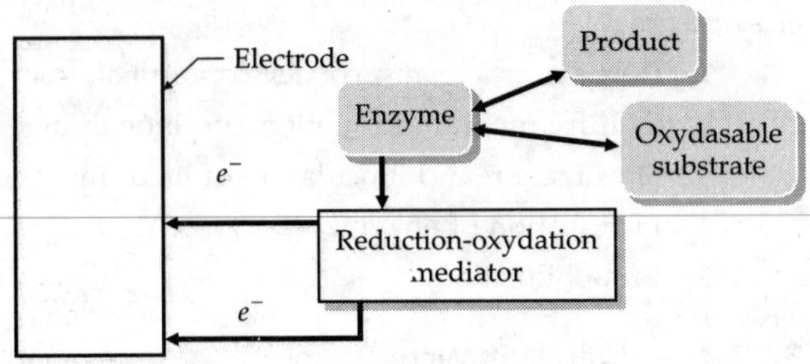


**Fig. 32.3**

The enzyme plays the part of catalyst in reduction oxidation reaction in presence of bio-product and substrate and coupled to the electrode. The electron transfer is from oxydasable substrate to electrode through enzyme. In this (enzymetic) type of biosensors, the oxygen requirement is replaced by oxydase.

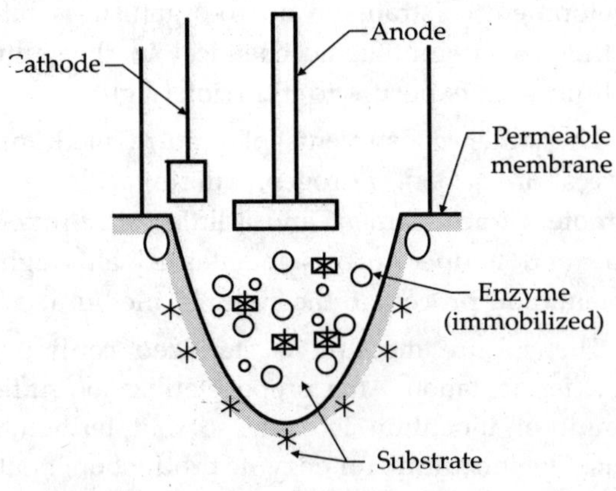


**Fig. 32.4**

In enzymes' electrode, the substrate which may be protein, amino acid, glucose, alcohol or lactic acid, reacts with a enzyme separated by thin permeable membrane. The reaction produce $O_2$, hydrogen ion, Ammonium ion, carbon dioxide or small molecules due to enzyme reaction. The potentiometric response from the electrode depends on the concentration of the substrate. The biochemical reaction between immobilized enzyme and bio-product sometimes generate heat which may be sensed by a thermistor and changed to electrical signal. As thermistor has a measurable sensitivity in the range of temperature 0.1–0.001°C this type of biosensors are widely used.

### 32.2.8 Electrochemical Electrode

The table show the substrate with its specific enzyme and response time for electro-chemical electrodes.

*Enzyme based biosensors*

| ***Substrate*** | ***Enzyme*** | ***Response time*** | ***Range*** |
|---|---|---|---|
| Amines | Monoamines oxidase | 4 min | 50 – 200μ mol m$^3$ |
| Cholesterol | Cholesterol oxidase | 2 min | $10^{-2} - 2 \times 10^{-5}$ mol dm$^{-3}$ |
| Glucose | Glucose oxidase | 20 sec | $2 \times 10^{-3} - 3 \times 10^{-6}$ mol dm$^{-3}$ |
| Penicillin | Penicillinases | 25 sec | 1-10 m mol dm$^{-3}$ |
| Sucrose | Invertage | 6 min | $10^{-2} - 2 \times 10^{-3}$ mol dm$^{-3}$ |
| Uric acid | Uricese | 30 min | $5 \times 10^{-3} - 5 \times 10^{-3}$ mol dm$^{-3}$ |

Other type of biosensor, recently developed is bio-affinity sensor. A charged radio labelled receptor is bound on the surface of the electrode reacting with the substrate. The immobilized substrate is called the determinate analogue. With the increase in substrate concentration, the receptor forms as bound complex with the determinant. This is measured by reducing signal from the bioreceptor. This type of bio affinity sensors are used for hormone analysis and sacchraride and drug analysis.

Whole cell biosensors are used with the addition of microbial cells or micro organism or their organelles which react with the immobilized substrate and show response. Though response is one hour, it can analyse the concentration of substrate at a very low level. Likewise microbe like acto fermentation reacts with assimilase sugar in nearly 10 minutes to measure sugar linearly above 1 m mole dm$^{-3}$.

***Colour changes due to biosensing.*** In these biosensor there is a coating of enzymes or antibodies on the membrane. The biosensor is also coated with dye on the membrane. The substrate reacts with enzyme or antibody and due to change in pH level of the yield there is a colour change is the dye-membrane coating. These changes in the colour of dye is sensed by light emitting diode.

***Use of biosensors.*** Biosensors have following applications in medicine, industry, pollution control, detergent industry etc. Glucose analysers are used to find the concentration of blood glucose level and the amount of insulin required. Detection of Mitomycin and aflatoxin, which causes cancer in infants may be detected by biosensors. The quality control of drinking water has been performed by monitoring with biosensors *i.e.,* measuring BOD (oxygen demand).

A microbial cell biosensor is used to detect the mutagenecity and carcinogenecity of some compounds. Biosensors are used in fermentation control to derive different industrial product.

The electrical circuitry has been replaced by biochips in modern days. The biosignal generated are is relation with electrical signal and these signals are tried to be processed by biochip. The main advantage of biochip is that the electrical element mean the reflected by a bio molecule (like protein) whose size is in nanometer order. The problem of electron tunneling is not a big disadvantage in biochip. As the bio molecule has less electrical resistance than Aluminium or Copper, less heat is dissipated in the biochip. Now-a-day efforts have been tried to develop bio-computers, which is presumably faster than its electrical counterpart.

## 32.3 NANOSENSORS

Before starting with nanosensors it is the need of the day to get accustomed with nanoscience. The basic particle with nanometer dimension is a molecule. As the size is reduced from micrometer to nanometer the technology which was designed with integrated electronics or integrated optics or microelectronic machines systems has moved apart in the new direction of nanotechnology. As nanotechnology concerns with atoms and molecules the concepts of quantization of the matter, wave particle duality matter waves Schrodinger equation and particularly the postulates of quantum mechanics and uncertainty principle are of prime interest.

**Problem with classical mechanics.** The basic assumption for classical mechanics is :

(*a*) There exist dynamical variable $q_i$ and a given time $t$ which is precisely $q_i(t) i = 1$ to $N$ for some integer $N$. Each of these variables take definite value at time $t$ *i.e.,* the state of the system is defined accurately at time $t$.

The development of $q_i(t)$ ($i = 1$ to N) *i.e.,* evolution of the system, with time is deterministic and at a particular time $t_1$ the set of values $q_i(t)$, $i = 1$ to $N$ gives the state of the system.

The differential equation with $q_i(t), i = 1 \ldots N$, was solved by system of particles by taking dynamical variable of position of the particle $q_i(t)$ and their velocities $q_i$ ($i = 1$ to $N$) though only obeyed by classical particles (upto μm in size). The force of law *i.e.*, gravitational force between two massive bodies and coulombs inverse square law, Maxwell's equation and Lorent'z force are well understood for a distance greater than a few micrometer ($10^{-6}$ m).

But at the end of nineteenth century certain phenomena involving the distance in range $10^{-8}$ m to $10^{-10}$ m could not be explained properly. Hence there is a contradiction with classical mechanics at the monolevel or at the molecular level.

One of the departure from classical mechanics is the structure of atoms. When the structure of atoms with orbital motion of electrons around nucleus, the classical picture, is analysed, it is found that a atomic structure is unstable and electrons are predicted to have collapse within the nucleus in much below 1 second. Other discrepancies in classical logic is found in case of black body radiator or harmonic oscillator.

Hence for nanoparticles another law is required to be formulated. The quantum mechanics has two stages of development. One concerns with Planck's introduction, to quantum of action which was not totally satisfactory and the second was the formulation of Schrodinger and Heisenberg, which gives birth to quantum mechanics which the nanoparticles follow. This new mechanics answer satisfactorily, the queries about structure of atoms and molecules, optical spectra from excited atoms, black body radiation and Compton effect.

The structure of atoms and their optical spectra has been solved for Hydrogen atom and Helium atom by Bohr's postulates which state that :

1. There exists a discrete set of stationary states (stable states) of an atom ; from each of these states no electromagnetic radiation is emitted due to accelerating motion of the electrons round the nucleus.
2. Transitions occur from one of these states to another with the emission of radiation. If the initial and final states have energies $E_n$ and $E_m$, the frequency of the emitted radiation is $\frac{E_n - E_m}{h}$ where $h$ is the Planck's constant.
3. The angular momentum of an electron in motion round the nucleus of hydrogen atom takes the value 0, $t$, $2t$, $3t$ ... only where $t = \frac{h}{2\pi}$ and the quantisation (discrete values) of angular momentum is the basic of quantum mechanics.

Though, from this Bohr's postulates the structure of hydrogen atom is somehow perceived, yet these postulates is unable to predict the relative intensities of the spectral lines of hydrogen atom. The postulates also fail to predict the actual line frequencies and for more complicated atoms or molecules.

The new turn in the Nanoscience is the wave particle duality. The wave property of matter is defined by the parameter wavelength (λ) whereas the particle property is defined by momentum $p$. The wavelength of an electron can be measured by the diffraction pattern when the electron passes through a crystal. If the diffraction of a beam of electrons with

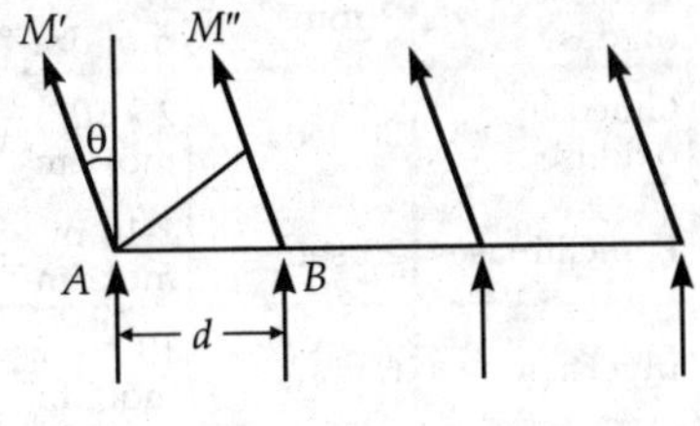


**Fig. 32.5**

wavelength λ at right angle to the atom in a plane (separation between atoms is $d$), the beam is deflected by an angle and the path difference between two adjacent paths ($AM'$, $BM''$) is BC is an integral number of wavelength. So if, the electron has a wave nature with wavelength λ, the maxima of Bragg's diffraction will be found as

$$BC = d \sin\theta = n\lambda \quad n = 1, 2, 3 \ldots$$

Hence this experiment reveals the relation between wavelength λ momentum $p$ of a nanoparticle by

$$\lambda = \frac{h}{p} \; h = \text{Planck's constant}$$

and this relation expresses the wave particle duality of matter and the equation is called de Broglie relation.

The immediate consequence of this wave particle dual nature of nanoparticle is the dynamics of the particle given by Schrödinger equation which is stated as

$$i\hbar\frac{\partial\psi}{\partial t} = \left[-\hbar^2\frac{\nabla^2}{2m} + V(r)\right]\psi$$

with the momentum operator as

$$\vec{p} \rightarrow \left(\frac{\hbar}{i}\right)\vec{\nabla}$$

and the Energy operator is

$$\vec{E} \rightarrow i\hbar\frac{\partial}{\partial t}$$

where $\psi$ is the wave function in view of probability of getting a particle or a wave is a region $\Delta x$ . $\Delta y$. $\Delta z$ at time $\Delta t$. This is connected to the basic Heigenbery uncertainty principle which states

$$\Delta x.\Delta p \geq \frac{h}{2}$$

This uncertainty principle is the limitation in accuracy of simultaneous measurement of position ($x$) and momentum ($p$) of a nanoparticle.

Nanoscience also includes the study of molecules. There is different chemical bonding between atoms which is basically caused by interactions between electrons and atoms or ions involved. The different molecular forces between electrons, atoms and ions include Coulomb's force, ionic force and also van-der-Walls force which is existent in the nanolevel.

Since electrons are responsible for making and breaking of bonds the chemical properties of atoms and molecules are due to electron transport. These bonds acts as mechanical device like hinges, bearings, structural members from nanoscale machines. In greater size (μm) the bonds are related to reaction between two molecules. Hence, bonds are tools in nanoscale. The change in phase, from vapour to liquid, water to ice, the way of packing of molecules is a matter of study in nanoscale. Like, $CO_2$ molecule is a gas but when many individual molecule cluster they form dry ice which is solid. The study of much larger chain of molecules like polymer, is the key to nanoscience. Most metals shine due to the fact that the electromagnetic radiation is scattered by free electrons on the metal. Some metals, though made up of same atoms, is not metallic but is a insulator as for example, coal, diamond, phosphorous, graphite etc. They are not shiny as they have no free surface elections.

Generally nanoscience discuss with four basic materials, metals, insulators, polymer and ceramic.

Polymers are macro molecules and generally a long carbon chain. As carbon has the unique property that a carbon atom may bond with other carbon atom, polymers are carbon compounds. The same carbon molecule are repeated in polymers forming a chain of different length. A polystyrene glass may have innumerable polymers molecules of different length. As electrons are localized in polymers chain, polymers generally act like insulator and this is why PVC (polyvinyl chloride) plastic is used for jacketing live electrical wires. In polymers carbon chains are cross linked Fig. 32.6 and they behave like non-metals and they are hard due to their rigid structure. Hence polymeric science is based on nanoscience.

**Fig. 32.6**

A molecular model of a segment of the polythelene chain. This segment contains 28 carbon atoms (dark), but in commercial polythene there are more than a thousand carbon atoms per strand.

Besides and enzymes are polymers which is responsible for blood oxygen carrier *i.e.*, hemoglobin. Another type of plant protein, nitrogense, is responsible for plant growth. Many types of protein is used for structure and growth of plant, animals and the most important one is Deoxyribonuclic acid (DNA) and ribonuclic acid (RNA) whose structure is responsible for animal's genetic codification. Cell fraction study reveals that the DNA is a part of cell nuclei whereas RNA is present in both nucleus and cytoplasm.

The fundamental part of DNA and RNA are nuclotide whose formation is given below.

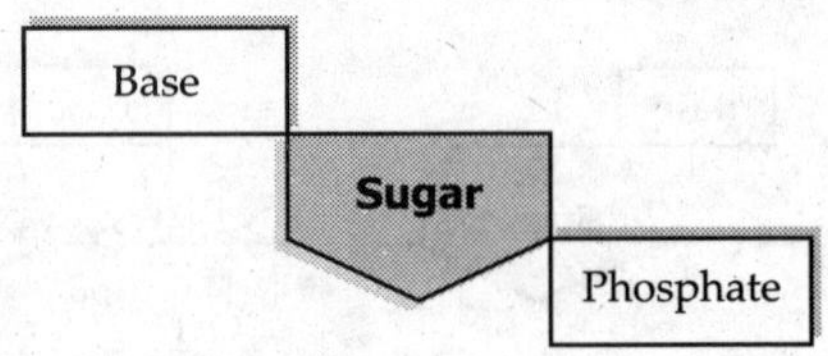


**Fig. 32.7**

The nucleotides are joined by diester link and they are of molecular weight 50000 units. From X-ray crystallgraphy, it is found that in the model of DNA, nitrogenous bases are at right angles to the axis of the chain and the planes of the sugar molecules and the nitrogenous bases are at right angle. Pauling and Corey postulated a helical structure of DNA from the results.

The structure of DNA has been shown in Fig. 32.8.

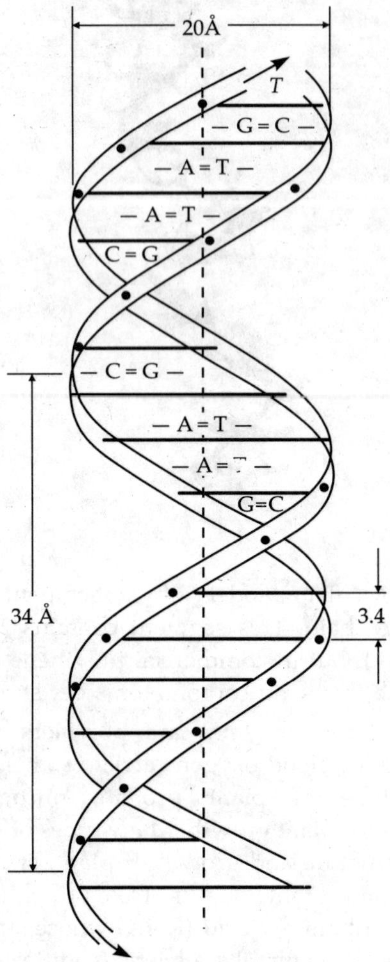


**Fig. 32.8** Model of DNA – The Biological Role of the Nucleic Acids, Arnold.

The nucleotides were joined by diester links between the phosphoric acid and the sugar molecule, *viz* :

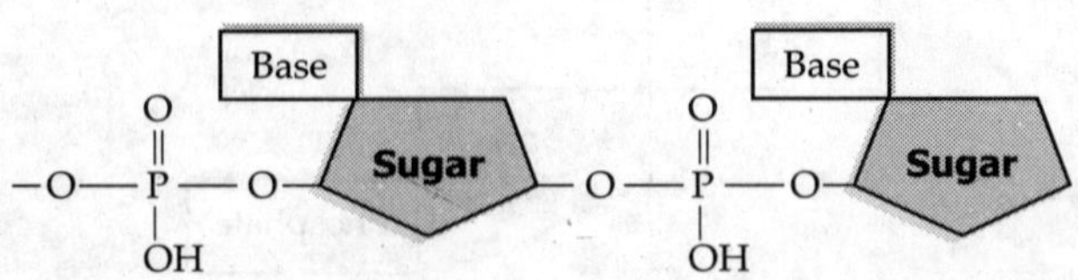


and such dinucleotides were first synthesized by Todd in 1955.

During this period it became clear from chemical analyses that both DNA and RNA were macro-molecules with molecular weights of the order of 50,000 units.

X-ray crystallography showed that the nitrogenous bases were at right angles to the axis of the chain and that the planes of the sugar molecules and the bases were mutually at right angles. Pauling and Corey (1953) then postulated a helical structure for the molecule which was, however, of incorrect form.

Crick and Watson's model of DNA consists of two coaxial polynucleotide right-handed helices, each of which were of equal pitch and subtended an angle of 120 degrees at the axis (Fig. 32.8).

If we temporarily forget about the helical twisting of the molecule, we may use the analogy of a chain ladder. The rungs of the ladder are formed by the hydrogen bonding of two nitrogenous compounds.

The DNA structure contains negative charge due to presence of phosphorus and oxygen. DNA structure is double helix with a pile of chips : Each base pair can have one of the four combinations namely *AT, TA,* (*G* or *GC*). For each position of the staircase the *two* planar molecules are any of the *four* molecules adenines Thymine, guanine and cytorine *i.e., A, T, G, C. A* and *T* bond between themselves and *G* and *C* can be bonded. So there are four combinations possible. The genetic code in the structure of DNA is read by proteins and RNA in the cell. The nanoscience and associated technology is a way to read the DNA code.

Another nanomaterial is ceramics. Ceramics, most of the time is oxide, but, it is not always oxide. In ceramic, a long chain has been produced with an extended structure of oxygen. Like clay is aluminium oxide, sand is silicon dioxide and fire brick is magnesium silicon oxide.

Unlike metals, ceramics are non-conductor of electricity and generally not glazy. They are hard and sometimes brittle.

***Molecular recognition.*** The bonding affinity between one large molecule with other molecule due to coulombic attraction is called to be molecular recognition. Like allergens is a large molecule which is recognised by large molecules in the body and the body molecule bind to the allergens and affected by foreign molecules. Our sensory experiences are also molecular recognition. The nasal bulb can, recognise large foreign molecules find with it and creates schration in our body. Due to molecular recognition insects attract at another and higher biological organisms is structured with proteins.

### 32.3.1 Measurement of Nanostructure

**Probe type sensing.** There are a few ways of measuring nanostructure. The probe type instruments are popular. Probe type instrument with microscopic observation type fitted with modern processor can measure upto one nanometer. The probe is generally made of tungsten wire whose tip is of nanoscale diameter and often one atomic diameter in thickness is used for scanning the surface with nanoscale irregularities. The nanoscale irregularities apply different pressure on the tip of the probe as the probe slides on the surface, which corresponds to different types of measurements. Atomic Force Microscopic (AFM) technique ensures the sensing of pressure exerted on the probe tip and as the probe scans the surface a composite signal is generated. This signal is analysed by a component to get the picture of irregularities on the surface in atomic scale.

The other way around is Scanning Tunneling Microscopy (STM). In this technique, the probe tip is fitted with a high frequency electrical circuit which generates different current at the touch of probe tip on different irregularities in surface and this technique test the local geometry of the surface *i.e.*, how much of the surface protrudes locally on measuring the local electrical conducting characteristics.

The other type of probe measurement is Magnetic Force Microscopy (MFM), in this technique, a magnetic tip or electromagnetic lines of force is used to get the local magnetic structure of the surface. It acts just like a proximity sensor whose magnetic field is distributed differently as the probe scans locally with difficult geometry protruding parts of the surface.

Figure 32.9 shows a nanoscale probe with dimension as seen under an electron microscope.

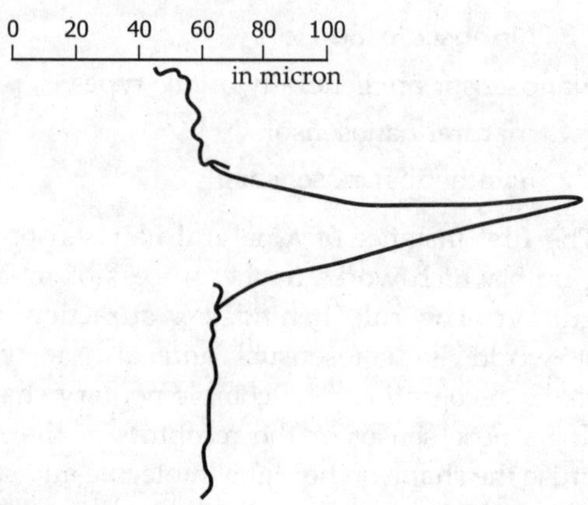


**Fig. 32.9**

The scanned data is fitted to the computer to get a human usable picture of the surface. The computer picture provides internal surface irregularities in micrometer length and breadth *i.e.*, the scanning probe makes it possible to view in atomic dimension, locally.

***Spectroscopic technique.*** The localised scanning of nanoscale structure can be made by spectroscopic technique. In spectroscopy a specific colour light is focused on the nanoscale surface and the information from the localised surface is gathered by observing the scattered or absorbed or refracted rays from the surface. Spectroscopic technique includes the processing of intensity or phase sensitive data from the incident and scattered or absorbed ray. X-ray spectroscopy is advantageous to use for nanostructure as wavelength of X-ray is as low as 1 μm or 10 nanometers whereas visible light spectroscopy provides a wavelength much greater than nanoscale irregularities. Now-a-days Magnetic Resonance Imaging (MRI) is used as nanoscale spectroscopic technique.

***Electrochemical probe.*** Electrochemical analysis surface with nanoscale irregularities is based on the principle of generation of charged layer at a specific permeable membrane due to chemical reaction of the probe tip with the surface. The membrane provided, is permeable to the surface atoms. The electrochemical analysis of the surface atoms in an array provides information about the nature of surface.

***Electron microscopy.*** Electron microscopy is one type of nano-surface scanning technique where the object is exposed to electron beam rather than visible light and the transmitted electron beam intensity is analysed to get idea about nanostructure. The popular Transmission Electron Microscopy (TEM) is what, the instrument is called.

The advantage of using TEM for nanomeasurement is that, it can measure local nanoscale physical structure rather than forces like electric or magnetic field. The TEM image is processed and viewed at length of few μm.

### 32.3.2 Generation of Nanostructure

Probe type nanostructure generation instruments are available now-a-days. Though expensive, and, to develop a nanostructure is time consuming, the probe type instrument is a way to create a nanostructure. The probe which is used to analyser nanostructure, is the same probe with its tip a few nanometer. The assembling of the structure on an atom-by-atom or molecule by molecule basis. The probe tip interacts with the surface by giving specific pressure or creating localized electric or magnetic field or by electro-chemical process. The nanoscale probe generation is a very slow process and is unable to satisfy mass demand.

32

***Nanoscale lithography.*** Lithography is the process to make a object from a single stone : In silicon technology lithographic image is produced on a silicon wafer by producing a mask on the wafer with a oxide coating on it and by exposing the wafer to UV ray or by chemical method eliminating the mask to get actual chip structure. The UV lithography and more modern lithographic technique limit the dimension of mask greater than 0.4 μm or 4 nanometer. Hence X-ray lithographic technique is used to generate nanoscale structure. The new lithographic technique that is accepted today is the imprint lithography. In this technique nanoink / molecular ink is used to develop a nano structure on a material rubber and press it on the target material so that nanoscale structure has been imprinted. This process of generating nanostructure is inexpensive and faster then other processes but the technology to create a nanostructure on a rubber by molecular ink is new.

***Dip pen technique.*** Atomic Force Microscopy (AFM) probe is used to read nanostructure with a dip pen. The molecular ink is used in the pen and as AFM scans and put force on the surface a nanoetching is developed on the surface with molecular ink. The Fig. 32.10 shows how Dip Pen nanolithography technique is used.

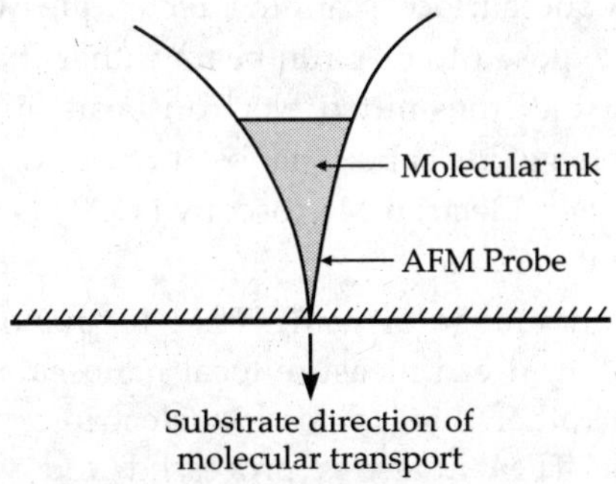


**Fig. 32.10** Dip pen technique.

Electron beam may be used to eliminate the mark on the substrate in a desired way and this type of nanolithography is called electron beam lithography. Electron beam lithography is an accepted technique and it is better than X-ray lithography in the sense the 1 Å X-ray is more particle in nature than wave nature and can damage the substrate molecule in a undesired way.

***Lift off nanolithography.*** In this process molecular ink is spray printed on the substrate and if an ink molecule is lifted off the surface, it produces a dot on the substrate. This process provides good edge accuracy at the edge of the drops. Figure 32.11 shows the lift off technique.

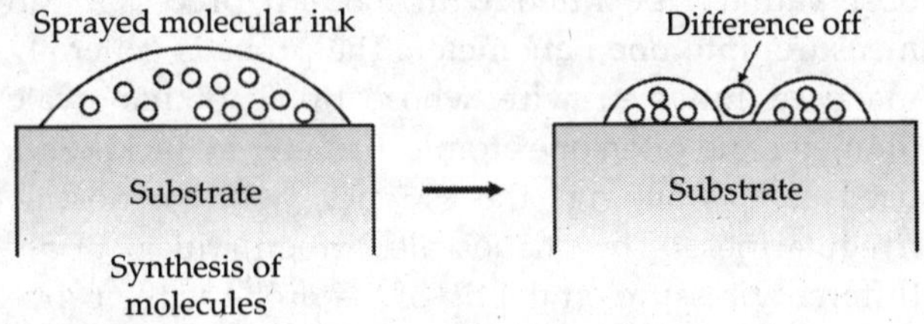


**Fig. 32.11** Lift off Lithography

**Synthesis of molecules.** The production of molecules to a target structure is called molecular synthesis. This includes synthesis of protein, catalyst, enzymes and last but not the least the DNA synthesis. Synthesis molecules are used for drug, detergent industries. This nanosynthesis can be realised by a process called self- assembly. As the molecule will seek the lowest energy level an atom or a molecule can be placed at a desired position by relating particular energy level diagram.

The other method for synthesis of particular molecular pattern is polymerization. A polymer like PVC contains million of molecular in chain. Modern day technology permits controlled polymerization *i.e.,* a monomar molecule is added to the polymeric growth at a time.

In DNA synthesis, when a synthetic DNA sample is added on to a bacterial DNA the DNA of micro-organism *i.e.,* bacteria, many copies of a particular protein or enzymes is produced. This bio-synthesis of nanoscale DNA is an important process in Drug industry.

### 32.3.3 Nanoscale Sensors

Nanosensor are generally of *two* types

1. natural nanosensors
2. synthetic nanosensors.

The first instance of a natural nanosensor is the nasal bulb which works in the process of molecular recognition. The rule behind sex attraction in the insect world is nanosensual and also a type of molecular recognition. The complementary shapes in the dog's nose sensor or the receptors of the insects recognise the shape of the signal molecule and sensors are actuated by particular distribution of electronic charge due to the generation of new polymeric molecule which is molecular recognition.

#### Synthetic Sensors : Synthesis Sensing of Biological Entities

One type of synthetic sensor is biosensor. When a animal body has been exposed to allergen or a microorganism like bacteria or fungi, the body is sentisized, *i.e.*, creates antibodies. The antibodies in view of the principle of molecular recognition, sense the allergic proteins or micro organism and release histamine. This histamine is a substance due to which body reacts in the way of itching, sensing or nansia. The blood glucose level has been identified with an electrode which contain one semi permeable membrane to separate glucose oxidase (an enzyme) from glucose solution and another membrane to separate glucose oxidase from electrode. The reaction of glucose with its enzyme activates the ion sensitive electrode.

## Nanoscale Biosensor : DNA Sensing

DNA sensing is another aspect of nanosensor technology. DNA hybridisation is defined as the ability of DNA to bind to a complementary strand (structure) and not binding to any other nano molecule *i.e.*, protein molecule. Like, a DNA sequence GCCGTTC could be sensed with a DNA sequence CGGCAAG obtained by translation. Different DNA sequences can be sensed using the complementary strand, therefore, the amino acid sequence of corresponding protein or enzyme is determined. The number of different combinations possible for 4 bases in a single strand combinations possible for 4 bases is a single strand is $4^6$. Hence if a particular biological molecule has a known DNA sequence, it is possible to get a short list of 10 to 15 sequence, which can sense it uniquely, without error. This type of biosensor with DNA sensing is used for finger print study or to predict a genetic disease. The probability of error in this type of biosensing is one in million for a 15-base strand to be tested.

## Micromachined Electromechanical Sensors (MEMS), Smart Sensors and Systems and System on Chip (SOC)

Silicon microsensors are used as an essential part of process control and analytical instrumentation in a wide application range like industrial monitoring and automation. Other uses are in the automobile engineering, transportation, telecommunication, environmental and biomedical engineering. The growth in micromachined electromechanical systems *i.e.*, MEM systems and application specific integrated circuit (ASIC) design, sensing and signal processing has become more and more precise and reliable. The integration of sensing part and signal processing part on the same integrated chip, the idea of SMART sensor has been developed. The same technology has been used for system on chip (SOC) in which the sensor, the logic to control the system, the micro actuator and also the dedicated display unit with a processor has been developed on the same chip. The display unit is preferably built with porous silicon or some other optoelectronic material. Silicon transducer are generally of *two* types. The first one is MEMs (Micromachined Electromechanical Systems) and the other porous silicon based mechanical structure.

Micromachined technology ensures the fabrication of mechanical devices on a silicon wafer of very small dimension (.1 – 10 μm) and these devices may be cantilevers, nozzles, cavities, diaphragms or membranes, small bridges, mechanical resonators etc. The silicon wafer is chosen for the development of mechanical devices as yield strength of silicon is double the yield strength of steel, whereas, Young's modulus is chosen to that of steel and density of silicon is one third of steel. As the yield strength is high and density of silicon is lower than steel it is easy to be machined in the micron order. Other advantage of silicon is that its plasticity is small and hence it yields less permanent deformation in stress and show very low hysteresis. Hence, the MEMS and porous silicon devices are ideal sensing device with high repeatability.

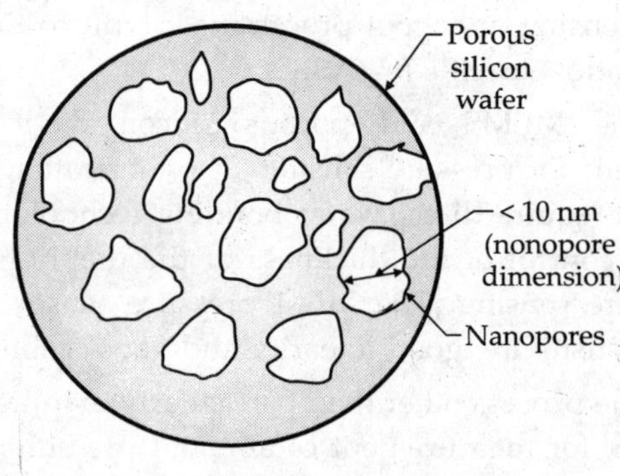


**Fig. 32.12** Porous silicon humidity or gas sensor.

32

The technology used for micromachining of silicon wafer is controlled etching in a preferential direction with anisotropic etchants like potassium hydroxide, TMAH, EDP. To develop microstructures like cantilever cavity, nozzle on the silicon chip, hydrazine and complexing agents like isopropyl alcohol are used in a controlled fashion. The crystal structure of the echant and controlled temperature used during etching process are the parameters on which etching rate is dependent. The other way of etching is electrochemical etching to develop the MEMS and this type of etching provides better control.

***Porous silicon humidity sensor, porous silicon devices.*** The porous silicon technology have delivered a convenient way to integrate sensor, filters, waveguides electroluminescent display on a single chip. The high surface to volume ratio of porous silicon ($> 500\,m^2/cm^3$) ensure a good sensing medium for pressure, humidity or as Bio-sensing. In a vapour sensor the small pores ($< 10$ nm) *i.e.*, nonpores of the porous silicon absorb the vapour or gas molecules and there is a change is permittivity or conductivity of the silicon chip. The nanopores are generally created by generating a current density on the silicon chip or by anodic etching of the silicon chip. The shape and position of the nanopores on silicon chip is controlled by the formation of current density or by properly composing the anodic etching solution. The porous silicon is a distributed type resistance capacitance network with high piezoresistive property. Hence, the magnitude and phase of the electrical signal both can be modulated during sensing by these devices. The other important criteria for porous silicon to be used as a system on chip is its lumniscence at room temperature. Hence LED display developed on chip of porous silicon can be integrated with sensing an signal processing circuit to develop application specific IC (ASIC).

The MEMS and porous silicon devices are excellent for pressure sensing. The fabrication of the sensor require direct wafer bonding technology and better control of the thickness of the diaphragm for pressure sensing. The MEM pressure sensor shows high sensitivity, good linearity and good stability.

The process other than particularly diaphragm as MEMS, for measurement of ambient pressure, is the use of a layer resulting in the formation of cavity which interact with ambient pressure.

## 32.4 BIOSENSORS

As silicon is bio-compatible, it can be embedded in the living cell, without toxic effect. A typical MEMS has been designed in which two electrodes on a free standing membrane has been developed to analyse the rate and amount and monitoring of fermentation process. Besides MEMS and porous silicon biosensors are used for blood pressure, blood flow, oxygen saturation and measurement of body temperature.

The humidity measurement scheme with fabrication of porous sensor and the system on chip (SOC) arrangement has been shown in Fig. 32.13.

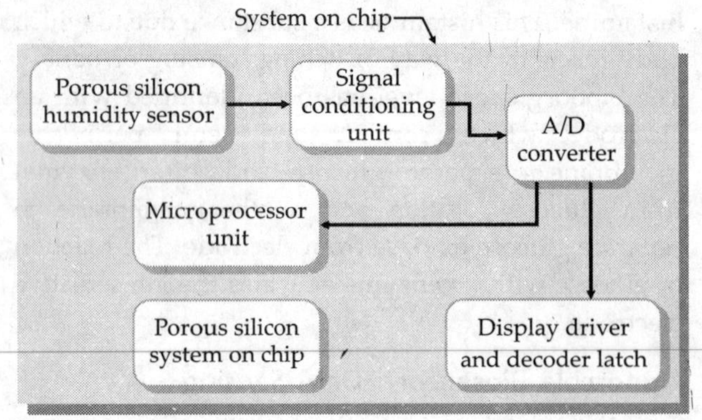


**Fig. 32.13** System on chip for humidity sensing.

## 32.5 ADVANCED INSTRUMENTATION TECHNIQUE

### 32.5.1 Virtual Instruments

Virtual instrument laboratories are associated with the modelling of instrument for a particular application with specified proper range resolution, accuracy, repeatability : A novel virtual measurement laboratory is based on a software support like internationally accepted software like CORBA or Jini which provides the tools and methodologies to assess the required specifications and deliver the inputs and output pattern of the sensor *i.e.*, instruments in a Java accepted objects. The applications software for virtual instrumentation is in a internet acceptable Java environment which can accept object *i.e.*, measuring instrument from the internet and place the developed instrument as an object in the internet service. Hence, the environment of the virtual instrumentation laboratory is a distributed system with networking facility. Like, a Jini virtual instrumentation laboratory, the measurement instrument concern the monitoring of the power network characteristic. Attention is focused on (1). the power network analyzer services and the monitoring services.

The work aims at experimenting in a distributed environment over the internet using industry based software techniques. This virtual instruments are developed in modular approach and the major idea is the modular design and exploitation of heterogeneous computing platforms for supporting a open and dynamically configurable measurement system.

The monitoring system of a smart power electrical network analyzer is shown in the following Fig. 32.14.

In the last few years a number of tools have been developed for designing virtual distributed system in a distributed way. CORBA software or Java related software JAVA-RMI, Java spaces, Jini and some mobile agent system are few of them. Like CORBA aims at developing interactions between object based, interoperable applications in different implementation language. JAVA technologies are able to transform a instrument developed of heterogeneous machine in a network of homogeneous virtual machine.

Figure 32.14 shows a distributed measurement example. In a reported work a series of measurement experiments were developed in JINI software. They include the instrument control, monitoring current and voltage at a remote load (Fig. 32.14) frequency response of an amplifier and calibration of a DAQ load.

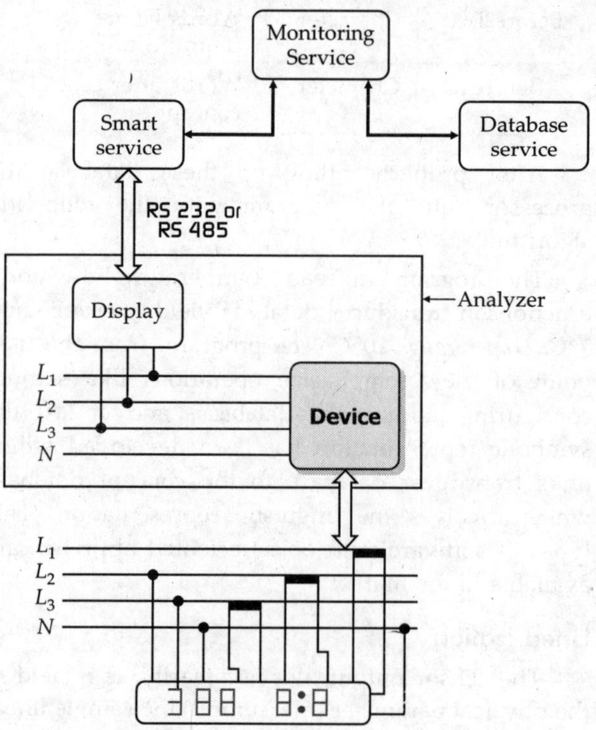


**Fig. 32.14** Typical virtual instrumentation power measurement object.

The measurement test in Fig. 32.14 is organised in *three* services :

(*a*) The power electrical network analyzer service

(*b*) Database service

(*c*) The monitoring service, which implements the measurement algorithm and controls the other two services.

At last all the measurement services in a modelling environment can be remotely controlled through distinct graphical user interface (GUI) which can be viewed in an acceptable way by the enduser.

## 32.6 INTELLIGENT INSTRUMENTATION

The manifestation of a measurement system or instrumentation in the real world so that human being can perceive the situation by high level of interaction preferably in English language is called intelligent instrumentation. The perception of the temperature in a room can be in form of a voltage level or symbolic like "too cold", "cold", "hot", "too hot" and warm. The symbolic representation provides a person newly acquinte with the instrument a idea about the room temperature. Similarly, for a level control system, the set point specification or the measurement output is of no meaning to a person who handles the instrument for the first time. Here another symbolic representation, which is called Man Machine Interface (MMI) which shows how level is getting changed in the reservoir with time helps a new user to adopt to the system quickly. Many software has been developed which includes the engine that converts the measurement results and a simulator that allows the user to verify the settings of the instrument. These software for intelligent measurement does *two* works

1. numeric data to symbolic data conversion and
2. hardware/software resources that will host and execute the program.

The symbolic transducer represents a processor base intelligent transducer that can talk to the user not in number but in English or process identifier graphics. The symbolic approach depends on a functional model based on the operator's qualitative observations about the measuring process. The measured quantity is represented by a qualitative model rather than numerical.

The software for a symbolic representation generator is shown below. This software requires *four* different databases to be filled up with proper data and diameter so that with the number of program on these databases yield representation of measurement.

## 1. Transducer Database Structure

| Field Name | Data Type | Explanation |
|---|---|---|
| Code | Numeric | Identify code for transducer |
| Name | Character | Transducer's name |
| Class | Character | Type of transducer |
| Lower range limit | Numeric | |
| Upper range limit | Numeric | |
| Accuracy | Numeric | |
| Measuring unit | Character | |

## 2. Application Database Structure

| Field Name | Data Type | Explanation |
|---|---|---|
| Transducer code | Numeric | Links to certain transducer |
| Application code | Numeric | Internal code for application |
| Name | Character | Application name |
| Lower range limit | Numeric | |
| Upper range limit | Numeric | APPLICATION |
| Central value | Numeric | RANGE LIMIT |
| Concepts | Numeric | Concept No. |
| Form code | Numeric | Code No. corresponding to concept form |
| Adoptation code | Numeric | The adoptation function code |

The application database structure contains a field created to link this database to transducer's database. This linking or relations is necessary because the applications are properties of the transducer.

## 3. Function Database

Each function has to satisfy several mathematical tricks so that the transducer output has been linked to the linguistic concept via functional output. Say, for examples, the lower limit is equal to lower range limit of transducer database + central value of application database.

**In this way the functional database structure has been grown as :**

| Field Name | Data Type | Explanation |
|---|---|---|
| Function code | Character | The name of |
| Name | Character | The function |
| Branch J | Numeric | The mathematical expression of 1st function branch |
| Lower 1 | Numeric | |
| Upper 1 | Numeric | |

Lower 1 and upper 1 is the first branch of the function which is defined on the interval determined by those two limits.

## 4. The last date base is the concept database

| Field Name | Data Type | Explanation |
|---|---|---|
| Code | Numeric | Central concept code |
| Text | Character | Words representing central concept |
| Before Text | Character | Words before central concept |
| After Text | Character | Words after central concept |

After properly filling up these database the processor runs the program with the following algorithm.

The program will read "Temperature Low" for a function on transducer database yields : *Lower range* 5°C, *Upper range* 10°C. The program, from the user point of view, emphasize operations like setting, configuring, adding the databases and at last the symbolic representation has been developed which links transducer database to the concept database which involves the linguistic representation. This types of software with object oriented approach are available in the market.

## Linearization

The sensor output, not necessarily, is related to the physical parameter (measurand) by a single linear relation. For such type of sensors where the output maintains nonlinear relation with the sensed variable (measurand) there is a need for prediction of the output for a particular input to draw the calibration curve. Again, for a unknown value of measurand (physical parameter) the experiment suggests a output value which is an indication of the measurand. When the input and output maintains a linear relation the

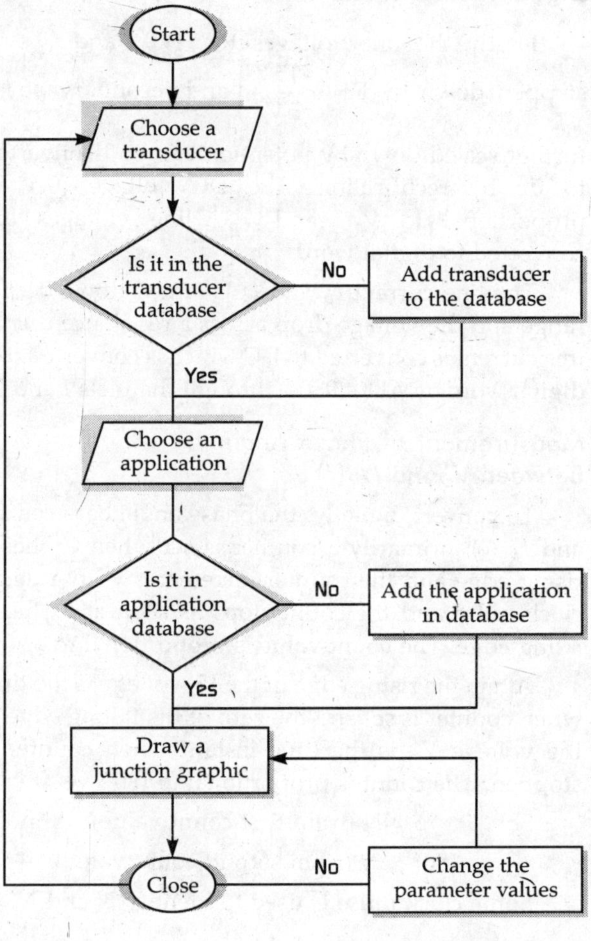


**Fig. 32.15** Flowchart for Intelligent Instrumentation.

prediction is easy, whereas, for nonlinear relation between input and output, there is a need for linearization. In a particular case, when the relation between sensor output and measurand is quadratic the output must be square rooted so that square root quantity makes an linear relation with the input and this linearlization technique ensures a linear calibration curve. Sometimes, the linearization technique is not as simple as the above mentioned example (Just square rooting the output). The input and output may be related by a more complex mathematical relation which needs more complicated linearization treatment to be processed. There are certain benchmark technique for complex linearization process. The Analog to Digital converters used now-a-days has a reference voltage input ($V_{REF}$) which coupled with some input resistances and capacitances gives linearization in case of a more complex process. The 0804 or 0808 A/D converter ICS have this facility. The physical logic behind this technique is that, the time constant (RC) of the voltage charging has been changed by changing the resistance and capacitance and the charged voltage is compared to the reference voltage $V_{REF}$ by a comparator whose output detects the proper linearization circuit.

The other method used as complex linearization technique is the Artificial Neural Network (ANN) based algorithm which seems to be very efficient. The ANN systems can stimulate any complex relation between input and output with a weight matrix. And the prediction is :

$$[\text{InputVector}] = [W] \times [\text{Output Vector}]$$

where [W] is the weight matrix. This weight matrix is obtained by supervised learning technique when a output-input pattern is presented as the learning pattern.

## 32.7 DIGITAL THREE PHASE REAL POWER AND ENERGY MEASUREMENT

There are few disadvantages of real power and energy measurement using power meter or energy meter. Due to reactive components *i.e.*, inductance or capacitance of coils the eddy current is induced in the metal parts of the instrument by varying magnetic field of the current coil. Other faults occur due to fault in potential transformer which cause under registration. This chapter in this book deals with some digital wattmeter or energy meter for power and energy measurement with the help of microprocessor or by optical means.

The schematic diagram of power measurement for three phase four wire system has been shown in Fig. 32.16.

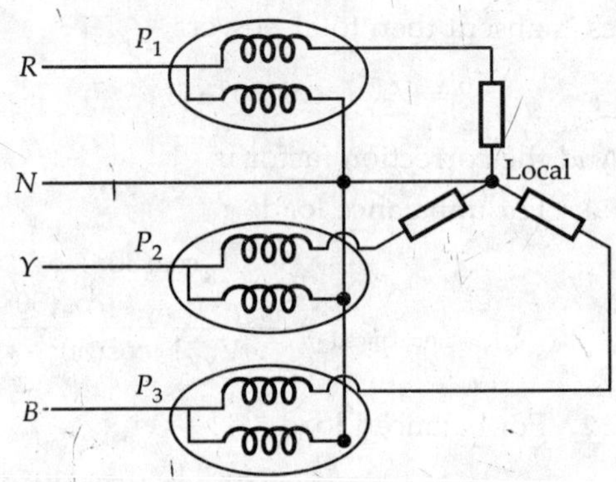


**Fig. 32.16**

The same diagram of measurement of power for three phase three-wire system has been given by Fig. 32.17.

The vector diagram corresponding to Fig. 32.18 is given by

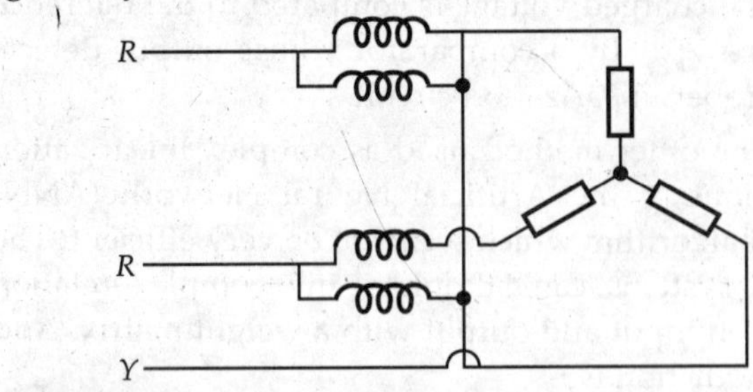


**Fig. 32.17**

The total power flowing through the load (Fig. 32.18).

$$P = V_{RY} I_R \cos(30°+\phi) + V_{BY} I_B \cos(30°-\phi) \quad ...(1)$$

If the load is balanced then

$$V_{RY} = V_{BY} = \sqrt{3}\, V$$

and $\quad I_R = I_B = I$

or, $\quad P = 3\, VI \cos\phi \quad ...(2)$

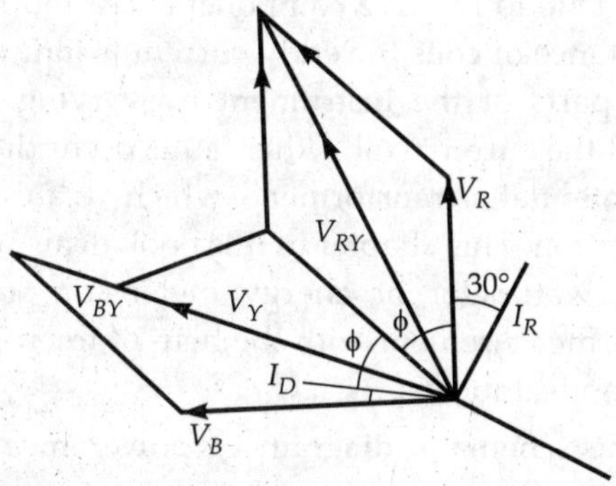


**Fig. 32.18**

If the potential coil between red and yellow phases is absent then total power

$$P = V_{BY} I_B \cos(30°-\phi)$$

And the correction factor is

1. For unbalance load

$$C_{RY-\text{unbalanced}} = \frac{V_{RY} I_R \cos(30°+\phi) + V_{BY} I_B \cos(30°-\phi)}{V_{BY} I_B \cos(30°-\phi)}$$

2. For balanced load

$$C_{RY-\text{balanced}} = \frac{\cos(30°+\phi) + \cos(30°-\phi)}{\cos(30°-\phi)}$$

Similarly, for fault in $PT$, between $Y$ and $B$ phases

$$C_{BY-\text{unbalanced}} = \frac{V_{RY} I_R \cos(30°+\phi) + V_{BY} I_B \cos(30°-\phi)}{V_{RY} I_R \cos(30°+\phi)}$$

## Digital Measurement Scheme with ADC

Firstly the line voltages $V_{RY}$, $V_{BY}$ and $V_{BY}$ is stepped down by PT $\left(\frac{415\,V}{5\,V}\right)$ and secondary of PT further scaled down by potentiometer and converted to dc by rectification. The DC voltage (0.5 V) proportional to $V_{RY}$ and $V_{BY}$ line voltages are converted to digital word.

The line current $I_R$ and $I_B$ is scaled down $0-1\ A$ range and the voltage drop across a resistance due to this current is converted to DC which is converted to a digital word by ADC (0804) through channels 1 and 2.

## Measurement of Phase Angle Between $V_R$ and $I_R$ (ϕ)

To convert digitally, the phase angle between $V_R$ and $I_R$ (ϕ), primarily a counter is set when $V_R$ has a rising edge and the counter increments with a delay clock pulse and the count stops as soon as $I_R$ has a rising edge. The count value is proportional to ϕ.

Again the rising edge of the $V_R$ voltage is the time when counter is set and the zero of the falling edge of the voltage $V_R$ is the time instant when counter is stopped. The count is proportional to 180°.

So $\quad \phi \propto N$ (from first count value)

$180° \propto N$ (from second count value)

Some clock pulse is used to identify $N$ and $N_1$.

$$\text{Now,}\ \cos\phi = \frac{3.14\,N}{N_1} = \frac{13\,A_H \times N}{N_1} \ \text{(in Hex)}$$

Changing to hexadecimal codes

$$\therefore\ 100\cos(30°+\phi) = \frac{1}{2}\left[C8_H - \frac{\left[34_H + \frac{13\,A_H \times N}{N_1}\right]}{64_H}\right]$$

[From $\cos x = 1 - \frac{x}{1}$ from (2)]

where, $\frac{\pi}{6}$ radian $= 34_H$ and $C8_H = 100_{10}$

Similarly,

$$100\cos(30°-\phi) = \frac{1}{2}\left[C8_H - \frac{\left[34_H - \frac{13\,A_H \times N}{N_1}\right]}{64_H}\right]$$

Hence the total three phase power calculated

$V_{RYH} \to V_{RY}$ Voltage in hex in hexadecimal

$I_{RH} \to I_R$ current

$$= \left[\frac{64_H \, mfV_{RY} V_{RYH} mf_{IR} I_{RH} \cos(30° + \phi)}{64_H}\right] + \frac{64_H \, mf_{VBY} V_{BYH} mf_{IB} I_{BH} \cos(30° + \phi)}{64_H} \times K$$

for line fault where,

$$V_{RYH} \neq 0 \quad \text{and} \quad Y_{BYH} \neq 0$$

for $\quad K = 1 \quad \text{and} \quad V_{RY} = 0$

then $\quad P = \dfrac{64_H [\cos(30° + \phi)_H + \cos(30° - \phi)_H]}{64_H \cos(30° - \phi)_H}$

Similarly, we can find an expression for $C_{RY-\text{unbalanced}}$.

The fault measurement algorithm is shown in the flow chart.

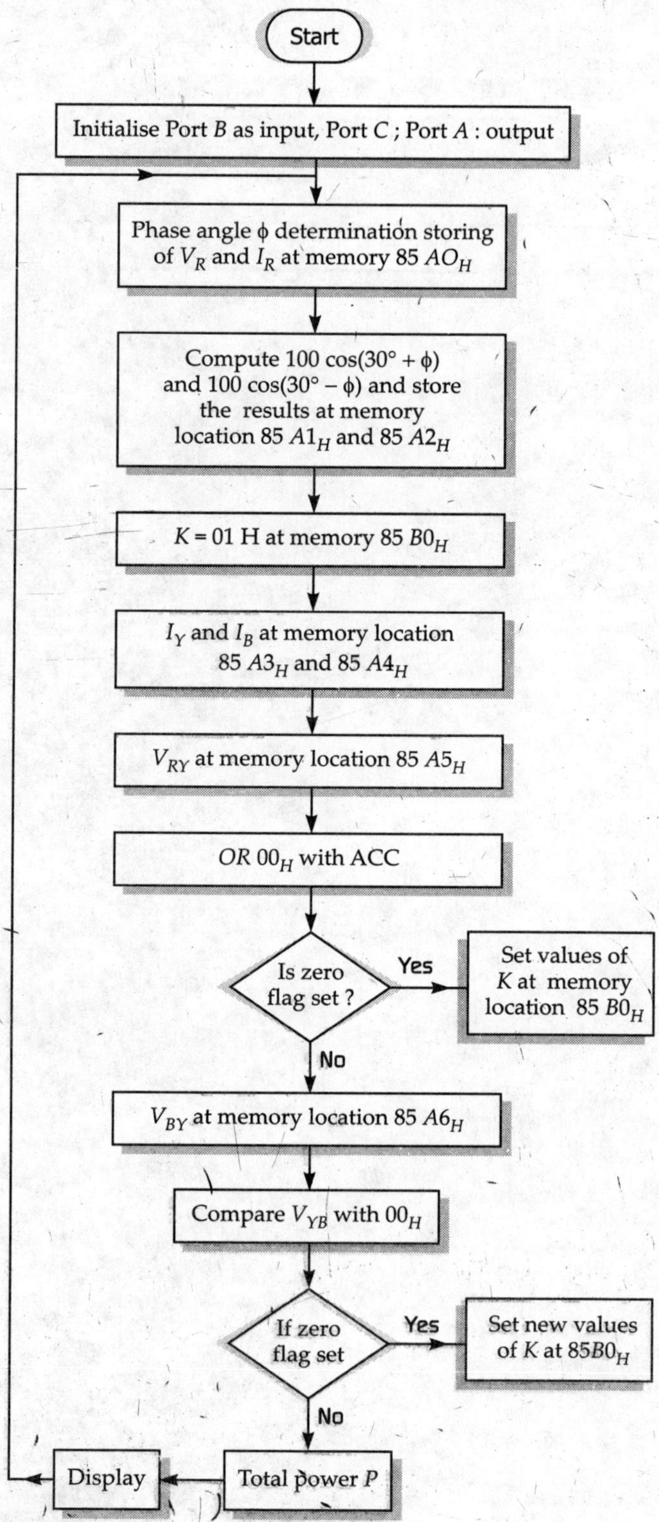


Fig. 32.19

# 33

# Digital Measurement Techniques

## 33.1 INTRODUCTION

The digital techniques used in Industrial/ Electronic Instrumentation enrich the user with high accuracy (5% to 0.05% of the range of Measurement) measurement scheme. Besides digital system provides high input impedance (~ 2 MΩ) to ensure less loading effect on the input circuit. The numerical readout of the digital system allows the worker to perform measurement with zero parallax error unlike analog system.

## 33.2 TYPES OF TOOLS USED IN DIGITAL SYSTEMS

The basic tools used in digital system are

1. Crystal oscillator (1 MHz) with high temperature stability ($10^{-8}/°C$) used as the main element in a LC circuit.
2. Decade (MOD 10, MOD 6, MOD 12) counters are used for realising clock pulses with different period.
3. The Analog to Digital (A/D) converters are used extensively in digital system. The main A/D converters used in digital technique are dual slope converter and successive approximation converter.
4. The voltage to frequency converters are used for high precision digital systems.
5. The Phase Lock Loop (PLL) is also a building block in digital measurement scheme.

The above mentioned tools in digital instrumentation are discussed hereby.

### 33.2.1 Crystal Oscillator

The inductor-capacitor (LC) oscillators generally produce higher frequency than resistance—capacitance oscillators. Whereas, the high frequency produced by LC circuit has low temperature stability and its operation is limited to temperature variation. But the digital systems need to maintain the timing sequence with a high accuracy. Hence the crystal (piezo electric) used in the LC circuit, ensures a clock pulse with high timing accuracy. Generally 1 MHz crystal is used in LC oscillators circuit which produce a clock pulse of 1 MHz and duty circle nearly 50%.

### 33.2.2 Counters

The counters are used in digital circuits to get a clock pulse of particular frequency (1 Hz – 1 MHz) by dividing the fixed frequency clock, (1 MHz), generated by the crystal oscillator circuit by 10, 100, 1000 and so on. The divide by 10 counter is achieved by a decade (MOD 10) counter.

Figure 33.1 shows a decade counter circuit with four RS flip flops.

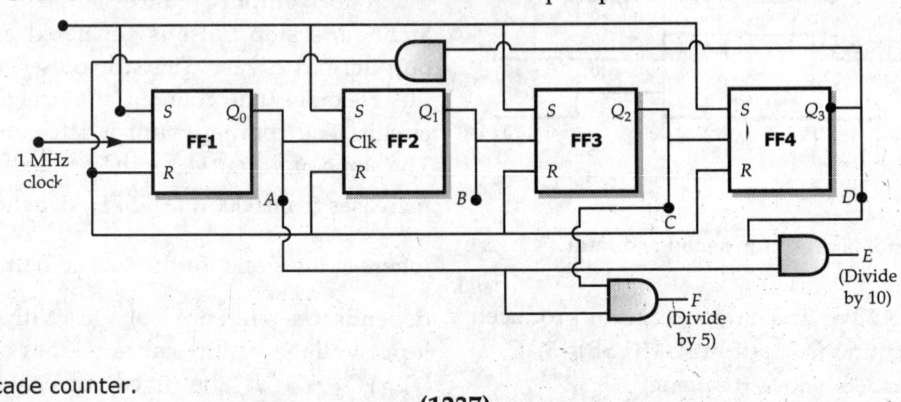


**Fig. 33.1** Decade counter.

Decade counter is a MOD 10 counter whose output (*ABCD*) changes state in the following way :

$$\begin{matrix} 0\ 0\ 0\ 0 & \rightarrow 0\ 0\ 0\ 1 & \rightarrow 0\ 0\ 1\ 0 \rightarrow 0\ 0\ 1\ 1 \rightarrow 0\ 1\ 0\ 0 \rightarrow 0\ 1\ 0\ 1 \rightarrow 0\ 1\ 1\ 0 \\ \uparrow\ \uparrow\ \uparrow\ \uparrow & \uparrow\ \uparrow\ \uparrow\ \uparrow & \\ D\ C\ B\ A & D\ C\ B\ A & \end{matrix}$$

$$0\ 0\ 0\ 0 \leftarrow 1\ 0\ 0\ 1 \leftarrow 1\ 0\ 0\ 0 \leftarrow 0\ 1\ 1\ 1$$

It is an asynchronous counter since different clock pulses has been used for different flip flops.

Now if the fixed stable frequency clock pulse of 1 MHz has been used for flip-flop 1 (FF1 in Fig. 33.1) the waveform from *E* is divide by 10. The AND operation of most significant bit *D* and least significant bit *A* is decade less than 1 MHz clock pulse *i.e.,* 100 kHz. Again the O/P of FF 3 and FF1, which is F, is a frequency which is divide by 5 of clock frequency *i.e.,* 200 kHz.

Similarly using MOD 12 counter we can generate a clock pulse which is of frequency 12 times less than master clock frequency and in this way the divide by 6 module of master clock has been achieved.

The circuit shown in Fig. 33.1 is an asynchronous circuit and hence delay of each flip-flop is supposed to be additive, but the use of AND gate *A*1 reduces total delay and for practical cases the delay is as small as 1 nanosecond.

The timing diagram for the decade counter is shown in Fig. 33.2 (taking the change in state to be negative edged triggered *i.e.,* change in state for $1 \rightarrow 0$ transition).

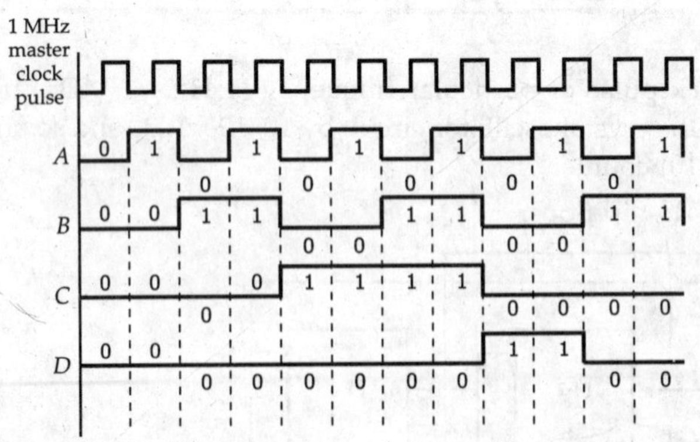


**Fig. 33.2** Timing diagram for decade counter.

From Fig. 33.2 we find the pulse train produced by the decade counter at outputs *A* (LSB), *B*, *C*, *D* (MSB) with the master clock frequency.

### 33.2.3 Converters

Generally two types of Analog to Digital converters (A/D Converter) are used in digital instrumentation. They are

(*a*) Dual Slope Converter

(*b*) Successive Approximation Converter.

**Dual slope converter.** The circuit shown in Fig. 33.3 is a dual slope converter. The RC integrator circuit integrates the input voltage. Voltage $V_{in}$ and the integrated output increases the count of a four

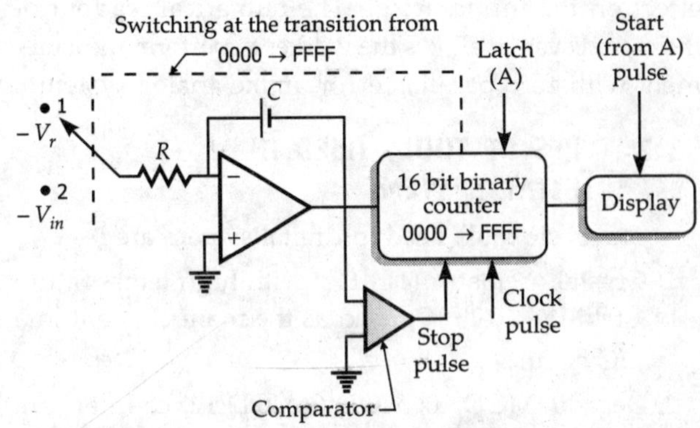


**Fig. 33.3** Circuit diagram of dual slope converter.

digit Hexadecimal Count (16 bit binary count). As the count reaches $FFFF_H$ starting from $0000_H$, with the next clock pulse the counter is reset ($FFFF \rightarrow 0000$). For this particular change of count value from *FFFF* to 0000 a switch is operated and negative reference voltage $V_r$ falls at the input of the integrator. As the integrator output reduces from the peak value attained, a stop pulse is generated as the integrator output reaches zero. This stop pulse stops the count of the Hexadecimal counter and triggers the display circuit. The counter count is latched and displayed. The slope of the curve for the first half when the count increases from $0000_H$ to $FFFF_H$ depends on $\frac{1}{RC}\int V_{in}\,dt$ whereas the slope for the second half is $\frac{1}{RC}\int V_r\,dt$ *i.e.,* depends on reference voltage. With the increase in input voltage $V_{in}$ the curve reaches at a higher level ($A' > A'' > A'''$) at the fixed time when the count increases from $0000_H - FFFF_H$ and for the second half all the negative going curve are of equal slope (As same reference voltage $-V_r$ falls at the input of the integrator) and thus $A'$ reaches $B'$, $A'' \to B''$, $A''' \to B'''$, where $B' > B'' > B'''$. Hence the counts for higher input voltage is higher.

The reference voltage ($V_r$) helps calibrating the converter as with the change in $-V_r$, the corresponding count of $B'$, $B''$, $B'''$ are changed.

This type of dual slope A/D converters also supply the user information about the polarity of $V_{in}$. As the polarity of $V_{in}$ goes from negative to positive the curves *I*, *II*, *III* in Fig. 33.4 moves from first quadrant to fourth quadrant, thus making it easier for the user to comment about the polarity of $V_{in}$.

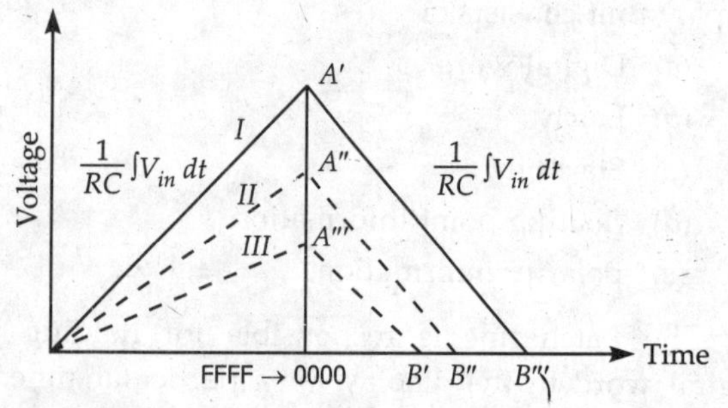


**Fig. 33.4** Voltage-time curve for dual slope converter.

Last but not the least, is the less noise corrupted voltage convertion achieved through this type of converters. This is because the high frequency noise has been averaged out at the integrated output and the systematic noise has been reduced due to the counting is *two* phase

1. $0000_H$ — $FFFF_H$
2. $0000_H$ — Hex count value corresponding to $V_{in}$.

**Successive approximation converter.** The conversion time for the dual slope converter depends on the count value *i.e.*, total number of clock pulses required from start count to stop count. Whereas, the conversion time of a successive approximation converter is fixed and does not depend on the final count value.

The successive approximation converter has a serial address register (SAR) whose flip-flops has been set or reset according to the ring counter. The SAR output has been converted to analog voltage and compared with the input voltage $V_{in}$ after adding a small bias. The comparator output in conjunction with Ring counter output changes Digital word at the output of the SAR. Hence, the SAR converts a analog signal to a digital word in a fixed clock pulse.

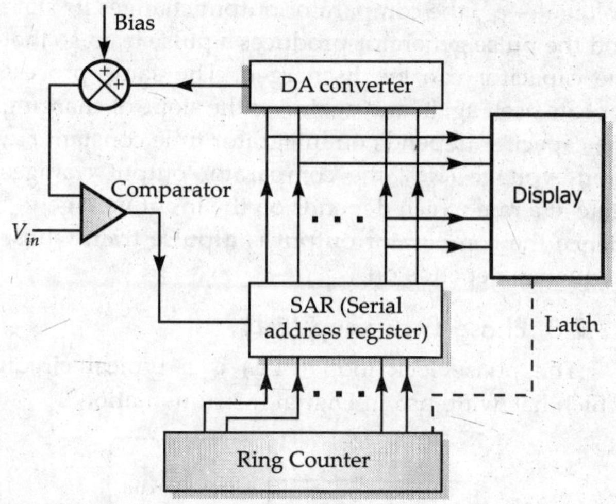


**Fig. 33.5** Successive approximation converter.

If, it is a 3 bit converter the steps followed for a count have been indicated below :

Let the count value (CV) is 011.

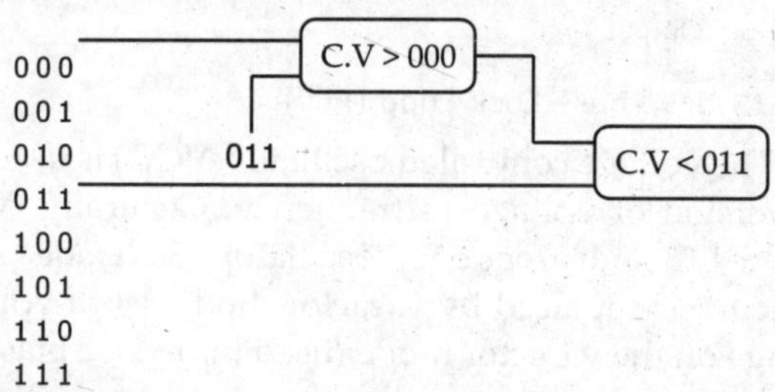


**Fig. 33.6**

Hence any count value can be reached with three or less than three clock-pulses. These type of converters has been used in digital instrumentation.

### 33.2.4 Voltage to Frequency Conversion

The circuit of voltage to frequency conversion has been shown in Fig. 33.7.

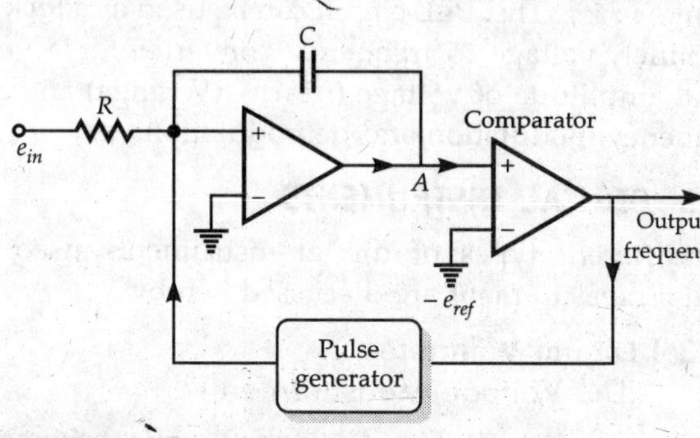


**Fig. 33.7** Voltage to frequency converter.

As the voltage $e_{in}$ charges the capacitor the negative voltage ($e_{in} > 0$) is increased at the output of the integrator. As the output is less than the reference voltage $- e_{ref}$ the comparator output changes its state and the pulse generator produces a pulse train so that the capacitor can be discharged. The same process repeats itself again and again. As the slope of charging the capacitor depends on integrator time constant *i.e.*, input voltage ($V_{in}$), the comparator output changes state at a rate which depends on the input voltage $V_{in}$. Hence the comparator output is a pulse train whose frequency depends on $V_{in}$.

### 33.2.5 Phase Lock Loop (PLL)

The phase lock loop (PLL) is a typical circuit which has wide use in Digital instrumentation.

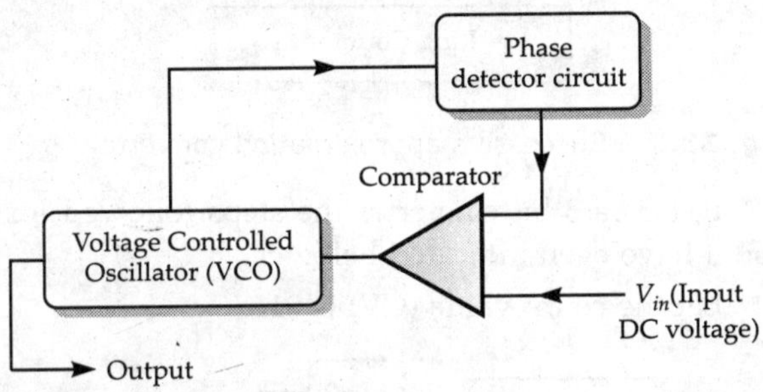


**Fig. 33.8** Phase Lock Loop (PLL).

The voltage controlled oscillator (VCO) is used for conversion of voltage to frequency. Generally, VCO uses a LC high frequency oscillator, with one of its capacitor is replaced by varactor diode. With voltage change on the varactor diode (used in reverse bias), its capacitance changes and the frequency of oscillation of the oscillator changes. The phase detector circuit shown in Fig. 33.8 detect the phase of the voltage controlled oscillator circuit and compares it with the input voltage and the comparator output changes the frequency of oscillation of voltage controlled oscillator (VCO). Hence, the phase of the output of the oscillator has been locked with the input voltage $V_{in}$, thus the VCO output frequency change is a replica of the change in $V_{in}$. This PLL circuit can be used as a lock in amplifier, voltage to frequency converter for a very small amplitude of voltage (nearly $n$V range) and for frequency modulation and demodulation.

## 33.3 DIGITAL INSTRUMENTS

Different types of digital instruments used in digital measurement are discussed hereby :

### 33.3.1 Digital Voltmeter (DC Voltage Measurement)

The circuit in Fig. 33.9 shows the schematic diagram of a digital voltmeter. The DC voltage to be measured in first attenuated and then converted to the digital word which is displayed in the display unit.

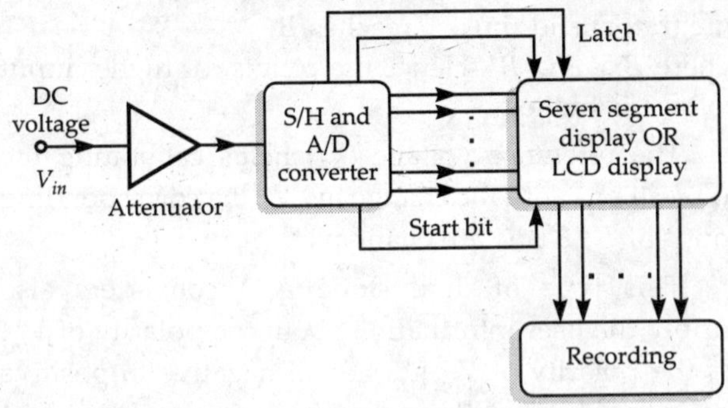


**Fig. 33.9**

The lines supplied by the A/D converter to the display unit consists of

(*a*) Digital word
(*b*) Latch
(*c*) Start bit
(*d*) floating point information
(*e*) polarity information.

The Latch line is responsible for holding the digital word in the display for a particular time, so that recording and the display for a particular time, so that recording and viewing of digital reading is possible for a specified time. The start bit provides the display, the information that the A/D conversion is over. The main *three* features of a DVM is its auto-ranging, auto-zeroing and polarity detection.

**1. Auto ranging**

The modern converter ICS are provided with this facility. The auto ranging circuit is shown in Fig. 33.10.

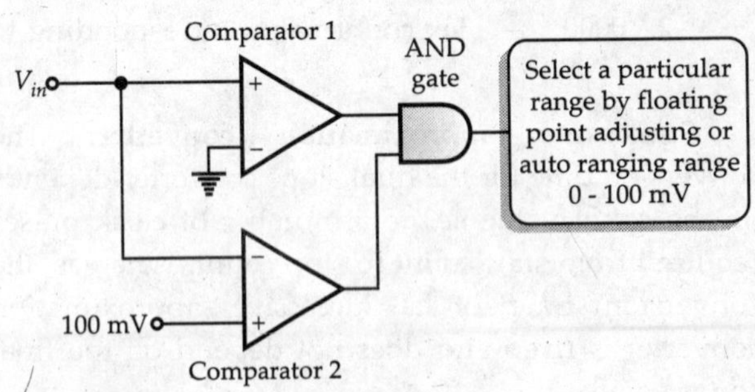


**Fig. 33.10**

The output of comparator 1 (COMP 1) changes state when the input voltage to be measured, is above zero volt whereas comparator 2 (COMP 2) is activated when $V_{in} < 100$ mV. Hence this circuit provides a range 0–100 mV.

Similar circuit may be configured for the range 100 mV – 1 Volt.

The schematic is shown in Fig. 33.11.

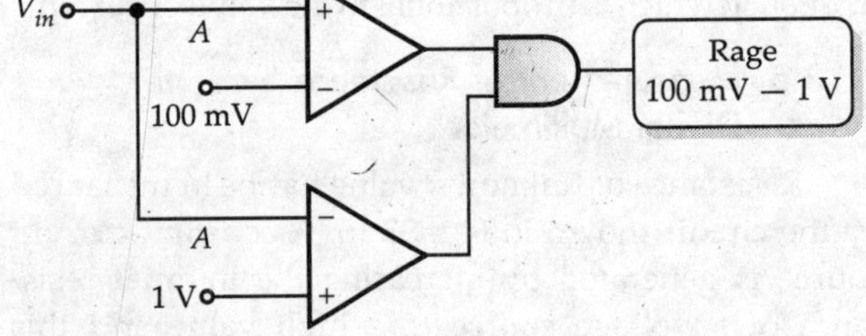


**Fig. 33.11**

Generally, the different range of voltage available for a auto ranging can be given as :

0 – 100 mV
100 mV – 1 V
1 V – 100 Volt
100 Volt – 1000 Volt.

**2. Polarity detection**

The information about polarity of the measurand voltage is given by the A/D converter. The dual-slope converter has the facility of extracting information about polarity. When the input DC voltage is positive the characteristic curve of the dual slope converter is in the first quadrant and when polarity of $V_{in}$ is negative the characteristic curve is in the fourth quadrant. This polarity indication line is supplied by the A/D converter to the display unit.

**3. Auto zeroing**

During start of conversion the zero indication is found on the display. It is called auto zeroing. This is achieved by properly discharging the integrator circuit capacitance to ground. The schematic diagram is shown in Fig. 33.12.

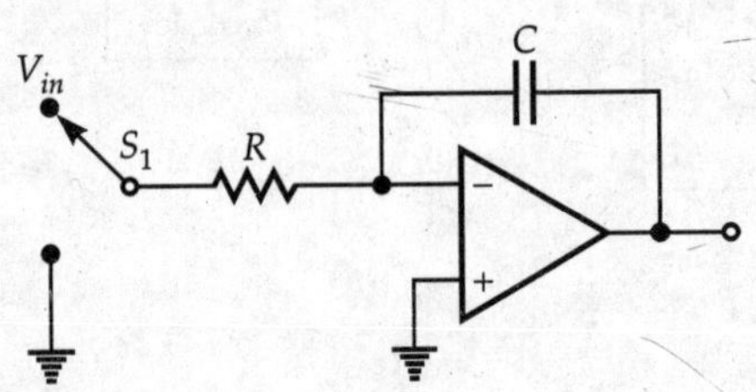


**Fig. 33.12**

During Auto-zeroing, $S_1$ switch from $V_{in}$ to ground and capacitor C of the dual slope converter discharges to the ground.

### 33.3.2 Digital Voltmeter (AC)

For AC input to Digital Voltmeter (DVM), the AC is first processed *i.e.*, phase compensated attenuation with rectification and filtering with realization of r.m.s. DC value of the AC. High frequency AC needs a

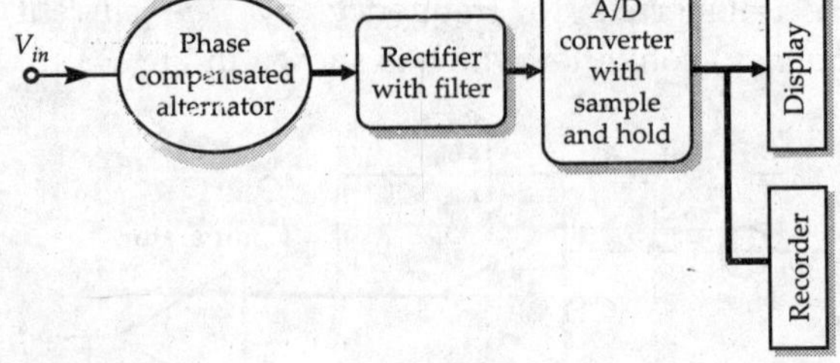


**Fig. 33.13**

precision diode with small reverse recovery *i.e.*, (the negative going edge of AC has been cut off properly) and proper use of capacitor for phase compensation. The DC produced after filtering from high frequency AC has been converted to digital word which is displayed and recorded.

### 33.3.3 DVM for Very Low Amplitude Signal

For small signal AC (range : 1 nanovolt – 10 nanovolt) the method used for conversion is different. The signal is first converted to frequency by voltage to frequency converter or phase lock loop (PLL) and the frequency has been measured by proper frequency determining network. This reduces the effect of noise for small signal measurement through analog to digital conversion. The circuit diagram has been shown in Fig. 33.14.

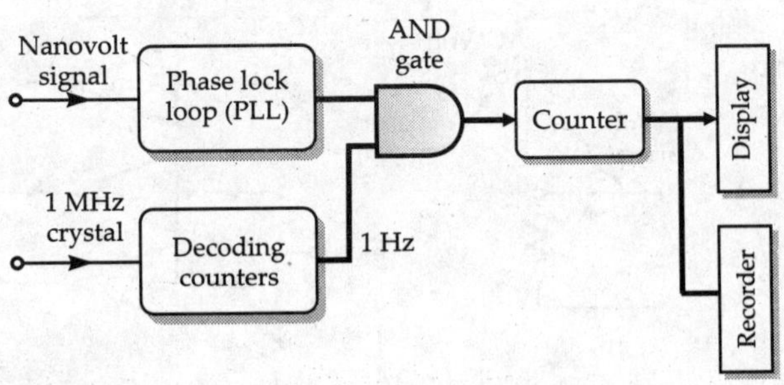


**Fig. 33.14**

### 33.3.4 Measurement of Current by Digital Multimeter

DMM can measure current in the range

(*a*) 0 – 10 mA,

(*b*) 10 mA – 100 mA,

(*c*) 100 mA – 1 A,

(*d*) 1 A – 10 A

with necessary facility of Auto ranging and Auto zeroing. The current may be measured by two different circuits. In one type of circuit the current is passed through constant resistance and the voltage drop across the resistance is measured using a Digital Voltmeter. Whereas, in other method the current has been transferred to frequency by *I-f* (current – frequency) converter which is shown in Fig. 33.15.

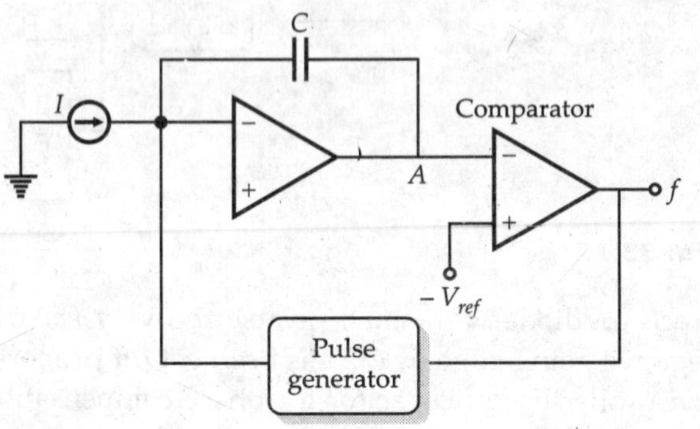


**Fig. 33.15**

In Fig. 33.15, the constant current (*I*) source charges up the capacitor (*C*) of the integrating circuit and the output of the integration has been compared with reference voltage ($-V_{ref}$). Due to the switching of output of the comparator, when $V_A < -V_{ref}$, the pulse generator is triggered and the capacitor *C* is discharged. The output of the comparator is a frequency which is proportional to the value of current.

### 33.3.5 Measurement of Resistance by Digital Multimeter

Resistance of unknown value has been measured by the circuit shown in Fig. 33.16. A constant current source is generated by increasing the internal resistance of a voltage source to a high value and this constant current has been passed through the unknown resistance, the drop across which is measured by a DVM.

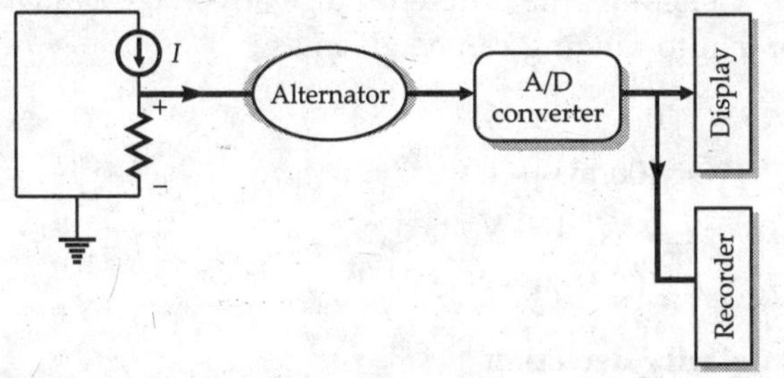


**Fig. 33.16**

### 33.3.6 Complete Circuit of Digital Multimeter (DMM)

Figure 33.17 is the schematic diagram of a complete Digital Multimetre (DMM). The change over switch $S_1$, $S_2$, and $S_3$ are coupled to each other *i.e.,* if one switch jumps to position 1, the other switches follow it. The individual circuits for AC voltage, DC voltage, Current, Resistance measurement has been already discussed. Digital multimeter display is generally $3\frac{1}{2}$ digit *i.e.,* range – 1999 to + 1999 (3 digits may vary 0 –9 and the first digit can be 0 or 1).

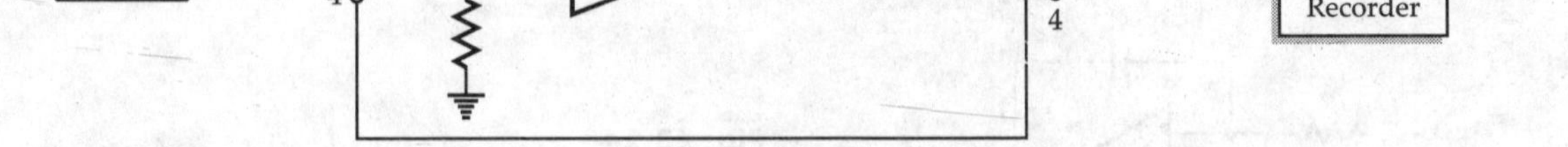


**Fig. 33.17**

### 33.3.7 Frequency Meter

The schematic block diagram of frequency meter has been shown in Fig. 33.18.

The 1 MHz clock pulse has been converted to 1 Hz, 10 Hz, 100 Hz, gating pulses by decade counter in cascade. The input high frequency AC has been amplitude limited and gated by AND operation with 1 Hz pulse. It is obvious that, as we choose 1 kHz or 100 Hz gating pulse (higher frequency gating pulse) the resolution suffers.

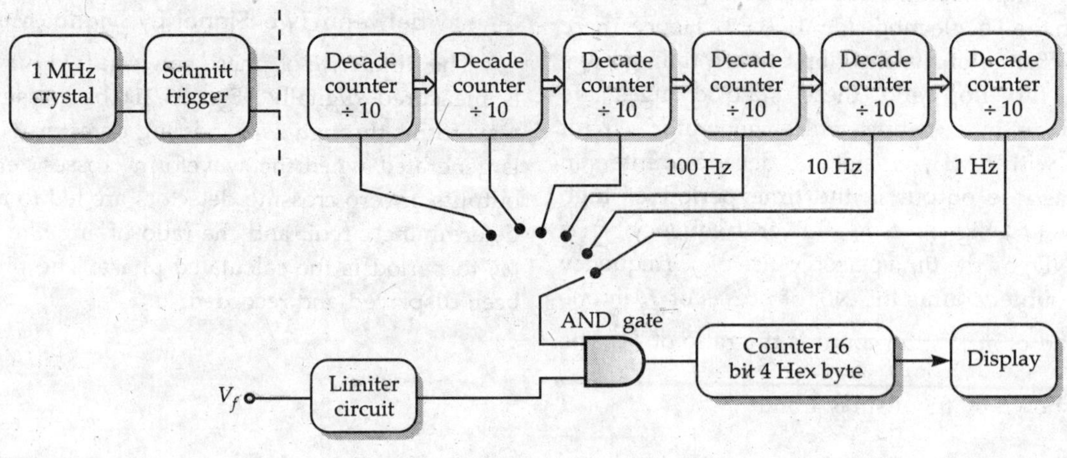


**Fig. 33.18**

This type of frequency metre has been used for a frequency range 10 Hz.

For higher frequency range 1 MHz (RF frequency range) the frequency has been heterodyned to a lower value of Intermediate frequency. The frequency of the *IF* is measured with precision. For this purpose zero bit frequency detection is sometimes used for zero beat, the *IF* is of frequency zero. Hence, a vertical straight line is found on CRO and at this period the high frequency local oscillator frequency is the signal frequency of interest. The heterodyne frequency detection scheme has been shown in Fig. 33.19.

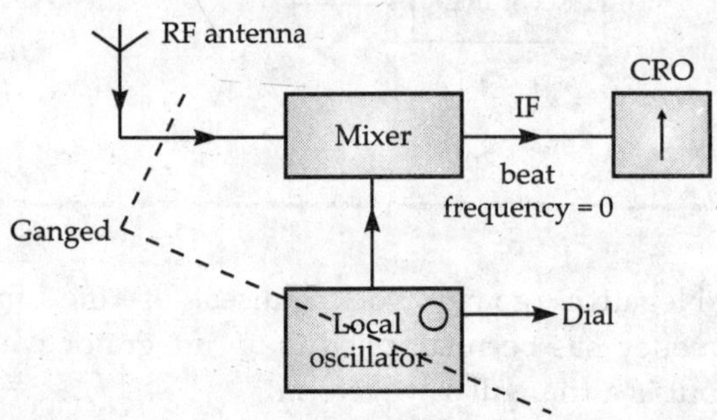


**Fig. 33.19**

The *RF* tuner and the local oscillator has been changed with the same gang capacitor like heterodyne receiver.

Higher range of frequency (100 MHz -- 50 MHz) can also be measured by frequency to voltage conversion by phase lock loop (PLL).

### 33.3.8 Ratio of Frequency Measurement

The ratio of frequency of two different high frequency signals has been measured by the circuit shown in Fig. 33.20.

In the schematic Fig. 33.20, the frequency $f_1$ is taken to be much larger than $f_2$ and we find the ratio of $\frac{f_1}{f_2}$.

The period triggering is used for high frequency signal $f_1$. The $f_1$ frequency signal is first attenuated

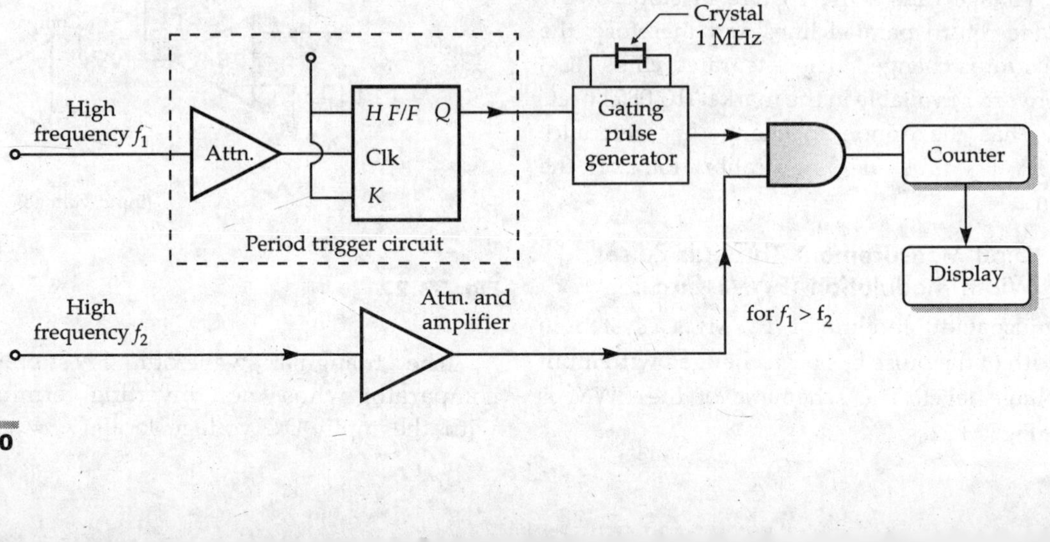


**Fig. 33.20**

and/ or preamplified and put as a clock pulse of a JK flip flop in the Toggle mode ($J = 1$, $K = 1$). Hence, there is a state change of the flip flop at each falling edge triggered flip flop and these state changes are processed with a gating pulse generator circuit (supplied with 1 MHz crystal and decade counters in cascade) whose output is the time period of high frequency ($f_1$) signal. Now, lower frequency $f_2$ is gated (AND) by the time period pulse of $f_1$ frequency and the counter counts the No. of pulses of $f_2$ in one time period of $f_1$ *i.e.*, the count is the ratio of $\frac{f_1}{f_2}$. The result is shown by the display unit.

### 33.3.9 Measurement of Phase Difference between Two Signal by Digital Method

The difference of phase between two waveforms is measured digitally (Fig. 33.21) by passing each waveforms through zero crossing detector (A trigger is generated when the waveform crosses zero). The outputs of zero crossing detectors are fed to a period determining circuit and the ratio of angular velocity (ω) to period is the calculated phase. The phase has been displayed and recorded.

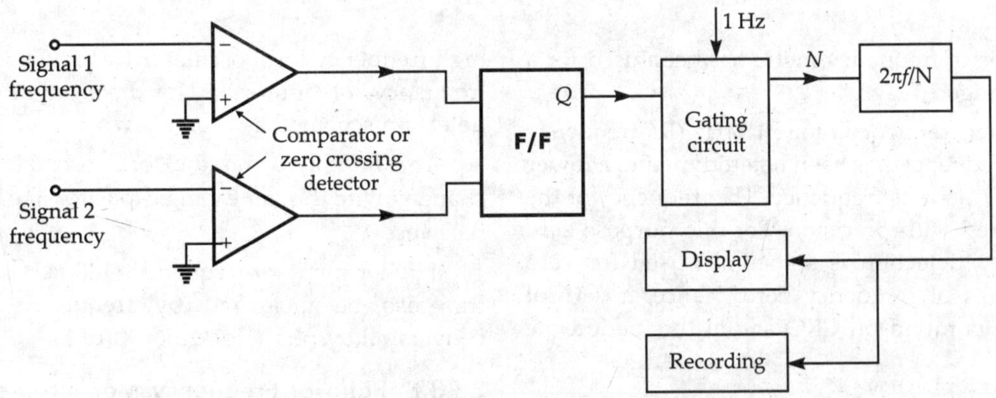


**Fig. 33.21**

### 33.3.10 Calibration Method for Digital Instruments

The calibration of digital instruments is normally done by changing the reference voltage input of the A/D converter. For dual slope converters the change in reference voltage causes the change in slope of the zero voltage return parallel lines and therefore, the ultimate count is changed. The integrated chips (ICs) of A/D converter available in the market (IC 0804 or IC 0808) also has the option to change the reference voltage levels, thus helping calibration of the instrument.

### 33.3.11 Digit Measurement Through Pulse Width Modulation (PWM) Circuit

The pulse width modulator (PWM) is a system in which width of the pulse has been changed with input direct voltage level. The schematic of the PWM is shown in Fig. 33.22.

Negative going clock pulses of the fixed frequency has been supplied to an integrator whose output is a triangular waveform.

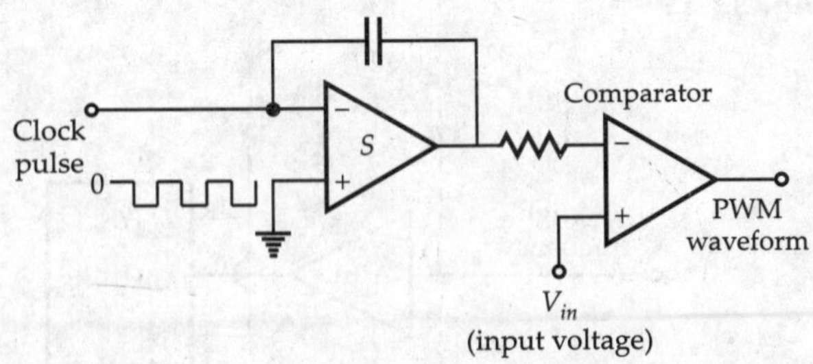


**Fig. 33.22**

The triangular waveform is supplied to a comparator whose non inverting terminal is fed with the input DC voltage local ($V_1$, $V_2$, $V_3$). From

Fig. 33.23, it is found that with increase of DC input voltage level ($V_1 > V_2 > V_3$), the ON time of the comparator output changes with the increase of duty cycle. For digital measurement of DC voltage level, the pulse ON-period gates a fixed frequency (from crystal oscillator) clock. Thus, the increase in ON-period of the pulse causes the final counter reading to increase.

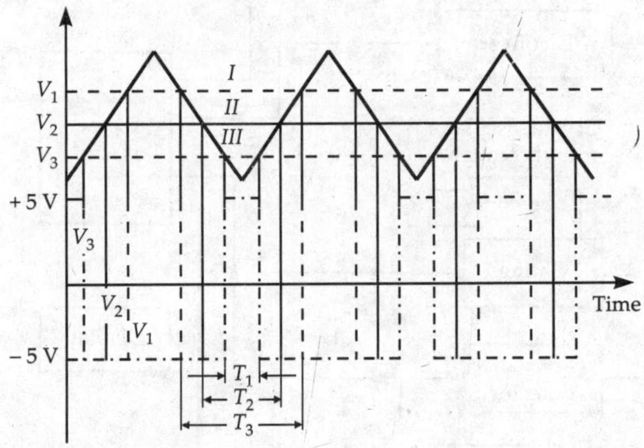


**Fig. 33.23**

Calibration in the PWM method of digital measurement is performed by changing the clock pulse frequency.

### 33.3.12 Digital Ph-Metre

The Ph-measurement is the measure of concentration of $H^+$ ions in a liquid sample. Its range is from 0-14. As discussed earlier the Ph measurement is performed by a glass electrode and a reference electrode. The EMF developed between inner electrode and outer electrode with the presence of semi-permeable membrane between them, is converted to digital count by a PWM converter. The circuit diagram is shown in Fig. 33.24.

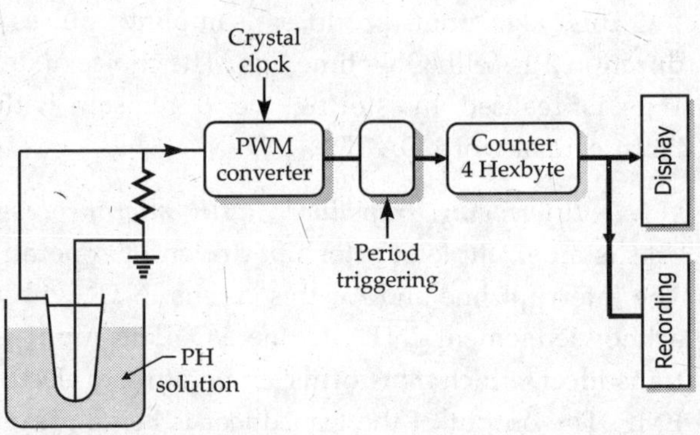


**Fig. 33.24**

This type of Ph metres are popular for their accuracy .005% of the range.

### 33.2.13 Digital Method of Measurement of Resistance, Capacitance and Inducted by DC and AC Bridges

**A.** ***DC bridge for measurement of resistance.*** The principle of working of whetstone bridge is that detector null has been obtained when the ratio $\frac{R_1}{R_2} = \frac{R_3}{R_4}$ is valid *i.e.,* balance of the bridge. Here $R_1$, $R_2$, $R_3$, $R_4$ are the resistances at the four arms of the bridge circuit. The digital circuit for measurement technique by DC bridge is shown in Fig. 33.25.

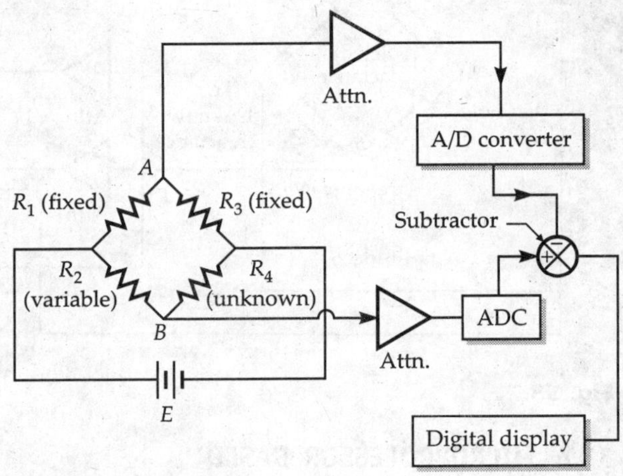


**Fig. 33.25**

The two points of the bridge $A$ and $B$ are of equal potential. Hence, the balance has been obtained by changing the variable resistance $R_2$ and thereby making the digital reading $0000_H$. As the relation between $R_2$ and $R_4$ are linear, hence calibration of the circuit is achieved by changing $R_2$.

This method of resistance measurement is highly sensitive to the change in resistance $R_2$.

**B.** ***AC bridge measurement of capacitance and inductance.*** The balance of AC bridges having reactance and resistance in the arm is balanced when the two points $A$ and $B$ (Fig. 33.26) has similar in amplitude (r.m.s. value), phase and frequency. 33

Hence measuring amplitude, phase, frequency at the points $A$ and $B$ and comparing them digitally, is the way to achieve null detection for AC bridges.

The digital words $A_1$ and $A_2$ are compared $B_1$ is compared with $B_2$ and $C_1$ is compared with $C_2$. AC bridge null is found when $A_1 = A_2$, $B_1 = B_2$ and $C_1 = C_2$ with the change of impedance $Z_2$.

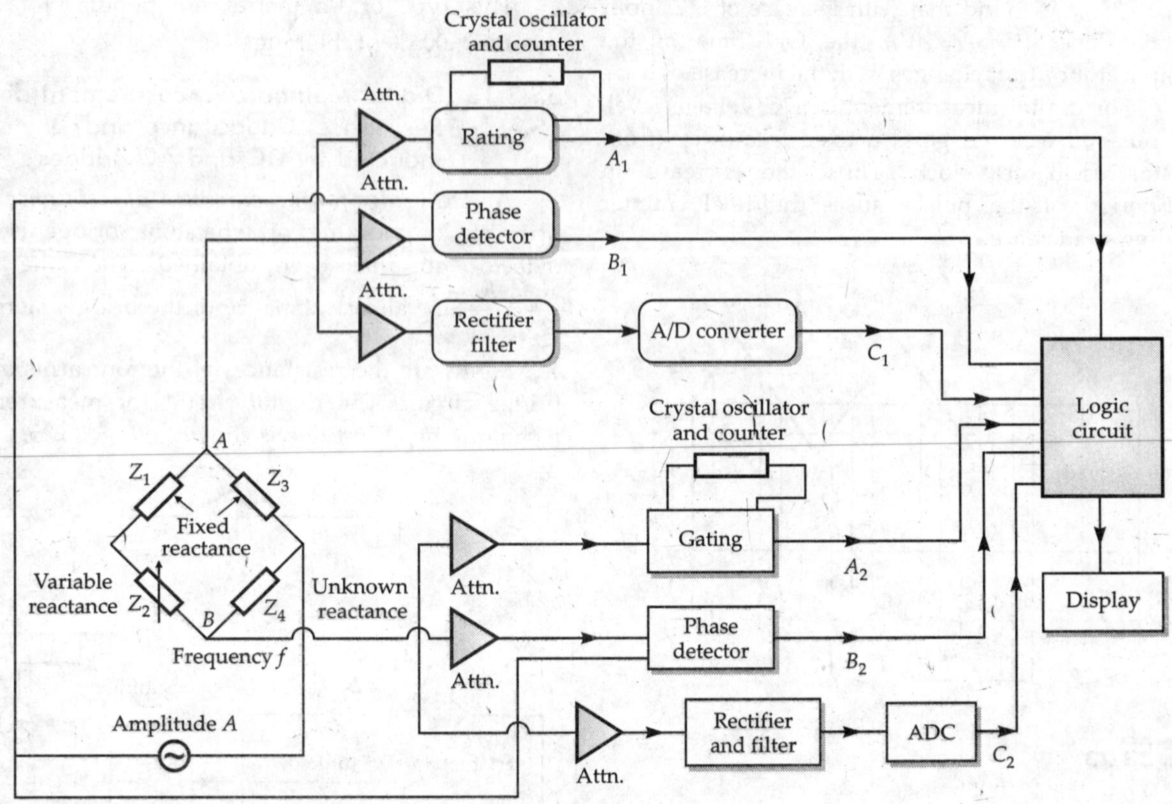


Fig. 33.26

## 33.4 MICROPROCESSOR BASED INSTRUMENTATION

**1. *Microprocessor based sampling or clock pulse generation.*** The microprocessor and associated circuit for sampling pulse generation has been shown in Fig. 33.27.

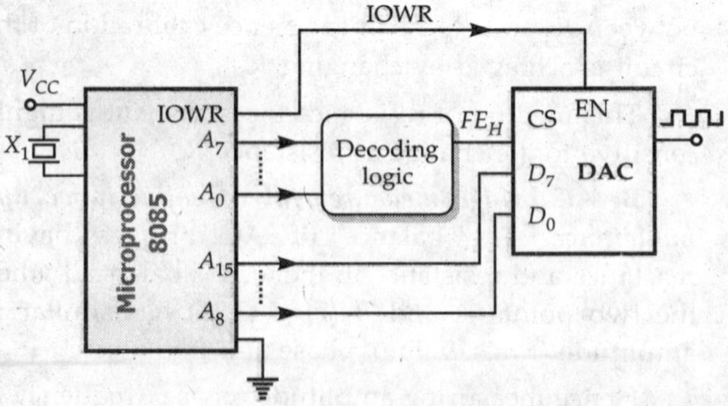


Fig. 33.27

The 8085 microprocessor has 8 data lines ($A_8 - A_{15}$) and 16. Address/Data lines of which 8 lines ($A_0 - A_7$) are dedicated address lines and $A_8 - A_{15}$ are address lines when ALE (Address Latch End) bit is high and data lines when ALE is low. Figure 33.28 shows input-output mapping (I/O mapped I/O) in which $A_8 - A_{15}$ lines are used as data lines and $A_0 - A_7$ are used as dedicated address lines. The address lines are decoded and select a digital to analog converter (8 bit DAC) as output device. The port address of DAC is $FE_H$ (Fig. 33.28). The control signal IOWR (Input — output write) has also been shown in Fig. 33.27.

The algorithm for sampling pulse generation is shown in Fig. 33.28.

This algorithm produces sampling pulses of duration 2 μs with ON - time 1 μs . The choice of delay 1 μs is realised by storing the $B$ register with a hexadecimal word $OACE_H$.

**2. *Interfacing transducer with microprocessor.*** This is an example of Interrupt driven I/O operation. The interrupt line and for this case is INTR with the acknowledgement $\overline{INTA}$ to the I/O. The input is a transducer, which transforms temperature to electrical EMF. The output of the transducer is being sampled and hold (S/H) and latched to the microprocessor data

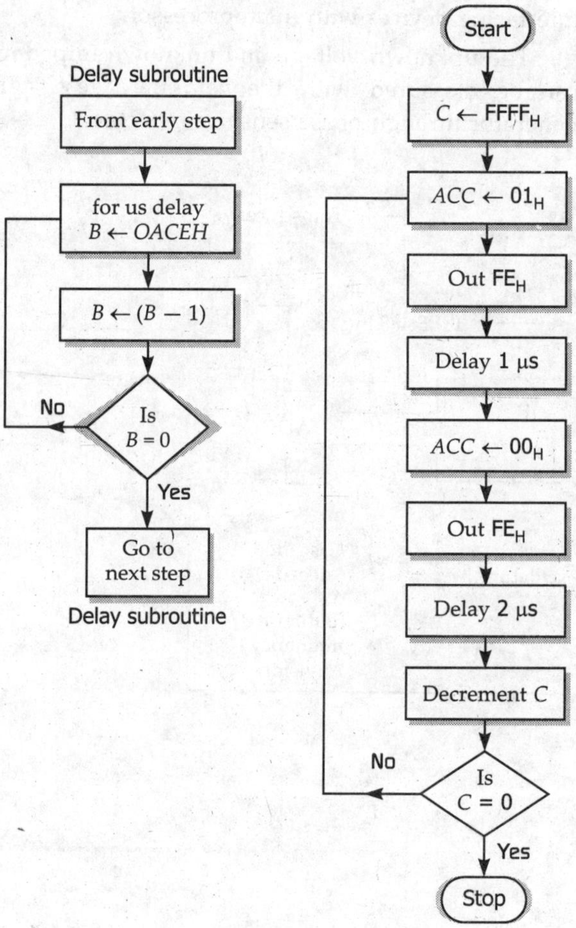


**Fig. 33.28** Sampling pulse generation subroutine.

bus for a predetermined time. The sequence of operation has been enumerated as follows :

1. Generation of sampling pulse (20 μs of period 50% duty cycle).
2. As the temperature transducer is ON the INTR input line to the $\mu P$ has been activated.
3. The microprocessor informs to the output device that it is ready to accept data through $\overline{INTA}$ (interrupt acknowledge line).
4. The microprocessor sends the sampling pulse to the transducer so that its output is sampled.
5. The sampled data has been converted to digital word with a conversion time $T$ so the $T < 5$ μs.
6. The digital word has been sent to the microprocessor register and the digital word has been displayed and recorded.
7. The steps 1 to 6 have been repeated.

The interfacing circuit is shown in Fig. 33.29.

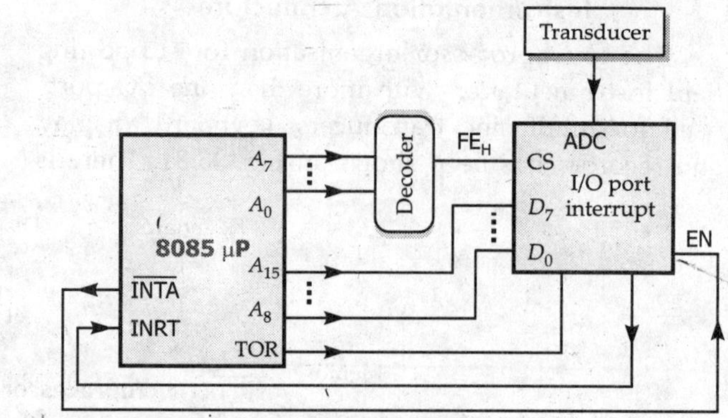


**Fig. 33.29**

The flow chart has been shown in Fig. 33.30.

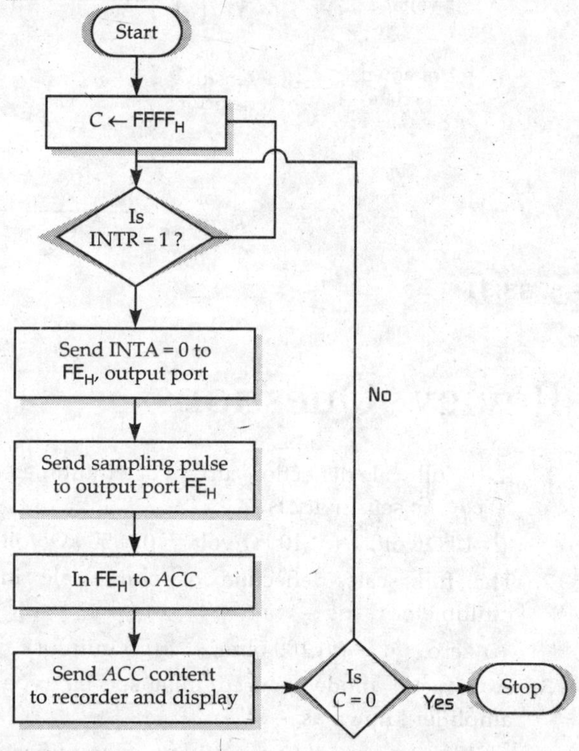


**Fig. 33.30**

### 33.4.1 Advantages of Microprocessor Based Instrumentation

The use of microprocessor in instrumentation enables the user to introduce software which is easy to commission and change. Besides a large number of input/output devices can be interfaced with the microprocessor at the same time with specific scanning algorithm. Again, the keyboard, display and recorder can be interfaced with microprocessor which helps to interact intelligently with the other I/O devices. Microprocessor is a digital device and hence accuracy reached for measurement is very high.

### 33.4.2 The Microprocessor Supported Instrumentation Architecture

The microprocessor organisation for supporting field instrumentation with more than one I/O ports used for interfacing transducers, keyboard, display and recorder has been shown in Fig. 33.31. There is also the presence of *IEEE* 488 bus technology for interfacing devices with microprocessor.

The unknown voltage and unknown impedance can be measured with the help of a test signal generator through phase sensitive detector.

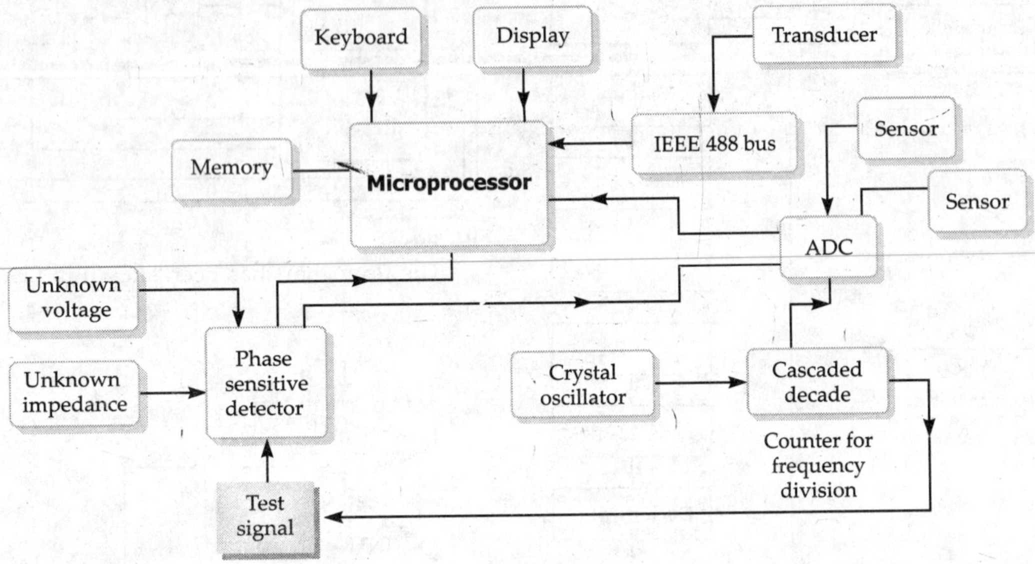


**Fig. 33.31**

# Review Questions

1. If the full scale deflection current of a multimeter is 10 μA, its sensitivity is —
   (*i*) 1 kΩ/volt, (*ii*) 10 kΩ/volt, (*iii*) 50 kΩ/volt
2. The full scale deflection of ohm scale in a multimeter reads —
   (*i*) zero, (*ii*) 100 ohm, (*iii*) infinity.
3. A varactor diode is used in a special type of amplifier known as —
   (*i*) high gain amplifier, (*ii*) tuned amplifier,
   (*iii*) parametric amplifier.
4. The input resistance and capacitance of a digital voltmeter is generally in the order of —
   (*i*) Megohms and picofarad,
   (*ii*) Megohms and microfarad,
   (*iii*) Kiloohm and picofarad.
5. A crystal oscillator is having stability of —
   (*i*) very high value, (*ii*) very low value,
   (*iii*) medium value.
6. One of the following A/D converters utilises a logic programmer of —
   (*i*) dual slope A/D converter,
   (*ii*) successive approximation type A/D converter,
   (*iii*) staircase ramp type A/D converter.
7. Which type of A/D converter is generally used in digital voltmeter for having high resolution and high speed —
   (*i*) ramp type, (*ii*) successive approximation type,
   (*iii*) dual slope type ?
8. Write a short note on Heterodyne frequency meter.
9. Enumerate with a block diagram the various elements involved in a digital data acquisition system. Why multiplexers or scanners are important in this type of data acquisition system ?
10. Explain the operation of (*a*) digital to analog multiplexing, and (*b*) analog to digital multiplexing.
11. Draw a basic circuit of a digital frequency meter using various sections. Explain the function of each section. Discuss the sources of errors in frequency meter.
12. Write explanatory notes on any two of the following :
    (*a*) Digital Ph meter. (*b*) PWM. (*c*) PLL.
    (*d*) Microprocessor based instrumentation.

# 34

# Chemical Sensors and Analytical Instruments

## 34.1 INTRODUCTION

Modern day industrial control instrumentation is largely dependent on the performance of chemical sensors. Chemical instrumentation includes the off line and on line analysis of composition of liquid and gas mixtures, detail analysis of chemical composition of gas and liquid, moisture content of sample, environmental analysis, dissolved oxygen measure, conductivity, residual chlorine and vapour pressure analysis, corrosion detection etc. Chemical sensors, now-a-days, provide the advantage of on-line quality control and measurement without the process being disturbed. The analyzers may be available in probe form or in the form of columns supported chromatograph (liquid or gas). The modern day fibre optical probes provide remote sensing facility, high immunity to EMI-EMC noise and performance less affected by humidity and dust, suitable in hazardous environmental conditions like sludge and slurry industrial application. These probe type sensors have most modern microprocessor based head cleaning arrangement and suitable for measurement of solid content in biological sludge. These fibre optic probes and chromatograph and spectroscopic analysis show high selectivity *i.e.*, characteristic of responding only to the property or component of interest. The aspect of calibration is also important for these type of sensors. Calibration is performed with respect to reliable reference sample. There may be interference between two different components of the same mass *i.e.*, the two components have a peak in the same wavelength in the spectroscopic measurement scheme or in chromatographic measurement scheme. The problem is solved by a compensation factor, if the effect of interfering substance is constant. The spectroscopic technique may be divided into groups depending on the wavelength of radiation used, such as Visible spectroscopy, Ultra-Violet (UV) spectroscopy and Infra-Red (IR) spectroscopy. There is also other type of characterization of spectroscopy depending on whether the sample absorbs or reflects or there is transmission of radiation through the sample. Fluorescence spectroscopy is also of major interest now-a-days. The most popular analytical instrument used recently is chromatograph. It separates the components of interest from the rest of the sample and then measurement is done. These are also used as binary sensors (presence of two components within the sample) through elution by a carrier gas. The chromatographs are on-stream analysers and they provide the widespread use of 30% of the total analyzers used in industry.

The chromatographic sensors are mainly thermal conductivity type, flame type, flame ionization detectors and provide superior sensitivity. Orifice type chromatograph are the most recent development in the history of chromatograph.

## 34.2 PROBE ANALYSERS

Analysis of the composition of a liquid or gas sample is performed with the help of the probe analysers. The probe is a small needle type of instrument with a hole in it for visible light region. For fibre optic probe, there is a variation of refractive index along the cross-section and specifically the refractive index decreases along the diameter of cross-section. For ultra violet and infra red probe the tip of the probe needle emits UV and IR radiation. To mention, the fibre optic probe works in the principle of

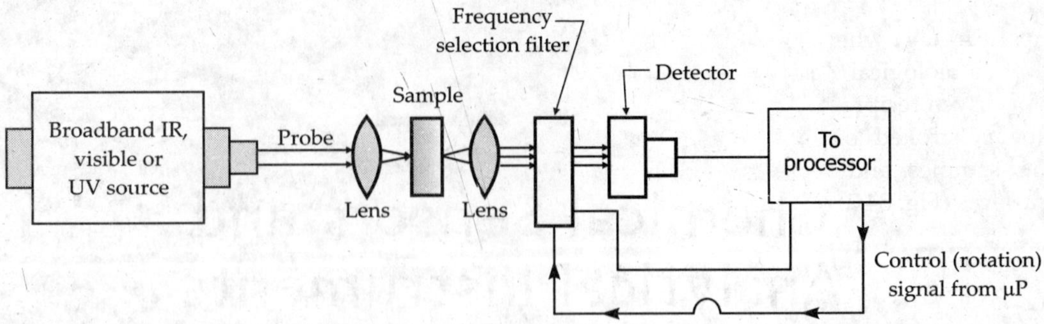


**Fig. 34.1** Probe type sensor setup (Absorption type).

total internal refraction. The ray emitted from the tip of the probe is reflected from the sample liquid or gas and the intensity of the reflected ray is determined with the help of a photo or UV and IR detector as the case may be. These types of probes are called reflective type probe. The reflected ray from the sample shows the variation of reflected intensity with different wavelength of visible or UV or IR light. The distribution of reflected intensity with wavelength helps to infer the composition of the sample in terms of quality (Chemical composition) or quantity (quantitative analysis).

Absorption type probes (Fig. 34.1) are also available in which case the detector in placed on the other side of the probe needle and in between the detector and probe the sample is placed in a transparent container (Fig. 34.2). The visible or UR or IR ray shows selective absorption lines depending on the composition of the sample. In between the sample and the detector and the probe a frequency selective filter (In the range of frequency of interest) is placed which is transparent to a particular frequency for a particular degree of rotation of the circular shaped filter along its axis. The detector is also a broadband one in a particular range of frequency UV, IR or visible

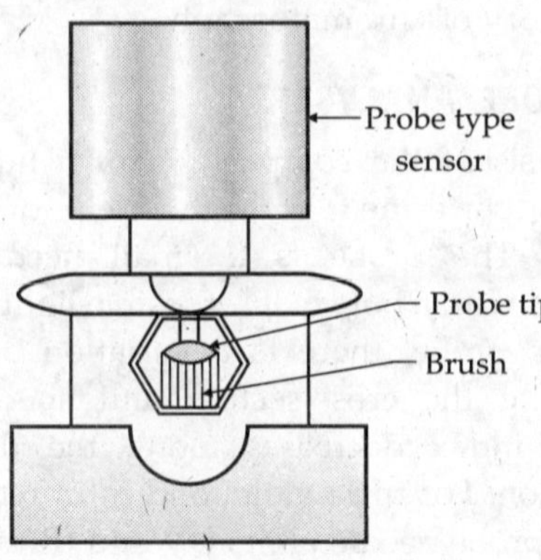


**Fig. 34.2** Probe cleaner.

region. Absorption or backscatter measurement is also extended in the case of Gamma ray radiation analysis. Because of Gamma ray can penetrate metallic walls, this type of probes give rise to a non-contact measurement scheme.

The probe type sensors for analysis of composition of liquid or gas, in the method of frequency selective absorption or reflectance, can be used for on line process measurement or for continuous quality control. The optical fibre (using multimode fibre) has the added advantage of its use in highly hostile industrial (High EMI, EMC ambiance) less disturbance in the process.

There probe type sensors have been calibrated with the help of a known sample composed of different chemical compounds of interest (two for binary compound) at different proportions. The sample used for calibration is called the reference sample. Two different chemical identities can show the same spectral. Intensity distribution and this case is called the interference. The way of calibration in such case is to determine whether effect of the interfering substance is constant and, if it is constant, addition of a compensation factor is done for proper calibration.

The extension of the probe type, analyser is microwave analyser which works on absorption or reflection mechanism, generally used for measurement of moisture in solids without requiring physical contact with the surface. Infra Red (IR) sensor is used for liquid or gas samples, generally used in non contact back scatter technology. Ultra violet probe type analysers are more sensitive than IR but less specific. UV can handle clean and single phase (all the components are either liquid or gas) sample. UV bands of some chemicals shift with the change of pH of the sample and hence disturb calibration process.

The probe tip may be opaque to the particular ray (visible or IR or UV) when the solid concentration is measured in a biological sludge. The head is cleaned by a motor driven reciprocating piston located within glass tube is stroked once every 15 seconds, for refreshing samples and deposits and wiping the optical surfaces (Fig. 34.2).

## 34.3 DIFFERENTIAL REFRACTOMETERS

Refractive index for chemical compounds is a unique property used for composition determination of binary samples with the help of differential refractometers.

When the ray passes from a chemical (with angle of incident $\alpha$) to air (with angle of refractance $\beta$) the RI of the chemical is $\frac{\sin \alpha}{\sin \beta}$. Differential refractometers utilise this relationship and keeping $\alpha$ constant the measurement of $\beta$ gives the RI.

$$\text{Now, } \alpha_{critical} = \text{arc sin}\left(\frac{\text{variable sample RI}}{\text{fixed prism RI}}\right).$$

As for binary chemical, the angle of incidence of visible light, is varied and the detector output shows peak at two different angles of incidence ($\alpha_1$ and $\alpha_2$) from which composition and quantitative analysis of the sample is done.

As, with refractometers, turbidity sensors work with visible light in absorption or reflectance mode and also for light scattering mode. The absorption sensor detects the sum of all light absorbing effects, including color variations, whereas reflectance measures only the concentration of suspended solids.

**Smart processing for probe type sensor.** Data gathered from the intensity transmitted to filter rotation (selection of frequency) of a particular sample used for processor based smart processing (Fig. 34.3) for analysis purpose. The quantitative and qualitative analysis are done with a perfection of 98-99% for the present day probe. These high accuracy probe type sensors are popular now-a-days.

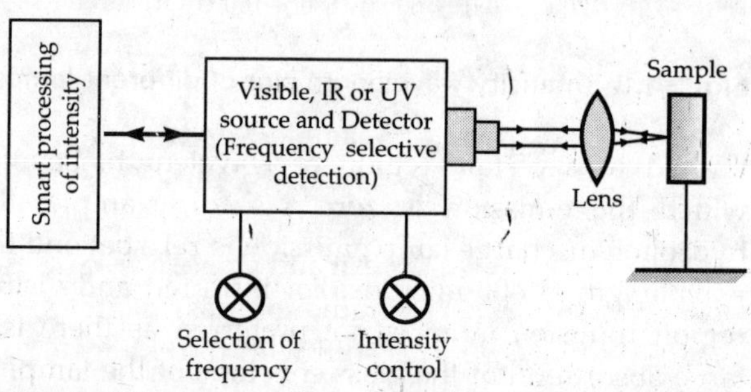


**Fig. 34.3** Reflection type probe with smart processing.

**Some compounds which absorb UV :**

- Acetic acid
- Acetone
- Ammonia
- Benzene
- Bromine
- Butadiene (1, 3 of) Carbon disulphate
- Carbon tetra chrode
- Chronine
- Chlorobenzene
- Cumene
- Formic acid
- Hydrogen peroxide
- Hydrogen sulphide
- Iodine, Isopreme
- Nitric acid.

Refractive index table for some common compounds :

| | |
|---|---|
| Acetic acid | 1.3718 |
| Acetone | 1.3558 |
| Acrylic acid | 1.4224 |
| Benzene | 1.5011 |
| Butyle acetate | 1.3951 |
| Carbon disulphide | 1.6225 |

## 34.4 SPECTROPHOTOMETERS

Selective absorption of electromagnetic radiation is of significance in a spectrometer (IR or visible) design. The amount of selective (frequency) absorption of electromagnetic radiation reflects the composition, concentration and analysing a particular sample. Special case is taken for the preparation of the sample. Absorption of selective frequency is narrated with quantum jump of electrons from valence band to the conduction band. The energy required for quantum jump of electrons from valence band to conduction band is supplied by the frequency of radiation absorbed and the particular frequency is determined by

$$E = h\nu$$

[$h$ = Planck's constant ($6.625 \times 10^{-27}$ ergs. s)]

where $E$ is the energy difference of the two bands and $\nu$ is frequency of absorbed energy. A molecule possesses *three* types of internal energy states : (1) electronic (2) vibrational and (3) radiational.

34

So, each type of quantized energy state, needs different frequency of electromagnetic radiation to change its internal energy. Hence there may be more than one characteristic frequency of absorption for a particular molecule or compound. Analysis of these selective absorption lines conclude, about the composition, concentration and analysis of the samples. Generally, the electronic absorption correspond to the UV spectroscopy, and also visible region calorimetry. whereas vibrational and rotational eigen states of a molecule contribute to infra red spectroscopy. Spectroscopic analysis is developed in such a level that only a trace of a molecule in a sample is detected with surety.

### 34.4.1 Different Aspects of Spectroscopy

#### Region of Electromagnetic Radiation of Spectroscopic Interest

| Region | Frequency | Wavelength |
|---|---|---|
| UV (ultra violet) spectroscopy | $7.5 \times 10^{-14}$ – $1.5 \times 10^{15}$ Hz | 200 nm – 400 nm |
| Visible region spectroscopy : or calorimetry | $4.4 \times 10^{14}$ – $7.5 \times 10^{14}$ Hz | 400 nm – 700 nm |
| Infrared region spectroscopy | $3 \times 10^{14}$ Hz – $4.4 \times 10^{14}$ Hz | 700 nm – 1000 nm |

Effect of radiation on matter is illustrated as incident, reflected, absorbed, scattered and transmitted (Fig. 34.4) ray analysis when the electromagnetic wave falls on the surface of the sample. There may also be excited fluorescence which is categorised as fluorescence spectroscopy.

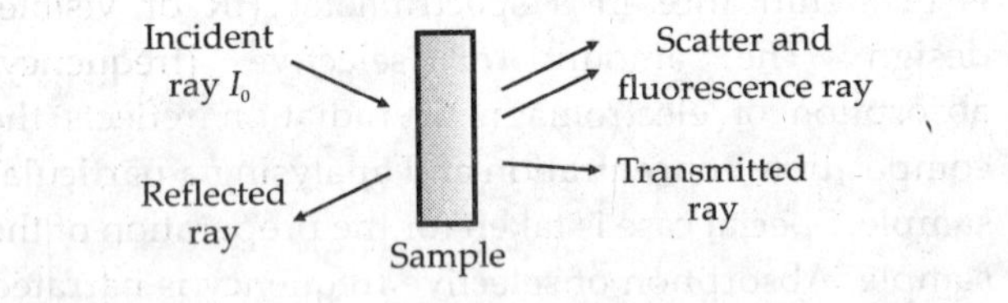


**Fig. 34.4**

For absorption spectroscopy, the absorbance ($A$) is defined in terms of molar absorption ($a$) to (sample path length) and absorber sample concentration ($c$). The definition is :

$$A = abc \qquad ...(i)$$

The Beer's law of absorbance is given as

$$A = abc = \ln \frac{I_0}{I}$$

where $I_0$ and $I$ are radiation intensity entering and leaving the sample. Figure 34.4 illustrates this.

Beer's law indicates the relationship between absorbance, the path length through the sample (*i.e.*, sample thickness) and concentration of different compounds in the sample. The absorbance is the logarithmic calculation of the incident to transmitted intensity and directly proportional to concentration. Though the relation between absorbance and concentration is not linear for high concentration. The relation can be plotted as (Fig. 34.5).

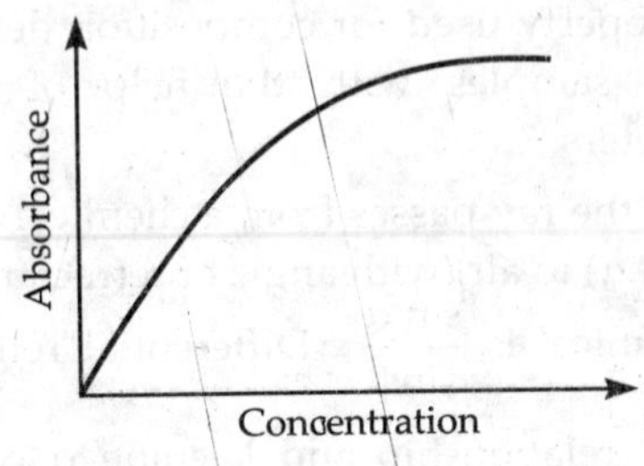


**Fig. 34.5** Absorbance-concentration curve.

Hence, it is transparent from the curve that absorbance is proportional to concentration at low concentration level of compound in sample.

### 34.4.2 Source of Radiation and Detectors for Spectrophotometers

The most important source for a spectrophotometer is the tungsten lamp. Generally tungsten lamp is used for work in the infrared and to some extent calorimetric (visible light) analysis. Hydrogen and deuterium discharge lamp is used for UV spectroscopy. The emission of these sources is

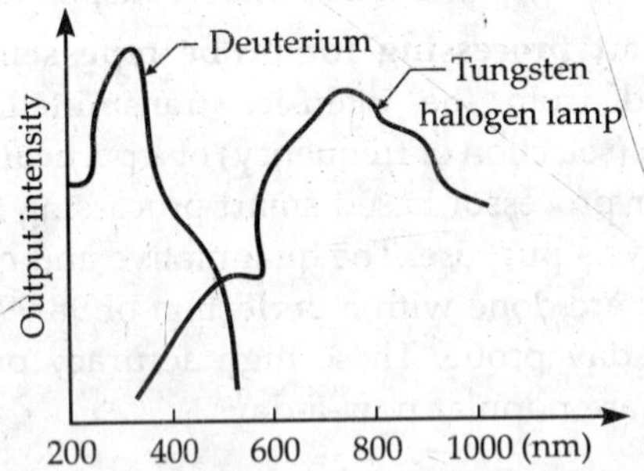


**Fig. 34.6** Intensity wavelength plot of different lamps.

limited to a certain region of wavelength, beyond which the emission is zero. As for example, the hydrogen discharge lamp emission is zero beyond the wavelength of 400 microns. For infra red and visible region tungsten lamp is not preferred as there is a large absorption of the glass envelope of the lamp. In recent development, Nernst sources or mercury lamps or tungsten halogen lamp has been introduced as sources in the region of UV or IR spectroscopy. Xenon arc lamps is also used is UV and visible region. Modern day lamps are fitted with the accessory like electronic ballast for ionising and running high pressure lamps like Mercury Argon or Tungsten halogen arc lamp.

Figure 34.6 shows the intensity or radiated energy to wavelength for different types of lamps.

## 34.5 DETECTORS

Like sources, the other important aspect of the spectrophotometry is detectors and filters. Generally photocells are used as detectors. There are different types of photocells like blue-photocell and red photocell. The other type is selenium cell (Sn cell). The red and blue photocells are of the types high vacuum photocells, gas filled phocell and photomultiplier tube. The phocell works as a transformation of EM radiation to high anode current. The ionisation of gas molecules at low pressure when electromagnetic radiation is incident causes an amplified electrical current at the output. The ionisation in a cell occurs at low pressure and high voltage (DC) between anode and cathode. The low pressure and presence of low content of inert gas like Argon in the photocell causes less saturation current. The transformation of electromagnetic radiation to current is high in photomultiplier tube and thus it is used to detect low intensity EM wave (Fig. 34.7). In photomultiplier tube there is the presence of more than one intermediary electrodes in between cathode and anode which are called the dynode. At each electrode the electromagnetic wave gets strengthened.

There are 8 – 16 dynodes in between cathode and anode for a particular photomultiplier tube. The power supply circuit for these tubes consists of Extra High-voltage Transformer (EHT) and high anode

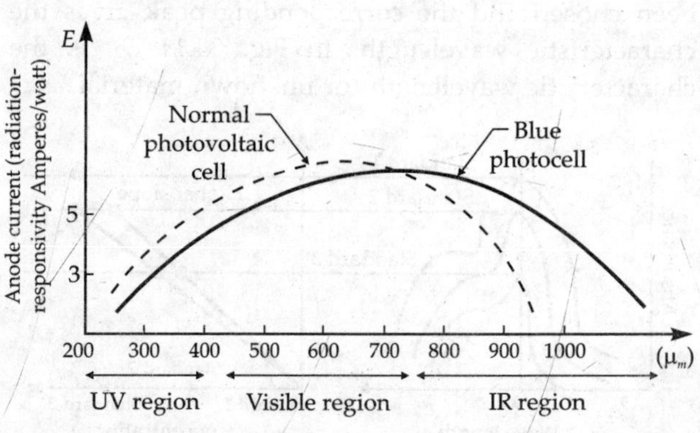


**Fig. 34.7** Responsivity vs. wavelength plot.

current may cause damage to the photomultiplier tube. The frequency response of photocells is shown in Fig. 34.7.

The semiconductor detectors, which are used extensively now-a-days are silicon diode detectors. These type of detectors consume less power and used extensively for UV and visible region. For longer responsivity (Current/EM radiation intensity) diodes in array maybe used.

For calorimetric detection, the other type of detector commonly used is photovoltaic cell. In this case a photosensitive material like selenium is coated on a semiconducting material. The cell is maintained at a low pressure and supply voltage is given on gold plated cathode and anode.

## 34.6 FILTERS

Filters are used in spectrophotometry to help transmitting the EM wave in a small spectral width. Coloured filters *i.e.*, use of colour gas, and film show absorption in a small spectral width of 50 micron (Fig. 34.8). The absorption type coloured filters are used in visible region spectrophotometry. For these coloured filters the 3dB transmission point of intensity, is 50 μm apart from the central frequency. Their efficiency of transmission is limited to 25%. To increase the selectivity of these colour filters, their thickness has been increased and also increasing their efficiency. The characteristic of the coloured filter is shown in Fig. 34.8.

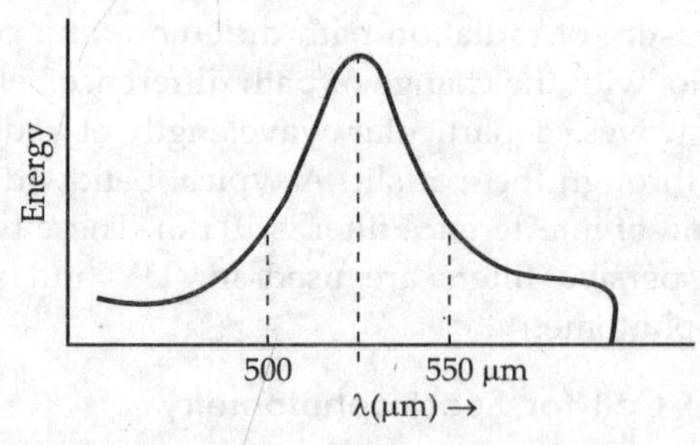


**Fig. 34.8** Filter transmittance.

The most commonly used filters for spectrophotometers are called monochromators. Monochromators enable single frequency measurement with a narrow pass band extending from 20 mm to 50 mm. The monochromators may be dispersive or non dispersive. The dispersive effect is defined as the spatial deviation of the beam when it passes through a monochromator (prism). The deviation cycle is a function of frequency. The two commonly used dispersive monochromators are grating structure and prism. The prism is a dispersive element and with the rotation of the prism (shown in Fig. 34.9) selective different frequency passes through the exist slit of the spectrophotometer (Fig. 34.9).

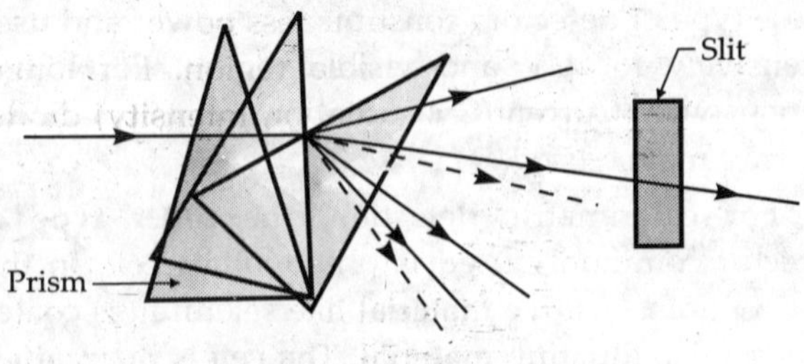


**Fig. 34.9** Prism monochromators.

The other type of dispersive monochromator is grating structure. Grating are made with scribing of parallel lines on a glass surface. The selected frequency passing through the exit slit depends on the distance (*d*) between the two parallel scribed lines and the angle of incident *EM* wave. This grating structure allows measurement of intensity of single wavelength with a narrow band of 20 nm.

The non-dispersive filters are categorised as photometers. Photometers use narrow band pass interference filters which are used to pass radiation at measurement frequency. Interference pattern is generated with two silts in which the input *EM* radiation cause secondary wavelet. The secondary wavelets emit coherent E.M. wave and the interference fringes are generated on the screen which is a measure of radiation path difference at a certain point. So, with the change of path difference between to *EM* waves, a particular wavelength of radiation passes through the exit slit. A typical bandwidth for passband of interference filter is 10 nm. These type of non dispersive filters are used for UV and visual spectrophotometry.

## Sample Cell for Spectrophotometry

The sample cell is the representation of the process stream. The cell material is made of stainless steel or monel or titanium or plastic body and the container is fitted with two transparent windows at the opposite side. Quartz, glass are the cell material or UV—IR—visual spectrophotometry. The sealing of the sample cell is made with a oring gasket. Important parameter for the cell is the optical path length between two of its windows which very from 1 mm to 1 m. As the photometric measurement depends on the path lengths for interference filter, the sensitivity of measurement increases with increase in path length.

Different types of spectrophotometers for UV—visual use have been configured and it is found that dual beam (one signal and the other reference beam) analysers are better than single beam spectrophotometry. It is because in the dual beam photometer, one beam passes through the sample and the other beam through the reference sample and the effects on the beam due to temperature, humidity and other physical parameters are connected out in the final measurement parameter, hence increasing the accuracy (Fig. 34.10).

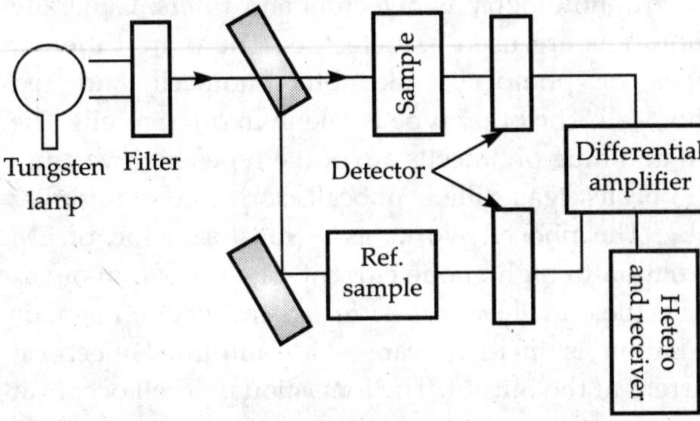


**Fig. 34.10** Dual beam analysers.

**Calibration of spectrophotometer.** Dealing with a unknown sample (material or compound) by a spectrophotometer, it is wise to choose a wavelength which is a characteristic to a unknown compound. To do this *i.e.*, plotting of calibration curve, a number of standard solutions of known concentration of the sample has been prepared. The spectrophotometer analysis reveals the variation of absorbance with change in wavelength, using one filter at a time. For all the peaks of the calibration curve the absorbance data is plotted against the known concentration of the standard samples (Fig. 34.11).

The maximum slope curve with zero intercept has been chosen and the corresponding peak gives the characteristic wavelength. In Fig. 34.11, $\lambda_c$ is the characteristic wavelength for unknown material.

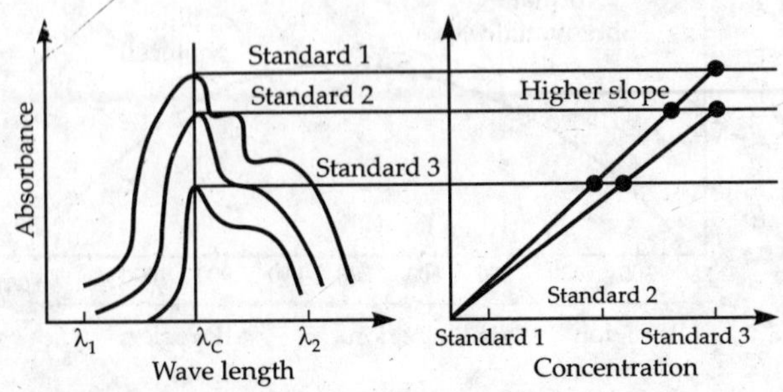


**Fig. 34.11** Calibration curve.

## 34.7 CHROMATOGRAPHY

Chromatographic technique is used to identify the component and their composition by separating different phase of a two phase (gas-liquid or gas-solid or liquid solid) mixture. It is urgently used as hydrocarbon analyser. The principle behind the chromatograph is that, a modulation in velocity of a gas-liquid mixture, during its passage through a chromatographic column, is created. After the column of different gas has been detected with detectors and plotted in a chromatogram whose time delay and amplitude are the characteristic of the composition and concentration of different components of the mixture. The column has been prepared in tabular form with finely divided solid in it. The columns are designed in such a way that the carrier gas (often Helium or Nitrogen) passes through it in steady flow. Generally the column types for liquid-solid mixture is gas adsorption type for gas-liquid mixture is gas permeation type and liquid-liquid mixture is liquid liquid partition type and for Ion exchange it is distribution type. The temperature controlled column is coated with non volatile liquid phase. Liquid chromatography generally use thin layer and paper chromatographic column. The constituents of the liquid pass through high pressure liquid chromatographic column (HPLC) and retarded with different delay due to selective absorption effect which is attributed by the column. The detectors may be fibre optic probe type or differential refractive index type, fixed or variable wavelength ultra violet type or conductivity type.

The block diagram of a gas chromatograph is shown in Fig. 34.12.

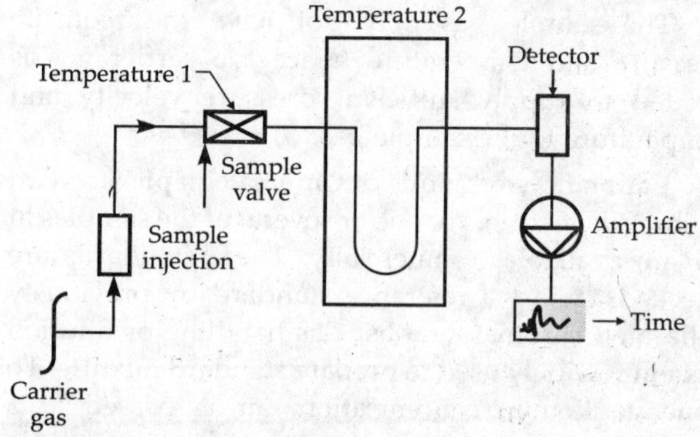


**Fig. 34.12** Gas chromatograph.

The liquid chromatograph operating conditions are different from gas chromatograph operating requirements. The carrier pressure is 1000 PSIG, its flow rate is 1 ml/mm and chromatographic circulation time is 3 – 20 minutes. The temperature $T_1$ in Fig. 34.12 is the temperature of vaporisation of different constituents of the sample. The chromatograph peaks show different delay which vary with the preparation of the column. For grains of solid with larger diameter the peaks are less separated (closed) and for finer graniules the peaks are widely separated. Hence the resolution, selectivity and efficiency of a chromatograph depends on the preparation of chromatographic column. The quality factor of a chromatograph also depends on the temperature distribution of different zones of the chromatographic column, though single temperature zone is frequently used. Now-a-days, temperature distribution in a column is programmed. It causes decreased analysis time for a sample of compounds of widely divergent boiling points and polarity. Temperature of column is constant for a fixed period while the low boiling point liquid elute from the column.

The on-line analysis by a chromatograph is shown in Fig. 34.13.

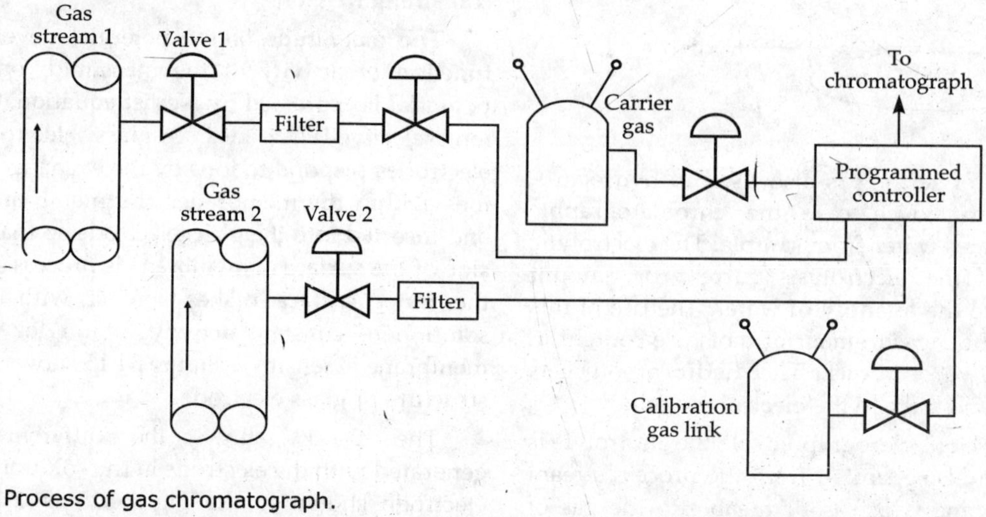


**Fig. 34.13** Process of gas chromatograph.

34

The sample gas may not have the requisite pressure and temperature hence the carrier gas is needed to supply sufficient pressure velocity and temperature to the sample gas.

**Calibration technique of Chromatograph.** Standard calibration mixtures of one or several of the samples in air are available commercially. These standards are checked against a reference standard by previously calibrated chromatographs. Gas handling or dilution system is widely used to prepare standard mixture. To generate known concentration, air is passed at a known rate over a permeation tube, in which sample is kept in a temperature controlled vessel. The other way of making standard for chromatorgraph calibration is the known sample gas added by a syringe into a carrier gas volume. The chromatograph is calibrated against the standard sample.

## 34.8 ELECTROCHEMICAL SENSORS

These are mainly, diffusion type sensors which comprise of two electrodes *i.e.,* measuring electron and a reference electrode, covered by an electrolytic solution and separated from the sample by a membrane. This membrane is chosen so that it exhibits permeability to the gas of interest (Fig. 34.14). These type of sensors are used for measurement application of toxic gases monitoring. In this system the toxic gas permeates through the meter and reacts with the electrolyte causing electric current or potential difference between the electrodes.

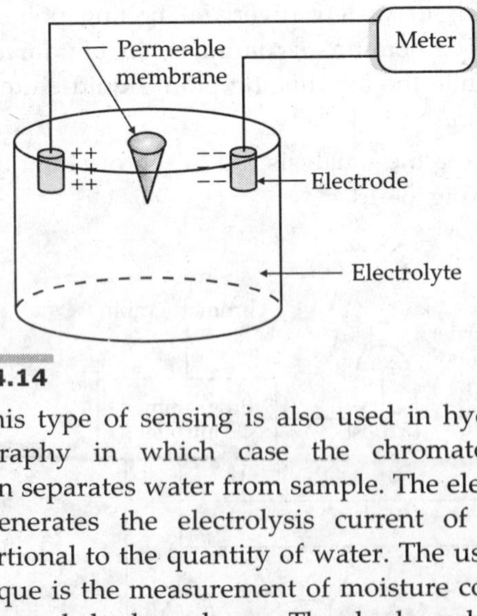


**Fig. 34.14**

This type of sensing is also used in hydrochromatography in which case the chromatographic column separates water from sample. The electrolytic cell generates the electrolysis current of amount proportional to the quantity of water. The use of this technique is the measurement of moisture content in unsaturated hydrocarbons. The hydrocarbon gets polymerized and stick to the electrolytic cell.

In Galvanic or polarographic cell, the electrolyte is in gel form and is separated from the process stream with a membrane which is permeable to the gas of interest. This helps the on stream measurement. This system performs well if the membrane is properly cleaned up. The cause behind the electrochemical reaction is the polarizing voltage and the resulting current flow is the indication of the composition.

The pH sensor is also an example of electrochemical sensor in which glass electrode has been used already indicated earlier in this book.

Other type of electrochemical sensor in use, is Oxidation Reduction potential sensors. These types of sensors detect the ratio of reducing agent to the oxidising agent which is an important parameter in effluent treatment control. These sensors are available in probe type.

## 34.9 NERNST EQUATION

The electric potential generated across an ion selective membrane is related to the ionic activity in response of the frequency of dissolved ionized solid in a solution is given by Nernst equation which states :

$$E = \frac{2.3\,RT}{nF}\log\frac{a_1}{a_2}$$

where

$E$ = Potential developed across the membrane

$a_1$ = activity of measured ion in the sample.

$a_2$ = the activity of the same ion in the internal solution.

$\frac{2.3\,RT}{nF}$ = slope of calibration curve *i.e.,* Nernst slope.

An ion selective electrode is used to obtain information about the chemical composition of process solution. They generate millivolt potential when they are dipped in a conducting solution containing free ions.

The magnitude of the potential developed is a function of activity of the measured ion and this potential is expressed by Nernst equation. One of the ion selective electrodes is glass electrode. These electrodes respond to ions by the exchange of mobile ions within the membrane. The membrane is fused and inserted into the glass electrode so that the outer side of the surface of membrane is process stream and the inner surface makes contact with the filling solution of constant activity of ion for which the membrane is sensitive. Figure 34.15 shows the typical structure of glass electrode.

The $E_1 - E_7$ shows the different potential generated with the electrode in the solution. The glass electrode shown in the Fig. 34.15 is used for pH measurement. In the pH measurement the buffer solution is at pH of 7 and contains a chloride level equal to that of external reference electrode. The glass electrode is inactive in case of sodium ion with the other principal inferences with silver, hydrogen, lithium, potassium, ions with lower detectable limit of 0.02 parts per milligram.

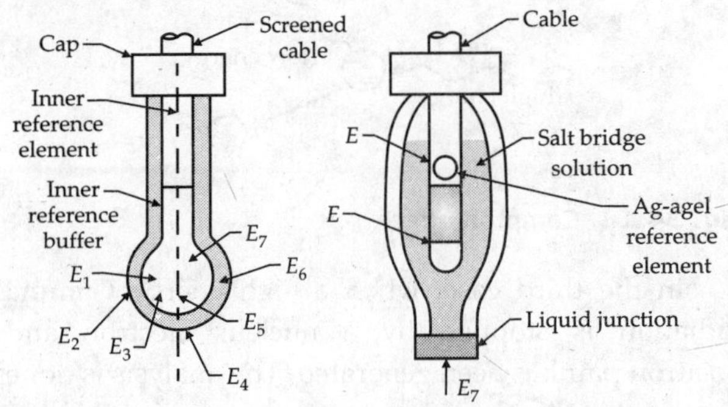


**Fig. 34.15**

Other popular ion exchange electrode used now-a-days is solid state electrode which shows activity in case of bromide, reference cadmium, silver, calcium, iodide, lead, copper, cyanide ions with a varying detection limit of 0.001 – 0.007 ppm. The solid state electrode is shown in Fig. 34.16.

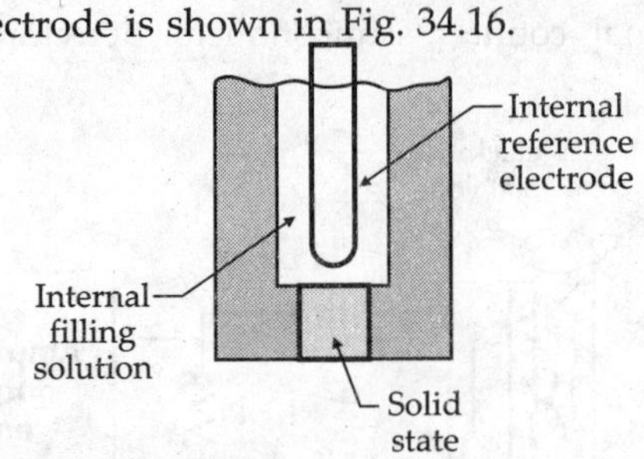


**Fig. 34.16** Solid state electrode.

All ion selective electrodes are similar in operation, only differ in the way the ion to be measured moves across the membrane and the other ions do not move through the membrane. Likewise, in the pH glass electrode does not respond to solium ion at pH 7 – 10 but respond to the $H^+$ ion.

The electrochemical instruments *i.e.* ion selective electrode measurement exhibits simple, rapid, non destructive, direct and continuous measurement scheme. The only disadvantage is in the form of measurement accuracy and the ion selective measurement shows relative errors of 14% to 18% rather than other analytical techniques discussed, whose relative error is ± 0.1%.

Ion selective electrodes are also used as conductivity sensor in which case the conductivity is correlated with the specific ions and its concentration in the process stream. Ideal reference electrode produces a constant potential independent of composition of the solution. A ion selective electrode gives 60 mV change in potential difference which is ten times the activity of monovalent ion.

**Calibration.** Like other analytical instruments, the calibration solution is used which are found to be affected by dilution, evaporation and contamination by foreign matter, and oxidation. So case should be taken to prepare and handling of these standard solutions. The reference solution is 0.01 M or 100 ppm which are diluted to get other solutions.

**Internal property sensor.** The detection of dielectric constant of a sample is sometimes used to measure the composition of a solid or liquid, or moisture content in samples or the measure of viscosity gives analytical indications of some liquid or gas. These types of sensors are called the internal property sensor. Instruments can also measure the dielectric loss of a moving material with the advantage that all components external to the piping with increased selectivity. The measurement of flow index intrinsic viscosity measurement for analytical significance is also called plasmometry. Determination of molecular weight distribution and plastic behaviour are included in the study of plasmometric analytical sensors. The capillary extension plasmometers with automated sampling system are used for on-stream measurement. The advantage in plasmometry is that data may be generated for molecular weight distribution in addition to molecular weight average. The draw back of plasmometry is the lengthy analysis behind the inferences drawn.

The block diagram of a differential pressure capillary viscometer is shown below.

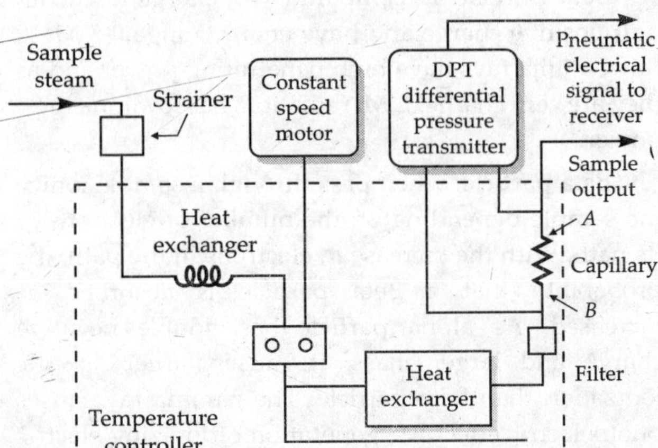


**Fig. 34.17** Differential pressure capillary viscometers.

34

Filter selects one particular velocity of the inflow gas stream at a particular temperature *i.e.*, viscosity is a function of temperature and due to capillary effect there is pressure head generated between two ends of the capillary ($A$, $B$ in Fig. 34.17) which is measured. The effect of viscosity on the capillary flow at a particular viscosity is considered to be one way of analytical instrumentation.

**Radio-chemical instruments.** Radio-isotopes emit radiation both in particle nature ($\alpha, \beta$ and positrons *i.e.*, equal in charge and mass with electron) and wave nature (Gamma ($\gamma$) ray). These radio-isotopes are possibly produced in artificial way with the help of cyclotron presently. Naturally found radio-isotopes were the base of radio chemical instrumentation upto 1980's. There are different ways to analyse a sample with radio-isotope. One particular way is to mix with the sample for analysis. After equilibrium is reached, *i.e.*, the mixture emits homogeneously, a part of the mixture sample is isolated and its activity has been analysed. This is because radio activity can be introduced in a sample, which is not radioactive, by adding to it to predetermined radio-isotope. One of the characteristics of radio isotope is its half life period *i.e.*, the period in which the activity of the isotope becomes 50% of its initial activity. The expression for half life period is

$$t_{1/2} = \frac{0.693}{\lambda}$$

where $\lambda$ is the decay constant.

The $\alpha$ particle emitted from a radio isotope (artificially produced or naturally found like uranium, thorium etc.) has energy 2–10 MeV (Mega electron volt) and their penetration power is low, in aluminium foils.

Beta particle is comparatively energetic carries unit positive charge and have energy range 0-3 MeV. But Gamma rays have high penetrating power and as they are zero charge, E.M. radiation, has low ionising power.

Beta particle, when pass through a sample ionize the sample depending on the number of electrons in its path. With the increase in electrons in the path, the probability that a Beta particle is absorbed, is increased. As alpha particle has double negative charge and larger mass, it causes larger specific ionization than Beta particle. The gamma ray causes photoelectric effect *i.e.*, generation of ionizing electric potential, due to absorption of Gamma rays by an atom. This photoelectric effect is generated by lower energy Gamma rays. The second ray is defined by Compton effect, when there is an elastic collision between Gamma ray and atom. Hence, with elastic collision lower energy Gamma ray has been produced absorption of low energy Gamma ray (Fig. 34.18).

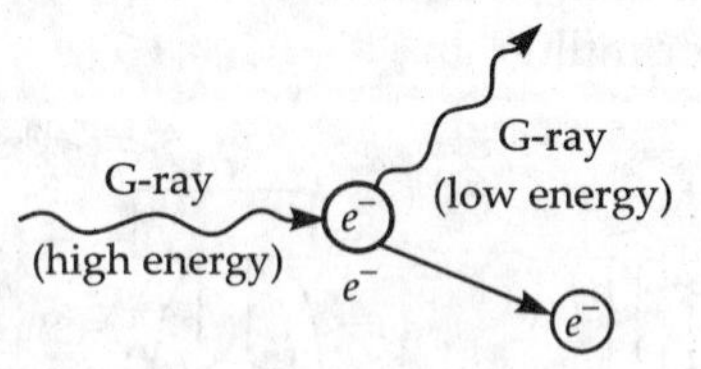


**Fig. 34.18** Compton effect.

In the third case, when a high energy Gamma radiation is stopped by a nucleus electron and positron pair has been generated. The analysis is done on the number of ion per centimeter for specific ionization. For this radiochemical instruments, no preparation of chemical standard is required.

**Detector.** Several detectors are used for counting of emission from a mixture of radio isotope with the sample in radioactive equilibrium. These are ionization chambers or Gaiger Müller counter and proportional counter, semiconductor detectors.

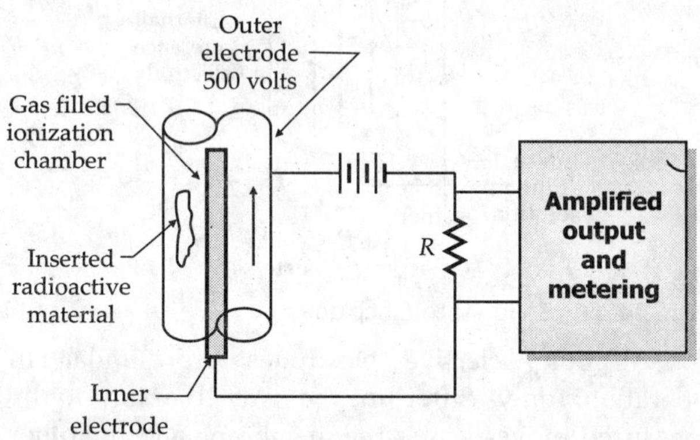


**Fig. 34.19** Ionization chamber.

The radioactive source placed in the chamber emit and the passage of the emission through the gas causes ionisation and ionised particles are channeled from inner electrode to the outer electrode due to high potential difference between them and the current produced is directly proportional to the activity of the radio isotope mixture with sample. The current produced is very small ($10^{-9}$ Ampere) and is detected by differential voltmeter which has high input impedance.

Different circuit is used, when the counting of ionized particle has been made with the counter. This is the way in which Gaiger Müller counter is built up. Here, due to radio-active ions in the gas are produced and as the voltage between anode and cathode is too high *i.e.*, 800-3000 volts, the gas ions collide with other gas molecules and an avalanche of production of gas discharge is produced. For each discharge there is a generation of voltage pulse of 1-5 volts which is counted by a counter. The relation with period of pulses with anode-cathode voltage is shown in Fig. 34.20.

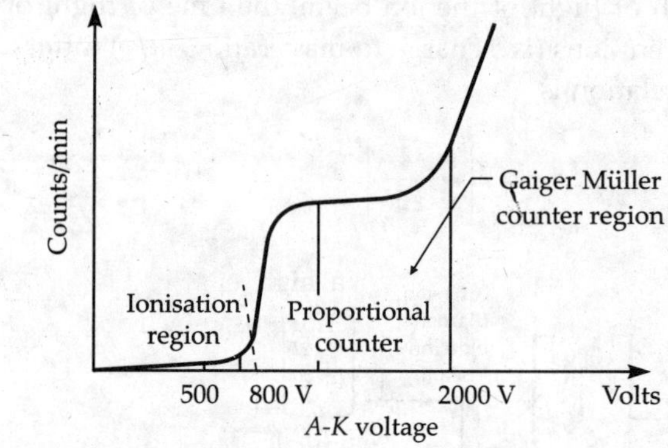


**Fig. 34.20** Period of pulses with anode cathode voltage.

**Dead time.** Positive ions are much less speedier than electrons, hence as the electrons move from cathode to anode the positive ions travel much less path due to the potential gradient between cathode to anode. Hence for a very small time the ionization chamber is inactive and this is the dead time *i.e.*, off period of pulse generation.

Hence, it is better to use integrating type of measurement for ionization chamber and counting of pulses for Gaiger-Müller counter. For proportional counter voltage between anode-Cathode is from 600 volts to 900 volts *i.e.*, below Gaiger Müllar region. Proportional counters behaves in a way that output pulse from the chamber starts to increase with voltage between anode and cathode, but proportional to the initial ionization.

Figure 34.21 shows a Gamma counting system.

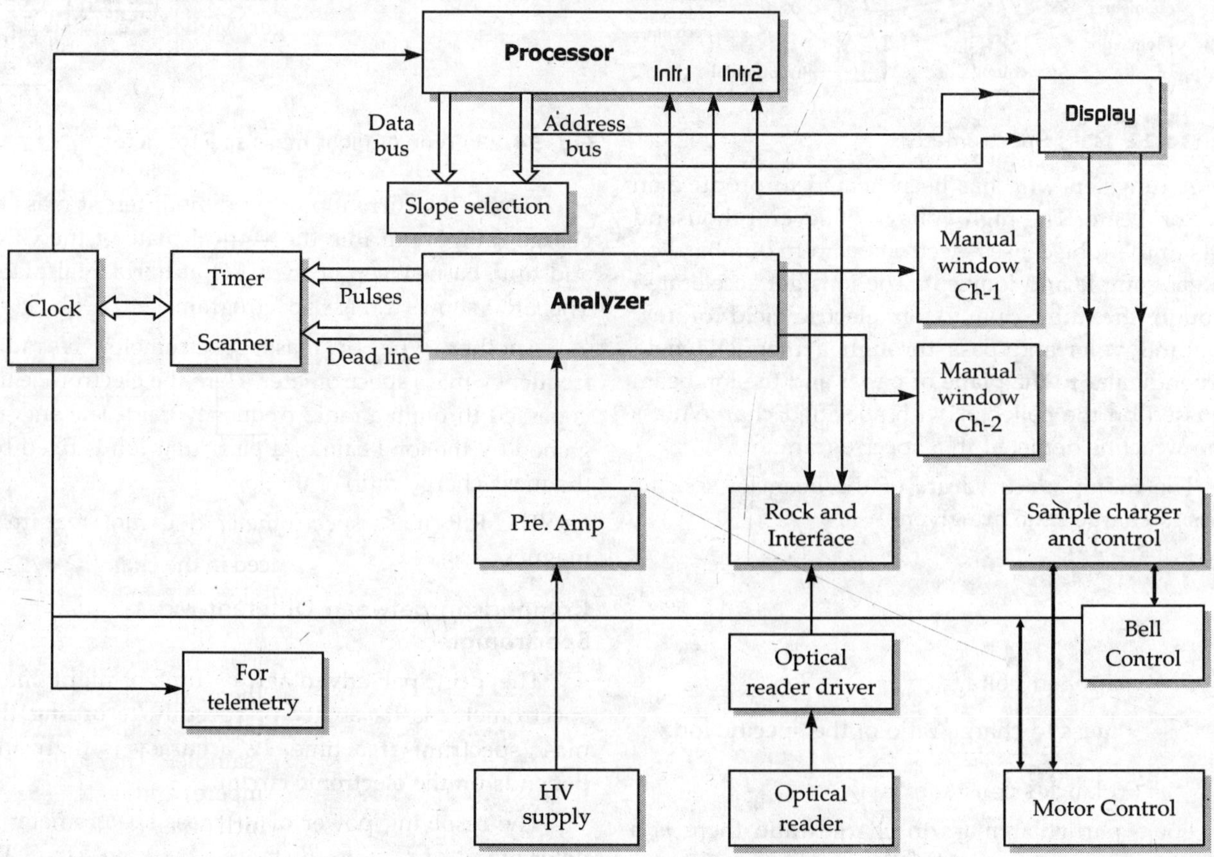


**Fig. 34.21**

**Mass spectrometer.** The mass spectrometry is a tool for direct determination of molecular weight of a sample, analysis of functional group of aromatic compounds, composition of compounds and to investigate reaction mechanism. In mass spectroscopy the sample is subjected to a phase change from solid to gas by chemical means. The gaseous form of the sample, when bombarded with electron beam, the gas sample gets ionized and different elemental ions have different mass to charge ratio. These ions are deflected in a semiconductor loop with high electric and magnetic field. The detector output is plotted in a mass spectrogram which is the analysis of the sample with proper calibration.

Due to the joint effect of electric and magnetic field the ions which come out from the slit gets defocussed and after traversing a circular path it is focused again to be collected and recorded in a mass (Fig. 34.22) spectrogram.

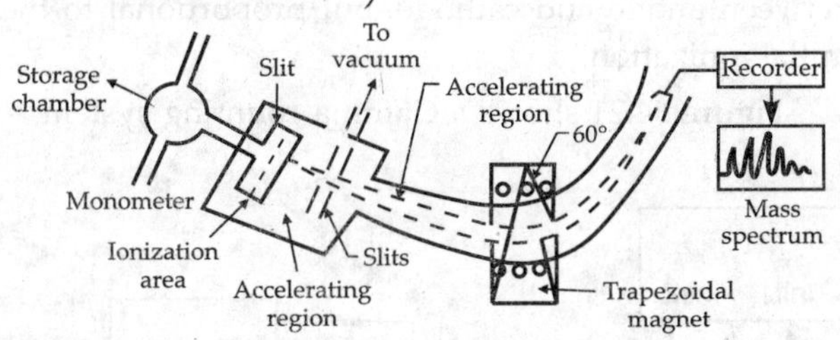


**Fig. 34.22** Mass Spectrometer.

A tungsten wire has been heated to produce an electron beam. The high voltage of several thousand volts in the tube causes electron beam to bombard on the gas sample and ionize it. The ions get accelerated through the tube due to an electric field of few thousand volts and pass through a magnetic field perpendicular to the plane of paper and the ion beam focussed on the collector with specified charge/mass ratio which is depicted in a spectrogram.

The radius of curvature of the beam of specific mass to charge ratio are given by

$$r = \frac{2\,V}{h^2} \times \left(\frac{m}{e}\right)$$

where

$V$ = applied voltage

$\frac{m}{e}$ = mass to charge ratio of the specific ion.

$h$ = Planck's constant.

For a particular mass to charge ratio there is a specific radius of curvature the ion beam follows, hence, each peak in the mass spectrogram is a characteristic of particular ion.

The main advantage of mass spectrometry is the requirement of very small aperture of sample (few tenths of milligrams) for the propose.

Other type of analyser is time-of-flight mass spectrometer. In this case the ion beam rather than traversing a curved path, traverses is a straight line between emitter and collector. Here, due to the same length of flight of the ion beam, the time of flight of different ions (*i.e.*, charge to mass ratio = $m/e$) differs. The relation is

$$t = L\left[\frac{m}{e}\frac{1}{2\,V}\right]^{1/2}$$

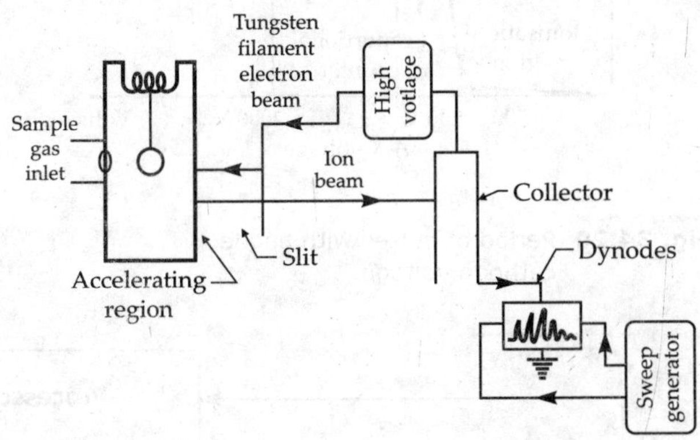


**Fig. 34.23** Time of flight mass spectrometer.

Hence, the current produced by different mass to charge ratio is put into the vertical plate of the CRO and time base sweep of the CRO at horizontal plate. The CRO shows a mass spectrogram.

Another type of mass spectrometer is radio frequency mass spectrometer. Here the electron beam is passed through a radio frequency field. The energy gained by the ion beam of a particular ion is fixed by the mass/charge ratio of the ion.

The R.F. mass spectrometer does not require a magnet.

### Comparison between Different Mass Spectrometers

The principal advantage of time-of-flight mass spectrometer is its ability of speedily recording the mass spectrum at a time. Its accuracy is high and depends on the electronic circuit.

The resolving power of RF mass spectrometer is determined by the distance between the RF accelerating stages. Particularly for RF (frequency of MHz order) mass spectrometer

$$\frac{m}{e} = \frac{0.26\,V}{d^2 f^2} \text{ (is CGS)}$$

where

$\frac{m}{e}$ = mass to charge ratio

$V$ = Voltage ≈ 50 – 250 Volts D.C.

$d$ = spacing between two R.F. coil grids.

$f$ = frequency of R.F.

For R.F. mass spectrometer, a low stage tube may be used but this may cause in between spurious lines.

## 34.9 NUCLEAR MAGNETIC RESONANCE (NMR)

NMR, though expensive, is used vastly as solid-moisture analyser and to distinguish the different atoms of a sample, based on specific absorption characteristics of the Nucleus. The absorption due to the nucleus of the atoms in sample is analysed at radio frequency energy, when the atoms are placed in a continuous magnetic field. NMR provides higher speed of analysis and higher accuracy with the inherent non destructive nature.

The magnetic property of the nucleus is exhibited due to spin angular momentum quantum number whose magnitude is $\frac{h}{2\pi}\sqrt{I(I+1)}$, where $I$ is the spin quantum number and $h$ is Planck's constant. Nuclei with $I = 0$ is non magnetic. The nuclei with $I = \frac{1}{2}$ gives the best resolved spectrum, as for example, $^1H$, $^{13}C$, $^{19}F$, $^{31}P$ nuclei. The other nuclei which show magnetic property for $I > \frac{1}{2}$ are $^2H(I=1)$, $^{14}N(I=1)$, $^{11}B(\neq\frac{3}{2})$. The magnetic axes of the nuclei have $2I+1$ different orientations with respect to external magnetic field $\vec{H}$ and each orientation link with a discrete energy level of $E = \frac{m\mu}{I}\beta H_0$, where, $m$ is magnetic quantum number, $\mu$ is the magnetic moment

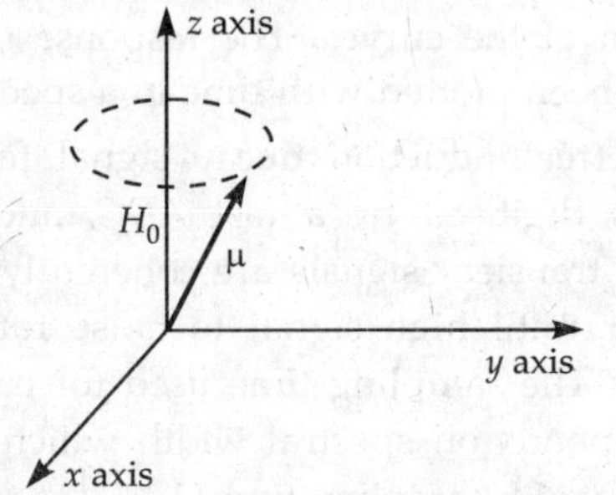


Fig. 34.24

of the nucleus expressed as Bohr magnetron, β is constant of value $5.049 \times 10^{-24}$ erg $G^{-1}$ and $H_0$ is the external magnetic field in gauss, when, $I$ is the spin quantum no. the spectrum assumes some allowed values for $I, I-I...-I-I-I$.

When the atom is subjected to an external magnetic field $\vec{H_0}$ the precession of the nuclear moment is shown in Fig. 34.24.

With the application a *RF* magnetic field $H$, applied along $y$-axis the precession of the nuclear moment is shown in Fig. 34.25.

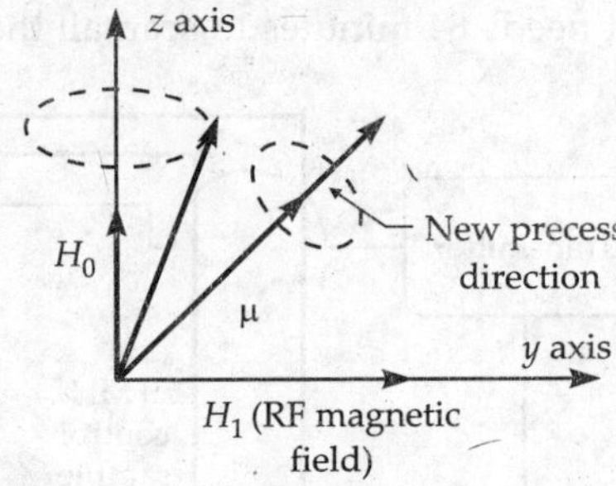


Fig. 34.25

The R.F. field is the resultant of two circularly varying fields one rotating clockwise and the other anticlockwise, whose vectorial addition is $\vec{H_1}$. The component which rotates in the same sense as the precessing nuclear magnetic moment has effect on the nucleus. When the R.F. field acquires a frequency of Nuclear magnetic resonance, absorption of energy due to iuclei jump from lower energy level to upper energy level occurs *i.e.*, the spin precessing with $H_0$, now precess against $\vec{H_0}$ (Fig. 34.26). So, the sweeping with the R.F. frequency from lower to higher valence state, the absorption peak occurs at a particular frequency which is Nuclear resonance frequency. This R.F. is a characteristic of the atom in a sample.

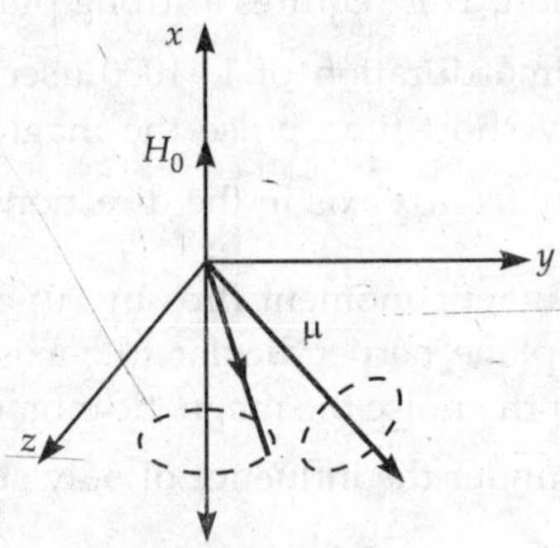


Fig. 34.26 Precession anti parallel to $\vec{H_0}$.

The characteristic RF frequency is given by

$$\nu = \frac{\Delta E}{h} = \frac{\mu \beta H_0}{h I}$$

Calculation shows that for $^1H$ isotope with magnetic moment of 2.7927 and $|H_0|$ is 14.09 kilo gauss the resonance *RF* frequency is 60 MHz. With other values, constant resonance frequency for R.F. is found to be 90 MHz for $|\vec{H}_0|$ as 21.14 *K* Gauss.

### Measurement Details

The continuous wave NMR has limitations of large scanning time. The R.F. frequency is scanned over a frequency range of25 MHz (Fig. 34.27). With a narrow band of 5 kHzs. at a rate 1 Hz. $sec^{-1}$. Thus this type of instrument needs 84 minutes to scan all the resonance lines in a frequency span of 25 MHz.

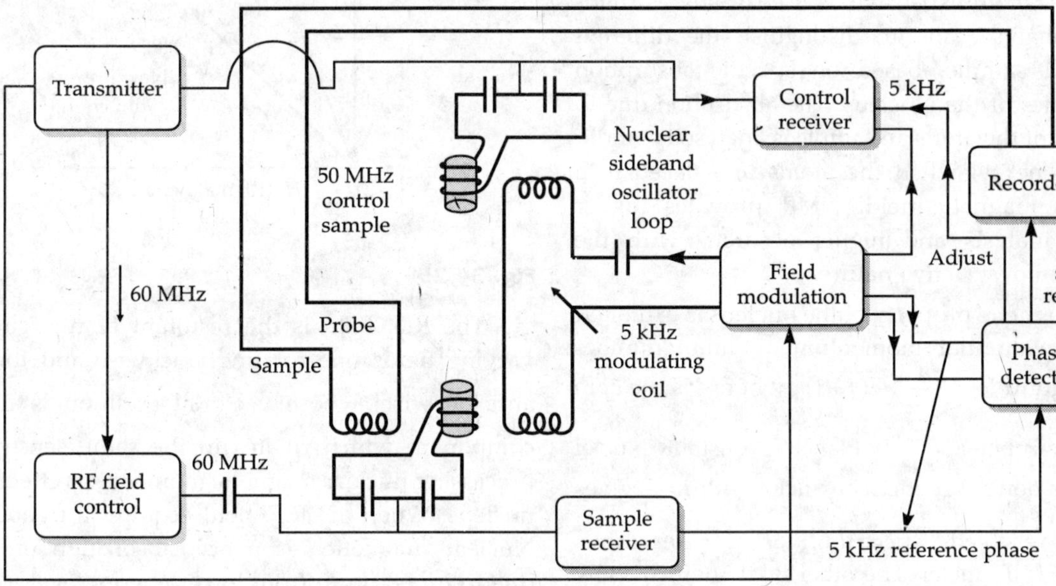


**Fig. 34.27** NMR spectrometer.

This problem has been solved with the pulsed (Fourier Transform) NMR (Fig. 34.28) which needs no scanning. This causes simultaneous excitation of all the resonance spectral lines with the help of a pulsed R.F. excitation. This requires a strong pulsed R.F. ($\vec{H}_1$) for short time duration of 1 – 1000 μsec. Due to the high energy short time pulse the magnetic moment spirals away from z axis in the direction of $\vec{H}_0$.

The magnetic moment flips by 90° and comes to rest in the plane perpendicular to z-axis. During the OFF time of the pulse the magnetic moment has a free precession under the influence of only static field $\vec{H}_0$. The free precession of the magnetic moment due to $\vec{H}_0$ there is a decaying induced sinusoidal voltage in the coil surrounding the sample. After several time constants ($3\,T_1$ to $5\,T_1$). The magnetic moment of the nuclei spirals back to its original position and the high energy pulse is again applied. The free induction decay (impulse response) and the continuous – wave display of the absorption lines can be analysed to plot the spin response with time which is the Fourier transform of the curves. The response of the entire spin has been plotted with time in a spectrometer.

The free induction decay signal for repetitive pulses is digitised by a fast *ADC*, and successive digitised transient signals are coherently added in a computer until high signal to noise ratio (SNR) is obtained. The sampling time used for convention to digits depends on spectral width which is typically $A = 5$ kHz with sampling time $(2\ A)^{-1}$ is first adjusted and then controlled.

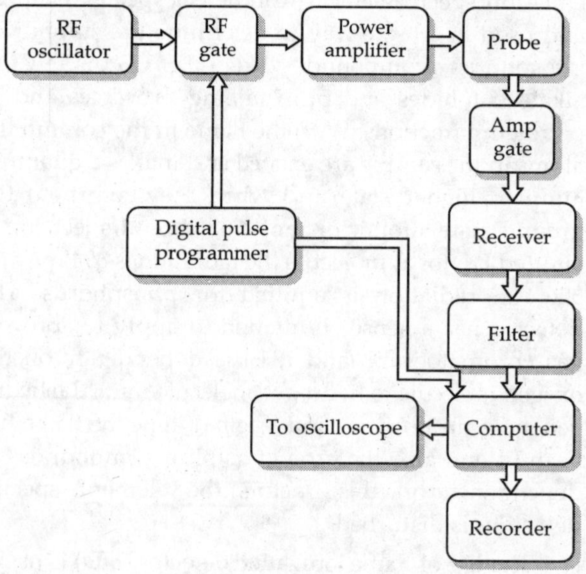


**Fig. 34.28** Pulse Fourier transform NMR spectrometer.

The phase of 5 kHz and 60 MHz has been adjusted and supplied to recorder. In control node the 60.004 MHz signal is supplied to field modulator.

A schematic block diagram of NMR instrument is shown in Fig. 34.27. Each of the R.F. frequency selected for magnetic nuclei is produced with the help of a 5 MHz crystal oscillator and mixed with output of low frequency oscillator (oscillation frequency 5 kHz) to generate the sweep. This part is shown in the figure as field modulator. The frequency synthesizer uses the twelfth harmonic of the base oscillator (59.9 MHz) a 0.1 MHz increased oscillator, and a single side band oscillator that mixes the two outputs and selects the upper side band (59.9 + 1.1) = 60 MHz. The final increment of 5 kHz is supplied by the 5 kHz audio-frequency oscillator and also modulates the $H_0$ field of permanent magnet. For other nuclei, plug in units provide a different multiplier circuit for selecting the desired harmonic incremental oscillator and sweep oscillator parameters.

## 34.10 X-RAY SPECTROMETER

The X-ray spectrometers are generally of three types. The sample analysis is done by the absorption to X-ray spectrometer while fluorescence type X-ray spectrometer gives induction of heavy metal presence and diffraction type is used to analyse the crystal structure. The X-ray spectrum is identified by *KX* unit, 1 *KX* unit is 100.202 picometers ($10^{-12}$ meter). For analytical purpose, the X-ray spectrum is useful in the wavelength range of 50 pm to 20 pm. The most important lines in X-ray spectrum are $K\alpha$ (70.9 pm for molybdenum X-ray target) and $K\beta$ (63.2 pm for molybdenum X-ray target). The second point of interest is the instrument for obtaining X-ray. Like other tubes, the cathode emits electron and the anode is the target and is its case a hard and heavy metal like molybdenum. The typical anode to cathode voltage for a X-ray tube is 60 kilo-volt. With the high energy electrode beam hits the target a spectrum of X-ray is generated so that the lowest wavelength is larger than a cut-off wavelength which is given by

$$\lambda_{min} = \frac{1240}{V}$$

The X-ray is a particle wave, though have the wave nature. The wave nature is utilised to realize the monochromator or wavelength selective device for X-ray spectrometry. The closely placed lines on the analysed crystal causes a diffraction grating for the X-ray. For the electromagnetic radiation quality inherent in X-ray the diffraction by grating produced by sample crystal follows the law :

$$n\lambda = d \sin \theta \qquad ...(1)$$

where, $\lambda$ is the X-ray wavelength in consideration
$d$ is the spacing between two lines in the grating.
and $\theta$ is the diffraction angle shown in Fig. 34.29.

Hence from the equation (1) for X-ray, smaller the spacings between grating lines the reasonable is $\theta$ value.

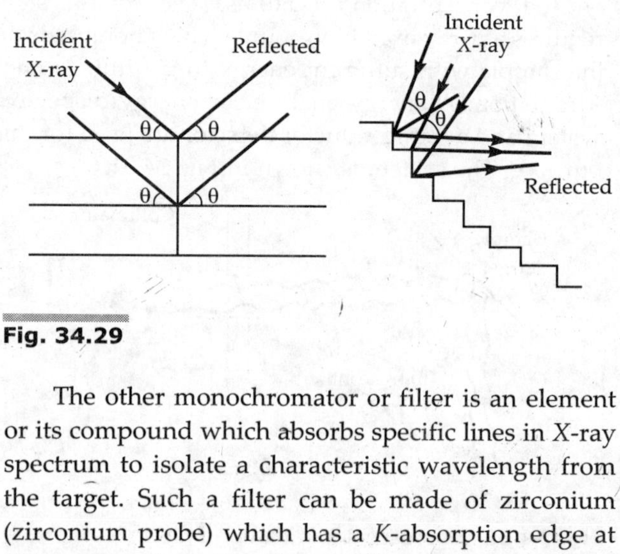


**Fig. 34.29**

The other monochromator or filter is an element or its compound which absorbs specific lines in X-ray spectrum to isolate a characteristic wavelength from the target. Such a filter can be made of zirconium (zirconium probe) which has a *K*-absorption edge at 68.9 pm and it absorbs molybdenum $K\beta$ line at 63.2 pm.

The detector for X-ray is the photographic film. The semiconductor detectors or solid state detectors are widely used now-a-days. It is found that

Germanium and Silicon detectors are used to detect X-radiation with addition of lithium. The proper way to detect X-ray is that, it is passed through a scintillation detector or gas. The ionized gas molecules due to X-radiation is sensed by semiconductor detector. Lastly, a collimating system is used for transit of X-radiation which is made of closely placed metal foils between the sample and analyser crystal and also between analyser and detector. Mere close are the foils, lesser X-radiation has been transmitted.

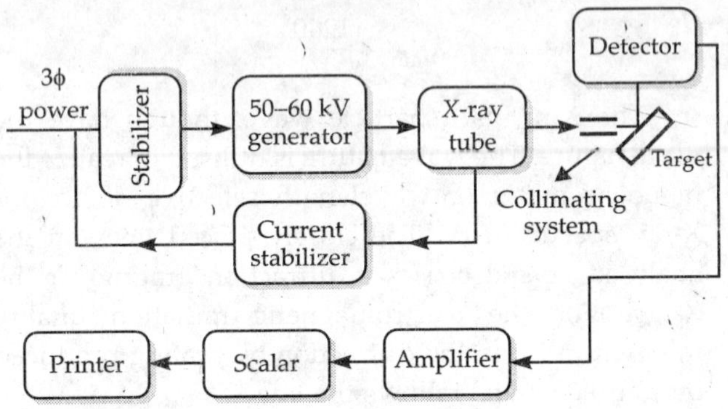


**Fig. 34.30** X-ray spectrometer.

The other type of X-ray spectroscopy is the Fluorescence spectrometry. It is used for amount detection and composition detection for heavy metals. In Fluorescence spectrometry, X radiation is generated by the irradiation of the sample by primary X-ray in tube and the energy of X-radiation is proportional to atomic number of sample (Z). The intensity depends on the concentration of atoms. The total system requires a X-ray excitation source which bombards on the sample with sufficient energy to get fluorescence X-radiation. A energy analyser or energy dispersive system and a semiconductor detector are used for this purpose. The system is shown in Fig. 34.31.

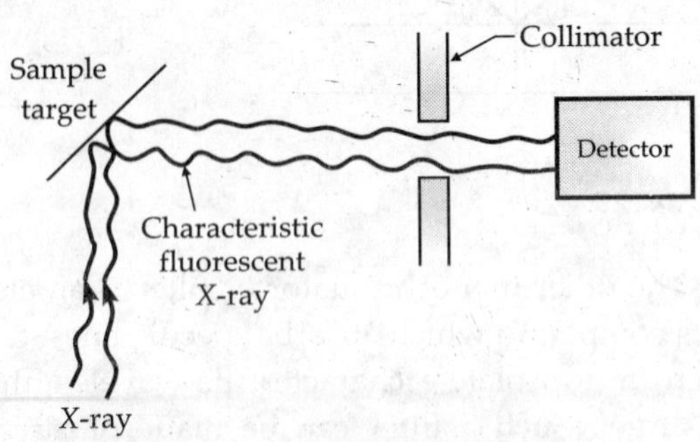


**Fig. 34.31** Fluorescent X-radiation system.

## 34.11 OTHER ANALYTICAL SENSORS

**Environment analysis-atomic absorption type photometric fluorometers.** The flame photometric detector is one type of flame analyser or a flame emission spectrometer. It can detect specific element and used in a large way to determine the sulphur or phosphorus compounds like $H_2S$, $CS_2$, $SO_2$, COS, alkale sulphides and pulp milling processes and in petroleum fractions. With the flame in the column the atoms in the sample are excited and makes a quantum jump to higher state and when they return to the ground state, atomic or molecular brand spectrum is emitted *i.e.*, for S, molecule the radiation is 394 nm and 526 nm radiation is emitted for phosphorus. The detector has a sensor limitation to apply for process gas chromatograph and it is its dependency on the oxygen molecule to hydrogen molecule ratio, the hydrogen molecule flow rate and length of time the flame has been lit and also the type of sulphur compound. For dependency on this, factors the element specific detection is disturbed.

The use of flame ionization detector (FID) is useful for its very large dynamic range ($10^{-7}$), high sensitivity, low detection limit ($10^{-2}$ gm/sec) and varnishingly small dead volume ($10^{-2}$ μl). The FID uses a hydrogen rich oxygen flame in which organic compounds are burnt producing ionized fragments. A high potential (150 – 300 volts) across the flame catch these ions at the collecting electrode *i.e.*, the collector electrode is positive. The FID process shown in Fig. 34.32.

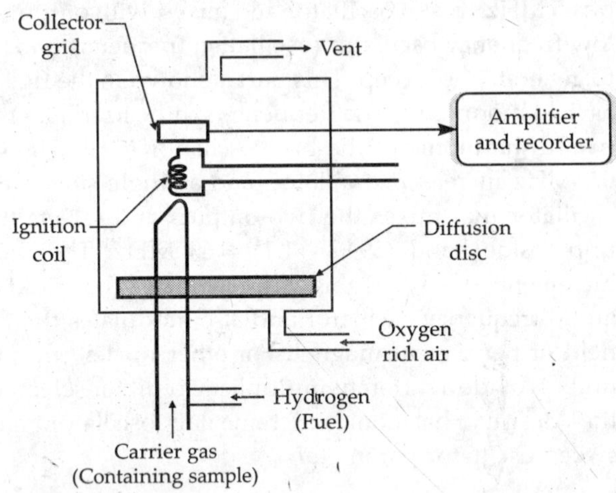


**Fig. 34.32** FID Chamber.

The FID response is different for different atoms and the analysis of no. of carbon atoms in aromatic compound is found from the FID response. This response is stated as ECN (Effective Carbon Number). The ECN is the addition of contribution of each carbon atom found in the organic compound modified by the functional group contributions.

The FID can't detect a number of compounds which are $H_2S$, nitrogen, $SO_2$, COS, CS, $CO_2$, $H_2O$,

$NH_3$ and other mobile gases. But, if, the carrier gas is doped with the ionic gas, these molecules can be detected. The response of the FID is very little noise prone ($10^{-14}$ Amps) and it is thus amplified with large gain amplifier.

**Flame and furnace atomic absorption (AA) Spectrometer.** This type of AA instrument is used for elemental analysis (Fig. 34.33).

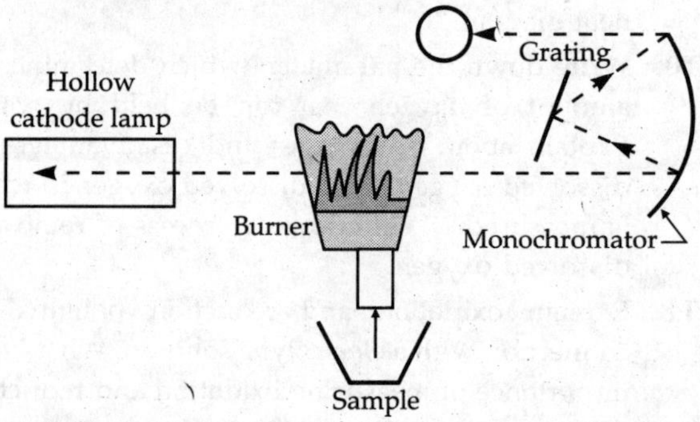


**Fig. 34.33** AA spectrometer.

In flame AA, the sample is neutralised and passed into the flame. The flame temperature is about 3000°C. As the radiation passes through the flame, the excited sample atoms absorb more than one particular lines from a continuous spectrum and these lines are characteristic of the sample atoms. The neutralised atoms in a flame temperature about 2000°C – 4000°C become plasma. In the furnace type AA spectrometer the flame is replaced by a certain cylinder. The cylinder is heated electrically to atomise the sample. For this type of AA the detection limit is improved by 100 – 1000 times, as compared to the flame.

**Dissolved oxygen and oxygen demand.** This type of analysis involves quantity up-take of oxygen by the sample. The dissolved oxygen concentration of water is a function of oxygen solubility at the existing temperature and barometer pressure. DO (dissolved oxygen) describes the oxygen of receiving water, while BOD (Biochemical $O_2$ demand), COD (chemical oxygen demand) and TOD (total oxygen demand) describes the pollution and pollutant content in waste water. Measurement of BOD provides necessary information about the quantity of oxygen required for biochemical degradation of organic matter in water. COD measures the quantity of oxygen that organic materials will consume in water when oxygen is released from reduction of a dichromate solution. TOD measures oxygen demand for both organic and inorganic compounds when they react and consume oxygen.

Dissolved oxygen concentration indicates the quality of water. It is measured by quality of water. It is measured by electrochemical probes. The principle is that the passage of dissolved oxygen through a membrane of chemical electrode is followed by the radiation of oxygen at the cathode. The anode current indicates the dissolved oxygen concentration.

**Fluorimeters and phosphorimeters.** The main difficulty which arises in the UV or visible or IR spectrometry is that, it is hard to distinguish between similar absorption lines, This problem has been addressed by fluorimeters and phosphorimeters. It is the property of some solutions, such that, when these solutions are irradiated with some Ultraviolet radiation or visible light, with specified spectrum, they emit light of different characteristic wavelength. This nature of radiation particularly found in phosphorous is called fluorescence or phosphorescence. This type of spectrometery is widely used in biomedical sample analysis particularly Diphosphonew-dolide (DAN) and transminase etc.

**Raman spectroscopy.** Raman effect is the root cause behind Raman spectroscopy. The Raman scattering effect states that a photon when bombards with the electron in an atom a few portion of the scattered radiation (1 in $10^5$) is perpendicular to the direction of the incident beams and the 1% of the total scattered intensity is of frequency that is different from incident monochromatic light intensity. A photomultiplier tube used as detector, is placed perpendicular to the direction of incident beam to detect radiation which is a characteristic to sample atom subjected to Raman scattering. Generally Raman spectrometery is used to monitor corrosion, medicinal preparation, ceramics industry and polymer industry.

**Pollution analysis in environment.** The chart below shows the different types of pollution present in environment air and the way to analyse the specific pollutions.

| ***Name of Pollutants*** | ***Presence in atmosphere*** | ***Way of analysis*** |
|---|---|---|
| CO | 0 – 300 ppm | Absorption spectrometer and Gas chromatography |
| $SO_2$ | 0 – 2 ppm | UV absorption, IR absorption, GC Flame detection |
| NO, $NO_2$ | 0 – 1 ppm | Visible spectroscopy chemiluniscence |
| Hydro carbons | 0 – 80 ppm | UV absorption GC Flame Ionisation detector (FID) |
| Oxidants | 0 – 500 ppm | Chemiluminiscence chemical analysis electro chemical analysis visible spectroscopy. |

# Review Questions

1. (*a*) Write down the Beer's Law for absorption of radiation by a gas medium. Draw the block diagram of an infrared spectrometer.
   (*b*) What are single channel and dual channel IR spectrometers?
2. (*a*) What are advantages of zirconia type oxygen analyser over paramagnetic type ?
   (*b*) Write down the principle of absorption type gas analyser.
   (*c*) Explain relative humidity and absolute humidity.
3. (*a*) State and explain Lambert and Beer's law regarding absorption of radiation in solution.
   (*b*) What are the basic features of fluorescence and phosphrescence ? Write the application of fluorescence in analysing different components.
4. (*a*) Write down the principle of liquid chromatography. Explain with schematic diagram.
   (*b*) State Bragg's equation of *x*-ray diffraction through a crystal.
5. (*a*) Draw a schematic diagram of a sampling system connected between the take off point of flue gas outlet and an infrared type co analyser and explain the system.
   (*b*) What is the function of air purging of the sample lines ?
   (*c*) Why pressure and temperature conditioning is required ?
6. Draw block diagram of an instrument for measurement of dissolved oxygen in a fluid. Explain functions of each block/component.
7. (*a*) Explain the basic principle of mass spectrometer.
   (*b*) Describe and explain a single focussing magnetic deflection type mass spectrometer.
8. Write short notes on ; (*i*) Mass spectroscopy, (*ii*) Gas chromatography.
9. Write down the working principle of Zirconic probe. For what type of gases Zirconic probe is used ? Explain the measuring procedure with a neat diagram.
10. Write down the parameters which determine the amount of oxygen that can be held by water. Explain about advantages and disadvantages of dissolved oxygen. How dissolved oxygen in water is measured ? Write down a process of removing dissolved oxygen.
11. Explain oxidation and reduction potential in connection with a electrolytic solution. What is the importance of measuring oxidation and reduction potentia ? Describe the process of measurement with necessary diagram.
12. What is characteristic *x*-rays ? Write down Mosley's law. Explain with diagram, how characteristics *x*-rays can be used for quantitative and qualitative micro analysis ?
13. Explain the theory of Gas chromatography Separation. Draw neat diagram of a modern gas chromatograph (GC) instrument and describe functions of each component of the system. Explain about flow rate and quality of carrier gases.
14. Write down about sampling system for gases. Draw schematic diagram of final sample system with analyzer. Explain function of each section.
15. What are the fundamental factor on which electrical safety depends ? Write down the factors that influence the degree of injury caused by electrical shock. Mention about precaution to be taken for avoiding electrical hazard.

# Objective Type Questions

1. Choose the correct answer from the given alternatives
   (*i*) Infrared radiation interacts with all molecules except :
   (*a*) carbon monoxide
   (*b*) Carbon dioxide
   (*c*) nitrogen (*d*) none of the above.
   (*ii*) The starting point for quantitative analysis is :
   (*a*) Bear Lambert law
   (*b*) Lambert Cosine Law.
   (*iii*) Which of the following wavelengths fall in the IR spectrum :
   (*a*) 110 mm (*b*) 800 mm
   (*c*) 200 mm (*d*) 10 cm ?

(*iv*) Oxygen has strong affinity for a :
(*a*) magnetic field
(*b*) electric field
(*c*) both magnetic and electric fields.

(*v*) The diffracted beam of X-rays can be detected :
(*a*) photo graphically
(*b*) electromagnetically
(*c*) electrostatically.

(*vi*) The wavelength :
(*a*) $\lambda = m / hv$
(*b*) $\lambda = hc / mv$
(*c*) $\lambda = h / mv$.

(*vii*) Filters in a gas sampling system remove :
(*a*) moisture
(*b*) toxic gases
(*c*) particulate water.

**2.** Write the correct answer out of the choices given

(*i*) For UV/visible Spectrophotometer sample cell is made up of :
(*a*) glass
(*b*) quartz
(*c*) sodium chloride..

(*ii*) When a current passes through a wire the final equilibrium temperature depends upon :
(*a*) thermal conductivity of the gas surrounding it
(*b*) material of the enclosed container
(*c*) none of these.

(*iii*) Paramagnetic gases have :
(*a*) low permeability and negative susceptibility
(*b*) high permeability and high susceptibility
(*c*) high permeability and low but +ve susceptibility.

(*iv*) The main interest of infrared region of the electro-magnetic spectrum is from :
(*a*) 4000 to 200 $cm^{-1}$
(*b*) 0.8 to 2.5 μm
(*c*) 200 to 10 $cm^{-1}$.

(*v*) IR transmission range for sodium chloride cell/window is :
(*a*) 2.5 – 59 μm
(*b*) 9.0 – 26.0 μm
(*c*) 2.0 – 15.4 μm.

(*vi*) The percentage of oxygen by weight in dry pure air is :
(*a*) 20.9%
(*b*) 20.2%
(*c*) 21%
(*d*) none of the these.

(*vii*) Roentgen, Wilhelm Conard was awarded Nobel Prize in Physics in the year :
(*a*) 1909
(*b*) 1901
(*c*) 1923 for discovery of X-rays.

(*viii*) In on-line sampling system the function of a analyzar is to :
(*a*) perform the measurement and the information at the desired time interval,
(*b*) perform the measurement as quick as possible,
(*c*) perform measurement as accurate as possible.

(*ix*) Dissolved oxygen is undesirable in boiles feed water because its presence increases the risk of :
(*a*) accident
(*b*) life of boiler
(*c*) corrosion.

(*x*) Chemical shift and spin-spin coupling are related to :
(*a*) G.C.
(*b*) EPR
(*c*) NMR.

(*xi*) The energy of a nucleus in the presence of an externally applied magnetic field is given by :
(*a*) $E = -\frac{m_1 \mu \beta B_0}{I}$ (*b*) $E = -\frac{\beta B_0}{hI}$
(*c*) $E = -\frac{\mu \beta B_0}{IKT}$

(*xii*) In gas chromatography the FID responds only to substances that can be ionized in :
(*a*) air acetylene flame
(*b*) air-hydrogen flame
(*c*) air nitrous oxide flame.

(*xiii*) Inventor of Electron Microscope is :
(*a*) C.W. Oatlay
(*b*) K. Knoil
(*c*) E. Ruska, who got Nobel Prize in the year 1986.

(*xiv*) Radiation which is not harmful of the following :
(*a*) *x*-rays
(*b*) UV
(*c*) γ-rays
(*d*) microwave.

# 35

# Automatic Process Control

## 35.1 INTRODUCTION

With the advancement of time, automatic process control has been introduced instead of manual process control. In manual process control the process variable has been measured with the help of process sensor not necessarily an electric transducer. The process variable has been compared with the set point value where human brain is the controller and control action has been taken with the help of manually operated switch or actuator. In automatic process control a controller, be it ON-OFF or proportional, is used to compare process variable with set point or any other control arithmetic than only comparison, and depending on the controller output a electrically actuated valve or servo motor or stepper motor does the function of final control element. With the modernisation of sensor technology most of the parameters treated for control, of a process, can be transducted to electrical signal and hence continuous process monitoring by automated control system is possible. Modern positioner and actuator technology has improved in such a way in the form of valve positioner, servo or AC or DC motor drivers, stepper motor drives that it is easy to implement the controller output. In this article, we discuss with the different controllers, their specification and mode of working of control process may be of *three* types :

- Open loop control,
- Feed forward control and
- Feed back or closed loop control.

## 35.2 OPEN LOOP CONTROL

The open loop control path diagram is shown in Fig. 35.1, where $r$ is the set point, $d$ is the load disturbance and $e$ is the error voltage. Here, $r$ and $d$ are unmeasured variables and hence $e$ is the unmeasured variable. Hence, if there is process disturbance $d_1$, it is difficult for error voltage $e$ to control the process.

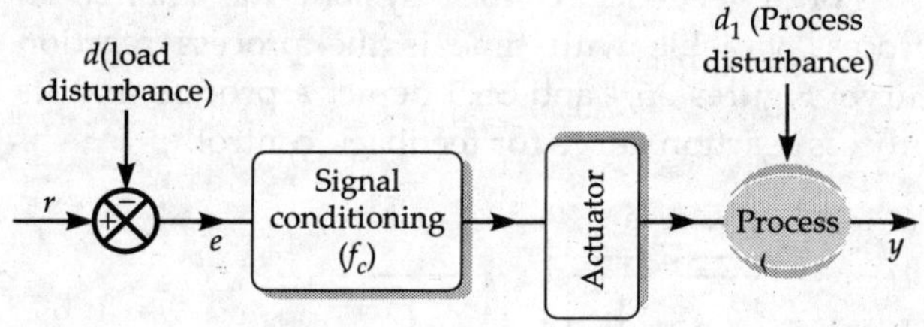


**Fig. 35.1**

## 35.3 FEED FORWARD CONTROL

Control by feedforward loop has been shown in Fig. 35.2. Feed forward control, though an open loop control, but it counteracts the anticipated effect of a measured load by error signal $e_M$ (measured) combined with feedback control to cancel the effect of unmeasured load $d_1$.

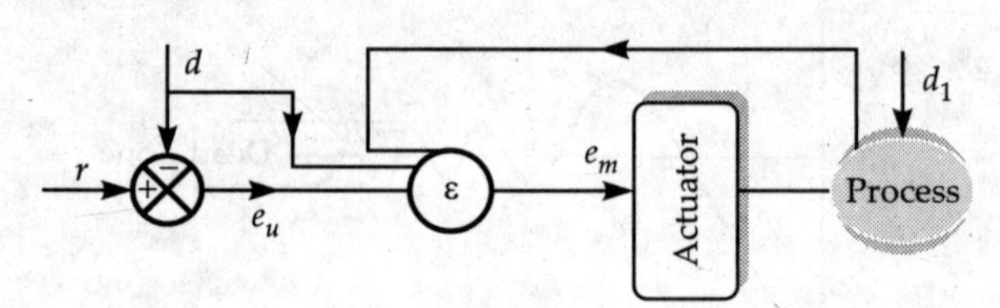


**Fig. 35.2**

## 35.3 FEEDBACK CONTROL

The block diagram of a feedback control loop is shown in Fig. 35.3.

In feedback loop the set point is compared to the feedback of measured value of process variable and control action is taken by the error signal $e_M$ through

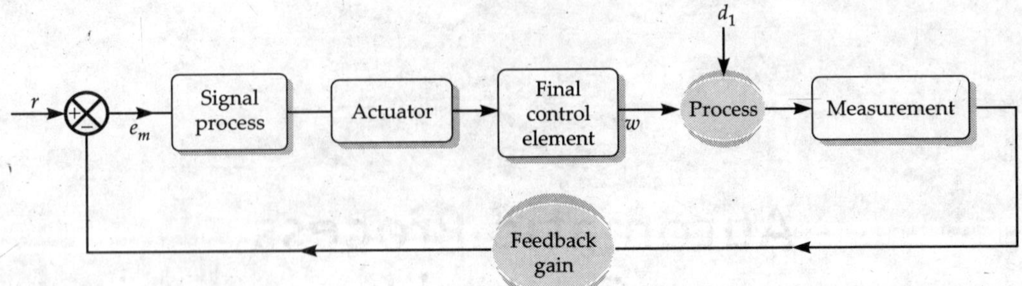


Fig. 35.3

the actuation of final control element. The measured variable is monitored continuously for modern automated controller. The load distribution is measured in the feedback controller and control action is instantaneous with little delay for any disturbance in the process.

## 35.5 PROCESS REACTION CURVE

For a feedback control system the change in process variable with time is the process reaction curve. Figures 35.4 and 35.5 depict a process and its process reaction curve for feedback control.

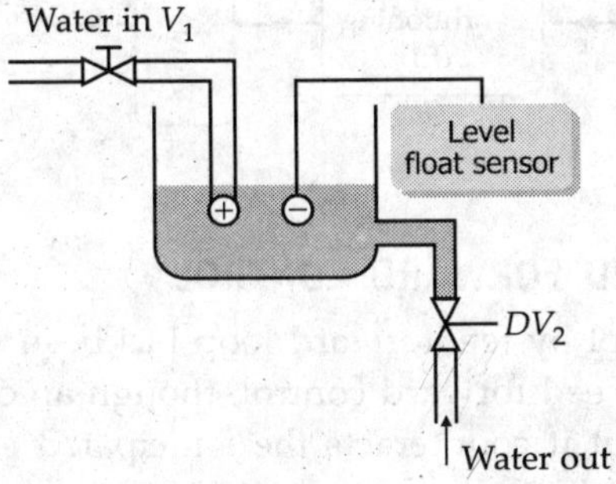


**Fig. 35.4** Level control process.

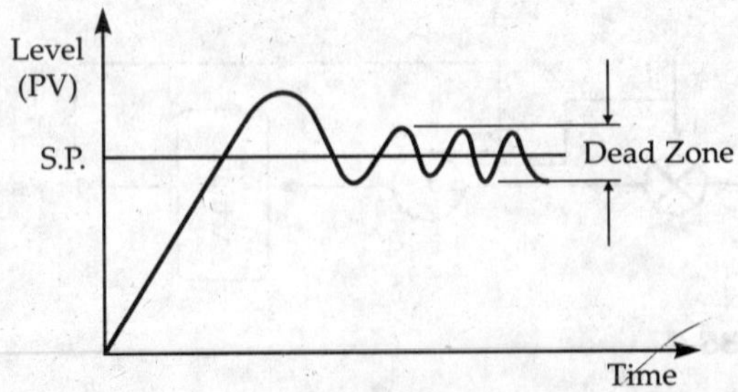


**Fig. 35.5** Process reaction curve.

The process shown in Fig. 35.6 is an experiment of level control of water in a bath where valves $V_1$ and $V_2$ are used for incoming and outgoing water and a level sensor is used to get the electrical signal corresponding to the present level *i.e.*, process variable (PV). The experiment can be viewed as an On-off control *i.e.*, when level goes below a certain level set point, $V_1$ is on and $V_2$ off and above the level set point $V_2$ on and $V_1$ off. The process reaction curve depicts relation between change in level with time.

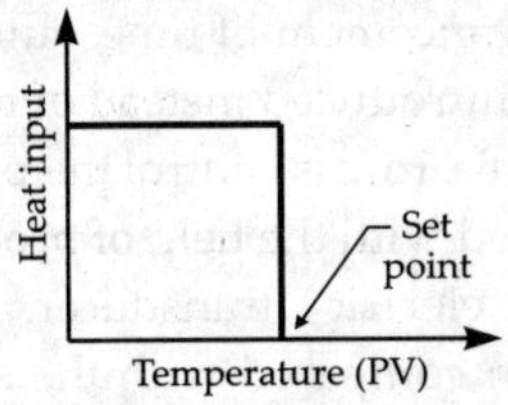


**Fig. 35.6** Controller transfer curve temperature (PV).

## 35.6 TRANSFER CURVE

Figures 35.6 and 35.7 depict a thermal process and its transfer curve respectively.

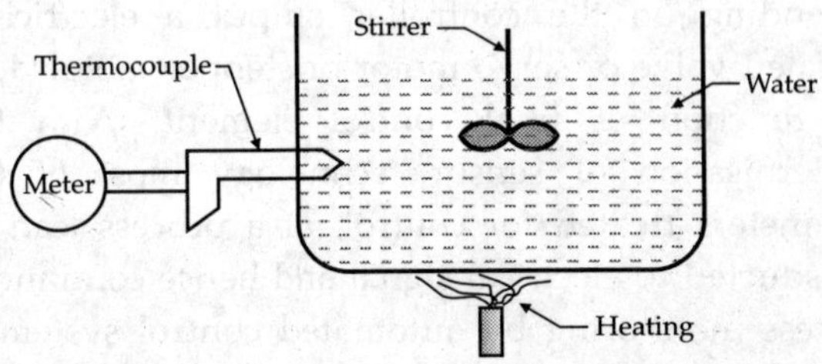


**Fig. 35.7**

In the process, Fig. 35.7, water has been heated and with its temperature has been monitored. The heat is well mixed with water with a stirrer. The transfer curve for the controller (ON-OFF) has been

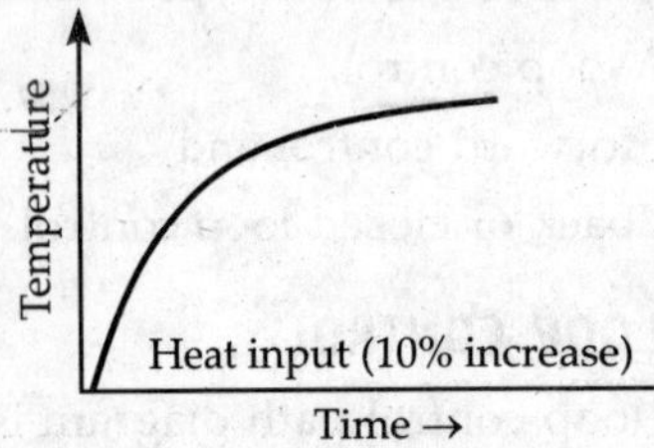


**Fig. 35.8** Process for endothermic process.

shown in Fig. 35.7 which depicts the heat input to the process with change in temperature. The process reaction curve also indicates the change is temperature for 10% increase in heat input. This is shown in Fig. 35.8 (Endothermic process) and Fig. 35.9 (exothermic process).

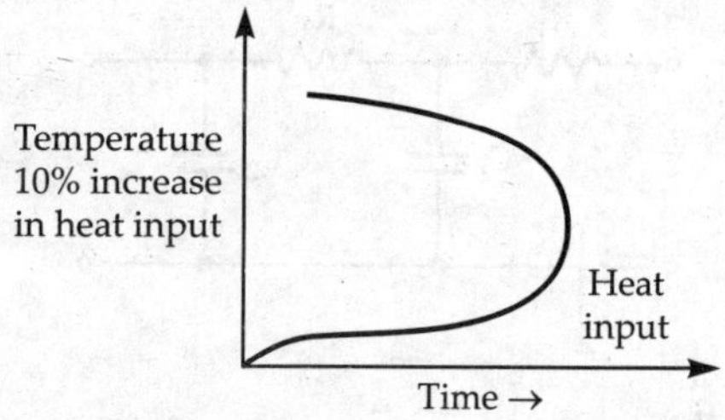


**Fig. 35.9** Process reaction curve for exothermic process.

Figure 35.8 shows that the process is endothermic. For an exothermic (process in which heat is produced) process, the curve is shown in Fig. 35.9.

**Capacity of process.** In a different context when the thermocouple is placed in pipe connecting to the thermal bath (Fig. 35.10), there is a lag between process temperature change and the temperature indicated by the thermocouple. If the process in Fig. 35.9 is called a single capacity process, Fig. 35.10 is a multi capacity process.

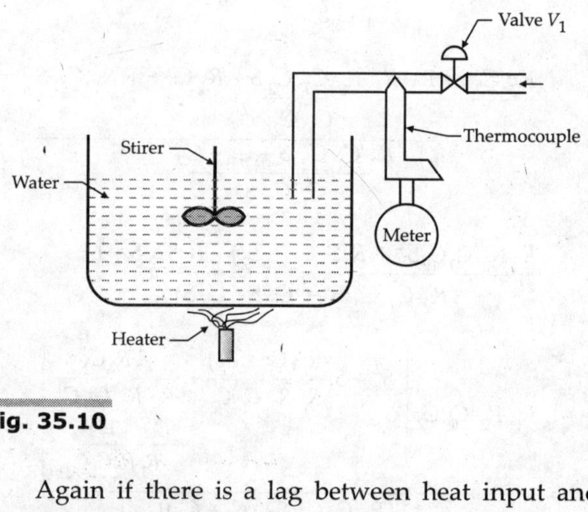


**Fig. 35.10**

Again if there is a lag between heat input and temperature rise localised to certain portion of water, it is a multiplicity process.

A single and two-capacity process can be configured by electrical circuits introducing a resistance capacitance lag. Figures 35.11 and 35.12 depict the equivalent electrical circuit for a single capacity and double capacity process.

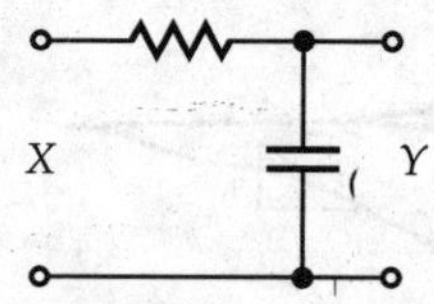


**Fig. 35.11** Single capacity process.

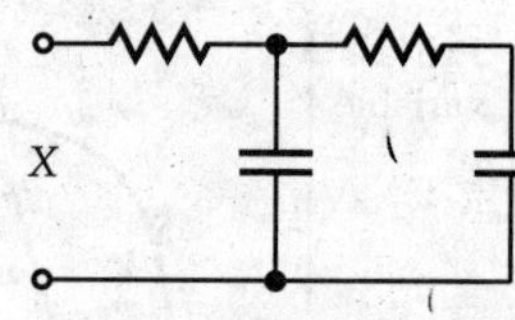


**Fig. 35.12** Double capacity process.

**Gain of a process.** The ideal transfer curve for a thermal process shown in Fig. 35.7 may be represented by Fig. 35.13.

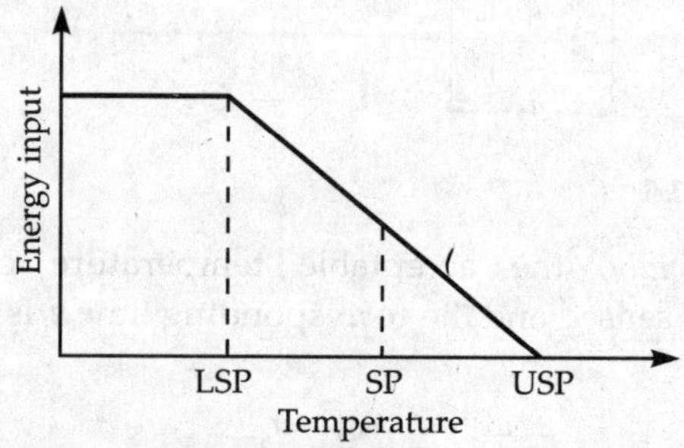


SP → Set point
LSP → Lower set point
USP → Upper set point.

**Fig. 35.13**

These curves assume that, the heat input changes proportionally with temperature the change in temperature about the set point, temperature.

And, gain of the controller is given by

$$\text{gain} = \frac{100}{\text{USP} - \text{LSP}} \%$$

where PB (Proportional band)

$$= \text{USP} - \text{LSP}$$

**Controllability and stability of process.** The controllability of a system is best understood from the process reaction curve which gives the change in process variable with time for 10% increase in energy input. Generally, the process reaction curve is a *S*-type curve. The process reaction curve for four different processes is shown in Fig. 35.14.

Two parameters for process reaction curves are important :

*Firstly*, the steepness of maximum rate of change of the curve.

It is found that curve 4 is the steepest with maximum rate of change ($b_4 = \tan\theta$) and curve 1 is least steep.

and $b_4 > b_3 > b_2 > b_1$

35

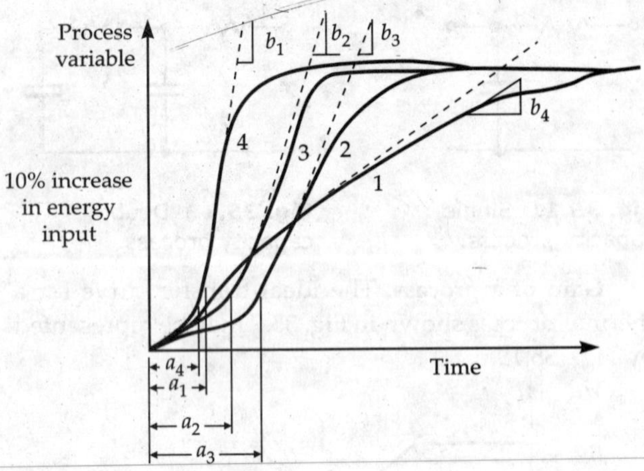


Fig. 35.14

*Secondly*, the acceptable temperature change which is sensed and the corresponding time $a_i$ is found as

$$a_3 > a_2 > a_1 > a_4$$

The product $a_i b_i$ suggests a measure to predict the controllability of the process. It is found that process 4 is most difficult to control and process 1 is least difficult to control and in between processes 2 and 3.

Stability of a process is judged by the analysis of the poles and zeros of the process. If the poles and zeros of the process are on the left half plane (LHP) of the real-imaginary plane the process is stable whereas if one of the poles or zeros is on the Right half plane (RHP) the process is not stable if not otherwise compensation is done for RHP poles or zeros.

Let us consider the stability of a single capacity process.

The transfer function for Fig. 35.15 in the Laplace domain is

$$H(S) = \frac{\frac{1}{RC}}{\left(S + \frac{1}{RC}\right)} \quad ...(i)$$

$$H(S) = \frac{y(s)}{x(s)}$$

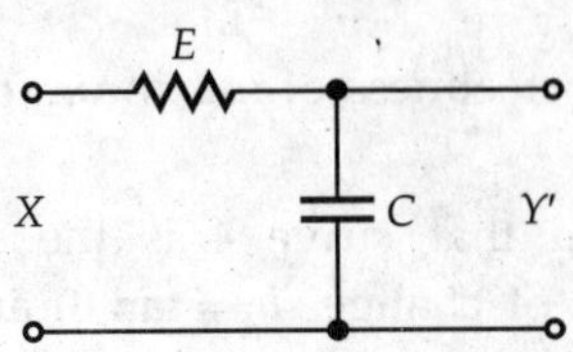


Fig. 35.15

Hence from Eqn. (*i*) it can be inferred that

1. The only pole is at $-\frac{1}{RC}$
2. Gain $\frac{1}{RC}$

So this single capacity process is a stable process.

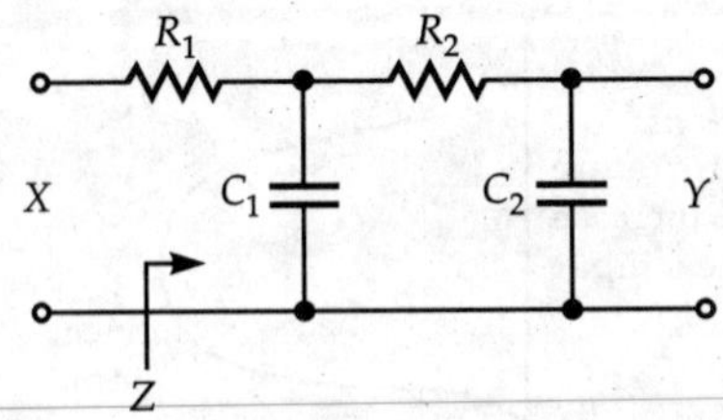


Fig. 35.16

For 2-capacity process,

$$Z = \left\{\left(R_2 + \frac{1}{C_2 S}\right) 11 \frac{1}{C_1 S}\right\} + R_1$$

or
$$Z = \frac{\left(R_2 + \frac{1}{C_2 S}\right)\frac{1}{C_1 S}}{R_2 + \frac{1}{C_2 S} + \frac{1}{C_1 S}} + R_1$$

$$= \frac{\frac{R_2}{C_1 S} + \frac{1}{C_1 C_2 S^2} + R_1 R_2 + \frac{R_1}{C_1 S} + \frac{R_1}{C_2 S}}{\frac{R_2 C_1 C_2 S + C_1 + C_2}{C_1 C_2 S}}$$

$$= \frac{\frac{R_2 C_2 S + 1 + R_1 R_2 C_1 C_2 S + R_1 C_2 S + R_1 C_1 S}{C_1 C_2 S^2}}{\frac{R_2 C_1 C_2 S + C_1 + C_2}{C_1 C_2 S}}$$

$$= \frac{R_1 R_2 C_1 C_2 S^2 + R_1 S(C_1 + C_2) + R_2 C_2 S + 1}{(R_2 C_1 C_2 S^2 + C_1 S + C_2 S)}$$

$$= \frac{R_1 R_2 C_1 C_2 \left[S^2 + S.\frac{R_1(C_1 + C_2)}{R_1 R_2 C_1 C_2} + \frac{S.R_2 C_2}{R_1 R_2 C_1 C_2} + \frac{1}{R_1 R_2 C_1 C_2}\right]}{S[R_2 C_1 C_2 S + (C_1 + C_2)]}$$

$$= \frac{\left[S^2 + S\left(\frac{1}{R_2}\left(\frac{1}{C_1} + \frac{1}{C_2}\right) + \frac{1}{R_1 C_1}\right) + \frac{1}{R_1 R_2 C_1 C_2}\right]}{S\left[S + \left(\frac{1}{R_2 C_1} + \frac{1}{R_2 C_2}\right)\right]}$$

$$H(S) = \frac{X}{Z} \times \frac{\frac{1}{C_1 S}}{\frac{1}{C_1 S} + \left[R_2 + \frac{1}{C_2 S}\right]} \times \frac{1}{C_2 S}$$

$$= \frac{X}{Z} \times \frac{\frac{1}{C_1 C_2 S^2}}{\frac{C_2 + C_1 + R_2 C_1 C_2 S}{C_1 C_2 S}}$$

$$= \frac{X}{Z} \times \frac{1}{(C_2 + C_2 + C_1 C_2 R_2 S)}$$

or $$H(S) = \frac{1}{R_1 R_2 C_1 C_2 \left[S^2 + S\left(\frac{1}{R_2}\left(\frac{1}{C_1} + \frac{1}{C_2}\right) + \frac{1}{R_1 C_1}\right) + R_1 R_2 C_1 C_2\right]}$$

$\therefore$ Gain is $\frac{1}{R_1 R_2 C_1 C_2}$

and the poles are

$$(i)\ -\frac{1}{2}\left(\frac{1}{R_2}\left(\frac{1}{C_1} + \frac{1}{C_2}\right) + \frac{1}{R_1 C_1}\right),$$

$$\pm \frac{\sqrt{\left(\frac{1}{R_2}\left(\frac{1}{C_1} + \frac{1}{C_2}\right) + \frac{1}{R_1 C_1}\right)^2 - 4 R_1 R_2 C_1 C_2}}{2}$$

which are given by $-\frac{1}{2}b \pm \frac{\sqrt{b^2 - 4ac}}{2}$

where $b = \frac{1}{R_2}\left(\frac{1}{C_1} + \frac{1}{C_2}\right) + \frac{1}{R_1 C_1}$; $\quad a = 1$

$c = 4 R_1 R_2 C_1 C_2$

Hence the transfer function has two poles of which one pole $\left(-\frac{1}{2a}b - \frac{\sqrt{b^2 - 4c}}{2a}\right)$ is a LHP whereas the other pole $-\frac{1}{2a}b + \frac{\sqrt{b^2 - 4c}}{2a}$ is a LHP if and only if

$b > (b^2 - 4c)^{1/2}$

or $\quad b^2 > b^2 - 4c$

or $\quad -c < 0$

or $\quad c > 0$

or $\quad c = R_1 R_2 C_1 C_2$ and $c$ is the gain.

And for real LHP,

$b^2 - 4c > 0$

or $\quad b > 2\sqrt{c}$

or $\quad c < \frac{b^3}{4}$

Hence the system is stable *i.e.*, all poles are LHP if

$$\frac{b^2}{4} > R_1 R_2 C_1 C_2 > 0$$

or gain is positive and less than $\frac{b^2}{4}$ for meeting stability condition of the system.

Controllers are categorised on the basis of their functioning and controller used in a process depends on the process reaction, capacity and the placing of poles and zeros in the complex plane. Extending the mathematics for multiplicity control the fixation of gain is important and more complex.

## 35.6 ON-OFF CONTROL

For a control of an ON-OFF controller is active only when process variable crosses the set point in the positive or negative direction. When the process variable is greater than set point the controller is OFF and when PV is less than set point the controller is ON. The transfer curve for the controller is given in Fig. 35.17.

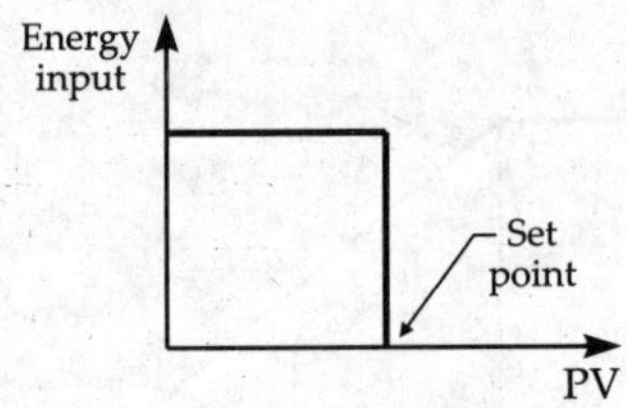


**Fig. 35.17** Transfer curve for ON-OFF control.

The figure suggests that correcting action is taken by the controller only when manipulated or error *i.e.*, (PV — set point), is non zero. The gain of these types of controller is infinity for ideal case and greater than 1000 for practical case. The process reaction curve of ON-Off controller is shown in Fig. 35.18.

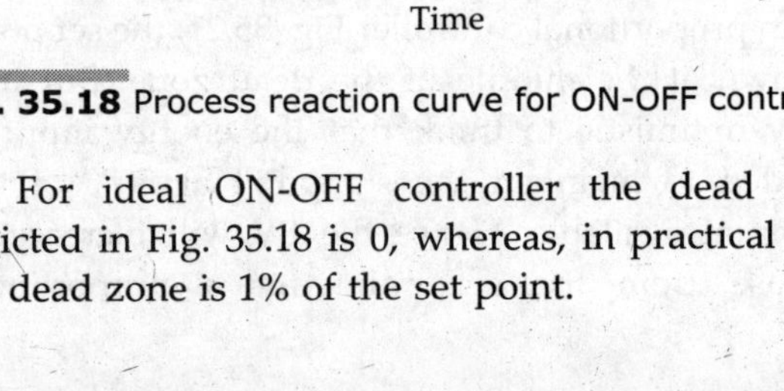


**Fig. 35.18** Process reaction curve for ON-OFF control.

For ideal ON-OFF controller the dead zone indicted in Fig. 35.18 is 0, whereas, in practical base the dead zone is 1% of the set point.

## 35.7 PROPORTIONAL CONTROL

A proportional controller provides energy input proportional to the manipulated variable so that the process, to an extent is balanced in energy demand. The transfer curve for a proportional controller is shown in Fig. 35.19, where LSP is the lower set point and USP is the upper set point.

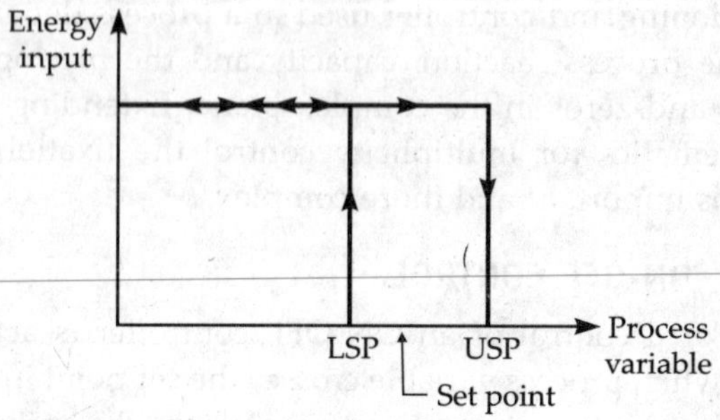


**Fig. 35.19** Impact side of the process.

The process demand side transfer curve of the process is given by

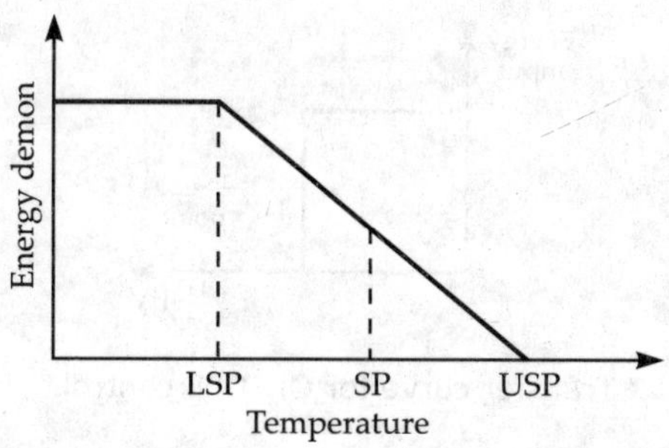


**Fig. 35.20**

Where the proportional band *i.e.,* (USP — LSP) limits the gain of the process. For wide PB the gain is lower and for narrow band PB the gain is high.

The advantage of using proportional controller than ON-OFF controller is the energy input is approximately equal to the process demand of energy to maintain a stable process variable. The proportional controller is designed in such a way that its gain is limited within the range which is permitted by poles and zeros of the process transfer function $H(S)$.

### Offset

In proportional controller Fig. 35.21, the set point is shown at the middle of the dead zone. But it is highly optimistic to think that the energy input is provided to maintain the set point at 50% of the maximum available. Hence the PV will increase or decrease from the set point for a proportional controller. The difference of stabilized process variable from the set point is the offset. Offset of a proportional controller action is shown in Fig. 35.21.

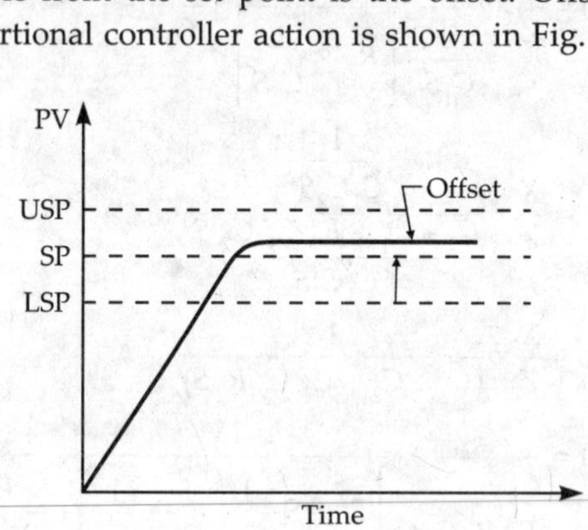


**Fig. 35.21**

**Nulling of offset.** Offset null can be achieved by manually or automatically. The process of offset nulling is to move the proportional band parallely so that set point and operating point are at the middle of the proportional band *i.e.,* S.P. and O.P. is equal to $\frac{1}{2}$(USP – LSP). This is shown in Fig. 35.22.

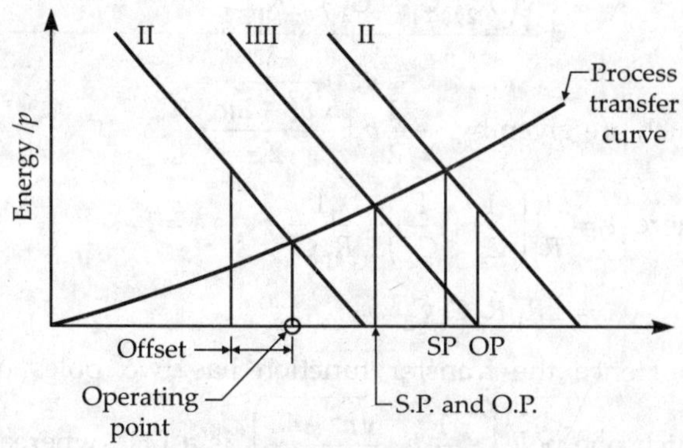


**Fig. 35.22**

Hence for curve II, the amount of *PB* shifting is done either automatically or manually by the operator in small increment until that controller output matches the process heat demand at set point *i.e.,* the set point is stabilised. For manual offset nulling *i.e.,* reset a potentionmeter is used to offset the proportional band electrically.

Automatic reset action uses an integrator with the proportional controller. The error signal is integrated with respect to time and the integrated output is added to error signal to move the proportional band. The PI controller *i.e.,* proportional—Integrated controller circuit is presented in Fig. 35.23.

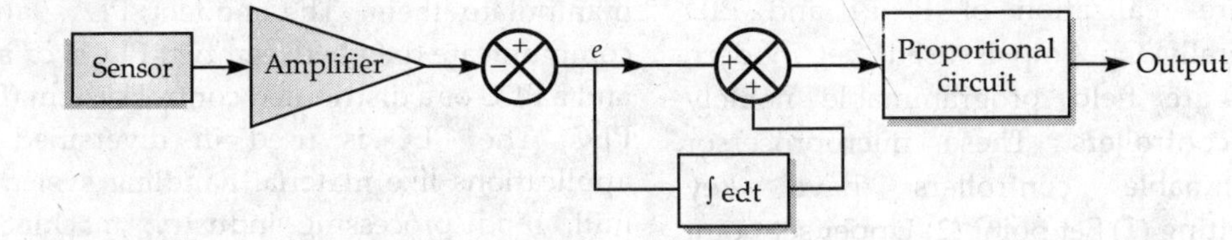


**Fig. 35.23** Shifting of proportional band–integrating action.

### Integral Saturation

In the case of integral saturation, the integrator integrates the error signal with respect to time when the process variable is outside the proportional band. The comparatively large integrated value causes the set point to move outside the proportional level. For this, the integration output starts to desturate and the PV—time curve shows large over shoot and there is detonation of controllability.

Another problem for PI controller is that large over-shoot is found in the controller action during start up. This is because, as the PV moves within the proportional band and the controller action starts, the reset action causes the proportional band to shift causing excess energy input. As the PV moves above the set point the deviation error signal reverses and the integrator brings the PB at a lower value, hence the overshoot is formed at starting.

## 35.8 PROPORTIONAL-DERIVATIVE CONTROLLER

Derivation of a quickly varying error signal generates a varying signal which leads the error signal in time. Hence derivative action is an anticipatory action. The derivative action has the ability to shift proportional band up and down for a quickly varying process variable. The amount of shift of the proportional band is proportional to the rate at which error signal vary. Hence it is also called the rate action controller. A schematic of the proportional derivative controller is shown in Fig. 35.24.

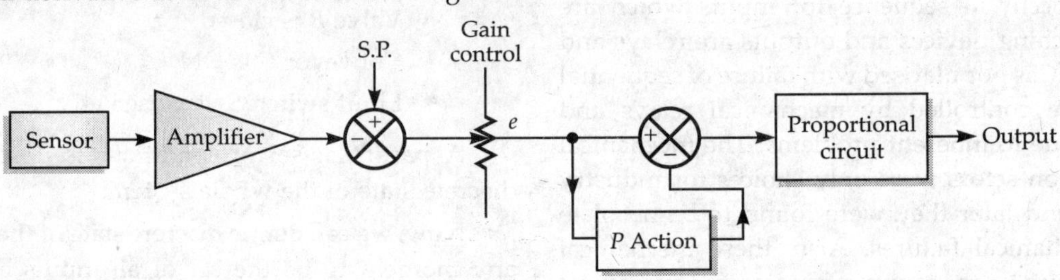


**Fig. 35.24**

As the derivative action works in a way to collaborate with the rapidly changing process variable, it is used to compensate the start up overshoot of the PI controller and also acts properly in the case of step change in set point.

## 35.9 PROPORTIONAL-INTEGRAL DERIVATION (PID) CONTROL

PID controllers are used in case of difficult processes where there is rapid change in process variable due to improper mixing of heat and localised heating in the water bath or very sluggish process. The *P*-action and *D*-action shift the proportional band to accommodate the PV in vicinity of the set point, when the *P*-action takes charge of rapidly varying PV and the starting up overshoot problem. The PID controller is shown in Fig. 35.25.

35

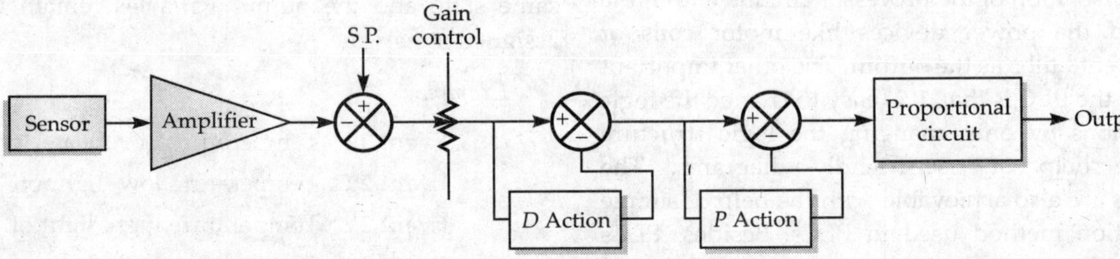


**Fig. 35.25**

For hardware realization of P, PI and PID controllers, generally Op-Amp is used, yet modern day controllers are field programmable namely programmable controllers. These microprocessor based programmable controllers have key orientations of setting (1) Set point (2) Upper set point (UJP) (3) Lower set point (LSP) (4). Proportional band (5) D-action set point whose value could be chosen between 0 sec to 3600 sec. (*i.e.*, 0-1 hour 6) I-action set point in the span 0-1 hour. Besides they are programmable to specify a particular input from a large number (4–16 inputs) of inputs. For input configuration of a controller the standard is followed *i.e.*, 0–10 volt standard or 4–20 mA standard. Similarly, for output specification same standards are followed *i.e.*, 0–10 volt or 4–20 mA.

## 35.10 PROGRAMMABLE LOGIC CONTROLLER (PLC)

Programmable logic controller (PLC) is a processor oriented digital device which contains that much intelligence to run a discrete logic system performing logical operations. The discrete logic system refers to a industrial application where operations occur in sequence for inputs which are mainly switching devices and outputs are relays and coils. The PLC is popularised with failure of sequential events to be controlled by mechanical relays and contactors due to inherent problems. The mechanical relays and contactors were only choices for industry upto 1960, and later they were found to be obsolete due to mechanical failures. Also, these mechanical devices require large amount of energy for operations and are used to generate high mechanical noise due to chattering. The advent of power electronics devices like silicon control rectifiers (SCR), triac, diac, insulated gate bipolar transistors (IGBT) and MOSFET, the mechanical relays and contactors are replaced by these solid state relays and contactors. The second advantage of the PLC is their installation case *i.e.*, the easy implementation of logical relations between different switching inputs and the relay and coils at the outputs. The opto coupler does the necessary isolation of the processor circuits within the PLCs and the power devices like motor coils or incandescent bulbs as the output. The other important feature of the PLC is that, it is easy to change the logic of operations by only changing the logic structure with the help of user-friendly diagrams. The contactors are also achievable with the help of simple identification method used in PLCs. Besides, PLCs have the ability to deal with large amount of data and manipulate them. The modem PLCs are used to communicate data between two PLCs, or a computer and a PLC or a distributed control system (DCS) and a PLC. The PLC is used in diversified industrial applications like material handling system, weaving mill, food processing industry, machine handling systems, robotics and also as a local controller, which is a part of DCS.

### 35.10.1 Discrete State Process Control Systems

The majority of industrial process control installations involve more than simply regulating a controlled variable. The requirement of regulation means that some variable tends to vary in a continuous fashion because of external influences. But there are many processes in industry, in which it is not a variable that is controlled but a sequence of events. This sequence of events leads to production of final products from raw materials. Discrete state means that each event in the sequence can be described by specification of the condition of all operating units in the process. Such conditions might be described by expressions

- Valve $A$ is open.
- Valve $B$ is closed.
- Conveyor $C$ is ON.
- Limit switch $S_1$ is closed etc.

A particular set of conditions is described as a discrete state of the whole system.

Now, we can define discrete state of the process at any moment to be the set of all inputs and output values. Each state is discrete in the sense that there are only a discrete number of possible states. If there are three input variables and three output variables, then a state consists of specification of six values. Because each variable can have two values (ON and OFF) the total number of possible discrete states are $2^6 = 64$.

An event in the system is defined by a particular state of the system, *i.e.*, particular assignment of output values for particular set of input values. The event lasts as long as the input variable remains in the same state and the output variables remain in the assigned state.

*Example* :

**Event 1 :** Temperature low, heater off
**Event 2 :** Temperature low, light on
**Event 3 :** Temperature high, light off
**Event 4 :** Temperature high, heater on.

These types of discrete state process control (sequential event control) are taken care of by the PLCs.

### 35.10.2 Popularity of Programmable Logic Controllers

The control relay was the most widely used device for controlling discrete manufacturing process upto 1960. Although these earlier devices are still being used today and many of the problems associated with using them, have been eliminated due to technological advances, in their design. But that approach continue to suffer from inherent problems. Relays are susceptible to mechanical failure, they require large amount of energy to operate and relays generate large amount of electrical noise (chattering). With the invention of solid state relays and devices, the mechanical relays are not in use now-a-days for discrete controllers. Later PLC was introduced to replace the sophisticated control system.

### 35.10.3 Use of PLC

In late 1970's the microprocessor became a reality and greatly enhanced the role of PLC, permitting it to replace mechanical relays. PLCs now have the ability to manipulate large amounts of data, perform mechanical calculations and communicate with other intelligent devices like Distributed Control System (DCS), computers etc. The PLC is used in many of the industrial applications including the control of machine tools, material handling system, food processing operations, and the continuous process control field.

### 35.10.4 Characteristic Function of PLC

PLC is a digital electronic device that uses a programmable memory to store instructions and to implement specific functions such as logic sequence, timing, counting and arithmetic to control machines and processes.

The characteristics are enumerated as :

- PLC is field programmable by the user. This characteristic allows the user to write and change programmes in the field without rewiring or sending the unit back to manufacturer.
- PLC contains programmed functions. PLCs contain logic, timing, counting and memory functions that the user can access through some type of control-oriented programming language.
- It scans memory and input/output (I/O), in a deterministic manner. This critical feature allows the control engineer to determine precisely how the machine or process will respond to the program.
- PLC provides error checking or diagnostic. A PLC will periodically run internal tests of its memory, processor and I/O systems to ensure that, what it is doing to the machine or process is what it is programmed to do.
- PLC provides some monitoring capability, either through indicating light or through external display.
- It is packaged appropriately.
- Generally, a PLC is not designed for a specific application but it can handle a wide variety of control tasks effectively.

### 35.10.5 Block Diagram of PLC

The block diagram of a PLC is shown in Fig. 35.26. The logic solver is the processor which interacts with ROM/RAM and these constitute the building block of main PLC functions. The program is written in popular ladder diagram and it is downloaded to the memory (ROM) of the PLC. The analysis of the program *i.e.,* to open different normally open (NO),

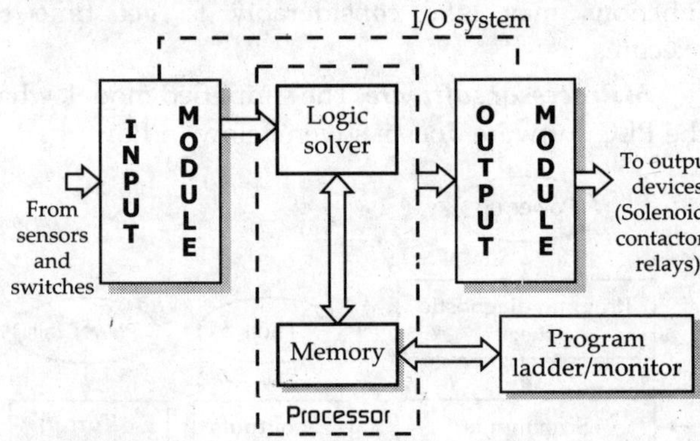


Fig. 35.26

normally closed (NC) contacts is performed within the PLC processor. The NC/NO contacts are identified in such notation that use of internal contactors are met. The external input devices are connected to the input module which is generally implemented in rack structure *i.e.,* as a card. The different external inputs generally come from switches like push button switches, limit switches, proximity switches, temperature, pressure and level switches etc. The external outputs are solenoid values, motor coils, contactors, relays and values.

35

**1. *Processor section : Main units are***

(*a*) Power supply
(*b*) Memory
(*c*) Central processing unit
(*d*) Input/output interface.

**2. *Power supply section***

(*a*) DC 5 volts and 12 volts.

**3. *Memory section***

(*a*) Battery backup CMOS RAM
(*b*) EPROM memory (Electrically programmable ROM)
(*c*) EEPROM Memory (Erasable Electrically Programmable ROM)

**4. *CPU (Central Processing Unit).*** The operational flexibility of a PLC is determined by the construction of CPU, particularly, whether the PLC can be modified for future expansion. In this context, the overall speed of the PLC is important. The speed of the CPU means how fast the PLC will scan the memory. This measure is called the scan rate, typically expressed in millisecond (ms) for thousand words scanned memory.

PLCs specify their scan rate time using contacts and coils only. A real program that uses other functions, such as timers, counters and mathematical functions may take considerably to get time to execute.

**5. *Processor software.*** The simplified model, what the PLC software does is given below in flow chart :

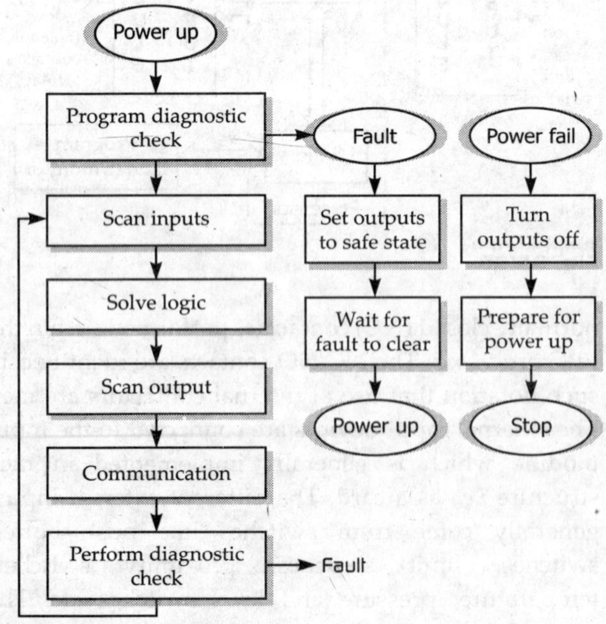


**Fig. 35.27**

**Multitasking**

PLCs are capable of executing multiple tasks with a single process. In particular, PLCs that support sequential function charts (SFC), tend to support some limited multitasking capability.

*Multitasking takes several forms*

- Time driven
- Event driven

**Time driven multitasking.** In some time driven multitasking systems, the user writes programs and assigns I/O for each task separately. The user then may be allowed to configure the processor to run each task on periodic time intervals. This feature allows the fine-critical portion of the control system, such that the portion that controls high speed motions or machine fault detection, to run many times per second while allowing the non critical portions such as servicing indicator, lights, to run much more slowly.

**Event driven multitasking.** It is also called interrupt driven. In this case the user defines a particular event, such as pressing a push bottom, that causes each task to run. Generally PLCs allow both time driven multitasking and event driven multitasking.

***User software.*** This software that the *control engineers* write and store in user memory in order to perform the required control over the machine or process. User software can contain both configuration data and language programs. The configuration data contain information that tells the processor what its environmental is and how it should execute the language program.

***Configuration.*** The configuration process typically consists of assigning I/O points to particular I/O racks, telling the processor how much memory and I/O it has, assigning specific memory for task. The configuration software also performs fatal and non fatal fault checking and assigning many other items interactively on a program loader.

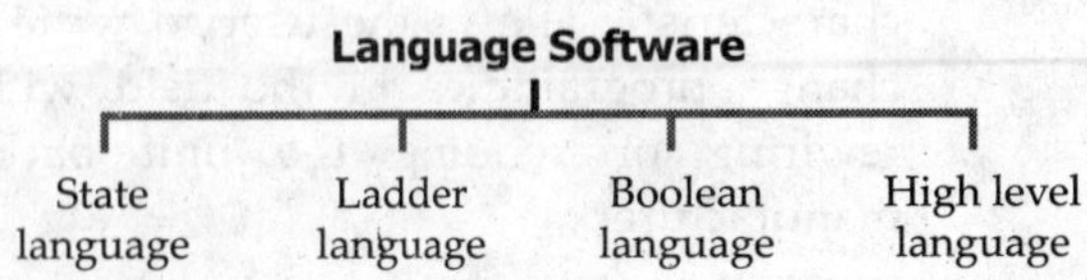


***Ladder Language.*** The advantage of using this language is :

1. It is readily understood and maintained by skilled workers.

2. It provides graphic display of program execution by sharing power flow through the ladder design.
3. Program is fast.
4. It generates more readable programs for sequence control.

***Boolean language.*** This language generally is used in every PLC. Boolean AND, OR, NOT, STORE. RECALL instructions (logical) are used in order to describe the program logic. This language is not easy to debug unless a single-stepping feature is provided that slows down program execution so that the effects on the I/O can be observed originally.

***Ladder diagram.*** The field programmable, PLCs generally use ladder diagram for representing a logic or sequence of events. The advantage of ladder programming for PLCs over boolean language programming is that, ladder programming explains the states closer to the field instrumentation. Some field instrumentation objects used in ladder diagram are shown below.

**6. *Different inputs and outputs (external and internal) used in ladder diagram***

(*a*) Internal inputs and switches :

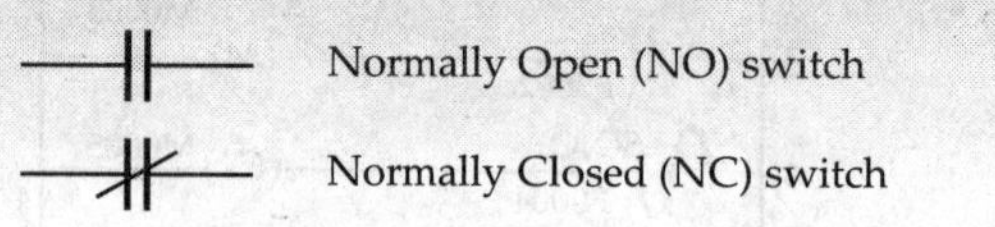


(*b*) Devices :

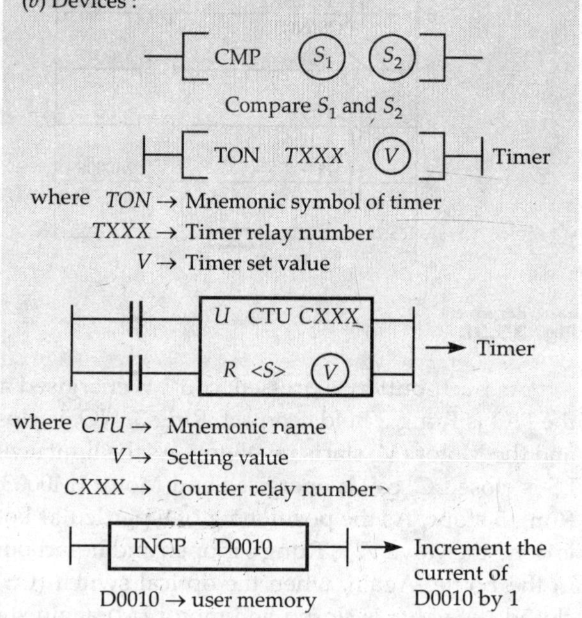


(*c*) Data Transfer Switches in Ladder

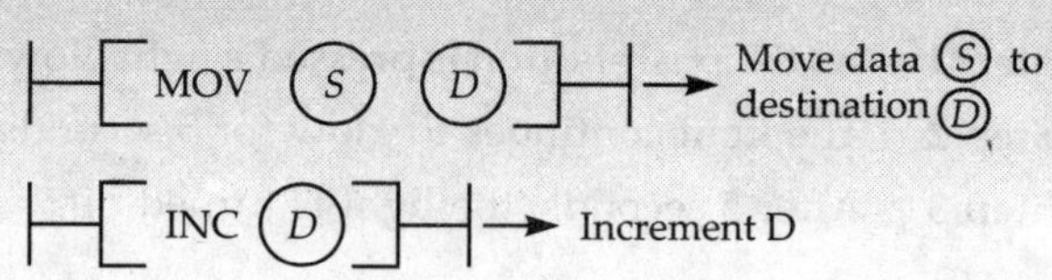


(*d*) External Switches and Buttons

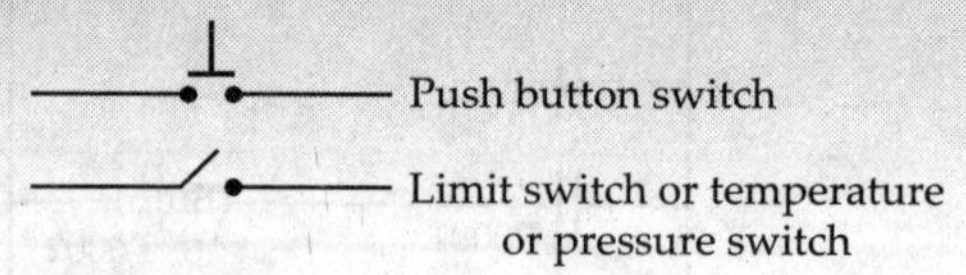


(*e*) Internal Output Devices

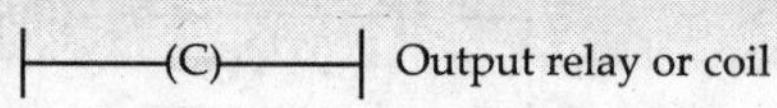


(*f*) External Output Devices

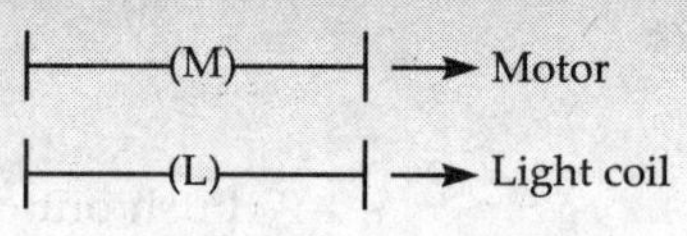


The objects are used in ladder programming to represent the discrete state of logic *i.e.*, switching logic in a batch processing mode. The ladder programming is converted to machine language bit level program which is downloaded from the computer to the PLC. The PLC is provided with specified inputs and outputs with identification as described in the ladder diagram for example, an external push-button input switch with identity P0001 in ladder program has been connected to specified input socket reserved in *P* rack with number 0001. Similar nomenclature is used for output device identification which is a link between ladder diagram and actual output device connections. Also, two different classes of identification number are reserved for external input and internal logic input. Like, a push-button switch which is an external input device is named as K 0001 and an NO switch which is an internal logic device, is named P0002. The PLC also uses the programmable relays and contacts and this time also, the identification number of the coil and its contactors matter. In the case of a coil (M 0003) has two contactors are normally open (NO) and the other normally closed (NC), both the contactors are identified by the device name (P0003), corresponding to the mother coil identification. Hence, when the coil M0003 is energised, the NO contact P0003 closes and NC contact trips. Instances of logic representation of a process by ladder diagram follows.

***Problem 1.** Draw the ladder diagram of the following process :*

**Step 1** As the push-button is pressed a light glows.

**Step 2** The light continues to glow for 5 seconds.

**Step 3** After 5 seconds, the light is turned off.

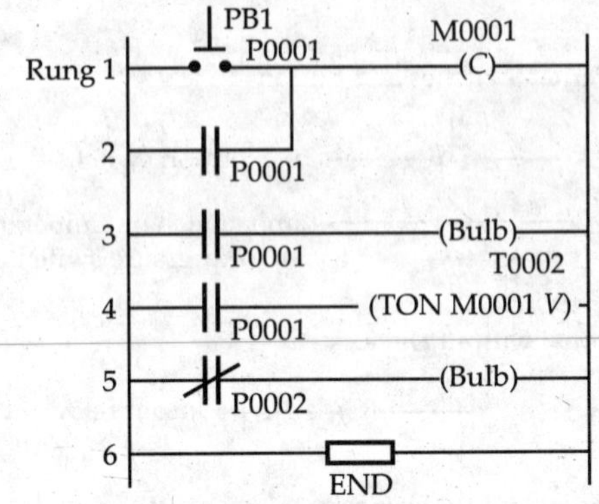


**Fig. 35.28**

where,

| | | |
|---|---|---|
| PB1 | → | Push button switch |
| —\| \|— | → | Normally open switch |
| —\|/\|— | → | Normally closed switch |
| —(TON M 0001 V)— | → | Timer |

where (V) is set for 5 seconds.

When PB1 is pressed, the coil M0001 is energised. As soon as, the coil is energised the normally open switch P 0001 is closed. This NO holds the contact as Push button switch is OFF and the coil remains in the energised state through the closed NO contact (Rung 2). The P0001 NO switch in Rung 3 is closed as soon as the coil is energised and the bulb glows. The NO in Rung 4 is also closed as the coil M0001 is energised and the tinier T0002 starts counting. As soon as timer count is equal to 5 second the timer coil trips and the NO, P 0002 in Rung 5 in open and the bulb is turned off.

***Problem 2.** (Liquid Bottling Plant)*

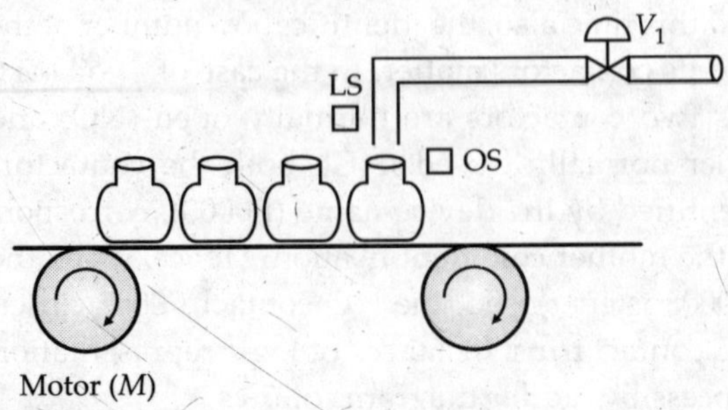


**Fig. 35.29**

| | | |
|---|---|---|
| LS | → | Limit switch (normally open) |
| OS | → | Optical switch (normally open) |
| $V_1$ | → | Valve |
| $M$ | → | Motor |

Time Sequences

**Step 1** The Motor $M$ rotates and the conveyor belt moves.

**Step 2** As soon as, the first bottle crosses the limit switch (LS), the motor stops, and bottle is under the pipe.

**Step 3** As the bottle is under the pipe, valve $V$ opens.

**Step 4** The bottle is filled up and when full, optical switch (OS) is closed and valve $V$ is closed.

**Step 5** The motor again rotates and the steps 1 to step 4 are repeated.

The ladder diagram for water bottling plant is shown in Fig. 35.30.

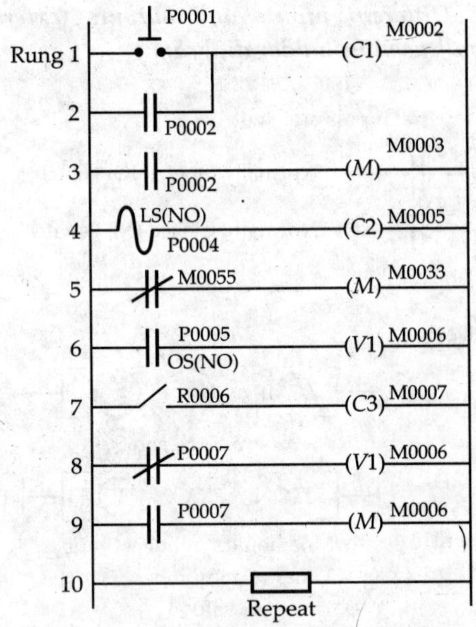


**Fig. 35.30**

As push-button is pressed, coil 1 is energised and the NO is Rung 2 holds contact. Ring 3, NO, is closed and the Motor ($M$) starts revolving. As the limit switch LS is closed $C_2$ coil is energised and Motor (M0003) is Rung 5 stops. As the positioning of a particular bottle is done the valve $V_1$ in Rung 6 is open and liquid pours in the bottle. Again, when the optical switch (OS) is closed the valve is closed and motor ($M$) again starts and steps from Rung 1 to Rung 9 are repeated.

***Problem 3.*** *Street lighting*

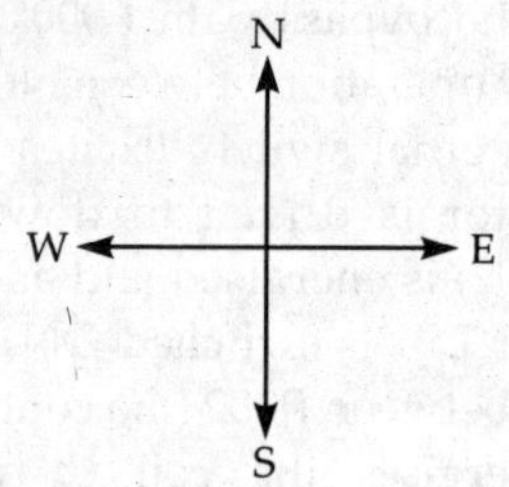


**Fig. 35.31**

The street lighting is configured as follows :

$R_N$ → Red light at north
$R_E$ → Red light at East
$R_W$ → Red light at West
$R_S$ → Red light at South

$G_N$ → Green light at North
$G_E$ → Green light at East
$G_W$ → Green light at West
$G_S$ → Green light at South.

The truth table for street lighting is shown below :

**Table 35.1**

| | $R_N$ | $R_S$ | $R_E$ | $R_W$ | $G_N$ | $G_S$ | $G_E$ | $G_W$ |
|---|---|---|---|---|---|---|---|---|
| 1. | × | × | √ | √ | √ | √ | × | × |
| 2. | √ | √ | × | × | × | × | √ | √ |

The logic given in Table 35.1 is simple and shows the north-south path is open (As given light $G_N$ $G_S$) glows and at the same time East-West red signal glows. For 2nd step the East-West path is open.

The ladder diagram (Fig. 35.32) explains a simple street lighting logic whose truth table is given in Table 35.1. As push button (NO) PB1 is closed in the ladder diagram the coil $C_1$ is energised. The NO (P0001) holds the contact as push button is released. As the coil $C_1$ (M0001) is energised, the NO for Green (North-South) and Red (East-West) closes and $G_{NS}$ and $R_{EW}$ bulbs glow. By similar reasoning the Red (North-South) and Green (East-West) are put off.

With the counter coil C0006 is set, the contactor P0006 is energised. Rung 11 – Rung 15 NOs and NCs are operated which set the respective bulbs $R_{NS}$, $G_{EW}$ glow, whereas, $G_{NS}$, $R_{ES}$ are put off.

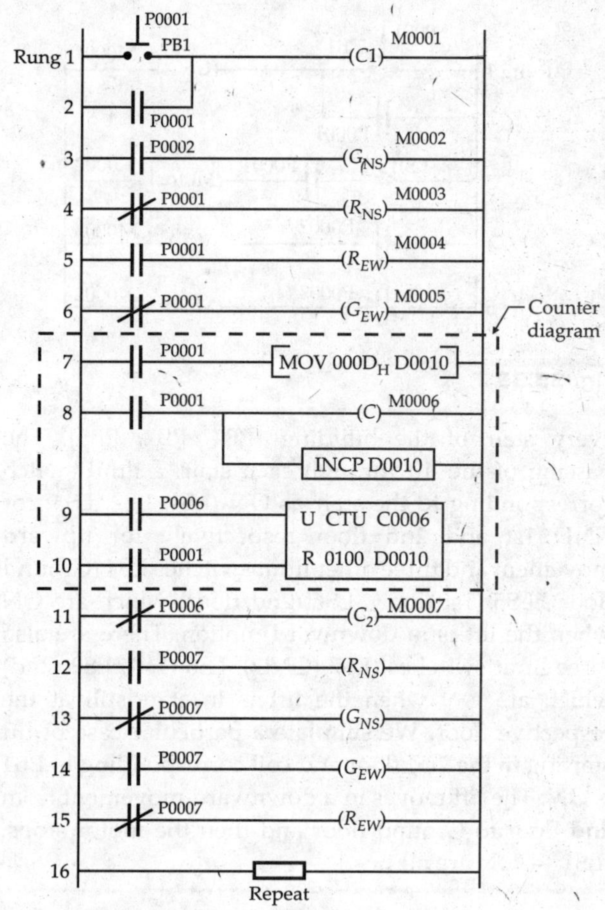


**Fig. 35.32**

**Counter Circuit**

The counter circuit (Rung 7 to Rung 10) is shown in the ladder diagram. The counter is a up-counter. At first, the user defined data register D0010 is set to $000_H$ (Hex value). As the coil M0001 is closed and coil M0006 is energised. The (P0001) in Rung 10 is closed and the counter is reset (*R*-mode). With M0006 is energised, the up-counter starts counting in up-mode. The user defined register (D 0010) content is compared to the set value $0100_H$ and, if, they are not equal, the counter coil C0006 remains in deenergised state, and hence M0006 is deenergised. The control then passes in parallel path in Rung 8 and the content of D0010 is increased by 1. Again the counter is reset and the same process continues until D0010, content is equal to $0100_H$, and as they are equal, counter coil C0006 is energised. Hence the counter causes a delay of $0100_H$ × pulse duration time.

***Problem 4.*** *(Elevator Problem)*

In this problem we treat up and down movement of an elevator in a three storeyed building. Generally, the lift is fitted with the push button corresponding to

35

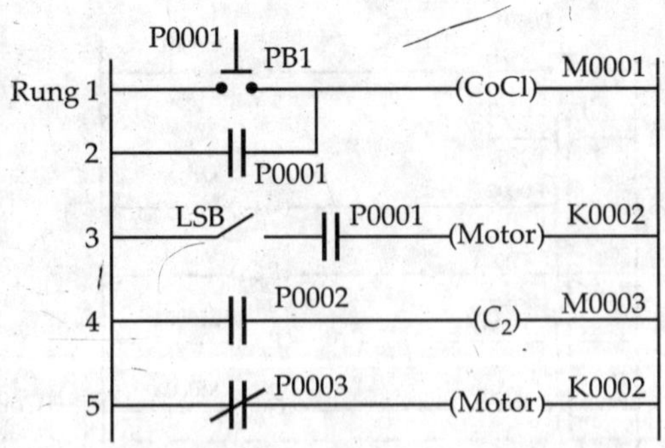


**Fig. 35.33**

every stair of the building (PB1, PB2, PB3). The passing of the lift through each stair, a limit switch corresponding to the stair as ON (LS1, LS2, LS3) for GND 1st and 2nd floor respectively for upward movement and three other limit switches LS4 (Ground floor), LS5 (1st floor), LS6 (2nd floor) which are ON when the lift is in downward motion. There are also three limit switches (LS7 (GND), LS8 (1st), LS9 (2nd) which are ON when the lift is in standstill at the respective floor. We simulate a particular case of lift starting in the 2nd floor and coil corresponding to PB1 is ON. The lift moves in a downward movement from 2nd floor to Ground floor and then the motor stops. (LS1 — LS9 are all nos.).

The first Rung starts action as soon as the push button for ground floor (PB1) is ON. The coil $C_1$ is energised and the bypass path P0001 holds contact. The logical AND operation between NO contact P0001 and LS8, the external switch, excites the coil of the motor. The motor is driven to downward motion (motor coil K0002) is energised and as the motor is at the ground floor LS4 is switched ON and the logical AND operation between P002 (the contactor of K0002) and LS4, deenergises the coil $C_2$ which in turn deenergises the motor coil K0002.

In this method, we can simulate all the possibilities of lift movement by using the ladder diagram.

7. ***Input/Output devices.*** The external input/output devices of the PLC are wired to the contacts exposed outside the racks. Generally there are 2-3 no. of input modules whose specifications being 0-12 V DC or 0-24 V AC or 0-24 V DC. The external output devices also occupy 2-3 no. of racks which are categorised by AC/DC voltage and the magnitude of the voltage. The input and output devices are isolated from the processor circuit by optical coupling *i.e.*, use of opto coupler. A rack generally contains 16 or 32 or 64 number of input/output connections.

One external input and one external output connection is shown in Figure 35.34 and 35.35.

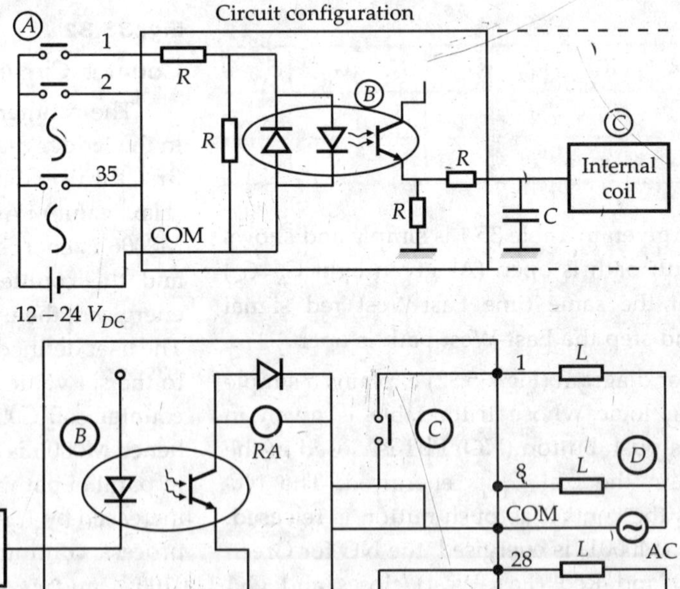


**Fig. 35.34**

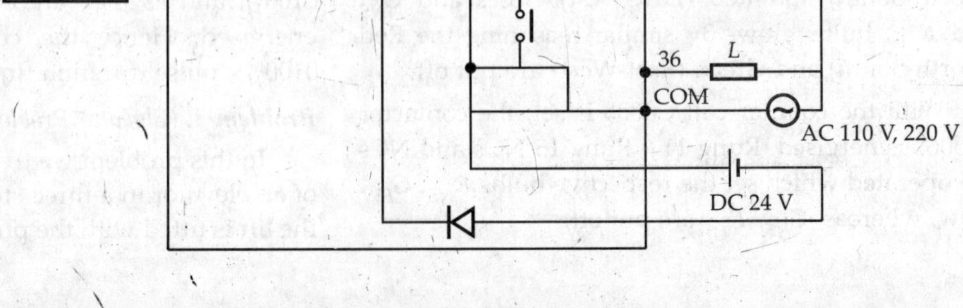


**Fig. 35.35**

### Basic External I/O Modules

(*a*) ***A 64 point 12/24 VDC input module.*** In Fig. 35.34, the basic input module (*A*) shows the input switch configuration, (*B*) shows the optocoupler and (*C*) shows the internal coil. As soon as push-button 1 is pressed the input circuit is closed and the current flows through the optocoupler, as the photo transistor is switched on. The optocoupler causes isolation between input and output. As the photo-transistor is switched on the internal coil is energised. Input voltage may vary 12/24 $V_{DC}$ to 220 $V_{DC}$.

(*b*) ***32 Point relay output module.*** In Fig. 35.35, the basic output module (*A*) shows the internal circuit, (*B*) shows the optocoupler, (*C*) shows the output contactor and (*D*) is output circuit. The internal logic circuit activates the optocoupler and current through the photo transmission activates the contactors which closes the 220 V AC output circuit. The optocoupler ensures isolation between input and output circuits.

## 35.11 DISTRIBUTED CONTROL SYSTEM

With the evolution of the major idea of using intelligent processor in loop control and PLCs, the new idea of integrating a process industry within a intelligent network has been undertaken and this new network has been named Distributed Control System. The major task for a process engineer is to commission, calibration and maintaining a process plant with panel board mounting which is rigorous. This has been replaced by a comparatively simple task of formulating a control strategy and implement it with designed programmable intelligent network so that local controls can feed data to a central control through some intermediate controllers and in this hierarchy each controller performs its duty of computing and control. Each part of the network can be calibrated with its instruments *i.e.,* sensors, converters, relays contactors, and logically linked to other parts of the network generate control functions and stabilize the network. Like a power plant uses nearly 4500 no. of sensors and that amount of input and output devices, it is hard to commission and maintain with panelboard mounting. The modern DCS and its counterpart Total Plant Solution (TPS) works with 20-50 sensors and 1/0 devices with a dedicated controller at the lowest level and with integrating the control function generated by the higher level of intermediate controllers and in this way advancing to the central controller. This strategic method reduces operating cost and implement easy commissioning by optimizing the control strategies. There is also a major improvement in the monitoring of the product path as failure of any local loop control can be measured instantly. With the advent of Supervisory Control and Data Accusation System (SCADA) software, the central operator is presented with the total plant picture and gets the history of last one hour to manipulate the plant control strategies for improvement in production with higher responsivity. To implement a production strategy abiding by the new Government regulations, product specifications and also addition of new products in a particular control loop, is easier. As the whole DCS is software controlled, and maintain a hierarchical network topology, it is easy to make a contingency strategy for any plant failure so that production is not upset. After the rapid development of computer network technologies it is easier to relate between DCS and a open system of computer network. With this advancement, the DCS becomes more and more intelligent, such as, in case of addition of new instruments with DCS, virtual processing of data *i.e.,* data communication to large distance apart from the plant and organising the operation of the plant with coordinated control strategies. Hence, the DCS becomes a platform for control engineers to support large scale real time process applications.

DCS, generally, has been categorised into five subsystems. These are

- Operations Workstation,
- Controller Subsystem,
- Data Collection Subsystem,
- Process Computing Subsystem
- Communication Subsystem.

### 35.11.1 Operations Workstation

One of the most important parts of the DCS architecture is the operations workstation. The central control room workstation of the DCS uses a mainframe/computer, the monitor of which shows the total plant architecture on the screen. It is difficult to resolute all the controls starting from lower level to higher level on the same screen, hence a display management is required. The higher level figure shows a few sections and double clicking on each section we can get control diagram of that section. In this way, with the help of modem display management we can roam from higher level to the lowest level of the DCS architecture.

Next important issue for operations workstation is Alarm management. It is to configure the Alarms for different sections of the plant by generating control

35

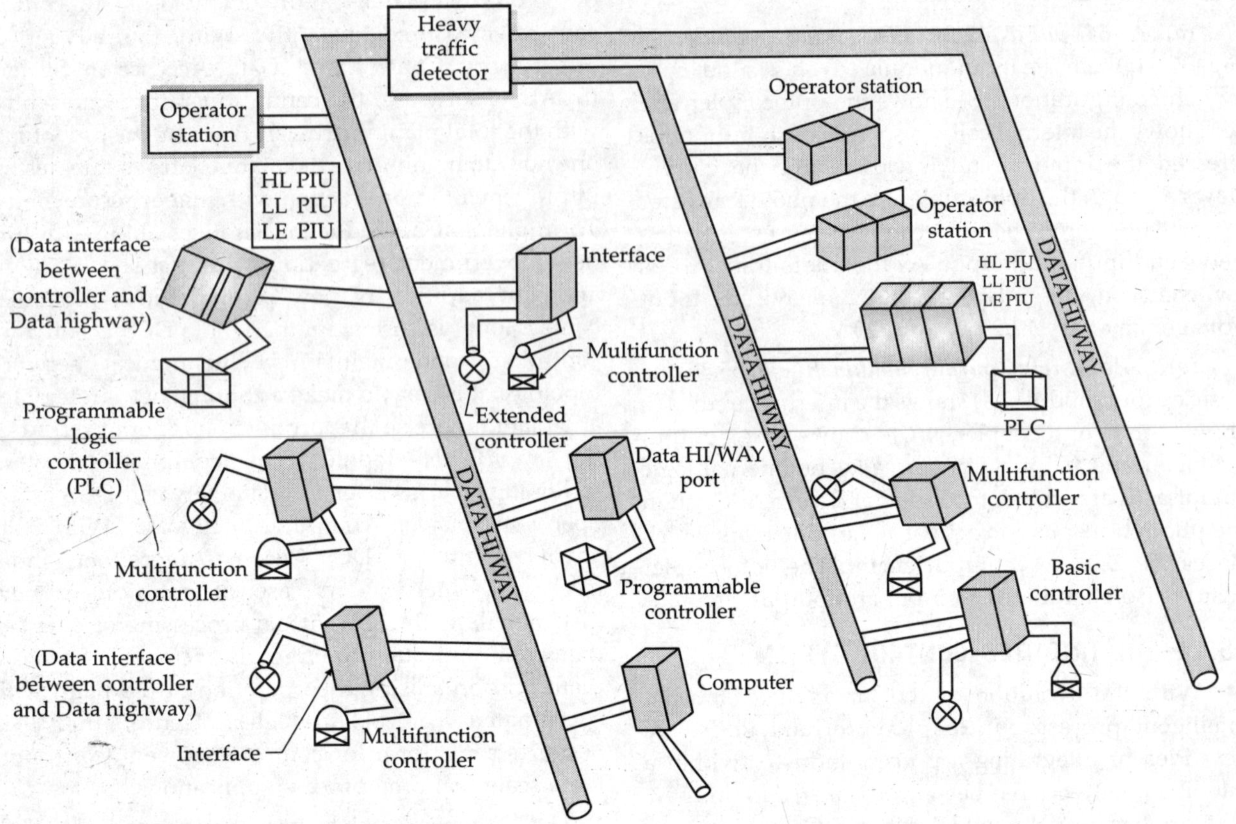


**Fig. 35.36** Honeywell's 2000 architecture (DCS).

function for each alarm from the respective section. The workstation also shows the current trend process condition by the way of information management. Besides central workstation there are other intermediatory workstations controlling different parts of the process and all these workstations are integrated in a network.

### 35.11.2 Controller Subsystem

The controller subsystem is used to integrate large numbers of local control loops (single or multiloop control) and PLCs all over the plant in a fashion, so that, they can be programmed and act as a part of the DCS network. With the advent of shared microprocessor controllers *i.e.,* more than one controller can share a single processor or memory, the importance of controller subsystem in DCS architecture has been enhanced. In short, the controller subsystem replaces the wiring of the panelboard mounting with software control. The controller subsystem also takes care of abnormal process conditions at any part of the plant and alert the operation with a alarm. Besides one of the major work done by controller subsystem is synchronization which is the proper sequencing or timing of digital algorithm running all over the plant. The data acquisition system with supervisory control or SCADA software is a part of modern controller subsystem procedural language.

### 35.11.3 Data Collection Subsystem

The collection of data and information from local loops to higher process elements *i.e.,* loops higher in DCS hierarchy for the purpose of recording, memorising, indicating and the initiating of alarm for exception in process condition is the work performed by data collection subsystem. This is a modern day technique which replaces the panelboard meters and recorders. This section is also responsible for data distribution throughout the DCS network. The data corresponds to input-output and the sensors used in DCS network the source of data which these subsystems collect in digital formats and use the data for the purpose of metering and recording.

### 35.11.4 Process Computing Subsystem

The process computing subsystem is an important part of DCS architecture whose input is from data collection subsystem and different controllers and it presents an intelligent device like computer for a computation handling large amount of data, to report advanced alert *i.e.,* alarm and for supervisory control optimization of production. The data highway which is the path between the controllers and process computing subsystem of DCS is a high speed double line digital data path unlike panel board mounting which takes 15 to 20 lines for data path. Modern DCS was coaxial cable or a backbone fiber optic cable for data path. The main outputs of information processing by a process computing subsystem are :

Production and maintenance scheduling, supervisory control optimization, expert system application and creation and maintenance of the database for expert system applications and also for statistical analysis of the data.

Some modem data collection subsystems also use relation database in the back end and graphical user interface in the front end which is easy for the maintenance, query and computation processes on the database. The modern DCS has been discussed in the next section.

### 35.11.5 Communication Subsystem

This subsystem is fed with data from all the four subsystems discussed earlier so that the total plant area has been integrated and dispose of timely information from a common information base. These informations help the central computer to initiate early action for any fault occurring in the process. The networking of different controllers and computers has been supervised by this communication subsystem. The network is open and allows modification of local loop with modern devices.

## 35.12 FUTURE TREND FOR DCS MODULE

The workstations attached to the real *time-information network* will be advanced versions of the distributed computing work stations. They will comply with the Graphic User Interface (GUI) standards emerging today and will be capable display information equally well from process and business application. The DCS protocol for a real-time-information network is shown as an example.

### 35.12.1 The DCS Protocol (MODERN DCS)

Three components comprise the DCS architecture :

- Distributed Hardware Server (DHS)
- GUI clients (Graphical user Interface clients)
- DCS Server (DCSS)

GUI clients and Hardware server (DHS) communicate solely with DCS server (DCSS); and DCSS operates as the only listening sockets (Network server) in the DCS system. Consequently hardware servers are in a sense network clients of DCSS and sometimes referred to as clients of DCSS in the DCS message protocol documentation. DCS hardware servers (DHS) encapsulate low level control of physical devices such as motors, shutters, detectors,

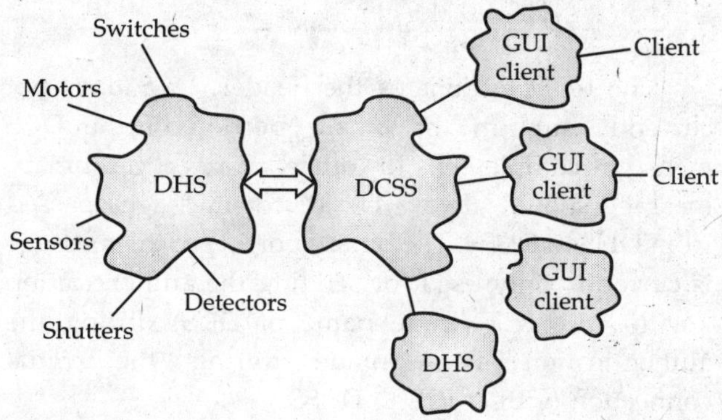


**Fig. 35.37**

switches and so on. DHS simply connects to DCSS and do whatever DCSS tells them to do. DCS GUI clients are the ultimate source of these instructions to the Hardware servers (DHS). DCSS passes the requests from GUI clients down to the appropriate DHS and broadcasts all replies from DHS are back to all of the GUI clients, thus keeping GUI clients in complete synchronization.

By convention the first four letters of the command in a DCS message specify the source and destination of the message in the DCS architecture. The five possibilities are :

- Stog : **S**erver to **g**ui client
- Stoh : **S**erver to **H**ardware clients
- gtos : **g**ui to **S**erver
- htos : **H**ardware client to **S**erver
- stoc : **S**erver to **C**lient (GUI) or **H**ardware.

Hardware client → switches, motors, sensors etc.

All communication between the components of the DCS architecture uses message passing protocol which runs on top of TCP/IP. The protocol is completely asynchronous (*i.e.,* not strictly client server. The network message is split into three distinct sections :

1. the heater
2. the text section, and
3. the binary section.

The DCS Message header : Each DCS message must start with 26 bytes of text message. These 26 bytes must contain 2 numbers in ASCII format. The two numbers indicate the size of the text section (in bytes) and binary section (in bytes). The text message must contain a terminating 0 at the end of the two numbers. This allows a library to simply use the "scanf" function to obtain the two numbers.

An example header may look like this, the first number in this example indicates that there are 1235 bytes of text following the header. The second number in this string indicates that there are 23.35 bytes of binary data following the 12.35 bytes of test.

| | | | | | | | | | | | | | | | | | | | | | | | | | |
|---|---|---|---|---|---|---|---|---|---|---|---|---|---|---|---|---|---|---|---|---|---|---|---|---|---|
| | | | | | | | | | | 1 | 2 | 3 | 5 | | | | | | | 2 | 3 | 3 | 5 | 0 | × 00 |

The 0X00 terminates the header. DCS does not currently support the binary portion of the DCS message completely. In other words, the binary message should always be "0" for messages passed from DHS to DCSS. The binary portion of the message is currently only used for sending the authentication key to BLU-ICE (trade name of DCS) during the authentication stage made during the initial connection of BLU-ICE to DCSS.

### 35.12.2 The Connection Protocol

The connection protocol has been simplified for the first two releases of DCS. The GUI and hardware clients connection to DCSs on different ports are used.

**Connection module for modern DCS** : For Hardware servers (DHS)

1. The new DHS opens a socket connection to the hardware listening port on DCSs.
2. DHS begins reading and handling messages sent from DCSs.
3. DCSS will send a stoc-send client type to DHS.
4. The DHS has 1 second to respond with the following message :

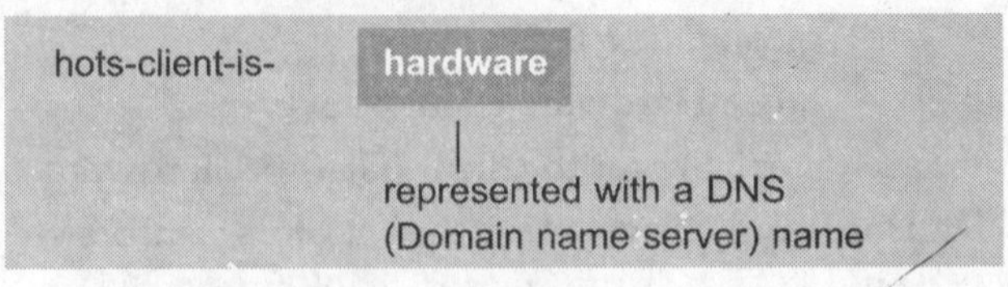


At all these points DCSs will disconnect a DHS for the following reasons :

- The DHS does not respond within 1 second
- The DHS does not respond with the name of a DHS listed within the database data file.
- There is already DHS connected with the same name as the DHS that is currently trying to connect.

For GUI clients :

1. The new GUI client opens a socket connection to the GUI listening port on DCSS.
2. DCSs will send a stoc-send-client type to the GUI.
3. The GUI has 1 second to respond with the following message :

   htos-client-is-gui *username hostnone display* where username is the unix account name of the user that initiated the GUI.

   At this point DCSs has the option to disconnect the GUI for the reasons :

   (*a*) The GUI does not respond within 1 second

   (*b*) The GUI responds with a username that is not listed within database users table.
4. After DCSs confirm that the username is listed in mysql database, DCSs will send the following message to GUI with 200 bytes of trailing binary data :

   Stog-respond-to-challenge
5. GUI must read the 200 bytes of binary data and respond to DCSs appropriately.

   At this point DCSs will disconnect the GUI for the following reasons :

   - The GUI client does not respond within 1 second.
   - The GUI does not send the appropriate response.
6. DCSs send message

   stog_login_appropriate.
7. DCSs send another message.

   Stog_set_permission_level permission level.

   Where permission level is a number attained from the mysql's user table that can be used by the GUI to disable or enable certain buttons and tabs. DCSs will also use this value to prevent the user from doing what they are not allowed to do.

8. DCSs will then proceed to update GUI client with the complete state of beam line. This will include motor position.
9. DCSs will then begin processing message from GUI client as they come in and will send messages to the GUI client to keep the GUI upto-date with the latest status of motor positions and operation activity.
10. DCSs will disconnect the GUI client under the following circumstances :
    - ▲ The GUI client is too slow in processing the message in its TCP/IP input buffer and DCSs TCP/ IP output buffer becomes full.
    - ▲ The GUI socket has an error.

Installation of DCS has following steps :

1. Installing the libraries
2. Installing DCSs
3. Configuration DCSs
4. Installing and configuration hardware simulator
5. Installing DHS
6. Installing the detector simulator
7. Installing GUI clients
8. The MySql database

Maintaining the database .dat file

1. Dumping the database .dat file
2. Recreating the database .dat file
3. Format the database .dat file.

Writing scripted devices and operations

- ▲ General DCS scripting commands
- ▲ Scripted Device family relationship
- ▲ Exception handling

Adding scripts to the scripting Engine :

- ▲ Adding new scripted device
- ▲ Adding new scripted operations.

### Example of some commands

**1.** gtos_abort_all [hard/soft]

This command requests that all operations either cease immediately or halt as soon as possible. All motors are stopped and data collection begins shutting down and stopping detector activity.

**2.** *Command used for motor control by program of GUI* :

gto S_configure_device motor Name : position : upper limit : Lower Limit : lower limit on : upper limit on : motor lock on :

- ▲ motor name is the name of the motor to configure.
- ▲ position is the scaled position of the motor.
- ▲ upper limit is the upper limits for the motor scaled units.
- ▲ lower limit is the lower limit for motor scaled units.
- ▲ lower limit on is a boolean (0 or 1) indicating if the lower limit is enabled.
- ▲ upper limit on is a boolean (0 or 1) indicating if the upper limit is enabled.
- ▲ motor lock on is a boolean (0 or 1) indicating if the motor is software locked.

**3.** *Command used for a particular operation by GUI programming.*

Similar message

Gtos_start_operation : Operation Name : Operation_Handle [arg 1 [arg 2 [arg 3 [....]]]]

- ▲ Operation_Name is the name of the operation to be started.
- ▲ Operation_Handle is a unique handle.

arg 1, arg 2, arg 3 are the list of arguments that should be passed to the operation.

These are the examples of GUI client to server messages. Similarly, there are Hardware to server message, Server to GUI messages and server to Hardware message.

The binary section of message is currently used only during the authentication stage between DCSs to GUI clients.

## 35.13 SMART INTELLIGENT TRANSMITTERS

The transmitters are devices which convert the sensor output data to a electrical signal according to standards. The common output standard followed for voltage transmission is 0-10 and for current transmitter 4-20 mA and for pneumatic transmitter is 3-15 Psi (Pound per square inch). Hence the variation of voltage or current at the output of the transducer has been converted to varying voltage current in the standard range of voltage/current, indicated earlier. The transmitter input and output need four wires :

1. the supply line
2. the ground,
3. the input line from sensor
4. the output line.

The adjustment of the transmitter for changing span *i.e.*, range, of sensed voltage, is called turndown and this adjustment is performed by a knob on the transmitter panel. Semiconductor transmitters are in all, accurate and less expensive. Also, the intelligent *i.e.,* smart transmitters can be built with semiconductor devices. These smart transmitters can/store data digitally for purpose of linearization, correction for sensor non-linearities and for temperature and pressure compensation. These smart transmitters can work with large input spans (spans can be 50 times of the standard output 0 – 10 off or 4 – 20 mA). They can also output the process variable dimension and unit like/°C, Psi, and can perform for outline programming. Due to the fact that, the calibration curve of the transmitter is in the memory of the processor the zero setting and span setting can be set with the keyboard of the transmitter. The other advantage of smart transmitters is the automatic relating at the start of the control and they can be used for two way communication between the field transducer and the controller. They can also provide damping for moving coil instruments and temperature compensation and can make a pointer fixed to a particular position or switch between the direct and reverse action.

1. ***Pneumatic controllers.*** Pneumatic controllers use high pressure for transmission and control of a process as transmission medium. The controlled pressure fluid (air) medium replaces control through processed electrical signal. These controllers are popular for their innate safety but its acceptance is diminished due to worse intrinsic safety. The intrinsic safe factor *i.e.,* the presence of high pressure air in a explosive environment, is poor for pneumatic controllers and hence they are used in metallurgical and steel industry where safe operation in explosive atmosphere is not the main problem. The coupling of electronic equipments with pneumatic controller are still popular. Though most of the pneumatic control has been replaced by electronic control in modern times, yet the pneumatically operated final control elements *i.e.,* pneumatic control valves, are still popular, now-a-days.

2. ***Pneumatic devices.*** Pneumatic devices for control purpose include transmitters, controllers and pneumatic signal processing elements. These include baffle-nozzle amplifier and the idea is extended to Baffle-Nozzle controllers pneumatic relays like pilot relays, pressure divider circuits and pneumatic controllers in the form of proportioned and proportional plus reset and proportional plus derivative controllers.

## 35.14 BAFFLE-NOZZLE AMPLIFIER

It is the pneumatic transmitter which converts mechanical motion into pneumatic signal. Figure 35.38 shows a Baffle-Nozzle-amplifiers.

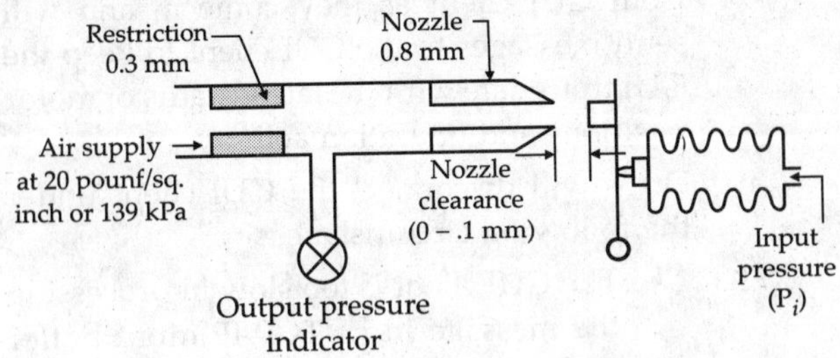


**Fig. 35.38**

The above figure shows that the nozzle clearance is changed with the input pressure ($P_i$) and the output pressure is a function of input pressure and the supply pressure. Generally pneumatic transmitters supply a output pressure variation of 3 – 15 psig (20-100 kilo pascal) like standard for current transmitter it is 4 – 20 mA. The celebration curve of a Baffle-Nozzle amplifier or transmitter has be shown in the figure for a fixed supply pressure 20 kPa.

From Fig. 35.39, it is clear that as the nozzle clearance decreases the output pressure increases with the supply pressure fixed at 20 kPa. Hence the acceptable nozzle clearance for a variation of output pressure from 20 kPa – 100 kPa, which is the standard, is .0016 – .0035 mm. For 20 kPa supply air pressure, the output pressure for a change in nozzle clearance from .0016mm to .0035 mm, varies linearly as found from Fig. 35.39. The gain of the transmitter is defined by the ratio $\Delta P / \Delta x$ and the gain is found to be 1000, for this particular design.

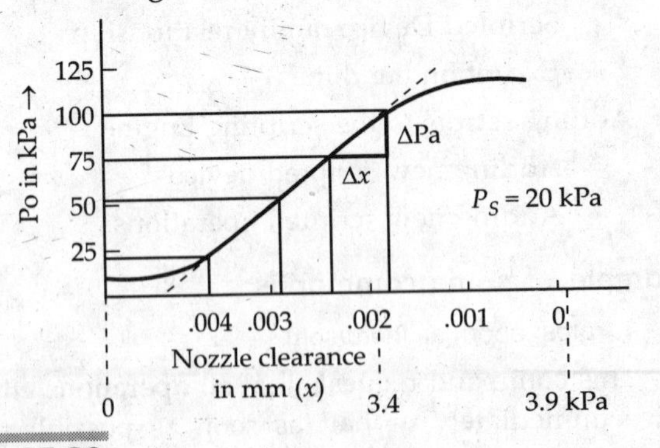


**Fig. 35.39**

### Pilot Relays

Pilot relays are also called booster relays or pilot valves. These are used as a part of Baffle-Nozzle assembly for good flow capacity and gain during transmission. A relay is used to increase the gain of a controller and a baffle nozzle fitted with a pilot relay increases the gain by a factor of 100. The figure of a pilot relay is shown in Fig. 35.40.

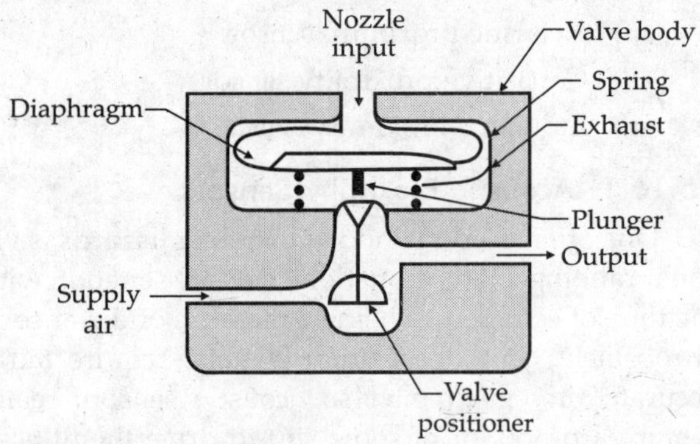


**Fig. 35.40**

This figure shows, as the nozzle pressure *i.e.*, input/air pressure is high, the pressure increases on the diaphragm and the valve positioner goes down due to spring action to open the supply port and the output pressure increases. With the nozzle pressure goes low the supply port is closed and exhaust port is open and the output pressure decreases. This type of pilot valve is used to achieve a pressure gain of 10. This type of relay as shown in Fig. 35.40 is a bleed type relay which causes the supply port and exhaust port to be open simultaneously and there is a continuous supply of air to the atmosphere. But stiffer positioner structure and longer travel of the positioner decreases the gain. The pressure gain is also a factor of the stiffness in input diaphragm or spring. As, a Baffle-nozzle output is linear the pilot relay output air has been fed back to the baffle nozzle and hence to maintain exact gain and linearity for a pilot relay is not essential. These feed back type pilot relays are non-bleed type, in which, there is no continuous flow of air to the atmosphere. They are of lower gain than bleed type but has ability to shut down input port and exhaust port at the sometime. Their pressure gain ranges from 2 to 6.

**Pneumatic Proportional Control Action.** The use of non-bleed type pilot relay, which is in the feedback path of a baffle nozzle controller is the heart of a pneumatic proportional controller, as shown in Fig. 35.41.

This is a force-balance type pneumatic proportional controller. The feed back bellow measures the error pressure which is the difference between supplied set point input pressure ($P_i$) and the output pressure which is transmitted to the feedback bellow through Baffle-nozzle amplifier and pilot relay. The difference is calculated by a mechanical

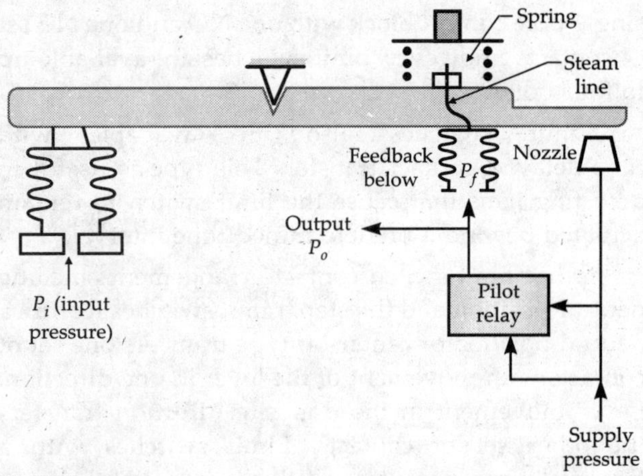


**Fig. 35.41**

arrangement of lever with a fulcrum. The input set point pressure and the feedback bellow pressure which is a measure of output pressure forces the mechanical rod on the two sides of the fulcrum. The gain of the controller may be changed by changing the fulcrum position and zerosetting is done by the zerosetting spring.

Another type of the pneumatic proportional controller is the motion balance type which uses the change in baffle-level by a bounden tube with input pressure and due to the motion of baffle the nozzle clearance changes. This has been illustrated in Fig. 35.42.

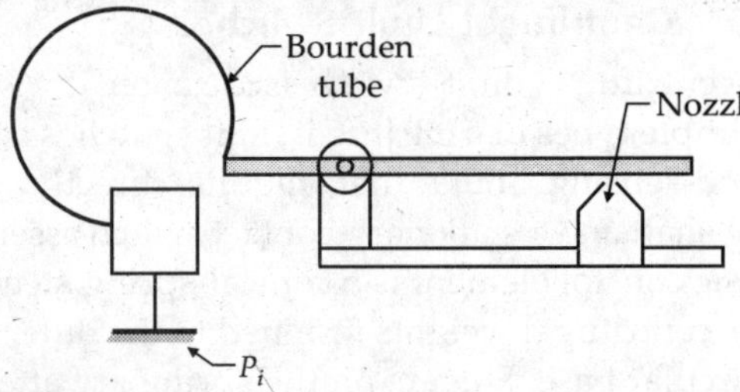


**Fig. 35.42**

This is an instance of the motion balance proportional controller.

**Proportional plus reset controller.** For a force balance proportional controller, the reset action (*i.e.*, integral action is generated by using a separate bellow whose pressure is changed by a needle reset valve and a reset volume ($P_r$) in the feedback path. The pressure in the reset volume is proportional to the integrated airflow into it.

## 35.15 LIMIT SWITCHES

A mechanical limit switch interlocks a mechanical motion or position with an electrical circuit. A good starting point for limit-switch selection is contact arrangement. The most common limit switch is the single-pole contact block with one NO and one NC set of contacts ; however, limit switches are available up to four poles.

Limit switches also are available with time-delayed contact transfer. This type is useful in detecting jams that cause the limit switch to remain actuated beyond a predetermined time interval.

Other limit switch contact arrangements include neutral-position and the step. Limit switches feature a neutral-position or center-off type transfers one set of contacts with movement of the lever in one direction. Lever movement in the opposite direction transfers the other set of contacts. Limit switches with a two-step arrangement, a small movement of the lever transfers one set of contacts, and further lever movement in the same direction transfers the other set of contacts.

Maintained-contact limit switches require a second definite reset motion. These limit switches are primarily used with reciprocating actuators, or where position memory or manual reset is required. Spring-return limit switches automatically reset when actuating force is removed.

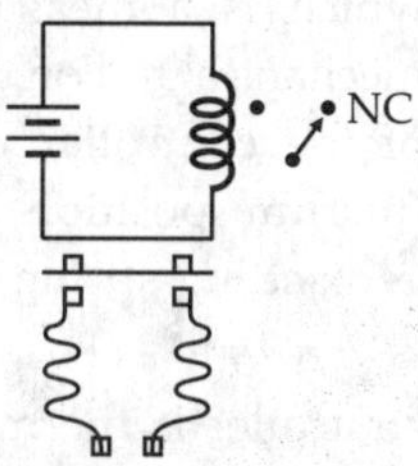


**Fig. 35.43** Pressure switch.

### 35.15.1 Centrifugal Limit Switches

A centrifugal limit switch is actuated by speed only. Simple types of centrifugal limit switches consist of speed-sensing units that mount directly on a rotating shaft and a stationary-contact switch assembly. The basic control element is a conical-spring steel disc that has centrifugal weights fastened to the outer edge of its circular base. Fingers on the spring are attached to an insulating spool that rides free of the shaft and actuates the movable switch contact. As the rotating sensing unit reaches switching speed, the centrifugal force of the calibrated weights overcomes spring force, resulting in an instantaneous axial displacement of the spring and the contact-actuating spool.

The contacts switch at one speed as speed increases from zero to operating speed, and at a lower speed as rotation slows from operating speed toward zero. The spring decreasingly opposes centrifugal force as rotational speed increases from standstill until the snap-over point is reached. Then, spring force adds to centrifugal force to axially snap the spool and actuate the contacts. As rotational speed decreases from operating speed, spring force overcomes the centrifugal force of the weights at a lower speed where snapback begins.

## 35.16 PROXIMITY SENSORS

These types of sensors are operated in proximity to a particular object. These sensors are of three types :

(*i*) Acoustic proximity sensor

(*ii*) Capacitive proximity sensor

(*iii*) Inductive proximity sensor.

### 35.16.1 Acoustic Proximity Sensors

One often used method of avoiding hazards is via sonar ranging. Here, acoustic signals are sent out, with the time of echo return being a measure of distance to an obstacle. This does, unfortunately, require fairly accurate timing circuitry-so acoustic sensors really require a processor of some sort to drive them. Also note that acoustic sensing essentially requires the use of commercial sensors. The most common acoustic proximity sensor is the kind used in Polaroid cameras.

### 35.16.2 Capacitive Proximity Sensors

When power is applied to the sensor, an electrostatic field is generated and reacts to changes in capacitance cause by the presence of a target. The main disadvantage to this sensor (often called a capaciflector) is that its usefulness is dependent on properties of the obstacles it is sensing (namely, their dielectric constant). The higher the dielectric constant (say, water), the more sensitive a capacitive sensor is to that target. The sensing distance depends on the dielectric constant of the target and the surface areas of the probe and the target.

### 35.16.3 Inductive Proximity Sensors

Another method for sensing distance to objects is through the use of induced magnetic fields. The primary problem with this method is that it is largely confined to sensing metallic objects.

## 35.17 TURBINE METERS

A bladed turbine, when in rotation, driven by a fluid can measure the flow of liquids and gases for very low flow rate. The rotation is sensed by optical arrangement *i.e.*, a mirror fitted to the blade reflects the optical signal which is sensed by a photo detector. The optical reflected pulse occurs once in each rotation of the blade and the counter counts the number of pulses in 1 sec. which is calibrated with the flow rate. The transmitted signal is through a hole on the blade is also a used practice. The other type of turbine flowmeter uses a toothed ferromagnetic wheel radiation system when a toothed ferromagnetic wheel rotates in a magnetic field there is temporal change in magnetic field in the vicinity of a fixed coil and change in magnetic field can be sensed by a fixed coil due to induction of voltage. All these types of sensing technique is a case of remote in magnetic field, sensing which does not disturb the process. Turbine flow meter are the not accurate meters for low and medium viscosity fluid. These meters often offer good rangeability and also for high and low temperature use. Again these types of flowmeters are suitable for high pressure operation. Ready installation small size and weight are also added advantages for these types of meters. But these meters are limited in the use of high viscosity fluid and also for dirty liquids. Sometimes filtration is used.

## Review Questions

1. What do you mean by the transfer function of a system? Derive the transfer function of the electrical system shown in the following figure and also obtain the output for unit step input.

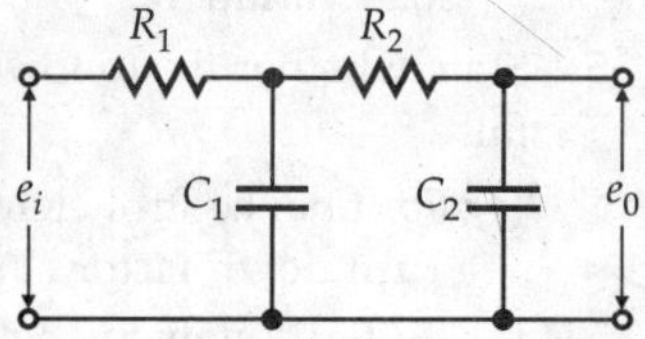


2. Draw the schematic diagram of a practical closed loop temperature control system and describe the function of each component-present in the loop.
3. What do you mean by transient and steady state part of process response ? Draw a neat diagram of time response of a first order system and define : (*i*) peak over shoot, (*ii*) settling time, (*iii*) rise time.
4. For which types of processes ON-OFF control is recommended ? Describe a practical ON-OFF type level control scheme for overhead tank where water is pumped from underground tank with the help of a pump.
5. Using operational amplifier realise a PID controller with separated P, I and D blocks. How the proportional, integral and derivative gains can be adjusted ?
6. Describe the construction of a pneumatic relay. What do you mean by the term 'derivative over run' and 'integral windup' ?
7. What is a positioner and why is it used ? Describe the construction of an electro-pneumatic converter (I/P converter).
8. Write the mathematical expression for a PID controller.
9. When an ON-OFF controller is selected ?
10. Draw and explain the operation of an I to P converter (current to pneumatic).
11. What are the basic components of a Final control Elements ?
12. Draw and explain the architecture of a PLC.
13. Differentiate between centralised and distributed control systems.
14. What is PID control ? What are the tuneable parameters of a PID controller ? What are the changes in the overall system dynamics when a derivative action is plugged in ?
15. With a schematic diagram describe the operation of a pneumatic flapper-nozzle amplifier. Why is a pneumatic relay an integral part of a pneumatic controller ?
16. What is meant by 'offset' ? How does it develop ? How can it be eliminated ?
17. Give the schematic design of a practical automatic level control system of an overhead tank where water is pumped from an underground reservoir.
18. Describe the construction of a pneumatic PID controller and obtain its transfer function.
19. What are the advantages of feedback compensation ? Discuss with a typical example.
20. What is a DCS ? Draw the basic structure of a DCS with a brief discussion of the different components of the same.
21. What are the basic modules of a PLC ? Discuss the inter connections amongst modules. How is the ladder diagram translated into a storable program in such a controller ?
22. Draw the driving circuitry of a four-stack variable reluctance type stepper motor. Why are diodes placed across the excitation coils ? Sketch the stack excitation waveforms.



## Objective Type Questions

1. Find the appropriate answer :

(*i*) Laplace transform of an impulse input is :

(*a*) 1, (*b*) 1/s, (*c*) $1/s^2$, (*d*) s

(*ii*) Pneumatic relay is an integral part of pneumatic controller to :

(*a*) increase gain (*b*) decrease gain

(*c*) increase flow rate

(*d*) decrease flow rate

(*iii*) Electronic controller produces an output of :

(*a*) 1 – 5 volt (*b*) 0 – 10 volt

(*c*) 4 – 20 mA (*d*) none of these

(*iv*) On applying negative feedback to a system, its gain bandwidth product will be :

(*a*) larger (*b*) smaller

(*c*) similar (*d*) none of these

(*v*) The relation between peak overshoot and rise time of a system is :

(*a*) linear (*b*) reciprocal

(*c*) square law (*d*) cubic

(*vi*) ON-OFF controller is basically a proportional controller with :

(*a*) large PB (*b*) medium PB

(*c*) zero PB (*d*) none of these

(*vii*) Reset wind-up problem is associated with :

(*a*) P control (*b*) PD control

(*c*) ON-OFF control (*d*) PI control

(*viii*) Offset can be eliminated with the help of :

(*a*) D-action (*b*) P-action

(*c*) I-action (*d*) none of these

(*ix*) If the process capacitance becomes large, its time constant will be :

(*a*) smaller (*b*) larger

(*c*) unchanged (*d*) none of these

(*x*) Solenoid values are operated by :

(*a*) pneumatic signal (*b*) hydraulic signal

(*c*) electrical signal (*d*) none of these

**2.** State whether the following statements are true or false :

(*i*) For a stable system any bounded input produces bounded output

(*ii*) Safety valve is a type of control valve

(*iii*) Regenerative feedback can be used in closed loop control

(*iv*) Transfer function of a system is unique.

**3.** State whether the following statements are true or false :

(*i*) Transfer function of a system is a function of time.

(*ii*) A system is said to be stable, if the output is bounded for a bounded output.

(*iii*) Integral control action improves the stability of a control system.

(*iv*) A traffic control system is an example of open loop system.

(*v*) In control system, positive feedback is used.

(*vi*) A system is said to be unstable, if at least one pole lies on the right of imaginary axis.

**4.** State whether the following statements are true or false :

(*i*) Proportional sensitivity and proportional band are directly related.

(*ii*) An ON-OFF controller has an infinite proportional band.

(*iii*) Derivative control action is not suitable for step excitation.

(*iv*) Positioner improves the response of a valve.

(*v*) Off-set in a control system is generated due to load variation.

**5.** State whether the following statements are true or false :

(*a*) For final control element, Rangeability ($R$) and Turn down factor ($T$) are related as $T = 0.7$ R.

(*b*) Butterfly valve is actually a sliding stem type control valve.

(*c*) PLC is an acronym for process logic controller.

(*d*) DCS has a layered structure.

(*e*) PLC is an integral part of DCS

(*f*) Solenoid valve is a digital final control element.

(*g*) Laplace transform transfers a linear differential equation into an algebraic equation.

(*h*) A control system that is open-loop unstable is always closed-loop unstable.

(*i*) Transfer function of a physical system is unique.

(*j*) Derivative control is also called an anticipatory control.

(*k*) Output offset can be eliminated by introducing integral action.

(*l*) Integral error compensation increases the order of the system.

(*m*) Feed forward control technique is suitable for a system having unpredictable disturbances.

(*n*) Bump-less transfer function can be realised by using PID algorithm in velocity form.

(*o*) For a cascade control system to function properly, the outer loop should be slower than the inner loop.

(*p*) In a position control system, output rate feedback can be incorporated by connecting a tachogenerator to the load shaft.

(*q*) On-off controllers can be used in processes that have high demand side capacity.

(*r*) An electronic PID controller can be made completely non-interactive by using three operational amplifiers.

# 36

# Biomedical Instrumentation

## 36.1 INTRODUCTION

After 1970, a steady growth has been reported in the field of biomedical instrumentation or engineering. The reported electro-physical investigation helps to observations related to cardio-vascular system. The cardiovascular system of study concerns with the activity of heart and the rhythmatic blood flow all over the body. The interrelation between cardiac activity, the blood flow and the electro physical responses of cell are found to be interlinked. The electro physical responses of cell, which is the biopotential generated in the cells throughout the body, can be considered to be the major break-through in Biomedical engineering. The measurement of bioelectric potential through electrodes and the relation between the bioelectric responses from different parts of the body has helped in the development of Electro-Cardiogram (ECG). The placing of silver electrode at different parts of the body, thus developing unipolar leads, bipolar leads to upto 9 channel leads, is a subject of interest to modern day engineering. Monitoring of blood pressure and the plot of blood pressure, systole and diastole and electro cardiogram (ECG) is also a subject of interest to engineers now-a-days. The subject of electro-cardiography has been extended to the study of Holter monitoring *i.e.,* 24 hours ECG monitoring, or monitoring of ECG at stressed physical condition (treadmill ECG). The other topic of interest is neural control of cardio-vascular system or the pacemaker technology. The chapter includes the physiological aspects concerning cardio vascular system and the measurements made to diagnose different physical disorder.

The similar type for measurement of recording of the activity of brain is called Electroencephalogram (EEG), and that of functioning of muscle is Electro-Myogram (EMG). These are of more complex pattern than ECG to recognise. The placing of electrode, for EEG pattern, is on the scalp and that of EMG is on the muscle.

The other important activity of the physiological system is neuronal activity. The neuronal system with the study of central nervous system, the transmission of responses through neurone is dealt with in this article.

## 36.2 BIOMETRICS

The subject concerning with the measurement of physiological variables of the human body is called *Biometrics.* The measurement of body temperature blood pressure, blood flow, plethysnography are the important topics in biometrics. The characteristics of bioengineering sensors are range, selectivity *i.e.,* low small variation of the variable the sensor can reliably measure, linearity *i.e.,* the degree of accuracy with which the input may be predicted from the output voltage of the sensor, repeatability of the sensor and the measure of variation of measurement within wide range of frequency. There are also different types of errors involved in bio-measurement. These include errors due to tolerances in electronic components, mechanical error, component error, poor frequency response, errors due to ambient temperature and pressure and other instrumental and systematic errors. The other important feature of instrumentation is the signal to noise ratio of the instrument. Due to interference of EMI—EMC noise (Electromagnetic noise or electrostatic noise) or interference due to diathermy equipment is also possible.

The ability of the control instrument not to be driven to uncontrollable state is called stability of the instrument. This factor is dependent on amplification, feedback and other features in a control loop. Degree of isolation between patient and ground is also a figure of merit of the biomedical instrumentation.

In human body, the transport system that exchanges products and oxygen between specialized cell and organs is blood. The culture of blood is an important measure by which the physiological activity of human body can be assessed. So chemical analysis of blood can be used for clear physiological inferences. Like blood, the study of urine of human body is also an important topic.

Human beings like other animals, respond to the environment and these responsiveness is reflected in their body functions. The body functions are controlled by the nervous system is a function of human psychology. The aspect of human responsiveness is called behaviour which is dependent on neurological aspects. Different types of instrumentation have been grown for sensory or psychological requirements and thus the study of behaviour. Particular interesting addition to this subject is the Bio feedback generation and mechanism. Some of the body functions of human being are perceived in an unconscious state or these controls can be stated as non-voluntary control. It has been found that if these body functions are measured and fedback to some senses of human being these uncontrollable states of human body can be controllable in a conscious or voluntary way. For example, the relaxation training of human being is a laboratory way to control anxiety.

X-ray imaging is also an important tool for medical diagnosis. X-ray can be used for imaging due to its high penetration power except hard bones but also X-ray can be used for therapeutic purpose. This subject is named as radiology. Like X-ray some other radioactive particles may be used for this purpose. Treatment of cancer uses the cobalt and other ionising radiation to make the tumour inactive and closely monitoring its temperature with the help of a fibre-optic temperature sensor. In modern diagnostic laboratory, the imaging quality of X-ray has been used extensively for three-dimensional imaging by movement of an X-ray source in a solid angle of 360° about the subject and for each position, detection has been done. Computer has been vastly used to process the information to get a 3-D imaging and this process is called the CT-Scan (Computed-Tomography Scan). Other modern scanners use the MRI scanning *i.e.,* magnetic resonance scanning which is the resonating current in a coil due to an atom subject to high DC magnetic field with R.F. magnetic field. With change in radio frequency the resonance lines have been found which is a characteristic to the atom.

The ultrasonic detection process which is a non-invasive diagnostic process *i.e.,* not getting inside the body, is an important testing tool now-a-days. Ultrasound is ionic energy with a frequency much above (1-10 MHz) audible range of frequency. Due to longitudinal nature, it has high penetrating power with a velocity of sound and extensively used for Imaging purpose. This imaging technique needs a proper ultrasonic source and an ultrasonic receiver. The imaging by ultrasonic is termed as echocardiogram or ultrasonogram. A well known characteristic of ultrasound used in bioengineering is Doppler effect which is the change in frequency of reflected untrasonic energy due to the moving interface. This is termed as Doppler and colour-Doppler. The frequency increases when the interface moves towards the sensor and decreases when moves away. A 40 Hz change in frequency for 1 cm/sec interface velocity is reported for the ultrasonic frequency of 3 MHz.

## 36.3 BIOPOTENTIAL

The physiological activity of a cell is defined by the gain of materials by a cell or losing material to the body fluid through a semi-permeable membrane. The transport of materials from the cell to body fluid or from body fluid to membrane happens in way of passive transport or active transport. If the hydrostatic pressure is equal on both sides of the membrane the materials or ions move due to concentration gradient across the membrane. This type of transport is called passive transport or diffusion. The transport phenomena of sodium ion ($Na^+$), potassium ion ($K^+$) and chromine ion ($Cl^-$) through the membrane are of importance to study biopotential or cell potential. The passive transport is found to support the movement of potassium and chlorine ion through membrane but does not allow sodium ions pass through the membrane. But active transport, refers to the movement of $Na^+$, $K^+$ $Cl^-$ and ions through the membrane irrespective of concentration gradient. Such transport requires expenditure of chemical energy and it may happen that, even when $Na^+$ concentration outside the cell is much more than intercellular $Na^+$ concentration, there may be transport of $Na^+$ through the membrane. So active transport supports movement of ions across the gradient and with the expenditure of the chemical energy the creation of selective permeable membrane *i.e.,* $Na^+$ permeability has been grown with the cells which actively maintain concentration gradient by expenditure of work.

At first, the membrane separates the body fluid and the intercellular materials. At rest, the membrane allows $K^+$ ions and $Cl^-$ ions to diffuse along the concentration gradient. Hence potassium ions move inside the cell through the membrane to restore neutrality of the cell. But as $Na^+$ ions cannot move through the membrane the concentration of $Na^+$ ion much exceeds the $K^+$ ion within the cell. Hence a potential develops across the membrane due to unbalance of ions. This potential is measured to be nearly – 60 mV to 100 mV and it is called the resting potential (Fig. 36.1).

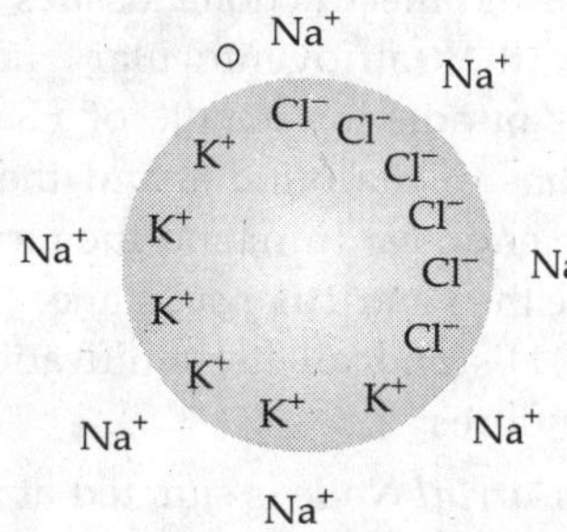


**Fig. 36.1**

Due to ionic current flow through the body fluid due to expenditure of energy the cell membrane becomes selectively permeable for $Na^+$ and $Na^+$ ions pump into the cell through the cell membrane and this physiological incident is called *Depolarization*. As the $Na^+$ ions move inside the cell the bioelectric potential is generated between inside and outside of the cell and this potential is called action potential (Fig. 36.2). This potential is nearly + 20 mV.

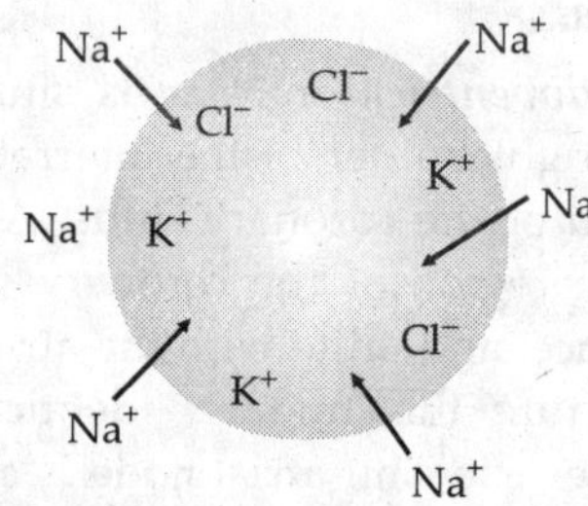


**Fig. 36.2**

The ionic current which is generated due to expenditure of energy by burning the phosphate molecule decreases and active transport through the cell membrane stops and cell recovers its state of rest potential. But this recovery according to our physical interpretation should be very slow, though it is much faster. And the $Na^+$ moves outside the cell through a type of active transport which is called sodium pump.

The theory has been framed so that sodium ions move from inside to outside of the cell against charge neutrality and concentration gradient by an active transport supported by the burning of high energy phosphate compound. This withdrawal of $Na^+$ from the cell is called repolarization. Hence the potential to time curve may be given as (Fig. 36.3).

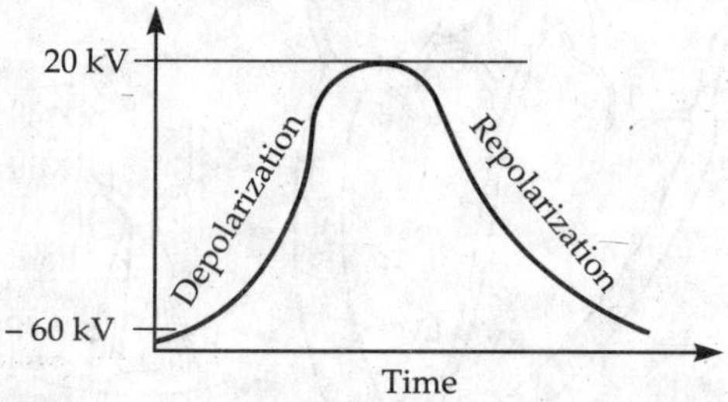


**Fig. 36.3**

## 36.4 CARDIO VASCULAR SYSTEM

Cardio vascular system *i.e.*, heart, consists of four chambers. Two of which are thin walled atria separated from each other by interrarial septum. The other two chambers are thick walled ventricules [Figs. 36.4(*a*) and (*b*)]. The common walls between two ventricules are inner ventricular septum. Atria and ventricles are connected by fibrous Atria-Ventricular (*A-V* ring). The opening between Right-Atria and right-ventricule through fibrous *A-V* ring is called tricuspid valve and that between left Atria and left ventricle is mitral valve or bicupsid valve. These two valves consist of flaps and cusps at the periphery of A-V ring. The right ventricle is connected to pulmonary artery by semilunar valve which has three flaps. Again the left ventricle is opened to Aorta by aortic orifice. There are some tissues which link with the initiation and propagation of heart beat. The excitation in heart is initiated in the sinoatrial (SA) node rather than starting from the central nervous system. These SA nodes are also characterised by pacemaker cell. The SA node is situated at the junction of superior vena cava and the right atrium. Hence, when the impulse created by SA node has been spread over to the heart from right to left atria and this stimulus impulse causes atria to contract and then stimulates the *A-V* node. This impulse, after a short time spreads over the ventricles through bundle of his and through Purbinje fibres, to the myocardium. The myocardium supplies the force for contraction of the ventricles to pump oxygen rich blood in the circulating system consisting of aorta, pulmonary artery.

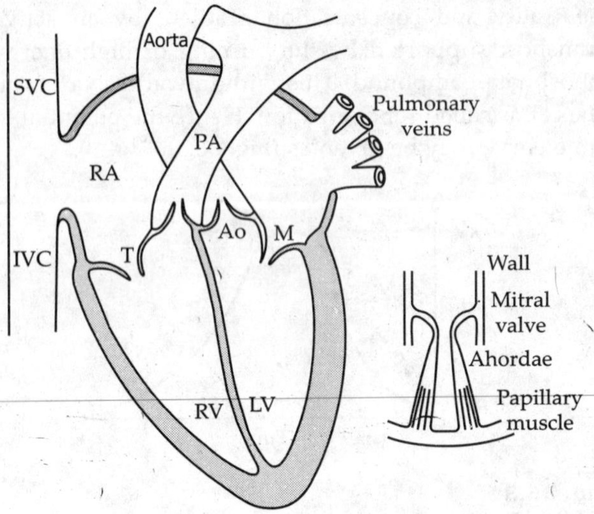


SVC = Superior vena cava
IVC = Inferior vena cava
RA = Right atrium
T = Tricuspid valve
RV = Right ventricle
PA = Pulmonary artery
M = Mitral valve
LV = Left ventricle
Ao = Aorta

**Fig. 36.4** (*a*) Direction of blood flow through the heart.

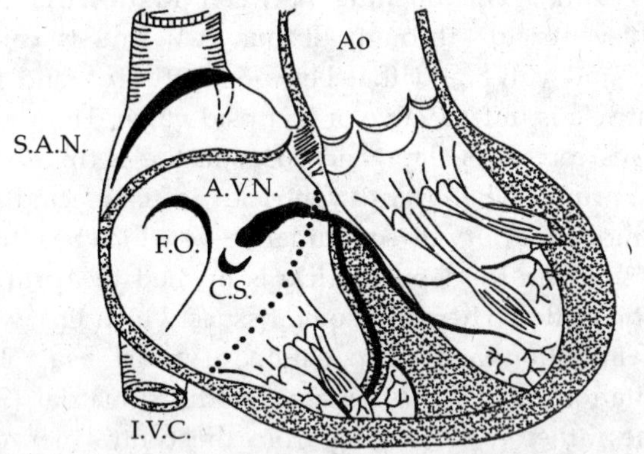


**Fig. 36.4** (*b*) The specialized tissue of the mammalian heart. The pulmonary trunk and part of the right ventricle have been removed.

The blood poor in oxygen, enters the atria through pulmonary veins and as the pressure of atria chamber increases, the bicupsid and tricupsid valve opens and blood is pumped to the ventricle and a new cycle starts.

The heart rate is controlled by the impulse from the SA node, the heart rate is increased or decreased by the action of sympathetic nerve and parasympathetic nervous system.

Ventricular ejection land close when the relevant arterial pressure exceeds that of the corresponding ventricle when it begins to relax.

Closure of both atrioventricular valves causes the first heart sound and closure of the semilunar valves causes the second heart sound.

## 36.5 SPECIAL JUNCTIONAL TISSUES OF THE HEART

This term is employed to describe certain tissues in the heart which are concerned with the initiation and propagation of the heart beat. They include the sinu-atrial node and the junctional tissues — the latter consisting of the atrioventricular node ; the atrioventricular bundle, or bundle of His ; the right and left divisions of the bundle, and their arborizations under the endo-cardium and the terminal fibres which penetrate the ventricular substance (Fig. 36.4(*b*)). The bundle of His and all its ramifications may be called Purkinje fibres.

1. ***The Sinu-atrial Node*** is situated at the junction of the superior vena cava and the free border of the right atrial appendix, and extends down along the sulcus terminals for a distance of 2 cm. It is 2 mm in width and has a rich capillary blood supply. It consists essentially of thin, elongated muscle fibres (about one-third the size of heart-muscle fibres), fusiform in shape and longitudinally striated, which interlace with one another in a plexiform manner. These fibres normally initiate the heart beat ; for this reason the sinu-atrial node is called the *pacemaker*. Nerve cells and fibres forming the excitor relay of the vagus nerve and excitor (postganglionic) fibres of the sympathetic are also present.

2. ***The Atrioventricular Node*** is situated at the posterior and right border of the interratrial septum near the mouth of the coronary sinus. Atrial muscle fibres from the region of the coronary sinus collect fanwise, interlace, and unite with the atrioventricular node. In structure this node is identical with the sinu-atrial node. The sinu-atrial node is a right-sided structure developing from tissue which lies at the entrance of the primitive right great vein (which later becomes the superior vena cava). This explains the position of the node and its supply by the right vagus nerve. The atrioventricular node is a left-sided structure, supplied therefore by the left vagus nerve, and developed from tissue in the vicinity of the entrance of the left great veins which becomes the coronary sinus.

3. ***The Bundle of His.*** Runs upwards to the posterior margin of the membranous part of the interventricular septum and then forwards below it,

unsheathed and isolated in a canal. At the interior part of the membranous septum, in front of the attachment of the septal cusp of the tricuspid valve to the a.v. ring, the bundle forks. The left division pierces the membrane and then lies on the upper border of the muscular septum to enter the subendocardial space of the left ventricle beneath the union of the anterior and right posterior cusps of the aortic valve.

## 36.6 PRESSURE PULSES IN THE CARDIAC CHAMBER

The atria and vertical pressure is 10 mm of mercury at the beginning of cardiac cycle as atria and vertricle are released and filled with blood from vanes. The *A-V* valves are open and pressure in each cavity is nearly equal. With the impulse of the S.A. node, atrial muscle contracts and pressure of atria ($P_A$) rises. The volume of ventricle during atrial contraction is the diastalic volume. The atrial contraction lasts for 0.1 sec. At this time pressure is both atria and verticle falls. With the impulse from S.A. node spreads to the ventricle by the bundle of His and purbinje tissues the ventricular contraction begins and $P_V > P_A$ and *A-V* valve closes with generation of first heart sound. $P_V$ rises sharply in closed ventricle chamber and $P_V$ rise exceeds the pressure in artery and semilunar valve

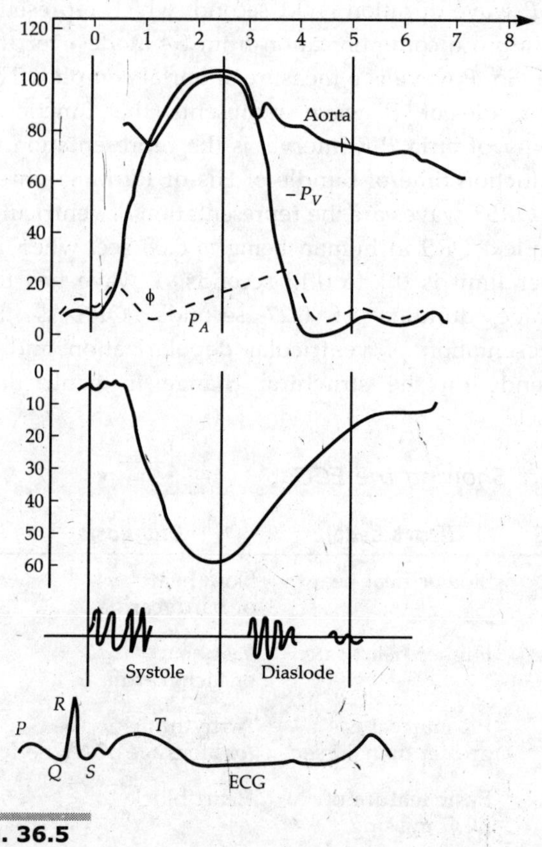


**Fig. 36.5**

opens, with the starting of ejection phase. During this phase, arterial and ventricular pressure are nearly equal and ventricular volume decreases. The arterial pressure rises as blood enters the vessel much faster than escapes through peripheral branches. The $P_V$ curve with time reaches its peak when aortic entry of blood begins and the $P_V$ declines as the ventricular volume increases as flow from artery and other branches continues to be high. This is ventricular systole and total systolic dilation is for 0.3 second. With the end of systole, the ventricular pressure drops sharply and at the same time $P_A > P_V$ as semilunar valve closes with a sharp heart sound. The curve showing the variation of $P_A$, $P_V$ and aorta pressure with time and the ventricular volume with systolic and diastolic heart sound and their time mapping with the electrocardiogram plot has been shown in Fig. 36.5.

## 36.7 ELECTRO CARDIOGRAM (ECG)

The electro chemical electrode theory has also been discussed earlier. The interface of metal ions in the solution with a metal electrode of principal interferences with the metal ions cause the generation of electrode potential. This potential is for the difference in diffusion rates into the metal and out of the metal. Non metals like hydrogen also show electrode potential when interfaced with associated ions. The difference in rate of flow of + ve and – ve ions from a two larger barrier at the interface, of metal and solution which acts as electrode potential.

A different source of electro chemical potential is the separation of liquid solution with different concentration of a particular ion by a membrane which is semi-permeable to the ion. To measure bioptential for electro cardiogram (ECG) skin surface electrodes are generally used, whereas, some needle electrode which penetrate the skin is generally used for Electroencephalogram (EEG) and electromyogram. Some Microelectrode has been developed to measure the biopotential of a single cell.

All the electrodes use a metal electrolyte interface and an electrode potential is developed at the interface due to exchange of ion through the interface. Generally used electrode is silver-silver chloride electrode *i.e.*, silver as metal and silver chloride as electrolyte solution. A coating of pure silver with silver chloride is the way how these electrodes are made. Generally more twan one electrodes are used and if the two electrodes are of same type, the difference of instantaneous ionic potential between the points of the body, where they are placed, has been measured.

Hence there are a some specific rules, how surface electrodes are placed on the body. Two kinds of indirect leads are applied *i.e.,* unipolar leads and bipolar leads.

### 36.7.1 Unipolar Lead

The exploring electrode *i.e.,* principal electrode is placed on the body surface of any one of the positions at Right arm (VR) or Left arm (VL) or Left leg (VE) or chest (VC). The zero potential is supplied by the indifferent electrode which is the superposition of the

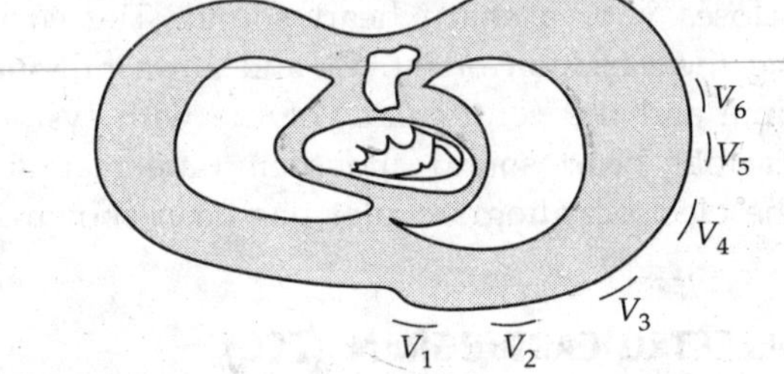


**Fig. 36.6** Cross-section of the chest and the position of exploring electrode.

signal from three electrodes at right arm, left arm and left leg, through the central terminal through 5 kΩ resistance. The position of the exploring electrode may be one of the six positions given in Figs. 36.6 and 36.7.

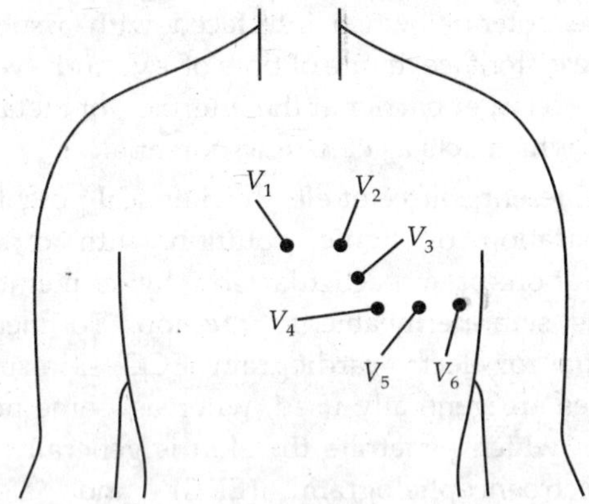


**Fig. 36.7** Placement of bipolar limb.

The VC or exploring electrode on chest is called precordial leads. The different position on which the exploring electrode is placed on the chest gives the characteristic signed of heart nearest to it. The leads $V_1$ and $V_2$ shows right ventricular activity whereas $V_5$ and $V_6$ show left ventricular activity.

The wiring of unipolar leads is such that, as the exploring electrode signal is negative with respect to zero terminal, the record is deflected downwards and when exploring electrode is positive it is deflected upwards.

In case of unipolar leads if the exploring electrode is placed on left leg, the electrical activity of the inferior surface of heart is found for the electrode on left arm *i.e.,* the activity of left upper side of the heart and for VR position the activity at the cavity of the ventricles. The leads are called limb leads.

### 36.7.2 Bipolar Leads

In this type of leads there is no central zero but the bipolar limb lead is the algebraic sum of two constituents leads at a particular time. The bipolar limb leads are described as

Lead *I* = Lead VL – VR

Lead *II* = Lead VF – VR

Lead *III* = Lead VF – VL

This type of bipolar leads supply more information about the axis of heart. Figure 36.5 shows the *PQRST* waves of the ECG. If the heart is vertical the predominant *R* wave is found in Lead *III* and predominant *S* wave in Lead *I*. This is the right axis deviation. For heart, to be more vertical, there is predominant *R* wave in *I* and *S* wave in lead *III*. This position indicates left ventricular hypertrophy.

### 36.7.3 Interpretation of ECG

*P* wave duration is 0.1 second, which represents the impulse communication from SA node over the atria. So, *P* wave is a measure of aortal complex. The magnitude of *P* wave represents the functional activity of atria *P-R* interval is the representation of conduction time of Bundle of His or Purkinje fibres. The *QRST* waves are the representation of ventricular complex. *QRS* in human being is 0.08 sec. when its upper limit is 0.1 to 0.12 seconds. *T* wave has the average duration of 0.27 seconds. *QRS* is the representation of ventricular depolarization and it depends on the structural change in ventricular muscle.

*Chart Showing the ECG*

| ***ECG*** | ***Heart beat*** | ***Diagnosis*** |
|---|---|---|
| 1. | Slower heat beat | Slow heart or bardycardia |
| 2. | Higher heart rate | Fast heart or tachycardia |
| 3. | PR interval is greater than 0.2 sec. | Arrhythmic (or blockage of AV node) |
| 4. | Basic feature of ECG missing | Heart block |

*Other problems are*

1. different path of excitation in the heart
2. changed origin of waves
3. altered relationship of features
4. altered magnitude of one or more features
5. different duration of waves.

### 36.7.4 ECG Recording and Monitoring

Figure 36.8 shows the assembled block for ECG recording.

The figure shows the ECG recording arrangement. The leads from the electrodes pass through the buffer amplifier so that proper impedance matching occurs between the electrodes and ECG recorder. In another scheme which is called driven right lead the right leg lead is not connected to the ECG chasis but it is provided with addition voltages from four leads and supply it to the fifth node to maintain the high voltage DC balance. The capacitors $C_1$, $C_2$, $C_3$, $C_4$ are used for high frequency grounding.

The ECG recorder switch connects one lead to the preamplifier. The preamplifier is a AC coupled different amplifier with high common node rejection ratio (CMRR) to eliminate the DC common node signal. The gain of the preamplifier can be changed by a switch. The pen control circuit is coupled with the ECG recorder switch and for each position of the recorder switch the pen is deflected from its original position to make available more than one channels.

The paper used for marking is a heat sensitive paper and the stylus temperature is controlled to realize the marking on the paper. The marker pen is required for writing the additional information on one side of the paper. The paper speed varies from 20 mm/sec. to 50 mm/sec by a selector switch.

The power switch, when in ON position, the buffers and amplifiers get supply voltage but the motor drive does not get supply whereas in RUN position the motor drive is ON and the paper moves. The 1 mV standard switch in ECG recorder supplies a standard voltage for positioning of the pen on the paper.

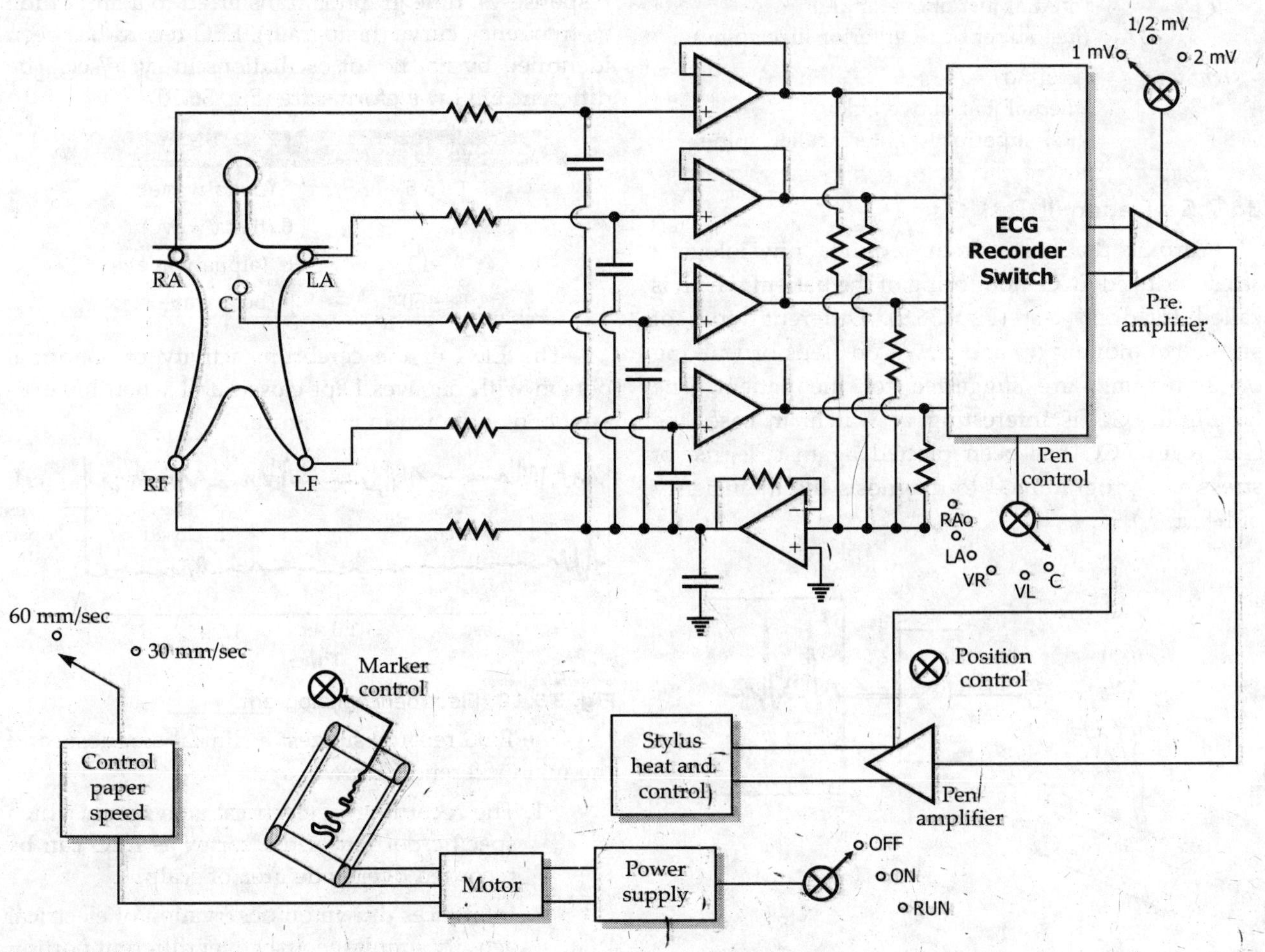


**Fig. 36.8** ECG Recording.

The ECG recorder must be provided with high voltage protection circuit where over voltage is caused by defibrillation *i.e.,* electrical shock treatment due to recovery of patient from losing the rhythm of heart.

The ECG recorders are of different types. Generally single channel recorder, or three channel recorders *i.e.,* simultaneously record three leads and has the capability to switch to other three leads automatically. Now-a-days for lead selection and switching between leads is much easier with the introduction of computer processing of electrocardiograms. The computer processing also provides more leads to be monitored at the same time.

| Electrode Placement | |
|---|---|
| *Limb Leads* | (Electrode placement stress for extremely leads) |
| *Left arm* | (medial wrist)<br>(distal end of clavicle at infrascapular fossa) |
| *Right arm* | (medial wrist) |
| *Left leg* | (medial just above ankle)<br>(just superior to anterior iliac spine) |
| *Right leg* | (ground)<br>(medial just above ankle)<br>(just superior to anterior iliac spine) |

### 36.7.5 Treadmill Test

Cardiac diagnosis often requires physiological stress induced ECG monitoring of the patient which is called treadmill test (Fig. 36.9). Different types of stress like moving up and down on steps or walking on a treadmill are suggested for the patient. The treadmill ECG is interesting to watch, in case, the change in ECG has been plotted against degree of stress and programmed to diagnosis the inefficiency of heart.

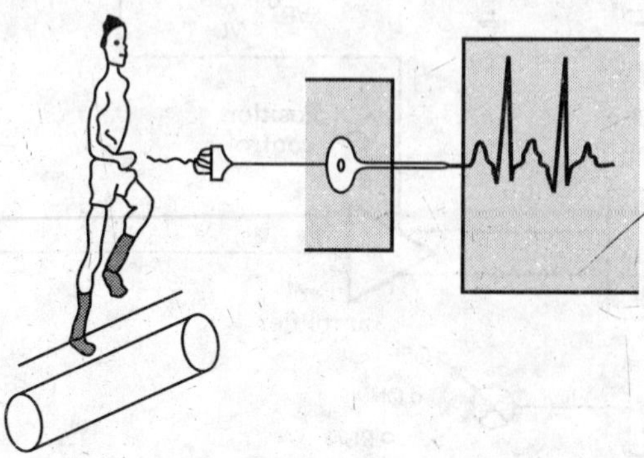

**Fig. 36.9** treadmill test.

### 36.7.6 Continuous ECG Recording (Holter Recording)

It may happen for some heart diseases like arrhythmia, that the missing of any of the *P, Q, R, S* wave or destructing of the wave may happen intermittently for emotional or physiological stress. This type of short time effect on ECG is difficult to diagnose with normal, ECG. Hence 24 hours of ECG monitoring is suggested *i.e.,* the Holter test. In Holter test, the cardiac activity for a long time has been recorded on a magnetic tape recorder. The device is small and can be placed on the breast of the patients. The recorded tape has been played back to monitor the 24 hour ECG.

### EEG and EMG

Electro-encephalogram (EEG) is the recording of activity of the cerebrum or brain for human being. EEG has been recorded with electrodes place on different portion of scalps cerebral hemisphere and amplified by an a.c. coupled differential amplifier. The output is recorded by a pen recorder like ECG or the response vs. time graph is transferred to a amplitude vs frequency curve (histogram). EEG record has been identified by the no. of oscillations in cycle/sec. The different EEG waveforms are (Fig. 36.10).

| Frequency cycle/sec. | Name |
|---|---|
| 1 – 3.5 | δ (delta wave) |
| 4 – 7 | θ (theta wave) |
| 8 – 13 | α (alpha wave) |
| 14 – 30 | β (beta wave) |

The EEG of the cerebrum activity of a normal person with his eyes kept closed and when his eyes are open as shown in the figure.

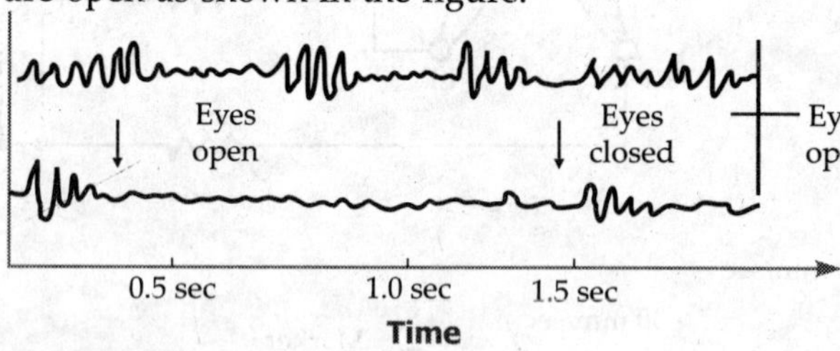


**Fig. 36.10** Electroencephalogram.

The EEG records suggest a clinical diagnosis and the main inferences drawn are :

1. The recording of electrical activity on which specific portion of the cortex as EEG can be recorded over wide area of scalp.
2. Inferences drawn on observation of electrical activity simultaneously over different portion of scalp.

Generally 4 to 6 simultaneous tracing of complex rythmitic wave have been taken. Each tracing never repeats itself. An analyser is used, which plot the histograms amplitude vs. frequency analysis, of the wave components. The sine wave classified to the frequency are one of the δ, θ, α or β wave. The histogram is shown in Fig. 36.11.

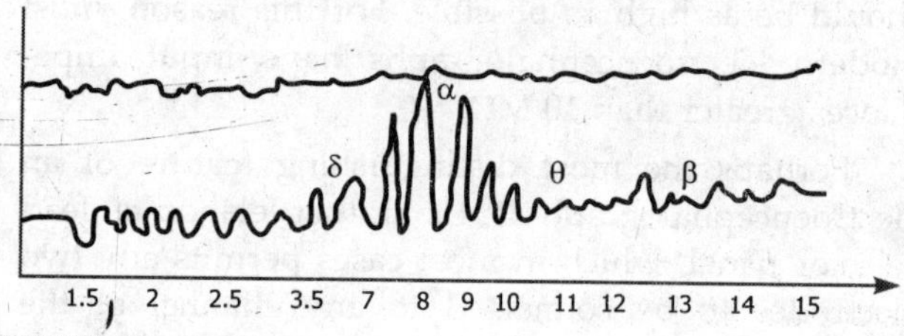


**Fig. 36.11** Histogram.

It is found from the figure that a wave is dominant. For EEG of a normal person the α wave is most dominant and δ, β and θ waves are of small amplitude. The chart shows the interpretation of EEG for human being with disorder and in different state.

| ***Wave*** | ***Area on scalp*** | ***Activity*** | ***Amplitude*** | ***Diagnosis*** |
|---|---|---|---|---|
| α | Occipital and paril to occipital area | Visual or mental activity | | |
| θ | parietal area or temporal area | | 10 μV | |
| δ | | (*a*) During sleep<br>(*b*) Over breathing | 100 μV | |
| β | | Mechanical stress on neighbourhood of tumour | | Incremental tumour |
| Waves occurring in groups separated by period of inactivity | thelamous | Patient, given anaesthesia, | | |

There is provision to record the electrical activity of gray matter of the brain by putting a needle electrode at different depth of the brain. Thermo-coagulation of the four layers outside the brain does not alter the pattern of EEG. Again, if the intralaminar thelemic nucleus are stimulated the response in EEG is

- bilateral and widespread
- of longer latency and of longer duration.

The different portions of the scalp and their nodal name are shown in the figure.

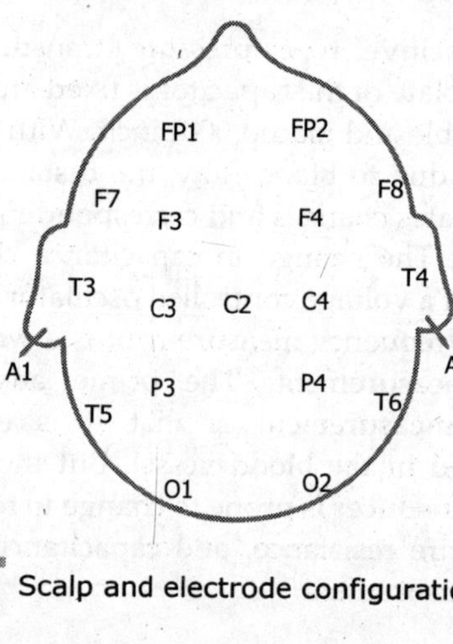


**Fig. 36.12** Scalp and electrode configuration.

For each signal, a complete channel of instrumentation is required. Thus, electroencephalographs having as many as 16 channels are available. A clinical instrument with eight channels and a portable unit are shown in the Fig.

Because of the low-level input signals, the electroencephalograph must have high-quality differential amplifiers with good common-mode rejection. The differential preamplifier is generally followed by a power amplifier to drive the pen mechanism for each channel. In nearly all clinical instruments, the amplifiers are ac-coupled with low-frequency cutoff below 1 Hz and a bandwidth extending to somewhere between 50 and 100 Hz. Stable dc amplifiers can be used, but possible variations in the dc electrode potentials are often bothersome. Most modern electroencephalographs include adjustable upper and lower-frequency limits to allow the operator to select a bandwidth suitable for the conditions of the measurement. In addition, some instruments include a fixed 60-Hz rejection filter to reduce powerline interference.

To reduce the effect of electrode resistance changes, the input impedance of the EEG amplifier should be as high as possible. For this reason, most modern electroencephalographs have input impedances greater than 10 MΩ.

Perhaps the most distinguishing feature of an electroencephalograph is the rather elaborate lead selector panel, which in most cases permits any two electrodes to be connected to any channel of the instrument. Either a bank of rotary switches or a panel of pushbuttons is used. The switch panel also permits one of several calibration signals to be applied to any desired channel for calibration of the entire instrument.

## 36.8 BLOOD PRESSURE MEASURING INSTRUMENTS

The blood pressure can be measured in two ways, indirect measurement and direct measurement.

### 36.8.1 Indirect Measurement

In indirect form, with application of pressure on the upper arm in between systolic and diastolic pressure, causes spurts in blood flow in the brachial artery and Korotkoff sound is heard by a stethoscope to identify the systolic and diastolic pressure. In the other method, a pressure transducer has been placed in the blood vessel and the recording of blood pressure is done on a recorder. The indirect method provides information only about systolic and diastolic pressure and gives no indication about the pressure variation with time, whereas, direct method provides continuous blood pressure monitoring. Though indirect method can be automatic with the instrument electro sphygmomanometre.

#### Sphygmomanometre

In this instrument a inflatable cuff has been used which can be inflated by a rubber bladder and the pressure in the cuff is measured by the manometer. The inflated cuff contains a rubber bladder and it can be deflated slowly by a needle valve.

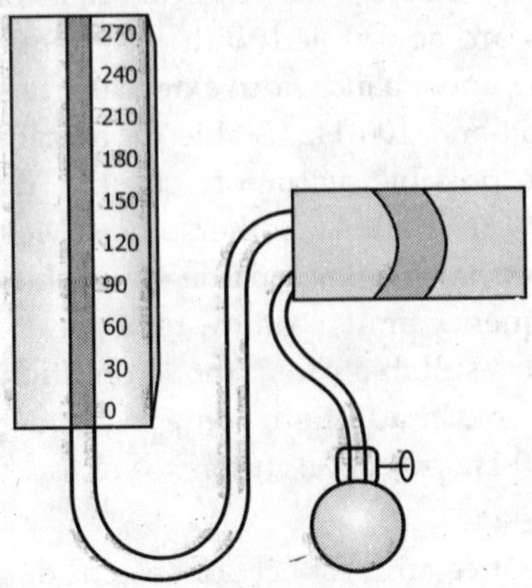


**Fig. 36.13** Sphygmomanometre.

The cuff is put on upper arm and fastened. With the pumping of rubber ball the pressure has been increased to 220 mm of Hg much about the brachial artery blood pressure. The slow release of pressure at the systole there is spurt of blood due to constriction and Korotkoff sound is heard. The pressure, as read from the manometer is systolic pressure. With more deflation of the cuff, the Korotkoff sound is heard and it continues to be heard upto diastolic pressure is reached when the sound stops. The arterial blood circulation is normal below the cuff pressure is diastolic pressure. The sphygmomanometre has been auto- mated with the help of a programmable device called electro-sphygmomanometer. In this instrument, a pressure transducer is placed below the cuff *i.e.,* in between artery and cuff surface and a microphone has been placed over it. The increase and slow decrease of pressure by the instrument can be programmed with specific rate. The systolic point pressure and diastolic point pressure has been recorded and can be played back.

### 36.8.2 Direct Measurement

In direct measurement a surgical operation is needed to place a catheter within the blood vessel. The method of measurement is of two type.

- A sterile saline solution is introduced in the catheter so that fluid pressure is transmitted to a transducer outside the body.
- A pressure transducer is used within the catheter so that continuous recording of blood pressure may be achieved.

In the second case the pressure transducer used in catheter is of three types :

- capacitive type
- inductive type
- linear variable differential transformer (LVDT) type.

In capacitive type pressure transducer metal membrane plate of the capacitor is fixed and the other one is movable and placed 0001 inch. With the change in pressure due to blood flow, the distance between the metal plates changes and corresponding change in capacitance. The change in capacitance changes the frequency of a voltage controlled oscillator (VCO) and the altered frequency measurement is a way of direct pressure measurement. The prime advantage of capacitive measurement is that it is easy to be implemented in the blood vessel. But the capacitive pressure transducer is prone to change in temperature and lead wire resistance, and capacitance introduce error in measurement.

The other direct method pressure transducer is LVDT. The spring coupled core within the primary and secondary winding of a transformer changes its position with change in pressure due to blood flow. This change in position causes flux linkage to the secondary winding vary and change the voltage output at the secondary. Generally, a high frequency low voltage source is used to supply the current in the primary. This type of measurement system is advantageous for their high sensitivity and linearity.

## 36.9 BLOOD FLOW MEASUREMENT

Blood supply lower than required limit to a organ of human body is a diagnosis criteria of different diseases. Blood flow in an organ may be indirectly measured by Plethismograph, in which the measurement of change in volume of particular organ has been done which is indication of blood flow in that organ.

Although there are different direct methods of blood flow measurement, in which the popular methods are :

- Electromagnetic induction
- Utilising Doppler effect
- Thermal convection
- Radiography
- Dilution method.

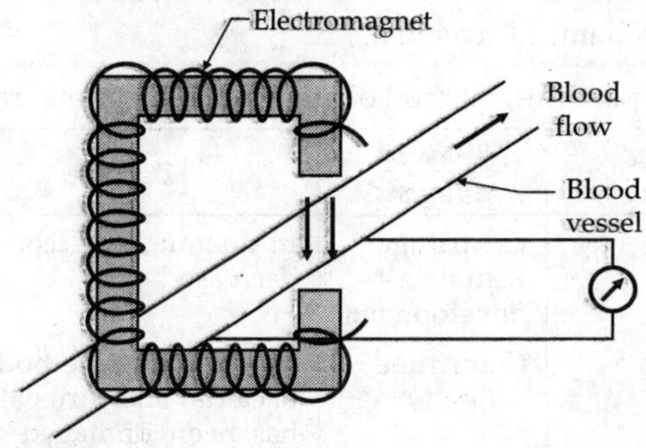


Fig. 36.14

### 36.9.1 Electromagnetic Induction

The electromagnet or permanent magnet is used to generate a magnetic field pattern and as blood is a conductive medium, there is a potential difference generated perpendicular to both blood flow and magnetic field due to flow of blood. This potential difference is generated due to magnetic induction and its magnitude is given by

$$\vec{E} = \vec{V} \times \vec{B}$$

When the potential difference is $e = \int_{l=0}^{d} \vec{E}.\vec{dl}$ where $d$ is the thickness of the blood vessel. The blood tube is designed by squeezing a blood vessel and passing needle to the two ends of the vessel. Hence this type of measurement needs surgery for catheterisation.

### 36.9.2 Blood Flow Measurement Utilising Ultrasonic Doppler Effect.

A ultrasonic metre is designed with an oscillator which operates at frequency 1–10 MHz and this oscillation excites a piezo electric crystal which generates a mechanical vibration at the frequency. This ultrasonic wave, when falls on the flow of blood at an angle, the reflected wave frequency has been increased or decreased by an amount, which depends on the flow direction and flow rate. Figure 36.15 shows the Doppler shift in frequency with direction. The Doppler signal

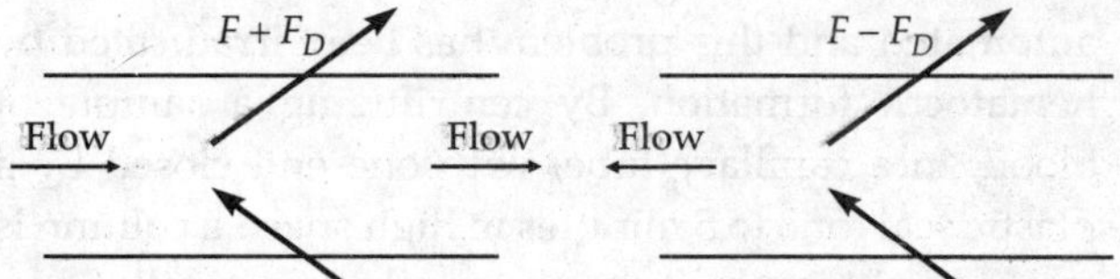


Fig. 36.15

$F_D$ is in the lower audio range of frequency and can be detected by a earphone. The Doppler metre generally can't discriminate between the direction of flow, but with modern circuitry which use two quadrature carrier ultrasonic wave, it can be discriminated.

### 36.9.3 Thermal Convection Measurement Method of Blood Flow

A lead thermistor is placed in the blood flow, the temperature of which is kept constant by servo controlled temperature controller. The thermistor temperature is fixed at 60°C and cooling due to blood flow happens. The electrical signal required to maintain the temperature of the thermistor at 60°C is the measure of blood flow.

### 36.9.4 Radiography

The blood velocity can be measured by X-ray imaging blood. This is possible by injecting a colouring medium into the blood vessel.

### 36.9.5 Dilution Method

The last method of blood flow measurement is Dilution method. This is the prime method to measure the blood speed rather than velocity. Visible spectroscopy is used to analyse the concentration of dye which is injected into the blood vessel.

## 36.10 PATHOLOGICAL TESTS

The clinical laboratory test for blood is henceforth discussed. Blood contains a serum (colourless liquid) in which different corpascular elements like WBC

(White blood cell) ; RBC (Red blood cell), Blood platelets or thermocytes of different count and concentration with different structure of the cells are present. There are a few basic ways to analyse blood :

1. by centrifuging the blood,
2. bacteriological test 3. Serological test and
4. Histological test.

### 36.10.1 Centrifugal Test

Centrifugal spinning is done with a blood sample so that blood plasma and blood cells are separated. Blood plasma or serum is separated in a yellowish viscous fluid. The blood plasma contains protein molecules and also fibrinogen which helps blood clotting. Centrifuging the blood sample is hard to be automated and this problem has been irradicated by hematocrit formation. By centrifuging a sample of blood, in a capillary tube, with one end closed by a plastic seal for 3 to 5 minutes in high speed a column is formed within the tube. The column is called the hematocrit. This colourwise identification of different blood cells and serums is one way of getting red or white corpasulor blood counts.

### 36.10.2 Bacteriological Test

The other ways of automated blood counts is the selective resistivity of different cell growth in the hematocrit. There is also way to find haemoglobin concentration in the blood. The membrane of red blood cell has been destroyed and haemoglobin coming out of cell has been subjected to chemical conversion and producing a coloured compound whose concentration has been analyzed by photo-spectro-photometry or colorimetry. The hematocrit colour has been shown in Fig. 36.16.

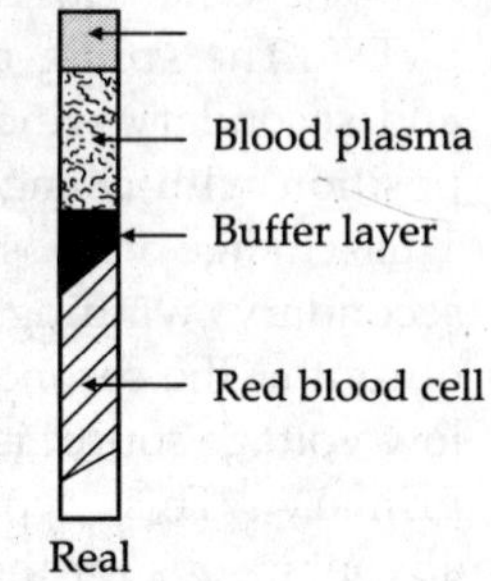


**Fig. 36.16**

### 36.10.3 Serological Test

If a living organism is exposed to either bacteria or virus or some infectious disease, antibody is formed in blood which defend the body against further infection of the same disease. If the antigen of the organism is added to the blood sample there is agglutination *i.e.*, clumping of blood which is visible under a microscope or send for analysis. The concentration of the antibody gives an idea of the disease and its diagnosis.

### 36.10.4 Histological Tests

Microscopical observation of the tissues of blood, with precise slicing, is named as Histology.

The blood serum has been analysed in a colorimetric detection after the serum has been treated with chemical reaction.

A chart has been presented which gives the different blood particles, method of analysis their normal range.

| | Blood particle | Structure (dia.) | Normal range (count) | Way of analysis | Disease |
|---|---|---|---|---|---|
| 1. | RBC | Round disk indented in the centre 8 μm dia | 4.5 – 5.5 million | Centrifuge, hematocrit development | In anemia RBC count is decreased |
| 2. | WBC | dia avg. 10 μm nucleus | 6000 – 10,000 | Centrifuge | Immunity of the body to bacterial and virus attack has been damaged with reducing of WBC count. |
| 3. | Thrombocycles | | 2 lack to 8 lack | Centrifuge | Blood clotting disturbed |
| 4. | Serum | | | | |
| | (*a*) Blood Urea nitrogen (BUN) | | 8 – 16 mg N/100 ml | | |
| | (*b*) Glucose | | 70 – 9 mg/100 ml | Chemical reaction and colorimetry Eq./litre | |
| | (*c*) Phosphate | | 3.4 – 4.5 D | | |
| | (*d*) Na, K, Cl | | | | |
| | (*e*) $CO_2$ | | 24 – 32 mg/litre → nm | | |
| | (*f*) Ca | | 0 – 11.5 mg/100 ml | | |
| | (*g*) Uric acid | | 3 – 6 mg/100 ml | | |
| | (*h*) Protein | | 6 – 8 g/100 ml | | |
| | (*i*) Albumin | | 4 – 6 g/100 ml | | |
| | Cholesterol | | 160 – 200 ng/100 ml | | |
| | Bilirrulein | | 2 – 1 mg/100 ml | | |

## 36.11 EMG (ELECTROMYOGRAM)

Electromyogram is the way to measure the action potential developed in the muscle fibre of the human body. The characteristic of muscle fibre is that the action potential generated in one fibre is not transmitted to other fibre like action potential in neurone but the action potential developed in neurone is of same magnitude of that generated in muscle fibre. The electrical activity of the whole muscle rather than a single muscle fibre is of interest. As the total electrical activity of the muscle strain is the weighted average of the electrical activity of all the fibre in strain weighted by the distance of the fibre from the electrode placing. Hence there is a relation between Electromyogram and the total muscle contraction. The Electromyographic voltage developed when integrated over time gives the isometric voluntary tension in the muscle. The EMG waveform has a peak amplitude in the range 50 μV to 1 mV which depends on the position of the electrode placing. The EMG activity looks like noise whose frequency ranges from about 100 Hz to 3 kHz.

The electrodes used for muscle fibre electrical activity measurement may be surface or needle electrode or wire electrode. For localised measurement needle or wire electrode is used, whereas, for the measurement of gross electrical activity of the muscle the surface electrode is used. The electromyogram display is generally shown by electromyograph on the oscilloscope screen. The pen or chart recorders are generally not used for EMG display. Also, the used equipments to monitor EMG are speakers.

For qualitative study of EMG the amplitude and the no. of repeatations of electrical activity of muscle is noted and the muscle activity can be measured by these parameters. The amplitude of the EMG is not a clear indication of the muscle activity as the amplitude also depends on the position of electrode placing of the muscle. The other type of measurement of muscle activity done by EMG is done integrating the Electromyogram over time. In this method the integrated value of EMG is recorded and plotted.

## 36.12 SPIROMETER

A spirometer is an apparatus for measuring the volume of air inspired and expired by the lungs. It is a precision differential pressure transducer for the measurement of respiration flow rates. The spirometer records the amount of air and the rate of air that is breathed in and out over a specified time. The Spirometer and attached flow head function together as a pneumotachometer, with an output signal proportional to airflow. It was invented by John Hutchinson in 1846.

The output produced by a spirometer is called a kymograph trace. From this, vital capacity, tidal volume, breathing rate and ventilation rate (= tidal volume × breathing rate) can be calculated. From the overall decline on the graph, the oxygen uptake can also be measured.

In humans and other mammals, the respiratory system consists of the airways, the lungs, and the respiratory muscles that mediate the movement of air into and out of the body. Within the alveolar system of the lungs, molecules of oxygen and carbon dioxide are passively exchanged, by diffusion, between the gaseous environment and the blood. Thus, the respiratory system facilitates oxygenation of the blood with a concomitant removal of carbon dioxide and other gaseous metabolic wastes from the circulation. The system also helps to maintain-the acid-base balance of the body through the efficient removal of carbon dioxide from blood.

## 36.13 ANATOMY

The respiratory system can be conveniently subdivided into an upper respiratory tract and lower respiratory tract, trachea and lungs. These sections are also called the conducting zone and a respiratory zone.

The conducting zone starts with the nares (nostrils) of the nose, which open into the nanospharynx (nasal cavity). The primary functions of the nasal passages are to :

- filter,
- warm,
- moisten, and
- provide resonance in speech.

The nasophamyx opens into the oropharynx (behind the oral cavity). The oropharynx leads to the laryngopharyn, and empties into the larynx (voicebox), which contains the vocal cords, passing through the glottis, connecting to the trachea (wind pipe) which leads down to the thoracic cavity (chest) where it divides into the right and left "main stem" bronchi. The subdivision of the bronchus are : primary, secondary, and tertiary divisions (first, second and third levels). In all, they divide 16 more times into even smaller bronchioles.

The bronchioles lead to the respiratory zone of the lungs which consists of respiratory bronchioles, alveolar ducts and the alveoli, the multi-lobulated sacs in which most of the gas exchange occurs.

Ventilation of the lungs is carried out by the muscles of respiration. Inhalation is initiated by the diaphragm and supported by the external intercostal muscles. Normal resting respirations are 10 to 18 breaths per minute. During vigorous inhalation (at rates exceeding 35 breaths per minute), or in approaching respiratory failure, accessory muscles of respiration are recruited for support. These consist of sternocleidomastoid, platysma, and the strap muscles of the neck. Exhalation is generally a passive process, however active or forced exhalation is achieved by the abdominal and the internal intercostal muscles.

Air moves through the body in the following order :

- Nostrils
- Nasal cavity
- Pharynx (naso-, oro-, laryngo-)
- Laryr.. (voice box)
- Trachea (wind pipe)
- Thoracic cavity (chest)
- Bronchi (right and left)
- Alveoli (site of gas exchange)

Ventilation occurs under the control of the autonomic nervous system from the part of the brain stem, the medulla oblongata and the pons. This area of the brain forms the respiration regulatory centre, a series of interconnected neurons within the lower and middle brain stem which coordinate respiratory movements. The sections are the pneumotaxic center, the appneustic center, and the dorsal and ventral respiratory groups. This section is especially sensitive during infancy, and the neurous can be destroyed if the infant is dropped or shaken violently.

## Circulation

The right side of the heart pumps blood from the right ventricle through the pulmonary semilunar valve into the pulmonary trunk. The trunk branches into right and left pulmonary arteries to the pulmonary blood vessels. The vessels generally accompany the airways and also undergo numerous branchings. Once the gas exchange process is complete in the pulmonary capillaries, blood is returned to the left side of the heart through four pulmonary veins, two from each side. The pulmonary circulation has a very low resistance, due to the short distance within the lungs, compared to the systemic circulation, and for this reason, all the pressures within the pulmonary blood vessels are normally low as compared to the pressure of the systemic circulation loop.

## Functions

The major function of the respiratory system is gas exchange. Respiration consists of a mechanical cycle of inhalation and exhalation, with gaseous exchange occurring in between. As gas exchange occurs, the acid-base balance of the body is maintained as part of homeostasis. If proper ventilation is not maintained two opposing conditions could occur :

1. respiratory acidosis, a life threatening condition, and
2. respiratory alkalosis.

Inhalation is driven primarily by the diaphragm. When the diaphragm contracts, the ribcage expands and the contents of the abdomen are moved downward. This results in a larger thoracic volume, which in turn causes a decrease in intrathoracic pressure. As the pressure in the chest falls, air moves into the conducting zone. Here, the air is filtered, warmed, and humidified as it flows to the lungs.

Exhalation, on the other hand, is typically a passive process. The lungs have a natural elasticity ; as they recoil from the stretch of inhalation, air flows back out until the pressures in the chest and the atmosphere reach equilibrium.

A simple model of how the lungs are inflated can be built from a bell jar.

During forced inhalation, as when taking a deep breath, the external intercostal muscles and accessory muscles further expand the thoracic cavity.

During forced exhalation, as when blowing out a candle, expiratory muscles including the abdominal muscles and internal intercostal muscles, generate abdominal and thoracic pressure, which forces air out of the lungs.

Upon inhalation, gas exchange occurs at the alveoli, the tiny sacs which are the basic functional components of the lungs. The alveolar walls are extremely thin (approx. 0.2 micrometers), and are permeable to gases. The alveoli are lined with pulmonary capillaries, the walls of which are also thin enough to permit gas exchange. All gases diffuse from the alveolar air to the blood in the pulmonary capillaries, as carbon dioxide diffuses in the opposite direction, from capillary blood to alveolar air. At this point, the pulmonary blood is oxygen-rich, and the lungs are holding carbon dioxide. Exhalation follows, thereby ridding the body of the carbon dioxide and completing the cycle of respiration.

## 36.14 OTHER TESTS : SPIROMETRY

Spirometer (meaning the measuring of breath) is the most common of the Pulmonary Function Tests (PFTs), measuring lung function, specifically the measurement of the amount (volume) and/or speed (flow) of air that can be inhaled and exhaled. Spirometry is an important tool used for assessing conditions such as asthma, cystic fibrosis, and COPD.

Spirometer maneuvres include slow vital capacity (SVC), forced vital capacity (FVC), tidal volume (TV) and maximum voluntary ventilation (MVV). The measurement of forced vital capacity is the most commonly performed maneuvre, sometimes in conjunction with slow vital capacity and/or tidal volume. A plethysmograph can be used to measure functional residual capacity (FRC).

Results are usually given in the both raw data (liters, liters per second) and percent predicted — the test result as a percent of the "predicted values" for the patients of similar characteristics (height, weight, age, sex, and sometimes race). The interpretation of the results can vary depending on the physician and source of the predicted values, but generally speaking, result to 100% predicted are the most normal, and generally results over 80% are often considered normal. However, review by a doctor is necessary for accurate diagnosis of any individual situation. The spirometry test is performed using a device called a spirometer, which comes in many different varieties. Many produce a graph called a Flow-Volume Loop, which graphically depicts the flow of air compared to the total volume inspired or expired. Flow is listed along the Y-axis and volume along the X-axis. Most spirometers also display a Volume—Time curve, showing volume (liters) along the Y-axis and time (seconds) along the X-Axis (not shown).

Flow—Volume loop showing successful FVC maneuver. Positive values represent expiration, negative values represent inspiration. The trace moves clockwise for expiration followed by inspiration.

The basic FVC test varies slightly depending on the equipment used. Generally, the patient is asked to take the deepest breath they can, and then exhale into the pressure sensor as hard as possible, for as long as possible. It is sometimes directly followed by a rapid inhalation (inspiration), in particular when assessing possible inspiratory obstruction or restriction. Sometimes, the test will be preceded by a period of quiet breathing in and out from the sensor (tidal volume), or the rapid breath in (forced inspiratory part) will come before the forced exhalation. During the test, soft nose clips may be used to prevent air escaping through the nose. Filter mouthpieces may be used to prevent the spread of germs, particularly for inspiratory maneuvers.

Sometimes, to assess the reversibility of a particular condition, a bronchodilator is administered before performing another round of tests for comparison. This is commonly referred to as a reversibility test, or a post bronchodilation test (Post BD), and is an important part in diagnosis asthma versus COPD.

Spirometery can also be part of a bronchial challenge test, used to determine bronchial hyper-responsiveness to either rigorous exercise, inhalation of cold/dry air, or with a pharmaceutical agent such as metacholine or histamine.

Due to the patient cooperation required, spirometry can only be used on children old enough to comprehend and follow the instructions given (typically about 4-5 years old), and only on patients that are able to understand and follow instructions — thus, this test is not suitable for patients that are able to understand and follow instructions — thus, this test is not suitable for patients that are unconscious, heavily sedated, or have limitations that would interfere with vigorous respiratory efforts. Other types of PFTs are available for infants and unconscious persons.

### 36.14.1 Basic Spirometer Sensors

- Volumetric Spirometers
  - Water bell
  - Bellows wedge
- Flow Measuring Spirometers
  - Fleisch – pneumotach
  - Lilly (screen) pneumotach
  - Turbine
  - Pilot tube
  - Hot-wire anemometer
  - Ultrasound

In an average resting adult, the lungs take up about 250 ml of oxygen every minute while excreting about 200 ml of carbon dioxide. During an average breath, an adult will exchange from 500 ml to 700 ml of air. This average breath capacity is called **tidal volume.**

## 36.15 DISEASES OF THE RESPIRATORY SYSTEM

Diseases of the respiratory system can be classified into *four* general areas :

- Obstructive Diseases
  (*e.g.*, Emphysema, Bronchitis, Asthma)
- Restrictive Diseases
  (*e.g.*, Fibrosis, Sarcoidosis, Alveolar Damage, Pleural Effusion)
- Vascular Diseases
  (*e.g.*, Pulmonary Edema, Pulmonary Embolism, Pulmonary Hypertension)
- Infectious, Environmental and Other Diseases (*e.g.*, Pneumonia, Tuberculosis, Asbestosis, Particulate Pollutants).

## 36.16 RESPIRATORY SYSTEM IN PLANTS

Plant respiration is limited by the process of diffusion. As well as using photosynthesis, plants take in oxygen (although very little of it) through their stomata, holes on the undersides of their leaves. However, even a baobab tree is mostly dead because air can penetrate only skin deep. However, most plants are not involved in highly metabolic activities like hunting, *i.e.*, they do not need the energy necessary for predators, and thus their breathing is limited.

## 36.17 TISSUE ENGINEERING

In tissue engineering, respiration is an essential problem. The small depth of diffusion respiration sufficient to support the metabolism of an average human cell is less than a millimeter (0.04 in). However, various substances can be used to enhance this depth.

## 36.18 DOPPLER SONOGRAPHY

**Colour Doppler of Common Cartoid Artery.** Sonography can be enhanced with Doppler measurements, which employ the Doppler effect to assess whether structures (usually blood) are moving towards or away from the probe, and its relative velocity. By calculating the frequency shift of a particular sample volume, for example, a jet of blood flow over a heart valve, its speed and direction can be determined and visualised. This is particularly useful in cardiovascular studies (sonography of the vasculature system and heart) and essential in many areas such as determining reverse blood flow in the liver vasculate in portal hypertension. The Doppler information is displayed graphically using spectral Doppler, or as an image using colour Doppler (directional Doppler) or power-Doppler (non directional Doppler). This Doppler shift falls in the audible range and is often presented audibly using stereo speakers : this produces a very distinctive, although synthetic, pulsing sound.

Strictly speaking, most modern ultrasound machines do not use the Doppler effect to measure velocity, as they rely on pulsed wave Doppler (PW). Pulsed wave machines transmit pulses of ultrasound, and then switch to receive mode. As such, the reflected pulse that they receive is not subject to a frequency shift, as the isolation is not continuous. However, by making several measurements, the phase change in subsequent measurements can be used to obtain the frequency shift (since frequency is the rate of change of phase). To obtain the phase shift between the received and transmitted signals, one of two algorithms is typically used the Kasai algorithm or the cross correlation. Older machines, that use continuous wave (CW) Doppler, exhibit the Doppler effect as described above. To do this, they must have separate transmission and reception transducers. The major drawback of CW machines, is that no distance information can be obtained (this is the major advantage of PW systems — the time between the transmitted and received pulses can be converted into a distance with knowledge of the speed of sound).

### 36.18.1 Strength of Sonography

- It images muscle and soft tissue very well and is particularly useful for delineating the interfaces between solid and fluid-filled spaces.
- It renders "live" images, where the operator can dynamically select the most useful section for diagnosing and documenting changes, often enabling rapid diagnosis.
- It shows the structure of organs.
- It has no known long-term side effects and rarely causes any discomfort to the patient.
- Equipment is widely available and comparatively flexible.
- Small, easily carried scanners are available ; examinations can be performed at the bedside.
- Relatively inexpensive compared to other modes of investigation (*e.g.*, computed X-ray tomography, DEXA or magnetic resonance imaging).

### 36.18.2 Weaknesses of Ultrasound Imaging

- Large body habitus, obese patients limit image quality as the overlying adipose tissue (fat) scatters the sound and greater depth the sound waves need to travel attenuate or weaken the signal on transmission and reflection back to the transducer. A fetus close to the surface will be imaged at a higher resolution than those at greater distance to the skin surface.
- Ultrasound devices have trouble-penetrating bone. For example, ultrasound imaging of the brain is very limited.
- Ultrasound can detect fluid surrounding the lung (pleural effusion) but the high impedance mismatch between the solid tissues and the air filled lungs limits image.
- Ultrasound performs very poorly when there is a gas between the scan head and the organ of interest, due to the extreme differences in acoustical impedance. For example, overlying gas in the gastro-intestinal tract often makes ultrasound scanning of the pancreas difficult, and lung imaging is not possible.
- Even in the absence of bone or air, the depth penetration of ultrasound is limited, making it difficult to image structures deep in the body, especially in obese patients.
- The method is operator-dependent. A high level of skill and experience is needed to acquire good quality images and make accurate diagnoses.
- There is no scout image as there is with CT and MRI. Once an image is acquired there is no exact way to tell which part of the body was imaged.

### 36.18.3 Dangers of Ultrasound Imaging

There have been disputes whether ultrasound is safe. Since ultrasound is energy, there are questions such as "What are the energy waves doing to the tissue". A meta-analysis of several ultrasonography studies were performed showing that there were no statistical significant harmful effects from ultrasonography. This however does not rule out the possibility that harmful effects are present, although they must be so small as not to show up in the sample sizes of choice in the studies.

There may be a number of side-effects, including but not necessarily limited to :

- Heat development : Local tissue absorbs the ultrasound energy and increases the temperature of those tissues.
- Bubble formation : dissolved gases come out of the solution due to local heat increases.

### 36.18.4 Uses of Sonography

Sonography (ultrasonography) is widely utilized in medicine. It is possible to perform diagnosis or therapeutic procedures with the guidance of sonography (for instance, biopsies or drainage of fluid collections). Sonographers are medical professionals who perform scans for diagnostic purposes, they work with specialized doctors called sonologists who provide reports of the image obtained. Sonographers typically used a hand-held probe (called a transducer) that is placed directly on and moved over the patient, a water-based gel ensures good coupling between the patient and scan head. This eliminates the air between the transducer head and the skin. Ultrasound image quality is limited by the amount of overlying adipose (fat) tissue, as fatty tissue tends to scatter the sound and greater depth leads to attenuation of the sound beam. Superficial structures such as muscles, tendons, testes, breast and neonatal brain are imaged at a higher frequency (7 – 15 MHz), which provides better axial and lateral resolution. Deeper structures such as liver and kidney are imaged at a lower frequency 1–6 MHz with lower arial and lateral resolution but greater pevetration. Medical sonography is used in, for example :

- Cardiology
- Endocrinology
- Gastroenetrology
- Gynaecology
- Obstetrics
- Ophthalmology
- Urology

### 36.18.5 Instrumentation of Sonography

Ultrasonography (sonography) uses a probe containing one or more acoustic transducers to send pulses or sound into a material (1–15 MHz). Whenever a sound wave encounters a material with a different acoustical impedance, part of the sound wave is reflected, which the probe detects as an echo. The time it takes for the echo or travel back to the probe is measured and used to calculate the depth of the tissue interface causing the echo. The greater the difference between acoustic impedances, the larger the echo is. The difference between gases and solids is so great that most of the acoustic energy is reflected, and so imaging of objects beyond that region is not possible.

The speed of sound is different in different materials, and is dependent on the acoustical impedance of the material. However, the ultrasound scanner assumes that the acoustic velocity is constant at 1540 m/s. Although part of the acoustic energy is lost every time an echo is formed, this effect is small compared to the attenuation of sound due to absorption.

To generate a 2D-image, the ultrasound beam is swept, either mechanically, or electronically using a phased array of acoustic transducers. The received data are processed and used to construct the image.

The frequencies used for medical imaging are generally in the range of 1 to 13 MHz. Higher frequencies have a correspondingly shorter wavelength, and so images can have a greater resolution. However, the attenuation of the sound wave is increased at higher frequencies, so in order to have better penetration of deeper tissues, a lower frequency (3 – 5 MHz) is used.

Most sonography machines can also produce colour images of sorts. This is done simply by assigning a range of colours to encode the amplitudes of the received echoes. In addition, 3-D images can be generated by acquiring a series of 2D images, often using a specialised probe.

## 36.19 MICROBUBBLES

The use of microbubble contrast media in medical sonography to improve ultrasound signal backscatter is known as contrast enhanced ultrasound. This technique is currently used in echocardiography, and may have future applications in molecular imaging and drug delivery.

## 36.20 COMPUTED TOMOGRAPHY IMAGING (CT SCAN, CAT SCAN)

### 36.20.1 Use of CT Scan

Computed Tomography (CT) imaging, also known as "CAT scanning" (Computed Axial Tomography), combines the use of a digital computer together with a rotating X-ray device to create exposed X-radiation detailed cross sectional images of the different organs and body parts such as the lungs, liver, kidneys, pancreas, pelvis, extremities, brain and blood vessels. For many patients, CAT can be performed on an outpatient basis.

Among the various techniques such as MR and X-ray, CAT has the unique ability to image a combination of soft tissue, bone and blood vessels. For example, X-ray imaging of the head can only show the intense bone structures of the skull. X-ray angiography of the head only depicts the blood vessels of the head and neck and not the soft brain-tissue. Magnetic resonance (MRI) imaging does an excellent job of showing soft tissue and blood vessels, but MRI does not give as much detail of bony structures such as the skull. CAT images of the head allow physicians to see soft-tissue a structures like the brain's ventricles or gray and white matter. Experts selectively "window" the digital CT images on the computer monitor to look at tissue, then the bone and then the blood vessels.

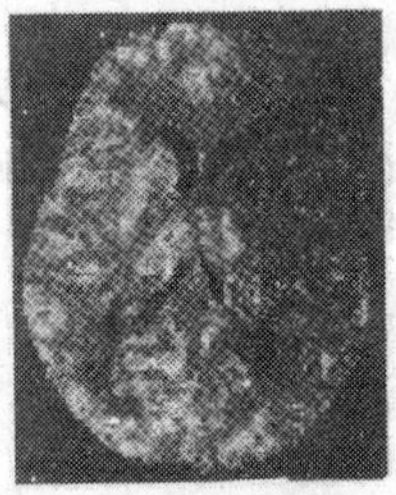

**Fig. 36.17** CT image of the brain.

CT is one of the best tools for studying the lungs and abdomen. CT is used in the cancer diagnosis process and is often the preferred method to diagnose liver and pancreas cancer. CT imaging is finding its use in the detection, diagnosis and treatment of heart disease, acute stroke and vascular diseases which can lead to stroke, gangrene or kidney failure. Additionally, CT Scan is used to measure bone mineral density for the detection of osteoporosis. CT is applied in trauma cases. All dedicated shock-traur have a CT scanner in the trauma department.

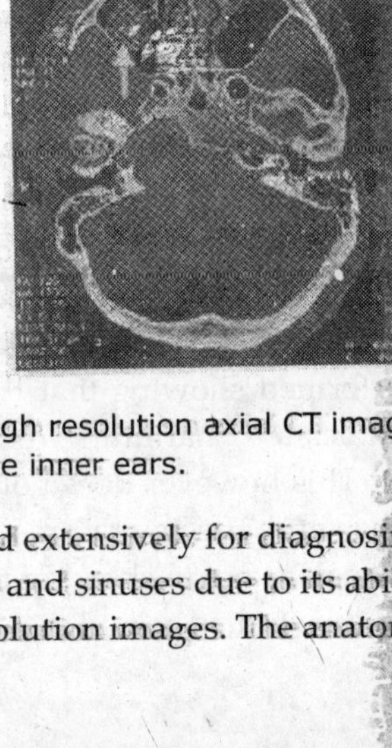

**Fig. 36.18** High resolution axial CT image of the inner ears.

CT is used extensively for diagnosing problems of the inner ears and sinuses due to its ability to generate very high resolution images. The anatomy of the inner ear sinuses is made up of delicate soft tissue structures and very fine bones. CT is used for imaging tumours in the sinuses and diseases that cause degeneration of small bones in the inner ear.

CT can provide detailed cross-sectional images and diagnostic information for every part of the body including :

- the neck, shoulders, cervical spine and blood vessels of the neck.
- the brain, vessels of the brain, eyes, inner ear, sinuses
- the chest, heart, aorta, lungs, mediastinum
- the upper abdomen, liver, kidney, spleen, pancreas and other abdomen
- the thoracic and lumbar spine
- the skeletal system including bones of the hands, feet, ankles, legs and jaws
- the pelvis and hips, male and female reproductive system, bladder,

## 36.20.2 Working Principle of CT Scan

Computed Tomography is based on the X-ray principle : as X-rays pass through the body they are absorbed or attenuated at differing levels creating a matrix or profile of X-ray beams of different strength. This X-ray profile is registered on film, thus creating an image in a computer. In the case of CT, the film is replaced by a banana shaped detector which measures the X-ray profile.

**Fig. 36.19** Inside view of modern CT system, the X-ray tube is on the top at the 1 o'clock position and the arc-shaped CT detector is on the bottom at the 7 o'clock position. The frame holding the X-ray tube and detector rotate around the patient as the data is gathered.

A CT scanner looks like a big square. The patient aperture (opening) is 60 cm to 70 cm (24" to 28") in diameter. Inside the covers of the CT scanner is a rotating frame which has an X-ray tube mounted on one side and detector mounted on the opposite side. A fan beam of X-ray is created as the rotating frame spins the X-ray tube and detector around the patient (Fig. 36.20). Each time the X-ray tube and detector make a 360° rotation, an image has been acquired. This image is collimated (focused) to a thickness between 1 mm and 10 mm using lead shutters in front of the X-ray tube and X-ray detector.

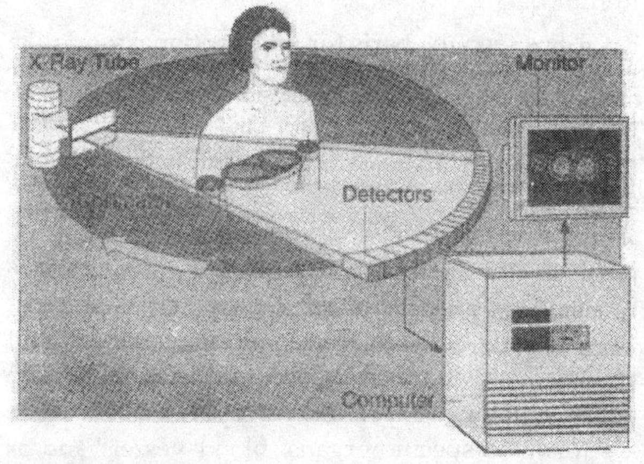


**Fig. 36.20** Showing relationship of X-ray tube, patient, detector, and image reconstruction computer and display monitor.

As the X-ray tube and detector make this 360° rotation, the detector takes numerous snapshots (called profiles) of the attenuated X-ray beam. Typically, in one 360 lap, about 1,000 profiles are sampled. Each profile is subdivided spatially by the detectors and fed into about 700 individual channels. Each profile is then backward projected by a dedicated computer into a two-dimensional image of the slice that was scanned.

Sometimes it is necessary to not drink anything for an hour to several hours before the exam. The preparation time varies depending on the actual exam as well as the imaging center's requirements.

## 36.20.3 Applications and Clinical Benefits of CT Imaging

Unlike other medical imaging techniques, such as conventional X-ray imaging (radiography), CT enables direct imaging and differentiation of soft tissue structures, such as liver, lung tissue, and fat. CT is especially useful in searching for large space occupying lesions, tumours and metastasis and can not only reveal their presence, but also the size, spatial location and extent of a tumour.

CT imaging of the head and brain can detect tumours, show blood clots and blood vessel defects, show enlarged ventricles (caused by a build up of cerebrospinal fluid) and image other abnormalities such as those of the nerves or muscles of the eye.

Due to the short scan times of 500 milliseconds to a few seconds, CT can be used for all anatomic regions, including those susceptible to patient motion and breathing. For example, in the thorax CT can be used for visualization of modular structures, infiltrations of fluid, fibrosis (for example from asbestos fibres), and effusions (filling of an air space with fiuid).

CT has seen the basis for interventional work like CT guided biopsy and minimally invasive therapy. CT images are also used as basis for planning radiotherapy cancer treatment. CT is also often used to follow the course of cancer treatment to determine how the tumour is responding to treatment.

For many computer tomography examinations patients may be asked to take a special contrast agent (orally, rectally or via injection). Intravenous, oral and rectal CT contrast are pharmaceutical agents (liquids) and are sometimes referred to as "dye". CT contrast is used to make specific organs, blood vessels and/or tissue types "stand out" with more image contrast to better show the presence of disease or injury. Thus CT contrast highlights specific areas of the resultant CT image or "dyes" it.

There are four types of contrast agents used in CT :

1. The type that is given via intravenous (through a vein) injection
2. The type that is given orally
3. The type that is given rectally
4. A much less common type of contrast used in CT is inhaled as a gas and used for special lung and brain imaging. This technique (called Xenon CT) is only available at a small number of locations throughout the world and is only performed for rare cases.

### CT Contrast Given via Intravenous Injection

Intravenous contrast is used in CT to help highlight blood vessels and to enhance the tissue structure of various organs such as the brain, spine, liver and kidneys. "Intraveneous" means that the contrast is injected into a vein using a small needle. Some imaging exams of the abdomen and gastrointestinal system use both the intraveneous iodine and orally administered barium contrast for maximum sensitivity. The intraveneous CT contrast is clear like water and has a similar consistency. It is typically packaged in glass bottle or vial. A sterile syringe is used to draw it from the bottle or a power injector is used to administer the contrast. Typically between 75 cc to 150 cc (about 2.5 oz. to 5 oz) of contrast is used depending upon the patient's age, weight, area being imaged and cardiovascular health.

### 36.20.4 Intravenous CT Contrast

A small needle is first placed into a vein in the hand or arm by the radiologist, technologist or a nurse and held in place with tape or a strap. Once the needle is in place, the vein is flushed with saline solution. Typically the contrast is loaded into a power-assisted injector, which injects the CT contrast using tubing through the needle into the body during a specific period in the CT exam. The injection is fully under the control of the technologist or radiologist. The injector is either mounted on a small trolley or hung from a ceiling mounted suspension next to the CT scanner. The contrast may also be hand injected using a large syringe connected to the needle via tubing.

Once the iodine contrast has been injected into the blood stream, it circulates through the heart and passes into the arteries, through the body's capillaries and then into the veins and back to the heart. As CT images are being acquired, the CT's X-ray beam is attenuated (weakened) as they pass through the blood vessels and organs flush with the contrast. This causes the blood vessels and organs filled with the contrast to "enhance" and show up as white areas on the X-ray or CT images. The kidneys and liver eliminate the contrast from the blood.

### 36.20.5 Use of Computers in CT Scan

Multiple computers are used to control the entire CT system. The main computer that combines the operation of the entire system is called the "host computer". There is also a dedicated computer that reconstructs the "raw CT data" into an image. A workstation with a mouse, keyboard and other dedicated controls allows the technologist to control and monitor the exam. The CT gantry and table have multiple microprocessors that control the rotation of the gantry, movement of the table (up/down and in/out), tilting of the gentry for angled images, and other functions such as turning the X-ray beam on and off.

## 36.21 MAGNETIC RESONANCE IMAGING (MRI)

Magnetic resonance imaging was developed from knowledge gained in the study of nuclear magnetic resonance. The original name for the medical technology is nuclear magnetic resonance imaging (NMRI), but the word nuclear is almost universally dropped. This is done to avoid the negative connotations of the word nuclear, and to prevent patients from associating the examination with radiation exposure, which is not one of the safety concerns for MRI.

### Difference between MRI and CT

A CT scanner uses ionizing radiation, X-rays, to acquire its images, making it a good tool for dense tissue (bone) exams. MRI, on the other hand, uses radio frequency signals to acquire its images, and is best suited for soft (non-calcified) tissue exams.

### 36.21.1 MRI-Technique

Medical MRI most frequently relies on the relaxation properties of excited hydrogen nuclei in water. When the object to be imaged is placed in a powerful, uniform magnetic field the spins of the atomic nuclei with non-zero spin numbers (essentially, an unpaired proton or neutron) within the tissue all align in one of two opposite directions parallel to the magnetic field or antiparallel. Common magnetic field strengths range from 0.3 to 3 teslas, although research instruments range as high as 20 teslas, and commercial suppliers are investing in 7 tesla platforms. (For comparison, average Earth's magnetic field is around 50 μT, that is more than 100000 times lower.)

An excess of only one in a million nuclei align themselves with the magnetic field since the thermal energy far exceeds the difference between the parallel and antiparallel states. Yet the vast quantity of nuclei in a small volume sum to produce a detectable change in field. Most basic explanations of NMR and MRI will say that the nuclei align parallel or anti-parallel with the static magnetic field though, because of quantum mechanical reasons beyond the scope of this article, the individual nuclei are actually set off an angle from the direction of the static magnetic field. The bulk collection of nuclei can be partitioned into a set whose sum spin are aligned parallel and a set whose sum spin are anti-parallel.

The magnetic dipole moment of the nuclei then precesses around the axial field. While the proportion is nearly equal, slightly more are oriented at the low energy angle. The frequency with which the dipole moments precess is called the Larmor frequency. The tissue is then briefly exposed to pulses of electromagnetic energy (RF pulses) in a plane perpendicular to the magnetic field, causing some of the magnetically aligned hydrogen nuclei to assume a temporary non-aligned high-energy state. The frequency of the pulses is governed by the Larmor equation.

In order to selectively image different voxels (volume picture elements) of the subject, orthogonal magnetic gradients are applied. Although it is relatively common to apply gradients in the principal axes of a patient (so that the patient is imaged in $x$, $y$, and $z$ from head to toe), MRI allows completely flexible orientations for images. All spatial encoding is obtained by applying magnetic field gradients which encode position within the phase of the signal. In one dimension, a linear phase with respect to position can be obtained by collecting data in the presence of a magnetic field gradient. In three dimensions (3D), a plane can defined by "slice selection", in which an RF pulse of defined bandwidth is applied in the presence of a magnetic field gradient in order to reduce spatial encoding to two dimensions (2D). Spatial encoding can then be applied in 2D after slice selection, or in 3D without slice selection. In either case, a 2D or 3D matrix of spatially-encoded phases is acquired, and these data represent the spatial frequencies of the images object. Images can be created from the acquired data using the discrete Fourier transform (DFT).

In order to understand MRI contrast, it is important to have some understanding of the time constants involved in relaxation processes that establish equilibrium following RF excitation. As the high-energy nuclei relax and realign, they emit energy at rates which are recorded to provide information about their environment. The realignment of nuclear spins with the magnetic field is termed longitudinal relaxation and the time (typically about 1 sec) required for a certain percentage of the tissue nuclei to realign is termed "Time 1" or $T_1$. $T_2$ weighed imaging relies upon local dephasing of spins following the application of the transverse energy pulse ; the transverse relaxation time (typically < 100 ms for tissue) is termed "Time 2" or $T_2$. A subtle but important variant of the $T_2$ technique is called $T_2{}^*$ imaging. $T_2$ imaging employs a spin echo technique, in which spins are refocused to compensate for local magnetic field inhomogeneities. $T_2{}^*$ imaging is performed without refocusing. This sacrifices some image integrity in order to provide additional sensitivity to relaxation processes that cause incoherence of transverse magnetization.

Image contrast is created by using a selection of image acquisition parameters that weights signal by $T_1$, $T_2$ or $T_2{}^*$, or no relaxation time ("proton-density images"). In the brain, $T_1$-weighting causes fibre tracts (nerve connections) to appear white, congregations of neurons to appear gray, and cerebrospinal fluid to appear dark. The contrast of "white matter", "gray

36

matter" and "cerebrospinal field" is reversed using $T_2$ or $T_2$ imaging, whereas proton-weighted imaging provides little contrast in normal subjects. Additionally, functional information (CBF, CBV, blood oxygenations) can be encoded within $T_1$, $T_2$ or $T_2$*, see functional MRI (fMRI) and the section below.

Diffusion Weighted Imaging (DWI) uses very fast scans with an additional series of gradients (diffusion gradients) rapidly turned on and off. Products from water diffusing randomly within the brain, via Brownian motion, lose phase coherence and , thus, signal during application of diffusion gradients. Within acutely infarcted brain, water diffusivity is impaired, and signal loss on DWI sequences is less than in normal brain. DWI is the most sensitive method of detecting cerebral infraction (stroke) and can identify an infarct within 30 minutes of ictus.

Typical medical resolution is about 1 $mm^3$, while research models can exceed 1 $\mu m^3$.

### 36.21.2 Contrast Enhancement

Both $T_1$ and $T_2$-weighted images are acquired for most medical examinations. However, these 2 sets of images are not always sufficient to adequately show anatomy or pathology. One option is to use a more sophisticated image acquisition technique *i.e.,* fast suppression, chemical-shift imaging. The other is to administer a contrast agent to delineate areas of interest.

A contrast agent may be as simple as water, taken orally, for imaging the stomach and small bowel. Alternatively, substances with specific magnetic properties may be used.

Most commonly, a paramagnetic contrast agent (usually a gadolinium compound) is given. Gadolinium-enhanced tissues and fluids appear extremely bright on $T_1$-weighted images. This provides high sensitivity for detection of vascular tissues (*e.g.,* tumours) and permits assessment of brain perfusion (*e.g.,* in stroke).

More recently, superparamagnetic contrast agents (*e.g.,* iron oxide nanoparticles) have become available. These agents appear very dark on $T_2$* weighted images. These agents maybe used for liver imaging - normal liver tissues retains the agent, but abnormal areas (*e.g.,* scans, tumours) do not. They can also be taken orally, to improve visualisation of the gastrointestinal tract, and to prevent water in the gastrointestinal tract from obscuring other organs (*i.e.,* pancreas).

Diamagnetic agents *i.e.,* barium sulphate have been studied for potential use in the Gl tract, but are less frequently used.

### 36.21.3 The *k*-Space formalism

In 1983 Ljunggren and Tweig independently introduced the *k*-space formalism, a technique that proved invaluable in unifying different MR imaging techniques. They showed that the demodulated MR signal $S(r)$ generated by spins freely processing in the presence of a linear magnetic field gradient $G$ equals the Fourier transform of the effective spin density $\rho_{eff}$ *i.e.,*

$$S(t) = \tilde{\rho}_{\text{effective}}(\vec{k}(t)) \equiv \int d^3x\, \rho(\vec{x}).e^{2\pi_2 \vec{k}(t).\vec{x}}$$

where, $\vec{k}(t) \equiv \int_0^t \vec{G}(t')dt'$

In other words, as time progresses the signal traces out a trajectory in *k*-space with the velocity vector of the trajectory proportional to the vector of the applied magnetic field gradient. By the term *effective spin density* we mean the true spin density $\rho(\vec{x})$ connected for the effects of $T_1$ preparation, $T_2$ decay, dephasing due to field inhomogeneity, flow, diffusion, etc. and any other phenomena that affect that amount of transverse magnetization available to induce signal in the antenna.

From the basic *k*-space formula, it follows immediately that we reconstructed an image $I(\vec{x})$ simply by taking the inverse Fourier transform of the sampled data :

$$I(\vec{x}) = \int d^3x\, S(\vec{k}(t)).e^{2\pi_2 \vec{k}(t).\vec{x}}$$

Using the *k*-space formalism, a number of seemingly complex ideas become simple. For example, it becomes very easy to understand the role of phase encoding (the so-called spin-warp method). In a standard spin echo or gradient echo scan, where the readout (or view) gradient is constant (*e.g.,* $G_x$), a single line of *k*-space is scanned per *RF* excitation. When a non-zero phase-encoding pulse encoding gradient is zero, the line scanned is the axis. When phase is added in between the *RF* excitation and the commencement of the readout gradient, this line moves up or down is *k*-space *i.e.,* we scan the line $k_y$ = constant. The *k*-space formalism also makes it very easy to compare different scanning techniques. In single-shot EPI, all of *k*-space is scanned in a single shot, following either a sinusoidal or zig-zag trajectory. Since alternate lines of *k*-space are scanned in opposite directions, this must be taken into account in the reconstruction. Multi-shot EPI and fast spin echo techniques acquire only part of *k*-space per excitation. In each shot, a different interleaved segment is acquired, and the shots are repeated until *k*-space is sufficiently well-covered. Since the data at the centre of *k*-space represent lower spatial frequencies than the data at the edges of *k*-space, the $T_E$ value for the centre of *k*-space determines the images $T_2$ contrast.

The importance of the centre of *k*-space in determining image contrast can be exploited in more advanced imaging techniques. One such technique is spiral acquisition–a rotating magnetic field gradient is applied, carrying the trajectory in *k*-space to trace out spiral out from the centre to the edge. Due to $T_7$ and $T_2$* decay the signal is greatest at the start of the acquisition, hence acquiring the centre of *k*-space first improves contrast to noise ratio (CNR) when compared to conventional zig-zag acquisitions, especially in the presence of rapid movement.

### 36.21.4 Application of MRI

In critical practice, MRI is used to distinguish pathologic tissue (such as a brain tumour) from normal tissue. One advantage of an MRI scan is that it is harmless to the patient. It uses strong magnetic fields and non-ionizing radiation in the radio frequency range. Compare this to CT scans and traditional X-rays which involve doses of ionizing radiation and may increase the chance of malignancy, especially in a fetus.

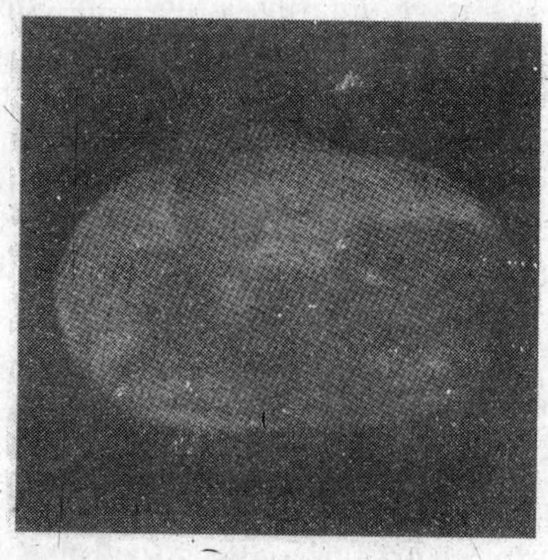

**Fig. 36.21** 3D MRI.

While CT provides good spatial resolution (the ability to distinguish two structures an arbitrarily small distance from each other as separate), MRI provides comparable resolution with far better contrast resolution (the ability to distinguish the differences between two arbitrarily similar but not identical tissues). The basis of this ability is the complex library of pulse sequences that the modern medical MRI scanner includes, each of which is optimized to provide image contrast based on the chemical sensitivity of MRI.

For example, with particular values of the echo time ($T_E$) and the repetition time ($T_R$), which are basic parameters of image acquisition, a sequence will take on the property of $T_2$-weighting. On a $T_2$-weighted scan, fat, water-and fluid-containing tissues are bright (most modern $T_2$ sequences are actually fast $T_2$ sequences). Damaged tissue tends to develop edema, which makes a $T_2$-weighted sequences sensitive for pathology, and generally able to distinguish pathologic tissue from normal tissue. With the addition of an additional radio frequency pulse and additional manipulation of the magnetic gradients, a $T_2$-weighted sequence can be converted to a FLAIR (Fluid Light Attenuation Inversion Recovery) sequence, in which free water is now dark, but edematous tissues remain bright. This sequence in particular is currently the most sensitive way to evaluate the brain.

## 36.22 SPECIALIZED MRI SCANS

### 36.22.1 Diffusion MRI

Diffusion MRI measures the diffusion of water molecules in biological tissues. In an isotropic medium (inside a glass of water for example) water molecules naturally move randomly according to Brownian motion. In biological tissues however, the diffusion may be anisotropic. For example a molecule inside the axon of a neuron has a low probability of crossing the myelin membrane. Therefore, the molecule will move principally along the axis of the neural fibre. If we know that molecules in a particular voxel diffuse principally in one direction we can make the assumption that the majority of the fibres in this area are going parallel to that direction.

The recent development of Diffusion Tensor Imaging (DTI) enables diffusion to be measured in multiple directions and the Fractional Anisotropy in each direction to be calculated for each voxel. This enables researchers to make brain maps of fibre directions to examine the connectivity of different regions in the brain (using tractography) or to examine areas of neural degeneration and demyelination in diseases like Multiple Sclerosis.

Finally, it has been proposed that diffusion MRI may be able to detect minute changes in extracellular

36

water diffusion and therefore could be used as a tool for FMRI. The nerve cell body enlarges when it conducts an action potential, hence restricting extracellular water molecules from diffusing naturally. Although this process works in theory, evidence is only moderately convincing.

Like many other specialized applications, this technique is usually coupled with a fast image acquisition sequence, such as Echo Planar Imaging sequence.

### 36.22.2 Magnetic Resonance Angiography

Magnetic resonance angiography (MRA) is used to generate pictures of the arteries in order to evaluate them for stenosis (abnormal narrowing) or aneuysms (vessel wall dilatations, at risk of rupture). MRA is often used to evaluate the arteries of the neck and brain, the thoracic and abdonominal aorta, the renal arteries, and the legs (called a "run-off"). A variety of techniques can be used to generate the pictures, such as administration of a paramagnetic contrast agent (gadolinium) or using a technique known as "flow-related enhancement" (*e.g.*, 2D and 3D time-of-flight sequences), where most of the signal on an image is due to blood which has recently moved into that lane. Magnetic Resonance Venography (MRV) is a similar procedure that is used to image veins. In this method the tissue is now excited inferiorily while signal is gathered in the plane immediately superior to the excitation plane, and this imaging the venous blood which has recently moved from the excited plane.

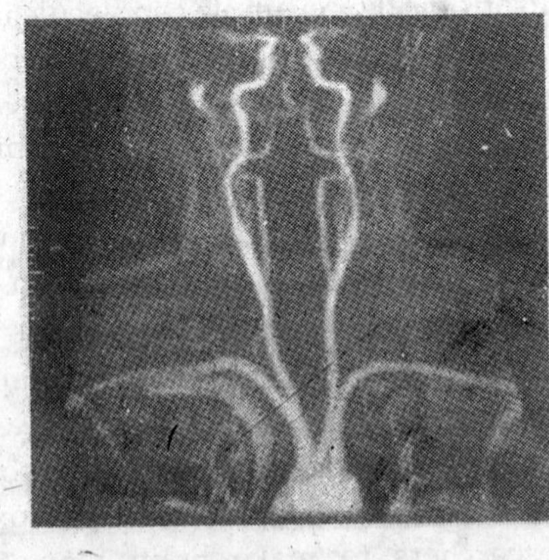

**Fig. 36.22** Magnetic resonance angiography.

## 36.23 BIOFEEDBACK

Biofeedback operates on the notion that we have innate ability and potential to influence the automatic functions of our bodies through the exertion of will and mind. Biofeedback has recently been shown to give us what had previously seemed an impossible degree of control over a variety of physiologic events.

For example, a person can be trained in a matter of days to cause the temperature of one hand to rise five to ten degrees higher than that of the other hand, while not contracting the hand muscles. What is amazing is that even animals can be trained. In one experiment, researchers trained a laboratory rat to produce a differential in the temperature of its two ears in order to receive a food reward.

Using a special machine and sensors to record muscle contractions and skin temperature, one can learn to control normally involuntary processes such as heart rate and blood pressure that increase under stress. The machine "feeds back" the efforts and eventually one can recognize and control facets of the stress response by himself. Once viewed with skepticism, the control of "involuntary" responses is now seen to be effective in the treatment of migraine headaches, asthma and other disorders in certain individuals.

The electroymogram (EMG) measures muscle tension. Two electrodes (or sensors) are placed on the skin over the muscle to be monitored. The most common muscles that biofeedback practitioners will use – are the frontalis (the "frowning" muscle in your forehead), the masseter (jaw muscle), and the trapezium (the shoulder muscles that hunch when you're stressed). This machine had been utilized to rehabilitate patients paralysed by stroke. Even when a person has no sensation in a paralysed limb and cannot move it voluntarily, EMG can often detect some electrical activity in the muscles. The EMG machine amplifies the electrical sound emitting from the paralysed limb, and as the patient becomes aware of the activity, his nervous system may stimulate more muscle activity. Eventually, new nerve endings may grow in the affected muscles and the patient may regain some mobility.

More often EMG is used to promote relaxation in muscles that have become tense in response to stress. When the electrodes pick up on muscle tension, the machine gives one a signal, such as a colored light or sound. In this way, one can see or hear continuous monitoring of his muscle activity and begin to focus on what the activity (or tension) feels like. As one become more aware of this internal processes, he will begin to recognize in his daily life when tension starts to build. He can then use the techniques he learn in the biofeedback training to control the tension before it gets worse or causes other physical problems.

EMG had been used for the treatment of tension headache, backache, neck pain, as well as in the stress related illnesses such as asthma and ulcers.

### 36.23.1 Electroencephalogram (EEG)

This device monitors brain wave activity. The method is relatively imprecise. The brain emits many electrical signals of various frequencies. Only a very few of these waves have been directly tied to specific disorders of mental states. The following are classification of brain waves

- Beta (awake)
- Alpha (calm relaxation)
- Theta (light sleep)
- Delta (deep sleep)

Since alpha waves are commonly observed during relaxation, the researchers thought that patients could relief from anxiety, insomnia and perhaps epilepsy by learning their alpha wave activity. But further research showed that alpha training is only useful if it is combined with other therapies. These days, insomnias are taught to control theta waves, and epileptics have found relief by monitoring waves produced during seizures. As technology advances, EEG may be used for other neurological diseases.

### 36.23.2 Other Types of Biofeedback Instruments

Other biofeedback machines can monitor heart rate and blood pressure, both of which change in response to stress, arrhythmia and hypertension. One can purchase biofeedback instruments to use at home, though the most affordable are the type that monitor only one system, such as temperature. Biofeedback is more effective if one combines it with relaxation techniques, self hypnotism and psychotherapy. In this way one not only learns how to control his reactions to stress but one can explore the causes of the stress and your thoughts and behaviour that contribute to it.

### 36.23.2A Temperature Biofeedback

Temperature biofeedback device monitors skin temperature and can be helpful in certain circulatory disorders. Reynolds disease is an example that can be benefited by this technique. Usually, a sensor is attached to ones foot or to the middle or small finger of ones dominant hand. When one is tense or anxious, his skin temperature drops as blood is redirected inward to muscles and internal organs.

Like monitoring muscle tension, measuring skin temperature is a useful tool in learning how to manage stress. This method may also reduce the frequency of migraine headaches, and is also used to promote relaxation.

### 36.23.2B Galvanic Skin Response (GSH) or Electrodermal Response (EDR)

Also known as Electrodermal Response (EDR), Galvanic skin response measures electrical conductance in the skin, which is associated with the activity of the sweat glands. A very slight electrical current (unnoticeable to one) is run through his skin. The machine measures changes in the salt and water in his sweat gland ducts. The more emotionally aroused he is, the more active his sweat glands are and the greater the electrical conductivity of your skin. GSR is effective in treating phobias, anxiety, excessive sweating, and, at times, stuttering since these emotions will affect his skin's conductivity. It is used as a lie detention test. Athletes use this technique to prepare for games — to make sure that they are not too anxious or have pre-game jitters.

The electrodes are attached to a machine which converts the electrical information into an easily observable form, such as a light or a buzzing noise. The machine can be adjusted, so that the buzzing sound is moderately audible at the beginning of the session. As the device picks up more perspiration, meaning more tension, the noise gets louder. If the person becomes calmer an [illegible]ere is less perspiration, the noise becomes lower and is finally extinguished.

Usually, the person hooked up to the biofeedback machine and told extinguished the buzz or the light. Since the person has no idea what to do, he or she will start experimenting to stop the annoying sound. If he tenses his muscles, for example, he will find that the noise is getting louder. Then may be he figures that if he relaxes, the buzz will go softer. So, he relaxes and the buzz does get softer. But it is not extinguished.

The person now will start putting himself/herself in various frames of mind that he believes will do the trick. There is a delay of several seconds between the feeling and the buzz, because it takes that long for the perspiration to appear on the skin, but he will soon enough find out if the machine is doing what he wants it to do. He tries other frames of mind. He imagines different scenes, different people, may be different colors. Then, quite suddenly, he discovers that the sound is no longer there. He will start mentally examine what he did to get to that condition. In practice, he will recall it and keeps it up. The therapist will now readjust the machine so that it has greater sensitivity. In other words, the buzz is going to sound when smaller amounts of perspiration are detected. In another session or two the person would probably learn how to counter this, and the process is continued until a satisfactory degree of relaxation is obtained. Once the person figures out how to do it with the help of machine, he can accomplish relaxation without the help of the machine by doing what he had to do as learned from the biofeedback techniques.

The same technique would be used to teach someone how to warm his hands, such as when we want him to control his migraine headaches. Here, instead of measuring perspiration, skin temperature would be measured. The person would imagine whatever he would imagine whatever he found necessary to do the trick. Incredibly, some people can not only boost the temperature of one hand over the other, but also to make one part of their palm warmer than the adjacent part.

### Number of Sessions

Biofeedback sessions usually last thirty to sixty minutes. The number of sessions each week and the duration of the training depend on your condition and the progress you make. It generally should not exceed 15 sessions. The crucial last step in biofeedback training is taking what you have learned and applying it in your everyday life without the help of the machines.

## Review Questions

1. What is the importance of Biomedical Instrumentation ?
2. State in brief the progresses of Biomedical Instrumentation from 1970.
3. What is biopotential ? How it is measured ?
4. Give the schematic of human heart and state its physiological aspects.
5. Describe in brief, the three leading systems used in ECG.
6. Describe how the action potential develop in a human cell. Give description of the activity of a cell. How human cell is polarised and depolarised ? What is the relation between cell firing activity and ECG ?
7. What is EEG and what are the different types of leads used in EEG ? What is θ, δ, β waves in EEG ? What is the difference between time domain EEG curve and EEG Histogram ?
8. Give the instrumentation diagram of a ECG monitoring system.
9. Give a chart of different diseases and their diagnoses from ECG.
10. Write short notes on :
    (*i*) Stressed ECG or treadmill
    (*ii*) Holter monitoring
    (*iii*) EEG at different conditions of human body.
    (*iv*) EMG.
11. Describe the physiological aspects of breathing.
12. What are the different transducers used in spirometry ?
13. What is the technique used in monitoring breathing by spirometer ?
14. In what different ways blood pressure may be monitored ?
15. Give a chart of diseases in relation to blood culture.
16. What are the differences between CT Scan and MRI ?
17. Write the principle of CT Scan.
18. What is the principle of MR imaging ?
19. What is the K-space formation in MRI ?
20. What is bio-feedback ? How it is used for relaxation of patient ?

# APPENDIX

## CONVERSION FACTORS

| To Convert | Into SI | Multiply by | Conversely multiply by |
|---|---|---|---|
| **Length** | | | |
| inches | cm | 2.540 | 0.3937 |
| feet | m | 0.3048 | 3.2808 |
| miles | km | 1.6093 | 0.6214 |
| micron | m | $10^{-6}$ | $10^{6}$ |
| **Area** | | | |
| inch$^2$ | cm$^2$ | 6.452 | 0.1550 |
| feet$^2$ | m$^2$ | 0.0929 | 10.764 |
| **Volume** | | | |
| inch$^3$ | cm$^3$ | 16.387 | 0.061 |
| feet$^3$ | m$^3$ | 0.02832 | 35.31 |
| gallon | m$^3$ | $4.546 \times 10^{-3}$ | 219.9 |
| litre | m$^3$ | $10^{-3}$ | $10^{3}$ |
| pint | m$^3$ | $0.568 \times 10^{-3}$ | 1760.5 |
| Quart | m$^3$ | $1.136 \times 10^{-3}$ | 879.9 |
| **Velocity** | | | |
| feet/second | m/s | 0.3048 | 3.281 |
| miles/hour | km/h | 1.609 | 0.6215 |
| knots | m/s | 0.5144 | 1.943 |
| **Mass** | | | |
| gram | kg | $10^{-3}$ | $10^{3}$ |
| pound (lbm) | kg | 0.4536 | 2.2046 |
| slug | kg | 14.59 | 0.0685 |
| ton | kg | 1016 | $0.09842 \times 10^{-3}$ |
| tonne (metric) | kg | $10^{3}$ | $10^{-3}$ |
| **Density** | | | |
| gram/cm$^3$ | kg/m$^3$ | $10^{3}$ | $10^{-3}$ |
| pound/foot$^3$ | kg/m$^3$ | 16.02 | 0.0624 |
| pound/inch$^3$ | kg/m$^3$ | $27.679 \times 10^{9}$ | $36.123 \times 10^{-12}$ |
| slug/foot$^3$ | kg/m$^3$ | 515.4 | $1.94 \times 10^{-3}$ |
| **Force** | | | |
| dyne | N | $10^{-5}$ | $10^{5}$ |
| $kg_f$ | N | 9.807 | 0.102 |
| $lb_f$ | N | 4.448 | 0.2248 |

| To Convert | Into SI | Multiply by | Conversely multiply by |
|---|---|---|---|
| **Pressure** | | | |
| dyne/cm$^2$ | N/m$^2$ | 0.1 | 10 |
| kgf/cm$^2$ | kN/m$^2$ | 98.07 | 0.0102 |
| psi(lb$_f$/inch)$^2$ | kN/m$^2$ | 6.895 | 0.145 |
| inches of water | N/m$^2$ | 249.5 | $4.008 \times 10^{-3}$ |
| feet of water | kN/m$^2$ | 2.989 | 0.3345 |
| inches of mercury | kN/m$^2$ | 3.386 | 0.2953 |
| mm of mercury | kN/m$^2$ | 0.1333 | 7.502 |
| 1 torr = 1 mm of Hg | | | |
| Pascal (Pa) | N/m$^2$ | 1 | 1 |
| bar | kN/m$^2$ | 100 | 0.01 |
| atm | kN/m$^2$ | 101.333 | $98.69 \times 10^{-3}$ |
| **Energy** | | | |
| erg | J | $10^{-7}$ | $10^7$ |
| foot pound | J | 1.356 | 0.7375 |
| calory | J | 4.187 | 0.2388 |
| Btu | kJ | 1.055 | 0.9478 |
| **Power** | | | |
| erg/second | W = Nm/s = J/s | $10^{-7}$ | $10^7$ |
| foot pound/second | W | 1.356 | 0.7376 |
| horse power | kW | 0.7457 | 1.341 |
| horse power (metric) | kW | 0.7355 | 1.360 |
| **Moment of Inertia** | | | |
| pound foot$^2$ | kg-m$^2$ | 0.0421 | 23.75 |
| pound inch$^2$ | kg-m$^2$ | $0.2926 \times 10^{-3}$ | $3.418 \times 10^3$ |
| slug foot$^2$ | kg-m | 1.355 | 0.7375 |
| **Discharge** | | | |
| cusecs (ft$^3$/s) | m$^3$/s | 0.02832 | 35.33 |
| gpm (gallons/min) | m$^3$/s | $3.0758 \times 10^3$ | $0.3251 \times 10^{-3}$ |
| litre/second | m$^3$/s | $10^{-3}$ | $10^3$ |
| pound/second | kg/s | 0.4536 | 2.204 |
| **Viscosity** | | | |
| Poise (P) | Ns/m$^2$ | 0.1 | 10 |
| centipoise (cP) | Ns/m$^2$ | 0.001 | 1000 |
| Stoke (St) | m$^2$/s | $10^{-4}$ | $10^4$ |
| centistock (cSt) | m$^2$/s | $10^{-6}$ | $10^6$ |
| **Thermal Conductivity** | | | |
| Btu/ft-hour-°F | W/m – °C | 1.7307 | 0.5778 |
| kCal/m-hour-°C | W/m – °C | 1.163 | 0.86 |
| **Entropy, Specific Heat, Latent Heat** | | | |
| Btu/lb – °R | kJ/kg – °K | 4.187 | 0.2388 |
| Btu/lb | kJ/kg | 2.326 | 0.423 |
| kcal/kg | kJ/kg | 4.187 | 0.2388 |
| **Heat Transfer coefficients** | | | |
| Btu/ft$^2$-hour-°F | W/m$^2$–°C | 5.678 | 0.1760 |
| kCal/m$^2$-hour-°C | W/m$^2$–°C | 1.163 | 0.86 |

# INDEX

## A



## B

## C



## D

# E

## F



## G



## H

# I



# J

# L

## N



## O

# P

## Q



## R

## S

## T



## U

## V



## W